CHILTON®

EUROPEAN MECHANICAL SERVICE 2006 EDITION

Audi
BMW
Jaguar
Land Rover
Mercedes
MINI
Saab
Volkswagen
Volvo

Australia • Canada • Mexico • Singapore • Spain • United Kingdom • United States

Chilton®

European
Mechanical Service
2006 Edition

Audi, BMW, Jaguar, Land Rover, Mercedes, MINI, Saab, Volkswagen, Volvo

Vice President, Technology Professional Business Unit:
Gregory L. Clayton

Publisher, Technology Prosfessional Business Unit:
David Koontz

Director of Marketing:
Beth A. Lutz

Production Director:
Patty Stephan

Editorial Assistant:
Rebecca Rokitowski

Production Manager:
Andrew Crouth

Marketing Manager:
Brian McGrath

Marketing Coordinator:
Jennifer Stall

Publishing Coordinator:
Paula Baillie

Sr. Content Project Manager:
Elizabeth C. Hough

Managing Editor:
Terry Blomquist

Senior Editor:
Rich Rivele

Editors:
Nick D'Andrea
Wayne Eiffes
Matt Frederick
Thomas A. Mellon
Rich Rivele
Jon Wallace
Lance Williams

Graphical Designer:
Melinda Possinger

Printed in the United States of America
1 2 3 4 5 XX 10 09 08 07 06

For more information contact: Thomson Delmar Learning
Executive Woods
5 Maxwell Drive, PO Box 8007,
Clifton Park, NY 12065-8007
Or find us on the World Wide Web at
www.delmarlearning.com

ISBN: 1-4180-0604-1

ISSN: 1548-0887

NOTICE TO THE READER

Publisher does not warrant or guarantee any of the products described herein or perform any independent analysis in connection with any of the product information contained herein. Publisher does not assume, and expressly disclaims, any obligation to obtain and include information other than that provided to it by the manufacturer.

The reader is expressly warned to consider and adopt all safety precautions that might be indicated by the activities herein and to avoid all potential hazards. By following the instructions contained herein, the reader willingly assumes all risks in connection with such instructions.

The publisher makes no representation or warranties of any kind, including but not limited to, the warranties of fitness for particular purpose or merchantability, nor are any such representations implied with respect to the material set forth herein, and the publisher takes no responsibility with respect to such material. The publisher shall not be liable for any special, consequential, or exemplary damages resulting, in whole or part, from the readers' use of, or reliance upon, this material.

Table of Contents

Sections

Model Index

USING THIS INFORMATION

Organization

To find where a particular model section or procedure is located, look in the Table of Contents. Main topics are listed with the page number on which they may be found. Following the main topics is an alphabetical listing of all of the procedures within the section and their page numbers.

Manufacturer and Model Coverage

This product covers 2002–2006 European models that are produced in sufficient quantities to warrant coverage, and which have technical content available from the vehicle manufacturers before our publication date. Although this information is as complete as possible at the time of publication, some manufacturers may make changes which cannot be included here. While striving for total accuracy, the publisher cannot assume responsibility for any errors, changes, or omissions that may occur in the compilation of this data.

Part Numbers & Special Tools

Part numbers and special tools are recommended by the publisher and vehicle manufacturer to perform specific jobs. Before substituting any part or tool for the one recommended, you must be completely satisfied that neither your personal safety, nor the performance of the vehicle will be endangered.

ACKNOWLEDGEMENT

The publisher would like to express appreciation to the following vehicle manufacturers for their assistance in producing this publication. No further reproduction or distribution of the material in this manual is allowed without the expressed written permission of the vehicle manufacturers and the publisher. Audi of America, Inc., BMW of North America, LLC, Jaguar Cars North America, Land Rover North America, Inc., Mercedes-Benz USA, LLC, MiniUSA, Volkswagen of America, Inc., Volvo Cars of North America, LLC

PRECAUTIONS

Before servicing any vehicle, please be sure to read all of the following precautions, which deal with personal safety, prevention of component damage, and important points to take into consideration when servicing a motor vehicle:

- Always wear safety glasses or goggles when drilling, cutting, grinding or prying.
- Steel-toed work shoes should be worn when working with heavy parts. Pockets should not be used for carrying tools. A slip or fall can drive a screwdriver into your body.
- Work surfaces, including tools and the floor should be kept clean of grease, oil or other slippery material.
- When working around moving parts, don't wear loose clothing. Long hair should be tied back under a hat or cap, or in a hair net.
- Always use tools only for the purpose for which they were designed. Never pry with a screwdriver.
- Keep a fire extinguisher and first aid kit handy.
- Always properly support the vehicle with approved stands or lift.
- Always have adequate ventilation when working with chemicals or hazardous material.
- Carbon monoxide is colorless, odorless and dangerous. If it is necessary to operate the engine with vehicle in a closed area such as a garage, always use an exhaust collector to vent the exhaust gases outside the closed area.
- When draining coolant, keep in mind that small children and some pets are attracted by ethylene glycol antifreeze, and are quite likely to drink any left in an open container, or in puddles on the ground. This will prove fatal in sufficient quantity. Always drain the coolant into a sealable container.
- To avoid personal injury, do not remove the coolant pressure relief cap while the engine is operating or hot. The cooling system is under pressure; steam and hot liquid can come out forcefully when the cap is loosened slightly. Failure to follow these instructions may result in personal injury. The coolant must be recovered in a suitable, clean container for reuse. If the coolant is contaminated it must be recycled or disposed of correctly.
- When carrying out maintenance on the starting system be aware that heavy gauge leads are connected directly to the battery. Make sure the protective caps are in place when maintenance is completed. Failure to follow these instructions may result in personal injury.
- Do not remove any part of the engine emission control system. Operating the engine without the engine emission control system will reduce fuel economy and engine ventilation. This will weaken engine performance and shorten engine life. It is also a violation of Federal law.
- Due to environmental concerns, when the air conditioning system is drained, the refrigerant must be collected using refrigerant recovery/recycling equipment. Federal law requires that refrigerant be recovered into appropriate recovery equipment and the process be conducted by qualified technicians who have been certified by an approved organization, such as MACS, ASI, etc. Use of a recovery machine dedicated to the appropriate refrigerant is necessary to reduce the possibility of oil and refrigerant incompatibility concerns. Refer to the instructions provided by the equipment manufacturer when removing refrigerant from or charging the air conditioning system.
- Always disconnect the battery ground when working on or around the electrical system.
- Batteries contain sulfuric acid. Avoid contact with skin, eyes, or clothing. Also, shield your eyes when working near batteries to protect against possible splashing of the acid solution. In case of acid contact with skin or eyes, flush immediately with water for a minimum of 15 minutes and get prompt medical attention. If acid is swallowed, call a physician immediately. Failure to follow these instructions may result in personal injury.
- Batteries normally produce explosive gases. Therefore, do not allow flames, sparks or lighted substances to come near the battery. When charging or working near a battery, always shield your face and protect your eyes. Always provide ventilation. Failure to follow these instructions may result in personal injury.
- When lifting a battery, excessive pressure on the end walls could cause acid to spew through the vent caps, resulting in personal injury, damage to the vehicle or battery. Lift with a battery carrier or with your hands on opposite corners. Failure to follow these instructions may result in personal injury.
- Observe all applicable safety precautions when working around fuel.

Whenever servicing the fuel system, always work in a well-ventilated area. Do not allow fuel spray or vapors to come in contact with a spark, open flame, or excessive heat (a hot drop light, for example). Keep a dry chemical fire extinguisher near the work area. Always keep fuel in a container specifically designed for fuel storage; also, always properly seal fuel containers to avoid the possibility of fire or explosion. Do not smoke or carry lighted tobacco or open flame of any type when working on or near any fuel-related components.

- Fuel injection systems often remain pressurized, even after the engine has been turned OFF. The fuel system pressure must be relieved before disconnecting any fuel lines. Failure to do so may result in fire and/or personal injury.
- The evaporative emissions system contains fuel vapor and condensed fuel vapor. Although not present in large quantities, it still presents the danger of explosion or fire. Disconnect the battery ground cable from the battery to minimize the possibility of an electrical spark occurring, possibly causing a fire or explosion if fuel vapor or liquid fuel is present in the area. Failure to follow these instructions can result in personal injury.
- The EPA warns that prolonged contact with used engine oil may cause a number of skin disorders, including cancer! You should make every effort to minimize your exposure to used engine oil. Protective gloves should be worn when changing oil. Wash your hands and any other exposed skin areas as soon as possible after exposure to used engine oil. Soap and water, or waterless hand cleaner should be used.
- Some vehicles are equipped with an air bag system, often referred to as a Supplemental Restraint System (SRS) or Supplemental Inflatable Restraint (SIR) system. The system must be disabled before performing service on or around system components, steering column, instrument panel components, wiring and sensors. Failure to follow safety and disabling procedures could result in accidental air bag deployment, possible personal injury and unnecessary system repairs.
- Always wear safety goggles when working with, or around, the air bag system. When carrying a non-deployed air bag, be sure the bag and trim cover are pointed away from your body. When placing a non-deployed air bag on a work surface, always face the bag and trim cover upward, away from the surface. This will reduce the motion of the module if it is accidentally deployed.
- Electronic modules are sensitive to electrical charges. The ABS module can be damaged if exposed to these charges.
- Brake pads and shoes may contain asbestos, which has been determined to be a cancer-causing agent. Never clean brake surfaces with compressed air. Avoid inhaling brake dust. Clean all brake surfaces with a commercially available brake cleaning fluid.
- When replacing brake pads, shoes, discs or drums, replace them as complete axle sets.
- When servicing drum brakes, disassemble and assemble one side at a time, leaving the remaining side intact for reference.
- Brake fluid often contains polyglycol ethers and polyglycols. Avoid contact with the eyes and wash your hands thoroughly after handling brake fluid. If you do get brake fluid in your eyes, flush your eyes with clean, running water for 15 minutes. If eye irritation persists, or if you have taken brake fluid internally, immediately seek medical assistance.
- Clean, high quality brake fluid from a sealed container is essential to the safe and proper operation of the brake system. You should always buy the correct type of brake fluid for your vehicle. If the brake fluid becomes contaminated, completely flush the system with new fluid. Never reuse any brake fluid. Any brake fluid that is removed from the system should be discarded. Also, do not allow any brake fluid to come in contact with a painted or plastic surface; it will damage the paint.
- Never operate the engine without the proper amount and type of engine oil; doing so will result in severe engine damage.
- Timing belt maintenance is extremely important! Many models utilize an interference-type, non-freewheeling engine. If the timing belt breaks, the valves in the cylinder head may strike the pistons, causing potentially serious (also time-consuming and expensive) engine damage.
- Disconnecting the negative battery cable on some vehicles may interfere with the functions of the on-board computer system (s) and may require the computer to undergo a relearning process once the negative battery cable is reconnected.
- Steering and suspension fasteners are critical parts because they affect performance of vital components and systems and their failure can result in major service expense. They must be replaced with the same grade or part number or an equivalent part if replacement is necessary. Do not use a replacement part of lesser quality or substitute design. Torque values must be used as specified during reassembly to ensure proper retention of these parts.

AUDI

1

A4 • A4 Avant • A6 • A6 Avant • A6 Avant Allroad • A8 • A8 Quattro • TT • TT Quattro

SPECIFICATION CHARTS

ENGINE AND VEHICLE IDENTIFICATION

Engine						
Code	Liters (cc)	Cu. In.	Cyl.	Fuel Sys.	Engine Type	Eng. Mfg.
AMB	1.8 (1781)	107	4	MFI-Turbo	DOHC	Audi
AMU	1.8 (1781)	107	4	MFI-Turbo	DOHC	Audi
ATC	1.8 (1781)	107	4	MFI-Turbo	DOHC	Audi
AWP	1.8 (1781)	107	4	MFI-Turbo	DOHC	Audi
BEA	1.8 (1781)	107	4	MFI-Turbo	DOHC	Audi
AXX	2.0 (1998)	121	4	FSI-Turbo	DOHC	Audi
BPY	2.0 (1998)	121	4	FSI-Turbo	DOHC	Audi
BPG	2.0 (1998)	121	4	FSI-Turbo	DOHC	Audi
APB	2.7 (2671)	169	6	MFI-Turbo	DOHC	Audi
AHA	2.8 (2771)	169	6	MFI	DOHC	Audi
AVK	3.0 (2976)	183	6	MFI	DOHC	Audi
BGN	3.0 (2976)	183	6	MFI	DOHC	Audi
BHE	3.2 (3200)	195	6	MFI	DOHC	Audi
BUB	3.2 (3200)	195	6	MFI	DOHC	Audi
BKH	3.2 (3200)	190	6	FSI	DOHC	Audi
AWN	4.2 (4172)	255	8	MFI	DOHC	Audi
ABZ	4.2 (4172)	255	8	MFI	DOHC	Audi
BNK	4.2 (4172)	255	8	MFI	DOHC	Audi

Model Year	
Code ①	Year
2	2002
3	2003
4	2004
5	2005
6	2006

MFI: Multi-point Fuel Injection

FSI: Direct Fuel Injection

DOHC: Double Overhead Camshaft

① 10th digit of the Vehicle Identification Number (VIN)

06041_AUDI_C0001

GENERAL ENGINE SPECIFICATIONS

Year	Model	Engine Displacement Liters	Engine ID	Net Horsepower @ rpm	Net Torque@rpm (ft. lbs.)	Bore x Stroke (in.)	Com-pression Ratio	Oil Pressure @ rpm
2002	A4	1.8	AMB	150@5700	155@1750	3.18x3.40	9.5:1	72-99@3000
	A4	3.0	AVK	217@6300	221@3200	3.24x3.65	10.3:1	44@2000
	A6	2.8	AHA	200@6000	207@3200	3.25x3.40	10.3:1	29@2000
	A8	3.7	AEW	230@5800	229@2300	3.33x3.24	10.8:1	44-72@3000
	A8 Quattro	4.2	ABZ	300@6000	295@3300	3.33x3.66	10.8:1	44-72@3000
	TT	1.8	AWP	177@5700	173@1750	3.18x3.45	9.5:1	51-65@2000
	TT Quattro	1.8	AMU	225@5900	207@3900	3.18x3.45	9.0:1	51-65@2000
2003	A4	1.8	AMB	167@5700	155@1750	3.18x3.40	9.5:1	51-65@2000
	A4	3.0	AVK	217@6300	221@3200	3.24x3.65	10.3:1	44@2000
	A6 Quattro	2.7	APB	250@5800	258@3200	3.18x3.40	9.3:1	29@2000
	A6	3.0	AVK	217@6300	221@3200	3.25x3.65	10.3:1	44@2000
	A6 Quattro	4.2	AWN	310@6000	295@3300	3.33x3.66	11.0:1	44-72@3000
	A8 Quattro	4.2	AYS	360@7000	302@3300	3.33x3.66	11.0:1	44-72@3000
	TT	1.8	AWP	177@5700	173@1750	3.18x3.45	9.5:1	51-65@2000
	TT Quattro	1.8	AMU	225@5900	207@3900	3.18x3.45	9.0:1	51-65@2000
2004	A4	1.8	AMB	167@5700	155@1750	3.18x3.40	9.5:1	51-65@2000
	A4	3.0	AVK	217@6300	221@3200	3.24x3.65	10.3:1	44@2000
	A6	3.0	AVK	217@6300	221@3200	3.25x3.65	10.3:1	44@2000
	A6 Quattro	4.2	AWN	310@6000	305@3300	3.33x3.66	11.0:1	44-72@3000
	A8 Quattro	4.2	AYS	360@7000	302@3300	3.33x3.66	11.0:1	44-72@3000
	TT	1.8	AWP	132@5500	235@1950	3.18x3.48	9.5:1	51-65@2000
	TT	1.8	BEA	225@5900	220@5900	3.18x3.40	9.0:1	51-65@3000
	TT Quattro	3.2	BHE	184@6300	320@2800	3.31x3.78	11.3:1	44@2000
2005-06	A3	2.0	AXX	200@5,100-600	207@1800-5000	3.25x3.65	10.3:1	39-65@2000
	A3	2.0	BPY	200@5,100-600	207@1800-5000	3.25x3.65	10.3:1	39-65@2000
	A3	3.2	BUB	250@6,300	236@2800-3200	3.31x3.78	11.3:1	39-65@2000
	A4	1.8	AMB	167@5700	155@1750	3.18x3.40	9.5:1	51-65@2000
	A4	2.0	BPG	200@5,100-600	207@1800-5000	3.25x3.65	10.5:1	43@2000
	A4	3.0	BGN	220@6300	221@3200	3.25x3.65	10.3:1	43@2000
	A4	3.2	BKH	255@6500	243@3250	3.33x3.65	12.5:1	43@2000
	A6	3.2	BKH	255@6500	243@3250	3.33x3.65	12.5:1	43@2000
	A6	4.2	BNK	335@6500	310@3500	3.33x3.65	11.0:1	29@2000
	TT	1.8	AWP	132@5500	235@1950	3.18x3.48	9.5:1	51-65@2000
	TT	1.8	BEA	225@5900	220@5900	3.18x3.40	9.0:1	51-65@3000
	TT Quattro	3.2	BHE	184@6300	320@2800	3.31x3.78	11.3:1	44@2000

MFI: Multi-point Fuel Injection

06041_AUDI_C0002

ENGINE TUNE-UP SPECIFICATIONS

Year	Engine Displacement Liters	Engine ID/VIN	Spark Plug Gap (in.)	Ignition Timing (deg.) MT	Ignition Timing (deg.) AT	Fuel Pump (psi)	Idle Speed (rpm) MT	Idle Speed (rpm) AT	Valve Clearance In.	Valve Clearance Ex.
2002	1.8	AMB	0.039	①	①	50-58	820-900	820-900	HYD	HYD
	1.8	AMU	0.039	①	①	36	700-820	700-820	HYD	HYD
	1.8	AWP	0.039	①	①	36	700-820	700-820	HYD	HYD
	2.8	AHA	0.039	①	①	55-61	700-800	700-800	HYD	HYD
	3.0	AVK	0.039	①	①	46-55	700-800	670-770	HYD	HYD
	3.7	AEW	0.039	①	①	52	700-800	700-800	HYD	HYD
	4.2	ABZ	0.039	①	①	52	700-800	700-800	HYD	HYD
2003	1.8	AMB	0.039	①	①	58	700-860	700-860	HYD	HYD
	1.8	AMU	0.039	①	①	36	700-820	700-820	HYD	HYD
	1.8	AWP	0.039	①	①	36	700-820	700-820	HYD	HYD
	2.7	APB	0.039	①	①	50	650-750	650-750	HYD	HYD
	3.0	AVK	0.039	①	①	46-55	700-800	670-770	HYD	HYD
	4.2	AWN	0.039	①	①	50	650-720	650-720	HYD	HYD
2004	1.8	AMB	0.039	①	①	58	700-860	700-860	HYD	HYD
	1.8	BEA	0.039	①	①	36	700-820	700-820	HYD	HYD
	1.8	AWP	0.039	①	①	36	700-820	700-820	HYD	HYD
	3.0	AVK	0.039	①	①	46-55	700-800	670-770	HYD	HYD
	3.2	BHE	0.039	①	①	46-55	650-750	650-750	HYD	HYD
	4.2	AWN	0.039	①	①	50	650-720	650-720	HYD	HYD
2005-06	1.8	AMB	0.039	①	①	58	700-860	700-860	HYD	HYD
	1.8	BEA	0.039	①	①	36	700-820	700-820	HYD	HYD
	1.8	AWP	0.039	①	①	36	700-820	700-820	HYD	HYD
	2.0	AXX	0.039	①	①	87	640-800	640-800	HYD	HYD
	2.0	BPG	0.039	①	①	87	640-800	640-800	HYD	HYD
	2.0	BPY	0.039	①	①	87	640-800	640-800	HYD	HYD
	3.0	BGN	0.039	①	①	46-55	700-800	700-800	HYD	HYD
	3.2	BUB	0.039	①	①	40	600-700	600-700	HYD	HYD
	3.2	BHE	0.039	①	①	46-55	650-750	650-750	HYD	HYD
	3.2	BKH	0.039	①	①	87	650-750	650-750	HYD	HYD
	4.2	BNK	0.039	①	①	58	650-720	650-720	HYD	HYD

NOTE: The Vehicle Emission Control Information label reflects specification changes made during production and must be used if different from this chart.

NOTE: Fuel pump pressure specifications with the fuel pressure regulator vacuum hose attached.

HYD: Hydraulic

① The basic setting is controlled by the ECU and is not adjustable

06041_AUDI_C0003

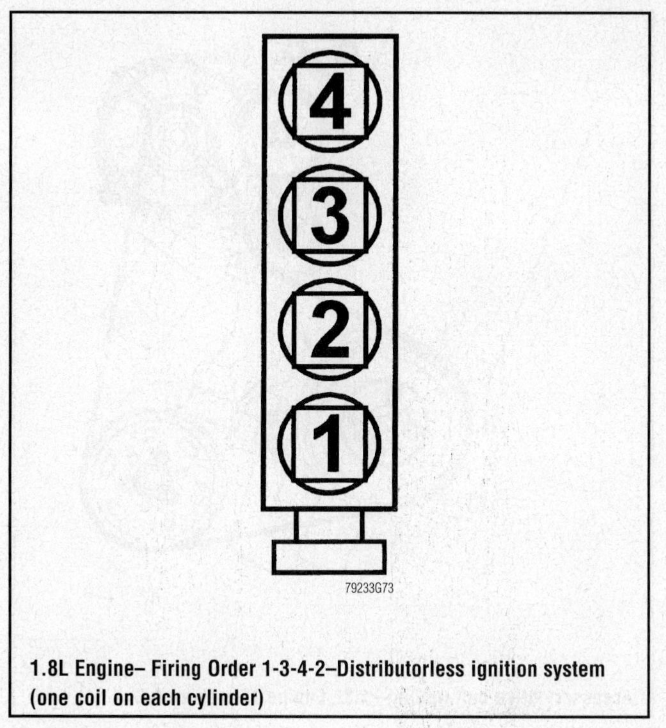

1.8L Engine– Firing Order 1-3-4-2–Distributorless ignition system (one coil on each cylinder)

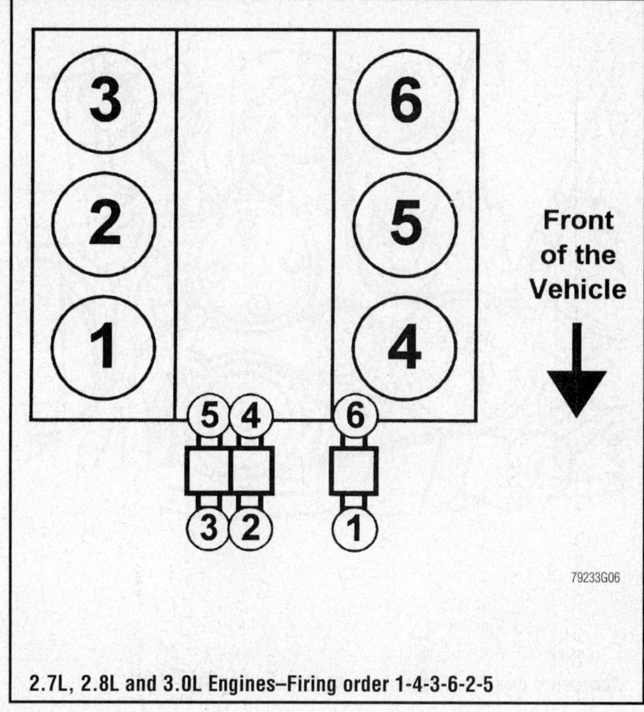

2.7L, 2.8L and 3.0L Engines–Firing order 1-4-3-6-2-5

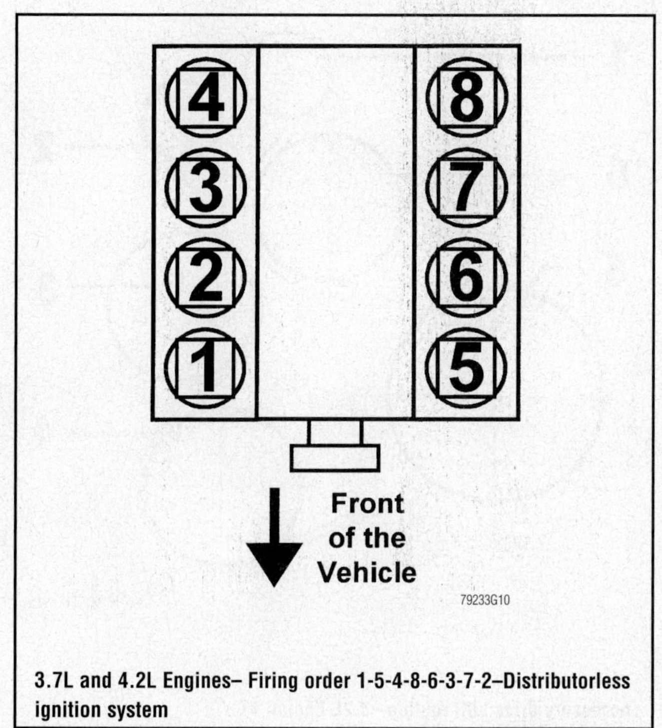

3.7L and 4.2L Engines– Firing order 1-5-4-8-6-3-7-2–Distributorless ignition system

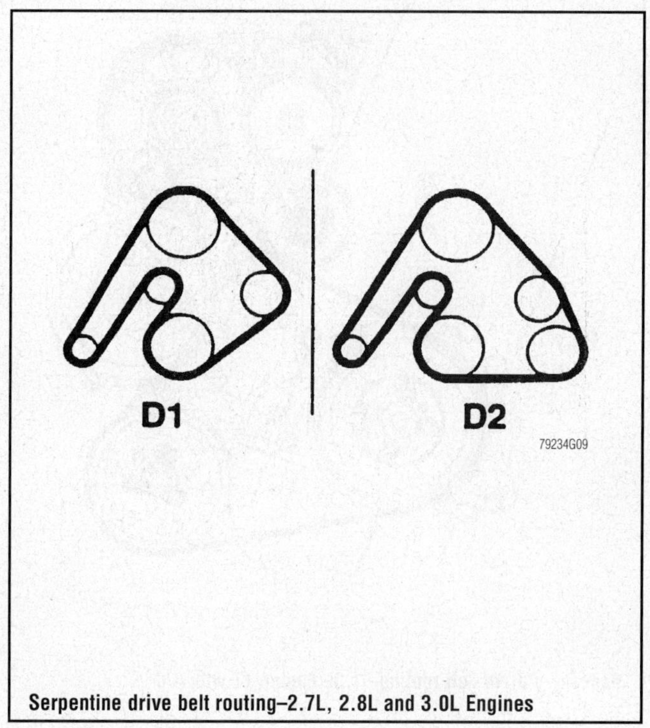

Serpentine drive belt routing–2.7L, 2.8L and 3.0L Engines

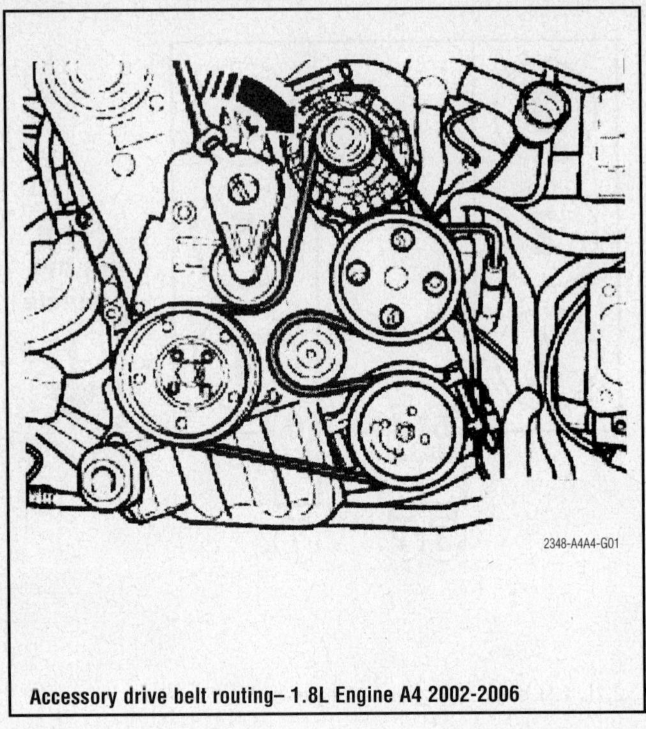

Accessory drive belt routing– 1.8L Engine A4 2002-2006

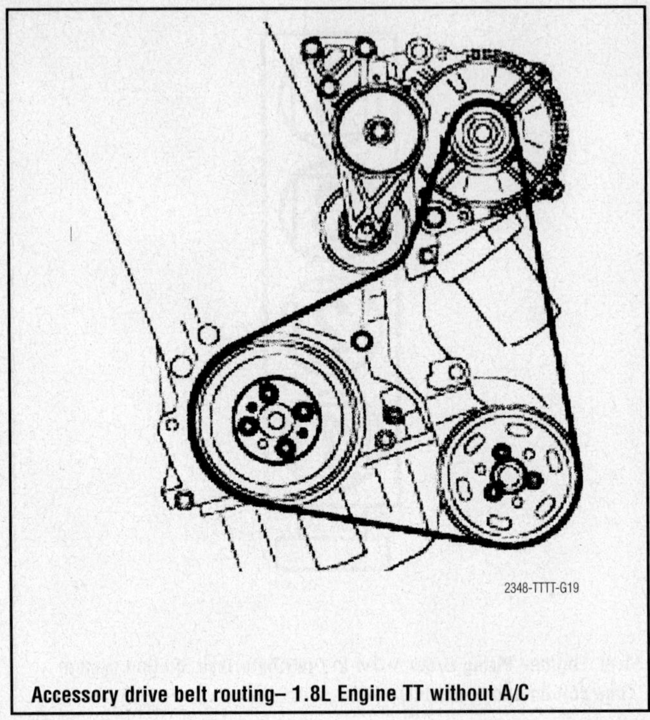

Accessory drive belt routing– 1.8L Engine TT without A/C

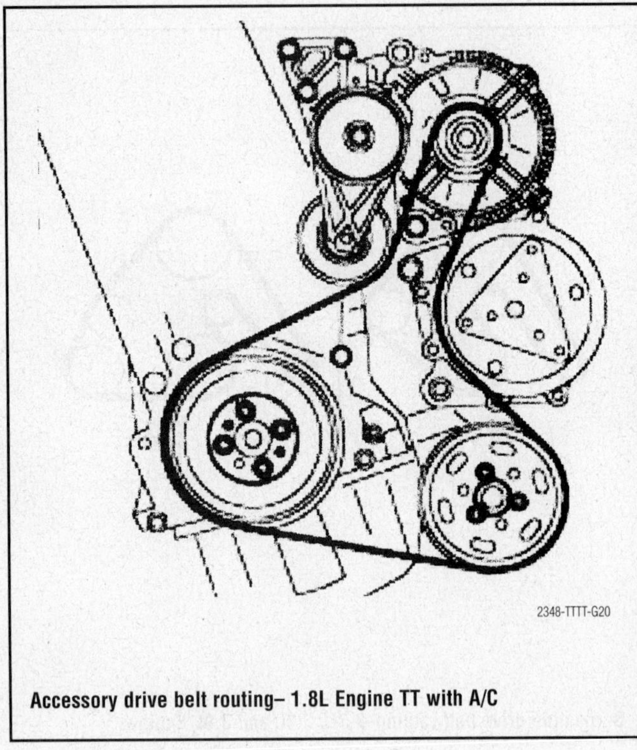

Accessory drive belt routing– 1.8L Engine TT with A/C

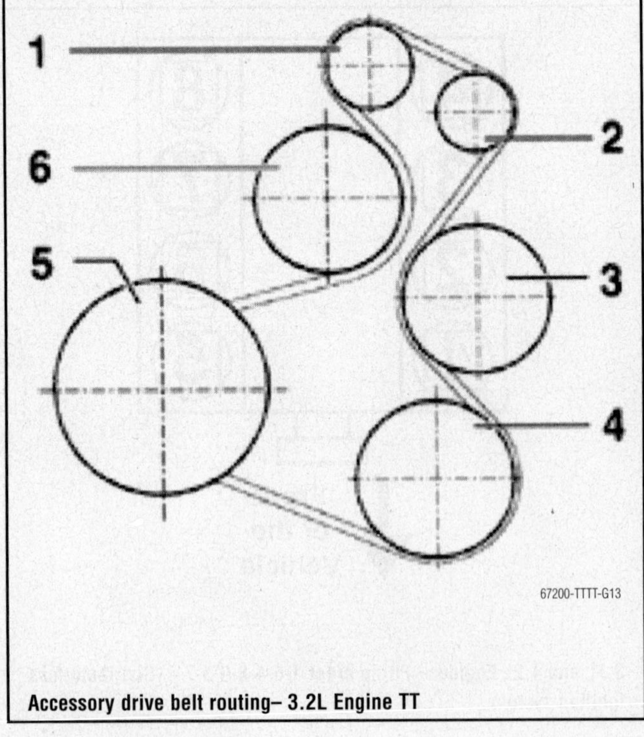

Accessory drive belt routing– 3.2L Engine TT

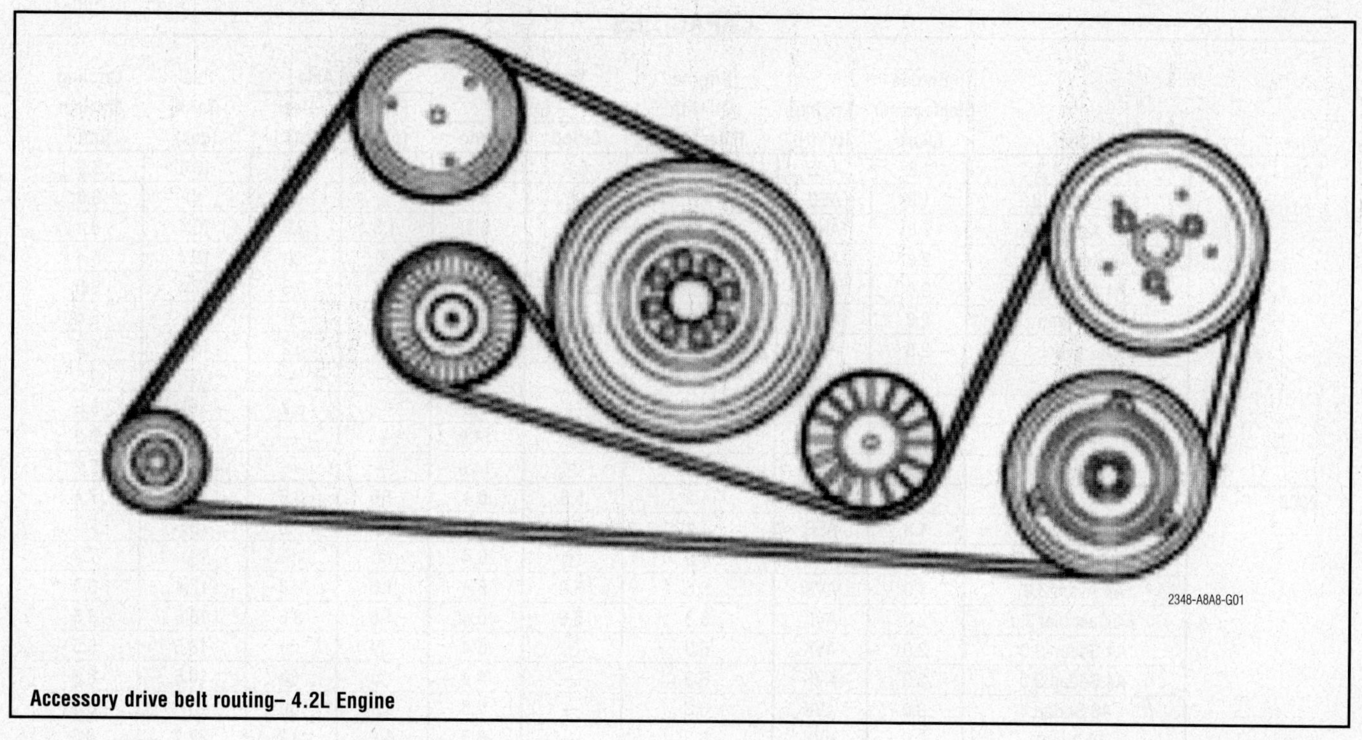

Accessory drive belt routing– 4.2L Engine

CAPACITIES

Year	Model	Engine Displacement Liters	Engine ID/VIN	Engine Oil with Filter (qts)	Transmission (pts.) 5-Spd	Transmission (pts.) Auto	Drive Axle Front (pts.)	Drive Axle Rear (pts.)	Fuel Tank (gal.)	Cooling System (qts.)
2002	A4 Avant 1.8 T	1.8	AMB	4.2	5.8	5.4	1.6	3.8	16.4	6.9
	A4 Sedan 1.8 T	1.8	AEB	4.2	①	5.4	②	③	④	6.6
	A4 Avant 2.8	2.8	AHA	5.3	5.8	5.4	1.6	3.8	16.4	6.4
	A4 Sedan 2.8	2.8	AHA	5.3	①	5.4	②	③	16.4	6.4
	A4 Avant 3.0	3.0	AVK	5.3	5.8	5.4	1.6	3.8	17.4	8.5
	A6 Sedan	2.8	AHA	6.9	—	5.5	⑤	⑥	⑦	8.4
	A6 Avant	2.8	AHA	6.9	—	5.5	1.9	3.2	18.5	8.4
	A8	3.7	AEW	8.0	—	5.5	—	—	⑧	9.5
	A8 Quattro	4.2	ABZ	8.0	—	8.0	—	3.2	⑧	9.5
	TT 1.8T	1.8	AMU	4.8	⑨	14.8	—	—	14.5	5.2
	TT Quattro 1.8T	1.8	AWP	4.8	⑨	14.8	—	—	16.3	5.2
2003	A4 Avant 1.8 T	1.8	AEB	4.2	5.8	6.4	1.6	3.8	17.4	7.4
	A4 Cabriolet 1.8 T	1.8	AEB	4.2	5.8	6.4	1.6	3.8	18.5	7.4
	A4 Sedan 1.8 T	1.8	AEB	4.2	①	6.4	②	③	⑨	7.4
	A4 Avant 3.0	3.0	AVK	5.3	5.8	5.4	1.6	3.8	17.4	8.5
	A4 Cabriolet 3.0	3.0	AVK	6.3	5.8	5.4	1.6	3.8	18.5	8.5
	A4 Sedan 3.0	3.0	AVK	6.3	①	5.4	②	③	18.5	8.5
	A4 Sedan 3.0	3.0	AVK	6.3	—	5.4	②	③	18.5	8.5
	A6 Sedan	3.0	AVK	6.3	—	5.5	⑤	⑥	⑦	8.5
	A6 Avant	3.0	AVK	6.3	—	5.5	1.9	3.2	21.7	8.5
	A8	4.2	AWN	8.0	—	5.5	—	—	⑧	9.5
	A8 Quattro	4.2	AWN	8.0	—	8.0	—	3.2	⑧	9.5
	TT Coupe 1.8T	1.8	AMU	4.8	⑩	14.8	—	—	14.5	5.2
	TT Roadster 1.8T	1.8	AMU	4.8	⑩	14.8	—	—	14.5	5.2
	TT Quattro 1.8T	1.8	AWP	4.8	⑩	14.8	—	—	16.3	5.2
2004	A4 Avant 1.8 T	1.8	AMB	4.2	5.8	6.4	1.6	3.8	17.4	7.4
	A4 Cabriolet 1.8 T	1.8	AMB	4.2	5.8	6.4	1.6	3.8	18.5	7.4
	A4 Sedan 1.8 T	1.8	AMB	4.2	①	6.4	②	③	⑨	7.4
	A4 Avant 3.0	3.0	AVK	5.3	5.8	5.4	1.6	3.8	17.4	8.5
	A4 Cabriolet 3.0	3.0	AVK	6.3	5.8	5.4	1.6	3.8	18.5	8.5
	A4 Sedan 3.0	3.0	AVK	6.3	①	5.4	②	③	18.5	8.5
	A4 Sedan 3.0	3.0	AVK	6.3	—	5.4	②	③	18.5	8.5
	A6 Sedan	3.0	AVK	6.3	—	5.5	⑤	⑥	⑦	8.5
	A6 Avant	3.0	AVK	6.3	—	5.5	1.9	3.2	21.7	8.5
	A8	4.2	AWN	8.0	—	5.5	—	—	⑧	9.5
	A8 Quattro	4.2	AWN	8.0	—	8.0	—	3.2	⑧	9.5
	TT Quattro 1.8T	1.8	AWP	4.8	⑩	14.8	—	—	16.3	5.2
	TT Coupe 1.8T	1.8	BEA	4.8	⑩	14.8	—	—	14.5	5.2
	TT Quattro 1.8T	3.2	BHE	5.8	⑩	⑪	—	—	—	—

NOTE: All capacities are approximate. Add fluid gradually and ensure a proper fluid level is obtained.

① All Wheel Drive: 5.8 pts.
Front Wheel Drive: 4.8 pts.

② Automatic Transmission, Front Wheel Drive:
Front Differential: 1.6 pts.
Automatic Transmission, All Wheel Drive:
Front Differential: 1.6 pts.
Center Differential: 1.6 pts.

③ All Wheel Drive: 3.8 pts.

④ Front Wheel Drive: 16.6 gal.
All Wheel Drive: 16.5 gal.

⑤ Automatic Transmission, Front Wheel Drive:
Front Differential: 1.9 pts.
Automatic Transmission, All Wheel Drive:
Front Differential: 1.9 pts.
Center Differential: 1.9 pts.

⑥ All Wheel Drive: 3.2 pts.
Front Differential: 1.9 pts.

⑦ All Wheel Drive: 21.1 gallons
Front Wheel Drive: 18.5 gallons

⑧ Either 21 or 24 gallon tank

⑨ Front Wheel Drive: 18.5
All Wheel Drive: 17.4 gal.

⑩ Front Wheel Drive: 4.2 pts.
All Wheel Drive: 5.4 pts.

⑪ Initial Fill: 15,2 pts.
Refill/Change: 12.7 pts.

06041_AUDI_C0004

CAPACITIES

Year	Model	Engine Displacement Liters	Engine ID/VIN	Engine Oil with Filter (qts)	Transmission (pts.) 5-Spd	Transmission (pts.) Auto	Drive Axle Front (pts.)	Drive Axle Rear (pts.)	Fuel Tank (gal.)	Cooling System (qts.)
2005-06	A3	2.0	AXX	4.8	4.8	12.6	NA	NA	14.5	8.5
	A3	2.0	BPY	4.8	4.8	12.6	NA	NA	14.5	8.5
	A3	3.2	BUB	NA	NA	NA	NA	NA	14.5	7.4
	A4	1.8	AMB	4.3	6.4	6.4	1.6	3.8	17.4	7.4
	A4	2.0	BPG	4.8	7.4	6.4	1.6	3.8	16.5	7.4
	A4	3.2	BKH	8.6	7.4	6.4	1.6	NA	16.6	—
	A6	3.2	BKH	8.6	7.4	4.6	NA	—	31.1	8.5
	A6	4.2	BNK	8.0	—	8.0	—	3.2	21.1	9.5
	TT	1.8	AWP	4.8	⑩	14.8	—	—	16.3	5.2
	TT Coupe 1.8T	1.8	BEA	4.8	⑩	14.8	—	—	14.5	5.2
	TT	3.2	BHE	5.8	⑩	⑪	—	—	—	—

NOTE: All capacities are approximate. Add fluid gradually and ensure a proper fluid level is obtained.

NA: Not Available

① All Wheel Drive: 5.8 pts.
Front Wheel Drive: 4.8 pts.

② Automatic Transmission, Front Wheel Drive:
Front Differential: 1.6 pts.
Automatic Transmission, All Wheel Drive:
Front Differential: 1.6 pts.
Center Differential: 1.6 pts.

③ All Wheel Drive: 3.8 pts.

④ Front Wheel Drive: 16.6 gal.
All Wheel Drive: 16.5 gal.

⑤ Automatic Transmission, Front Wheel Drive:
Front Differential: 1.9 pts.
Automatic Transmission, All Wheel Drive:
Front Differential: 1.9 pts.
Center Differential: 1.9 pts.

⑥ All Wheel Drive: 3.2 pts.
Front Differential: 1.9 pts.

⑦ All Wheel Drive: 21.1 gallons
Front Wheel Drive: 18.5 gallons

⑧ Either 21 or 24 gallon tank

⑨ Front Wheel Drive: 18.5
All Wheel Drive: 17.4 gal.

⑩ Front Wheel Drive: 4.2 pts.
All Wheel Drive: 5.4 pts.

⑪ Initial Fill: 15,2 pts.
Refill/Change: 12.7 pts.

06041_AUDI_C0005

CRANKSHAFT AND CONNECTING ROD SPECIFICATIONS

All measurements are given in inches.

Year	Engine Size Liters	Engine ID/VIN	Crankshaft: Main Brg. Journal Dia.	Crankshaft: Main Brg. Oil Clearance	Crankshaft: Shaft End-play	Crankshaft: Thrust on No.	Connecting Rod: Journal Diameter	Connecting Rod: Oil Clearance ①	Connecting Rod: Side Clearance
2002	1.8	AMB	2.1267-2.1275	0.0008-0.0016	0.0030-0.009	3	1.8811-1.8837	0.0004-0.002	0.0040-0.014
	1.8	AMU	2.1251-2.1259	0.0004-0.0015	0.0028-0.0083	3	1.8802-1.8819	0.0004-0.002	0.0040-0.014
	1.8	AWP	2.1251-2.1259	0.0004-0.0015	0.0028-0.0083	3	1.8802-1.8819	0.0004-0.002	0.0040-0.014
	2.8	AHA	2.5573-2.5598	0.0007-0.0018	0.0028-0.0091	3	2.1243-2.1268	0.0006-0.0024	0.0060-0.014
	3.0	AVK	2.5573-2.559	0.0007-0.0018	0.0028-0.009	3	2.1243-2.126	0.0006-0.0024	NA
	3.7	AEW	NA	NA	NA	NA	NA	NA	NA
	4.2	ABZ	NA	NA	NA	NA	NA	NA	NA
2003	1.8	AMB	1.8883-1.8897	0.0008-0.0016	0.0028-0.009	3	1.6518-1.6535	0.0004-0.002	0.0040-0.014
	1.8	AMU	2.1251-2.1259	0.0004-0.0015	0.0028-0.0083	3	1.8802-1.8819	0.0004-0.002	0.0040-0.014
	1.8	AWP	2.1251-2.1259	0.0004-0.0015	0.0028-0.0083	3	1.8802-1.8819	0.0004-0.002	0.0040-0.014
	2.7	APB	NA	0.0007-0.0018	0.0035-0.0098	4	2.1243-2.126	0.0006-0.0024	NA
	3.0	AVK	2.5573-2.559	0.0007-0.0018	0.0028-0.009	3	2.1243-2.126	0.0006-0.0024	NA
	4.2	AWN	NA	NA	NA	NA	NA	NA	NA
2004	1.8	AMB	1.8883-1.8897	0.0008-0.0016	0.0028-0.009	3	1.6518-1.6535	0.0004-0.002	0.0040-0.014
	1.8	AWP	2.1251-2.1259	0.0004-0.0015	0.0028-0.0083	3	1.8802-1.8819	0.0004-0.002	0.0040-0.014
	1.8	BEA	2.1251-2.1259	0.0004-0.0015	0.0028-0.0083	3	1.8802-1.8819	0.0004-0.002	0.0040-0.014
	3.0	AVK	2.5573-2.559	0.0007-0.0018	0.0028-0.009	3	2.1243-2.126	0.0006-0.0024	NA
	3.2	BHE	2.3606-2.3613	0.0008-0.0024	0.0003-0.009	5	2.1243-2.1251	0.0008-0.0028	NA
	4.2	AWN	NA	NA	NA	NA	NA	NA	NA
2005-06	1.8	AMB	1.8883-1.8897	0.0008-0.0016	0.0028-0.009	3	1.6518-1.6535	0.0004-0.002	0.0040-0.014
	1.8	AWP	2.1251-2.1259	0.0004-0.0015	0.0028-0.0083	3	1.8802-1.8819	0.0004-0.002	0.0040-0.014
	1.8	BEA	2.1251-2.1259	0.0004-0.0015	0.0028-0.0083	3	1.8802-1.8819	0.0004-0.002	0.0040-0.014
	2.0	AXX	NA	NA	NA	NA	NA	NA	NA
	2.0	BPY	NA	NA	NA	NA	NA	NA	NA
	2.0	BPG	NA	NA	NA	NA	NA	NA	NA
	3.2	BUB	2.3606-2.3613	0.0008-0.0024	0.0003-0.009	5	2.1243-2.1251	0.0008-0.0028	NA

06041_AUDI_C0006

CRANKSHAFT AND CONNECTING ROD SPECIFICATIONS

All measurements are given in inches.

	Engine		Crankshaft				Connecting Rod		
Year	Size Liters	Engine ID/VIN	Main Brg. Journal Dia.	Main Brg. Oil Clearance	Shaft End-play	Thrust on No.	Journal Diameter	Oil Clearance ①	Side Clearance
2005-06	3.0	BGN	NA	NA	NA	NA	NA	NA	NA
	3.2	BHE	2.3606-2.3613	0.0008-0.0024	0.0003-0.009	5	2.1243-2.1251	0.0008-0.0028	NA
	3.2	BKH	2.3606-2.3613	0.0008-0.0024	0.0003-0.009	5	2.1243-2.1251	0.0008-0.0028	NA
	4.2	BNK	NA	NA	NA	NA	NA	NA	NA

NA: Not Available

① To measure oil clearance torque as follows:

AFC and AHA engine connecting rods: 15 ft. lbs.

AEB engine connecting rods: 22 ft. lbs.

06041_AUDI_C0007

VALVE SPECIFICATIONS

Year	Engine Displacement Liters	Engine ID/VIN	Seat Angle (deg.)	Face Angle (deg.)	Spring Test Pressure (lbs. @ in.)	Spring Installed Height (in.)	Stem-to-Guide Clearance (in.)		Stem Diameter (in.)	
							Intake	Exhaust	Intake	Exhaust
2002	1.8	AMB	45	45	NA	NA	0.031 ①	0.031 ①	0.2339-0.2350	0.2339-0.2343
	1.8	AMU	45	45	NA	NA	0.031 ①	0.031 ①	NA	NA
	1.8	AWP	45	45	NA	NA	0.031 ①	0.031 ①	NA	NA
	2.8	AHA	45	45	NA	NA	0.031 ①	0.031 ①	0.2339-0.2350	0.2339-0.2343
	3.0	AVK	45	45	NA	NA	0.031 ①	0.031 ①	0.2338-0.2346	0.2338-0.2343
	3.7	AEW	45	45	NA	NA	0.039 ①	0.051 ①	NA	NA
	4.2	ABZ	45	45	NA	NA	0.039 ①	0.051 ①	NA	NA
2003	1.8	AMB	45	45	NA	NA	0.031 ①	0.031 ①	0.2343-0.2350	0.2339-0.2343
	1.8	AMU	45	45	NA	NA	0.031 ①	0.031 ①	NA	NA
	1.8	AWP	45	45	NA	NA	0.031 ①	0.031 ①	NA	NA
	2.7	APB	45	45	NA	NA	0.031 ①	0.031 ①	0.2339-0.2350	0.2339-0.2343
	3.0	AVK	45	45	NA	NA	0.031 ①	0.031 ①	0.2338-0.2346	0.2338-0.2343
	4.2	AWN	45	45	NA	NA	0.031 ①	0.031 ①	0.2346-0.2350	0.2339-0.2343
2004	1.8	AMB	45	45	NA	NA	0.031 ①	0.031 ①	0.2343-0.2350	0.2339-0.2343
	1.8	AMU	45	45	NA	NA	0.031 ①	0.031 ①	NA	NA
	1.8	BEA	45	45	NA	NA	0.031 ①	0.031 ①	NA	NA
	3.0	AVK	45	45	NA	NA	0.031 ①	0.031 ①	0.2338-0.2346	0.2338-0.2343
	3.2	BHE	45	45	NA	NA	0.031 ①	0.031 ①	0.2346	0.2343
	4.2	AWN	45	45	NA	NA	0.031 ①	0.031 ①	0.2346-0.2350	0.2339-0.2343
2005-06	1.8	AMB	45	45	NA	NA	0.031 ①	0.031 ①	0.2343-0.2350	0.2339-0.2343
	1.8	AWP	45	45	NA	NA	0.031 ①	0.031 ①	NA	NA
	1.8	BEA	45	45	NA	NA	0.031 ①	0.031 ①	NA	NA
	2.0	AXX	45	45	NA	NA	0.031 ①	0.031 ①	NA	NA
	2.0	BPY	45	45	NA	NA	0.031 ①	0.031 ①	NA	NA
	2.0	BPG	45	45	NA	NA	0.031 ①	0.031 ①	NA	NA
	3.0	BGN	45	45	NA	NA	0.031 ①	0.031 ①	0.2338-0.2346	0.2338-0.2343
	3.2	BUB	45	45	NA	NA	0.031 ①	0.031 ①	0.2346	0.2343
	3.2	BHE	45	45	NA	NA	0.031 ①	0.031 ①	0.2346	0.2343
	3.2	BKH	45	45	NA	NA	0.031 ①	0.031 ①	0.2346	0.2343
	4.2	BNK	45	45	NA	NA	0.031 ①	0.031 ①	0.2346-0.2350	0.2339-0.2343

NA: Not Available

① To measure: Insert a new valve into guide with end of valve flush with end of guide. Use a dial indicator to measure axial valve head movement.

06041_AUDI_C0008

PISTON AND RING SPECIFICATIONS

All measurements are given in inches

Year	Engine Size Liters	Engine ID/VIN	Piston Clearance	Ring Gap			Ring Side Clearance		
				Top Compression	Bottom Compression	Oil Control	Top Compression	Bottom Compression	Oil Control
2002	1.8	AEB	0.0014-0.0022	0.0078-0.0157	0.0078-0.0157	0.0098-0.0197	0.0035	0.0019-0.0315	0.0011-0.0023
	1.8	AMU	0.0005-0.0011	0.006-0.0157	0.006-0.0157	0.0098-0.0197	0.0028	0.0008-0.0028	0.0008-0.0023
	1.8	AWP	0.0005-0.0011	0.006-0.0157	0.006-0.0157	0.0098-0.0197	0.0028	0.0008-0.0028	0.0008-0.0023
	2.8	AHA	0.0008-0.0012	0.0140-0.0200	0.0200-0.0280	0.0100-0.0200	0.0010-0.0030	0.0100-0.0030	0.0010-0.0300
	3.0	AVK	0.0018-0.0012	0.0138-0.0200	0.0236-0.0315	0.0100-0.0200	0.0031	0.0008-0.0031	0.0008-0.0031
	3.7	AEW	NA	NA	NA	NA	NA	NA	NA
	4.2	ABZ	NA	NA	NA	NA	NA	NA	NA
2003	1.8	AMB	0.0017-0.0023	0.0078-0.0157	0.0078-0.0157	0.0098-0.0197	0.0035	0.0019-0.0315	0.0011-0.0023
	1.8	AMU	0.0005-0.0011	0.006-0.0157	0.006-0.0157	0.0098-0.0197	0.0028	0.0008-0.0028	0.0008-0.0023
	1.8	AWP	0.0005-0.0011	0.006-0.0157	0.006-0.0157	0.0098-0.0197	0.0028	0.0008-0.0028	0.0008-0.0023
	2.7	AHA	0.0008-0.0012	0.0140-0.0200	0.0200-0.0280	0.0100-0.0200	0.0010-0.0030	0.0100-0.0030	0.0010-0.0300
	3.0	AVK	0.0018-0.0012	0.0138-0.0200	0.0236-0.0315	0.0100-0.0200	0.0031	0.0008-0.0031	0.0008-0.0031
	4.2	ABZ	NA	NA	NA	NA	NA	NA	NA
2004	1.8	AMB	0.0017-0.0023	0.0078-0.0157	0.0078-0.0157	0.0098-0.0197	0.0035	0.0019-0.0315	0.0011-0.0023
	1.8	AWP	0.0005-0.0011	0.006-0.0157	0.006-0.0157	0.0098-0.0197	0.0028	0.0008-0.0028	0.0008-0.0023
	1.8	BEA	0.0005-0.0011	0.006-0.0157	0.006-0.0157	0.0098-0.0197	0.0028	0.0008-0.0028	0.0008-0.0023
	3.0	AVK	0.0018-0.0012	0.0138-0.0200	0.0236-0.0315	0.0100-0.0200	0.0031	0.0008-0.0031	0.0008-0.0031
	3.2	BHE	NA	0.0078-0.0157	0.0078-0.0157	0.0098-0.0197	0.0016-0.0035	0.0012-0.0023	0.0008-0.0023
	4.2	ABZ	NA	NA	NA	NA	NA	NA	NA
2005-06	1.8	AMB	0.0017-0.0023	0.0078-0.0157	0.0078-0.0157	0.0098-0.0197	0.0035	0.0019-0.0315	0.0011-0.0023
	1.8	AWP	0.0005-0.0011	0.006-0.0157	0.006-0.0157	0.0098-0.0197	0.0028	0.0008-0.0028	0.0008-0.0023
	1.8	BEA	0.0005-0.0011	0.006-0.0157	0.006-0.0157	0.0098-0.0197	0.0028	0.0008-0.0028	0.0008-0.0023
	2.0	AXX	NA	NA	NA	NA	NA	NA	NA
	2.0	BPY	NA	NA	NA	NA	NA	NA	NA
	2.0	BPG	NA	NA	NA	NA	NA	NA	NA
	3.0	BGN	0.0018-0.0012	0.0138-0.0200	0.0236-0.0315	0.0100-0.0200	0.0031	0.0008-0.0031	0.0008-0.0031
	3.2	BUB	NA	0.0078-0.0157	0.0078-0.0157	0.0098-0.0197	0.0016-0.0035	0.0012-0.0023	0.0008-0.0023
	3.2	BHE	NA	0.0078-0.0157	0.0078-0.0157	0.0098-0.0197	0.0016-0.0035	0.0012-0.0023	0.0008-0.0023
	3.2	BKH	NA	0.0078-0.0157	0.0078-0.0157	0.0098-0.0197	0.0016-0.0035	0.0012-0.0023	0.0008-0.0023
	4.2	ABZ	NA	NA	NA	NA	NA	NA	NA

NA - Not Available.

06041_AUDI_C0009

TORQUE SPECIFICATIONS

All readings in ft. lbs.

Year	Engine Displacement Liters	Engine ID/VIN	Cylinder Head Bolts	Main Bearing Bolts	Rod Bearing Bolts	Crankshaft Damper Bolts	Flywheel Bolts	Manifold Intake	Manifold Exhaust	Spark Plugs	Oil Drain Plug
2002	1.8	AEB	①	②	③	④	⑤	7	18	22	22
	1.8	AMU	⑥	⑦	③	④	⑧	7	18	22	22
	1.8	AWP	⑥	⑦	③	④	⑧	7	18	22	22
	2.8	AHA	①	⑦	③	④	⑤	15	18	22	22
	3.0	AVK	⑪	⑫	③	NA	⑬	7	18	22	22
	3.7	AEW	⑨	NA	NA	332	⑧	15	18	22	NA
	4.2	ABZ	⑨	NA	NA	332	⑧	15	18	22	26
2003	1.8	AEB	①	②	③	④	⑤	7	18	22	22
	1.8	AMU	⑥	⑦	③	④	⑧	7	18	22	22
	1.8	AWP	⑥	⑦	③	④	⑧	7	18	22	22
	2.7	APB	②	⑩	③	④	②	NA	NA	22	22
	3.0	AVK	⑪	⑫	③	NA	⑬	7	18	22	22
	4.2	AWN	⑨	NA	NA	332	⑧	15	18	22	26
2004	1.8	AEB	①	②	③	④	⑤	7	18	22	22
	1.8	AWP	⑥	⑦	③	④	⑧	7	18	22	22
	1.8	BEA	⑰	⑦	③	⑱	⑲	7	18	22	22
	3.0	AVK	⑬	⑫	③	NA	⑬	7	18	22	22
	3.2	BHE	⑭	⑮	③	⑯	⑧	10	17	22	22
	4.2	AWN	⑨	NA	NA	332	⑧	15	18	22	26
2005-06	1.8	AEB	①	②	③	④	⑤	7	18	22	22
	1.8	AWP	⑥	⑦	③	④	⑧	7	18	22	22
	1.8	BEA	⑰	⑦	③	⑱	⑬	7	18	22	22
	2.0	AXX	⑰	⑦	③	NA	⑬	9	18	22	22
	2.0	BPY	⑰	⑦	③	NA	⑬	9	18	22	22
	2.0	BPG	⑰	⑦	③	NA	⑬	9	18	22	22
	3.0	BGN	⑰	⑫	③	NA	⑬	7	18	22	22
	3.2	BUB	⑭	⑮	③	NA	⑬	9	18	22	22
	3.2	BHE	⑭	⑮	③	⑯	⑧	10	17	22	22
	3.2	BKH	⑰	⑮	③	⑯	—	10	17	22	22
	4.2	BNK	⑲	NA	NA	30	⑧	7	18	22	36

NA: Not Available

① Step 1: 44 ft. lbs.
Step 2: 90 degrees
Step 3: 90 degrees

② Step 1: 44 ft. lbs.
Step 2: 180 degrees

③ Step 1: 22 ft. lbs.
Step 2: 90 degrees

④ Center Bolt, installed with oil:
Step 1: 148 ft. lbs.
Step 2: 180 degrees
Damper Bolts: 15 ft. lbs.

⑤ Flywheel MT:
Step 1: AFC engine: 30 ft. lbs. ;
all others 44 ft. lbs.
Step 2: 90 degrees
Step 3: 90 degrees
Flexplate AT:
Step 1: 44 ft. lbs.
Step 2: 90 degrees

⑥ Step 1: 30 ft. lbs.
Step 2: Plus 90 degrees

⑦ Step 1: 48 ft. lbs.
Step 2: Plus 90 degrees

⑧ Step 1: 44 ft. lbs.
Step 2: Plus 90 degrees

⑨ Step 1: 30 ft. lbs.
Step 2: 44 ft. lbs.
Step 3: Plus 180 degrees

⑩ Step 1: 22 ft. lbs.
Step 2: 44 ft. lbs.
Step 3: Plus 90 degrees

⑪ Step 1: 30 ft. lbs.
Step 2: Plus 180 degrees

⑫ Cap Nuts: 26 ft. lbs. Plus 90 degrees
Side Bolts: 15 ft. lbs. Plus 180 degrees

⑬ 22.5 mm Bolt: 44 ft. lbs. Plus 90 degrees
35 & 43 mm Bolts: 44 ft.lbs. Plus 180 degrees

⑭ Step 1: 22 ft. lbs.
Step 2: 37 ft. lbs.
Step 3: Plus 90 degrees
Step 4: Plus 90 degrees

⑮ Step 1: 22 ft. lbs.
Step 2: Plus 90 degrees

⑯ Step 1: 74 ft. lbs.
Step 2: Plus 90 degrees

⑰ Step 1:29 ft. lbs.
Step 2: Plus 90 degrees
Step 3: Plus 90 degrees

⑱ Step 1:74 ft. lbs.
Step 2: Plus 90 degrees

⑲ Step 1: 22 ft. lbs.
Step 2: 44 ft. lbs.
Step 3: Plus 90 degrees
Step 4: Plus 90 degrees

06041_AUDI_C0010

WHEEL ALIGNMENT

Year	Model		Caster Range (+/-Deg.)	Caster Preferred Setting (Deg.)	Camber Range (+/-Deg.)	Camber Preferred Setting (Deg.)	Toe Setting (in.)
2002	A4 Standard Suspension	F	—	—	0.42	-0.50	0.16 +/- 0.03
		R	—	—	0.50	-1.17	0.50 +/- 0.16
	A4 Sport Suspension	F	—	—	0.42	-0.78	0.16 +/- 0.03
		R	—	—	0.50	-1.17	0.50 +/- 0.16
	A4 Heavy Duty Suspension	F	—	—	0.42	-0.35	0.16 +/- 0.03
		R	—	—	0.42	-0.35	0.50 +/- 0.16
	A6 ③ FWD	F	—	—	0.42	-0.83	0.18 +/- 0.03
		R	—	—	0.33	-1.50	0.13 +/- 0.08
	A6 ④ FWD	F	—	—	0.42	-1.09	0.18 +/- 0.03
		R	—	—	0.33	-1.50	0.13 +/- 0.08
	A6 ⑤ FWD	F	—	—	0.42	-0.58	0.18 +/- 0.03
		R	—	—	0.33	-1.50	0.18 +/- 0.03
	A6 ③ AWD	F	—	—	0.42	-0.83	0.16 +/- 0.03
		R	—	—	0.50	-0.67	0.18 +/- 0.03
	A6 ④ AWD	F	—	—	0.42	-0.83	0.16 +/- 0.03
		R	—	—	0.50	-0.67	018 +/- 0.03
	A6 ⑤ AWD	F	—	—	0.42	-0.83	0.16 +/- 0.03
		R	—	—	0.50	-0.67	0.18 +/- 0.03
	A6 ③ Quattro	F	—	—	0.42	-1.09	0.18 +/- 0.03
		R	—	—	0.50	-0.67	0.33 +/- 0.21
	A6 ④ Quattro	F	—	—	0.42	-1.08	0.18 +/- 0.03
		R	—	—	0.50	-1.00	0.33 +/- 0.21
	A6 ⑤ Quattro	F	—	—	0.42	-0.83	0.18 +/- 0.03
		R	—	—	0.50	-0.67	0.23 +/- 0.21
	A8	F	—	—	0.50	-0.83	018 +/- 0.03
		R	—	—	0.50	-0.67	0.10 +/- 0.08
	A8 Quattro	F	—	—	0.50	-0.75	018 +/- 0.03
		R	—	—	0.50	-0.67	0.10 +/- 0.08
	TT ③ FWD	F	—	—	0.50	-0.75	0.06 +/- 0.02
		R	—	—	0.33	-2.00	0.23 +/- 0.08
	TT ③ ⑥ AWD	F	—	—	0.50	-0.75	0.06 +/- 0.02
		R	—	—	0.33	-2.16	0.12 +/- 0.09
	TT ④ FWD	F	—	—	0.50	-0.97	0.06 +/- 0.02
		R	—	—	0.33	-2.00	0.32 +/- 0.24
	TT ④ ⑥ AWD	F	—	—	0.50	-0.97	0.06 +/- 0.02
		R	—	—	0.33	-2.46	0.12 +/- 0.02

① With aluminum mounting brackets

② Without aluminum mounting brackets

③ With standard suspension

④ With sport suspension

⑤ With heavy duty suspension

⑥ Rear camber at standing vehicle height of 13.8", measured from centerline of rear wheel to bottom of rear wheel opening.

⑦ Range is -0.75 to +0.50 degrees for FWD and AWD models.

⑧ With standard suspension, rear camber at standing vehicle height of 13.9", measured from centerline of rear wheel to bottom of rear wheel opening.

⑨ With sport suspension, rear camber at standing vehicle height of 13.1", measured from centerline of rear wheel to bottom of rear wheel opening.

⑩ Preferred setting is 0.075 in., with a range of +0.075 in. to -0.05 in.

06041_AUDI_C0011

WHEEL ALIGNMENT

Year	Model		Caster Range (+/-Deg.)	Caster Preferred Setting (Deg.)	Camber Range (+/-Deg.)	Camber Preferred Setting (Deg.)	Toe Setting (in.)
2003	A4	F	—	—	0.42	-0.50	0.16 +/- 0.03
	Standard Suspension	R	—	—	0.50	-1.17	0.50 +/- 0.16
	A4	F	—	—	0.42	-0.78	0.16 +/- 0.03
	Sport Suspension	R	—	—	0.50	-1.17	0.50 +/- 0.16
	A4	F	—	—	0.42	-0.35	0.16 +/- 0.03
	Heavy Duty Suspension	R	—	—	0.42	-0.35	0.50 +/- 0.16
	A6 ③	F	—	—	0.42	-0.83	0.18 +/- 0.03
	FWD	R	—	—	0.33	-1.50	0.13 +/- 0.08
	A6 ④	F	—	—	0.42	-1.09	0.18 +/- 0.03
	FWD	R	—	—	0.33	-1.50	0.13 +/- 0.08
	A6 ⑤	F	—	—	0.42	-0.58	0.18 +/- 0.03
	FWD	R	—	—	0.33	-1.50	0.18 +/- 0.03
	A6 ③	F	—	—	0.42	-0.83	0.16 +/- 0.03
	AWD	R	—	—	0.50	-0.67	0.18 +/- 0.03
	A6 ④	F	—	—	0.42	-0.83	0.16 +/- 0.03
	AWD	R	—	—	0.50	-0.67	018 +/- 0.03
	A6 ⑤	F	—	—	0.42	-0.83	0.16 +/- 0.03
	AWD	R	—	—	0.50	-0.67	0.18 +/- 0.03
	A6 ③	F	—	—	0.42	-1.09	0.18 +/- 0.03
	Quattro	R	—	—	0.50	-0.67	0.33 +/- 0.21
	A6 ④	F	—	—	0.42	-1.08	0.18 +/- 0.03
	Quattro	R	—	—	0.50	-1.00	0.33 +/- 0.21
	A6 ⑤	F	—	—	0.42	-0.83	0.18 +/- 0.03
	Quattro	R	—	—	0.50	-0.67	0.23 +/- 0.21
	A8	F	—	—	0.50	-0.83	018 +/- 0.03
		R	—	—	0.50	-0.67	0.10 +/- 0.08
	A8 Quattro	F	—	—	0.50	-0.75	018 +/- 0.03
		R	—	—	0.50	-0.67	010 +/- 0.08
	TT ③	F	—	—	0.50	-0.75	0.06 +/- 0.02
	FWD	R	—	—	0.33	-2.00	0.23 +/- 0.08
	TT ③ ⑥	F	—	—	0.50	-0.75	0.06 +/- 0.02
	AWD	R	—	—	0.33	-2.16	0.12 +/- 0.09
	TT ④	F	—	—	0.50	-0.97	0.06 +/- 0.02
	FWD	R	—	—	0.33	-2.00	0.32 +/- 0.24
	TT ④ ⑥	F	—	—	0.50	-0.97	0.06 +/- 0.02
	AWD	R	—	—	0.33	-2.46	0.12 +/- 0.02

① With aluminum mounting brackets

② Without aluminum mounting brackets

③ With standard suspension

④ With sport suspension

⑤ With heavy duty suspension

⑥ Rear camber at standing vehicle height of 13.8", measured from centerline of rear wheel to bottom of rear wheel opening.

⑦ Range is -0.75 to +0.50 degrees for FWD and AWD models.

⑧ With standard suspension, rear camber at standing vehicle height of 13.9", measured from centerline of rear wheel to bottom of rear wheel opening.

⑨ With sport suspension, rear camber at standing vehicle height of 13.1", measured from centerline of rear wheel to bottom of rear wheel opening.

⑩ Preferred setting is 0.075 in., with a range of +0.075 in. to -0.05 in.

06041_AUDI_C0012

WHEEL ALIGNMENT

Year	Model		Caster Range (+/-Deg.)	Caster Preferred Setting (Deg.)	Camber Range (+/-Deg.)	Camber Preferred Setting (Deg.)	Toe Setting (in.)
2004	A4	F	—	—	0.42	-0.50	0.16 +/- 0.03
	Standard Suspension	R	—	—	0.50	-1.17	0.50 +/- 0.16
	A4	F	—	—	0.42	-0.78	0.16 +/- 0.03
	Sport Suspension	R	—	—	0.50	-1.17	0.50 +/- 0.16
	A4	F	—	—	0.42	-0.35	0.16 +/- 0.03
	Heavy Duty Suspension	R	—	—	0.42	-0.35	0.50 +/- 0.16
	A6 ③	F	—	—	0.42	-0.83	0.18 +/- 0.03
	FWD	R	—	—	0.33	-1.50	0.13 +/- 0.08
	A6 ④	F	—	—	0.42	-1.09	0.18 +/- 0.03
	FWD	R	—	—	0.33	-1.50	0.13 +/- 0.08
	A6 ⑤	F	—	—	0.42	-0.58	0.18 +/- 0.03
	FWD	R	—	—	0.33	-1.50	0.18 +/- 0.03
	A6 ③	F	—	—	0.42	-0.83	0.16 +/- 0.03
	AWD	R	—	—	0.50	-0.67	0.18 +/- 0.03
	A6 ④	F	—	—	0.42	-0.83	0.16 +/- 0.03
	AWD	R	—	—	0.50	-0.67	018 +/- 0.03
	A6 ⑤	F	—	—	0.42	-0.83	0.16 +/- 0.03
	AWD	R	—	—	0.50	-0.67	0.18 +/- 0.03
	A6 ③	F	—	—	0.42	-1.09	0.18 +/- 0.03
	Quattro	R	—	—	0.50	-0.67	0.33 +/- 0.21
	A6 ④	F	—	—	0.42	-1.08	0.18 +/- 0.03
	Quattro	R	—	—	0.50	-1.00	0.33 +/- 0.21
	A6 ⑤	F	—	—	0.42	-0.83	0.18 +/- 0.03
	Quattro	R	—	—	0.50	-0.67	0.23 +/- 0.21
	A8	F	—	—	0.50	-0.83	018 +/- 0.03
		R	—	—	0.50	-0.67	0.10 +/- 0.08
	A8	F	—	—	0.50	-0.75	0.18 +/- 0.03
	Quattro	R	—	—	0.50	-0.67	0.10 +/- 0.08
	TT	F	—	—	⑦	—	0.08 +/- 0.08
	FWD Std. Suspension	R	—	—	0.33	-0.20	0.20 +/- 0.01
	TT	F	—	—	0.50	-0.75	0.08 +/- 0.08
	FWD Sport Suspension	R	—	—	0.20	-2.00	0.20 +/- 0.05
	TT	F	—	—	⑧	—	0.08 +/- 0.08
	AWD	R	—	—	⑧⑨	⑧⑨	⑩

① With aluminum mounting brackets

② Without aluminum mounting brackets

③ With standard suspension

④ With sport suspension

⑤ With heavy duty suspension

⑥ Rear camber at standing vehicle height of 13.8", measured from centerline of rear wheel to bottom of rear wheel opening.

⑦ Range is -0.75 to +0.50 degrees for FWD and AWD models.

⑧ With standard suspension, rear camber at standing vehicle height of 13.9", measured from centerline of rear wheel to bottom of rear wheel opening.

⑨ With sport suspension, rear camber at standing vehicle height of 13.1", measured from centerline of rear wheel to bottom of rear wheel opening.

⑩ Preferred setting is 0.075 in., with a range of +0.075 in. to -0.05 in.

06041_AUDI_C0013

WHEEL ALIGNMENT

			Caster		Camber		
Year	Model		Range (+/-Deg.)	Preferred Setting (Deg.)	Range (+/-Deg.)	Preferred Setting (Deg.)	Toe Setting (in.)
2005-06	A3	F	—	—	0.30	0.30	0.16 +/- 0.03
	Standard Suspension	R	—	—	0.20	0.20	0.50 +/- 0.16
	A3	F	—	—	0.41	-0.78	0.16 +/- 0.03
	Sport Suspension	R	—	—	0.20	-1.17	0.50 +/- 0.16
	A3	F	—	—	0.42	-0.35	0.16 +/- 0.03
	Heavy Duty Suspension	R	—	—	0.42	-0.35	0.50 +/- 0.16
	A4	F	—	—	0.42	-0.50	0.16 +/- 0.03
	Standard Suspension	R	—	—	0.50	-1.17	0.50 +/- 0.16
	A4	F	—	—	0.42	-0.78	0.16 +/- 0.03
	Sport Suspension	R	—	—	0.50	-1.17	0.50 +/- 0.16
	A4	F	—	—	0.42	-0.35	0.16 +/- 0.03
	Heavy Duty Suspension	R	—	—	0.42	-0.35	0.50 +/- 0.16
	A6 ③	F	—	—	0.42	-0.83	0.18 +/- 0.03
	FWD	R	—	—	0.33	-1.50	0.13 +/- 0.08
	A6 ④	F	—	—	0.42	-1.09	0.18 +/- 0.03
	FWD	R	—	—	0.33	-1.50	0.13 +/- 0.08
	A6 ⑥	F	—	—	0.42	-0.58	0.18 +/- 0.03
	FWD	R	—	—	0.33	-1.50	0.18 +/- 0.03
	A6 ③	F	—	—	0.42	-0.83	0.16 +/- 0.03
	AWD	R	—	—	0.50	-0.67	0.18 +/- 0.03
	A6 ④	F	—	—	0.42	-0.83	0.16 +/- 0.03
	AWD	R	—	—	0.50	-0.67	018 +/- 0.03
	A6 ⑤	F	—	—	0.42	-0.83	0.16 +/- 0.03
	AWD	R	—	—	0.50	-0.67	0.18 +/- 0.03
	A6 ③	F	—	—	0.42	-1.09	0.18 +/- 0.03
	Quattro	R	—	—	0.50	-0.67	0.33 +/- 0.21
	A6 ④	F	—	—	0.42	-1.08	0.18 +/- 0.03
	Quattro	R	—	—	0.50	-1.00	0.33 +/- 0.21
	A6 ⑤	F	—	—	0.42	-0.83	0.18 +/- 0.03
	Quattro	R	—	—	0.50	-0.67	0.23 +/- 0.21
	TT	F	—	—	⑦	—	0.08 +/- 0.08
	FWD Std. Suspension	R	—	—	0.33	-0.20	0.20 +/- 0.01
	TT	F	—	—	0.50	-0.75	0.08 +/- 0.08
	FWD Sport Suspension	R	—	—	0.20	-2.00	0.20 +/- 0.05
	TT	F	—	—	⑧	—	0.08 +/- 0.08
	AWD	R	—	—	⑧⑨	⑧⑨	⑩

① With aluminum mounting brackets

② Without aluminum mounting brackets

③ With standard suspension

④ With sport suspension

⑤ With heavy duty suspension

⑥ Rear camber at standing vehicle height of 13.8", measured from centerline of rear wheel to bottom of rear wheel opening.

⑦ Range is -0.75 to +0.50 degrees for FWD and AWD models.

⑧ With standard suspension, rear camber at standing vehicle height of 13.9", measured from centerline of rear wheel to bottom of rear wheel opening.

⑨ With sport suspension, rear camber at standing vehicle height of 13.1", measured from centerline of rear wheel to bottom of rear wheel opening.

⑩ Preferred setting is 0.075 in., with a range of +0.075 in. to -0.05 in.

06041_AUDI_C0014

TIRE, WHEEL AND BALL JOINT SPECIFICATIONS

Year	Model	OEM Tires		Tire Pressures (psi)		Wheel	Ball Joint	Lug Nut
		Standard	Optional	Front	Rear	Size	Inspection	(ft. lbs.)
2002	A8	225/60HR16	225/55WR17	34	34	Std: 7-J Opt: 8-J	NS	120
	A6	195/65HR15	205/55HR16	34	34	7-J	NS	120
	A4	205/60HR15	205/55ZR16	34	34	7-J	NS	120
	TT	205/55WR16	225/45YR17	34	34	7-J	NS	120
2003	A8	225/60HR16	225/45R18 245/45R18	34	34	Std: 7-J Opt: 8-J	NS	120
	A6	205/55R16	235/45R17	34	34	7-J	NS	120
	A4	205/65HR15	215/55HR16	34	34	7-J	NS	120
	TT	205/55WR16	225/45YR17	34	34	7-J	NS	120
2004	A8	225/60HR16	225/45R18 245/45R18	34	34	Std: 7-J Opt: 8-J	NS	120
	A6	205/55R16	235/45R17	34	34	7-J	NS	120
	A4	205/65HR15	215/55HR16	34	34	7-J	NS	120
	TT	205/55WR16	225/45YR17	34	34	7-J	NS	120
2005-06	A8	225/60HR16	225/45R18 245/45R18	34	34	Std: 7-J Opt: 8-J	NS	120
	A6	205/55R16	235/45R17	34	34	7-J	NS	120
	A4	205/65HR15	215/55HR16	34	34	7-J	NS	120
	A3	205/65HR15	215/55HR16	34	34	7-J	NS	120
	TT	205/55WR16	225/45YR17	34	34	7-J	NS	120

OEM: Original Equipment Manufacturer

PSI: Pounds Per Square Inch

STD: Standard

OPT: Optional

NS: Not Specified by manufacturer

06041_AUDI_C0015

BRAKE SPECIFICATIONS

All measurements in inches unless noted

			Brake Disc			Minimum Lining Thickness		Brake Caliper	
Year	Model		Original Thickness	Minimum Thickness	Maximum Runout	Front	Rear	Bracket Bolts (ft. lbs.)	Mounting Bolts (Nm.)
2002	A4 Sedan	F	①	②	0.002	③	—	170	30
	and Avant	R	0.390	0.310	0.002	—	0.080	75	35
	A6 Sedan	F	0.984	0.906	0.002	0.078	—	89	18
	and Avant	R	0.394	0.315	0.002	—	0.079	70	26
	A8	F	0.984	0.905	0.002	0.078	—	—	18
		R	0.390	0.315	0.002	—	0.079	⑥	26
	A8 Quattro	F	1.181	1.102	0.002	0.078	—	—	18
		R	0.866	0.787	0.002	—	0.079	⑥	26
	TT	F	0.984	0.905	0.002	0.078	—	92	21
		R	0.354	0.275	0.002	—	0.079	50	26
2003	A4 Sedan	F	①	②	0.002	③	0.275	170	30
	and Avant	R	0.394	0.315	0.002	—	0.275	75	35
	A6 Sedan	F	0.984	0.906	0.002	0.078	—	96	18
	and Avant	R	0.394	0.315	0.002	—	0.079	70	26
	A8	F	0.984	0.905	0.002	0.078	—	—	18
		R	0.390	0.315	0.002	—	0.079	⑥	26
	TT	F	0.984	0.905	0.002	0.078	—	92	21
		R	0.354	0.275	0.002	—	0.079	50	26
2004	A4 Sedan	F	①	②	0.002	③	0.275	144	18
	and Avant	R	0.394	0.315	0.002	—	0.275	103	25
	A6 Sedan	F	0.984	0.906	0.002	0.078	—	96	18
	and Avant	R	0.394	0.315	0.002	—	0.079	70	26
	A8	F	0.984	0.905	0.002	0.078	—	—	18
		R	0.390	0.315	0.002	—	0.079	⑥	26
	TT	F	⑦	⑧	0.002	0.551	—	92	21
		R	⑨	⑩	0.002	—	0.079	50	26

06041_AUDI_C0016

BRAKE SPECIFICATIONS

All measurements in inches unless noted

			Brake Disc			Minimum Lining Thickness		Brake Caliper	
Year	Model		Original Thickness	Minimum Thickness	Maximum Runout	Front	Rear	Bracket Bolts (ft. lbs.)	Mounting Bolts (Nm.)
2005-06	A4 Sedan	F	①	②	0.002	③	0.275	144	18
	and Avant	R	0.394	0.315	0.002	—	0.275	103	25
	A3 Sedan	F	⑪	⑬	0.002	⑮	0.023	144	22
	and Avant	R	L	N	0.002	—	0.023	66	25
	A6 Sedan	F	0.984	0.906	0.002	0.078	—	144	18
	and Avant	R	0.394	0.315	0.002	—	0.079	103	25
	A8	F	0.984	0.905	0.002	0.078	—	—	18
		R	0.390	0.315	0.002	—	0.079	⑥	26
	TT	F	⑦	⑧	0.002	0.551	—	92	20
		R	⑨	⑩	0.002	—	0.079	47	25

① Teves/Ate Calipers:
Ventilated Disc: 0.984 inches
Non-ventilated Disc: 0.590 inches
Lucas Calipers:
Ventilated Disc: 0.870 inches
Non-ventilated Disc: 0.590 inches
Double Piston Calipers:
Ventilated Disc: 1.180 inches

③ Teves/Ate and Lucas Calipers: 0.080 inches
Double Piston Calipers: 0.12 inches

④ Teves/ATE: 18 ft. lbs.
Lucas: 22 ft. lbs.
Double Piston Calipers: 148 ft. lbs.

⑤ FWD Models (Ribbed Bolt): 70 ft. lbs.
AWD Models (Socket-head Bolt): 44 ft. lbs.

⑥ Steel wheel carrier: 44 ft. lbs.
Aluminum wheel carrier: 51 ft. lbs. Plus 90 deg.

⑦ Exc. P/N 1ZT: 0.984 inches
P/N 1ZT: 1.26 inches

⑧ Exc. P/N 1ZT: 0.905 inches
P/N 1ZT: 1.181 inches

⑨ Non-ventilated Disc: 0.354 inches
Ventilated Disc: 0.866 inches

⑩ Non-ventilated Disc: 0.276 inches
Ventilated Disc: 0.787 inches

⑪ C54 15 inch :
FN 3 15 inch
FN 3 16 inch
FNR-G60 16 inch
FNR-G60 17 inch
Brembo 18 inch

⑫ C38 15 inch
C38 16 inch
C41 16 inch
C43 16 inch
C43 18 inch

⑬ C54: 0.790 inch
FN 3 15in.: 0.980 inch
FN 3 16 in: 0.980 inch
FNR-G60 16 in: 1.180 inc
FNR-G60 17 in: 1.180 inch
Brembo 18 inch: 1.340 inch

⑭ C38 15 inch: 0.31 inch
C38 16 inch: 0.39 inch
C41 16 inch: 0.39 inch
C43 16 inch: 0.79 inch
C43 18 inch: 0.79 inch

⑮ C54: 0.063 inch
FN 3 15in.: 0.066 inch
FN 3 16 in: 0.066 inch
FNR-G60 16 in: 0.066 inc
FNR-G60 17 in: 0.066 inch
Brembo 18 inch: not avaialable

06041_AUDI_C0017

SCHEDULED MAINTENANCE INTERVALS

Audi—TT, A3, A4, A6 and A8

TO BE SERVICED	TYPE OF SERVICE	VEHICLE MILEAGE INTERVAL (x1000)												
		5	10	20	30	40	50	60	70	80	90	100	105	120
Engine oil & filter ①	R	✓	✓	✓	✓	✓	✓	✓	✓	✓	✓	✓	✓	✓
Service reminder reset	S/I	✓	✓	✓	✓	✓	✓	✓	✓	✓	✓	✓	✓	✓
Rotate wheels	S/I	✓												
Fluid levels	S/I	✓	✓	✓	✓	✓	✓	✓	✓	✓	✓	✓	✓	✓
Auto shift lock	S/I	✓	✓	✓	✓	✓	✓	✓	✓	✓	✓	✓	✓	✓
Brake system	S/I	✓	✓	✓	✓	✓	✓	✓	✓	✓	✓	✓	✓	✓
Brake fluid	R	①	①	①	①	①	①	①	①	①	①	①	①	①
M/T shift & clutch interlock	S/I	✓	✓	✓	✓	✓	✓	✓	✓	✓	✓	✓	✓	✓
Cooling system	S/I		✓	✓	✓	✓	✓	✓	✓	✓	✓	✓	✓	✓
Exhaust system	S/I		✓	✓	✓	✓	✓	✓	✓	✓	✓	✓	✓	✓
ODB check for codes	S/I		✓	✓	✓	✓	✓	✓	✓	✓	✓	✓	✓	✓
Door hinges - lubricate	S/I		✓	✓	✓	✓	✓	✓	✓	✓	✓	✓	✓	✓
Battery level	S/I		✓	✓	✓	✓	✓	✓	✓	✓	✓	✓	✓	✓
Windshield washer fluid	S/I		✓	✓	✓	✓	✓	✓	✓	✓	✓	✓	✓	✓
Tire condition & pressure	S/I		✓	✓	✓	✓	✓	✓	✓	✓	✓	✓	✓	✓
Drive shaft boots	S/I		✓	✓	✓	✓	✓	✓	✓	✓	✓	✓	✓	✓
Road test	S/I		✓	✓	✓	✓	✓	✓	✓	✓	✓	✓	✓	✓
Lighting	S/I		✓	✓	✓	✓	✓	✓	✓	✓	✓	✓	✓	✓
Engine for leaks	S/I			✓		✓		✓		✓		✓		✓
Front axle dust seals on ball joints & tie rod ends	S/I			✓		✓		✓		✓		✓		✓
Haldex clutch oil & filter ②	R			✓		✓		✓		✓		✓		✓
Transmission for leaks	S/I			✓		✓		✓		✓		✓		✓
Multronic trans fluid	R					✓				✓				✓
MT final drive fluid	S/I			✓		✓		✓		✓		✓		✓
AT final drive fluid	S/I					✓				✓				✓
Dust/pollen filter	R			✓		✓		✓		✓		✓		✓
Sliding roof rails	S/I			✓		✓		✓		✓		✓		✓
PS fluid	S/I					✓				✓				✓
Air cleaner element	R					✓				✓				✓
Spark plugs	R					✓				✓				✓
Serpentine belt	R									✓				
Timing belt 1.8L Turbo ③	R													✓
Timing belt 2.0 2.7L & 4.2L	R												✓	
Timing belt & tensioner roller 2.8L	R												✓	

R: Replace S/I: Service or Inspect

① Replace fluid every two years, regardless of mileage.

② TT Quattro 2001 only.

③ 1.8L Turbo, exc. TT models; on TT models with 1.8L Turbo & 3.2L V6, replace every 105K miles.

06041_AUDI_C0018

ENGINE REPAIR

➡Disconnecting the negative battery cable on some vehicles may interfere with the functions of the on-board computer systems and may require the computer to undergo a relearning process, once the negative battery cable is disconnected.

Alternator

REMOVAL

A3

2.0L ENGINE

1. Before servicing the vehicle, refer to the precautions section.
2. Disconnect the negative battery cable.
3. Disconnect the electrical connection at Mass Air Flow (MAF) sensor.
4. Remove the engine cover.
5. Lift the EVAP canister with the hoses still connected out of bracket and set aside.

➡Before removing accessory drive belt, mark direction of rotation. Belt damage will result if not reinstalled in proper direction.

6. Remove the accessory drive belt tension by turning tensioner in direction of.

Lock tensioner in place using mandrel T10060 A as illustrated.

7. Remove the accessory drive belt.
8. Remove the bolts and remove tensioner.
9. Remove the bolts and remove alternator from bracket.
10. Swing alternator to right.
11. Disconnect the electrical connection.
12. Remove the B+ terminal nut and terminal eyelet.
13. Remove the cable retainer.
14. Remove the alternator.

3.2L ENGINE

1. Before servicing the vehicle, refer to the precautions section.
2. Disconnect the negative battery cable.
3. Remove the right front wheel.
4. Remove the noise insulation.
5. Remove the cover for air duct and the air duct.
6. Pull off vacuum hose at air intake hose.
7. Remove the air hose from Throttle Valve Control Module.
8. Disconnect the electrical connection at Mass Air Flow (MAF) sensor.
9. Remove the top section of air cleaner housing.
10. Remove the filter element.
11. Remove the screws and remove bottom section of air cleaner housing.
12. Separate secondary air hose at point indicated.
13. Set aside air hose to secondary air pump.
14. Remove the top screws at radiator cowl.

➡Before removing accessory drive belt, mark direction of rotation. Belt damage will result if not reinstalled in proper direction.

15. Remove the accessory drive belt.
16. Move aside and retain lower coolant hose at radiator cowl.
17. Disconnect the electrical connection at Secondary Air Injection (AIR) Pump Motor set aside harness.
18. Remove the bolt and remove bracket for coolant pipe.
19. Remove the bolts.
20. Remove the bolt and remove secondary air pump with bracket.
21. Disconnect the electrical connections for coolant fan.
22. Remove the lower screws for radiator cowl.
23. Pull out radiator cowl with both fans from below.
24. Remove the electrical connection for a/c clutch and set aside wiring.
25. Remove the bolts for air conditioner compressor and remove compressor with the hoses still attached from bracket.
26. Disconnect the electrical connection.
27. Remove the B+ terminal nut and terminal eyelet.
28. Remove the upper idler roller.
29. Loosen the bolts and.

➡The alternator can only be removed with upper bolt still in place. If alternator is difficult to remove from mounting, reinstall the mounting bolts loosely and carefully strike bolt heads with a rubber mallet in order to loosen the mounting threaded bushings.

30. Pull alternator away from bracket.
31. Remove the alternator from lower left engine compartment.

Except A6 2.7L and 3.0L, A6 and A8 4.2L and TT

1. Before servicing the vehicle, refer to the precautions section.
2. Remove or disconnect the following:
   - Negative battery cable
   - All covers necessary for access
   - Alternator drive belt
   - Alternator wiring harness connectors
   - Alternator

A6

2.7L AND 3.0L

1. Before servicing the vehicle, refer to the precautions section.
2. Remove or disconnect the following:
   - Air intake channel
   - Engine undercover
   - Accessory drive belt
   - Hose to charge air cooler
   - Alternator wiring harness
   - A/T and A/C lines
   - Alternator bolts
   - Alternator

A6 and A8

4.2L

1. Before servicing the vehicle, refer to the precautions section.
2. Remove or disconnect the following:
   - Negative battery cable
   - Air cleaner assembly (A8)
   - Engine undercover
   - Alternator air duct hose
   - Alternator wiring harness
   - Accessory drive belt
   - Alternator bolts
   - Alternator

TT

2002–03 1.8L ENGINE

1. Before servicing the vehicle, refer to the precautions section.
2. Remove or disconnect the following:
   - Negative battery cable
   - Electrical connections at throttle valve control module and charge pressure sender
   - Air and vacuum hoses at throttle valve control
   - Carbon canister
   - Alternator wiring harness
   - Accessory drive belt
   - Alternator bolts
   - Alternator

2004–06 1.8L ENGINE

1. Before servicing the vehicle, refer to the precautions section.

⁂ CAUTION

Before beginning repairs on the electrical system:

- Obtain the anti-theft radio security code.
- Switch off all electrical components.
- Switch ignition off and remove the ignition key.
- Disconnect negative battery terminal.

2. Remove the engine cover and cover in front of the intake manifold.
3. Release the fasteners and remove the left lock carrier cover.
4. Release the fasteners and remove the right lock carrier cover and expansion tank cover.
5. Before removing drive belt, mark direction of rotation. Belt damage will result if not reinstalled in proper direction.
6. Remove the drive belt.
7. Remove the secondary air system pipe from brackets at engine.
8. Remove the bolts and pull off holder from guide tube for dipstick.
9. On the underside of the holder, disconnect all electrical connections.
10. Disconnect the vacuum hose to solenoid valves at the intake manifold.
11. Set aside the holder with hoses still connected.
12. Pull the air hoses off of the secondary air pump.
13. Disconnect electrical connector.
14. Remove the nut from alternator terminal B+ and remove the connector eyelet from terminal.
15. Remove the harness retainer at alternator.
16. Remove the bolts and the alternator.

3.2L

1. Before servicing the vehicle, refer to the precautions section.

⁂ CAUTION

Before beginning repairs on the electrical system:

- Obtain the anti-theft radio security code.
- Switch off all electrical components.
- Switch ignition off and remove the ignition key.
- Disconnect negative battery terminal.

2. Disconnect the negative battery cable.
3. Release the fastener and remove the cover.
4. Before removing drive belt, mark direction of rotation. Belt damage will result if not reinstalled in proper direction.
5. Thread M8 x 40 bolt into the tensioner enough to relieve drive belt tension.
6. Remove the drive belt.
7. Remove the upper bolt.
8. Remove the screws and the center noise insulation panel.
9. Remove the screws and the right and left noise insulation.
10. Remove the bolts and the cross brace.
11. Remove the bolts from the power steering pump and position the pump aside with the lines attached.
12. Disconnect A/C compressor electrical connector, remove the bolts and position aside with the hoses attached.
13. Remove the alternator lower bolt.
14. Remove the alternator from bracket leaving the wiring connected.
15. Remove the harness retainer at alternator.
16. Remove the nut from alternator terminal B+ and the connector eyelet from terminal.
17. Disconnect the electrical connector and remove the alternator.

INSTALLATION

A3

2.0L ENGINE

1. Installation is the reverse order of removal.
2. Tighten the alternator to bracket bolts and belt tensioner to bracket bolts to 23 Nm.

3.2L ENGINE

1. Installation is the reverse order of removal.
2. Tighten the alternator to bracket bolts to 23 Nm.
3. Tighten the tensioning roller bracket bolt to 50 Nm.
4. Tighten the A/C compressor bracket bolt to 25 Nm.

Except A6 2.7L and 3.0L, A6, A8 4.2L and TT

1. Before servicing the vehicle, refer to the precautions section.
2. Install or connect the following:
   - Alternator. Torque the bolt to 18 ft. lbs. (25 Nm).
   - Alternator wiring harness connectors
   - Alternator drive belt
   - Adjust the alternator belt
   - All covers removed for access
   - Negative battery cable

A6

2.7L AND 3.0L ENGINE

1. Before servicing the vehicle, refer to the precautions section.
2. Install or connect the following:
   - Alternator. Tighten the 8mm bolt to 16 ft. lbs. (22 Nm) and the 10mm bolts to 33 ft. lbs. (45 Nm).
   - A/T and A/C lines
   - Alternator wiring harness
   - Hose to charge air cooler
   - Accessory drive belt
   - Engine undercover
   - Air intake channel
   - Negative battery cable

A6 and A8

4.2L ENGINE

1. Before servicing the vehicle, refer to the precautions section.
2. Install or connect the following:
   - Alternator. Tighten the alternator to engine bolts to 22 Nm, tighten the tensioner roller to bracket bolts to 22 Nm.
   - Accessory drive belt
   - Alternator wiring harness. Tighten the **B** terminal to 12 ft. lbs. (16 Nm) and the **D** to 9 Nm.
   - Alternator air duct hose
   - Engine undercover
   - Air cleaner assembly
   - Negative battery cable

TT

2002–03 1.8L ENGINE

1. Before servicing the vehicle, refer to the precautions section.
2. Install or connect the following:
   - Alternator. Torque the bolt to 18 ft. lbs. (25 Nm).
   - Alternator wiring harness
   - Accessory drive belt
   - Carbon canister
   - Air and vacuum hoses at throttle valve control
   - Electrical connections at throttle valve control module and charge pressure sender
   - Negative battery cable

2004–06 1.8L ENGINE

1. Installation is the reverse of removal, please note the following:

2. Press the sleeves into alternator mounting to aid in installation.

3. Before installing the drive belt, make sure that all subassemblies (alternator, a/c compressor etc.) are securely mounted and turn freely. When installing belt, ensure correct seating in the belt pulleys!

4. Note running direction of drive belt as previously marked and route as illustrated.

5. Tighten the alternator mounting bolts to 16 ft. lbs. (23 Nm).

6. Start engine and check for proper operation.

3.2L ENGINE

1. Installation is the reverse of removal, please note the following:

2. Before installing the drive belt, make sure that all subassemblies (alternator, a/c compressor etc.) are securely mounted and turn freely. When installing belt, ensure correct seating in the belt pulleys!

3. Note running direction of drive belt as previously marked and route as illustrated.

4. Tighten the alternator mounting bolts to 16 ft. lbs. (23 Nm).

5. Start engine and check for proper operation.

Ignition Timing

All vehicles in this section are equipped with distributorless ignition systems. No adjustments are possible.

Engine Assembly

REMOVAL & INSTALLATION

Except TT

1.8L ENGINE

➡To allow clearance for removal of the engine assembly, the front bumper and the hood lock carrier assembly must be removed from the front of the vehicle.

1. Before servicing the vehicle, refer to the precautions section.

2. Turn the ignition switch to the **OFF** position, and then disconnect the negative battery cable.

3. Position the wipers to the vertical position.

4. Properly relieve the fuel system pressure.

5. Drain the engine coolant.

6. Remove or disconnect the following:

- Negative battery cable
- Lower engine splash shield
- Water pump housing drain plug, drain coolant and reinstall
- Left and right lower bumper air grilles
- Inner fender lining
- Front bumper
- Power steering cooling coil
- Transaxle oil cooling lines
- Electric cooling fan thermal switch
- Air intake duct/assembly
- Headlight height adjuster wiring harness
- Turn signal bulb sockets
- Power steering fluid reservoir cap/dipstick
- Anti-lock Brake System (ABS) hydraulic unit wiring harness
- Horn electrical connectors
- Air guides at the left and right sides of the radiator
- Wires or connectors that would inhibit hood lock carrier assembly removal
- Hood release cable
- Air guide between the lock carrier and air filter
- Outside temperature electrical connector
- Upper and lower radiator hoses from the radiator
- Outside air temperature sensor/cooling line
- Power steering hydraulic cooling line bracket without disconnecting the fluid lines
- Air conditioning condenser assembly without disconnecting the pressure hoses

⁂ CAUTION

Use care to not kink the A/C lines. Do not allow the condenser to hang on the lines.

- Outside air temperature sensor/cooling line bracket
- Power steering hydraulic cooling line bracket from bottom of the radiator. Do not open the fluid lines
- Hood weather seal
- Front hood lock carrier
- Air conditioning low-pressure switch
- Green harness connector from the air conditioning compressor magnetic clutch
- Engine covers
- Wiring harness connectors for the wastegate bypass regulator valve
- EVAP canister purge regulator valve
- Power output stage

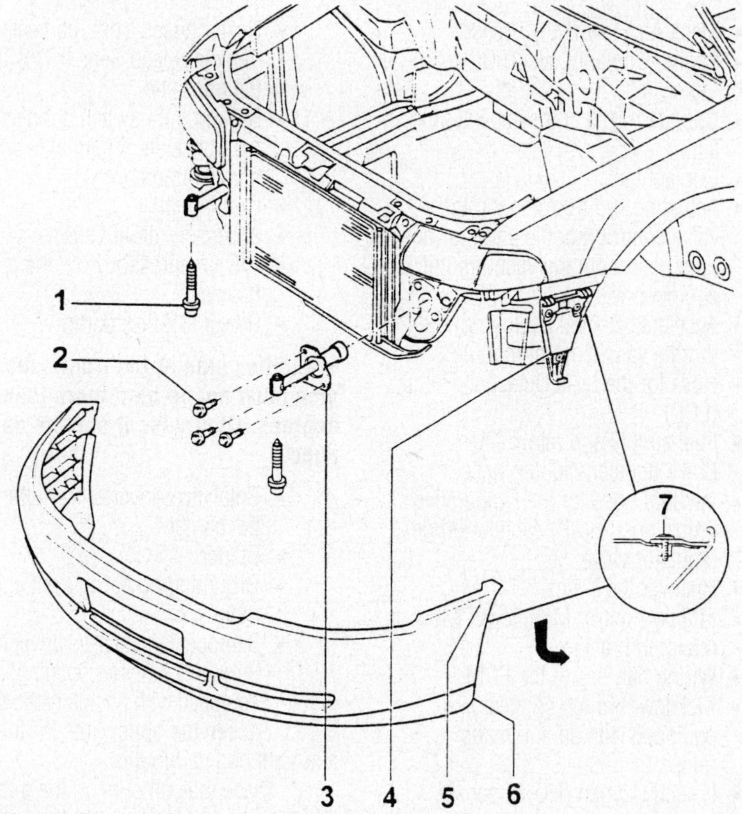

Bumper mounting fastener (1), energy absorbing strut fasteners (2), strut assembly (3), saddle (4), bumper cover (5), spoiler (6) and liner fastener (7)—A4 models

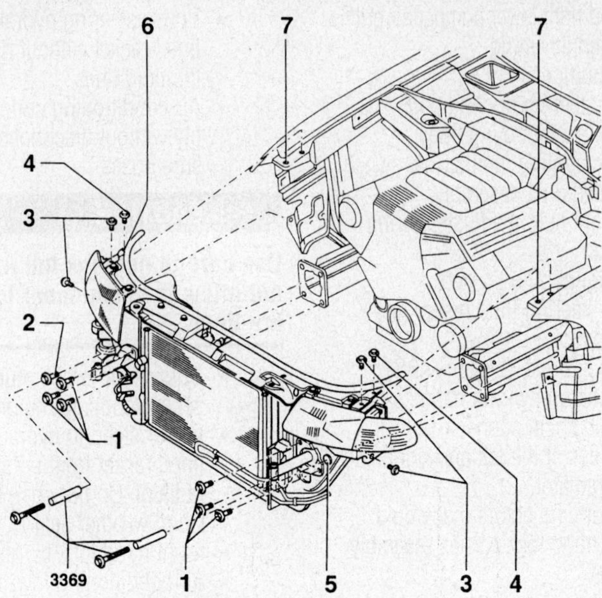

1. Bolts 33 ft. lbs. (45 Nm)
2. Bolts 33 ft. lbs. (45 Nm)
3. Bolts 7 ft. lbs. (10 Nm)
4. Bolts 7 ft. lbs. (10 Nm)
5. Bore for support tool
6. Lock carrier bore
7. Fender bore

Exploded view of the hood lock carrier assembly—A4 models

- Mass Air Flow (MAF) sensor
- Engine Coolant Level (ECL) warning switch
- Coolant hoses at the expansion tank
- Coolant tank
- Actuating rod from the throttle valve control module and the vacuum hose from the vacuum unit, if equipped with cruise control
- Accelerator pedal cable from the throttle valve control module
- Hose for the Leak Detection Pump (LDP)
- Fuel supply and return lines
- Brake booster vacuum hose
- Vacuum hose for the Evaporative Emissions (EVAP) canister purge regulator valve
- Uncover the E-box
- Engine Control Module (ECM) retaining bracket
- Wiring harness to the ECM
- Kickdown switch connector, if equipped with an automatic transaxle
- Heated Oxygen (HO_2S) sensor wiring harness
- Ground connection at the plenum chamber
- Heater hoses from the heater core
- Vehicle Speed Sensor (VSS) from the transaxle
- Backup light switch connector from the transaxle, if equipped with a manual transaxle
- Cooling fan
- Accessory drive belts
- A/C compressor from the mounting bracket
- Power steering pump

➡The flex pipe at the front exhaust pipe must not be bent more than 10 degrees. Otherwise it may be damaged.

- Catalytic converter from the turbocharger
- Starter
- Ground strap at the right engine mount
- 3 torque converter-to-driveplate bolts through the starter opening, if equipped with an automatic transaxle

7. Loosen the upper nuts for the left and right engine mounts.

8. Place matchmarks on the threaded bolt and centering sleeves at the bottom of the left and right engine mounts, then remove the mounting nuts.

9. Remove or disconnect the following:
   - Lower engine-to-transaxle mounting bolts
   - Cooler line bracket form the left side of the engine, if equipped with an automatic transaxle
   - Upper nuts from the engine mounts

10. Position an Engine Support Bridge tool 10-222A to the bolted flanges of the fenders with the spindle facing forward.

11. Attach the Support Adapter tool 3147 to the bolt hole above the starter area in the transaxle bell housing.

12. Connect the Engine Support Adapter tool 3147 to the Engine Support Bridge tool 10-222A using Adapter tool 2024A/1 and Extension tool 2024A/2 and support the transaxle.

13. Attach an engine sling between the engine and the hoist.

14. Remove or disconnect the following:
   - Upper engine-to-transaxle bolts
   - Engine from the transaxle, then out the front of the engine compartment

15. If equipped with an automatic transaxle, secure the torque converter to prevent it from falling out.

To install:

16. Installation is the reverse of the removal procedure, while using the following torque values:
   - Transaxle flange M12 bolts: 48 ft. lbs. (65 Nm)
   - Transaxle flange M10 bolts: 33 ft. lbs. (45 Nm)
   - Engine mounting fasteners: 18 ft. lbs. (25 Nm)
   - Torque converter bolts: 63 ft. lbs. (85 Nm), if equipped with an automatic transaxle
   - Catalytic converter bolts: 22 ft. lbs. (30 Nm)

TT

1.8L ENGINE

1. Before servicing the vehicle, refer to the precautions section.

2. Turn the ignition switch to the **OFF** position, then disconnect the negative battery cable.

3. Properly relieve the fuel system pressure.

4. Drain the engine coolant.

5. Remove or disconnect the following:
   - Negative battery cable
   - Engine cover from cylinder head cover
   - Cover in front of intake manifold
   - Air duct at bottom of right-hand longitudinal member

- Noise insulation in center and on left and right sides
- Pendulum support
- Both drive axles from transmission flange
- Move left-hand drive axle toward rear of vehicle and secure in place by tying to anti-roll bar
- Front exhaust pipe

➡**Do not bend the flexible connection on front exhaust pipe more than 10 degrees or it may be damaged**

- Secure support rails to subframe with M8 x 25 bolt. Also secure engine support device to support rails
- Press engine forward
- Loosen right-hand 12-point nut for drive axle. When doing this, vehicle must be on its wheels
- Unbolt wheel
- Press out and remove drive axle
- Air hose and wiring harness from air mass meter
- Vacuum hose from air recirculation valve
- Hose from pressure control valve for crankcase breather
- Connector from charge pressure control valve
- Hose from charge pressure control valve
- Hose from solenoid valve to turbocharger
- Solenoid charge pressure control valve from air intake hose and place it to one side on engine
- Hose between activated charcoal filter and turbocharger
- Retainer off turbocharger connection and detach air intake hose
- Battery and platform
- Ground wire on engine/transmission flange
- Wires from starter and retainer on starter and move wires to one side
- Connector for electrical changeover valve on underside of bracket
- Bracket from intake manifold and disengage dipstick tube
- Alternator wiring harness
- A/C compressor wiring harness
- Coolant hoses
- Connectors from knock sensors, speed sensor and oil pressure switch
- Vacuum pipe from intake manifold.
- Connector for intake air temperature sensor and throttle valve control part (under throttle valve control part)
- Air hose from throttle valve control part
- Vacuum hose (for activated charcoal filter) from throttle valve control part
- Connector from Hall sensor
- Vacuum pipe from vacuum reservoir
- Vacuum reservoir and detach it from the bracket
- Connectors from injectors and unclip support bar from fuel rail
- Ground wire between ignition coils 1 and 2
- Connectors from ignition coils
- Connector from coolant temperature sensor and speedometer sensor
- Mark threaded sections of both selector cables with a permanent marker pen so that they can be installed in the same positions later
- Pull collar toward ball joint to compress selector cable spring, then turn collar clockwise to lock in position
- Disengage both threaded sections
- Unclip clutch slave cylinder hose from selector cable support bracket

➡**Do not depress the clutch or open the pipe system.**

- Fuel supply line and fuel return line at connection point

➡**Check the colors of connectors when connecting fuel supply and return lines**

➡**Before removing ribbed belt, mark direction of rotation with chalk or a felt pen. If a used belt rotates in the wrong direction when reinstalled, this can result in breakage. When installing belt, ensure it is correctly seated in the pulleys.**

- Loosen ribbed belt by turning tensioning element clockwise and remove belt

➡**Tensioning element can be locked in position with a suitable punch.**

- Connecting pipe between both left and right crossmembers
- Right cross member
- Connectors at both starter and transmission and starter positive cable
- Power steering fluid cooling line from transmission
- Open hose clamp at lower left of radiator and completely remove cooling line
- Pulley from power steering pump (counterhold with socket wrench)
- Place clean container under pump. Open spring clips from intake hose and remove hose.
- Pump bolts from bracket
- Connector and hoses at secondary air pump
- Secondary air pump
- Two radiator fan connectors on the radiator cowl (bottom left)
- Cowl from radiator (4 bolts) together with the two fans and remove by pulling down
- Unbolt A/C compressor, lift clear together with refrigerant hoses (do not disconnect) and tie in place on hood lock

6. Loosen engine/transmission mounting bolts (about two turns only) on engine side. Do not remove the bolts completely. Loosen engine/transmission mounting bolts (about two turns only) on transmission side. Do not remove the bolts completely.
7. Raise vehicle
8. Bolt engine bracket T10012, "or equivalent" to cylinder block with securing nut and bolt (M10 x 25/8.8). Tighten to about 15 ft. lb. (20 Nm).
9. Install engine/transmission jack to engine bracket and raise engine/transmission slightly
10. Unbolt engine/transmission mounting on the engine side
11. Unbolt engine/transmission mounting on the transmission side
12. Carefully lower the engine/transmission assembly

To install:

13. Installation is the reverse of the removal procedure, while using the torque values below. Note the following engine mounting adjustment procedure.
14. Adjust the engine mounting so that distance "a" (on the right-hand mounting) is about 0.51" (13 mm). It should be possible

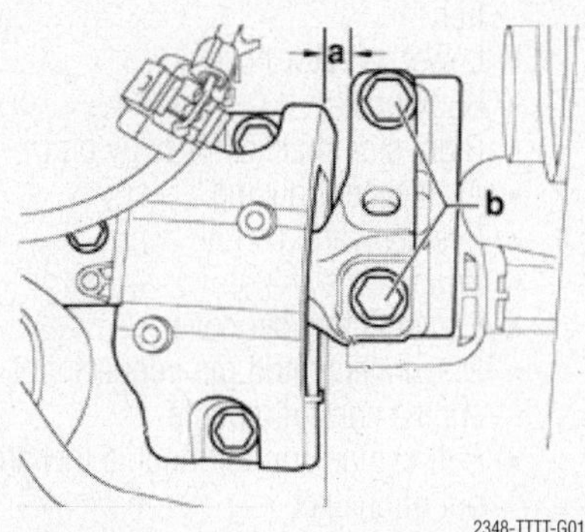

Adjusting engine mounting—1.8L engine TT

to insert a 0.47" (12 mm) flat metal bar without any difficulty.

15. To move the engine within the engine console, first loosen bolts "b" on the left and right engine supports by about 2 turns each

16. If the gap is too narrow or too wide, proceed as follows: Remove air hose from air mass meter and disconnect electrical connectors.

17. Remove air cleaner housing

18. Raise engine slightly, taking up the weight evenly on both sides

19. Loosen bolts "b" on left and right-hand support arms by about two turns each

20. Insert a metal bar between engine console and support arm, then move engine until gap "a" measures 0.51" (13 mm)

21. Tighten (by hand) all four bolts and left and right-hand support arms, then final tighten all four bolts to 63 ft. lbs. (85 Nm.)

- M8 bolts: 15 ft. lbs. (205 Nm)
- M10 bolts: 33 ft. lbs. (45 Nm)
- M12 bolts: 48 ft. lbs. (65 Nm)
- Drive shaft-to-transmission bolts: 30 ft. lbs. (40 Nm)
- Engine support-to-engine console bolts: 63 ft. lbs. (85 Nm)
- Transmission support-to-transmission console bolts: 44 ft. lbs. (60 Nm), plus 90 degrees

3.2L ENGINE

1. Before servicing the vehicle, refer to the precautions section.

2. Turn the ignition switch to the **OFF** position, then disconnect the negative battery cable.

3. Properly relieve the fuel system pressure.

4. Drain the engine coolant.

5. Remove or disconnect the following:

- Negative battery cable
- Both front wheels
- Unbolt bracket for headlight range control in left wheel housing
- Center, left and right sound insulation
- Lower radiator hose
- Body brace
- Right side bracket for body brace
- Air cleaner housing
- Fuse box and carrier
- Wiper arms
- Plenum chamber cover
- Disconnect multi pin connector at engine control module
- Pull engine control module forward out of bracket
- Disconnect electrical connector from bracket near left side headlight
- Place engine control module with wiring harness on top of engine

➡Mark installed position of hoses at hose bracket behind intake manifold

- Disconnect hoses from hose bracket and remove bracket
- Fuel lines

➡Before removing ribbed belt, mark direction of rotation with chalk or a felt pen. If a used belt rotates in the wrong direction when reinstalled, this can result in breakage. When installing belt, ensure it is correctly seated in the pulleys.

- Release tension on belt by screwing a bolt into bore on tensioning element and remove belt
- Upper radiator hose
- Upper securing bolts for radiator cowl
- Bracket and selector cable from transmission
- Loosen, but do not remove bolts on mounting assembly on engine and transmission
- Cross piece
- Power steering pump
- Secondary air injection pump and bracket
- Lower securing bolts for radiator cowl
- Radiator cowl with both fans

➡The air conditioning compressor can only be removed from the bracket with the upper securing bolt still inserted. To accomplish this, press the engine simultaneously to the left side. If necessary, loosen all bolts of the assembly mounting.

- Mounting bolt for air conditioning compressor
- Secure the air conditioning compressor together with the hoses at front longitudinal member (hoses remain connected)
- Alternator
- Catalytic converter
- Front exhaust pipe

➡When loosening and removing the axle shaft, the wheel bearing must NOT be under load (vehicle must not be standing on its wheels)

- Right and left axle shafts
- Unbolt flexible coupling for driveshaft, push driveshaft as far back as possible

➡The oil seal in the flange of the driveshaft must NOT be damaged. The driveshaft must be replaced if the oil seal is damaged

- Pendulum support

6. Attach engine bracket T10102 or equivalent to cylinder block

7. Install engine/transmission jack to engine bracket and raise assembly slightly

8. Remove engine/transmission mounting bolts

9. Pull engine/transmission assembly as far forward as possible and carefully lower

To install:

10. Installation is the reverse of the removal procedure, while using the torque values below. Note the following engine mounting adjustment procedure.

11. Adjust the engine mounting so that distance "X" (on the right-hand mounting) is about 0.51" (13 mm). It should be possible to insert a 0.47" (12 mm) flat metal bar without any difficulty.

12. To move the engine within the engine console, first loosen bolts on the left and right engine supports by about 2 turns each

13. If the gap is too narrow or too wide, proceed as follows: Remove air hose from air mass meter and disconnect electrical connectors.

14. Remove air cleaner housing

15. Raise engine slightly, taking up the weight evenly on both sides

16. Loosen bolts on left and right-hand support arms by about two turns each

17. Insert a metal bar between engine console and support arm, then move engine until gap "a" measures 0.51" (13 mm)

18. Tighten (by hand) all four bolts and left and right-hand support arms, then final tighten all four bolts to 63 ft. lbs. (85 Nm)

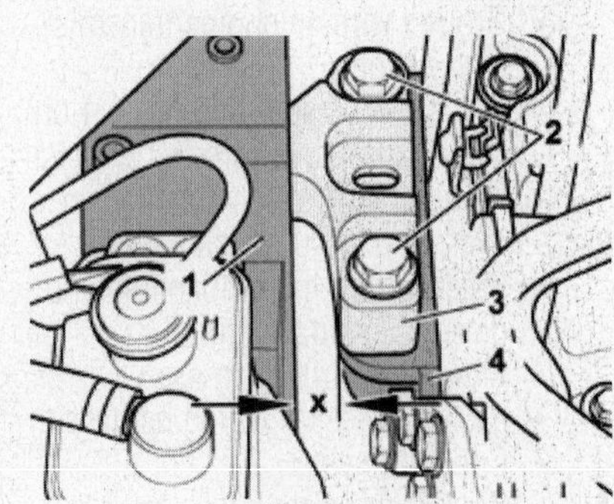

1. Engine mount
2. Bolts
3. Support arm
4. Engine support

67200-TTTT-G11

Adjusting engine mounting—3.2L engine TT

19. Transmission to engine block
    - M12 bolts 59 ft. lbs (80 Nm)
    - M10 bolts 30 ft. lbs (40 Nm)
20. Other bolts/nuts
    - M6 7 ft. lbs (10 Nm)
    - M8 15 ft. lbs (20 Nm)
    - M10 33 ft. lbs (45 Nm)
    - M12 48 ft. lbs (65 Nm)
    - Flexible coupling to bevel box 44 ft. lbs (60 Nm)
    - Exhaust pipe to manifold 30 ft. lbs (40 Nm)
    - Exhaust pipe to catalytic converter 18 ft. lbs (25 Nm)
    - Pendulum support to transmission 30 ft. lbs (40 Nm) + 90° (1/4 turn)
    - Pendulum support to subframe 15 ft. lbs (20 Nm) + 90° (1/4 turn)

A3

2.0L ENGINE

1. Before servicing the vehicle, refer to the precautions section.

➡The engine is removed downward together with the transmission.

2. Disconnect the electrical connector and unclip intake connection.
3. Remove the engine cover upward and out in direction of.
4. Remove the electrical connector for oxygen sensor before catalytic converter from the bracket, disconnect it and lay it aside.
5. Remove the coolant expansion tank cap.
6. Remove the cover from above battery, by pressing release button.
7. Disconnect the negative battery cable.
8. Remove the battery.
9. Remove the battery carrier.
10. Pry off caps on wiper arms with a screwdriver and remove nuts.
11. Pull off and remove wiper arms from wiper axles.
12. Unclip spray nozzles.
13. Push spray nozzles, with the lines still attached, back through the opening and into the plenum chamber.
14. Remove the rubber seal for plenum chamber cover.
15. Remove the plenum chamber cover.
16. Loosen up the rear engine wiring harness at partition for plenum chamber.
17. Remove the partition for plenum chamber.
18. Remove the Engine Control Module (ECM).
19. Disconnect the electrical connector for the engine wiring harness.
20. Pull up guide for engine wiring harness.
21. Open the wiring guide bracket.
22. Remove the engine wiring harness to control module from the wiring guide.
23. Disconnect the electrical connector for back-up light switch.
24. Remove the ground wire.
25. Push the cover towards the rear.
26. Remove the wires at the starter.
27. Remove the cover from above the battery, by pressing the buttons.
28. Remove the electrical harness from alternator.
29. Remove the front wheels.
30. Remove the center sound insulation fasteners.
31. Remove the left and right sound insulation fasteners.
32. Drain the coolant.
33. Disconnect the upper coolant hose from the radiator.
34. Disconnect the coolant hoses to the heat exchanger.
35. Remove the nuts for front exhaust pipe/turbocharger that can be reached from above.
36. Disconnect the vacuum hose from brake booster.
37. Unclip lock washers and at both cables. Remove the cable retainers from transmission selector lever and relay lever.
38. Remove the cable mounting bracket from the transmission and lay aside.

⁂ CAUTION

Do not operate the clutch pedal anymore after the hose for the slave cylinder has been disconnected.

39. Pull out the clip to stop and pull off the slave cylinder hose line.
40. Relieve the fuel system pressure.
41. Disconnect the fuel line and lay it aside.
42. Disconnect the EVAP line and lay it aside.
43. Disconnect the vacuum line to EVAP canister.
44. Remove the EVAP canister upwards.
45. Disconnect the coolant hoses.
46. Discharge air conditioning system
47. Remove the refrigerant line at compressor.
48. Unscrew bolts of engine mountings about 2 turns.
49. Unscrew bolts of transmission mountings about 2 turns.
50. Disconnect the electrical connector.
51. Open cable retainer. and loosen up any electrical wiring.
52. Remove the bolts from above.
53. Remove the air duct. for the charge air cooler.
54. Disconnect the electrical connector.
55. Remove the bolts and remove fan shroud.
56. Disconnect the connector at left longitudinal member and loosen up cable.
57. Pull air duct off from charge air cooler.
58. Disconnect the refrigerant line from condenser.
59. Disconnect the connector from Charge Air Pressure sensor
60. Disconnect the electrical connector at oil level sensor.
61. Disconnect the electrical connector at after-run water pump motor and loosen up wiring.
62. Remove the heat shield for right drive axle.
63. Remove the left and right drive axle.
64. Remove the right cover for vehicle floor.
65. Disconnect the connector for oxygen sensor after catalytic converter and loosen up cable.
66. Remove the remaining nuts for front exhaust pipe/turbocharger from below.
67. Remove the supports for front exhaust pipe with catalytic converter, by unscrewing bolts.

➡Decoupling elements in front exhaust pipes must not be bent more than 10° as damage may occur.

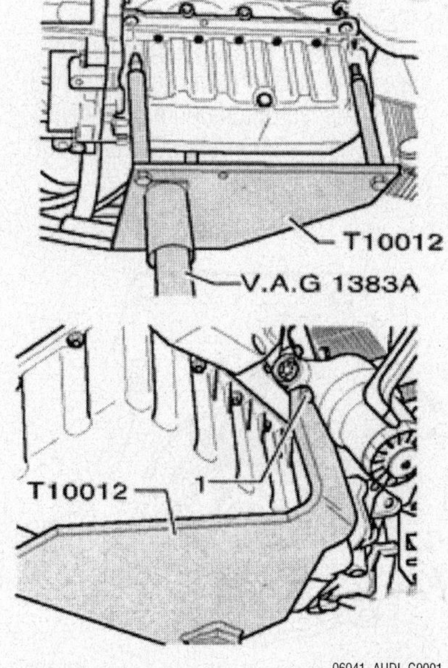

Fasten engine/transmission assembly support T10012 to cylinder block—A3 with 2.0L engine

68. Disconnect the exhaust system at clamping sleeve.
69. Remove the cross member for vehicle floor.
70. Remove the bracket for exhaust system.
71. Remove the catalytic converter with front exhaust pipe.
72. Remove the bolts and remove pendulum support.
73. Remove the auxiliary water pump bolts.
74. Fasten engine/transmission assembly support T10012 to cylinder block with bolt and nut to about 20 Nm.
75. Insert an engine/transmission assembly jack into engine/transmission assembly support T10012 and lift slightly.
76. Remove the bolts at engine assembly mounting.
77. Remove the bolts at transmission assembly mounting.
78. Verify that all hose and line connections between engine, transmission and body have been disconnected.
79. While lowering, carefully guide the engine/transmission assembly in order to prevent damage.
80. Pull engine/transmission assembly as far forward as possible and slowly lower it.

To install:

81. Installation is in the reverse order of removal, please note the following:

a. Check whether dowel sleeves for centralizing engine/transmission assembly are in cylinder block, install if necessary.

b. Make sure that the intermediate plate is hooked in at the sealing flange and is pushed onto alignment sleeves.

c. Check clutch release bearing for wear and replace if necessary.

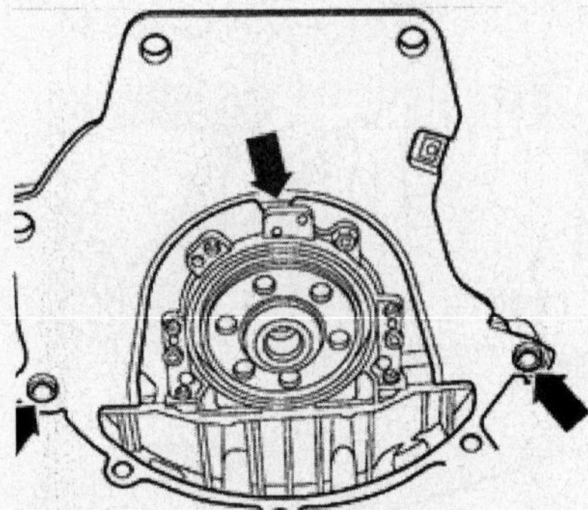

06041_AUDI_G0002

Check whether dowel sleeves for centralizing engine/transmission assembly are in cylinder block—A3 with 2.0L engine

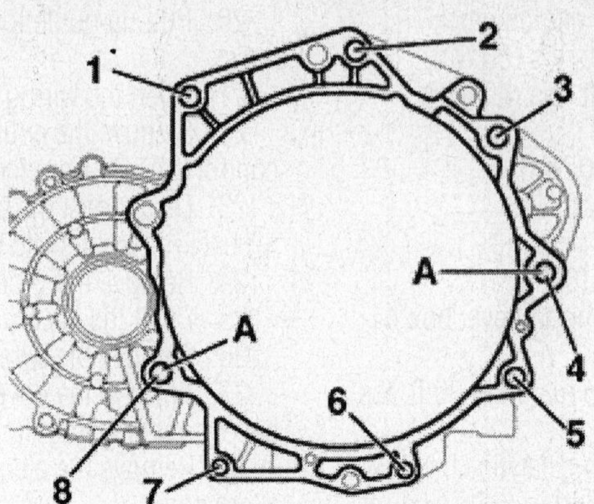

Engine/transmission, fastening

Pos.	Bolt	Nm
1, 2	M12×65	80
3 *, 4 *	M12×135	80
5 to 7	M10×60	40
8	M12×90	65

* Bolt with threaded pin M8

06041_AUDI_G0003

Transmission to engine bolts torque values—A3 with 2.0L engine

➡Torque specifications only apply to lightly greased, oiled, phosphated or blackened nuts and bolts. Additional lubricants, such as engine or transmission oil are permissible, although lubricants containing graphite are not.

d. Do not use any degreased parts.

e. Tighten the M12 X 65 and M12 X 135 bolts to 80 Nm, the M12 X 90 bolts to 65 Nm and the M10 X 60 bolts 40 Nm.

➡The bolts are first tightened to final torque after the engine mounts have been adjusted.

f. Remove the engine/transmission assembly support T10012 from engine.

g. Fasten pendulum supports to transmission and to sub frame. Tighten pendulum to transmission bolts to 40 Nm plus an additional 90 degrees. Tighten the pendulum to sub frame bolts to 100 Nm plus 90 degrees.

⁂ CAUTION

Do not use a battery charger for starting assistance! There is the risk that the vehicle control modules could be damaged.

h. Tighten all remaining M6 bolts to 10 Nm, M8 bolts to 20 Nm, M10 bolts to 45 Nm and M12 bolts to 65 Nm.

i. Tighten the A/C compressor to accessory assembly bracket to 25 Nm.

j. Tighten the ground strap to transmission to 23 Nm.

3.2L ENGINE

1. Before servicing the vehicle, refer to the precautions section.

➡The engine is removed downward together with the transmission.

2. Remove the tool formed insert under luggage compartment floor covering.
3. Remove the cover for battery compartment
4. Remove the formed insert from over battery.
5. Disconnect the negative battery cable.
6. Remove the coolant expansion tank cap.
7. Remove the both front wheels.
8. Remove the center noise insulation fasteners.
9. Remove the left and right noise insulation fasteners.
10. Disconnect the electrical connector at Engine Coolant Temperature (ECT) sensor.

11. On vehicles with drain plug, turn the drain plug left on radiator, place assisting hose on supports if necessary.

12. Remove the lower coolant hose from radiator by removing retaining clamps and drain the coolant.

13. If equipped, remove vacuum hose to air guide hose.

14. Disconnect the air guide hose at Throttle Valve Control Module.

15. Disconnect the Mass Air Flow (MAF) sensor connector.

16. Remove the upper part of air filter housing.

17. Remove the filter element.

18. Remove the air duct cover by releasing retaining clasps on sides.

19. Remove the lower part of air filter housing.

20. Remove the air filter housing bracket.

21. Disconnect the electrical connector for Heated Oxygen sensors (HO2S)

22. Move the electrical wiring to oxygen sensors aside.

23. Pry off the wiper arms caps using a screwdriver.

24. Loosen the wiper arm nuts using several turns.

25. Loosen the wiper arms by gently rocking wiper arm.

26. Remove the nuts and remove wiper arms.

27. Unclip spray nozzles. Push the spray nozzle with lines still attached back through opening and into plenum chamber.

28. Remove the rubber seal for plenum chamber cover.

29. Remove the plenum chamber cover.

30. Loosen up the rear engine wiring harness at the partition for plenum chamber.

31. Remove the partition for plenum chamber.

32. Remove the Engine Control Module (ECM)

33. Separate electrical connector for engine wiring harness.

34. Release pass-through for engine wiring harness and pull off upward.

35. On vehicles with the older with E-box version, press the retaining clamps in and remove the engine compartment E-box cover.

36. On vehicles with the newer with E-box version, press in both latches and remove engine compartment E-box cover.

37. Remove the terminal 30 wiring at engine compartment E-box. Open the wiring bracket.

38. Loosen up the electrical connector and disconnect, open underlying wiring guide bracket.

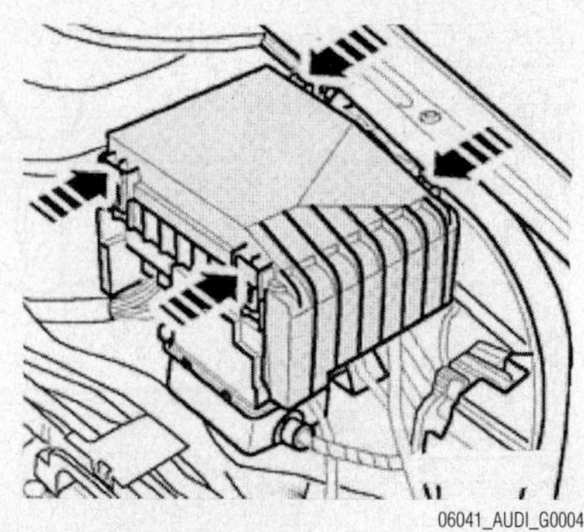

View of the older E-box version—A3 with 3.2L engine

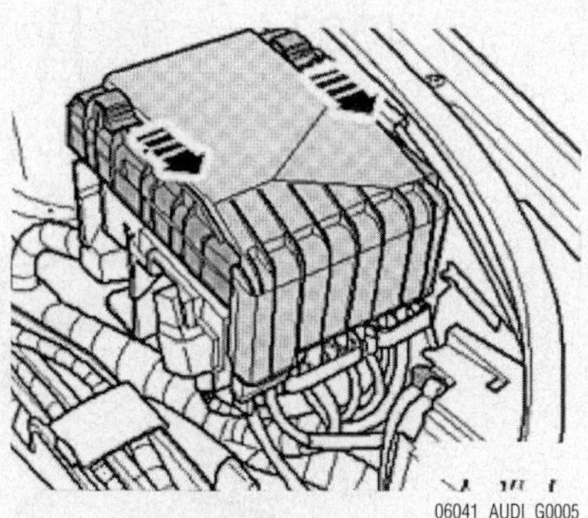

View of the newer E-box version—A3 with 3.2L engine

39. Remove the engine wiring harness to control module from wiring.

40. Disconnect the Secondary Air Injection hose at position indicated by.

41. Loosen up the air hose to the secondary air injection pump.

42. Disconnect the electrical harness connectors.

43. Remove the protective boot and remove electrical wiring at starter solenoid switch.

44. Remove the ground wire.

45. Remove the bolts for bracket.

46. Remove the retaining clip upward and remove selector lever cable from transmission.

47. Remove the coolant hoses to heater core on bulkhead.

48. Disconnect the coolant hoses.

49. Disconnect the electrical connection at throttle valve control module.

50. Remove the Throttle Valve Control Module from the intake manifold.

51. Properly relieve the fuel system pressure.

52. Disconnect the following hose connections:

a. Coolant hose to reservoir.

b. Vacuum hose to Evaporative Emission (EVAP) canister .

c. Vacuum hose to leak detection pump.

d. Fuel supply line.

e. Coolant hose to reservoir.

53. Remove the following hoses and electrical connections::

a. Electrical connector to Evaporative Emission (EVAP) canister Purge Regulator Valve.

b. Vacuum hose to brake booster.

c. Vacuum hose to exhaust flap, also to leak detection pump.

d. Vacuum hose to evaporative emission (EVAP) canister purge regulator valve.

54. Remove the wiring connection.

55. Remove the wiring connections with coolant and vacuum hoses.

56. Remove the Heated Oxygen sensors.

57. Remove the bolts and remove front exhaust pipe support.

58. Remove the front exhaust pipe nuts and exhaust manifolds.

59. Remove the fan shroud upper screws—upper arrows.

➡Before removing accessory drive belt, mark the turning direction on it with chalk or a felt tip pen. A reversed direction of rotation can cause damage to the belt under operating conditions.

60. Remove the accessory drive belt.

61. Loosen the subframe mounting bolts to engine about 2 turns.

62. Loosen the subframe mounting bolts to transmission about 2 turns.

63. Remove the cover for vehicle floor on the right.

64. Disconnect the electrical connector to Oxygen sensors (O2S) behind the Three Way Catalytic Converter (TWC).

65. Remove the connector coupling from bracket and loosen up wiring to the oxygen sensors.

66. Remove the front vehicle floor cross member by removing the retainers.

➡Flex joints in front exhaust pipe must not be bent more than 30 degrees, otherwise they may be damaged.

67. Separate exhaust system at clamping sleeve.

68. Remove the exhaust system bracket and remove exhaust system.

69. Loosen up the lower coolant hose at fan shroud.

70. Disconnect the electrical connector at Secondary Air Injection (AIR) Pump Motor and loosen up electrical wiring.

71. Remove the coolant pipe bracket to transmission oil cooler.

72. Remove the bolts.

73. Loosen the bolt and remove Secondary Air Injection (AIR) pump with bracket.
74. Disconnect the lower coolant fan electrical connectors at fan shroud.
75. Remove the lower fan shroud screws.
76. Remove the fan shroud with both fans downward.
77. Disconnect the solenoid clutch electrical connector at A/C compressor and loosen up electrical wiring.
78. Remove the A/C compressor bolts.
79. Secure A/C compressor with coolant hoses attached at front of longmember.
80. Remove the electrical wire on alternator.
81. Disconnect the electrical connector.
82. Remove the upper idler roller.
83. Remove the alternator bolts and.

➡Alternator can only be removed from bracket with upper securing bolt still in place.

84. Remove the alternator from accessory assembly bracket.
85. Remove the alternator toward lower left.
86. Disconnect the electrical connector to Secondary Air Injection (AIR) Solenoid Valve.
87. Have a helper press foot brake and loosen the collar bolt at right drive axle.
88. Remove the right drive axle heat shield from the bevel box.
89. Remove the right drive axle.
90. Unbolt left drive axle from transmission flange shaft.
91. Remove the drive shaft flexible disc at bevel box.

➡To loosen bolts, counter hold at drive shaft flange/final drive with a lever. Press drive shaft horizontally as far back as possible.

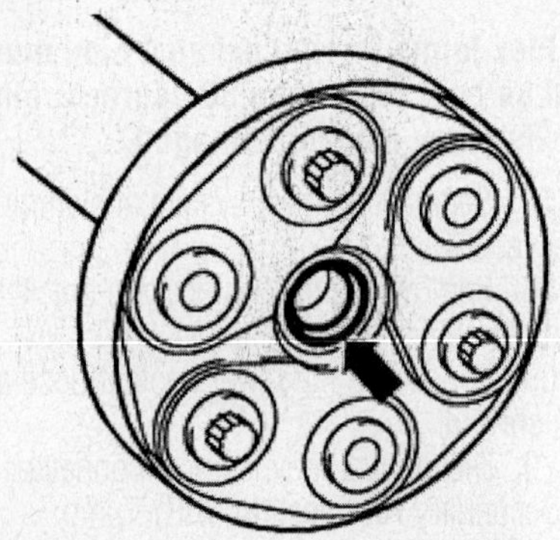

To loosen bolts, counter hold at drive shaft flange/final drive with a lever—A3 with 3.2L engine

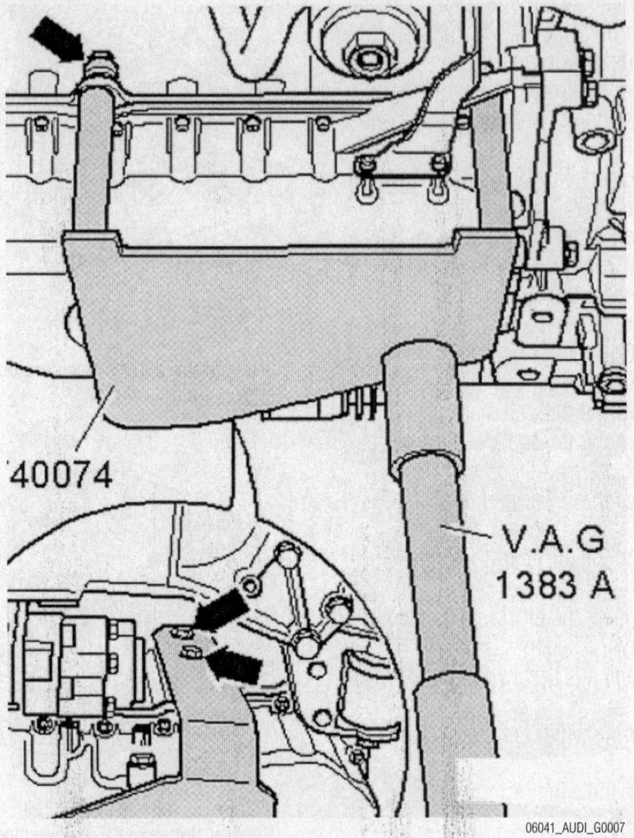

Bolt engine mount T40074 to cylinder block and tighten to 20 Nm—A3 with 3.2L engine

⁂ CAUTION

The seal in drive axle flange must not be damaged when removing engine/transmission assembly. If seal is damaged, drive axle must be replaced. Be aware that the engine/transmission assembly swings forward slightly when pendulum support/subframe bolts are loosened.

92. Remove the bolts and remove pendulum support.
93. Disconnect the electrical connector at Oil Level Thermal sensor.
94. Unclip bracket for electrical wire to Oil Level Thermal sensor at subframe.
95. Disconnect the electrical connector at After-Run Coolant Pump.
96. Remove the After-Run Coolant Pump downward with coolant lines attached from rubber bracket loops. Spray rubber loops with silicone-free lubricant, if necessary.

➡The water pump remains with coolant hoses connected at engine.

97. Remove the bolts and accessory assembly bracket.
98. Bolt engine mount T40074 to cylinder block and tighten to 20 Nm.
99. Insert engine and transmission jack at engine mount T40074 and raise engine/transmission assembly slightly.
100. Remove bolts for the assembly mounting.
101. Remove the subframe mounting bolts at the engine.
102. Remove the subframe mounting bolts at the transmission.

➡Verify that all hose and line connections between engine, transmission and body have been disconnected.

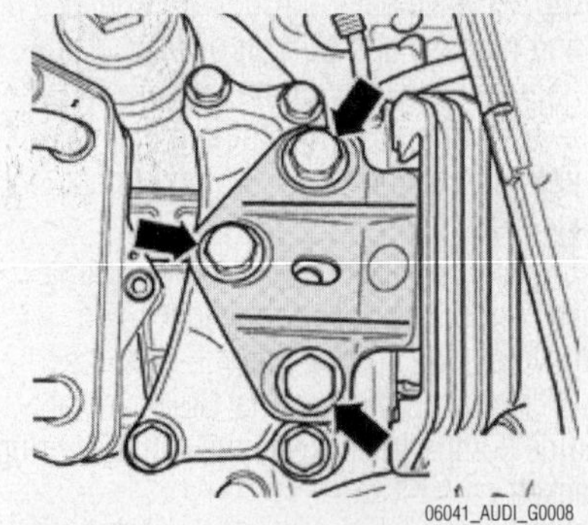

Remove the sub frame mounting bolts at the transmission—A3 with 3.2L engine

103. While lowering, carefully guide the engine/transmission assembly in order to prevent damages.

104. Pull engine/transmission assembly as far forward as possible while slowly lowering it.

To install:

105. Installation is the reverse of removal, please note the following:

a. Always replace self-locking nuts, bolts that have been tightened to tightening torque as well as gaskets and O-rings.

b. Make sure alignment sleeves for engine to transmission are installed in cylinder block.

c. A pilot needle bearing must be installed in the crankshaft in engines for vehicles with DSG transmission.

d. Use new bolts when securing the transmission to engine.

➡Torque specifications only apply to lightly greased, oiled, phosphated or blackened nuts and bolts. Additional lubricants, such as engine or transmission oil are permissible, although lubricants containing graphite are not.

e. Tighten bolts position 1 on the transmission to engine illustration to 80 Nm.

f. Tighten bolts position 2 on the transmission to engine illustration to 40 Nm.

g. Tighten bolts position 3 and 6 on the transmission to engine illustration to 80 Nm.

h. Tighten bolts position 4 and 9 on the transmission to engine illustration to 40 Nm.

i. Tighten bolts position 5 on the transmission to engine illustration to 80 Nm.

j. Tighten bolts position 7 and 8 on the transmission to engine illustration to 40 Nm.

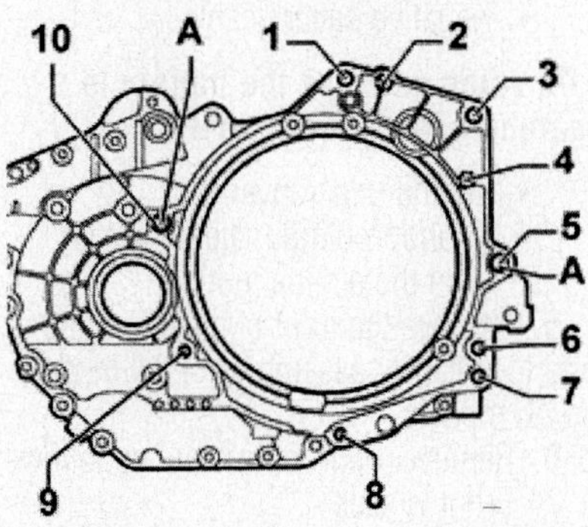

Transmission to engine bolt locations—A3 with 3.2L engine

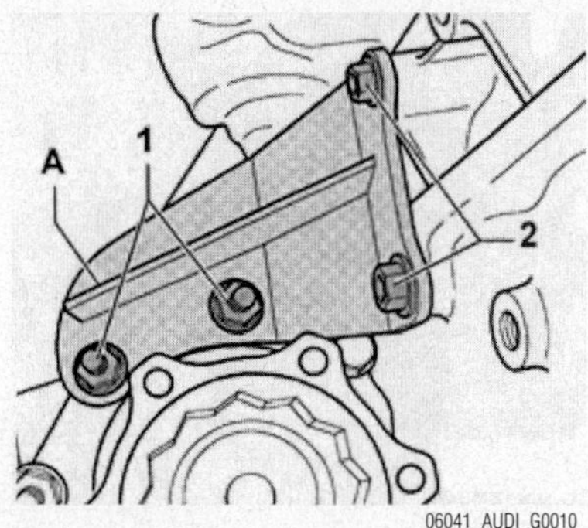

Bevel box bolt locations—A3 with 3.2L engine

k. Tighten bolts position 10 on the transmission to engine illustration to 80 Nm.

l. Install the bevel box bracket, tighten the bolts (1) to 3 Nm, then bolts (2) to 35 Nm and finally bolts (1) to 45 Nm. Refer to the illustration for bolt locations.

m. Guide engine/transmission assembly into the body and hand tighten engine support bolts to engine console.

n. Tighten transmission support bolts to transmission console.

o. Remove the engine/transmission assembly support from engine.

p. Press engine/transmission assembly toward bulkhead, when doing so bevel box pins must be carefully guided into drive shaft flange.

q. Fasten pendulum supports to transmission and to sub frame. Tighten pendulum to transmission bolts to 40 Nm plus an additional 90 degrees. Tighten the pendulum to sub frame bolts to 100 Nm plus 90 degrees.

r. Fasten drive shaft with flexible disc to bevel box flange to 60 Nm.

s. To make it easier to install the alternator, drive the bushings for the retaining bolts back slightly.

⁂ CAUTION

Do not use a battery charger for starting assistance, there is the risk that the vehicle control modules could be damaged.

t. Tighten the front exhaust pipe to exhaust manifold to 40 Nm with new fasteners.

⁂ CAUTION

Do not use a battery charger for starting assistance! There is the risk that the vehicle control modules could be damaged.

u. Tighten all remaining M6 bolts to 10 Nm, M7 bolts to 15 Nm, M8 bolts to 20 Nm, M10 bolts to 45 Nm and M12 bolts to 65 Nm.

v. Tighten the ground strap to transmission to 22 Nm.

2.7L Engine

➡The engine is removed without the transaxle, through the top of the engine compartment. The front bumper and the front lock carrier assembly must be removed.

1. Before servicing the vehicle, refer to the precautions section.
2. Relieve the fuel system pressure.
3. Drain the engine coolant.
4. Remove or disconnect the following:
   - Negative battery cable
   - Plenum cover
   - Engine covers
   - Engine undercover
   - Air filter cover
   - Soundproofing material holder and bracket from the engine mount
5. Remove the front bumper as follows:
   - Front wheels
   - Fasteners attaching the wheel housing lining to the front bumper
   - Bumper cover flanged nuts
   - Lower air grille assemblies
   - Front bumper mounting bolts
   - Front bumper
6. Remove or disconnect the following:
   - Hood release cable
   - Air intake duct for the air cleaner housing
   - Wring connectors for the horns, headlights and electrical equipment mounted on the front hood lock carrier assembly
   - Hood lock carrier assembly to the front fenders fasteners
   - Radiator coolant hoses
   - A/C condenser from the brackets and move it aside with the hoses attached

⁂ CAUTION

The condenser lines must not be bent or kinked.

   - Power steering cooling line brackets located on the engine support and the transmission and move them aside with the power steering hoses attached
   - Cooling lines, if equipped with an automatic transmission
   - Hood seal at the front fenders
   - Energy absorbing bumper struts to the carrier fasteners

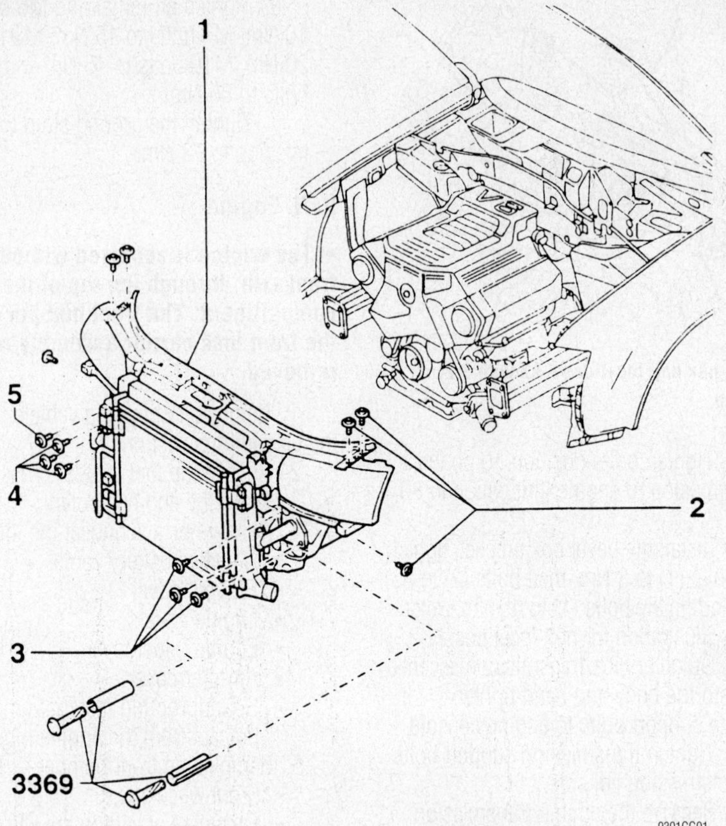

The hood lock carrier assembly (1), fender fasteners (2) and energy absorbing strut fasteners (3). Tool 3369 is used when placing the carrier in the service position—A6 models

- Lock carrier assembly
- Viscous fan left hand threaded nut
- Accessory drive belt
- Coolant reservoir
- Valve cover for cylinder bank 4-6
- Coolant Hoses
- Wiring harnesses
- Air distributor
- Fuel lines
- Air filter
- Pressure hoses from charge air cooler
- Power steering hose
- Heat sensor from right hand turbocharger
- Heat shield from both turbochargers
- Wiper arms and water deflector trim
- ECM retaining bracket and ECM
- Brake booster hose
- Firewall ground
- Bulkhead connectors
- Heat exchanger coolant hoses
- Turbocharger and charge air cooler pressure hoses
- Starter cable clip
- Ground at engine support
- Torque support below crank pulley
- Oil filter
- Oil cooler
- Heat shield above both drive axles
- Left and right exhaust pipes from turbocharger
- Air conditioning condenser brackets and move the condenser aside with the hoses attached
- Alternator
- Air conditioning compressor and move it aside with the hoses connected
- Air intake pipe
- Both wheels
- Starter
- Flexplate to torque converter bolts through the starter opening, if equipped with an automatic transmission
- Exhaust system from the manifolds
- Mark positions of engine mountings on both sides
- Nuts at engine mountings
- Engine-to-transmission bolts

7. Position an Engine Support Bridge tool 10-222A to the bolted flanges of the fenders with the spindle facing forward.

8. Connect the Engine Support Bridge tool 10-222A using Adapter tool 10-222A1 and Extension tool 10-222A3 and support the transaxle.

9. Attach a Engine Sling tool 2024A to the right rear and left front of the engine.

10. Attach an engine hoist to the sling.

11. Support transmission with transmission jack Lift the engine slightly.

12. Check that all hoses, wires, cables and mounts have been disconnected and carefully lift the engine in conjunction with the transmission to clear the right engine mount.

13. Pull the engine forward until it is separated from the transmission.

14. After the engine has been separated from the transmission, lift the engine up and out of the vehicle.

To install:

15. Installation is the reverse of the removal procedure, while using the following torque values:
    - Engine mount bolts: 33 ft. lbs. (45 Nm).
    - Flexplate-to-torque converter bolts: 63 ft. lbs. (85 Nm)
    - Engine-to-transaxle bolts: 33 ft. lbs. (45 Nm) for the M10x 60 and M10x70 bolts, 48 ft. lbs. (65 Nm) for the M10x100 and all the M12 bolts
    - General bolts: 18 ft. lbs. (24 Nm) for the M8 bolts, 33 ft. lbs. (45 Nm) for the M10 bolts and 44 ft. lbs. (60 Nm) for the M12 bolts

2.8L Engine

➡The engine is removed without the transaxle, through the top of the engine compartment. On all A4 and A6 models, the front bumper and the front lock carrier assembly must be removed.

1. Before servicing the vehicle, refer to the precautions section.
2. Relieve the fuel system pressure.
3. Drain the engine coolant.
4. Remove or disconnect the following:
    - Negative battery cable

➡On some vehicles the battery is located under the rear seat.

- Engine undercover
- Soundproofing material holder from the engine mount

5. Remove the front bumper and the hood lock carrier assembly, for A4 models and A6 models.
6. Remove the front bumper as follows:
7. Front wheels
8. Fasteners attaching the wheel housing lining to the front bumper
9. Bumper cover flanged nuts

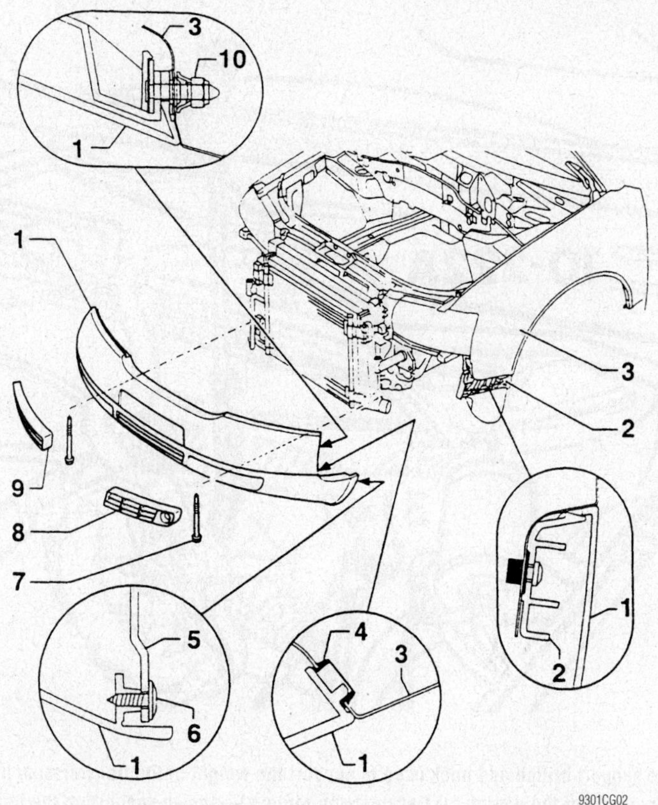

Bumper cover (1), saddle (2), fender (3), alignment tab (4), inner fender liner (5), liner fastener (6), bumper mounting fastener (7), air grilles (8, 9) and flanged nut (10)—A6 models

10. Lower air grille assemblies
11. Front bumper mounting bolts
12. Front bumper
13. Remove or disconnect the following:
    - Hood release cable
    - Air intake duct for the air cleaner housing
    - Wring connectors for the horns, headlights and electrical equipment mounted on the front hood lock carrier assembly
    - Hood lock carrier assembly to the front fenders fasteners
    - Radiator coolant hoses
    - A/C condenser from the brackets and move it aside with the hoses attached

⁂ CAUTION

The condenser lines must not be bent or kinked.

- Power steering cooling line brackets located on the engine support and the transmission and move them aside with the power steering hoses attached
- Cooling lines, if equipped with an automatic transmission
- Hood seal at the front fenders
- Energy absorbing bumper struts to the carrier fasteners
- Carrier assembly
- Stabilizer brace from the right rear of the engine compartment
- Wiper arms and water deflector trim, on A4 and A6 models
- Hoses
- Wiring harnesses
- Lines and cables for engine removal, as necessary
- Accessory drive belt guard
- Accessory drive belt
- Front engine mount at the cross member
- Hydraulic lines that route above the cylinder head covers from the power steering pump
- Ground strap from the right side engine support
- Air intake for the alternator
- Alternator
- Air conditioning condenser brackets and move the condenser aside with the hoses attached
- Air conditioning compressor and move it aside with the hoses connected
- Exhaust system from the manifolds
- Crossmember to access the cat-

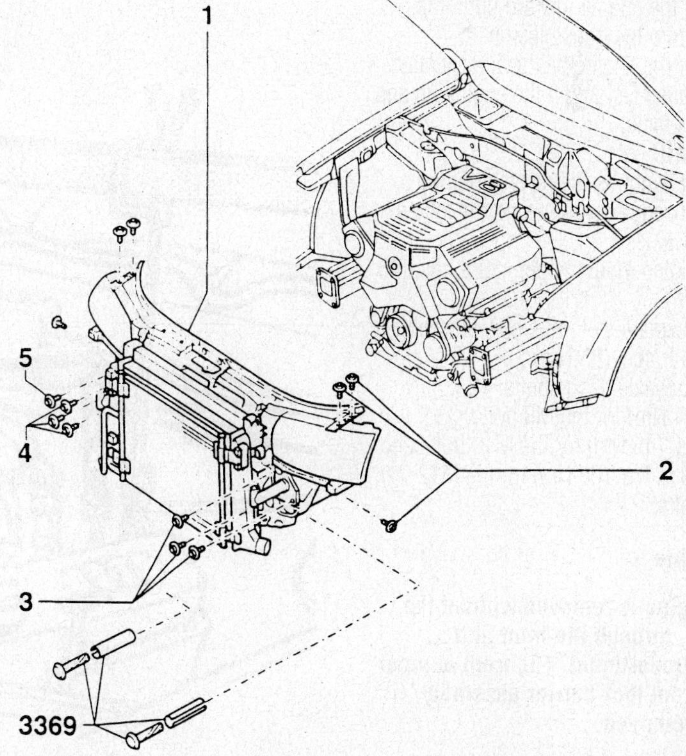

The hood lock carrier assembly (1), fender fasteners (2) and energy absorbing strut fasteners (3). Tool 3369 is used when placing the carrier in the service position—A6 models

alytic converters and the exhaust pipe
- Front exhaust pipes with the catalytic converters
- Starter
- Oil filter
- Oil cooler
- Flexplate to torque converter bolts through the starter opening, if equipped with an automatic transmission
- Engine-to-transmission bolts

14. Position an Engine Support Bridge tool 10-222A to the bolted flanges of the fenders with the spindle facing forward.
15. Attach the Support Adapter tool 3147 to the bolt hole above the starter mounting area in the transaxle bell housing.
16. Connect the Engine Support Adapter tool 3147 to the Engine Support Bridge tool 10-222A using Adapter tool 2024A/1 and Extension tool 2024A/2 and support the transaxle.
17. Attach a Engine Sling tool 2024A to the right rear and left front of the engine.
18. Attach an engine hoist to the sling.
19. Lift the engine slightly.
20. Check that all hoses, wires, cables and mounts have been disconnected and carefully lift the engine in conjunction with the transmission to clear the right engine mount.
21. Pull the engine forward until it is separated from the transmission.
22. After the engine has been separated from the transmission, lift the engine up and out of the vehicle.

To install:

23. Installation is the reverse of the removal procedure, while using the following torque values:
    - Engine mounts bolts: 33 ft. lbs. (45 Nm).
    - Flexplate-to-torque converter bolts: 63 ft. lbs. (85 Nm)
    - Transaxle flange bolts: 18 ft. lbs. (24 Nm) for the M8 bolts, 33 ft. lbs. (45 Nm) for the M10 bolts and 48 ft. lbs. (65 Nm) for the M12 bolts

3.0L Engine

➡The engine is removed without the transaxle, through the front of the engine compartment. The front bumper and the front lock carrier assembly must be removed.

1. Before servicing the vehicle, refer to the precautions section.
2. Relieve the fuel system pressure.
3. Drain the engine coolant.

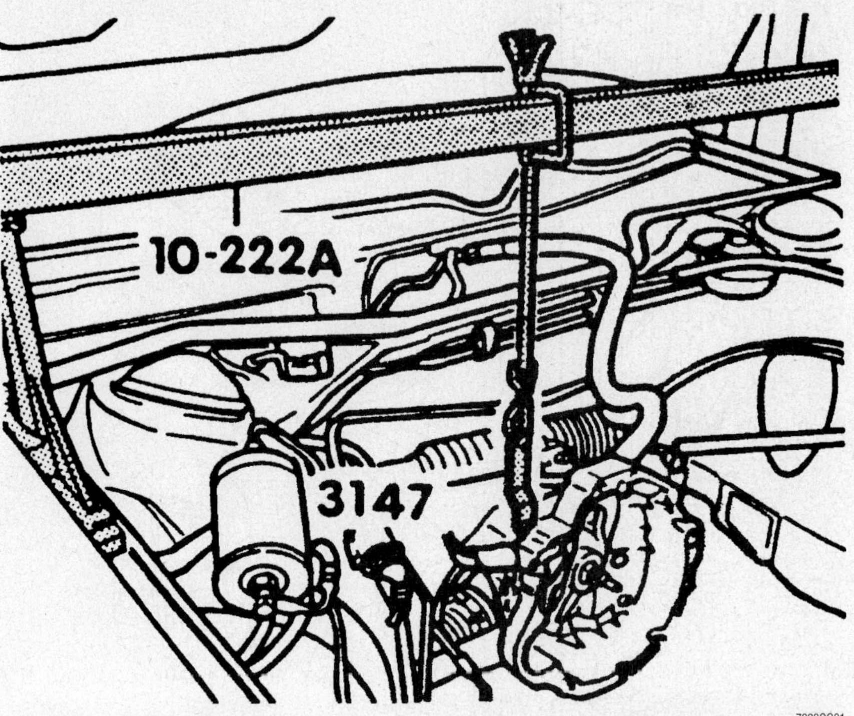

The engine support bridge and hook used to support the weight of the transmission for engine removal or weight of the engine for transmission removal—shown supporting the transmission with the engine removed

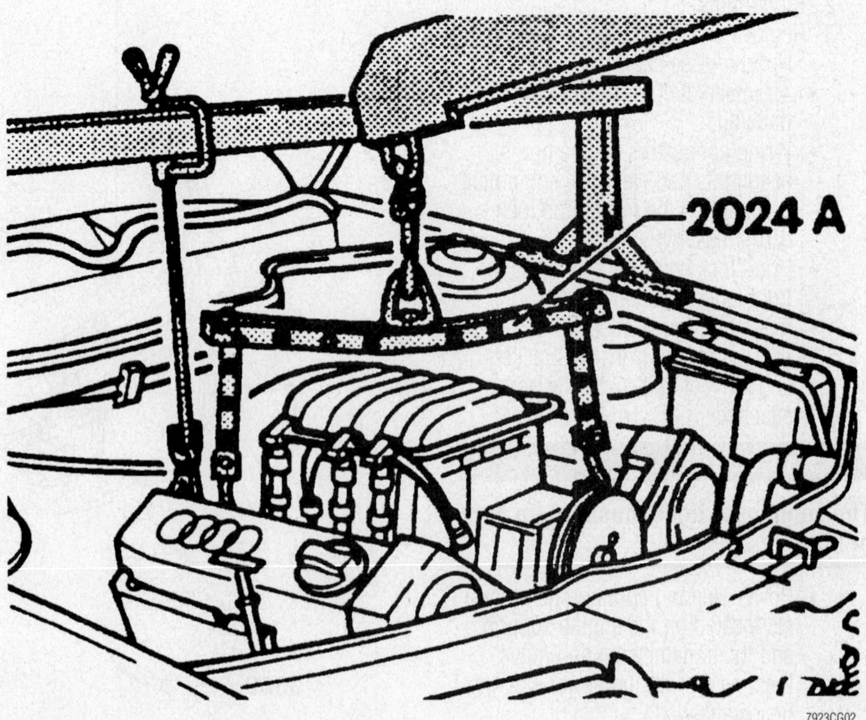

Be sure to attach the engine sling properly

4. Remove or disconnect the following:
- Negative battery cable
- Plenum cover
- Both front wheels
- Vehicles with auxiliary heater, remove bolts for exhaust pipe of parking heater/auxiliary heater at noise insulation
- Carefully open cap of expansion tank
- Engine undercover
- Soundproofing material holder and bracket from the engine mount

5. Remove the front bumper as follows:
- Front wheels
- Fasteners attaching the wheel housing lining to the front bumper
- Bumper cover flanged nuts
- Lower air grille assemblies
- Front bumper mounting bolts
- Front bumper

6. Drain the engine coolant, then remove the lower radiator hose from the radiator.

7. Remove or disconnect the following, as equipped:
- Remaining coolant hoses
- Hood release cable
- Air intake duct for the air cleaner housing
- Wring connectors for the horns, headlights and other necessary electrical equipment that might obstruct removal
- Hood lock carrier assembly to the front fenders fasteners
- A/C condenser from the brackets and move it aside with the hoses attached

⁂ CAUTION

The condenser lines must not be bent or kinked.

- Power steering cooling line brackets located on the engine support and the transmission and move them aside with the power steering hoses attached
- Cooling lines, if equipped with an automatic transmission
- Hood seal at the front fenders
- Air guides from LH and RH of radiator
- Temperature sensor, PS cooling coil, and connectors from in front of radiator
- A/C pressure sensor connector at side of radiator
- A/C condenser from the brackets and move it aside with the hoses attached

⁂ CAUTION

The condenser lines must not be bent or kinked.

- Energy absorbing bumper struts to the carrier fasteners
- Hood lock and lock carrier assembly (use assistant for this step)
- Rear engine cover
- Solenoid holding plate
- MAF sensor intake hose
- Air guide hose
- Throttle valve control module air hose
- Coolant reservoir
- Vacuum hoses
- Power steering hoses from reservoir
- Connector, cable and tube from AIR pump
- EVAP canister hoses and wiring
- Fuel line
- PS pressure line at rear coolant pipe
- Air filter housing
- Brake booster connector
- Wiring harness at ECM and place ECM aside
- Wiring harnesses from bulkhead
- Exhaust pipes
- Heater hoses
- Upper mounting bolts for engine/transmission
- Front engine cover
- Accessory drive belt
- Engine oil and filter
- Oil cooler
- Air conditioning compressor and move it aside with the hoses connected
- Alternator
- Heat shields, as equipped
- Starter
- Flexplate to torque converter bolts through the starter opening, if equipped with an automatic transmission
- Exhaust system from the manifolds, including oxygen sensor wiring
- Engine speed sensor
- Back-up light harness and speedometer connector from transmission
- Other transmission connectors and harnesses, as equipped
- Mark positions of connections and sleeves on engine mounts
- Engine mount nuts
- Torque support below crank pulley
- Install T10062 support on threaded bores of torque support

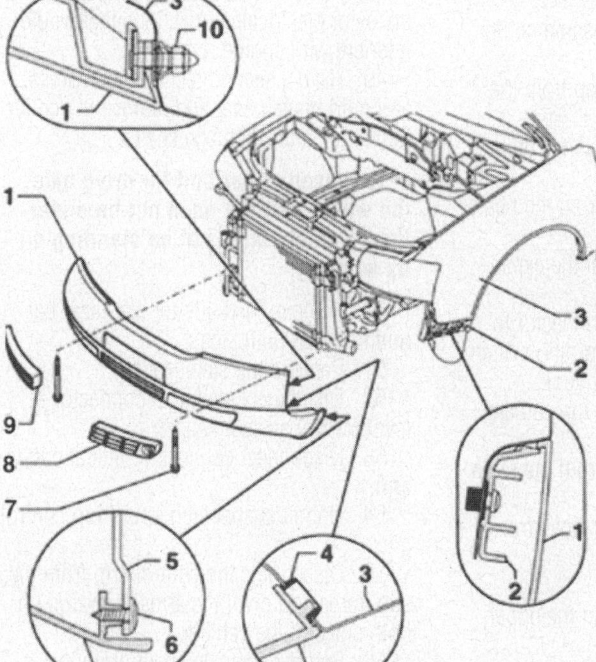

Bumper cover (1), guide (2), fender (3), clip (4), inner fender liner (5), screw/washer (6), socket head bolt (7), air grilles (8, 9) and flanged nut (10)—A6 models

- Raise engine and remove connecting bolts at engine/transmission from below
- Lower engine

8. Position an Engine Support Bridge tool 10-222A to the bolted flanges of the fenders with the spindle facing forward.
9. Attach the Support Adapter tool 3147 to the bolt hole above the starter mounting area in the transaxle bell housing.
10. Connect the Engine Support Adapter tool 3147 to the Engine Support Bridge tool 10-222A using Adapter tool 10-222A/2 and Extension tool 2024A/2 and support the transaxle.
11. Attach an Engine Sling tool 2024A to the right rear and left front of the engine.
12. Remove final mounting bolt
13. Support transmission with transmission jack
14. Lift the engine slightly.
15. Check that all hoses, wires, cables and mounts have been disconnected and carefully lift the engine in conjunction with the transmission to clear the right engine mount.
16. Pull the engine forward until it is separated from the transmission.
17. After the engine has been separated from the transmission, lift the engine up and out of the vehicle.

To install:

18. Installation is the reverse of the removal procedure, while using the following torque values:
    - Engine mount bolts: 33 ft. lbs. (45 Nm).
    - Flexplate-to-torque converter bolts: 63 ft. lbs. (85 Nm)
    - Engine-to-transaxle bolts: 33 ft. lbs. (45 Nm) for the M10x 60 and M10x70 bolts, 48 ft. lbs. (65 Nm) for the M10x100 and all the M12 bolts
    - General bolts: 18 ft. lbs. (24 Nm) for the M8 bolts, 33 ft. lbs. (45 Nm) for the M10 bolts and 44 ft. lbs. (60 Nm) for the M12 bolts

A6

3.2L ENGINE

1. Before servicing the vehicle, refer to the precautions section.

⁂ CAUTION

Before removing the engine, secure the vehicle to prevent it from tipping over. The luggage compartment must be empty.

2. Remove the luggage compartment floor trim.

➡So that the front wheels can still be turned with the battery disconnected, the battery must only be disconnected with the ignition key inserted.

3. Disconnect the negative battery cable.
4. Remove the front engine cover.
5. Remove the rear engine cover.
6. Open cap of coolant expansion tank after the engine has fully cooled.
7. Remove the drive belt.
8. Remove the both front wheels.

➡Secure the brake rotors with wheel bolts.

9. Loosen quick-release fasteners and remove sound insulation.
10. Place a drip tray under the engine.
11. Remove the drain plug on coolant thermostat housing and drain coolant from engine.
12. Disconnect lower coolant hose from radiator and drain radiator coolant.
13. Disconnect the A/C compressor clutch connector.
14. Unbolt the A/C compressor and move to one side with the lines attached.
15. Disconnect the power steering hoses, unbolt the pump and move aside.
16. If equipped with an auxiliary heater, remove the left front fender liner.
17. Remove the muffler and corrugated exhaust pipe.
18. Remove the bolts and the torque support.
19. Disconnect the transmission cooler lines.
20. Remove the ground strap from the longitudinal member.
21. Disconnect vacuum hose for brake booster at bulkhead.
22. Disconnect the hoses from the heater core.
23. Disconnect the hoses at the expansion tank and remove the tank.
24. Disconnect electrical connector to Engine Coolant Level (ECL) warning switch at bottom of coolant expansion tank.
25. Disconnect the hose at the coolant line.
26. Disconnect the hoses from the radiator.
27. Disconnect the vacuum hose from the leak detection pump.
28. Disconnect fuel line.
29. Remove the air duct and the upper air filter housing.
30. Remove the lower air filter housing.
31. Disconnect vacuum hose to EVAP canister.
32. Remove the nuts and the right bracket for harness connectors from bulkhead.
33. Remove the ground wire on left strut tower.
34. Remove the nuts and left bracket for harness connectors from bulkhead.
35. Remove the rubber seal for plenum chamber cover.
36. Remove the plenum chamber cover
37. Remove the wiper arms and the cowl grill
38. Remove the strut tower brace if equipped.
39. Remove the E box cover and box electrical connections.
40. Remove the bolts and remove cover from electronics box at left in engine compartment.
41. Release retaining tabs and remove the fuse holder and the relay carrier.
42. Remove the bolt for electrical wiring connection
43. Disconnect all electrical connections at rear on connector strip.
44. Disengage engine wiring harness at electronics box and move the wiring aside.
45. Remove the bolts and remove cover from electronics box at right in engine compartment.
46. Carefully pry off retaining clip with a screwdriver and remove ECM from electronics box.
47. Leave the ECM connected at wiring harness.
48. Remove or disconnect any additional hoses or electrical connections that would interfere with engine removal.
49. Have a second technician depress and hold brake pedal and remove the collar bolt at left and right axle shafts.

➡To loosen collar bolt for drive axle, the wheel bearing must not be under load (vehicle must not be standing on its wheels).

50. Remove the bolts for stabilizer bar uniformly on both sides.
51. Remove the stabilizer bar.
52. Disconnect electrical connector at level control system sensor.
53. Disconnect connecting link at control arm.
54. Unbolt suspension strut from control arm.
55. Disconnect the control arm from the sub frame and pivot the arms outwards on both sides of the vehicle.
56. Remove the axle shafts from the steering knuckle.
57. Unbolt the transverse beam from the sub frame.
58. Unhook shift cable.

59. Remove the transverse beam.
60. Remove the bolts at transmission/driveshaft flange.
61. Remove the left and right axle shafts.
62. Position a lift table such as VAS 6131 with support set VAS 6131/10 and VAS 6131/11.
63. Remove the bolts at tunnel cross member.
64. Remove the bolts at the sub frame, mark installation position of sub frame and of both engine bearing consoles to long members using a felt-tip marker.
65. Remove the bolts in a diagonal sequence using multiple passes.
66. Make sure that all hoses and lines between engine, transmission, sub frame and body have been disconnected.
67. While lowering, carefully guide engine/transmission subassembly with sub frame out of engine compartment in order to prevent damage.
68. Slowly lower engine/transmission subassembly.

To install:

69. Tighten the engine to transmission bolts as follows:
    a. M12 bolts to 65 Nm.
    b. M10 X 150 bolts to 65 Nm.
    c. M10 X 70 bolts to 45 Nm.
70. Tighten the sub frame bolts as follows, refer to the illustration for bolt locations:
71. Position 1 : 50 Nm.
72. Position 2 : 150 Nm.
73. Position 3 : 150 Nm.
74. Position 4 : 75 Nm.
75. Install the remaining components in the reverse order of removal, please note the following specifications:
    a. Tighten all M6 bolts to 9 Nm.
    b. Tighten all M8 bolts to 20 Nm.
    c. Tighten all M10 bolts to 40 Nm.
    d. Tighten all M12 bolts to 65 Nm.
    e. Tighten the drive plate to torque to 85 Nm using new bolts.

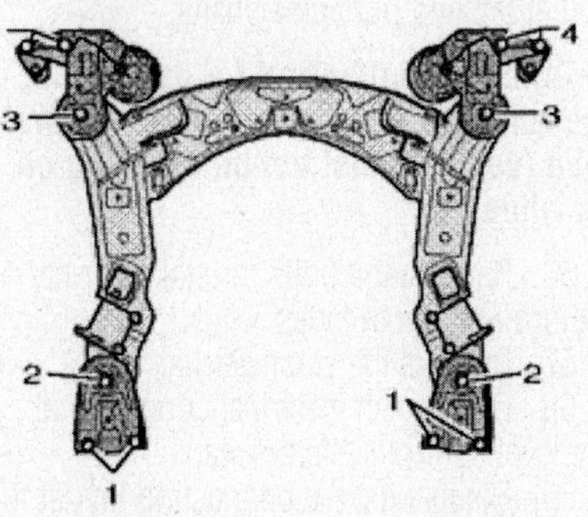

Sub frame bolt locations—2005–06 A6 with the 3.2L and 4.2L engines

    f. Tighten the engine support to cylinder block to 40 Nm.
    g. Tighten the engine mount console to longitudinal member to 75 Nm.
    h. Tighten the tunnel cross member to body to 40 Nm.
    i. Tighten the pressure line to the power steering pump to 47 Nm.
    j. Tighten the fuel supply line to 22 Nm.
    k. Tighten the torque bracket to engine to 40 Nm.
    l. Tighten the torque support stop to lock carrier to 23 Nm.

2002-04 4.2L ENGINE

➡The engine is removed from the front together with the transmission.

➡All tie wraps that are loosened or removed in order to remove the engine must be replaced or reinstalled to the same place when the engine is reinstalled.

➡The battery is located in the luggage compartment to the right and under the cover.

1. Before servicing the vehicle, refer to the precautions section.
2. Drain the cooling system.
3. Relieve the fuel system pressure.
4. Drain the transaxle fluid.
5. Remove or disconnect the following:
    - Negative battery cable
    - Intake air grille
    - Noise insulation panel
    - Both front wheels and wheel housings
    - Front bumper
    - Washer hoses and jets for headlight washers, if equipped
    - Intake air duct between air cleaner and throttle body
    - Engine cover
    - Coolant hose clamps
    - Radiator shroud
    - Hydraulic line brackets at bottom of radiator
    - Temperature sensor and pressure switch at radiator
    - Cooling loop for power steering fluid and place to side
    - A/C condenser and tie up to side (to relieve strain on lines and radiator)
    - Left and right headlight connectors
    - Left and right indicator lights connector
    - ATF lines for automatic transmission
    - Coolant hose at upper right radiator
    - Thermal switch
    - Horn connectors
    - ABS unit cover
    - Connectors in front of ABS unit
    - Left and right impact absorbers
    - Hood release cable
    - Radiator shroud with radiator
    - PCV and throttle body hoses
    - Air cleaner hose
    - Frequency vale bracket from air cleaner
    - MAF sensor connector
    - Air cleaner housing with mass air flow sensor and air intake duct
    - Oil level sensor connector
    - Viscous fan (left hand thread)
    - Accessory drive belt
    - Hydraulic pump belt pulley (save washers and reinstall)
    - Hydraulic pump from bracket and move it aside
    - A/C refrigerant line clamp
    - A/C compressor, lines remain connected
    - ATF lines at engine
    - Torque support with bonded rubber bushing at left-front of engine
    - Left and right side oxygen sensor connectors
    - Fuel supply and return lines
    - Coolant expansion tank hoses and tank
    - Coolant level sensor at expansion tank
    - Coolant hose, connectors and ground wire at sides and back of cylinder head
    - Ignition coil connectors
    - Knock sensor connectors
    - Fuel injector connectors for cylinder bank 1-4
    - Upper and lower exhaust pipe bolts
    - Exhaust pipe from retainer

➡Do not bend the flexible connection on exhaust pipe more than 10 degrees or it may be damaged

    - Double clamp at Three Way Catalytic Converter (TWC)
    - Exhaust pipe with TWC
6. For all-wheel drive vehicles, mark alignment of driveshaft to transmission output, disconnect the driveshaft, and install support for the driveshaft.
7. For all vehicles, remove or disconnect the following:
    - Left and right drive axles
    - Alternator wiring
    - Bracket for protective cover under vehicle
    - Oil filter
    - Ground cable and starter bracket
    - Starter wiring and starter

➡Before unbolting torque converter, mark alignment with drive plate

- Torque converter
- Left and right engine mounts

8. Install brackets VAG 3180 on both rear engine lifting points.
9. Attach lifting device 3033 to both brackets evenly.
10. Slide in assembly crane VAG 1202 A (500 kg), and attach lifting device 3033.
11. Raise engine carefully and pull out toward front.

To install:

12. Installation is the reverse of the removal procedure, while using the following torque values:
    - A/C compressor bracket to engine: 18 ft. lbs. (25 Nm)
    - A/C compressor to bracket: 18 ft. lbs. (25 Nm)
    - Drive axles to flange shafts: 33. ft. lbs. (45 Nm)
    - Engine support (right front) to engine: 33 ft. lbs. (45 Nm)
    - Engine support to body: 37 ft. lbs. (50 Nm)
    - Generator to engine 8mm bolt: 18 ft. lbs. (25 Nm)
    - Generator to engine 10mm bolt: 30 ft. lbs. (40 Nm)
    - Exhaust manifold to cylinder head: 18 ft. lbs. (25 Nm)
    - Exhaust pipe to exhaust manifold: 30 ft. lbs. (40 Nm)
    - Torque converter to drive plate: 63 ft. lbs. (85 Nm)

2005-06 4.2L ENGINE

1. Before servicing the vehicle, refer to the precautions section.

CAUTION

Before removing the engine, secure the vehicle to prevent it from tipping over. The luggage compartment must be empty.

2. Remove the luggage compartment floor trim.

➡So that the front wheels can still be turned with the battery disconnected, the battery must only be disconnected with the ignition key inserted.

3. Disconnect the negative battery cable.
4. Remove the front engine cover.
5. Remove the rear engine cover.
6. Open cap of coolant expansion tank after the engine has fully cooled.
7. Remove the both front wheels.

➡Secure the brake rotors with wheel bolts.

8. Loosen quick-release fasteners and remove sound insulation.
9. Place a drip tray under the engine.
10. Remove the drain plug on coolant thermostat housing and drain coolant from engine.
11. Disconnect lower coolant hose from radiator and drain radiator coolant.
12. Remove the bumper.
13. Remove the lock carrier and attachments.
14. Disconnect connector from the secondary air injection pump and move the wiring aside.
15. Disconnect the following hoses:
    a. Ventilation hose.
    b. Variable intake manifold hose.
    c. Secondary Air Injection (AIR) pump.
    d. Disconnect coolant hose from the thermostat housing.
    e. Coolant expansion tank.
    f. Coolant line.
16. Remove the coolant expansion tank.
17. Disconnect electrical connector to Engine Coolant Level (ECL) warning switch at the bottom of coolant expansion tank.
18. Clamp off hose of power steering fluid reservoir using hose clamps up to 25mm diameter.
19. Disconnect hose for power steering .
20. Unclip fuel hose.
21. Disconnect electrical connector from mass air flow (MAF) sensor and separate the sensor from air intake hose.
22. Remove the upper part of air filter housing with mass air flow (MAF) sensor.
23. Separate hose for secondary air injection (AIR) system.
24. Disconnect electrical connector from the air flap positioner.
25. Remove the pin from spreader clip and remove lower part of air filter housing.
26. Pull resonator up and unclip it.
27. Properly relieve the fuel system pressure.
28. Disconnect fuel line.
29. Disconnect vacuum hose for brake booster at bulkhead.
30. Remove the intake hose.
31. Disconnect vacuum hose to EVAP canister.
32. Remove the ground wire on right strut tower.
33. Remove the nuts and the right bracket for harness connectors from bulkhead.
34. Remove the ground wire on left strut tower.
35. Remove the nuts and left bracket for harness connectors from bulkhead.
36. Remove the rubber seal for plenum chamber cover.
37. Remove the plenum chamber cover.
38. Remove the wiper arms and the cowl grill.
39. Remove the bolts and remove cover from electronics box at left in engine compartment.
40. Release retaining tabs and remove the fuse holder and the relay carrier.
41. Remove the bolt for electrical wiring connection.
42. Disconnect all electrical connections at rear on connector strip.
43. Disengage engine wiring harness at electronics box and move the wiring aside.
44. Remove the bolts and remove cover from electronics box at right in engine compartment.
45. Carefully pry off retaining clip with a screwdriver and remove ECM from electronics box.
46. Leave the ECM connected at wiring harness.
47. Remove or disconnect any additional hoses or electrical connections that would interfere with engine removal.
48. Remove the bolts and remove front transverse beam.
49. Remove the heat shield for the driveshaft.
50. Remove the bolts at transmission/driveshaft flange.
51. Push driveshaft toward rear final drive the Constant velocity (CV) joints can move axially.
52. Tie driveshaft up and to side against heat shield
53. Remove the heat shield for left driveshaft.
54. Remove the left and right axle shafts from transmission.
55. Have a second technician depress and hold brake pedal and remove the collar bolt at left and right axle shafts.

➡To loosen collar bolt for drive axle, the wheel bearing must not be under load (vehicle must not be standing on its wheels).

56. Remove the bolts for stabilizer bar uniformly on both sides.
57. Remove the stabilizer bar.
58. Disconnect electrical connector at the level control system sensor.
59. Disconnect the connecting link at the control arm.
60. Unbolt the suspension strut from the control arm.

61. Loosen the nut of the guide link ball joint until it is flush with end of thread.

➡Ensure that the ball joint protective boot does not become damaged when pressing off. Make sure that both lever arms of ball joint puller are positioned parallel to one another while increasing force.

62. Secure ball joint puller against falling.
63. Using ball joint puller T40043 press guide link ball joint off of tapered seat.
64. Remove the nut and remove ball joint from guide link.
65. Loosen nut of ball joint at control arm until it is flush with end of thread.

➡Make sure that the ball joint protective boot does not become damaged when pressing off. Make sure that both lever arms of ball joint puller are positioned parallel to one another while increasing force.

66. Using ball joint puller T40042 press control arm ball joint off from tapered seat.
67. Remove the nut and remove ball joint from control arm.
68. Unbolt guide link and control arm at sub frame and remove them.
69. Perform these removal procedures on both sides of the vehicle.
70. Remove the bolts for support bracket from transmission.
71. Unhook shift cable.
72. Position a lift table such as VAS 6131 with support set VAS 6131/10 and VAS 6131/11.
73. Remove the bolts at tunnel cross member.
74. Remove the bolts at the sub frame, mark installation position of sub frame and of both engine bearing consoles to long members using a felt-tip marker.
75. Remove the bolts in a diagonal sequence using multiple passes.
76. Make sure that all hoses and lines between engine, transmission, sub frame and body have been disconnected.
77. While lowering, carefully guide engine/transmission subassembly with sub frame out of engine compartment in order to prevent damage.
78. Slowly lower engine/transmission subassembly.

To install:

79. Tighten the engine to transmission bolts as follows:
   a. M12 bolts to 65 Nm.
   b. M10 X 150 bolts to 65 Nm.
   c. M10 X 70 bolts to 45 Nm.

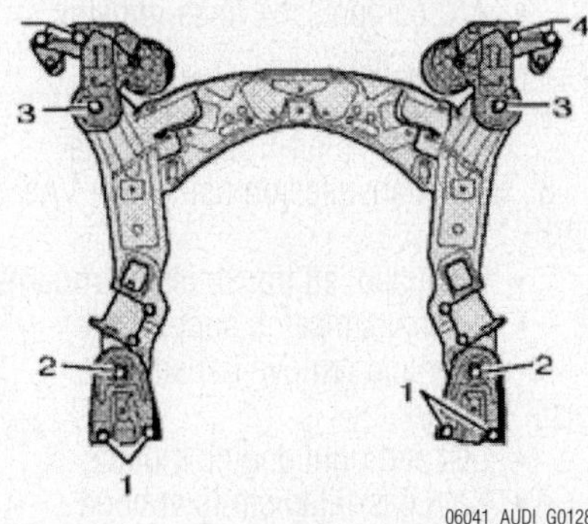

Sub frame bolt locations—2005–06 A6 with the 3.2L and 4.2L engines

80. Tighten the sub frame bolts as follows, refer to the illustration for bolt locations:
81. Position 1 : 50 Nm.
82. Position 2 : 150 Nm.
83. Position 3 : 150 Nm.
84. Position 4 : 75 Nm.
   a. Tighten all M6 bolts to 9 Nm.
   b. Tighten all M8 bolts to 20 Nm.
   c. Tighten all M10 bolts to 40 Nm.
   d. Tighten all M12 bolts to 65 Nm.
   e. Tighten the right engine support to cylinder block to 42 Nm.
   f. Tighten the engine mount to the engine support and to the engine console to 23 Nm.
   g. Tighten the pressure line to the power steering pump to 47 Nm.
   h. Tighten the transmission mount to sub frame to 23 Nm.
   i. Tighten the selector lever heat shield to 22 Nm.
   j. Tighten the exhaust pipe straps to 25 Nm.
   k. Tighten the heat shield to the transmission to 25 Nm.
   l. Tighten the fuel supply line to 25 Nm.
   m. Tighten the chassis ground wire to 25 Nm.
   n. Tighten the torque support stop to lock carrier to 28 Nm.

A8

➡The engine is removed from the front together with the transmission.

➡All tie wraps that are loosened or removed in order to remove the engine must be replaced or reinstalled to the same place when the engine is reinstalled.

➡The battery is located in the luggage compartment to the right and under the cover.

1. Before servicing the vehicle, refer to the precautions section.
2. Drain the cooling system.
3. Relieve the fuel system pressure.
4. Drain the transaxle fluid.
5. Remove or disconnect the following:
   - Negative battery cable
   - Intake air grille
   - Noise insulation panel
   - Front bumper
   - Intake air duct between air cleaner and throttle body
   - Engine cover
   - Coolant hose clamps at right-side belt guard
   - Electronics box cover by removing seven bolts
   - All connectors from Engine Control Module (ECM) and Transmission Control Module (TCM) at bulkhead
   - Cruise control module with relay and fuse bracket
   - Sealing strip between engine compartment and plenum chamber
   - Wiring harness from plenum panel, remove spacer sleeves, expose wiring and place down on engine
   - Vacuum line to brake booster unit
   - Vacuum line at cruise control module
   - Accelerator pedal cable
   - Coolant lines to heater core (supply and return) at vent valves
   - Heated Oxygen (HO_2S) sensor connectors on left and right sides and push out of holder
   - Ignition power output stage
   - Top part of air cleaner.
   - Harness connector at Mass Air Flow (MAF) sensor
   - Coolant vent line to expansion tank at radiator
   - Coolant supply line at expansion tank
   - Hydraulic reservoir
   - Vacuum line for intake manifold change-over valve, located at front by left headlight
   - Left and right knock sensor connectors at fuel rail
   - Harness connectors at fuel injectors
   - Fuel rail with fuel injectors, and place down to side
   - Exhaust manifold to front exhaust pipe on left and right sides
   - Crossmember
   - Three Way Catalytic Converter (TWC)
   - Exhaust system at retaining loop and remove
   - Exhaust pipes (left and right) from exhaust manifold

- TWC bracket along with spring
- TWC along with front pipe
- Heat shield above three way catalytic converter
- Heat shield for selector lever cable at transmission pan

6. For all-wheel drive vehicles, mark alignment of driveshaft to transmission output, disconnect the driveshaft, and install support for the driveshaft.
7. For all vehicles, remove or disconnect the following:
   - Selector lever cable at bracket
   - Left and right drive axles
   - Left drive axle shield
   - Starter cable connector at junction box on right side long member
   - Junction box cover
   - Electrical connectors
   - Starter cable in junction box and at bracket
   - Generator air guide
   - Transmission oil cooler
   - Lower radiator hose
   - Engine Coolant Temperature (ECT) sensor above hose
   - Torque support (note washers)
   - Radiator air guide on left and right sides
   - High pressure switch connector
   - Left and right headlight connectors
   - Left headlight
   - Outside temperature sensor at bottom in front of radiator, cut open cable clip, and expose cable
   - Cooling loop for power steering fluid and place to side
   - A/C condenser and tie up to side (to relieve strain on lines and radiator)

✲✲ CAUTION

Disconnect A/C refrigerant line brackets and support points only. DO NOT open the air conditioning refrigerant circuit. Avoid damage from bending. Refrigerant lines kink easily.

- Coolant fan wiring
- Hood supports at both front fenders
- Lock carrier bolts located below hood support (note washers)
- Impact absorbers for bumper on left and right sides
- Lock carrier along with radiator
- Accessory drive belt
- Hydraulic pump belt pulley (note washers)
- Hydraulic pump from mounting bracket and place down at side member
- A/C compressor, lines remains connected
- Torque support with bonded rubber bushing at left-front of engine

8. Raise transmission using tool VAG 1383.
   - Right and left transmission mounts
   - Left transmission support
9. Lower and remove transmission hoist.
   - Left and right engine mounts
   - Both gas-filled struts at hood
10. Position hood in vertical position and support with auxiliary tool.
11. Install brackets VAG 3180 on both rear engine lifting points.
12. Attach lifting device 3033 to both brackets evenly.
13. Slide in assembly crane VAG 1202 A (500 kg), and attach lifting device 3033.
14. Raise engine carefully and pull out toward front.

To install:

15. Installation is the reverse of the removal procedure, while using the following torque values:
   - A/C compressor bracket to engine: 18 ft. lbs. (25 Nm)
   - A/C compressor to bracket: 18 ft. lbs. (25 Nm)
   - Drive axles to flange shafts: 33. ft. lbs. (45 Nm)
   - Engine support (right front) to engine: 33 ft. lbs. (45 Nm)
   - Engine support to body: 37 ft. lbs. (50 Nm)
   - Generator to engine 8mm bolt: 18 ft. lbs. (25 Nm)
   - Generator to engine 10mm bolt: 30 ft. lbs. (40 Nm)
   - Exhaust manifold to cylinder head: 18 ft. lbs. (25 Nm)
   - Exhaust pipe to exhaust manifold: 30 ft. lbs. (40 Nm)
   - Hydraulic pump bracket to engine: 18 ft. lbs. (25 Nm)

Water Pump

REMOVAL & INSTALLATION

A4

1.8L ENGINE

➡The coolant pump is bolted to the brackets for the alternator, power steering pump and cooling fan. To gain access to the front of the engine, the front bumper must be removed and the hood lock carrier assembly moved forward to the service position.

1. Before servicing the vehicle, refer to the precautions section.
2. Drain the engine coolant.
3. Turn the ignition switch to the **OFF** position.
4. Remove or disconnect the following:
   - Negative battery cable
   - Front bumper assembly and move the front inner fender lock carrier assembly forward to the lock carrier service position, on A4 models
5. With the front bumper removed, move the lock carrier assembly into the service position as follows:
   a. Release the 3 quick-release screws on the front of the lower engine splash shield.
   b. Unbolt the air guide between the lock carrier and the air filter assembly.
   c. Remove the 2 bolts that attach the lock carrier assembly to the side of the front fender and the 2 front bolts that mount the carrier to the top of the fender.
   d. Remove the right side top outer bumper energy absorbing strut mounting bolt and install support tool 3369 into the top outer threaded holes on both the left and right sides of the bumper energy absorbing strut mounting surface.

➡The 2 front bumper-to-bumper energy absorbing strut fasteners can be substituted for support tool 3369.

   e. Remove the remaining bumper energy absorbing strut mounting bolts.
   f. Remove the remaining 2 bolts that attach the lock carrier assembly to the top of the front fenders. Pull the lock carrier assembly forward until the rearmost bolt holes of the carrier align with the front-most threaded mounting points on the top of the front fender. Install the 2 carrier mounting bolts through the carrier into threaded mounting points of front fender to secure the lock carrier assembly in this position.

➡The 2 front bumper mounting fasteners can be substituted for Support tool 3369.

   g. Remove the remaining bumper energy absorbing strut bolts and pull the lock carrier out to the stop.
   h. To secure the lock carrier, install the appropriate M6 bolts into the rear of the lock carrier and fender.
6. Remove or disconnect the following:
   - Accessory drive belt
   - Cooling fan
   - Lower engine slash shield
7. Loosen the clamps for the coolant hoses at the water pump.

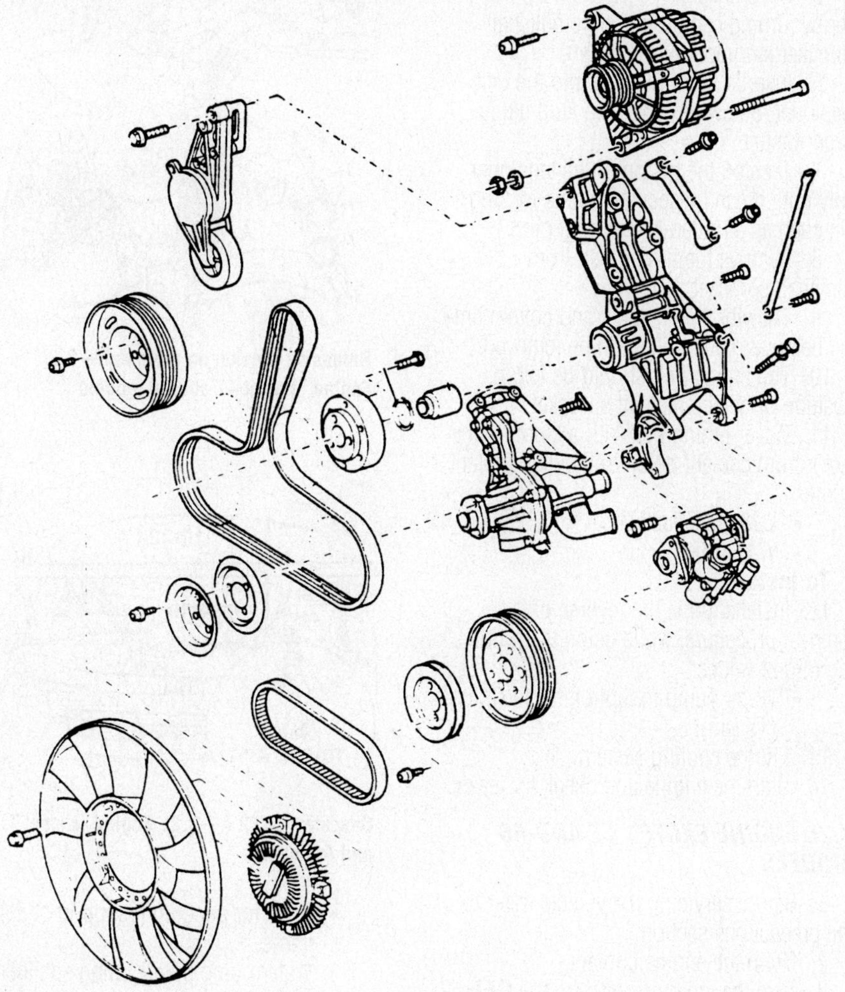

The water pump, alternator and power steering pump all mount to the same engine bracket—1.8L engines

8. Remove or disconnect the following:
   - Intake air duct between the intake manifold and the charge air cooler
   - Alternator mounting bolts and slide it forward
   - Alternator wiring

9. Unbolt the following supports and brackets for the alternator, power steering pump and engine cooling fan:
   - Intake manifold support
   - Support for the engine torque bracket
   - Brace to the cylinder block
   - Alternator brackets
   - Power steering pump brackets
   - Cooling fan brackets. Position the brackets for the alternator, power steering pump and engine cooling fan to the left side using a piece of wire
   - Coolant hoses from the pump
   - Coolant hoses from the thermostat housing
   - Coolant pump housing from the timing belt cover
   - Coolant pump
   - Impeller housing from the pump housing

10. Clean all gasket and O-ring sealing surfaces.

To install:

11. Install or connect the following:
    - Coolant pump to the pump housing, using a new gasket. Torque the bolts to 84 inch lbs. (10 Nm).
    - Coolant pump, using a new gasket and O-ring. Torque the bolts in sequence to 18 ft. lbs. (25 Nm).
    - Coolant pump housing to the timing belt cover. Torque it to 84 inch lbs. (10 Nm).
    - Coolant hoses to the pump and thermostat housing
    - Brackets. Torque the bolts to 18 ft. lbs. (25 Nm)
    - Alternator and wire connectors
    - Air intake duct between the intake manifold and the charge air cooler

12. The remaining steps are the reverse of the removal procedure noting the following items:
    - Fill the engine with coolant
    - Verify that the key is in the **OFF** position before connecting the battery
    - Fully close all power windows to stop, operate all window switches for at least 1 second in the close

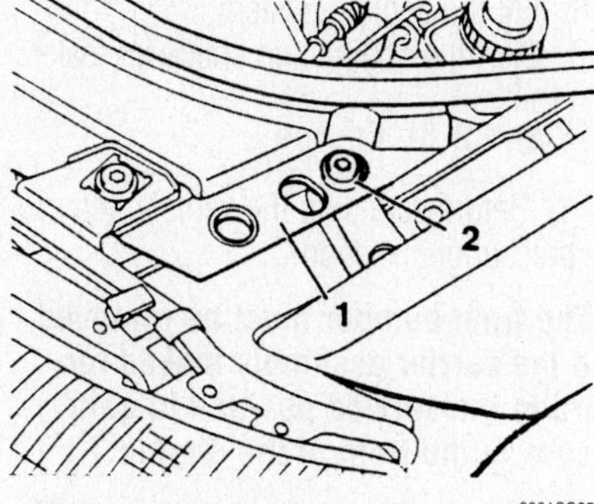

The hood lock carrier assembly (1) secured in the service position with the mounting fastener (2) installed—A4 models

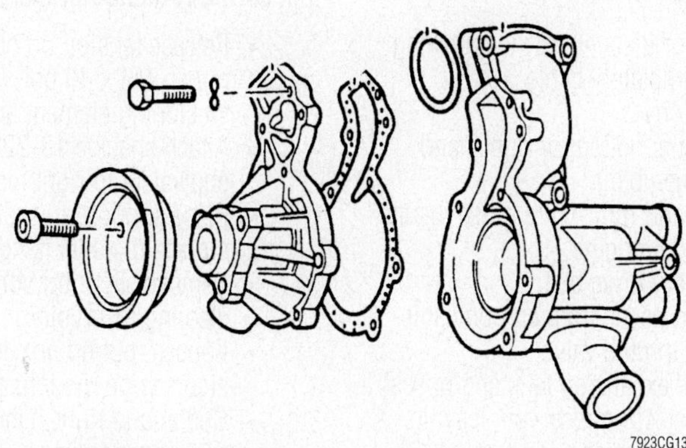

Exploded view of the water pump, housing and related components—1.8L Engine

- direction to activate the one touch opening/closing function
- After installing the lock carrier, check the wiring for proper routing near the cooling fan

TT

2002–03 MODELS

1. Before servicing the vehicle, refer to the precautions section.
2. Drain the engine coolant.
3. Turn the ignition switch to the **OFF** position.
4. Remove or disconnect the following:
   - Negative battery cable
   - Engine cover
   - Air duct at bottom of right-hand cross member
   - Center and right-hand noise insulation under engine
   - Accessory drive belt
   - Vacuum hoses from carbon canister and throttle valve
   - Coolant expansion tank and hoses
   - Power steering reservoir, leaving hoses connected
   - Upper and center timing belt covers
   - Timing belt from camshaft sprocket, but leave in position on crankshaft sprocket
   - Lower air duct
   - Water pump

To install:

5. Installation is the reverse of the removal procedure, while using the following torque values:
   - Water pump mount bolts: 11 ft. lbs. (15 Nm).
6. Fill the cooling system.
7. Start the engine and check for leaks.

2004–06 MODELS

1. Before servicing the vehicle, refer to the precautions section.
2. Drain the engine coolant.
3. Turn the ignition switch to the **OFF** position.
4. Remove or disconnect the following:
   - Negative battery cable
   - Engine cover
   - Air duct at bottom of right-hand cross member
   - Center and right-hand noise insulation under engine
   - Accessory drive belt
   - Vacuum hoses from carbon canister and throttle valve
   - Coolant expansion tank and hoses
   - Power steering reservoir, leaving hoses connected
   - Upper and center timing belt covers
5. Set crankshaft to TDC of No. 1 cylinder by turning central bolt on crankshaft sprocket in direction of rotation.
6. Insert a M5 x 55 stud into the belt tensioner. Screw hex nut onto stud using large washer.
7. Tension the piston of the tensioner only until it can be secured with a locking pin such as an Allen key or split pin.
8. Remove the timing belt from camshaft sprocket.
9. The vibration damper and bottom timing belt guard do not have to be removed.
10. The timing belt should be left in position on the crankshaft sprocket.
11. Cover timing belt with a cloth to protect it from coolant before removing coolant pump.
    - Lower air duct
    - Water pump

To install:

12. Installation is the reverse of the removal procedure, while using the following torque values:
    - Water pump mount bolts: 11 ft. lbs. (15 Nm).
13. Fill the cooling system.
14. Start the engine and check for leaks.

3.2L ENGINE EXCEPT TT AND A6 MODELS

1. Before servicing the vehicle, refer to the precautions section.
2. Drain the engine coolant.
3. Turn the ignition switch to the **OFF** position.
4. Remove or disconnect the following:
   - Negative battery cable
   - Air hose at throttle valve
   - Covers (right side and at lock carrier above coolant tank)

➡Before removing ribbed belt, mark direction of rotation with chalk or a felt pen. If a used belt rotates in the wrong direction when reinstalled, this can result in breakage. When installing belt, ensure it is correctly seated in the pulleys.

- Release tension on belt by screwing an M8 x 40 bolt into bore on tensioning element and remove belt
- Attach shackle 10-222 A/12 or equivalent to right rear lifting eye
- Place support bracket 10-222 A on surface of panel flange
- Remove securing bolt for power steering reservoir
- Loosen, but do not remove, left side engine mounting bolts
- Center and right-hand noise insulation under engine
- Lower radiator hose

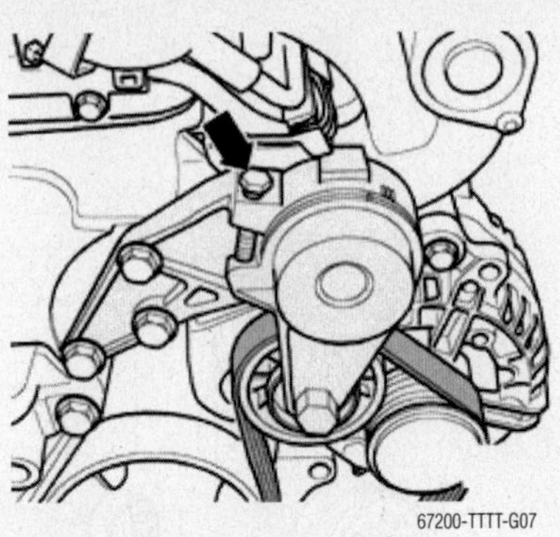

Releasing tension on drive belt—3.2L Engine, except TT and A6 models

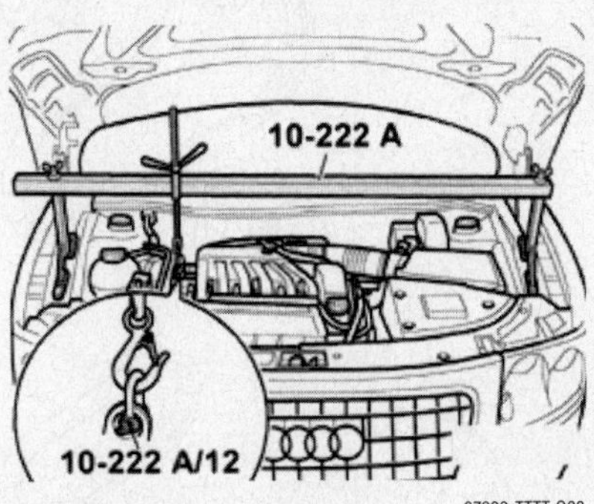

Shackle 10-222 A—3.2L Engine, except TT and A6 models

- Bolts for pendulum support at subframe
- Water pump pulley using spanner VAG 1590 or equivalent
- Water pump mounting bolts
- Lower engine approximately 65 mm
- Swing coolant pump forward and remove

To install:

5. Installation is the reverse of the removal procedure, while using the following torque values:
   - Water pump mount bolts: 6 ft. lbs. (8 Nm)
   - Pulley to pump 15 ft. lbs (20 Nm)
6. Fill the cooling system.
7. Start the engine and check for leaks.

2.7L and 2.8L Engine

1. Before servicing the vehicle, refer to the precautions section.

➡The front bumper must be removed and the carrier assembly moved forward to the service position to gain access to the front of the engine.

2. Turn the ignition switch to the **OFF** position.
3. Remove or disconnect the following:
   - Negative battery cable

- Front bumper assembly and move the hood lock carrier assembly forward and lock the carrier in the service position, on A4 and A6 models

4. On A6 models, to move the hood lock carrier into the service position, perform the following:
   a. Tag and remove any wiring or connector that would inhibit moving the carrier.
   b. Remove the 3 quick-release screws on the front noise insulation panel.
   c. Unbolt the air guide between the lock carrier and the air filter.
   d. If installed, remove the retaining clamps for the wiring harness at the left side of the radiator frame.
   e. Remove the right side top outer bumper energy absorbing strut mounting bolt and install support tool 3369 into the top outer threaded holes on both the left and right side frame rails.

➡The 2 front bumper mounting fasteners can be substituted for the Support tool 3369.

   f. Remove the remaining bumper energy absorbing strut bolts and pull the lock carrier out to the stop.
   g. To secure the lock carrier, install the appropriate M6 bolts into the rear of the lock carrier and fender.
5. Drain the cooling system.
6. Remove or disconnect the following
   - V-belts and the timing belt covers
   - Timing belt.
   - Water pump and discard the gasket or O-ring

To install:

7. Installation is the reverse of the removal procedure.
8. Tighten the water pump retaining bolts to 89 inch lbs. (10 Nm).
9. Reinstall the timing belt.
10. Properly tension the belt with the water pump. Refer to the necessary service procedures.
11. Fill and bleed the cooling system.

3.0L Engine

1. Before servicing the vehicle, refer to the precautions section.
2. Turn the ignition switch to the **OFF** position.
3. Remove or disconnect the following:
   - Negative battery cable
   - Engine coolant
   - Timing belt
   - Camshaft gears from right cylinder head
   - Idler pulley and retaining bracket

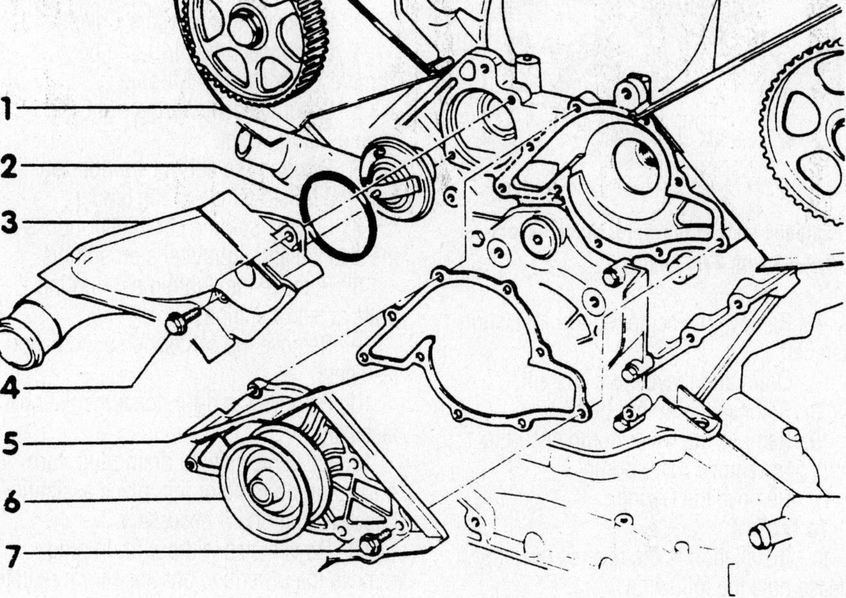

Exploded view of the water pump and related components—2.8L engine

   - Timing belt rear guard
   - Eccentric pulley for timing belt
   - Water pump

To install:

4. Install or connect the following:
   - Water pump. Tighten the bolts to 89 inch lbs. (10 Nm).
   - Eccentric pulley for timing belt
   - Timing belt rear guard. Tighten the bolts to 89 inch lbs. (10 Nm).
   - Idler pulley and retaining bracket. Tighten the bolts to 89 inch lbs. (10 Nm).
   - Camshaft gears from right cylinder head
   - Timing belt
   - Negative battery cable
5. Fill the cooling system.
6. Start the engine and check for leaks.

A6 and A8

3.2L ENGINE

1. Before servicing the vehicle, refer to the precautions section.
2. Drain the cooling system.
3. Remove or disconnect the following:
   - Negative battery cable
   - Water pump

To install:

4. Install or connect the following:
   - Water pump. Tighten the bolts to 89 inch lbs. (10 Nm).
   - Negative battery cable
5. Fill the cooling system.
6. Start the engine and check for leaks.

4.2L ENGINE

1. Before servicing the vehicle, refer to the precautions section.
2. Drain the cooling system.
3. Remove or disconnect the following:
   - Negative battery cable
   - Timing belt
   - Water pump

To install:

4. Install or connect the following:
   - Water pump. Tighten the bolts to 89 inch lbs. (10 Nm).
   - Timing belt
   - Negative battery cable
5. Fill the cooling system.
6. Start the engine and check for leaks.

A3

2.0L ENGINE

1. Before servicing the vehicle, refer to the precautions section.
2. Remove the coolant expansion tank cap.
3. Disconnect the electrical connector.
4. Disconnect the intake connection using spring-type clip pliers.
5. Remove the engine cover.

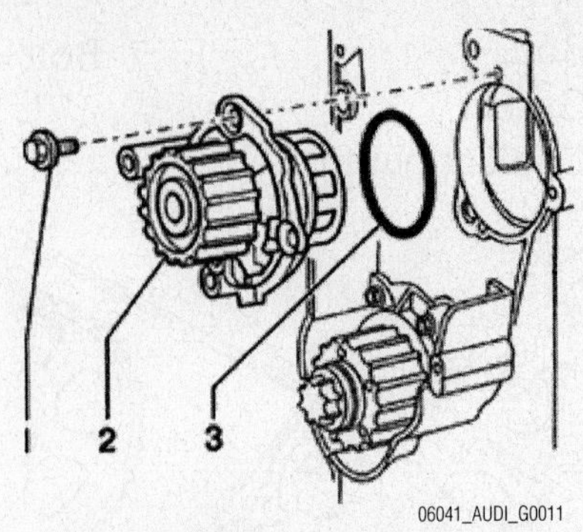

Exploded view of the water pump assembly—A3 with 2.0L engine

6. Remove the center sound insulation fasteners.
7. Drain and recycle the coolant.
8. Remove the timing belt.
9. Remove the water pump securing bolts and remove water pump.
10. Remove the O-ring.

To install:

11. Installation is the reverse of removal, please note the following:
   a. Clean and/or smooth O-ring sealing surface.
   b. Moisten new O-ring with coolant.
   c. Insert water pump.
   d. Make sure the sealing plug in housing points downward.
   e. Tighten bolts of water pump to 15 Nm.
   f. Install the timing belt.
   g. Install accessory drive belt
12. Fill the cooling system.

3.2L ENGINE

1. Before servicing the vehicle, refer to the precautions section.
2. Remove the coolant expansion tank cap.

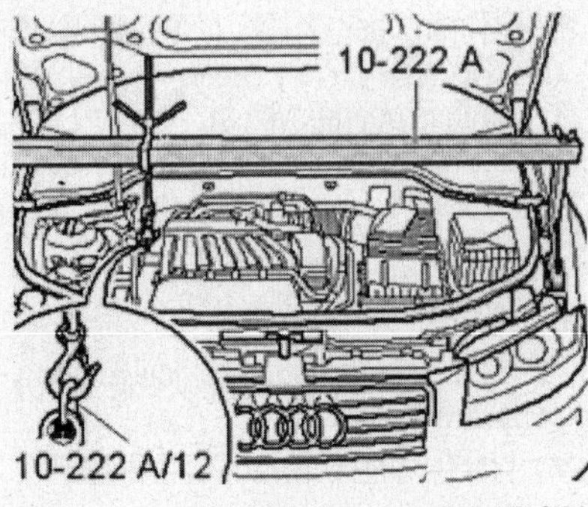

Secure shackle to the right rear engine lifting eyelet, then position the engine support tool onto the bolted flange of the fenders—A3 with 3.2L engine

➡Before removing accessory drive belt, mark the turning direction on it with chalk or a felt tip pen. A reversed direction of rotation can cause damage to the belt under operating conditions.

3. Remove the accessory drive belt.
4. Disconnect the air guide hose at Throttle Valve Control Module.
5. Secure shackle to the right rear engine lifting eyelet.
6. Position the engine support tool onto the bolted flange of the fenders.
7. Engage spindle carabineer hooks at the shackle and lightly tension spindle.
8. Remove the subframe mounting bolts at left of engine.
9. Remove the center noise insulation fasteners.
10. Remove the right noise insulation fasteners.
11. On vehicles with drain plug, turn drain plug on radiator left, place assisting hose on supports if necessary.
12. On vehicles without drain plug, remove the lower coolant hose from radiator by removing retaining clamps.
13. Disconnect the electrical connector at Engine Coolant Temperature (ECT) sensor.

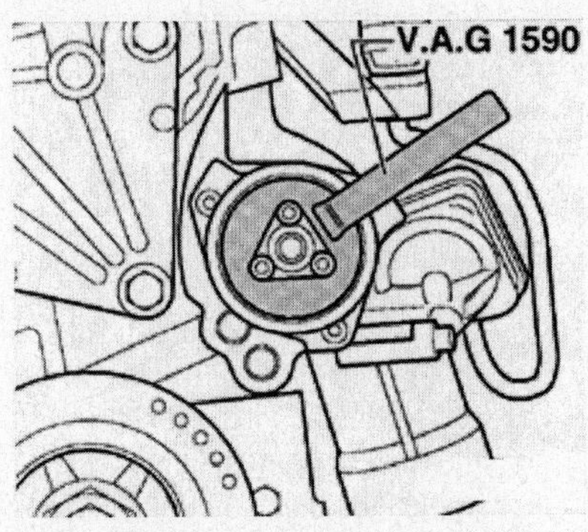

Remove the belt pulley while securing pulley with the water pump wrench—A3 with 3.2L engine

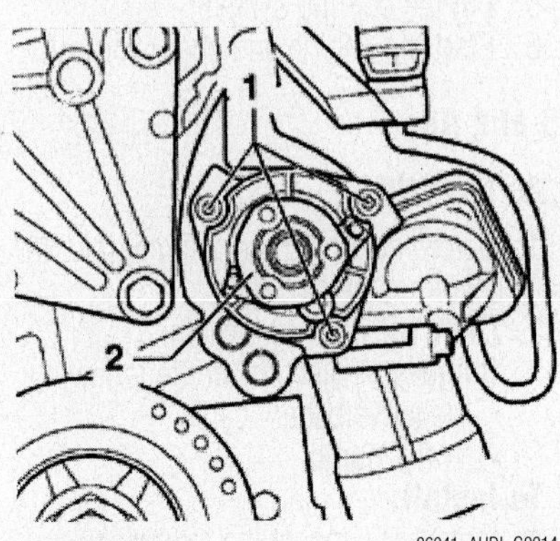

Water pump mounting—A3 with 3.2L engine

14. Remove the lower coolant hose to After-Run Coolant Pump and allow remaining coolant to drain.
15. Remove the pendulum support by removing bolts.
16. Unclip electrical wiring bracket to Oil Level Thermal sensor.
17. Lower the engine at the spindle about 60 mm.
18. Remove the belt pulley while securing pulley with the water pump wrench.

➡If engine is removed, the belt pulley must not be removed. Water pump bolts are then accessible through belt pulley holes.

19. Remove the bolts on water pump and swing the water pump forward.

To install:

20. Installation is the reverse of removal, please note the following:

➡Replace O-ring.

   a. Clean or smooth O-ring sealing surface.
   b. Coat new O-ring with coolant additive.
   c. Install the water pump, tighten bolts to 8 Nm.
21. Install the belt pulley, tighten the bolts to 20 Nm.
22. Install the accessory drive belt .
23. Fill the cooling system.
   a. Adjust engine mounts.

Heater Core

REMOVAL & INSTALLATION

A4 and TT

1. Perform the Output Diagnostic Test Mode (DTM) using the VAG 1551 function 03 by performing the following procedure:
   - Once the air flow flap closes, cancel the Output DTM by pressing the "C" button.

➡If equipped with power seats, move the seats as far rearward as possible. Also, obtain the anti-theft radio coding and the preset radio stations from the owner.

2. Discharge and recover the air conditioning system refrigerant.
3. Remove or disconnect the following:
   - Air plenum cover, the water guide and the dust/pollen filter, located at the right side

- Negative battery cable
- Positive battery cable and remove the battery
- Coolant recovery tank cap to relieve the pressure from the system

4. Drain the cooling system into a clean container for reuse.
5. Label and disconnect or clamp off the heater hoses from (at) the heater core.
6. Place a container under the right heater core tube, induce pressurized air to the left tube and blow compressed into the tube to drain excess coolant.
7. Remove or disconnect the following:
   - Air conditioning system vacuum supply hose and attach it at the heater core inlet/outlet
   - Heater core-to-chassis boot
   - Refrigerant lines-to-evaporator core clamp bolt, disconnect the refrigerant line clamp and discard the O-rings. Plug the openings to prevent contamination.
   - Evaporator core-to-chassis boot
   - Low pressure switch electrical connector and secure it to the evaporator fixture
8. Remove the glove box, the driver's side lower shelf and the instruments panel center section by removing or disconnecting the following:
   - 5 glove box-to-instrument panel bolts and the glove box
   - 3 clips, remove the 3 stowage compartment bolts and the stowage compartment at the driver's side
   - Rear console
   - 3 the knobs, remove the 4 center instrument panel trim bolts, the 2 screws and the trim
   - Radio
   - 4 front console-to-instrument panel bolts
   - Trim cover and the console-to-chassis nut at the left side of the front console
   - Pedal assembly at the driver's side
9. Place the front wheels in the straight-ahead position.
10. Remove the driver's side air bag module by removing or disconnecting the following:
    - Air bag module-to-steering wheel screws using a T30 Torx bit; the screws are located on both sides of the steering wheel
    - Air bag module and disconnect the electrical connector
    - Place the air bag module in a safe place with the front facing upward.
11. Remove the steering wheel by removing or disconnecting the following:
    - Steering wheel nut
    - Horn and air bag electrical connectors
    - 4 carrier unit-to-steering wheel bolts and the carrier unit
    - Steering wheel
12. At the steering column-to-instrument panel connection, secure the steering column wire to keep the steering column from sliding apart.
13. Remove or disconnect the following:
    - Steering column-to-steering gear bolt
    - Electronic box electrical connectors located in the left air plenum chamber
    - All instrument panel-to-chassis electrical connectors
14. Remove the instrument panel by removing or disconnecting the following:
    - Instrument panel-to-chassis bolts
    - Any necessary electrical connectors
    - Instrument panel
15. Remove or disconnect the following:
    - Ducts, heater/air conditioning housing assembly-to-chassis bolts and the assembly
    - Heater core-to-heater/air conditioning housing screws
    - Press the heater core-to-heater housing catches and remove the

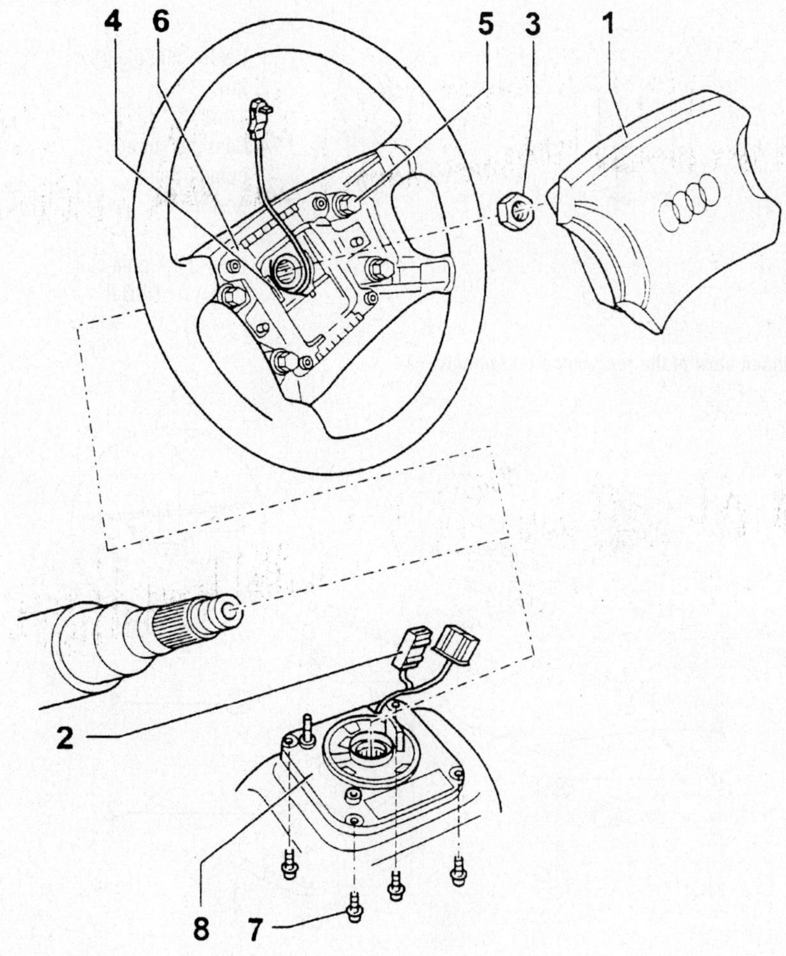

Exploded view of the air bag module, steering wheel and related components—A4

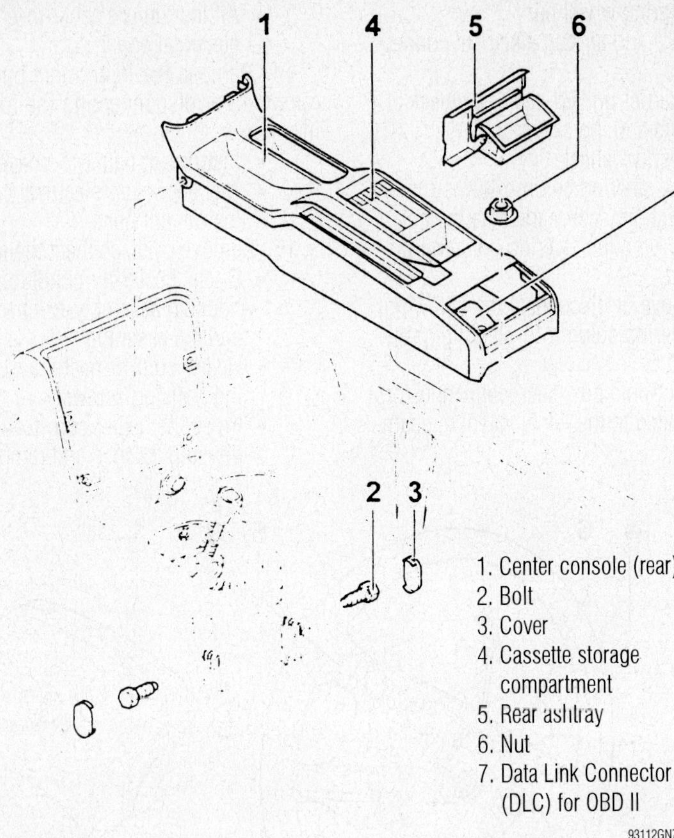

Exploded view of the rear console assembly—A4

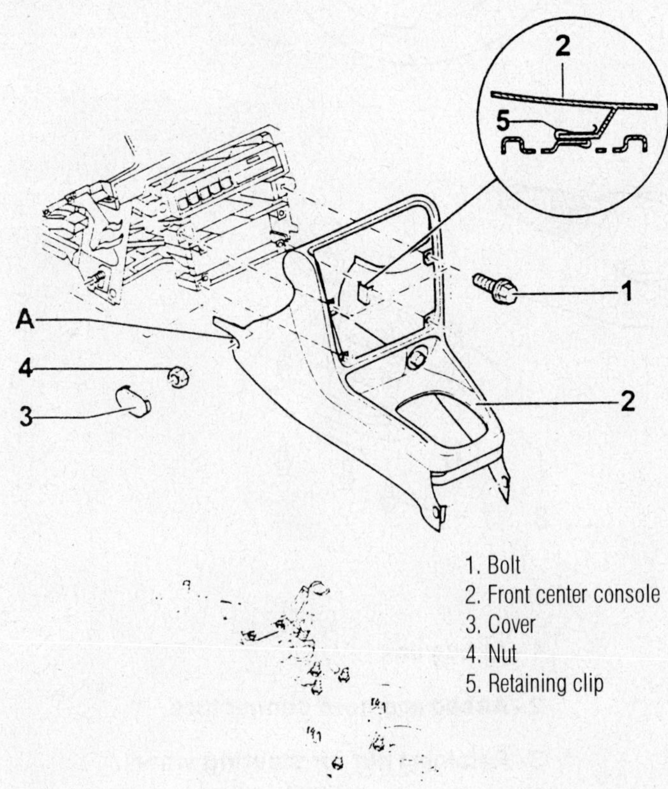

Exploded view of the front console assembly—A4

heater core from the heater/air conditioning housing

To install:

16. Install or connect the following:
    - Heater core to the heater/air conditioning housing and press the heater core into heater housing until the latches catch
    - Heater core-to-heater/air conditioning housing screws
    - Heater/air conditioning housing assembly and connect the ducts

17. Install the instrument panel by installing or connecting the following:
    - Instrument panel
    - Any necessary electrical connectors
    - Instrument panel-to-chassis bolts and torque the bolts to 44 inch lbs. (5 Nm)
    - All instrument panel-to-chassis electrical connectors
    - Electronic box electrical connectors located in the left air plenum chamber
    - Steering column-to-steering gear bolt

18. Install the steering wheel by installing or connecting the following:
    - Steering wheel
    - Carrier unit and the 4 carrier unit-to-steering wheel bolts
    - Horn and air bag electrical connectors
    - Steering wheel nut

19. Install the driver's side air bag module by installing or connecting the following:
    - Air bag module and connect the electrical connector
    - Torque the air bag module-to-steering wheel screws to 53 inch lbs. (6 Nm) using a T30 Torx bit; the screws are located on both sides of the steering wheel
    - Pedal assembly at the driver's side

20. Install the glove box, the driver's side lower shelf and the instruments panel center section by installing or connecting the following:
    - Console-to-chassis nut and the trim cover at the left side of the front console. Torque the nut to 31 inch lbs. (3.5 Nm)
    - 4 front console-to-instrument panel bolts and torque the bolts to 44 inch lbs. (5 Nm)
    - Radio
    - 4 center instrument panel trim, the 4 center instrument panel trim bolts, the 2 screws and the 3 the knobs. Torque the bolts to 44 inch lbs. (5 Nm) and the screws to 31 inch lbs. (3.5 Nm)

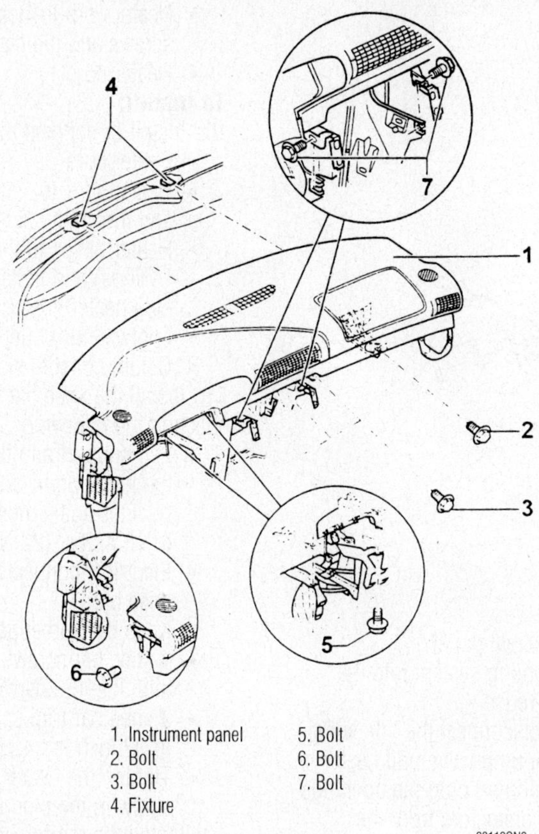

Exploded view of the instrument panel assembly—A4

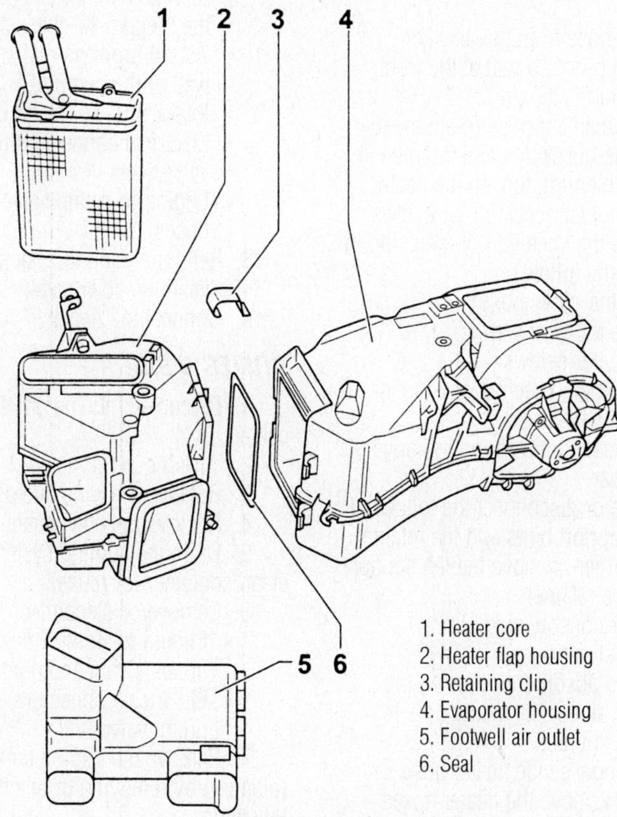

Exploded view of the heater core, heater/air conditioning housing and related components—A4

- Rear console
- Stowage compartment, the 3 stowage compartment bolts and engage the 3 clips. Torque the bolts to 44 inch lbs. (5 Nm)
- Glove box and the 5 glove box-to-instrument panel bolts; then, torque the bolts to 44 inch lbs. (5 Nm)

21. Install or connect the following:
    - Low pressure switch electrical connector
    - Evaporator core-to-chassis boot.
    - Refrigerant lines-to-evaporator core clamp and connect the refrigerant line clamp bolt making sure to use new O-rings
    - Heater core-to-chassis boot
    - Air conditioning system vacuum supply hose
    - Heater hoses to the heater core
22. Refill the cooling system.
23. Install the coolant recovery tank cap.
24. Install the battery and connect the positive (+) battery cable.
25. Connect the negative (−) battery cable.
26. At the right side, install the dust/pollen filter, the water guide and the air plenum cover.
27. Evacuate, charge and leak test the air conditioning system.
28. Operate the engine to normal operating temperature; then, check the climate control operation and check for leaks.

A6

➡If equipped with power seats, move the seats as far rearward as possible. Also, acquire the anti-theft radio coding from the owner.

1. Remove or disconnect the following:
    - Negative battery cable
    - Positive battery cable and remove the battery
    - Coolant recovery tank cap to relieve the pressure from the system
    - Coolant
    - Heater hoses from (at) the heater core
    - Heater core tubes-to-chassis grommet

➡If desired, instead of draining cooling system, disconnect heater hoses from tubes in engine compartment and apply compressed air to upper tube and catch coolant from heater core coming out of bottom tube.

2. Remove the storage compartment on driver's side.
3. From inside passenger compartment, remove bolts and brackets from heater core tubes.

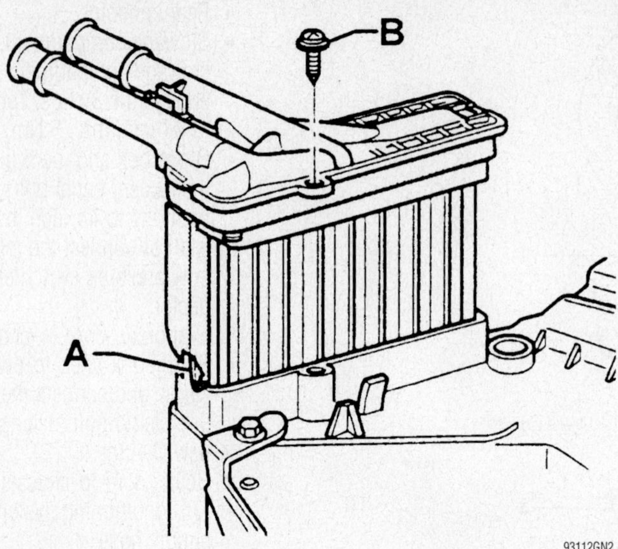

Removing the heater core—A4

4. Slightly pull the heater core from the A/C-heater housing.
5. Pull two coolant lines out of the heater core.
6. Remove the heater core.
7. If the heater core cannot be pulled out of the A/C housing, loosen the instrument panel cross member by removing or disconnecting the following:
- Bolt at base of windshield (above meter cluster)
- Loosen RH panel cross member bolts
- Two bolts from bottom of instrument panel brace
- ECM cover
- Pedal bracket from cross member
- LH instrument panel cover (push in steering column as far as possible)
- Loosen 2 LH instrument panel nuts

8. Pull back instrument panel and cross member enough to allow heater core to be removed.

To install:

9. Installation is the reverse of the removal procedure, noting the following:
- Install new clamps and O-rings at all connections
- Before installing storage compartment on driver's side, check for cooling system leaks
- Ensure WOT stop of accelerator pedal is incorrect position

A8

PASSENGER'S SIDE

1. Disconnect the negative battery cable.
2. Turn the ignition switch OFF.
3. Remove the cowl panel.
4. Drain the cooling system into the clean container for reuse.
5. Remove or disconnect the following:
- Engine-to-pump valve unit heater hoses. Drain and plug the openings
- Electrical connectors from the pump valve unit

6. Remove the reinforcement plate (plenum) by removing or disconnecting the following:
- Intake hose from the air filter
- Sound proofing mat at the front wall of the plenum
- Reinforcement plate (plenum)-to-chassis nuts/bolts and the plenum
- In the plenum, loosen the heater hose holder screw about 2 turns

7. Remove the knee bar by removing or disconnecting the following:
- Open the glove box
- 2 knee bar trim screws
- 4 knee bar screws
- Pull out the knee bar and disengage the latch
- Electrical connector and remove the knee bar

8. Remove or disconnect the following:
- Left support bolts and the retainer bolts; then, remove the left support and the retainer
- Center console side trim.
- Footwell air outlet

9. Place an absorbent cover on the floor of the car, under the heater core to catch any spilt coolant.
- Passenger's side heater hose clamps, move the heater hoses toward the plenum and discard the O-rings
- Heater core-to-heater case clamp screws and the clamp
- Heater core

To install:

10. Install or connect the following:
- Heater core
- Heater core-to-heater case clamp and the clamp screws
- Heater hoses to the plenum (using O-rings) and install the passenger's side heater hose clamps
- Footwell air outlet
- Center console side trim

11. Install the knee bar by installing or connecting the following:
- Left support and the retainer and the left support bolts and the retainer bolts; then, torque the bolts to 16 ft. lbs. (22 Nm)
- Electrical connector and install the knee bar
- Knee bar and engage the latch
- 4 knee bar screws and torque to 22 inch lbs. (2.5 Nm)
- 2 knee bar trim screws and torque to 22 inch lbs. (2.5 Nm)
- Tighten the heater hose holder screw in the plenum

12. Install the reinforcement plate (plenum) by installing or connecting the following:
- Reinforcement plate (plenum) and the plenum-to-chassis nuts/bolts
- Sound proofing mat at the front wall of the plenum
- Intake hose to the air filter
- Electrical connectors to the pump valve unit
- Engine-to-pump valve unit heater hoses

13. Refill the cooling system.
14. Install the cowl panel.
15. Connect the negative battery cable.

DRIVER'S SIDE

1. Disconnect the negative battery cable.
2. Turn the ignition switch OFF.
3. Remove the windshield.
4. Remove the cowl panel.
5. Drain the cooling system into the clean container for reuse.
6. Remove or disconnect the following:
- Engine-to-pump valve unit heater hoses. Drain and plug the openings
- Electrical connectors from the pump valve unit

7. Remove the reinforcement plate (plenum) by removing or disconnecting the following:
- Remove the intake hose from the air filter.

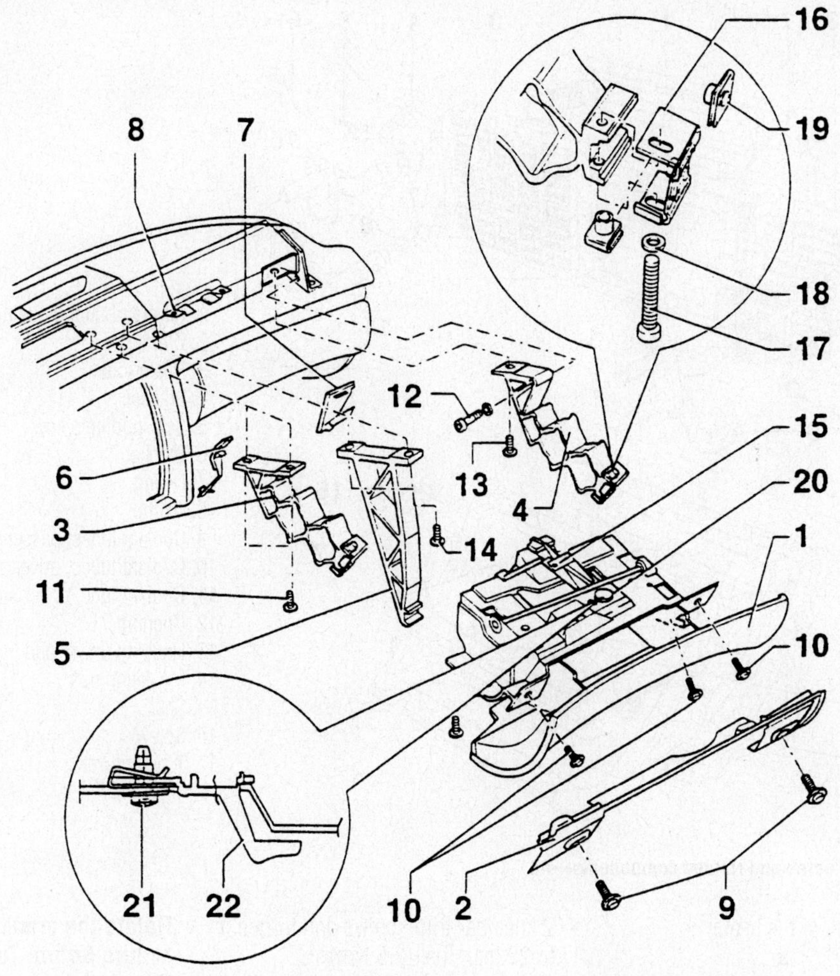

1. Knee bar
2. Knee bar trim
3. Support, left
4. Support, right
5. Retainer
6. Knee bar bracket, left
7. Knee bar bracket, right
8. Latch
9. Screws
10. Screws
11. Bolts
12. Bolt
13. Bolt
14. Bolts
15. Knee bar support
16. Mounting bracket
17. Screw
18. Washer
19. Clip
20. Adjusting eccentric
21. Screw
22. Eccentric arm

93112GN0

Exploded view of the glove box, knee bar assembly and related components—A8

- At the front wall of the plenum, remove the sound proofing mat.
- Remove the reinforcement plate (plenum)-to-chassis nuts/bolts and the plenum

8. Remove or disconnect the following:
   - Loosen the heater hose holder screw in the plenum about 2 turns
9. Remove the knee bar by removing or disconnecting the following:
   - Open the glove box
   - 2 knee bar trim screws
   - 4 knee bar screws
   - Pull out the knee bar and disengage the latch
   - Electrical connector and remove the knee bar
10. Remove or disconnect the following:
   - Left support bolts and the retainer bolts; then, remove the left support and the retainer
   - Center console side trim
   - Passenger's footwell air outlet
   - Passenger's side heater hose clamps, move the heater hoses toward the plenum and discard the O-rings
   - Passenger's side heater core-to-heater case clamp screws and the clamp
   - Passengers side heater core
   - Driver's side shelf and center console's side trim
   - Loosen the heater hose holder screws about 2 turns at the driver's side heater core
   - Driver's side footwell air outlet
   - Driver's side heater hose clamps, move the heater hoses toward the plenum and discard the O-rings
   - Driver's side heater core through the passenger's side of the heater housing

To install:

11. Install or connect the following:
   - Driver's side heater core through

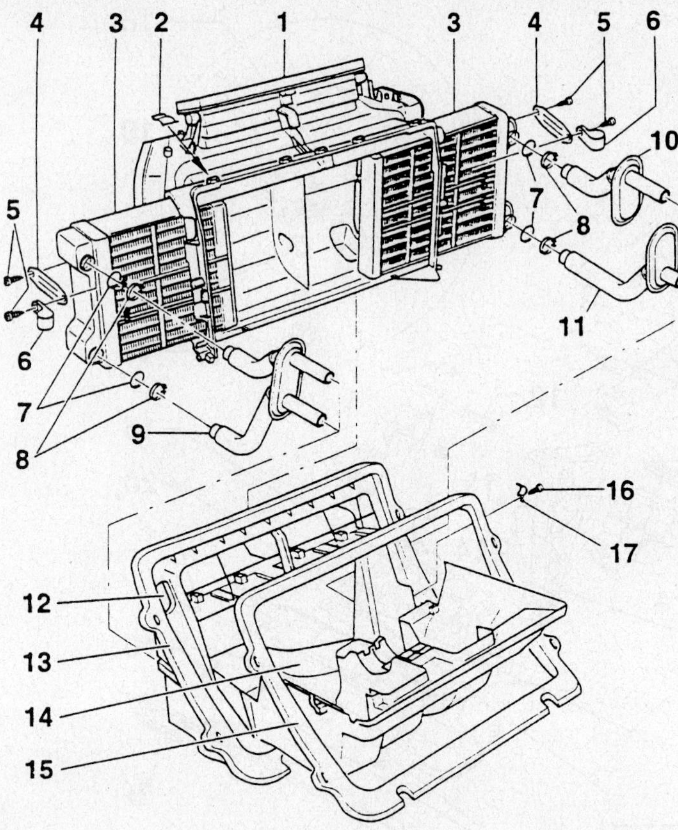

Exploded view of the heater cores and related components—A8

the passenger's side of the heater housing

- Move the heater hoses away from the plenum and install the driver's side heater hose clamps using new O-rings
- Driver's side footwell air outlet
- Heater hose holder screws at the driver's side heater core
- Driver's side shelf and center console's side trim
- Passenger's side heater core
- Heater core-to-heater case clamp and the clamp screws
- Heater hoses to the plenum using O-rings and install the passenger's side heater hose clamps
- Footwell air outlet
- Center console side trim

12. Install the knee bar by installing or connecting the following:
    - Left support and the retainer and the left support bolts and the retainer bolts; then, torque the bolts to 16 ft. lbs. (22 Nm)
    - Electrical connector and install the knee bar
    - Knee bar and engage the latch
    - 4 knee bar screws and torque to 22 inch lbs. (2.5 Nm)
    - 2 knee bar trim screws and torque to 22 inch lbs. (2.5 Nm)
13. In the plenum, tighten the heater hose holder screw.
14. Install the reinforcement plate (plenum) by installing or connecting the following:
    - Reinforcement plate (plenum) and the plenum-to-chassis nuts/bolts
    - Sound proofing mat at the front wall of the plenum
    - Intake hose to the air filter.
15. Install or connect the following:
    - Electrical connectors to the pump valve unit
    - Engine-to-pump valve unit heater hoses
16. Refill the cooling system.
    - Cowl panel
    - Windshield
17. Connect the negative battery cable.
18. Operate the engine to normal operating temperature; then, check the climate control operation and check for leaks.

Cylinder Head

➡Before removing or installing the cylinder head, align the engine timing marks at Top Dead Center (TDC). Rotate the crankshaft mark away about ¼ turn Before Top Dead Center (BTDC). This will prevent the valves from hitting the piston heads. Be sure to turn the crankshaft to the proper position after cylinder head installation.

REMOVAL & INSTALLATION

⁂ CAUTION

Cylinder head removal should not be attempted unless the engine is cold.

A4

1.8L ENGINE

1. Before servicing the vehicle, refer to the precautions section.
2. Remove the front bumper if necessary for access.
3. Place the hood lock carrier into the service position.
4. Turn the ignition switch to the **OFF** position.
5. Remove or disconnect the following:
    - Negative battery cable
    - Accessory drive belt
    - Cooling fan
6. Drain the engine coolant.

7. Remove or disconnect the following:
   - Intake manifold
   - Accessory drive belts
   - Wastegate bypass regulator valve
   - Evaporative Emissions (EVAP) canister purge regulator valve
   - Power outage stage
   - Mass Air Flow (MAF) sensor
   - Air cleaner housing
   - Engine Temperature Control (ETC) and the temperature II sensor harness connector
   - All connections from the cylinder head
   - Crankcase breather line
   - Oil supply line at the cylinder head
   - Exhaust manifold heat shield
   - Turbocharger from the exhaust manifold
   - Coolant hose to the heat exchanger at the rear of the cylinder head
   - Upper timing belt cover
8. Turn the crankshaft, in the direction of rotation (clockwise), until the No. 1 cylinder is at TDC.
9. Using Torx® wrench T45, loosen the timing belt tensioner.
10. Push down on the tensioner and remove the belt from the camshaft gear.
11. Remove or disconnect the following:
   - Torx® bolt and swing the tensioner assembly bracket forward
   - Valve cover
   - Cylinder head bolts, in sequence, as shown
   - Cylinder head
12. Clean the gasket mating surfaces.
13. Clean and dry out the cylinder head bolt holes.

To install:

➡Always replace the cylinder head bolts, self-locking nuts, bolts, gaskets and O-rings.

14. Before installing the cylinder head, set the crankshaft and camshaft to TDC for the No. 1 cylinder.
15. Loosen the turbocharger support bracket to reduce the likelihood of any tension while installing the cylinder head.
16. Install or connect the following:
   - Head gasket with the part number visible from the intake side
   - Cylinder head
   - New cylinder head bolts, tighten by hand
17. Tighten the new cylinder head bolts in sequence in 2 steps:
   a. Step 1: Exc. 2004, 44 ft. lbs. (60 Nm); on 2004, 30 ft. lbs. (40 Nm)
   b. Step 2: Plus 180 degrees
18. Install or connect the following:
   - Turbocharger to the exhaust manifold using new gaskets and the bolts coated with Hot Bolt Paste G 052 112 A3. Torque the bolts to 26 ft. lbs. (35 Nm).
   - Turbo support bracket. Torque the bolts to 33 ft. lbs. (40 Nm).
   - Valve cover
   - Timing belt
   - Accessory drive belts
   - Exhaust manifold heat shield
   - Oil supply lines to the cylinder head. Torque the retaining straps to 15 ft. lbs. (20 Nm).
   - Crankcase breather
   - Coolant temperature sensors
   - Air cleaner housing
19. Fill the engine with coolant and bleed, if necessary.
20. Connect the negative battery cable.
21. Fully close all power windows to stop, operate all window switches for at least 1 second in the close direction to activate the one touch opening/closing function.
22. Check the oil level before starting the engine and top off, as necessary.
23. Install the hood lock carrier assembly and front bumper.
24. Adjust the headlights.
25. Start the vehicle, check for leaks and repair if necessary.

TT

1.8L ENGINE

1. Before servicing the vehicle, refer to the precautions section.
2. Turn the ignition switch to the **OFF** position.
3. Drain the engine coolant.
4. Remove or disconnect the following:
   - Negative battery cable
   - Intake manifold
   - Coolant line at thermostat
   - Connectors on ignition coils
   - All connections from the cylinder head
   - Oil return line retainer
   - Upper air line to turbocharger
   - Exhaust manifold heat shield
   - Turbocharger from the exhaust manifold
   - Accessory drive belt
   - Coolant expansion tank with hoses
   - Power steering reservoir, leaving hoses attached
   - Upper timing belt cover
5. Turn the crankshaft, in the direction

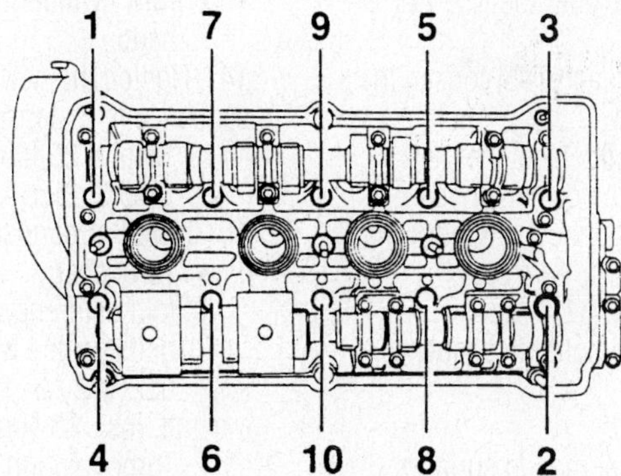

Cylinder head bolt removal sequence—A4 1.8L engine

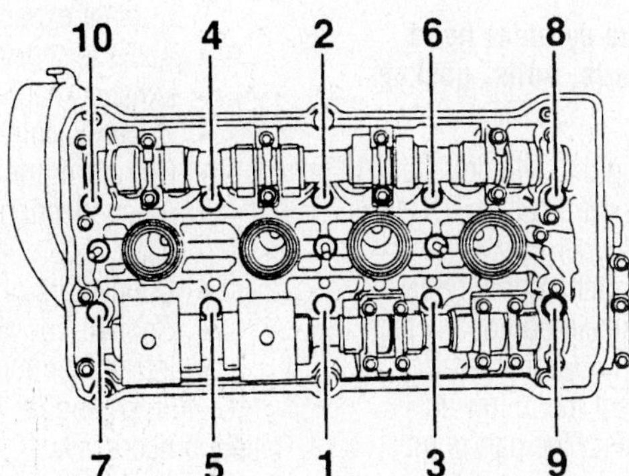

Cylinder head torque sequence—A4 1.8L engine

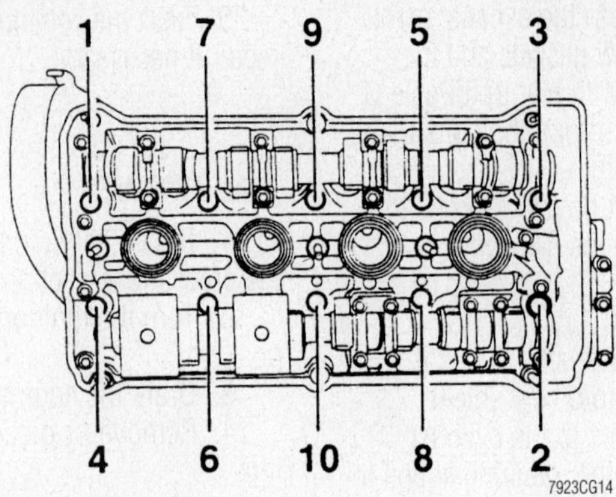

Cylinder head bolt removal sequence—TT 1.8L engine

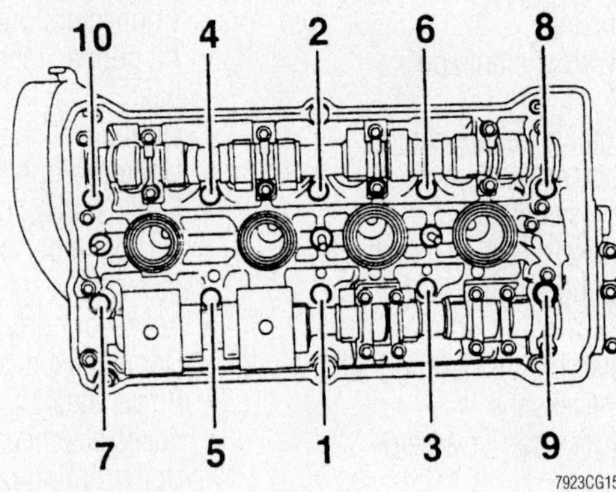

Cylinder head torque sequence—TT 1.8L engine

of rotation (clockwise), until the No. 1 cylinder is at TDC.

6. Using Torx® wrench T45, loosen the timing belt tensioner.

7. Push down on the tensioner and remove the belt from the camshaft gear.

8. Remove or disconnect the following:
   - Ignition coils
   - Valve cover
   - Cylinder head bolts, in sequence, as shown
   - Cylinder head

9. Clean the gasket mating surfaces.

10. Clean and dry out the cylinder head bolt holes.

To install:

➡Always replace the cylinder head bolts, self-locking nuts, bolts, gaskets and O-rings.

11. Before installing the cylinder head, set the crankshaft and camshaft to TDC for the No. 1 cylinder.

12. Loosen the turbocharger support bracket to reduce the likelihood of any tension while installing the cylinder head.

13. Install or connect the following:
   - Head gasket with the part number visible from the intake side
   - Cylinder head
   - New cylinder head bolts, tighten by hand

14. Tighten the new cylinder head bolts in sequence in 2 steps:
   a. Step 1: 30 ft. lbs. (40 Nm)
   b. Step 2: Plus 180 degrees

15. Install or connect the following:
   - Turbocharger to the exhaust manifold using new gaskets and the bolts coated with Hot Bolt Paste G 052 112 A3. Torque the bolts to 26 ft. lbs. (35 Nm).
   - Turbo support bracket. Torque the bolts to 33 ft. lbs. (40 Nm).
   - Timing belt.
   - Power steering reservoir
   - Coolant expansion tank
   - Accessory drive belt
   - Exhaust manifold heat shield
   - Upper air line to turbocharger
   - Oil return line retainer
   - All connections from the cylinder head
   - Connectors on ignition coils
   - Coolant line at thermostat
   - Intake manifold

16. Fill the engine with coolant and bleed, if necessary.

17. Connect the negative battery cable.

18. Fully close all power windows to stop, operate all window switches for at least 1 second in the close direction to activate the one touch opening/closing function.

19. Check the oil level before starting the engine and top off, as necessary.

20. Start the vehicle, check for leaks and repair if necessary.

3.2L ENGINE

1. Before servicing the vehicle, refer to the precautions section.

2. Turn the ignition switch to the **OFF** position.

3. Drain the engine coolant.

4. Remove or disconnect the following:
   - Negative battery cable

➡Before removing ribbed belt, mark direction of rotation with chalk or a felt pen. If a used belt rotates in the wrong direction when reinstalled, this can result in breakage. When installing belt, ensure it is correctly seated in the pulleys.

   - Release tension on belt by screwing an M8 x 40 bolt into bore on tensioning element and remove belt
   - Upper mounting bolt for the alternator
   - Upper mounting bolts for radiator cowl
   - Secondary air hose
   - Center and right-hand noise insulation under engine
   - Cross piece
   - Power steering pump from bracket, hang pump from lock carrier with hoses still attached
   - Secondary air injection pump with bracket
   - Lower radiator cowl bolts and remove the radiator cowl with fans

➡The air conditioning compressor can only be removed from the bracket with the upper securing bolt still inserted. To accomplish this, press the engine simultaneously to the left side. If necessary, loosen all bolts of the assembly mounting.

   - Mounting bolt for air conditioning compressor
   - Secure the air conditioning compressor together with the hoses at front longitudinal member (hoses remain connected)
   - Alternator
   - Oil dipstick tube
   - Vacuum reservoir
   - Mounting bracket for alternator and air conditioning compressor

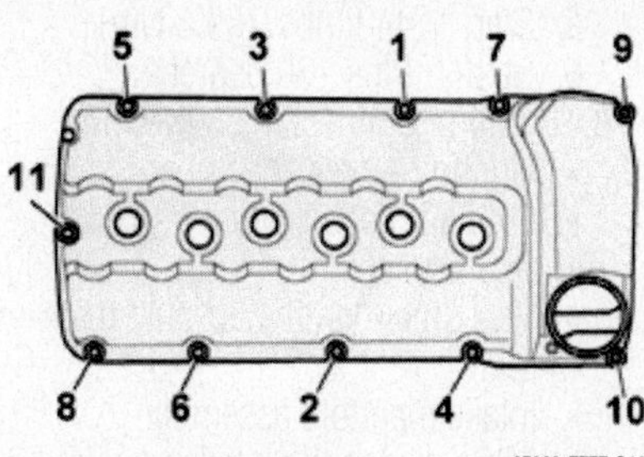

Cylinder head cover bolt removal sequence—TT 3.2L engine

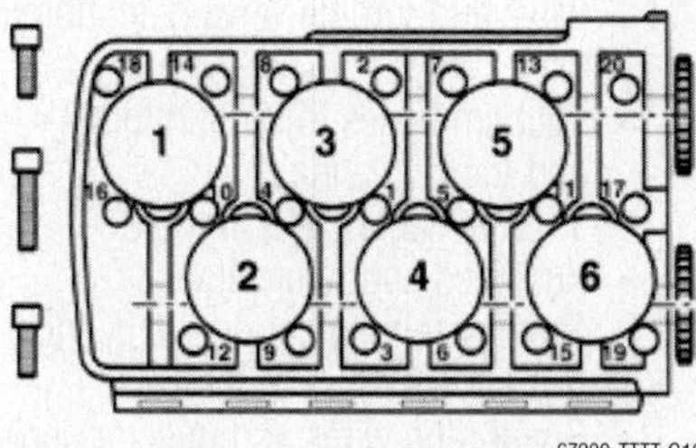

Cylinder head bolt removal and tightening sequence—TT 3.2L engine

- Bolts at front of intake manifold
- Attach puller T10095A or equivalent to ignition coils and remove coils
- Body brace
- Air intake hose
- Crankcase breather hose
- Throttle body
- Hose bracket
- Rear intake mounting bolts, swing intake manifold forward
- Cylinder head cover, remove bolts in sequence
- Timing chain from camshafts
- Exhaust pipe from manifold
- Remove cylinder head bolts in sequence indicated
- Attach lifting shackle 3033 or equivalent and carefully lift off cylinder head

➡Replace cylinder head bolts. Replace self locking nuts and bolts that are tightened by turning to a specified angle

5. Set camshafts to TDC for number 1 cylinder
   - It must be possible to insert camshaft plate T10068 in both camshaft grooves
6. Install cylinder head gasket and cylinder head

➡The long cylinder head bolts are inserted in the middle bores of the cylinder head

7. Tighten cylinder head bolts in 4 stages using the following sequence:
   - 1st stage 22 ft. lbs (30 Nm)
   - 2nd stage 37 ft. lbs (50 Nm)
   - 3rd stage 90¯ (1/4 turn) further
   - 4th stage 90¯ (1/4 turn) further
8. Install or connect the following
   - Timing chain
   - Cylinder head cover
   - Intake manifold
9. Fill the engine with coolant and bleed, if necessary.
10. Connect the negative battery cable.
11. Check the oil level before starting the engine and top off, as necessary.
12. Start the vehicle, check for leaks and repair if necessary.

2.7L Engine

➡On A6 models, the front lock carrier assembly must be placed in the service position. The front bumper must be removed before the lock carrier can be placed in the service position

1. Drain engine coolant.
2. Properly relieve fuel pressure.
3. Before servicing the vehicle, refer to the precautions section.
4. Remove or disconnect the following:
   - Negative battery cable
   - Camshaft timing control connector
   - Injector connectors
   - Ignition coil connectors
   - Crankcase breather
   - Ignition coils
   - Turbocharger intake hose
   - Solenoid valve for charge air pressure control
   - EVAP valve connector
   - Throttle unit connector
   - Charge air sensor connector
   - Intake air temperature sensor connector
   - Accessory belt
   - Accessory belt tensioner
   - Timing belt front covers
5. Place crankshaft in TDC position for no. 1 cylinder.
6. Screw clamping bolt 3242 for crankshaft into sealing plug hole and tighten.
7. Remove or disconnect the following:
   - Crankshaft damper
   - Idler wheel
   - Belt guard behind vibration damper
   - Timing belt
   - Camshaft sprockets
   - Intake manifold
   - Coolant brackets and water lines
   - Heat sensor
   - Turbocharger from exhaust manifold
   - Valve covers
   - Cylinder head bolts by loosening them in the reverse of the tightening sequence
   - Cylinder head

To install:

8. Clean all sealing surfaces. Check cylinder head for distortion. Measure at several locations. The maximum permissible distortion is 0.0039 in. (0.1mm).
9. Before installing cylinder head, place camshaft and crankshaft sprockets in TDC position of no. 3 cylinder.
10. Install or connect the following:
    - New cylinder head gasket with the lettering facing toward cylinder head
    - Cylinder head
11. Tighten the cylinder head bolts following the torque tightening sequence in 2 steps as follows:
    a. Step 1: 44 ft. lbs. (60 Nm)

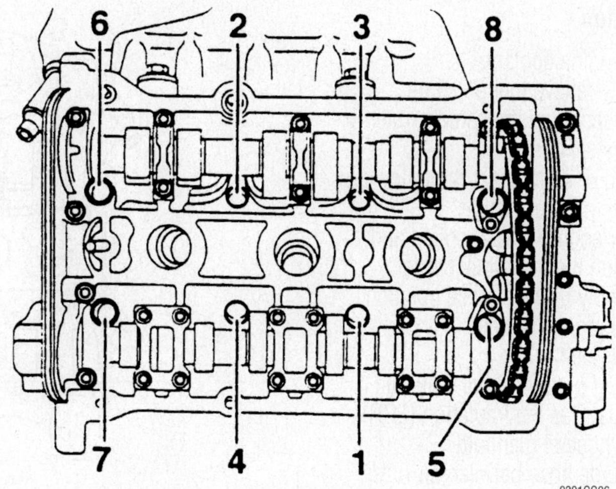

Cylinder head torque sequence—2.7L and 2.8L DOHC engine

b. Step 2: Plus 180 degrees

12. Install or connect the following:
- Valve covers
- Turbocharger
- Heat sensor
- Coolant brackets and water lines
- Intake manifold
- Camshaft sprockets
- Timing belt
- Belt guard behind vibration damper
- Idler wheel
- Crankshaft damper
- Timing belt front covers
- Accessory belt tensioner
- Accessory belt
- Intake air temperature sensor connector
- Charge air sensor connector
- Throttle unit connector
- EVAP valve connector
- Solenoid valve for charge air pressure control
- Turbocharger intake hose
- Ignition coils
- Crankcase breather
- Ignition coil connectors
- Injector connectors
- Camshaft timing control connector

13. Fill and bleed the cooling system, as necessary.
14. Connect the negative battery cable.
15. Check all fluid levels and top off as necessary.
16. Operate the engine and check for leaks.
17. Road test the vehicle for proper operation.

2.8L Engine

➡On A4 models and A6 models, the front lock carrier assembly must be placed in the service position. The front bumper must be removed before the lock carrier can be placed in the service position

1. Drain engine coolant.
2. Properly relieve fuel pressure.
3. Before servicing the vehicle, refer to the precautions section.
4. Remove or disconnect the following:
- Negative battery cable
- Lower engine splash shield and position the front lock carrier assembly in the service position
- Accessory drive belt
- Timing belt
- Exhaust pipe from the manifold
- Exhaust Gas Recirculation (EGR) valve hose at manifold
- Air guide hose between air mass sensor and intake manifold
- Spark plug wires
- Injector connectors
- Crankcase breathers on the left and right cylinder head covers
- Fuel feed and return lines
- Left side cover for fuel line
- Throttle cable
- Vacuum hoses from vacuum pump and intake manifold
- Idling stabilization valve and throttle valve potentiometer
- Vacuum hose on vacuum control unit
- Oil pressure sender
- Oil pressure switch
- Hall sender sensor
- EGR valve from the intake manifold
- Intake manifold assembly
- Coolant pipe at the rear of the cylinder head
- Oxygen (O_2S) sensor
- Heat shield on the exhaust manifold
- Cylinder head cover
- Timing belt rear belt guard
- Hydraulic line from reservoir to pump
- Cylinder head bolts by loosening them in the reverse of the torque sequence
- Cylinder head

To install:

5. Clean all sealing surfaces. Check cylinder head for distortion. Measure at several locations. The maximum permissible distortion is 0.0039 in. (0.1mm).
6. Install or connect the following:
- New cylinder head gasket with the lettering facing upwards
- Cylinder head

7. Tighten the cylinder head bolts following the torque tightening sequence in 2 steps as follows:

a. Step 1: 44 ft. lbs. (59.8 Nm)
b. Step 2: Plus 180 degrees

8. Install or connect the following:
- Timing belt rear belt guard
- Cylinder head cover
- O_2S sensor
- Heat shield on the exhaust manifold
- Intake manifold assembly
- EGR valve to the intake manifold
- Hall sender sensor
- Oil pressure sender
- Oil pressure switch
- Connectors on idling stabilization valve and throttle valve potentiometer
- Vacuum hoses to vacuum pump and intake manifold
- Throttle cable
- Fuel feed and return lines
- Crankcase breather on the cylinder head cover
- Spark plug wires and injector connectors
- Exhaust manifold
- Timing belt and accessory drive belt

9. Fill and bleed the cooling system, as necessary.
10. Connect the negative battery cable.
11. Check all fluid levels and top off as necessary.
12. Operate the engine and check for leaks.
13. Road test the vehicle for proper operation.

3.0L Engine

➡On 3.0L models, the following procedure is with the engine installed in the vehicle.

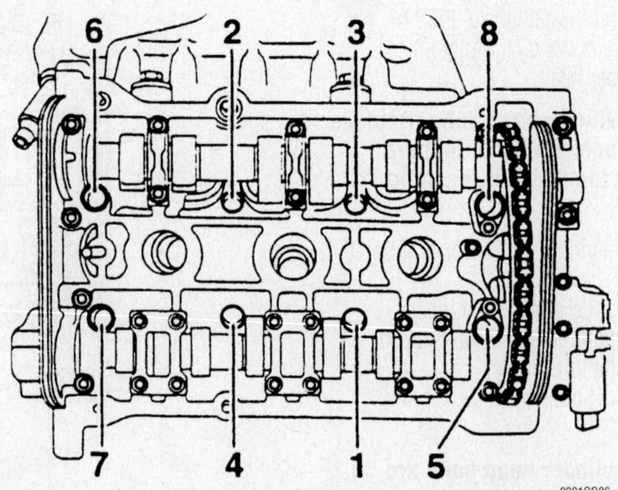

Cylinder head torque sequence—2.7L and 2.8L DOHC engine

1. Drain engine coolant.
2. Properly relieve fuel pressure.
3. Before servicing the vehicle, refer to the precautions section.
4. Remove or disconnect the following:
   - Negative battery cable
   - Intake manifold
   - Engine covers
   - Coolant reservoir
   - Resonator
   - Fuel rail
   - Ducting
   - Solenoid valve holding plate
   - Connectors from CMP sensors
   - Ground wire from rear of cylinder head
   - PS pressure line from pump at RH cylinder head (position aside)
   - Timing belt
   - Camshaft gears from both heads
   - Belt guard
   - Idler roller and bracket from right head
   - Timing belt brackets from both heads
   - Coolant pipes from both heads
   - Vacuum hose to brake booster
   - Secondary air injection combination valves
   - Exhaust manifold to exhaust pipe nuts on both sides

➡The flex pipes at the front exhaust pipes must not be bent more than 10 degrees. Otherwise they may be damaged.

   - Exhaust pipes from manifold
   - Cylinder head bolts in proper sequence
   - Cylinder head

To install:

5. Clean all sealing surfaces. Check cylinder head for distortion. Measure at several locations. The maximum permissible distortion is 0.0039 in. (0.1mm).

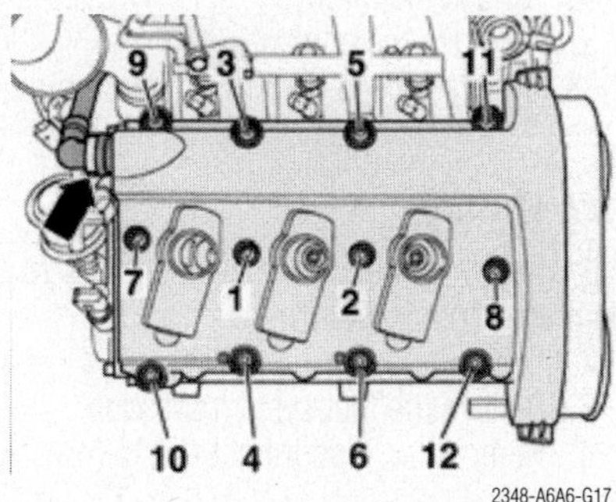

Cylinder head cover removal sequence—Audi 3.0L engine

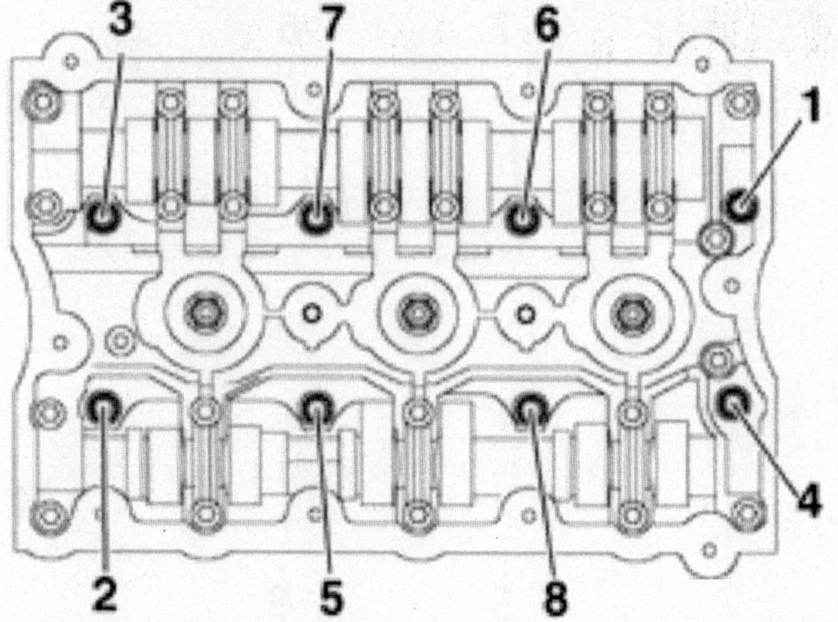

Cylinder head removal sequence—3.0L engine

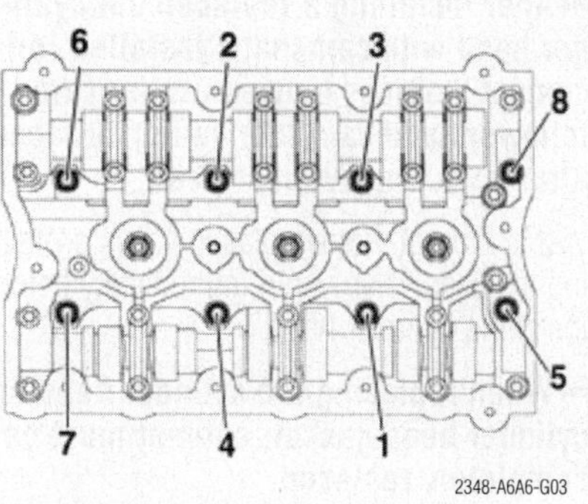

Cylinder head torque sequence—3.0L engine

6. Install new cylinder head gasket with the lettering facing toward cylinder head.
7. Install the cylinder head.
8. Tighten the cylinder head bolts following the torque tightening sequence in 2 steps as follows:
   a. Step 1: 30 ft. lbs. (40 Nm)
   b. Step 2: Plus 180 degrees
9. Install or connect the following:
   - Exhaust pipes
   - Exhaust pipes to exhaust manifold
   - Secondary air injection combination valves
   - Vacuum hose to brake booster
   - Coolant pipes
   - Timing belt brackets
   - Idler roller and bracket to right head
   - Camshaft gears
   - Timing belt
   - Power steering pressure line
   - Ground wire on left head
   - Crankshaft and camshaft position sensor connectors
   - Coolant level warning switch connector at bottom of expansion tank
   - Coolant hoses at reservoir and reservoir
   - Solenoid valve holding plate
   - Ducting
   - Fuel rail
   - Resonator
   - Exhaust manifold
   - Engine cover
10. Fill and bleed the cooling system, as necessary.
11. Connect the negative battery cable.
12. Check all fluid levels and top off as necessary.
13. Operate the engine and check for leaks.
14. Road test the vehicle for proper operation.

A6 and A8

3.2L ENGINE

1. Before servicing the vehicle, refer to the precautions section.
2. Drain the cooling system.
3. Remove the front exhaust pipe.
4. Remove the drive belt.
5. Remove the front coolant pipe.
6. Remove the power-steering pump.
7. To remove the right cylinder head, the vacuum pump for brake booster must be removed.
8. Remove the intake manifold.

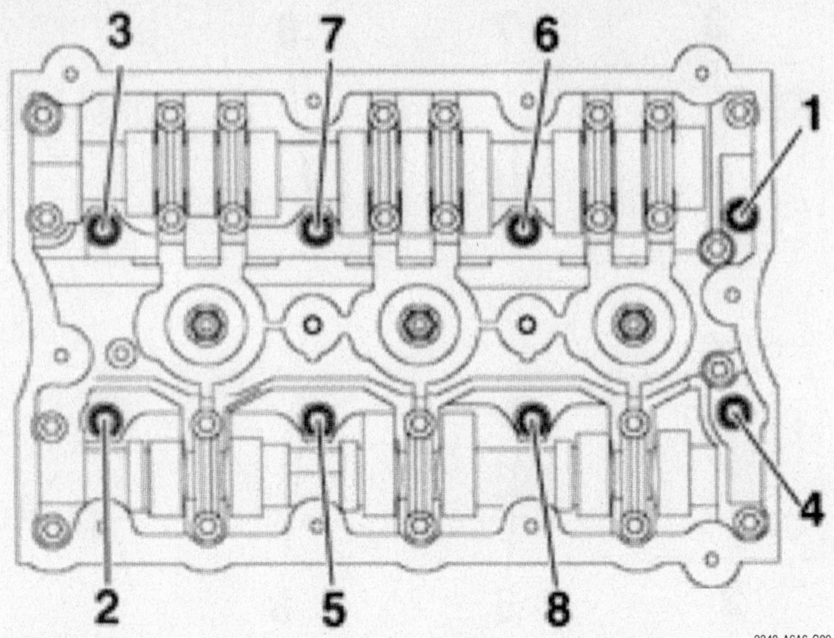

Cylinder head removal sequence—A6 with 3.2L engine

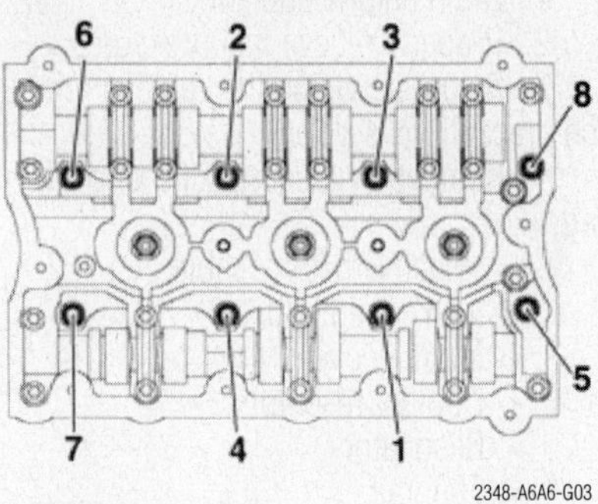

Cylinder head torque sequence—A6 with 3.2L engine

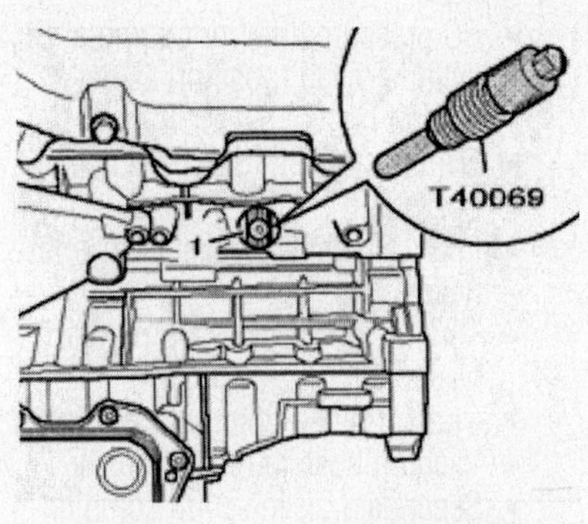

Sub frame bolt locations The camshaft locating tool T40070 is correctly positioned when the holes for cylinder head bolts remain free—A6 with the 3.2L engine

9. Remove the oil filter housing.
10. Disconnect the fuel line.
11. Remove the low-pressure line.
12. If equipped, remove the large lifting eye.
13. Disconnect the electrical connectors.
14. If necessary, remove high pressure pump from left cylinder head.
15. Remove the bolt and the oil dipstick tube.
16. Remove the cylinder head cover.
17. Remove the left and right timing chain covers.
18. Remove the camshaft timing chains.
19. Remove the bolts at rear of cylinder head. There are 3 bolts on the left side head and four on the right side head.
20. Remove the head bolts in sequence illustrated.
21. Carefully remove cylinder head.

To install:

➡Replace cylinder head bolts. During assembly, replace self-locking nuts and bolts. Always replace bolts that are tightened to torque as well as O-rings and gaskets.

22. Clean the gasket mating surfaces.

➡Only unpack new cylinder head gasket immediately prior to installation. Handle gasket carefully. Damages to the silicone layer and in areas of recesses may result in leaks.

23. Install the cylinder head gasket onto guide sleeves. Marking "oben" (top) or part number must face toward cylinder head.

➡After installing a replacement cylinder head with camshafts installed, oil contact surfaces between roller cam followers and cam lubricating surfaces after installing cylinder head.

24. Do not remove plastic bases protecting exposed valves until immediately before installing cylinder head.

➡When replacing the cylinder head or cylinder head gasket, coolant must be completely replaced.

➡After working on the valvetrain, carefully rotate engine by hand at least 2 full revolutions to ensure that valves do not strike the pistons when starting. Change contaminated engine oil.

25. Before installing cylinder head, set crankshaft and camshafts to TDC setting, mounting camshaft locating tool T40070 on both cylinder heads and tightening to 20 Nm.
26. The camshaft locating tool T40070 is correctly positioned when the holes for cylinder head bolts remain free. Crankshaft holder T40069 must be screwed into crankshaft.
27. Position cylinder head gasket.
28. Pay close attention to centering pins in cylinder block.
29. Pay attention to installation position of cylinder head gasket, marking "oben" (top) or part number must face toward cylinder head.
30. Install the cylinder head.
31. Insert new cylinder head bolts and tighten by hand.
32. Tighten cylinder head in sequence in 3 steps as follows:
33. Step1: tighten to 40 Nm.
   a. Step 2: Tighten an additional 90 degree.
   b. Step 3: Tighten an additional 90 degree.
34. Tighten the bolts at the rear of the heads to 9 Nm.
35. Installation of the remaining components is the reverse order of removal.
36. Check the oil level.
37. Replace the coolant.
38. Fill and bleed the power steering system.

4.2L ENGINE

1. Before servicing the vehicle, refer to the precautions section.
2. Drain the cooling system.
3. Relieve the fuel system pressure.
4. Remove or disconnect the following:
   - Negative battery cable
   - Dipstick from LH cylinder head

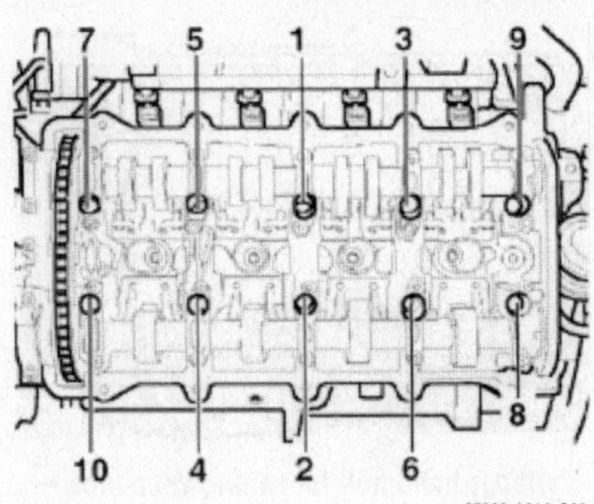

Cylinder head torque sequence—4.2L Engines

- Intake duct for air cleaner
- Fuel supply and return lines
- Upper bolt between exhaust manifold and exhaust pipe
- Oxygen sensor connectors
- Exhaust pipe from LH manifold (loosen double clamps rearward)
- Upper LH and RH timing belt guards
- Timing belt from camshafts (without moving position)
- Lower timing belt guard
- Connectors, pipes and coolant lines (mark for position reference)
- Cylinder head covers
- Left and right knock sensor connectors
- Fuel injector harness connectors
- Fuel rail, carefully remove along with fuel injectors, and place down to one side
- Coolant line at rear between cylinder heads
- Intake manifold
- Cylinder head bolts, in sequence, following reverse order of tightening sequence
- Cylinder head

To install:

5. Install the cylinder head with a new gasket and tighten the bolts in sequence as follows:
    a. Step 1: 26 ft. lbs. (35 Nm)
    b. Step 2: 44 ft. lbs. (60 Nm)
    c. Step 3: Plus 90 degrees
    d. Step 4: Plus 90 degrees
6. The remainder of the installation is the reverse of the removal procedure.

A3

2.0L ENGINE

1. Before servicing the vehicle, refer to the precautions section.
2. Disconnect the electrical connector.
3. Remove the engine cover upward and out in direction of
4. Remove the coolant expansion tank cap.
5. Remove the center sound insulation fasteners.
6. Remove the right sound insulation fasteners.
7. Drain and recycle the coolant.
8. Remove the wiper arms.
9. Unclip spray nozzles and push spray nozzles, with lines still attached, back through opening and into plenum chamber.
10. Remove the rubber seal for plenum chamber cover.
11. Remove the plenum chamber cover.
12. Loosen up the rear engine wiring harness at partition for plenum chamber.
13. Remove the electrical connector for oxygen sensor before catalytic converter from bracket, disconnect it and lay it aside.
14. Remove the partition for plenum chamber.
15. Disconnect the coolant line to coolant expansion tank.
16. Disconnect the coolant hose.
17. Unscrew the ground wire.
18. Disconnect the electrical connector.
19. Disconnect the coolant hoses.
20. Remove the intake manifold
21. Remove the catalytic converter with front exhaust pipe
22. Remove the right driveshaft
23. Remove the heat shield for right driveshaft.
24. Pull air duct off from charge air cooler.
25. Unfasten the air duct from turbocharger.
26. Disconnect the connector and loosen up cable.
27. Unscrew bolts and remove air duct.
28. Disconnect the oil supply line for turbocharger at cylinder block.
29. Remove the oil supply line for turbocharger.
30. Disconnect the coolant supply line for turbocharger at cylinder block.
31. Unscrew coolant supply line to turbocharger.
32. Remove the coolant supply line for turbocharger at cylinder block.
33. Remove the oil return line at turbocharger.
34. Remove the bolts and remove turbocharger support.
35. Remove the timing belt.
36. Loosen the bolt for camshaft sprocket using retainer 3036.
37. Pull off camshaft sprocket using two-arm puller T40001 and claws T40001/6 and T40001/7.
38. Remove the rear the timing belt housing from cylinder head.
39. Remove the cylinder head cover.

Loosen the bolt for camshaft sprocket using retainer 3036—A3 with 2.0L engine

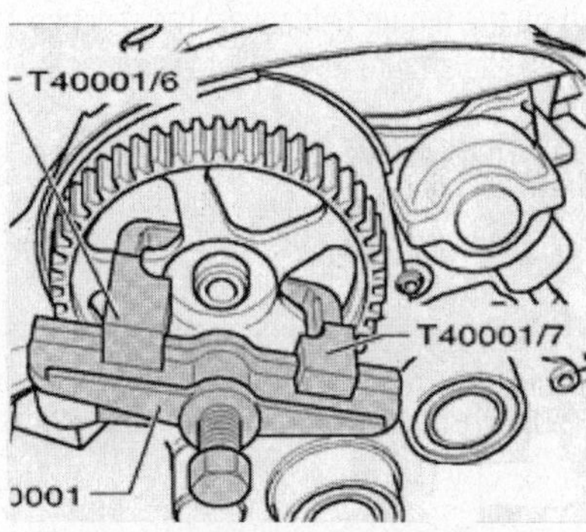

Pull off camshaft sprocket using two-arm puller T40001 and claws T40001/6 and T40001/7—A3 with 2.0L engine

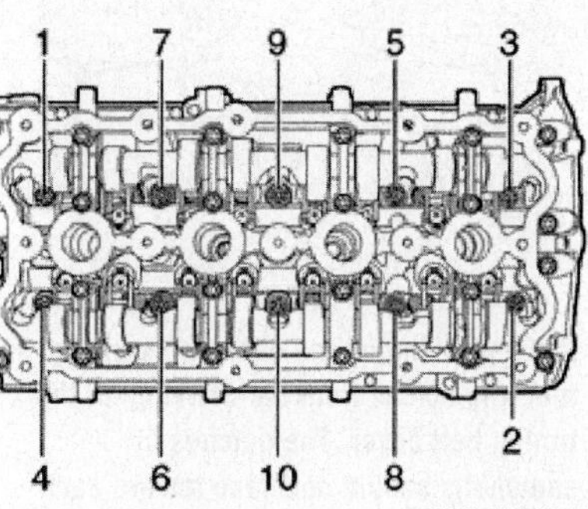

Cylinder head bolt loosening sequence—A3 with 2.0L engine

40. Loosen the cylinder head bolts in the sequence illustrated.

➡Verify that all hose and line connections between engine, transmission and body have been disconnected.

41. Remove the cylinder head.

To install:

➡Use new cylinder head bolts.

42. Always replace self-locking nuts, bolts that have been tightened to tightening torque as well as gaskets and O-rings.

43. Clean all gasket mating surfaces.

44. Only unpack new cylinder head gasket immediately prior to installation. Handle gasket carefully. Damages to the silicone layer and in areas of recesses may result in leaks.

45. There must be no oil or coolant in the blind holes for the cylinder head bolts in the cylinder block.

➡Only turn over the engine at the crankshaft in direction of engine rotation (clockwise) using the crankshaft bolt.

46. Move marking on the camshaft sprocket so that it is across from the marking on the timing belt cover. The notches of camshafts should now face toward each other vertically.

47. If the crankshaft was turned in the meantime: set cyl1 to TDC and turn crankshaft back again slightly.

48. Set cylinder head gasket in place.

49. Pay attention to centering pins in cylinder block.

50. Observe the installed position of the

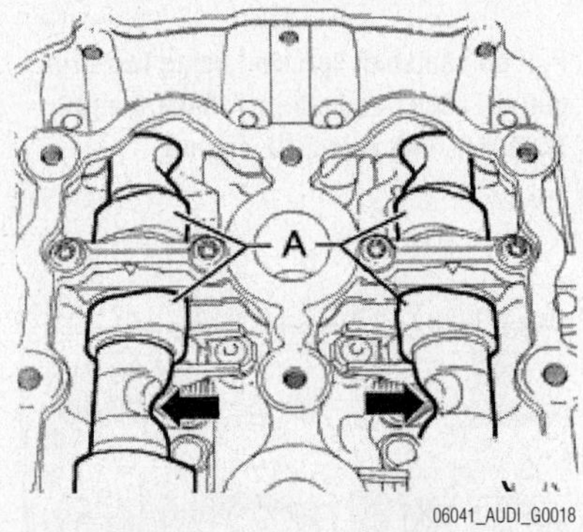

Move marking on the camshaft sprocket so that it is across from the marking on the timing belt cover. The notches of camshafts should now face toward each other vertically—A3 with 2.0L engine

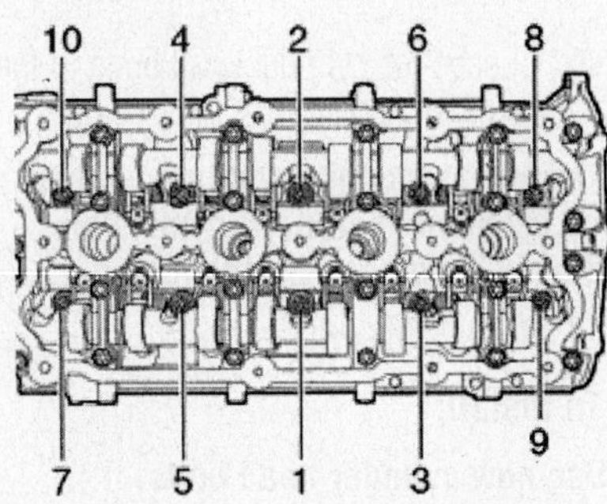

Cylinder head bolt tightening sequence—A3 with 2.0L engine

cylinder head gasket marking. The replacement part numbers must be legible from the intake side.

51. Install the cylinder head in place.

52. Insert cylinder head bolts and tighten by hand.

53. Tighten the cylinder head in the sequence illustrated as follows:

a. Step 1: tighten to 40 Nm.

b. Step 2: tighten and additional 180 degrees.

54. Install the cylinder head cover and tighten to 10 Nm.

55. Install the wiper arms.

56. Install the timing belt.

57. Install the accessory drive belt

58. Install the intake manifold

59. Connect the negative battery cable.

60. Check the engine oil level and replenish as needed.

61. Fill the cooling system.

⁂ CAUTION

Do not use a battery charger for starting assistance! There is the risk that the vehicle control modules could be damaged.

3.2L ENGINE

1. Before servicing the vehicle, refer to the precautions section.

2. Drain the cooling system.

3. Remove the intake manifold

4. Remove the coolant hoses to heater core at the bulkhead.

5. Disconnect the following hose connections:

a. Coolant hose to reservoir.

b. Vacuum hose to Evaporative Emission (EVAP) canister.

c. Vacuum hose to leak detection pump

d. Fuel supply line.

e. Coolant hose to reservoir.

f. Disconnect the coolant hoses.

6. Remove the wiring connections with coolant and vacuum hoses.

7. Remove the camshaft timing chain from camshafts.

➡Flex joints in front exhaust pipe must not be bent more than 30 degrees, otherwise they may be damaged.

8. Remove the front exhaust pipe from exhaust manifolds.

9. Loosen and remove cylinder head bolts in sequence shown.

10. Install lifting tackle 3033 as illustrated.

11. Carefully raise cylinder head.

12. Place clean cloth in cylinders so that

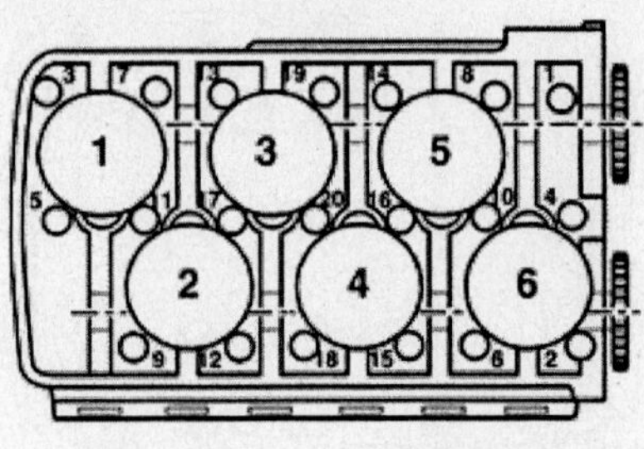

Cylinder head bolt loosening sequence—A3 with 3.2L engine

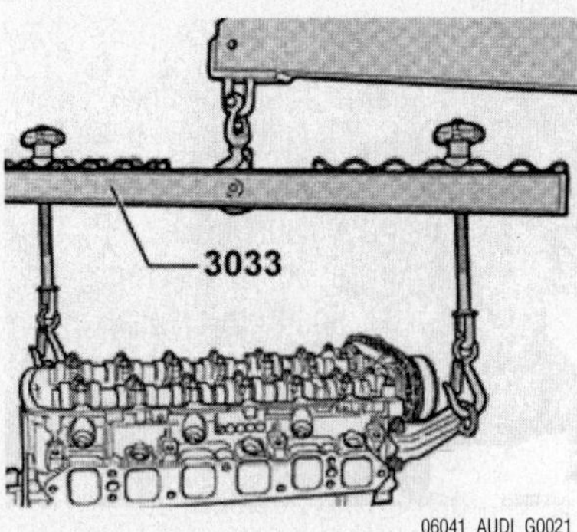

Install lifting tackle 3033 as shown—A3 with 3.2L engine

no dirt can enter between cylinder wall and pistons.

To install:

➡Use new cylinder head bolts.

➡Always replace self-locking nuts, bolts that have been tightened to tightening torque as well as gaskets and O-rings.

13. Clean the gasket mating surfaces.

14. Only unpack new cylinder head gasket immediately prior to installation. Handle gasket carefully. Damage in silicon layer and recessed area lead to leakage. The plastic protectors installed to protect the open valves must only be removed immediately before installing the cylinder head.

15. When installing a cylinder head other than the original cylinder head, the contact surfaces between bearing elements, roller rocker levers and cam contact surfaces must be oiled before installation.

➡There must be no oil or coolant in the blind holes for the cylinder head bolts in the cylinder block.

16. Position camshafts at TDC marking before positioning cylinder head. It must be possible to insert camshaft bar T10068 into both shaft grooves.

17. Make sure the crankshaft is positioned at TDC.

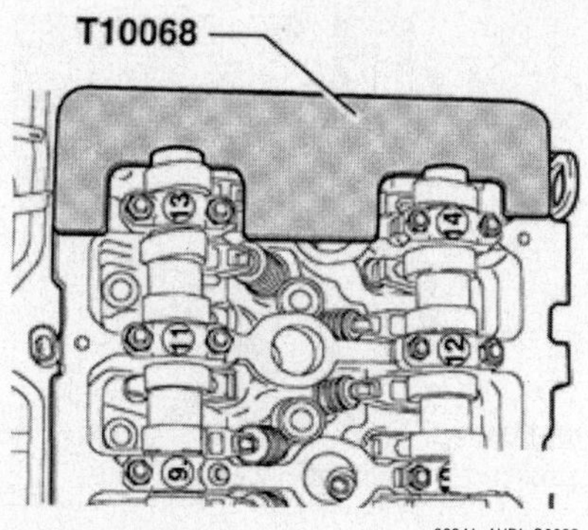

Position camshafts at TDC marking before positioning cylinder head. It must be possible to insert camshaft bar T10068 into both shaft grooves—A3 with 3.2L engine

18. Install cylinder head gasket. Check the centering pins in cylinder block. Check the cylinder head seal location marks, the part number must be visible from intake side.
19. Install the cylinder head and hand tighten the bolts.

➡The longer cylinder head bolts must be inserted in the middle holes of cylinder head.

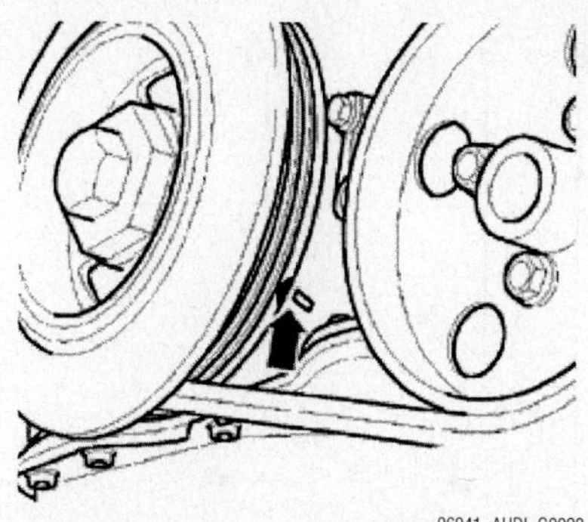

Make sure the crankshaft is positioned at TDC—A3 with 3.2L engine

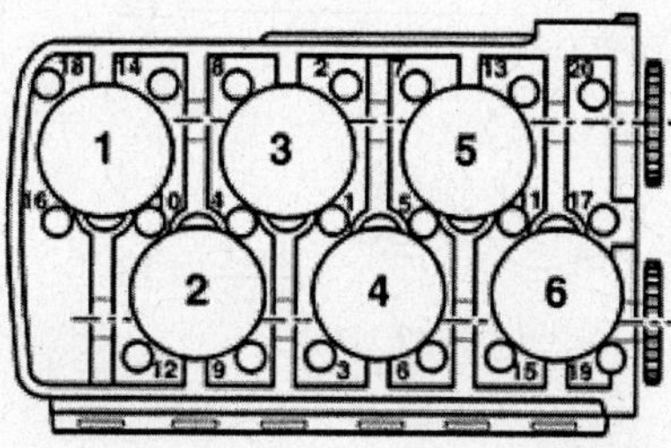

Cylinder head torque sequence—A3 with 3.2L engine

20. Tighten the bolts in sequence as follows:
    a. Step 1: tighten to 30 Nm.
    b. Step 2: tighten to 50 Nm.
    c. Step 3: Using a Torx key tighten an additional 90 degrees.
    d. Step 4: Using a Torx key tighten an additional 90 degrees.
21. Install the camshaft timing chain.
22. Install the cylinder head cover
23. Install the intake manifold
24. Fill the cooling system.
25. Connect the negative battery cable.

Turbocharger

REMOVAL & INSTALLATION

A4

1.8L ENGINE

1. Before servicing the vehicle, refer to the precautions section.
2. Remove or disconnect the following:
    - Negative battery cable
    - Engine undercover
    - Air conditioning compressor and move it aside with the lines attached

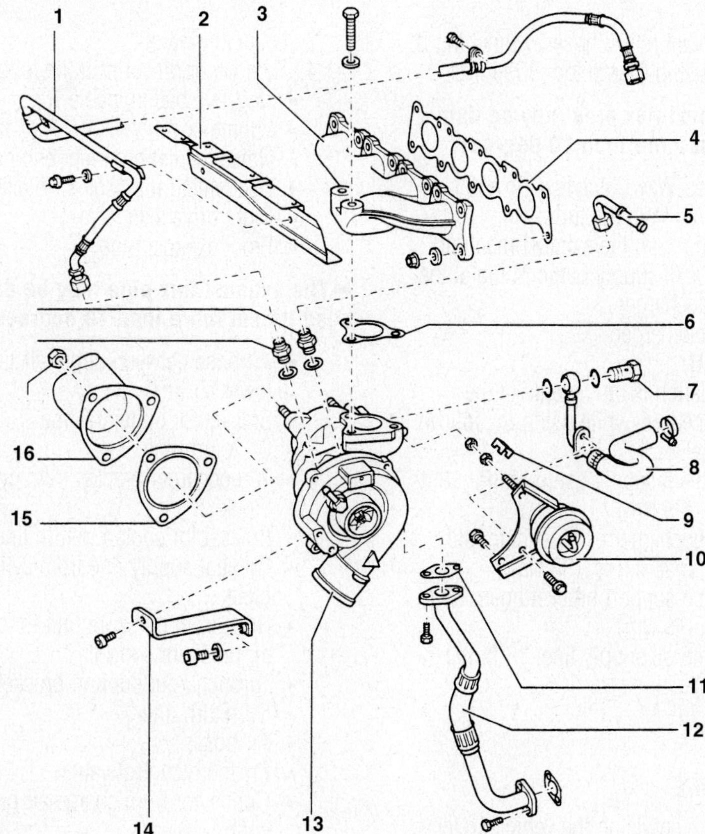

1. Oil supply line
2. Heat shield
3. Exhaust manifold
4. Exhaust manifold gasket
5. Coolant return line
6. Exhaust manifold-to-turbo gasket
7. Banjo bolt
8. Coolant supply hose
9. Fuse
10. Vacuum diaphragm for the wastegate
11. Gasket
12. Oil return line
13. Turbocharger
14. Support bracket
15. Gasket
16. Three Way Catalytic Converter (TWC)

7923CG17

Exploded view of the turbocharger and related components—1.8L engine

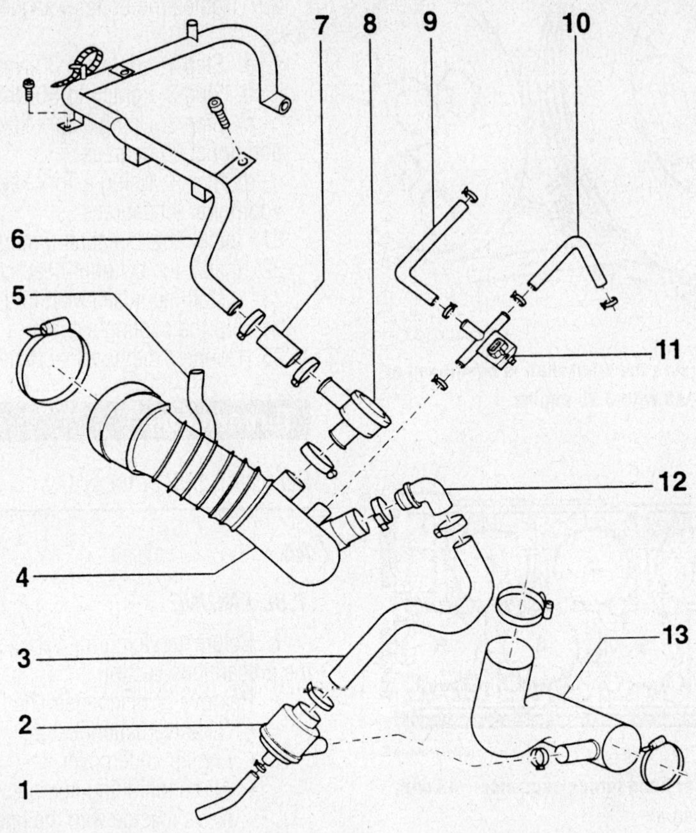

Exploded view of the vacuum hoses related to the turbocharger—1.8L engine

- Turbocharger support bracket
- Oil return line at the turbocharger
- Air hoses from the turbocharger
- Oil feed line at the turbocharger
- Hose for the boost pressure regulation valve vacuum diaphragm
- Bracket for the coolant supply line at the boost pressure regulation valve vacuum diaphragm
- Coolant supply hose by pinching it off using Clamp tool 3094
- Intake air duct between the cowl and the air cleaner housing
- Air cleaner housing cover

3. Label and detach the following lines and electrical connectors:
   - Wastegate bypass regulator valve
   - Evaporative Emissions (EVAP) canister purge regulator valve
   - Power outage stage
   - Mass Air Flow (MAF) sensor
4. Remove or disconnect the following:
   - Air cleaner housing and the engine cover
   - Crankcase breather hose at the valve cover
   - Oil supply line at the turbocharger
   - Heat shield and sleeve from the coolant return hose
   - Coolant return hose by pinching it off using Clamp tool 3094 first

➡The exhaust flex pipe may be damaged if bent more than 10 degrees.

- Three-Way Catalytic Converter (TWC) from the turbo
- Turbo from the exhaust manifold
- Coolant supply banjo fitting at the turbocharger
- Turbocharger

To install:

5. Installation is the reverse of the removal procedure, while using the following torque values:
   - Coolant supply banjo fitting: 18 ft. lbs. (25 Nm)
   - Turbocharger exhaust manifold bolts: 26 ft. lbs. (35 Nm)
   - Turbo support bracket bolts: 33 ft. lbs. (45 Nm)
   - Turbo oil supply line: 18 ft. lbs. (25 Nm)

TT

1.8L ENGINE

1. Before servicing the vehicle, refer to the precautions section.
2. Drain coolant
3. Drain engine oil
4. Remove or disconnect the following:
   - Negative battery cable
   - Engine cover
   - Noise insulation under engine
   - Heat shield for right drive axle
   - Right drive axle
   - Front exhaust pipe

➡The exhaust flex pipe may be damaged if bent more than 10 degrees.

- Air hose between upper air line and lower air line
- Bracket for upper air line
- Oil return line
- Turbocharger bracket from cylinder block
- Bracket for coolant return line
- Coolant supply line from cylinder block
- Hose from pressure unit for charge air pressure control
- Turbocharger support bracket
- Oil return line
- Air hoses
- Pressure control valve
- Connector from charge air pressure control valve
- Vacuum line from air recirculation valve

- Air intake hose from air cleaner
- Pull locking clip out of turbocharger connection and take off air intake hose
- Coolant return hose from the Y connection on the right next to the cylinder head
- Heat shield on back of cylinder head
- Turbocharger from exhaust manifold
- Exhaust manifold
- Retainer piece for oil supply line from turbocharger
- Oil supply line and coolant return line
- Turbocharger

To install:

5. Installation is the reverse of the removal procedure, while using the following notes and torque values:
6. Bolt turbocharger bracket onto turbocharger but do not tighten bolts
7. Position turbocharger against engine from below, then tighten bolts by hand to secure bracket to cylinder block
8. Tighten bolts for oil supply line
9. Install coolant return line with spacer sleeve and bolt it to the turbocharger
10. Unbolt turbocharger bracket from cylinder block again
11. Install exhaust manifold
12. Bolt turbocharger to exhaust manifold
13. Install or connect the following:
    - Coolant supply banjo fitting: 26 ft. lbs. (35 Nm)
    - Turbocharger exhaust manifold bolts: 18 ft. lbs. (25 Nm)
    - Turbo support bracket bolts: 22 ft. lbs. (30 Nm)
    - Turbo return line: 26 ft. lbs. (35 Nm)

2.7L Engine

➡Turbocharger replacement procedure is given with the engine removed. The 2.7L engine uses turbochargers on both sides of the engine.

1. Before servicing the vehicle, refer to the precautions section.

RIGHT SIDE TURBOCHARGER

1. Remove or disconnect the following:
    - Cover panel from cylinder head cover
    - Remove hoses
    - Upper section of intake line
    - Plug lower section of intake line, then remove line
    - Vacuum hoses

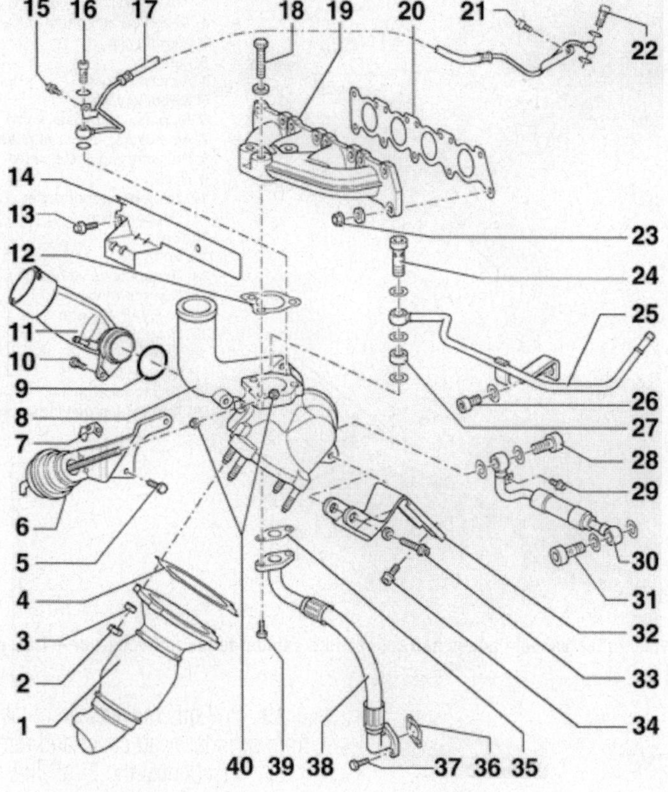

1. Front exhaust pipe
2. Nut
3. Spacer sleeve
4. Gasket
5. Bolt
6. Pressure unit
7. Securing clip
8. Turbocharger
9. O-ring
10. Bolt
11. Connection
12. Gasket
13. Bolt
14. Heat shield
15. Bolt
16. Banjo bolt
17. Oil supply line
18. Bolt
19. Exhaust manifold
20. Gasket
21. Bolt
22. Banjo bolt
23. Nut
24. Banjo bolt
25. Coolant return line
26. Bolt
27. Spacer sleeve
28. Banjo bolt
29. Bolt
30. Coolant supply line
31. Banjo bolt
32. Bracket
33. Bolt
34. Bolt
35. Gasket
36. Gasket
37. Bolt
38. Oil return line
39. Bolt
40. Nut

2348-TTTT-G04

Exploded view of the turbocharger and related components—1.8L engine TT

- Bracket with transmission oil lines
- Pressure hose from turbocharger
- Oil return and water lines
- Hose from cylinder head
- Turbocharger

LEFT SIDE TURBOCHARGER

1. Remove or disconnect the following:
    - Vacuum hoses
    - Lower section of pressure line
    - Upper section of intake line
    - Pressure hose from turbocharger
    - Hose to turbocharger
    - Oil return and water lines
    - Turbocharger

To install:

2. Installation is the reverse of the removal procedure, using the following torque values:
    - Intake lines: 84 inch lbs. (10 Nm)
    - Heat shield: 84 inch lbs. (10 Nm)
    - Oil line to turbocharger: 11 ft. lbs. (15 Nm)
    - Oil line to intake manifold: 15 ft. lbs. (20 Nm)
    - Turbocharger: 15 ft. lbs. (20 Nm

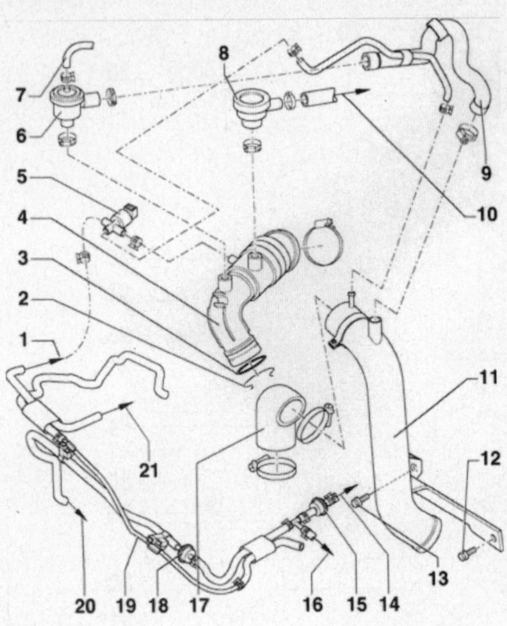

Exploded view of the vacuum hoses and connections related to the turbocharger—1.8L engine TT

A3

2.0L ENGINE

1. Before servicing the vehicle, refer to the precautions section.
2. Disconnect the intake connection using spring-type clip pliers.
3. Remove the engine cover upward and out in direction of.
4. Remove the coolant expansion tank cap.
5. Drain and recycle the coolant.
6. Remove the wiper arms.
7. Unclip spray nozzles and push spray nozzles, with lines still attached, back through opening and into plenum chamber.
8. Remove the rubber seal for plenum chamber cover.
9. Remove the plenum chamber cover.
10. Loosen up the rear engine wiring harness at partition for plenum chamber.
11. Remove the electrical connector for the Heated Oxygen sensor (HO2S) and Oxygen sensor (O2S) located before the catalytic converter from the bracket, disconnect it and lay it aside.
12. Remove the partition for plenum chamber.
13. Disconnect the connectors from ignition coils.
14. Remove the heat shield.
15. Disconnect the coolant line to coolant expansion tank.
16. Disconnect the coolant hoses.
17. Unfasten the coolant line.
18. Unfasten the crankcase ventilation line with heat shield from turbocharger.
19. Pull off crankcase ventilation line from cylinder head cover and remove it.
20. Disconnect the EVAP line to turbocharger from cylinder head cover.
21. Unscrew oil supply line from turbocharger.

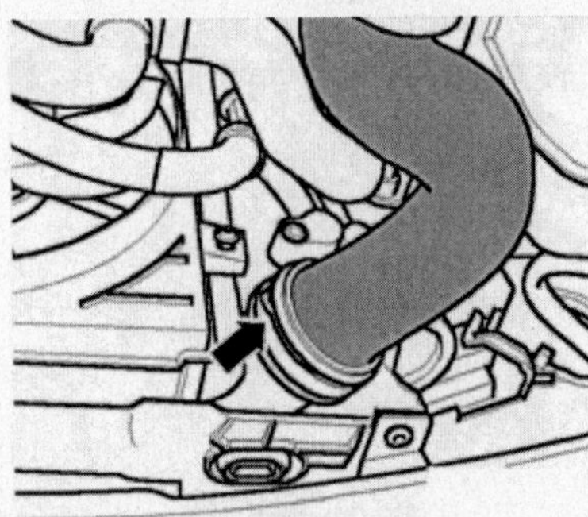

Pull air duct off from charge air cooler—A3 with 2.0L engine

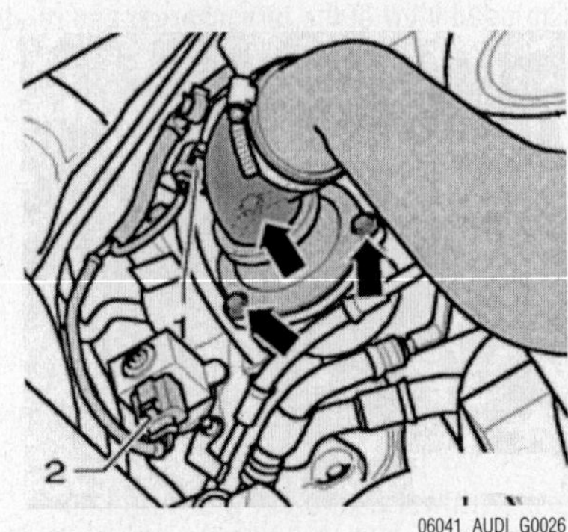

Unfasten the air duct from turbocharger—A3 with 2.0L engine

22. Remove the catalytic converter with front exhaust pipe
23. Remove the center sound insulation fasteners.
24. Remove the right sound insulation fasteners.
25. Remove the right drive axle.
26. Remove the heat shield for right drive axle.
27. Pull air duct off from charge air cooler.
28. Unfasten the air duct from turbocharger.
29. Unscrew bolts and remove air duct.
30. Remove the oil supply line from turbocharger.
31. Unscrew coolant supply line to turbocharger.
32. Remove the oil return line at turbocharger.
33. Remove the bolts and remove turbocharger support.
34. Disconnect the coolant pipe.
35. Remove the coolant pipe from cylinder head.
36. Remove the upper nuts.
37. Remove the turbocharger/exhaust manifold upward and out.

To install:

38. Installation is the reverse of removal, please note the following:

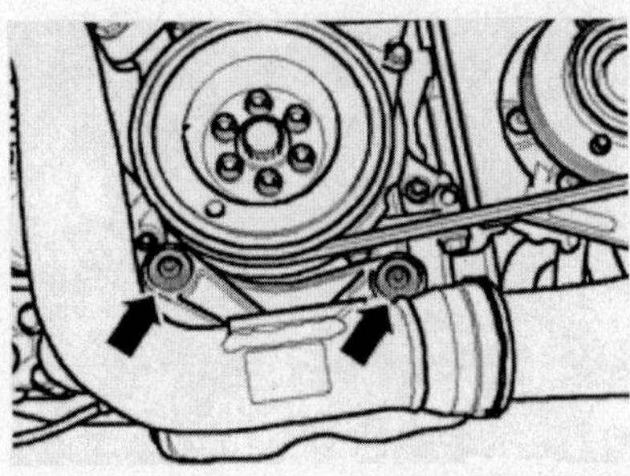

Unscrew bolts and remove air duct—A3 with 2.0L engine

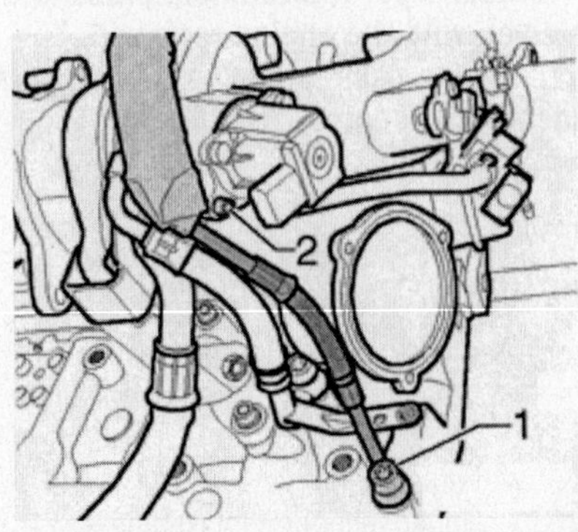

Remove the oil supply line from turbocharger—A3 with 2.0L engine

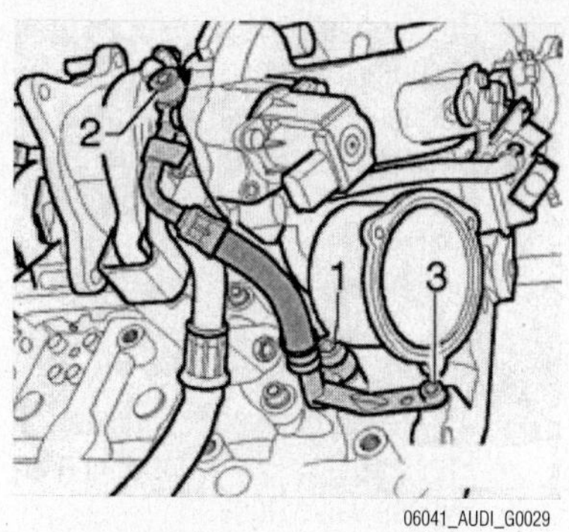

Unscrew coolant supply line to turbocharger—A3 with 2.0L engine

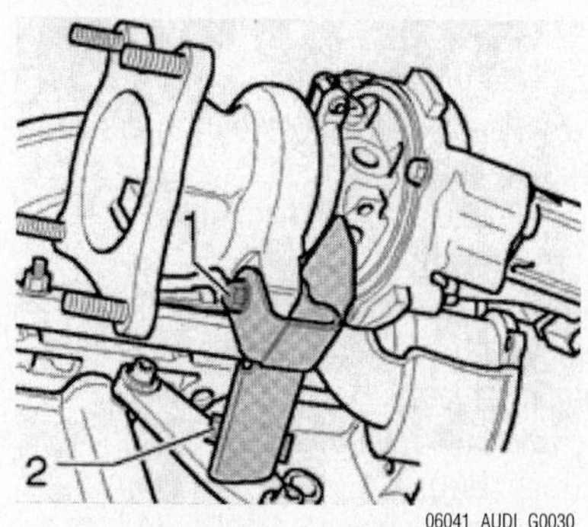

Remove the turbocharger support—A3 with 2.0L engine

➡**Replace gaskets, seals and self-locking nuts.**

a. Tighten the exhaust manifold/turbocharger to cylinder head nuts to 21 Nm using new fasteners.

b. Tighten the heat shield to exhaust manifold/turbocharger nuts to 20 Nm using new fasteners.

c. Tighten the oil supply line to turbocharger to 30 Nm.

d. Tighten the bracket for oil supply line to exhaust manifold to 20 Nm using new fasteners.

e. Tighten the bracket for turbocharger to cylinder block to 30 Nm using new fasteners.

f. Tighten the bracket for turbocharger to turbocharger to 30 Nm using new fasteners.

g. Tighten the right charge air tube to oil pan and rear charge air tube to bracket to 10 Nm.

h. Tighten the oil supply line to oil filter bracket to 25 Nm.

i. Fill the turbocharger with engine oil at the connection for oil supply line.

j. Hose unions and hoses for charge air system must be free of oil and grease before installing.

k. Install the exhaust system and align it free of tension

l. Fill the cooling system.

m. Check the engine oil level and replenish as needed.

➡**After installing turbocharger, let engine idle for about 1 minute without increasing engine speed. This ensures adequate oil supply to the turbocharger.**

Intake Manifold

REMOVAL & INSTALLATION

A4

1.8L ENGINE

1. Before servicing the vehicle, refer to the precautions section.
2. Drain the engine coolant.
3. Properly relieve the fuel system pressure.
4. Turn the ignition switch to the **OFF**.
5. Remove or disconnect the following:
    - Negative battery cable
    - Engine covers
    - Hose for the Leak Detection Pump (LDP)
    - Accelerator pedal cable from the throttle valve control module
    - Air guide hose from the throttle valve control module
    - Vacuum line from the Evaporative Emissions (EVAP) canister
    - Brake booster vacuum hose
    - Intake Air Temperature (IAT) sensor and the throttle valve control module
    - Camshaft Position (CMP) sensor wiring harness connector
    - Fuel rail with the injectors
    - Coolant hoses from to the intake manifold
    - Crankcase breather hose at the intake manifold
    - Intake manifold brace and the oil dipstick
    - Manifold at the mounting flange
6. Clean all gasket surfaces.

To install:

7. Install or connect the following:
    - Intake manifold using new gaskets. Torque the fasteners to 89 inch lbs. (10 Nm).
    - Manifold brace. Torque the bolts to 15 ft. lbs. (20 Nm).
    - Dipstick
    - Crankcase breather and coolant hoses to the intake manifold
    - Fuel injector sealing O-ring
    - Fuel rail with the injectors. Torque the bolts to 89 inch lbs. (10 Nm).
    - CMP and the IAT sensors to the throttle valve control module
    - Brake booster and EVAP canister vacuum hoses
    - Air guide hose and the accelerator pedal cable to the throttle valve control module
    - Hose for the LDP
8. Top off the engine coolant and bleed, if necessary.
9. Connect the negative battery cable.
10. Fully close all power windows to stop, operate all window switches for at least 1 second in the close direction to activate the one-touch opening/closing function.
11. Check the oil level before starting the engine and top off, as necessary.

TT

1.8L ENGINE

1. Before servicing the vehicle, refer to the precautions section.
2. Drain the engine coolant.
3. Properly relieve the fuel system pressure.
4. Turn the ignition switch to the **OFF**.
5. Remove or disconnect the following:
    - Negative battery cable
    - Engine cover
    - IAT and throttle control valve connectors
    - Throttle control valve air hose
    - Vacuum hoses
    - Hall sensor connector
    - Fuel injector connectors
    - Support bar from fuel rail
    - Electrical change over valve connector
    - Dipstick tube bracket
    - Release tabs from fuel supply and return lines
    - PCV hose
    - Intake manifold support
    - Intake manifold

To install:

6. Installation is the reverse of the removal procedure, using the following torque values:
    - Fuel rail bolts: 84 inch lbs. (10 Nm)
    - Intake manifold bolts: 84 inch lbs. (10 Nm)

3.2L ENGINE

1. Before servicing the vehicle, refer to the precautions section.
2. Drain the engine coolant.

3. Properly relieve the fuel system pressure.
4. Turn the ignition switch to the **OFF**.
5. Remove or disconnect the following:
   - Negative battery cable
   - Right and left covers at lock carrier, above main fuse box and coolant expansion tank

➡Before removing ribbed belt, mark direction of rotation with chalk or a felt pen. If a used belt rotates in the wrong direction when reinstalled, this can result in breakage. When installing belt, ensure it is correctly seated in the pulleys.

   - Release tension on belt by screwing an M8 x 40 bolt into bore on tensioning element and remove belt
   - Upper mounting bolt for the alternator
   - Upper mounting bolts for radiator cowl
   - Secondary air hose
   - Center and right-hand noise Insulation under engine
   - Cross piece
   - Power steering pump from bracket, hang pump from lock carrier with hoses still attached
   - Secondary air injection pump with bracket
   - Lower radiator cowl bolts and remove the radiator cowl with fans

➡The air conditioning compressor can only be removed from the bracket with the upper securing bolt still inserted. To accomplish this, press the engine simultaneously to the left side. If necessary, loosen all bolts of the assembly mounting.

   - Mounting bolt for air conditioning compressor
   - Secure the air conditioning compressor together with the hoses at front longitudinal member (hoses remain connected)
   - Alternator
   - Oil dipstick tube
   - Vacuum reservoir
   - Mounting bracket for alternator and air conditioning compressor
   - Bolts at front of intake manifold
   - Attach puller T10095A or equivalent to ignition coils and remove coils
   - Body brace
   - Air intake hose
   - Crankcase breather hose
   - Throttle body
   - Hose bracket
   - Rear intake mounting bolts, swing intake manifold forward
   - Fuel rail with fuel injectors
   - Swing back to installed position and remove manifold

To install:

6. Installation is the reverse of the removal procedure, using the following torque values:
   - Intake manifold to cylinder head 115 inch lbs. (13 Nm)
   - Bracket for water pump to cylinder block 15 ft. lbs. (20 Nm)
   - Intake manifold support to intake manifold 15 ft. lbs. (20 Nm)
   - Throttle body to intake manifold 88 inch lbs. (10 Nm)

2.7L Engine

1. Before servicing the vehicle, refer to the precautions section.
2. Drain the engine coolant.
3. Properly relieve the fuel system pressure.
4. Turn the ignition switch to the **OFF**.
5. Remove or disconnect the following:
   - Negative battery cable
   - Wastegate bypass regulator from fuel rail
   - Connector from EVAP canister purge valve
   - Connector at turbocharger recirculating valve
   - Recirculating valve hose clamps
   - Fuel pressure regulator hose clamps
   - Exhaust temperature sensors, (leave connected)
   - Brake booster hose
   - Vacuum reservoir hose
   - Connectors for fuel injectors
   - Suction hose at turbocharger
   - Hose clamps at bypass valve, turbocharger intake line and fuel vapor line
   - Fuel rail (4) bolts
   - Throttle valve connector
   - Air pressure sensor connector
   - Pressure hoses at front of engine
   - Bypass valve hose
   - Intake air temperature sensor connector
   - PCV hose
   - Fuel return and supply lines to fuel rail
   - EVAP valve hose
6. Remove fuel rail and injectors and support fuel rail
7. Remove intake manifold

To install:

8. Installation is the reverse of the removal procedure, using the following torque values:
   - Fuel rail bolts: 84 inch lbs. (10 Nm)
   - Intake manifold bolts: 84 inch lbs. (10 Nm)

2.8L Engine

1. Before servicing the vehicle, refer to the precautions section.
2. Disconnect the negative battery cable.
3. Properly relieve the fuel system pressure.
4. Remove or disconnect the following:
   - Fuel supply and return lines

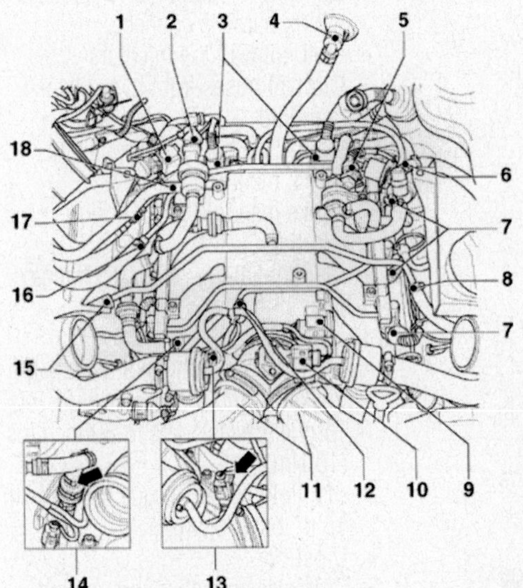

Component view of the intake manifold assembly and related components—2.7L engine

- Fuel injector electrical connectors and the retainers for the fuel injectors
- Fuel pressure regulator clamp and the regulator
- Fuel manifold
- Injectors
- Any hoses or connectors associated with the intake manifold
- Upper intake manifold mounting bolts
- Upper intake manifold and discard the gasket
- Idle Air Control (IAC) valve
- Exhaust Gas Recirculation (EGR) valve
- Accelerator cable form the throttle body
- Lower intake manifold mounting bolts
- Lower intake manifold

5. Clean all gasket mating surfaces.

To install:

6. Install or connect the following:
   - Lower intake manifold using a new gasket. Torque the bolts to 15 ft. lbs. (20 Nm).
   - EGR valve
   - Accelerator cable to the throttle body
   - IAC valve. Torque the bolts to 89 inch lbs. (10 Nm).
   - Upper intake manifold. Torque the short bolts to 89 inch lbs. (10 Nm) and the long bolts to 15 ft. lbs. (20 Nm).
   - All hoses or connectors that were removed
   - Fuel injectors using new O-rings
   - Fuel manifold. Torque the bolts to 89 inch lbs. (10 Nm).
   - Pressure regulator and clamp
   - Fuel injector retainers
   - Electrical connectors
   - Fuel return and supply lines
   - Negative battery cable

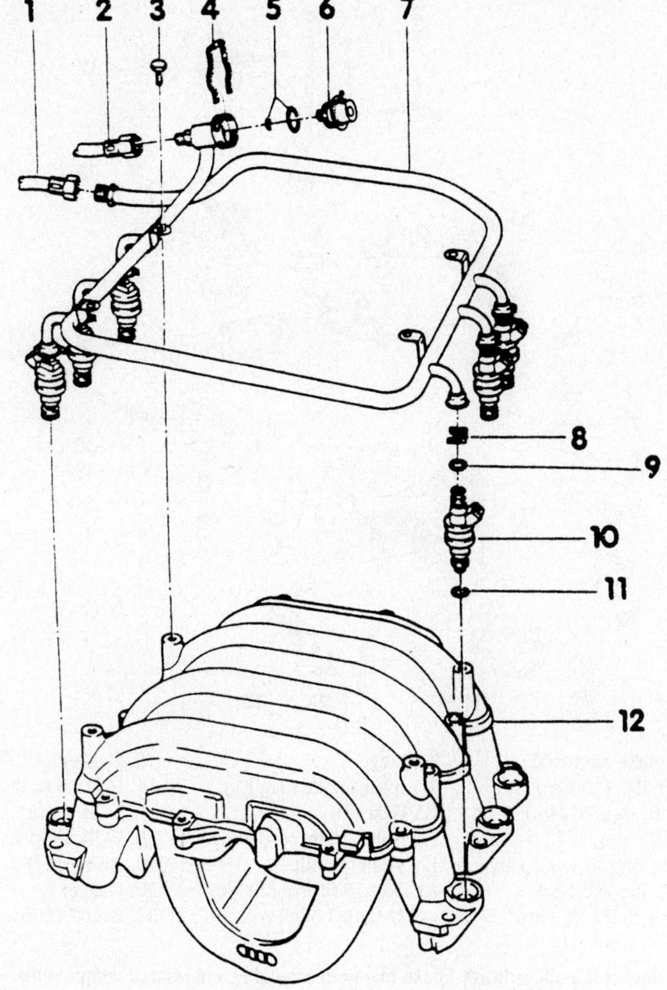

Exploded view of the fuel injector assembly—2.8L engine

3.0L Engine

1. Before servicing the vehicle, refer to the precautions section.
2. Drain the engine coolant.
3. Properly relieve the fuel system pressure.
4. Turn the ignition switch to the **OFF**.
5. Remove or disconnect the following:
   - Negative battery cable
   - Engine covers
   - Air filter induction tube
   - EVAP canister purge valve at air filter housing
   - Connector at Mass Air Flow (MAF) sensor
   - Air guide hose with MAF sensor
   - Air filter housing
   - Vacuum hoses at intake manifold change-over valve and check valve
   - PCV hose
   - Solenoid valve holding plate
   - Fuel injector connectors
   - Fuel pressure regulator
   - Fuel distributor with fuel injectors
   - All remaining vacuum and electrical connectors
   - Remaining air ducting from throttle body
   - Dipstick tube
   - Secondary air injection system
   - Coolant hoses
   - Intake manifold

To install:

6. Installation is the reverse of the removal procedure, using the following torque values:
   - Fuel rail bolts: 84 inch lbs. (10 Nm)
   - Intake manifold bolts: 84 inch lbs. (10 Nm)
   - All remaining bolts, except M8 on AIR: 84 inch lbs. (10 Nm)
   - M8 bolts on AIR system: 17 ft. lbs. (23 Nm)

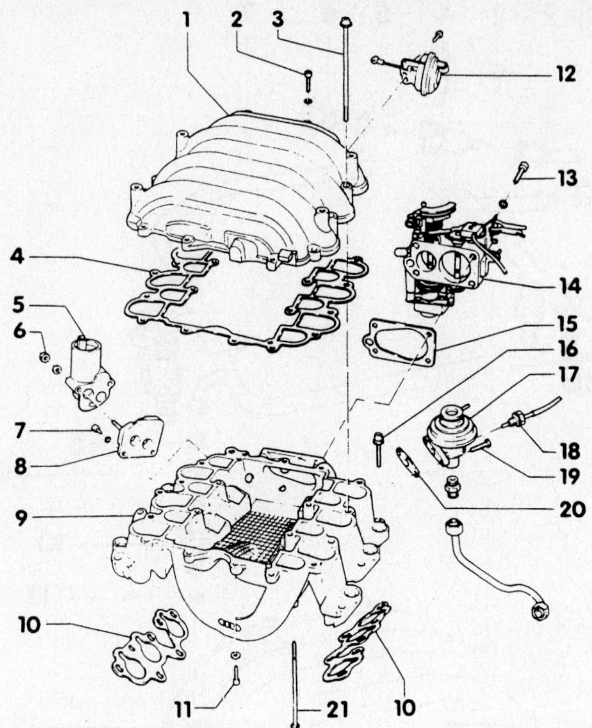

1. Upper intake manifold
2. Bolt 7 ft. lbs. (10 Nm)
3. Bolt 15 ft. lbs. (20 Nm)
4. Gasket
5. Idle Air Control (IAC) valve
6. Bolt 7 ft. lbs. (10 Nm)
7. Bolt 53 inch lbs. (6 Nm)
8. Flange
9. Lower intake manifold
10. Gasket
11. Bolt 7 ft. lbs. (10 Nm)
12. Vacuum unit
13. Bolt 15 ft. lbs. (20 Nm)
14. Throttle body
15. Gasket
16. Bolt 15 ft. lbs. (20 Nm)
17. EGR valve
18. EGR temp sensor
19. Bolt 7 ft. lbs. (10 Nm)
20. Gasket
21. Bolt 7 ft. lbs. (10 Nm)

7923CG08

Exploded view of the upper/lower intake manifold assembly and related components—2.8L engine

A6 and A8

3.2L ENGINE

1. Before servicing the vehicle, refer to the precautions section.
2. Remove the engine covers.
3. Disconnect the check valve from connection at air duct hose.
4. Remove the air duct hose.
5. Disconnect the electrical harness connectors from the change-over valve for intake manifold flap and the Intake Manifold Tuning (IMT) valve position sensor.
6. Remove and necessary vacuum lines and hoses .
7. Disconnect the following electrical connectors.
   a. Camshaft adjustment valve.
   b. Throttle valve control module.
   c. Evaporative emission (EVAP) canister purge regulator valve.
   d. Intake Air Temperature (IAT) sensor G42 / Manifold Absolute Pressure (MAP) sensor.
   e. Intake Manifold Runner Control (IMRC) valve.
8. Remove the Evaporative Emission (EVAP) canister purge regulator valve from the throttle valve control module.
9. Disconnect the crankcase breather hose at pressure regulator valve.
10. Remove the bolts and the upper intake manifold.

➡Plug the intake ports of the cylinder head with clean rags.

11. Remove the high-pressure line, thereby removing bolts and union nuts—1 to 7—.
12. Disconnect the following electrical connectors:
    a. Intake Manifold Runner Position sensor.
    b. Fuel Pressure sensor.
13. Remove the nuts and bolts and remove lower intake manifold with fuel rail.

➡Plug the intake ports of the cylinder head with clean rags.

To install:

14. Replace the lower manifold gaskets and O-rings.
15. Tighten the lower manifold bolts to 9 Nm using several passes in a diagonal sequence.
16. Replace the upper manifold gaskets and O-rings.
17. Tighten the upper manifold bolts to 6 Nm using several passes in a diagonal sequence.
18. Tighten the high pressure line to 25 Nm at the pump and fuel rail.
19. Install the remaining components in the reverse order of removal.

4.2L ENGINE

1. Before servicing the vehicle, refer to the precautions section.
2. Relieve the fuel system pressure.
3. Remove or disconnect the following:
   - Negative battery cable
   - Noise insulation panel
   - Intake air grille in bumper
   - Accessory drive belt
   - Timing belt
   - Vacuum line to brake booster unit
   - Vacuum line at cruise control module
   - Accelerator pedal cable
   - Coolant lines to heater core (supply and return to vent valves)
   - Left and right knock sensor connectors at fuel rail and remove cable ties (4x)
   - Fuel injector harness connectors
   - Fuel rail, carefully remove along with fuel injectors, and place down to one side
   - Intake manifold

To install:

4. Install or connect the following:
   - Intake manifold. Tighten the fasteners to 15 ft. lbs. (20 Nm).
   - Fuel rail. Tighten the bolts to 84 inch lbs. (10 Nm).
   - Fuel injector harness connectors
   - Left and right knock sensor connectors at fuel rail
   - Coolant lines to heater core (supply and return to vent valves)
   - Accelerator pedal cable
   - Vacuum line at cruise control module
   - Vacuum line to brake booster unit
   - Timing belt
   - Accessory drive belt
   - Intake air grille in bumper
   - Noise insulation panel
   - Negative battery cable

A3

2.0L ENGINE

1. Before servicing the vehicle, refer to the precautions section.
2. Relieve the fuel system pressure.
3. Remove the engine cover.
4. Disconnect the all necessary electrical harness connectors.
5. Disconnect the vacuum hose between intake manifold and vacuum pump from intake manifold.
6. Disconnect the hose connections from valve cover.
7. Loosen the fuel supply line and hose connection from EVAP canister and cap off open lines.
8. Open both fuel lines at high-pressure pump.
9. Disconnect the electrical connector from Fuel Pressure sensor.
10. Remove the oil dipstick and remove bolt for oil dipstick guide tube.
11. Remove the bolt for oil line from intake manifold.
12. Open hose clamps and pull intake hose off from Throttle Valve Control Module and intake manifold.
13. Remove the nut from intake manifold bracket, if necessary, remove Throttle Valve Control Module.
14. Separate oil dipstick guide tube at separation point and remove bolt from intake manifold bracket.
15. Remove the bolts from intake manifold.
16. Carefully pull intake manifold with fuel rail off from cylinder head.

➡The fuel injectors could remain stuck in the fuel rail.

To install:

17. Installation is the reverse of removal, please note the following:
    a. Intake manifold bolts are tightened to 9 Nm
    b. Intake manifold support bolts are tightened to 23 Nm
    c. Intake manifold support nuts are tightened to 10 Nm
    d. Fuel supply line to pump are tightened to 27 Nm
    e. New fuel return line to pump are tightened to 17 Nm

3.2L ENGINE

1. Remove the molded insert for tools under luggage compartment floor cover.
2. Unscrew cover for battery compartment.
3. Remove the molded insert over battery.
4. Disconnect the negative battery cable.
5. Remove the noise insulation.
6. Disconnect the vacuum hose from air duct hose.
7. Disconnect the air duct hose from Throttle Valve Control Module.
8. Disconnect the electrical harness connector at Mass Air Flow (MAF) sensor.
9. Unbolt upper section of air filter housing.
10. Remove the filter element.
11. Unbolt lower section of air duct.
12. Unbolt lower section of air filter housing.
13. Disconnect the secondary air hose at position indicated by.
14. Loosen up the air hose to Secondary Air Injection (AIR) pump.
15. Disconnect the electrical harness connectors and.
16. Disconnect the crankcase ventilation hose.
17. Unscrew Throttle Valve Control Module from the intake manifold and lay it aside with coolant hoses connected.
18. Disconnect the electrical connectors from the ignition coils.
19. Attach Puller for ignition coil tool T10095A on ignition coils as shown in the illustration and pull out ignition coils in succession.

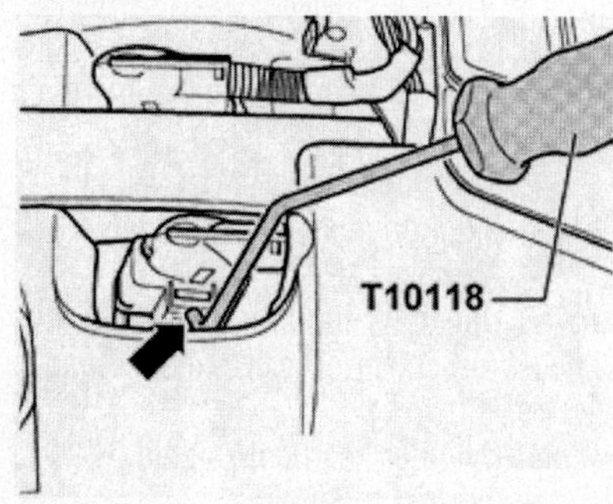

Disconnect the electrical connectors from the ignition coils—A3 with 3.2L engine

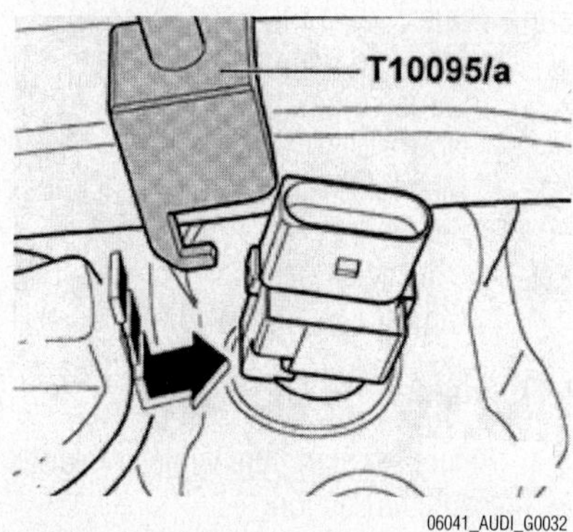

Attach Puller for ignition coil tool T10095A on ignition coils and pull out ignition coils in succession —A3 with 3.2L engine

20. Remove the ignition wiring harness strip and set it aside.
21. Disconnect the vacuum hoses at rear from intake manifold.
22. Remove the line union from intake manifold and place it in the rear with hoses connected.
23. Remove the bolts for rear intake manifold mount.
24. Unscrew top mounting bolts for fan shroud.
25. Remove the dipstick.

➡Mark direction of rotation of accessory drive belt using chalk or felt-tip marker before removing. Reversing the direction of rotation of a run-in belt can destroy the belt

26. Remove the accessory drive belt.
27. Loosen up the coolant hose at bottom on fan shroud.
28. Disconnect the electrical harness connector at Secondary Air Injection (AIR) pump motor.
29. Unscrew bracket for coolant pipe.
30. Remove the bolts.
31. Loosen the bolt and remove Secondary Air Injection (AIR) pump with bracket.
32. Disconnect the electrical harness connectors for coolant fans at bottom on fan shroud.
33. Unscrew mounting bolts for fan shroud at the bottom.
34. Pull out fan shroud downward with both fans.
35. Disconnect the electrical harness connector for A/C clutch on A/C compressor and loosen up electrical wire.
36. Remove the bolts for A/C compressor. Securely tie the A/C compressor with connected coolant hoses at front on the longitudinal member.
37. Unscrew electrical wire on alternator.
38. Disconnect the electrical harness connector.
39. Remove the upper idler roller.
40. Remove the mounting bolts and for alternator.

➡The alternator can be removed from bracket only with the upper mounting bolt still installed.

41. Remove the alternator with electrical wires connected from bracket for assemblies.
42. Remove the alternator downward and to the left.
43. Disconnect the electrical harness connector at the after-run coolant pump.
44. Remove the bolts and remove bracket for assemblies.
45. Pull the after-run coolant pump with coolant hoses connected, downward out of

rubber loops of retainer. Spray rubber loops with silicon-free lubricant if necessary.

➡**The water pump remains on the engine with the coolant hoses connected.**

46. Disengage harness connector for Engine Speed (RPM) sensor from bracket on oil dipstick guide tube.
47. Unscrew bolt and pull out oil dipstick guide tube.
48. Unscrew bolt and pull off vacuum reservoir from intake manifold.
49. Set aside vacuum reservoir with lines still connected.
50. Remove the bolts at front on intake manifold.
51. Disconnect the vacuum line from actuator for intake manifold change-over.

➡**Protect intake manifold from damage with a clean cloth.**

52. Swivel intake manifold forward and then pull slightly toward left.

➡**Seal intake channels in cylinder head with clean rags or foam pieces so that small pieces cannot fall in.**

53. Remove the bracket for right fuel line at cylinder head cover.
54. Expose electrical wiring harness on cylinder head cover.
55. Remove the bolts.
56. Pull off fuel rail pipe with fuel injectors from cylinder head and set aside with fuel line connected.
57. Swivel intake manifold back into installation position and remove.

To install:

58. Installation is the reverse of removal, please note the following:

➡**Replace all gaskets.**

a. Intake manifold to cylinder head retainers are tightened to 13 Nm.
b. Intake manifold support to intake manifold are tightened to 20 Nm.
c. Throttle Valve Control Module to intake manifold retainers are tightened to 10 Nm.

Exhaust Manifold

REMOVAL & INSTALLATION

A4

1.8L ENGINE

1. Before servicing the vehicle, refer to the precautions section.
2. Remove or disconnect the following:
   - Cylinder head cover
   - Intake manifold cover
   - Turbocharger to charcoal filter hose
   - Upper air pipe and heat shield
   - Turbocharger
   - Exhaust manifold nuts
   - Exhaust manifold
3. Clean the gasket mounting surfaces.

To install:

4. Install or connect the following:
   - Exhaust manifold using a new gasket. Torque the nuts to 18 ft. lbs. (25 Nm).
   - Turbocharger. Torque the bolts for 2001-03 to 22 ft. lbs. (30 Nm) and for 2004 to 25 ft. lbs. (35 Nm)
   - Upper air pipe and heat shield. Torque the bolts to 89 inch lbs. (10 Nm).
   - Turbocharger to charcoal filter hose
   - Intake manifold cover
   - Cylinder head cover

TT

1.8L ENGINE

1. Before servicing the vehicle, refer to the precautions section.
2. Remove or disconnect the following:
   - Negative battery cable.
   - Cylinder head and intake manifold covers
   - Noise insulation under vehicle
   - Right drive axle heat shield
   - Air hoses
   - Upper air line bracket
   - Turbocharger from exhaust manifold
   - Cylinder head heat shield
   - Exhaust manifold
3. Clean all gasket mating surfaces.

To install:

4. Installation is the reverse of the removal procedure, using the following torque values:
   - Manifold using a new gasket. Torque the nuts to 18 ft. lbs. (25 Nm)
   - Manifold heat shield: 84 inch lbs. (10 Nm)
   - Drive shaft heat shield: 26 ft. lbs. (35 Nm)

2.7L Engine

1. Before servicing the vehicle, refer to the precautions section.
2. Remove engine
3. Remove turbocharger
4. Remove exhaust manifold

To install:

5. Install exhaust manifold and tighten to 18 ft. lbs. (25 Nm)

2.8L Engine

RIGHT MANIFOLD

1. Before servicing the vehicle, refer to the precautions section.
2. Remove or disconnect the following:
   - Negative battery cable
   - Heated Oxygen (HO_2S) sensor
   - Exhaust system from the manifold
   - Heat shield
   - Manifold nuts
   - Exhaust manifold
3. Clean all gasket mating surfaces.

To install:

4. Install or connect the following:
   - Exhaust manifold using a new gasket. Torque the nuts to 18 ft. lbs. (24 Nm)
   - Heat shield
   - Exhaust system to the manifold using a new gasket
   - HO_2S sensor
   - Negative battery cable
5. Start the vehicle, check for leaks and repair if necessary.

LEFT MANIFOLD

1. Before servicing the vehicle, refer to the precautions section.
2. Disconnect the negative battery cable.
3. Drain the engine coolant.
4. Remove or disconnect the following:
   - Heated Oxygen (O_2S) sensor
   - Exhaust system from the manifold
   - Coolant tube from the cylinder head
   - Heat shield
   - Exhaust Gas Recirculation (EGR) tube from the rear of the manifold
   - Exhaust manifold nuts
   - Exhaust manifold
5. Clean all gasket mating surfaces.

To install:

6. Install or connect the following:
   - Manifold using a new gasket. Torque the nuts to 18 ft. lbs. (24 Nm)
   - Exhaust system to the manifold using a new gasket
   - HO_2S sensor
   - EGR tube to the rear of the manifold
   - Heat shield
   - Coolant tube to the cylinder head
   - Negative battery cable
7. Fill the engine with coolant.
8. Start the vehicle, check for leaks and repair if necessary.

3.0L Engine

LEFT MANIFOLD

1. Before servicing the vehicle, refer to the precautions section.
2. Remove or disconnect the following:
   - Negative battery cable
   - Covers, components and electrical connections for access
   - Exhaust system from the manifold
   - Heat shield
   - Manifold nuts
   - Exhaust manifold
3. Clean all gasket mating surfaces.

To install:

4. Install or connect the following:
   - Exhaust manifold using a new gasket. Torque the nuts to 18 ft. lbs. (25 Nm)
   - Heat shield. Torque M6 bolts to 84 inch lbs. (10 Nm) and M8 bolts to 18 ft. lbs. (25 Nm)
   - Exhaust system to the manifold using a new gasket
   - Covers, components and electrical connections
   - Negative battery cable
5. Start the vehicle, check for leaks and repair if necessary.

RIGHT MANIFOLD

1. Before servicing the vehicle, refer to the precautions section.
2. Remove or disconnect the following:
   - Negative battery cable.
   - Covers, components and electrical connections for access (mark connections for reinstallation)
   - Right front exhaust pipe with catalytic converter
   - Secondary air injection valve bolts (resonator remains installed)
   - Dipstick guide tube
   - Heat shield bracket
   - Heat shield
   - Exhaust manifold nuts
   - Exhaust manifold
3. Clean all gasket mating surfaces.

To install:

4. Install or connect the following:
   - Manifold using a new gasket. Torque the nuts to 18 ft. lbs. (25 Nm)
   - Heat shield and bracket. Torque M6 bolts to 84 inch lbs. (10 Nm) and M8 bolts to 18 ft. lbs. (25 Nm)
   - Dipstick guide tube
   - Secondary air injection line bolts
   - Covers, components and electrical connections
   - Negative battery cable
5. Fill the engine with coolant.
6. Start the vehicle, check for leaks and repair if necessary.

3.2L Engine, except TT and A6

1. Before servicing the vehicle, refer to the precautions section.
2. Remove or disconnect the following:
   - Negative battery cable
   - Intake manifold
   - Heat shield
   - Remove 7 nuts and washers on both exhaust manifolds
   - Pull exhaust manifolds towards the rear and remove

To install:

3. Installation is the reverse of the removal procedure, using the following torque values:
   - Exhaust manifold to cylinder head 17 ft. lbs (23 Nm)
   - Exhaust pipe to manifold 29 ft. lbs (40 NM)

A6 and A8 4.2L Engine

BOTH MANIFOLDS

➡This procedure requires engine removal. See ENGINE ASSEMBLY section.

1. Before servicing the vehicle, refer to the precautions section.
2. Remove or disconnect the following:
   - Negative battery cable
   - Dipstick tube (LH manifold only)
   - Exhaust system from the manifold
   - Heat shield
   - Manifold nuts
   - Exhaust manifold
3. Clean all gasket mating surfaces.

To install:

4. Install or connect the following:
   - Exhaust manifold using a new gasket. Torque the nuts to 18 ft. lbs. (25 Nm)
   - Heat shield
   - Exhaust system to the manifold using a new gasket
   - Dipstick tube
   - Negative battery cable
5. Reinstall engine assembly.
6. Start the vehicle, check for leaks and repair if necessary.

Front Crankshaft Seal

REMOVAL & INSTALLATION

A4

1.8L ENGINE

1. Before servicing the vehicle, refer to the precautions section.

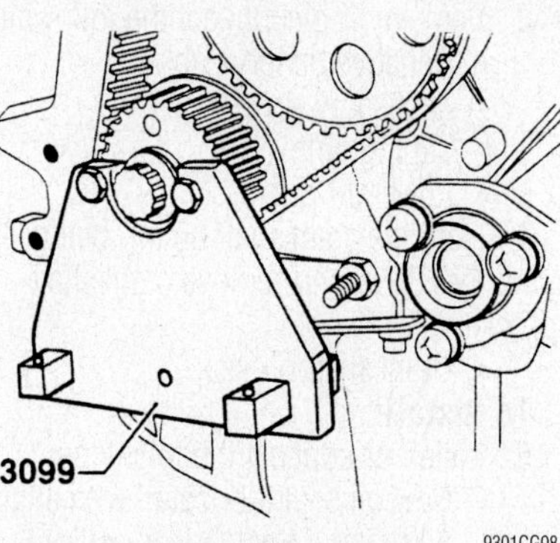

Tool 3099 is used to hold the crankshaft toothed gear, allowing removal of the center bolt—1.8L engines

2. Place the lock carrier into the service position.
3. Turn the ignition switch to the **OFF** position.
4. Remove or disconnect the following:
   - Negative battery cable
   - Accessory drive belts
   - Timing belt
   - Torque arm mounting bracket
5. Hold the crankshaft timing belt gear using tool 3099.
6. Remove or disconnect the following:
   - Crankshaft timing belt gear retaining bolt
   - Crankshaft timing belt gear
   - Crankshaft oil seal

To install:

7. Install or connect the following:
   - New oil seal using a Seal Driver until it is flush. Do NOT oil sealing lip of oil seal.
   - Timing belt gear. Torque the new bolt to 66 ft. lbs. (90 Nm) plus ¼ (90 degree) turn.
   - Torque arm mounting bracket. Torque the bolts to 18 ft. lbs. (25 Nm)
   - Timing belt
   - Accessory drive belts
   - Negative battery cable
8. Fully close all power windows to stop, operate all window switches for at least 1 second in the close direction to activate the one-touch opening/closing function.
9. Check the oil level before starting the engine.

TT

1.8L ENGINE

1. Before servicing the vehicle, refer to the precautions section.
2. Turn the ignition switch to the **OFF** position.

3. Remove or disconnect the following:
   - Negative battery cable
   - Accessory drive belt
   - Timing belt
   - Lower air duct
4. Hold the crankshaft timing belt gear using tool 3099 and remove crankshaft sprocket
   - Crankshaft oil seal

To install:

5. Install or connect the following:
   - New oil seal lubricated with engine oil using a Seal Driver until it is flush
   - Crankshaft sprocket: Torque to 66 ft. lbs. (90 Nm) plus 90 degrees
   - Lower air duct
   - Timing belt
   - Accessory drive belt
   - Negative battery cable

2.7L Engine

1. Before servicing the vehicle, refer to the precautions section.
2. On all A6 models, place the lock carrier into the service position.
3. Turn the ignition switch to the **OFF** position.
4. Remove or disconnect the following:
   - Negative battery cable
   - Timing belt
   - Crankshaft sprocket
   - Crankshaft oil seal

To install:

5. Install or connect the following:
   - New oil seal lubricated with engine oil using a Seal Driver until it is flush
   - Crankshaft sprocket: Torque to 148 ft. lbs. (200 Nm) plus 90 degrees
   - Timing belt
   - Negative battery cable

2.8L Engine

1. Before servicing the vehicle, refer to the precautions section.
2. On all A4 and A6 models, place the lock carrier into the service position.
3. Remove the accessory drive belt.
4. Rotate the engine to Top Dead Center (TDC).
5. Remove or disconnect the following:
   - Sealing plug from the left side of the engine
   - Crankshaft by locking it in position using tool 3242
   - Timing belt
   - Crankshaft timing belt sprocket from the crankshaft
   - Seal using a Seal remover
6. Clean the running and sealing surfaces.

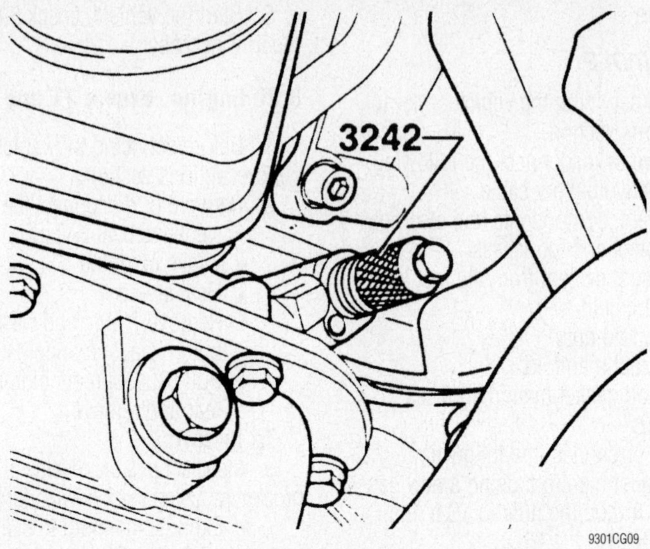

The crankshaft is locked in the TDC position by installing tool 3242 through the sealing plug on the left side of the engine block—2.8L engines

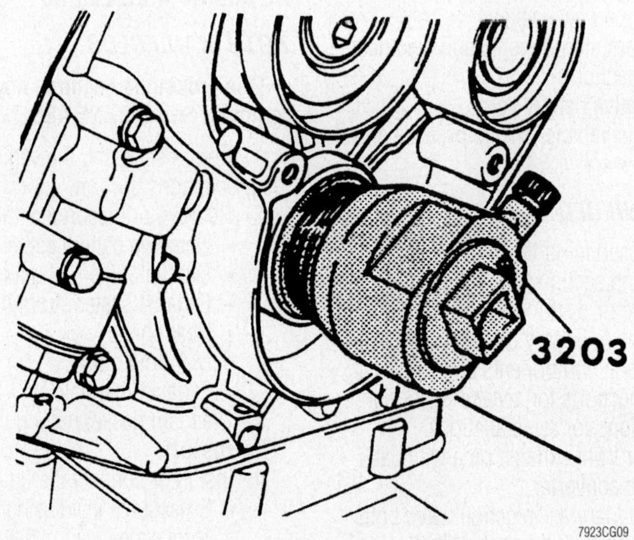

Removing the seal using the seal remover—2.8L and 3.0L engines

To install:

7. Slide the new seal over the Installing Sleeve tool 3202.
8. Install or connect the following:
   - New oil seal using a Seal Installer tool 3265 until it is flush
   - Center crankshaft bolt
   - Timing belt sprocket. Torque the center crankshaft bolt to 148 ft. lbs. (200 Nm), plus an additional ½ (180 degree) turn
   - Timing belt
   - Accessory drive belt

3.0L Engine

1. Before servicing the vehicle, refer to the precautions section.

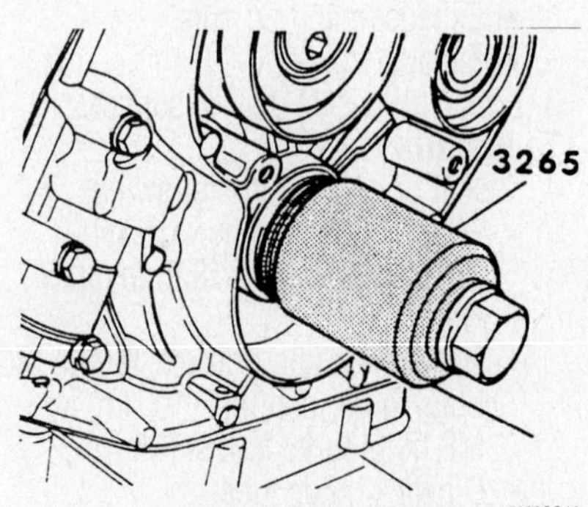

Installing the seal using the seal installer and the crankshaft center bolt—2.8L and 3.0L engines

2. On all A4 and A6 models, place the lock carrier into the service position.
3. Turn the ignition switch to the **OFF** position.
4. Remove or disconnect the following:
   - Negative battery cable
   - Timing belt
   - Crankshaft sprocket bolt, sprocket and washer
   - Crankshaft oil seal using Seal Remover tool (3203 and 2085)
5. Clean oil seal and crankshaft washer running surfaces.

To install:

6. Install or connect the following:
   - New oil seal lubricated with engine oil using a Seal Driver (3202/1 and 3265) until it is flush

⁂ WARNING

2004 models. Do not oil sealing lip or outer edge of sealing ring before pressing in.

- Crankshaft sprocket, using new bolt; torque to 148 ft. lbs. (200 Nm) plus 180 degrees
- Timing belt
- Negative battery cable

3.2L Engine, except TT and A6

1. Before servicing the vehicle, refer to the precautions section.
2. Turn the ignition switch to the **OFF** position.
3. Remove or disconnect the following:
   - Negative battery cable
   - Cover over coolant expansion tank

➡Before removing ribbed belt, mark direction of rotation with chalk or a felt pen. If a used belt rotates in the wrong direction when reinstalled, this can result in breakage. When installing belt, ensure it is correctly seated in the pulleys.

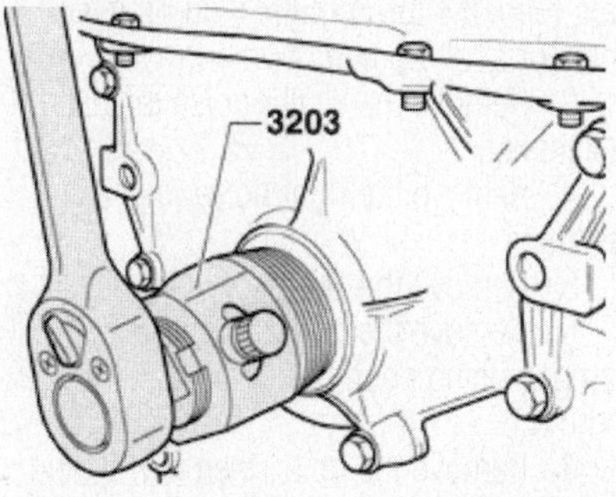

Removing the seal using the seal remover—3.2L engine, except TT and A6

- Release tension on belt by screwing an M8 x 40 bolt into bore on tensioning element and remove belt
- Center and right sound insulation from under vehicle
- Remove bolt for vibration damper using counter hold tool T10069 or equivalent
- Unscrew inner section of oil seal extractor 3203 3 turns and lock with knurled screw
- Lubricate the threaded head of oil seal extractor, put into position, exert firm pressure and screw into oil seal as far as possible
- Loosen the knurled screw and turn inner part against the crankshaft until the oil seal is pulled out

To install:

➡Do not lubricate the oil seal before installing

4. Place guide sleeve T10053/1 or equivalent on crankshaft journal
5. Push oil seal carefully over the guide sleeve onto the crankshaft journal as far as possible and remove the guide sleeve
6. Press the oil seal in as far as possible using the old securing bolt for the vibration damper and installation sleeve from 3266 or equivalent
7. Install or connect the following:
   - Vibration damper with new securing bolt and tighten to 74 ft. lbs (100 Nm) + 90 degrees (1/4 turn)
   - Accessory drive belt
   - Sound insulation
   - Cover over coolant expansion tank
   - Negative battery cable

A6 and A8

3.2L ENGINE

1. Before servicing the vehicle, refer to the precautions section.
2. Remove the transmission
3. Remove the drive plate
4. Pry out sealing ring using extractor lever T20143/2.
5. Clean operating and sealing surfaces.
6. Insert assembly device T10122/1 onto pull sleeve T10122/2 and slide seal on to sleeve.
7. Remove the assembly device T10122/1.

To install:

8. Install pull sleeve T10122/2 with sealing ring onto crankshaft.
9. Press sealing ring in evenly and flush, using pressure sleeve T10122/3.
10. Install drive plate.
11. Install transmission.

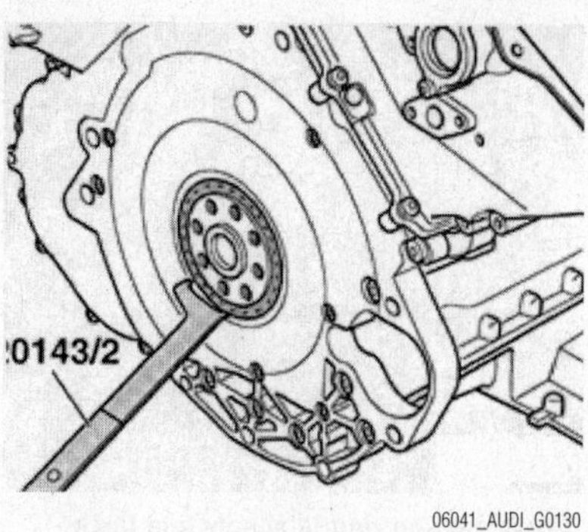

Pry out sealing ring using extractor lever T20143/2—A6 with 3.2L engine

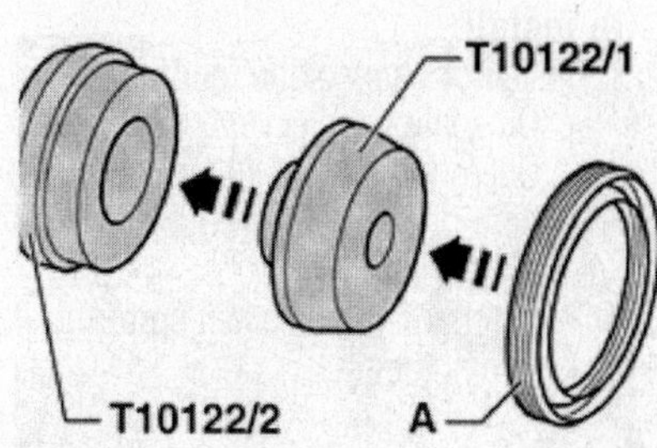

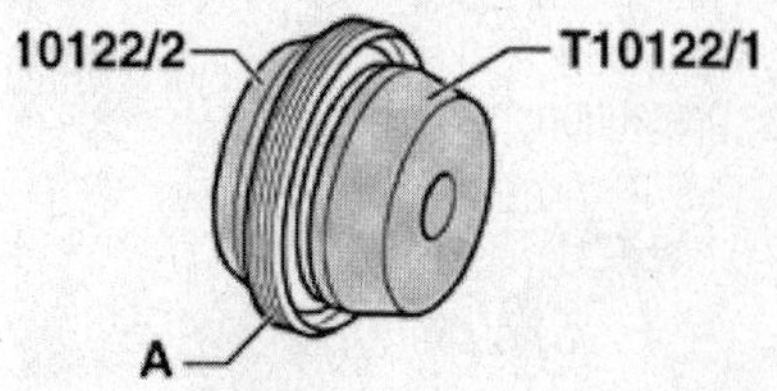

Insert assembly device T10122/1 onto pull sleeve T10122/2 and slide seal Bonto pull sleeve—A6 with 3.2L engine

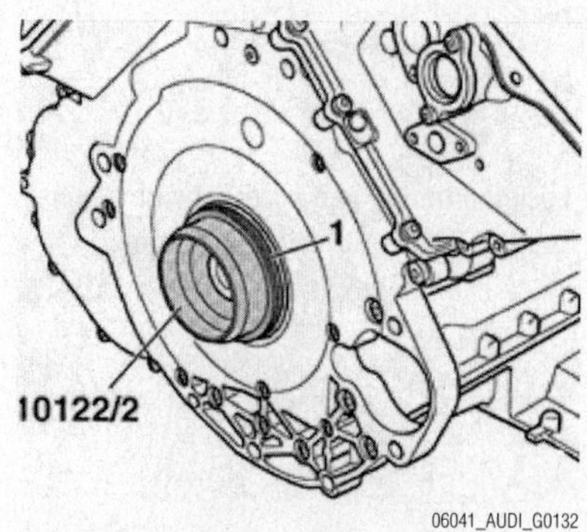

Install pull sleeve T10122/2 with sealing ring onto crankshaft—A6 with 3.2L engine

4.2L ENGINE

1. Before servicing the vehicle, refer to the precautions section.
2. Remove or disconnect the following:
   - Timing belt
   - Timing belt crankshaft sprocket
   - Oil seal using seal remover 3203

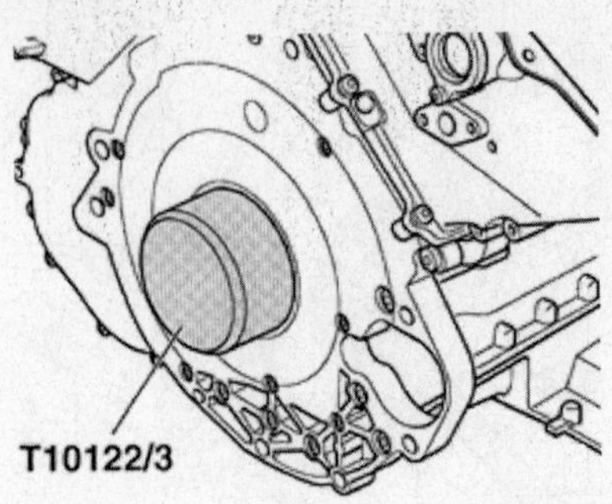

Press sealing ring in evenly and flush, using pressure sleeve T10122/3—A6 with 3.2L engine

To install:

3. Install or connect the following:
   - Oil seal over sleeve 3202
   - Using vibration damper mounting bolt, press in seal with sleeve 3202 until flush
   - Timing belt crankshaft sprocket
   - Timing belt

A3

2.0L ENGINE

1. Before servicing the vehicle, refer to the precautions section.

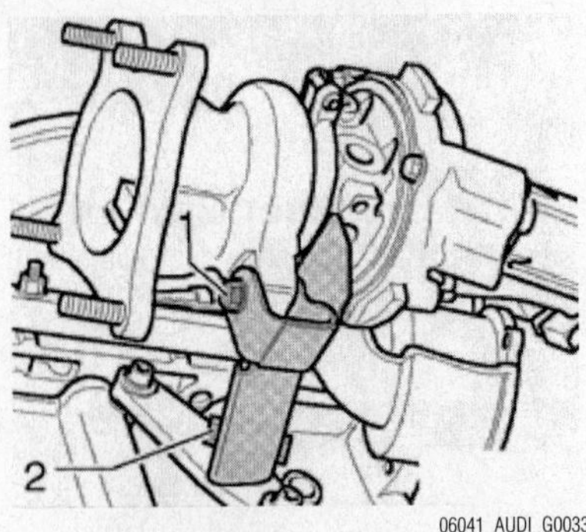

Lock the timing belt sprocket with counter support 3415—A3 with 2.0L engine

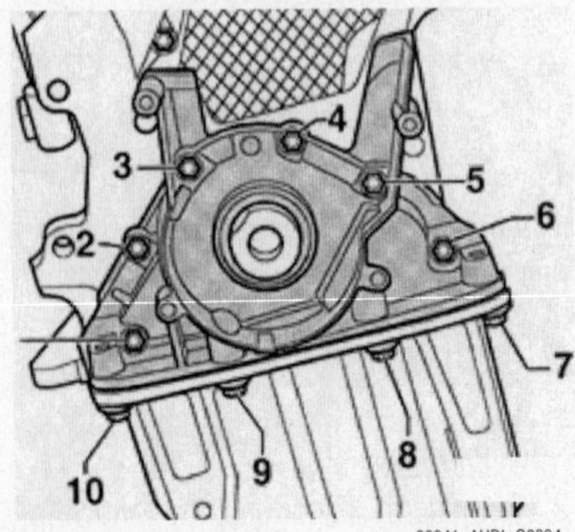

Location of the sealing flange bolts—A3 with 2.0L engine

2. Remove the accessory drive belt.
3. Remove the timing belt..
4. Remove the timing belt crankshaft sprocket. To do this, lock the timing belt sprocket with counter support 3415.
5. Remove the bolts.
6. Pry off and remove front sealing flange.
7. Drive out seal from removed flange.

To install:

➡Lay a rag over the open portion of the oil pan. Carefully remove any residual sealant from cylinder block and oil pan.

Apply a 2 to 3mm bead of sealant to edge between cylinder block and oil pan—A3 with 2.0L engine

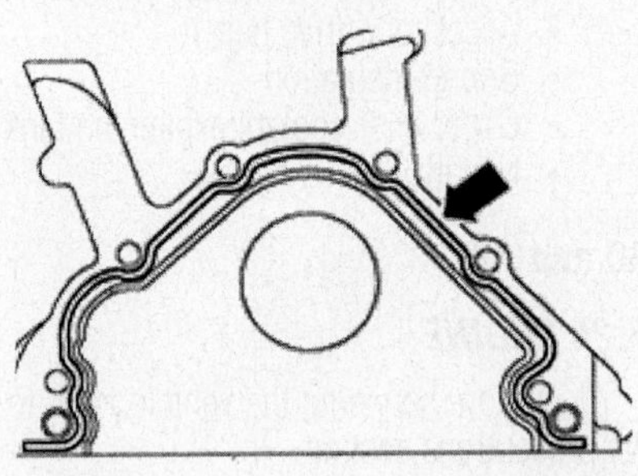

Apply a 2 to 3mm bead of sealant to sealing surface of sealing flange—A3 with 2.0L engine

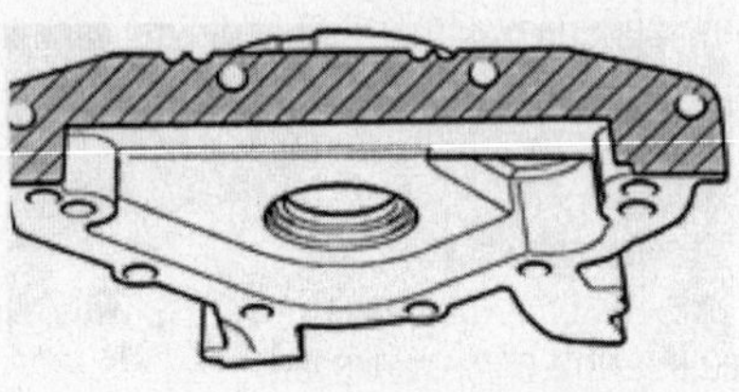

Lightly coat lower sealing surface of sealing flange with sealant —A3 with 2.0L engine

8. Remove the residual sealant from sealing flange e.g. with a rotating plastic brush.
9. Clean sealing surfaces, they must be free of oil and grease.
10. Apply a 2 to 3mm bead of sealant to edge between cylinder block and oil pan.
11. Apply a 2 to 3mm bead of sealant to sealing surface of sealing flange.

➡The sealant bead may not be thicker than 3 mm, otherwise excess sealant could enter the oil pan and clog the oil intake tube.

12. Lightly coat lower sealing surface of sealing flange with sealant.

➡The sealing flange must be installed within 5 minutes of being applied with sealant.

13. Carefully push sealing flange onto locating pins on cylinder block.

➡To install the sealing flange with installed seal, use the guide sleeve T10053/1.

14. Tighten bolts 1 to 6 first to 15 Nm and then bolts 7 through 10 to 15 Nm.
15. Install the crankshaft seal on the belt pulley side.
16. Replace central bolt for the timing belt sprocket.

➡There must be no oil on contact surface between the timing belt sprocket and crankshaft. Do not oil bolt for crankshaft the timing belt sprocket.

17. Install the crankshaft the timing belt sprocket, counter-holding the timing belt sprocket with counter support 3415.
18. Install the remaining components in the reverse order of removal.

3.2L ENGINE

1. Before servicing the vehicle, refer to the precautions section.
2. Before removing accessory drive belt, mark the turning direction on it with chalk or a felt tip pen. A
3. Remove the center noise insulation fasteners.
4. Remove the right noise insulation fasteners.
5. Remove the accessory drive belt.
6. Remove the bolt from vibration damper, using counter-holder tool T10069 to hold it.
7. Remove the bolts, then pry off and remove front sealing flange.
8. Drive out the seal from the removed flange.

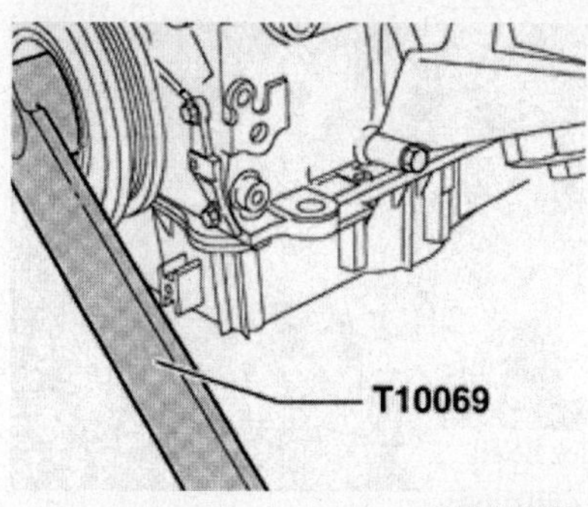

Remove the bolt from vibration damper, using counter-holder tool T10069 hold it—A3 with 3.2L engine

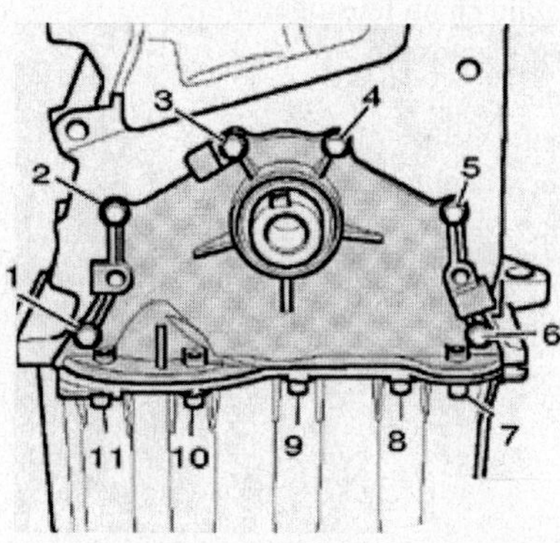

Front sealing flange bolt locations—A3 with 3.2L engine

To install:

➡Lay a rag over the open portion of the oil pan. Carefully remove any residual sealant from cylinder block and oil pan.

9. Remove the residual sealant from sealing flange e.g. with a rotating plastic brush.
10. Clean sealing surfaces, they must be free of oil and grease.
11. Apply a 2 to 3mm bead of sealant to edge between cylinder block and oil pan.
12. Apply a 2 to 3mm bead of sealant to surface of sealing flange.

➡The sealant bead may not be thicker than 3 mm, otherwise excess sealant could enter the oil pan and clog the oil intake tube.

13. Lightly coat lower sealing surface of sealing flange with sealant.

➡The sealing flange must be installed within 5 minutes of being applied with sealant.

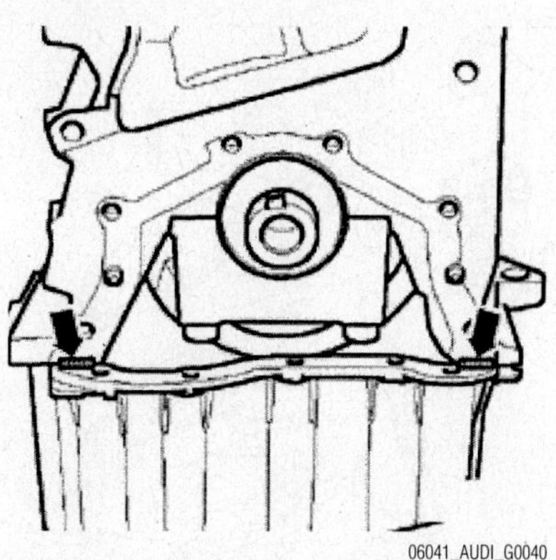

Apply a 2 to 3mm bead of sealant to edge between cylinder block and oil pan (indicated by arrows in illustration)—A3 with 3.2L engine

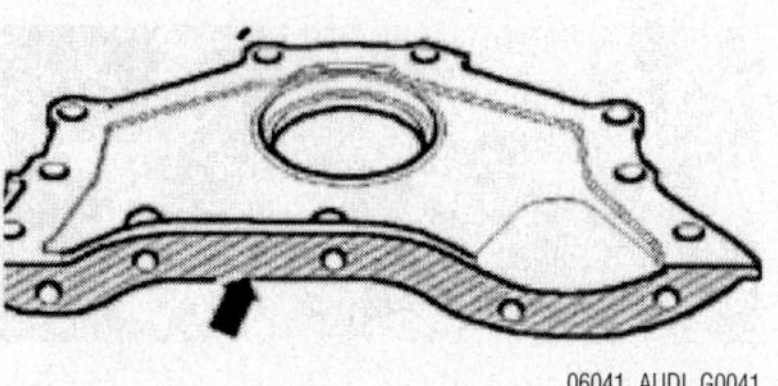

Apply a 2 to 3mm bead of sealant to surface of sealing flange—A3 with 3.2L engine

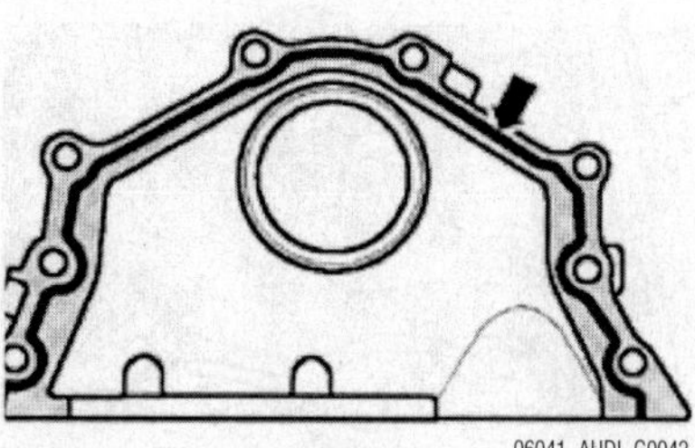

Lightly coat lower sealing surface of sealing flange with sealant—A3 with 3.2L engine

14. Carefully push sealing flange onto locating pins on cylinder block.

➡To install the sealing flange with installed seal, use the guide sleeve T10053/1.

15. Tighten bolts 1 to 6 first to 10 Nm and then bolts 7 through 10 to 10 Nm.
16. Install the crankshaft seal on the belt pulley side.
17. Replace central bolt for the timing belt sprocket.

➡There must be no oil on contact surface between the timing belt sprocket and crankshaft. Do not oil bolt for crankshaft the timing belt sprocket.

18. Install the crankshaft the timing belt sprocket, counter-holding the timing belt sprocket with counter support 3415.
19. Install the remaining components in the reverse order of removal.

Camshaft

REMOVAL & INSTALLATION

1.8L Engine

1. Before servicing the vehicle, refer to the precautions section.
2. Turn the ignition switch to the **OFF** position.
3. Disconnect the negative battery cable.
4. On A4 models, place the lock carrier into the service position.
5. Remove or disconnect the following:
   - Accessory drive belts
   - Engine covers
   - Timing belt upper cover
6. Turn the crankshaft, in the direction of rotation (clockwise), until the No. 1 cylinder is at Top Dead Center (TDC).
7. Remove or disconnect the following:
   - Timing belt tensioner by loosening it using Torx® wrench T45
   - Belt from the camshaft gear by pushing the tensioner downward
   - Torx® bolt and swing the tensioner assembly bracket forward
   - Timing belt
   - Valve cover
   - Cam gear retaining bolt by loosening it using retainer tool 3036
   - Camshaft gear
   - Camshaft Position (CMP) housing sensor and shutter wheel
   - Hydraulic chain tensioner by securing it with bracket tensioner tool 3366
8. Verify that the camshafts are at TDC for the No. 1 cylinder. Both camshaft markings must align with arrows on the bearing caps.
9. Clean the drive chain and the cam chain gears opposite both arrows on the bearing caps. Matchmark the installed position using paint.

➡The distance between the 2 arrows/paint marks is equivalent to 16 drive chain rollers and the notch on the exhaust camshaft is slightly offset inward toward the drive chain roller.

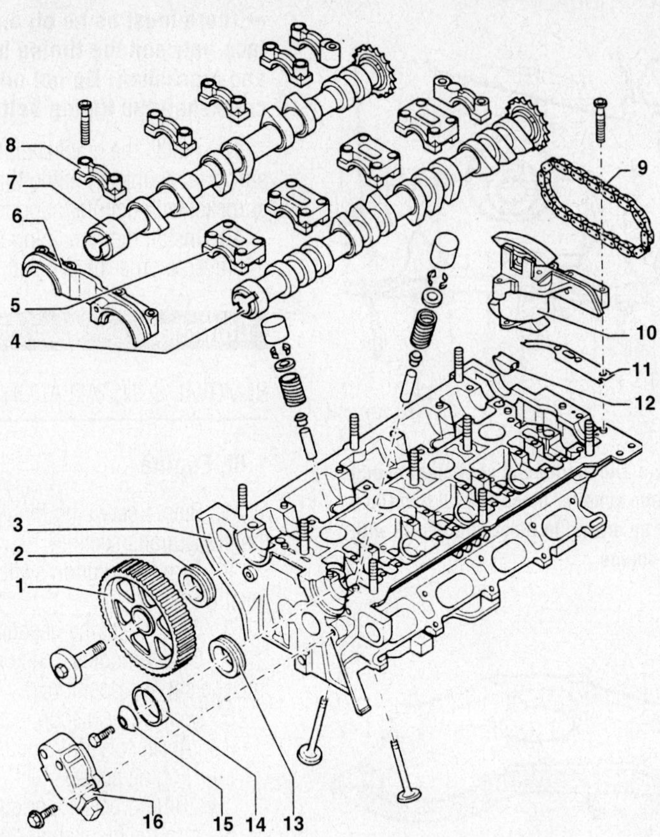

1. Camshaft gear
2. Oil seal
3. Cylinder head
4. Intake camshaft
5. Intake camshaft bearing cap
6. Double bearing cap
7. Exhaust camshaft
8. Exhaust camshaft bearing cap
9. Drive chain
10. Hydraulic chain tensioner
11. Rubber/metal seal
12. Gasket
13. Oil seal
14. Shutter wheel for the CMP
15. Washer
16. CMP sensor housing

7923CG20

Exploded view of the camshaft mounting and related components—1.8L engine

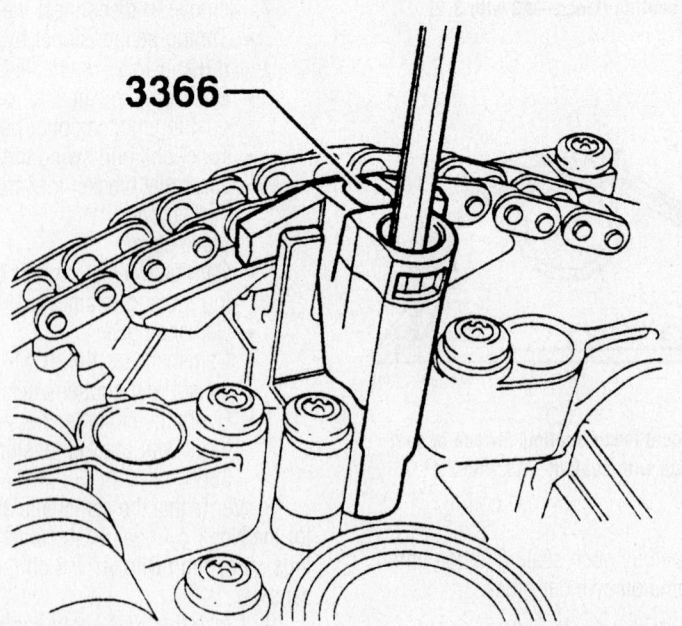

7923CG21

Do not overtighten the chain tensioner tool 3366, it can be damaged—1.8L and 2.7L and 2.8L DOHC engines

10. Remove or disconnect the following:
    - Bearing caps No. 3 and 5 from the intake and exhaust camshafts
    - Double bearing cap
    - Both bearing caps from the chain gears on the intake and exhaust camshafts
    - Hydraulic chain tensioner retaining bolts
    - Intake and exhaust manifold bearing caps No. 2 and 4 by loosening them in an alternating and diagonal sequence
    - Camshafts with the hydraulic chain tensioner

To install:

11. Replace the rubber/metal chain tensioner gasket and apply sealant to the hatched area, as shown.
12. Install or connect the following:
    - Drive chain on the camshaft

➡If installing the old chain, align the paint marks with the camshaft marks. If installing a new chain, the distance between the notches A and B on the camshafts must equal the distance between 16 drive chain rollers.

- Hydraulic chain tensioner by sliding it between the drive chains
- Camshafts with the chain tensioner lubricated with engine oil into the cylinder head

➡When installing the bearing caps, verify the markings on the caps are readable from the intake side of the cylinder head.

- Intake and exhaust camshafts bearing caps No. 2 and 4. Torque them

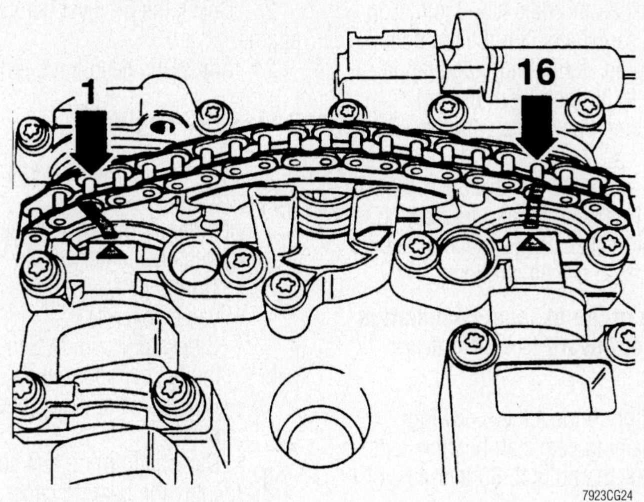

To ensure proper installation, matchmark the chain-to-camshaft position—1.8L and 2.7L and 2.8L DOHC engines

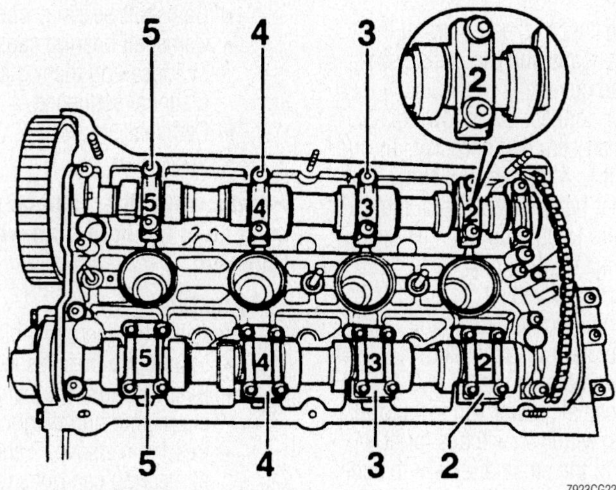

Camshaft bearing cap identification—1.8L engine

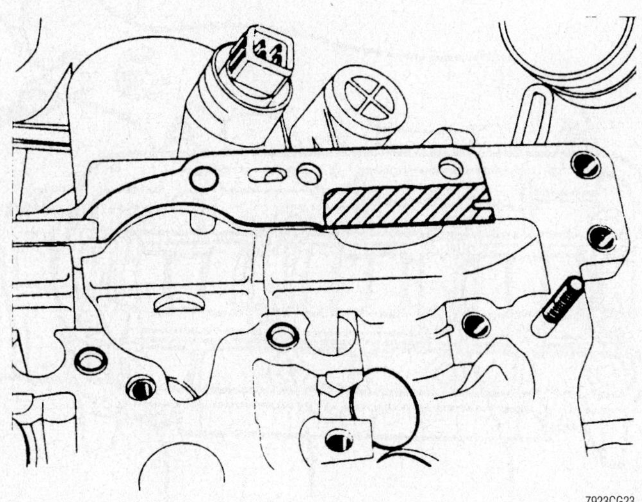

To ensure a proper seal, be sure to apply sealant to the hatched area—1.8L and 2.7L and 2.8L DOHC Engines

in an alternating diagonal sequence to 89 inch lbs. (10 Nm).

- Both the intake and exhaust camshafts bearing caps on the chain sprockets. Torque the bolts to 89 inch lbs. (10 Nm).

13. Verify the correct positions of the camshafts.
14. Remove the bracket tensioner.
15. Install or connect the following:
    - Cylinder head-to-double bearing cap mating surface by lightly coating it with sealant. Torque the remaining bearing caps to 89 inch lbs. (10 Nm)
    - Camshaft gear. Torque the bolt to 48 ft. lbs. (65 Nm)
    - CMP shutter wheel and housing cover
    - Valve cover
16. Align the camshaft gear and the vibration damper with the TDC markings.
17. Install or connect the following:
    - Timing belt
    - Accessory drive belts and the engine cover
    - Lock carrier
    - Negative battery cable
18. Fully close all power windows to stop, operate all window switches for at least 1 second in the close direction to activate the one-touch opening/closing function

✲✲ CAUTION

After installing the lifters or the camshaft(s), the engine must NOT be started for at least 30 minutes. Otherwise the valves could strike the pistons. Rotate the engine by hand, at least 2 revolutions, to ensure that the valves do not strike the pistons.

19. Check the oil level before starting the engine.
20. Adjust the headlights.

2.7L Engine

1. Before servicing the vehicle, refer to the precautions section.
2. Properly relieve the fuel system pressure.
3. Remove the engine undercover.
4. Remove the front bumper.
5. Place the hood lock carrier assembly in the service position.
6. Disconnect the battery negative cable.
7. Remove or disconnect the following:
    - Air cleaner assembly
    - Accessory drive belt
    - Tooted cam belt

- Cylinder head covers
- Camshaft Position (CMP) sensor housing and shutter wheel

8. Install the Camshaft Locator Bar tool 3391 onto the camshaft locking plates.
9. Remove or disconnect the following:
   - Camshaft bolts by loosening them 5 revolutions
   - Camshaft locator bar
   - Camshaft gears using tool T40001
   - Oil supply lines for the camshaft bearings

➡Use care to not damage the positioning clips.

10. Secure the camshaft adjuster using the bracket tensioner tool 3366. Use care to not over tighten the bracket adjuster as the camshaft adjuster could be damaged.
11. Verify that both camshafts are at TDC. The notch on the camshaft behind the cam sprocket should align with the arrows on the camshaft bearing caps.
12. Clean the camshaft sprockets and cam drive chain and make alignment marks on the sprocket and chain in alignment with the top dead center marks.
13. Remove or disconnect the following:
   - Camshaft adjuster retaining bolts, leaving the adjuster in place
   - Camshaft journals No. 1, 3 and 5 from both cams and journal No. 7 from the exhaust camshaft
   - Camshaft journal caps No. 2 and 4 by working in a diagonal sequence
   - Camshafts with the camshaft adjuster assembly

To install:

14. Install or connect the following:
   - Rubber/metal gasket for the camshaft adjuster assembly by applying a thin layer of sealant, as illustrated
   - Drive chain on the camshaft sprockets by aligning the matchmarks

➡If a new chain is installed or the marks are no longer visible, install the chain on the camshaft sprockets such that the distance between the 2 camshaft sprockets is 16 drive chain rollers.

➡Use an assistant to place the camshaft adjuster between the camshaft drive chain.

15. Install or connect the following:
   - Camshafts and adjuster into the cylinder head by lubricating them with engine oil. Torque the camshaft adjuster to 89 inch lbs. (10 Nm)
   - Camshaft journals No. 2 and 4 on both camshafts. Torque the bolts, using an alternating diagonal pattern, to 89 inch lbs. (10 Nm)
16. Align both camshafts with the TDC marks.
17. Apply sealant to the double camshaft bearing cap (No. 1) and the outer bearing cap (No. 7) at their mating surfaces. Refer to the shaded area in the illustration.

➡The notch of the exhaust camshaft is slightly offset inward from the drive chain roller.

18. Install or connect the following:
   - Remaining camshaft bearing caps. Torque the bolts to 89 inch lbs. (10 Nm)
   - Camshaft bearing oil lines
   - New camshaft oil seal
19. Remove the cam chain bracket tensioner tool 3366.
20. Apply sealant to the corners of the camshaft journal caps and the gasket surface for the head covers.
21. Install or connect the following:
   - Camshaft cam belt sprocket. Torque the bolt to 41 ft. lbs. (55 Nm)
   - Cylinder head covers. Torque the fasteners to 89 inch lbs. (10 Nm)
   - Camshaft toothed belt
   - Accessory drive belt
   - Remaining equipment in the reverse order of removal
   - Negative battery cable
22. Fully close all power windows to stop, operate all window switches for at least 1 second in the close direction to activate the one-touch opening/closing function
23. Check the oil level before starting the engine.
24. Adjust the headlights if necessary.

2.8L SOHC Engine

1. Before servicing the vehicle, refer to the precautions section.
2. Remove or disconnect the following:
   - Negative battery cable
   - Timing belt
   - Valve cover(s)
   - Camshaft Position Sensor (CMP) at the left cylinder head
   - Plug/cover at the right cylinder head
   - Camshaft timing belt sprocket
3. Identify the bearing caps.

➡DO NOT allow the bearing caps to become mixed up.

4. Remove or disconnect the following:
   - Camshaft bearing caps No. 2 and 3
   - Camshaft bearing caps No. 1 and 4 by loosening them gradually, in a diagonal sequence
   - Camshaft
   - Valve lifter

➡If the valve lifter is to be reused, it must go in the bore from which it was removed.

To install:

5. Install or connect the following:
   - Lifters into their respective bore
   - Bearing caps No. 1 and 4 in an alternating and diagonal sequence
   - Bearing caps No. 2 and 3. Torque all bearing cap bolts to 15 ft. lbs. (20 Nm)

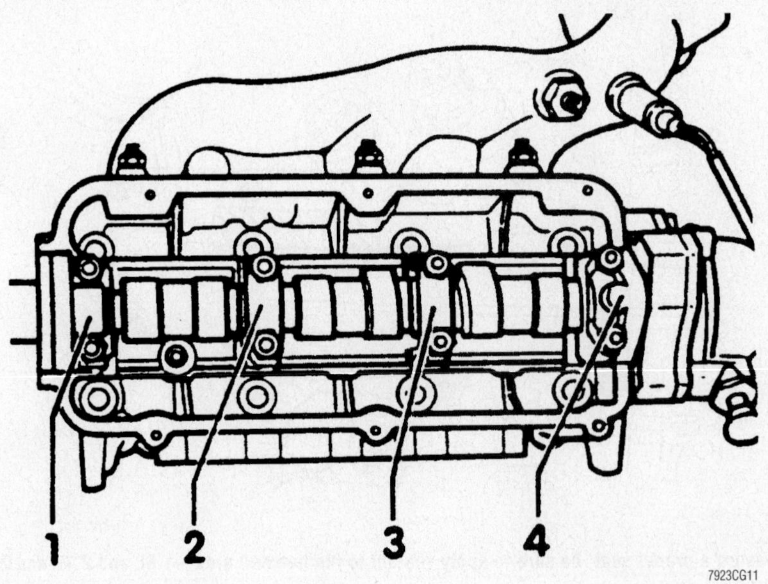

Camshaft bearing cap identification—2.8L SOHC engine

✲✲ CAUTION

After installing the lifters or the camshaft(s), the engine must NOT be started for at least 30 minutes. Otherwise the valves could strike the pistons. Rotate the engine by hand, at least 2 revolutions, to ensure that the valves do not strike the pistons.

- Camshaft timing belt sprocket. Torque the bolt to 52 ft. lbs. (71 Nm)
- Plug/cover, on the right cylinder head
- CMP on the left cylinder head. Torque the bolts to 89 inch lbs. (10 Nm)
- Valve cover
- Timing belt
- Negative battery cable

6. Fully close all power windows to stop, operate all window switches for at least 1 second in the close direction to activate the one-touch opening/closing function.
7. Check the oil level before starting the engine.
8. Check and adjust the headlights.

2.8L DOHC Engine

1. Before servicing the vehicle, refer to the precautions section.
2. Properly relieve the fuel system pressure.
3. On A4 and A6 models, place the hood lock carrier assembly in the service position.
4. Disconnect the battery negative cable.
5. Rotate the engine to Top Dead Center (TDC).
6. Remove or disconnect the following:
   - Accessory drive belt
   - Toothed cam belt
   - Engine cosmetic covers
   - Engine lower splash shield
   - Crankshaft housing ventilation line from the cylinder head cover
   - Intake air duct
   - Coolant expansion tank and place aside
   - Mass Air Flow (MAF) connector
   - Evaporative Emissions (EVAP) purge regulator and valve
7. Rotate the secondary air injection pump duct 45 degrees clockwise.
8. Remove or disconnect the following:
   - Air cleaner assembly
   - Camshaft Position (CMP) sensor housing and shutter wheel
   - Cylinder head covers
9. Install the Camshaft Locator Bar tool 3391 onto the camshaft locking plates.
10. Remove or disconnect the following:
    - Camshaft bolts by loosening them 5 revolutions
    - Camshaft locator bar
    - Camshaft gears using tool T40001
    - Oil supply lines for the camshaft bearings

➡Use care to not damage the positioning clips.

11. Secure the camshaft adjuster using the bracket tensioner tool 3366. Use care to not over tighten the bracket adjuster as the camshaft adjuster could be damaged.
12. Verify that both camshafts are at TDC. The notch on the camshaft behind the cam sprocket should align with the arrows on the camshaft bearing caps.
13. Make sure the camshafts are at TDC.
14. Clean the camshaft sprockets and cam drive chain and make alignment marks on the sprocket and chain in alignment with the top dead center marks.
15. Remove or disconnect the following:
    - Camshaft adjuster retaining bolts, leaving the adjuster in place
    - Camshaft journals No. 1, 3 and 5

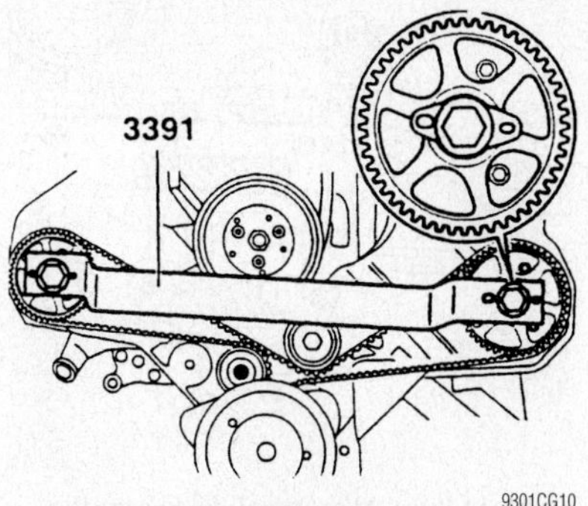

The camshaft sprockets are held using tool 3391—2.7L and 2.8L DOHC engine

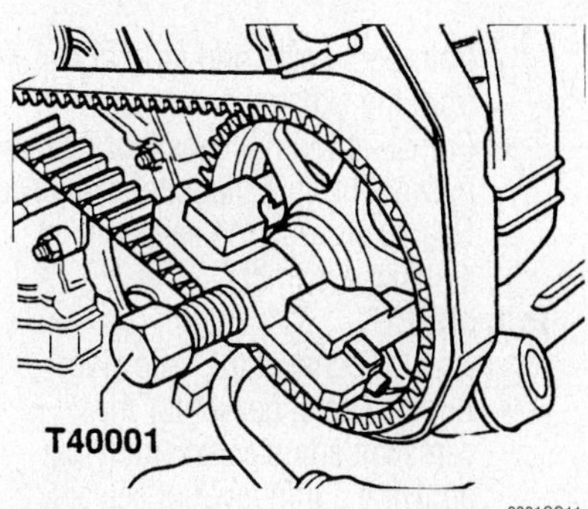

The camshaft gears are removed using tool T40001—2.7L and 2.8L DOHC engine

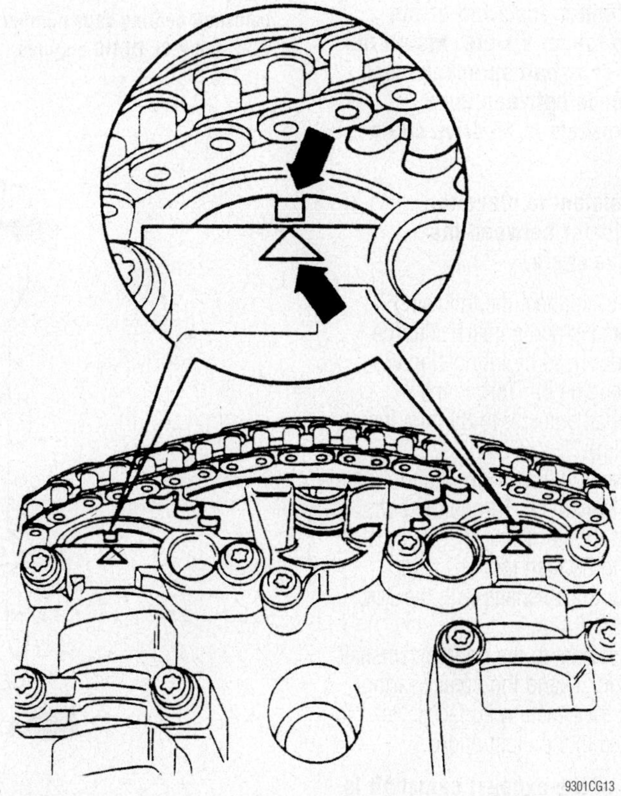

The notches in the camshaft align with the arrows on the cam journal caps when the camshafts are at TDC—2.7L and 2.8L DOHC engine

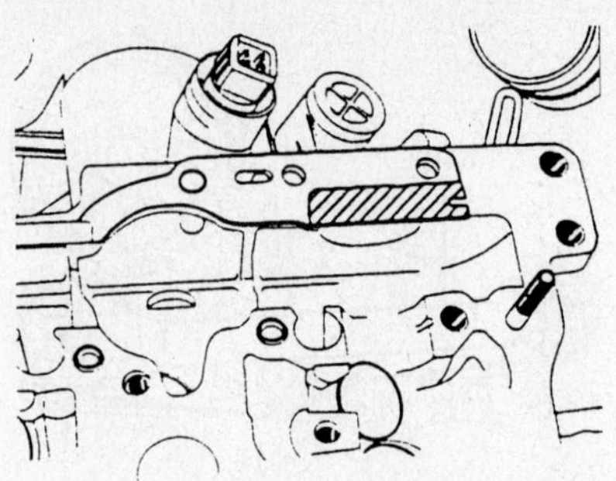

Replace the rubber/metal gasket for the camshaft adjuster assembly, and apply a thin layer of sealant on the hatched surface as illustrated

from both cams and journal No. 7 from the exhaust camshaft
- Camshaft journal caps No. 2 and 4 by working in a diagonal sequence
- Camshafts with the camshaft adjuster assembly

To install:

16. Install or connect the following:
  - Rubber/metal gasket for the camshaft adjuster assembly by applying a thin layer of sealant, as illustrated
  - Drive chain on the camshaft sprockets by aligning the matchmarks

➡If a new chain is installed or the marks are no longer visible, install the chain on the camshaft sprockets such that the distance between the 2 camshaft sprockets is 16 drive chain rollers.

➡Use an assistant to place the camshaft adjuster between the camshaft drive chain.

17. Install or connect the following:
  - Camshafts and adjuster into the cylinder head by lubricating them with engine oil. Torque the camshaft adjuster to 89 inch lbs. (10 Nm)
  - Camshaft journals No. 2 and 4 on both camshafts. Torque the bolts, using an alternating diagonal pattern, to 89 inch lbs. (10 Nm)

18. Align both camshafts with the TDC marks.

19. Apply sealant to the double camshaft bearing cap (No. 1) and the outer bearing cap (No. 7) at their mating surfaces. Refer to the shaded area in the illustration.

➡The notch of the exhaust camshaft is slightly offset inward from the drive chain roller.

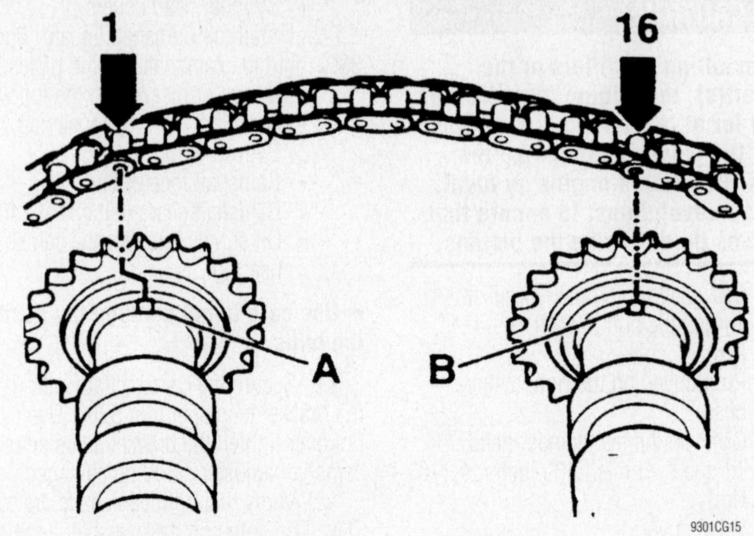

The camshaft drive chain is installed on the camshaft sprockets 16 drive chain rollers apart—1.8L and 2.7L and 2.8L DOHC engines

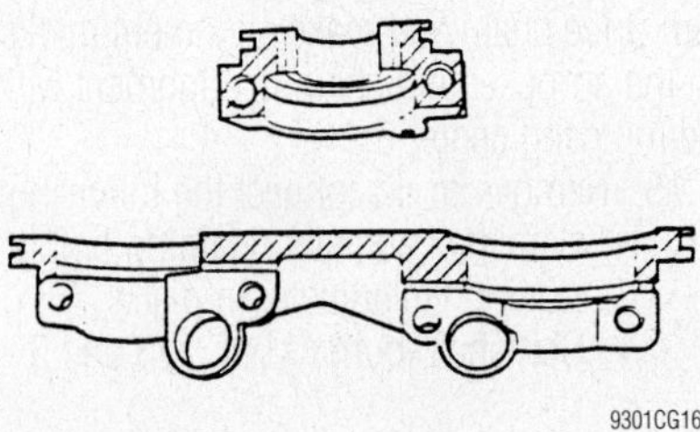

Apply sealant to the shaded areas for camshaft bearing caps numbers 1 and 7—2.7L and 2.8L DOHC engines

20. Install or connect the following:
  - Remaining camshaft bearing caps. Torque the bolts to 89 inch lbs. (10 Nm)
  - Camshaft bearing oil lines

21. Remove the cam chain bracket tensioner tool 3366.

22. Apply sealant to the corners of the camshaft journal caps and the gasket surface for the head covers.

23. Install or connect the following:
  - New camshaft oil seal

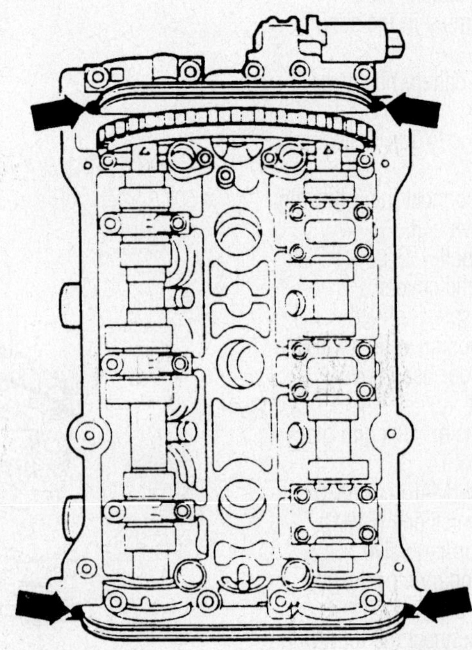

Apply a thin bead of sealant at the corners of the outer camshaft journals—2.7L and 2.8L DOHC engines

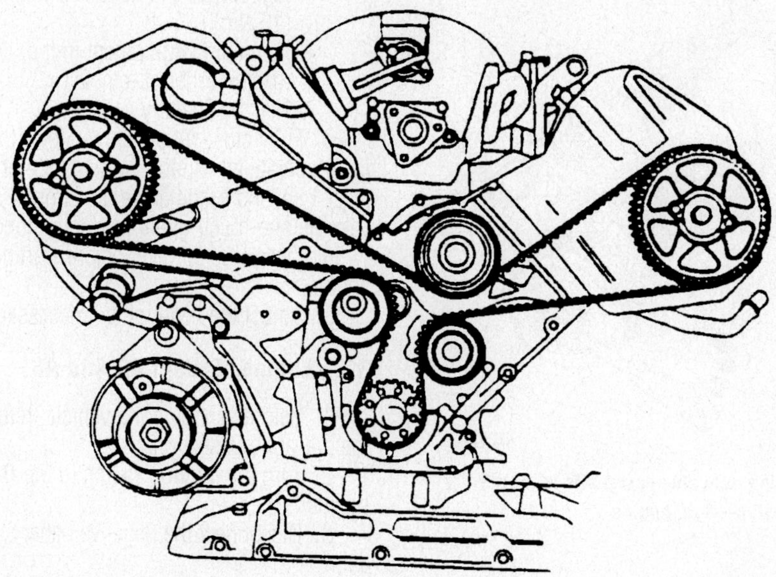

Camshaft belt routing—2.7L and 2.8L DOHC engines

- Camshaft cam belt sprocket. Torque the bolt to 41 ft. lbs. (55 Nm)
- Cylinder head covers. Torque the fasteners to 89 inch lbs. (10 Nm)
- Camshaft toothed belt
- Accessory drive belt
- Remaining equipment in the reverse order of removal
- Negative battery cable

24. Fully close all power windows to stop, operate all window switches for at least 1 second in the close direction to activate the one-touch opening/closing function
25. Check the oil level before starting the engine.
26. Adjust the headlights if necessary.

3.0L Engine

1. Before servicing the vehicle, refer to the precautions section.
2. Properly relieve the fuel system pressure.
3. Remove the engine undercover.
4. Place the hood lock carrier assembly in the service position.
5. Disconnect the battery negative cable.
6. Remove or disconnect the following:
    - Accessory drive belt
    - Timing belt
    - CMP sensor connectors
    - RH idler pulley
    - Timing belt guards at rear
    - Housing for camshaft adjustment valves
    - Camshaft Position (CMP) sensor housing

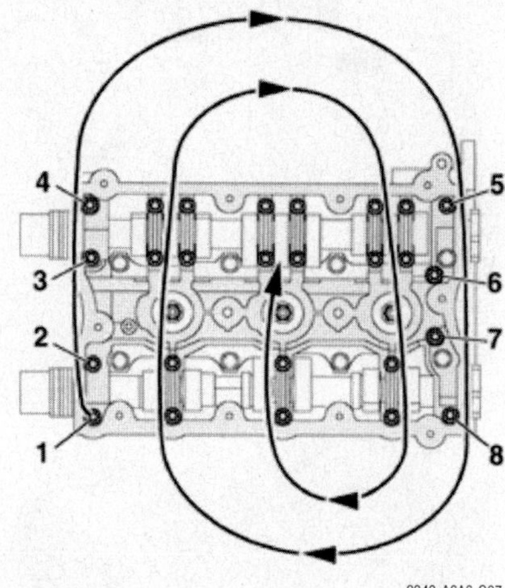

Camshaft guide frame bolt loosening sequence—3.0L engine

7. Loosen camshaft guide frame bolt in sequence shown.
8. Remove the guide frame.
9. Remove both camshafts.

To install:

10. Install the crankshaft alignment pin (T40026) to lock the crankshaft in position.
11. Oil the camshaft journal surfaces.
12. Insert left side camshafts into cylinder head so cam lobes at cylinder no. 5 point downward.
13. Insert right side camshafts into cylinder head so cam lobes at cylinder no. 3 point upward.
14. Apply sealant evenly to the camshaft guide frame, avoiding bolt holes
15. Set the guide frame in place.

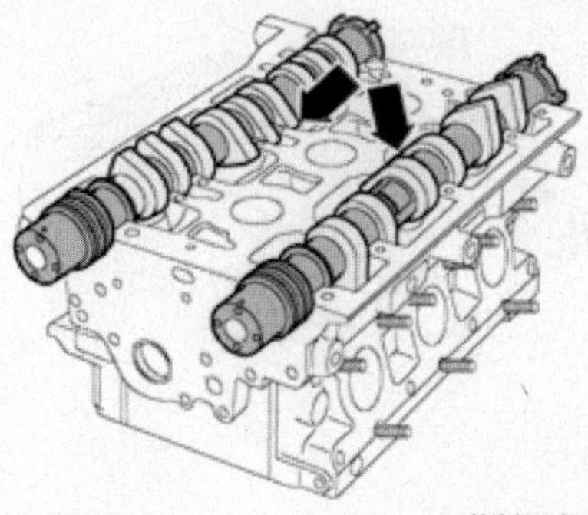

Left side camshafts installation position—3.0L engine

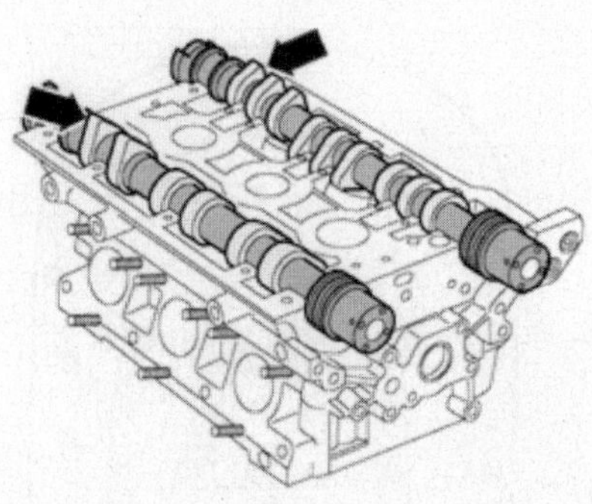

Right side camshafts installation position—3.0L engine

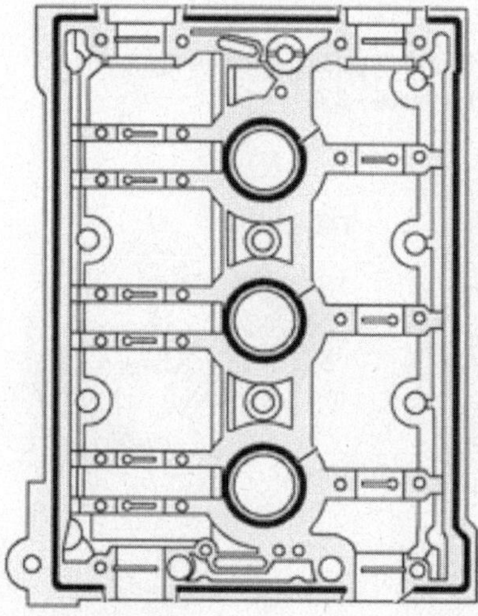

Applying sealant to camshaft guide frame—3.0L engine

16. Install the T40029 guide frame alignment pin.
17. Torque the guide frame bolts in sequence shown to 84 inch lbs. (10 Nm).
18. On left side camshafts, remove T40029 alignment pin.
19. Install T40030 camshaft adjuster gauge to camshafts.

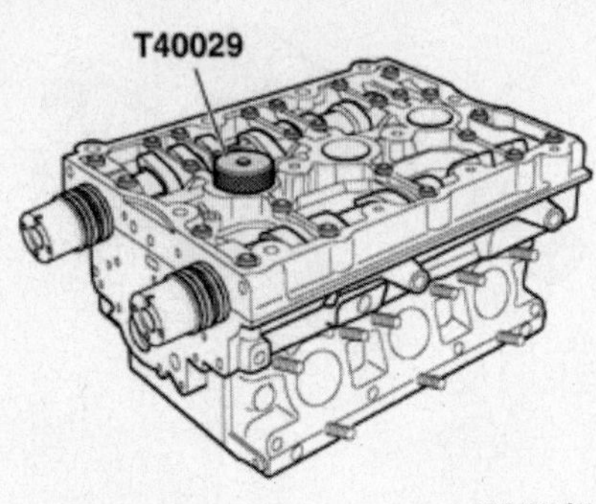

Installing guide frame aligning pin—3.0L engine

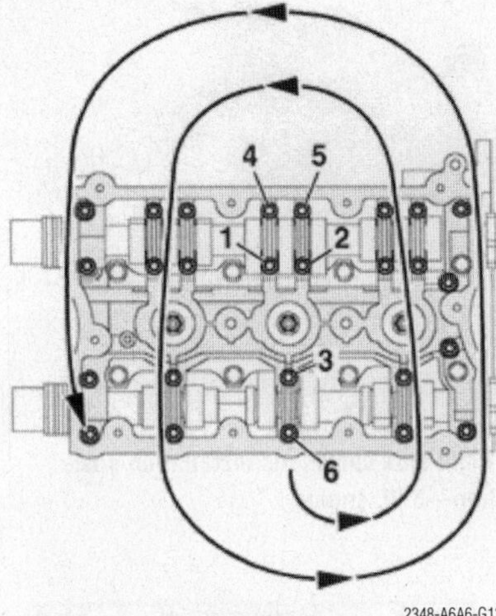

Camshaft guide frame bolt tightening sequence—3.0L engine

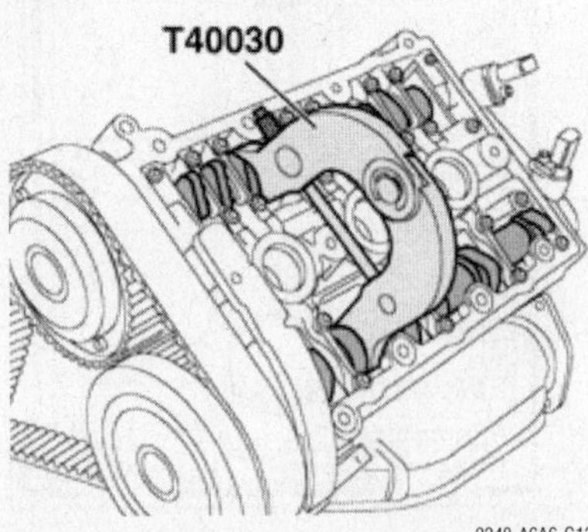

Installing camshaft adjuster gauge T40030 on left side camshafts—3.0L engine

20. Spread adjuster gauge with threaded shaft until it seats without end play.
21. On right side camshafts, install exhaust camshaft gear.
22. Turn exhaust camshaft and intake camshaft in direction shown until cam lobes of no. 3 cylinder are at position shown.
23. Install T40030 camshaft adjuster gauge to camshafts.

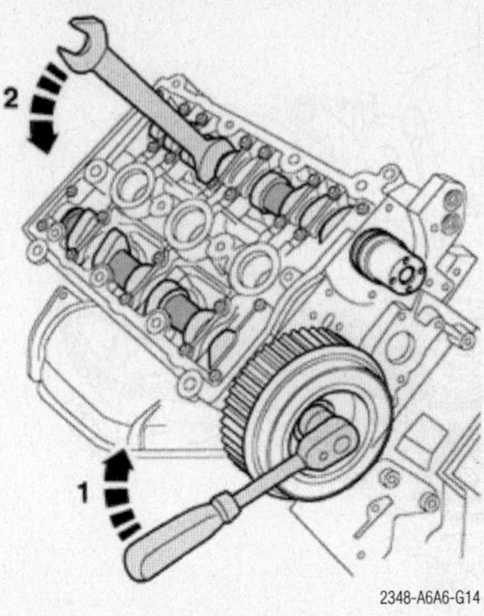

Turning right side camshafts in proper direction—3.0L engine

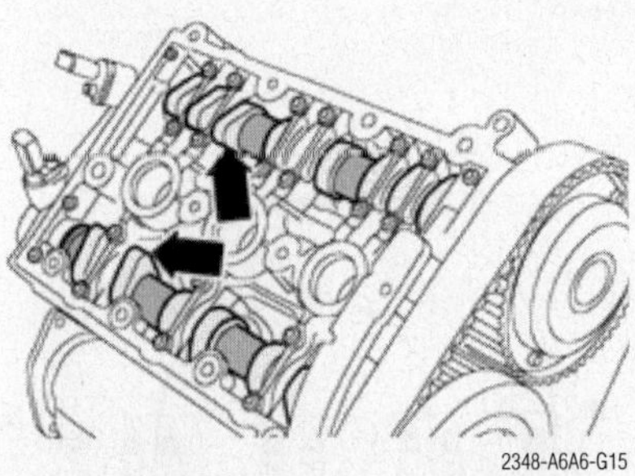

Proper position of right side camshaft lobes —3.0L engine

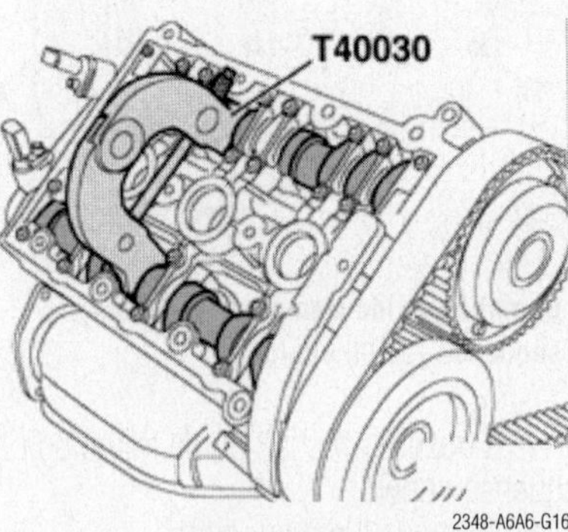

Installing camshaft adjuster gauge T40030 on right side camshafts—3.0L engine

24. Spread adjuster gauge with threaded shaft until it seats without end play.
25. Remove the exhaust cam gear.
26. Remove the adjuster gauge.
27. Ensure the camshaft sealing rings are in place.
28. Install or connect the following to both cylinder heads:
- Camshaft position sensor housing: Torque the bolts to 84 inch lbs. (10 Nm).
- Camshaft adjustment valve housing: Torque the bolts to 84 inch lbs. (10 Nm).
- Remaining equipment in the reverse order of removal
- Negative battery cable

29. Fully close all power windows to stop, operate all window switches for at least 1 second in the close direction to activate the one-touch opening/closing function
30. Check the oil level before starting the engine.
31. Adjust the headlights if necessary.

3.2L Engine, Except A3 And A6

1. Before servicing the vehicle, refer to the precautions section.
2. Turn the ignition switch to the **OFF** position.
3. Disconnect the negative battery cable.
4. Drain coolant
5. Remove or disconnect the following:
- Intake manifold
- Air cleaner housing
- Fuse box
- Carrier
- Coolant hoses
- Bracket for secondary air injection hose

➡When removing the thermostat housing, push the coolant pipe toward the coolant pump to prevent it from also being removed

- Thermostat housing
- Throttle valve control module
- Camshaft position sensors
- Camshaft adjustment valves
- Fuel lines
- Cylinder head cover removing bolts in sequence
- Insert camshaft plate T10068 or equivalent into grooves on both camshafts
- Bolt for timing chain tensioner
- Upper timing chain cover.

➡Be careful not to damage the head gasket when removing the upper timing chain cover

- Guide rail
- Remove camshaft plate T10068

➡The camshaft plate must not be in place when loosening or tightening the bolts for the camshaft adjuster

- Camshaft adjuster, intake side
- Camshaft adjuster, exhaust side
- Timing chain from camshafts and set aside
- Valve timing housing

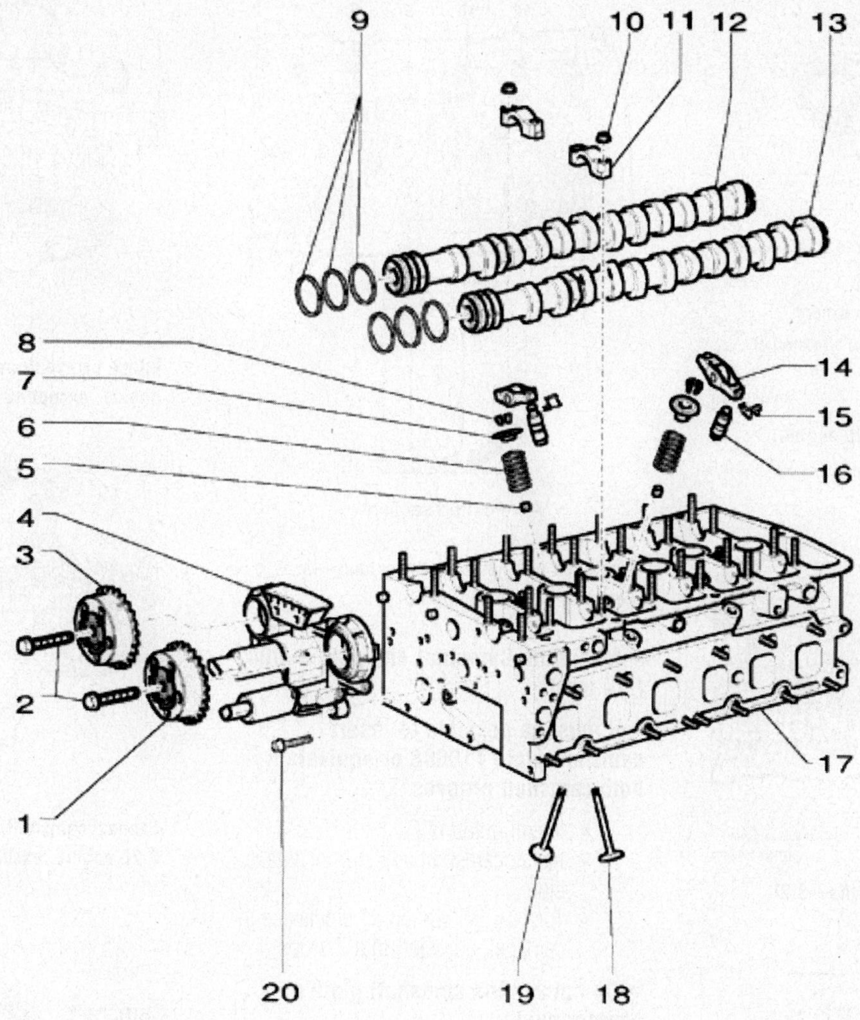

1	Camshaft adjuster-exhaust	11	Bearing cap
2	60 Nm + 90° further	12	Intake camshaft
3	Camshaft adjuster-intake	13	Exhaust camshaft
4	Valve timing housing	14	Roller rocker finger
5	Valve stem oil seal	15	Securing clip
6	Valve spring	16	Supporting element
7	Valve spring plate	17	Cylinder head
8	Valve keepers	18	Valves
9	Camshaft seals	19	8 Nm
10	5 Nm + 45°		

67200-TTTT-G01

Exploded view of the camshaft mounting and related components—3.2L engine, except A3 and A6

6. Remove intake camshaft as follows
   - Unbolt bearing caps 1 and 13
   - Unbolt bearing caps 3 and 11
   - Unbolt bearing cap 7
   - Loosen bearing caps 5 and 9 alternately in a diagonal sequence and remove
7. Remove exhaust camshaft as follows
   - Unbolt bearing caps 2 and 14
   - Unbolt bearing caps 4 and 12
   - Unbolt bearing cap 8
   - Loosen bearing caps 6 and 10 alternately in a diagonal sequence and remove
8. Remove camshafts

To install:

➡Pay attention to the camshaft identification located between cam pairs for cylinder number 4 and 5.

- Exhaust camshaft identification 022—index 101
- Intake camshaft identification 022—index 102

9. Install camshafts so that they are located at TDC for cylinder number 1
   - It must be possible to insert camshaft plate T10068 or equivalent in both camshaft grooves
10. Install bearing caps, pointed section "A" of bearing cap faces outward
11. Tighten intake camshaft as follows:
    - Caps 5 and 9 alternately in a diagonal sequence and then tighten to final torque
    - Caps 1 and 13

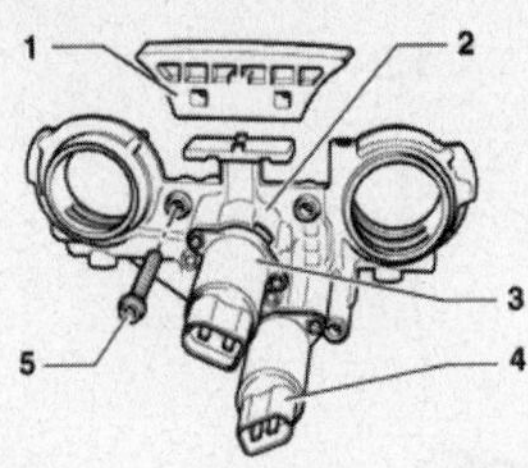

1 - Guide rail; clipped onto valve timing housing
2 - Valve timing housing
3 - Valve -1- for camshaft adjustment
4 - Camshaft adjustment valve 1 (exhaust)
5 - 8 Nm

67200-TTTT-G03

Valve timing housing—3.2L engine, except A3 and A6

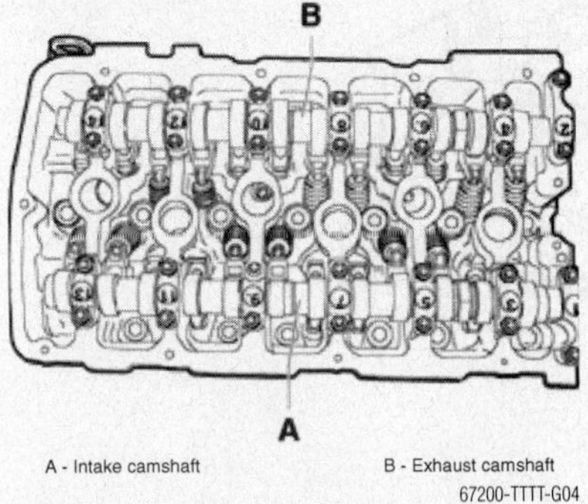

A - Intake camshaft B - Exhaust camshaft

67200-TTTT-G04

Intake and exhaust camshafts—3.2L engine, except A3 and A6

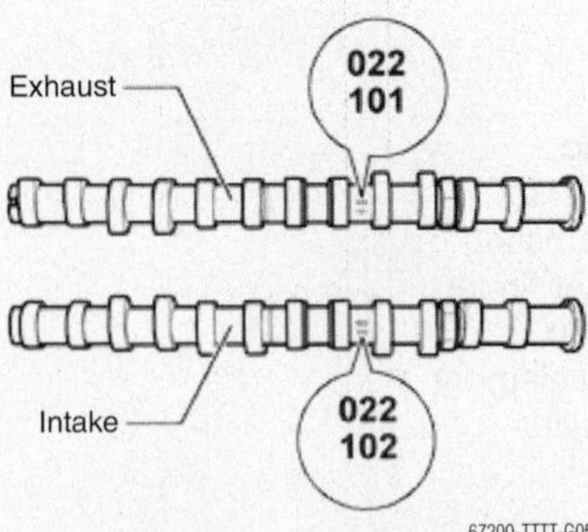

67200-TTTT-G05

Intake and exhaust camshaft identification—3.2L engine, except A3 and A6

- Cap 7
- Caps 3 and 11

12. Tighten exhaust camshaft as follows:
    - Caps 6 and 10 alternately in a diagonal sequence and then tighten to final torque
    - Caps 2 and 14
    - Cap 8
    - Caps 4 and 12
    - Bearing cap to cylinder head 44 in. lbs (5Nm) + 45° (45° = 1/8 turn)
13. Install valve timing housing
    - Timing housing 71 in. lbs (8 Nm)

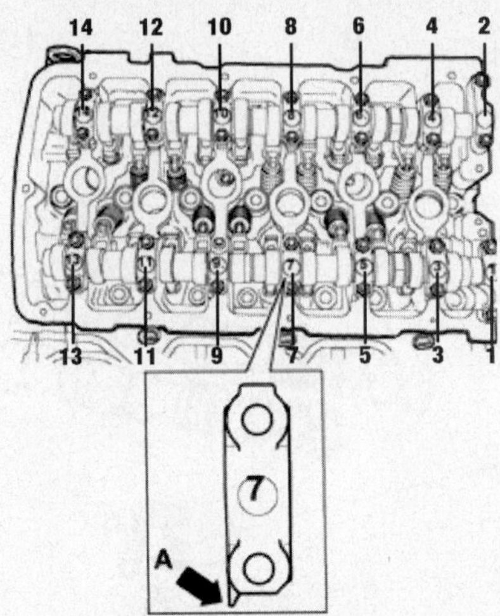

A = Pointed section

67200-TTTT-G06

Installed position of bearing caps—3.2L engine, except A3 and A6

➡Maximum Camshaft end play = .004 (.10 mm)

➡It must be possible to insert camshaft plate T10068 or equivalent in both camshaft grooves

- Install guide rail
- Install camshaft adjuster for intake side
- Turn sender wheel "1" clockwise in camshaft adjuster until it stops

➡Do not use the camshaft plate as a counter hold

- Tighten intake camshaft adjuster to intake camshaft in this position finger tight
- Inscription "24E" and tooth just behind it must align with notch "3" in valve timing housing
- Lay timing chain in this position tautly onto the intake camshaft adjuster
- Count exactly 16 rollers of timing chain starting from "24E" on the camshaft adjuster moving towards the exhaust side

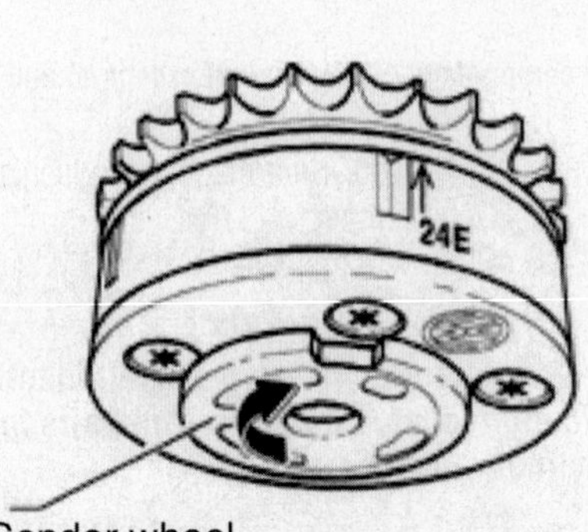

67200-TTTT-G14

Intake camshaft adjuster—3.2L engine, except A3 and A6

67200-TTTT-G15

Intake camshaft adjuster alignment—3.2L engine, except A3 and A6

67200-TTTT-G17

Exhaust camshaft adjuster alignment—3.2L engine, except A3 and A6

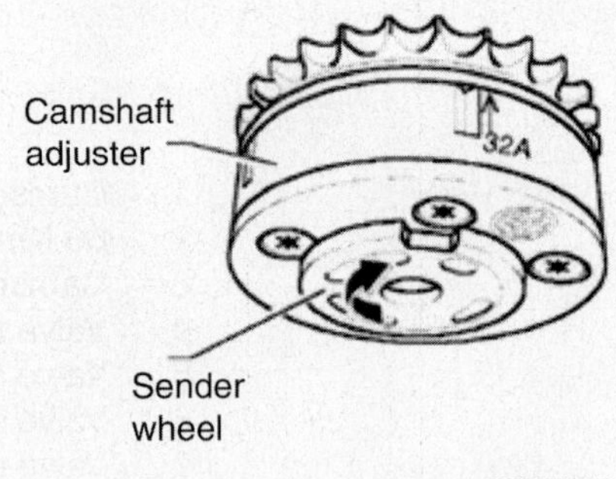

67200-TTTT-G16

Exhaust camshaft adjuster—3.2L engine, except A3 and A6

- Place tooth on "32A" on exhaust camshaft adjuster exactly behind the 16th roller of the chain
- Inscription "32A" on exhaust camshaft adjuster must align with the notch in the valve timing housing
- Turn sender wheel "2" clockwise in exhaust camshaft adjuster until it stops
- Tighten exhaust camshaft adjuster

to exhaust camshaft in this position finger tight

➡When turning the crankshaft it is necessary to press the guide rail firmly against the timing chain to prevent the timing chain from slipping

- Turn the crankshaft 2 revolutions in normal direction and return to TDC
- Insert camshaft plate T10068 in both camshaft grooves. If camshaft plate cannot be inserted, repeat adjustment of valve timing

➡If valve timing is correct, torque camshaft adjuster bolts to 44 ft. lbs (60 Nm) + 90 degrees (1/4 turn) install new bolts

- Install upper timing chain cover
- Tighten bolts in following sequence
- Timing cover to engine block initially to 44 inch lbs (5 Nm)
- Bottom of timing cover at head gasket 17 ft. lbs (23 Nm)
- Timing cover to engine block to 88 inch lbs (10 Nm)

14. The remainder of the installation is the reverse of the removal procedure

➡The engine must not be started for 30 minutes after installing the camshafts. The hydraulic valve compensation elements have to settle otherwise valves will strike pistons. After 30 minutes, turn the engine carefully at least 2 rotations to ensure that none of the valves make contact when the starter is operated

3.2L Engine–A6 Models

1. Before servicing the vehicle, refer to the precautions section.
2. Remove the cylinder head covers.
3. Remove the left and right timing chain covers.
4. Remove the camshaft timing chains.
5. Loosen bearing bracket bolts in sequence.
6. Remove the bearing bracket.
7. Mark camshafts and their related components and remove.

To install:

8. Replace all gaskets and seals
9. Secure crankshaft using crankshaft holder T40069 .
10. Carefully clean all gasket mating surfaces.
11. Oil the camshaft journal surfaces.
12. Place bearing bracket onto a soft cloth on bench.
13. Set the camshafts into bearing bracket.

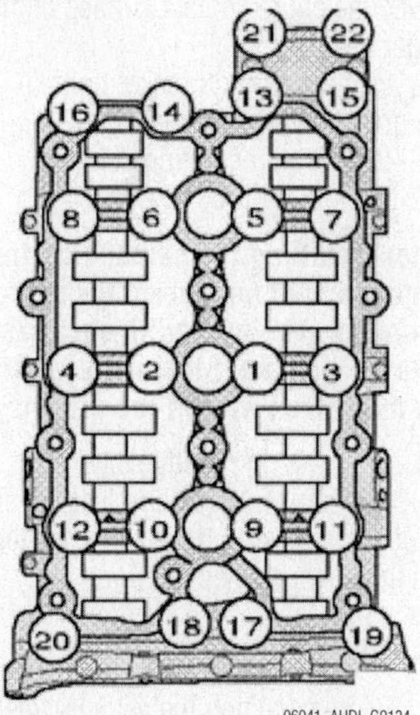

Bearing bracket loosening sequence—A6 with 3.2L engine

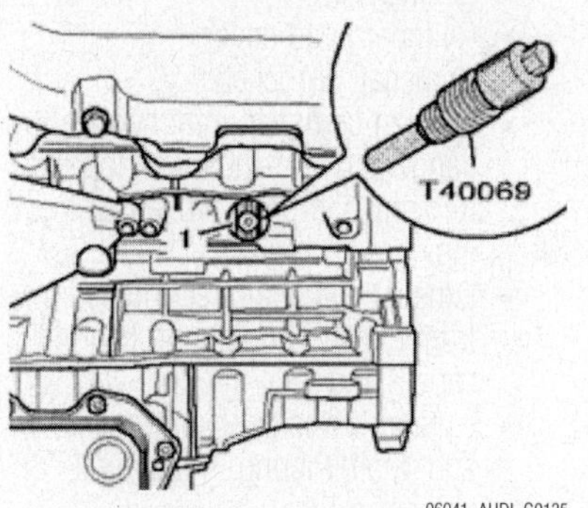

Secure crankshaft using crankshaft holder T40069—A6 with 3.2L engine

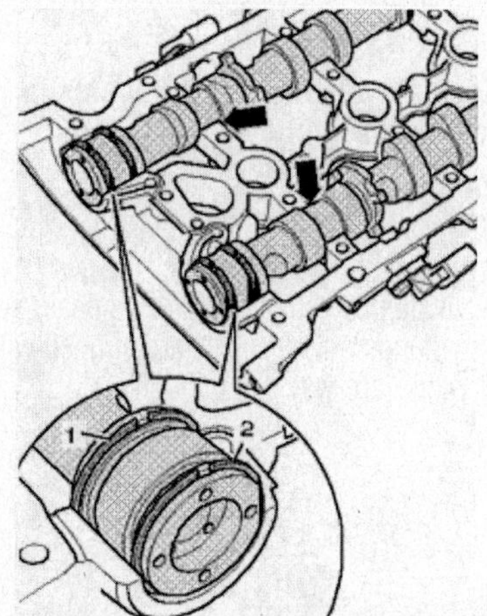

The placement of camshafts must be exactly within the axial bearings (arrows) of the bearing bracket. The ends of the piston rings (1 and 2) must face upward or downward, never sideways—A6 with 3.2L engine

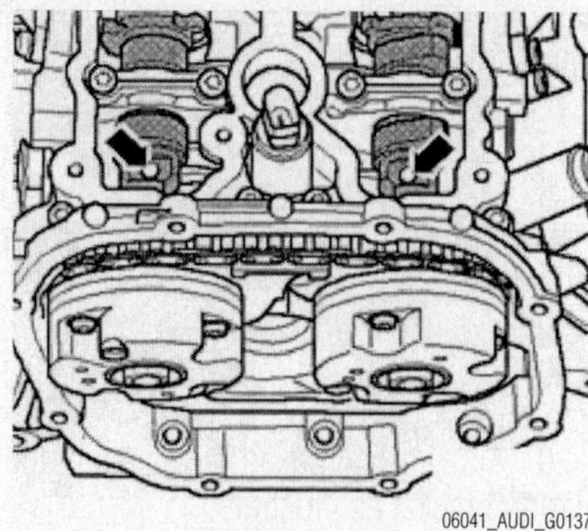

Turn camshafts until threaded holes face upward—A6 with 3.2L engine

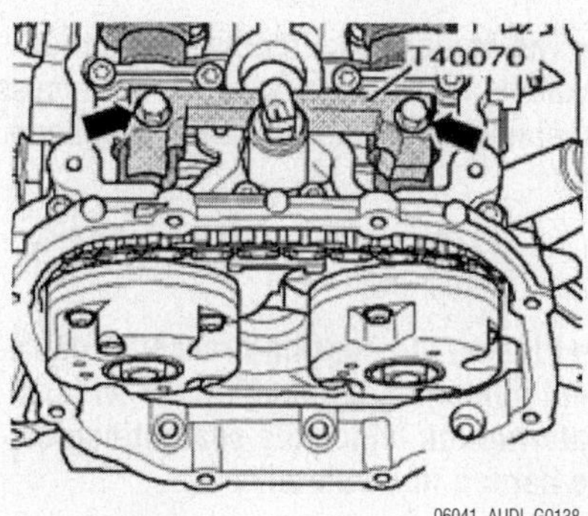

Mount camshaft locating tool T40070—A6 with 3.2L engine

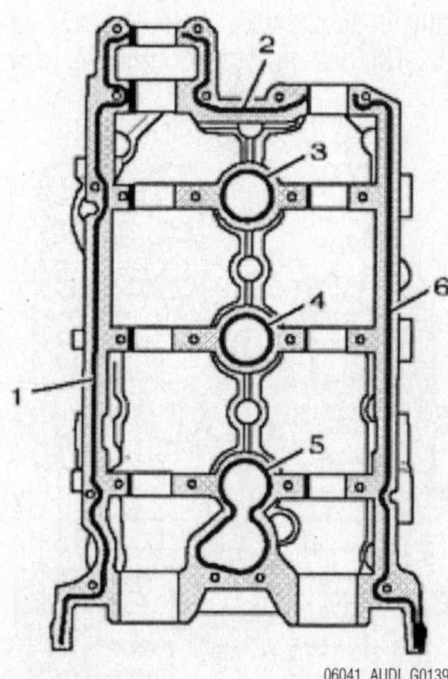

Apply a 1mm bead of sealant of the bearing bracket at the locations indicated—A6 with 3.2L engine

14. The placement of camshafts must be exactly within the axial bearings of the bearing bracket.
15. The ends of the piston rings must face upward or downward, and must never face sideways.
16. Turn over the bearing bracket with

installed camshafts, holding camshafts tight within bearing bracket.

17. Turn camshafts until threaded holes face upward.

18. Make sure that the camshafts are still positioned exactly within the axial bearings of the bearing bracket.

19. Mount camshaft locating tool T40070 as shown and tighten bolts to 20 Nm.

20. Apply a 1mm bead of sealant of the bearing bracket as illustrated.

21. The grooves of sealing surface must be completely filled with sealant.

22. The sealant beads must be 1.5 to 2.0 mm above sealing surface.

➡The sealant beads must be applied exactly as instructed, otherwise excess sealant could enter the camshaft bearings.

23. Place bearing bracket immediately onto cylinder head.

➡Placing the bearing bracket in place and tightening it should occur without interruption, since the sealant begins to harden immediately.

➡After installing bearing bracket, sealant must dry for approx. 30 minutes.

24. Hand-tighten bearing bracket bolts equally, in sequence.

25. The bearing bracket must be in contact with the entire contact surface of the cylinder head.

26. Fasten bearing bracket bolts in sequence to 9 Nm until they stop.

27. Install the remaining components in the reverse order of removal.

➡After installing the camshafts, the engine may not be started for approx. 30 minutes. The hydraulic equalization elements must seat themselves (otherwise the valves will hit the pistons).

28. After working on the valve gear, carefully rotate engine by hand at least 2 full revolutions to ensure that valves do not strike the pistons when starting.

Bearing bracket tightening sequence—A6 with 3.2L engine

4.2L Engine

1. Before servicing the vehicle, refer to the precautions section.
2. Remove or disconnect the following:
   - Negative battery cable
   - Timing belt
   - Cylinder head cover
   - Camshaft sprockets
   - Exhaust camshaft intermediate flange and rear-most bearing cap for camshaft position sensor housing
   - Exhaust camshaft bearing cap in front of drive chain, and bearing caps 2 and 3
   - Exhaust camshaft bearing caps 1 and 4, alternating in diagonal sequence
   - Intake camshaft bearing cap 6 and 7
   - Intake camshaft bearing caps 5 and 8, alternating in diagonal sequence
   - Intake and exhaust camshafts
   - Hydraulic valve lifters

➡Keep all valvetrain components in order for assembly.

To install:

3. Install the hydraulic valve lifters in their original positions.
4. Install camshafts with drive chain so that markings on sprockets are aligned with each other (arrows).

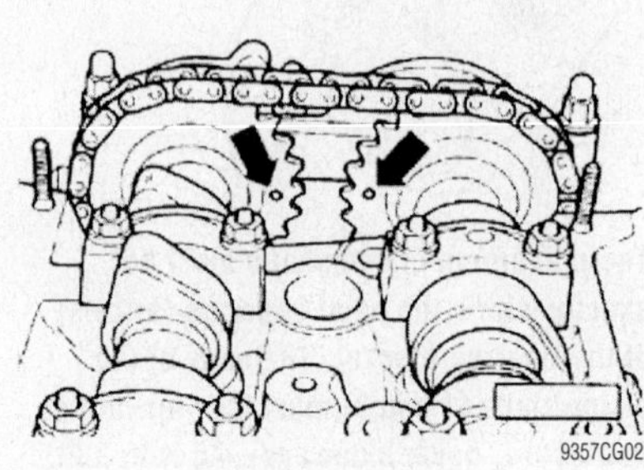

Camshaft alignment—A8

➡When installing the bearing caps, make sure that the stamped numbers appear upright when viewed from the intake side.

➡The contact surface of both outer bearing caps should be coated lightly with AMV 174 003 01.

5. Install the intake camshaft bearing caps as follows:
   a. Step 1: Tighten bearing caps 5 and 8 evenly, alternating in diagonal sequence to 11 ft. lbs. (15 Nm).
   b. Step 2: Tighten the remaining bearing caps to 11 ft. lbs. (15 Nm).
6. Install the exhaust camshaft bearing caps as follows:
   a. Step 1: Tighten bearing caps 1 and 4 evenly, alternating in diagonal sequence to 11 ft. lbs. (15 Nm).
   b. Step 2: Tighten the remaining bearing caps to 11 ft. lbs. (15 Nm).
7. Install the camshaft sprockets and tighten the bolts to 41 ft. lbs. (55 Nm).
8. The remainder of the installation is the reverse of the removal procedure.

A3

2.0L ENGINE

1. Before servicing the vehicle, refer to the precautions section.

➡The camshaft bearings are integrated into the cylinder head and into the bearing bracket. Before removing the bearing bracket, the timing belt must be relieved of tension. If the bearing bracket was loosened, the camshaft seal and the sealing cap must be replaced.

2. Remove the cylinder head cover
3. Remove the camshaft adjuster as follows:
4. Remove the high-pressure pump
5. Remove the cylinder head cover

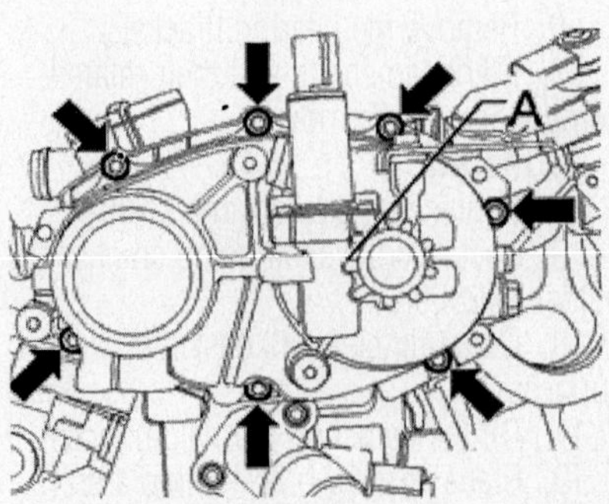

Remove the camshaft adjuster cover—A3 with 2.0L engine

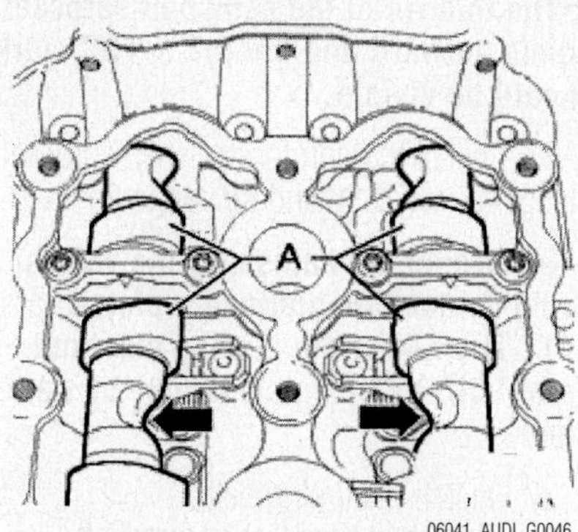

Move marking on camshaft sprocket so that it is across from the marking on the timing belt cover. The notches of camshafts now face toward each other vertically—A3 with 2.0L engine

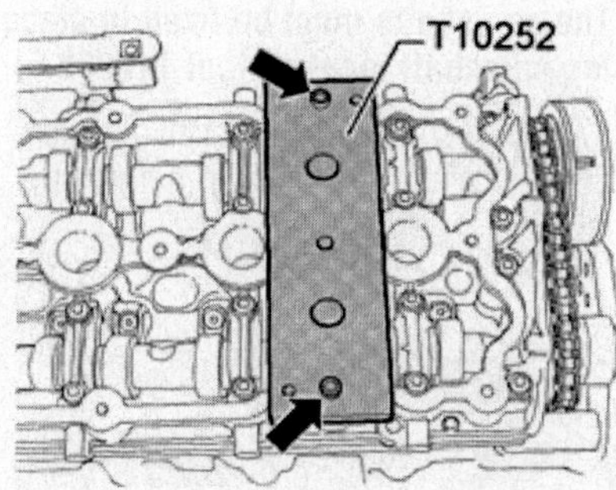

Install the camshaft locating tool T10252 and secure it—A3 with 2.0L engine

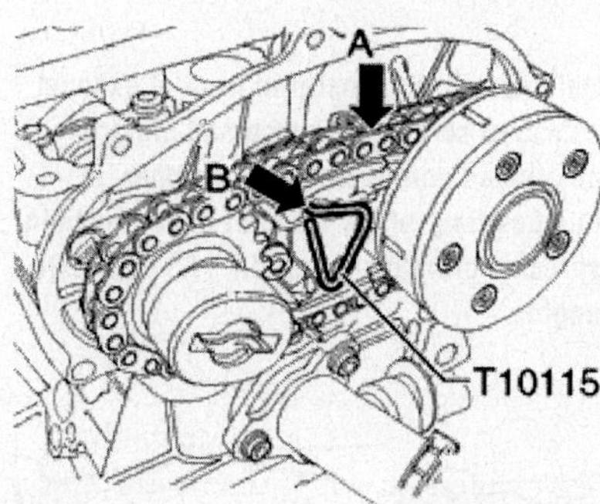

Compress chain tensioner and secure it with locking pin T10115—A3 with 2.0L engine

6. Remove the camshaft adjuster cover.

7. Move marking on camshaft sprocket so that it is across from the marking on the timing belt cover. The notches of camshafts now face toward each other vertically.

8. Install the camshaft locating tool T10252 and secure it.

9. Loosen the bolt from camshaft adjuster using socket T40080.

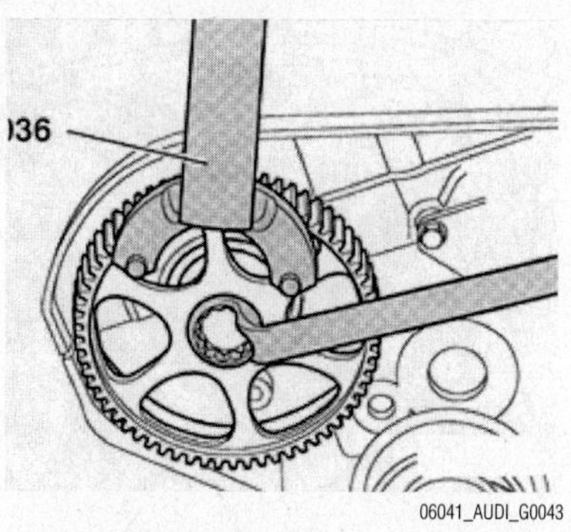

Loosen the camshaft adjuster—A3 with 2.0L engine

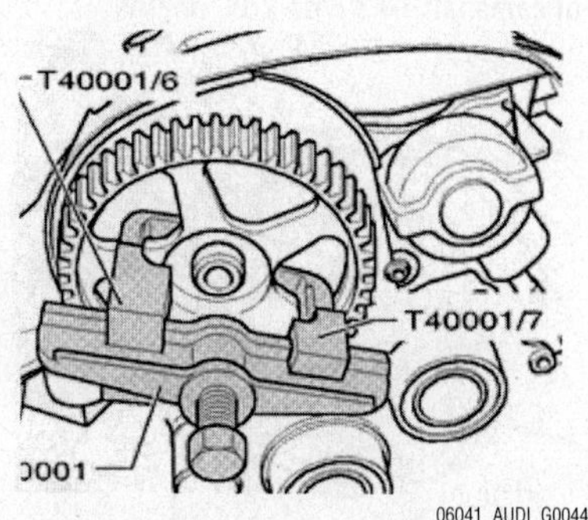

Pull off camshaft sprocket using puller T40001 , claw T40001/6 and claw T40001/7—A3 with 2.0L engine

10. Compress chain tensioner and secure it with locking pin T10115.

11. Remove the bolt from camshaft adjuster and remove it together with chain.

12. Remove the timing belt.

13. Loosen the camshaft adjuster.

14. Pull off camshaft sprocket using puller T40001 , claw T40001/6 and claw T40001/7.

15. Remove the rear the timing belt cover from cylinder head.

16. Loosen the bearing bracket bolts evenly from outside to inside and remove bearing bracket.

17. Carefully remove camshafts upward and place them on a clean surface.

18. Remove the old sealant from bearing bracket groove and from sealing surfaces.

19. Do not allow dirt or sealant residue to get into cylinder head.

To install:

➡The sealing surfaces must be free of oil and grease.

➡The pistons must not be positioned at TDC.

20. Make sure that all roller rocker levers are positioned correctly on the valve stem ends.

21. Oil the journal surfaces of camshafts.

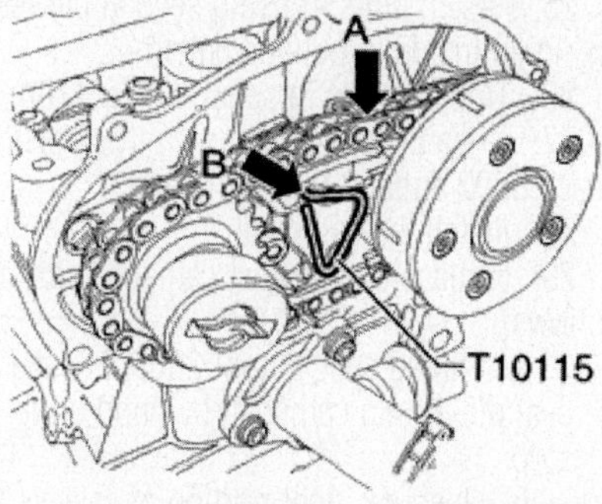

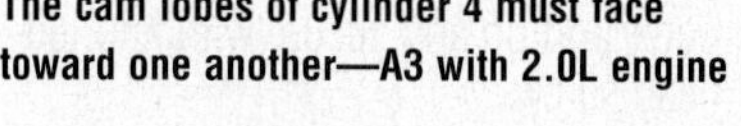

The cam lobes of cylinder 4 must face toward one another—A3 with 2.0L engine

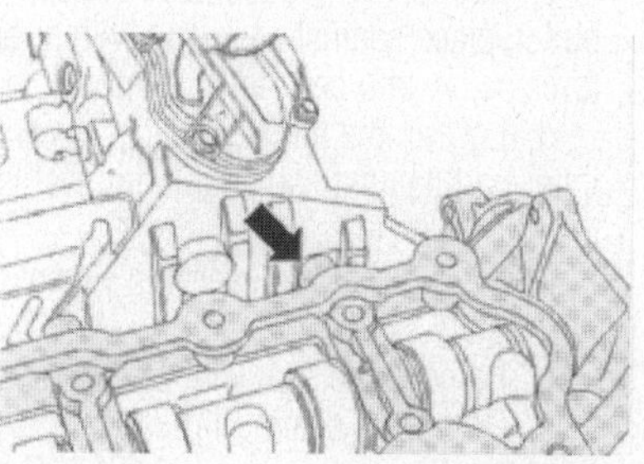

Set bearing bracket in place so that it can get by the EGR valve—A3 with 2.0L engine

22. Carefully place camshafts into camshaft bearings of cylinder head. The cam lobes of cylinder 4 must face toward one another.

23. Apply an even, light sealant bead into clean bearing bracket groove. Spread sealant evenly on sealing surface.

➡The sealant must not be applied too thickly. Wipe off excess sealant with a lint-free rag, if necessary.

➡Install the bearing bracket and tighten it right away, since the sealant begins to harden immediately when the sealing surfaces come into contact.

24. Set bearing bracket in place so that it can get by the EGR valve.

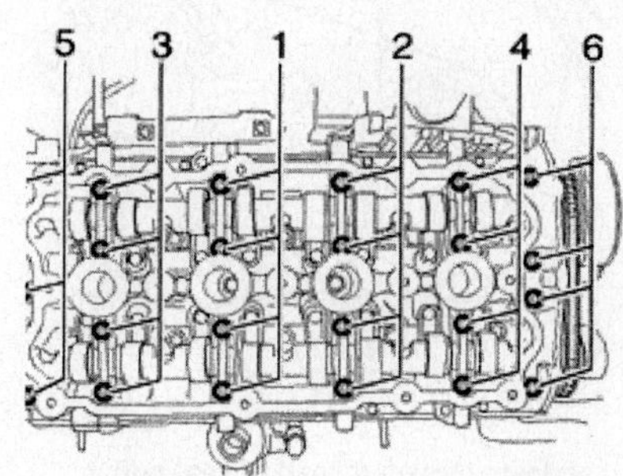

Camshaft bolt tightening sequence—A3 with 2.0L engine

25. Fasten bolts in using several passes using an inside to outside pattern.
26. Tighten bolts in sequence to 8 Nm.
27. Drive in sealing cap with the thrust piece 3202 about 1 to 2 mm deep.
28. Install the camshaft.
29. Replace the exhaust camshaft seal as follows:
   a. Thread special tool 2085/2 of the seal puller into camshaft by hand up to stop.
   b. Unscrew inner portion of seal puller two rotations or about 3 mm from outer portion and secure with knurled-head screw.
   c. Lubricate threaded head of seal puller, place against seal, and with strong force screw into seal as far as possible.
   d. Loosen the knurled bolt and turn inner part against camshaft until seal is removed.
   e. Secure seal remover in a vise at the flat spots.
   f. Remove the seal with pliers
   g. Do not oil sealing lip of seal.
   h. Position guide sleeve T10070/1 of polydrive bit and drive socket T10070 on end of camshaft.

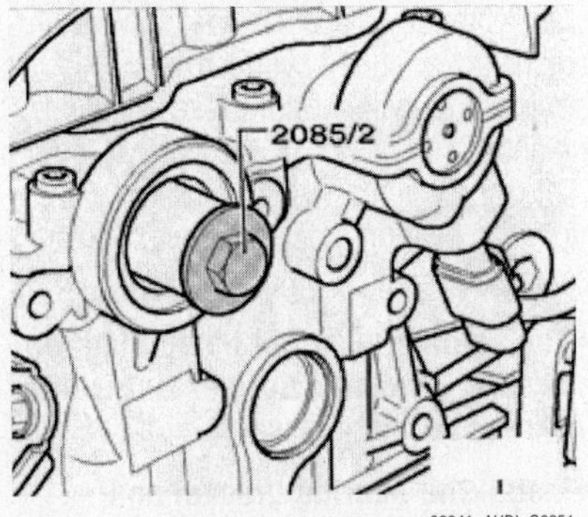

Thread special tool 2085/2 of the seal puller into camshaft by hand up to stop—A3 with 2.0L engine

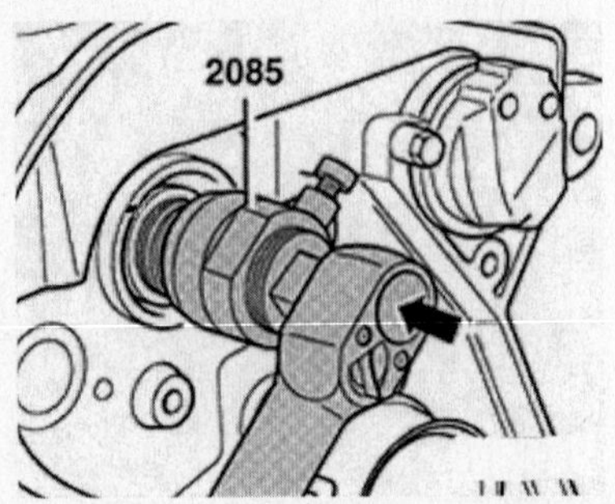

Lubricate threaded head of seal puller, place against seal, and with strong force screw into seal as far as possible—A3 with 2.0L engine

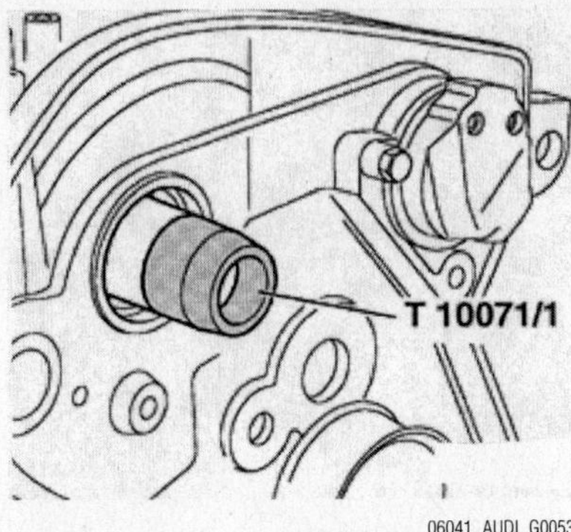

Position guide sleeve T10070/1 of poly-drive bit and drive socket T10070 on end of camshaft—A3 with 2.0L engine

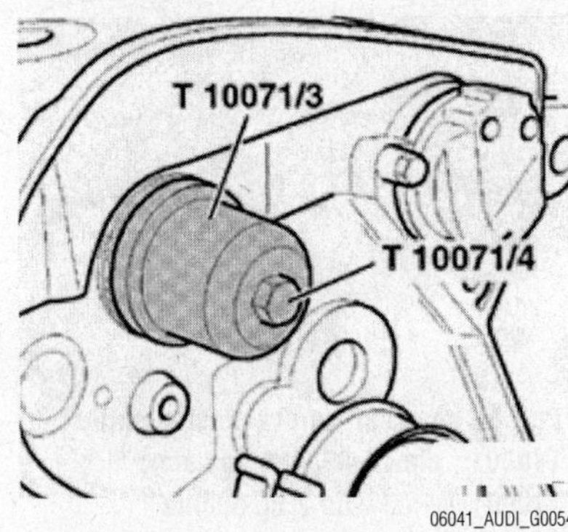

Press in seal with pressure sleeve T10071/3 and bolt T10071/4 from Poly-drive Bit And Drive Socket T10070 up to stop—A3 with 2.0L engine

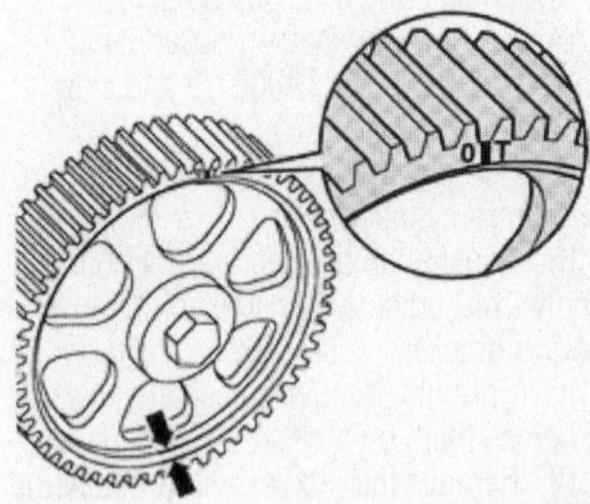

The thin rib of the camshaft sprocket points outward and the cyl. 1 TDC mark should be visible—A3 with 2.0L engine

   i. Slide seal over guide sleeve onto camshaft end.
   j. Remove the guide sleeve.
   k. Press in seal with pressure sleeve T10071/3 and bolt T10071/4 from Poly-drive Bit And Drive Socket T10070 up to stop.
30. Install the camshaft sprocket.

➡The thin rib of the camshaft sprocket points outward and the cyl. 1 TDC mark should be visible.

31. Install the bolt for camshaft sprocket using retainer 3036 and tighten to 65 Nm.

➡By turning the camshaft, the valves could contact the stationary pistons at TDC. Therefore, the pistons must not be at TDC. Valves/pistons may be damaged.

32. Install the timing belt.
33. Camshaft sprocket to camshaft.
34. Install the rear the timing belt cover.
35. Insert fitted key into camshaft.

➡Make sure fitted key is properly seated.

36. Install the camshaft adjuster

➡The camshafts must be fixed in place using camshaft locating tool T10252.

37. Place chain onto camshaft adjuster.
38. Hold camshaft adjuster in front of

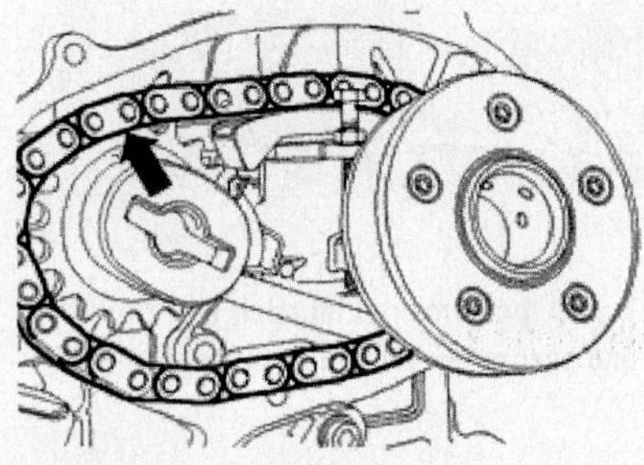

Hold camshaft adjuster in front of exhaust camshaft so that the notch and pin are positioned across from one another. In this position, place chain at top onto chain sprocket of intake manifold—A3 with 2.0L engine

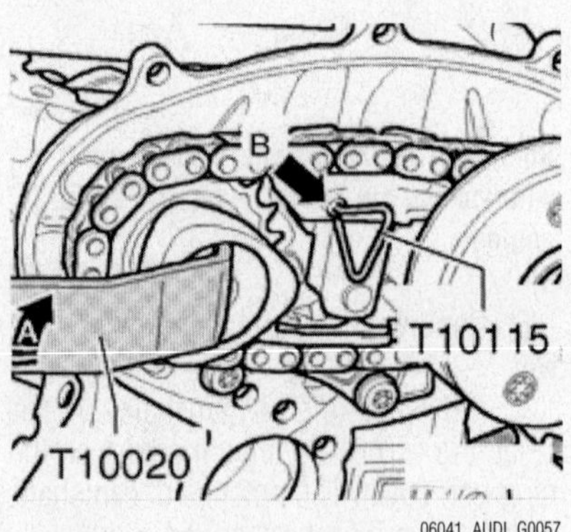

Slowly turn intake camshaft using the timing belt tensioner tool T10020 until camshaft adjuster fits onto camshaft—A3 with 2.0L engine

exhaust camshaft so that the notch and pin are positioned across from one another. In this position, place chain at top onto chain sprocket of intake manifold.

39. Slowly turn intake camshaft using the timing belt tensioner tool T10020 until camshaft adjuster fits onto camshaft.

➡**If the pin does not fit into the notch, remove the chain and then re-install the chain again.**

40. Tighten bolt of camshaft adjuster to 100 Nm.

41. Remove the locking pin T10115.

42. Install the remaining components in the reverse of removal.

3.2L ENGINE

1. Before servicing the vehicle, refer to the precautions section.
2. Remove the camshaft timing chain from camshafts.
3. Remove the control housing.
4. Carefully pull control housing off camshaft seals.
5. Remove the intake camshaft as follows:
   a. Remove the bearing caps 1 and 13.
   b. Remove the bearing caps 3 and 11.
   c. Remove the bearing cap 7.
   d. Loosen the bearing caps 5 and 9 alternately and in diagonal sequence, and remove.
6. Remove the exhaust camshaft as follows:
   a. Remove the bearing caps 2 and 14.
   b. Remove the bearing caps 4 and 12.
   c. Remove the bearing cap 8.
   d. Loosen the bearing caps 6 and 10 alternately and in diagonal sequence, and remove.
7. Carefully remove camshafts and place on a clean surface.

To install:

➡**When installing camshafts, crankshaft must not be at TDC with any piston. Valves and/or pistons may be damaged.**

➡**After installing the camshafts, the engine may not be started for about 30 minutes. The hydraulic equalization elements must seat themselves (otherwise the valves will crash into the pistons). After working on the valvetrain and lifters, carefully rotate the crankshaft by hand at least 2 full revolutions before starting to be sure that valves do not strike the pistons.**

8. Insert support element in cylinder head.
9. Install the rocker lever so that it rests against valve stem ends and is clipped into support element.
10. Oil the journal surfaces of camshafts.
11. Place respective camshaft into camshaft bearings of cylinder head.
12. Note the camshaft identification between cylinder 4 and cylinder 5 cam pair:
    a. A—Exhaust camshaft—Identification 022—Index 101.

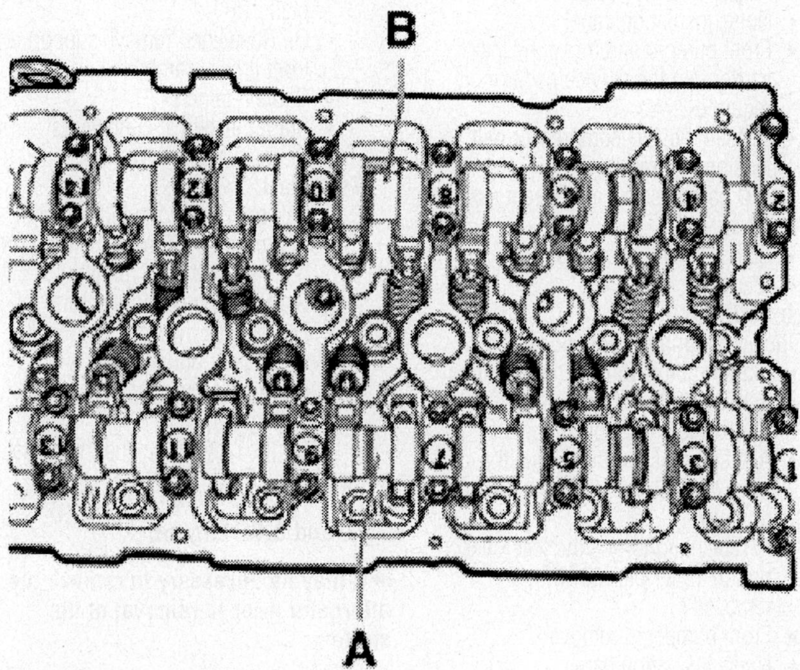

Camshaft bearing cap location—A3 with 3.2L engine

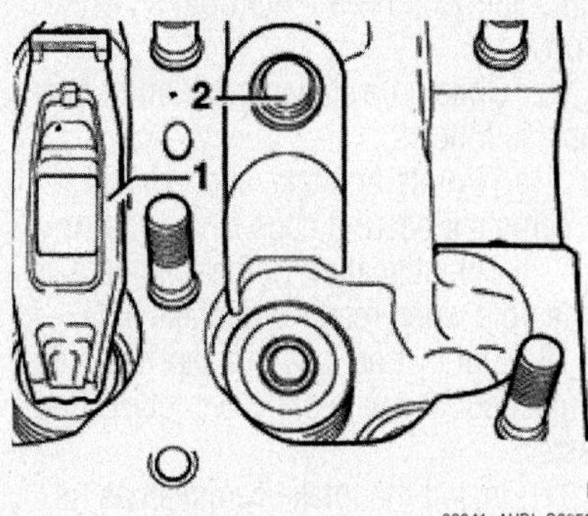

Install the rocker lever so that it rests against valve stem ends and is clipped into support element—A3 with 3.2L engine

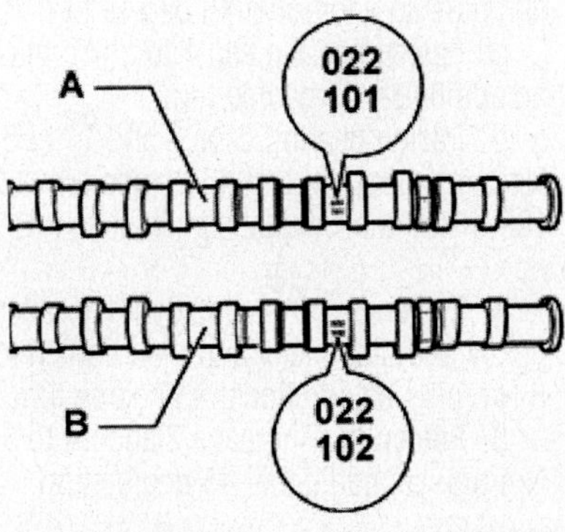

Note the camshaft identification between cylinder 4 and cylinder 5 cam pair—A3 with 3.2L engine

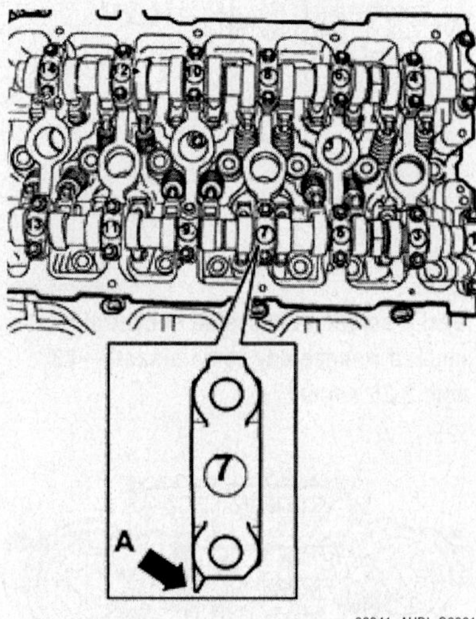

Points of intake and exhaust camshaft bearing caps face outward—A3 with 3.2L engine

    b. B—Intake camshaft—Identification 022—Index 102.
13. Insert camshafts into cylinder head in such a way that they are set at TDC.
14. It must be possible to insert

camshaft bar T10068 into both shaft grooves.

15. Observe installed position of bearing caps as follows:
   a. Points of intake and exhaust camshaft bearing caps face outward.
   b. Identification on bearing caps is legible when read from intake side.

16. Lightly coat bearing cap 7 and 8 contact surface with adhesive lubricating paste.

17. Install the intake camshaft as follows:
   a. Fasten bearing caps 5 and 9 alternately and diagonally and then tighten to 5 Nm plus an additional 45 degree turn.
   b. Fasten bearing caps 1 and 13 to 5 Nm plus an additional 45 degree turn.
   c. Fasten bearing cap 7 to 5 Nm plus an additional 45 degree turn.
   d. Fasten bearing caps 3 and 11 to 5 Nm plus an additional 45 degree turn.

18. Install the exhaust camshaft as follows:
   a. Fasten bearing caps 6 and 10 alternately and diagonally and then tighten to 5 Nm plus an additional 45 degree turn.
   b. Fasten bearing caps 2 and 14 to 5 Nm plus an additional 45 degree turn.

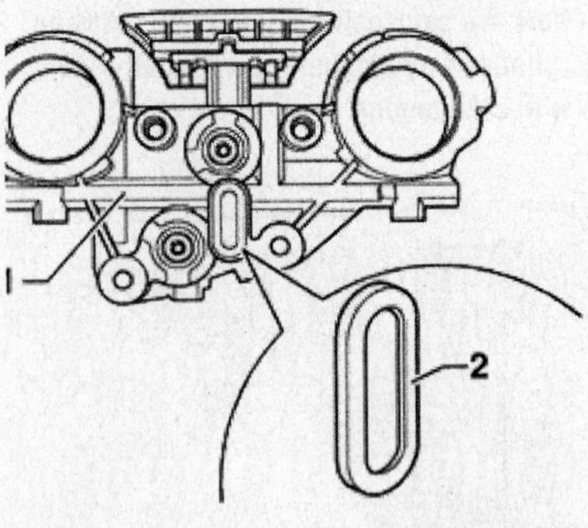

Unclip screen at backside of control housing and remove any contaminants—A3 with 3.2L engine

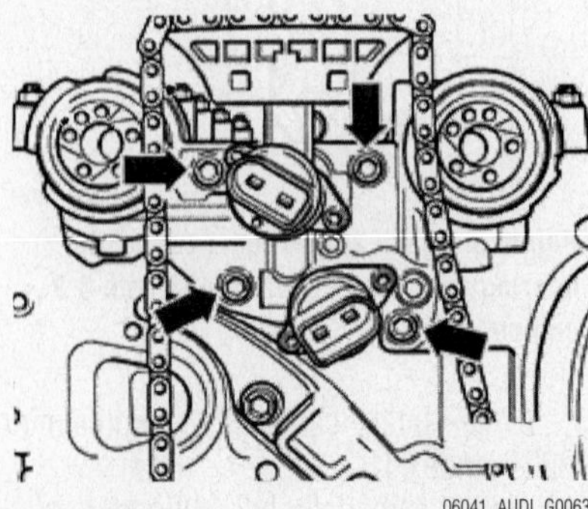

Tighten the control housing(indicated by arrows)—A3 with 3.2L engine

   c. Fasten bearing cap 8 to 5 Nm plus an additional 45 degree turn.
   d. Fasten bearing caps 4 and 12 to 5 Nm plus an additional 45 degree turn.

19. Position camshafts in cylinder head to TDC. It must be possible to insert camshaft bar T10068 into both shaft grooves.

20. Unclip screen at backside of control housing and remove any contaminants.

21. Lightly oil camshaft seal contact surfaces in control housing.

22. Lightly oil sealing ring contact surfaces of camshafts and carefully slide control housing over camshaft sealing rings.

23. Tighten control housing.

24. Install the camshaft timing chain.

Valve Lash

ADJUSTMENT

Audi engines are equipped with hydraulic lash adjusters. No adjustment is possible.

Starter Motor

REMOVAL & INSTALLATION

A4

1.8L ENGINE

1. Before servicing the vehicle, refer to the precautions section.
2. Remove or disconnect the following:
   - Negative battery cable
   - Noise insulation panel
   - Front bumper and move the lock carrier into the service position, if required
   - Loosen the A/C compressor belt tensioner, if required
   - A/C compressor and move it aside, if required
   - Starter electrical connectors
   - Starter

To install:

3. Install or connect the following:
   - Starter motor. Torque the mounting bolts to 48 ft. lbs. (65 Nm).
   - Starter electrical connectors
   - A/C compressor and torque the bolts to 18 ft. lbs. (25 Nm), if removed
   - A/C belt and torque the tensioner bolt to 15 ft. lbs. (20 Nm), if removed
   - Front bumper, if removed
   - Noise insulation panel
   - Negative battery cable
4. Enter the radio code and the preset frequencies.

TT

1.8L ENGINE

1. Before servicing the vehicle, refer to the precautions section.
2. Remove or disconnect the following:
   - Battery
   - Battery platform
   - Air cleaner housing bolts on battery platform
   - On Roadsters, remove support at front cross member
   - Cable retainers
   - Starter electrical connectors
   - Starter

To install:

3. Install or connect the following:
   - Starter motor. Torque the mounting bolts to 48 ft. lbs. (65 Nm)
   - Starter electrical connectors
   - Front cross member support
   - Air cleaner housing
   - Battery platform
   - Battery cables
4. Enter the radio code and the preset frequencies.

3.2L ENGINE

1. Before servicing the vehicle, refer to the precautions section.
2. Remove or disconnect the following:
   - Battery
   - Battery platform
   - Air cleaner housing bolts on battery platform
   - On Roadsters, remove support at front cross member
   - Cable retainers
   - Starter electrical connectors
   - Starter

To install:

3. Install or connect the following:
   - Starter motor. Torque the mounting bolts to 48 ft. lbs. (65 Nm)
   - Starter electrical connectors
   - Front cross member support
   - Air cleaner housing
   - Battery platform
   - Battery cables
4. Enter the radio code and the preset frequencies.

2.7L and 3.0L Engine

➡It may be necessary to remove the alternator prior to removal of the starter.

1. Before servicing the vehicle, refer to the precautions section.

2. Remove or disconnect the following:
   - Negative battery cable
   - Right front wheel, if necessary
   - Alternator, if necessary
   - Starter wiring connectors
   - Ground wire at engine mount
   - Starter

To install:

3. Install or connect the following:
   - Starter and torque the bolts to 48 ft. lbs. (65 Nm)
   - Starter wiring connectors
   - Ground wire
   - Alternator, if removed
   - Right front wheel, if removed
   - Negative battery cable

2.8L Engine

➡It may be necessary to remove the alternator prior to removal of the starter.

1. Before servicing the vehicle, refer to the precautions section.
2. Remove or disconnect the following:
   - Negative battery cable
   - Air duct
   - Starter wiring connectors
   - Rear noise insulation panel, if equipped with a manual transmission
   - Right front wheel , if equipped with an automatic transmission
   - Starter

To install:

3. Install or connect the following:
   - Starter and torque the bolts to 48 ft. lbs. (65 Nm)
   - Right front wheel, if equipped with an automatic transmission
   - Rear noise insulation panel, if equipped with a manual transmission
   - Starter wiring connectors
   - Air duct
   - Negative battery cable
   - Alternator, if removed

A6

1. Before servicing the vehicle, refer to the precautions section.
2. Remove or disconnect the following:
   - Negative battery cable
   - Noise insulation panel
   - Starter retainer at engine mount
   - Electrical connection at engine mount
   - Starter ground wire
   - Starter wiring connectors
   - Rotate starter downward while turning clockwise to remove

To install:

3. Install or connect the following:
   - Starter and torque the bolts to 48 ft. lbs. (65 Nm)
   - Starter wiring connectors
   - Ground wire
   - Electrical connection
   - Starter retainer
   - Noise insulation panel
   - Negative battery cable

A8

4.2L ENGINE

1. Before servicing the vehicle, refer to the precautions section.
2. Relieve the fuel system pressure.
3. Remove or disconnect the following:
   - Negative battery cable
   - Engine cover
   - Intake air duct between the air filter and the throttle valve part
   - Air-induction ports between the air filter housing and the metering unit
   - Fan shroud and the electric fan
   - Engine undercover
   - Alternator air duct hose
   - Alternator wiring harness
   - Accessory drive belt
   - Alternator bolts
   - Alternator
   - The coverings over the strut tower and the air filter housing
4. Place the support device 10-222A with adapter for support device 10-222A/4 onto the screws of the strut tower.
5. Unscrew the fuel line from the pressure regulator.
6. Insert support device 3180 on the transmission side into the engine hoisting ring and screw it in securely.
7. Insert additional hook A (10-222A/2) into the hoisting ring on the pressure regulator.
8. Remove or disconnect the following:
   - Engine mount bolts
   - Attachment screws from the torque converter bearing on the right front
9. Use the spindles to hoist the engine until the engine mounts stand out from the subframe.

➡Be careful not to let the throttle valve part damage the soundproofing material on the bulkhead.

10. Remove or disconnect the following:
    - Harness connector at the engine mount bracket
    - Engine mount bracket
    - Starter harness connectors
    - Starter motor mounting bolts
    - Starter motor

To install:

11. Installation is the reverse of the removal procedure, while using the following torque values:
    - Starter motor bolts: 48 ft. lbs. (65 Nm)
    - Starter motor **B+** terminal: 12 ft. lbs. (16 Nm)

A3

2.0L ENGINE

1. Before servicing the vehicle, refer to the precautions section.
2. Disconnect the negative battery cable.
3. Disconnect the electrical connection at Mass Air Flow (MAF) sensor.
4. Remove the engine cover.
5. Remove the cable tie for protective sleeve.
6. Disconnect the electrical connections.
7. Remove the top starter bolt.
8. Remove the bottom starter bolt and remove starter.

To install:

9. Installation is the reverse of removal, please note the following:
   a. Tighten the starter to transmission bolt to 80 Nm.
   b. Tighten the ground cable to starter bolt to 23 Nm.
   c. Tighten the wiring bracket bolt to 23 Nm.

3.2L ENGINE

1. Before servicing the vehicle, refer to the precautions section.
2. Disconnect the negative battery cable.
3. Pull off vacuum hose at air intake hose.
4. Remove the air hose from Throttle Valve Control Module.
5. Disconnect electrical connection at Mass Air Flow (MAF) sensor.
6. Remove the top section of air cleaner housing.
7. Remove the filter element.
8. Pull off cover for air duct.
9. Unclip air duct.
10. Remove the screws and remove bottom section of air cleaner housing.
11. Remove the screws and remove bracket for air cleaner housing.
12. Disconnect the electrical connections.
13. Remove the top starter bolt.
14. Remove the bottom starter bolt and remove starter.

To install:

15. Installation is the reverse of removal, please note the following:

a. Tighten the starter to transmission bolt to 40 Nm.

b. Tighten the ground cable bolt to 23 Nm.

c. Tighten the wiring bracket bolt to 23 Nm.

Oil Pan

REMOVAL & INSTALLATION

A4

1.8L ENGINE

1. Before servicing the vehicle, refer to the precautions section.
2. Drain the engine oil.
3. Remove or disconnect the following:
   - Negative battery cable
   - Engine undercover
   - Accessory drive belts and the air conditioning belt tension pulley
   - Torquc support stop and side brace
   - Starter wiring
   - Hose from the turbocharger at the air guide tube in the lock carrier
   - Bottom nuts from the lower engine mount
   - Top engine cover
4. Install the Engine Support Bridge tool 10-222A and the Engine Sling tool 2024A; then, lift the engine as far a possible.
5. Support the subframe.
6. Remove or disconnect the following:
   - Front bolts No. 2 and 3
   - Bolt No. 1 from the subframe
7. Slowly lower the subframe.
8. If equipped with a manual transaxle, loosen the left transaxle mount nut until it is aligned with the lower edge of the bolt (approx. 4 turns).
9. If equipped with an automatic transaxle, loosen the rear bolt for the left transaxle mount several turns, then remove the front bolt for the transaxle mount.
10. At the right transaxle mount, loosen the rear bolt mount several turns and remove the front bolt.

➡If equipped with a manual transaxle, both of the rear bolts on the oil pan can be accessed through the opening on the flywheel. Turn the flywheel as needed.

11. Remove the oil pan.
12. Clean all gasket mating surfaces.

To install:

13. Apply sealant to the front and rear contact areas of the oil seal carriers.
14. Install or connect the following:
   - New oil pan gasket
   - Oil pan
15. Tighten the oil pan bolts as follows:
   - Oil pan-to-engine bolts: 44 inch lbs. (5 Nm)
   - M10 bolts between the oil pan and engine: 33 ft. lbs. (45 Nm)
   - M6 bolts between the oil pan and engine: 84 inch lbs. (10 Nm)
16. Tighten the subframe bolts/nuts using the illustration as follows:
   - Bolts 2 and 5: 81 ft. lbs. (110 Nm) plus a ¼ (90 degree) turn
   - Bolt 6: 55 ft. lbs. (75 Nm)
   - Bolt 1: 17 ft. lbs. (23 Nm)

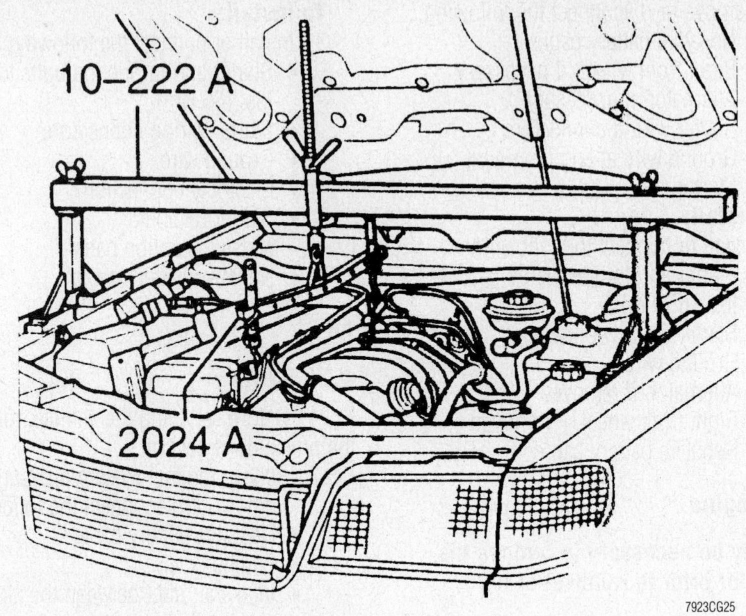

The engine must be supported, because the subframe mounting bolts must be loosened—1.8L engine

If equipped with a manual transaxle, align the flywheel as shown to remove the rear oil pan bolts—1.8L engine

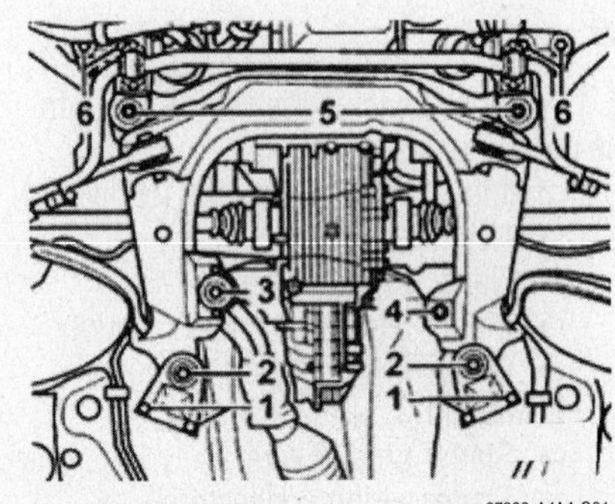

Subframe bolt tightening sequence—1.8L engine

- Nuts 3 and 4: 30 ft. lbs. (40 Nm)

17. Install or connect the following:
    - Subframe-to-transaxle. Torque the nuts to 17 ft. lbs. (23 Nm)
    - Engine mount-to-subframe. Torque the nuts to 18 ft. lbs. (25 Nm)
    - Turbocharger air hose
    - Starter wiring
    - Torque support stop and brace. Torque the fasteners to 18 ft. lbs. (25 Nm)
    - Air conditioning belt tensioner
    - Accessory drive belts
    - Negative battery cable
18. Fill the engine with oil and check the level.
19. Start the vehicle and check for leaks, then recheck the engine oil level.
20. Install or connect the following:
    - Engine cover
    - Undercover

TT

1.8L ENGINE

1. Before servicing the vehicle, refer to the precautions section.
2. Disconnect the battery negative cable.
3. Drain engine oil
4. Remove or disconnect the following:
    - Engine undercover
    - Oil return line
    - Oil pan

To install:

5. Apply silicone sealant to oil pan sealing surface
6. Install or connect the following:
    - Oil pan
    - Tighten the oil pan bolts to engine diagonally to 84 inch lbs., (10 Nm)
    - Tighten the oil pan bolts to transmission to 33 ft. lbs. (45 Nm)
    - Oil return line
    - Engine undercover
    - Negative battery cable

2.7L Engine

1. Before servicing the vehicle, refer to the precautions section.
2. Properly relieve the fuel system pressure.
3. On A6 models, place the hood lock carrier assembly in the service position.
4. Disconnect the battery negative cable.
5. Drain engine oil
6. Remove or disconnect the following:
    - Engine undercover
    - Accessory drive belt
    - Air conditioning lines from oil pan
    - Air conditioning compressor lines
    - Oil pan

To install:

7. Install or connect the following:
    - New oil pan gasket
    - Oil pan
    - Tighten the oil pan bolts to 84 inch lbs. (10 Nm) working from the center outward
    - Air conditioning compressor lines
    - Air conditioning lines
    - Accessory drive belt
    - Engine undercover
    - Install lock carrier in normal position
    - Negative battery cable

2.8L Engine

1. Before servicing the vehicle, refer to the precautions section.
2. Drain the engine oil.
3. Remove or disconnect the following:
    - Oil pan bolts
    - Oil pan from the engine

To install:

4. Be sure the gasket surface is flat.
5. Install or connect the following:
    - New oil pan gasket
    - Oil pan. Torque the bolts in a crisscross pattern to 11 ft. lbs. (15 Nm)
6. Fill the engine with oil.
7. Start the engine, check for leaks and repair if necessary.

3.0L Engine

1. Before servicing the vehicle, refer to the precautions section.
2. For vehicles with auxiliary heater, remove bolts for exhaust pipe of parking heater/auxiliary heater at engine undercover.
3. Properly relieve the fuel system pressure.
4. Disconnect the battery negative cable.
5. Drain engine oil
6. Remove or disconnect the following:
    - Engine undercover
    - Accessory drive belt
    - Bracket for automatic transmission fluid lines
    - Bracket for air conditioning lines
    - Connector at oil level sensor
    - Oil pan

To install:

7. Apply silicone sealant to oil pan sealing surface
8. Install or connect the following:
    - Oil pan
    - Tighten the oil pan bolts diagonally to 84 inch lbs., (10 Nm)
    - Air conditioning lines bracket
    - Automatic transmission fluid lines bracket
    - Accessory drive belt
    - Engine undercover
    - Negative battery cable

3.2L Engine, Except A3 And A6

1. Before servicing the vehicle, refer to the precautions section.
2. Disconnect the battery negative cable.
3. Drain engine oil
4. Remove or disconnect the following:
    - Center and right sound insulation
    - Oil thermal sensor
    - Secondary air injection pump with bracket
    - Oil pan

To install:

5. Installation is the reverse of the removal procedure, while using the following torque values:
    - Tighten the oil pan bolts in diagonal sequence initially to 44 inch lbs. (5 Nm)
    - Tighten bolts for sump/transmission to 30 ft. lbs. (40 Nm) Use new bolts
    - Tighten the oil pan bolts in diagonal sequence to 106 inch lbs. (12 Nm)

A6 and A8

3.2L ENGINE

1. Before servicing the vehicle, refer to the precautions section.
2. Disconnect the battery negative cable.
3. Remove the engine undercover sound insulation.
4. Drain engine oil
5. Remove or disconnect the following:
    - Oil thermal sensor
    - Oil pan

To install:

6. Installation is the reverse of the removal procedure, while using the following torque values:
    - Tighten the oil pan bolts in diagonal sequence initially to 9 Nm
    - Tighten the oil pan drain plug to 30 Nm.

4.2L ENGINE

1. Before servicing the vehicle, refer to the precautions section.
2. Remove the engine from the vehicle and mount it on a suitable workstand.
3. Remove or disconnect the following:
    - Lower oil pan
    - Upper oil pan
    - Honeycomb baffle

To install:

4. Installation is the reverse of the removal procedure, while using the following torque values:

- Upper oil pan bolts: 89 inch lbs. (10 Nm)
- Lower oil pan bolts: 89 inch lbs. (10 Nm)

A3

2.0L ENGINE

1. Before servicing the vehicle, refer to the precautions section.
2. Disconnect the negative battery cable.
3. Disconnect the electrical connector.
4. Disconnect the intake connection using spring-type clip pliers.
5. Remove the engine cover upward and out in direction of.
6. Remove the center sound insulation fasteners.
7. Remove the right sound insulation fasteners.
8. Remove the air duct from charge air cooler.
9. Unfasten the air duct from turbocharger.
10. Remove the bolts and remove air duct.
11. Remove the air duct for charge air cooler.
12. Disconnect the charge air pressure sensor connector.
13. Remove the air duct.
14. Disconnect the oil Level thermal sensor connector.
15. Remove the oil return line from turbocharger.
16. Drain the engine oil.
17. Remove the bolts for oil pan/transmission.
18. Loosen the bolts in a diagonal sequence.
19. Remove the oil pan, if necessary loosen by applying light strikes with a rubber mallet.

To install:

➡The oil pan must be installed within 5 minutes after applying the silicone sealant.

20. Clean the oil pan mating surfaces.
21. Apply 2 to 3mm bead of silicone sealant as illustrated to the sealing surface of oil pan.

➡The sealant bead may not be thicker than specified, otherwise excess sealant could enter the oil pan and clog the oil intake tube.

22. Apply a bead of sealant at the rear area of the sealing flange as shown in the illustration.

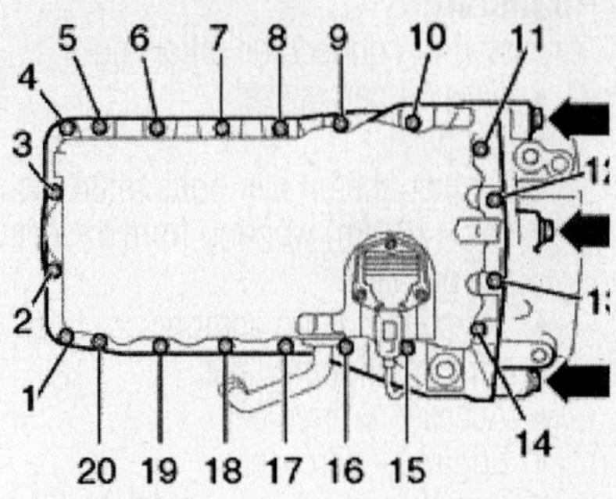

Oil pan bolt loosening/tightening sequence—A3 with 2.0L engine

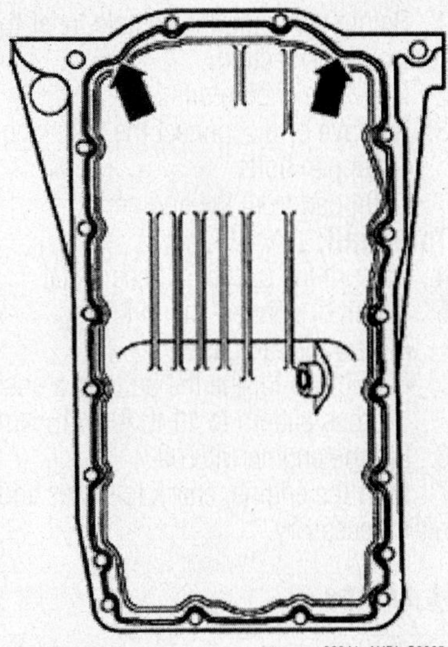

Apply 2 to 3mm bead of silicone sealant to the sealing surface of oil pan —A3 with 2.0L engine

23. Install the oil pan immediately and tighten bolts in sequence as follows:
    a. Step 1: Bolts 1 through 20 to 5 Nm.
    b. Step 2: Oil pan to transmission bolts to 40 Nm.
    c. Step 3: Bolts 1 through 20 to 15 Nm.

➡Make sure that the oil pan is positioned flush with the cylinder block on the flywheel side.

➡After installing oil pan, allow sealant to dry for about 30 minutes, then fill the crankcase with oil.

24. Check and adjust the oil level.
25. Install the remaining components in the reverse order of removal.

3.2L ENGINE

1. Before servicing the vehicle, refer to the precautions section.
2. Disconnect the negative battery cable.
3. Disconnect the secondary air injection hose.
4. Remove the center noise insulation fasteners.
5. Remove the right noise insulation fasteners.
6. Drain the engine oil.
7. Disconnect the lower coolant hose at fan shroud.
8. Disconnect the electrical connector at Secondary Air Injection (AIR) pump.
9. Remove the coolant pipe bracket to transmission oil cooler.
10. Remove the bolts.
11. Loosen the bolt and remove Secondary Air Injection (AIR) pump with bracket.
12. Disconnect the oil level thermal sensor connector.
13. Remove the bolts for oil pan/transmission.
14. Loosen the oil pan bolts diagonally and remove.
15. Remove the oil pan, and if necessary loosen by applying light strikes with a rubber mallet.

To install:

➡The oil pan must be installed within 5 minutes after applying the silicone sealant.

16. Clean the oil pan mating surfaces.
17. Apply 2 to 3mm bead of silicone sealant as illustrated to the sealing surface of oil pan.

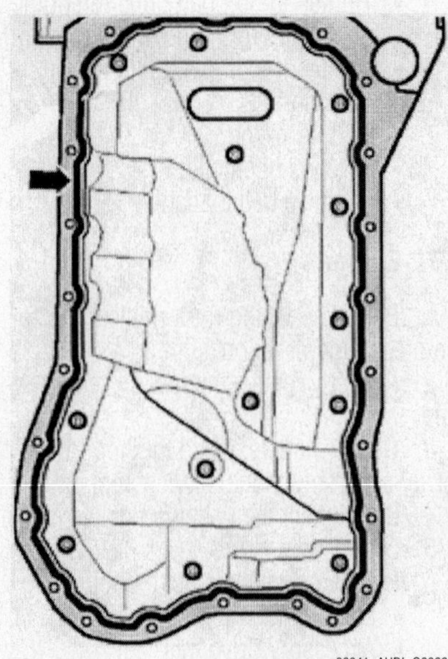

Apply 2 to 3mm bead of silicone sealant to the sealing surface of oil pan—A3 with 3.2L engine

Oil pan bolt location—A3 with 3.2L engine

➡The sealant bead may not be thicker than specified, otherwise excess sealant could enter the oil pan and clog the oil intake tube.

18. Install the oil pan immediately and tighten bolts in sequence as follows:
    a. Step 1: Oil pan bolts to 5 Nm.
    b. Step 2: New oil pan to transmission bolts to 40 Nm.
    c. Step 3: Oil pan bolts to 12 Nm.

➡Make sure that the oil pan is positioned flush with the cylinder block on the flywheel side.

➡After installing oil pan, allow sealant to dry for about 30 minutes, then fill the crankcase with oil.

19. Check and adjust the oil level.
20. Install the remaining components in the reverse order of removal.

Oil Pump

REMOVAL & INSTALLATION

A4

1.8L ENGINE

1. Before servicing the vehicle, refer to the precautions section.
2. Drain the engine oil.
3. Remove or disconnect the following:
    - Oil pan
    - Baffle plate
    - Oil pump-to-engine bolts
    - Oil pump by pressing down on the subframe

To install:

4. Install or connect the following:
    - Oil pump by pressing down on the subframe. Torque the Oil pump-to-engine bolts to 18 ft. lbs. (25 Nm)
    - Baffle plate
    - Oil pan
5. Fill the engine with clean oil.
6. Start the vehicle, check for leaks and repair if necessary.

TT

1.8L ENGINE

1. Before servicing the vehicle, refer to the precautions section.
2. Drain the engine oil.
3. Remove or disconnect the following:
    - Oil pan
    - Oil pump
    - Chain sprocket

To install:

4. Install or connect the following:
    - Dowel sleeves at top of oil pump
    - Chain sprocket. Torque the sprocket bolts to 11 ft. lbs. (15 Nm).
    - Oil pump. Torque the Oil pump-to-engine bolts to 18 ft. lbs. (25 Nm).
    - Oil pan
5. Fill the engine with clean oil.
6. Start the vehicle, check for leaks and repair if necessary.

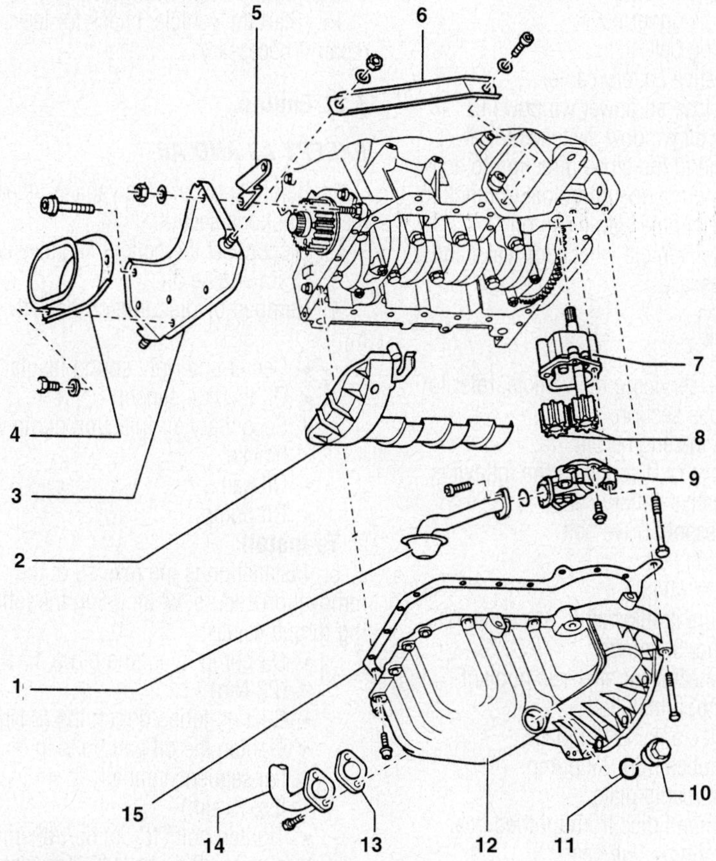

Exploded view of the oil pan and pump—1.8L engine A4

2.7L Engine

1. Before servicing the vehicle, refer to the precautions section.
2. Drain the engine oil.
3. Remove or disconnect the following:
   - Negative battery cable
   - Engine
   - Accessory drive belt
   - Timing belt
   - Engine oil dipstick tube
   - Lower oil pan
   - Oil supply lines
   - Hex head bolts retaining upper oil pan
   - Upper oil pan
   - Oil pump
   - Chain sprocket from oil pump

To install:

4. Clean the gasket mating surfaces.
5. Install or connect the following:
   - Chain sprocket. Torque the bolts to 18 ft. lbs. (25 Nm).
   - Oil Pump. Torque the bolts to 18 ft. lbs. (25 Nm).
   - Upper oil pan. Torque the hex head bolts to 15 ft. lbs. (20 Nm).
   - Oil supply lines
   - Lower oil pan. Torque the bolts to 84 inch lbs. (10 Nm).
   - Engine oil dipstick tube
   - Timing belt
   - Accessory drive belt
   - Engine
   - Negative battery cable
6. Fill the engine with clean oil.
7. Start the vehicle, check for leaks and repair if necessary.

2.8L Engine

The oil pump is part of the engine front cover, and the cooling system does not have to be opened during this procedure.

1. Before servicing the vehicle, refer to the precautions section.
2. Drain the engine oil.
3. Remove or disconnect the following:
   - Negative battery cable
   - Timing belt
   - Engine under cover
   - Engine oil dipstick tube
   - Engine oil tube from the crankcase
   - Crankshaft vibration damper and sprocket
   - Timing belt idler and tensioner pulleys
   - Any wiring, hoses, lines and cables that interfere with oil pump removal
   - Oil filter
   - Oil cooler line from the filter housing
   - Engine support front bolts

⁂ CAUTION

Be prepared for the engine support to drop 3/8 inch.

   - Starter bolts
   - Upper and lower oil pans
   - Oil suction tube
   - Oil pump

To install:

4. Clean the gasket mating surfaces.
5. Install or connect the following:
   - Oil pump and verify the oil pump drive is properly engaged
   - Upper and lower oil pans
   - Starter bolts
   - Engine support bolts
   - Oil cooler line to the oil filter housing
   - New oil filter
   - Wiring, hoses, lines and cables that were removed
   - Timing belt idler pulley. Torque the bolt to 15 ft. lbs. (20 Nm)
   - Tensioner pulleys. Tighten after adjusting the timing belt.
   - Crankshaft sprocket and vibration damper. Torque the center bolt to 148 ft. lbs. (200 Nm) plus an additional ½ (180 degree) turn
   - Dipstick tube and dipstick
   - Engine under cover
   - Timing belt
   - Negative battery cable
6. Fully close all power windows to stop, operate all window switches for at least 1 second in the close direction to activate the one-touch opening/closing function
7. Fill the engine with clean oil.
8. Start the vehicle, check for leaks and repair if necessary.

3.0L Engine

1. Before servicing the vehicle, refer to the precautions section.
2. Drain the engine oil.
3. Remove or disconnect the following:
   - Negative battery cable
   - Accessory drive belt
   - Timing belt
   - Lower oil pan
   - Ensure crankshaft is placed at TDC on no. 3 cylinder
   - Connector for after-run coolant pump, if installed
   - Lock crankshaft in place
   - Oil tubes from oil pump
   - Front baffle plate
   - Diamond disc from toothed gear
   - Oil pump drive gear
   - Oil pump

To install:

4. Clean the gasket mating surfaces.
5. Replace all seals and gaskets and O-rings
6. Coat oil pump mounting nuts with locking fluid. Use new bolts on oil pump.
7. Install oil pump uniformly at all mounting points, so that shear pin does not cant. Torque oil pump nuts to 15 ft. lbs. (20 Nm), plus 120 degrees, torque bolts to 16 ft. lbs. (22 Nm).
8. Install diamond disc on toothed gear.
9. When installing drive gear, press piston of chain tensioner back using a screwdriver.
10. Turn the drive gear until a pin 5 mm pin can be slid into alignment hole at bottom side of oil pump.
11. Tighten drive gear bolt to 15 ft. lbs. (20 Nm), plus 90 degrees.
12. Remove 5 mm pin.
13. Install or connect the following:
    - Front baffle plate. Torque bolts to 84 inch lbs. (10 Nm).
    - Oil tubes. Torque bolts to 84 inch lbs. (10 Nm).
    - Lower oil pan
    - Timing belt
    - Accessory drive belt
    - Negative battery cable
14. Fill the engine with clean oil.
15. Start the vehicle, check for leaks and repair if necessary.

3.2L Engine

EXCEPT A3 AND A6

1. Before servicing the vehicle, refer to the precautions section.
2. Disconnect the battery negative cable.
3. Drain engine oil
4. Remove or disconnect the following:
   - Center and right sound insulation
   - Oil thermal sensor
   - Secondary air injection pump with bracket
   - Oil pan
   - Oil pump

To install:

5. Installation is the reverse of the removal procedure, while using the following torque values:
   - Oil pump to engine block 17 ft. lbs (23 Nm)
   - Suction tube 70 inch lbs (8 Nm)
   - Tighten the oil pan bolts in diagonal sequence initially to 44 inch lbs. (5 Nm)
   - Tighten bolts for sump/transmission to 30 ft. lbs. (40 Nm) Use new bolts

- Tighten the oil pan bolts in diagonal sequence to 106 inch lbs. (12 Nm)

A6

1. Before servicing the vehicle, refer to the precautions section.
2. Drain the engine oil.
3. Remove the oil pan.
4. Remove the pump and pickup tube bolts and remove the pump assembly.

To install:

5. Installation is the reverse of removal, tighten the oil pump bolts to 20 Nm.

4.2L Engine

1. Before servicing the vehicle, refer to the precautions section.
2. Drain the engine oil.
3. Remove or disconnect the following:
   - Engine oil dipstick
   - Lower oil pan
   - Oil pump

To install:

4. Installation is the reverse of the removal procedure, while using the following torque values:
   - Oil pump bolts: 18 ft. lbs. (25 Nm)
   - Lower oil pan bolts: 89 inch lbs. (10 Nm)

A3

2.0L ENGINE

1. Before servicing the vehicle, refer to the precautions section.
2. Remove the oil pan

➡Only rotate the engine at the crankshaft in a clockwise direction of rotation.

3. Set camshaft gear to marking for TDC by turning crankshaft. The marking on the camshaft sprocket must align with marks on the timing belt cover.
4. Pull off the chain guard. If necessary, release retaining tabs through openings with a small screwdriver.
5. Loosen the bolt from oil pump chain sprocket while holding the central bolt for the vibration damper.
6. Relieve tension on chain rail with a screwdriver and secure it with a 3mm Allen wrench.
7. Remove the oil pump chain sprocket and unhook chain at balance shaft drive.

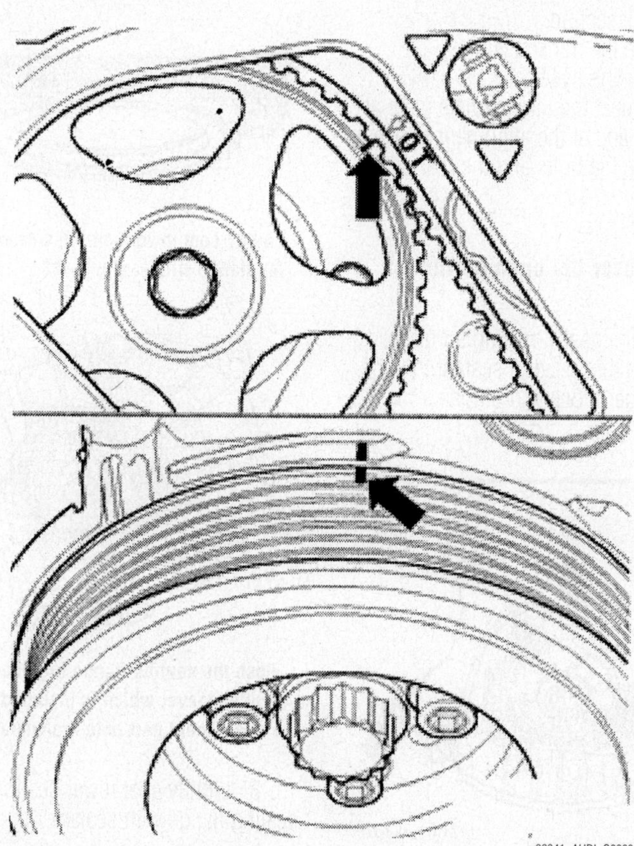

Oil pan bolt location Set camshaft gear to marking for TDC by turning crankshaft. The marking on the camshaft sprocket must align with marks on the timing belt cover—A3 with 2.0L engine

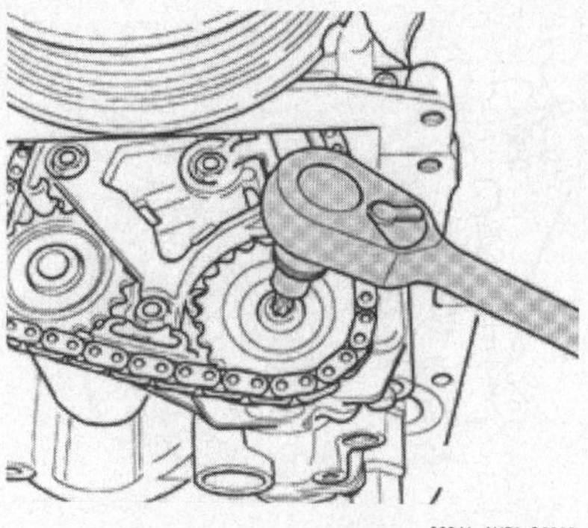

Loosen the bolt from oil pump chain sprocket while holding the central bolt for the vibration damper—A3 with 2.0L engine

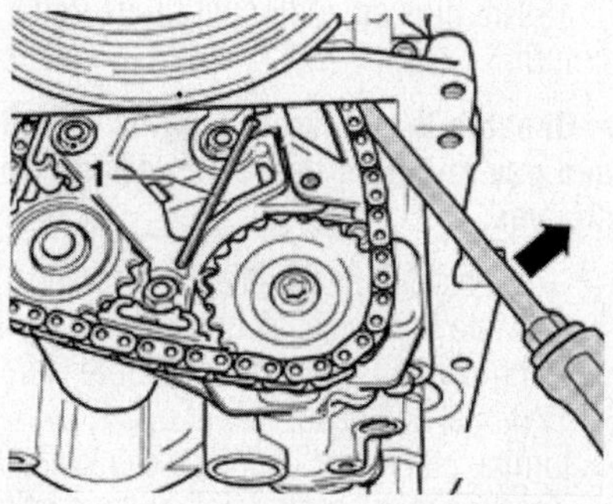

Relieve tension on chain rail with a screwdriver and secure it with an 3mm Allen wrench—A3 with 2.0L engine

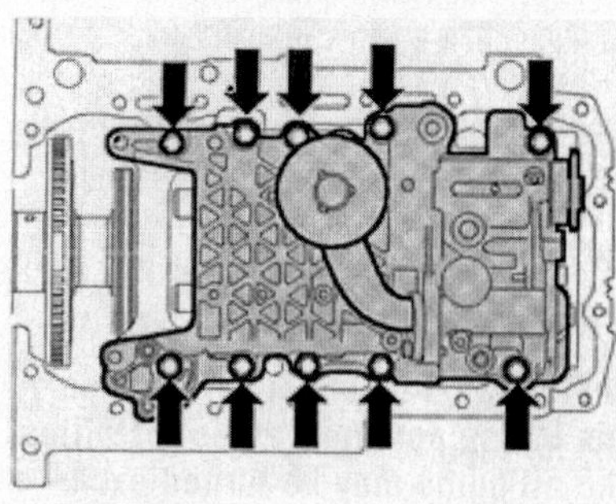

Loosen the bolts from balance shaft housing from outside to inside and remove it—A3 with 2.0L engine

8. Remove the oil baffle.
9. Loosen the bolts from balance shaft housing from outside to inside and remove it.

To install:

10. Installation is the reverse of removal, please note the following:
    a. Replace all bolts of balance shaft housing
    b. Install the balance shaft housing with oil pump and intermediate plate. Fasten bolts from inside to outside to

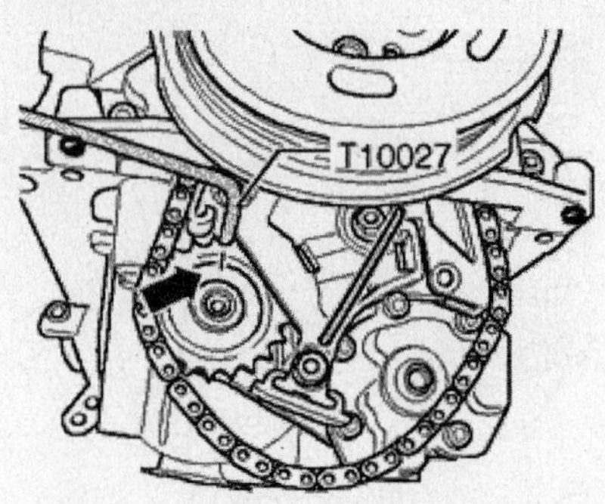

Turn the crankshaft clockwise the marking on camshaft sprocket and marking on crankshaft must be positioned at TDC for cyl. 1—A3 with 2.0L engine

15 Nm plus an additional 90 degree turn.

➡Observe the different lengths of bolts and pay attention to the centering sleeves.

c. Install the oil battle, by inserting tabs into balance shaft housing and tightening bolt to 21 Nm and bolts to 9 Nm.

d. Set engine to TDC at cyl. 1, by turning crankshaft in engine clockwise direction at the central bolt of the crankshaft the timing belt sprocket. Marking on camshaft sprocket and marking on crankshaft must be positioned at TDC for cyl. 1.

e. Position the marking on the balance shaft chain sprocket across from locating hole. Secure the chain sprocket in this position using connecting pin T10027. Place the chain onto chain sprocket of balance shaft.

f. Install the oil pump chain sprocket and tighten bolt by hand.

➡The oil pump chain sprocket only fits in one position. When installing, the oil pump may be turned exclusively.

g. Remove the connecting pin T10027 and Allen wrench. Fasten the oil pump chain sprocket. To 20 Nm plus and additional 90 degrees.

3.2L ENGINE

1. Before servicing the vehicle, refer to the precautions section.
2. Drain the engine oil.
3. Remove the oil pan.
4. Remove the pump and pickup tube bolts and remove the pump assembly.

To install:

5. Installation is the reverse of removal, tighten the oil pump bolts to 23 Nm.

Rear Main Seal

REMOVAL & INSTALLATION

Except A3 and A8

1. Before servicing the vehicle, refer to the precautions section.
2. Remove or disconnect the following:
   - Negative battery cable
   - Transaxle
   - Oil pan if needed to access seal
   - Flywheel/flexplate assembly
   - Oil seal by prying it out of the housing.

To install:

3. Install or connect the following:
   - New oil seal by coating it with engine oil and press it into place

➡Be careful not to damage the seal or score the crankshaft.

   - Flywheel/flexplate
   - Oil pan if removed
   - Transaxle
   - Negative battery cable

A3

1. Before servicing the vehicle, refer to the precautions section.
2. Remove the transmission.
3. Remove the flywheel.
4. Disconnect the intermediate plate at sealing flange and at the alignment sleeves.
5. Remove the bolts and the rear sealing flange.

To install:

➡Lay a rag over the open portion of the oil pan.

6. Clean the gasket mating surfaces.
7. Apply a thin bead of sealant to edge between cylinder block and oil pan.

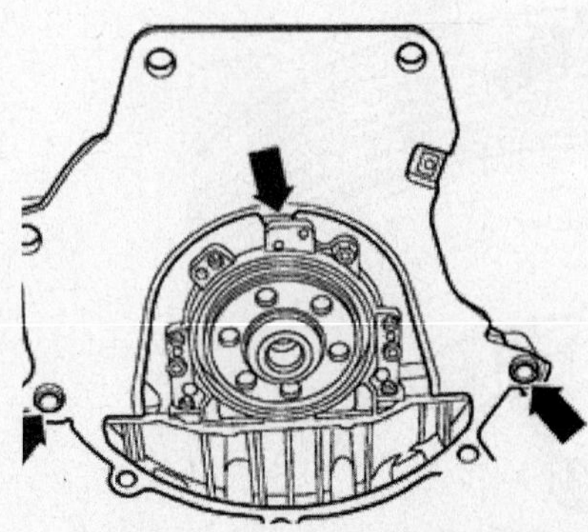

Disconnect the intermediate plate at sealing flange and at the alignment sleeves—A3 engine

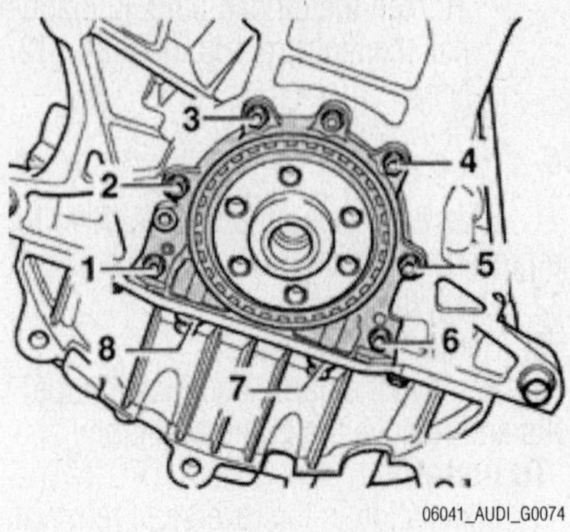

Rear sealing flange bolt location—A3

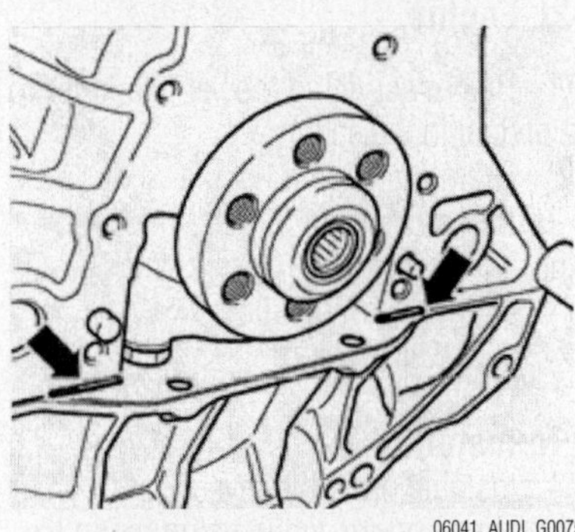

Apply a thin bead of sealant to edge between cylinder block and oil pan—A3

Lightly coat lower sealing surface of sealing flange with sealant—A3

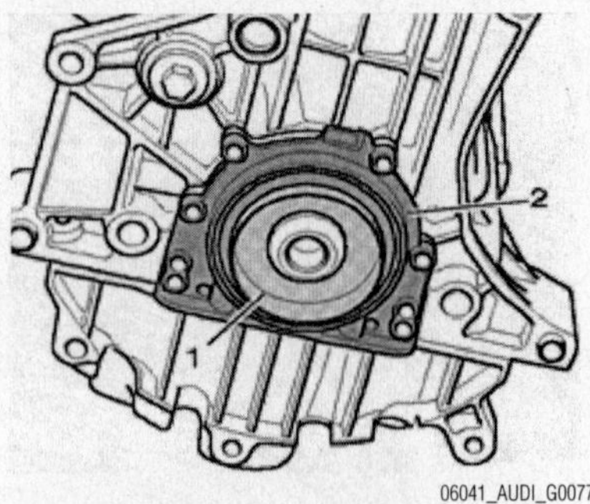

Push the sealing flange together with the guide sleeve, which is included with replacement part onto crankshaft—A3

8. Lightly coat lower sealing surface of sealing flange with sealant.

➡The sealing flange must be installed within 5 minutes of sealant being applied.

9. Push the sealing flange together with the guide sleeve, which is included with replacement part onto crankshaft.

10. Carefully push the sealing flange onto the locating pins on cylinder block.

11. Fasten the bolts to 15 Nm.

12. Install the remaining components in the reverse order of removal.

A8

1. Before servicing the vehicle, refer to the precautions section.

2. Remove the transaxle from the vehicle.

3. Remove the flexplate from the crankshaft.

4. Use extraction hook 10-221 and spacer, approx. 6 mm (1/4 in.) thick, and pry out oil seal.

To install:

5. Using tool 2003 and drive plate mounting bolts, press in oil seal until fully seated.

6. Install the flexplate. Tighten the bolts to 44 ft. lbs. Plus 90 degrees.

7. Install the transaxle to the vehicle.

8. Start the engine and check for leaks.

Timing Belt, Front Cover and Seal

REMOVAL & INSTALLATION

1.8L Engine

A4

1. Disconnect the negative battery cable.

2. Raise and safely support the vehicle.

3. From under the vehicle, remove the splash shield.

4. Remove the front bumper.

5. Remove the intake air duct between the grille/front end assembly and the air cleaner housing.

6. Remove the grille/front end assembly-to-chassis bolts and the grille/front end assembly-to-vehicle fasteners.

7. If installed, remove the wiring harness retaining clamps from the left side of the radiator frame.

8. Install Support tools 3369 bolts into the grille/front end assembly-to-chassis holes; then, pull the grille/front end assembly forward until it hits the stops.

➡If necessary to secure the grille/front end assembly, install M6 bolts into the rear bored holes of the grille/front end assembly and the fender.

9. Loosen the A/C compressor's serpentine belt tensioner bolts; release the belt tension and remove the belt.

10. Place an open-end wrench on the alternator belt tensioner and rotate it clockwise toward the alternator to release the belt's tension. Remove the alternator serpentine drive belt and release the tensioner.

➡If necessary to lock the alternator tensioner in position, align the housing holes and insert an Allen wrench into the holes to secure its movement.

11. Using a 5 x 60mm bolt, secure the viscous fan pulley. Using a hex wrench, remove the viscous fan-to-pulley bolts. Remove the viscous fan assembly.

12. Remove the upper timing belt cover.

➡If reusing the timing belt, mark its rotational direction so it may be installed in its original position.

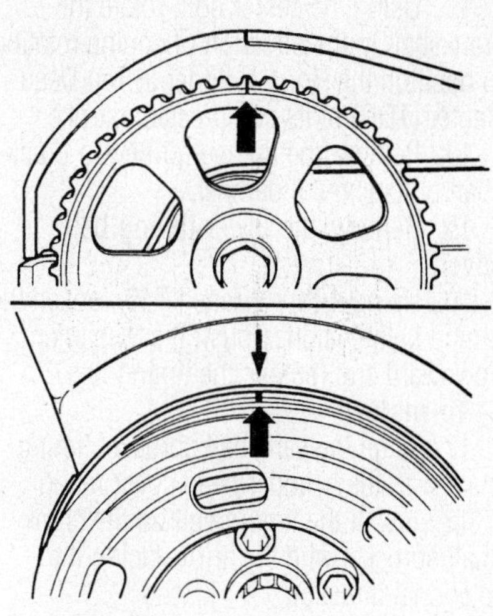

Crankshaft pulley and camshaft sprocket alignment locations—Audi A4 1.8L 4-Cyl engine

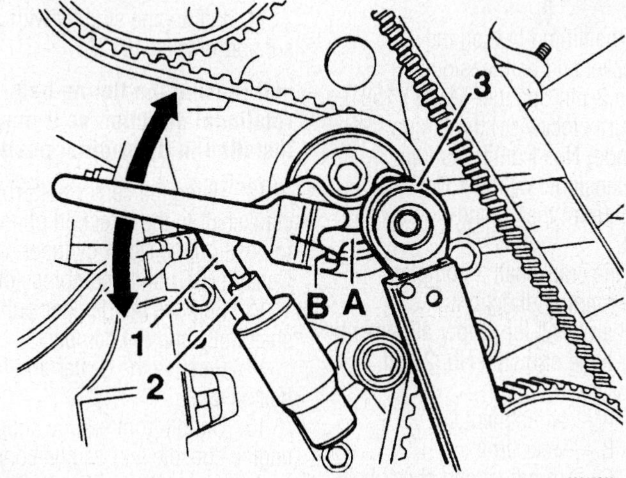

Timing belt tension adjustment—Audi A4 1.8L 4-Cyl engine

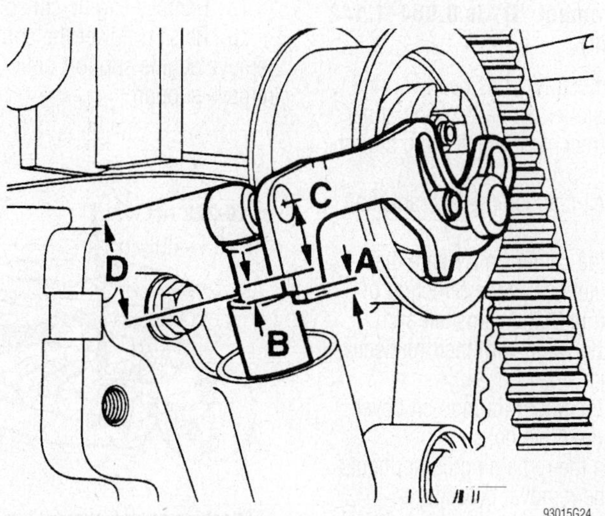

Timing belt tension wear limits—Audi A4 1.8L 4-Cyl engine

13. Using the center bolt, rotate the crankshaft in the direction of engine rotation to position the No. 1 cylinder at Top Dead Center (TDC) of its compression stroke.

14. Remove the damper pulley-to-crankshaft bolts and the damper.

15. Remove the lower timing belt cover.

16. Using a Torx Wrench T45, loosen the timing belt tensioner, push the tensioner downward and remove the timing belt.

To install:

17. Align the camshaft sprocket timing mark with the cylinder head cover mark.

18. Install the timing belt on the crankshaft sprocket with the arrow facing the rotational direction.

19. Install the lower timing belt cover.

20. Using a bolt, secure the damper/belt pulley on the crankshaft.

21. Align the crankshaft damper/belt pulley with the housing timing mark so that the No. 1 cylinder is at TDC of its compression stroke.

22. Install the timing belt on the camshaft sprocket and belt tensioner.

23. Using a 2-pin Spanner Matra V159 Wrench, lift (turn clockwise) the timing belt tensioner cylinder No. 1 until it is fully extended and tensioner cylinder No. 2 is raised approx. 1mm; then, hand-tighten the mounting bolt.

24. Rotate the crankshaft 2 complete rotation in the running direction.

25. Inspect area "A" for proper alignment with the upper edge of piston No. 2 and adjust if necessary.

- Area "A"—adjustment OK
- Area "B"—wear limit
- Area "C"—re-adjust and check belt drive including tensioner for wear.

➡If the piston edge is located in area "A", measurement "D" is 0.984–1.142 in. (25–29mm).

26. After adjustment has been verified, secure the tensioner with a 2-pin Spanner Matra V159 Wrench and tighten the mounting bolt.

27. Complete the damper to crankshaft installation.

28. Using the center bolt, rotate the crankshaft 2 rotations in the direction of engine rotation until the camshaft and crankshaft marks align with their respective reference points.

29. Install the upper timing belt cover.

30. Install the drive belts.

31. Replace the remaining components by reversing the removal procedures.

32. Install the negative battery cable last.

33. Test drive the vehicle.

TT

1. Disconnect the negative battery cable.

2. From under the vehicle, remove the splash shield and right side noise insulation panel.

3. Remove the intake air duct at the bottom of the right cross member.

4. Remove the accessory drive belt and belt tensioner.

5. Remove coolant expansion tank and power steering reservoir with hoses attached and place to one side.

6. Disconnect throttle valve and charcoal canister vacuum hoses.

7. Disconnect electrical connectors from coolant tank and charcoal canister.

8. Remove the upper timing belt cover.

9. Install Engine Support bracket 10-222A and supports 10-222A1.

10. Install Retainer 3180 to right engine lifting eye and attach it to support bracket 10-222A.

11. Lift engine slightly with spindle of support bracket 1022A.

➡If reusing the timing belt, mark its rotational direction so it may be installed in its original position.

12. Using the center bolt, rotate the crankshaft in the direction of engine rotation to position the No. 1 cylinder at Top Dead Center (TDC) of its compression stroke.

13. Remove the damper pulley-to-crankshaft bolts and the damper.

14. Remove the center and lower timing belt covers.

15. Unbolt front engine support from engine console and engine console from body.

16. Disconnect connector between engine console and engine support.

17. Remove engine console.

18. Raise or lower the engine enough to remove engine support bolts. Do not remove support.

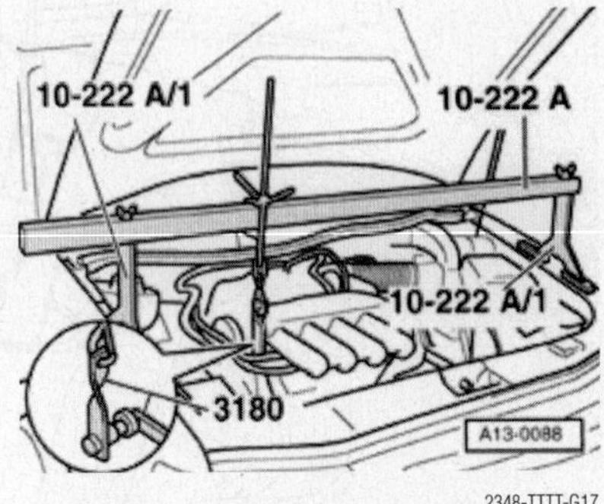

Attaching engine support and lifting tools—Audi TT 1.8L 4-Cyl engine

19. Using a Torx Wrench T45, loosen the timing belt tensioner, push the tensioner downward and remove timing belt.

To install:

20. Align the camshaft sprocket timing mark with the cylinder head cover mark.

21. Install the timing belt on the crankshaft sprocket with the arrow facing the rotational direction.

22. Install the lower timing belt cover. Torque the bolts to 84 inch lbs. (10 Nm).

23. Using a bolt, secure the damper/belt pulley on the crankshaft. Torque the bolt to 18 ft. lbs. (25 Nm).

24. Align the crankshaft damper/belt pulley with the housing timing mark so that the No. 1 cylinder is at TDC of its compression stroke.

25. Install the timing belt on the water pump, tensioning pulley and finally the camshaft sprocket.

26. Rotate the crankshaft 2 complete rotations in the running direction, and then set to TDC on no. 1 cylinder.

27. Check that timing belt is properly aligned.

28. Tighten engine support bolts to engine. Torque the bolts to 33 ft. lbs. (45 Nm).

29. Install engine console to body. Torque the bolts to 40 ft. lbs. (30 Nm) plus 90 degrees.

30. Install engine console to engine support. Torque the bolts to 44 ft. lbs. (60 Nm) plus 90 degrees.

31. Install the center and upper timing belt covers. Torque the bolts to 84 inch lbs. (10 Nm).

32. Replace the remaining components by reversing the removal procedures.

33. Install the negative battery cable last.

34. Test drive the vehicle.

2.7L Engine

1. Before servicing the vehicle, refer to the precautions section.

2. Disconnect negative battery cable .

3. Remove the front bumper.

4. Place the hood lock carrier in the service position.

5. Remove the serpentine drive belt by performing the following procedure:

6. Using Spanner Wrench No. 3212, secure the viscous fan pulley. Using an Open-end Spanner Wrench 3312, remove the viscous fan bearing housing by turning it clockwise.

➡The viscous fan is mounted with a left-handed thread; turn it clockwise to loosen it.

7. Place a 17mm box wrench on the serpentine drive belt tensioner and turn the tensioner clockwise until the 2 holes align; insert drift 3204 into the holes to secure the tensioner in place.
8. Mark the running direction of the serpentine drive belt and remove it from the pulleys.
9. Remove the top engine covers.
10. Remove the pressure hoses between charge air coolers and pressure lines.
11. Remove the pressure hoses located at the front of the engine.
12. Remove the timing belt tensioner.
13. Remove all three timing belt covers.
14. Rotate the crankshaft by hand to align the crankshaft pulley mark with the arrow on the engine housing and the large hole in each camshaft sprocket must face inward and must align; this should be Top Dead Center (TDC) of the No. 1 cylinder's compression stroke. If these conditions are not correct, rotate the crankshaft one complete revolution and realign.
15. On the left side of the cylinder block near the crankshaft, remove the sealing plug.
16. Insert Crankshaft Holder tool No. 3242 into the sealing plug hole to secure the crankshaft.
17. Remove the damper-to-crankshaft bolts and the damper.

➡It is not necessary to remove the center bolt when removing the crankshaft damper.

18. Remove the serpentine belt idler and the crankshaft damper guard.
19. Using a 8mm Allen® wrench, rotate the timing belt tensioner roller clockwise until the tensioner is compressed; then, insert a 2mm spring pin through the tensioner housing and tensioner plunger to secure it in place. When the plunger is secure, release the wrench tension.
20. Mark the running direction of the timing belt and remove it from the pulleys.

To install:

21. Insert camshaft clamp 3391 in the securing plates of both camshaft sprockets.
22. Loosen both camshaft bolts and remove approximately 5 turns.
23. Remove the camshaft clamp.
24. Pull off both camshaft sprockets with special tool T40001.
25. Reinstall both camshaft sprockets with securing plates and tighten hand-tight again.

➡The camshaft sprockets should be just tight enough on the camshaft tapers so that they can still be turned but do not move axially.

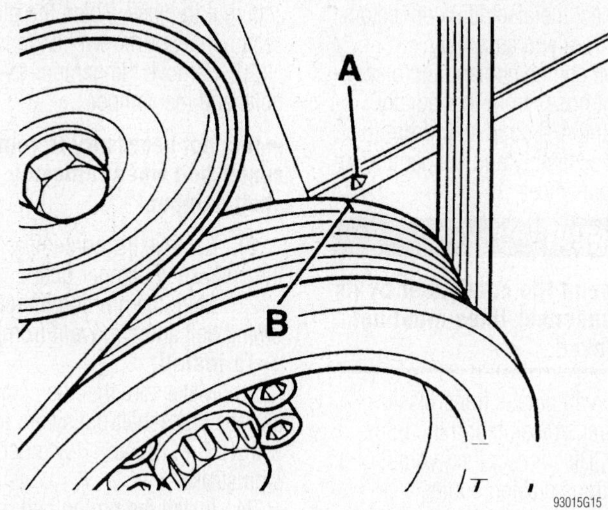

Crankshaft pulley alignment location for TDC—Audi 2.7L V6 engine

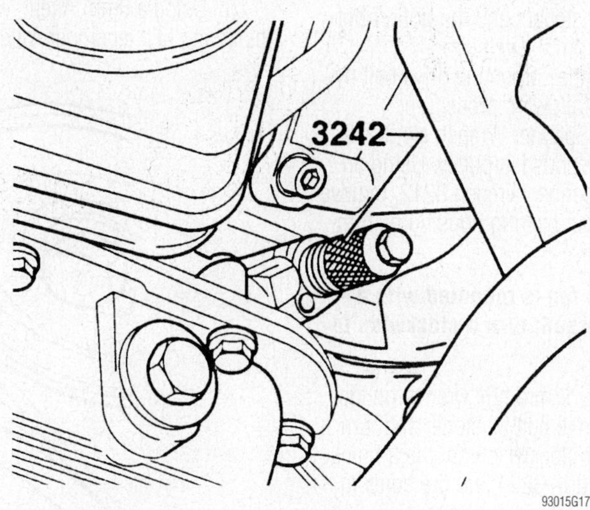

View of crankshaft holding tool installed—Audi 2.7L V6 engine

26. Install the timing belt; make sure the timing belt is installed in the correct running direction from which it was removed.
27. Reinstall the camshaft clamp.
28. Using a 8mm Allen® wrench, rotate the timing belt tensioner roller clockwise until the tensioner is compressed; then, remove the 2mm spring pin from the tensioner housing. Using a torque wrench, turn tensioner roller counterclockwise to 11 ft. lbs. (15 Nm) to put pressure on the timing belt.
29. Tighten he camshaft sprocket bolts. Torque the bolts to 41 ft. lbs. (55 Nm).
30. Install the crankshaft damper guard and the serpentine belt idler pulley; torque the idler pulley bolts to 33 ft. lbs. (45 Nm).
31. Install the crankshaft damper and torque the damper-to-crankshaft bolts to 15 ft. lbs. (20 Nm). If the damper-to-crankshaft center bolt was removed, torque it to 147 ft. lbs. (200 Nm) plus 180°.
32. Remove the Crankshaft Holder tool No. 3242 and install the sealing plug.
33. Replace the remaining components by reversing the removal procedures.
34. Connect the electrical connectors. Install the negative battery cable last.
35. Test drive the vehicle.

2.8L Engines

1. Disconnect the negative battery cable.
2. Remove the upper engine cover.
3. Raise and safely support the vehicle.
4. From under the vehicle, remove the splash shield.
5. Remove the front bumper.
6. Disengage the hood lock cable
7. Remove the intake air duct between the lock carrier and the air cleaner housing at the grille/front end assembly.
8. Remove the grille/front end assembly-to-chassis bolts.

9. Disconnect the electrical connectors from the grille/front end assembly.

10. Drain the engine coolant and disconnect the coolant hoses from the radiator.

11. Detach the A/C condenser from the grille/front end assembly and suspend it on a wire at the front wheel.

⁂ WARNING

DO NOT suspend the condenser by its lines. The condenser lines must not be bent or kinked.

12. Drain the automatic transmission fluid from the transmission and the transmission cooler. Disconnect the hydraulic lines from the transmission cooler.

13. If equipped, remove the charge air cooler.

14. Remove the grille/front end assembly-to-vehicle fasteners and the grille/front end assembly from the vehicle.

15. Remove the serpentine drive belt by performing the following procedure:

a. Using Spanner Wrench No. 3212, secure the viscous fan pulley. Using an Open-end Spanner Wrench 3212, remove the viscous fan bearing housing by turning it clockwise.

➡The viscous fan is mounted with a left-handed thread; turn it clockwise to loosen it.

b. Place a 17mm box wrench on the serpentine drive belt tensioner and turn the tensioner clockwise until the 2 holes align; insert drift 3204 into the holes to secure the tensioner in place.

c. Mark the running direction of the serpentine drive belt and remove it from the pulleys.

16. Rotate the crankshaft by hand to align the crankshaft pulley mark with the arrow on the engine housing and the large hole in each camshaft sprocket must face inward and must align; this should be Top Dead Center (TDC) of the No. 1 cylinder's compression stroke. If these conditions are not correct, rotate the crankshaft one complete revolution and realign.

17. On the left side of the cylinder block near the crankshaft, remove the sealing plug.

18. Insert Crankshaft Holder tool No. 3242 into the sealing plug hole to secure the crankshaft.

19. Using a 8mm Allen® wrench, rotate the timing belt tensioner roller clockwise until the tensioner is compressed; then, insert a 2mm spring pin through the tensioner housing and tensioner plunger to secure it in place. When the plunger is secure, release the wrench tension.

20. Remove the damper-to-crankshaft bolts and the damper.

➡It is not necessary to remove the center bolt when removing the crankshaft damper.

21. Remove the serpentine belt idler and the crankshaft damper guard.

22. Mark the running direction of the timing belt and remove it from the pulleys.

To install:

23. Make sure that the camshaft pulleys and the crankshaft pulley are in alignment with TDC of the No. 1 cylinder's compression stroke.

24. Install the timing belt; make sure the timing belt is installed in the correct running direction from which it was removed.

25. Using a 8mm Allen® wrench, rotate the timing belt tensioner roller clockwise until the tensioner is compressed; then, remove the 2mm spring pin from the tensioner housing. Slowly, release the tensioner's spring pressure to put pressure on the timing belt.

26. Install the crankshaft damper guard and the serpentine belt idler pulley; torque the idler pulley bolts to 33 ft. lbs. (45 Nm).

27. Install the crankshaft damper and torque the damper-to-crankshaft bolts to 15 ft. lbs. If the damper-to-crankshaft center bolt was removed, torque it to 147 ft. lbs. (200 Nm) plus 180° ½ turn).

28. Remove the Crankshaft Holder tool No. 3242 and install the sealing plug.

29. Replace the remaining components by reversing the removal procedures.

30. Refill the cooling system and the automatic transaxle. Connect the electrical connectors. Install the negative battery cable last.

31. Test drive the vehicle.

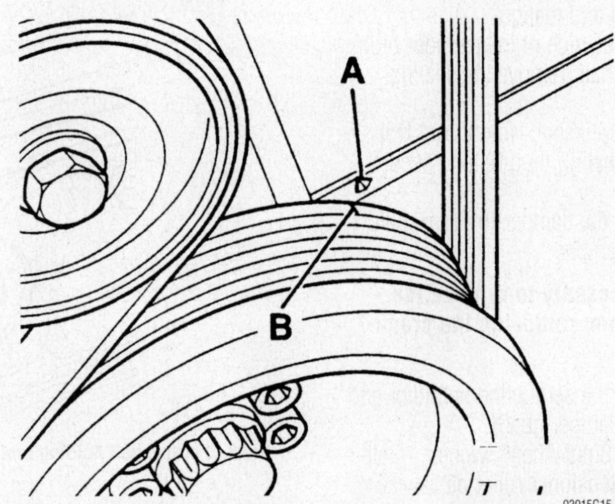

Crankshaft pulley alignment location for TDC—Audi 2.8L V6 engine

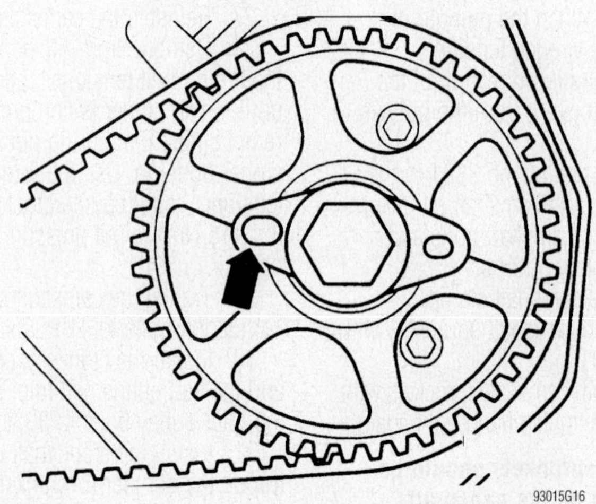

Left camshaft sprocket alignment position for TDC; right camshaft position is similar—Audi 2.8L V6 engine

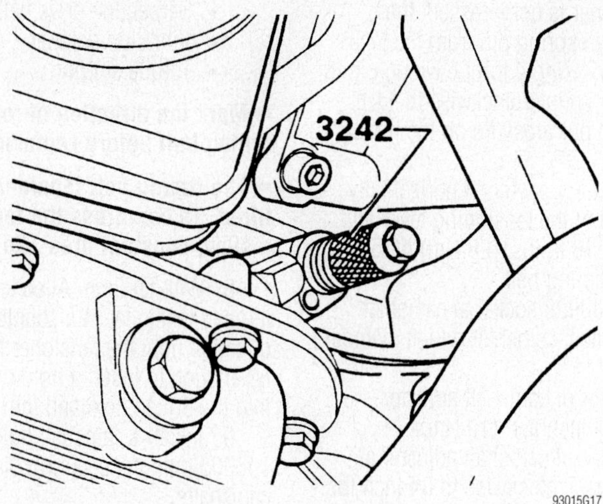

View of crankshaft holding tool installed—Audi 2.8L V6 engine

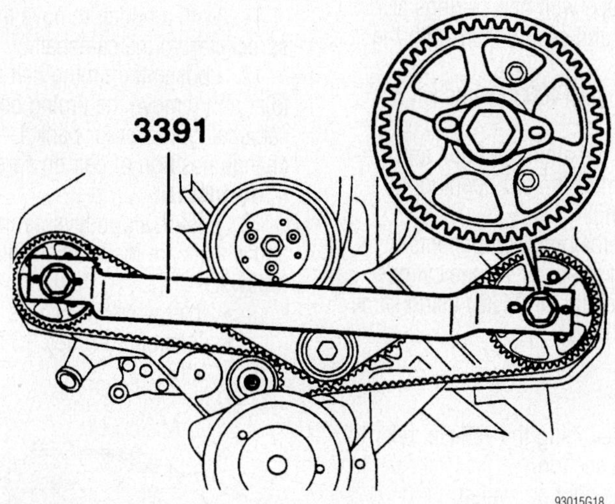

View of camshaft locator bar installed—Audi 2.8L V6 engine

3.0L Engine

1. Before servicing the vehicle, refer to the precautions section.
2. Disconnect negative battery cable.
3. Place the hood lock carrier in the service position.
4. Remove the serpentine drive belt by turning belt tensioner clockwise using tensioner tool 3299.
5. Mark the running direction of the serpentine drive belt and remove it from the pulleys.
6. Remove the serpentine belt tensioner.
7. Remove power steering pump pulley.
8. Remove the harmonic balancer and thrust washer.
9. Remove 12 bolts and the timing belt covers.
10. Remove the rear engine cover.
11. Remove the coolant expansion tank and set to one side with hoses attached.
12. Remove all necessary wiring connectors and hoses, and remove air filter housing.
13. Remove the oil dipstick guide tube.
14. Disconnect all ignition coil electrical connectors.

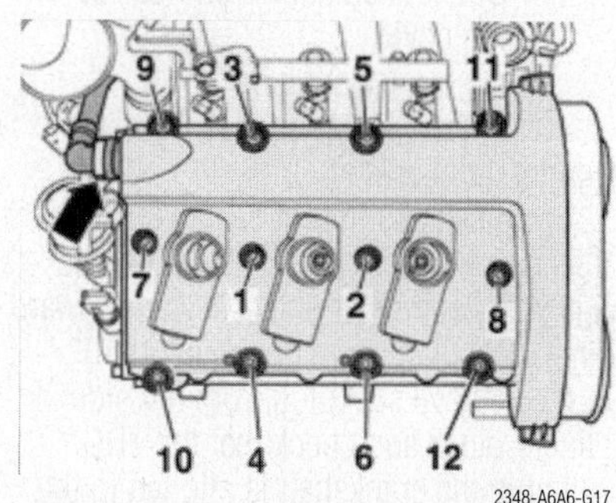

Cylinder head cover removal sequence—Audi 3.0L engine (same for both head covers)

15. Remove both cylinder head covers by removing screws in the sequence shown.
16. Turn crankshaft until camshaft lobes at cylinder no. 3 on right cylinder head point upward.
17. Install camshaft adjuster gauge T40030 on right cylinder head camshafts.
18. Spread adjuster gauge using wrench on threaded shaft until gauge is seated without end play.
19. Install camshaft adjuster gauge T4030 on left cylinder head camshafts.
20. Spread adjuster gauge using wrench on threaded shaft until gauge is seated without end play.

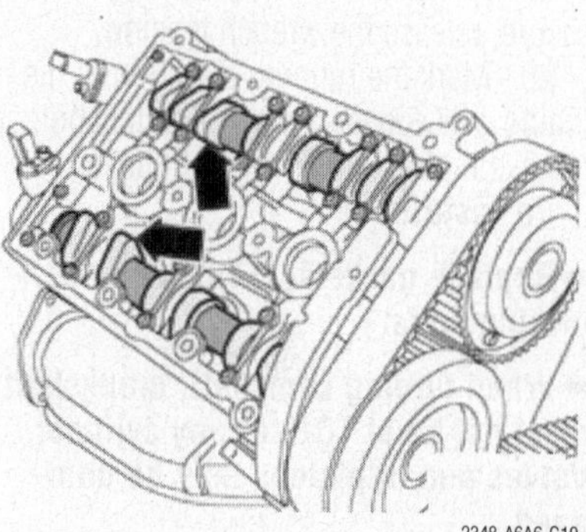

Right side camshaft lobes positioning—Audi 3.0L engine

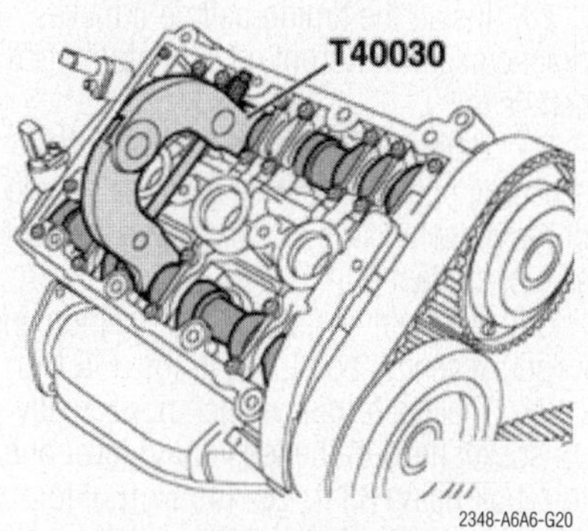

Installing camshaft adjuster gauge on right cylinder head—Audi 3.0L engine

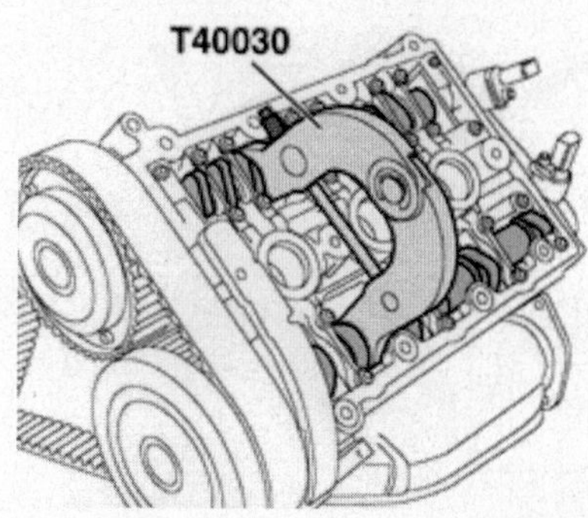

Installing camshaft adjuster gauge on left cylinder head—Audi 3.0L engine

21. On the left side of the cylinder block near the crankshaft, remove the sealing plug.

22. Insert Crankshaft Holder tool No. 3242 into the sealing plug hole to secure the crankshaft.

23. Remove securing ring from all camshaft gears.

24. Loosen, but do not remove all camshaft gear mounting bolts.

25. Using a 8mm Allen® wrench, rotate the timing belt tensioner roller clockwise until the tensioner is compressed; then, insert a 2mm spring pin through the tensioner housing and tensioner plunger to secure it in place. When the plunger is secure, release the wrench tension.

26. Mark the running direction of the timing belt, loosen eccentric pulley bolt and remove timing belt from the pulleys.

To install:

➡Replace gaskets, O-rings and self-locking bolts.

➡When turning camshaft, crankshaft must not be at TDC for any cylinder. Valves and/or pistons may be damaged.

27. Bolt in the camshaft gears 1 to 4 far enough so they can still just be turned without canting.

28. Install the timing belt so it makes exact contact with front edge of all toothed belt gears.

29. Using 3387 pin wrench, turn the eccentric pulley clockwise until the handle of the pin wrench is exactly above the center axle of coolant pump gear.

30. Hold pin wrench in this position and torque eccentric bolt to 33 ft. lbs. (45 Nm).

31. Place a torque wrench horizontally on socket head of the tensioning lever and turn bolt using 33 ft. lbs. (45 Nm) of torque, to pretension the timing belt.

32. Using a 8mm Allen® wrench, rotate the timing belt tensioner roller clockwise until the tensioner is compressed; then, remove the 2mm spring pin from the tensioner housing. Using a torque wrench, turn tensioner roller counterclockwise to 11 ft. lbs. (15 Nm) to put pressure on the timing belt.

33. Place a torque wrench horizontally on socket head of the tensioning lever and turn bolt using 18 ft. lbs. (25 Nm) of torque, to tension the timing belt.

34. Place T40028 socket at camshaft adjuster of exhaust camshaft on left cylinder bank.

35. Turn rotor of camshaft adjuster clockwise to 84 inch lbs. (10 Nm).

36. Turn rotor of camshaft adjuster at right cylinder bank clockwise to 84 inch lbs. (10 Nm).

37. Torque bolts for all four camshaft gears to 74 ft. lbs. (100 Nm).

38. Install caps with new O-rings at camshaft gears and secure them with the securing ring.

39. Remove the T40030 camshaft adjuster gauge.

40. Remove the alignment bolt and thread sealing plug of TDC opening in using new O-ring.

41. Install remaining components in reverse order of removal. Torque cylinder covers to 84 inch lbs. (10 Nm) using same sequence as removal.

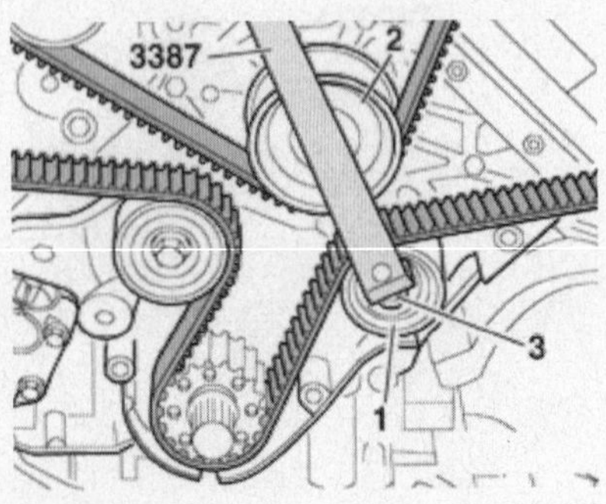

Positioning eccentric pulley for timing belt installation—Audi 3.0L engine

4.2L Engine

1. Before servicing the vehicle, refer to the precautions section.

2. Drain the cooling system.

➡The battery is located in the luggage compartment, right side, under a cover.

3. Remove or disconnect the following:
   - Negative battery cable
   - Engine cover
   - Cylinder head covers
   - Spring clips at timing belt cover
   - Timing belt cover
   - Sound insulation from rear of timing cover
   - Accessory drive belt

4. Turn crankshaft to TDC by hand to align timing marks

5. Check position of camshafts; ensure larger holes in securing plates on sprockets align with each other and are facing inward (if not, turn crankshaft one revolution).

6. Remove seal plug from left side of cylinder block and check that the TDC drilling in the crankshaft is aligned in the hole. Then screw in a clamping bolt (3242) to hold this position.

7. Remove or disconnect the following:
   - Serpentine drive belt tensioner
   - Vibration damper
   - Torque support

➡Mark the direction of rotation of the timing belt before removing.

➡The timing belt tensioner is oil-filled. To compress the tensioner, apply a slow, constant pressure.

8. With an 8mm Allen key, turn the timing belt tensioner in a counterclockwise direction until the tensioner is compressed far enough to insert a holding pin T40011 into the drilled hole and into the plunger.

9. Install a camshaft holding bar T40005 onto the securing plated of the two camshafts.

10. Loosen two camshaft bolts about five turns, and then remove the camshaft holding bar.

11. With a puller, remove the camshaft sprocket from the camshaft.

12. Loosen the timing belt tensioning idler and remove the timing belt with the camshaft sprocket for bank 1—4. Do not change position of belt on sprocket.

To install:

13. Ensure all pulleys, timing belt idlers and wheels are installed. Torque bolts as follows:
   - Idler wheel: 33 ft. lbs. (45 Nm)
   - Tensioning roller: 15 ft. lbs. (20 Nm)

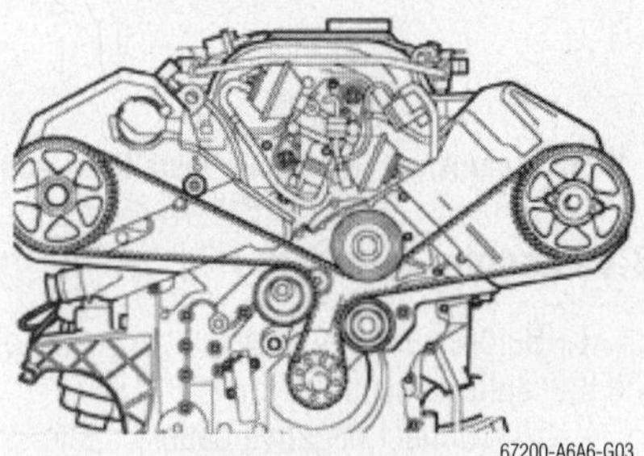

Showing timing belt routing—4.2L engines

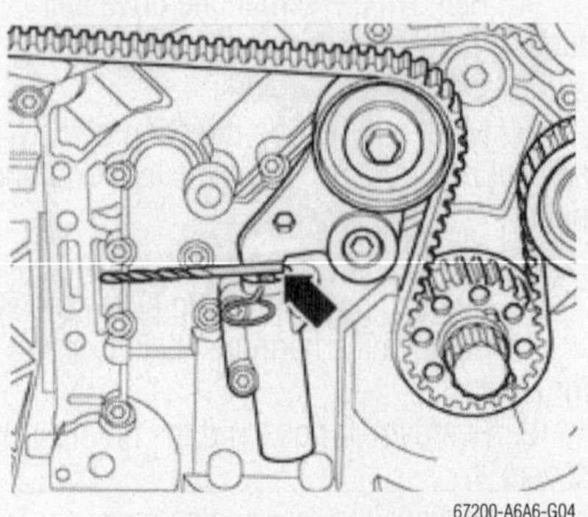

Inserting drill bit between tensioner and hydraulic piston—4.2L engines

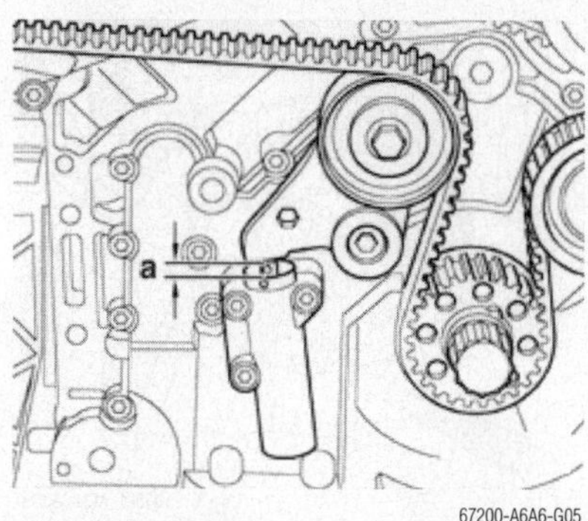

Measuring dimension "a" between timing belt tensioning lever and hydraulic tensioning element—4.2L engines

- Pulley to crankshaft: 15 ft. lbs. (20 Nm)
- Timing belt tensioner: 7 ft. lbs. (10 Nm)
- Eccentric wheel: 16 ft. lbs. (22 Nm)
- Center bolt to crankshaft: 148 ft. lbs. (200 Nm), plus 180 degrees

14. First, install the timing belt onto the crankshaft sprocket, idler wheel for tensioner, tensioning roller, camshaft sprocket for bank 5—8, coolant pump, and damper wheel.

15. Then, take camshaft sprocket for cylinder bank 1—4 and install timing belt and bolt sprocket onto camshaft.

16. Lightly secure both camshaft sprockets with securing plates by tightening hand tight.

➡Camshaft sprockets must be tight enough on the camshaft tapers to be turned, but not to be moveable axially.

17. Reinstall the camshaft holding bar T40005.

18. Insert a 5mm drill bit between the tensioner and hydraulic piston.

19. Tension the timing belt initially to 35 inch lbs. (4 Nm), using tensioning key T40009 and torque wrench with socket attachment VAS 5122 by turning in a counterclockwise direction. Then, tighten the eccentric wheel to 16 ft. lbs. (22 Nm).

20. Using the 8mm Allen key, turn the timing belt tensioning lever to the left until the insert pin T40011 can be removed.

21. Remove the camshaft holding bar T40005, 5mm drill bit, and crankshaft clamping bolt 3242 and reinstall the sealing plug in the hole.

22. Tighten the camshaft sprocket bolts to 41 ft. lbs. (55 Nm).

23. Turn the crankshaft two revolutions.

24. Measure the dimension between the timing belt tensioning lever and the hydraulic tensioning element (dimension "a"). If dimension is not 5mm, readjust the tensioning roller as needed.

25. Measure the clearance between the timing belt guard and outer surface of timing belt (at both cam sprockets). The clearance should be 1mm. If not, adjust cover until clearance is set.

26. Install all remaining components in reverse of removal procedure.

A8

1. Before servicing the vehicle, refer to the precautions section.

2. Drain the cooling system.

➡The battery is located in the luggage compartment, right side, under a cover.

3. Remove or disconnect the following:
   - Negative battery cable
   - Noise insulation panel
   - Intake air grille from bumper
   - Accessory drive belt
   - Engine cover
   - Coolant hose clamp on right belt guard
   - Air intake duct between air cleaner and throttle body
   - Air shroud for viscous fan and electric coolant fan on radiator (upper-left)
   - Viscous fan. Counterhold with 3212 two-hole nut driver.

➡The viscous fan has a left-hand thread; loosen by turning clockwise.

   - Viscous fan with air shroud, from above
   - Engine support mount (right front)
   - Coolant hoses at engine
   - Loosen coolant hose on upper-right of radiator, and turn hose to right
   - Loosen center bolt for vibration damper approx. 1 turn. Counterhold using special tool 3197.

➡Loosening and removing the vibration damper with the camshaft drive belt sprocket is only necessary if the belt is to be replaced, or if other engine disassembly requires it.

➡When the center bolt has been loosened, it must be replaced.

4. Rotate crankshaft to align TDC marks on vibration damper.

5. Remove Camshaft Position (CMP) sensor housing at rear of left cylinder head.

6. If camshaft position sensor is not positioned behind sensor plate window, rotate crankshaft 360.

7. Remove camshaft flange at rear of right cylinder head.

8. Remove or disconnect the following:
   - Harness connector from switch for intake manifold change-over valve -N156-
   - Belt guard
   - Cap at guide pulley for ribbed belt at right belt guard
   - Top part of air cleaner housing
   - Right belt guard mounting bolts (6x)
   - Right belt guard
   - Hold camshaft sprocket with holder 3036 and loosen camshaft sprockets bolts
   - Camshaft Position (CMP) sensor plate and housing

9. Install camshaft locking device 3341 at rear of each cylinder head, and tighten. If necessary, rotate camshaft slightly to allow locking device to engage.

⁂ CAUTION

The camshaft locking devices 3341 are not to be used as counterholds when loosening or tightening the camshaft sprocket bolts.

10. Using two-hole nut turner, e.g. Matra V/159, loosen eccentric tensioning roller and turn to lowest point.

11. Compress drive belt damper by hand and remove tensioning roller.

➡Mark the running direction of the belt with a felt pen or equivalent. Reversed running direction can lead to damage.

12. Remove belt from camshaft sprockets.

13. Remove 4 screws fastening vibration damper to drive belt sprocket, remove center bolt, and remove vibration damper.

14. Remove camshaft drive belt.

To install:

15. Place belt over crankshaft sprocket and install vibration damper on crankshaft.

16. By hand, screw in 4 bolts that connect vibration damper to sprocket.

17. Lightly oil thread and bolt contact surfaces of center bolt.

18. Thread in new center bolt and tighten by hand. Counterhold with special tool 3197.

➡Always replace center bolt.

19. Place belt in position over all sprockets, guide pulleys and water pump pulley.

20. Push belt over tensioning roller with eccentric insert and tighten nut.

➡Eccentric insert must still be able to turn.

21. Tighten 4 mounting bolts connecting vibration damper and crankshaft sprocket to 18 ft. lbs. (25 Nm).

22. Turn eccentric insert of tensioning roller clockwise using commercial two-hole nut turner (e.g. Matra V/159).

23. Adjust position of tensioning roller until damper length is within specified range.

24. Install or connect the following:

- Engine cold: 5.35-5.47 inches (136-139 mm)
- Engine warm: 4.96-5.08 inches (126-129 mm)

25. Tighten tensioning roller bolt to 18 ft. lbs. (25 Nm).

26. Remove camshaft locking device from each camshaft (rear of cylinder head.

27. Install housing behind camshaft position sensor and tighten to 15 ft. lbs. (20 Nm).

28. Crank engine at least two revolutions by hand and check timing belt alignment.

29. Tighten crankshaft center bolt as follows:

- With special tool 2079: 258 ft. lbs. (350 Nm)
- Without special tool 2079: 332 ft. lbs. (450 Nm)

30. The remaining installation of the camshaft drive belt is the reverse of removal.

A3

2.0L ENGINE

1. Remove the engine cover.
2. Open the coolant expansion tank cap after the engine is cooled.
3. Remove the center sound insulation.
4. Remove the right sound insulation.
5. Drain the cooling system.
6. Relieve the fuel system pressure.
7. Disconnect the fuel line.
8. Disconnect the EVAP lines .
9. Remove the EVAP canister.
10. Loosen coolant hose from retainer.
11. Remove the bracket for EVAP the filter.
12. Unbolt filler tube for windshield washer fluid reservoir.
13. Remove the retaining clip, disconnect the electrical connector and remove the coolant expansion tank.
14. Disconnect the coolant pipe.
15. Remove the accessory drive belt and remove locking pin T10060A .
16. Install an engine support tool such as 10—222 A onto fender bolting edge. The spindles should be positioned forward. Hook the carabiner hooks of spindles with shackles 10—222 A /12 into lifting eyes.
17. Remove the bolts from assembly mount/engine support remove the assembly mount completely.

⁂ CAUTION

When lifting the engine using the engine support tool , make sure that no components/hoses are damaged, overstressed or torn off.

18. Lift engine using engine support tool far enough so that both upper bolts of engine support can be loosened and removed.

Set camshaft gear to marking for TDC by turning crankshaft. The marking on the camshaft sprocket must align with the mark on timing belt cover.

19. Remove the vibration damper/belt pulley.

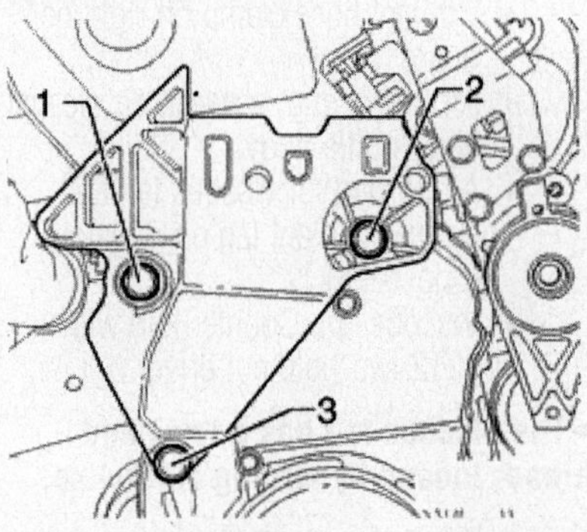

Lift the engine using the support tool and remove the upper engine support bolts—A3 with 2.0L engine

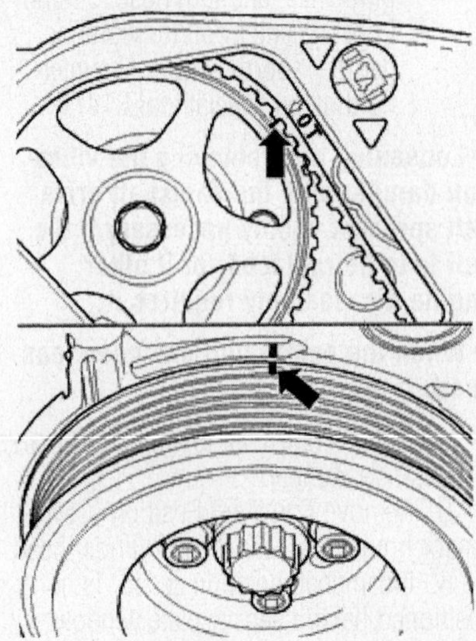

Align the marks as shown by setting the engine to TDC—A3 with 2.0L engine

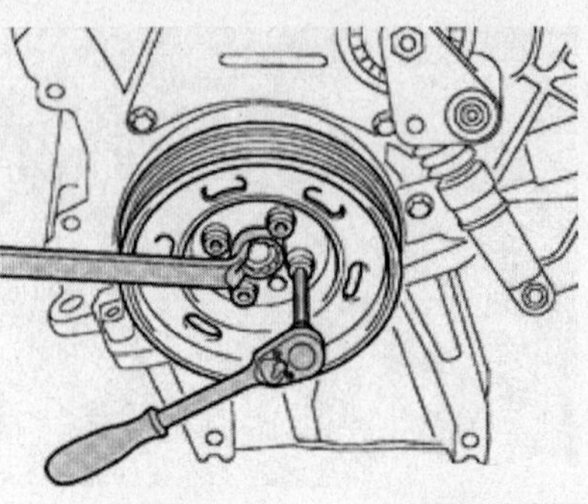

Remove the vibration damper pulley—A3 with 2.0L engine

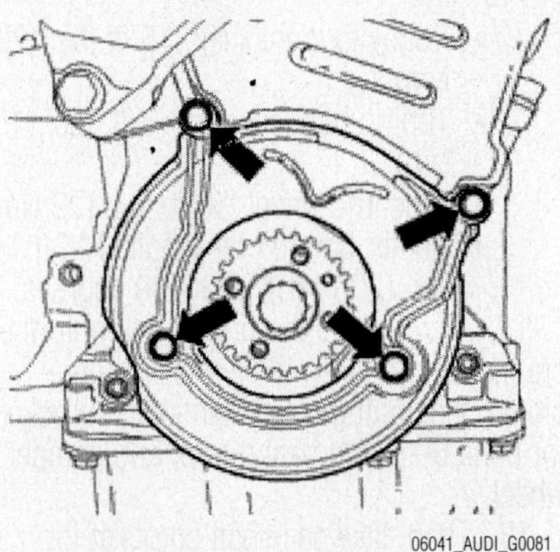

Remove the lower timing belt cover—A3 with 2.0L engine

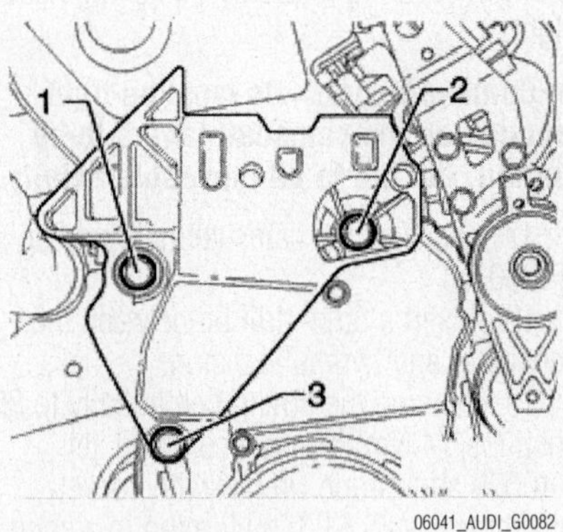

Remove the lower bolts and the engine support—A3 with 2.0L engine

20. Remove the bolts from lower timing belt cover.

21. Loosen the lower bolts and remove the engine support.

22. Remove the remaining bolts from the timing belt cover and remove timing belt cover from engine.

23. Mark the direction of rotation of timing belt.

24. Loosen tensioning roller and remove the timing belt.

25. Turn crankshaft back slightly.

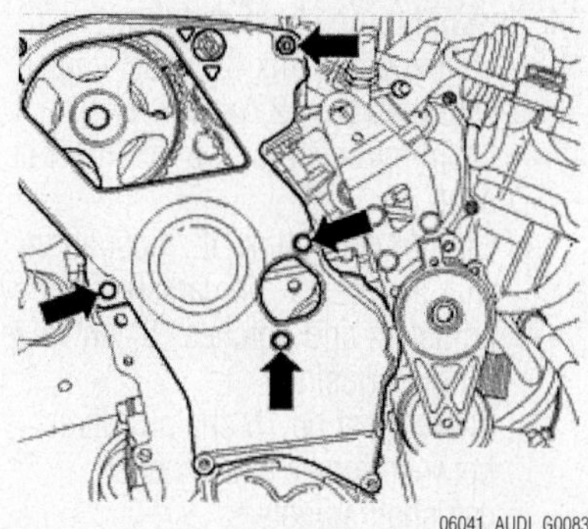
06041_AUDI_G0083

Remove the timing belt cover—A3 with 2.0L engine

To install:

➡When turning the camshaft, the crankshaft must not be at TDC as the valves and/or pistons may be damaged.

26. Place the timing belt onto the crankshaft sprocket making sure of the belt direction of rotation.
27. Secure lower timing belt cover with two lower bolts.
28. Install the vibration damper/belt pulley with new bolts. Tighten to 10 Nm plus an additional 90 degrees.
29. Set the crankshaft and camshaft to TDC at cylinder 1.
30. Route the timing belt as follows:
    a. Tension roller.
    b. Camshaft sprocket.
    c. Coolant pump.
    d. Idler pulley.

➡Make sure the tensioning roller is correctly seated in the cylinder head.

31. Tension timing belt by turning hex key on the eccentric toward right in a clockwise direction until notch is positioned above tab.
32. Relieve tension on timing belt again.
33. Now tension timing belt until notch and tab are positioned across from each other and tighten the nut to 25 Nm.
34. Turn the crankshaft 2 revolutions further in direction of engine rotation until engine is positioned at TDC again. When doing so, it is important that the last 45 degrees is turned without interruption.
35. Check timing belt tension again. The tab and notch are positioned across from each other.
36. Check valve timing again.
37. If the marks do not align, repeat the valve timing adjustment procedure.
38. If the marks align, install timing belt guard.

Note: Observe the different lengths of bolts. The lower bolt is approx. 25 mm shorter than the upper bolts.

39. Insert the lower bolt into the engine support.
40. Install engine support from below onto cylinder block and hand-tighten bolt.
41. Lift engine using engine support tool so that both upper bolts can be inserted. Tighten the bolts to 45 Nm.
42. Lower engine to installed position.
43. Install assembly mount for engine completely. The clearance between engine support and right longitudinal member should be 16mm. The casting edge of engine support must be parallel to the support arm.
44. Install the engine assembly mount to engine support bringing contact surfaces into contact using engine support tool. Tighten to 60 Nm plus an additional 90 degrees.
45. Remove the engine support bridge 10-222A .
46. Install the remaining components in the reverse order of removal, please note the following:

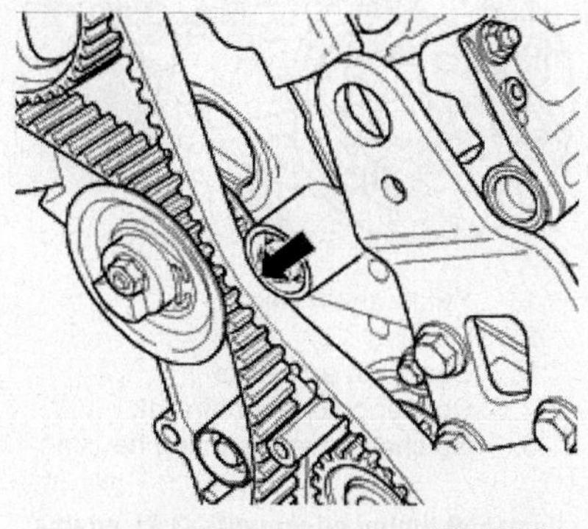
06041_AUDI_G0084

Install the belt over the tension roller, camshaft sprocket, water pump and idler puller—A3 with 2.0L engine

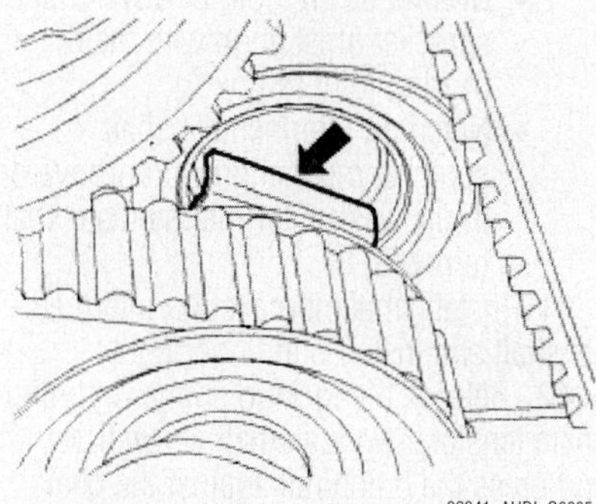
06041_AUDI_G0085

Make sure the tension roller is seated correctly in the cylinder head—A3 with 2.0L engine

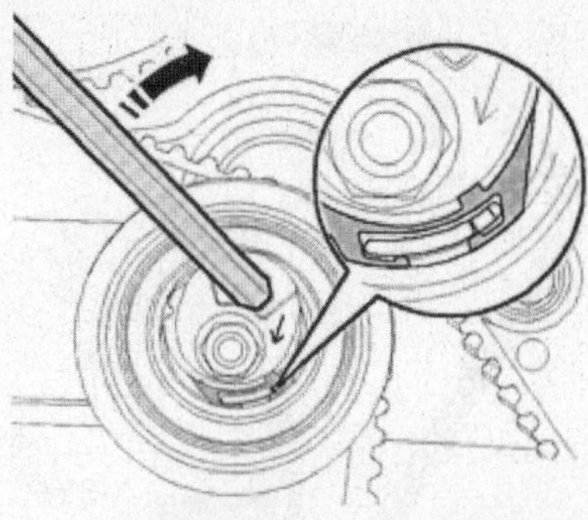
06041_AUDI_G0086

Tension the timing belt by turning the hex key clockwise until the notch is positioned above the tab—A3 with 2.0L engine

a. Tighten the timing belt tensioning roller to cylinder head to 23 Nm.
b. Tighten the lower and middle timing belt covers to cylinder block, use a locking compound and tighten to 10 Nm.
c. Tighten the engine support to cylinder block to 45 Nm.
d. Tighten the engine console to body, using new bolts to 40 Nm plus an additional 90 degrees.
e. Tighten the bracket to body/engine console, using new bolts to 20 Nm plus an additional 90 degrees.

Timing Chain

REMOVAL & INSTALLATION

3.2L Engine

TT

1. Disconnect the negative battery cable.
2. Remove engine
3. Remove flywheel
4. Remove intake manifold
5. Remove cylinder head cover
6. Remove thermostat housing
7. Remove oil pan
8. Remove or disconnect the following:
    - Bolt for timing chain tensioner
    - Upper timing chain cover. Be careful not to damage the head gasket when removing
    - Lower timing chain cover. Be careful not to damage the head gasket when removing
    - Crankshaft oil seal (rear main)

➡Mark rotation direction of timing chains with a colored arrow before removing. Do not mark the chains using a center punch, notch, or anything similar

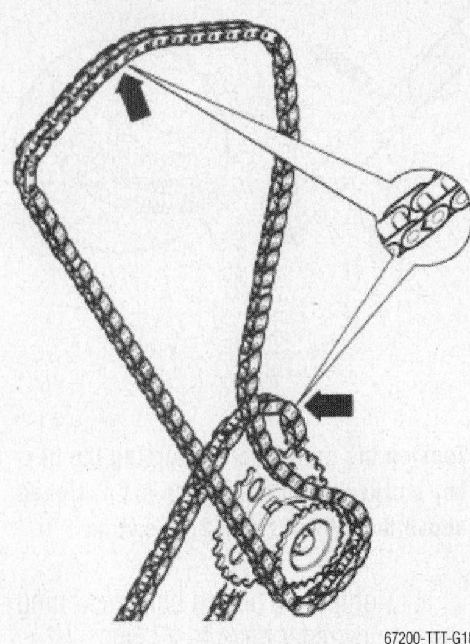

Mark timing chain rotation—3.2L engine

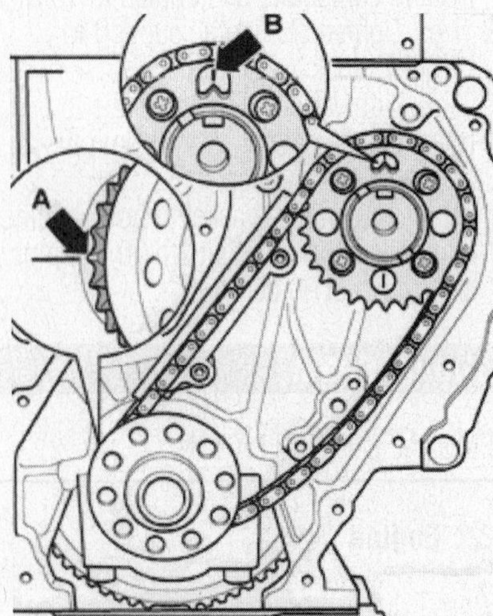

A. Main bearing joint
B. Thrust washer groove

Timing gear alignment—3.2L engine

- Guide rail
- Timing chain from sprocket for intermediate shaft and from behind projection of head gasket
- Loosen bolt on intermediate shaft sprocket approximately 1 turn by counter holding the vibration damper with T10069 or equivalent
- Loosen valve gear chain tensioner
- Remove bolt from intermediate shaft gear and removing chain together with gears

To install:

9. Install chain together with intermediate shaft sprocket
   - Sprocket can only be installed on intermediate shaft in one position.

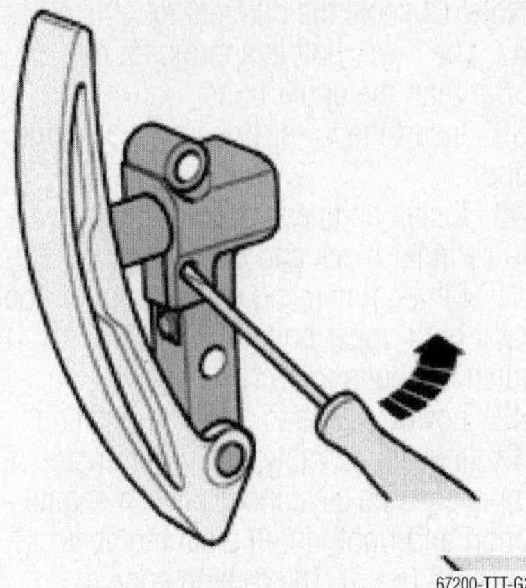

Release locking spline—3.2L engine

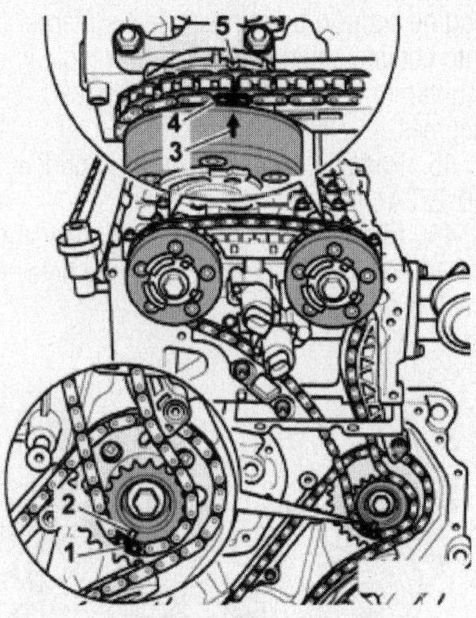

1. Yellow chain link
2. Marking
3. Camshaft adjusters
4. Copper colored chain link
5. Notches on valve timing housing

Camshaft timing alignment—3.2L engine

Turn intermediate shaft slightly if necessary

10. Check TDC position of crankshaft
    - Ground down tooth of drive chain sprocket must align with main bearing joint (arrow A)
    - Notch on intermediate shaft sprocket must align with groove on thrust washer for intermediate shaft (arrow B)
11. Install small sprocket on intermediate shaft and install bolt finger tight
12. Release the locking spline in timing chain tensioner with a small screwdriver and press the guide rail against the chain tensioner. Secure in this position
13. Tighten intermediate shaft securing bolt by counter holding at vibration damper with T10069
14. Install timing chain as follows:
    - Yellow chain link "1" must align with marking "2" on the small chain sprocket for the intermediate shaft
    - Copper chain link "4" must align with arrows "3" on camshaft adjusters and notches "5" on valve timing housing
    - There must be 16 chain rollers between intake and exhaust camshaft adjusters
15. Install guide rail
16. Remove camshaft plate T10068

➡When turning the crankshaft you must press the guide rail firmly against the timing chain to prevent the chain from slipping

17. Turn the crankshaft 2 revolutions in normal direction until timing mark is at TDC

➡The marked chain links will now be in an undefined position

18. Insert camshaft plate T10068 in both camshaft grooves. If camshaft plate cannot be inserted, repeat adjustment of valve timing
19. Install lower timing cover. Tighten bolts in diagonal sequence and in stages
20. Install upper timing cover and tighten bolts in following sequence:
    - Timing cover to engine block initially to 44 inch lbs (5 Nm)
    - Bottom of timing cover at head gasket 17 ft. lbs (23 Nm)
    - Timing cover to engine block to 88 inch lbs (10 Nm)
21. Secure timing chain tensioner
22. The remainder of the installation is the reverse of the removal procedure

➡Torque Specs

- Pins for guide rail to cylinder block 88 inch lbs (10 Nm)
- Chain tensioner to cylinder block 71 inch lbs (8 Nm)
- Chain sprockets to intermediate shaft 44 ft. lbs (60 Nm) + 90 degrees (1/4/ turn) Install new bolt
- Camshaft adjusters to camshaft 44 ft. lbs (60 Nm) + 90 degrees (1/4/ turn) Install new bolt
- Lower timing chain cover to cylinder block 88 inch lbs (10 Nm)

A3

1. Before servicing the vehicle, refer to the precautions section.
2. Remove the engine.
3. Remove the transmission and flywheel.

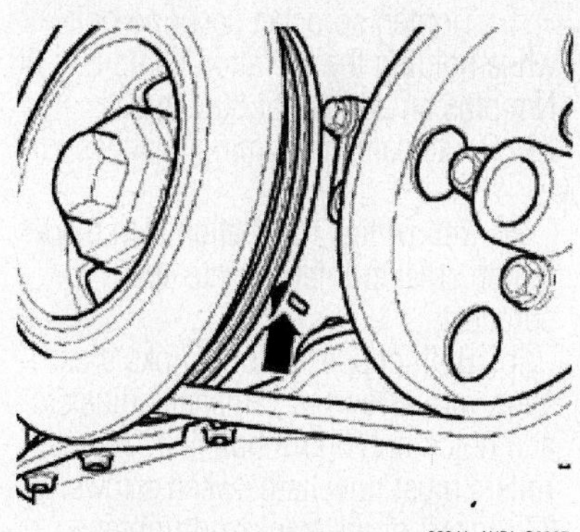

Set the engine to TDC—A3 with 3.2L engine

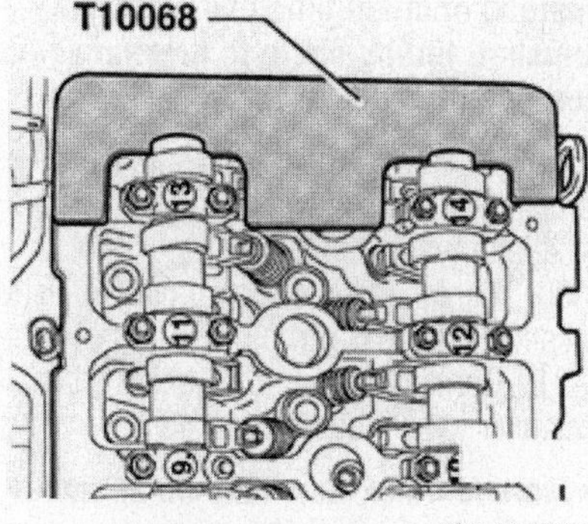

When at TDC the camshaft bar T10068 must engage in both shaft grooves—A3 with 3.2L engine

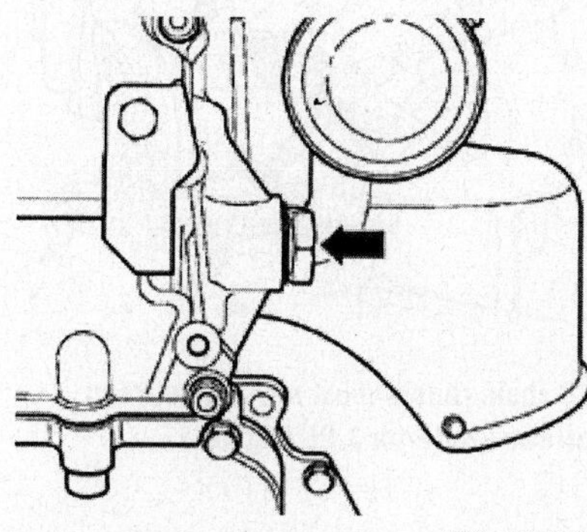

Remove the camshaft timing chain tensioner—A3 with 3.2L engine

4. Remove the intake manifold.
5. Remove the valve cover.
6. Remove the coolant thermostat housing.
7. Remove the oil pan
8. Remove the transmission.
9. Set crankshaft to TDC mark by turning crankshaft on vibration damper bolt in a clockwise direction of engine rotation. At the same time, camshaft bar T10068 must engage in both shaft grooves. Turn crankshaft 1 additional turn if necessary.
10. Disconnect all necessary electrical connections.
11. Remove the camshaft timing chain tensioner.
12. Remove the bolts and upper timing chain cover horizontally from cylinder head. When doing so, ensure cylinder head bolts are not damaged.
13. Remove the bolts and remove lower timing chain cover horizontally from cylinder block. When doing so, ensure cylinder head bolt is not damaged.
14. Press crankshaft seal on the timing chain side out of the cover.
15. Identify timing chain direction of rotation with arrow mark before removing, using paint.

➡Do not mark chain with punch, notch or something similar!

16. Remove the bolts and remove guide track.

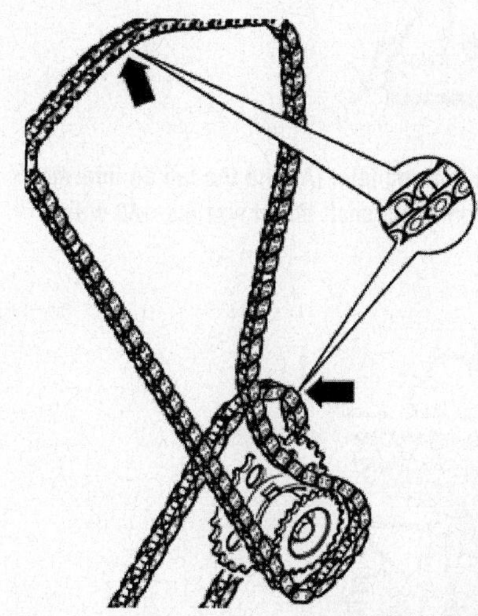

Identify timing chain direction of rotation with arrow mark before removing, using paint—A3 with 3.2L engine

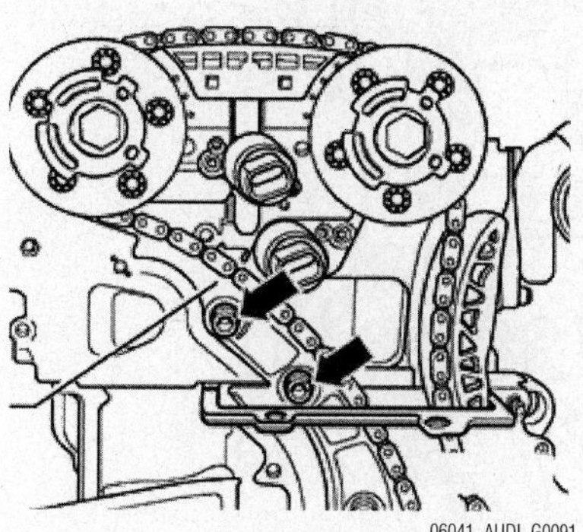

Remove the bolts and remove guide track—A3 with 3.2L engine

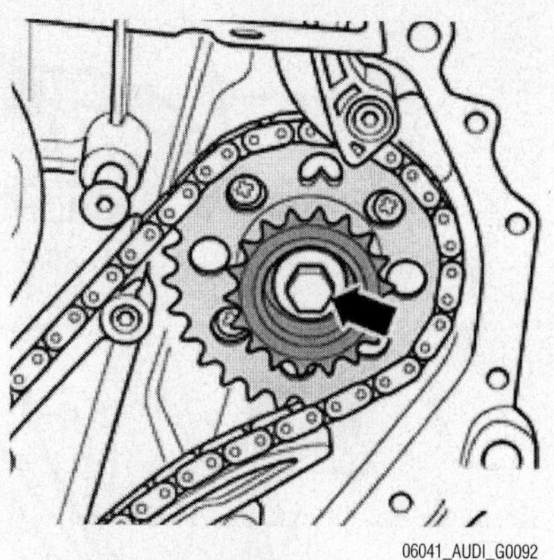

Remove the camshaft timing chain from intermediate shaft sprocket—A3 with 3.2L engine

17. Remove the camshaft timing chain from intermediate shaft sprocket.
18. Carefully remove camshaft timing chain behind cylinder head seal rib.
19. Loosen the chain sprocket bolt on intermediate shaft about 1 turn while holding the vibration damper.
20. Remove the timing drive chain tensioner.
21. Remove the bolt and remove drive chain together with chain sprockets from intermediate shaft.

To install:

➡When reinstalling a used timing mechanism drive chain, note running direction identification mark.

22. Install the drive chain together with large sprocket.
23. Note the location of sprocket, it can only be installed in one position in intermediate shaft. Turn intermediate shaft slightly if necessary.
24. Verify the crankshaft is at the TDC position as follows:
    a. The milled drive chain sprocket tooth must align with bearing joint.
    b. The tab on intermediate shaft sprocket must align with notch on intermediate shaft thrust washer.
25. Install the small sprocket on intermediate shaft as follows:
    a. It can only be installed in one position on intermediate shaft.
    b. Replace intermediate shaft sprocket securing bolt.
    c. Hand tighten bolt.
    d. Release locking splines in chain tensioner with a small screwdriver and press tensioning rail against chain tensioner.
    e. Tighten chain tensioner in this position.

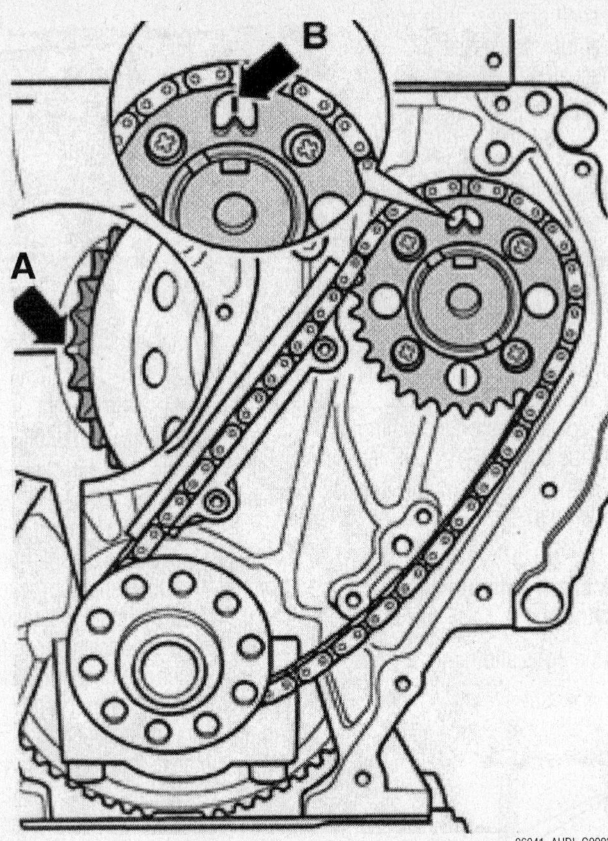

The milled drive chain sprocket tooth must align with bearing joint (A) and the tab on intermediate shaft sprocket (B) must align with notch on intermediate shaft thrust washer—A3 with 3.2L engine

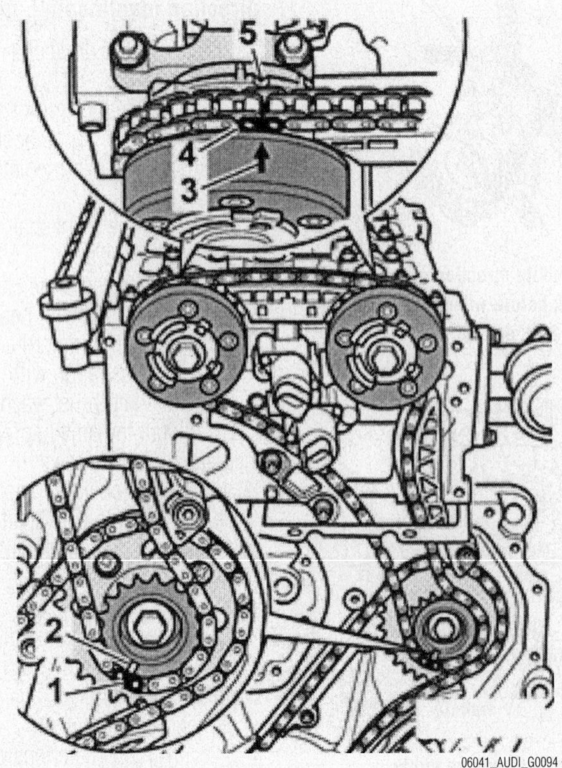

Timing chain routing—A3 with 3.2L engine

f. Tighten sprocket securing bolt while holding the vibration damper to 60 Nm plus an additional 60 degrees.

26. Route camshaft timing chain as follows:

a. Yellow link must align with marking on smaller intermediate chain sprocket.

b. Both copper-colored links must align with arrows at camshaft adjusters and notch on control housing. 16 chain rollers must now lie between arrows.

27. Insert guide track and tighten bolts.

28. Remove the camshaft bar T10068.

➡When turning crankshaft, tensioning rail must be pressed strongly with hand (instead of chain tensioner) against camshaft timing chain to prevent chain from jumping off.

29. Set crankshaft to TDC and at the same time, the camshaft bar T10068 must engage in both shaft grooves.

30. Tighten the camshaft timing adjuster to camshafts to 60 Nm plus 90 degrees

31. Tighten the camshaft timing chain tensioner to cylinder head to 40 Nm

➡Marked links, now take on an undefined position.

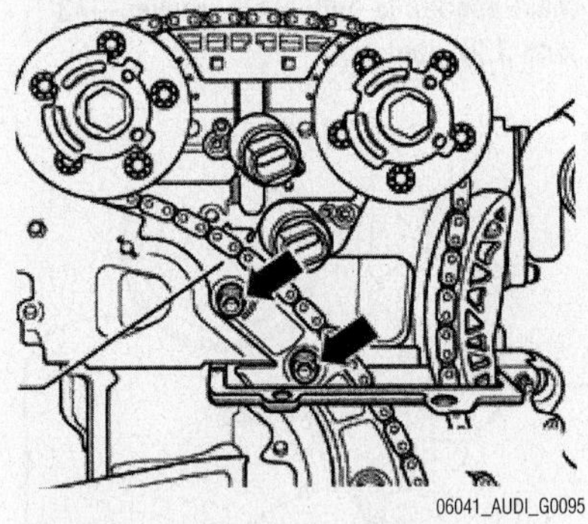

16 chain rollers must now lie between arrows—A3 with 3.2L engine

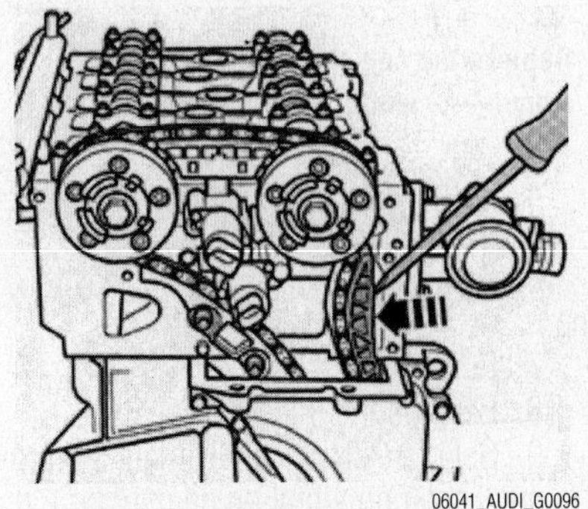

Insert guide track and tighten bolts—A3 with 3.2L engine

32. If camshaft bar cannot be inserted, repeat valve timing adjustment.

33. Clean old sealant from the 3mm holes in cylinder head seal.

➡With the cylinder head installed only half of the holes in the cylinder head gasket are visible.

34. Clean sealing surfaces at engine and both covers; they must be free of oil and grease.

35. Check whether lower timing chain cover alignment pins are inserted in cylinder block.

36. Lightly coat the lower timing chain cover sealing surfaces with sealing paste AMV 188 001 02.

➡The lower timing chain cover must be installed within 5 minutes of applying sealing paste.

37. Tighten the lower timing chain cover bolts diagonally in stages.

38. Fill the 3mm holes in cylinder head seal using black sealant AMV 174 04 01.

39. Replace seal in upper timing chain cover

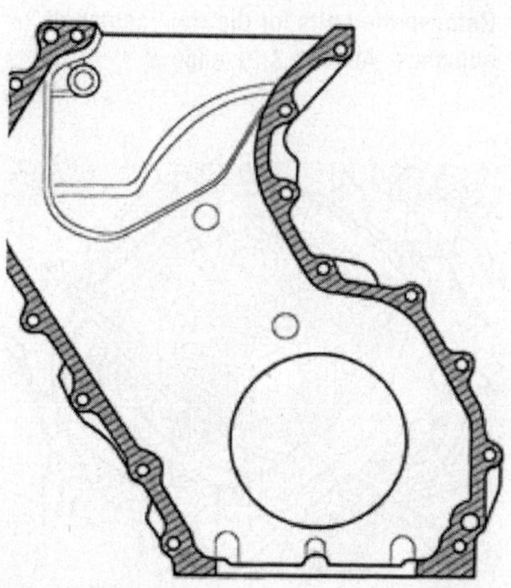

06041_AUDI_G0097

Lightly coat the lower timing chain cover sealing surfaces with sealing paste—A3 with 3.2L engine

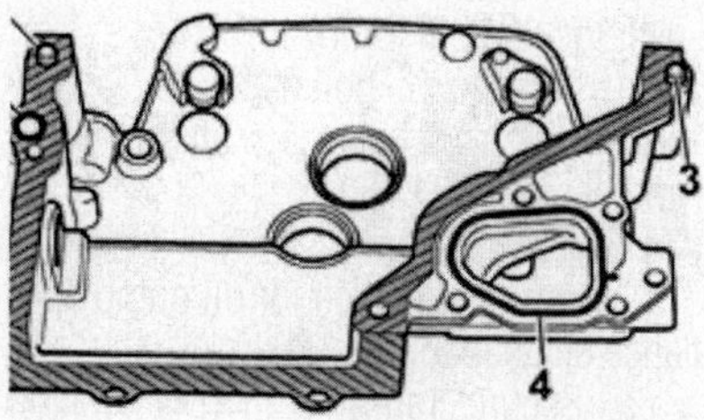

06041_AUDI_G0098

Lightly coat the upper timing chain cover sealing surfaces with sealing paste—A3 with 3.2L engine

40. Check whether alignment sleeves and are inserted in upper timing chain cover.

41. Insert a new gasket and a new seal.

42. Lightly coat clean upper timing chain cover sealing surfaces with sealing paste AMV 188 001 02.

43. Set upper timing chain cover in place and tighten bolts as follows:

a. Step 1: Tighten bolts 1 and 2 to 5 Nm.

b. Step 2: Tighten bolts 3 to 23 Nm.

c. Step 3: Tighten bolts 1 and 2 to 10 Nm.

d. Step 1: Tighten camshaft timing chain tensioner.

e. Step 1: Turn crankshaft 2 times in direction of engine rotation and check valve timing again.

44. Install the sealing ring for crankshaft on the timing belt side.

45. Install the cylinder head cover.

46. Install the coolant thermostat housing

47. Install the oil pan.

48. Install the flywheel.

49. Install the transmission and engine.

A6

1. Before servicing the vehicle, refer to the precautions section.
2. Drain engine oil.
3. Remove the transmission
4. Remove the drive plate.
5. Remove the left and right timing chain covers as follows:
6. Remove the rear engine covers as follows:

a. Remove the coolant expansion tank.

b. Disconnect electrical connection from Engine Coolant Level (ECL) switch at bottom of coolant reservoir and set aside coolant reservoir with coolant hoses still connected.

c. Disconnect electrical harness connectors.

d. Remove the nut.

e. Remove the retainer for connection.

f. Remove the double-bolt lying beneath.

g. Remove the retainer for connections.

h. Remove the bolts and separate electrical connections at ignition coils.

i. Remove the bolts and remove the left timing chain cover.

j. Disconnect check valve from connection at air duct hose.

k. Remove the air duct hose.

l. Disconnect electrical connector.

m. Remove the bolt and remove retainer for connection.

n. Remove the bolts and remove right timing chain cover.

7. Remove the lower timing chain cover as follows:

a. Remove the cap for oil filter housing.

b. Remove the oil filter element.

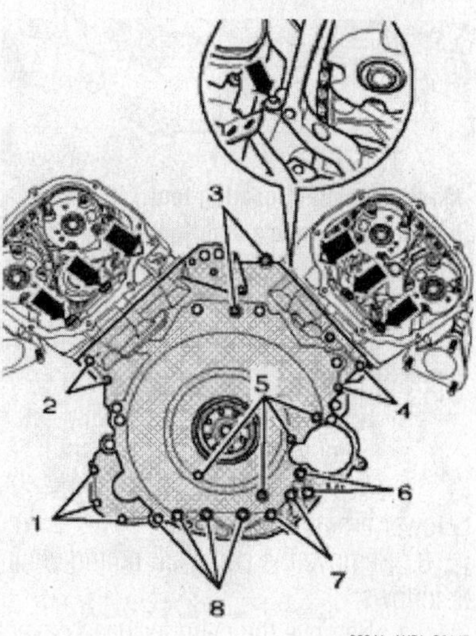

06041_AUDI_G0145

Remove the bolts and remove lower timing chain cover—A6 with 3.2L engine

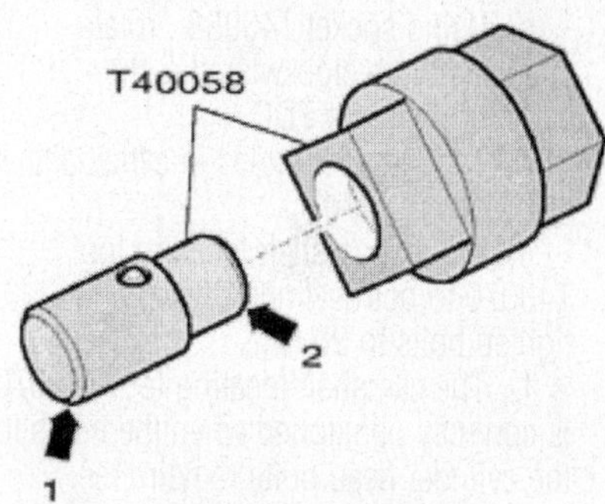

06041_AUDI_G0150

Insert guide pin of adapter T40058 so that large diameter points to engine the small diameter should point to the adapter—A6 with 3.2L engine

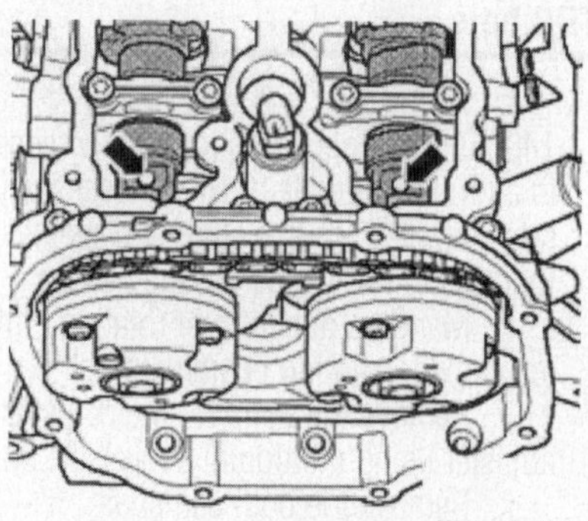

06041_AUDI_G0151

The threaded holes in camshafts must face upward—A6 with 3.2L engine

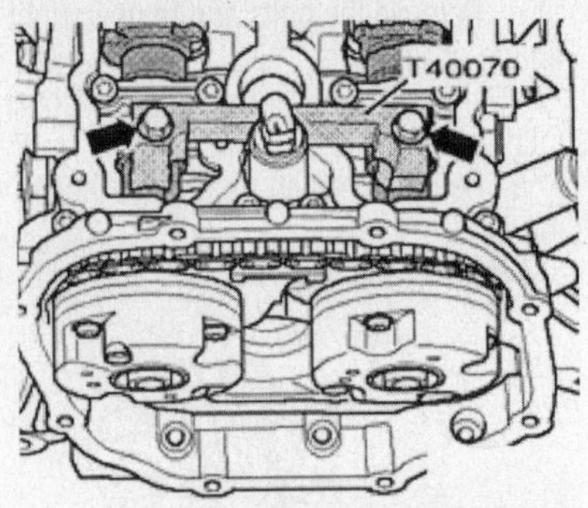

Mount camshaft locating tool T40070 to both cylinder heads and tighten bolts to 20 Nm—A6 with 3.2L engine

c. Disconnect electrical harness connector from oil pressure switch.
d. Remove the oil pressure switch.
e. Remove the oil filter housing.
f. Remove the bolts and remove lower timing chain cover.

8. Remove the camshaft timing chains as follows:
a. Remove the cylinder head cover.
b. Insert guide pin of adapter T40058 so that large diameter points to engine the small diameter should point to the adapter.
c. Using socket T40058 , rotate crankshaft in a clockwise direction of engine rotation to TDC.
d. The threaded holes in camshafts must face upward.
e. Mount camshaft locating tool T40070 to both cylinder heads and tighten bolts to 20 Nm.
f. The camshaft locating tool T40070 is correctly positioned when the holes for the cylinder head bolts remain free.
g. Remove the sealing plug from the cylinder block.

CAUTION

Due to a risk of injury, do not to turn the crankshaft while touching the TDC hole.

h. Screw the crankshaft holder T40069 into hole using a torque wrench to a torque value of 10 Nm, if necessary rotate crankshaft very slightly back and forth to completely center the holder.
i. Mark the direction of rotation of the left camshaft timing chain with paint.
j. Remove the bolts for the camshaft adjuster using multipoint socket T10035.
k. Remove the both camshaft adjusters.
l. Remove the bolts and remove the chain tensioner.

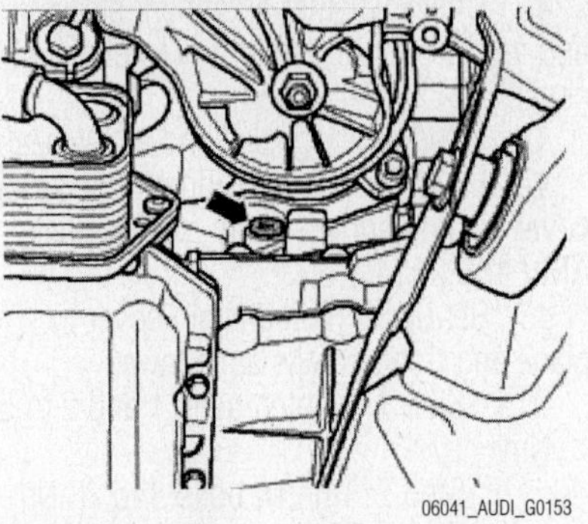

Remove the sealing plug from the cylinder block—A6 with 3.2L engine

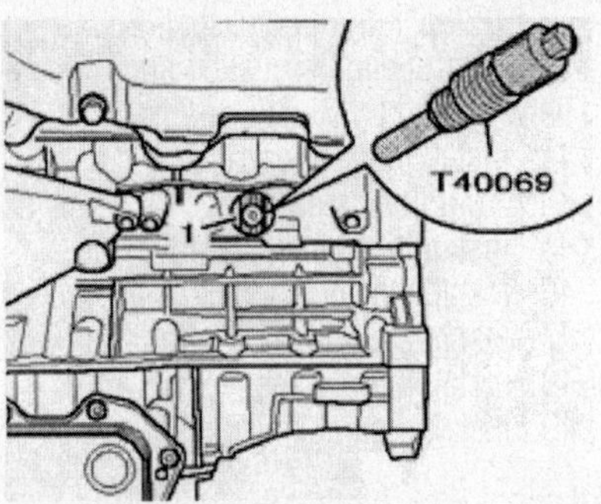

Screw the crankshaft holder T40069 into hole using a torque wrench to a torque value of 10 Nm—A6 with 3.2L engine

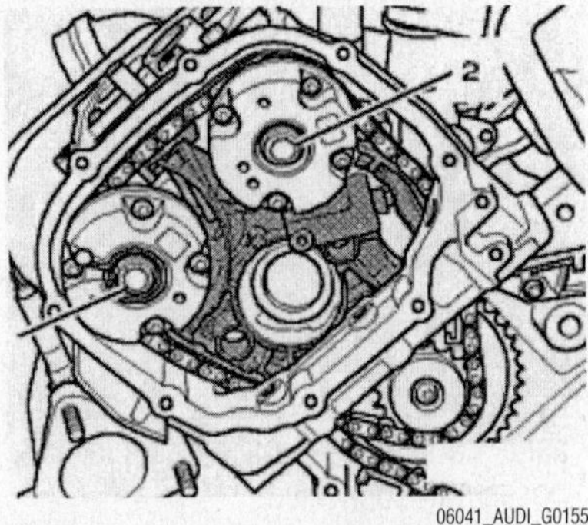

Remove the bolts for the left camshaft adjuster—A6 with 3.2L engine

m. Mark the direction of rotation of the right camshaft timing chain with paint.
n. Remove the bolts for the camshaft adjuster using multipoint socket T10035.
o. Remove the both camshaft adjusters.
p. Remove the bolts and remove the chain tensioner.

9. Remove the drive chain for oil pump and balance shaft as follows:
a. Mark running direction of drive chain for oil pump and balance shaft with paint.

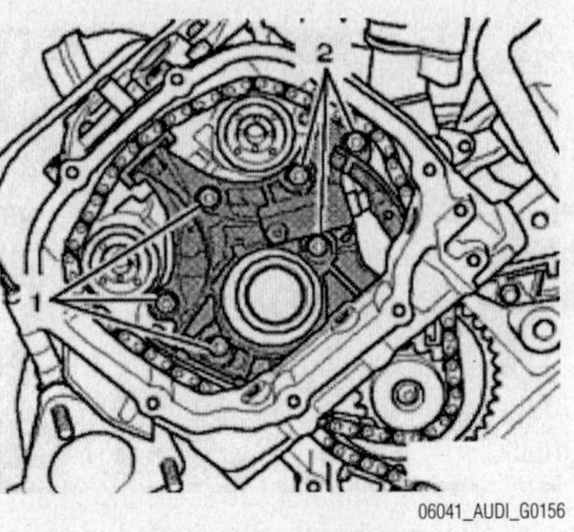

Remove the bolts for the left chain tensioner—A6 with 3.2L engine

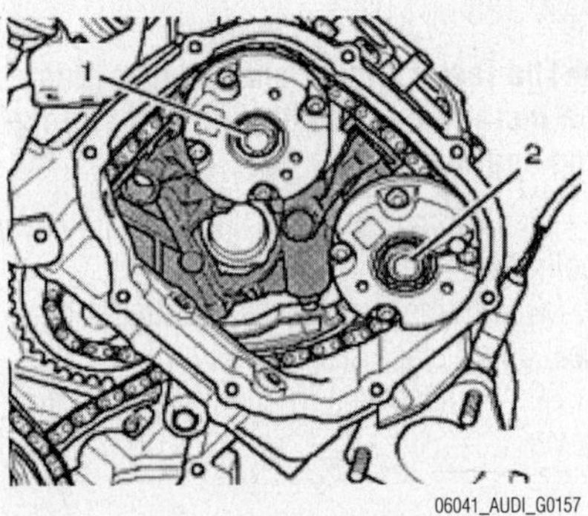

Remove the bolts for the right camshaft adjuster—A6 with 3.2L engine

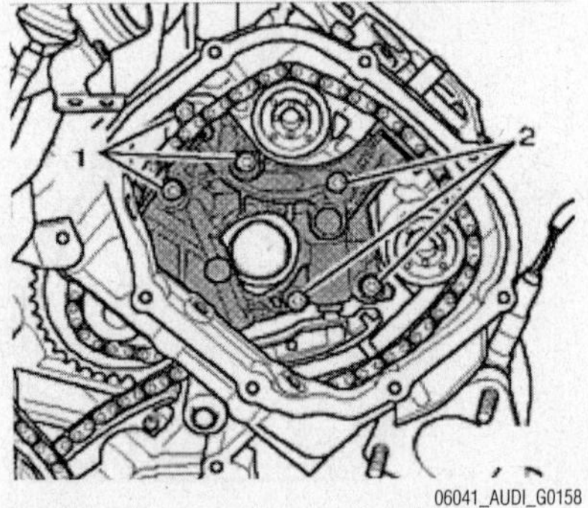

Remove the bolts for the right chain tensioner—A6 with 3.2L engine

b. Press chain tensioner guide rail in downwards and secure chain tensioner with locking pin T40071.
c. Remove the bolts and remove chain tensioner, balance shaft sprocket and chain.

To install:

10. Install drive chain for oil pump and balance shaft as follows:
a. Secure crankshaft in TDC position using crankshaft holder T40069.
b. Mount chain tensioner with chain and balance shaft sprocket.
c. Secure balance shaft with 8 mm

Secure balance shaft with 8 mm diameter drill bit in TDC position—A6 with 3.2L engine

diameter drill bit in TDC position. The slots in balance shaft sprocket must be at the middle position in relation to the threaded holes of the balance shaft. If necessary, adjust chain by one tooth.

d. Tighten chain tensioner bolts.

e. Loosely screw in bolts for sprocket.

f. The chain sprocket must still be able to be rotated on balance shaft and must not tip.

g. Remove the locking pin T40071 to release the chain tensioner.

h. Press against chain tensioner guide rail with a screwdriver and fasten the bolts for sprocket.

i. Tighten the chain tensioner to cylinder block to 6 Nm plus an additional 45 degrees and the balance shaft sprocket to balance weight to 15 Nm plus an additional 90 degrees.

j. Remove the drill from the balance shaft.

k. Install a new sealing plug in the block and tighten to 14 Nm.

11. Install the camshaft timing chains as follows:

➡Always replace bolts that are tightened to torque as well as O-rings and gaskets.

➡When turning camshaft, crankshaft must not be at TDC for any cylinder. Valves and/or pistons may be damaged.

a. The drive chain for the timing mechanism installed.

b. Secure crankshaft in TDC position using crankshaft holder T40069 and the camshaft locating tool T40070 must be mounted on both cylinder heads and fastened to 20 Nm.

c. Fully relieve tension of guide rail for left and right camshaft timing chain

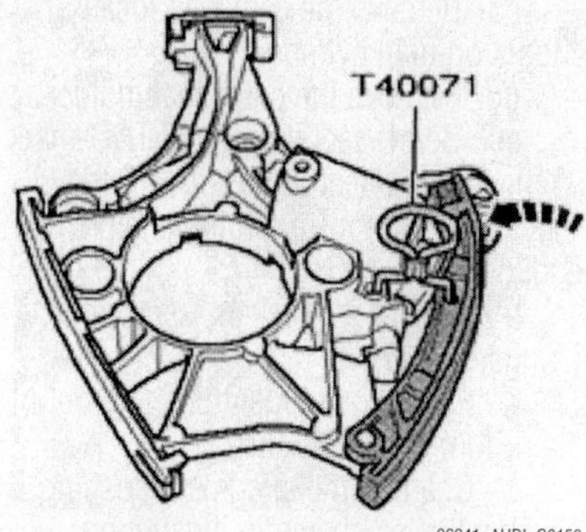

Press guide rail of the left and right camshaft timing chain inward up to the stop and secure chain tensioner with the locking pin T40071—A6 with 3.2L engine

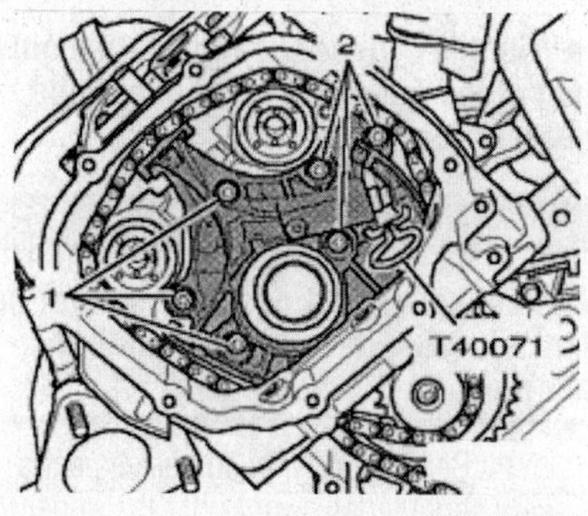

Install the set chain tensioner in place on the left cylinder head and install camshaft timing chain. Install the bolts indicated by the numbers 1 and 2—A6 with 3.2L engine

tensioner. The chain tensioner piston must be driven out completely, releasing the retainer.

d. Press guide rail of the left and right camshaft timing chain inward up to the stop and secure chain tensioner with the locking pin T40071.

e. Place a new gasket onto the rear of the chain tensioner.

f. Install the set chain tensioner in place on the left cylinder head and install camshaft timing chain, as illustrated.

g. Install the bolts indicated by the numbers 1 and 2 on the illustration. Replace the camshaft bolts.

h. Both camshaft adjusters must be able to rotate on camshaft and must not tip.

i. Set chain tensioner in place on right cylinder head and install camshaft timing chain, as illustrated.

j. Install the bolts indicated by the numbers 1 and 2 on the illustration. Replace the camshaft bolts.

k. Both camshaft adjusters must be

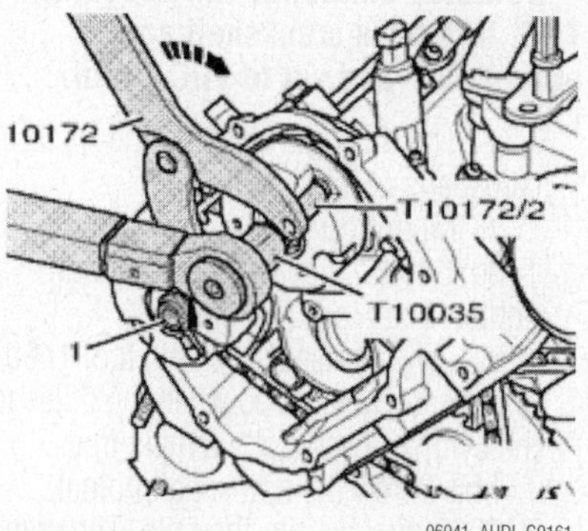

Set counter-holder T10172 with pin T10172/2 in place on camshaft adjuster of the left intake camshaft—A6 with 3.2L engine

able to rotate on camshaft and must not tip.

l. Set counter-holder T10172 with pin T10172/2 in place on camshaft adjuster of the left intake camshaft.

m. Hold camshaft timing chain pre-tensioned by pressing on counter-holder in direction of—arrow -.

n. Simultaneously, pre-torque camshaft bolt using multipoint socket T10035 and torque wrench. Tighten to 40 Nm.

o. Continue holding the pretension on intake camshaft and pre-torque bolt on the exhaust camshaft and tighten to 40 Nm.

p. Tighten the camshaft bolts on left cylinder head to 80 Nm plus an additional 90 degrees

q. Set counter-holder T10172 with pin T10172/2 in place on camshaft adjuster of the right exhaust camshaft.

r. Hold camshaft timing chain pre-tensioned by pressing on counter-holder in clockwise direction and pre-torque camshaft bolt using multipoint socket T10035 and torque wrench to 40 Nm.

s. Continue holding the pretension on the exhaust camshaft and pre-torque bolt on intake camshaft to 40 Nm.

t. Tighten camshaft bolts on right cylinder head to final torque specification to 80 Nm plus an additional 90 degrees.

u. Remove the locking pin T40071 to release chain tensioner

v. Remove the camshaft locators T40070 on both cylinder heads.

w. Remove the crankshaft holder T40069.

x. Using key T40058 turn crankshaft two complete rotations in clockwise direction of engine rotation until crankshaft stands at TDC again.

➡If rotated unintentionally beyond TDC, turn back crankshaft again approx. 30° and set to TDC again.

y. The threaded holes in the camshafts must face upward.

z. Mount camshaft locating tools T40070 to both cylinder heads and tighten bolts to 20 Nm.

aa. The camshaft locating tool T40070 is correctly positioned when the holes for the cylinder head bolts remain free.

bb. Screw the crankshaft holder T40069 directly into the hole. The crankshaft holder T40069 must engage in locating hole of crankshaft, otherwise repeat adjustment.

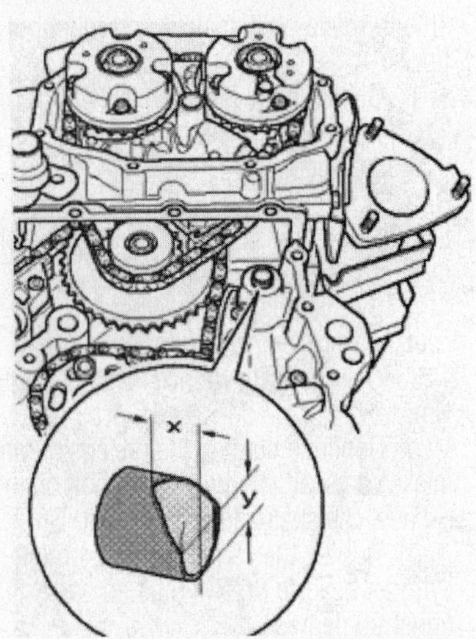

Chamfer the alignment bushing with a file, as shown. Dimension X should be 6.5 mm and dimension Y should be 8 mm—A6 with 3.2L engine

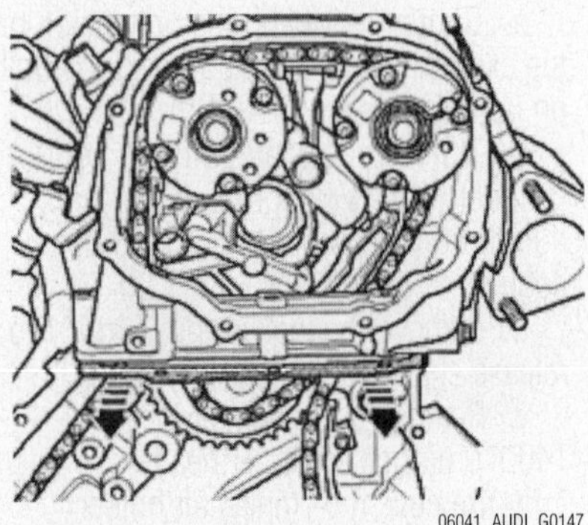

Bend cylinder head gaskets ends very slightly downward until the upper sealing surface of the gasket and cylinder head can be cleaned—A6 with 3.2L engine

cc. Remove the camshaft locating tools on both cylinder heads.

dd. Remove the crankshaft holder.

ee. Screw sealing plug of TDC-marking with new seal into cylinder block.

12. Install the lower timing chain cover as follows:

a. Replace all gaskets, seals and O-rings.

b. Pull the alignment bushing out of top right of cylinder block.

c. Chamfer the alignment bushing with a file, as shown in illustration. Dimension **X** should be 6.5 mm and dimension **Y** should be 8 mm.

d. Install alignment bushing into cylinder block so that the chamfered side faces upward.

e. Clean all gasket mating surfaces.

➡With the cylinder head installed only half of the holes in the cylinder head gasket are visible.

CAUTION

The cylinder head gasket must not be kinked. A kinked cylinder head gasket must be replaced.

f. Bend cylinder head gaskets ends very slightly downward until the upper sealing surface of the gasket and cylinder head can be cleaned.

g. Clean both cylinder head gaskets, top and bottom, so they are completely free of any oil or grease.

h. Coat sealing surfaces of the cylinder head gaskets, top and bottom, with a 2mm bead of sealant, slightly bending cylinder head gaskets downward again. A feeler gauge can be used to coat the surface between the cylinder head and gasket.

➡The covers for timing chain must be installed within 5 minutes after applying sealant.

i. Apply a 2mm bead of sealant as shown to the lower timing chain cover. The groove of sealing surface must be completely filled with sealant and the sealant bead must be 1.5 to 2.0 mm above sealing surface.

j. Install the lower timing chain cover in place, guiding cover at an angle from below onto sealing surface of the cylinder block and cylinder head.

k. Insert bolts indicated by arrows on the illustration with locking compound and tighten to 5 Nm.

l. Tighten bolts 1 through 8 using a diagonal sequence to 10 Nm.

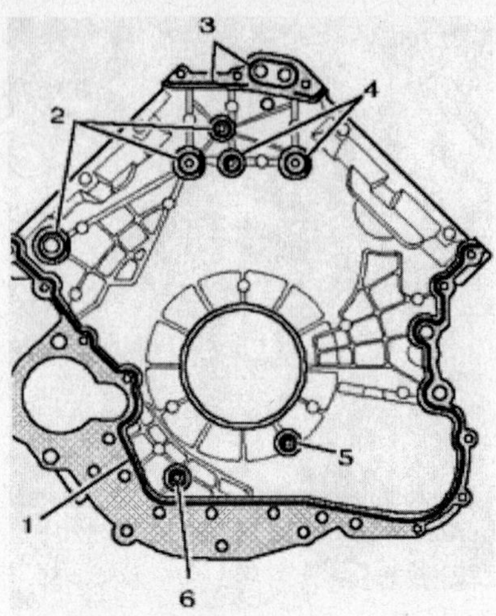

Apply a 2mm bead of sealant to the locations illustrated on the lower timing chain cover—A6 with 3.2L engine

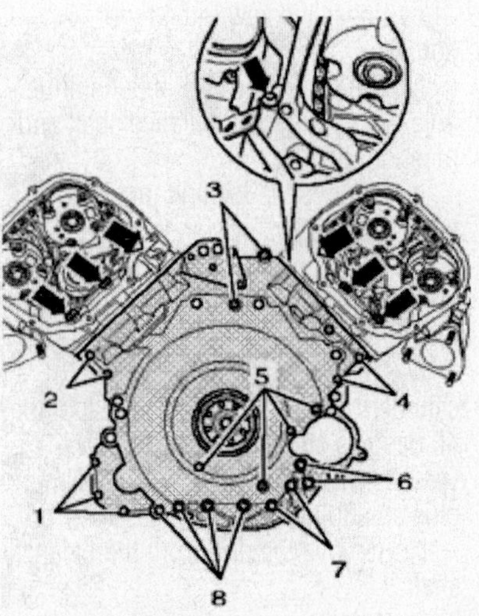

Lower timing chain cover bolt locations—A6 with 3.2L engine

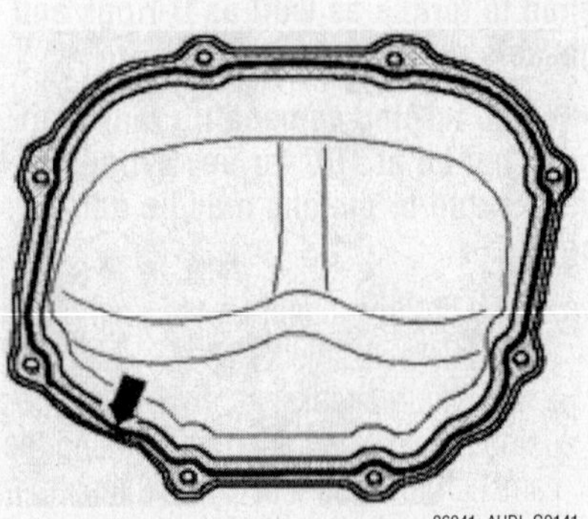

Apply a 1mm bead of sealant as shown to the left timing chain cover—A6 with 3.2L engine

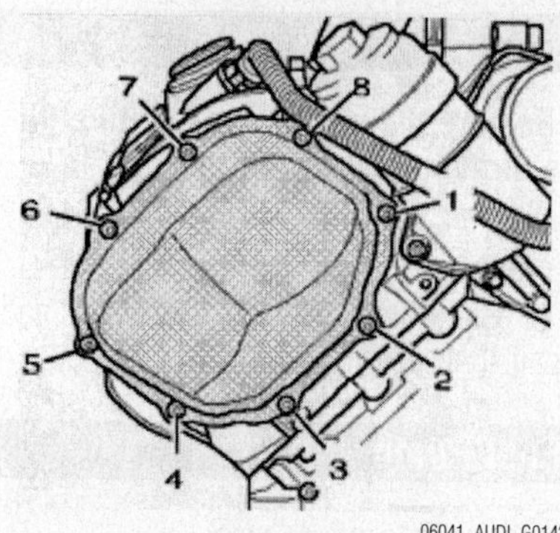

Left timing chain cover bolt locations—A6 with 3.2L engine

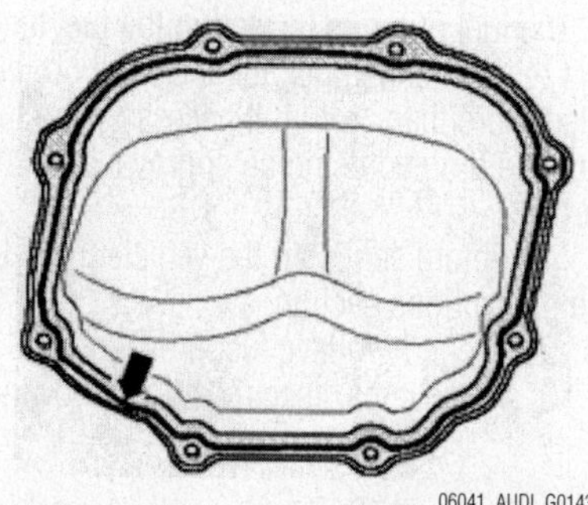

Apply a 1mm bead of sealant as shown to the right timing chain cover—A6 with 3.2L engine

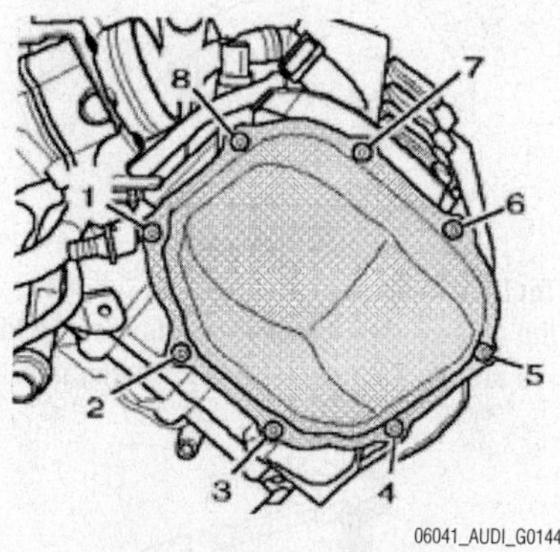

Right timing chain cover bolt locations—A6 with 3.2L engine

m. Tighten the bolts indicated by arrows in the illustration to 10 Nm.

n. Tighten bolts 6, 7 and 8 to 22 Nm.

13. Install the rear engine covers as follows:

a. Clean all gasket mating surfaces.

➡Covers for the timing chain must be installed within 5 minutes after applying sealant.

b. Apply a 1mm bead of sealant as illustrated to the left timing chain cover. The groove of sealing surface must be completely filled with sealant and the sealant bead must be 1.5 to 2.0 mm above sealing surface.

c. Install the left timing chain cover and tighten bolts in the sequence illustrated to 9 Nm.

d. Apply a 1mm bead of sealant as illustrated to the right timing chain cover. The groove of sealing surface must be completely filled with sealant and the sealant bead must be 1.5 to 2.0 mm above sealing surface.

e. Install the right timing chain cover and tighten bolts in the sequence illustrated to 9 Nm.

Piston and Ring

POSITIONING

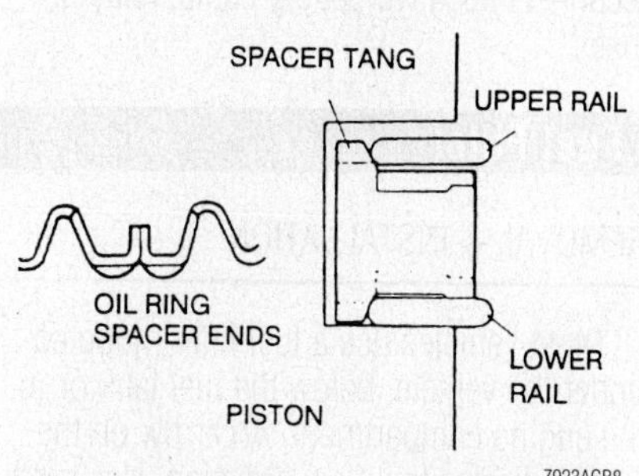

Piston ring positioning mark and location—Audi engines

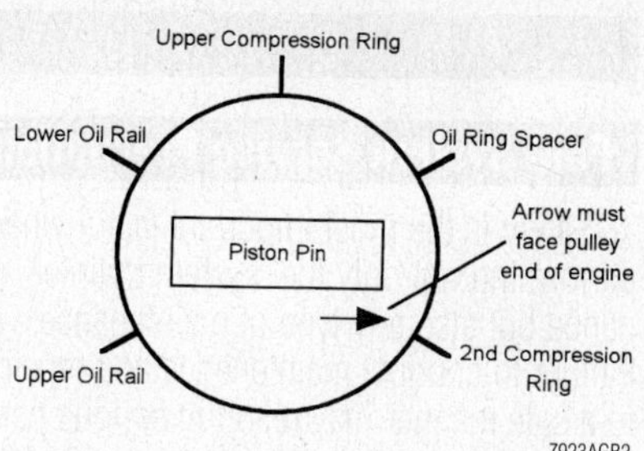

Piston ring and end-gap spacing—Audi engines

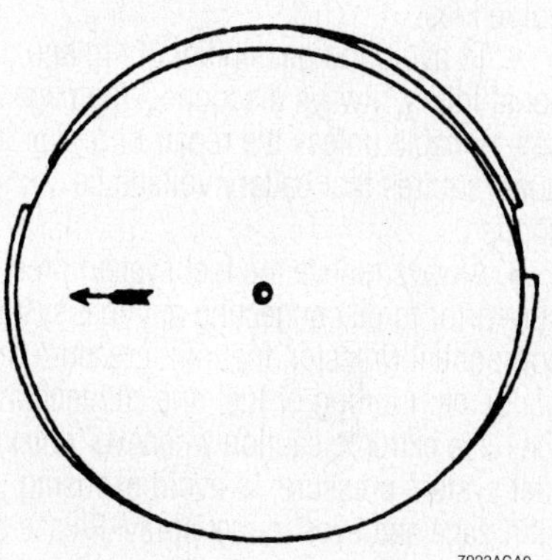

Arrow on the piston crown must face the front of the engine—Audi engines

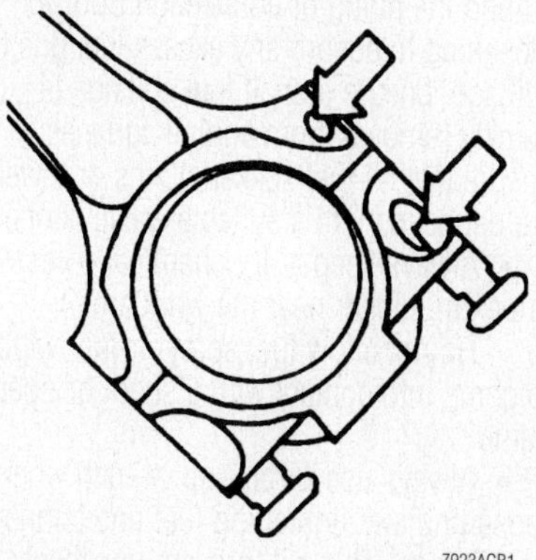

Connecting rod to bearing cap assembly—Audi engines

FUEL SYSTEM

Fuel System Service Precautions

Safety is the most important factor when performing not only fuel system maintenance but also any type of maintenance. Failure to conduct maintenance and repairs in a safe manner may result in serious personal injury or death. Maintenance and testing of the vehicle's fuel system components can be accomplished safely and effectively by adhering to the following rules and guidelines.

- To avoid the possibility of fire and personal injury, always disconnect the negative battery cable unless the repair or test procedure requires that battery voltage be applied.
- Always relieve the fuel system pressure prior to disconnecting any fuel system component (injector, fuel rail, pressure regulator, etc.), fitting or fuel line connection. Exercise extreme caution whenever relieving fuel system pressure, to avoid exposing skin, face and eyes to fuel spray. Please be advised that fuel under pressure may penetrate the skin or any part of the body that it contacts.
- Always place a shop towel or cloth around the fitting or connection prior to loosening to absorb any excess fuel due to spillage. Ensure that all fuel spillage is quickly removed from engine surfaces. Ensure that all fuel soaked cloths or towels are deposited into a suitable waste container.
- Always keep a dry chemical (Class B) fire extinguisher near the work area.
- Do not allow fuel spray or fuel vapors to come into contact with a spark or open flame.
- Always use a back-up wrench when loosening and tightening fuel line connection fittings. This will prevent unnecessary stress and torsion to fuel line piping. Always follow the proper torque specifications.
- Always replace worn fuel fitting O-rings with new. Do not substitute fuel hose where fuel pipe is installed.

Fuel System Pressure

RELIEVING

The fuel injection system operates under high pressure. This makes it necessary to first relieve the system of pressure before servicing. The pressurized fuel, when released, may ignite or cause personal injury.

**** CAUTION**

The fuel injection system remains under pressure after the engine has been turned OFF. Properly relieve fuel pressure before disconnecting any fuel lines. Failure to do so may result in fire or personal injury.

1. Before servicing the vehicle, refer to the precautions section.
2. Remove or disconnect the following:
   - Power to the fuel pump by removing the relay or the fuel pump fuse. The fuse can be removed to stop the fuel pump from running. With the engine operating at idle, wait until the engine stalls from fuel starvation.
3. Switch the ignition **OFF** and remove the negative battery cable.
4. Slowly and carefully open the fuel tank filler cap for a brief moment and reinstall.
5. Carefully loosen the fuel line on the control pressure regulator or component to be serviced.
6. Wrap a clean rag around the connection, while loosening, to catch any fuel.
7. After service is complete, discard the fuel soaked rag in the proper manner and reconnect negative battery cable, relay or fuses.

Fuel Filter

REMOVAL & INSTALLATION

Most vehicles use a fuel filter mounted under the vehicle, below the fuel tank or in the engine compartment. An arrow on the filter indicates fuel flow direction. Use care not to mix up fuel supply or return lines. Fuel pressure applied to the return side of the system will cause damage.

1. Before servicing the vehicle, refer to the precautions section.
2. Properly relieve the residual fuel pressure.
3. Remove or disconnect the following:
   - Negative battery cable
   - Fuel lines leading into and out of the fuel filter.
   - Filter retaining bracket
   - Fuel filter

To install:

4. Install or connect the following:
   - New fuel filter in the bracket
   - Fuel filter bracket

➡Be sure the arrows are pointing in the direction of the fuel flow

   - Fuel lines
   - Negative battery cable
5. Start the engine, check for leaks and repair if necessary.

Fuel Pump

REMOVAL & INSTALLATION

Except A3

The fuel pump is located in the fuel tank. It is recommended that the fuel tank not be filled more than ⅓ full. If necessary the fuel must be drained using an approved fuel cart.

1. Before servicing the vehicle, refer to the precautions section.
2. Properly relieve fuel pressure.
3. Remove or disconnect the following:

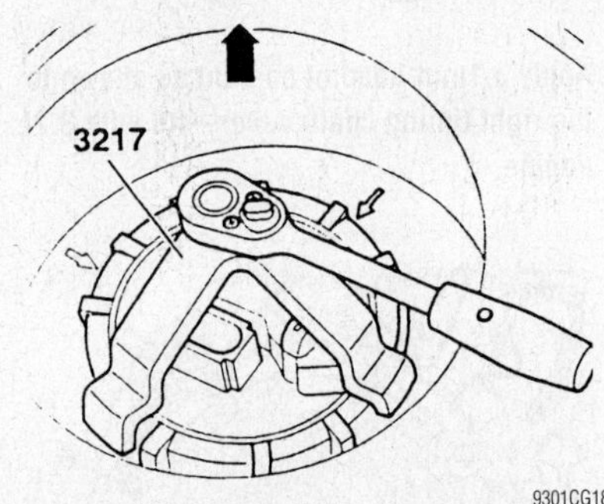

Tool 3217 is used to loosen and tighten the union nut on the fuel tank for access to the fuel pump and fuel level sending unit

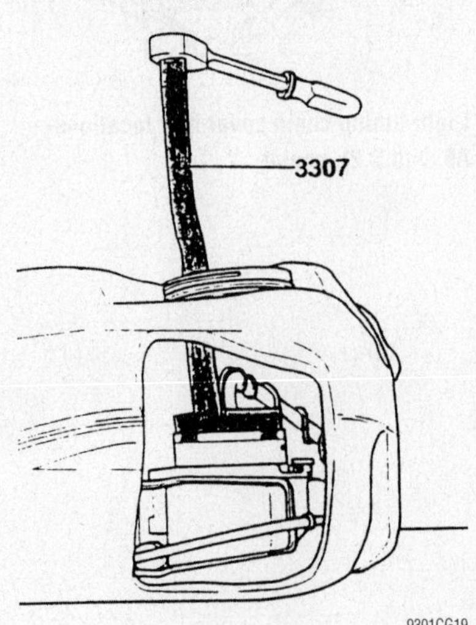

Tool 3307 is used to remove and install the fuel pump in the fuel tank

- Negative battery cable
- Inspection cover by lifting the cargo area trim

4. Mark the fuel pump/gauge sending unit assembly supply and return lines for reassembly.
5. Remove or disconnect the following:
   - Electrical connector
   - Union nut by loosening it using tool 3217
   - Fuel pump/gauge sending unit assembly by matchmarking it
   - Level connector and the fuel return line
6. Using tool 3307, turn the fuel pump module to the left (counterclockwise) about 15 degrees.
7. Remove or disconnect the following:
   - Fuel pump supply hose
   - Electrical connectors
   - Fuel pump

To install:

8. The installation is reverse of the removal procedure.
9. Install or connect the following:
   - Flange cover using a new O-ring lubricated the O-ring with fuel
   - Union nut using tool 3217. Torque it to 59 ft. lbs. (80 Nm)
   - Negative battery cable
10. Start the vehicle, check for leaks and repair if necessary.

A3

1. Before servicing the vehicle, refer to the precautions section.

➡To remove fuel pump, the fuel tank may only be a maximum of one third filled, drain the tank as needed.

2. Remove the spare wheel and formed insert for tool kit from under luggage compartment floor covering.
3. Disconnect the negative battery cable.
4. Remove the rear seat bench.
5. Unclip cover retainers for right sealing flange.
6. Disconnect the electrical harness connector at locking flange.
7. Mark fuel supply line and fuel return line.
8. Pull both lines from locking flange, by pressing release buttons.
9. Remove the locking ring using wrench 3087.
10. Pull the right sealing flange and fuel pump seal out from opening in fuel tank and lay aside with lines still attached.
11. Through the fuel tank opening, separate suction jet pipe from suction jet pump, by pressing release button.

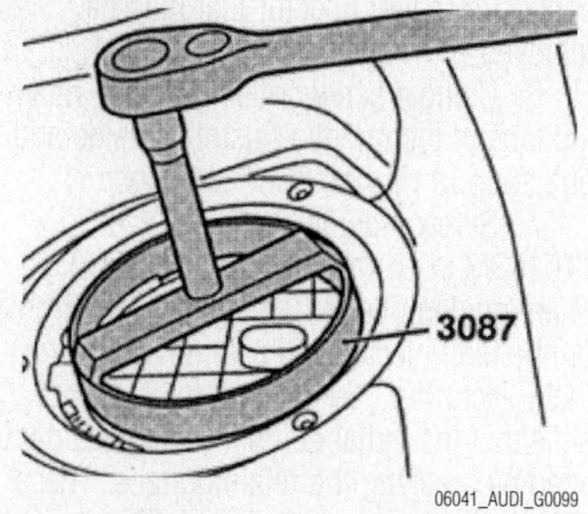

Remove/tighten the locking ring using wrench 3087—A3

12. Disconnect the fuel delivery line by pressing the release.
13. Remove the fuel pump.

➡When removing fuel pump, be sure not to bend floater arm of fuel level sensor.

To install:

14. Installation is the reverse of removal, please note the following:
    a. When inserting fuel pump, be sure not to bend fuel level sensor.
    b. Ensure fuel hoses are seated securely.
    c. Insert fuel pump into fuel tank with locking flange set aside.
    d. Connect the fuel lines until a click is heard..
    e. Install the new locking flange seal dry.
    f. Insert guide of locking flange into guide at fuel pump, observing spring while doing so.
    g. Press locking flange down against spring pressure and bring it into installation position.

➡The tab on locking flange faces toward the arrow marking and must lie between tabs and on fuel tank.

    h. Tighten locking ring to 145 Nm.
    i. Install the rear seat bench.

Fuel Injector

REMOVAL & INSTALLATION

Except A3

1. Before servicing the vehicle, refer to the precautions section.
2. Relieve the fuel system pressure.
3. Remove or disconnect the following:
   - Negative battery cable

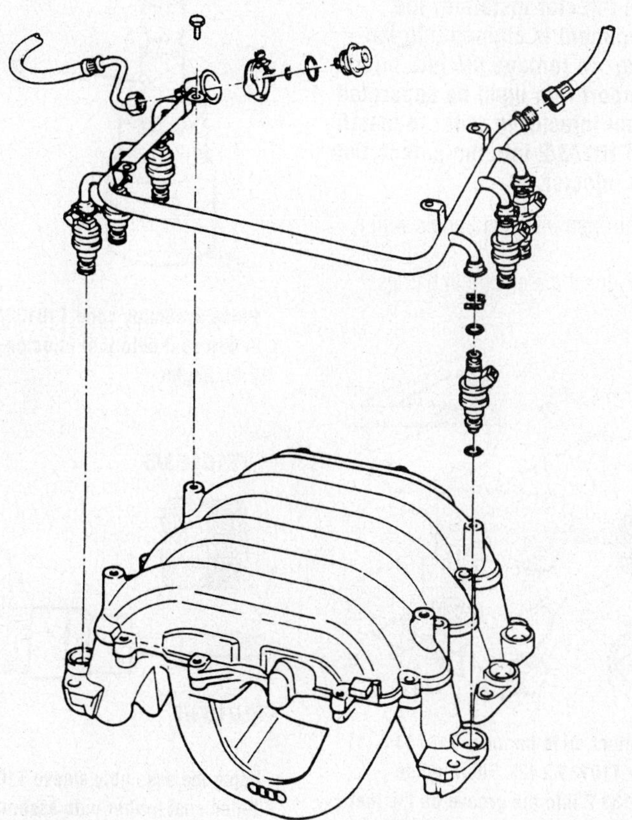

Fuel injectors and rail—2.8L V6 engine

- Engine under cover, if applicable
- Fuel lines
- Fuel injector connectors
- Fuel pressure line
- Fuel supply rail with injectors attached
- Fuel injectors from the supply rail

To install:

4. Install or connect the following:
   - Fuel injectors to the fuel supply rail using new O-rings
   - Fuel supply rail with injectors attached. Tighten the bolts to 89 inch lbs. (10 Nm)
   - Fuel injector connectors
   - Fuel pressure line
   - Fuel lines
   - Engine under cover, if applicable
   - Negative battery cable
5. Start the engine, check for leaks and repair if necessary.

A3

2.0L ENGINE

1. Before servicing the vehicle, refer to the precautions section.
2. Remove the intake manifold and fuel rail .
3. Carefully pull fuel injectors out of fuel rail.

➡With fuel injector installed, the radial adjustment is clipped into the support ring. To remove the fuel injector, the support ring must be separated from the fuel injector in order to install the puller T10133/2 into the indentation on the fuel injector.

4. Cover open intake channels with a clean rag.
5. Disconnect the electrical harness connector at fuel injector that is to be removed.
6. Using a screwdriver, bend the retaining tabs of the radial adjustment aside and pull support ring off from fuel injector.
7. Screw together slide hammer T10133/3 (1) with puller T10133/2 (2). Then, guide puller T10133/2 into the groove on the fuel injector and carefully drive the fuel injector out. By doing this, it is possible that the radial adjustment will be damaged by breaking the retaining tabs. These must be replaced when re-installing the fuel injector.

➡The combustion chamber seal must always be replaced before re-installing the high-pressure fuel injector.

8. Clean off the old Teflon seal, make sure not to damage the groove and rib in groove base.

➡If the groove is damaged, the fuel injector must be replaced.

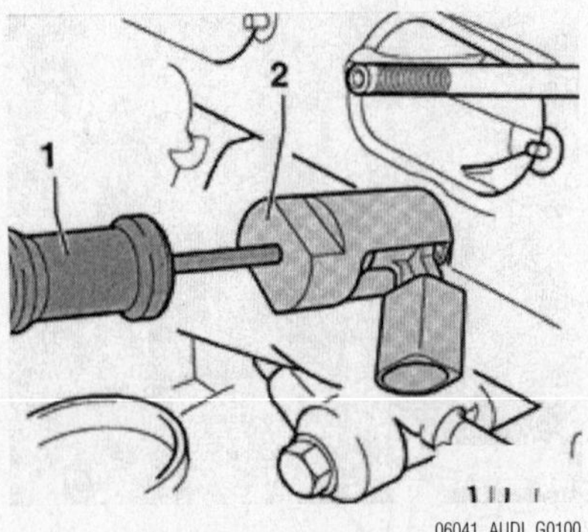

Screw together slide hammer T10133/3 (1) with puller T10133/2 (2). Then, guide puller T10133/2 into the groove on the fuel injector and carefully drive the fuel injector out—A3 with 2.0L engine

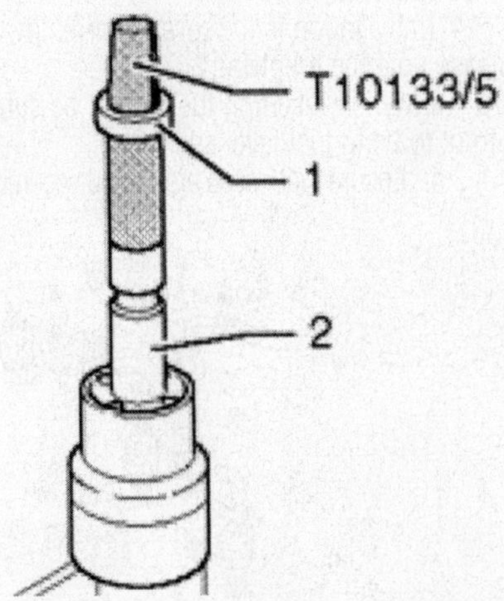

Place assembly cone T10133/5 with new Teflon seal onto fuel injector—A3 with 2.0L engine

T10133/6

T10133/5

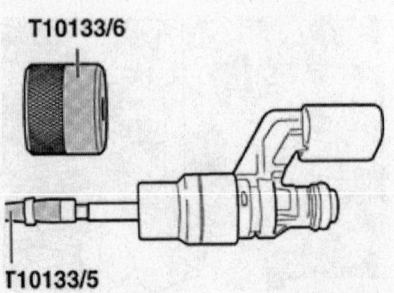

06041_AUDI_G0102

Using the assembly sleeve T10133/6, push Teflon seal further onto assembly cone T10133/5 until the Teflon seal is engaged in seal groove—A3 with 2.0L engine

To install:

9. Before installing new Teflon seal, the seal groove and shaft of the fuel injector must be clean.
10. Place assembly cone T10133/5 with new Teflon seal onto fuel injector.
11. Using the assembly sleeve T10133/6, push Teflon seal further onto assembly cone T10133/5 until the Teflon seal is engaged in seal groove. Do not use any lubricants.
12. By pushing the Teflon seal onto the fuel injector, the Teflon seal is expanded. For this reason, the Teflon seal must be contracted again, once it has been pushed on as follows:
    a. Using calibration sleeve T10133/7, a 180 degree rotating motion and light pressure, push calibration sleeve T10133/7 over the fuel injector up to stop. Pull calibration sleeve T10133/7 off again, rotating it in the opposite direction.
    b. Using calibration sleeve T10133/8 , 180 degree rotating motion and light pressure, push calibration sleeve T10133/8 over the fuel injector up to stop. Pull calibration sleeve T10133/8 off again, rotating it in the opposite direction.
13. Replace O-ring on the fuel injector and on spacer sleeve. Moisten O-ring with clean engine oil before installing.

➡The Teflon seal must not get oil on it.

14. Clean the bores for the high-pressure fuel injectors in the cylinder head before installing the fuel injectors.

➡An open intake valve could possibly hinder the cleaning process. In this case, turn the engine further by hand using a socket wrench at the crankshaft.

15. Equip the fuel injector again with support ring and clip the radial adjustment in at the support ring.

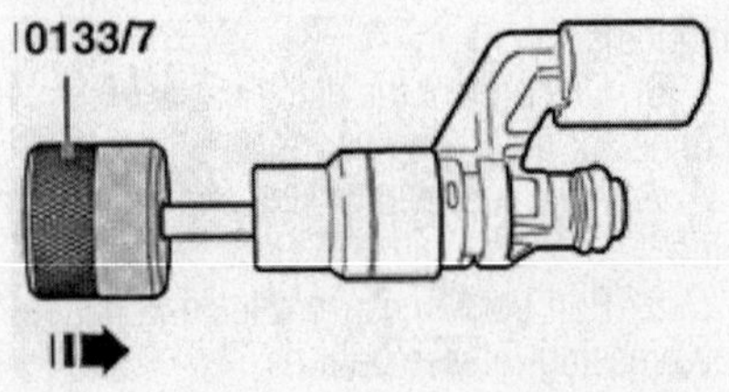

Using calibration sleeve T10133/7, a 180 degree rotating motion and light pressure, push calibration sleeve T10133/7 over the fuel injector up to a stop—A3 with 2.0L engine

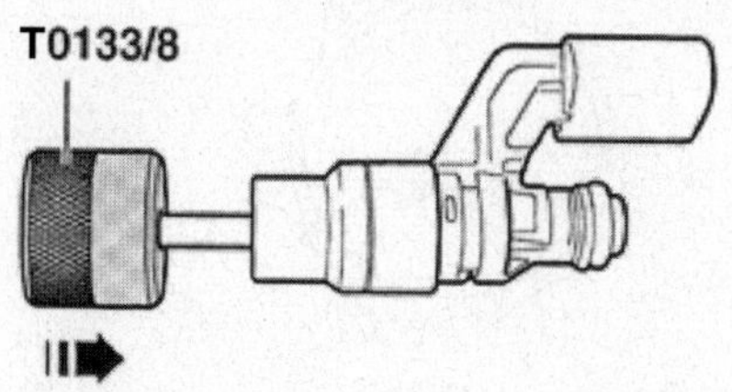

Using calibration sleeve T10133/8, 180 degree rotating motion and light pressure, push calibration sleeve T10133/8 over the fuel injector up to a stop—A3 with 2.0L engine

16. Using the assembly drift T10133/9, press the fuel injector into its bore in the cylinder head until it stops. Make sure fuel injectors are correctly positioned in cylinder head.
17. Install the remaining components in the reverse order of removal.

3.2L ENGINE

1. Before servicing the vehicle, refer to the precautions section.
2. Relieve the fuel system pressure.
3. Disconnect the fuel line.
4. Remove the intake manifold and fuel rail.
5. Pull off retaining clip and remove the relevant fuel injector.

To install:

6. Installation is the reverse of removal, please note the following:
   a. Replace O-rings at all opened connections. Never pull off the plastic cap from valve head to replace the front O-ring always lift O-ring over the plastic cap.
   b. Lightly lubricate O-rings with clean engine oil.
   c. Make sure fuel injectors are in place and seated correctly
   d. Check clip for unobstructed fitting.
   e. Place fuel rail with secured fuel injectors onto intake manifold, and apply uniform pressure to press it in and tighten the rail bolts to 10 Nm.

DRIVE TRAIN

Transaxle Assembly

REMOVAL & INSTALLATION

Manual

A3

1. Before servicing the vehicle, refer to the precautions section.
2. Obtain the radio anti-theft code.
3. Disconnect the negative battery cable.
4. Remove the engine cover.
5. Remove the battery and battery carrier.
6. Remove the circlips and at both cables.
7. Pull cable retainers with cables off from transmission shift lever and relay lever.
8. Remove the cable mounting bracket from transmission and tie it up off to side with cables.
9. Unfasten the ground cable from upper starter bolt.
10. Remove the connector and wiring from starter.

➡During the following procedures, make sure that brake fluid does not leak onto the starter or the transmission. If this does occur, these areas must be cleaned thoroughly.

11. Pull clamp for line/hose assembly out to the stop.
12. Pull pressure line off from bleeder/slave cylinder and seal it off.

⁂ CAUTION

Do not press clutch pedal after line/hose assembly has been removed.

13. Remove the bolts at engine/transmission assembly flange.
14. Remove the upper bolt on starter.
15. Disconnect the any hoses and cables in the area of engine lifting eyes on engine.
16. Install engine support tool.
17. Pretension engine with the support tool, but do not lift.
18. With the vehicle weight still on the wheels, loosen left front collar (drive axle) bolt a maximum of 90 degrees, otherwise wheel bearing will be damaged.
19. Loosen the left front wheel bolts.
20. Remove the left front wheel.
21. Remove the sound insulation.
22. Remove the front part of left front wheel housing liner.
23. If equipped, remove left front level control system sensor.
24. Disconnect the electrical connector for back-up light switch.
25. Remove the nut and bracket with electrical lines from lower starter bolt.
26. Remove the starter motor.
27. Separate air duct hose/pressure pipe connection.
28. Remove the exhaust system bracket from subframe.
29. Separate exhaust system at clamping sleeve and tie up front exhaust pipe.

➡To prevent damage, decoupling element of the front exhaust pipe must never be kinked by more than 10 degrees.

30. Disconnect the oil level thermal sensor.
31. Remove the shield for right drive axle.
32. Unbolt right drive axle from flange shaft on transmission.
33. Completely remove left drive axle

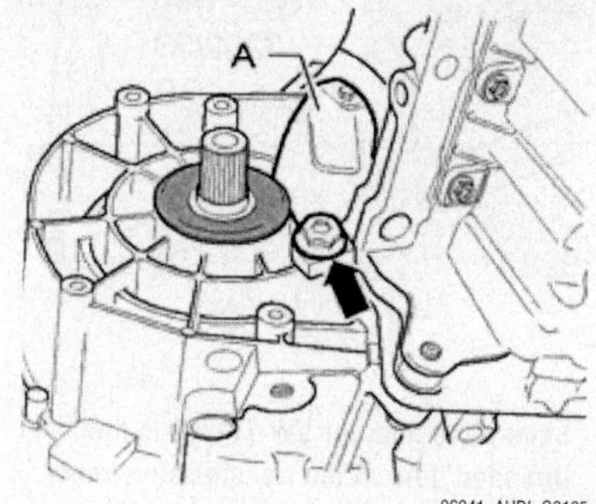

Remove the bolt and small cover plate—A3 2.0L engine with manual transmission

34. Remove the bolt and small cover plate.
35. Remove the bolts and remove pendulum support.
36. Before removing subframe, match mark its position.
37. Remove the subframe.
38. Remove the left console with the control arm.
39. Remove the bolts from assembly mounting on transmission.
40. Lower transmission the engine support tool by 60 mm.
41. Unfasten the console from transmission.
42. Pull off circlip and remove relay lever.
43. Remove the lower engine/transmission assembly bolts.
44. Loosen the engine transmission bolts and leave them screwed in hand-tight.
45. Attach transmission support 3282 for removal of the transmission using adjustment plate 3282/33.
46. Place transmission support 3282 in the engine/transmission assembly jack V.A.G 1383A.

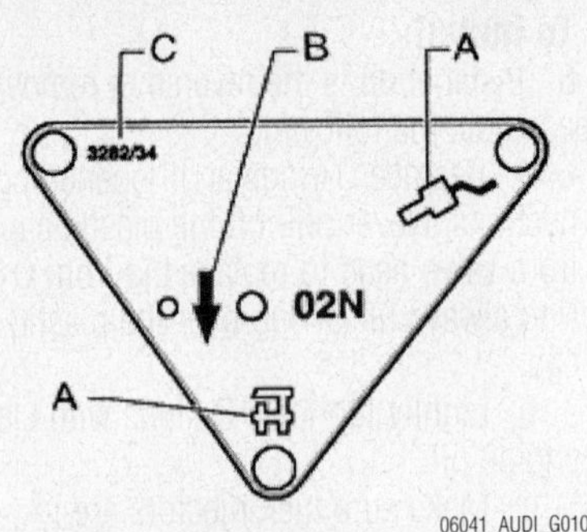

Attach transmission support 3282 for removal of the transmission using adjustment plate 3282/33—A3 2.0L engine with manual transmission

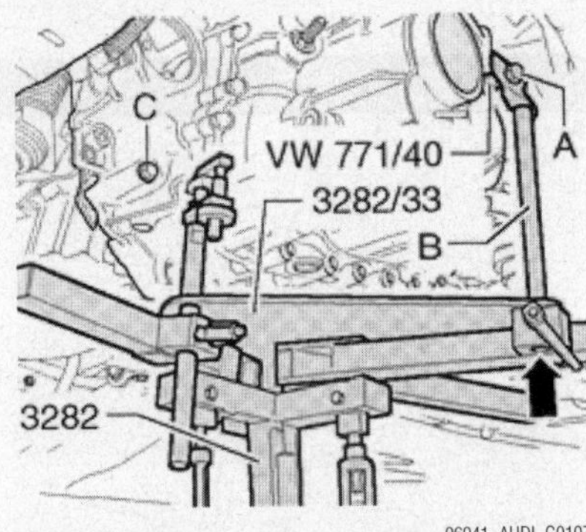

Secure the adapter VW 771/40 in the threaded bore of the transmission housing—A3 2.0L engine with manual transmission

47. Align the arms of transmission support according to holes in adjustment plate.
48. Screw the support elements onto adjustment plate as shown.
49. Instead of the support element, screw in the stud 3282/29.
50. Place engine/transmission assembly jack under vehicle, with arrow symbol on adjustment plate pointing to front of vehicle.
51. Align adjustment plate and transmission parallel to one another.
52. Secure the adapter VW 771/40 in the threaded bore of the transmission housing, as shown in the accompanying illustration.

Screw stud 3282/29 into bore for securing bolt of pendulum support on transmission.

53. Secure transmission on transmission support 3282 using an M10X M20 bolt. The drift should thereby sit flush at bottom with guide of transmission support 3282.
54. Remove the last connecting bolt for engine/transmission assembly.
55. Press transmission off from engine locating sleeves.
56. Using spindles of the transmission support 3282, screw up the transmission in the area of the differential.
57. Tilt transmission toward left via the spindles of the transmission support 3282.
58. Carefully lower transmission.

To install:

59. Replace self-locking nuts and bolts.
60. Replace bolts that are tightened to torque as well as O-rings and gaskets.
61. Clean the input shaft splines and on a used clutch plate, the hub splines, remove corrosion and apply only a very thin coating of grease onto the splines. Then move clutch plate back and forth on input shaft until the hub moves freely on the shaft. Excess grease must be removed.
62. If the transmission is being replaced, transfer over the transmission shift lever.
63. Check if the alignment sleeves for centering engine/transmission assembly are in the cylinder block, and install if necessary. If alignment sleeves are not installed, difficulty in shifting, clutch problems and possibly noises from the transmission (loose gear chatter) may develop.
64. Make sure that the intermediate plate is hooked into sealing flange and is pushed onto alignment sleeves.
65. Check release bearing for wear. If necessary, replace slave cylinder with release
66. Carefully raise transmission and move it into installation position.
67. Align transmission to engine and install. Tighten M12 bolts to 80 Nm and the M10 bolts to 40 Nm.
68. Insert relay lever into selector mechanism and clip in circlip.
69. Install the console onto transmission with new hex bolts to 60 Nm plus 90 degrees
70. Align engine/transmission assembly in installed position. To do this, lift until console is in complete contact with left assembly mounting.

➡To prevent the threads in the console from being damaged, the transmission mount and console must be parallel to one another.

71. Fasten console to left assembly mounting , using new hex bolts to 60 Nm plus 90 degrees.

⁂ CAUTION

Do not remove the engine support tool until all bolts for left assembly mounting have been tightened to torque specification.

72. Install the small cover plate for flywheel.
73. Install left and right drive axles to transmission

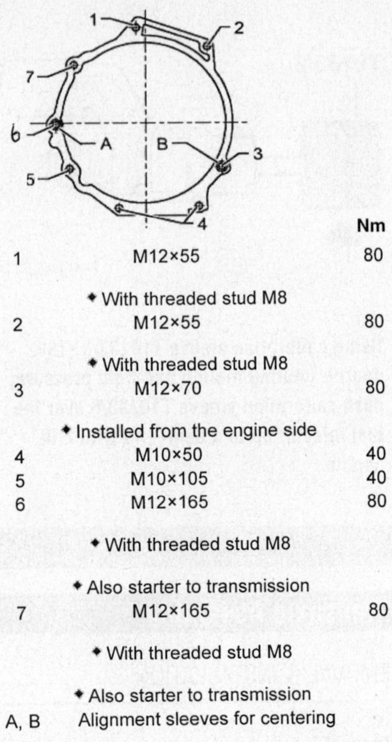

		Nm
1	M12×55	80
	◆ With threaded stud M8	
2	M12×55	80
	◆ With threaded stud M8	
3	M12×70	80
	◆ Installed from the engine side	
4	M10×50	40
5	M10×105	40
6	M12×165	80
	◆ With threaded stud M8	
	◆ Also starter to transmission	
7	M12×165	80
	◆ With threaded stud M8	
	◆ Also starter to transmission	
A, B	Alignment sleeves for centering	

06041_AUDI_G0108

Transmission bolt location and their torque values—A3 2.0L engine with manual transmission

74. Install the shield for right drive axle.
75. Install the subframe/control arm.
76. Attach steering gear to sub frame.
77. Fasten pendulum supports to transmission first and then to subframe bolts , and use new bolts to fasten. Replace the bolts, tighten the outer bolt to 100 Nm plus 90 degrees and the two inner bolts to 40 Nm plus 90 degrees.
78. Check adjustment of subframe mounting, adjust if necessary
79. Install the remaining components in the reverse order of removal.
80. Bleed the clutch system .
81. Check the gear oil level in manual transmission.

A4 AND A6

1. Before servicing the vehicle, refer to the precautions section.
2. Remove or disconnect the following:
   - Negative battery cable
   - Intake hose
   - Expansion tank (move aside)
   - Oxygen sensors
   - Upper engine-to-transaxle mounting bolts
   - Engine undercover and bracket
   - Front exhaust pipe with the catalytic converter

- Driveshaft for All Wheel Drive (AWD) models only
- Starter
- Shift rod and joint bolt at the transaxle and separate from the rear of the shift rod
- Shift rod
- Vehicle Speed Sensor (VSS) connector
- Backup light switch connector
- Driveshaft from transaxle and rest on heat shield

3. Support the transaxle, using a transmission/transaxle jack.
4. Remove or disconnect the following:
   - Transaxle mount heat shield
   - Right mount at the transaxle
   - Left mount with the bushings
   - Left and right halfshafts
   - Axle heat shield
   - Remaining engine-to-transaxle mounting bolts
5. Pry the transaxle off the dowel sleeves and carefully lower the transaxle until the slave cylinder is accessible, approx. 6 inches. (15cm).
6. Remove or disconnect the following:
   - Clutch slave cylinder from the transaxle with the hydraulic line attached
   - Transaxle

To install:

7. Installation is the reverse of the removal procedure, while using the following torque values:
   - Transaxle flange M12 bolts: 48 ft. lbs. (65 Nm)
   - Transaxle flange M10 bolts: 33 ft. lbs. (45 Nm)
   - Transaxle flange M8 bolts: 18 ft. lbs. (25 Nm)
   - Slave cylinder bolts: 15 ft. lbs. (20 Nm)
   - Transaxle mount bolts: 30 ft. lbs. (40 Nm)
   - Shift rod bolts: 15 ft. lbs. (20 Nm)
   - Pushrod to transaxle: 30 ft. lbs. (40 Nm)
   - Driveshaft to transaxle: 41 ft. lbs. (55 Nm)
   - Exhaust pipe clamp: 30 ft. lbs. (40 Nm)
   - Axle shaft to drive flange: M8 bolts 30 ft. lbs. (40 Nm); M10 bolts 52 ft. lbs. (70 Nm)

TT

1. Before servicing the vehicle, refer to the precautions section.
2. Place gearshift in neutral.
3. Remove or disconnect the following:
   - Negative battery cable
   - Engine covers
   - Engine top cover over cylinder head
   - Hoses and connectors from air filter housing
   - Air filter housing
   - Battery and battery carrier
   - Cable mounting bracket at transaxle
   - Shift cables at transaxle
   - Speed sensor connector
   - Ground cable
   - Cables and connectors from starter
   - Upper starter mounting bolts
   - Mark rotation direction of accessory belt, then remove belt
   - Air duct at lower right cross member
   - Front Wheels
   - Engine undercover
   - Charge air cooler lines
   - Power steering pump pulley
   - Clamp off power steering hoses
   - Pressure pipe at power steering pump
   - Power steering pipe from transaxle
   - Starter
   - Reverse light connector
   - Clutch cylinder hydraulic line
   - Exhaust pipe clamps
   - Loosen exhaust system and suspend under vehicle
   - Mark position of driveshaft relative to flexible coupling
   - Driveshaft from flexible coupling
   - Press front driveshaft tube as far back as possible
   - Sway brace support under vehicle
   - Steering box bolts
4. Place a transaxle jack under transaxle.
5. Remove the subframe bolts.
6. Lower the subframe and leave it resting on the ball joints.
7. Suspend steering box up out the way.
8. Disconnect the drive axles and place on subframe.
9. Pull the subframe back and tie it under vehicle.
10. Remove right drive axle heat shield bolted to bevel box.
11. Remove the bevel box.
12. Remove the flywheel heat shield behind the bevel box.
13. Install Engine Support bracket 10-222A and supports 10-222A1.
14. Install Retainer 3180 to right engine lifting eye and attach it to support bracket 10-222A.
15. Lift engine slightly with spindle of support bracket 1022A.
16. Remove the left side transaxle support bolts.

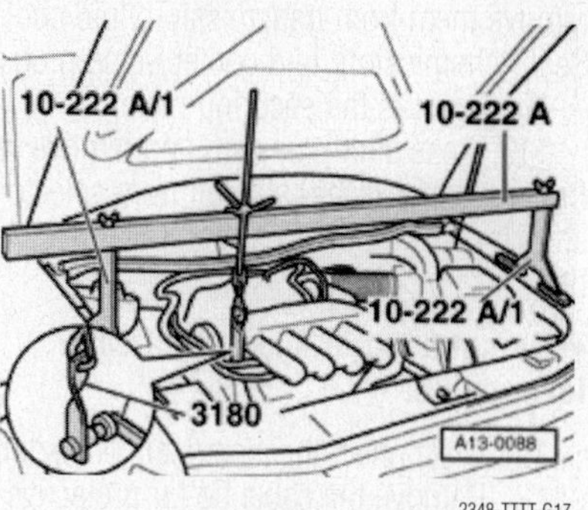

Attaching engine support and lifting tools—Audi TT 1.8L 4-Cyl engine

17. Lower the engine slightly.
18. Position transaxle jack so it can support full weight of transaxle.
19. Remove the engine to transaxle mounting bolts.
20. Slide transaxle back and then lower away from vehicle.

To install:

21. Installation is the reverse of the removal procedure, while using the following torque values:
    - Cable mounting bracket: 17 ft. lbs. (23 Nm)
    - Transaxle flange M12 bolts: 59 ft. lbs. (80 Nm)
    - Transaxle flange M10 bolts: 30 ft. lbs. (40 Nm)
    - Slave cylinder bolts: 18 ft. lbs. (25 Nm)
    - Starter bolts: 48 ft. lbs. (65 Nm)
    - Steering box bolts: 15 ft. lbs. (20 Nm) plus 90°
    - Subframe bolts: 74 ft. lbs. (100 Nm) plus90°
    - Sway brace side bolts: 15 ft. lbs. (20 Nm) plus 90°
    - Sway brace bottom bolts: 30 ft. lbs. (40 Nm) plus 90°

Automatic

A3 FRONT WHEEL DRIVE

1. Before servicing the vehicle, refer to the precautions section.
2. Place selector lever in position "P".
3. Disconnect the electrical harness connector at Mass Air Flow (MAF) sensor.
4. Unclip air intake nozzle.
5. Remove the engine cover.
6. Disconnect the battery cables, negative cable first.
7. Remove the battery.
8. Remove the battery carrier.
9. Clamp off coolant hoses of transmission fluid cooler using hose clamps and

remove them from transmission fluid cooler. Seal transmission fluid cooler using plugs.

10. Remove the securing plate.

11. Press circlip upward and remove it. Press selector lever cable off from selector shaft lever and pull it out of mounting bracket.

➡Do not bend or kink the selector lever cable.

12. Disconnect the electrical connector.

13. Remove the cable tie for protective sleeve.

14. Flip protective sleeve back and unscrew B+ cable from solenoid switch.

CAUTION

Never touch the contacts in the transmission connector, under any circumstances, because static discharge could destroy the control module along with the mechatronics unit. Touch a vehicle ground with your hand to discharge static.

15. Twist bayonet connection of 20-pin connector to release it from transmission, and disconnect.

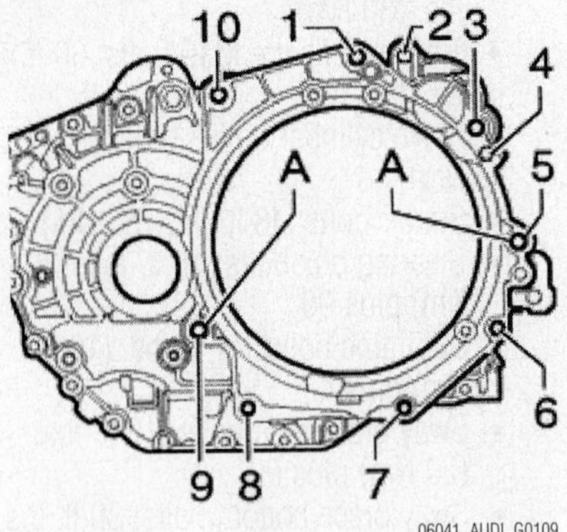

Transmission bolt locations—Front wheel drive A3 with automatic transmission

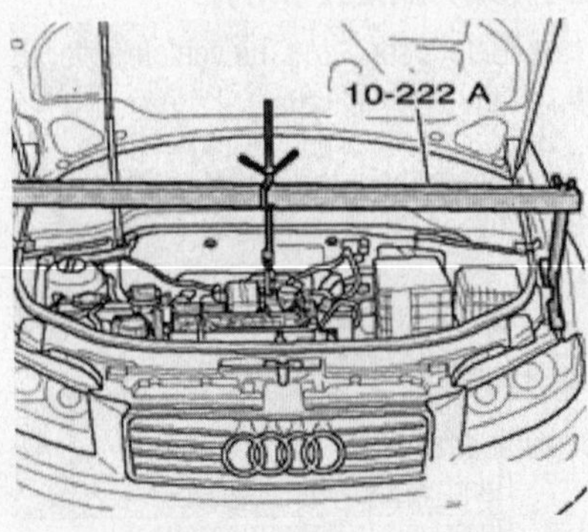

Attach an engine support tool—Front wheel drive A3 with automatic transmission

16. Unbolt ground cable.

17. Unhook connector and wiring harnesses from bracket on transmission.

18. Remove the bolts starter.

19. Remove the engine/transmission assembly bolts which are accessible from above.

20. Set engine support tool onto bolted flanges of fender.

21. Hook spindle of engine support tool into left engine lifting eye.

22. Pretension engine using spindle, but do not lift.

23. Remove the both front wheels.

24. Remove the center sound insulation fasteners.

25. Remove the left sound insulation fasteners.

26. Remove the air duct hose to charge air cooler, if equipped, remove shield at bottom of transmission.

27. Remove the both nuts that secure bracket on front side of transmission oil pan.

➡The threaded studs are welded to the front of the transmission oil pan.

28. Press off the bracket from threaded studs of transmission oil pan and lay it aside up top, so that it will not get caught up when lowering transmission.

29. Remove the heat shield for right drive axle.

➡To prevent damage, decoupling element of the front exhaust pipe must never be angled by more than 10 degrees.

30. Loosen the nuts of clamping sleeve.

31. To disconnect exhaust system, slide the clamping sleeve towards the rear.

32. Remove the bracket for exhaust system.

33. Remove the bolt from exhaust system bracket.

34. Remove the bolts and remove pendulum support.

35. Disconnect the oil Level thermal sensor connector and unclip the bracket at the sub frame.

36. Unbolt the drive axles from the transmission flange and support them with wire.

37. Remove the bolt and pull right flange shaft off from transmission.

38. Disconnect the left front level control system sensor connector.

39. Remove the bolt and the sub frame.

➡When removing the control arm console, the locating fixture T10096 must be used, otherwise a vehicle alignment will need to be performed afterwards.

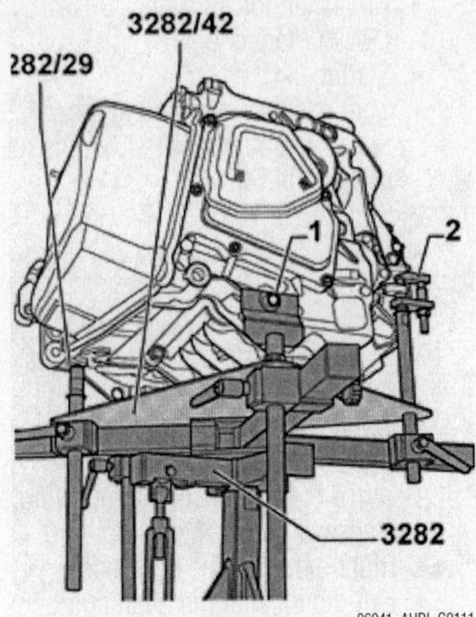

Remove the transmission with a suitable jack and attachments—Front wheel drive A3 with automatic transmission

40. Remove the left control arm console from body.

41. Remove the bolts and remove transmission support.

42. Lower the engine/transmission assembly, using spindle of engine support tool 100 to 110mm between transmission housing and transmission mounting.

43. To remove transmission, transmission support 3282 is equipped with adjustment plate 3282/42 and positioned on engine/transmission assembly jack .

44. Align arms of transmission support to correspond with holes in adjustment plate 3282/42.

45. Screw in support elements as shown on adjustment plate 3282/42.

46. Place engine/transmission assembly jack under vehicle, with the arrow on adjustment plate 3282/42 pointing to front of vehicle.

47. Align transmission support 3282 parallel to transmission.

48. Secure support element to transmission with bolt.

49. Screw stud 3282/29 into transmission.

50. Insert support element into transmission and secure by tightening nut.

51. Support transmission from underneath by raising engine/transmission assembly jack .

52. Remove the remaining engine/transmission assembly bolts.

53. Press transmission off from alignment sleeves.

54. Pull transmission off from engine.

55. Carefully lower transmission using engine/transmission assembly jack.

56. Use spindles of transmission support 3282 to adjust position of transmission while lowering.

To install:

57. Replace self-locking nuts and bolts.
58. Replace bolts that are tightened to torque as well as O-rings and gaskets.
59. Clean the input shaft splines and the hub splines, remove corrosion and apply only a very thin coating of grease onto the splines. Excess grease must be removed.
60. Check if the alignment sleeves for centering engine/transmission assembly are in cylinder block; if necessary, install alignment sleeves.
61. Replace seal for right flange shaft
62. Make sure that the intermediate plate is hooked into sealing flange of engine and is pushed onto alignment sleeves.
63. Carefully raise transmission using engine/transmission assembly jack and move it into installation position using transmission support 3282.
64. Fasten transmission to engine. Tighten the M12 bolts to 80 Nm and the M10 bolts to 40 Nm.
65. Install the mountings as follows:
    a. Position transmission support between transmission and support arm of transmission mount.
    b. Fasten transmission support to transmission using new bolts and tighten to 40 Nm plus 90 degrees.
    c. Use spindle of engine support tool to lift transmission up to support arm of transmission mounting.
    d. Temporarily screw in bolts hand-tight.
    e. Remove the transmission support 3282 from transmission.
    f. Install the left control arm console.
    g. Install the sub frame and tighten the bolts to 110 Nm plus 90 degrees.
66. Install the left front level control system sensor, screw in bolt and connect electrical connector.
67. Install the flange.
68. Install the drive axles.
69. Fasten pendulum supports to transmission first and then to subframe bolts.
70. Install the exhaust system and align it free of tension.
71. Install the starter
72. Check adjustment of assembly mounting. Fasten bolts of assembly mounting for transmission to 60 Nm plus an additional 90 degrees.
73. Remove the engine support tool from engine.

➡Do not bend or kink the selector lever cable.

74. Install the selector lever cable to mounting bracket and to selector shaft lever.
75. Install the remaining components in the reverse order of removal.

A3 ALL WHEEL DRIVE

1. Before servicing the vehicle, refer to the precautions section.
2. Remove the molded tool insert under luggage compartment floor covering.
3. Remove the battery compartment cover.
4. Remove the molded insert over battery.
5. Disconnect the negative battery cable.
6. Remove the vacuum hose on air duct hose.
7. Disconnect the air guide hose at throttle valve control module.
8. Disconnect the electrical harness connector at Mass Air Flow (MAF) sensor.
9. Remove the upper part of air filter housing.
10. Remove the filter insert.
11. Remove the air duct cover.
12. Unclip air duct.
13. Unscrew bottom part of air filter housing.
14. Remove the air filter housing bolts.
15. Place selector lever in position "P".
16. Press lock washer upward and remove it. Press off selector lever cable from lever/selector shaft in direction of and set it on top. Do not loosen bolt.
17. Remove the both bolts and remove the selector cable. Do not kink selector lever cable.
18. Remove the starter.

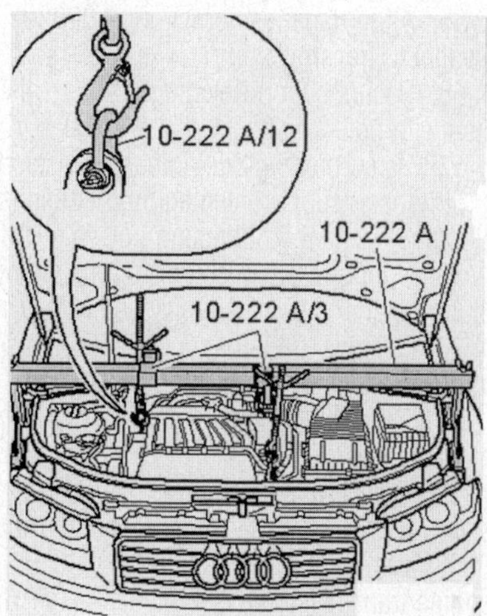

Attaching the engine support device to the engine—All wheel drive A3 with automatic transmission

19. Mark coolant hoses to transmission fluid cooler to prevent mix-ups when installing.
20. Disconnect the transmission fluid cooler hoses with hose clamps from transmission fluid cooler. Plug transmission fluid cooler.
21. Remove the all engine/transmission assembly bolts that can be reached from top.
22. Secure a shackle at right rear engine lifting eye.
23. Position engine support tool with adapters on to bolted flanges of fender.
24. Disconnect the any hoses and cables in area of engine support tool lifting eyes on engine.
25. Engage spindles in lifting eyes or at shackles.
26. Slightly pre-tension engine/transmission assembly on spindle.
27. Remove the center noise insulation fasteners.
28. Remove the left and right noise insulation fasteners.
29. Remove the both front wheels.
30. If installed, remove guard plate at bottom of transmission.
31. Remove the bracket for exhaust system.
32. Unclip bracket for electrical wiring to oil level thermal sensor at subframe. Remove the bolts and bracket.
33. Remove the nut and press bracket upward so that it does not hang when transmission is lowered.

➡Bracket is secured at front of transmission fluid pan.

34. Disconnect the exhaust system at double clamp. Push clamping sleeve toward rear.
35. Tie exhaust system in position.

➡Flex joint of exhaust system must not be bent more than 10 degrees, otherwise it may be damaged.

36. Remove the right axle shaft heat shield from bevel box.
37. Remove the drive axles from transmission flange.
38. Secure the axles with wire.
39. Mark the position of flexible disc and bevel box flange in relation to each other.
40. Remove the drive shaft flexible disc at bevel box.

➡When loosening bolts, provide support at drive shaft flange/final drive with a suitable lever. Press drive axle horizontally toward rear as far as possible.

CAUTION

The sealing ring in drive shaft flange must not be damaged when removing transmission. If the seal is damaged, the drive axle must be replaced. Be aware that loosening pedal support/subframe bolts, the engine/transmission assembly moves forward slightly.

41. Remove the screws and press assembly forward slightly and then detach drive shaft from bevel box.
42. Set down drive shaft above bevel box and secure it in position.
43. If equipped, remove the left front level control system sensor.
44. Remove the sub frame.
45. Remove the left control arm console from the body.
46. Remove the bolts and at bevel gear transfer case bracket.
47. Remove the bracket for bevel gear transfer case.
48. Remove the bolts and remove transmission console.

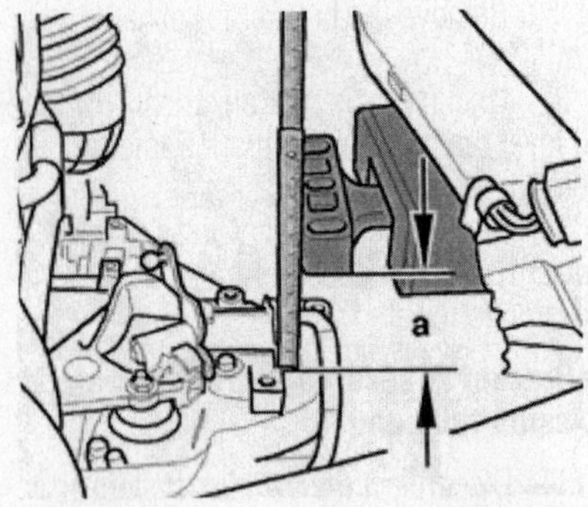

The distance between transmission housing and transmission bearing must be 100 to 110 mm—All wheel drive A3 with automatic transmission

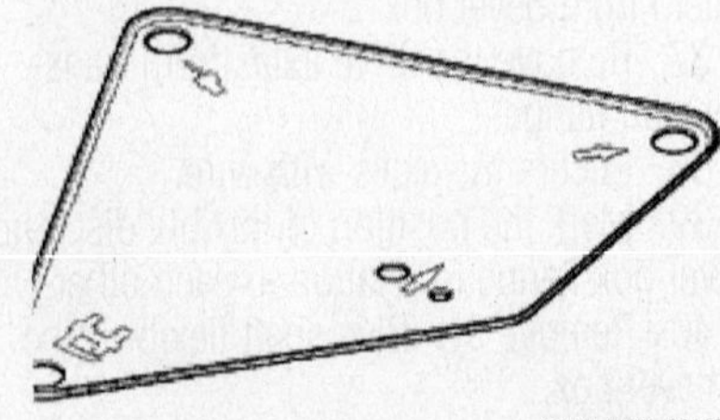

To remove transmission, the transmission support 3282 is equipped with adjustment plate 3282/42 and positioned on engine/transmission assembly jack—All wheel drive A3 with automatic transmission

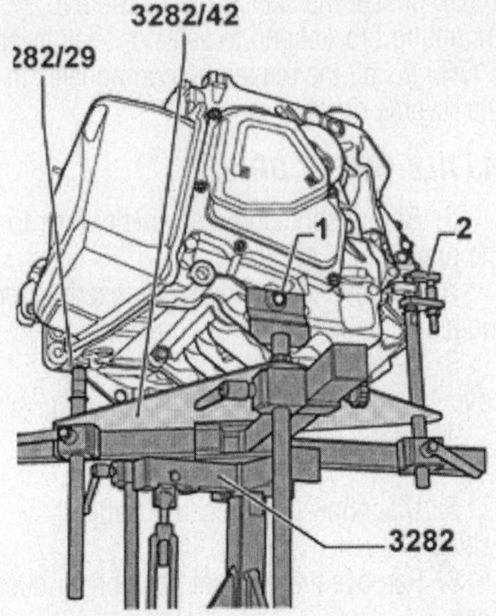

Screw in support elements as shown on adjustment plate 3282/42—All wheel drive A3 with automatic transmission

49. Lower engine/transmission assembly at spindle a short distance.
50. The distance between transmission housing and transmission bearing must be 100 to 110 mm.
51. To remove transmission, the transmission support 3282 is equipped with adjustment plate 3282/42 and positioned on engine/transmission assembly jack .
52. Align arms of transmission support to correspond with holes in adjustment plate 3282/42.
53. Screw in support elements as shown on adjustment plate 3282/42.
54. Place engine/transmission assembly jack under vehicle, arrow symbol on adjustment plate 3282/42 points in direction of travel/vehicle.
55. Align transmission support 3282 parallel to transmission.
56. Secure support element to transmission with bolt.
57. Screw bolt 3282/29 into transmission.
58. Insert support element into transmission and secure by tightening nut.
59. Support transmission from below by lifting engine/transmission assembly jack .
60. Remove the remaining engine/transmission assembly connecting bolts.
61. Press off transmission from alignment sleeves.
62. Detach transmission from engine, starting on bevel box side and then at front.
63. Carefully lower transmission over engine/transmission assembly jack , making sure no wires are pinched.
64. Use spindles of transmission support 3282 to adjust position of transmission while lowering.

To install:

65. Replace self-locking nuts and bolts.
66. Replace bolts that are tightened to torque as well as O-rings and gaskets.
67. Clean the input shaft splines and on a used clutch plate, the hub splines, remove corrosion and apply only a very thin coating of grease onto the splines. Then move clutch plate back and forth on input shaft until the hub moves freely on the shaft. Excess grease must be removed.
68. If the transmission is being replaced, transfer over the transmission shift lever.
69. Check if the alignment sleeves for centering engine/transmission assembly are in the cylinder block, and install if necessary. If alignment sleeves are not installed, difficulty in shifting, clutch problems and possibly noises from the transmission (loose gear chatter) may develop.
70. Make sure that the intermediate plate is hooked into sealing flange and is pushed onto alignment sleeves.
71. Check release bearing for wear. If necessary, replace slave cylinder with release
72. Carefully raise transmission and move it into installation position.
73. Align transmission to engine and install. Tighten M12 bolts to 80 Nm and the M10 bolts to 40 Nm.
   a. Use spindle of engine support tool to lift transmission up to support arm of transmission mounting.
   b. Temporarily screw in bolts hand-tight.
   c. Remove the transmission support 3282 from transmission.
74. Tighten the mounting bracket to transmission to 20 Nm plus 90 degrees.
75. Tighten the transmission to console to 40 Nm plus 90 degrees.

CAUTION

Before installing the bolts, the transmission console and transmission mount support arm must be absolutely parallel, otherwise the threads will be damaged. If necessary, use a floor jack to raise the rear of the transmission. Remove the engine support tool only when all sub frame mounting bolts are tightened to the torque specifications and the sub frame is installed.

76. Install the bolts and hand-tight for securing bevel box bracket as follows:.
   a. Step 1: Tighten nuts (1) first to 3 Nm.
   b. Step 1: Tighten nuts (2) to 35 Nm.

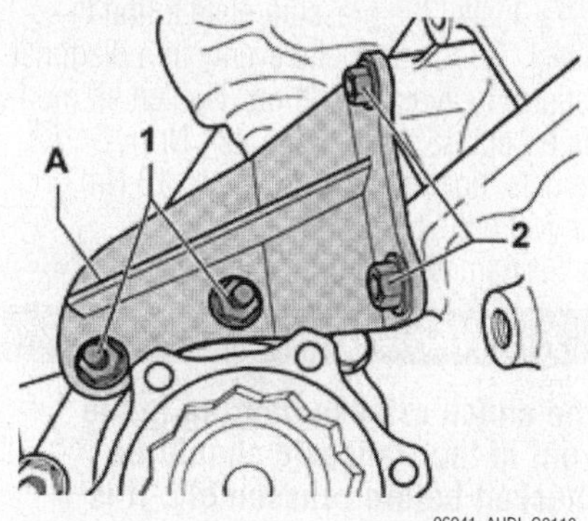

Bevel box bracket bolt locations—All wheel drive A3 with automatic transmission

c. Step 1: Tighten nuts to (1) 45 Nm.

77. Install the pendulum support to transmission bolts to 40 Nm plus 90 degrees.

78. Install the pendulum support to sub frame bolts to 110 Nm plus 90 degrees.

79. Install the transmission mount to transmission console bolts to 60 Nm plus 90 degrees.

80. Install the left front level control system sensor, screw in bolt and connect electrical connector.

81. Install the flange.

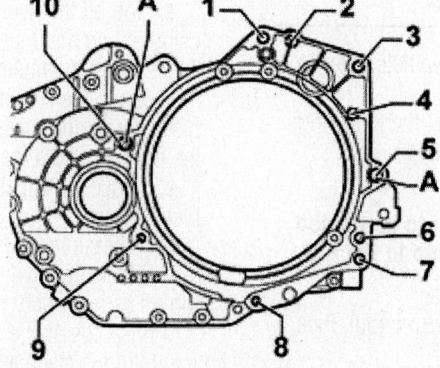

82. Install the drive axles.

83. Fasten pendulum supports to transmission first and then to sub frame bolts.

84. Install the exhaust system and align it free of tension.

85. Install the starter

86. Check adjustment of assembly mounting. Fasten bolts of assembly mounting for transmission to 60 Nm plus an additional 90 degrees.

87. Remove the engine support tool from engine.

➡Do not bend or kink the selector lever cable.

88. Install the selector lever cable to mounting bracket and to selector shaft lever.

89. Install the remaining components in the reverse order of removal.

A4 AND A6

1. Before servicing the vehicle, refer to the precautions section.

2. Remove or disconnect the following:

- Negative battery cable
- Engine cover
- Air cleaner duct
- Air cleaner assembly
- Upper engine-to-transaxle bolts

3. Using an engine support tool, secure it to the engine and the vehicle.

4. Remove or disconnect the following:

- Front wheels
- Transmission range switch, connector and bracket
- Heat shields for drive shafts
- Drive shaft from transmission flanges
- Speed sensor connectors (mark for reinstallation)
- Both oxygen sensors (mark left and right for reinstallation)
- Both top bolts at the front of the engine
- Starter
- Torque converter-to-driveplate bolts through the starter opening
- Torque converter cover plate
- Coolant hoses at the transmission cooler by clamping them off

5. Support the halfshafts

6. Remove or disconnect the following:

- Inner halfshaft-to-transaxle bolts
- Remove the ball joint and the support
- Oil filler tube from the oil pan and drain the fluid
- Exhaust pipe-to-transaxle bracket
- Selector cable bracket from the transaxle
- Selector cable circlip and the cable at the transaxle shift lever
- Accelerator cable bracket and the cable from the operating lever
- Center bolt, from the transaxle mount.

7. Using the engine support tool, lift the engine slightly.

8. Remove the throttle cable bracket bolts and the bracket.

9. Support the transaxle and lift it slightly.

10. Remove or disconnect the following:

- Lower transaxle-to-engine bolts
- Transaxle from the engine

➡Be sure to secure the torque converter.

To install:

11. Installation is the reverse of the removal procedure, while using the following torque values:

- Transaxle flange bolts: 41 ft. lbs. (56 Nm)
- Subframe bolts: 52 ft. lbs. (71 Nm)
- Transaxle mount center bolt: 30 ft. lbs. (40 Nm)
- Torque converter bolts: 63 ft. lbs. (85 Nm)
- Halfshaft bolts: 33 ft. lbs. (45 Nm)
- Ball joint bolts: 48 ft. lbs. (65 Nm)

Securing engine/transmission

		Nm
1, 3, 6	M12x55	80
5	M12x65	80
7, 8	M10x50	40
9	M10x45	40
10	M12x70	80
A	Alignment sleeves for centering	

06041_AUDI_G0117

Transmission bolt locations and torque values—All wheel drive A3 with automatic transmission

- Starter motor bolts: 48 ft. lbs. (65 Nm)

TT

1. Before servicing the vehicle, refer to the precautions section.
2. Place gearshift in neutral.
3. Before servicing the vehicle, refer to the precautions section.
4. Place transaxle in neutral.
5. Remove or disconnect the following:
   - Negative battery cable
   - Engine covers
   - Engine top cover over cylinder head
   - Hoses and connectors from air filter housing
   - Air filter housing
   - Battery and battery carrier
   - Selector lever off selector lever shaft
   - Selector lever cable
   - Multi-function switch connector
   - Cables and connectors from starter
   - Starter
   - ATF cooler hoses after clamping off
6. Install Engine Support bracket 10-222A and supports 10-222A1.
7. Install Retainer 3180 to right engine lifting eye and attach it to support bracket 10-222A.
8. Lift engine slightly with spindle of support bracket 1022A.
9. Remove or disconnect the following:
   - Ground cable at transaxle
   - Engine to transaxle mounting bolts accessible from above
   - Engine undercover
   - Front wheels
   - Left drive shaft
   - Left side sound insulator
   - Power steering line at transmission
   - Front tubular cross member (if equipped)
   - Cross brace at subframe
   - Sway brace
10. Press out the right ball joint.
11. Remove right drive shaft heat shield.
12. Disconnect right drive shaft from transaxle, but not from wheel hub
13. Tie the right drive shaft up out of the way.
14. Remove six torque converter nuts, turning crankshaft 60° each time to access nuts.
15. Remove exhaust pipe clamps and suspend exhaust pipe under vehicle.
16. Remove electrical connectors from transaxle.
17. Remove transaxle console support bracket.
18. Lower engine and transaxle assembly approximately 2 inches (50 mm).
19. Position transaxle jack so it can support full weight of transaxle.
20. Remove the engine to transaxle mounting bolts.
21. With the aid of an assistant slide transaxle back slightly until torque converter can be secured in place, then lower transaxle and move away from vehicle.

To install:

22. Installation is the reverse of the removal procedure, while using the following torque values:
    - Transaxle to engine M12 bolts: 59 ft. lbs. (80 Nm)
    - Transaxle to engine M10 bolts: 33 ft. lbs. (45 Nm)
    - Starter bolts: 59 ft. lbs. (80 Nm)
    - Subframe bolts: 74 ft. lbs. (100 Nm) plus90°
    - Sway brace side bolts: 15 ft. lbs. (20 Nm) plus 90°
    - Sway brace bottom bolts: 30 ft. lbs. (40 Nm) plus 90°

Clutch

ADJUSTMENT

All vehicles use a hydraulic clutch release mechanism. No free-play adjustment is required or possible.

REMOVAL & INSTALLATION

1. Before servicing the vehicle, refer to the precautions section.
2. Remove or disconnect the following:
   - Negative battery cable
   - Transaxle

➡If the pressure plate is to be reused, matchmark its relationship to the flywheel.

3. Using a Flywheel Locking tool, lock the flywheel.
4. Remove or disconnect the following:
   - Pressure plate from the flywheel by loosening the bolts alternately, a little at a time, to prevent warpage
   - Clutch disc

To install:

5. Install or connect the following:
   - Clutch with the driven plate on the pressure plate so the spring cage is facing the pressure plate.
6. Hold the clutch assembly against the flywheel, aligning the matchmarks and the dowel pins on the flywheel with the pressure plate. Insert an alignment shaft tool through the pressure plate and the driven plate into the crankshaft pilot bearing.
7. Install the pressure plate to the flywheel. Torque the bolts evenly in a diagonal pattern, to avoid distortion. Tighten all models except A3 to 18 ft. lbs. (24 Nm). On A3 models, tighten the M6 bolts to 13 Nm or the M8 bolts to 20 Nm.
8. Remove the alignment shaft.

⁂ WARNING

The clutch release bearing in the front of the transaxle should be checked before reassembly. It is retained by 2 springs.

9. Install or connect the following:
   - Transaxle
   - Negative battery cable

Hydraulic Clutch System

BLEEDING

The clutch system can be bled using a pressure bleeder. Follow the instructions that come with the pressure bleeder for the proper pressure bleeding procedure. The maximum line pressure while pressure bleeding must not exceed 36 psi (248 kPa).

1. Before servicing the vehicle, refer to the precautions section.
2. To bleed the system perform the following:
   a. Top off the hydraulic fluid reservoir using a fluid that meets the standards of the vehicle's hydraulic system.
   b. Open the clutch slave cylinder bleed screw and press the clutch pedal to the floor and hold the pedal down.
   c. Close the clutch slave cylinder bleed screw.
   d. Release the clutch pedal.
   e. Check the hydraulic fluid level and top off as necessary.
3. Repeat the above steps until the discharged fluid is clean and no air bubbles appear during the bleeding process.

Halfshaft

REMOVAL & INSTALLATION

Front

A3

1. Before servicing the vehicle, refer to the precautions section.
2. Remove the sound insulation.
3. Remove the drive axle bolts on wheel side.
4. Remove the wheel.

5. Remove the hex flange bolts.
6. Remove the hex nuts on ball joint.
7. Remove the hex nut from bracket for level control system sensor if equipped with automatic vertical headlight aim control on left side of vehicle.
8. Disengage control arm from ball joint.
9. Swing suspension strut outward, at the same time press drive axle out of wheel bearing unit using a brass drift if necessary
10. Support the axle with and pull drive axle out of splines and remove.

To install:

11. Installation is the reverse of removal, please note the following:
    a. Replace gasket at transmission bolt.
    b. Install the drive axle on transmission side first. Tighten hex flange bolts to 20 Nm.

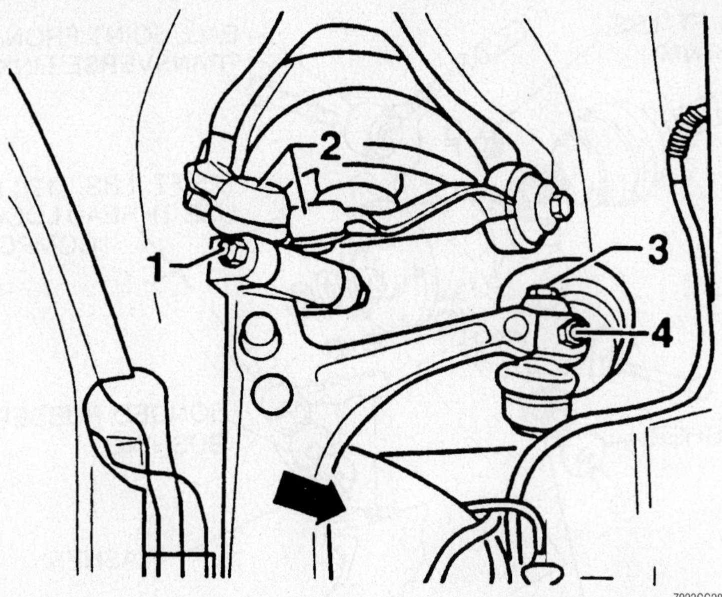

Loosen nut (1), remove the hex bolt and pull both arms (2) upward and out—A4 models

A6 MODELS

1. Before servicing the vehicle, refer to the precautions section.
2. Remove or disconnect the following:
    - Halfshaft nut
    - Front wheels
    - Hex bolt for outer joint of drive axle
    - Anti-lock Brake System (ABS) speed sensor by sliding it partly out of its mount
    - Halfshaft from the transaxle flange
3. Press the halfshaft upward toward the front of the vehicle.
4. Turn the steering to full lock.
5. Remove the halfshaft.

To install:

➡If equipped, replace the inner CV-joint gasket.

- Halfshaft into the wheel hub
- Halfshaft-to-transaxle flange. Torque the bolts to 33 ft. lbs. (45 Nm) for the M8 bolts and to 59 ft. lbs. (80 Nm) for the M10 bolts
- ABS speed sensor
- Halfshaft nut. Torque it to 148 ft. lbs. (200 Nm) plus an additional 1/4 (90 degree) turn
- Hex bolt for outer joint of drive axle. Torque M14 bolts to 85 ft. lbs. (115 Nm) and M16 bolts to (190 Nm).
- Front wheels

A4 MODELS

1. Before servicing the vehicle, refer to the precautions section.
2. Remove or disconnect the following:
    - Hub cap or center cap
    - Hex collar bolt by loosening it
    - Front wheels
    - Halfshaft-to-transaxle flange bolts
    - Hex collar bolt
    - Anti-lock Brake System (ABS) wheel speed sensor cable from the brake caliper bracket
    - ABS speed sensor by sliding it partly out of its mount
    - Remove nut/bolt No. 1, as shown
3. Pull both arms up and out of the swing arm

⁂ WARNING

The slots in the swing arm must not be widened. Do not loosen the bolts No. 3 and 4, otherwise the axle geometry must be checked.

4. Tilt the swing arm out and to the rear of the vehicle
5. Remove the halfshaft

To install:

6. Install or connect the following:
    - Halfshaft into the wheel hub
    - Swing arm bolt. Torque it to 30 ft. lbs. (40 Nm)
    - Halfshaft-to-transaxle flange. Torque the bolts to 30 ft. lbs. (40 Nm) for the M8 bolts and to 57 ft. lbs. (77 Nm) for the M10 bolts
    - ABS wheel speed sensor
    - Sensor cable into the caliper bracket
    - Front wheels
7. Tighten the axle bolt as follows:
    - M14 bolt: 85 ft. lbs. (115 Nm) plus an additional 1/4 (90 degree) turn
    - M16 bolt: 140 ft. lbs. (190 Nm) plus an additional 1/4 (90degree) turn

Rear

A6 MODELS

1. Before servicing the vehicle, refer to the precautions section.
2. Remove or disconnect the following:
    - Halfshaft bolt
    - Rear wheel
    - Brake caliper without disconnecting the hydraulic line and support it on a wire
    - Brake rotor
    - Inner halfshaft flange bolts
3. Support the halfshaft.
4. Remove or disconnect the following:
    - Fuel tank cover plate and/or inner CV-joint heat shield
    - Anti-lock Brake System (ABS) speed sensor by sliding it partly out of its mount
    - Lower strut bolt
    - Transverse link from the wheel bearing housing
    - Halfshaft by pressing down on wheel bearing housing
5. Clean the halfshaft splines of any grease, dirt or locking compound.

To install:

6. Use a new inner flange gasket and reverse the removal procedures.
7. Torque the halfshaft flange bolts as follows:
    - M8 bolts: 33 ft. lbs. (45 Nm)
    - M10 bolts: 59 ft. lbs. (80 Nm)
8. Install or connect the following:
    - Brake caliper. Torque the bolts to 48 ft. lbs. (65 Nm)

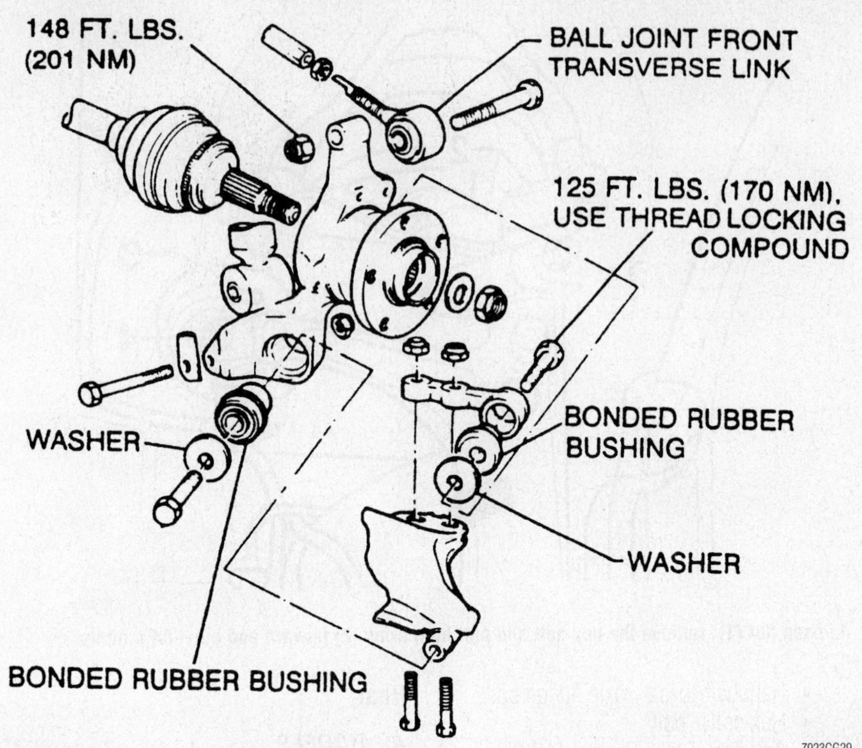

Exploded view of the rear suspension and halfshaft—A6 Quattro models

➡**Adjustment of parking brake may be necessary.**

- Halfshaft bolt and washer assembly. Torque the bolt to 148 ft. lbs. (200 Nm) plus an additional ¼ (90 degree) turn
- ABS speed sensor
- Rear wheels

A4 MODELS

1. Before servicing the vehicle, refer to the precautions section.
2. Remove or disconnect the following:
   - Halfshaft bolt
   - Rear wheel
   - Anti-lock Brake System (ABS) speed sensor by sliding it partly out of its mount
   - CV-joint from the final drive
3. Loosen the sway bar link mounting bolt at the wheel bearing housing.
4. Loosen the upper control arm mounting bolt at the wheel bearing housing.
5. Remove or disconnect the following:
   - The center and rear mufflers, if servicing the left halfshaft
   - Halfshaft from the final drive and the wheel bearing housing

To install:

6. Install or connect the following:
   - Halfshaft into the wheel bearing and the final drive
   - Upper control arm. Torque the bolt to 52 ft. lbs. (70 Nm) plus a ¼ (90 degree) turn
   - Sway bar link. Torque the bolt to 37 ft. lbs. (50 Nm)
   - Center and rear mufflers, if removed
   - CV-joint to the final drive. Torque the bolts to 30 ft. lbs. (40 Nm)
   - ABS wheel speed sensor
   - Halfshaft bolt. Torque the bolt to 85 ft. lbs. (115 Nm) plus ¼ (90 degree) turn
   - Rear wheel

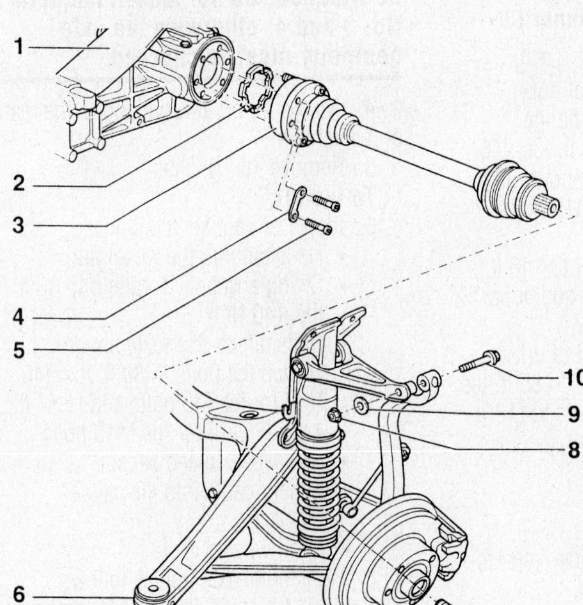

Exploded view of the rear halfshaft and related component mounting—A4 Quattro models

CV-Joints

OVERHAUL

Outer CV-Joint

The outer CV-joint is serviced with the axle halfshaft as an assembly. The outer CV-joint boot may be serviceable by first removing the inner joint.

Inner CV-Joint

TRI-POD JOINT

1. Before servicing the vehicle, refer to the precautions section.
2. Remove or disconnect the following:
   - Axle halfshaft
   - Inner boot clamps and push the boot back
   - CV-joint housing
   - Snapring
   - Spider and rollers
   - CV-joint boot

➡**Do not disassemble the spider and rollers.**

To install:

3. Install or connect the following:
   - CV-joint boot

- Spider and rollers
- Snapring

4. Apply clean grease to the CV-joint housing and the boot.
5. Install or connect the following:
   - CV-joint housing and tighten the boot clamps
   - Axle to the vehicle

DOUBLE OFFSET JOINT

1. Before servicing the vehicle, refer to the precautions section.
2. Remove or disconnect the following:
   - Axle halfshaft
   - Inner boot clamps and push the boot back
   - Circlip
   - CV-joint housing
   - Snapring
   - Double offset joint inner race, cage and ball assembly
   - CV-joint boot

➡Do not disassemble the inner race, cage and ball assembly.

To install:

3. Install or connect the following:
   - CV-joint boot
   - Double offset joint inner race, cage and ball assembly
   - Snapring
4. Apply clean grease to the CV-joint housing and the boot.
5. Install or connect the following:
   - CV-joint housing
   - Circlip
   - Boot clamps by tightening them
   - Axle to the vehicle

STEERING AND SUSPENSION

Air Bag

⁂ CAUTION

Some vehicles are equipped with an air bag system. The system must be disabled before performing service on or around system components, steering column, instrument panel components, wiring and sensors. Failure to follow safety and disabling procedures could result in accidental air bag deployment, possible personal injury and unnecessary system repairs.

PRECAUTIONS

Several precautions must be observed when handling the inflator module to avoid accidental deployment and possible personal injury.

- Never carry the inflator module by the wires or connector on the underside of the module.
- When carrying a live inflator module, hold securely with both hands and ensure that the bag and trim cover are pointed away from you.
- Place the inflator module on a bench or other surface with the bag and trim cover facing up.
- With the inflator module on the bench, never place anything on or close to the module that may be thrown in the event of an accidental deployment.
- Before installing a computer memory saver on vehicles with electronic radio lock, detach the air bag voltage connector.
- DO NOT use air bag components that have been dropped from heights of 18 in. (45cm) or higher.
- Disable the SRS before performing electric welding on the vehicle.
- SRS can only be tested using Diagnostic Tester (VAG 1551) and Adapter Test Harness (VAG 1551/1) or their equivalents. DO NOT use Air Bag Tester (VAG 1619). Never use a test light, ohmmeter or voltmeter to test the air bag system, except when testing the clockspring.

DISARMING

1. Before servicing the vehicle, refer to the precautions section.
2. Disconnect and shield the negative battery cable.
3. Wait at least 5 minutes before servicing the vehicle.

ARMING

1. Before servicing the vehicle, refer to the precautions section.
2. Remove the shield.
3. Connect the negative battery cable.

Power Rack and Pinion Steering Gear

REMOVAL & INSTALLATION

A3

1. Before servicing the vehicle, refer to the precautions section.
2. Disconnect the battery.
3. Remove the nuts and remove foot well trim.
4. Remove the bolt and pull universal joint from steering gear.
5. Remove the front wheels.
6. Loosen the nut from tie rod end but do not remove completely.
7. Press off tie rod end from wheel bearing housing using Ball Joint Puller 3287A.
8. Remove the sound insulation at bottom.
9. Remove the coupling rod from stabilizer bar.
10. Remove the bolts.
11. Remove the bolts and remove pendulum support from transmission.
12. Remove the bracket for exhaust system from subframe.
13. Remove the bolts for heat shield.
14. Remove the heat shield from subframe.
15. Loosen the bolts for steering gear and stabilizer bar.
16. Secure sub frame and consoles.
17. Position engine/transmission assembly jack under subframe.
18. Place wooden block or similar between engine/transmission assembly jack and subframe.
19. Loosen the bolts and lower sub frame and consoles slightly. Make sure all electrical connections are out of the way.
20. Remove the heat shield above steering gear.
21. Remove the bolts.
22. Remove the cable guide from subframe.
23. Unclip all other cable fasteners from steering gear.
24. Disconnect the all electrical connections at steering gear.
25. Remove the stabilizer bar from subframe, then remove steering gear.

To install:

➡Threaded sleeves of steering gear must sit in holes of subframe.

➡Coat seal on steering gear with lubricant, e.g. lubricating soap, before installing.

26. After attaching steering gear to drive axle, make sure that seal on steering gear makes contact with mounting plate without kinking and seals opening to foot well properly. Otherwise, this may result in water leaks and/or noise.

27. Before positioning bolts for subframe, position steering gear on subframe and fasten bolts for steering gear and stabilizer bar.

28. Bolt on heat shields with steering gear.

29. Bolt on sub frame and consoles with body.

30. Bolt on steering gear with sub frame.

31. Bolt on ball joint with control arm. Always use new bolts.

➡Make sure that boot is not damaged or twisted.

32. Bolt on universal joint with steering gear.

33. After installation, position of steering wheel must be checked with a road test.

A6

1. Before servicing the vehicle, refer to the precautions section.
2. Remove or disconnect the following:
   - Battery
   - Driver's side storage compartment
   - Set wheels straight ahead
   - Nut and bolt at universal joint at lower end of steering column
3. Pull the steering column from the steering column.
4. With the steering wheel centered, remove the ignition key.
5. Pinch off the intake and return power steering hoses.
6. Remove the front wheels.
7. Remove the tie rod end from the steering knuckle connection.
8. Remove the steering column opening cover from the dash panel.
9. If necessary, remove the left exhaust pipe.
10. Drain the power steering fluid by removing the banjo bolts from the steering gear.
11. Remove the steering gear mounting bolts from the cross member and remove the steering gear through the left wheel housing.

To install:

12. Steering gear into the vehicle. Temporarily install mounting bolts from above, but do not tighten.
13. Install the single mounting bolt from below and tighten.
14. Install the return hose and tighten the banjo bolt to 37 ft. lbs. (50 Nm).
15. Screw of the pressure hose and tighten the banjo fitting to 30 ft. lbs. (40 Nm).
16. Install the ignition key and release the steering lock.
17. Center the steering wheel, then install the universal joint to the steering gear pinion.
18. Install the eccentric bolt and tighten in a counterclockwise direction.
19. Install and tighten a new self-locking nut on the eccentric bolt. Torque the nut to 22 ft. lbs. (30 Nm).
20. Install the driver's side storage compartment.
21. Install the dash panel cover.
22. Fully tighten steering gear mounting bolts in cross member.
23. Remove the power steering hose pinchers.
24. Install the front wheels and battery.
25. Refill the power steering fluid.
26. Start the engine, cycle the steering wheel from lock to lock. Turn the engine off and check and top up the power steering fluid.

A4

1. Before servicing the vehicle, refer to the precautions section.
2. Remove or disconnect the following:
   - Battery
   - Battery box
   - Steering column U-joint bolt

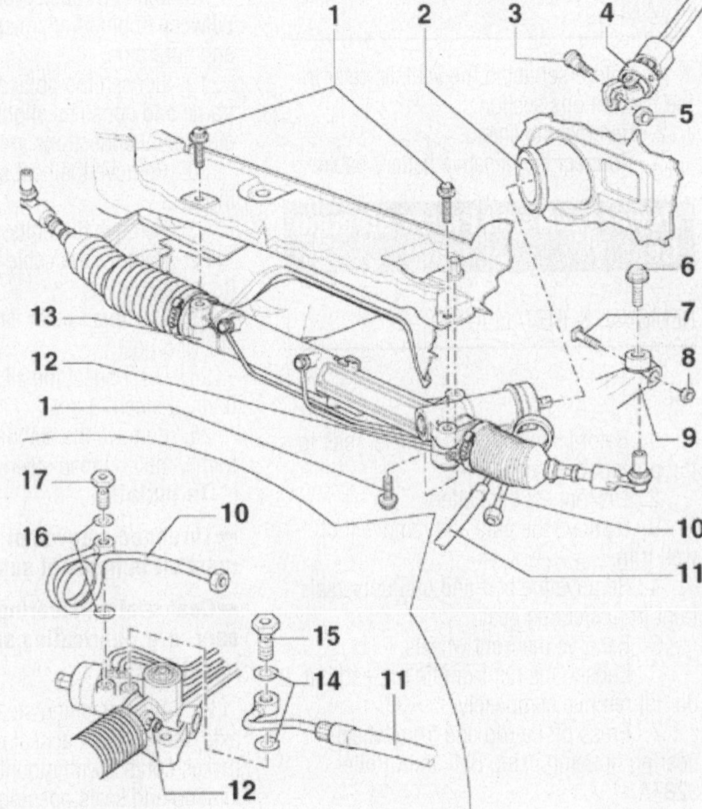

Exploded view of the steering rack assembly—A6

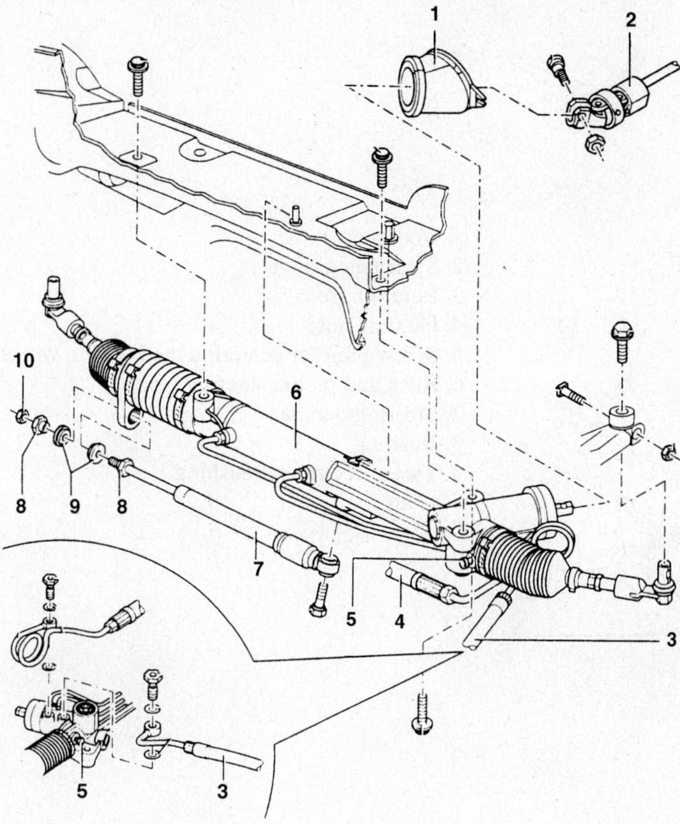

Exploded view of the steering gear mounting—A4 models

3. Release the eccentric by turning the Torx® T50 bolt clockwise, then remove the bolt.

➡**Before removing the steering column from the steering gear, secure the steering column with safety wire.**

⁂ WARNING

Be sure to lock the steering wheel, otherwise the air bag unit coil spring may be damaged.

4. Lock the steering wheel in the center position and do not move during the repairs.

➡**The splines between the top and bottom part of the steering column must not be separated.**

5. Move the U-joint down and out of the way.
6. Using hose clamps tool 3094, pinch off the suction and return lines to the steering gear.
7. Remove or disconnect the following:
   - Front wheels
   - Left and right tie rods
   - Tie rod opening cover

➡**Place a drip tray under the vehicle to catch any residual power steering fluid.**

   - Banjo bolts for the steering gear suction and return hydraulic hoses
   - Steering gear mounting bolts
   - Steering gear through the left side wheel opening

To install:

8. Remove the screw plug to lock the steering gear in the center position with locking tool VAG 1907 and torque to 13 ft. lbs. (18 Nm)
9. Install or connect the following:
   - Steering gear through the left side wheel opening. Torque bolt No. 3 to 48 ft. lbs. (65 Nm) and bolts No. 1 and 2 to 48 ft. lbs. (65 Nm).
   - Power steering gear hoses using new sealing gaskets. Torque the return hose banjo bolt to 37 ft. lbs. (50 Nm) and the suction hose banjo bolt to 30 ft. lbs. (40 Nm).
   - Left and right tie rods. Torque the bolts to 33 ft. lbs. (45 Nm).
   - Tie rod opening cover
   - U-joint to the steering gear and the Torx® adjusting bolt by turning it clockwise
10. Remove the locking tool VAG 1907.
11. Install or connect the following:
   - Screw plug. Torque it to 13 ft. lbs. (18 Nm)
   - Adjusting bolt. Torque the nut to 30 ft. lbs. (40 Nm)
12. Remove the steering wheel lock
13. Remove the Hose Clamp tools 3094 and check the hydraulic fluid
14. Install or connect the following:
   - Battery tray
   - Battery

⁂ WARNING

If the hydraulic fluid requires being topped of, use only an approved fluid, otherwise internal damage may occur.

15. Start the vehicle and check for leaks.
16. Check and/or adjust the wheel alignment.

TT

1. Before servicing the vehicle, refer to the precautions section.
2. Secure the front wheel in a straight ahead position.
3. Remove or disconnect the following:
   - Cover in back of brake pedal
   - Steering column U-joint bolt
   - Engine undercover
   - Hydraulic hose bracket on left subframe

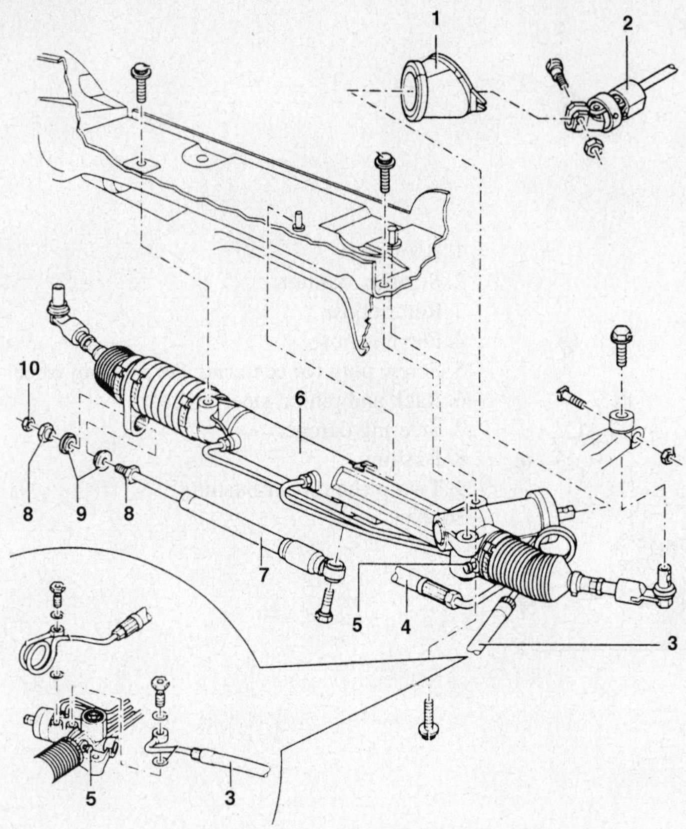

Exploded view of the steering gear mounting—A4 models

- Servo reservoir suction hose
- Servo reservoir return line
- Front wheels
- Left and right tie rods

➡Place a drip tray under the vehicle to catch any residual power steering fluid.

- Banjo bolts for the steering gear suction and return hydraulic hoses
- Sway brace support on transaxle side

4. Place a suitable jack under steering gear subframe.
5. Remove the bolts and screws from subframe.
6. Pry off subframe using a tire iron and carefully lower it.
7. Unbolt steering gear return line from clamp and rotary slide valve housing.
   - Steering gear mounting bolts
   - Steering gear through the left side wheel opening

To install:

8. Remove the screw plug to lock the steering gear in the center position with locking tool VAG 1907 and torque to 13 ft. lbs. (18 Nm)
9. Install or connect the following:
   - Steering gear with guide sleeve into subframe and hand-tighten the bolts.
   - Return line to clamp on steering gear
   - Return line with new gaskets to rotary slide valve housing
10. Check sealing cuff at the steering pinion for proper seating.
11. Raise the subframe and insert steering pinion into hole on underside of vehicle. Bolt on subframe with old bolts. Torque subframe bolts to: 74 ft. lbs. (100 Nm) plus90°. Torque steering gear bolts to 15 ft. lbs. (20 Nm).

➡Replace old bolts of subframe when aligning the vehicle.

12. Install or connect the following:
   - Expansion hose with new gaskets to rotary slide valve housing
   - Sway brace support and exhaust system bracket. Torque sway brace side bolts: 15 ft. lbs. (20 Nm) plus 90°, sway brace bottom bolts: 30 ft. lbs. (40 Nm) plus 90°
   - Tie rods in steering knuckle and bolt on. If joint bolt also turns when tightening, counterhold with internal Torx screw (T40). Torque the bolts to 33 ft. lbs. (45 Nm).
   - Return line to servo fluid reservoir
   - Engine undercover.
13. Slide universal joint onto steering pinion and secure with new bolts. Torque the bolts to: 22 ft. lbs. (30 Nm).
14. Install cover behind brake pedal.

⁂ WARNING

If the hydraulic fluid requires being topped of, use only an approved fluid, otherwise internal damage may occur.

15. Start the vehicle and check for leaks.
16. Check and/or adjust the wheel alignment.

Strut

REMOVAL & INSTALLATION

Front

A3

1. Before servicing the vehicle, refer to the precautions section.
2. Loosen the hex-combination bolt for drive axle.
3. Remove the wheel.
4. Remove the upper coupling rod hex nut from suspension strut.

5. Disengage wheel speed sensor wire from suspension strut.
6. Remove the bolts.
7. Pull out wheel bearing housing with lower ball joint out of control arm.
8. Fasten drive axle to body with wire.

➡Drive axle must not hang down, otherwise inner joint will be damaged by over bending.

9. Bolt ball joint to the control arm.
10. Secure Engine/transmission jack with wheel hub support T10149 to wheel hub using a wheel bolt.
11. Remove the wheel bearing housing/suspension strut bolt connection
12. Insert Spreader 3424 into slot of wheel bearing housing.
13. Turn ratchet 90 degrees and remove from spreader 3424.
14. Press brake disc by hand in direction of suspension strut. Otherwise strut tube may be canted in hole of wheel bearing housing.
15. Pull off wheel bearing housing downward from strut tube using engine/transmission jack and lower until strut tube hangs free.

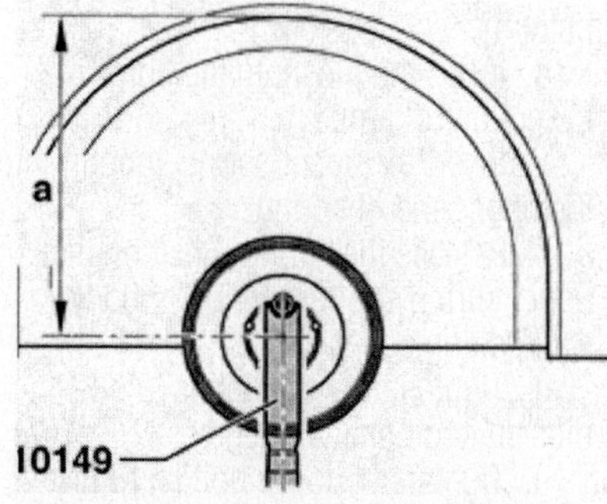

Secure Engine/transmission jack with wheel hub support T10149 to wheel hub using a wheel bolt—A3

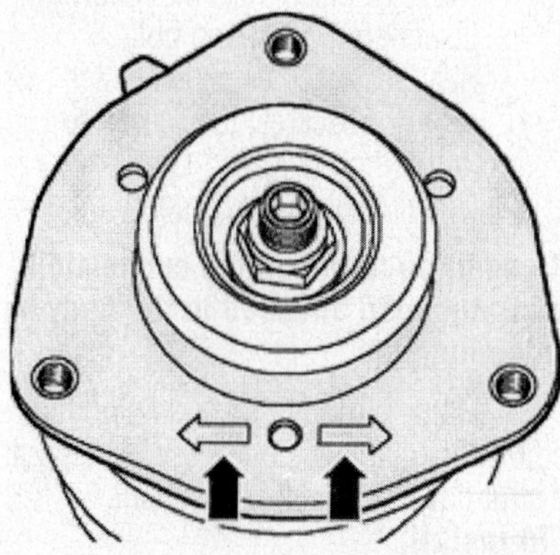

Insert strut, one of two markings must point in direction of travel—A3

16. Tie wheel bearing housing with wire at console/subframe.
17. Remove the Engine/transmission jack under wheel bearing housing.
18. Remove the wiper arms and plenum chamber cover
19. Remove the hex bolts for top strut mounting and remove strut.

To install:

20. Insert strut, one of two markings must point in direction of travel.
21. Tighten hex bolts for top strut mounting to 15 Nm plus 90 degrees.
22. Secure Engine/transmission jack with Wheel hub support T10149 to wheel hub using wheel bolt.
23. Place suspension strut on wheel bearing housing.
24. Remove the wire on wheel bearing housing.
25. Install the remaining components in the reverse order of removal, please note the following:
    a. Tighten the strut to wheel bearing housing to bolt/nut to 70 Nm plus an additional 90 degrees.
    b. Tighten the drive axle bolt to 200 Nm plus an additional 180 degrees.

A4 MODELS

1. Before servicing the vehicle, refer to the precautions section.
2. Remove or disconnect the following:
    - Front wheels
    - Rubber grommets from the plenum chamber
    - Upper strut-to-body mounting nuts
    - Anti-lock Brake System (ABS) wheel speed sensor wire from the brake caliper bracket
    - Upper control arm pinch bolt and both upper control links
    - Guide link ball joint, by swiveling the wheel bearing housing aside
    - Lower strut mounting bolt

➡When removing the strut, be sure not to damage the CV-joint boot.

- Strut

To install:

➡The bonded rubber bushing can only turned to a limited extent. The bolted connections between the suspension strut and the lower track control links should therefore only be tightened when the vehicle is standing on the ground.

3. Install or connect the following:
    - Strut by positioning it so that the spring hole plate faces the middle of the vehicle. Torque the bolt to 66 ft. lbs. (90 Nm)
    - Upper control links to the wheel bearing housing. Torque the pinch bolt to 30 ft. lbs. (40 Nm)

➡It may be necessary to hold the ball joint stud with a 4mm hex wrench.

- Ball joint. Torque the nut to 74 ft. lbs. (100 Nm)
- ABS wheel speed sensor wire into the brake caliper holder
- Upper strut-to-body nuts. Torque them to 15 ft. lbs. (20 Nm)
- Rubber grommets into the plenum chamber
- Front wheels

4. Test drive the vehicle.
5. Check and/or adjust the front alignment.

A6 MODELS

1. Before servicing the vehicle, refer to the precautions section.
2. Remove or disconnect the following:
    - Plenum chamber cover
    - Battery
    - Front wheels
    - Rubber grommets in plenum
    - Support lower control arm with a strong jack
    - Top mounting nuts for strut through plenum openings
    - ABS sensor cable from bracket on caliper
    - Steering knuckle upper mounting nut (innermost nut only; do not loosen outer bolts)
    - Guide link from wheel bearing housing

⁂ WARNING

Do not damage CV boot during this process.

- Lower strut mounting bolt
- Strut assembly from vehicle.

To install:

3. Installation is the reverse of the removal procedures.
4. Install or connect the following:
    - Lower strut mounting nut to 66 ft. lbs. (90 Nm)
    - Upper links in wheel bearing housing nuts to 30 ft. lbs. (40 Nm)
    - Joint bolt nut to 74 ft. lbs. (100 Nm)
    - Strut upper mounting bolts (new) to 15 ft. lbs. (20 Nm)

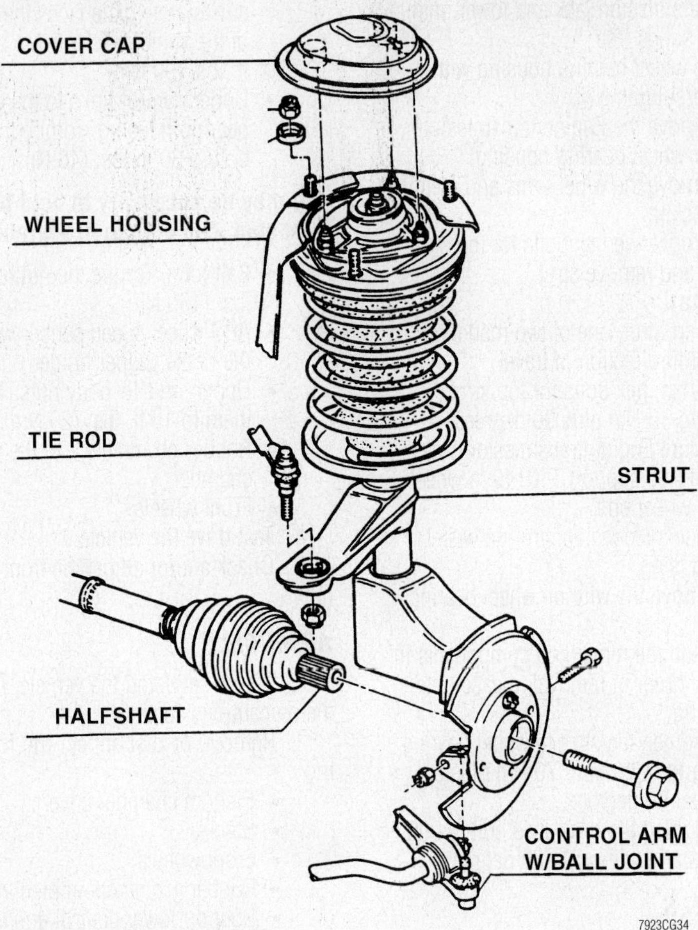

Exploded view of the front strut mounting—A6

TT MODELS

1. Before servicing the vehicle, refer to the precautions section.
2. Remove or disconnect the following:
   - Front wheels
   - Coupling rod from suspension strut on both sides
   - Spring clip from bracket and detach brake hose
   - Speed sensor wiring
   - Noise insulation panel
   - Drive axle from flange shaft/transmission
   - Secure drive axle with wire
   - Wheel bearing housing/suspension strut bolt connection
3. Insert special tool 3424 into slot on wheel bearing housing and ratchet around 90°.
4. Press the brake disc by hand in direction of suspension strut
5. Remove the wheel bearing housing from strut tube downward.
6. Tie wheel bearing housing to subframe with wire
7. Remove hex-nut for top strut mounting .

➡When removing the strut, be sure not to damage the CV-joint boot.

To install:

➡Before inserting suspension strut, coat strut mounting with installation lubricant G 294 421 A1.

8. Install the strut to the upper mounting. Torque the bolt to 44 ft. lbs. (60 Nm).
9. Place suspension strut on wheel bearing housing.
10. Carefully lift wheel bearing housing using transmission jack far enough until bolt for suspension strut/wheel bearing housing can be inserted.
11. Press the brake disc by hand in direction of the suspension strut.
12. Remove spreader 3424 from wheel bearing housing.
13. Strut must be installed up to stop in wheel carrier. Torque the bolt to 44 ft. lbs. (60 Nm) plus 90°.
14. Install the ball joint. Torque the bolt to 55 ft. lb. (75) Nm).
15. Reverse the removal procedure for the remaining components.
16. Test drive the vehicle.
17. Check and/or adjust the front alignment.

Rear

A3

1. Before servicing the vehicle, refer to the precautions section.
2. Remove the wheel.
3. Measure dimension from center of wheel to lower edge of wheel housing
4. Remove the wheel housing liner.
5. Remove the coil spring as follows:
   a. Compress the coil spring until it can be removed. Using spring compressor V.A.G 1752/1 and spring holders V.A.G 1752/3.
   b. Remove the spring.
6. Remove the upper strut bolts.
7. Remove the lower bolt.
8. Remove the shock absorber.

To install:

9. Installation is the reverse order of removal, please note the following:
   a. The end of the spring must rest against stop of lower spring support.
   b. Install the spring together with spring support.
   c. The spring support has two pins at bottom.
   d. Insert the pins into the bores of lower control arm.
   e. Insert the upper spring support into the upper end of spring.
   f. Release the spring. Position the upper spring support on lug of body while doing so.
   g. Tighten the upper bolts to 50 Nm plus an additional 45 degrees.
   h. Tighten the lower bolt to 180 Nm.

A4 MODELS

1. Before servicing the vehicle, refer to the precautions section.
2. Support the trailing arms.
3. Remove or disconnect the following:
   - Lower strut mounting bolt
   - Rear seat backrest side bolster cover or backrest to access the upper mounting
   - Upper strut-to-body mounting nuts

➡In addition to the bolted connection, the strut is also attached to the body by 4 retaining lugs.

4. Turn the strut until the retaining lugs are positioned above the recesses, then pull the strut downward out of its mount.

To install:

➡The bonded rubber bushing can only turned to a limited extent. The bolted

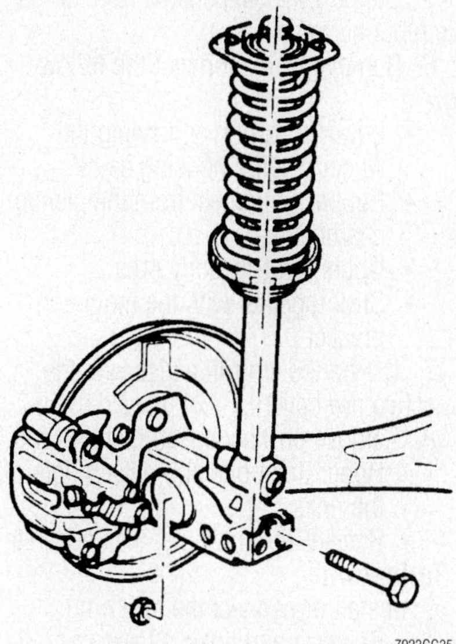
7923CG35

Be sure to support the trailing arm before removing the lower strut mounting bolt—A4 models

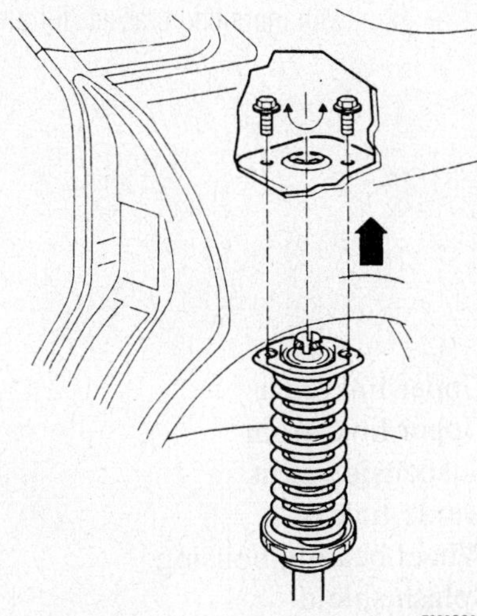
7923CG36

After loosening the 2 attaching bolts, rotate the upper strut mount to disengage the strut from the vehicle—A4 models

connections between the suspension strut and the rear axle should therefore only be tightened when the vehicle is standing on the ground.

5. Install or connect the following:
   - Strut by engaging the retaining lugs
   - Upper strut-to-body. Torque the nuts to 18 ft. lbs. (25 Nm)
   - Lower strut. Torque the bolt to 37 ft. lbs. (50 Nm) plus a ¼ (90 degree) turn with the suspension loaded
   - Rear seat backrest side bolster cover or backrest

Coil Spring

REMOVAL & INSTALLATION

All Models

1. Before servicing the vehicle, refer to the precautions section.
2. Remove the strut from the vehicle.
3. Clamp the spring compressor tool VAG 1752/2, in a vise.
4. Install the strut into the spring compressor.
5. Pry off the mounting bolt cap.
6. Compress the coil spring and

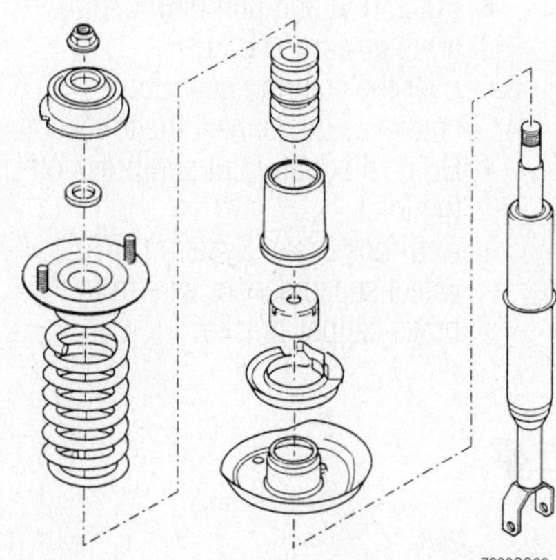
7923CG38

Exploded view of the front strut—A4

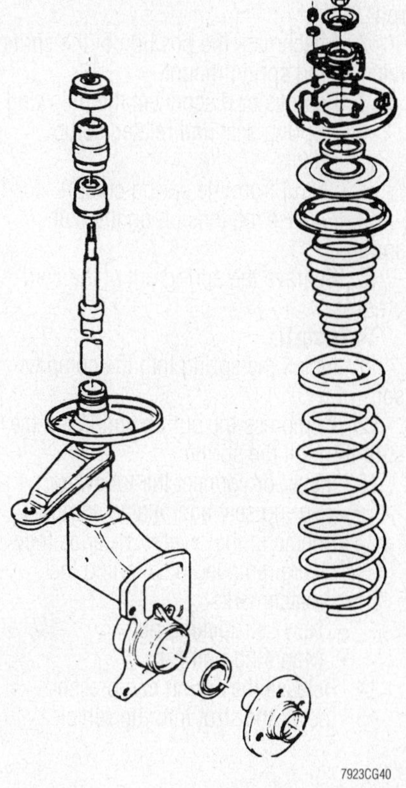
7923CG40

Exploded view of the front strut—except A4 model

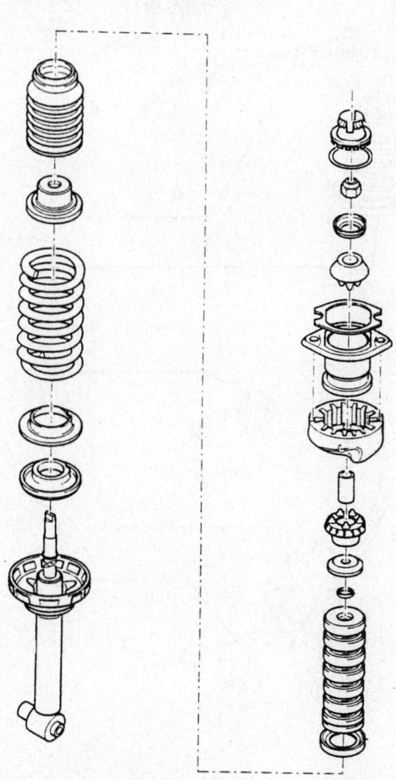
7923CG37

Exploded view of the rear strut—A4 model

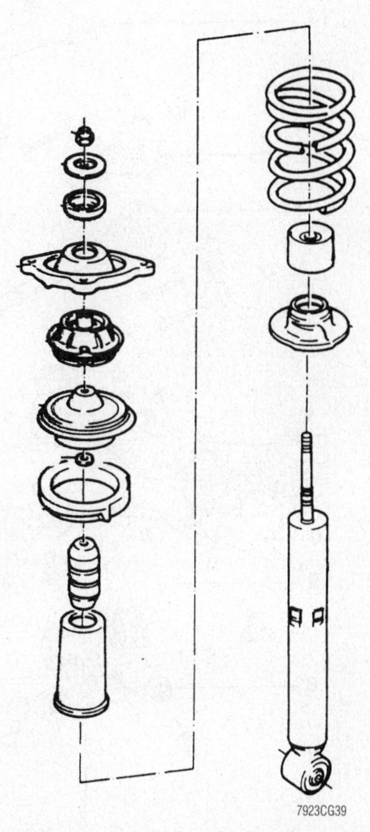
7923CG39

Exploded view of the rear strut—except A4 model

remove the self-locking nut from the piston rod.

7. Matchmark the position of the spring retainer and spring mount.

8. Remove or disconnect the following:
- Spring seat and related components
- Strut from the spring compressor

9. Release the tension on the coil spring.

10. Remove the spring out of the compressor.

To install:

11. Install the spring into the compressor.

12. Compress the spring and insert the strut through the spring.

13. Install or connect the following:
- Spring seat and related components in the reverse order as they were removed by aligning the matchmarks
- New self-locking nut
- Mounting bolt cap

14. Release the spring compressor.

15. Install the strut into the vehicle.

Upper Ball Joint

REMOVAL & INSTALLATION

A4

The Audi A4 front suspension is equipped with 2 separate upper ball joints that are not replaceable, therefore the upper link (front or rear) must be replaced. To remove this link, perform the following procedures.

1. Before servicing the vehicle, refer to the precautions section.

2. Remove or disconnect the following:
- Front wheels
- Pinch bolt and pull both control arms upward and out

3. Cover the steering gear boot.

4. Remove or disconnect the following:
- Guide link ball joint and press off the joint
- Anti-lock Brake System (ABS) wheel speed sensor wire from the brake caliper bracket

5. Support the suspension from excessive rebound travel.

6. Remove or disconnect the following:
- Lower strut bolt and swing the wheel bearing housing aside
- Rubber grommets from the plenum chamber
- Upper strut-to-body nuts
- Strut together with the mounting bracket

7. Clamp the strut in a vise with the protective jaw covers.

8. Remove or disconnect the following:
- Upper link bolts and detach both of the links
- Bracket-to-strut nuts, then separate

To install:

9. Install or connect the following:
- Brackets and links, as shown. Torque the bracket-to-strut mounting nuts to 15 ft. lbs. (20 Nm)
- Links by aligning them, as shown. Torque to 37 ft. lbs. (50 Nm) plus a ¼ (90 degree) turn
- Strut with mounting bracket. Torque

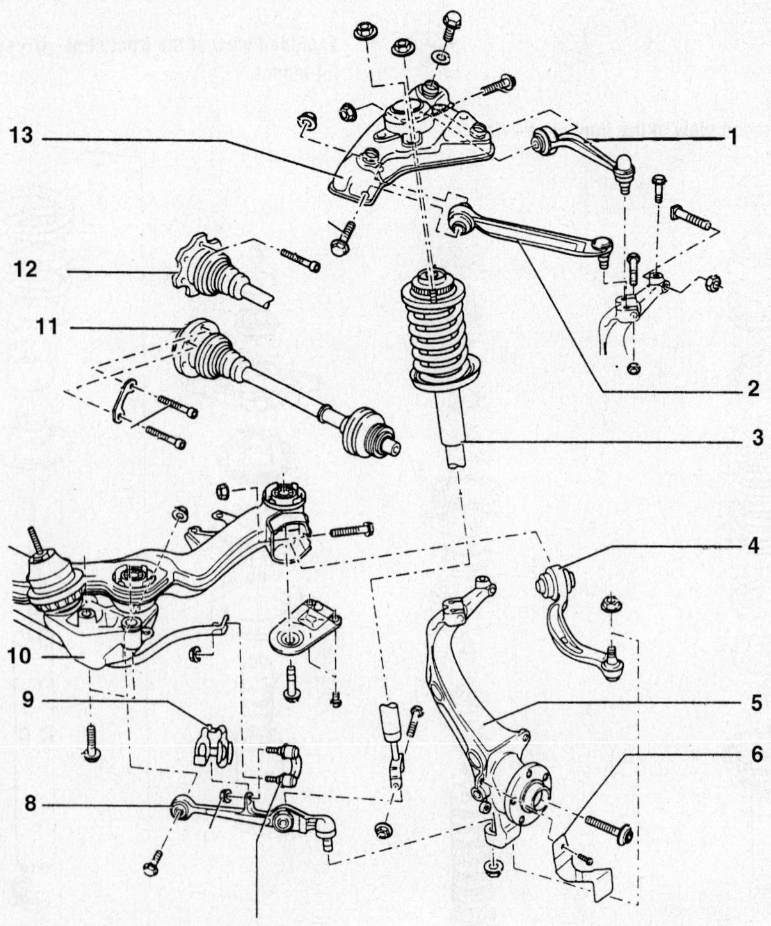

1. Upper link, rear
2. Upper link, front
3. Suspension strut
4. Guide link
5. Wheel bearing housing
6. Splash shield
7. Connecting link
8. Lower track control link
9. Clamp
10. Subframe
11. Halfshaft w/CV joint
12. Halfshaft w/triple-rotor joint
13. Mounting bracket

Exploded view of the front suspension—A4 and A6 models

the upper strut-to-body nuts to 48 ft. lbs. (75 Nm)

- Lower strut bolt. Torque it to 66 ft. lbs. (90 Nm)
- Nut on the ball joint. Torque to 74 ft. lbs. (100 Nm)
- Upper links to the wheel bearing housing. Torque the pinch bolt to 30 ft. lbs. (40 Nm)
- ABS wiring to the brake caliper bracket
- Wheels

10. Check the front suspension alignment.

Lower Ball Joint

REMOVAL & INSTALLATION

A3

1. Before servicing the vehicle, refer to the precautions section.
2. Remove the hex-combination bolt for drive axle
3. Remove the wheel.
4. Remove the hex nuts from ball joint.
5. Remove the hex nut from bracket for level control system sensor, if equipped on left side of vehicle.
6. Disengage control arm from ball joint.
7. Press constant velocity joint out of wheel bearing unit using a brass drift.
8. Swing suspension strut outward, at the same time guide drive axle out of wheel bearing.

➡Drive axle must not hang down, otherwise inner joint will be damaged by over bending. Tie up drive axle at body using wire.

9. Position ball joint puller 3287 A at ball joint as shown in the accompanying illustration.
10. Press ball joint out of wheel bearing housing.

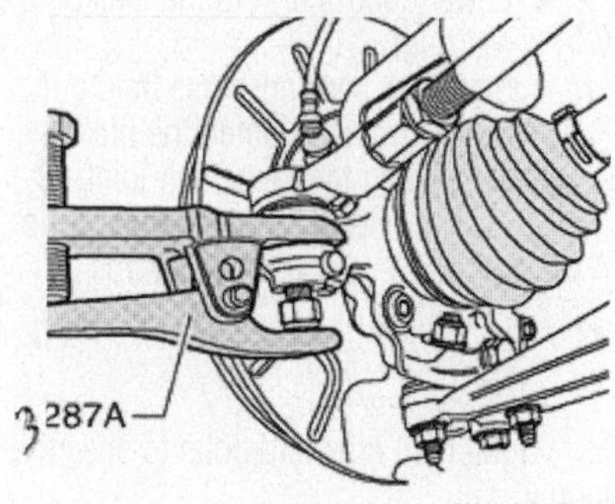

Press ball joint out of wheel bearing housing using puller 3287 —A3

To install:

11. Installation is the reverse order of removal, please note the following:
   a. Install the new self-locking nut, while holding with internal Torx T40. Tighten the ball joint nuts to 75 Nm.

➡Instead of insertion tool V.A.G 1332/10 other commercially available 18mm ring-insertion tools may be used.

   b. Guide drive axle into wheel hub splines, tighten the bolt to 200 Nm plus 180 degrees.

A4 And A6

➡The A4 and A6 models are equipped with 2 lower ball joints that are not serviceable. The control arms must be replaced if a joint is worn. The lower track control link ball joint stud faces down, and the guide link ball joint stud faces up.

LOWER TRACK CONTROL LINK

1. Before servicing the vehicle, refer to the precautions section.
2. Remove or disconnect the following:
   - Front wheels
   - Nut from the lower track control link
3. Press the ball joint out of the tapered seat.
4. Support the wheel bearing housing to prevent excessive rebound travel in the suspension.
5. Remove or disconnect the following:
   - Stabilizer link and lower strut mounting bolt
   - Lower track control link-to-subframe attaching bolt
   - Lower track control link

To install:

6. Install or connect the following:
   - Lower track control link
   - Subframe attaching bolt. Torque the bolt to 74 ft. lbs. (100 Nm)
7. Install or connect the following:
   - Stabilizer link. Torque the upper bolt to 30 ft. lbs. (40 Nm) plus ¼ (90 degree) turn and the lower bolt to 74 ft. lbs. (100 Nm)
8. Load the suspension and torque the subframe bolt to 59 ft. lbs. (80 Nm) plus ¼ (90 degree) turn
9. Front wheels
10. Check and/or adjust the front suspension alignment.

LOWER GUIDE LINK

1. Before servicing the vehicle, refer to the precautions section.
2. Remove or disconnect the following:
   - Front wheels
   - Nut from the lower guide link joint and press the joint from the wheel bearing housing
3. Loosen lower guide link-to-subframe attaching bolt

➡The subframe must be lowered at the rear to remove the lower guide link-to-subframe attaching bolt.

4. Loosen the rear subframe support plate bolts and subframe bolts.
5. Remove or disconnect the following:
   - Lower guide link-to-subframe bolt
   - Link from the vehicle

To install:

6. Install or connect the following:
   - Link into the vehicle
   - Guide link-to-subframe mounting bolt
7. Torque the support plate bolts as follows:
   - Bolt type **A**: 18 ft. lbs. (25 Nm)
   - Bolt type **B**: 55 ft. lbs. (75 Nm)
8. Install or connect the following:
   - New subframe bolts. Torque the bolts to 81 ft. lbs. (110 Nm) plus a ¼ (90 degree) turn
   - Joint end into the wheel bearing housing. Torque the nut to 74 ft. lbs. (100 Nm)
9. Load the suspension. Torque the lower guide link-to-subframe attaching bolt to 66 ft. lbs. (90 Nm) plus ¼ (90 degree) turn.
10. Install the front wheels.
11. Check and/or adjust the front suspension alignment.

TT

1. Before servicing the vehicle, refer to the precautions section.
2. Remove appropriate wheel.
3. Mark the installation positions of the ball joint nuts.

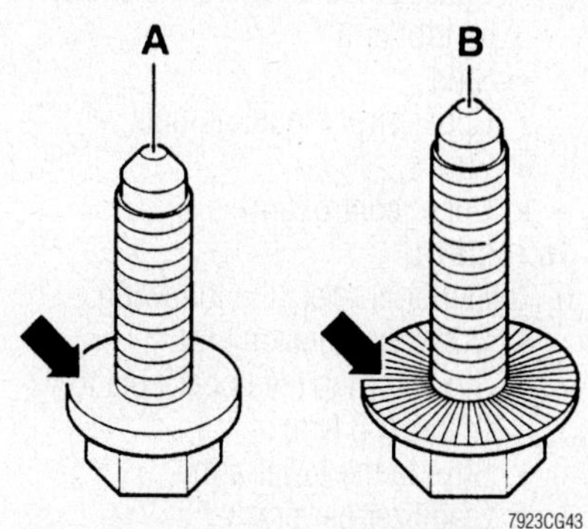

Subframe support bracket bolt identification—A4 model

4. Remove upper and lower ball joint nuts.
5. Press ball joint from control arm.

To install:

6. Install ball joint into control arm and install retaining nuts to marked positions as removed. Torque the bolts to 55 ft. lbs. (75 Nm).
7. Install ball joint into wheel bearing housing using new self locking nut. Torque the bolts to 55 ft. lbs. (75 Nm).
8. Install wheel.

Upper Control Arm

REMOVAL & INSTALLATION

Except A4 and A6 Models

1. Before servicing the vehicle, refer to the precautions section.
2. Remove or disconnect the following:
   - Negative battery cable
   - Wheel
3. Loosen the upper strut mounting nuts.
4. Loosen, but do not remove, the upper strut rod nut.

**** CAUTION**

DO NOT completely remove the upper strut nut at this time.

5. Remove or disconnect the following:
   - Brake caliper, leaving the line attached and secure it out of the way
   - Anti-lock Brake System (ABS) speed sensor and harness, if applicable
   - Cotter pin and nut from the upper control arm
   - Upper control arm from the steering knuckle
   - Stabilizer bar from the link, if applicable
   - Cotter pin and nut from the lower control arm
   - Strut
   - Upper strut mounting nuts
   - Strut
   - Upper control arm

To install:

6. Install or connect the following:
   - Upper suspension arm
   - Strut. Torque the upper nuts to 42 ft. lbs. (56 Nm)
   - Strut to the lower arm
   - Stabilizer bar bracket
   - Stabilizer bar to the link
   - Upper suspension arm to the steering knuckle. Torque the nut to 64 ft. lbs. (87 Nm)
   - New cotter pin
   - ABS speed sensor. Torque the bolt to 69 inch lbs. (8 Nm)
   - Brake caliper
   - Front wheel
7. Bounce the vehicle several times to stabilize the suspension.
8. Tighten the lower strut bolt
9. Check and/or adjust the front wheel alignment.

CONTROL ARM BUSHING REPLACEMENT

The upper control arm bushings are serviced with the control arm as an assembly.

Lower Control Arm

REMOVAL & INSTALLATION

A3

1. Before servicing the vehicle, refer to the precautions section.
2. Remove the wheel.
3. Remove the lower sound insulation
4. Remove the coupling rod of left front level control system sensor from control arm, if installed.
5. Remove the bolts.
6. Pull out wheel bearing housing with lower ball joint out of control arm.
7. Set position for mounting bracket.
8. Replace bolt for left side and the right side by using locating pins T10096 and tighten locating pins to 20 Nm.

**** CAUTION**

Locating pins T10096 must not be tightened to more than 20 Nm, otherwise thread of locating pins can be damaged.

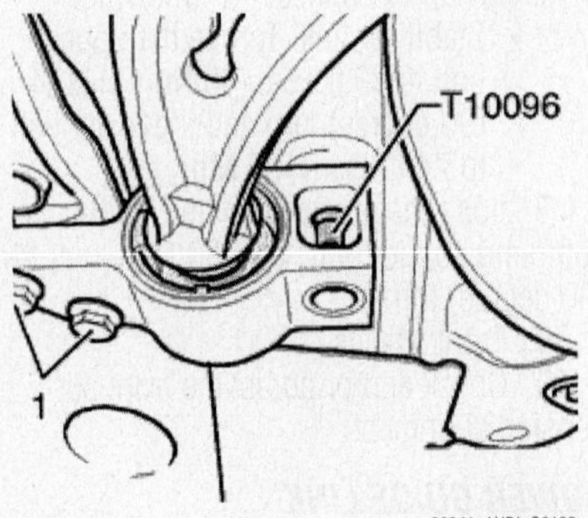

Replace bolt for left side and the right side by using locating pins T10096—A3

9. Remove the item for left vehicle side and item for right vehicle side.
10. Remove the bolts.
11. Remove the control arms.

To install:

12. Installation is the reverse order of removal, please note the following:
    a. Insert control arm with mounting bracket into subframe.
    b. Position but do not tighten yet.
    c. Tighten the mounting bracket bolts to 50 Nm plus 90 degrees.
    d. Tighten the lower control arm to spindle bolts to 75 Nm
    e. Replace locating pins T10096 with new bolt and tighten to 70 Nm plus 90 degrees.
13. Bolt the control arm to the ball joint.
14. Tighten control arm to the console in the curb weight position.
15. Install the remaining components in the reverse order of removal.

Except A4 and A6 Models

1. Remove or disconnect the following:
   - Front wheels
   - Cotter pin and castle nut from the tie rod end
   - Tie rod end from the steering knuckle
   - Cotter pin and castle nut from the ball joint stud
   - Knuckle from the ball joint
   - Stabilizer bar link from the lower arm
   - Nuts and bolts that connect the lower control arm
   - Lower control arm from the vehicle

To install:

➡Tightening of the suspension component fasteners should be performed with the full weight of the vehicle resting on the suspension.

2. Install or connect the following:
   - Control arm to the cross member and install the through bolt, lockwasher and nut
   - Lower control arm to the tension rod nuts
   - Stabilizer bar transverse link to the lower arm and tighten the link
   - Knuckle to the lower ball joint.
   - Tie rod end to the knuckle. Torque the castle nut to 22–36 ft. lbs. (29–49 Nm)
   - New cotter pin
   - Front wheels
3. Tighten all nuts and bolts to specification
   - Check and/or adjust the wheel alignment.

CONTROL ARM BUSHING REPLACEMENT

Except A3

FRONT BUSHINGS

1. Before servicing the vehicle, refer to the precautions section.
2. Remove the lower control arm from the vehicle.
3. Press the front bushing out of the control arm.

To install:

4. Lubricate the front bushing with soap and press into the control arm.
5. Install the control arm to the vehicle.
6. Check and/or adjust the wheel alignment.

REAR BUSHINGS

1. Before servicing the vehicle, refer to the precautions section.
2. Remove or disconnect the following:
   - Rear wheel
   - Rear control arm
   - Press the rear control arm bushing out

To install:

3. Lube the rear bushing with soap
4. Install or connect the following:
   - Rear bushing
   - Rear control arm
   - Front wheel
5. Check and/or adjust the wheel alignment.

A3

1. Before servicing the vehicle, refer to the precautions section.
2. Press out bonded rubber bushing, as shown in the accompanying illustration.
3. Pressing in bonded rubber bushing

➡Bonded rubber bushing must be positioned at an angle when pressed in to avoid damage. When pressing in, bonded rubber bushing will straighten.

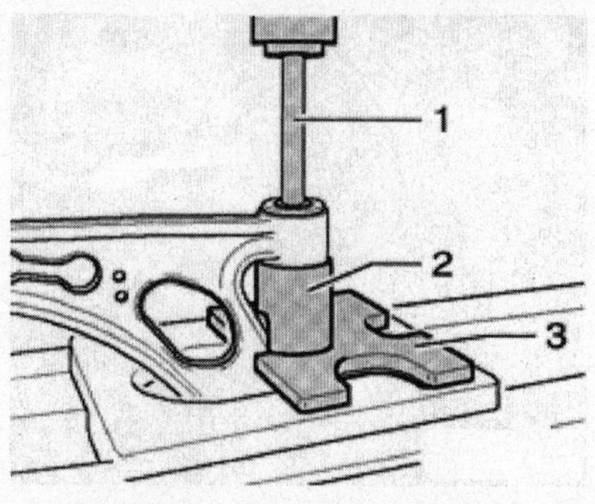

Press out bonded rubber bushing—A3

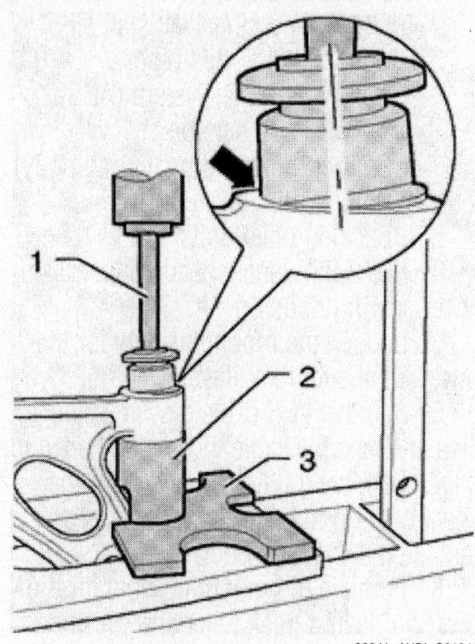

Position bonded rubber bushing at an angle in direction of control arm. The lip (arrow) must slide into hole—A3

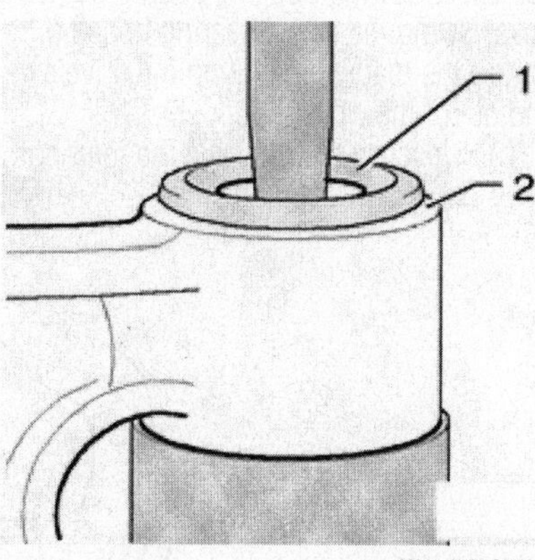

Press in bonded rubber bushing until core and hole of control arm are at same level—A3

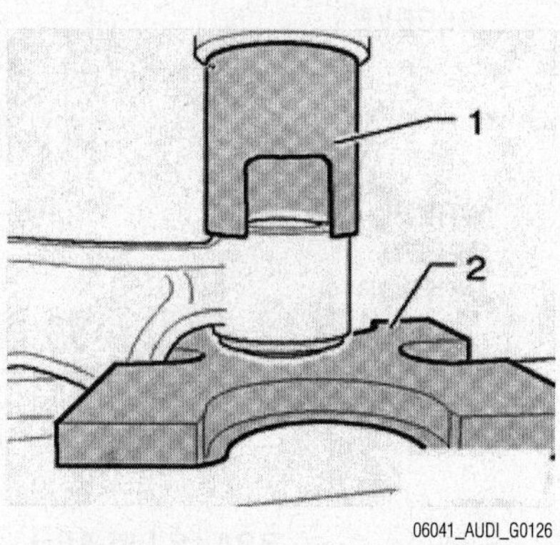

Press back bushing slightly in control arm—A3

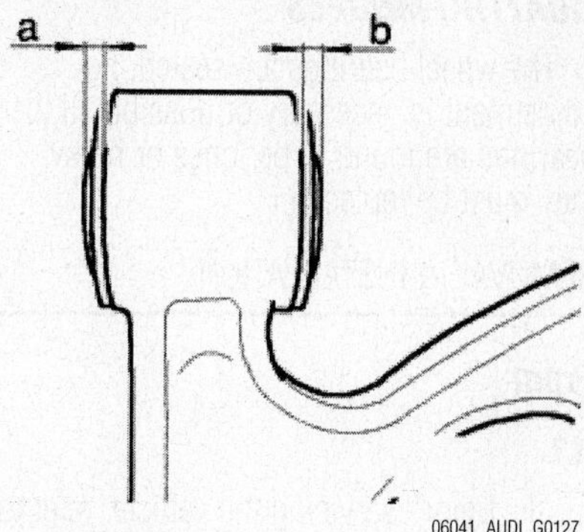

The dimensions and must be the same as shown—A3

4. Apply a suitable lubricant on outside of bonded rubber bushing.
5. Position bonded rubber bushing at an angle in direction of control arm. The lip must slide into hole as shown in the accompanying illustration.
6. Press in bonded rubber bushing until core and hole of control arm are at same level.
7. Press back bushing slightly in control arm.
8. The dimensions and must be the same as shown.

Wheel Bearings

ADJUSTMENT

Front

The front wheel bearings are sealed and no adjustment is necessary or possible. If the bearings are found to be loose or noisy, they must be replaced.

Rear

NON-QUATTRO MODELS

1. Before servicing the vehicle, refer to the precautions section.
2. Remove or disconnect the following:
   - Grease cap
   - Cotter pin and the locking nut
3. While turning the wheel, so the wheel bearing does not jam, tighten the adjusting nut firmly.
4. Back the nut off slightly. The nut is properly adjusted when it is possible to pry the thrust washer side to side with some drag but using light pressure on the tool.
5. Install the locking nut and a new cotter pin.
6. When installing the cap, be sure it is securely in place.

QUATTRO MODELS

The wheel bearings are sealed; no adjustment is necessary or possible. If the bearings are found to be loose or noisy, they must be replaced.

REMOVAL & INSTALLATION

Front

A3

1. Before servicing the vehicle, refer to the precautions section.
2. Remove the hex-combination bolt for drive axle.
3. Remove the wheel.
4. Remove the brake caliper and hang on body using tie wire.
5. Remove the ABS wheel speed sensor
6. Remove the brake disc.
7. Loosen the nut from tie rod end but do not remove completely. Press off tie rod end from wheel bearing housing using Ball Joint Puller 3287A.
8. Remove the wheel bearing housing/suspension strut bolt connection
9. Insert Spreader 3424 into slot of wheel bearing housing.
10. Turn ratchet around 90 degrees and remove from the ratchet from spreader 3424.
11. Remove the lower control arm to spindle bolts.
12. Guide control arm out of wheel bearing housing with ball joint.
13. Remove the wheel bearing housing from strut tube and pull driveshaft out of wheel hub.

➡Drive axle must not hang down, otherwise inner joint will be damaged by over bending.

14. Fasten drive axle to body with wire.
15. Remove the wheel bearing housing with ball joint.

To install:

16. Installation is the reverse order of removal, please note the following:
    a. Tighten the NEW wheel bearing housing bolts to 70 Nm plus 90 degrees.
    b. Tighten the tie rod end nut to 20 Nm plus 90 degrees.
    c. Tighten the lower control arm to spindle bolts to 75 Nm
    d. Tighten the drive axle bolts to 200 Nm plus 180 degrees.

A4 AND A6 MODELS

1. Before servicing the vehicle, refer to the precautions section.
2. Loosen the halfshaft retaining bolt.
3. Remove or disconnect the following:
   - Front wheel, then reinstall all 5 wheel bolts at this time
   - Anti-lock Brake System (ABS) wheel speed sensor
   - Caliper (suspend by wire to body)
   - Rotor
   - Backing plate bolts
4. Pull ABS wheel speed sensor from wheel bearing housing
5. Loosen the mounting nuts for the lower guide and track links.
6. Remove ABS cable grommet and clips and feed the cable through the opening.
7. Remove the tie rod end from the wheel bearing housing (use care to not damage the CV joint boot).
8. Remove the mounting nuts for the lower guide and track links and press out the joints

➡If equipped with self-leveling control (for headlight aiming), disconnect or remove actuating arm.

9. Remove the upper mounting nuts for the wheel bearing housing and lift out both links. Swing the wheel bearing housing toward the rear while pulling the drive axle end out of the wheel hub.
10. Unscrew the nut from the joint bolt of the track control link and remove the wheel bearing housing.
11. Place the wheel bearing housing on a press.
12. Drive out the hub with the wheel bearing.
13. Using a bearing separator and press, drive hub out of the bearing.

To install:

14. Press the new wheel bearing into the bearing housing using the appropriate bearing driver.
15. Press the hub into the wheel bearing using the appropriate bearing driver.
16. Install or connect the following:
    - CV-joint by sliding it through the wheel hub and hand-tighten the new nut
    - Lower track control and guide link. Torque the new self-locking nut to 74 ft. lbs. (100 Nm)
    - Both of the upper link ball joints into the wheel bearing. Torque the pinch bolt to 30 ft. lbs. (40 Nm)
    - Socket-head bolt and self-locking nut for self-leveling control arm. Torque to 7 ft. lbs. (10 Nm)
    - Tie rod end. Torque the new self-locking nut to 37 ft. lbs. (50 Nm) and the bolt to 44 inch lbs. (5 Nm)

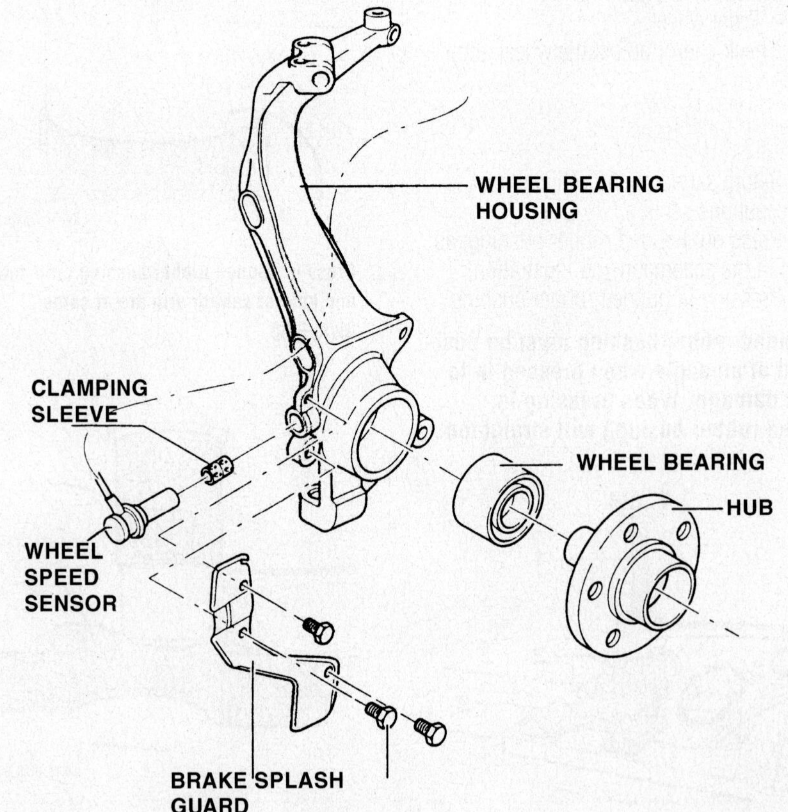

Exploded view of the front wheel bearing housing—A4 models

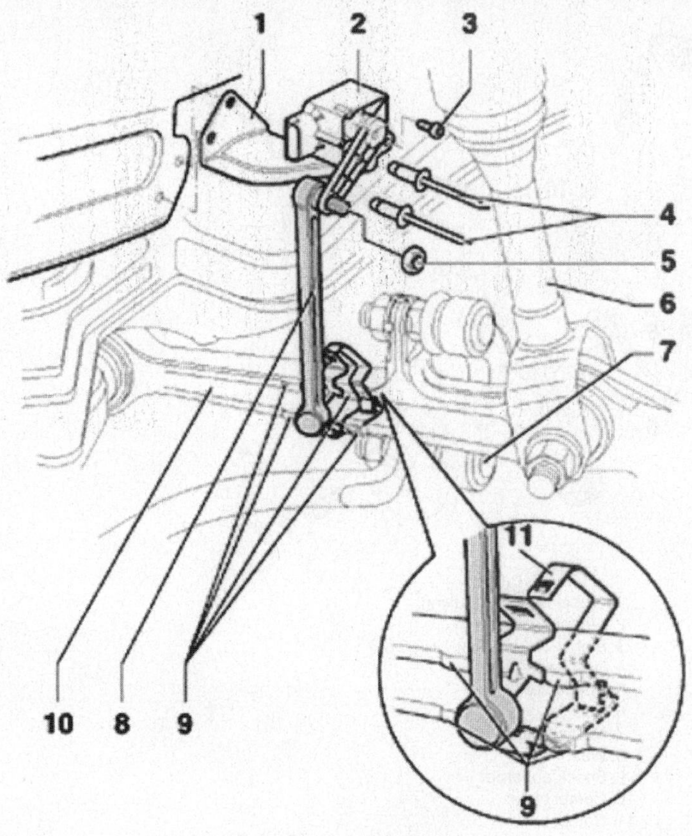

1. Bracket
2. LF level control system sensor
3. Bolt
4. Rivets
5. Nut
6. Front strut
7. Stabilizer bar link
8. Actuating arm
9. Alignment marks for mounting
10. Lower front link
11. Clip

Showing the self-leveling control assembly on front suspension—A6 models (optional equipment)

- ABS wheel speed sensor
- Brake splash guard. Torque the bolts to 84 inch lbs. (10 Nm)
- Brake rotor
- Brake caliper. Torque the bolt to 89 ft. lbs. (120 Nm)
- Front wheel

17. Tighten the halfshaft retaining bolt as follows:
    - M14 bolt: 85 ft. lbs. (115 Nm) plus ½ (180 degree) turn
    - M16 bolt: 140 ft. lbs. (190 Nm) plus ½ (180 degree) turn
18. Check and/or adjust the front alignment, if necessary.

Rear

A3

1. Before servicing the vehicle, refer to the precautions section.
2. Remove the wheel.
3. Remove the dust cap.
4. Remove the brake carrier caliper and support to one side with a piece of wire.
5. Remove the brake rotor.
6. Remove the nut and wheel bearing.

To install:

7. Installation is the reverse order of removal, tighten the bearing nut to 180 Nm plus 180 degrees:

A4 ALL MODELS

1. Before servicing the vehicle, refer to the precautions section.
2. Remove or disconnect the following:
    - Rear wheel
    - Brake caliper, without disconnecting the hydraulic line and suspend it on a wire.
    - Brake rotor
    - Dust cap
    - Center hub bolt
    - Speed sensor
    - Collared bolt securing wheel bearing housing to rear shock absorber (mark position for proper reinstallation)
    - Shock absorber, if necessary
    - Wheel bearing housing
    - Wheel hub

➡The wheel hub assembly and wheel bearing are an integral unit. The wheel bearing cannot be replaced separately. If the wheel hub assembly can be pulled from the wheel bearing housing by hand, it is okay and can be reinstalled. If not okay, wheel hub assembly must be replaced.

To install:

3. Clean and inspect mating surfaces for bearing races.
4. Install in reverse of the removal procedure, noting the following:

➡If shock absorber was removed, torque upper mounting bolts to 27 ft. lbs. (36 Nm).

- Center hub bolt to 140 ft. lbs. (190 Nm), plus an additional 180° with wheel raised off ground; when vehicle is returned to ground, loosen bolt by 90°
- Wheel bearing housing to shock absorber new bolt. Torque self-locking nut to 118 ft. lbs. (160 Nm), plus an additional 90°, with vehicle standing on ground.

5. Check the brakes for proper operation.

A6 FRONT WHEEL DRIVE MODELS

1. Before servicing the vehicle, refer to the precautions section.
2. Remove or disconnect the following:
    - Wheel(s)
    - Brake caliper (suspend by wire from body)
    - Rotor
    - Wheel speed sensor bracket
    - Speed sensor
    - Wheel hub/bearing assembly bolts

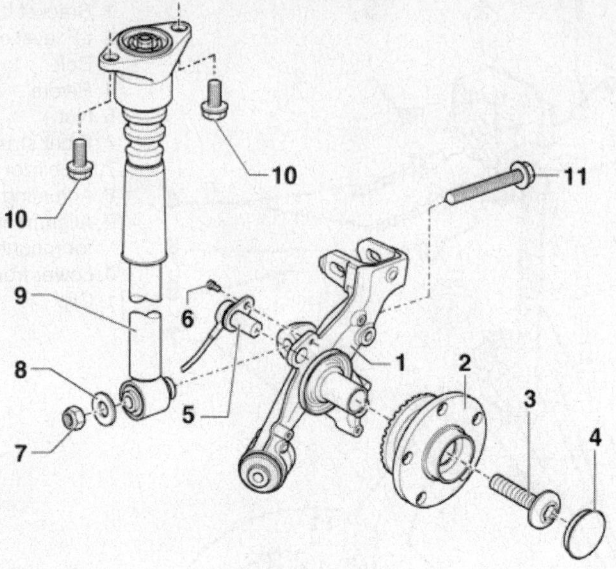

Exploded view of the rear wheel bearing—A4 with front wheel drive vehicles

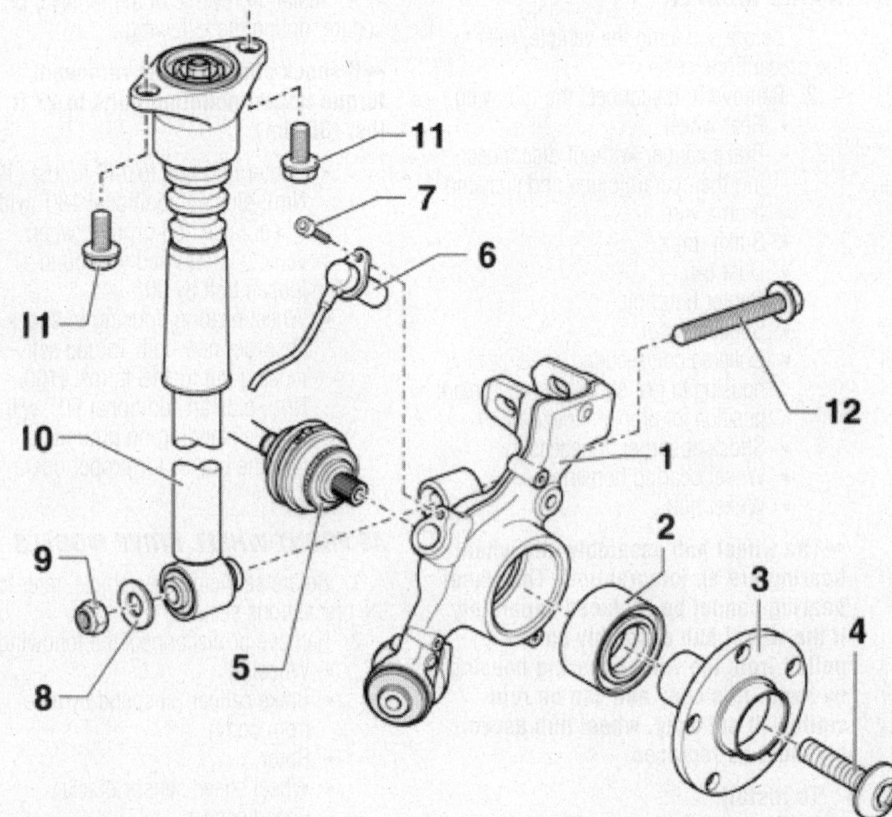

1. Wheel bearing housing
2. Wheel bearing
3. Hub
4. Center bolt
5. Drive axle
6. Speed sensor
7. Bolt
8. Washer
9. Nut
10. Shock absorber
11. Bolts
12. Bolt

Exploded view of the rear wheel bearing housing—A4 Quattro models

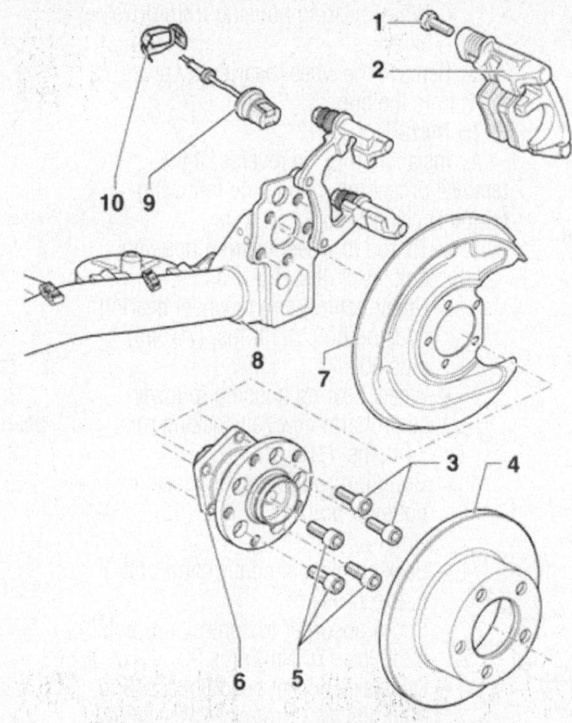

Exploded view of rear wheel hub/bearing assembly—A6 models with front wheel drive

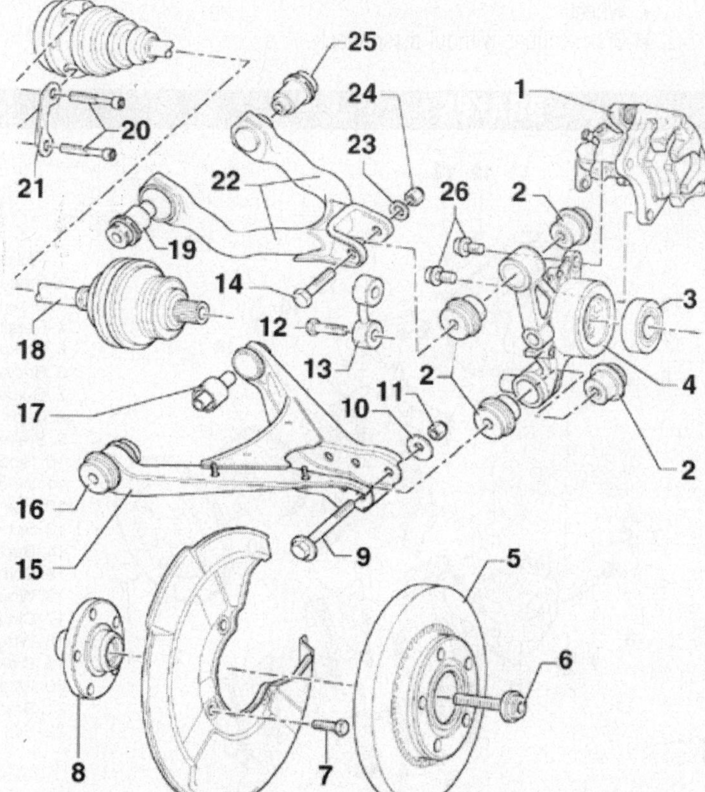

Exploded view of the rear wheel bearing housing and related components—A6 Quattro models

- Wheel hub/bearing assembly with backing plate

To install:

➡Wheel hub and bearing assembly is an integral unit. The wheel bearing cannot be replaced separately. If the wheel hub and bearing assembly is replaced, also replace the wheel speed sensor.

3. Installation is the reverse of the removal procedure, noting the following:
   - Wheel hub/bearing bolts: torque to 44 ft. lbs. (60 Nm)
   - Brake caliper bolts: torque to 26 ft. lbs. (35 Nm)
   - Wheel bolts: torque to 89 ft. lbs. (120 Nm)

A6 QUATTRO MODELS

1. Before servicing the vehicle, refer to the precautions section.
2. Remove or disconnect the following:
   - Wheel assembly
   - Caliper (hang by wire from body)
   - Backing plate
   - Brake rotor
   - ABS wheel speed sensor from the wheel bearing housing
   - Coupling rod bolt
   - Tie rod bolt (mark position of eccentric bolt before removal)
   - Control arm bolt
   - Wheel bearing housing from drive axle
3. Remove the wheel bearing by pressing it from the hub.

To install:

4. Installation is the reverse of the removal procedure, noting the following torque specifications:
   - Tie rod to wheel bearing housing bolt: 70 ft. lbs. (95 Nm)
   - Upper control arm to wheel bearing housing bolt: 52 ft. lbs. (70 Nm) plus 90°
   - Wheel bearing housing to lower control arm new self-locking nut: 74 ft. lbs. (100 Nm)
   - Coupling rod to wheel bearing housing bolt: 41 ft. lbs. (55 Nm)
   - Backing plate to control arm bolt: 7 ft. lbs. (10 Nm)
   - Shock absorber to control arm bolt: 52 ft. lbs. (70 Nm) plus 90°
   - Drive axle socket head bolts: M8 to 30 ft. lbs. (40 Nm) or M10 to 52 ft. lbs. (70 Nm)
5. Check and/or adjust the wheel alignment.

TT MODELS

1. Before servicing the vehicle, refer to the precautions section.
   - Wheel
   - Brake caliper, without disconnecting the hydraulic line and suspend it on a wire.
   - Grease cap
   - Cotter pin, nut and washer
   - Outer bearing
   - Brake rotor
   - Bearing inner bearing and seal from the rotor hub, using a soft drift or press
   - Bearing inner and outer race(s) from the rotor, using a soft drift or press

To install:

2. Clean and inspect mating surfaces for bearing races.
3. Install or connect the following:
   - New races using soft drift or press
   - New bearing packed with grease and set it into the inner race
   - Seal, making sure it is square in the rotor hub
   - Rotor, outer bearing, washer, and nut and adjust bearing play
   - Cotter pin and dust cap
   - Brake caliper
4. If hydraulic lines were removed, install and bleed brakes.
5. If parking brake cable has been remove, install and adjust as necessary.
6. Install the wheel .
7. Check the brakes for proper operation.

BRAKES

Brake Caliper

REMOVAL & INSTALLATION

All Models

FRONT

➡Procedures may vary slightly, depending on vehicle or brake type. Refer to appropriate illustration for additional assistance.

1. Before servicing the vehicle, refer to the precautions section.
2. Remove the wheels.
3. Loosen the hydraulic line at the caliper, then remove the caliper from the carrier. With guide pin calipers, be sure to hold the pin with a back-up wrench when removing the caliper bolts.
4. Remove the caliper from the hydraulic line.
5. The carrier can be removed by removing the 2 bolts.

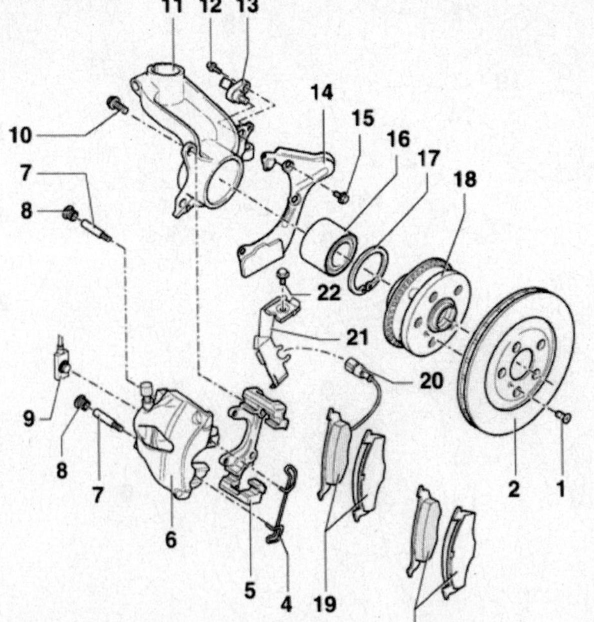

Exploded view of front brake components—TT models

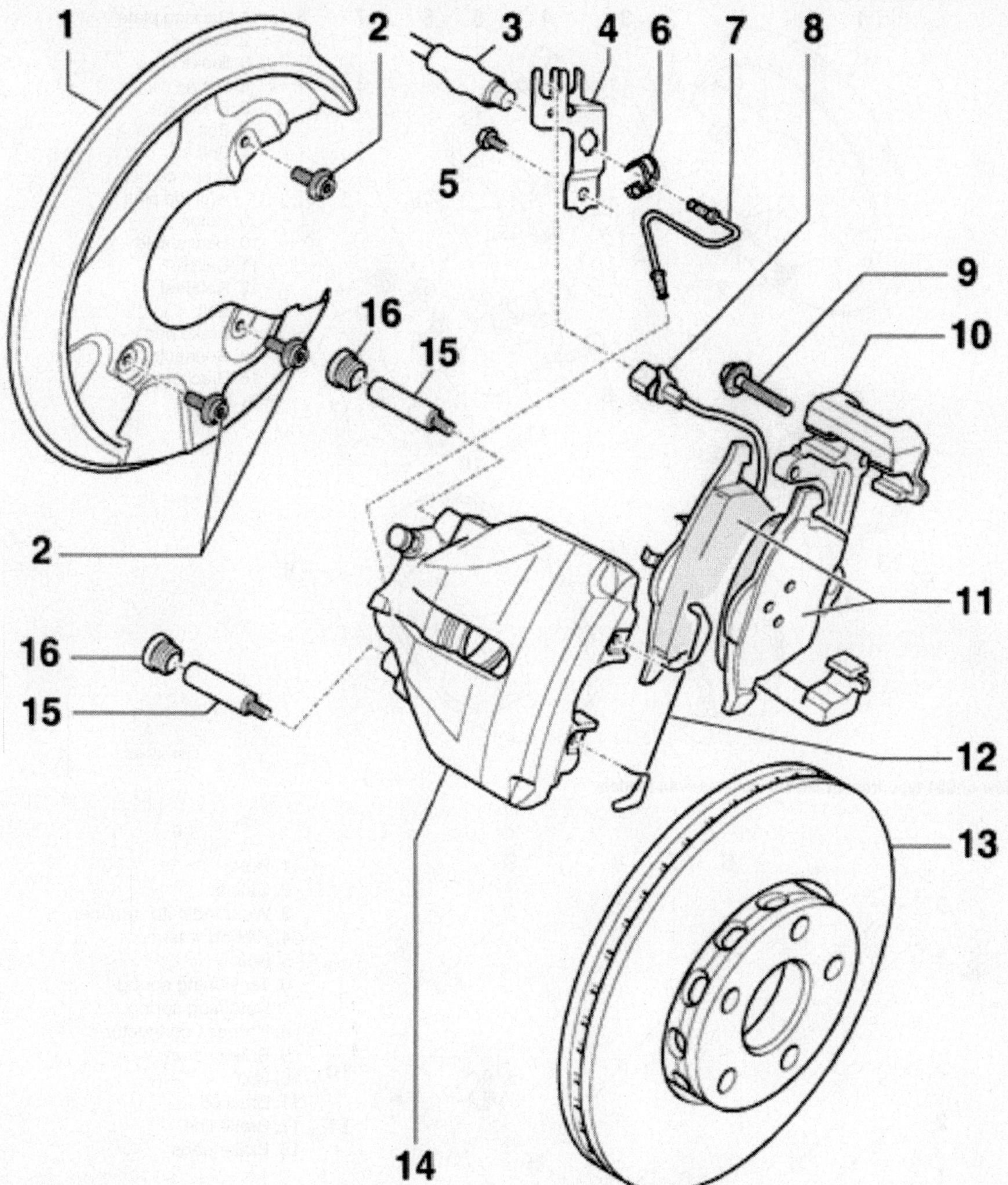

1. Brake disc cover
2. Flange bolt
3. Brake hose
4. Brake hose mount
5. Hex bolt
6. Spring clip
7. Brake line
8. Connector
9. Ribbed bolt
10. Brake carrier
11. Upper air line
12. Brake pads
13. Retention spring
14. Brake caliper housing
15. Guide pin
16. Trim cap

2348-A4A4-G23

Exploded view of FN3 type front brake components—A4 and A6 shown

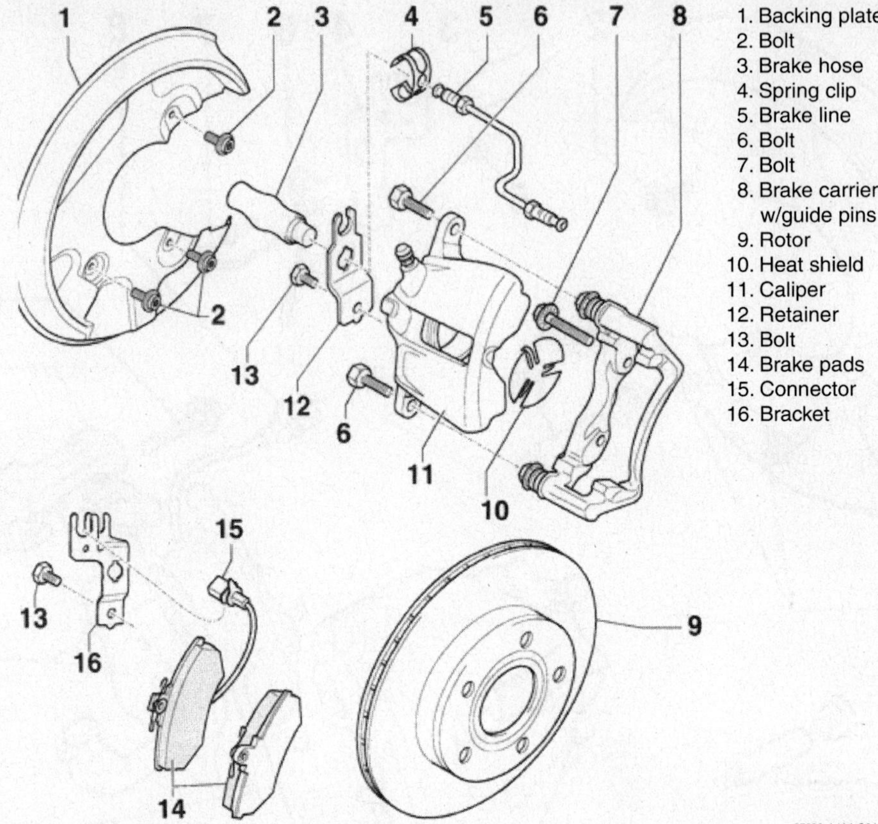

1. Backing plate
2. Bolt
3. Brake hose
4. Spring clip
5. Brake line
6. Bolt
7. Bolt
8. Brake carrier w/guide pins
9. Rotor
10. Heat shield
11. Caliper
12. Retainer
13. Bolt
14. Brake pads
15. Connector
16. Bracket

Exploded view of C54 type front brake components—A4 models

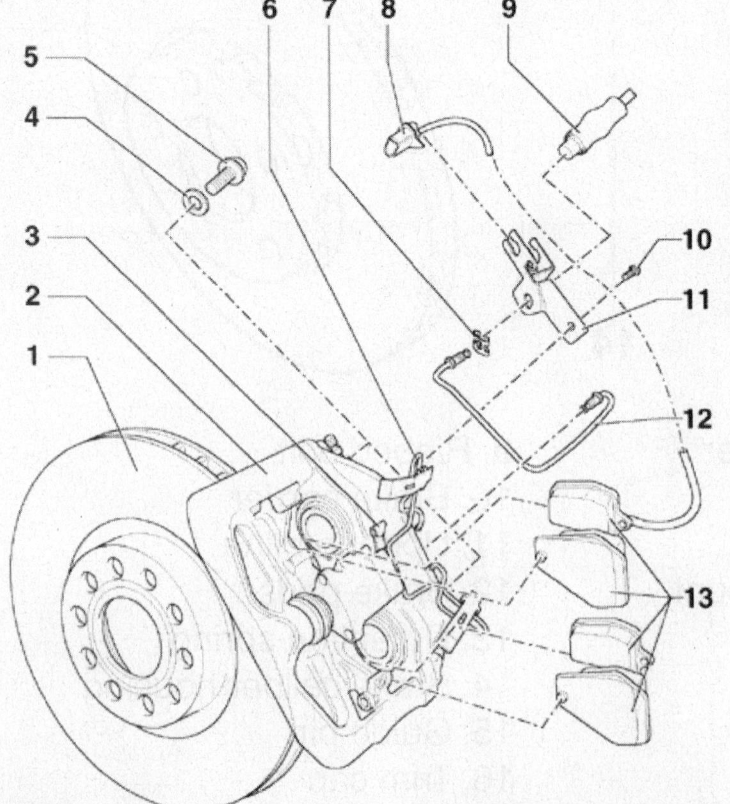

1. Rotor
2. Caliper
3. Wear indicator retainer
4. Ribbed washer
5. Bolt
6. Tensioning spring
7. Retaining spring
8. Harness connector
9. Brake hose
10. Bolt
11. Bracket
12. Brake line
13. Brake pads

Exploded view of HP-2 type (aluminum caliper) front brake components—A6 shown; others similar

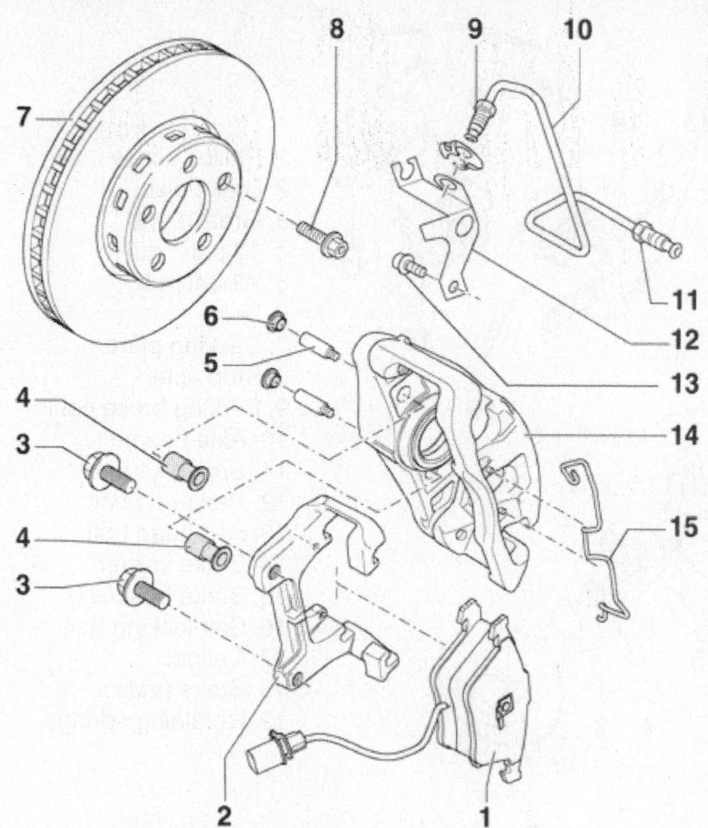

1. Brake pads
2. Carrier
3. Bolt
4. Bushing
5. Guide pin
6. Trim cap
7. Rotor
8. Wheel bolts
9. Line-to-hose connection
10. Brake line
11. Line-to-caliper connection
12. Bracket
13. Bolt
14. Caliper
15. Retaining spring

67200-A6A6-G11

Exploded view of FNR-60 type front brake components—A4 and A6 shown

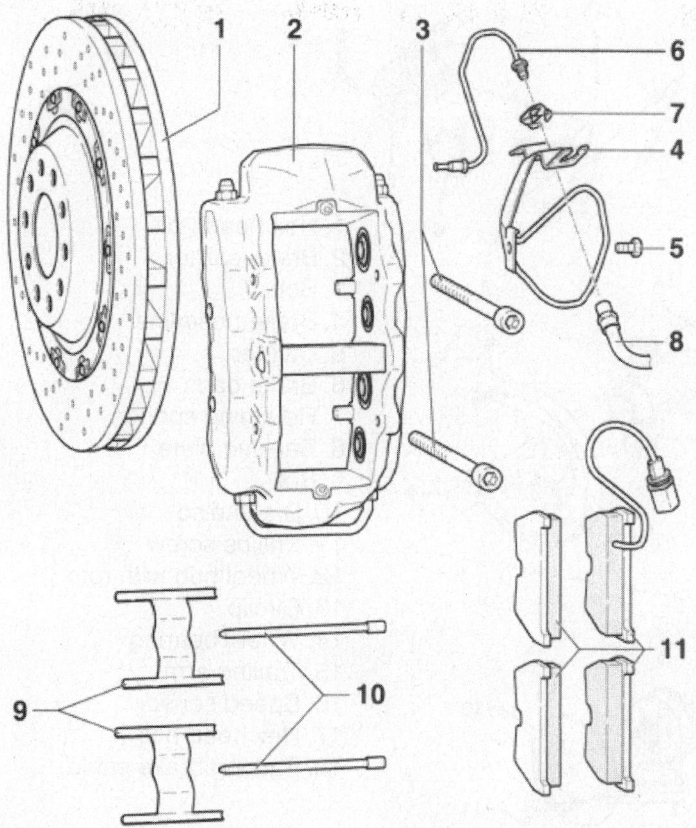

1. Rotor
2. Caliper
3. Guide pins
4. Bracket
5. Bolt
6. Brake line
7. Spring clip
8. Brake hose
9. Retaining springs
10. Pad retaining pins
11. Brake pads

67200-A6A6-G12

Exploded view of Brembo 18” type front brake components—A6 shown; others similar

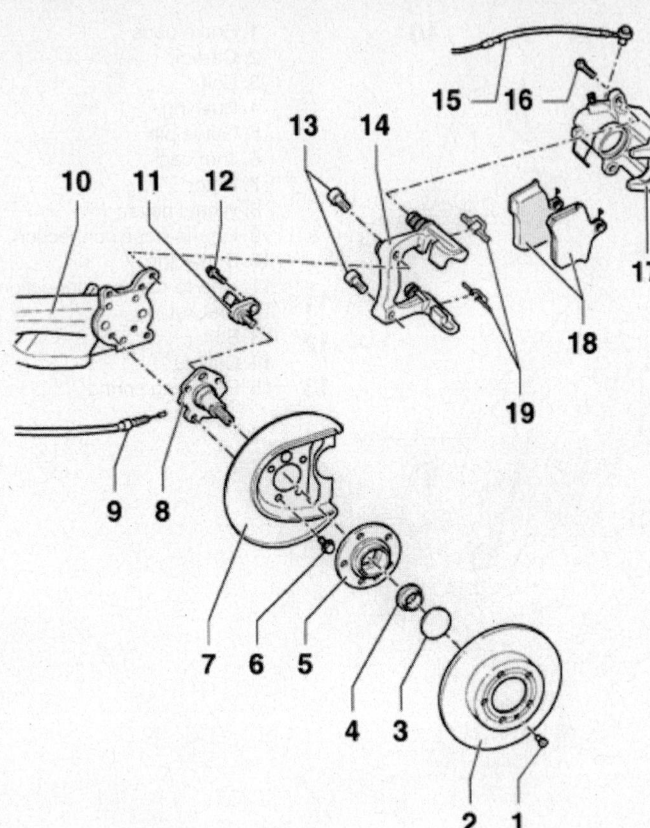

Exploded view of rear disc brakes—TT with front wheel drive

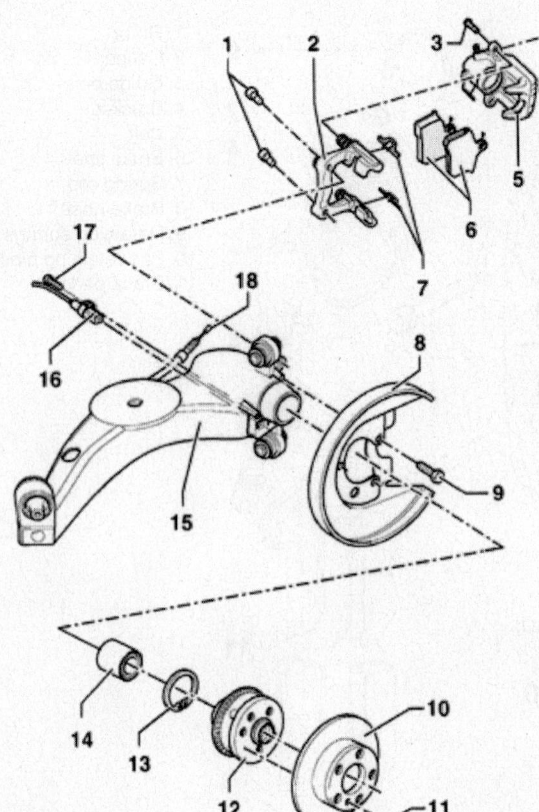

Exploded view of rear disc brakes—TT with all wheel drive

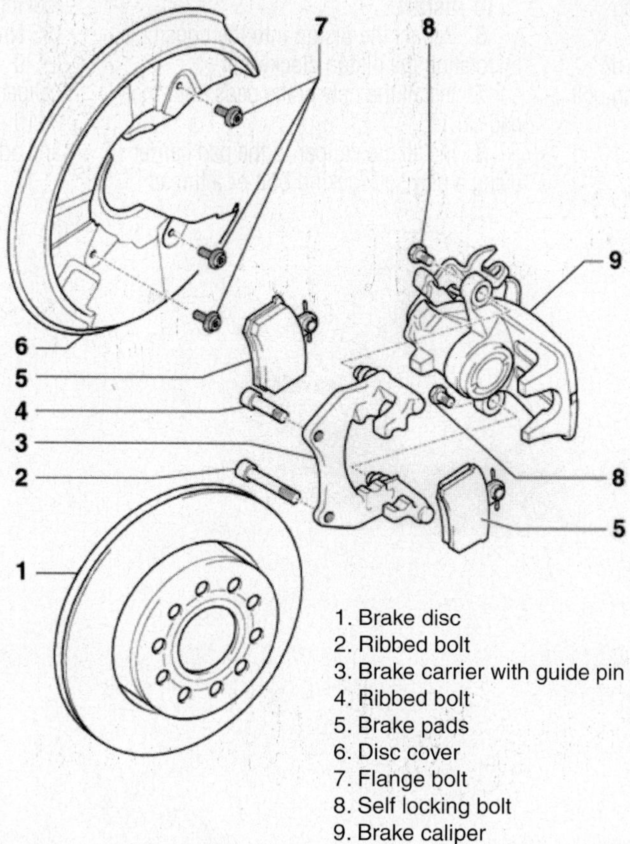

Exploded view of rear disc brakes—A4 shown; A6 and A8 similar

To install:

6. If removed, install the carrier. Torque the carrier bolts to 92 ft. lbs. (125 Nm).
7. Thread the caliper onto the hydraulic line and hand-tighten it. Fit the caliper into place on the carrier.
8. Torque the caliper bolts (guide pins) as follows:
   - On TT models: 21 ft. lbs. (28 Nm)
   - On A4 and A6 models with FN3 type brakes: (25 Nm)
   - On All A3 models, A4 and A6 models with FNR-G60 type brakes: (30 Nm)
   - On A6 models with HP2 (aluminum caliper) type brakes: (190 Nm)
   - On A6 models with 18" Brembo type brakes: (110 Nm)
9. Torque the brake line connection to (15 Nm).
10. Tighten the hydraulic line and bleed the brakes.

REAR

1. Before servicing the vehicle, refer to the precautions section.
2. Turn the ignition switch **OFF** and pump the brake pedal 25–35 times to relieve the system pressure.
3. Remove the wheels.
4. Disconnect the parking brake cable.
5. Loosen the hydraulic line.
6. Use a back-up wrench to hold the guide pins and remove the caliper bolts.
7. Lift the caliper off the carrier and unscrew it from the hydraulic line.

To install:

8. Thread the caliper onto the hydraulic line and hand-tighten it. Fit the caliper into place on the carrier. Torque the bolts to 26 ft. lbs. (35 Nm).
9. Bleed the brakes.
10. Install the wheels.

Brake Pads

REMOVAL & INSTALLATION

All Models

FRONT

1. Before servicing the vehicle, refer to the precautions section.
2. Remove the front wheels.
3. Hold the lower guide pin with an open wrench and remove the bolt securing the caliper to the guide pin.
4. Pivot the caliper up on the upper guide pin and slide the pads straight out to remove them.

To install:

5. Compress the caliper piston into the bore.
6. Fit the new pads into the carrier and pivot the caliper into place.
7. The original bolts are micro-encapsulated with a thread locking compound. Install a new bolt or clean the old bolt and apply a thread-locking compound.
8. When tightening the caliper bolts (guide pins) torque them as follows:
   - On TT models: 21 ft. lbs. (28 Nm)
   - On all A3 models, A4 and A6 models with FN3 type brakes: (25 Nm)
   - On A4 and A6 models with FNR-G60 type brakes: (30 Nm)
   - On A6 models with HP2 (aluminum caliper) type brakes: (190 Nm)
   - On A6 models with 18" Brembo type brakes: (110 Nm)
9. Install the wheels.

REAR

1. Before servicing the vehicle, refer to the precautions section.
2. Remove the rear wheels.
3. Remove the parking brake cable clip

from the caliper. Disconnect the parking brake cable.

4. Hold the guide pin with a back-up wrench and remove the upper mounting bolt from the brake caliper.

5. Swing the caliper downward and remove the brake pads.

To install:

6. Retract the piston into the housing by rotating the piston clockwise.

7. Install the new brake pads onto the pad carrier.

8. Install the caliper to the pad carrier using a new self locking bolt or a thread locking compound and torque to 26 ft. lbs. (35 Nm).

9. Attach the hand brake cable to the caliper.

10. Check the parking brake operation and adjust the cable if necessary.

11. Install the wheels.

BMW

M3 • Z3 • Z4 • 3 • 5 • 7 SERIES

2

SPECIFICATIONS AND MAINTENANCE CHARTS

ENGINE AND VEHICLE IDENTIFICATION

Engine						
Code	Liters (cc)	Cu. In.	Cyl.	Fuel Sys.	Engine Type	Eng. Mfg.
M54B25	2.5 (2494)	152	6	②	DOHC	BMW
M54B30	3.0 (2979)	182	6	②	DOHC	BMW
N52B30	3.0 (2996)	183	6	②	DOHC	BMW
S54B32	3.2 (3246)	192	6	③	DOHC	BMW
M62B44	4.4 (4398)	268	8	④	DOHC	BMW
N62B44	4.4 (4398)	268	8	⑤	DOHC	BMW
N62B48	4.8 (4841)	292	8	⑥	DOHC	BMW
S62B50	4.9 (4398)	300	8	⑥	DOHC	BMW

Model Year	
Code ①	Year
2	2002
3	2003
4	2004
5	2005
6	2006

DOHC: Double Overhead Camshaft

① 10th digit of the Vehicle Identification Number (VIN)

② Siemens DME MS 43.0

③ Siemens DME

④ Bosch DME 7.2

⑤ Bosch DME ME 9.2

⑥ Bosch DME

06041_BMWC_C0001

GENERAL ENGINE SPECIFICATIONS

Year	Body Type	Model	Engine Displacement Liters	Engine ID	Net Horsepower @ rpm	Net Torque @ rpm (ft. lbs.)	Bore x Stroke (in.)	Com-pression Ratio	Oil Pressure @ rpm
2002	E46	325i Sedan	2.5	M54B25	184@6000	175@3500	3.31x2.95	10.5:1	7.3@idle
	E46	325xi Sedan	2.5	M54B25	184@6000	175@3500	3.31x2.95	10.5:1	7.3@idle
	E46	325Ci Coupe	2.5	M54B25	184@6000	175@3500	3.31x2.95	10.5:1	7.3@idle
	E46	325Ci Cabriolet	2.5	M54B25	184@6000	175@3500	3.31x2.95	10.5:1	7.3@idle
	E46	325i Wagon	2.5	M54B25	184@6000	175@3500	3.31x2.95	10.5:1	7.3@idle
	E46	325xi Wagon	2.5	M54B25	184@6000	175@3500	3.31x2.95	10.5:1	7.3@idle
	E46	330i Sedan	3.0	M54B30	225@4900	214@3500	3.31x3.53	10.2:1	7.3@idle
	E46	330xi Sedan	3.0	M54B30	225@4900	214@3500	3.31x3.53	10.2:1	7.3@idle
	E46	330Ci Coupe	3.0	M54B30	225@4900	214@3500	3.31x3.53	10.2:1	7.3@idle
	E46	330Ci Cabriolet	3.0	M54B30	225@4900	214@3500	3.31x3.53	10.2:1	7.3@idle
	E46	M3	3.2	S54B32	333@7900	262@4900	3.43x3.58	11.5:1	10@idle
	E36/7	Z3 2.5	2.5	M54B25	184@6000	175@3500	3.31x2.95	10.5:1	7.3@idle
	E36/7	Z3 3.0	3.0	M54B30	225@4900	214@3500	3.31x3.53	10.2:1	7.3@idle
	E36/7	Z3 3.2	3.2	S54B32	333@7900	262@4900	3.43x3.58	11.5:1	7.3@idle
	E39	525i	2.5	M54B25	184@6000	175@3500	3.31x2.95	10.5:1	7.3@idle
	E39	525i Wagon	2.5	M54B25	184@6000	175@3500	3.31x2.95	10.5:1	7.3@idle
	E39	530i	3.0	M54B30	225@4900	214@3500	3.31x3.53	10.2:1	7.3@idle
	E39	540i	4.4	M62B44	290@5400	324@3600	3.62x3.26	10.0:1	7.3@idle
	E39	540i Wagon	4.4	M62B44	290@5400	324@3600	3.62x3.26	10.0:1	7.3@idle
	E39	M5	4.9	S62B50	394@6600	368@3800	3.70x3.50	11.0:1	7.3@idle
	E65	745i	4.4	N62B44	325@6100	330@3600	3.62x3.26	10.5:1	14.5@idle
	E66	745Li	4.4	N62B44	325@6100	330@3600	3.62x3.26	10.5:1	14.5@idle
2003	E46	325i Sedan	2.5	M54B25	184@6000	175@3500	3.31x2.95	10.5:1	7.3@idle
	E46	325xi Sedan	2.5	M54B25	184@6000	175@3500	3.31x2.95	10.5:1	7.3@idle
	E46	325Ci Coupe	2.5	M54B25	184@6000	175@3500	3.31x2.95	10.5:1	7.3@idle
	E46	325Ci Cabriolet	2.5	M54B25	184@6000	175@3500	3.31x2.95	10.5:1	7.3@idle
	E46	325i Wagon	2.5	M54B25	184@6000	175@3500	3.31x2.95	10.5:1	7.3@idle
	E46	325xi Wagon	2.5	M54B25	184@6000	175@3500	3.31x2.95	10.5:1	7.3@idle
	E46	330i Sedan	3.0	M54B30	225@4900	214@3500	3.31x3.53	10.2:1	7.3@idle
	E46	330xi Sedan	3.0	M54B30	225@4900	214@3500	3.31x3.53	10.2:1	7.3@idle
	E46	330Ci Coupe	3.0	M54B30	225@4900	214@3500	3.31x3.53	10.2:1	7.3@idle
	E46	330Ci Cabriolet	3.0	M54B30	225@4900	214@3500	3.31x3.53	10.2:1	7.3@idle
	E46	M3	3.2	S54B32	333@7900	262@4900	3.43x3.58	11.5:1	10@idle
	E85	Z4 2.5	2.5	M54B25	170@5500	181@3500	3.31x2.95	10.5:1	7.3@idle
	E85	Z4 3.0	3.0	M54B30	193@5500	206@3500	3.31x3.31	10.2:1	7.3@idle
	E39	525i	2.5	M54B25	170@5500	181@3500	3.31x2.95	10.5:1	7.3@idle
	E39	530i	3.0	M54B30	225@4900	214@3500	3.31x3.53	10.2:1	7.3@idle
	E39	540i	4.4	M62B44	325@6100	330@3600	3.62x3.26	10.5:1	7.3@idle
	E39	M5	4.9	S62B50	394@6600	368@3800	3.70x3.50	10.5:1	7.3@idle
	E65	745i	4.4	N62B44	325@6100	330@3600	3.62x3.26	10.5:1	14.5@idle
	E66	745Li	4.4	N62B44	325@6100	330@3600	3.62x3.26	10.5:1	14.5@idle
2004	E46	325i Sedan	2.5	M54B25	184@6000	175@3500	3.31x2.95	10.5:1	7.3@idle
	E46	325xi Sedan	2.5	M54B25	184@6000	175@3500	3.31x2.95	10.5:1	7.3@idle
	E46	325Ci Coupe	2.5	M54B25	184@6000	175@3500	3.31x2.95	10.5:1	7.3@idle
	E46	325Ci Cabriolet	2.5	M54B25	184@6000	175@3500	3.31x2.95	10.5:1	7.3@idle
	E46	325i Wagon	2.5	M54B25	184@6000	175@3500	3.31x2.95	10.5:1	7.3@idle

06041_BMWC_C0002

GENERAL ENGINE SPECIFICATIONS

Year	Body Type	Model	Engine Displacement Liters	Engine ID	Net Horsepower @ rpm	Net Torque @ rpm (ft. lbs.)	Bore x Stroke (in.)	Com-pression Ratio	Oil Pressure @ rpm
2004 Cont.	E46	325xi Wagon	2.5	M54B25	184@6000	175@3500	3.31x2.95	10.5:1	7.3@idle
	E46	330i Sedan	3.0	M54B30	225@4900	214@3500	3.31x3.53	10.2:1	7.3@idle
	E46	330xi Sedan	3.0	M54B30	225@4900	214@3500	3.31x3.53	10.2:1	7.3@idle
	E46	330Ci Coupe	3.0	M54B30	225@4900	214@3500	3.31x3.53	10.2:1	7.3@idle
	E46	330Ci Cabriolet	3.0	M54B30	225@4900	214@3500	3.31x3.53	10.2:1	7.3@idle
	E46	M3	3.2	S54B32	333@7900	262@4900	3.43x3.58	11.5:1	10@idle
	E85	Z4 2.5	2.5	M54B25	170@5500	181@3500	3.31x2.95	10.5:1	7.3@idle
	E85	Z4 3.0	3.0	M54B30	193@5500	206@3500	3.31x3.31	10.2:1	7.3@idle
	E60	525i	2.5	M54B25	170@5500	181@3500	3.31x2.95	10.5:1	7.3@idle
	E60	530i	3.0	M54B30	225@4900	214@3500	3.31x3.53	10.2:1	7.3@idle
	E60	545i	4.4	N62B44	325@6100	330@3600	3.62x3.26	10.5:1	14.5@idle
	E65	745i	4.4	N62B44	325@6100	330@3600	3.62x3.26	10.5:1	14.5@idle
	E66	740Li	4.4	N62B44	325@6100	330@3600	3.62x3.26	10.5:1	14.5@idle
2005	E46	325i Sedan	2.5	M54B25	184@6000	175@3500	3.31x2.95	10.5:1	7.3@idle
	E46	325xi Sedan	2.5	M54B25	184@6000	175@3500	3.31x2.95	10.5:1	7.3@idle
	E46	325Ci Coupe	2.5	M54B25	184@6000	175@3500	3.31x2.95	10.5:1	7.3@idle
	E46	325Ci Cabriolet	2.5	M54B25	184@6000	175@3500	3.31x2.95	10.5:1	7.3@idle
	E46	325i Wagon	2.5	M54B25	184@6000	175@3500	3.31x2.95	10.5:1	7.3@idle
	E46	325xi Wagon	2.5	M54B25	184@6000	175@3500	3.31x2.95	10.5:1	7.3@idle
	E46	330i Sedan	3.0	M54B30	225@4900	214@3500	3.31x3.53	10.2:1	7.3@idle
	E46	330xi Sedan	3.0	M54B30	225@4900	214@3500	3.31x3.53	10.2:1	7.3@idle
	E46	330Ci Coupe	3.0	M54B30	225@4900	214@3500	3.31x3.53	10.2:1	7.3@idle
	E46	330Ci Cabriolet	3.0	M54B30	225@4900	214@3500	3.31x3.53	10.2:1	7.3@idle
	E46	M3	3.2	S54B32	333@7900	262@4900	3.43x3.58	11.5:1	10@idle
	E85	Z4 2.5	2.5	M54B25	170@5500	181@3500	3.31x2.95	10.5:1	7.3@idle
	E85	Z4 3.0	3.0	M54B30	193@5500	206@3500	3.31x3.31	10.2:1	7.3@idle
	E60	525i	2.5	M54B25	170@5500	181@3500	3.31x2.95	10.5:1	7.3@idle
	E60	530i	3.0	M54B30	225@4900	214@3500	3.31x3.53	10.2:1	7.3@idle
	E60	545i	4.4	N62B44	325@6100	330@3600	3.62x3.26	10.5:1	14.5@idle
	E65	745i	4.4	N62B44	325@6100	330@3600	3.62x3.26	10.5:1	14.5@idle
	E66	740Li	4.4	N62B44	325@6100	330@3600	3.62x3.26	10.5:1	14.5@idle
2006	E46	325i Sedan	3.0	M54B30	215@6250	185@2750	3.35x3.46	10.7:1	7.3@idle
	E46	325xi Sedan	3.0	M54B30	215@6250	185@2750	3.35x3.46	10.7:1	7.3@idle
	E46	325Ci Coupe	2.5	M54B25	184@6000	175@3500	3.31x2.95	10.5:1	7.3@idle
	E46	325Ci Cabriolet	2.5	M54B25	184@6000	175@3500	3.31x2.95	10.5:1	7.3@idle
	E46	325xi Wagon	3.0	M54B30	215@6250	185@2750	3.35x3.46	10.7:1	7.3@idle
	E46	330i Sedan	3.0	M54B30	225@6600	220@2750	3.35x3.46	10.7:1	7.3@idle
	E46	330xi Sedan	3.0	M54B30	225@6600	220@2750	3.35x3.46	10.7:1	7.3@idle
	E46	330Ci Coupe	3.0	M54B30	225@5900	214@3500	3.31x3.53	10.2:1	7.3@idle
	E46	330Ci Cabriolet	3.0	M54B30	225@5900	214@3500	3.31x3.53	10.2:1	7.3@idle
	E46	M3	3.2	S54B32	333@7900	262@4900	3.43x3.58	11.5:1	10@idle
	E85	Z4 Coupe	3.0	M54B30	255@6600	220@2750	3.35x3.46	10.7:1	7.3@idle
	E85	Z4 3.0i	3.0	M54B30	215@6248	185@2748	3.35x3.46	10.7:1	7.3@idle
	E85	Z4 3.0si	3.0	N52B30	255@6600	220@2750	3.35x3.46	10.7:1	7.3@idle
	E60	525i	3.0	N52B30	215@6250	185@2750	3.35x3.46	10.7:1	7.3@idle
	E60	525xi	3.0	N52B30	215@6250	185@2750	3.35x3.46	10.7:1	7.3@idle
	E60	530i	3.0	N52B30	215@6250	185@2750	3.35x3.46	10.7:1	7.3@idle
	E60	530xi	3.0	N52B30	215@6250	185@2750	3.35x3.46	10.7:1	7.3@idle
	E60	550i	4.8	N62B48	360@6300	360@3400	3.66x3.48	10.5:1	14.5@idle

06041_BMWC_C0003

GENERAL ENGINE SPECIFICATIONS

Year	Body Type	Model	Engine Displacement Liters	Engine ID	Net Horsepower @ rpm	Net Torque @ rpm (ft. lbs.)	Bore x Stroke (in.)	Com-pression Ratio	Oil Pressure @ rpm
2006	E65	750i	4.8	N62B48	360@6300	360@3400	3.66x3.48	10.5:1	14.5@idle
Cont.	E65	750Li	4.8	N62B48	360@6300	360@3400	3.66x3.48	10.5:1	14.5@idle

NA: Not Available

① Siemens DME MS 43.0

② Siemens DME

③ Bosch DME 7.2

④ Bosch DME

⑤ Bosch DME ME 9.2

06041_BMWC_C0004

ENGINE TUNE-UP SPECIFICATIONS

Year	Engine Displacement Liters	Engine ID	Spark Plug Gap (in.)	Ignition Timing (deg.) MT	Ignition Timing (deg.) AT	Fuel Pump (psi)	Idle Speed (rpm) MT	Idle Speed (rpm) AT	Valve Clearance In.	Valve Clearance Ex.
2002	2.5	M54B25	①	②	②	48-54 ③	NA	NA	HYD	HYD
	3.0	M54B30	①	②	②	48-54 ③	NA	NA	HYD	HYD
	3.2	S54B32	①	②	②	72-73 ③	NA	NA	HYD	HYD
	4.4	M62B44	①	②	②	48-54 ③	NA	NA	HYD	HYD
	4.4	N62B44	①	②	②	48-54 ③	NA	NA	HYD	HYD
	4.9	S62B50	①	②	②	72-73 ③	NA	NA	HYD	HYD
2003	2.5	M54B25	①	②	②	48-54 ③	NA	NA	HYD	HYD
	3.0	M54B30	①	②	②	48-54 ③	NA	NA	HYD	HYD
	3.2	S54B32	①	②	②	72-73 ③	NA	NA	HYD	HYD
	4.4	M62B44	①	②	②	48-54 ③	NA	NA	HYD	HYD
	4.4	N62B44	①	②	②	48-54 ③	NA	NA	HYD	HYD
	4.9	S62B50	①	②	②	72-73 ③	NA	NA	HYD	HYD
2004	2.5	M54B25	①	②	②	48-54 ③	NA	NA	HYD	HYD
	3.0	M54B30	①	②	②	48-54 ③	NA	NA	HYD	HYD
	3.2	S54B32	①	②	②	72-73 ③	NA	NA	HYD	HYD
	4.4	N62B44	①	②	②	48-54 ③	NA	NA	HYD	HYD
2005	2.5	M54B25	①	②	②	48-54 ③	NA	NA	HYD	HYD
	3.0	M54B30	①	②	②	48-54 ③	NA	NA	HYD	HYD
	3.2	S54B32	①	②	②	72-73 ③	NA	NA	HYD	HYD
	4.4	N62B44	①	②	②	48-54 ③	NA	NA	HYD	HYD
2006	2.5	M54B25	①	②	②	48-54 ③	NA	NA	HYD	HYD
	3.0	M54B30	①	②	②	48-54 ③	NA	NA	HYD	HYD
	3.0	N52B30	①	②	②	48-54 ③	NA	NA	HYD	HYD
	3.2	S54B32	①	②	②	72-73 ③	NA	NA	HYD	HYD
	4.8	N62B48	①	②	②	48-54 ③	NA	NA	HYD	HYD

NOTE: The Vehicle Emission Control Information label reflects specification changes during production and must be used if they differ from this chart.

NA: Not Available

B: Before Top Dead Center

HYD: Hydraulic

① Three mass and four-mass electrodes cannot be adjusted
  Dual mass electrodes: 0.035-0.039 inches
  All models except M models: 0.028-0.031 inches
  M models: 0.024-0.028 inches

② Controlled by the Engine Control Module (ECM) and cannot be adjusted

③ At idle, pressure measured at injectors

06041_BMWC_C0005

CAPACITIES

Year	Body Type	Model	Engine Displacement Liters	Engine ID	Engine Oil with Filter (qts.)	Transmission (pts.) 5-Spd	Transmission (pts.) 6-Spd	Transmission (pts.) Auto.	Drive Axle Front (pts.)	Drive Axle Rear (pts.)	Fuel Tank (gal.)	Cooling System (qts.)
2002	E46	325i Sedan	2.5	M54B25	6.9	3.2	—	6.4	—	3.6	16.4	8.9
	E46	325xi Sedan	2.5	M54B25	6.9	3.2	—	6.4	1.48	3.6	16.4	8.9
	E46	325Ci Coupe	2.5	M54B25	6.9	3.2	—	6.4	—	3.6	16.4	8.9
	E46	325Ci Cabriolet	2.5	M54B25	6.9	3.2	—	6.4	—	3.6	16.4	8.9
	E46	325i Wagon	2.5	M54B25	6.9	3.2	—	6.4	—	3.6	16.4	8.9
	E46	325xi Wagon	2.5	M54B25	6.9	3.2	—	6.4	1.48	3.6	16.4	8.9
	E46	330i Sedan	3.0	M54B30	6.9	3.2	—	6.4	—	3.6	16.4	8.9
	E46	330xi Sedan	3.0	M54B30	6.9	3.2	—	6.4	1.48	3.6	16.4	8.9
	E46	330Ci Coupe	3.0	M54B30	6.9	3.2	—	6.4	—	3.6	16.4	8.9
	E46	330Ci Cabriolet	3.0	M54B30	6.9	3.2	—	6.4	—	3.6	16.4	8.9
	E46	M3	3.2	S54B32	5.8	3.2	—	6.4	—	3.8	16.4	9.8
	E36/7	Z3 2.5	2.5	M54B25	6.9	3.2	—	6.4	—	3.6	13.5	8.9
	E36/7	Z3 3.0	3.0	M54B30	6.9	3.2	—	6.4	—	3.6	13.5	8.9
	E36/7	Z3 3.2	3.2	S54B32	5.8	3.2	—	6.4	—	3.8	13.5	9.8
	E39	525i	2.5	M54B25	6.9	3.2	—	6.4	—	3.4	18.5	11.1
	E39	525i Wagon	2.5	M54B25	6.9	3.2	—	6.4	—	3.4	18.5	11.1
	E39	530i	3.0	M54B30	6.9	3.2	—	6.4	—	3.4	18.5	11.1
	E39	540i	4.4	M62B44	7.9	—	4.8	7.0	—	3.4	18.5	13.2
	E39	540i Wagon	4.4	M62B44	7.9	—	4.8	7.0	—	3.4	18.5	13.2
	E39	M5	4.9	S62B50	7.9	—	4.8	7.0	—	3.4	18.5	13.2
	E65	745i	4.4	N62B44	8.4	—	—	NA	—	3.4	23.2	NA
	E66	745Li	4.4	N62B44	8.4	—	—	NA	—	3.4	23.2	NA
2003	E46	325i Sedan	2.5	M54B25	6.9	3.2	—	6.4	—	3.6	16.4	8.9
	E46	325xi Sedan	2.5	M54B25	6.9	3.2	—	6.4	1.48	3.6	16.4	8.9
	E46	325Ci Coupe	2.5	M54B25	6.9	3.2	—	6.4	—	3.6	16.4	8.9
	E46	325Ci Cabriolet	2.5	M54B25	6.9	3.2	—	6.4	—	3.6	16.4	8.9
	E46	325i Wagon	2.5	M54B25	6.9	3.2	—	6.4	—	3.6	16.4	8.9
	E46	325xi Wagon	2.5	M54B25	6.9	3.2	—	6.4	1.48	3.6	16.4	8.9
	E46	330i Sedan	3.0	M54B30	6.9	3.2	—	6.4	—	3.6	16.4	8.9
	E46	330xi Sedan	3.0	M54B30	6.9	3.2	—	6.4	1.48	3.6	16.4	8.9
	E46	330Ci Coupe	3.0	M54B30	6.9	3.2	—	6.4	—	3.6	16.4	8.9
	E46	330Ci Cabriolet	3.0	M54B30	6.9	3.2	—	6.4	—	3.6	16.4	8.9
	E46	M3	3.2	S54B32	5.8	3.2	—	6.4	—	3.8	16.4	9.8
	E85	Z4 2.5	2.5	M54B25	6.9	3.2	—	6.4	—	3.6	13.5	9.3
	E85	Z4 3.0	3.0	M54B30	6.9	3.2	—	6.4	—	3.6	13.5	9.3
	E39	525i	2.5	M54B25	6.9	3.2	—	6.4	—	3.4	18.5	11.1
	E39	530i	3.0	M54B30	6.9	3.2	—	6.4	—	3.4	18.5	11.1
	E39	540i	4.4	M62B44	7.9	—	4.8	7.0	—	3.4	18.5	13.2
	E39	M5	4.9	S62B50	7.9	—	4.8	7.0	—	3.4	18.5	13.2
	E65	745i	4.4	N62B44	8.4	—	—	NA	—	3.4	23.2	NA
	E66	745Li	4.4	N62B44	8.4	—	—	NA	—	3.4	23.2	NA
2004	E46	325i Sedan	2.5	M54B25	6.9	3.2	—	6.4	—	3.6	16.4	8.9
	E46	325xi Sedan	2.5	M54B25	6.9	3.2	—	6.4	1.48	3.6	16.4	8.9
	E46	325Ci Coupe	2.5	M54B25	6.9	3.2	—	6.4	—	3.6	16.4	8.9
	E46	325Ci Cabriolet	2.5	M54B25	6.9	3.2	—	6.4	—	3.6	16.4	8.9
	E46	325i Wagon	2.5	M54B25	6.9	3.2	—	6.4	—	3.6	16.4	8.9
	E46	325xi Wagon	2.5	M54B25	6.9	3.2	—	6.4	1.48	3.6	16.4	8.9
	E46	330i Sedan	3.0	M54B30	6.9	3.2	—	6.4	—	3.6	16.4	8.9
	E46	330xi Sedan	3.0	M54B30	6.9	3.2	—	6.4	1.48	3.6	16.4	8.9

06041_BMWC_C0006

CAPACITIES

Year	Body Type	Model	Engine Displacement Liters	Engine ID	Engine Oil with Filter (qts.)	Transmission (pts.) 5-Spd	6-Spd	Auto.	Drive Axle Front (pts.)	Rear (pts.)	Fuel Tank (gal.)	Cooling System (qts.)
2004	E46	330Ci Coupe	3.0	M54B30	6.9	3.2	—	6.4	—	3.6	16.4	8.9
Cont.	E46	330Ci Cabriolet	3.0	M54B30	6.9	3.2	—	6.4	—	3.6	16.4	8.9
	E46	M3	3.2	S54B32	5.8	3.2	—	6.4	—	3.8	16.4	9.8
	E85	Z4 2.5	2.5	M54B25	6.9	3.2	—	6.4	—	3.6	13.5	9.3
	E85	Z4 3.0	3.0	M54B30	6.9	3.2	—	6.4	—	3.6	13.5	9.3
	E60	525i	2.5	M54B25	NA	NA	NA	NA	NA	NA	NA	NA
	E60	530i	3.0	M54B30	NA	NA	NA	NA	NA	NA	NA	NA
	E60	545i	4.4	N62B44	NA	NA	NA	NA	NA	NA	NA	NA
	E65	745i	4.4	N62B44	8.4	—	—	NA	—	3.4	23.2	NA
	E66	740Li	4.4	N62B44	8.4	—	—	NA	—	3.4	23.2	NA
2005	E46	325i Sedan	2.5	M54B25	6.9	3.2	—	6.4	—	3.6	16.4	8.9
	E46	325xi Sedan	2.5	M54B25	6.9	3.2	—	6.4	1.48	3.6	16.4	8.9
	E46	325Ci Coupe	2.5	M54B25	6.9	3.2	—	6.4	—	3.6	16.4	8.9
	E46	325Ci Cabriolet	2.5	M54B25	6.9	3.2	—	6.4	—	3.6	16.4	8.9
	E46	325i Wagon	2.5	M54B25	6.9	3.2	—	6.4	—	3.6	16.4	8.9
	E46	325xi Wagon	2.5	M54B25	6.9	3.2	—	6.4	1.48	3.6	16.4	8.9
	E46	330i Sedan	3.0	M54B30	6.9	3.2	—	6.4	—	3.6	16.4	8.9
	E46	330xi Sedan	3.0	M54B30	6.9	3.2	—	6.4	1.48	3.6	16.4	8.9
	E46	330Ci Coupe	3.0	M54B30	6.9	3.2	—	6.4	—	3.6	16.4	8.9
	E46	330Ci Cabriolet	3.0	M54B30	6.9	3.2	—	6.4	—	3.6	16.4	8.9
	E46	M3	3.2	S54B32	5.8	3.2	—	6.4	—	3.8	16.4	9.8
	E85	Z4 2.5	2.5	M54B25	6.9	3.2	—	6.4	—	3.6	13.5	9.3
	E85	Z4 3.0	3.0	M54B30	6.9	3.2	—	6.4	—	3.6	13.5	9.3
	E60	525i	2.5	M54B25	NA	NA	NA	NA	NA	NA	NA	NA
	E60	530i	3.0	M54B30	NA	NA	NA	NA	NA	NA	NA	NA
	E60	545i	4.4	N62B44	NA	NA	NA	NA	NA	NA	NA	NA
	E65	745i	4.4	N62B44	8.4	—	—	NA	—	3.4	23.2	NA
	E66	740Li	4.4	N62B44	8.4	—	—	NA	—	3.4	23.2	NA
2006	E46	325i Sedan	3.0	M54B30	6.9	3.2	—	6.4	—	3.6	16.4	8.9
	E46	325xi Sedan	3.0	M54B30	6.9	3.2	—	6.4	—	3.6	16.4	8.9
	E46	325Ci Coupe	2.5	M54B25	6.9	3.2	—	6.4	—	3.6	16.4	8.9
	E46	325Ci Cabriolet	2.5	M54B25	6.9	3.2	—	6.4	—	3.6	16.4	8.9
	E46	325xi Wagon	3.0	M54B30	6.9	3.2	—	6.4	—	3.6	16.4	8.9
	E46	330i Sedan	3.0	M54B30	6.9	3.2	—	6.4	—	3.6	16.4	8.9
	E46	330xi Sedan	3.0	M54B30	6.9	3.2	—	6.4	—	3.6	16.4	8.9
	E46	330Ci Coupe	3.0	M54B30	6.9	3.2	—	6.4	—	3.6	16.4	8.9
	E46	330Ci Cabriolet	3.0	M54B30	6.9	3.2	—	6.4	—	3.6	16.4	8.9
	E46	M3	3.2	S54B32	5.8	3.2	—	6.4	—	3.8	16.4	9.8
	E85	Z4 Coupe	3.0	M54B30	6.9	3.2	—	6.4	—	3.6	13.5	9.3
	E85	Z4 3.0i	3.0	M54B30	6.9	3.2	—	6.4	—	3.6	13.5	9.3
	E85	Z4 3.0si	3.0	N52B30	6.9	3.2	—	6.4	—	3.6	13.5	9.3
	E60	525i	3.0	N52B30	NA	NA	NA	NA	NA	NA	18.5	NA
	E60	525xi	3.0	N52B30	NA	NA	NA	NA	NA	NA	18.5	NA
	E60	530i	3.0	N52B30	NA	NA	NA	NA	NA	NA	18.5	NA
	E60	530xi	3.0	N52B30	NA	NA	NA	NA	NA	NA	18.5	NA
	E60	550i	4.8	N62B48	8.4	NA	NA	NA	NA	NA	18.5	NA
	E65	750i	4.8	N62B48	NA	NA	NA	9	NA	NA	23.3	NA
	E65	750Li	4.8	N62B48	NA	NA	NA	9	NA	NA	23.3	NA

NA: Information not available

NOTE: All capacities are service, not overhaul and are approximate. Add fluid gradually and ensure a proper fluid level is obtained.

06041_BMWC_C0007

VALVE SPECIFICATIONS

Year	Engine Displacement Liters	Engine ID	Seat Angle (deg.)	Face Angle (deg.)	Spring Test Pressure (lbs. @ in.)	Spring Installed Height (in.)	Stem-to-Guide Clearance (in.)		Stem Diameter (in.)	
							Intake	Exhaust	Intake	Exhaust
2002	2.5	M54B25	①	45	NA	NA	② 0.0197	② 0.0197	0.0234-0.0235	0.0234-0.0235
	3.0	M54B30	①	45	NA	NA	② 0.0197	② 0.0197	0.0234-0.0235	0.0234-0.0235
	3.2	S54B32	①	45	NA	NA	② 0.0197	② 0.0197	0.0234-0.0235	0.0234-0.0235
	4.4	M62B44	①	45	NA	NA	② 0.0197	② 0.0197	0.0234-0.0235	0.0234-0.0235
	4.4	N62B44	①	45	NA	NA	② 0.0197	② 0.0197	0.0234-0.0235	0.0234-0.0235
	4.9	S62B50	①	45	NA	NA	② 0.0197	② 0.0197	0.0234-0.0235	0.0234-0.0235
2003	2.5	M54B25	①	45	NA	NA	② 0.0197	② 0.0197	0.0234-0.0235	0.0234-0.0235
	3.0	M54B30	①	45	NA	NA	② 0.0197	② 0.0197	0.0234-0.0235	0.0234-0.0235
	3.2	S54B32	①	45	NA	NA	② 0.0197	② 0.0197	0.0234-0.0235	0.0234-0.0235
	4.4	M62B44	①	45	NA	NA	② 0.0197	② 0.0197	0.0234-0.0235	0.0234-0.0235
	4.4	N62B44	①	45	NA	NA	② 0.0197	② 0.0197	0.0234-0.0235	0.0234-0.0235
	4.9	S62B50	①	45	NA	NA	② 0.0197	② 0.0197	0.0234-0.0235	0.0234-0.0235
2004	2.5	M54B25	①	45	NA	NA	② 0.0197	② 0.0197	0.0234-0.0235	0.0234-0.0235
	3.0	M54B30	①	45	NA	NA	② 0.0197	② 0.0197	0.0234-0.0235	0.0234-0.0235
	3.2	S54B32	①	45	NA	NA	② 0.0197	② 0.0197	0.0234-0.0235	0.0234-0.0235
	4.4	N62B44	①	45	NA	NA	② 0.0197	② 0.0197	0.0234-0.0235	0.0234-0.0235
2005	2.5	M54B25	①	45	NA	NA	② 0.0197	② 0.0197	0.0234-0.0235	0.0234-0.0235
	3.0	M54B30	①	45	NA	NA	② 0.0197	② 0.0197	0.0234-0.0235	0.0234-0.0235
	3.2	S54B32	①	45	NA	NA	② 0.0197	② 0.0197	0.0234-0.0235	0.0234-0.0235
	4.4	N62B44	①	45	NA	NA	② 0.0197	② 0.0197	0.0234-0.0235	0.0234-0.0235
2006	2.5	M54B25	①	45	NA	NA	② 0.0197	② 0.0197	0.0234-0.0235	0.0234-0.0235
	3.0	M54B30	①	45	NA	NA	② 0.0197	② 0.0197	0.0234-0.0235	0.0234-0.0235
	3.0	N52B30	①	45	NA	NA	② 0.0197	② 0.0197	0.0234-0.0235	0.0234-0.0235
	3.2	S54B32	①	45	NA	NA	② 0.0197	② 0.0197	0.0234-0.0235	0.0234-0.0235

06041_BMWC_C0008

VALVE SPECIFICATIONS

Year	Engine Displacement Liters	Engine ID	Seat Angle (deg.)	Face Angle (deg.)	Spring Test Pressure (lbs. @ in.)	Spring Installed Height (in.)	Stem-to-Guide Clearance (in.)		Stem Diameter (in.)	
							Intake	Exhaust	Intake	Exhaust
2006 Cont.	4.8	N62B48	①	45	NA	NA	② 0.0197	② 0.0197	0.0234-0.0235	0.0234-0.0235

NA: Information not available

① Valve seat angle: 45 degrees
Correction angle outside: 15 degrees
Correction angle inside: 60 degrees

② To measure: Insert a new valve into guide with the end of the valve flush with the end of the guide.
Use a dial indicator to measure axial valve head movement.

06041_BMWC_C0009

CRANKSHAFT AND CONNECTING ROD SPECIFICATIONS

All measurements are given in inches.

Year	Engine Displacement Liters	Engine ID	Crankshaft: Main Brg. Journal Dia.	Crankshaft: Main Brg. Oil Clearance	Crankshaft: Shaft End-play	Crankshaft: Thrust on No.	Connecting Rod: Journal Diameter	Connecting Rod: Oil Clearance	Connecting Rod: Side Clearance
2002	2.5	M54B25	①	0.0007-0.0023	0.0031-0.0064	NA	1.7706-1.7712	0.0007-0.0022	NA
	3.0	M54B30	①	0.0007-0.0023	0.0031-0.0064	NA	1.7706-1.7712	0.0007-0.0022	NA
	3.2	S54B32	②	0.0007-0.0020	0.0055-0.0099	NA	1.9281-1.9287	0.0015-0.0027	NA
	4.4	M62B44	③	0.0007-0.0018	0.0033-0.0101	NA	1.8901-1.8887	0.0007-0.0022	NA
	4.4	N62B44	⑤	0.0009-0.0020	0.0031-0.0096	NA	2.1249-2.1256	0.0011-0.0027	NA
	4.9	S62B50	④	0.0009-0.0019	0.0033-0.0101	NA	1.9281-1.9287	0.0011-0.0026	NA
2003	2.5	M54B25	①	0.0007-0.0023	0.0031-0.0064	NA	1.7706-1.7712	0.0007-0.0022	NA
	3.0	M54B30	①	0.0007-0.0023	0.0031-0.0064	NA	1.7706-1.7712	0.0007-0.0022	NA
	3.2	S54B32	②	0.0007-0.0020	0.0055-0.0099	NA	1.9281-1.9287	0.0015-0.0027	NA
	4.4	M62B44	③	0.0007-0.0018	0.0033-0.0101	NA	1.8901-1.8887	0.0007-0.0022	NA
	4.4	N62B44	⑥	0.0009-0.0020	0.0031-0.0096	NA	2.1249-2.1256	0.0011-0.0027	NA
	4.9	S62B50	④	0.0009-0.0019	0.0033-0.0101	NA	1.9281-1.9287	0.0011-0.0026	NA
2004	2.5	M54B25	①	0.0007-0.0023	0.0031-0.0064	NA	1.7706-1.7712	0.0007-0.0022	NA
	3.0	M54B30	①	0.0007-0.0023	0.0031-0.0064	NA	1.7706-1.7712	0.0007-0.0022	NA
	3.2	S54B32	⑤	0.0007-0.0020	0.0055-0.0099	NA	1.9281-1.9287	0.0015-0.0027	NA
	4.4	N62B44	⑥	0.0009-0.0020	0.0031-0.0096	NA	2.1249-2.1256	0.0011-0.0027	NA
2005	2.5	M54B25	①	0.0007-0.0023	0.0031-0.0064	NA	1.7706-1.7712	0.0007-0.0022	NA
	3.0	M54B30	①	0.0007-0.0023	0.0031-0.0064	NA	1.7706-1.7712	0.0007-0.0022	NA
	3.2	S54B32	⑤	0.0007-0.0020	0.0055-0.0099	NA	1.9281-1.9287	0.0015-0.0027	NA
	4.4	N62B44	⑥	0.0009-0.0020	0.0031-0.0096	NA	2.1249-2.1256	0.0011-0.0027	NA
2006	2.5	M54B25	①	0.0007-0.0023	0.0031-0.0064	NA	1.7706-1.7712	0.0007-0.0022	NA
	3.0	M54B30	①	0.0007-0.0023	0.0031-0.0064	NA	1.7706-1.7712	0.0007-0.0022	NA
	3.0	N52B30	①	0.0007-0.0023	0.0031-0.0064	NA	1.7706-1.7712	0.0007-0.0022	NA
	3.2	S54B32	⑤	0.0007-0.0020	0.0055-0.0099	NA	1.9281-1.9287	0.0015-0.0027	NA
	4.8	N62B48	⑥	0.0009-0.0020	0.0031-0.0096	NA	2.2677-2.8740	0.0011-0.0027	NA

NA: Information not available

① Standard yellow 2.3615-2.3618 inches
Standard green: 2.3613-2.3615 inches
Standard white: 2.3611-2.3613 inches

② Standard yellow 2.3607-2.3610 inches for bearing 1, 2.3615-2.3618 inches for other bearing positions
Standard green: 2.3605-2.3610 inches for bearing 1, 2.3605-2.3607 inches for other bearing positions
Standard white: 2.3602-2.3604 inches for bearing 1, 2.3610-2.3612 inches for other bearing positions

③ Standard yellow 2.7553-2.7555 inches
Standard green: 2.7550-2.7552 inches
Standard white: 2.7548-2.76550 inches

④ Standard yellow 2.7544-2.7547 inches for bearing 1, 2.7555 inches for other bearing positions
Standard green: 2.7542-2.7544 inches for bearing 1, 2.7549-2.7552 inches for other bearing positions
Standard white: 2.7539-2.7541 inches for bearing 1, 2.7547-2.7549 inches for other bearing positions

⑤ Standard yellow 2.9521-2.9523 inches
Standard green: 2.9518-2.952 inches
Standard white: 2.9516-2.9518 inches

⑥ Standard yellow 2.7548-2.7551 inches for bearing 1, 2.7552-2.7555 inches for other bearing positions
Standard green: 2.7545-2.7548 inches for bearing 1, 2.7549-2.7552 inches for other bearing positions
Standard violet: 2.75436-2.7545 inches for bearing 1, 2.7547-2.7549 inches for other bearing positions

06041_BMWC_C0010

PISTON AND RING SPECIFICATIONS

All measurements are given in inches

	Engine			Ring Gap			Ring Side Clearance		
Year	Size Liters	Engine ID	Piston Clearance	Top Compression	Bottom Compression	Oil Control	Top Compression	Bottom Compression	Oil Control
2002	2.5	M54B25	0.0004-0.0016	0.0078-0.0157	0.0078-0.0157	0.0078-0.0177	0.0008-0.0024	0.0008-0.0024	0.0005-0.0023
	3.0	M54B30	0.0004-0.0016	0.0078-0.0157	0.0078-0.0157	0.0078-0.0177	0.0008-0.0024	0.0008-0.0024	0.0005-0.0023
	3.2	S54B32	0.0010-0.0022	0.0078-0.0137	0.0137-0.0236	0.0098-0.0196	0.0011-0.0027	0.0005-0.0023	0.0008-0.0024
	4.4	M62B44	0.0002-0.0014	0.0039-0.0118	0.0078-0.0157	0.0078-0.0354	0.0012-0.0026	0.0012-0.0026	NA
	4.4	N62B44	0.0010-0.0028	0.0039-0.0118	0.0078-0.0157	0.0078-0.0354	0.0007-0.0027	0.0007-0.0023	NA
	4.9	S62B50	0.0001-0.0014	0.0059-0.0098	0.0078-0.0157	0.0078-0.0157	0.0005-0.0023	0.0007-0.0026	NA
2003	2.5	M54B25	0.0004-0.0016	0.0078-0.0157	0.0078-0.0157	0.0078-0.0177	0.0008-0.0024	0.0008-0.0024	0.0005-0.0023
	3.0	M54B30	0.0004-0.0016	0.0078-0.0157	0.0078-0.0157	0.0078-0.0177	0.0008-0.0024	0.0008-0.0024	0.0005-0.0023
	3.2	S54B32	0.0010-0.0022	0.0078-0.0137	0.0137-0.0236	0.0098-0.0196	0.0011-0.0027	0.0005-0.0023	0.0008-0.0024
	4.4	M62B44	0.0002-0.0014	0.0039-0.0118	0.0078-0.0157	0.0078-0.0354	0.0012-0.0026	0.0012-0.0026	NA
	4.4	N62B44	0.0010-0.0028	0.0039-0.0118	0.0078-0.0157	0.0078-0.0354	0.0007-0.0027	0.0007-0.0023	NA
	4.9	S62B50	0.0001-0.0014	0.0059-0.0098	0.0078-0.0157	0.0078-0.0157	0.0005-0.0023	0.0007-0.0026	NA
2004	2.5	M54B25	0.0004-0.0016	0.0078-0.0157	0.0078-0.0157	0.0078-0.0177	0.0008-0.0024	0.0008-0.0024	0.0005-0.0023
	3.0	M54B30	0.0004-0.0016	0.0078-0.0157	0.0078-0.0157	0.0078-0.0177	0.0008-0.0024	0.0008-0.0024	0.0005-0.0023
	3.2	S54B32	0.0010-0.0022	0.0078-0.0137	0.0137-0.0236	0.0098-0.0196	0.0011-0.0027	0.0005-0.0023	0.0008-0.0024
	4.4	N62B44	0.0010-0.0028	0.0039-0.0118	0.0078-0.0157	0.0078-0.0354	0.0007-0.0027	0.0007-0.0023	NA
2005	2.5	M54B25	0.0004-0.0016	0.0078-0.0157	0.0078-0.0157	0.0078-0.0177	0.0008-0.0024	0.0008-0.0024	0.0005-0.0023
	3.0	M54B30	0.0004-0.0016	0.0078-0.0157	0.0078-0.0157	0.0078-0.0177	0.0008-0.0024	0.0008-0.0024	0.0005-0.0023
	3.2	S54B32	0.0010-0.0022	0.0078-0.0137	0.0137-0.0236	0.0098-0.0196	0.0011-0.0027	0.0005-0.0023	0.0008-0.0024
	4.4	N62B44	0.0010-0.0028	0.0039-0.0118	0.0078-0.0157	0.0078-0.0354	0.0007-0.0027	0.0007-0.0023	NA
2006	2.5	M54B25	0.0004-0.0016	0.0078-0.0157	0.0078-0.0157	0.0078-0.0177	0.0008-0.0024	0.0008-0.0024	0.0005-0.0023
	3.0	M54B30	0.0004-0.0016	0.0078-0.0157	0.0078-0.0157	0.0078-0.0177	0.0008-0.0024	0.0008-0.0024	0.0005-0.0023
	3.0	N52B30	0.0004-0.0016	0.0078-0.0157	0.0078-0.0157	0.0078-0.0177	0.0008-0.0024	0.0008-0.0024	0.0005-0.0023
	3.2	S54B32	0.0010-0.0022	0.0078-0.0137	0.0137-0.0236	0.0098-0.0196	0.0011-0.0027	0.0005-0.0023	0.0008-0.0024

06041_BMWC_C0011

PISTON AND RING SPECIFICATIONS

All measurements are given in inches

	Engine			Ring Gap			Ring Side Clearance		
Year	Size Liters	Engine ID	Piston Clearance	Top Compression	Bottom Compression	Oil Control	Top Compression	Bottom Compression	Oil Control
2006 Cont.	4.8	N62B48	0.0001-0.0014	0.0059-0.0098	0.0078-0.0136	0.0078-0.0157	0.0005-0.0023	0.0007-0.0026	0.0007-0.0026

NA: Information not available

06041_BMWC_C0012

TORQUE SPECIFICATIONS

All measurements in ft. Lbs.

Year	Engine Displacement Liters	Engine ID	Cylinder Head Bolts	Main Bearing Bolts	Rod Bearing Bolts	Crankshaft Damper Bolts	Flywheel Bolts	Manifold		Spark Plugs	Wheel Lug Nut
								Intake	Exhaust		
2002	2.5	M54B25	①	②	③	302	④	⑤	⑥	⑦	80
	3.0	M54B30	①	②	③	302	④	⑤	⑥	⑦	80
	3.2	S54B32	①	②	③	302	④	⑤	⑥	⑦	80
	4.4	M62B44	⑧	⑨	⑩	⑪	④	⑤	⑫	⑦	100
	4.4	N62B44	⑧	⑨	⑩	⑪	④	⑤	⑫	⑦	100
	4.9	S62B50	⑧	⑨	⑩	⑪	④	⑤	⑫	⑦	100
2003	2.5	M54B25	①	②	③	302	④	⑤	⑥	⑦	80
	3.0	M54B30	①	②	③	302	④	⑤	⑥	⑦	80
	3.2	S54B32	①	②	③	302	④	⑤	⑥	⑦	80
	4.4	M62B44	⑧	⑨	⑩	⑪	④	⑤	⑫	⑦	100
	4.4	N62B44	⑧	⑨	⑩	⑪	④	⑤	⑫	⑦	100
	4.9	S62B50	⑧	⑨	⑩	⑪	④	⑤	⑫	⑦	100
2004	2.5	M54B25	①	②	③	302	④	⑤	⑥	⑦	80
	3.0	M54B30	①	②	③	302	④	⑤	⑥	⑦	80
	3.2	S54B32	①	②	③	302	④	⑤	⑥	⑦	80
	4.4	N62B44	⑧	⑨	⑩	⑪	④	⑤	⑫	⑦	100
2005	2.5	M54B25	①	②	③	302	④	⑤	⑥	⑦	80
	3.0	M54B30	①	②	③	302	④	⑤	⑥	⑦	80
	3.2	S54B32	①	②	③	302	④	⑤	⑥	⑦	80
	4.4	N62B44	⑧	⑨	⑩	⑪	④	⑤	⑫	⑦	100
2006	2.5	M54B25	①	②	③	302	④	⑤	⑥	⑦	80
	3.0	M54B30	①	②	③	302	④	⑤	⑥	⑦	80
	3.0	N52B30	①	②	③	302	④	⑤	⑥	⑦	80
	3.2	S54B32	①	②	③	302	④	⑤	⑥	⑦	80
	4.8	N62B48	①	⑬	③	302	④	⑤	⑥	⑦	88

① Cast iron block. Replace, wash and oil bolts
Step 1: 22 ft. lbs.
Step 2: 90 degrees
Step 3: 90 degrees

② Cast iron block. Replace, wash and oil bolts
Step 1: 14.8 ft. lbs.
Step 2: 50 degrees

③ Replace, wash and oil connecting rod bolts
Step 1: 14.8 ft. lbs.
Step 2: 70 degrees

④ New micro-encapsulated screws:
Automatic transmission: 88 ft. lbs.
Manual transmission: 77.4 ft. lbs.

⑤ All M6 fasteners: 88 inch lbs.
All M7 fasteners: 11 ft. lbs.
All M8 fasteners: 16 ft. lbs.

⑥ Coat with Molykkote HSC compound or equivalent
All M6 fasteners: 88 inch lbs.
All M7 fasteners: 14.8 ft. lbs.

⑦ M12x1.25:14.8-19.1 ft. lbs.
M14x1.25: 21.4-24.3 ft. lbs.

⑧ Aluminum block. Replace bolts. Do not remove coating
Step 1: M52: 29.5 ft. lbs. M62, M73: 22 ft. lbs.
Step 2: M52: 90 degrees. M62: 80 degrees. M73: 60 degrees
Step 2: M52: 90 degrees. M62: 80 degrees. M73: 60 degrees

⑨ Aluminum block. Replace bolts. Do not remove coating
Step 1: 14.8 ft. lbs.
Step 2: 70 degrees

⑩ Replace, wash and oil connecting rod bolts
Step 1: 14.8 ft. lbs.
Step 2: 80 degrees

⑪ Step 1: 73.7 ft. lbs.
Step 2: 60 degrees
Step 3: 60 degrees
Step 4: 30 degrees

⑫ Coat with Molykkote HSC compound
All M6 fasteners: 88 inch lbs.
All M7 fasteners: 11 ft. lbs.

⑬ Replace bolts. Do not remove coating
Step 1: 14.75 ft. lbs.
Step 2: 100 degrees

06041_BMWC_C0013

WHEEL ALIGNMENT

Year	Model		Caster Range (+/-Deg.)	Caster Preferred Setting (Deg.)	Camber Range (+/-Deg.)	Camber Preferred Setting (Deg.)	Toe-in (Deg.)
2002	3 Series ①	F	0.50	5.43	0.33	-0.33	0.23 +/- 0.13
		R	—	—	0.25	1.50	0.26 +/- 0.11
	3 Series ②	F	0.50	5.6	0.33	-0.71	0.23 +/- 0.13
		R	—	—	0.25	-2.05	0 +/- 0.06
	3 Series ③	F	0.50	5.28	0.33	0.13	0.23 +/- 0.13
		R	—	—	0.25	-0.76	0.26 +/- 0.1
	3 Series ④	F	0.50	7.41	0.33	-1.0	0.26 +/- 0.11
		R	—	—	0.25	-1.75	0.36 +/- 0.1
	5 Series ①	F	0.50	6.30	0.50	-0.21	0.23 +/- 0.16
		R	—	—	0.08	-2.06	0.26 +/- 0.16
	5 Series ②	F	0.50	6.56	0.50	-0.61	0.23 +/- 0.16
		R	—	—	0.08	-2.06	0.26 +/- 0.20
	5 Series ③	F	0.50	6.08	0.50	-0.21	0.23 +/- 0.16
		R	—	—	0.08	-2.06	0.26 +/- 0.20
	5 Series ④	F	0.50	6.45	0.50	0.50	0.16 +/- 0.16
		R	—	—	0.08	-1.81	0.16 +/- 0.16
	5 Series ⑤	F	0.50	6.30	0.50	-0.21	0.23 +/- 0.16
		R	—	—	0.08	-2.06	0.26 +/- 0.16
	7 Series	F	0.33	7.65	0.33	-0.1	0.16 +/- 0.13
		R	—	—	0.33	0.5	0.16 +/- 0.16
2003	3 Series ①	F	0.50	5.43	0.33	-0.33	0.23 +/- 0.13
		R	—	—	0.25	1.50	0.26 +/- 0.11
	3 Series ②	F	0.50	5.6	0.33	-0.71	0.23 +/- 0.13
		R	—	—	0.25	-2.05	0 +/- 0.06
	3 Series ③	F	0.50	5.28	0.33	0.13	0.23 +/- 0.13
		R	—	—	0.25	-0.76	0.26 +/- 0.1
	3 Series ④	F	0.50	7.41	0.33	-1.0	0.26 +/- 0.11
		R	—	—	0.25	-1.75	0.36 +/- 0.1
	5 Series ①	F	0.50	6.30	0.50	-0.21	0.23 +/- 0.16
		R	—	—	0.08	-2.06	0.26 +/- 0.16
	5 Series ②	F	0.50	6.56	0.50	-0.61	0.23 +/- 0.16
		R	—	—	0.08	-2.06	0.26 +/- 0.20
	5 Series ③	F	0.50	6.08	0.50	-0.21	0.23 +/- 0.16
		R	—	—	0.08	-2.06	0.26 +/- 0.20
	5 Series ④	F	0.50	6.45	0.50	0.50	0.16 +/- 0.16
		R	—	—	0.08	-1.81	0.16 +/- 0.16
	5 Series ⑤	F	0.50	6.30	0.50	-0.21	0.23 +/- 0.16
		R	—	—	0.08	-2.06	0.26 +/- 0.16
	7 Series	F	0.33	7.65	0.33	-0.1	0.16 +/- 0.13
		R	—	—	0.33	0.5	0.16 +/- 0.16
2004	3 Series ①	F	0.50	5.43	0.33	-0.33	0.23 +/- 0.13
		R	—	—	0.25	1.50	0.26 +/- 0.11
	3 Series ②	F	0.50	5.6	0.33	-0.71	0.23 +/- 0.13
		R	—	—	0.25	-2.05	0 +/- 0.06
	3 Series ③	F	0.50	5.28	0.33	0.13	0.23 +/- 0.13
		R	—	—	0.25	-0.76	0.26 +/- 0.1
	3 Series ④	F	0.50	7.41	0.33	-1.0	0.26 +/- 0.11
		R	—	—	0.25	-1.75	0.36 +/- 0.1
	5 Series	F	0.50	7.51	0.33	-0.2	0.13 +/- 0.16
		R	—	—	0.33	-2	0.3 +/- 0.16
	7 Series	F	0.33	7.65	0.33	-0.1	0.16 +/- 0.13
		R	—	—	0.33	0.5	0.16 +/- 0.16

06041_BMWC_C0014

WHEEL ALIGNMENT

			Caster		Camber		
Year	Model		Range (+/-Deg.)	Preferred Setting (Deg.)	Range (+/-Deg.)	Preferred Setting (Deg.)	Toe-in (Deg.)
2005	3 Series ①	F	0.50	5.43	0.33	-0.33	0.23 +/- 0.13
		R	—	—	0.25	1.50	0.26 +/- 0.11
	3 Series ②	F	0.50	5.6	0.33	-0.71	0.23 +/- 0.13
		R	—	—	0.25	-2.05	0 +/- 0.06
	3 Series ③	F	0.50	5.28	0.33	0.13	0.23 +/- 0.13
		R	—	—	0.25	-0.76	0.26 +/- 0.1
	3 Series ④	F	0.50	7.41	0.33	-1.0	0.26 +/- 0.11
		R	—	—	0.25	-1.75	0.36 +/- 0.1
	5 Series	F	0.50	7.51	0.33	-0.2	0.13 +/- 0.16
		R	—	—	0.33	-2	0.3 +/- 0.16
	7 Series	F	0.33	7.65	0.33	-0.1	0.16 +/- 0.13
		R	—	—	0.33	0.5	0.16 +/- 0.16
2006	3 Series ①	F	0.50	5.43	0.33	-0.33	0.23 +/- 0.13
		R	—	—	0.25	1.50	0.26 +/- 0.11
	3 Series ②	F	0.50	5.6	0.33	-0.71	0.23 +/- 0.13
		R	—	—	0.25	-2.05	0 +/- 0.06
	3 Series ③	F	0.50	5.28	0.33	0.13	0.23 +/- 0.13
		R	—	—	0.25	-0.76	0.26 +/- 0.1
	3 Series ④	F	0.50	7.41	0.33	-1.0	0.26 +/- 0.11
		R	—	—	0.25	-1.75	0.36 +/- 0.1
	5 Series	F	0.50	7.51	0.33	-0.2	0.13 +/- 0.16
		R	—	—	0.33	-2	0.3 +/- 0.16
	7 Series ①	F	0.33	7.65	0.33	-0.1	0.16 +/- 0.13
		R	—	—	0.33	0.5	0.16 +/- 0.16
	7 Series ②	F	0.50	+0.20	0.50	-0.60	0.12 +/- 0.08
		R	—	—	0.08	+0.45	0.15 +/- 0.08

① Standard suspension
② Sport suspension
③ Rough road package
④ M series
⑤ Air suspension

06041_BMWC_C0015

BRAKE SPECIFICATIONS

All measurements in inches unless noted

				Brake Disc			Parking Brake Drum		Minimum Lining Thickness		Brake Caliper	
Year	Body Type	Model		Original Thickness	Minimum Thickness	Maximum Runout	Original Inside Diameter	Max. Runout	Front	Rear	Bracket Bolts (ft. lbs.)	Mounting Bolts (ft. lbs.)
2002	E46	325i Sedan	F	0.803	①	0.007	—	—	0.118	—	81	26
			R	②	①	0.007	6.299	0.004	—	③	48	26
	E46	325xi Sedan	F	0.803	①	0.007	—	—	0.118	—	81	26
			R	②	①	0.007	6.299	0.004	—	③	48	26
	E46	325Ci Coupe	F	0.803	①	0.007	—	—	0.118	—	81	26
			R	②	①	0.007	6.299	0.004	—	③	48	26
	E46	325Ci Cabriolet	F	0.803	①	0.007	—	—	0.118	—	81	26
			R	②	①	0.007	6.299	0.004	—	③	48	26
	E46	325i Wagon	F	0.803	①	0.007	—	—	0.118	—	81	26
			R	②	①	0.007	6.299	0.004	—	③	48	26
	E46	325xi Wagon	F	0.803	①	0.007	—	—	0.118	—	81	26
			R	②	①	0.007	6.299	0.004	—	③	48	26
	E46	330i Sedan	F	0.803	①	0.007	—	—	0.118	—	81	26
			R	②	①	0.007	6.299	0.004	—	③	48	26
	E46	330xi Sedan	F	0.803	①	0.007	—	—	0.118	—	81	26
			R	②	①	0.007	6.299	0.004	—	③	48	26
	E46	330Ci Coupe	F	0.803	①	0.007	—	—	0.118	—	81	26
			R	②	①	0.007	6.299	0.004	—	③	48	26
	E46	330Ci Cabriolet	F	0.803	①	0.007	—	—	0.118	—	81	26
			R	②	①	0.007	6.299	0.004	—	③	48	26
	E46	M3	F	0.803	①	0.007	—	—	0.118	—	81	26
			R	②	①	0.007	—	—	—	③	48	26
	E36/7	Z3 2.5	F	0.803	①	0.007	—	—	0.118	—	81	26
			R	②	①	0.007	6.299	0.004	—	③	48	26
	E36/7	Z3 3.0	F	0.803	①	0.007	—	—	0.118	—	81	26
			R	②	①	0.007	6.299	0.004	—	③	48	26
	E36/7	Z3 3.2	F	0.803	①	0.007	—	—	0.118	—	81	26
			R	②	①	0.007	—	—	—	③	48	26
	E39	525i	F	0.803	①	0.007	—	—	0.118	—	81	26
			R	②	①	0.007	—	—	—	③	48	26
	E39	525i Wagon	F	0.803	①	0.007	—	—	0.118	—	81	26
			R	②	①	0.007	—	—	—	③	48	26
	E39	530i	F	0.803	①	0.007	—	—	0.118	—	81	26
			R	②	①	0.007	—	—	—	③	48	26
	E39	540i	F	1.118	①	0.007	—	—	0.118	—	81	26
			R	④	①	0.007	7.283	0.004	—	③	48	26
	E39	540i Wagon	F	1.118	①	0.007	—	—	0.118	—	81	26
			R	④	①	0.007	7.283	0.004	—	③	48	26
	E39	M5	F	1.118	①	0.007	—	—	0.118	—	81	26
			R	⑥	①	0.007	—	—	—	③	48	26
	E65	745i	F	1.118	①	0.007	—	—	0.118	—	81	26
			R	0.409	①	0.007	7.086	0.004	—	③	48	26
	E66	745Li	F	1.118	①	0.007	—	—	0.118	—	81	26
			R	0.409	①	0.007	7.086	0.004	—	③	48	26
2003	E46	325i Sedan	F	0.803	①	0.007	—	—	0.118	—	81	26
			R	②	①	0.007	6.299	0.004	—	③	48	26
	E46	325xi Sedan	F	0.803	①	0.007	—	—	0.118	—	81	26
			R	②	①	0.007	6.299	0.004	—	③	48	26

06041_BMWC_C0016

BRAKE SPECIFICATIONS

All measurements in inches unless noted

				Brake Disc			Parking Brake Drum		Minimum Lining Thickness		Brake Caliper	
Year	Body Type	Model		Original Thickness	Minimum Thickness	Maximum Runout	Original Inside Diameter	Max. Runout	Front	Rear	Bracket Bolts (ft. lbs.)	Mounting Bolts (ft. lbs.)
2003 Cont.	E46	325Ci Coupe	F	0.803	①	0.007	—	—	0.118	—	81	26
			R	②	①	0.007	6.299	0.004	—	③	48	26
	E46	325Ci Cabriolet	F	0.803	①	0.007	—	—	0.118	—	81	26
			R	②	①	0.007	6.299	0.004	—	③	48	26
	E46	325i Wagon	F	0.803	①	0.007	—	—	0.118	—	81	26
			R	②	①	0.007	6.299	0.004	—	③	48	26
	E46	325xi Wagon	F	0.803	①	0.007	—	—	0.118	—	81	26
			R	②	①	0.007	6.299	0.004	—	③	48	26
	E46	330i Sedan	F	0.803	①	0.007	—	—	0.118	—	81	26
			R	②	①	0.007	6.299	0.004	—	③	48	26
	E46	330xi Sedan	F	0.803	①	0.007	—	—	0.118	—	81	26
			R	②	①	0.007	6.299	0.004	—	③	48	26
	E46	330Ci Coupe	F	0.803	①	0.007	—	—	0.118	—	81	26
			R	②	①	0.007	6.299	0.004	—	③	48	26
	E46	330Ci Cabriolet	F	0.803	①	0.007	—	—	0.118	—	81	26
			R	②	①	0.007	6.299	0.004	—	③	48	26
	E46	M3	F	0.803	①	0.007	—	—	0.118	—	81	26
			R	②	①	0.007	—	—	—	③	48	26
	E85	Z4 2.5	F	0.803	①	0.007	—	—	0.118	—	81	26
			R	②	①	0.007	—	—	—	③	48	26
	E85	Z4 3.0	F	0.803	①	0.007	—	—	0.118	—	81	26
			R	②	①	0.007	—	—	—	③	48	26
	E39	525i	F	0.803	①	0.007	—	—	0.118	—	81	26
			R	②	①	0.007	—	—	—	③	48	26
	E39	530i	F	0.803	①	0.007	—	—	0.118	—	81	26
			R	②	①	0.007	—	—	—	③	48	26
	E39	540i	F	1.118	①	0.007	—	—	0.118	—	81	26
			R	④	①	0.007	7.283	0.004	—	③	48	26
	E39	M5	F	1.118	①	0.007	—	—	0.118	—	81	26
			R	④	①	0.007	—	—	—	③	48	26
	E65	745i	F	1.118	①	0.007	—	—	0.118	—	81	26
			R	0.409	①	0.007	7.086	0.004	—	③	48	26
	E66	745Li	F	1.118	①	0.007	—	—	0.118	—	81	26
			R	0.409	①	0.007	7.086	0.004	—	③	48	26
2004	E46	325i Sedan	F	0.803	①	0.007	—	—	0.118	—	81	26
			R	②	①	0.007	6.299	0.004	—	③	48	26
	E46	325xi Sedan	F	0.803	①	0.007	—	—	0.118	—	81	26
			R	②	①	0.007	6.299	0.004	—	③	48	26
	E46	325Ci Coupe	F	0.803	①	0.007	—	—	0.118	—	81	26
			R	②	①	0.007	6.299	0.004	—	③	48	26
	E46	325Ci Cabriolet	F	0.803	①	0.007	—	—	0.118	—	81	26
			R	②	①	0.007	6.299	0.004	—	③	48	26
	E46	325i Wagon	F	0.803	①	0.007	—	—	0.118	—	81	26
			R	②	①	0.007	6.299	0.004	—	③	48	26
	E46	325xi Wagon	F	0.803	①	0.007	—	—	0.118	—	81	26
			R	②	①	0.007	6.299	0.004	—	③	48	26
	E46	330i Sedan	F	0.803	①	0.007	—	—	0.118	—	81	26
			R	②	①	0.007	6.299	0.004	—	③	48	26
	E46	330xi Sedan	F	0.803	①	0.007	—	—	0.118	—	81	26
			R	②	①	0.007	6.299	0.004	—	③	48	26

06041_BMWC_C0017

BRAKE SPECIFICATIONS

All measurements in inches unless noted

Year	Body Type	Model		Brake Disc Original Thickness	Brake Disc Minimum Thickness	Brake Disc Maximum Runout	Parking Brake Drum Original Inside Diameter	Parking Brake Drum Max. Runout	Minimum Lining Thickness Front	Minimum Lining Thickness Rear	Brake Caliper Bracket Bolts (ft. lbs.)	Brake Caliper Mounting Bolts (ft. lbs.)
2004	E46	330Ci Coupe	F	0.803	①	0.007	—	—	0.118	—	81	26
Cont.			R	②	①	0.007	6.299	0.004	—	③	48	26
	E46	330Ci Cabriolet	F	0.803	①	0.007	—	—	0.118	—	81	26
			R	②	①	0.007	6.299	0.004	—	③	48	26
	E46	M3	F	0.803	①	0.007	—	—	0.118	—	81	26
			R	②	①	0.007	—	—	—	③	48	26
	E85	Z4 2.5	F	0.803	①	0.007	—	—	0.118	—	81	26
			R	②	①	0.007	—	—	—	③	48	26
	E85	Z4 3.0	F	0.803	①	0.007	—	—	0.118	—	81	26
			R	②	①	0.007	—	—	—	③	48	26
	E60	525i	F	NA	①	NA	—	—	0.118	—	81	22
			R	NA	①	NA	—	—	—	③	81	22
	E60	530i	F	NA	①	NA	—	—	0.118	—	81	22
			R	NA	①	NA	—	—	—	③	81	22
	E60	545i	F	NA	①	NA	—	—	0.118	—	81	22
			R	NA	①	NA	—	—	—	③	81	22
	E65	745i	F	1.118	①	0.007	—	—	0.118	—	81	26
			R	0.409	①	0.007	7.086	0.004	—	③	48	26
	E66	740Li	F	1.118	①	0.007	—	—	0.118	—	81	26
			R	0.409	①	0.007	7.086	0.004	—	③	48	26
2005	E46	325i Sedan	F	0.803	①	0.007	—	—	0.118	—	81	26
			R	②	①	0.007	6.299	0.004	—	③	48	26
	E46	325xi Sedan	F	0.803	①	0.007	—	—	0.118	—	81	26
			R	②	①	0.007	6.299	0.004	—	③	48	26
	E46	325Ci Coupe	F	0.803	①	0.007	—	—	0.118	—	81	26
			R	②	①	0.007	6.299	0.004	—	③	48	26
	E46	325Ci Cabriolet	F	0.803	①	0.007	—	—	0.118	—	81	26
			R	②	①	0.007	6.299	0.004	—	③	48	26
	E46	325i Wagon	F	0.803	①	0.007	—	—	0.118	—	81	26
			R	②	①	0.007	6.299	0.004	—	③	48	26
	E46	325xi Wagon	F	0.803	①	0.007	—	—	0.118	—	81	26
			R	②	①	0.007	6.299	0.004	—	③	48	26
	E46	330i Sedan	F	0.803	①	0.007	—	—	0.118	—	81	26
			R	②	①	0.007	6.299	0.004	—	③	48	26
	E46	330xi Sedan	F	0.803	①	0.007	—	—	0.118	—	81	26
			R	②	①	0.007	6.299	0.004	—	③	48	26
	E46	330Ci Coupe	F	0.803	①	0.007	—	—	0.118	—	81	26
			R	②	①	0.007	6.299	0.004	—	③	48	26
	E46	330Ci Cabriolet	F	0.803	①	0.007	—	—	0.118	—	81	26
			R	②	①	0.007	6.299	0.004	—	③	48	26
	E46	M3	F	0.803	①	0.007	—	—	0.118	—	81	26
			R	0.803	①	0.007	—	—	—	③	48	26
	E85	Z4 2.5	F	0.803	①	0.007	—	—	0.118	—	81	26
			R	②	①	0.007	—	—	—	③	48	26
	E85	Z4 3.0	F	0.803	①	0.007	—	—	0.118	—	81	26
			R	②	①	0.007	—	—	—	③	48	26
	E60	525i	F	NA	①	NA	—	—	0.118	—	81	22
			R	NA	①	NA	—	—	—	③	81	22
	E60	530i	F	NA	①	NA	—	—	0.118	—	81	22
			R	NA	①	NA	—	—	—	③	81	22

06041_BMWC_C0018

BRAKE SPECIFICATIONS

All measurements in inches unless noted

				Brake Disc			Parking Brake Drum		Minimum Lining Thickness		Brake Caliper	
Year	**Body Type**	**Model**		**Original Thickness**	**Minimum Thickness**	**Maximum Runout**	**Original Inside Diameter**	**Max. Runout**	**Front**	**Rear**	**Bracket Bolts (ft. lbs.)**	**Mounting Bolts (ft. lbs.)**
2005 Cont.	E60	545i	F	NA	①	NA	—	—	0.118	—	81	22
			R	NA	①	NA	—	—	—	③	81	22
	E65	745i	F	1.118	①	0.007	—	—	0.118	—	81	26
			R	0.409	①	0.007	7.086	0.004	—	③	81	26
	E66	740Li	F	1.118	①	0.007	—	—	0.118	—	81	26
			R	0.409	①	0.007	7.086	0.004	—	③	81	26
2006	E46	325i Sedan	F	0.803	①	0.007	—	—	0.118	—	81	26
			R	②	①	0.007	6.299	0.004	—	③	48	26
	E46	325xi Sedan	F	0.803	①	0.007	—	—	0.118	—	81	26
			R	②	①	0.007	6.299	0.004	—	③	48	26
	E46	325Ci Coupe	F	0.803	①	0.007	—	—	0.118	—	81	26
			R	②	①	0.007	6.299	0.004	—	③	48	26
	E46	325Ci Cabriolet	F	0.803	①	0.007	—	—	0.118	—	81	26
			R	②	①	0.007	6.299	0.004	—	③	48	26
	E46	325xi Wagon	F	0.803	①	0.007	—	—	0.118	—	81	26
			R	②	①	0.007	6.299	0.004	—	③	48	26
	E46	330i Sedan	F	0.803	①	0.007	—	—	0.118	—	81	26
			R	②	①	0.007	6.299	0.004	—	③	48	26
	E46	330xi Sedan	F	0.803	①	0.007	—	—	0.118	—	81	26
			R	②	①	0.007	6.299	0.004	—	③	48	26
	E46	330Ci Coupe	F	0.803	①	0.007	—	—	0.118	—	81	26
			R	②	①	0.007	6.299	0.004	—	③	48	26
	E46	330Ci Cabriolet	F	0.803	①	0.007	—	—	0.118	—	81	26
			R	②	①	0.007	6.299	0.004	—	③	48	26
	E46	M3	F	0.803	①	0.007	—	—	0.118	—	81	26
			R	②	①	0.007	—	—	—	③	48	26
	E85	Z4 Coupe	F	0.803	①	0.007	—	—	0.118	—	81	26
			R	②	①	0.007	—	—	—	③	48	26
	E85	Z4 3.0i	F	0.803	①	0.007	—	—	0.118	—	81	26
			R	②	①	0.007	—	—	—	③	48	26
	E85	Z4 3.0si	F	0.803	①	0.007	—	—	0.118	—	81	26
			R	②	①	0.007	—	—	—	③	48	26
	E60	525i	F	NA	①	NA	—	—	0.118	—	81	22
			R	NA	①	NA	—	—	—	③	48	22
	E60	525xi	F	NA	①	NA	—	—	0.118	—	81	22
			R	NA	①	NA	—	—	—	③	48	22
	E60	530i	F	NA	①	NA	—	—	0.118	—	81	22
			R	NA	①	NA	—	—	—	③	48	22
	E60	530xi	F	NA	①	NA	—	—	0.118	—	81	22
			R	NA	①	NA	—	—	—	③	48	22
	E60	550i	F	NA	①	NA	—	—	0.118	—	81	22
			R	NA	①	NA	—	—	—	③	48	22
	E65	750i	F	1.118	①	0.007	—	—	0.118	—	81	26
			R	0.409	①	0.007	7.086	0.004	—	③	48	26
	E65	750Li	F	1.118	①	0.007	—	—	0.118	—	81	26
			R	0.409	①	0.007	7.086	0.004	—	③	48	26

NA: Information not available
F: Front
R: Rear
① Minimum thickness is stamped in the brake disk shell
Maximum machining limit per side: 0.031 inches

② Solid brake rotor: 0.331 inches
Ventilated brake rotor: 0.685 inches
③ Rear brake pad wear limit: 0.118 inches
Parking brake shoe wear limit: 0.059 inches
④ Solid brake rotor: 0.331 inches
Ventilated brake rotor: 0.724 inches

SCHEDULED MAINTENANCE INTERVALS

BMW—3 SERIES, 5 SERIES, 7 SERIES (EXCEPT V12), Z3, Z4

TO BE SERVICED	TYPE OF SERVICE	SERVICE INTERVALS			
		INITIAL 1200 MILES	OIL SERVICE	INSPECTION I	INSPECTION II
Oil level	S/I	✓			
Engine oil	R	①			
Engine oil & filter	R ②		✓	✓	✓
Engine air cleaner element	R ③				✓
Spark plugs	R				✓
Fuel filter	R ④				✓
Fuel, vapor lines & fuel cap	S/I	✓		✓	✓
Cooling system	S/I	✓		✓	✓
Exhaust pipe & muffler	S/I	✓		✓	✓
Catalytic converter & shielding	S/I	✓		✓	✓
Throttle linkage	S/I			✓	✓
Engine (check for leakage)	S/I	✓			
Engine drive belts	S/I				✓
Maintenance Indicators	RE		⑤	✓	✓
Engine coolant	R			⑥	⑥
Oxygen sensor	R ⑦				
Intake air dust separators	S/I ⑧				✓
Brake & clutch fluids ⑥	S/I			✓	✓
Brake pads & discs	S/I			✓	✓
Parking brake system	S/I			✓	✓
Power steering system	S/I			✓	✓
Rear axle fluid	S/I			✓	✓
Steering play, suspension track rods, front axle joints, steering linkage & joint disc	S/I			✓	✓
Transmission fluid/oil	S/I			✓	⑨
Wheel centering hubs	S/I			✓	✓
Rear axle fluid ⑩	R		✓		✓
OBD system for codes	S/I	✓		✓	✓

R: Replace S/I: Service or Inspect RE: Reset

Note: BMW does not rely solely on vehicle mileage to determine service intervals. An on-oboard diagnostic center, monitors engine operating conditions, along with mileage, to determine the most effective maintenance intervals. The information is then conveyed to the driver through the service indicator lights, located in the center of the instrument panel.

① Service is not required for 325 models.

② On vehicles operated less than 6200 miles per year, more frequent service may be required.

③ Replace more frequently if vehicle is operated in dusty conditions.

④ Recommended service for California models, required for all other models.

⑤ Reset the oil service indicator lights only.

⑥ Replace every 2 years with inspection service.

⑦ Replace every 100,000 miles on all models.

⑧ Required service for manual transmission models only.

⑨ Change fluid (A/T) or oil (M/T) at inspection.

⑩ Check at first oil service, then at each inspection.

FREQUENT OPERATION MAINTENANCE (SEVERE SERVICE)

If a vehicle is operated under any of the following conditions it is considered severe service

- **Extremely dusty areas.**
- **50% or more of the vehicle operation is in 32°C (90°F) or higher temperatures, or constant operation in temperatures below 0°C (32°F).**
- **Prolonged idling (vehicle operation in stop and go traffic).**
- **Frequent short running periods (engine does not warm to normal operating temperatures).**
- **Police, taxi, delivery usage or trailer towing usage.**

06041_BMWC_C0020

ENGINE REPAIR

➡Disconnecting the negative battery cable on some vehicles may interfere with the operation of the on-board computer system. The computer to undergo a relearning process once the negative battery cable is reconnected.

Alternator

REMOVAL & INSTALLATION

During installation of the alternator, refer to the following information for proper tightening of the fasteners. Alternator fastener torque specifications for all models:

- Alternator terminal M6 fasteners: 61 inch lbs. (7 Nm)
- Alternator terminal M8 fasteners: 115 inch lbs. (13 Nm)
- Alternator bearing block fasteners: 31 ft. lbs. (43 Nm)
- Alternator crankcase fasteners (M8 thread): 15 ft. lbs. (21 Nm)
- Alternator V-belt type pulley nut: 33 ft. lbs. (45 Nm)
- Alternator ribbed belt pulley nut (with cooling jacket): 59 ft. lbs. (80 Nm)
- Alternator ribbed belt pulley nut (without cooling jacket): 51 ft. lbs. (70 Nm)

M54 and S54 Engines

1. Before servicing the vehicle, refer to the Precautions Section.
2. Remove or disconnect the following:
   - Negative battery cable
   - Suction filter housing
   - Fan clutch
   - Alternator drive belt
   - Power steering pump supply tank, secure aside with hoses attached
   - Alternator air hose, if equipped
   - Electrical connectors
   - Idler pulley, if equipped
   - Bolts and remove the alternator

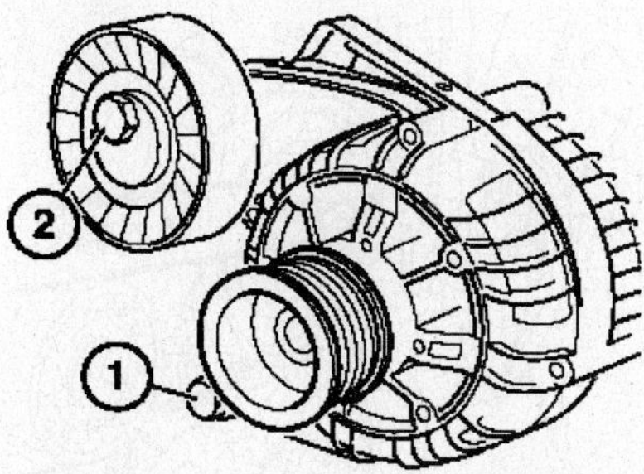

Mounting bolt (1) and idler pulley (2) used on some models

Mounting bolt on models without idler pulley

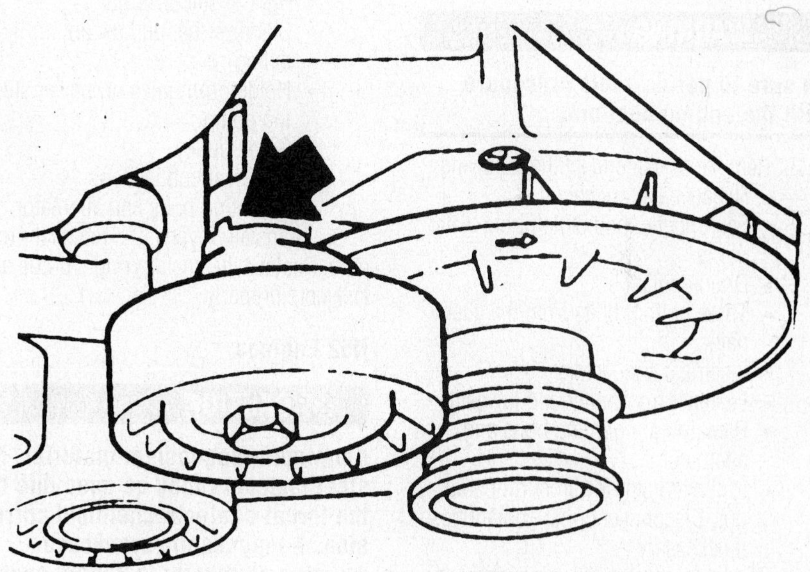

When installing the belt idler tensioner roller, make sure the alignment tab is seated into the slot

3. To install, reverse the removal procedure. Torque all fasteners properly. When installing the belt idler tensioner roller, make sure the alignment tab is seated into the slot

M62 and S62 Engines

AIR COOLED ALTERNATOR

1. Before servicing the vehicle, refer to the Precautions Section.
2. Remove or disconnect the following:
   - Negative battery cable
   - Fasteners from oil filter housing, place housing aside
   - Electrical connectors
   - Cooling air guide
   - Bolts and remove the alternator
3. To install, reverse the removal procedure. Torque all fasteners properly.

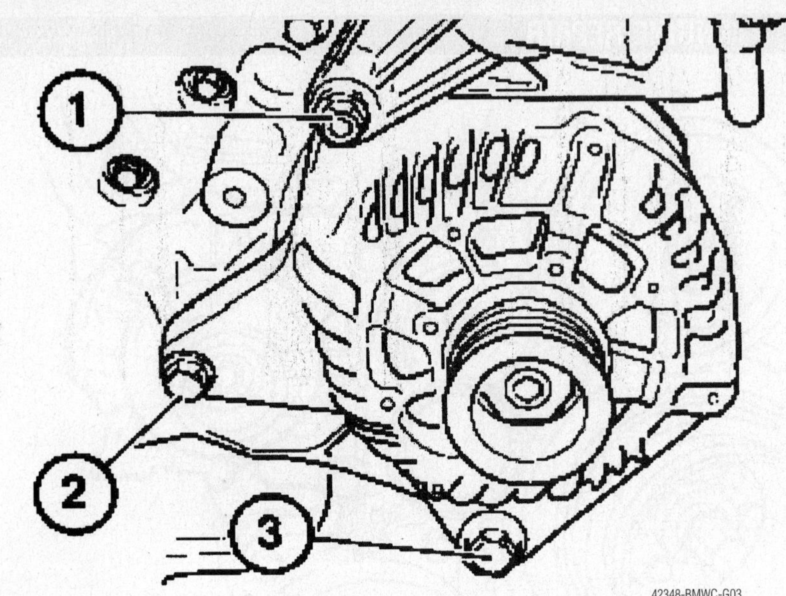

Alternator mounting bolts on M62/S62 air cooled alternator

LIQUID COOLED ALTERNATOR

1. Before servicing the vehicle, refer to the Precautions Section.

WARNING

Be sure to perform this procedure with the engine cool only.

2. Remove or disconnect the following:
   - Negative battery cable
   - Fan impeller and clutch from water pump
   - Drain coolant
   - Three bottom bolts from the alternator
   - Electrical connectors
   - Fastener from roller and lay aside
   - Remaining mounting bolts and alternator
   - Pull alternator towards front and out. Disconnect upper coolant hose if necessary
3. To install, reverse the removal procedure. Replace the sealing ring. Torque all fasteners properly.

N62 Engines

1. Before servicing the vehicle, refer to the Precautions Section.

WARNING

Be sure to perform this procedure with the engine cool only.

2. Remove or disconnect the following:
   - Negative battery cable
   - Drain coolant
   - Radiator
   - Alternator drive belt
   - A/C compressor drive belt
   - Antifriction bearings on left and right of stabilizer bar
   - Stabilizer bar and secure from falling
   - Holder from lines on power steering pump
   - Guide pulley
   - Electrical connections
   - Mounting bolts and alternator
3. To install, reverse the removal procedure. Replace the sealing ring. Torque all fasteners properly.

N52 Engines

CAUTION

Aluminum-magnesium material. No steel fasteners may be used due to the threat of electrochemical corrosion. A magnesium crankcase requires aluminum fasteners exclusively. Aluminum fasteners must be replaced each time they are removed. The end faces of aluminum fasteners are painted blue for purposes of identification. Torque specifications and torque angles must be observed for risk of damage.

1. Before servicing the vehicle, refer to the Precautions Section.
2. Remove or disconnect the following:
   - Negative battery cable
   - Drain coolant
   - Radiator
   - Intake filter housing
   - Alternator drive belt
   - Electrical connections and slacken nut
   - Remove bolts and remove alternator
3. To install, reverse the removal procedure. Replace all aluminum bolts. Torque all fasteners properly.

Ignition Timing

ADJUSTMENT

The ignition timing is controlled by the Digital Motor Electronics (DME). No adjustments are necessary.

Engine Assembly

REMOVAL & INSTALLATION

M54 and S54 Engines

1. Before servicing the vehicle, refer to the Precautions Section.
2. Drain the cooling system.
3. Relieve the fuel system pressure.
4. Raise the hood all the way up into the service position. This usually requires disconnecting the gas struts and securing the hood hinges with bolts.
5. Remove or disconnect the following:
   - Negative battery terminal
   - Cables from the duct on the filter housing
   - Filter housing
   - Engine splash guard
   - Transmission assembly
   - A/C compressor drive belt
   - A/C compressor and secure aside
   - Upper and lower radiator hoses
   - Radiator
   - Cover from fuel injectors
   - Both Oxygen (O_2) sensor monitors
   - Intake manifold
   - Hoses from water pipe and heating valve
   - Engine wiring harness and secure aside

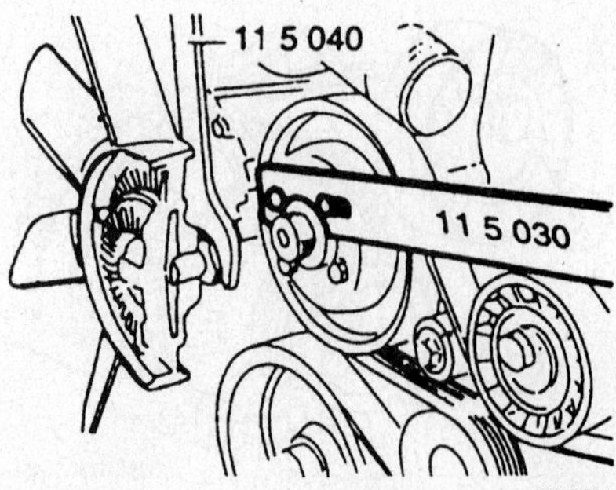

Cooling fan removal tools

Engine mount, showing the ground cable attachment

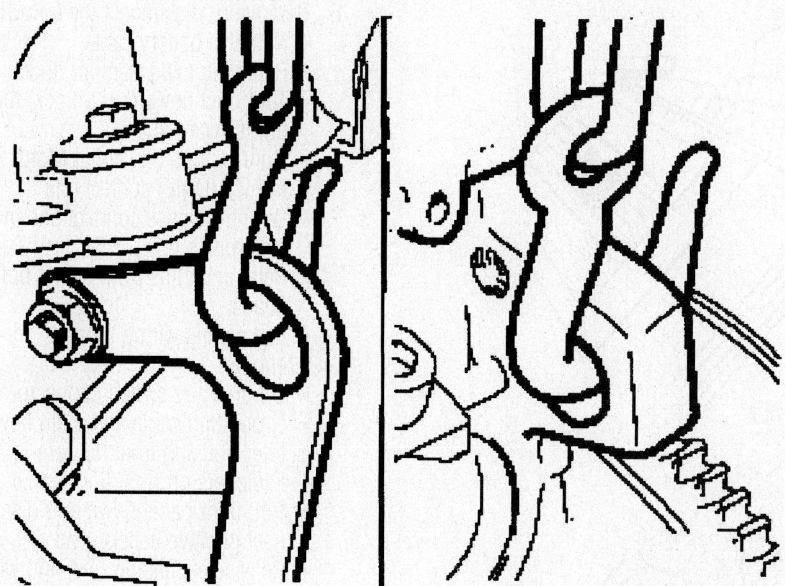

Engine front and rear lift points

- Power steering reservoir and secure aside
- Fan and fan clutch assembly
- Power steering pump and secure aside

6. Carefully support the engine and remove the engine mounts and ground strap.
7. Slowly and carefully lift the engine out of the vehicle being careful of the painted surfaces for the front radiator mount.
8. Installation is the reverse of the removal procedure, while using the following torque values:

- 6mm fasteners: 61 inch lbs. (7 Nm)
- 7mm fasteners: 10 ft. lbs. (13 Nm)
- 8mm fasteners: 16 ft. lbs. (22 Nm)
- 10mm fasteners: 31 ft. lbs. (42 Nm)

M62 Engines

1. Before servicing the vehicle, refer to the Precautions Section.
2. Drain the cooling system.
3. Relieve the fuel system pressure.
4. Raise the hood all the way up into the service position. This usually requires disconnecting the gas struts and securing the hood hinges with bolts.
5. Remove or disconnect the following:
   - Negative battery cable
   - Engine splash shield
   - Radiator
   - Nut holding the transmission oil cooler lines to the engine oil pan, if equipped
   - Transmission assembly
   - Drive belts for the power steering pump and the air conditioner compressor, then remove the bolts holding the pump and compressor to the engine and remove them from the engine, keeping the high pressure hoses connected. Secure the pump and compressor out of the way and without any tension or binding on the hoses.
   - Hoses from the coolant reservoir expansion tank, then remove the fasteners on the side of the expansion tank and remove the expansion tank from the engine compartment
   - Heater hoses from the heater control valve and the heater core inlet pipe
   - Electrical connections to the ignition coil, then the air cleaner housing assembly
   - Idle speed control valve from the intake duct
   - Wire harness connectors for the air flow meter, then the ducting to the air flow meter and crankcase breather vacuum line
   - With a cable operated throttle, cruise control cable and the throttle cable at the throttle, then remove the cable mounting bracket
   - With an electronically controlled throttle, electrical connector to the throttle control unit
   - Wiring harness from the starter, then the electrical connectors located near the starter motor assembly
   - Oil level sender connector and the wiring harness from the alternator
   - Air duct to the alternator
   - Fuel tank venting valve and the hose to the canister
   - Fuels lines for the fuel rail
   - Vacuum line from the brake booster and plug the opening
   - Wiring harness connections to the temperature sensors and the Digital Motor Electronics (DME)
   - Engine ground cable and verify that all remaining fluid lines or electrical leads have been disconnected and properly placed aside

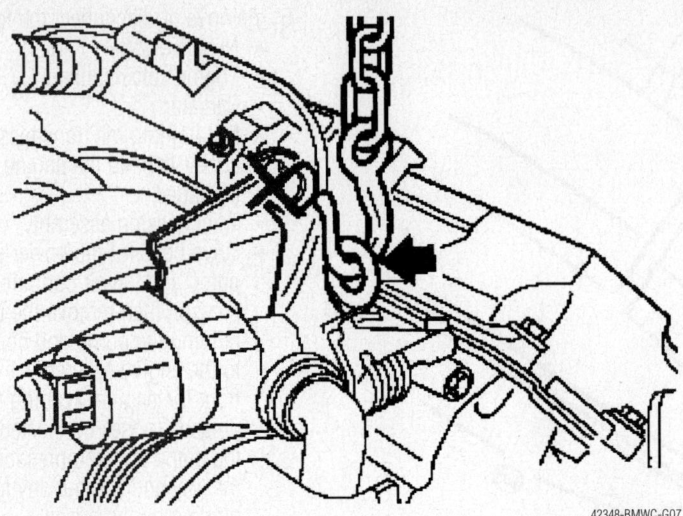

Engine front lift point

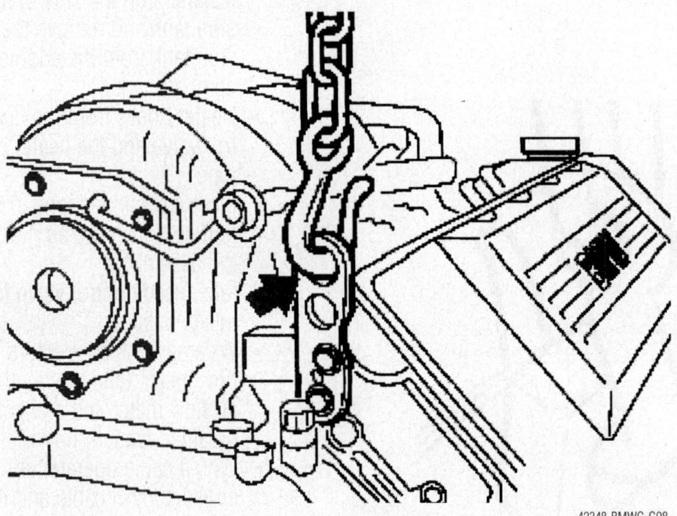

Engine rear lift point

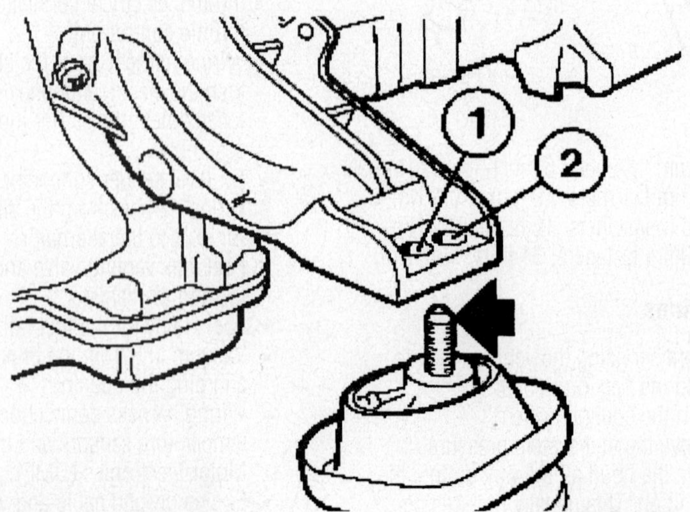

During installation on M62 and S62 engines, place mount stud in hole (1) of the support bracket

- Engine mount nuts and bolts and carefully lift the engine from the vehicle

6. Installation is the reverse of the removal procedure. When installing the engine, place the mount stud in hole (1) of the support bracket. Use the following torque values:
    - 6mm fasteners: 61 inch lbs. (7 Nm)
    - 7mm fasteners: 10 ft. lbs. (13 Nm)
    - 8mm fasteners: 16 ft. lbs. (22 Nm)
    - 10mm fasteners: 31 ft. lbs. (42 Nm)

S62 Engines

1. Before servicing the vehicle, refer to the Precautions Section.
2. Properly discharge the A/C system.
3. Drain the cooling system.
4. Relieve the fuel system pressure.
5. Raise the hood all the way up into the service position. This usually requires disconnecting the gas struts and securing the hood hinges with bolts.
6. Remove or disconnect the following:
    - Negative battery cable
    - Retaining clips from air ducts
    - Turn duct upward to unlock, then pull forward and out
    - Connections on Digital Motor Electronics (DME) control unit
    - Wiring harness connections in electronics box
    - Retaining clips from heater bulkheads
    - Lay engine wiring harness on engine
    - Oil pressure switch connector
    - Grounding cable and main flow oil filter housing from holder
    - Fasteners on right floor panel
    - Pull floor panel down and disconnect positive battery lead
    - Release expansion rivet and withdraw positive battery lead from holders
    - Remove transmission
    - Alternator drive belt
    - A/C compressor drive belt and compressor
    - Power steering pump bolts and secure pump aside
    - Left side filter housing
    - Low pressure lead from brake booster
    - Locks and clips from fuel hose, then the fuel hose
    - Vacuum line and cooling system hoses
    - Right exhaust manifold
    - Ground strap from right engine support

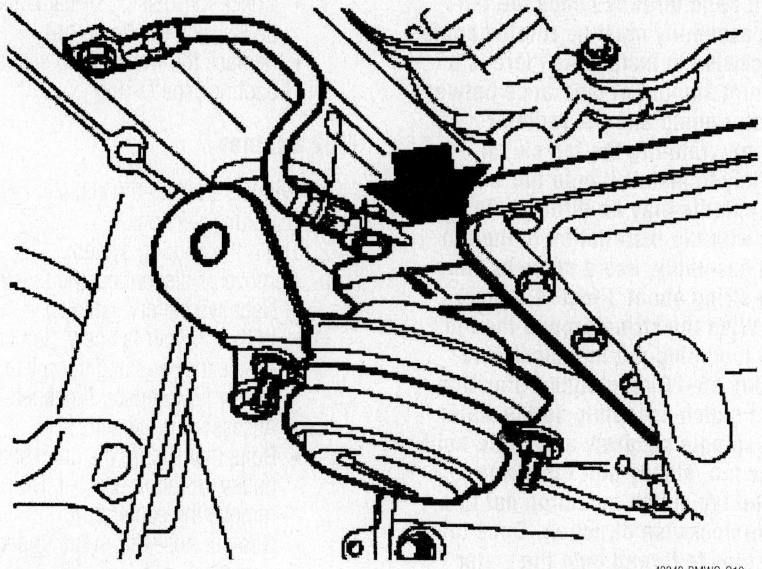

Engine ground strap and mount

- Top left and right nuts from engine mount
- Intake manifold cover
- Verify that all remaining fluid lines or electrical leads have been disconnected and properly placed aside
- Carefully lift the engine from the vehicle

7. Installation is the reverse of the removal procedure. When installing the engine, place the mount stud in hole (1) of the support bracket. Use the following torque values:
   - 6mm fasteners: 61 inch lbs. (7 Nm)
   - 7mm fasteners: 10 ft. lbs. (13 Nm)
   - 8mm fasteners: 16 ft. lbs. (22 Nm)
   - 10mm fasteners: 31 ft. lbs. (42 Nm)

N62 Engines

1. Before servicing the vehicle, refer to the Precautions Section.
2. Drain the cooling system.
3. Relieve the fuel system pressure.
4. Raise the hood all the way up into the service position. This usually requires disconnecting the gas struts and securing the hood hinges with bolts.
5. Secure the engine with a holding tool to prevent it from tilting.
6. Remove or disconnect the following:
   - Negative battery cable
   - Transmission assembly
   - Radiator and hoses
   - Expansion tank
   - Intake manifold
   - Heater hoses
   - Engine wiring harness from the control unit
   - Spark plug wiring harness
   - Negative lead from spring strut dome
   - Vacuum line and hose from vacuum pump
   - Line on tank venting valve
   - Both oxygen monitor and control sensors
   - Negative lead from engine support arm
   - Vibration damper and tensioning pulley fro A/C compressor drive belt
   - Fasteners from power steering pump, secure pump aside
   - Fasteners from A/C compressors, secure compressor aside
   - Fasteners from ABS control unit, secure unit aside
   - Lower universal joint from the steering spindle
   - If equipped with Dynamic Drive, unlock and detach the hydraulic line
   - Verify that all remaining fluid lines or electrical leads have been disconnected and properly placed aside
   - Engine mount nuts and bolts and carefully lift the engine from the vehicle
7. Installation is the reverse of the removal procedure. Use the following torque values:
   - 6mm fasteners: 61 inch lbs. (7 Nm)
   - 7mm fasteners: 10 ft. lbs. (13 Nm)
   - 8mm fasteners: 16 ft. lbs. (22 Nm)
   - 10mm fasteners: 31 ft. lbs. (42 Nm)

N52 Engines

⁂ CAUTION

Aluminum-magnesium material. No steel fasteners may be used due to the threat of electrochemical corrosion. A magnesium crankcase requires aluminum fasteners exclusively. Aluminum fasteners must be replaced each time they are removed. The end faces of aluminum fasteners are painted blue for purposes of identification. Torque specifications and torque angles must be observed for risk of damage.

1. Before servicing the vehicle, refer to the Precautions Section.
2. Drain the cooling system and engine oil.
3. Relieve the fuel system pressure.
4. Raise the hood all the way up into the service position. This usually requires disconnecting the gas struts and securing the hood hinges with bolts.
5. Secure the engine with a holding tool to prevent it from tilting.
6. Remove or disconnect the following:
   - Negative battery cable
   - Exhaust system
   - Transmission assembly
   - Intake filter housing
   - Fan cowl with electric fan
   - Radiator and hoses
   - Water pump
   - Thermostat
   - All coolant hoses from engine
   - Intake manifold
   - Vacuum line and hose from brake booster
   - Ignition wiring harness, secure and set aside.
   - Engine wiring harness, secure and set aside
   - Fuel injector rail, secure and set aside

⁂ CAUTION

Do not disconnect coolant pipe from crankcase.

   - A/C compressor, set down and secure on front axle carrier

➡Do not disconnect hydraulic lines.

   - Power steering pump, set down and secure on front axle carrier
   - If equipped with Dynamic Drive, remove bracket
   - Verify that all remaining fluid lines or electrical leads have been dis-

connected and properly placed aside
- Engine mount nuts and bolts and carefully lift the engine from the vehicle

7. Installation is the reverse of the removal procedure. Check DME function.

Replace all aluminum bolts. Use the following torque values:

- 6mm fasteners: 61 inch lbs. (7 Nm)
- 7mm fasteners: 10 ft. lbs. (13 Nm)
- 8mm fasteners: 16 ft. lbs. (22 Nm)
- 10mm fasteners: 31 ft. lbs. (42 Nm)

Water Pump

REMOVAL & INSTALLATION

M54 Engines

1. Before servicing the vehicle, refer to the Precautions Section.
2. Drain the cooling system.
3. Remove or disconnect the following:
   - Negative battery cable
   - With an upper fan cowl, fan cowl
   - Air cleaner housing complete with the air flow sensor, if necessary
   - Fan clutch mounting nut from the water pump shaft. The shaft uses left-hand threads; thus the nut is turned clockwise to remove.
   - Accessory drive belt
   - Fasteners securing the pulley to the water pump
   - Water pump mounting bolts
   - Water pump assembly

➡The water pump can be separated from the engine by using 2 M6 bolts threaded into the tapped bores of the water pump housing. To press the water pump away from the engine thread the bolts evenly.

To install:

4. Clean and remove any residual debris or gasket material from the engine mounting surface for the water pump.
5. Lubricate and install a new O-ring.
6. Install the water pump and tighten the bolts as follows:
   - M6 bolts: 78 inch lbs. (9 Nm)
   - M8 bolts: 16 ft. lbs. (22 Nm)
7. Install or disconnect the following:
   - Water pump pulley onto the water pump
   - Cooling fan and fan clutch assembly onto the threaded water pump spindle

➡The water pump spindle threads for the cooling fan and fan clutch assembly are left hand threads hence the fan clutch assembly must be rotated counterclockwise to install. As there is a minimum amount of clearance between the water pump and the radiator assembly, rotating the fan clutch assembly to install it onto the water pump spindle may be difficult. To assist with the installation of the fan clutch assembly, use a piece of thin, sturdy string about 3 feet (1 meter) long. Wind the string around the fan clutch mounting nut in a clockwise direction 15–20 revolutions. Position the fan clutch assembly on the water pump spindle squarely and while holding the fan, slowly pull the string to spin the fan clutch mounting nut in a counterclockwise direction. Once the nut begins to thread onto the water pump spindle, remove the remainder of the string.

   - Fan clutch assembly using tool No. 11–5–040. Nut: 29 ft. lbs. (40 Nm). If using the fan tool, torque the nut to 22 ft. lbs. (30 Nm); as the additional length of the tool multiples the torque to achieve 29 ft. lbs. (40 Nm) at the nut.
   - The remaining components are installed in the reverse order from which they were removed
8. Fill the cooling system.
9. Start the engine and check for leaks.

➡It may be necessary to bleed the cooling system a second time once the engine has been started.

S54 Engines

1. Before servicing the vehicle, refer to the Precautions Section.
2. Drain the cooling system.
3. Remove or disconnect the following:
   - Negative battery cable
   - Water pump and alternator drive belts
   - Pulley from water pump
   - Thermostat
   - Screws and coolant pipe from the exhaust side of the cylinder head, on 3 series
   - Water pump fasteners and water pump
4. Installation is the reverse of the removal procedure. Note the following:
   - Clean and remove any residual debris or gasket material from the engine mounting surface for the water pump
   - Check for missing or damaged dowel pins on the pump
   - Replace the water pump seal and cooling pipe O-ring

M62 Engines

1. Before servicing the vehicle, refer to the Precautions Section.
2. Drain the cooling system.
3. Remove or disconnect the following:
   - Negative battery cable
   - With an upper fan cowl, fan cowl
   - Air cleaner housing complete with the air flow sensor, if needed
   - Accessory drive belts
   - Bolts that secure the crankshaft pulley vibration damper. Do not remove the center bolt.
   - Coolant hoses from the upper water pump housing
   - Fan clutch mounting nut from the water pump shaft. The shaft uses left-hand threads; thus the nut is turned clockwise to remove.
   - Fasteners securing the pulley to the water pump
   - Thermostat housing from the water pump assembly
   - Water pump mounting bolts
   - Water pump assembly

To install:

4. Clean and remove any residual debris or gasket material from the engine mounting surface for the water pump.
5. Install or connect the following:
   - Water pump with a new gasket. Tighten the M6 bolts to 78 inch lbs. (9 Nm), and the M8 bolts to 16 ft. lbs. (22 Nm).
   - Thermostat and thermostat housing onto the water pump assembly using a new housing gasket sealing ring
   - Crankshaft pulley vibration damper using new fasteners. Be sure to align the damper with the dowel.
   - Water pump pulley onto the water pump
   - Accessory drive belts
   - Cooling fan and fan clutch assembly onto the threaded water pump spindle

➡The water pump spindle threads for the cooling fan and fan clutch assembly are left hand threads hence the fan clutch assembly must be rotated counterclockwise to install. As there is a minimum amount of clearance between the water pump and the radiator assembly, rotating the fan clutch assembly to install it onto the water

pump spindle may be difficult. To assist with the installation of the fan clutch assembly, use a piece of thin, sturdy string about 3 feet (1 meter) long. Wind the string around the fan clutch mounting nut in a clockwise direction 15–20 revolutions. Position the fan clutch assembly on the water pump spindle squarely and while holding the fan, slowly pull the string to spin the fan clutch mounting nut in a counterclockwise direction. Once the nut begins to thread onto the water pump spindle, remove the remainder of the string.

6. Install or connect the following:
   - Fan clutch assembly using tool No. 11–5–040. Nut: 29 ft. lbs. (40 Nm). If using the fan tool, torque the nut to 22 ft. lbs. (30 Nm); as the additional length of the tool multiplies the torque to achieve 29 ft. lbs. (40 Nm) at the nut.
   - The remaining components are installed in the reverse order from which they were removed
7. Fill the cooling system.
8. Start the engine and check for leaks.

➡It may be necessary to bleed the cooling system a second time once the engine has been started.

N62 Engines

1. Before servicing the vehicle, refer to the Precautions Section.
2. Drain the cooling system.
3. Remove or disconnect the following:
   - Negative battery cable
   - Fan cowl
   - Alternator drive belt
   - Connectors and vacuum line holder
   - Hoses and pulley from water pump
   - Vibration damper
   - Water pump fasteners and water pump
4. Installation is the reverse of the removal procedure. Note the following:
   - Clean and remove any residual debris or gasket material from the engine mounting surface for the water pump
   - Check for missing or damaged adapter sleeve on the pump
   - Replace the water pump seals and coolant pipes
   - Coat the sealing points of the coolant pipes with a rubber coating

N52 Engines

⁂ CAUTION

Aluminum-magnesium material. No steel fasteners may be used due to the threat of electrochemical corrosion. A magnesium crankcase requires aluminum fasteners exclusively. Aluminum fasteners must be replaced each time they are removed. The end faces of aluminum fasteners are painted blue for purposes of identification. Torque specifications and torque angles must be observed for risk of damage.

1. Before servicing the vehicle, refer to the Precautions Section.
2. Drain the cooling system.
3. Remove or disconnect the following:
   - Negative battery cable
   - Front splash guard
   - Coolant thermostat
   - Water hose and plug connector
   - Water pump fastener and water pump
4. Installation is the reverse of the removal procedure. Tighten all fasteners to correct torque.

Note the following:
   - Replace all aluminum fasteners.
   - If the water pump is to be reused, it must be mechanically rotated once (breakaway torque at impellers). One water pump rotation will be sufficient.
   - Vent cooling system and check for leaks
   - Clean and remove any residual debris or gasket material from the engine mounting surface for the water pump
   - Replace the water pump seals and coolant pipes

Heater Core

REMOVAL & INSTALLATION

3 Series

➡The heater case assembly must be removed to remove the heater core on all 3 Series vehicles.

1. Disconnect the negative battery cable.
2. Drain the cooling system into a clean container for reuse.
3. Discharge and recover the air conditioning system refrigerant.

Heater bulkhead retaining screws

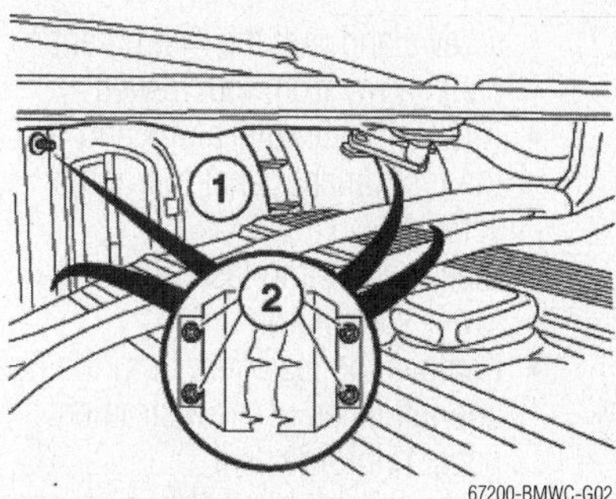

Heater case retaining screws

4. Remove or disconnect the following:
   - Remove the heater bulkhead from the engine compartment
   - Release hose clamps and detach the water hoses
   - Release heater case screws on the firewall
   - Steering wheel and column switches

⁂ CAUTION

When removing the steering wheel on airbag equipped vehicles, use extreme caution handling the airbag assembly. Store the airbag with the airbag facing up, to avid injury in case of accidental deployment. Replace damaged airbag assemblies. Do not try to repair or reuse deployed airbag assemblies.

   - Center console retaining screws
   - All electrical leads and remove the console
   - Open the glove box and remove the glove box door assembly
   - Instrument panel trim
   - Left and right "A" pillar trim
   - Headlight switch assembly
   - Instrument cluster
   - Radio assembly and the radio trim
   - Ventilation control head
   - Passenger side airbag
   - Left and right defroster vents
   - Upper instrument panel retaining

67200-BMWC-G03

Instrument panel reinforcement side fasteners and head airbag assembly

screws and pull the instrument panel away from the firewall
- All electrical multi-plugs and pull the instrument panel out of the vehicle
- Steering column fasteners and lower the column
- Remove the fasteners from the right and left sides of the instrument panel reinforcement
- Right and left head airbag assemblies if equipped
- Electrical leads from the heater
- Right and left center console reinforcements
- Cut the cable ties from the fusebox. Release the locks and pivot assembly downward
- Fusebox and the transverse reinforcement
- Heater case retaining bolts and heater case assembly from the vehicle
- Remove the core from the case assembly

To install:

5. Install or connect the following:
   - Heater core into the case. Make sure the gasket is positioned properly.
   - Heater case in position in the vehicle
   - Fusebox and the transverse reinforcement
   - Pivot fusebox assembly upward and engage locks. Install new cable ties
   - Right and left center console reinforcements
   - Electrical leads from the heater
   - Right and left head airbag assemblies if equipped
   - Right and left side fasteners of the instrument panel reinforcement
   - Raise the steering column and install new fasteners
   - All instrument panel electrical multi-plugs and the instrument panel into the vehicle
   - Left and right defroster vents
   - Passenger side airbag
   - Ventilation control head
   - Radio assembly and the radio trim
   - Instrument cluster
   - Headlight switch assembly
   - Left and right "A" pillar trim
   - Instrument panel trim
   - The glove box door assembly
   - Center console
   - Column switches and steering wheel
   - Heater case screws on the firewall
   - Water hoses and clamps
   - The heater bulkhead in the engine compartment
6. Refill the cooling system.
7. Evacuate, charge and leak test the air conditioning system.
8. Connect the negative battery cable.
9. Operate the engine to normal operating temperatures; then, check the climate control operation and check for leaks.

5 and 7 Series

2002–03 5 SERIES AND 2001 7 SERIES

1. Disconnect the negative battery cable.
2. Drain the cooling system into a clean container for reuse.
3. Remove or disconnect the following:
   - Screws retaining the center console
   - Ashtray assembly

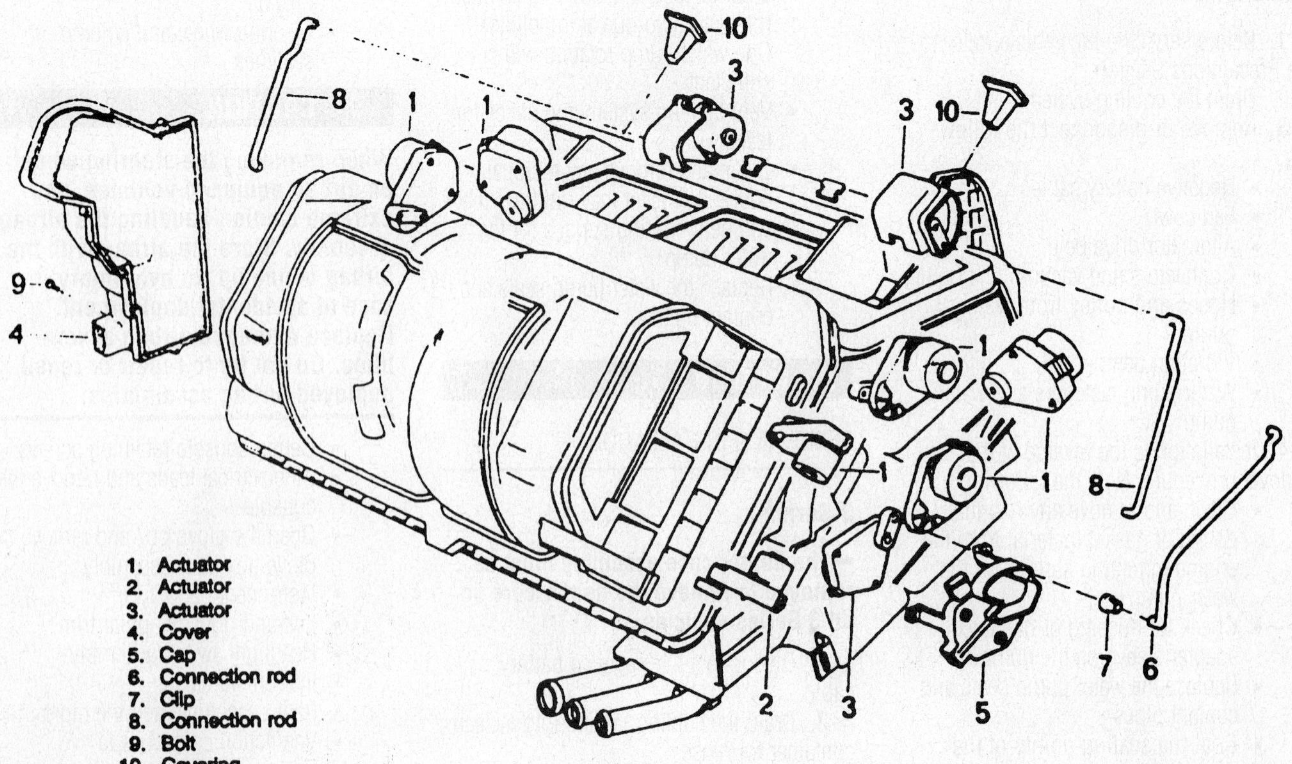

View of the heater/blower housing assembly

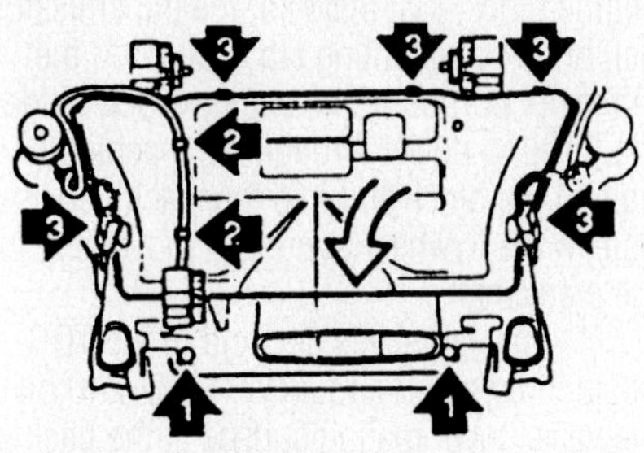

Removing the heater core cover

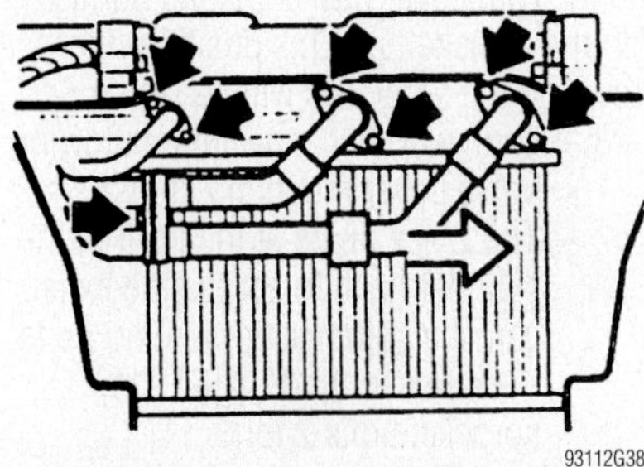

Removing the heater water pipes

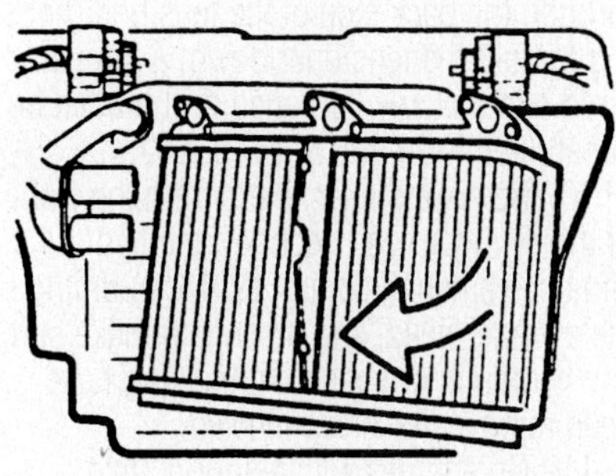

Remove the heater by pulling to the right

- Glove box door assembly
- Right side heater case cover screws and remove the cover
- Front vent drive motor
- Plug from the inside temperature sensor
- Heater core cover retaining screws
- Cover clips and straps
- Cover and gasket
- Coolant hoses from the heater core
- Heater pipe-to-heater core retaining bolts
- Heater pipes
- Heater core retaining bolts
- Heater core by tilting it to the side

To install:

4. Install or connect the following:
   - Heater core in the heater case
   - Heater core retaining bolts
   - Heater pipes to the heater core. Tighten the M6 bolts to 72–84 inch lbs. (8.1–9.5 Nm) and the M8 bolts to 16–17 ft. lbs. (21.7–23 Nm). Connect the heater hoses. Always use new O-rings on the heater pipes.
   - Heater case cover and retaining clips
   - Cover retaining screws. Make sure the cover gasket is properly positioned.
   - Front vent drive motor
   - Right heater case cover
   - Inside temperature sensor
   - Glove box door assembly
   - Center console and the ashtray

5. Refill the cooling system and connect the negative battery cable.

6. Operate the engine to normal operating temperatures; then, check the climate control operation and check for leaks.

EXCEPT 2002–03 5 SERIES AND 7 SERIES

1. Disconnect the negative battery cable.

2. Discharge and recover the air conditioning system refrigerant.

3. Drain the cooling system into a clean container for reuse.

4. Remove or disconnect the following:

5. Move the front seats completely back

6. Move steering column into the lower and extended position
   - Pillar trim
   - Instrument cluster
   - Headlight switch
   - On-board monitor assembly
   - Trim panel above the pedal assembly
   - Glovebox and instrument panel trim
   - Center speaker connection
   - Solar sensor connection
   - Fasteners securing the instrument panel
   - Instrument panel
   - A/C and coolant lines from the heater case
   - Fasteners securing the heater case
   - Heater case assembly
   - Heater core cover from the case
   - Coolant pipe retainers and pipes from the core
   - Heater core

To install:

7. Install or connect the following:
   - Heater core in the heater case
   - Coolant pipes
   - Heater core cover
   - Heater case assembly and fasteners
   - A/C and coolant lines and fasteners
   - Instrument panel and fasteners
   - Solar sensor connection
   - Center speaker connection
   - Glovebox and instrument panel trim
   - Trim panel above the pedal assembly
   - On-board monitor assembly
   - Headlight switch
   - Instrument cluster
   - Pillar trim

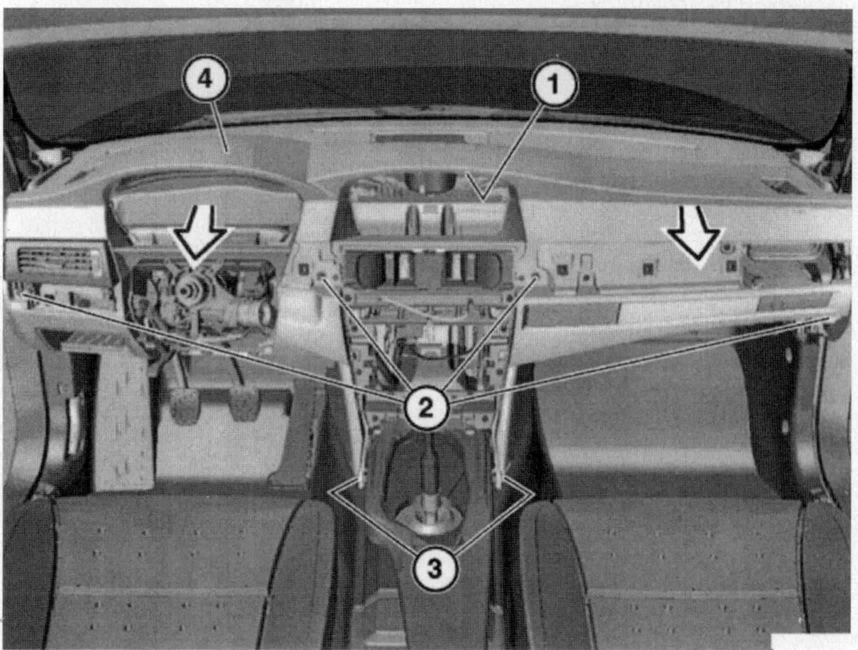

Remove the center speaker connector (1), instrument panel fasteners (2, 3), then pull the instrument panel (4) back

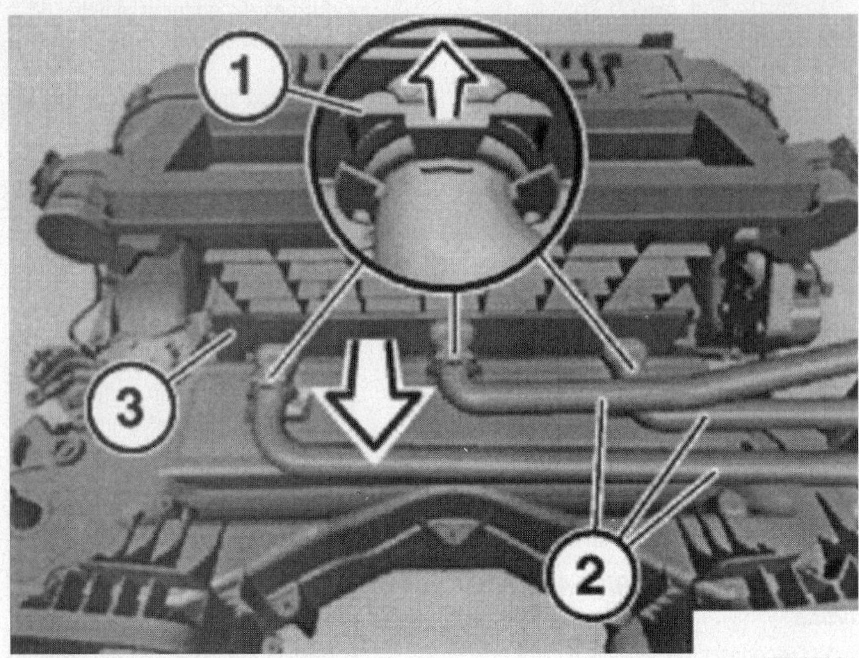

Remove retainers (1) and coolant pipes (2) from the core (3)

8. Refill the cooling system and connect the negative battery cable.
9. Recharge the A/C system with refrigerant
10. Operate the engine to normal operating temperatures; then, check the climate control operation and check for leaks.

Cylinder Head

REMOVAL & INSTALLATION

M54 Engines

1. Before servicing the vehicle, refer to the Precautions Section.
2. Drain the cooling system.
3. Relieve the fuel system pressure.
4. Remove or disconnect the following:
   - Negative battery cable
   - Lower engine splash shield
   - Exhaust manifolds from the exhaust pipes
   - Rear hood seal, then remove the fasteners securing the sealed plastic housing for the engine wiring harness from the passenger compartment fresh air intake shroud
   - Fresh air intake shroud
   - If equipped, plastic trim covers from the top of the engine
   - Mass Air Flow (MAF) sensor and remove intake manifold air duct along with the air filter housing assembly
   - Throttle cable from the throttle body linkage, coolant hoses, the Throttle Position (TP) sensor electrical connector, the throttle body mounting fasteners, and remove the throttle body assembly
   - Intake manifold support bracket fasteners from the intake manifold and loosen the support bracket fasteners on the engine block
   - Fuel feed and return lines
   - Vacuum hoses from the intake manifold, and loosen the clamps for the idle speed stabilizer underneath the manifold, and disconnect the hoses
   - Intake manifold intake port fasteners and carefully lift the manifold away from the engine
   - Ground cable at the front of the cylinder head, then remove the 2 radiator hoses from the thermostat housing, and remove the housing
   - Exhaust manifolds
   - Wire connectors for the ignition coils and cylinder head sensors and remove the coils, then remove the cylinder head cover
5. Rotate the engine in the direction of rotation to Top Dead center (TDC) for cylinder number one. Cylinder number 1 will be at TDC when the intake and exhaust camshaft peaks for cylinder number 1 face each other
6. Lock the engine in the TDC position by placing the holding dowel tool No. 11–2–300 through the machined hole in the engine block, just inside of the transmission bell housing mounting tab located on the left lower portion of the engine block. Slide the locating dowel through the machined hole in the block and into the machined hole in the flywheel to prevent movement of the crankshaft.
7. The camshafts are held in the TDC position by placing tool No. 11–3–240 on the valve cover mating surface at the back of the cylinder head and onto the squared ends of the camshafts. Secure the camshafts so that 2 sides of the squared ends are parallel with the cam cover gasket mating surface. With the camshafts in this position, the arrows on the sprockets will be facing up.
8. Remove or disconnect the following:
   - Valve cover mounting studs
   - The 2 hex plugs at the front of the cylinder head to access the exhaust camshaft sprocket mounting bolts, then loosen the exhaust cam sprocket bolts 2 turns
9. Press down on the secondary cam chain tensioner between the 2 camshaft sprockets and install tool No. 11–3–292 through the back side of the tensioner housing to hold the tensioner down. A similar sized and suitably hardened drill bit can be substituted for tool No. 11–3–292.
10. Remove the fasteners from the front of the cylinder head securing the hydraulic variable camshaft control (VANOS) unit to the cylinder head, and inspect to make sure any hydraulic or sensor connectors have been removed or disconnected.
11. On engines with a spring plate installed on the intake camshaft, place tool No. 11–5–490, onto the exhaust camshaft sprocket. Carefully rotate the sprocket clockwise to allow the helical gear of the VANOS unit to release the intake camshaft and to allow the VANOS unit to be pulled away from the front of the cylinder head.
12. If tool No. 11–5–490 is not available, move the camshaft sprockets to release the VANOS by using a suitable drift and soft faced mallet and lightly tapping on a sprocket tooth of the intake cam sprocket to rotate both cam sprockets clockwise, while alternately pulling on the VANOS unit to release it. This procedure may need to be repeated several times to fully release the VANOS unit, and must be performed very carefully, in such a manner to not distort or damage the teeth of the cam sprocket.
13. With the VANOS assembly removed, remove the intake and exhaust camshaft sprockets, the hydraulic cam chain tensioner, and the cam chain guide.
14. From the side of the right front area of the engine, remove the cap nut for the

cam chain tensioner for the cam chain that runs between the crankshaft and the exhaust camshaft sprocket. Use care when removing the tensioner cap nut as the cam tensioner spring applies pressure to the cap nut.

15. Remove the exhaust cam tensioner, and then release the sprocket from the cam chain.

16. Attach a wire tie or mechanics wire to the cam chain and temporarily secure the chain fully extended.

17. Remove the dowel from the engine and flywheel locking the engine in the TDC position.

18. While holding the crankshaft to exhaust camshaft camchain, rotate the engiune 30 degrees counterclockwise to avoid damaging the valves during cylinder head removal and reinstallation.

19. With the attached wire tie or mechanics wire secured to the cam chain, carefully lower the chain downward making sure there is enough exposed wire to retrieve the chain for reinstallation.

20. Remove the fasteners located in a recessed area near the front of the camshafts securing the front of the cylinder head to the engine block.

21. In the reverse order of the tightening sequence, using a proper sized Torx® bit or tool No. 11–2–250, loosen the cylinder head mounting bolts in 3 steps as follows:

a. Step 1: 90 degrees.
b. Step 2: 90 degrees.
c. Step 3: Completely remove the bolt.

22. With all of the cylinder head bolts removed, verify that all electrical connectors and fluid lines have been removed and lift the cylinder head off the engine block.

➡It is not recommended to mill the cylinder head surface of the S52 engine. If milling the cylinder head of an M52 engine, a thicker head gasket is available for reassembly.

To install:

23. Thoroughly clean the deck surface of the engine cylinder block.

24. If the camshafts have been removed and reinstalled a waiting period dependent on the ambient temperature is necessary before mounting the cylinder head on the engine. At room temperature wait 4 minutes to allow the lifters to compress fully. At temperatures down to 50°F (10°C), wait 11 minutes. At temperatures lower than 50°F (10°C) wait 30 minutes. This is to prevent contact between the valves and the piston tops. The engine may not be cranked under the same condition for a period of 10 minutes at room temperature; 30 minutes for

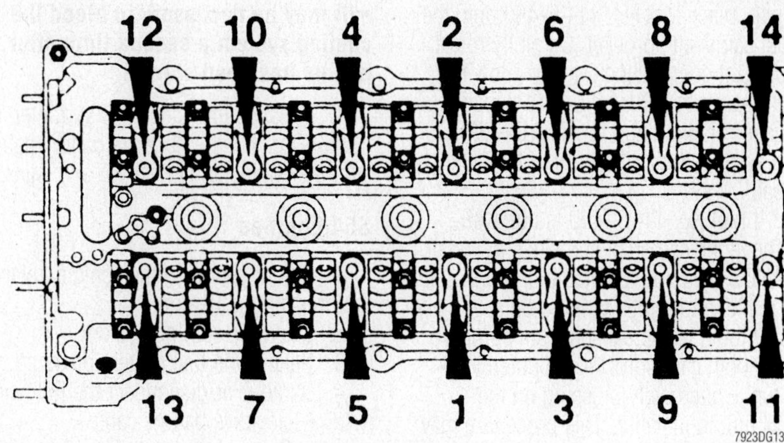

Cylinder head torque sequence—M54 engines

temperatures down to 50°F (10°C); 75 minutes for temperatures below 50°F (10°C).

25. Make sure the cylinder head has been checked for warpage and coolant leakage.

26. Do not to drop any pieces of gasket or debris into the oil or coolant passages. Check the condition of the head locating dowel sleeves.

27. Place a new head gasket on the engine block over the locating dowels and gently place the head on the engine with the dowel sleeves properly aligned and check that the head sits flat on the engine.

➡The cylinder head bolts may not be reused. The bolts are a different length and have a different torque specification for aluminum engine blocks than those for the cast iron engine blocks.

28. Install and torque the new cylinder head bolts in 3 steps following the tightening sequence.

29. On cast iron engine blocks, wash the bolts in a cleaning solvent, then apply oil to the threads of the bolts and torque in sequence as follows:

a. Step 1: 22 ft. lbs. (30 Nm).
b. Step 2: 90 degrees.
c. Step 3: 90 degrees.

30. On aluminum engine blocks do not remove the coating on the head bolts, apply oil to the threads and torque in sequence as follows:

a. Step 1: 30 ft. lbs. (40 Nm).
b. Step 2: 90 degrees.
c. Step 3: 90 degrees.

31. Install the fasteners located in the recessed area near the front of the camshafts securing the front of the cylinder head to the engine block and torque to 96 inch lbs. (10 Nm).

32. With the attached wire secured to the crankshaft cam chain, carefully raise the chain upward and apply a light tension to the chain, then while holding the chain, carefully rotate the engine clockwise to TDC on cylinder number one.

33. Lock the engine in the TDC position by placing the holding dowel tool No. 11–2–300 through the machined hole in the engine block, just inside of the transmission bell housing mounting tab located on the left lower portion of the engine block. Slide the locating dowel through the machined hole in the block and into the machined hole in the flywheel to prevent movement of the crankshaft.

34. Hold the camshafts in the TDC position by placing tool No. 11–3–240, on the valve cover mating surface at the back of the cylinder head and onto the squared ends of the camshafts, securing the camshafts so that 2 sides of the squared ends are parallel with the cam cover gasket mating surface. With the camshafts in this position, the arrows on the sprockets will be facing up.

35. Position the crankshaft cam chain over the exhaust cam chain sprocket and install the sprocket on the exhaust cam such that the slotted holes are centered with the fastener bores in the camshaft.

36. Install the hydraulic cam chain tensioner, and the cam chain guide, then the intake and exhaust camshaft sprockets along with the cam chain and remove the tool used to hold the hydraulic cam chain tensioner collapsed.

37. Install the crankshaft cam chain tensioner assembly, spring and cap nut.

38. Apply sealant to the upper corners of the cylinder head where the VANOS housing mounts and install a new VANOS housing gasket.

39. Press the helical gear of the VANOS assembly toward the housing and install the VANOS assembly. To do so on engines with a spring plate installed on the intake

camshaft, place tool No. 11–5–490 onto the exhaust camshaft sprocket. Carefully rotate the sprocket counterclockwise to allow the helical gear of the VANOS unit to thread into the intake camshaft and to allow the VANOS unit to be pulled into the front of the cylinder head.

40. If tool No. 11–5–490 is not available, move the camshaft sprockets to install the VANOS by using a suitable drift and soft faced mallet and lightly tapping on a sprocket tooth of the exhaust cam sprocket to rotate both cam sprockets counterclockwise, while alternately pressing on the VANOS unit to install it. This procedure may need to be repeated several times to fully install the VANOS unit, and must be performed very carefully, in such a manner to not distort or damage the teeth of the cam sprocket.

➡Make sure when assembling the VANOS unit is able to rest on the front of the cylinder head without being forced or without binding. If the VANOS unit does not fully seat it may be necessary to reposition the camshaft sprockets such that the slots in the camshaft sprockets allow enough movement of the sprockets for the helical gear of the VANOS assembly to be fully seated during assembly.

41. Tighten the VANOS unit fastener and then torque the camshaft sprocket bolts of both the intake and exhaust cams to 16 ft. lbs. (22 Nm).

42. Remove the crankshaft TDC holding dowel tool No. 11–3–240 from the engine, and remove the camshaft TDC positioning tool No. 11–3–240 from the valve cover mating surface at the back of the cylinder head.

43. Slowly and carefully rotate the engine clockwise 4 complete revolutions bringing cylinder number 1 to TDC. If the engine binds for any reason stop immediately to evaluate and rectify the cause of the binding.

44. With cylinder number 1 at TDC slide the camshaft TDC positioning tool No. 11–3–240 over the ends of the cams and onto the valve cover mating surface. If the tool slides over the cams and is flush with the mating surface, the camshafts are properly timed. If the tool does not slide easily over the ends of the cams, or if the tool is not flush with the valve cover mating surface, the camshaft timing must be repeated until the tool fits squarely.

45. The balance of installation is the reverse of the removal procedure.

46. Top off the engine cooling system and bleed as necessary.

➡It may be necessary to bleed the cooling system a second time after the engine has been started.

47. Change the engine oil and filter.

48. Connect the negative battery cable, start the engine and check for any leaks.

S54 Engines

1. Before servicing the vehicle, refer to the Precautions Section.
2. Drain the cooling system.
3. Relieve the fuel system pressure.
4. Remove or disconnect the following:
   - Negative battery cable
   - Both exhaust manifolds
   - Intake filter housing and mass air sensor
   - Cylinder head cover
   - All spark plugs
   - Intake manifold
   - Camshafts
   - Raise rocker arms
   - Adjustment plates using magnetic tool No. 11–4–400, set all aside in order
   - Connector strip in sequence 1 to 6 with tool No. 12–1–120

⁂ WARNING

The connector strip may break if tool No 12–1–120 is not used.

   - Press lock and detach supplementary air line
   - Bracket for idle speed control valve from return valve and secure aside
   - Temperature sensor connector
   - Thermostat housing
   - Electrical and vacuum connections from the end of the cylinder head
   - Pull rod and fasteners securing the pipe under it
   - Coolant return hose
   - Fasteners from fuel feed line on the engine block
   - Fuel feed line retaining clips
   - Pull fuel feed line downward
   - Thrust bearing flange on the end of the cylinder head
   - Sliding rail in the cylinder head
   - Two bolts between the cylinder head and timing case cover, and the cable holder
   - Cylinder head bolts from outside to inside in sequence, 14 to 1
   - Cylinder head

➡The cylinder head bolts must be replaced.

To install:

5. Thoroughly clean all mounting surfaces and check the head for warpage. Take care not to drop any pieces of gasket or debris into the oil or coolant passages. Check the condition of the head locating dowel sleeves and clean out the bolt threads with a tap.

6. Mount the cylinder head on the block and use new bolts. Do not remove the coating on the head bolts. Apply oil to the threads and washer contact area. Torque following the tightening sequence as follows:

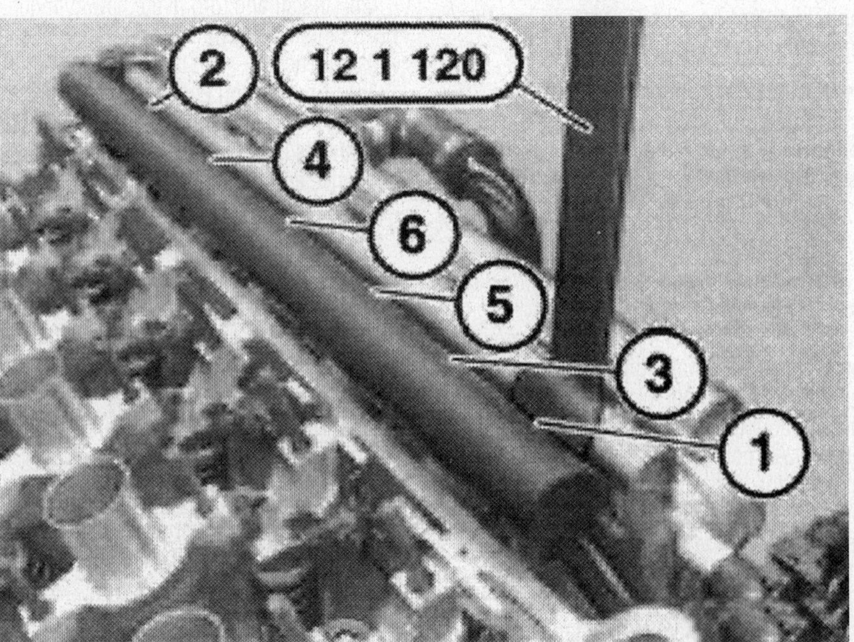

Remove the connector strip in sequence 1 to 6

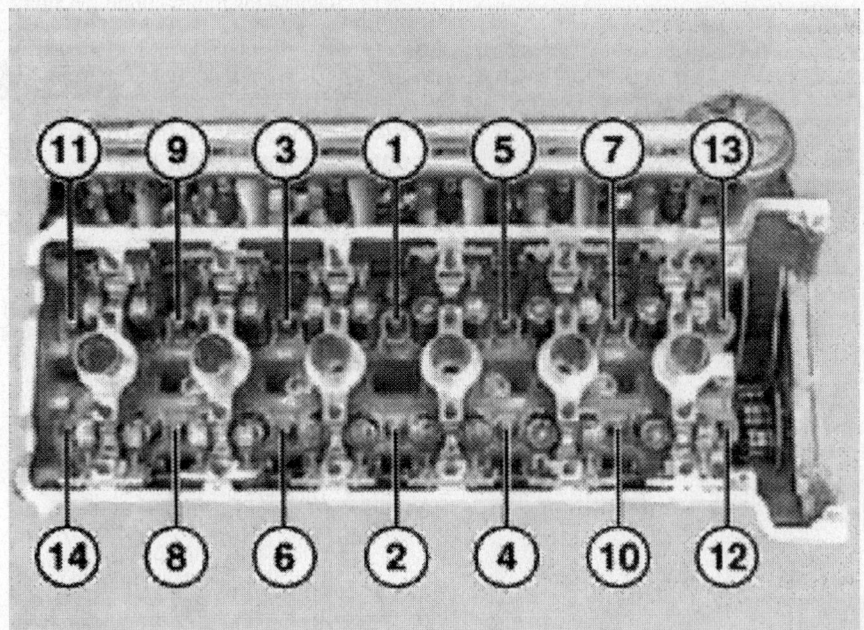

Cylinder head bolt sequence on S54 engines

a. Step 1: 22 ft. lbs. (30 Nm).
b. Step 2: 90 degrees.
c. Step 3: 90 degrees.

7. The balance of installation is the reverse of the removal procedure. Be sure to replace all seals and O-rings.
8. Fill the cooling system.
9. Start the engine and check for leaks.

➡It may be necessary to bleed the cooling system a second time after the engine has been started.

M62 Engines

1. Before servicing the vehicle, refer to the Precautions Section.
2. Drain the cooling system.
3. Relieve the fuel system pressure.
4. Remove or disconnect the following:
   - Negative battery cable
   - Both exhaust manifolds from each side of the engine, then remove the heat shields from the front axle carrier
   - Coolant expansion tank
   - Upper timing case cover
   - Oil pipes from the cylinder head
   - Intake manifold
   - Cylinder head cover
   - Engine vent pipe together with the O-ring, disconnect the coolant hoses on the coolant collector, remove the coolant collector mounting bolts, and then remove the coolant collector
   - Spark plugs
   - Camshaft sprockets, and the timing chain tensioner
   - Bolts retaining the guide rail on the cylinder head's left-hand side
   - Cylinder head bolts in the reverse order of the tightening sequence
   - Cylinder head

➡The cylinder head bolts must be replaced.

To install:

5. Thoroughly clean all mounting surfaces and check the head for warpage. Take care not to drop any pieces of gasket or debris into the oil or coolant passages. Check the condition of the head locating dowel sleeves and clean out the bolt threads with a tap.
6. Mount the cylinder head on the block and use new bolts. Do not remove the coating on the head bolts. Apply oil to the threads and torque following the tightening sequence in 3 steps as follows:
   a. Step 1: 22 ft. lbs. (30 Nm).
   b. Step 2: 80 degrees.
   c. Step 3: 80 degrees.
7. The balance of installation is the reverse of the removal procedure.
8. Fill the cooling system.
9. Start the engine and check for leaks.

➡It may be necessary to bleed the cooling system a second time after the engine has been started.

S62 Engines

LEFT SIDE

1. Before servicing the vehicle, refer to the Precautions Section.
2. Drain the cooling system.
3. Relieve the fuel system pressure.
4. Remove or disconnect the following:
   - Negative battery cable
   - Left exhaust manifold
   - Intake filter housing and mass air sensor
   - Intake manifold
   - Both cylinder head covers
   - All spark plugs
   - Fan clutch with impeller and cowl
   - Pulse generator connectors
   - Cable channel and bracket
   - Throttle actuator
   - Hose from supplementary air distributor
   - Knock sensor cable guide from the return line
   - Coolant hoses from the block and both return hoses on the heads
   - Brake booster hose

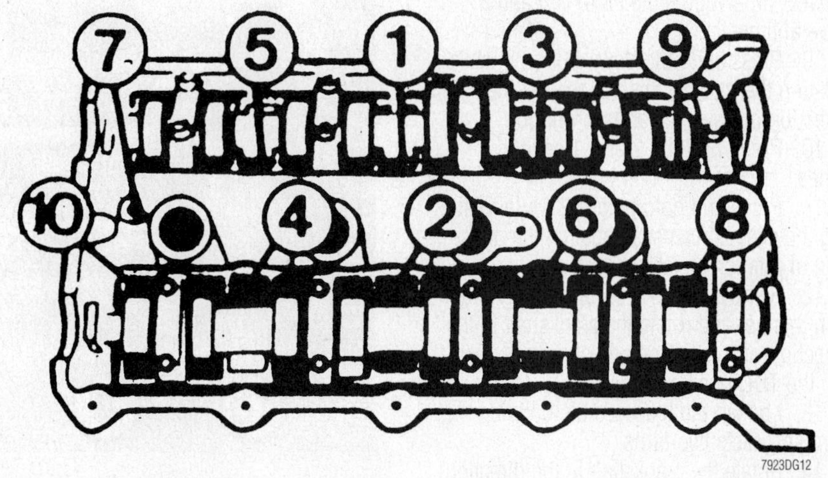

Cylinder head torque sequence—M62 engines

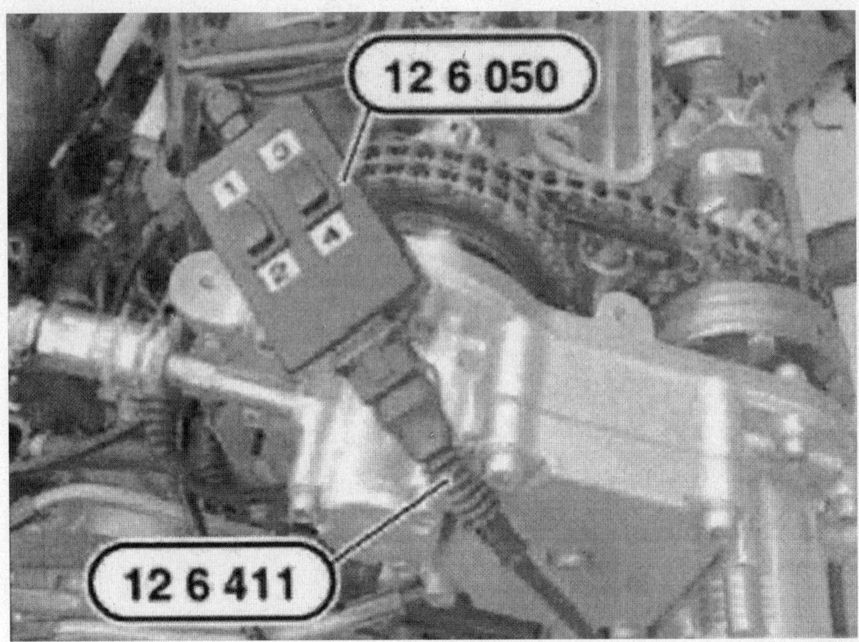

Special tools connected to the VANOS unit

- Vacuum hose from pressure regulator
- Oil lines on both cylinder heads
- Pulse generators on the end of the cylinder head

5. Hold the camshafts on the hexagon heads provided and remove the banjo bolts securing the signal rings.
6. Rotate the crankshaft in the direction of rotation to the Top Dead Center (TDC) firing position of cylinder 1. Secure the vibration damper in place. This can be done with tool No. 11–2–300.
7. Remove the oil line from the VANOS unit. Fit an air line adapter to the hole and apply 30–115 psi (2–8 bar) of compressed air.
8. Unplug the solenoid and connect tool No. 12–6–050 and 12–6–411 to the VANOS unit. Toggle the buttons 4 and 3 several times.
9. Press and hold down button 4 while rotating the intake camshaft against the direction of travel as far as it will go.
10. Press buttons 2 and 1 several times.
11. Press and hold button 2 while rotating the exhaust camshaft against the direction of rotation as far as it will go.
12. Remove the tool securing the vibration damper and rotate the crankshaft in the direction of rotation one revolution to overlap the TDC position.
13. Loosen the 6 accessible bolts on the cam sprockets two turns.
14. Rotate the crankshaft in the direction of rotation until cylinder 1 is in the TDC position. Secure the vibration damper from rotating again.
15. Loosen the 6 remaining bolts on the cam sprockets two turns.
16. Align the camshaft opening to the locating bores in the bearing caps and secure in position. Tool No. 11 7 120 can be used for this.
17. Remove the fasteners securing the VANOS unit and remove it from the cylinder head. Be sure to remove the compressed air first.
18. Remove the tool securing the vibration damper.

⁂ WARNING

No piston must be in the TDC position when the cylinder head is removed.

➡When the VANOS unit is removed, the chain sprockets of the camshafts are not positively connected to the camshafts. Camshafts on cylinder bank 5 to 8 do not rotate.

19. Rotate the crankshaft against the direction of rotation to 45 degrees BTDC.
20. Remove or disconnect the following:
    - Thermostat
    - Left timing case cover.
    - Chain tensioning piston
    - The slackened bolts from the spacer ring and remove the ring
    - Spline hub with plate spring and supporting ring. Keep note of the direction the plate spring faces
    - Intake chain sprocket from the centering sleeve. Keep the chain tensioned and secure from slipping into the lower timing cover
    - Sliding rail from the cylinder head
    - Cylinder head bolts in sequence from 10 to 1
    - Cylinder head

To install:

21. Thoroughly clean all mounting surfaces and check the head for warpage. Take care not to drop any pieces of gasket or

Cylinder head bolt sequence on S62 engines

debris into the oil or coolant passages. Check the condition of the head locating dowel sleeves and clean out the bolt threads with a tap.

22. Coat the joint between the engine block and timing case cover with Drei Bond 1209.

23. Mount the cylinder head on the block and use new bolts. Do not remove the coating on the head bolts. Apply oil to the threads and washer contact area. Torque following the tightening sequence as follows:

a. Step 1: 22 ft. lbs. (30 Nm).
b. Step 2: 80 degrees.
c. Step 3: 80 degrees.

24. Install the slide rail. Install the chain sprocket and chain on the centering sleeve. Do not secure it with fasteners. Note that the bores of the chain sprocket do not have to line up with the thread holes yet.

25. Install the chain tensioning piston. If the old one is being reused, drain the oil from it.

26. Rotate the crankshaft in the direction of rotation to the TDC position again. Secure the vibration damper. Be sure to keep the intake sprocket from slipping off the centering sleeve.

27. Check the position of the chain sprocket. If the threads are not visible through the holes, pull the sprocket from the sleeve (while keeping the chain tensioned), and refit. Fine adjustment can be made by rotating the exhaust side.

➡On engines manufactured up to 9/2001, the plate spring points in the direction of the supporting ring. On engine built after 9/2001, it faces towards the spline hub.

28. Install the spline hub with the plate spring and supporting ring. Align the hub so the oil pump is positioned as pictured.

29. Fit the spacer ring and install the bolts snug. Slacken the bolts until the spline hub can be moved.

30. Install the left timing case cover.

31. Replace the seals on the VANOS unit. Coat the seals and gears with oil.

32. Connect tool No. 12–6–050 and 12–6–411 to the VANOS unit. Press buttons 2 and 4. Push the splined shafts by hand into the VANOS unit.

33. Turn the splined hubs of the intake and exhaust camshafts clockwise to the right limit positon.

34. Attach the VANOS unit. Rotate the splined shafts until the spur tooting is engaged. Push the VANOS unit into the gear until the helical cut splines are positioned shortly before meshing with the spline hub.

35. If the gears on the exhaust side cannot be pushed into the splined hub, rotate the hub against the direction of rotation until correctly positioned. Do not rotate any further. Do the same for the intake side.

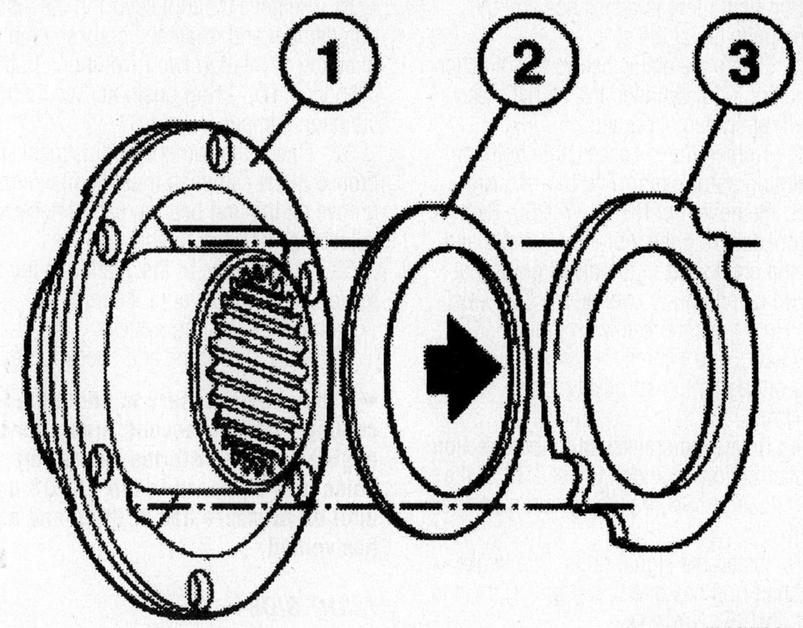

Spline hub (1), plate spring (2), and supporting ring (3)—version up to 9/2001

Spline hub (1) position

36. Align the oil pump to the holes in the spline hub. Push the VANOS unit in until the seals just rest against the timing cover. If necessary, realign the oil pump drive to the spline hub.

37. Insert one bolt on each side finger tight.

38. Use tool No. 11 7 200 to secure the timing sprockets. Tighten the six accessible bolts to 7 ft. lbs. (10 Nm). Then slacken all the bolts a quarter of a turn.

39. Install all the bolts on the VANOS unit. Alternately tighten them in half turn increments evenly until the unit rests against the timing cover.

40. Remove the tool securing the vibration damper. Make sure the timing sprockets are still secured to allow the camshafts to remain locked in position. Carefully rotate the engine against the direction of

rotation until there is resistance and the spline hubs are at the stop.

41. Rotate the crankshaft in the direction of rotation until cylinder 1 is at TDC. Secure the vibration damper again.

42. Tighten the six accessible bolts on the timing sprockets to 7 ft. lbs. (10 Nm).

43. Remove tool No. 11–7–200. Remove the tool securing the vibration damper and turn the crankshaft in the direction of rotation on revolution to overlap the TDC position. Hold the sprockets with tool No. 11–7–200, and tighten the six remaining bolts on the timing sprockets to 7 ft. lbs. (10 Nm).

44. Rotate the crankshaft in the direction of rotation until cylinder 1 is at TDC of the firing position. Secure the vibration damper again.

45. Install the signal rings. The intake camshaft ring has one dowel pin. Tighten to 37 ft. lbs. (50 Nm).

46. Remove the oil line from the VANOS unit on cylinder bank 1 to 4. Fit an air line adapter to the hole and apply 30–115 psi (2–8 bar) of compressed air.

47. Unplug the solenoid and connect tool No. 12–6–050 and 12–6–411 to the VANOS unit. Toggle the buttons 1 and 2 several times.

48. Press and hold down button 1 while rotating the intake camshaft against the direction of travel as far as it will go.

49. Press buttons 3 and 4 several times.

50. Press and hold button 3 while rotating the exhaust camshaft against the direction of rotation as far as it will go.

51. Remove the tool securing the vibration damper and rotate the crankshaft in the direction of rotation two revolutions to the cylinder 1 TDC firing position. Secure the vibration damper.

52. Check the camshaft adjustment. The groove in the camshaft must be inside the groove on the first bearing cap. If necessary, adjust the camshaft timing.

53. The balance of installation is the reverse of the removal procedure.

54. Fill the cooling system.

55. Start the engine and check for leaks.

➡It may be necessary to bleed the cooling system a second time after the engine has been started. A rattling noise may be heard in the VANOS unit until oil pressure has built up and air has vented.

RIGHT SIDE

1. Before servicing the vehicle, refer to the Precautions Section.
2. Drain the cooling system.
3. Relieve the fuel system pressure.
4. Remove or disconnect the following:
   - Negative battery cable
   - Right exhaust manifold
   - Intake filter housing and mass air sensor
   - Intake manifold
   - Wiring harness from heads and secure aside
   - Fuel line with injectors
   - Idle speed control valve
   - Hoses from supplementary air distributor to tank venting valve and throttle
   - Hoses to solenoid valve
   - Coolant hoses from the block and both return hoses on the heads
   - Brake booster hose
   - Pull rod from throttle actuator
   - Both cylinder head covers
   - All spark plugs
   - Fan clutch with impeller and cowl
   - Thermostat
   - Cable channel and bracket
   - Throttle actuator
   - Hose from supplementary air distributor
   - Knock sensor cable guide from the return line
   - Vacuum hose from pressure regulator
   - Oil lines on both cylinder heads
   - Pulse generators on the end of the cylinder head
5. Hold the camshafts on the hexagon heads provided and remove the banjo bolts securing the signal rings.
6. Rotate the crankshaft in the direction of rotation to the TDC firing position of cylinder 1. Secure the vibration damper in place. This can be done with tool No. 11–2–300.
7. Remove the oil line from the VANOS unit. Fit an air line adapter to the hole and apply 30–115 psi (2–8 bar) of compressed air.
8. Unplug the solenoid and connect tool No. 12–6–050 and 12–6–411 to the VANOS unit. Toggle the buttons 1 and 2 several times.
9. Press and hold down button 1 while rotating the intake camshaft against the direction of travel as far as it will go.
10. Press buttons 3 and 4 several times.
11. Press and hold button 3 while rotating the exhaust camshaft against the direction of rotation as far as it will go.
12. Remove the tool securing the vibration damper and rotate the crankshaft in the direction of rotation one revolution to overlap the TDC position.
13. Loosen the 6 accessible bolts on the cam sprockets two turns.
14. Rotate the crankshaft in the direction of rotation until cylinder 1 is in the TDC firing position. Secure the vibration damper from rotating again.
15. Loosen the 6 remaining bolts on the cam sprockets two turns.
16. Align the camshaft opening to the locating bores in the bearing caps and secure in position. Tool No. 11 7 120 can be used for this.
17. Remove the fasteners securing the

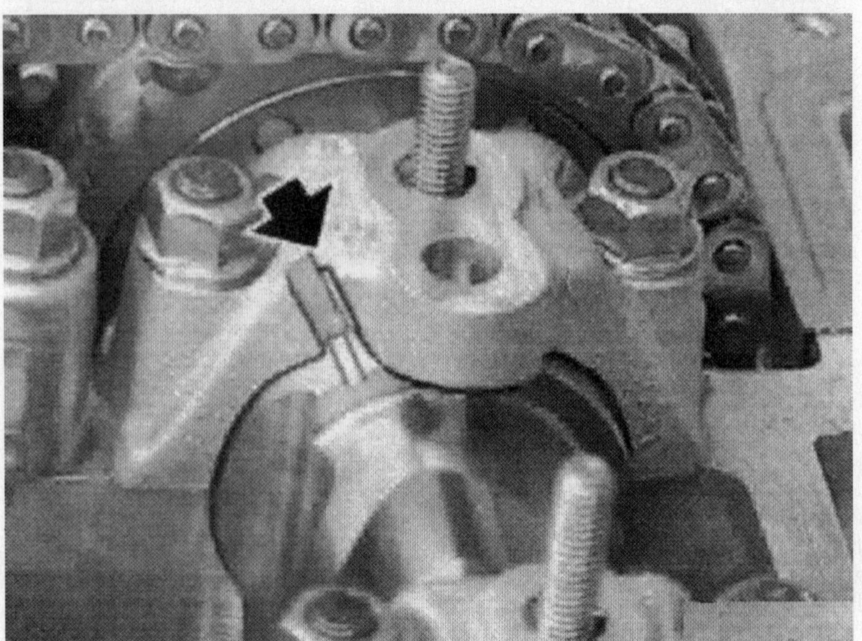

42348-BMWC-G17

The groove in the camshaft must be inside the groove on the first bearing cap

VANOS unit and remove it from the cylinder head. Be sure to remove the compressed air first.

18. Remove the tool securing the vibration damper.

⁂ WARNING

No piston must be in the TDC position when the cylinder head is removed.

➡When the VANOS unit is removed, the chain sprockets of the camshafts are not positively connected to the camshafts. Camshafts on cylinder bank 1 to 4 do not rotate.

19. Rotate the crankshaft against the direction of rotation to 45 degrees BTDC.

20. Remove or disconnect the following:
- Chain tensioning piston
- Left timing case cover.
- The slackened bolts from the spacer ring and remove the ring
- Spline hub with plate spring and supporting ring. Keep note of the direction the plate spring faces
- Intake chain sprocket from the centering sleeve. Keep the chain tensioned and secure from slipping into the lower timing cover
- Cylinder head bolts in sequence from 10 to 1
- Cylinder head

To install:

21. Thoroughly clean all mounting surfaces and check the head for warpage. Take care not to drop any pieces of gasket or debris into the oil or coolant passages. Check the condition of the head locating dowel sleeves and clean out the bolt threads with a tap.

22. Coat the joint between the engine block and timing case cover with Drei Bond 1209.

23. Mount the cylinder head on the block and use new bolts. Do not remove the coating on the head bolts. Apply oil to the threads and washer contact area. Torque following the tightening sequence as follows:

a. Step 1: 22 ft. lbs. (30 Nm).
b. Step 2: 80 degrees.
c. Step 3: 80 degrees.

24. Install the chain sprocket and chain on the centering sleeve. Do not secure it with fasteners. Note that the bores of the chain sprocket do not have to line up with the thread holes yet.

25. Rotate the crankshaft in the direction of rotation to the TDC position again. Secure the vibration damper. Be sure to keep the intake sprocket from slipping off the centering sleeve.

26. Check the position of the chain sprocket. If the threads are not visible through the holes, pull the sprocket from the sleeve (while keeping the chain tensioned), and refit. Fine adjustment can be made by rotating the exhaust side.

➡The plate spring faces towards the spline hub.

27. Install the spline hub with the plate spring and supporting ring. Align the hub so the oil pump is positioned as pictured.

28. Fit the spacer ring and install the bolts snug. Slacken the bolts until the spline hub can be moved.

29. Install the right timing case cover.

30. Replace the seals on the VANOS unit. Coat the seals and gears with oil.

31. Connect tool No. 12–6–050 and 12–6–411 to the VANOS unit. Press buttons 1 and 3. Push the splined shafts by hand into the VANOS unit.

32. Turn the splined hubs of the intake and exhaust camshafts clockwise to the right limit positon.

33. Attach the VANOS unit. Rotate the splined shafts until the spur tooting is engaged. Push the VANOS unit into the gear until the helical cut splines are positioned shortly before meshing with the spline hub.

34. If the gears on the exhaust side cannot be pushed into the splined hub, rotate the hub against the direction of rotation until correctly positioned. Do not rotate any further. Do the same for the intake side.

35. Align the oil pump to the holes in the spline hub. Push the VANOS unit in until the seals just rest against the timing cover. If necessary, realign the oil pump drive to the spline hub.

36. Insert one bolt on each side finger tight.

37. Use tool No. 11 7 200 to secure the timing sprockets. Tighten the six accessible bolts to 7 ft. lbs. (10 Nm). Then slacken all six a quarter turn.

38. Install all the bolts on the VANOS unit. Alternately tighten them in half turn increments evenly until the unit rests against the timing cover.

39. Remove the tool securing the vibration damper. Make sure the timing sprockets are still secured to allow the camshafts to remain locked in position. Carefully rotate the engine against the direction of rotation until there is resistance and the spline hubs are at the stop.

40. Rotate the crankshaft in the direction of rotation until cylinder 1 is at TDC. Secure the vibration damper again.

41. Tighten the six accessible bolts on the timing sprockets to 7 ft. lbs. (10 Nm).

42. Remove tool No. 11–7–200. Remove the tool securing the vibration damper and turn the crankshaft in the direction of rotation on revolution to overlap the TDC position. Hold the sprockets with tool No. 11–7–200, and tighten the six remaining bolts on the timing sprockets to 7 ft. lbs. (10 Nm).

43. Rotate the crankshaft in the direction of rotation until cylinder 1 is at TDC of the firing position. Secure the vibration damper again.

44. Install the signal rings. The intake camshaft ring has one dowel pin. Tighten to 37 ft. lbs. (50 Nm).

45. Remove the oil line from the VANOS unit on cylinder bank 5 to 8. Fit an air line adapter to the hole and apply 30–115 psi (2–8 bar) of compressed air.

46. Unplug the solenoid and connect tool No. 12–6–050 and 12–6–411 to the VANOS unit. Toggle the buttons 4 and 3 several times.

47. Press and hold down button 4 while rotating the intake camshaft against the direction of travel as far as it will go.

48. Press buttons 2 and 1 several times.

49. Press and hold button 2 while rotating the exhaust camshaft against the direction of rotation as far as it will go.

50. Remove the tool securing the vibration damper and rotate the crankshaft in the direction of rotation two revolutions to the cylinder 1 TDC firing position. Secure the vibration damper.

51. Check the camshaft adjustment. The groove in the camshaft must be inside the groove on the first bearing cap. If necessary, adjust the camshaft timing.

52. The balance of installation is the reverse of the removal procedure.

53. Fill the cooling system.

54. Start the engine and check for leaks.

➡It may be necessary to bleed the cooling system a second time after the engine has been started. A rattling noise may be heard in the VANOS unit until oil pressure has built up and air has vented.

N62 Engines

LEFT SIDE

1. Before servicing the vehicle, refer to the Precautions Section.
2. Drain the cooling system.
3. Relieve the fuel system pressure.
4. Remove or disconnect the following:
   - Negative battery cable
   - Left exhaust manifold

- Ignition coils
- Eccentric shaft servo motor
- Cylinder head cover
- Intake manifold
- Timing case cover
- Engine wiring harness and secure aside
- Spark plugs
- Vent hose from cylinder head
- Hose and check valve. Do not remove the guide tube for the dipstick.

5. Rotate the eccentric shaft at the dihedral head to reduce tension on the torsion spring. Pull on the torsion spring with a strap and rotate it out from the roller.
6. Rotate the dihedral head eccentric shaft back to the minimum stroke position.
7. Remove or disconnect the following:
   - Mounting screw and spring mount from locating pin
   - Eccentric shaft sensor
   - Guide rail from cylinder head
   - Screws between cylinder head and timing case cover
   - Slacken cylinder head bolts in sequence one turn from 10 to 1
   - Rotate the eccentric shaft and remove bolt 7
   - Rotate the eccentric shaft back and remove the remaining cylinder head bolts
   - Cylinder head

To install:

8. Thoroughly clean all mounting surfaces and check the head for warpage. Take care not to drop any pieces of gasket or debris into the oil or coolant passages. Check the condition of the head locating dowel sleeves and clean out the bolt threads with a tap.
9. Coat the joint between the engine block and timing case cover with Drei Bond 1209.
10. Mount the cylinder head on the block and use new bolts. Do not remove the coating on the head bolts. Apply oil to the threads and washer contact area. Rotate the eccentric shaft to insert bolt 7. Torque following the tightening sequence as follows:
    a. Step 1: 22 ft. lbs. (30 Nm).
    b. Step 2: 90 degrees.
    c. Step 3: 90 degrees.
11. The balance of installation is the reverse of the removal procedure.
12. Fill the cooling system.
13. Start the engine and check for leaks.

Cylinder head bolt sequence on N62 engines

RIGHT SIDE

1. Before servicing the vehicle, refer to the Precautions Section.
2. Drain the cooling system.
3. Relieve the fuel system pressure.
4. Remove or disconnect the following:
   - Negative battery cable
   - Right exhaust manifold
   - Ignition coils
   - Eccentric shaft servo motor
   - Cylinder head cover
   - Intake manifold
   - Timing case cover
   - Engine wiring harness and secure aside
   - Spark plugs
   - Vent hose from cylinder head
   - Hose and check valve
5. Slacken chain tensioning piston by one turn. Rotate the eccentric shaft at the dihedral head to reduce tension on the torsion spring. Pull on the torsion spring with a strap and rotate it out from the roller.
6. Rotate the dihedral head eccentric shaft back to the minimum stroke position.
7. Remove or disconnect the following:

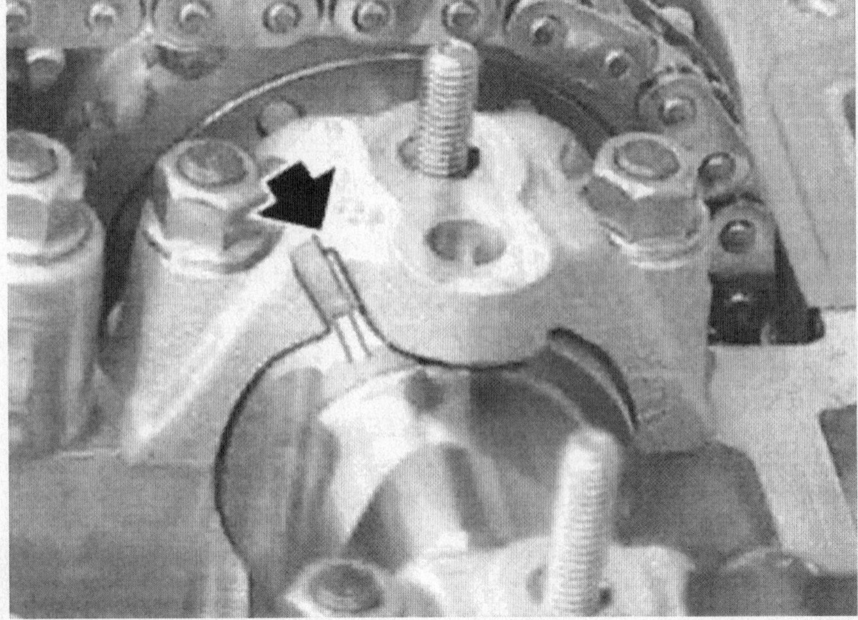

Rotate the eccentric shaft (1), pull on the torsion spring and rotate out from roller (3)

- Mounting screw and spring mount from locating pin
- Eccentric shaft sensor
- Guide rail from cylinder head
- Screws between cylinder head and timing case cover
- Slacken cylinder head bolts in sequence one turn from 10 to 1
- Rotate the eccentric shaft and remove bolt 7
- Rotate the eccentric shaft back and remove the remaining cylinder head bolts
- Cylinder head
- Chain tensioning piston

To install:

8. Place piston on a level surface and compress slowly. Repeat twice.
9. Thoroughly clean all mounting surfaces and check the head for warpage. Take care not to drop any pieces of gasket or debris into the oil or coolant passages. Check the condition of the head locating dowel sleeves and clean out the bolt threads with a tap.
10. Coat the joint between the engine block and timing case cover with Drei Bond 1209.
11. Replace sealing ring on piston. Install piston and tighten unit there is no play.
12. Mount the cylinder head on the block and use new bolts. Do not remove the coating on the head bolts. Apply oil to the threads and washer contact area. Rotate the eccentric shaft to insert bolt 7. Torque following the tightening sequence as follows:
    a. Step 1: 22 ft. lbs. (30 Nm).
    b. Step 2: 90 degrees.
    c. Step 3: 90 degrees.
13. The balance of installation is the reverse of the removal procedure.
14. Fill the cooling system.
15. Start the engine and check for leaks.

N52 Engines

⁂ CAUTION

Aluminum-magnesium material. No steel fasteners may be used due to the threat of electrochemical corrosion. A magnesium crankcase requires aluminum fasteners exclusively. Aluminum fasteners must be replaced each time they are removed. The end faces of aluminum fasteners are painted blue for purposes of identification. Torque specifications and torque angles must be observed for risk of damage.

1. Before servicing the vehicle, refer to the Precautions Section.
2. Drain the cooling system and engine oil.
3. Relieve the fuel system pressure.
4. Remove or disconnect the following:
    - Exhaust system
    - Both exhaust manifolds
    - Intake manifold
    - Coolant hoses from cylinder head
    - Cylinder head cover
    - Inlet and exhaust adjustment unit

⁂ CAUTION

If the timing chain is stored in the gearcase, the crankshaft must not be rotated. Only during assembly can the timing chain be lifted out.

- Timing chain module and fasteners
- Eccentric shaft sensor and fasteners towards front

⁂ CAUTION

After removing, secure magnet wheel in a plastic bag. Magnet wheel must be protected against metal chips.

- Magnet wheel and fastener towards front

5. Pretension eccentric shaft upward. Remove mini stop screw between first and second cylinders
6. Use special tool Torx Socket T60 (No. 115190) to remove M10 fasteners. Use special tool Torx Socket T50 (No. 114420) to remove M9 fasteners.

⁂ CAUTION

Do not set cylinder head down on sealing face, risk of damage to engine valves.

7. Remove cylinder head. Install seal plugs into bores.

⁂ CAUTION

Do not use any metal cutting tools for cleaning.

8. Thoroughly clean all mounting surfaces of gasket remnants and check the head for warpage. Take care not to drop any pieces of gasket or debris into the oil or coolant passages. Check the condition of the head locating dowel.

➡The cylinder head bolts must be replaced.

To install:

Install bolts 1–10 with special tool No. 115190. Install bolts 11–14 with special tool No. 114420.

Torque following the tightening sequence as follows:

a. Step 1:Bolts 1–14: 22 ft. lbs. (30 Nm).
b. Step 2: Bolts 1–14: 90 degrees.
c. Step 3: Bolts 1–10: 90 degrees.
d. Step 4: Bolts 1–14: 42 degrees

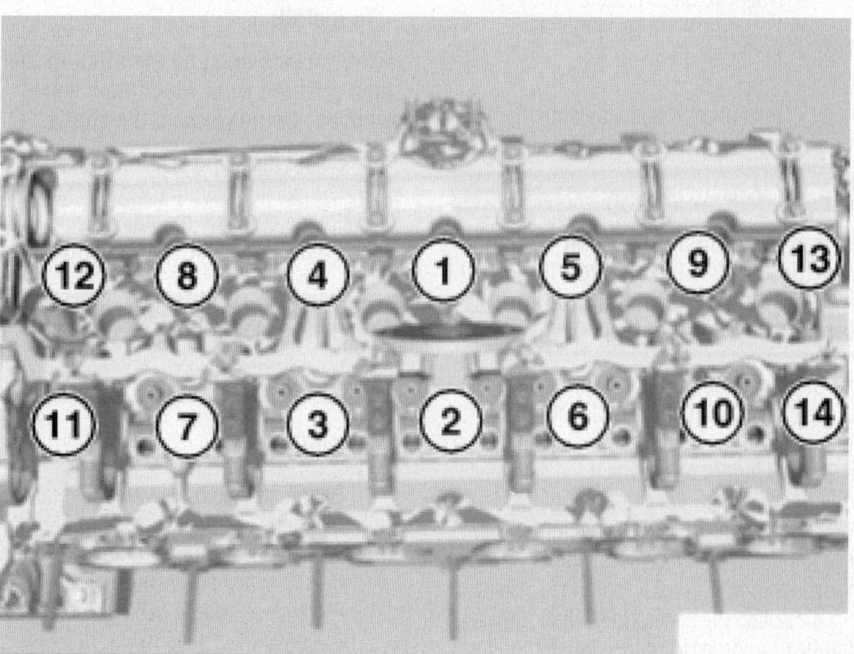

Cylinder head bolt sequence on N52 engines

9. The balance of installation is the reverse of the removal procedure.
10. Fill the cooling system.
11. Start the engine and check for leaks.

Rocker Arms/Shafts

REMOVAL & INSTALLATION

S54 Engines

1. Before servicing the vehicle, refer to the Precautions Section.
2. Remove or disconnect the following:
   - Negative battery cable
   - Cylinder head
   - Raise rocker arms and remove adjustment plates; keep aside in order
   - Spring clip
   - Camshaft sensor
   - Rocker shaft securing screw
   - Rocker shaft

To install:

➡The rockers must be installed in the exact position from which they were removed. Do not mix up the shafts. The intake shaft is marked with a groove between cylinders 5 and 6.

3. Installation is the reverse of removal.

N62 Engines

1. Before servicing the vehicle, refer to the Precautions Section.
2. Remove or disconnect the following:
   - Negative battery cable
   - Eccentric shaft servomotor
   - Ignition coils
   - Spark plugs
   - Inlet and exhaust adjustment units
   - Cylinder head cover

⁂ WARNING

Rotate the camshaft so that when the bearing bracket is removed the intermediate levers do not slip out and damage the camshaft.

3. If working on the left inlet side, rotate intake camshaft against the direction of rotation until the lettering on the camshaft at cylinder 8 points upward and the camshaft lobe is horizontal. For the right inlet side, rotate the camshaft in the direction of rotation until camshaft lobe for cylinder one is horizontal pointing away from the rockers.
4. Rotate the eccentric shaft at the dihedral head in order to reduce tension on torsion spring. Pull on the torsion spring with a strap and feed it out from the roller.
5. Rotate the eccentric shaft back to the minimum stroke position.
6. Remove the bearing cap marked R E1 (for the left inlet side) or L E1 (for the right inlet side). Do not mix up the camshaft bearing covers.
7. Remove the eight nuts securing the bearing bracket assembly. Start from the outside and work to the inside.
8. Clamp special tool No. 11 9 470 in a vice. This will support the bearing bracket assembly.
9. Carefully lift out the bearing bracket assembly. Do not tilt the bracket; secure it to the tool with a nut.
10. Remove the rocker arms. Do not mix them up; they must be returned to their original positions.

To install:

11. Installation is the reverse of the removal procedure, while using the following torque values:
    - M6 fasteners: 88 inch lbs. (10 Nm)
    - M7 fasteners: 10 ft. lbs. (15 Nm)
    - M8 fasteners: 15 ft. lbs. (20 Nm)

N52 Engines

1. Before servicing the vehicle, refer to the Precautions Section.
2. Remove or disconnect the following:
   - Negative battery cable
   - Cylinder head
   - Intermediate lever
   - Exhaust camshaft
   - Rocker shaft
   - Roller cam from HVC element
   - HVC element

To install:

➡The rockers must be installed in the exact position from which they were removed. Do not mix up the shafts.

3. Before installing exhaust camshaft and intermediate lever, make sure roller cam followers are correctly seated.
4. Installation is the reverse of removal.

Intake Manifold

REMOVAL & INSTALLATION

M54 Engines

1. Before servicing the vehicle, refer to the Precautions Section.
2. Remove or disconnect the following:
   - Fuel system pressure
   - Negative battery cable
   - Coolant
   - Throttle cable cover and throttle cable with the rubber holder
   - Vacuum fitting from the brake booster and plug the openings
   - Engine and intake manifold covers, remove the bolt holding the ground strap on the front lifting eye and reinstall the bolt
   - Bolts holding the ignition coil wire harness plug plate. Be careful not to damage the rubber seals. Take off the ignition coil electrical plugs, then remove the plug plate complete with the electrical leads.
   - Cylinder head vent hose and remove the air temperature sensor plug. Remove the tank venting hose and the coolant hoses from the throttle body. Remove the throttle valve switch connector.
   - Idle speed control valve mounted on the manifold, then disconnect the fuel hoses from the feed and return lines
   - Hardware holding the intake manifold to the cylinder head
   - Intake manifold taking care not to drop anything into the exposed ports

To install:

3. Installation is the reverse of the removal procedure, while using the following torque values:
   - 6mm fasteners: 88 inch lbs. (10 Nm)
   - 7mm fasteners: 11 ft. lbs. (15 Nm)
   - 8mm fasteners: 16 ft. lbs. (22 Nm)

S54 Engines

1. Before servicing the vehicle, refer to the Precautions Section.
2. Remove or disconnect the following:
   - Negative battery cable
   - On SMG transmissions, unplug the hydraulic pump relay from the electronics box. Disconnect the line between the expansion tank and hydraulic unit
   - Microfilter assembly
   - Hoses and positive cable lead from the intake manifold
   - Oil dipstick guide tube from the manifold
   - Manifold support fasteners
   - Clamps from the throttle assemblies
   - Intake manifold

To install:

3. Installation is the reverse of the removal procedure.

M62 Engines

1. Before servicing the vehicle, refer to the Precautions Section.

2. Remove or disconnect the following:
   - Fuel system pressure
   - Negative battery cable
   - Center cover from the cylinder head cover
   - Hose clamps on the idle speed control and the throttle valve assembly
   - Plug on the mass air flow sensor
   - Upper section of the air cleaner assembly along with the MAF sensor
   - Right-hand cover from the cylinder head cover
   - Plug for the oil level switch
   - Connectors for the ignition coils
   - Both KS, for cylinders 1 and 2 along with 3 and 4, and the pulse sensor
   - Intake Air Temperature (IAT) sensor, Throttle Valve (TVS) sensor and the Idle Speed Control (ISC) valve
   - Diagnostic connector from the mounting bracket and disconnect the engine wiring connector
   - Ignition coil ground wire, located near the rear engine lifting eye, then the temperature sensor (black) for the temperature gauge, and the temperature sensor (white) for the Digital Motor Electronics (DME)
   - 4 bolts for the intake manifold cover, and remove the holder
   - Throttle cable
   - Left-hand cover from the cylinder head cover
   - Ignition coil electrical connectors
   - Both KS and the camshaft reference sender
   - Coolant expansion tank plug and overflow hose, then remove the 2 mounting bolts, and move the tank aside
   - Oil pressure switch electrical connector and remove the wiring
   - Fasteners for the wiring ducts on the cylinder heads
   - Vacuum hoses on the radiator, and loosen the hose clamp. Pull the vacuum hose off of the brake booster.
   - Tank vapor venting hose from the throttle valve assembly
   - Fuel feed and return lines
   - Hose off of the end cover on the back of the manifold
   - Mounting bolts, and remove the end cover together with the pressure regulating valve straight back to prevent damaging the vent pipe
   - Intake manifold

To install:

3. Installation is the reverse of the removal procedure, while using the following torque values:

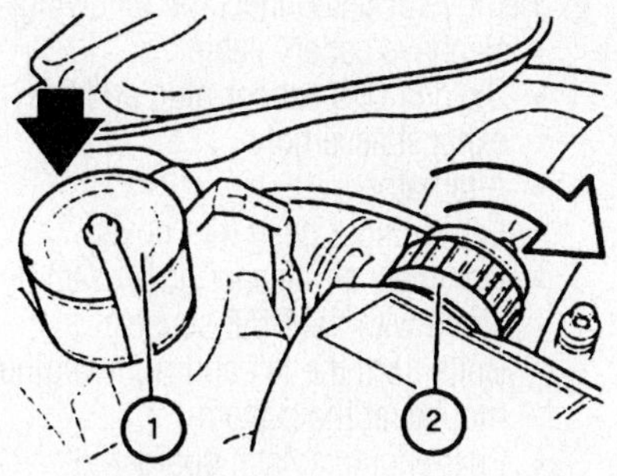

Diagnosis connector (1) and engine connector (2) locations on M62 engines

   - 6mm fasteners: 88 inch lbs. (10 Nm)
   - 7mm fasteners: 11 ft. lbs. (15 Nm)
   - 8mm fasteners: 16 ft. lbs. (22 Nm)

S62 Engines

1. Remove or disconnect the following:
   - Negative battery cable
   - Intake filter housing and mass air flow sensor
   - Left and right air ducts
   - Engine cover
   - Intake funnels

➡Note that the funnels for cylinders 1 and 5 are different from the others; mark them and set aside.

2. Slacken the bolts one turn starting from the outside and work towards the center. Start with the outer two rows, then do the inner rows near the center of the manifold. Once all are loosened one turn, remove all bolts following the same sequence. Disconnect any hoses and remove the manifold.

To install:

3. Installation is the reverse of the removal procedure. Tighten all manifold fasteners in the reverse procedure of removal, first to (5 Nm) then to 88 inch lbs. (10 Nm).

N62 Engines

1. Remove or disconnect the following:
   - Negative battery cable
   - Engine covers
   - Intake hose
   - Injection pipe
   - Differential pressure sensor and servomotor connectors
   - Vent hose and cable holder
   - Intake manifold fasteners and the intake manifold

To install:

2. Installation is the reverse of the removal procedure, while using the following torque values:
   - 6mm fasteners: 88 inch lbs. (10 Nm)
   - 7mm fasteners: 11 ft. lbs. (15 Nm)
   - 8mm fasteners: 16 ft. lbs. (22 Nm)

N52 Engines

1. Before servicing the vehicle, refer to the Precautions Section.
2. Remove or disconnect the following:
   - Negative battery cable
   - Tension strut

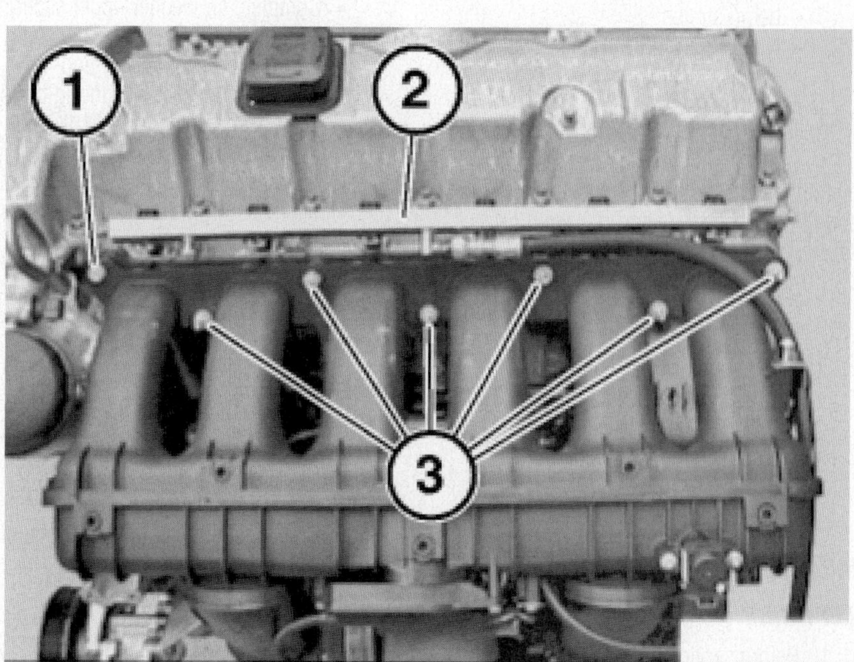

Intake manifold fastener locations (1, 3) and fuel rail (2) on N52 engines

- Intake filter housing and mass air flow sensor
- Engine covers
- Vent hose and cable holder
- Electrical connectors and engine wiring harness, set aside and secure.
- Oil pressure switch plug connector

➡**Do not detach fuel line.**

- Fuel rail, set aside and secure
- Intake manifold fasteners
- Tank vent line behind throttle valve assembly
- Intake manifold

To install:

3. Replace all seals and gaskets.
4. Installation is the reverse of the removal procedure, while using the following torque values:
   - 7mm fasteners: 11 ft. lbs. (15 Nm)

Exhaust Manifold

REMOVAL & INSTALLATION

M54 and S54 Engines

1. Before servicing the vehicle, refer to the Precautions Section.
2. Remove or disconnect the following:
   - Negative battery terminal
   - Mounting nuts on each flange connection and separate the exhaust pipes from the manifold. Support the exhaust system. Be sure the Oxygen (O_2S) sensor wire is not being stretched.
   - Nuts securing the manifold to the cylinder head
   - Manifolds

To install:

3. Clean the mounting surfaces on the manifolds and the cylinder head. Check the condition of the studs and replace if necessary.
   - New exhaust manifold gaskets with the graphite side towards the cylinder head and install the manifolds. Nuts, with exhaust studs coated with an anti-seize compound: 14 ft. lbs. (19 Nm).
   - Exhaust pipe to the manifolds using new mounting nuts
   - Negative battery terminal

M62 Engine

LEFT SIDE

1. Before servicing the vehicle, refer to the Precautions Section.
2. Remove or disconnect the following:
   - Negative battery cable
   - Oxygen O_2S sensor plug and the exhaust assembly
   - Alternator
   - Left cylinder head trim cover
   - Complete air cleaner upper section along with the MAF sensor
   - Bolts from the left and right engine mounts at the bottom
   - Rear engine splash guard
   - Bolts of the center of gravity mount to front axle carrier
   - Left heat shields on the front axle carrier
3. Support the engine and ensure clearance between the engine and the firewall.
4. Remove the manifold fasteners and remove the manifold downward.

To install:

5. Remove the old gasket off of the cylinder head and exhaust manifold and replace the gasket. The gasket beads face the exhaust manifolds.
6. Install or connect the following:
   - Exhaust manifold, with upper row of the exhaust fasteners bolts coated with a locking agent. Mounting fasteners: 16 ft. lbs. (22 Nm).
   - Left heat shields and the center of gravity mount-to-front axle carrier bolts
   - Rear engine splash guard
   - Bolts for the left and right engine mounts at the bottom. 10mm bolts: 31 ft. lbs. (42 Nm). 8mm bolts: 16 ft. lbs. (22 Nm).
   - Complete air cleaner upper section along with the mass air flow sensor
   - Left cylinder head cover and replace the gasket. Mounting bolts, in a crisscross pattern: 11 ft. lbs. (15 Nm)
   - Alternator
   - O_2S sensor plug
   - Exhaust assembly
   - Negative battery cable

RIGHT SIDE

1. Before servicing the vehicle, refer to the Precautions Section.
2. Remove or disconnect the following:
   - Negative battery cable
   - Oxygen (O_2S) sensor plug and remove the exhaust assembly
   - Right heat shields on the front axle carrier
   - Rear engine splash guard
   - Windshield washer fluid tank

➡**Remove the manifold for cylinders number 2 and 4 first.**

   - Manifold fasteners and remove the manifold upward

To install:

3. Remove the old gasket from the cylinder head and exhaust manifold and replace the gasket. The gasket beads face the exhaust manifolds.
4. Install or connect the following:
   - Exhaust manifold, with upper row of the exhaust fasteners coated with a locking agent. Mounting fasteners: 16 ft. lbs. (22 Nm).
   - Washer fluid tank
   - Rear engine splash guard
   - Right heat shields
   - O_2S sensor plug and install the exhaust assembly
   - Negative battery cable

N52 Engines

1. Before servicing the vehicle, refer to the Precautions Section.
2. Remove or disconnect the following:
   - Negative battery cable
   - Engine splash guard
   - Reinforcement plate
   - Oxygen (O_2S) sensor plug from cylinders number 4 and 6 and remove the exhaust assembly
   - Manifold for cylinders number 1 and 3 downward
   - Manifold for cylinders number 4 and 6 downward

To install:

3. Remove the old gasket from the cylinder head and exhaust manifold and replace the gasket. The gasket beads face the exhaust manifolds.
4. Installation is reverse of removal.

Camshaft and Valve Lifters

REMOVAL & INSTALLATION

M54 and S54 Engines

1. Before servicing the vehicle, refer to the Precautions Section.
2. Remove or disconnect the following:
   - Fuel pressure
   - Negative battery cable
   - Cylinder head, if necessary. If the fresh air intake shroud on the firewall is removable, removing the shroud should allow enough room to remove the camshafts. If the body of the vehicle interferes with the camshaft removal, removing the cylinder head is necessary

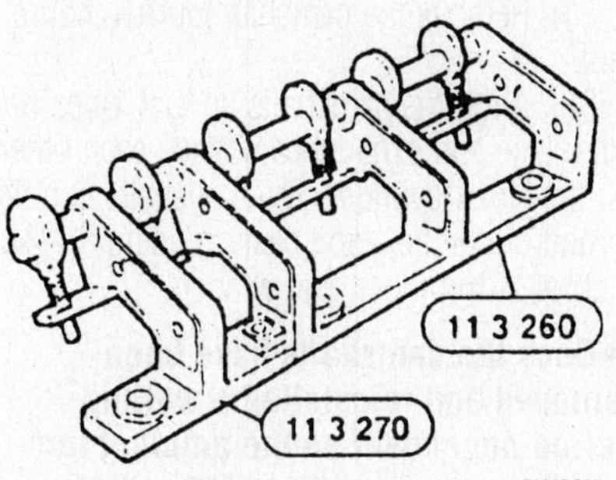

Camshaft removal tools 11–3–260 and 11–3–270 used to hold the camshaft journals in place during camshaft removal

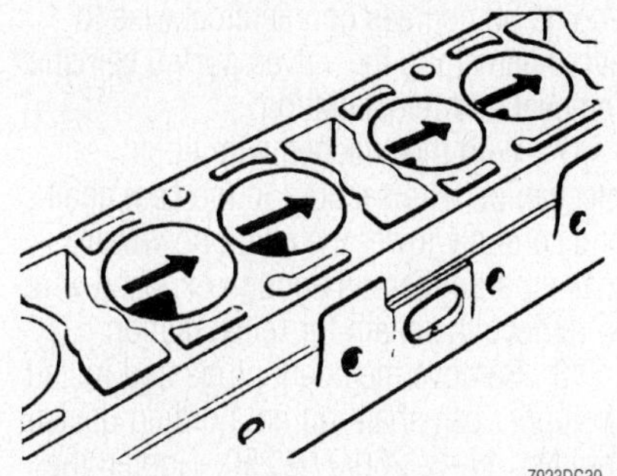

Check the bore surfaces of the valve clearance compensators for scoring

The lower camshaft journal and tappet bore assemblies are marked A for exhaust and E for intake

Camshaft bearing journal bolt locations

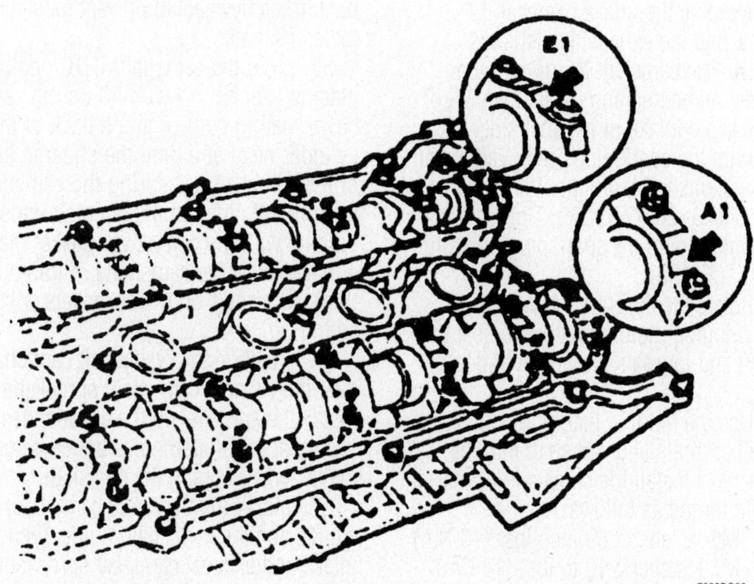

Bearing cap ID markings. The intake camshaft journals are indicated by the letter E, the exhaust journals are indicated by the letter A

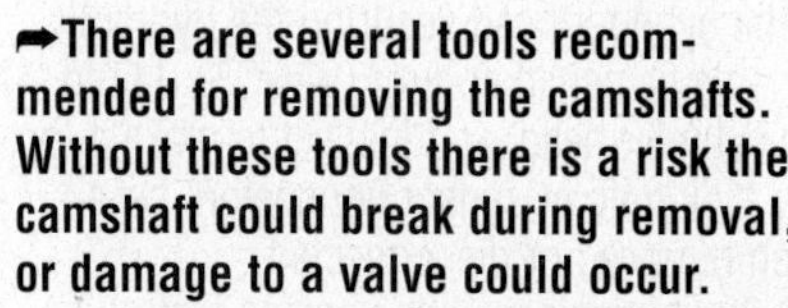

➡There are several tools recommended for removing the camshafts. Without these tools there is a risk the camshaft could break during removal, or damage to a valve could occur.

- Rear hood seal, then remove the fasteners securing the sealed plastic housing for the engine wiring harness from the passenger compartment fresh air intake shroud.
- Fasteners securing the fresh air intake shroud to the firewall, and remove the shroud
- If equipped, plastic trim covers from the top of the engine
- Wire connectors for the ignition coils and cylinder head sensors and remove the coils, then remove the cylinder head cover

3. Rotate the engine in the direction of rotation to TDC for cylinder number one. Cylinder number 1 will be at Top Dead center (TDC) when the intake and exhaust camshaft peaks for cylinder number 1 face each other

4. Lock the engine in the TDC position by placing the holding dowel tool No. 11–2–300 through the machined hole in the engine block, just inside of the transmission bell housing mounting tab located on the left lower portion of the engine block. Slide the locating dowel through the machined hole in the block and into the machined hole in the flywheel to prevent movement of the crankshaft.

5. The camshafts are held in the TDC position by placing tool No. 11–3–240 on the valve cover mating surface at the back of the cylinder head and onto the squared ends of the camshafts, securing the camshafts such that 2 sides of the squared ends are parallel with the cam cover gasket mating surface. With the camshafts in this position, the arrows on the sprockets will be facing up.

6. Remove or disconnect the following:
   - Valve cover mounting studs
   - 2 hex plugs at the front of the cylinder head to access the exhaust camshaft sprocket mounting bolts, then loosen the exhaust cam sprocket bolts 2 turns

7. Press down on the secondary cam chain tensioner between the 2 camshaft sprockets and install tool No. 11–3–292 through the back side of the tensioner housing to hold the tensioner down. A similar sized and suitably hardened drill bit can be substituted for tool No. 11–3–292.

8. Remove the fasteners from the front of the cylinder head securing the hydraulic variable camshaft control (VANOS) unit to the cylinder head, and inspect to make sure any hydraulic or sensor connectors have been removed or disconnected.

9. On engines with a spring plate installed on the intake camshaft, place tool No. 11–5–490 onto the exhaust camshaft sprocket and carefully rotate the sprocket clockwise to allow the helical gear of the VANOS unit to release the intake camshaft and to allow the VANOS unit to be pulled away from the front of the cylinder head.

10. If tool No. 11–5–490 is not available, move the camshaft sprockets to release the VANOS by using a suitable drift and soft faced mallet and lightly tapping on a sprocket tooth of the intake cam sprocket to rotate both cam sprockets clockwise, while alternately pulling on the VANOS unit to release it. This procedure may need to be repeated several times to fully release the VANOS unit, and must be performed very carefully, in such a manner to not distort or damage the teeth of the cam sprocket.

11. With the VANOS assembly removed, remove the intake and exhaust camshaft sprockets, the hydraulic cam chain tensioner, and the cam chain guide.

12. From the side of the right front area of the engine, remove the cap nut for the cam chain tensioner for the cam chain that runs between the crankshaft and the exhaust camshaft sprocket. Use care when removing the tensioner cap nut as the cam tensioner spring applies pressure to the cap nut.

13. Remove the exhaust cam tensioner, and then release the sprocket from the cam chain.

14. Attach a wire tie or mechanic's wire to the cam chain and temporarily secure the chain fully extended.

15. Remove the dowel from the engine and flywheel locking the engine in the TDC position.

16. While holding the crankshaft to exhaust camshaft cam chain, rotate the engine 30 degrees counterclockwise to avoid damaging the valves during camshaft removal and reinstallation.

17. With the attached wire tie or mechanic's wire secured to the cam chain, and carefully lower the chain downward making sure there is enough exposed wire to retrieve the chain for reinstallation.

18. Remove the spark plugs and install the upper camshaft journal holding device, tool No. 11–3–260/270/250. Tighten the hold down bolts in the spark plug bores to 17 ft. lbs. (23 Nm).

19. Apply a load to the bearing caps by rotating the eccentric shaft. This relieves the tension on the bearing cap bolts. Loosen and remove the bearing cap bolts.

20. Slowly and carefully rotate the eccentric of the cam cap holding fixture to release the camshaft.

21. Remove and rotate the camshaft fixture tool 180 degrees and repeat the above procedures to remove the other camshaft.

22. Remove the camshafts and the bearing caps. Note that the intake camshaft is marked **E** and the exhaust camshaft is marked **A**. The camshaft bearing are consecutively numbered and lettered with **A** or **E** to designate intake or exhaust side.

23. Hold the valve lash compensators in place using tool 11–3–250 and remove the bearing plate along with the valve plungers.

To install:

24. Apply fresh engine oil to the camshaft journals and lobes.

25. Place the camshaft in cam journals of the cylinder head.

26. Place the upper camshaft journal bearing caps on the camshaft in the correct order. Note that the intake camshaft is marked **E** and the exhaust camshaft is marked **A**. The camshaft bearing are consecutively numbered and lettered with **A** or **E** to designate intake or exhaust side.

27. With the spark plugs removed install the upper camshaft journal holding device, tool No. 11–3–260/270/250. Tighten the hold down bolts in the spark plug bores to 17 ft. lbs. (23 Nm).

28. Rotate the eccentric shaft of the camshaft journal holding fixture to seat the camshaft and journals into the cylinder head.

29. Use the holding fixture to secure the camshaft journals and camshaft into the cylinder head Install the camshaft journal bolts and torque as follows:
   - M6 fasteners: 89 inch lbs. (10 Nm)
   - M7 fasteners: 10 ft. lbs. (14 Nm)
   - M8 fasteners: 14 ft. lbs. (19 Nm)

30. Repeat the above procedures to install the second camshaft.

31. Remove the camshaft journal fixture tool.

32. Align the camshafts so that lobes of the intake and exhaust cams face each other for the No. 1 cylinder. The camshafts can be turned on the hexagon casting using a 1$\frac{1}{16}$ inch or 27mm open end wrench.

➡Once the camshafts have been removed and reinstalled a waiting period dependent on the ambient temperature is necessary before rotating the engine. At room temperature wait 4 minutes to allow the lifters to compress fully. At temperatures down to 50°F (10°C) wait 11 minutes. At temperatures lower than 50°F (10°C) wait 30 minutes. This is to prevent contact between the valves and the piston tops. The engine may not be cranked under the same condition for a period of 10 minutes at room temperature; 30 minutes for temperatures down to 50°F (10°C); 75 minutes for temperatures below 50°F (10°C).

33. With the attached wire secured to the crankshaft cam chain, carefully raise the chain upward and apply a light tension to the chain, then while holding the chain, carefully rotate the engine clockwise to TDC on cylinder number one.

34. Lock the engine in the TDC position by placing the holding dowel tool No. 11–2–300 through the machined hole in the engine block, just inside of the transmission bell housing mounting tab located on the left lower portion of the engine block. Slide the locating dowel through the machined hole in the block and into the machined hole in the flywheel to prevent movement of the crankshaft.

35. Hold the camshafts TDC position by placing tool No. 11–3–240 on the valve cover mating surface at the back of the cylinder head and onto the squared ends of the camshafts, securing the camshafts such that 2 sides of the squared ends are parallel with the cam cover gasket mating surface. With the camshafts in this position, the arrows on the sprockets will be facing up.

36. Position the crankshaft cam chain over the exhaust cam chain sprocket and install the sprocket on the exhaust cam such that the slotted holes are centered with the fastener bores in the camshaft.

37. Install the hydraulic cam chain tensioner, and the cam chain guide, then the intake and exhaust camshaft sprockets along with the cam chain and remove the tool used to hold the hydraulic cam chain tensioner collapsed.

38. Install the crankshaft cam chain tensioner assembly, spring and cap nut.

39. Apply sealant to the upper corners of the cylinder head where the VANOS housing mounts and install a new VANOS housing gasket.

40. Press the helical gear of the VANOS assembly toward the housing and install the VANOS assembly. To do so on engines with a spring plate installed on the intake camshaft, place tool No. 11–5–490 onto the exhaust camshaft sprocket and carefully rotate the sprocket counterclockwise to allow the helical gear of the VANOS unit to thread into the intake camshaft and to allow the VANOS unit to be pulled into the front of the cylinder head.

41. If tool No. 11–5–490 is not available, move the camshaft sprockets to install the VANOS by using a suitable drift and soft faced mallet and lightly tapping on a sprocket tooth of the exhaust cam sprocket to rotate both cam sprockets counterclockwise, while alternately pressing on the VANOS unit to install it. This procedure may need to be repeated several times to fully install the VANOS unit, and must be performed very carefully, in such a manner to not distort or damage the teeth of the cam sprocket.

➡Make sure when assembling the VANOS unit is able to rest on the front of the cylinder head without being forced or without binding. If the VANOS unit does not fully seat it may be necessary to reposition the camshaft sprockets such that the slots in the camshaft sprockets allow enough movement of the sprockets for the helical gear of the VANOS assembly to be fully seated during assembly.

42. Tighten the VANOS unit fastener and then torque the camshaft sprocket bolts of both the intake and exhaust cams to 16 ft. lbs. (22 Nm).

43. Remove the crankshaft TDC holding dowel tool No. 11–3–240 from the engine, and remove the camshaft TDC positioning tool No. 11–3–240 from the valve cover mating surface at the back of the cylinder head.

44. Slowly and carefully rotate the engine clockwise 4 complete revolutions bringing cylinder number 1 to TDC. If the engine binds for any reason stop immediately to evaluate and rectify the cause of the binding.

45. With cylinder number 1 at TDC slide the camshaft TDC positioning tool No. 11–3–240 over the ends of the cams and onto the valve cover mating surface. If the tool slides over the cams and is flush with the mating surface, the camshafts are properly timed. If the tool does not slide easily over the ends of the cams, or if the tool is not flush with the valve cover mating surface, the camshaft timing must be repeated until the tool fits squarely.

46. The balance of the assembly is in reverse order of disassembly.

47. Check all fluid levels as necessary.

48. Connect the negative battery cable.

M62 Engines

LEFT CAMSHAFT (CYLINDER BANK 5–8)

1. Before servicing the vehicle, refer to the Precautions Section.

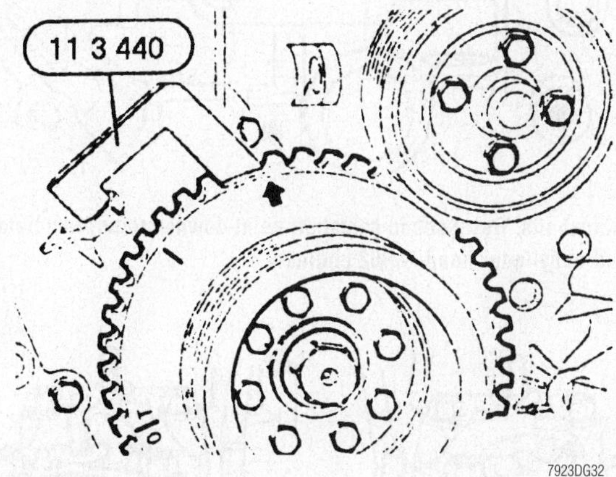

Gap in increment gear must fit in tool No. 11–3–440—M62 engines

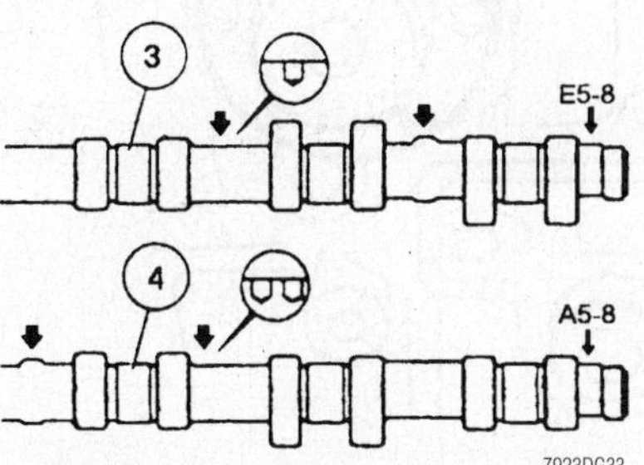

Left side camshaft identification (cylinder bank 5 to 8): hex head (3) on intake camshaft between cylinders 7 and 8, hex head (4) on exhaust camshaft between cylinders 5 and 6—M62 engines

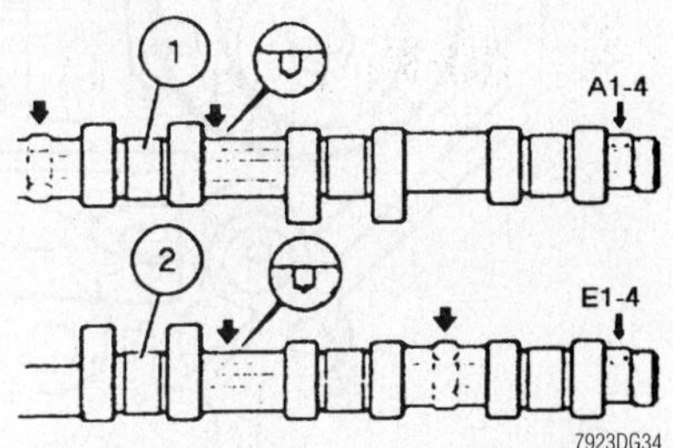

Right side camshaft identification (cylinder bank 1 to 4): hex-head (2) on intake camshaft between cylinders 3 and 4, hex-head (1) on exhaust camshaft between cylinders 1 and 2—M62 engines

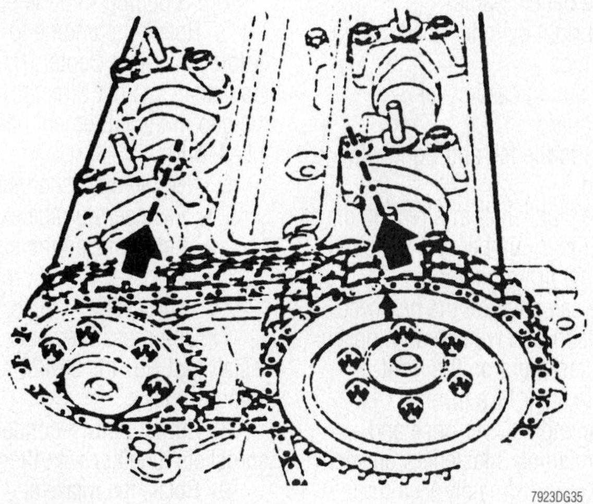

Camshaft positioning—M62 engines

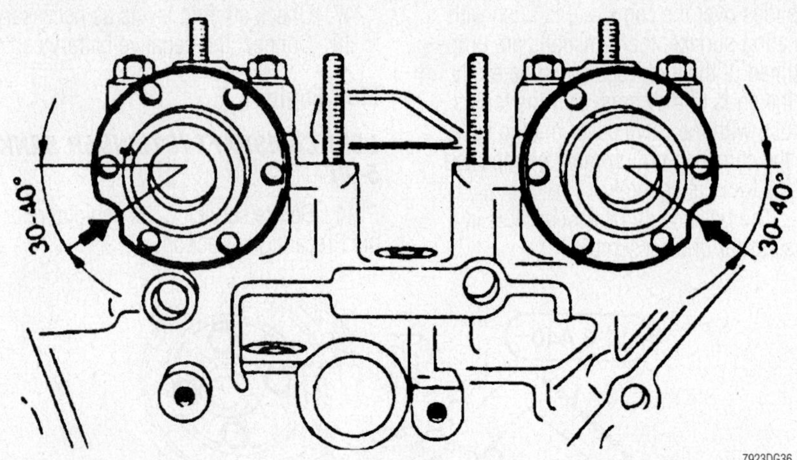

Install left-hand camshafts: Recesses in camshaft point downward approximately 30–40 degrees from plane of cylinder head—M62 engines

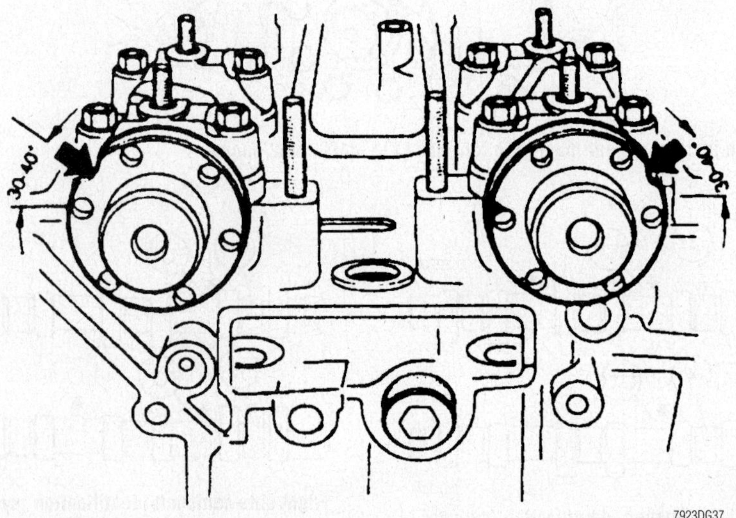

Install right-hand camshafts: Recesses in camshaft point upward approximately 30–40 degrees from plane of cylinder head—M62 engines

2. Remove or disconnect the following:

- Negative battery cable
- Left and right cylinder head covers
- Spark plugs
- Top left timing case cover
- Splash guard
- Oil lines to the left and right cylinder head

3. Rotate the crankshaft in direction of rotation until cylinder number 1 is in Top Dead Center (TDC) firing position.

4. Brace the camshaft on the hex head with a suitable open-end wrench and loosen the 3 accessible fasteners on the right sprocket approximately ½ a turn.

a. Turn the engine over once and loosen the remaining 3 fasteners on each right sprocket approximately ½ a turn.

b. Remove the primary sprocket from the left-hand intake camshaft (cylinder bank 5–8). Secure the chain to prevent it from dropping into the engine.

5. Rotate the engine to 45 degrees Before Top Dead Center (BTDC). Rotate the crankshaft against direction of rotation until the gap in the increment gear fits in tool No. 11–3–440.

6. Remove or disconnect the following:

- Fasteners on the exhaust camshaft
- Fasteners on the exhaust camshaft sprocket. Do not remove the sprocket.

7. Compress the chain tensioner and install tool No. 11–3–420 to lock the tensioner in place.

8. Lift off both secondary camshaft sprockets together with the chain.

9. Rotate the intake and exhaust camshafts to the installed position as follows:

a. Using tool No. 11–3–430, rotate the camshafts until the recess in both camshaft flange points approximately 30–40 degrees downward from the plane of the cylinder head.

b. Check the installed position by installing tool No. 11–2–430 to the camshafts. The cylinder designation of the tool must point upward.

10. Loosen the both camshaft bearing caps uniformly from outside to inside ½ turn.

11. Remove or disconnect the following:

- All bearing caps. Label each bearing cap to facilitate reassembly and place aside.
- Camshafts noting their locations
- Hydraulic valve lifters. To remove, use tool No. 11–3–250 to pull them out of the cylinder head. Be sure that no damage occurs to the guides in the head. Inspect the bearing surfaces of the tappets for wear and scoring.

To install:

12. If the lifters were removed, install them with tool No. 11–32–250.

13. Lubricate and install the camshafts in their correct position.

➡The intake camshaft will have a hexagon between cylinders 7 and 8. The exhaust camshaft will have the hexagon between cylinders 5 and 6.

14. Rotate the intake and exhaust camshafts to the installed position as follows:

a. Using tool No. 11–3–430, rotate the camshafts until the recess in both camshaft flange points approximately 30–40 degrees downward from the plane of the cylinder head.

b. Check the installed position by installing tool No. 11–2–430 to the camshafts. The cylinder designation of the tool must point upward.

15. Install the bearing caps, tightened from outside to inside in 1/2 turn increments to 9–13 ft. lbs. (12–17 Nm).

➡Do not confuse camshaft bearing caps of cylinders No. 1–4 and 5–8. The exhaust camshaft bearing caps are marked A1 through A5. The intake camshaft bearing caps are marked E1 through E5 from intake end.

16. Fit tool No. 11–3–430 to the camshaft. Rotate the camshaft until the marker bores face upward.

17. Install tool Nos. 11–2–442/446 to the camshaft on cylinder bank 5–8.

18. Install tool Nos. 11–2–441/445 to the camshaft on cylinder bank 1–4.

19. Using a suitable open-end wrench,

align all camshafts in such a way that the tools fit on the cylinder heads without any gaps.

20. Fit tool Nos. 11–2–443 to tool Nos. 11–2–441/442/445/446 and secure them with tool No. 11–2–444 using the spark plug threads.

21. Install:
- Secondary sprockets together with chain to the camshafts on cylinder bank 5–8
- Fasteners on the exhaust camshaft sprocket and tighten

22. Remove the tool used to lock the chain tensioner in position.

23. Rotate the engine from 45 degrees BTDC in direction of rotation as far as TDC. Install tool No. 11–2–300 at the flywheel to lock the crankshaft in TDC position.

24. Assemble the primary sprocket and chain to the intake camshaft with the arrow pointing upward (in cylinder axis) and the long bores centrally aligned. Install the fasteners snuggly.

25. Install tool No. 11–3–390 in the right timing case cover and with a suitable torque wrench, tension the tool to 11 inch lbs. (1.2 Nm).

26. Tighten the sprockets to 11 ft. lbs. (15 Nm) in the following order:
- All fasteners on the left exhaust camshaft
- The 3 fasteners in the right exhaust camshaft
- All fasteners on the left intake camshaft
- The 3 fasteners in the right intake camshaft

27. Remove:
- Tool Nos. 11–2–444/443/441/445 or their equivalents
- Tool Nos. 11–2–444/443/442/446 or their equivalents
- Tool No. 11–2–300 or the equivalent used to locked the crankshaft in TDC position

28. Turn the engine over once.

29. Tighten the remaining 3 fasteners on right exhaust camshaft and remaining 3 fasteners on the right intake camshaft to 11 ft. lbs. (15 Nm).

30. Relieve the load and remove tool No. 11–3–390 from the right timing case cover.

31. The balance of installation is the reverse of the removal procedure.

32. Check and top off all fluid levels as necessary.

RIGHT CAMSHAFT (CYLINDER BANK 1–4)

1. Before servicing the vehicle, refer to the Precautions Section.

2. Remove or disconnect the following:
- Negative battery cable
- Left and right cylinder head covers
- Spark plugs
- Fan assembly
- Top right timing case cover
- Splash guard
- Oil lines to the left and right cylinder head

3. Rotate the crankshaft in direction of rotation until the first cylinder is in Top Dead Center (TDC) firing position.

4. Brace the camshaft on the hex head with a suitable open-end wrench and loosen the 3 accessible fasteners on each left sprocket approximately ½ a turn.

5. Turn the engine over once and loosen the remaining 3 fasteners on each left sprocket approximately ½ turn.

6. Remove the primary sprocket from the right-hand intake camshaft (cylinder bank 1–4). Secure the chain to prevent it from dropping.

7. Rotate the engine to 45 degrees Before Top Dead Center (BTDC) setting position. Rotate the crankshaft against direction of rotation until the gap in the increment gear fits in the tool No. 11–3–440.

8. Remove the fasteners on the exhaust camshaft sprocket. Do not remove the sprocket.

9. Compress the chain tensioner and install tool No. 11–3–420 to lock the tensioner in place.

10. Lift off both secondary camshaft sprockets together with the chain.

11. Rotate the intake and exhaust camshafts to the installed position.

12. Using tool No. 11–3–430, rotate the camshafts until the recess in both camshaft flange points approximately 30–40 degrees upward from the plane of the cylinder head.

13. Check the installed position by installing tool No. 11–2–430 to the camshafts. The cylinder designation of the tool must point upward.

14. Loosen the both camshaft bearing caps uniformly from outside to inside ½ turn.

15. Remove or disconnect the following:
- Bearing journal caps. Label each bearing cap to facilitate reassembly and position aside.
- Camshafts noting their locations
- Hydraulic valve lifters. To remove, use tool No. 11–3–250 to pull them out of the cylinder head. Be sure that no damage occurs to the guides in the head. Inspect the bearing surfaces of the tappets for wear and scoring.

To install:

16. If the tappets were removed, lubricate and install them with tool No. 11–32–250.

17. Lubricate and install the camshafts in their correct position.

➡The intake camshaft will have a hexagon between cylinders 3 and 4. The exhaust camshaft will have the hexagon between cylinders 1 and 2.

18. Rotate the intake and exhaust camshafts to the installed position and perform the following:
   a. Using tool No. 11–3–430, rotate the camshafts until the recess in both camshaft flange points approximately 30–40 degrees upward from the plane of the cylinder head.
   b. Check the installed position by installing tool No. 11–2–430 to the camshafts. The cylinder designation of the tool must point upward.

19. Install the bearing caps. Tighten the bearing caps from outside to inside in ½ turn increments. Tighten the bolts to 10–13 ft. lbs. (13–17 Nm).

➡Do not confuse camshaft bearing caps of cylinders No. 1–4 and 5–8. The exhaust camshaft bearing caps are marked A1 through A5 counting from intake side. The intake camshaft bearing caps are marked E1 through E5 counting from the intake end.

20. Fit tool No. 11–3–430 to the camshaft. Rotate the camshaft until the marker bores face upward, then proceed as follows:
   a. Install tool Nos. 11–2–442/446 to the camshaft on cylinder bank 5–8.
   b. Install tool Nos. 11–2–441/445 to the camshaft on cylinder bank 1–4.
   c. Using a suitable open-end wrench, align all camshafts in such a way that the tools fit on the cylinder heads without any gaps.
   d. Fit tool Nos. 11–2–443 to tool Nos. 11–2–441/442/445/446 and secure them with tool No. 11–2–444 using the spark plug threads.

21. Install:
- Secondary sprockets together with chain to the camshafts on cylinder bank 1–4
- Fasteners on the exhaust camshaft sprocket and tighten

22. Remove the tool used to lock the chain tensioner in position.

23. Rotate the engine from 45 degrees BTDC in direction of rotation as far as TDC setting. Install tool No. 11–2–300 or

equivalent at the flywheel to lock the crankshaft in TDC position.

24. Assemble the primary sprocket and chain with sensor pin to the intake camshaft with the arrow pointing upward (in cylinder axis) and the long bores centrally aligned. Install the fasteners.

25. Install tool No. 11–2–400 to the right cylinder head (cylinder bank 1–4). Install tool No. 11–3–390 to tool No. 11–2–400. Using a suitable torque wrench, tension the tool to 11.5 inch lbs. (1.3 Nm).

26. Tighten the sprockets to 11 ft. lbs. (15 Nm) in the following order:
- 3 fasteners on the left exhaust camshaft
- All fasteners on the right exhaust camshaft
- The 3 fasteners on the left intake camshaft
- All fasteners in the right intake camshaft

27. Remove:
- Tool Nos. 11–2–444/443/441/445 or the equivalent
- Tool Nos. 11–2–444/443/442/446 or the equivalent
- Tool No. 11–2–300 used to locked the crankshaft in TDC position

28. Turn the engine over once.

29. Tighten the remaining 3 fasteners on left exhaust camshaft and remaining 3 fasteners on left intake camshaft to 11 ft. lbs. (15 Nm).

30. Relieve the load and remove the tool Nos. 11–3–390 and 11–2–400 or their equivalents.

31. Install the remaining components in the reverse order of removal.

32. Check and top off the fluid levels as necessary.

Valve Lash

ADJUSTMENT

All engines are equipped with hydraulic valve lash adjusters. No adjustments are possible.

Oil Pan

REMOVAL & INSTALLATION

M54 and S54 Engines

3 SERIES

1. Before servicing the vehicle, refer to the Precautions Section.

2. Remove or disconnect the following:
- Negative battery cable
- Front lower splash guard, if necessary
- Oil
- Holding bolt for the oil dipstick guide pipe and remove the clamp. Pull the guide tube free of the pan.
- Electrical terminal from the oil level sending unit
- Power steering box, leaving the hoses attached from the front axle carrier

3. Support the engine, then remove the engine mount fasteners.

4. Support the crossmember and remove the fasteners securing the front cross member to the vehicle's chassis. Carefully lower the jack.

5. Lift the engine slightly.

6. Remove or disconnect the following:
- Flywheel cover
- Oil pan bolts and lower the oil pan away from the engine and over the cross member

To install:

7. Before installing the oil pan, clean the gasket surfaces and install a new gasket on the oil pan.

8. Coat the joints on the ends of the front engine cover with a universal sealing compound.

9. Install or connect the following:
- Oil pan and tighten the fasteners
- Flywheel cover

10. Raise the crossmember into position and install the mounting bolts.

11. Slowly lower then engine onto the engine mounts and install the engine mount fasteners.
- Power steering box
- Dipstick guide tube using a new base seal and tighten the holding bolt
- Electrical wiring harness to oil sending unit
- Front lower splash guard, if removed
- Oil pan drain plug, then fill the engine with oil
- Negative battery cable

12. Start the engine and check that oil pressure is present; if the oil pressure lamp does not turn off within 5–7 seconds of starting the engine, turn the engine **OFF**. Check for any oil leaks.

5 SERIES

1. Before servicing the vehicle, refer to the Precautions Section.

2. Install or connect the following:
- Negative battery terminal
- Engine oil

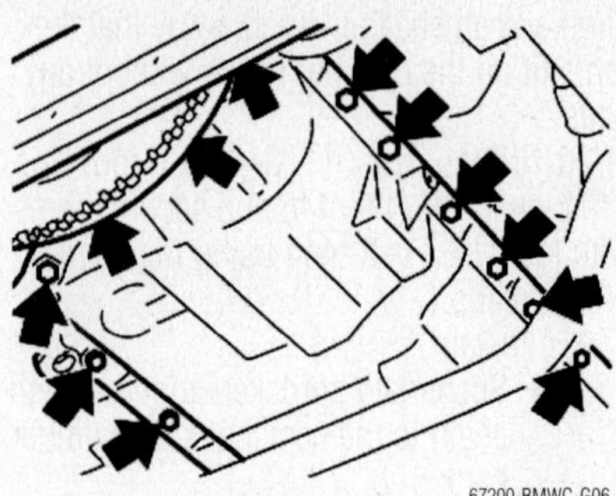

5 series oil pan fasteners

- Intake air filter assembly
- Automatic transmission lines
- Holding bolt for the oil dipstick guide pipe and remove the clamp. Pull the guide tube free of the pan.
- All oil pan bolts and remove the pan. Raise the engine slightly if needed for clearance.

To install:

3. Apply sealer to the joint between the pan, front cover and block.

4. Install or connect the following:
- Oil pan, with new gaskets. Mounting bolts: 78–96 inch lbs. (9–11 Nm).
- Dipstick guide tube using a new base seal and tighten the holding bolt
- Automatic transmission lines
- Intake air filter assembly
- Engine oil
- All remaining fluids as necessary

M62 Engines

1. Before servicing the vehicle, refer to the Precautions Section.

2. Remove or disconnect the following:
- Negative battery cable
- Intake manifold cover
- Top clips on the radiator
- Cooling fan
- Guide tube for the oil dipstick
- Engine splash guards
- Cover for the oil filter so the oil will run back to the pan
- Engine oil, then disconnect the plug for the level switch
- Lower oil pan
- Gasket and clean the mounting surfaces
- Left and right engine mounts at the bottom
- Unbolt the power steering pump at the holder
- Oil pipes at the power steering pump, on automatic transmissions
- Banjo bolt for the oil return pipe from the oil filter at the oil pan

3. Attach a suitable engine hoist and lift the engine by the front eye hook. Observe the distance between the engine and the firewall while lifting the engine.

4. Remove the oil pan bolts and remove the oil pan.

To install:

5. Clean the mounting surfaces and install a new gasket.

6. Install upper oil pan. Bolts: 84–96 inch lbs. (9–11 Nm). Lower the engine.

7. Check the seals on the oil pipes and replace it if necessary. Lubricate the seals with oil.

8. Install or connect the following:
- Banjo bolt for the oil return pipe from the oil filter at the oil pan
- Power steering pump and connect the oil lines, if equipped
- Ground strap, then connect the left and right engine mounts at the bottom and tighten to 32 ft. lbs. (43 Nm).

➡If replacing the lower oil pan, remove the level switch from the old pan and install it in the new pan with a new O-ring.

- Lower oil pan with a new gasket. Bolts, beginning in the middle and working to the outside: 84–96 inch lbs. (9–11 Nm).
- Plug for the level switch, making sure to replace the O-ring
- Engine splash guards
- Oil dipstick guide tube, making sure to replace the O-ring
- Cooling fan
- Intake manifold cover
- Top clips on the radiator
- Engine oil
- Negative battery cable

N52 Engines

⁂ CAUTION

Aluminum-magnesium material. No steel fasteners may be used due to the threat of electrochemical corrosion. A magnesium crankcase requires aluminum fasteners exclusively. Aluminum fasteners must be replaced each time they are removed. The end faces of aluminum fasteners are painted blue for purposes of identification. Torque specifications and torque angles must be observed for risk of damage.

➡This removal and installation procedure is for AWD vehicles only.

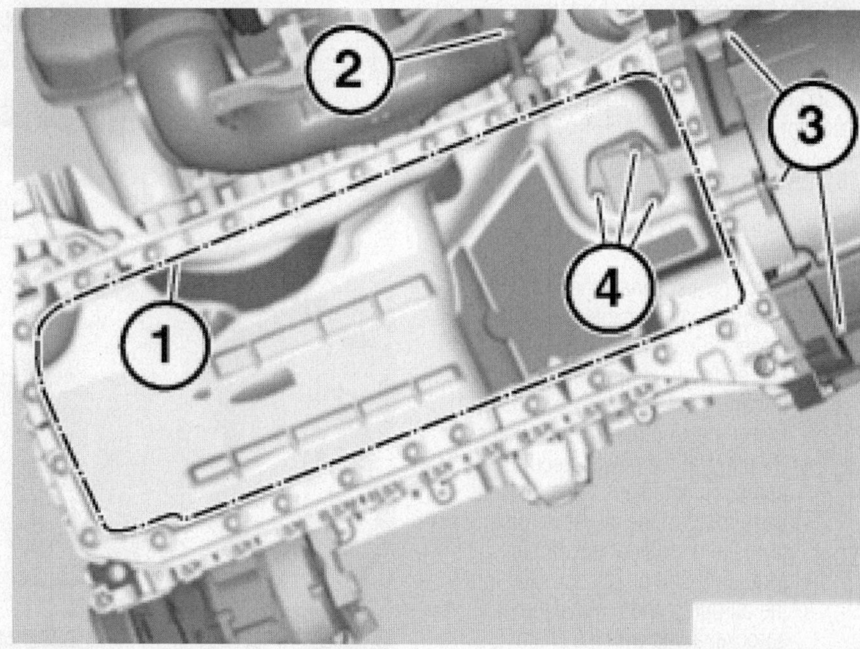

Oil pan bolts (1), oil return hose (2), transmission bolts (3) and oil level sensor (4)—N52 Engine

1. Before servicing the vehicle, refer to the Precautions Section.

2. Install engine support tool or equivalent.

3. Remove or disconnect the following:
- Negative battery cable
- Lower front axle
- Left drive shaft
- Right drive shaft
- Front axle differential
- Engine oil

➡Vehicles equipped with automatic transmission, oil lines must be detached from the engine oil pan. If necessary, remove vane pump and set it aside.

- Two bolts securing oil pan to transmission
- Oil return hose

4. Remove oil pan bolts and remove the oil pan. If necessary, remove oil level sensor bolts and oil level sensor.

To install:

5. Clean the mounting surfaces and install a new gasket and all seals. Replace all aluminum fasteners.

6. Oil pan and mounting bolts. Torque following the tightening sequence as follows:
   a. Step 1: 70 inch lbs (8 Nm)
   b. Step 2: 90 degrees

7. Reverse remaining removal procedure.
- Engine oil
- All remaining fluids as necessary

Oil Pump

REMOVAL & INSTALLATION

M54 and S54 Engines

1. Before servicing the vehicle, refer to the Precautions Section.

2. Remove or disconnect the following:
- Negative battery cable
- Oil
- Oil pan, to access the oil pump drive sprocket
- Oil pump drive sprocket nut. Note that it is a left-hand thread
- Oil pump drive sprocket from the oil pump shaft. Check the shaft splines.
- Oil pump body. Check the condition of the dowel sleeves.

To install:

3. Clean the oil pump mounting surfaces.

4. Install or connect the following:
- Oil pump on the engine. Mounting bolts: 16 ft. lbs. (22 Nm).
- Oil pump drive sprocket onto the oil pump shaft. Oil pump drive sprocket nut: 18 ft. lbs. (25 Nm). The sprocket nut must be tightened in a counterclockwise direction; it has a reverse, or left hand thread.
- Oil pan
- Oil pan drain plug, then fill the engine with oil
- Negative battery cable

5. Start the engine and check that oil pressure is present; if the oil pressure lamp does not extinguish within 5–7 seconds of starting the engine, turn the engine **OFF**. Check for any oil leaks.

M62 Engines

1. Before servicing the vehicle, refer to the Precautions Section.
2. Remove or disconnect the following:
   - Negative battery cable
   - Lower oil pan
   - Left and right engine mounts at the bottom
   - Power steering pump at the holder
   - Oil pipes at the power steering pump, on automatic transmissions
   - Banjo bolt for the oil return pipe from the oil filter at the oil pan
   - Mounting bolt on the sprocket for the oil pump and remove the sprocket along with the chain
   - 3 oil pump mounting bolts and remove the oil pump. Remove the oil pipes out of the crankcase.

To install:

3. Check the seals on the oil pipes and replace it if necessary. Lubricate the seals with oil and the oil pipes.
4. Check the seal in the oil pump and replace it if necessary. Screw the hexagon adapter back into the oil pump until it stops.
5. Install or connect the following:
   - Oil pump and 2 right side oil pump mounting bolts. Bolts: 14–17 ft. lbs. (20–24Nm).
6. Position the chain on the pump and the sprocket and install the sprocket. Bolt: 35 ft. lbs. (47 Nm). Verify that the chain is positioned correctly.
7. Adjust the chain sag to 0.315–0.472 inches (8–12mm) by turning the hexagon adapter in the oil pump, then install the left side mounting bolt.
8. Install or connect the following:
   - Banjo bolt for the oil return pipe from the oil filter at the oil pan

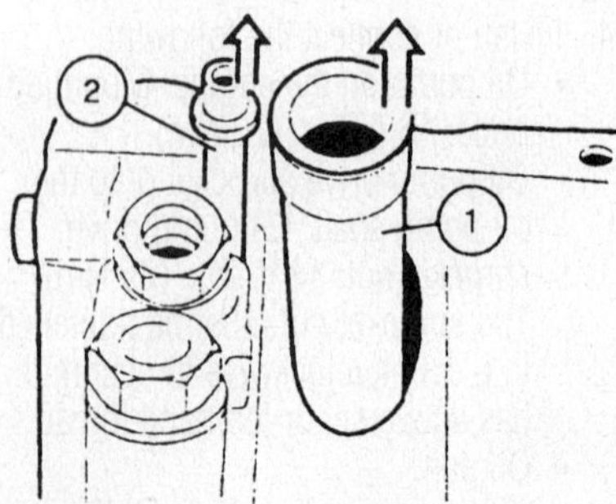

Fresh oil pipe (1), and pure oil pipe (2) locations—M62 engines

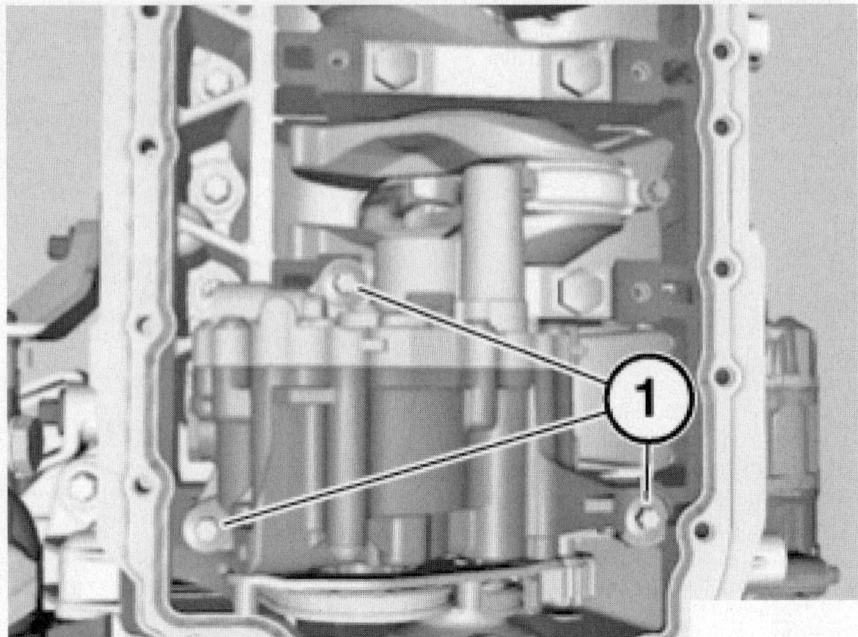

Oil pump mounting bolts (1)—N52 Engine

   - Power steering pump and connect the oil lines, if equipped
   - Lower oil pan
   - Engine oil
   - Negative battery cable

N52 Engines

1. Before servicing the vehicle, refer to the Precautions Section.
2. Remove or disconnect the following:
   - Negative battery cable
   - Oil pan
   - Oil pump intake pipe fasteners and oil pump intake pipe, pull towards transmission
   - Oil pump pulley bolt
   - Oil pump mounting bolts

➡Timing chain of triangular drive is pressed upward by chain tensioner.

Do not remove pulley from assembly.
   - Oil pump pulley, pull towards front of engine bay
   - Oil pump

To install:

3. Check the seals on the oil pipes and replace it if necessary. Lubricate the seals with oil and the oil pipes.
4. Check the seal in the oil pump and replace it if necessary.
5. Align twin surface on oil pump to sprocket wheel.
6. Replace all aluminum fasteners.
7. Install or connect the following:
   - Oil pump to bedplate:
   a. Step 1: 25 ft. lbs. (34 Nm).
   b. Step 2: 180 degrees
   - Chain module to crankcase and oil pump:
   c. Step 1: 35 inch lbs. (4 Nm).
   d. Step 2: 45 degrees
   - Pulley to oil pump:
   e. Step 1: 15 ft. lbs. (20 Nm)
   f. Step 2: 45 degrees
   - Intake pipe to bedplate:
   g. Step 1: 35 inch lbs. (4 Nm)
   h. Step 2: 100 degrees
   - Engine oil
8. Reverse remaining removal procedure.
9. Start engine and check for oil leaks.

Rear Main Seal

REMOVAL & INSTALLATION

The rear main bearing oil seal can be replaced after the transmission and flywheel have been removed from the engine.

1. Before servicing the vehicle, refer to the Precautions Section.
2. Remove or disconnect the following:
   - Transmission
   - Flywheel assembly
   - Oil seal, using a suitable seal removal tool

To install:

3. Coat the sealing lips of the new seal with oil and install the new seal into the end cover housing with a suitable seal installation tool. On the 6-cylinder engines, press the seal in until it is about 0.039–0.079

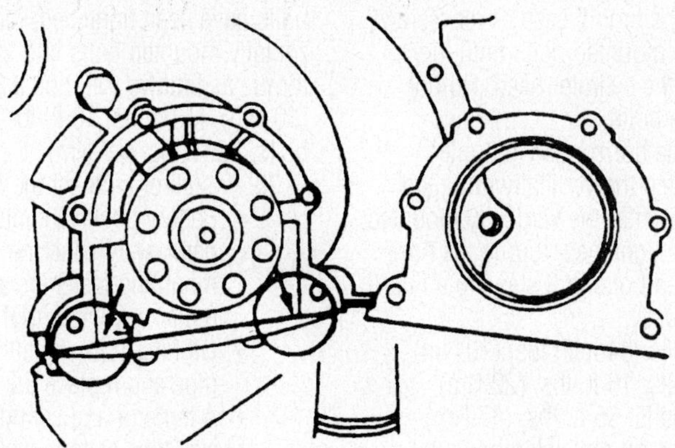

When installing, apply sealer to the joints (as marked) of the rear main seal housing if it has been removed

inches (0.991–2.070mm) deeper than the standard seal, which was installed flush with the housing.

4. Install the remaining components in the opposite order from which they were removed.
5. Connect the negative battery cable.
6. Start the engine and check that oil pressure is present; if the oil pressure lamp does not extinguish within 5–7 seconds of starting the engine, turn the engine **OFF**.
7. Check and top off all fluid levels.

Timing Chain, Sprockets, Front Cover and Seal

REMOVAL & INSTALLATION

M54 and S54 Engines

1. Before servicing the vehicle, refer to the Precautions Section.
2. Drain the cooling system.
3. Remove or disconnect the following:
   - Negative battery cable
   - Radiator and fan assembly
   - Drive belts and any accessories that block access to the timing cover
   - Engine splash shield
   - Vibration damper using a suitable tool, then remove the central bolt and remove the vibration damper hub
   - Timing case cover bolts and timing cover

➡The timing case cover can be removed without removing the water pump.

   - Upper chain guide and top bolt on the right chain guide
   - Timing chain sprockets and the lift out the chain
   - Timing chain guide
   - Tensioning rail, if necessary.
   - Crankshaft sprocket with a suitable tool and lift out the Woodruff key
   - Reversing roller, if needed

➡The reversing roller can only be replaced complete with bearings.

To install:

4. Install the Woodruff key into the channel in the crankshaft. Slide the crankshaft sprocket over the end of the crankshaft with the Woodruff key aligning with the channel in the crankshaft. Use the central mounting bolt to draw the sprocket entirely into position.
5. Apply sealant at the intersections of the timing cover and the oil pan.
6. The remaining components are installed in the reverse order from which they were removed.
7. Tighten the remaining fasteners as follows:
   - Camshaft sprocket bolts: 16 ft. lbs. (22 Nm)
   - Timing cover M6 bolts: 78–96 inch lbs. (9–11 Nm)
   - Timing cover M8 bolts: 16 ft. lbs. (22 Nm)
   - Vibration damper central bolt: M44 engines: 243 ft. lbs. (330 Nm)
   - Vibration damper central bolt: M52 and S52 engines: 302 ft. lbs. (410 Nm)
   - Vibration damper pulley bolts: 17 ft. lbs. (23 Nm)
8. Connect the negative battery cable.

M62 Engines

1. Before servicing the vehicle, refer to the Precautions Section.
2. Remove or disconnect the following:
   - Negative battery cable
   - Accessory drive belt
   - front engine splash shield
   - Left cylinder head cover
   - Alternator
   - Oil filter housing
   - Left upper timing cover
   - Right cylinder head cover
   - Mass Air Flow (MAF) sensor
   - Timing chain tensioner and mount
   - Camshaft Position (CMP) sensor
   - Oil dipstick tube
   - Right timing cover
   - Engine cooling fan
   - Water pump pulley
   - Crankshaft damper pulley and hub
   - Front crankshaft seal using tool No. 11-2-380 and 11-2-383
   - Lower timing cover
3. Turn the engine in the direction of rotation and set cylinder number 1 to TDC firing. The arrows on the sprockets should face up in the cylinder axis. Use a crankshaft holder to keep the TDC position.
4. Loosen and remove the camshaft sprocket bolts from both banks of cylinders. Compress the hydraulic tensioning element to loosen the timing chain. Lock the element with tool No. 11-3-420, or equivalent and remove the sprockets with the chain. Do not rotate the engine with the timing chain removed.
5. Guide the chain out of the tensioner rails and off the lower sprocket.
6. To remove the guide rail:
   a. On the left side (cylinder bank 1–4), remove the lower mounting bolt and remove the tensioning rail. Remove the spacer for the tensioning rail.
   b. On the left side, remove the 2 mounting bolts for the guide rail. Do not mix the 2 bolts, it is important to install the same bolt in the same hole. Remove the sliding rail.
   c. On the right side (cylinder bank 5–8), remove the 2 mounting bolts on the tensioning rail, and the 2 bolts on the guide rail. Remove the rails.

To install:

7. On the right cylinder bank, position the guide rail and install the mounting bolts. Position the tensioning rail, and install the mounting bolts.
8. On the left cylinder bank, check the seal for the spacer. Position the guide rail, and install the mounting bolts in the correct holes. Install the spacer. Position the tensioning rail, and install the lower mounting bolt.
9. Inspect the sprockets for wear and replace if necessary.
10. Install the chain in position.
11. Be sure No. 1 piston remains at the TDC on the firing stroke and the key on the crankshaft is in the 12 o'clock position.

12. Position the chain on the guide rail and swing the chain inward and to the left.

13. Engage the chain on the crankshaft gear and install the camshaft sprockets into the chain.

14. The sprocket, on the intake camshaft for cylinder bank 1–4, has a sender pin. With the arrow pointing up, align the pin in the middle of the slots, then install the camshaft sprockets. Remove tool No. 11–3–420 and remove the crankshaft holder.

15. Install the chain tensioner piston, spring and cap plug, but do not tighten.

16. To bleed the chain tensioner, fill the oil pocket located on the upper timing housing cover with engine oil and move the tensioner back and forth with a suitable pry tool until oil is expelled at the cap plug, then tighten the cap plug securely.

17. Check for the correct seating of the dowel sleeves. Clean the sealing surfaces thoroughly, then place a new gasket on the lower cover.

18. Trim the protruding ends of the gasket, making sure the cutting tool is level. Do not allow the pieces to fall into the engine.

19. Position the lower cover and install the mounting bolts with an even distribution of pressure. Tighten the 6mm bolts to 84 inch lbs. (10 Nm), 8mm bolts to 16 ft. lbs. (22 Nm) and 10mm bolts to 35 ft. lbs. (47 Nm)

20. Install the oil seal in the timing case cover using tool No. 11 1 220 or an equivalent seal driver. Make sure the seal is flush with the cover.

21. Install the vibration damper hub and install the mounting bolt. Tighten the hub bolt as follows:
   a. Step 1: 74 ft. lbs. (100 Nm).
   b. Step 2: Plus 60 degrees.
   c. Step 3: Plus 60 degrees.
   d. Step 4: Plus 30 degrees.

22. Raise and safely support the vehicle. Install the cooling air guide for the alternator, located on the engine carrier. Install the front engine splash shield and lower the vehicle.

23. Position the vibration damper pulleys and install the mounting bolts for the damper.

24. Install the pulley on the water pump.

25. Install the drive belt and the cooling fan. Rotate the fan counterclockwise to install.

26. Install the intake hose between the throttle body and the air volume meter, then install the manifold cover.

27. Replace the hydraulic tensioner oil seal in the timing case cover.

28. Check for the correct seating of the dowel sleeves. Clean the sealing surfaces to remove any oil or old gasket material and install a new gasket.

29. Mount the timing case cover. Screw in the vertically mounted bolts until the cover contacts the cylinder head. Do not fully tighten the bolts.

30. Install the horizontally mounted bolts, then tighten the vertically mounted bolts in 2 steps. After the vertically mounted bolts have been tightened, torque the horizontally mounted bolts in 2 steps to a final torque as follows:
- M6 bolts: 84 inch lbs. (10 Nm)
- M8 bolts: 16 ft. lbs. (22 Nm)
- M10 bolts: 35 ft. lbs. (47 Nm)

31. Install the oil dipstick guide tube, making sure to replace and lubricate the O-ring.

32. Install the camshaft sender fastener and the chain tensioner.

33. Install the air cleaner upper section along with the MAF sensor.

34. Install the right cylinder head cover.

35. Check for the correct seating of the dowel sleeves. Clean the sealing surfaces to remove any oil or old gasket material and install a new gasket.

36. Mount the timing case cover together with the inserted bolt. This bolt cannot be installed with the cover in place. Install the rest of the mounting bolts and fasten the vertically mounted bolts until the cover just contacts the cylinder head. Do not fully tighten the bolts.

37. Install the horizontally mounted bolts, then tighten the vertically mounted bolts in 2 steps. After the vertically mounted bolts have been tightened, torque the horizontally mounted bolts in 2 steps to a final torque as follows: M6 bolts: 84 inch lbs. (10 Nm), M8 bolts: 16 ft. lbs. (22 Nm), M10 bolts: 35 ft. lbs. (47 Nm)

38. Install or connect the following:
- Battery positive cable for the alternator and install the protective tube mounting fasteners and connect the remaining wires to the alternator
- Oil filter housing and the return pipe and replace the housing cover
- Alternator and cylinder head cover
- Negative battery cable

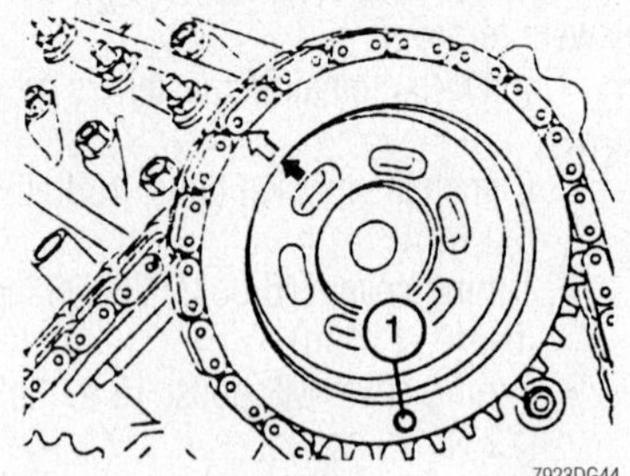

Camshaft sprocket sender pin—M62 engines

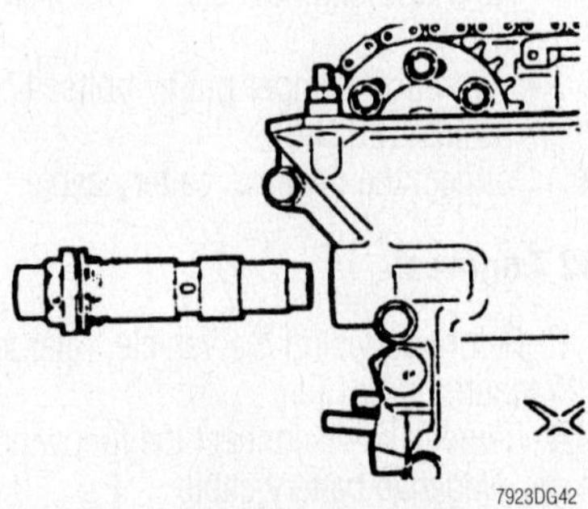

The timing chain tensioner slides out of the side of the front cover—M62 engines

Piston and Ring

POSITIONING

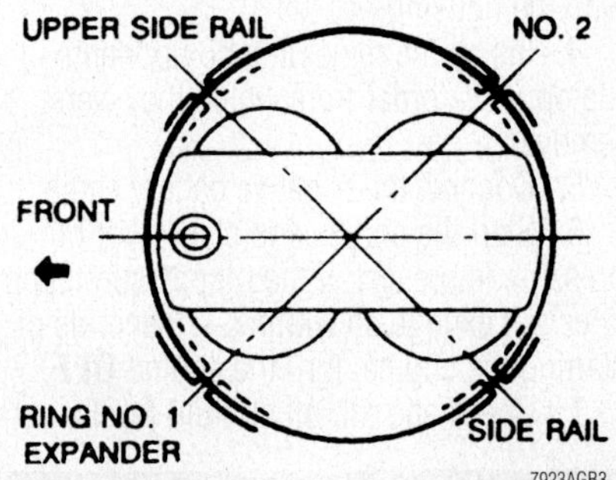

Piston ring end-gap spacing—all engines

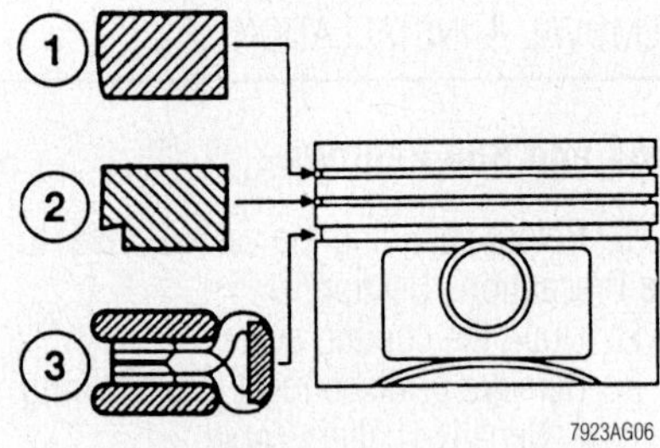

Compression and oil control ring locations—all engines

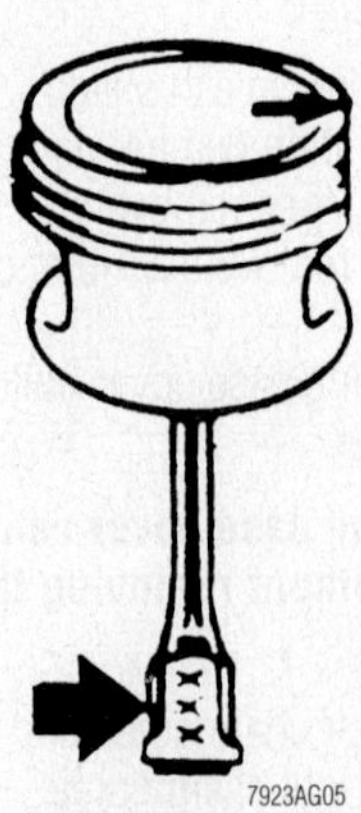

Connecting rod-to-piston positioning—all engines

FUEL SYSTEM

Fuel System Service Precautions

Safety is the most important factor when performing not only fuel system maintenance but also any type of maintenance. Failure to conduct maintenance and repairs in a safe manner may result in serious personal injury or death. Maintenance and testing of the vehicle's fuel system components can be accomplished safely and effectively by adhering to the following rules and guidelines.

- To avoid the possibility of fire and personal injury, always disconnect the negative battery cable unless the repair or test procedure requires that battery voltage be applied.
- Always relieve the fuel system pressure prior to disconnecting any fuel system component (injector, fuel rail, pressure regulator, etc.), fitting or fuel line connection. Exercise extreme caution whenever relieving fuel system pressure, to avoid exposing skin, face and eyes to fuel spray. Fuel under pressure may penetrate the skin or any part of the body that it contacts.
- Always place a shop towel or cloth around the fitting or connection prior to loosening to absorb any excess fuel due to spillage. Ensure that all fuel spillage (should it occur) is quickly removed from engine surfaces. Ensure that all fuel soaked cloths or towels are deposited into a suitable waste container.
- Always keep a dry chemical (Class B) fire extinguisher near the work area.
- Do not allow fuel spray or fuel vapors to come into contact with a spark or open flame.
- Always use a back-up wrench when loosening and tightening fuel line connection fittings. This will prevent unnecessary stress and torsion to fuel line piping. Always follow the proper torque specifications.
- Always replace worn fuel fitting O-rings with new. Do not substitute fuel hose where fuel pipe is installed.

Relieving Fuel System Pressure

To relieve the pressure in the system, locate fuel pump relay located on the cowl. The relay can sometimes be distinguished by the orange color of the housing. Unplug and remove the relay, and place it in a safe location. With the fuel pump relay removed, start the engine and operate it until it stalls. Crank the engine for 10 seconds after it stalls to remove any residual pressure.

Fuel Filter

REMOVAL & INSTALLATION

On filters that are located near the fuel tank, it is necessary to clamp the fuel lines closed before disconnecting them, or fuel will run out continuously.

1. Before servicing the vehicle, refer to the Precautions Section.
2. Relive the fuel system pressure and disconnect the negative battery cable.
3. Clamp the fuel lines closed if the filter is mounted low, near the fuel tank or underneath the vehicle. Then, loosen the clamps and disconnect the inlet and outlet hoses. Remove the hose clamps or slide them back, well off the connections to make it easier to remove the hoses, if necessary.
4. The filters are usually attached to a bracket on the frame, floor pan or wheel well. Loosen the bracket and remove the filter. Note the direction of flow for reinstallation or replacement of the filter.

To install:

5. Position the new fuel filter and install the fuel lines onto the correct fuel filter fittings. Tighten the fuel line clamps until tight, but not to the point where the fuel lines become excessively pinched or damaged, then tighten the mounting bracket until snug.
6. Connect negative battery cable. Cycle the ignition **ON** and **OFF** several times to build fuel pressure.
7. Inspect the fuel filter and fuel lines for any fuel leaks.

Fuel Pump

REMOVAL & INSTALLATION

The fuel pump is mounted through the top of the fuel tank along with the fuel level sending unit. The fuel tank should not be filled more than ⅓ of the total fuel tank capacity to prevent fuel leakage during fuel pump removal. If the fuel tank is filled beyond this level, the fuel level must be reduced using an approved fuel removal device.

1. Before servicing the vehicle, refer to the Precautions Section.
2. Relive the fuel system pressure and disconnect the negative battery cable.
3. Drain the fuel, if filled beyond ⅓ of the capacity of the fuel tank. Drain the fuel tank enough to prevent spillage when removing the pump using an approved fuel removal device.

➡The fuel pump must be removed through the top of the fuel tank, thus the location of the fuel tank determines whether the fuel pump is accessed by removal of the rear seat, or removal of the trim panels in the trunk.

4. Remove or disconnect the following:
   - Rear seat, or the trim panels in the trunk, depending on fuel tank location to access the top of the fuel tank

➡On models which require removal of the rear seat, the insulation mat under the seat must be cut in a "U" shape to allow the insulation to be folded up to access the top of the fuel tank.

   - Fasteners securing the metal cover located above the fuel tank, and remove the cover
   - Electrical connector at the top of the combination fuel pump and fuel level sending unit assembly
   - Fuel feed and return lines
5. Match mark the combination fuel pump and fuel level sending unit assembly to the fuel tank to ensure proper installation during reassembly.
   - Fasteners or fastener securing the combination fuel pump and fuel level sending unit assembly to the fuel tank. The fasteners are one of 2 types.
6. If the fuel pump assembly is fastened to the fuel tank with a series of 6 mm nuts:
   a. Loosen the nuts evenly using a crisscross sequence and carefully lift the cover and place aside.
   b. Compress the large plastic tongue to release the fuel pump, and lift the pump along with the fuel sending unit out of the fuel tank.
7. If the fuel pump assembly is fastened to the fuel tank with a large sealing ring:
   a. Use tool No. 16–1–020 to loosen the sealing ring in a counterclockwise direction.
   b. With the seal ring removed, lift the fuel pump assembly out of the fuel tank.

To install:

➡Always use a new seal or gasket when installing the fuel pump or fuel level gauge sending unit assembly.

8. Install the fuel pump into the fuel tank taking care not to bend or damage the fuel sending unit assembly.

9. If the fuel pump is held in place by a plastic bracket in the fuel tank perform the following:
   a. Make sure the fuel pump is fully snapped in place.
   b. Install the fuel tank cover plate with a new gasket and torque the fasteners using a crisscross pattern to 57 inch lbs. (6.5 Nm).
10. If the fuel pump is held in place with a sealing ring perform the following:
   a. Ensure the pump is properly aligned with the fuel tank matchmarks made during disassembly.
   b. Install a new seal and torque the sealing ring using tool No. 16–1–020 as follows:
      - Metal sealing rings: 26 ft. lbs. (35 Nm)
      - Plastic sealing rings: 41 ft. lbs. (55 Nm)
11. The balance of the assembly is in reverse order of disassembly.
12. Connect the negative battery cable.
13. Once the vehicle is started, check for leaks. If a strong fuel odor is present, or any fuel leakage is noted, stop the engine immediately and repair as necessary.

Fuel Injector

REMOVAL & INSTALLATION

1. Remove the upper intake manifold.
2. Disconnect the electrical connectors for the fuel injectors and place the harness aside, then remove the dampening valve vacuum hose at the front of the fuel rail.
3. The two hose clamps for the fuel rail return line are stamped to be tamper proof. To remove, use a locking pliers and attach then onto the clamp's adjustment screw head securely. Turn the adjustment screw counterclockwise until the clamp is loosened. Loosen both clamps and slide to the middle of the fuel line.
4. Using a 12 mm line wrench, loosen the fuel rail feel line flare nut and once loose, slide the flare nut away from the fuel rail.
5. To avoid debris from entering the injector ports, If compressed air is available, using an air nozzle, and wearing eye protection, carefully spray compressed air around the base of the injectors. If excessive debris exists, spray the base of the injectors with an evaporating brake cleaner and spray again with compressed air. If compressed air is not available, spray the brake cleaner to clean as best as possible.
6. Using a 10 mm socket, short extension and a ratchet, remove the two fuel rail mounting bolts.
7. Use a clean shop towel folded over several times to form a small pad, and place on the corner of the valve cover. Then using a suitable and sturdy pry tool, slowly and carefully lift the fuel rail until the injectors are just removed from the manifold, then remove the fuel feed and fuel return lines away from the fuel rail and remove the entire assembly with the large vent hoses still attached and set aside.

DRIVE TRAIN

Transmission Assembly

REMOVAL & INSTALLATION

Manual

3 SERIES

1. Before servicing the vehicle, refer to the Precautions Section.
2. Remove or disconnect the following:
   - Negative battery cable
   - Exhaust system. Remove the cross brace and heat shield.
   - Flexible coupling at the front of the driveshaft. Some vehicles may also have a vibration damper located at this point in the drive train. This damper is mounted on the transmission output flange with bolts that are pressed into the damper. On these vehicles, remove the nuts located behind the damper.
3. Loosen the threaded sleeve on the driveshaft. Use a suitable tool to hold the splined portion of the shaft while turning the sleeve.
   - Mounting bolts and center driveshaft mount. Then, bend the driveshaft downward at the center and pull it off the transmission output flange. Keep the sections of the driveshaft from pulling apart and suspend it from the vehicle.
   - Retainer and washer and pull out the shift selector rod
   - Self-locking bolts that retain the shift rod bracket at the rear of the transmission, then remove the bracket. If equipped with a shift arm, use a suitable pry tool to pry the spring clip up off the boss on the transmission case and swing it upward. Then, pull out the shift shaft pin.
   - Clutch slave cylinder and support it so the hydraulic line can remain connected
4. The transmission incorporates sending units for flywheel rotating speed and position. Remove the heat shield that protects these from the exhaust heat, then remove the retaining bolt for each sending unit. Note that the speed sending unit, which has no identifying ring is located on the right, and that the reference mark sending unit, which has a marking ring, is located on the left. If the sending units are installed in reverse positions, the engine will not run. Remove both units from the flywheel housing.
5. Detach the wiring connector going to the back-up light switch and place the wires aside.
6. Support the transmission. Remove the mounting bolts and crossmember holding the rear of the transmission to the body. Then, lower the transmission onto the front axle carrier.
7. Using the proper tool, remove the bolts holding the transmission flywheel housing to the engine. Be sure to retain the washers with the bolts. Pull the transmission rearward to slide the input shaft out of the clutch disc, then lower the transmission and carefully remove it from the vehicle.

To install:

8. Installation is the reverse of the removal procedure, while using the following torque values:
   - Front mounting bolts: 46–58 ft. lbs. (62–80 Nm)
   - Shift rod bracket bolts: 16 ft. lbs. (22 Nm)
   - Driveshaft center bearing bolts: 16–17 ft. lbs. (21–23 Nm)
   - Flexible coupling bolts: 83–94 ft. lbs. (114–129 Nm)

5 SERIES

1. Before servicing the vehicle, refer to the Precautions Section.
2. Remove or disconnect the following:
   - Negative battery cable
   - Exhaust system, to provide clearance for transmission removal
   - Heat shield brace and transmission heat shield

- Driveshaft coupling at the rear of the transmission
- Screw-on ring type connection that attaches the driveshaft to the center bearing. Then, unbolt the center bearing mount. Bend the driveshaft downward and pull it off the centering pin. If equipped with a vibration damper, turn it and pull it back over the output flange before pulling the driveshaft off the guide pin. Suspend it from the vehicle.
- The wires for the back-up light switch
- Passenger compartment console to disconnect it from the top of the transmission by removing the self-locking bolts. Discard and replace the bolts.
- Shift rod at the rear of the transmission, by pulling out the locking clip

3. If the transmission is linked to the shift lever with an arm, use a small pry tool to lift the spring out of the holder on the bracket, then raise the arm, then remove the shift shaft bolt.

- With a flywheel housing cover (semi-circular in shape), the mounting bolts and cover
- The speed sensor and reference mark sensor. Note their locations. The speed sensor is located in the upper bore, marked **D**. The reference mark sensor, which has a ring, is located in the lower bore, marked **B**. Check the O-rings for the sensors and install new ones if they are damaged.
- Rear transmission crossmember, with transmission supported
- Upper and lower attaching nuts and remove the clutch slave cylinder, supporting it so the hydraulic line need not be disconnected
- Reverse gear back-up light switch and pull the wires out of the holders and place aside
- The bolts fastening the transmission to the bell housing, using the proper tool. Pull the transmission rearward until the input shaft has disengaged from the clutch disc, then lower and remove it.

To install:

4. Installation is the reverse of the removal procedure, while using the following torque values:

➡**All aluminum fasteners must be replaced**

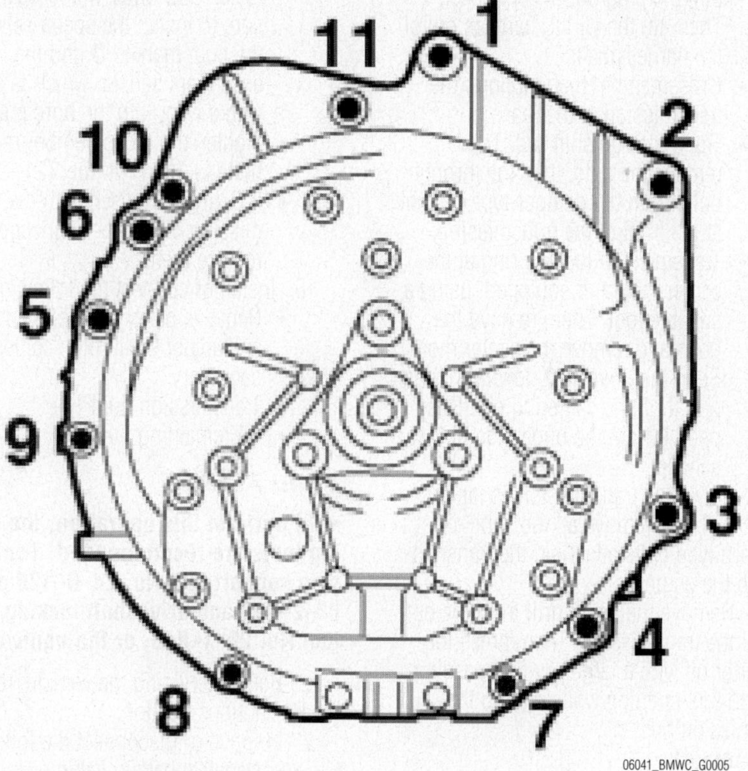

Manual and automatic transmission tightening sequence

- Transmission-to-bell housing: 52–58 ft. lbs. (70–80 Nm)
- Rear/top transmission Torx® bolts: 46–58 ft. lbs. (62–80 Nm)
- Center mount-to-body: 16–17 ft. lbs. (21–23 Nm)
- Front joint-to-transmission: 83–94 ft. lbs. (114–129 Nm)

Automatic Transmission

3 SERIES

➡**To perform this operation, a support for the transmission, tool Nos. 24-0-120 and 00-2-020 or the equivalents and a tool for tightening the driveshaft locking ring, tool No. 26-1-040, are recommended.**

1. Before servicing the vehicle, refer to the Precautions Section.
2. Remove or disconnect the following:

- Negative battery cable
- Throttle cable adjusting nuts, release the cable tension and disconnect the cable at the throttle lever. Then, remove and retain the nuts and pull the cable housing out of the bracket.
- Exhaust system at the manifold and hangers and lower it aside
- Hanger that runs across under the driveshaft
- Exhaust heat shield from under the center of the vehicle
- Transmission oil
- Oil filler neck
- Oil cooler lines at the transmission by removing the flare nuts and plug the open connections

3. Support the transmission. Separate the torque converter housing from the transmission by removing the Torx® bolts with the proper tool, then remove the bolts from underneath. Retain the washers used with the Torx® bolts.

- Bolts attaching the torque converter housing to the engine, making sure to retain the spacer used behind one of the bolts. Then, loosen the mounting bolts for the oil level switch just enough so the plate can be removed while pushing the switch mounting bracket to one side.
- Bolts attaching the torque converter to the driveplate. Turn the flywheel as necessary to gain access to each of the bolts. Be sure to re-use the same bolts and retain the washers.
- Speed and reference mark sensors. Remove the attaching bolt for each and remove each sensor.
- Bayonet type electrical connector, turning it counterclockwise, then

pull the plug out of the socket. Then, lift the wiring harness out of the harness bails
- Crossmember that supports the transmission at the rear
- Transmission shift rod. Then, remove the nuts, then the through-bolts from the damper-type U-joint at the front of the transmission.
- Transmission locking ring at the center mount, if equipped, using a suitable tool. Then, remove the bolts and remove the center mount. Bend the driveshaft downward and pull it off the centering pin. Suspend it from the underside of the vehicle.

4. Lower the transmission as far as possible. Then, remove all the Torx® or standard type bolts attaching the transmission to the engine.

5. Remove the small grill from the bottom of the transmission. Then, press the converter off with a large pry tool passing through this opening while sliding the transmission out.

To install:

6. Position the transmission under the vehicle and install the torque converter onto the transmission.

7. Be sure the converter is fully seated onto the transmission, such that the ring on the front is inside the edge of the case. Install the small air grille onto the bottom of the transmission.

8. Install or connect the following:
   - Engine-to-transmission attaching bolts
   - The remaining components in the reverse order from which they were removed

9. Make note of the following points during reassembly:
   - When reinstalling the driveshaft, tighten the lockring with the proper tool. Be sure to replace the self-locking nuts on the driveshaft flexible joint and to hold the bolts while tightening the nuts to keep from distorting the joint.
   - When installing the center mount, preload it forward from its most natural position 0.157–0.236 inches (3.98–5.99mm).
   - When reconnecting the bayonet type electrical connector, be sure the alignment marks are aligned after the plug it twisted into its final position.
   - When reinstalling the speed and reference mark sensors, inspect the O-rings used on the sensors and install new ones, if necessary. Be sure to install the speed sensor into the bore marked **D** and the reference mark sensor, which is marked with a ring, into the bore marked **B**.
   - Tighten the crossmember mounting bolts to 16–17 ft. lbs. (21–23 Nm).
   - If O-rings are used with the transmission oil cooler connections, replace them

10. Install or connect the following:
   - Negative battery cable, then check and adjust the throttle cables as necessary
   - Transmission fluid
   - All remaining fluids

5 AND 7 SERIES

➡To perform this operation, the following tools are recommended. Transmission support tool No. 24–0–120 and 00–2–020 and driveshaft locking ring tool No. 26–1–040, or the equivalents.

1. Before servicing the vehicle, refer to the Precautions Section.

2. Remove or disconnect the following:
   - Negative battery cable
   - Throttle cable adjusting nuts, release the cable tension and disconnect the cable at the throttle lever. Then, remove the nuts and pull the cable housing out of the bracket.
   - Exhaust system at the manifold and hangers and lower it out of the way
   - Hanger that runs across under the driveshaft
   - Exhaust heat shield from under the center of the vehicle
   - Crossmember that supports the transmission at the rear, with transmission supported
   - Driveshaft coupling through-bolts and nuts or the CV-joint through bolts and nuts. Either type is located right at the rear of the transmission. Discard and replace the self-locking coupling nuts. Keep the CV-joint clean and replace its gasket upon reassembly.
   - Transmission locking ring at the center mount, if equipped. Then, remove the bolts and remove the center mount. Bend the driveshaft downward and pull it off the centering pin. Suspend it from the underside of the vehicle.
   - Transmission oil
   - Oil filler neck
   - Oil cooler lines at the transmission by removing the flare nuts and plug the open connections

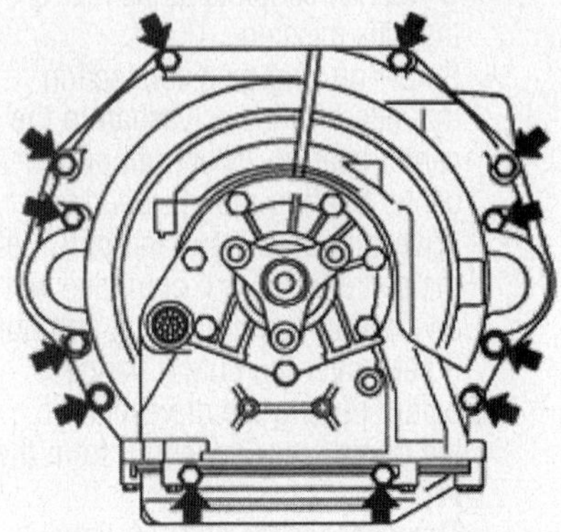

Automatic transmission mounting bolt locations—5 and 7 series

   - Converter cover by removing the Torx® bolts from behind and the regular bolts from underneath, if equipped
   - Bolts fastening the torque converter to the driveplate, turning the flywheel as necessary to gain access from below
   - Guard for the speed and reference mark sensors, if equipped
   - Attaching bolt for each sensor and remove. Keep the sensors clean.
   - Shift cable by loosening the locknut fastening it to the shift lever and disconnecting the cable at the cable housing bracket

3. If the transmission has an electrical connection, turn the bayonet fastener to the left to release the connection, disconnect it and pull the wire out of the brackets and place aside.

4. Lower the transmission as far as possible. Then, remove all the fasteners attaching the transmission to the engine.

5. Remove the small grill from the bottom of the transmission. Then, press the converter off with a large pry tool through this opening while sliding the transmission out.

To install:

6. Place the transmission under the vehicle and raise it into position. Slide the torque converter and the transmission together before installing the transmission completely to the engine. Make sure the converter is fully seated onto the transmission, such that the ring on the front is inside the edge of the case.

7. Install or disconnect the following:
   - Small grill onto the transmission
   - Transmission to the engine
   - The remaining components in the reverse order from which they were removed

8. Make note of the following points during reassembly:
   a. Driveshaft, as follows: Tighten the lockring with a suitable tool. If the driveshaft has a simple coupling, rather than a CV-joint, be sure to replace the self-locking nuts and to hold the bolts still while tightening the nuts to keep from distorting the coupling.
   b. Center mount. Preload it forward from its most natural position 0.157–0.236 inches (3.98–5.99mm).
   c. Transmission fluid.
   d. All remaining fluids as necessary.
   e. Negative battery cable, then check and adjust the throttle cables as necessary.

SHIFT LINKAGE ADJUSTMENT

1. Before servicing the vehicle, refer to the Precautions Section.
2. Move the selector lever to **P** position. Loosen the nut until the cable is free.
3. Push the transmission lever to the **D** or **P** position. Then, push the cable rod in the opposite direction.
4. Clamp down the cable rod without tension.
5. Tighten the nut to 84–102 inch lbs. (9–11 Nm).

➡Do not bend the cable.

THROTTLE LINKAGE ADJUSTMENT

1. Before servicing the vehicle, refer to the Precautions Section.
2. On the injection system throttle body, loosen the 2 locknuts at the end of the throttle cable and adjust the cable until there is a play of 0.010–0.030 inches (0.254–0.762mm).
3. Loosen the locknut and lower the kickdown stop under the accelerator pedal. Have someone depress the accelerator pedal until the transmission detent can be felt. Then, back the kickdown stop back out until it just touches the pedal.
4. Check that the distance from the seal at the throttle body end of the cable housing is at least 1.732 inches (43.9mm) from the rear end of the threaded sleeve and tighten the locknuts.

Clutch

ADJUSTMENT

These vehicles are equipped with a hydraulic clutch actuating system. No adjustment is possible.

REMOVAL & INSTALLATION

1. Before servicing the vehicle, refer to the Precautions Section.
2. Remove or disconnect the following:
   - Negative battery cable
   - Heat shield, then the mounting bolts
   - Speed and reference mark sensors at the flywheel housing. Mark the plugs for reinstallation.
   - Transmission and clutch housing
   - With a 265/6 transmission (without an integral clutch housing), clutch housing. On vehicles with 6-cylinder engines, a Torx® socket is required.
3. Prevent the flywheel from turning, using a suitable locking tool.
4. Loosen the mounting bolts one after another gradually, 1–1 ½ turns at a time, to relieve tension from the clutch.
5. Remove the mounting bolts, clutch and driveplate. Coat the splines of the transmission input shaft with Molykote® Long-term 2, Microlube® GL 2611. Be sure the clutch pilot bearing, located in the center of the crankshaft, turns easily.
6. Check the clutch driven disc for excess wear or cracks. Check the integral torsional damping springs, used with lighter flywheels only, for tight fit. Inspect the rivets to be sure they are all tight. Inspect the flywheel to be sure it is not scored, cracked, or burned. Use a straightedge to verify the contact surface is true. Replace any defective parts.

To install:

➡On vehicles with 6-cylinder engines, the clutch pressure plate must fit over dowel pins.

7. To install, fit the new clutch disk on a suitable clutch alignment tool and center the tool with the flywheel by centering it with the clutch pilot bearing. Mount the clutch pressure plate and install the mounting bolts. Clutch mounting bolts: 16–19 ft. lbs. (21–26 Nm).
8. When installing the clutch retaining bolts turn them in gradually to evenly tighten the clutch pressure plate to prevent warpage.
9. Install or connect the following:
   - Clutch housing, if equipped with separate housing, then the transmission. If the clutch housing is part of the transmission, install the transmission.
   - Speed and reference mark sensors, if equipped; then the heat shield

Clutch pressure plate mounting bolts

   - The remaining components in the reverse order from which they were removed
   - All remaining fluids as necessary
   - Negative battery cable

Hydraulic Clutch System

BLEEDING

The clutch system can be bled using a pressure bleeder. Follow the instructions that come with the pressure bleeder for the proper pressure bleeding procedure. The maximum line pressure while pressure bleeding must not exceed 36 psi (248 pa).

1. Before servicing the vehicle, refer to the Precautions Section.
2. To bleed a clutch manually requires the following:
   - The assistance of a second person
   - A section of hose that is compatible with brake fluid (preferably clear) and fits the slave cylinder bleeder valve snugly
   - A container to catch the fluid that is bled through the system.

➡Cleanliness is of the utmost importance. Always be sure the fluid reservoir is as clean as possible to avoid bleeding debris or contaminated fluid into the system check valves and seals. As brake hydraulic fluid easily absorbs moisture, always use fresh fluid when bleeding a clutch hydraulic system. If the clutch hydraulic system has been run dry, or if the hydraulic fluid line, clutch slave or master cylinder has just been replaced, it may be necessary to repeat the bleeding procedure a number of times to remove all of the air. Make sure the reservoir does not run dry during bleeding. If the reservoir runs dry the bleeding process must be repeated until no air is present.

3. To bleed the system perform the following:

a. Top off the clutch hydraulic fluid reservoir using a fluid that meets the standards of the vehicle's clutch hydraulic system.

b. With one end of a section of hose attached to the slave cylinder bleed valve and the other end of the hose submerged into a container of fresh brake hydraulic fluid, open the clutch slave cylinder bleed valve. Slowly press the clutch pedal to the floor and hold the pedal down.

c. Close the clutch slave cylinder bleed valve.

d. Slowly release the clutch pedal.

e. Check the hydraulic fluid level and top off as necessary.

4. Repeat the above steps until the discharged fluid is clean and no air bubbles appear during the bleeding process.

Halfshafts

REMOVAL & INSTALLATION

3 Series

1. Before servicing the vehicle, refer to the Precautions Section.
2. Remove or disconnect the following:
   - Rear tire and wheel assembly
   - Output shaft at the outer flange and suspend it
   - Caliper, and suspend it with the brake line connected. Unbolt and remove the rear disc.
   - Large nut and lock plate. If equipped with ABS, disconnect, then remove the ABS speed sensor.
   - Collar nut. Then, remove the drive flange with a suitable press tool or install the collar nut until it is just flush with the end of the shaft and use a suitable soft faced hammer to knock out the shaft.
   - Snapring. Pull out the wheel bearings, using a suitable tool.
3. Pull the inner bearing race off the axle shaft with tool No. 00–7–500.

To install:

4. Install or connect the following:
   - New bearing assembly, using suitable bearing driver tools. Install the snapring.
   - Rear axle shaft with tool Nos.: 23–1–300, 33–4–080 and 33–4–020 or their equivalents.
   - Collar nut, after lubricating, and drive in the lock plate using a suitable tool. Collar nut: 148 ft. lbs. (200 Nm).

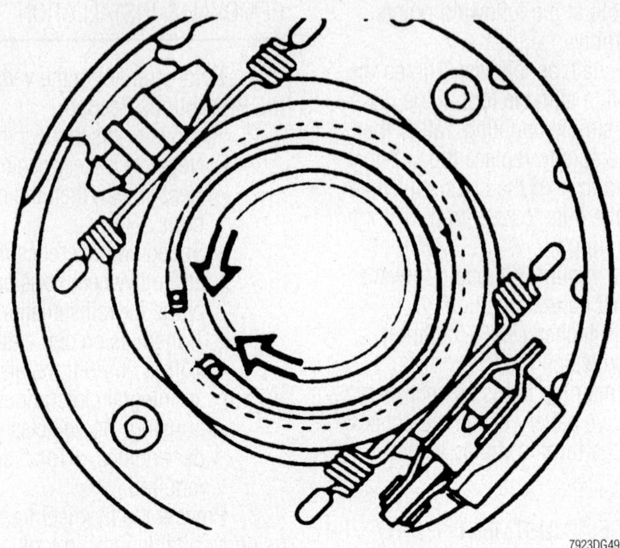

Remove the snapring from the hub—3 Series

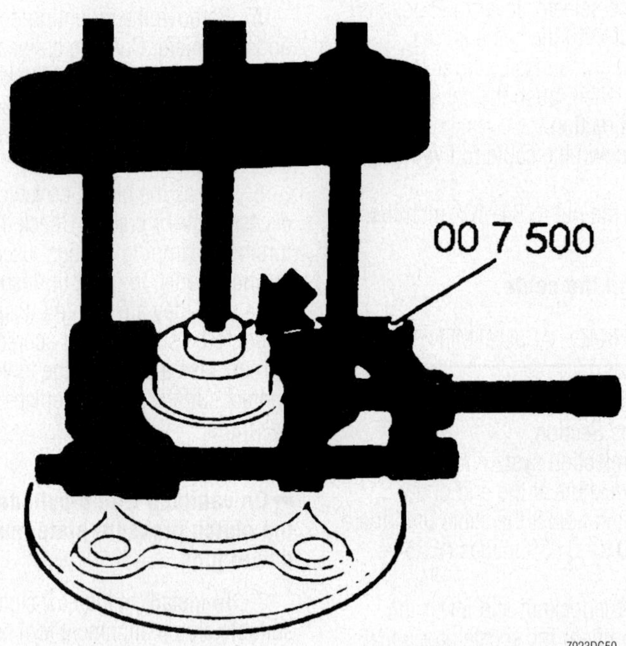

Tool No. 00–7–500 bearing puller used to remove the inner bearing race—3 Series

   - ABS speed sensor
   - Brake disc and caliper
   - Output shaft
   - Rear tire and wheel assembly

5 and 7 Series

1. Before servicing the vehicle, refer to the Precautions Section.
2. Remove or disconnect the following:
   - Rear tire and wheel assembly
   - Lock plate and if equipped, the ABS sensor
   - Retaining nut from the output flange
   - Flange
   - Output half shaft from the final drive (differential carrier) by pressing out with a suitable tool and suspend it
   - Output shaft from the drive flange hub using a suitable tool
   - Rear axle shaft with a suitable tool
   - Snapring
   - Pull out the wheel bearings, using a suitable tool
   - Pull out the seal with a suitable tool
3. If the inner bearing shell is damaged, pull it off with a suitable puller and thrust pad.

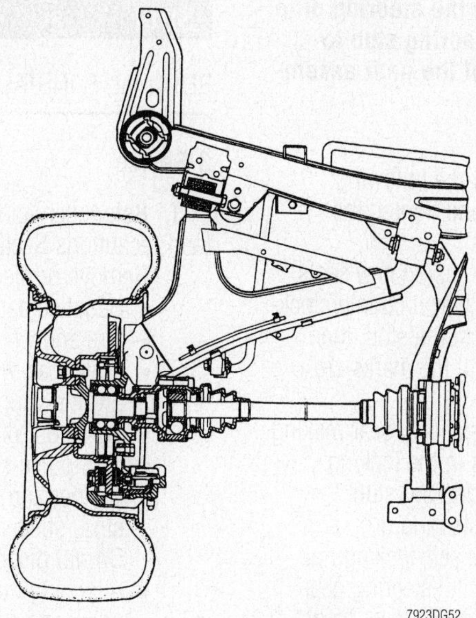

Cut-away schematic of the rear halfshaft and suspension system—5 Series

To install:

4. Using an appropriate bearing installer, install the wheel bearing assembly, then install the seal, and insert the snapring, then install the rear axle shaft, all in reverse of the removal procedure.
5. Install or connect the following:
   - Axle shaft seal
   - Output shaft, as follows: screw the threaded spindle into the shaft all the way, then use the nut and washer against the outside of the bridge
   - Output shaft to the final drive. Mounting bolts: 42–46 ft. lbs. (58–63 Nm).
   - Outer nut, bearing surface lubricated. Nut: 169–188 ft. lbs. (234–260 Nm).
   - ABS sensor, if removed
   - Rear tire and wheel assembly

STEERING AND SUSPENSION

Air Bag

⁂ CAUTION

The vehicles are equipped with an air bag system. The system must be disarmed before performing service on, or around, system components, the steering column, instrument panel components, wiring and sensors. Failure to follow the safety precautions and the disarming procedure could result in accidental air bag deployment, possible personal injury and unnecessary system repairs.

PRECAUTIONS

Several precautions must be observed when handling the inflator module to avoid accidental deployment and possible personal injury.

- Never carry the inflator module by the wires or connector on the underside of the module.
- When carrying a live inflator module, hold securely with both hands, and ensure that the bag and trim cover are pointed away.
- Place the inflator module on a bench or other surface with the bag and trim cover facing up.
- With the inflator module on the bench, never place anything on or close to the module that may be thrown in the event of an accidental deployment.

DISARMING & ARMING

1. Before servicing the vehicle, refer to the Precautions Section.
2. Place the ignition switch in the **OFF** position.
3. Disconnect the negative battery terminal and cover the battery terminal to prevent accidental contact.
4. Once the battery has been disconnected, wait for a period of approximately 10 minutes allowing the capacitor in the control unit to discharge.
5. When repairs are completed, connect the negative battery cable.

Power Rack and Pinion Steering Gear

REMOVAL & INSTALLATION

3 Series

1. Before servicing the vehicle, refer to the Precautions Section.
2. Use a siphon tool to empty the power steering fluid reservoir.
3. Remove or disconnect the following:
   - Front wheels
   - Intermediate shaft pinch bolt
   - Steering shaft spindle off the steering gear
   - Hydraulic fluid return line from the power steering unit. Drain the fluid into a sealable container.
   - Pressure line
   - Left and right side nuts and press off the tie rods where they connect to the spring struts
   - Bolts attaching the steering unit to the front crossmember and remove it

To install:

4. Install the steering unit to the front crossmember, then install and tighten the mounting bolts.
5. Install remaining components in reverse order of disassembly.
6. Note the following:
   - The steering unit bolts to the rear holes of the axle carrier. Always use new self-locking nuts and tighten them to 29–34 ft. lbs. (40–46 Nm).
   - When reconnecting tie rods to the spring struts, be sure tie rod pins and strut bores are clean. Replace self-locking nut and tighten to 40–48 ft. lbs. (54–66 Nm).
   - Replace the seals on the power steering pump connection and tighten the bolt to 29–32 ft. lbs. (40–43 Nm).
7. Connect the negative battery cable.
8. Refill the fluid reservoir with specified fluid. Idle the engine and turn the steering wheel back and forth until it has reached full lock both right and left twice in each direction. Then, turn the engine **OFF** and refill the reservoir as necessary.

Power Recirculating Ball Steering Gear

REMOVAL & INSTALLATION

5 and 7 Series

1. Before servicing the vehicle, refer to the Precautions Section.
2. Remove or disconnect the following:
   - Negative battery cable
   - Steering wheel
3. Discharge the pressure reservoir by pushing in on the brake pedal about 10 times. Draw off hydraulic fluid in the supply tank.
   - Bolt and press the tie rod off the steering drop arm with the proper tool
   - Heat shield on the steering gear and disconnect the ride level height control pipes, on 750iL models
   - U-joint from the steering gear
   - Hydraulic lines and plug
   - Steering gear mounting bolts and the steering gear

➡If necessary, move the steering drop arm by turning the steering stub to enable the removal of the gear assembly.

To install:

4. Install or connect the following:
   - Steering gear and tighten the mounting bolts
   - Hydraulic lines, using new seals
5. Turn the steering wheel counterclockwise or clockwise against the stop, then back about 1¾ turn until the marks are aligned.
   - U-joint to the steering gear making sure the bolt is in the locking groove of the steering stub
   - Tie rod to the steering drop arm and replace the self-locking nut
   - Heat shield on the steering gear and connect the ride level height control pipes, on 750iL models
   - Hydraulic fluid
   - Steering wheel
   - Negative battery cable

Strut

REMOVAL & INSTALLATION

3 Series

1. Before servicing the vehicle, refer to the Precautions Section.
2. Remove or disconnect the following:
   - Negative battery cable
   - Tire and wheel assembly
   - Brake pad wear indicator plug and ground wire
   - Wires out of the holder on the strut
   - ABS pulse sender, if equipped
   - Caliper and pull it away from the strut, suspending it from the body. Do not disconnect the brake line.
   - Attaching nut, then detach the pushrod on the stabilizer bar at the strut
   - Attaching nut and press off the guide joint with the proper tool
   - Nut and press off the tie rod joint
3. Press the bottom of the strut outward

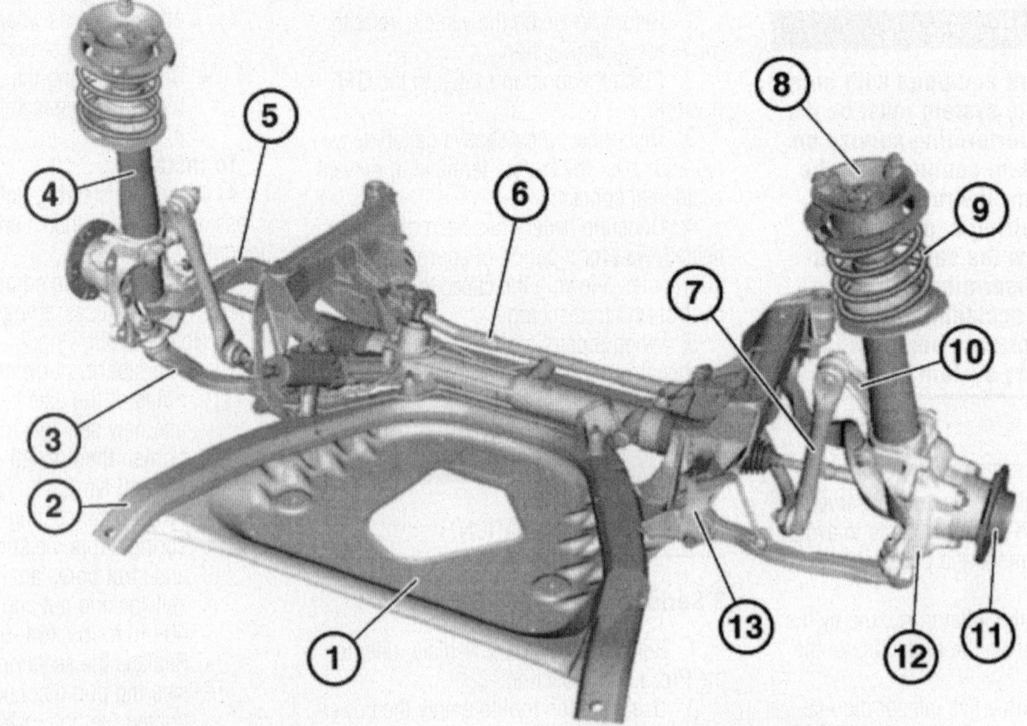

1	Reinforcement plate
2	Front axile support
3	Control arm
4	Spring strut shock absorber
5	Tension strut
6	Stabilizer/stabilizer suspension
7	Stabilizer link
8	Thrust bearing
9	Coil spring
10	Stabilizer link holder
11	Wheel bearing/drive flange
12	Swivel bearing
13	Ride level sensor

67200-BMWC-G09

5 Series front suspension and steering components

and push it over the guide joint pin, using the proper tool. Support the bottom of the strut.

4. Remove the nuts at the top of the strut from inside the engine compartment, then remove the strut.

To install:

5. Position the strut in the vehicle. Upper strut mounting nuts: 16–17 ft. lbs. (21–23 Nm). The upper strut mounting nuts must be replaced with new self-locking nuts.

6. The remaining components are installed in the reverse order from which they were removed. Tie rod and guide joints must have both pins and bores clean for reassembly. Replace both self-locking nuts. Tighten the control arm to spring strut attaching nut to 43–51 ft. lbs. (59–69 Nm).

7. Install the front tire and wheel assemblies.

8. Connect the negative battery cable.

5 and 7 Series

1. Before servicing the vehicle, refer to the Precautions Section.
2. Remove or disconnect the following:
   - Negative battery cable
   - Tire and wheel assembly
   - Brake pad wear indicator plug and ground wire
   - Wires out of the holder on the strut
   - ABS pulse sender, if equipped
   - Stabilizer pushrod with the proper tool
   - Lower strut bolts at the control arm
3. Support the bottom of the strut and remove the nuts at the top of the strut, from inside the engine compartment. Remove the strut.

To install:

4. The installation is the reverse of the removal procedure.

Shock Absorber

REMOVAL & INSTALLATION

3 Series

1. Before servicing the vehicle, refer to the Precautions Section.
2. Remove the trunk trim panel to expose the upper shock mounts.
3. Support the trailing arm and remove the lower mounting bolt.

WARNING

The support must not be removed until the new shock absorber is installed, and the vehicle must not be raised since this could damage the halfshafts.

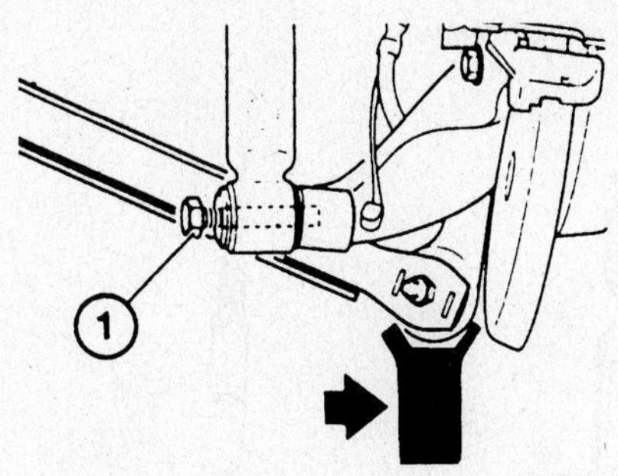

Support the trailing arm and remove the bolt (1)—3 Series

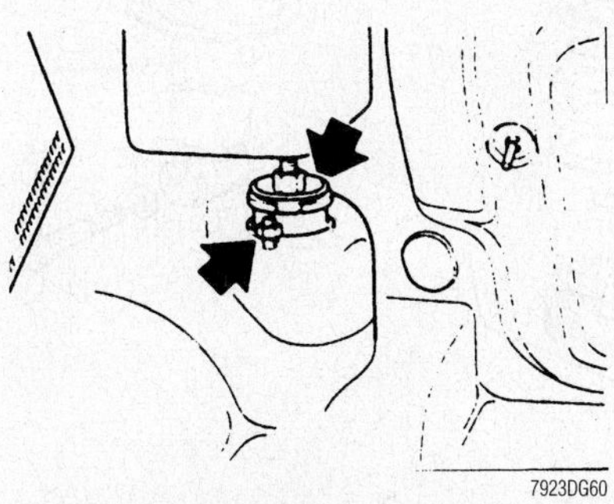

Upper mounting nut locations—3 Series

4. Remove the cap and remove the upper mounting nuts and remove the shock from the vehicle.

To install:

5. Place the shock into position with new seals fitted between the shock absorber and body. Renew the upper self-locking nuts: 11 ft. lbs. (15 Nm) for the Z3, 16 ft. lbs. (22 Nm) for all other 3 series vehicles.
6. Install the trunk trim panel.
7. Install the lower shock mounting to the rear axle assembly. The thrust washer on the rubber mount must face the screw head.
8. With the vehicle resting at standard ride height, tighten the mounting bolt to 63 ft. lbs. (87 Nm), or to 94 ft. lbs. (130 Nm) if marked with 10.9, for the Z3, or to 74 ft. lbs. (100 Nm) for all other 3 series models.

5 and 7 Series

STANDARD SUSPENSION

1. Before servicing the vehicle, refer to the Precautions Section.
2. Remove or disconnect the following:
   - Rear seat cushion and back rest
   - Trim panel over the strut mount
   - With the control arm supported, rubber cap and remove the nuts at the top of the strut mount
   - Lower mounting bolt and lower the spring/shock assembly. Remove the assembly from the vehicle.
3. Use a spring compressor and compress the spring. Remove the top nut and pull the top mount off. Remove the spring.

To install:

4. Compress the new spring or replace the old spring on the shock. Install the mount and washers. Use a new locknut and tighten to 18 ft. lbs. (25 Nm). Release the spring.
5. Install the shock. Upper mount nuts: 16 ft. lbs. (21.5 Nm). Loosely install the lower mounting bolt.
6. With the vehicle lowered to the ground and at standard riding height, tighten the lower mount to 94 ft. lbs. (130 Nm).
7. Install the trim and seat cushions.

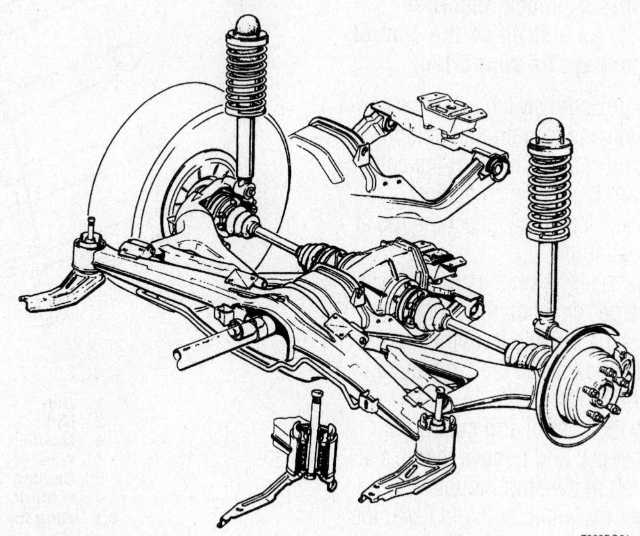

Rear suspension system—5 Series

REAR SPRING STRUT LAYOUT DRAWING

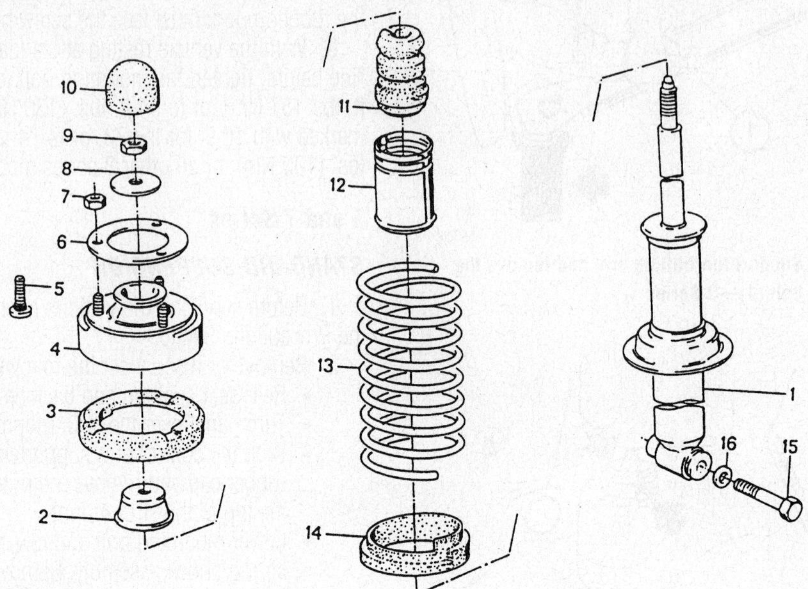

3 Upper spring ring
4 Mount
5 Bolt
6 Insulator
7 Collar nut M 8
8 Disc
9 Hexagon nut M 10 x 1.8 ZN
12 Protective tube
13 Coil spring
14 Lower spring ring
15 Bolt M 14 x 1.5 x 85
16 Washer

7923DG62

Exploded view of the shock and coil spring assembly—5 and 7 Series

RIDE LEVEL HEIGHT CONTROL SUSPENSION

1. Before servicing the vehicle, refer to the Precautions Section.
2. Remove or disconnect the following:
   - Rear seat cushion and back rest
   - Trim panel over the strut mount

➡The coil spring, shock absorber assembly acts as a strap so the control arm should always be supported.

   - Low pressure switch electrical connection and turn on the ignition
   - Control rod nut, holding the collar with an 8mm wrench against torque. Don't disconnect the rod at the ball joint.
3. Operate the lever on the control switch in the "discharge" direction for about 20 seconds to discharge fluid from the lines.
   - Hydraulic line on the strut and turn off the ignition
   - With the control arm supported, rubber cap and remove the nuts at the top of the strut mount
   - Lower mounting bolt and lower the spring strut assembly. Remove the assembly from the vehicle.
4. Use a spring compressor and compress the spring. Remove the top nut and pull the top mount off. Remove the spring.

To install:

5. Compress the new spring or replace the old spring on the strut. Install the mount and washers. Use a new locknut and tighten to 18 ft. lbs. (25 Nm). Release the spring.
6. Install or connect the following:
   - Spring strut. Upper mount nuts: 16 ft. lbs. (21.5 Nm). Loosely install the lower mounting bolt.
   - Hydraulic line on the strut
   - Control rod nut, holding the collar with an 8mm wrench against torque
   - Low pressure switch electrical connection
7. With the vehicle lowered to the ground and at standard riding height, tighten the lower mount to 94 ft. lbs. (130 Nm).
8. Install the trim and seat cushions.

Coil Spring

REMOVAL & INSTALLATION

Front

⁂ CAUTION

This procedure calls for the spring to be compressed. A compressed spring

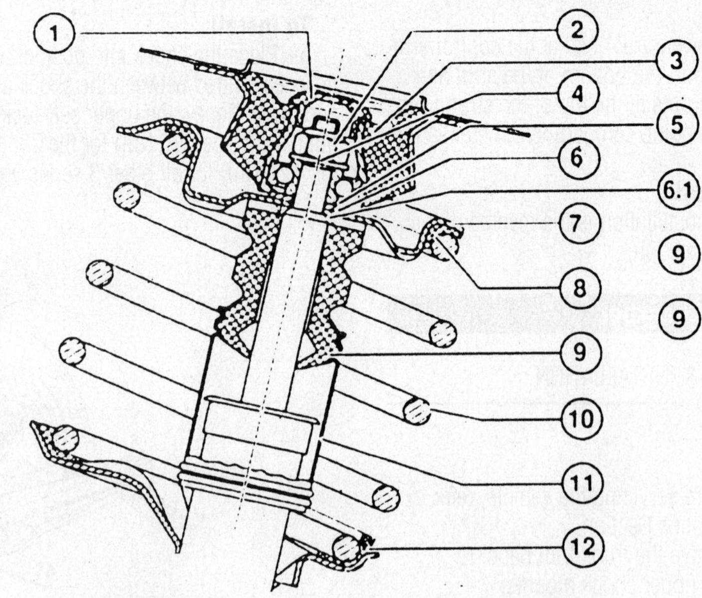

1 Cap
2 Nut
3 Mount
4 Washer
5 Sealing ring
6 Washer
6.1 Ring for hollow piston rod
7 Upper plate spring
8 Upper spring ring
9 Rubber damper
10 Coil spring
11 Protective tube
12 Lower spring ring

7923DG66

Cut away view of the strut mount and related components—3 Series models

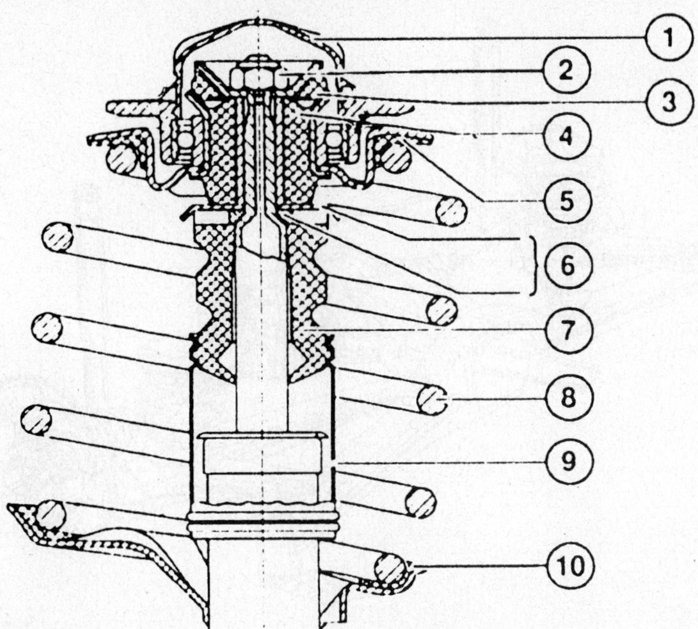

Cut away view of the strut mounting with a separate support bearing and related components—3 Series models

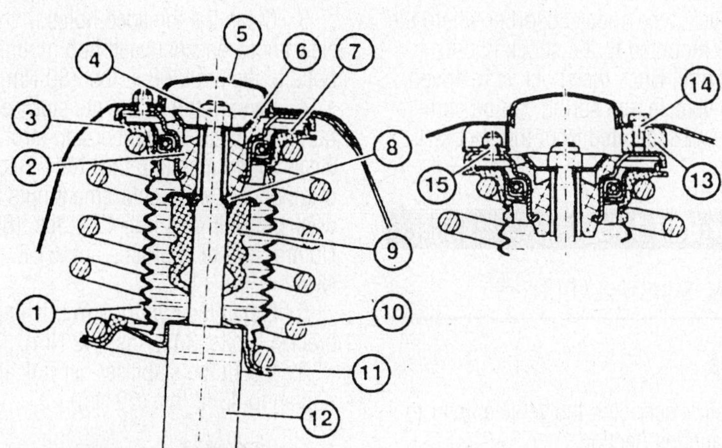

Cut away view of the strut mount and related components—5 and 7 Series

has high potential energy and if released suddenly can cause severe damage and personal injury.

1. Before servicing the vehicle, refer to the Precautions Section.
2. Remove the strut from the vehicle and mount in a vise using a strut holder. This will prevent damage to the strut tube.
3. Using a proper spring compressor, compress the spring and lock into place.
4. Remove the top nut of the strut mount. Counterhold the strut rod during removal.
5. Pull the strut mount off the strut rod. Note the positioning of the spacers and washer for replacement.
6. Pull the spring off the strut and place aside in a safe area.
7. Slowly release the compression of the spring.

To install:

8. Install the spring in the compressor and compress.
9. Install the spring and strut mount with all the spacers and washers in their original positions. New strut rod nut: 47 ft. lbs. (65 Nm).
10. Release the spring slowly and check that it seats in the spring holders. Install the strut in the vehicle.

Rear

3 SERIES—EXCEPT Z3 AND Z4

1. Before servicing the vehicle, refer to the Precautions Section.
2. Remove the tire and wheel assembly.
3. Support the lower trailing arm at the hub with a suitable jack and disconnect the stabilizer bar at the control arm and the subframe.
4. Remove the shock absorber lower mounting bolt.
5. Lower the trailing arm slowly and remove the spring to the side.

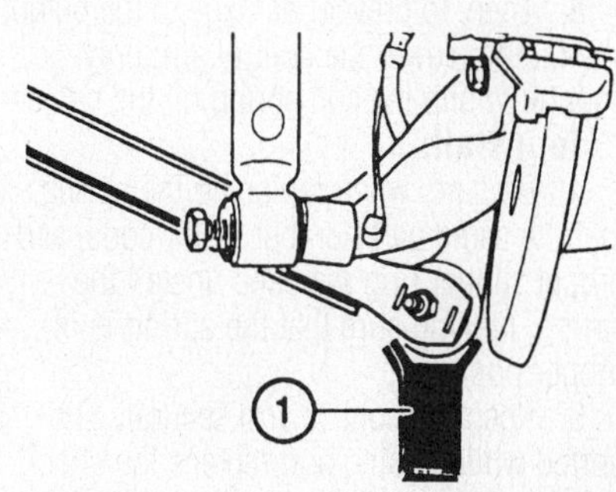

Support the trailing arm (1)—3 Series, except Z3 and Z4

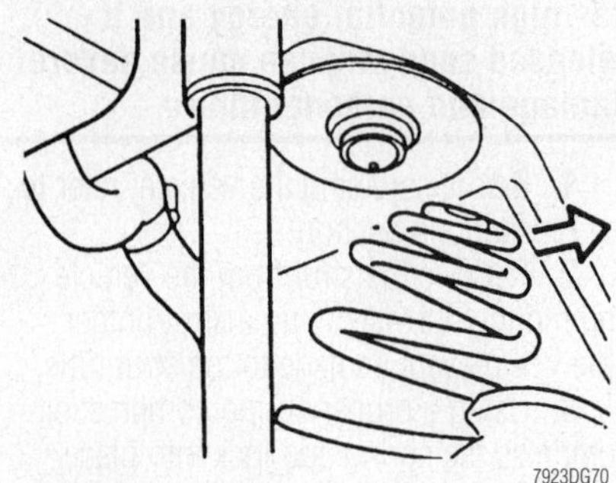

Remove the coil spring—3 Series, except Z3 and Z4

To install:

6. Install the spring with the bushing in place and the top of the upper spring ring lubricated.
7. Raise the trailing arm to a level where the bolt can be replaced in the lower shock mount. Connect the stabilizer bar. Do not fully tighten the bolts at this time.
8. Install the tire and wheel assembly.
9. Tighten the stabilizer bolt to 16 ft. lbs. (21.5 Nm) and the shock bolt to 63 ft. lbs. (87 Nm) with the control arm in the normal ride position.

Z3 AND Z4

1. Before servicing the vehicle, refer to the Precautions Section.
2. Remove or disconnect the following:
   - Rear portion of the exhaust system and support it from the body
   - Final drive rubber mount, push it downward, and hold it down with a suitable wedge
   - Bolt that connects the rear stabilizer bar to the strut on the side being worked on. Be careful not to damage the brake line.

➡Support the lower control arm securely with a suitable jack or other device that will permit it to be lowered gradually, while maintaining a secure support.

3. Then, to prevent damage to the output shaft joints, lower the control arm only enough to slip the coil spring off the retainer.

To install:

4. Be sure, when replacing the spring, that the same part number, color code, and proper rubber ring are used. Install the spring, making sure that the spring is in proper position.
5. Keep the control arm securely supported while raising and replace the shock bolt. Install the bolts in the final drive rubber mount and tighten to 69 ft. lbs. (95 Nm).

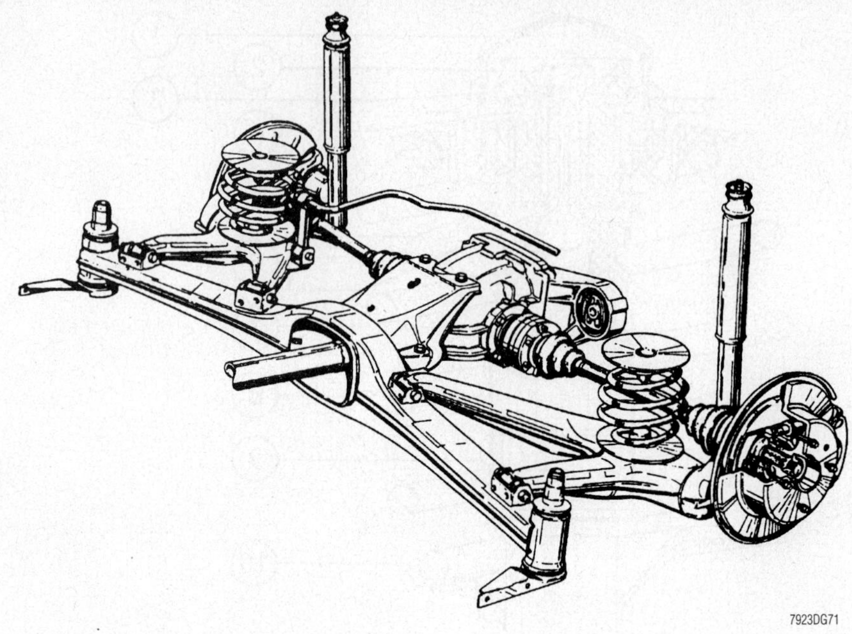

Rear suspension setup on Z3 and Z4 models

6. Tighten the stabilizer bolt to 16 ft. lbs. (21.5 Nm), and the shock bolt to 63 ft. lbs. (87 Nm) with the control arm in the normal ride position. Install the exhaust system.

5 AND 7 SERIES

The coil spring is removed along with the shock absorber. The 5 and 7 Series use a "coil over" type shock absorber where the spring is mounted to the shock in one compact unit. Once the shock is removed from the vehicle, the spring can be compressed and separated from the shock absorber.

Lower Ball Joint

REMOVAL & INSTALLATION

3 Series

1. Before servicing the vehicle, refer to the Precautions Section.
2. Remove or disconnect the following:
   - Front tire and wheel assembly
   - Rear control arm bushing bracket where it connects to the body by removing the bolts
   - Nut and disconnect the link on the front stabilizer bar where it connects to the control arm
   - Nut that attaches the control arm to the crossmember, and remove the nut from above the crossmember. Then, use a soft faced hammer to knock the stud out of the crossmember.
   - Nut to the point it contacts the strut housing
   - Bolts connecting the hub to the struts. Press off the ball joint where the control arm attaches to the lower end of the strut, using a suitable tool.

To install:

3. Clean the threaded holes in the hub. New micro-encapsulated hub mounting bolts to the strut: 58 ft. lbs. (80 Nm).
4. Be sure the ball joints studs and the bores in the crossmember and strut are clean before inserting the studs. Replace the original nuts with replacement nuts and washers. Ball joint nut: 47 ft. lbs. (65 Nm). Control arm to subframe nut: 61 ft. lbs. (85 Nm).
5. Install the control arm bushing bracket. Bolts: 30 ft. lbs. (42 Nm).
6. Install the stabilizer bar link and tighten to 43 ft. lbs. (59 Nm).

5 and 7 Series

The lower ball joint is serviced with the lower control arm as an assembly.

Lower Control Arms

REMOVAL & INSTALLATION

3 Series

1. Before servicing the vehicle, refer to the Precautions Section.
2. Raise and safely support the vehicle. Remove the front tire and wheel assembly.

Use a piece of wire to prevent the strut from extending to far and damaging the brake hose.

3. Disconnect the rear control arm bushing bracket where it connects to the body by removing the bolts.

4. Remove the nut and disconnect the link on the front stabilizer bar where it connects to the control arm.

5. Unscrew the nut which attaches the control arm to the crossmember and remove the nut from above the crossmember. Then, use a plastic hammer to knock the stud out of the crossmember.

6. Unscrew the nut to the point it contacts the strut housing. Remove the bolts connecting the hub to the struts. Press off the ball joint where the control arm attaches to the lower end of the strut, using the proper tool.

To install:

7. Clean the threaded holes in the hub. Install new micro-encapsulated hub mounting bolts to the strut and torque to 58 ft. lbs. (80 Nm).

8. Make sure the ball joints studs and the bores in the crossmember and strut are clean before inserting the studs. Replace the original nuts with replacement nuts and washers. Torque the ball joint nut to 47 ft. lbs. (65 Nm) for 2 wheel drive. Torque the control arm to subframe nut to 61 ft. lbs. (85 Nm) for 2 wheel drive.

9. Install the control arm bushing bracket and torque the bolts to 30 ft. lbs. (42 Nm).

10. Install the stabilizer bar link and torque to 43 ft. lbs. (59 Nm).

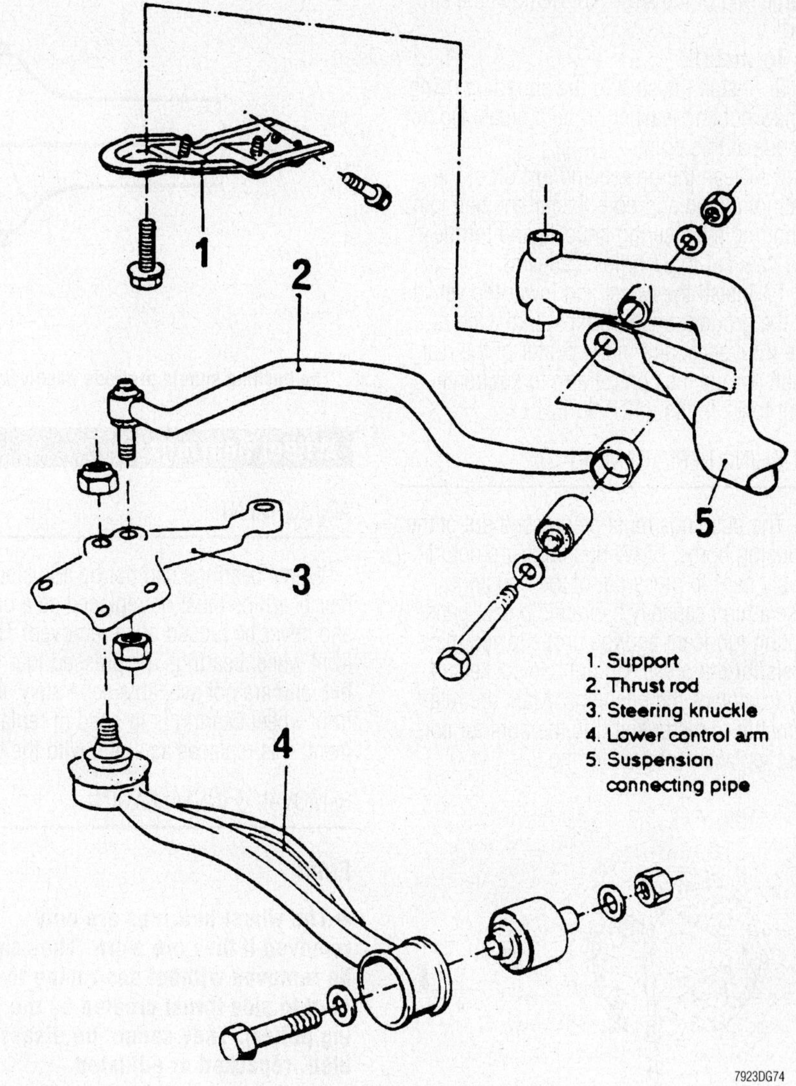

Exploded view of the front suspension—5 and 7 Series

5 and 7 Series

1. Before servicing the vehicle, refer to the Precautions Section.

2. Raise and safely support the front end. Do not place the jackstands under any suspension parts. Remove the wheel.

3. Remove the 3 bolts holding the steering knuckle to the bottom of the strut.

4. Remove the ball joint nut and press the stud out of the steering knuckle with a ball joint remover tool.

5. Remove the nut and bolt at the subframe end of the control arm. Remove the control arm.

To install:

6. Install the control arm to the subframe using a new nut and washer on both sides. Do not torque at this point.

7. Clean the grease and dirt off of the ball joint stud and bore. Install the ball joint stud into the steering knuckle and torque the new nut to 67 ft. lbs. (93 Nm).

8. Clean the threads and bores of the steering knuckle mounting bolts and the strut housing. Install the bolts, using threadlocker, and torque to 80 ft. lbs. (110 Nm). There is a groove that will align the strut and knuckle.

9. Install the wheel and lower the vehicle to the ground. Load 150 lbs. into each of the front seats and in the center of the rear seat. Torque the control arm to subframe bolt to 56 ft. lbs. (77.5 Nm).

BUSHING REPLACEMENT

The bushings must be pressed out of the housing bores. BMW bushings are notoriously hard to press out of the housings. Use a high capacity hydraulic press, penetrating lubricant and the proper sized mandrels for the press. Do not use sockets to try to replace the bushings. Mark the relationship of the bushing to the bore for correct replacement positioning.

Thrust Rod

REMOVAL & INSTALLATION

5 and 7 Series

Always replace the strut rods in pairs. If the strut rods are not replaced in pains, uneven driving response may result.

1. Before servicing the vehicle, refer to the Precautions Section.

2. Raise and safely support the front end. Do not place the jackstands under any suspension parts. Remove the wheel.

3. Remove the thrust rod ball joint nut and press the stud out of the steering knuckle with a ball joint remover tool.

4. Remove the nut and bolt at the sub-

frame end of the strut rod. Remove the strut rod.

To install:

5. Install the strut to the subframe using a new nut and washer on both sides. Do not torque at this point.
6. Clean the grease and dirt off of the ball joint stud and bore. Install the ball joint stud into the steering knuckle and torque the new nut to 67 ft. lbs. (93 Nm).
7. Install the wheel and lower the vehicle to the ground. Load 150 lbs. into each of the front seats and in the center of the rear seat. Torque the control arm to subframe bolt to 92 ft. lbs. (127 Nm).

BUSHING REPLACEMENT

The bushings must be pressed out of the housing bores. BMW bushings are notoriously hard to press out of the housings. Use a high capacity hydraulic press, penetrating lubricant and the proper sized mandrels for the press. Do not use sockets to try to replace the bushings. Mark the relationship of the bushing to the bore for correct replacement positioning.

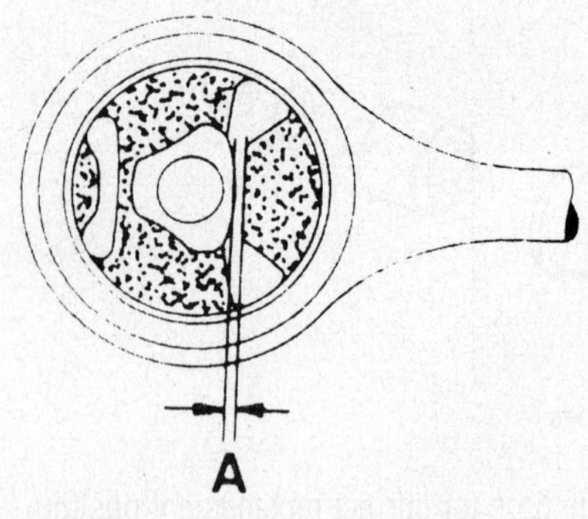

With the vehicle loaded and at normal ride height, the gap at A in the lower control arm bushing should be 0.039–0.079 in. (1.0–2.0 mm)

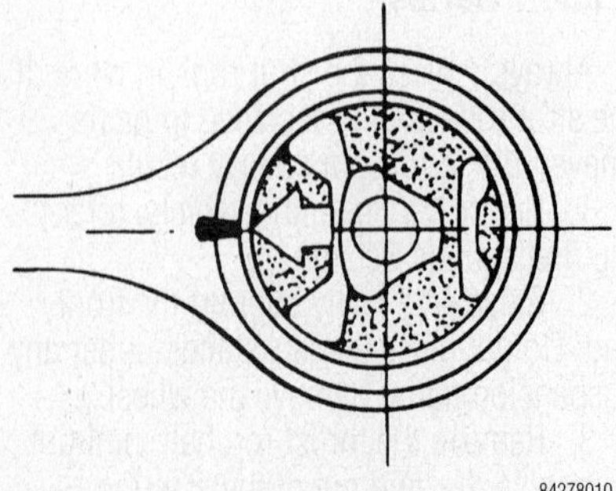

The lower control arm bushing arrow should match the cast boss on the bracket

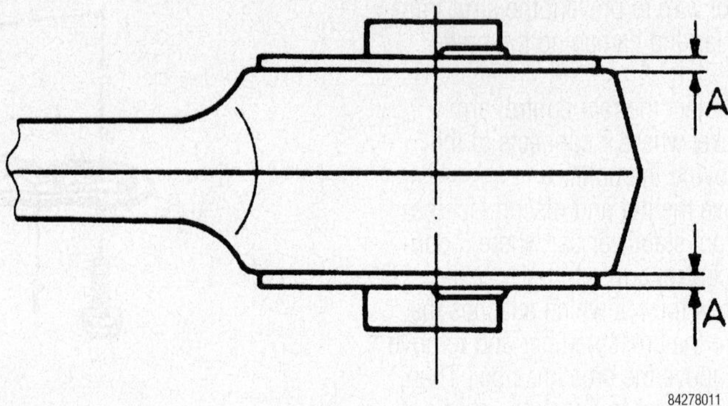

The bushing should protrude evenly when installing a replacement

Wheel Bearings

ADJUSTMENT

Wheel bearings can not be adjusted. The rear bearings must be replaced as a unit and never be reused once removed. The front wheel bearings are pressed into the hub and are not available separately. If a front wheel bearing is in need of replacement, it is replaces as a unit with the hub.

REMOVAL & INSTALLATION

Front

➡The wheel bearings are only removed if they are worn. They cannot be removed without destroying them (due to side thrust created by the bearing puller). They cannot be disassembled, repacked or adjusted.

1. Before servicing the vehicle, refer to the Precautions Section.
2. Remove or disconnect the following:
   - Front wheel
   - Caliper, then suspend the brake caliper to avoid putting stress on the brake line
   - Setscrew, with a suitable hex wrench
   - Brake disc, then the dust cover suing a suitable prytool or chisel to remove the pressed on cover
3. Using a suitable chisel, knock the tab on the collared nut away from the shaft. Remove and discard the nut.
   - The bearing with a puller set tool Nos. 31–2–101/102/104 or their equivalents and discard it. On M3 models, use a puller set tool No. 31–2–102/105/106 or their equivalents. On M3 models, install the main bracket of the puller with 3 wheel bolts.
4. If the inside bearing inner race remains

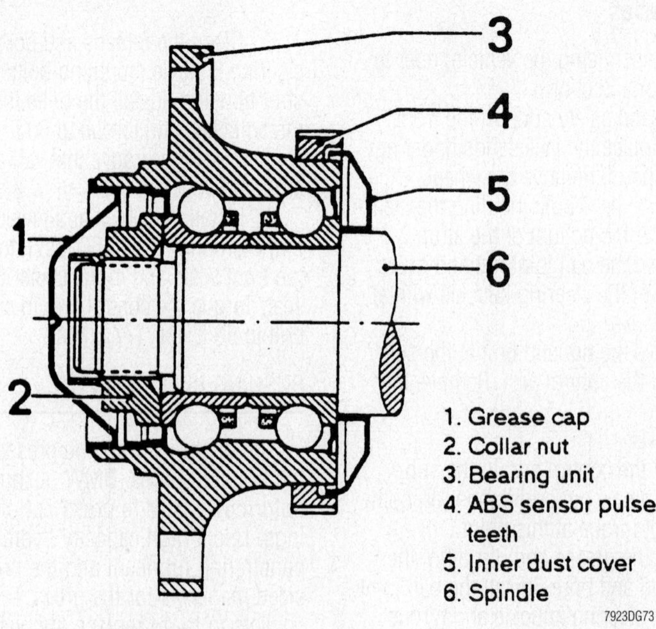

Cut away view of the front wheel bearing

on the stub axle, unbolt and remove the dust guard. Bend back the inner dust guard and pull the inner race off using tool No. 00–7–500 and 33–1–309 or their equivalents, which are capable of getting under the race to remove it. Reinstall the dust guard.

To install:

5. If the dust guard has been removed, install a new one. Install tool No. 31–2–120; except the M3 models, which require tool No. 31–2–110. Place the tool over the stub axle and screw it in the entire length of the guide sleeve's threads, then press the bearing on.
6. Reverse the remaining removal procedures to install the disc and caliper. Using a new collared mounting nut, tighten the wheel hub collar nut to 210 ft. lbs. (290 Nm). Lock the collar nut by bending over the tab into the groove of the spindle.

Rear

1. Before servicing the vehicle, refer to the Precautions Section.
2. Remove or disconnect the following:
    - Negative battery cable
    - Wheel assembly
    - Half shaft assembly
    - Brake caliper assembly and support it such that there is no tension on the brake hose
    - Rear brake caliper and disconnect the parking brake cable
    - Brake dust cover along with the parking brake assembly
3. Matchmark the mounting brackets for the trailing arm
4. Support the trailing arm with a suitable jack and remove the fasteners that attach the rear shock assembly to the trailing arm.
5. Remove or disconnect the following:
    - Stabilizer bar, and if installed, the upper control arm
    - Trailing arm assembly
    - Large snapring in front of the wheel bearing
6. Using suitable press tools, support the trailing arm in a suitable hydraulic press and press the bearing from the trailing arm assembly.

To install:

7. Inspect the trailing arm for damage and thoroughly clean the area where the wheel bearing is installed. Make sure there are no sharp edges on the entrance of the bore where the bearing is to be installed. If necessary, chamfer the edges of the housing to avoid binding when pressing the bearing in place.
8. Using a suitable hydraulic press and press tools, press the bearing into the trailing arm.
9. Install or connect the following:
    - Bearing snapring into the trailing arm
    - Trailing arm onto the vehicle, making sure to align the match marks made during disassembly
10. The balance of the assembly procedure is in reverse of the removal procedure.
11. Install the road wheel and torque the wheel bolts as follows:
    - All models except the 740i with sport package 18 inch wheels: 80 ft. lbs. (108 Nm)
    - 740i models with sport package 18 inch wheels: 100 ft. lbs. (136 Nm)
12. Check and adjust the rear wheel alignment if necessary.

BRAKES

Brake Caliper

REMOVAL & INSTALLATION

All Models

FRONT

1. Before servicing the vehicle, refer to the Precautions Section.
2. Draw off brake fluid with a syringe.
3. Remove or disconnect the following:
    - Hydraulic brake lines
    - Wheels
    - Caliper mounting bolts and disconnect the brake pad wear indicator plug
    - Caliper assembly

To install:

4. Install or connect the following:
    - Brake caliper onto the steering knuckle. Caliper mounting bolt torque for all 3 Series is 63–79 ft. lbs. (85–108 Nm). Caliper mounting bolt torque is 80–89 ft. lbs. (109–121 Nm) for all other models.
    - Guide bolts to 22–25 ft. lbs. (30–34 Nm)

➡Make sure the brake wear indicator wire is held in the correct position by the tab of the dust cap.

- Front wheels
- Hydraulic brake system lines and bleed the brake system

REAR

1. Before servicing the vehicle, refer to the Precautions Section.
2. Draw off brake fluid with a syringe.
3. Remove or disconnect the following:
    - Hydraulic brake lines
    - Rear wheels
    - Caliper mounting bolts and disconnect the brake pad wear indicator plug
    - Caliper assembly by pulling to the rear

To install:

4. Install or connect the following:
    - Brake caliper onto the steering knuckle. Tighten the caliper mounting bolts to 42–48 ft. lbs. (56–66 Nm).
    - Guide bolts to 22–25 ft. lbs. (30–34 Nm)

➡Make sure the brake wear indicator wire is held in the correct position by the tab of the dust cap.

- Wheels
- Hydraulic brake system lines and bleed the brake system

Disc Brake Pads

REMOVAL & INSTALLATION

All Models

FRONT

1. Before servicing the vehicle, refer to the Precautions Section.
2. Remove or disconnect the following:
    - Wheels
    - Brake pad wear indicator connector
    - Caliper guide bolts and the spring clamp
3. Turn up the caliper and remove the brake pads. The inner pad is located with a spring in the piston.

To install:

4. Lubricate the mounting pads with suitable grease.
5. Install or connect the following:
    - Brake pads onto the brake caliper, then swing the caliper down until the lower mounting bolt holes are aligned
    - Mounting bolts and spring clamp. Caliper mounting bolt torque for all 3 Series is 63–79 ft. lbs. (85–108 Nm). Caliper mounting bolt torque

is 80–89 ft. lbs. (109–121 Nm) for all other models.
- Guide bolts to 22–25 ft. lbs. (30–34 Nm).
- Wheels

6. Bleed the brake system if any of the brake system lines were disconnected.

REAR

1. Before servicing the vehicle, refer to the Precautions Section.
2. Remove or disconnect the following:
   - Wheels
   - Plug for the brake pad wear indicator
   - Caliper guide bolts and the spring clamp
3. Turn up the caliper and remove the brake pads. The inner pad is located with a spring in the piston.

To install:

4. Lubricate the mounting pads with suitable grease.
5. Install or connect the following:
   - Brake pads onto the brake caliper, then swing the caliper down until the lower mounting bolt holes are aligned
   - Mounting bolts and spring clamp. Tighten the caliper mounting bolts to 42–48 ft. lbs. (56–66 Nm), and the guide bolts to 22–25 ft. lbs. (30–34 Nm).
   - Wheels
6. Bleed the brake system if any of the brake system lines were disconnected.

BMW

X3

3

SPECIFICATIONS AND MAINTENANCE CHARTS

ENGINE AND VEHICLE IDENTIFICATION

Engine							Model Year	
Code	Liters (cc)	Cu. In.	Cyl.	Fuel Sys.	Engine Type	Eng. Mfg.	Code ①	Year
M54B30	3.0 (2979)	182	6	SMPI	DOHC	BMW	6	2006
							7	2007

SMPI: Sequencial Multi-port Fuel Injection

DOHC: Double Overhead Camshaft

① 10th digit of the Vehicle Identification Number (VIN)

06041_BMX3_C0001

GENERAL ENGINE SPECIFICATIONS

Year	Body Type	Model	Engine Displacement Liters	Engine ID	Net Horsepower @ rpm	Net Torque @ rpm (ft. lbs.)	Bore x Stroke (in.)	Compression Ratio	Oil Pressure @ rpm
2006	E83	X3	3.0	M54B30	225@5900	214@3500	3.31x3.53	10.2:1	7.4@700

06041_BMX3_C0002

ENGINE TUNE-UP SPECIFICATIONS

Year	Engine Displacement Liters	Engine ID	Spark Plug Gap (in.)	Ignition Timing (deg.)		Fuel Pump (psi)	Idle Speed (rpm)		Valve Clearance	
				MT	AT		MT	AT	In.	Ex.
2006	3.0	M54B30	0.024-0.028	①	①	48-54	②	②	HYD	HYD

NOTE: The Vehicle Emission Control Information label reflects specification changes during production and must be used if they differ from this chart.

HYD: Hydraulic

① Ignition timing is regulated by the Engine Control Module (ECM) and cannot be adjusted.

② Idle speed is controlled by the Engine Control Module (ECM) and cannot be adjusted.

06041_BMX3_C0003

CAPACITIES

Year	Model	Engine Displacement Liters	Engine ID	Engine Oil with Filter (qts.)	Transmission (pts.)		Drive Axle		Fuel Tank (gal.)	Cooling System (qts.)
					6-Spd	Auto.	Front (pts.)	Rear (pts.)		
2006	X3	3.0	M54B30	8.0	3.2	7.0	—	3.4	24.6	11.1

NOTE: Capacities given are service, not overhaul. All capacities are approximate. Add fluid gradually and ensure a proper fluid level is obtained.

06041_BMX3_C0004

VALVE SPECIFICATIONS

Year	Engine Displacement Liters	Engine ID	Seat Angle (deg.)	Face Angle (deg.)	Spring Test Pressure (lbs. @ in.)	Spring Installed Height (in.)	Stem-to-Guide Clearance (in.)		Stem Diameter (in.)	
							Intake	Exhaust	Intake	Exhaust
2006	3.0	M54B30	①	45	NA	NA	②	②	0.2372-0.2340	0.278-0.2384

NA: Information not available

① Valve seat angle: 45 degrees
Correction angle outside: 15 degrees
Correction angle inside: 60 degrees

② To measure: Insert a new valve into guide with the end of the valve flush with the end of the guide.
Use a dial indicator to measure axial valve head movement.

06041_BMX3_C0005

CRANKSHAFT AND CONNECTING ROD SPECIFICATIONS

All measurements are given in inches.

Year	Engine Displacement Liters	Engine ID	Crankshaft				Connecting Rod		
			Main Brg. Journal Dia.	Main Brg. Oil Clearance	Shaft End-play	Thrust on No.	Journal Diameter	Oil Clearance	Side Clearance
2006	3.0	M54B30	①	0.0007-0.0023	0.0031-0.0064	5	1.7720-1.7706	0.0007-0.0022	0.0060-0.0160

① Standard yellow 2.3615-2.3618 inches
Standard green: 2.3613-2.3615 inches
Standard white: 2.3611-2.3613 inches

06041_BMX3_C0006

PISTON AND RING SPECIFICATIONS

All measurements are given in inches

	Engine			Ring Gap			Ring Side Clearance		
Year	Size Liters	Engine ID	Piston Clearance	Top Compression	Bottom Compression	Oil Control	Top Compression	Bottom Compression	Oil Control
2006	3.0	M54B30	0.0004-0.0016	0.0078-0.0157	0.0078-0.0157	0.0098-0.0197	0.0008-0.0024	0.00012-0.0026	0.0007-0.0024

06041_BMX3_C0007

TORQUE SPECIFICATIONS

All readings in ft. lbs

Year	Engine Displacement Liters	Engine ID	Cylinder Head Bolts	Main Bearing Bolts	Rod Bearing Bolts	Crankshaft Damper Bolts	Flywheel Bolts	Manifold Intake	Manifold Exhaust	Spark Plugs	Wheel Lug Nut
2006	3.0	M54B30	①	②	③	302	④	⑤	⑥	⑦	80

① Cast iron block. Replace, wash and oil bolts
Step 1: 22 ft. lbs.
Step 2: +90 degrees
Step 3: +90 degrees

② Cast iron block. Replace, wash and oil bolts
Step 1: 14.8 ft. lbs.
Step 2: +50 degrees

③ Replace, wash and oil connecting rod bolts
Step 1: 14.8 ft. lbs.
Step 2: +70 degrees

④ New micro-encapsulated screws:
Automatic transmission: 88 ft. lbs.
Manual transmission: 77.4 ft. lbs.

⑤ All M6 fasteners: 88 inch lbs.
All M7 fasteners: 11 ft. lbs.
All M8 fasteners: 16 ft. lbs.

⑥ Coat with Molykkote HSC compound or equivalent
All M6 fasteners: 88 inch lbs.
All M7 fasteners: 14.8 ft. lbs.

⑦ M12x1.25:14.8-19.1 ft. lbs.
M14x1.25: 21.4-24.3 ft. lbs.

06041_BMX3_C0008

WHEEL ALIGNMENT

			Caster		Camber			Steering
Year	Model		Range (+/-Deg.)	Preferred Setting (Deg.)	Range (+/-Deg.)	Preferred Setting (Deg.)	Toe-in (Deg.)	Axis Inclination (Deg.)
2006	X3	F	0.50	+0.83	0.42	-0.20	0.1 +/- 0.16	—
		R	—	—	0.33	+0.83	0.26 +/- 0.13	0.06

F: Front
R: Rear

06041_BMX3_C0009

BRAKE SPECIFICATIONS

All measurements in inches unless noted

			Brake Disc			Parking Brake Drum				Brake Caliper	
Year	Model		Original Thickness	Minimum Thickness	Maximum Runout	Original Inside Diameter	Max. Wear Limit	Maximum Machine Diameter	Minimum Lining Thickness	Bracket Bolts (ft. lbs.)	Mounting Bolts (ft. lbs.)
2006	X3	F	NA	0.921 ①	0.003 ②	—	—	—	0.118	81	23
		R	NA	0.803 ①	0.003 ②	7.28	NA	0.031	0.118	49	21

NA: Information not available

F: Front

R: Rear

① Minimum thickness is stamped in the brake disk shell

② Only with precision turned brake disc

06041_BMX3_C0010

SCHEDULED MAINTENANCE INTERVALS

BMW—X3

TO BE SERVICED	TYPE OF SERVICE	SERVICE INTERVALS			
		INITIAL 1200 MILES	OIL SERVICE	INSPECTION I	INSPECTION II
Oil level	S/I	✓			
Engine oil & filter	R ②		✓	✓	✓
Engine air cleaner element	R ③				✓
Spark plugs	R				✓
Fuel filter	R ④				✓
Fuel, vapor lines & fuel cap	S/I	✓		✓	✓
Cooling system	S/I	✓		✓	✓
Exhaust pipe & muffler	S/I	✓		✓	✓
Catalytic converter & shielding	S/I	✓		✓	✓
Throttle linkage	S/I			✓	✓
Engine (check for leakage)	S/I	✓			
Engine drive belts	S/I				✓
Maintenance Indicators	RE		⑤	✓	✓
Engine coolant	R			⑥	⑥
Oxygen sensor	R ⑦				
Intake air dust separators	S/I ⑧				✓
Brake & clutch fluids ⑥	S/I			✓	✓
Brake pads & discs	S/I			✓	✓
Parking brake system	S/I			✓	✓
Power steering system	S/I			✓	✓
Rear axle fluid	S/I			✓	✓
Steering play, suspension track rods, front axle joints, steering linkage & joint disc	S/I			✓	✓
Transmission fluid/oil	S/I			✓	⑨
Wheel centering hubs	S/I			✓	✓
Rear axle fluid ⑩	R		✓		✓
OBD system for codes	S/I	✓		✓	✓

R: Replace S/I: Service or Inspect RE: Reset

Note: BMW does not rely solely on vehicle mileage to determine service intervals. An on-oboard diagnostic center, monitors engine operating conditions, along with mileage, to determine the most effective maintenance intervals. The information is then conveyed to the driver through the service indicator lights, located in the center of the instrument panel.

② On vehicles operated less than 6200 miles per year, more frequent service may be required.

③ Replace more frequently if vehicle is operated in dusty conditions.

④ Recommended service for California models, required for all other models.

⑤ Reset the oil service indicator lights only.

⑥ Replace every 2 years with inspection service.

⑦ Replace every 100,000 miles on all models.

⑧ Required service for manual transmission models only.

⑨ Change fluid (A/T) or oil (M/T) at inspection.

⑩ Check at first oil service, then at each inspection.

FREQUENT OPERATION MAINTENANCE (SEVERE SERVICE)

If a vehicle is operated under any of the following conditions it is considered severe service

- Extremely dusty areas.

- 50% or more of the vehicle operation is in 32°C (90°F) or higher temperatures, or constant operation in temperatures below 0°C (32°F).

- Prolonged idling (vehicle operation in stop and go traffic).

- Frequent short running periods (engine does not warm to normal operating temperatures).

- Police, taxi, delivery usage or trailer towing usage.

06041_BMX3_C0011

ENGINE REPAIR

➡**Disconnecting the negative battery cable on some vehicles may interfere with the functions of the on-board computer systems and may require the computer to undergo a relearning process.**

Ignition Timing

ADJUSTMENT

The Digital Motor Electronics (DME) control, unit controls all ignition and fuel injection functions. Ignition timing is fully electronically controlled; there is no vacuum advance or manual adjustment. Ignition functions are calculated from internal maps and from the same sensors used for the fuel injection system. On vehicles with an automatic transmission, the control unit will retard ignition timing briefly when the transmission is about to shift up or down. For this reason, there is a data link between the DME control unit and the transmission control unit.

Since the ignition timing is controlled by the DME, checking and adjusting the timing is impossible. There is no method of setting dynamic or static timing.

Alternator

REMOVAL & INSTALLATION

➡**When the battery is disconnected the radio code, on-board computer and clock settings will be lost. The radio code should be obtained before disconnecting the battery or radio. Once the battery has been reconnected, the radio will not function unless the code is keyed in.**

1. Before servicing the vehicle, refer to the Precautions Section.
2. Read out stored fault codes (if applicable).
3. Switch off ignition.
4. Disconnect negative battery cable.
5. Remove or disconnect the following:
   - Suction filter housing
   - Fan clutch
   - Alternator drive belt
   - Power steering pump from mounting (move aside)
   - Alternator air hose (if equipped)
   - Electrical connections from alternator

Showing idler pulley bolt and alternator bolt

   - Alternator mounting bolts (versions without idler pulley)
   - Idler pulley cover and idler pulley bolt (versions with idler pulley)
   - Alternator

To install:

6. To install, reverse removal procedure.
7. If equipped with idler pulley, turn lock of tensioning roller to engage alternator groove.
8. With scan tool, clear fault code memory.

Engine Assembly

REMOVAL & INSTALLATION

1. Before servicing the vehicle, refer to the Precautions Section.
2. Disconnect the negative battery cable.
3. Remove or disconnect the following:
4. Remove the heater bulkhead by pulling off the sealing strips from the firewall, turning the toggle retainers about 90 degrees, lifting out the cover, and removing the heater bulkhead.
   - Engine cover
   - Both oxygen sensors
   - Hood to full open (assembly) position
   - Front splash guard
   - Reinforcement plate from undercarriage
   - Vacuum line from vacuum accumulator

➡**On automatic transmissions, remove transmission fluid line bolt on engine oil sump.**

   - Transmission
   - A/C compressor from mounting (move aside, with lines connected)

5. Drain the cooling system (drain plug is located on exhaust side of cylinder number 2 in engine block). Reinstall drain plug with new seal ring.
6. Pull the locks and disconnect the coolant hoses from the water pump.
7. Remove or disconnect the following:
   - Radiator
   - Water hoses from pipe and from heating valve
   - Tension strut
   - Wiring harness section for engine (lay aside)
8. Siphon off some fluid from power steering reservoir. Unbolt reservoir and tie out of the way, leaving lines connected.
9. Remove or disconnect the following:
   - Fan clutch with water pump impeller
   - Alternator drive belt
   - Vane pump for power steering (tie out of way, leaving lines connected)
10. Install an engine lift and chains, attaching to eye hooks.
11. Disconnect the right ground wire.

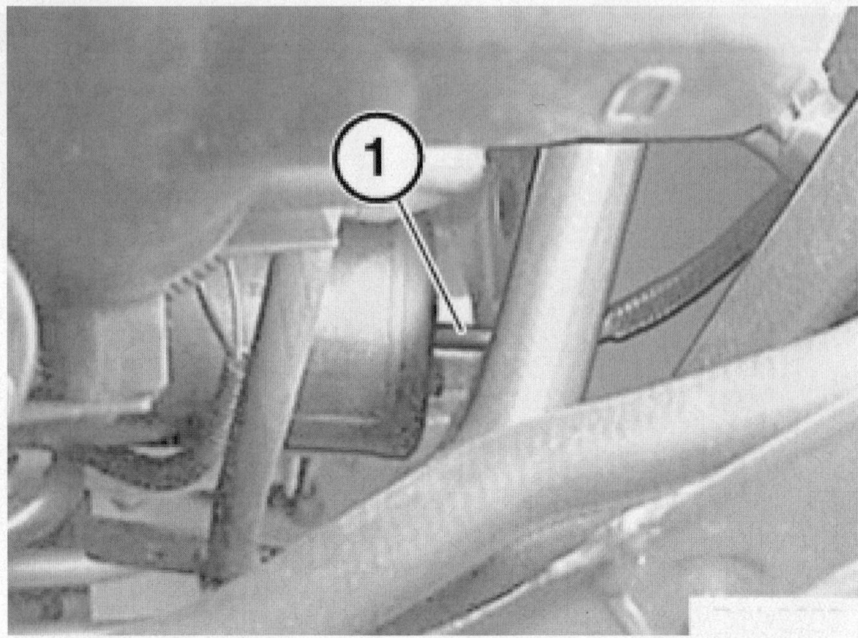

Showing vacuum line and vacuum accumulator—3.0L engine

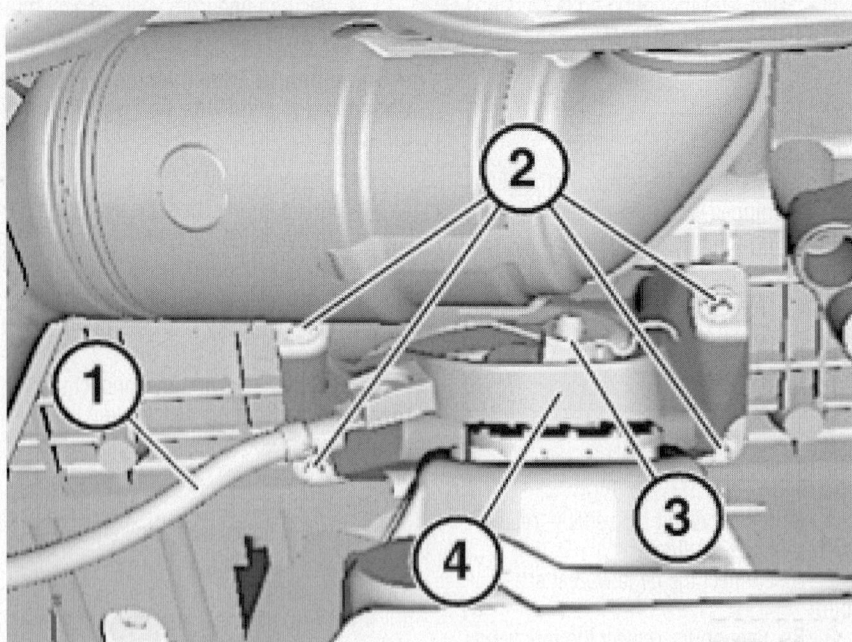

Showing right ground wire and nut on right engine mount

12. Unbolt the left and right engine mounts.
13. Carefully lift out the engine assembly.

To install:

14. Install in reverse of removal procedure.
15. Tighten components to the following torque settings:
    a. Radiator bolts: 6–7 ft. lbs. (9–10 Nm)
    b. Power steering pump bracket bolts: 16 ft. lbs. (22 Nm)
16. Tighten the automatic transmission–to–engine bolts to the following specifications:
    a. M8 hex bolts: 18 ft. lbs. (24 Nm).
    b. M10 hex bolts: 33 ft. lbs. (45 Nm).
    c. M12 hex bolts: 60 ft. lbs. (82 Nm).
    d. M8 Torx bolts: 15 ft. lbs. (21 Nm).
    e. M10 Torx bolts: 31 ft. lbs. (42 Nm).
    f. M12 Torx bolts: 53 ft. lbs. (72 Nm).

Water Pump

REMOVAL & INSTALLATION

1. Before servicing the vehicle, refer to the Precautions Section.
2. Remove the alternator drive belt.
3. Drain the cooling system. Drain plug is located on exhaust side of block, next to cylinder number 2.
4. Remove the water pump pulley.
5. Remove the 4 water pump retaining bolts. Use 2 M6 bolts in holes next to mounting bolt holes and screw in until water pump releases from timing cover.

To install:

6. When installing, use a new O-ring.
7. Tighten mounting bolts as follows:
    a. M6 bolts: 7 ft. lbs. (10 Nm).
    b. M7 bolts: 11 ft. lbs. (15 Nm).
    c. M8 bolts: 16 ft. lbs. (22 Nm).
8. Install water pump pulley.
9. Refill cooling system.
10. Install the alternator drive belt.
11. Start the vehicle, check for leaks and repair as necessary.

Heater Core

REMOVAL & INSTALLATION

1. Before servicing the vehicle, refer to the Precautions Section.
2. Disconnect the negative battery cable.
3. Remove the heater bulkhead as follows:
    a. Pull off the weatherstrips on top of the bulkhead.
    b. Release the cowl retaining toggle clips.
    c. Lift out the cover.
    d. Remove the screws and remove the heater bulkhead.
4. Remove or disconnect the following:
    - Heater hoses from firewall connections
    - Air duct for left footwell from mountings
    - Left footwell vent
    - Heater pipes from under dash
    - Heater duct for right footwell
    - Air duct for rear footwell on right

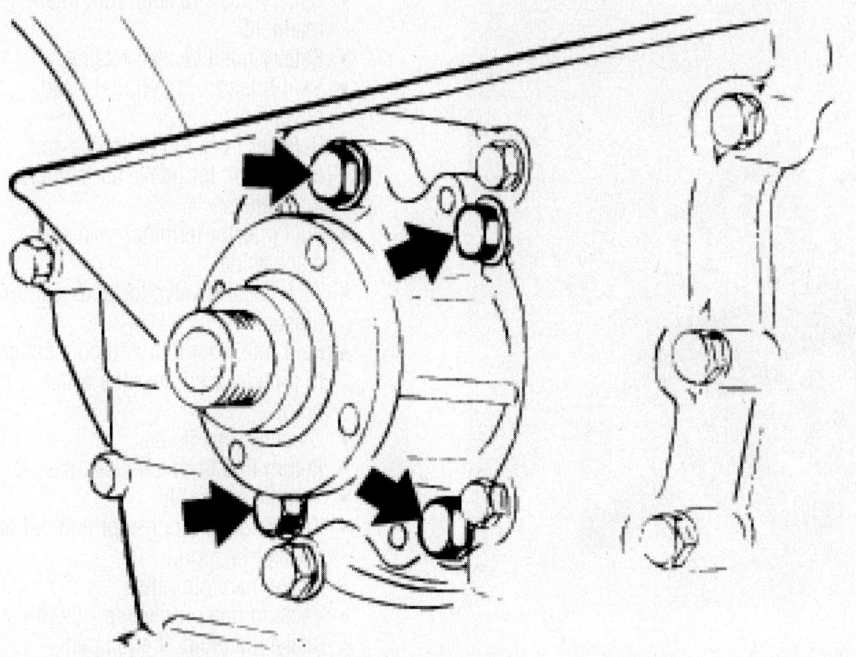

Water pump bolt location

- Cover from heater housing
- Evaporator temperature sensor (pull out to one side)
- Heater core

To install:

5. Install or connect the following:
    - Heater core
    - Evaporator temperature sensor
    - Heater housing cover
    - Air ducts in footwells
    - Heater pipes under dash
    - Heater hoses at firewall connections
    - Heater bulkhead
6. Reconnect the negative battery cables
7. Check coolant level

Cylinder Head

REMOVAL & INSTALLATION

1. Before servicing the vehicle, refer to the Precautions Section.
2. Properly relieve the fuel system pressure.
3. Remove or disconnect the following:
    - Negative battery cable
    - Ignition coils
    - Left side exhaust manifold
    - Cylinder head cover
    - Spark plugs
    - Intake manifold
    - Drain cooling system
    - Thermostat housing
    - Coolant pipe from side of block
    - VANOS adjustment unit and camshafts with bearings
    - Timing case cover–to– cylinder head bolts
    - Timing chain guide
4. Remove the cylinder head bolts in reverse of sequence shown and remove the cylinder head.

To install:

5. Clean all sealing material from mating faces

⁂ CAUTION

There must be no oil in the cylinder head bolt holes in the block and timing case cover or there is a possibility of cracking and distorting torque values.

6. Be sure that dowel sleeve are correctly positioned in block.
7. Apply permanently elastic sealing compound, Drei Bond 1209, at joints to timing case cover.
8. Install a new cylinder head gasket.
9. Position the cylinder head onto the block and install new cylinder head bolts, applying a light coat of oil to the washer contact area and the threads of the new bolts.
10. Tighten cylinder head bolts, following the sequence shown, in 3 steps, to the following:
    a. Step 1: 30 ft. lbs. (40 Nm).
    b. Step 2: Additional 90 degrees.
    c. Step 3: Additional 90 degrees.
11. Install or connect the following:
    - Timing chain guide
    - Timing case cover–to–cylinder head bolts
    - Camshafts
    - VANOS adjustment unit

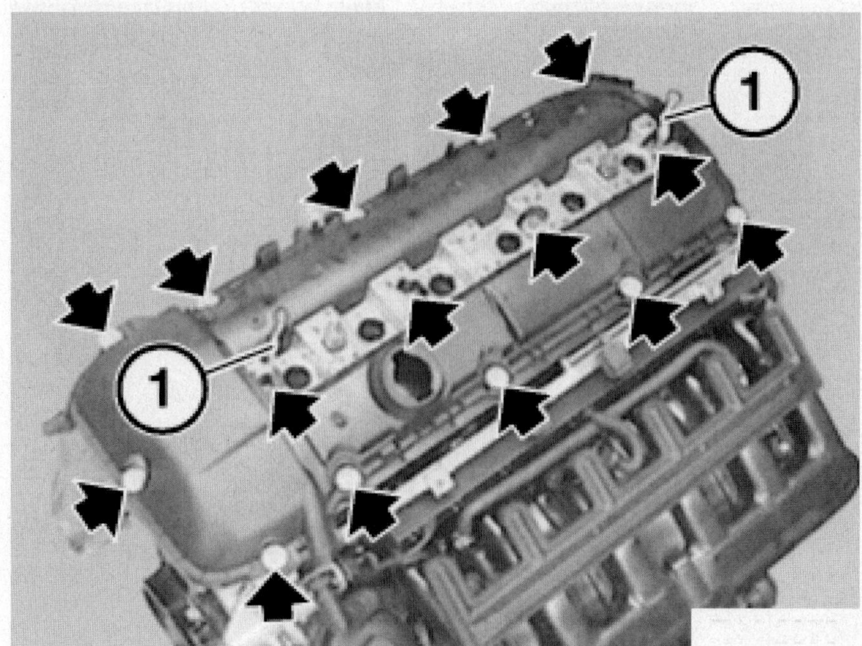

Removing grounding straps and cylinder head bolts

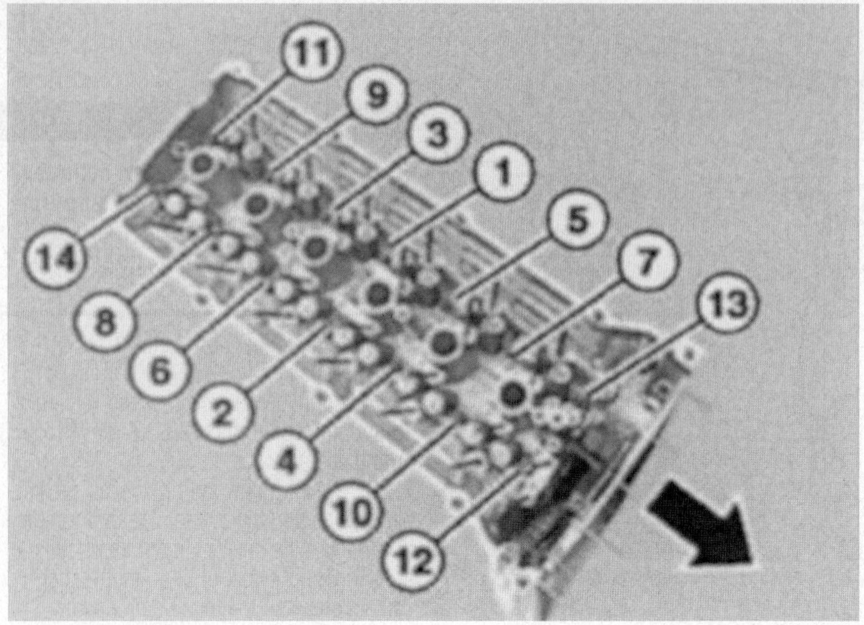

Cylinder head bolt tightening sequence

12. Complete installation in reverse of the removal procedure.

Intake Manifold

REMOVAL & INSTALLATION

1. Before servicing the vehicle, refer to the Precautions Section.
2. Properly relieve the fuel system pressure.
3. Remove or disconnect the following:
   - Negative battery cable
   - Engine cover
   - Oxygen (O_2S) sensor connectors (mark connector locations before removing)
   - Wiring from clips on intake manifold
   - Battery positive lead from intake manifold
   - Battery positive lead retainer
   - Vent hose from cylinder head cover
   - Retainer from top of engine and connection for intake air temperature sensor
   - Fuel injector terminal strip and place aside
   - Tank venting valve located near the dipstick
   - Fuel line from clip and connection to pipe located near the water pump
   - Dipstick tube guide
   - Return line from dipstick tube
   - Throttle assembly
   - Knock sensor connector and nut on manifold support
   - Intake manifold nuts
   - Vacuum line on reverse side of intake manifold, if equipped
   - Intake manifold

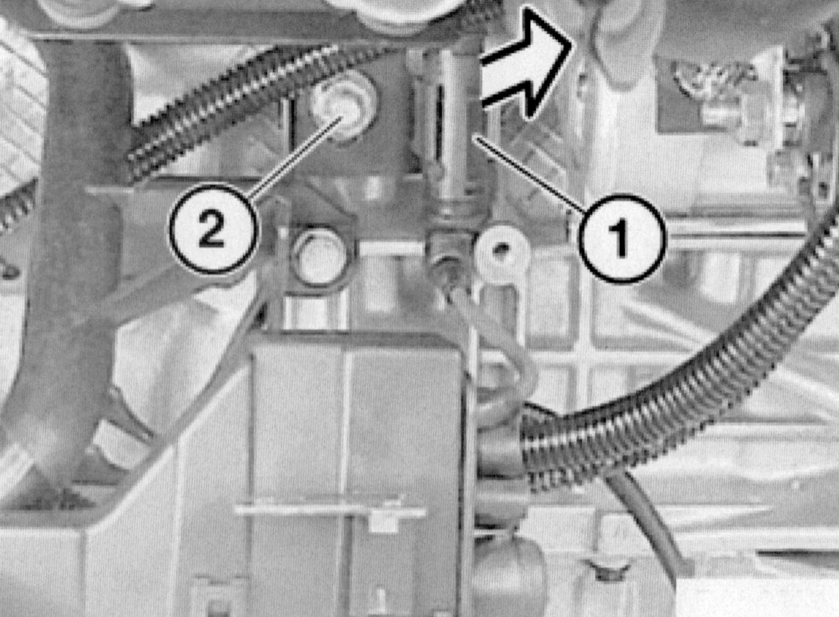

Showing knock sensor connector and nut on manifold support

To install:

4. Check all intake manifold gaskets and replace if necessary.
5. Inspect the rubber dampers at manifold connection and replace if needed.
6. Install intake manifold and torque bolts as follows:
   a. M6 bolts: 7 ft. lbs. (10 Nm).
   b. M7 bolts: 11 ft. lbs. (15 Nm).
   c. M8 bolts: 16 Nm (22 Nm).
7. Install or connect the following:
   - Knock sensor connector and manifold support nut
   - Throttle assembly
   - Return line to dipstick tube
   - Dipstick tube guide
   - Fuel line to clip and connection to pipe located near the water pump
   - Tank venting valve located near the dipstick
   - Fuel injector terminal strip
   - Retainer to top of engine and connection for intake air temperature sensor
   - Vent hose to cylinder head cover
   - Battery positive lead retainer
   - Battery positive lead to intake manifold
   - Wiring to clips on intake manifold
   - O_2S sensor connectors
   - Engine cover
   - Negative battery cable

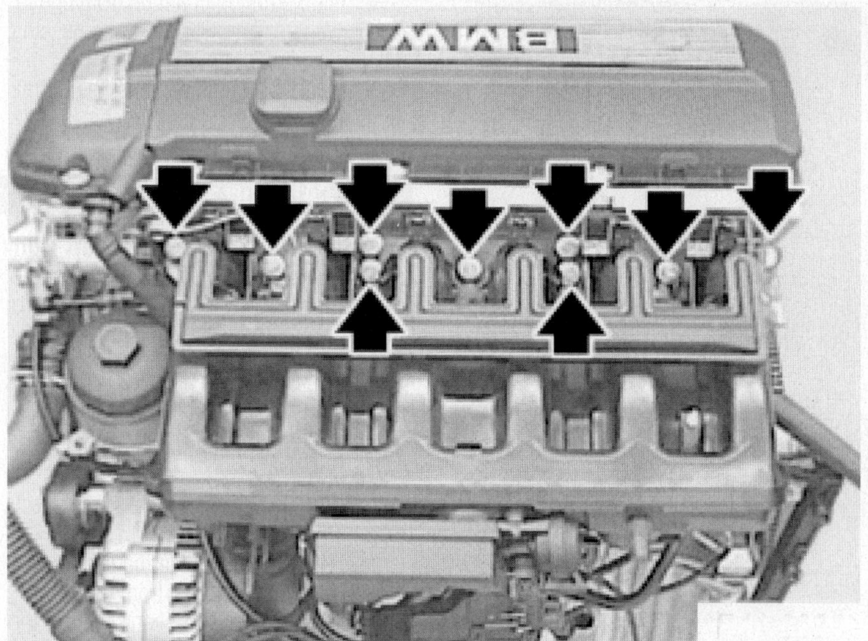

Showing intake manifold nuts

Exhaust Manifold

REMOVAL & INSTALLATION

1. Before servicing the vehicle, refer to the Precautions Section.
2. Remove or disconnect the following:
   - Negative battery cable
   - Intermediate muffler
   - Reinforcement plate from undercarriage
   - Bracket, rubber and both stabilizers
   - Oxygen control sensors from intake manifold, make sure to mark locations for reinstallation purposes
   - Fuel injector cover
   - Plug connections from intake manifold, make sure to mark all connection locations for reinstallation purposes
   - Oxygen (O_2S) sensors
   - Lower front axle
   - Front exhaust manifold with catalytic converter
   - Rear exhaust manifold with catalytic converter

To install:

3. Clean all mating faces.
4. Install new gaskets.
5. Position the rear exhaust manifold and front exhaust manifold and install new retaining nuts as follows:
   a. M6 nuts: 7 ft. lbs. (10 Nm).
   b. M7 nuts: 15 ft. lbs. (20 Nm).
6. Install or connect the following:
   - O_2S monitor sensors on exhaust pipes; torque to 37 ft. lbs. (50 Nm)
   - Connections on intake manifold and components
   - Fuel injector cover
   - Oxygen control sensors on intake manifold; torque to 37 ft. lbs. (50 Nm)
   - Bracket, rubber and both stabilizers from struts; torque nuts to 74 ft. lbs. (100 Nm)
   - Reinforcement plate from undercarriage; torque bolts to 74 ft. lbs. (100 Nm)
   - Negative battery cable

Camshaft and Valve Lifters

REMOVAL & INSTALLATION

1. Before servicing the vehicle, refer to the Precautions Section.
2. Disconnect the negative battery cable.
3. Remove the camshaft (Double VANOS) adjustment unit as follows:
   a. Remove the intake filter housing with Mass Air Flow (MAF) sensor.
   b. Remove the fan impeller with fan clutch and cowl.
   c. Remove the cylinder head covers.
   d. Remove the spark plugs.
   e. Remove the plastic cover for intake camshaft.
   f. Remove the oil pressure pipe.
   g. Install a special tool, 11–3–450, with banjo bolt on oil pressure connection unit.
   h. Cover the camshaft adjustment unit.
   i. Connect a compressed air hose to the special tool fitting on the oil line connection.
   j. Rotate the engine, a least 2 turns in direction of rotation, to return the camshafts to TDC position (front cam lobes pointing to 10 o'clock and 2 o'clock positions).
   k. Install a special locking tool, 11–2–300, into hole to lock flywheel in TDC position.
   l. Remove the 2 studs from the rear of the cylinder head outer edge. Secure the camshafts with locking tools, 11–2–240.
   m. Disconnect the compressed air hose from the special tool.
   n. Using care so no oil drips onto belt drive, remove the screw plugs from the camshaft adjustment unit cover corresponding to the forward ends of the camshafts. Have a container to catch oil that runs out.
   o. Using short, flat nose pliers, remove sealing caps through screw plug openings, then remove the fitting screws (left-hand thread) from both camshaft forward ends.
   p. Detach the plug connection from the camshaft sensor and solenoid valves on both camshafts.
   q. Remove the engine lifting eye from the camshaft adjustment unit cover.
   r. Remove nuts and remove the camshaft adjustment unit

⁂ WARNING

Once camshaft adjust unit is removed, DO NOT crank engine. The toothed shaft on the intake side camshaft may slip out of the spline teeth and valves could rest of the piston.

4. Remove the cylinder for the chain tensioning piston. Use caution as piston is under spring pressure.
5. Press down on the secondary chain tensioner at the top and lock the chain in place with a special tool, 11–3–292.
6. Remove or disconnect the following:
   - Sensor gear from exhaust camshaft
   - Spring plate from behind sensor gear
   - Nuts on intake camshaft and remove corrugated washer

42348-BMX5-G06

Attaching compressed air hose to oil line connector

42348-BMX5-G07

Showing location of retaining nuts on camshaft adjustment unit (Double VANOS)

- Screws from chain gear on exhaust camshaft
- Toothed shaft with sleeve
- Secondary (upper) chain tensioner
- Release screw-in pins from exhaust camshaft chain gear
- Exhaust camshaft chain gear (leave chain on end of camshaft)
- Thrust washer on intake camshaft
- Intake camshaft sensor gear

✲✲ CAUTION

DO NOT release screws from front ends of either camshaft.

- Studs from between both camshafts

7. Release the special locking tool so flywheel is free.
8. Lift the timing chain off from end of exhaust camshaft and hold it under tension by lifting it straight up.
9. Rotate the engine against the normal direction of rotation about 30 degrees, using the crankshaft bolt.

✲✲ CAUTION

To prevent the intake camshaft from moving the bearing inserts, remove the camshaft no. 1 bearing cap nuts and remove the bearing cap.

10. Install a special tool, 11–2–260, to the cylinder head and install retaining bolts into tool and into spark plug holes for cylinders no.1 and no. 4.
11. Turn the eccentric shaft of the special tool to pre-tension the bearing caps. Remove the nuts on all bearing caps.
12. Relieve the tension from the eccentric shaft of the special tool and remove the tool.
13. Remove the bearing caps and keep in order as removed.
14. Remove the camshaft. Repeat the process for the other camshaft.

➡If cylinder head is to be removed, remove the complete bearing strip with the bucket tappets. Keep in order of original locations.

To install:

15. Check bearing points of bucket tappets for signs of scoring.

➡Bearing strips are marked "A" for exhaust side and "E" for intake side.

16. Be sure centering dowels are in place on retaining pins at bearing points 2 and 7.
17. Install the bearing strips.
18. Oil all teeth, camshafts, bearings, bearing caps, and friction washers before installation.

✲✲ CAUTION

Bucket tappets expand when not subjected to load by the camshaft and therefore require some time before they can be pushed back down. During a rapid assembly sequence, the "closed" valves may still be open and therefore be in contact with the piston. After assembly, wait at least 30 minutes before cranking the engine back to the TDC position.

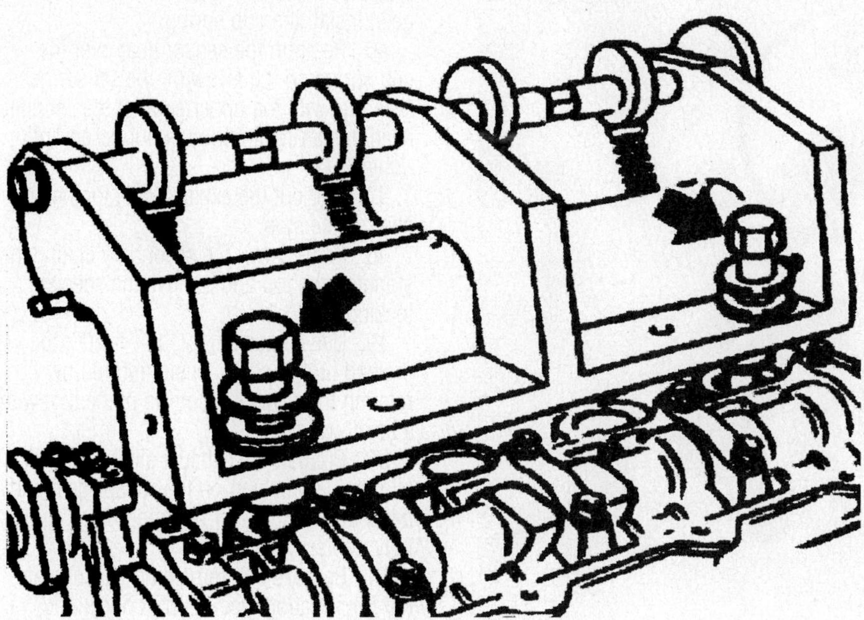

42348-BMX5-G08

Installing special tool to pre-tension camshaft bearing caps for removal

19. Pull up the timing chain and feed in the exhaust camshaft. Position the timing chain onto the end of the exhaust camshaft.

CAUTION

To prevent damaging valves when fitting camshafts, no pistons should be in TDC position.

20. Install the camshafts so the cam tips on the intake and exhaust valves on no. 1 cylinder face each other at about the 10 o'clock and 2 o'clock positions.

21. Install the bearing caps (caps are marked from the exhaust side, A1 through A7 for the exhaust side, and E1 through E7 for the intake side).

22. Install the special tool, 11–3–260, onto the cylinder head as during removal.

23. Turn the eccentric shaft to pre-tension the bearing caps. Tighten the bearing cap bolts as follows:

a. M6 bolts: 7 ft. lbs. (10 Nm).

b. M7 bolts: 10 ft. lbs. (14 Nm).

c. M8 bolts: 15 ft. lbs. (20 Nm).

24. Remove the special tool.

25. Install camshaft locking tools, 11–3–240 on back end of camshafts. Use an open-end wrench to align camshafts (if necessary, machine wrench head so it does not contact cylinder head) so there is no gap at the locking tools.

26. Install the middle locking tool, 11–3–244 so it adjoins other tools and is bolted into spark plug hole.

27. Lift the timing chain straight up off the exhaust camshaft and hold it under tension.

28. Rotate the engine from 30 degrees BTDC to TDC, in normal direction of rotation.

29. Using special tool, 11–2–300, lock flywheel in this position.

30. Install sensor gear onto intake camshaft. Install the thrust washer over the sensor gear and torque the gear screw-in pins to 15 ft. lbs. (20 Nm).

31. Feed the chain wheel onto the timing chain so the arrow on the chain wheel faces the upper edge of the cylinder head.

32. Install a special tool, 11–4–220 into the cylinder head in the chain tensioning piston bore and bring tool adjustment screw into contact with the timing chain tensioning rail, but no further at this time.

33. Check the arrow mark on the chain wheel and adjust if needed so it still points to upper edge of the cylinder head.

34. Install and tighten the retaining screw-in pins in the exhaust chain wheel to 15 ft. lbs. (20 Nm).

35. Install the secondary (upper) chain tensioner.

36. Position the toothed sleeve into the exhaust camshaft to the toothed gaps are opposed. Secure the toothed shaft. Insert the pin of the toothed shaft into the tooth gaps of the splines on the camshaft and toothed sleeve.

37. Push in the toothed shaft on the exhaust camshaft until the elongated holes in the tooth sleeve wheel are centered over the bolt holes.

38. Place forward chain wheels onto special tool, 116–6–180, and position tooth gap on intake chain wheel as shown as feed on the timing chain.

CAUTION

DO NOT alter position of chain wheels and chain when removing the special tool.

39. Remove the special tool and position chain wheels on camshafts, so that tooth

42348-BMX5-G09

Setting timing chain and chain wheel mark in position during installation of chain

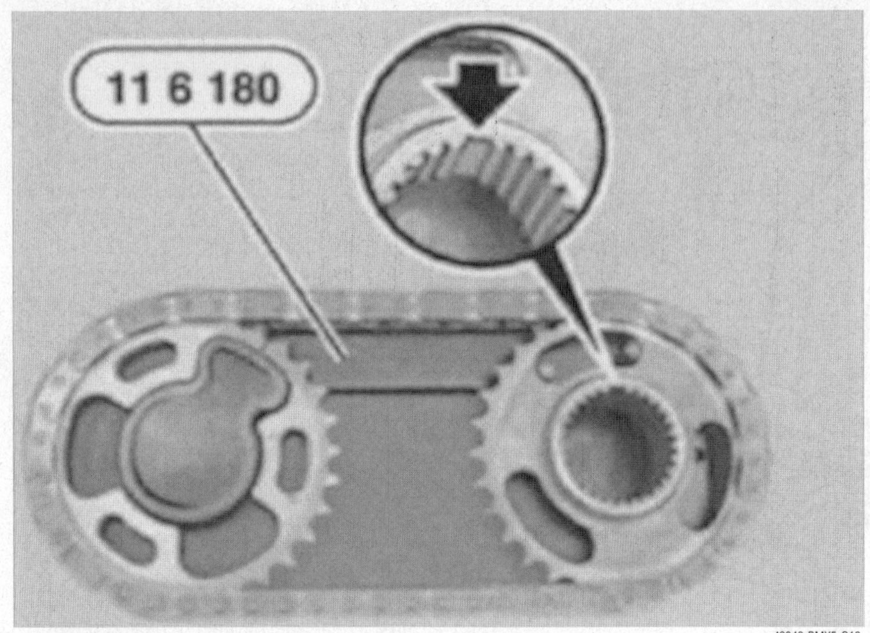

Aligning forward timing chain on chain wheels with special tool

spaces oppose each other on intake camshaft.

40. Align chain with sprocket wheels so tooth spaces are positioned exactly over each other on intake side.

41. Install and secure the toothed shaft into the tooth gaps of the splines of the camshaft and chain wheel. Push in the toothed shaft until about 0.4 inch (1 mm) of splines can still be seen.

42. Note the installation direction of the corrugated washed so "FRONT" is visible. Install the washer and tighten retaining nuts snug only at this time.

43. Install 4 bolts on exhaust side to retain chain wheel. Initially tighten to about 3 ft. lbs. (5 Nm), then slack off one-half turn.

44. Install the thrust washer over the exhaust side chain wheel.

45. Note the installation direction of the cup spring (2) so the "F" stamp is forward. If "F" is no longer visible on used engine, install the cup spring so the small locating

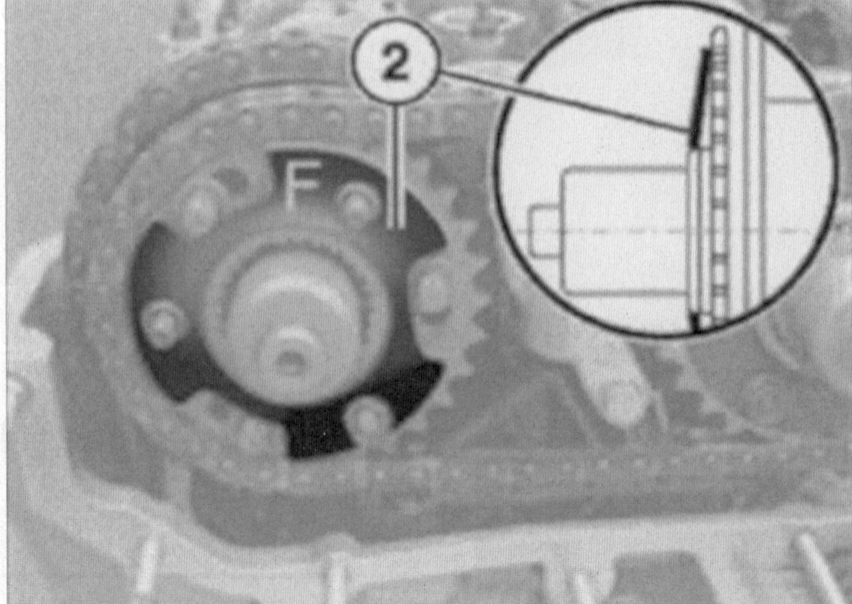

Installing the cup spring on the exhaust camshaft chain wheel

diameter of the spring points to the sensor gear. Install the cup spring.

46. Position the sensor gear over the cup spring so the arrow on the sensor gear is in line with the upper edge of the cylinder head. Install retaining nuts but do not fully tighten.

47. Pull out the exhaust side toothed shaft to the stop.

48. Press down the secondary chain tensioner at the top and remove the special locking tool.

49. Use a special tool, 11–4–200, to preload timing chain tensioning rail by rotating the adjusting screws of the tool with a small wrench.

50. Preload the exhaust side cup spring slightly be pressing on the sensor gear and hand-tighten the gear retaining nuts (do not fully tighten yet).

51. Remove any remaining gasket material from mating face on front of cylinder head.

52. Check dowel sleeves for damage and for correct installation position.

53. Make sure sealing face is clean and free of oil.

54. Fit special tool, 11–6–150 without gasket over both toothed shafts. Tighten retaining nuts to hold tool in place. Hand-tighten only until special tool is uniformly in contact with cylinder head.

55. Insert bolts on exhaust side chain wheel and tighten to about 3 ft. lbs. (5 Nm).

56. Initially tighten nuts on both exhaust and intake chain wheels to about 3 ft. lbs. (5 Nm).

57. Torque bolts on exhaust chain wheel to 15 ft. lbs. (20 Nm).

58. Torque nuts on exhaust and intake chain wheels screw-in pins to 7 ft. lbs. (10 Nm).

59. Pull back the special tool from the flywheel.

60. Remove the special tools from the rear of the camshafts.

61. Crank the engine twice in the direction of rotation until the cam lobe tips on the front of the camshafts point inward at about the 10 o'clock and 2 o'clock positions.

62. Lock the flywheel with the special tool to the crankshaft is at Top Dead Center (TDC) position.

WARNING

DO NOT turn the engine against normal rotation.

63. Install the special camshaft locking tool, 11–3–240, onto rear end of camshafts to hold its position.

➡The camshaft timing is correctly set if the special tool rests on the camshafts without a gap on the cylinder head or any protrusion up to 0.04 inch (1 mm) on the intake side.

64. Remove the special tool from the camshaft forward ends.
65. Install the camshaft adjustment (Double VANOS) unit as follows:
   a. Ensure dowel sleeves on forward mating face of cylinder head are not damaged and are correctly installed.
   b. Keep sealing faces clean and free of oil.
   c. Apply a thin and even coat of sealing compound, Drei Bond 1209, to contact surface edges of the separating face between the cylinder head and the camshaft adjustment unit.
   d. Install or connect the following:
   - New seal on mating face
   - Camshaft adjustment unit
   - Screws for hydraulic pistons on toothed shaft end on both camshafts (through front openings); torque to 7 ft. lbs. (10 Nm). Screws have left-hand threads.
   - Sealing caps into camshaft adjustment unit openings
   - Screw plugs, with new seal rings; torque to 37 ft. lbs. (50 Nm)
   - Engine lifting eye
66. Set camshaft timing, if necessary.
67. Complete installation in reverse of removal procedure.

Valve Lash

ADJUSTMENT

All engines are equipped with hydraulic valve lash adjusters. This design does not permit adjustments nor are adjustments possible.

Starter

REMOVAL & INSTALLATION

➡When the battery is disconnected, the radio code, on-board computer and clock settings will be lost. The radio code should be obtained before disconnecting the battery or radio. Once the battery has been reconnected, the radio will not function unless the code is keyed in.

1. Before servicing the vehicle, refer to the Precautions Section.
2. If needed, read the stored fault memories from the control module.

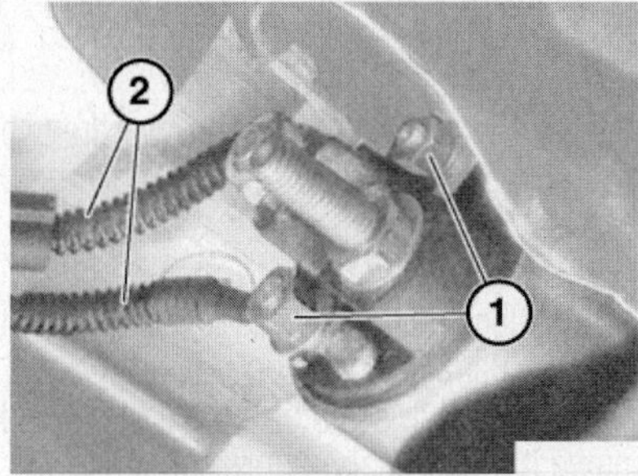

Showing control lead, nuts and starter

3. Relieve the fuel system pressure.
4. Set the ignition switch to the **OFF** position.
5. Remove or disconnect the following:
   - Negative battery cable
   - Pressure accumulator
   - Positive battery cable from the starter
   - Starter electrical connectors
   - Starter from the transmission mount
   - Control leads
   - Starter

To install:

6. Check starter pinion and starter ring gear for damage
7. Install or connect the following:
   - Starter to transmission mount
   - Control leads
   - Starter electrical connectors
   - Positive battery cable to the starter
   - Pressure accumulator
   - Negative battery cable

Oil Pan

REMOVAL & INSTALLATION

1. Before servicing the vehicle, refer to the Precautions Section.
2. Drain the engine oil.
3. Remove or disconnect the following:
4. Attach an engine lift and hoist to engine.
5. Remove top left and right nuts on engine mounts
6. Raise the engine about 0.15 inch (5mm).
7. Remove and disconnect the following:
   - Front splash guard
   - Left and right swivel bearings
   - Output shafts
   - Propeller shaft at front
   - Alternator drive belt from belt pulley for vane pump
   - Power steering pump and move aside with the lines still connected
   - Return hose from oil separator
   - Dipstick tube
   - Steering spindle from steering gear tie rods
8. Release the screws on the oil pan.
9. Move the pan backward when removing.
10. Remove transmission bolts

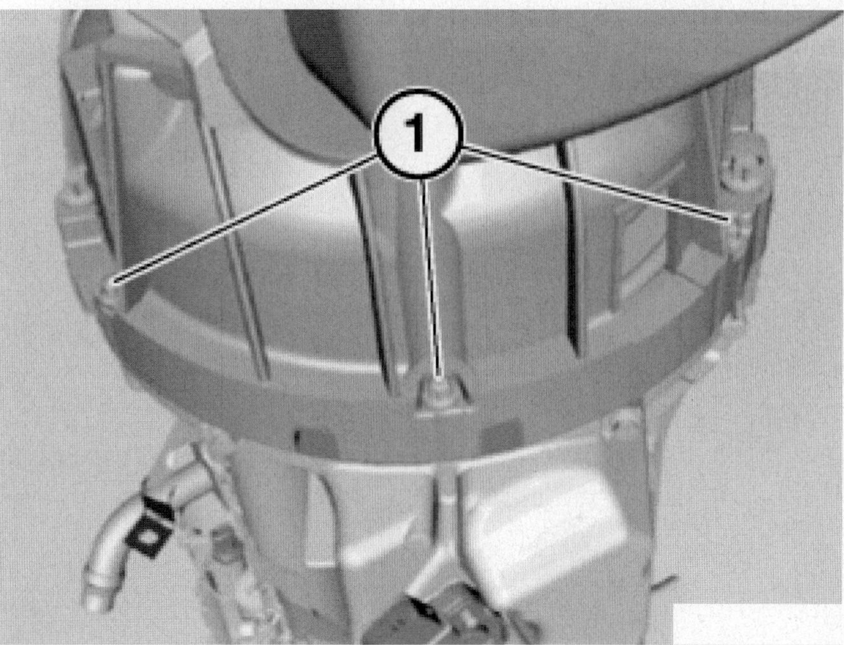

Showing transmission bolts

To install:

11. Ensure mating surfaces are clean.
12. Install a new gasket and position the oil pan in place.
13. Install and tighten the oil pan bolts, but do not fully tighten the transmission end bolts.
14. Tighten the engine end bolts, then tighten the transmission end bolts.
15. Restore the front axle assembly to position.
16. Install or connect the following:
    - Dipstick tube
    - Return hose from oil separator
    - Power steering pump
    - Alternator drive belt
    - Propeller shaft
    - Output shafts
    - Swivel bearings
    - Splash guard
17. Reposition the engine onto the mount and torque the upper nuts.
18. Remove the engine hoist
19. Reconnect the negative battery cable.
20. Refill the engine oil.

Oil Pump

REMOVAL & INSTALLATION

1. Before servicing the vehicle, refer to the Precautions Section.
2. Drain the engine oil.
3. Remove or disconnect the following:
    - Negative battery cable
    - Oil pan
    - Oil pump sprocket wheel and chain
    - If the oil pump is integrated with the oil scraper, remove oil pump with the scraper
    - If the oil pump and scraper are two parts, remove oil pump intake pipe
    - Oil pump

To install

4. Check dowel sleeves for damage and correct installation position.
5. Check the seals on the oil pipes and replace it if necessary. Lubricate the seals with oil.
6. Check the seal in the oil pump and replace it if necessary.
7. If the oil pump is integrated with the oil scraper, install oil pump with scraper
8. If the oil pump and scraper are two parts, install oil pump intake pipe
9. Install or connect the following:
    - Oil pump. Torque the bolts to 17 ft. lbs. (22 Nm).
    - Oil pump sprocket wheel and chain. Torque the nut to 35 ft. lbs. (47 Nm).
    - Oil pan
    - Negative battery cable
10. Fill the engine with clean oil.
11. Start the vehicle and check for leaks, repair if necessary.

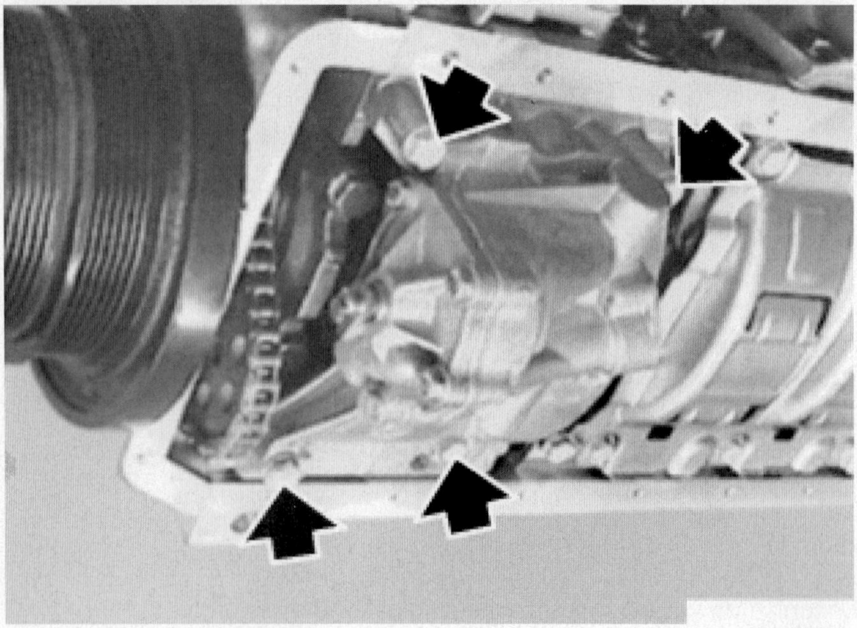

06041_BMX3_G0010

Showing oil pump mounting bolts

Rear Main Seal

REMOVAL & INSTALLATION

The rear main bearing oil seal can be replaced after the transmission and flywheel has been removed from the engine.

1. Before servicing the vehicle, refer to the Precautions Section.
2. Drain the transmission fluid.
3. Remove or disconnect the following:
    - Negative battery cable
    - Transmission
    - Flywheel assembly
    - Oil seal, using a suitable seal removal tool

To install:

4. Coat the sealing lips of the new seal with oil.
5. Install or connect the following:
    - New seal into the end cover housing with a suitable seal installation tool
    - Flywheel
    - Transmission
    - Negative battery cable
6. Fill the transmission with new fluid.
7. Start the engine and check that oil pressure is present; if the oil pressure lamp does not extinguish within 5–7 seconds of starting the engine, turn the engine **OFF**.
8. Check and top off all fluid levels.

Timing Chain, Sprockets, Front Cover and Seal

REMOVAL & INSTALLATION

➡This procedure covers lower timing case cover removal only. Procedure for removal of timing chain is covered under "Camshaft" for this engine.

1. Before servicing the vehicle, refer to the Precautions Section.
2. Drain the engine oil.
3. Remove or disconnect the following:
    - Negative battery cable
    - Camshaft adjustment unit (Double VANOS), as described under "Camshaft."
    - Alternator drive belt tensioner and belt
    - Water pump pulley
    - Vibration damper and hub, if equipped
    - Oil pan
4. Drive out the dowel pins from the timing case cover toward the rear.
5. Remove the lower timing case cover bolts from cylinder head.
6. Remove the timing case cover bolts and cover.

To install:

7. Check cylinder head and gasket for possible damage.

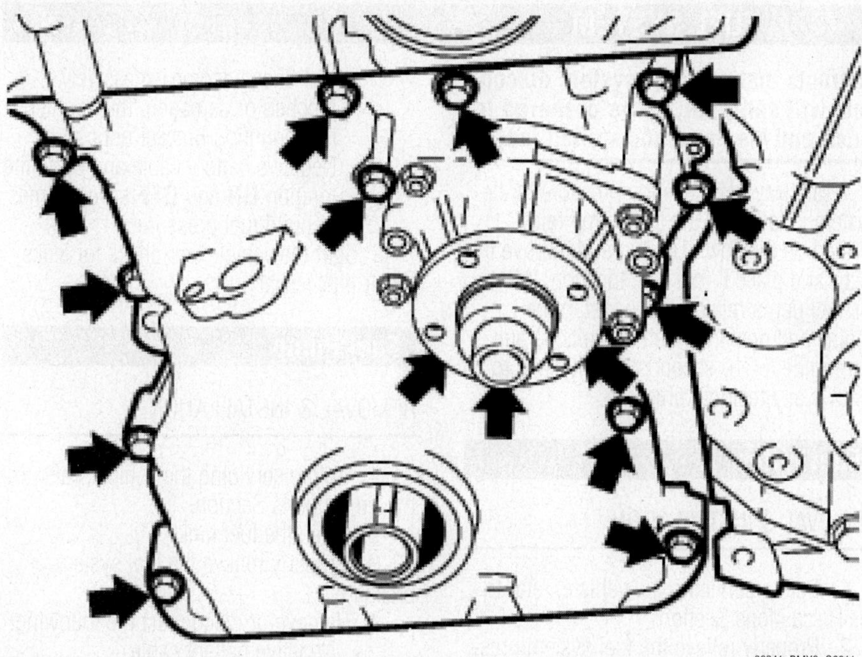
Showing the timing case cover bolts and cover

8. Clean all mating surfaces.
9. Drive dowel pins into timing case cover until they protrude about 0.08-0.12 inch (2-3mm).
10. Use grease to hold timing case cover in place during installation.
11. Apply sealing compound, Drei Bond 1209, to joints at cylinder head gasket left and right, and thinly and evenly to entire timing case cover sealing surface adjoining the cylinder head.
12. Install all timing cover retaining bolts to about 3 ft. lbs. (5 Nm).
13. Drive in the dowel pins until flush.
14. Torque all timing cover retaining screws, in an alternating sequence, to the following:
    a. M6 bolts: 7 ft. lbs. (10 Nm).
    b. M7 bolts: 11 ft. lbs. (15 Nm).
    c. M8 bolts: 16 ft. lbs. (22 Nm).
    d. M10 bolts: 35 ft. lbs. (47 Nm).
    e. Repeat the torque sequence a second time.
15. Install and tighten the bolts connecting the timing case cover to the cylinder head.
16. Replace the crankshaft oil seal.
17. Install or connect the following:
    - Oil pan
    - Vibration damper and hub
    - Water pump pulley
    - Drive belt tensioner and belt
    - Camshaft (Double VANOS) adjustment unit
    - Negative battery cable.
18. Refill the engine oil.

Piston and Ring Positioning

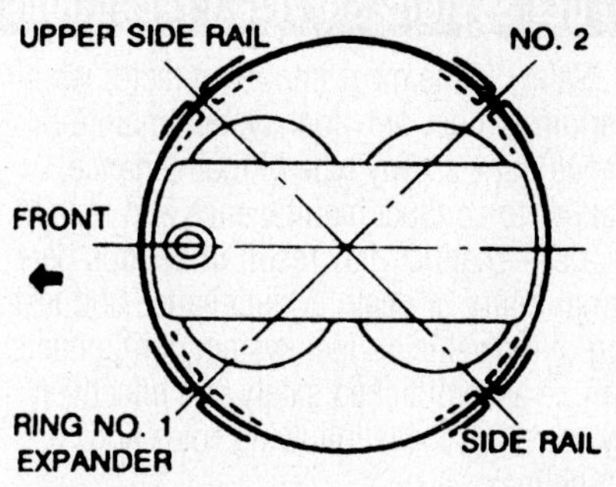

Piston ring end-gap spacing

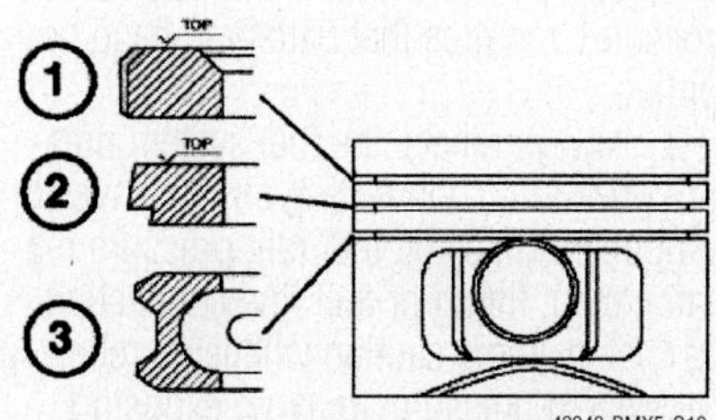

Piston ring locations and identification

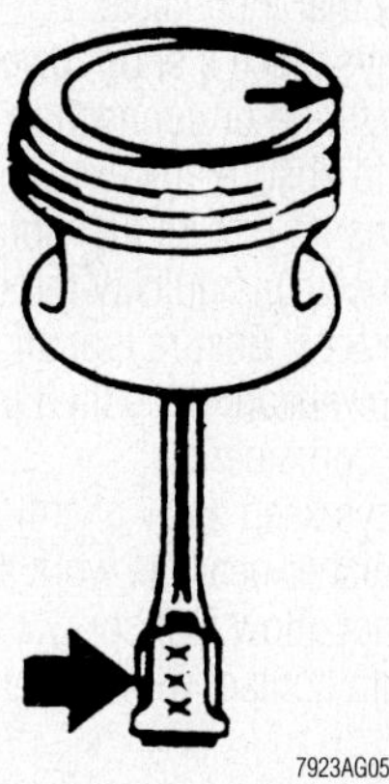

Showing connecting rod-to-piston positioning

FUEL SYSTEM

Fuel System Service Precautions

Safety is the most important factor when performing not only fuel system maintenance but also any type of maintenance. Failure to conduct maintenance and repairs in a safe manner may result in serious personal injury or death. Maintenance and testing of the vehicle's fuel system components can be accomplished safely and effectively by adhering to the following rules and guidelines.

1. To avoid the possibility of fire and personal injury, always disconnect the negative battery cable unless the repair or test procedure requires that battery voltage be applied.
2. Always relieve the fuel system pressure prior to disconnecting any fuel system component (injector, fuel rail, pressure regulator, etc.), fitting or fuel line connection. Exercise extreme caution whenever relieving fuel system pressure, to avoid exposing skin, face and eyes to fuel spray. Fuel under pressure may penetrate the skin or any part of the body that it contacts.
3. Always place a shop towel or cloth around the fitting or connection prior to loosening to absorb any excess fuel due to spillage. Ensure that all fuel spillage (should it occur) is quickly removed from engine surfaces. Ensure that all fuel soaked cloths or towels are deposited into a suitable waste container.
4. Always keep a dry chemical (Class B) fire extinguisher near the work area.
5. Do not allow fuel spray or fuel vapors to come into contact with a spark or open flame.
6. Always use a back-up wrench when loosening and tightening fuel line connection fittings. This will prevent unnecessary stress and torsion to fuel line piping. Always follow the proper torque specifications.
7. Always replace worn fuel fitting O-rings with new. Do not substitute fuel hose or equivalent where fuel pipe is installed.

Fuel System Pressure

RELIEVING

✲✲ WARNING

Fuel lines are under about 43.5–72.5 psi (3–5 bar) of pressure. Manufacturer does not provide a specific pressure relieving procedure, but instructs that any fuel system disconnect will spill fuel, so be prepared to catch and clean up any spilled fuel.

A safe way to relieve the pressure in the system is to locate the fuel pump relay located on the cowl. Unplug and remove the relay, and place it in a safe location. With the fuel pump relay removed, start the engine and operate it until it stalls. Crank the engine for 10 seconds after it stalls to remove any residual pressure.

Fuel Filter

REMOVAL & INSTALLATION

1. Before servicing the vehicle, refer to the Precautions Section.
2. Properly relieve the fuel system pressure.
3. Remove or disconnect the following:
   - Negative battery cable
   - Fuel pressure regulator and seal the fuel line before and after the filter with special tool 16–1–020
   - Clips and fuel line from the filter
   - Fuel filter

To install:

   - New fuel filter
   - New seal
   - Fuel lines onto the correct fittings. Tighten the fuel line clamps until tight, but not to the point where the fuel lines become excessively pinched or damaged, then tighten the mounting bracket until snug.
   - Negative battery cable and cycle the ignition **ON** and **OFF** several times to build fuel pressure
4. Start the vehicle and check for leaks, repair if necessary.

Fuel Pump

REMOVAL & INSTALLATION

1. Before servicing the vehicle, refer to the Precautions Section.
2. Drain the fuel tank.
3. Properly relieve the fuel system pressure.
4. Remove or disconnect the following:
   - Negative battery cable
   - Rear seat bench
   - Rubber plug above the sender unit and fold the rubber mat back
   - Metal cover
   - Fuel gauge level sending unit electrical connector
   - Fuel lines and service vent line
   - Rotary connection with special tool 16–1–020
5. Raise the fuel level sensor and expose the spiral hose.
6. Remove the fuel level sensor and fuel pump from the tank.

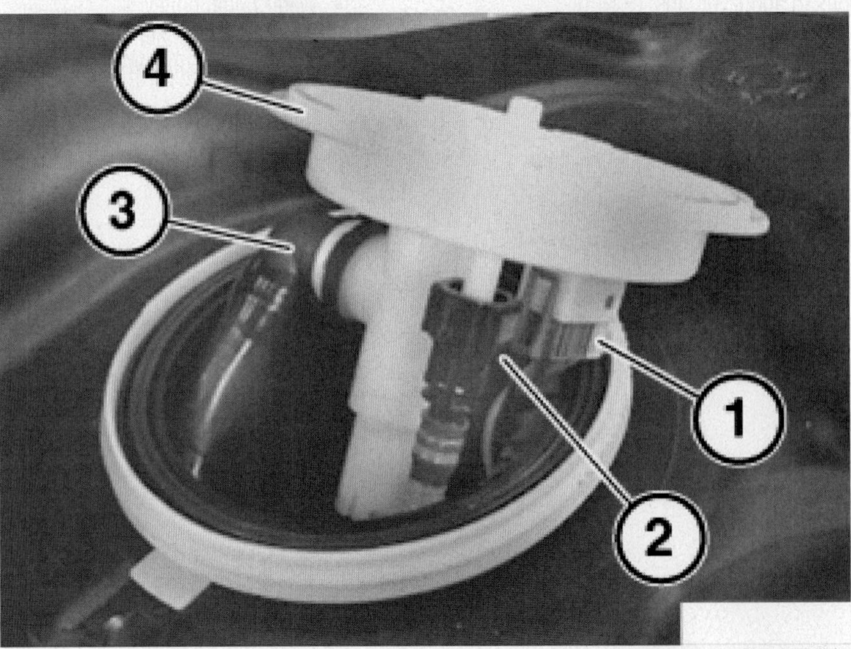

Showing service cap, fuel line, vent line and electrical connector on fuel pump

To install:

➡**Always use a new seal or gasket when installing the fuel pump or fuel level gauge sending unit assembly.**

7. Install the fuel pump into the fuel tank, taking care not to bend or damage the fuel sending unit assembly.
8. Install a new seal and torque the sealing ring using tool No. 16-1-020 as follows:
   a. Metal sealing rings: 26 ft. lbs. (35 Nm).
   b. Plastic sealing rings: 41 ft. lbs. (55 Nm).
9. Install or connect the following:
   - Fuel lines and service vent line
   - Fuel gauge level sending unit electrical connector
   - Metal cover
   - Rubber plug above the sender unit
   - Rear seat bench
   - Negative battery cable
10. Start the vehicle and check for leaks, repair if necessary.

Fuel Injector(s)

REMOVAL & INSTALLATION

3.0L Engines

➡**This procedure involves removing the complete injection pipe with injectors.**

1. Before servicing the vehicle, refer to the Precautions Section.
2. Properly relieve the fuel system pressure.
3. Mark the locations of each oxygen sensor connector for proper reinstallation.
4. Remove or disconnect the following:
   - Negative battery cable
   - Connectors from clips on injection pipe
   - Intake Air Temperature (IAT) sensor connector
   - Plug connection on solenoid valve for camshaft adjustment (VANOS) unit
   - Terminal strip for fuel injectors from injection pipe
5. Unclip the fuel line from its holder, then unlock the fuel line from the cylinder head at the rear (be prepared to catch any fuel).
6. Remove the retaining bolts and remove the injection pipe with the fuel injectors.

To install:

7. Coat the sealing rings of the injectors with anti-seize compound prior to fitting.
8. Install the fuel injectors and fuel pipe to the engine.
9. Install or connect the following:
   - Fuel line
   - Terminal strip to fuel injectors
   - Connectors as removed
   - Refill the tank with gas
   - Negative battery cable

DRIVE TRAIN

Transmission Assembly

REMOVAL & INSTALLATION

Manual Transmission

1. Before servicing the vehicle, refer to the Precautions Section.
2. Remove or disconnect the following:
   - Negative battery cable
   - Grounding strap to body
   - Heat shield
   - Reinforcement plate from undercarriage
   - Front splash shield
   - Front propeller shaft
   - Exhaust hangar bracket
   - 3 transmission–to–engine lower bolts
3. Support the transmission with special cradle, 23–3–130/1/2/3.
   - Cable from transmission
   - Crossmember
   - Driveshaft–to–transmission flange nuts and discard the nuts
   - Propeller shaft nuts (bend shaft downward at center bearing)
   - Propeller shaft from transmission output flange and tie aside
4. Slightly lower the transmission, slowly releasing tension on the clutch slave cylinder.
5. Remove the nuts and remove the clutch slave cylinder, leaving pressure lines connected.
6. Detach the reverse light switch connector and the wiring cables.
7. Detach servomotor electrical connectors.
8. Support the engine with a proper jack.
9. Disconnect the gear selector rod from the joint. Pry off the spring from the bearing pin.

⁂ CAUTION

Be sure the transmission does not put any weight on the input shafts or clutch plate will be distorted.

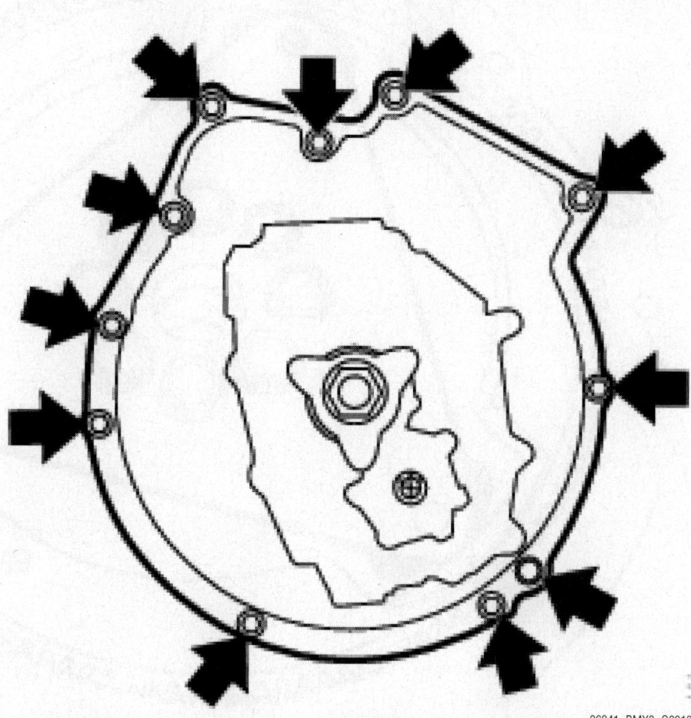

Showing transmission housing bolts—manual transmission

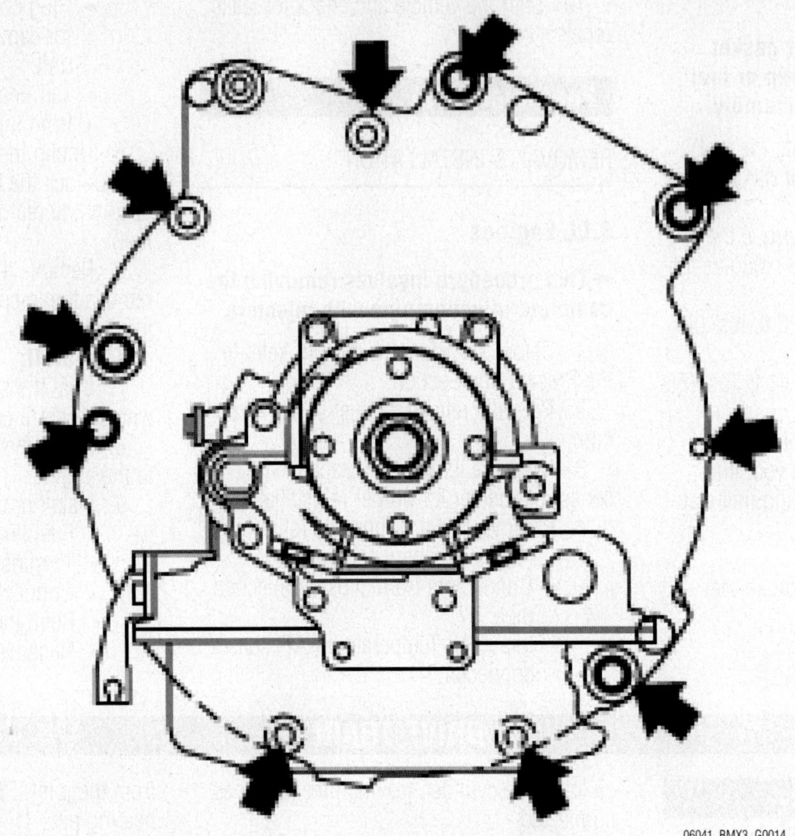

Showing transmission housing bolts—Automatic transmission

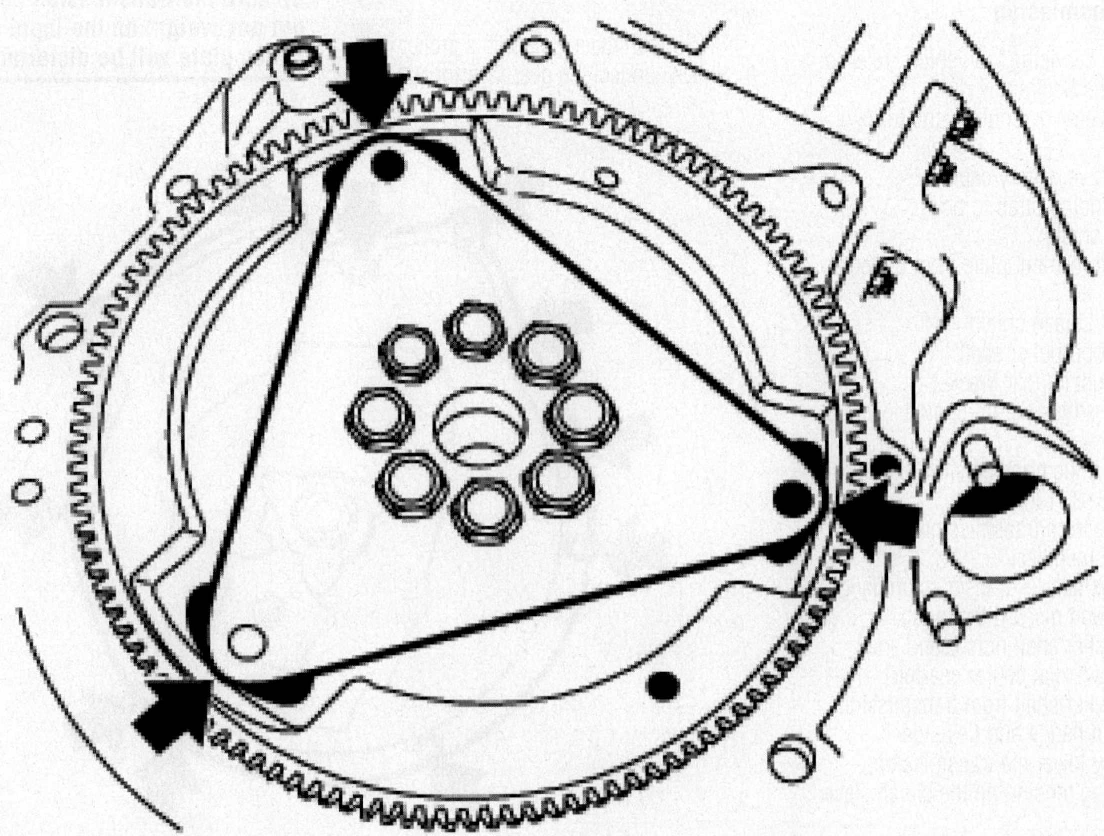

Showing three recesses for the torque converter retaining tabs—Automatic transmission

10. Remove the transmission housing bolts and pull the transmission toward the rear as far as possible, turn it about 10 degrees counterclockwise and remove from the vehicle.

To install:

11. Clean all mating surfaces.
12. Grease the bearing pin
13. Check the condition of the adapter sleeves on either side of the transmission at the flywheel. Replace if necessary.
14. Position the transmission into the vehicle.
15. Install and torque the transmission–to–engine housing bolts to the following:
    a. M8 hex bolts: 18 ft. lbs. (25 Nm).
    b. M10 hex bolts: 36 ft. lbs. (49 Nm).
    c. M12 hex bolts: 55 ft. lbs. (74 Nm).
    d. M8 Torx bolts: 16 ft. lbs. (22 Nm).
    e. M10 Torx bolts: 32 ft. lbs. (43 Nm).
    f. M12 Torx bolts: 53 ft. lbs. (72 Nm).
16. Install or connect the following:
    - Spring clip at bearing pin
    - Gear selector rod
    - Reverse light switch and wiring cables
    - Clutch slave cylinder; torque to 16 ft. lbs. (22 Nm)
    - Propeller shaft to output flange new nuts; torque to 47 ft. lbs. (64 Nm)
17. Install the crossmember and torque the bolts as follows:
    a. To rubber mounts: 55 ft. lbs. (74 Nm).
    b. To body, M8 bolts: 16 ft. lbs. (21 Nm).
    c. To body, M10 bolts: 30 ft. lbs. (41 Nm).
18. Install the cable to the transmission.
19. Install and torque the lower transmission–to–engine bolts to the following:
    a. M8 hex bolts: 18 ft. lbs. (25 Nm).
    b. M10 hex bolts: 36 ft. lbs. (49 Nm).
    c. M12 hex bolts: 55 ft. lbs. (74 Nm).
    d. M8 Torx bolts: 16 ft. lbs. (22 Nm).
    e. M10 Torx bolts: 32 ft. lbs. (43 Nm).
    f. M12 Torx bolts: 53 ft. lbs. (72 Nm).
20. Install or connect the following:
    - Front propeller shaft
    - Stabilizer bar
    - Front splash guard
    - Reinforcement plate
    - Heat shields
    - Grounding strap to body
    - Negative battery cable
21. Refill the transmission and check operation.

Automatic Transmission

1. Before servicing the vehicle, refer to the Precautions Section.
2. Remove or disconnect the following:
    - Negative battery cable
    - Engine under guard at front
    - Reinforcement plate from undercarriage
    - Exhaust system
    - Cables for oxygen sensors, mark the locations for installation purposes
    - Front propeller shaft
    - Vent line from transmission at bracket
    - Exhaust bracket
    - Nut and bushing from shift lever
    - Shift control cable from body bracket
    - Electrical connection and cable from transmission
    - Hydraulic cooler lines from transmission
3. Position the proper transmission cradle, 00–2–030, 24–5–301/305.
4. Support the transmission and transfer case with transmission cradle.
5. Remove or disconnect the following:
    - Wiring cables from transmission
    - Crossmember
    - Propeller shaft flange nuts
    - Center bearing support and support the propeller shaft
    - Propeller shaft from transmission and tie aside
    - Hex bolt near protective cap on transmission flange
    - Protective cap
6. Protect torque converter from slipping out. Insert special tools 24–4–131/137 in opening in transmission housing. Clamp torque converter.
7. Remove or disconnect the following:
    - Bolts from torque converter through protective cap opening using special tool 24–1–110
    - Transmission housing–to–engine bolts
8. Pull the transmission, with transfer case, out toward rear.

To install:

➡The sheet metal flywheel is equipped with three recesses for the torque converter retaining tabs. When guiding the engine and transmission together, the three mounting tabs on the torque converter must be aligned with the three indentations in the sheet metal flywheel. Turning on the torque converter, or engine, is no longer possible after guiding the engine and transmission together and can lead to damage.

9. Note the position and condition of dowel sleeves on flywheel housing. Replace dowels if damaged.
10. Remove special tools 24–4–131/137 before install transmission to engine.
11. Position the transmission to the engine and install the housing bolts. Torque the bolts as follows:
    a. M8 hex bolts: 18 ft. lbs. (24 Nm).
    b. M10 hex bolts: 33 ft. lbs. (45 Nm).
    c. M12 hex bolts: 60 ft. lbs. (82 Nm).
    d. M8 Torx bolts: 15 ft. lbs. (21 Nm).
    e. M10 Torx bolts: 31 ft. lbs. (42 Nm).
    f. M12 Torx bolts: 53 ft. lbs. (72 Nm).
12. Install and torque the torque converter–to–flywheel bolts to the following:
    a. M8 hex bolts: 19 ft. lbs. (26 Nm).
    b. M10 hex bolts: 33 ft. lbs. (45 Nm).
13. Install or connect the following:
    - Protective cap over flywheel bolt opening
    - Hex bolt near protective cap
    - Propeller shaft into center support; torque mount–to–body bolts to 15 ft. lbs. (21 Nm).
    - Propeller shaft flange new nuts; torque to 74 ft. lbs. (100 Nm)
14. Install the transmission crossmember. Torque the bolts as follows:
    a. To rubber mounts: 55 ft. lbs. (74 Nm).
    b. To body: 30 ft. lbs. (41 Nm).
15. Install or connect the following:
    - Transmission cables
    - Transmission cooler lines with new sealing ring
    - Wiring cable to transmission
    - Shift control cable to body bracket
    - Shift selector lever bushing and nut
    - Exhaust hanger bracket
    - Vent line to bracket
    - Engine guard
    - Reinforcement plate
    - Oxygen sensor connectors
    - Heat shield
    - Exhaust system
    - Negative battery cable
16. Refill transmission and check operation, then recheck fluid level.

Transfer Case

REMOVAL & INSTALLATION

1. Before servicing the vehicle, refer to the Precautions Section.
2. Remove or disconnect the following:

- Negative battery cable
- Exhaust system
- Heat shield
- Front propeller shaft set aside and secure
- Propeller shaft nuts, bend shaft downward at center bearing
- Propeller shaft from transmission and tie aside

➡Automatic transmissions: Use special tools 00–2–030, 24–5–301, 24–5–305. Manual transmission: Support with hydraulic lifter.

3. Support the transmission
   - Crossmember
   - Electrical connectors to servomotor
4. Support the transfer case. Remove the retaining bolts and remove the transfer case.

To install:

5. Check the condition of the dowel pins and replace if necessary.
6. Be sure the mating surfaces are clean.
7. Apply a thin coat of anti-seize grease to transfer case splines
8. Replace the sealing ring of the driveshaft of the transfer case.
9. Install the transfer case and torque the retaining bolts to 32 ft. lbs. (43 Nm).
10. Reposition the propeller shaft in the center support bearing and to the flange. Torque the center support bearing nuts to 15 ft. lbs. (21 Nm) and the flange nuts to 24 ft. lbs. (32 Nm).
11. Install or connect the following:
    - Cable to transfer case
12. Install the transmission crossmember. Torque the bolts as follows:
    a. To rubber mounts: 55 ft. lbs. (74 Nm).
    b. To body: 30 ft. lbs. (41 Nm).
13. Install or connect the following:
    - Electrical connectors to servomotor
    - Propeller shaft and center bearing
    - Reinforcement plate
    - Heat shields
    - Exhaust system
14. Refill the transfer case.

Halfshafts

REMOVAL & INSTALLATION

Front

1. Before servicing the vehicle, refer to the Precautions Section.
2. Remove or disconnect the following:
   - Negative battery cable
   - Front wheel
   - Reinforcement plate
   - Front splash guard
   - ABS pulse generator
   - Brake caliper from disc and tie out of the way with the hose still connected
   - Steering gear tie rod from swivel bearing
   - Tension strut, with guide joint, from swivel bearing
   - Control arm from swivel bearing
   - Collar nut on halfshaft, press halfshaft out of drive flange
   - Halfshaft

To install:

3. Push in output shaft over the resistance of the retaining ring until it snaps in place.

➡Use a new snap ring on halfshaft spline

4. Install or connect the following:
   - Collar nut
   - Control arm to swivel bearing
   - Tension strut to swivel bearing
   - Tie rod to swivel bearing
   - Brake caliper to disc
   - ABS pulse generator
   - Front wheel
   - Splash guard
   - Reinforcement plate
5. Check the front axle differential fluid level.

Rear

1. Before servicing the vehicle, refer to the Precautions Section.
2. Remove or disconnect the following:
   - Rear wheel
   - Rear muffler from center muffler
   - Flange bolts to differential
   - Depressurize ride control system, if equipped
3. Raise the rear wheel carrier with a proper jack and support.
4. Press the halfshaft from the wheel carrier.
5. Remove the halfshaft.

To install:

6. Position the halfshaft and press into the wheel carrier.
7. Fit the shims, then install the bolts on the inner halfshaft flange to 61 ft. lbs. (83 Nm).
8. Attach the rear muffler to the center muffler.
9. Install the rear wheel.

STEERING AND SUSPENSION

Air Bag

⁂ CAUTION

The vehicles are equipped with an air bag system. The system must be disarmed before performing service on, or around, system components, the steering column, instrument panel components, wiring and sensors. Failure to follow the safety precautions and the disarming procedure could result in accidental air bag deployment, possible personal injury and unnecessary system repairs.

PRECAUTIONS

Several precautions must be observed when handling the inflator module to avoid accidental deployment and possible personal injury.

1. Never carry the inflator module by the wires or connector on the underside of the module.
2. When carrying a live inflator module, hold securely with both hands, and ensure that the bag and trim cover are pointed away.
3. Place the inflator module on a bench or other surface with the bag and trim cover facing up.
4. With the inflator module on the bench, never place anything on or close to the module that may be thrown in the event of an accidental deployment.

DISARMING

1. Before servicing the vehicle, refer to the Precautions Section.

2. Place the ignition switch in the **OFF** position.
3. Disconnect the negative battery terminal and cover the battery terminal to prevent accidental contact.
4. Once the battery has been disconnected, wait for a period of approximately 5 seconds allowing the capacitor in the control unit to discharge. Once the capacitor is discharged, a trigger pulse cannot be generated inadvertently.

REARMING

1. Before servicing the vehicle, refer to the Precautions Section.
2. Place the ignition switch in the **OFF** position.
3. Attach the sensors, the steering column connector and the seat belt tensioner connectors.
4. Connect the negative battery terminal.
5. Place the ignition switch in the **ON** position. Check that the SRS light illuminates for 6 seconds and extinguishes. If it illuminates in any other pattern, check the components and their connections for proper operation and recheck operation of the warning light.

Power Rack and Pinion Steering Gear

REMOVAL & INSTALLATION

1. Before servicing the vehicle, refer to the Precautions Section.
2. Set the steering gear in the straight ahead position by aligning the marks on the steering gear and spindle.
3. Drain the power steering fluid.
4. Remove or disconnect the following:
   - Negative battery cable
   - Front wheels
   - Reinforcement plate
   - Nuts from the left and right engine support arms and raise the engine slightly
   - 2 bolts accessible through holes in frame
   - Lower clamping screw
   - Steering gear clamps (after locking wheels in straight-ahead position)
   - Tie rod by pressing it off with special tool 32–3–090
   - Stabilizer end-links
   - Heat shield from power steering gear
   - Hydraulic lines, set aside and secure
   - Electrical connector on EH converter
   - If necessary, pressure line from front axle carrier
   - Self-locking nuts and brace the front axle support
   - Banjo bolts and slide the steering gear out through the left wheel opening

To install:

5. Install the steering gear through the left side wheel opening
6. Install new sealing rings and banjo bolts. Torque the bolts as follows:
   a. M10: 7 ft. lbs. (12 Nm).
   b. M14: 25 ft. lbs. (35 Nm).
   c. M16: 29 ft. lbs. (40 Nm).
   d. M18: 34 ft. lbs. (45 Nm).
7. Install or connect the following:
   - Front axle support screws. Torque the screws to 74 ft. lbs. (100 Nm).
   - Self-locking nuts. Torque the nuts to 74 ft. lbs. (100 Nm).
   - Tie rod. Torque the castle nut to 58 ft. lbs. (80 Nm).
   - Steering gear to the spindle. Torque the fastener to 18 ft. lbs. (24 Nm).
   - Steering gear clamp
   - Engine support arm nuts. Torque the nuts to 60 ft. lbs. (85 Nm).
   - Reinforcement plate
   - Splash guard
   - Both front wheels
   - Negative battery cable
8. Fill and bleed the power steering system.
9. Start the vehicle and check for leaks, repair if necessary.

Strut

REMOVAL & INSTALLATION

1. Before servicing the vehicle, refer to the Precautions Section.
2. Mark the position of the threaded pin to the wheel arch to retain the camber setting when installed.
3. Remove or disconnect the following:
   - Negative battery cable
   - Tire and wheel assembly

➡Carefully mark position of threaded pin to strut tower to ensure original camber is retained during installation.

   - Two of the nuts on the spring strut support bearing
   - Center strut bracket nut
   - Speed sensor/brake wear cable and disconnect the plug housing
   - Swivel bearing and tie it aside
   - Remaining nut on the spring strut support bearing
   - Strut assembly

To install:

4. Install or connect the following:
   - Strut assembly
   - One nut to the spring strut support bearing and hand tighten at this time
   - Swivel bearing. Torque the new self-locking nut to 176 ft. lbs. (250 Nm).
   - Speed sensor/brake wear cable and connect the housing plug
   - Center strut bracket. Torque the nut to 74 ft. lbs. (100 Nm).
5. Align the three upper spring strut support bearing nuts and match the threaded pin with the mark made during the removal procedure. When aligned properly torque the nuts to 25 ft. lbs. (34 Nm).
6. Install the tire and wheel assembly.
7. Reconnect the negative battery cable.

Shock Absorber

REMOVAL & INSTALLATION

1. Before servicing the vehicle, refer to the Precautions Section.
2. Remove or disconnect the following:
   - Negative battery cable
   - Wheel
   - Brake caliper
   - Stabilizer link from spring/strut
   - Tie rod end from swivel bearing
   - Line for pulse generator from spring/strut
   - If necessary, brake pad wear sensor
   - Swivel bearing from spring/strut, set aside and secure

➡If the centering pin missing, mark the position of studs in relation to wheel arch.

3. Secure spring so that it doesn't fall out. Remove mounting nuts.

Remove shock absorber downwards out of wheel arch.

To install:

4. Installation is reverse of removal procedure.
5. Replace all bolts and self-locking nuts
6. Tighten all following fasteners to correct torque specifications:
   - Spring/strut support: 25 ft. lbs. (30 Nm)
   - Shock absorber piston rod on thrust bearing: 47 ft. lbs. (64 Nm)
   - Spring strut shock absorber to pivot mount: 73 ft. lbs. (100 Nm)

Coil Spring

REMOVAL & INSTALLATION

Front

CAUTION

This procedure calls for the spring to be compressed. A compressed spring has high potential energy and if released suddenly can cause severe damage and personal injury.

1. Before servicing the vehicle, refer to the Precautions Section.
2. disconnect the negative battery cable.
3. Remove the strut from the vehicle and mount in a vise using a strut holder. This will prevent damage to the strut tube
4. Using a proper spring compressor, compress the spring until the stress on the thrust bearing is released.
5. Remove the top nut of the strut mount. Counterhold the strut rod during removal.
6. Pull the strut mount off the strut rod. Note the positioning of the spacers and washer for replacement.
7. Pull the spring off the strut and place aside in a safe area.
8. Slowly release the compression of the spring.

To install:

9. Install or connect the following:
   - Spring in the compressor and compress
   - Spring and strut mount with all the spacers and washers in their original positions. Torque the new strut rod nut: 47 ft. lbs. (65 Nm).
10. Release the spring slowly and check that it seats in the spring holders. Install the strut in the vehicle.
11. Connect the negative battery cable.

Lower Ball Joint

REMOVAL & INSTALLATION

1. Before servicing the vehicle, refer to the Precautions Section.
2. Remove or disconnect the following:
   - Negative battery cable
   - Coil spring
   - Upper control arm from trailing arm
   - Shock absorber from the swinging arm
3. Using special tool 33–3–331/2/3 and pull the ball joint out of the steering knuckle.

To install:

4. Install or connect the following:
   - Ball joint into the steering knuckle with special tools 33–3–332/3/4
   - Shock absorber to the swinging arm
   - Upper control arm to trailing arm
   - Coil spring
   - Negative battery cable

Lower Control Arm

REMOVAL & INSTALLATION

1. Before servicing the vehicle, refer to the Precautions Section.
2. Remove or disconnect the following:
   - Negative battery cable
   - Front wheel
   - Control arm from the front axle support and loosen the nut from the control arm to swivel bearing
   - Control arm from the swivel bearing by pressing it off with special tool 31–2–240

To install:

3. Install or connect the following:
   - Lower control arm to the swivel bearing. Torque the nut to 58 ft. lbs. (80 Nm).
   - Lower control arm to the front axle support. Torque the nut to 74 ft. lbs. (100 Nm) plus an additional 90 degrees.
   - Front wheel
   - Negative battery cable
4. Check and adjust the front end alignment as needed.

BUSHING REPLACEMENT

1. Before servicing the vehicle, refer to the Precautions Section.
2. Remove or disconnect the following:
   - Negative battery cable
   - Control arm and tie it back to prevent damage to the ball joint
   - Control arm bushing by installing special tools 31–1–331/2/3

To install:

3. Install or connect the following:
   - Control arm bushing by installing special tools 31–1–331/2/3
   - Control arm
   - Negative battery cable

Upper Control Arm

REMOVAL & INSTALLATION

1. Before servicing the vehicle, refer to the Precautions Section.
2. Remove or disconnect the following:
   - Negative battery cable
   - Wheel assembly
   - Coil spring
   - Upper control arm from the stabilizer link from upper control arm
   - If necessary, jointed rod of ride-height sensor from upper control arm
3. Support rear differential and release bearing bolts on rear differential. Halfshalfts do not have to be removed.
4. Remove or disconnect the following:
   - Upper control arm through bolts towards rear
   - Upper control arm

To install:

5. Installation is reverse of removal.
6. Install and torque the following to correct torque specifications:
   - Upper control arm. Torque the bolt to 78 ft. lbs. (106 Nm).
   - Upper control arm to the steering knuckle. Torque the bolt to 74 ft. lbs. (100 Nm).
   - Wheel assembly
   - Negative battery cable

Wheel Bearings

ADJUSTMENT

Wheel bearings cannot be adjusted and must be replaced as a unit and never be reused once removed.

REMOVAL & INSTALLATION

Front

➡The wheel bearings are only removed if they are worn. They cannot be removed without destroying them (due to side thrust created by the bearing puller). They cannot be disassembled, repacked or adjusted.

1. Before servicing the vehicle, refer to the Precautions Section.
2. Remove or disconnect the following:
   - Negative battery cable

⁂ CAUTION

To avoid damaging the dust sleeve, use special tool tools 33–2–160 to drive out and draw in the drive flange after removing the brake disk. Rounded inside edge of special tool must point to drive flange.

   - Front brake disc
   - Drive flange by installing special tools 33–2–116, 160 and 33–4–200
   - Bearing inner race from the flange
   - Swivel bearing
   - Bearing by installing special tools 31–2–113, 33–4–266, 33–3–266 and 33–4–266/261

To install:

3. Install or connect the following:
   - Wheel bearing with the wider chamfer facing the swivel bearing to the drive flange with special tools 31–2–113, 33–4–266, 33–3–266 and 33–4–266/261
   - Snap ring and circlip
   - Bearing inner race to the drive flange
   - Drive flange to the swivel bearing by using special tool 33–2–116, 160 and 33–4–200
   - Swivel bearing
   - Front brake disc
   - Negative battery cable

Rear

1. Before servicing the vehicle, refer to the Precautions Section.
2. Remove or disconnect the following:
   - Negative battery cable
   - Wheel assembly
   - Collar nut
   - Brake disc
   - Rear axle shaft drive flange
   - Wheel bearing inner race from the drive flange
   - Retaining ring
   - Wheel bearing with special tools 33–3–261/2/3

To install:

3. Install or connect the following:
   - New wheel bearing with special tools 33–3–261/4/5
   - Retaining ring
   - Wheel bearing inner race to the drive flange.
   - Drive flange to the axle shaft
   - Brake disc
   - Collar nut
   - Wheel assembly
   - Negative battery cable

BRAKES

Brake Caliper

REMOVAL & INSTALLATION

Front

1. Before servicing the vehicle, refer to the Precautions Section.
2. Remove or disconnect the following:
   - Negative battery cable
   - Wheel assembly
3. Apply the brake pedal slightly with a brake clamp.
   - Brake pipe from the connection with the brake hose
   - Connector for the wear indicator on the left side
   - Caliper guide bolts
   - Brake caliper

To install:

4. Install or connect the following:
   - Caliper and torque the guide bolts to 22 ft. lbs. (30 Nm)
   - Brake hose to the brake pipe to 13 ft. lbs. (18 Nm)
5. Set the wheel in a straight-ahead position.
6. Be sure brake hose is positively attached to the mounting fixture.
   - Wear indicator on the left side
   - Brake clamp
   - Front wheels
7. Bleed the brakes.

Rear

1. Before servicing the vehicle, refer to the Precautions Section.
2. Remove or disconnect the following:
   - Negative battery cable
   - Wheel assembly
3. Apply the brake pedal slightly with a brake clamp.
   - Slacken union nut
   - Brake hose from the caliper fitting
   - Connector for the wear indicator on the right side
   - Caliper guide bolts and remove the brake caliper

To install:

4. Install or connect the following:
   - Caliper and torque the guide bolts to 21 ft. lbs. (28 Nm)
   - Brake hose to the brake pipe and torque to 14 ft. lbs. (19 Nm)
   - Wear indicator
5. Remove the brake clamp.
6. Install the rear wheels.
7. Bleed the brakes.

Disk Brake Pads

REMOVAL & INSTALLATION

Front

1. Before servicing the vehicle, refer to the Precautions Section.
2. Remove the front wheels.
3. Remove the disk pad retaining spring from the caliper.
4. Remove the calipers from the disk.
5. Use a special tool, 34–1–050, to force piston back into caliper.
6. Remove the outer brake pad (the inner pad is held in place with a spring in the piston).

To install:

7. Be sure the pads marked "L" and "R" are inserted properly on left and right sides, respectively.
8. Apply anti-squeak compound to all mounting surfaces.
9. Install the calipers.
10. Reposition the retaining spring.
11. Install the front wheels.
12. Fully depress brake pedal several times to set proper contact of pads with rotor.

13. Hold ignition key for at least 30 seconds in position “1” without starting engine. This clear any fault codes stored in system and prevent the wear indicator light from coming on.
14. Bleed brake system. If necessary.

Rear

1. Before servicing the vehicle, refer to the Precautions Section.
2. Remove the rear wheels.
3. Remove the plastic plugs from the inside of the caliper.
4. Disconnect the plug connection for the wear indicator.
5. Remove the calipers from the disk.
6. Lift out the pad retaining spring from the caliper.
7. Use a special tool, 34–1–050, to force piston back into caliper.
8. Remove the outer brake pad (the inner pad is held in place with a spring in the piston).

To install:

9. Apply anti-squeak compound to all mounting surfaces.
10. Reposition retaining spring.
11. Install calipers.
12. Install rear wheels.
13. Fully depress brake pedal several times to set proper contact of pads with rotor.
14. Hold ignition key for at least 30 seconds in position “1” without starting engine. This will clear any fault codes stored in system and prevent the wear indicator light from coming on.
15. Bleed brake system. If necessary.

BMW

X5

4

SPECIFICATION AND MAINTENANCE CHARTS

ENGINE AND VEHICLE IDENTIFICATION

Engine						
Code ①	Liters (cc)	Cu. In.	Cyl.	Fuel Sys.	Engine Type	Eng. Mfg.
M54	3.0 (2979)	182	6	SMPI	DOHC	BMW
M62	4.4 (4398)	268	8	SMPI	DOHC	BMW
M62	4.6 (4619)	282	8	SMPI	DOHC	BMW

Model Year	
Code ②	Year
2	2002
3	2003
4	2004
5	2005
6	2006

DOHC: Double Overhead Camshaft

SMPI: Sequential Multi-Port Injection

① 8th position of VIN

② 10th position of VIN

06041_BMW_X5_C0001

GENERAL ENGINE SPECIFICATIONS

Year	Model	Engine Displacement Liters (VIN)	Net Horsepower @ rpm	Net Torque @ rpm (ft. lbs.)	Bore x Stroke (in.)	Com-pression Ratio	Oil Pressure @ rpm
2002	X5	3.0 (M54/5)	225@5900	214@3500	3.31x3.53	10.2:1	7.4@700
	X5	4.4 (M62/5)	290@5400	324@3600	3.62x3.26	10.0:1	7.4@580
	X5	4.6 (M62) ①	340@5700	350@3700	3.66x3.35	10.5:1	7.4@580
2003	X5	3.0 (M54/5)	225@5900	214@3500	3.31x3.53	10.2:1	7.4@700
	X5	4.4 (M62/5)	290@5400	324@3600	3.62x3.26	10.0:1	7.4@580
	X5	4.6 (M62) ①	340@5700	350@3700	3.66x3.35	10.5:1	7.4@580
2004	X5	3.0 (M54/5)	225@5900	214@3500	3.31x3.53	10.2:1	7.4@700
	X5	4.4 (M62/5)	290@5400	324@3600	3.62x3.26	10.0:1	7.4@580
	X5	4.6 (M62) ①	340@5700	350@3700	3.66x3.35	10.5:1	7.4@580
2005-06	X5	3.0 (M54/5)	225@5900	214@3500	3.31x3.53	10.2:1	7.4@700
	X5	4.4 (M62/5)	290@5400	324@3600	3.62x3.26	10.0:1	7.4@580
	X5	4.6 (M62) ①	340@5700	350@3700	3.66x3.35	10.5:1	7.4@580

SMPI: Sequential Multi-port Fuel Injection

① Specific VIN information not available.

06041_BMW_X5_C0002

ENGINE TUNE-UP SPECIFICATIONS

Year	Engine Displacement Liters (VIN)	Spark Plug Gap (in.)	Ignition Timing (deg.)	Fuel Pump (psi)	Idle Speed (rpm)	Valve Clearance In.	Valve Clearance Ex.
2002	3.0 (M54/5)	0.024-0.028	①	48-54	②	HYD	HYD
	4.4 (M62/5)	0.024-0.028	①	48-54	②	HYD	HYD
	4.6 (M62) ③	0.024-0.028	①	48-54	②	HYD	HYD
2003	3.0 (M54/5)	0.024-0.028	①	48-54	②	HYD	HYD
	4.4 (M62/5)	0.024-0.028	①	48-54	②	HYD	HYD
	4.6 (M62) ③	0.024-0.028	①	48-54	②	HYD	HYD
2004	3.0 (M54/5)	0.024-0.028	①	48-54	②	HYD	HYD
	4.4 (M62/5)	0.024-0.028	①	48-54	②	HYD	HYD
	4.6 (M62) ③	0.024-0.028	①	48-54	②	HYD	HYD
2005-06	3.0 (M54/5)	0.024-0.028	①	48-54	②	HYD	HYD
	4.4 (M62/5)	0.024-0.028	①	48-54	②	HYD	HYD
	4.6 (M62) ③	0.024-0.028	①	48-54	②	HYD	HYD

NOTE: The Vehicle Emission Control Information label often reflects specification changes made during production. The label figures must be used if they differ from those in this chart.

HYD: Hydraulic

① Ignition timing is regulated by the Electronic Control Module (ECM), and cannot be adjusted.

② Idle speed is controled by the Electronic Control Module (ECM), and cannot be adjusted.

③ Specific VIN information not available.

06041_BMW_X5_C0003

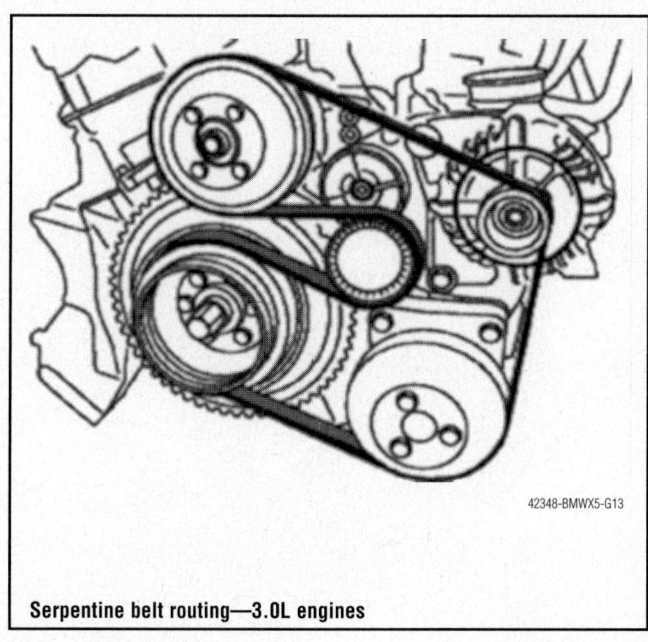

Serpentine belt routing—3.0L engines

Serpentine belt routing—4.4L and 4.6L engines

CAPACITIES

Year	Model	Engine Displacement Liters (VIN)	Engine Oil with Filter (qts.)	Automatic Transaxle (qts.)	Manual Transaxle (qts.)	Rear Drive Axle (pts.)	Fuel Tank (gal.)	Cooling System (qts.)
2002	X5	3.0 (M54/5)	8.0	7.0	3.2	3.4	24.6	11.1
		4.4 (M62/5)	8.5	11.7	NA	3.4	24.6	13.7
		4.6 (M62) ①	8.5	11.7	NA	3.4	24.6	13.7
2003	X5	3.0 (M54/5)	8.0	7.0	3.2	3.4	24.6	11.1
		4.4 (M62/5)	8.5	11.7	NA	3.4	24.6	13.7
		4.6 (M62) ①	8.5	11.7	NA	3.4	24.6	13.7
2004	X5	3.0 (M54/5)	8.0	7.0	3.2	3.4	24.6	11.1
		4.4 (M62/5)	8.5	11.7	NA	3.4	24.6	13.7
		4.6 (M62) ①	8.5	11.7	NA	3.4	24.6	13.7
2005-06	X5	3.0 (M54/5)	8.0	7.0	3.2	3.4	24.6	11.1
		4.4 (M62/5)	8.5	11.7	NA	3.4	24.6	13.7
		4.6 (M62) ①	8.5	11.7	NA	3.4	24.6	13.7

NOTE: All capacities are approximate. Add fluid gradually and check to be sure a proper fluid level is obtained.

NA: Not available.

① Specific VIN information not available.

06041_BMW_X5_C0004

VALVE SPECIFICATIONS

Year	Engine Displacement Liters (VIN)	Seat Angle (deg.)	Face Angle (deg.)	Spring Test Pressure (lbs. @ in.)	Spring Installed Height (in.)	Stem-to-Guide Clearance (in.)		Stem Diameter (in.)	
						Intake	Exhaust	Intake	Exhaust
2002	3.0 (M54/5)	45	45	NA	NA	0.0197	0.0197	0.2372-0.2340	0.2378-0.2384
	4.4 (M62/5)	45	45	NA	NA	0.0197	0.0197	0.2156-0.2159	0.2146-0.2150
	4.6 M62 ①	45	45	NA	NA	0.0016	0.0016	0.1998-0.2378	0.1992-0.2384
2003	3.0 (M54/5)	45	45	NA	NA	0.0197	0.0197	0.2372-0.2340	0.2378-0.2384
	4.4 (M62/5)	45	45	NA	NA	0.0197	0.0197	0.2156-0.2159	0.2146-0.2150
	4.6 M62 ①	45	45	NA	NA	0.0016	0.0016	0.1998-0.2378	0.1992-0.2384
2004	3.0 (M54/5)	45	45	NA	NA	0.0197	0.0197	0.2372-0.2340	0.2378-0.2384
	4.4 (M62/5)	45	45	NA	NA	0.0197	0.0197	0.2156-0.2159	0.2146-0.2150
	4.6 M62 ①	45	45	NA	NA	0.0016	0.0016	0.1998-0.2378	0.1992-0.2384
2005-06	3.0 (M54/5)	45	45	NA	NA	0.0197	0.0197	0.2372-0.2340	0.2378-0.2384
	4.4 (M62/5)	45	45	NA	NA	0.0197	0.0197	0.2156-0.2159	0.2146-0.2150
	4.6 M62 ①	45	45	NA	NA	0.0016	0.0016	0.1998-0.2378	0.1992-0.2384

NA: Not available

① Specific VIN information not available

06041_BMW_X5_C0005

CRANKSHAFT AND CONNECTING ROD SPECIFICATIONS

All measurements are given in inches.

	Engine	Crankshaft				Connecting Rod		
Year	Displacement Liters (VIN)	Main Brg. Journal Dia.	Main Brg. Oil Clearance	Shaft End-play	Thrust on No.	Journal Diameter	Oil Clearance	Side Clearance
2002	3.0 (M54/5)	①	0.0007-0.0029	0.0031-0.0064	5	1.7720-1.7706	0.0007-0.0022	0.0060-0.0160
	4.4 (M62/5)	①	0.0007-0.0018	0.0033-0.0101	3	1.8901-1.8887	0.0007-0.0022	0.0060-0.0196
	4.6 M62 ②	①	0.0007-0.0018	0.0033-0.0101	3	1.8901-1.8904	0.0007-0.0022	0.0060-0.0196
2003	3.0 (M54/5)	①	0.0007-0.0029	0.0031-0.0064	5	1.7720-1.7706	0.0007-0.0022	0.0060-0.0160
	4.4 (M62/5)	①	0.0007-0.0018	0.0033-0.0101	3	1.8901-1.8887	0.0007-0.0022	0.0060-0.0196
	4.6 M62 ②	①	0.0007-0.0018	0.0033-0.0101	3	1.8901-1.8904	0.0007-0.0022	0.0060-0.0196
2004	3.0 (M54/5)	①	0.0007-0.0029	0.0031-0.0064	5	1.7720-1.7706	0.0007-0.0022	0.0060-0.0160
	4.4 (M62/5)	①	0.0007-0.0018	0.0033-0.0101	3	1.8901-1.8887	0.0007-0.0022	0.0060-0.0196
	4.6 M62 ②	①	0.0007-0.0018	0.0033-0.0101	3	1.8901-1.8904	0.0007-0.0022	0.0060-0.0196
2005-06	3.0 (M54/5)	①	0.0007-0.0029	0.0031-0.0064	5	1.7720-1.7706	0.0007-0.0022	0.0060-0.0160
	4.4 (M62/5)	①	0.0007-0.0018	0.0033-0.0101	3	1.8901-1.8887	0.0007-0.0022	0.0060-0.0196
	4.6 M62 ②	①	0.0007-0.0018	0.0033-0.0101	3	1.8901-1.8904	0.0007-0.0022	0.0060-0.0196

① Standard green: 2.3613-2.3615 inches

Standard white: 2.3611-2.3613 inches

② No specific VIN information available

06041_BMW_X5_C0006

PISTON AND RING SPECIFICATIONS

All measurements are given in inches.

Year	Engine Displacement Liters (VIN)	Piston Clearance	Ring Gap Top Compression	Ring Gap Bottom Compression	Ring Gap Oil Control	Ring Side Clearance Top Compression	Ring Side Clearance Bottom Compression	Ring Side Clearance Oil Control
2002	3.0 (M54/5)	0.0004-0.0016	0.0039-0.0118	0.0078-0.0157	0.0098-0.0197	0.0008-0.0024	0.0012-0.0026	0.0007-0.0024
	4.4 (M62/5)	0.0002-0.0015	0.0039-0.0118	0.0078-0.0157	0.0078-0.0354	0.0008-0.0024	0.0008-0.0024	①
	4.6 M62 ②	0.0002-0.0015	0.0039-0.0118	0.0078-0.0157	0.0078-0.0354	0.0008-0.0022	0.0008-0.0022	①
2003	3.0 (M54/5)	0.0004-0.0016	0.0039-0.0118	0.0078-0.0157	0.0098-0.0197	0.0008-0.0024	0.0012-0.0026	0.0007-0.0024
	4.4 (M62/5)	0.0002-0.0015	0.0039-0.0118	0.0078-0.0157	0.0078-0.0354	0.0008-0.0024	0.0008-0.0024	①
	4.6 M62 ②	0.0002-0.0015	0.0039-0.0118	0.0078-0.0157	0.0078-0.0354	0.0008-0.0022	0.0008-0.0022	①
2004	3.0 (M54/5)	0.0004-0.0016	0.0039-0.0118	0.0078-0.0157	0.0098-0.0197	0.0008-0.0024	0.0012-0.0026	0.0007-0.0024
	4.4 (M62/5)	0.0002-0.0015	0.0039-0.0118	0.0078-0.0157	0.0078-0.0354	0.0008-0.0024	0.0008-0.0024	①
	4.6 M62 ②	0.0002-0.0015	0.0039-0.0118	0.0078-0.0157	0.0078-0.0354	0.0008-0.0022	0.0008-0.0022	①
2005-06	3.0 (M54/5)	0.0004-0.0016	0.0039-0.0118	0.0078-0.0157	0.0098-0.0197	0.0008-0.0024	0.0012-0.0026	0.0007-0.0024
	4.4 (M62/5)	0.0002-0.0015	0.0039-0.0118	0.0078-0.0157	0.0078-0.0354	0.0008-0.0024	0.0008-0.0024	①
	4.6 M62 ②	0.0002-0.0015	0.0039-0.0118	0.0078-0.0157	0.0078-0.0354	0.0008-0.0022	0.0008-0.0022	①

① Does not require measurement.

② No specific VIN information available.

06041_BMW_X5_C0007

TORQUE SPECIFICATIONS

All readings in ft. lbs.

Year	Engine Displacement Liters (VIN)	Cylinder Head Bolts	Main Bearing Bolts	Rod Bearing Bolts	Crankshaft Damper Bolts	Flywheel Bolts	Manifold Intake	Manifold Exhaust	Spark Plugs	Oil Pan Drain Plug
2002	3.0 (M54/5)	①	②	③	NS	77	④	15	⑤	⑥
	4.4 (M62/5)	①	NS	⑧	NS	77	④	7	⑤	⑥
	4.6 (M62) ⑦	①	NS	⑧	NS	77	④	7	⑤	⑥
2003	3.0 (M54/5)	①	②	③	NS	77	④	15	⑤	⑥
	4.4 (M62/5)	①	NS	⑧	NS	77	④	7	⑤	⑥
	4.6 (M62) ⑦	①	NS	⑧	NS	77	④	7	⑤	⑥
2004	3.0 (M54/5)	①	②	③	NS	77	④	15	⑤	⑥
	4.4 (M62/5)	①	NS	⑧	NS	77	④	7	⑤	⑥
	4.6 (M62) ⑦	①	NS	⑧	NS	77	④	7	⑤	⑥
2005-06	3.0 (M54/5)	①	②	③	NS	77	④	15	⑤	⑥
	4.4 (M62/5)	①	NS	⑧	NS	77	④	7	⑤	⑥
	4.6 (M62) ⑦	①	NS	⑧	NS	77	④	7	⑤	⑥

NS: Not specified by manufacturer

① See repair procedure for torque information.

② Step 1: 15 ft. lbs.
Step 2: Additional 70 degrees.

③ Step 1: 4 ft. lbs.
Step 2: 15 ft. lbs.
Step 3: Additional 70 degrees.

④ M6 bolts: 7 ft. lbs.
M7 bolts: 11 ft. lbs.
M8 bolts: 16 ft. lbs.

⑤ With thread M12x1.25: 15-19 ft. lbs.
With thread M14x1.25: 20-24 ft. lbs.

⑥ M12 X 1.5 bolts: 18 ft. lbs.
M18 X 1.5 bolts: 22 ft. lbs.
M22 X 1.5 bolts: 44 ft. lbs.

⑦ No specific VIN information available.

⑧ Step 1: 4 ft. lbs.
Step 2: 15 ft. lbs.
Step 3: Additional 80 degrees.

06041_BMW_X5_C0008

TIRE, WHEEL AND BALL JOINT SPECIFICATIONS

Year	Model	Engine	OEM Tires Standard	OEM Tires Optional	Tire Pressures (psi) Front	Tire Pressures (psi) Rear	Wheel Size	Ball Joint Inspection	Lug Nuts
2002	X5	3.0L	235/65R17	255/55R18	32	32	7.5J/8.5J	NA	81-96
		4.4L	255/55R18	255/50R19 ①	32 ①	32 ①	9J ①	NA	81-96
				285/45R19 ②	32 ②	32 ②	10J ②	NA	
2003	X5	3.0L	235/65R17	255/55R18	32	32	7.5J/8.5J	NA	81-96
		4.4L	255/55R18	255/50R19 ①	32 ①	32 ①	9J ①	NA	81-96
				285/45R19 ②	32 ②	32 ②	10J ②	NA	
		4.6L	275/40R20 ①	NA	32 ①	32 ①	9J ①	NA	81-96
			315/35R20 ②	NA	32 ②	32 ②	10J ②	NA	
2004	X5	3.0L	235/65R17	255/55R18	32	32	7.5J/8.5J	NA	81-96
		4.4L	255/55R18	255/50R19 ①	32 ①	32 ①	9J ①	NA	81-96
				285/45R19 ②	32 ②	32 ②	10J ②	NA	
		4.6L	275/40R20 ①	NA	32 ①	32 ①	9J ①	NA	81-96
			315/35R20 ②	NA	32 ②	32 ②	10J ②	NA	
2005-06	X5	3.0L	235/65R17	255/55R18	32	32	7.5J/8.5J	NA	81-96
		4.4L	255/55R18	255/50R19 ①	32 ①	32 ①	9J ①	NA	81-96
				285/45R19 ②	32 ②	32 ②	10J ②	NA	
		4.6L	275/40R20 ①	NA	32 ①	32 ①	9J ①	NA	81-96
			315/35R20 ②	NA	32 ②	32 ②	10J ②	NA	

OEM: Original Equipment Manufacturer

PSI: Pounds Per Square Inch

NA: Not Available

① Front

② Rear

06041_BMW_X5_C0009

WHEEL ALIGNMENT

Year	Model		Caster Range (+/-Deg.)	Caster Preferred Setting (Deg.)	Camber Range (+/-Deg.)	Camber Preferred Setting (Deg.)	Toe-in (in.)
2002	X5	F	0.50	+0.83	0.42	-0.20	0.30+/-0.13
		R	—	—	0.33	+0.83	0.30+/-0.13
2003	X5	F	0.50	+0.83	0.42	-0.20	0.30+/-0.13
		R	—	—	0.33	+0.83	0.30+/-0.13
2004	X5	F	0.50	+0.83	0.42	-0.20	0.30+/-0.13
		R	—	—	0.33	+0.83	0.30+/-0.13
2005-06	X5	F	0.50	+0.83	0.42	-0.20	0.30+/-0.13
		R	—	—	0.33	+0.83	0.30+/-0.13

NA: Not available.

06041_BMW_X5_C0010

BRAKE SPECIFICATIONS

All measurements in inches unless noted

			Brake Disc			Brake Drum Diameter			Min.	Brake Caliper	
Year	Model		Original Thickness	Minimum Thickness	Maximum Run-out	Original Inside Diameter	Max. Wear Limit	Maximum Machine Diameter	Lining Thick-ness	Bracket Bolts (ft. lbs.)	Mounting Bolts (ft. lbs.)
2002	M54	F	1.118	①	0.005	NA	NA	NA	0.118	81	23
		R	0.470	①	0.005	8.63-8.65	NA	NA	①	49	21
	M62	F	1.118	①	0.007	NA	NA	NA	0.118	81	23
		R	0.470 ②	①	0.007	8.63-8.65	NA	NA	①	49	21
2003	M54	F	1.118	①	0.005	NA	NA	NA	0.118	81	23
		R	0.470	①	0.005	8.63-8.65	NA	NA	①	49	21
	M62	F	1.118	①	0.007	NA	NA	NA	0.118	81	23
		R	0.470 ②	①	0.007	8.63-8.65	NA	NA	①	49	21
2004	M54	F	0.803	①	0.005	NA	NA	NA	0.118	81	23
		R	0.470	①	0.005	8.63-8.65	NA	NA	①	49	21
	M62	F	1.118	①	0.007	NA	NA	NA	0.118	81	23
		R	0.470 ②	①	0.007	8.63-8.65	NA	NA	①	49	21
2005-06	M54	F	0.803	①	0.005	NA	NA	NA	0.118	81	23
		R	0.470	①	0.005	8.63-8.65	NA	NA	①	49	21
	M62	F	1.118	①	0.007	NA	NA	NA	0.118	81	23
		R	0.470 ②	①	0.007	8.63-8.65	NA	NA	①	49	21

NA: Not Available

① Minimum thickness is stamped in the brake disc shell

② Figure shown is for 4.4L engine models; with 4.6L engine, thickness is 1.118 inches.

06041_BMW_X5_C0011

SCHEDULED MAINTENANCE INTERVALS

BMW—X5

TO BE SERVICED	TYPE OF SERVICE	SERVICE INTERVALS			
		INITIAL 1200 MILES	OIL SERVICE	INSPECTION I	INSPECTION II
Oil level	S/I	✓			
Engine oil	R	✓			
Engine oil & filter	R ①		✓	✓	✓
Engine air cleaner element	R ②				✓
Spark plugs	R				✓
Fuel filter	R ③				✓
Fuel, vapor lines & fuel cap	S/I	✓		✓	✓
Cooling system	S/I	✓		✓	✓
Exhaust pipe & muffler	S/I	✓		✓	✓
Catalytic converter & shielding	S/I	✓		✓	✓
Throttle linkage	S/I			✓	
Engine (check for leakage)	S/I	✓			
Engine drive belts	S/I				✓
Maintenance Indicators	RE		④	✓	✓
Engine coolant	R			⑤	⑤
Oxygen sensor	R ⑥				
Brake & clutch fluids ⑥	S/I			✓	✓
Brake pads & discs	S/I			✓	✓
Parking brake system	S/I			✓	✓
Power steering system	S/I			✓	✓
Rear axle fluid	S/I			✓	✓
Steering play, suspension track rods, front axle joints, steering linkage & joint disc	S/I			✓	✓
Transmission fluid/oil	S/I			✓	✓
Wheel centering hubs	S/I			✓	✓
Rear axle fluid	R		✓		✓
OBD system for codes	S/I	✓		✓	

R: Replace S/I: Service or Inspect RE: Reset

Note: BMW does not rely solely on vehicle mileage to determine service intervals. An on-oboard diagnostic center, monitors engine operating conditions, along with mileage, to determine the most effective maintenance intervals. The information is then conveyed to the driver through the service indicator lights, located in the center of the instrument panel.

Note: Maintenance and most wear items are covered by the manufacturer. Refer to the operator's manual for additional information.

① On vehicles operated less than 6200 miles per year, more frequent service may be required.

② Replace more frequently if vehicle is operated in dusty conditions.

③ Recommended service for California models, required for all other models.

④ Reset the oil service indicator lights only.

⑤ Replace every 2 years with inspection service.

⑥ Replace every 100,000 miles on all models.

FREQUENT OPERATION MAINTENANCE (SEVERE SERVICE)

If a vehicle is operated under any of the following conditions it is considered severe service

- Extremely dusty areas.

- 50% or more of the vehicle operation is in 32°C (90°F) or higher temperatures, or constant operation in temperatures below 0°C (32°F).

- Prolonged idling (vehicle operation in stop and go traffic).

- Frequent short running periods (engine does not warm to normal operating temperatures).

- Police, taxi, delivery usage or trailer towing usage.

06041_BMW_X5_C0012

ENGINE REPAIR

➡Disconnecting the negative battery cable on some vehicles may interfere with the functions of the on-board computer systems and may require the computer to undergo a relearning process.

Ignition Timing

ADJUSTMENT

The Digital Motor Electronics (DME) control, unit controls all ignition and fuel injection functions. Ignition timing is fully electronically controlled; there is no vacuum advance or manual adjustment. Ignition functions are calculated from internal maps and from the same sensors used for the fuel injection system. On vehicles with an automatic transmission, the control unit will retard ignition timing briefly when the transmission is about to shift up or down. For this reason, there is a data link between the DME control unit and the transmission control unit.

Since the ignition timing is controlled by the DME, checking and adjusting the timing is impossible. There is no method of setting dynamic or static timing.

Alternator

REMOVAL & INSTALLATION

➡When the battery is disconnected the radio code, on-board computer and clock settings will be lost. The radio code should be obtained before disconnecting the battery or radio. Once the battery has been reconnected, the radio will not function unless the code is keyed in.

3.0L Engines

1. Before servicing the vehicle, refer to the precautions section.
2. Read out stored fault codes (if applicable).
3. Switch off ignition.
4. Disconnect negative battery cable.
5. Remove or disconnect the following:
   - Suction filter housing
   - Fan clutch
   - Alternator drive belt
   - Power steering pump from mounting (move aside)
   - Alternator air hose (if equipped)
   - Electrical connections from alternator
   - Alternator mounting bolts (versions without idler pulley)
   - Idler pulley cover and idler pulley bolt (versions with idler pulley)
   - Alternator

To install:

6. To install, reverse removal procedure.
7. If equipped with idler pulley, turn lock of tensioning roller to engage alternator groove.
8. With scan tool, clear fault code memory.

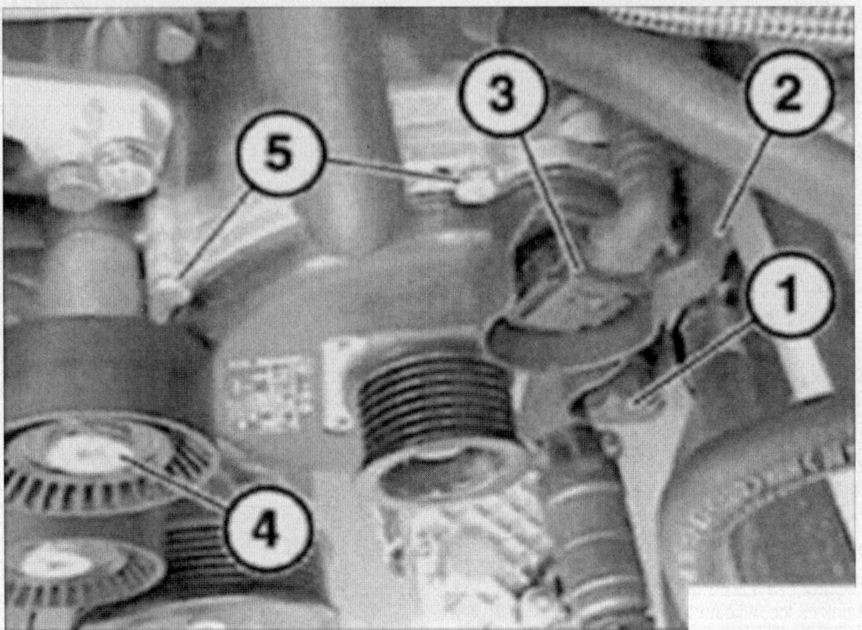

Showing electrical and other components to release in order to remove the alternator (4.4L and 4.6L engines)

4.4L & 4.6L Engines

1. Switch off ignition.
2. Disconnect negative battery cable.
3. Remove or disconnect the following:
   - Alternator drive belt
   - Fan cowl
   - Drain cooling system
   - Alternator lower mounting bolts
   - Nut on front of alternator (1)
   - Electrical connections from alternator (2, 3)
   - Tensioning roller (4)
   - Alternator upper mounting bolts (5)
   - Alternator

To install:

4. To install, reverse removal procedure, using new sealing ring on alternator.

Engine Assembly

REMOVAL & INSTALLATION

3.0L Engines

1. Before servicing the vehicle, refer to the precautions section.
2. Disconnect the negative battery cable.
3. Remove or disconnect the following:
4. Remove the heater bulkhead by pulling off the sealing strips from the firewall, turning the toggle retainers about 90 degrees, lifting out the cover, and removing the heater bulkhead.
   - Engine cover
   - Both oxygen sensors
   - Hood to full open (assembly) position
   - Front splash guard
   - Reinforcement plate from undercarriage
   - Transmission
   - A/C compressor from mounting (move aside, with lines connected)
5. Drain the cooling system (drain plug is located on exhaust side of cylinder number 2 in engine block). Reinstall drain plug with new seal ring.
6. Pull the locks and disconnect the coolant hoses from the water pump.

7. Remove or disconnect the following:
   - Radiator
   - Water hoses from pipe and from heating valve
   - Wiring harness section for engine (lay aside)

8. Siphon off some fluid from power steering reservoir. Unbolt reservoir and tie out of the way, leaving lines connected.

9. Remove or disconnect the following:
   - Fan clutch with water pump impeller
   - Alternator drive belt
   - Vane pump for power steering (tie out of way, leaving lines connected)

10. Install an engine lift and chains, attaching to eye hooks.

11. Disconnect the right ground wire.

12. Unbolt the left and right engine mounts.

13. Carefully lift out the engine assembly.

To install:

14. Install in reverse of removal procedure.

15. Tighten components to the following torque settings:
   a. Radiator bolts: 6–7 ft. lbs. (9–10 Nm)
   b. Power steering pump bracket bolts: 16 ft. lbs. (22 Nm)

16. Tighten the automatic transmission–to–engine bolts to the following specifications:
   a. M8 hex bolts: 18 ft. lbs. (24 Nm).
   b. M10 hex bolts: 33 ft. lbs. (45 Nm).
   c. M12 hex bolts: 60 ft. lbs. (82 Nm).
   d. M8 Torx bolts: 15 ft. lbs. (21 Nm).
   e. M10 Torx bolts: 31 ft. lbs. (42 Nm).
   f. M12 Torx bolts: 53 ft. lbs. (72 Nm).

4.4L and 4.6L Engines

1. Before servicing the vehicle, refer to the precautions section.

2. Disconnect negative battery cable.

3. Remove the heater bulkhead by pulling off the sealing strips from the firewall, turning the toggle retainers about 90 degrees, lifting out the cover, and removing the heater bulkhead.

4. Fully open the hood and properly secure it in place.

5. Remove or disconnect the following:
   - Engine cover
   - Throttle bellows
   - Intake filter housing
   - Mass Air Flow (MAF) sensor
   - Windshield washer reservoir
   - Brake booster hose from the suction jet pump
   - Fuel feed line from the injection pipe
   - Engine splash shield and reinforcement plate
   - Drain cooling system (drain plug on right side of block)
   - Alternator drive belt
   - Power steering pump (move aside, with lines connected)
   - Evacuate and recover A/C refrigerant
   - A/C system lines between the compressor and condenser
   - A/C suction line
   - Transmission
   - Starter electrical connectors and heat shield
   - Starter
   - Oil lines to the transmission on the heat exchanger
   - Radiator
   - Coolant hoses on the alternator and thermostat housing
   - Coolant hoses from the coolant manifold
   - Heating valve and hoses
   - Fuel tank vent hose
   - Engine wire harness from the control unit box
   - Transmission wire harness from the control unit box
   - Positive lead from support point (place on engine)
   - Oxygen sensor wiring and place all wires on top of the engine
   - Expansion tank
   - Supply reservoir from the carrier
   - Ground strap from the oil filter housing
   - Left and right swivel bearings and output shafts
   - Propeller shaft
   - Partition wall
   - Ground tape from the right side engine support
   - Upper nuts from the left and right side engine mounts and install an engine removal tool to the locating lugs
   - Engine from the vehicle

To install:

6. Carefully lower the engine into the engine compartment.

7. Install or connect the following:
   - Engine mounts. Torque the bolts to 32 ft. lbs. (45 Nm).
   - Ground tape
   - Partition wall
   - Propeller shafts
   - Output shafts
   - Left and right swivel bearings
   - Ground strap to the oil filter housing
   - Supply reservoir
   - Expansion tank
   - Oxygen sensor electrical connector
   - Engine and transmission wire harness to the control unit box
   - Fuel tank vent hose
   - Heater valve and hoses
   - Coolant hoses to the manifold and thermostat housing
   - Radiator
   - Transmission
   - Oil lines to the transmission. Torque the nuts to 25 ft. lbs. (34 Nm).
   - Starter and electrical connectors
   - A/C lines
   - Power steering pump
   - Drive belt
   - Engine splash shield and reinforcement plate
   - Fuel feed line to the injection pipe
   - Brake booster vacuum hose
   - Windshield washer reservoir
   - MAF sensor
   - Intake filter housing
   - Throttle cable to the filter housing
   - Engine cover
   - Negative battery cable

8. Fill and bleed the power steering system.

9. Fill and bleed the coolant system.

10. Recharge the A/C system.

11. Fill the engine with clean oil.

12. Start the vehicle and check for leaks, repair if necessary.

Water Pump

REMOVAL & INSTALLATION

3.0L Engines

1. Before servicing the vehicle, refer to the precautions section.

2. Remove the alternator drive belt.

3. Drain the cooling system. Drain plug is located on exhaust side of block, next to cylinder number 2.

4. Remove the water pump pulley.

5. Remove the 4 water pump retaining bolts. Use 2 M6 bolts in holes next to mounting bolt holes and screw in until water pump releases from timing cover.

To install:

6. When installing, use a new O-ring.

7. Tighten mounting bolts as follows:
   a. M6 bolts: 7 ft. lbs. (10 Nm).
   b. M7 bolts: 11 ft. lbs. (15 Nm).
   c. M8 bolts: 16 ft. lbs. (22 Nm).

8. Install water pump pulley.

9. Refill cooling system.

10. Install the alternator drive belt.

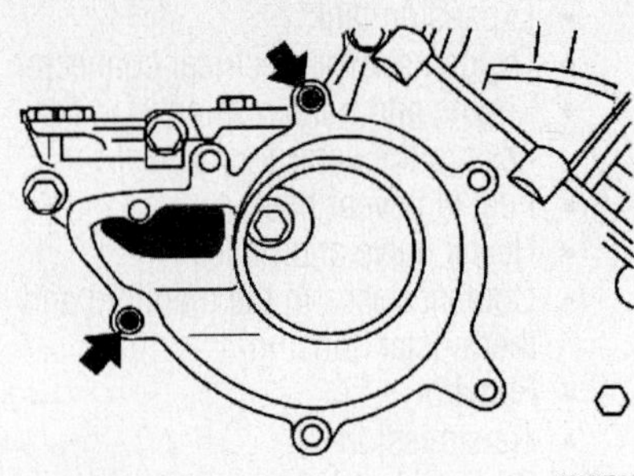

Showing locating sleeve position on the water pump—4.4L engine

11. Start the vehicle, check for leaks and repair as necessary.

4.4L & 4.6L Engines

1. Before servicing the vehicle, refer to the precautions section.
2. Drain the cooling system.
3. Remove or disconnect the following:
   - Negative battery cable
   - Vibration damper
   - Thermostat housing
   - Water pump pulley
   - Coolant hoses
   - Water pump and discard the seal

To install:

4. Clean and remove any residual debris or gasket material from the engine mounting surface for the water pump.
5. Install the water pump with a new gasket. Torque the bolts as follows:
   a. M6 bolts: 78 inch lbs. (9 Nm).
   b. M7 bolts: 89 inch lbs. (10 Nm).
   c. M8 bolts: 16 ft. lbs. (22 Nm).
6. Install or connect the following:
   - Coolant hoses to the water pump
   - Water pump pulley
   - Thermostat housing
   - Vibration damper
   - Negative battery cable
7. Fill and bleed the cooling system.
8. Start the vehicle, check for leaks and repair as necessary.

Heater Core

REMOVAL & INSTALLATION

1. Before servicing the vehicle, refer to the precautions section.
2. Disconnect the negative battery cable.
3. On 3.0L engines, remove the heater bulkhead as follows:
   a. Pull off the weatherstrips on top of the bulkhead.
   b. Release the cowl retaining toggle clips.
   c. Lift out the cover.
   d. Remove the screws and remove the heater bulkhead.
4. Remove or disconnect the following:
   - Heater hoses from firewall connections
   - Air duct for left footwell from mountings
   - Left footwell vent
   - Heater pipes from under dash
   - Heater duct for right footwell
   - Air duct for rear footwell on right
   - Cover from heater housing
   - Evaporator temperature sensor (pull out to one side)
   - Heater core

To install:

5. Install or connect the following:
   - Heater core
   - Evaporator temperature sensor
   - Heater housing cover
   - Air ducts in footwells
   - Heater pipes under dash
   - Heater hoses at firewall connections
   - Heater bulkhead (3.0L)
6. Reconnect the negative battery cables
7. Check coolant level

Cylinder Head

REMOVAL & INSTALLATION

3.0L Engines

1. Before servicing the vehicle, refer to the precautions section.
2. Properly relieve the fuel system pressure.
3. Remove or disconnect the following:
   - Negative battery cable
   - Left side exhaust manifold
   - Cylinder head cover
   - Spark plugs
   - Intake manifold
   - Drain cooling system
   - Thermostat housing
   - Coolant pipe from side of block
   - Double VANOS adjustment unit and camshafts with bearings
   - Timing case cover–to– cylinder head bolts
   - Timing chain guide
4. Remove the cylinder head bolts in reverse of sequence shown and remove the cylinder head.

To install:

5. Clean all sealing material from mating faces

⁂ CAUTION

There must be no oil in the cylinder head bolt holes in the block and timing case cover or there is a possibility of cracking and distorting torque values.

6. Be sure that dowel sleeve are correctly positioned in block.
7. Apply permanently elastic sealing compound, Drei Bond 1209, at joints to timing case cover.
8. Install a new cylinder head gasket.

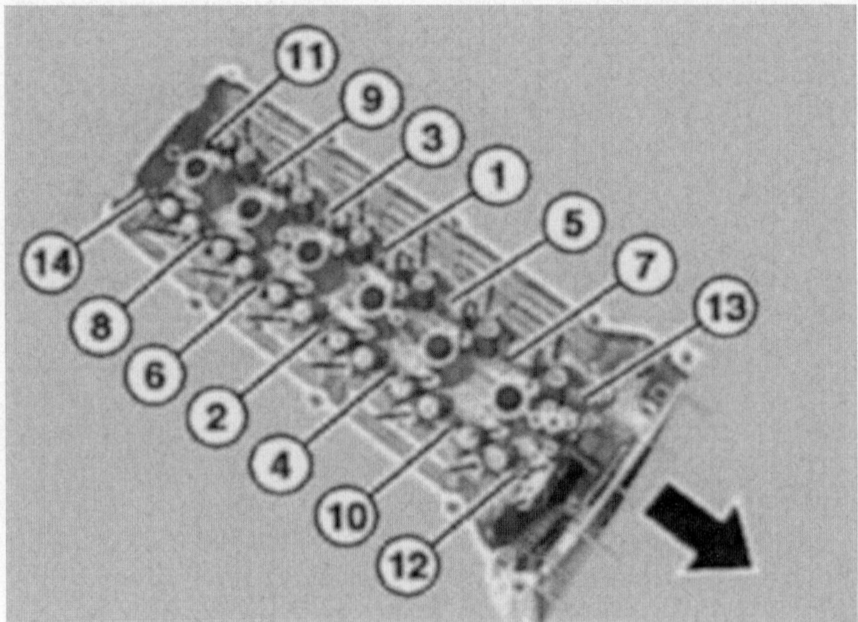

Cylinder head bolt tightening sequence—3.0L engines

9. Position the cylinder head onto the block and install new cylinder head bolts, applying a light coat of oil to the washer contact area and the threads of the new bolts.

10. Tighten cylinder head bolts, following the sequence shown, in 3 steps, to the following:
   a. Step 1: 30 ft. lbs. (40 Nm).
   b. Step 2: Additional 90 degrees.
   c. Step 3: Additional 90 degrees.

11. Install or connect the following:
   - Timing chain guide
   - Timing case cover–to–cylinder head bolts
   - Camshafts
   - Double VANOS adjustment unit

12. Complete installation in reverse of the removal procedure.

4.4L & 4.6L Engines

LEFT SIDE

1. Before servicing the vehicle, refer to the precautions section.
2. Properly relieve the fuel system pressure.
3. Drain the cooling system.
4. Remove or disconnect the following:
   - Negative battery cable
   - Left side exhaust manifold
   - Engine cover
   - Throttle bellows
   - Upper section of suction filter housing with Mass Air Flow (MAF) sensor
   - Lower section of suction filter housing
   - Battery positive lead from firewall connection and lay aside
   - Left cylinder head cover
   - Washer fluid reservoir
   - Right cylinder head cover
   - Water hoses from heater valve
   - Vent line for front axle differential
   - Positive lead
   - Ignition coils
   - Vent hose from cylinder head
   - Left and right cable ducts and fold inward
   - Both cylinder head covers
   - Spark plugs
   - Intake manifold
   - Coolant manifold
   - Left side camshaft VANOS adjustment unit
5. Install special tool, 11–5–180, and pull back until the flywheel is no longer secured.
6. Lift the timing chain and hold it under tension.
7. Crank the engine, at the crank pulley bolt against the direction of rotation, to 45 degrees Before Top Dead Center (BTDC).
8. Remove the special tools.
9. Remove or disconnect the following:
   - Guide rail from the cylinder head
   - Cylinder head bolts in reverse of the tightening sequence illustrated
   - Cylinder head

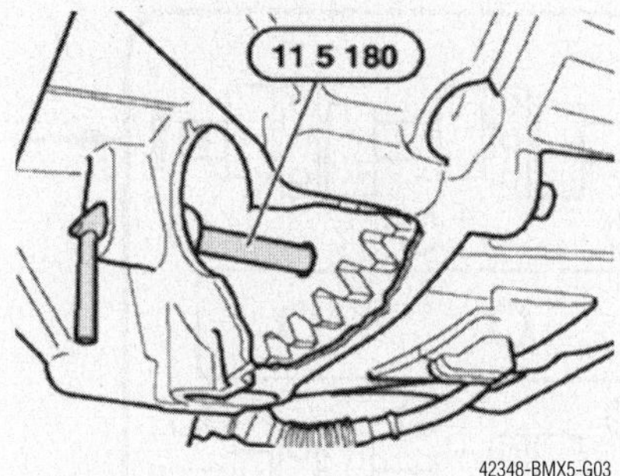

Using a special tool to release the flywheel—4.4L & 4.6L engines

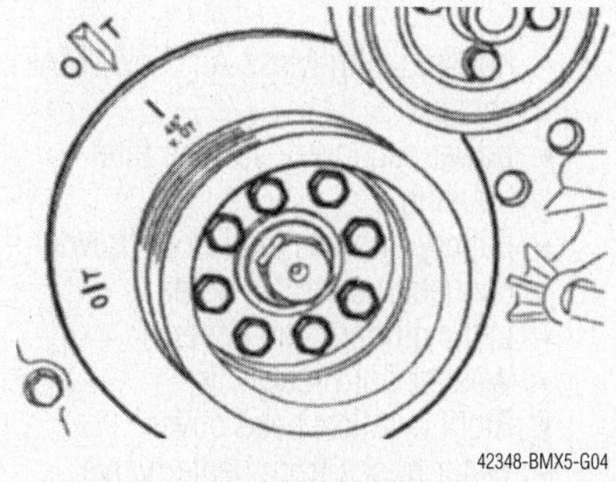

Showing the crankshaft pulley 45° BTDC mark—4.4L & 4.6L engines

To install:

10. Thoroughly clean all mounting surfaces and check the head for warpage. Take care not to drop any pieces of gasket or debris into the oil or coolant passages. Check the condition of the head locating dowel sleeves and clean out the bolt threads with a tap.
11. Apply a permanently elastic sealant, Drei Bond 1209, to joint between engine block and timing case cover.
12. Install a new cylinder head gasket.
13. Mount the cylinder head on the block and install new bolts. Apply a light coat of clean oil to the threads and washer area.
14. Torque the bolts in the sequence illustrated in 3 steps as follows:
   a. Step 1: 22 ft. lbs. (30 Nm).
   b. Step 2: 80 degrees.
   c. Step 3: 80 degrees.
15. Install and tighten the guide rail screw.

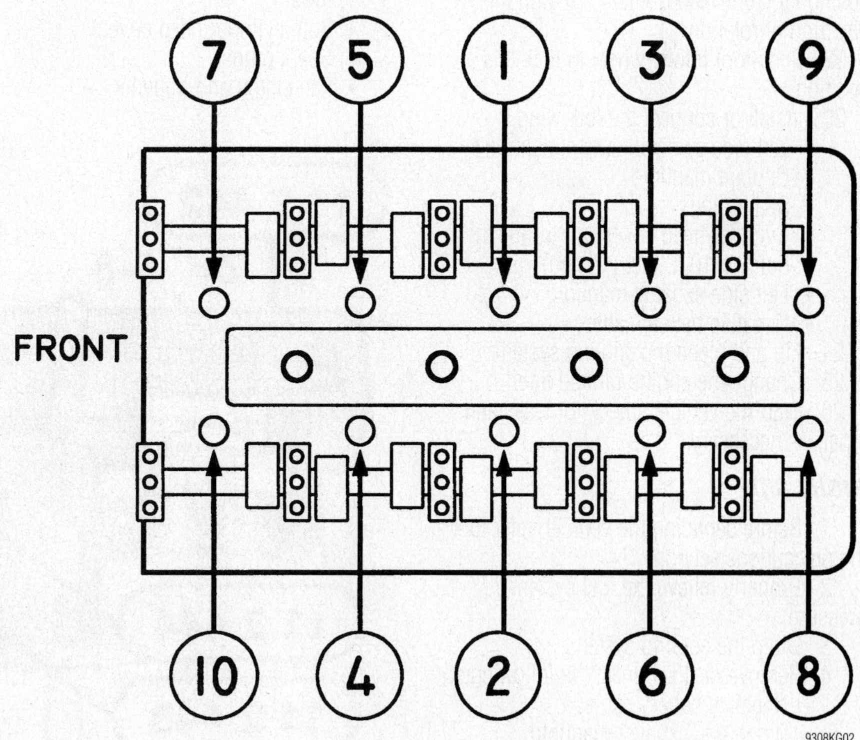

Cylinder head bolt tightening sequence—4.4L & 4.6L engines

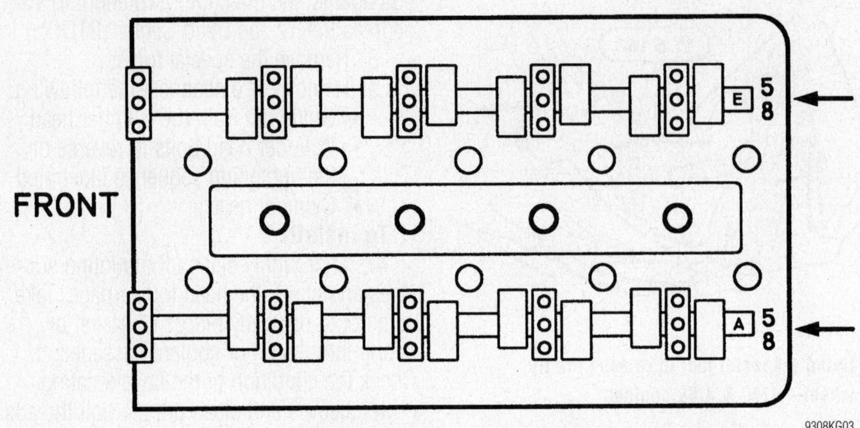

Rotate the camshafts until the markings face upward as shown—4.4L & 4.6L engines

16. Carefully rotate the camshafts until the markings on end of camshafts face upward.
17. Install special tools, 11–2–446/442, and set camshaft position.
18. Align camshafts, using an open-end wrench on camshaft flats, until there is no gap between special tools and cylinder head.
19. Install a locking tool, 11–2–443, in front of other camshaft tools and bolt into position.
20. Lift the timing chain and hold it under tension.
21. Crank the engine from the 45° BTDC position to Top Dead Center (TDC), in the direction of rotation.
22. Push tool into flywheel to lock this position.
23. Install or connect the following:
    - Left side camshaft adjustment unit
    - Coolant manifold
    - Spark plugs
    - Cylinder head cover and torque the bolts to 10 ft. lbs. (15 Nm)
    - Left side exhaust manifold
    - Negative battery cable
24. Fill and bleed the cooling system.
25. Change the engine oil and filter.
26. Start the vehicle, check for leaks and repair as necessary.

RIGHT SIDE

1. Before servicing the vehicle, refer to the precautions section.
2. Properly relieve the fuel system pressure.
3. Drain the cooling system.
4. Remove or disconnect the following:
    - Negative battery cable
    - Right side exhaust manifold
    - Engine cover
    - Throttle bellows
    - Upper section of suction filter housing with Mass Air Flow (MAF) sensor
    - Lower section of suction filter housing
    - Battery positive lead from firewall connection and lay aside
    - Left cylinder head cover
    - Washer fluid reservoir
    - Right cylinder head cover
    - Water hoses from heater valve
    - Vent line for front axle differential
    - Positive lead
    - Ignition coils
    - Vent hose from cylinder head
    - Left and right cable ducts and fold inward
    - Both cylinder head covers
    - Spark plugs
    - Fan clutch and impeller
    - Intake manifold
    - Coolant manifold
    - Right side camshaft VANOS adjustment unit
5. Install special tool 11–5–180 and pull back until the flywheel is no longer secured.
6. Lift the timing chain and hold it under tension.
7. Crank the engine at the central bolt against the direction of rotation to 45° Before Top Dead Center (BTDC).
8. Remove the special tools from camshafts.
9. Remove the cylinder head bolts in reverse of the tightening sequence shown.
10. Remove the cylinder head.

To install:

11. Thoroughly clean all mounting surfaces and check the head for warpage. Take care not to drop any pieces of gasket or debris into the oil or coolant passages. Check the condition of the head locating dowel sleeves and clean out the bolt threads with a tap.
12. Apply permanently elastic sealant, Drei Bond 1209, to joint between block and timing case cover.
13. Install a new cylinder head gasket.
14. Mount the cylinder head on the block and use new bolts. Do not remove the coating on the head bolts, apply oil to the threads and torque the bolts in 3 steps in the tightening sequence illustrated as follows:
    a. Step 1: 22 ft. lbs. (30 Nm).
    b. Step 2: 80 degrees.

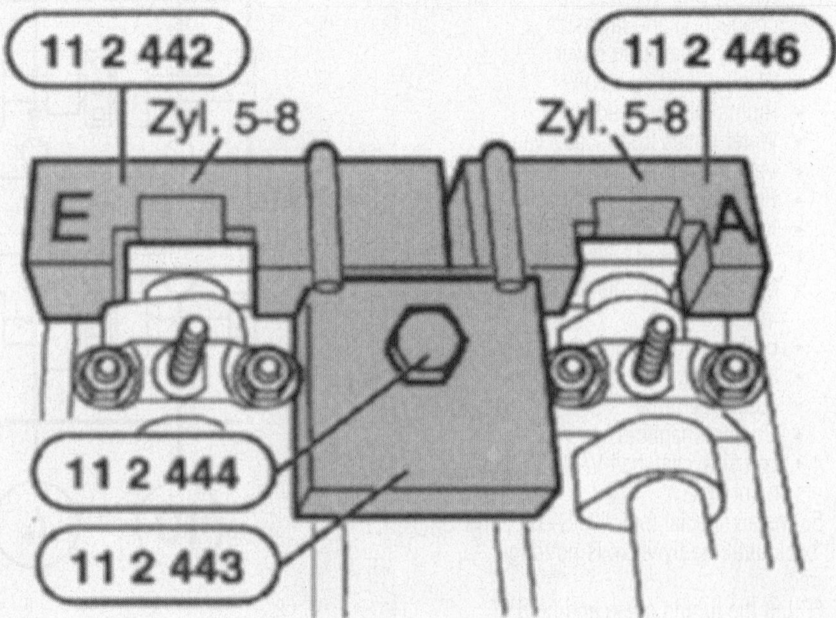

Showing camshaft holding tools in position—4.4L engines

c. Step 3: 80 degrees.

15. Rotate the camshafts until markings are facing upward.

16. Install special tools, 11–2–441/445, and set camshaft position.

17. Align camshafts, using an open-end wrench on camshaft flats, until there is no gap between special tools and cylinder head.

18. Install a locking tool, 11–2–443, in front of other camshaft tools and bolt into position.

19. Lift the timing chain and hold it under tension.

20. Crank the engine in direction of rotation to Top Dead Center (TDC).

21. Push tool into flywheel to lock this position.

22. Install or connect the following:

- Left side camshaft adjustment unit
- Coolant manifold
- Spark plugs
- Cylinder head cover. Torque the bolts to 10 ft. lbs. (15 Nm).
- Left side exhaust manifold
- Negative battery cable

23. Fill and bleed the cooling system.

24. Change the engine oil and filter.

- Start the vehicle and check for leaks, repair if necessary.

Intake Manifold

REMOVAL & INSTALLATION

3.0L Engines

1. Before servicing the vehicle, refer to the precautions section.
2. Properly relieve the fuel system pressure.
3. Remove or disconnect the following:

- Negative battery cable
- Engine cover
- Oxygen (O_2S) sensor connectors (mark connector locations before removing)
- Wiring from clips on intake manifold
- Battery positive lead from intake manifold
- Battery positive lead retainer
- Vent hose from cylinder head cover
- Retainer from top of engine and connection for intake air temperature sensor
- Fuel injector terminal strip and place aside
- Tank venting valve located near the dipstick
- Fuel line from clip and connection to pipe located near the water pump
- Dipstick tube guide
- Return line from dipstick tube
- Throttle assembly
- Knock sensor connector and nut on manifold support
- Intake manifold nuts
- Vacuum line on reverse side of intake manifold, if equipped
- Intake manifold

To install:

4. Check all intake manifold gaskets and replace if necessary.

5. Inspect the rubber dampers at manifold connection and replace if needed.

6. Install intake manifold and torque bolts as follows:

a. M6 bolts: 7 ft. lbs. (10 Nm).

b. M7 bolts: 11 ft. lbs. (15 Nm).

c. M8 bolts: 16 Nm (22 Nm).

7. Install or connect the following:

- Knock sensor connector and manifold support nut
- Throttle assembly
- Return line to dipstick tube
- Dipstick tube guide
- Fuel line to clip and connection to pipe located near the water pump
- Tank venting valve located near the dipstick
- Fuel injector terminal strip
- Retainer to top of engine and connection for intake air temperature sensor
- Vent hose to cylinder head cover
- Battery positive lead retainer
- Battery positive lead to intake manifold
- Wiring to clips on intake manifold
- O_2S sensor connectors
- Engine cover
- Negative battery cable

4.4L & 4.6L Engines

1. Before servicing the vehicle, refer to the precautions section.
2. Properly relieve the fuel system pressure.
3. Remove or disconnect the following:

- Negative battery cable
- Heater bulkhead
- Engine cover
- Ignition coil covers (pry plug from cylinder head cover) and electrical connectors
- Throttle bellows
- Intake filter housing
- Mass Air Flow (MAF) sensor
- Wiring harness from the intake manifold
- Air injection vacuum control hoses
- Throttle body vacuum hose
- Fuel feed line from the injection pipe
- Engine ventilation hose from the cylinder head cover
- Engine ventilation hose from the oil separator
- Brake booster vacuum hose
- Oil separator from the cover
- Decoupling elements from under the intake manifold
- Oil drain hose from the rear cover after raising the intake manifold slightly
- Intake manifold
- Vent pipe from intake manifold, if necessary

➡The intake manifold is vibrationally separated from the cylinder head by decoupling elements and gaskets.

To install:

4. Install the decoupling elements to the cylinder head

5. Position the intake manifold

6. Install the remaining components of the decoupling elements. Torque the nuts as follows:

a. M6 nuts: 89 inch lbs. (10 Nm).

b. M7 nuts: 10 ft. lbs. (15 Nm).

c. M8 nuts: 16 ft. lbs. (22 Nm).

7. Install or connect the following:

- Oil drain hose to the rear cover
- Oil separator
- Brake booster vacuum hose

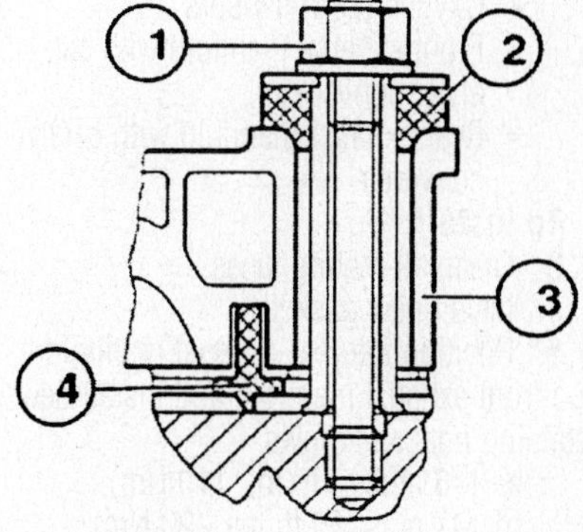

Separating the intake manifold from the cylinder head

- Engine vent hose to the oil separator and the cylinder head
- Fuel feed line to the injection pipe
- Throttle body vacuum hose
- Air injection vacuum control hoses
- Wiring harness to the intake manifold
- MAF sensor
- Intake filter housing
- Throttle bellows
- Ignition coil electrical connectors
- Engine cover
- Heater bulkhead
- Negative battery cable

Exhaust Manifold

REMOVAL & INSTALLATION

3.0L Engines

1. Before servicing the vehicle, refer to the precautions section.
2. Remove or disconnect the following:
   - Negative battery cable
   - Intermediate muffler
   - Reinforcement plate from undercarriage
   - Bracket, rubber and both stabilizers
   - Oxygen control sensors from intake manifold, make sure to mark locations for reinstallation purposes
   - Fuel injector cover
   - Plug connections from intake manifold, make sure to mark all connection locations for reinstallation purposes
   - Oxygen (O_2S) sensors
   - Front exhaust manifold with catalytic converter
   - Rear exhaust manifold with catalytic converter

To install:

3. Clean all mating faces.
4. Install new gaskets.
5. Position the rear exhaust manifold and front exhaust manifold and install new retaining nuts as follows:
   a. M6 nuts: 7 ft. lbs. (10 Nm).
   b. M7 nuts: 15 ft. lbs. (20 Nm).
6. Install or connect the following:
   - O_2S monitor sensors on exhaust pipes; torque to 37 ft. lbs. (50 Nm)
   - Connections on intake manifold and components
   - Fuel injector cover
   - Oxygen control sensors on intake manifold; torque to 37 ft. lbs. (50 Nm)
   - Bracket, rubber and both stabilizers from struts; torque nuts to 74 ft. lbs. (100 Nm)
   - Reinforcement plate from undercarriage; torque bolts to 74 ft. lbs. (100 Nm)
   - Negative battery cable

4.4L & 4.6L Engines

1. Before servicing the vehicle, refer to the precautions section.
2. Remove or disconnect the following:
   - Negative battery cable
   - Exhaust system
   - Reinforcement plate
   - Propeller shaft, left side manifold only
   - Screw connection at the exhaust manifold
   - Exhaust manifold downward and discard the gaskets

To install:

3. Remove the old gasket off of the cylinder head and exhaust manifold and replace the gasket. The gasket beads face the exhaust manifolds.
4. Install or connect the following:
   - Exhaust manifold with new gaskets. Torque the bolts to 10 ft. lbs. (15 Nm).
   - Screw connection at the exhaust manifold
   - Propeller shaft, left side manifold only
   - Reinforcement plate
   - Exhaust system
   - Negative battery cable

Camshaft and Valve Lifters

REMOVAL & INSTALLATION

3.0L Engines

1. Before servicing the vehicle, refer to the precautions section.
2. Disconnect the negative battery cable.
3. Remove the camshaft (Double VANOS) adjustment unit as follows:
   a. Remove the intake filter housing with Mass Air Flow (MAF) sensor.
   b. Remove the fan impeller with fan clutch and cowl.
   c. Remove the cylinder head covers.
   d. Remove the spark plugs.
   e. Remove the plastic cover for intake camshaft.
   f. Remove the oil pressure pipe.
   g. Install a special tool, 11–3–450, with banjo bolt on oil pressure connection unit.
   h. Cover the camshaft adjustment unit.
   i. Connect a compressed air hose to the special tool fitting on the oil line connection.
   j. Rotate the engine, a least 2 turns in direction of rotation, to return the camshafts to TDC position (front cam lobes pointing to 10 o'clock and 2 o'clock positions).
   k. Install a special locking tool, 11–2–300, into hole to lock flywheel in TDC position.
   l. Remove the 2 studs from the rear of the cylinder head outer edge. Secure the camshafts with locking tools, 11–2–240.
   m. Disconnect the compressed air hose from the special tool.
   n. Using care so no oil drips onto belt drive, remove the screw plugs from the camshaft adjustment unit cover corresponding to the forward ends of the camshafts. Have a container to catch oil that runs out.
   o. Using short, flat nose pliers, remove sealing caps through screw plug openings, then remove the fitting screws (left-hand thread) from both camshaft forward ends.
   p. Detach the plug connection from the camshaft sensor and solenoid valves on both camshafts.
   q. Remove the engine lifting eye from the camshaft adjustment unit cover.
   r. Remove nuts and remove the camshaft adjustment unit

⁂ WARNING

Once camshaft adjust unit is removed, DO NOT crank engine. The toothed shaft on the intake side camshaft may slip out of the spline teeth and valves could rest of the piston.

4. Remove the cylinder for the chain tensioning piston. Use caution as piston is under spring pressure.
5. Press down on the secondary chain tensioner at the top and lock the chain in place with a special tool, 11–3–292.
6. Remove or disconnect the following:
   - Sensor gear from exhaust camshaft
   - Spring plate from behind sensor gear
   - Nuts on intake camshaft and remove corrugated washer

42348-BMX5-G06

Attaching compressed air hose to oil line connector—3.0L engines

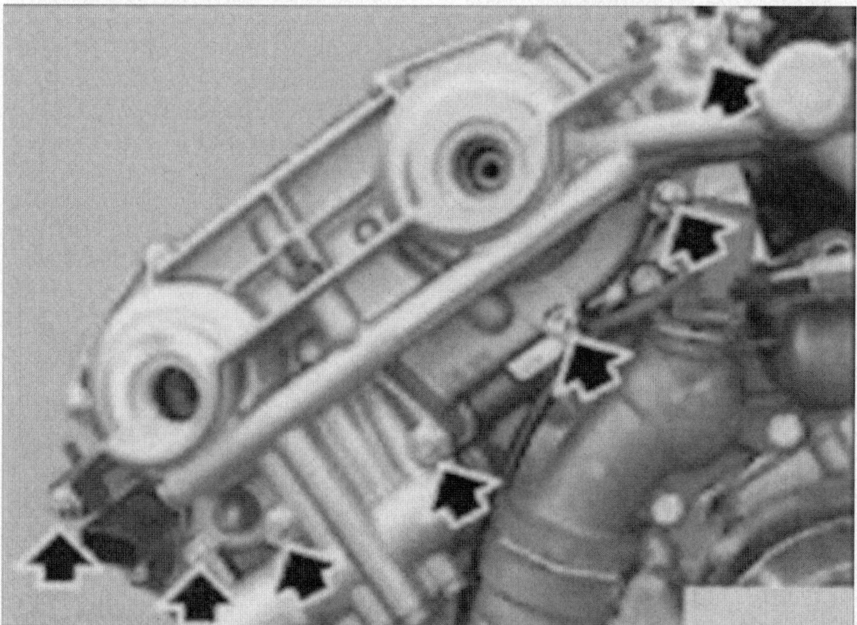
42348-BMX5-G07

Showing location of retaining nuts on camshaft adjustment unit (Double VANOS)—3.0L engines

- Screws from chain gear on exhaust camshaft
- Toothed shaft with sleeve
- Secondary (upper) chain tensioner
- Release screw-in pins from exhaust camshaft chain gear
- Exhaust camshaft chain gear (leave chain on end of camshaft)
- Thrust washer on intake camshaft
- Intake camshaft sensor gear

⁂ CAUTION

DO NOT release screws from front ends of either camshaft.

- Studs from between both camshafts

7. Release the special locking tool so flywheel is free.
8. Lift the timing chain off from end of exhaust camshaft and hold it under tension by lifting it straight up.
9. Rotate the engine against the normal direction of rotation about 30 degrees, using the crankshaft bolt.

⁂ CAUTION

To prevent the intake camshaft from moving the bearing inserts, remove the camshaft no. 1 bearing cap nuts and remove the bearing cap.

10. Install a special tool, 11–2–260, to the cylinder head and install retaining bolts into tool and into spark plug holes for cylinders no.1 and no. 4.
11. Turn the eccentric shaft of the special tool to pre-tension the bearing caps. Remove the nuts on all bearing caps.
12. Relieve the tension from the eccentric shaft of the special tool and remove the tool.
13. Remove the bearing caps and keep in order as removed.
14. Remove the camshaft. Repeat the process for the other camshaft.

➡If cylinder head is to be removed, remove the complete bearing strip with the bucket tappets. Keep in order of original locations.

To install:

15. Check bearing points of bucket tappets for signs of scoring.

➡Bearing strips are marked "A" for exhaust side and "E" for intake side.

16. Be sure centering dowels are in place on retaining pins at bearing points 2 and 7.
17. Install the bearing strips.
18. Oil all teeth, camshafts, bearings, bearing caps, and friction washers before installation.

⁂ CAUTION

Bucket tappets expand when not subjected to load by the camshaft and therefore require some time before they can be pushed back down. During a rapid assembly sequence, the "closed" valves may still be open and therefore be in contact with the piston. After assembly, wait at least 30 minutes before cranking the engine back to the TDC position.

19. Pull up the timing chain and feed in the exhaust camshaft. Position the timing chain onto the end of the exhaust camshaft.

⁂ CAUTION

To prevent damaging valves when fitting camshafts, no pistons should be in TDC position.

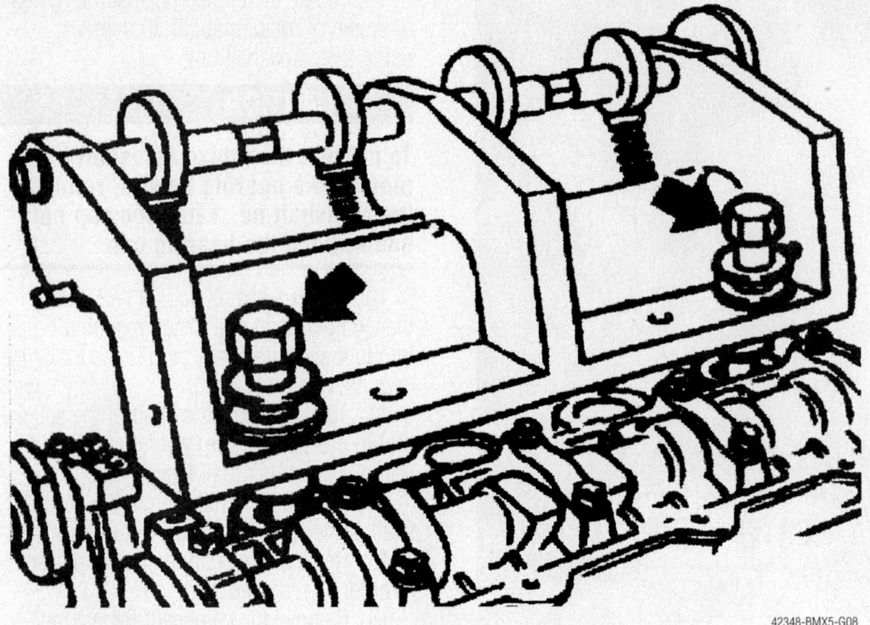

42348-BMX5-G08

Installing special tool to pre-tension camshaft bearing caps for removal—3.0L engines

20. Install the camshafts so the cam tips on the intake and exhaust valves on no. 1 cylinder face each other at about the 10 o'clock and 2 o'clock positions.

21. Install the bearing caps (caps are marked from the exhaust side, A1 through A7 for the exhaust side, and E1 through E7 for the intake side).

22. Install the special tool, 11–3–260, onto the cylinder head as during removal.

23. Turn the eccentric shaft to pre-tension the bearing caps. Tighten the bearing cap bolts as follows:

a. M6 bolts: 7 ft. lbs. (10 Nm).
b. M7 bolts: 10 ft. lbs. (14 Nm).
c. M8 bolts: 15 ft. lbs. (20 Nm).

24. Remove the special tool.

25. Install camshaft locking tools, 11–3–240 on back end of camshafts. Use an open-end wrench to align camshafts (if necessary, machine wrench head so it does not contact cylinder head) so there is no gap at the locking tools.

26. Install the middle locking tool, 11–3–244 so it adjoins other tools and is bolted into spark plug hole.

27. Lift the timing chain straight up off the exhaust camshaft and hold it under tension.

28. Rotate the engine from 30 degrees BTDC to TDC, in normal direction of rotation.

29. Using special tool, 11–2–300, lock flywheel in this position.

30. Install sensor gear onto intake camshaft. Install the thrust washer over the sensor gear and torque the gear screw-in pins to 15 ft. lbs. (20 Nm).

31. Feed the chain wheel onto the timing chain so the arrow on the chain wheel faces the upper edge of the cylinder head.

32. Install a special tool, 11–4–220 into the cylinder head in the chain tensioning piston bore and bring tool adjustment screw into contact with the timing chain tensioning rail, but no further at this time.

33. Check the arrow mark on the chain wheel and adjust if needed so it still points to upper edge of the cylinder head.

34. Install and tighten the retaining screw-in pins in the exhaust chain wheel to 15 ft. lbs. (20 Nm).

35. Install the secondary (upper) chain tensioner.

36. Position the toothed sleeve into the exhaust camshaft to the toothed gaps are opposed. Secure the toothed shaft. Insert the pin of the toothed shaft into the tooth gaps of the splines on the camshaft and toothed sleeve.

37. Push in the toothed shaft on the exhaust camshaft until the elongated holes in the tooth sleeve wheel are centered over the bolt holes.

38. Place forward chain wheels onto special tool, 116–6–180, and position tooth gap on intake chain wheel as shown as feed on the timing chain.

**** CAUTION**

DO NOT alter position of chain wheels and chain when removing the special tool.

39. Remove the special tool and position chain wheels on camshafts, so that tooth spaces oppose each other on intake camshaft.

40. Align chain with sprocket wheels so tooth spaces are positioned exactly over each other on intake side.

41. Install and secure the toothed shaft into the tooth gaps of the splines of the camshaft and chain wheel. Push in the toothed shaft until about 0.4 inch (1 mm) of splines can still be seen.

42348-BMX5-G09

Setting timing chain and chain wheel mark in position during installation of chain—3.0L engines

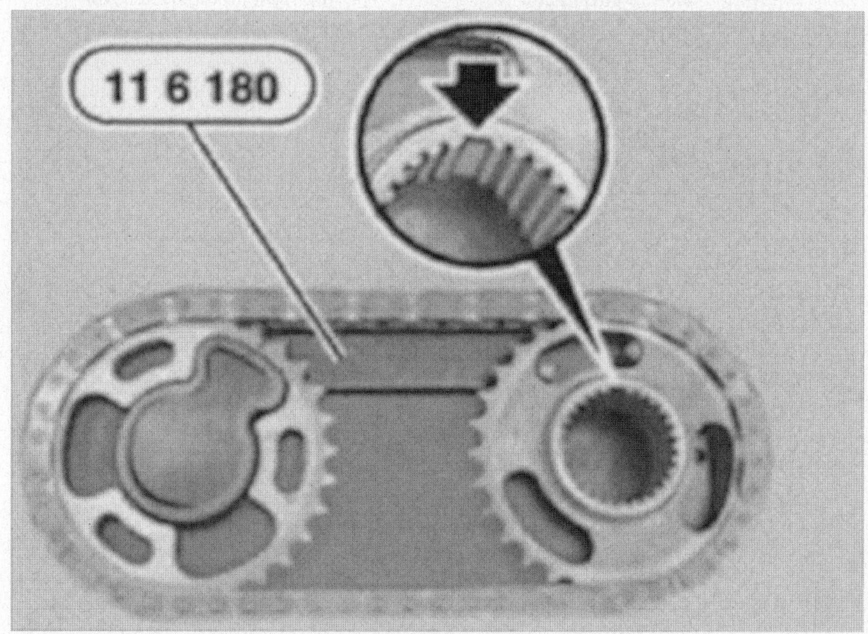

42348-BMX5-G10

Aligning forward timing chain on chain wheels with special tool—3.0L engines

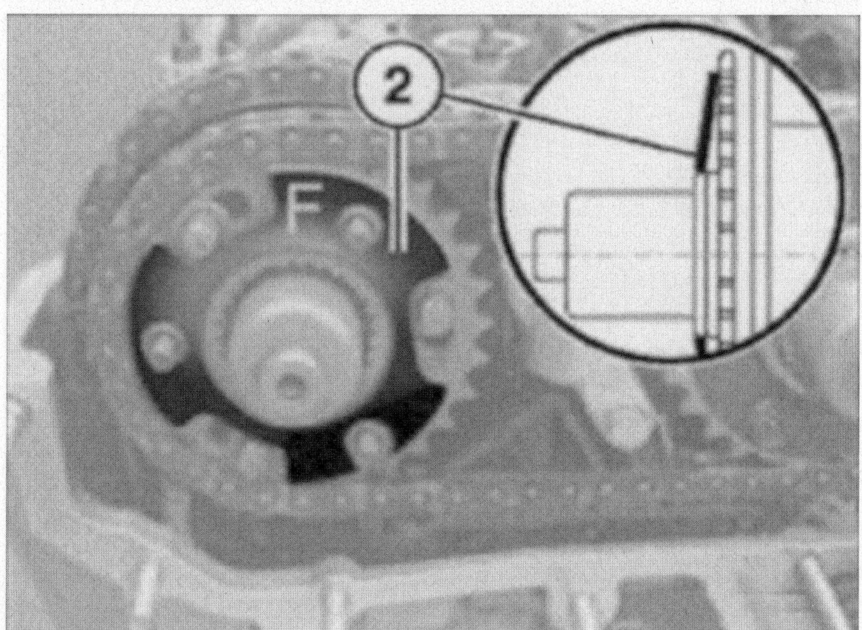

42348-BMX5-G11

Installing the cup spring on the exhaust camshaft chain wheel—3.0L engines

42. Note the installation direction of the corrugated washed so "FRONT" is visible. Install the washer and tighten retaining nuts snug only at this time.
43. Install 4 bolts on exhaust side to retain chain wheel. Initially tighten to about 3 ft. lbs. (5 Nm), then slack off one-half turn.
44. Install the thrust washer over the exhaust side chain wheel.
45. Note the installation direction of the cup spring (2) so the "F" stamp is forward. If "F" is no longer visible on used engine, install the cup spring so the small locating diameter of the spring points to the sensor gear. Install the cup spring.
46. Position the sensor gear over the cup spring so the arrow on the sensor gear is in line with the upper edge of the cylinder head. Install retaining nuts but do not fully tighten.
47. Pull out the exhaust side toothed shaft to the stop.
48. Press down the secondary chain tensioner at the top and remove the special locking tool.
49. Use a special tool, 11–4–200, to preload timing chain tensioning rail by rotating the adjusting screws of the tool with a small wrench.
50. Preload the exhaust side cup spring slightly be pressing on the sensor gear and hand-tighten the gear retaining nuts (do not fully tighten yet).
51. Remove any remaining gasket material from mating face on front of cylinder head.
52. Check dowel sleeves for damage and for correct installation position.
53. Make sure sealing face is clean and free of oil.
54. Fit special tool, 11–6–150 without gasket over both toothed shafts. Tighten retaining nuts to hold tool in place. Hand-tighten only until special tool is uniformly in contact with cylinder head.
55. Insert bolts on exhaust side chain wheel and tighten to about 3 ft. lbs. (5 Nm).
56. Initially tighten nuts on both exhaust and intake chain wheels to about 3 ft. lbs. (5 Nm).
57. Torque bolts on exhaust chain wheel to 15 ft. lbs. (20 Nm).
58. Torque nuts on exhaust and intake chain wheels screw-in pins to 7 ft. lbs. (10 Nm).
59. Pull back the special tool from the flywheel.
60. Remove the special tools from the rear of the camshafts.
61. Crank the engine twice in the direction of rotation until the cam lobe tips on the front of the camshafts point inward at about the 10 o'clock and 2 o'clock positions.
62. Lock the flywheel with the special tool to the crankshaft is at Top Dead Center (TDC) position.

WARNING

DO NOT turn the engine against normal rotation.

63. Install the special camshaft locking tool, 11–3–240, onto rear end of camshafts to hold its position.

➡The camshaft timing is correctly set if the special tool rests on the camshafts without a gap on the cylinder head or any protrusion up to 0.04 inch (1 mm) on the intake side.

64. Remove the special tool from the camshaft forward ends.

65. Install the camshaft adjustment (Double VANOS) unit as follows:

a. Ensure dowel sleeves on forward mating face of cylinder head are not damaged and are correctly installed.

b. Keep sealing faces clean and free of oil.

c. Apply a thin and even coat of sealing compound, Drei Bond 1209, to contact surface edges of the separating face between the cylinder head and the camshaft adjustment unit.

d. Install or connect the following:

- New seal on mating face
- Camshaft adjustment unit
- Screws for hydraulic pistons on toothed shaft end on both camshafts (through front openings); torque to 7 ft. lbs. (10 Nm). Screws have left-hand threads.
- Sealing caps into camshaft adjustment unit openings
- Screw plugs, with new seal rings; torque to 37 ft. lbs. (50 Nm)
- Engine lifting eye

66. Set camshaft timing, if necessary.

67. Complete installation in reverse of removal procedure.

Left Camshaft (Cylinder Bank 5–8)—4.4L & 4.6L Engines

1. Before servicing the vehicle, refer to the precautions section.

2. Remove or disconnect the following:

- Negative battery cable
- Splash guard
- Left and right cylinder head covers
- Spark plugs
- Timing chain tensioning piston
- Top left timing case cover
- Left camshaft adjustment unit and distributor
- Oil lines to the left and right cylinder head

3. Install special tool 11–2–300 and pull back on it until the flywheel is no longer secured in position.

4. Lift the timing chain and hold it under tension.

5. Crank the engine counter-engine wise on the central screw into the 45° Before Top Dead Center (BTDC) position.

6. Rotate the exhaust camshaft at the hex head until the cam at cylinder No. 6 faces upward.

7. Rotate the inlet camshaft at the hex head until the cam at cylinder No. 8 faces upward.

8. Evenly release the bearing covers on the exhaust and inlet camshafts in ½ turn steps from the outside working in.

9. Remove the bearing covers and remove the camshafts.

10. Remove the hydraulic valve lifters. To remove, use tool No. 11-3-250 to pull them out of the cylinder head. Be sure that no damage occurs to the guides in the head. Inspect the bearing surfaces of the tappets for wear and scoring.

To install:

11. If the lifters were removed, install them with tool No. 11-32-250.

12. Lubricate and install the camshafts in their correct position.

13. Install or connect the following:

- Exhaust and inlet camshafts and rotate them until exhaust cam at cylinder No. 6 faces up and inlet cam at cylinder No. 8 faces up
- Bearing caps. The exhaust camshaft bearing covers are marked A1 to A5 and the inlet covers are marked E1 to E5. Evenly tighten the bearing covers in ½ turn steps from the outside working in.

14. Torque the camshaft bearing covers as follows:

a. M6 to 89 inch lbs. (10 Nm).

b. M7 to 9 ft. lbs. (14 Nm).

c. M8 to 15 ft. lbs. (20 Nm).

15. Rotate the camshafts until the markings face upwards.

16. Using special tools 11–2–446 and 11–2–442 align the inlet and exhaust camshafts with an open-end wrench so the tools rest without a gap on the cylinder head.

17. Crank the engine from the 45 degrees BTDC position in the direction of rotation up to the TDC position.

18. Hold the crankshaft and install the distributor on the camshaft adjustment unit.

19. Install or connect the following:

- Camshaft adjustment unit
- Left timing case cover
- Timing chain tensioning piston
- Spark plugs
- Cylinder head covers
- Splash guard
- Negative battery cable

Right Camshaft (Cylinder Bank 1–4)

1. Before servicing the vehicle, refer to the precautions section.

2. Remove or disconnect the following:

- Negative battery cable
- Splash guard
- Left and right cylinder head covers
- Spark plugs
- Timing chain tensioning piston
- Top right timing case cover
- Right camshaft adjustment unit and distributor
- Oil lines to the left and right cylinder head

3. Install special tool 11–2–300 and pull back on it until the flywheel is no longer secured in position.

4. Lift the timing chain and hold it under tension.

5. Crank the engine counter-engine wise on the central screw into the 45 degrees Before Top Dead Center (BTDC) position.

6. Rotate the exhaust camshaft at the hex head until the cam at cylinder No. 6 faces upward.

7. Rotate the inlet camshaft at the hex head until the cam at cylinder No. 8 faces upward.

8. Evenly release the bearing covers on the exhaust and inlet camshafts in ½ turn steps from the outside working in.

9. Remove the bearing covers and remove the camshafts.

10. Remove the hydraulic valve lifters. To remove, use tool No. 11-3-250 to pull them out of the cylinder head. Be sure that no damage occurs to the guides in the head. Inspect the bearing surfaces of the tappets for wear and scoring.

To install:

11. If the lifters were removed, install them with tool No. 11-32-250.

12. Lubricate and install the camshafts in their correct position.

13. Install or connect the following:

- Exhaust and inlet camshafts and rotate them until exhaust cam at cylinder No. 6 faces up and inlet cam at cylinder No. 8 faces up
- Bearing caps. The exhaust camshaft bearing covers are marked A1 to A5 and the inlet covers are marked E1 to E5. Evenly tighten the bearing covers in ½ turn steps from the outside working in.

14. Torque the camshaft bearing covers as follows:

a. M6 to 89 inch lbs. (10 Nm).

b. M7 to 9 ft. lbs. (14 Nm).

c. M8 to 15 ft. lbs. (20 Nm).

15. Rotate the camshafts until the markings face upwards.

16. Using special tools 11–2–446 and 11–2–442 align the inlet and exhaust camshafts with an open-end wrench so the tools rest without a gap on the cylinder head.

17. Crank the engine from the 45 degrees BTDC position in the direction of rotation up to the TDC position.

18. Hold the crankshaft and install the distributor on the camshaft adjustment unit.
19. Install or connect the following:
    - Camshaft adjustment unit
    - Both timing case covers
    - Timing chain tensioning piston
    - Spark plugs
    - Cylinder head covers
    - Splash guard
    - Negative battery cable

Valve Lash

ADJUSTMENT

All engines are equipped with hydraulic valve lash adjusters. This design does not permit adjustments nor are adjustments possible.

Starter

REMOVAL & INSTALLATION

➡When the battery is disconnected, the radio code, on-board computer and clock settings will be lost. The radio code should be obtained before disconnecting the battery or radio. Once the battery has been reconnected, the radio will not function unless the code is keyed in.

1. Before servicing the vehicle, refer to the precautions section.
2. If needed, read the stored fault memories from the control module.
3. Relieve the fuel system pressure.
4. Set the ignition switch to the **OFF** position.
5. Remove or disconnect the following:
    - Negative battery cable
    - Reinforcement plate
    - Positive battery cable from the starter
    - Starter electrical connectors
    - Heat shield (4.4L and 4.6L)
    - Starter from the transmission mount
    - Control leads (3.0L)
    - Starter

To install:

6. Install or connect the following:
    - Starter to transmission mount
    - Control leads (3.0L)
    - Heat shield (4.4L and 4.6L); torque bolts to 38 ft. lbs. (47 Nm).
    - Starter electrical connectors
    - Positive battery cable to the starter
    - Reinforcement plate
    - Negative battery cable

Oil Pan

REMOVAL & INSTALLATION

3.0L Engines

1. Before servicing the vehicle, refer to the precautions section.
2. Drain the engine oil.
3. Remove or disconnect the following:
    - Bulkhead heater
    - Cover over fuel injectors
    - Upper section of suction filter housing with Mass Air Flow (MAF) sensor
4. Attach an engine lift and hoist to engine.
5. Remove top left and right nuts on engine mounts
6. Raise the engine about 0.15 inch (5mm).
7. Remove and disconnect the following:
    - Front splash guard
    - Reinforcement plate
    - Left and right swivel bearings
    - Output shafts
    - Propeller shaft at front
    - Alternator drive belt from belt pulley for vane pump
    - Power steering pump and move aside with the lines still connected
    - Return hose from oil separator
    - Dipstick tube
    - Steering spindle from steering gear tie rods
8. Temporarily reinstall the left and right swivel bearings and connect with a bolt to the spring strut.
9. Support the entire front axle assembly.

➡The steering gear will remain bolted to the front axle support. Return hose must not be over-stretched when front axle is lowered.

10. Detach front axle support from engine carrier and lower about 3-4 inches (90-100mm).
11. Remove the front axle differential.
12. Release the screws on the oil pan.
13. Move the pan backward when removing.

To install:

14. Ensure mating surfaces are clean.
15. Install a new gasket and position the oil pan in place.
16. Install and tighten the oil pan bolts, but do not fully tighten the transmission end bolts.
17. Tighten the engine end bolts, then tighten the transmission end bolts.
18. Restore the front axle assembly to position.
19. Reconnect the steering gear tie rods.
20. Install or connect the following:
    - Dipstick tube
    - Return hose from oil separator
    - Power steering pump
    - Alternator drive belt
    - Propeller shaft
    - Output shafts
    - Swivel bearings
    - Reinforcement plate
    - Splash guard
21. Reposition the engine onto the mount and torque the upper nuts.
22. Remove the engine hoist
23. Install the upper section of suction filter housing with MAF sensor
24. Install the cover of the fuel injectors.
25. Install the bulkhead heater.
26. Reconnect the negative battery cable.
27. Refill the engine oil.

4.4L & 4.6L Engines

1. Before servicing the vehicle, refer to the precautions section.
2. Drain the engine oil.
3. Remove or disconnect the following:
    - Negative battery cable
    - Reinforcement plate
    - Oil level switch plug
    - Cable guide clips
    - Lower oil pan section

➡To remove the upper section of the oil pan, proceed with the following steps.

4. Remove or disconnect the following:
    - Upper nuts on the left and right engine mounts
    - Front splash guard
    - Positive battery cable from the starter
    - Left and right swivel bearings
    - Output shafts
    - Bearing pedestal from the right output shaft
    - Propeller shaft
    - Steering spindle from the steering gear and support the front axle
    - Front axle support from the engine carrier and slightly lower the axle support
    - Drive belt
    - Vane pump from the oil pan
    - Adjustable plate from the oil pan after releasing the tension from the A/C compressor belt
    - Guide tube for the oil dipstick

- Oil return line from the oil separator to the oil pan
- Oil pump snorkel
- Cable guide for the positive lead
- Upper oil pan section towards the rear of the vehicle

To install

5. Clean the mounting surfaces.
6. Check the seals on the oil pipes and replace it if necessary. Lubricate the seals with oil.
7. Install or connect the following:
    - Install upper oil pan. Torque the bolts to 89 inch lbs. (10 Nm) and lower the engine.
    - Cable guide for the positive lead
    - Banjo bolt for the oil return pipe from the oil filter at the oil pan
    - Drive belt
    - Left and right engine mounts Torque the bottom bolts to 32 ft. lbs. (43 Nm).
    - Lower oil pan with a new gasket. Torque the bolts, beginning in the middle and working to the outside to 89 inch lbs. (10 Nm).
    - Plug for the level switch, making sure to replace the O-ring
    - Steering spindle to the steering gear
    - Propeller shaft
    - Bearing pedestal
    - Output shafts
    - Positive battery cable
    - Engine splash guards
    - Oil dipstick guide tube, making sure to replace the O-ring
    - Reinforcement plate
    - Negative battery cable
8. Fill the engine with clean oil.
9. Start the vehicle and check for leaks, repair if necessary.

Oil Pump

REMOVAL & INSTALLATION

1. Before servicing the vehicle, refer to the precautions section.
2. Drain the engine oil.
3. Remove or disconnect the following:
    - Negative battery cable
    - Oil pan
    - Oil pump sprocket wheel and chain
    - Oil pump

To install

4. Check the seals on the oil pipes and replace it if necessary. Lubricate the seals with oil.
5. Check the seal in the oil pump and replace it if necessary. Screw the hexagon adapter back into the oil pump until it stops.
6. Install or connect the following:
    - Oil pump. Torque the bolts to 17 ft. lbs. (22 Nm).
    - Oil pump sprocket wheel and chain. Torque the nut to 35 ft. lbs. (47 Nm).
    - Oil pan
    - Negative battery cable
7. Fill the engine with clean oil.
8. Start the vehicle and check for leaks, repair if necessary.

Rear Main Seal

REMOVAL & INSTALLATION

The rear main bearing oil seal can be replaced after the transmission and flywheel has been removed from the engine.

1. Before servicing the vehicle, refer to the precautions section.
2. Drain the transmission fluid.
3. Remove or disconnect the following:
    - Negative battery cable
    - Transmission
    - Flywheel assembly
    - Oil seal, using a suitable seal removal tool

To install:

4. Coat the sealing lips of the new seal with oil.
5. Install or connect the following:
    - New seal into the end cover housing with a suitable seal installation tool
    - Flywheel
    - Transmission
    - Negative battery cable
6. Fill the transmission with new fluid.
7. Start the engine and check that oil pressure is present; if the oil pressure lamp does not extinguish within 5–7 seconds of starting the engine, turn the engine **OFF**.
8. Check and top off all fluid levels.

Timing Chain, Sprockets, Front Cover and Seal

REMOVAL & INSTALLATION

3.0L Engines

➡This procedure covers lower timing case cover removal only. Procedure for removal of timing chain is covered under "Camshaft" for this engine.

1. Before servicing the vehicle, refer to the precautions section.
2. Drain the engine oil.
3. Remove or disconnect the following:
    - Negative battery cable
    - Camshaft adjustment unit (Double VANOS), as described under "Camshaft."
    - Alternator drive belt tensioner and belt
    - Water pump pulley
    - Vibration damper and hub, if equipped
    - Oil pan
4. Drive out the dowel pins from the timing case cover toward the rear.
5. Remove the lower timing case cover bolts from cylinder head.
6. Remove the timing case cover bolts and cover.

To install:

7. Check cylinder head and gasket for possible damage.
8. Clean all mating surfaces.
9. Drive dowel pins into timing case cover until they protrude about 0.08-0.12 inch (2-3mm).
10. Use grease to hold timing case cover in place during installation.
11. Apply sealing compound, Drei Bond 1209, to joints at cylinder head gasket left and right, and thinly and evenly to entire timing case cover sealing surface adjoining the cylinder head.
12. Install all timing cover retaining bolts to about 3 ft. lbs. (5 Nm).
13. Drive in the dowel pins until flush.
14. Torque all timing cover retaining screws, in an alternating sequence, to the following:
    a. M6 bolts: 7 ft. lbs. (10 Nm).
    b. M7 bolts: 11 ft. lbs. (15 Nm).
    c. M8 bolts: 16 ft. lbs. (22 Nm).
    d. M10 bolts: 35 ft. lbs. (47 Nm).
    e. Repeat the torque sequence a second time.
15. Install and tighten the bolts connecting the timing case cover to the cylinder head.
16. Replace the crankshaft oil seal.
17. Install or connect the following:
    - Oil pan
    - Vibration damper and hub
    - Water pump pulley
    - Drive belt tensioner and belt
    - Camshaft (Double VANOS) adjustment unit
    - Negative battery cable.
18. Refill the engine oil.

4.4L & 4.6L Engines

1. Before servicing the vehicle, refer to the precautions section.

2. Drain the engine oil.
3. Remove or disconnect the following:
   - Negative battery cable
   - Spark plugs
   - Cylinder head covers

➡In the Top Dead Center (TDC) firing position, the inlet camshaft twists in the splines of the camshaft adjustment unit.

   - Oil lines from the cylinder head
   - Vibration damper and rotate the engine at the central bolt so that the first cylinder is at the TDC position.
   - Timing chain tensioning piston
   - Both top timing case covers
   - Left hand threaded nut from the sensor gear on cylinder bank 1–4
   - Left hand threaded nut from the sensor gear on cylinder bank 5–8
4. Slacken the screw connection for the exhaust camshaft on cylinder bank 5–8 by ½ turn.
5. Slacken the screw connection for the exhaust camshaft on cylinder bank 1–4 by ½ turn.
6. Slacken the screw connection for the inlet camshaft on cylinder bank 5–8 by ½ turn.
7. Slacken the screw connection for the inlet camshaft on cylinder bank 1–4 by ½ turn.
8. Align the camshafts and install special tool 11–2–445/441 to the camshafts on cylinder back 1–4.
9. Align the camshafts and install special tool 11–2–446/442 to the camshafts on cylinder back 5–8.
10. Remove or disconnect the following:
   - Central bolt and hub from the vibration damper
   - Oil pump
   - Water pump and thermostat housing
   - Bottom timing case cover
   - Tensioning rail and oil guide
   - Timing chain from the camshaft adjustment unit

To install:

11. Install or connect the following:
   - Timing chain over the reversing rail, camshaft adjustment unit and crankshaft sprocket wheel for cylinder bank 5–8
   - Timing chain inside screw-in pin
   - Timing chain onto the camshaft adjustment unit for cylinder bank 1–4
12. Raise the timing chain slightly by the guide rail and slide the rail over the pin until the retaining lug can be heard snapping into place on the lower guide pin.
13. Align the timing chain to the guide rail.
14. Install or connect the following:
   - Oil guide for the bow cover in the pivot rail
   - Tensioning rail screw. Press the cover against the timing chain and secure the cover with the plastic strap.
   - Upper oil pan section and secure the crankshaft in the TDC position with special tool 11–5–180
   - Special tool 11–7–380 to the right side cylinder bank and install special tool 11–4–230
   - Adjustment screw into the tensioning rail and hand tighten
   - Special tool 11–6–440 to the camshaft adjustment unit on cylinder bank 5–8 and move it 31 ft. lbs. (40 Nm) to the left hand stop.
15. Tighten the screw connection on the inlet camshaft on cylinder bank 5–8 to 10 ft. lbs. (15 Nm) and back off by ¼ turn.
16. Tighten the screw connection on the exhaust camshaft on cylinder bank 5–8 to 10 ft. lbs. (15 Nm) and back off by ¼ turn.
17. Special tool 11–6–440 to the camshaft adjustment unit on cylinder bank 1–4 and move it 31 ft. lbs. (40 Nm) to the left hand stop.
18. Tighten the screw connection on the inlet camshaft on cylinder bank 1–4 to 10 ft. lbs. (15 Nm) and back off by ¼ turn.
19. Tighten the screw connection on the exhaust camshaft on cylinder bank 1–4 to 10 ft. lbs. (15 Nm) and back off by ¼ turn.
20. Tighten the tensioning rail by turning the adjusting screw on special tool 11–4–230.

➡When the timing chain is pretensioned, the camshaft adjustment unit moves and must be reset to the left hand stop.

21. Tighten the inlet camshaft screw connection on cylinder bank 5–8 to 85 ft. lbs. (110 Nm).
22. Tighten the exhaust camshaft screw connection on cylinder bank 5–8 to 85 ft. lbs. (110 Nm).
23. Install special tool 11–6–451 to the camshaft adjustment unit on cylinder bank 1–4 and move it 31 ft. lbs. (40 Nm) to the left hand stop.
24. Tighten the inlet camshaft screw connection on cylinder bank 1–4 to 85 ft. lbs. (110 Nm).
25. Tighten the exhaust camshaft screw connection on cylinder bank 1–4 to 85 ft. lbs. (110 Nm).
26. Install the sensor gear to cylinder bank 1–4 and hand tighten the nut.
27. Align the locating bore on the sensor gear to the positioning pin on special tool 11–6–451. Press the tool downward and align it to the cylinder head. Torque the sensor gear screw to 30 ft. lbs. (40 Nm). Remove the special tool.
28. Install special tool 11–6–452 to the camshaft adjustment unit on cylinder bank 5–8 and move it 31 ft. lbs. (40 Nm) to the left hand stop.
29. Tighten the inlet camshaft screw connection on cylinder bank 5–8 to 85 ft. lbs. (110 Nm).
30. Tighten the exhaust camshaft screw connection on cylinder bank 5–8 to 85 ft. lbs. (110 Nm).
31. Install the sensor gear to cylinder bank 5–8 and hand tighten the nut.
32. Align the locating bore on the sensor gear to the positioning pin on special tool 11–6–452. Press the tool downward and align it to the cylinder head. Torque the sensor gear screw to 30 ft. lbs. (40 Nm). Remove the special tool.
33. Check for the correct seating of the dowel sleeves. Clean the sealing surfaces thoroughly, and then place a new gasket on the lower cover.
34. Trim the protruding ends of the gasket, making sure the cutting tool is level. Do not allow the pieces to fall into the engine.
35. Position the lower cover and install the mounting bolts with an even distribution of pressure. Tighten the 6mm bolts to 84 inch lbs. (10 Nm), 8mm bolts to 16 ft. lbs. (22 Nm) and 10mm bolts to 35 ft. lbs. (47 Nm).
36. Install the oil seal in the timing case cover using tool No. 11–1–220. Make sure the seal is flush with the cover.
37. Install the vibration damper hub and install the mounting bolt. Tighten the hub bolt to:
   a. Step 1: 74 ft. lbs. (100 Nm).
   b. Step 2: turn an additional 60 degrees.
   c. Step 3: turn an additional 60 degrees.
   d. Step 4: turn an additional 30 degrees.
38. Position the vibration damper pulleys and install the mounting bolts for the damper.
39. Install or connect the following:
   - Water pump pulley
   - Drive belt and the cooling fan. Rotate the fan counterclockwise to install.

- Intake hose between the throttle body and the air volume meter
- Battery positive cable for the alternator and install the protective tube mounting fasteners and connect the remaining wires to the alternator
- Oil filter housing and the return pipe and replace the housing cover
- Alternator and cylinder head cover
- Negative battery cable

40. Fill the engine with clean oil.
41. Start the vehicle and check for leaks, repair if necessary

Piston and Ring Positioning

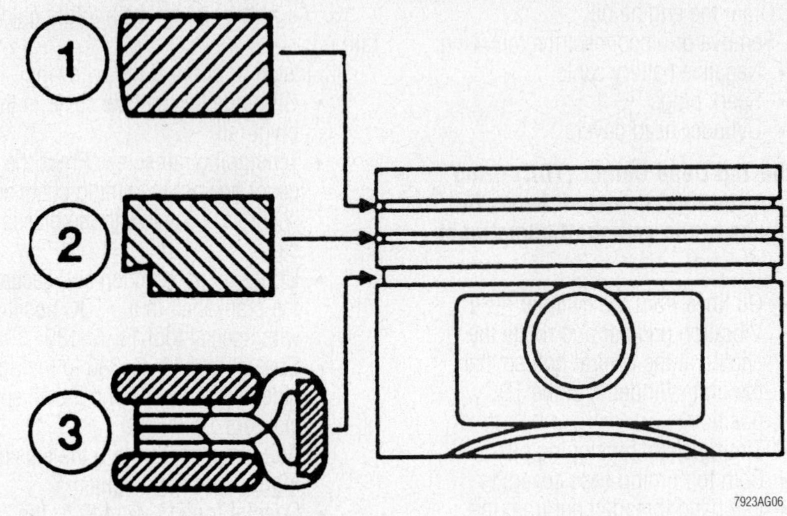

Compression and oil control ring locations—4.4L & 4.6L engines

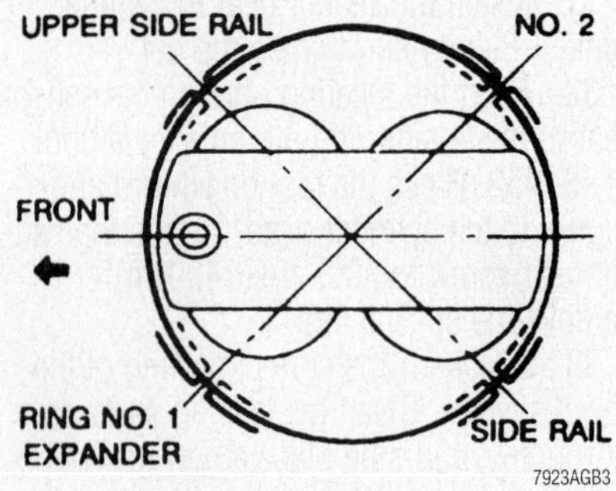

Piston ring end-gap spacing

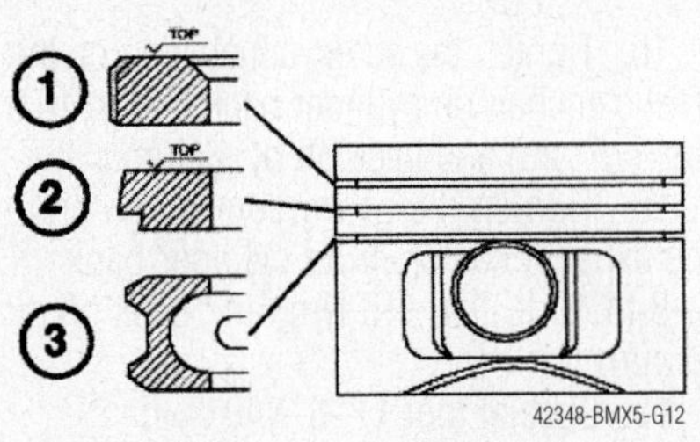

Piston ring locations and identification—3.0L engines

Showing connecting rod-to-piston positioning

FUEL SYSTEM

Fuel System Service Precautions

Safety is the most important factor when performing not only fuel system maintenance but also any type of maintenance. Failure to conduct maintenance and repairs in a safe manner may result in serious personal injury or death. Maintenance and testing of the vehicle's fuel system components can be accomplished safely and effectively by adhering to the following rules and guidelines.

1. To avoid the possibility of fire and personal injury, always disconnect the negative battery cable unless the repair or test procedure requires that battery voltage be applied.
2. Always relieve the fuel system pressure prior to disconnecting any fuel system component (injector, fuel rail, pressure regulator, etc.), fitting or fuel line connection. Exercise extreme caution whenever relieving fuel system pressure, to avoid exposing skin, face and eyes to fuel spray. Fuel under pressure may penetrate the skin or any part of the body that it contacts.
3. Always place a shop towel or cloth around the fitting or connection prior to loosening to absorb any excess fuel due to spillage. Ensure that all fuel spillage (should it occur) is quickly removed from engine surfaces. Ensure that all fuel soaked cloths or towels are deposited into a suitable waste container.
4. Always keep a dry chemical (Class B) fire extinguisher near the work area.
5. Do not allow fuel spray or fuel vapors to come into contact with a spark or open flame.
6. Always use a back-up wrench when loosening and tightening fuel line connection fittings. This will prevent unnecessary stress and torsion to fuel line piping. Always follow the proper torque specifications.
7. Always replace worn fuel fitting O-rings with new. Do not substitute fuel hose or equivalent where fuel pipe is installed.

Fuel System Pressure

RELIEVING

**** WARNING**

Fuel lines are under about 43.5–72.5 psi (3–5 bar) of pressure. Manufacturer does not provide a specific pressure relieving procedure, but instructs that any fuel system disconnect will spill fuel, so be prepared to catch and clean up any spilled fuel.

A safe way to relieve the pressure in the system is to locate the fuel pump relay located on the cowl. Unplug and remove the relay, and place it in a safe location. With the fuel pump relay removed, start the engine and operate it until it stalls. Crank the engine for 10 seconds after it stalls to remove any residual pressure.

Fuel Filter

REMOVAL & INSTALLATION

1. Before servicing the vehicle, refer to the precautions section.
2. Properly relieve the fuel system pressure.
3. Remove or disconnect the following:
   - Negative battery cable
   - Fuel pressure regulator and seal the fuel line before and after the filter with special tool 13–3–010
   - Clips and fuel line from the filter
   - Fuel filter

To install:

   - New fuel filter
   - Fuel lines onto the correct fittings. Tighten the fuel line clamps until tight, but not to the point where the fuel lines become excessively pinched or damaged, then tighten the mounting bracket until snug.
   - Negative battery cable and cycle the ignition **ON** and **OFF** several times to build fuel pressure
4. Start the vehicle and check for leaks, repair if necessary.

Fuel Pump

REMOVAL & INSTALLATION

1. Before servicing the vehicle, refer to the precautions section.
2. Drain the fuel tank.
3. Properly relieve the fuel system pressure.
4. Remove or disconnect the following:
   - Negative battery cable
   - Rear seat bench
   - Rubber plug above the sender unit and fold the rubber mat back
   - Metal cover
   - Fuel gauge level sending unit electrical connector
   - Fuel lines
   - Rotary connection with special tool 16–1–020
5. Raise the fuel level sensor and expose the spiral hose.
6. Remove the fuel level sensor and fuel pump from the tank.

To install:

➡Always use a new seal or gasket when installing the fuel pump or fuel level gauge sending unit assembly.

7. Install the fuel pump into the fuel tank, taking care not to bend or damage the fuel sending unit assembly.
8. Install a new seal and torque the sealing ring using tool No. 16-1-020 as follows:
   a. Metal sealing rings: 26 ft. lbs. (35 Nm).
   b. Plastic sealing rings: 41 ft. lbs. (55 Nm).
9. Install or connect the following:
   - Fuel lines
   - Fuel gauge level sending unit electrical connector
   - Metal cover
   - Rubber plug above the sender unit
   - Rear seat bench
   - Negative battery cable
10. Start the vehicle and check for leaks, repair if necessary.

Fuel Injector(s)

REMOVAL & INSTALLATION

3.0L Engines

➡This procedure involves removing the complete injection pipe with injectors.

1. Before servicing the vehicle, refer to the precautions section.
2. Properly relieve the fuel system pressure.
3. Mark the locations of each oxygen sensor connector for proper reinstallation.
4. Remove or disconnect the following:
   - Negative battery cable
   - Connectors from clips on injection pipe
   - Intake Air Temperature (IAT) sensor connector
   - Plug connection on solenoid valve for camshaft adjustment (VANOS) unit
   - Terminal strip for fuel injectors from injection pipe
5. Unclip the fuel line from its holder, then unlock the fuel line from the cylinder head at the rear (be prepared to catch any fuel).
6. Remove the retaining bolts and remove the injection pipe with the fuel injectors.

To install:

7. Coat the sealing rings of the injectors with anti-seize compound prior to fitting.
8. Install the fuel injectors and fuel pipe to the engine.
9. Install or connect the following:
   - Fuel line
   - Terminal strip to fuel injectors
   - Connectors as removed
   - Negative battery cable

4.4L & 4.6L Engines

1. Before servicing the vehicle, refer to the precautions section.
2. Properly relieve the fuel system pressure.
3. Remove or disconnect the following:
   - Negative battery cable
   - Acoustic cover
   - Knock Sensor (KS) electrical connector from the cable strips
   - Changeover valve electrical connector
   - Camshaft (CMP) sensor electrical connector
   - Left side cylinder head ignition coil cover
   - Ignition coil electrical connectors
   - Ignition coil cover
   - Cable strip from the fuel injectors
   - Vacuum accumulator lines
   - Fuel line
   - Both fuel injection pipe retaining brackets
   - Fuel injectors from the fuel pipe
4. Check the O-rings on the fuel injectors and replace if damaged.

To install:

5. If the O-rings are being replaced, coat the new O-ring with petroleum jelly.
6. Install or connect the following:
   - Fuel injectors to the fuel pipe
   - Fuel line
   - Vacuum lines to the vacuum accumulator
   - Cable strip to the fuel injectors
   - Ignition coil electrical connectors
   - Ignition coil cover
   - CMP sensor electrical connector
   - Changeover valve electrical connector
   - KS electrical connector
   - Acoustic cover
   - Negative battery cable
7. Start the vehicle and check for leaks, repair if necessary.

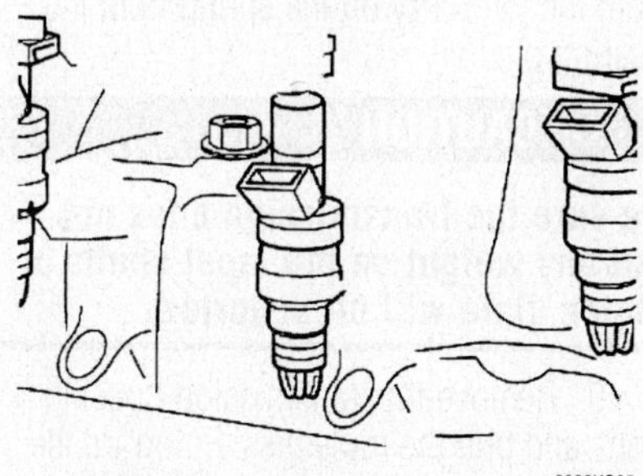

Remove the fuel injector from the injection pipe—4.4L & 4.6L engines

DRIVE TRAIN

Transmission Assembly

REMOVAL & INSTALLATION

3.0L Engines

MANUAL TRANSMISSION

1. Before servicing the vehicle, refer to the precautions section.
2. Remove or disconnect the following:
   - Negative battery cable
   - Grounding strap to body
   - Center muffler
   - Heat shield at right front
   - Heat shield at rear
   - Reinforcement plate from undercarriage
   - Front splash shield
   - Stabilizer bar (slide forward)
   - Front propeller shaft
   - Exhaust hangar bracket
   - 3 transmission–to–engine lower bolts
3. Support the transmission with special cradle, 23–3–130/1/2/3.
   - Cable from transmission
   - Crossmember
   - Driveshaft–to–transmission flange nuts and discard the nuts
   - Propeller shaft nuts (bend shaft downward at center bearing)
   - Propeller shaft from transmission output flange and tie aside
4. Slightly lower the transmission, slowly releasing tension on the clutch slave cylinder.
5. Remove the nuts and remove the clutch slave cylinder, leaving pressure lines connected.
6. Detach the reverse light switch connector and the wiring cables.
7. Support the engine with a proper jack.
8. Disconnect the gear selector rod from the joint. Pry off the spring from the bearing pin.

⁂ CAUTION

Be sure the transmission does not put any weight on the input shafts or clutch plate will be distorted.

9. Remove the transmission housing bolts and pull the transmission toward the rear as far as possible, turn it about 10 degrees counterclockwise and remove from the vehicle.

To install:

10. Clean all mating surfaces.
11. Grease the bearing pin
12. Check the condition of the adapter sleeves on either side of the transmission at the flywheel. Replace if necessary.
13. Position the transmission into the vehicle.
14. Install and torque the transmission–to–engine housing bolts to the following:
    a. M8 hex bolts: 18 ft. lbs. (25 Nm).
    b. M10 hex bolts: 36 ft. lbs. (49 Nm).
    c. M12 hex bolts: 55 ft. lbs. (74 Nm).
    d. M8 Torx bolts: 16 ft. lbs. (22 Nm).
    e. M10 Torx bolts: 32 ft. lbs. (43 Nm).
    f. M12 Torx bolts: 53 ft. lbs. (72 Nm).
15. Install or connect the following:
    - Spring clip at bearing pin
    - Gear selector rod
    - Reverse light switch and wiring cables
    - Clutch slave cylinder; torque to 16 ft. lbs. (22 Nm)
    - Propeller shaft to output flange new nuts; torque to 47 ft. lbs. (64 Nm)
16. Install the Crossmember and torque the bolts as follows:
    a. To rubber mounts: 55 ft. lbs. (74 Nm).
    b. To body, M8 bolts: 16 ft. lbs. (21 Nm).
    c. To body, M10 bolts: 30 ft. lbs. (41 Nm).
17. Install the cable to the transmission.
18. Install and torque the lower transmission–to–engine bolts to the following:
    a. M8 hex bolts: 18 ft. lbs. (25 Nm).
    b. M10 hex bolts: 36 ft. lbs. (49 Nm).
    c. M12 hex bolts: 55 ft. lbs. (74 Nm).
    d. M8 Torx bolts: 16 ft. lbs. (22 Nm).
    e. M10 Torx bolts: 32 ft. lbs. (43 Nm).
    f. M12 Torx bolts: 53 ft. lbs. (72 Nm).
19. Install or connect the following:
    - Exhaust hangar bracket
    - Front propeller shaft
    - Stabilizer bar
    - Front splash guard
    - Reinforcement plate
    - Heat shields
    - Grounding strap to body
    - Negative battery cable
20. Refill the transmission and check operation.

AUTOMATIC TRANSMISSION

1. Before servicing the vehicle, refer to the precautions section.
2. Remove or disconnect the following:
   - Negative battery cable
   - Exhaust system
   - Cables for oxygen sensors, mark the locations for installation purposes
   - Reinforcement plate from undercarriage
   - Engine under guard at front
   - Stabilizer bar and slide forward
   - Front propeller shaft
   - Vent line from transmission at bracket
   - Exhaust bracket
   - Nut and bushing from shift lever
   - Shift control cable from body bracket
   - Electrical connection and cable from transmission
   - Hydraulic cooler lines from transmission
3. Position the proper transmission cradle, 00–2–030, 24–5–301/305.
4. Support the transmission and transfer case with transmission cradle.
5. Remove or disconnect the following:
   - Oxygen (O_2S) sensor wiring cable at transmission crossmember
   - Wiring cable from transmission
   - Crossmember
   - Propeller shaft flange nuts
   - Center bearing support and support the propeller shaft
   - Propeller shaft from transmission and tie aside
   - Hex bolt near protective cap on transmission flange
   - Protective cap
   - Bolts from torque converter through protective cap opening
   - Transmission housing–to–engine bolts
6. Pull the transmission, with transfer case, out toward rear.

To install:

7. Note the position and condition of dowel sleeves on flywheel housing. Replace dowels if damaged.
8. Position the transmission to the engine and install the housing bolts. Torque the bolts as follows:
   a. M8 hex bolts: 18 ft. lbs. (24 Nm).
   b. M10 hex bolts: 33 ft. lbs. (45 Nm).
   c. M12 hex bolts: 60 ft. lbs. (82 Nm).
   d. M8 Torx bolts: 15 ft. lbs. (21 Nm).

e. M10 Torx bolts: 31 ft. lbs. (42 Nm).

f. M12 Torx bolts: 53 ft. lbs. (72 Nm).

9. Install and torque the torque converter–to–flywheel bolts to the following:

a. M8 hex bolts: 19 ft. lbs. (26 Nm).

b. M10 hex bolts: 33 ft. lbs. (45 Nm).

10. Install or connect the following:

- Protective cap over flywheel bolt opening
- Hex bolt near protective cap
- Propeller shaft into center support; torque mount–to–body bolts to 15 ft. lbs. (21 Nm).
- Propeller shaft flange new nuts; torque to 74 ft. lbs. (100 Nm)

11. Install the transmission crossmember. Torque the bolts as follows:

a. To rubber mounts: 55 ft. lbs. (74 Nm).

b. To body: 30 ft. lbs. (41 Nm).

12. Install or connect the following:

- Transmission cable
- O_2S sensor wiring cables
- Transmission cooler lines with new sealing ring
- Wiring cable to transmission
- Shift control cable to body bracket
- Shift selector lever bushing and nut
- Exhaust hanger bracket
- Vent line to bracket
- Stabilizer bar
- Engine guard
- Reinforcement plate
- Oxygen sensor connectors
- Heat shield
- Exhaust system
- Negative battery cable

13. Refill transmission and check operation, then recheck fluid level.

4.4L & 4.6L Engines

AUTOMATIC TRANSMISSION

1. Before servicing the vehicle, refer to the precautions section.

2. Drain the transmission fluid.

3. Remove or disconnect the following:

- Negative battery cable
- Exhaust system
- Front splash guard
- Reinforcement plate
- Heat shields
- Oxygen (O_2S) sensor connectors, mark the locations for installation purposes
- Stabilizer bar and slide it forward
- Rear heat shield
- Front propeller shaft and unclip the vent line from the transmission
- Retaining plate and brace the clamping bush
- Bracket for the oil lines at the oil pan
- Power steering pump oil line bracket
- Oil line and banjo bolt
- Union screw on the oil return line and properly support the transmission and transfer case assembly
- O_2S sensor cable from the transmission crossmember
- Transmission crossmember
- Nuts for the center bearing after bracing the propeller shaft

➡Do not allow the propeller shaft to damage the CV-joints.

- Transmission output flange by bending the propeller shaft downward at the center bearing
- Cable connector from the transmission case
- Impulse sensor electrical connector
- Torque converter retaining screws

4. Support the engine at the front housing and turn the front wheel to the right lock position.

5. Remove the remaining screws and remove the transmission and transfer case assembly.

To install:

6. Align the transmission/transfer case as an assembly.

7. Torque the transmission to engine screws as follows:

a. M8 Hex screws: 18 ft. lbs. (24 Nm).

b. M10 Hex screws: 33 ft. lbs. (45 Nm).

c. M12 Hex screws: 60 ft. lbs. (82 Nm).

d. M8 Torx bolts: 16 ft. lbs. (21 Nm).

e. M10 Torx bolts: 31 ft. lbs. (42 Nm).

f. M12 Torx bolts: 54 ft. lbs. (72 Nm).

8. Install or connect the following:

- Torque converter. Torque the bolts to 30 ft. lbs. (40 Nm).
- Impulse sensor electrical connector
- Transmission case cable connector
- Transmission output flange
- Center bearing. Torque the bolts to 16 ft. lbs. (21 Nm).
- Transmission crossmember. Torque the bolts to 16 ft. lbs. (21 Nm).
- O_2S sensor cable to the crossmember
- Union screw to the oil return line. Torque the screw to 21 ft. lbs. (28 Nm).
- Oil line and banjo bolt. Torque the bolt to 21 ft. lbs. (28 Nm).
- Oil line bracket for the power steering pump
- Retaining plate
- Front propeller shaft. Torque the bolts to 47 ft. lbs. (64 Nm) and clip the vent line to the transmission.
- Rear heat shield
- Stabilizer bar. Torque the bolts to 16 ft. lbs. (22 Nm).
- Front heat shields
- O_2S sensor connector
- Reinforcement plate
- Front splash guard
- Exhaust system
- Negative battery cable

9. Fill the transmission with the proper fluid to the proper level.

10. Start the vehicle and check for leaks, repair if necessary.

Transfer Case

REMOVAL & INSTALLATION

3.0L Engines

WITH MANUAL TRANSMISSION

1. Before servicing the vehicle, refer to the precautions section.

2. Remove or disconnect the following:

- Negative battery cable
- Exhaust system
- Heat shield at right front
- Heat shield at rear
- Reinforcement plate from undercarriage
- Stabilizer bar
- Front propeller shaft

3. Support the transmission

- Crossmember
- Vent line from transfer case
- Driveshaft–to–transmission flange nuts and discard the nuts
- Propeller shaft nuts, bend shaft downward at center bearing
- Propeller shaft from transmission and tie aside

4. Support the transfer case. Remove the retaining bolts and remove the transfer case.

To install:

5. Check the condition of the dowel pins and replace if necessary.

6. Be sure the mating surfaces are clean.

7. Apply a thin coat of grease to transfer case splines

8. Replace the sealing ring of the driveshaft of the transfer case.

9. Install the transfer case and torque the retaining bolts to 32 ft. lbs. (43 Nm).

10. Reposition the propeller shaft in the center support bearing and to the flange. Torque the center support bearing nuts to 15 ft. lbs. (21 Nm) and the flange nuts to 24 ft. lbs. (32 Nm).

11. Install or connect the following:
- Cable to transfer case
- Vent line

12. Install the transmission crossmember. Torque the bolts as follows:
    a. To rubber mounts: 55 ft. lbs. (74 Nm).
    b. To body: 30 ft. lbs. (41 Nm).

13. Install or connect the following:
- Propeller shaft
- Front stabilizer
- Reinforcement plate
- Heat shields
- Exhaust system

14. Refill the transfer case.

4.4L & 4.6L Engines

1. Before servicing the vehicle, refer to the precautions section.
2. Drain the transmission fluid.
3. Remove or disconnect the following:
- Negative battery cable
- Exhaust system
- Front splash guard
- Reinforcement plate
- Heat shields and unclip the oxygen sensor
- Stabilizer bar and slide it forward
- Rear heat shield
- Front propeller shaft and unclip the vent line from the transmission. Properly support the transmission.
- Oxygen (O_2S) sensor cable from the transmission crossmember
- Transmission crossmember
- O_2S sensor cable from the transfer case
- Center bearing after bracing the properller shaft. Do not allow the propeller shaft to damage the CV-joints.
- Transmission output flange by bending the propeller shaft downward at the center bearing
- Transfer case from the transmission

To install:

4. Connect the transfer case to the transmission. Torque the bolts as follows:
    a. M8 Hex screws: 18 ft. lbs. (24 Nm).
    b. M10 Hex screws: 33 ft. lbs. (45 Nm).
    c. M12 Hex screws: 60 ft. lbs. (82 Nm).
    d. M8 Torx bolts: 16 ft. lbs. (21 Nm).
    e. M10 Torx bolts: 31 ft. lbs. (42 Nm).
    f. M12 Torx bolts: 54 ft. lbs. (72 Nm).
5. Install or connect the following:
- Transmission output flange
- Center bearing. Torque the bolts to 16 ft. lbs. (21 Nm).
- Transmission crossmember. Torque the bolts to 16 ft. lbs. (21 Nm).
- O_2S sensor cable to the crossmember
- Union screw to the oil return line. Torque the screw to 21 ft. lbs. (28 Nm).
- Front propeller shaft. Torque the bolts to 47 ft. lbs. (64 Nm) and clip the vent line to the transmission.
- Rear heat shield
- Stabilizer bar. Torque the bolts to 16 ft. lbs. (22 Nm).
- Front heat shields
- O_2S sensor connector
- Reinforcement plate
- Front splash guard
- Exhaust system
- Negative battery cable

6. Fill the transmission with the proper fluid to the proper level.

7. Start the vehicle and check for leaks, repair if necessary.

Halfshafts

REMOVAL & INSTALLATION

3.0L Engines

FRONT

1. Before servicing the vehicle, refer to the precautions section.
2. Remove or disconnect the following:
- Negative battery cable
- Front wheel
- Reinforcement plate
- Front splash guard
- ABS pulse generator
- Brake caliper from disc and tie out of the way with the hose still connected
- Steering gear tie rod from swivel bearing
- Tension strut, with guide joint, from swivel bearing
- Control arm from swivel bearing
- Collar nut on halfshaft, press halfshaft out of drive flange
- Halfshaft

To install:

3. Push in output shaft over the resistance of the retaining ring until it snaps in place.

➡Use a new snap ring on halfshaft spline

4. Install or connect the following:
- Collar nut
- Control arm to swivel bearing
- Tension strut to swivel bearing
- Tie rod to swivel bearing
- Brake caliper to disc
- ABS pulse generator
- Front wheel
- Splash guard
- Reinforcement plate

5. Check the front axle differential fluid level.

REAR

1. Before servicing the vehicle, refer to the precautions section.
2. Remove or disconnect the following:
- Rear wheel
- Rear muffler from center muffler

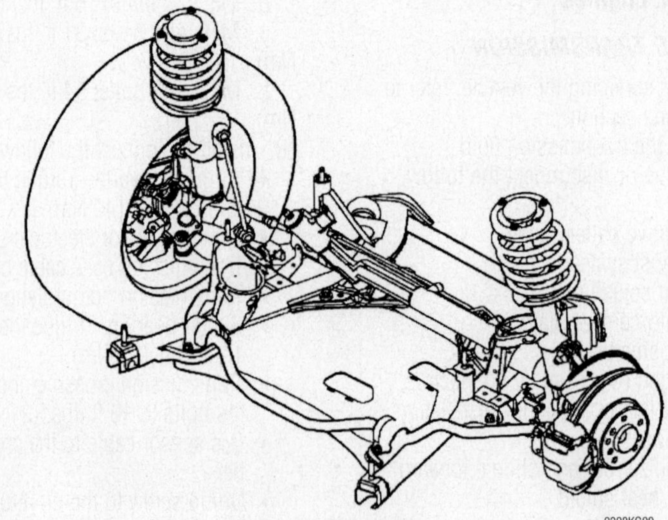

Assembled view of the front axle assembly—X5 Series

- Flange bolts to differential
- Depressurize ride control system, if equipped

3. Raise the rear wheel carrier with a proper jack and support.
4. Press the halfshaft from the wheel carrier.
5. Remove the halfshaft.

To install:

6. Position the halfshaft and press into the wheel carrier.
7. Fit the shims, then install the bolts on the inner halfshaft flange to 61 ft. lbs. (83 Nm).
8. Attach the rear muffler to the center muffler.
9. Install the rear wheel.

4.4L & 4.6L Engines

FRONT

1. Before servicing the vehicle, refer to the precautions section.
2. Remove or disconnect the following:
   - Negative battery cable
   - Front wheel
   - Reinforcement plate
   - Front splash guard
   - Swivel bearing
   - Output shaft from the differential by pressing it out with special tool 31–5–110

To install:

3. Install or connect the following:
   - New output shaft radial seal
   - New snap ring on the output shaft
   - Press the output shaft in by pushing it in over the resistance of the snap ring
   - Swivel bearing
   - Front splash guard
   - Reinforcement plate
   - Front wheel
   - Negative battery cable

REAR

1. Before servicing the vehicle, refer to the precautions section.
2. Remove or disconnect the following:
   - Negative battery cable
   - Rear tire and wheel assembly
   - Collar nut
   - Brake disc
   - ABS sensor
   - Retaining nut from the output flange and remove the drive flange hub

➡The wheel bearing will be destroyed when the flange is removed. The wheel bearing must be replaced.

   - Halfshaft from the vehicle by removing the shaft from the final drive output flange and by pressing

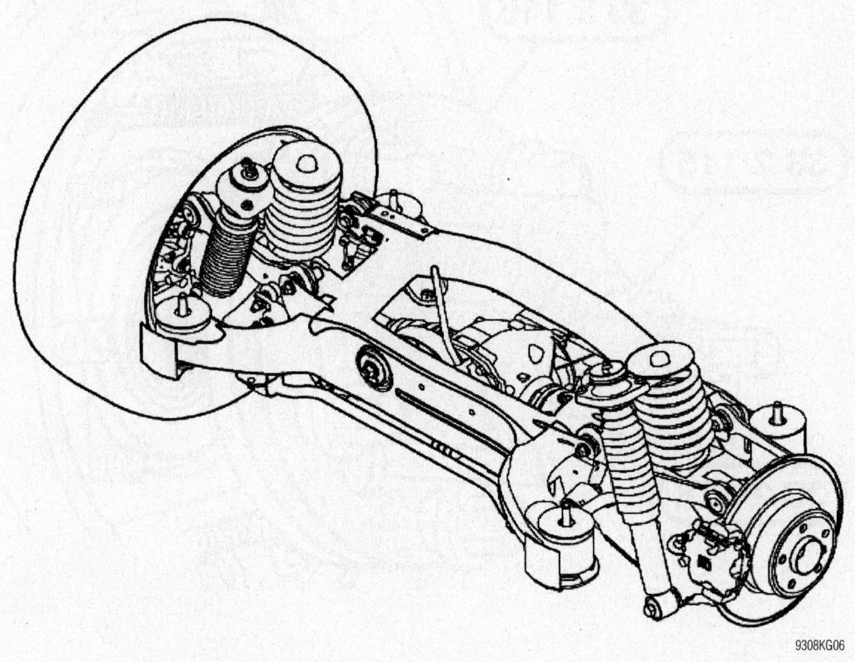

Assembled view of the rear halfshaft and suspension system—X5 Series

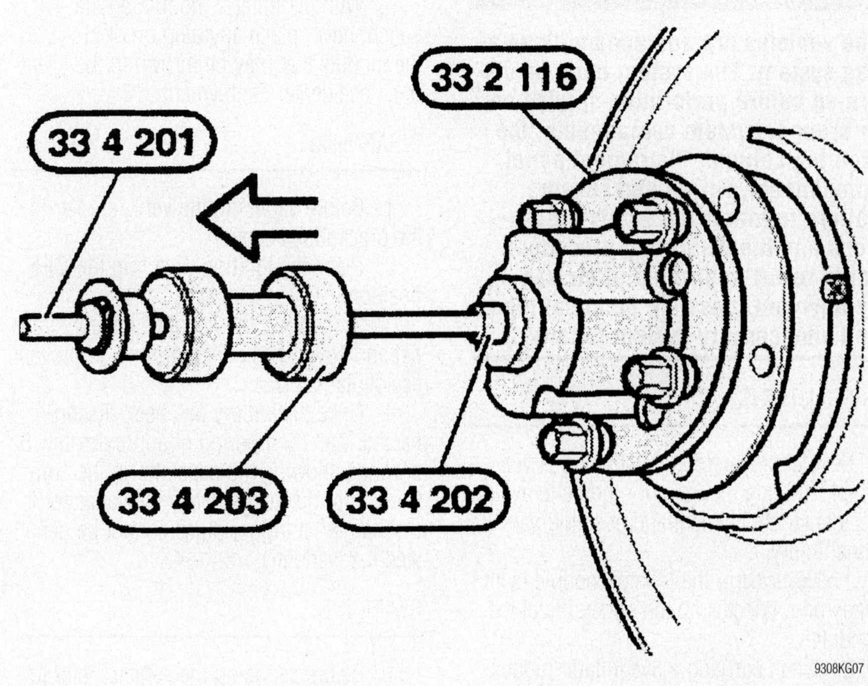

Halfshaft removal tools Nos. 33-2-116, 33-4-201, 201 and 203

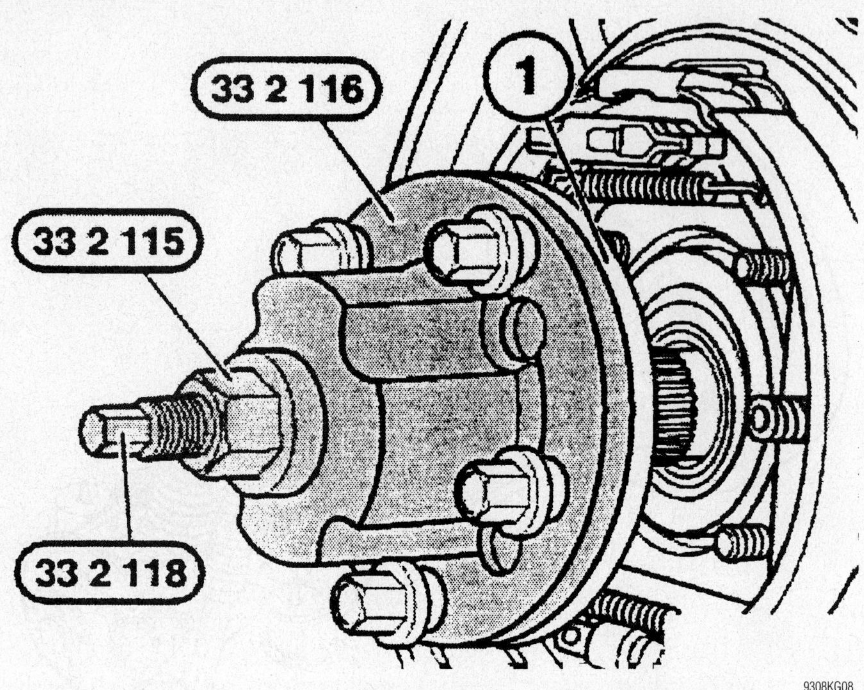

Drive flange hub tool Nos. 33-2-115, 116 and 118 used for drive flange installation

the halfshaft out of the drive flange hub using tool Nos. 33-2-116, 201, 202 and 203

3. Press out the drive flange hub.
4. Pull out the seal with a suitable tool.
5. If the bearing inner race is damaged, pull it off of the drive flange hub with tool No. 33–3–240.

To install:

6. Using an appropriate bearing installer, install the wheel bearing assembly, install the seal, then insert the snapring, and install the drive flange hub.
7. Install or connect the following:
   - Axle shaft seal
   - Output shaft, using tool Nos. 33-2-115, 116 and 118
   - Output shaft to the final drive. Torque the bolts to 61 ft. lbs. (83 Nm).
   - Outer nut with bearing surface lightly oiled. Torque the nut to 310 ft. lbs. (420 Nm).
   - Brake disc
   - ABS sensor
   - Rear tire and wheel assembly
   - Negative battery cable

STEERING AND SUSPENSION

Air Bag

CAUTION

The vehicles are equipped with an air bag system. The system must be disarmed before performing service on, or around, system components, the steering column, instrument panel components, wiring and sensors. Failure to follow the safety precautions and the disarming procedure could result in accidental air bag deployment, possible personal injury and unnecessary system repairs.

PRECAUTIONS

Several precautions must be observed when handling the inflator module to avoid accidental deployment and possible personal injury.

1. Never carry the inflator module by the wires or connector on the underside of the module.
2. When carrying a live inflator module, hold securely with both hands, and ensure that the bag and trim cover are pointed away.
3. Place the inflator module on a bench or other surface with the bag and trim cover facing up.
4. With the inflator module on the bench, never place anything on or close to the module that may be thrown in the event of an accidental deployment.

DISARMING

1. Before servicing the vehicle, refer to the precautions section.
2. Place the ignition switch in the **OFF** position.
3. Disconnect the negative battery terminal and cover the battery terminal to prevent accidental contact.
4. Once the battery has been disconnected, wait for a period of approximately 5 seconds allowing the capacitor in the control unit to discharge. Once the capacitor is discharged, a trigger pulse cannot be generated inadvertently.

REARMING

1. Before servicing the vehicle, refer to the precautions section.
2. Place the ignition switch in the **OFF** position.
3. Attach the sensors, the steering column connector and the seat belt tensioner connectors.
4. Connect the negative battery terminal.
5. Place the ignition switch in the **ON** position. Check that the SRS light illuminates for 6 seconds and extinguishes. If it illuminates in any other pattern, check the components and their connections for proper operation and recheck operation of the warning light.

Power Rack and Pinion Steering Gear

REMOVAL & INSTALLATION

1. Before servicing the vehicle, refer to the precautions section.
2. Set the steering gear in the straight ahead position by aligning the marks on the steering gear and spindle.
3. Drain the power steering fluid.
4. Remove or disconnect the following:
   - Negative battery cable
   - Front wheels
   - Reinforcement plate
   - Nuts from the left and right engine

support arms and raise the engine slightly
- 2 bolts accessible through holes in frame
- Lower clamping screw
- Steering gear clamps (after locking wheels in straight-ahead position)
- Tie rod by pressing it off with special tool 32–3–090
- Self-locking nuts and brace the front axle support
- Banjo bolts and slide the steering gear out through the left wheel opening

To install:

5. Install the steering gear through the left side wheel opening
6. Install new sealing rings and banjo bolts. Torque the bolts as follows:
   a. M10: 7 ft. lbs. (12 Nm).
   b. M14: 25 ft. lbs. (35 Nm).
   c. M16: 29 ft. lbs. (40 Nm).
   d. M18: 34 ft. lbs. (45 Nm).
7. Install or connect the following:
   - Front axle support screws. Torque the screws to 74 ft. lbs. (100 Nm).
   - Self-locking nuts. Torque the nuts to 74 ft. lbs. (100 Nm).
   - Tie rod. Torque the castle nut to 58 ft. lbs. (80 Nm).
   - Steering gear to the spindle. Torque the fastener to 18 ft. lbs. (24 Nm).
   - Steering gear clamp
   - Engine support arm nuts. Torque the nuts to 60 ft. lbs. (85 Nm).
   - Reinforcement plate
   - Splash guard
   - Both front wheels
   - Negative battery cable
8. Fill and bleed the power steering system.
9. Start the vehicle and check for leaks, repair if necessary.

Strut

REMOVAL & INSTALLATION

1. Before servicing the vehicle, refer to the precautions section.
2. Mark the position of the threaded pin to the wheel arch to retain the camber setting when installed.
3. Remove or disconnect the following:
   - Negative battery cable
   - Tire and wheel assembly

➡Carefully mark position of threaded pin to strut tower to ensure original camber is retained during installation.

- Two of the nuts on the spring strut support bearing
- Center strut bracket nut
- Speed sensor/brake wear cable and disconnect the plug housing
- Swivel bearing and tie it aside
- Remaining nut on the spring strut support bearing
- Strut assembly

To install:

4. Install or connect the following:
   - Strut assembly
   - One nut to the spring strut support bearing and hand tighten at this time
   - Swivel bearing. Torque the new self-locking nut to 176 ft. lbs. (250 Nm).
   - Speed sensor/brake wear cable and connect the housing plug
   - Center strut bracket. Torque the nut to 74 ft. lbs. (100 Nm).
5. Align the three upper spring strut support bearing nuts and match the threaded pin with the mark made during the removal procedure. When aligned properly torque the nuts to 25 ft. lbs. (34 Nm).
6. Install the tire and wheel assembly.
7. Reconnect the negative battery cable.

Shock Absorber

REMOVAL & INSTALLATION

1. Before servicing the vehicle, refer to the precautions section.
2. Remove or disconnect the following:
   - Fuse in the air supply system
   - Negative battery cable
   - Rear wheel
   - Nuts and expansion rivets and luggage compartment trim
   - 3 upper nuts after supporting the wheel carrier
   - Shock absorber from the swinging arm and insert a bushing into the bore of the arm
   - Thrust bearing after bracing the piston rod with a ring spanner
   - Upper nut and remove the shock absorber

To install:

3. Install or connect the following:
   - Shock absorber to the swinging arm. Torque the bolt to 41 ft. lbs. (56 Nm).
   - Thrust bearing. Torque the nut to 18 ft. lbs. (25 Nm).
   - Upper nuts. Torque the nuts to 41 ft. lbs. (56 Nm).
   - New expansion rivets and nuts
   - Luggage compartment trim
   - Rear wheel
   - Negative battery cable
   - Fuse for the air supply system

Coil Spring

REMOVAL & INSTALLATION

Front

⁂ CAUTION

This procedure calls for the spring to be compressed. A compressed spring has high potential energy and if released suddenly can cause severe damage and personal injury.

1. Before servicing the vehicle, refer to the precautions section.
2. disconnect the negative battery cable.
3. Remove the strut from the vehicle and mount in a vise using a strut holder. This will prevent damage to the strut tube
4. Using a proper spring compressor, compress the spring until the stress on the thrust bearing is released.
5. Remove the top nut of the strut mount. Counterhold the strut rod during removal.
6. Pull the strut mount off the strut rod. Note the positioning of the spacers and washer for replacement.
7. Pull the spring off the strut and place aside in a safe area.
8. Slowly release the compression of the spring.

To install:

9. Install or connect the following:
   - Spring in the compressor and compress
   - Spring and strut mount with all the spacers and washers in their original positions. Torque the new strut rod nut: 47 ft. lbs. (65 Nm).
10. Release the spring slowly and check that it seats in the spring holders. Install the strut in the vehicle.
11. Connect the negative battery cable.

Lower Ball Joint

REMOVAL & INSTALLATION

1. Before servicing the vehicle, refer to the precautions section.
2. Remove or disconnect the following:
   - Negative battery cable

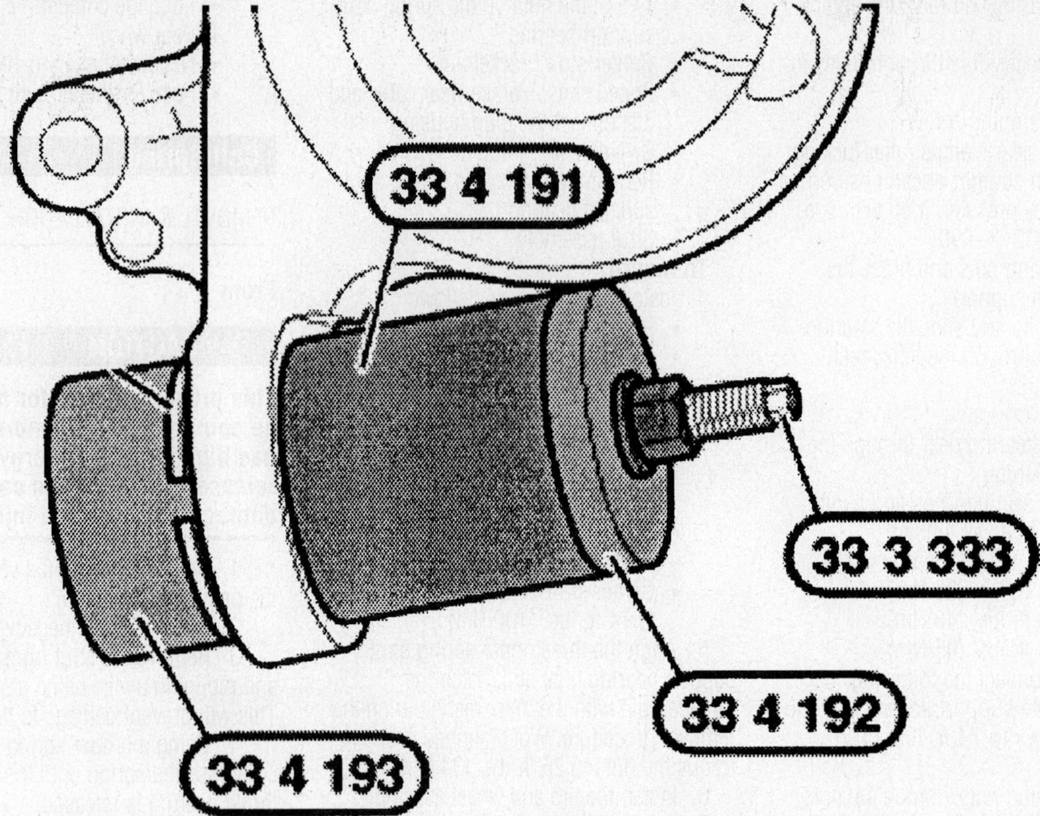

Remove the lower ball joint from the steering knuckle—X5 Series

- Push rod/integral link assembly and properly support the wheel carrier
- Shock absorber from the swinging arm
- Circlip

3. Using special tool 33–4–191, 192, 193 and 33–3–333 pull the ball joint out of the steering knuckle.

To install:

4. Install or connect the following:
   - Ball joint into the steering knuckle with special tools 33–4–191, 192, 194 and 33–3–333
   - New circlip
   - Shock absorber and remove the support from the wheel carrier
   - Push rod/integral link
   - Negative battery cable

Lower Control Arm

REMOVAL & INSTALLATION

1. Before servicing the vehicle, refer to the precautions section.
2. Remove or disconnect the following:
   - Negative battery cable
   - Front wheel
   - Control arm from the front axle support and loosen the nut from the control arm to swivel bearing
   - Control arm from the swivel bearing by pressing it off with special tool 31–2–240

To install:

3. Install or connect the following:
   - Lower control arm to the swivel bearing. Torque the nut to 58 ft. lbs. (80 Nm).
   - Lower control arm to the front axle support. Torque the nut to 74 ft. lbs. (100 Nm) plus an additional 90 degrees.
   - Front wheel
   - Negative battery cable
4. Check and adjust the front end alignment as needed.

BUSHING REPLACEMENT

1. Before servicing the vehicle, refer to the precautions section.
2. Remove or disconnect the following:
   - Negative battery cable
   - Control arm and tie it back to prevent damage to the ball joint
   - Control arm bushing by installing special tools 31–1–051, 052, 33–3–051, 052, 054 and 310

To install:

3. Install or connect the following:
   - Control arm bushing by installing special tools 31–1–051, 052, 33–3–051, 052, 054 and 310
   - Control arm
   - Negative battery cable

Upper Control Arm

REMOVAL & INSTALLATION

1. Before servicing the vehicle, refer to the precautions section.
2. Remove or disconnect the following:
   - Negative battery cable
   - Wheel assembly
   - Fuse for the air supply system, if equipped with air suspension and loosen the pipes on the distributor block
   - Upper control arm from the steering knuckle
   - Plastic shim and unhook the lines
   - Upper control arm

To install:

3. Install or connect the following:
   - Upper control arm. Torque the bolt to 74 ft. lbs. (100 Nm).

- Plastic shim and connect the lines
- Upper control arm to the steering knuckle. Torque the bolt to 122 ft. lbs. (165 Nm).
- Fuse for the air supply system and tighten the pipes on the distributor block
- Wheel assembly
- Negative battery cable

Wheel Bearings

ADJUSTMENT

Wheel bearings cannot be adjusted and must be replaced as a unit and never be reused once removed.

REMOVAL & INSTALLATION

Front

➡The wheel bearings are only removed if they are worn. They cannot be removed without destroying them (due to side thrust created by the bearing puller). They cannot be disassembled, repacked or adjusted.

1. Before servicing the vehicle, refer to the precautions section.
2. Remove or disconnect the following:
   - Negative battery cable
   - Swivel bearing and clamp it in a vise
   - Drive flange by installing special tools 33–2–116, 150 and 33–4–200
   - Bearing inner race from the flange
   - Circlip
   - Snap ring
   - Bearing by installing special tools 31–2–113, 33–3–261, 262 and 266

To install:

3. Install or connect the following:
   - Wheel bearing with the wider camfer facing the swivel bearing to the drive flange with special tools 33–2–261, 264, 268 and 31–2–113
   - Snap ring and circlip
   - inner race to the drive flange
   - Drive flange to the swivel bearing by using special tool 33–3–261, 266, 268 and 31–2–113
   - Swivel bearing
   - Negative battery cable

Rear

1. Before servicing the vehicle, refer to the precautions section.
2. Remove or disconnect the following:
   - Negative battery cable
   - Wheel assembly
   - Collar nut
   - Brake disc
   - Drive flange by installing special tool 33–2–116, 33–4–201, 202 and 203
   - Inner race from the drive flange
   - Wheel bearing

To install:

3. Install or connect the following:
   - Wheel bearing
   - Inner race to the drive flange. Torque the bolts to 74 ft. lbs. (100 Nm).
   - Drive flange to the axle shaft. Torque the collar nut to 310 ft. lbs. (420 Nm).
   - Brake disc
   - Wheel assembly
   - Negative battery cable

BRAKES

Brake Caliper

REMOVAL & INSTALLATION

Front

1. Before servicing the vehicle, refer to the precautions section.
2. Remove or disconnect the following:
   - Negative battery cable
   - Wheel assembly
3. Apply the brake pedal slightly with a brake clamp.
   - Brake pipe from the connection with the brake hose
   - Connector for the wear indicator on the left side
   - Caliper guide bolts
   - Brake caliper

To install:

4. Install or connect the following:
   - Caliper and torque the guide bolts to 81 ft. lbs. (110 Nm)
   - Brake hose to the brake pipe to 13 ft. lbs. (18 Nm)
5. Set the wheel in a straight-ahead position.
6. Be sure brake hose is positively attached to the mounting fixture.
   - Wear indicator on the left side
   - Brake clamp
   - Front wheels
7. Bleed the brakes.

Rear

1. Before servicing the vehicle, refer to the precautions section.
2. Remove or disconnect the following:
   - Negative battery cable
   - Wheel assembly
3. Apply the brake pedal slightly with a brake clamp.
   - Brake hose from the caliper fitting
   - Connector for the wear indicator on the right side
   - Caliper guide bolts and remove the brake caliper

To install:

4. Install or connect the following:
   - Caliper and torque the guide bolts to 21 ft. lbs. (28 Nm)
   - Brake hose to the brake pipe and torque to 14 ft. lbs. (19 Nm)
   - Wear indicator
5. Remove the brake clamp.
6. Install the rear wheels.
7. Bleed the brakes.

Disk Brake Pads

REMOVAL & INSTALLATION

Front

1. Before servicing the vehicle, refer to the precautions section.
2. Remove the front wheels.
3. Remove the disk pad retaining spring from the caliper.
4. Remove the calipers from the disk.
5. Use a special tool, 34–1–050, to force piston back into caliper.
6. Remove the outer brake pad (the inner pad is held in place with a spring in the piston).

To install:

7. Be sure the pads marked "L" and "R" are inserted properly on left and right sides, respectively.
8. Apply anti-squeak compound to all mounting surfaces.
9. Install the calipers.
10. Reposition the retaining spring.
11. Install the front wheels.
12. Fully depress brake pedal several times to set proper contact of pads with rotor.

13. Hold ignition key for at least 30 seconds in position "1" without starting engine. This clear any fault codes stored in system and prevent the wear indicator light from coming on.

14. Bleed brake system. If necessary.

Rear

1. Before servicing the vehicle, refer to the precautions section.
2. Remove the rear wheels.
3. Remove the plastic plugs from the inside of the caliper.
4. Disconnect the plug connection for the wear indicator.
5. Remove the calipers from the disk.
6. Lift out the pad retaining spring from the caliper.
7. Use a special tool, 34–1–050, to force piston back into caliper.
8. Remove the outer brake pad (the inner pad is held in place with a spring in the piston).

To install:

9. Apply anti-squeak compound to all mounting surfaces.
10. Reposition retaining spring.
11. Install calipers.
12. Install rear wheels.
13. Fully depress brake pedal several times to set proper contact of pads with rotor.
14. Hold ignition key for at least 30 seconds in position "1" without starting engine. This clear any fault codes stored in system and prevent the wear indicator light from coming on.
15. Bleed brake system. If necessary.

JAGUAR

S Type • S Type R • X Type 2.5 • X Type 3.0 • XJ • XJR • XK • XKR

5

SPECIFICATIONS AND MAINTENANCE CHARTS

ENGINE AND VEHICLE IDENTIFICATION

Engine						
Code ①	Liters (cc)	Cu. In.	Cyl.	Fuel Sys.	Type	Eng. Mfg.
AJ26	4.0 (3988)	243	8	MFI (N/A)	DOHC	Jaguar
AJ27	4.0 (3988)	243	8	MFI (N/A)	DOHC	Jaguar
AJ28	4.0 (3996)	244	8	MFI/PCM(SEFI)	DOHC	Jaguar
AJ34	4.2 (4195)	256	8	MFI/PCM(SEFI)	DOHC	Jaguar
AJV6	2.5 (2495)	152	6	MFI/PCM(SEFI)	DOHC	Jaguar
AJV6	3.0 (2967)	181	6	MFI/PCM (SEFI)	DOHC	Jaguar
AJV6	3.0 (2970)	181	6	MFI/PCM (SEFI)	DOHC	Jaguar
AJV8	4.2 (4196)	256	8	MFI/PCM(SEFI)	DOHC	Jaguar
A	3.0 (2970)	181	6	MFI/PCM (SEFI)	DOHC	Jaguar
B	4.2 (4196)	256	8	MFI/PCM(SEFI)	DOHC	Jaguar
C ③	4.2 (4196)	256	8	MFI/PCM(SEFI)	DOHC	Jaguar

Model Year	
Code ②	Year
2	2002
3	2003
4	2004
5	2005
6	2006

MFI: Multi-Port Fuel Injection

PCM: Powertrain Control Module maintains fuel pressure at injectors

SEFI: Sequential Electronic Fuel Injection

DOHC: Double Overhead Camshaft

NA: Not Available

① 8th digit of the VIN

② 10th digit of the VIN

③ Supercharged

06041_JAGU_C0001

GENERAL ENGINE SPECIFICATIONS

Year	Model	Engine Displacement Liters	Engine ID/VIN	Net Horsepower @ rpm	Net Torque @ rpm (ft. lbs.)	Bore x Stroke (in.)	Com-pression Ratio	Oil Pressure @ rpm
2002	S Type	3.0	AJV6	240@6800	221@4500	3.50x3.12	10.5:1	NA
	S Type	4.0	AJ28	281@6100	287@4300	3.39x3.39	10.75:1	NA
	X Type	2.5	AJV6	192@6600	178@3000	3.21x3.12	10.3:1	NA
	X Type	3.0	AJV6	227@6800	206@3000	3.50x3.12	10.5:1	NA
	XJ Super V8	4.0	AJ26 ①	365@6000	388@3500	NA	NA	NA
	XJ VP	4.0	AJ26	290@6100	290@4250	NA	NA	NA
	XJ Sport	4.0	AJ26	290@6100	290@4250	NA	NA	NA
	XJR	4.0	AJ26 ①	365@6000	388@3500	NA	NA	NA
	XJ8	4.0	AJ26	290@6100	290@4250	NA	NA	NA
	XK8	4.0	AJ27	290@6100	290@4250	NA	NA	NA
	XKR	4.0	AJ27 ①	365@6000	388@3500	NA	NA	NA
2003	S Type	3.0	AJV6	235@6800	216@4100	3.50x3.13	10.5:1	NA
	S Type	4.2	AJ34	293@6000	303@4100	3.39x3.56	11.0:1	NA
	S Type-R	4.2	AJ34 ①	390@6100	399@3500	3.39x3.56	9.1:1	NA
	X Type	2.5	AJV6	192@6600	178@3000	3.21x3.12	10.3:1	NA
	X Type	3.0	AJV6	227@6800	206@3000	3.50x3.12	10.5:1	NA
	XJ Super V8	4.0	AJ26 ①	365@6000	388@3500	NA	NA	NA
	XJ VP	4.0	AJ26	290@6100	290@4250	NA	NA	NA
	XJ Sport	4.0	AJ26	290@6100	290@4250	NA	NA	NA
	XJ8	4.0	AJ26	290@6100	290@4250	NA	NA	NA
	XJR	4.0	AJ26 ①	365@6000	388@3500	NA	NA	NA
	XK8	4.2	AJV8	294@6000	303@4100	3.38x3.56	11.0:1	NA
	XKR	4.2	AJV8 ①	387@6150	387@3600	3.38x3.56	9.0:1	NA
2004	S Type	3.0	AJV6	235@6800	216@4100	3.50x3.13	10.5:1	NA
	S Type	4.2	AJ34	293@6000	303@4100	3.39x3.56	11.0:1	NA
	S Type-R	4.2	AJ34 ①	390@6100	399@3500	3.39x3.56	9.1:1	NA
	X Type 2.5	2.5	AJV6	192@6600	178@3000	3.21x3.12	10.3:1	NA
	X Type 3.0	3.0	AJV6	227@6800	206@3000	3.50x3.12	10.5:1	NA
	X Type 3.0 Sport	3.0	AJV6	227@6800	206@3000	3.50x3.12	10.5:1	NA
	XJ VP	4.2	AJ34	293@6000	303@4100	3.39x3.56	11.0:1	NA
	XJ8	4.2	AJV8	294@6000	303@4100	3.38x3.56	11.0:1	NA
	XJR	4.2	AJV8 ①	387@6150	387@3600	3.38x3.56	9.0:1	NA
	XK8	4.2	AJV8	294@6000	303@4100	3.38x3.56	11.0:1	NA
	XKR	4.2	AJV8 ①	387@6150	387@3600	3.38x3.56	9.0:1	NA
2005	S Type	3.0	A	240@6800	221@4100	3.50x3.13	10.5:1	NA
	S Type	4.2	B	293@6000	303@4100	3.39x3.56	11.0:1	NA
	S Type-R	4.2	C	390@6100	399@3500	3.39x3.56	9.1:1	NA
	X Type 2.5	2.5	NA	192@6600	178@3000	3.21x3.12	10.3:1	NA
	X Type 3.0	3.0	A	227@6800	206@3000	3.50x3.12	10.5:1	NA
	XJ	4.2	B	293@6000	303@4100	3.39x3.56	11.0:1	NA
	XJ	4.2	C	387@6150	387@3600	3.38x3.56	9.0:1	NA
	XJR	4.2	C	387@6150	387@3600	3.38x3.56	9.0:1	NA
	XK	4.2	B	294@6000	303@4100	3.38x3.56	11.0:1	NA
	XKR	4.2	C	387@6150	387@3600	3.38x3.56	9.0:1	NA
2006	S Type	3.0	A	240@6800	221@4100	3.50x3.13	10.5:1	NA
	S Type	4.2	B	293@6000	303@4100	3.39x3.56	11.0:1	NA
	S Type-R	4.2	C	390@6100	399@3500	3.39x3.56	9.1:1	NA
	X Type 2.5	2.5	NA	192@6600	178@3000	3.21x3.12	10.3:1	NA
	X Type 3.0	3.0	A	227@6800	206@3000	3.50x3.12	10.5:1	NA
	XJ	4.2	B	293@6000	303@4100	3.39x3.56	11.0:1	NA

06041_JAGU_C0002

GENERAL ENGINE SPECIFICATIONS

Year	Model	Engine Displacement Liters	Engine ID/VIN	Net Horsepower @ rpm	Net Torque @ rpm (ft. lbs.)	Bore x Stroke (in.)	Com-pression Ratio	Oil Pressure @ rpm
2006	XJ	4.2	C	387@6150	387@3600	3.38x3.56	9.0:1	NA
Cont.	XJR	4.2	C	387@6150	387@3600	3.38x3.56	9.0:1	NA
	XK	4.2	B	294@6000	303@4100	3.38x3.56	11.0:1	NA
	XKR	4.2	C	387@6150	387@3600	3.38x3.56	9.0:1	NA

NA: Information not available

VP: Vanden Plas

① Supercharged

06041_JAGU_C0003

ENGINE TUNE-UP SPECIFICATIONS

Year	Engine Displacement Liters	Engine ID/VIN	Spark Plugs Gap (in.)	Ignition Timing (deg.)		Fuel Pump (psi)	Idle Speed (rpm)		Valve Clearance	
				MT	AT		MT	AT	In.	Ex.
2002	2.5	AJV6	0.054	①	①	②	NA	NA	0.009-0.026	0.014-0.032
	3.0	AJV6	0.054	—	①	②	—	NA	0.009-0.026	0.014-0.032
	4.0	AJ27	0.035-0.039	—	①	43	—	700	0.007-0.009	0.009-0.011
	4.0	AJ28	0.035-0.039	—	①	②	—	NA	0.009-0.023	0.012-0.026
2003	2.5	AJV6	0.054	①	①	②	NA	NA	0.009-0.026	0.014-0.032
	3.0	AJV6	0.054	—	①	②	—	NA	0.009-0.026	0.014-0.032
	4.0	AJ26	0.035-0.039	—	①	43	—	700	0.012-0.014	0.012-0.014
	4.0	AJ28	0.035-0.039	—	①	②	—	NA	0.009-0.023	0.012-0.026
	4.2	AJ34	0.035-0.039	—	①	②	—	NA	NA	NA
2004	2.5	AJV6	0.051-0.056	①	①	②	NA	NA	0.009-0.026	0.014-0.032
	3.0	AJV6	0.051-0.056	—	①	②	—	NA	NA	NA
	4.2	AJV8	0.035-0.039	—	①	②	—	NA	NA	NA
2005	2.5	NA	0.051-0.056	①	①	②	NA	NA	0.007-0.009	0.013-0.015
	3.0	A	0.051-0.056	—	①	②	—	NA	0.007-0.009	0.013-0.015
	4.2	B & C	0.035-0.039	—	①	②	—	NA	0.007-0.009	0.009-0.011
2006	2.5	NA	0.051-0.056	①	①	②	NA	NA	0.007-0.009	0.013-0.015
	3.0	A	0.051-0.056	—	①	②	—	NA	0.007-0.009	0.013-0.015
	4.2	B & C	NA	—	①	②	—	NA	0.007-0.009	0.009-0.011

The underhood specifications sticker often reflects specification changes in production. Sticker figures must be used if they disagree with those in this chart.

NA: Information not available.

① Ignition timing is controlled by the engine computer

② (Powertrain Control Module) & (Rear Electronic Module). The PCM and ECM calculates the
frequency and determines the current required by the fuel pump to maintain the correct fuel pressure at the fuel injectors.

06041_JAGU_C0004

S-TYPE up to 2002.5 MY

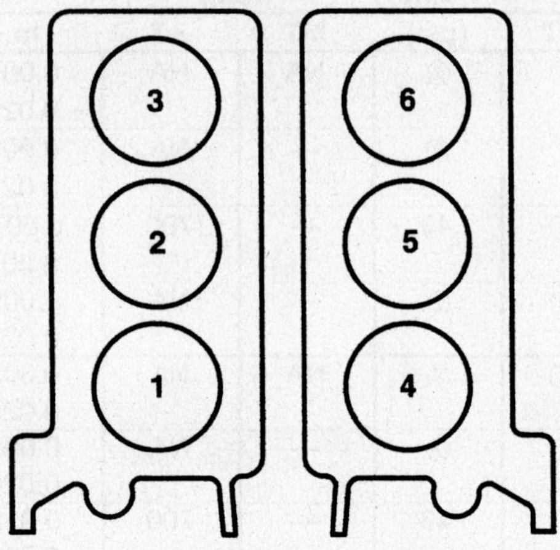

Front of Engine
Bank 1 Bank 2
Firing Order:
1 : 4 : 2 : 5 : 3 : 6

X-TYPE 2001 MY onwards
S-TYPE 2002.5 MY onwards
XJ 2003 MY onwards

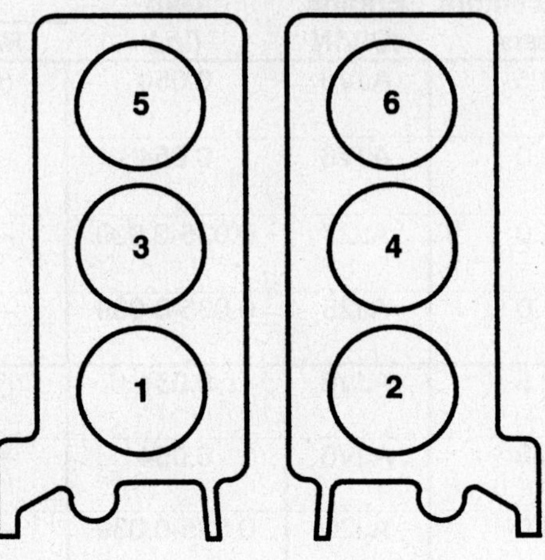

Front of Engine
Bank 1 Bank 2
Firing Order:
1 : 2 : 3 : 4 : 5 : 6

06041_JAGU_G0001

V6 firing orders

up to 2002.5 MY

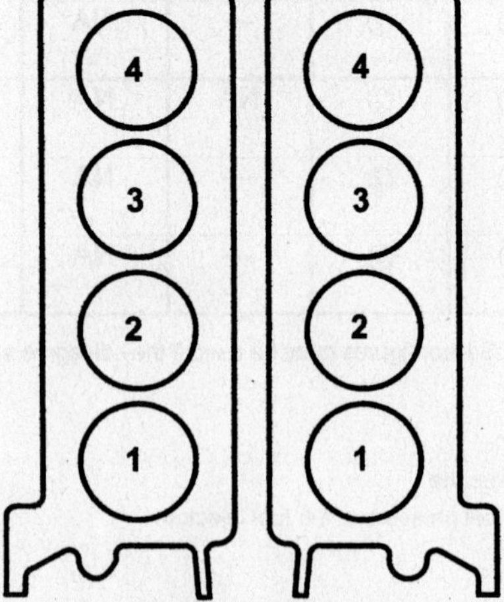

Front of Engine
Bank 1 Bank 2
Firing Order:
(Bank) – Cylinder
(1)-1 : (2)-1 : (1)-4 : (1)-2 : (2)-2 : (1)-3 : (2)-3 : (2)-4

2002.5 MY onwards

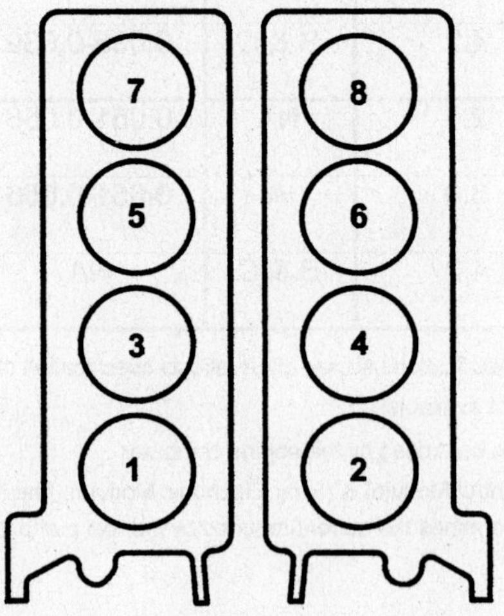

Front of Engine
Bank 1 Bank 2
Firing Order:
1 : 2 : 7 : 3 : 4 : 5 : 6 : 8

06041_JAGU_G0002

V8 firing orders

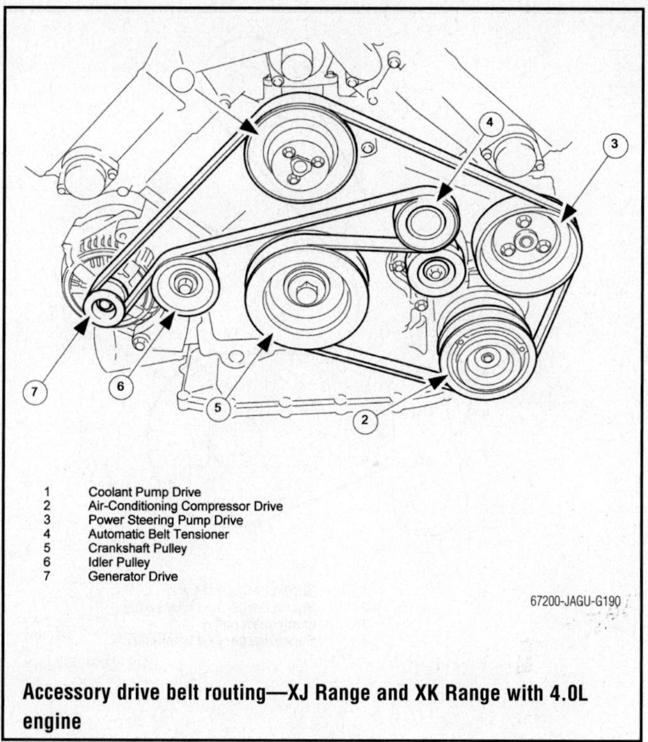

Accessory drive belt routing—XJ Range and XK Range with 4.0L engine

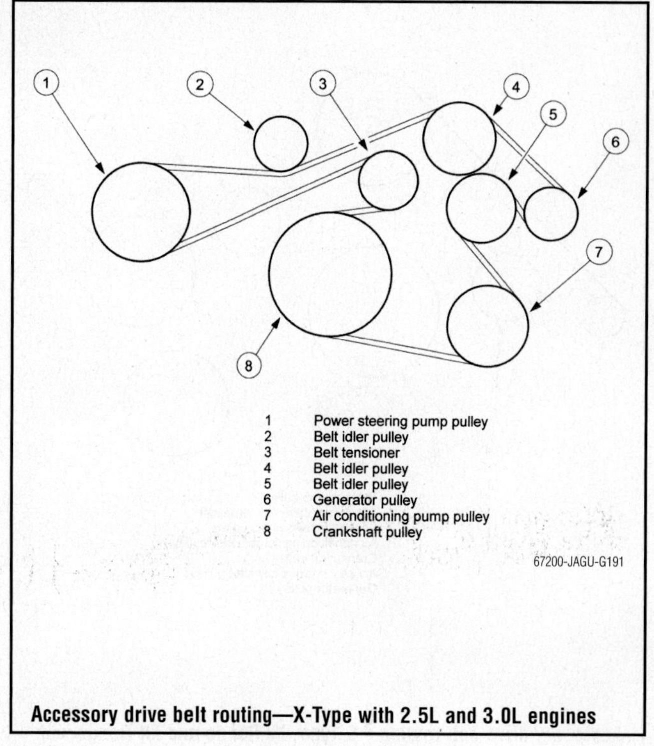

Accessory drive belt routing—X-Type with 2.5L and 3.0L engines

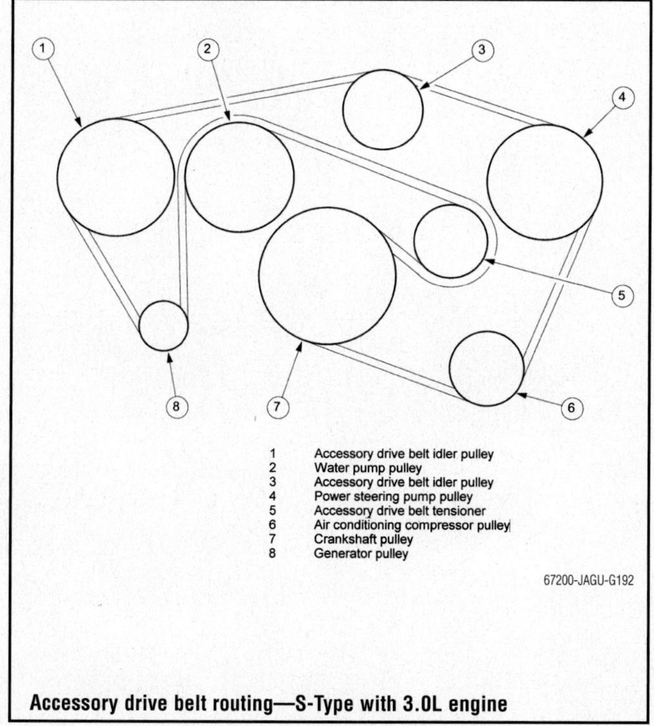

Accessory drive belt routing—S-Type with 3.0L engine

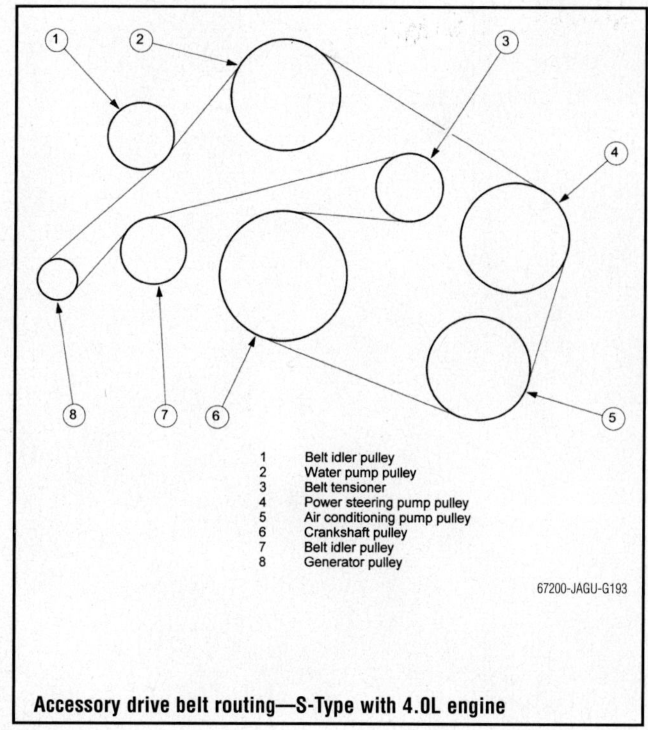

Accessory drive belt routing—S-Type with 4.0L engine

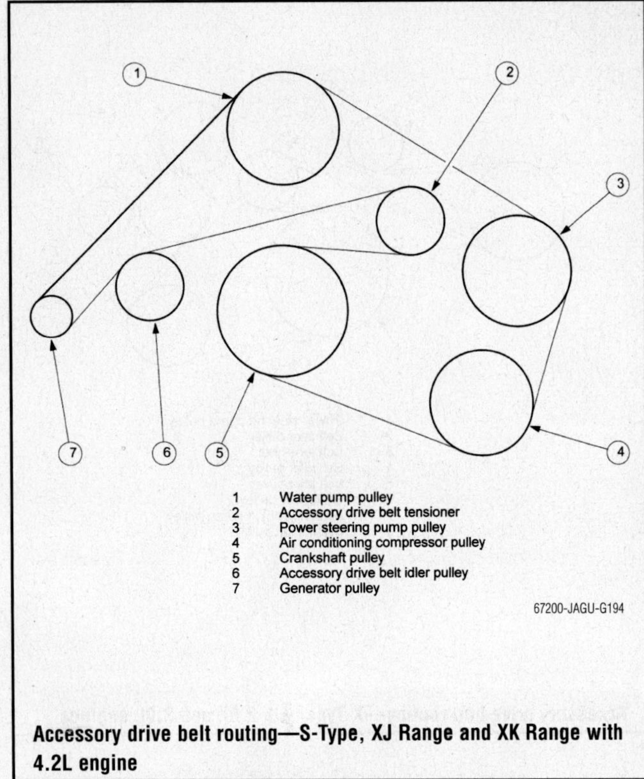

Accessory drive belt routing—S-Type, XJ Range and XK Range with 4.2L engine

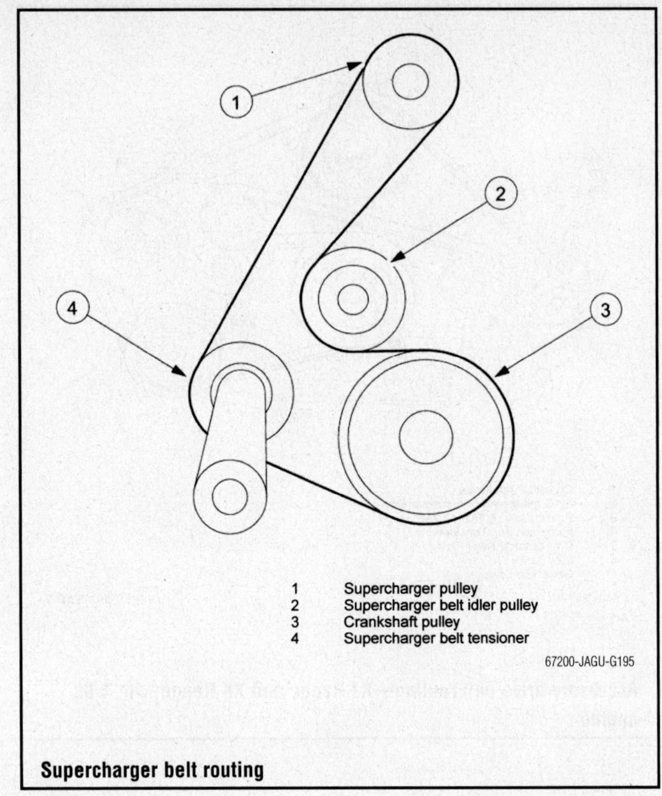

Supercharger belt routing

CAPACITIES

Year	Model	Engine Displacement Liters	Engine ID	Engine Oil with Filter (qts.)	Transmission (pts.) Man.	Transmission (pts.) Auto.	Transfer Case (pts.)	Drive Axle (pts.)	Fuel Tank (gal.)	Cooling System (qts.)
2002	S Type	3.0	AJV6	6.9	—	19.0	—	3.2	18.4	11.1
	S Type	4.0	AJ28	6.9	—	19.0	—	3.2	18.4	11.6
	X Type	2.5	AJV6	6.2	3.7	16.7	1.3	2.5	16.25	8.7
	X Type	3.0	AJV6	6.2	—	16.7	1.3	2.5	16.25	8.7
	XJ Super V8	4.0	AJ26 ①	NA	—	20.0	—	4.0	23.1	NA
	XJ VP	4.0	AJ26	NA	—	21.1	—	4.0	23.1	NA
	XJ Sport Sedan	4.0	AJ26	NA	—	21.1	—	4.0	23.1	NA
	XJ8	4.0	AJ26	7.3	—	21.1	—	2.7	23.1	10.6
	XJR	4.0	AJ26 ①	7.3	—	20.0	—	2.7	23.1	10.6
	XK8	4.0	AJ27	7.3	—	21.1	—	2.7	19.8	10.6
	XKR	4.0	AJ27 ①	8.8	—	20.0	—	2.7	19.8	12.2
2003	S Type	3.0	AJV6	6.2	—	21.1	—	2.8	18.4	9.9
	S Type	4.2	AJ34	6.2	—	21.1	—	2.8	18.4	14.1
	S Type R	4.2	AJ34 ①	6.2	—	21.1	—	2.8	18.4	14.1
	X Type	2.5	AJV6	6.2	3.7	16.7	1.3	2.5	16.25	8.7
	X Type	3.0	AJV6	6.2	—	16.7	1.3	2.5	16.25	8.7
	XJ Super V8	4.0	AJ26 ①	NA	—	20.0	—	4.0	23.1	NA
	XJ VP	4.0	AJ26	NA	—	21.1	—	4.0	23.1	NA
	XJ Sport Sedan	4.0	AJ26	NA	—	21.1	—	4.0	23.1	NA
	XJ8	4.0	AJ26	7.3	—	21.1	—	2.7	23.1	10.6
	XJR	4.0	AJ26 ①	7.3	—	20.0	—	2.7	23.1	10.6
	XK8 cpe	4.0	AJ26	7.3	—	21.1	—	2.7	19.9	10.6
	XK8 conv.	4.0	AJ26	7.3	—	21.1	—	2.7	19.9	NA
	XKR	4.0	AJ26 ①	8.8	—	20.0	—	2.7	19.9	NA
2004	S Type	3.0	AJV6	6.9	—	21.1	—	2.8	18.4	11
	S Type	4.2	AJ34	6.3	—	21.1	—	2.8	18.4	10.6
	S Type R	4.2	AJ34 ①	7.2	—	21.1	—	2.8	18.4	10.6
	X Type	2.5	AJV6	6.2	3.7	16.7	1.3	2.5	16.25	8.7
	X Type	3.0	AJV6	6.2	—	16.7	1.3	2.5	16.25	8.7
	X Type Sport	3.0	AJV6	6.2	—	16.7	1.3	2.5	16.25	8.7
	XJ VP	4.2	AJV8	6.9	—	NA	—	2.7	22.5	14.1
	XJ8	4.2	AJV8	6.9	—	21.1	—	2.7	22.5	10.6
	XJR	4.2	AJV8 ①	6.9	—	22.3	—	2.8	22.5	14.1
	XK8	4.2	AJV8	8.2	—	NA	—	3.8	19.8	10.0
	XKR	4.2	AJV8 ①	8.2	—	NA	—	3.8	19.8	12.2
2005	S Type	3.0	A	6.9	—	20.0	—	2.6	18.4	11
	S Type	4.2	B	6.4	—	20.0	—	2.6	18.4	10.6
	S Type R	4.2	C	6.4	—	21.6	—	2.6	18.4	10.6
	X Type	2.5	NA	6.2	3.7	18.6	1.0	2.5	16.25	8.7
	X Type	3.0	A	6.2	—	18.6	1.0	2.5	16.25	8.7
	X Type Sport	3.0	A	6.2	—	18.6	1.0	2.5	16.25	8.7
	XJ VP	4.2	B	6.4	—	20.0	—	2.6	22.5	14.1
	XJ8	4.2	B	6.4	—	20.0	—	2.6	22.5	10.6
	XJR	4.2	C	6.4	—	21.6	—	2.6	22.5	14.1
	XK8	4.2	B	7.4	—	20.0	—	4.0	19.8	10.0
	XKR	4.2	C	7.4	—	21.6	—	4.0	19.8	12.2
2006	S Type	3.0	A	6.9	—	20.0	—	2.6	18.4	10
	S Type	4.2	B	6.9	—	20.0	—	2.6	18.4	14
	S Type R	4.2	C	6.9	—	21.6	—	2.6	18.4	NA

06041_JAGU_C0005

CAPACITIES

Year	Model	Engine Displacement Liters	Engine ID	Engine Oil with Filter (qts.)	Transmission (pts.) Man.	Transmission (pts.) Auto.	Transfer Case (pts.)	Drive Axle (pts.)	Fuel Tank (gal.)	Cooling System (qts.)
2006 Cont.	X Type	2.5	NA	6.2	3.7	18.6	1.0	2.5	16.25	8.7
	X Type	3.0	A	6.2	—	18.6	1.0	2.5	16.25	8.7
	XJ	4.2	B	6.4	—	20.0	—	2.6	22.5	10
	XJ	4.2	C	6.4	—	20.0	—	2.6	22.5	10
	XJR	4.2	C	6.4	—	21.6	—	2.6	22.5	NA
	XK	4.2	B	7.1	—	20.0	—	4.0	19.8	10.0
	XKR	4.2	C	7.1	—	21.6	—	4.0	19.8	12.2

NA: Information not available

VP: Vanden Plas

SC: Supercharged

06041_JAGU_C0006

VALVE SPECIFICATIONS

Year	Engine Displacement Liters	Engine ID	Seat Angle (deg.)	Face Angle (deg.)	Spring Test Pressure (lbs. @ in.)	Spring Free Length (in.)	Stem-to-Guide Clearance (in.) Intake	Stem-to-Guide Clearance (in.) Exhaust	Stem Diameter (in.) Intake	Stem Diameter (in.) Exhaust
2002	2.5	AJV6	NA	45.25-45.75	NA	1.740	0.0009-0.0026	0.0014-0.0032	0.215-0.216	0.215-0.216
	3.0	AJV6	NA	45.25-45.75	NA	1.740	0.0009-0.0026	0.0014-0.0032	0.215-0.216	0.215-0.216
	4.0	AJ26	44.50-44.75	45.00-45.50-	NA	1.580	0.0012-0.0014	0.0012-0.0014	0.310-0.311	0.310-0.311
	4.0	AJ28	NA	45.00-45.25	NA	1.713	0.0009-0.0023	0.0012-0.0026	0.196-0.196	0.195-0.195
2003	2.5	AJV6	NA	45.25-45.75	NA	1.740	0.0009-0.0026	0.0014-0.0032	0.215-0.216	0.215-0.216
	3.0	AJV6	NA	45.25-45.75	NA	1.740	0.0009-0.0026	0.0014-0.0032	0.215-0.216	0.215-0.216
	4.0	AJ26	44.50-44.75	45.00-45.50-	NA	1.580	0.0012-0.0014	0.0012-0.0014	0.310-0.311	0.310-0.311
	4.2	AJ34	NA	NA	NA	NA	NA	NA	NA	NA
2004	2.5	AJV6	NA	45.25-45.75	NA	1.740	0.0009-0.0026	0.0014-0.0032	0.215-0.216	0.215-0.216
	3.0	AJV6	NA	①	NA	1.740	0.0009-0.0026	0.0014-0.0032	NA	NA
	4.2	AJV8	NA	NA	NA	NA	NA	NA	NA	NA
2005	2.5	NA	NA	45.25-45.75	NA	1.740	0.0009-0.0026	0.0014-0.0032	0.215-0.216	0.215-0.216
	3.0	A	NA	①	NA	1.740	0.0009-0.0026	0.0014-0.0032	NA	NA
	4.2	B&C	NA	NA	NA	NA	NA	NA	NA	NA
2006	2.5	NA	NA	45.25-45.75	NA	1.740	0.0009-0.0026	0.0014-0.0032	0.215-0.216	0.215-0.216
	3.0	A	NA	①	NA	1.740	0.0009-0.0026	0.0014-0.0032	NA	NA
	4.2	B & C	NA	NA	NA	NA	NA	NA	NA	NA

NA: Information not available

① Intake: 45.75; Exhaust: 45.25

06041_JAGU_C0007

CRANKSHAFT AND CONNECTING ROD SPECIFICATIONS

All measurements are given in inches.

Year	Engine Displacement Liters	Engine ID	Crankshaft Main Brg. Journal Dia.	Crankshaft Main Brg. Oil Clearance	Crankshaft Shaft End-play	Thrust on No.	Connecting Rod Journal Diameter	Connecting Rod Oil Clearance	Connecting Rod Side Clearance
2002	2.5	AJV6	2.2815-2.2820	0.0012-0.0018	0.0039-0.0098	3	1.8872-1.8884	0.0008-0.0020	NA
	3.0	AJV6	2.2815-2.2820	0.0012-0.0018	0.0039-0.0098	3	1.8872-1.8884	0.0008-0.0020	NA
	3.0	AJV6	NA	NA	NA	NA	NA	NA	NA
	4.0	AJ26	NA	NA	NA	NA	NA	NA	NA
	4.0	AJ26 ①	2.2811-2.2820	0.0012-0.0018	0.0039-0.0098	5	1.8872-1.8884	0.0008-0.0020	NA
	4.0	AJ28	2.5177-2.5183	0.0008-0.0018	0.0039-0.0087	3	1.8880-1.8884	0.0012-0.0022	0.0087-0.0149
2003	2.5	AJV6	2.2815-2.2820	0.0012-0.0018	0.0039-0.0098	3	1.8872-1.8884	0.0008-0.0020	NA
	3.0	AJV6	NA	NA	NA	NA	NA	NA	NA
	3.0	AJV6	2.2811-2.2820	0.0012-0.0020	NA	5	NA	NA	NA
	4.0	AJ26	NA	NA	NA	NA	NA	NA	NA
	4.0	AJ26 ①	NA	NA	NA	NA	NA	NA	NA
	4.2	AJ34	NA	NA	NA	NA	NA	NA	NA
	4.2	AJ34 ①	2.5177-2.5183	0.0008-0.0018	0.0039-0.0087	3	1.8880-1.8884	0.0012-0.0022	0.0087-0.0149
2004	2.5	AJV6	2.2815-2.2820	0.0012-0.0018	0.0039-0.0098	3	1.8872-1.8884	0.0008-0.0020	NA
	3.0	AJV6	2.2815-2.2820	0.0012-0.0018	0.0039-0.0098	3	1.8872-1.8884	0.0008-0.0020	NA
	4.2	AJV8	2.2811-2.2820	0.0012-0.0020	NA	5	NA	NA	NA
2005	2.5	NA	NA	NA	NA	NA	NA	NA	NA
	3.0	A	NA	NA	NA	NA	NA	NA	NA
	4.2	B & C	NA	NA	NA	NA	NA	NA	NA
2006	2.5	NA	NA	NA	NA	NA	NA	NA	NA
	3.0	A	NA	NA	NA	NA	NA	NA	NA
	4.2	B & C	NA	NA	NA	NA	NA	NA	NA

NA: Not Available

① Supercharged

06041_JAGU_C0008

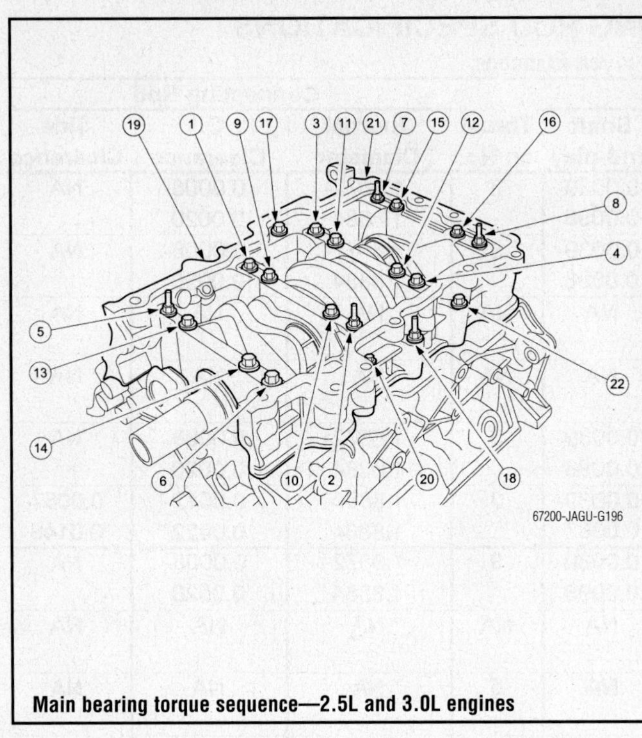

Main bearing torque sequence—2.5L and 3.0L engines

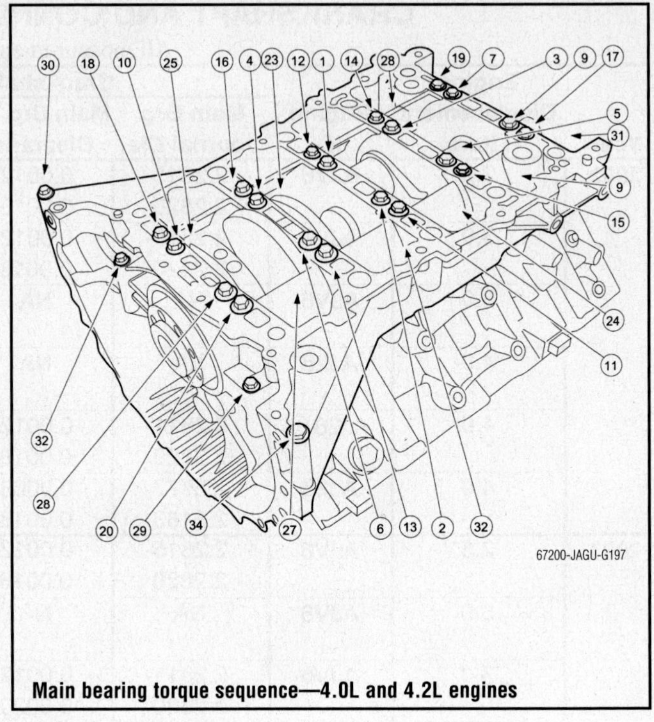

Main bearing torque sequence—4.0L and 4.2L engines

TORQUE SPECIFICATIONS

All readings in ft. lbs.

Year	Engine Displacement Liters	Engine ID	Cylinder Head Bolts	Main Bearing Bolts	Rod Bearing Bolts	Crankshaft Damper Bolts	Flywheel Bolts	Manifold		Spark Plugs	Oil Pan Drain Plug
								Intake**	Exhaust		
2002	2.5	AJV6	①	②	③	④	59	7	15	11	18
	3.0	AJV6	①	②	③	④	59	7	15	11	18
	4.0	AJ26	⑤	NA	NA	⑥	NA	NA	NA	21	25
	4.0	AJ28	⑦	⑧	⑨	278	⑩	15	13	21	NA
2003	2.5	AJV6	①	②	③	④	59	7	15	11	18
	3.0	AJV6	①	②	③	④	59	7	15	11	18
	4.0	AJ26	⑤	NA	NA	⑥	NA	NA	NA	21	25
	4.2	AJ34	⑪	⑧	⑨	278	⑩	16	15	20	18
2004	2.5	AJV6	①	②	③	④	59	7	15	11	18
	3.0	AJV6	⑤	②	③	④	59	7	15	11	18
	4.2	AJV8	⑪	⑧	⑨	278	⑩	16	28	20	18
2005	2.5	NA	⑬	②	③	④	59	7	15	11	18
	3.0	A	⑬	②	③	④	59	7	15	11	18
	4.2	B & C	⑫	⑧	⑨	236	⑩	16	15	20	18
2006	2.5	NA	⑬	②	③	④	59	7	15	11	18
	3.0	A	⑬	②	③	④	59	7	15	11	18
	4.2	B & C	⑫	⑧	⑨	278	⑩	16	18	20	18

NA: Information not available

** Lower manifold

① See illustration in the text:
Bolts 1-10:
Step 1: 22 ft. lbs.
Step 2: + 90 degrees.
Step 3: Loosen 360 degrees
Step 4: 22 ft. lbs.
Step 5: +90 degrees
Step 6: +90 degrees
Bolts 11-12: 17 ft. lbs.

② See accompanying illustration
Bolts 1-8: 19 ft. lbs.
Bolts 9-16: 30 ft. lbs.
Bolts 1-16: plus 90 degrees
Bolts 17-22: 19 ft. lbs.
Bolts 1-16 must be new

③ Step 1: 17 ft. lbs.
Step 2: 32 ft. lbs.
Step 3: plus 105 degrees

④ Step 1: 89 ft. lbs.
Step 2: Loosen (minimum one turn)
Step 3: 37 ft. lbs.
Step 4: Angle torque to 90 degrees

⑤ Step 1: Tighten M10 bolts to 15 ft. lbs.
Step 2: Tighten M10 bolts to 26 ft. lbs.
Step 3: Tighten M10 bolts + 90 degree turn
Step 4: Tighten M10 bolts + 90 degree turn
Step 5: Tighten M8 bolts to 20 ft. lbs.

⑥ With split locking ring: 268-285 ft. lbs.
Without split locking ring:
59 ft. lbs. + 80 degree turn.

⑦ Step 1: 15 ft. lbs.
Step 2: 26 ft. lbs.
Step 3: +90 degrees
Step 4: Tighten M10 +90 degrees
Step 5: Tighten bolts at front of cyl. head

⑧ See accompanying illustration
Tighten the bolts in the following steps:
Bolts 21-32: 11 ft. lbs.
Bolts 33 & 34: 11 ft. lbs.
Bolts 1-10: 18 ft. lbs.
Bolts 11-20: 11 ft. lbs.
Bolts 1-10:26 ft. lbs. plus 135 degrees
Bolts 11-20: 15 ft. lbs. plus 150 degrees
Bolts 21-32: 15 ft. lbs. plus 90 degrees
Bolts 33 & 34: 15 ft. lbs. plus 150 degrees

⑨ Step 1: 7 ft. lbs.
Step 2: 22 ft. lbs.
Step 3: plus 90 degrees

⑩ Step 1: 11 ft. lbs.
Step 2: 81 ft. lbs.

⑪ See illustration in text:
Bolts 1-10:
Step 1: 15 ft. lbs.
Step 2: 26 ft. lbs.
Step 3: +90 degrees
Step 4: +90 degrees
Bolts 11-12: 18 ft. lbs.

⑫ See illustration in text:
Bolts 1-10:
Step 1: 15 ft. lbs.
Step 2: 26 ft. lbs.
Step 3: +90 degrees
Step 4: +90 degrees
Bolts Y: 17 ft. lbs.

⑬ See illustration in text
Step 1: 22 ft. lbs.
Step 2: + 90 degrees.
Step 3: Loosen 360 degrees
Step 4: 22 ft. lbs.
Step 5: +90 degrees
Step 6: +90 degrees

06041_JAGU_C0009

WHEEL ALIGNMENT

Year	Model		Caster Range (+/-Deg.)	Caster Preferred Setting (Deg.)	Camber Range (+/-Deg.)	Camber Preferred Setting (Deg.)	Toe-in (in.)
2002	S Type ①	F	0.70	+7.62	0.50	-0.10	-0.10+/-0.83
		R	—	—	0.75	-0.75	0.12+/-0.10
	X Type 2.5	F	1.0	+2.48	1.0	-1.03	-0.10+/-0.27
		R	—	—	—	—	—
	X Type 3.0	F	1.0	+2.48	1.0	-1.03	-0.10+/-0.27
		R	—	—	—	—	—
	XJ Super V8	F	1.0	+4.00	1.0	-1.03	0.25+/-0.17
		R	—	—	1.0	-1.20	0.08+/-0.08
	XJ Vanden Plas	F	1.0	+4.00	1.0	-1.03	0.25+/-0.17
		R	—	—	1.0	-1.20	0.08+/-0.08
	XJ Sport Sedan	F	1.0	+4.00	1.0	-1.03	0.25+/-0.17
		R	—	—	1.0	-1.20	0.08+/-0.08
	XJR	F	1.0	+4.00	1.0	-1.03	0.25+/-0.17
		R	—	—	1.0	-1.20	0.08+/-0.08
	XJ8	F	NA	NA	NA	NA	NA
		R	—	—	NA	NA	NA
	XKR	F	1.0	②	0.60	③	0.15+/-0.10
		R	—	—	1.00	④	0.08+/-0.08
	XK8	F	1.0	②	0.60	③	0.15+/-0.10
		R	—	—	1.00	④	0.08+/-0.08
2003	S Type ⑤	F	0.70	+7.62	0.50	-0.10	-0.10+/-0.83
		R	—	—	0.75	-0.75	0.12+/-0.10
	S Type R ⑤	F	0.70	+7.62	0.50	-0.10	-0.10+/-0.83
		R	—	—	0.75	-0.75	0.12+/-0.10
	X Type 2.5	F	1.0	+2.48	1.0	-1.03	-0.10+/-0.27
		R	—	—	—	—	—
	X Type 3.0	F	1.0	+2.48	1.0	-1.03	-0.10+/-0.27
		R	—	—	—	—	—
	XJ Super V8	F	—	+4.0 to +8.0	—	⑥	+0.08 to +0.42
		R	—	—	—	-1.20 to -1.10	+0.17 to +0.52
	XJ Vanden Plas	F	—	+4.0 to +8.0	—	⑥	+0.08 to +0.42
		R	—	—	—	-1.20 to -1.10	+0.17 to +0.52
	XJ Sport Sedan	F	—	+4.0 to +8.0	—	⑥	+0.08 to +0.42
		R	—	—	—	-1.20 to -1.10	+0.17 to +0.52
	XJR	F	—	+4.0 to +8.0	—	⑥	+0.08 to +0.42
		R	—	—	—	-1.20 to -1.10	+0.17 to +0.52
	XKR	F	1.0	②	0.60	③	0.15+/-0.10
		R	—	—	1.00	④	0.08+/-0.08
	XK8	F	1.0	②	0.60	③	0.15+/-0.10
		R	—	—	1.00	④	0.08+/-0.08
2004	S Type	F	0.70	+7.62	0.50	-0.10	-0.10+/-0.83
		R	—	—	0.75	-0.75	0.12+/-0.10
	S Type R	F	0.70	+7.62	0.50	-0.10	-0.10+/-0.83
		R	—	—	0.75	-0.75	0.12+/-0.10
	X Type 2.5	F	1.0	+2.48	1.0	-1.03	-0.10+/-0.27
		R	—	—	—	—	—
	X Type 3.0	F	1.0	+2.48	1.0	-1.03	-0.10+/-0.27
		R	—	—	—	—	—

06041_JAGU_C0010

WHEEL ALIGNMENT

			Caster		Camber		
Year	Model		Range (+/-Deg.)	Preferred Setting (Deg.)	Range (+/-Deg.)	Preferred Setting (Deg.)	Toe-in (in.)
2004 Cont.	X Type 3.0 Sport	F	1.0	+2.48	1.0	-1.03	-0.10+/-0.27
		R	—	—	—	—	—
	XJ Vanden Plas	F	—	+4.0 to +8.0	—	⑥	+0.08 to +0.42
		R	—	—	—	-1.20 to -1.10	+0.17 to +0.52
	XJR	F	—	+4.0 to +8.0	—	⑥	+0.08 to +0.42
		R	—	—	—	-1.20 to -1.10	+0.17 to +0.52
	XJ8	F	—	+4.0 to +8.0	—	⑥	+0.08 to +0.42
		R	—	—	—	-1.20 to -1.10	+0.17 to +0.52
	XKR	F	1.0	②	0.60	③	0.15+/-0.10
		R	—	—	1.00	④	0.08+/-0.08
	XK8	F	1.0	②	0.60	③	0.15+/-0.10
		R	—	—	1.00	④	0.08+/-0.08
2005	S Type	F	—	⑫	—	⑪	-0.37 to -0.03
		R	—	—	—	⑮	⑯
	S Type R	F	—	⑫	—	⑪	-0.37 to -0.03
		R	—	—	—	⑮	⑯
	X Type 2.5	F	—	⑬	—	⑭	-0.27 to -0.07
		R	—	—	—	⑰	⑱
	X Type 3.0	F	—	⑬	—	⑭	-0.27 to -0.07
		R	—	—	—	⑰	⑱
	X Type 3.0 Sport	F	—	⑬	—	⑭	-0.27 to -0.07
		R	—	—	—	⑰	⑱
	XJ Vanden Plas	F	1.4	+7.48	—	⑧	-0.43 to -0.10
		R	—	—	0.50	-0.89	-0.02 to +0.18
	XJR	F	4.0	+6.0	—	⑦	0.08 to 0.42
		R	—	—	0.50	-0.89	-0.02 to +0.18
	XJ8	F	1.4	+7.48	—	⑧	-0.43 to -0.10
		R	—	—	0.50	-0.89	-0.02 to +0.18
	XKR	F	—	⑩	—	⑨	0.08 to 0.42
		R	—	—	0.40	-0.50	0.08 to 0.25
	XK8	F	—	⑩	—	⑨	0.08 to 0.42
		R	—	—	0.40	-0.50	0.08 to 0.25
2006	S Type	F	0.7	⑲	0.50	⑳	0.10+/- 0.17
		R	—	—	0.75	-0.54	㉑
	S Type R	F	0.7	⑲	0.50	⑳	0.10+/- 0.17
		R	—	—	0.75	-0.54	㉑
	X Type 2.5	F	—	⑬	—	⑭	-0.27 to -0.07
		R	—	—	—	NA	NA
	X Type 3.0	F	—	⑬	—	⑭	-0.27 to -0.07
		R	—	—	—	NA	NA
	XJ	F	0.70	+7.88	0.50	㉒	-0.20+/-0.17
		R	—	—	0.75	-0.75	-0.25 +/-to 0.17
	XJR	F	0.42	+7.53	0.30	㉓	-0.12+/-0.10
		R	—	—	0.45	-0.45	0.15+/-0.25
	XKR	F	—	⑩	—	⑨	0.08 to 0.42
		R	—	—	0.40	-0.50	0.08 to 0.25
	XK	F	—	⑩	—	⑨	0.08 to 0.42
		R	—	—	0.40	-0.50	0.08 to 0.25

06041_JAGU_C0011

WHEEL ALIGNMENT SPECIFICATIONS
Footnotes

NA: Not Available

SC: Supercharged

① Through VIN M45254, use 2001 figures

② Left: +5.87; right: +7.39

③ Left: -1.18; right -0.18

④ Left: -0.90; right; -0.10

⑤ Through VIN F59525, use 2002 figures

⑥ Left: -0.53 to + 0.47; Right: -1.03 to -0.03

⑦ Left: -0.53 to +0.47; right: -1.03 to -0.03

⑧ left -0.75 to +0.25; right -1.25 to -0.25

⑨ left -1.18 to -0.18; right -1.30 to -0.30

⑩ left +5.87 to -7.27; right +6.64 to +8.04

⑪ left -0.70 to +0.30; right -1.05 to -0.05

⑫ left +5.91 to +7.31; right +6.04 to +7.44

⑬ left and right: Min. +1.98, Max. +2.98

⑭ left and right -1.04 to -0.04

⑮ left & right -1.25 to -0.25

⑯ Naturally aspirated: -0.02 to +0.18
Supercharged: +0.03 to +0.23

⑰ left & right -1.20 to -0.20

⑱ +0.03 to +0.23

⑲ Left: 6.61, right: 6.74

⑳ Left: 0.20; right: 0.55

㉑ Naturally aspirated: 0.167 +/- 0.167
Supercharged: 0.25 +/- 0.167

㉒ Left: -0.10; Right: -0.60

㉓ Left: -0.60; Right: -0.36

06041_JAGU_C0012

TIRE, WHEEL AND BALL JOINT SPECIFICATIONS

Year	Model	OEM Tires Standard	OEM Tires Optional	Tire Pressures (psi) ① Front	Tire Pressures (psi) ① Rear	Wheel Size	Ball Joint Inspection	Lugnut Torque (ft. lbs.)
2002	S-Type	225/55ZR16	225/55R16	32	34	Std: 7.0/7.5x16 in.	②	94
			235/50ZR17	32	34	Opt: 7.5x17 in.		
	X-Type	205/55R16	225/45R17	32	32	Std: 6.5x16 in.	③	④
				30	30	Opt: 7x17 in.		
	XJ Super V8	225/60ZR16	225/40ZR18	32	34	NA	NS	88-102
	XJ Vanden Plas	225/60ZR16	225/40ZR18	32	34	NA	NS	88-102
	XJ Sport Sedan	225/60ZR16	225/40ZR18	32	34	NA	NS	88-102
	XJ8 Sedan	225/55R16	225/60ZR16	32	34	8-J	NS	88-102
			225/40ZR18	32	34			
	XJR	225/40ZR18	None	32	34	8-J	NS	88-102
	XK8 Series	245/50ZR17	245/45ZR18	32	34	8-J	NS	88-102
			255/45ZR18	32	34	8.5-J	NS	
			255/35ZR20	34	34	9-J	NS	
			285/30ZR20	34	34	10-J	NS	
2003	S-Type 3.0L	205/60R16	225/55R16	Std 32	Std 32	Std: 7.5x16 in.	②	94
			235/50R17	Opt 31	Opt 31	Opt: 7.5x17 in.		
			245/40ZR18			Opt: 7.5x17 in.		
	S-Type 4.2L	205/60R16	225/55R16	Std 32	Std 32	F: 8x18 in.	②	94
			235/50R17	Opt 31	Opt 31	R: 9.5x18 in.		
			245/40ZR18					
	S Type R	245/40ZR18	275/35ZR18	Std 36	Std 36	F: 8x18 in.	②	94
				Opt 40	Opt 40	R: 9.5x18 in.		
	X-Type	205/55R16	225/45R17	32	32	Std: 6.5x16 in.	③	④
				30	30	Opt: 7x17 in.		
	XJ Series	235/55R17	None	30	32	7.5-J	NS	88-102
		235/50R18	None	30	32	8-J	NS	
		255/40R19	None	32	34	8.5-J	NS	
		255/35R20	NA	34	36	10-J	NS	
	XK8 Series	245/50ZR17	245/45ZR18	32	34	8-J	NS	75
			255/45ZR18	32	34	8.5-J	NS	
			255/35ZR20	34	34	9-J	NS	
			285/30ZR20	34	34	10-J	NS	
2004	S Type 3.0L	235/50R17	245/40ZR18	⑤	⑤	7-J	②	94
	S Type 4.2L	235/50R17	245/40ZR18	⑤	⑤	7-J	②	94
	S Type R	245/40ZR18	None	⑤	⑤	8-J	②	94
	X Type 2.5L	205/55HR16	None	⑤	⑤	7-J	③	④
	X Type 3.0L	225/45HR17	225/45HR18	⑤	⑤	7-J	③	④
	X Type Sport	NA	NA	⑤	⑤	NA	③	④
	XJ Vanden Plas	235/50HR18	None	⑤	⑤	8-J	NS	92
	XJ8	235/55R17	235/50R18	⑤	⑤	Std. 7.5x17 in.	NS	92
			255/40R19	⑤	⑤	Opt. 8x18 in.	NS	
	XJR	235/55R17	235/50R18	⑤	⑤	Std. 7.5x17 in.	NS	92
			255/40R19	⑤	⑤	Opt. 8x18 in.	NS	
	XKR	245/45ZR18	255/45ZR18	⑤	⑤	Std. 8; Opt. 9	NS	75
	XK8	245/45ZR18	255/45ZR18	⑤	⑤	Std. 8; Opt. 9	NS	75
2005	S Type 3.0L	235/50R17	245/40ZR18	⑤	⑤	7-J	②	85-100
	S Type 4.2L	235/50R17	245/40ZR18	⑤	⑤	7-J	②	85-100
	S Type R	245/40ZR18	None	⑤	⑤	8-J	②	85-100
	X Type 2.5L	205/55HR16	None	⑤	⑤	7-J	③	④
	X Type 3.0L	225/45HR17	225/45HR18	⑤	⑤	7-J	③	④

06041_JAGU_C0013

TIRE, WHEEL AND BALL JOINT SPECIFICATIONS

Year	Model	OEM Tires		Tire Pressures (psi) ①		Wheel Size	Ball Joint Inspection	Lugnut Torque (ft. lbs.)
		Standard	Optional	Front	Rear			
2005	X Type Sport	NA	NA	⑤	⑤	NA	③	④
Cont.	XJ Vanden Plas	235/50HR18	None	⑤	⑤	8-J	NS	85-100
	XJ	235/55HR17	235/50HR18	⑤	⑤	Std. 7.5x17 in.	NS	85-100
			255/40R19	⑤	⑤	Opt. 8x18 in.	NS	85-100
	XJR	235/55R17	235/50HR18	⑤	⑤	Std. 7.5x17 in.	NS	85-100
			235/50ZR18	⑤	⑤	Opt. 8.5x19	NS	85-100
			255/40ZR19	⑤	⑤	Opt. 8x18 in.	NS	85-100
			235/55ZR20	⑤	⑤	Opt. 9x20	NS	85-100
	XK	245/45ZR18	255/45ZR18	⑤	⑤	Std. 8	NS	85-100
			245/40ZR19	⑤	⑤	Opt. 8.5, 9 and 10	NS	85-100
			255/40ZR19	⑤	⑤		NS	85-100
			255/35ZR20	⑤	⑤		NS	85-100
			285/30ZR20	⑤	⑤		NS	85-100
2006	S Type	235/50R17	245/40ZR18	⑤	⑤	7-J	②	85-100
	S Type R	245/40ZR18	None	⑤	⑤	8-J	②	85-100
	X Type 2.5L	205/55HR16	None	⑤	⑤	7-J	③	④
	X Type 3.0L	225/45HR17	225/45HR18	⑤	⑤	7-J	③	④
	XJ	235/55HR17	235/50HR18	⑤	⑤	Std. 7.5x17 in.	NS	85-100
			255/40R19	⑤	⑤	Opt. 8x18 in.	NS	85-100
	XJR	235/55R17	235/50HR18	⑤	⑤	Std. 7.5x17 in.	NS	85-100
			235/50ZR18	⑤	⑤	Opt. 8.5x19	NS	85-100
			255/40ZR19	⑤	⑤	Opt. 8x18 in.	NS	85-100
			235/55ZR20	⑤	⑤	Opt. 9x20	NS	85-100
	XK/XKR	245/45ZR18	255/45ZR18	⑤	⑤	Std. 8	NS	85-100
			245/40ZR19	⑤	⑤	Opt. 8.5, 9 and 10	NS	85-100
			255/40ZR19	⑤	⑤		NS	85-100
			255/35ZR20	⑤	⑤		NS	85-100
			285/30ZR20	⑤	⑤		NS	85-100

NA: Information not available

NS: Not specified by manufacturer

OEM: Original Equipment Manufacturer

PSI: Pounds Per Square Inch

STD: Standard

OPT: Optional

① Standard tire pressures are the first listed per model.

② Upper and lower: 0.8mm max. play

③ Lower: 0.8mm max.

④ Steel wheels: 59 ft. lbs.
Aluminum wheels: 76 ft. lbs.

⑤ See placard on vehicle

06041_JAGU_C0014

BRAKE SPECIFICATIONS

All measurements in inches unless noted

			Brake Disc				Brake Caliper	
Year	Model		Original Thickness	Machine Thickness	Maximum Runout	Minimum Lining Thickness	Bracket Bolts (ft. lbs.)	Mounting Bolts (ft. lbs.)
2002	S Type	F	①	⑧	NA	0.080	76	25 ② ⑲
		R	0.79	0.72	NA	0.080	76	25 ③ ㉑
	X Type ④	F	NA	NA	NA	NA	98	22
		R	NA	NA	NA	NA	52	24
	XJ Super V8	F	NA	NA	NA	NA	104 ⑤	18-22 ⑥
		R	NA	NA	NA	NA	40	18-22 ⑥
	XJ Vanden Plas	F	NA	NA	NA	NA	104 ⑤	18-22 ⑥
		R	NA	NA	NA	NA	40	18-22 ⑥
	XJ Sport Sedan	F	NA	NA	NA	NA	104 ⑤	18-22 ⑥
		R	NA	NA	NA	NA	40	18-22 ⑥
	XJ8	F	NA	NA	NA	NA	104 ⑤	18-22 ⑥
		R	NA	NA	NA	NA	40	18-22 ⑥
	XJR	F	NA	NA	NA	NA	104 ⑤	18-22 ⑥
		R	NA	NA	NA	NA	40	18-22 ⑥
	XK8	F	NA	NA	NA	NA	104 ⑤	18-22 ⑥
		R	NA	NA	NA	NA	40	18-22 ⑥
2003	S Type	F	①	⑧	NA	0.080	76	25 ② ⑲
		R	0.79	0.72	NA	0.080	76	25 ③ ㉑
	S Type R	F	①	⑧	NA	0.080	76	25 ② ⑲
		R	0.79	0.72	NA	0.080	76	25 ③ ㉑
	X Type ④	F	NA	NA	NA	NA	98	22
		R	NA	NA	NA	NA	52	24
	XJ Super V8 ⑦	F	NA	NA	NA	NA	76	25 ② ⑳
		R	NA	NA	NA	NA	76	25 ③ ㉒
	XJ Vanden Plas ⑧	F	NA	NA	NA	NA	76	25 ② ⑳
		R	NA	NA	NA	NA	76	25 ③ ㉒
	XJ Sport Sedan ⑧	F	NA	NA	NA	NA	76	25 ② ⑳
		R	NA	NA	NA	NA	76	25 ③ ㉒
	XJ8 ⑧	F	NA	NA	NA	NA	76	25 ② ⑳
		R	NA	NA	NA	NA	76	25 ③ ㉒
	XJR ⑧	F	NA	NA	NA	NA	76	25 ② ⑳
		R	NA	NA	NA	NA	76	25 ③ ㉒
	XK8	F	NA	NA	NA	NA	104 ⑤	18-22 ⑤
		R	NA	NA	NA	NA	40	18-22 ⑤
	XKR	F	NA	NA	NA	NA	104 ⑤	18-22 ⑤
		R	NA	NA	NA	NA	40	18-22 ⑤
2004	S Type 3.0L	F	①	⑧	NA	0.080	76	25 ② ⑲
		R	0.79	0.72	NA	0.080	76	25 ③ ㉑
	S Type 4.2L	F	①	⑧	NA	0.080	76	25 ② ⑲
		R	0.79	0.72	NA	0.080	76	25 ③ ㉑
	S Type R	F	①	⑧	NA	0.080	76	25 ② ⑲
		R	0.79	0.72	NA	0.080	76	25 ③ ㉑
	X Type 2.5L	F	NA	NA	NA	NA	98	22
		R	NA	NA	NA	NA	52	24
	X Type 3.0L	F	NA	NA	NA	NA	98	22
		R	NA	NA	NA	NA	52	24
	X Type Sport	F	NA	NA	NA	NA	98	22
		R	NA	NA	NA	NA	52	24

06041_JAGU_C0015

BRAKE SPECIFICATIONS

All measurements in inches unless noted

			Brake Disc				Brake Caliper	
Year	Model		Original Thickness	Machine Thickness	Maximum Runout	Minimum Lining Thickness	Bracket Bolts (ft. lbs.)	Mounting Bolts (ft. lbs.)
2004 Cont.	XJ Vanden Plas ⑨	F	NA	NA	NA	NA	76	25 ② ⑳
		R	NA	NA	NA	NA	76	25 ③ ㉒
	XJ8	F	NA	NA	NA	NA	76	25 ② ⑳
		R	NA	NA	NA	NA	76	25 ③ ㉒
	XJR	F	NA	NA	NA	NA	76	25 ② ⑳
		R	NA	NA	NA	NA	76	25 ③ ㉒
	XKR	F	NA	NA	NA	NA	104 ⑤	18-22 ⑥
		R	NA	NA	NA	NA	40	18-22 ⑥
	XK8	F	NA	NA	NA	NA	104 ⑤	18-22 ⑥
		R	NA	NA	NA	NA	40	18-22 ⑥
2005	S Type 3.0L	F	①	⑧	NA	0.080	76	25 ② ⑲
		R	⑨	⑩	NA	0.080	76	25 ③ ㉑
	S Type 4.2L	F	①	⑧	NA	0.080	76	25 ② ⑲
		R	⑨	⑩	NA	0.080	76	25 ③ ㉑
	S Type R	F	①	⑧	NA	0.080	76	25 ② ⑲
		R	⑨	⑩	NA	0.080	76	25 ③ ㉑
	X Type 2.5L	F	NA	NA	NA	NA	98	22
		R	NA	NA	NA	NA	52	24
	X Type 3.0L	F	NA	NA	NA	NA	98	22
		R	NA	NA	NA	NA	52	24
	X Type Sport	F	NA	NA	NA	NA	98	22
		R	NA	NA	NA	NA	52	24
	XJ Vanden Plas ⑨	F	NA	NA	NA	NA	76	25 ② ⑳
		R	NA	NA	NA	NA	76	25 ③ ㉒
	XJ	F	⑦	⑪	NA	0.080	76	25 ② ⑳
		R	0.79	0.72	NA	0.080	76	25 ③ ㉒
	XJR	F	⑦	⑪	NA	0.080	76	25 ② ⑳
		R	0.79	0.72	NA	0.080	76	25 ③ ㉒
	XKR	F	NA	NA	NA	NA	104 ⑤	18-22 ⑥
		R	NA	NA	NA	NA	40	18-22 ⑥
	XK8	F	NA	NA	NA	NA	104 ⑤	18-22 ⑥
		R	NA	NA	NA	NA	40	18-22 ⑥
2006	S Type	F	①	⑧	NA	0.080	76	25 ② ⑲
		R	⑨	⑩	NA	0.080	76	25 ③
	S Type R	F	①	⑧	NA	0.080	76	25 ② ⑲
		R	⑨	⑩	NA	0.080	76	25 ③
	X Type 2.5L	F	0.95	0.87	NA	0.080	98	22
		R	0.47	0.40	NA	0.080	52	24
	X Type 3.0L	F	0.95	0.87	NA	0.080	98	22
		R	0.47	0.40	NA	0.080	52	24
	XJ	F	⑦	⑪	NA	0.080	76	25 ② ⑳
		R	0.79	0.72	NA	0.080	76	25 ③ ㉒
	XJR	F	⑦	⑪	NA	0.080	76	25 ② ⑳
		R	0.79	0.72	NA	0.080	76	25 ③ ㉒
	XK	F	⑫	⑬	NA	0.080	140 ⑯	25-30 ⑰
		R	⑭	⑮	NA	0.080	60	25-30 ⑱
	XKR	F	⑫	⑬	NA	0.080	140 ⑯	25-30 ⑰
		R	⑭	⑮	NA	0.080	60	25-30 ⑱

06041_JAGU_C0016

BRAKE SPECIFICATIONS
Footnotes

NA: Information not available

NS: Not Specified by Manufacturer

F: Front

R: Rear

SC: Supercharged

① Standard: 1.20 in.; High performance: 1.28 in.

② With Brembo brakes: 156 ft. lbs.

③ With Brembo brakes: 51 ft. lbs.

④ Through VIN C59260

⑤ With Brembo brakes: 133 ft. lbs.

⑥ With Brembo brakes: Do not remove the caliper-to-anchor bolt

⑦ Standard: 1.18 in.; High performance: 1.28 in.

⑧ Std.: 1.12; high performance: 1.2 in.

⑨ Std.: 0.8; high performace: 0.6 in.

⑩ Std.: 0.73; high performance: 0.52 in.

⑪ Standard: 1.14 in.; high performance: 1.18 in.

⑫ Standard 1.12 in.; Brembo option: 1.26 in.

⑬ Standard 1.10 in.; Brembo option 1.18 in.

⑭ Standard: 0.79 0n.; Brembo option 1.10 in.

⑮ Standard 0.73 in.; Brembo option 1.02 in.

⑯ Standard brakes

⑰ Brembo option: 180 ft. lbs.

⑱ Brembo option: 60 ft. lbs.

⑲ VIN N52048-N99999: Standard brakes, 21 ft. lbs.; Brembo brakes, 43 ft. lbs.

⑳ VIN G45704-G99999 Standard brakes: 21 ft. lbs.; Brembo brakes, 43 ft. lbs.

㉑ 2002.5-06 VIN N52048-N99999 with or without Brembo brakes: 21 ft. lbs.

㉒ 2003.5-6 VIN G45708-G99999 with or without Brembo Brakes: 26 ft. lbs.

06041_JAGU_C0017

SCHEDULED MAINTENANCE INTERVALS
Jaguar S-Type 3.0L and 4.0L Engines, 2002 through VIN M44997

TO BE SERVICED	TYPE OF SERVICE	VEHICLE MILEAGE INTERVAL (x1000)												
		10	20	30	40	50	60	70	80	90	100	110	120	130
Engine oil & filter & drain plug washer	R	✓	✓	✓	✓	✓	✓	✓	✓	✓	✓	✓	✓	✓
Engine coolant	R	every 5 years or 150,000 miles												
Cooling system	S/I	✓	✓	✓	✓	✓	✓	✓	✓	✓	✓	✓	✓	✓
Windshield washer fluid	S/I	✓	✓	✓	✓	✓	✓	✓	✓	✓	✓	✓	✓	✓
Windshield wiper blade rubber insert	R	✓	✓	✓	✓	✓	✓	✓	✓	✓	✓	✓	✓	✓
Windshield wiper blade assembly	R				✓				✓				✓	
Passenger compartment air filter	R	✓	✓	✓	✓	✓	✓	✓	✓	✓	✓	✓	✓	✓
Battery electrolyte level	S/I	✓	✓	✓	✓	✓	✓	✓	✓	✓	✓	✓	✓	✓
Brake system (brake pads & fluid level)	S/I	✓	✓	✓	✓	✓	✓	✓	✓	✓	✓	✓	✓	✓
Engine (check for leaks)	S/I	✓	✓	✓	✓	✓	✓	✓	✓	✓	✓	✓	✓	✓
Fuel filter	R						✓						✓	
Air cleaner element ①	R			✓			✓			✓			✓	
Spark Plugs ①	R							✓						
Power steering fluid level	S/I	✓	✓	✓	✓	✓	✓	✓	✓	✓	✓	✓	✓	✓
Brake Fluid	R	every 2 years												
Accessory drive belt	S/I	✓	✓	✓	✓	✓	✓	✓	✓	✓	✓	✓	✓	✓
Accessory drive belt	R	every 100,000 miles												
Tire pressure and condition	S/I	✓	✓	✓	✓	✓	✓	✓	✓	✓	✓	✓	✓	✓

R: Replace S/I: Service or Inspect

① Naturally aspirated engines

06041_JAGU_C0018

SCHEDULED MAINTENANCE INTERVALS

Jaguar S-Type 3.0L and 4.2L Engines, 2002.5 onward from VIN M44998

TO BE SERVICED	TYPE OF SERVICE	VEHICLE MILEAGE INTERVAL (x1000)												
		10	20	30	40	50	60	70	80	90	100	110	120	130
Engine oil & filter & drain plug washer	R	✓	✓	✓	✓	✓	✓	✓	✓	✓	✓	✓	✓	✓
Engine coolant	R	every 5 years or 150,000 miles												
Cooling system	S/I	✓	✓	✓	✓	✓	✓	✓	✓	✓	✓	✓	✓	✓
Windshield washer fluid	S/I	✓	✓	✓	✓	✓	✓	✓	✓	✓	✓	✓	✓	✓
Windshield wiper blade rubber insert	R	✓	✓	✓	✓	✓	✓	✓	✓	✓	✓	✓	✓	✓
Windshield wiper blade assembly	R				✓				✓				✓	
Passenger compartment air filter	R		✓		✓		✓		✓		✓		✓	
Battery electrolyte level	S/I	✓	✓	✓	✓	✓	✓	✓	✓	✓	✓	✓	✓	✓
Brake system (brake pads & fluid level)	S/I	✓	✓	✓	✓	✓	✓	✓	✓	✓	✓	✓	✓	✓
Engine (check for leaks)	S/I	✓	✓	✓	✓	✓	✓	✓	✓	✓	✓	✓	✓	✓
Fuel filter	R						✓						✓	
Air cleaner element	R				✓				✓				✓	
Spark Plugs V6	R							✓						
Spark Plugs V8	R										✓			
Power steering fluid level	S/I	✓	✓	✓	✓	✓	✓	✓	✓	✓	✓	✓	✓	✓
Brake Fluid	R	every 2 years												
Accessory drive belt	S/I	✓	✓	✓	✓	✓	✓	✓	✓	✓	✓	✓	✓	✓
Accessory drive belt	R	every 150,000 miles												
Tire pressure and condition	S/I	✓	✓	✓	✓	✓	✓	✓	✓	✓	✓	✓	✓	✓

R: Replace S/I: Service or Inspect

06041_JAGU_C0019

SCHEDULED MAINTENANCE INTERVALS
Jaguar X-Type 2.5L and 3.0L Engines

TO BE SERVICED	TYPE OF SERVICE	VEHICLE MILEAGE INTERVAL (x1000)												
		10	20	30	40	50	60	70	80	90	100	110	120	130
Engine oil & filter & drain plug washer	R	✓	✓	✓	✓	✓	✓	✓	✓	✓	✓	✓	✓	✓
Engine coolant	R	every 5 years or 150,000 miles												
Cooling system	S/I	✓	✓	✓	✓	✓	✓	✓	✓	✓	✓	✓	✓	✓
Windshield washer fluid	S/I	✓	✓	✓	✓	✓	✓	✓	✓	✓	✓	✓	✓	✓
Windshield wiper blades	S/I	✓	✓	✓	✓	✓	✓	✓	✓	✓	✓	✓	✓	✓
Passenger compartment air filter	R		✓		✓		✓		✓		✓		✓	
Battery electrolyte level	S/I	✓	✓	✓	✓	✓	✓	✓	✓	✓	✓	✓	✓	✓
Brake system (brake pads & fluid level)	S/I	✓	✓	✓	✓	✓	✓	✓	✓	✓	✓	✓	✓	✓
Engine (check for leaks)	S/I	✓	✓	✓	✓	✓	✓	✓	✓	✓	✓	✓	✓	✓
Fuel filter	R							✓						
Air cleaner element ①	R			✓			✓			✓			✓	
Spark Plugs ①	R							✓						
Power steering fluid level	S/I	✓	✓	✓	✓	✓	✓	✓	✓	✓	✓	✓	✓	✓
Brake Fluid	R	every 2 years												
Accessory drive belts	S/I	✓	✓	✓	✓	✓	✓	✓	✓	✓	✓	✓	✓	✓
Accessory drive belts	R	every 100,000 miles												
Tire pressure and condition	S/I	✓	✓	✓	✓	✓	✓	✓	✓	✓	✓	✓	✓	✓

R: Replace S/I: Service or Inspect
① Naturally aspirated engines

06041_JAGU_C0020

SCHEDULED MAINTENANCE INTERVALS
Jaguar XJ Range 2002

		VEHICLE MILEAGE INTERVAL (x1000)												
TO BE SERVICED	TYPE OF SERVICE	10	20	30	40	50	60	70	80	90	100	110	120	130
Axle shaft U-joints	L	✓	✓	✓	✓	✓	✓	✓	✓	✓	✓	✓	✓	✓
Engine oil & filter & drain plug washer	R	✓	✓	✓	✓	✓	✓	✓	✓	✓	✓	✓	✓	✓
Engine coolant	R	every 5 years or 150,000 miles												
Cooling system	S/I	✓	✓	✓	✓	✓	✓	✓	✓	✓	✓	✓	✓	✓
Windshield washer fluid	S/I	✓	✓	✓	✓	✓	✓	✓	✓	✓	✓	✓	✓	✓
Windshield wiper blade rubber insert	R	✓	✓	✓	✓	✓	✓	✓	✓	✓	✓	✓	✓	✓
Windshield wiper blade assembly	R				✓				✓				✓	
Passenger compartment air filter	R		✓		✓		✓		✓		✓			
Battery electrolyte level	S/I	✓	✓	✓	✓	✓	✓	✓	✓	✓	✓	✓	✓	✓
Brake system (brake pads & fluid level)	S/I	✓	✓	✓	✓	✓	✓	✓	✓	✓	✓	✓	✓	✓
Engine (check for leaks)	S/I	✓	✓	✓	✓	✓	✓	✓	✓	✓	✓	✓	✓	✓
Fuel filter	R						✓						✓	
Air cleaner element ①	R			✓			✓			✓			✓	
Air cleaner element ②			✓		✓		✓		✓		✓		✓	
Spark Plugs ①	R							✓						
Spark Plugs ②	R			✓			✓			✓			✓	
Power steering fluid level	S/I	✓	✓	✓	✓	✓	✓	✓	✓	✓	✓	✓	✓	✓
Brake Fluid	R	every 2 years												
Accessory drive belt	S/I	✓	✓	✓	✓	✓	✓	✓	✓	✓	✓	✓	✓	✓
Accessory drive belt	R	every 100,000 miles												
Supercharger belt	R				✓				✓				✓	
Tire pressure and condition	S/I	✓	✓	✓	✓	✓	✓	✓	✓	✓	✓	✓	✓	✓

R: Replace S/I: Service or Inspect L: Lube

① Naturally aspirated engines

② Supercharged engines

06041_JAGU_C0021

SCHEDULED MAINTENANCE INTERVALS

Jaguar XJ Range 2003 Onward

		VEHICLE MILEAGE INTERVAL (x1000)												
TO BE SERVICED	TYPE OF SERVICE	10	20	30	40	50	60	70	80	90	100	110	120	130
Engine oil & filter & drain plug washer	R	✓	✓	✓	✓	✓	✓	✓	✓	✓	✓	✓	✓	✓
Engine coolant	R	every 5 years or 150,000 miles												
Cooling system	S/I	✓	✓	✓	✓	✓	✓	✓	✓	✓	✓	✓	✓	✓
Windshield washer fluid	S/I	✓	✓	✓	✓	✓	✓	✓	✓	✓	✓	✓	✓	✓
Windshield wiper blade rubber insert	I	✓	✓	✓	✓	✓	✓	✓	✓	✓	✓	✓	✓	✓
Exhaust system	S/I	✓	✓	✓	✓	✓	✓	✓	✓	✓	✓	✓	✓	✓
Radiator & condenser (debris)	Clean		✓		✓		✓		✓		✓		✓	
All hydraulic lines	S/I	✓	✓	✓	✓	✓	✓	✓	✓	✓	✓	✓	✓	✓
Headlamp alignment	S/I		✓		✓		✓		✓		✓		✓	
Passenger compartment air filter	R		✓		✓		✓		✓		✓		✓	
Windshield ①			✓		✓		✓		✓		✓		✓	
Shock absorbers	S/I	✓	✓	✓	✓	✓	✓	✓	✓	✓	✓	✓	✓	✓
Battery electrolyte level	S/I	✓	✓	✓	✓	✓	✓	✓	✓	✓	✓	✓	✓	✓
Steering rack boots	I	✓	✓	✓	✓	✓	✓	✓	✓	✓	✓	✓	✓	✓
Brake system (brake pads & fluid level)	S/I	✓	✓	✓	✓	✓	✓	✓	✓	✓	✓	✓	✓	✓
Engine (check for leaks)	S/I	✓	✓	✓	✓	✓	✓	✓	✓	✓	✓	✓	✓	✓
Transmission check for leaks)	S/I	✓	✓	✓	✓	✓	✓	✓	✓	✓	✓	✓	✓	✓
Final drive	S/I	✓	✓	✓	✓	✓	✓	✓	✓	✓	✓	✓	✓	✓
Fuel filter	R						✓						✓	
Fuel system (check for leaks)	S/I	✓	✓	✓	✓	✓	✓	✓	✓	✓	✓	✓	✓	✓
Air cleaner element	R				✓				✓				✓	
Spark Plugs	R										✓			
Power steering system	I	✓	✓	✓	✓	✓	✓	✓	✓	✓	✓	✓	✓	✓
Door check arm	L		✓		✓		✓		✓		✓		✓	
Power steering fluid level	S/I	✓	✓	✓	✓	✓	✓	✓	✓	✓	✓	✓	✓	✓
Brake Fluid	R	every 2 years												
Accessory drive belt	S/I	✓	✓	✓	✓	✓	✓	✓	✓	✓	✓	✓	✓	✓
Accessory drive belt	R	every 150,000 miles												
Supercharger belt	R	every 100,000 miles												
Tire pressure and condition	S/I	✓	✓	✓	✓	✓	✓	✓	✓	✓	✓	✓	✓	✓

R: Replace S/I: Service or Inspect

① Apply Jaguar Screen Clean paste to the exterior of the windshield

06041_JAGU_C0022

SCHEDULED MAINTENANCE INTERVALS

Jaguar XK Range 4.0L Engine

TO BE SERVICED	TYPE OF SERVICE	VEHICLE MILEAGE INTERVAL (x1000)												
		10	20	30	40	50	60	70	80	90	100	110	120	130
Engine oil & filter & drain plug washer	R	✓	✓	✓	✓	✓	✓	✓	✓	✓	✓	✓	✓	✓
Engine coolant	R	every 5 years or 150,000 miles												
Axle shaft U-joints	L	✓	✓	✓	✓	✓	✓	✓	✓	✓	✓	✓	✓	✓
Cooling system	S/I	✓	✓	✓	✓	✓	✓	✓	✓	✓	✓	✓	✓	✓
Windshield washer fluid	S/I	✓	✓	✓	✓	✓	✓	✓	✓	✓	✓	✓	✓	✓
Windshield wiper blade rubber insert	R	✓	✓	✓	✓	✓	✓	✓	✓	✓	✓	✓	✓	✓
Windshield wiper blade assembly	R				✓				✓				✓	
Passenger compartment air filter	R		✓		✓		✓		✓		✓		✓	
Battery electrolyte level	S/I	✓	✓	✓	✓	✓	✓	✓	✓	✓	✓	✓	✓	✓
Brake system (brake pads & fluid level)	S/I	✓	✓	✓	✓	✓	✓	✓	✓	✓	✓	✓	✓	✓
Engine (check for leaks)	S/I	✓	✓	✓	✓	✓	✓	✓	✓	✓	✓	✓	✓	✓
Fuel filter	R						✓						✓	
Air cleaner element ①	R			✓			✓			✓			✓	
Air cleaner element ②	R		✓		✓		✓		✓		✓		✓	
Spark Plugs ①	R							✓						
Spark Plugs ②	R			✓			✓			✓			✓	
Power steering fluid level	S/I	✓	✓	✓	✓	✓	✓	✓	✓	✓	✓	✓	✓	✓
Brake Fluid	R	every 2 years												
Accessory drive belt	S/I	✓	✓	✓	✓	✓	✓	✓	✓	✓	✓	✓	✓	✓
Accessory drive belt	R	every 100,000 miles												
Supercharger belt	R				✓				✓				✓	
Tire pressure and condition	S/I	✓	✓	✓	✓	✓	✓	✓	✓	✓	✓	✓	✓	✓

R: Replace S/I: Service or Inspect L: Lube

① Normally aspirated engine

② Supercharged engines

06041_JAGU_C0023

SCHEDULED MAINTENANCE INTERVALS
Jaguar XK Range 4.2L Engine

TO BE SERVICED	TYPE OF SERVICE	VEHICLE MILEAGE INTERVAL (x1000)												
		10	20	30	40	50	60	70	80	90	100	110	120	130
Axle shaft U-joints	L	✓	✓	✓	✓	✓	✓	✓	✓	✓	✓	✓	✓	✓
Engine oil & filter & drain plug washer	R	✓	✓	✓	✓	✓	✓	✓	✓	✓	✓	✓	✓	✓
Engine coolant	R	every 5 years or 150,000 miles												
Cooling system	S/I	✓	✓	✓	✓	✓	✓	✓	✓	✓	✓	✓	✓	✓
Windshield washer fluid	S/I	✓	✓	✓	✓	✓	✓	✓	✓	✓	✓	✓	✓	✓
Windshield wiper blade rubber insert	R	✓	✓	✓	✓	✓	✓	✓	✓	✓	✓	✓	✓	✓
Windshield wiper blade assembly	R				✓				✓				✓	
Passenger compartment air filter	R		✓		✓		✓		✓		✓		✓	
Battery electrolyte level	S/I	✓	✓	✓	✓	✓	✓	✓	✓	✓	✓	✓	✓	✓
Brake system (brake pads & fluid level)	S/I	✓	✓	✓	✓	✓	✓	✓	✓	✓	✓	✓	✓	✓
Engine (check for leaks)	S/I	✓	✓	✓	✓	✓	✓	✓	✓	✓	✓	✓	✓	✓
Fuel filter	R						✓						✓	
Air cleaner element ①	R			✓			✓			✓			✓	
Air cleaner element ②	R		✓		✓		✓		✓		✓		✓	
Spark Plugs	R										✓			
Power steering fluid level	S/I	✓	✓	✓	✓	✓	✓	✓	✓	✓	✓	✓	✓	✓
Brake Fluid	R	every 2 years												
Accessory drive belt	S/I	✓	✓	✓	✓	✓	✓	✓	✓	✓	✓	✓	✓	✓
Accessory drive belt	R	every 100,000 miles												
Supercharger belt	R				✓				✓				✓	
Tire pressure and condition	S/I	✓	✓	✓	✓	✓	✓	✓	✓	✓	✓	✓	✓	✓

R: Replace S/I: Service or Inspect L: Lube

① Naturally aspirated engines

② Supercharged engines

06041_JAGU_C0024

ENGINE REPAIR

➡Disconnecting the negative battery cable on some vehicles may interfere with the functions of the on board computer system. The computer may undergo a relearning process once the negative battery cable is reconnected.

Distributor

The vehicles covered in this section are equipped with a Distributorless Ignition System (DIS).

Alternator

REMOVAL & INSTALLATION

All Engines Except 4.2L

1. Before servicing the vehicle, refer to the Precautions Section.
2. Remove or disconnect the following:
   - Negative battery cable
   - Air intake assembly
   - Cooling fan assembly
   - Supercharger drive belt, if equipped
   - Accessory drive belt
   - Alternator air intake ducting
   - Alternator wiring connectors
   - Supercharger water pump, if equipped
   - Alternator

To install:

3. Before servicing the vehicle, refer to the Precautions Section.
4. Install or connect the following:
   - Alternator. Tighten the mounting bolts to 28–35 ft. lbs. (38–48 Nm).
   - Supercharger water pump, if equipped
   - Alternator wiring connectors
   - Alternator air intake ducting
   - Accessory drive belt
   - Supercharger drive belt, if equipped
   - Cooling fan assembly
   - Air intake assembly
   - Negative battery cable

4.2L Engines

1. Before servicing the vehicle, refer to the Precautions Section.
2. Remove or disconnect the following:
   - Negative battery cable
   - Accessory drive belt
   - Alternator cooling duct
   - Splash shield
   - Right side lower engine mount bolt
3. Using suitable lifting device, raise engine slightly.
4. Remove the steering gear shaft pinch bolt.
5. Remove the steering gear mounting bolts and suspend gear out of the way.
6. Remove the engine mounting bracket assembly.
7. Remove the alternator wiring connectors and alternator.

To install:

8. Before servicing the vehicle, refer to the Precautions Section.
9. Install or connect the following:
   - Alternator. Tighten the upper bolt to 15 ft. lbs. (21 Nm) and the lower bolt to 30 ft. lbs. (40 Nm).
   - Alternator wiring connectors
   - Engine mounting bracket. Tighten bolts to 18 ft. lbs. (25 Nm).
   - Steering gear. Tighten bolts to 74 ft. lbs. (100 Nm).
   - Steering gear pinch bolt. Tighten bolts to 26 ft. lbs. (35 Nm).
   - Right side lower engine mount bolt. Tighten bolt to 46 ft. lbs. (63 Nm).
   - Splash shield
   - Alternator cooling duct
   - Accessory drive belt
   - Negative battery cable

Ignition Timing

ADJUSTMENT

The ignition timing is not adjustable. It is controlled by the PCM.

Engine Assembly

REMOVAL & INSTALLATION

S-Type 2.5L and 3.0L

➡The engine is removed from the vehicle after the transmission is removed.

1. Before servicing the vehicle, refer to the Precautions Section.
2. Drain the cooling system.
3. Recover the A/C refrigerant.
4. Remove or disconnect the following:
   - Negative battery cable
   - Transmission
   - Hood
   - Air cleaner
   - Pollen air filter
   - Engine compartment panel
   - Engine cover
   - Passenger side wheel and tire
   - Engine harness ground
   - Powertrain control module electrical connector
   - Engine harness electrical connector
   - Charcoal canister vacuum hose
   - Fuel supply manifold coupling
   - Cooling module
   - Brake booster vacuum hose
   - Power steering lines and cap openings
   - Valve cover wiring harnesses
   - Ignition coil connectors
   - Dipstick tube bolt
   - One exhaust manifold retaining nut on each side
5. Install engine lifting bracket and support bars 303-661 to exhaust manifold.
6. Install engine lifting bracket 303-021 to support bars.
7. Raise and support vehicle.
8. Rotate the accessory drive belt tensioner counter-clockwise using 3/8 inch square drive bar.
9. Remove the accessory drive belt.
10. Remove or disconnect the following:
    - Alternator wiring
    - Air conditioning compressor and wire aside
    - Power steering hose
    - Lower engine mount bolts
    - Transmission cooler line bracket and position lines aside
11. Lower the vehicle and remove lifting bracket.
12. Attach suitable engine hoist to support bars and carefully remove engine from vehicle.

To install:

13. With the aid of an assistant, guide the engine into the vehicle.
14. Install or connect the following:
    - Transmission cooler line bracket
    - Lower engine mount bolts.
    - Power steering hose
    - Air conditioning compressor
    - Alternator wiring
    - Accessory drive belt.
    - Exhaust manifold retaining nut on each side. Tighten to 18 ft. lbs. (25 Nm).
    - Dipstick tube bolt
    - Ignition coil connectors
    - Valve cover wiring harnesses
    - Power steering lines and cap openings
    - Brake booster vacuum hose
    - Cooling module
    - Fuel supply manifold coupling
    - Charcoal canister vacuum hose

- Engine harness electrical connector
- Powertrain control module electrical connector
- Engine harness ground
- Passenger side wheel and tire and tighten to 95 ft. lbs. (128 Nm)
- Engine cover
- Engine compartment panel
- Pollen air filter
- Air cleaner
- Hood, tighten bolts to 18 ft. lbs. (25 Nm).
- Transmission
- Negative battery cable

15. Charge the A/C refrigerant.
16. Refill the cooling system.

XJR 4.2L

➡The engine is removed from the vehicle after the transmission is removed.

1. Before servicing the vehicle, refer to the Precautions Section.
2. Drain the cooling system.
3. Recover the A/C refrigerant.
4. Remove or disconnect the following:
   - Negative battery cable
   - Transmission
   - Engine mount retaining nut
   - On models with supercharger, remove oil cooler lines
   - Starter motor positive cables
   - Air conditioning compressor lines
   - Power steering lines
   - Hood
   - Radiator
   - Engine compartment braces
   - Pollen air filter
   - Engine compartment cover
   - Engine harness electrical connector
   - Powertrain control module electrical connector
   - Power steering fluid reservoir
   - Fuel line retaining bracket
   - On models with supercharger, remove air intake tube
   - Brake booster vacuum pipe
   - Fuel purge line, discard retaining clip
5. Attach suitable engine hoist and carefully remove engine from vehicle.

To install:

6. With the aid of an assistant, guide the engine into the vehicle.
7. Install new O rings on all fuel, oil and hydraulic lines.
8. Install or connect the following:
   - Fuel purge line
   - Brake booster vacuum pipe
   - On models with supercharger, air intake tube
   - Fuel line retaining bracket
   - Power steering fluid reservoir, tighten bolt to 18 ft. lbs. (25 Nm)
   - Powertrain control module electrical connector
   - Engine harness electrical connector
   - Engine compartment cover
   - Pollen air filter
   - Engine compartment braces
   - Radiator
   - Hood, tighten bolts to 18 ft. lbs. (25 Nm).
   - Power steering lines
   - Air conditioning compressor lines
   - Starter motor positive cables
   - On models with supercharger, oil cooler lines
   - Engine mount retaining nut
   - Transmission
   - Negative battery cable
9. Charge the A/C system.
10. Fill engine with coolant.

XJR and XJ8 4.0L

➡The engine is removed from the vehicle with the transmission as an assembly.

1. Before servicing the vehicle, refer to the Precautions Section.
2. Drain the cooling system.
3. Recover the A/C refrigerant.
4. Remove or disconnect the following:
   - Negative battery cable
   - Hood
   - Engine covers
   - Mass Air Flow (MAF) meter
   - Air cleaner assembly
   - Center trim panel
   - Coolant recovery tank
   - Cooling fan assembly
   - Radiator hoses
   - Radiator
   - Oil pressure switch connector
   - Power steering hoses
   - Heated Oxygen (HO_2S) sensor connectors
   - Exhaust front pipes
   - Driveshaft
   - Transmission ground cable
   - Transmission control wiring harness
   - Transmission cooler lines
   - Engine control wiring harness
   - Fuel lines
   - Accelerator cable
   - Heater hoses
   - Brake booster vacuum hose
   - Left and right motor mounts
   - Transmission mount
   - Transmission crossmember
5. With the aid of an assistant, lower the engine/transmission assembly from the vehicle.

To install:

6. With the aid of an assistant, raise the engine/transmission assembly into the vehicle.
7. Install or connect the following:
   - Transmission crossmember. Tighten the bolts to 16–21 ft. lbs. (22–28 Nm).
   - Transmission mount. Tighten the bolts to 22–30 ft. lbs. (30–40 Nm).
   - Left and right motor mounts
   - Brake booster vacuum hose
   - Heater hoses
   - Accelerator cable
   - Fuel lines
   - Engine control wiring harness
   - Transmission cooler lines
   - Transmission control wiring harness
   - Transmission ground cable
   - Driveshaft. Tighten the bolts to 55–65 ft. lbs. (75–88 Nm).
   - Exhaust front pipes
   - HO_2S sensor connectors
   - Power steering hoses
   - Oil pressure switch connector
   - Radiator
   - Radiator hoses
   - Cooling fan assembly
   - Coolant recovery tank
   - Center trim panel
   - Air cleaner assembly
   - MAF meter
   - Engine covers
   - Hood
   - Negative battery cable
8. Fill the cooling system.
9. Recharge the A/C system.
10. Start the vehicle and check for leaks.

XK8 (2002)

➡The engine and transmission are removed as an assembly.

1. Before servicing the vehicle, refer to the Precautions Section.
2. Place the gear selector in **N**.
3. Drain the cooling system.
4. Recover the A/C refrigerant.
5. Remove or disconnect the following:
   - Negative battery cable
   - Hood
   - Engine covers
   - Mass Air Flow (MAF) meter
   - Air cleaner assembly
   - Center trim panel
   - Left side enclosure panel
   - Coolant level sensor connector
   - Coolant recovery tank
   - Radiator hoses

- Cooling fans
- Transmission cooler lines
- Upper radiator retainer
- A/C condenser lines
- Fan switch connector
- Radiator
- Oil pressure sender connector
- Power steering hoses
- Heated Oxygen (HO_2S) sensor connectors
- Front muffler
- Catalytic converters
- Driveshaft
- Transmission ground cable
- Generator suppression module harness
- Starter motor wiring connectors
- P.I. harness connector
- Transmission rotary switch connector
- Fuel lines
- Purge valve pipes
- Accelerator cable
- Heater hoses
- Brake booster vacuum line
- Left and right motor mounts
- Transmission selector cable and arm
- Transmission mount
- Transmission crossmember

6. With the aid of an assistant, lower the engine/transmission assembly from the vehicle.

To install:

7. With the aid of an assistant, raise the engine/transmission assembly into the vehicle.
8. Install or connect the following:
    - Transmission crossmember
    - Transmission mount
    - Transmission selector cable and arm
    - Left and right motor mounts
    - Brake booster vacuum line
    - Heater hoses
    - Accelerator cable
    - Purge valve pipes
    - Fuel lines
    - Transmission rotary switch connector
    - P.I. harness connector
    - Starter motor wiring connectors
    - Generator suppression module harness
    - Transmission ground cable
    - Driveshaft
    - Catalytic converters
    - Front muffler
    - HO_2S sensor connectors
    - Power steering hoses
    - Oil pressure sender connector
    - Radiator
    - Fan switch connector
    - A/C condenser lines
    - Upper radiator retainer
    - Transmission cooler lines
    - Cooling fans
    - Radiator hoses
    - Coolant recovery tank
    - Coolant level sensor connector
    - Left side enclosure panel
    - Center trim panel
    - Air cleaner assembly
    - MAF meter
    - Engine covers
    - Hood
    - Negative battery cable
9. Fill the cooling system.
10. Recharge the A/C system.
11. Start the vehicle and check for leaks.

S-Type 4.0L

➡The engine is removed from the vehicle after the transmission is removed.

1. Before servicing the vehicle, refer to the Precautions Section.
2. Drain the cooling system.
3. Recover the A/C refrigerant.
4. Remove or disconnect the following:
    - Negative battery cable
    - Transmission
    - Radiator
    - Evaporative emission canister purge valve
    - Wiper arms
    - Cowl grille
    - Engine compartment braces
    - Power steering reservoir
    - Air cleaner
    - Power steering pump supply hose.
    - Fuel supply pipe
    - Fuel supply pipe retaining bolt
    - Engine harness support bracket
    - Engine harness electrical connectors
    - Powertrain control module
5. Install a lifting eye at rear of engine.
6. Raise and support the vehicle.
7. Disconnect air conditioning compressor tubes.
8. Remove the left and right engine mounting nuts.
9. Lower the vehicle.
10. Remove the engine using suitable hydraulic lift.

To install:

11. With the aid of an assistant, install the engine into the vehicle.
12. Install new O rings on all fuel, oil and hydraulic lines.
13. Install or connect the following:
14. Left and right engine mounting nuts. Tighten to 46 ft. lbs. (62 Nm).
15. Air conditioning compressor tubes.
16. Lifting eye at rear of engine.
    - Powertrain control module
    - Engine harness electrical connectors
    - Engine harness support bracket
    - Fuel supply pipe retaining bolt
    - Fuel supply pipe
    - Power steering pump supply hose
    - Air cleaner
    - Power steering reservoir
    - Engine compartment braces
    - Cowl grille
    - Wiper arms
    - Evaporative emission canister purge valve
    - Radiator
    - Transmission
    - Negative battery cable
17. Charge the A/C system.
18. Fill engine with coolant.

S-Type 4.2L

➡The engine is removed from the vehicle after the transmission is removed.

1. Before servicing the vehicle, refer to the Precautions Section.
2. Drain the cooling system.
3. Recover the A/C refrigerant.
4. Remove or disconnect the following:
    - Negative battery cable
    - Transmission
    - Engine mount nuts
    - On models with supercharger, disconnect oil cooler lines.
    - Starter wiring
    - Air conditioning compressor line
    - Remove the right hand wheel and tire assembly
    - Remove the right hand fender splash shield
    - Hood
    - Radiator
    - Pollen filter and filter housing
    - Engine compartment support brace
    - Engine compartment cover at firewall
    - Powertrain control module connector
    - Engine harness connector
    - Air cleaner
    - Power steering fluid reservoir
    - Supercharger air intake tube, if equipped
5. Install lifting eyes at front and rear of engine.
6. Disconnect brake vacuum booster hose.

7. Disconnect fuel lines and cap openings.

8. Remove the engine using suitable hydraulic lift.

To install:

9. With the aid of an assistant, install the engine into the vehicle.

10. Install new O rings on all fuel, oil and hydraulic lines.

11. With engine in place, remove the lifting eyes.

12. Install or connect the following:
- Fuel lines
- Brake vacuum booster hose
- Supercharger air intake tube, if equipped
- Power steering fluid reservoir
- Air cleaner
- Engine harness connector
- Powertrain control module connector
- Engine compartment cover at firewall
- Engine compartment support brace
- Pollen filter and filter housing
- Radiator
- Hood. Tighten bolts to 18 ft. lbs. (25 Nm).
- Right hand fender splash shield
- Right hand wheel and tire assembly
- Air conditioning compressor lines. Tighten to 15 ft. lbs. (20 Nm).
- Starter wiring
- On models with supercharger, oil cooler lines. Tighten connections to 9 ft. lbs. (12 Nm).
- Engine mount nuts. Tighten to 46 ft. lbs. (62 Nm).
- Transmission
- Negative battery cable

13. Charge the A/C system.

14. Fill engine with coolant.

X-Type All Engines

➡The engine and transmission are removed as an assembly. Powertrain Assembly JackHTJ1200-2 is required to remove the engine/transmission assembly.

1. Before servicing the vehicle, refer to the Precautions Section.

2. Drain the cooling system.

3. Recover the A/C refrigerant.

4. Remove or disconnect the following:
- Battery cables, battery and tray
- Steering column lower retaining bolt
- Front wheels and tires
- Wheel well splash shields
- Brake caliper and suspend aside
- ABS wheel speed sensor and suspend aside
- Cooling fan motor and shroud
- Halfshafts
- Front muffler from exhaust manifold
- A/C compressor tubes
- Transmission selector cables
- Transmission electrical connectors
- Air cleaner
- Air filter intake pipe
- Accelerator cable and position the accelerator lever to the fully open position.
- Throttle cable
- On manual transmissions, the slave clutch cylinder.
- Engine control module electrical connector
- Power steering hose and cap line
- Strut and spring top mounting bolts

5. Raise and support the vehicle.

6. Install powertrain assembly jack HTJ1200-2.

7. Position and adjust the powertrain jack rear height adjuster to the locating hole in the floor pan.

8. Position and adjust the powertrain jack front height adjuster.

9. Adjust the engine height adjusters until jack is tensioned.

10. Remove the engine support bar.

11. Remove the engine support bracket.

12. Remove the transmission mount.

13. Remove the front subframe reinforcement plate retaining bolts.

14. Remove the front sub frame retaining bolt and remove sub frame.

⁂ WARNING

Rotate the special tool height adjustment valve slowly. Failure to do so may result in personal injury.

15. Rotate the special tool height adjustment valve counter clockwise.

16. Carefully remove the engine and separate the transmission.

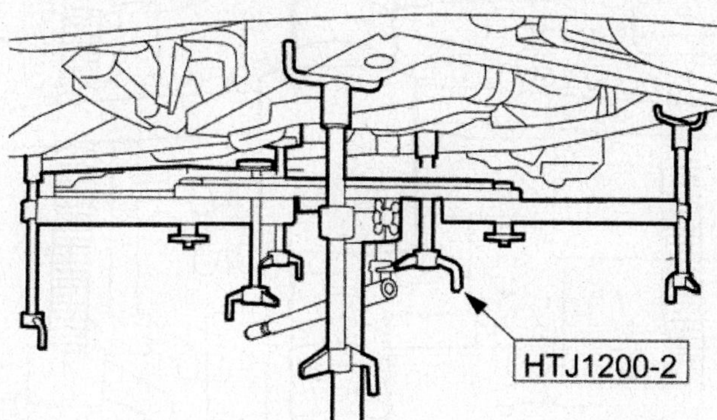

Installing powertrain assembly jack HTJ1200-2—X-Type all engines

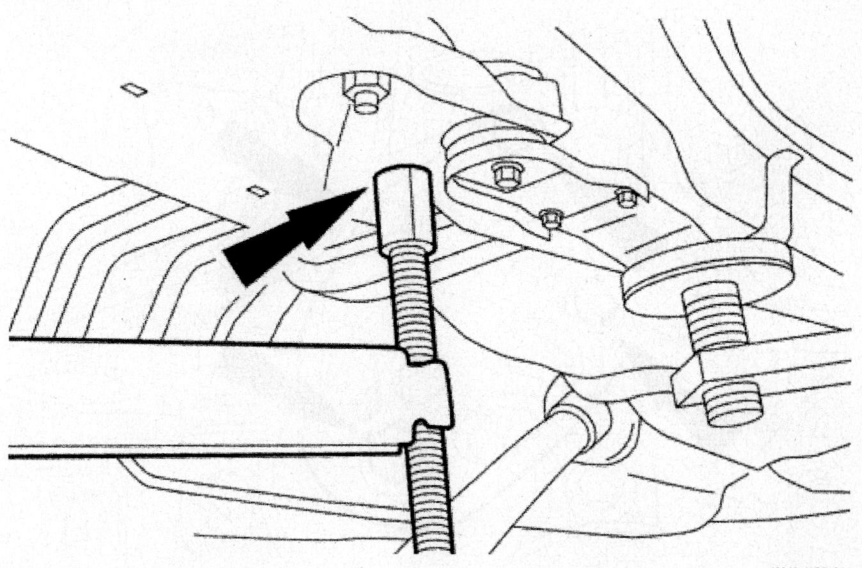

Positioning powertrain assembly jack rear height adjuster—X-Type all engines

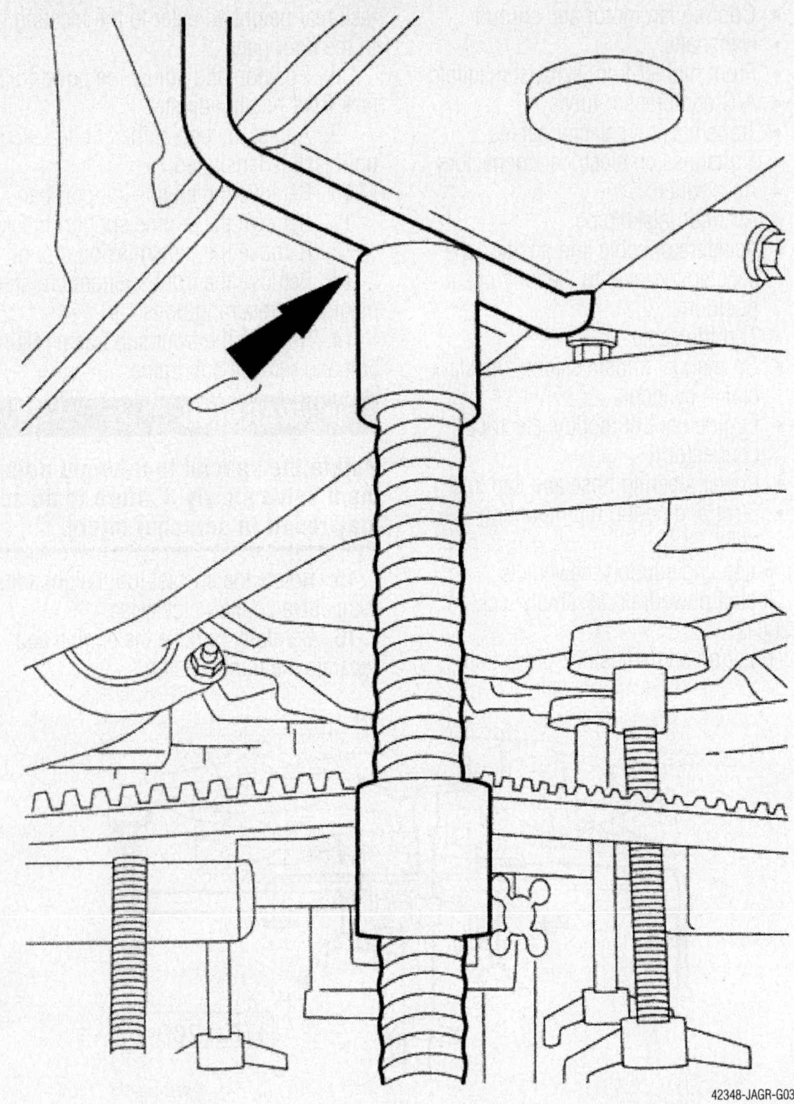

Positioning powertrain assembly jack front height adjuster—X-Type all engines

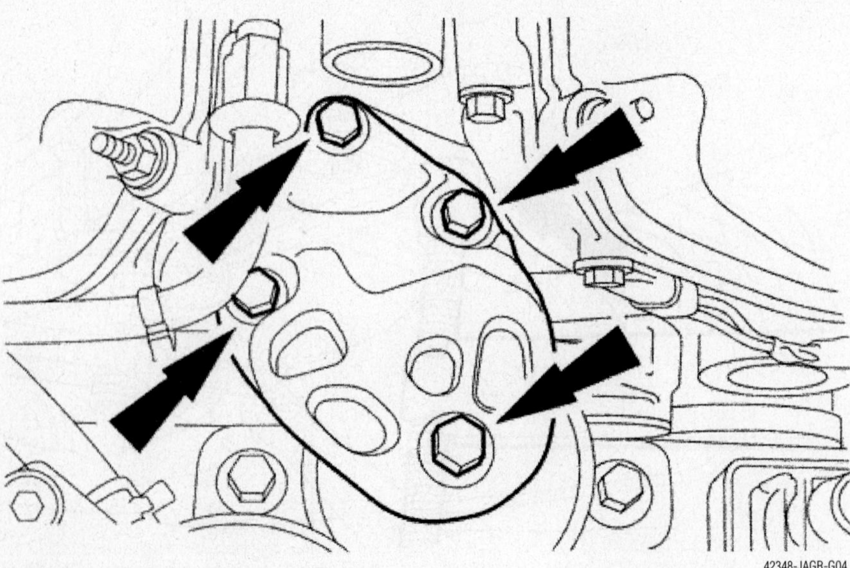

Removing engine support bracket—X-Type all engines

To install:

➡Install new O rings on all fuel, oil and hydraulic lines.

17. Connect the transmission to the engine.
18. Install powertrain assembly jack HTJ1200-2.
19. Rotate the special tool height adjustment valve clockwise and raise the engine/transmission assembly into the vehicle.
20. Install the front sub frame and the front subframe retaining bolt hand tight.
21. Install the front subframe reinforcement plate retaining bolts. Tighten to 16 ft. lbs. 22 (Nm).
22. Tighten front subframe retaining bolt to 105 ft. lbs. (142 Nm).
23. Install the transmission mount bolts. Tighten to 35 ft. lbs. (48 Nm).
24. Install the engine support bar bolts. Tighten bolts to 98 ft. lbs. (133 Nm).
25. Install the engine support bracket bolts. Tighten front bolt to 59 ft. lbs. (80 Nm), and rear bolts to 35 ft. lbs. (48 Nm).
26. Remove the special support tool.
27. Lower the vehicle.
28. Install or connect the following:
    - Strut and spring top mounting bolts. Tighten to 22 ft. lbs. (30 Nm).
    - Power steering hose
    - Engine control module electrical connector
    - On manual transmissions, the slave clutch cylinder.
    - Throttle cable
    - Accelerator cable.
    - Air filter intake pipe
    - Air cleaner
    - Transmission electrical connectors
    - Transmission selector cables
    - A/C compressor tubes
    - Front muffler to exhaust manifold. Tighten bolts to 41 ft. lbs. (55 Nm).

➡Driveshaft. Tighten bolts to 33 ft. lbs. (44 Nm).

- Cooling fan motor and shroud
- ABS wheel speed sensor
- Brake caliper. Tighten bolts to 98 ft. lbs. (133 Nm).
- Wheel well splash shields
- Front wheels and tires
- Steering column lower retaining bolt. Tighten to 18 ft. lbs. (25 Nm).
- Battery cables, battery and tray

29. Refill and bleed cooling system.
30. Bleed power steering system.
31. Recharge the A/C system.

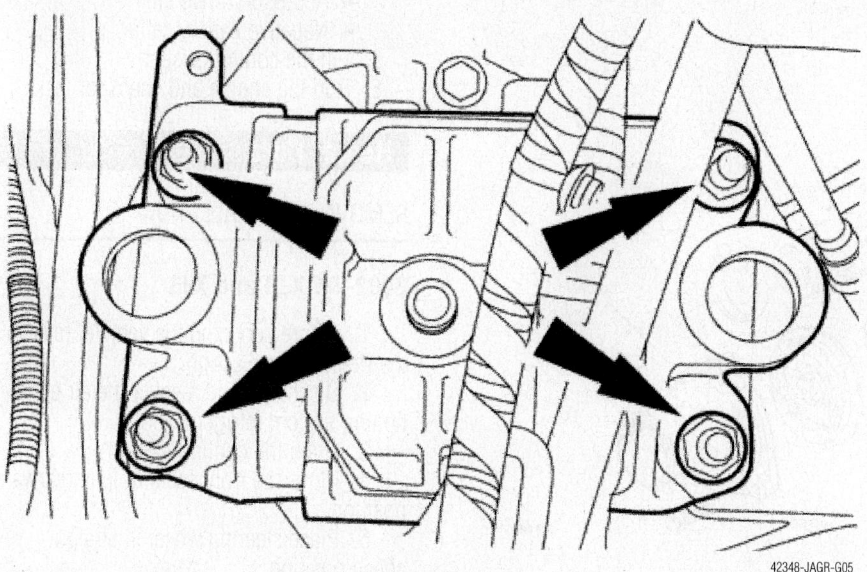

Removing transmission mount—X-Type all engines

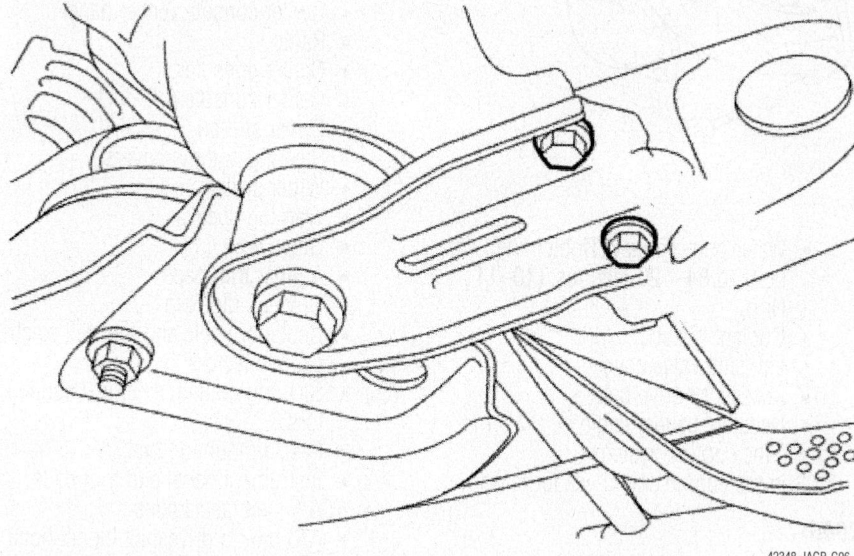

Removing front subframe reinforcement plate—X-Type all engines

Water Pump

REMOVAL & INSTALLATION

XJ Range and S-Type 4.0L and 4.2L Engines

1. Before servicing the vehicle, refer to the Precautions Section.
2. Drain the cooling system.
3. Remove or disconnect the following:
    - Negative battery cable
    - Supercharger drive belt, if equipped
    - Accessory drive belt
    - Thermostat housing, if necessary

Exploded view of the water pump mounting and related components

- Water pump pulley
- Water pump

To install:

4. Install or connect the following:
    - Water pump using a new gasket and O-ring. Tighten the bolts to 18 ft. lbs. (25 Nm) for the 4.0L; 72 inch lbs. (8 Nm), plus 90 degrees, for the 4.2L.
    - Water pump pulley. Tighten the bolts to 11 ft. lbs. (14 Nm) for the 4.0L; 96 inch lbs. (10 Nm) for the 4.2L.
    - Accessory drive belt
    - Negative battery cable
5. Fill the cooling system.
6. Start the engine and check for leaks.

XK Range

1. Before servicing the vehicle, refer to the Precautions Section.
2. Drain the cooling system.
3. Remove or disconnect the following:
    - Negative battery cable
    - Supercharger drive belt, if equipped
    - Accessory drive belt
    - Water pump pulley
    - Water pump

To install:

4. Install or connect the following:
    - Water pump using a new gasket and O-ring. Tighten the bolts to 96–108 inch lbs. (11–13 Nm) for 2002 models; 72 inch lbs. (8 Nm), plus 90 degrees, for 2003–06 models.
    - Water pump pulley. Tighten the bolts to 84–120 inch lbs. (10–14 Nm) for 2002 models; 96 inch lbs. (10 Nm) for 2003–06 models.
    - Accessory drive belt
    - Supercharger drive belt, if equipped
    - Negative battery cable
5. Fill the cooling system.
6. Run the engine and check for leaks

S-Type 3.0L

1. Before servicing the vehicle, refer to the Precautions Section.
2. Drain the cooling system.
3. Remove or disconnect the following:
    - Negative battery cable
    - Accessory drive belt
    - Upper shield
    - Throttle body
    - Coolant hoses
    - Upper radiator mounting bracket
    - Intake manifold support bracket
    - Water pump

To install:

4. Install or connect the following:
    - Water pump using a new gasket

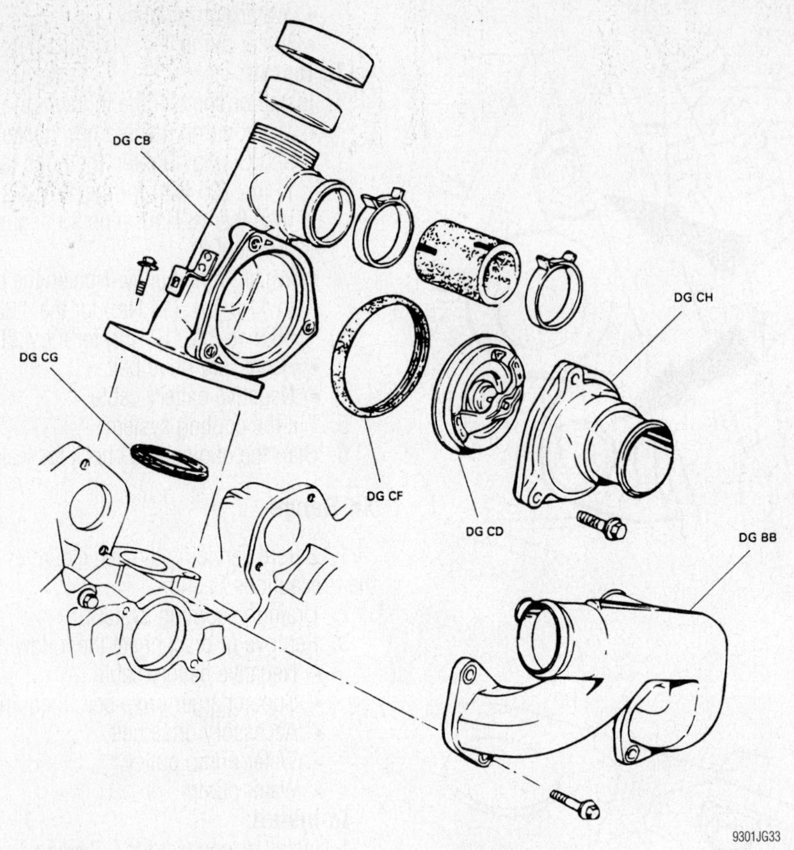

Water rail and thermostat—V8

and O-ring. Tighten the bolts to 18 ft. lbs. (25 Nm).
- Intake manifold support bracket
- Upper radiator mounting bracket
- Coolant hoses
- Throttle body
- Upper shield
- Accessory drive belt
- Negative battery cable

5. Fill the cooling system.
6. Run the engine and check for leaks.

X-Type 2.5L and 3.0L

2002–05

1. Before servicing the vehicle, refer to the Precautions Section.
2. Drain the cooling system.
3. Remove or disconnect the following:
   - Negative battery cable
   - Accessory drive belt
   - Left side valve cover
   - Coolant hoses
   - Water pump pulley
   - Water pump

To install:

4. Install or connect the following:
   - Water pump using a new gasket and O-ring. Tighten the bolts to 96–108 inch lbs. (11–13 Nm).
   - Water pump pulley. Tighten the bolts to 84–120 inch lbs. (10–14 Nm).
   - Coolant hoses
   - Left side valve cover
   - Accessory drive belt
   - Negative battery cable
5. Fill the cooling system.
6. Run the engine and check for leaks

2006

1. Before servicing the vehicle, refer to the Precautions Section.
2. Drain the cooling system.
3. Remove or disconnect the following:
   - Negative battery cable
   - Accessory drive belt
   - Left side valve cover
   - Coolant hoses
   - Water pump pulley
   - Water pump

To install:

4. Install or connect the following:
   - Water pump using a new gasket and O-ring. Tighten the bolts to 84 inch lbs. (10 Nm).
   - Water pump pulley. Tighten the bolts to 96 inch lbs. (11 Nm).
   - Coolant hoses
   - Left side valve cover
   - Accessory drive belt
   - Negative battery cable
5. Fill the cooling system.
6. Run the engine and check for leaks

Heater Core

REMOVAL & INSTALLATION

2002–03 XJR and XJ8

1. Before servicing the vehicle, refer to the Precautions Section.
2. Discharge and recover the air conditioning system refrigerant.
3. Drain the cooling system.
4. Move the front seats to fully rearward position.
5. Place steering wheel in straight ahead position.
6. Remove or disconnect the following:
   - Negative battery cable
   - Selector lever finish panel
   - Center console veneer panel
   - Radio
   - Center console
   - Center console A/C duct
   - Driver air bag
   - Steering column covers
   - Wiper switch
   - Warning speaker
   - Glove box
   - A pillar trim pads
   - Drive storage bin
   - Center console and footwell electrical connectors
   - Steering column electrical connectors
   - Fascia mounting bracket
   - Instrument panel end trim pads
   - A/C unit fascia plate
   - A/C unit to driveshaft tunnel bolts
   - Electrical connectors underneath and behind instrument panel
   - Instrument panel fascia panel retaining bolts
   - Instrument panel fascia panel, with assistance
   - Key transponder module
   - Light switch module
   - Control processor behind glove box opening
   - Instrument panel veneer
   - Trunk release switch
   - Left side ducts
   - Passenger side air bag
   - Right side ducts
   - Center vents
   - Clock
   - Defroster vents
   - Air vents

- Air distribution box
- Aspirator motor assembly
- Harness retaining clips from fascia
- Fascia assembly
- A/C lines at evaporator
- Heater hoses
- Steering column mountings and lay column on floor
- Remaining blower ducts
- All connectors and lines to heater core
- Heater core

To install:

7. To install, reverse the removal procedure.
8. Tighten heater core bolts to 62 inch lbs. (7 Nm).
9. Tighten steering column lower bolts to 15 ft. lbs. (20 Nm). Upper bolts to 18 ft. lbs. (25 Nm).
10. Tighten all instrument panel retaining bolts to 18 ft. lbs. (25 Nm).
11. Tighten air bag module to 62 inch lbs. (7 Nm).
12. Reconnect negative battery cable.
13. Refill the cooling system.
14. Recharge the air conditioning system.

2004–06 XJ Series

1. Before servicing the vehicle, refer to the Precautions Section.
2. Discharge and recover the air conditioning system refrigerant.
3. Drain the cooling system.
4. Remove or disconnect the following:
   - Negative battery cable
   - Cowl vent screen
   - Engine compartment brace
   - Pollen filter and housing
   - Engine compartment panel
   - Heater hoses
   - Cap the air conditioning ports
   - Expansion valve manifold and tube assembly
   - Instrument panel lower trim panel
   - Steering column lower shroud
   - Driver air bag module
   - Clockspring
   - Steering wheel rotation sensor
   - Multifunction switch wiring harnesses
   - Steering column electrical connector
   - Steering column pinch bolt
   - Steering column
   - Steering column lock actuator connector
   - Instrument panel temperature sensor hose
   - Instrument panel console
   - Ashtray
   - Floor console register duct
   - Transmission selector switch connector
   - Lateral/yaw rate sensor connector
   - Wiring harness ground connector
   - Instrument panel retaining bracket
   - Floor console duct retaining clip
   - Front door opening weather strip
   - Scuff plate trim panel
   - Cowl side trim panel
   - Footwell electrical connectors
   - Climate control electrical connectors
   - Instrument panel end panel
   - Sunload sensor
   - Defogger duct
   - Instrument panel retaining bolts
   - Instrument panel
   - Floor console ducts
   - Heater motor electrical connector
   - Rear footwell vent ducts
   - Heater and evaporator core housing

To install:

5. To install, reverse the removal procedure.
6. Tighten heater core bolts to 62 inch lbs. (7 Nm).
7. Use new O-ring seals on expansion valve manifold and tube assembly. Tighten to 71 inch lbs. (8 Nm).
8. Tighten all instrument panel retaining bolts to 15 ft. lbs. (20 Nm).
9. Tighten air bag module to 62 inch lbs. (7 Nm).
10. Reconnect negative battery cable.
11. Refill the cooling system.
12. Recharge the air conditioning system.

S-Type

1. Before servicing the vehicle, refer to the Precautions Section.
2. Discharge and recover the air conditioning system refrigerant.
3. Drain the cooling system.
4. Remove or disconnect the following:
   - Negative battery cable
   - Heater hoses
   - Throttle body on supercharged models
   - Expansion valve manifold and tube and cap ports
   - Heater core housing retaining nut.
   - Instrument panel lower trim panel
   - Steering column lower shroud
   - Driver air bag module
   - Clockspring
   - Steering wheel rotation sensor
   - Steering column connectors
   - Steering column pinch bolt
   - Steering column
   - Floor console register duct
   - Instrument panel multi-switch pack
   - Ashtray
   - Parking brake switch
   - Floor console bracket
   - Instrument panel support bracket connectors
   - Instrument panel support bracket
   - Hood release handle
   - A pillar trim panels
   - Instrument panel end caps
   - Door weather stripping
   - Defroster vent
   - Passenger side trim panels
   - Instrument panel connectors
   - Instrument panel bolts and screws
   - Instrument panel
   - Rear footwell ducts
   - PCM retaining bracket
   - Heater core

To install:

5. To install, reverse the removal procedure.
6. Tighten heater core bolts to 62 inch lbs. (7 Nm).
7. Use new O-ring seals on expansion valve manifold and tube assembly. Tighten to 71 inch lbs. (8 Nm).
8. Tighten all instrument panel retaining bolts to 15 ft. lbs. (20 Nm).
9. Tighten steering column bolts to 18 ft. lbs. (25 Nm).
10. Tighten air bag module to 62 inch lbs. (7 Nm).
11. Reconnect negative battery cable.
12. Refill the cooling system.
13. Recharge the air conditioning system.

2005 X-Type

1. Before servicing the vehicle, refer to the Precautions Section.
2. Discharge and recover the air conditioning system refrigerant.
3. Drain the cooling system.
4. Remove or disconnect the following:
   - Negative battery cable
   - Heater core drain tube
   - Heater hoses
   - Instrument panel console
   - Climate control panel
   - Navigation screen, if equipped
   - Instrument cluster
   - Passenger finish panel
   - Passenger air vent
   - Passenger air bag electrical connectors
   - Passenger air bag
   - Multi-function switch
   - Steering column electrical connectors

- Steering column
- Headlight switch
- Transmission gear selector
- Reverse gear interlock
- Right side end panel trim
- Support panel
- Lower vent
- Instrument panel retaining screws
- Instrument panel
- Center register air duct
- Glove box mounting frame
- Remaining electrical connectors
- Instrument panel cross beam
- Heater core

To install:

5. To install, reverse the removal procedure.
6. Tighten heater core bolts to 62 inch lbs. (7 Nm).
7. Tighten instrument panel cross beam bolts to 18 ft. lbs. (25 Nm).
8. Tighten all instrument panel retaining bolts to 15 ft. lbs. (20 Nm).
9. Tighten steering column bolts to 18 ft. lbs. (25 Nm).
10. Tighten air bag module to 80 inch lbs. (9 Nm).
11. Reconnect negative battery cable.
12. Refill the cooling system.
13. Recharge the air conditioning system.

2006 X-Type

➡If a leak is suspected, the heater core must be leak tested in the vehicle before removal.

1. Drain the cooling system.
2. Disconnect the drain tube from the heater core.
3. Disconnect the heater hose from the heater core.
4. Disconnect the heater hose from the heater core.
5. Vehicles with automatic transaxle, remove the J-gate surround.

➡To aid removal of the climate control assembly place the J-gate selector in the neutral position.

6. Vehicles with manual transaxle, detach the gear selector surround.
7. Disconnect the battery ground cable. For additional information, refer to <<414-01>>.
8. Remove the ashtray retaining screws.
9. Remove the ashtray. Disconnect the electrical connectors.
10. Remove the climate control assembly mounting bracket upper retaining screws cover.
11. Remove the climate control assembly mounting bracket upper retaining screws.
12. Remove the climate control assembly mounting bracket lower retaining screws.
13. Disconnect the hazard lamp switch assembly electrical connectors.
14. Remove the hazard lamp switch assembly. Remove the retaining screws.
15. Disconnect the climate control assembly electrical connectors.
16. Remove the climate control assembly retaining screws.
17. Remove the retaining screws. Remove the climate control assembly.
18. Position the heater core air duct up.
    a. Remove the heater core air duct retaining screw.
    b. Position the heater core air duct up.
19. Remove the heater core air duct housing.
    a. Remove the heater core air duct housing retaining screws.
    b. Remove the heater core air duct housing. Release the retaining tabs.
20. Remove the heater core housing retaining clips.
21. Remove the heater core housing. Release the heater core housing retaining tabs.
22. Remove the seal.
23. Remove the heater core retaining screw.
24. Remove the heater core.
25. Installation is the reverse of the removal procedure.

➡Make sure the climate control assembly is correctly located into the audio unit.

Cylinder Head

REMOVAL & INSTALLATION

XJ Range and S-Type 4.0L and 4.2L Engines

1. Before servicing the vehicle, refer to the Precautions Section.
2. Drain the cooling system.
3. Remove or disconnect the following:
    - Negative battery cable
    - Accessory drive belt
    - Valve cover
    - Timing chain cover
    - Intake manifold
    - Radiator and heater hoses
    - Engine Coolant Temperature (ECT) sensor connector
    - Coolant outlet pipe
    - Exhaust front pipes
    - Variable Valve Timing (VVT) sensor
    - Crankshaft Position (CKP) sensor
4. Rotate the crankshaft until the triangular arrow indent on the driveplate is visible through the access hole.
5. Install the Crankshaft Setting Peg JD 216 into the CKP sensor location.
6. Install the Camshaft Locking tool JD 215 on the camshafts.
7. Remove or disconnect the following:
    - Camshaft sprocket mounting bolt by loosening it
    - VVT mounting bolt by loosening it
    - Camshaft Locking tool JD 215
    - Primary timing chain tensioner and backing plate
    - Primary timing chain guide
    - VVT unit and exhaust camshaft sprocket
    - Secondary timing chain tensioner and guide

➡Keep all valvetrain components in order for assembly.

- Camshafts. Loosen the bearing cap bolts evenly and in several passes.
- Valve tappets and shims
- Cylinder head

To install:

8. Clean and inspect the cylinder head bolts. They may be reused twice. When reused, each bolt head should be marked with 1 dot from an automatic center punch.
9. Install the cylinder head with a new gasket. Tighten the bolts in sequence as follows:
    a. Bolts 1-10: 15 ft. lbs. (20 Nm).
    b. Bolts 1-10: 26 ft. lbs. (35 Nm).
    c. Bolts 1-10: plus 90 degrees.
    d. Bolts 1-10: plus 90 degrees.
    e. Bolts 11 & 12: 18 ft. lbs. (25 Nm).
10. Install or connect the following:
    - Tappets and shims into their original position
    - Camshaft bearing caps and tighten evenly, in stages, to 84–96 inch lbs. (9–11 Nm).
11. Install the Camshaft Locking tool JD 215.
12. Prepare the timing chain tensioners for installation by using a paperclip or other wire to unseat the check valves and compressing the pistons into their bores.
13. Install or connect the following:
    - Secondary timing chain guide. Tighten the bolt to 89–124 inch lbs. (10–14 Nm).
    - Secondary timing chain tensioner. Tighten the bolt to 89–124 inch lbs. (10–14 Nm).
    - VVT, secondary timing chain and exhaust cam sprocket
    - Primary timing chain
    - Primary timing chain guide.

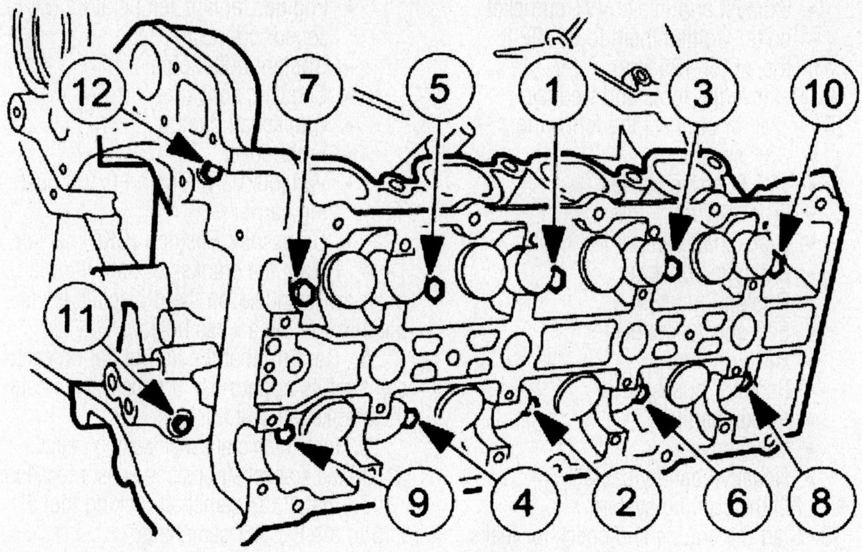

Cylinder head bolt torque sequence—4.0L and 4.2L engine

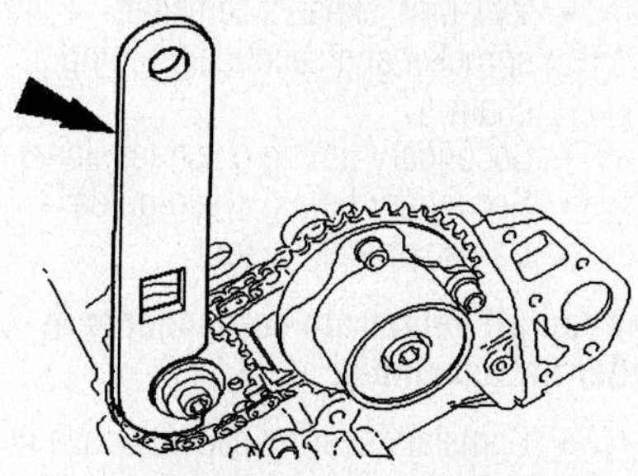

Apply force in a counterclockwise direction when tightening the sprocket mounting bolts—4.0 and 4.2L engines

Tighten the bolt to 10–12 ft. lbs. (13–16 Nm).

- Primary timing chain tensioner and backing plate. Tighten the bolts to 89–124 inch lbs. (10–14 Nm).
- Primary timing chain slack, eliminate it by placing a wedge between the primary timing chain tensioner and the guide shoe.
- Secondary timing chain by applying counterclockwise force to the exhaust camshaft sprocket.
- Exhaust and intake VVT sprocket bolts. Tighten them to 85–92 ft. lbs. (115–125 Nm).

14. Remove the tools and wedges.
15. Install or connect the following:
    - CKP sensor
    - VVT sensor
    - Exhaust front pipes
    - Coolant outlet pipe
    - ECT sensor connector
    - Radiator and heater hoses
    - Intake manifold
    - Timing chain cover
    - Valve cover
    - Accessory drive belt
    - Negative battery cable
16. Fill the cooling system.
17. Start the engine and check for leaks.

XK Range

A-BANK

1. Before servicing the vehicle, refer to the Precautions Section.
2. Drain the cooling system.
3. Remove or disconnect the following:
    - Negative battery cable
    - Valve cover
    - Intake manifold
    - Radiator hoses
    - Heater hoses
    - Engine Coolant Temperature (ECT) sensor connector
    - Coolant outlet pipe
    - Catalytic converter
    - Crankshaft damper
    - Front cover
    - Variable Valve Timing (VVT) bushing carrier
    - Crankshaft Position (CKP) sensor
4. Rotate the crankshaft until the triangular arrow indent on the driveplate is visible through the access hole.
5. Confirm that the timing flat on each camshaft is uppermost. If necessary, rotate the crankshaft one full turn.
6. Install the crankshaft setting peg JD 216 to the crankshaft position sensor location.
7. Install the camshaft locking tool JD 215 to the A-Bank camshafts.
8. Remove or disconnect the following:
    - Primary timing chain tensioner and backing plate
    - Primary timing chain
    - VVT unit, exhaust camshaft sprocket and secondary timing chain
    - Secondary timing chain tensioner
    - Secondary timing chain guide
    - Camshaft locking tool

➡Keep all valvetrain components in order for assembly.

- Camshaft bearing caps. Loosen the bolts evenly in stages.
- Camshafts
- Tappets and shims
- Cylinder head

To install:

9. Clean and inspect the cylinder head bolts. They may be reused twice. When reused, each bolt head should be marked with 1 dot from an automatic center punch.
10. Install the cylinder head with a new gasket. Tighten the bolts in sequence as follows:
    a. Step 1: M10 bolts to 15 ft. lbs. (20 Nm).
    b. Step 2: M10 bolts to 26 ft. lbs. (35 Nm).
    c. Step 3: M10 bolts plus 90 degrees.
    d. Step 4: M10 bolts plus 90 degrees.
    e. Step 5: M8 bolts to 17–20 ft. lbs. (23–27 Nm).
11. Install or connect the following:
    - Tappets and shims into their original position
    - Camshaft bearing caps. Tighten the bolts evenly, in stages, to 84–96 inch lbs. (9–11 Nm).

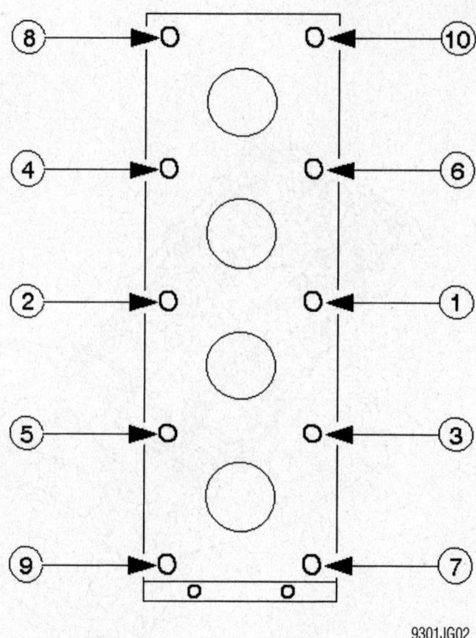

Cylinder head bolt torque sequence—XK V8 A-bank

12. Install the Camshaft Locking tool JD 215.

13. Prepare the timing chain tensioners for installation by using a paperclip or other wire to unseat the check valves and compressing the pistons into their bores.

14. Install or connect the following:

- Secondary timing chain guide. Tighten the bolt to 89–124 inch lbs. (10–14 Nm).
- Secondary timing chain tensioner. Tighten the bolt to 89–124 inch lbs. (10–14 Nm).
- VVT, secondary timing chain, and exhaust cam sprocket
- Primary timing chain
- Primary timing chain guide. Tighten the bolt to 10–12 ft. lbs. (13–16 Nm).
- Primary timing chain tensioner and backing plate. Tighten the bolts to 89–124 inch lbs. (10–14 Nm).
- Primary timing chain slack, eliminate it by placing a wedge between the primary timing chain tensioner and the guide shoe
- Secondary timing chain, tension it by applying counterclockwise force to the exhaust camshaft sprocket
- Exhaust and intake VVT sprocket bolts. Tighten them to 85–92 ft. lbs. (115–125 Nm).

15. Remove the tools and wedges.

16. Install or connect the following:

- CKP sensor
- VVT bushing carrier
- Front cover
- Crankshaft damper
- Catalytic converter
- Coolant outlet pipe
- ECT sensor connector
- Heater hoses
- Radiator hoses
- Intake manifold
- Valve cover
- Negative battery cable

17. Fill the cooling system.

18. Start the engine and check for leaks.

B-BANK

1. Before servicing the vehicle, refer to the Precautions Section.

2. Drain the cooling system.

3. Remove or disconnect the following:

- Negative battery cable
- Valve cover
- Intake manifold
- Radiator hoses
- Heater hoses
- Engine Coolant Temperature (ECT) sensor connector
- Coolant outlet pipe
- Catalytic converter
- Crankshaft damper
- Front cover
- Variable Valve Timing (VVT) bushing carrier
- Crankshaft Position (CKP) sensor

4. Rotate the crankshaft until the triangular arrow indent on the driveplate is visible through the access hole.

5. Confirm that the timing flat on each camshaft is uppermost. If necessary, rotate the crankshaft one full turn.

6. Install the crankshaft setting peg JD 216 to the crankshaft position sensor location.

7. Install the camshaft locking tool JD 215 to the B-Bank camshafts.

8. Remove or disconnect the following:

- Primary timing chain tensioner and backing plate
- Primary timing chain
- VVT unit, exhaust camshaft sprocket and secondary timing chain
- Secondary timing chain tensioner
- Secondary timing chain guide
- Camshaft locking tool

➡Keep all valvetrain components in order for assembly.

- Camshaft bearing caps. Loosen the bolts evenly in stages.
- Camshafts
- Tappets and shims
- Cylinder head

To install:

9. Clean and inspect the cylinder head bolts. They may be reused twice. When

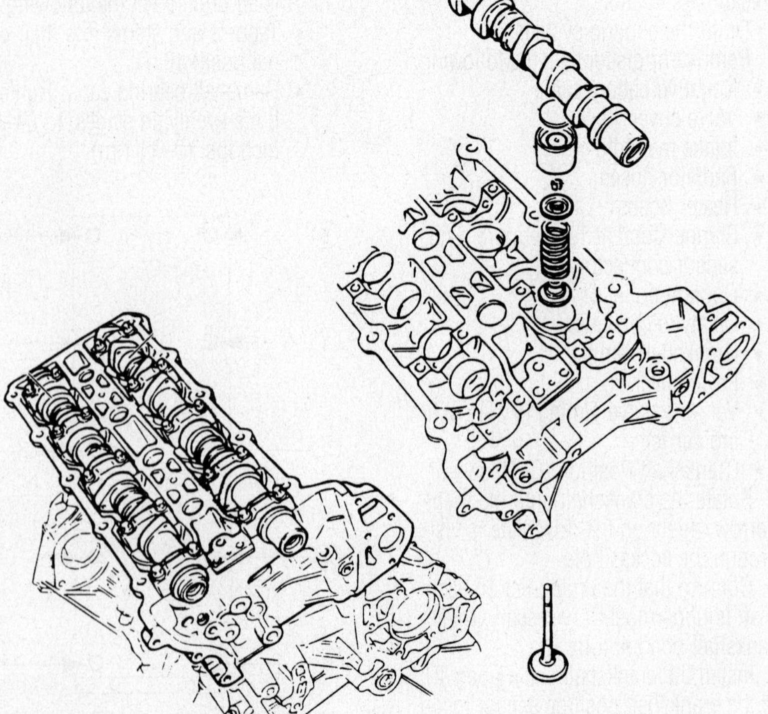

9301JG04

Exploded view of the camshaft and valve installation—V8 shown

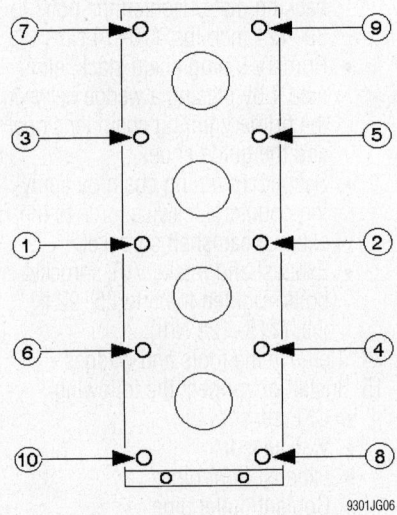

7 3 1 6 10 9 5 2 4 8

9301JG06

Cylinder head bolt torque sequence—XK V8 B-bank

reused, each bolt head should be marked with 1 dot from an automatic center punch.

10. Install the cylinder head with a new gasket. Tighten the bolts in sequence as follows:

a. Step 1: M10 bolts to 15 ft. lbs. (20 Nm).
b. Step 2: M10 bolts to 26 ft. lbs. (35 Nm).
c. Step 3: M10 bolts plus 90 degrees.
d. Step 4: M10 bolts plus 90 degrees.
e. Step 5: M8 bolts to 17–20 ft. lbs. (23–27 Nm).

11. Install or connect the following:

- Tappets and shims into their original position
- Camshaft bearing caps. Tighten the bolts evenly, in stages, to 84–96 inch lbs. (9–11 Nm).

12. Install the Camshaft Locking tool JD 215.

13. Prepare the timing chain tensioners for installation by using a paperclip or other wire to unseat the check valves and compressing the pistons into their bores.

14. Install or connect the following:

- Secondary timing chain guide. Tighten the bolt to 89–124 inch lbs. (10–14 Nm).
- Secondary timing chain tensioner. Tighten the bolt to 89–124 inch lbs. (10–14 Nm).
- VVT, secondary timing chain, and exhaust cam sprocket
- Primary timing chain
- Primary timing chain guide. Tighten the bolt to 10–12 ft. lbs. (13–16 Nm).
- Primary timing chain tensioner and backing plate. Tighten the bolts to 89–124 inch lbs. (10–14 Nm).
- Primary timing chain slack, eliminate it by placing a wedge between the primary timing chain tensioner and the guide shoe
- Secondary timing chain, tension it by applying counterclockwise force to the exhaust camshaft sprocket
- Exhaust and intake VVT sprocket bolts. Tighten them to 85–92 ft. lbs. (115–125 Nm).

15. Remove the tools and wedges.

16. Install or connect the following:

- CKP sensor
- VVT bushing carrier
- Front cover
- Crankshaft damper
- Catalytic converter
- Coolant outlet pipe
- ECT sensor connector
- Heater hoses
- Radiator hoses

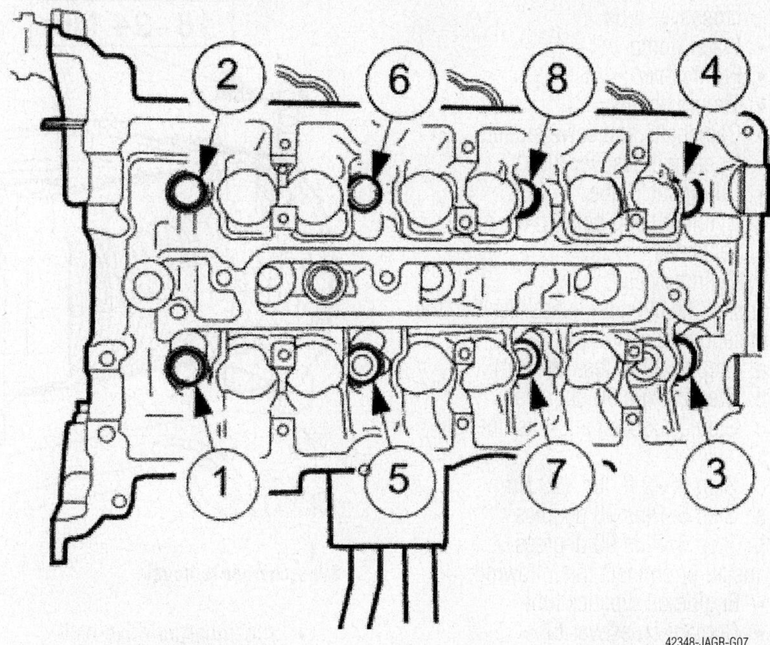

Cylinder head bolt removal sequence—2.5L and 3.0L engines—left side shown, right side same

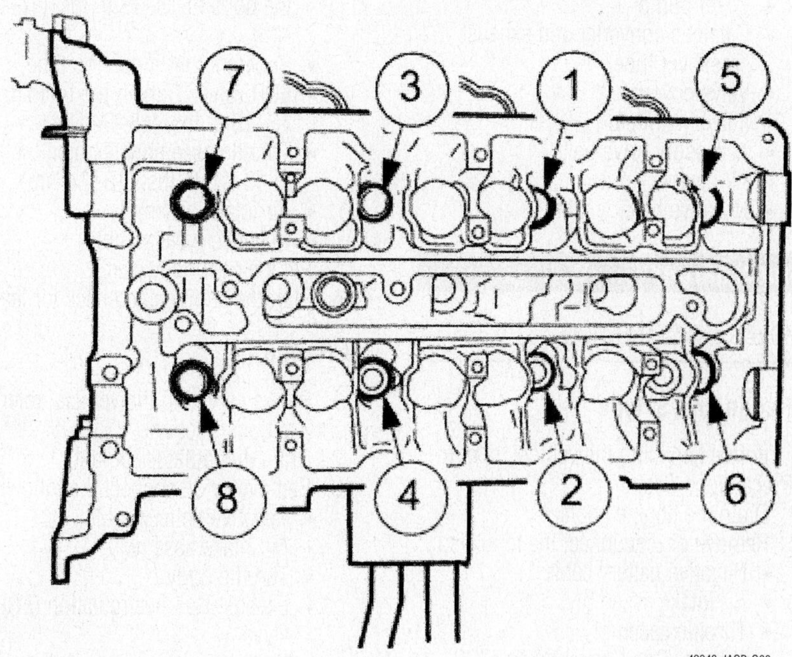

Cylinder head bolt installation sequence—2.5L and 3.0L engines—left side shown, right side same

- Intake manifold
- Valve cover
- Negative battery cable

17. Fill the cooling system.

18. Start the engine and check for leaks.

S-Type and X-Type 2.5L and 3.0L Engines

➡The cylinder head bolts are a torque-to-yield design and must be replaced.

1. Before servicing the vehicle, refer to the Precautions Section.

2. Drain the crankcase and cooling system.

3. Remove or disconnect the following:

- Negative battery cable
- Air cleaner housing
- Accessory drive belt
- Intake manifold
- Valve covers

- Catalytic converter and exhaust crossover tube
- Water pump
- Front cover
- Timing chains
- Camshafts and valve tappets
- Coolant crossover tube
- Oil dipstick tube
- Cylinder head bolts in the proper sequence
- Cylinder heads

4. Install the cylinder heads. Tighten the bolts in sequence as follows:
   a. Step 1: 22 ft. lbs. (30 Nm)
   b. Step 2: Plus 90 degrees
   c. Step 3: Loosen all bolts one full turn
   d. Step 4: 22 ft. lbs. (30 Nm)
   e. Step 5: Plus 90 degrees
   f. Step 6: Plus 90 degrees
5. Install or connect the following:
   - Engine oil dipstick tube
   - Coolant crossover tube
   - Camshafts and valve tappets
   - Timing chains
   - Front cover
   - Water pump
   - Catalytic converter and exhaust crossover tube
   - Valve covers
   - Intake manifold
   - Accessory drive belt
   - Air cleaner housing
   - Negative battery cable

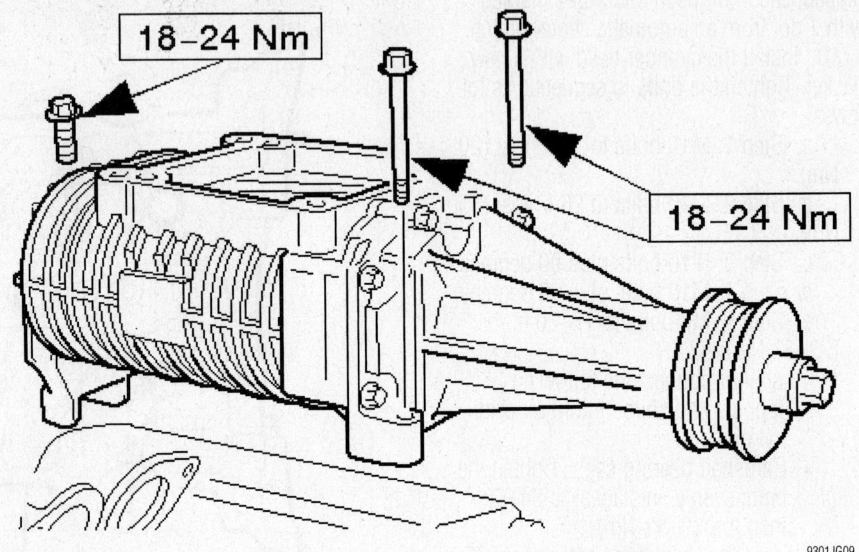

Supercharger removal

Supercharger

REMOVAL & INSTALLATION

XJ Range and S-Type

1. Before servicing the vehicle, refer to the Precautions Section.
2. Drain the engine coolant.
3. Remove or disconnect the following:
   - Negative battery cable
   - Air intake assembly
   - Throttle assembly
   - Exhaust Gas Recirculation (EGR) valve
   - Supercharger induction tube
   - Supercharger outlet duct
   - Coolant hoses
   - Coolant outlet
   - Supercharger drive belt
   - Supercharger

To install:

4. Install or connect the following:
   - Supercharger with a new gasket. Tighten the bolts to 13–18 ft. lbs. (18–24 Nm).
   - Supercharger drive belt
   - Coolant outlet
   - Coolant hoses
   - Supercharger outlet duct. Tighten the bolts to 13–18 ft. lbs. (18–24 Nm).
   - Supercharger induction tube
   - EGR valve. Tighten the bolts to 13–18 ft. lbs. (18–24 Nm).
   - Throttle assembly. Tighten the bolts to 13–18 ft. lbs. (18–24 Nm).
   - Air intake assembly
   - Negative battery cable
5. Fill the cooling system
6. Start the engine and check for leaks.

XK Range

1. Before servicing the vehicle, refer to the Precautions Section.
2. Drain the radiator coolant.
3. Remove or disconnect the following:
   - Negative battery cable
   - Air intake assembly
   - Throttle body
   - Exhaust Gas Recirculation (EGR) valve
   - Supercharger induction elbow
   - Supercharger outlet duct
   - Intercooler and hoses
   - Coolant outlet
   - Supercharger drive belt
   - Supercharger

To install:

4. Install or connect the following:
   - Supercharger with a new gasket. Tighten the bolts to 13–18 ft. lbs. (18–24 Nm).
   - Supercharger drive belt
   - Coolant outlet
   - Intercooler and hoses
   - Supercharger outlet duct
   - Supercharger induction elbow
   - EGR valve
   - Throttle body
   - Air intake assembly
   - Negative battery cable
5. Fill the cooling system
6. Start the engine and check for leaks.

Intake Manifold

REMOVAL & INSTALLATION

S-Type 3.0L

2002

1. Disconnect the battery ground cable.
2. Drain the cooling system.
3. Remove the throttle body.
4. Disconnect the EGR vacuum regulator (EVR) valve electrical connector.
5. Disconnect the vacuum hose connector from the EGR vacuum regulator valve.
6. Remove the intake manifold tuning valve cover.
7. Disconnect the electrical connector.
8. Remove the intake manifold tuning valve cover.
9. Disconnect the differential pressure feedback EGR sensor electrical connector.
10. Disconnect the hoses from the differential pressure feedback EGR sensor.
11. Disconnect the inlet and outlet intake manifold coolant hoses.
12. Disconnect the purge line tube.
13. Detach the EGR valve.
14. Remove the intake manifold bracket securing bolt.

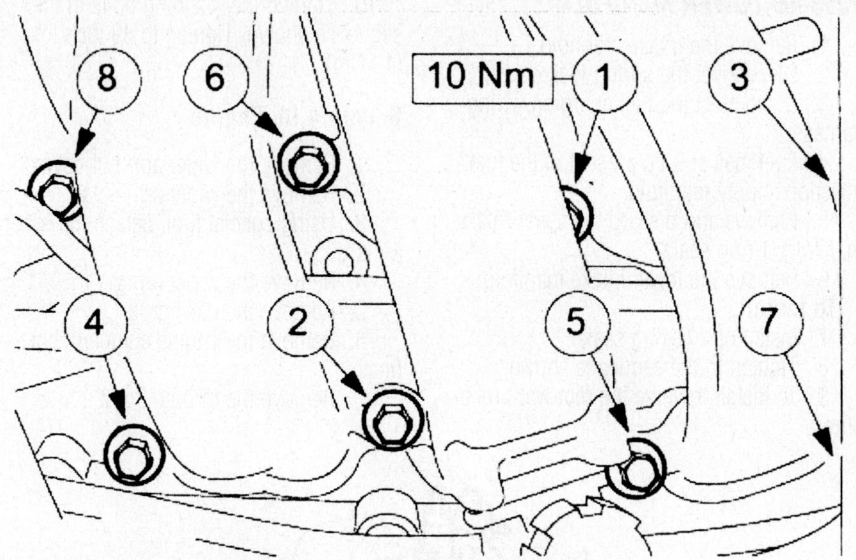

Intake manifold bolt identification—3.0L engine

15. Remove the intake manifold bracket securing bolt.
16. Remove the intake manifold securing bolts.
17. Remove the fuel pulse damper protective bracket.
18. Displace the intake manifold.
19. Disconnect the engine breather hose and the intake manifold tuning valve electrical connector.
20. Remove the intake manifold.
21. Discard the intake manifold gaskets.

➡The intake manifold retaining bolts in position 1,2,3 and 4 are longer than the retaining bolts in position 5,6,7 and 8.

22. To install, reverse the removal procedure. Install new intake manifold gaskets. Install a new EGR valve gasket.

➡When the battery has been disconnected and reconnected, some abnormal drive symptoms may occur while the vehicle relearns its adaptive strategy.

23. Connect the battery ground cable.
24. Fill and bleed the cooling system.

2003–06 UPPER MANIFOLD

1. Release the cooling system pressure.
2. Remove the coolant expansion tank pressure cap.
3. Remove the air cleaner outlet pipe.
4. Remove the engine cover.
5. Disconnect the throttle motor electrical connector.
6. Disconnect the throttle position sensor (TPS) electrical connector.
7. Disconnect the positive crankcase ventilation (PCV) hose.

➡Cap the coolant hoses to minimize coolant loss.

8. Detach the coolant hoses from the throttle body.
9. Detach the engine wiring harness from the intake manifold side retaining bracket.
10. Detach the intake manifold right-hand side retaining bracket.
11. Loosen the intake manifold right-hand side retaining bracket.
12. Detach the intake manifold front retaining bracket.
13. Loosen the intake manifold front retaining bracket.
14. Detach the coolant temperature sensor wiring harness.
15. Remove the engine cover front retaining bracket.
16. Detach the evaporative emission canister purge valve transfer pipe.
17. Remove the engine cover left-hand side retaining bracket.
18. Remove the fuel pressure sensor bracket retaining bolt.
19. Detach the fuel pressure sensor bracket.
20. Detach the intake manifold.

➡The evaporative emission canister purge valve transfer pipe is attached to the induction manifold by a quick release coupling.

21. Disconnect the evaporative emission canister purge valve transfer pipe.
22. Disconnect the manifold absolute pressure (MAP) sensor electrical connector.
23. Disconnect the fuel pressure sensor vacuum hose.
24. Disconnect the IMT valves electrical connectors.
25. Disconnect the fuel pressure regulator vacuum hose.
26. Disconnect the brake booster vacuum hose.
27. Remove the intake manifold.
28. Install new intake manifold gaskets.

To install:

➡The intake manifold retaining bolts in position 1, 4 and 5 are longer than the retaining bolts in position 2, 3 and 6. Do not fully tighten the retaining bolts.

29. Install the intake manifold.
30. Connect the brake booster vacuum hose.

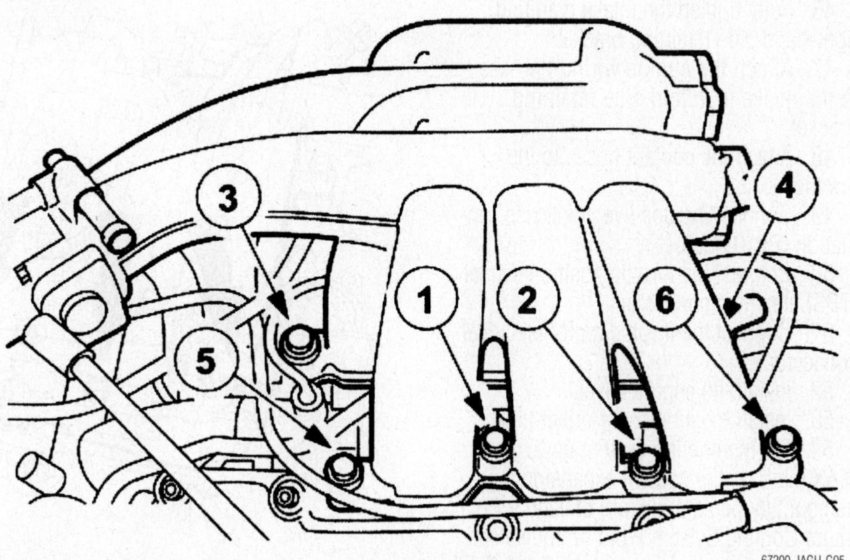

Upper intake manifold bolt identification—2003–06 S-Type 3.0L

31. Connect the fuel pressure regulator vacuum hose.
32. Connect the IMT valves electrical connectors.
33. Disconnect the fuel pressure sensor vacuum hose
34. Connect the manifold absolute pressure (MAP) sensor electrical connector.

➡The evaporative emission canister purge valve transfer pipe is attached to the induction manifold by a quick release coupling.

35. Connect the evaporative emission canister purge valve transfer pipe.

➡The intake manifold retaining bolts in position 1, 4 and 5 are longer than the retaining bolts in position 2, 3 and 6.

36. Reposition the intake manifold. Tighten to 15 ft. lbs. Complete the tightening sequence.
37. Install the fuel pressure sensor bracket retaining bolt.
38. Attach the fuel pressure sensor bracket.
39. Install the engine cover left-hand side retaining bracket.
40. Attach the evaporative emission canister purge valve transfer pipe.
41. Install the engine cover front retaining bracket.
42. Attach the coolant temperature sensor wiring harness.
43. Attach the intake manifold front retaining bracket.
44. Fully tighten the intake manifold front retaining bracket.
45. Fully tighten the intake manifold right-hand side retaining bracket.
46. Fully tighten the intake manifold right-hand side retaining bracket.
47. Attach the engine wiring harness to the intake manifold side retaining bracket.
48. Attach the coolant hoses to the throttle body.
49. Connect the positive crankcase ventilation (PCV) hose.
50. Connect the throttle position sensor (TPS) electrical connector.
51. Connect the throttle motor electrical connector.
52. Install the engine cover.
53. Install the air cleaner outlet pipe.
54. Fill the cooling system up to the MAX mark on the coolant expansion tank using a fifty percent mixture of Jaguar Premium Cooling System Fluid or equivalent, meeting Jaguar specification WSS M97B44-D and fifty percent water.

2003–06 LOWER MANIFOLD

1. Remove the intake manifold.
2. Disconnect the spring lock coupling.
3. Disconnect the fuel charging wiring harness.
4. Fuel may still be present in the fuel injection supply manifold.
5. Remove and discard the lower intake manifold O-ring seals.
6. Remove the lower intake manifold.

To install:

7. Install new O-ring seals.
8. Tighten in the sequence shown.
9. To install, reverse the removal procedure.
10. Tighten the retaining bolts in the sequence shown. Tighten to 89 inch lbs. (10 Nm).

S-Type 4.0L Engine

1. Remove the wiper arm nut covers.
2. Remove the wiper arm nuts.
3. Using special tool, detach the wiper arms.
4. Remove the wiper arms.
5. Remove the cowl grille.
6. Remove the engine compartment brace.
7. Remove the throttle body.

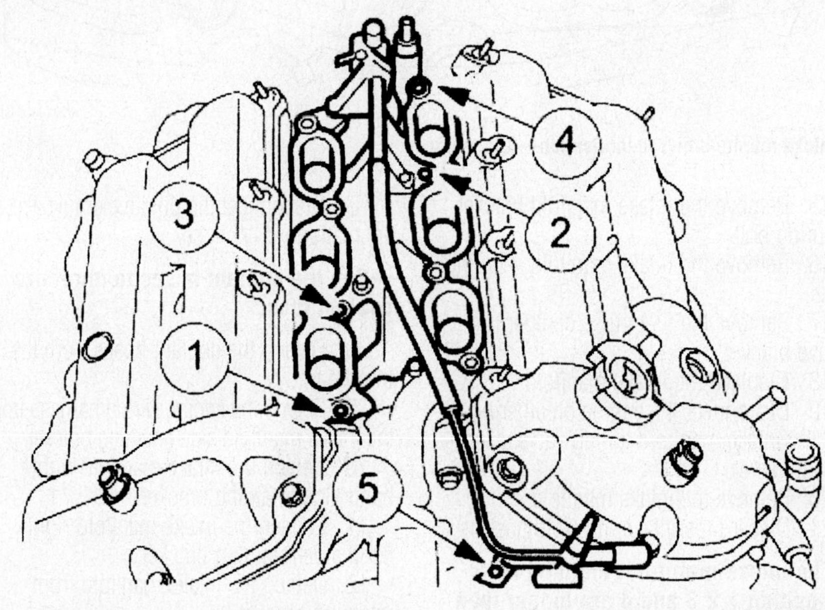

Lower manifold torque sequence—3.0L engine

4.0L engine intake manifold bolts—right side

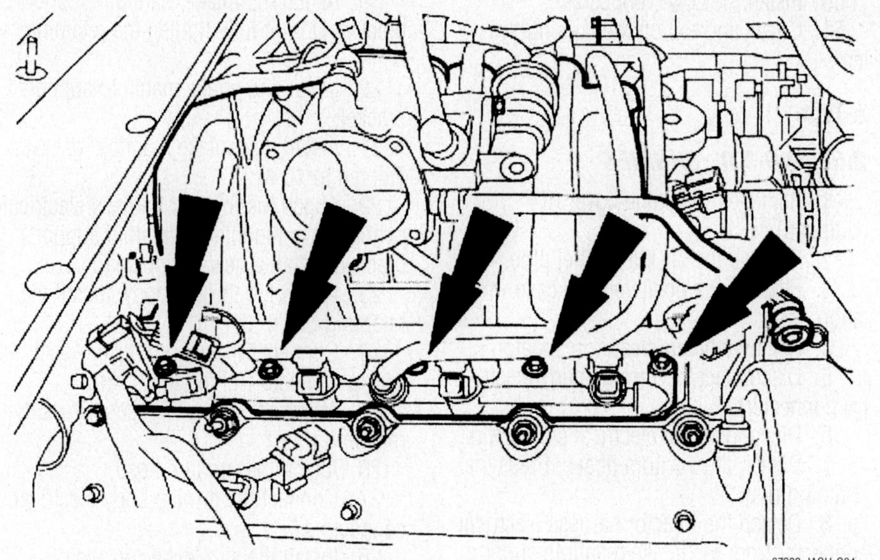

4.0L engine intake manifold bolts—left side

8. Disconnect the right-hand bank camshaft position sensor.
9. Disconnect the right-hand bank injector electrical connectors.
10. Disconnect the fuel temperature sensor electrical connector.
11. Detach the engine harness.
12. Disconnect the knock sensor electrical connector.
13. Detach the engine harness.
14. Disconnect the idle speed control valve electrical connector.
15. Disconnect the fuel pressure regulator electrical connector.
16. Remove the engine harness bracket.
17. Disconnect the camshaft position sensor electrical connector.
18. Disconnect the left-hand bank fuel injector electrical connectors.
19. Disconnect the brake booster vacuum supply hose.
20. Disconnect the evaporative emission canister purge valve vacuum supply hose.
21. Remove the engine harness bracket.
22. Disconnect the intake manifold heater coolant tube.
23. Detach the intake manifold.
24. Remove the intake manifold.
25. To install, reverse the removal procedure. Torque the manifold bolts to 13 ft. lbs. (18 Nm).

S-Type 4.2L Engine

1. Remove the cowl vent screen.
2. Remove the air cleaner outlet pipe.
3. Remove the engine compartment brace.
4. Remove the on-plug coil covers.
5. Disconnect the throttle potentiometer (TP) electrical connector.
6. Disconnect the throttle motor electrical connector.

➡Cap the coolant system hose to minimize coolant loss.

7. Disconnect the throttle body coolant hose.
8. Disconnect the knock sensors (KS) electrical connectors.
9. Remove the left-hand engine cover bracket.
10. Detach the left-hand engine wiring harness.
11. Disconnect the left-hand fuel injector electrical connectors.
12. Detach the fuel line.
13. Detach the left-hand camshaft position sensor electrical connector.
14. Disconnect the coolant temperature sensor electrical connector.
15. Remove the right-hand engine cover bracket.
16. Remove the positive crankcase ventilation (PCV) valve.
17. Disconnect the fuel pressure regulator electrical connector.
18. Disconnect the exhaust gas recirculation (EGR) valve electrical connector.

➡Cap the coolant hose.

19. Disconnect the exhaust gas recirculation (EGR) valve coolant hose.
20. Detach the right-hand engine wiring harness.
21. Disconnect the right-hand fuel injector electrical connectors.
22. Disconnect the fuel temperature sensor electrical connector.
23. Disconnect the right-hand camshaft position (CMP) sensor.
24. Detach the brake booster vacuum pipe.
25. Disconnect the manifold absolute pressure (MAP) sensor electrical connector.
26. Detach the engine wiring harness.

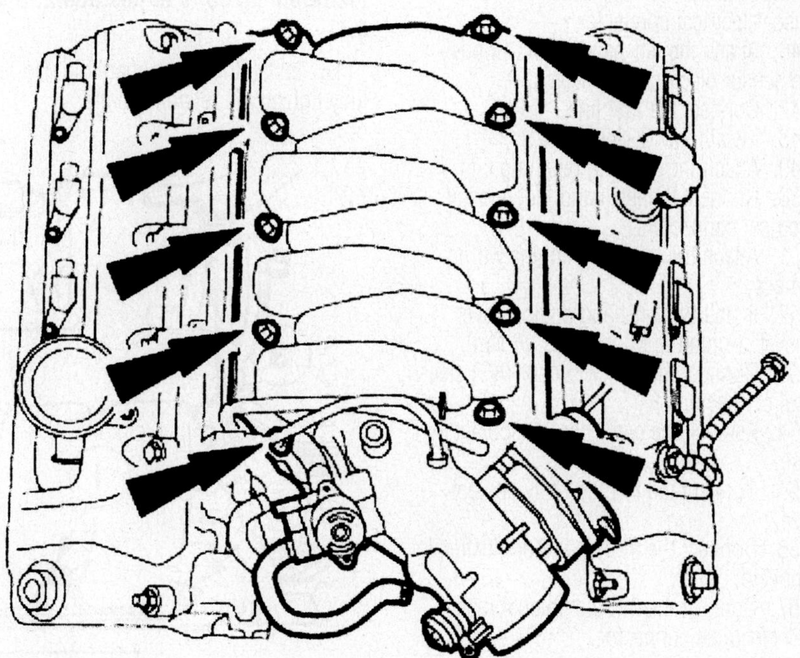

Intake manifold bolt locations—S-Type 4.2L engine

27. Detach the exhaust gas recirculation (EGR) valve tube.
28. Remove and discard the gasket. Remove the intake manifold.
29. Remove and discard the gaskets.

To install:

30. Install the intake manifold. Install new gaskets. Tighten to 16 ft. lbs. (22 Nm).
31. Attach the exhaust gas recirculation (EGR) valve tube. Install a new gasket.
32. Attach the engine wiring harness. Tighten to 84 inch lbs. (10 Nm).
33. Connect the manifold absolute pressure (MAP) sensor electrical connector.
34. Connect the right-hand camshaft position (CMP) sensor.
35. Connect the fuel temperature sensor electrical connector.
36. Attach the brake booster vacuum pipe.
37. Connect the right-hand fuel injector electrical connectors.
38. Attach the right-hand engine wiring harness.
39. Remove the cap from the coolant hose.
40. Connect the exhaust gas recirculation (EGR) valve coolant hose.
41. Connect the exhaust gas recirculation (EGR) valve electrical connector.
42. Connect the fuel pressure regulator electrical connector.
43. Install the positive crankcase ventilation (PCV) pipe.
44. Install the right-hand engine cover bracket. Tighten to 50 inch lbs. (6 Nm).
45. Connect the coolant temperature sensor electrical connector.
46. Attach the left-hand camshaft position sensor electrical connector.
47. Connect the fuel line.
48. Install new O-ring seals.
49. Attach the fuel line retaining clip.
50. Connect the left-hand fuel injector electrical connectors.
51. Attach the left-hand engine wiring harness.
52. Install the left-hand engine cover bracket. Tighten to 50 inch lbs. (6 Nm).
53. Connect the knock sensor (KS) electrical connectors.
54. Remove the cap from the coolant hose.
55. Connect the throttle body coolant hose.
56. Connect the throttle motor electrical connector.
57. Connect the throttle potentiometer (TP) electrical connector.
58. Install the engine compartment brace. Tighten to 13 ft. lbs. (18 Nm).
59. Install the air cleaner outlet pipe.
60. Install the cowl vent screen.
61. Check and top up the cooling system.

X-Type

2.5L AND 3.0L ENGINES

1. Remove the engine cover (if equipped).
2. Remove the air filter outlet pipe.
3. Remove the positive crankcase ventilation hose.
4. Remove the air cleaner resonator.
5. Disconnect the throttle body electrical connectors.
6. Disconnect the electrical connectors.
7. Detach the vacuum hoses. Press the retaining ring.
8. Detach the injector harness electrical connector from the intake manifold support bracket.
9. Remove the intake manifold support bracket.
10. Detach the wiring harness.
11. Detach the intake manifold support bracket.
12. Remove and discard the O-ring seal.
13. Detach the throttle body.
14. Remove the retaining bolts.
15. Remove the intake manifold.
16. Remove and discard the intake manifold gaskets.

To install:

17. Install new intake manifold gaskets.

➡The intake manifold retaining bolts in position 1, 4 and 5 are longer than the retaining bolts in position 2, 3 and 6.

18. Install the intake manifold. Do not fully tighten the retaining bolts.
19. Attach the intake manifold support bracket. Do not fully tighten the retaining bolts.
20. Install the intake manifold support bracket.
21. Install new intake manifold gaskets. Tighten to 10 Nm.
22. Attach the injector harness electrical connector to the intake manifold support bracket. Install a new O-ring seal.
23. Attach the throttle body. Install the retaining bolts.
24. Connect the vacuum hose. Attach the wiring harness. Tighten to 10 Nm.
25. Attach the vacuum hoses. Press the retaining ring.
26. Attach the vacuum hose.
27. Connect the throttle body electrical connectors.
28. Install the air cleaner resonator.
29. Install the positive crankcase ventilation hose.
30. Install the air filter outlet pipe.
31. Install the engine cover (if equipped).

XJ Range

2002

1. Before servicing the vehicle, refer to the Precautions Section.
2. Drain the cooling system.
3. Remove or disconnect the following:
   - Negative battery cable
   - Mass Air Flow (MAF) meter
   - Air intake assembly
   - Throttle body
   - Fuel lines
   - Fuel injector wiring connectors
   - Intake manifold

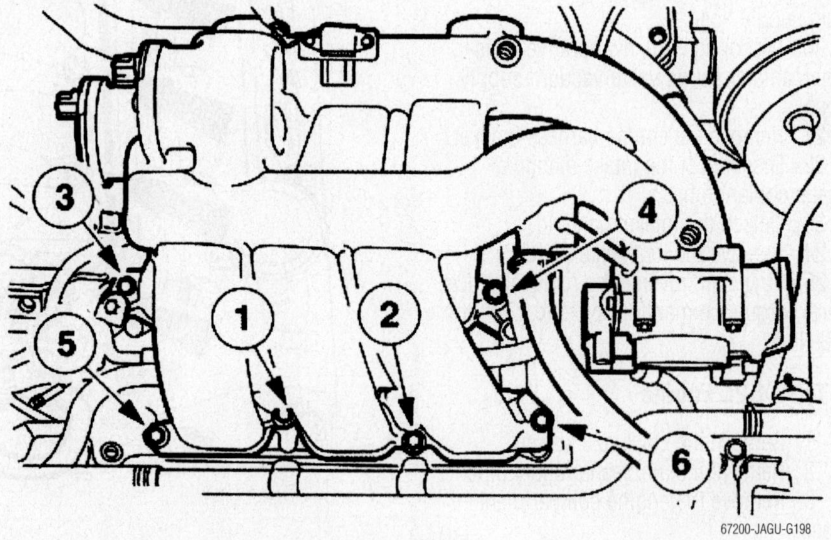

Upper intake manifold torque sequence—2.5L and 3.0L engines

To install:

4. Install or connect the following:
   - Intake manifold with a new gasket. Tighten the mounting bolts to 13–18 ft. lbs. (18–24 Nm).
   - Fuel injector wiring connectors
   - Fuel lines
   - Throttle body. Tighten the mounting bolts to 13–18 ft. lbs. (18–24 Nm).
   - Air intake assembly
   - MAF meter
   - Negative battery cable
5. Fill the cooling system.
6. Start the engine and check for leaks.

2003–06

1. Remove the cowl vent screen.
2. Remove the air cleaner outlet pipe.
3. Remove the engine compartment support.
4. Remove the on-plug coil covers.
5. Disconnect the throttle potentiometer (TP) electrical connector.
6. Disconnect the throttle motor electrical connector.

⁂ CAUTION

To avoid having scalding hot coolant or steam blow out of the cooling system, use extreme care when removing the coolant pressure cap from a hot cooling system. Wait until the engine has cooled, then wrap a thick cloth around the coolant pressure cap and turn it slowly until the pressure begins to release. Step back while the pressure is released from the system. When certain all the pressure has been released (still with a cloth) turn and remove the coolant pressure cap from the coolant expansion tank. Failure to follow these instructions may result in personal injury.

7. Remove the coolant expansion tank cap. Cap the coolant hose.
8. Disconnect the throttle body coolant hose.
9. Disconnect the knock sensors (KS) electrical connectors.
10. Remove the left-hand engine cover bracket.
11. Detach the engine wiring harness.
12. Disconnect the fuel injector electrical connectors.
13. Detach the fuel line.
14. Detach the left-hand camshaft position sensor electrical connector.
15. Disconnect the coolant temperature sensor electrical connector.
16. Remove the right-hand engine cover bracket.
17. Remove the positive crankcase ventilation (PCV) valve.
18. Disconnect the fuel pressure regulator electrical connector.
19. Disconnect the exhaust gas recirculation (EGR) valve electrical connector.
20. Cap the coolant hose.
21. Disconnect the exhaust gas recirculation (EGR) valve coolant hose.
22. Detach the right-hand engine wiring harness.
23. Disconnect the right-hand fuel injector electrical connectors.
24. Disconnect the fuel temperature sensor electrical connector.
25. Disconnect the right-hand camshaft position (CMP) sensor.
26. Detach the brake booster vacuum pipe.
27. Disconnect the manifold absolute pressure (MAP) sensor electrical connector.
28. Detach the engine wiring harness.
29. Detach the exhaust gas recirculation (EGR) valve tube.
30. Remove and discard the gasket.
31. Remove the intake manifold.
32. Remove and discard the gaskets.

To install:

33. Make sure that all gasket mating faces are clean and dry.
34. Install the intake manifold.
35. Install new gaskets. Tighten to 16 ft. lbs. (22 Nm).
36. Attach the exhaust gas recirculation (EGR) valve tube. Install a new gasket. Tighten to 16 ft. lbs. (22 Nm).
37. Attach the engine wiring harness. Tighten to 10 Nm.
38. Connect the manifold absolute pressure (MAP) sensor electrical connector.
39. Connect the right-hand camshaft position (CMP) sensor.
40. Connect the fuel temperature sensor electrical connector.
41. Attach the brake booster vacuum pipe.
42. Connect the right-hand fuel injector electrical connectors.
43. Attach the engine wiring harness.
44. Remove the cap from the coolant hose.
45. Connect the exhaust gas recirculation (EGR) valve coolant hose.
46. Connect the exhaust gas recirculation (EGR) valve electrical connector.
47. Connect the fuel pressure regulator electrical connector.
48. Install the positive crankcase ventilation (PCV) pipe.
49. Install the right-hand engine cover bracket. Tighten to 50 inch lbs. (6 Nm).
50. Connect the coolant temperature sensor electrical connector.
51. Attach the left-hand camshaft position sensor electrical connector.
52. Connect the fuel line. Install new O-ring seals.
53. Attach the fuel line retaining clip.
54. Connect the fuel injector electrical connectors.
55. Attach the engine wiring harness.
56. Install the left-hand engine cover bracket. Tighten to 6 Nm.
57. Connect the knock sensor (KS) electrical connectors.
58. Remove the cap from the coolant hose.
59. Connect the throttle body coolant hose.
60. Connect the throttle motor electrical connector.
61. Connect the throttle potentiometer (TP) electrical connector.
62. Install the on-plug coil covers.
63. Install the engine compartment support. Tighten to 13 ft. lbs. (18 Nm).
64. Install the air cleaner outlet pipe.
65. Install the cowl vent screen.
66. Check and top up the cooling system.

XK Range

4.0L ENGINE

1. Disconnect the battery ground cable.
2. Remove the battery cover.
3. Release the fuel system pressure.

⁂ CAUTION

Never remove the coolant expansion tank pressure cap under any circumstances while the engine is operating. Failure to follow this instruction may result in personal injury. To avoid hot coolant or steam blowing out of the cooling system, use extreme care when removing the coolant expansion tank pressure cap. Wait until the engine has cooled down, then insulate the coolant pressure cap with a suitable cloth and slowly loosen the coolant expansion tank pressure cap until the cooling system pressure is released. Do not remove the coolant expansion tank pressure cap. Step back while the pressure is released from the system. When all of the pressure has been released slowly remove the coolant expansion tank pressure cap (still with the suitable cloth in position) from the coolant expansion tank. Failure to follow this instruction may result in personal injury.

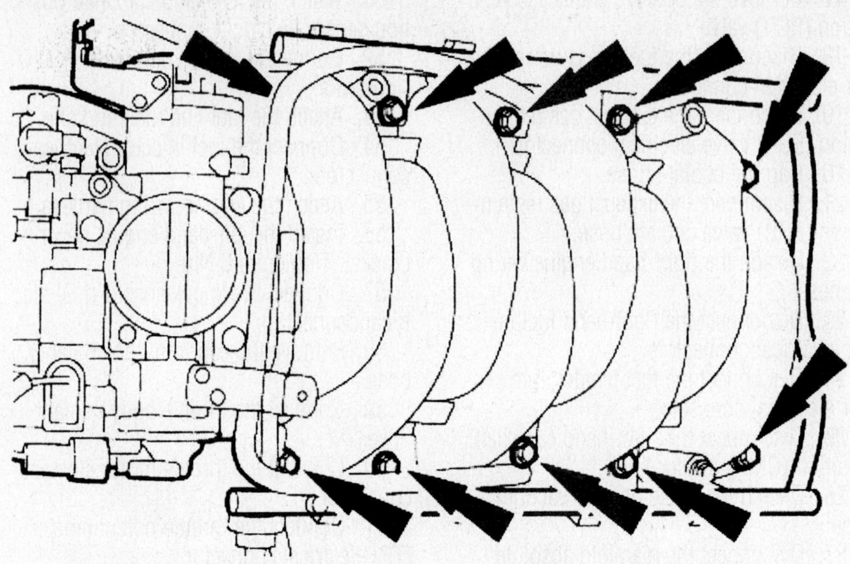

Intake manifold bolt locations—4.0L engine XK Range

4. Release the cooling system pressure.
5. Remove the coolant expansion tank pressure cap.
6. Disconnect the intake tube from the throttle body.
7. Remove the two bolts which secure the intake tube to the throttle body.
8. Disconnect the intake tube from the throttle body.
9. Disconnect the air cleaner cover.
10. Release the five clips which secure the air cleaner cover.
11. Disconnect the mass air flow meter electrical connector.
12. Disconnect the breather pipe from the camshaft cover.
13. Disconnect the mass air flow meter electrical connector.
14. Remove the intake tube/air flow meter/air cleaner cover assembly.
15. Remove the intake tube/air flow meter/air cleaner cover assembly.
16. Remove and discard the rubber seal; air intake tube to throttle body.
17. Detach the coolant pipe.
18. Remove the engine left-hand cover.
19. Disconnect the throttle body electrical connectors.
20. Detach the right-hand camshaft position (CMP) sensor wiring harness.
21. Disconnect the brake booster vacuum pipe.

➡Cap the coolant hoses to minimize coolant loss.

22. Disconnect the throttle body coolant hose.
23. Remove the throttle body elbow lower retaining bolts.

➡Cap the coolant hoses to minimize coolant loss.

24. Disconnect the throttle body coolant hose.
25. Disconnect the accelerator inner cable.
26. Reposition the accelerator pulley to the full throttle position.
27. Disconnect the accelerator inner cable.
28. Disconnect the accelerator outer cable.
29. Remove and discard the retaining strap.
30. Disconnect the accelerator outer cable.
31. Disconnect the throttle body electrical connectors.
32. Detach the left-hand CMP sensor wiring harness from the throttle body elbow.
33. Disconnect the knock sensor (KS) electrical connectors.
34. Disconnect the fuel injectors electrical connectors.

➡Place an absorbent cloth underneath fuel connection to absorb any spillages.

➡Cap the fuel line and fuel injection supply manifold to prevent contamination.

35. Disconnect the fuel feed hose.
36. Disconnect the fuel feed hose at the Quick Fit connector, using special tool 310-044, or equivalent.

➡Place an absorbent cloth underneath fuel connection to absorb any spillages.

➡Cap the fuel line and fuel injection supply manifold to prevent contamination.

37. Disconnect the fuel return hose.
38. Disconnect the fuel return hose at the Quick Fit connector, using special tool 310-054, or equivalent.
39. Disconnect the engine breather hose from the throttle body elbow.
40. Disconnect the vacuum hose from the throttle body elbow.
41. Disconnect the engine breather hose from the throttle body elbow.
42. Remove the intake manifold.
43. Remove and discard the eight seals from the intake manifold ports.
44. Clean all parts and relevant mating faces.

To install:

45. Install eight new seals to the intake manifold ports.
46. Install the intake manifold. Tighten to 16 ft. lbs. (21 Nm).
47. Connect the engine breather hose to the throttle body elbow.
48. Connect the vacuum hose to the throttle body elbow.
49. Connect the engine breather hose from the throttle body elbow.
50. Uncap the fuel line and fuel injection supply manifold.
51. Connect the fuel feed hose.
52. Ensure that the hose is fully seated to the feed pipe.
53. Uncap the fuel line and fuel injection supply manifold.
54. Connect the fuel return hose.
55. Connect the fuel return hose at the Quick Fit connector. Ensure that the hose is fully seated to the feed pipe.
56. Connect the fuel injectors electrical connectors.
57. Connect the KS electrical connectors.
58. Attach the right-hand camshaft sensor wiring harness to the throttle body elbow.
59. Connect the throttle body electrical connectors.
60. Connect the accelerator inner cable.
61. Reposition the accelerator pulley to the full throttle position.
62. Connect the accelerator inner cable.
63. Uncap the coolant hose.
64. Connect the coolant hose to the throttle body.
65. Install the intake elbow lower retaining bolts. Tighten to 13 ft. lbs. (18 Nm).
66. Connect the coolant hose to the throttle body.
67. Attach the brake servo vacuum pipe to the throttle body elbow.

68. Attach the left-hand CMP sensor wiring harness to the throttle body elbow.

69. Connect the throttle body electrical connectors.

70. Install the engine right-hand cover.

71. Install the engine left-hand cover.

72. Attach the coolant pipe.

73. Install the intake tube/air flow meter/air cleaner cover assembly.

74. Install a new rubber seal; air intake tube to throttle body.

75. Install the intake tube/air flow meter/air cleaner cover assembly.

76. Connect the mass air flow meter electrical connector.

77. Connect the breather pipe from the camshaft cover.

78. Connect the mass air flow meter electrical connector.

79. Reconnect the air cleaner cover.

80. Install the five clips which secure the air cleaner cover.

81. Connect the intake tube to the throttle body.

82. Install the two bolts which secure the intake tube to the throttle body.

83. Connect the intake tube from the throttle body.

84. Fill the cooling system to the correct level.

85. Install the coolant expansion tank pressure cap.

86. Connect the battery.

87. Adjust the accelerator cable.

4.2L ENGINE

1. Release the fuel system pressure.

2. Remove the air cleaner outlet pipe.

3. Remove the engine cover retaining brackets.

4. Detach the knock sensor (KS) electrical connectors.

5. Detach the engine wiring harness.

6. Disconnect the hose.

7. Disconnect the pipe.

8. Disconnect the exhaust gas recirculation (EGR) valve electrical connector.

9. Disconnect the fuel temperature sensor electrical connector.

10. Disconnect the right-hand camshaft position (CMP) sensor.

11. Detach the wiring harness.

12. Disconnect the electrical connectors.

13. Detach the wiring harness.

14. Disconnect the fuel pressure regulator electrical connector.

CAUTION

To avoid hot coolant or steam blowing out of the cooling system, use extreme care when removing the coolant expansion tank pressure cap. Wait until the engine has cooled down, then insulate the coolant pressure cap with a suitable cloth and slowly loosen the coolant expansion tank pressure cap until the cooling system pressure is released. Do not remove the coolant expansion tank pressure cap. Step back while the pressure is released from the system. When all of the pressure has been released slowly remove the coolant expansion tank pressure cap (still with the suitable cloth in position) from the coolant expansion tank. Failure to follow this instruction may result in personal injury.

15. Release the cooling system pressure.

16. Disconnect the coolant hose from the expansion tank.

17. Detach the wiring harness.

18. Disconnect the electrical connectors.

19. Detach the wiring harness.

20. Disconnect the electrical connector.

21. Disconnect the throttle body electrical connectors.

22. Disconnect the manifold absolute pressure (MAP) sensor electrical connector.

23. Disconnect the coolant hose.

24. Disconnect the coolant hose.

25. Disconnect the hose.

26. Remove the exhaust manifold to exhaust gas recirculation (EGR) valve tube heat shield retaining nut.

27. Remove the exhaust manifold to EGR valve tube heat shield.

28. Remove the exhaust manifold to EGR valve tube retaining bolts.

29. Remove and discard the gasket.

30. Remove the intake manifold.

31. Remove and discard the gaskets.

To install:

32. Install the intake manifold. Install new gaskets. Tighten to 16 ft. lbs. (22 Nm).

33. Install the exhaust manifold to EGR valve tube retaining bolts. Install a new gasket. Tighten to 16 ft. lbs. (21 Nm).

34. Install the exhaust manifold to EGR valve tube heat shield. Tighten to 80 inch lbs. (9 Nm).

35. Install the exhaust manifold to EGR valve tube heat shield retaining nut. Tighten to 44 ft. lbs. (60 Nm). Connect the hose.

36. Connect the coolant hose.

37. Connect the electrical connector.

38. Connect the manifold absolute pressure (MAP) sensor electrical connector.

39. Connect the throttle body electrical connectors.

40. Connect the CMP electrical connector.

41. Attach the wiring harness.

42. Connect the electrical connectors.

43. Attach the wiring harness.

44. Connect the hose to the expansion tank.

45. Connect the fuel pressure regulator electrical connector.

46. Attach the wiring harness.

47. Connect the electrical connectors.

48. Attach the wiring harness.

49. Connect the right-hand CMP sensor.

50. Disconnect the fuel temperature sensor electrical connector.

51. Connect the EGR valve electrical connector.

52. Connect the pipe. Connect the hose.

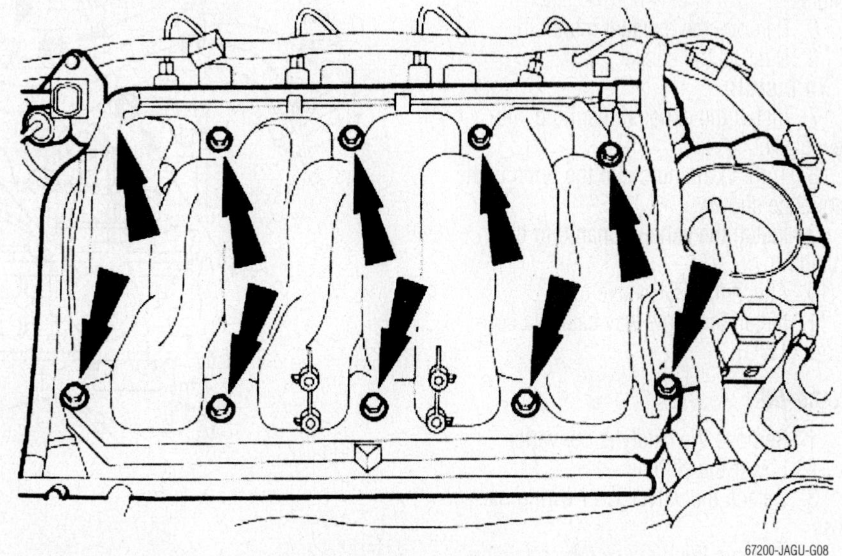

Intake manifold bolt locations—XK Range 4.2L engine

53. Attach the engine wiring harness.
54. Install the engine cover retaining brackets.
55. Detach the knock sensor (KS) electrical connectors. Tighten to 53 inch lbs. (6 Nm).
56. Install the air cleaner outlet pipe.
57. Check and top up the cooling system.

Exhaust Manifolds

REMOVAL & INSTALLATION

S-Type 3.0L Engine

2002 LEFT SIDE

1. Remove the three way catalytic converter (TWC).
2. Lower the vehicle.
3. Remove the exhaust manifold heat shield.
4. Remove the exhaust manifold.
5. Discard the gasket.

To install:

6. Install the exhaust manifold with a new gasket.
7. Tighten the nuts and the bolts in the sequence shown.
8. Install the exhaust manifold heat shield.
9. Install the three way catalytic converter (TWC).

2002 RIGHT SIDE

1. Remove the three way catalytic converter (TWC).
2. Lower the vehicle.
3. Detach the EGR valve tube.
4. Remove the exhaust manifold heat shield.
5. Remove the exhaust manifold.
6. Discard the gasket.

To install:

7. Install the exhaust manifold with a new gasket.
8. Tighten the nuts and the bolts in the sequence shown.
9. Install the exhaust manifold heat shield.
10. Attach the EGR valve tube.
11. Install the three way catalytic converter (TWC).

2003–06

1. Remove the catalytic converter.
2. Lower the vehicle.
3. Detach the valve cover wiring harness.
4. Remove the dipstick tube retaining bolt.

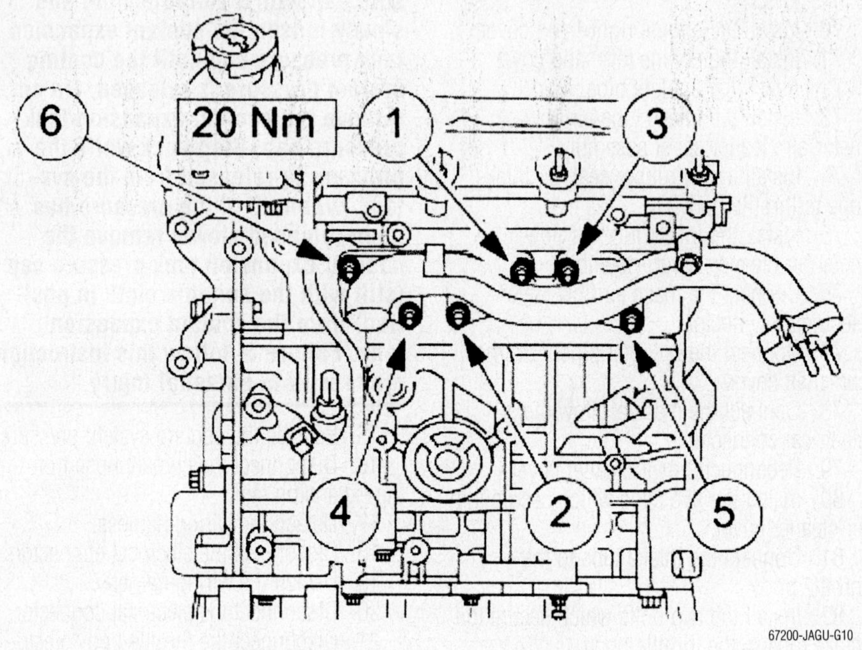

Left side exhaust manifold torque sequence—S-Type 3.0L engine

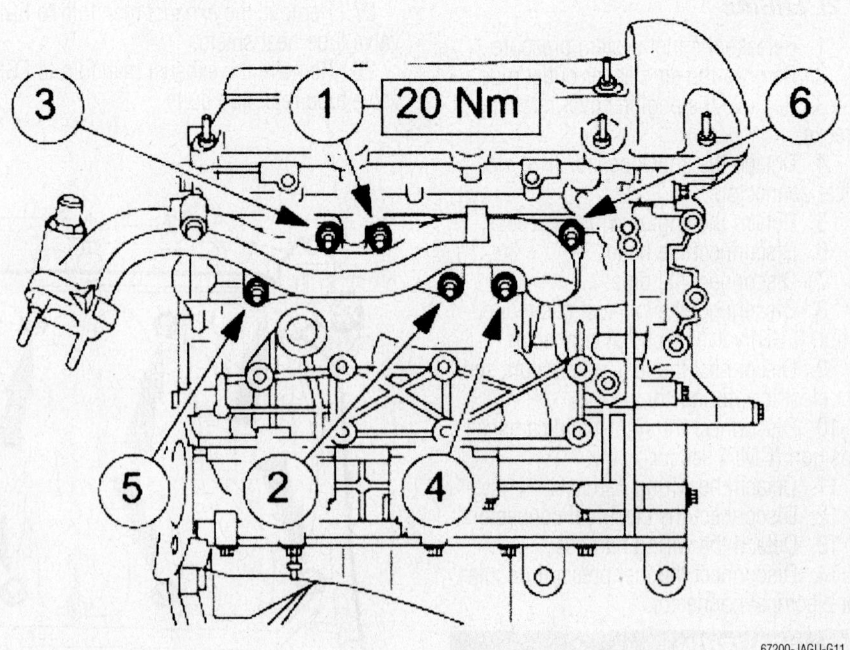

Right side exhaust manifold torque sequence—S-Type 3.0L engine

5. Reposition the dipstick tube (left side).
6. Remove the exhaust manifold heat shield.
7. Raise the vehicle.
8. Remove the exhaust manifold lower retaining nuts.
9. Lower the vehicle.
10. Remove the exhaust manifold.
11. Remove and discard the exhaust gasket.
12. To install, reverse the removal procedure.
13. Install a new gasket. Tighten to 15 ft. lbs. (25 Nm).

S-Type 4.0L

LEFT SIDE

1. Detach the power assisted steering reservoir.
2. Remove the air cleaner. Remove the oil deposit tube.
3. Remove the left-hand three way catalytic converter.
4. Remove the exhaust manifold.
5. To install, reverse the removal procedure.

RIGHT SIDE

1. Remove the air intake tube.
2. Remove the right-hand three way catalytic converter.
3. Remove the exhaust manifold.

S-Type 4.2L

1. Remove the exhaust system.
2. Remove the exhaust manifold heat shield.
3. Detach the exhaust gas recirculation (EGR) valve.
4. Remove and discard the retaining nuts and gasket.
5. Remove the exhaust manifold heat shield/gasket retaining bolt.

➡New exhaust manifold retaining studs must be fitted if the old studs are removed.

6. Remove the exhaust manifold.
7. Remove and discard the retaining nuts and heat shield/gasket.

To install:

➡When all retaining nuts have been torqued, re-torque all retaining nuts to 18 ft. lbs. (25 Nm).

➡Ensure exhaust manifold and heat shield/gasket is correctly aligned to cylinder head and studs.

8. Install the exhaust manifold.
9. Install new retaining nuts and a new heat shield/gasket. Tighten to 18 ft. lbs. (25 Nm).
10. Install the exhaust manifold heat shield/gasket retaining bolt. Tighten to 50 Nm.
11. Attach the exhaust gas recirculation (EGR) valve.
12. Install new retaining nuts and a new gasket. Tighten to 16 ft. lbs. (22 Nm).
13. Install the exhaust manifold heat shield. Tighten to 27 inch lbs. (3 Nm).
14. Install the exhaust system.

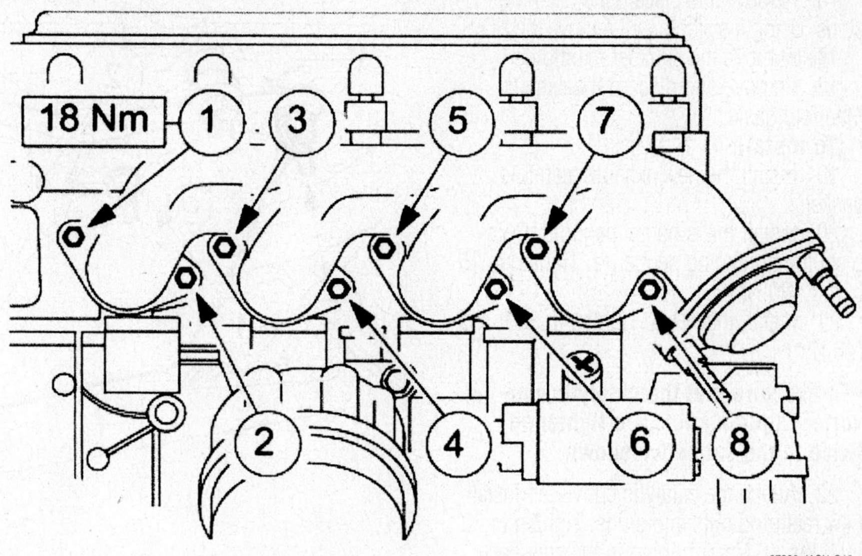

Left exhaust manifold bolt torque sequence—2002 S-Type 4.0L

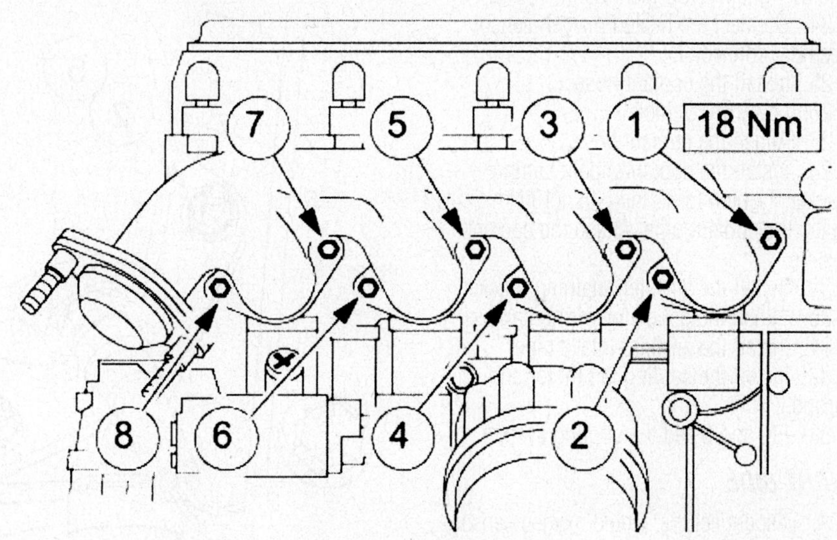

Right exhaust manifold bolt torque sequence—2002 S-Type 4.0L

X-Type 2.5L and 3.0L

LEFT SIDE

1. Drain the cooling system.
2. Remove the cooling fan motor and shroud.
3. Lower the vehicle.
4. Remove the air filter intake pipe.
5. Detach the generator wiring harness.
6. Remove the air filter retaining bracket.
7. Detach the coolant pipe.
8. Detach the coolant hose.
9. Raise the vehicle.
10. Remove the coolant hose.
11. Disconnect the heated oxygen sensor electrical connector.
12. Remove the exhaust manifold heat shield retaining bolt.
13. Lower the vehicle.
14. Remove the exhaust manifold heat shield.
15. Detach the catalytic converter. Discard the retaining nuts.

16. Remove and discard the retaining studs. Using a suitable tool remove the HO_2S.
17. Remove the exhaust manifold.
18. Remove and discard the exhaust manifold gasket.

To install:

19. Install the new exhaust manifold gasket.
20. Install the exhaust manifold. Complete the tightening sequence. Tighten to 15 ft. lbs. (20 Nm).
21. Install the HO_2S. Tighten to 30 ft. lbs. (40 Nm).

➡**Make sure that the catalytic converter retaining nuts are tightened twice in the sequence shown.**

22. Attach the catalytic converter. Install new retaining nuts and studs. Tighten to 80 inch lbs. (9 Nm). Tighten in the sequence shown to 19 ft. lbs. (25 Nm).
23. Install the exhaust manifold heat shield. Tighten to 98 inch lbs. (11 Nm).
24. Connect the heated oxygen sensor electrical connector.
25. Install the coolant hose.
26. Lower the vehicle.
27. Attach the coolant hose.
28. Install the coolant pipe retaining bracket. Tighten to 89 inch lbs. (10 Nm). Attach the coolant pipe. Attach the coolant hose.
29. Install the air filter retaining bracket.
30. Attach the generator wiring harness.
31. Install the air filter intake pipe.
32. Install the cooling fan motor and shroud.
33. Fill and bleed the cooling system.

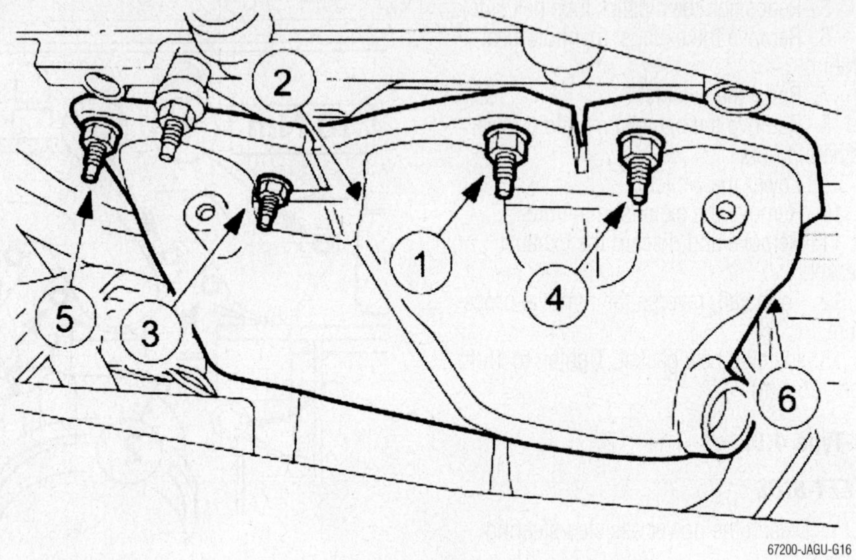

Left side exhaust manifold torque sequence—X-Type 2.5L and 3.0L engine

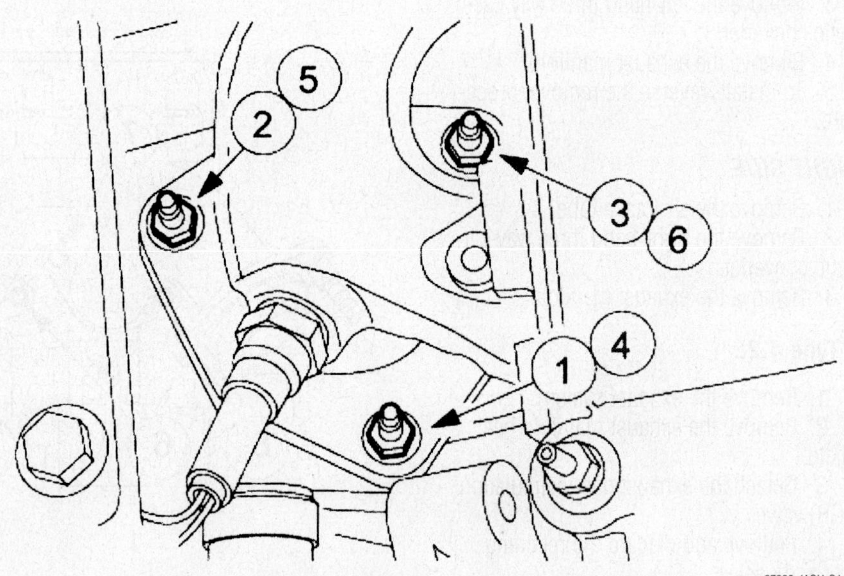

Catalytic converter torque sequence—2.5L and 3.0L

RIGHT SIDE

1. Disconnect the heated oxygen sensor electrical connector. Remove the three way catalytic converter (TWC).
2. Using a suitable tool remove the HO_2S.
3. Remove the exhaust manifold.
4. Remove and discard the exhaust manifold gasket.

To install:

5. Install the exhaust manifold.
6. Install the new exhaust manifold gasket. Complete the tightening sequence. Tighten to 15 ft. lbs. (20 Nm).
7. Install the three way catalytic converter (TWC). Tighten to 30 ft. lbs. (40 Nm).

XJ Range

1. Remove the exhaust system.
2. Remove the right-hand exhaust manifold heat shield.
3. Detach the exhaust gas recirculation valve.
4. Remove and discard the retaining nuts and gasket.
5. Remove the exhaust manifold heat shield/gasket retaining bolt.

➡**New exhaust manifold retaining studs must be fitted if the old studs are removed.**

6. Remove the exhaust manifold.
7. Remove and discard the retaining nuts and heat shield/gasket.

To install:

➡**Ensure exhaust manifold and heat shield/gasket is correctly aligned to cylinder head and studs.**

8. Install the exhaust manifold.

➡**When all retaining nuts have been torqued, re-torque all retaining nuts to 18 ft. lbs. (25 Nm).**

9. Install new retaining nuts and a new heat shield/gasket. Tighten to 18 ft. lbs. (25 Nm).
10. Install the exhaust manifold heat shield/gasket retaining bolt. Tighten to 37 ft. lbs. (50 Nm).
11. Attach the exhaust gas recirculation valve. Install new retaining nuts and a new gasket. Tighten to 16 ft. lbs. (22 Nm).
12. Install the exhaust manifold heat shield. Tighten to 26 inch lbs. (3 Nm).
13. Install the exhaust system.

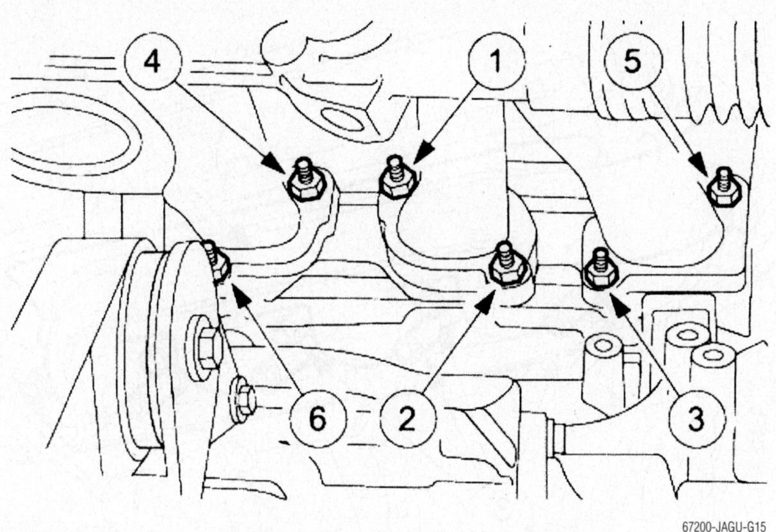

Right side exhaust manifold torque sequence—X-Type 2.5L and 3.0L engine

3. Remove the down pipe catalytic converter, right hand side.
4. From above the vehicle, disconnect the EGR pipe from the exhaust manifold.
5. Remove the two nuts which secure the EGR flange to the right hand exhaust manifold.
6. Reposition the EGR pipe from the manifold.
7. Remove the bolts which secure the front section of the exhaust manifold to the cylinder head.
8. Working from below the vehicle, remove the manifold bolts.
   a. Remove the bolts which secure the rear section of the exhaust manifold to the cylinder head.
   b. Remove the manifold and the manifold gasket from the cylinder head.

XK Range

LEFT SIDE

1. Remove the front silencer.
2. Remove the down pipe catalytic converter, left hand side.
3. Working from above the vehicle, remove the bolts which secure the front section of the exhaust manifold to the cylinder head.
4. Working from below the vehicle, reposition the steering rack for access.
   a. Remove the two bolts which secure the steering rack at the driver's side.
   b. Loosen, but do not remove, the two bolts which secure the steering rack at the passenger side.
   c. Reposition the steering rack for access.
5. Working from below the vehicle, remove the manifold.
   a. Remove the bolts which secure the rear section of the exhaust manifold to the cylinder head.
   b. Remove the manifold and the manifold gasket.
6. Clean the manifold faces (if refitting) and the mating faces on the cylinder head.
7. Use a thread tap (plug) to clean out the threads of the manifold securing bolt holes, in the cylinder head.
8. Installation is the reverse of removal. Use new gaskets and new securing bolts for the manifold.

RIGHT SIDE

2002

1. Remove the air cleaner assembly.
2. Remove the front silencer.

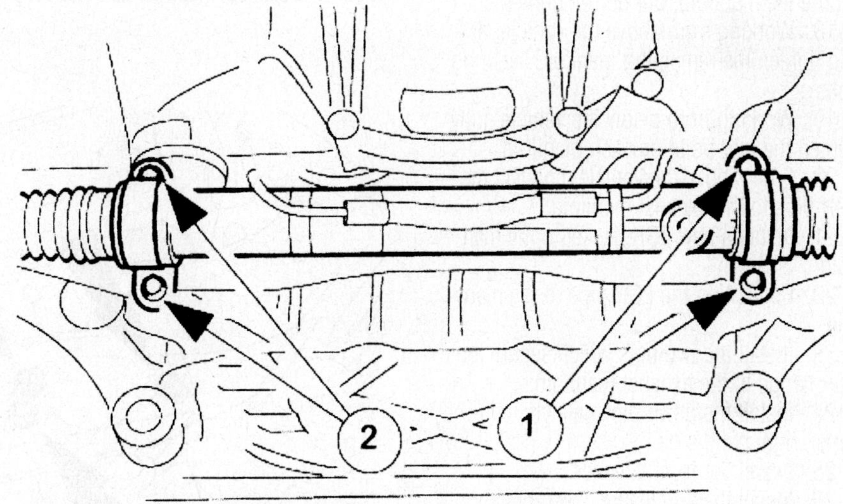

Steering rack bolt locations—XK Range

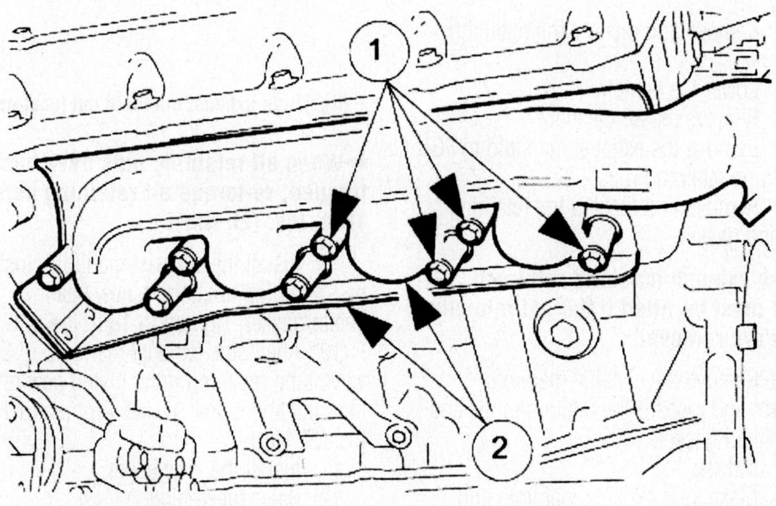

Exhaust manifold bolt locations—XK Range

9. Remove the gasket from the EGR flange.
10. If fitting a new manifold, remove the two studs from the EGR flange of the manifold, and transfer them to the new manifold.
11. Clean the manifold faces (if refitting) and the mating faces on the cylinder head.
12. Use a thread tap (plug) to clean out the threads of the manifold securing bolt holes, in the cylinder head.
13. Install a spacer to each new manifold securing bolt (8 bolts).
14. Install the manifold and manifold gasket.
15. Install and align the new gasket to the cylinder head.
16. Install and align the manifold to the cylinder head.
17. Install the rear four bolts which secure the manifold, but do not tighten.
18. Working from above the vehicle, fit and tighten the remaining manifold securing bolts.
19. Working from below the vehicle, fully tighten the rear bolts on each manifold.
20. From above the vehicle, connect the EGR pipe to the exhaust manifold.
21. Install a new EGR gasket to the manifold.
22. Reposition the EGR pipe to the manifold.
23. Install the two nuts which secure the EGR flange to the exhaust manifold.
24. Install the down pipe catalytic converter, right hand side.
25. Install the front silencer.
26. Install the air cleaner assembly.

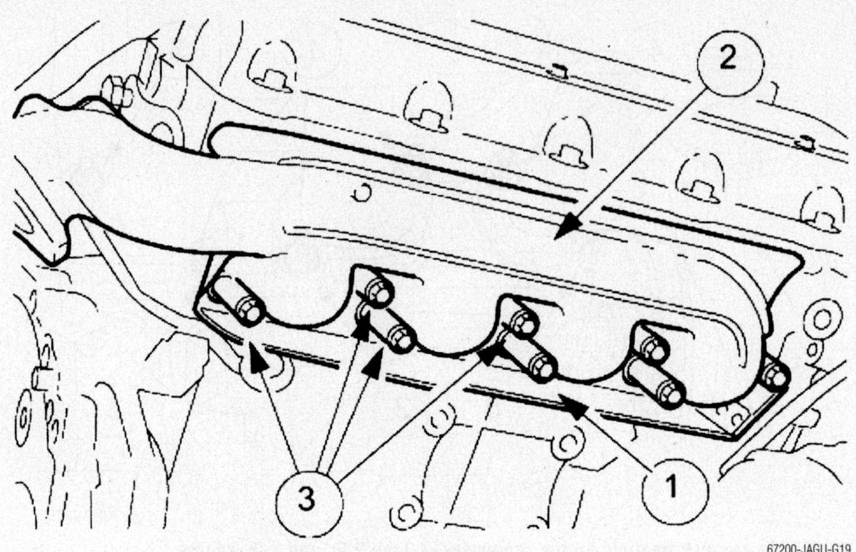

Right side exhaust manifold bolt locations—2002 XR Range

Right side exhaust manifold nut locations—2003–04 XK Range

2003–06

1. Remove the right-hand catalytic converter.
2. Remove and discard the retaining nuts.
3. Lower the vehicle.
4. Remove the air cleaner.
5. Remove the exhaust manifold to EGR valve tube retaining nuts.
6. Remove and discard the retaining nuts and gasket.

➡New exhaust manifold retaining studs must be fitted if the old retaining studs are removed.

7. Remove the exhaust manifold. Remove and discard the retaining nuts and heat shield/gasket.

To install:

8. Make sure exhaust manifold and heat shield/gasket is correctly aligned to cylinder head and studs.

➡When all retaining nuts have been torqued, re-torque all retaining nuts to 18 ft. lbs. (25 Nm).

9. Install the exhaust manifold. Install new retaining nuts and a new heat shield/gasket. Tighten to 18 ft. lbs. (25 Nm).
10. Attach the exhaust manifold to EGR valve tube retaining nuts. Install new retaining nuts and a new gasket. Tighten to 16 ft. lbs. (22 Nm).
11. Install the air cleaner.
12. Raise the vehicle.
13. Install the right-hand catalytic converter.

Camshaft

REMOVAL & INSTALLATION

XJ Range and S-Type 4.0L and 4.2L Engines

1. Before servicing the vehicle, refer to the Precautions Section.
2. Drain the cooling system.
3. Remove or disconnect the following:
   - Negative battery cable
   - Accessory drive belt

- Valve cover
- Timing chain cover
- Intake manifold
- Radiator and heater hoses
- Engine Coolant Temperature (ECT) sensor connector
- Coolant outlet pipe
- Exhaust front pipes
- Variable Valve Timing (VVT) sensor
- Crankshaft Position (CKP) sensor

4. Rotate the crankshaft until the triangular arrow indent on the driveplate is visible through the access hole.
5. Install the Crankshaft Setting Peg JD 216 into the CKP sensor location.
6. Install the Camshaft Locking tool JD 215 on the camshafts.
7. Remove or disconnect the following:
   - Camshaft sprocket mounting bolt by loosening it
   - VVT mounting bolt by loosening it
   - Camshaft Locking tool JD 215.
   - Primary timing chain tensioner and back plate
   - Primary timing chain guide
   - VVT unit and exhaust camshaft sprocket
   - Secondary timing chain tensioner and guide
   - Camshafts. Loosen the bearing cap bolts evenly and in several passes.

To install:

8. Install or connect the following:
   - Camshafts. Tighten the bearing cap bolts evenly, in stages, to 84–96 inch lbs. (9–11 Nm).
9. Install the Camshaft Locking tool JD 215.
10. Prepare the timing chain tensioners for installation by using a paperclip or other wire to unseat the check valves and compressing the pistons into their bores.
11. Install or connect the following:
    - Secondary timing chain guide. Tighten the bolt to 89–124 inch lbs. (10–14 Nm).
    - Secondary timing chain tensioner. Tighten the bolt to 89–124 inch lbs. (10–14 Nm).
    - VVT, secondary timing chain, and exhaust cam sprocket
    - Primary timing chain
    - Primary timing chain guide. Tighten the bolt to 10–12 ft. lbs. (13–16 Nm).
    - Primary timing chain tensioner and backing plate. Tighten the bolts to 89–124 inch lbs. (10–14 Nm).
    - Primary timing chain slack, eliminate it by placing a wedge between the primary timing chain tensioner and the guide shoe.

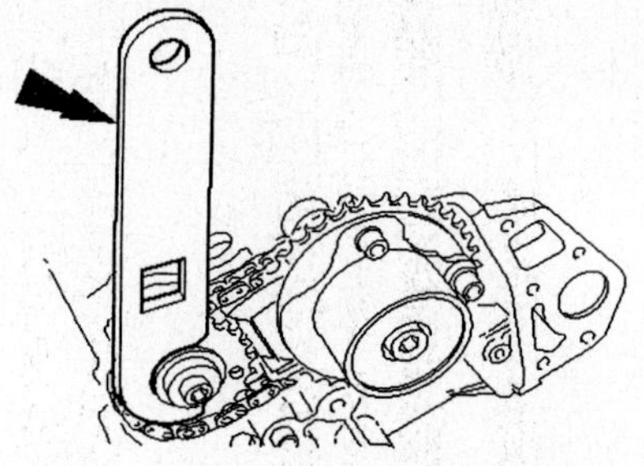

Apply force in a counterclockwise direction when tightening the sprocket mounting bolts

- Secondary timing chain, tension it by applying counterclockwise force to the exhaust camshaft sprocket
- Exhaust and intake VVT sprocket bolts. Tighten them to 85–92 ft. lbs. (115–125 Nm).

12. Remove the tools and wedges.
13. Install or connect the following:
    - CKP sensor
    - VVT sensor
    - Exhaust front pipes
    - Coolant outlet pipe
    - ECT sensor connector
    - Radiator and heater hoses
    - Intake manifold
    - Timing chain cover
    - Valve cover
    - Accessory drive belt
    - Negative battery cable
14. Fill the cooling system.
15. Start the engine and check for leaks.

XK

A-BANK

1. Before servicing the vehicle, refer to the Precautions Section.
2. Drain the cooling system.
3. Remove or disconnect the following:
   - Negative battery cable
   - Valve cover
   - Intake manifold
   - Radiator hoses
   - Heater hoses
   - Engine Coolant Temperature (ECT) sensor connector
   - Coolant outlet pipe
   - Catalytic converter
   - Crankshaft damper
   - Front cover
   - Variable Valve Timing (VVT) bushing carrier
   - Crankshaft Position (CKP) sensor
4. Rotate the crankshaft until the triangular arrow indent on the driveplate is visible through the access hole.
5. Confirm that the timing flat on each camshaft is uppermost. If necessary, rotate the crankshaft one full turn.
6. Install the crankshaft setting peg JD 216 to the crankshaft position sensor location.
7. Install the camshaft locking tool JD 215 to the A-Bank camshafts.
8. Remove or disconnect the following:
   - Primary timing chain tensioner and backing plate
   - Primary timing chain
   - VVT unit, exhaust camshaft sprocket and secondary timing chain
   - Secondary timing chain tensioner
   - Secondary timing chain guide
   - Camshaft locking tool
   - Camshaft bearing caps. Loosen the bolts evenly in stages.
   - Camshafts

To install:

9. Install or connect the following:
   - Camshafts. Tighten the bearing cap bolts evenly, in stages, to 84–96 inch lbs. (9–11 Nm).
10. Install the Camshaft Locking tool JD 215.
11. Prepare the timing chain tensioners for installation by using a paperclip or other wire to unseat the check valves and compressing the pistons into their bores.
12. Install or connect the following:
    - Secondary timing chain guide. Tighten the bolt to 89–124 inch lbs. (10–14 Nm).
    - Secondary timing chain tensioner. Tighten the bolt to 89–124 inch lbs. (10–14 Nm).
    - VVT, secondary timing chain, and exhaust cam sprocket
    - Primary timing chain
    - Primary timing chain guide. Tighten the bolt to 10–12 ft. lbs. (13–16 Nm).
    - Primary timing chain tensioner and backing plate. Tighten the bolts to 89–124 inch lbs. (10–14 Nm).
    - Primary timing chain slack, eliminate it by placing a wedge between the primary timing chain tensioner and the guide shoe.
    - Secondary timing chain, tension it by applying counterclockwise force to the exhaust camshaft sprocket.
    - Exhaust and intake VVT sprocket bolts. Tighten them to 85–92 ft. lbs. (115–125 Nm).
13. Remove the tools and wedges.
14. Install or connect the following:
    - CKP sensor
    - VVT bushing carrier

- Front cover
- Crankshaft damper
- Catalytic converter
- Coolant outlet pipe
- ECT sensor connector
- Heater hoses
- Radiator hoses
- Intake manifold
- Valve cover. Tighten the bolts in sequence to 84–96 inch lbs. (9–11 Nm).
- Negative battery cable

15. Fill the cooling system.
16. Start the engine and check for leaks.

B-BANK

1. Before servicing the vehicle, refer to the Precautions Section.
2. Drain the cooling system.
3. Remove or disconnect the following:
   - Negative battery cable
   - Valve cover
   - Intake manifold
   - Radiator hoses
   - Heater hoses
   - Engine Coolant Temperature (ECT) sensor connector
   - Coolant outlet pipe
   - Catalytic converter
   - Crankshaft damper
   - Front cover
   - Variable Valve Timing (VVT) bushing carrier
   - Crankshaft Position (CKP) sensor
4. Rotate the crankshaft until the triangular arrow indent on the driveplate is visible through the access hole.
5. Confirm that the timing flat on each camshaft is uppermost. If necessary, rotate the crankshaft one full turn.
6. Install the crankshaft setting peg JD 216 to the crankshaft position sensor location.
7. Install the camshaft locking tool JD 215 to the B-Bank camshafts.
8. Remove or disconnect the following:
   - Primary timing chain tensioner and backing plate
   - Primary timing chain
   - VVT unit, exhaust camshaft sprocket and secondary timing chain
   - Secondary timing chain tensioner
   - Secondary timing chain guide
   - Camshaft locking tool

➡Keep all valvetrain components in order for assembly.

- Camshaft bearing caps. Loosen the bolts evenly in stages.
- Camshafts

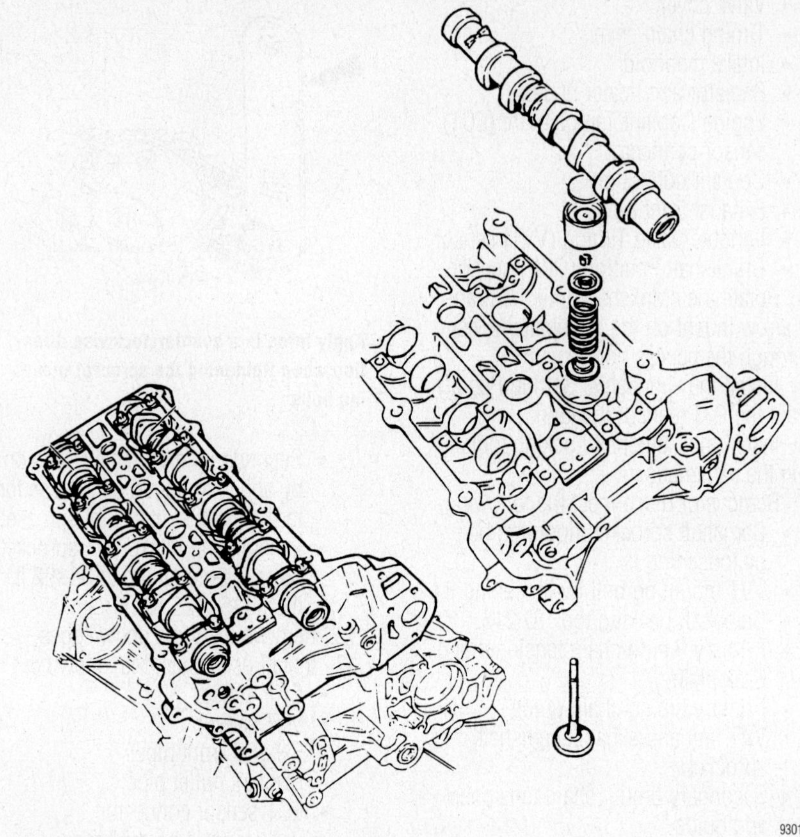

Exploded view of the camshaft and valve installation—V8 shown

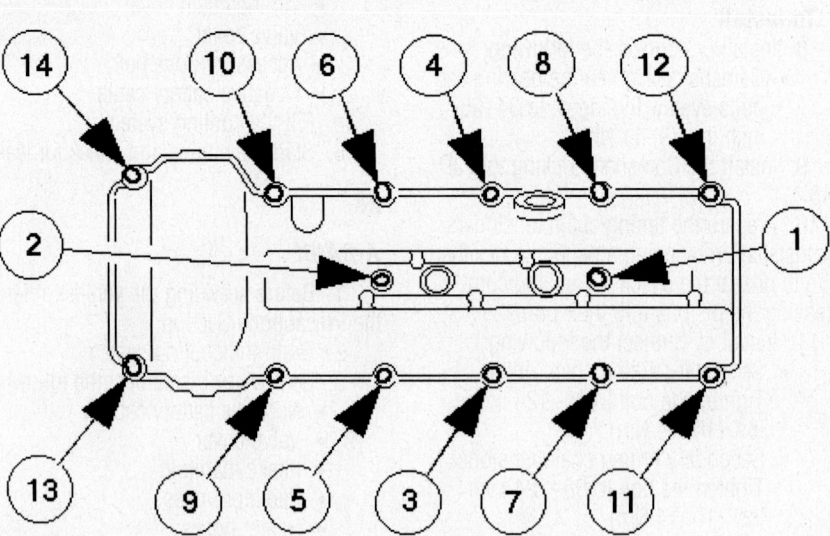

14 10 6 4 8 12
2 1
13 9 5 3 7 11
9301JG07

Camshaft cover bolt torque sequence—V8 A-bank

To install:

9. Install or connect the following:
   - Camshafts. Tighten the bearing cap bolts evenly, in stages, to 84–96 inch lbs. (9–11 Nm).
10. Install the Camshaft Locking tool JD 215.
11. Prepare the timing chain tensioners for installation by using a paperclip or other wire to unseat the check valves and compressing the pistons into their bores.
12. Install or connect the following:
    - Secondary timing chain guide. Tighten the bolt to 89–124 inch lbs. (10–14 Nm).
    - Secondary timing chain tensioner. Tighten the bolt to 89–124 inch lbs. (10–14 Nm).

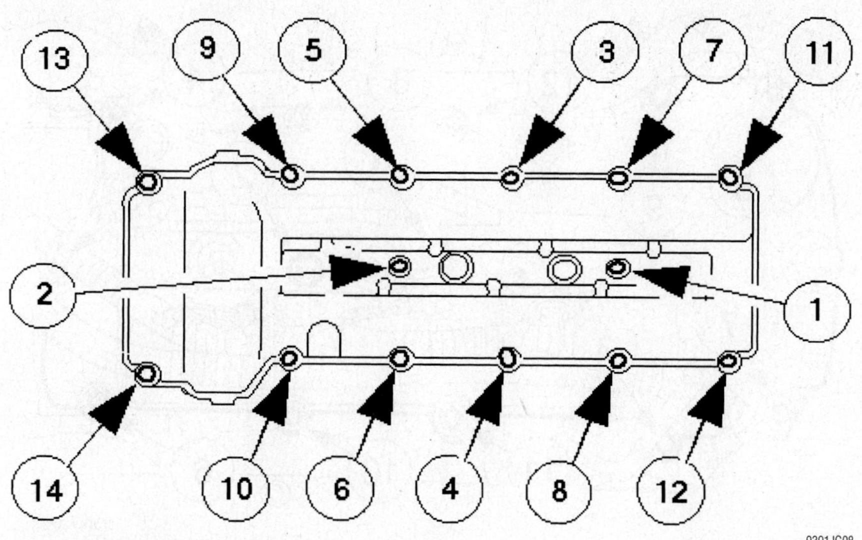

Camshaft cover bolt torque sequence—V8 B-bank

- VVT, secondary timing chain, and exhaust cam sprocket
- Primary timing chain
- Primary timing chain guide. Tighten the bolt to 10–12 ft. lbs. (13–16 Nm).
- Primary timing chain tensioner and backing plate. Tighten the bolts to 89–124 inch lbs. (10–14 Nm).
- Primary timing chain slack, eliminate it by placing a wedge between the primary timing chain tensioner and the guide shoe.
- Secondary timing chain, tension it by applying counterclockwise force to the exhaust camshaft sprocket.
- Exhaust and intake VVT sprocket bolts. Tighten them to 85–92 ft. lbs. (115–125 Nm).

13. Remove the tools and wedges.
14. Install or connect the following:
    - CKP sensor
    - VVT bushing carrier
    - Front cover
    - Crankshaft damper
    - Catalytic converter
    - Coolant outlet pipe
    - ECT sensor connector
    - Heater hoses
    - Radiator hoses
    - Intake manifold
    - Valve cover. Tighten the bolts in sequence to 84–96 inch lbs. (9–11 Nm).
    - Negative battery cable
15. Fill the cooling system.
16. Start the engine and check for leaks.

S-Type and X-Type 2.5L and 3.0L Engines

1. Before servicing the vehicle, refer to the Precautions Section.
2. Remove or disconnect the following:
    - Negative battery cable
    - Valve covers
    - Front cover
    - Timing chains

⁂ WARNING

The camshaft journal thrust caps must be removed before loosening the remaining camshaft journal cap bolts to ensure that the camshaft journal thrust caps are not damaged.

➡Keep all valvetrain components in order for assembly.

- Camshaft journal thrust caps
- Remaining camshaft journal caps
- Camshafts
- Valve tappets and shims

To install:

3. Install or connect the following:
    - Valve tappets and shims in their original locations
    - Camshafts
    - Camshaft journal caps in their original positions. Install the thrust journal caps last. Tighten the bolts in sequence to 89 inch lbs. (10 Nm).
    - Timing chains
    - Front cover
    - Valve covers
    - Negative battery cable
4. Start the engine and check for leaks.

Valve Lash

ADJUSTMENT

V6

1. Remove the left hand valve cover.
    a. Remove the ignition coils.
    b. Remove the air filter intake pipe.
    c. Detach the wiring harness.
    d. Remove the air filter retaining bracket.
    e. Detach the positive crankcase ventilation hose.
    f. Disconnect the variable camshaft timing (VCT) solenoid electrical connector.
    g. Detach the coolant hose.

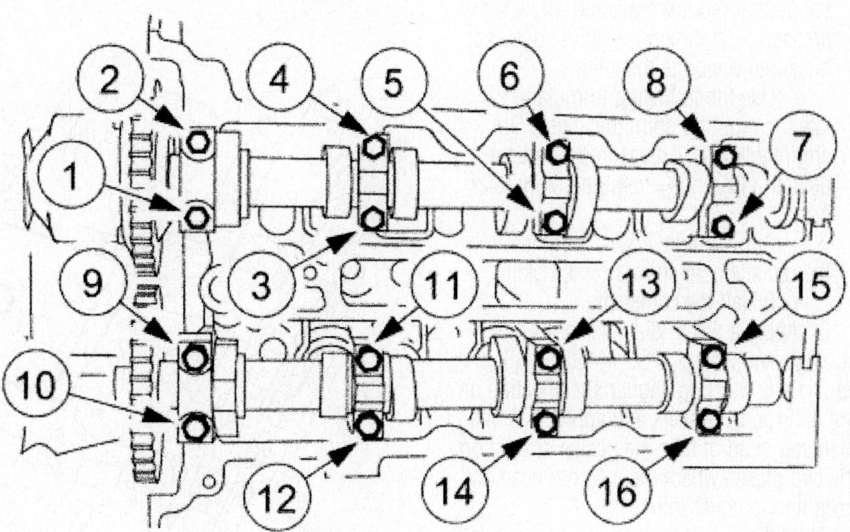

Camshaft journal cap torque sequence—2.5L and 3.0L engine

h. Remove the valve cover.
i. Remove and discard the valve cover gaskets.
2. Remove the right hand valve cover.
a. Remove the ignition coils.
b. Detach the wiring harness.
c. Disconnect the variable camshaft timing (VCT) solenoid electrical connector.
d. Remove the engine cover retaining bracket.
e. Remove the intake manifold support bracket.
f. Remove the valve cover.
g. Remove and discard the valve cover gaskets.

⁂ WARNING

Rotating the crankshaft in a counterclockwise direction may cause engine damage. Crankshaft journals are directionally machined. Rotating the crankshaft counterclockwise can raise burrs on bearing surfaces, reducing engine life.

➡Camshaft lobes must be 180 degrees away from each valve shim or valve clearance measurement will be incorrect.

3. Rotate the engine clockwise to position the camshaft lobe away from the shim surface.
4. Using a feeler gauge set, measure the clearance between the camshaft and the shim surface. Record and check the readings.
5. Adjust the clearances as necessary.
a. Remove the camshafts.
b. Use compressed air to remove the shims that require replacing. Blow compressed air between the shim edge and bucket to dislodge the shim.
c. Use the following formula to calculate the required shim thickness. Original shim thickness + measured clearance – desired clearance = required shim thickness.
d. Apply a light coat of engine oil to the replacement shim(s) and install.
e. Install the camshafts.
6. Apply a 5 mm diameter bead of silicone gasket sealant WSE-M4G323-A6, or equivalent, meeting Jaguar specification on the half round gaskets and apply an 8 mm diameter bead of silicone gasket sealant on the two places where the cylinder head and front timing cover join.
7. Make sure that the valve cover isolator mounts are correctly installed to the new valve cover gaskets.

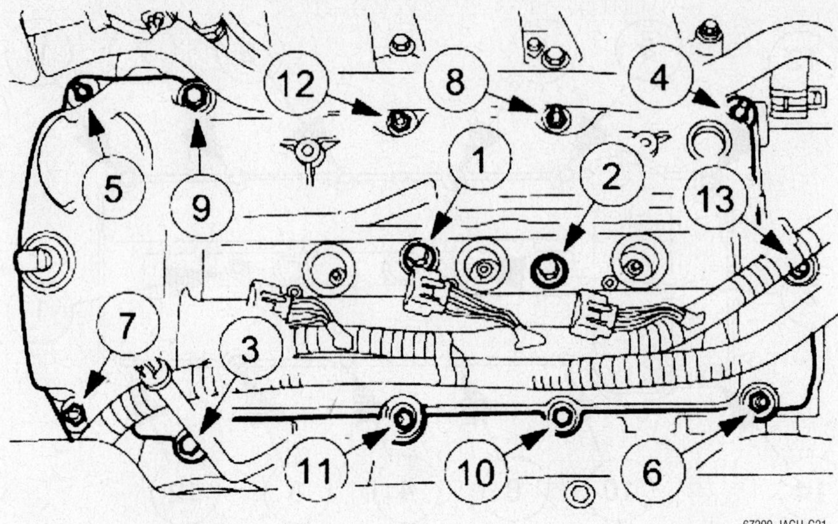

Left side valve cover torque sequence

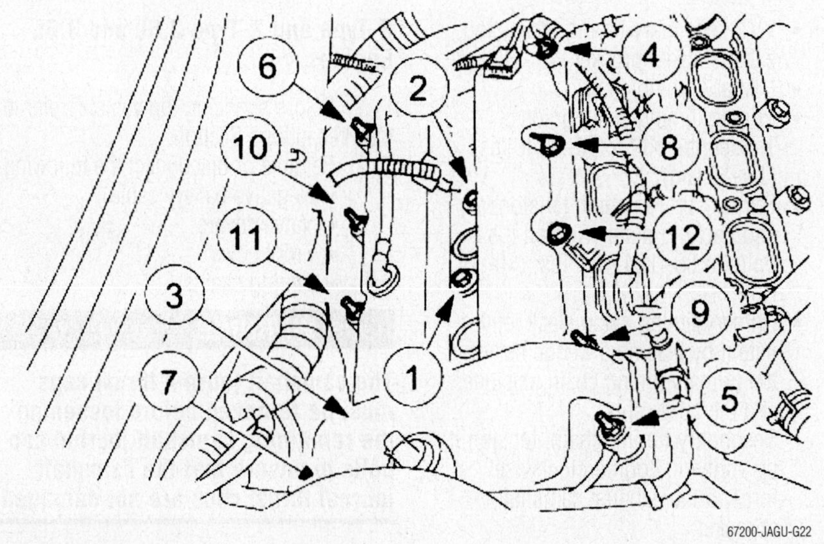

Right side valve cover torque sequence

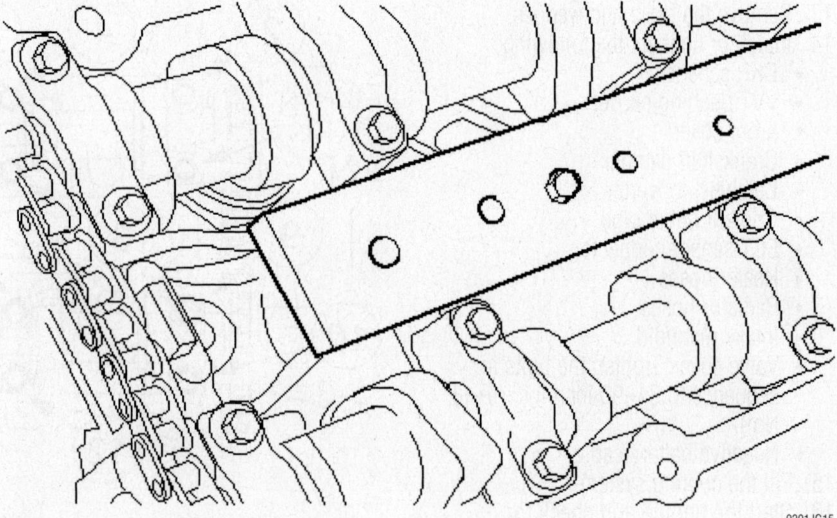

Valve adjustment tool base plate—V8

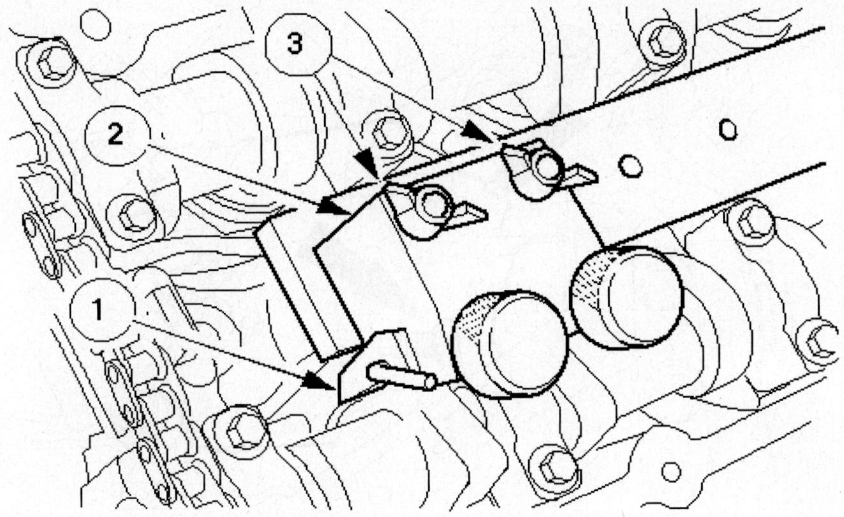

Valve adjustment tool attachment—V8

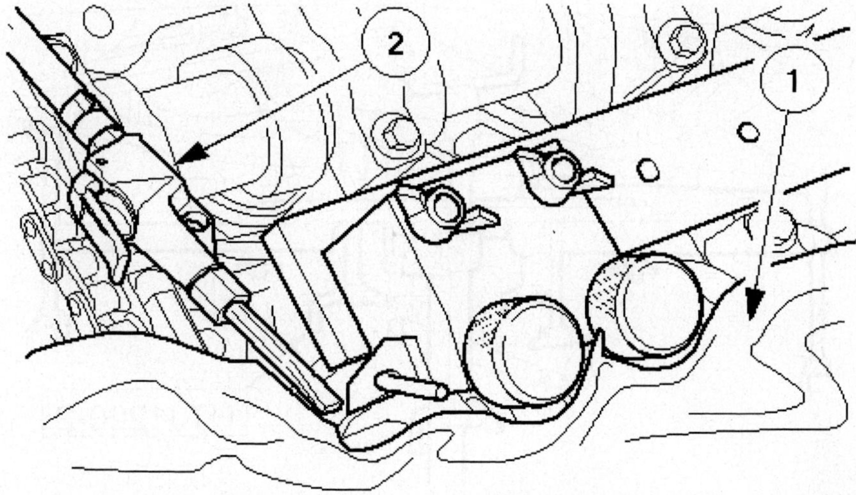

Remove the shims with compressed air—V8

8. To install, reverse the removal procedure. Install new valve cover gaskets. Install the valve cover. Torque the bolts, in the sequence shown, to 89 inch lbs. (10 Nm).

V8

1. Before servicing the vehicle, refer to the Precautions Section.
2. Remove or disconnect the following:
   - Negative battery cable
   - Engine appearance covers
   - Ignition coils
   - Valve covers
3. Measure the valve clearance while the camshaft lobe is pointed away from the valve shim. Rotate the crankshaft as necessary for each valve to be measured.
4. Valve clearance should be 0.012–0.014 inch.
5. If adjustment is necessary, compress the valves with the special tools and remove the shim with compressed air as shown. Repeat for each valve to be adjusted.
6. Install or connect the following:
   - Valve covers
   - Ignition coils
   - Engine appearance covers
   - Negative battery cable

Starter Motor

REMOVAL & INSTALLATION

2002 S-Type 3.0L and 4.0L Engines

1. Disconnect the battery ground cable.
2. Raise and support the vehicle.
3. Remove the electrical connector protective cover.
4. Disconnect the starter motor electrical connectors.
5. Remove the oxygen sensor.
6. Disconnect the electrical connector.
7. Remove the oxygen sensor.
8. Disconnect the ground cable.
9. Remove the harness retaining bracket.
10. Detach the harness.
11. Remove the harness retaining bracket.
12. Remove the starter motor bolts.
13. To install, reverse the removal procedure. Torque the mounting bolts to 18 ft. lbs. (25 Nm).

XJ Range and 2003–06 S-Type 3.0L and 4.0L Engines

1. Disconnect the battery ground cable.
2. Remove the air deflector.

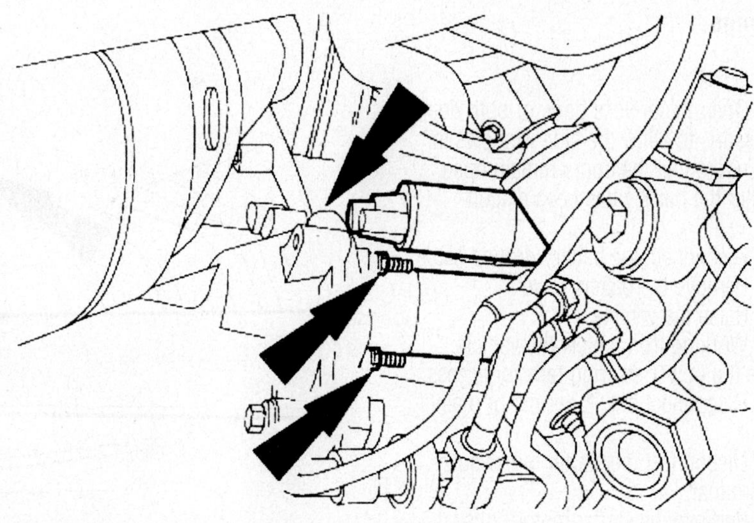

Starter mounting bolts—3.0L engines

3. Disconnect the engine ground cable.
4. Detach the starter motor positive cables.
5. Remove the starter motor retaining bolts.
6. Remove the starter motor.
7. To install, reverse the removal procedure. With the V6, tighten the mounting bolts to 18 ft. lbs. (25Nm). With 4.2L engine, tighten the mounting bolts to 33 ft. lbs. (45 Nm).

X-Type 2.5L and 3.0L Engines

1. Disconnect the battery ground cable.
2. Raise and support the vehicle.
3. Install the special tool.
4. Position and adjust the special tool rear height adjuster. Left-hand shown, right-hand similar.
5. Position and adjust the special tool front height adjuster. Right-hand shown, left-hand similar.
6. Position and adjust the special tool engine height adjusters.
7. Detach the support bar center fasteners. Remove the support bar end bolts.
8. Detach the starter motor solenoid cover.
9. Disconnect the starter motor electrical connectors.
10. Remove the starter motor rear retaining bolt.
11. Remove the starter motor front retaining bolt.
12. Remove the starter motor.
13. To install, reverse the removal procedure. Tighten the starter mounting bolts to the mounting bolts to 26 ft. lbs. (35 Nm). Tighten to support bar end bolts to 98 ft. lbs. (133 Nm); the center bolts to 18 ft. lbs. (25 Nm).

XK Range

2002

1. Open the driver's door, or both doors if necessary, to allow the side windows to drop. Ensure that the doors remain open until after the battery has been disconnected.
2. Disconnect the battery ground cable.
3. Remove the battery cover.
4. Raise the vehicle.
5. Working from below the vehicle, remove the power steering rack for access.
6. Disconnect the starter motor main terminal.
7. Disconnect the starter motor solenoid terminal.
8. Remove the starter motor bolts.
9. Carefully remove the starter motor.

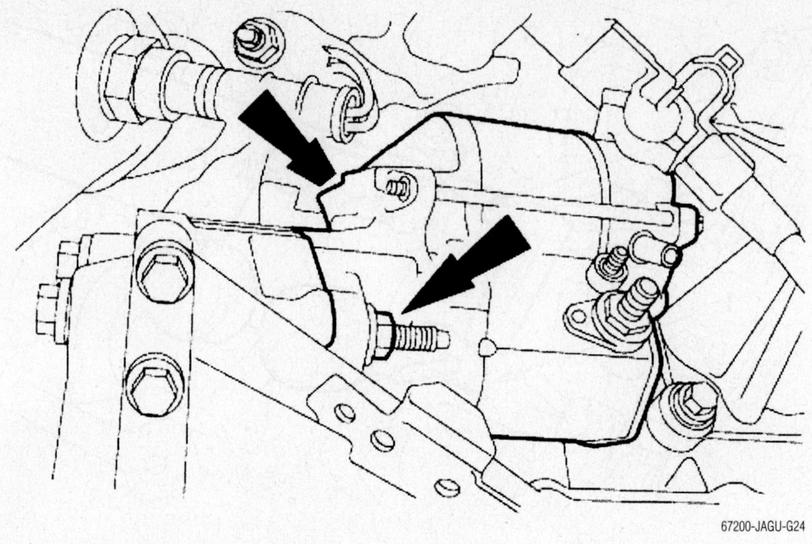

Starter mounting bolts—4.2L engine

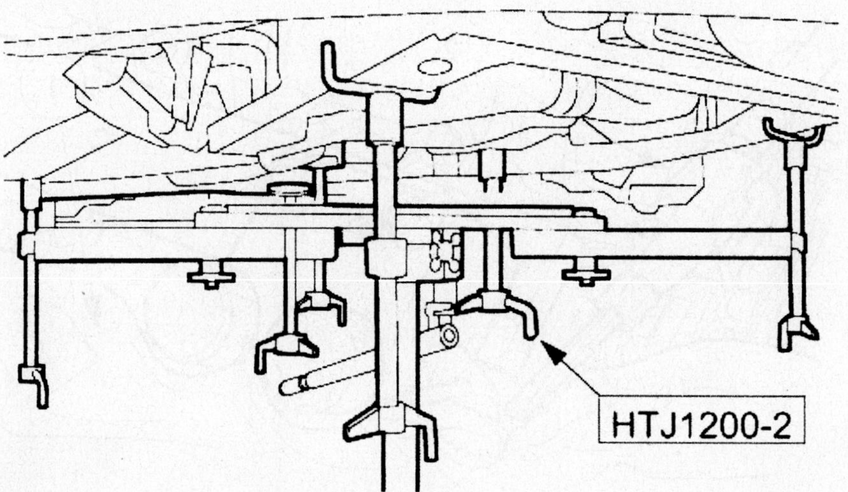

Power train assembly jack

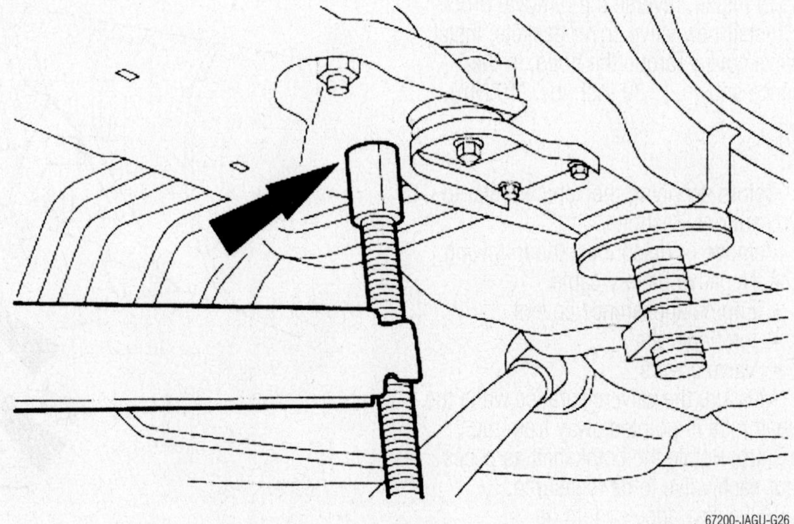

Position and adjust the special tool rear height adjuster. Left-hand shown, right-hand similar

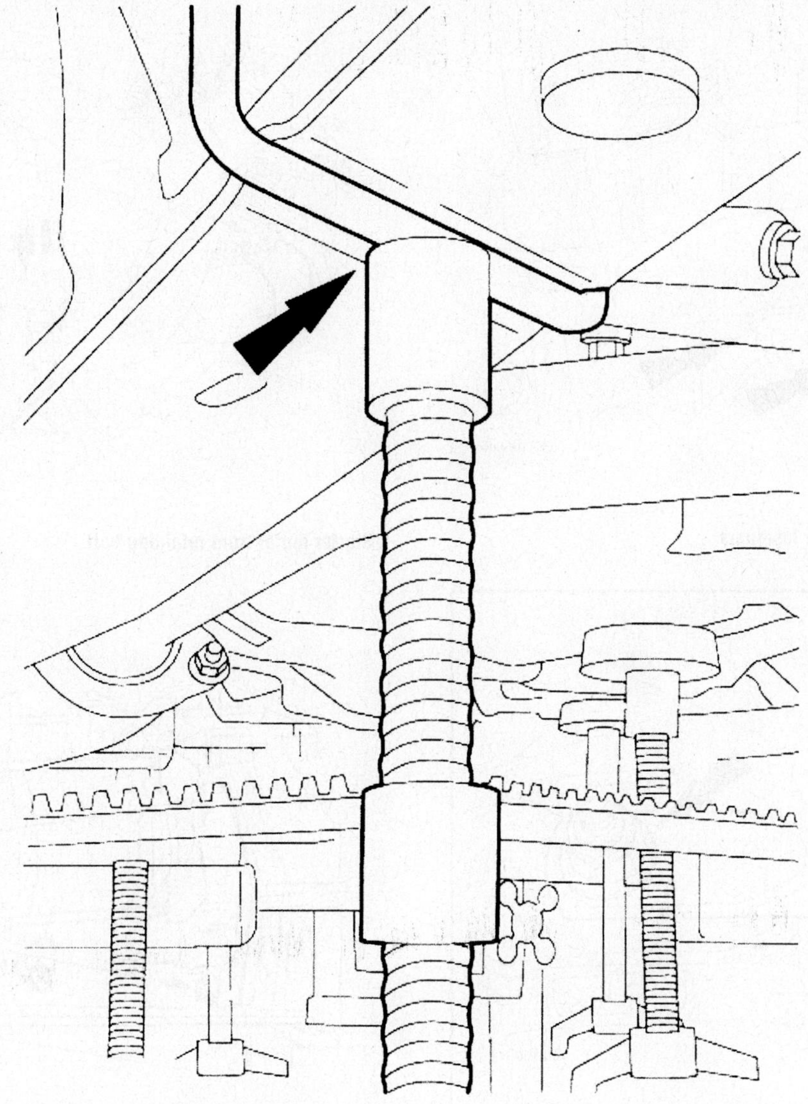

67200-JAGU-G27

Position and adjust the special tool front height adjuster. Right-hand shown, left-hand similar

10. Installation is the reverse of the removal procedure. Tighten the starter motor securing bolts to 38–48 Nm.
11. Install the steering rack. Refill the reservoir with the recommended fluid and bleed the steering system.
12. Lower the car.
13. Connect the battery and install the battery cover.

2003–06

1. Disconnect the battery ground cable.
2. On coupes, raise and support the vehicle.
3. On convertibles:
    a. Remove the radiator splash shield.
    b. Remove the underbody front cross brace.
4. Remove the generator lower cooling duct.

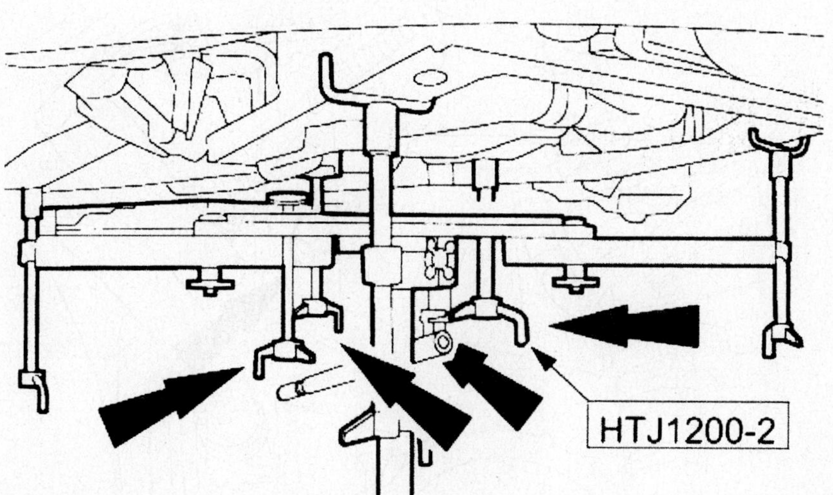

67200-JAGU-G28

Position and adjust the special tool engine height adjusters

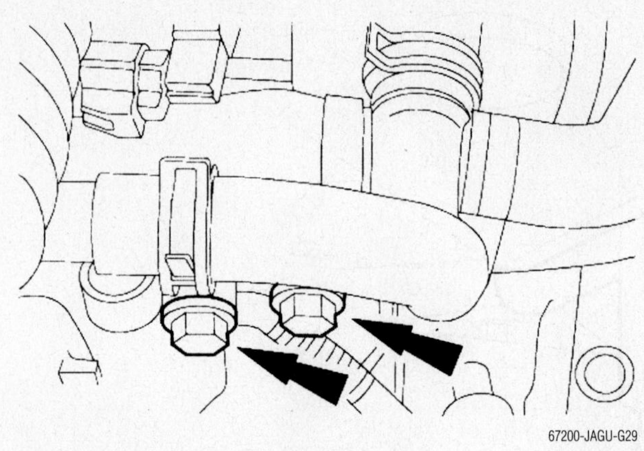

Detach the support bar center fasteners

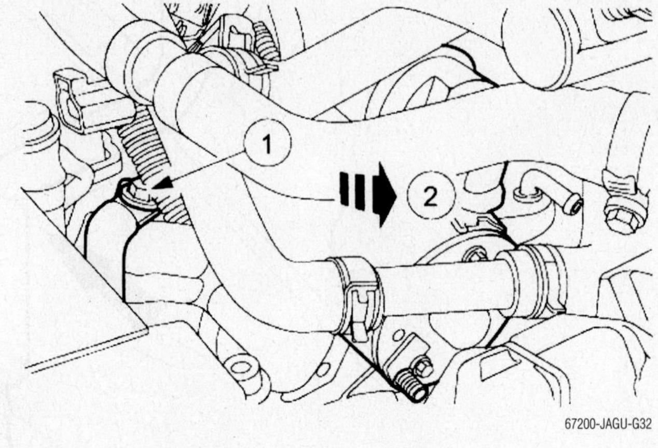

Starter motor front retaining bolt

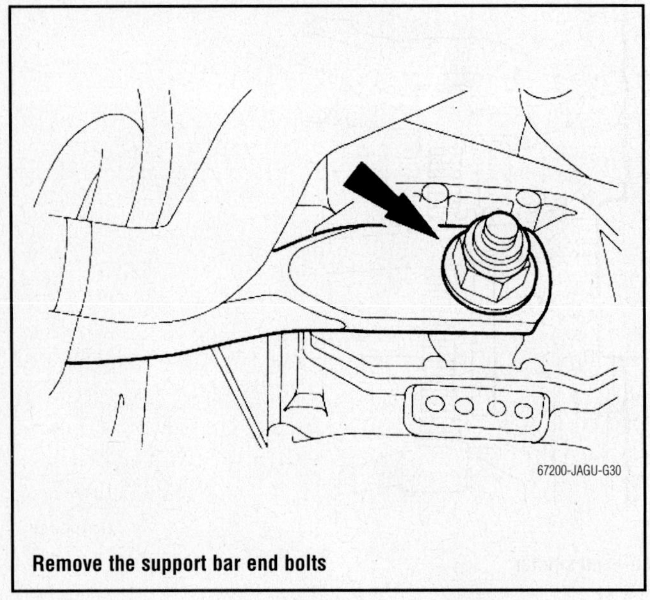

Remove the support bar end bolts

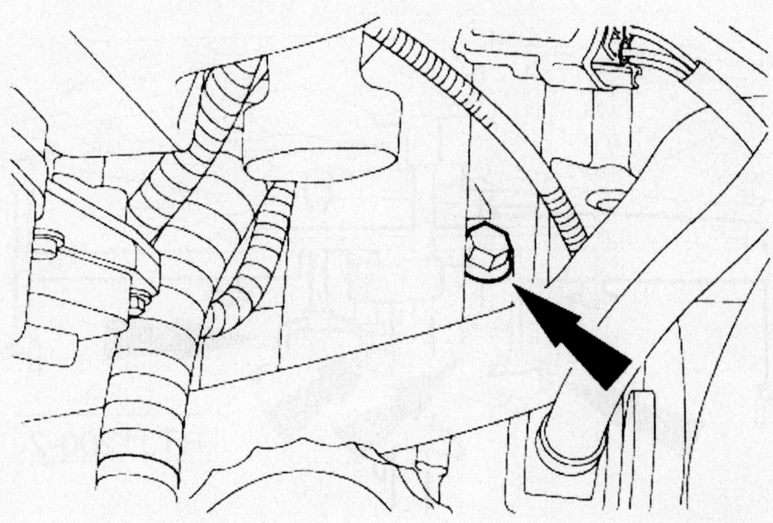

Starter motor rear retaining bolt

Starter motor bolt locations—2002 XK Range

5. Remove the generator upper cooling duct.
6. Disconnect the starter cables.
7. Remove the starter motor.
8. To install, reverse the removal procedure. Tighten the mounting bolts to 33 ft. lbs. (45 Nm).

Oil Pan

REMOVAL & INSTALLATION

2002 S-Type 3.0L Engine through VIN M45254

1. Remove the front axle crossmember.
2. Remove the generator.
3. Drain the engine oil.

4. Disconnect the air conditioning compressor electrical connector.
5. Support the air conditioning compressor using tie straps.
6. Detach the air conditioning compressor.
7. Detach the power steering fluid tube mounting bracket.
8. Detach the wiring harness.
9. Detach the transmission fluid cooler tubes mounting bracket.
10. Remove the oil pan to transmission retaining bolts.
11. Remove the oil pan.
12. Remove and discard the gasket.
13. Clean and inspect the oil pan and cylinder block sealing surfaces.

To install:

14. Apply an 8mm dot of silicone gasket, or equivalent, meeting Jaguar specification WSE-M4G323-A6 to the sealing surface on the engine block and engine front cover joint.

➡Install the oil pan and tighten the bolts within 6 minutes of applying sealant.

15. Using a new gasket, position the pan on the block.
16. Loosely install the oil pan to cylinder block retaining bolts. Tighten to 18 ft. lbs. (25 Nm).
17. Loosely install the oil pan to transmission retaining bolts. Tighten to 35 ft. lbs. (47 Nm).
18. Attach the transmission fluid cooler tubes mounting bracket. Tighten to 89 inch lbs. 10 Nm.
19. Attach the wiring harness.
20. Attach the power steering fluid tube mounting bracket. Tighten to 89 inch lbs. (10 Nm).
21. Attach the air conditioning compressor. Tighten to 18 ft. lbs. (25 Nm).
22. Connect the air conditioning compressor electrical connector.
23. Refill the engine with oil.
24. Install the generator.
25. Install the front axle crossmember.

2002.5–04 S-Type 3.0L Engine from VIN M45255

1. Remove the cowl vent screen.
2. Remove the engine compartment support.
3. Detach the right-hand valve cover wiring harness.
4. Disconnect the left-hand ignition coils electrical connectors.
5. Detach the left-hand valve cover wiring harness.

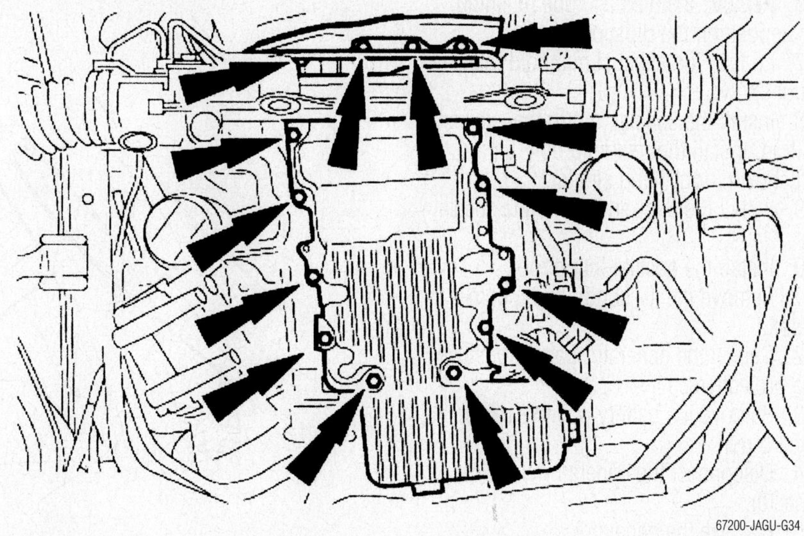

Oil pan bolt locations—2002 S-Type 3.0L Engine through VIN M45254

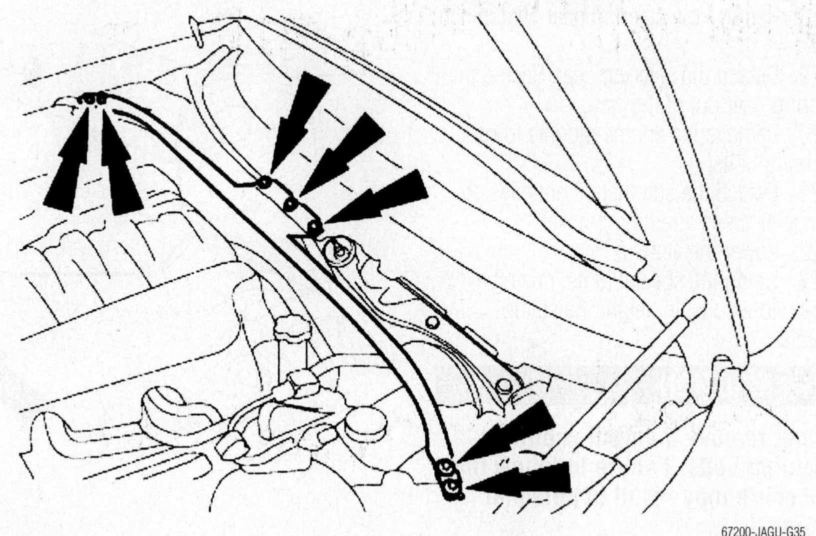

Remove the engine compartment support—2002.5–04 S-Type 3.0L Engine from VIN M45255

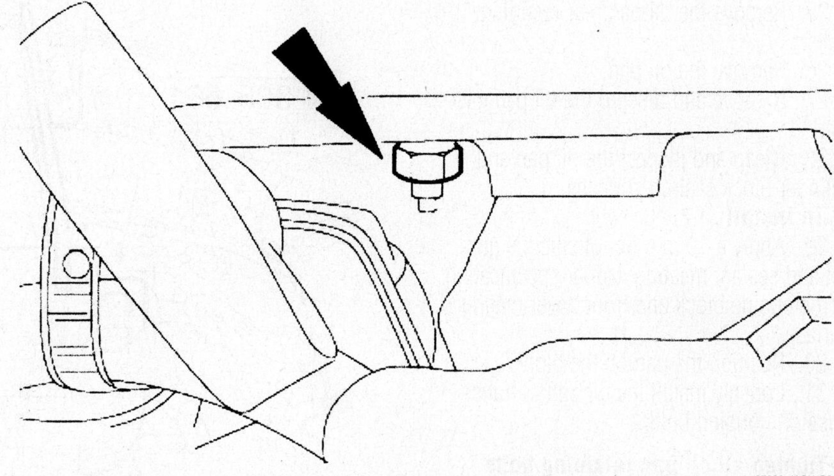

Loosen the exhaust manifold retaining nut shown—2002.5–04 S-Type 3.0L Engine from VIN M45255

6. Remove the dipstick tube retaining bolt. Reposition the dipstick tube.
7. Loosen the exhaust manifold retaining nut shown.
8. Install the special tool to the exhaust manifold. Install the retaining bolt. Left-hand shown; right-hand similar.
9. Install the support bars to the special tool.
10. Install the special lifting tool.
11. Remove the air conditioning compressor.
12. Detach the generator battery positive cable protective cover.
13. Remove the battery positive cable retaining nut.
14. Disconnect the generator electrical connector.
15. Remove the generator.
16. Drain the engine oil.
17. Detach the power steering hose.
18. Remove the steering gear shaft pinch bolt.
19. Detach the steering gear. Secure the steering gear out of the way.
20. Remove the engine mounts lower retaining bolts.
21. Detach the automatic transmission oil cooler tubes retaining bracket.
22. Lower the vehicle.
23. Using the special tools, raise the engine to a suitable height. Raise the vehicle.

CAUTION

Do not remove the front subframe retaining bolts. Failure to follow this procedure may result in personal injury.

24. Loosen the front subframe retaining bolts to enable the oil pan to be removed.
25. Remove the oil pan rear retaining bolts.
26. Remove the oil pan.
27. Remove and discard the oil pan gasket.
28. Clean and inspect the oil pan and cylinder block sealing surfaces.

To install:

29. Apply a 10 mm dot of silicone gasket and sealant meeting Jaguar specification to the engine block and front cover mating surface.
30. Position the pan on the block.
31. Loosely install the oil pan to transmission housing bolts.

➡Tighten all oil pan retaining bolts with in six minutes of applying the sealer.

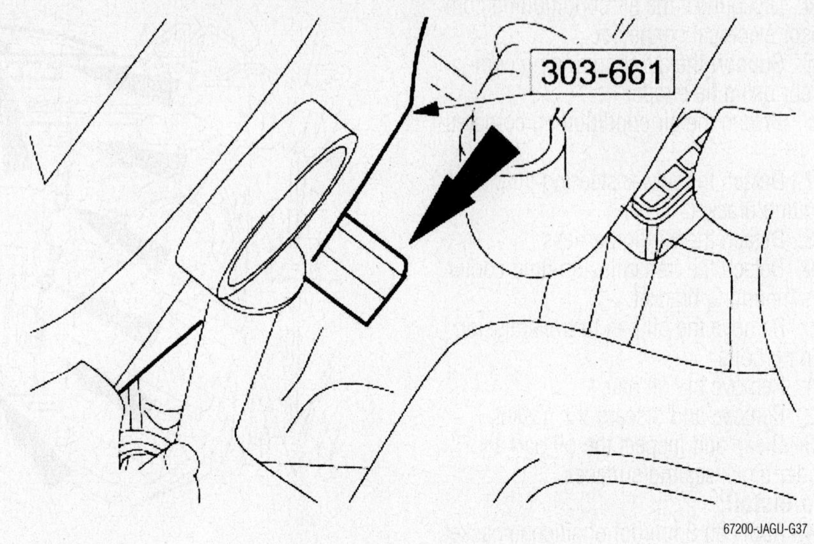

Install the special tool to the exhaust manifold—2002.5–04 S-Type 3.0L Engine from VIN M45255

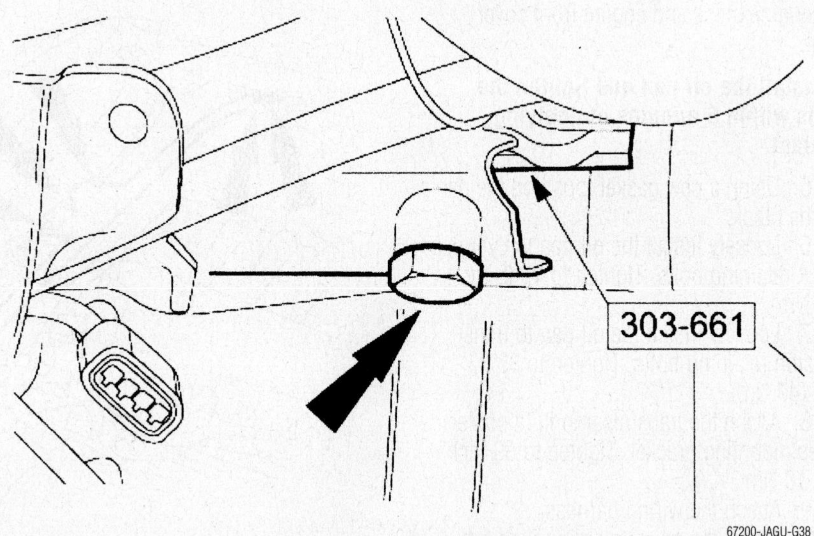

Install the special tool retaining bolt—2002.5–04 S-Type 3.0L Engine from VIN M45255

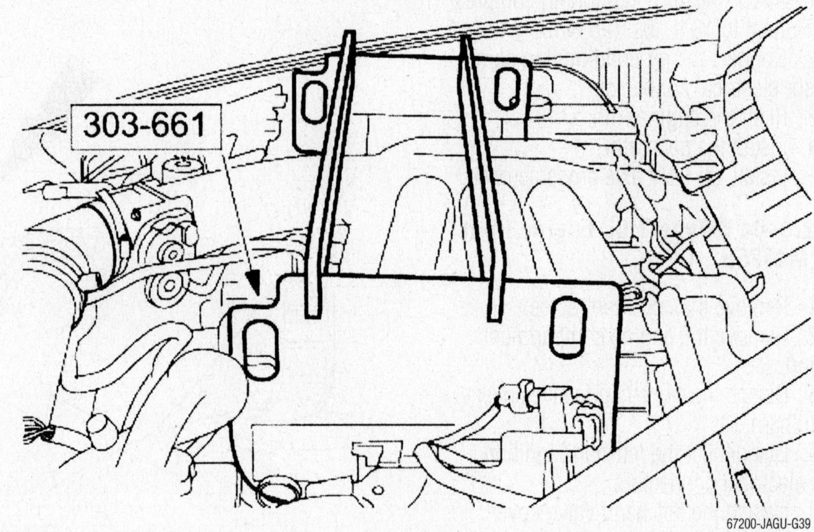

Install the support bars to the special tool—2002.5–04 S-Type 3.0L Engine from VIN M45255

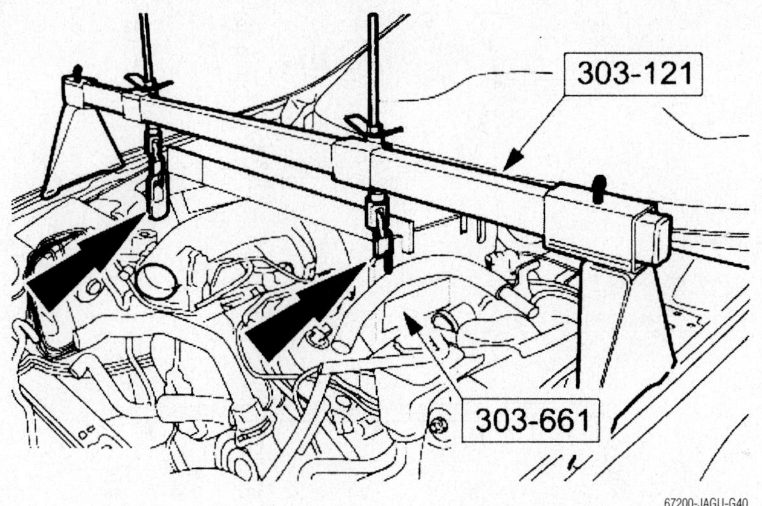

Install the special lifting tool—2002.5–04 S-Type 3.0L Engine from VIN M45255

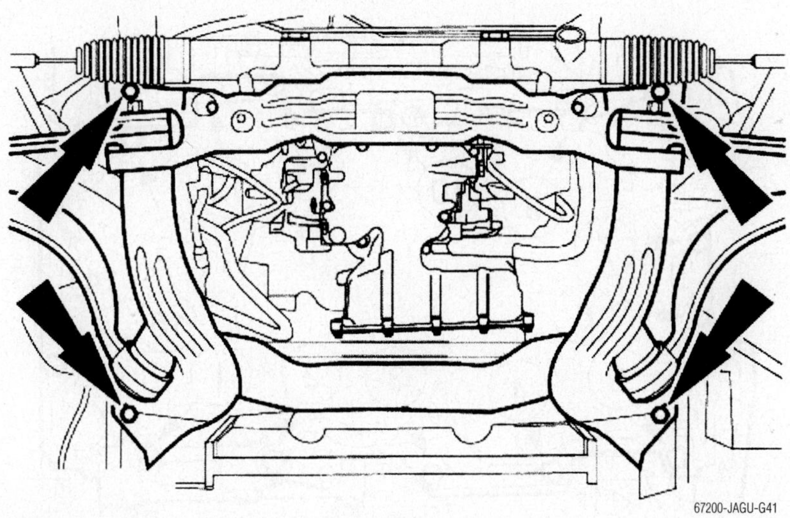

Loosen the front subframe retaining bolts to enable the oil pan to be removed—2002.5–04 S-Type 3.0L Engine from VIN M45255

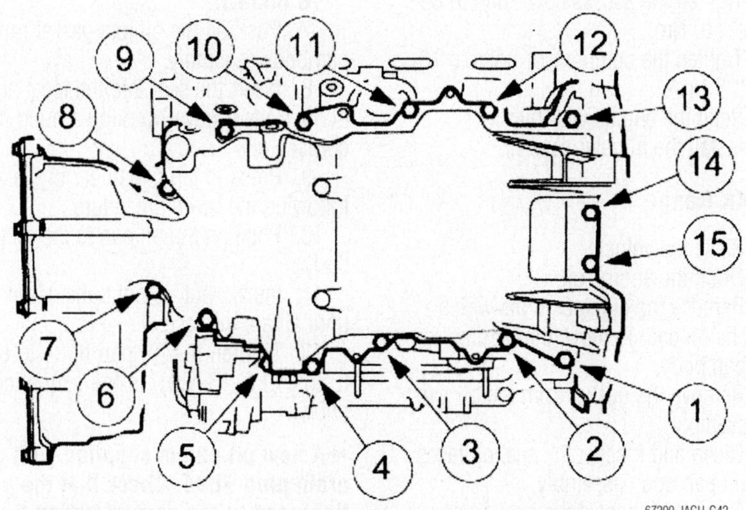

Oil pan bolt tightening sequence—2002.5–04 S-Type 3.0L Engine from VIN M45255

32. Install the oil pan retaining bolts. Tighten in the sequence shown to 18 ft. lbs. (25 Nm).
33. Tighten the rear bolts to 33 ft. lbs. (45 Nm).
34. Lower the vehicle.

➡**Make sure the engine mounts locate into the correct position when the engine is repositioned.**

35. Using the special tools, reposition the engine.
36. Raise the vehicle.
37. Attach the automatic transmission oil cooler tubes retaining bracket. Tighten to 89 inch lbs. (10 Nm).
38. Install the engine mounts lower retaining bolts. Tighten to 47 ft. lbs. (63 Nm).
39. Install the steering gear. Tighten to 74 ft. lbs. (100 Nm).
40. Install the steering gear shaft pinch bolt. Tighten to 26 ft. lbs. (35 Nm).
41. Attach the power steering hose. Tighten to 89 inch lbs. (10 Nm).
42. Install a new sealing washer to oil pan drain plug. Install the oil pan drain plug. Tighten to 18 ft. lbs. (24 Nm).
43. Install the generator. Tighten to 35 ft. lbs. (48 Nm). Connect the generator electrical connector.
44. Install the battery positive cable retaining nut. Tighten to 106 inch lbs. (12 Nm).
45. Attach the generator battery positive cable protective cover.
46. Install the air conditioning compressor.
47. Remove the special tool.
48. Install the exhaust manifold retaining nut. Tighten to 18 ft. lbs. (25 Nm).
49. Install the dipstick tube retaining bolt. Tighten to 89 inch lbs. (10 Nm).
50. Attach the left-hand valve cover wiring harness.
51. Connect the left-hand ignition coils electrical connectors.
52. Attach the right-hand valve cover wiring harness.
53. Install the engine compartment support. Tighten to 18 ft. lbs. (25 Nm).
54. Install the cowl vent screen.
55. Refill the engine with oil.

S-Type 4.0L and 4.2L Engine through VIN M45254; XJ Range, 4.0L and 4.2L Engines

1. Raise and support the vehicle.
2. With the 4.2L engine, remove the air deflector.
3. Drain the engine oil.

4. Remove the seventeen retaining bolts. Remove the oil pan.
5. Remove the gasket.
6. Make sure all surfaces are clean.

To install:

7. Install the oil pan gasket to the oil pan assembly.
8. Insert the seal into the locating positions in the groove, working around from a corner.
9. Press in the remainder of the seal, following the same procedure.
10. Install the oil pan to the block.
11. Install, but do not fully tighten the retaining bolts. When all bolts are installed, tighten the oil pan bolts to 8–10 ft. lbs. (11–13 Nm), in the sequence shown.
12. Install the air deflector.

➡A new oil pan is supplied with the drain plug installed. Check that the drain plug is tightened to the correct torque figure.

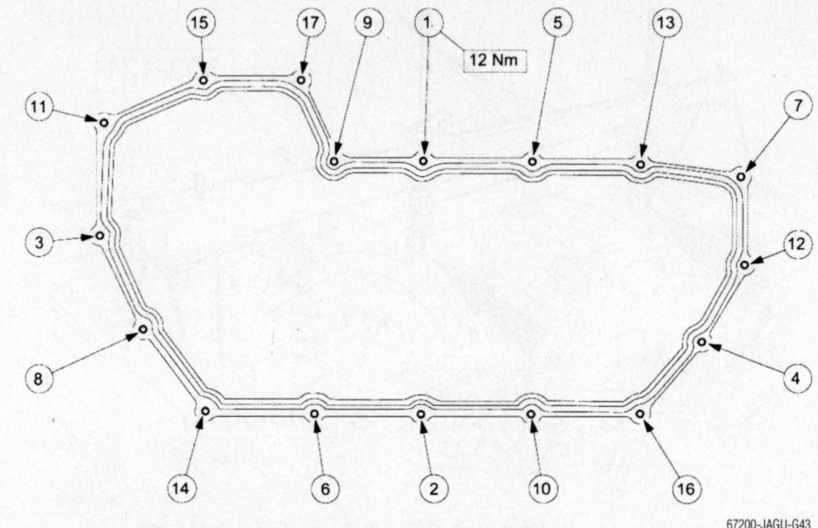

Oil pan bolt torque sequence—2002 S-Type 4.0L Engine through VIN M45254

X-Type

1. Remove the transfer case.
2. Drain the engine oil.
3. Remove the accessory drive belt.
4. Remove the A/C compressor lower retaining bolts. Loosen the A/C compressor upper retaining bolts, and swing it out of the way.
5. Remove the lower access cover.
6. Remove the oil pan-to-transmission retaining bolts.
7. Remove the oil pan.
8. Remove and discard the gasket.
9. Clean and inspect the oil pan and the cylinder block sealing surfaces.

To install:

10. Apply a 10 mm diameter dot of silicone sealant WSE-M4G323-A6, or equivalent, meeting Jaguar specification to the engine front cover to the cylinder block mating joints.
11. Loosely install the oil pan to transmission housing bolts.

➡The oil pan retaining bolts numbered 1 and 2 are longer than the retaining bolts numbered 3 through 15.

➡Tighten the oil pan bolts within six minutes of applying sealer.

12. To install, reverse the removal procedure.
13. Use a new oil pan gasket. Tighten the bolts, in the sequence shown, to 18 ft. lbs. (25 Nm).
14. Tighten the pan-to-transmission bolts to 33 ft. lbs. (45 Nm).

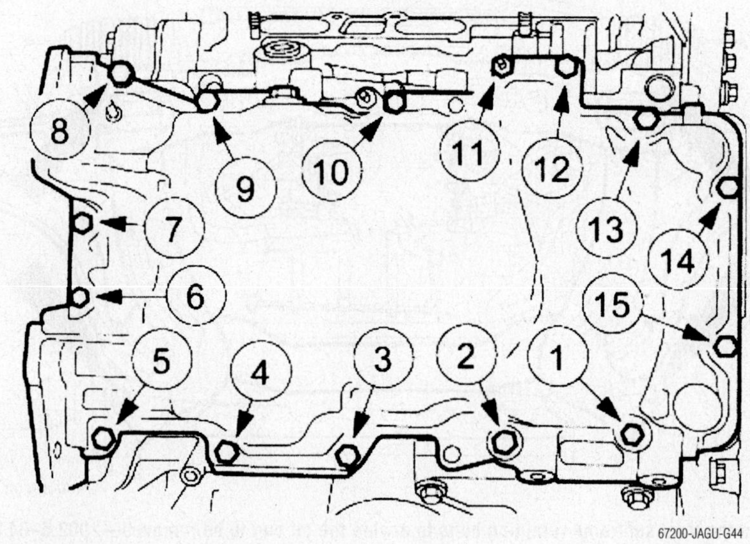

Oil pan bolt torque sequence—X-Type

15. Tighten the access cover bolt to 89 inch lbs. (10 Nm).
16. Tighten the compressor bolts to 18 ft. lbs. (25 Nm).
17. Refill the engine with oil.
18. Install the transfer case.

2002 XK Range

1. Raise the vehicle.
2. Drain the engine oil.
3. Remove the eighteen bolts which secure the oil pan. Remove the oil pan from the oil pan body.
4. Remove the gasket from the oil pan body assembly.
5. Clean and inspect the sealing faces on the oil pan body assembly.
6. Clean and inspect the new oil pan mating face.

To install:

7. Position the oil pan gasket to the oil pan body assembly.
8. Insert the seal into the locating positions in the groove, working around from a corner.
9. Press in the remainder of the seal, following the same procedure.
10. Position the oil pan to the oil pan body.
11. Install, but do not tighten, the bolts until all are in place.
12. Tighten the oil pan bolts to 8–10 ft. lbs. (11–13 Nm), in the sequence shown.

➡A new oil pan is supplied with the drain plug fitted. Check that the plug is tightened to the correct torque figure of 22–29 ft. lbs. (30–40 Nm).

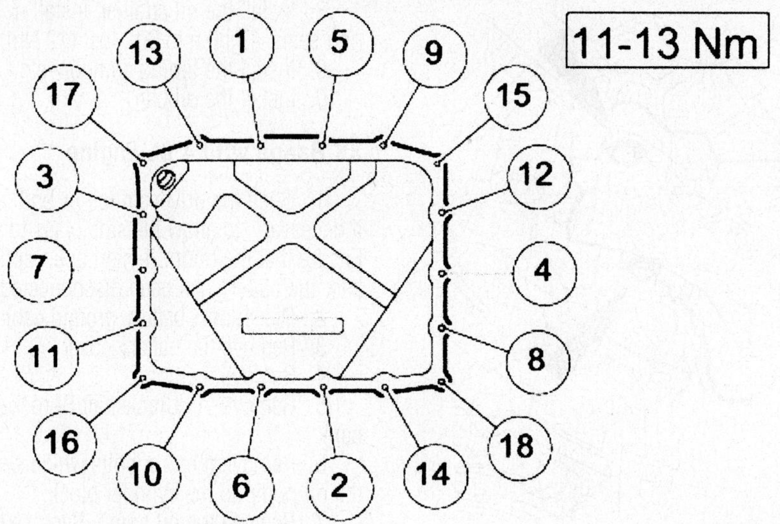

Oil pan bolt torque sequence—XK Range

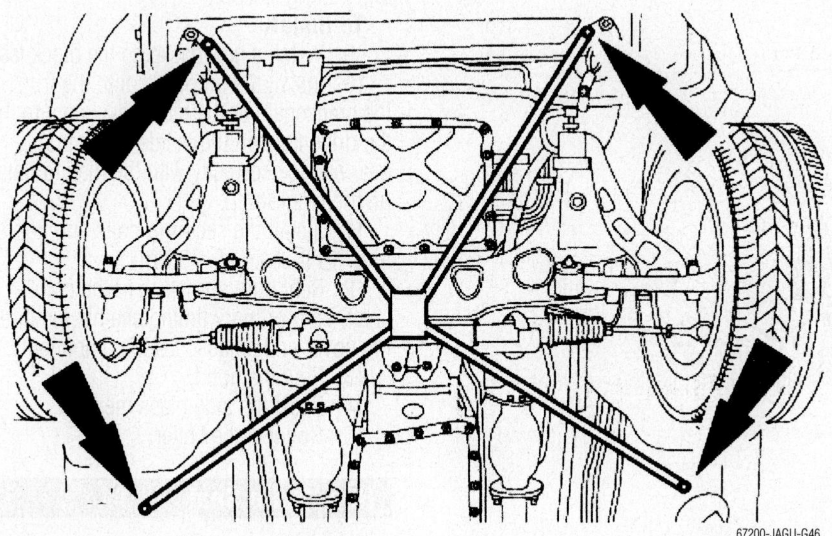

Underbody front cross brace—XK Range

➡When the engine is started from cold and run for a short period only, before being switched off, a quantity of oil will be hung up within the engine because it is not up to operating temperature.

13. Should the oil level subsequently be topped-up, the level could become too high, causing oil burning. Typically, about a quart (0.875L) is hung up.

14. Lower the car. Refill the engine with the correct specification engine oil.

2003–06 XK Range

1. On coupes, raise and support the vehicle.
2. On convertibles:
   a. Remove the radiator splash shield.
   b. Remove the underbody front cross brace.
3. Drain the engine oil.
4. Remove the oil pan from the oil pan body.
5. Remove the gasket from the oil pan body assembly.
6. Make sure all surfaces are clean.

To install:

7. Install the oil pan gasket to the oil pan body assembly.
8. Position the oil pan.
9. Install, but do not fully tighten the retaining bolts until all bolts are in place. Tighten to 89 inch lbs. (12 Nm), in the sequence shown.

➡A new oil pan is supplied with the drain plug installed. Check that the drain plug is tightened to 18 ft. lbs. (25 Nm).

10. On convertibles, install the underbody front cross brace.
11. Install the radiator splash shield.
12. On coupes, lower the vehicle.
13. Fill the engine with engine oil to the correct level.

Oil Pump

REMOVAL & INSTALLATION

S-Type 3.0L Engine, X-Type 2.5L and 3.0L Engines

1. Remove the timing drive components.
2. Remove the oil pump tube.
3. Remove the oil pump.
4. Remove and discard the O-ring seal.
5. Inspect the oil pump for damage and wear.
6. Install the oil pump flush to the cylinder block for correct sealing.
7. Rotate the inner rotor of the oil pump to align with the flats on the crankshaft before installation.
8. Install the oil pump. Tighten all fasteners to 89 inch lbs. (10 Nm).

S-Type and XJ Range with 4.0L Engine

1. Remove the left-hand primary timing chain.
2. Remove the four bolts which secure the oil pump to the cylinder block.
3. Remove the oil pump.
4. Remove the gasket.
5. Make sure all surfaces are clean.
6. Install the new gasket to the block face.
7. Install the oil pump. Make sure that the drive flats on the inside of the pump gear engage correctly with the mating flats on the crankshaft.
8. Install the retaining bolts and tighten to 9 ft. lbs. (12 Nm).
9. Install the left-hand primary timing chain.

S-Type, XJ Range and XK Range with 4.2L Engine

1. Remove the oil pan.
2. Remove the oil strainer.
3. Remove and discard the O-ring seal.
4. Remove the timing chains and drive sprockets.
5. Remove the oil pump. Remove and discard the gasket.
6. Make sure that all mating surfaces are clean.

To Install:

7. Install a new gasket. Install the oil pump. Tighten to 9 ft. lbs. (12 Nm).

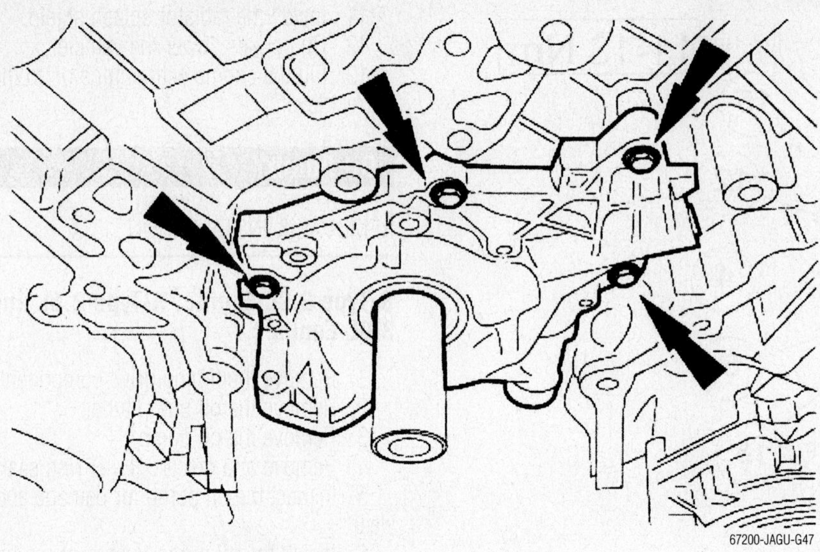

Oil pump bolt locations—2.5L and 3.0L engines

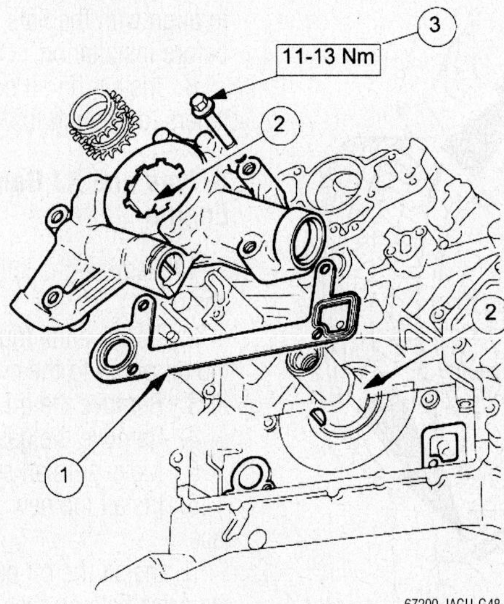

Oil pump—4.0L engine

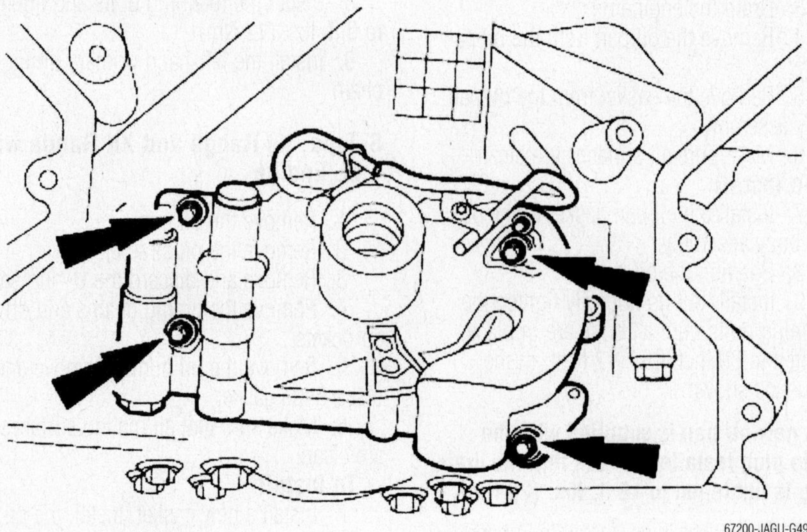

Oil pump—4.2L engine

8. Install the oil strainer. Install new O-ring seals. Tighten to 9 ft. lbs. (12 Nm).
9. Install the timing components.
10. Install the oil pan.

XK Range with 4.0L Engine

1. Open the driver's door, or both doors if necessary, to allow the side glass to drop. Ensure that the doors remain open until after the battery has been disconnected.
2. Disconnect battery ground cable.
3. Remove the battery cover.
4. Remove the hood.
5. Remove the Crankshaft Sprocket B-bank.
6. Remove the four bolts which secure the oil pump to the cylinder block.
7. Remove the oil pump. Remove the gasket.
8. Clean and inspect all components and mating faces.

To install:

9. Install a new gasket to the block face.
10. Install the oil pump onto the front of the crankshaft, to the block face; ensure that the drive flats on the inside of the pump gear engage correctly with the mating flats on the crankshaft.
11. Install the securing bolts and tighten to 8–10 ft. lbs. (11–13 Nm).
12. Refit both sprockets to the crankshaft, both primary timing chains, the timing cover and all associated components.
13. Refit the hood.
14. Reset the side glass memory.
15. Connect the battery.

Rear Main Seal

REMOVAL & INSTALLATION

2.5L and 3.0L Engine

1. Remove the flexplate.

➡Avoid scratching or damaging the oil seal sealing surfaces on the crankshaft and cylinder block.

2. Using the special tools, remove the crankshaft rear main oil seal.
3. Clean and inspect the crankshaft rear oil seal sealing surfaces.
4. Lubricate the crankshaft flange, the crankshaft rear oil bore seal, and the oil seal lip. Use engine assembly lubricant or equivalent meeting Jaguar specification ESR-M99C80-A.

➡Alternate bolt tightening to correctly seat the crankshaft rear oil seal until it is flush with the cylinder block.

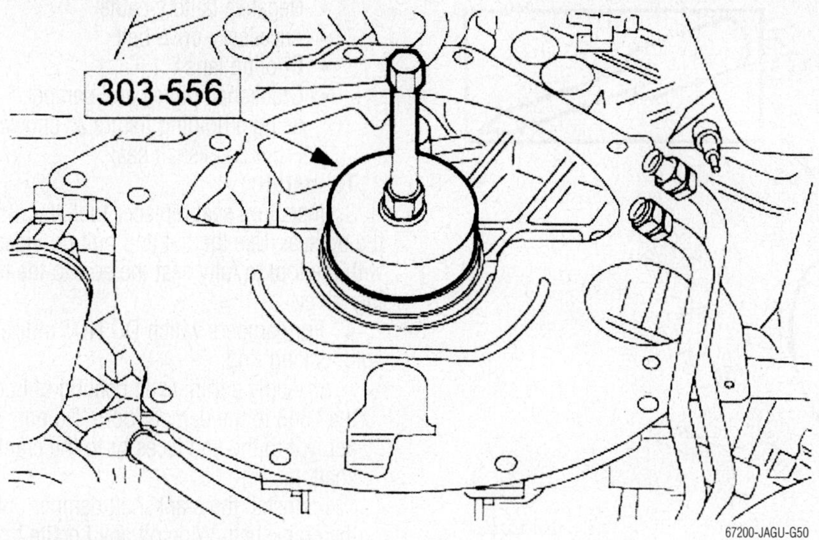

Rear main seal removal tool—3.0L engine

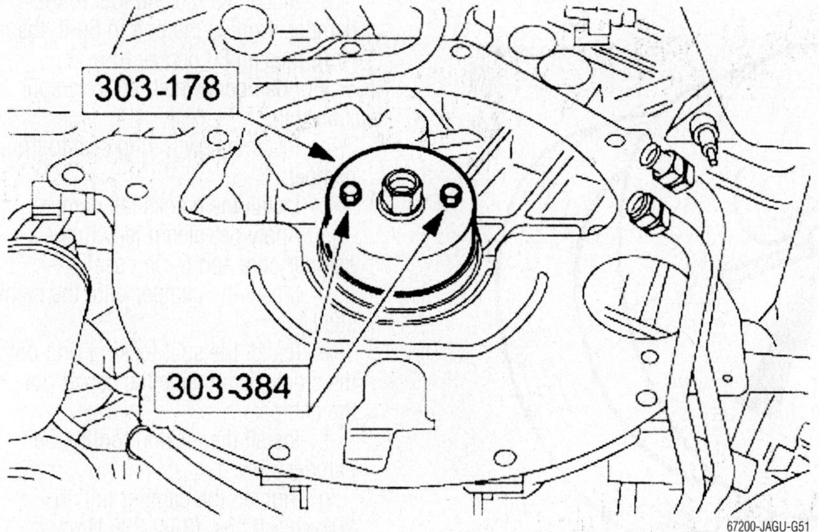

Rear main seal installation tool—2.5L and 3.0L engine

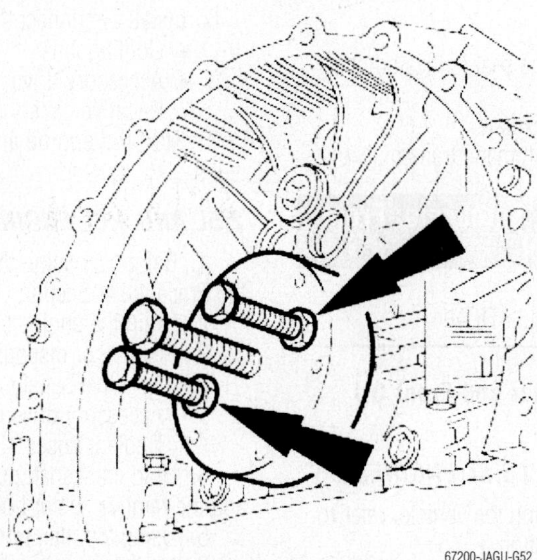

Rear main seal removal/installation tool—4.0L and 4.2L engines

5. Using the special tools, install the crankshaft rear oil seal.
6. Install the flexplate.

4.0L and 4.2L Engines

1. Remove the automatic transmission.
2. Remove the flexplate.
3. Install seal removal tool 303538 to the crankshaft.
4. Unscrew the two nuts towards the head of each bolt to allow the bolts to install fully into the crankshaft.
5. Unscrew the center bolt until it is below the face of the main tool and position the tool to the rear of the crankshaft.
6. Lightly tighten the two bolts so that they fully locate in the crankshaft.
7. Hold the tool against the crankshaft and lightly and evenly tighten the two nuts to the face of the tool to secure the tool and to hold it parallel to the face of the crankshaft.
8. Secure the oil seal to the tool using the three self-tapping screws spaced evenly around the face of the tool.
9. Use the punch (provided) and a hammer to pierce the seal face and provide a pilot hole for one of the self-tapping screws provided.
10. Install a self-tapping screw and tighten by 1.5 turns only.

**** WARNING**

DO NOT TIGHTEN BY MORE THAN TWO TURNS OR THE SCREW MAY BREAK.

11. Repeat the above steps for the other two screws.
12. Hold the two bolts stationary and loosen fully the two nuts.
13. Tighten the center bolt.
14. Remove the two bolts to release the tool and seal assembly from the crankshaft.
15. Remove the three self-tapping screws.
16. Remove and discard the seal.
17. Unscrew the center bolt until it is below the face of the tool.
18. Make sure all components are clean.

To install:

19. Make sure the transit sleeve is correctly in place and install the new seal over the crankshaft.

➡Do not use any lubricant on the seal, the transit sleeve or the crankshaft.

20. Carefully remove the transit sleeve, leaving the seal in place.
21. Install the tool to the crankshaft.
22. Make sure that the center bolt is

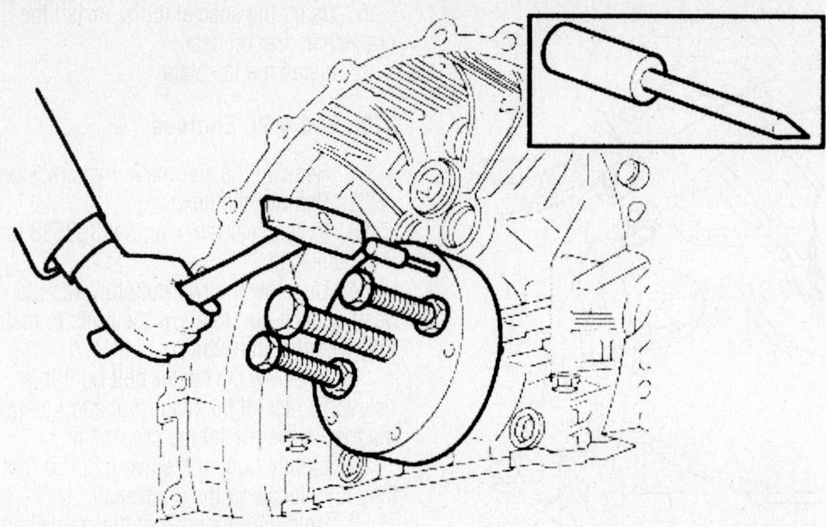

Use the punch (provided) and a hammer to pierce the seal face and provide a pilot hole for one of the self-tapping screws provided—4.0L and 4.2L engines

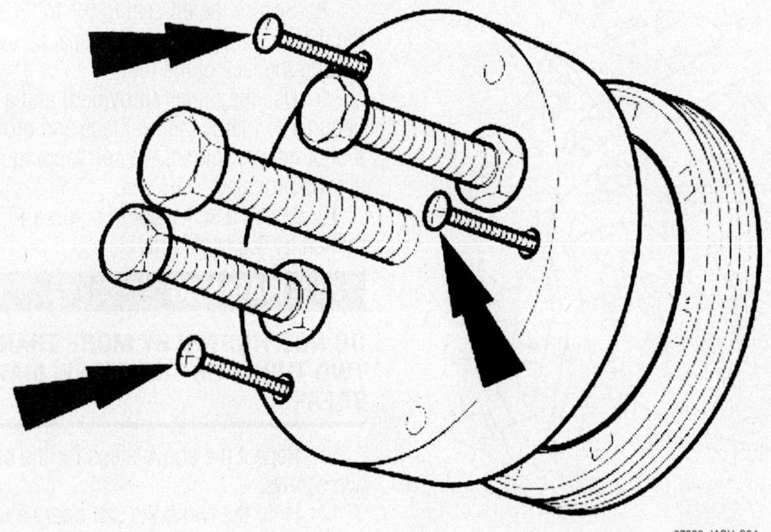

Self-tapping screws installed—4.0L and 4.2L engines

unscrewed to below the face which is to fit to the crankshaft and position the tool to the crankshaft.

23. Check that the nuts are fully unscrewed and install the two bolts to the crankshaft threads and lightly tighten.

24. Tighten the nuts, finger tight, to hold the tool assembly against the seal. Check that the seal and the tool are approximately parallel to the rear of the engine.

25. Press the seal into the housing.

26. Hold each bolt stationary, in turn, and tighten its nut a small amount (approximately two to three flats) in turn, to press the seal fully into the housing, until the tool is in contact with the cylinder block.

27. Remove the tool from the crankshaft.

28. Check that the seal is located correctly.

29. Install the flexplate.

30. Install the automatic transmission.

Timing Chain, Sprockets, Front Cover and Seal

REMOVAL & INSTALLATION

Crankshaft Damper and Front Oil Seal

EXCEPT 2.5L AND 3.0L ENGINES

1. Before servicing the vehicle, refer to the Precautions Section.

2. Remove or disconnect the following:
   - Negative battery cable
   - Accessory drive belt
   - Cooling fans
   - Crankshaft pulley and damper, using a holding fixture as shown
   - Front crankshaft seal

To install:

3. Install oil seal replacer tool JD-235 to the oil seal. Use the nut and bolt provided with the tool to fully seat the seal to the timing cover.

4. For dampers which DO NOT utilize a split locking ring:
   a. Apply a thin, even coating of Loctite® 648 to the damper bore. Do not apply it to the end faces or to the crankshaft.
   b. Install the crankshaft damper onto the crankshaft. Wipe off any Loctite that has squeezed out from the front of the damper.
   c. Install the locking tool to the damper. Tighten the bolt to 59 ft. lbs. (80 Nm), plus an 80 degree turn.

5. For dampers which utilize a split locking ring:
   a. Install a new O-ring seal to the damper.
   b. Install the crankshaft damper.
   c. Apply petroleum jelly to the damper bore and O-ring seal.
   d. Install the damper onto the crankshaft.
   e. Install the split locking ring onto the crankshaft, inside the center bore of the damper.
   f. Install the locking tool to the damper.
   g. Tighten the damper bolt to 266–285 ft. lbs. (364–386 Nm).
   h. Remove the locking tool from the damper.

6. Install or connect the following:
   - Cooling fans
   - Accessory drive belt
   - Negative battery cable

7. Start the engine and check for leaks.

2.5L AND 3.0L ENGINES

1. Before servicing the vehicle, refer to the Precautions Section.

2. Drain the engine coolant.

3. Remove or disconnect the following:
   - Negative battery cable
   - Accessory drive belt
   - Coolant hoses

4. Hold crankshaft pulley from turning and the remove center bolt.

5. Using a puller, remove pulley.

6. Using special tool 303-700, remove the crankshaft seal.

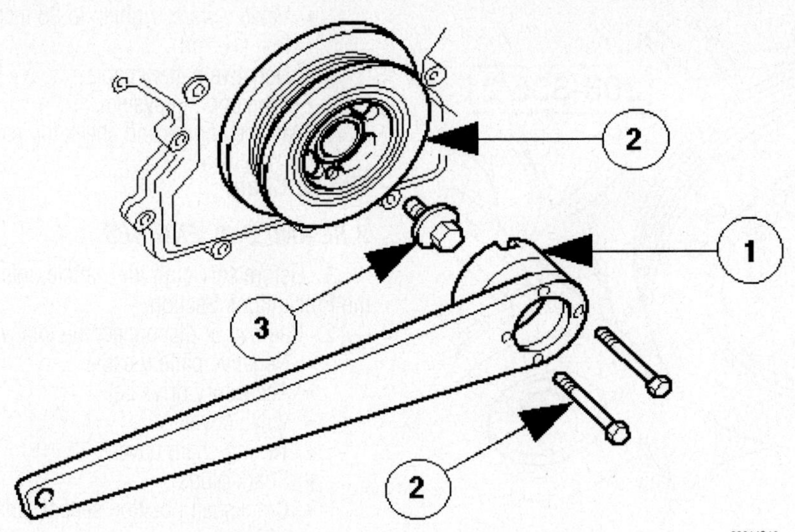

Crankshaft pulley removal tool—V8

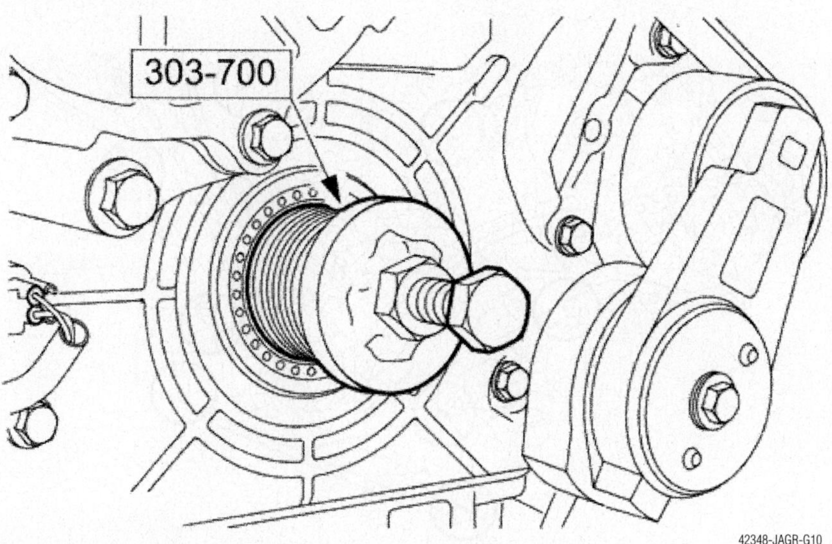

Crankshaft seal removal tool—2.5L and 3.0L engines

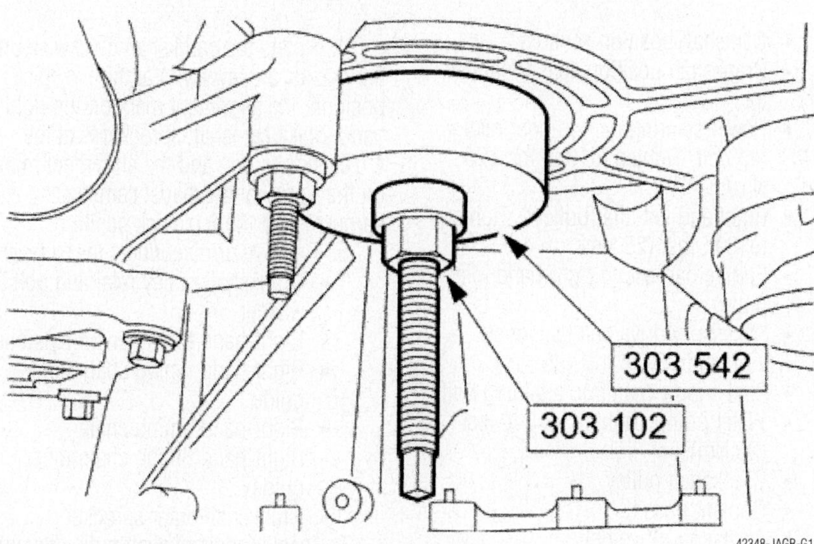

Crankshaft seal installation tool—2.5L and 3.0L engines

To install:

7. Coat new seal with clean engine oil.
8. Install seal using special tools 303-102 and 303-542.

➡Coat the crankshaft pulley keyway and sealing surfaces with silicone gasket sealant.

9. Install the crankshaft pulley using special tools 303-102 and 303-335/2.
10. Tighten the damper bolt to 89 ft. lbs. (120 Nm). Loosen the bolt 1 turn. Tighten bolt to 37 ft. lbs. (50 Nm). Tighten an additional 90°.
11. Install or connect the following:
    - Coolant hoses
    - Accessory drive belt
    - Negative battery cable
12. Refill the engine with coolant.
13. Start the engine and check for leaks

Timing Cover

EXCEPT 2.5L AND 3.0L ENGINES

1. Before servicing the vehicle, refer to the Precautions Section.
2. Drain the cooling system.
3. Remove or disconnect the following:
    - Negative battery cable
    - Upper radiator hose
    - Accessory drive belt
    - Water pump pulley
    - Accessory drive belt tensioner
    - Idler pulley
    - Crankshaft damper
    - Engine appearance covers
    - Mass Air Flow (MAF) meter
    - Air intake assembly
    - Ignition coils
    - Canister purge valve
    - Valve covers
    - Variable Valve Timing (VVT) solenoids
    - Engine harness retaining clips
    - Timing cover

To install:

4. Apply sealant to the 8 joints on the engine face as shown.
5. Install or connect the following:
    - Timing cover with new seals. Tighten the bolts in sequence to 96–120 inch lbs. (11–13 Nm).
    - Engine harness retaining clips
    - VVT solenoids
    - Valve covers
    - Canister purge valve
    - Ignition coils
    - Air intake assembly
    - MAF meter
    - Engine appearance covers
    - Crankshaft damper
    - Idler pulley

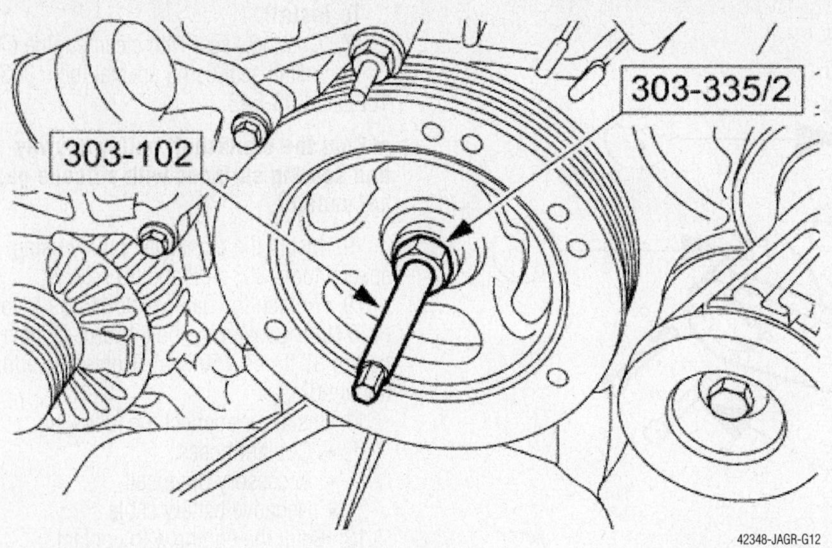

Crankshaft pulley installation tools—2.5L and 3.0L engines

- Accessory drive belt tensioner
- Water pump pulley
- Accessory drive belt
- Upper radiator hose
- Negative battery cable

6. Fill the cooling system.
7. Start the engine and check for leaks.

2.5L AND 3.0L ENGINES

1. Before servicing the vehicle, refer to the Precautions Section.
2. Drain the cooling system.
3. Remove or disconnect the following:
   - Negative battery cable
   - Valve covers
   - Oil pan
   - Accessory drive belt
   - Coolant hoses
   - Crankshaft pulley
   - Water pump
   - Fuel supply manifold retaining bolt
   - Accessory drive belt tensioner
   - Engine harness from right-hand idler pulley
   - Right and left idler pulleys
   - Power steering pump upper retaining bolt
   - Crankshaft position sensor connector
   - Camshaft position sensors
   - Front cover

To install:

4. Apply sealant to the flanges on the engine face as shown.
5. Install or connect the following:
   - Timing cover with new seals. Tighten the bolts in sequence to 18 ft. lbs. (25 Nm).

➡Bolts 3, 4, 10 and 11 are longer. Bolt 14 has a stud head.

- Camshaft position sensors
- Crankshaft position sensor connector
- Power steering pump upper retaining bolt: Tighten to18 ft. lbs. (25 Nm).
- Right and left idler pulleys: Tighten to18 ft. lbs. (25 Nm).
- Engine harness to right-hand idler pulley
- Accessory drive belt tensioner: Tighten to 33 ft. lbs. (45 Nm).
- Fuel supply manifold retaining bolt
- Water pump: Tighten to18 ft. lbs. (25 Nm).
- Crankshaft pulley
- Coolant hoses
- Accessory drive belt
- Oil pan
- Valve covers: Tighten to 88 inch lbs. (10 Nm).
- Negative battery cable

6. Fill the cooling system.
7. Start the engine and check for leaks.

Timing Chain

2.5L AND 3.0L ENGINES

1. Before servicing the vehicle, refer to the Precautions Section.
2. Remove or disconnect the following:
   - Negative battery cable
   - Accessory drive belt
   - Valve cover
   - Timing chain cover
   - Spark plugs
   - Crankshaft position sensor pulse wheel
3. Install the crankshaft pulley retaining bolt and washer

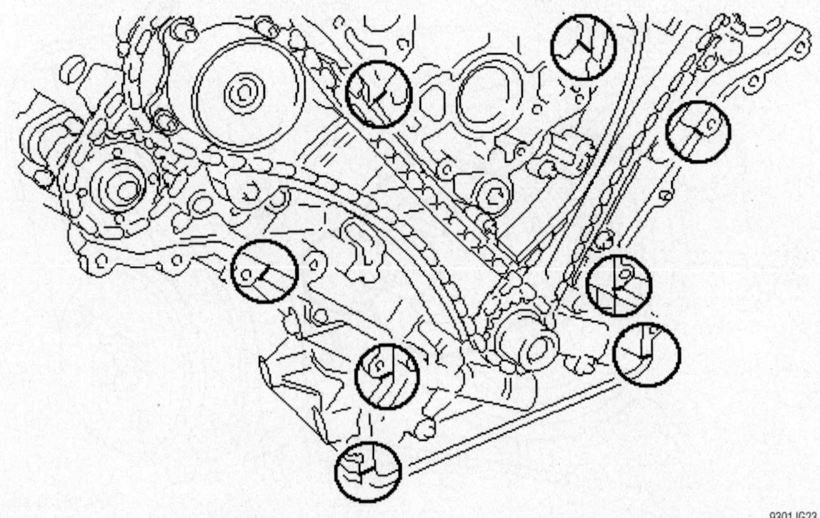

Sealant application points—V8

4. Rotate the crankshaft clockwise until the crankshaft keyway is at the 7 o'clock position, the alignment mark on the right-hand intake camshaft sprocket is at the 1 o'clock position and the alignment mark on the right-hand exhaust camshaft sprocket is at the 8 o'clock position.
5. Remove or disconnect the following:
   - Crankshaft pulley retaining bolt and washer
   - Right-hand timing chain tensioner
   - Right-hand timing chain outer guide
   - Right-hand timing chain
   - Right-hand timing chain inner guide
   - Crankshaft outer sprocket
6. Install the crankshaft pulley retaining bolt and washer.

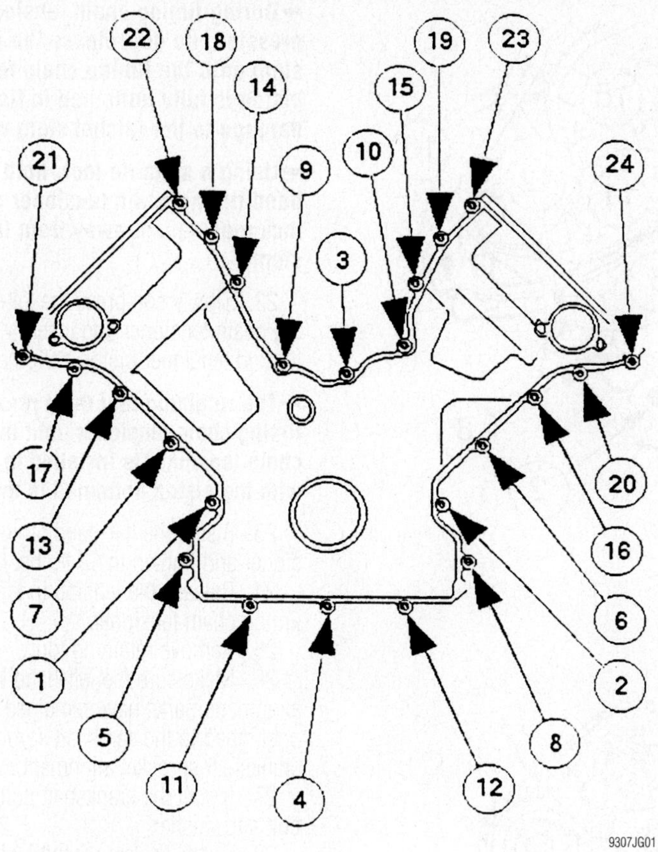

Timing cover torque sequence—V8

7. Rotate the crankshaft clockwise until the crankshaft keyway is at the 11 o'clock position, the alignment mark on the left-hand intake camshaft sprocket is at the 9 o'clock position and the alignment mark on the left-hand exhaust camshaft sprocket is at the 2 o'clock position.

8. Remove or disconnect the following:
- Crankshaft pulley retaining bolt and washer
- Left-hand timing chain tensioner
- Left-hand timing chain inner guide
- Left-hand timing chain
- Left-hand timing chain outer guide
- Crankshaft inner sprocket

To install:

9. Remove the right-hand intake camshaft and right-hand exhaust camshaft bearing caps evenly.

10. Rotate the right-hand intake camshaft clockwise until the camshaft sprocket alignment mark is at the 5 o'clock position and rotate the right-hand exhaust camshaft sprocket clockwise until the camshaft sprocket alignment mark is at the 12 o'clock position.

➡Do not install the cylinder head camshaft journal thrust caps until the camshaft journal caps are installed or damage to the thrust caps may occur.

11. Lubricate the camshafts and the camshaft bearing caps.

12. Install the right-hand intake camshaft and right-hand exhaust camshaft bearing caps in their original position.

13. Install the camshaft bearing cap retaining bolts evenly and tighten in sequence to 88 inch lbs. (10 Nm).

14. Install the crankshaft inner sprocket.

15. Install the left-hand timing chain outer guide. Tighten bolts starting with inner bolt to 18 ft. lbs. (25 Nm).

16. Make sure the crankshaft keyway is at the 11 o'clock position, the alignment mark on the left-hand intake camshaft sprocket is at the 9 o'clock position and the alignment mark on the left-hand exhaust camshaft sprocket is at the 2 o'clock position.

17. The timing chain alignment marks should be correctly positioned to the crankshaft sprocket and camshaft sprocket alignment marks.

18. The timing chain slack is on the tensioned side of the timing chain.

19. Install the left-hand timing chain.

20. Install the left-hand timing chain inner guide.

21. Place the left-hand timing chain tensioner in a vise.

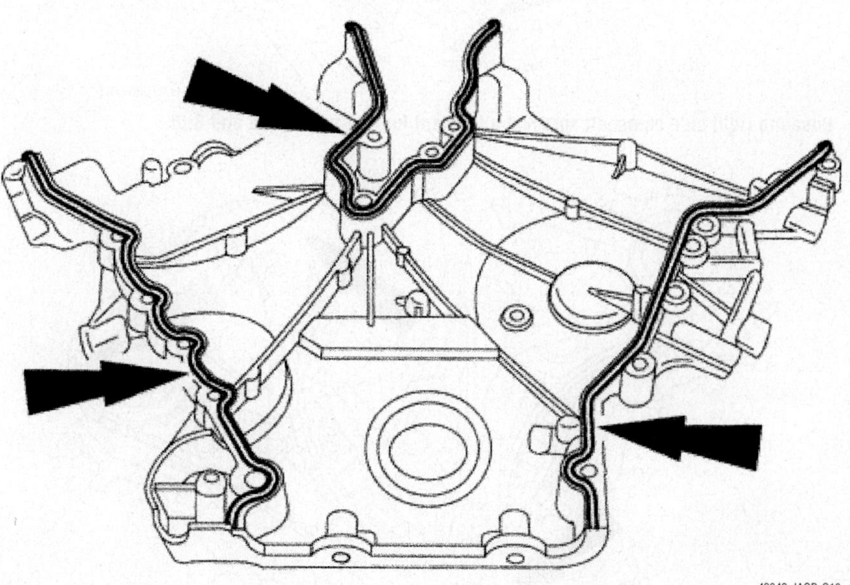

Sealant application points—2.5L and 3.0L

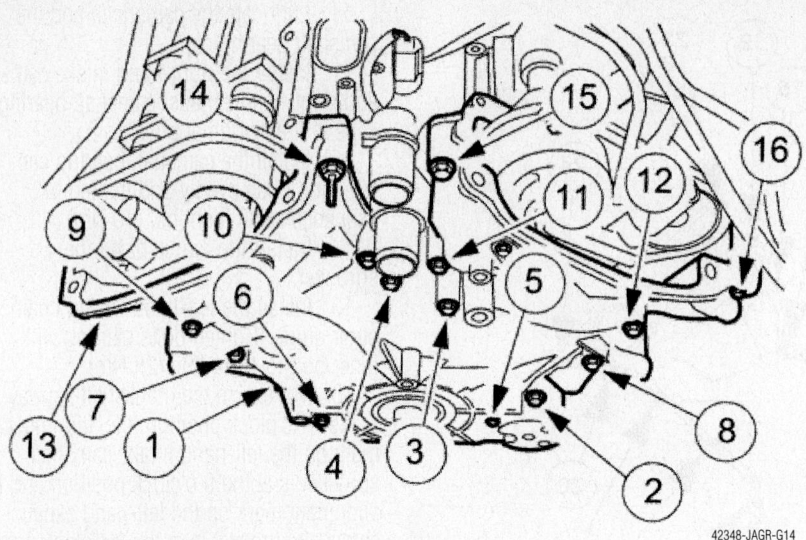

Timing cover torque sequence—2.5L and 3.0L

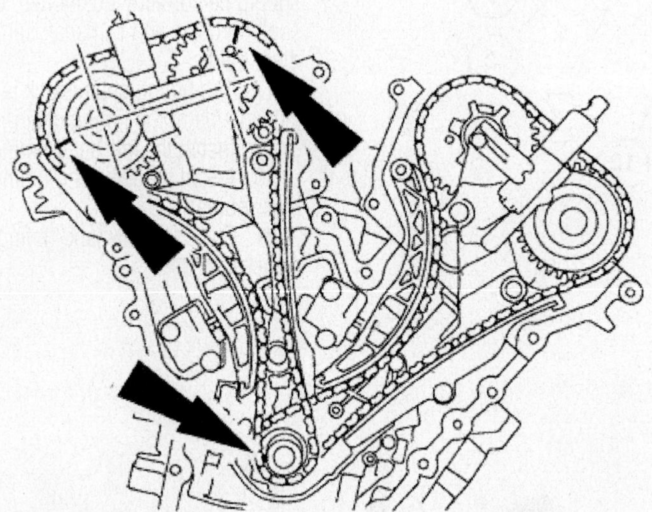

Locating right side camshaft sprocket alignment for removal—2.5L and 3.0L

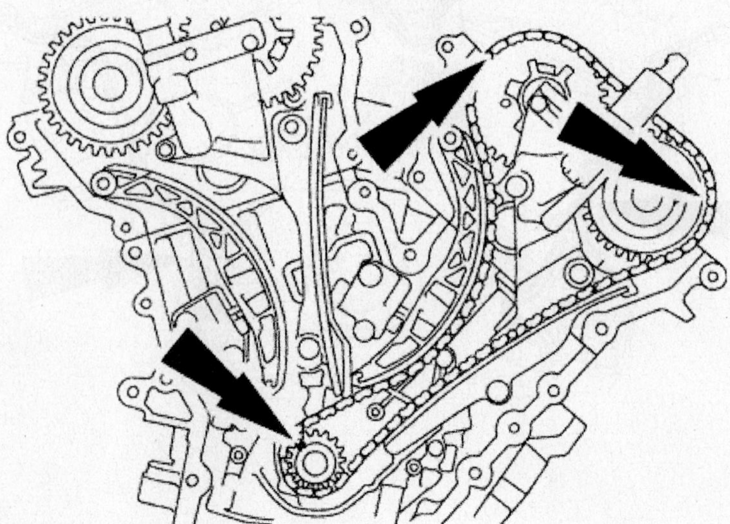

Locating left side camshaft sprocket alignment for removal—2.5L and 3.0L

➡During timing chain tensioner compression, do not release the ratchet stem until the timing chain tensioner piston is fully bottomed in its bore or damage to the ratchet stem will result.

➡Using a suitable tool, hold the left-hand timing chain tensioner ratchet lock mechanism away from the ratchet stem.

22. Slowly compress the left-hand timing chain tensioner and insert a paper clip to hold tensioner piston bottomed in bore.

➡The retaining tool must remain in the timing chain tensioner until the timing chain tensioner is installed to the engine with the piston bottomed in the bore.

23. Install the left-hand timing chain tensioner and tighten to 18 ft. lbs. (25 Nm).
24. Release the tension in the left-hand timing chain tensioner.
25. Remove retaining tool.
26. Make sure the left-hand timing chain alignment marks have remained correctly positioned to the camshaft sprocket and crankshaft sprocket alignment marks.
27. Install the crankshaft pulley retaining bolt and washer.
28. Rotate the crankshaft clockwise until the crankshaft keyway is at the 3 o'clock position.
29. Remove the crankshaft pulley retaining bolt and washer.
30. Install the crankshaft outer sprocket.
31. Install the right hand timing chain outer guide. Tighten bolts starting with inner bolt to 18 ft. lbs. (25 Nm).
32. Make sure the crankshaft keyway is at the 3 o'clock position, the alignment mark on the right-hand intake camshaft sprocket is at the 5 o'clock position and the alignment mark on the right-hand exhaust camshaft sprocket is at the 12 o'clock position.
33. Make sure the timing chain alignment marks are correctly positioned to the crankshaft sprocket and camshaft sprocket alignment marks.
34. The timing chain slack is on the tensioned side of the timing chain.
35. Install the right-hand timing chain.
36. Install the right-hand timing chain outer guide.
37. Place the left-hand timing chain tensioner in a vise.

➡During timing chain tensioner compression, do not release the ratchet stem until the timing chain tensioner piston is fully bottomed in its bore or damage to the ratchet stem will result.

Locating right side camshaft sprocket alignment for installation—2.5L and 3.0L

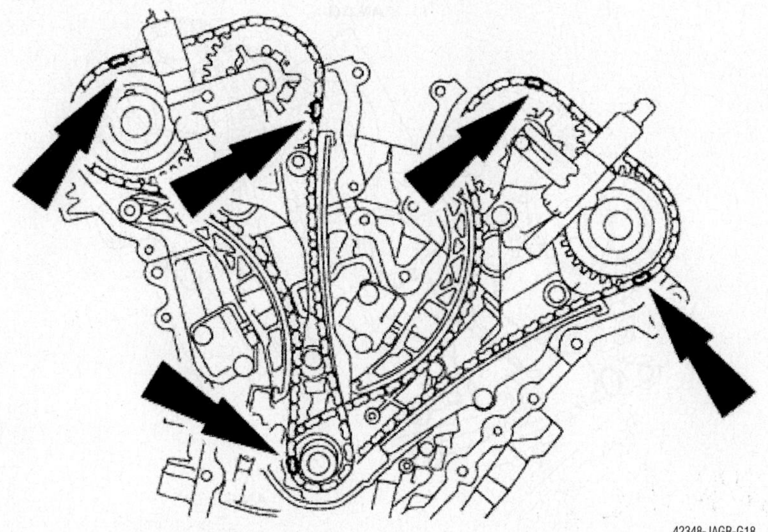

Correct installation position for timing chain—2.5L and 3.0L

➡Using a suitable tool, hold the left-hand timing chain tensioner ratchet lock mechanism away from the ratchet stem.

38. Slowly compress the left-hand timing chain tensioner and insert a paper clip to hold tensioner piston bottomed in bore.

➡The retaining tool must remain in the timing chain tensioner until the timing chain tensioner is installed to the engine with the piston bottomed in the bore.

39. Install the left-hand timing chain tensioner and tighten to 18 ft. lbs. (25 Nm).
40. Release the tension in the left-hand timing chain tensioner.
41. Remove retaining tool.
42. Make sure the right-hand timing chain alignment marks have remained correctly positioned to the camshaft sprocket and crankshaft sprocket alignment marks.
43. Make sure all the timing chain alignment marks are in the positions shown.
44. Install the crankshaft pulley retaining bolt and washer.
45. Check the engine valve timing is correctly set.
46. Rotate the crankshaft two complete turns clockwise.
47. Remove the crankshaft pulley retaining bolt and washer.
48. Install the CKP sensor pulse wheel with the teeth pointing outwards.
49. Install the spark plugs.
50. Install the engine front cover, valve covers and accessory drive belt.
51. Connect the negative battery cable.
52. Fill the cooling system.
53. Start the engine and check for leaks.

V8—A-BANK

1. Before servicing the vehicle, refer to the Precautions Section.
2. Drain the cooling system.
3. Remove or disconnect the following:
   - Negative battery cable
   - Valve cover
   - Intake manifold
   - Radiator hoses
   - Heater hoses
   - Engine Coolant Temperature (ECT) sensor connector
   - Coolant outlet pipe
   - Catalytic converter
   - Crankshaft damper
   - Front cover
   - Variable Valve Timing (VVT) bushing carrier
   - Crankshaft Position (CKP) sensor
4. Rotate the crankshaft until the triangular arrow indent on the driveplate is visible through the access hole.
5. Confirm that the timing flat on each camshaft is uppermost. If necessary, rotate the crankshaft one full turn.
6. Install the crankshaft setting peg JD 216 to the crankshaft position sensor location.
7. Install the camshaft locking tool JD 215 to the A-Bank camshafts.
8. Remove or disconnect the following:
   - Primary timing chain tensioner and backing plate
   - Primary timing chain
   - VVT unit, exhaust camshaft sprocket and secondary timing chain
   - Secondary timing chain tensioner
   - Secondary timing chain guide

To install:

9. Install or connect the following:
   - Timing chain tensioners, prepare them for installation by using a paperclip or other wire to unseat the check valves and compressing the pistons into their bores
   - Secondary timing chain guide. Tighten the bolt to 89–124 inch lbs. (10–14 Nm).
   - Secondary timing chain tensioner. Tighten the bolt to 89–124 inch lbs. (10–14 Nm).
   - VVT, secondary timing chain, and exhaust cam sprocket
   - Primary timing chain
   - Primary timing chain guide. Tighten the bolt to 10–12 ft. lbs. (13–16 Nm).
   - Primary timing chain tensioner and

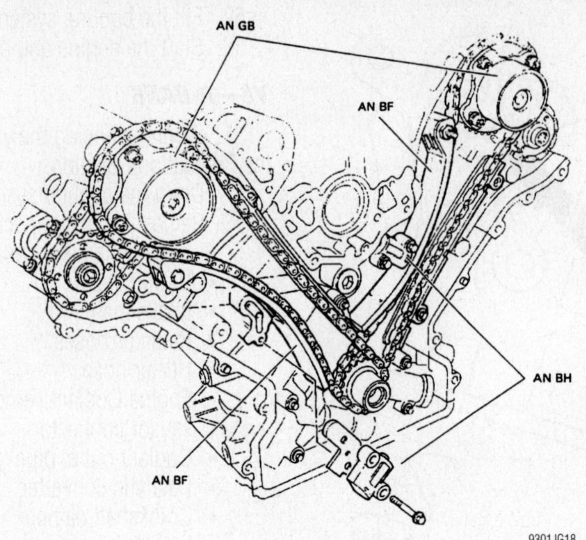

Timing chain arrangement—V8

Variable Valve Timing (VVT) solenoid—V8

backing plate. Tighten the bolts to 89–124 inch lbs. (10–14 Nm).
- Primary timing chain slack, eliminate it by placing a wedge between the primary timing chain tensioner and the guide shoe
- Secondary timing chain, tension it by applying counterclockwise force to the exhaust camshaft sprocket
- Exhaust and intake VVT sprocket bolts. Torque them to 85–92 ft. lbs. (115–125 Nm).

10. Remove the tools and wedges.
11. Install or connect the following:
    - CKP sensor
    - VVT bushing carrier
    - Front cover
    - Crankshaft damper
    - Catalytic converter
    - Coolant outlet pipe
    - ECT sensor connector
    - Heater hoses
    - Radiator hoses
    - Intake manifold
    - Valve cover
    - Negative battery cable
12. Fill the cooling system.
13. Start the engine and check for leaks.

V8—B-BANK

1. Before servicing the vehicle, refer to the Precautions Section.
2. Drain the cooling system.
3. Remove or disconnect the following:
    - Negative battery cable
    - Valve cover
    - Intake manifold
    - Radiator hoses
    - Heater hoses
    - Engine Coolant Temperature (ECT) sensor connector
    - Coolant outlet pipe
    - Catalytic converter
    - Crankshaft damper
    - Front cover
    - Variable Valve Timing (VVT) bushing carrier
    - Crankshaft Position (CKP) sensor

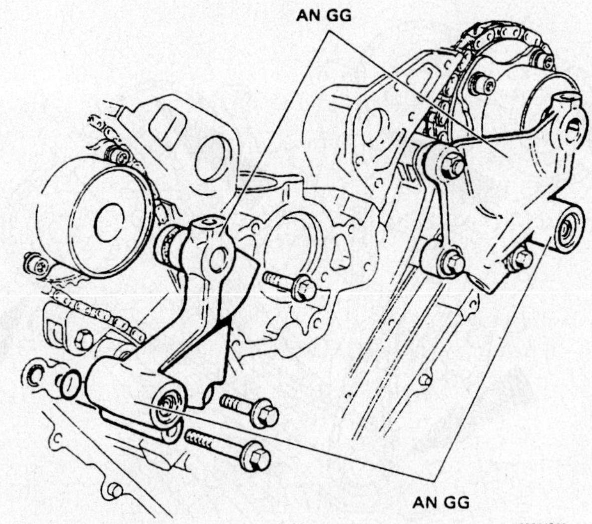

Variable Valve Timing (VVT) bushing carrier assembly—V8

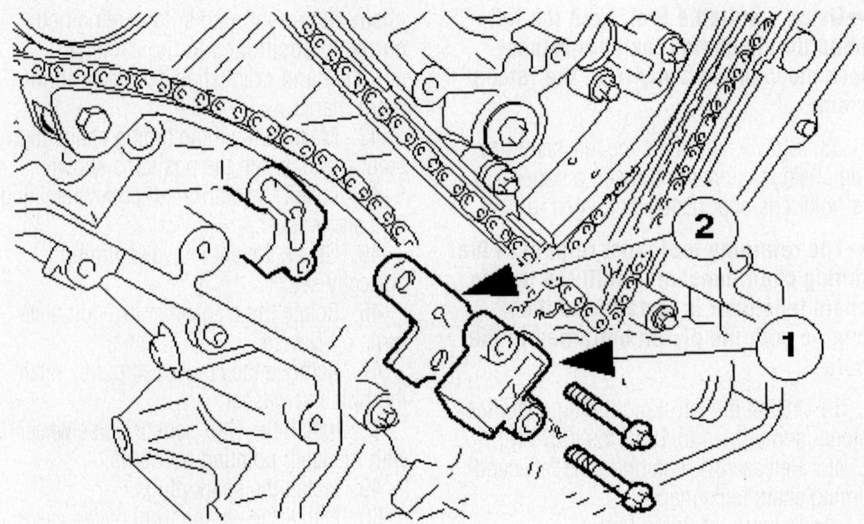

Timing chain tensioner—V8

4. Rotate the crankshaft until the triangular arrow indent on the driveplate is visible through the access hole.

5. Confirm that the timing flat on each camshaft is uppermost. If necessary, rotate the crankshaft one full turn.

6. Install the crankshaft setting peg JD 216 to the crankshaft position sensor location.

7. Install the camshaft locking tool JD 215 to the B-Bank camshafts.

8. Remove or disconnect the following:
   - Primary timing chain tensioner and backing plate
   - Primary timing chain
   - VVT unit, exhaust camshaft sprocket and secondary timing chain
   - Secondary timing chain tensioner
   - Secondary timing chain guide

To install:

9. Install or connect the following:
   - Timing chain tensioners, prepare them for installation by using a paperclip or other wire to unseat the check valves and compressing the pistons into their bores
   - Secondary timing chain guide. Tighten the bolt to 89–124 inch lbs. (10–14 Nm).
   - Secondary timing chain tensioner. Tighten the bolt to 89–124 inch lbs. (10–14 Nm).
   - VVT, secondary timing chain, and exhaust cam sprocket
   - Primary timing chain
   - Primary timing chain guide. Tighten the bolt to 10–12 ft. lbs. (13–16 Nm).
   - Primary timing chain tensioner and backing plate. Tighten the bolts to 89–124 inch lbs. (10–14 Nm).
   - Primary timing chain slack, eliminate it by placing a wedge between the primary timing chain tensioner and the guide shoe
   - Secondary timing chain, tension it by applying counterclockwise force to the exhaust camshaft sprocket
   - Exhaust and intake VVT sprocket bolts. Torque them to 85–92 ft. lbs. (115–125 Nm).

10. Remove the tools and wedges.

11. Install or connect the following:
    - CKP sensor
    - VVT bushing carrier
    - Front cover
    - Crankshaft damper
    - Catalytic converter
    - Coolant outlet pipe
    - ECT sensor connector
    - Heater hoses
    - Radiator hoses
    - Intake manifold
    - Valve cover
    - Negative battery cable

12. Fill the cooling system.

13. Start the engine and check for leaks.

FUEL SYSTEM

Fuel System Service Precautions

Safety is the most important factor when performing not only fuel system maintenance but any type of maintenance. Failure to conduct maintenance and repairs in a safe manner may result in serious personal injury or death. Maintenance and testing of the vehicle's fuel system components can be accomplished safely and effectively by adhering to the following rules and guidelines.

- To avoid the possibility of fire and personal injury, always disconnect the negative battery cable unless the repair or test procedure requires that battery voltage be applied.
- Always relieve the fuel system pressure prior to disconnecting any fuel system component (injector, fuel rail, pressure regulator, etc.), fitting or fuel line connection. Exercise extreme caution whenever relieving fuel system pressure, to avoid exposing skin, face and eyes to fuel spray. Please be advised that fuel under pressure may penetrate the skin or any part of the body that it contacts.
- Always place a shop towel or cloth around the fitting or connection prior to loosening to absorb any excess fuel due to spillage. Ensure that all fuel spillage (should it occur) is quickly removed from engine surfaces. Ensure that all fuel soaked cloths or towels are deposited into a suitable waste container.
- Always keep a dry chemical (Class B) fire extinguisher near the work area.
- Do not allow fuel spray or fuel vapors to come into contact with a spark or open flame.
- Always use a back-up wrench when loosening and tightening fuel line connection fittings. This will prevent unnecessary stress and torsion to fuel line piping.
- Always replace worn fuel fitting O-rings with new. Do not substitute fuel hose or equivalent where fuel pipe is installed.

Fuel System Pressure

RELIEVING

S-TYPE 2002–06

1. Disconnect the battery ground cable.
2. Remove the fuel system pressure relief valve cap.

⁂ WARNING

Make sure the tap of the special tool is rotated fully clockwise before installing the special tool 310-012 to the fuel system pressure relief valve. Failure to follow these instructions may result in personal injury.

3. Install the special tool to the fuel system pressure relief valve.

➡When relieving the fuel system pressure, catch any displaced fuel in a suitable container.

4. Rotate the tap of the special tool fully counterclockwise.
5. Remove the special tool from the fuel system pressure relief valve.
6. Install the fuel system pressure relief valve cap.
7. Connect the battery ground cable.

2002–03 X-TYPE

1. Disconnect the battery ground cable.
2. Remove the Schrader valve cap.

⁂ WARNING

Make sure that the tap is turned fully clockwise before fitting the fuel pressure gauge to the Schrader valve. Failure to follow these instructions may result in personal injury.

3. Connect the fuel pressure gauge JD 209 to the Schrader valve.

➡When relieving the fuel pressure, catch any displaced fuel in a suitable container.

4. Turn the tap on the fuel pressure gauge fully counterclockwise to relieve the fuel pressure.
5. Remove the fuel pressure gauge and install the Schrader valve cap.

2004–06 X-TYPE

1. Remove the fuel pump fuse.
2. Start the engine and allow to idle until the engine stalls.

3. Crank the engine for approximately five seconds to make sure the fuel injection supply manifold pressure has been released.
4. Install the fuel pump fuse.

2002–03 XJ

1. Make sure the vehicle is in Park for vehicles with automatic transmission, Neutral for vehicles with manual transmission.
2. Apply the parking brake.
3. Place suitable absorbent material around the Schrader valve to absorb any spillage when connecting the gauge set.
4. Remove the protective cap from the Schrader valve.

➡Depending on the design of the gauge set, there may be a drain valve and tube. Make sure this valve is closed, and the drain tube placed in a suitable container BEFORE connecting the gauge to the Schrader valve.

5. Connect the gauge, using adaptors if required.
6. Remove the absorbent material and dispose of safely.
7. Disconnect and plug the vacuum hose from the pressure regulator.
8. Start the engine and record the fuel pressure reading. As a guide, the fuel pressure should be approximately 3.0 bar (44.1 psi) with the regulator disconnected.
9. With the engine still running, reconnect the vacuum hose to the pressure regulator and record the reading. The pressure should drop to approximately 2.6 bar (38 psi) when the vacuum hose is reconnected.
10. Switch off the engine.
11. Place more absorbent material around the Schrader valve.
12. Where fitted, open the drain valve and allow the fuel from the gauge and line to flow into the container to depressurize the system.
13. Disconnect and remove the gauge set and any adaptors.
14. Clean up any fuel which may have been spilt and remove the absorbent material. Dispose of safely.
15. Refit and secure the protective cap to the Schrader valve.
16. Reconnect the fuel pressure regulator vacuum hose.
17. Start the engine and make sure there are no leaks.

2003.5–06 XJ

1. Disconnect the battery ground cable.
2. Remove the fuel system pressure relief valve cap.

✲✲ WARNING

Make sure the tap of the special tool is rotated fully clockwise before installing the special tool to the fuel system pressure relief valve. Failure to follow these instructions may result in personal injury.

3. Install the special tool JD 209 to the fuel system pressure relief valve.

➡When relieving the fuel system pressure, catch any displaced fuel in a suitable container.

4. Rotate the tap of the special tool fully counterclockwise.
5. Remove the special tool from the fuel system pressure relief valve.
6. Install the fuel system pressure relief valve cap.
7. Connect the battery ground cable.

2002 XK

1. Make sure the vehicle is in Park for vehicles with automatic transmission, Neutral for vehicles with manual transmission.
2. Apply the parking brake.
3. Place suitable absorbent material around the Schrader valve to absorb any spillage when connecting the gauge set.
4. Remove the protective cap from the Schrader valve.

➡Depending on the design of the gauge set, there may be a drain valve and tube. Make sure this valve is closed, and the drain tube placed in a suitable container BEFORE connecting the gauge to the Schrader valve.

5. Connect the gauge, using adaptors if required.
6. Remove the absorbent material and dispose of safely.
7. Disconnect and plug the vacuum hose from the pressure regulator.
8. Start the engine and record the fuel pressure reading. As a guide, the fuel pressure should be approximately 3.0 bar (44.1 psi) with the regulator disconnected.
9. With the engine still running, reconnect the vacuum hose to the pressure regulator and record the reading. The pressure should drop to approximately 2.6 bar (38 psi) when the vacuum hose is reconnected.
10. Switch off the engine.
11. Place more absorbent material around the Schrader valve.
12. Where fitted, open the drain valve and allow the fuel from the gauge and line to flow into the container to depressurize the system.
13. Disconnect and remove the gauge set and any adaptors.
14. Clean up any fuel which may have been spilt and remove the absorbent material. Dispose of safely.
15. Refit and secure the protective cap to the Schrader valve.
16. Reconnect the fuel pressure regulator vacuum hose.
17. Start the engine and make sure there are no leaks.

2003–06 XK

1. Disconnect the battery ground cable.
2. Remove the oil filler cap.
3. Remove the engine cover.
4. Remove the fuel system pressure release valve cap.

✲✲ WARNING

Make sure the tap of the special tool is rotated fully clockwise before installing the special tool to the fuel system pressure release valve. Failure to follow these instructions may result in personal injury.

5. Install the special tool 310-012 to the fuel system pressure relief valve.

➡When releasing the fuel system pressure, catch any displaced fuel in a suitable container.

6. Rotate the tap of the special tool fully counterclockwise.
7. Remove the special tool from the fuel system pressure release valve.
8. Install the fuel system pressure relief valve cap.
9. Install the engine cover.
10. Install the oil filler cap.
11. Connect the battery ground cable.

Fuel Filter

REMOVAL & INSTALLATION

1. Before servicing the vehicle, refer to the Precautions Section.
2. Relieve the fuel system pressure.
3. Remove or disconnect the following:
   - Negative battery cable
   - Fuel filter bracket cover
   - Fuel lines
   - Fuel filter

To install:

4. Install or connect the following:
   - Fuel filter into the bracket making sure the flow direction is correct. Tighten the clamp to 15–25 inch lbs. (2–3 Nm).

- Fuel lines. Tighten the fittings to 22 ft. lbs. (30 Nm).

5. Start the engine and check for leaks.

Fuel Pump

REMOVAL & INSTALLATION

1. Before servicing the vehicle, refer to the Precautions Section.
2. Relieve the fuel system pressure.
3. Drain the fuel tank.
4. Remove or disconnect the following:
   - Negative battery cable
   - Trunk liner
   - Trunk seal retainer
   - Rear lamp assembly interior trim finisher
   - Left and right side liners
   - Fuel feed and return lines
   - Fuel filler and vent hoses
   - Fuel tank wiring connectors
   - Fuel filler cap
   - Fuel tank retaining straps
   - Fuel tank
   - Fuel pump module

To install:

5. Install or connect the following:
   - Fuel pump module
   - Fuel tank
   - Fuel tank retaining straps
   - Fuel filler cap
   - Fuel tank wiring connectors
   - Fuel filler and vent hoses
   - Fuel feed and return lines
   - Left and right side liners
   - Rear lamp assembly interior trim finisher
   - Trunk seal retainer

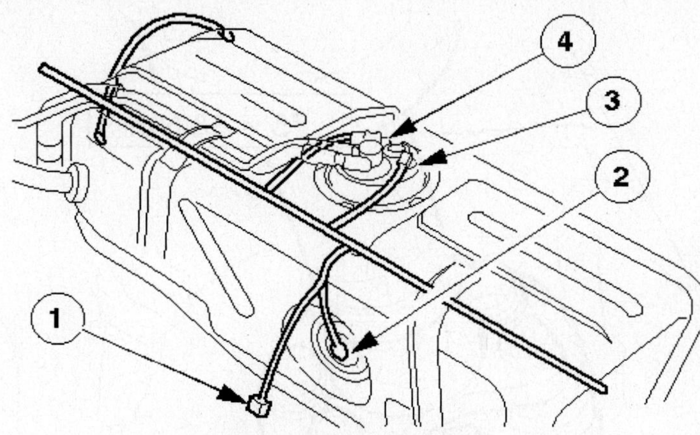
4
3
2
1
1 If fitted, disconnect the wiring to the accelerometer.
2 Disconnect the wiring to the fuel level sensor harness.
3 Disconnect the wiring to the fuel pump connector.
4 If fitted, disconnect the wiring to the pressure sensor connector.
9301JG24

XK8 fuel tank

   - Trunk liner
   - Negative battery cable

6. Fill the fuel tank with at least 10 gallons (38L) of fuel.
7. Start the engine and check for leaks.

Fuel Injector

REMOVAL & INSTALLATION

1. Before servicing the vehicle, refer to the Precautions Section.
2. Relieve fuel system pressure.
3. Remove or disconnect the following:
   - Negative battery cable
   - Engine appearance covers
   - Fuel lines
   - Fuel pressure regulator
   - Fuel cross over elbow
   - Fuel injector connectors
   - Fuel injector clamping plates
   - Fuel injectors

To install:

4. Install or connect the following:
   - Fuel injectors. Use new O-ring seals.
   - Fuel injector clamping plates
   - Fuel injector connectors
   - Fuel cross over elbow
   - Fuel pressure regulator
   - Fuel lines
   - Engine appearance covers
   - Negative battery cable
5. Start the engine and check for leaks.

DRIVE TRAIN

Transmission Assembly

REMOVAL & INSTALLATION

Automatic

ALL MODELS EXCEPT X-TYPE

1. Before servicing the vehicle, refer to the Precautions Section.
2. Install a support fixture to the engine lifting eyes.
3. Drain the transmission fluid.
4. Remove or disconnect the following:
   - Negative battery cable
   - Engine appearance covers
   - Mass Air Flow (MAF) meter
   - Air intake assembly
   - Coolant recovery tank
   - Exhaust front pipes
   - Starter
   - Driveshaft
   - Shift selector cable
   - Transmission electrical connectors
   - Transmission oil cooler lines
   - Torque converter
   - Transmission mount and bracket
   - Transmission flange bolts
   - Transmission

To install:

5. Install or connect the following:
   - Transmission to the engine. Tighten the flange bolts to 32–42 ft. lbs. (43–57 Nm).
   - Transmission mount and bracket. Tighten the mount bolts to 22–30 ft. lbs. (30–40 Nm) and the bracket bolts to 16–21 ft. lbs. (22–28 Nm).
   - Torque converter. Tighten the bolts to 32–42 ft. lbs. (43–57 Nm).
   - Transmission oil cooler lines
   - Transmission electrical connectors
   - Shift selector cable
   - Driveshaft. Tighten the bolts to 55–65 ft. lbs. (75–88 Nm).
   - Exhaust front pipes
   - Coolant recovery tank
   - Air intake assembly
   - MAF meter
   - Engine appearance covers
   - Negative battery cable
6. Fill the transmission to the correct level with the proper fluid. Do not overfill.
7. Start the engine and check for leaks.

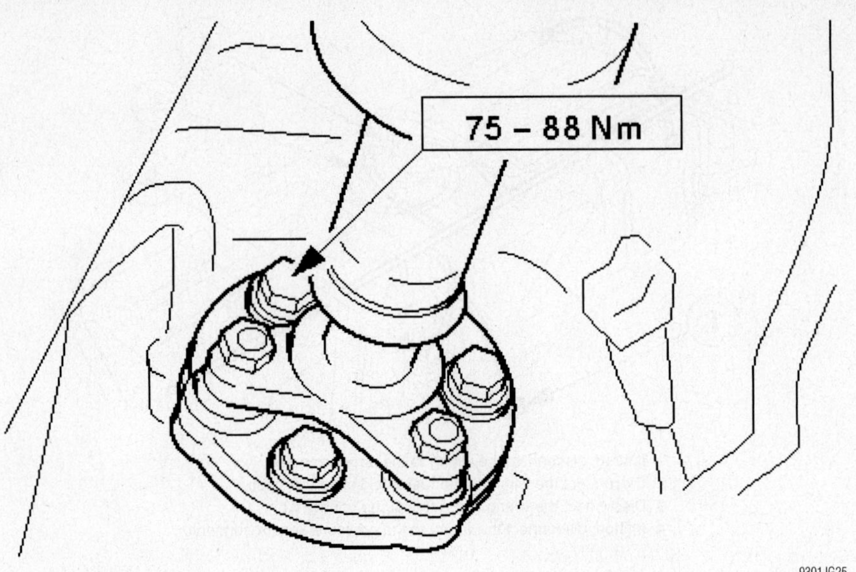

Driveshaft flange bolts—except X-Type

X-TYPE

1. Before servicing the vehicle, refer to the Precautions Section.
2. Install a support fixture to the engine lifting eyes.
3. Drain the transmission fluid.
4. Remove or disconnect the following:
   - Negative battery cable
   - Engine/transaxle assembly
   - Transfer case
   - Transfer case link shaft
   - Torque converter access cover
   - Transaxle dust cover
   - Torque converter bolts
   - Transaxle bolts
5. Support the torque converter and remove the transaxle from the engine.

To install:

6. Install or connect the following:
   - Transmission to the engine. Tighten the bolts to 35 ft. lbs. (48 Nm).
   - Torque converter. Tighten the bolts to 41 ft. lbs. (55 Nm).
   - Transfer case. Tighten the bolts to 66 ft. lbs. (90 Nm). Tighten the nuts to 18 ft. lbs. (25 Nm).
   - Engine/transaxle assembly
   - Negative battery cable
7. Fill the transmission to the correct level with the proper fluid. Do not over-fill.
8. Start the engine and check for leaks.

Manual

X-TYPE

1. Before servicing the vehicle, refer to the Precautions Section.
2. Install a support fixture to the engine lifting eyes.
3. Drain the transmission fluid.
4. Remove or disconnect the following:
   - Negative battery cable
   - Engine/transaxle assembly
   - Transfer case
   - Transfer case link shaft
   - Transaxle dust cover
   - Transaxle

To install:

5. Install or connect the following:
   - Transmission to the engine. Tighten the bolts to 35 ft. lbs. (48 Nm).
   - Transfer case. Tighten the bolts to 66 ft. lbs. (90 Nm). Tighten the nuts to 18 ft. lbs. (25 Nm).
   - Engine/transaxle assembly
   - Negative battery cable
6. Fill the transmission to the correct level with the proper fluid. Do not over-fill.
7. Bleed the clutch system.
8. Start the engine and check for leaks.

Clutch

ADJUSTMENT

All vehicles use a hydraulic clutch release mechanism. No free-play adjustment is required or possible.

REMOVAL & INSTALLATION

1. Before servicing the vehicle, refer to the Precautions Section.
2. Remove or disconnect the following:
   - Negative battery cable
   - Transaxle

➡If the pressure plate is to be reused, matchmark its relationship to the flywheel.

3. Using a Flywheel Locking tool, lock the flywheel.
4. Remove or disconnect the following:
   - Pressure plate from the flywheel by loosening the bolts alternately, a little at a time, to prevent warpage
   - Clutch disc

To install:

5. Install or connect the following:
   - Clutch with the driven plate on the pressure plate so the spring cage is facing the pressure plate.
6. Hold the clutch assembly against the flywheel, aligning the matchmarks and the dowel pins on the flywheel with the pressure plate. Insert an alignment shaft tool through the pressure plate and the driven plate into the crankshaft pilot bearing.
7. Install the pressure plate to the flywheel. Torque the bolts evenly to 18 ft. lbs. (24 Nm), in a diagonal pattern, to avoid distortion.
8. Remove the alignment shaft.

⁂ WARNING

The clutch release bearing in the front of the transaxle should be checked before reassembly. It is retained by 2 springs.

9. Install or connect the following:
   - Transaxle
   - Negative battery cable

Hydraulic Clutch System

BLEEDING

The clutch system can be bled using a pressure bleeder. Follow the instructions that come with the pressure bleeder for the proper pressure bleeding procedure.

1. Before servicing the vehicle, refer to the Precautions Section.
2. To bleed the system perform the following:
   a. Top off the hydraulic fluid reservoir using a fluid that meets the standards of the vehicle's hydraulic system.
   b. Open the clutch slave cylinder bleed screw and press the clutch pedal to the floor and hold the pedal down.
   c. Close the clutch slave cylinder bleed screw.
   d. Release the clutch pedal.
   e. Check the hydraulic fluid level and top off as necessary.

3. Repeat the above steps until the discharged fluid is clean and no air bubbles appear during the bleeding process.

Halfshaft

REMOVAL & INSTALLATION

Front

X-TYPE

1. Before servicing the vehicle, refer to the Precautions Section.
2. Drain the transmission fluid.
3. Remove or disconnect the following:
   - Negative battery cable
   - Wheel and tire
   - Wheel hub nut
   - Tie rod end
   - Lower ball joint nut
   - Subframe reinforcing plate
   - Subframe mounting bolts
   - Wheel knuckle
   - Halfshaft
   - Halfshaft snap ring

To install:

4. Install or connect the following:
   - New halfshaft snap ring
   - Halfshaft
   - Wheel knuckle
   - Lower ball joint: Tighten nut to 61 ft. lbs. (83 Nm).
   - Subframe: Tighten bolts to 105 ft. lbs. (142 Nm).
   - Subframe reinforcing plate: Tighten bolts to 26 ft. lbs. (35 Nm).
   - Tie rod end: Tighten bolt to 26 ft. lbs. (35 Nm).
   - Loosely install wheel hub nut
   - Wheel and tire
5. Lower vehicle so suspension is loaded, and tighten the wheel hub nut to 199 ft. lbs. (270 Nm).
6. Fill the transmission with fluid and road test vehicle.

Rear

ALL MODELS EXCEPT S-TYPE AND X-TYPE

1. Before servicing the vehicle, refer to the Precautions Section.
2. Remove or disconnect the following:
   - Negative battery cable
   - Rear wheel
   - Brake caliper
   - Wheel speed sensor
   - Hub carrier pivot bolt
   - Hub retaining nut
3. Press the stub shaft out of the hub and remove the halfshaft.

To install:

4. Install or connect the following:
   - 4 axle shaft-to-differential output shaft bolts. Tighten the bolts 60–73 ft. lbs. (81–99 Nm).
   - Halfshaft in the wheel hub by applying Loctite® 270 thread locking compound to the splines
   - Hub carrier pivot bolt by aligning the bolt head matchmarks. Tighten it to 66–81 ft. lbs. (90–110 Nm).
   - Hub retaining nut. Use a new nut and tighten it to 224–248 ft. lbs. (304–337 Nm).
   - Wheel speed sensor
   - Brake caliper
   - Rear wheel
   - Negative battery cable
5. Check the wheel alignment and adjust as necessary.

➡If the hub is removed for any reason, a new bearing assembly must be installed. Never attempt to re-use a bearing.

S-TYPE

1. Before servicing the vehicle, refer to the Precautions Section.
2. Remove or disconnect the following:
   - Negative battery cable
   - Wheel and tire
   - Loosen the wheel hub nut
   - Brake disc
   - Wheel speed sensor
   - Outer tie rod nut
   - Lower arm from wheel knuckle
   - Wheel hub nut
   - Disconnect halfshaft
   - Wheel knuckle
   - Halfshaft
   - Halfshaft retaining clip
   - Halfshaft seal

To install:

3. Install or connect the following:
   - Halfshaft seal
   - New halfshaft retaining clip
4. Apply Loctite® to the halfshaft splines and install the halfshaft, making sure the retaining clip is correctly seated.

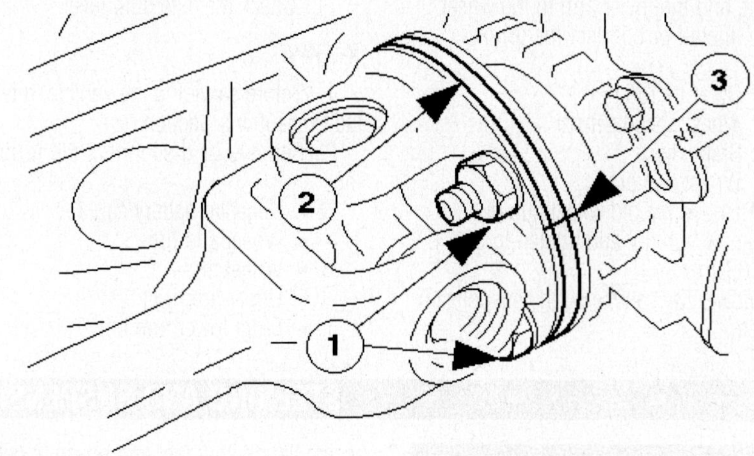

Matchmark the flange

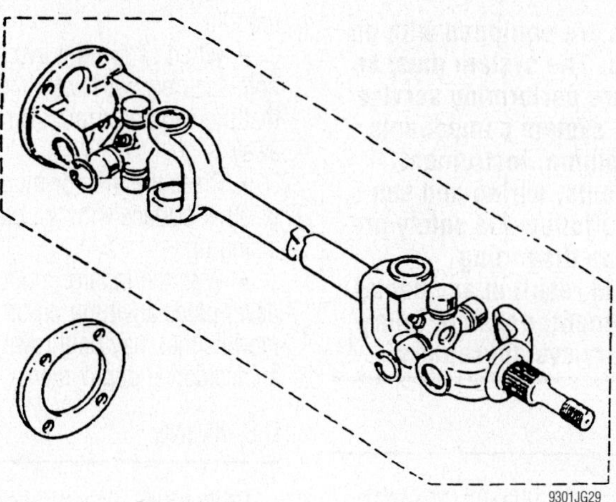

Halfshaft assembly

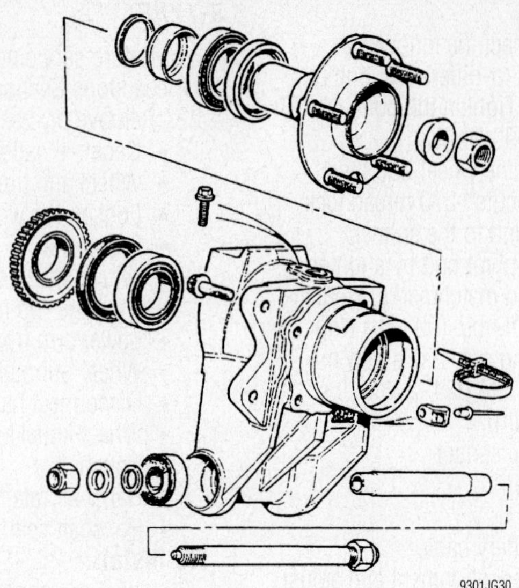

Rear hub exploded view

5. Install the wheel knuckle assembly, then install the old wheel hub nut and tighten to 111 ft. lbs. (150 Nm).

6. Attach the wheel knuckle to the halfshaft, and the lower arm to the wheel knuckle. Install or connect the following:
   - Outer tie rod nut
   - Wheel speed sensor
   - Brake disc
   - Wheel and tire

7. Remove the old wheel hub nut. Install a new hub nut and tighten to 222 ft. lbs. (300 Nm).

8. Lower the vehicle so the suspension is loaded.

9. Tighten the lower arm to the wheel knuckle bolt to 111 ft. lbs. (150 Nm).

10. Tighten the outer tie rod nut to 41 ft. lbs. (55 Nm).

11. Check the axle fluid level.

X-TYPE

1. Before servicing the vehicle, refer to the Precautions Section.

2. Remove or disconnect the following:
   - Negative battery cable
   - Wheel and tire
   - Wheel hub
   - Upper arm bolt
   - Front lower arm bolt
   - Support rear lower arm with jack stand
   - Rear lower arm bolt
   - Halfshaft

To install:

3. Reverse the removal procedure to install. Tighten all control arm bolts to 89 ft. lbs. (120 Nm).

CV-Joints

OVERHAUL

1. Before servicing the vehicle, refer to the Precautions Section.

2. Remove or disconnect the following:
   - Axle halfshaft
   - Inner boot clamps and push the boot back
   - CV-joint housing
   - Snapring
   - Spider and rollers
   - CV-joint boot

➡Do not disassemble the spider and rollers.

To install:

3. Install or connect the following:
   - CV-joint boot
   - Spider and rollers
   - Snapring

4. Apply clean grease to the CV-joint housing and the boot.

5. Install or connect the following:
   - CV-joint housing and tighten the boot clamps
   - Axle to the vehicle

STEERING AND SUSPENSION

Air Bag

⁂ CAUTION

Some vehicles are equipped with an air bag system. The system must be disarmed before performing service on, or around, system components, the steering column, instrument panel components, wiring and sensors. Failure to follow the safety precautions and the disarming procedure could result in accidental air bag deployment, possible injury and unnecessary system repairs.

PRECAUTIONS

Several precautions must be observed when handling the inflator module to avoid accidental deployment and possible personal injury.

- Never carry the inflator module by the wires or connector on the underside of the module.
- When carrying a live inflator module, hold securely with both hands, and ensure that the bag and trim cover are pointed away.
- Place the inflator module on a bench or other surface with the bag and trim cover facing up.
- With the inflator module on the bench, never place anything on or close to the module that may be thrown in the event of an accidental deployment.

DISARMING

Proper SRS disarming can be obtained by disconnecting and isolating the negative battery cable. Allow the air bag system capacitor at least 2 minutes to discharge before removing any air bag system components.

Power Rack and Pinion Steering Gear

REMOVAL & INSTALLATION

S-Type

2002 S-Type through VIN M45254

1. Disconnect the battery ground cable.

➡Install the special tool to the lower steering column to maintain a straight ahead steering wheel position. Failure to do so may cause damage to the steering wheel clock-spring.

2. Install the special tool to the lower steering column.

3. Loosen the lower steering column slider pinch bolt.

4. Remove both front wheels and tires.

5. Disconnect the actuator electrical connector.

6. Detach the lower steering column.

7. Remove the steering gear shaft pinch bolt.

➡Whenever the steering gear unions and hoses are disconnected the steering gear valve body ports and hoses should be plugged to prevent dirt ingress. Failure to follow this instruction may result in damage to the vehicle.

8. Disconnect the power steering supply and return tubes.

9. Drain the fluid into a suitable container.

10. Remove and discard the O-ring seals.

11. Prevent the tie-rod end ball joint ball pin hexagon from rotating. Failure to do so may result in damage to the tie-rod end ball joint boot.

12. Detach both of the tie rod ends. Remove and discard the retaining nuts.

13. Vehicles with 4.0L engine, detach the steering gear supply and return tubes.

14. Remove the steering gear.

To install:

15. Install the steering gear. Tighten to 76 ft. lbs. (103 Nm).

16. Vehicles with 4.0L engine, attach the steering gear supply and return tubes.

➡Prevent the tie-rod end ball joint ball pin from rotating. Failure to do so may result in damage to the tie-rod end ball joint boot. Make sure the mating surfaces are free from dirt and lubricant. Failure to do so may result in damage to the vehicle.

17. Attach both of the tie-rod ends to the wheel knuckles. Install new retaining nuts. Tighten to 56 ft. lbs. (75 Nm).

18. Install new O-ring seals. Connect the power steering supply and return tubes. Tighten to 89 inch lbs. (10 Nm).

19. Install the steering gear shaft pinch bolt. Tighten to 26 ft. lbs. (35 Nm).

20. Connect the actuator electrical connector.

21. Install both front wheels and tires.

22. Tighten the lower steering column slider pinch bolt. Tighten to 26 ft. lbs. (35 Nm).

23. Remove the special tool from the lower steering column.

Steering wheel locking tool installed—2002 S-Type through VIN M45254

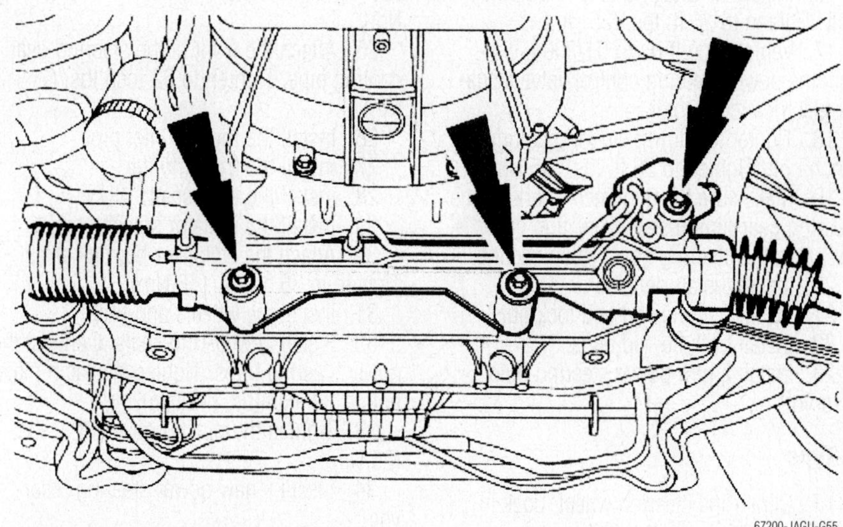

Steering gear mounting points—S-Type

➡When the battery has been disconnected and reconnected, some abnormal driving conditions may occur while the vehicle relearns its driving strategy.

24. Connect the battery ground cable.

25. Install a new power steering fluid reservoir.

26. Check and adjust front wheel alignment.

2002.5–2004 S-Type

1. Center the steering wheel. Lock in position, remove the ignition key.

2. Remove both tie-rod ends. Remove both tie-rod end lock nuts.

3. Remove the air deflector.

4. Make sure the alignment mark on the steering gear pinion seal protection cover is central to the steering gear pinion casting.

5. Loosen the lower steering column pinch bolt.

6. Vehicles built up to 01/2004, disconnect the power steering control valve actuator electrical connector.

7. Detach the lower steering column.

8. Remove the steering gear shaft pinch bolt.

➡Cap the power steering lines to prevent loss of fluid and dirt ingress. If power steering fluid comes into contact with the paintwork, the affected area must be immediately washed down with cold water. Drain the fluid into a suitable container.

9. Disconnect the power steering supply and return lines. Remove and discard the O-ring seals.
10. Vehicles built up to 01/2004, remove the steering gear.
11. Vehicles built 02/2004 onwards, remove the steering gear. Disconnect the power steering control valve actuator electrical connector.
12. Remove the steering gear.

To install:

13. Vehicles built 02/2004 onwards, install the steering gear. Connect the power steering control valve actuator electrical connector. Tighten to 74 ft. lbs. (100 Nm).
14. Vehicles built up to 01/2004, install the steering gear. Tighten to 74 ft. lbs. (100 Nm).
15. Install new O-ring seals. Connect the power steering supply and return lines. Tighten to 15 Nm.
16. Install the steering gear shaft pinch bolt. Tighten to 26 ft. lbs. (35 Nm).
17. Vehicles built up to 01/2004, connect the power steering control valve actuator electrical connector.
18. Tighten the lower steering column pinch bolt. Tighten to 26 ft. lbs. (35 Nm).
19. Make sure the alignment mark on the steering gear pinion seal protection cover is central to the steering gear pinion casting.
20. Install the air deflector.
21. Install both tie-rod end lock nuts.
22. Install both tie-rod ends.
23. Install a new power steering fluid reservoir.

X-Type

1. Center the steering wheel. Lock in position and remove the ignition key.
2. Remove the extension shaft lower retaining bolt.
3. Remove the wheels and tires.
4. Detach the stabilizer bar link arms.
5. Remove the radiator splash shield.
6. Remove the driveshaft.
7. Remove the front muffler.
8. Remove the muffler inlet pipe.
9. Detach the rear of the steering gear heat shield.
10. Detach the right-hand front of the steering gear heat shield.
11. Detach the left-hand front of the steering gear heat shield.
12. Remove the engine roll restrictor.
13. Detach the front of the steering gear cooling pipe.
14. Detach the rear of the steering gear cooling pipe.
15. Remove both tie rod ends from the wheel hub assemblies.
16. Disconnect the actuator electrical connector.
17. Remove the latch plate bolt. Rotate latch plate counter clockwise. Disconnect the high and low pressure pipes.
18. Remove the steering gear.

To install:

19. Install the steering gear. Tighten to 98 ft. lbs. (133 Nm).
20. Attach the left-hand front of the steering gear heat shield. Tighten to 62 inch lbs. (7 Nm).
21. Attach the right-hand front of the steering gear heat shield. Tighten to 62 inch lbs. (7Nm).
22. Attach the rear of the steering gear heat shield. Tighten to 62 inch lbs. (7 Nm).
23. Install the engine roll restrictor. Tighten to 59 ft. lbs. (80 Nm).
24. Attach the rear of the steering gear cooling pipe. Tighten to 62 inch lbs. (7 Nm).
25. Attach the front of the steering gear cooling pipe. Tighten to 62 inch lbs. (7 Nm).
26. Install the muffler inlet pipe.
27. Install the front muffler.
28. Install the driveshaft.
29. Install the radiator splash shield.
30. Attach the stabilizer bar link arms. Tighten to 36 ft. lbs. (48 Nm).
31. Install the wheels and tires.
32. Install new O-ring seals. Connect the power steering lines. Tighten the latch plate bolt to 89 inch lbs. (10 Nm).
33. Tighten the pinch bolt to 18 ft. lbs. (25 Nm).
34. Install a new power steering reservoir.

2002–03 XJ Range through VIN F59525 and all XK Range

1. Centralize steering the wheel.
2. Raise and support the vehicle on a hoist.
3. Place a suitable drain pan under the steering gear.
4. From below the vehicle, check that steering gear is centralized by aligning the centralizing notch on the dirt protection cover with the location hole in the pinion housing casting.
5. Remove and discard both tie rod end nuts. Detach both of the tie rod ends.
6. Loosen the steering column lower shaft retaining bolt.
7. Remove and discard the steering column lower shaft retaining bolt.
8. Detach the steering column lower shaft.
9. Disconnect the steering gear transducer electrical connector.
10. Cut and discard the power steering return pipe tie-strap.
11. Detach the power steering return pipe from the mounting rubber.
12. Remove the power steering supply and return pipe retaining plate bolt and withdraw the plate.
13. Drain the power steering fluid.
14. Disconnect the power steering supply and return pipes and allow the fluid to drain.
15. Remove and discard the O-ring seals from the pipe connectors.
16. Install caps to the pipe connectors and pinion housing ports.

➡The steering gear mounting bushings must be replaced as a pair.

17. Remove the steering gear mounting bracket bushings.

➡When removing the steering gear, maneuver the power steering return pipe past the pinion housing to steering gear pipes.

18. Remove the steering gear.
19. Using a tap (M10 x 1.5 mm), clean the steering gear to crossbeam bolt hole threads.

To install:

➡If a new steering gear assembly is to be fitted, make sure steering gear is centralized before installation. A new steering gear assembly is supplied with a dirt protection cover/centralizing cap fitted.

20. Position the steering gear to vehicle.
21. Install the steering gear mounting bracket bushings. Torque to 33 ft. lbs. (45 Nm).
22. Connect the power steering supply and return pipes to the pinion housing.
23. Install new O-ring seals in pipe connectors. Lubricate the O-ring seals with clean power steering fluid.
24. Position the power steering supply and return pipes to the pinion ports and seat O-rings. Attach the power steering supply and return pipes to the pinion housing.
25. Ensure the power steering supply and return pipes and the retaining plate are correctly seated. Install the bolt. Tighten to 9 ft. lbs. (12 Nm).
26. Attach the power steering return pipe to the steering gear.
27. Attach the rubber support to the power steering return pipe.
28. Install a new power steering return pipe tie-strap.

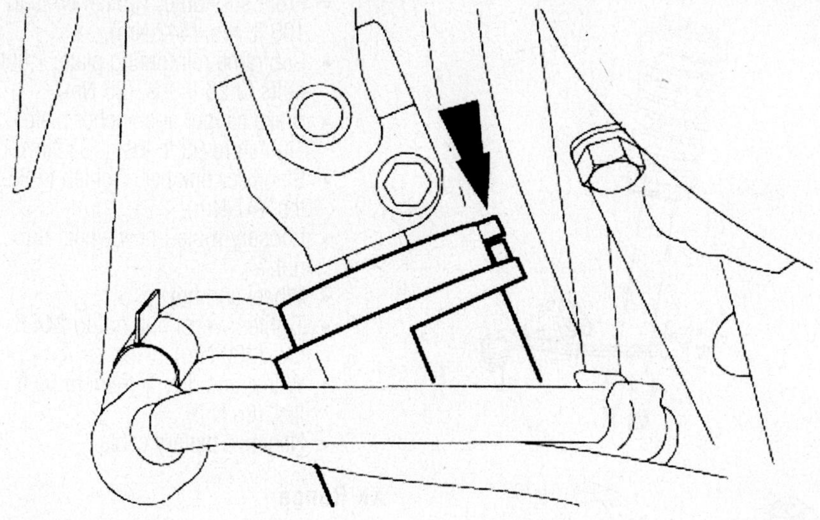

Centering the steering shaft

29. Install a new steering gear transducer harness tie-strap.
30. Connect the steering gear transducer electrical connector.
31. Locate the tie rod end into the steering arm. Install a new nut. Tighten the nut to 21 ft. lbs. (28 Nm).

➡Make sure the steering wheel and wheel and tire assemblies are in the straight ahead position before fitting the steering column lower shaft.

32. Attach the steering column lower shaft.
33. Attach the steering column lower shaft.
34. Install the steering column lower shaft retaining bolt. Tighten to 25 ft. lbs. (34 Nm).
35. Install a new power steering fluid reservoir, refill with new power steering fluid and bleed air from the system.
36. Check and adjust the front toe.

2003.5–06 XJ Range from VING00442

1. Center the steering wheel. Lock in position, remove the ignition key.
2. Remove both tie-rod ends. Remove both tie-rod end lock nuts.
3. Remove the air deflector.
4. Disconnect the power steering control valve actuator electrical connector.
5. Remove the steering gear shaft pinch bolt.
6. Cap the power steering lines to prevent loss of fluid and prevent dirt ingress.

➡If power steering fluid comes into contact with the paintwork, the affected area must be immediately washed down with cold water.

7. Disconnect the power steering supply and return lines. Remove and discard the O-ring seals.
8. Remove the steering gear.

To install:

9. Install the steering gear. Tighten to 100 Nm.
10. Install new O-ring seals. Connect the power steering supply and return lines. Tighten to 15 Nm.
11. Attach the lower steering column. Install the steering gear shaft pinch bolt. Tighten to 35 Nm.
12. Connect the power steering control valve actuator electrical connector.
13. Install the air deflector.
14. Install both tie-rod ends.
15. Install both tie-rod end lock nuts.
16. Install a new Power Steering Fluid Reservoir.

Front Strut and Spring Assembly

REMOVAL & INSTALLATION

S-Type

1. Before servicing the vehicle, refer to the Precautions Section.
2. Remove or disconnect the following:
   - Negative battery cable
   - Wheel and tire
   - Stabilizer bar link
   - Support wheel knuckle
   - Upper arm
   - Shock absorber and spring cover plate
   - Shock lower bolt
   - Electrical connector
   - Shock absorber and spring upper bolts
   - Shock absorber and spring

To install:

3. Install or connect the following:
   - Shock absorber and spring upper bolts: Tighten to 20 ft. lbs. (27 Nm).
   - Electrical connector
   - Shock absorber and spring cover plate
   - Upper arm: Tighten to 66 ft. lbs. (90 Nm).
   - Shock absorber lower bolt: Tighten to 129 ft. lbs. (175 Nm).
   - Stabilizer bar link: Tighten to 35 ft. lbs. (47 Nm).
   - Wheel and tire: Tighten to 52 ft. lbs. (70 Nm).
   - Negative battery cable

X-Type

1. Before servicing the vehicle, refer to the Precautions Section.
2. Loosen the wheel hub nut.
3. Remove or disconnect the following:
   - Negative battery cable
   - Wheel and tire
   - Stabilizer bar link
   - Brake caliper and anchor plate and suspend aside
   - Subframe reinforcement plate
   - Loosen subframe mounting bolts and lower subframe
   - Lower ball joint
   - Wheel knuckle
   - Wheel hub nut
   - Halfshaft
   - Support spring and shock

➡Top inner spring retaining stud is marked for orientation reference.

   - Strut and spring assembly
   - Wheel knuckle pinch bolt
   - Wheel knuckle

To install:

4. Install or connect the following:
   - Wheel knuckle: Tighten to 63 ft. lbs. (85 Nm).

➡Make sure orientation arrow on the strut top rubber mount faces from front to rear of the vehicle. Also make sure the stud with the paint spot on the strut top mount is located in the inner most fixing.

   - Strut and spring upper nuts: Tighten to 18 ft. lbs. (25 Nm).
   - Lower ball joint: Tighten to 63 ft. lbs. (83 Nm).

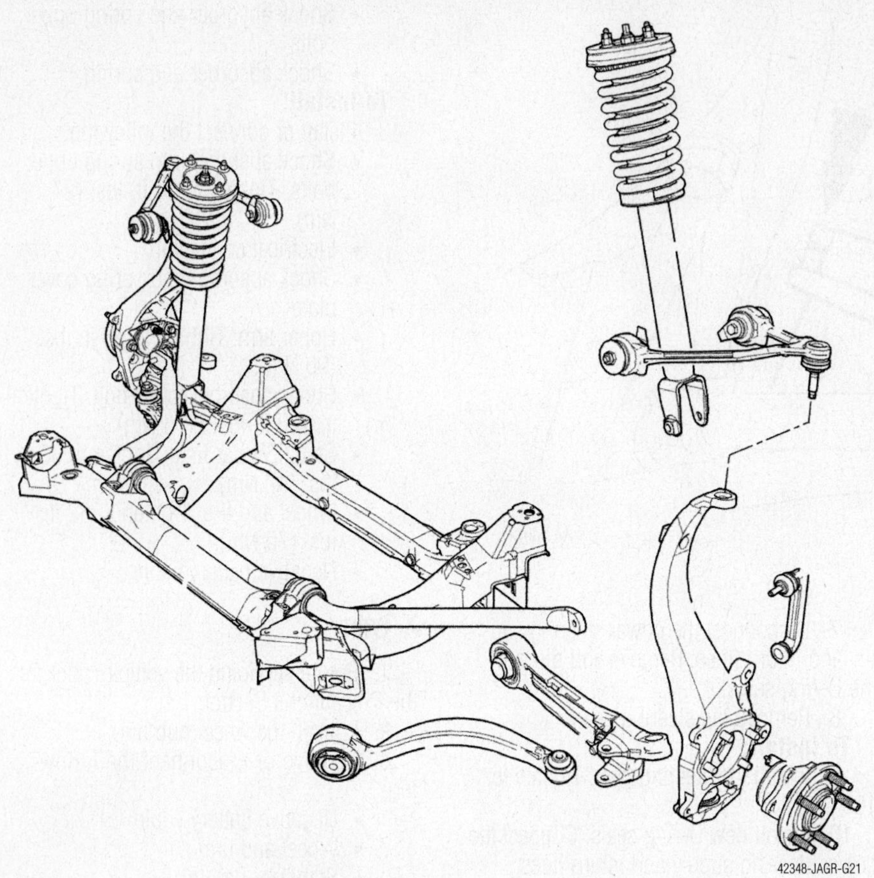

Front suspension assembly—S-Type

- Front subframe: Tighten bolts to 105 ft. lbs. (142 Nm).
- Subframe reinforcing plate: Tighten bolts to 26 ft. lbs. (35 Nm).
- Brake caliper and anchor plate: Tighten to 98 ft. lbs. (133 Nm).
- Stabilizer link bar: Tighten to 35 ft. lbs. (47 Nm).
- Loosely install new wheel hub nut.
- Wheel and tire
- Tighten wheel hub nut to 244 ft. lbs. (330 Nm).
- Wheel and tire: Tighten to 59 ft. lbs. (80 Nm).
- Negative battery cable

XK Range

➡Various thread-locking devices are used on nuts and bolts throughout the vehicle. These devices restrict the number of times a nut or bolt can be used.

1. Raise front of vehicle and support on stands.
2. Remove front wheel.

➡The suspension must remain supported by the jack throughout this operation, to prevent damage to the upper and lower ball-joints.

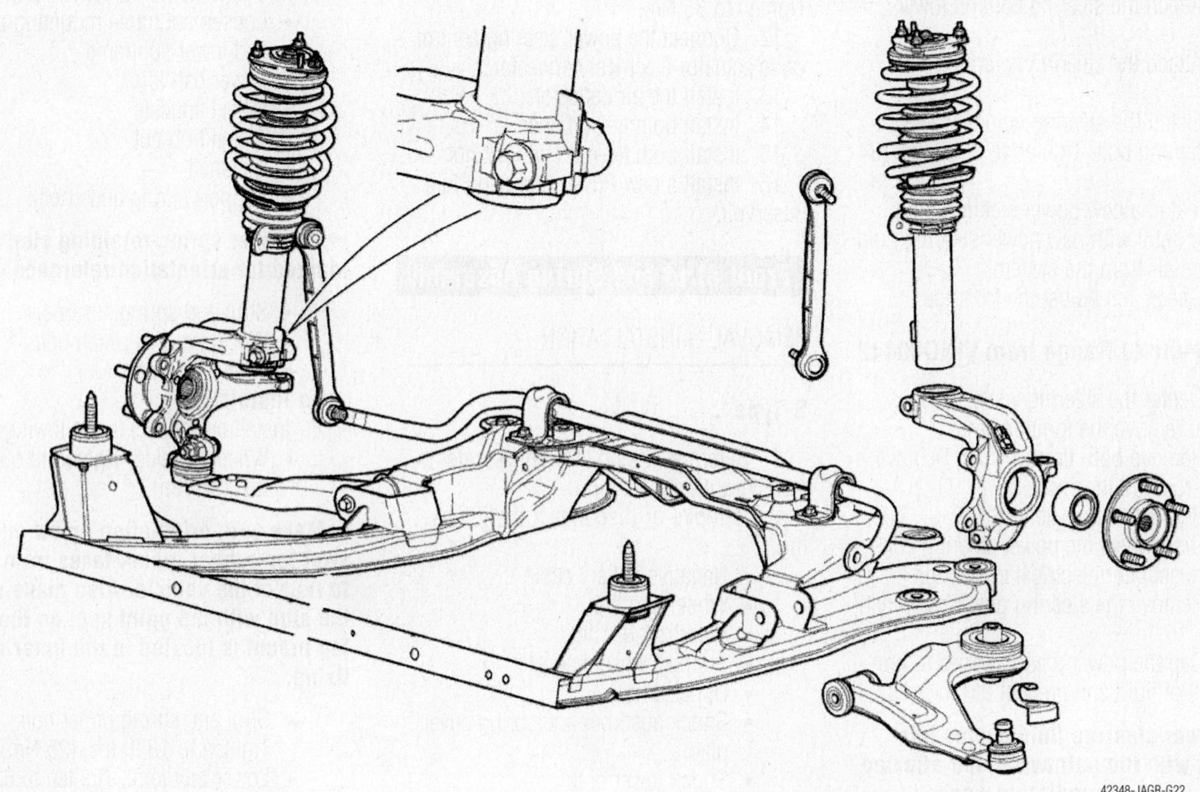

Front suspension assembly—X-Type

3. Place a jack under the lower control arm, raise the jack until the weight of the suspension is supported.
4. Remove shock absorber upper-mounting nuts.
5. Vehicles fitted with adaptive damping: disconnect connector.
6. Remove nut securing shock absorber to lower mounting, DO NOT remove bolt at this stage.
7. Carefully lower the jack supporting the suspension, while guiding the shock absorber, upper mounting studs through the body.
8. Stop lowering the jack when the studs have passed through the body.

➡For installation reference, note position of washers and shims before removing fulcrum shaft.

9. Remove nut. Remove fulcrum shaft. Collect washers and shims. Release upper control arm from crossbeam.

➡Do not stretch the brake hose when removing shock absorber and spring assembly.

10. With assistance, remove shock absorber and spring assembly.
11. Remove bolt from shock absorber lower mounting.
12. Remove shock absorber and spring assembly from vehicle.

To install:

13. Reposition upper control arm.
14. Align upper control arm to crossbeam.
15. Temporarily fit fulcrum shaft.
16. Clean relevant parts.
17. Install jaw protectors to vice.
18. Secure assembly in the vice by clamping across two of the upper mounting studs.
19. Align the lower bush bore so that it is parallel with two of the upper mounting studs.
20. Remove assembly from vice.
21. Release upper control arm to provide access for the fitting of the shock absorber and spring assembly.
22. Remove fulcrum shaft.
23. Release upper control arm.

➡Do not stretch brake-hose when fitting shock absorber and spring assembly.

24. With assistance, fit shock absorber and spring assembly.
25. Position shock absorber to lower mounting.
26. Install bolt to lower mounting, DO NOT fit nut at this stage.
27. Install upper control arm.
28. Position control arm to crossbeam.
29. Install fulcrum shaft; washers and shims in the positions noted in removal.
30. Install and tighten nut to 98 ft. lbs. (132 Nm).

➡Make sure the adaptive damping electrical connector is correctly aligned. Failure to follow these instructions may cause damage to the electrical connector.

31. Install shock absorber and spring assembly upper-mounting to body.

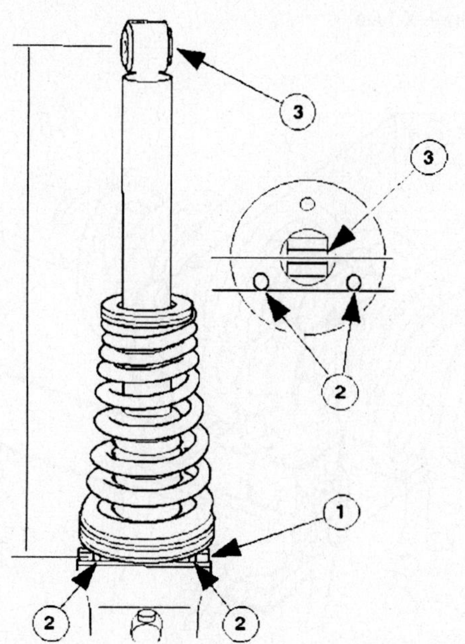

Final alignment of shock absorber with upper mounting studs

32. Raise suspension carefully on jack, and guide upper mounting studs into body stud-holes.
33. Install and tighten nuts to 25 ft. lbs. (34 Nm).
34. Vehicles fitted with adaptive damping, connect connector.
35. Install nut to shock absorber lower mounting.
36. Install and tighten nut to 59 ft. lbs. (80 Nm).
37. Lower and remove jack supporting suspension.
38. Install wheel.
39. Remove stands and lower vehicle.

DISASSEMBLY AND ASSEMBLY

Front and rear strut disassembly is the same for each model.

S-Type

➡Various thread-locking devices are used on nuts and bolts throughout the vehicle. These devices restrict the number of times a nut or bolt can be used.

1. Remove the shock absorber and spring assembly.

✻✻ CAUTION

Make sure the shock absorber is secured by fully inserting the locking pin in to the special tool. Failure to follow these instructions may result in personal injury.

2. Install the shock absorber and spring assembly to the special tool as shown.

✻✻ CAUTION

As the spring is under extreme tension care must be taken at all times. Failure to follow these instructions may result in personal injury.

➡The shock absorber assembly has a unique upper isolator orientation requirement to minimize the side-load effects of the spring on the damper. Therefore make a note of this orientation before disassembling the shock absorber.

3. Clamp the road spring.
4. Remove the shock absorber retaining nut.
5. Release the road spring.
6. Remove the shock absorber.

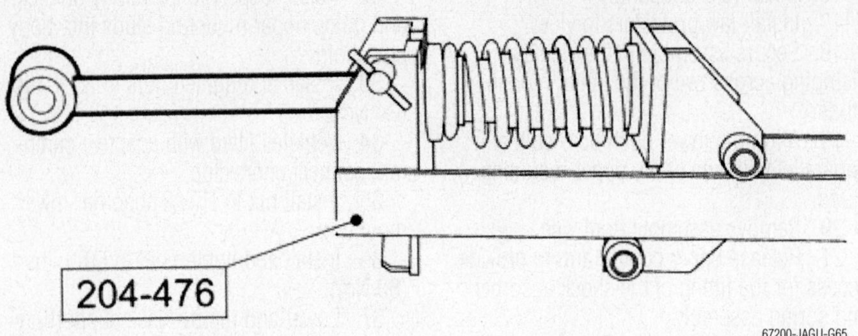

Strut installed in a spring compressor—S-Type and XK Range

To assemble:

➡**Make sure the spring ends locate correctly against the spring seats.**

7. Install the shock absorber and spring assembly to the special tool.
8. Clamp the road spring.
9. Vehicles without adaptive damping, install a new retaining nut. Tighten to 37 ft. lbs. (50 Nm).
10. Vehicles with adaptive damping, install a new retaining nut. Tighten to 20 ft. lbs. (27 Nm).
11. Release the road spring.
12. Remove the shock absorber and spring assembly from the special tool.
13. Install the shock absorber and spring assembly.

X-Type

1. Remove the strut and spring assembly.

⁂ CAUTION

As the spring is under extreme tension care must be taken at all times. Failure to follow these instructions may result in personal injury.

2. Using a spring compressor, compress the spring.
3. Loosen the thrust bearing retaining nut.
4. Disassemble the strut and spring assembly:
   a. Remove the spacer.
   b. Remove the top mount.
   c. Remove the thrust bearing.
   d. Remove the spring seat.
   e. Remove the spring.
   f. Remove the boot.
   g. Remove the bump stop.

➡**Make sure the spring locates correctly against the spring seats.**

5. To assemble, reverse the disassembly procedure. Tighten the nut to 47 Nm.

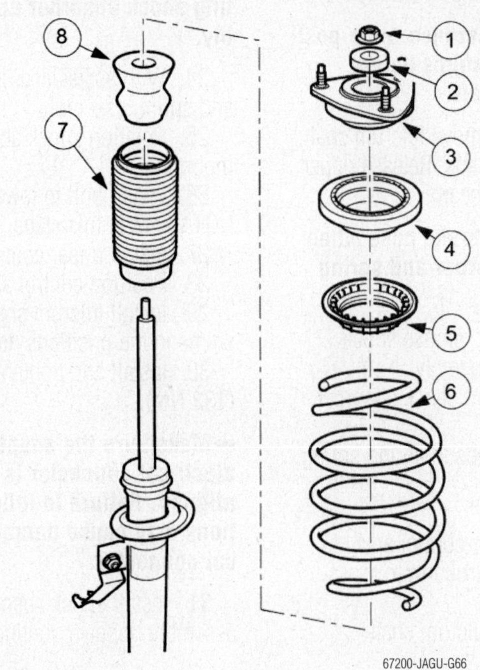

Front strut exploded view—X-Type

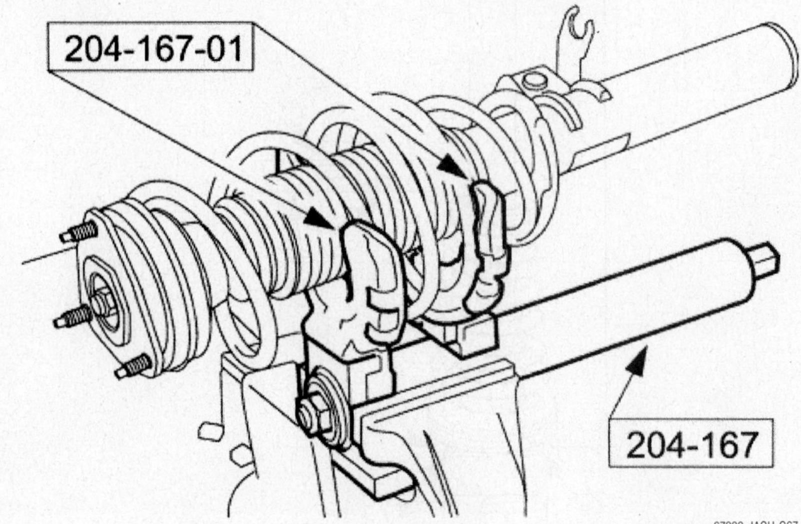

Using a spring compressor, compress the spring—X-Type

XK Range

➡**Various thread-locking devices are used on nuts and bolts throughout the vehicle. These devices restrict the number of times a nut or bolt can be used.**

1. Remove the front shock absorber and spring assembly.

⁂ CAUTION

Make sure the shock absorber is secured by fully inserting the locking pin in to the special tool. Failure to

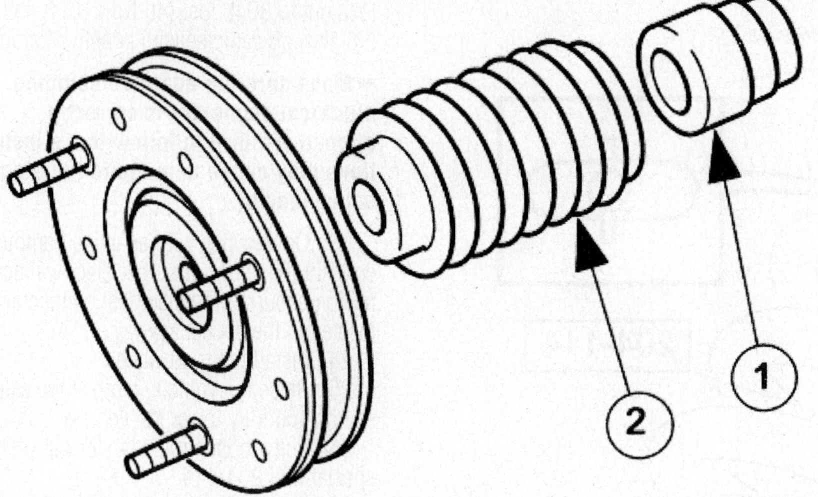

Strut upper end parts—XK Range

follow these instructions may result in personal injury.

2. Install the shock absorber and spring assembly to the special tool as shown.

⁂ CAUTION

As the spring is under extreme tension care must be taken at all times. Failure to follow these instructions may result in personal injury.

3. Clamp the road spring.
4. Rotate the handle until spring pressure is released.
5. Remove the shock absorber retaining nut.
6. Remove and discard the retaining nut.
7. Release the road spring.
8. Remove the shock absorber.
9. Disengage the boot from the upper mounting.
10. Remove the spring assister from the boot.
11. Clean the relevant parts.

To install:

12. To ease fitting apply a suitable lubricant to the spring assister and boot.
13. Make sure the boot assembly is seated securely and squarely into the upper mounting.
14. Insert the spring assister fully into the boot.
15. Insert the boot into upper mounting location.
16. Make sure circlip on the shock absorber abuts with the lower spring pan.
17. Make sure the spring ends butt correctly against the spring seats.
18. Install the shock absorber and spring assembly to the special tool.
19. Clamp the road spring.
20. Vehicles without adaptive damping. Install a new retaining nut. Tighten to 30 ft. lbs. (40 Nm).
21. Vehicles with adaptive damping. Install a new retaining nut. Tighten to 13 ft. lbs. (18 Nm).
22. Release the road spring.
23. Remove the shock absorber and spring assembly from the special tool.
24. Install the front shock absorber and spring assembly.

Front Shock Absorber

REMOVAL & INSTALLATION

2002–03 XJ Range

➡Various thread-locking devices are used on nuts and bolts throughout the vehicle. These devices restrict the number of times a nut or bolt can be used.

1. Raise the front of the vehicle.
2. Fully loosen the adjuster nut of the special tool 204-114 and position in the center of the road spring.

➡Ensure that the dowel of the special tool 204-114 is correctly positioned across the suspension turret slot.

3. Locate the special tool in the suspension turret slot.
4. Pass the upper dowel through the slot in the suspension turret and rotate to locate the dowel.
5. Lower the front of the vehicle.

➡When tightening the tool nut, use the lower dowel as a guide to ensure that the tool stem does not rotate and disturb the position of the dowel in the suspension turret.

6. Tighten the nut of the special tool 204-114 to its full extent, to achieve a suspension mid-laden position.

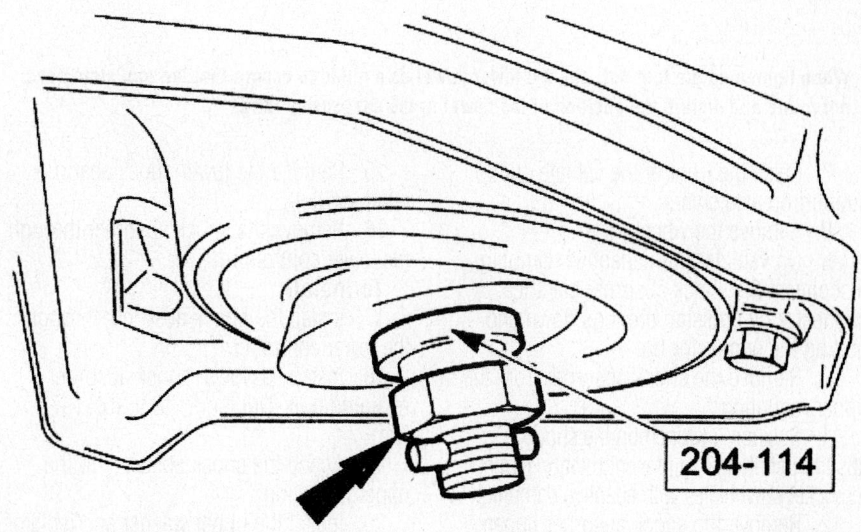

Fully loosen the adjuster nut of the special tool 204-114 and position in the center of the road spring

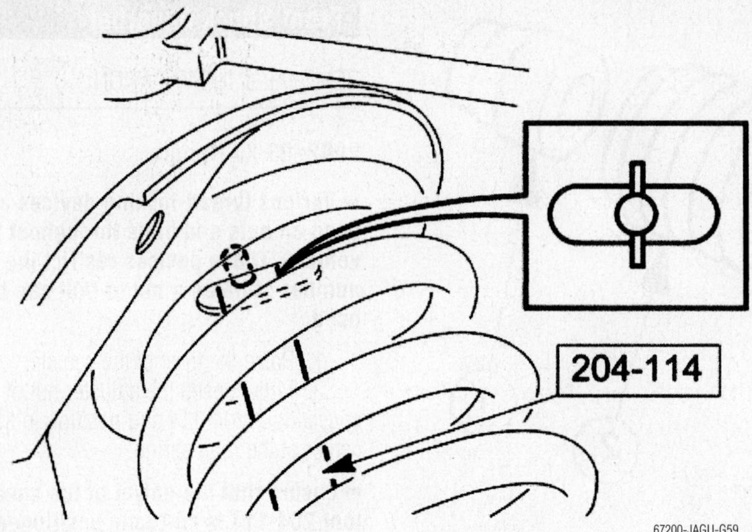

Ensure that the dowel of the special tool 204-114 is correctly positioned across the suspension turret slot

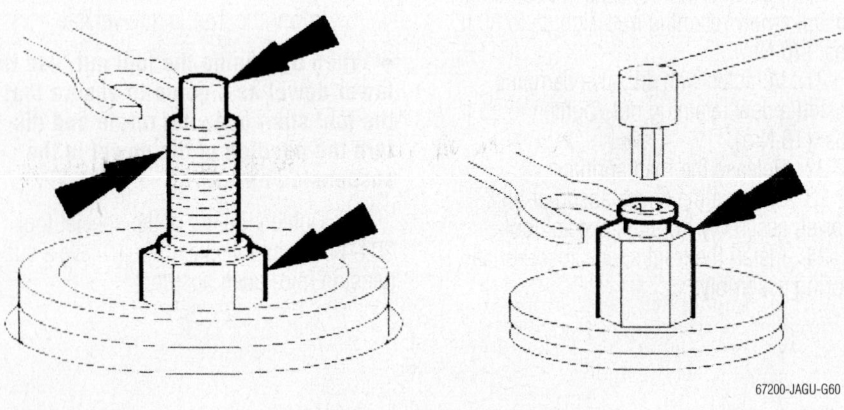

When tightening the tool nut, use the lower dowel as a guide to ensure that the tool stem does not rotate and disturb the position of the dowel in the suspension turret

7. Raise the front of the vehicle and support on axle stands.
8. Remove the wheel.
9. On vehicles with adaptive damping, disconnect the shock absorber electrical connector by pressing the lugs down and pulling the connector up.
10. Remove the shock absorber from the upper mounting.
11. Position a wrench on the shock absorber shaft flats to prevent rotation (wrench in socket on vehicles with adaptive damping).
12. Remove the shock absorber upper retaining nut.
13. Remove the upper washer and isolator.
14. Remove the lower shock absorber retaining nut.
15. Remove the lower shock absorber retaining bolt.
16. Remove the shock absorber through the lower control arm.

To install:

17. Install the shock absorber through the lower control arm.
18. Install the lower shock absorber retaining bolt. Tighten to 42 ft. lbs. (57 Nm).
19. Attach the shock absorber to the upper mounting.
20. Install the upper washer and isolator.
21. Install the shock absorber upper retaining nut.
22. Position a wrench on the shock absorber shaft flats to prevent rotation (wrench in socket on vehicles with adaptive damping). Tighten the shock absorber upper lock nut to 30 ft. lbs. (40 Nm); 13 ft. lbs. (18 Nm) on vehicles with adaptive damping.

➡Make sure the adaptive damping electrical connector is correctly aligned. Failure to follow these instructions may cause damage to the electrical connector.

23. On vehicles with adaptive damping, connect the shock absorber electrical connector, ensuring the electrical connector key locates in the socket keyway.
24. Install the road wheel.
25. Raise the vehicle, remove the axle stands and fully lower the vehicle.
26. Fully loosen the adjuster nut of the special tool 204-114.
27. Rotate the special tool 204-114 to release the dowel from the suspension turret slot.
28. Raise the vehicle for access and remove the special tool 204-114 from the suspension.
29. Fully lower the vehicle.

Rear Strut and Spring Assembly

REMOVAL & INSTALLATION

S-Type

1. Remove the luggage compartment side trim panel.
2. Vehicles with Dynamic Stability Control, disconnect the DSC electrical connector.
3. Remove the shock absorber and spring assembly upper retaining nuts.
4. Remove the wheel.
5. Support the shock absorber and spring assembly.

⁂ CAUTION

Failure to follow these instructions may result in personal injury. It is essential that the springs are correctly positioned in the vehicle. The upper mount housing contains a painted mark that corresponds with the coil spring upper wire end. If the mark is not evident it is necessary to paint a new mark in the required location. If the upper mount housing is replaced, the painted mark must be copied across from the previous mount.

6. Remove the lower bolt. Remove the shock absorber and spring assembly.
7. Make sure the dynamic stability control electrical connector is correctly aligned.

⁂ WARNING

Failure to follow these instructions may cause damage to the electrical connector.

➡The final tightening of the suspension components must be carried out with the vehicle on its wheels.

8. To install, reverse the removal procedure. Torque the lower bolt to 98 ft. lbs. (133 Nm); the upper nuts to 41 ft. lbs. (50 Nm) without DSC; 20 ft. lbs. (27 Nm) with DSC.

XJ Range and XK Range

➡Various thread-locking devices are used on nuts and bolts throughout the vehicle. These devices restrict the number of times a nut or bolt can be used.

1. Raise the rear of the vehicle and support on axle stands.
2. Remove the wheel.

➡Do not suspend the caliper from the brake hose as this will damage the hose.

3. Remove the caliper from the carrier and secure to the suspension mounting bracket using a tie strap.
4. Disconnect the wheel speed sensor electrical connector.
5. Remove and discard the tie strap securing the wheel speed sensor harness to the brake hose clip.
6. Mark the control arm at the pivot pin eccentric flange marker to aid installation.

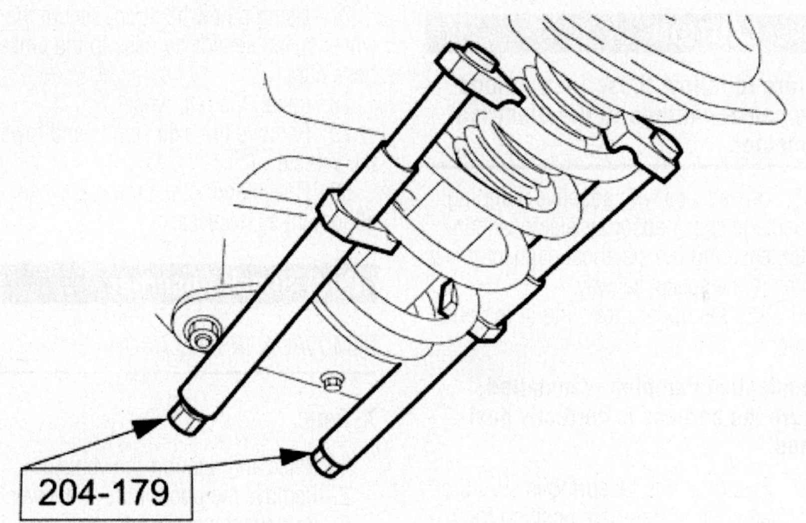

Install a spring compressor set on the spring— XJ Range and XK Range

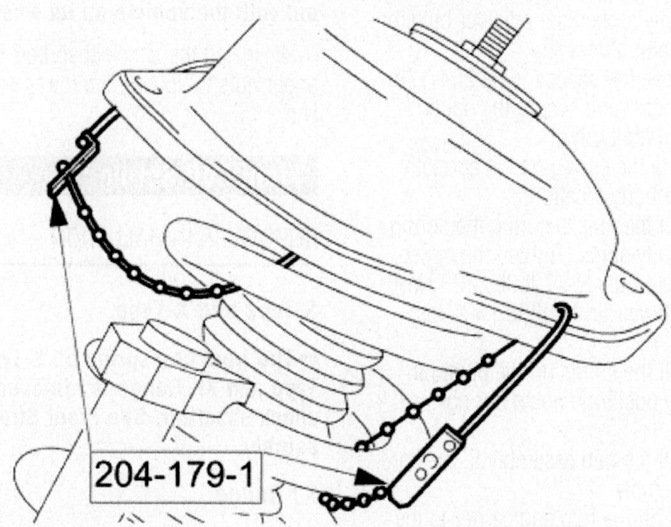

Using the special tools 204-179-1, secure the upper mounting to the spring—XJ Range

⁂ WARNING

To prevent damage to the parking brake cable, the hub carrier must be supported on removal from the control arm.

7. Remove the hub carrier from the control arm.
8. Remove the pivot pin from the hub carrier.
9. Remove the hub carrier and support on a block of wood.

➡Note the position of the shims during removal. They must be returned to their original position during installation.

10. Noting their positions, remove the shim from each end of the pivot pin sleeve.

⁂ WARNING

Ensure the special tools are positioned diametrically opposite on the spring and the arms are correctly seated. When compressing the spring, tighten the special-tool adjustment bolts evenly.

11. Install a spring compressor set on the spring. Evenly tighten the special tool tensioning bolts to compress the spring.
12. Remove the shock absorber to control arm mounting nut and bolt.
13. Remove the nuts securing the shock absorber and spring upper mounting to the body.
14. Using the special tools 204-179-1, secure the upper mounting to the spring.
15. Using a jack, retract the shock absorber through the control arm aperture.
16. Position a bolt through the spring coils and shock absorber lower bushing, to temporarily locate the shock absorber.
17. Partially lower the jack for access to the top of the shock absorber.
18. On vehicles with adaptive damping, disconnect the shock absorber electrical connector by pressing the lugs down and pulling the connector up.
19. Lower and remove the jack.
20. Remove the shock absorber and spring assembly from the vehicle and place the assembly on a work bench.
21. Clean the upper mounting spring location.
22. Remove the pigtail seat and clean the control arm lower spring location.
23. Compress the shock absorber and withdraw the bolt from the spring coils.
24. Remove the shock absorber from the upper mounting.

25. Position a wrench on the shock absorber shaft flats to prevent rotation (wrench in socket on vehicles with adaptive damping).
26. Remove the shock absorber upper retaining nut.
27. Remove the upper washer and isolator.
28. Remove the special tools 204-179-1 and remove the upper mounting from the spring.
29. Remove the shock upper mounting assembly from the spring.
30. Remove the spring upper isolator.
31. Remove the upper spring pan.
32. Remove the packing ring.

To install:

33. Install the spring pigtail lower seat on to the control arm.
34. Install the spring in the upper mounting.
35. Install the spring upper isolator.
36. Install the upper spring pan.
37. Install the packing ring.
38. Ensuring the correct orientation, secure the upper mounting to the spring using the special tools 204-179-1.
39. Position the spring upper mounting assembly on to the shock absorber. Apply a thin coat of grease to the upper mounting bush.
40. Ensure that the washer, lower isolator and bush are correctly positioned on the shock absorber shaft.
41. Position the shock absorber shaft in the upper mounting.
42. Secure the top of the shock absorber to the upper mounting.
43. Install the spring upper isolator.
44. Install the upper washer.
45. Install but do not tighten the locknut on to the shock absorber.
46. Manually retract the shock absorber and place a bolt through the spring coils and the shock absorber lower-bush. Secure the shock absorber to the upper mounting.
47. Position a wrench on the shock absorber shaft flats to prevent rotation (wrench in socket on vehicles with adaptive damping).
48. Tighten the shock absorber upper lock nut to 30 ft. lbs. (40 Nm); 13 ft. lbs. (18 Nm) on vehicles with adaptive damping.
49. Position the shock absorber and spring assembly to the vehicle.
50. Manually retract the shock absorber and remove the temporary securing bolt.
51. Remove the special tools 204-179-1.
52. Align the shock absorber lower bush to the control arm and install the mounting bolt.
53. Install but do not tighten the nut.
54. Make sure the adaptive damping electrical connector is correctly aligned.

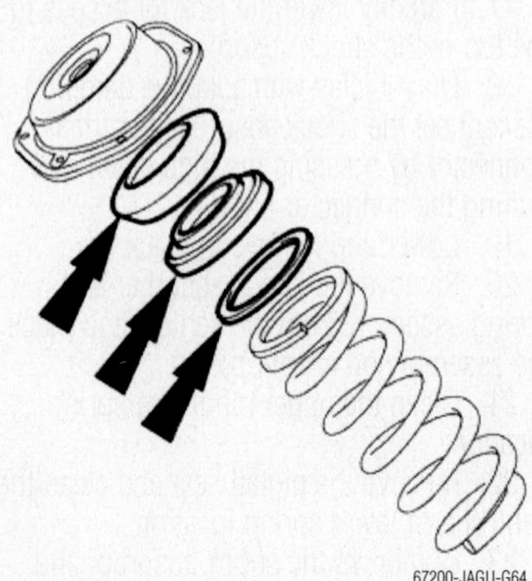

Assembling the upper strut parts— XJ Range and XK Range

⁂ CAUTION

Failure to follow these instructions may cause damage to the electrical connector.

55. On vehicles with adaptive damping, connect the shock absorber electrical connector, ensuring the electrical connector key locates in the socket keyway.
56. Seat the upper mounting in the body location.

➡If adaptive damping is installed, ensure the harness is correctly positioned.

57. Ensuring that the spring is seated on the isolator and spring pan, position the upper spring pan on to the body studs.
58. Release the spring tension.
59. Ensuring that the spring is correctly located, evenly loosen the adjuster bolts on the special tools 204-179.
60. Remove the special tools 204-179.
61. Fully seat and secure the upper spring pan to the body.
62. Ensure the spring pan is correctly seated in the body location.
63. Install the nuts securing the spring pan to the body studs. Tighten the upper strut mounting nuts to 17 ft. lbs. (23 Nm). Tighten the lower strut bolt to 74 ft. lbs. (100 Nm).
64. Install the shims on the pivot pin sleeve in the positions noted during removal.
65. Install the hub assembly to the control arm assembly.
66. Position the hub and carrier to the control arm.
67. Install the pivot pin ensuring that the eccentric flange correctly seats in the control arm groove.
68. Align the marker on the pivot pin flange with the mark made on control arm.
69. Install the nut on to the pivot pin and, ensuring the pin position is not disturbed, tighten the nut to 87 ft. lbs. (118 Nm).
70. Release the tie strap securing the brake caliper to the suspension mounting bracket.
71. Install the brake caliper on to the carrier.
72. Position the caliper assembly on to the carrier.
73. Install the guide pins and tighten to 22 ft. lbs. (30 Nm).
74. Install the dust caps.
75. Install the brake caliper spring clip.
76. Connect the wheel speed sensor electrical connector.
77. Using a new tie strap, secure the wheel speed sensor harness to the brake hose clip.
78. Install the rear wheel.
79. Remove the axle stands and lower the vehicle.
80. Check and adjust the rear wheel alignment as necessary.

Rear Shock Absorber

REMOVAL & INSTALLATION

X-Type

1. Raise and support the vehicle.
2. Remove the upper nuts and lower bolt.
3. Remove the shock absorber.

➡The final tightening of the rear suspension components must be carried out with the vehicle on its wheels.

4. Install the shock absorber. Torque the upper nuts to 25 Nm; the lower bolt to 130 Nm.

Front Coil Spring

REMOVAL & INSTALLATION

S-Type and X-Type

➡The front coil spring on S-Type, X-Type and XK Range is removed with the shock absorber. See Front Strut Disassembly.

XJ Range

1. Before servicing the vehicle, refer to the Precautions Section.

2. Remove the front wheel(s).
3. Install the Special Tool 204-111 in road spring.
4. Slacken the tool adjuster to suit spring length.
5. Install the adaptor and thrust collar.
6. Position the stem of tool in center of spring passing dowel through slot in suspension turret.
7. Rotate the stem of to position dowel across slot.

➡When the turning the tool adjuster, ensure tool stem does not rotate and change dowel position.

8. Rotate the adjuster of Special Tool to tension spring.
9. Remove spring pan to lower control arm.
10. Rotate adjuster of Special Tool to release tension from spring.
11. Remove spring assembly and special tool.

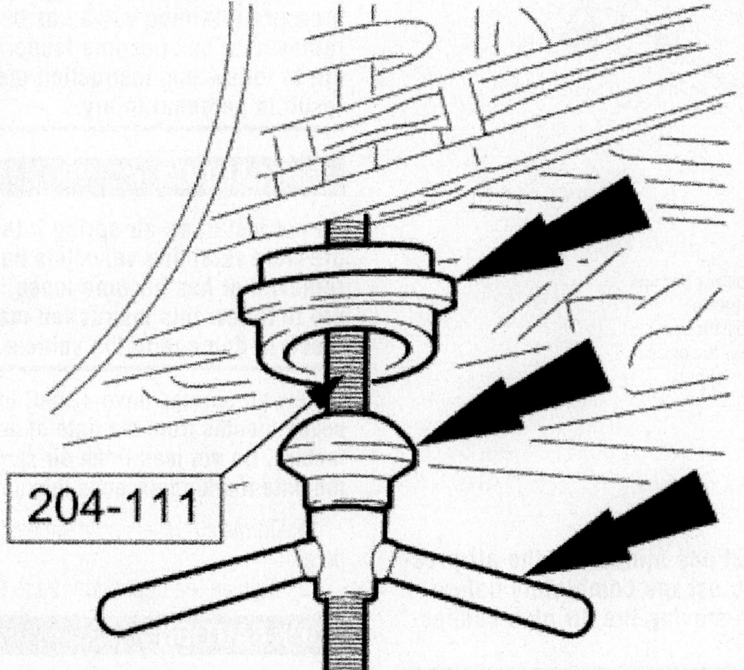

Installing spring compressor tool 204-111—XJ Range

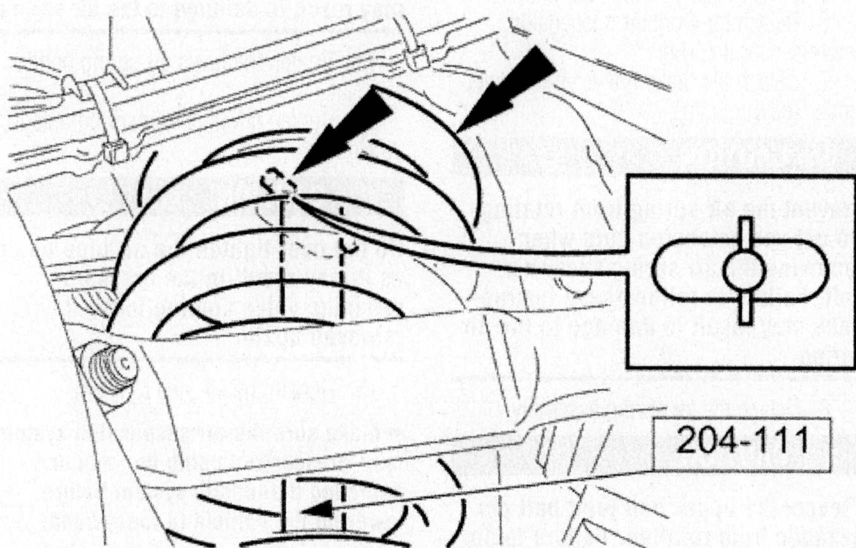

Positioning spring compressor dowel across slot—XJ Range

To install:

12. Install spring pan, new spring and new spring rubber upper and lower seats on tool 204-111.
13. Position spring assembly and tool in suspension turret.
14. Ensure spring upper rubber seat is correctly positioned and fully seated.
15. Engage tool dowel in turret slot.
16. Position spring pan on lower control arm.
17. Rotate adjuster of tool to lightly tension spring.
18. Align spring pan with lower control arm bolt locations.
19. Install suitable guide pins in spring pan/control arm bolt locations.

➡When the turning the tool adjuster, ensure tool stem does not rotate and change dowel position.

20. Install spring pan to control arm securing bolts.
21. Rotate tool adjuster to gradually increase spring tension.
22. Progressively substitute spring pan securing bolts for guide pins.
23. Tighten spring pan securing bolts to 52–66 (70–90 Nm).
24. Remove Special Tool 204-111 from suspension.
25. Ensure spring lower rubber seat is correctly positioned and fully seated.
26. Rotate adjuster of Special Tool to release tension on spring.
27. Rotate tool stem to release dowel from turret and withdraw tool assembly.
28. Install front wheel(s).

Rear Coil Spring

REMOVAL & ILLUSTRATION

For all except the X-Type, the rear coil spring is serviced with the strut assembly.

X-Type

1. Before servicing the vehicle, refer to the Precautions Section.
2. Remove the rear wheels.
3. Using a suitable jack, support the rear lower arm.
4. Remove or disconnect the following:
   - Rear wheels
   - Stabilizer bar link
   - Rear lower arm
   - Jack
   - Lower arm from subframe
   - Spring
   - Isolator

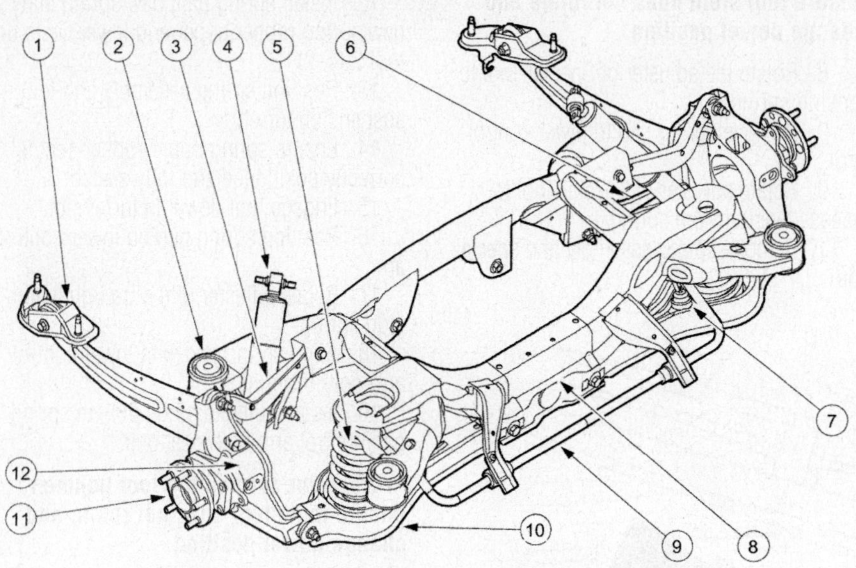

Rear suspension assembly—S-Type

To install:

5. Install or connect the following:
   - Isolator
   - Spring
   - Lower arm: Tighten bolt to 85 ft. lbs. (115 Nm).
   - Stabilizer bar link: Tighten bolt to 26 ft. lbs. (35 Nm).
   - Rear wheels
6. Check rear wheel alignment and adjust as necessary.

Front Air Spring

REMOVAL & INSTALLATION

XJ Series

1. Using the Jaguar approved diagnostic system depressurize the air suspension system.
2. Remove the wheel and tire assembly.

➡Disconnect the air spring electrical connector if fitted.

⁂ WARNING

Loosen the air pipe until the stored air pressure starts to vent. Wait for at least one minute for the air pressure to escape completely before fully removing the air pipe connector.

3. Detach the air pipe.
4. Remove the air spring retaining nuts.
5. Remove the stabilizer bar link.
6. Detach the front air suspension height sensor link rod.
7. Detach the brake line from the inner fender securing clips

⁂ CAUTION

Prevent the air spring from rotating. Do not use extension bars when removing the air spring securing bolt. Failure to follow these instructions may result in damage to the air spring.

8. Detach the air spring assembly.

⁂ CAUTION

Prevent the upper ball joint ball pin hexagon from rotating. Failure to do so may result in damage to the upper ball joint boot.

⁂ CAUTION

Make sure the wheel knuckle is supported. Failure to follow these instructions may result in damage to the vehicle.

9. Detach the upper arm.
10. Remove the air spring.

To install:

⁂ WARNING

Do not install an air spring if the pressure retaining valve has been removed or has become loose. Failure to follow this instruction may result in personal injury.

⁂ CAUTION

Do not install an air spring if the pressure retaining valve has been removed or has become loose. Failure to follow this instruction may result in damage to the vehicle.

➡New air springs have a shelf life of seven months from the date of manufacture. Do not install the air spring if the date marking exceeds this date.

11. To install, reverse the removal procedure.
12. Tighten the upper arm nut to 90 Nm.

⁂ CAUTION

Prevent the air spring from rotating. Do not use extension bars when tightening the air spring securing bolt. Failure to follow these instructions may result in damage to the air spring.

13. Tighten the lower air spring bolt to 175 Nm.
14. Tighten the upper air spring nuts to 25 Nm.

⁂ CAUTION

Do not over tighten the air pipe union as it may result in the pressure retaining valve coming loose if removed again.

15. Tighten the air pipe to 5 Nm.

➡Make sure the air suspension system is re-pressurized using the Jaguar approved diagnostic system before lowering the vehicle to the ground.

16. Re-pressurize the air suspension system. For further information, refer to the Jaguar approved diagnostic system.

Rear Air Spring

REMOVAL & INSTALLATION

VIN G00442–G45703

1. Disconnect the battery ground cable.
2. Remove the luggage compartment floor covering securing screws.
3. Remove the luggage compartment floor covering.
4. Remove the luggage compartment side trim panel.
5. Remove the air spring trim panel.
6. Disconnect the air spring electrical connector.

⁂ WARNING

Loosen the pipe no more than one full turn to allow the stored air pressure to vent for a minimum of one minute.

7. Detach the air pipe.
8. Remove the air spring retaining nuts.
9. Remove the rear wheel and tire assembly.
10. Detach the rear fender splash shield.
11. Remove the rear fender splash shield.
12. Remove the air spring bolt.
13. Displace the air spring lower mounting from the lower arm.
14. Displace the air spring from the upper mounting.
15. Position the air spring to the outside of the lower arm.
16. Remove the air spring.

To install:

⁂ WARNING

Do not install an air spring if the pressure retaining valve has been removed or has become loose. Failure to follow this instruction may result in personal injury.

⁂ CAUTION

Do not install an air spring if the pressure retaining valve has been removed or has become loose. Failure to follow this instruction may result in damage to the vehicle.

➡New air springs have a shelf life of seven months from the date of manufacture. Do not install the air spring if the date marking exceeds this date.

17. Install the air spring.
18. Position the air spring to the front of the lower arm.
19. Locate the air spring into the upper mount.
20. Locate the lower air spring mounting into the lower arm.
21. Install the air spring bolt. Tighten to 133 Nm.
22. Attach the rear fender splash shield.
23. Install the rear fender splash shield.
24. Install the rear wheel and tire assembly.
25. Install the rear air spring retaining nuts. Tighten to 25 Nm.
26. Attach the air pipe. Tighten to 5 Nm.
27. Connect the air spring electrical connector.
28. Install the air spring trim panel.
29. Install the luggage compartment side trim panel.
30. Install the luggage compartment floor covering.
31. Install the luggage compartment floor covering securing screws.
32. Connect the battery ground cable.
33. Start and run the engine for 2 minutes to allow the air spring to recharge and to allow the compressor to replenish the reservoir.

VIN G45704–G99999

1. Disconnect the battery ground cable.
2. Remove the luggage compartment floor covering securing screws.
3. Remove the luggage compartment floor covering.
4. Remove the luggage compartment side trim panel.
5. Remove the air spring trim panel.
6. Disconnect the air spring electrical connector.

⁂ WARNING

Loosen the pipe no more than one full turn to allow the stored air pressure to vent for a minimum of one minute.

7. Detach the air pipe.
8. Remove the air spring retaining nuts.
9. Remove the rear wheel and tire assembly.
10. Reposition the stabilizer bar link cover.
11. Detach the stabilizer bar link.
12. Remove the air spring retaining bolt.
13. Remove the air spring.

To install:

⁂ WARNING

Do not install an air spring if the pressure retaining valve has been removed or has become loose. Failure to follow this instruction may result in personal injury.

⁂ CAUTION

Do not install an air spring if the pressure retaining valve has been removed or has become loose. Failure to follow this instruction may result in damage to the vehicle.

➡New air springs have a shelf life of seven months from the date of manufacture. Do not install the air spring if the date marking exceeds this date.

14. Install the air spring.
15. Tighten the lower end to 133 Nm.
16. Tighten the stabilizer link to 48 Nm.
17. Reposition the stabilizer bar link cover.
18. Install the rear wheel and tire assembly.
19. Tighten the upper spring nuts to 25 Nm.
20. Tighten air pipe to 5 Nm.
21. Connect the air spring electrical connector.
22. Install the air spring trim panel.
23. Install the luggage compartment side trim panel.
24. Install the luggage compartment floor covering.
25. Install the luggage compartment floor covering securing screws.
26. Connect the battery ground cable.
27. Start and run the engine for 2 minutes to allow the air spring to recharge and to allow the compressor to replenish the reservoir.

Upper Ball Joint

REMOVAL & INSTALLATION

XJ Range and XK Range

1. Remove the control arm.
2. Secure upper control arm in a vise.
3. Using a hacksaw, partially cut through two diametrically opposite segments of ball joint flange.
4. Carefully chisel through and remove the two ball joint flange segments to permit installation between press bed plates.

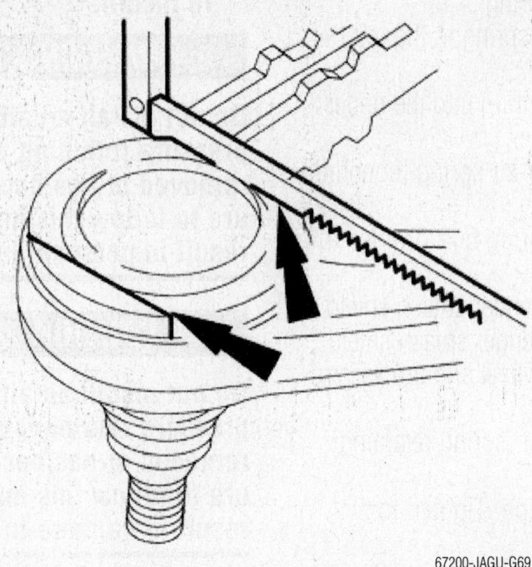

Using a hacksaw, partially cut through two diametrically opposite segments of ball joint flange— XJ Range and XK Range

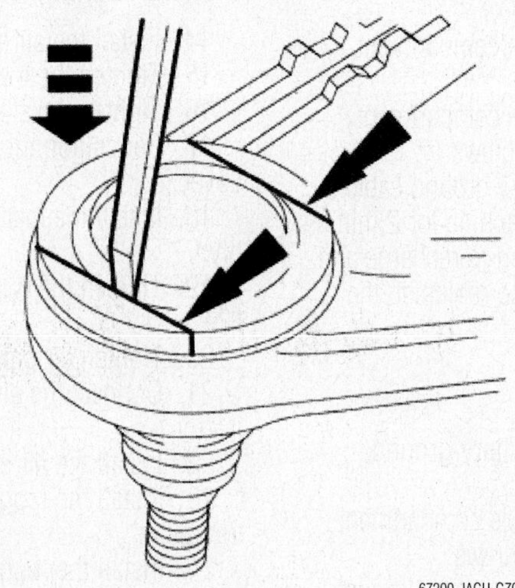

Carefully chisel through and remove the two ball joint flange segments— XJ Range and XK Range

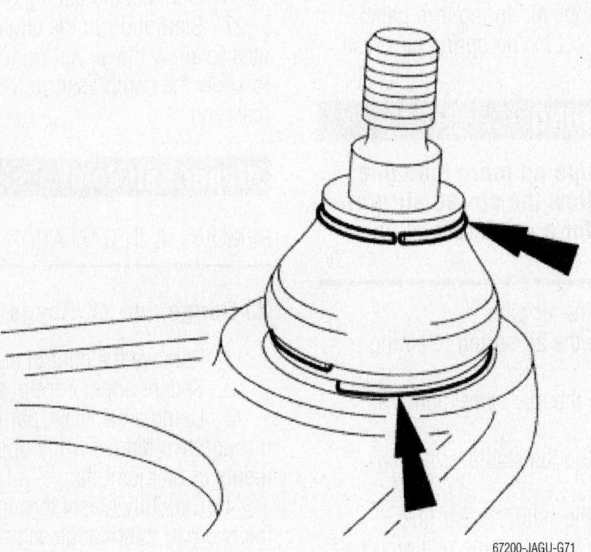

Remove retaining clips and boot— XJ Range and XK Range

5. Remove control arm from vise and thoroughly clean to remove road deposits.
6. Remove retaining clips and boot.
7. Exercising care to avoid splitting or otherwise damaging boot, remove retaining clips.
8. Remove boot and retain for warranty purposes.
9. Invert control arm and position on press with ball joint flange between press bed plates. Install special tool 204-200 on ball joint and operate press to release joint from control arm.
10. Remove special tool from ball joint.
11. Remove ball joint and control arm from press.
12. Thoroughly clean control arm paying particular attention to ball joint and pivot bolt locations. Install ball joint in control arm.
13. Position support of special tool 204-199 on press bed and place control arm on tool.
14. Using ball joint plastic cover to align, position mandrel of special tool 204-199 on ball joint
15. Align assembly with press ram and operate press to fully seat ball joint in control arm.

Upper Control Arm

REMOVAL & INSTALLATION

S-Type

2002 THROUGH VIN M45254

1. Raise and support the vehicle.
2. Remove the wheel.
3. Detach the upper arm from the wheel knuckle.
4. Remove the upper arm-to-frame bolts.
5. Remove the upper arm.
6. To install, reverse the removal procedure. Torque all fasteners to 90 Nm.

2002.5–06 S-TYPE FROM VIN M45255

LEFT SIDE

1. Remove the shock absorber and spring assembly.
2. Remove the cowl vent screen.
3. Disconnect the electrical connector.
4. Remove the brake pedal travel sensor.
5. Remove the brake pedal travel sensor retaining clip.
6. Remove the brake pedal travel sensor.

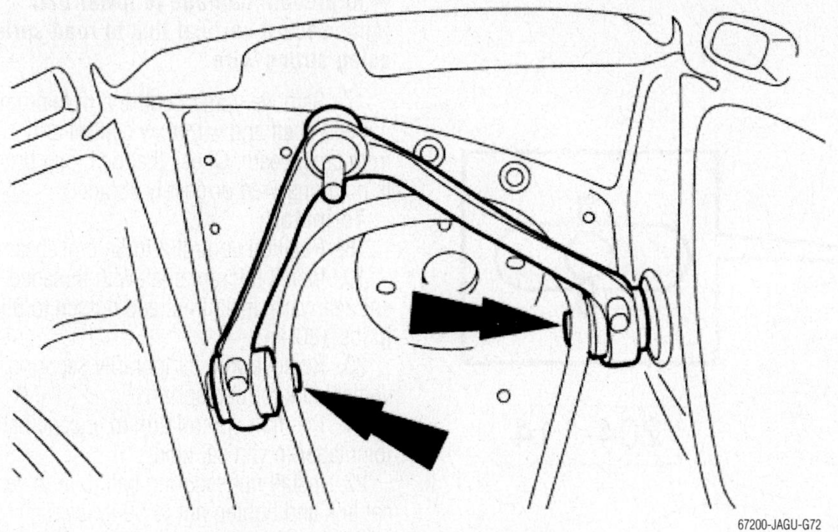

Upper arm-to-frame nuts—2002.5–06 S-Type from VIN M45255

7. Remove the upper arm ball stud nut. Separate the upper arm from the knuckle.
8. Remove the air cleaner assembly.
9. Remove the upper arm-to-frame retaining nuts.
10. Remove the upper arm.

To install:

11. Install the upper arm. Install the upper arm-to-frame retaining nuts. Tighten to 35 ft. lbs. (47 Nm).
12. Install the upper arm-to-knuckle stud nut retaining nut. Use a new cotter pin. Torque to 67 ft. lbs. (90 Nm).
13. Install the shock absorber and spring assembly.

➡The final tightening of the upper arm-to-frame retaining nuts must be carried out with the vehicle on its wheels.

14. Install the air cleaner assembly.
15. Install the brake pedal travel sensor.
16. Install the brake pedal travel sensor retaining clip.
17. Connect the electrical connector.
18. Install the cowl vent screen.

RIGHT SIDE

1. Remove the shock absorber and spring assembly.
2. Remove the cabin air filter.
3. Remove the cabin air filter housing retaining nut.
4. Remove the cabin air filter housing.
5. Remove the engine compartment support.
6. Remove the engine compartment panel.
7. Detach the wiring harness mounting bracket.
8. Remove the upper arm-to-knuckle ball stud nut.
9. Detach the engine compartment battery junction box.
10. Release the engine compartment battery junction box retaining clips.
11. Detach the engine compartment battery junction box.
12. Detach the hydraulic control unit.
13. Remove the upper arm-to-frame retaining nuts.
14. Remove the upper arm.

To install:

15. Install the upper arm.
16. Install the upper arm-to-knuckle ball stud retaining nut. Use a new cotter pin. Torque to 67 ft. lbs. (90 Nm).
17. Install the upper arm-to-frame retaining nuts. Torque to 35 ft. lbs. (47 Nm).
18. Install the shock absorber and spring assembly.

➡The final tightening of the upper arm retaining nut must be carried out with the vehicle on its wheels.

19. Attach the hydraulic control unit.
20. Attach the engine compartment battery junction box.
21. Attach the engine compartment battery junction box.
22. Attach the wiring harness mounting bracket.
23. Install the engine compartment panel.
24. Install the engine compartment support.
25. Install the cabin air filter housing.
26. Install the cabin air filter housing retaining nut.
27. Install the cabin air filter.

XJ Range

2002–03 THROUGH VIN F59525

1. Raise front of vehicle.
2. Position camber tie down special tool 204-114 on suspension.
3. Fully loosen tool adjuster nut.
4. Position tool in road spring and pass upper dowel through slot in suspension turret.
5. Lower front of vehicle.
6. Rotate stem of special tool 204-114 to locate dowel in suspension turret slot.

➡When tightening tool nut, use lower dowel as a guide to ensure that the tool stem does not rotate and disturb position of dowel in suspension turret.

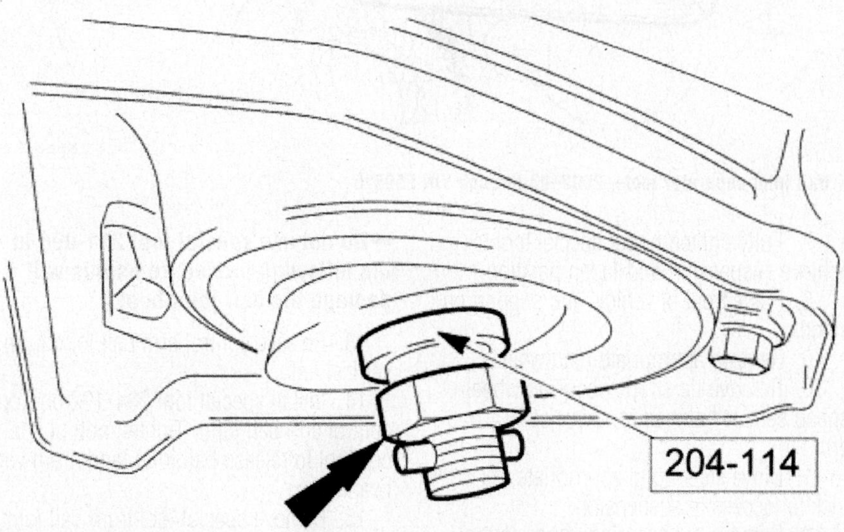

Position camber tie down special tool 204-114 on suspension—2002–03 through VIN F59525

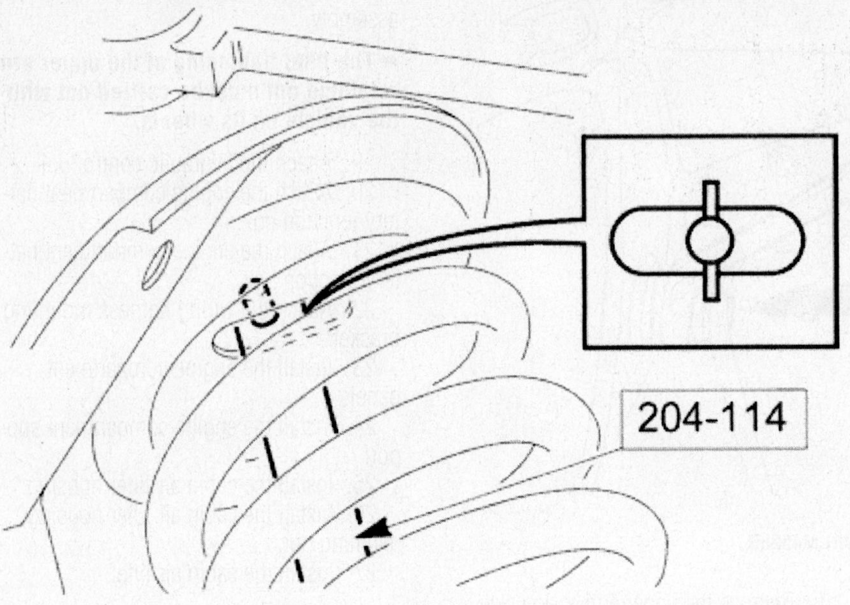

Rotate stem of special tool 204-114 to locate dowel in suspension turret slot—2002–03 through VIN F59525

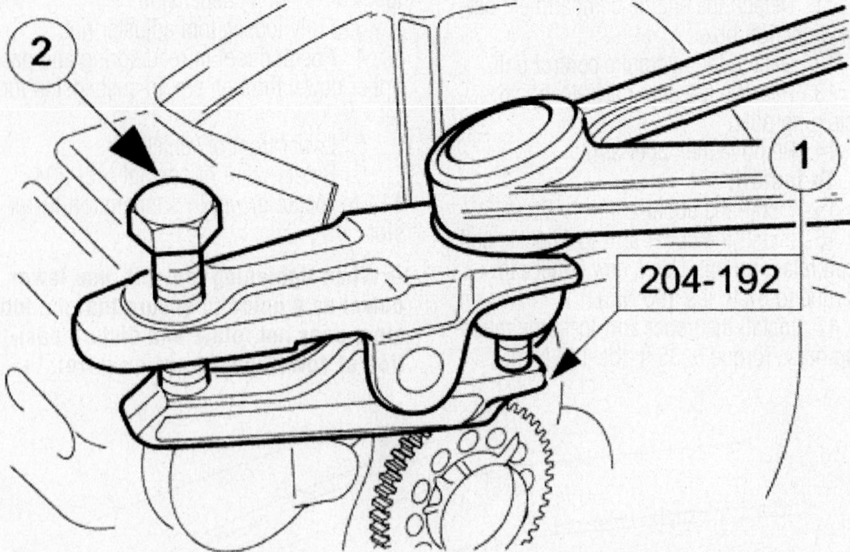

Ball joint separator tool—2002–03 through VIN F59525

7. Fully tighten nut of special tool to achieve suspension mid-laden position.
8. Raise front of vehicle and support on stands.
9. Remove appropriate road wheel.
10. Remove tie straps securing wheel speed sensor harness to upper control arm.
11. Move steering to appropriate full lock for access to suspension.
12. Remove nut securing upper control arm ball joint to vertical link.

➡Do not use special tool 211-098 in the following procedure as this will damage the ball joint boot.

13. Release control arm ball joint from vertical link.
14. Install special tool 204-192 on upper control arm ball joint. Tighten bolt of special tool to release ball joint taper from vertical link.
15. Remove special tool from ball joint.
16. Lift upper control arm to withdraw ball joint from vertical link.

➡To prevent damage to lower ball joint, secure vertical link to road spring using strong wire.

17. Remove upper control arm fulcrum nut and shaft and withdraw control arm from crossbeam. Check that ball joint boot is not punctured or deeply abraded.

To install:

18. Position control arm on crossbeam.
19. Install fulcrum shaft with threaded end rearward. Install nut and tighten to 59 ft. lbs. (80Nm).
20. Remove wire temporarily securing vertical link to road spring.
21. Position control arm to engage ball joint taper in vertical link.
22. Install nut securing ball joint to vertical link and tighten nut to 59 ft. lbs. (80Nm).
23. Reposition wheel speed sensor harness on control arm and secure with two tie straps.
24. Install road wheel.
25. Raise vehicle, remove stands and fully lower vehicle.
26. Fully loosen nut on special tool. Release dowel of special tool 204-114 from suspension turret slot.
27. Raise vehicle for access and remove special tool from suspension.
28. Fully lower vehicle.
29. Carry out steering geometry and wheel alignment checks and if necessary adjust.

2003.5–04 FROM VIN G00442

1. Remove the air spring assembly.
2. Remove the cowl vent screen access panel.
3. Remove the upper arm rear retaining nut.
4. Remove the air cleaner assembly.
5. Remove the upper arm front retaining nut.
6. Note the position of the upper arm retaining bolt locating tangs to aid installation.
7. Remove the upper arm.

To install:

8. Make sure the upper arm retaining bolt locating tangs are correctly located.

⁂ WARNING

Failure to follow this instruction may result in damage to the vehicle.

9. Position the upper arm.
10. Install the upper arm front retaining nut.
11. Install the upper arm rear retaining nut.

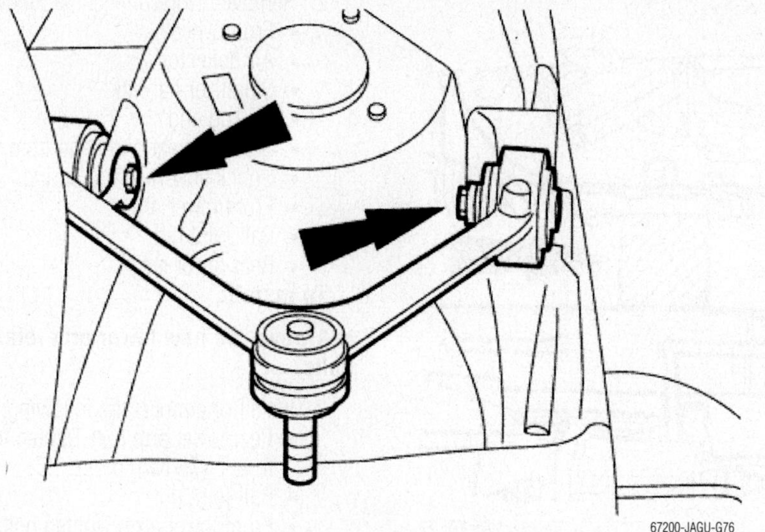

Make sure the upper arm retaining bolt locating tangs are correctly located—2003.5–04 FROM VIN G00442

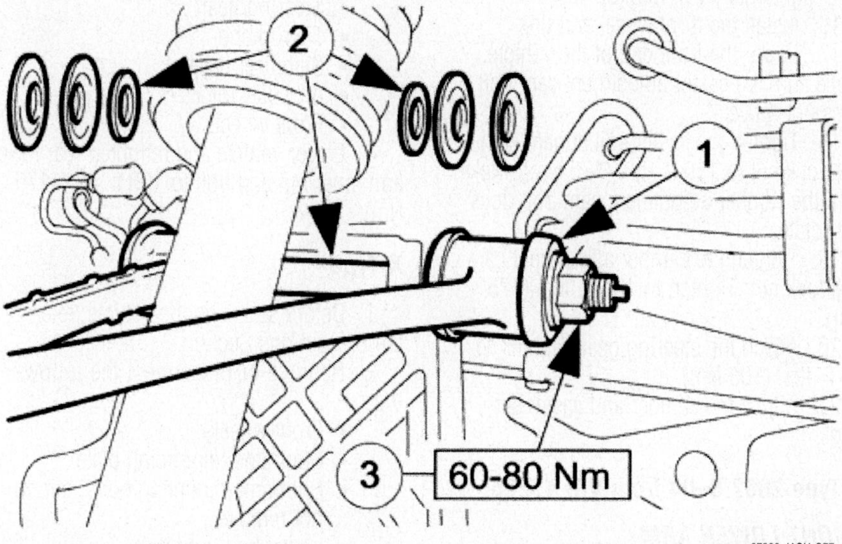

Shim and washer positioning— XK Range

12. Connect the arm to the knuckle. Torque the nut to 66 ft. lbs. (90 Nm).
13. Install the air spring assembly.

➡The final tightening of the upper arm-to-frame retaining nuts must be carried out with the vehicle on its wheels.

14. Tighten the arm-to-frame nuts to 35 ft. lbs. (47 Nm).
15. Install the cowl vent screen access panel.
16. Install the air cleaner assembly.

XK Range

➡Various thread-locking devices are used on nuts and bolts throughout the vehicle. These devices restrict the number of times a nut or bolt can be used.

1. Raise front of vehicle and support on stands.
2. Remove front wheel.
3. Release ABS speed sensor harness from upper control arm.
4. Remove tie straps. Reposition harness.
5. Remove nut from upper ball joint.

➡Do not allow the weight of the vertical link to hang on the brake hose.

6. Release upper ball-joint from vertical link.
7. Using a strong length of wire secure the vertical link to the shock absorber.
8. Install special tool to ball joint. Tighten tool bolt to release ball joint pin-taper.
9. Remove tool and disengage ball joint from vertical link.

➡For installation reference, note the position of washers and shims before removing fulcrum bolt.

10. Remove nut. Remove fulcrum bolt. Collect washers and shims. Remove control arm.

To install:

11. Align control arm to crossbeam.
12. Install fulcrum bolt, shims and washers to their original positions as noted in the removal section. Install and tighten nuts to (80 Nm).
13. Install upper ball joint to vertical link.
14. Engage ball joint stud into vertical link. Install and tighten nut to (80 Nm).
15. Release vertical link from shock absorber.
16. Secure ABS harness to upper control arm.
17. Position harness and secure with tie clips.
18. Install wheel.
19. Remove stands and lower vehicle.

Lower Control Arm

REMOVAL & INSTALLATION

S-Type 2002 through VIN M45254

➡This procedure must be carried out using a 4-post ramp.

1. Raise and support the vehicle.
2. Detach and reposition the steering gear.
3. Raise and support the front end of the vehicle on the No. 2 crossmember.
4. Position two 4 in. x 4 in. pieces of wood between the No. 2 crossmember and the lifting device.
5. Remove the front wheel and tire.

➡Prevent the stabilizer bar ball joint ball pin hexagon from rotating. Failure to do so may result in damage to the stabilizer bar ball joint boot.

6. Detach the stabilizer bar connecting link.
7. Detach the strut and spring assembly.

➡Prevent the lower ball joint ball pin hexagon from rotating. Failure to do so may result in damage to the lower ball joint boot.

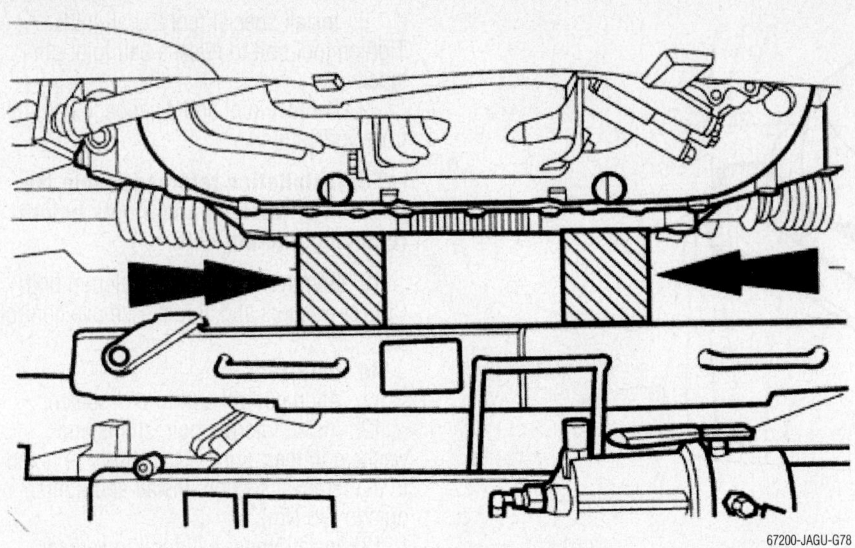

Position two 4 in. x 4 in. pieces of wood between the No. 2 crossmember and the lifting device

8. Detach the lower arm.
9. Vehicles without camber and caster adjustment cam bolts, remove the lower arm. Discard the lower arm front and rear retaining bolts and lock nuts.
10. Vehicles with camber and caster adjustment cam bolts, remove the lower arm. Discard the lower arm front and rear lock nuts.

To install:

11. Vehicles with camber and caster adjustment cam bolts, do not tighten the lock nuts. Install new lower arm front and rear lock nuts. Install the lower arm.
12. Vehicles without camber and caster adjustment cam bolts, do not tighten the lock nut. Install a caster adjustment cam bolt and a new lock nut. Install the lower arm. Install the bolt from the rear. Install the bolt with the cam lobe down. The cam must be seated between the cam guides on the No. 1 crossmember. Do not tighten the lock nut. Install a camber adjustment cam bolt and a new lock nut. Install the bolt from the rear. Install the bolt with the cam lobe down. The cam must be seated in the groove in the No. 2 crossmember.

➡Prevent the lower ball joint ball pin hexagon from rotating. Failure to do so may result in damage to the lower ball joint boot.

13. Attach the lower arm. Tighten the ball stud nut to 111 ft. lbs. (150 Nm).
14. Attach the strut and spring assembly. Tighten to 130 ft. lbs. (175 Nm).

➡Prevent the stabilizer bar ball joint ball pin hexagon from rotating. Failure to do so may result in damage to the stabilizer bar ball joint boot.

15. Attach the stabilizer bar connecting link. Tighten to 41 ft. lbs. (55 Nm).
16. Install the front wheel and tire.
17. Lower the front end of the vehicle. Make sure the caster adjustment cam bolt does not rotate.
18. Tighten the caster adjustment cam bolt lock nut to 130 ft. lbs. (175 Nm). Make sure the camber adjustment cam bolt does not rotate.
19. Tighten the camber adjustment cam bolt lock nut. Tighten to 130 ft. lbs. (175 Nm).
20. Attach the steering gear. Tighten to 76 ft. lbs. (103 Nm).
21. Adjust the camber and caster settings.

S-Type 2002.5–04 from VIN 45255

FRONT LOWER ARM

1. Before servicing the vehicle, refer to the Precautions Section.
2. Remove or disconnect the following:
   - Front wheels
   - Air deflector
   - Front lower arm

To install:

➡Always use new lower arm retaining bolts.

3. Install or connect the following:
   - Lower arm: Tighten bolts to 44 ft. lbs. (60 Nm) plus 135°.
   - Air deflector
   - Front wheels

REAR LOWER ARM

1. Before servicing the vehicle, refer to the Precautions Section.
2. Remove or disconnect the following:
   - Front wheels
   - Air deflector
   - Stabilizer bar link
   - Tie rod end
   - Steering gear and hang from frame
   - Shock absorber and spring
   - Front lower arm
   - Ball joint nut
   - Rear lower arm

To install:

➡Always use new lower arm retaining bolts.

3. Install or connect the following:
   - Rear lower arm nut: Tighten to 55 ft. lbs. (75 Nm) plus 135°.
   - Ball joint nut
   - Front lower arm: Tighten bolts to 44 ft. lbs. (60 Nm) plus 135°.
   - Rear lower arm inner nut and bolt: Tighten to 129 ft. lbs. (175 Nm).
   - Shock absorber and spring
   - Steering gear
   - Tie rod end
   - Stabilizer bar link
   - Air deflector
   - Front wheels
4. Lower vehicle and retighten rear lower arm inner nut and bolt to 129 ft. lbs. (175 Nm).

X-Type

1. Before servicing the vehicle, refer to the Precautions Section.
2. Remove or disconnect the following:
   - Front wheels
   - Subframe reinforcing plate
   - Subframe mounting bolts, but do not remove
   - Lower ball joint bolt
   - Wheel knuckle
   - Lower arm

To install:

➡Always use new lower arm retaining bolts.

3. Press the bearing into the wheel knuckle.
4. Install or connect the following:
   - Lower arm: Tighten bolts to 66 ft. lbs. (90 Nm) plus 60 degrees.
   - Wheel knuckle
   - Lower ball joint: Tighten to 61 ft. lbs. (83 Nm).
   - Subframe: Tighten bolts to 105 ft. lbs. (142 Nm).
   - Subframe retaining plate: Tighten bolts to 25 ft. lbs. (35 Nm).
   - Front wheels

XJ Range

2002–03 THROUGH VIN F59525

➡Various thread-locking devices are used on nuts and bolts throughout the vehicle. These devices restrict the number of times a nut or bolt can be used.

1. Raise front of vehicle and support on stands.
2. Remove appropriate road wheel.

⁂ WARNING

Do not suspend caliper from brake hose as this will damage the hose.

3. Remove brake calipers from carriers and tie to suspension mounting bracket.
4. Raise vehicle further to provide access for Special Tool 204-111 and support on stands.
5. Install Special Tool 204-111 in road spring.
6. Slacken tool adjuster to suit spring length.
7. Install adaptor and thrust collar.
8. Position stem of tool in center of spring passing dowel through slot in suspension turret. Rotate stem of Special Tool to position dowel across slot.

➡When turning tool adjuster, ensure that tool stem does not rotate and disturb position of dowel in suspension turret.

9. Rotate adjuster of Special Tool to tension spring.
10. Remove spring pan to lower control arm securing bolts.
11. Rotate adjuster of Special Tool to release tension from spring. It is not necessary to remove Special Tool from spring at this stage.
12. Remove spring assembly and Special Tool 204-111 from suspension turret.

Lower vehicle to working height and support on stands.

13. Remove and discard tie straps securing wheel speed sensor harness to upper control arm.
14. Remove wheel speed sensor securing screw and withdraw sensor from vertical link.
15. Remove nut securing tie rod outer ball joint to steering arm.
16. Install special tool 211-098 on tie rod ball joint. Tighten tool bolt to release taper pin from vertical link.
17. Remove nut securing upper control arm ball joint to vertical link. Do not use special tool 211-098 in the following procedure as this will damage the ball joint boot.
18. Install special tool 204-192 on ball joint. Tighten tool bolt to release ball joint

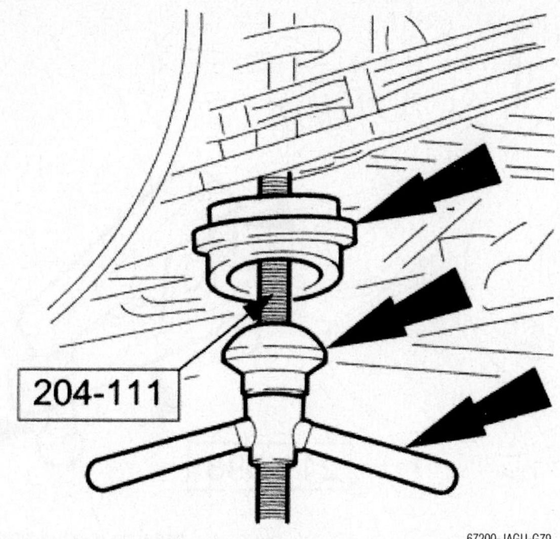

Install Special Tool 204-111 in road spring—XJ Range 2002–03 through VIN F59525

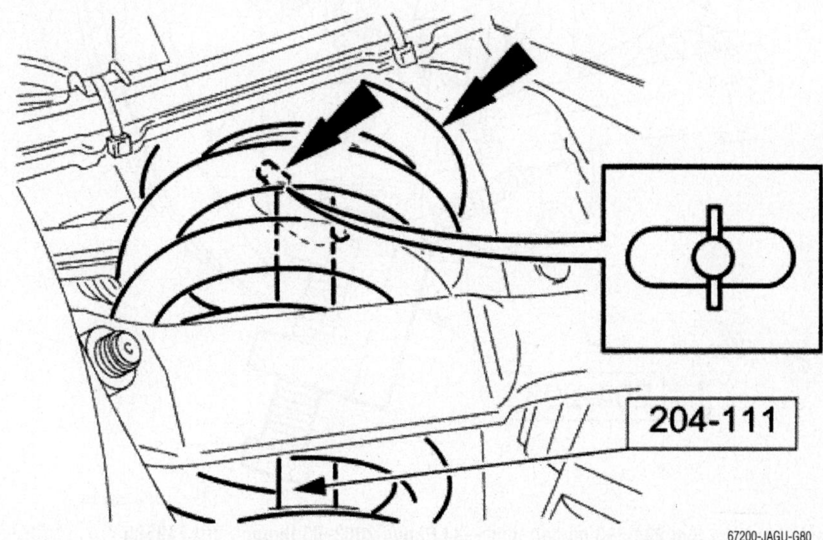

Rotate stem of Special Tool to position dowel across slot—XJ Range 2002–03 through VIN

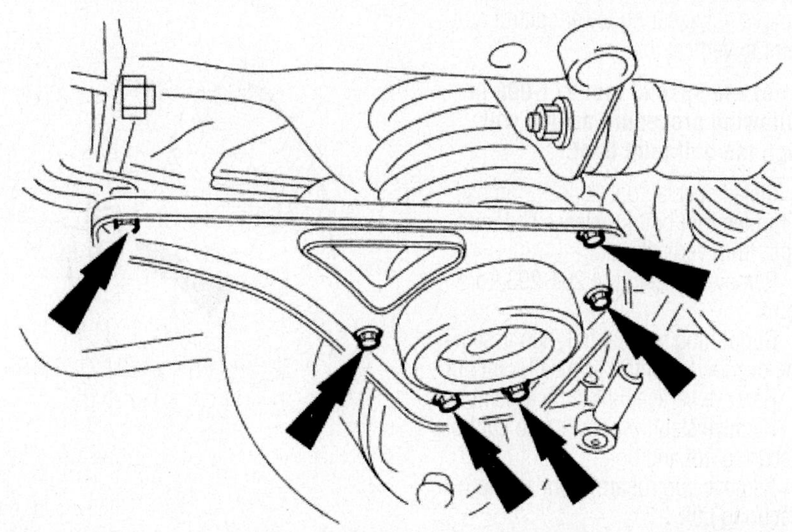

Spring pan to lower control arm securing bolts—XJ Range 2002–03 through VIN F59525

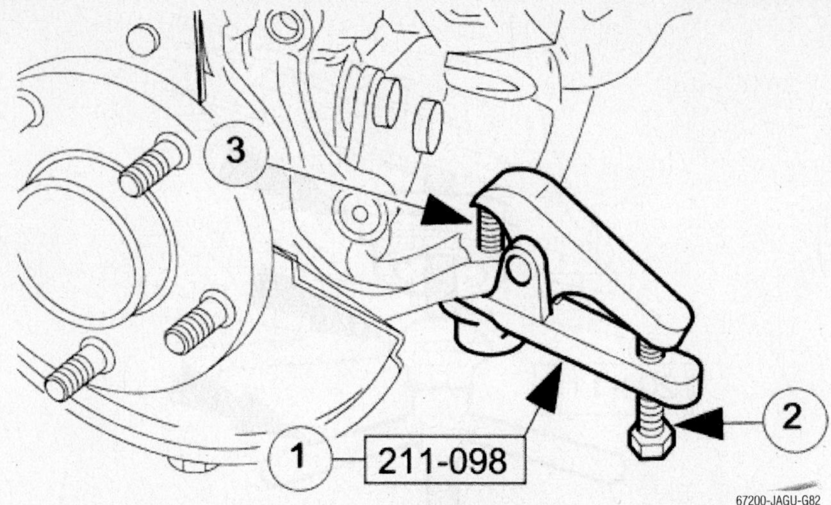

Install special tool 211-098 on tie rod ball joint—XJ Range 2002–03 through VIN F59525

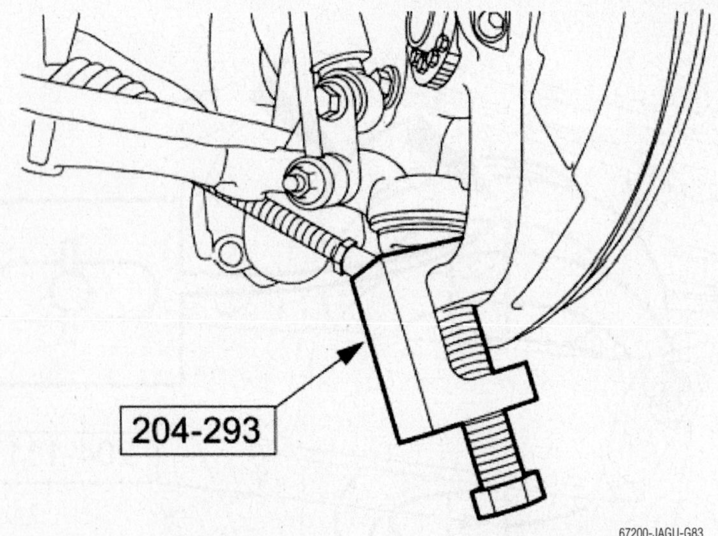

Install special tool 204-293 on ball joint—XJ Range 2002–03 through VIN F59525

taper pin from vertical link. Slacken but do not remove nut securing lower control arm ball joint to vertical link.

➡Do not use special tool 211-098 in the following procedure as this will damage the ball joint boot.

19. Install special tool 204-293 on ball joint. Tighten tool bolt to release ball joint taper pin from vertical link.
20. Remove special tool 204-293 from ball joint.
21. Supporting vertical link/hub assembly, remove lower control arm ball joint nut.
22. Remove vertical link /hub assembly.
23. Remove stabilizer bar link to control arm securing nut and bolt.
24. Remove control arm front arm to rear arm securing bolt.
25. Slacken but do not remove shock absorber lower nut and bolt.
26. Release lower steering column from pinion shaft.
27. Remove lower steering column nut and bolt.
28. Move column upwards to release it from pinion shaft.
29. Disconnect harness connector from steering transducer.
30. Remove tie strap securing return pipe and rubber support to steering rack.
31. Do not allow steering pipes to support weight of steering rack.
32. Pivot steering arm fully downwards.
33. Remove steering rack securing bolts.
34. Carefully move steering rack downwards for access to control arm eccentric bolt.
35. Remove lower control arm rear arm eccentric bolt and nut.
36. Mark position of eccentric bolt large graduation mark on crossbeam.
37. Remove control arm rear eccentric bolt and nut.
38. Remove lower control arm rear arm/ball joint assembly from crossbeam.

To install:

39. Position control arm rear arm on crossbeam and install but do not tighten eccentric bolt.
40. Install steering rack on crossbeam.
41. Ensure that bushes are correctly positioned on steering rack.
42. Position steering rack on crossbeam.
43. Position support brackets over bushes.
44. Install mounting bracket bolts and tighten to 30–40 ft. lbs. (41–54Nm).
45. Install lower steering column on pinion shaft.
46. Move column downwards to engage coupling splines on pinion shaft.

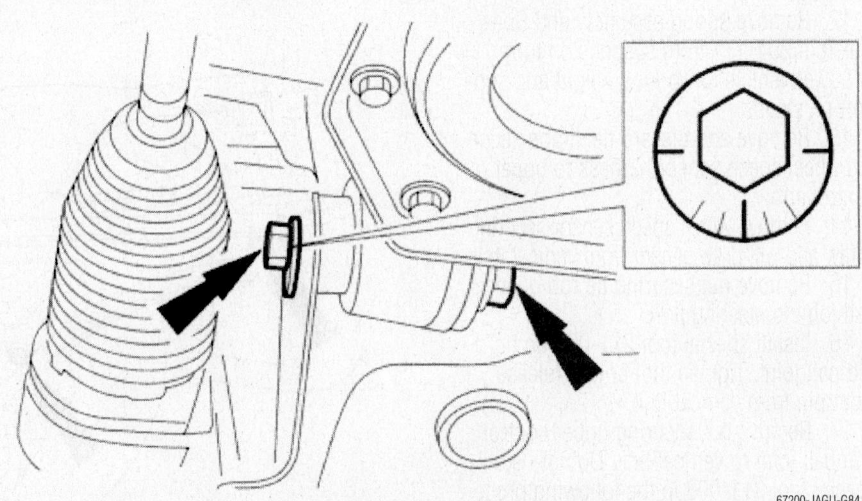

Mark position of eccentric bolt large graduation mark on crossbeam—XJ Range 2002–03 through VIN F59525

47. Install nut and bolt and tighten to 16–21 ft. lbs. (22–28Nm).
48. Position and connect harness to steering transducer.
49. Connect harness to transducer.
50. Secure return pipe on steering rack.
51. Position pipe rubber support on steering rack.
52. Position pipe on support.
53. Secure pipe and harness to support with new tie strap.
54. Install shock absorber and mounting bracket.
55. Position shock absorber mounting bracket on control arm rear arm.
56. Install control arm front arm to rear arm bolt and tighten to 57–76 ft. lbs. (77–103Nm).
57. Tighten shock absorber lower bolt and nut to 32–42 ft. lbs. (43–57Nm).
58. Position and partially tighten control arm rear eccentric bolt.
59. Rotate eccentric bolt to align large graduation with mark made during removal.
60. Ensuring bolt remains in position, partially tighten nut.
61. Install stabilizer bar link to control arm securing nut and bolt and tighten to 30–40 ft. lbs. (41–54Nm).
62. Position vertical link/hub assembly on lower ball joint taper pin, install nut and tighten to 44–59 ft. lbs. (60–80Nm).
63. Position vertical link/hub assembly on upper ball joint taper pin.
64. Install nut securing upper ball joint to vertical link and tighten to 44–59 ft. lbs. (60–80 Nm).
65. Position tie rod ball joint taper pin on steering arm and install and tighten nut to 71–85Nm.
66. Raise vehicle and support on stands to provide access for Special Tool 204-111 and road spring.
67. Install spring pan, rubber lower seat, spring and rubber upper seat on Special Tool 204-111.
68. Install road spring in suspension turret.
69. Install spring pan/spring assembly and Special Tool 204-111 in suspension turret.
70. Ensure spring upper seat is correctly positioned in turret.
71. Engage tool dowel in turret slot.
72. Rotate adjuster of Special Tool 204-111 to lightly tension spring.
73. Install guide pins (supplied with Special Tool 204-111) in spring pan/control arm bolt locations.

➡**When rotating tool adjuster, ensure that tool stem does not turn and disturb position of dowel in suspension turret.**

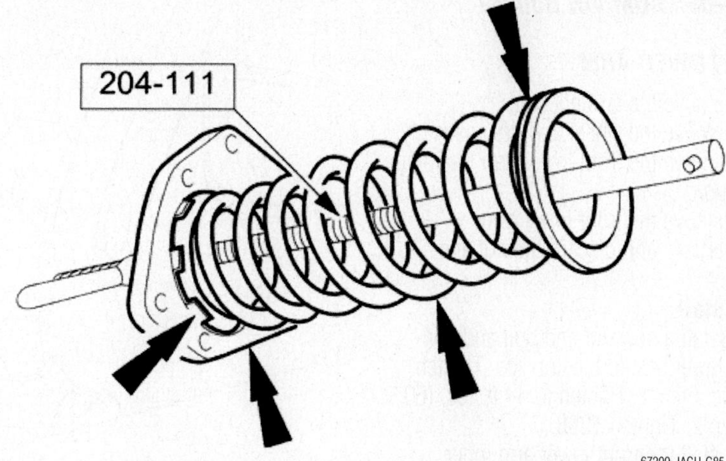

Install spring pan, rubber lower seat, spring and rubber upper seat on Special Tool 204-111—XJ Range 2002–03 through VIN F59525

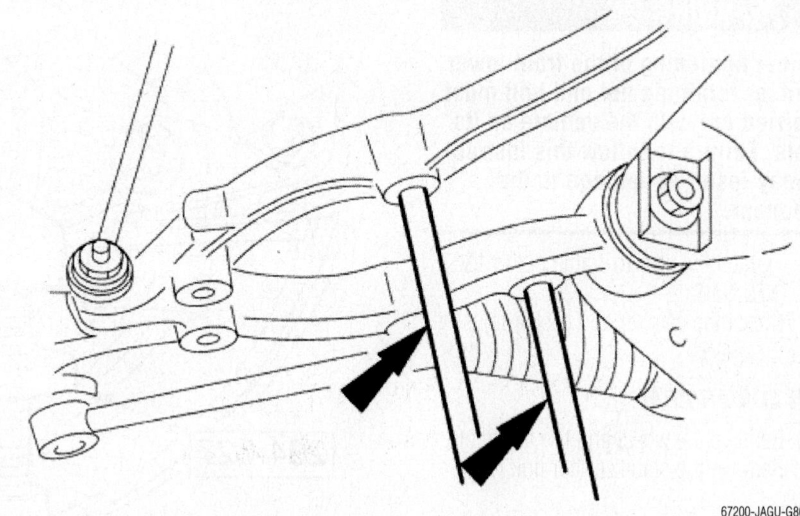

Install guide pins (supplied with Special Tool 204-111) in spring pan/control arm bolt locations—XJ Range 2002–03 through VIN F59525

74. Install spring pan to control arm securing bolts.
75. Rotate tool adjuster to gradually increase spring tension.
76. Progressively install but do not tighten spring pan securing bolts. Tighten spring pan bolts to 52–67 ft. lbs. (70–90Nm).
77. Ensure spring lower seat is correctly positioned in spring pan.

➡**When rotating tool adjuster, ensure that tool stem does not turn and disturb position of dowel in suspension turret.**

78. Remove Special Tool 204-111 from suspension.
79. Rotate adjuster of Special Tool to release tension on spring.
80. Rotate tool stem to release dowel from turret.
81. Remove tool and adaptor assembly.
82. Lower vehicle to working height and support on stands.
83. Position wheel speed sensor on vertical link.
84. Install sensor bolt and tighten to 89 inch lbs. (10Nm).
85. Connect wheel speed sensor harness.
86. Using new tie straps secure wheel speed sensor harness to upper control arm.
87. Install brake calipers.
88. Install road wheel.
89. Raise vehicle, remove stands and lower vehicle onto road wheels.
90. Ensuring graduation mark remains aligned, finally tighten control arm rear eccentric bolt to 84–112 ft. lbs. (113–153Nm).
91. Check wheel alignment and if necessary adjust.

2003.5–06 FROM VIN G00442

FRONT LOWER ARM

1. Remove the air deflector.
2. Remove the wheel and tire assembly.
3. Remove the front lower arm retaining nut and bolt.
4. Remove the front lower arm.
5. Remove and discard the nut and bolt.

To install:

6. Install a new nut and bolt and connect the lower arm to the knuckle. Tighten in 2 steps: Step 1: Tighten to 44 ft. lbs. (60 Nm). Step 2: Tighten 135ß.
7. Install the front lower arm inner retaining nut and bolt.
8. Install the wheel and tire assembly.
9. Install the air deflector.

WARNING

The final tightening of the front lower arm inner retaining nut and bolt must be carried out with the vehicle on its wheels. Failure to follow this instruction may result in damage to the component.

10. Tighten the arm-to-frame nut to 130 ft. lbs. (175 Nm).
11. Check the camber and caster adjustment.

REAR LOWER ARM

1. Remove the wheel and tire assembly.
2. Remove the stabilizer bar link retaining nut.
3. Remove the stabilizer bar link.
4. Detach the headlamp leveling sensor retaining clip.
5. Detach the tie rod end.
6. Detach the air spring assembly.
7. Detach the rear lower arm.
8. Detach the front lower arm.
9. Remove and discard the retaining nut and bolt.

WARNING

Prevent the rear lower arm ball joint ball pin hexagon from rotating. Failure to do so may result in damage to the lower ball joint boot.

10. Remove the rear lower arm ball joint retaining nut.

WARNING

Make sure the special tool is correctly located and the lower ball joint boot is not damaged during opera-

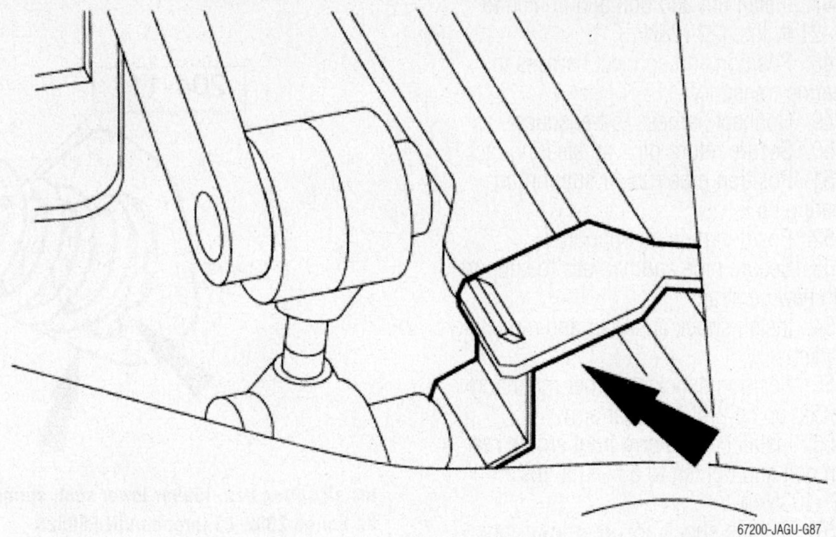

Detach the headlamp leveling sensor retaining clip—XJ 2003.5–06 from VIN G00442

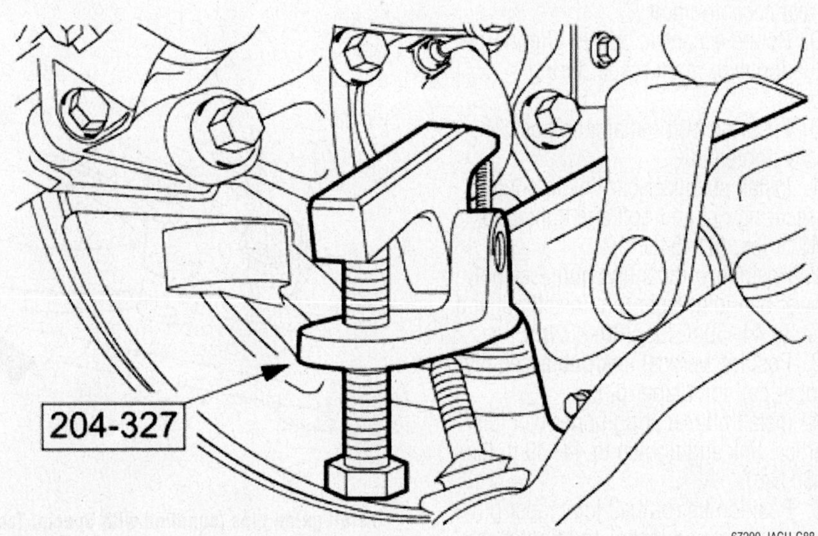

Using the special tool, remove the rear lower arm—XJ 2003.5–06 from VIN G00442

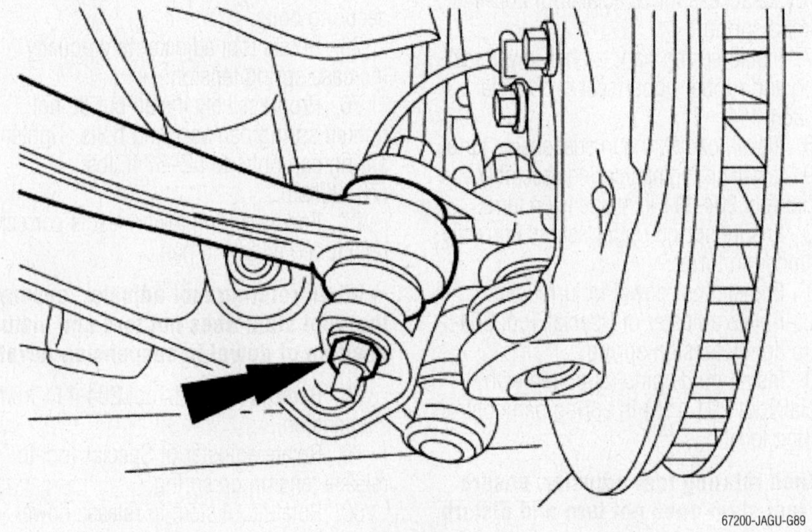

Front lower arm-to-knuckle attachment—XJ 2003.5–05 from VIN G00442

tion. Failure to follow this instruction may result in damage to the component.

11. Using the special tool, remove the rear lower arm.

To install:

12. Install the rear lower arm. Tighten the ball joint stud nut to 56 ft. lbs. (75 Nm).
13. Install a new retaining nut and bolt. Attach the front lower arm to the knuckle. Tighten in 2 steps. Step 1: 44 ft. lbs. (60 Nm). Step 2: Tighten 135°.
14. Attach the rear lower arm. Install the nut loosely. Tighten the nut completely with the wheels on the ground.
15. Attach the air spring assembly. Tighten to 130 ft. lbs. (175 Nm).
16. Attach the tie rod end. Tighten to 56 ft. lbs. (75 Nm).
17. Attach the head lamp leveling sensor retaining clip.
18. Attach the stabilizer bar link. Tighten to 32 ft. lbs. (43 Nm).
19. Install the wheel and tire assembly.
20. The final tightening of the rear lower arm inner retaining nut and bolt must be carried out with the vehicle on its wheels. Tighten to 130 ft. lbs. (175 Nm).
21. Tighten the stabilizer link to 52 ft. lbs. (70 Nm).
22. Check the camber and caster adjustment.

XK Range

➡Various thread-locking devices are used on nuts and bolts throughout the vehicle. These devices restrict the number of times a nut or bolt can be used.

1. Raise front of vehicle and support on stands.
2. Remove appropriate front road wheel.
3. Remove nut and bolt securing stabilizer bar link-rod to lower control arm.

➡To prevent damage to upper ball joint, secure vertical link to road spring using strong wire.

4. Remove nut securing lower ball joint to vertical link.

⁂ WARNING

Do not use special tool 211-098 in the following procedure as this will damage the ball joint boot.

5. Install special tool 204-293 on ball joint.

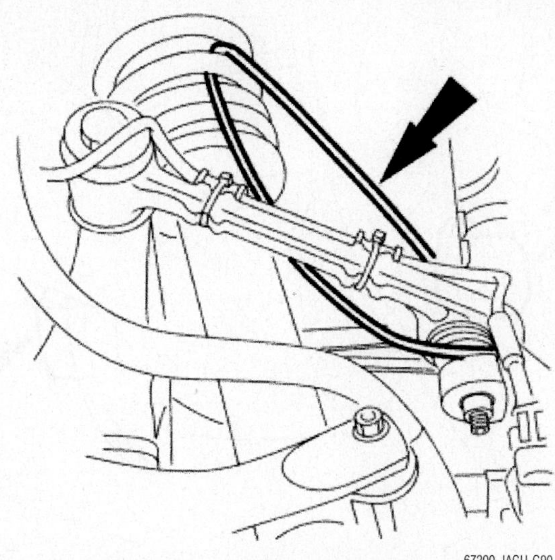

To prevent damage to upper ball joint, secure vertical link to road spring using strong wire

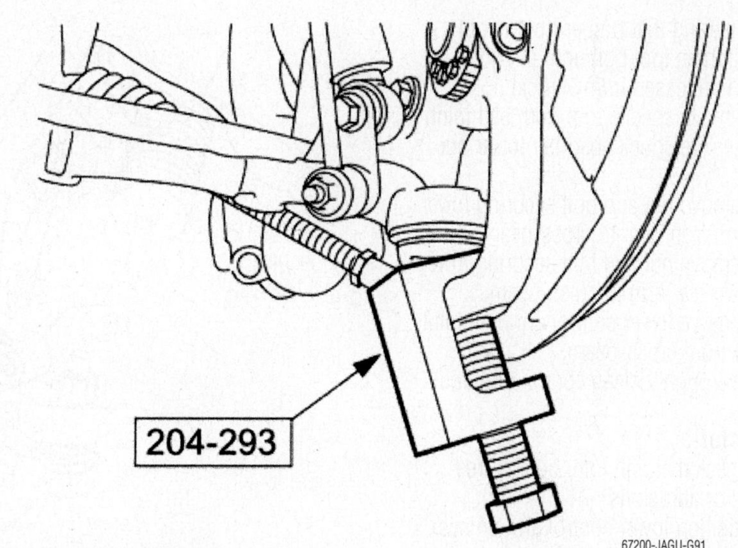

Install special tool 204-293 on ball joint

Remove nut and bolt securing lower control arm front arm to cross-beam

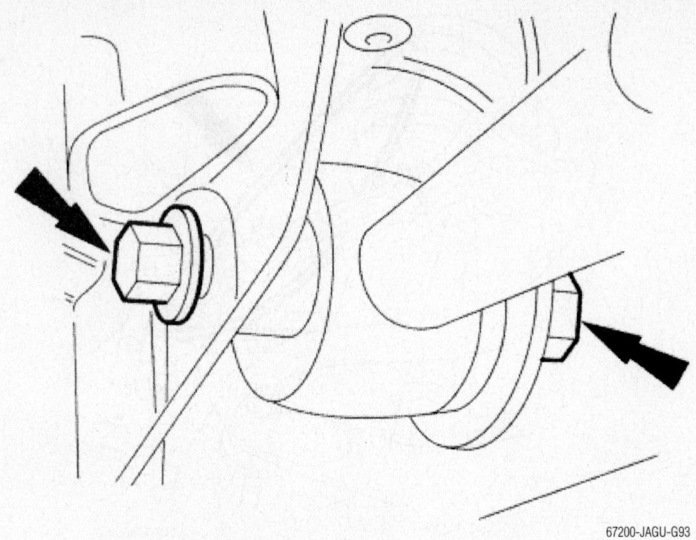
67200-JAGU-G93

Remove nut and bolt securing lower control arm rear arm to cross-beam

6. Ensuring that rubber boot is not trapped, tighten tool bolt until ball joint taper pin is released from vertical link.
7. Remove special tool from ball joint.
8. Remove shock absorber lower nut and bolt.
9. Remove nut and bolt securing lower control arm front arm to cross-beam.
10. Remove nut and bolt securing lower control arm rear arm to cross-beam.
11. Remove lower control arm/ball joint assembly from cross-beam.
12. Thoroughly clean control arm and ball joint.

To install:

13. Check that ball joint boot is free from cuts or abrasions.
14. Position lower control arm on suspension cross-beam and vertical link.
15. Install but do not tighten control arm rear arm to cross beam pivot nut and bolt.
16. Install but do not tighten control arm front arm to cross beam pivot nut and bolt.
17. Install nut securing lower ball joint to vertical link and tighten to 44–59 ft. lbs. (60–80 Nm).
18. Align lower end of shock absorber with control arm, install nut and bolt and tighten to 44–59 ft. lbs. (60–80 Nm).
19. Remove wire temporarily securing vertical link to road spring.
20. Tighten control arm rear arm pivot nut and bolt to 84–113 ft. lbs. (113–153 Nm).
21. Tighten control arm front arm pivot nut and bolt to 84–113 ft. lbs. (113–153 Nm).
22. Align stabilizer bar link rod with lower control arm, install nut and bolt and tighten to 29–41 ft. lbs. (39–55Nm).
23. Install road wheel.
24. Raise vehicle, remove stands and lower vehicle.

Front Wheel Bearings

ADJUSTMENT

The front wheel bearings are sealed and no adjustment is necessary or possible. If the bearings are found to be loose or noisy, they must be replaced.

REMOVAL & INSTALLATION

S-Type

1. Remove the brake disc.
2. Remove and discard the retaining clip. Detach the ABS sensor.

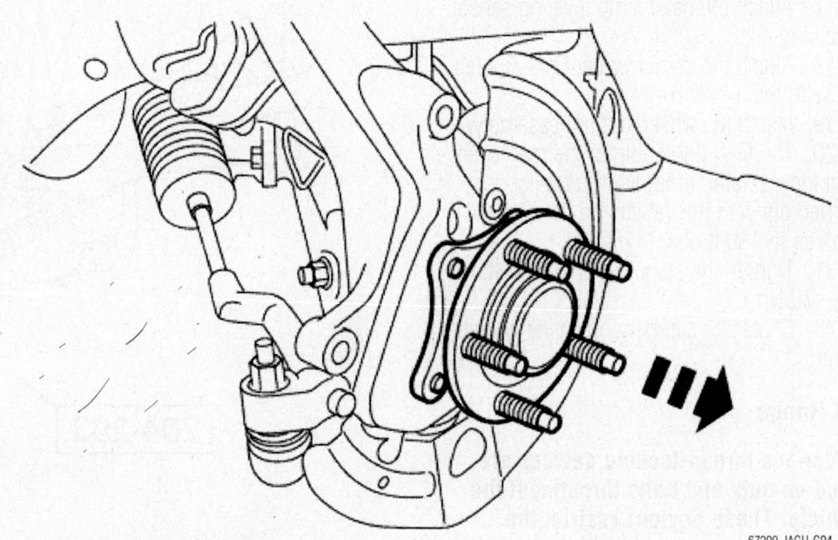
67200-JAGU-G94

Front hub/bearing assembly—S-Type

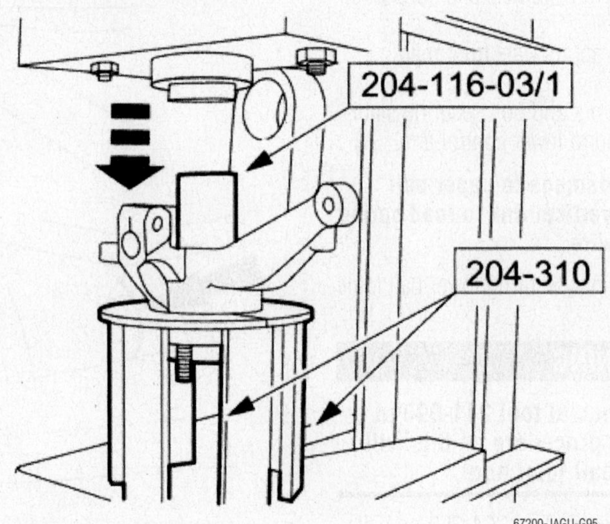

67200-JAGU-G95

Using special tools, remove the wheel hub—X-Type

3. Remove and discard the O-ring seal.
4. Remove the wheel hub and bearing retaining bolts.
5. Remove the wheel hub and bearing.
6. Prior to applying sealer, clean the ABS sensor sealing surface and the ABS sensor aperture sealing surface.
7. Install a new O-ring seal.
8. Apply a 2.5mm bead of Loctite Ultra Blue 587 Silicone (RTV) sealer to the ABS sensor sealing surface.
9. Install a new ABS sensor retaining clip.
10. Install the hub/bearing assembly. Tighten to 90 Nm.

X-Type

1. Remove wheel knuckle.

➡When removing the wheel hub from the wheel knuckle the bearing will be destroyed and the bearing inner ring will remain on the wheel hub.

2. Using special tools, remove the wheel hub.
3. Using a suitable tool, remove the bearing inner ring.
4. Using special tools, remove the wheel bearing.

To install:

5. Insert special tool 204-312 in to 204-309.
6. Support the wheel knuckle on special tools.

➡Make sure wheel bearing is correctly orientated with the colored side facing inboard.

7. Align the bearing over special tool.
8. Using special tools, press bearing into wheel knuckle.
9. Change special tool 204-312 to 204-313.
10. Align the bearing and wheel knuckle over special tool.
11. Slide wheel hub over the guide and align to the bearing.
12. Using special tools install the wheel hub.
13. Install the wheel knuckle.

2002–03 XJ Range and 2002–06 XK Range

1. Raise front of vehicle and support on stands.
2. Remove appropriate road wheel.
3. Using wire or string, secure brake caliper to road spring to take weight off brake hose.
4. Raise vehicle sufficient to provide access for special tool 204-111.

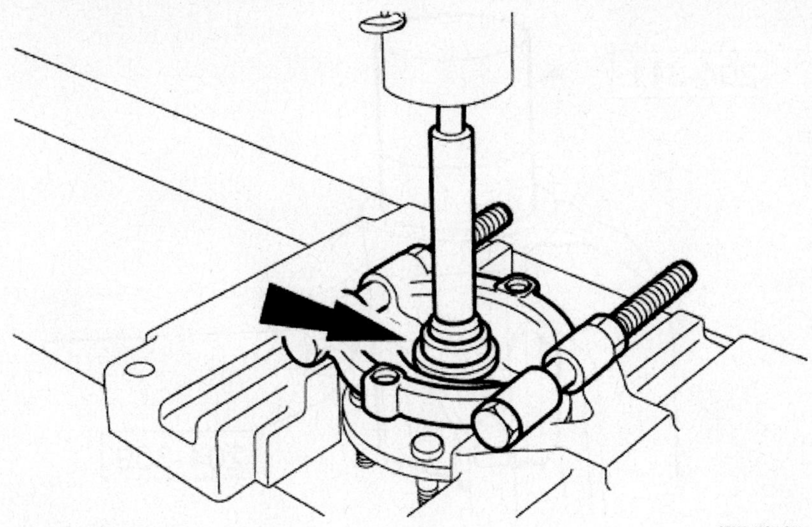

Using a suitable tool, remove the bearing inner ring—X-Type

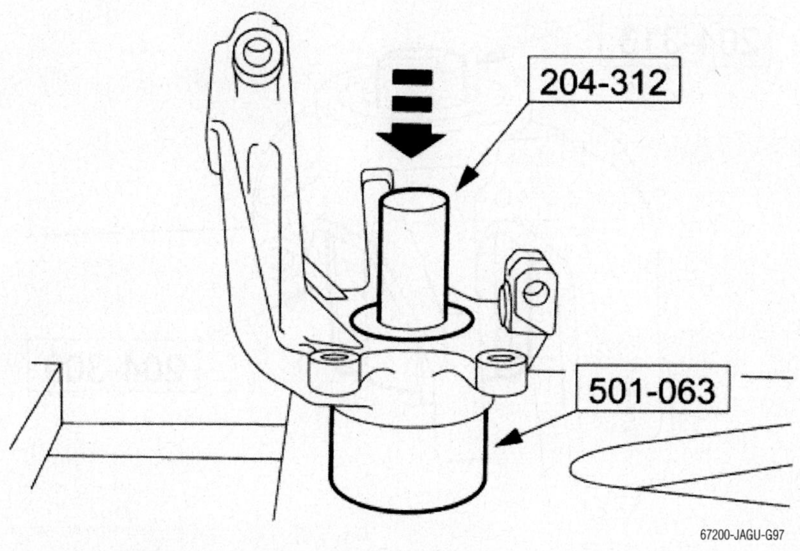

Using special tools, remove the wheel bearing—X-Type

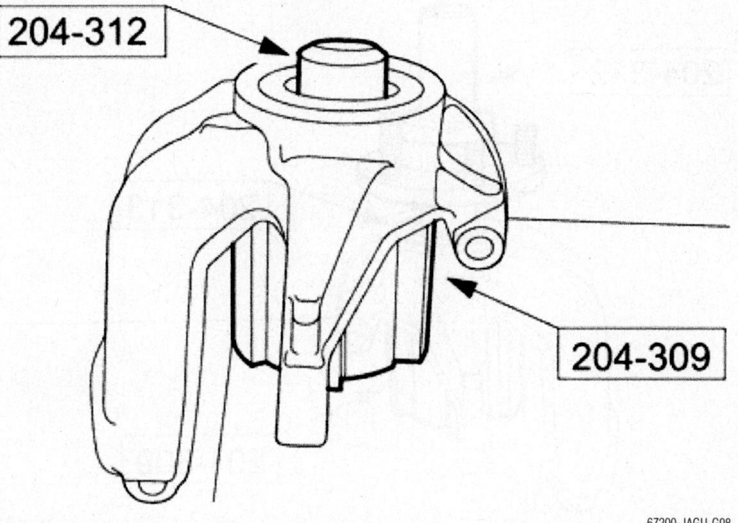

Insert special tool 204-312 in to 204-309—X-Type

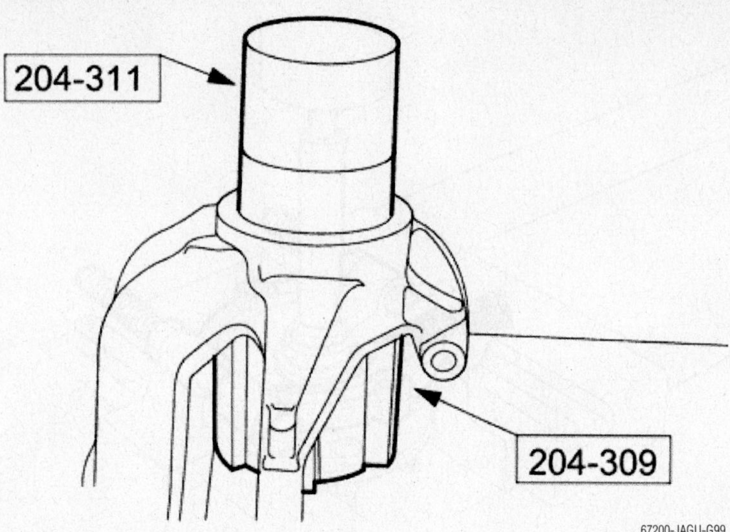

Using special tools, press bearing into wheel knuckle—X-Type

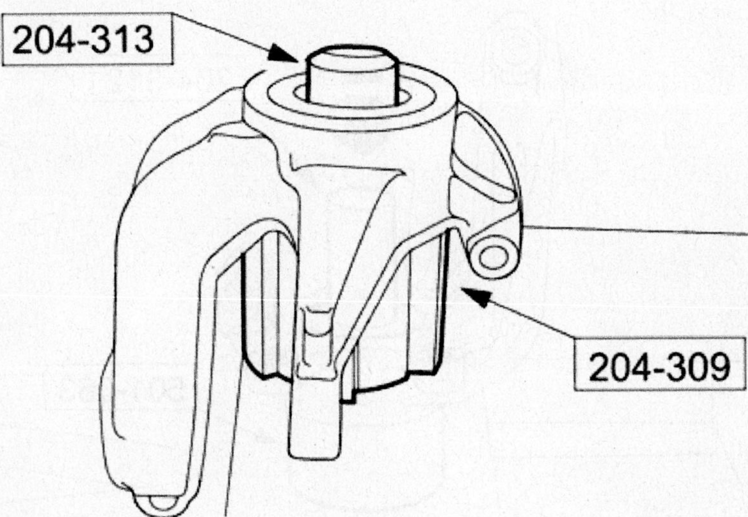

Change special tool 204-312 to 204-313—X-Type

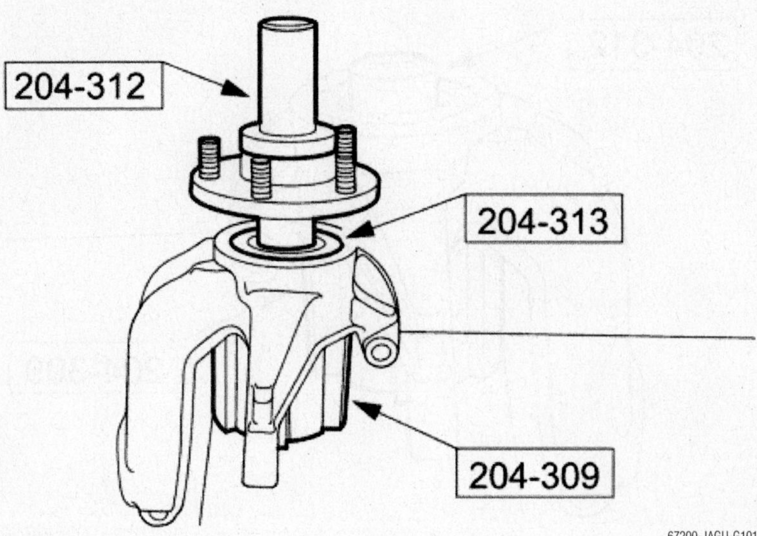

Align the bearing and wheel knuckle over special tool—X-Type

5. Install special tool 204-111 in road spring.
6. Slacken tool adjuster to suit spring length.
7. Install adaptor and thrust collar.
8. Position stem of tool in center of spring passing dowel through slot in suspension turret.

➡When turning tool adjuster, ensure that tool stem does not rotate and disturb position of dowel in suspension turret.

9. Engage dowel of special tool 204-111 in suspension turret and rotate to position across slot.
10. Lower vehicle onto stands.
11. Rotate tool adjuster to tension spring.
12. Remove wheel speed sensor securing screw and withdraw sensor from vertical link.
13. Remove nut securing tie rod outer ball joint to steering arm.
14. Release ball joint from vertical link.
15. Install special tool 211-098 on tie rod outer ball joint.
16. Tighten tool bolt to release taper pin from vertical link.
17. Remove special tool from ball joint and position tie rod for access.
18. Remove nut securing upper control arm ball joint to vertical link.

⁂ WARNING

Do not use special tool 211-098 in the following procedure as this will damage the ball joint boot.

19. Release upper control arm ball joint from vertical link/hub assembly.
20. Install special tool 204-192 on ball joint.
21. Tighten tool bolt to release ball joint taper pin from vertical link.
22. Raise upper control arm to release ball joint pin from vertical link.
23. Remove nut securing lower control arm ball joint to vertical link.
24. To prevent damage to lower ball joint, secure vertical link to road spring using strong wire.
25. Do not use special tool 211-098 in the following procedure as this will damage the ball joint boot.
26. Release lower control arm ball joint from vertical link/hub assembly.
27. Install special tool 204-293 on ball joint.
28. Tighten tool bolt to release ball joint taper pin from vertical link.

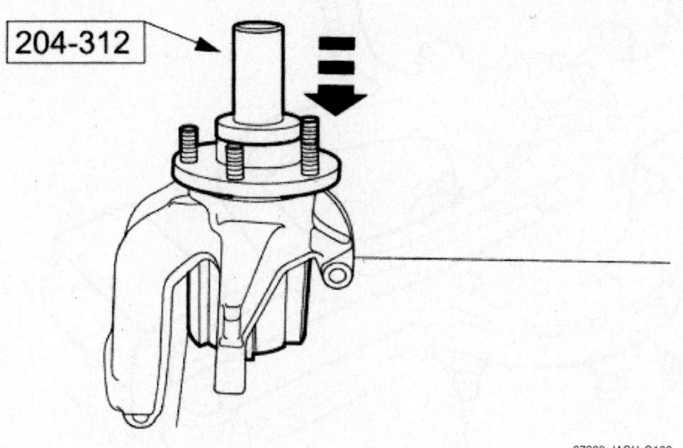

Slide wheel hub over the guide and align to the bearing—X-Type

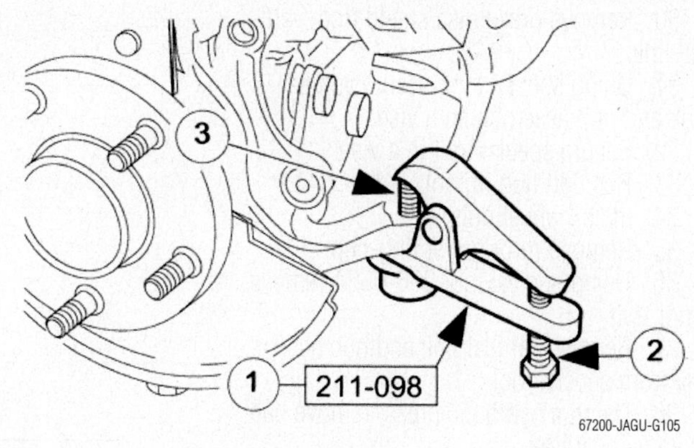

Install special tool 211-098 on tie rod outer ball joint—2002–03 XJ Range and 2002–06 XK Range

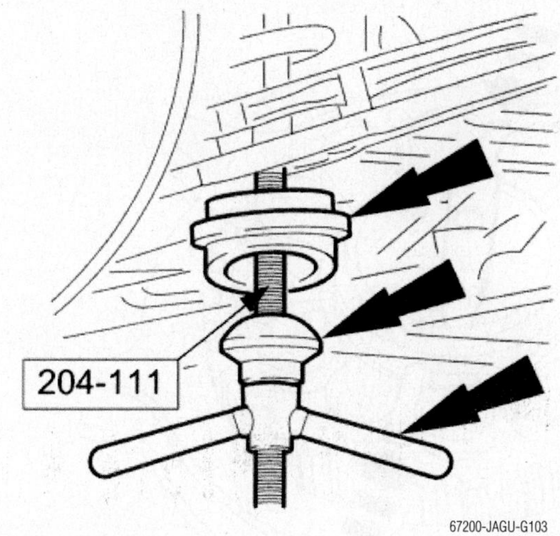

Install special tool 204-111 in road spring—2002–03 XJ Range and 2002–06 XK Range

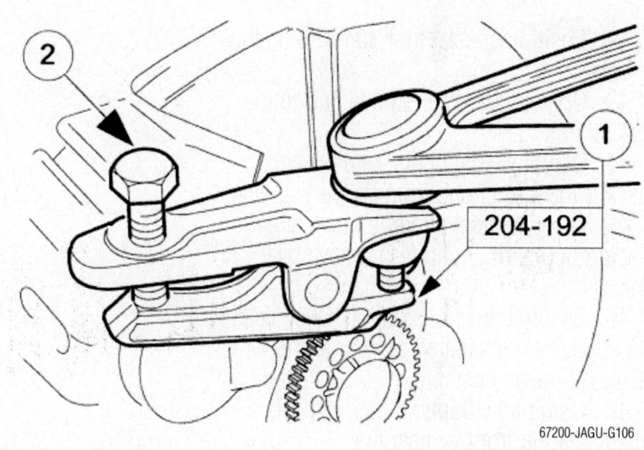

Install special tool 204-192 on ball joint—2002–03 XJ Range and 2002–06 XK Range

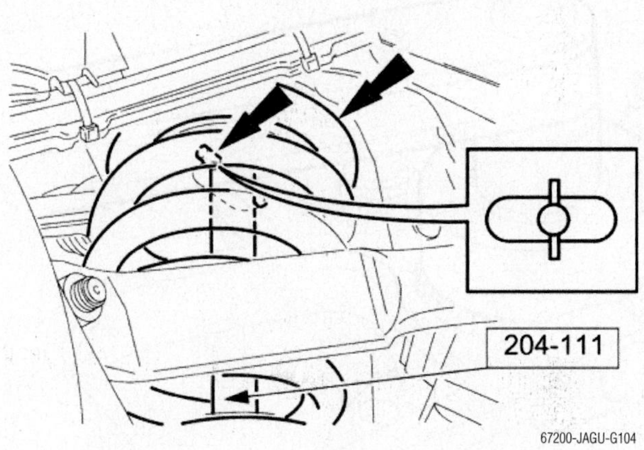

Position stem of tool in center of spring passing dowel through slot in suspension turret—2002–03 XJ Range and 2002–06 XK Range

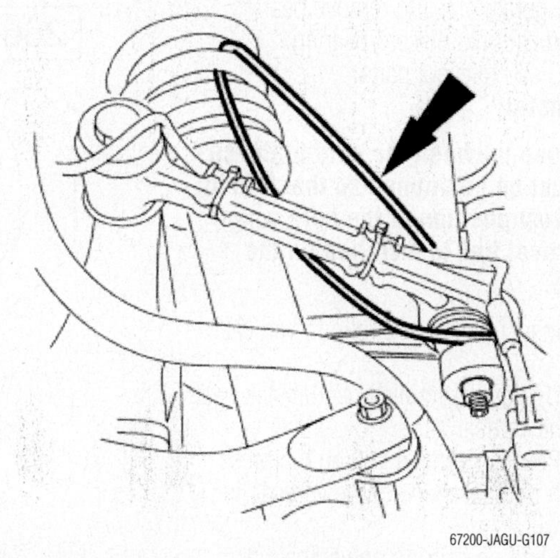

Secure vertical link to road spring using strong wire—2002–03 XJ Range and 2002–06 XK Range

29. Remove special tool 204-293 from ball joint.

30. Remove brake disc shield from vertical link.

31. Using special tool, secure vertical link and hub assembly in a vise.

32. Secure special tool in a vise.

33. Position hub in tool.

34. Install wheel nuts.

35. Remove rotor nut spring-clip.

36. Using special tool 206-066A remove rotor nut.

37. Remove vertical link and hub assembly from special tool.

38. Using a hydraulic press, remove hub from vertical link.

39. Place special tool 204-194 on press bed.

40. Position vertical link and hub assembly into tool collets.

41. Ensure one of the disc shield screw-bosses is in the center of one of the tool collets.

42. Position special tool 204-193 on top of hub.

43. Operate press to remove hub from vertical link.

44. Using special pliers, remove and discard inboard circlip from the vertical link.

45. Using special pliers, remove and discard outboard circlip from the vertical link.

46. Ensure the vertical link is level, and the supports are situated as near to the bearing bore as possible.

47. Using a hydraulic press, remove wheel bearing from vertical link.

48. Position the vertical link so that inboard side of the bearing bore is resting on suitable supports on the press-bed.

49. Position special tool 204-197 on top of bearing.

50. Operate press to remove bearing from vertical link, discard bearing.

51. Clean relevant parts.

To install:

➡The gap between the ears of the circlip must be positioned so that it is in the lowest position of the bore when the vertical link is installed on the vehicle.

52. Install outboard circlip in vertical link.

53. Using a hydraulic press, install bearing in the vertical link.

54. Position vertical link on the press bed with outboard side of the vertical link face down.

55. Position bearing on vertical link.

56. Position special tool on top of wheel bearing.

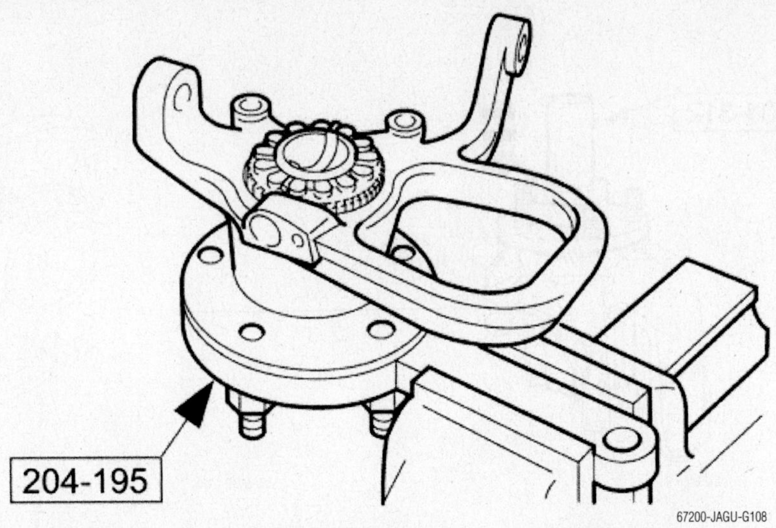

Using special tool, secure vertical link and hub assembly in a vise—2002–03 XJ Range and 2002–06 XK Range

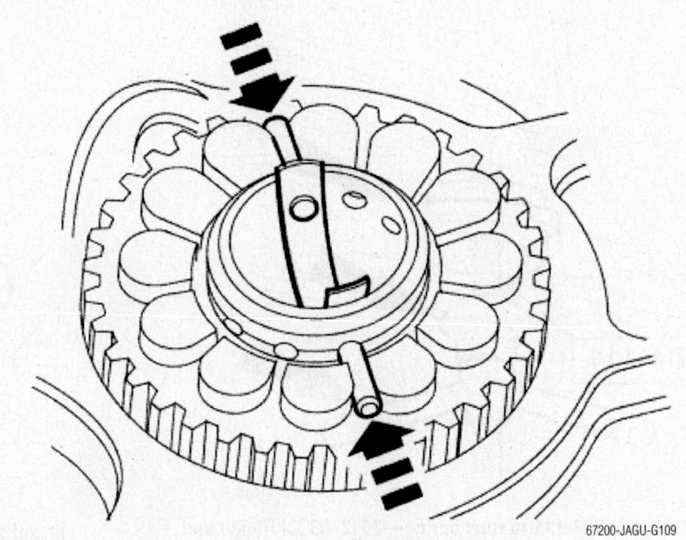

Remove rotor nut spring-clip—2002–03 XJ Range and 2002–06 XK Range

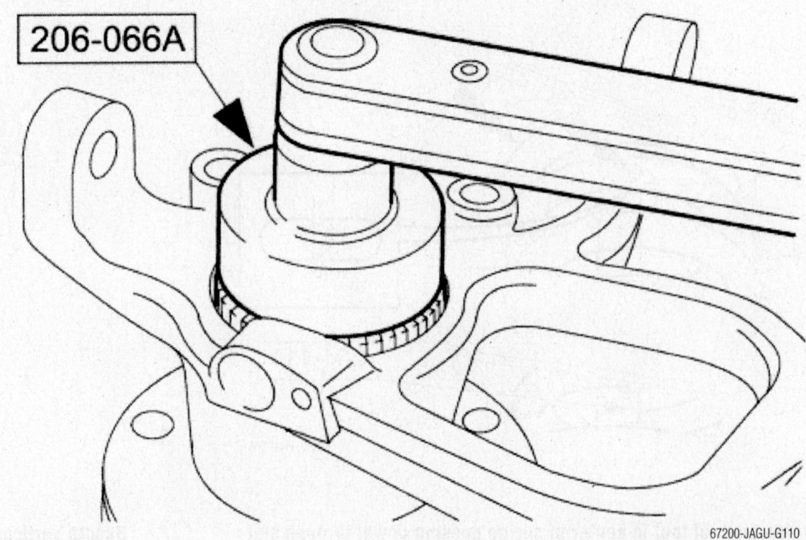

Using special tool 206-066A remove rotor nut—2002–03 XJ Range and 2002–06 XK Range

57. Operate press to install bearing, applying a three-ton load to ensure that the bearing is fully seated.

58. To prevent damage to the integral wheel-bearing seal, select circlip pliers ends that do not protrude through the circlip ears.

59. The gap between the ears of the circlip must be positioned so that it is in the lowest position of the bore, when the vertical link is installed on the vehicle.

60. Install inboard circlip in vertical link.

61. Position special tool on press bed.

62. Position the vertical link on the tool, with the inboard-side face down.

63. Position hub on bearing.

64. Position a flat steel plate across the face of the hub, DO NOT place plate across wheel studs.

65. Operate press to install the hub, applying a three-ton load to ensure hub is fully seated in bearing.

66. Using special tool, secure vertical link and hub assembly in a vise.

67. Secure special tool in a vise.

68. Position hub in tool.

69. Install wheel nuts.

70. Install rotor nut on hub. Using special tool tighten rotor nut. Do not loosen the rotor nut to engage the spring clip. If the castellated slots of the rotor nut are not aligned, tighten the rotor nut further until the spring clip engages the slots.

71. Install disc shield.

72. Position disc shield on vertical link.

73. Install bolts.

74. Clean all suspension to vertical link/hub assembly mating surfaces.

75. Position vertical link/hub assembly on upper ball joint taper pin.

76. Position vertical link/hub assembly on lower ball joint taper pin, install securing nut and tighten to 44–59 ft. lbs. (60–80Nm).

77. Install nut securing upper ball joint to vertical link and tighten to 44–59 ft. lbs. (60–80Nm).

78. Position tie rod ball joint taper pin on steering arm and install and tighten securing nut to 53–63 ft. lbs. (71–85Nm).

79. Install wheel speed sensor.

80. Position wheel speed sensor on vertical link.

81. Install sensor securing bolt and tighten to 89 inch lbs. (10Nm).

82. Remove wire temporarily securing vertical link to road spring.

83. Install brake disc.

84. Release tension on road spring.

85. Rotate adjuster of special tool 204-111 to release spring tension.

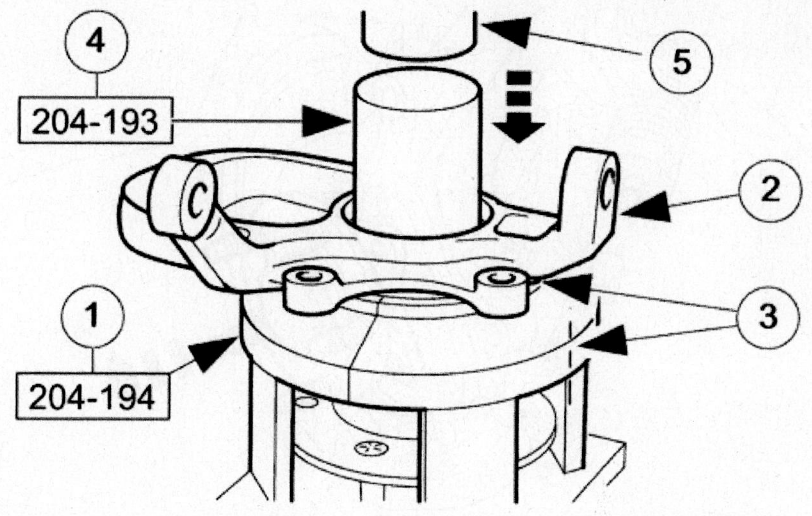

Place special tool 204-194 on press bed—2002–03 XJ Range and 2002–06 XK Range

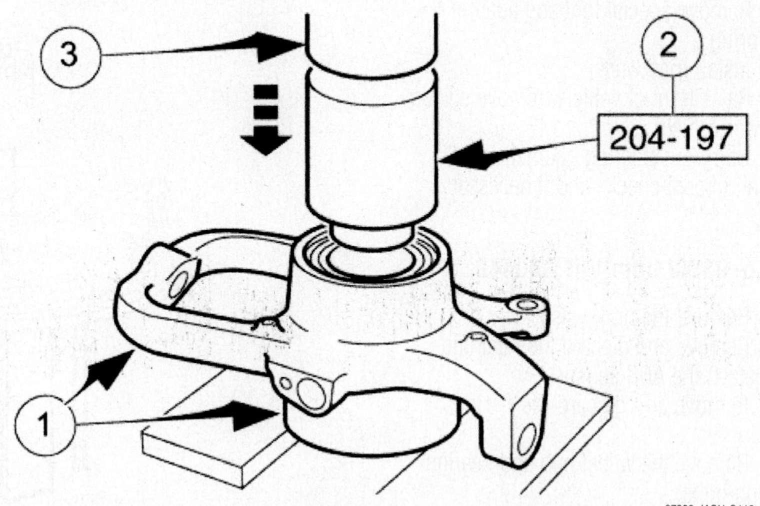

Position special tool 204-197 on top of bearing—2002–03 XJ Range and 2002–06 XK Range

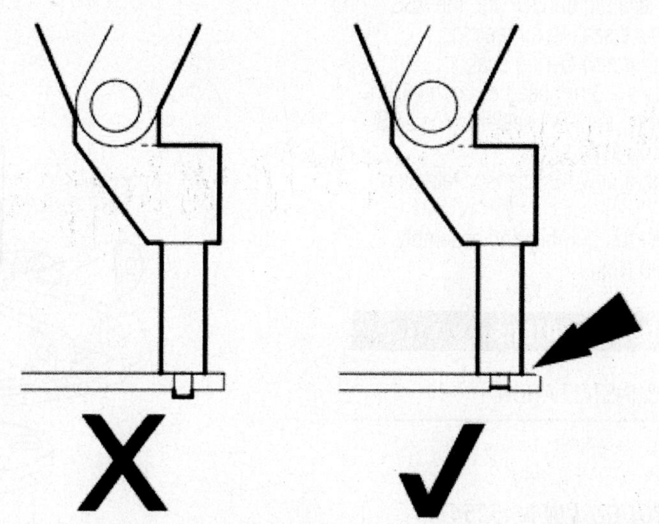

To prevent damage to the integral wheel-bearing seal, select circlip pliers ends that do not protrude through the circlip ears—2002–03 XJ Range and 2002–06 XK Range

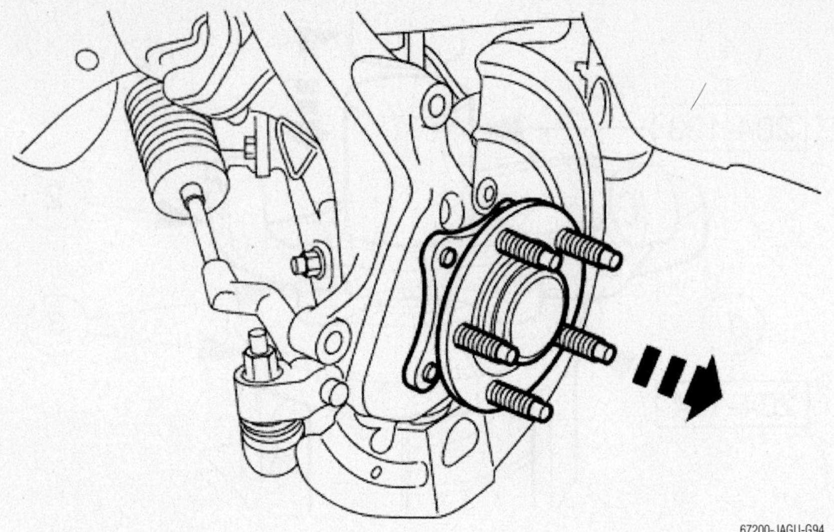

Front hub/bearing assembly—S-Type

86. Remove special tool and adaptor from spring.
87. Install road wheel.
88. Raise front of vehicle, remove stands and lower vehicle.
89. Carry out steering geometry and wheel alignment checks and if necessary adjust.

2003.5–06 XJ from VIN G00442

1. Remove the brake disc.
2. Remove and discard the retaining clip. Detach the ABS sensor.
3. Remove and discard the O-ring seal.
4. Remove the wheel hub and bearing retaining bolts.
5. Remove the wheel hub and bearing.
6. Prior to applying sealer, clean the ABS sensor sealing surface and the ABS sensor aperture sealing surface.
7. Install a new O-ring seal.
8. Apply a 2.5mm bead of Loctite Ultra Blue 587 Silicone (RTV) sealer to the ABS sensor sealing surface.
9. Install a new ABS sensor retaining clip.
10. Install the hub/bearing assembly. Tighten to 90 Nm.

Rear Wheel Bearings

REMOVAL & INSTALLATION

S-Type

2002 THROUGH VIN M45254

1. Loosen the wheel nuts.
2. Raise and support the vehicle.
3. Remove the wheel.
4. Remove the axle shaft hub nut.
5. Detach the brake caliper.
6. Unclip the ABS sensor wiring harness.
7. Detach the ABS sensor.
8. Detach the outer tie rod.
9. Detach the lower arm from the hub assembly.
10. Remove the brake disc.
11. Detach the hub assembly from the driveshaft.
12. Using the special tools, remove the hub from the bearing.
13. Using the special tools, remove the inner bearing race from the hub.
14. Remove the circlip.
15. Press the bearing from the hub carrier.

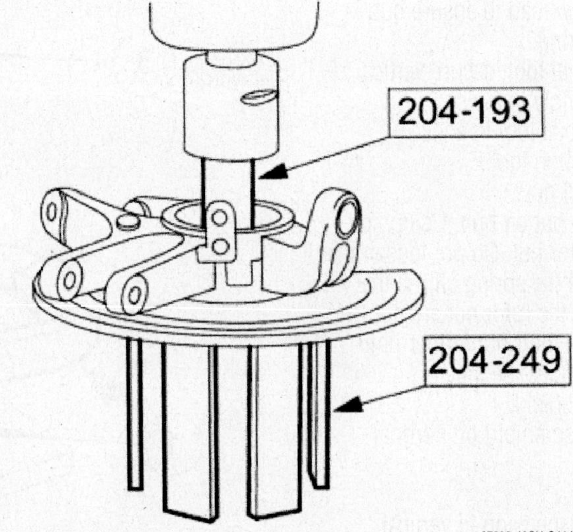

Using the special tools, remove the hub from the bearing

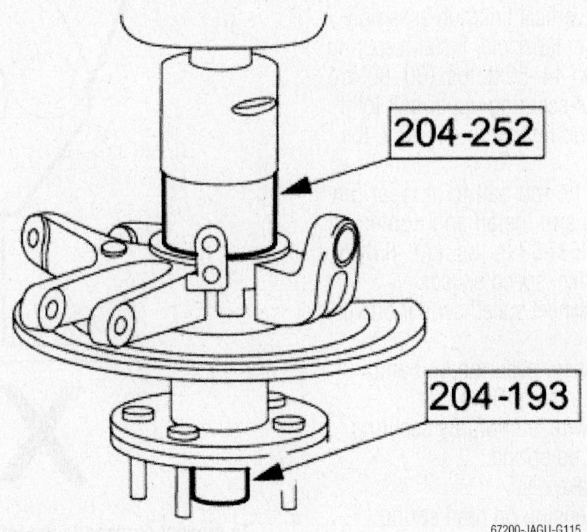

Using the special tools, remove the inner bearing race from the hub

To install:

➡The final tightening of the rear suspension components must be carried out with the vehicle on its wheels.

16. Press the hub bearing to the hub carrier.
17. Install the circlip.
18. Using the special tools, install the hub to the bearing.
19. Install the hub assembly to the upper arm. Torque to 67 ft. lbs. (90 Nm).
20. Attach the hub assembly to the driveshaft.
21. Install the brake disc.
22. Install the lower arm to the hub assembly. Torque to 111 ft. lbs. (150 Nm).
23. Install the outer tie rod. Torque to 74 ft. lbs. (100 Nm).
24. Install the ABS sensor.
25. Attach the ABS sensor wiring harness.
26. Install the brake caliper.
27. Install the axle shaft hub nut. Torque to 303 ft. lbs. (410 Nm).
28. Install the wheel.
29. Lower the vehicle.
30. Tighten the wheel nuts.

2002.5–06 FROM VIN M45255

1. Loosen the wheel hub nut.
2. Remove the brake disc.
3. Detach the anti-lock brake system (ABS) sensor.
4. Detach the outer tie rod.
5. Detach the lower arm from the wheel knuckle.
6. Remove the wheel hub nut.
7. Using the special tools, detach the halfshaft.
8. Remove the wheel knuckle assembly
9. Using the special tools, remove the wheel hub from the bearing.
10. Using the special tools, remove the inner bearing race from the hub.
11. Remove the circlip.
12. Press the bearing from the wheel knuckle.

To install:

➡The final tightening of the rear suspension components must be carried out with the vehicle on its wheels.

13. Press the bearing to the wheel knuckle.
14. Install the circlip.
15. Using the special tools, install the wheel hub to the bearing.
16. Install the wheel knuckle assembly. Tighten to 67 ft. lbs. (90 Nm).
17. Using the old wheel hub nut tighten to 111 ft. lbs. (150 Nm).

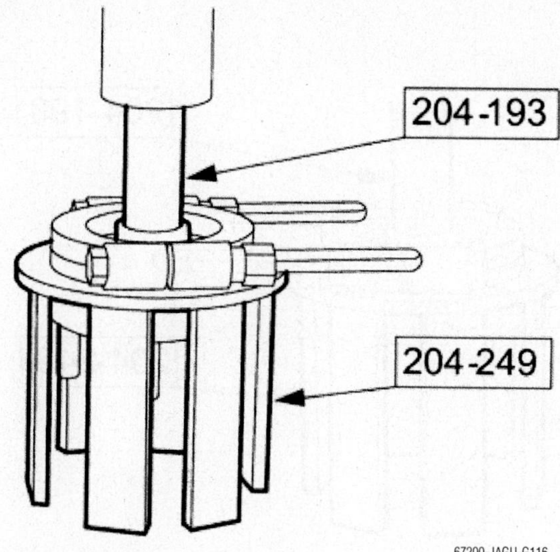

Using the special tools, install the hub to the bearing

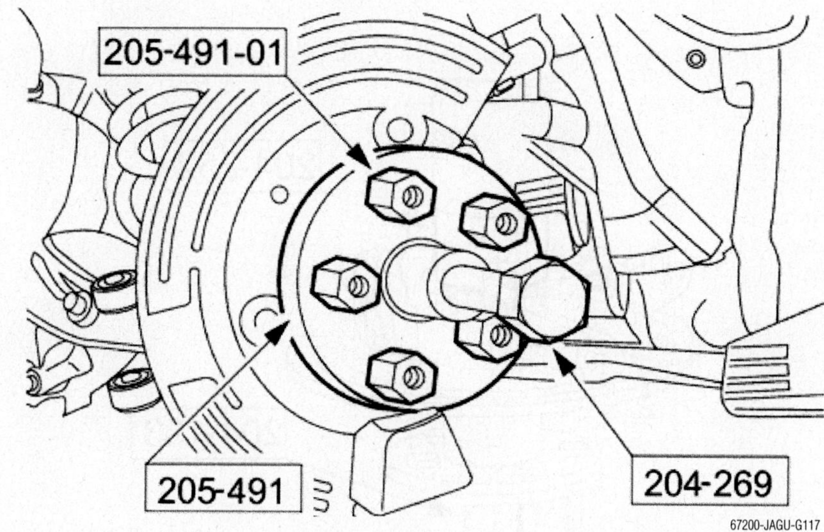

Using the special tools, detach the halfshaft—2002.5–06 from VIN M45255

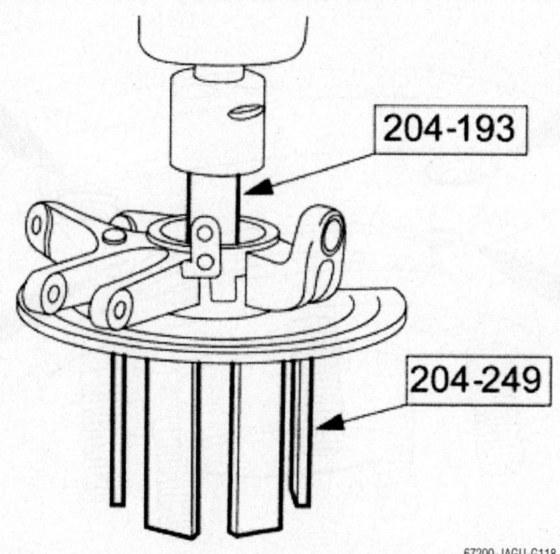

Using the special tools, remove the wheel hub from the bearing—2002.5–06 from VIN M45255

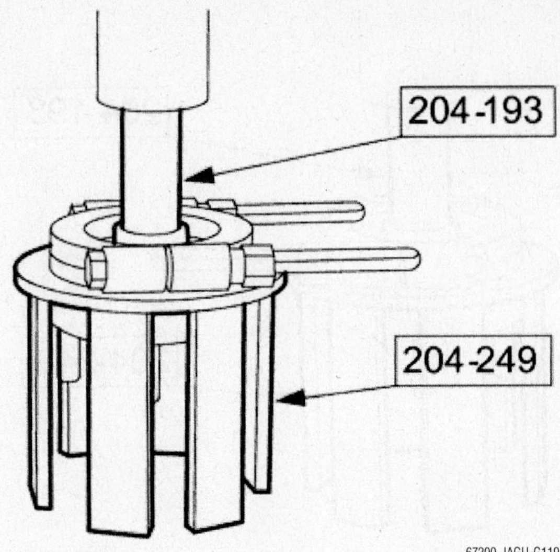

Using the special tools, remove the inner bearing race from the hub—2002.5–06 from VIN M45255

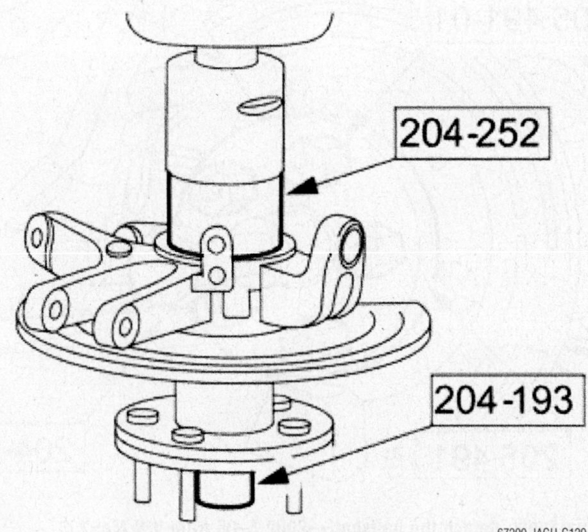

Using the special tools, install the wheel hub to the bearing—2002.5–06 from VIN M45255

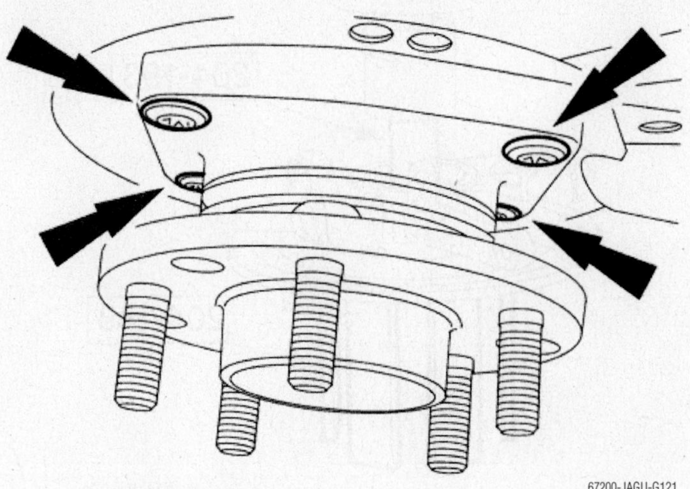

Wheel hub retaining bolts—X-Type

18. Attach the wheel knuckle to the half-shaft.
19. Install the lower arm to the wheel knuckle retaining bolt.
20. Install the outer tie rod retaining bolt.
21. Install the ABS sensor. Tighten to 89 inch lbs. (10 Nm).
22. Install the brake disc.
23. Lower the vehicle.
24. Remove and discard the old wheel hub nut, install a new wheel hub nut.
25. Tighten the wheel hub nut. Tighten to 222 ft. lbs. (300 Nm).

➡The final tightening of the rear suspension components must be carried out with the vehicle on its wheels.

26. Tighten the lower arm-to-knuckle to 111 ft. lbs. (150 Nm).
27. Tighten the tie rod end to 41 ft. lbs. (55 Nm).

X-Type

1. Remove the rear brake disc.
2. Remove and discard the rear halfshaft retaining nut.
3. Remove the wheel hub retaining bolts.
4. Using the special tools, remove the wheel hub.

To install:

5. Install the wheel hub. Tighten to
   - 2002 through VIN C59260: 52 ft. lbs. (70 Nm)
   - 2002.25–05 from VIN C59261: 47 ft. lbs. (63 Nm)
6. Install a new rear halfshaft retaining nut. Tighten to: 200 ft. lbs. (270 Nm)
7. Install the rear brake disc.

XJ Range and XK Range

XJ RANGE 2002–03 THROUGH VIN F59525 AND 2002–06 XK RANGE

➡Various thread-locking devices are used on nuts and bolts throughout the vehicle. These devices restrict the number of times a nut or bolt can be used.

1. Release handbrake.
2. Raise the vehicle on a four-post lift.
3. Fully loosen but do not remove parking brake cable adjuster nut.
4. Release appropriate side parking brake cable from connecting clip.
5. Remove parking brake cable through equalizer (LH side cable only).
6. Lower four-post lift.
7. Raise rear of vehicle to working height and support on stands.
8. Remove rear wheel.

9. Position special tool 204-195 on hub studs and install securing nuts.
10. Slacken but do not remove hub to axle shaft securing nut.
11. Remove securing nuts and withdraw special tool 204-195 from hub.
12. Remove rear brake disc.
13. Remove parking brake shoes.
14. Remove wheel speed sensor.
15. Disconnect wheel speed sensor harness from sensor.
16. Remove wheel speed sensor securing bolt and withdraw sensor from hub carrier.
17. Sever and discard wheel speed sensor harness tie strap and position harness for access.
18. Remove hub nut and collar.
19. Install a thread protector on axle shaft.
20. Release hub from axle shaft using a hub puller.
21. Remove hub puller and thread protector.
22. Remove securing nuts and release puller from hub.
23. Remove thread protector from axle shaft.
24. Release parking brake cable from suspension mounting bracket.
25. Remove cable retaining clip.
26. Draw cable through suspension mounting bracket aperture.
27. Remove parking brake cable to hub carrier grommet.
28. Remove back plate.
29. Remove two bolts and two screws securing back plate to hub carrier.
30. Remove back plate complete with parking brake cable.
31. Mark control arm at pivot pin eccentric flange marker to aid installation.
32. Remove nut and washer from pivot pin.
33. Remove hub and carrier from control arm and axle shaft.
34. Remove pivot pin and remove hub and carrier assembly.

➡Note position of shims during removal. They must be returned to their original positions during installation.

35. Noting positions, remove shim from each end of pivot pin sleeve.
36. Remove pivot pin sleeve from hub carrier.
37. Using a hydraulic press, remove hub from carrier.
38. Place hub carrier on a work bench.
39. Remove ABS rotor from hub carrier.
40. Remove outer seal and bearing assembly from hub carrier.

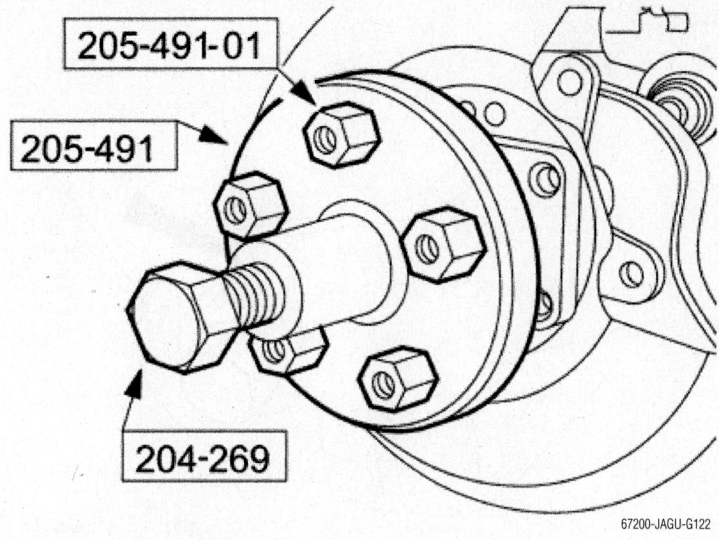

Using the special tools, remove the wheel hub—X-Type

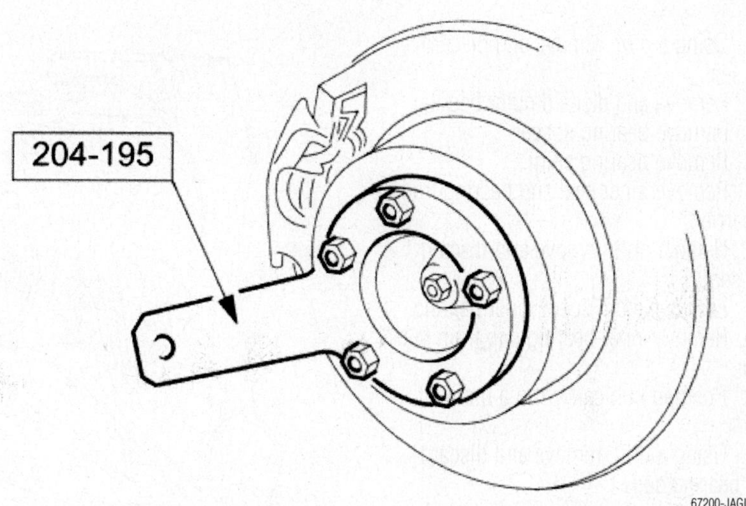

Position special tool 204-195 on hub studs and install securing nuts—2002–03 through VIN F59525

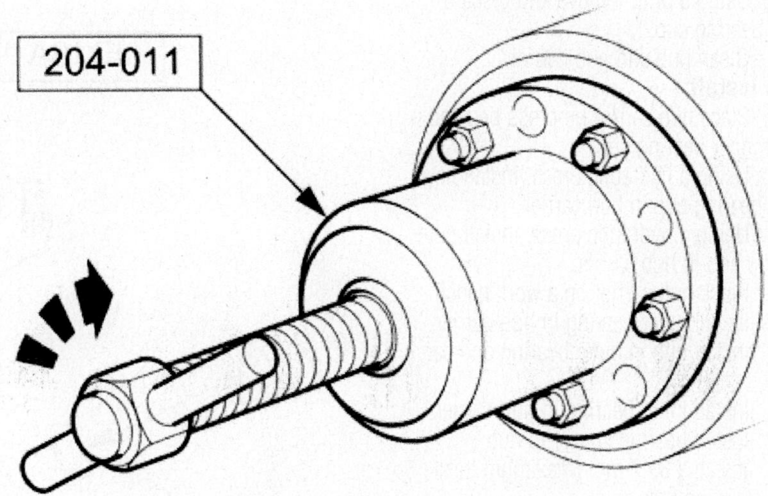

Release hub from axle shaft using a hub puller—2002–03 through VIN F59525

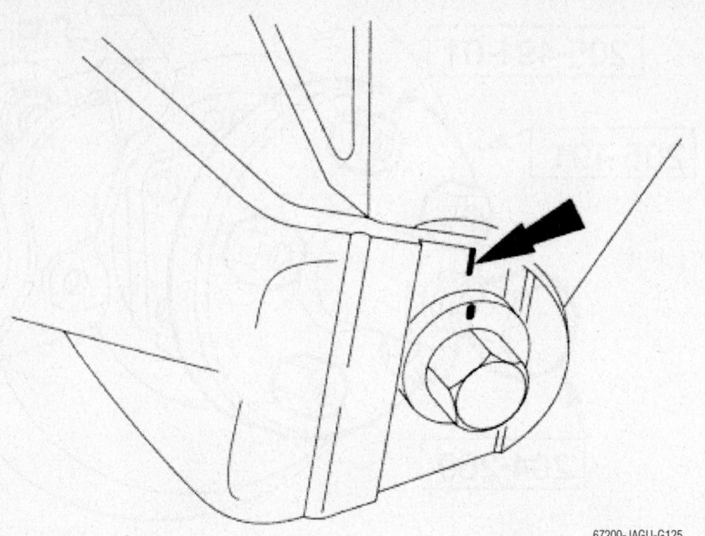

Mark control arm at pivot pin eccentric flange marker to aid installation—2002–03 through VIN F59525

41. Using a drift, remove and discard outer seal.
42. Remove and discard outer bearing.
43. Remove bearing spacer.
44. Remove bearing shim.
45. Remove inner seal and bearing from hub carrier.
46. Using a drift, remove and discard inner seal.
47. Remove and discard inner bearing.
48. Remove inner bearing cup from hub carrier.
49. Position hub carrier on a block of wood.
50. Using a drift, remove and discard inner bearing cup.
51. Remove outer bearing cup from hub carrier.
52. Position hub carrier on a block of wood.
53. Using a drift, remove and discard outer bearing cup.
54. Clean hub and hub carrier.

To install:

55. Place hub carrier on press bed and insert inner bearing cup.
56. Using a hydraulic press, install new inner bearing cup in hub carrier.
57. Using a hydraulic press, install outer bearing cup in hub carrier.
58. Place hub carrier on a work bench.
59. Install outer bearing in hub carrier.
60. Install hub in outer bearing and carrier.
61. Install inner bearing in hub carrier.
62. Install bearing spacer to hub.
63. Install a 3.47 mm (maximum thickness) shim on hub.
64. Install inner bearing in hub carrier.

➡Do not allow press load to be directed on to hub studs.

65. Using a hydraulic press, install ABS-sensor rotor to hub.

➡Do not use excessive force when levering hub.

66. Measure hub end-float.
67. Secure hub carrier in vice.
68. Install a dial indicator on the hub.
69. Using two levers, measure and record hub end-float.
70. Remove special tools from hub carrier.
71. Using hydraulic press, remove hub and outer bearing assembly from carrier.
72. Remove ABS rotor and inner bearing from hub carrier.
73. Remove ABS rotor.
74. Remove inner bearing.
75. Remove outer bearing assembly from hub.

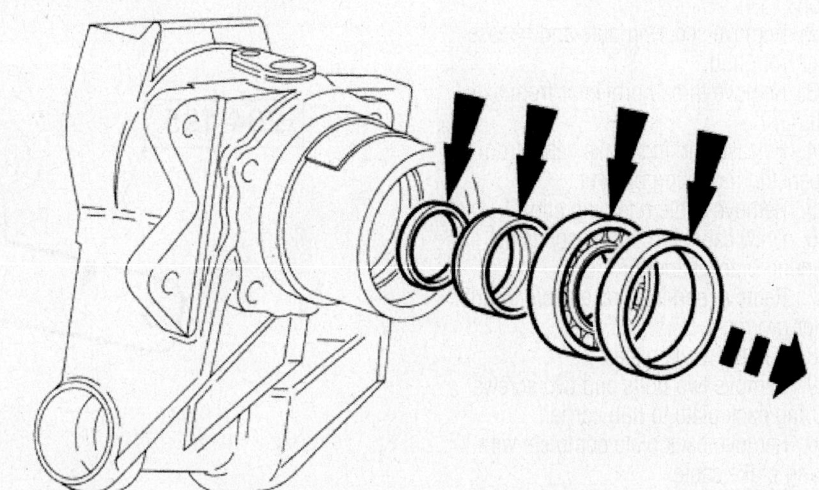

Seal, outer bearing, spacer and shim—2002–03 through VIN F59525

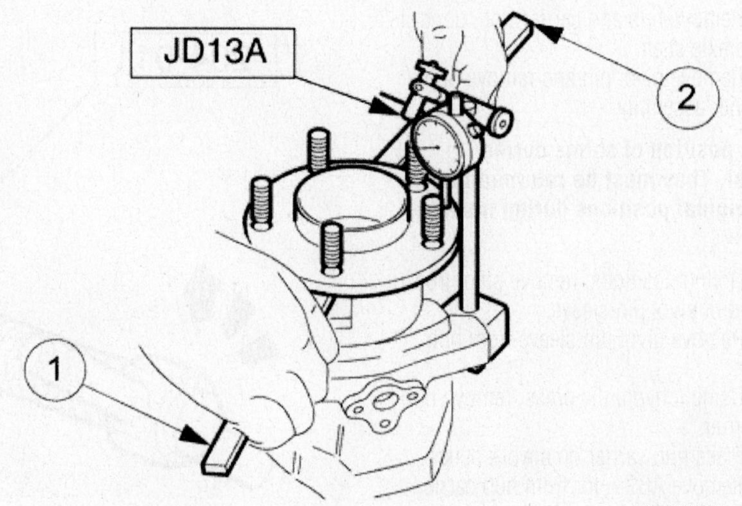

Install a dial indicator on the hub—2002–03 through VIN F59525

76. Remove bearing shim.

77. Remove bearing spacer.

78. Remove outer bearing.

79. Pack new inner and outer bearings with Shell Retinax 'A', or equivalent, grease.

80. Install outer bearing and using special tool 205-234, fully seat seal in hub carrier.

81. Using a press, install hub in carrier/bearing assembly.

82. Install inner bearing assembly.

83. Install bearing spacer.

84. Install bearing shim.

85. Install inner bearing.

86. Determine hub end float. Measure and record end-float as a datum. Calculating from datum, select a shim that will achieve a 0.076 mm end-float.

87. Install new bearing shim.

88. Install selected shim.

89. Install inner bearing.

90. Using special tool 205-234, install inner bearing seal.

91. Place block of wood on hydraulic press-bed and position hub carrier on block.

➡Do not allow press load to be directed on to hub studs.

92. Align ABS rotor to hub and operate press to fully seat rotor on hub.

93. Install pivot pin sleeve in carrier bearings.

94. Install shims on pivot pin sleeve in positions noted during removal.

95. Position hub carrier assembly on control arm.

96. Ensuring shims remain positioned on sleeve, position hub carrier assembly on control arm.

97. Install hub assembly to control arm and axle assembly.

98. Position hub and carrier to control arm and axle shaft.

99. Install pivot pin ensuring that eccentric flange correctly seats in control arm groove.

100. Install washer and nut on pivot pin, but DO NOT tighten nut.

101. Install hub on axle shaft.

102. Route parking brake cable through hub carrier aperture and install grommet.

103. Apply Loctite 270 to axle shaft splines and thread, and fully seat hub on shaft.

104. Install locating collar and hub nut on axle shaft, but DO NOT tighten hub nut.

105. Tighten pivot pin nut.

106. Align marker on pivot pin flange with mark made on control arm.

107. Ensuring pin position is not disturbed, tighten nut to 65–87 ft. lbs. (88–118 Nm).

108. Install wheel speed sensor on hub carrier.

109. Position sensor on hub carrier and install and tighten bolt to 89 inch lbs. (10 Nm).

110. Connect wheel speed sensor harness to sensor.

111. Using new tie strap, secure wheel speed sensor harness to hub carrier.

112. Install handbrake shoes.

113. Install brake disc.

114. Position special tool 204-195 on hub studs and install securing nuts.

115. Tighten hub nut to 225–249 ft. lbs. (304–336 Nm).

116. Remove securing nuts and withdraw special tool 204-195 from hub.

117. Install rear wheel.

118. Remove stands and lower vehicle onto wheels.

119. Raise vehicle on four-post lift.

120. Route parking brake cable through suspension mounting bracket and secure with clip. Route parking brake LH cable through equalizer.

121. Install LH cable in connecting clip.

122. Adjust parking brake cable.

123. Lower four-post lift.

124. Apply parking brake.

125. Carry out rear wheel alignment check and adjust as necessary.

XJ RANGE 2003.5–06 FROM VIN G00442

1. Loosen the wheel hub nut.
2. Remove the brake disc.

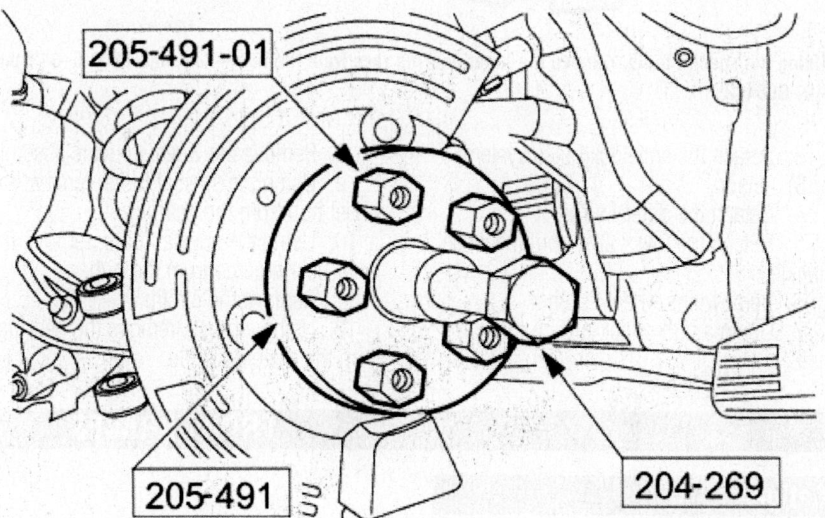

Using a hub puller, detach the halfshaft—XJ range 2003.5–04 from VIN G00442

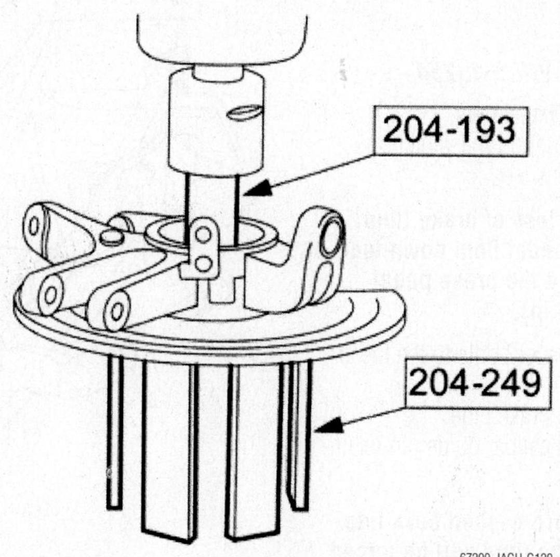

Using the special tools, remove the wheel hub from the bearing—XJ range 2003.5–04 from VIN G00442

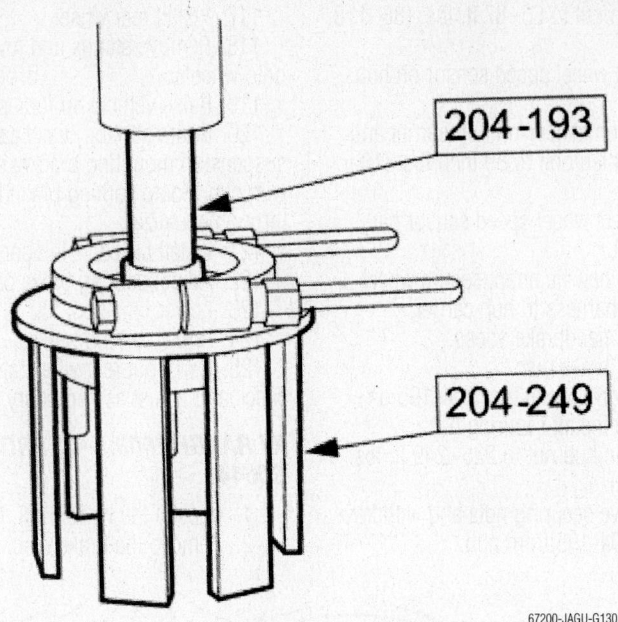

Using the special tools, remove the inner bearing race from the hub—XJ range 2003.5–04 from VIN G00442

3. Detach the anti-lock brake system (ABS) sensor.
4. Detach the outer tie rod.
5. Detach the lower arm from the wheel knuckle.
6. Remove the wheel hub nut.
7. Using a hub puller, detach the half-shaft.
8. Remove the wheel knuckle assembly.
9. Using the special tools, remove the wheel hub from the bearing.
10. Using the special tools, remove the inner bearing race from the hub.
11. Remove the circlip.
12. Using a press, remove the bearing from the wheel knuckle.

To install:

➡The final tightening of the rear suspension components must be carried out with the vehicle on its wheels.

13. Using the special tools, install the bearing to the wheel knuckle.
14. Install the circlip.
15. Using the special tools, install the wheel hub to the bearing.
16. Install the wheel knuckle assembly. Tighten to 67 ft. lbs. (90 Nm).
17. Using the old wheel hub nut tighten to 111 ft. lbs. (150 Nm).
18. Remove and discard the old wheel hub nut.
19. Install a new wheel hub nut tighten to 111 ft. lbs. (150 Nm).
20. Attach the wheel knuckle to the half-shaft.
21. Install the lower arm to the wheel knuckle.
22. Install the outer tie rod retaining bolt.
23. Install the ABS sensor. Tighten to 89 inch lbs. (10 Nm).
24. Install the brake disc.
25. Lower the vehicle.
26. Tighten the wheel hub nut to 222 ft. lbs. (300 Nm).
27. Tighten the lower arm to 111 ft. lbs. (150 Nm).
28. Tighten the tie rod end to 41 ft. lbs. (55 Nm).

FRONT BRAKES

Caliper

REMOVAL & INSTALLATION

S-Type

2002 THROUGH VIN M45254

1. Loosen the wheel nuts.
2. Raise and support the vehicle.
3. Remove the wheel.

➡To prevent the loss of brake fluid, apply the brake pedal hold down tool and adjust to hold the brake pedal down 40mm (1.6 in).

4. Install a bleeder bottle to the bleed nipple and loosen.
5. Remove the brake pipe.
6. Remove the caliper guide pin retaining bolts.

➡If the pistons are pushed back into the housing, brake fluid will be forced out of the inlet port of the brake caliper.

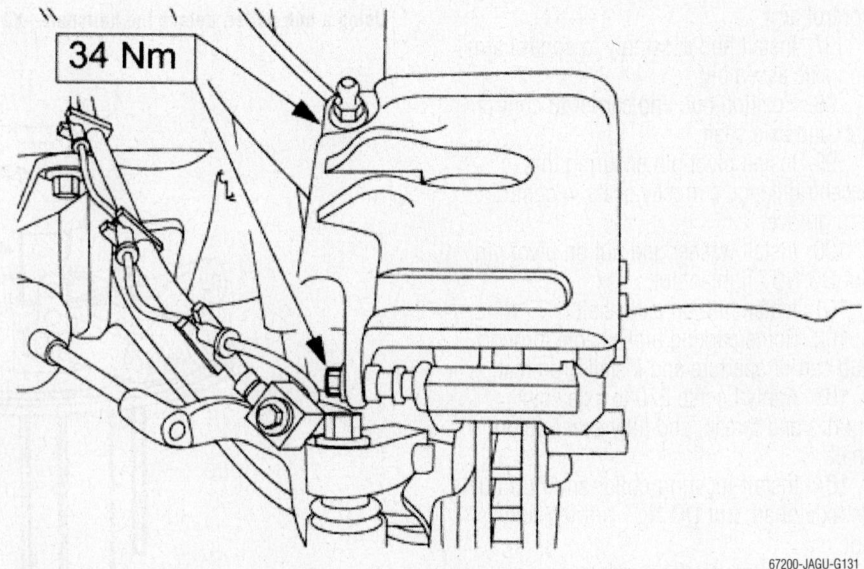

Caliper mounting bolts—S-Type 2002 through VIN M45254

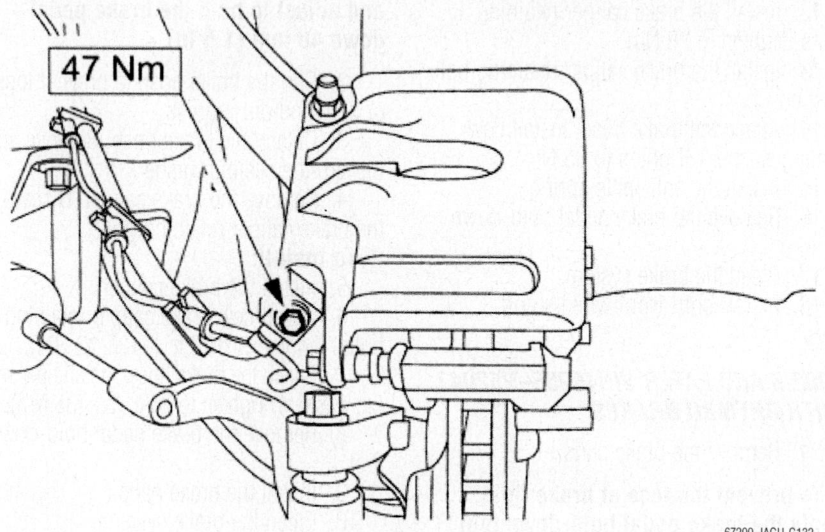

Brake line union bolt—S-Type 2002 through VIN M45254

7. Using the a piston tool fully retract the brake caliper pistons.
8. Remove the caliper.

To install:

9. Install the caliper.
10. Install new caliper guide pin retaining bolts. Torque to 25 ft. lbs. (34 Nm).
11. Install the brake line, using new washers. Torque to 35 ft. lbs. (47 Nm).
12. Bleed the brake system.

2002.5 AND LATER VIN N45255–52047 WITHOUT BREMBO BRAKES

1. Remove the brake pads.

➡**The brake caliper must be supported at all times.**

2. Cap the brake hose to prevent loss of fluid and dirt ingress.
3. To prevent the loss of brake fluid, apply the brake pedal hold-down tool and adjust to hold the brake pedal down 40 mm (1.6 in).
4. Detach the brake hose.
5. Remove and discard the sealing washers.
6. Remove the brake caliper.
7. Remove and discard the brake caliper upper retaining bolt.

To install:

8. Install the brake caliper.
9. Install a new brake caliper upper retaining bolt. Tighten to 25 ft. lbs. (34 Nm).
10. Attach the brake hose. Install new sealing washers. Tighten to 26 ft. lbs. (35 Nm).
11. Remove the brake pedal hold-down tool.
12. Install the brake pads.
13. Bleed the brake system.

2002.5 AND LATER VIN N52048–99999 WITHOUT BREMBO BRAKES

1. Remove both front wheels and tires.
2. Remove the anti-rattle spring.

✲✲ WARNING

Brake fluid contains polyglycol ethers and polyglycols. Avoid contact with the eyes. Wash hands thoroughly after handling, as prolonged contact may cause irritation and dermatitis. If brake fluid contacts the eyes, flush the eyes with cold water or eyewash solution and seek medical attention. If taken internally do not induce vomiting, seek immediate medical attention. Failure to follow these instructions may result in personal injury.

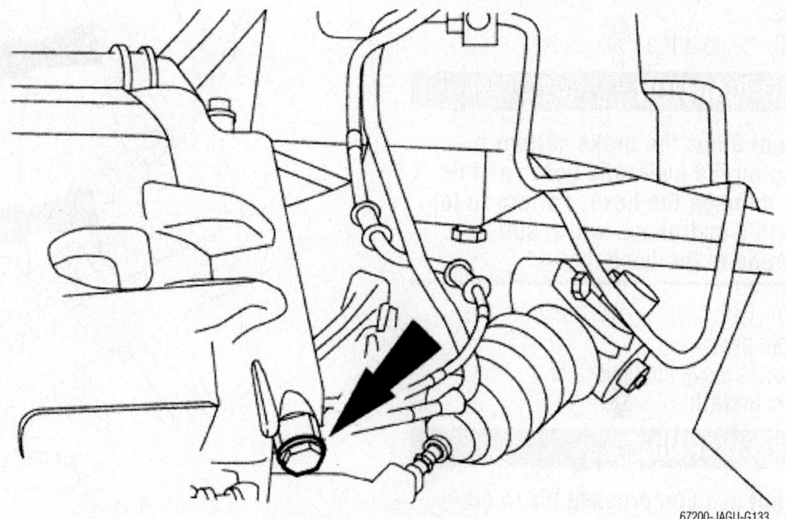

Brake union bolt—S-Type 2002.5 and later VIN 45255–52047 without Brembo brakes

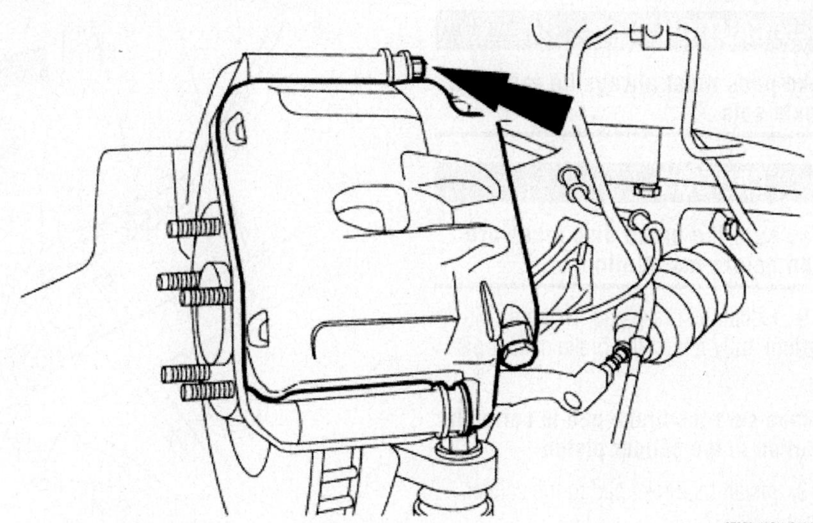

Caliper mounting bolts—S-Type 2002.5 and later VIN 45255–52047 without Brembo brakes

CAUTION

If brake fluid comes into contact with the paintwork, the affected area must be immediately washed down with cold water.

CAUTION

Cap the brake hose to prevent loss of fluid and dirt ingress.

➡To prevent the loss of brake fluid, apply the brake pedal hold-down tool and adjust to hold the brake pedal down 40 mm (1.6 in).

3. Detach the brake hose. Remove and discard the sealing washers.
4. Remove the brake caliper retaining bolt dust covers.
5. Remove the brake caliper retaining bolts.
6. Remove the brake caliper.

CAUTION

Do not allow the brake caliper to hang on the hydraulic hose, as this will damage the hose. Failure to follow this instruction will result in damage to the hose.

7. Remove the brake pad from the caliper piston.
8. Remove the brake pad.

To install:

WARNING

Do not use compressed air to disperse brake dust into the atmosphere.

CAUTION

Brake pads must always be replaced in axle sets.

CAUTION

Make sure the brake disc faces are clean before installation.

9. Install the brake pad. Using a suitable tool, fully retract the brake caliper piston.

➡Make sure the brake pad is correctly installed to the caliper piston.

10. Install the brake pad to the caliper piston.
11. Install the brake caliper.
12. Install the brake caliper retaining bolts. Tighten to 28 Nm.
13. Install the brake caliper retaining bolt dust covers.
14. Attach the brake hose. Install new sealing washers. Tighten to 35 Nm.
15. Install the anti-rattle spring.
16. Remove the brake pedal hold-down tool.
17. Bleed the brake system.
18. Install both front wheels and tires.

2002.5 AND LATER VIN45255–N52047 WITH BREMBO BRAKES

1. Remove the brake pads.

➡To prevent the loss of brake fluid, apply the brake pedal hold-down tool and adjust to hold the brake pedal down 40 mm (1.6 in).

2. Cap the brake hose to prevent loss of fluid and dirt ingress.
3. Detach the brake hose. Remove and discard the sealing washers.
4. Remove the brake caliper. Discard the brake caliper retaining bolts.

To install:

5. Install the brake caliper.
6. Install new brake caliper retaining bolts. Tighten to 157 ft. lbs. (212 Nm).
7. Attach the brake hose. Install new sealing washers. Tighten to 26 ft. lbs. (35 Nm).
8. Remove the brake pedal hold-down tool.
9. Install the brake pads.
10. Bleed the brake system.

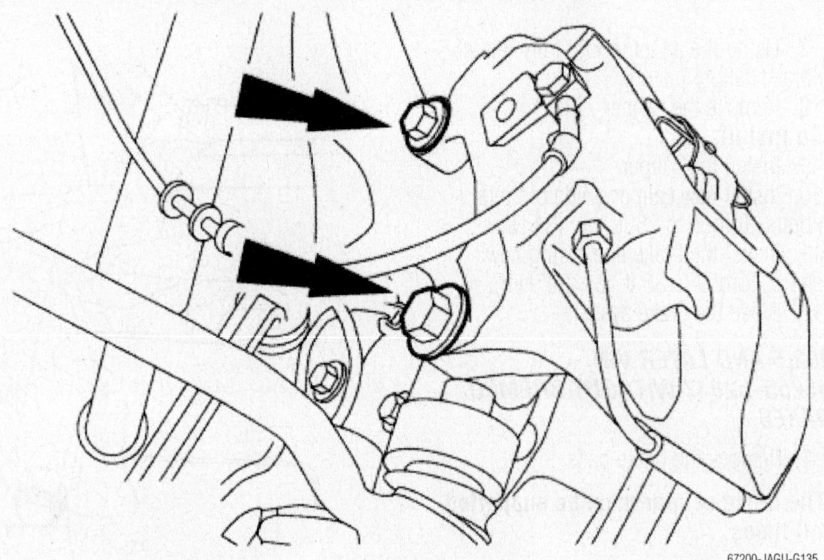

Front caliper bolts—S-Type 2002.5 and later VIN45255–N52047 with Brembo brakes

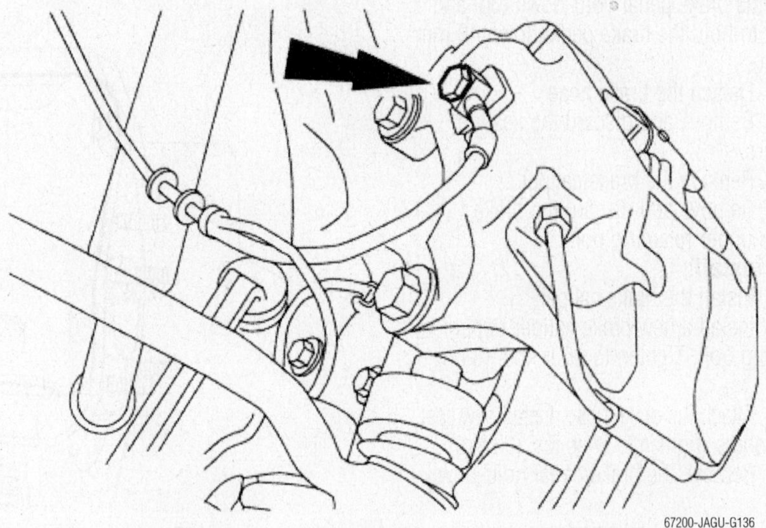

Brake hose union bolt—S-Type 2002.5 and later VIN45255–N52047 with Brembo brakes

2002.5 AND LATER VIN N52048–99999 WITH BREMBO BRAKES

1. Remove both front wheels and tires.
2. Remove the anti-rattle spring.

WARNING

Brake fluid contains polyglycol ethers and polyglycols. Avoid contact with the eyes. Wash hands thoroughly after handling, as prolonged contact may cause irritation and dermatitis. If brake fluid contacts the eyes, flush the eyes with cold water or eyewash solution and seek medical attention. If taken internally do not induce vomiting, seek immediate medical attention. Failure to follow these instructions may result in personal injury.

CAUTION

If brake fluid comes into contact with the paintwork, the affected area must be immediately washed down with cold water.

CAUTION

Cap the brake hose to prevent loss of fluid and dirt ingress.

➡To prevent the loss of brake fluid, apply the brake pedal hold-down tool and adjust to hold the brake pedal down 40 mm (1.6 in).

3. Detach the brake hose. Remove and discard the sealing washers.
4. Remove the brake caliper retaining bolt dust covers.
5. Remove the brake caliper retaining bolts.
6. Remove the brake caliper.

CAUTION

Do not allow the brake caliper to hang on the hydraulic hose, as this will damage the hose. Failure to follow this instruction will result in damage to the hose.

7. Remove the brake pad from the caliper piston.
8. Remove the brake pad.

To install:

WARNING

Do not use compressed air to disperse brake dust into the atmosphere.

CAUTION

Brake pads must always be replaced in axle sets.

CAUTION

Make sure the brake disc faces are clean before installation.

9. Install the brake pad. Using a suitable tool, fully retract the brake caliper piston.

➡Make sure the brake pad is correctly installed to the caliper piston.

10. Install the brake pad to the caliper piston.
11. Install the brake caliper.
12. Install the brake caliper retaining bolts. Tighten to 58 Nm.
13. Install the brake caliper retaining bolt dust covers.
14. Attach the brake hose. Install new sealing washers. Tighten to 35 Nm.
15. Install the anti-rattle spring.
16. Remove the brake pedal hold-down tool.
17. Bleed the brake system.
18. Install both front wheels and tires.

X-Type

1. Remove the wheel and tire.
2. To prevent the loss of brake fluid, apply the brake pedal hold down tool and adjust to hold the brake pedal down 40mm.
3. Install a bleeder bottle to the bleed nipple and loosen.

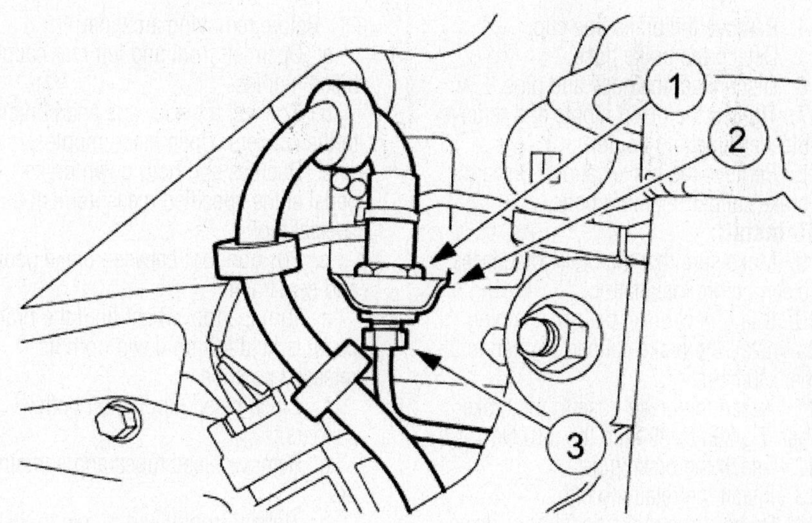

Brake hose retaining clip—X-Type

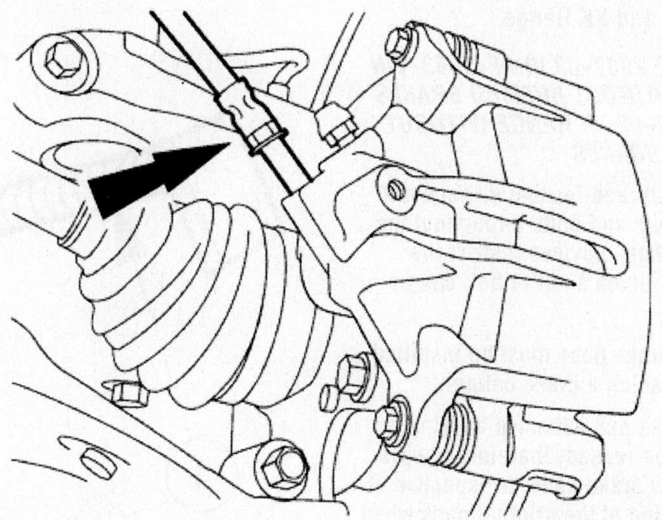

Brake hose attachment—X-Type

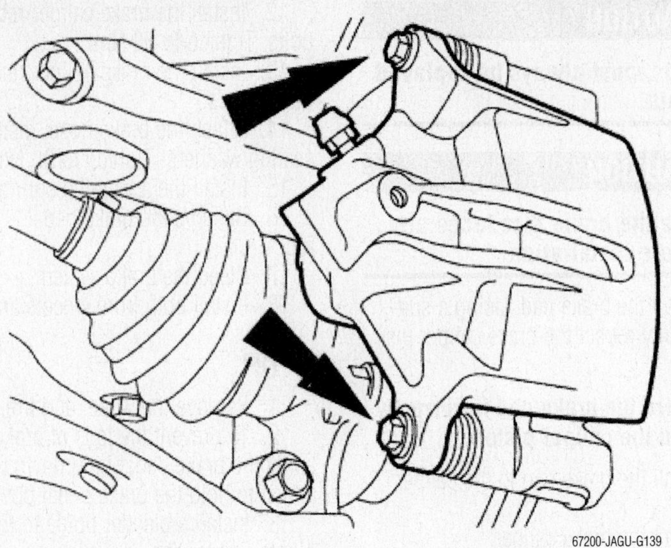

Brake caliper bolts—X-Type

4. Remove the brake line clip.
5. Detach the brake hose.
6. Disconnect the hose and pipe.
7. Tighten the bleed nipple and remove the bleeder tube and bottle.
8. Remove the brake caliper. Discard the brake caliper retaining bolts.

To install:

9. Make sure that the brake disc faces are clean before installation.
10. Using new brake caliper retaining bolts, install the brake caliper. Tighten to 22 ft. lbs. (30 Nm).
11. Attach the brake hose to the brake caliper. Tighten to 89 inch lbs. (10 Nm).
12. Attach the brake pipe.
13. Install the retaining clip.
14. Remove the brake pedal hold-down tool.
15. Bleed the brake system.

XJ Range and XK Range

XJ RANGE 2002–03 VIN F41863–VIN F59525 WITHOUT BREMBO BRAKES AND 2002–06 XK RANGE WITHOUT BREMBO BRAKES

➡Various thread-locking devices are used on nuts and bolts throughout the vehicle. These devices restrict the number of times a nut or bolt can be used.

➡A new brake hose must be installed when replacing a brake caliper.

➡The brake pad adhesive backing must not be reused, therefore always install new brake pads, irrespective of the condition of the original pads when replacing a brake caliper. Always replace brake pads as an axle set.

1. Before removing a caliper:
   a. Open left front and left rear caliper bleed nipples.
   b. Connect a bleed tube and container to the calipers. Open bleed nipples.
   c. Depress and hold down brake pedal at the specified measurement using special tool.
   d. Position tool between brake pedal and seat frame.
   e. Turn the tool wheel until the brake pedal is held 60mm down from its released position.
   f. Close bleed nipples on both calipers.
   g. Remove bleed tubes and containers.
   h. Tighten front bleed nipple to 36–53 inch lbs. (4–6 Nm). Tighten rear bleed nipple to 71–97 inch lbs. (8–11 Nm). A new brake hose must be installed when replacing a brake caliper.
2. Disconnect brake hose from brake tube.
3. Unscrew brake tube union nut.
4. Remove nut from hose.
5. Remove hose from bracket.
6. Remove brake hose from caliper. Discard brake hose. Plug the caliper port.
7. Remove caliper housing from caliper carrier.
8. Remove anti-rattle spring.
9. Remove dust covers.
10. Remove guide pins.
11. Remove housing from carrier.

➡The inner brake pad is fitted with a clip which secures the pad into the caliper piston.

12. Remove and discard brake pads.
13. Remove brake carrier from vertical link.
14. Remove bolts.
15. Remove carrier.
16. Remove brake dust and clean caliper mating surfaces.

To install:

17. Align carrier to vertical link. Install and tighten bolts to 104 ft. lbs. (140 Nm).
18. Install brake pads into caliper housing.
19. Remove plug from port and slowly retract piston into caliper housing.
20. Remove paper backing from pads to reveal adhesive shim.
21. Install brake pads, insert clip located on back of inner pad into caliper piston.
22. Install caliper housing to caliper carrier.
23. Align caliper to carrier. Install and tighten guide pins to 18–22 ft. lbs. (25–30 Nm).

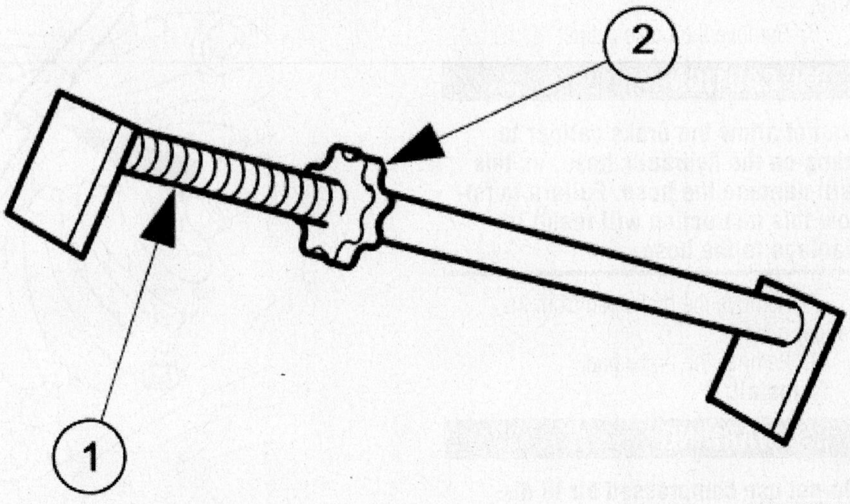

Brake pedal holding tool

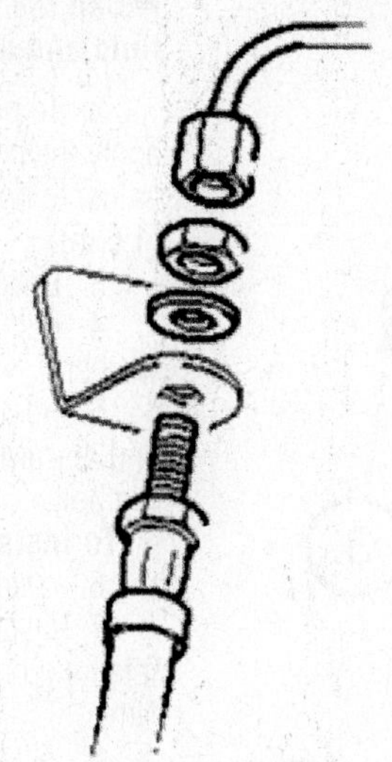

Disconnect brake hose from brake tube—XJ Range 2002–03 VIN F41863–VIN F59525 without Brembo brakes; 2002–06 XK Range without Brembo brakes

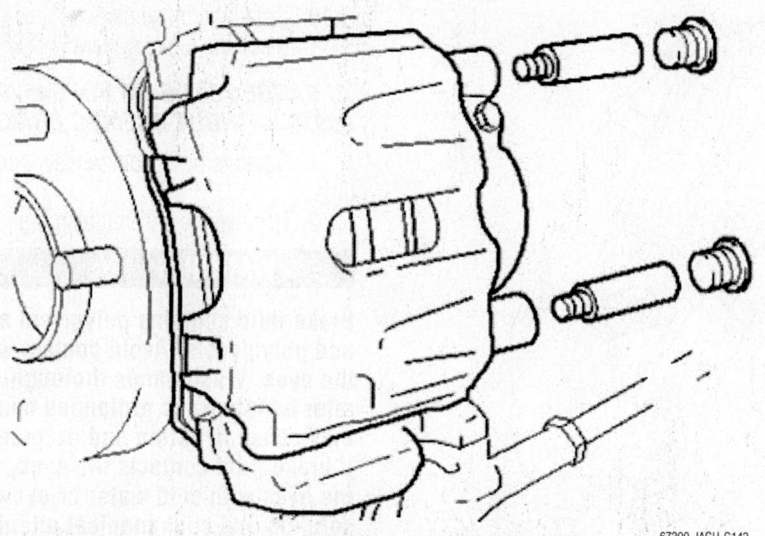

Brake caliper and related parts—XJ Range 2002–03 VIN F41863–VIN F59525 without Brembo brakes; 2002–06 XK Range without Brembo brakes

24. Install dust covers.
25. Install anti-rattle spring.

➡The brake hose must only be installed with the vehicle's body supported and the suspension hanging.

26. Install and tighten brake hose to caliper housing to 10 ft. lbs. (14 Nm).

➡Make sure the brake hose does not twist when tightening the hose locknut.

27. Connect brake hose to brake tube.
28. Position hose thread through bracket.
29. Install and tighten locking nut.
30. Install and tighten brake-tube union nut.
31. Remove special tool from brake pedal.
32. Install wheels.

➡Make sure the brake hose is routed correctly.

33. Remove stands and lower vehicle.
34. Bleed brake hydraulic system.

XJ RANGE 2002–03 VIN F41863–VIN F59525 WITH BREMBO BRAKES

XK RANGE 2002–04 WITH BREMBO BRAKES

➡Various thread-locking devices are used on nuts and bolts throughout the vehicle. These devices restrict the number of times a nut or bolt can be used.

⁂ WARNING

The Brembo caliper is aligned to the brake disc when it is first installed to the vehicle, therefore, care must be taken not to disturb this alignment. When removing the caliper; remove the bolts that secure the anchor bracket to the vertical link only. DO NOT loosen any other caliper bolts.

1. Raise front of vehicle and support on stands.
2. Remove front wheels.
3. Open left front and left rear caliper bleed nipples.
4. Connect a bleed tube and container to the calipers.
5. Open bleed nipples.
6. Depress and hold down brake pedal at the specified measurement using the special tool. Position tool between brake pedal and seat frame. Turn the tool wheel until the brake pedal is held 60mm down from its released position.
7. Close bleed nipples on both calipers. Tighten bleed nipples to 9–12 ft. lbs. (12–16 Nm).
8. Remove bleed tubes and containers.
9. Disconnect brake hose from brake tube. Unscrew brake tube union nut.
10. Remove nut from hose.
11. Remove hose from bracket.
12. Remove union bolt. Release brake hose from caliper. Remove and discard sealing washers.
13. Install plugs into caliper and hose.
14. Remove R-clips. Remove pins from caliper.
15. Remove anti-rattle spring.
16. Remove brake pads.

⁂ WARNING

Only remove the bolts securing the caliper anchor bracket to the vertical link.

17. Remove bolts. Remove caliper from vertical link.
18. Remove brake dust and clean mating surfaces.

To install:

19. Install caliper to vertical link. Install and tighten bolts to 133 ft. lbs. (180 Nm).

➡Retracting the caliper piston may cause the fluid reservoir to overflow. Remove brake fluid spillage immediately from paint work, with clean water.

➡Note the condition of pins and R-clips and replace if necessary.

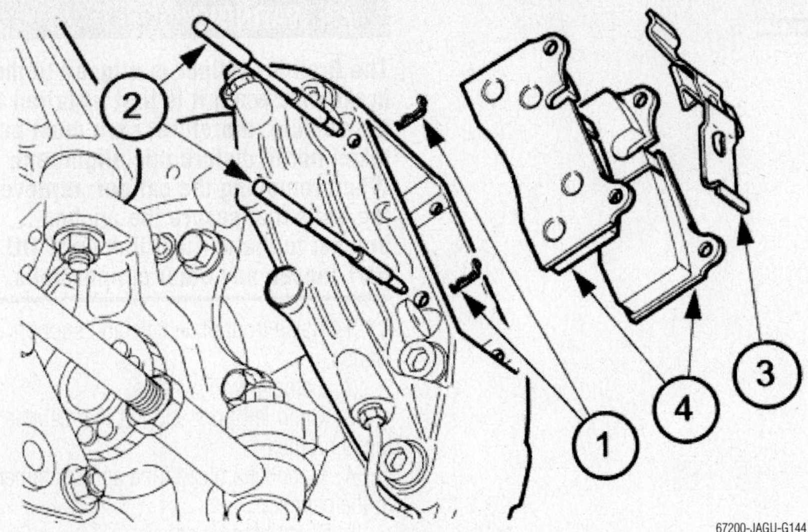

Brake pads and related parts. (1) R-clips (2) pins (3) anti-rattle spring (4) pads—XJ Range 2002–03 VIN F41863–VIN F59525 with Brembo brakes; XK Range 2002–04 with Brembo brakes

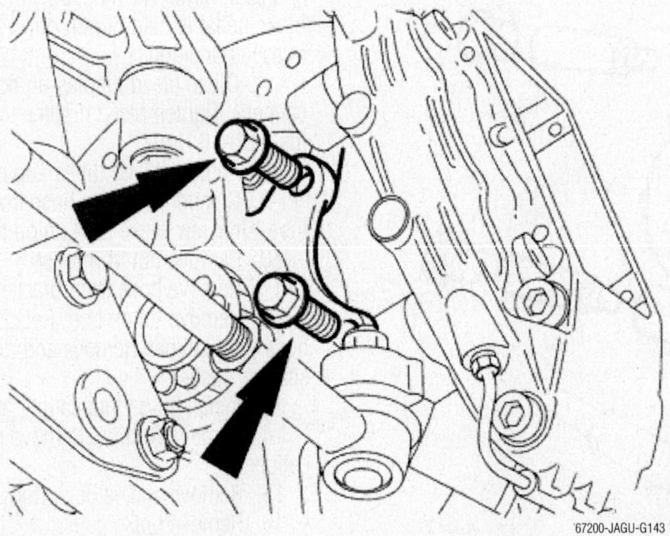

Remove only these bolts—XJ Range 2002–03 VIN F41863–VIN F59525 with Brembo brakes; XK Range 2002–04 with Brembo brakes

20. Install brake pads into caliper.
21. Slowly retract caliper pistons.
22. Install pads.
23. Position anti-rattle spring. Install pins. Install R-clips.

➡**The brake hose must only be installed with the vehicle's body supported and the suspension hanging.**

24. Using new sealing washers, install and tighten union bolt to 16–19 ft. lbs. (22–26 Nm). Make sure the brake hose does not twist when tightening the hose lock-nut.
25. Connect brake hose to brake tube. Position hose thread through bracket. Install and tighten locking nut to 11–15 ft. lbs. (15–20 Nm.)
26. Install and tighten brake-tube union nut to 16–19 ft. lbs. (22–26 Nm).
27. Remove special tool from brake pedal.
28. Install wheels.

➡**Make sure the brake hose is routed correctly.**

29. Remove stands and lower vehicle.
30. Bleed brake hydraulic system.

XJ RANGE 2003.5–06 VIN G00442–45703 WITHOUT BREMBO BRAKES

1. Remove the brake pads.

➡**The brake caliper must be supported at all times.**

➡**Cap the brake hose to prevent loss of fluid and dirt ingress.**

2. To prevent the loss of brake fluid, apply the brake pedal hold-down tool and adjust to hold the brake pedal down 40 mm (1.6 in).
3. Detach the brake hose.
4. Remove and discard the sealing washers.
5. Remove the brake caliper. Remove and discard the brake caliper upper retaining bolt.

To install:

6. Install the brake caliper.
7. Install a new brake caliper upper retaining bolt. Tighten to 25 ft. lbs. (34 Nm).
8. Attach the brake hose. Install new sealing washers. Tighten to 26 ft. lbs. (35 Nm).
9. Remove the brake pedal hold-down tool.
10. Install the brake pads.
11. Bleed the brake system.

XJ RANGE 2003.5–06 VIN G45703–99999 WITHOUT BREMBO BRAKES

1. Remove both front wheels and tires.
2. Remove the anti-rattle spring.

⁂ WARNING

Brake fluid contains polyglycol ethers and polyglycols. Avoid contact with the eyes. Wash hands thoroughly after handling, as prolonged contact may cause irritation and dermatitis. If brake fluid contacts the eyes, flush the eyes with cold water or eyewash solution and seek immediate medical attention. If taken internally do not induce vomiting, seek immediate medical attention. Failure to follow these instructions may result in personal injury.

3. If brake fluid comes into contact with the paintwork, the affected area must be immediately washed down with cold water.
4. Cap the brake hose to prevent loss of fluid and dirt ingress.

➡**To prevent the loss of brake fluid, apply the brake pedal hold-down tool and adjust to hold the brake pedal down 40 mm (1.6 in).**

5. Detach the brake hose. Remove and discard the sealing washers.
6. Remove the brake caliper retaining bolt dust covers.

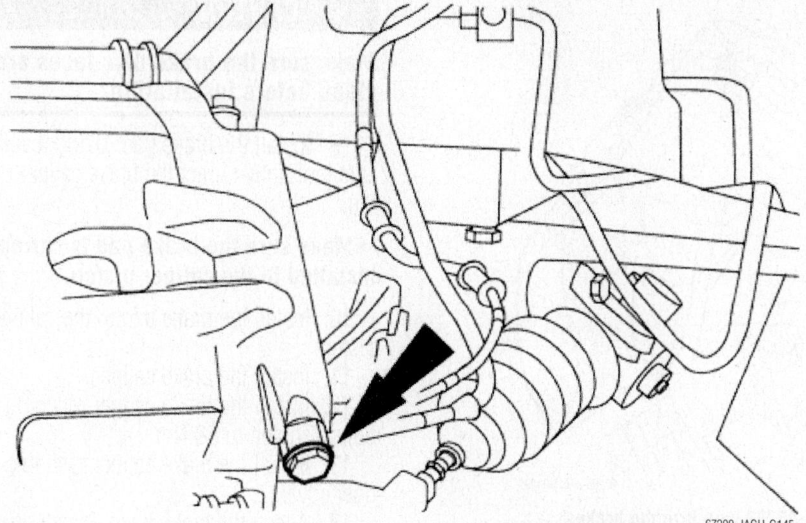

Brake line union bolt—XJ Range 2003.5–04 from VIN G00442 without Brembo brakes

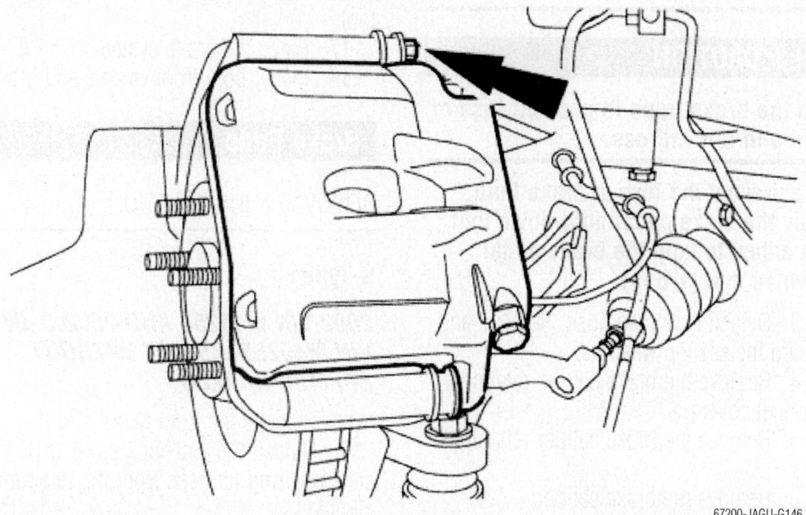

Brake caliper upper bolt—XJ Range 2003.5–04 from VIN G00442 without Brembo brakes

7. Remove the brake caliper retaining bolts.
8. Remove the brake caliper.

⁂ CAUTION

Do not allow the brake caliper to hang on the hydraulic hose, as this will damage the hose. Failure to follow this instruction will result in damage to the hose.

9. Remove the brake pad from the caliper piston.
10. Remove the brake pad.

To install:

⁂ WARNING

Do not use compressed air to disperse brake dust into the atmosphere.

⁂ CAUTION

Brake pads must always be replaced in axle sets.

⁂ CAUTION

Make sure the brake disc faces are clean before installation.

11. Install the brake pad. Using a suitable tool, fully retract the brake caliper piston.

➡Make sure the brake pad is correctly installed to the caliper piston.

12. Install the brake pad to the caliper piston.
13. Install the brake caliper.
14. Install the brake caliper retaining bolts. Tighten to 28 Nm.
15. Install the brake caliper retaining bolt dust covers.
16. Attach the brake hose. Install new sealing washers. Tighten to 35 Nm.
17. Install the anti-rattle spring.
18. Remove the brake pedal hold-down tool.
19. Bleed the brake system.
20. Install both front wheels and tires.

XJ RANGE 2003.5–06 VIN G00442–45703 WITH BREMBO BRAKES

1. Remove the brake pads.
2. To prevent the loss of brake fluid, apply the brake pedal hold-down tool and adjust to hold the brake pedal down 40 mm (1.6 in).
3. Cap the brake hose to prevent loss of fluid and prevent dirt ingress.

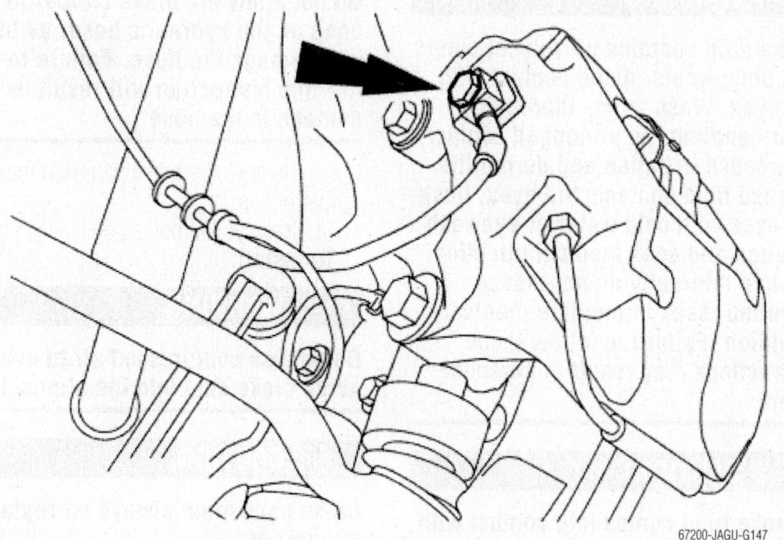

Brake line union bolt—XJ Range 2003.5–06 VIN G00442–45703 with Brembo brakes

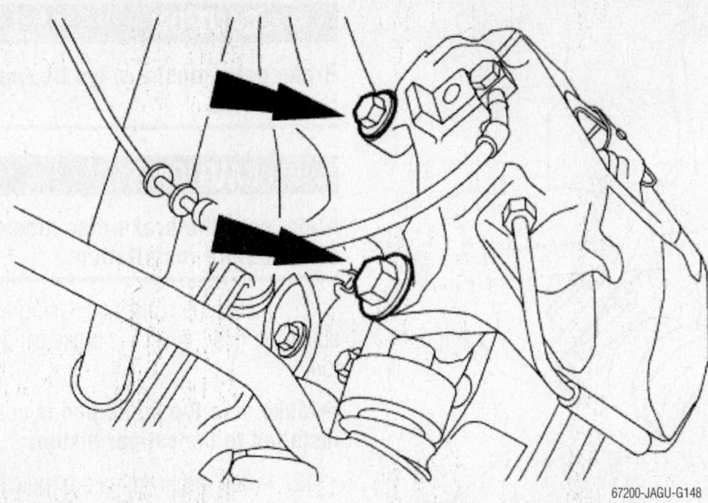

Brake caliper bolts—XJ Range 2003.5-06 VIN G00442-45703 with Brembo brakes

4. Detach the brake hose. Remove and discard the sealing washers.
5. Remove the brake caliper. Discard the brake caliper retaining bolts.

To install:

6. Install the brake caliper.
7. Install new brake caliper retaining bolts. Tighten to 157 ft. lbs. (212 Nm).
8. Attach the brake hose. Install new sealing washers. Tighten to 26 ft. lbs. (35 Nm).
9. Remove the brake pedal hold-down tool.
10. Install the brake pads.
11. Bleed the brake system.

XJ RANGE 2003.5–06 VIN G457043–99999 WITH BREMBO BRAKES

1. Remove both front wheels and tires.
2. Remove the anti-rattle spring.

WARNING

Brake fluid contains polyglycol ethers and polyglycols. Avoid contact with the eyes. Wash hands thoroughly after handling, as prolonged contact may cause irritation and dermatitis. If brake fluid contacts the eyes, flush the eyes with cold water or eyewash solution and seek medical attention. If taken internally do not induce vomiting, seek immediate medical attention. Failure to follow these instructions may result in personal injury.

CAUTION

If brake fluid comes into contact with the paintwork, the affected area must be immediately washed down with cold water.

CAUTION

Cap the brake hose to prevent loss of fluid and dirt ingress.

➡**To prevent the loss of brake fluid, apply the brake pedal hold-down tool and adjust to hold the brake pedal down 40 mm (1.6 in).**

3. Detach the brake hose. Remove and discard the sealing washers.
4. Remove the brake caliper retaining bolt dust covers.
5. Remove the brake caliper retaining bolts.
6. Remove the brake caliper.

CAUTION

Do not allow the brake caliper to hang on the hydraulic hose, as this will damage the hose. Failure to follow this instruction will result in damage to the hose.

7. Remove the brake pad from the caliper piston.
8. Remove the brake pad.

To install:

WARNING

Do not use compressed air to disperse brake dust into the atmosphere.

CAUTION

Brake pads must always be replaced in axle sets.

CAUTION

Make sure the brake disc faces are clean before installation.

9. Install the brake pad. Using a suitable tool, fully retract the brake caliper piston.

➡**Make sure the brake pad is correctly installed to the caliper piston.**

10. Install the brake pad to the caliper piston.
11. Install the brake caliper.
12. Install the brake caliper retaining bolts. Tighten to 58 Nm.
13. Install the brake caliper retaining bolt dust covers.
14. Attach the brake hose. Install new sealing washers. Tighten to 35 Nm.
15. Install the anti-rattle spring.
16. Remove the brake pedal hold-down tool.
17. Bleed the brake system.
18. Install both front wheels and tires.

Brake Pads

REMOVAL & INSTALLATION

S-Type

2002 VIN M45254 AND 2002.5–06 VIN M45255–N52047 WITHOUT BREMBO BRAKES

1. Remove the front wheel and tire.
2. Detach the anti-lock brake (ABS) sensor wiring harness from the retaining clips.
3. Remove and discard the brake caliper lower retaining bolt.
4. Pivot the brake caliper upwards.
5. Remove the brake pads.

➡**If the pistons are pushed back into the housing, brake fluid will be forced back into the fluid reservoir.**

6. Make sure that the brake disc faces are clean before installation.
7. To install, reverse the removal procedure. Install a new brake caliper lower retaining bolt. Tighten to 25 ft. lbs. (34 Nm).

2002.5–06 VIN N52048–99999 WITHOUT BREMBO BRAKES

1. Remove both front wheels and tires.
2. Remove the anti-rattle spring.
3. Remove the brake caliper retaining bolt dust covers.

CAUTION

Do not allow the brake caliper to hang on the hydraulic hose, as this will damage the hose. Failure to follow this instruction will result in damage to the hose.

4. Remove the brake caliper retaining bolts.

CAUTION

Do not allow the brake caliper to hang on the hydraulic hose, as this will damage the hose. Failure to follow this instruction will result in damage to the hose.

5. Detach the brake caliper.

CAUTION

Do not allow the brake caliper to hang on the hydraulic hose, as this will damage the hose. Failure to follow this instruction will result in damage to the hose.

6. Remove the brake pad.
7. Remove the brake pad from the caliper piston.

To install:

WARNING

Do not use compressed air to disperse brake dust into the atmosphere.

CAUTION

If the pistons are pushed back into the housing, brake fluid will be forced back into the brake fluid reservoir.

CAUTION

Make sure the brake disc faces are clean before installation.

CAUTION

Brake pads must always be replaced in axle sets.

➡Make sure that the inboard brake pad is correctly installed to the caliper piston.

8. To install, reverse the removal procedure. Using a suitable tool, fully retract the brake caliper pistons. Tighten to 28 Nm.

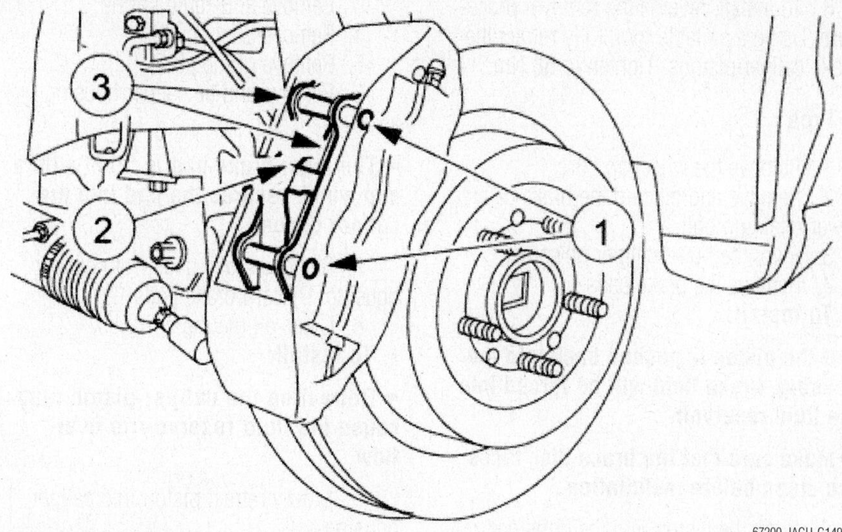

Brake pads and related parts (1) pins (2) anti-rattle spring (3) pads—S-Type 2002.5–04 from VIN M45255 with Brembo brakes

2002.5–04 FROM VIN M45255 WITH BREMBO BRAKES

1. Remove both front wheels and tires.
2. Remove the brake pad retaining pins.
3. Remove the brake pad anti-rattle spring plate.
4. Remove the brake pads.

➡If the pistons are pushed back into the housing, brake fluid will be forced back into the brake fluid reservoir.

5. Make sure the brake disc faces are clean before installation.
6. Brake pads must always be replaced in axle sets.
7. To install, reverse the removal procedure.
8. Using a suitable tool, fully retract the brake caliper pistons.

2002.5–06 VIN N52048–99999 WITH BREMBO BRAKES

1. Remove both front wheels and tires.
2. Remove the anti-rattle spring.
3. Remove the brake caliper retaining bolt dust covers.
4. Remove the brake caliper retaining bolts.

CAUTION

Do not allow the brake caliper to hang on the hydraulic hose, as this will damage the hose. Failure to follow this instruction will result in damage to the hose.

5. Detach the brake caliper.

CAUTION

Do not allow the brake caliper to hang on the hydraulic hose, as this will damage the hose. Failure to follow this instruction will result in damage to the hose.

6. Remove the brake pad from the caliper piston.
7. Remove the brake pad.

To install:

WARNING

Do not use compressed air to disperse brake dust into the atmosphere.

CAUTION

If the pistons are pushed back into the housing, brake fluid will be forced back into the brake fluid reservoir.

CAUTION

Make sure the brake disc faces are clean before installation.

CAUTION

Brake pads must always be replaced in axle sets.

➡Make sure that the inboard brake pad is correctly installed to the caliper piston.

8. To install, reverse the removal procedure. Using a suitable tool, fully retract the brake caliper pistons. Tighten to 58 Nm.

X-Type

1. Remove the wheel and tire.
2. Remove and discard the brake caliper lower retaining bolt.
3. Pivot the brake caliper upwards.
4. Remove the brake pads.

To install:

➡If the piston is pushed back into the housing, brake fluid will be forced into the fluid reservoir.

➡Make sure that the brake disc faces are clean before installation.

5. Push the piston into the housing.
6. Install the brake pads.
7. Pivot the brake caliper downwards.
8. Install a new brake caliper lower retaining bolt. Tighten to 22 ft. lbs. (30 Nm).
9. Install the wheel.

XJ 2002–03 through VIN F59525 without Brembo Brakes and 2002–06 XK Without Brembo Brakes

1. Raise front of vehicle and support on stands.
2. Loosen brake fluid reservoir-cap. Position a cloth around the reservoir to collect any fluid spillage.

➡Do not allow the caliper housing to hang on the hydraulic hose, as this will damage the hose.

3. Remove anti-rattle spring.
4. Remove dust covers.
5. Remove guide pins.
6. Remove and tie caliper housing aside.

➡The inner brake pad is fitted with a clip which secures the pad into the caliper piston.

7. Remove brake pads from caliper housing. Discard brake pads.
8. Clean all mating surfaces.

To install:

➡Retracting the caliper piston may cause the fluid reservoir to overflow.

9. Slowly retract piston into caliper housing.
10. Remove paper backing from pads to reveal adhesive shim.
11. Install brake pads, insert clip located on back of inner pad into caliper piston.
12. Install caliper housing to caliper carrier.
13. Install and tighten guide pins to 18–22 ft. lbs. (25–30 Nm).
14. Install dust covers.
15. Install anti-rattle spring.
16. Install wheels.
17. Remove stands and lower vehicle.
18. Check brake fluid level is at the maximum mark.
19. Start the engine and repeatedly press the brake pedal until brake pressure is evident.

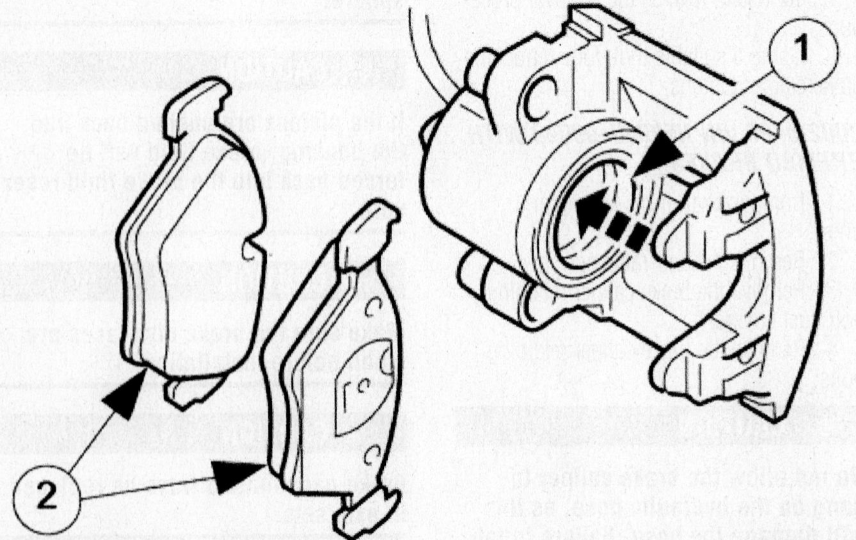

Brake pads—XJ Range 2002–03 through VIN F59525 without Brembo brakes; XK Range without Brembo brakes

XJ Range 2003.5–06 VIN G00442–G45703 Without Brembo Brakes

1. Remove both front wheels and tires.
2. Detach the anti-lock brake (ABS) sensor wiring harness from the retaining clips.
3. Remove and discard the brake caliper lower retaining bolt.
4. Rotate the brake caliper upwards.
5. Remove the brake pads.

➡If the pistons are pushed back into the housing, brake fluid will be forced back into the brake fluid reservoir.

6. Make sure the brake disc faces are clean before installation.

✲✲ CAUTION

Brake pads must always be replaced in axle sets.

7. To install, reverse the removal procedure.
8. Using a suitable tool, fully retract the brake caliper pistons.
9. Install a new brake caliper lower retaining bolt. Tighten to 25 ft. lbs. (34 Nm).

XJ Range 2003.5–06 VIN G45704–G99999 Without Brembo Brakes

1. Remove both front wheels and tires.
2. Remove the anti-rattle spring.
3. Remove the brake caliper retaining bolt dust covers.

✲✲ CAUTION

Do not allow the brake caliper to hang on the hydraulic hose, as this will damage the hose. Failure to follow this instruction will result in damage to the hose.

4. Remove the brake caliper retaining bolts.

✲✲ CAUTION

Do not allow the brake caliper to hang on the hydraulic hose, as this will damage the hose. Failure to follow this instruction will result in damage to the hose.

5. Detach the brake caliper.

✲✲ CAUTION

Do not allow the brake caliper to hang on the hydraulic hose, as this will damage the hose. Failure to follow this instruction will result in damage to the hose.

6. Remove the brake pad.
7. Remove the brake pad from the caliper piston.

To install:

⁂ WARNING

Do not use compressed air to disperse brake dust into the atmosphere.

➡If the pistons are pushed back into the housing, brake fluid will be forced back into the brake fluid reservoir.

8. Make sure the brake disc faces are clean before installation.
9. Brake pads must always be replaced in axle sets.

➡Make sure that the inboard brake pad is correctly installed to the caliper piston.

10. To install, reverse the removal procedure. Using a suitable tool, fully retract the brake caliper pistons. Tighten to 28 Nm.

2002 XJ Range, 2003–06 XJ VIN G00442–G45703 and XK Range With Brembo Brakes

➡The Brembo caliper is aligned to the brake disc when it is first installed to the vehicle, therefore, care must be taken not to disturb this alignment.

⁂ WARNING

When removing the caliper; remove the bolts that secure the anchor bracket to the vertical link only. DO NOT loosen any other caliper bolts.

1. Remove R-clips. Remove pins from caliper.
2. Remove anti-rattle spring.
3. Discard brake pads.
4. Clean all mating surfaces.
5. Note condition of pins and R-clips and replace if necessary.

To install:

6. Install brake pads into caliper.
7. Slowly retract caliper pistons into caliper.
8. Install pads, making sure that arrow faces the direction of rotation.
9. Position anti-rattle spring.
10. Install pins.
11. Install R-clips.

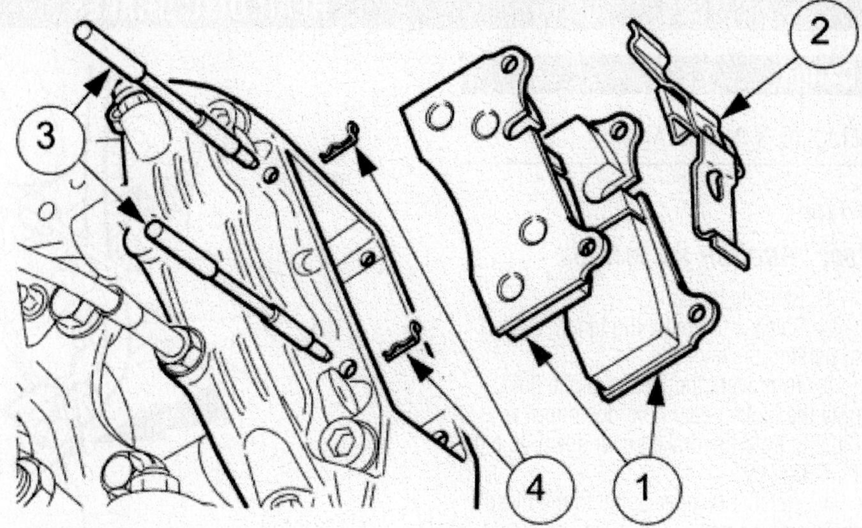

Brake pads and related parts—XJ Range and XK Range with Brembo brakes. (1) pads (2) anti-rattle spring (3) pins (4) R-clips

12. Install wheels.
13. Remove stands and lower vehicle.
14. Check brake fluid level is at the maximum mark.
15. Install cap.
16. Start the engine and repeatedly press the brake pedal until brake pressure is evident.

XJ Range 2003.5–06 VIN G45704–G99999 with Brembo Brakes

1. Remove both front wheels and tires.
2. Remove the anti-rattle spring.
3. Remove the brake caliper retaining bolt dust covers.
4. Remove the brake caliper retaining bolts.

⁂ CAUTION

Do not allow the brake caliper to hang on the hydraulic hose, as this will damage the hose. Failure to follow this instruction will result in damage to the hose.

5. Detach the brake caliper.

⁂ CAUTION

Do not allow the brake caliper to hang on the hydraulic hose, as this will damage the hose. Failure to follow this instruction will result in damage to the hose.

6. Remove the brake pad from the caliper piston.
7. Remove the brake pad.

To install:

⁂ WARNING

Do not use compressed air to disperse brake dust into the atmosphere.

⁂ CAUTION

If the pistons are pushed back into the housing, brake fluid will be forced back into the brake fluid reservoir.

⁂ CAUTION

Make sure the brake disc faces are clean before installation.

⁂ CAUTION

Brake pads must always be replaced in axle sets.

➡Make sure that the inboard brake pad is correctly installed to the caliper piston.

8. To install, reverse the removal procedure. Using a suitable tool, fully retract the brake caliper pistons. Tighten to 58 Nm.

REAR DISC BRAKES

Caliper

REMOVAL & INSTALLATION

S-Type

2002 THROUGH VIN M45254

1. Remove the wheel.
2. Release the retaining tangs. Detach the parking brake cable.
3. To prevent the loss of brake fluid, apply the brake pedal hold down tool and adjust to hold the brake pedal down 40 mm (1.6 in).
4. Install a bleeder bottle to the bleed nipple and loosen.
5. Remove the brake pipe.
6. Remove the caliper guide pin retaining bolts.

➡If the pistons are pushed back into the housing, brake fluid will be forced out of the inlet port of the brake caliper.

7. Using the Snap On tool KS326, or equivalent, fully retract the brake caliper piston.
8. Make sure the piston location mark is in the correct position so that the locator pin on the pad backplate locates correctly into the piston.

➡Caliper guide pin bolts must be replaced.

9. Install the caliper. Torque the new bolts to 24 ft. lbs. (33 Nm).
10. Install the new brake hose. Torque to 35 ft. lbs. (47 Nm).
11. Remove the brake hold down tool.
12. Bleed the brake system.
13. Install the wheel.

2002.5–06 FROM VIN M45255–N52047 WITHOUT BREMBO BRAKES

1. Remove the luggage compartment floor covering.
2. Remove the luggage compartment trim panel.
3. Remove the luggage compartment side trim panel.
4. Disconnect the parking brake module electrical connector.
5. Connect the special tool to the parking brake module wiring harness electrical connector.
6. Operate the parking brake release tool until an audible noise is heard from the parking brake release actuator.

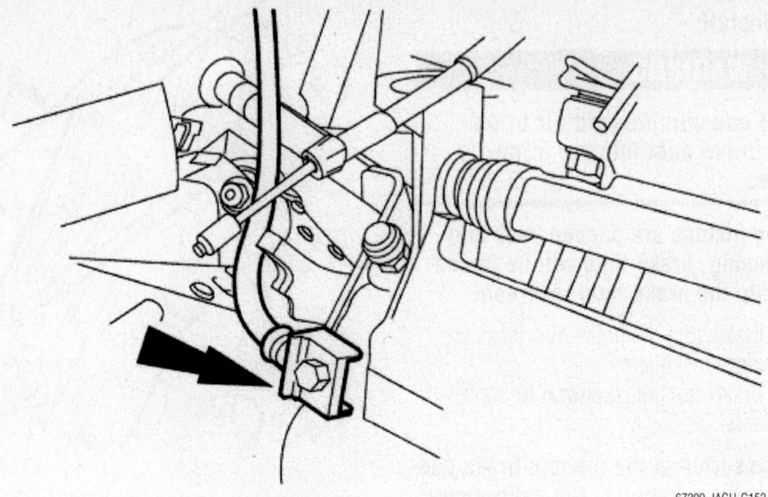

Brake pipe—S-Type 2002 through VIN M45254

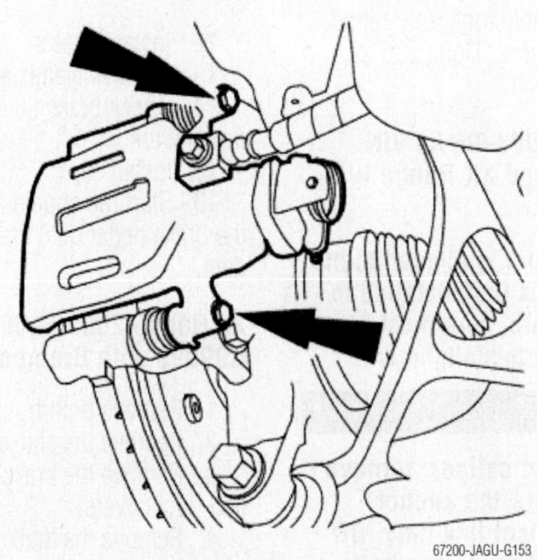

Rear caliper retaining bolts—S-Type 2002 through VIN M45254

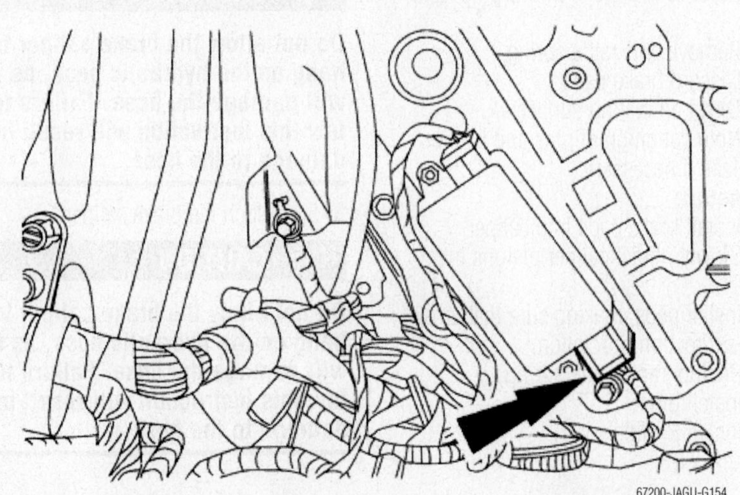

Disconnect the parking brake module electrical connector—S-Type 2002.5–06 VIN M45255–N52047 without Brembo brakes

7. Release the parking brake cable tension.
8. Remove both rear wheels and tires.
9. Reposition the brake caliper lever.
10. Detach the parking brake cable.
11. Release the retaining tangs.
12. Detach the parking brake cable from the bracket.

➡The brake caliper must be supported at all times.

13. Cap the brake hose to prevent loss of fluid and dirt ingress.
14. To prevent the loss of brake fluid, apply the brake pedal hold-down tool and adjust to hold the brake pedal down 40 mm (1.6 in).
15. Detach the brake hose. Remove and discard the sealing washers.
16. Remove and discard the brake caliper retaining bolts.

To install:

17. Install the brake caliper.
18. Install new brake caliper upper retaining bolts. Tighten to 25 ft. lbs. (34 Nm).
19. Attach the brake hose. Install new sealing washers. Tighten to 26 ft. lbs. (35 Nm).
20. Attach the parking brake cable to the bracket.
21. Reposition the brake caliper lever. Attach the parking brake cable.
22. Remove the brake pedal hold-down tool.
23. Bleed the brake system.
24. Install both rear wheels and tires.
25. Remove the special tool.
26. Connect the parking brake module electrical connector.
27. Install the luggage compartment side trim panel.
28. Install the luggage compartment side trim panel retaining clips.
29. Install the luggage compartment trim panel.
30. Install the luggage compartment floor covering.
31. Apply the parking brake.

2002.5–06 VIN M45255–N52047 WITH BREMBO BRAKES

1. Remove the brake pads.
2. To prevent the loss of brake fluid, apply the brake pedal hold-down tool and adjust to hold the brake pedal down 40 mm (1.6 in).
3. Cap the brake hose to prevent loss of fluid and dirt ingress.
4. Detach the brake hose. Remove and discard the sealing washers.
5. Remove and discard the brake caliper retaining bolts.

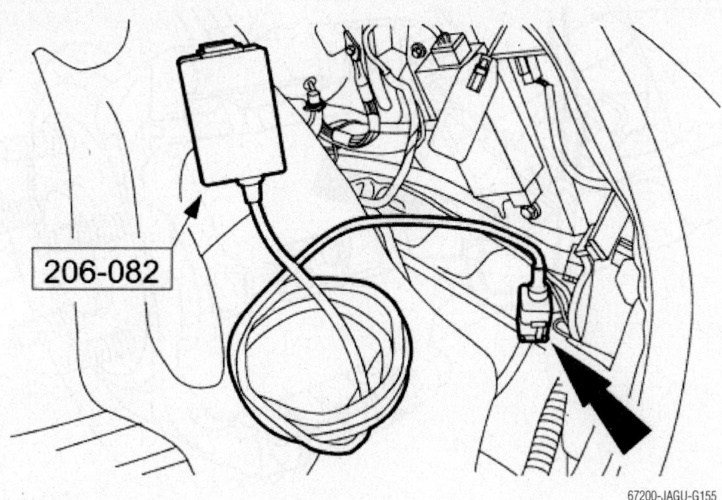

Connect the special tool to the parking brake module wiring harness electrical connector—S-Type 2002.5–06 VIN M45255–N52047 without Brembo brakes

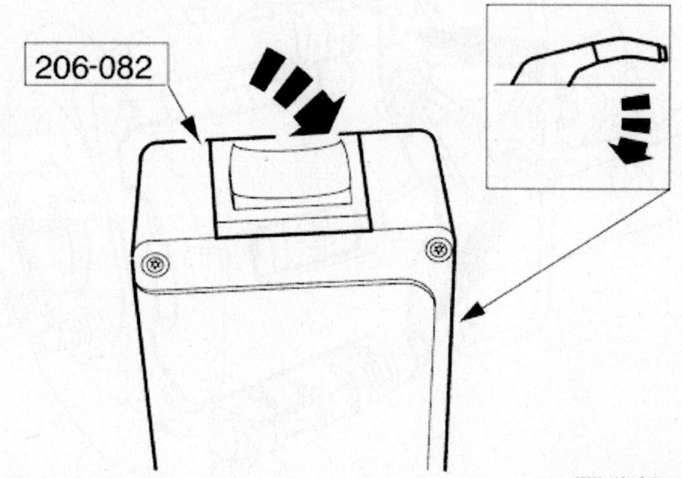

Operate the parking brake release tool until an audible noise is heard from the parking brake release actuator—S-Type 2002.5–06 VIN M45255–N52047 without Brembo brakes

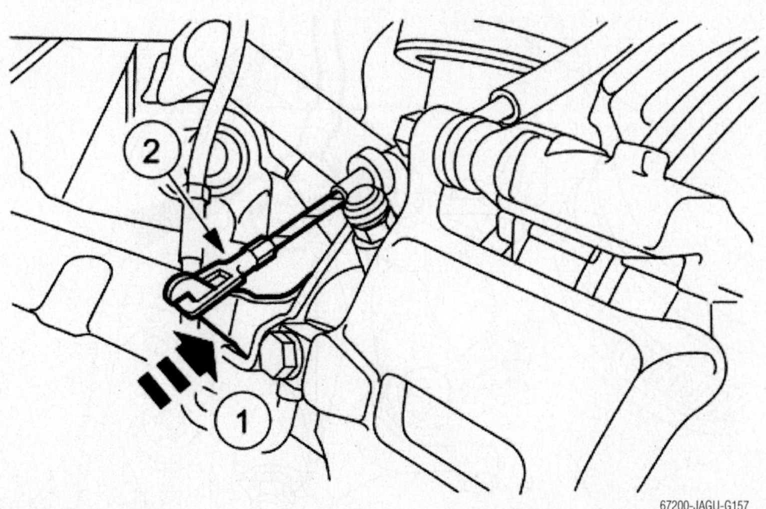

Detach the parking brake cable from the lever—S-Type 2002.5–06 VIN M45255–N52047 without Brembo brakes

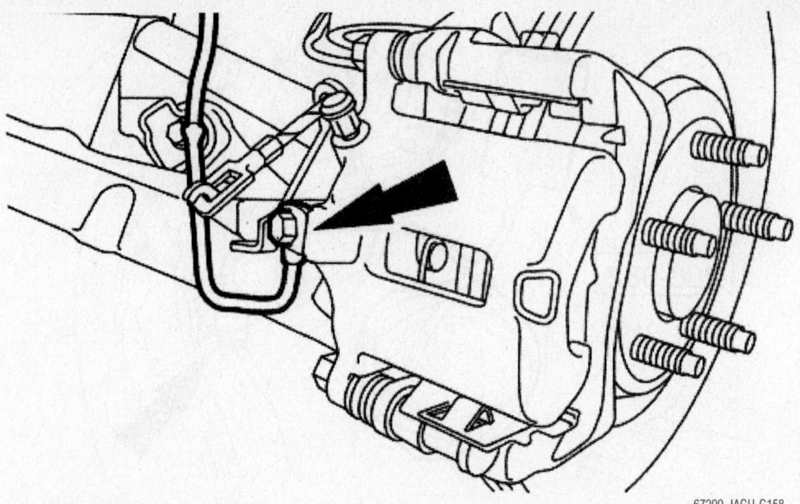

Detach the parking brake cable from the bracket—S-Type 2002.5–06 VIN M45255–N52047 without Brembo brakes

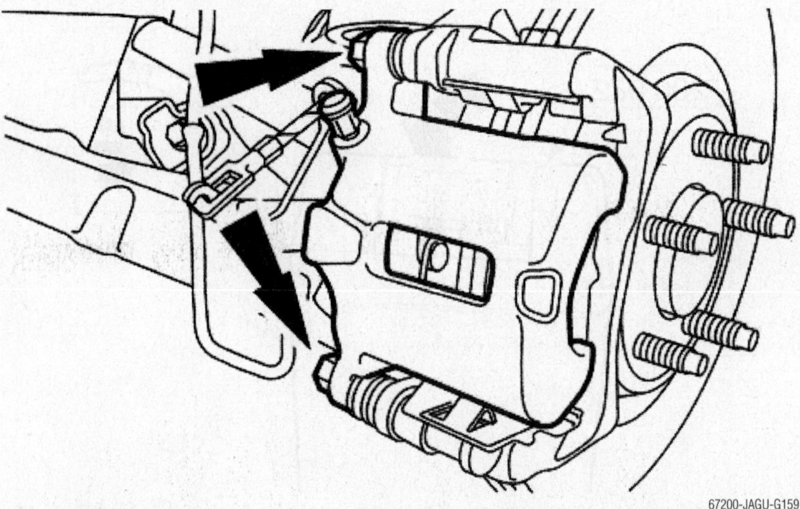

Remove and discard the brake caliper retaining bolts—S-Type 2002.5–06 VIN M45255–N52047 without Brembo brakes

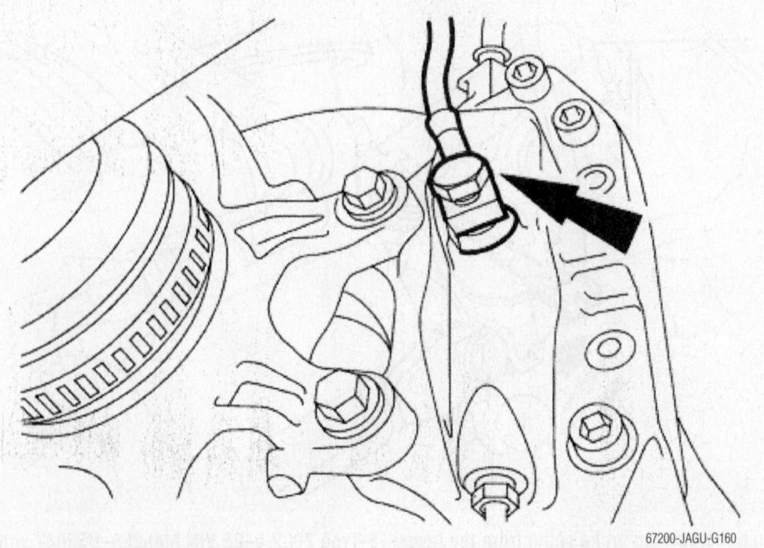

Brake line bolt—S-Type 2002.5–06 VIN M45255–N52047 with Brembo brakes

To install:

6. Install the brake caliper. Install new brake caliper retaining bolts. Tighten to 52 ft. lbs. (70 Nm).
7. Attach the brake hose. Install new sealing washers. Tighten to 26 ft. lbs. (35 Nm).
8. Remove the brake pedal hold-down tool.
9. Install the brake pads.
10. Bleed the brake system.

2002.5–06 VIN N52048–N99999 WITH OR WITHOUT BREMBO BRAKES

1. All vehicles:
   a. Carry out the parking brake cable tension release procedure.
   b. Remove both rear wheels and tires.
   c. Detach the parking brake cable.
2. Vehicles with high performance brakes: Remove the logo badge.
3. All vehicles:
   a. Remove the anti-rattle spring.

⁂ CAUTION

Brake fluid contains polyglycol ethers and polyglycols. Avoid contact with the eyes. Wash hands thoroughly after handling, as prolonged contact may cause irritation and dermatitis. If brake fluid contacts the eyes, flush the eyes with cold water or eyewash solution and seek medical attention. If taken internally do not induce vomiting, seek immediate medical attention. Failure to follow these instructions may result in personal injury.

⁂ CAUTION

If brake fluid comes into contact with the paintwork, the affected area must be immediately washed down with cold water.

⁂ CAUTION

Cap the brake hose to prevent loss of fluid and prevent dirt ingress.

➡To prevent the loss of brake fluid, apply the brake pedal hold-down tool and adjust to hold the brake pedal down 40 mm (1.6 in).

   b. Detach the brake hose. Remove and discard the sealing washers.
   c. Remove the brake caliper retaining bolt dust covers.
   d. Remove the brake caliper retaining bolts.

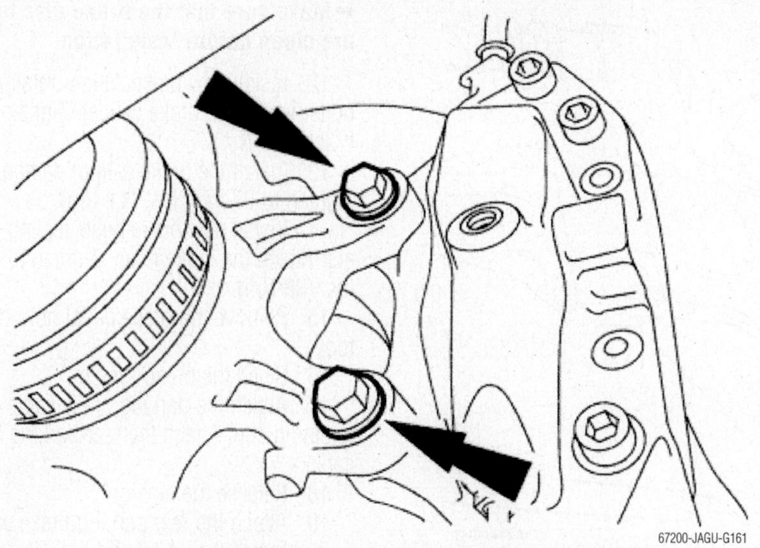

Brake caliper bolts—S-Type 2002.5–06 VIN M45255–52047 with Brembo brakes

e. Remove the brake caliper.

f. Remove the brake pads.

To install:

4. All vehicles:

WARNING

Do not use compressed air to disperse brake dust into the atmosphere.

CAUTION

Brake pads must always be replaced in axle sets.

CAUTION

Make sure the brake disc faces are clean before installation.

➡Make sure that the brake pad with the anti-rattle spring is installed to the inboard side of the brake disc.

a. Install the brake pads.

CAUTION

As the piston is pushed back into the caliper housing, the brake fluid level in the reservoir will rise. Do not allow the reservoir to overflow.

CAUTION

If brake fluid comes into contact with the paintwork, the affected area must be immediately washed down with cold water.

➡Only the extraction bolt from special tool 303-588 is used.

➡Only the forcing plate from special tool 206-080 is used.

b. Using the special tools, fully retract the brake caliper piston.

c. Install the brake caliper. Tighten to 21 ft. lbs. (28 Nm).

d. Install the brake caliper retaining bolt dust covers.

e. Attach the brake hose. Install new sealing washers. Tighten to 26 ft. lbs. (35 Nm).

f. Install the anti-rattle spring.

5. Vehicles with high performance brakes: Install the logo badge. Tighten to 44 inch lbs. (5 Nm).

6. All vehicles:

a. Attach the parking brake cable.

b. Remove the brake pedal hold-down tool.

c. Bleed the brake system.

d. Install both rear wheels and tires.

X-Type

2002–03 FROM VIN C00344 THROUGH VIN D55321

1. Remove the wheel and tire.
2. Release the parking brake tension. Detach the rear parking brake cable. Remove the parking brake cable retaining clip.
3. To prevent the loss of brake fluid, apply the brake pedal hold down tool and adjust to hold the brake pedal down 40 mm.
4. Install a bleeder bottle to the bleed nipple and loosen.
5. Remove and discard the brake hose sealing washers.
6. Detach the brake hose.
7. Tighten the bleed nipple and remove the bleeder tube and bottle.
8. Remove the brake caliper damper.
9. Remove and discard the brake caliper retaining bolts. Remove the brake caliper.

To install:

10. If the piston is pushed back into the housing, brake fluid will be forced out of the inlet port of the brake caliper.

➡The left-hand brake caliper piston retracts counter clockwise and the right-hand retracts clockwise.

11. Using the special tools, fully retract the brake caliper piston. Make sure the piston location mark is in the correct position so that the locator pin on the brake pad backing plate locates correctly into the piston.

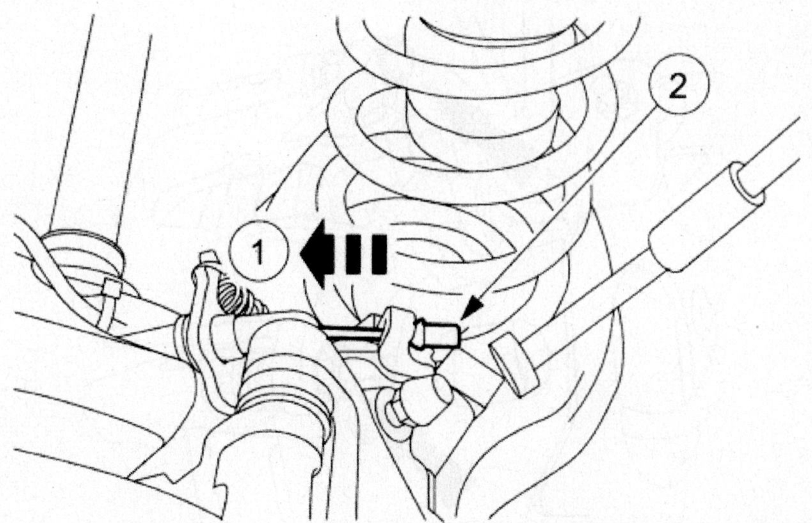

Detach the rear parking brake cable—2002–03 from VIN C00344 through VIN D55321

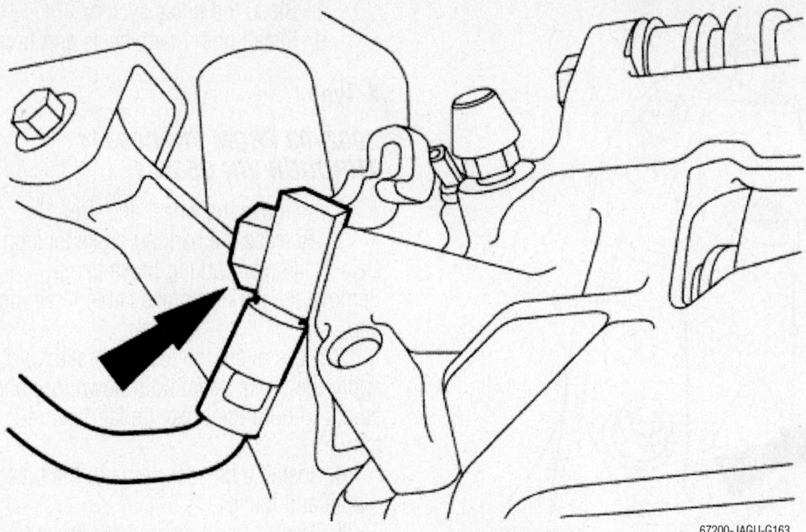

Brake hose bolt—2002–03 from VIN C00344 through VIN D55321

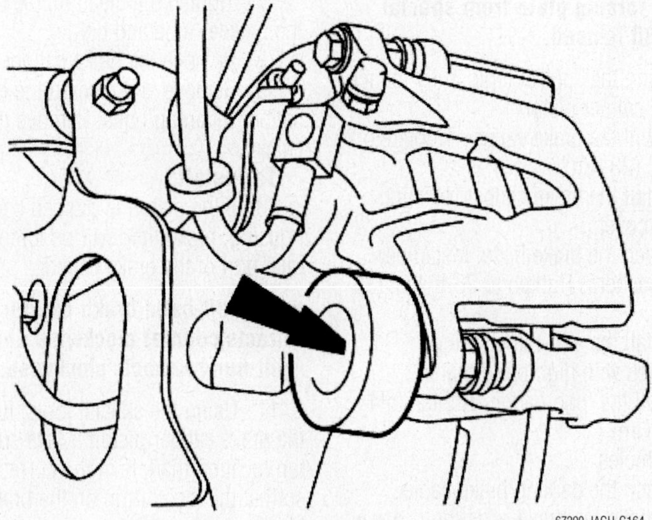

Brake caliper damper—X-Type 2002–03 from VIN C00344 through VIN D55321

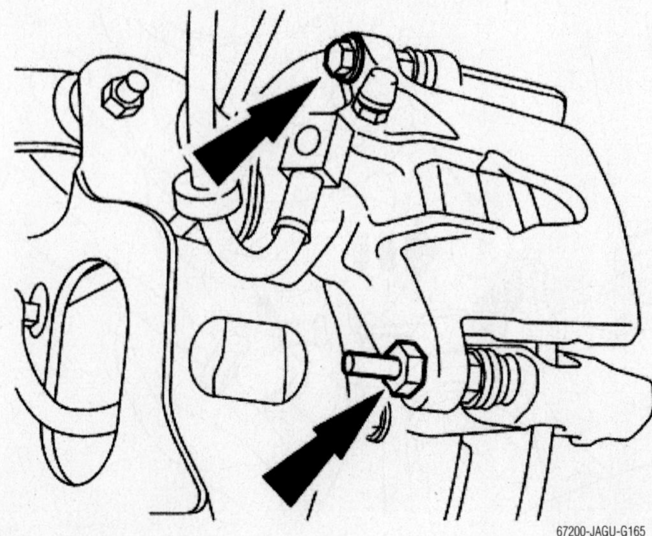

Brake caliper retaining bolts—X-Type 2002–03 from VIN C00344 through VIN D55321

➡Make sure that the brake disc faces are clean before installation.

12. Install new brake caliper retaining bolts. Install the brake caliper. Tighten to 24 ft. lbs. (32 Nm).
13. Install the brake caliper damper. Tighten to 97 inch lbs. (11 Nm).
14. Install new brake hose sealing washers. Attach the brake hose. Tighten to 28 ft. lbs. (38 Nm).
15. Remove the brake pedal hold down tool.
16. Bleed the brake system.
17. Attach the parking brake cable retaining clip. Attach the rear parking brake cable.
18. Release the tension.
19. Attach the rear parking brake cable.
20. Install the wheel and tire.

2004–06 FROM VIN D55322

1. Remove the wheel and tire.
2. Release the tension.
3. Detach the parking brake cable.
4. Remove the parking brake cable retaining clip.
5. To prevent the loss of brake fluid, apply the brake pedal hold down tool and adjust to hold the brake pedal down 40mm.
6. Install a bleeder bottle to the bleed nipple and loosen.
7. Remove and discard the brake hose sealing washers.
8. Detach the brake hose.
9. Tighten the bleed nipple and remove the bleeder tube and bottle.
10. Remove the brake caliper. Discard the brake caliper retaining bolts.

➡If the piston is pushed back into the housing, brake fluid will be forced out of the inlet port of the brake caliper.

➡The left-hand brake caliper piston retracts counter clockwise and the right-hand retracts clockwise.

11. Using the special tools, fully retract the brake caliper piston. Make sure the piston location mark is in the correct position so that the locator pin on the brake pad backing plate locates correctly into the piston.
12. Make sure that the brake disc faces are clean before installation.

To install:

13. Install new brake caliper retaining bolts. Install the brake caliper. Tighten to 24 ft. lbs. (32 Nm).
14. Install new brake hose sealing washers. Attach the brake hose. Tighten to 28 ft. lbs. (38 Nm).

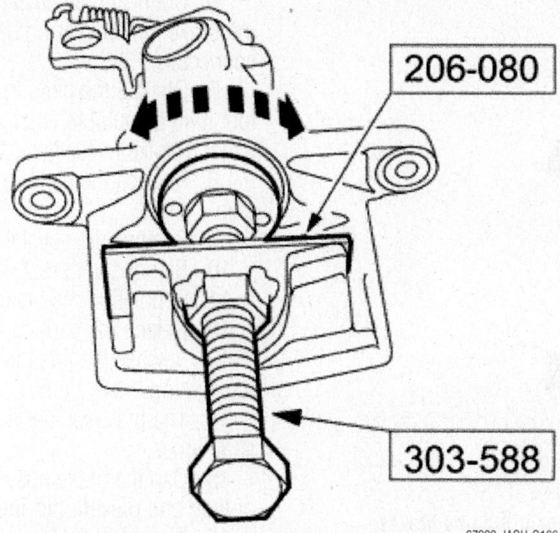

The left-hand brake caliper piston retracts counter clockwise and the right-hand retracts clockwise—2002–03 from VIN C00344 through VIN D55321

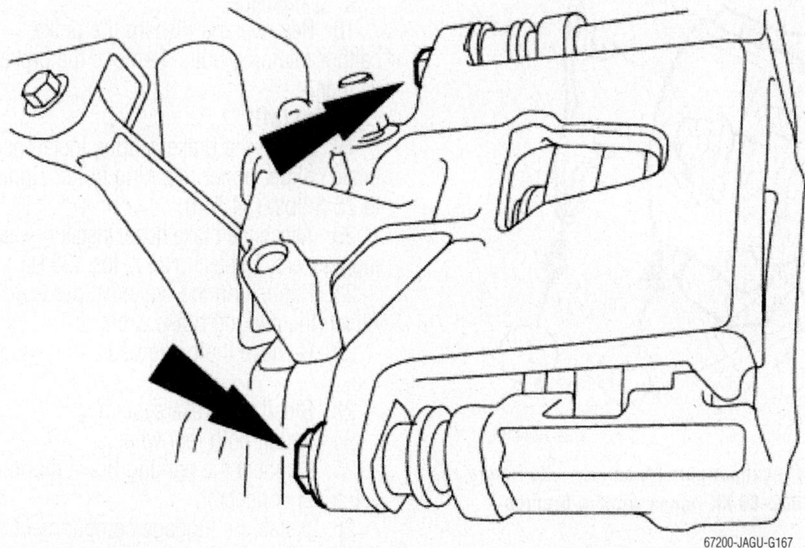

Caliper mounting bolts—X-Type 2004–06

15. Remove the brake pedal hold-down tool.
16. Bleed the brake system.
17. Attach the parking brake cable.
18. Install the wheel and tire.

2002–03 Through XJ Range VIN F59525 without Brembo Brakes

2002–06 XK Range without Brembo Brakes

➡The brake pad adhesive backing must not be reused, therefore always Install new brake pads, irrespective of the condition of the original pads when replacing a brake caliper. Always replace brake pads as an axle set.

1. Raise vehicle on a four-post lift.
2. Raise rear of vehicle and support on stands.
3. Remove rear wheel.
4. Open left front and left rear caliper bleed nipples.
5. Connect a bleed tube and container to the calipers.
6. Open bleed nipples. Depress and hold down brake pedal at the specified measurement using the special tool.
7. Position tool between brake pedal and seat frame. Turn the tool wheel until the brake pedal is held 60mm down from its released position.
8. Close bleed nipples on both calipers. Remove bleed tubes and containers.
9. Tighten front bleed nipple to 53 inch lbs. (6 Nm).
10. Tighten rear bleed nipple to 97 inch lbs. (11 Nm).
11. Disconnect brake hose from rear caliper housing.
12. Install plugs to caliper and brake hose.
13. Remove anti-rattle spring. Remove dust covers. Remove guide pins. Remove caliper from carrier.

➡The inner brake pad is fitted with a clip which secures the pad into the caliper piston.

14. Remove and discard brake pads.
15. Remove bolts. Remove carrier.
16. Clean all mating surfaces and Remove brake dust.

To install:

17. Install caliper carrier to hub carrier. Install and tighten bolts to 44 ft. lbs. (60 Nm).

➡Make sure piston is fully retracted into caliper housing.

18. Remove plug from caliper housing.
19. Retract piston.
20. Install brake pads into caliper housing.
21. Remove paper backing from pads to reveal adhesive shim.
22. Install brake pads, insert clip located on back of inner pad into caliper piston.
23. Install caliper housing to caliper carrier.
24. Install and tighten guide pins to 18–22 ft. lbs. (25–30 Nm).
25. Install dust covers.
26. Install anti-rattle spring.
27. Connect brake hose to caliper housing.
28. Install and tighten bolt to 22–29 ft. lbs. (30–40 Nm).
29. Remove special tool from brake pedal.
30. Install wheels.
31. Remove stands and lower vehicle.
32. Bleed brake hydraulic system.

2003.5–06 XJ Range VIN G00442–G45703 without Brembo Brakes

1. Move the selector lever to the "P" position.
2. Release the parking brake.
3. Remove the luggage compartment floor covering.
4. Remove the luggage compartment side trim panel.
5. Disconnect the parking brake module electrical connector.

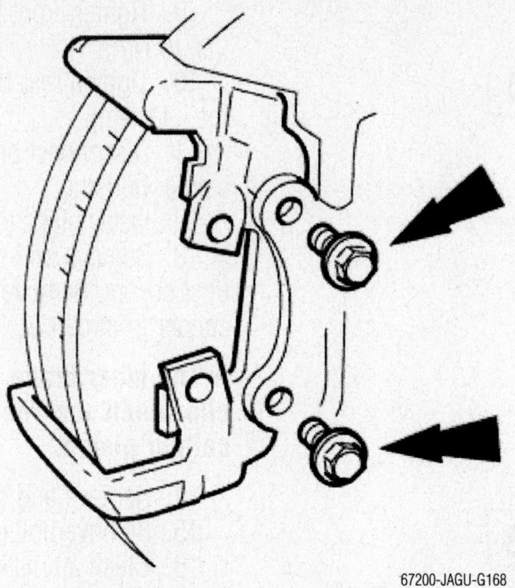

Rear caliper carrier bolts—XJ Range 2002–03 through VIN F59525 without Brembo brakes; 2002–06 XK Range without Brembo brakes

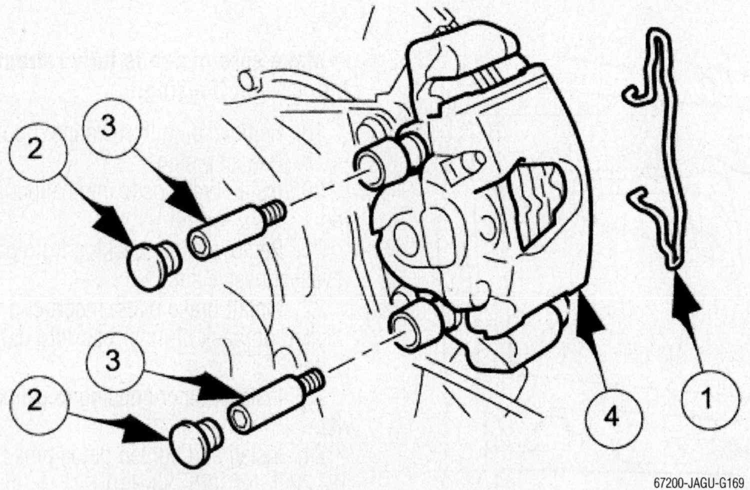

Rear caliper parts (1) anti-rattle spring (2) dust covers (3) caliper pins (4) caliper—XJ Range 2002–03 through VIN F59525 without Brembo brakes; 2002–06 XK Range without Brembo brakes

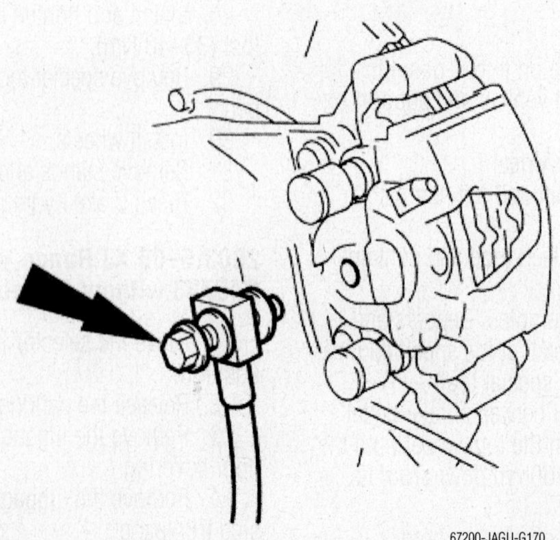

Rear caliper brake line union bolt—XJ Range 2002–03 through VIN F59525 without Brembo brakes; 2002–06 XK Range without Brembo brakes

6. Connect the special tool to the parking brake module wiring harness electrical connector.
7. Operate the parking brake release tool until an audible noise is heard from the parking brake release actuator.
8. Release the parking brake cable tension.
9. Remove the special tool.
10. Remove both rear wheels.
11. Reposition the brake caliper lever.
12. Detach the parking brake cable.
13. Release the retaining tangs. Detach the parking brake cable.
14. The brake caliper must be supported at all times.
15. Cap the brake hose to prevent loss of fluid and prevent dirt ingress.
16. To prevent the loss of brake fluid, apply the brake pedal hold-down tool and adjust to hold the brake pedal down 40 mm (1.6 in).
17. Detach the brake hose. Discard the sealing washers.
18. Remove and discard the brake caliper retaining bolts. Remove the brake caliper.

To install:

19. Install the brake caliper. Install new brake caliper upper retaining bolts. Tighten to 25 ft. lbs. (34 Nm).
20. Attach the brake hose. Install new sealing washers. Tighten to 26 ft. lbs. (35 Nm).
21. Reposition the brake caliper lever. Attach the parking brake cable.
22. Remove the brake pedal hold-down tool.
23. Bleed the brake system.
24. Install both rear wheels.
25. Connect the parking brake module electrical connector.
26. Install the luggage compartment side trim panel.

➡The brake pedal must be applied when calibrating the parking brake switch.

27. Apply the parking brake.

2002–03 XJ Range through VIN F59525 with Brembo Brakes

2002–06 XK Range with Brembo Brakes

➡The Brembo caliper is aligned to the brake disc when it is first installed to the vehicle, therefore, care must be taken not to disturb this alignment. When removing the caliper, remove the bolts that secure the anchor bracket to the hub carrier only. DO NOT loosen any other caliper bolts.

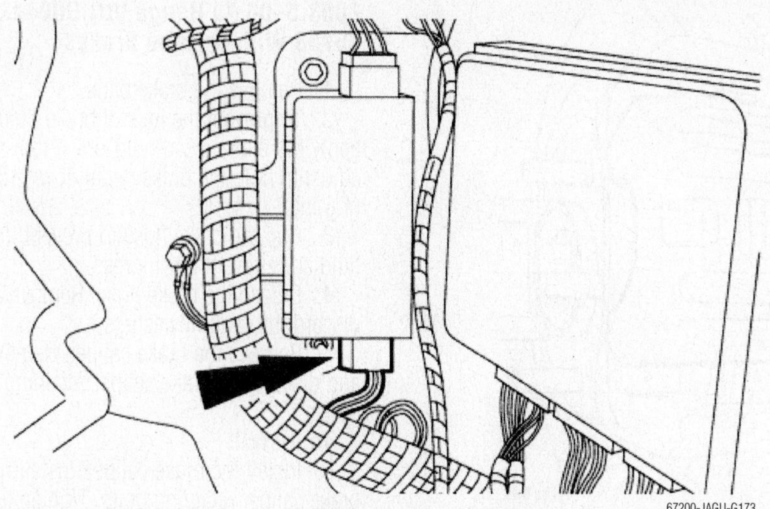

Parking brake module electrical connector—XJ Range 2003.5–06 VIN G00442–45703 without Brembo brakes

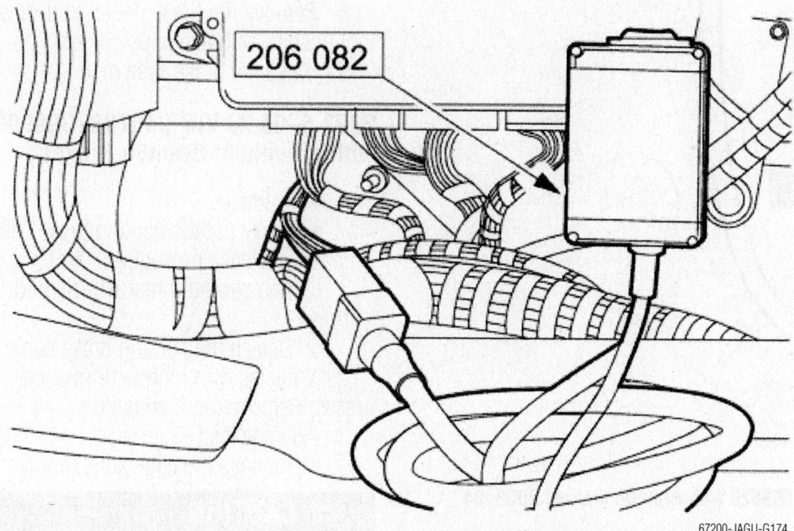

Parking brake release tool—XJ Range 2003.5–06 VIN G00442–45703 without Brembo brakes

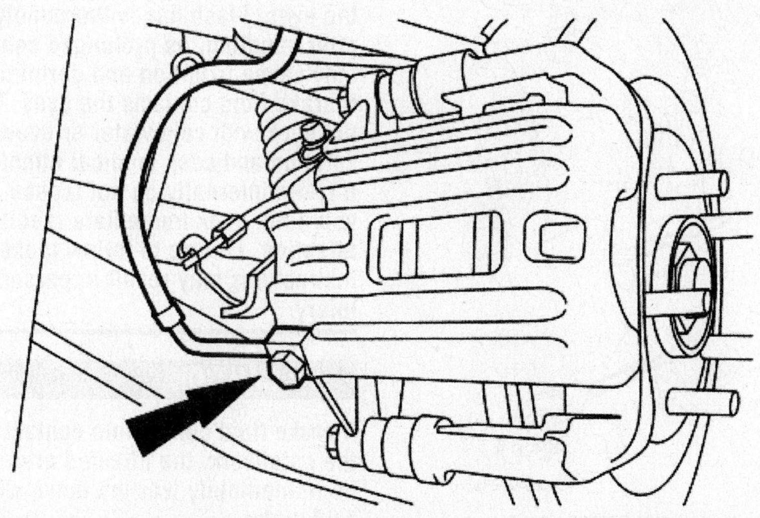

Brake pipe union bolt—XJ Range 2003.5–06 VIN G00442–45703 without Brembo brakes

1. Raise front of vehicle and support on stands.
2. Open left front and left rear caliper bleed nipples.
3. Connect a bleed tube and container to the calipers.
4. Open bleed nipples.
5. Depress and hold down brake pedal at the specified measurement using the special tool. Position tool between brake pedal and seat frame.
6. Turn the tool wheel until the brake pedal is held 60mm down from its released position.
7. Close bleed nipples on both calipers. Tighten bleed nipples to 12–16 Nm.
8. Remove bleed tubes and containers.
9. Remove union bolt and brake hose. Remove and discard sealing washers. Install plugs into caliper and hose.
10. Remove pins from caliper. Remove anti-rattle spring. Remove brake pads.
11. Remove bolts and remove caliper from hub carrier.
12. Remove brake dust and clean mating surfaces.

To install:

13. Install caliper to hub carrier. Install and tighten bolts to 44 ft. lbs. (60 Nm).
14. Retracting the caliper piston may cause the fluid reservoir to over-flow.
15. Note the condition of the pins and replace if necessary.
16. Slowly retract caliper pistons. Install pads. Position anti-rattle spring. Install pins. Make sure pin collets are fully engaged into caliper.

⁂ WARNING

The brake hose must only be installed with the vehicle's body supported and the suspension hanging. It is important that the brake hose is aligned correctly when installed to prevent it fouling on the steering and suspension components. After installation the brake hose must be aligned with the center line as shown.

17. Install brake hose to caliper.
18. Remove plugs from caliper and hose. Install new sealing washers. Install and tighten union bolt to 16–19 ft. lbs. (22–26 Nm).
19. Remove special tool from brake pedal.
20. Install wheels.
21. Make sure the brake hose is routed correctly.
22. Remove stands and lower vehicle

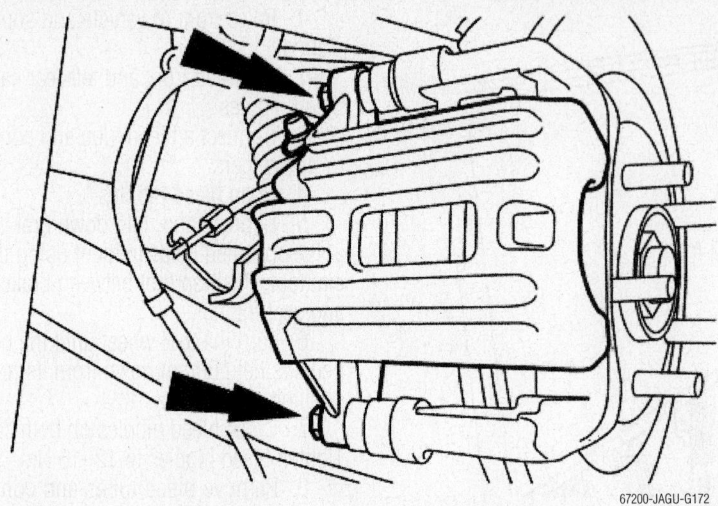

Caliper mounting bolts—XJ Range 2003.5–06 VIN G00442–45703 without Brembo brakes

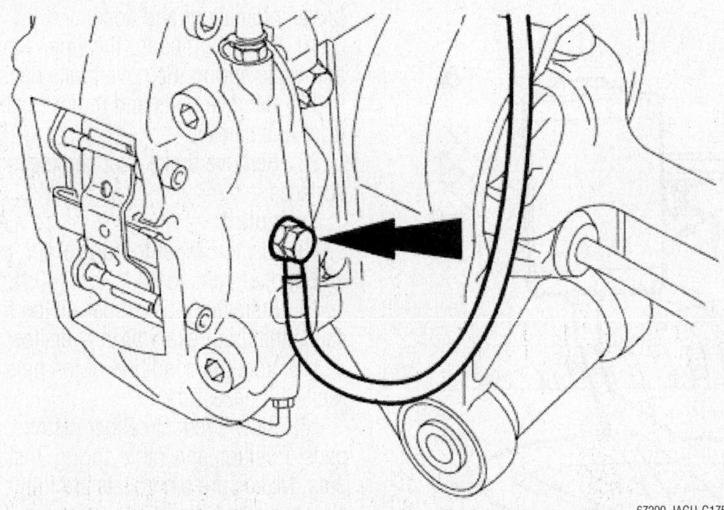

Brake hose union bolt—XJ Range 2002–03 through VIN F59525 with Brembo brakes; 2002–04 XK Range with Brembo brakes

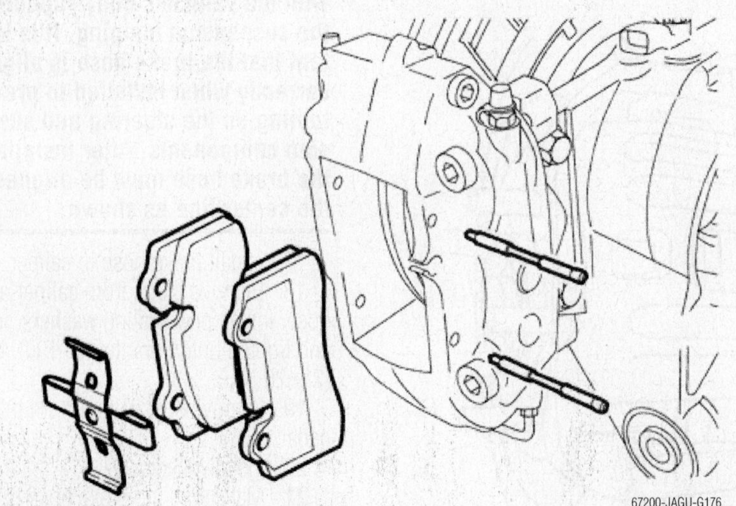

Brake pads, pins and anti-rattle spring—XJ Range 2002–03 through VIN F59525 with Brembo brakes; 2002–04 XK Range with Brembo brakes

2003.5–06 XJ Range VIN G00442–45703 With Brembo Brakes

1. Remove the brake pads.
2. To prevent the loss of brake fluid, apply the brake pedal hold-down tool and adjust to hold the brake pedal down 40 mm (1.6 in).
3. Cap the brake hose to prevent loss of fluid and prevent dirt ingress.
4. Detach the brake hose. Remove and discard the sealing washers.
5. Remove the brake caliper. Remove and discard the brake caliper retaining bolts.

To install:

6. Install the brake caliper. Install new brake caliper retaining bolts. Tighten to 52 ft. lbs. (70 Nm).
7. Attach the brake hose. Install new sealing washers. Tighten to 26 ft. lbs. (35 Nm).
8. Remove the brake pedal hold-down tool. Install the brake pads.
9. Bleed the brake system.

2003.5–06 XJ VIN G45708–G99999 with or without Brembo Brakes

1. All vehicles:
   a. Carry out the parking brake cable tension release procedure.
   b. Remove both rear wheels and tires.
   c. Detach the parking brake cable.
2. Vehicles with high performance brakes: Remove the logo badge.
3. All vehicles:
   a. Remove the anti-rattle spring.

CAUTION

Brake fluid contains polyglycol ethers and polyglycols. Avoid contact with the eyes. Wash hands thoroughly after handling, as prolonged contact may cause irritation and dermatitis. If brake fluid contacts the eyes, flush the eyes with cold water or eyewash solution and seek medical attention. If taken internally do not induce vomiting, seek immediate medical attention. Failure to follow these instructions may result in personal injury.

CAUTION

If brake fluid comes into contact with the paintwork, the affected area must be immediately washed down with cold water.

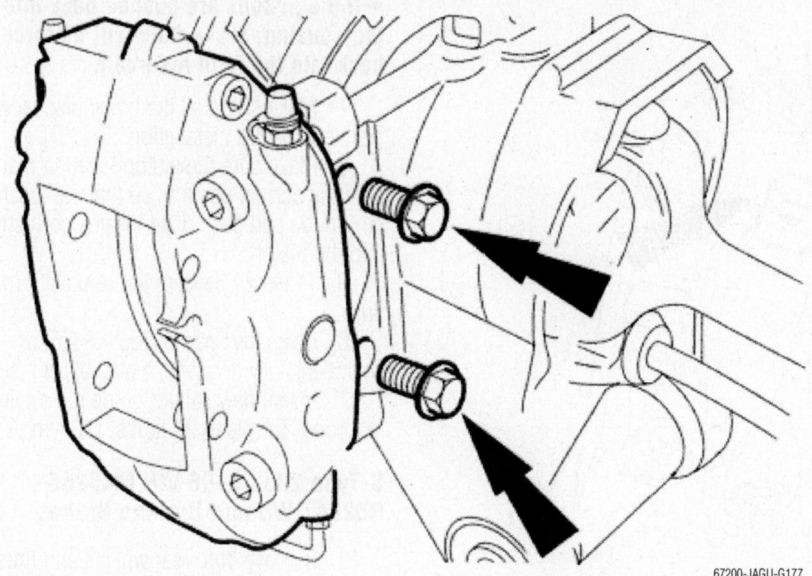

Removing the caliper—XJ Range 2002–03 through VIN F59525 with Brembo brakes; 2002–04 XK Range with Brembo brakes

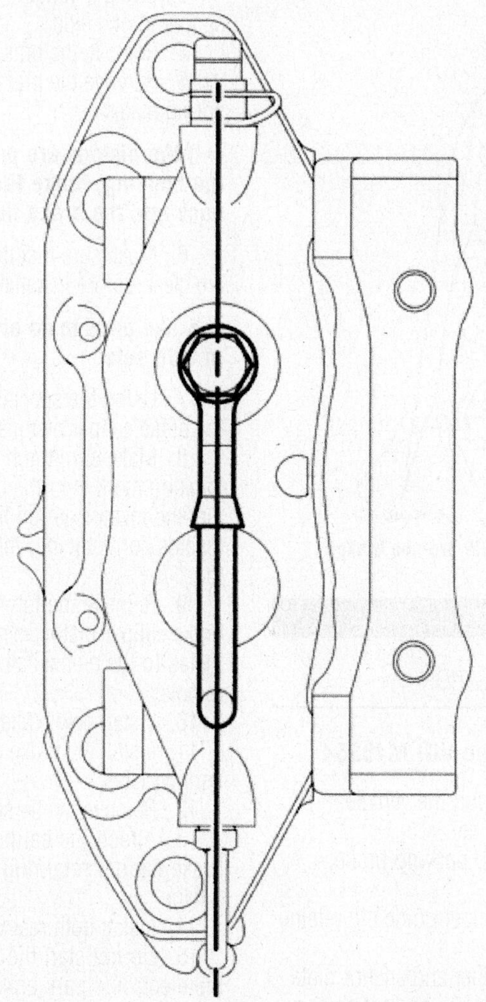

The brake hose must be aligned as shown—XJ Range 2002–03 through VIN F59525 with Brembo brakes; 2002–04 XK Range with Brembo brakes

CAUTION

Cap the brake hose to prevent loss of fluid and prevent dirt ingress.

➡To prevent the loss of brake fluid, apply the brake pedal hold-down tool and adjust to hold the brake pedal down 40 mm (1.6 in).

b. Detach the brake hose. Remove and discard the sealing washers.
c. Remove the brake caliper retaining bolt dust covers.
d. Remove the brake caliper retaining bolts.
e. Remove the brake caliper.
f. Remove the brake pads.

To install:

4. All vehicles

WARNING

Do not use compressed air to disperse brake dust into the atmosphere.

CAUTION

Brake pads must always be replaced in axle sets.

CAUTION

Make sure the brake disc faces are clean before installation.

➡Make sure that the brake pad with the anti-rattle spring is installed to the inboard side of the brake disc.

a. Install the brake pads.

CAUTION

As the piston is pushed back into the caliper housing, the brake fluid level in the reservoir will rise. Do not allow the reservoir to overflow.

CAUTION

If brake fluid comes into contact with the paintwork, the affected area must be immediately washed down with cold water.

➡Only the extraction bolt from special tool 303-588 is used.

➡Only the forcing plate from special tool 206-080 is used.

b. Using the special tools, fully retract the brake caliper piston.

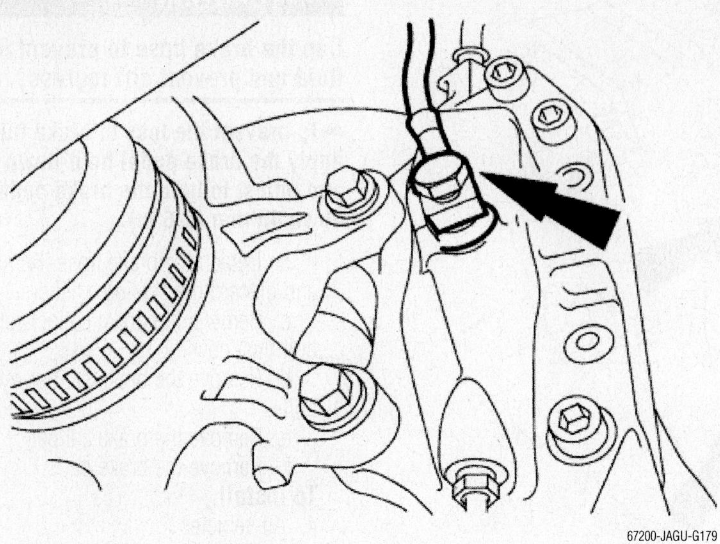

Brake pipe union bolt—XJ Range 2003.5–06 VIN G00442–45703 with Brembo brakes

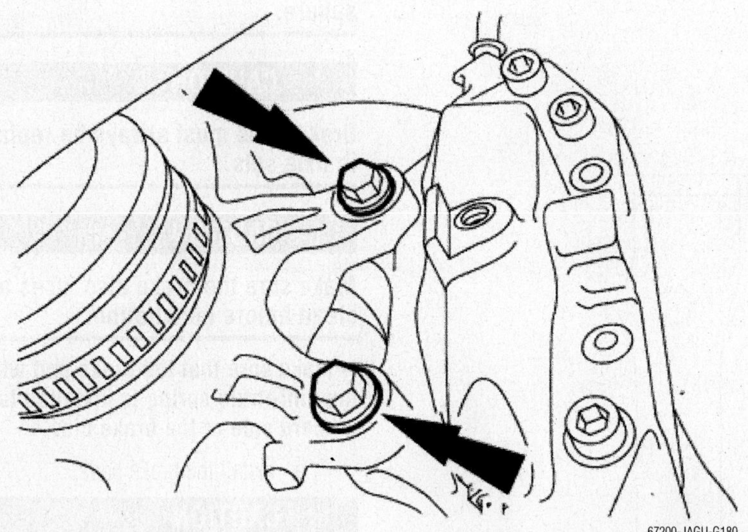

Brake caliper mounting bolts—XJ Range 2003.5–06 VIN G00442–45703 with Brembo brakes

c. Install the brake caliper. Tighten to 28 Nm.

d. Install the brake caliper retaining bolt dust covers.

e. Attach the brake hose. Install new sealing washers. Tighten to 28 ft. lbs. (35 Nm).

f. Install the anti-rattle spring.

5. Vehicles with high performance brakes: Install the logo badge. Tighten to 44 inch lbs. (5 Nm).

6. All vehicles:

a. Attach the parking brake cable.

b. Remove the brake pedal hold-down tool.

c. Bleed the brake system.

d. Install both rear wheels and tires.

Brake Pads

REMOVAL & INSTALLATION

S-Type 2002 through VIN M45254

1. Raise and support the vehicle.
2. Remove the wheel.
3. Disconnect the parking brake cable.
4. Remove the caliper guide pin retaining bolts.
5. Secure the caliper and anchor plate assembly out of the way to prevent damage to the brake hose.
6. Remove the brake pads.

➡If the pistons are pushed back into the housing, brake fluid will be forced back into the fluid reservoir.

7. Make sure that the brake disc faces are clean before installation.
8. Make sure the piston location mark is in the correct position so that the locator pin on the pad back plate locates correctly into the piston.
9. To install, reverse the removal procedure.
10. Using the Snap On tool KS326, or equivalent, fully retract the brake caliper piston.
11. Install new caliper guide pin retaining bolts. Torque to 24 ft. lbs. (33 Nm).

S-Type 2002.5–06 VIN M45255–N52047 Without Brembo Brakes

1. Remove both rear wheels and tires.
2. Detach the brake caliper.

➡The brake caliper must be supported at all times.

3. Using a suitable tie strap, secure the caliper to one side.
4. Remove the brake pads.
5. Remove the brake pad anti-rattle spring plates.

➡If the pistons are pushed back into the housing, brake fluid will be forced back into the brake fluid reservoir.

6. Make sure that the brake disc faces are clean before installation.

➡Brake pads must always be replaced in axle sets.

7. Using the special tools, fully retract the brake caliper pistons.
8. Make sure the brake caliper piston location mark is in the correct position so that the locator pin on the brake pad locates correctly into the brake caliper piston.
9. Remove the forcing screw from the brake caliper piston retractor tool and install to the crankshaft pulley/damper remover.
10. Install new brake pads.
11. Install the brake pad anti-rattle spring plates.
12. Remove the tie strap.
13. Attach the caliper. Install new brake caliper retaining bolts. Tighten to 34Nm.
14. Install both rear wheels and tires.
15. Do not start the engine or operate the electronic park brake before depressing the brake pedal. Repeatedly depress the brake pedal until brake pressure is evident.

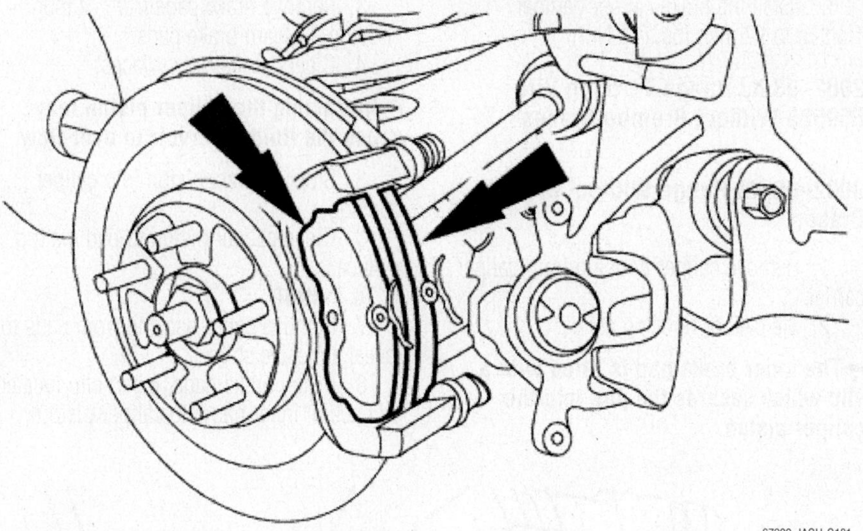

Brake pads—S-Type 2002 through VIN M45254

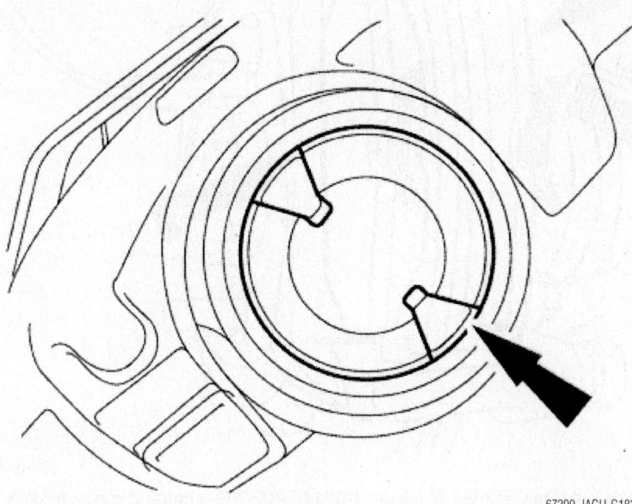

Make sure the brake caliper piston location mark is in the correct position so that the locator pin on the brake pad locates correctly into the brake caliper piston—S-Type 2002.5–06 VIN M45255–N52047 Without Brembo Brakes

2002.5–06 VIN M45255–N52047 with Brembo Brakes

1. Remove both rear wheels and tires.
2. Remove the brake pad retaining pins. Remove the brake pad anti-rattle spring plate. Remove the brake pads.

➡If the pistons are pushed back into the housing, brake fluid will be forced back into the brake fluid reservoir.

➡Brake pads must always be replaced in axle sets. Make sure the brake pads are installed with the direction arrow facing downwards.

3. To install, reverse the removal procedure. Using a suitable tool, fully retract the brake caliper pistons.

S-Type 2002.5–06 VIN N52048–N99999 with or without Brembo Brakes

1. All vehicles
   a. Remove both rear wheels and tires.
2. Vehicles with high performance brakes: Remove the logo badge.
3. All vehicles:
   a. Remove the anti-rattle spring.
   b. Remove the brake caliper retaining bolt dust covers.

CAUTION

The brake caliper must be supported at all times.

   c. Remove the brake caliper retaining bolts.
   d. Detach the brake caliper.
   e. Remove the brake pads.

To installation:

4. All vehicles:

CAUTION

As the piston is pushed back into the caliper housing, the brake fluid level in the reservoir will rise. Do not allow the reservoir to overflow.

CAUTION

If brake fluid comes into contact with the paintwork, the affected area must be immediately washed down with cold water.

➡Only the extraction bolt from special tool 303-588 is used.

➡Only the forcing plate from special tool 206-080 is used.

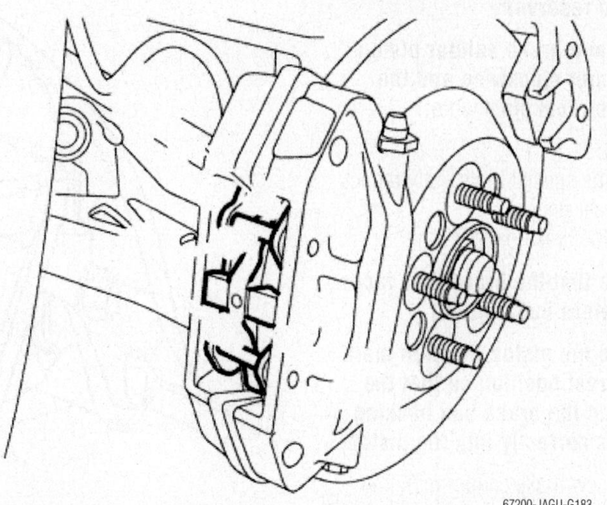

Rear brake pads—S-Type 2002.5–06 VIN M45255–N52047 with Brembo brakes

a. Using the special tools, fully retract the brake caliper piston.

⁂ WARNING

Do not use compressed air to disperse brake dust into the atmosphere.

⁂ CAUTION

Make sure the brake disc faces are clean before installation.

⁂ CAUTION

Brake pads must always be replaced in axle sets.

➡Make sure that the brake pad with the anti-rattle spring is installed to the inboard side of the brake disc.

b. To install, reverse the removal procedure. Tighten the caliper pins to 21 ft. lbs. (28 Nm).

5. Vehicles with high performance brakes: Install the logo badge. Tighten to 44 inch lbs. (5 Nm).

6. All vehicles:

➡Do not start the engine or operate the electronic park brake before depressing the brake pedal.

a. Repeatedly depress the brake pedal until brake pressure is evident.

X-Type

1. Detach the brake caliper and secure to one side. Remove and discard the brake caliper retaining bolts.
2. Remove the brake pads.

➡If the piston is pushed back into the housing, brake fluid will be forced back into the fluid reservoir.

➡The left-hand brake caliper piston retracts counter clockwise and the right-hand retracts clockwise.

To install:

3. Using the special tools, fully retract the brake caliper piston.
4. Install the brake pads.

➡Make sure that the brake disc faces are clean before installation.

➡Make sure the piston location mark is in the correct position so that the locator pin on the brake pad backing plate locates correctly into the piston.

5. Install new brake caliper retaining bolts. Attach the brake caliper. Tighten to 27 ft. lbs. (32 Nm).
6. Install the brake caliper damper. Tighten to 97 inch lbs. (11 Nm).

2002–03 XJ Range Through VIN F59525 Without Brembo Brakes

2002–06 XK Range Without Brembo Brakes

1. Remove caliper housing from caliper carrier.
2. Tie caliper housing aside.

➡The inner brake pad is fitted with a clip which secures the pad into the caliper piston.

3. Remove brake pads from caliper housing. Discard brake pads.
4. Clean all mating surfaces.

➡Retracting the caliper piston may cause the fluid reservoir to over-flow.

5. Slowly retract piston into caliper housing.
6. Clean caliper piston to pad mating surface.

To install:

7. Remove paper backing from pads to reveal adhesive shim.
8. Install brake pads, insert clip located on back of inner pad into caliper piston.

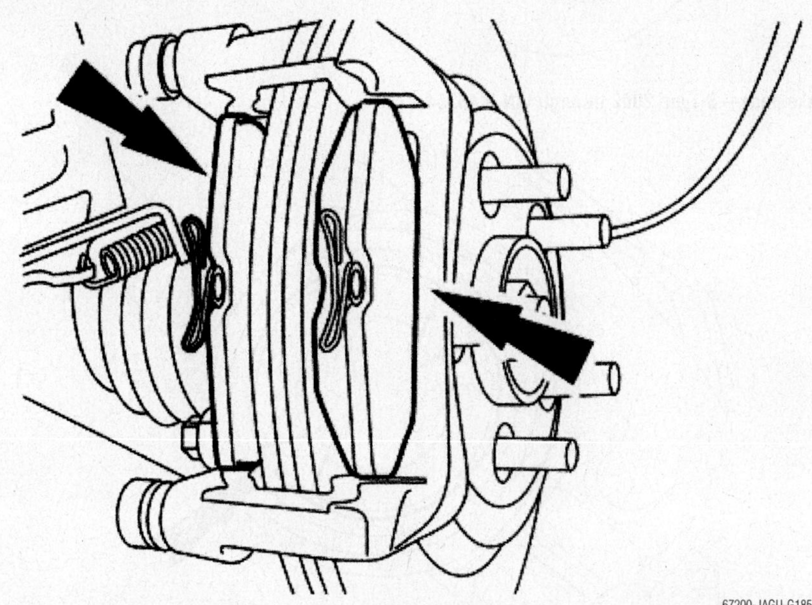

Rear brake pads—XJ Range 2003.5–06 VIN G00442–G45703 without Brembo brakes

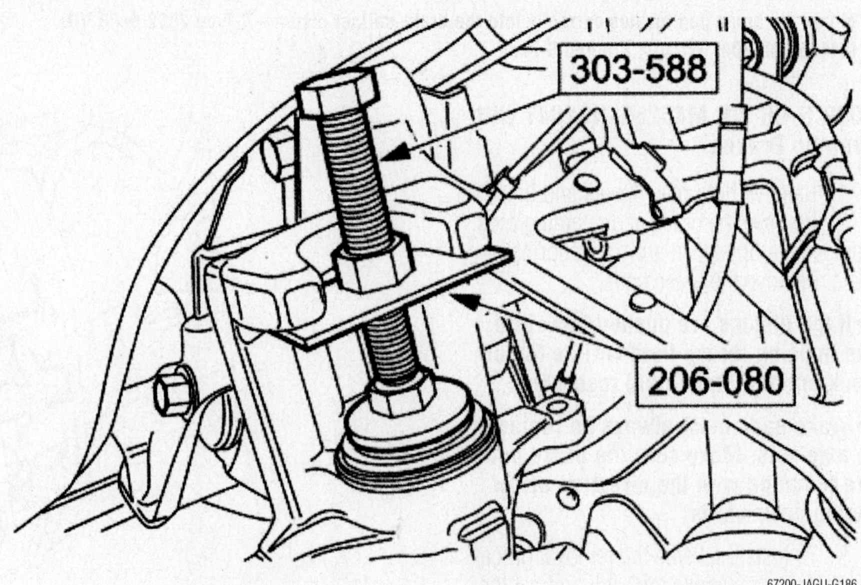

Brake caliper piston tool—XJ Range 2003.5–06 VIN G00442–G45703 without Brembo brakes

9. Align housing to carrier.
10. Install and tighten guide pins to 18–22 ft. lbs. (25–30 Nm).
11. Install dust covers.
12. Install anti-rattle spring.

2003.5–06 XJ Range VIN G00442–G45703 Without Brembo Brakes

1. Remove both rear wheels and tires. Detach the brake caliper.

➡The brake caliper must be supported at all times.

2. Remove the brake pads. Remove the brake pad anti-rattle spring plates.

➡If the pistons are pushed back into the housing, brake fluid will be forced back into the brake fluid reservoir.

3. Make sure that the brake disc faces are clean before installation.

WARNING

Brake pads must always be replaced in axle sets.

4. Using the special tools, fully retract the brake caliper piston.

➡Make sure the brake caliper piston location mark is in the correct position so that the locator pin on the brake pad locates correctly into the brake caliper piston.

5. Install new brake pads. Install the brake pad anti-rattle spring plates.
6. Attach the caliper. Install new brake caliper lower retaining bolts. Tighten to 25 ft. lbs. (34 Nm).
7. Install both rear wheels and tires.

WARNING

Do not start the engine or operate the electronic park brake before depressing the brake pedal.

8. Repeatedly depress the brake pedal until brake pressure is evident.

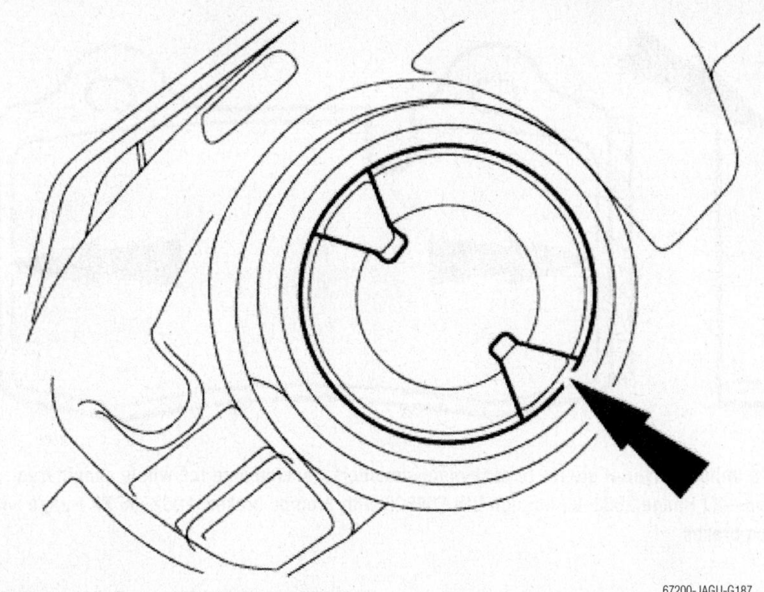

Make sure the brake caliper piston location mark is in the correct position so that the locator pin on the brake pad locates correctly into the brake caliper piston—XJ Range 2003.5–06 VIN G00442–G45703 without Brembo brakes

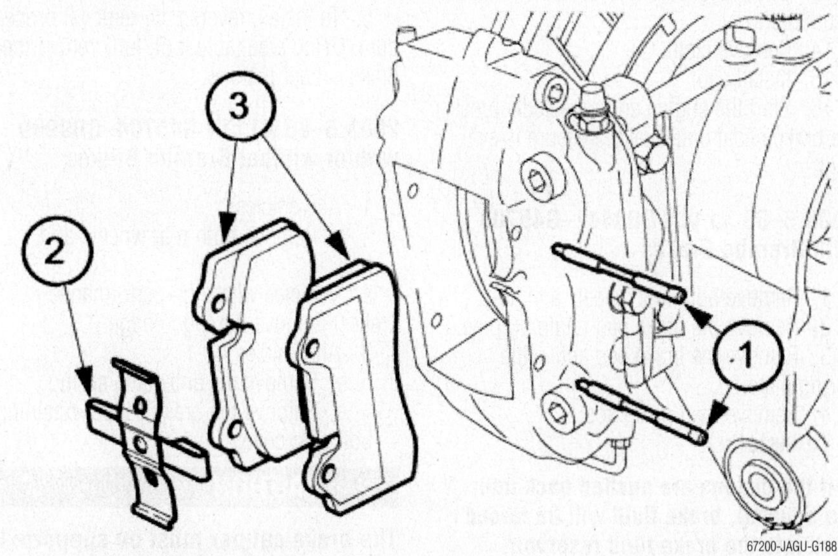

Rear brake pads (3), anti-rattle spring (2) and retaining pins (1)—XJ Range 2002–03 through VIN F59525 with Brembo brakes; 2002–06 XK Range with Brembo brakes

2002–03 XJ Range through VIN F59525 with Brembo Brakes

2002–06 XK Range with Brembo Brakes

1. Raise rear of vehicle and support on stands.
2. Remove rear wheels.
3. Loosen brake fluid reservoir-cap.
4. Position a cloth around the reservoir to collect any fluid spillage.
5. Remove pins from caliper. Remove anti-rattle spring. Discard brake pads.
6. Clean all mating surfaces.

To install:

CAUTION

Make sure grease does not contact brake-pad friction surface or brake disc.

➡To prevent brake squeal apply disc brake grease to the shoulders of the brake-pad back plate. Apply a uniform layer of grease to back plate shoulders, making sure the whole shoulder is covered.

➡Retracting the caliper piston may cause the fluid reservoir to over-flow.

7. Note condition of pins and replace if necessary.
8. Slowly retract pistons into caliper. Install pads carefully, making sure that the grease on the shoulders is not disturbed. This is particularly relevant to the grease on the pad's trailing shoulder.

Apply a uniform layer of grease to back plate shoulders, making sure the whole shoulder is covered—XJ Range 2002–03 through VIN F59525 with Brembo brakes; 2002–06 XK Range with Brembo brakes

9. Position anti-rattle spring.
10. Install pins. Make sure pin collets are fully engaged into caliper.
11. Install wheels.
12. Remove stands and lower vehicle.
13. Check brake fluid level is at the maximum mark.
14. Remove cloth.
15. Install cap.
16. Start the engine and repeatedly press the brake pedal until brake pressure is evident.

2003.5–06 XJ VIN G00442–G45703 with Brembo Brakes

1. Remove both rear wheels and tires.
2. Remove the brake pad retaining pins.
3. Remove the brake pad anti-rattle spring plate.
4. Remove the brake pads.

To install:

➡If the pistons are pushed back into the housing, brake fluid will be forced back into the brake fluid reservoir.

➡Make sure the brake disc faces are clean before installation.

⁂ WARNING

Brake pads must always be replaced in axle sets.

⁂ WARNING

Make sure the brake pads are installed with the direction arrow facing downwards.

5. To install, reverse the removal procedure. Using a suitable tool, fully retract the brake caliper pistons.

2003.5–06 XJ VIN G45704–G99999 with or without Brembo Brakes

1. All vehicles:
   a. Remove both rear wheels and tires.
2. Vehicles with high performance brakes: Remove the logo badge.
3. All vehicles:
   a. Remove the anti-rattle spring.
   b. Remove the brake caliper retaining bolt dust covers.

⁂ CAUTION

The brake caliper must be supported at all times.

   c. Remove the brake caliper retaining bolts.
   d. Detach the brake caliper.
   e. Remove the brake pads.

To install:

4. All vehicles:

⁂ CAUTION

As the piston is pushed back into the caliper housing, the brake fluid level in the reservoir will rise. Do not allow the reservoir to overflow.

⁂ CAUTION

If brake fluid comes into contact with the paintwork, the affected area must be immediately washed down with cold water.

➡Only the extraction bolt from special tool 303-588 is used.

➡Only the forcing plate from special tool 206-080 is used.

   a. Using the special tools, fully retract the brake caliper piston.

⁂ WARNING

Do not use compressed air to disperse brake dust into the atmosphere.

⁂ CAUTION

Make sure the brake disc faces are clean before installation.

⁂ CAUTION

Brake pads must always be replaced in axle sets.

➡Make sure that the brake pad with the anti-rattle spring is installed to the inboard side of the brake disc.

   b. To install, reverse the removal procedure. Tighten the caliper pins to 21 ft. lbs. (28 Nm).
5. Vehicles with high performance brakes: Install the logo badge: Tighten to 44 inch lbs. (5 Nm).
6. All vehicles:

➡Do not start the engine or operate the electronic park brake before depressing the brake pedal.

   a. Repeatedly depress the brake pedal until brake pressure is evident.

LAND ROVER

Discovery • Freelander • Range Rover

6

SPECIFICATION AND MAINTENANCE CHARTS

ENGINE AND VEHICLE IDENTIFICATION

Engine						
Code ①	Liters (cc)	Cu. In.	Cyl.	Fuel Sys.	Engine Type	Eng. Mfg.
G	2.5 (2497)	152	6	MFI	DOHC	Land Rover
2	2.5 (2497)	152	6	SFI	DOHC	Land Rover
2	4.0 (3950)	241	8	MFI	OHV	Land Rover
V	4.0 (3950)	241	8	MFI	OHV	Land Rover
1	4.4 (4398)	268	8	MFI	DOHC	Land Rover
4	4.4 (4398)	268	8	MFI	OHV	Land Rover
J	4.6 (4554)	278	8	MFI	OHV	Land Rover
4	4.6 (4554)	278	8	MFI	OHV	Land Rover

Model Year	
Code ②	Year
1	2001
2	2002
3	2003
4	2004
5	2005

MFI: Multi-port Fuel Injection

SFI: Sequential Fuel Injection

OHV: Overhead Valve

DOHC: Dual Overhead Camshaft

① 8th position of VIN

② 10th position of VIN

67200-RANG-C01

GENERAL ENGINE SPECIFICATIONS

Year	Model	Engine Displacement Liters	Engine ID/VIN	Net Horsepower @ rpm	Net Torque @ rpm (ft. lbs.)	Bore x Stroke (in.)	Com-pression Ratio	Oil Pressure @ rpm
2001	Discovery Series II	4.0	2	188@4750	250@2600	3.7x2.8	9.34:1	35@2400
	Range Rover	4.6	J	222@4750	300@2600	3.7x3.2	9.34:1	35@2400
2002	Freelander	2.5	G	175@6250	177@3000	3.2x3.3	10.5:1	44@3000
	Discovery Series II	4.0	2	188@4750	250@2600	3.7x2.8	9.34:1	35@2400
	Range Rover	4.6	J	222@4750	300@2600	3.7x3.2	9.34:1	35@2400
2003	Freelander	2.5	2	177@6500	177@4000	3.2x3.3	10.5:1	44@3000
	Discovery Series II	4.6	4	217@4750	177@4000	3.2x3.3	10.5:1	50@2000
	Range Rover	4.4	4	282@5400	325@3600	3.6X3.3	10.0:1	50@2000
2004	Freelander	2.5	2	177@6500	177@4000	3.2x3.3	10.5:1	44@3000
	Discovery Series II	4.6	4	217@4750	300@2600	3.7x3.3	9.35:1	50@2000
	Range Rover	4.4	1	282@5400	325@3600	3.6X3.3	10.0:1	44@3500

67200-RANG-C02

GASOLINE ENGINE TUNE-UP SPECIFICATIONS

Year	Engine Displacement Liters	Engine ID/VIN	Spark Plug Gap (in.)	Ignition Timing (deg.) MT	Ignition Timing (deg.) AT	Fuel Pump (psi)	Idle Speed (rpm) MT	Idle Speed (rpm) AT	Valve Clearance Intake	Valve Clearance Exhaust
2001	4.0	2	0.035	—	①	34-37	—	675-725	HYD	HYD
	4.6	J	0.038	—	①	34-37	—	680-720	HYD	HYD
2002	2.5	G	0.039	—	①	54.7	—	700-800	HYD	HYD
	4.0	2	0.035	—	①	34-37	—	675-725	HYD	HYD
	4.6	J	0.038	—	①	34-37	—	680-720	HYD	HYD
2003	2.5	2	0.039	—	①	43.5	—	700-800	HYD	HYD
	4.4	4	0.040	—	①	50.75	—	610-720	HYD	HYD
	4.6	4	0.040	—	①	50.75	—	610-720	HYD	HYD
2004	2.5	2	0.039	—	①	54.7	—	700-800	HYD	HYD
	4.4	1	0.040	—	①	50.75	—	600-750	HYD	HYD
	4.6	4	0.040	—	①	50.75	—	610-710	HYD	HYD

HYD: Hydraulic

① Automatically controlled by the Powertrain Control Module (PCM).

67200-RANG-C03

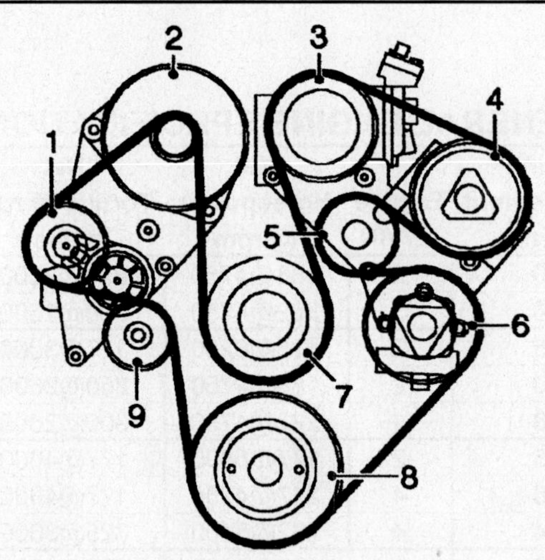

V8 auxiliary belt

1 Tensioner pulley
2 Alternator
3 A/C compressor
4 ACE pump
5 Idler pulley - V8 only
6 PAS pump
7 Viscous fan pulley
8 Crankshaft pulley
9 Idler pulley - V8 only

42348-RANG-G37

Accessory drive belt routing—Discovery II and Range Rover 2001–02

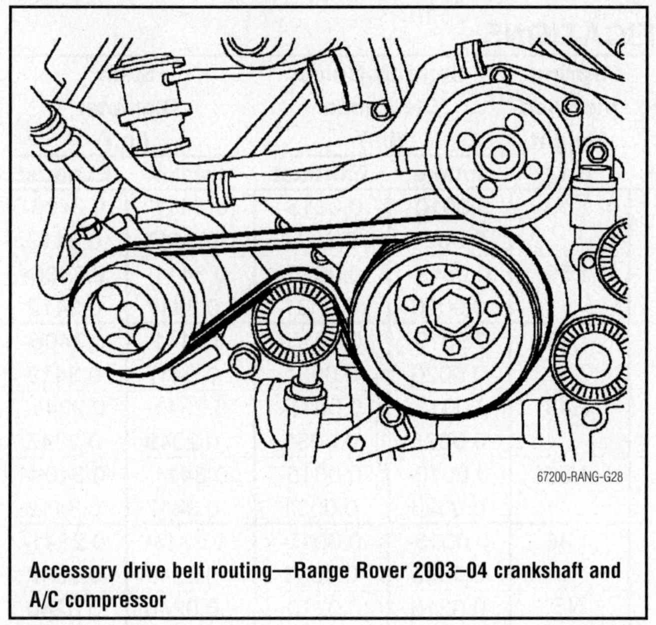

Accessory drive belt routing—Range Rover 2003–04 crankshaft and A/C compressor

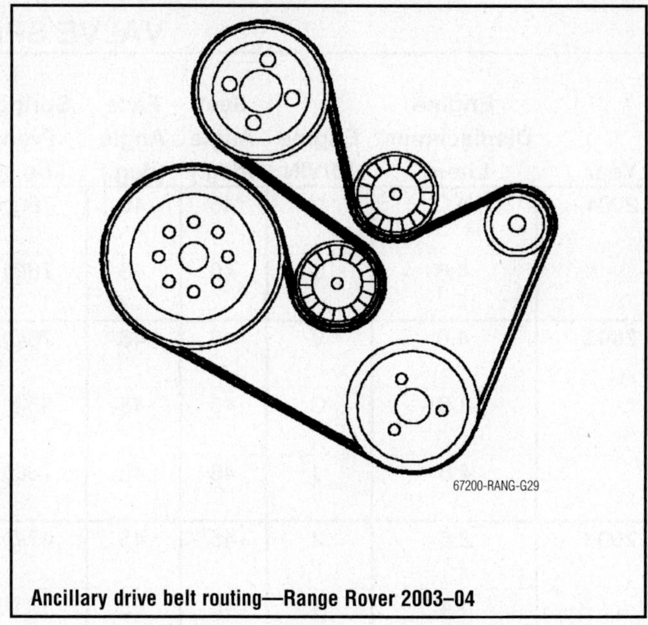

Ancillary drive belt routing—Range Rover 2003–04

CAPACITIES

Year	Model	Engine Displacement Liters	Engine ID/VIN	Engine Oil with Filter (qts.)	Transmission (pts.) 5-Spd	Transmission (pts.) Auto.	Transfer Case (pts.)	Drive Axle Front (pts.)	Drive Axle Rear (pts.)	Fuel Tank (gal.)	Cooling System (qts.)
2001	Discovery Series II	4.0	V	5.6	—	19.2	4.9	3.6	3.6	24.6	24.0
	Range Rover	4.6	J	7.0	—	23.2	5.0	3.6	3.6	24.6	24.0
2002	Discovery Series II	4.0	V	5.6	—	19.2	4.9	3.6	3.6	24.6	24.0
	Freelander	2.5	G	6.3	—	18.0	2.3	—	1.75	15.6	16.5
	Range Rover	4.6	J	7.0	—	20.5	5.0	3.6	3.6	24.6	24.0
2003	Discovery Series II	4.6	4	6.6 ①	—	19.7	4.1	3.4	3.4	25.5	8.1
	Freelander	2.5	2	6.3	—	18.2	—	—	②	17.2	8.6
	Range Rover	4.4	1	9.0	—	12.6	2.0	1.6	2.1	25.8	13.7
2004	Discovery Series II	4.6	4	6.6 ①	—	19.7	4.1	3.4	3.4	25.5	8.1
	Freelander	2.5	2	6.3	—	18.2	—	—	②	17.2	8.6
	Range Rover	4.4	1	9.0	—	12.6	2.0	1.6	2.1	25.8	13.7

NOTE: All capacities are approximate. Add fluid gradually and check to be sure a proper fluid level is obtained.

① With oil cooler: 6.6 qts.
Without oil cooler: 6.1 qts.

② Maximum differential refill is 28.1 fl. oz.; no specifications given for front and rear axle capacities.

67200-RANG-C04

VALVE SPECIFICATIONS

Year	Engine Displacement Liters	Engine ID/VIN	Seat Angle (deg.)	Face Angle (deg.)	Spring Test Pressure (lbs. @ in.)	Spring Installed Height (in.)	Stem-to-Guide Clearance (in.) Intake	Stem-to-Guide Clearance (in.) Exhaust	Stem Diameter (in.) Intake	Stem Diameter (in.) Exhaust
2001	4.0	V	46	46	76@1.59	1.59	0.0010-0.0026	0.0015-0.0031	0.3411-0.3417	0.3406-0.3412
	4.6	J	46	46	76@1.59	1.59	0.0010-0.0026	0.0015-0.0031	0.3411-0.3417	0.3406-0.3412
2002	4.0	V	46	46	76@1.59	1.59	0.0010-0.0026	0.0015-0.0031	0.3411-0.3417	0.3406-0.3412
	2.5	G	45	45	55@1.48	1.48	0.0013-0.0025	0.0015-0.0031	0.2343-0.2349	0.2341-0.2347
	4.6	J	46	46	76@1.59	1.59	0.0010-0.0026	0.0015-0.0031	0.3411-0.3417	0.3406-0.3412
2003	2.5	2	45	45	47@1.46	1.46	0.0013-0.0025	0.0015-0.0031	0.2343-0.2349	0.2341-0.2347
	4.4	4	45	45	NS	NS	0.0210	0.0210	0.0240	0.0240
	4.6	4	46	46	NS	1.61	NS	NS	0.3411-0.3417	0.3400-0.3410
2004	2.5	2	45	45	47@1.46	1.46	0.0013-0.0025	0.0015-0.0031	0.2343-0.2349	0.2341-0.2347
	4.4	1	45	45	NS	NS	0.0210	0.0210	0.0240	0.0240
	4.6	4	46	46	NS	1.61	NS	NS	0.3411-0.3417	0.3400-0.3410

NS: Not specified by manufacturer.

67200-RANG-C05

CRANKSHAFT AND CONNECTING ROD SPECIFICATIONS

All measurements are given in inches.

	Engine		Crankshaft				Connecting Rod		
Year	Displacement Liters	Engine ID/VIN	Main Brg. Journal Dia.	Main Brg. Oil Clearance	Shaft End-play	Thrust on No.	Journal Diameter	Oil Clearance	Side Clearance
2001	4.0	V	2.4995-2.5000	0.0004-0.0019	0.0040-0.0080	3	2.1850-2.1856	0.0006-0.0022	0.006-0.014
	4.6	J	2.4995-2.5000	0.0004-0.0019	0.0040-0.0080	3	2.1850-2.1856	0.0006-0.0022	0.006-0.014
2002	4.0	V	2.4995-2.5000	0.0004-0.0019	0.0040-0.0080	3	2.1850-2.1856	0.0006-0.0022	0.006-0.014
	2.5	G	2.6670-2.6673	0.0008-0.0015	0.0040-0.0080	4	2.1279-2.1281	0.0009-0.0016	0.007-0.011
	4.6	J	2.4995-2.5000	0.0004-0.0019	0.0040-0.0080	3	2.1850-2.1856	0.0006-0.0022	0.006-0.014
2003	2.5	2	①	0.0008-0.0015	0.0040-0.0120	NS	NS	NS	NS
	4.4	4	②	0.0008-0.0020	0.0033-0.0100	NS	NS	NS	NS
	4.6	4	2.4995-2.5200	0.00059-0.00063	0.0030-0.0100	3	NS	NS	NS
2004	2.5	2	①	0.0008-0.0015	0.0040-0.0120	NS	NS	NS	NS
	4.4	4	②	0.0008-0.0020	0.0033-0.0100	NS	NS	NS	NS
	4.6	4	2.4995-2.5200	0.00059-0.00063	0.0030-0.0100	3	NS	NS	NS

NS: Not specified by manufacturer.

① Grade 1: 2.6670-2.6673
Grade 2: 2.6668-2.6670
Grade 3: 2.6666-2.6668

② Standard Yellow: 2.755
Standard Green: 2.754
Standard White: 2.753

67200-RANG-C06

PISTON AND RING SPECIFICATIONS

All measurements are given in inches.

Year	Engine Displacement Liters	Engine ID/VIN	Piston Clearance	Ring Gap: Top Compression	Ring Gap: Bottom Compression	Ring Gap: Oil Control	Ring Side Clearance: Top Compression	Ring Side Clearance: Bottom Compression	Ring Side Clearance: Oil Control
2001	4.0	V	0.0010-0.0020	0.010-0.020	0.016-0.030	0.014-0.050	0.0020-0.0040	0.0020-0.0040	SNUG
	4.6	J	0.0010-0.0020	0.010-0.020	0.016-0.030	0.014-0.050	0.0020-0.0040	0.0020-0.0040	SNUG
2002	4.0	V	0.0010-0.0020	0.010-0.020	0.016-0.030	0.014-0.050	0.0020-0.0040	0.0020-0.0040	SNUG
	2.5	G	0.0013-0.0014	0.008-0.014	0.011-0.018	0.010-0.039	0.0020-0.0031	0.0012-0.0024	0.0004-0.0071
	4.6	J	0.0010-0.0020	0.010-0.020	0.016-0.030	0.014-0.050	0.0020-0.0040	0.0020-0.0040	SNUG
2003	2.5	2	0.0013-0.0014	0.008-0.014	0.011-0.018	0.010-0.039	0.0020-0.0031	0.0012-0.0024	0.0004-0.0071
	4.4	4	0.0002-0.0014	0.003-0.011	0.008-0.016	0.008-0.035	0.0008-0.0023	0.0008-0.0023	SNUG
	4.6	4	0.0010-0.0020	0.012-0.020	0.016-0.026	0.015-0.055	0.0020-0.0040	0.0020-0.0040	SNUG
2004	2.5	2	0.0013-0.0014	0.008-0.014	0.011-0.018	0.010-0.039	0.0020-0.0031	0.0012-0.0024	0.0004-0.0071
	4.4	1	0.0002-0.0014	0.003-0.011	0.008-0.016	0.008-0.035	0.0008-0.0023	0.0008-0.0023	SNUG
	4.6	4	0.0010-0.0020	0.012-0.020	0.016-0.026	0.015-0.055	0.0020-0.0040	0.0020-0.0040	SNUG

67200-RANG-C07

TORQUE SPECIFICATIONS

All readings in ft. lbs.

Year	Engine Displacement Liters	Engine ID/VIN	Cylinder Head Bolts	Main Bearing Bolts	Rod Bearing Bolts	Crankshaft Damper Bolts	Flywheel Bolts	Manifold		Spark Plugs	Oil Drain Plug
								Intake	Exhaust		
2002	4.0	V	①	②	③	200	58	④	40	15	24
	2.5	G	⑤	⑥	⑦	118	55	18	18	18	18
	4.6	J	①	②	③	200	58	④	40	15	24
2002	4.0	V	①	②	③	200	58	④	40	15	24
	2.5	G	⑤	⑥	⑦	118	55	18	18	18	18
	4.6	J	①	②	③	200	58	④	40	15	24
2003	2.5	2	⑧	⑨	⑩	118	⑪	18	33	18	18
	4.4	4	⑫	NS	NS	⑬	NS	11	17	23	13
	4.6	4	①	⑭	⑮	200	58	⑯	40	15	24
2004	2.5	2	⑧	⑨	⑩	118	⑪	18	33	18	18
	4.4	4	⑫	⑬	⑭	⑮	NS	11	17	23	13
	4.6	4	①	⑯	⑰	200	58	⑱	40	15	24

NS: Not specified by manufacturer.

① Step 1: 15 ft. lbs.
Step 2: additional 90 degrees
Step 3: additional 90 degrees

② Bolts 1-8:
Step 1: 10 ft. lbs.
Step 2: 53 ft. lbs.
Bolts 9 & 10:
Step 1: 10 ft. lbs.
Step 2: 68 ft. lbs.
Bolts 11-20:
Step 1: 10 ft. lbs.
Step 2: 33 ft. lbs.

③ Step 1: 15 ft. lbs.
Step 2: additional 80 degrees

④ Step 1: 84 inch lbs.
Step 2: 38 ft. lbs.

⑤ Step 1: 18 ft. lbs.
Step 2: retighten to 18 ft. lbs.
Step 3: Retighten to 18 ft. lbs.
Step 4: Tighten all an additional 180 degrees

⑥ Bearing cradle
Step 1: 15 ft. lbs.
Step 2: plus 90 degrees

⑦ Step 1: 15 ft. lbs.
Step 2: plus 70 degrees

⑧ Step 1: 18 ft. lbs.
Step 2: 18 ft. lbs.
Step 3: 18 ft. lbs.
Step 4: additional 180 degrees

⑨ M6 Bolts: 7 ft. lbs.
M8 Bolts: 18 ft. lbs.

⑩ Step 1: 15 ft. lbs.
Step 2: additional 45 degrees

⑪ Driveplate to crankshaft
Step 1: 18 ft. lbs.
Step 2: 74 ft. lbs.

⑫ Step 1: 22 ft. lbs.
Step 2: additional 80 degrees
Step 3: additional 80 degrees

⑬ Step 1: 15 ft. lbs.
Step 2: additional 100 degrees

⑭ Step 1: 4 ft. lbs.
Step 2: 15 ft. lbs.
Step 3: additional 70 degrees

⑮ Step 1: 74 ft. lbs.
Step 2: additional 60 degrees
Step 3: additional 60 degrees
Step 4: additional 30 degrees

⑯ Initial Torque:
All bolts: 10 ft. lbs.
Final Torque:
Cap bolts 1-8: 53 ft. lbs.
Cap bolts 9,10: 68 ft. lbs.
Cap side bolts 11-15: 33 ft. lbs.
Cap bolts 16-20: 33 ft. lbs.

⑰ Step 1: 15 ft. lbs.
Step 2: additional 80 degrees

⑱ Step 1: 7 ft. lbs.
Step 2: 38 ft. lbs.

67200-RANG-C08

WHEEL ALIGNMENT

			Caster		Camber		
Year	Model		Range (+/-Deg.)	Preferred Setting (Deg.)	Range (+/-Deg.)	Preferred Setting (Deg.)	Toe-in (in.)
2001	Discovery	F	0.10	+3.70	NS	0	0.16+/-0.83
	Range Rover	F	NS	+4.00	NS	0	0.05+/-0.03
2002	Discovery	F	0.10	+3.70	NS	0	0.16+/-0.83
	Freelander	F	1.00	+3.42	0.75	-0.25	0.23+/-0.25 (out)
		R	NS	NS	0.75	+0.50	0.30+/-0.50
	Range Rover	F	NS	+4.00	NS	0	0.05+/-0.03
2003	Discovery	F	0.75	+3.75	0.50	+0.40	0.1+/-0.1 (out)
	Freelander	F	1.00	+3.50	0.75	-0.25	0.23+/-0.23
		R	1.00	NS	0.75	-0.50	NS
	Range Rover	F	NS	+6.70	0.50	-0.40	0.0+/-0.25
		R	NS	NS	NS	-1.00	0.40+/-0.25
2004	Discovery	F	1.00	+3.34	0.75	+0.34	0.1+/-0.1 (out)
	Freelander	F	1.00	+3.50	1.00	-0.25	0.23+/-0.23
		R	NS	NS	1.00	-0.50	NS
	Range Rover	F	NS	+6.41	0.50	-0.34	0.0+/-0.23
		R	NS	NS	NS	-1.34	0.30+/-0.20

NS: Not specified by manufacturer.

67200-RANG-C09

TIRE, WHEEL AND BALL JOINT SPECIFICATIONS

Year	Model	OEM Tires		Tire Pressures (psi)		Wheel	Lug Nut
		Standard	Optional	Front	Rear	Size	(ft. lbs.)
2001	Discovery	255/65R16	255/55R18	28	46	8J	103
	Range Rover SE	255/65R16	None	28	38	8J	103
	Range Rover HSE	255/55HR18	None	28	38	8J	103
2002	Discovery SD	255/65HR16	None	28	46	8J	103
	Discovery SE	P255/55HR18	None	28	46	8J	103
	Freelander S	P215/65R16	None	26 ②	26 ②	6J	85
	Freelander SE	P225/55HR17	None	26 ②	26 ②	7J	85
	Freelander HSE	P225/55HR17	None	26 ②	26 ②	7J	85
	Range Rover HSE	P225/65HR18	None	28	38	8J	103
2003	Discovery S	235/70R16	None	30 ③	38 ③	8J	103
	Discovery SE	P255/65R16	None	30 ③	38 ③	8J	103
	Discovery HSE	P255/55R18	None	30 ③	38 ③	8J	103
	Freelander S	P215/65R15	None	30	30	6J	85
	Freelander SE/HSE	P225/55R17	None	30	30	7J	85
	Range Rover HSE	P255/55HR19	None	33 ④	36 ④	8J	103
2004	Discovery S	P255/65HR16	None	30 ③	38 ③	8J	103
	Discovery SE/HSE	P255/55HR18	None	30 ③	38 ③	8J	103
	Freelander SE	P215/65HR16	None	30	30	6J	85
	Freelander SE3/HSE	P225/55HR17	None	30	30	7J	85
	Range Rover HSE	P255/55HR19	None	33 ④	36 ④	8J	103

OEM: Original Equipment Manufacturer

PSI: Pounds Per Square Inch

STD: Standard

OPT: Optional

NS: Not Specified By Manufacturer

① Ft. lbs.

② For towing or max GVW: 30 psi

③ For towing or max GVW:
Front: 38 psi
Rear: 46 psi

④ For towing or max GVW:
Front: 36 psi
Rear: 44 psi

67200-RANG-C10

BRAKE SPECIFICATIONS

All measurements in inches unless noted

		Front Brake Disc			Rear Brake Disc			Minimum Lining Thickness		Brake Caliper	
Year	Model	Original Thickness	Minimum Thickness	Maximum Run-out	Original Thickness	Minimum Thickness	Maximum Run-out	Front	Rear	Bracket Bolts (ft. lbs.)	Mounting Bolts (ft. lbs.)
2001	Discovery	0.984	0.866	0.006	0.500	0.461	0.006	0.079	0.079	①	②
	Range Rover	1.000	0.870	0.006	0.500	0.460	0.006	0.080	0.080	③	④
2002	Discovery	0.984	0.866	0.006	0.500	0.461	0.006	0.079	0.079	①	②
	Freelander	0.822	0.708	0.0016	⑤	⑥	⑦	0.118	0.079	20	70
	Range Rover	1.000	0.870	0.006	0.500	0.460	0.006	0.080	0.080	③	④
2003	Discovery	0.988	0.866	0.006	0.500	0.461	0.006	0.079	0.079	22	⑧
	Freelander	0.826	0.708	0.0016	⑤	⑥	⑦	0.118	0.079	20	74
	Range Rover	1.180	1.120	0.0023	0.470	0.410	0.003	0.012	0.012	24	⑨
2004	Discovery	0.988	0.866	0.006	0.500	0.461	0.006	0.079	0.079	22	⑧
	Freelander	0.826	0.708	0.0016	⑤	⑥	⑦	0.118	0.079	20	74
	Range Rover	1.180	1.120	0.0028	0.470	0.410	0.003	0.012	0.012	24	⑨

① Front: 129 ft. lbs.
Rear: 70 ft. lbs.

② Both front and rear calipers: 22 ft. lbs.

③ Front: 122 ft. lbs.
Rear: 74 ft. lbs.

④ Front: 19 ft. lbs.
Rear: 26 ft. lbs.

⑤ Drum original diameter: 10.0 in.

⑥ Drum discard diameter: 10.059 in.

⑦ Drum out of round limit: 0.0005 in.

⑧ Front: 129 ft. lbs.
Rear: 70 ft. lbs.

⑨ Front: 81 ft. lbs.; Rear: 48 ft. lbs.

67200-RANG-C11

SCHEDULED MAINTENANCE INTERVALS
Land Rover—Discovery, Range Rover & Freelander

TO BE SERVICED	TYPE OF SERVICE	VEHICLE MILEAGE INTERVAL (x1000)															
		7.5	15	22.5	30	37.5	45	52.5	60	67.5	75	82.5	90	97.5	105	112.5	120
Air cleaner filter	R				✓				✓				✓				✓
Brake pads, calipers & rotors	S/I	✓	✓	✓	✓	✓	✓	✓	✓	✓	✓	✓	✓	✓	✓	✓	✓
Coolant hoses	S/I	✓	✓	✓	✓	✓	✓	✓	✓	✓	✓	✓	✓	✓	✓	✓	✓
Driveshafts & U-joints	L		✓		✓		✓		✓		✓		✓		✓		✓
Engine coolant	R				✓				✓				✓				✓
Engine oil & filter	R	✓	✓	✓	✓	✓	✓	✓	✓	✓	✓	✓	✓	✓	✓	✓	✓
Exhaust system	S/I	✓	✓	✓	✓	✓	✓	✓	✓	✓	✓	✓	✓	✓	✓	✓	✓
Front and rear axle oil	R				✓				✓				✓				✓
Fuel filter	R								✓								✓
Oxygen sensors	R											✓					
Parking brake	S/I		✓		✓		✓		✓		✓		✓		✓		✓
Power steering fluid	S/I		✓		✓		✓		✓		✓		✓		✓		✓
Radiator and A/C condenser	S/I		✓		✓		✓		✓		✓		✓		✓		✓
Serpentine drive belt	R										✓						
Spark plugs	R				✓				✓				✓				✓
Steering linkage	S/I		✓		✓		✓		✓		✓		✓		✓		✓
Air bag (SRS)	S/I	Every 10 years															
Suspension links & mountings	S/I		✓		✓		✓		✓		✓		✓		✓		✓
Tires	S/I	✓	✓	✓	✓	✓	✓	✓	✓	✓	✓	✓	✓	✓	✓	✓	✓
Transfer gearbox oil	R				✓				✓				✓				✓
Transmission fluid	R				✓				✓				✓				✓
Transmission fluid filter	R				✓								✓				

R: Replace S/I: Service or Inspect L: Lubricate A: Adjust

FREQUENT OPERATION MAINTENANCE (SEVERE SERVICE)

If a vehicle is operated under any of the following conditions it is considered severe service:

- Towing a trailer or using a camper or car-top carrier.

- Repeated short trips of less than 5 miles in temperatures below freezing, or trips of less than 10 miles in any temperature.

- Extensive idling or low-speed driving for long distances as in heavy commercial use, such as delivery, taxi or police cars.

- Operating on rough, muddy or salt-covered roads.

- Operating on unpaved or dusty roads.

- Frequent operation in temperatures above 90°F.

Air cleaner element: replace every 15,000 miles

Brake fluid: replace every 15,000 miles

Driveshafts & U-joints: lubricate every 7,500 miles

Engine coolant: replace every 15,000 miles

Engine oil & filter: replace every 3,750 miles.

Front & rear axle oil: replace every 15,000 miles.

Fuel filter: replace every 30,000 miles.

Power steering fluid level: inspect every 7,500 miles.

Serpentine drive belt: inspect every 15,000 miles and replace every 30,000 miles.

Shock absorbers: inspect every 7,500 miles.

Spark plugs: replace every 15,000 miles.

Tires: inspect every 3,750 miles.

Transfer gearbox oil: replace every 15,000 miles.

Transmission fluid & filter: replace every 15,000 miles.

67200-RANG-C12

PRECAUTIONS

Before servicing any vehicle, please be sure to read all of the following precautions, which deal with personal safety, prevention of component damage, and important points to take into consideration when servicing a motor vehicle:

- Never open, service or drain the radiator or cooling system when the engine is hot; serious burns can occur from the steam and hot coolant.
- Observe all applicable safety precautions when working around fuel. Whenever servicing the fuel system, always work in a well-ventilated area. Do not allow fuel spray or vapors to come in contact with a spark, open flame, or excessive heat (a hot drop light, for example). Keep a dry chemical fire extinguisher near the work area. Always keep fuel in a container specifically designed for fuel storage; also, always properly seal fuel containers to avoid the possibility of fire or explosion. Refer to the additional fuel system precautions later in this section.
- Fuel injection systems often remain pressurized, even after the engine has been turned **OFF**. The fuel system pressure must be relieved before disconnecting any fuel lines. Failure to do so may result in fire and/or personal injury.
- Brake fluid often contains polyglycol ethers and polyglycols. Avoid contact with the eyes and wash your hands thoroughly after handling brake fluid. If you do get brake fluid in your eyes, flush your eyes with clean, running water for 15 minutes. If eye irritation persists, or if you have taken brake fluid internally, IMMEDIATELY seek medical assistance.
- The EPA warns that prolonged contact with used engine oil may cause a number of skin disorders, including cancer! You should make every effort to minimize your exposure to used engine oil. Protective gloves should be worn when changing oil. Wash your hands and any other exposed skin areas as soon as possible after exposure to used engine oil. Soap and water, or waterless hand cleaner should be used.
- All new vehicles are now equipped with an air bag system. The system must be disabled before performing service on or around system components, steering column, instrument panel components, wiring and sensors. Failure to follow safety and disabling procedures could result in accidental air bag deployment, possible personal injury and unnecessary system repairs.
- Always wear safety goggles when working with, or around, the air bag system. When carrying a non-deployed air bag be sure the bag and trim cover are pointed away from your body. When placing a non-deployed air bag on a work surface, always face the bag and trim cover upward, away from the surface. This will reduce the motion of the module if it is accidentally deployed. Refer to the additional air bag system precautions later in this section.
- Clean, high quality brake fluid from a sealed container is essential to the safe and proper operation of the brake system. You should always buy the correct type of brake fluid for your vehicle. If the brake fluid becomes contaminated, completely flush the system with new fluid. Never reuse any brake fluid. Any brake fluid that is removed from the system should be discarded. Also, do not allow any brake fluid to come in contact with a painted surface; it will damage the paint.
- Never operate the engine without the proper amount and type of engine oil; doing so WILL result in severe engine damage.
- Timing belt maintenance is extremely important! Many models utilize an interference-type, non-freewheeling engine. If the timing belt breaks, the valves in the cylinder head may strike the pistons, causing potentially serious (also time-consuming and expensive) engine damage. Refer to the maintenance interval charts in the front of this chapter for the recommended replacement interval for the timing belt.
- Disconnecting the negative battery cable on some vehicles may interfere with the functions of the on-board computer system(s) and may require the computer to undergo a relearning process once the negative battery cable is reconnected.
- When servicing drum brakes, only disassemble and assemble one side at a time, leaving the remaining side intact for reference.
- Only an MVAC-trained, EPA-certified automotive technician should service the air conditioning system or its components.

ENGINE REPAIR

Distributor

The vehicles covered in this section are equipped with DIS.

Alternator

REMOVAL & INSTALLATION

Freelander (2001–04)

1. Before servicing the vehicle, refer to the precautions in the beginning of this section.
2. Disconnect the battery ground cable.
3. Remove the top arm.
4. Remove the drive belt.
5. Disconnect the wiring from the alternator.
6. Remove the lower bolt, upper nut and front plate bolt
7. Remove the alternator.
8. Installation is the reverse of removal. Tighten all mounting bolts/nut to 33 ft. lbs. (45 Nm).

Discovery (2001–02)

1. Before servicing the vehicle, refer to the precautions in the beginning of this section.
2. Disconnect the battery ground cable.
3. Remove the drive belt.
4. Remove the 2 mounting bolts.
5. Lift the alternator from the bracket and disconnect the wires.
6. Installation is the reverse of removal. If the pulley was removed, tighten the pulley bolt to 30 ft. lbs. (40 Nm). Torque the mounting bolts to 18 ft. lbs. (25 Nm).

Discovery (2003–04)

1. Before servicing the vehicle, refer to the precautions in the beginning of this section.
2. Disconnect the battery ground cable.
3. Remove the upper fan cowl. Remove viscous fan.
4. Loosen the belt tensioner and remove drive belt.
5. Remove the 2 mounting bolts.
6. Disconnect alternator wiring, then remove the alternator from its mounting bracket.
7. Installation is the reverse of removal. Torque alternator mounting bolts to 33 ft. lbs. (45 Nm). Hold tensioner fully clockwise

in order to fit the drive belt around the pulley during installation.

Range Rover (2001–02)

1. Before servicing the vehicle, refer to the precautions in the beginning of this section.
2. Disconnect the battery ground cable.
3. Remove the drive belt.
4. Remove the 2 mounting bolts.
5. Lift the alternator from the bracket and disconnect the wires.
6. Installation is the reverse of removal. Tighten B+ wire nut to 13 ft. lbs. (18 Nm). Tighten D+ nut to 3.5 ft. lbs. (5 Nm). Tighten the alternator mounting bolts to 33 ft. lbs. (45 Nm). Install alternator belt.

Rover (2003–04)

1. Before servicing the vehicle, refer to the precautions in the beginning of this section.
2. Disconnect the battery ground cable.
3. Drain the cooling system.
4. Remove the upper radiator hose.
5. Remove the cooling fan assembly.
6. Position the charcoal canister purge valve aside.
7. Remove the drive belt.
8. Remove the drive belt tensioner.
9. Disconnect the alternator wiring.
10. Remove the 6 mounting bolts.
11. Lift the alternator from the bracket and disconnect the wires.
12. Installation is the reverse of removal. Tighten the alternator wire nuts to 10 ft. lbs. (13 Nm). Tighten the alternator mounting bolts to 10 ft. lbs. (13 Nm). Tighten the drive belt tensioner to 22 ft. lbs. (30 Nm). Install alternator belt.

Ignition Timing

➡The ignition timing is not adjustable. It is controlled by the PCM.

Engine Assembly

REMOVAL & INSTALLATION

Freelander (2001–02)

➡The engine and transaxle are removed as an assembly.

1. Before servicing the vehicle, refer to the precautions in the beginning of this section.
2. Remove the hood.
3. Remove the battery and battery box.
4. Drain the cooling system and engine oil.
5. Drain the transaxle fluid.
6. Release the fuel system pressure.
7. Remove or disconnect the following:
   - Acoustic cover
   - ECM
   - Engine wiring harness connectors
   - Air box
   - Fuse box cover
   - Battery and starter leads from the fuse box
   - All wiring from the fuse box
   - Engine ground strap
   - Wiring from the starter
   - Wiring from the transaxle
   - Fuel line from the fuel rail
   - Hose from the purge valve
   - Throttle cable
   - Heater hoses
   - Radiator hoses
   - All remaining wiring and hoses
   - Front wheels
   - Halfshaft hub nuts
   - Front suspension rear beam
   - Exhaust pipe from the manifold
8. Matchmark the driveshaft–to–IRD unit. Remove the 6 nuts and bolts and disconnect the driveshaft from the IRD. Tie it aside.

WARNING

Do not allow the driveshaft to hang. Damage to the Tripod joint will occur.

9. Drain the IRD fluid.
10. Remove the right and left splash shields.
11. Remove or disconnect the following:
    - Clips securing the brake hoses to the struts
    - ABS sensor wiring
    - Track rod ball studs from the steering arms
    - Halfshafts
    - Selector lever from the gearbox
    - Accessory drive belts
    - A/C compressor and tie it out of the way
12. Attach an engine crane to the engine using an equalizer.
13. Raise the engine just far enough to take the weight off the mounts.
14. Remove the left side engine mount and left transaxle mount.
15. Secure a lifting bracket to the transaxle using the mount bolts.
16. Place a wood block on a jack and raise the transaxle just enough to take its weight. Unhook the chain from the rear engine lift bracket and hook it to the transaxle bracket.
17. Remove the right side engine brackets.
18. Disconnect the power steering pressure hose from the front plate and top arm.
19. Remove the top arm.
20. Remove the power steering pulley.
21. Remove the power steering pump and position it out of the way.
22. Lift the engine from the vehicle.

To install:

23. With assistance, raise engine and gearbox into engine compartment.
24. Position PAS pump to front mounting plate, fit and tighten bolts to 18 ft. lbs. (25 Nm).
25. Position PAS pump pulley, fit and tighten Torx screws to 7 ft. lbs. (9 Nm).
26. Position top arm to engine mounting bracket and right side hydramount, fit and tighten bolts to 74 ft. lbs. (100 Nm).
27. Position PAS pipe support bracket to right side hydramount, fit and tighten nut to 63 ft. lbs. (85 Nm).
28. Place a wooden block on jack, position jack under gearbox and raise jack sufficient to support weight of gearbox. Release lifting hook from lifting bracket on gearbox and connect to rear engine lifting bracket.
29. Lower and remove jack supporting gearbox.
30. Remove bolts securing lifting bracket to gearbox and remove bracket.
31. Position left side mounting bracket to gearbox, fit and tighten bolts to 63 ft. lbs. (85 Nm).
32. Position left side mounting to body, fit and tighten bolts to 35 ft. lbs. (48 Nm).
33. Align gearbox bracket to left side body mounting, fit and tighten through bolt to 74 ft. lbs. (100 Nm).
34. Position upper right side engine steady to top arm, fit and tighten bolt to 74 ft. lbs. (100 Nm).
35. Tighten bolt securing upper right side engine steady to body to 74 ft. lbs. (100 Nm).
36. Position PAS pipe to engine front mounting plate, fit and tighten bolt to 18 ft. lbs. (25 Nm).
37. Lower hoist, disconnect and remove lifting bracket.
38. Raise vehicle on ramp.
39. Position A/C compressor to front mounting plate and cylinder block, align heat shield, fit and tighten bolts to 18 ft. lbs. (25 Nm).
40. Connect multiplug to A/C compressor.
41. Using a 3⁄8 square drive socket bar, raise auxiliary drive belt tensioner and fit drive belt to pulleys.
42. Position selector cable to gearbox bracket and secure with clip.

43. Position selector lever to selector shaft, fit and tighten nut to 18 ft. lbs. (25 Nm).

44. Clean splines and seal areas on each driveshaft and mating faces in front hubs.

45. Install new circlips to right side and left side driveshaft inner joint splines.

46. Install driveshafts to IRD and gearbox, ensuring that the circlip on each driveshaft is fully engaged.

47. Engage left side and right side driveshafts into front hubs.

48. Install new driveshaft flange nuts but do not tighten at this stage.

49. Clean ball joint tapers and taper seats.

50. Position left side and right side track rod ends to steering arms, fit new nuts and tighten to 40 ft. lbs. (55 Nm).

51. Clean ABS sensors and mating faces.

52. Apply anti-seize grease to both ABS sensors and position sensors in front hubs.

➡Ensure ABS sensor is fully located into hub, so that sensor touches pole wheel teeth.

53. Position left side and right side brake hoses to front shock absorber brackets and secure with clips.

54. Position left side and right side splash shields, fit and tighten bolts.

55. Ensure mating face of propeller shaft and IRD drive flange are clean.

56. Install propeller shaft to IRD flange and align marks. Tighten nuts and bolts to 31 ft. lbs. (42 Nm).

57. Install rear beam.

58. Fill IRD to correct level with fluid.

59. Install exhaust front pipe.

60. With assistant depressing the brake pedal, tighten front hub nuts to 295 ft. lbs. (400 Nm).

61. Stake nut to shaft.

62. Install front road wheels, fit and tighten nuts to 85 ft. lbs. (115 Nm).

63. Lower vehicle on ramp.

64. Connect brake servo vacuum hose to intake manifold chamber.

65. Connect coolant hose to underside of expansion tank and secure with clip.

66. Connect expansion tank hose to intake manifold and secure clip.

67. Connect top hose to radiator and secure with clip. Position hose in bracket.

68. Connect heater feed and return hoses and secure with clips.

69. Connect throttle inner cable to throttle cam and secure outer cable in abutment bracket, if fitted.

70. Secure throttle cable in clips on harness brackets, if fitted.

71. Adjust throttle cable, if fitted.

72. Connect hose to purge control valve.

73. Connect fuel hose to fuel rail pipe, fit rubber sleeve over hose connector.

74. Connect gearbox harness multiplugs and secure multiplugs in mounting bracket clips.

75. Connect Lucar connector to starter solenoid.

76. Position ground lead to gearbox housing, fit and tighten bolt to 18 ft. lbs. (25 Nm).

77. Position engine harness to air box mounting bracket and secure with clips.

78. Connect ground header multiplug.

79. Connect multiplug to underhood fuse box.

80. Position battery and starter motor lead to underhood fuse box, fit and tighten bolts to 6 ft. lbs. (8 Nm).

81. Install underhood fuse box cover.

82. Position air box, secure in retaining clip, fit and tighten nut to 7 ft. lbs. (9 Nm).

83. Position carrier in air box and secure with clips.

84. Position and secure air duct and harness rubber sleeve in air box.

85. Connect multiplugs securing main harness to engine harness.

86. Position ECM harness and multiplug to air box, align harness clamp and secure screws to air box.

87. Install engine ECM.

88. Install battery carrier.

89. Fill cooling system.

90. Connect battery ground lead.

91. Fill gearbox with fluid.

92. Install engine acoustic cover.

Freelander (2003–04)

➡The engine and automatic transaxle are removed as an assembly.

1. Before servicing the vehicle, refer to the precautions in the beginning of this section.

2. Disconnect the battery ground.

3. Open the hood and tie it back in an upright position.

4. Drain the cooling system.

5. Remove the engine acoustic cover by releasing clips and retaining straps and disconnecting hoses.

6. Remove the battery and carrier.

7. Remove the ECM by removing electrical box cover and disconnecting plugs, releasing 2 clips and removing the ECM.

8. Remove the bolt retaining the heated front screen harness to the battery positive cable. Remove the clamp.

9. Disconnect the body harness from the engine harness at the electrical box. Move the harness aside.

10. Drain the transaxle fluid.

11. Mark and disconnect remaining multiplugs at electrical box. Release the air duct from the electrical box.

12. Rotate the heated front windshield relay counterclockwise and release it from the electrical box. Then remove the electrical box.

13. Remove the engine compartment fuse box cover. Remove the 2 bolts holding the battery and starter leads to the fuse box.

14. Disconnect the multiplug connector from the fuse box.

15. Disconnect the ground multiplug from the shock tower, and release 3 clips holding the engine harness to the electrical box mounting bracket. Lay the harness over the engine.

16. Release the fuel system pressure.

17. Remove or disconnect the engine ground strap, starter solenoid connector, and the transmission harness multiplug connectors.

18. Position an absorbent cloth around the fuel pipe connections. Pull the back rubber sleeve on the fuel pipe connector. Release the connector and disconnect the fuel pipe from the fuel rail.

19. Remove or disconnect the following:

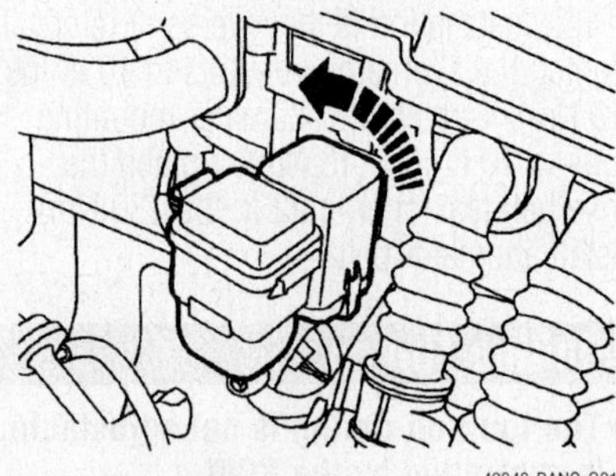

Removing heated front windshield relay—Freelander

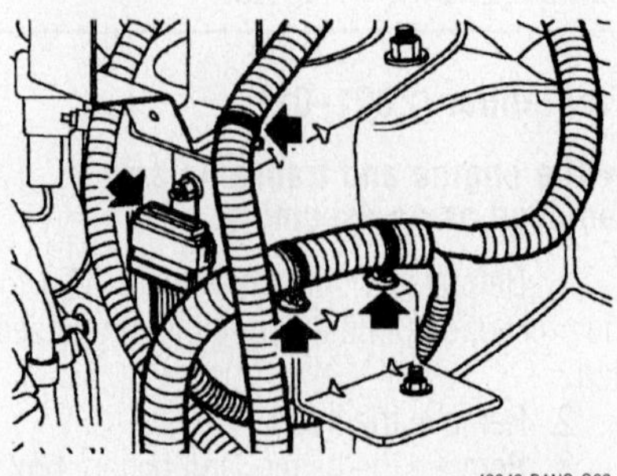

Showing location of ground multiplug and engine harness clips—Freelander

- Canister purge hose from clip and from purge control valve
- Heater hoses
- Upper radiator hose from radiator and retainers
- Expansion tank hose from intake manifold
- Hose from lower side of expansion tank
- Brake servo hose from intake manifold chamber, then plug openings

20. Raise the vehicle on a lift. Remove the front wheels.
21. Release the stake in the driveshaft (hub) nut, hold brake and remove and discard left and right driveshaft nuts.
22. Remove the front suspension rear crossmember by removing the under-vehicle panel, rear engine mount bolt, anti-roll bar clamp bolts (support anti-roll bar), lower control arm rear nuts and bushing housing bolts.
23. Release lower control arms from the crossmember. Support the crossmember and remove 2 remaining crossmember bolts. Remove crossmember.
24. Remove the front exhaust pipes from the manifolds.
25. In order to access the propeller shaft retaining bolts, raise one rear wheel for rotation of the propeller shaft.
26. Reference mark the position of coupling flange to Intermediate Reduction Drive (IRD) unit flange, then remove 6 nuts and bolts holding propeller shaft to IRD drive flange. Release propeller shaft and tie out of way.

CAUTION

Use care to support the U-joint when it is removed from the IRD drive flange coupling. Do not allow unit to fully extend or drop, or damage to boot or component will result.

27. Drain the transmission fluid. Drain the fluid from the IRD axle.
28. Remove the right and left splash shields.
29. Remove the clips securing the brake hoses to their brackets. Release the ABS sensor leads from the brackets. Remove the ABS sensors from the front hubs.
30. Separate the right and left ball joints from the steering arms.
31. Pull the right and left hubs outward and release the driveshafts from the hubs. Use care so driveshafts do not fall.
32. Using a special clamping tool (LRT-54-026), release the driveshafts from the IRD unit and transmission.

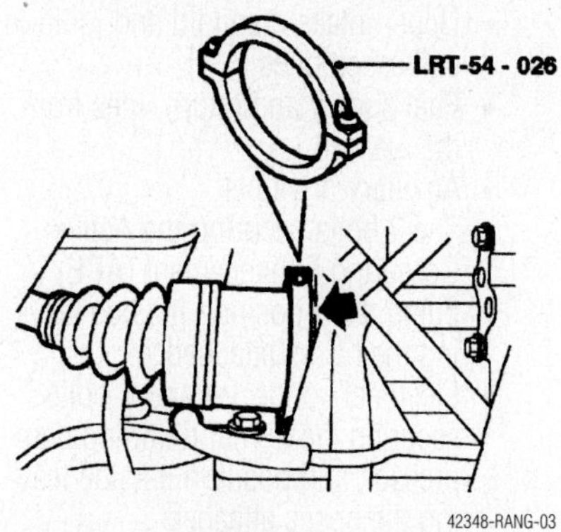

Releasing driveshafts from intermediate reduction drive unit—Freelander

33. Remove the selector lever from the selector shaft on the transmission. Release the cable–to–transmission clip. Move cable aside.
34. Release tension from the accessory drive belt tensioner by loosening the tensioner nut and moving the tensioner clockwise, then retighten the nut in this position.
35. Disconnect the connector from the A/C compressor. Remove compressor bolts and move the compressor aside, with the heat shield attached.
36. Lower the vehicle.
37. With a hoist and lifting bracket attached to engine, raise hoist to remove weight from engine mounts. Remove engine mount bolts.
38. Using transmission bracket bolts, attach transmission-lifting bracket (LRT-44-026). Position a transmission jack under the transmission to support its weight.
39. Release the lifting hook from the rear lift bracket and connect it to the transmission-lifting bracket.
40. Loosen bolts holding the upper left and right engine stays. Position the stay out of the way.
41. Remove or disconnect the following:
    - Power steering pipe from engine front mounting plate
    - Engine top arm bracket from right hydra-mount and front mounting bracket
    - Power steering pipe support bracket from hydra-mount (position the pipe aside)
    - Power steering pipe from engine front mounting plate
    - Top arm
    - Power steering pump from front mounting plate (position pump aside)
42. With an assistant, carefully lower engine and transmission assembly from the vehicle.

To install:

43. With an assistant, raise engine and transmission into engine compartment.
44. Position the power steering pump to its front mounting plate. Install and torque bolts to 18 ft. lbs. (25 Nm).
45. Install the top arm to the engine mounting bracket and right hydra-mount. Torque bolts to 74 ft. lbs. (100 Nm).
46. Install power steering pipe support bracket to right hydra-mount. Torque retaining nut to 63 ft. lbs. (85 Nm).
47. Place a transmission jack under the transmission and raise the jack enough to support the transmission weight. Release the lifting hook from the transmission-lifting bracket and connect it to the rear engine-lifting bracket.
48. Remove the transmission jack.
49. Remove the lifting bracket from the transmission.
50. Install the left transmission mount bracket. Torque bolts to 63 ft. lbs. (85 Nm).
51. Install the left mount to the body and torque bolts to 35 ft. lbs. (48 Nm).
52. Align the transmission mount bracket to the left body mount. Install and torque the bolt to 74 ft. lbs. (100 Nm).
53. Install the upper right engine stay to the top arm. Install and torque the bolt to 74 ft. lbs. (100 Nm).
54. Torque the bolt securing the upper right engine stay to body to 74 ft. lbs. (100 Nm).
55. Install the power steering pipe to the engine front mounting plate and torque the bolt to 18 ft. lbs. (25 Nm).
56. Lower the engine hoist to release pressure, then remove the lifting bracket.
57. Raise the vehicle on the lift.
58. Position the A/C compressor to the front mounting plate and cylinder block. Align the heat shift. Torque the retaining bolts to 18 ft. lbs. (25 Nm).
59. Clean the splines and seal areas on each driveshaft, and the mating faces in the front hubs. Install new circlips to the inner joint splines.
60. Install the driveshaft to the IRD unit and to the transmission. Be sure the circlip on each driveshaft is fully engaged.
61. Position the driveshafts into the front hubs. Install new driveshaft flange nuts, but do not tighten nuts at this stage.
62. Reconnect the ball joints and track rod ends to steering arms. Tighten new retaining nuts to 41 ft. lbs. (55 Nm).
63. Clean the ABS sensors and mating faces. Apply anti-seize grease to both sensors and position into the front hubs. Ensure both sensors are fully seated and touching the pole wheel teeth.

64. Reconnect the brake hoses to their brackets and clips. Install the splash shields.

65. Be sure the mating faces of the propeller shaft and IRD drive flange are clean. Position the shaft to the IRD flange and align the marks made during removal. Torque the retaining nuts and bolts to 31 ft. lbs. (41 Nm).

66. Install the rear crossmember in reverse of the removal procedure.

67. Fill the IRD unit with proper fluid.

68. Install the front exhaust pipes.

69. Have an assistant hold the brake pedal, then install the front hub nuts. Torque nuts to 295 ft. lbs. (400 Nm), and stake the nuts to the shaft.

70. Install front wheels and torque the wheel nuts to 85 ft. lbs. (115 Nm).

71. Lower the vehicle on the lift.

72. Reconnect or install the following:
- Brake servo vacuum hose to intake manifold chamber
- Coolant hoses to expansion tank
- Upper radiator hose
- Heater hoses
- Throttle inner cable to throttle cam and outer cable in bracket
- Canister purge hose
- Fuel hose to rail pipe, then rubber sleeve over hose connector
- Transmission electrical connectors
- Connector to starter solenoid
- Ground strap to transmission; torque bolt to 18 ft. lbs. (25 Nm)
- Engine harness to electrical box
- Multiplug to underhood fuse box
- Battery and starter motor led to underhood fuse box
- Fuse box cover
- Electrical box (E–box) and multiplugs
- Heated front screen harness to battery cable
- ECM
- Battery carrier

73. Refill cooling system.

74. Connect battery cables to battery.

75. Refill transmission.

76. Install engine sound shield.

Discovery (2001–04)

1. Before servicing the vehicle, refer to the precautions in the beginning of this section.

2. Remove or disconnect the following:
- Negative battery cable
- Coolant
- Fuel system pressure
- Engine oil and filter
- Radiator hoses
- Radiator
- Upper intake manifold and ignition coil assemblies
- Fuel supply and return lines from the fuel rail
- Auxiliary drive belt
- The 3 bolts securing the Active Cornering Enhancement (ACE) pump, then position it aside leaving the hoses attached.
- Electrical connector and 4 bolts securing the air conditioning compressor, and position it aside leaving the hoses attached.

⁂ CAUTION

Always plug any line or connections openings immediately.

- Power steering hoses from the pump, then position hoses out of the way
- Coolant hoses at the water pump and rail
- Bolt and harness clips securing the coolant rail and position the rail aside
- Engine ground and power supply cables
- Starter wiring harness
- The 2 engine wiring harness connectors from the fuse box
- EVAP solenoid connector
- Nut mounting the engine harness ground–to–body and detach the engine harness–to–main harness connector.
- Right side interior kick panel
- The 5 connectors attaching the engine harness to the ECM

3. Release the engine wiring harness, pull it into the engine bay and coil on top of engine.

4. Raise and safely support the vehicle.

5. Remove the 3 bolts securing the oil cooler lines to the engine block.

6. Detach the engine oil cooler lines and tie the lines aside.

7. Disconnect the front exhaust pipes from the manifolds.

8. Remove the torque converter access plug, matchmark the torque converter–to–driveplate relationship, then remove the 4 bolts securing the torque converter to the driveplate.

9. Remove the 12 engine–to–transmission mounting bolts (rotating engine as needed for access).

10. Collect support brackets from bell housing.

11. Attach suitable lifting equipment to the engine.

12. Remove the 4 nuts securing the engine mounts, raise the engine and remove the engine mounts.

13. Support the transmission on a jack.

14. Separate the engine from the transmission dowels.

15. With the aid of an assistant, remove the engine from the engine bay.

To install:

16. Clean the mating faces of the engine and transmission, dowel and dowel holes.

17. Lubricate the splines and bearing surface of first motion shaft.

18. With the aid of an assistant, position the engine in the engine bay, align to gearbox and locate on dowels.

19. Install the engine–to–transmission bolts and tighten to 37 ft. lbs. (50 Nm).

20. Install the engine mounts and tighten the fasteners to 63 ft. lbs. (85 Nm).

21. Remove the engine lifting equipment.

22. Align the torque converter to the driveplate, install the bolts and tighten to 37 ft. lbs. (50 Nm).

23. Install or connect the following:
- Torque converter access plug
- Front exhaust pipes to the manifolds
- All electrical connectors and replace any ties.
- Power steering hoses and check the fluid level
- Coolant hoses
- Radiator
- Coolant
- Air conditioning compressor mounting bolts to 16 ft. lbs. (22 Nm)
- ACE pump mounting bolts to 16 ft. lbs. (22 Nm)
- Accessory drive belt
- Fuel supply and return lines to the fuel rail
- Ignition coils
- Spark plug wires
- Upper intake manifold
- Oil filter
- Engine oil
- Transmission fluid level
- Negative battery cable

Range Rover (2001–02)

The engine and transmission are removed and installed, as an assembly, from the vehicle.

1. Before servicing the vehicle, refer to the precautions in the beginning of this section.

2. Remove or disconnect the following:
- Battery
- Fuel system pressure
- Coolant

- ECM located next to the battery
- Wiring harness from the starter and alternator
- Fuel supply and return lines from the fuel rail, then plug the openings to prevent contaminants from entering.
- Purge valve
- Intake hose/air flow meter assembly
- Throttle and cruise control cables from the throttle linkage
- All coolant hoses related to engine/transmission service
- Battery tray
- The 2 fuse box mounting bolts and pivot the fuse box aside.
- Engine wiring harness connector from the base of the fuse box
- Ground cable from the valance stud
- The 2 engine wiring harness connectors from the main harness
- Hood struts from the body locations. Raise the hood in the vertical position and stabilize.

3. Properly discharge the air conditioning system.
4. Remove or disconnect the following:
   - Cooling fan and viscous coupling
   - Grille
   - Hood release cable strap from the upper radiator support
   - Radiator upper support
   - Left and right radiator air deflectors
   - Washer bottle filler neck
   - Engine and transmission oil coolers
   - Coolant hoses from the radiator and thermostat housing
   - The 2 fog lamp breather hoses from the clips on either side of the radiator
   - Power steering fluid
   - Transmission oil temperature sensor connector
   - Refrigerant lines from the air conditioning condenser
   - The 2 nuts and bolts securing the radiator mountings to chassis
5. With the aid of an assistant, raise the radiator assembly for access to the condenser cooling fan connectors.
6. Remove or disconnect the following:
   - The 2 condenser cooling fan connectors
   - Radiator/condenser/oil cooler assembly
   - Window switch pack
   - Handbrake and cable clevis pin
   - Transmission fluid
   - Transfer case fluid
   - Engine oil
   - Exhaust pipes from the manifolds
   - Hand brake cable from the grommet in tunnel
   - Rear driveshaft shaft guard
   - Driveshafts
   - Gear selector cable from the transmission lever
   - The 2 bolts securing selector cable bracket to the transmission
   - Transfer case fluid temperature sensor
   - High/Low motor and output shaft speed sensor connectors
   - Gear selection position switch and transmission speed sensor connectors
   - Engine wiring harness–to–transmission harness connector
   - The 4 nuts securing the engine mounts to the chassis and engine brackets

➡When attaching the engine lifting chain and hoist, it may be necessary to remove the oil filler cap to prevent it from being damaged. Place a cloth over plenum chamber to protect from damage during lifting.

7. Install a suitable engine lifting chain and hoist.
8. Raise the engine slightly. Ensure that lifting bracket does not foul the bulkhead. Remove both engine mountings.

➡It may be necessary to lower the transmission support slightly during the above operation.

9. Raise the engine/transmission and pull forward, while lowering the transmission support.

➡The engine/transmission must be tilted at an angle of approximately 45 degrees before it can be removed from the engine compartment.

10. Remove the engine/transmission assembly.

To install:

11. Guide the engine into position.
12. With the aid of an assistant, raise the transmission and lower the engine until the engine mountings can be installed.
13. Attach the engine mounts to the chassis with new flange nuts, but do not tighten at this stage.
14. Lower and guide the engine onto the mounting studs.
15. Remove the engine lifting chain and hoist.
16. Route the transmission wiring harness.
17. The remaining step of the installation is the reverse of the removal procedure while keeping the following items in mind:
    - Attach all wiring harness and electrical connectors.
    - Align the driveshaft matchmarks and tighten the mounting nuts and bolts to 35 ft. lbs. (48 Nm).
    - Attach all cables to the transmission.
    - Tighten the engine mounting nuts to 33 ft. lbs. (45 Nm).
    - Connect the handbrake cable and associated components.
    - Install and connect the radiator/condenser/oil cooler assembly, using new sealing rings.
    - Tighten the fittings on the air conditioning compressor to 17 ft. lbs. (23 Nm) and condenser to 11 ft. lbs. (15 Nm).
    - Tighten the fittings for the power steering lines 12 ft. lbs. (16 Nm).
    - Tighten the oil cooler lines to 22 ft. lbs. (30 Nm).
18. Evacuate and recharge the air conditioning system.
19. Connect the fuel supply and return lines to the fuel rail and tighten to 12 ft. lbs. (16 Nm).
20. Check and correct the fluid levels in the engine, transmission and transfer case.
21. Install and connect the battery.
22. Start the engine. Check for fuel, coolant and oil leaks.

Range Rover (2003–04)

The engine and transmission are removed and installed, as an assembly, from the vehicle.

1. Before servicing the vehicle, refer to the precautions in the beginning of this section.
2. Properly discharge the air conditioning system.
3. Remove or disconnect the following:
   - Negative battery cable
   - Fuel system pressure
   - Coolant
   - Air intake plenum
   - Mass airflow (MAF) sensor
   - Cooling fan
   - Battery positive lead on shock tower
   - Ignition coil covers
   - Camshaft cover ground wires
   - Ignition coil connectors
   - Throttle body, knock sensors, Camshaft Position (CMP) sensors, thermostat heater and coolant temperature sensor connectors

- Variable Camshaft Control (VCC) connectors
- Alternator connector
- Vacuum reservoir mounting bracket
- Fuel injector connectors
- Engine wiring harness from camshaft cover
- Air injection valve
- Charcoal canister connector
- Canister purge valve pipe and plug opening
- Transmission breather pipes from water pump
- Heater hoses
- Transmission connector bracket from bellhousing
- Starter connector
- Fuel feed line
- Necessary vacuum hoses
- Radiator hoses
- A/C connector
- A/C compressor and condenser lines and discard O-rings
- Oil pressure switch lines
- Oil filter housing and place aside
- Drain power steering fluid and remove hoses from power steering pump
- Remove power steering reservoir
- Sub-frame air deflector
- Exhaust pipes
- Front wheels
- Right halfshaft
- Front differential
- Lower engine ground
- Crankshaft Position (CKP) sensor
- Transmission cooler lines
- Torque converter bolts
- Starter

4. Place transmission jack under transmission.
5. Remove transmission-to-engine bolts
6. Install suitable engine lifting device to engine.
7. Remove engine mount attaching nuts.
8. Remove 2 top bellhousing bolts.
9. Raise engine clear of mounts and dowels and carefully remove the engine.

To install:

10. Install new o-ring to differential and lubricate with clean engine oil.
11. With the aid of an assistant, guide the engine into position onto the transmission, ensuring the torque converter is engaged properly.
12. Install the transmission-to-engine bolts and tighten to 33 ft. lbs. (45 Nm).
13. Install starter and tighten bolts to 33 ft. lbs. (45 Nm).
14. Install top 2 belhousing bolts and tighten to 18 ft. lbs. (25 Nm).
15. Remove transmission jack and lower engine onto mounts. Tighten mount nuts to 74 ft. lbs. (100 Nm).
16. Remove engine lifting device.
17. Install or connect the following:
    - Drive plate bolts and tighten to 33 ft. lbs. (45 Nm).
    - Exhaust pipes and tighten nuts to 33 ft. lbs. (45 Nm).
    - Differential and tighten bolts to 81 ft. lbs. (110 Nm).
    - Differential breathe hose
    - Lower engine ground
    - Sub-frame air deflector and tighten bolts to 33 ft. lbs. (45 Nm)
    - Right halfshaft
    - Power steering lines, using new O-rings. Tighten nuts to 18 ft. lbs. (25 Nm).
    - Oil filter housing
    - Oil pressure switch connector
    - A/C lines using new o-rings. Tighten fittings to 18 ft. lbs. (25 Nm).
    - A/C connector
    - Upper and lower radiator hoses
    - Vacuum lines
    - Fuel feed line
    - Starter connector
    - Transmission connector bracket
    - Heater hoses
    - Attach all wiring harness and electrical connectors
    - Ignition coil covers
    - Cooling fan
    - Wheels and tighten to 103 ft. lbs. (140 Nm).
    - Air intake plenum
18. Evacuate and recharge the air conditioning system.
19. Check and correct the fluid levels in the engine, power steering, transmission and transfer case.
20. Install and connect the battery.
21. Start the engine. Check for fuel, coolant and oil leaks.

Water Pump

REMOVAL & INSTALLATION

Freelander (2001–04)

1. Before servicing the vehicle, refer to the precautions in the beginning of this section.
2. Disconnect battery ground lead.
3. Drain cooling system.
4. Remove camshaft timing belt.
5. Remove and discard 7 bolts securing coolant pump to cylinder block.

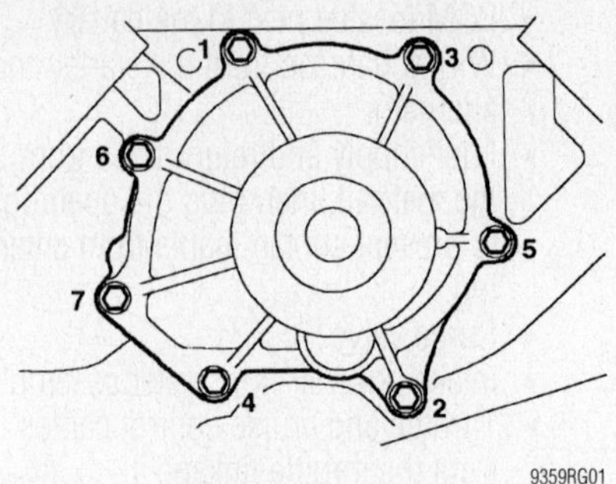

Water pump mounting bolt location torque sequence—Freelander

6. Release coolant pump from cylinder block and remove.
7. Remove and discard O-ring from coolant pump.

To install:

8. Clean coolant pump and mating face of cylinder block.
9. Lubricate new O-ring with rubber grease and fit to coolant pump.
10. Fit coolant pump to cylinder block. Fit new bolts and tighten progressively in the sequence illustrated to 7 ft. lbs. (9 Nm).
11. Fit camshaft timing belt.
12. Fill cooling system.
13. Connect battery ground lead.

Discovery (2001–04) and Range Rover (2001–02)

1. Before servicing the vehicle, refer to the precautions in the beginning of this section.
2. Disconnect the negative battery cable.
3. Drain and recycle the engine coolant.
4. Remove the accessory drive belt.
5. Remove the 3 bolts securing the pulley to the coolant pump, and remove the pulley.
6. Release the clip and disconnect the feed hose from the coolant pump.

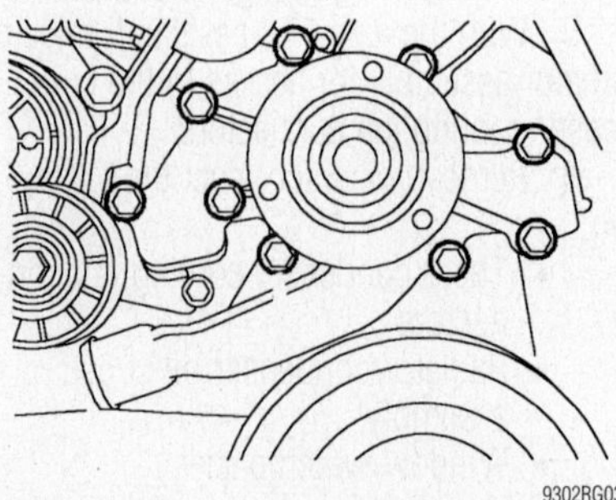

Water pump mounting bolt locations—Discovery II (2001–04) and Range Rover (2001–02)

7. Remove the 9 bolts securing the coolant pump, remove the pump and discard the gasket.

To install:

8. Clean the coolant pump and mating face.
9. Install new gasket and coolant pump to the cylinder block.
10. Install the mounting bolts and tighten to 18 ft. lbs. (24 Nm).
11. Connect the feed hose to the coolant pump, and secure with clip.
12. Ensure the mating faces of coolant pump pulley and flange are clean.
13. Install the pulley, and tighten the bolts to 16 ft. lbs. (22 Nm).
14. Install the auxiliary drive belt.
15. Connect the negative battery cable.

Range Rover (2003–04)

1. Before servicing the vehicle, refer to the precautions in the beginning of this section.
2. Disconnect the negative battery cable.
3. Drain and recycle the engine coolant.
4. Remove the accessory drive belt.
5. Remove the cooling fan.
6. Remove the upper and lower radiator hoses.
7. Remove the thermostat.
8. Disconnect secondary air pipe and move aside.
9. Remove the 4 bolts securing the pulley to the coolant pump, and remove the pulley.
10. Disconnect the engine coolant temperature sensor connector.
11. Remove the 7 bolts securing the coolant pump, remove the pump and discard the gasket.

To install:

12. Clean the coolant pump and mating face.
13. Install new o-rings and gasket and coolant pump to the cylinder block.
14. Install the mounting bolts and tighten to 7 ft. lbs. (10 Nm).
15. Connect temperature sensor connector.
16. Install the pulley, and tighten the bolts to 7 ft. lbs. (10 Nm).
17. Reconnect secondary air pipe.
18. Install the thermostat.
19. Install the auxiliary drive belt.
20. Install the cooling fan.
21. Fill cooling system.
22. Connect the negative battery cable.
23. Start the engine and check for leaks.

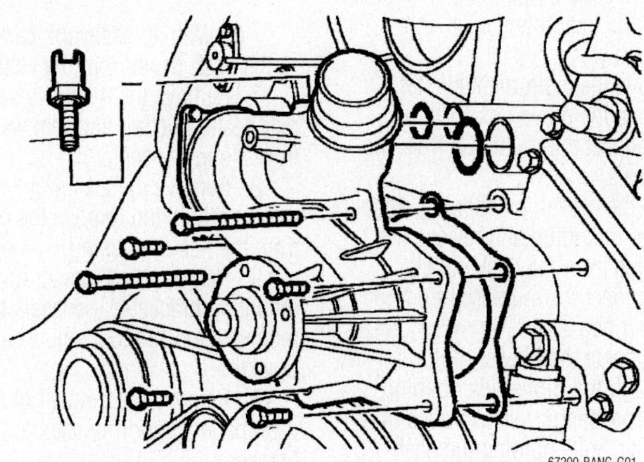

Water pump mounting bolt locations—Range Rover (2003–04)

Heater Core

REMOVAL & INSTALLATION

Freelander (2001–04)

1. Drain cooling system.
2. Release 2 clips securing heater hoses to heater and disconnect hoses.
3. Remove dash panel by the following steps:
   a. Remove front console
   b. Remove rotary couple for air bag system:
      - Steering wheel centered and road wheels straight ahead
      - Remove steering wheel
      - Remove steering column covers
      - Disconnect 2 multiplugs from rotary coupler
      - Remove 4 screws and remove rotary coupler
   c. Remove the 4 multiplugs from the wiper/indicator switch.
   d. Loosen the clamping screw and remove the wiper/indicator switch.
   e. Remove the 2 screws from the steering column lower finish panel and remove panel.
   f. Open the driver side glove box lid and remove the fuse box cover.
   g. Disconnect 2 multiplugs from the fuse box. Close the glove box.
   h. Remove the clock.
   i. Remove the upper trim panels from both A-pillars.
   j. Remove the passenger side airbag module by removing or disconnecting the following:
      - 2 screws holding airbag lower finish panel
      - 4 Torx bolts hold airbag module to dash panel
      - Airbag connector from airbag module
      - Remove airbag module and store properly
   k. Disconnect the 2 main harness–to–dash multiplugs through the center console opening.
   l. Remove the 12 dash panel retaining bolts.
   m. With an assistant, carefully remove the dash panel.
4. Remove the 2 bolts from the console support bracket, release the 2 relays and remove the brackets from in front of the shift lever.
5. Disconnect the 4 multiplugs from the heater controls.
6. Disconnect the multiplug from the heater assembly and release the diagnostic connector.
7. Disconnect the multiplug from the evaporator.
8. Disconnect the ducts from the left and right outer face level vents.
9. Detach the left and right defogger vents.
10. Release the air inlet connector hose.
11. Disconnect the harness from the heater housing.
12. Remove 2 nuts and 1 bolt and remove the heater assembly.

➡Do not carry out further dismantling if component is removed for access only.

13. Disconnect the 2 multiplugs from the heater controls.
14. Release the air blend control cable from the lever and abutment.
15. Release the air distribution cable from the lever and abutment.
16. Remove the heater controls.
17. Remove screw securing pipe clamp to the heater casing and remove clamp.
18. Remove 2 screws from core cover and remove cover.
19. Remove the heater core.

To install:

20. Fit the heater core to the heater assembly.
21. Fit the core cover and secure with screws.
22. Fit pipe clamp and secure with screw.
23. Install the heater assembly.
24. Position the heater controls to replacement heater, if necessary.
25. Connect the air distribution cable to lever and abutment.
26. Connect the air blend cable to lever and abutment.
27. Connect the multiplugs to heater controls.
28. Position the heater and secure with nuts and bolt.
29. Position harness and connect to clips on heater.
30. Fit the air inlet connector hose.
31. Fit the right side defogger vent duct to heater.
32. Fit the right side outer face level vent duct to heater and secure to body with bolt.
33. Fit the left side defogger vent duct to heater.
34. Fit left side outer face level vent duct to heater and secure to body with bolt.
35. Connect the multiplug to evaporator.
36. Connect the multiplug to the heater and secure the diagnostic socket.
37. Connect the multiplugs to heater controls.
38. Fit console support bracket and secure with bolts.
39. Fit relays to console support bracket.
40. Install the dash panel in reverse of the removal procedure.

⁂ CAUTION

Use special care to ensure the airbag module is properly restored.

41. Connect the heater hoses and secure with clips.
42. Refill the cooling system.
43. Operate the engine to normal operating temperatures. Test the controls and check for leaks.

Discovery (2001–04)

1. Disconnect the negative (–) battery cable, then disconnect the positive (+) battery cable.

➡Remove the key from the ignition switch and wait 10 minutes before starting work to allow system airbags to fully discharge.

2. Drain the cooling system into a clean container for reuse.

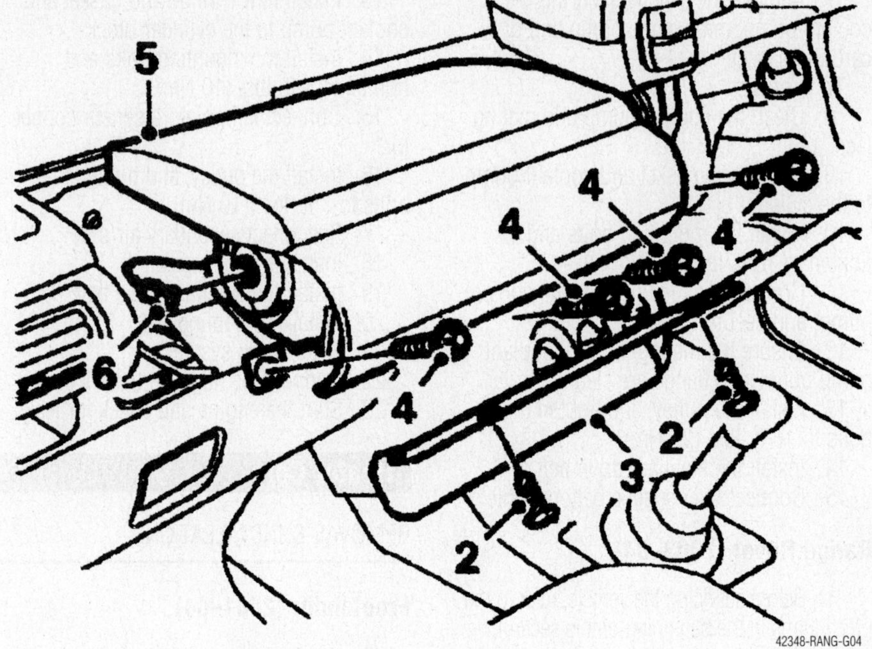

Removing passenger airbag module—Freelander

3. Discharge and recover the air conditioning system refrigerant.
4. Detach the heater hoses from the firewall connections.
5. Disconnect A/C refrigerant pipes from the evaporator fittings at the firewall and discard the O-rings.
6. Remove the dash panel assembly by performing the following procedure:
   a. Remove the stereo (radio/cassette/CD) player, using special tools as shown.
   b. Remove the steering wheel.
   c. Remove the steering column cover.
   d. Remove the center console assembly.
   e. Remove the A-pillar trim panels.
   f. Remove the passenger and driver side lower trim panels.
   g. Disconnect the multiplugs from the switches in the instrument cowl and remove the cowl.
   h. Remove the instrument cluster.
   i. Remove the glove box.
   j. Disconnect the multiplug from the heater controls.
   k. From the glove box opening, separate the Blue multiplug and release the coaxial cables from the dash panel.
   l. Disconnect the multiplug from the passenger air bag.
   m. Disconnect the blower motor.
   n. Through the glove box opening, remove the passenger air bag assembly.
   o. Remove bolts holding the dash panel at the transmission tunnel and at the A-pillars.

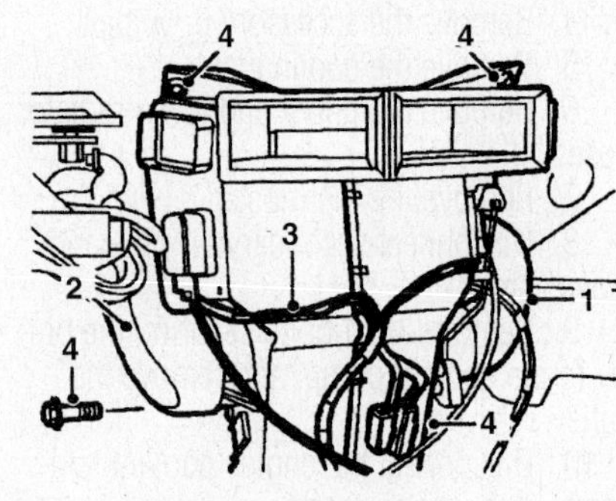

Identifying key components on the heater assembly—Freelander

   p. With an assistant, carefully remove the dash panel from the vehicle.
7. Remove the 4 screws securing the rear heater ducting and remove the ducting from along the center.
8. Remove the 2 blower motor housing nuts and bolt and remove the blower motor from the heater housing.
9. Remove the 2 center console front mounting bracket–to–chassis bolts and remove the center console front mounting bracket.
10. Remove the 2 nuts holding the right dash panel support bracket and remove the bracket.
11. Disconnect both drain tubes.

12. Remove the front heater ducting.

13. Remove the 4 nuts and bolt holding the heater assembly. Remove the heater by detaching it from the bulkhead grommet.

To install:

14. Position the heater assembly to bulkhead, fixing the heater tubes into the bulkhead grommet.

15. Install the heater assembly nuts and bolt and torque to 12 ft. lbs. (16 Nm).

16. Position right dash panel support bracket and install and tighten nuts.

17. Install the front heater ducting.

18. Position the center console support bracket and tighten retaining screws.

19. Connect both drain tubes.

20. Install the heater motor to its casing. Install and tighten the nuts and bolt to 14 ft. lbs. (19 Nm).

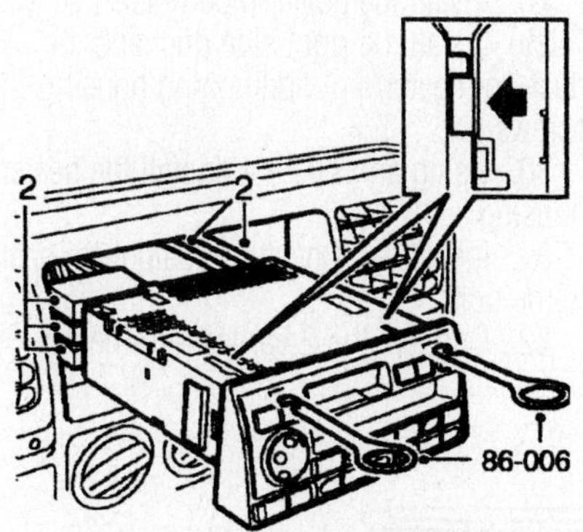

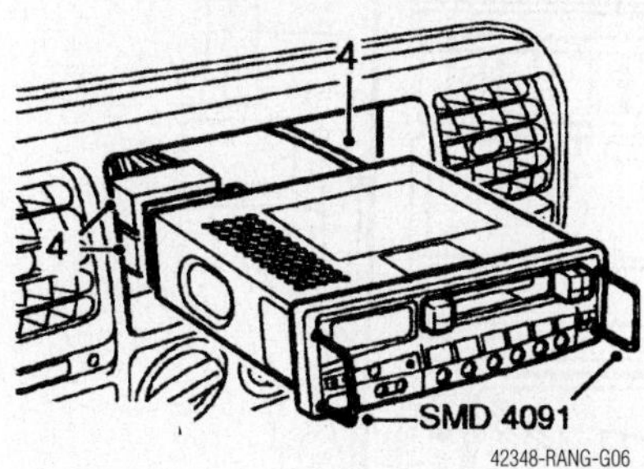

Removing stereo system from dash—Discovery

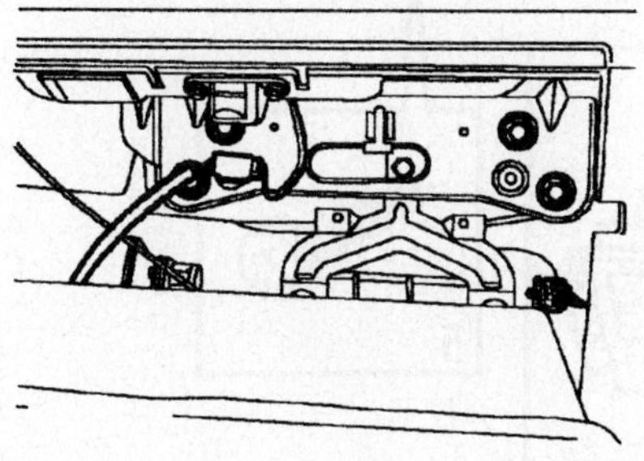

Disconnecting blue multiplug and coaxial cables—Discovery

21. Connect the multiplug for the heater motor.

22. Install the rear heater ducting and secure with screws.

23. Install the dash panel assembly by performing the following procedure:

a. Using an assistant, install the dash panel assembly to the vehicle. Install and tighten the lower edge bolts to 19 ft. lbs. (26 Nm).

b. Install and tighten the bolts retaining the instrument panel to the frame to 19 ft. lbs. (26 Nm).

c. Install the instrument panel–to–steering column bracket nuts and tighten to 8 ft. lbs. (11 Nm).

d. Secure the coaxial cables to the instrument panel and secure the blue multiplug to the main ICE multiplug, working through the glove box opening.

e. Connect the heater control multiplug connector.

f. Connect the multiplugs for the blower motor and for the passenger air bag.

g. Install the glove box and secure with bolts.

h. Connect the dash multiplugs to the main harness and to the fuse box.

i. Position to instrument cluster and connect the multiplugs, then install the cluster assembly with its screws.

j. Install the instrument cluster cowl, connecting the multiplugs to the cowl switches.

k. Working below the steering column, install the driver's side access panel.

l. Install both lower closing panels and secure with their clips.

m. Install both dash panel mats.

n. Install the center console assembly.

o. Install the steering column covers, then install the steering wheel, using a new self-locking nut torqued to 32 ft. lbs. (43 Nm).

p. Install the radio assembly.

24. Use new O-rings and insert the A/C pipes into the evaporator. Tighten the bolts to 2.7 ft. lbs. (5 Nm).

25. Connect the heater hoses and install the clamps at the firewall.

26. Evacuate and charge the air conditioning system.

27. Refill the cooling system.

28. Connect the positive (+) battery cable and then the negative (–) battery cable.

29. Run the engine to normal operating temperatures; then, check the climate control operation and check for leaks.

Range Rover (2001–02)

1. Disconnect the negative (–) battery cable; then the positive (+) battery cable.

➡Wait at least 20 minutes for the air bag(s) to discharge before performing any work on the system(s).

2. If equipped with SRS, remove the battery.

3. Drain the cooling system into a clean container for reuse.

4. Loosen the hose clips and disconnect the heater hoses from the heater tubes.

5. If equipped with air conditioning, perform the following procedure:

a. Discharge and recover the air conditioning refrigerant.

b. Remove the refrigerant pipes mount–to–evaporator bolt

c. Disconnect the pipes and discard the O-rings.

d. Plug the openings to prevent contamination.

6. Remove the dash panel assembly by performing the following procedures:

a. Remove the center console.

b. Remove the wiper motor and linkage.

c. Remove the steering column.

d. At the passenger's side, remove the clip and disconnect the heated front screen multiplug connector.

e. Remove the 6 scuttle side panel–to–chassis bolts and the side panel.

f. Remove the heater intake pollen filters.

g. Remove the 8 pollen filter housing screws from each housing and remove both housings.

h. Remove the radio.

i. Near the A-post lower trim panels, remove the door aperture seal.

j. If equipped with a footrest on the driver's side, remove the 3 foot rest–to–A-post lower trim bolts and remove the foot rest.

k. At each A-post's lower trim panel, remove the screw and release the spring clip and remove both trim panels.

l. At the driver's seat base trim, remove the fuse cover.

m. Remove the screw, the 2 trim studs and the seat base trim.

n. At the driver's side, release the 4 spring clips and remove the carpet retainer.

o. Remove the 2 lower closing panel–to–passenger side scrivet fasteners and the panel.

p. Release the closing panel, disconnect the footwell lamp and the diagnostic multiplug connector; then, remove the closing panel.

q. Remove the 4 dash center bracket bolts and the bracket.

r. Disconnect the 4 multiplug connectors from the Body Control Module (BCM).

s. At the base of the A-post on the driver's side, remove the ground wires from the stud.

t. Disconnect the multiplug electrical connectors at the base of each A-post.

u. Disconnect the BCM electrical harness from the sill and move it into the dash panel so it will not hamper removal of the dash panel.

v. At the brake and clutch switches, disconnect the multiplug electrical connectors and vacuum hose.

w. If equipped with SRS, disconnect the following items:

- The SRS harness connector from the main wiring harness
- The SRS harness connector from the control module

x. Remove both front wheel arch liners.

y. At the left wheel arch, remove the 2 air cleaner baffle scrivet fasteners and the baffle.

z. If equipped with SRS, disconnect both SRS crash sensor electrical connectors.

aa. Remove the 4 battery tray bolts and the 2 air cleaner–to–valance bolts.

bb. Raise the air cleaner and battery tray to access the crash sensor harness clips.

cc. Disconnect the crash sensor harness–to–valance clips; then, move the harnesses into the wheel arches.

dd. Disconnect the 3 crash sensor harness–to–underside wheel arches; then, move the harness through the bulkhead and into the dash.

ee. At the top of the dash panel, remove the 4 dash panel–to–scuttle panel tube bolts.

ff. Remove the dash panel–to–chassis bolts. Pull the panel rearward and support it on 50mm deep wooden blocks.

7. With the dash panel supported on 50mm deep wooden blocks, remove the face level vent ducts–to–dash screws; there is a vent duct located on both sides of the dash.

8. Remove the face level vent ducts inserts from the heater unit.

9. Remove the passenger's side blower duct.

10. Remove the 4 heater control panel–to–dash panel screws and remove the panel.

11. Disconnect the 4 multiplug connectors from the heater control panel and remove the control panel.

12. Remove the 5 center switch assembly–to–dash screws and remove the center switch assembly.

13. Disconnect the multiplug connectors from the solar sensor and the alarm LED; then push the leads into the dash ducting.

14. Disconnect the harness–to–dash ducting clip; then, position the solar sensor/LED harness aside.

15. Disconnect the water temperature sensor–to–heater core inlet pipe clip and position the sensor aside.

16. Disconnect the multiplug connector from the evaporator sensor.

17. Disconnect the 2 harness–to–heater base clips.

18. Remove the 4 heater housing–to–dash frame bolts.

19. Using an assistant, hold the harness away and remove the heater housing.

20. Remove the right side duct–to–heater/air conditioning housing screws and the duct.

21. Remove the heater pipe bracket screw.

22. Remove the 2 right side servo–to–heater/air conditioning housing screws and the servo.

23. Remove the heater core/pipe assembly–to–heater/air conditioning housing clips and the heater core/pipe assembly.

24. If installing a new heater core, remove the 2 heater core–to–heater pipe assembly screws and separate the heater pipe assembly. Discard the O-rings.

To install:

25. If installing a new heater core, install new O-rings, the heater pipe assembly and the 2 heater core–to–heater pipe assembly screws.

26. Install the heater core/pipe assembly and the heater core/pipe assembly–to–heater/air conditioning housing clips.

27. Install the right side servo and the 2 servo–to–heater/air conditioning housing screws.

28. Install the heater pipe bracket screw.

29. Install the right side duct and the duct–to–heater/air conditioning housing screws.

30. Using an assistant, install the heater housing.

31. Install the 4 heater housing–to–dash frame bolts.

32. Connect the 2 harness–to–heater base clips.

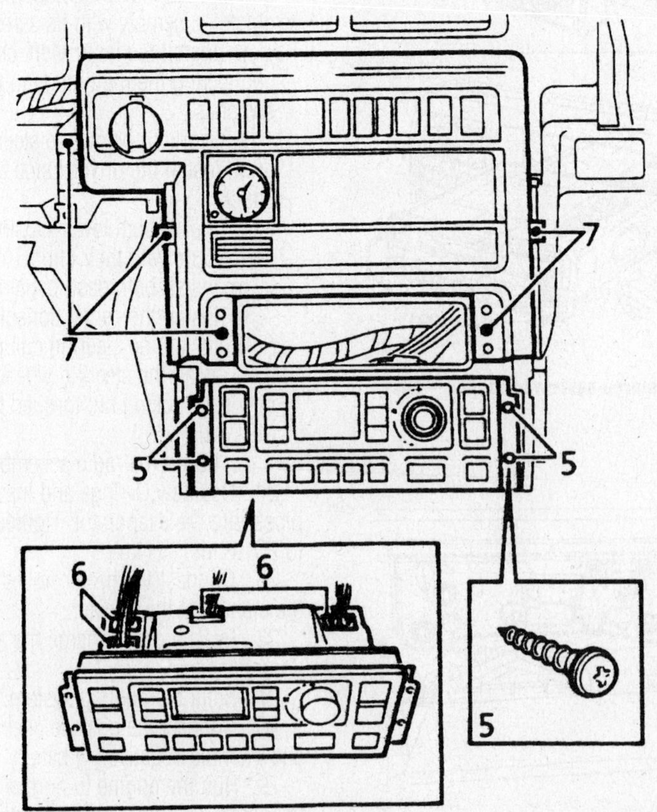

View of the center switch assembly—Range Rover (2001–02)

33. Connect the multiplug connector to the evaporator sensor.
34. Connect the water temperature sensor–to–heater core inlet pipe clip.
35. Connect the harness–to–dash ducting clip.
36. Connect the multiplug connectors to the solar sensor and the alarm LED.
37. Install the center switch assembly and the 5 center switch assembly–to–dash screws.
38. Install the control panel and connect the 4 multiplug connectors to the heater control panel.
39. Install the heater control panel and the 4 panel–to–dash panel screws.
40. Install the passenger's side blower duct.
41. Install the face level vent ducts inserts to the heater unit.
42. With the dash panel supported on 50mm deep wooden blocks, install the face level vent ducts–to–dash screws; there is a vent duct located on both sides of the dash.
43. Install the dash panel assembly by performing the following procedures:
  a. Push the panel forward and support it on 50mm deep wooden blocks. Install the dash panel–to–chassis bolts.
  b. At the top of the dash panel, install the 4 dash panel–to–scuttle panel tube bolts.
  c. Connect the 3 crash sensor harness–to–underside wheel arches.
  d. Connect the crash sensor harness–to–valance clips.
  e. Install the 4 battery tray bolts and the 2 air cleaner–to–valance bolts.
  f. If equipped with SRS, connect both SRS crash sensor electrical connectors.
  g. At the left wheel arch, install the air cleaner baffle and the 2 baffle scrivet fasteners.
  h. Install both front wheel arch liners.
  i. If equipped with SRS, connect the following items:
    - The SRS harness connector to the main wiring harness
    - The SRS harness connector to the control module
  j. At the brake and clutch switches, connect the multiplug electrical connectors and vacuum hose.
  k. Connect the BCM electrical harness to the sill.
  l. Connect the multiplug electrical connectors at the base of each A-post.
  m. At the base of the A-post on the driver's side, install the ground wires to the stud.
  n. Connect the 4 multiplug connectors to the Body Control Module (BCM).
  o. Install the dash center bracket and the 4 bracket bolts.
  p. Install the closing panel; then, connect the footwell lamp and the diagnostic multiplug connector.
  q. Install the lower closing panel and the 2 panel–to–passenger side scrivet fasteners.
  r. At the driver's side, install the carpet retainer and the 4 spring clips.
  s. Install the seat base trim, the 2 trim studs and the screw.
  t. At the driver's seat base trim, install the fuse cover.
  u. At each A-post's lower trim panel, install both trim panels, the screw and the spring clip.
  v. If equipped with a foot rest on the driver's side, install the foot rest and the 3 foot rest–to–A-post lower trim bolts.
  w. Near the A-post lower trim panels, install the door aperture seal.
  x. Install the radio.
  y. Install the both pollen filter housings and the 8 housing screws to each housing.
  z. Install the heater intake pollen filters.
  aa. Install the scuttle side panel and the 6 side panel–to–chassis bolts.
  bb. At the passenger's side, connect the heated front screen multiplug connector and install the clip.
  cc. Install the steering column.
  dd. Install the wiper motor and linkage.
  ee. Install the center console.
44. If equipped with air conditioning, perform the following procedure:
  a. Lubricate and install new O-rings and connect the pipes.
  b. Install the refrigerant pipes mount–to–evaporator bolt.
45. Connect the heater hoses to the heater tubes and install the hose clips.
46. Refill the cooling system.
47. If the battery was removed, install it.
48. Connect the positive (+) battery cable; then, the negative (–) battery cable.
49. Evacuate and charge the air conditioning system.
50. Run the engine to normal operating temperatures; then, check the climate control operation and check for leaks.

Range Rover (2003–04)

1. Before servicing the vehicle, refer to the precautions in the beginning of this section.
2. Disconnect the negative (–) battery cable; then the positive (+) battery cable.

➡Wait at least 20 minutes for the air bag(s) to discharge before performing any work on the system(s).

➡The instrument panel fascia is released from the passenger side and swung out for access with the main harness still attached.

3. Drain the cooling system into a clean container for reuse.
4. Discharge and recover the air conditioning refrigerant.
5. Remove the air intake plenum.
6. Release the driver side air bag by inserting a 2 inch (50 mm) rod through the steering wheel access hole on each side and push on the rod until the air bag clips are released.
7. Disconnect the electrical connectors and remove air bag.
8. With the front wheels in the straight ahead position, note the position of the steering wheel to column alignment marks, then remove the steering wheel.
9. Remove the passenger side air bag.
10. Remove the front seat belt anchor bolts.
11. Remove the front seats.
12. Remove the center console footwell panels.
13. Remove the driver side upper finish panel.
14. Remove the passenger side lower finish panel.
15. Remove 7 screws attaching the glove box.
16. Disconnect lamp connector, glove box latch cable and 2 screws attaching CD player to glove box.
17. Remove CD player and glove box assembly.
18. Remove center console finish panel.
19. Remove center console side panels.
20. Remove storage box from center console, along with 2 screws and washer at base of storage box opening.
21. Remove 2 screws attaching ignition switch to center console.
22. Remove 1 screw attaching center console to instrument panel.
23. Remove center console rear closing panel.
24. Raise the rear of the center console and disconnect 3 electrical connectors and air duct.
25. Remove 3 connectors from the accessory power socket and then remove center console.

26. Release 4 clips attaching the heater control panel, and remove the control panel.
27. Remove the air suspension control panel, located below the heater control panel.
28. Remove the handbrake lever assembly.
29. Remove the air ducts from the heater-A/C assembly.
30. Remove the driver's foot rest.
31. Remove the upper and lower A pillar finish panels.
32. Remove instrument panel side fascia from both sides.
33. Remove the gearshift selector assembly.
34. Remove 4 nuts attaching the gearshift selector assembly carrier and remove the carrier.

➡Cover the selector assembly mounting studs to prevent damage to the instrument panel.

35. Remove 4 bolts attaching the handbrake mounting bracket and remove the bracket.
36. Remove 4 bolts attaching the center console mounting bracket and remove the bracket.
37. Mark the location of the steering column lower shaft to the intermediate shaft, then remove the clamp securing them together.
38. Remove 2 nuts attaching steering support bearing to the body, release the bearing and disconnect the shaft.
39. Remove the wiper arm, disconnect the wiper motor and position aside.
40. From inside the engine compartment, remove 2 bolts attaching the instrument panel to the firewall.
41. Disconnect the A/C lines from the firewall, discard the o-rings and cap the line openings.
42. Disconnect the heater hoses at the firewall.
43. Remove 3 nuts attaching the heater assembly to the firewall.
44. From inside the vehicle, remove 4 nuts and 2 bolts attaching the instrument panel to the body.
45. Disconnect electrical connectors and release wiring harnesses from both A pillars.
46. Pull out instrument panel enough to disconnect airbag connectors.
47. Remove the transfer case Electronic Control Unit (ECU).
48. From inside the engine compartment, disconnect the power lead, relay and connector from firewall fuse box.
49. Disconnect headlight check control module from A pillar and position aside.

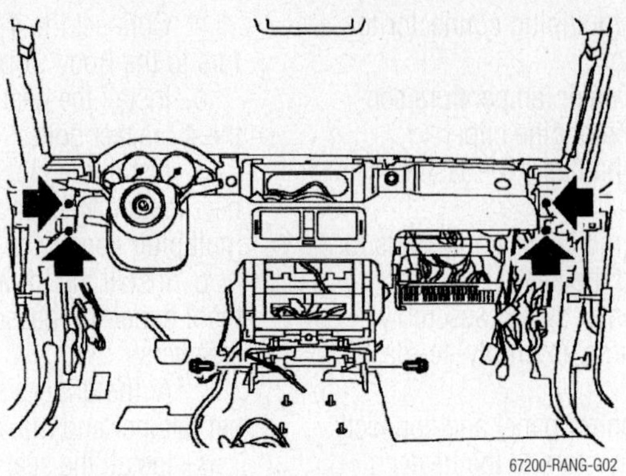

Instrument panel mounting—Range Rover (2003–04)

50. Remove door harness connector block from A pillar.
51. Remove ground header from A pillar.
52. Raise carpet and remove 3 bolts attaching wiring harness cover panel and remove panel.
53. Disconnect wiring harness connector block from Body Control Unit (BCU) carrier.
54. Remove 2 nuts and position BCU carrier aside.
55. Release the main wiring harness from the channel and withdraw the harness through the firewall.
56. With the aid of an assistant, rotate the instrument panel outward and support it on suitable stands to protect trim and carpet.
57. Disconnect air blend motor connector and remove motor.
58. Disconnect screws and nut from rear of instrument panel tray and release clip to allow wiring harness to move.
59. Remove 2 nuts attaching HVAC assembly.
60. With an assistant, remove the HVAC assembly from vehicle.
61. Turn HVAC assembly upside down and remove left side heat duct.
62. Remove heater pipes from assembly and discard o-rings.

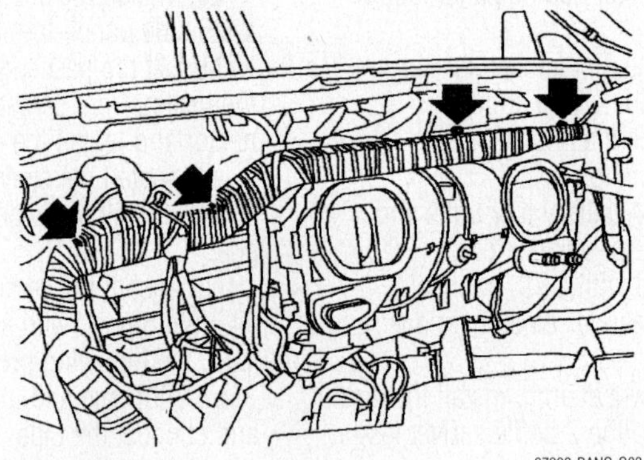

Rear of instrument panel wiring harness mounting—Range Rover (2003–04)

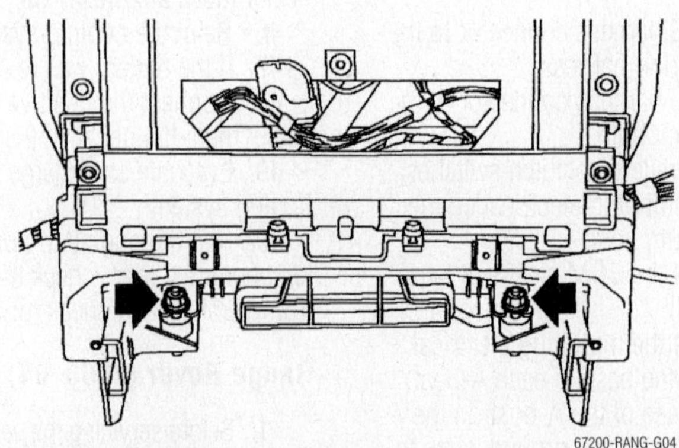

HVAC assembly mounting bolts—Range Rover (2003–04)

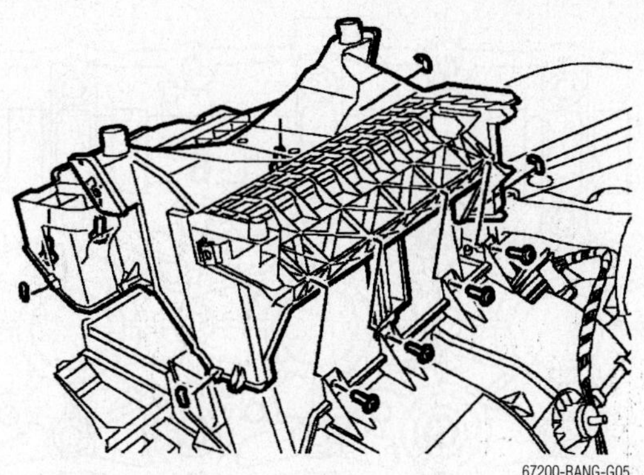

Removing HVAC lower access panel—Range Rover (2003–04)

63. Remove right side heat duct.
64. Disconnect all electrical connectors.
65. Remove rear air vent motor.
66. Remove thermal expansion valve cover.
67. Remove 7 screws and 3 clips from lower HVAC access panel and remove panel.
68. Remove heater core from HVAC assembly.

To install:

69. Install heater core to HVAC assembly.
70. Install lower access panel.
71. Install thermal expansion valve cover.
72. Install rear air vent motor.
73. Connect all electrical connectors.
74. Install right side heat duct
75. Install heater pipes to HVAC assembly using NEW o-rings
76. Install left side heat duct.
77. With an assistant, install the HVAC assembly into the vehicle and tighten the nuts.
78. Place wiring harness into position in rear of instrument panel tray and clips and screws.
79. Connect air blend motor connector and install motor.
80. With the aid of an assistant, rotate the instrument panel inward and place it in position.
81. Feed the main wiring harness through the firewall.
82. Raise carpet and place the main harness in the channel and install wiring harness cover panel.
83. Install the BCU carrier.
84. Connect wiring harness connector block to BCU carrier.
85. Install ground header to A pillar.
86. Install door harness connector block to A pillar.
87. Install headlight check control module to A pillar.
88. From inside the engine compartment, install the power lead, relay and connector to firewall fuse box.
89. Install the transfer case Electronic Control Unit (ECU).
90. Pull out instrument panel enough to reconnect airbag connectors.
91. Connect electrical connectors and attach wiring harnesses to both A pillars.
92. From inside the vehicle, install 4 nuts and 2 bolts attaching the instrument panel to the body. Tighten the fasteners to 18 ft. lbs. (25 Nm).
93. Install 3 nuts attaching the heater assembly to the firewall and tighten to 7 ft. lbs. (10 Nm).
94. Connect the heater hoses at the firewall.
95. Connect the A/C lines to the firewall, using NEW the o-rings.
96. From inside the engine compartment, install 2 bolts attaching the instrument panel to the firewall. Tighten the bolts to 18 ft. lbs. (25 Nm).
97. Install the wiper motor and tighten to 7 ft. lbs. (10 Nm).
98. Install the wiper arm and tighten to 18 ft. lbs. (25 Nm).
99. Install 2 nuts attaching steering support bearing to the body, and tighten to 18 ft. lbs. (25 Nm).
100. Using marks made on removal, position the steering column lower shaft to the intermediate shaft, and tighten the bolt to 18 ft. lbs. (25 Nm).
101. Install 4 bolts attaching the center console mounting bracket and tighten to 7 ft. lbs. (10 Nm).
102. Install 4 bolts attaching the handbrake mounting bracket and tighten to 18 ft. lbs. (25 Nm).
103. Install 4 nuts attaching the gearshift selector assembly carrier and tighten to 18 ft. lbs. (25 Nm).
104. Install the gearshift selector assembly.
105. Install instrument panel side fascia to both sides.
106. Install the upper and lower A pillar finish panels.
107. Install the driver's foot rest.
108. Install the air ducts to the heater-A/C assembly.
109. Install the handbrake lever assembly.
110. Install the air suspension control panel.
111. Install the heater control panel.
112. Install the center console.
113. Connect 3 connectors to accessory power socket.
114. Raise the rear of the center console and connect 3 electrical connectors and air duct.
115. Install center console rear closing panel.
116. Install 1 screw attaching center console to instrument panel.
117. Install 2 screws attaching ignition switch to center console.
118. Install 2 screws and washer at base of storage box opening.
119. Install storage box.
120. Install center console side panels.
121. Install center console finish panel.
122. Install CD player and glove box assembly.
123. Install the passenger side lower finish panel.
124. Install the driver side upper finish panel.
125. Install the center console footwell panels.
126. Install the front seats and tighten the bolts to 33 ft. lbs. (45 Nm).
127. Install the seal belt anchor bolts and tighten to 35 ft. lbs. (48 Nm).
128. Install the passenger side air bag and tighten the bolts to 16 ft. lbs. (22 Nm).
129. Noting the position of the steering wheel to column alignment marks, install the steering wheel and tighten the bolt to 46 ft. lbs. (63 Nm).
130. Release the driver side air bag by inserting a 2 inch (50 mm) rod through the steering wheel access hole on each side and push on the rod until the air bag clips are released.
131. Connect the electrical connectors and install driver side air bag. Tighten the screws to 22 inch lbs. (2.5 Nm).
132. Install the air intake plenum.
133. Charge the air conditioning refrigerant.
134. Fill the cooling system.
135. Connect the battery cables.

136. Start the vehicle and check for leaks and proper heater operation.

Cylinder Head

REMOVAL & INSTALLATION

Freelander (2001–04)

LEFT SIDE

1. Before servicing the vehicle, refer to the precautions in the beginning of this section.
2. Disconnect the battery ground lead.
3. Drain cooling system.
4. Remove the intake manifold chamber.
5. Remove left side exhaust manifold gasket.
6. Remove the camshaft timing belt.
7. Remove 4 bolts securing left side front camshaft timing belt cover backplate to cylinder head and then remove the backplate.
8. Remove the left camshaft cover gasket
9. Disconnect the ignition coils from the left intake manifold, and disconnect the coil ground lead.
10. Remove the ignition coils and position them aside.
11. Disconnect the coolant bleed hose from the intake manifold and move it aside.
12. Disconnect the PCV breather hose from the left intake manifold.
13. Remove special bolt securing high tension lead guide bracket to left side camshaft cover and remove bracket.
14. Disconnect camshaft position (CMP) sensor multiplug and release male multiplug from mounting bracket.
15. Remove the bolt securing the multiplug bracket, remove bracket.
16. Disconnect clips securing right side injector harness to injector protection cover.

➡**Protection covers are NOT fitted to vehicles.**

17. Remove 2 bolts securing protection cover and fuel rail to right side intake manifold, remove cover.
18. Remove 2 bolts securing fuel rail to left side intake manifold.
19. Release the injectors from the manifolds and carefully position the fuel rail and injectors aside.

CAUTION

Always fit plugs to open connections to prevent contamination.

20. Remove 14 bolts and remove left side camshaft cover and gasket.

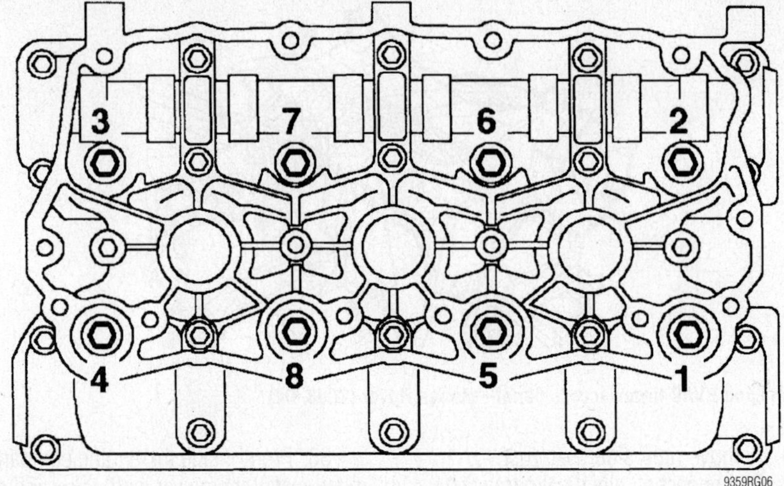

Cylinder head bolt loosening sequence—Freelander

21. Progressively loosen and remove 8 cylinder head bolts.
22. With assistance, remove cylinder head assembly.

CAUTION

Support both ends of cylinder head on blocks of wood.

23. Remove and discard cylinder head gasket and the cylinder head locating dowels.
24. Install cylinder liner clamps LRT-12-144 and secure with bolts. Ensure that feet of clamps do not protrude over bores.

To install:

25. Clean cylinder head face.
26. Remove bolts securing cylinder liner clamps LRT-12-144 to cylinder block and remove clamps.
27. Clean cylinder block face, dowels and dowel holes.
28. Clean cylinder head bolts and wipe dry.
29. Lightly lubricate threads and beneath heads of cylinder head bolts with clean engine oil.
30. Install 2 new locating dowels, supplied with the cylinder head gasket.
31. Check that the installed height of the dowels is 0.40–0.43 inch (10–11mm) above the cylinder block top face.
32. Install new cylinder head gasket onto cylinder block with the word 'TOP' uppermost.
33. With assistance, fit cylinder head and carefully position left side inlet to right side intake manifold, cylinder head to dowels.
34. Carefully place bolts into the cylinder head. DO NOT DROP. Screw bolts into place by hand.
35. Tighten cylinder head bolts progressively in the sequence illustrated to:
   - Step 1: 18 ft. lbs. (25 Nm)
   - Step 2: Tighten in the same sequence to 18 ft. lbs. (25 Nm)
   - Step 3: Tighten in the same sequence to 18 ft. lbs. (25 Nm)
   - Step 4: In the same sequence to plus 180 degrees.

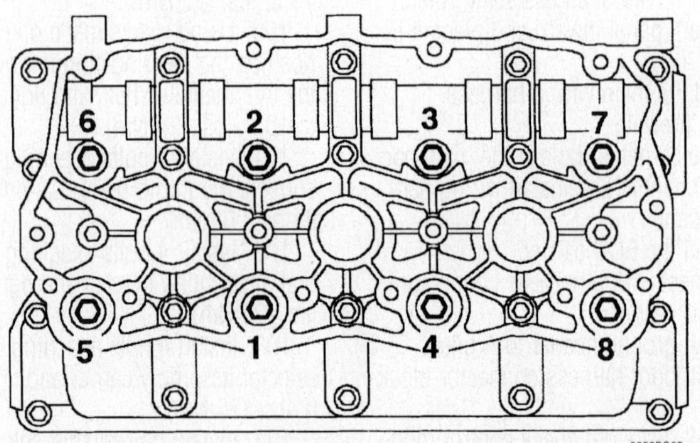

Cylinder head bolt torque sequence—Freelander

36. Clean mating surfaces of camshaft cover and carrier.

37. Clean inside of camshaft cover. If necessary, wash oil separator gauze and blow dry.

38. Install gasket and position camshaft cover to carrier.

CAUTION

When fitting camshaft cover gasket, ensure arrows on gasket point towards intake manifold.

39. Install bolts and tighten in sequence illustrated to 7 ft. lbs. (9 Nm).

40. Remove and discard O-rings from injectors.

41. Clean injectors and injector locations in fuel rail.

42. Lubricate new O-rings with castor oil and fit to injectors.

43. Position fuel rail assembly and secure injectors to intake manifolds.

44. Position injector protection cover to right side fuel rail and secure injector harness to protection cover with clips.

45. Install bolts securing fuel rail to intake manifolds and tighten to 7 ft. lbs. (9 Nm).

46. Position CMP sensor bracket, fit and tighten bolt to 7 ft. lbs. (9 Nm).

47. Connect CMP sensor multiplug and secure to bracket.

48. Connect breather hose to left side intake manifold.

49. Connect coolant bleed hose to intake manifold and secure with clip.

50. Position ignition coils to left side intake manifold.

51. Position ground lead, fit nuts and bolts securing ignition coils to left side intake manifold and tighten to 7 ft. lbs. (9 Nm).

52. Position high tension lead guide bracket, fit and tighten special bolt.

53. Connect high tension leads to spark plugs and secure high tension leads in guide clips.

54. Clean camshaft timing belt cover backplate bolts and apply Loctite 242 to the first 3 threads.

55. Position backplate, fit and tighten bolts to 7 ft. lbs. (9 Nm).

56. Install camshaft timing belt.

57. Install left side exhaust manifold gasket.

58. Install intake manifold chamber.

59. Fill cooling system.

RIGHT SIDE

1. Before servicing the vehicle, refer to the precautions in the beginning of this section.

2. Disconnect battery ground lead.

3. Drain cooling system.

4. Remove intake manifold chamber.

5. Remove right side exhaust manifold gasket.

6. Remove camshaft timing belt.

7. Remove 4 bolts securing right side front timing belt cover backplate to cylinder head and remove the backplate.

8. Depress the locking collar and release the breather hose from right side intake manifold.

9. For 2003–04 models, remove the right side camshaft cover gasket.

10. Release the 3 locking clips and disconnect 3 multiplugs from plug top coils.

11. Remove the bolt securing ground lead to right side camshaft cover and release ground lead.

12. Remove 6 bolts securing plug top coils to camshaft cover and remove coils.

13. Remove 3 nuts and 3 bolts securing ignition coils to left side intake manifold and release coil ground lead.

14. Remove ignition coils and position aside.

15. Release clip and disconnect coolant bleed hose from right side intake manifold. Position the hose aside.

16. Release the injector harness clips from injector protection cover or from the bosses on the fuel rail.

17. If so equipped, remove 2 bolts securing the injector protection cover and the right side fuel rail to right side intake manifold. Remove the cover.

18. Remove 2 bolts securing fuel rail to left side intake manifold.

19. Release the injectors from the manifolds and carefully position the fuel rail and injectors aside.

CAUTION

Always plug all open connections to prevent contamination.

20. For 2000–02 models, remove the 14 bolts and remove right side camshaft cover and gasket.

21. Progressively loosen and remove 8 cylinder head bolts, using the sequence shown.

22. With assistance, remove cylinder head assembly.

CAUTION

Carefully support both ends of the cylinder head on blocks of wood.

23. Remove and discard the right cylinder head gasket.

CAUTION

DO NOT rotate the crankshaft with one cylinder head installed.

24. For 2003–04 models, remove and discard both cylinder head locating dowels.

25. Install cylinder liner clamps LRT-12-144 and secure with bolts. Ensure that feet of clamps do not protrude over the bores.

To install:

26. Clean the cylinder head face, using gasket removal spray and a plastic scraper.

27. Remove bolts securing cylinder liner clamps LRT-12-144 to cylinder block and remove clamps, without rotating the crankshaft.

28. Clean the cylinder block face, dowels and dowel holes.

29. Clean cylinder head bolts and wipe dry.

30. Lightly lubricate threads and beneath heads of cylinder head bolts with clean engine oil.

31. For 2000–02 models, ensure the locating dowels are secured in the cylinder block.

32. For 2003–04 models, install new dowels supplied with the cylinder head gasket kit. Ensure dowels are installed so 0.40–0.43 inch (10–11mm) of the dowel is above the cylinder block top face.

33. Install the new cylinder head gasket onto the cylinder block with the word 'TOP' uppermost.

34. With assistance, fit cylinder head and carefully position right side inlet to left side intake manifold and to the cylinder head dowels.

35. Carefully install cylinder head bolts. DO NOT DROP. Screw bolts into place by hand.

36. Tighten cylinder head bolts progressively in the sequence illustrated to:

- Step 1: 18 ft. lbs. (25 Nm)
- Step 2: Tighten in the same sequence to 18 ft. lbs. (25 Nm)
- Step 3: Tighten in the same sequence to 18 ft. lbs. (25 Nm)
- Step 4: In the same sequence to plus 180 degrees.

37. Clean mating surfaces of camshaft cover and carrier.

38. Clean inside of camshaft cover. If necessary, wash oil separator gauze and blow dry.

39. Install new gasket and position camshaft cover to carrier.

CAUTION

When fitting camshaft cover gasket, ensure arrows on gasket point towards intake manifold.

40. Install bolts and tighten in sequence illustrated to 7 ft. lbs. (9 Nm).

41. Remove and discard lower O-rings from injectors.

42. Clean injectors and injector locations in fuel rail.

43. Lubricate new O-rings with castor oil and install them on the injectors.

44. Position fuel rail assembly and secure injectors to intake manifolds.

45. If so equipped, position injector protection cover to right side fuel rail and secure injector harness to protection cover with clips.

46. Install bolts securing fuel rail to intake manifolds and tighten to 7 ft. lbs. (9 Nm).

47. Connect coolant bleed hose to intake manifold and secure with clip.

48. Connect the breather hoses to the right side camshaft cover and intake manifold.

49. Position ignition coils to left side intake manifold.

50. Position ground lead, fit nuts and bolts securing ignition coils to left side intake manifold and tighten to 7 ft. lbs. (9 Nm).

51. Position plug top coils to spark plugs and right side camshaft cover, fit and tighten bolts to 7 ft. lbs. (9 Nm).

52. If so equipped, position the ground lead to the right side camshaft cover. Install the bolt and tighten to 7 ft. lbs. (9 Nm).

53. Connect multiplugs to the plug top coils and secure with locking clips.

54. Clean camshaft timing belt cover backplate bolts and apply Loctite 242 to the first 3 threads.

55. Position the backplate, fit and tighten bolts to 7 ft. lbs. (9 Nm).

56. Install the camshaft timing belt.

57. Install right side exhaust manifold gasket.

58. Install the intake manifold chamber.

59. Fill the cooling system.

Discovery (2001–02) and Range Rover (2001–02)

1. Before servicing the vehicle, refer to the precautions in the beginning of this section.

2. Remove or disconnect the following:
- Negative battery cable
- Coolant
- Intake manifold
- Alternator
- Air conditioning compressor and position it aside leaving the hoses attached.
- Rocker arm cover
- Rocker shafts and pushrods
- Front exhaust pipes from the manifolds
- Exhaust manifolds
- Air cleaner and intake air duct assembly

3. On the left cylinder head, remove the ground cable attached to the left-hand cylinder head.

4. On the right cylinder head, remove the breather pipe from the engine lifting bracket.

5. Loosen the cylinder head bolts, reversing the tightening sequence.

6. Lift the cylinder head off the engine block.

To install:

7. Remove the cylinder head gaskets, then clean the gasket mating surfaces.

8. Clean the exhaust gasket mating faces.

9. Ensure that the bolt holes in cylinder block are clean and dry.

10. Install new cylinder head gaskets with the word 'TOP' facing out.

11. Clean, then lightly oil the threads of the head bolts.

12. Install the cylinder heads on the block.

13. Install the cylinder head bolts in the positions as illustrated.
- 96mm long bolts: 2, 4, 6, 7, 8, 9, 10
- 66mm long bolts: 1, 3, 5

➡There are no bolts fitted in the 4 lower holes in each cylinder head.

14. Tighten the cylinder head bolts progressively in sequence, as shown, as follows:
- 15 ft. lbs. (20 Nm)
- + ¼ turn (90°)
- + ¼ (90°)

15. The remaining steps are the reverse of the removal procedure.

16. Connect the negative battery cable.

17. Fill and bleed the cooling system. Run the engine and check for leaks.

Discovery (2003–04)

1. Before servicing the vehicle, refer to the precautions in the beginning of this section.

2. Disconnect the negative battery cable.

3. Remove the intake manifold and gasket.

4. For right side intake manifold, perform the following steps:

a. Remove the auxiliary drive belt tensioner bolt. Remove the tensioner.

b. Remove the alternator mounting bracket.

5. Noting the installed order, disconnect the high tension leads from the spark plugs.

6. For left side intake manifold, perform the following steps:

a. Remove the bolt securing the engine harness to the read of the cylinder head.

b. Remove the brake vacuum servo heat shield.

7. For right side intake manifold, remove the bolt holding the ground connection.

8. Remove 8 bolts and remove the exhaust manifold from the cylinder head. Remove the gaskets.

9. Progressively, remove the 4 bolts securing the rocker shaft, and remove the rocker shaft.

10. Remove the push rods and store them in order as removed.

11. If equipped with secondary air injection (SAI), remove the 2 air injection adapters from the cylinder head and discard.

12. In the sequence shown, remove 10 cylinder head bolts. Discard the bolts.

➡Loosening sequence is the same for both cylinder heads. Right cylinder head shown in illustration.

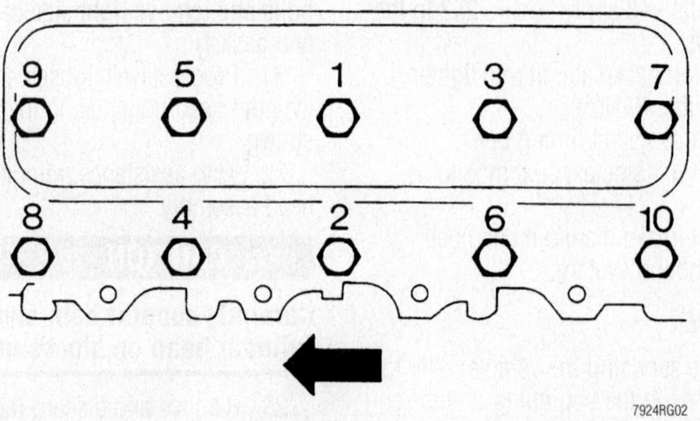

Cylinder head bolt torque sequence—4.0L and 4.6L engines (2000–02)

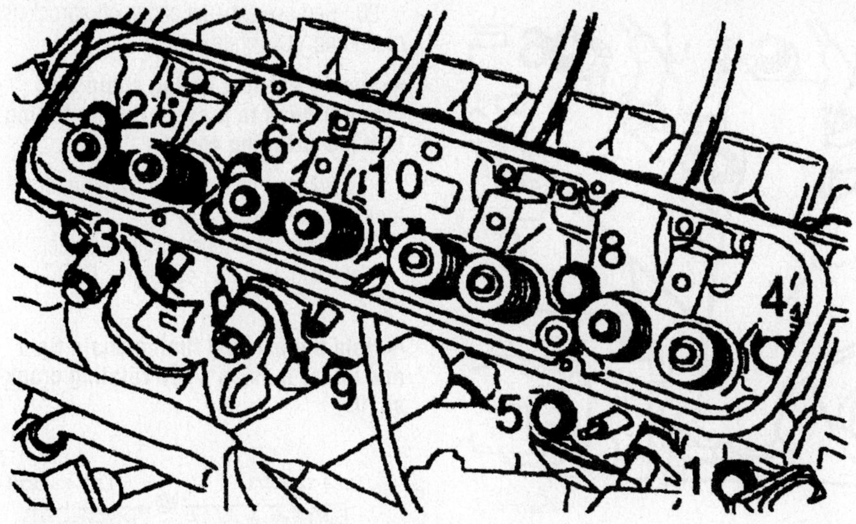

Cylinder head bolt loosening sequence—Discovery (2003–04)

13. Remove the cylinder head.

⁂ CAUTION

Carefully support both ends of the cylinder head on blocks of wood.

14. Remove the cylinder head gasket.

To install:

15. Clean mating faces of cylinder block and head, using gasket removal spray and plastic scraper. Ensure bolt holes in block are clean and dry.

16. Check the head face for warping, pitting or other damage. If out of specification, repair or replace head. Maximum allowable warping is 0.002 inch (0.05mm).

➡Not more than 0.02 inch (0.50mm) material can be removed from cylinder head surface during refacing.

17. Install a new cylinder head gasket (without sealant), with the word 'TOP' uppermost.

18. Carefully install cylinder head, locating head on dowels.

19. Lightly lubricate the new cylinder head bolt threads with clean engine oil.

20. Noting that bolts 1, 3 and 5 are longer than the others, be sure bolts are installed in their proper locations.

21. Tighten bolts, in sequence shown, to the following specifications:

- First step: 15 ft. lbs. (20 Nm)
- Second step: 90 degrees further
- Third step: 90 degrees further

⁂ CAUTION

DO NOT tighten bolts 180 degrees in one operation.

22. If equipped with SAI, install new air injection adapters and tighten to 24 ft. lbs. (33 Nm).

23. Clean the push rods and lubricate ends with clean engine oil.

24. Install push rods in same order as removed.

25. Clean the bases of the rocker pillars and mating faces on the cylinder head.

26. Clean the contact surfaces on the rockers, valves and push rods.

27. Lubricate the contact surfaces and rocker shaft with clean engine oil.

28. Install the rocker shaft assembly and engage the push rods.

29. Install the rocker shaft bolts and progressively tighten to 30 ft. lbs. (40 Nm).

30. On left side cylinder head, perform the following:

a. Install alternator mounting bracket and torque bolts to 30 ft. lbs. (40 Nm).

b. Install the auxiliary drive belt tensioner and torque the bolt to 33 ft. lbs. (45 Nm).

31. On right side cylinder heads, connect the high tension spark plug leads to their original locations.

32. Install new gaskets with the exhaust manifold and tighten the bolts, in the sequence shown, to the following specifications:

- First step: 11 ft. lbs. (15 Nm)
- Second step: 28 ft. lbs. (36 Nm)

➡Bolt tightening sequence is the same for both exhaust manifolds. Illustration shows left side exhaust manifold.

33. On right side cylinder head, reconnect the ground strap and tighten the bolts to 16 ft. lbs. (22 Nm).

34. On left side cylinder head, perform the following steps:

a. Install the brake vacuum servo heat shield.

b. Install the engine harness and tighten the retaining bolts to 16 ft. lbs. (22 Nm).

35. On left side cylinder head, reconnect the high tension leads to the spark plugs in their original order.

36. Install the intake manifold gasket.

37. Reconnect the negative battery cable.

Range Rover (2003–04)

1. Before servicing the vehicle, refer to the precautions in the beginning of this section.

2. Release fuel system pressure.

Cylinder head bolt torque sequence—Discovery (2003–04)

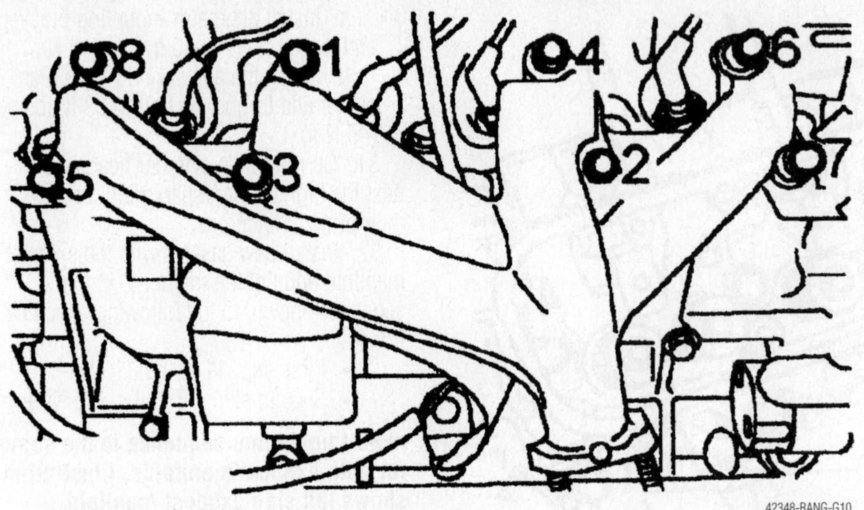

Exhaust manifold bolt torque sequence—Discovery (2003–04)

3. Remove or disconnect the following:
    - Negative battery cable
    - Coolant
    - Engine oil and oil filter
    - Air intake hose
    - Radiator hoses
    - Front exhaust pipes from the manifolds
    - Exhaust manifolds
    - Water pump-to-coolant manifold pipes.
    - Cooling fan cowl
    - Engine appearance cover
    - Valve cover grounds
    - Ignition coil covers
    - Ignition coil connectors
    - Ignition coils
    - Dipstick tube
    - Valve covers.
    - Spark plugs
    - Engine lifting eyes at rear of cylinder heads
    - 20 nuts attaching 4 camshaft oil lines from each head.
4. Rotate crankshaft until cylinder no. 1 camshaft lobes are at TDC.

➡The rear of the camshafts may appear twisted and out of alignment when cylinder no. 1 is at TDC. This is normal.

5. Place locking tool LRT-12-227 through timing tool access hole to lock flywheel in position.
6. On each cylinder head, remove the Variable Camshaft Control (VCC) solenoid connector.
7. Remove 2 bolts securing each VCC seal and remove seals.
8. Using socket LRT-12-226, remove the VCC solenoids.
9. Disconnect coolant hose from alternator.
10. On the right cylinder head, disconnect the engine wiring harness from the upper timing cover and remove the harness bracket.
11. Remove the Camshaft Position Sensor (CMP) harness ties and connector.
12. Remove Secondary Air Intake (SAI) pipes from each cylinder head and position aside.
13. Remove upper timing gear covers.
14. Remove CMP sensor rings.

➡Sensor ring attaching bolt is a left hand thread.

15. Loosen both VCC units and exhaust camshaft sprocket bolts ½ turn.
16. Position camshaft locking tools LRT-12-223/3 and 223/4 to rear of camshafts and tighten in position.
17. Position tools LRT-12-223/5 and 223/6 over locking tools and tighten into spark plug threads.
18. Compress exhaust timing chain tensioner and install tool LRT-12-220 to chain tensioner.
19. Remove VCC unit and exhaust camshaft sprocket bolts.

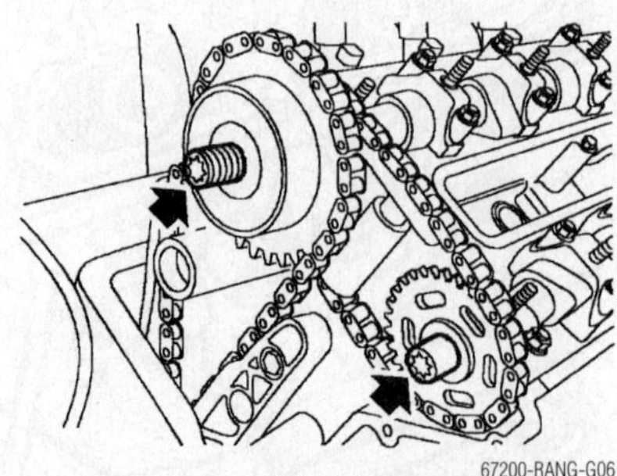

VCC unit and exhaust camshaft bolt locations—Range Rover (2003–04) left side shown; right side similar

20. Remove exhaust camshaft sprockets, chain and VCC units.

➡Secure exhaust timing chain the cylinder head to prevent it from falling into lower timing cover.

21. Remove locking tool from timing tool access hole.
22. Rotate crankshaft pulley counterclockwise to align the 45 degree BTDC mark.

➡Hold the exhaust timing chain tight and under tension when rotating crankshaft.

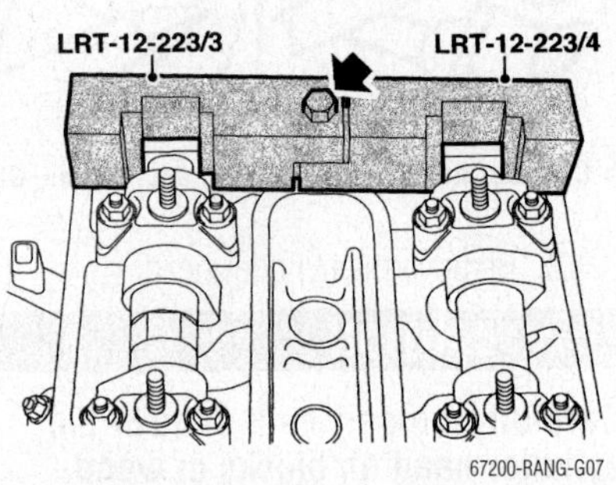

Installing locking tools to rear of camshafts—Range Rover (2003–04) left side shown; right side similar

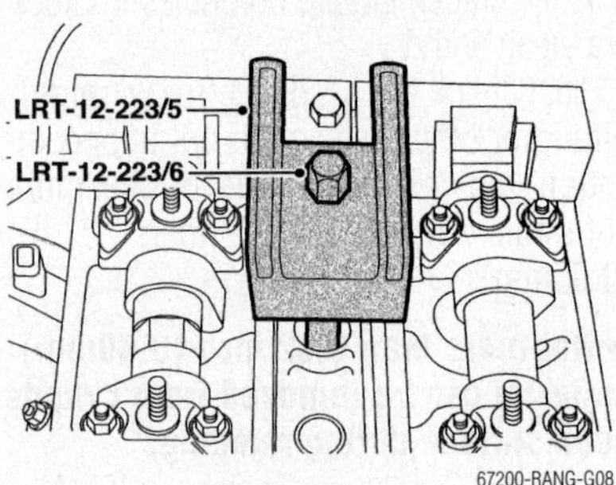

Installing locking tool to cylinder head—Range Rover (2003–04) left side shown; right side similar

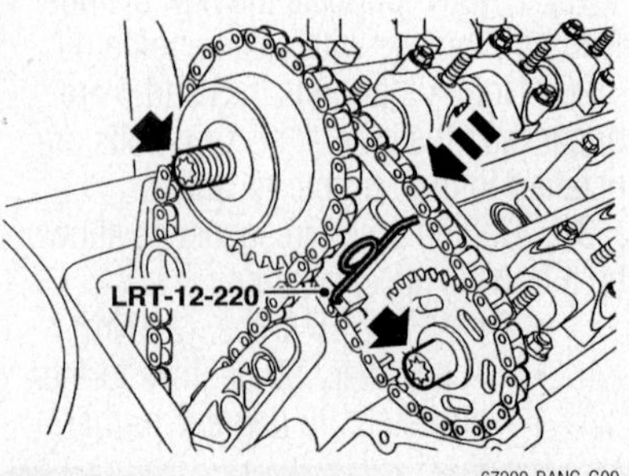

Installing tool to exhaust timing chain tensioner—Range Rover (2003–04) left side shown; right side similar

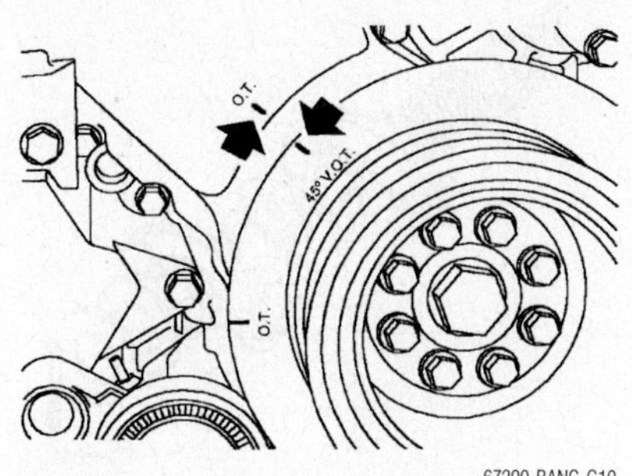

Crankshaft pulley timing marks—Range Rover (2003–04)

23. Remove tools from camshaft sprockets and cylinder head.

24. Remove bolt attaching timing chain guide rail to cylinder head.

25. Remove cylinder heads bolts in reverse of tightening sequence.

26. Remove cylinder head.

27. Remove o-ring from top of cylinder block.

To install:

28. Clean mating faces of cylinder block and head, using gasket removal spray and plastic scraper. Ensure bolt holes in block are clean and dry.

29. Check the head face for warping, pitting or other damage. If out of specification, repair or replace head. Maximum allowable warping is 0.002 inch (0.05mm).

➡Not more than 0.02 inch (0.50mm) material can be removed from cylinder head surface during refacing.

30. Install a new o-ring into the cylinder block.

31. Apply sealant to cylinder block as shown.

32. Install a new cylinder head gasket to locating dowels on cylinder block.

33. Place cylinder head on block.

34. Install washers and the new cylinder head bolts.

35. Tighten bolts, in sequence shown, to the following specifications:

- First step: 22 ft. lbs. (30 Nm)
- Second step: 80 degrees further
- Third step: 80 degrees further

36. Install bolt attaching timing chain guide rail to cylinder head.

37. Carefully rotate the camshafts until the letter and cylinder number markings on the rear of the camshafts are in an upright position.

38. Install the camshaft locking tools to the rear of the camshafts.

39. Install the locking plate over locking tools and tighten into spark plug threads.

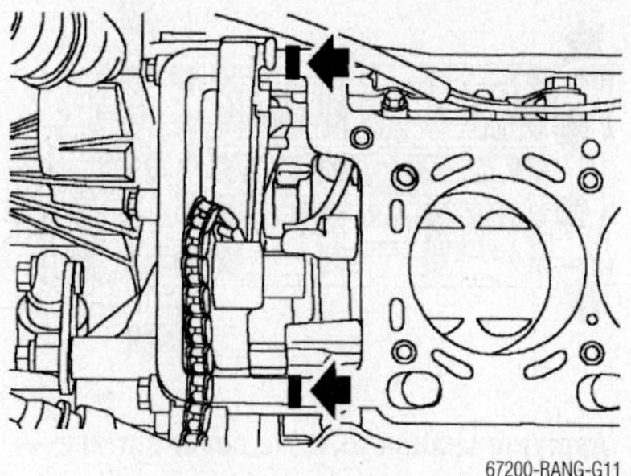

Applying sealant to cylinder block—Range Rover (2003–04) left side shown; right side similar

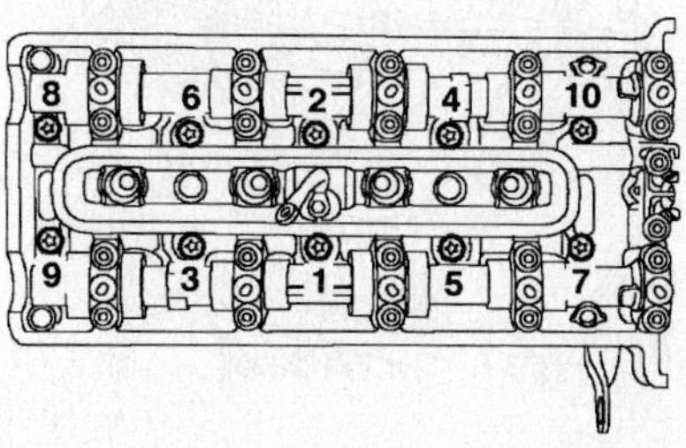

Cylinder head bolt tightening sequence—Range Rover (2003–04)

40. Rotate crankshaft pulley clockwise to align the 45 degree BTDC mark.

➡Hold the exhaust timing chain tight and under tension when rotating crankshaft.

41. Place locking tool LRT-12-227 through timing tool access hole to lock flywheel in position.

42. Position exhaust timing chains on VCC and exhaust camshaft sprockets.

43. Install VCC unit and exhaust camshaft sprocket bolts and hand tighten.

44. Compress exhaust timing chain tensioners and remove tool LRT-12-220 from chain tensioners.

45. Install Special tools LRT-12-221 and 222 to the right side cylinder head and hand tighten the bolt.

46. Install special nut LRT-12-224 to the left side VCC unit.

47. Connect a multimeter set to the ohms scale between the pin of the left side VCC unit and the stud on the oil line.

48. Rotate the VCC unit using a wrench on the special nut and 30 ft. lbs (40 Nm) of torque, to remove the oil cushion and ensure the VCC is at the full stop position. When the unit reaches the full stop position, continuity should be present.

49. Tighten the left side VCC unit and exhaust camshaft sprockets bolts to 11 ft. lbs. (15 Nm). Loosen the bolts ¼ turn.

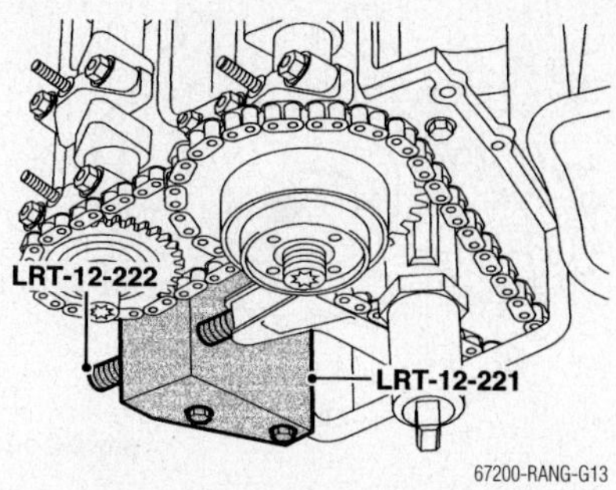

Installing special tools to cylinder head—Range Rover (2003–04) right side shown; left side similar

50. Repeat this procedure on the right side VCC unit and exhaust camshafts.

51. Tighten the bolt on tool LRT-12-222 to 6.3 inch lbs. (0.7 Nm) to set the chain tension.

52. Rotate the left side VCC unit using a wrench on the special nut and 30 ft. lbs (40 Nm) of torque, to remove the oil cushion and ensure the VCC is at the full stop position. When the unit reaches the full stop position, continuity should be present.

53. Tighten the left side VCC unit bolt to 81 ft. lbs. (110 Nm). Tighten the exhaust camshaft sprockets bolts to 92 ft. lbs. (125 Nm).

54. Rotate the right side VCC unit using a wrench on the special nut and 30 ft. lbs (40 Nm) of torque, to remove the oil cushion and ensure the VCC is at the full stop position. When the unit reaches the full stop position, continuity should be present.

55. Tighten the right side VCC unit bolt to 81 ft. lbs. (110 Nm). Tighten the exhaust camshaft sprockets bolts to 92 ft. lbs. (125 Nm).

56. Install the CMP sensor ring on the right side and hand tighten the nut.

57. Install special tool LRT-12-228/1 to the locating dowel on the CMP ring and tighten the tool to the cylinder head.

➡To ensure correct CMP alignment, make sure the tool is aligned flush with the cylinder head and lower timing gear cover before tightening the bolts.

58. Tighten the right side CMP bolt to 30 ft. lbs. 40 Nm).

59. Repeat the procedure on the left side CMP sensor ring, except use tool LRT-12-228/2.

60. Remove all special tools from all locations.

61. Clean upper timing gear covers and install new gaskets.

62. Install new chain tensioner o-ring and apply sealant to areas shown.

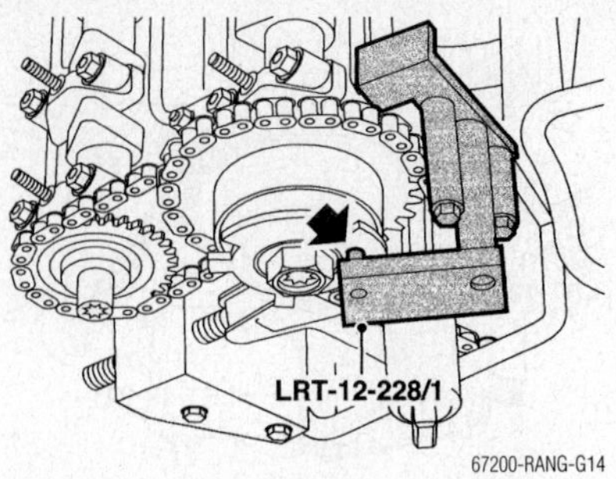

Installing special tool to VCC unit—Range Rover (2003–04) right side shown; left side similar

Applying sealant to upper timing cover surface—Range Rover (2003–04) right side shown; left side similar

63. Install new gaskets to upper timing belt gear covers and install the covers and bolts. Tighten the bolts hand tight.

64. Install new valve cover gaskets and hand tighten 4 bolts.

65. Tighten upper timing gear cover bolts to 11 ft. lbs. (15 Nm).

66. Remove valve covers.

67. Install new sealing washer and install timing chain tensioner and tighten plug to 30 ft. lbs. (40 Nm).

68. Install the VCC solenoids and tighten to 18 ft. lbs. (25 Nm).

69. Install VCC seal and tighten the bolts.remove seals.

70. Install Secondary Air Intake (SAI) pipes to each cylinder head.

71. Install the CMP harness ties and connectors.

72. On the right cylinder head, install the engine wiring harness bracket and harness to the upper timing cover.

73. Connect coolant hose to alternator.

74. Install 20 nuts securing 4 camshaft oil lines to each head.

75. Install engine lifting eyes at rear of cylinder heads.

76. Apply sealant to cylinder head as shown.

77. Install new gaskets and sealing

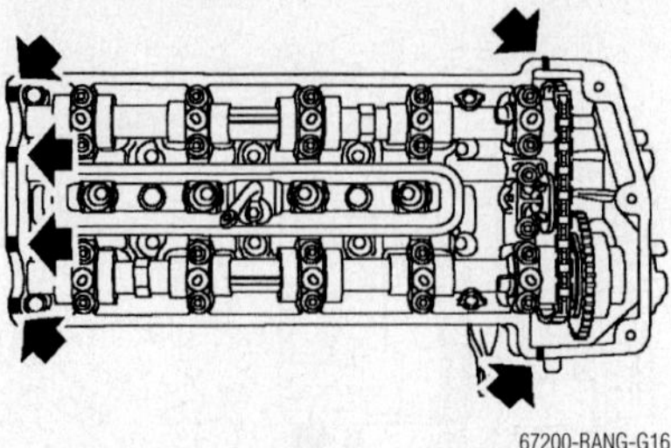

Applying sealant to valve cover surface—Range Rover (2003–04) right side shown; left side similar

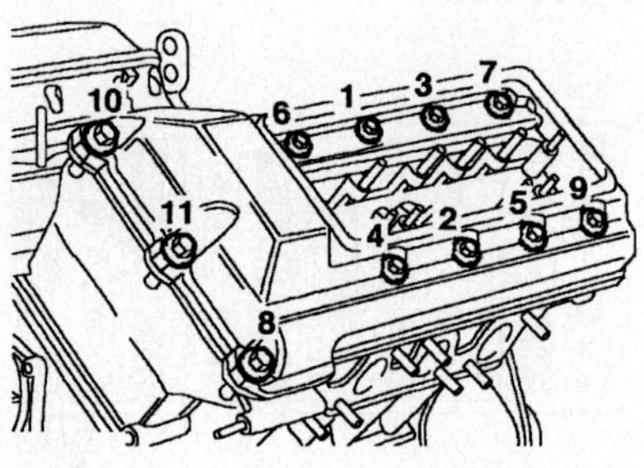

Valve cover bolt tightening sequence—Range Rover (2003–04)

washers to valve covers and tighten nuts and bolts in sequence to 7 ft. lbs. (10 Nm).

78. Install or connect the following:
- Spark plugs
- Dipstick tube
- Ignition coils
- Ignition coil connectors
- Ignition coil covers
- Valve cover grounds
- Engine appearance cover
- Cooling fan cowl
- Water pump-to-coolant manifold pipes.
- Exhaust manifolds
- Front exhaust pipes from the manifolds
- Air intake hose
- Radiator hoses
- Engine oil and oil filter
- Coolant

79. Reconnect the negative battery cable.

80. Start the engine and check for leaks.

Rocker Arms/Shafts

REMOVAL & INSTALLATION

Discovery (2001–04) and Range Rover (2001–02)

1. Before servicing the vehicle, refer to the precautions in the beginning of this section.

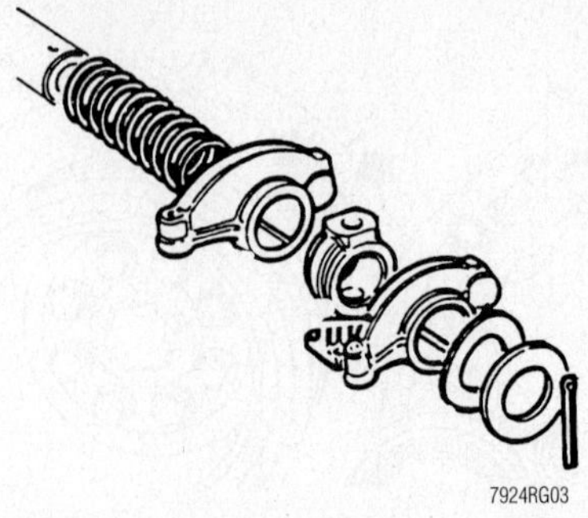

Exploded view of the rocker arm shaft components—Discovery II (2001–04) and Range Rover (2000–02)

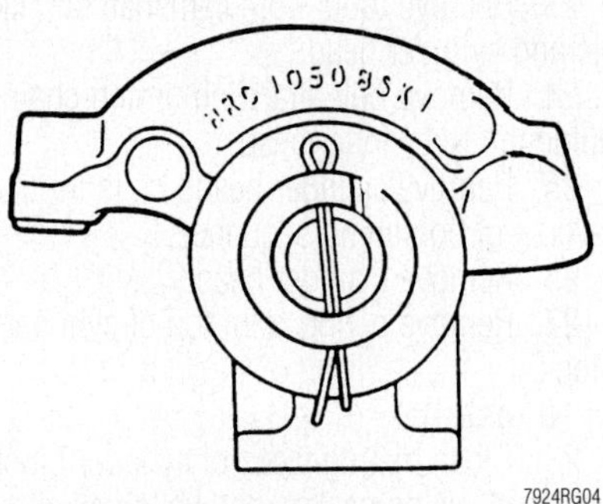

The end notches of the rocker shaft must face up—Discovery II (2000–02) and Range Rover (2000–02)

2. Disconnect the negative battery cable.

3. Relieve the fuel system pressure.

4. Drain the cooling system.

5. Remove the rocker arm covers.

6. Remove the 4 rocker arm shaft bolts, in progressive steps. Release the rocker shaft from the push rods and remove the rocker arm shaft assemblies.

➡Be sure push rods remain located in their tappets when the rocker shaft is removed. Retain all components in their removed sequence for proper reassembly.

7. Remove the cotter pin from either end of the shaft and slide the components from the shaft, keeping them in order for reassembly.

8. Inspect the parts for wear or damage, and replace any suspect parts. Discard any weak springs.

To install:

9. Reassemble the shaft and components. Note the position of the oil feed holes. Use new cotter pin(s).

10. Position the rocker shaft assembly on the head. Be sure that the shaft is installed with the notches or identification groove positioned at the one o'clock position at each end on the upper side, with

push rod locations of rocker arms to the right. Be sure that each rocker ball stud engages its respective pushrod.

11. Install the bolts and tighten them, gradually, to 30 ft. lbs. (40 Nm), starting with the 2 inner, then the 2 outer bolts.

12. Complete installation of rocker cover and components in reverse of removal procedure.

Intake Manifold

REMOVAL & INSTALLATION

Freelander (2001–04)

LEFT SIDE

1. Before servicing the vehicle, refer to the precautions in the beginning of this section.
2. Disconnect battery ground lead.
3. Drain cooling system.
4. Remove fuel rail.
5. Depress locking collar and release breather hose from left side intake manifold.
6. Release clip and disconnect coolant hose from left side intake manifold.
7. Progressively loosen and remove 7 bolts securing left side intake manifold to cylinder head.
8. Remove intake manifold and discard gasket.

To install:

9. Clean intake manifold and cylinder head mating faces.
10. Install new intake manifold gasket to cylinder head, position intake manifold, fit and tighten bolts to 18 ft. lbs. (25 Nm).
11. Connect breather hose to left side intake manifold.
12. Connect coolant hose to intake manifold and secure clip.
13. Install fuel rail.
14. Fill cooling system.
15. Connect battery ground lead.

RIGHT SIDE

1. Before servicing the vehicle, refer to the precautions in the beginning of this section.
2. Disconnect battery ground lead.
3. Drain cooling system.
4. Remove fuel rail.
5. Release clip and disconnect coolant hose from right side intake manifold.
6. Depress locking collar and release breather hose from right side intake manifold.
7. Progressively loosen and remove 7 bolts securing right side intake manifold to cylinder head.
8. Remove intake manifold and gasket.
9. Remove bolt securing coolant breather hose mounting bracket to intake manifold and remove bracket.

To install:

10. Clean intake manifold and cylinder head face.
11. Position coolant/breather hose bracket to intake manifold, fit and tighten bolt to 7 ft. lbs. (9 Nm).
12. Install new intake manifold gasket to cylinder head, position intake manifold, fit and tighten bolts to 18 ft. lbs. (25 Nm).
13. Connect coolant hose to intake manifold and secure clip.
14. Connect breather hose to manifold and secure into collar.
15. Install fuel rail.
16. Fill cooling system.
17. Connect battery ground lead.

Discovery (2001–02) and Range Rover (2001–02)

1. Before servicing the vehicle, refer to the precautions in the beginning of this section.
2. Remove or disconnect the following:
   - Negative battery cable
   - Coolant
   - Fuel system pressure
   - Air intake hose from the plenum chamber
   - Throttle and cruise control cables from the throttle linkage and mounting brackets
   - Breather hose from the plenum chamber
   - Purge hose from plenum chamber
   - TP sensor
   - Stepper motor electrical connector
   - The 6 bolts securing the plenum chamber to the ram housing
   - Coolant hoses attached to the plenum chamber and intake manifold
   - Plenum chamber assembly
   - Breather hoses, as necessary
   - ECT and temperature gauge sensors
   - The 8 fuel injector connectors
   - Fuel temperature sensor connector
   - Fuel supply and return lines
   - The 6 nuts securing fuel rail and ignition coil bracket to the intake manifold
3. Lift the fuel rail slightly to remove the ignition coil bracket from the intake manifold studs and place aside.
4. Using the sequence shown, remove the 12 bolts securing the intake manifold to the cylinder heads, then remove the manifold.
5. Remove the bolts and clamps securing the manifold gasket to the cylinder block.
6. Remove the intake manifold gasket and discard.
7. Remove the gasket seals and discard.

To install:

8. Clean all gasket mating faces.
9. Apply a thin bead of Loctite® Superflex (black) sealant to the 4 notches between cylinder head and block.
10. Install a new gasket seals. Be sure the ends engage correctly in the notches.
11. Install a new intake manifold gasket and tighten the manifold gasket clamps to 6 inch lbs. (0.7 Nm).
12. With the aid of an assistant, hold the harness and ignition coils aside, position the intake manifold assembly.

➡Tighten the manifold bolts in the reverse order of removal.

13. Install the intake manifold bolts and tighten as follows:
    - 84 inch lbs. (10 Nm).
    - 37 ft. lbs. (50 Nm).
14. Tighten the gasket clamp bolts to 13 ft. lbs. (17 Nm).
15. Install or connect the following:
    - Ignition coil bracket on the intake manifold studs, and tighten the mounting bolts to 72 inch lbs. (8 Nm).
    - Fuel lines
    - Connectors to the fuel injectors and fuel temperature sensor
    - ECT sensor and temperature gauge sensor
    - Breather hoses as removed
16. Clean the mating faces of the plenum chamber.
17. Connect any coolant hoses that were removed.
18. Apply a thin, uniform coating of Loctite® 577 sealant to the mating face of the plenum chamber.
19. Install or connect the following:
    - Plenum chamber to the ram housing and tighten to 18 ft. lbs. (24 Nm)
    - Stepper motor and TP sensor connectors
    - Purge hose to the plenum chamber
    - Breather hose to the plenum chamber
    - Throttle and cruise control cables
    - Air intake hose to the plenum chamber and secure with clip
    - Coolant
    - Alternator
    - Negative battery cable

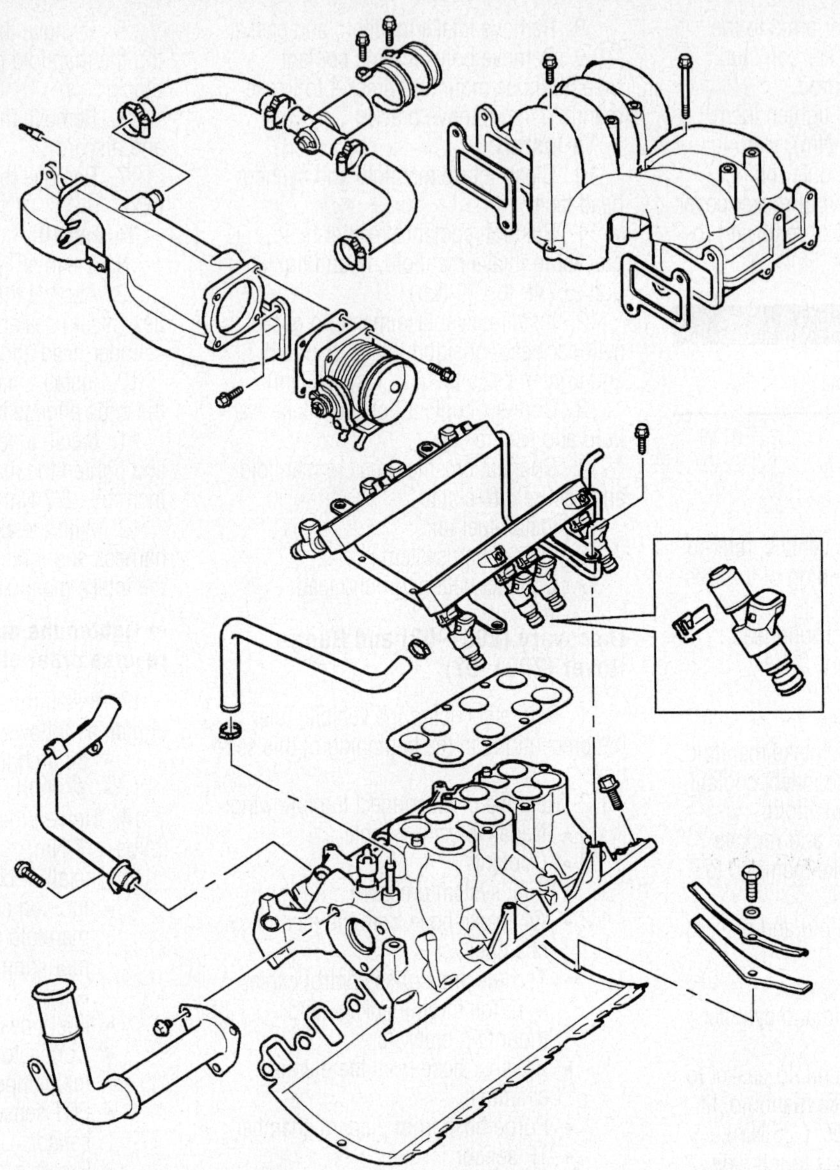

Intake manifold components—Discovery (2001–02) and Range Rover (2001–02)

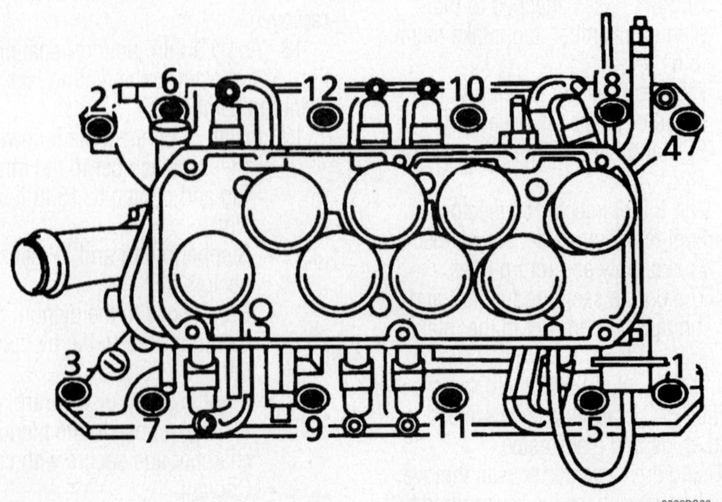

Intake manifold bolt removal sequence—Discovery II (2001–02) and Range Rover (2001–02)

Discovery With Secondary Air Injection (2003–04)

UPPER INTAKE MANIFOLD

1. Before servicing the vehicle, refer to the precautions in the beginning of this section.
2. Disconnect the negative battery cable.
3. Disconnect or remove the following:
   - Multiplug from the MAF sensor
   - Clip on the intake hose and release the harness
   - IACV hose
   - Air intake hose
   - Vacuum hose from secondary air valve
   - 2 nuts holding hose to air valve
   - Air manifold unions from adapters in cylinder head

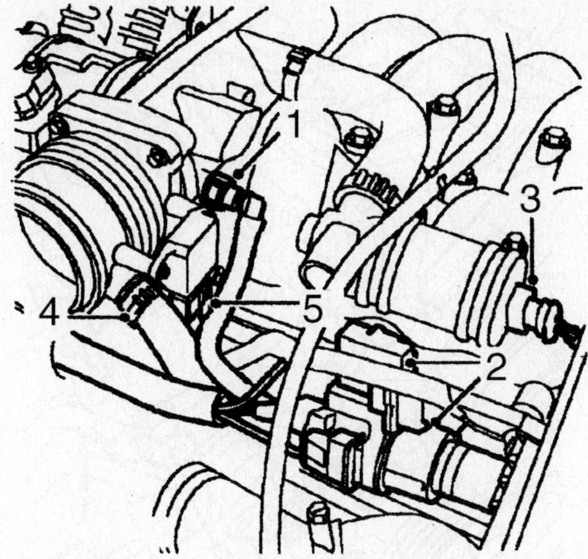

Identifying the location of the purge valve, IACV, breather hose and TP sensor—Discovery (2003–04)

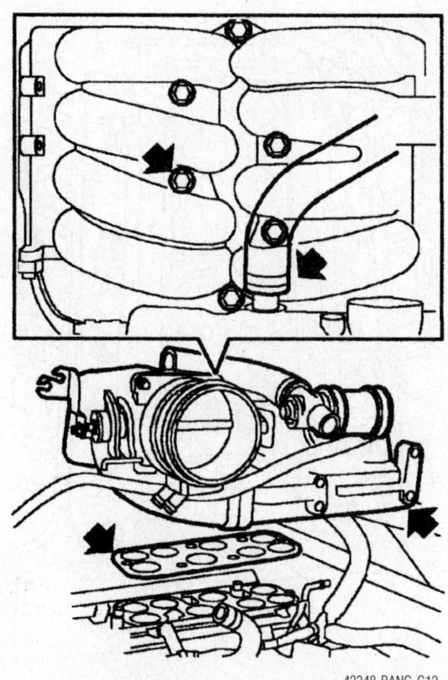

Identifying location of upper intake manifold bolts—Discovery (2003–04)

- Left side air manifold
- Purge valve hose from the intake manifold
- Purge valve and air control valve from bracket (lay it aside)
- Multiplug from the IACV
- Breather hose from the throttle housing
- Multiplug from TP sensor
- 2 coolant hoses from throttle housing
- Throttle and cruise control cables from throttle body (lay cables aside)

4. Depress the plastic collar and disconnect the brake servo hose from the intake manifold.

5. Disconnect the vacuum hose from the manifold.

6. Release the engine harness from the bulkhead clips and lay the harness aside.

7. Remove the hood seal from the rear of the engine compartment.

8. Remove the 2 bolts holding the air pipe and coil bracket to the manifold.

9. Loosen 2 lower bolts enough to allow the coil bracket to clear the manifold.

10. Release plug leads from upper manifold clips, near bulkhead.

11. Disconnect the air hose from the secondary air bypass valve.

12. Remove the 2 nuts securing the right side air manifold support bracket to the upper intake manifold.

13. Disconnect the IACV hose from the throttle housing.

14. Remove the 6 bolts securing the upper intake manifold and remove the manifold. Remove and discard the gasket.

To install:

15. Clean upper and lower intake manifold mating faces, dowels and dowel holes.

16. Using a new gasket, position the upper intake manifold into place. Install the retaining bolts. Working in a diagonal sequence, tighten bolts to 16 ft. lbs. (22 Nm).

17. Complete installation by reconnecting or installing components in reverse of removal procedure.

Discovery Without Secondary Air Injection (2003–04)

UPPER INTAKE MANIFOLD

1. Before servicing the vehicle, refer to the precautions in the beginning of this section.

2. Disconnect the negative battery cable.

3. Disconnect or remove the following:
   - MAF sensor harness from clip
   - Air intake hose
   - Throttle and cruise control cables from throttle body (lay cables aside)

4. Disconnect the EVAP pipe from the intake manifold.

5. Disconnect the multiplug from the throttle body.

6. Remove the clip securing the breather hose to the throttle body and release the hose.

7. Disconnect the coolant hoses from the throttle body.

8. Disconnect the brake servo vacuum pipe and the breather hose from the intake chamber

9. Remove the bolt holding the coolant rails.

10. Release the engine harness from the bulkhead clips and position the harness aside.

11. Remove the 4 bolts securing the ignition coils. Position the coils aside.

12. Detach the plug leads from the upper manifold clips. Disconnect the multiplug from the IACV. Release and disconnect the IACV hose.

13. Remove the 6 bolts securing the upper intake manifold and remove the manifold. Remove and discard the gasket.

To install:

14. Clean upper and lower intake manifold mating faces, dowels and dowel holes.

15. Using a new gasket, position the upper intake manifold into place. Install the retaining bolts. Working in a diagonal

sequence, tighten bolts to 16 ft. lbs. (22 Nm).

16. Complete installation by reconnecting or installing components in reverse of removal procedure.

Discovery (2003–04)

LOWER INTAKE MANIFOLD

1. Before servicing the vehicle, refer to the precautions in the beginning of this section.
2. Relieve fuel system pressure.
3. Disconnect the negative battery cable.
4. Drain the cooling system.
5. Remove both rocker arm covers.
6. Disconnect the left side injector harness and connectors.
7. Remove the upper radiator hose.
8. Remove the auxiliary drive belt.
9. Unbolt active cornering enhancement (ACE) pump and move out of the way.
10. Remove the alternator.
11. Remove the power steering pump pulley and the adjoining auxiliary drive belt pulley.
12. Release power steering pump high pressure pipe. Plug the pipe and pump openings immediately.
13. Remove the bolt holding the oil cooling pipe and release the bracket from the auxiliary housing.
14. Remove 2 bolts holding the power steering pump.
15. Remove 5 bolts and one nut securing the auxiliary housing. Pull the housing forward, remove the power steering pump and the auxiliary housing.
16. Remove 4 bolts securing top cooling hose outlet and remove outlet pipe. Remove and discard the O-ring.
17. Disconnect the fuel pipe from the intake manifold. Plug the openings immediately. Wipe up any spilled fuel.
18. Using the sequence shown, remove the 12 lower intake manifold retaining bolts. Remove the lower intake manifold.
19. Remove the 2 bolts retaining the intake manifold gasket and collect the gasket end clamps. Remove the gasket and seals.

To install:

20. Clean all gasket mating faces on cylinder head, block and lower intake manifold.
21. Apply sealant STC 50550 to cylinder head and cylinder block notches.
22. Install new gasket seals, making sure ends of seals engage correctly in notches.
23. Install a new intake manifold gasket. Position the gasket end clamps and install the bolts, but do not tighten at this stage.
24. Position the lower intake manifold to the engine. Install the retaining bolts. Working in the sequence shown, tighten the bolts in 2 steps:
    a. Step 1: 7 ft. lbs. (10 Nm)
    b. Step 2: 38 ft. lbs. (51 Nm)
25. Tighten the gasket end clamp bolts to 13 ft. lbs. (18 Nm).
26. Reconnect the fuel pipe to the intake manifold.
27. Clean the top hose outlet pipe mating faces, install a new O-ring, position the pipe and tighten the retaining bolts to 16 ft. lbs. (22 Nm).
28. Install the alternator and tighten the bolts to 33 ft. lbs. (45 Nm).
29. Install the power steering pump to the auxiliary housing, position the auxiliary housing on the engine, and tighten the retaining bolts to 30 ft. lbs. (40 Nm). Tighten the nut to 7 ft. lbs. (10 Nm).
30. Install and tighten the power steering pump bolts to 16 ft. lbs. (22 Nm).
31. Install the oil cooling pipe bracket bolt and tighten to 16 ft. lbs. (22 Nm).
32. Position the lower auxiliary drive belt pulley and tighten the bolt to 37 ft. lbs. (50 Nm).
33. Clean the power steering pump pulley mating faces. Install the pulley and tighten the bolts to 16 ft. lbs. (22 Nm).
34. Install the auxiliary drive belt.
35. Reconnect the injector harness and connectors.
36. Install the upper radiator hose.

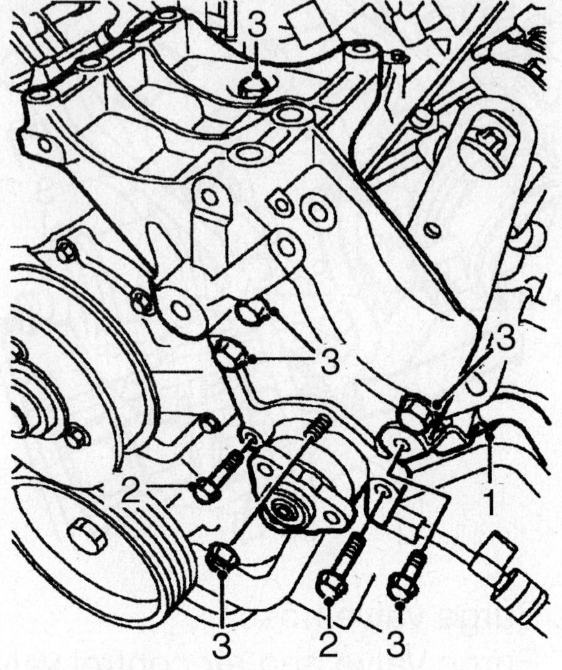

Identifying auxiliary housing and related retaining bolts—Discovery (2003–04)

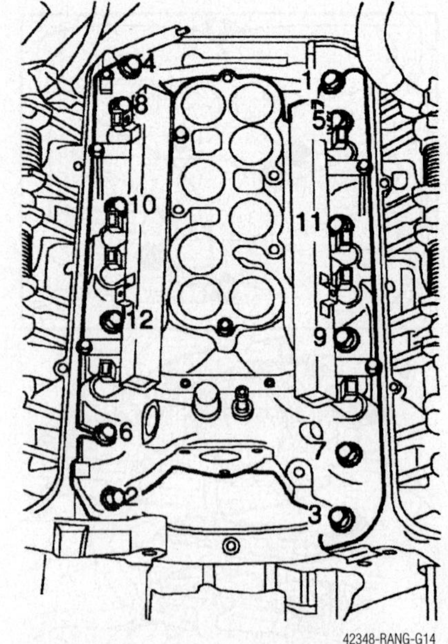

Lower intake manifold bolt removal sequence—Discovery (2003–04)

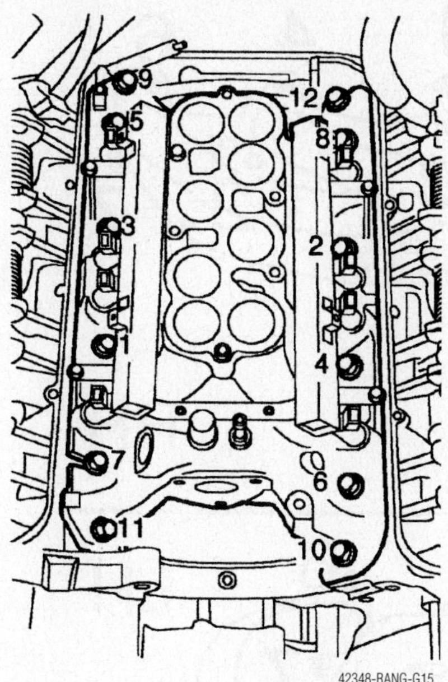

Lower intake manifold bolt torque sequence—Discovery (2003–04)

37. Install the rocker covers.
38. Check and refill all fluids.
39. Reconnect the negative battery cable.

Range Rover (2003–04)

1. Before servicing the vehicle, refer to the precautions in the beginning of this section.
2. Relieve fuel system pressure.
3. Disconnect the negative battery cable.
4. Drain the cooling system.
5. Remove the air intake hose.
6. Remove both ignition coil covers.
7. Disconnect all ignition coil connectors.
8. Disconnect and label all intake manifold electrical connectors.
9. Remove 2 nuts attaching fuel injector harness to fuel rail.
10. On the left side, remove vacuum reservoir and mounting bracket and fuel hose and mounting bracket.
11. Disconnect alternator harness clip from engine cover bracket.
12. Disconnect all fuel injector connectors.
13. Disconnect main engine harness and move aside.
14. Disconnect charcoal canister connector.
15. Disconnect all vacuum and breather hoses.
16. Remove the secondary air solenoid valve.
17. Position the fuel injector harness aside.
18. Remove the air intake plenum.
19. Remove the terminal cover, attaching nuts and disconnect 2 cables from the terminal post.
20. Disconnect fuel feed hose from fuel rail.
21. Remove the oil separator.
22. Remove the 10 intake manifold attaching nuts.
23. Position aside the brake servo vacuum pipe.
24. Lift the intake manifold, disconnect the breather pipe and remove manifold.

To install:

25. Clean all gasket mating faces.
26. Install new o-rings to the fuel injectors.
27. Install 4 new intake manifold gaskets.
28. Place the intake manifold over the mounting position and connect the lower breather pipe.
29. Lower the intake manifold and install the 10 spacers and attaching nuts. Tighten the nuts to 11 ft. lbs. (15 Nm).
30. Install the brake servo vacuum pipe, front cover and brake vacuum hoses.
31. Connect the 2 breather hoses.
32. Clean the fuel hose connections, then connect fuel feed line to fuel rail.
33. Connect 2 cables to terminal post, tighten nut and install cover.
34. Install the air intake plenum.
35. Position the fuel injector harness, install the washer, place the vacuum switching valve on the right side injector mounting stud and install the 2 nuts and tighten.
36. Connect charcoal canister solenoid connector.
37. Position and secure main engine harness to mounting bracket.
38. Connect fuel injector connectors.
39. Install alternator harness on front cover mounting bracket.
40. Place fuel hose and mounting bracket to left side injector stud.
41. Install vacuum reservoir and mounting bracket to left side injector mounting stud.
42. Install and tighten 2 nuts to left side injector mounting stud.
43. Connect all labeled electrical connectors to appropriate locations.
44. Connect ignition coil connectors.
45. Install ignition coil covers.
46. Connect air intake hose.
47. Connect the negative battery cable.
48. Fill the engine with coolant.
49. Start engine and check for leaks.

Exhaust Manifold

REMOVAL & INSTALLATION

Freelander (2001–04)

RIGHT SIDE

1. Before servicing the vehicle, refer to the precautions in the beginning of this section.
2. Disconnect battery ground lead.
3. Remove right side catalytic converter.
4. Remove 4 nuts securing exhaust manifold to cylinder head, remove manifold and discard gasket.

To install:

5. Clean exhaust manifold and mating face.
6. Using new gasket, fit exhaust manifold and tighten nuts to 45 Nm (33 ft. lbs.).
7. Fit right side catalytic converter.
8. Connect battery ground lead.

LEFT SIDE

1. Before servicing the vehicle, refer to the precautions in the beginning of this section.
2. Position vehicle on 4 post ramp.
3. Disconnect battery ground lead.
4. Remove engine acoustic cover.
5. Remove underbelly panel.
6. Using a ⅜ in. socket bar raise auxiliary drive belt tensioner and release drive belt from alternator and PAS pump pulleys.
7. Remove 3 bolts securing A/C compressor to fixing brackets and position compressor aside.
8. Remove compressor heat shield.
9. Release and disconnect HO_2S multiplug.
10. Release HO_2S harness from clip.
11. Release and disconnect post catalyst HO_2S multiplug.
12. Release HO_2S harness from clips.
13. Remove 2 nuts securing exhaust front pipe to exhaust manifold.
14. Remove 4 nuts securing manifold and remove exhaust manifold.
15. Remove and discard manifold and down pipe gaskets.

To install:

16. Clean exhaust manifold and mating faces.
17. Using new gaskets, fit exhaust manifold and tighten nuts to cylinder head to 33 ft. lbs. (45 Nm) and nuts to front pipe to 37 ft. lbs. (50 Nm).
18. Connect HO_2S multiplugs, secure multiplugs to brackets and harness to clips.
19. Clean compressor and mounting bracket mating faces.
20. Fit compressor heat shield.

21. Position compressor to mounting and align heat shield. Fit and tighten bolts to 18 ft. lbs. (25 Nm).

22. Using a 3/8 in. square drive socket bar, raise auxiliary drive belt tensioner and fit drive belt to pulleys.

23. Fit underbelly panel.

24. Fit engine acoustic cover.

25. Connect battery ground lead.

Discovery (2001–04)

➡This procedure is the same for both exhaust manifolds.

1. Before servicing the vehicle, refer to the precautions in the beginning of this section.
2. Raise vehicle on a 4-post lift or raise front of vehicle and support on safety stands.
3. Disconnect the negative battery cable.
4. Remove the exhaust manifold–to–exhaust pipe retaining nuts.
5. Remove the 8 manifold bolts. Remove the exhaust manifold and discard the 2 gaskets.

To install:

6. Clean all mating surfaces.
7. Place the exhaust manifold in position on the cylinder head along with new gaskets.
8. Install spacers and the exhaust manifold bolts. Working from the center outward, tighten bolts to 28 ft. lbs. (38 Nm).
9. Install the exhaust pipe to the exhaust manifold. With a new gasket in place, tighten the exhaust pipe retaining nuts to 22 ft. lbs. (30 Nm).
10. Lower the vehicle.
11. Connect the negative battery cable.
12. Start the engine and check for exhaust leaks.
13. Road test the vehicle and check for proper engine operation.

Range Rover (2001–02)

1. Before servicing the vehicle, refer to the precautions in the beginning of this section.

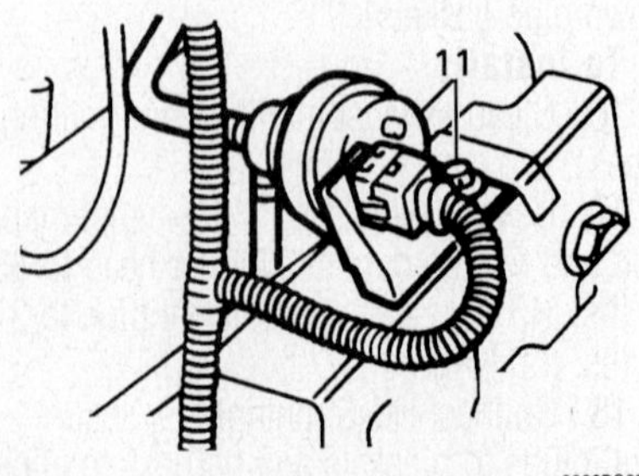

Purge valve mounting bolt location—Range Rover (2001–02)

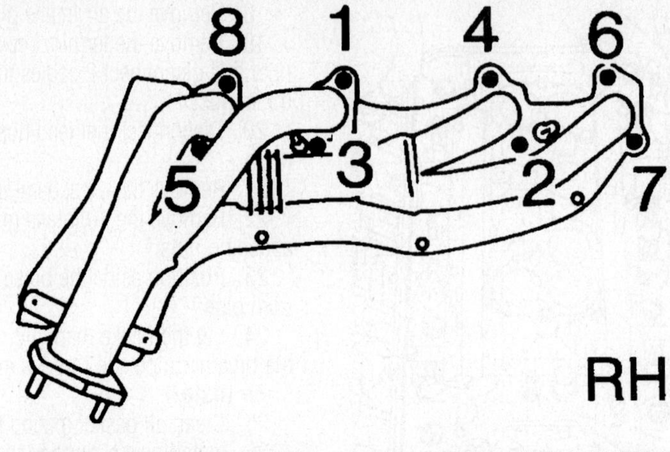

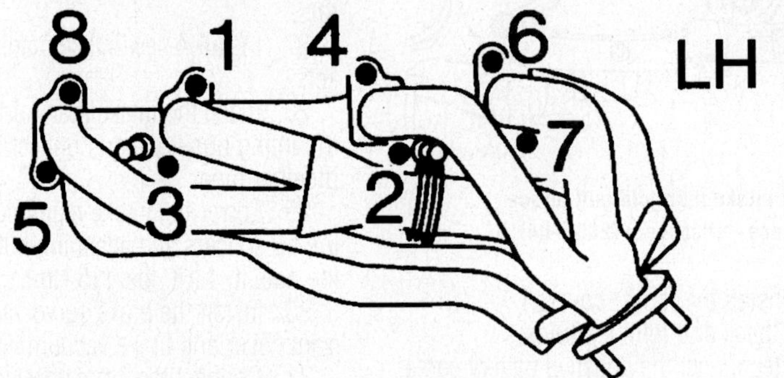

Exhaust manifold torque sequence—Range Rover (2001–02)

2. Remove or disconnect the following:
   - Negative battery cable
   - Front exhaust pipes from the exhaust manifolds
3. Lower the vehicle.
4. For the left manifold, perform the following:
   a. Release the intake hose from the plenum chamber.
   b. Release the harness from the intake hose clip.
   c. Remove the air flow meter.
   d. Disconnect the purge valve (11) and position the valve aside.
5. For the right manifold, perform the following:
   a. Disconnect the spark plug wires and position them out of the way.
   b. Unscrew the right shock absorber top mounting bolt to provide additional clearance for the removal of the heat shield.
6. Remove or disconnect the following:
   - The 8 bolts (for the right manifold) or 7 bolts (for the left manifold) securing the outer heat shield to the manifold.
   - The 8 exhaust manifold–to–cylinder head mounting bolts, then remove manifolds.

To install:

7. Ensure all gasket mating surfaces are clean.
8. Install or connect the following:
   - Manifolds using new gaskets. Tighten the bolts 40 ft. lbs. (55 Nm), in the sequence as shown.
   - Heat shields and tighten the mounting bolts to 72 inch lbs. (8 Nm).
   - Purge valve on the shock absorber turret
   - Air flow meter
   - Right shock absorber top mounting bolt to 63 ft. lbs. (85 Nm).
   - Spark plug wires
   - Front exhaust pipes to the manifolds and tighten the mounting nuts to 37 ft. lbs. (50 Nm).
   - Negative battery cable

Range Rover (2000–02)

1. Before servicing the vehicle, refer to the precautions in the beginning of this section.

➡The following procedure applies to either side exhaust manifold

2. Remove or disconnect the following:
   - Negative battery cable

- Oxygen sensor connectors from ahead and behind the catalytic converter
- Support exhaust system with a suitable jack
- 4 nuts attaching the front exhaust rubber mountings to body
- Front exhaust pipes from the exhaust manifolds
- Lower the exhaust system with the jack
- 8 nuts attaching the exhaust manifold
- Exhaust manifold and discard the gaskets

To install:

3. Ensure all gasket mating surfaces are clean.
4. Install or connect the following:
   - Apply copper grease to exhaust manifold mounting studs
   - Manifolds using new gaskets and nuts. Tighten the nuts 17 ft. lbs. (23 Nm).
   - Raise exhaust system with the jack
   - Exhaust pipes using new mounting nuts. Tighten the nuts to 33 ft. lbs. (45 Nm).
   - Exhaust rubber mounting nuts and tighten to 18 ft. lbs. (25 Nm)
   - Lower the jack
   - Connect oxygen sensor connectors
   - Negative battery cable

Camshafts and Valve Lifters

REMOVAL & INSTALLATION

Freelander (2001–04)

LEFT SIDE

1. Before servicing the vehicle, refer to the precautions in the beginning of this section.
2. Disconnect battery ground lead.
3. Remove camshaft timing belt.
4. Remove the left SIDE camshaft rear timing belt.
5. For 2000–02 models, remove the left SIDE camshaft cover gasket.
6. Remove the front camshaft drive belt cover backplate from the cylinder head.
7. Remove the left SIDE rear camshaft drive belt cover backplate from the rear of the cylinder head.
8. Using the sequence shown, progressively loosen the 22 bolts securing the camshaft carrier to the cylinder head. Loosen bolts until valve spring pressure is released, then remove the bolts.
9. Release the camshaft carrier from the dowels and remove the carrier.
10. Remove the camshaft and remove and discard the oil seals.
11. Disconnect CMP sensor multiplug and release male multiplug from mounting bracket.
12. Using a stick magnet, remove the 12 hydraulic tappets from the cylinder head, keeping tappets in order as removed for installation in original positions.

⁂ CAUTION

Maintain absolute cleanliness when handling hydraulic tappets. Failure to do so can result in engine damage.

13. Clean camshafts and bearing running surfaces in camshaft carrier and cylinder head.
14. Inspect camshafts and replace camshafts if scored, pitted or excessively worn.

➡Camshafts are color-coded blue for exhaust and orange for intake.

To install:

15. Thoroughly clean and lubricate hydraulic tappets with clean engine oil. Fit hydraulic tappets to original bores in cylinder head.
16. Ensure that mating faces of camshaft carrier and cylinder head are clean and dry.
17. Lubricate camshafts and bearing journals with clean engine oil.
18. Position camshafts in cylinder head with rear timing gear drive slots in each camshaft facing towards the center as shown.
19. Apply continuous thin beads of sealant STC 4600 to paths on camshaft carrier as shown. Spread sealant to an even film using a roller.

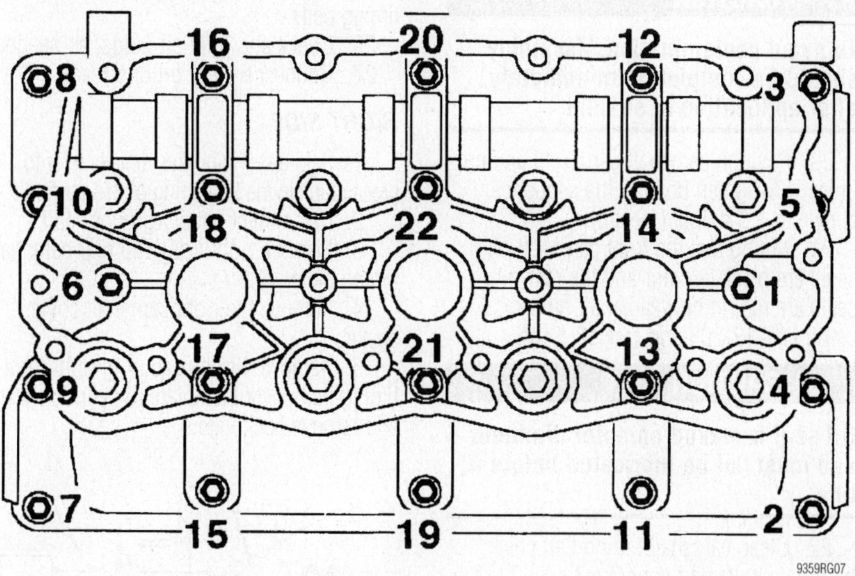

Camshaft carrier bolt loosening sequence—Freelander (2001–04)

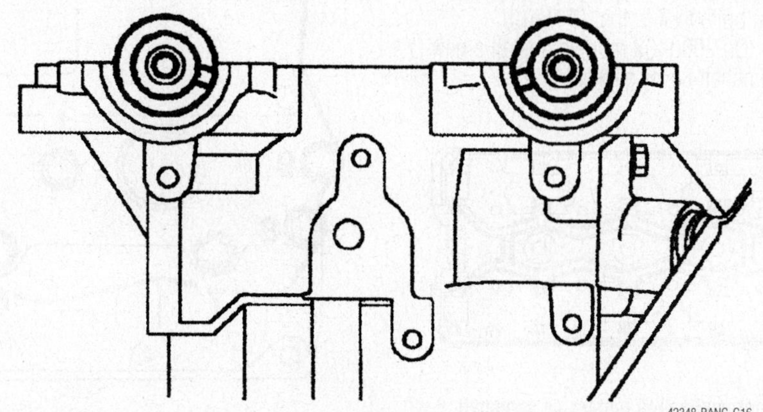

Camshaft installed positions—Freelander (2001–04)

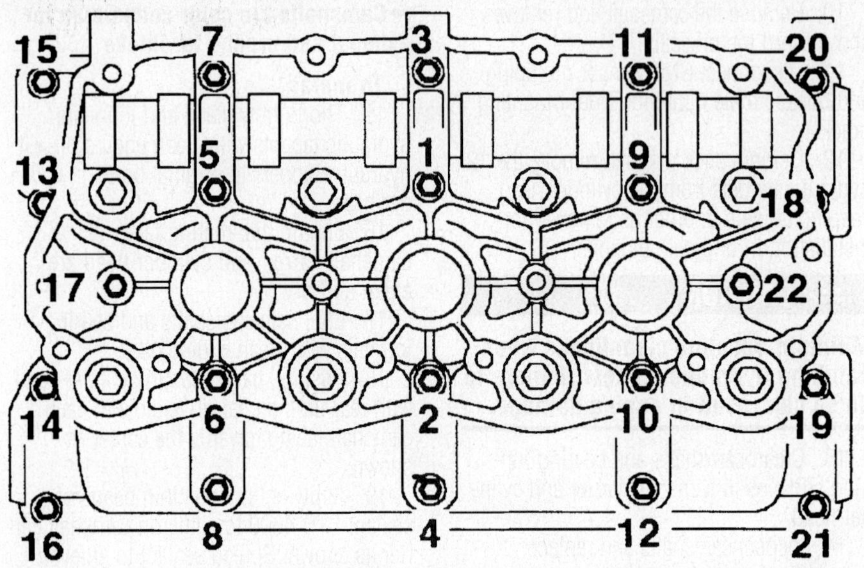

Camshaft carrier bolt torque sequence—Freelander (2001–04)

CAUTION

To avoid contamination, assembly should be completed immediately after application of sealant.

20. Position camshaft carrier, fit and progressively tighten bolts in the sequence shown to 7.5 ft. lbs. (10 Nm).
21. Noting that the front camshaft oil seals are black in color and the rear oil seals are red, fit new camshaft oil seals using LRT-12-203 and LRT 12-148A

CAUTION

Oil seal is waxed on outer diameter and must not be lubricated before fitting.

22. Clean camshaft timing belt cover backplate bolts and apply Loctite 242 to the first 3 threads.
23. Position camshaft timing belt rear cover backplates to cylinder head, fit and tighten bolts to 7 ft. lbs. (9 Nm).
24. On 2000–02 models, install a new left camshaft cover gasket.
25. Install the front and rear camshaft timing belts.
26. Check and refill all fluids, as needed.
27. Connect battery ground lead.

RIGHT SIDE

1. Before servicing the vehicle, refer to the precautions in the beginning of this section.
2. Disconnect battery ground lead.
3. Remove right front and rear camshaft timing belts.
4. Remove the right camshaft cover gasket.
5. Remove 4 bolts securing right front timing belt cover backplate to cylinder head and remove backplate.
6. Remove 2 bolts securing right rear timing belt cover backplate and remove backplate.
7. Using sequence shown, progressively loosen 22 bolts securing camshaft carrier to cylinder head until valve spring pressure is released and remove bolts.
8. Release camshaft carrier from dowels and remove carrier.
9. Remove both camshafts and discard oil seals.
10. Using a stick magnet, remove 12 hydraulic tappets from cylinder head.

CAUTION

Store hydraulic tappets in their fitted order and store upright. Maintain absolute cleanliness when handling hydraulic tappets. Failure to observe these precautions can result in engine failure.

To install:

11. Clean camshafts and bearing running surfaces in camshaft carrier and cylinder head.
12. Inspect camshafts and replace camshafts if scored, pitted or excessively worn.
13. Use a lint-free cloth and solvent, clean sealing surfaces on cylinder head and camshaft carrier.

➡Camshafts are color-coded blue for exhaust and orange for intake.

14. Thoroughly clean and lubricate hydraulic tappets with clean engine oil. Install tappets to original bores is cylinder head.

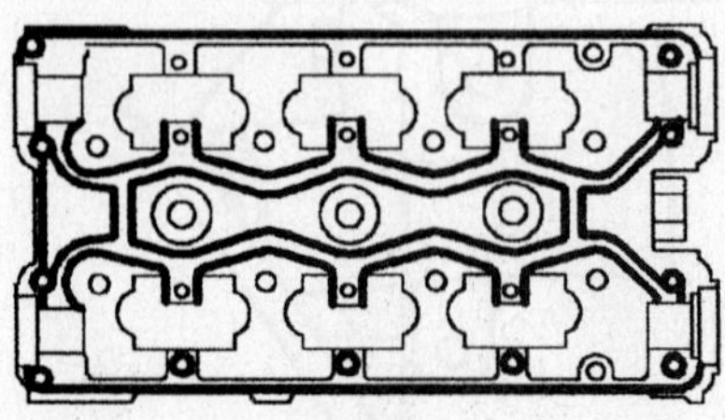

Sealant application pattern on camshaft carrier—Freelander (2001–04)

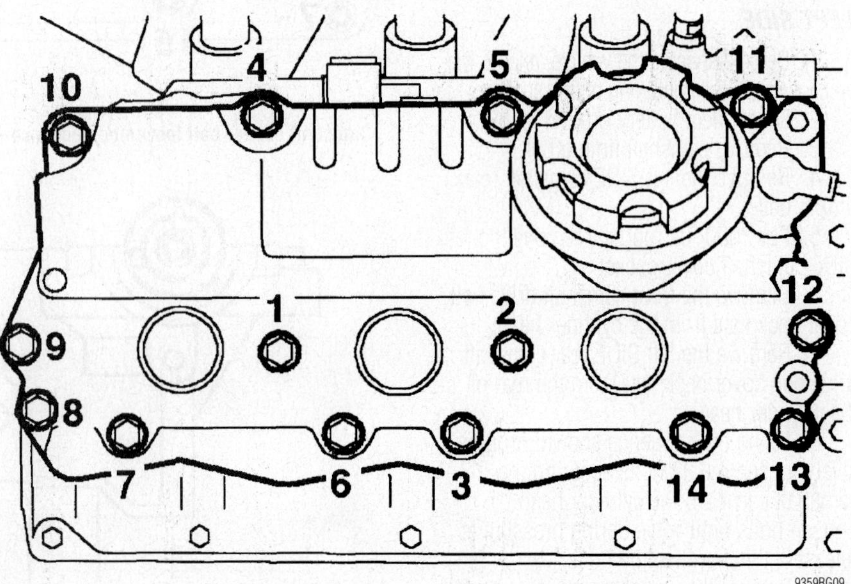

Camshaft cover bolt torque sequence—Freelander (2001–04)

15. Ensure mating faces of camshaft carrier and cylinder head are clean and dry.

16. Lubricate camshafts and bearing journals with clean engine oil.

17. Position camshafts in cylinder head with rear timing gear drive slots in each camshaft facing towards the center as shown.

18. Apply continuous thin beads of sealant STC 4600 to paths on camshaft carrier as shown. Spread sealant to an even film using a roller.

⁂ CAUTION

To avoid contamination, assembly should be completed immediately after application of sealant.

19. Position camshaft carrier, fit and progressively tighten bolts in the sequence shown to 7.5 ft. lbs. (10 Nm).

20. Noting that the front camshaft oil seals are black in color and the rear oil seals are red, fit new camshaft oil seals using LRT-12-203 and LRT -12-148A

⁂ CAUTION

Oil seal is waxed on outer diameter and must not be lubricated before fitting.

21. Clean camshaft timing belt cover backplate bolts and apply Loctite 242 to the first 3 threads.

22. Position camshaft timing belt rear cover backplates to cylinder head, fit and tighten bolts to 7 ft. lbs. (9 Nm).

23. Clean mating surfaces of camshaft cover and carrier.

24. Clean inside of camshaft cover. If necessary, wash oil separator elements in solvent and blow dry.

25. Position new gasket to camshaft carrier with arrows on gasket pointing towards inlet manifold.

26. Fit front and rear camshaft timing belts.

27. Check and refill fluids as necessary.

28. Connect battery ground lead.

Discovery (2001–02) and Range Rover (2001–02)

1. Before servicing the vehicle, refer to the precautions in the beginning of this section.

2. Remove or disconnect the following:
   - Engine from the vehicle
   - Intake manifold and gasket
   - Both rocker arm shaft assemblies

➡Identify each rocker shaft assembly to ensure installation on original cylinder bank.

   - Pushrods and place them in order
   - Valve lifters and place them in order
   - Front cover and timing chain
   - Camshaft thrust plate
   - Camshaft

➡The camshaft installed in the 4.0L engines is color-coded orange, while the camshaft in the 4.6L engines is color-coded red.

To install:

3. Lubricate the camshaft journals with engine oil.

4. Install or connect the following:
   - Camshaft into cylinder block. Tighten the thrust plate mounting bolts to 18 ft. lbs. (25 Nm).
   - Timing chain and front cover

5. Soak the lifters in engine oil. Before installing each lifter, pump the inner sleeve of the lifter several times using a pushrod to prime the lifter; this will reduce lifter noise when the engine is first started.

6. Lubricate the lifter bores with engine oil, then install the lifters.

➡Some lifter noise may still be heard on initial start-up. If necessary, run the engine at 2500 rpm for a few minutes until the noise clears.

7. Install or connect the following:
   - Pushrods
   - Rocker arm shaft assemblies
   - Intake manifold
   - Engine

Discovery (2003–04)

1. Before servicing the vehicle, refer to the precautions in the beginning of this section.

2. Remove intake manifold.

3. Remove rocker shaft assemblies as follows:
   a. Progressively loosen and remove 8 bolts securing rocker shaft assemblies.
   b. Mark each rocker shaft in relation to original position and cylinder head.
   c. Remove rocker shaft assemblies.

4. Remove push rods and store in their order as removed.

5. Remove tappets. Keep tappets in order and orientation as they are removed.

6. Remove timing chain and gears.

7. Before removing the camshaft, check camshaft end play as follows:
   a. Temporarily fit camshaft gear and lightly tighten the retaining bolt.
   b. Attach a dial indicator to front of cylinder block with tip of dial indicator contacting surface of camshaft gear.
   c. Push camshaft rearward and zero the dial indicator.
   d. Using the camshaft gear, pull the camshaft forward and note the dial indicator reading.
   e. End play should be within 0.003-0.010 inch (0.075-0.25mm).

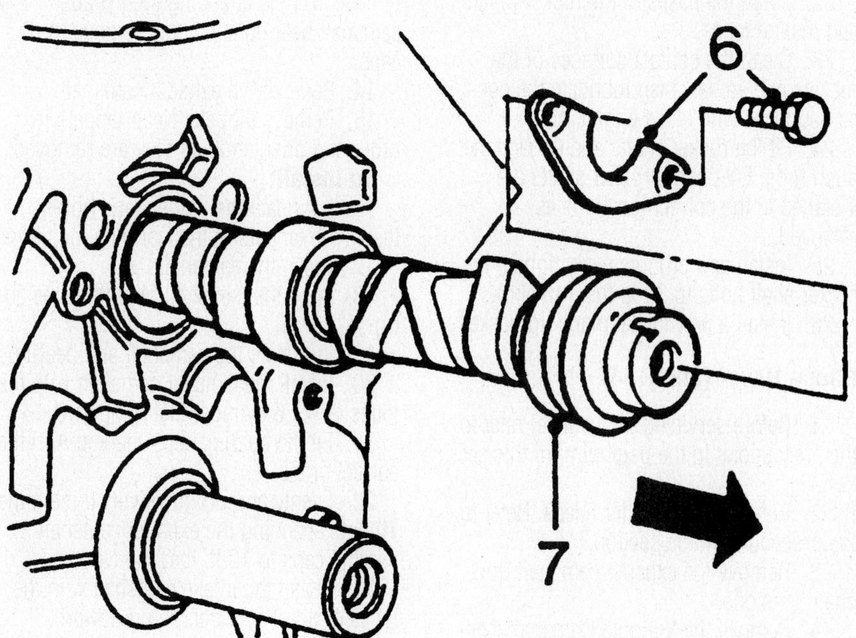

Exploded view of the camshaft (7) and thrust plate (6) mounting—Discover Series II (2001–04) and Range Rover (2001–02)

f. If end play is not within specified range, install a new thrust plate. If end play is still not within range, a new camshaft and/or gear must be installed.

g. Remove dial indicator, camshaft gear bolt and gear.

8. Remove the 2 bolts holding the camshaft thrust plate, and remove the thrust plate.

9. Carefully remove the camshaft, avoiding damage to camshaft bearings.

To install:

10. Clean camshaft bearings in cylinder block.

11. Clean the camshaft.

12. Wipe the camshaft bearing faces and lobes.

13. Clean the thrust plate and mating face.

14. Lubricate the camshaft bearings with clean engine oil.

15. Position the thrust plate. Install and tighten the bolts to 17 ft. lbs. (22 Nm).

16. Install the timing chain and gears.

17. Immerse the tappets in engine oil before installation. Pump the inner sleeve of the tappet several times, using a push rod to prime the tappets. Clean the tappet bores.

18. Lubricate the tappets and tappet bores with clean engine oil.

19. Install the tappets in their original positions.

20. Clean the pushrods. Lubricate the tappet end of the push rods with clean engine oil.

21. Install the push rods in their original locations.

22. Clean the bases of the rocker pillars and mating faces.

23. Clean the contact surfaces of the rockets and valves, then lubricate the contact surfaces with clean engine oil.

24. Fit the rocker shafts and engage the push rods. Ensure the rocker shafts are installed in the correct locations as removed.

25. Install and progressively tighten the rocker shaft bolts to 30 ft. lbs. (40 Nm).

26. Install a new intake manifold gasket.

Range Rover (2003–04)

1. Before servicing the vehicle, refer to the precautions in the beginning of this section.

2. Remove the cylinder heads. Refer to the procedure in this section.

3. Remove the exhaust camshaft drive chain tensioner.

4. Remove the Variable Camshaft Control (VCC) oil distribution housing. Remove and discard sealing washer from the long bolt.

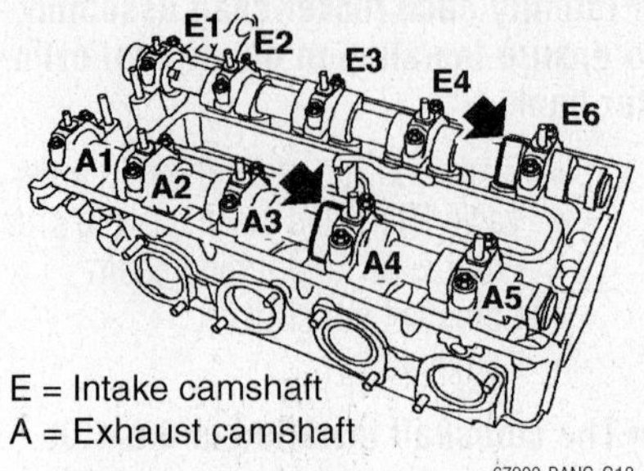

Identifying camshaft bearing caps—Range Rover (2003–04)

5. Remove 3 VCC oil non-return rings from the intake camshaft.

6. Fit an M10 bolt into the VCC non-return valve and by hand, remove the valve. Discard the o-ring.

7. Check camshaft end play. End play should be 0.08–0.014 inch (0.20–0.36 mm).

8. Using a wrench on the intake camshaft hexagon, rotate the camshaft until the lobes of cylinder no. 8 face upward.

9. Using a wrench on the exhaust camshaft hexagon, rotate the camshaft until the lobes of cylinder no. 6 face upward.

10. Identify the camshaft bearing caps for installation reference as shown.

11. Evenly and progressively loosen and remove 10 nuts attaching the intake camshaft bearing caps, and remove the caps.

12. Remove the intake camshaft.

13. Evenly and progressively loosen and remove 10 nuts attaching the exhaust camshaft bearing caps, and remove the caps.

14. Remove the exhaust camshaft.

15. Remove tappets. Keep tappets in order and orientation as they are removed.

To install:

16. Lubricate the tappets and bores. Install the tappets in their original positions.

17. Clean the camshaft.

18. Wipe the camshaft bearing faces and caps.

19. Lubricate the camshaft and bearings.

20. Install the exhaust camshaft with the lobes of no. 6 cylinder facing upward.

21. Fit the bearing caps marked A in the correct order.

22. Evenly and progressively tighten the 10 nuts attaching the exhaust camshaft bearing caps to 10 ft. lbs. (14 Nm).

23. Install the intake camshaft with the lobes of no. 8 cylinder facing upward.

24. Fit the bearing caps marked E in the correct order.

25. Evenly and progressively tighten the 10 nuts attaching the intake camshaft bearing caps to 10 ft. lbs. (14 Nm).

26. Clean the VCC non-return valve, and using a new o-ring, install the valve.

27. Lubricate the VCC rings and carefully fit the rings in the intake camshaft grooves with the ring gaps facing upward.

28. Clean the VCC oil distribution housing and cylinder head mating surfaces.

29. Using a new gasket, install the VCC housing.

30. Install a new sealing washer to the long bolt, then install VCC housing bolts and tighten to 7 ft. lbs. (10 Nm).

31. Lubricate the exhaust camshaft drive chain tensioner.

32. Using a new o-ring, install the chain tensioner and tighten the bolts to 7 ft. lbs. (10 Nm).

33. Install the cylinder heads.

Starter Motor

REMOVAL & INSTALLATION

Freelander (2001–04)

1. Before servicing the vehicle, refer to the precautions in the beginning of this section.

2. Remove battery carrier.

3. Release starter motor solenoid terminal cover, remove nut securing battery lead to solenoid and disconnect battery lead.

4. Disconnect connector from starter solenoid.

5. Remove 3 bolts securing starter motor to gearbox noting that the left-hand bolt also secures the mounting bracket for the CKP sensor multiplug.

6. Maneuver and remove starter motor.

To install:

7. Clean starter motor and mating face on gearbox.

8. Position starter motor to gearbox, align CKP sensor multiplug bracket, install and tighten bolts to 33 ft. lbs. (45 Nm).

9. Connect battery lead to solenoid terminal, install nut and tighten to 9.5 ft. lbs. (13 Nm). Secure terminal cover.

10. Connect connector to starter solenoid.

11. Install battery carrier.

Discovery (2001–02) and Range Rover (2001–02)

1. Before servicing the vehicle, refer to the precautions in the beginning of this section.

2. Disconnect the battery ground cable.

3. Remove the right side transmission sound shield.

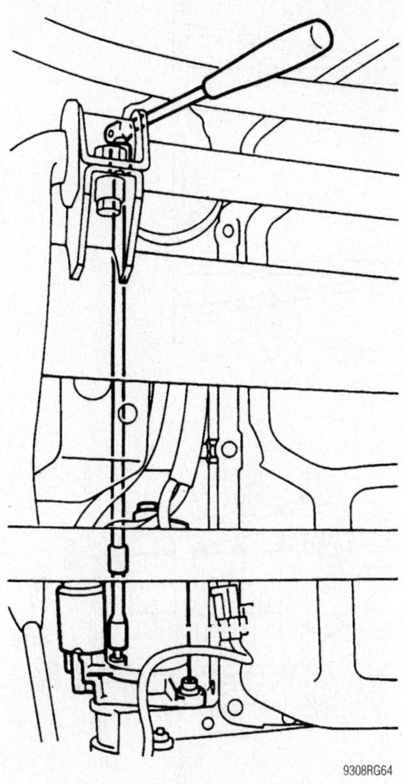

Accessing the starter mounting bolts—Discovery (2001–02) and Range Rover (2001–02)

4. Disconnect the wiring from the solenoid.
5. Remove the 2 starter mounting bolts.
6. Installation is the reverse of removal. Torque the mounting bolts to 33 ft. lbs. (45 Nm).

Discovery (2003–04)

1. Before servicing the vehicle, refer to the precautions in the beginning of this section.
2. Remove the battery cover, then disconnect the ground cable.
3. Raise and support the front of the vehicle on jack stands.
4. Remove the 3 nuts holding the exhaust flange to the front pipe. Disconnect the pipe and collect the gasket.
5. Remove the heat shield from the engine mounting bracket.
6. Remove the battery lead from the starter solenoid.
7. Disconnect the starter solenoid connector.
8. Remove the 2 Allen bolts and remove the starter.

To install:

9. Clean the starter and engine block mating faces.
10. Install the starter and tighten the Allen bolts to 32 ft. lbs. (44 Nm).
11. Reattach the connector to the starter solenoid.
12. Connect the battery lead to the solenoid and tighten the nut.
13. Position the heat shield over the starter motor, locating the tag in the engine mounting bracket. Install and tighten the bolt to 7 ft. lbs. (10 Nm).
14. Clean the exhaust manifold and front pipe mating faces. Install a new gasket and tighten nuts to 22 ft. lbs. (30 Nm).
15. Lower the vehicle.
16. Reconnect the battery ground cable.
17. Install the battery cover and attachments.

Range Rover (2003–04)

1. Before servicing the vehicle, refer to the precautions in the beginning of this section.
2. Disconnect the negative battery cable.
3. Remove the air intake plenum.
4. Raise and support the front of the vehicle on jack stands.
5. On the intake manifold, remove the terminal post cover, post nut and disconnect the 2 cables.
6. Remove the right side exhaust manifold.
7. Remove the heat shield.
8. Remove the 2 bolts and move the starter enough to disconnect the electrical connectors.
9. Disconnect the starter breather pipe.
10. Remove the starter.

To install:

11. Clean the starter and engine block mating faces.
12. Connect the breather pipe.
13. Position the starter enough to connect the electrical leads.
14. Install the starter and tighten the bolts to 35 ft. lbs. (47 Nm).
15. Install the heat shield.
16. Install the right exhaust manifold.
17. Connect the 2 cables to the intake manifold.
18. Install the air intake plenum.
19. Lower the vehicle and connect the negative battery cable.

Oil Pan

REMOVAL & INSTALLATION

Freelander (2001–04)

1. Before servicing the vehicle, refer to the precautions in the beginning of this section.
2. Disconnect battery ground.
3. Remove engine acoustic cover.
4. Drain engine oil.
5. Remove 3 bolts securing right side splash shield to body and remove shield.
6. Remove 4 nuts securing engine oil cooler to oil pan and position oil cooler aside.
7. Remove bolt securing dipstick tube to cylinder block.

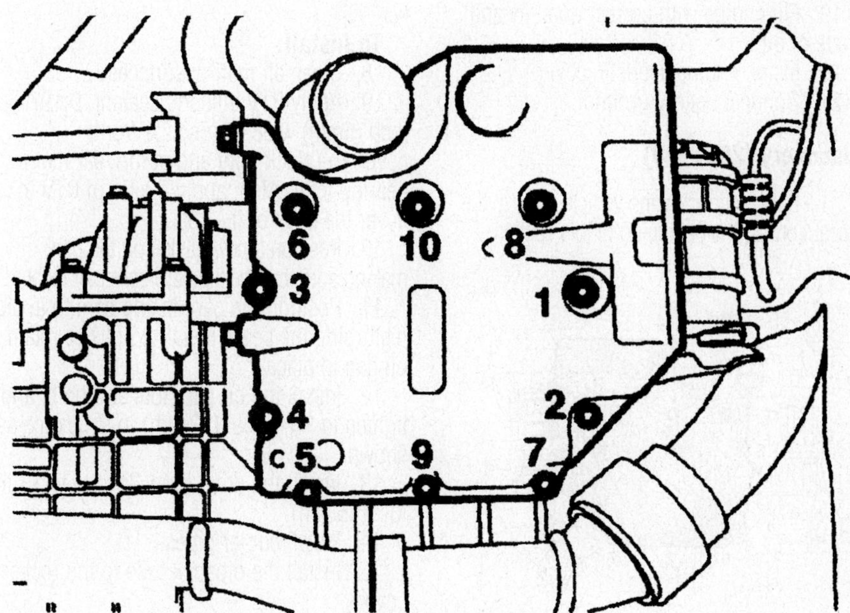

Oil pan mounting bolt loosening sequence—Freelander (2003–04)

8. Depress collar and remove dipstick tube from oil pan.

9. Remove 3 bolts securing IRD support bracket to oil pan.

10. Using sequence shown, and noting each bolt's installed position, remove 10 oil pan bolts.

11. Release and remove oil pan from lower crankcase.

To install:

12. Using and suitable cleaning solvent, clean oil pan and mating face on lower crankcase. DO NOT use a metal scraper on sealing surfaces.

13. Apply a 0.1 inch (2mm) bead of sealant STC 4600 or equivalent, along center of oil pan flange, then spread to an even film using a roller. Do not allow sealant into bolt holes.

CAUTION

To avoid contamination, assembly should be completed immediately after application of sealant.

14. Position oil pan, install bolts and tighten progressively in the sequence shown to 35 Nm (26 ft. lbs.).

15. Fit bolts securing IRD support bracket to oil pan and tighten to 33 ft. lbs. (45 Nm).

16. Position engine oil cooler to oil pan mounting bracket, install and tighten nuts to 18 ft. lbs. (25 Nm).

17. Position dipstick tube to oil pan and cylinder block, install bolt and tighten to 7 ft. lbs. (9 Nm).

18. Install splash shield and secure with bolts.

19. Fill engine with correct quantity and grade of oil.

20. Install engine acoustic cover.

21. Connect battery ground.

Discovery (2001–04)

1. Before servicing the vehicle, refer to the precautions in the beginning of this section.

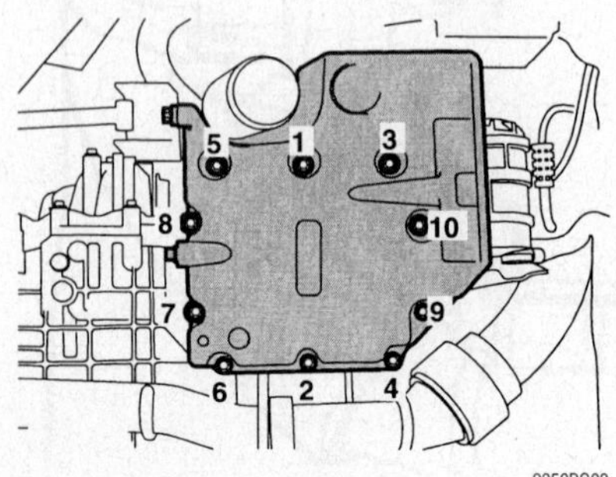

Oil pan mounting bolt torque sequence—Freelander (2001–02)

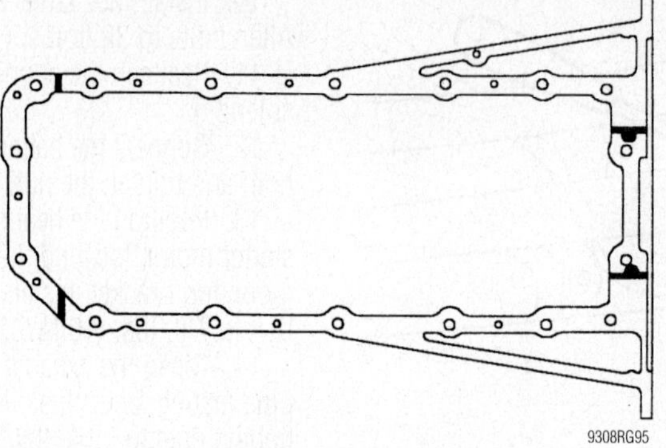

Oil pan sealant application—Discovery (2001–04)

2. Lift the vehicle on a 4-post lift, or, on jackstands under the chassis with a jack under the axle.

3. Remove or disconnect the following:
- Battery ground cable
- Engine oil dipstick
- Engine oil
- Front crossmember (8 bolts)

4. Raise the vehicle front end in order to gain clearance between engine and front axle.

5. Remove or disconnect the following:
- Transmission cooler lines from connection and clips (discard O-ring)
- The 2 forward-facing and 4 rear-facing bolts securing the oil pan to the bell housing
- The 2 bolts in the oil pan recess
- The 3 nuts and 12 bolts securing the pan

6. Remove the oil pan by maneuvering it over the front axle.

7. Remove and discard the oil pan gasket.

To install:

8. Clean all mating surfaces.

9. Apply RTV silicone sealant, 0.197 inch (5mm) wide, across the front cover–to–block joint and at the rear main bearing joint. Also, apply a glob of RTV to cover the ends of the seal.

10. Position a new gasket on the pan, making sure that the tabs are correctly located.

11. Position the pan, taking care to avoid disturbing the bead. Install 2 bolts to retain oil pan in place.

12. Install the oil pan nuts and bolts and tighten to 16 ft. lbs. (22 Nm) in sequence as shown.

13. Install the drain plug. Torque to 33 ft. lbs. (45 Nm).

14. Install cooler pipes.

15. Install the dipstick tube to the rocker cover.

16. Install the crossmember. Torque the bolts to 18 ft. lbs. (25 Nm).

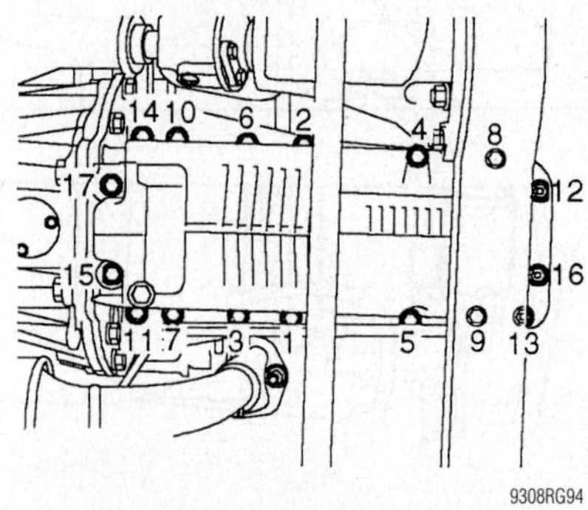

Oil pan fastener torque sequence—Discovery (2001–04)

17. Lower vehicle. Connect negative battery cable.

18. Fill the engine with oil, start the engine and check for leaks.

Range Rover (2001–02)

1. Before servicing the vehicle, refer to the precautions in the beginning of this section.

2. Lift the vehicle on a 4-post lift, or, on jackstands under the chassis with a jack under the axle.

3. Remove or disconnect the following:
- Battery ground cable
- Engine acoustic cover
- Transmission acoustic cover
- Engine oil dipstick
- Engine oil

4. Position a support under the front crossmember.

5. Lower the axle for clearance.

6. Remove or disconnect the following:
- O_2 sensor connectors
- The 3 nuts and 14 bolts securing the pan
- Oil Pan

To install:

7. Clean all mating surfaces.

8. Apply RTV Hylosil 101 or 106, or equivalent sealant, to the oil pan. Using the

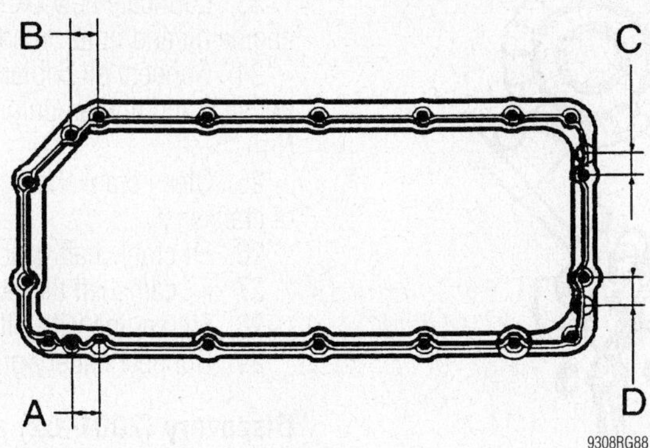

Sealant bead application—Range Rover (2001–02)

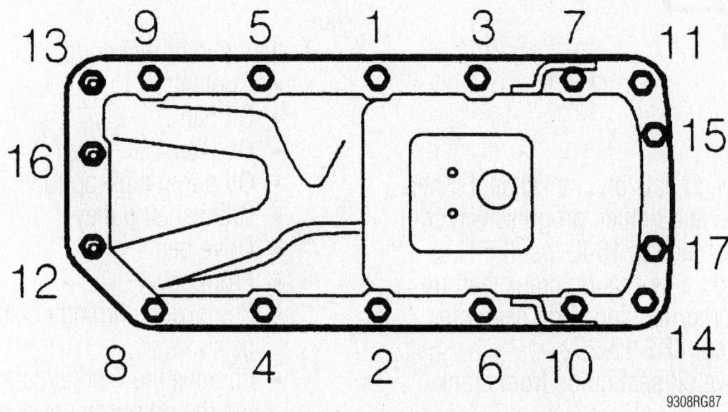

Oil pan fastener torque sequence—Range Rover (2001–02)

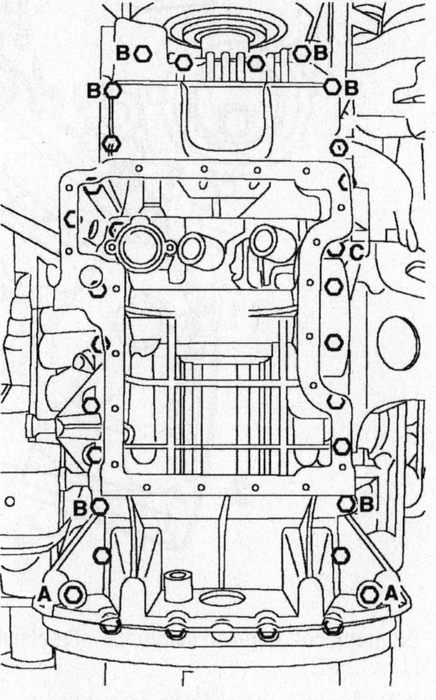

Oil pan bolt locations—Range Rover (2002–04)

illustration, the bead width and length should be:

- Width at A, B, C and D: 12mm
- Width at remaining areas: 5mm
- Length at A and B: 32mm
- Length at remaining areas: 19mm

➡Do not spread the bead. Install the pan immediately!

9. Position the pan, taking care to avoid disturbing the bead.
10. Install the oil pan and tighten the nuts bolts to 17 ft. lbs. (23 Nm) in sequence as shown.
11. Install the drain plug. Torque to 33 ft. lbs. (45 Nm).
12. Connect the sensors.
13. Install the dipstick tube to the rocker cover.
14. Install the covers.
15. Fill the engine with oil, start the engine and check for leaks.

Range Rover (2003–04)

1. Before servicing the vehicle, refer to the precautions in the beginning of this section.
2. Lift the vehicle on a 4-post lift, or, on jackstands under the chassis with a jack under the axle.
3. Remove or disconnect the following:
   - Battery ground cable
   - Engine oil
   - Oil dipstick tube
   - Power steering pump
   - Front differential
   - Oil pickup strainer
   - Banjo bolt from oil return line to oil pan and discard washers
   - Banjo bolt from oil filter return line to oil pan and discard washers
   - 4 bolts attaching oil pan to bell-housing
   - 31 bolts attaching oil pan, noting the bolt lengths and locations

To install:

4. Clean all mating surfaces.
5. Install a new gasket and fit oil pan to the block.
6. Note the bolt positions as shown, and install in the correct locations hand tight.
7. Install oil pan-to-bellhousing bolts and tighten the bolts to 39 ft. lbs. (53 Nm).
8. Using new washers, connect both banjo bolts and tighten to 22 ft. lbs. (30 Nm).
9. Install the differential.
10. Install the oil pickup strainer.
11. Install the power steering pump.
12. Install the dipstick tube.
13. Connect the battery ground cable.
14. Fill the engine with oil.
15. Start the engine and check for leaks.

Oil Pump

REMOVAL & INSTALLATION

Freelander (2001–04)

1. Before servicing the vehicle, refer to the precautions in the beginning of this section.
2. Disconnect battery ground lead.
3. Drain engine oil.
4. Remove camshaft timing belt.
5. Remove crankshaft gear.
6. Disconnect multiplug from oil pressure switch.
7. Using sequence shown, loosen and remove oil cooler pipe unions from oil filter housing, remove and discard 2 O-rings.
8. Using strap wrench, remove and discard oil filter.
9. Remove and discard 16 bolts securing oil pump to cylinder block.
10. Release oil pump from locating dowels and remove oil pump.
11. Remove and discard oil pump gasket.

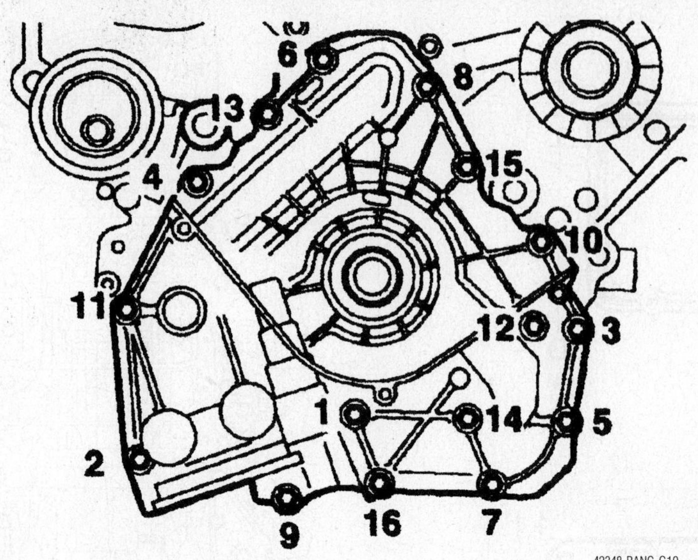

Oil pump bolt removal sequence—Freelander (2001–04)

12. Remove and discard crankshaft front oil seal from oil pump housing.

To install:

13. Using a lint free cloth and a suitable cleaning solvent, clean oil pump and mating face on cylinder block.

14. Using a lint free cloth, thoroughly clean oil seal recess in oil pump and running surface on crankshaft.

15. Fit new oil pump gasket, dry, to cylinder block.

16. Fit oil seal guide, from seal kit, over end of crankshaft.

17. Position oil pump, aligning flats on oil pump drive to flats on crankshaft. Fit new Patchlok bolts and tighten progressively in sequence illustrated to 18 ft. lbs. (25 Nm).

18. Position new seal on crankshaft up against oil pump housing. Drift seal into place using tool LRT-12-202.

19. Remove oil seal guide from crankshaft.

20. Connect multiplug to oil pressure switch.

21. Lubricate oil filter sealing ring with clean engine oil.

22. Fit oil filter and tighten by hand until it seats then tighten a further half turn.

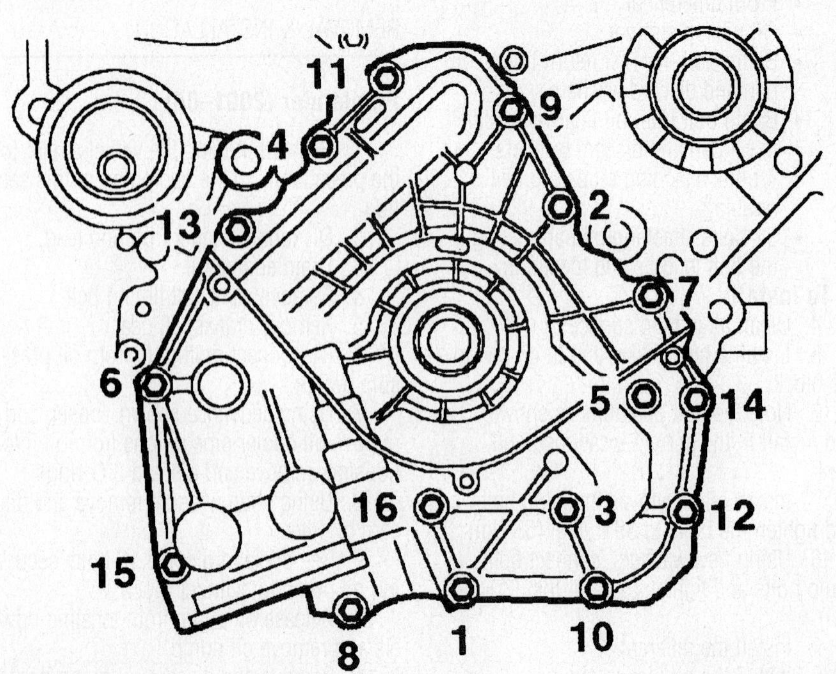

Oil pump bolt torque sequence—Freelander (2001–04)

23. Lubricate new O-rings with clean engine oil and fit to oil cooler pipe unions.

24. Connect oil cooler pipes to oil filter housing and tighten unions to 19 ft. lbs. (26 Nm).

25. Clean crankshaft gear and wipe end of crankshaft.

26. Fit crankshaft gear.

27. Fit camshaft timing belt.

28. Fill engine with oil.

29. Connect battery ground lead.

Discovery (2001–02) and Range Rover (2001–02)

1. Before servicing the vehicle, refer to the precautions in the beginning of this section.

2. Remove or disconnect the following:
   - Coolant
   - Oil pan
   - Oil filter
   - Oil pump pick-up tube
   - Crankshaft pulley
   - Drive belt
   - Front cover
   - Remove the timing chain and gears.
   - Remove the 7 bolts/screws mounting the oil pump cover plate.
   - Matchmark the inner and outer oil pump rotors, then remove the rotors and oil pump drive gear as an assembly.

To install:

3. Lubricate the rotors, oil pump drive gear, cover plate and housing with engine oil.

4. Align the rotor matchmarks, then install the oil pump drive gear and rotors as an assembly.

5. Apply Loctite® 222 to the mounting bolts/screws for the oil pump cover plate, then install the plate and tighten the bolts to 72 inch lbs. (8 Nm) and the screws to 36 inch lbs. (4 Nm).

6. Install the timing chain and gears.

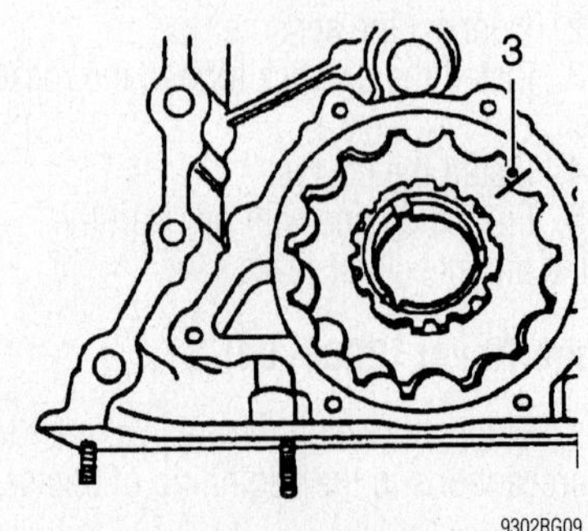

Place matchmarks on the rotors (3) before removing—Discovery (2001–02) and Range Rover (2001–02)

7. Connect the oil pump pick-up tube and install the oil pan.

8. Install a new oil filter and fill the engine with oil.

9. Start the vehicle and check for leaks.

Discovery (2003–04)

1. Before servicing the vehicle, refer to the precautions in the beginning of this section.

2. Remove the timing gear cover (oil pump is integral with timing gear cover).

3. Remove the camshaft position (CMP) sensor.

4. Remove 6 bolts securing the water pump. Remove the water pump.

5. Remove the oil pressure switch. Discard the sealing washer.

6. Remove the 4 bolts hold the oil filter head. Remove the oil filter head and discard the 2 O-rings.

To install:

7. Ensure the oil filter head adapter is clean.

8. Install the filter head and tighten the bolts to 9 ft. lbs. (13 Nm).

9. Clean the oil filter head and mating face.

10. Install new O-rings to oil filter head and adapter. Install and tighten filter head bolts to 6 ft. lbs. (8 Nm).

11. Clean oil pressure switch mating faces. Install new sealing washer and tighten switch to 11 ft. lbs. (15 Nm).

12. Clean the water pump and cover mating faces. Install water pump with a new gasket. Tighten water pump bolts to 19 ft. lbs. (25 Nm).

13. Clean CMP sensor and tighten the retaining bolt to 6 ft. lbs. (8 Nm).

14. Install the camshaft timing gear cover with a new gasket. Tighten bolts to 17 ft. lbs. (22 Nm), following illustrated sequence.

Range Rover (2003–04)

1. Before servicing the vehicle, refer to the precautions in the beginning of this section.

2. Drain the engine oil

3. Remove the oil pan.

4. Remove the oil pump sprocket.

5. Remove 2 nuts and 1 bolt attaching oil pump and remove pump. Discard o-rings.

To install:

6. Clean the oil pump mating surface, install new o-rings and lubricate with clean engine oil.

7. Install the oil pump and tighten the fasteners.

8. Install the oil pump sprocket and tighten the nut.

➡The oil pump chain is adjusted by turning the Allen bolt spacer. The spacer is accessed through the oil pump securing bolt hole.

9. Adjust the oil pump chain by turning the Allen bolt spacer until the chain deflection is 0.314–0.470 inch (8–12 mm).

10. Install the oil pan.

11. Fill the engine with oil.

12. Start the engine and check for leaks.

Rear Main Seal

REMOVAL & INSTALLATION

Freelander (2001–04)

1. Before servicing the vehicle, refer to the precautions in the beginning of this section.

2. Remove automatic gearbox.

3. Assemble LRT-12-161 to LRT-12-199 and secure with clamp bolt.

4. Position LRT-12-161 and LRT-12-199 to crankshaft pulley to hold crankshaft from turning.

5. With assistance, remove and discard 6 bolts securing drive plate to crankshaft.

6. Remove drive plate (flywheel) from crankshaft.

7. Remove and discard 5 bolts securing crankshaft rear oil seal housing to cylinder block.

8. Remove crankshaft rear oil seal.

To install:

9. Clean cylinder block face and rear main oil seal running surface on crankshaft.

10. Position oil seal protector, LRT-12-061, to crankshaft.

11. Install the rear main oil seal onto the crankshaft. Seal must be fitted dry.

12. Install new Patchlok® bolts to rear main oil seal. Tighten bolts to 6 ft. lbs. (8 Nm).

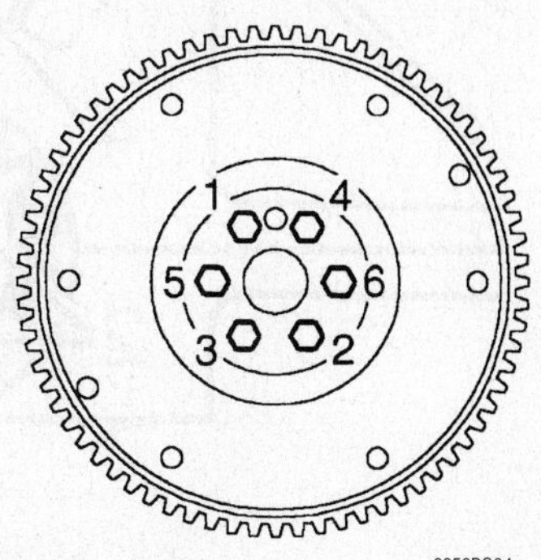

Drive plate bolt torque sequence—Freelander (2001–04)

13. Remove oil seal protector.

14. Clean bolt holes in crankshaft using an old drive plate bolt with two saw cuts at an angle of 45° to the bolt shank.

15. Clean drive plate and mating face of crankshaft.

16. Position drive plate to crankshaft and fit new Patchlok bolts, but do not tighten at this time.

17. Position LRT-12-161 and LRT-12-199 to crankshaft pulley to restrain crankshaft.

18. With assistance, tighten bolts in the following sequence:

- Step 1 — 18 ft. lbs. (25 Nm)
- Step 2 — 74 ft. lbs. (100 Nm)

19. Remove special tool from crankshaft pulley, loosen bolt and remove LRT-12-161 from LRT-12-199.

20. Fit automatic gearbox.

Discovery (2001–04) and Range Rover (2001–02)

1. Before servicing the vehicle, refer to the precautions in the beginning of this section.

2. Remove the transmission and drive-plate.

3. Using a suitable seal removal tool, remove the oil seal from the engine block.

To install:

4. Be sure both the seal location and running surface on the crankshaft are clean.

5. Fit a seal guide, LRT-12-095, onto crankshaft.

6. Install the crankshaft rear main oil seal over the end of the crankshaft. Use installer tools LRT-12-091 and LRT-99-003 to install seal squarely onto crankshaft. Install seal dry.

7. Remove installer tools and seal guide.

8. Install the driveplate and tighten the mounting bolts as follows:

- For 2001–02 models: 63 ft. lbs. (85 Nm)
- For 2003–04 models: 58 ft. lbs. (78 Nm)

9. Install the transmission.

Range Rover (2003–04)

1. Before servicing the vehicle, refer to the precautions in the beginning of this section.

2. Remove the transmission and drive-plate.

3. Remove 12 bolts and rear oil seal housing

4. Remove the oil seal from the hosing.

To install:

5. Be sure the seal housing, locating dowels and mating surface on the block are clean.

6. Fit seal guides, LRT-12-179 and 217 onto a new seal and install into the housing.
7. Check that the cylinder head dowels are in correct position.
8. Install new gasket to oil seal housing.
9. Apply sealant to oil pan upper contact surface.
10. Install crankshaft guide tool LRT-12-218 to crankshaft.
11. Apply a thin coat of oil to seal inner lip and install seal and housing over the tool and onto the locating dowels.
12. Using new washers on the Torx bolts, install 6 Torx bolts and 6 standard bolts to housing and tighten to 9 ft. lbs. (12 Nm).
13. Install the drive plate and transmission.
14. Fill the engine with oil.
15. Connect battery ground cable.
16. Start the engine and check for leaks.

Timing Chain, Sprockets, Front Cover and Seal

REMOVAL & INSTALLATION

Discovery (2001–02) and Range Rover (2001–02)

FRONT COVER WITH SEAL

➡For seal replacement, see the next procedure.

1. Before servicing the vehicle, refer to the precautions in the beginning of this section.
2. Remove or disconnect the following:
   - Negative battery cable
   - Coolant
   - Oil pan
   - Oil pick up strainer
   - Crankshaft pulley and drive belt tensioner
   - Hose from the water pump
   - Oil cooler hoses from the front cover, then plug the hoses and connections to prevent dirt from entering.
   - Oil pressure switch and CMP sensor electrical connectors
   - Timing cover mounting bolts, then remove the cover.
3. Clean the gasket mating surfaces and drive out the crankshaft seal.
4. Clean the timing gears and turn the crankshaft until the timing mark on the crankshaft and camshaft face each other (camshaft at 6 o'clock and crankshaft at twelve o'clock).
5. Remove the camshaft timing gear mounting bolt.
6. Remove the timing gears and chain as an assembly.

To install:

7. Assemble the timing chain and gears on a work bench, with the timing marks aligned.
8. Install the timing chain and gear assembly onto the engine with the timing marks facing outwards.
9. Install the camshaft timing gear mounting bolt and tighten to 37 ft. lbs. (50 Nm).
10. Lubricate the new timing cover oil seal with Shell Retinax LX, or equivalent grease, ensuring that the space between seal lips is filled with grease.
11. Install or connect the following:
    - Seal with a suitable seal driver
    - Front cover with a new gasket and tighten the cover bolts, in the sequence shown, to 16 ft. lbs. (22 Nm).
    - Oil pressure switch and CMP sensor electrical connectors
    - Plugs from the oil cooler hoses
    - Oil cooler hoses using new O-ring seals and tighten to 11 ft. lbs. (15 Nm)
    - Drive belt tensioner and tighten the bolt to 37 ft. lbs. (50 Nm)
    - Hose to the water pump
    - Oil filter
    - Crankshaft pulley and tighten the bolt 200 ft. lbs. (270 Nm).
    - Oil pump with a new O-ring and tighten the mounting bolt to 72 inch lbs. (8 Nm).
    - Oil pan
    - Engine oil
    - Coolant
    - Negative battery cable

SEAL REPLACEMENT ONLY

1. Before servicing the vehicle, refer to the precautions in the beginning of this section.
2. Remove or disconnect the following:
   - Battery ground cable
   - Cooling fan
   - Water pump pulley bolts
   - Drive belt
   - Engine under-cover
3. Install a holding tool, such as LRT-12-080, on the pulley
4. Remove the pulley bolt and pulley.
5. Using a seal remover, such as LRT-12-088, remove the seal.

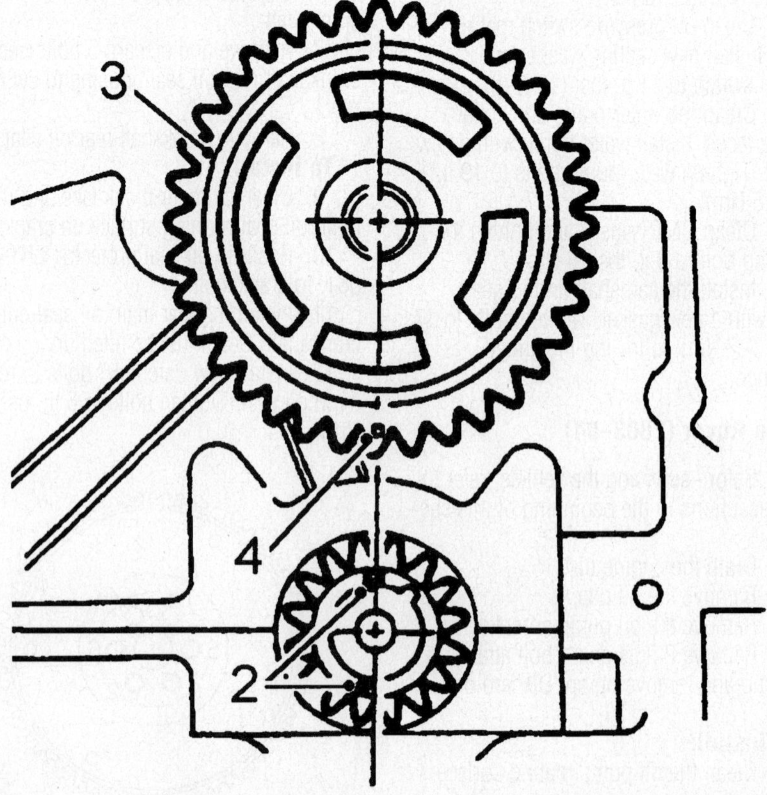

Be sure to align the cam gear (3) timing mark (4) with the crankshaft gear (2), as shown—Discovery (2001–02) and Range Rover (2001–02)

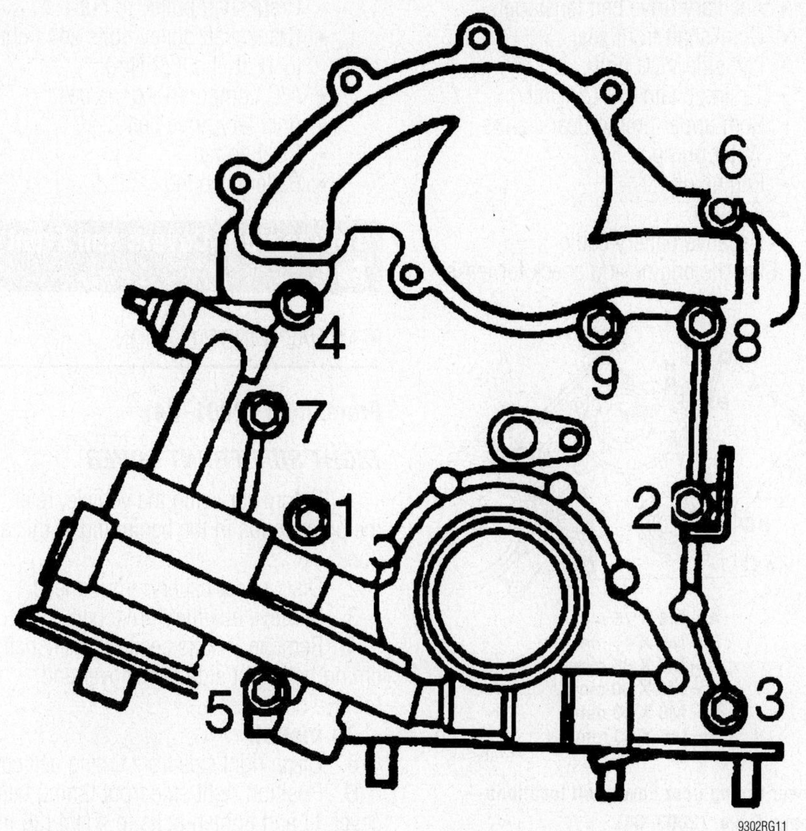

Front timing chain cover bolt torque sequence — Discovery (2001–04) and Range Rover (2001–02)

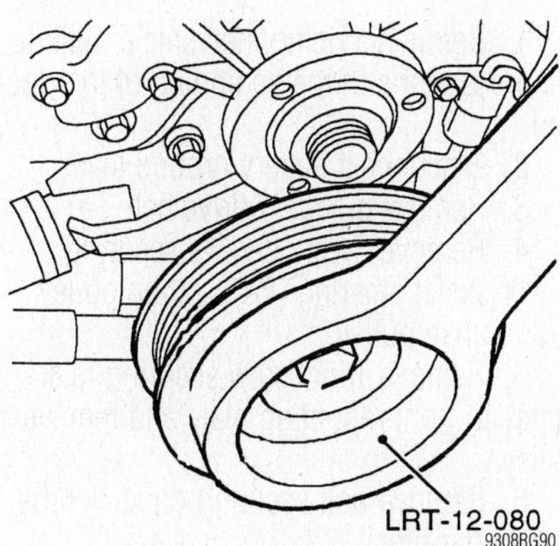

Crankshaft pulley holding tool

To install:

6. Clean all sealing surfaces.
7. Coat the outer edge of the seal with engine oil.
8. Using an installer, such as LRT-12-089, install the seal.
9. Coat the seal lip with engine oil.
10. Install the holding tool, position the pulley on the crankshaft and install the bolt. Torque the bolt to 200 ft. lbs. (270 Nm).
11. The remainder of installation is the reverse of removal.

Discovery (2003–04)

1. Before servicing the vehicle, refer to the precautions in the beginning of this section.
2. Remove auxiliary drive belt.
3. Remove oil pan.
4. Remove 2 bolts and 1 nut and remove oil pickup strainer.
5. Remove and discard 3 water pump pulley bolts. Remove the pulley.
6. Use a tool, LRT-12-080, to hold crankshaft pulley, then remove the pulley bolt and pulley.
7. Remove and discard the oil filter.
8. Remove the 9 bolts from the timing gear cover, noting position of the longer bolts. Remove the cover.
9. Remove and discard the cover gasket and seal.
10. Temporarily install crankshaft pulley bolt and rotate engine to align timing marks on both sprockets. Remove the crankshaft pulley bolt.
11. Hold camshaft gear and remove gear retaining bolt.
12. Remove the timing chain and both gears as an assembly.
13. Remove the gears from the timing chain.
14. If necessary, remove the key from the crankshaft keyway.

To install:

15. Clean the timing chain, gears, camshaft and crankshaft mating surfaces.
16. Temporarily fit gears to camshaft and crankshaft, rotating shafts as needed to line up timing marks.
17. When aligned (camshaft mark at 6 o'clock and crankshaft mark at 12 o'clock), remove gears from shafts and fit timing chain, retaining timing mark alignment.
18. With timing marks aligned, install timing chain and gears as an assembly.
19. Hold camshaft gear from turning and tighten camshaft retaining bolt to 37 ft. lbs. (50 Nm).
20. Clean timing gear cover and mating face.
21. Clean oil seal register in timing gear cover.
22. Install a new timing gear cover gasket (dry), align oil pump drive gear to key in crankshaft, and fit timing gear cover.
23. Align 2 camshaft sensor harness clips to timing gear cover bolt holes. Noting positions of the longer bolts, install timing gear cover bolts. Working in sequence shown, tighten bolts to 16 ft. lbs. (22 Nm).
24. Carefully apply grease, Retinax LX, to inner recess of a new crankshaft oil seal until recess is half-filled with grease. Do not allow grease to coat any other part of the seal. Do not apply oil to the seal lips.
25. Apply a smear of grease, Retinax LX, to oil seal running surface on crankshaft.
26. Install the seal, using seal installer LRT-12-089.
27. Clean and install crankshaft pulley. Tighten bolt to 200 ft. lbs. (270 Nm).
28. Clean oil filter mating face and install a new oil seal after coating filter seal with clean engine oil.
29. Clean water pump and pulley mating faces. Remove all traces of thread locking material from water pump pulley drive flange bolt holes.
30. Install water pump pulley and tighten new Patchlok bolts to 16 ft. lbs. (22 Nm).
31. Install oil pickup strainer as follows:
    a. Clean oil pickup strainer and O-ring recess.
    b. Lubricate and install a new O-ring.
    c. Locate washer on stud and position oil pickup strainer.
    d. Install and tighten bolts to 8 ft. lbs. (10 Nm).
    e. Install and tighten nut to 16 ft. lbs. (22 Nm).
32. Install oil pan with new gasket.
33. Install auxiliary drive belt.

Range Rover (2003–04)

➡For seal replacement, see the next procedure.

1. Before servicing the vehicle, refer to the precautions in the beginning of this section.

➡To remove the timing gear upper covers, upper timing chains, sprockets and Variable Camshaft Control (VCC) unit, see the cylinder head procedure in this section.

2. Remove or disconnect the following:
   - Negative battery cable
   - Coolant
   - Engine oil
   - Water pump
   - Both upper timing gear covers
   - Left side VCC unit
   - Crankshaft front seal
   - Ancillary drive belt tensioner
   - Alternator battery cable
   - Alternator
   - Power steering pump and move aside
   - 6 bolts attaching lower timing cover to oil pan
   - 15 lower timing cover bolts and timing cover
   - Timing chain

To install:

3. Clean the timing chain guides and sprockets.
4. Lubricate the timing chain with clean engine oil.
5. Install the timing chain to the crankshaft sprocket and right side VCC unit.

➡The timing chain may be installed in any position to the crankshaft and VCC sprocket.

6. Secure the timing chain to the left cylinder head until the left side VCC is installed.
7. Clean the lower timing gear cover and mating surfaces.
8. Apply sealant to the oil pan upper contact surface.
9. Using new gaskets, install the timing gear cover and install the bolts in the correct locations as shown.
10. Tighten all the bolts to 44 inch lbs. (5 Nm).
11. Tighten the M6 bolts to 7 ft. lbs. (10 Nm), and the M8 bolts to 16 ft. lbs. (2 Nm).
12. Tighten the 6 oil pan-to-lower cover bolts to 9 ft. lbs. (12 Nm).
13. Recheck the torque of all the bolts.
14. Install or connect the following:
    - Power steering pump
    - Alternator
    - Alternator battery cable
    - Ancillary drive belt tensioner
    - Crankshaft front seal
    - Left side VCC unit
    - Timing chain to VCC unit
    - Both upper timing gear covers
    - Water pump
    - Engine oil
    - Coolant
    - Negative battery cable
15. Start the engine and check for leaks.

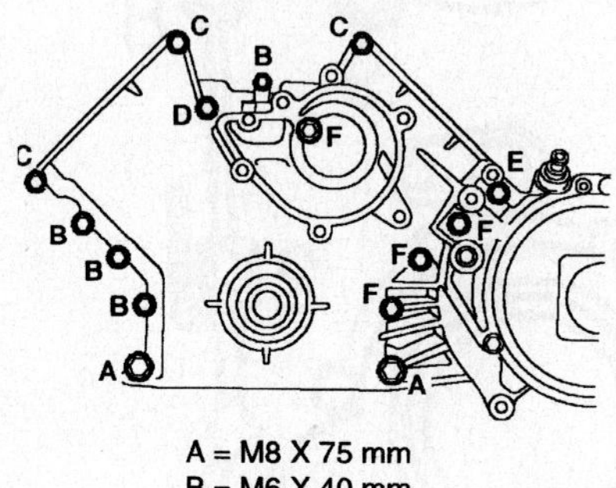

Lower timing gear cover bolt locations—Range Rover (2003–04)

SEAL REPLACEMENT ONLY

1. Before servicing the vehicle, refer to the precautions in the beginning of this section.
2. Remove or disconnect the following:
   - Battery ground cable
   - Cooling fan
   - Ancillary drive belt
   - A/C compressor drive belt
   - Crankshaft pulley
3. Install a holding tool LRT-12-229 on the pulley hub bolt.
4. Remove the hub bolt and hub.
5. Use a suitable puller and remove the crankshaft seal.

To install:

6. Clean the seal mating surfaces.
7. Using a seal installer, fit a new seal into the front cover until it is flush with the cover.
8. Clean crankshaft pulley hub and mating surfaces.
9. Install the hub, lock the crankshaft and using a new hub bolt, initially tighten the bolt to 74 ft. lbs. (100 Nm).
10. Tighten the hub bolt in the following sequence:
    - Step 1 an additional 60 degrees
    - Step 2 an additional 60 degrees
    - Step 3 an additional 30 degrees
11. Install or connect the following:
    - Crankshaft pulley to hub
    - Crankshaft pulley bolts and tighten to 16 ft. lbs. (22 Nm).
    - A/C compressor drive belt
    - Ancillary drive belt
    - Cooling fan
    - Battery ground cable

Timing Belt, Covers and Front Seal

REMOVAL & INSTALLATION

Freelander (2001–04)

RIGHT SIDE FRONT COVER

1. Before servicing the vehicle, refer to the precautions in the beginning of this section.
2. Disconnect battery ground lead.
3. Remove auxiliary drive belt.
4. Remove 3 bolts securing camshaft timing belt right side front cover and remove cover.

To install:

5. Clean right side front timing belt cover.
6. Position right side front timing belt cover, fit and tighten bolts to 4 Nm (36 inch lbs.).
7. Fit auxiliary drive belt.
8. Connect battery ground lead.

LEFT SIDE FRONT COVER

1. Before servicing the vehicle, refer to the precautions in the beginning of this section.
2. Disconnect battery ground lead.
3. Remove auxiliary drive belt.
4. Remove 3 Torx screws securing power-assist steering (PAS) pump pulley, and remove pulley.
5. Remove Torx screw securing idler pulley to front mounting plate and remove pulley.
6. Remove bolt securing dipstick tube to cylinder block.
7. Remove 3 bolts securing camshaft timing belt left side front cover and remove cover.

To install:

8. Clean left side front timing belt cover.
9. Position left side front timing belt cover, fit and tighten bolts to 3 ft. lbs. (4 Nm).
10. Position dipstick tube bracket to cylinder block and tighten bolt to 7 ft. lbs. (9 Nm).
11. Clean idler pulley, position pulley to engine front mounting plate and tighten Torx bolt to 33 ft. lbs. (45 Nm).
12. Position PAS pump pulley, fit and tighten Torx screws to 7 ft. lbs. (9 Nm).

13. Fit auxiliary drive belt.
14. Connect battery ground lead.

RIGHT SIDE REAR COVER

1. Before servicing the vehicle, refer to the precautions in the beginning of this section.
2. Disconnect battery ground lead.
3. Remove intake manifold chamber.
4. Remove 2 bolts securing right side rear timing belt cover heat shield to cylinder head and remove heat shield.
5. Remove 3 bolts securing right side rear cover and remove cover.

To install:

6. Clean right side rear timing belt cover.
7. Position right side rear cover, fit bolts and tighten to 3 ft. lbs. (4 Nm).
8. Position heat shield, fit bolts and tighten M6 bolt to 7 ft. lbs. (9 Nm) and M8 bolt to 18 ft. lbs. (25 Nm).
9. Fit intake manifold chamber.
10. Connect battery ground lead.

LEFT SIDE REAR COVER

1. Before servicing the vehicle, refer to the precautions in the beginning of this section.
2. Disconnect battery ground lead.
3. Remove engine acoustic cover.
4. Remove 3 bolts securing camshaft timing belt left side rear cover and remove cover.

To install:

5. Clean left side rear timing belt cover.
6. Position left side rear cover, fit bolts and tighten to 3 ft. lbs. (4 Nm).
7. Fit the engine acoustic cover.
8. Connect battery ground lead.

FRONT TIMING BELT

1. Before servicing the vehicle, refer to the precautions in the beginning of this section.
2. Disconnect battery ground lead.
3. Remove auxiliary drive belt.
4. Remove right side front road wheel.
5. Remove 3 bolts securing right side splash shield to body and remove shield.
6. Remove 3 bolts securing camshaft timing belt left side rear cover and remove cover.
7. Remove intake manifold chamber.
8. Remove 2 bolts from right side rear timing belt cover heat shield and remove shield.
9. Remove 3 bolts from right side rear timing belt cover and remove the cover.
10. Using a socket, rotate crankshaft pulley bolt in a clockwise direction and align the engine 'SAFE' position, notch on crankshaft pulley aligned to the 'ARROW' on front mounting plate and the timing marks aligned on the rear camshaft gears as illustrated.
11. Insert a timing pin, LRT-12-232, through hole in lower crankcase (hole immediately adjacent to side of lower crankcase). Ensure pin is located in hole in drive plate.
12. Remove bolt securing PAS pipe to engine front mounting plate.
13. Remove 3 Torx screws securing PAS pump pulley and remove pulley.
14. Remove 3 bolts securing PAS pump to front mounting plate and tie pump aside.
15. Release alternator battery lead terminal cover, loosen terminal nut and disconnect lead from terminal.
16. Disconnect multiplug from alternator.
17. Remove lower bolt and upper nut and bolt securing alternator to front mounting plate.
18. Remove alternator.
19. Remove Torx screw securing idler pulley to front mounting plate and remove pulley.
20. Loosen bolt securing lower engine steady to front sub-frame.
21. Remove bolt securing lower engine steady to sub-frame.
22. Remove bolt securing engine steady to oil pan mounting, release lower engine steady from oil pan mounting.
23. Remove 3 bolts securing camshaft timing belt right side front cover and remove cover.
24. Remove 3 bolts securing camshaft timing belt left side front cover and remove cover.
25. Assemble LRT-12-161 to LRT-12-199 and secure with clamp bolt.
26. Insert LRT-12-161 with LRT-12-199 into crankshaft pulley, loosen and remove pulley bolt. Ensure crankshaft does not rotate during this operation.
27. Remove special tool from crankshaft pulley and remove crankshaft pulley.
28. Remove 3 bolts securing camshaft timing belt lower cover to cylinder block and remove cover.
29. Remove 2 bolts securing auxiliary drive belt tensioner and remove tensioner.
30. Note installed position of heat shield on A/C compressor and remove 3 bolts securing A/C compressor to front mounting plate and cylinder block. Release A/C compressor and heat shield and position aside.
31. Drain engine oil.
32. Remove bolt securing dipstick tube to cylinder block.
33. Depress locking collar, release and remove dipstick tube from engine oil pan.
34. Remove rubber blanking plug from around camshaft timing belt tensioner.
35. Remove 3 bolts securing engine front mounting plate and lifting bracket to cylinder block.
36. Remove bolt securing lifting bracket to right side cylinder head and remove lifting bracket.
37. Remove 2 bolts securing IRD support bracket to engine front mounting plate.
38. Remove 5 bolts and 2 pillar bolts securing engine front mounting plate to cylinder block.
39. With care, release and remove front mounting plate from engine.
40. Insert a suitable 1.5mm diameter pin through the hole in the tensioner body and into the hole in the plunger. If holes in body and plunger are not aligned, move tensioner backplate, using a suitable bread-bladed screwdriver; this will move plunger enough to enable pin to be inserted.
41. Remove and discard 2 bolts securing tensioner to cylinder block, and remove tensioner. DO NOT apply excessive force during removal; if bolts are stuck, apply anti-seize lubricant.

CAUTION

DO NOT loosen Allen screw securing tensioner pulley.

42. If camshaft timing belt is to be refitted, mark direction of rotation on timing belt before removal.
43. With care, ease camshaft timing belt from gears using fingers only and remove timing belt.

CAUTION

Camshaft timing belt must be replaced if cylinder head is to be removed or new drive gears, tensioner or coolant pump are to be installed. Camshaft timing belts must be stored and handled with care. Always store a camshaft timing belt on its edge with a bend radius greater than 2.0 inches (50mm). Do not use a camshaft timing belt that has been twisted or bent double as this will damage the reinforcing fibers. Do not use a camshaft timing belt if debris other than belt dust is found in timing belt covers. Do not use a camshaft timing belt if partial engine seizure has occurred. Do not use a camshaft timing belt if mileage exceeds 48,000 miles (77, 000 km). Do not use an oil or coolant contaminated timing belt, cause of contamination must be rectified.

44. Remove left side and right side exhaust camshaft cap seals.

45. Position tools LRT-12-196 to left side and right side front inlet camshaft gears and into the end of each exhaust camshaft.

⁂ CAUTION

Special tools must be installed when tightening or loosening gear retaining bolts; otherwise, damage to camshafts may occur.

46. Remove and discard bolts retaining front inlet camshaft gears to camshafts.

47. Remove tools LRT-12-196 from both inlet camshaft gears and exhaust camshafts.

48. Remove camshaft drive gears and hub assemblies.

To install:

➡After the front or rear timing belt installation is complete, it is possible that, after rotating the engine and positioning the crankshaft pulley to the 'SAFE' position, the timing marks on the rear timing gears may be misaligned. This misalignment is acceptable, provided that the timing belt installation procedure is carried out correctly.

49. Clean camshaft gears and hubs, crankshaft gear, tensioner and water pump pulleys.

⁂ CAUTION

If the sintered metal gears have been subjected to prolonged oil contamination, they must be soaked in a solvent bath and then thoroughly washer in clean solvent before refitting. Because of the porous construction of the sintered material, oil impregnated in the gears will emerge and contaminate the belt.

50. Install hubs to camshaft gears and install gears to camshafts. Install new bolts and tighten sufficiently to allow gears to rotate without tipping.

51. Position timing belt to gears.

➡To prevent timing belt from disengaging from crankshaft gear when installing, place a suitable wedge between the belt and oil pump belt guard.

52. Install tools LRT-12-196 to both front inlet camshaft gears and into the end of each exhaust camshaft.

53. With assistance, rotate each exhaust camshaft slightly and align timing marks on left and right rear camshaft gears.

54. Remove tool LRT-12-197 from exhaust camshaft.

55. Install tools LRT-12-175 to left and right rear camshaft gears.

56. Rotate both front intake camshaft gears fully clockwise, as viewed from front of engine.

57. Using fingers only, install timing belt to gears. Start at the crankshaft gear and work in an counterclockwise direction, keeping the belt run as taut as possible and turning the camshaft gears only a minimum amount counterclockwise to install timing belt.

⁂ CAUTION

Gears must not be rotated counterclockwise more than 1 tooth.

58. Position an Allen key in the tensioner backplate and hold tensioner pulley against timing belt.

59. With assistance, position tensioner, install new bolts and tighten to 18 ft. lbs. (25 Nm).

60. Remove tools LRT-12-175 from rear timing belt gears.

61. Install tools LRT-12-196 to both front intake camshaft gears, and into the end of each exhaust camshaft.

⁂ CAUTION

Special tools must be installed when tighten or loosening gear retaining bolts; otherwise, damage to camshafts may occur.

62. Tighten front intake camshaft gear bolts to the following:

- Step 1: 20 ft. lbs. (27 Nm)
- Step 2: Tighten bolts 90° further

63. Remove tools LRT-12-196 from both intake camshaft gears and exhaust camshafts.

64. Remove wedge from between drive belt and oil pump belt guard.

65. Remove 1.5mm diameter pin from tensioner.

66. Remove timing pin LRT-12-232.

67. Clean exhaust camshaft front cap seal locations and install new cap seals.

⁂ CAUTION

The sealing edge of the cap seal and mating face must be clean and dry.

68. Position engine front mounting plate and maneuver plate into position. Install bolts, but do not tighten at this time.

69. Install bolts securing IRD support bracket to front mounting plate. Do not tighten bolts at this time.

70. Install and tighten bolts in the sequence shown, to the following specifications:

- Bolt No. 1: 18 ft. lbs. (25 Nm)
- Bolt No 5: 33 ft. lbs. (45 Nm)
- Bolts Nos. 2, 3, 4, 6 and 7: 63 ft. lbs. (85 Nm)

71. Position engine lifting bracket, install bolts securing lifting bracket and front mounting plate to cylinder block and right side cylinder head. Tighten bolts as follows:

- M10 bolts: 33 ft. lbs. (45 Nm)
- M8 bolt: 18 ft. lbs. (25 Nm)

72. Clean end of dipstick tube.

73. Position dipstick tube to oil pan and cylinder block, install bolt and tighten to 7 ft. lbs. (9 Nm).

74. Position A/C compressor to front mounting plate and cylinder block, align heat shield, install and tighten bolts to 18 ft. lbs. (25 Nm).

75. Clean lower timing belt cover.

76. Position lower timing belt cover, install and tighten bolts to 7 ft. lbs. (9 Nm).

77. Clean crankshaft pulley.

78. Install crankshaft pulley to crankshaft gear and ensure that the indent on pulley locates over the lug on crankshaft gear.

79. Install crankshaft pulley bolt and washer, position LRT-12-161 with LRT-12-199 into crankshaft pulley. Tighten pulley bolt to 118 ft. lbs. (160 Nm).

80. Remove LRT-12-161 and LRT-12-199 from crankshaft pulley.

81. Clean auxiliary drive belt tensioner.

82. Position auxiliary belt tensioner, install bolts and tighten to 18 ft. lbs. (25 Nm).

83. Clean left side and right side front timing belt covers.

84. Position left side and right side front timing belt covers, install and tighten bolts to 3 ft. lbs. (4 Nm).

85. Clean idler pulley, position pulley to engine front mounting plate and tighten Allen bolt to 33 ft. lbs. (45 Nm).

86. Position alternator to front mounting plate.

87. Install bolt and nut and bolt securing alternator to front mounting plate and tighten to 33 ft. lbs. (45 Nm).

88. Connect alternator multiplug.

89. Connect battery lead to alternator, install and tighten nut to 6 ft. lbs. (8 Nm), install terminal cover.

90. Position PAS pump to front mounting plate, install and tighten bolts to 18 ft. lbs. (25 Nm).

91. Position PAS pipe to engine front

mounting plate, install and tighten bolt to 18 ft. lbs. (25 Nm).

92. Clean PAS pump pulley mating faces.

93. Position PAS pump pulley, install and tighten Torx screws to 7 ft. lbs. (9 Nm).

94. Position lower engine steady to oil pan mounting, install and tighten bolt to 74 ft. lbs. (100 Nm).

95. Tighten bolt securing lower engine steady to sub-frame to 74 ft. lbs. (100 Nm).

96. Clean left side rear timing belt cover.

97. Position left side rear timing belt cover, install and tighten bolts to 3 ft. lbs. (4 Nm).

98. Install heat shield. Install and tighten bolts as follows:

- M6 bolt: 7 ft. lbs. (9 Nm)
- M8 bolt: 18 ft. lbs. (25 Nm)

99. Install intake manifold chamber.

100. Install right side road wheel and tighten nuts to 85 ft. lbs. (115 Nm).

101. Install auxiliary drive belt.

102. Fill engine with oil.

RIGHT SIDE REAR TIMING BELT

1. Before servicing the vehicle, refer to the precautions in the beginning of this section.

2. If necessary for access, remove intake manifold chamber.

3. Remove underbelly panel.

4. Remove right side front road wheel.

5. Remove 3 bolts securing right side splash shield to body and remove shield.

6. If equipped, remove 2 bolts securing right side rear timing belt cover heat shield. Remove heat shield.

7. Remove 3 bolts securing right side rear timing belt cover. Remove cover.

8. Rotate crankshaft in a clockwise direction and align the engine 'SAFE' position, notch on crankshaft pulley aligned to the 'ARROW' on front mounting plate and the timing marks aligned on the rear camshaft gears as illustrated.

9. Remove and discard right side front exhaust camshaft cap seal from cylinder head.

10. Position LRT-12-175 to rear camshaft gears as illustrated, remove and discard bolts securing gears to camshafts.

11. Remove rear camshaft gears, timing belt and special tool as an assembly.

⁂ CAUTION

Do not turn the crankshaft or the camshafts while the timing belt is removed.

12. If camshaft timing belt is to be refitted, mark direction of rotation on timing belt.

13. Remove LRT-12-175 from camshaft gears and remove timing belt from gears.

⁂ CAUTION

Camshaft timing belts must be stored and handled with care. Always store belt on its edge and with a bend radius of more than 2 inches (20mm). Do not reuse a belt that has been twisted or bent double. Do not use a belt if debris other than belt dust is found in timing belt covers, or if belt has been contaminated with coolant or oil. Do not reuse belt if partial engine seizure has occurred. Do not reused a belt if mileage is more than 45,000 miles (72,000 km).

Aligning rear camshaft timing marks and crankshaft pulley marks (left side camshaft shown) — Freelander (2001–04)

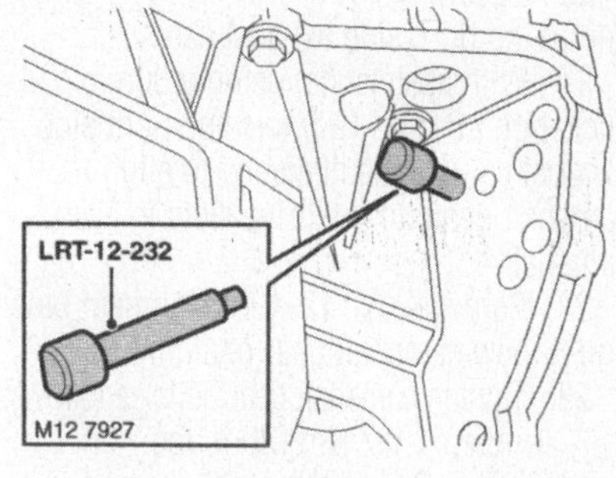

Inserting timing pin LRT-12-232 in lower crankcase—Freelander (2001–04)

To install:

➡After the front or rear timing belt installation is complete, it is possible that, after rotating the engine and positioning the crankshaft pulley to the 'SAFE' position, the timing marks on the rear timing gears may be misaligned. This misalignment is acceptable, provided that the timing belt installation procedure is carried out correctly.

⁂ CAUTION

If the sintered metal gears have been subjected to prolonged oil contamination, they must be soaked in a solvent bath and then thoroughly washer in clean solvent before refitting. Because of the porous construction of the sintered material, oil impregnated in the gears will emerge and contaminate the belt.

14. Clean camshaft gears and mating faces on camshafts.

15. Place gears inverted on a flat surface, with the locating lugs on the gears positioned so they are facing each other.

16. Keeping the timing marks aligned, position timing belt onto gears.

17. Position LRT-12-195 between the gears, turn center nut sufficiently to spread drive belt and position LRT-12-175 to camshaft gears, remove LRT-12-195 from between camshaft gears.

18. Fit LRT-12-198 alignment pins into the end of each camshaft.

19. Position timing belt and gears over LRT-12-198 and locate gears onto camshafts.

20. Position LRT-12-197 into the front end of the right side exhaust camshaft.

21. With assistance, using a 30mm socket on LRT-12-197, turn the right side exhaust camshaft sufficiently to align camshaft gears to the drive slots in each camshaft.

22. Remove LRT-12-198 alignment pins and fit new camshaft gear retaining bolts.

23. Tighten camshaft gear bolts as follows:

- Step 1: 27 Nm (20 ft. lbs.)
- Step 2: Additional 90 degrees

24. Remove LRT-12-175 from camshaft gears.

25. Remove LRT-12-197 from front end of exhaust camshaft.

26. Clean right side exhaust camshaft cap seal recess and fit new cap seal.

27. Fit splash shield and secure with bolts.

28. Fit right side road wheel and tighten nuts to 85 ft. lbs. (115 Nm).

29. Fit right side rear timing belt cover.

30. Fit underbelly panel.

31. Connect battery ground lead.

LEFT SIDE REAR TIMING BELT

1. Before servicing the vehicle, refer to the precautions in the beginning of this section.

2. Disconnect battery ground lead.

3. Remove underbelly panel.

4. Remove left side rear timing belt cover.

5. Remove right side front road wheel.

6. Remove 3 bolts securing right side splash shield to body and remove shield.

7. Rotate crankshaft in a clockwise direction and align the engine 'SAFE' position, notch on crankshaft pulley aligned to the 'ARROW' on front mounting plate and the timing marks aligned on the rear camshaft gears as illustrated.

8. Insert timing pin LRT-12-232 through hole in lower crankcase, ensuring pin is located in hole in flywheel. Hole immediately adjacent to side of lower crankcase must be used.

9. Remove and discard left side front exhaust camshaft cap seal from cylinder head.

10. Position LRT-12-175 to rear camshaft gears as illustrated, remove and discard bolts securing gears to camshafts.

11. Remove rear camshaft gears, timing belt and special tool as an assembly.

⁂ CAUTION

Do not turn the crankshaft or the camshafts while the timing belt is removed.

12. If camshaft timing belt is to be refitted, mark direction of rotation on timing belt.

13. Remove LRT-12-175 from camshaft gears and remove timing belt from gears.

⁂ CAUTION

Camshaft timing belts must be stored and handled with care. Always store belt on its edge and with a bend radius of more than 2 inches (20mm). Do not reuse a belt that has been twisted or bent double. Do not use a belt if debris other than belt dust is found in timing belt covers, or if belt has been contaminated with coolant or oil. Do not reuse belt if partial engine seizure has occurred. Do not reused a belt if mileage is more than 45,000 miles (72,000 km).

To install:

➡After the front or rear timing belt installation is complete, it is possible that, after rotating the engine and positioning the crankshaft pulley to the 'SAFE' position, the timing marks on the rear timing gears may be misaligned. This misalignment is acceptable, provided that the timing belt installation procedure is carried out correctly.

⁂ CAUTION

If the sintered metal gears have been subjected to prolonged oil contamination, they must be soaked in a solvent bath and then thoroughly washer in clean solvent before refitting. Because of the porous construction of the sintered material, oil impregnated in the gears will emerge and contaminate the belt.

14. Clean camshaft gears and mating faces on camshafts.

15. Place gears inverted on a flat surface, with the locating lugs on the gears positioned facing each other.

16. Keeping the timing marks aligned, position timing belt onto gears.

17. Position LRT-12-195 between the gears, turn center nut sufficiently to spread drive belt and position LRT-12-175 to camshaft gears, remove LRT-12-195 from between camshaft gears.

18. Fit LRT-12-198 alignment pins into the end of each camshaft.

19. Position timing belt and gears over LRT-12-198 and locate gears onto camshafts.

20. Position LRT-12-197 into the front end of the left side exhaust camshaft.

21. With assistance, using a 30mm socket on LRT-12-197, turn the left side exhaust camshaft sufficiently to align camshaft gears to the drive slots in each camshaft.

22. Remove LRT-12-198 alignment pins and fit new camshaft gear retaining bolts.

23. Tighten camshaft gear bolts to 20 ft. lbs. (27 Nm) then a further 90°. Remove LRT-12-175 from camshaft gears.

24. Remove LRT-12-197 from front end of exhaust camshaft.

25. Clean left side exhaust camshaft cap seal recess and fit new cap seal.

26. Fit splash shield and secure with bolts.

27. Fit right side road wheel and tighten nuts to 85 ft. lbs. (115 Nm).

28. Fit left side rear timing belt cover.

29. Fit underbelly panel.

30. Connect battery ground lead.

FRONT CRANKSHAFT SEAL

1. Before servicing the vehicle, refer to the precautions in the beginning of this section.

2. Disconnect battery ground lead.

3. Remove camshaft timing belt.

4. Remove crankshaft gear.

5. Fit thrust button, LRT-12-200/3 to end of crankshaft.

6. Screw LRT-12-200 into crankshaft front oil seal.

7. Tighten center bolt of LRT-12-200 to remove oil seal.

8. Remove and discard oil seal from special tool.

9. Remove thrust button from crankshaft.

To install:

10. Using a lint free cloth, thoroughly clean oil seal recess in oil pump and running surface on crankshaft.

11. Fit oil seal guide, from seal kit, over end of crankshaft.

12. Position new seal on crankshaft up against oil pump housing. Drift seal into place using tool LRT-12-202.

13. Remove LRT-12-202 and oil seal guide from crankshaft.

14. Fit gear to crankshaft.

15. Fit camshaft timing belt.

16. Connect battery ground lead.

Piston and Ring Positioning

Before servicing the vehicle, refer to the precautions in the beginning of this section.

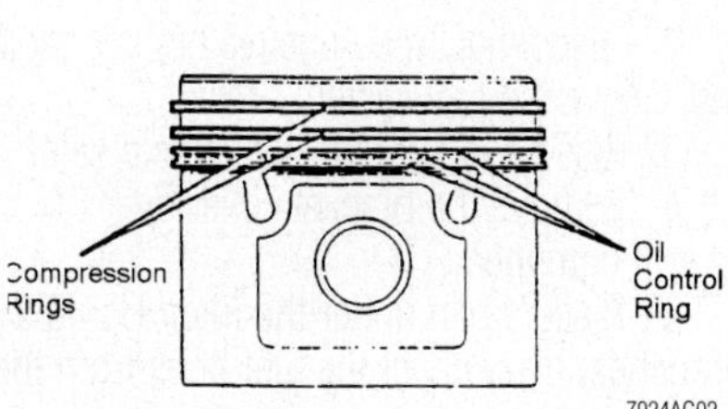

Piston ring positioning

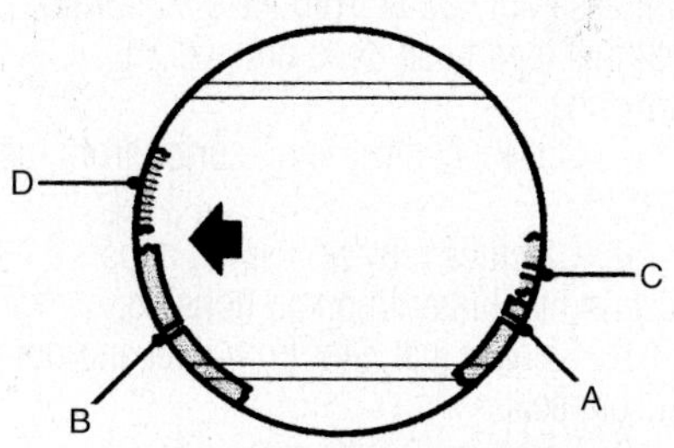

Piston ring end-gap spacing—Freelander shown

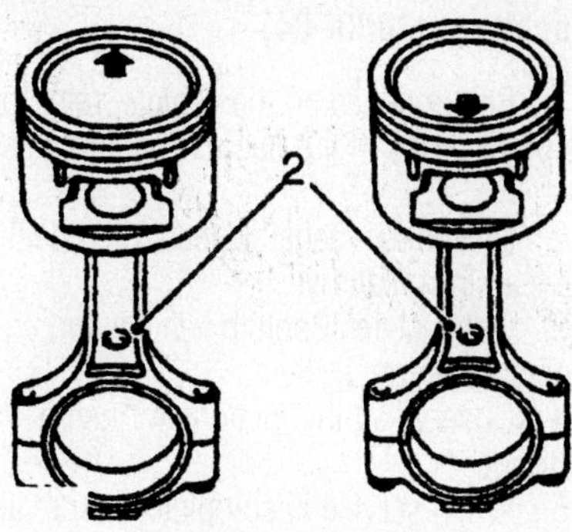

Connecting rod front mark location—Discovery shown (piston arrow same direction as domed-shaped connecting rod marks on right bank; away from domed-shaped connecting rod marks on for left bank)

FUEL SYSTEM

Fuel System Service Precautions

Safety is the most important factor when performing not only fuel system maintenance but any type of maintenance. Failure to conduct maintenance and repairs in a safe manner may result in serious personal injury or death. Maintenance and testing of the vehicle's fuel system components can be accomplished safely and effectively by adhering to the following rules and guidelines.

- To avoid the possibility of fire and personal injury, always disconnect the negative battery cable unless the repair or test procedure requires that battery voltage be applied.
- Always relieve the fuel system pressure prior to disconnecting any fuel system component (injector, fuel rail, pressure regulator, etc.), fitting or fuel line connection. Exercise extreme caution whenever relieving fuel system pressure, to avoid exposing skin, face and eyes to fuel spray. Please be advised that fuel under pressure may penetrate the skin or any part of the body that it contacts.
- Always place a shop towel or cloth around the fitting or connection prior to loosening to absorb any excess fuel due to spillage. Ensure that all fuel spillage (should it occur) is quickly removed from engine surfaces. Ensure that all fuel soaked cloths or towels are deposited into a suitable waste container.
- Always keep a dry chemical (Class B) fire extinguisher near the work area.
- Do not allow fuel spray or fuel vapors to come into contact with a spark or open flame.
- Always use a back-up wrench when loosening and tightening fuel line connection fittings. This will prevent unnecessary stress and torsion to fuel line piping.
- Always replace worn fuel fitting O-rings with new. Do not substitute fuel hose or equivalent where fuel pipe is installed.

Before servicing the vehicle, be sure to also refer to the precautions in the beginning of this section as well.

Fuel System Pressure

RELIEVING

The fuel injection system operates under high pressure. This makes it necessary to first relieve the system of pressure before servicing. The pressurized fuel, when released, may ignite or cause personal injury.

Freelander (2001–04)

1. Before servicing the vehicle, refer to the precautions in the beginning of this section.
2. Disconnect negative battery cable.
3. Remove acoustic cover.
4. If coil is obstructing access to Schrader valve, unbolt coil and position aside.
5. Remove Schrader valve cap.
6. Position absorbent cloth around fuel feed pipe connection to collect spillage.

⁂ WARNING

The spilling of fuel is unavoidable during this operation. Ensure that all necessary precautions are taken to prevent fire and explosion.

7. Connect adapter LRT-19-006 to Schrader valve.
8. Position opposite end of adapter LRT-19-006 into container, turn tap to release pressure.
9. Remove adapter LRT-19-006 from Schrader valve.
10. Fit cap to Schrader valve.
11. Install coil to mounting, if removed.
12. Fit acoustic cover.

Discovery (2001–02) and Range Rover (2001–02)

1. Before servicing the vehicle, refer to the precautions in the beginning of this section.
2. Disconnect the power to the fuel pump by removing the relay or the fuel pump fuse. Check the list on the fuse box lid to be sure. The fuse can be removed to stop the fuel pump from running. With the engine operating at idle, wait until the engine stalls from fuel starvation.
3. Switch the ignition **OFF** and remove the negative battery cable.
4. Carefully loosen the fuel line on the control pressure regulator or component to be serviced.
5. Wrap a clean rag around the connection, while loosening, to catch any fuel.
6. After service is complete, discard the fuel soaked rag in the proper manner and reconnect negative battery cable, relay or fuses.

Discovery (2003–04)

1. Before servicing the vehicle, refer to the precautions in the beginning of this section.
2. Disconnect negative lead, then positive lead from battery.
3. Connect TestBook to vehicle and depressurize fuel system.
4. Observe all instructions for using TestBook.
5. When service is complete, ensure all fuel connections are properly restored and all fuel spillage is thoroughly cleaned before starting vehicle.

Range Rover (2003–04)

1. Before servicing the vehicle, refer to the precautions in the beginning of this section.
2. Disconnect negative lead from battery.
3. Remove engine appearance cover.
4. Remove cap from Schraeder valve on fuel rail.
5. Position absorbent cloth around fuel feed pipe connection to collect spillage.

✻✻ WARNING

The spilling of fuel is unavoidable during this operation. Ensure that all necessary precautions are taken to prevent fire and explosion.

6. Connect adapter LRT-19-006 to Schrader valve.
7. Position opposite end of adapter LRT-19-006 into container, turn tap to release pressure.
8. Remove adapter LRT-19-006 from Schrader valve.
9. Fit cap to Schrader valve.
10. Install coil to mounting, if removed.
11. Fit acoustic cover.

Fuel Filter

REMOVAL & INSTALLATION

Freelander (2001–04)

1. Before servicing the vehicle, refer to the precautions in the beginning of this section.
2. Disconnect battery ground lead.
3. Properly release fuel system pressure.
4. Remove fuel pump, disconnect 2 connectors from top of fuel pump, release sprag clip holding sender to pump.
5. Release 3 slots in top of pump unit assembly from lugs in base.
6. Carefully maneuver top of pump unit assembly away from base, ensuring that fuel feed hose does not become strained.
7. Collect compression spring from fuel filter.
8. Carefully release 3 sprag clips securing fuel filter in pump housing.
9. Release fuel filter from inlet and outlet connections.
10. Remove fuel filter.
11. Collect O-rings.

To install:

12. Lubricate NEW O-rings with silicone grease and fit to ports.
13. Carefully fit fuel filter to ports and push fully home, ensuring that sprag clips engage fully.

✻✻ WARNING

Ensure ground spring fitted to fuel pressure regulator is correctly located.

14. Position spring to fuel filter recess and engage in top location.

✻✻ WARNING

Ensure that filter ground tag is correctly located to contact the base of fuel filter.

15. Engage pump top to base, ensuring that slots engage correctly with lugs.

✻✻ WARNING

During refit, ensure that all electrical connections are made correctly. Earth tag on fuel pump negative terminal must not become distorted.

16. Fit sender unit.
17. Connect battery ground lead.

Discovery (2001–04) and Range Rover (2000–02)

1. Before servicing the vehicle, refer to the precautions in the beginning of this section.
2. Disconnect the negative battery cable and relieve the fuel system pressure.
3. Raise and support the vehicle safely.
4. Remove the bracket cover over the filter, if equipped.
5. Place a pan under the filter. Using 2 wrenches, disconnect the fuel lines from the filter.
6. Remove the fuel filter from the bracket and retainer, if equipped. Note the direction of the flow arrow so the replacement filter can be installed correctly.

To install:

7. Install the fuel filter into the bracket making sure the flow direction is correct. Tighten the clamp to 15–25 inch lbs. (2–3 Nm).
8. Connect the fuel lines, using new O-rings. Tighten the fittings to 15 ft. lbs. (20 Nm).
9. Lower the vehicle.
10. Start the engine and check for leaks.

Range Rover (2003–04)

1. Before servicing the vehicle, refer to the precautions in the beginning of this section.
2. Disconnect the negative battery cable and relieve the fuel system pressure.
3. Raise and support the vehicle safely.
4. Under the right front of the vehicle, remove the 2 screws securing the fuel filter access panel, and remove the panel.
5. Remove the fuel filter from the mounting bracket.
6. Disconnect 3 fuel lines and 1 reference pipe and remove fuel filter.

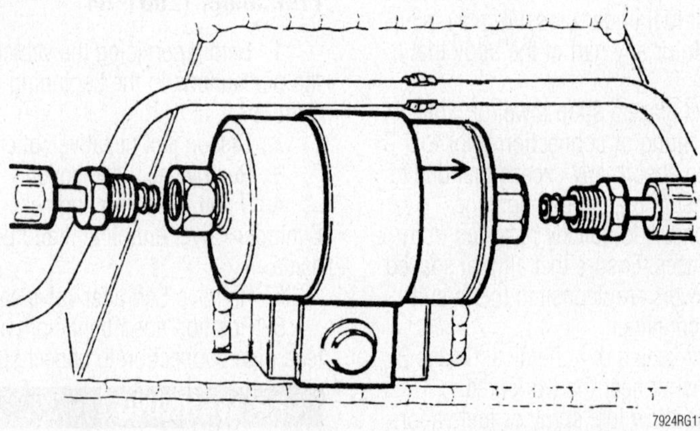

Exploded view of the fuel filter mounting—Discover Series II (2001–04) and Range Rover (2001–02)

To install:

7. Connect the fuel lines and the reference pipe.
8. Install the fuel filter into the bracket making sure the flow direction is correct.
9. Lower the vehicle.
10. Connect the battery ground cable.
11. Start the engine and check for leaks.
12. Install the fuel filter access cover.

Fuel Pump

REMOVAL & INSTALLATION

Freelander (2001–04)

1. Before servicing the vehicle, refer to the precautions in the beginning of this section.
2. Disconnect battery ground lead.
3. Properly depressurize fuel system before proceeding.

⁂ WARNING

Depressurize the system before disconnecting any components. Fuel pressure will be present in the system even if the ignition has been switched off for some time.

4. Open right side rear and tail doors.
5. Fold right side rear seat forward.
6. Remove 2 fasteners securing front and rear carpets.
7. Pull back load space carpet from fuel pump access panel.
8. Remove 6 screws securing access panel.
9. Remove access panel.

⁂ WARNING

The spilling of fuel is unavoidable during this operation. Ensure that all necessary precautions are taken to prevent fire and explosion.

10. Disconnect multiplug and fuel hose from fuel pump housing.

⁂ CAUTION

Always fit plugs to open connections to prevent contamination.

11. Use LRT-19-009 to remove locking ring from fuel pump housing.
12. Remove fuel pump housing and discard sealing ring.

To install:

13. Clean fuel pump housing and mating face on fuel tank.
14. Fit new seal to fuel pump housing.
15. Fit fuel pump housing to fuel tank, fit and tighten locking ring to 26 ft. lbs. (35 Nm) using LRT-19-009.
16. Connect multiplug and fuel hose to fuel pump housing.
17. Fit access panel and secure with screws.
18. Position carpet and secure with fasteners.
19. Reposition rear seat.
20. Close rear and tail doors.
21. Connect battery ground lead.

Discovery (2001–02) and Range Rover (2001–02)

1. Before servicing the vehicle, refer to the precautions in the beginning of this section.
2. Relieve the fuel system pressure.
3. Remove or disconnect the following:
   - Negative battery cable
   - Fuel from the tank
   - Fuel filler tube from the tank
   - Feed pipe at the rear of the filter
   - Return line
4. Support the fuel tank with a jack.
5. Remove the 3 nuts and 2 bolts securing the fuel tank cradle–to–the floor pan.
6. Lower the tank about 6 inches (150mm) and unplug the wiring connectors.
7. Remove the tank/cradle assembly.
8. Remove the tank from the cradle.
9. Remove any dirt that has accumulated around the fuel pump flange so it will not enter the fuel tank during removal and installation.

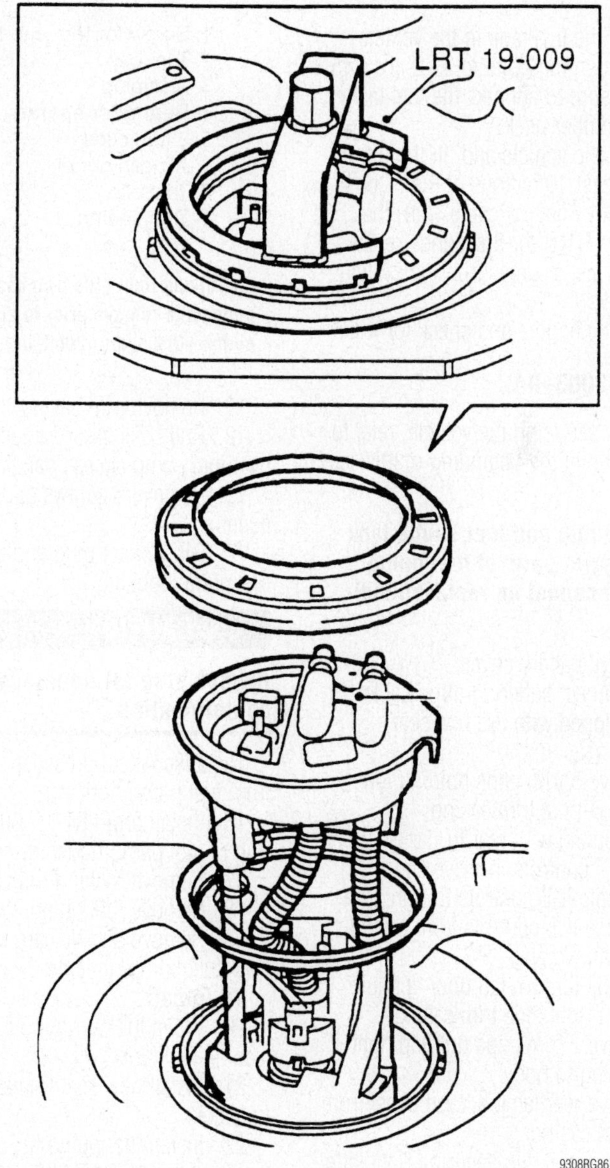

Fuel pump mounting—Discovery (2001–02) and Range Rover (2001–02)

10. Disconnect the breather hose from the pressure sensor.
11. Disconnect the hoses from the pump.
12. Using a locking ring tool, such as LRT 19-009, unscrew the retaining ring from the pump flange.

➡A lifting ring is provided to pull the unit out. Don't pull on any other parts!

13. Remove the fuel pump and discard the seal ring. Separate the fuel pump from the sending unit, if required.

To install:

14. Clean the fuel pump mounting flange and tank mounting surface.
15. Apply a light coating of sealer on a new seal ring.
16. Install the fuel pump on the sending unit, if removed. Install the fuel pump assembly in the tank and tighten it to 26 ft. lbs. (35 Nm).
17. Install the fuel tank in the vehicle, tighten the mounting bolts to 33 ft. lbs. (45 Nm), and be sure to connect the fuel tank vent hose and filler neck.
18. Lower the vehicle and fill the fuel tank with at least 10 gallons of fuel. Connect the negative battery cable. Turn the ignition key to **RUN** for 3 seconds repeatedly, 5–10 times, to pressurize the system. Check for leaks.
19. Start the engine and check for leaks.

Discovery (2003–04)

1. Before servicing the vehicle, refer to the precautions in the beginning of this section.

➡The fuel pump and fuel gauge tank unit are integral parts of fuel pump housing and cannot be replaced individually.

2. Remove battery cover.
3. Disconnect battery ground cable.
4. If equipped with 3rd row seat, remove right seat.
5. Remove 5 trim clips holding lower edge of right D-post trim casing.
6. If equipped with rear in-car entertainment (ICE) controls, remove screw securing remote ICE controls to right trim casing. Release ICE controls from casing and detach connectors.
7. Partially remove tail door aperture seal in area of right side trim casing.
8. Remove 2 trim clips holding right side trim casing to body.
9. Remove rear lamp access panel from right side trim casing.
10. Remove trim clip securing right side trim casing to lower E-post.

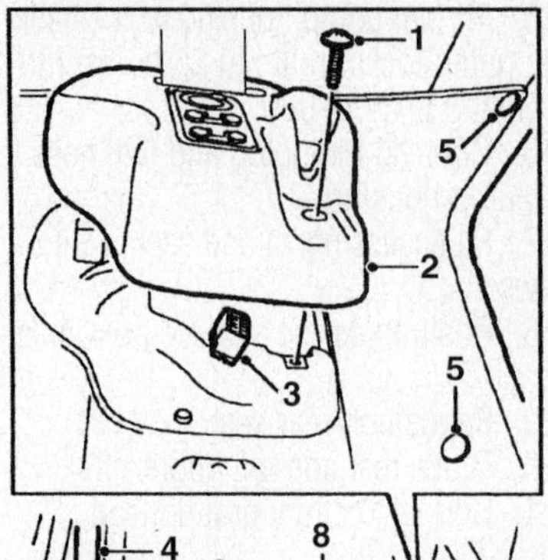

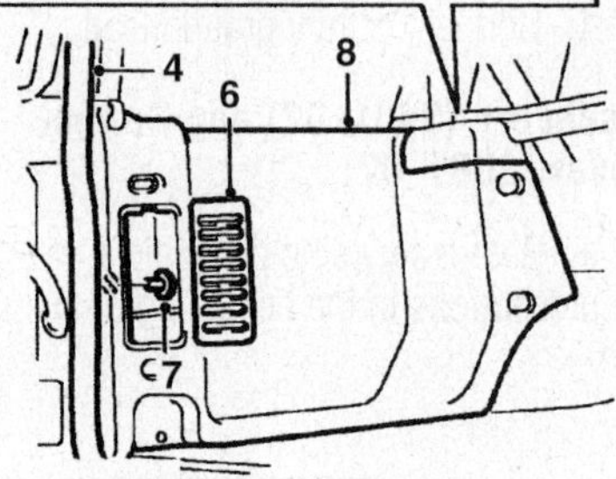

1. Screw for trim cap to casing
2. Trim cap
3. Multiplug
4. Rear door aperature seal
5. 2 trim clips
6. Access panel
7. Trim clip
8. Trim casing

42348-RANG-G24

Identifying right side trim casing and components to remove prior to access to fuel pump—Discovery (2003–04)

11. Remove trim casing.
12. Pull back cargo space carpet from over fuel pump access panel.
13. Remove 6 screws and remove access panel.
14. Disconnect plug and fuel hose from fuel pump housing.

⁂ CAUTION

Always plug all openings to prevent contamination.

15. Disconnect pressure sensor pipe from fuel pump housing.
16. Install tool LRT-19-009 onto locking ring of fuel pump housing.
17. Remove locking ring.
18. Remove fuel pump housing.
19. Remove and discard sealing ring from fuel pump housing.

To install:

20. Clean fuel pump housing and mating face on fuel tank.
21. Install new seal to mating face on fuel tank.
22. Install fuel pump housing to fuel tank and use LRT-19-009 to install and set locking ring.
23. Connect multiplug and fuel hose to fuel pump housing.
24. Connect pressure sensor pipe to fuel pump housing.
25. Install access panel and secure with screws.
26. Reposition cargo space carpet.
27. If equipped with 3rd row seats, install bolt securing lower mounting of 3rd row seat belt to bolt and tighten to 37 ft. lbs. (50 Nm).
28. Install trim casing and components in reverse of removal procedure.
29. Install 3rd row right side seat.
30. Reconnect battery ground cable.
31. Install battery cover.

Range Rover (2003–04)

1. Before servicing the vehicle, refer to the precautions in the beginning of this section.

➡The fuel pump and fuel gauge tank unit are integral parts of fuel pump housing and cannot be replaced individually.

➡Before disconnecting the battery, remove the ignition key. If the key is not removed, the fuel gauge will remember its previous position and result in inaccurate readings.

2. Disconnect battery ground cable.
3. Relieve the fuel system pressure.
4. Drain the fuel tank.
5. At the rear of the vehicle, remove both D pillar trim finish panels.
6. Remove the spare wheel hold down straps.
7. Pull back cargo space carpet from over fuel pump access panel.
8. Remove 4 nuts and remove right side fuel pump access panel.
9. Disconnect fuel pump connector.
10. Disconnect fuel feed and return pipes from pump.
11. Remove left side fuel pump access panel and from the opening, disconnect the fuel gauge sender connector from the fuel pump body.
12. Install tool LRT-19-009 onto locking ring of fuel pump housing.
13. Remove locking ring.
14. Remove fuel pump assembly.
15. Remove and discard sealing ring from fuel pump housing.

To install:

16. Clean fuel pump housing and mating face on fuel tank.
17. Install new seal to mating face on fuel tank.
18. Install the fuel pump.

19. Connect the fuel gauge sender to the pump.
20. Use LRT-19-009 to install and set locking ring.
21. Connect the fuel feed and return lines.
22. Connect the fuel pump connector.
23. Install access panel and secure with screws.
24. Reposition cargo space carpet.
25. Install spare wheel hold down straps.
26. Install D pillar trim finish panels.
27. Fill the tank with at least 10 gallons of gasoline.
28. Reconnect battery ground cable.
29. Start the vehicle and check for leaks.

Fuel Rail and Injectors

REMOVAL & INSTALLATION

Freelander (2001–04)

1. Before servicing the vehicle, refer to the precautions in the beginning of this section.
2. Disconnect battery ground lead.
3. Remove intake manifold chamber.
4. Remove underbelly panel.
5. Using a 3/8 in. socket bar raise auxiliary drive belt tensioner and release drive belt from alternator and PAS pump pulleys.
6. Remove nuts and bolts securing alternator.
7. Release alternator and place aside.
8. Note position of coil ground lead, remove 3 nuts and 3 bolts securing ignition coils to left side intake manifold.
9. Remove ignition coils ground. Release and position coil aside.
10. Disconnect clips securing right side injector harness to injector protection cover, or to bosses on fuel rail.
11. Remove 2 bolts securing right side fuel rail to right side intake manifold.
12. Release clips and disconnect injector multiplugs.
13. Remove 2 bolts securing fuel rail to left side intake manifold.
14. Release clips and disconnect injector multiplugs.
15. Release injectors from manifolds, raise fuel rail assembly, disconnect 4 clips securing injector harness to fuel rails and move harness aside.
16. Remove fuel rail and discard O-rings.

⁂ CAUTION

Always fit plugs to open connections to prevent contamination.

17. Remove clip securing injector to fuel rail.
18. Remove injector from fuel rail, remove and discard O-ring from injector.

⁂ CAUTION

Always fit plugs to open connections to prevent contamination.

To install:

19. Clean injector and mating face in fuel rail.
20. Remove cap from new injector and fit to old injector.
21. Lubricate new O-ring with castor oil and fit to injector.
22. Install injector to fuel rail and secure with clip.
23. Lubricate new O-rings with castor oil and fit to injectors.
24. Remove plugs from intake manifolds.
25. Position injector harness to fuel rail and secure with clips.
26. Position injectors and fuel rail to intake manifolds.
27. Connect multiplugs to injectors.
28. Install and tighten 4 bolts securing fuel rail to intake manifold to 7 ft. lbs. (9 Nm).
29. Position ignition coils to left side intake manifold.
30. Position ground lead, fit nuts and bolts securing ignition coils to left side intake manifold and tighten to 7 ft. lbs. (9 Nm).
31. Position alternator and secure with nuts and bolts, tighten to 45 Nm (33 ft. lbs.).
32. Using a 3/8 square drive socket bar, raise auxiliary drive belt tensioner and fit drive belt to pulleys.
33. Install underbelly panel.
34. Install intake manifold chamber.
35. Connect battery ground lead.

Discovery (2001–02) and Range Rover (2001–02)

1. Before servicing the vehicle, refer to the precautions in the beginning of this section.
2. Remove or disconnect the following:
   - Battery ground cable
   - Fuel system pressure
   - Throttle cable
   - Cruise control cable
   - Harness clip from the throttle linkage bracket
   - Breather hose form the plenum
   - IAC connector
   - TPS connector
   - Plenum chamber
   - Purge hose
   - Crankcase breather hose
   - Pressure regulator vacuum hose
   - Ram housing from the intake manifold

➡It may be necessary to pry on the housing to break the seal. If so, use a small block of wood as a pry point between the manifold and ram housing. NEVER pry on the fuel rail!

   - The 8 injector plugs
   - Fuel temperature sensor connection
   - Fuel feed hose form the fuel rail
   - Fuel return hose from the pressure regulator

➡Advanced EVAPS vehicles have a threaded connector.

   - Fuel rail and ignition coil bracket from the manifold
   - Coil bracket
   - Fuel rail and injectors
   - Injector retaining clips and injectors
   - Two O-rings from each injector

To install:

3. Install or connect the following:
   - New O-rings coated with silicone grease
   - Injectors and clips on the rail
   - Fuel rail and injectors on the manifold
   - Ignition coil and bracket
   - Return hose
   - Fuel feed line. Torque the union to 12 ft. lbs. (16 Nm).
   - All injector wiring
   - Ram housing. Use Loctite 577, or equivalent, as a sealant. Torque the bolts to 18 ft. lbs. (24 Nm).
   - All vacuum hoses
   - All remaining wiring
   - Plenum chamber. Use Loctite 577, or equivalent, as a sealant. Torque the bolts to 18 ft. lbs. (24 Nm).
   - Coolant hoses
   - All remaining hoses and wires
   - Battery ground

Discovery (2003–04)

1. Before servicing the vehicle, refer to the precautions in the beginning of this section.
2. Remove upper intake manifold.
3. Carefully maneuver ignition coil assembly from between intake manifold and bulkhead.
4. Position absorbent cloth beneath fuel pipe to catch spillage.
5. Disconnect fuel feed hose from fuel rail.
6. Release injector harness from fuel rail and disconnect injector multiplugs.
7. Remove 4 bolts securing fuel rail to lower intake manifold.

8. Release injectors from intake manifold and remove fuel rail and injectors.

⁂ CAUTION

Always plug all openings immediately to prevent contamination.

9. Remove and discard 2 O-rings from each injector.
10. Fit protective caps to each engine of injectors.

To install:

11. Clean injectors and recesses in fuel rail and intake manifold.
12. Lubricate new O-rings with silicone grease and fit to each end of injectors.
13. Fit injectors to fuel rail and secure with spring clips.
14. Position fuel rail assembly and push-fit each injector into intake manifold.
15. Position bolts securing fuel rail to intake manifold and tighten bolts to 7 ft. lbs. (9 Nm).
16. Connect fuel feed hose to fuel rail.
17. Connect injector harness multiplugs and secure to fuel rail.
18. Carefully position ignition coil assembly between lower intake manifold and bulkhead.
19. Install upper intake manifold.

Range Rover (2003–04)

1. Before servicing the vehicle, refer to the precautions in the beginning of this section.
2. Relieve fuel system pressure.
3. Disconnect the negative battery cable.
4. Remove the air intake hose.
5. Remove both ignition coil covers.
6. Disconnect all ignition coil connectors.
7. Disconnect and label all intake manifold electrical connectors.
8. Remove 2 nuts attaching fuel injector harness to fuel rail.
9. On the left side, remove vacuum reservoir and mounting bracket and fuel hose and mounting bracket.
10. Disconnect left side fuel injector connectors.
11. Disconnect main engine harness and move aside.
12. Remove the secondary air solenoid valve from the right side.
13. Disconnect right side fuel injector connectors.
14. Position the fuel injector harness aside.
15. Remove the air intake plenum.
16. Disconnect the fuel feed line from the fuel rail.
17. Disconnect air injection vacuum hose.
18. Remove 4 bolts attaching engine cover brackets to intake manifold.
19. Remove bolt attaching fuel rail to intake manifold, and remove fuel rail with injectors.

To install:

20. Clean injectors and recesses in fuel rail and intake manifold.
21. Lubricate new O-rings with silicone grease and fit to each end of injectors.
22. Fit injectors to fuel rail and secure with spring clips.
23. Position fuel rail assembly and push-fit each injector into intake manifold.
24. Install bolt attaching fuel rail to intake manifold.
25. Connect air injection vacuum hose.
26. Connect the fuel feed line from the fuel rail.
27. Clean the fuel line connections.
28. Install the air intake plenum.
29. Reposition the fuel injector harness.
30. Connect right side fuel injector connectors.
31. Install the secondary air solenoid valve to the right side.
32. Connect the main engine harness.
33. Connect left side fuel injector connectors.
34. On the left side, install vacuum reservoir and mounting bracket and fuel hose and mounting bracket.
35. Connect fuel injector harness to fuel rail.
36. Connect all labeled intake manifold electrical connectors.
37. Connect all ignition coil connectors.
38. Install both ignition coil covers.
39. Connect the air intake hose.
40. Connect the negative battery cable.
41. Start the vehicle and check for leaks.

DRIVE TRAIN

Transmission Assembly

REMOVAL & INSTALLATION

Freelander (2001–04)

1. Before servicing the vehicle, refer to the precautions in the beginning of this section.
2. Tie hood back in upright position.
3. Disconnect battery ground lead.
4. Remove engine acoustic cover.
5. Remove intermediate reduction drive (IRD).
6. Loosen gear selector cable trunnion nut.
7. Release clip securing selector cable to gearbox bracket, remove selector cable and collect trunnion from selector lever.
8. Remove 3 screws securing left side splash shield and remove shield.
9. Secure LRT-54-026 to drive shaft inboard joint. Using a suitable lever, release inboard joint from gearbox.
10. With assistance pull hub outwards and release drive shaft from gearbox.

⁂ CAUTION

Care must be taken not to damage oil seal when removing drive shaft from gearbox

11. Remove and discard circlip from end of drive shaft.
12. Remove 2 bolts securing torque converter access plate.
13. Remove access plate.
14. Mark drive plate to torque converter, for refit purposes.
15. Remove 4 bolts securing drive plate to converter.
16. Remove bolt securing IRD cooling hose retainer. Remove retainer.
17. Remove 2 bolts securing gearbox to engine.
18. Release HO_2S multiplug from support bracket on left side camshaft cover, disconnect multiplug.
19. Remove 4 nuts securing left side exhaust manifold to cylinder head.
20. Remove exhaust manifold, remove and discard gasket.
21. Position container to collect fluid spillage.
22. Disconnect 2 fluid cooler hose unions and discard O-rings.

⁂ CAUTION

Always fit plugs to open connections to prevent contamination.

23. Remove 3 bolts securing fluid cooler bracket.
24. Move fluid cooler aside.
25. Remove bolt securing CKP sensor

to gearbox, release sensor and position aside.

26. Remove nut and bolt, adjacent to CKP sensor, securing gearbox to engine.

27. Release throttle cable from abutment bracket and disconnect cable from throttle body cam, if fitted.

28. Depress collars and release 2 breather pipes from throttle housing, if fitted.

29. On vehicles with cruise control, disconnect vacuum hose from cruise control actuator.

30. Remove 4 Torx screws securing throttle housing, release throttle housing and position aside.

31. Remove and discard O-ring from throttle housing.

32. Remove starter motor.

33. Remove bolt securing engine ground lead.

34. Release multiplugs from clips attached to fluid pan.

35. Disconnect 2 gearbox harness to main harness multiplugs.

36. Using a hoist, connect adjustable lifting bracket, LRT-12-138 to engine.

37. Raise hoist to take weight without exerting any load on the engine mountings.

38. Install suitable lifting brackets to gearbox and secure with nuts and bolts.

39. Connect lifting equipment to brackets.

40. Remove through bolt securing left side engine mount to gearbox bracket.

41. Remove 4 bolts securing left side mount to body and remove mount.

42. Remove 4 bolts securing left side mounting bracket to gearbox and remove bracket.

43. Remove 2 top bolts securing gearbox to engine.

44. Release gearbox from 2 dowels.

45. Maneuver and lower gearbox to floor.

46. Install converter retaining plate and secure with bolts.

To install:

47. Remove torque converter retaining plate.

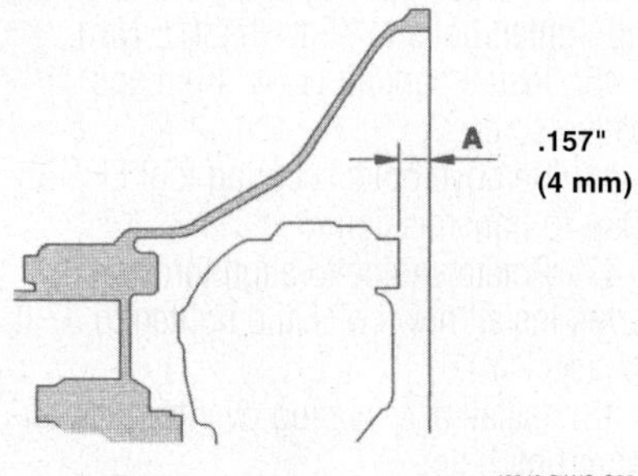

Measuring converter depth in oil pump drive—Freelander (2001–04)

48. Ensure converter is fully located in oil pump drive by checking depth 'A' as illustrated. Depth A = 0.157 inch (4mm).

49. Clean gearbox to engine mating faces, dowels and dowel holes.

50. Install gearbox assembly.

51. Install bolts securing gearbox and tighten to 63 ft. lbs. (85 Nm).

52. Position left side mounting bracket to gearbox, fit and tighten bolts to 63 ft. lbs. (85 Nm).

53. Position left side mounting to body, fit and tighten bolts to 35 ft. lbs. (48 Nm).

54. Align gearbox bracket to left side body mounting, fit and tighten through bolt to 74 ft. lbs. (100 Nm).

55. Disconnect lifting equipment.

56. Remove nuts and bolts securing lifting brackets to gearbox and remove brackets.

57. Connect engine and gearbox harness multiplugs to main harness.

58. Secure multiplugs to clips.

59. Position engine ground lead and secure with bolt.

60. Install starter motor.

61. Clean throttle housing and manifold chamber mating faces.

62. Install new seal to intake manifold chamber.

63. Position throttle housing to manifold chamber, fit Torx screws and tighten to 7 ft. lbs. (9 Nm).

64. Connect throttle inner cable to throttle cam and secure outer cable in abutment bracket, if fitted.

65. Connect hose to cruise control actuator.

66. Secure breather hoses to throttle housing, if fitted.

67. Adjust throttle cable, if fitted.

68. Clean CKP sensor and mating face.

69. Install CKP sensor, fit bolt and tighten to 7 ft. lbs. (9 Nm).

70. Position fluid cooler, tighten M12 bolts to 63 ft. lbs. (85 Nm) and M8 bolt to 18 ft. lbs. (25 Nm).

71. Clean fluid cooler unions.

72. Lubricate new O-rings with clean transmission fluid and fit O-rings to fluid cooler hoses.

73. Connect fluid cooler hoses to gearbox and tighten unions to 13 ft. lbs. (18 Nm).

74. Clean exhaust manifold and mating face on cylinder head.

75. Install exhaust manifold gasket.

76. Position exhaust manifold, fit nuts and progressively tighten, from center outwards to 33 ft. lbs. (45 Nm).

77. Connect HO_2S multiplug and secure to support bracket.

78. Position IRD cooling hose retainer, fit bolt and tighten to 18 ft. lbs. (25 Nm).

79. Align marks on drive plate to torque converter.

80. Install bolts securing drive plate to torque converter and tighten bolts to 45 Nm (33 ft. lbs.).

81. Clean torque converter access plate.

82. Position access plate, fit bolts and tighten to 7 ft. lbs. (9 Nm).

83. Clean end of drive shaft and mating splines in gearbox.

84. Install new circlip to left side drive shaft.

85. With assistance pull hub outwards, align drive shaft and fit to gearbox, taking care not to damage drive shaft oil seal.

**** CAUTION**

Pull the drive shaft to ensure the circlip is fully engaged and retains the shaft.

86. Install splash shield and secure with bolts.

87. Position trunnion to selector lever, locate inner cable through trunnion, do not tighten nut at this stage.

88. Position selector cable to gearbox bracket and secure with clip.

89. Adjust selector cable.

90. Install IRD.

91. Connect battery ground lead.

92. Install engine acoustic cover.

93. Untie and close hood.

Discovery (2001–02)

1. Before servicing the vehicle, refer to the precautions in the beginning of this section.

2. Remove or disconnect the following:
   - Negative battery cable
   - Fan shroud from the radiator
   - Transmission breather pipes from the right cylinder head at the rear and the dipstick
   - Kickdown cable from the throttle linkage
   - Shift boot knob and boot from the center console
   - Fluid from the transmission and the transfer case
   - Exhaust system, from the manifolds back
   - Speedometer from the transfer case
   - Driveshafts at the transmission and transfer case and support them out of the way.
   - Transmission oil cooler lines and secure them out of the way.
   - Transmission shift cable and the

wiring harness from the transmission

3. Secure a transmission jack to the transmission.

4. Remove or disconnect the following:

- Transmission crossmember
- Transfer case side mounts and mounting brackets
- Parking brake cable from the lever
- Driveplate inspection cover and matchmark the torque converter to the driveplate.
- Torque converter bolts and the fill tube from the transmission
- Bell housing to engine bolts

5. Pull the transmission rearward, slightly, secure the torque converter to the transmission.

6. Remove the transmission from the vehicle.

To install:

7. Install or connect the following:

- Transmission to the vehicle
- Bell housing–to–engine bolts and tighten to 31 ft. lbs. (42 Nm)

8. Align the matchmarks for the torque converter to the driveplate.

9. Apply Loctite® to the torque converter bolts and tighten to 29 ft. lbs. (39 Nm).

10. Install or connect the following:

- Fill tube to the transmission
- Driveplate inspection cover
- Parking brake cable to the lever
- Transfer case side mounts and tighten the bolts to 33 ft. lbs. (45 Nm).
- Transmission crossmember
- Transmission shift cable and the wiring harness to the transmission
- Transmission oil cooler lines
- Driveshafts at the transmission and transfer case
- Speedometer to the transfer case
- Exhaust system
- Transmission and transfer case with Dexron®II ATF
- Shift boot and knob
- Kickdown cable to the throttle linkage
- Transmission breather pipes to the right cylinder head at the rear
- Fan shroud to the radiator
- Negative battery cable

Discovery (2003–04)

1. Before servicing the vehicle, refer to the precautions in the beginning of this section.

2. Disconnect battery ground cable.

3. Raise vehicle on ramp.

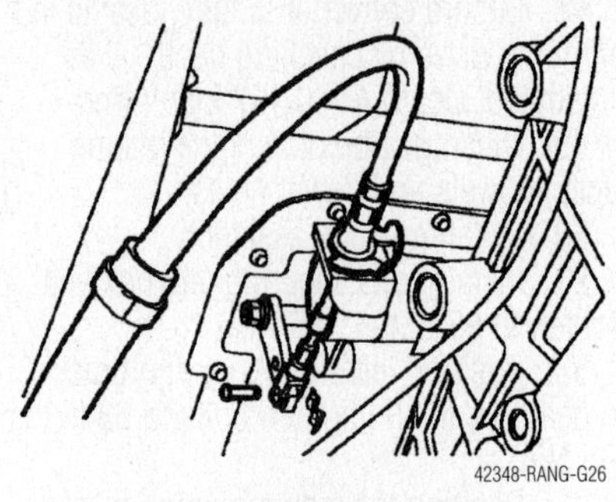

Showing the location of high/low ratio selector cable—Discovery (2003–04)

4. Remove 8 bolts securing center crossmember and remove crossmember.

5. Remove front exhaust pipe.

6. Remove intermediate muffler from tail pipe and rubber mountings.

7. Drain automatic transmission fluid.

8. Drain transfer case oil.

9. Remove front and rear propeller shafts.

10. Remove plug from torque converter housing to access torque converter bolts.

11. Remove and discard 4 bolts securing torque converter to drive plate.

12. Remove parking brake drum.

13. Remove parking brake backing plate from mounting and tie aside.

14. Remove retaining clip holding high/low ratio selector cable to transfer case lever. Remove C-clip and remove outer cable from gearbox, as shown.

15. Disconnect 2 connectors from differential lock switch and multiplugs from high/low ratio switch and neutral sensor.

16. Remove cable tie and multiplug from bracket on transfer case.

17. Release transfer case and automatic transmission breather pipes from clip of rear of cylinder head.

18. Position transmission support jack and secure tool LRT-99-008A to support plate on transmission jack.

19. Position a second support jack under engine, using wooden block between oil pan and jack to prevent damage to oil pan.

20. Remove left side rear engine mount.

21. Remove right side rear engine mount.

22. Remove transmission oil cooler pipes from engine clips.

23. Unscrew transmission cooler pipe connections from transmission and remove O-rings.

⁂ CAUTION

Always fit plugs to open connections to prevent contamination.

24. Remove C-clip securing gear selector cable to transmission bracket.

25. Remove nut holder gear selector lever to inhibitor switch (neutral safety switch). Release lever.

26. Disconnect multiplug from inhibitor switch.

27. Loosen transmission multiplug locking ring and disconnect plug.

28. Lower transmission enough to access bell housing bolts. Use care not to trap any pipes or cables when lowering transmission.

29. Remove 14 transmission–to–engine bolts. With assistance, remove transmission from engine.

30. Attach a suitable strap to retain torque converter in place.

To install:

31. Clean transmission and engine mating faces.

32. Remove torque converter retaining strap.

33. With assistance, position transmission to engine. Install and tighten bell housing bolts to 34 ft. lbs. (46 Nm).

34. Connect multiplug to transmission and tighten locking ring.

35. Connect inhibitor switch plug. Install switch and tighten nut to 18 ft. lbs. (25 Nm).

36. Attach gear selector cable to transmission with C-clip.

37. Clean oil cooler pipe connections at transmission and install new O-rings. Tighten pipe fittings.

38. Attach oil cooler pipes to retaining clamps on engine.

39. Install both rear engine mounts. Tighten bolts and nuts to 63 ft. lbs. (85 Nm) .

40. Attach transmission breather pipes to bulkhead clip.

41. Connect multiplugs to high/low ratio switch and neutral sensor. Attach connectors to differential lock switch.

42. Position cable tie and multiplug to transfer case bracket.

43. Connect high/low ratio cable to transfer case and secure with retaining clip.

44. Position parking brake backing plate and tighten bolts to 55 ft. lbs. (75 Nm).

45. Install parking brake drum and tighten screw.

46. Remove bolts securing tool LRT-99-008A to transmission.

47. Rotate engine to align torque converter. Install new bolts and tighten to 37 ft. lbs. (50 Nm).

48. Install access plug over torque converter bolt hole.

49. Install rear, then front, propeller shafts.

50. Fill transfer case with oil.

51. Fill automatic transmission with fluid.

52. Position intermediate muffler to rubber mounts and to tail pipe, with new gasket, tightening nuts to 30 ft. lbs. (40 Nm).

53. Install front exhaust pipe.

54. Install center crossmember. Tighten bolts to 19 ft. lbs. (26 Nm).

55. Connect battery ground cable.

56. Install battery cover.

57. Check and adjust gear selector cable.

Range Rover (2001–02)

1. Before servicing the vehicle, refer to the precautions in the beginning of this section.
2. Disconnect the negative battery cable.
3. Remove or disconnect the following:
   - Fan shroud
   - Transmission filler tube from the engine
   - Window switch pack
   - Handbrake cable clevis pin
   - Handbrake cable from the tunnel grommet
   - Transmission and transfer case fluid
   - Exhaust front pipe
   - Transmission mount
4. Secure a transmission jack to the transmission.
5. Remove or disconnect the following:
   - Front and rear driveshafts
6. Place a block of wood between the axle housing and the oil pan for support, and slightly lower the assembly.
7. Remove or disconnect the following:
   - Gear selector cable from the transmission lever
   - Selector cable
   - Transmission temperature sensor wiring
   - High/low motor wiring
   - Speed sensor wiring
   - Position sensor wiring
   - Wiring harness from the clips
   - Cooler pipe clamp bolt
   - Cooler pipes
   - Filler pipe from the transmission
   - Breather pipes
   - Converter housing cover
8. Matchmark the drive plate and converter.
9. Remove or disconnect the following:
   - Drive plate–to–converter bolts
   - Converter housing–to–engine bolts
10. Pull the transmission back, making sure that you don't drop the converter.
11. Lower and remove the transmission/transfer case assembly.
12. Support the assembly, nose down, on a bench.
13. Remove the 6 transfer case–to–transmission bolts.
14. Release the 2 ring dowels and separate the transfer case from the transmission.

To install:

15. Clean all mating surfaces.
16. Support the transmission, nose down, on a bench.
17. Position the transfer case and ring dowels. Install the bolts and tighten them to 33 ft. lbs. (45 Nm).
18. Raise the assembly into position, aligning the torque converter and drive plate. Tighten the transmission–to–engine bolts to 31 ft. lbs. (42 Nm).
19. Install the converter–to–flexplate bolts and torque to 33 ft. lbs. (45 Nm).
20. Install or connect the following:
   - Converter cover plate, with a new gasket
   - Breather pipes
   - Filler pipe
   - Cooler lines
   - Cooler line clamp
   - Wiring
   - Selector cable
   - Transmission mount. Torque the bolts to 33 ft. lbs. (45 Nm).

Range Rover (2003–04)

1. Before servicing the vehicle, refer to the precautions in the beginning of this section.
2. Disconnect the negative battery cable.
3. Remove or disconnect the following:
   - Transmission fluid
   - Cooling fan
   - Air intake plenum
   - Inhibitor switch connector
   - Oxygen sensor connectors
   - 2 bolts at top of bellhousing
   - Front and rear driveshafts
   - Left side heat shield
   - Place support tool LRT-44-032 on a suitable jack and support the transmission
   - Crossmember
   - Snubbers
   - Lower transmission on jack
   - High/low shift motor connector
   - Transmission housing multi-plug connector
   - Oxygen sensor leads at transfer case
   - Selector lever from selector shaft
   - Selector cable support bracket
   - Crankshaft position sensor
   - Front air dam
   - Front splash shield
   - Torque converter cover
   - 4 drive plate-to-torque converter bolts
   - Starter motor and wire aside
   - Oil cooler clamp
   - Cooler lines from both sides of transmission
   - Transmission mounting bolts
4. With the aid of a helper, support torque converter and carefully lower transmission from vehicle.

To install:

5. Clean all mating surfaces.
6. Carefully raise transmission into position, ensuring torque converter spigot and transmission engages the mounting dowels.
7. Install transmission bolts and tighten to 33 ft. lbs. (45 Nm).
8. Install new o-rings and connect the transmission cooler lines.
9. Install or connect the following:
   - Oil cooler clamp
   - Starter motor and tighten bolts to 33 ft. lbs. (45 Nm)
   - 4 drive plate-to-torque converter bolts and tighten to 33 ft. lbs. (45 Nm)
   - Torque converter cover
   - Front splash shield
   - Front air dam
   - Crankshaft position sensor
   - Selector cable support bracket
   - Selector lever to selector shaft
   - Oxygen sensor leads at transfer case
   - Transmission housing multi-plug connector
   - High/low shift motor connector
   - Snubbers
   - Crossmember and tighten bolts to 50 ft. lbs. (68 Nm)
   - Remove transmission jack
   - Left side heat shield
   - Front and rear driveshafts
   - 2 bolts at top of bellhousing and tighten bolts to 18 ft. lbs. (25 Nm)
   - Oxygen sensor connectors
   - Inhibitor switch connector
   - Air intake plenum
   - Cooling fan
   - Negative battery cable
   - Transmission fluid
   - Start the engine and check for leaks.

Transfer Case Assembly

REMOVAL & INSTALLATION

Discovery (2001–02)

1. Before servicing the vehicle, refer to the precautions in the beginning of this section.

2. Remove or disconnect the following:
- Negative battery cable
- Radiator fan shroud
- Transfer case shift knob and boot
- Transfer case fluid
- Heat shield from the front exhaust pipe
- Catalytic converter assembly
- Crossmember from under the transfer case
- Speedometer cable and tie it out of the way

3. Matchmark the front and rear driveshafts–to–flanges relation.
4. Disconnect the front and rear driveshafts and tie them out of the way.
5. Secure an adapter plate LRT-99-010.
6. Place 4 1³⁄₁₆ in. (30mm) spacers between the top of the hoist and the adapter plate, then secure the plate to the hoist.
7. Remove the 4 central bolts from the transfer case and secure the hoist with the adapter plate to the unit.
8. Raise the hoist to take the weight off the transfer case.
9. Remove the left and right transfer case rubber mounts.
10. Slowly lower the hoist until the park brake drum clears the passenger footwell. Be sure the engine does not crush any components.
11. Loosen the park brake adjustment nut and remove the park brake drum assembly from the rear output flange.
12. Label and unplug all sensors and switches from the transfer case.
13. Remove the transfer case breather banjo bolt and position it aside.
14. Disconnect the differential lock engaging rod.
15. Place the transfer case in low range.
16. Remove the range selector rod lower nut, then the rod from the yoke.
17. Support the transmission with a wooden block.
18. Remove the upper and lower transfer case mounting bolts.
19. Fit guide studs 18G 1425 to the transmission and move the transfer case rearward to remove.

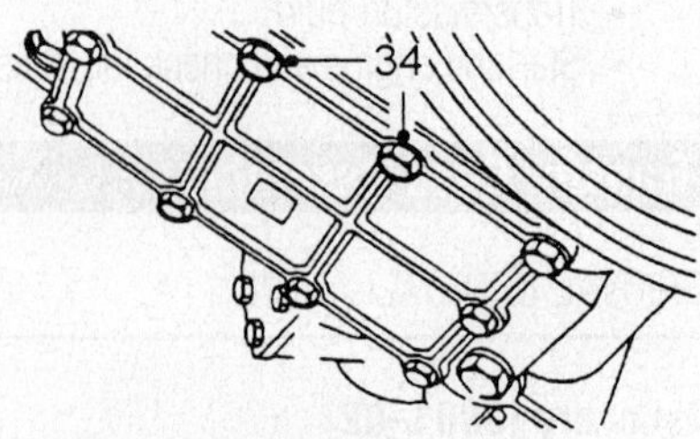

Location of the 4 central bolts—mount the adapter plate after removing these bolts—Discovery (2001–04)

To install:

20. Be sure the mating surface of the transmission and transfer case is free from dirt and debris.
21. Raise the transfer case until it is located over the guide studs 18G 1425, then slide it forward onto the transmission.
22. Remove the guide studs and bolt the transfer case to the transmission.
23. Complete the installation procedure in the reverse order of the removal procedure, noting the following items:
- After removing the adapter plate, clean the 4 bottom cover bolts and coat them with Loctite® 290, then tighten to 19 ft. lbs. (25 Nm).
- Fill the transfer case with 90W oil.
- Check and adjust the park brake cable.

Discovery (2003–04)

1. Before servicing the vehicle, refer to the precautions in the beginning of this section.
2. Disconnect battery ground cable.
3. Remove front exhaust pipe.
4. Drain transfer case oil.
5. Remove 8 bolts securing rear crossmember and remove crossmember.
6. Remove intermediate muffler from tail pipe and from rubber mounts.
7. Remove front and rear propeller shafts.
8. Remove parking brake drum.
9. Remove parking brake backing plate and tie aside.
10. On models with high/low shift interlock solenoid, disconnect solenoid connector from main harness. Also, remove sleeve retaining rings and remove sleeve from high/low selector cable.
11. Release and remove clevis pin holding differential lock selector cable to transfer case.
12. Loosen locknuts holding differential lock selector cable to bracket, and remove cable.
13. Remove clevis pin securing high/low ratio selector cable to selector lever. Remove C-clip and release outer cable from bracket.
14. Remove clevis pin and C-washer holding high/low ratio selector cable to selector lever. Release cable from bracket.
15. Remove 2 cable ties securing cable to fuel pipes.
16. If fitted, disconnect 2 Lucar connectors from oil temperature sensor and disconnect reverse lamp switch connector.
17. Remove banjo bolt securing breather pipe and discard washers.
18. If fitted, disconnect 2 Lucar connectors from differential lock switch.
19. Disconnect multiplug from reverse light switch.
20. If fitted, disconnect both differential lock warning lamp switch connectors from main harness.
21. Release and disconnect transfer case neutral switch connector.
22. Remove 4 bolts from transfer case bottom plate, then position support plate LRT-99-010 to transfer case and tighten bolts.
23. Position transmission support jack and secure support plate LRT-99-010 to jack with 4 bolts.
24. Remove 3 transfer case–to–transmission bolts and install 3 guide studs LRT-41-009 through transfer case bolt holes for support during removal.

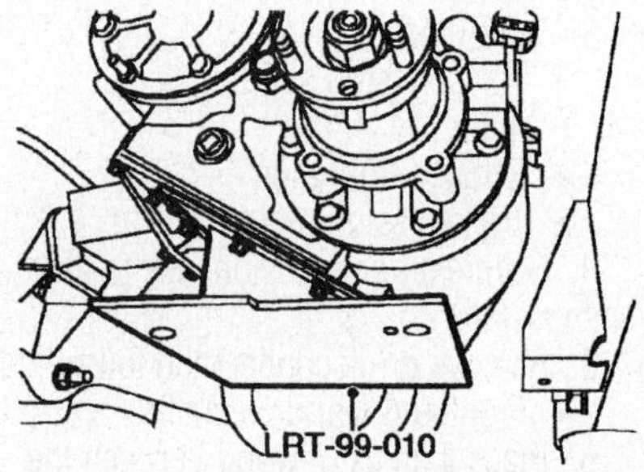

Showing transfer case support plate LRT-99-010 installed—Discovery (2003–04)

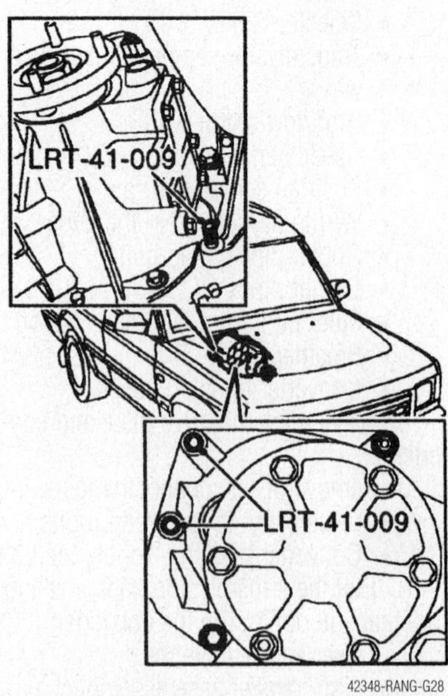

Installed positions of guide studs—Discovery (2003–04)

25. Remove 2 bolts and 1 nut securing transfer case to transmission.

CAUTION

If securing stud is removed during this operation, it must be discarded and a new stud fitted.

26. Position a jack to support transmission and engine.
27. Remove 4 bolts and nut from right side transfer case mount.
28. Remove nut from left side transfer case mount.
29. Raise transmission slightly and remove right side mount.
30. Remove transfer case input shaft oil seal.

To install:

31. Install a new transfer case input shaft oil seal.
32. Clean transfer case–to–transmission mating faces.
33. Raise transfer case on jack and align to guide studs LRT-41-099.
34. Clean transfer case mounting bolt threads and apply sealant to bolt and stud threads.
35. Fit and lightly tighten 2 bolts and nut securing transfer case to transmission.

➡If a new stud is fitted, apply sealant STC 50552 to stud threads and tighten to 33 ft. lbs. (45 Nm).

36. Remove guide studs, a replace them with 3 bolts, lightly tightened.
37. Tighten all transfer case–to–transmission bolts, in a diagonal pattern, to 33 ft. lbs. (45 Nm).
38. Position transmission right side mount and tighten bolts to 63 ft. lbs. (85 Nm).
39. Tighten right side mount nut to 35 ft. lbs. (48 Nm).
40. Remove 4 bolts securing support plate to transmission jack. Remove support plate–to–transfer case bolts and remove plate.
41. Clean threads of bottom plate bolts. Apply sealant STC 50552 to threads.
42. Install and tighten bottom plate bolts to 18 ft. lbs. (25 Nm).
43. Using new washers, position breather pipe and tighten banjo bolt.
44. Connect multiplug to transfer case neutral switch and secure harness.
45. If fitted, connect Lucar connectors to differential lock switch.
46. If fitted, connect differential lock warning lamp connectors and secure harness.
47. Attach connector to reverse lamp switch.
48. If fitted, connect Lucar connectors to oil temperature switch.
49. Position high/low ratio selector cable and secure with C-washer and clevis pin.
50. Position differential lock selector cable to bracket and tighten locknuts sufficiently to retain cable.
51. Fit selector cable to transfer case and secure with clevis pin. Adjust cable.
52. If fitted with high/low shift interlock solenoid, position solenoid harness to shift cable, fit sleeve around cable and harness, and secure sleeve, then connect multiplug to main harness and secure to bracket.
53. With new cable ties, attach cable to fuel pipes.
54. Clean parking brake backing plate and mating face. Install plate and tighten bolts to 55 ft. lbs. (75 Nm).
55. Clean parking brake drum, then install and tighten retaining screw.
56. Install rear then front propeller shafts.
57. Install intermediate muffler, using new gasket. Tighten nuts to 18 ft. lbs. (25 Nm).
58. Install rear crossmember and tighten bolts to 19 ft. lbs. (26 Nm).
59. Refill transfer case.
60. Install front exhaust pipe.
61. Reconnect battery ground cable.

Range Rover (2001–02)

1. Before servicing the vehicle, refer to the precautions in the beginning of this section.
2. Remove or disconnect the following:
   - Negative battery cable
   - Window switch pack
   - Hand brake cable clevis pin
   - Transmission and transfer case fluids
   - Front exhaust pipes from the manifolds
   - Hand brake cable from the grommet in the tunnel
   - Rear driveshaft shaft guard
   - Driveshafts
   - Gear selector cable from the transmission lever, then from the cable bracket.
   - Transfer case fluid temperature sensor electrical connector
   - High/Low motor and output shaft speed sensor electrical connectors
3. Attach a suitable transmission jack to the transfer case.
4. Remove the 6 transfer case–to–transmission mounting bolts.

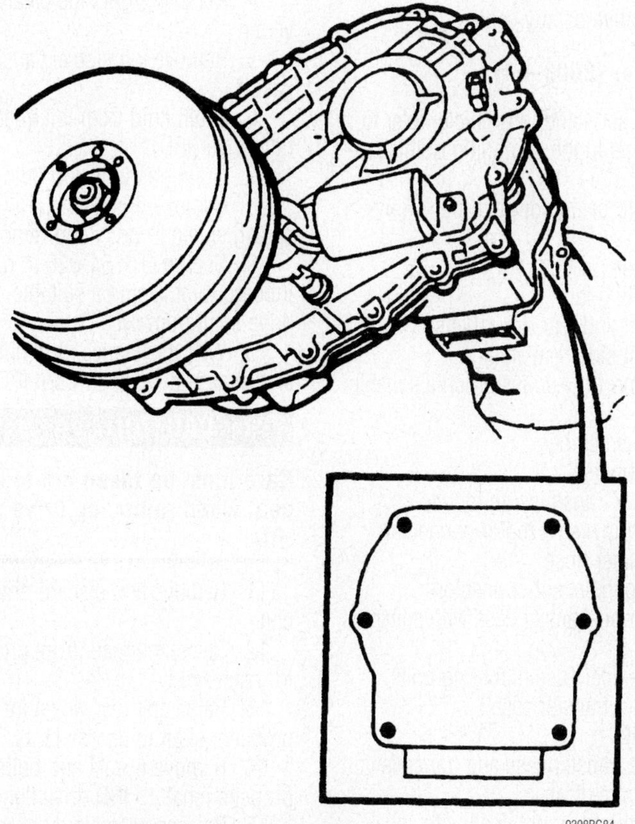

Transfer case bolt locations—Range Rover (2001–02)

5. Lower the transfer case form the vehicle.
6. Using a suitable seal removal tool, remove the transmission output seal.

To install:

7. Install a new transmission output seal, using a suitable seal installer.
8. Ensure transfer case and transmission mating faces are clean.
9. Lubricate the input shaft with transmission fluid.
10. Install the transfer case and tighten the mounting bolts to 33 ft. lbs. (45 Nm).
11. Remove the transmission jack.
12. Install or connect the following:
   - High/Low motor and output shaft speed sensor electrical connectors
   - Transfer case fluid temperature sensor electrical connector
   - Gear selector cable
   - Driveshafts, align the matchmarks and tighten the mounting bolts to 35 ft. lbs. (48 Nm).
   - Driveshaft guard
   - Hand brake cable through the grommet in the transmission tunnel
   - Exhaust system
   - Transmission and transfer case fluid
   - Handbrake cable
   - Window switch pack
   - Negative battery cable

Range Rover (2003–04)

1. Before servicing the vehicle, refer to the precautions in the beginning of this section.
2. Remove or disconnect the following:
   - Transfer case fluid
   - Cooling fan
   - Front and rear driveshafts
   - Right side heat shield
   - Support transmission on a suitable jack
   - Crossmember
   - Snubbers
   - Lower transmission for access
   - High/low shift motor connector
   - Breather hose
   - Oxygen sensor connectors
   - Support transfer case with suitable jack
   - 9 transfer case mounting bolts
   - Lower transfer case

To install:

3. Ensure transfer case and transmission mating faces are clean.
4. Lubricate the input shaft with anti-sieze.
5. Install the transfer case and tighten the mounting bolts to 32 ft. lbs. (43 Nm).
6. Remove the transmission jack.
7. Install or connect the following:
   - Oxygen sensor connectors
   - Breather hose
   - High/low shift motor connector
   - Snubbers
   - Crossmember, but do not tighten bolts
   - Raise transmission
   - Align crossmember and tighten bolts to 50 ft. lbs. (68 Nm).
   - Remove jack
   - Right side heat shield
   - Front and rear driveshafts
   - Cooling fan
   - Transfer case fluid
8. Start vehicle and check for leaks.

Intermediate Reduction Drive

REMOVAL & INSTALLATION

Freelander (2001–04)

1. Before servicing the vehicle, refer to the precautions in the beginning of this section.
2. Disconnect battery ground lead.
3. Remove rear beam.
4. Remove right side catalytic converter.
5. Remove left side exhaust manifold gasket.
6. Drain fluid from intermediate reduction drive (IRD).
7. Drain transmission fluid.
8. Remove 3 bolts securing right side splash shield to body and remove shield.
9. Secure LRT-54-026 to drive shaft inboard joint. Using a suitable lever, release drive shaft from IRD.
10. With assistance, pull hub outwards and release drive shaft from IRD.

CAUTION

Care must be taken not to damage oil seal when removing drive shaft from IRD.

11. Remove and discard drive shaft circlip.
12. Reference mark front propeller shaft for reassembly.
13. Raise one rear wheel for rotation of propeller shaft to access bolts.
14. Remove 6 nuts and bolts securing propeller shaft to IRD drive flange.
15. Release propeller shaft from IRD drive flange and tie shaft aside.

CAUTION

Care must be taken to support the Tripod joint when removed from the IRD unit. To avoid damage to boot or steel can, the joint should not be allowed to fully extend or be dropped.

16. Remove nut securing manifold heat shield to IRD unit.
17. Remove nut securing heat shield to IRD pinion housing.
18. Remove 2 bolts securing heat shield and remove heat shield.
19. Disconnect breather hose from IRD housing.
20. Position container to collect coolant spillage.
21. Release clips and disconnect coolant hoses from IRD.
22. Remove bolt securing engine lower steady to IRD support bracket.
23. Remove lower engine steady noting that 'TOP' mark on engine steady faces uppermost.
24. Remove 3 bolts securing IRD support bracket to oil pan.
25. Remove 2 bolts securing IRD support bracket to engine front mounting plate.
26. Remove 5 bolts securing support bracket to IRD.
27. Remove support bracket.
28. Remove 4 bolts securing IRD.
29. With assistance, release IRD from gearbox and remove.
30. Remove and discard O-ring from IRD.

To install:

31. Clean mating faces of IRD and gearbox.
32. Lubricate and fit new O-ring
33. With assistance, fit IRD.
34. Install bolts securing IRD to gearbox and tighten sufficiently only to pull mating faces of IRD and gearbox together at this stage.
35. Install IRD support bracket and tighten bolts sufficiently only to pull mating faces together.
36. Final tighten bolts securing IRD to gearbox to 59 ft. lbs. (80 Nm).
37. Final tighten bolts securing IRD support bracket in following sequence:
   - 5 bolts securing support bracket to IRD 37 ft. lbs. (50 Nm)
   - 2 bolts securing support bracket to engine front mounting bracket 37 ft. lbs. (50 Nm)
   - 3 bolts securing support bracket to oil pan 33 ft. lbs. (45 Nm)

38. Position lower engine steady, 'TOP' mark uppermost. Install bolt, but do not tighten at this stage.

39. Connect coolant hoses and secure with clips.

40. Connect breather hose to IRD housing.

41. Install manifold heat shield and fit nut securing heat shield to pinion housing finger tight.

42. Install bolts securing manifold heat shield to IRD support bracket and tighten to 7 ft. lbs. (9 Nm).

43. Install nut securing heat shield to IRD and tighten to 33 ft. lbs. (45 Nm).

44. Final tighten nut securing manifold heat shield to IRD pinion housing to 18 ft. lbs. (25 Nm).

45. Clean propeller shaft flange and mating face.

46. Install propeller shaft to IRD flange and align marks. Tighten nuts and bolts to 30 ft. lbs. (40 Nm).

47. Inspect drive shaft oil seal, renew if worn or damaged.

48. Clean drive shaft and flange splines.

49. Install new circlip to driveshaft.

50. With assistance, pull hub outwards, align drive shaft and fit to IRD taking care not to damage oil seal.

CAUTION

Pull the drive shaft to ensure the circlip is fully engaged and retains the shaft.

51. Install splash shield and secure with bolts.

52. Install exhaust front pipe.

53. Install rear beam.

54. Final tighten bolt securing lower engine steady to IRD support bracket to 74 ft. lbs. (100 Nm).

55. Fill IRD to correct level with fluid.

56. Install right side catalytic converter.

57. Install left side exhaust manifold gasket.

58. Fill gearbox with fluid.

59. Connect battery ground lead.

60. Refill cooling system.

Front Axle Swivel Hubs

REMOVAL & INSTALLATION

Range Rover (2001–02)

CAUTION

Before beginning, depressurize the air suspension.

1. Before servicing the vehicle, refer to the precautions in the beginning of this section.

2. Remove or disconnect the following:
   - Drive hub
   - Axle shaft
   - Track rod from the swivel hub
   - Drag link from the swivel hub
   - Swivel hub. Use a forcing tool, such as the one illustrated. If the joint pin turns in its socket, use a 6mm Allen wrench to hold it.
   - Adjusting collar from the hub

To install:

3. Install or connect the following:
   - New adjusting collar into the hub. Install the collar until a 4mm gap exists between the shoulder of the collar and the top of the hub.
   - Hub onto the axle. Install the upper nut, only. Torque the nut to 81 ft. lbs. (110 Nm).

4. Clean the seal surface in the axle.

5. Turn the clamp screw of tool LRT-54-006/1 fully counterclockwise. Make sure that the clamp toggle rotates freely. Position the tool in the axle with the **TOP** mark upwards.

6. Make sure that the tool is located squarely on the sealing surface. Use a plastic or brass mallet to tap the end of the screw. This will ensure that the tool is properly positioned.

7. Install and tighten the lower nut, until the ball stud is squarely seated in the joint but the collar can still turn. Adjust the height of the collar until the tool is a slide-fit in the hub.

8. Remove the tool. Tighten the collar 1¼ turn while tightening the lower nut to 100 ft. lbs. (135 Nm). The hub should turn smoothly and evenly. If not, repeat the assembly procedure.

9. Install or connect the following:
   - Drag link. Torque the nut to 59 ft. lbs. (80 Nm).
   - Track rod. Torque the nut to 59 ft. lbs. (80 Nm).
   - Axle shaft
   - Drive hub

Front Axle Shaft CV-Joints

REMOVAL & INSTALLATION

Freelander (2001–04)

1. Before servicing the vehicle, refer to the precautions in the beginning of this section.

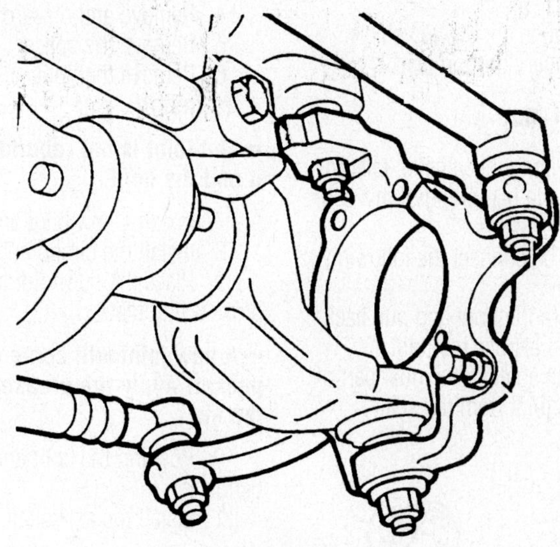

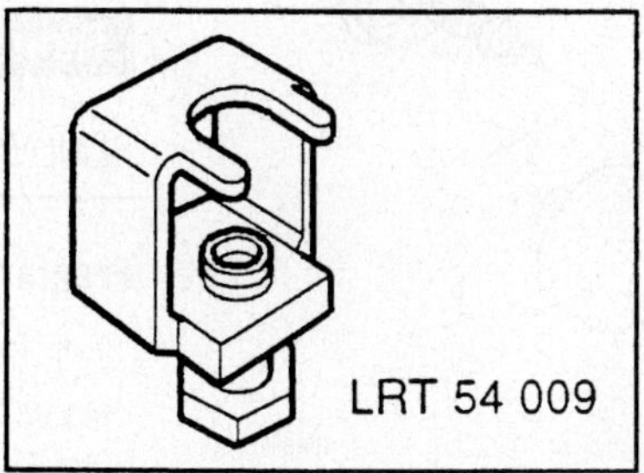

9308RG74

Front axle swivel hub. The inset shows the ball stud removal tool—Range Rover (2001–02)

2. Disconnect battery ground lead.
3. Raise front of vehicle.

⁂ WARNING

Do not work on or under a vehicle supported only by a jack. Always support the vehicle on safety stands.

4. Remove road wheel.
5. Remove drive shaft.
6. Place drive shaft in vise.
7. Release both boot clips and discard.
8. Slide boot along shaft to gain access to outer joint.
9. Using a suitable drift against the inner part of the joint, remove joint from shaft.
10. Inspect boot for damage and renew if necessary.

To install:

11. Clean drive shaft and boot.
12. Fit new circlip to drive shaft.
13. Position outer joint to shaft, use a screwdriver to press circlip into its groove and push joint fully onto shaft.
14. Apply grease from the sachet to the joint.
15. Position boot to joint and use a 'Band-it Thriftool' LRT-99-019 to secure the 2 new clips.
16. Fit drive shaft.
17. Fit front road wheels, fit and tighten nuts to 85 ft. lbs. (115 Nm).
18. Remove stands and lower vehicle.
19. Connect battery ground lead.

Discovery (2001–04)

1. Before servicing the vehicle, refer to the precautions in the beginning of this section.
2. Remove or disconnect the following:
   - Axle shaft
   - Bands from the boot and pull back the boot to expose the joint.
3. Using a drift against the inner part of the joint, drive the joint from the shaft.

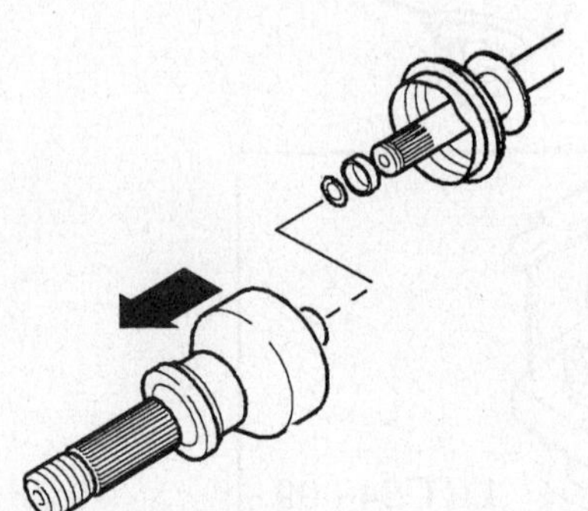

Front axle shaft CV-joint—Discovery (2001–04)

4. Remove and discard the circlip.
5. Remove the spacer.
6. Remove the boot.

To install:

➡The joint is not rebuildable. Replace it and the boot.

7. Install a new boot and bands.
8. Install the circlip and spacer.
9. Place the new joint into position and press it into place.

➡A new joint will come with a grease packet. Apply the grease as directed in the kit.

10. Position the boot and secure the bands.
11. Install the axle shaft.

Range Rover (2003–04)

➡The same procedure is used for the inner and outer joints.

1. Before servicing the vehicle, refer to the precautions in the beginning of this section.
2. Remove or disconnect the following:
   - Axle shaft
   - Place axle shaft in a vise
   - Bands from the boot and pull back the boot to expose the joint.
3. Using a drift against the inner part of the joint, drive the joint from the shaft.
4. Remove and discard the circlip.
5. Remove the spacer.
6. Remove the boot.

To install:

➡The joint is not rebuildable. Replace it and the boot.

7. Install a new boot and bands.
8. Install the circlip and spacer.
9. Place the new joint into position and press it into place.

➡A new joint will come with a grease packet. Apply the grease as directed in the kit.

10. Position the boot and secure the bands.
11. Install the axle shaft.

Rear Axle Shaft CV-Joints

REMOVAL & INSTALLATION

Freelander (2001–04)

1. Before servicing the vehicle, refer to the precautions in the beginning of this section.
2. Disconnect battery ground lead.
3. Raise rear of vehicle.

⁂ WARNING

Do not work on or under a vehicle supported only by a jack. Always support the vehicle on safety stands.

4. Remove road wheel.
5. Remove drive shaft.
6. Place drive shaft in vise.
7. Release both boot clips and discard.
8. Slide boot along shaft to gain access to outer joint.
9. Using a suitable drift against the inner part of the joint, remove joint from shaft.
10. Inspect boot for damage and renew if necessary.
11. Remove and discard circlip from drive shaft.

To install:

12. Clean drive shaft and boot.
13. Fit new circlip to drive shaft.
14. Position outer joint to shaft, use a screwdriver to press circlip into its groove and push joint fully onto shaft.
15. Apply grease from the sachet to the joint.
16. Position boot to joint and use a 'Band-it Thriftool' LRT-99-019 to secure the 2 new clips.
17. Fit drive shaft.
18. Fit road wheel(s) and tighten nuts to 85 ft. lbs. (115 Nm).
19. Remove stands and lower vehicle.
20. Connect battery ground lead.

Range Rover (2003–04)

➡The same procedure is used for the inner and outer joints.

1. Before servicing the vehicle, refer to the precautions in the beginning of this section.
2. Remove or disconnect the following:
   - Axle shaft
   - Place axle shaft in a vise
   - Bands from the boot and pull back the boot to expose the joint.
3. Using a drift against the inner part of the joint, drive the joint from the shaft.
4. Remove and discard the circlip.
5. Remove the spacer.
6. Remove the boot.

To install:

➡The joint is not rebuildable. Replace it and the boot.

7. Install a new boot and bands.
8. Install the circlip and spacer.
9. Place the new joint into position and press it into place.

➡A new joint will come with a grease packet. Apply the grease as directed in the kit.

10. Position the boot and secure the bands.
11. Install the axle shaft.

Front Axle Drive Hub

REMOVAL & INSTALLATION

Freelander (2001–04)

1. Before servicing the vehicle, refer to the precautions in the beginning of this section.
2. Raise front of vehicle.

WARNING

Do not work on or under a vehicle supported only by a jack. Always support the vehicle on safety stands.

3. Remove road wheel.
4. Release stake in drive shaft nut.
5. With an assistant applying brakes, remove and discard drive shaft hub nut.
6. Remove clip securing right side brake hose to support bracket, release hose from bracket.
7. Release ABS sensor and pad wear sensor harnesses from bracket.
8. Remove 2 bolts securing brake caliper to hub. Release caliper from hub and tie aside.

CAUTION

Do not allow caliper to hang on brake hose.

9. Mark brake disc to hub relationship.
10. Remove 2 screws securing brake disc and remove brake disc.
11. Remove 2 nuts and bolts securing hub to shock absorber.
12. Release hub from shock absorber.
13. Release drive shaft from hub.
14. Restrain hub from rotating and remove nut from lower swivel joint.
15. Break taper joint using LRT-57-043.
16. Remove swivel hub.

To install:

17. Clean drive shaft.
18. Fit hub assembly to lower joint, fit new nut and tighten to 48 ft. lbs. (65 Nm).
19. Fit drive shaft to hub.
20. Fit hub to shock absorber, fit nuts and bolts and tighten to 151 ft. lbs. (205 Nm).
21. Clean brake disc to drive flange mating faces.
22. Fit disc to drive flange, align reference marks, fit screws and tighten to 4 ft. lbs. (5 Nm).
23. Clean mating faces of caliper and hub.
24. Position caliper to brake disc fit bolts and tighten to 74 ft. lbs. (100 Nm).
25. Fit brake hose to abutment bracket and fit clip.
26. Clean ABS sensor, smear sensor with an anti-seize grease and fit sensor to hub.

CAUTION

Ensure ABS sensor is fully located into hub, so that sensor touches pole wheel teeth.

27. Fit ABS sensor lead to bracket.
28. Fit new drive shaft nut and tighten to 295 ft. lbs. (400 Nm). Stake nut to shaft.
29. Stake drive shaft hub nut.
30. Fit road wheel(s) and tighten nuts to 85 ft. lbs. (115 Nm).
31. Remove stands and lower vehicle.

Discovery (2001–02)

1. Before servicing the vehicle, refer to the precautions in the beginning of this section.

CAUTION

If so equipped, before beginning, depressurize the air suspension.

2. Remove the center cap from the wheel.
3. Loosen the axle shaft nut.
4. Remove or disconnect the following:
   - Wheel
   - Brake rotor shield
   - Brake rotor
   - ABS sensor harness from the brackets
   - Sensor from the hub
   - Sensor bushing
   - 4 bolts securing the hub to the carrier
   - Axle shaft nut
   - Drive hub

To install:

5. Apply anti-seize compound to the mating surfaces of the hub and knuckle.
6. Install or connect the following:
   - Hub onto the shaft. Install, but don't tighten, the nut.
   - Hub–to–carrier bolts. Torque to 74 ft. lbs. (100 Nm).
   - ABS sensor bushing and sensor, coated with silicone grease
   - Wiring harness to brackets
   - Rotor
   - Rotor shield
   - Wheel
7. Lower the vehicle and torque the axle shaft nut to 360 ft. lbs. (490 Nm). Stake the nut.

Discovery (2003–04)

1. Before servicing the vehicle, refer to the precautions in the beginning of this section.
2. Remove ABS sensor harness grommet from inner fender valance and disconnect multiplug.
3. Raise front of vehicle.

WARNING

Do not work on or under a vehicle supported only by a jack. Always support vehicle on safety stands.

4. Remove front wheel.
5. Pull ABS sensor harness up through opening in wheel well.
6. Release harness from brackets on inner fender panel, shock tower, and front hub.
7. Remove drive shaft nut. Discard nut.
8. Remove front brake caliper and rotor.
9. Remove 4 bolts securing front drive hub to steering knuckle, separate hub from knuckle, then remove front wheel hub and drive shaft assembly from vehicle.

CAUTION

Do not remove ABS sensor from hub. Hub and sensor are supplied as an assembly.

10. Position wheel hub and drive shaft on a press. Place support beneath wheel stud heads and press drive shaft from wheel hub.
11. Remove drive shaft oil seal from axle casing.

To install:

12. Clean drive shaft oil seal recess in axle casing. Clean drive shaft splines, oil seal running surface, wheel hub, and steering knuckle mating faces. Clean ABS sensor and sensor recess.
13. Lubricate lip of new drive shaft oil seal and running surface of drive shaft with clean differential oil.
14. Use seal installer LRT-54-012 to install new oil seal to axle casing.
15. Position drive shaft in axle casing.
16. Apply anti-seize compound to wheel hub and steering knuckle mating faces.
17. Apply a 0.125 inch (3mm) wide bead of sealant STC 50554 around drive shaft

circumference at middle beveled edge out from flange.

18. Ensure ABS harness is located in cutout in steering knuckle.

19. Fit wheel hub to drive shaft and align steering knuckle. Sealant will smear along length of splines as wheel hub is fitted to drive shaft.

20. Install wheel hub bolts and tighten to 74 ft. lbs. (100 Nm).

21. Install new drive shaft nut and tighten lightly.

22. Install front brake rotor and caliper.

23. With assistance, final-tighten drive shaft nut to 360 ft. lbs. (490 Nm). Stake nut.

➡Drive shaft nut MUST be tightened before sealant has cured.

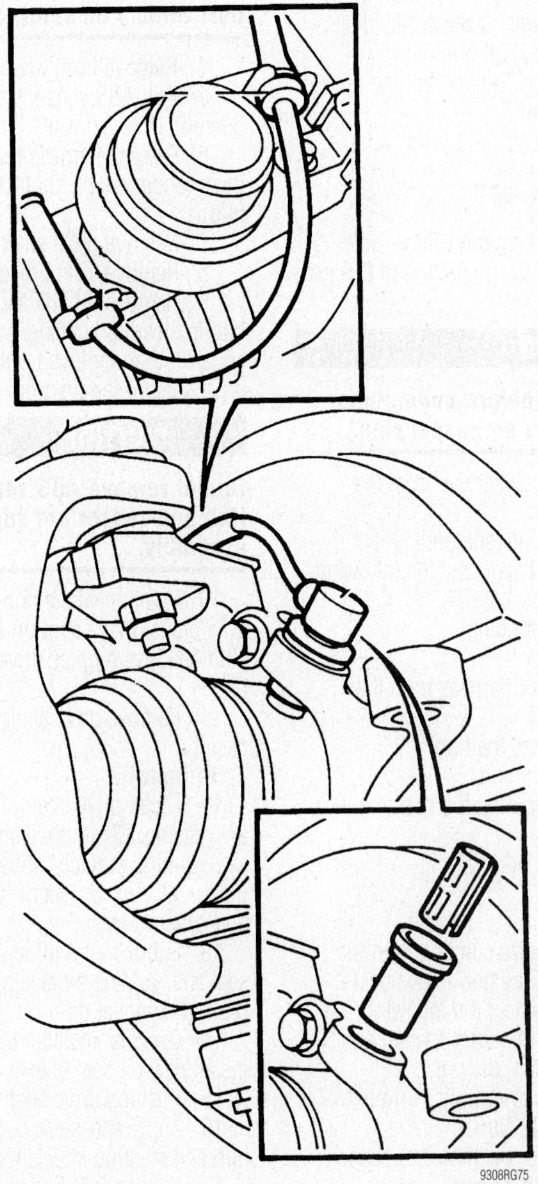

Front axle ABS sensor installation—Range Rover

24. Secure ABS harness to brackets and install grommet to inner fender panel.

25. Install road wheel and tighten nuts to 103 ft. lbs. (140 Nm).

26. Remove safety stands and lower vehicle.

27. Connect ABS sensor connector.

Range Rover (2001–02)

✻✻ CAUTION

If so equipped, before beginning, depressurize the air suspension.

1. Before servicing the vehicle, refer to the precautions in the beginning of this section.

2. Remove the center cap from the wheel.

3. Loosen the axle shaft nut.

4. Remove or disconnect the following:
   - Wheel
   - Brake rotor shield
   - ABS sensor harness from the brackets
   - Sensor from the hub
   - Sensor bushing
   - 4 bolts securing the hub to the carrier
   - Axle shaft nut
   - Drive hub

To install:

5. Install or connect the following:
   - Hub onto the shaft. Install, but don't tighten, the nut.
   - Hub–to–carrier bolts. Torque to 100 ft. lbs. (135 Nm).
   - ABS sensor bushing and sensor, coated with silicone grease
   - Wiring harness to brackets
   - Rotor shield
   - Wheel

6. Lower the vehicle and torque the axle shaft nut to 192 ft. lbs. (260 Nm). Stake the nut.

Rear Axle Drive Hub

REMOVAL & INSTALLATION

Freelander (2001–04)

1. Before servicing the vehicle, refer to the precautions in the beginning of this section.

2. Raise rear of vehicle.

✻✻ WARNING

Do not work on or under a vehicle supported only by a jack. Always support the vehicle on safety stands.

3. Remove road wheel.

4. With assistant depressing the brake pedal, remove and discard drive shaft nut.

5. Remove brake shoe assembly.

6. Using LRT-70-007 release handbrake cable from backplate and remove from backplate.

7. Remove clip securing brake hose to bracket on shock absorber.

8. Disconnect brake pipe union from wheel cylinder.

✻✻ CAUTION

Always fit plugs to open connections to prevent contamination.

9. Release ABS sensor from hub.
10. Remove nut and bolt securing trailing link to hub.
11. Remove nut, bolt and washers securing transverse links to hub.
12. Remove 2 nuts and bolts securing hub to shock absorber.
13. Release shock absorber from hub.
14. Remove hub assembly from drive shaft.

To install:

15. Install hub assembly to drive shaft.
16. Install hub to shock absorber and tighten bolts to 151 ft. lbs. (205 Nm).

CAUTION

Nuts and bolts must be tightened with weight of vehicle on suspension.

17. Install nut, bolt and washers securing transverse links to hub and tighten nut to 89 ft. lbs. (120 Nm).
18. Install trailing link to hub and tighten nut and bolt to 89 ft. lbs. (120 Nm).

➡Ensure that washers are fitted to both ends of bolts

19. Clean ABS sensor, smear sensor with an anti-seize grease and fit sensor to hub.

CAUTION

Ensure ABS sensor is fully located into hub, so that sensor touches pole wheel teeth.

20. Install brake pipe to wheel cylinder and tighten union to 10 ft. lbs. (14 Nm).
21. Install clip securing brake pipe to bracket.
22. Install and secure handbrake cable to backplate.
23. Install brake shoes.
24. Install new drive shaft nut and tighten to 295 ft. lbs. (400 Nm). Stake nut to shaft.
25. Bleed brakes.
26. Install road wheel(s) and tighten nuts to 85 ft. lbs. (115 Nm).
27. Remove stands and lower vehicle.

Discovery (2003–04)

1. Before servicing the vehicle, refer to the precautions in the beginning of this section.

WARNING

If model is equipped with SLS, ensure air suspension system is made safe before commencing work; otherwise, chassis may lower onto bump stops during repair.

2. Raise rear of vehicle.

WARNING

Do not work on or under a vehicle supported only by a jack. Always support vehicle on safety stands.

3. Remove road wheel.
4. Remove drive shaft nut.
5. Remove rear brake rotor.
6. Disconnect ABS sensor multiplug.
7. Release harness from brake hose and bracket.
8. Remove 4 bolts holding wheel hub to axle flange.
9. Remove wheel hub and drive shaft from axle housing.
10. Remove and discard O-ring from wheel hub.
11. Position wheel hub and drive shaft on a press. Place support beneath wheel stud heads and press drive shaft from wheel hub.

To install:

12. Clean drive shaft splines, wheel hub and axle mating faces, ABS sensor and sensor recess.
13. Install drive shaft partially into axle housing.
14. Lubricate new bearing hub O-ring with clean differential oil. Install O-ring to wheel hub.
15. Apply a 0.125 inch (3mm) wide bead of sealant STC 50554 around drive shaft circumference at middle beveled edge out from flange.
16. Position wheel hub to drive shaft and align to axle. Sealant will smear along length of splines as wheel hub is fitted to drive shaft.
17. Install wheel hub–to–axle bolts and tighten to 74 ft. lbs. (100 Nm).
18. Install new drive shaft nut. Lightly tighten at this time.
19. Install rear brake rotor.
20. With an assistant, tighten drive shaft nut to 360 ft. lbs. (490 Nm). Stake nut.

➡Drive shaft nut must be tightened before sealant is cured.

21. Connect ABS sensor multiplug and secure harness to bracket and brake hose.
22. Install road wheel. Tighten nuts to 103 ft. lbs. (140 Nm).
23. Remove safety stands and lower vehicle.

Front Axle Halfshafts

REMOVAL & INSTALLATION

Freelander (2001–04)

LEFT SIDE

1. Before servicing the vehicle, refer to the precautions in the beginning of this section.
2. Disconnect battery ground lead.
3. Raise front of vehicle.

WARNING

Do not work on or under a vehicle supported only by a jack. Always support the vehicle on safety stands.

4. Remove road wheel.
5. Remove underbelly panel.
6. Remove 3 bolts and remove splash shield.
7. Release stake from drive shaft nut.
8. With assistant depressing the brake pedal, remove and discard the drive shaft nut.
9. Remove clip securing brake hose and release hose from shock absorber bracket.
10. Release ABS sensor harness and brake hose from shock absorber.
11. Release ABS sensor from hub.
12. Remove 2 bolts securing brake caliper to hub.

Release caliper from hub and tie aside.

CAUTION

Do not allow caliper to hang on brake hose.

13. Remove 2 bolts securing shock absorber to hub.
14. Release hub from shock absorber.
15. Release drive shaft from hub.
16. Position container to catch oil spillage
17. Secure LRT-54-026 to drive shaft inboard joint. Using a suitable lever, release drive shaft from gearbox.
18. Remove drive shaft.
19. Remove and discard circlip from drive shaft.

To install:

20. Inspect gearbox seal for signs of wear or damage.
21. Wipe drive shaft ends, gearbox oil seal and hub.
22. Lubricate oil seal running surfaces.
23. Install new circlip to drive shaft.
24. Install drive shaft ensuring circlip is fully engaged.

CAUTION

Drive shaft must be fitted with care to prevent damage to gearbox oil seal.

25. Position drive shaft in hub.
26. Install new hub nut but do not tighten at this stage.
27. Install hub to shock absorber, fit nuts and bolts and tighten to 150 ft. lbs. (205 Nm).
28. Position caliper to brake disc fit bolts and tighten to 74 ft. lbs. (100 Nm).
29. Clean ABS sensor and mating face.
30. Lubricate ABS sensor with anti-seize grease.
31. Install ABS sensor .

CAUTION

Ensure ABS sensor is fully located into hub, so that sensor touches reluctor ring.

32. Position ABS harness and brake hose in bracket and secure with clip.
33. Install splash shield and secure with bolts.
34. Tighten front hub nut to 295 ft. lbs. (400 Nm).
35. Stake front hub nut.
36. Install road wheel(s) and tighten nuts to 85 ft. lbs. (115 Nm).
37. Check and top up oil level as required.
38. Install underbelly panel.
39. Remove stands and lower vehicle.
40. Connect battery ground lead.

RIGHT SIDE

1. Before servicing the vehicle, refer to the precautions in the beginning of this section.
2. Disconnect battery ground lead.
3. Raise front of vehicle.

WARNING

Do not work on or under a vehicle supported only by a jack. Always support the vehicle on safety stands.

4. Remove road wheel.
5. Remove underbelly panel.
6. Remove 3 bolts securing right side splash shield to body and remove shield.
7. Release stake from drive shaft nut.
8. With assistant depressing the brake pedal, remove and discard the drive shaft nut.
9. Remove clip securing brake hose and release hose from shock absorber bracket.
10. Release ABS sensor harness and brake hose from shock absorber.
11. Release ABS sensor from hub.
12. Remove 2 bolts securing brake caliper to hub.

Release caliper from hub and tie aside.

CAUTION

Do not allow caliper to hang on brake hose.

13. Remove 2 bolts securing shock absorber to hub.
14. Release hub from shock absorber.
15. Release drive shaft from hub.
16. Position container to catch oil spillage.
17. Secure LRT-54-026 to drive shaft inboard joint.

Using a suitable lever, release drive shaft from gearbox.

18. Remove drive shaft.

CAUTION

Care must be taken not to damage oil seal when removing drive shaft from gearbox

19. Remove and discard circlip from drive shaft.

To install:

20. Inspect gearbox seal for signs of wear or damage.
21. Wipe drive shaft ends, gearbox oil seal and hub.
22. Lubricate oil seal running surfaces.
23. Install new circlip to drive shaft.
24. Install drive shaft ensuring circlip is fully engaged.

CAUTION

Drive shaft must be fitted with care to prevent damage to gearbox oil seal.

25. Position drive shaft in hub.
26. Install new hub nut but do not tighten at this stage.
27. Install hub to shock absorber, fit nuts and bolts and tighten to 150 ft. lbs. (205 Nm).
28. Position caliper to brake disc fit bolts and tighten to 74 ft. lbs. (100 Nm).
29. Clean ABS sensor and mating face.
30. Lubricate ABS sensor with anti-seize grease.
31. Install ABS sensor .

CAUTION

Ensure ABS sensor is fully located into hub, so that sensor touches reluctor ring.

32. Position ABS harness and brake hose in bracket and secure with clip.
33. Install splash shield and secure with bolts.
34. Tighten front hub nut to 295 ft. lbs. (400 Nm).
35. Stake front hub nut.
36. Install road wheel(s) and tighten nuts to 85 ft. lbs. (115 Nm).
37. Check and top up oil level as required.
38. Install underbelly panel.
39. Remove stands and lower vehicle.
40. Connect battery ground lead.

Range Rover (2003–04)

FRONT

1. Before servicing the vehicle, refer to the precautions in the beginning of this section.
2. Raise and safely support the vehicle and remove the front wheels.
3. Remove the axle hub nut.
4. Unbolt the brake caliper and position it aside, leaving the hose attached.
5. Remove the brake rotor.
6. Disconnect the ABS sensor lead, then remove the ABS sensor from the hub.
7. Remove and discard the nut attaching the track rod ball joint to the steering arm.
8. Install an M14 nut to the ball pin so it is flush with the end of the pin.
9. Using a separator, disconnect the ball pin from the steering arm.
10. Remove the M14 nut.
11. Disconnect the height sensor link.
12. Loosen the lower arm-to-subframe bolt a quarter turn.
13. Remove the lower arm ball joint-to-hub nut.
14. Release the lower ball joint and disconnect the lower arm from the hub.
15. Loosen the tie rod-to-subframe bolt one quarter turn.
16. Remove the tie rod-to-ball joint nut.
17. Disconnect the tie rod from the ball joint.
18. Install tool set LRT-60-030 to the front hub.

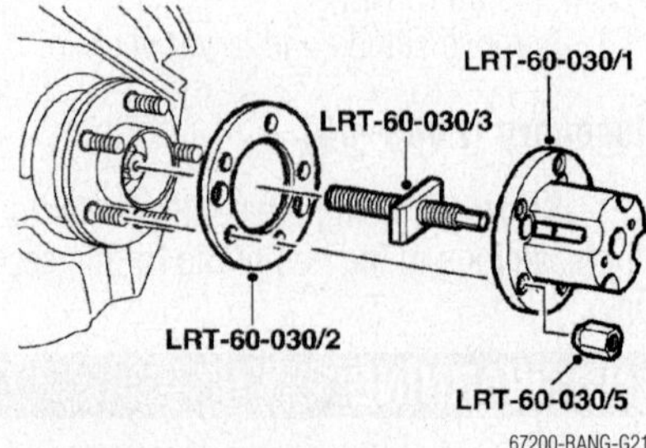

Install tool set to the front hub—Range Rover (2003–04)

19. Tighten the center screw on the tool to press the driveshaft from the flange.
20. Remove the tool set.
21. Install collar LRT-54-026 to the driveshaft inside joint.
22. Using levers, release the driveshaft from the hub.
23. Remove the driveshaft.

To install:

24. Clean driveshaft and differential opening.
25. Install new circlip to the driveshaft.
26. Lubricate the oil seal lip with clean differential oil.
27. Install the driveshaft ensuring the circlip is engaged.
28. Clean the driveshaft and flange splines and lightly lubricate the splines.
29. Position the drive shaft in the hub.
30. Reinstall the tool set to the front hub.
31. Tighten the tool until the driveshaft is seated in the drive flange.
32. Using new nuts, install the ball joint to steering arm, tie rod to ball joint and ball joint to hub. Tighten all nuts to 59 ft. lbs. (80 Nm).
33. Install new bolts in lower arm to subframe and tie rod to subframe, but do not tighten.
34. Connect height sensor link.
35. Install ABS sensor after coating it with anti-sieze.
36. Attach ABS sensor lead.
37. Install the brake rotor.
38. Install a new axle hub nut and tighten to 311 ft. lbs. (420 Nm).
39. Stake the nut to the shaft.
40. Install the brake caliper.
41. Install the road wheels and lower the vehicle.
42. Tighten the tie rod-to-subframe and lower arm-to-subframe bolts to 121 ft. lbs. (165 Nm), plus an additional 90 degrees.
43. Check the front wheel alignment.

Rear Axle Halfshafts

REMOVAL & INSTALLATION

Freelander (2001–04)

1. Before servicing the vehicle, refer to the precautions in the beginning of this section.
2. Disconnect battery ground lead.
3. Raise rear of vehicle.

⁂ WARNING

Do not work on or under a vehicle supported only by a jack. Always support the vehicle on safety stands.

4. Remove road wheel.
5. Release stake from drive shaft nut.
6. With assistant depressing the brake pedal, remove and discard drive shaft nut.
7. Remove nut and bolt securing trailing link to rear hub, collect spacer from under bolt head.
8. Remove nut and bolt securing fixed transverse link to subframe. Collect dynamic shock absorber.
9. Remove nut and bolt securing adjustable transverse link to subframe.
10. Position container to catch oil spillage.
11. With assistance pull hub assembly outwards and release drive shaft outer joint from hub assembly.
12. Taking care not to damage 'Flinger', release drive shaft inner joint from differential using LRT-51-014 and remove drive shaft.
13. Remove and discard drive shaft circlip.

To install:

14. Inspect differential seal, renew if worn or damaged.
15. Clean ends of drive shaft and locations in hub and differential.
16. Lubricate oil seal running surface with transmission oil.
17. Install new circlip to drive shaft.
18. Install drive shaft to differential and push fully home.

⁂ CAUTION

Pull the drive shaft to ensure the circlip is fully engaged and retains the shaft.

19. With assistance fit drive shaft to hub.
20. Install new drive shaft nut but do not tighten at this stage.
21. Install nut, bolt and dynamic shock absorber to adjustable transverse link and tighten to 89 ft. lbs. (120 Nm).

⁂ CAUTION

Nuts and bolts must be tightened with the weight of the vehicle on the suspension.

22. Install nut and bolt to fixed transverse link and tighten to 89 ft. lbs. (120 Nm).
23. Install spacer, nut and bolt to trailing link and tighten to 89 ft. lbs. (120 Nm).
24. Install new drive shaft nut and tighten to 295 ft. lbs. (400 Nm). Stake nut to shaft.
25. Check and top up oil level.
26. Install road wheel(s) and tighten nuts to 85 ft. lbs. (115 Nm).
27. Remove stands and lower vehicle.
28. Connect battery ground lead.

Range Rover (2003–04)

1. Before servicing the vehicle, refer to the precautions in the beginning of this section.
2. Disconnect battery ground lead.
3. Raise rear of vehicle.

⁂ WARNING

Do not work on or under a vehicle supported only by a jack. Always support the vehicle on safety stands.

4. Remove road wheel.
5. Remove 6 bolts attaching the driveshaft to differential.
6. Remove the axle shaft nut.
7. Remove the brake caliper and wire aside.
8. Remove the brake rotor.
9. Install tool set LRT-60-030 to the front hub. See Front Axle Shaft procedure for instruction.
10. Press the driveshaft from the hub.
11. Support the lower control arm with a jack.
12. Remove the lower control arm-to-hub bolt.
13. Lower the jack.
14. Pull the driveshaft outward from the differential.
15. Remove the driveshaft.

To install:

16. Inspect differential seal, renew if worn or damaged.
17. Clean driveshaft and differential opening.
18. Clean the driveshaft and flange splines and lightly lubricate the splines.
19. Position the drive shaft in the differential and the hub.
20. Align the hub to the lower arm, install and lightly tighten the bolt.
21. Reinstall the tool set to the rear hub.
22. Tighten the tool until the driveshaft is seated in the drive flange.
23. Remove the tool set.
24. Install the brake rotor and the caliper.
25. Install a new axle hub nut and tighten to 311 ft. lbs. (420 Nm).
26. Stake the nut to the shaft.
27. Using new bolts, attach the plates and bolts attaching the driveshaft to differential and tighten to 30 ft. lbs. (40 Nm), plus an additional 60 degrees.
28. Install the wheels.
29. Lower the vehicle and tighten the

lower control arm-to-hub bolt to 184 ft. lbs. (250 Nm).

Front Axle Shaft and Seal

REMOVAL & INSTALLATION

Discovery (2001–02)

1. Before servicing the vehicle, refer to the precautions in the beginning of this section.

CAUTION

If so equipped, before beginning, depressurize the air suspension.

2. Remove the center cap from the wheel.
3. Loosen the axle shaft nut.
4. Remove or disconnect the following:
   - Wheel
   - Brake rotor shield
   - Brake rotor
   - ABS sensor harness from the brackets
   - Sensor from the hub
   - Sensor bushing
   - 4 bolts securing the hub to the carrier
   - Axle shaft nut
   - Drive hub
   - Axle shaft
   - Oil Seal

To install:

5. Clean all mating surfaces thoroughly.
6. Apply gear oil to the new seal's lip and outer edge.
7. Drive the new seal into place.
8. Slide the axle shaft into place.
9. Apply anti-seize compound to the mating surfaces of the hub and knuckle.
10. Install or connect the following:
    - Hub onto the shaft. Install nut, but do not tighten at this time.
    - Hub–to–carrier bolts. Torque to 74 ft. lbs. (100 Nm).
    - ABS sensor bushing and sensor, coated with silicone grease
    - Wiring harness to brackets
    - Rotor
    - Rotor shield
    - Wheel
11. Lower the vehicle and torque the axle shaft nut to 360 ft. lbs. (490 Nm). Stake the nut.

Discovery (2003–04)

1. Before servicing the vehicle, refer to the precautions in the beginning of this section.
2. Raise and support front of vehicle.

WARNING

Never work on or under vehicle supported only by a jack. Always use safety stands.

3. Remove road wheel.
4. Remove front brake rotor.
5. Disconnect ABS sensor in-line connector.
6. Release ABS harness from brake hose and bracket.
7. Remove 4 bolts holding wheel hub to steering knuckle.
8. Remove wheel hub and drive axle from axle housing.
9. With a suitable pry tool, carefully remove oil seal from axle housing.

To install:

10. Clean drive shaft oil seal recess, oil seal running surface, wheel hub and swivel hub mating faces.
11. Lubricate oil seal lip and running surface with clean differential oil.
12. Using seal installer tools LRT-54-012 and LRT-99-003, install new oil seal into axle housing.
13. Apply anti-seize compound to wheel hub and steering knuckle mating faces.
14. Ensure ABS harness is properly located in steering knuckle cutout.
15. Install drive shaft and wheel hub into axle housing. Align wheel hub with steering knuckle. Install hub bolts and tighten to 74 ft. lbs. (100 Nm).
16. Connect ABS sensor multiplug and secure harness to bracket and brake hose.
17. Install front brake rotor.
18. Install road wheel and tighten nuts to 103 ft. lbs. (140 Nm).
19. Remove stands and lower vehicle.
20. Check differential oil level.

Range Rover (2001–02)

1. Before servicing the vehicle, refer to the precautions in the beginning of this section.

CAUTION

Before beginning, depressurize the air suspension.

➡If you're not separating the hub and shaft, don't loosen the axle shaft nut.

2. Remove the center cap from the wheel.
3. Loosen the axle shaft nut.
4. Remove or disconnect the following:
   - Wheel
   - Brake rotor shield
   - ABS sensor harness from the brackets
   - Sensor from the hub
   - Sensor bushing
   - 4 bolts securing the hub to the carrier
   - Axle shaft nut, if loosened
   - Drive hub and axle shaft
   - Seal from the axle tube

To install:

5. Clean all surfaces.
6. Lubricate the new seal lip and running surface with silicone grease. Drive a new seal into place.
7. Install or connect the following:
   - Axle shaft into the axle tube
   - Hub onto the shaft. Install, but don't tighten, the nut.
   - Hub–to–carrier bolts. Torque to 100 ft. lbs. (135 Nm).
   - ABS sensor bushing and sensor, coated with silicone grease
   - Wiring harness to brackets
   - Rotor shield
   - Wheel
8. Lower the vehicle and torque the axle shaft nut to 192 ft. lbs. (260 Nm). Stake the nut.

Rear Axle Shaft and Bearing

REMOVAL & INSTALLATION

Discovery (2001–02)

1. Before servicing the vehicle, refer to the precautions in the beginning of this section.

CAUTION

If so equipped, before beginning, depressurize the air suspension.

2. Remove the center cap from the wheel.
3. Loosen the axle shaft nut.
4. Remove or disconnect the following:
   - Wheel
   - Brake rotor shield
   - Brake rotor
   - ABS sensor harness from the brackets
   - Sensor from the hub
   - Sensor bushing
   - 4 bolts securing the hub to the carrier
   - Axle shaft nut
   - Hub/bearing assembly
   - Axle shaft
   - Oil Seal

To install:

5. Clean all mating surfaces thoroughly.
6. Apply gear oil to the new seal's lip and outer edge.

7. Drive the new seal into place.
8. Slide the axle shaft into place.
9. Apply anti-seize compound to the mating surfaces of the hub and knuckle.
10. Install or connect the following:
  - Hub onto the shaft. Install, but don't tighten, the nut.
  - Hub–to–axle carrier bolts. Torque to 74 ft. lbs. (100 Nm).
  - ABS sensor bushing and sensor, coated with silicone grease
  - Wiring harness to brackets
  - Rotor
  - Rotor shield
  - Wheel
11. Lower the vehicle and torque the axle shaft nut to 360 ft. lbs. (490 Nm). Stake the nut.

Range Rover (2001–02)

1. Before servicing the vehicle, refer to the precautions in the beginning of this section.

CAUTION

Before beginning, depressurize the air suspension.

➡If you're not separating the hub and shaft, don't loosen the axle shaft nut.

2. Remove the center cap from the wheel.
3. Loosen the axle shaft nut.
4. Remove or disconnect the following:
  - Wheel

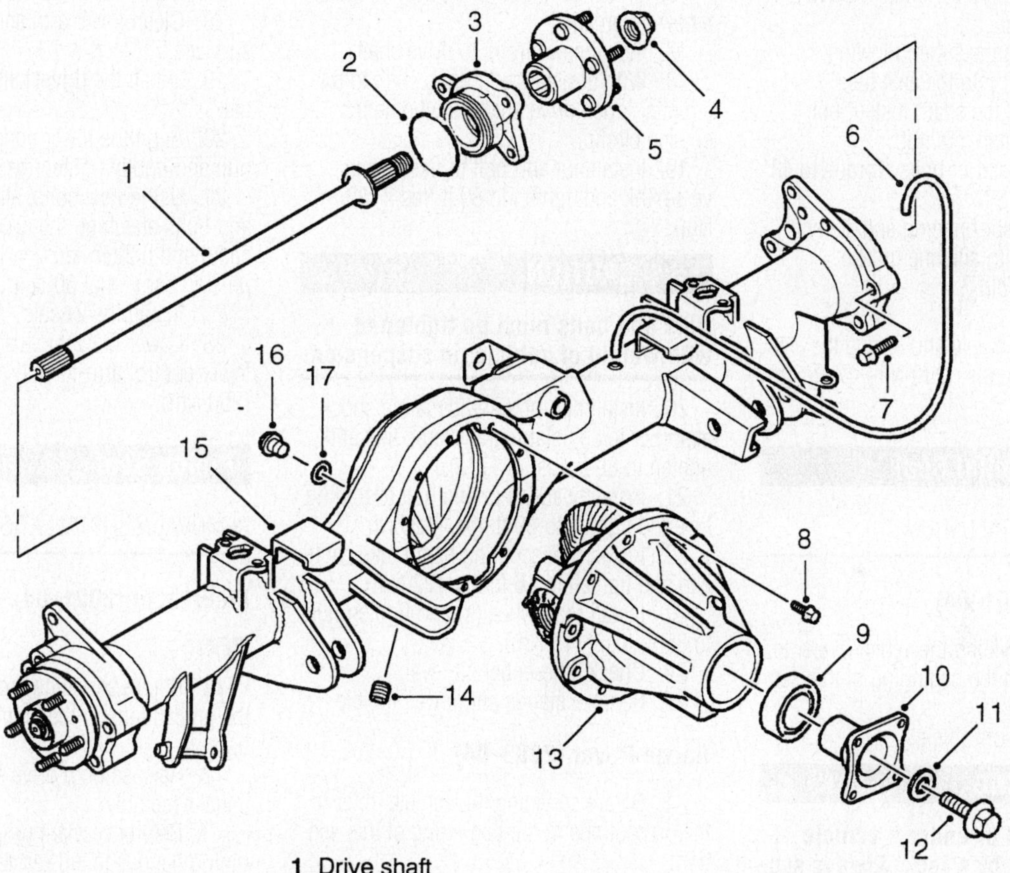

Rear axle components—Discovery II (2001–02) and Range Rover (2001–02)

- Brake rotor shield
- 2 bolts and remove the backstrap from the hub.
- Sensor from the hub
- Sensor bushing
- 6 bolts securing the hub to the axle carrier
- Drive hub and axle shaft
- Axle shaft nut, if loosened
- Seal from the axle tube

To install:

5. Clean all splines and surfaces.
6. Lubricate the new seal lip and running surface with silicone grease. Drive a new seal into place.
7. Install or connect the following:
   - Axle shaft into the axle tube
   - Hub onto the shaft. Install, but don't tighten, the nut.
   - Hub–to–carrier bolts. Torque to 48 ft. lbs. (65 Nm).
   - ABS sensor bushing and sensor, coated with silicone grease
   - Rotor shield
   - Wheel
8. Lower the vehicle and torque the axle shaft nut to 192 ft. lbs. (260 Nm). Stake the nut.

Rear Differential Seal

REMOVAL & INSTALLATION

Freelander (2001–04)

1. Before servicing the vehicle, refer to the precautions in the beginning of this section.
2. Raise rear of vehicle.

⁂ WARNING

Do not work on or under a vehicle supported only by a jack. Always support the vehicle on safety stands.

3. Remove road wheel.
4. Remove bolt securing handbrake cable to subframe.
5. Remove nut and bolt securing trailing link to rear hub, collect spacer from under bolt head.
6. Remove nut and bolt securing fixed transverse link to subframe. Collect dynamic shock absorber.
7. Remove nut and bolt securing adjustable transverse link to subframe.
8. Position drain tin to catch oil spillage.
9. With assistance pull hub assembly outwards.
10. Taking care not damage oil seal 'flinger', release drive shaft from differential using LRT-51-014 and position shaft aside.
11. Remove and discard circlip from drive shaft.
12. Remove differential oil seal.

To install:

13. Clean drive shaft oil seal recess in axle casing.
14. Install new oil seal using LRT-51-012.
15. Clean end of drive shaft and location in differential.
16. Check condition of oil seal slinger, renew if damaged.
17. Install new circlip to drive shaft.
18. With assistance fit drive shaft to differential, push drive shaft fully home to engage circlip.
19. Install nut and bolt to fixed transverse link and tighten to 89 ft. lbs. (120 Nm).

⁂ CAUTION

Nuts and bolts must be tightened with weight of vehicle on suspension.

20. Install nut, bolt and dynamic shock absorber to adjustable transverse link and tighten to 89 ft. lbs. (120 Nm).
21. Install spacer, nut and bolt to trailing link and tighten to 89 ft. lbs. (120 Nm).
22. Install bolt securing handbrake cable clip and tighten to 16 ft. lbs. (22 Nm).
23. Install road wheel(s) and tighten nuts to 85 ft. lbs. (115 Nm).
24. Check differential oil level.
25. Remove stands and lower vehicle.

Range Rover (2003–04)

1. Before servicing the vehicle, refer to the precautions in the beginning of this section.
2. Disconnect battery ground lead.
3. Raise rear of vehicle.

⁂ WARNING

Do not work on or under a vehicle supported only by a jack. Always support the vehicle on safety stands.

4. Remove road wheel.
5. Remove 6 bolts attaching the driveshaft to differential.
6. Support the lower control arm with a jack.
7. Remove the lower control arm-to-hub bolt.
8. Lower the jack.
9. Pull the driveshaft outward from the differential.
10. Using a drift, remove the drive flange.
11. Remove the circlip from the driveshaft and discard.
12. Remove the oil seal and discard.

To install:

13. Clean drive shaft oil seal recess in axle casing.
14. Install new oil seal using LRT-51-021.
15. Install a new circlip to the driveshaft.
16. Lubricate driveshaft splines and oil seal mating surface.
17. Install the drive flange.
18. Clean lower arm and hub mating surfaces.
19. Install the driveshaft to the differential.
20. Align the lower control arm to the hub and lightly tighten the bolt.
21. Using new bolts, attach the plates and bolts attaching the driveshaft to differential and tighten to 30 ft. lbs. (40 Nm), plus an additional 60 degrees.
22. Install the wheels.
23. Lower the vehicle and tighten the lower control arm-to-hub bolt to 184 ft. lbs. (250 Nm).

Pinion Seal

REMOVAL & INSTALLATION

Freelander (2001–04)

REAR

1. Before servicing the vehicle, refer to the precautions in the beginning of this section.
2. Release both drive shafts from differential assembly.
3. Reference mark propeller shaft and pinion flanges to aid reassembly.
4. Remove 4 nuts and bolts securing propeller shaft to differential. Release propeller shaft and tie aside.
5. Check and record the torque required to rotate the pinion and differential.

⁂ CAUTION

Drive shafts must be removed to obtain correct torque to turn figure.

6. Using LRT-51-003 to restrain differential flange, remove nut and washer securing pinion flange. Discard nut.
7. Remove pinion flange.
8. Carefully remove and discard oil seal, take care not to damage oil seal recess.
9. Remove oil thrower.

10. Remove pinion bearing inner race.
11. Remove and discard collapsible spacer.

To install:

12. Fit new collapsible spacer.
13. Fit pinion bearing and oil thrower.
14. Clean pinion flange and oil seal recess.
15. Fit new oil seal using LRT-51-010.
16. Fit pinion flange and washer.
17. Restrain pinion flange using LRT-51-003.
18. Fit new pinion nut and tighten to 140 ft. lbs. (190 Nm).
19. Check for end float on pinion. If end float exists continue to tighten pinion nut until end float is removed.
20. Continue to tighten pinion nut until correct preload is obtained.
21. Pinion preload is 1.2–2.1 ft. lbs. (1.7–2.8 Nm), if higher replace collapsible spacer.

⁂ CAUTION

Do not tighten pinion nut to more than 275 ft. lbs. (373 Nm), or the collapsible spacer will compress too far.

22. Clean propeller shaft flange and mating face.
23. Position propeller shaft to rear axle and align reference marks.
24. Tighten propeller shaft nuts and bolts to 48 ft. lbs. (65 Nm).
25. Fit drive shafts.
26. Check differential oil level.

Discovery (2001–02) and Range Rover (2001–02)

FRONT OR REAR

1. Before servicing the vehicle, refer to the precautions in the beginning of this section.

Pinion seal and flange—Discovery II (2001–02) and Range Rover (2001–02)

2. Matchmark the driveshaft and flange and disconnect the driveshaft.
3. Hold the flange and remove the nut.
4. Remove the flange.
5. Remove the seal.

To install:

6. Clean all mating surfaces.
7. Lubricate the oil seal lips with gear oil.
8. Drive the seal into place.
9. Position the flange and install the holding tool.
10. Install the nut or bolt. Torque the nut to 100 ft. lbs. (135 Nm); the bolt to 74 ft. lbs. (100 Nm).
11. Connect the driveshaft. Torque the nuts to 35 ft. lbs. (48 Nm).

Discovery (2003–04)

FRONT OR REAR

1. Before servicing the vehicle, refer to the precautions in the beginning of this section.
2. Raise front or rear of vehicle, as needed.

⁂ WARNING

Do not work on or under a vehicle supported only by a jack. Always use safety stands.

3. Reference mark rear propeller shaft for reassembly.
4. Remove nuts and bolts securing flexible coupling to pinion flange. Tie propeller shaft aside.
5. If working on rear differential, use tool LRT-51-008 to extract the centering peg from the rear pinion flange, then remove the peg from the removal tool.
6. Using tool LRT-51-003 to hold pinion flange, remove pinion flange bolt and washer. Remove flange.
7. Position a container to catch oil spillage.
8. Using a suitable pry tool, carefully remove the pinion oil seal. Avoid damaging oil seal recess.

To install:

9. Clean pinion oil seal recess and pinion flange. Lubricate oil seal lip with clean differential oil.
10. Using tool LRT-51-010, install a new pinion oil seal.
11. Position pinion flange and bolt and washer in place. Use tool LRT-51-003 to hold pinion flange and tighten bolt to 74 ft. lbs. (100 Nm).
12. On rear differential, use a tubular drift, install the centering peg to the pinion flange. Ensure the large diameter part of the centering peg is below the pinion flange mounting surface.
13. Position propeller shaft and align reference marks.
14. Install nuts and bolts holding flexible coupling to pinion flange. Tighten to 55 ft. lbs. (76 Nm).
15. Remove safety stands and lower vehicle.
16. Check differential oil.

Range Rover (2003–04)

FRONT OR REAR

1. Before servicing the vehicle, refer to the precautions in the beginning of this section.
2. Raise and support the chassis.
3. Disconnect the exhaust system.
4. Remove the center heat shield.
5. Remove the fuel tank heat shield.
6. Reference mark the propeller shaft to differential and remove the propeller shaft nuts.
7. Remove the center support bearing nuts and lower and support the propeller shaft.
8. Remove the differential locking plate.
9. Reference mark the pinion nut to pinion shaft and remove the pinion nut.
10. Using a puller, remove the pinion flange.
11. Remove the pinion seal.

To install:

12. Clean pinion oil seal recess and pinion flange. Lubricate oil seal lip with clean differential oil.
13. Using tool LRT-12-179 and LRT 51-019, install a new pinion oil seal.
14. Clean the pinion flange.
15. Install the pinion flange and tighten the nut to match the reference marks made previously.
16. Install a new locking plate.
17. Clean the differential and propeller shaft mating surfaces.
18. Install the propeller shaft to the differential flange and locate the support bearing.
19. Install the support bearing nuts but do not tighten.
20. Install the propeller shaft to the differential and tighten the nuts to 52 ft. lbs. (70 Nm).
21. Tighten the support bearing nuts to 15 ft. lbs. (21 Nm).
22. Install the fuel tank heat shield.
23. Install the center heat shield.
24. Install the exhaust system.
25. Check the differential oil and fill as needed.

Rear Axle Housing Assembly

REMOVAL & INSTALLATION

Discovery (2001–04) and Range Rover (2001–02)

1. Before servicing the vehicle, refer to the precautions in the beginning of this section.

**** CAUTION**

Before beginning, depressurize the air suspension.

2. Raise and support the chassis.
3. Support the axle assembly.
4. Matchmark the driveshaft and flange. Disconnect the driveshaft.
5. Remove or disconnect the following:
   - Wheels
   - Shock absorbers from the axle
   - Clips securing the air springs to the axle
   - Panhard rod from the axle
6. Disconnect the brake pipes from the body bracket. Plug the pipes and connections.
7. Remove or disconnect the following:
   - 2 clips securing the brake pipes from the body bracket
   - Banjo bolt and strap securing the breather hose to the axle
   - Height sensors from the trailing arms
   - Trailing arms from the body
   - Trailing arms from the axle
8. Lower the axle from the vehicle.

To install:

9. Install or connect the following:
   - Axle into position
   - Trailing arms to the axle. M16 8.8 grade bolts are torqued to 118 ft. lbs. (160 Nm); M16 10.9 grade bolts are torqued to 177 ft. lbs. (240 Nm); M12 bolts are torqued to 92 ft. lbs. (125 Nm).
   - Trailing arms to the chassis. Torque the bolts to 118 ft. lbs. (160 Nm).
   - Air spring clips
   - Height sensors
   - Shock absorbers to the axle. Torque the nuts to 33 ft. lbs. (45 Nm).
   - Breather hose
   - Brake pipes
   - ABS wiring
   - Panhard rod. Torque the bolt to 148 ft. lbs. (200 Nm).
   - Driveshaft. Torque the nuts to 35 ft. lbs. (48 Nm).
10. Refill the axle.
11. Bleed the brakes.

Front Drive Axle Housing Assembly

REMOVAL & INSTALLATION

Discovery (2001–04) and Range Rover (2001–02)

1. Before servicing the vehicle, refer to the precautions in the beginning of this section.

**** CAUTION**

Before beginning, depressurize the air suspension.

2. Remove or disconnect the following:
   - Brake pads
   - Caliper (tie it out of the way)
   - ABS sensors
   - Brake hoses from the knuckles
   - Drag link from the knuckle
   - Panhard rod from the axle
   - Sway bar
   - Track rod
   - Axle oil
   - Driveshaft
   - Height sensors
   - Breather hose
3. Support the axle
4. Remove or disconnect the following:
   - Air spring retaining clips
   - Air springs from the axle
   - Shock absorbers from the axle
   - Radius arms–to–chassis nuts
5. Move the axle forward and release the radius arms from the chassis brackets.
6. Remove the axle/radius arms assembly.
7. Remove the radius arms from the axle.

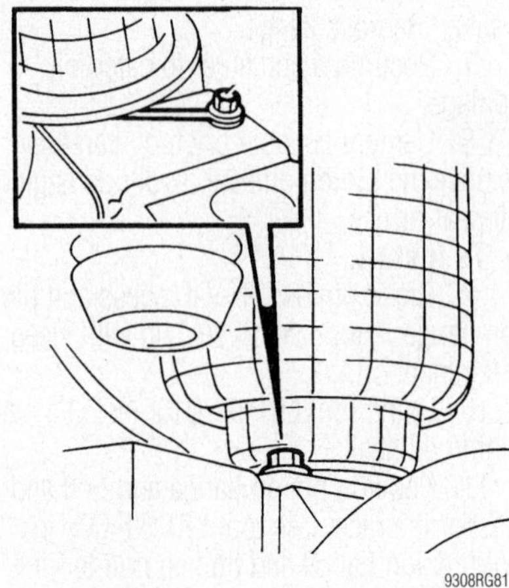

Lower air spring mounting bolt—Range Rover (2001–02)

To install:

8. Clean all mating surfaces.
9. Install or connect the following:
   - Radius arms to the axle. Torque the nuts to 92 ft. lbs. (125 Nm).
   - Axle into position. Locate the radius arms, with bushings, into the chassis brackets. Tighten the nuts to 118 ft. lbs. (160 Nm).
   - Shock absorbers. Torque the nuts to 33 ft. lbs. (45 Nm).
   - Air springs
   - Air spring retaining pins and bolts. Torque the bolts to 15 ft. lbs. (20 Nm).
   - Breather hose
   - Height sensors
   - Driveshaft. Torque the nuts to 35 ft. lbs. (48 Nm).
   - Track rod. Torque the nuts to 59 ft. lbs. (80 Nm).
   - Sway bar
   - Panhard rod. Torque the bolt to 148 ft. lbs. (200 Nm).
   - Drag link. Torque the nut to 59 ft. lbs. (80 Nm).
   - ABS sensors, coated with silicone grease
   - Brake hoses
   - Calipers. Torque the bolts to 162 ft. lbs. (220 Nm).
   - Brake pads
   - Axle fluid

Front Differential Assembly

REMOVAL & INSTALLATION

Range Rover (2003–04)

1. Before servicing the vehicle, refer to the precautions in the beginning of this section.
2. Disconnect the negative battery cable.
3. Drain the differential oil.
4. Remove the air intake plenum.
5. Remove the cooling fan.
6. Disconnect the power steering pump feed line.
7. Remove 2 nuts attaching engine brackets to their mountings.
8. Connect an engine lift to the lifting brackets and take the weight off the engine.
9. Remove the front propeller shaft.
10. Remove the right driveshaft seal.
11. Remove the differential seal.

➡Place the front wheels in the straight ahead position and ensure the ignition key is removed.

12. Remove the lower steering column clamp bolt.
13. Disconnect the pinion from the column.
14. Disconnect height sensor harness from the subframe.
15. Disconnect the power steering gear connector.
16. Remove the sway bar heat shields.
17. Place a transmission jack under the subframe to support it.
18. Remove the subframe.
19. Disconnect the differential breather hose.
20. Remove 4 differential bolts and remove the differential.

To install:

21. Clean all mating surfaces.
22. Install a new o-ring to the differential housing after coating it with petroleum jelly.
23. Install the differential but do not tighten the bolts.
24. Align the differential to transfer case front output flange.
25. Connect the breather hose.
26. Install the subframe and tighten the bolts to 122 ft. lbs. (165 Nm), plus an additional 90 degrees.
27. Install the sway heat shields.
28. Lower the engine onto the mounts and tighten the nuts to 74 ft. lbs. (100 Nm).
29. Connect the power steering gear connector.
30. Connect height sensor harness to the subframe.
31. Install the power steering feed line using a new o-ring.
32. Connect the pinion to the steering column.
33. Install the lower steering column clamp bolt and tighten to 18 ft. lbs. (25 Nm).
34. Install the right driveshaft seal.
35. Connect the front propeller shaft.
36. Install the cooling fan.
37. Install the air intake plenum.
38. Fill the differential with oil.
39. Connect the negative battery cable.
40. Bleed the power steering system.
41. Check the front wheel alignment.

Rear Differential Assembly

REMOVAL & INSTALLATION

Freelander (2001–04)

1. Before servicing the vehicle, refer to the precautions in the beginning of this section.
2. Remove both drive shafts.
3. Reference mark rear propeller shaft for reassembly.
4. Position container to catch oil spillage.
5. Remove 4 nuts and bolts securing propeller shaft to differential. Release propeller shaft and tie aside.
6. Support weight of differential assembly on a jack.
7. Remove 2 bolts securing differential to front mounting.
8. Depress red locking collar and disconnect breather pipe from differential casing.
9. Remove 4 bolts securing differential assembly to rear mountings.
10. With assistance, rotate differential assembly through 90 degrees and remove from subframe.

To install:

11. With assistance position differential assembly to subframe and locate in mountings, fit bolts but do not tighten at this stage.
12. Position centralizing jig LRT-51-013 to align differential assembly.
13. Tighten forward bolts to 48 ft. lbs. (65 Nm).
14. Tighten rearward bolts to 48 ft. lbs. (65 Nm).
15. Remove LRT-51-013.
16. Connect breather pipe.
17. Position propeller shaft to rear axle and align reference marks.
18. Install and tighten nuts and bolt securing propeller shaft to rear axle to 48 ft. lbs. (65 Nm).
19. Install drive shafts.
20. Check differential oil level.

Discovery (2003–04)

1. Before servicing the vehicle, refer to the precautions in the beginning of this section.
2. Raise and support rear of vehicle.

⁂ WARNING

Do not work on or under a vehicle supported only by a jack. Always use safety stands.

3. Drain oil from differential.
4. If same components are to be reinstalled, reference-mark propeller shaft and mating components.
5. Remove flexible coupling from pinion flange. Release propeller shaft and tie aside.
6. Using tool LRT-51-008, extract centering peg from pinion flange. Remove peg from tool.
7. Remove right and left side rear brake rotors.
8. Disconnect ABS sensor connectors.
9. Release each harness from brake hose and hose bracket.
10. Remove 4 bolts holding each rear wheel hub to axle flange.
11. Remove wheel hubs and drive shafts from rear axle. Remove and discard O-rings from hubs.
12. Remove 10 bolts securing differential to axle. Remove differential from axle.

To install:

13. Clean drive shafts, wheel hubs and hub locations in rear axle.
14. Remove all traces of sealant from differential and axle mating faces.
15. Apply sealant STC 3811 to differential or axle mating faces (except bolt holes).
16. Apply sealant STC 50552 to threads of differential bolts.
17. Position differential to axle and tighten bolts to 41 ft. lbs. (55 Nm).
18. Clean pinion flange and centering peg.
19. With a tubular drift, install centering peg into pinion flange. Ensure large diameter of peg is below pinion flange mounting surface.
20. Position propeller shaft and align reference marks.
21. Fit nuts and bolts holding flexible coupling to pinion flange. Tighten to 56 ft. lbs. (76 Nm).
22. Lubricate 2 new wheel hub O-rings with clean differential oil.
23. Install O-rings on hubs.
24. Fit both drive shafts and wheel hubs to rear axle housing. Install and tighten hub bolts to 74 ft. lbs. (100 Nm).
25. Connect each ABS sensor multiplug and secure each harness to brake hose and hose bracket.
26. Install both brake rotors.
27. Remove stands and lower vehicle.
28. Refill differential with clean differential oil.

Range Rover (2003–04)

1. Before servicing the vehicle, refer to the precautions in the beginning of this section.
2. Raise and support the chassis.
3. Disconnect the exhaust system.
4. Remove the center heat shield.
5. Remove the fuel tank heat shield.
6. Reference mark the propeller shaft to differential and remove the propeller shaft nuts.
7. Remove the center support bearing nuts and lower and support the propeller shaft.
8. Raise and support the vehicle and remove the wheels.

9. Disconnect the breather hose from the differential.
10. Drain the differential oil.
11. Reference mark the driveshaft-to-differential flanges and remove the attaching bolts and plates.
12. Position a jack under the lower control arm.
13. Remove the lower control arm-to-hub bolt.
14. Remove the jack and separate the lower arm from the hub.
15. Disconnect the driveshaft from the differential.
16. Repeat the procedure on the opposite side.
17. Place a suitable jack under the differential.
18. Remove 2 bolts and 1 nut and remove the differential.

To install:

19. Raise the differential into position.
20. Using new bolts, attach the differential to the subframe and tighten the front bolts to 74 ft. lbs. (100 Nm).
21. Install a new nut and bolt and tighten the rear of the differential to the subframe to 121 ft. lbs. (165 Nm).
22. Connect the driveshaft to the differential.
23. Align the lower control arm-to-hub and lightly tighten the bolt.
24. Using the reference mark attach the driveshaft-to-differential flange and tighten the bolts to 30 ft. lbs. (40 Nm) plus an additional 60 degrees. Do this on both sides.
25. Connect breather hose.
26. Install the wheels.
27. Lower the vehicle.
28. Tighten the lower control arm-to-hub bolts to 184 ft. lbs. (250 Nm).
29. Install the propeller shaft and support bearing.
30. Install fuel tank heat shield.
31. Install the center heat shield.
32. Connect the exhaust system.
33. Fill the differential with oil.

STEERING AND SUSPENSION

Air Suspension

DEPRESSURIZING

The air suspension is pressurized up to 150 psi (1034 kPa). Before beginning work on suspension or steering components, the system should be depressurized. A special tool called a TestBook is required to depressurize the system properly and safely. The tool is self-guiding. Depressurizing will lower the suspension to the bump stops. Before beginning any work, make sure that all air springs are deflated. A spring that remains inflated is due to a stuck solenoid valve. In that case, it will be necessary to disconnect the pipe at that air spring.

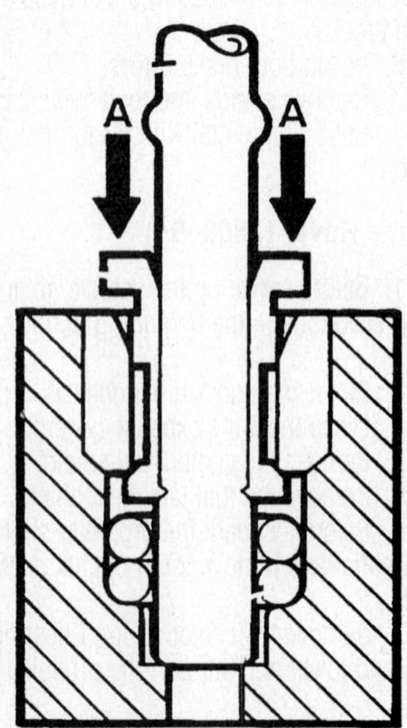

Air pipe disconnection

PIPE DISCONNECT

1. Before servicing the vehicle, refer to the precautions in the beginning of this section.
2. If it is necessary to disconnect a pipe:
    a. **Wear hand, ear and eye protection. Wrap a cloth around the connection.**
    b. Clean the connection with a stiff wire brush.
    c. Peel back the boot.
    d. Apply equal downward pressure on the collet flange (A) as shown.
    e. Pull the pipe firmly through the center of the collet.
    f. To connect the pipe, push it firmly through the 2 O-rings until it contacts the base of the housing. Gently pull back on the pipe. Some slight movement will be noticed.
    g. Reposition the boot.

When the job is done, run the engine to repressurize the system. The system must be recalibrated using the TestBook tool. Recalibration must be done on a floor surface that is level and smooth in all directions.

Air Bag

⁂ CAUTION

Air bag system must be disabled before performing service on or around system components, steering column, instrument panel components, wiring and sensors. Failure to follow safety and disabling procedures could result in accidental air bag deployment, possible personal injury and unnecessary system repairs.

PRECAUTIONS

Several precautions must be observed when handling the inflator module to avoid accidental deployment and possible personal injury.

- Never carry the inflator module by the wires or connector on the underside of the module.
- When carrying a live inflator module, hold securely with both hands, and ensure that the bag and trim cover are pointed away.
- Place the inflator module on a bench or other surface with the bag and trim cover facing up.
- With the inflator module on the bench, never place anything on or close to the module which may be thrown in the event of an accidental deployment.

Before servicing the vehicle, be sure to also refer to the precautions in the beginning of this section as well.

DISARMING

1. Before servicing the vehicle, refer to the precautions in the beginning of this section.
2. Remove the key from the ignition.
3. Disconnect the negative battery cable first, then the positive cable.
4. Wait 10-20 minutes for the back-up power to discharge.

ARMING

1. After performing the required service, rearm the SRS by reconnecting the battery.
2. Start the vehicle and the SRS service light should go OFF after 5 seconds.

Power Steering Pump

REMOVAL & INSTALLATION

Freelander (2001–04)

1. Before servicing the vehicle, refer to the precautions in the beginning of this section.
2. Disconnect battery ground lead.
3. Remove engine mounting top arm.
4. Loosen power-assisted steering (PASP pump pulley bolts.
5. Using a ⅜ in. socket bar, raise auxiliary drive belt tensioner and release drive belt from alternator and PAS pump pulleys.
6. Position container to collect PAS fluid spillage.
7. Release clip and disconnect fluid inlet hose from PAS pump.
8. Remove banjo bolt securing fluid outlet hose to PAS pump, release hose, remove and discard sealing washers.

⁂ CAUTION

To prevent damage to components, use two spanners when loosening or tightening unions.

9. Remove bolt securing PAS outlet pipe support bracket to cylinder head and move pipe aside.
10. Remove 3 bolts securing PAS pump pulley to PAS pump and remove pulley.

⁂ CAUTION

Always fit plugs to open connections to prevent contamination.

11. Remove 3 bolts securing PAS pump to mounting bracket.
12. Remove PAS pump.

To install:

13. Position PAS pump and align to mounting bracket.
14. Fit bolts securing PAS pump to mounting bracket and tighten to 18 ft. lbs. (25 Nm).
15. Fit PAS pump pulley to PAS pump and tighten bolts to finger tight.
16. Remove plug from fluid outlet hose.
17. Clean banjo bolt and mating faces.
18. Using new sealing washers, fit PAS outlet hose to pump and tighten banjo bolt to 15 ft. lbs. (20 Nm).

⁂ CAUTION

To prevent damage to components, use two spanners when loosening or tightening unions.

19. Align outlet pipe clip. Fit bolt and tighten to 16 ft. lbs. (22 Nm).
20. Clean elbow on PAS pump.
21. Remove plug from fluid inlet hose, fit new clip and connect hose to PAS pump.
22. Remove container.
23. Clean pulley 'V's and tensioner pulley running surface.
24. Using a ⅜ in. square drive socket bar, raise auxiliary drive belt tensioner and fit drive belt to pulleys.
25. Tighten PAS pump pulley bolts to 7.5 ft. lbs. (Nm).
26. Fit engine mounting top arm.
27. Connect battery ground lead.
28. Bleed PAS system.

Discovery (2001–02) and Range Rover (2001–02)

1. Before servicing the vehicle, refer to the precautions in the beginning of this section.
2. Remove the drive belt.
3. Remove the 3 pulley bolts.
4. Disconnect the return hose.
5. Disconnect the high pressure hose.
6. Remove the 4 bolts securing the pump/compressor bracket to the engine.
7. Remove the 3 bolts securing the mounting plate to the pump.
8. Remove the pump.
9. Installation is the reverse of removal. Tighten as follows:
   - Lift bracket to the pump: 13 ft. lbs. (18 Nm).
   - Bracket to the engine: 30 ft. lbs. (40 Nm).
   - Pump mounting bolts: 13 ft. lbs. (18 Nm).
   - High pressure fitting: 12 ft. lbs. (16 Nm).

Discovery (2003–04)

1. Before servicing the vehicle, refer to the precautions in the beginning of this section.
2. Remove auxiliary belt.
3. Remove cable tie securing harness to air intake hose.
4. Loosen 3 clips and remove air intake hose.
5. On models with ACE, remove Active Cornering Enhancement (ACE) pump from its mounting and position aside.
6. Disconnect A/C compressor connector, remove mounting bolts and position compressor aside.
7. Remove power assist steering (PAS) pump pulley.
8. Position container to catch fluid, and release PAS pump pressure pipe.

⁂ CAUTION

Always fit plugs to open connections to prevent contamination.

9. Remove the jockey pulley.
10. Remove oil cooling pipe bracket from auxiliary housing.
11. Remove PAS pump.
12. Remove 4 bolts and 1 nut securing auxiliary housing. Pull housing forward and remove PAS pump.

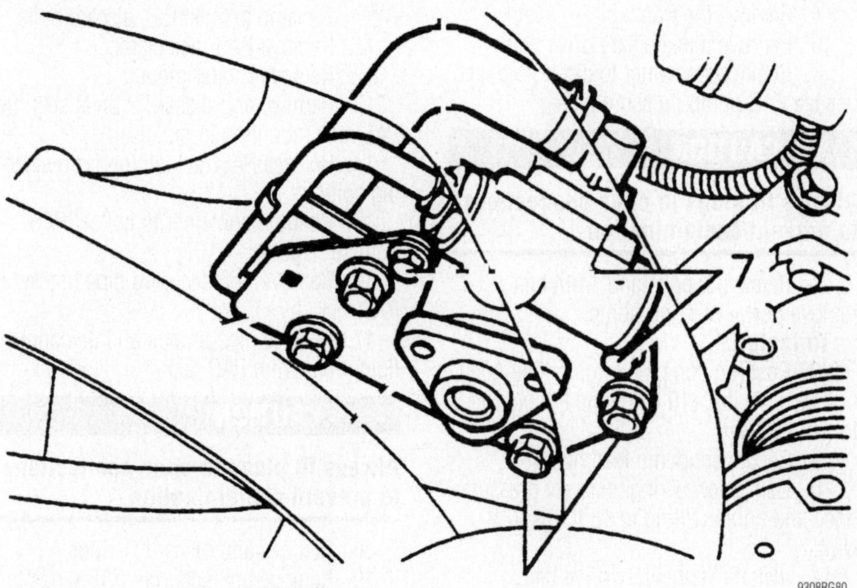

Power steering pump/compressor mounting bracket—Discovery II (2001–02) and Range Rover (2001–02)

To install:

13. Position PAS pump to auxiliary housing and reposition auxiliary housing to engine. Install and tighten auxiliary housing bolts to 30 ft. lbs. (40 Nm).
14. Tighten auxiliary housing nut to 7 ft. lbs. (10 Nm).
15. Reconnect oil cooling pipe bracket to auxiliary housing.
16. Install PAS pump mounting bolts and tighten to 16 ft. lbs. (22 Nm).
17. Reconnect PAS pump pressure pipe.
18. Install PAS pump inlet hose and secure with clip.
19. Position jockey pulley and tighten pulley bolt to 37 ft. lbs. (50 Nm).
20. Clean PAS pump pulley mating faces. Install pulley and tighten bolts to 16 ft. lbs. (22 Nm).
21. On models with ACE, clean ACE pump dowels and dowel holes. Position pump, install bolts, and tighten to 18 ft. lbs. (25 Nm).
22. Clean A/C compressor dowels and dowel holes. Install compressor and tighten bolts to 16 ft. lbs. (22 Nm).
23. Install air intake hose and secure with clips. Attach harness to air intake hose with new cable tie.
24. Install auxiliary drive belt.
25. Bleed PAS system.

Range Rover (2003–04)

1. Before servicing the vehicle, refer to the precautions in the beginning of this section.
2. Disconnect negative battery cable.
3. Remove the front air dam.
4. Remove the front splash shield.
5. Remove the ancillary drive belt.
6. Position container to catch fluid, and release PAS pump pressure pipe.

**** CAUTION**

Always fit plugs to open connections to prevent contamination.

7. Remove 2 bolts and 1 nut and remove power steering pump.

To install:

8. Position pump and tighten the Allen bolts to 7 ft. lbs. (10 Nm), and the nut to 18 ft. lbs. (25 Nm).
9. Reconnect pump feed hose.
10. Using new o-ring, connect pressure hose and tighten fitting to 18 ft. lbs. (25 Nm).
11. Install the ancillary drive belt.
12. Install the front splash shield
13. Install the air dam.
14. Connect the negative battery cable.
15. Fill system with clean power steering fluid and then bleed the system by starting the engine and turning the steering wheel lock to lock several times. Recheck the fluid level.

Power Steering Gear

REMOVAL & INSTALLATION

Freelander (2001–04)

1. Before servicing the vehicle, refer to the precautions in the beginning of this section.
2. Raise front of vehicle.

**** WARNING**

Do not work on or under a vehicle supported only by a jack. Always support the vehicle on safety stands.

3. Remove front road wheels.
4. Remove and discard nuts securing track rod ball joints to steering arms.
5. Fit an M12 nut to each ball pin, flush with end of each pin.
6. Using LRT-57-043, separate ball pins from right side and left side steering arms. Remove M12 nuts and release ball pins from steering arms.
7. Remove pinch bolt securing steering column to PAS rack pinion.
8. Remove 2 nuts securing steering rack heat shield and remove heat shield.
9. Remove 2 bolts securing coolant rail to cylinder block.
10. Remove 2 bolts and washers securing PAS rack clamp to bulkhead, discard bolts.
11. Remove PAS rack clamp.
12. Remove rubber mount.
13. Remove and discard 2 bolts securing PAS rack mounting to bulkhead.
14. Release PAS rack pinion from steering column.
15. Position container to collect PAS fluid spillage.
16. Remove bolt securing pipe bracket to PAS rack.
17. Release pipe unions and disconnect fluid pipes from PAS rack.

**** CAUTION**

Always fit plugs to open connections to prevent contamination.

18. Remove and discard O-rings.
19. Remove bolt securing PAS pipes to clamp and loosen clamp bolt.
20. With assistance remove PAS rack from passenger side of vehicle.
21. Remove dust seal from pinion housing.

To install:

22. Fit PAS rack to vehicle from passengers side.
23. Fit dust shield to pinion housing.
24. Ensure pipe unions are clean.
25. Fit new O-rings to fluid pipes.
26. Fit fluid pipes to PAS rack but do not tighten at this stage.
27. Align fluid pipe bracket to PAS rack, fit bolt but do not tighten at this stage.
28. With assistance fit PAS rack pinion to steering column, ensuring column coupling is aligned with gear input flag.
29. Fit washers and new bolts securing steering rack mounting to bulkhead, but do not tighten at this stage. Ensure large washer is fitted to lower bolt.
30. Fit rubber mount and clamp to PAS rack.
31. Fit bolts securing clamp to bulkhead but do not tighten at this stage.
32. Tighten PAS rack mounting bolts to 33 ft. lbs. (45 Nm).
33. Tighten PAS rack clamp bolts to 33 ft. lbs. (45 Nm).
34. Tighten PAS rack fluid feed pipe union to 13 ft. lbs. (18 Nm).
35. Tighten PAS rack fluid return pipe union to 16 ft. lbs. (22 Nm).
36. Tighten fluid pipe bracket to 7.5 ft. lbs. (10 Nm).
37. Align PAS rack clamp and tighten bolt.
38. Align pipes to clamp fit bolt and tighten to 7.5 ft. lbs. (10 Nm).
39. Fit pinch bolt to steering column and tighten to 24 ft. lbs. (32 Nm).

**** CAUTION**

Nuts and bolts must be tightened with the weight of the vehicle on the suspension.

40. Ensure tapers in track rod end and steering arm are clean and rubber boot is not damaged.
41. Fit ball joints to steering arms, fit new nuts and tighten to 41 ft. lbs. (55 Nm).
42. Align coolant rail to cylinder block, fit bolts and tighten to 7 ft. lbs. (9 Nm).
43. Position heat shield, fit nuts and tighten to 18 ft. lbs. (25 Nm)
44. Fit road wheel(s) and tighten nuts to 85 ft. lbs. (115 Nm).
45. Remove stands and lower vehicle.
46. Bleed PAS system.
47. Check and adjust front wheel alignment.

Discovery (2001–02) and Range Rover (2001–02)

1. Before servicing the vehicle, refer to the precautions in the beginning of this section.
2. Disconnect the negative battery cable.
3. Drain the fluid from the power steering fluid reservoir.
4. Clean the steering box to prevent dirt from entering when the hoses are removed.
5. Disconnect the feed and return lines from the steering box.
6. Raise and safely support the vehicle.
7. Remove the under tray
8. Disconnect the drag link from the drop arm.
9. Remove the universal joint connecting the steering column to the steering box.
10. Remove the power steering box mounting bolts, then remove the box from the vehicle.

To install:

11. Install the steering box and tighten the mounting bolts as follows:
   - Discovery: 66 ft. lbs. (90 Nm)
   - Range Rover: 92 ft. lbs. (125 Nm)
12. Connect the universal joint and tighten the pinch bolts 18 ft. lbs. (25 Nm).
13. Connect the drag link to the drop arm and tighten the retaining nut as follows:
   - Discovery: 30 ft. lbs. (40 Nm)
   - Range Rover: 59 ft. lbs. (80 Nm)
14. Lower the vehicle.
15. Connect the hydraulic hoses to the steering box and tighten the 16mm thread to 37 ft. lbs. (50 Nm) and the 14mm thread to 22 ft. lbs. (30 Nm).
16. Fill the power steering fluid reservoir.
17. With engine running, test steering system for leaks by holding steering in both full lock directions.
18. Ensure steering wheel is correctly aligned when wheels are positioned straight ahead.
19. If necessary, reposition the steering wheel.
20. Road test vehicle.

Discovery (2003–04)

1. Before servicing the vehicle, refer to the precautions in the beginning of this section.
2. Raise and support front of vehicle.

⁂ CAUTION

Do not work on or under a vehicle supported only on a jack. Always use safety stands.

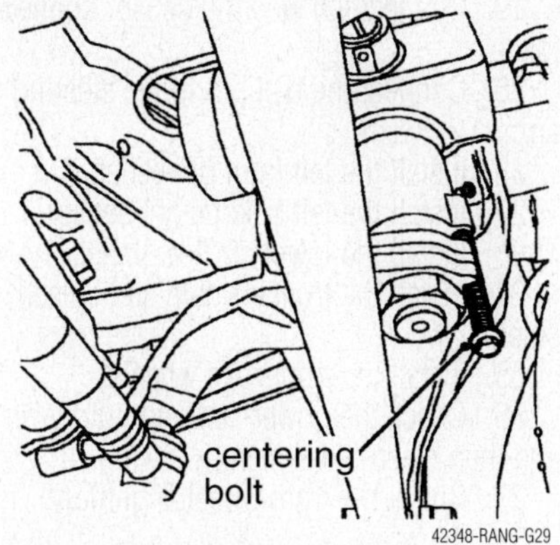

Steering box centering bolt installed—Discovery (2003–04)

3. Remove front road wheels.
4. Ensure steering wheel is centered, then fit centering bolt to steering box, as illustrated.
5. Remove key from steering lock and ensure column lock is engaged.
6. Remove 3 bolts securing intermediate shaft and U-joint.
7. Push shaft upward to release and remove U-joint.

⁂ CAUTION

Do not turn steering wheel with intermediate shaft or U-joint disconnected, as damage to rotary coupler and steering wheel switches may occur.

8. Position a container to catch steering fluid.
9. Remove bolt securing PAS pipe bracket to steering box. Release pipes and discard O-rings.

⁂ CAUTION

Always install plugs to open connections to prevent contamination.

10. Remove securing nut and bolt and release the panhard rod.
11. Remove the nut securing drag link to drop arm. Use LRT-57-036 to break taper joint and release drag link.
12. With assistance, remove 4 securing bolts and remove steering gear box.
13. Remove centering bolt from steering box.

To install:

14. Install centering bolt into steering box.
15. With assistance, position steering gear box and install and tighten bolts to 66 ft. lbs. (90 Nm).
16. Position drag link to drop arm. Fit nut and tighten to 59 ft. lbs. (80 Nm).
17. Position panhard rod. Install bolt and nut and tighten to 170 ft. lbs. (230 Nm).
18. Clean PAS pipe ends and O-ring recesses.
19. Fit new O-rings to pipes and attach to steering gear box. Position pip bracket and tighten bracket bolt to 16 ft. lbs. (22 Nm).
20. Ensure steering wheel is centered. Connect U-joint to intermediate shaft and tighten bolts to 18 ft. lbs. (25 Nm).
21. Remove centering bolt from steering gear box.
22. Install road wheels and tighten nuts to 103 ft. lbs. (140 Nm).
23. Remove safety stands and lower vehicle.
24. Check and refill PAS oil.
25. Bleed PAS system.
26. Center steering linkage.

Range Rover (2003–04)

1. Before servicing the vehicle, refer to the precautions in the beginning of this section.
2. Raise and support front of vehicle.

⁂ CAUTION

Do not work on or under a vehicle supported only on a jack. Always use safety stands.

3. Remove front road wheels.
4. Remove the front air dam and splash shield.
5. Remove the lower air flow panel.
6. Remove the left side height sensor.
7. Remove the left front driveshaft.
8. Separate the ball pin from the steering arm.
9. Disconnect the steering sensor connector.
10. Mark the position of the lower steering column to the gear, then remove the bolt from the steering column camp.

⁂ CAUTION

Do not turn steering wheel with shaft disconnected, as damage to rotary coupler and steering wheel switches may occur.

11. Disconnect the power steering feed line clamp.
12. Position a container to catch steering fluid.
13. Remove 2 bolts and 2 nuts and remove steering rack by rotating it out from the left side.

To install:

14. With assistance, position steering gear and install and tighten bolts to 74 ft.

lbs. (100 Nm), then tighten an additional 90 degrees.

15. Using a new nut, attach the track rod end ball joints to the steering arm and tighten the nuts to 59 ft. lbs. (80 Nm).

16. Using new washers, attach the banjo bolts for fluid feed and return lines and tighten the M14 bolt to 26 ft. lbs. (36 Nm), and the M16 bolt to 30 ft. lbs. (40 Nm).

17. Connect the feed line clamp.

18. Using the position marked on removal, attach the lower steering column to the rack and tighten the bolt to 18 ft. lbs. (25 Nm).

19. Connect the steering sensor connector.

20. Connect the ball pin to the steering arm.

21. Install the left front driveshaft.

22. Install the left side height sensor.

23. Install the lower air flow panel.

24. Install the front air dam and splash shield.

25. Install the front road wheels.

26. Check the power steering fluid level and then bleed the system.

27. Check the front wheel alignment.

Shock Absorbers

REMOVAL & INSTALLATION

Range Rover (2001–02)

FRONT

**** CAUTION**

Be sure to support the axle when the shock is removed, otherwise the pressurized air spring could fail and cause component damage and pos-

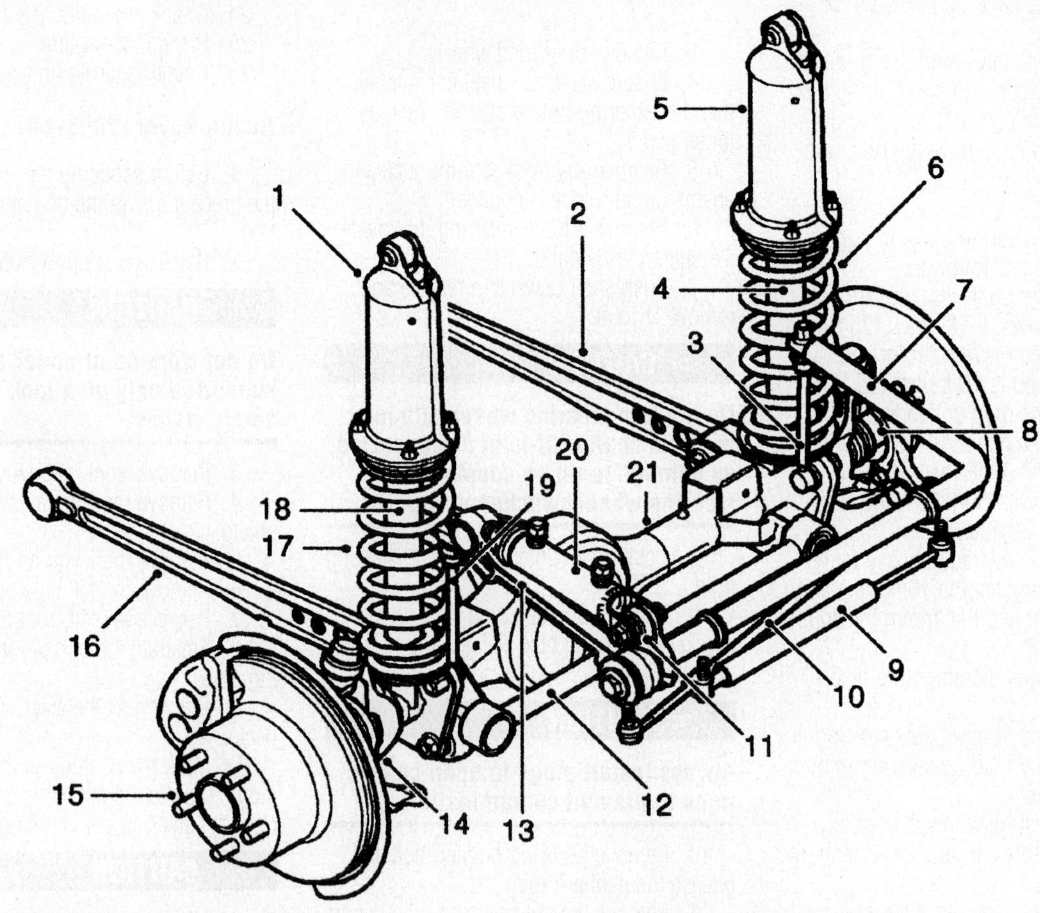

1 Turret RH
2 Radius arm LH
3 Anti-roll bar link LH
4 Damper LH
5 Turret LH
6 Coil spring LH
7 Torsion/anti-roll bar
8 Steering knuckle
9 Steering damper
10 Drag link
11 ACE short arm (if fitted)
12 Panhard rod
13 ACE long arm (if fitted)
14 Steering knuckle
15 Brake caliper and hub assembly
16 Radius arm RH
17 Coil spring RH
18 ACE actuator
19 Anti-roll bar link RH
20 ACE actuator
21 Anti-roll bar link RH

42348-RANG-G30

View of front suspension component layout (ACE torsion bar shown)—Discovery (2001–04)

sible personal injury. It is possible to remove the shock absorber without depressurizing air springs, BUT the distance between the axle and chassis must be held as if the shock absorber was still fitted. This is achieved by supporting the vehicle on supports, with a jack under the axle.

1. Before servicing the vehicle, refer to the precautions in the beginning of this section.
2. Raise and safely support the vehicle.
3. Support the front axle on a jack.
4. Remove front wheel.

⁂ WARNING

Do not lower axle when shock absorber is removed. This may result in air spring damage.

5. Remove the lower shock absorber retaining nut.
6. Remove the upper shock absorber retaining bolt.
7. Remove the shock absorber.

To install:

8. Install the shock absorber.
9. Install the upper mounting bolt and tighten to 92 ft. lbs. (125 Nm).
10. Install the lower mounting bolt and tighten to 33 ft. lbs. (45 Nm).
11. Install the front wheel and tighten the lug nuts to 80 ft. lbs. (108 Nm).
12. Remove the jack supporting the axle.
13. Remove the supports and lower vehicle.

REAR

⁂ CAUTION

Be sure to support the axle when the shock is removed. Otherwise, the pressurized air spring could fail and cause component damage and possible personal injury. It is possible to remove the shock absorber without depressurizing air springs, BUT the distance between the axle and chassis must be held as if the shock absorber was still fitted. This is achieved by supporting the vehicle with supports and a jack under the axle.

1. Before servicing the vehicle, refer to the precautions in the beginning of this section.
2. Raise and safely support the vehicle.
3. Support the axle on a jack.
4. Remove the rear wheels.

⁂ WARNING

Do not lower axle when shock absorber is removed. This may result in air spring damage.

5. Remove the lower shock absorber retaining nut.
6. Remove the upper shock absorber retaining bolt.
7. Remove the shock absorber.

To install:

8. Install the shock absorber.
9. Install the upper mounting bolt and tighten to 92 ft. lbs. (125 Nm).
10. Install the lower mounting bolt and tighten to 33 ft. lbs. (45 Nm).
11. Install the rear wheel and tighten the lug nuts to 80 ft. lbs. (108 Nm).
12. Remove the jack supporting the axle.
13. Remove the supports and lower vehicle.

Discovery (2001–04)

FRONT

1. Before servicing the vehicle, refer to the precautions in the beginning of this section.
2. Raise and support the front end securely.

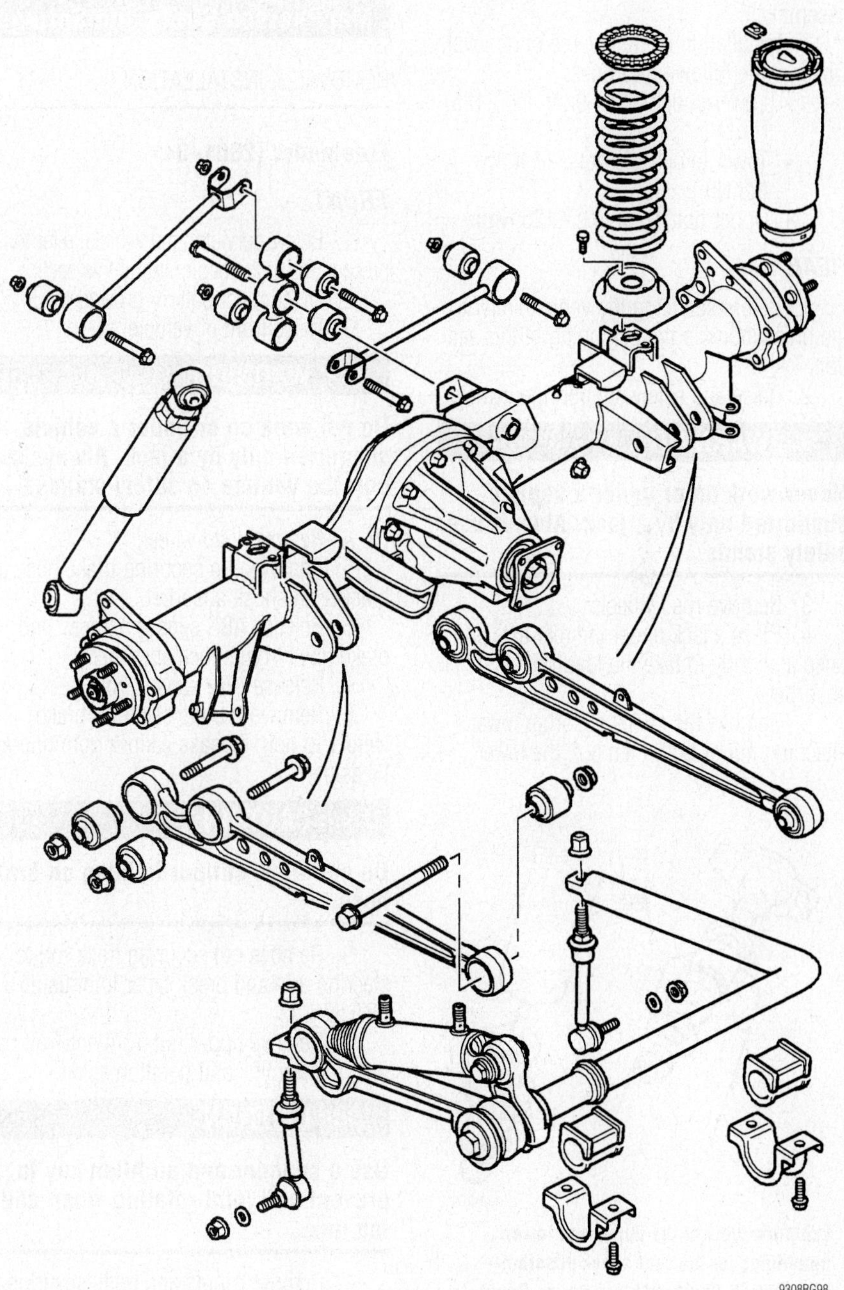

Exploded view of rear suspension components—Discovery (2001–04)

3. Remove the wheel.
4. To remove right side shock absorber, remove the coolant reservoir.
5. Support the weight of the axle with a jack.
6. Loosen the upper bolt securing the shock to the tower.
7. Remove the 4 nuts securing the shock tower to the chassis.
8. Remove the 2 lower shock absorber bolts.
9. On ACE equipped models, remove the ACE pipe clamp bolt and clamp.
10. Raise the tower and remove the upper bolt.
11. Compress the shock and remove the assembly.
12. Installation is the reverse of removal. Observe the following torques:
- Lower shock bolts: 92 ft. lbs. (125 Nm).
- Tower to chassis nuts: 17 ft. lbs. (23 Nm).
- Upper bolt: 92 ft. lbs. (125 Nm).

REAR

1. Before servicing the vehicle, refer to the precautions in the beginning of this section.
2. Raise and safely support the vehicle.

⁂ WARNING

Never work on or under a vehicle supported only by a jack. Always use safety stands.

3. Remove road wheels.
4. Place a jack under the rear axle and raise it slightly to take the load off the shock absorbers.
5. Remove the shock absorber lower attaching nut or bolt, then pull the lower end free of the mounting bracket on the axle housing.
6. Remove the upper attaching nut or bolt and remove the shock absorber.

To install:

7. Install the shock absorber onto the lower axle bracket. Install bolt or nut. Extend shock absorber and fix to top mount with bolt or nut.
8. Tighten the upper and lower attaching bolts or nuts to 92 ft. lbs. (125 Nm).
9. Install road wheels and tighten nuts to 103 ft. lbs. (140 Nm).
10. Remove safety stands and jack and lower vehicle.

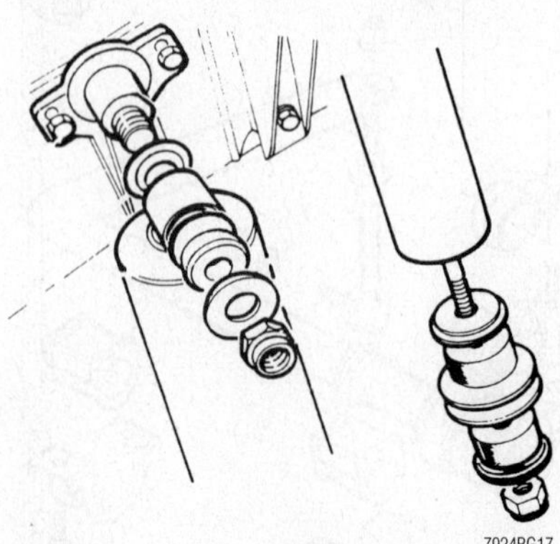

Exploded view of the upper and lower mountings for the rear shock absorber—Discovery II (2001–04) and Range Rover (2001–02)

MacPherson Struts and Springs

REMOVAL & INSTALLATION

Freelander (2001–04)

FRONT

1. Before servicing the vehicle, refer to the precautions in the beginning of this section.
2. Disconnect battery ground lead.
3. Raise front of vehicle.

⁂ WARNING

Do not work on or under a vehicle supported only by a jack. Always support the vehicle on safety stands.

4. Remove road wheel.
5. Remove clip securing brake hose to bracket on shock absorber.
6. Release ABS sensor harness and brake hose from shock absorber.
7. Release ABS sensor from hub.
8. Remove 2 bolts securing brake caliper to hub. Release caliper from hub and tie aside.

⁂ CAUTION

Do not allow caliper to hang on brake hose.

9. Remove nut securing track rod to steering arm and break taper joint using LRT-57-043.
10. Remove upper nut from anti-roll bar link, release link and position aside.

⁂ CAUTION

Use a spanner and an Allen key to prevent ball joint rotating when undoing link.

11. Remove 2 nuts and bolts securing hub to shock absorber.
12. Release shock absorber from hub.
13. On the left side: Remove 2 bolts securing positive and negative leads to fusebox.
14. On the left side: Disconnect multi-plug from engine compartment fuse box.
15. On the left side: Release leads and position aside.
16. On the left side: Remove 3 nuts securing fusebox and position fusebox aside.
17. On the right side: Remove bolt securing coolant reservoir and position reservoir aside.
18. Reference mark top mounting in relationship to body.
19. Remove 3 nuts from shock absorber top mounting and remove spring and shock absorber assembly.
20. Position a suitable spring compressor in vice.
21. Position spring and shock absorber assembly to spring compressor. Compress spring.

⁂ CAUTION

Note alignment of top mounting, spring and shock absorber dust cover.

22. Compress spring by 2 to 3 cm until loose, hold shock absorber shaft with Allen key, remove and discard mounting plate nut.
23. Remove rebound washer and mounting plate.
24. Remove spring aid and bump plate.
25. Remove spring seat, dust cover and bump stop cup.
26. Remove shock absorber from spring.
27. Release and remove spring from compressor.

To install:

28. Inspect shock absorber, spring mounting rubbers and bearing for deterioration and damage.
29. Clean mating faces of spring, mounting and mounting plate.
30. Clean shock absorber shaft and bump stop plate.
31. Position spring and shock absorber assembly to spring compressor. Compress spring.
32. Install shock absorber to spring, ensure spring locates in cut recess in shock absorber plate.
33. Install bump cup, bump stop and dust cover to shock absorber.
34. Install spring aid and bump plate.
35. Install mounting plate and rebound washer.

36. Using new nut, hold shock absorber shaft with Allen key and tighten nut to 42 ft. lbs. (57 Nm).

✲✲ CAUTION

Note alignment of top mounting, spring and shock absorber dust cover.

37. Release and remove spring from compressor.
38. Clean mating face of top mounting plate.
39. Position shock absorber assembly and align top mounting to body, fit nuts and tighten to 33 ft. lbs. (45 Nm).
40. On the left side: Position fusebox, fit nuts and tighten to 6 ft. lbs. (8 Nm).
41. On the left side: Connect positive and negative leads, fit bolts and tighten to 6 ft. lbs. (8 Nm).
42. On the left side: Connect multiplug to fusebox.
43. On the right side: Position coolant reservoir and secure with bolt.
44. Install hub to shock absorber, fit nuts and bolts and tighten to 151 ft. lbs. (205 Nm).
45. Clean anti-roll bar link taper and mating face.
46. Connect anti-roll bar link, fit new nut and tighten to 41 ft. lbs. (55 Nm).

✲✲ CAUTION

Use a spanner and an Allen key to prevent ball joint rotating when undoing link.

47. Clean track rod taper and mating face.
48. Connect track rod end to steering arm using new nut and tighten nut to 41 ft. lbs. (55 Nm).
49. Clean ABS sensor, smear sensor with an anti-seize grease and fit sensor to hub.

✲✲ CAUTION

Ensure ABS sensor is fully located into hub, so that sensor touches pole wheel teeth.

50. Position ABS harness and brake hose in bracket and secure with clip.
51. Install road wheel(s) and tighten nuts to 85 ft. lbs. (115 Nm).
52. Remove stand and lower vehicle.
53. Connect battery ground lead.

REAR

1. Before servicing the vehicle, refer to the precautions in the beginning of this section.
2. Raise rear of vehicle.

✲✲ WARNING

Do not work on or under a vehicle supported only by a jack. Always support the vehicle on safety stands.

3. Remove road wheel.
4. Clamp brake hose to prevent fluid loss.
5. Position absorbent cloth to catch spillage.
6. Loosen brake pipe union to hose and release union.

✲✲ CAUTION

Always fit plugs to open connections to prevent contamination.

7. Remove clip securing brake hose to bracket on shock absorber. Release brake hose from bracket.
8. Release ABS sensor harness and brake hose from shock absorber.
9. Release ABS sensor from hub.
10. Remove 2 nuts and bolts securing hub to shock absorber.
11. Release shock absorber from hub.
12. Remove rear quarter lower casing.
13. Remove 3 nuts from shock absorber top mounting and remove spring and shock absorber assembly.
14. Remove rubber seal from top mounting.
15. Position a suitable spring compressor in vice.

✲✲ CAUTION

Note alignment of top mounting, spring and shock absorber dust cover.

16. Position spring and shock absorber assembly to spring compressor. Compress spring.
17. Reference mark between top mounting and spring.
18. Remove cover from top mounting.
19. Compress spring by 2–3 cm until loose, hold shock absorber shaft with Allen key, remove and discard mounting plate nut.
20. Remove top mounting plate.
21. Remove rebound washer and mounting plate.
22. Remove spring aid and bump plate.
23. Remove spring seat, dust cover and bump stop cup.
24. Remove shock absorber from spring.
25. Release and remove spring from compressor.

To install:

26. Inspect shock absorber, spring mounting rubbers and bearing for deterioration and damage.
27. Clean mating faces of spring, mounting and mounting plate.
28. Clean shock absorber shaft and bump stop plate.
29. Position spring and shock absorber assembly to spring compressor. Compress spring.
30. Install shock absorber to spring, ensure spring locates in cut recess in shock absorber plate.
31. Install bump stop, bump stop cup and dust cover to shock absorber.
32. Install spring aid and bump plate.
33. Install mounting plate and rebound washer.
34. Using new nut, hold shock absorber shaft with Allen key and tighten nut to 42 ft. lbs. (57 Nm).

✲✲ CAUTION

Note alignment of top mounting, spring and shock absorber dust cover.

35. Install top mounting cover.
36. Release and remove spring from compressor.
37. Clean mating face of top mounting plate.
38. Install rubber seal to top mounting.
39. Position shock absorber assembly and align top mounting to body, fit nuts and tighten to 33 ft. lbs. (45 Nm).
40. Install rear quarter lower trim casings.
41. Install hub to shock absorber and tighten bolts to 151 ft. lbs. (205 Nm).
42. Clean ABS sensor, smear sensor with an anti-seize grease and fit sensor to hub.

✲✲ CAUTION

Ensure ABS sensor is fully located into hub, so that sensor touches pole wheel teeth.

43. Secure brake hose and ABS sensor harness to shock absorber.
44. Secure brake hose with C-clip.
45. Remove plugs and clean brake pipe male end.
46. Align hose to brake pipe and tighten union to 10 ft. lbs. (14 Nm).
47. Remove clamp from brake hose.
48. Bleed brake system.
49. Install road wheel(s) and tighten nuts to 85 ft. lbs. (115 Nm).
50. Remove stands and lower vehicle.

Coil Springs

REMOVAL & INSTALLATION

Discovery (2003–04)

FRONT

1. Before servicing the vehicle, refer to the precautions in the beginning of this section.
2. Raise and support front of vehicle, supporting vehicle under chassis.

**** WARNING**

Do not work on or under a vehicle supported only by a jack. Always use safety stands.

3. Remove front road wheels.
4. Support weight of front axle.
5. Remove 2 nuts securing anti-roll bar links to front axle and disconnect links from axle.
6. Remove 2 bolts securing each shock absorber to front axle.

**** WARNING**

Ensure axle cannot move when shock absorber is disconnected. Shock absorbers limit downward motion of the axle. If axle is not restrained, disconnecting shock absorber will allow unrestricted movement which may cause personal injury or damage to vehicle or equipment.

7. Low the front axle.

**** CAUTION**

Ensure brake hoses and ABS sensor harnesses are not damaged when front axle is lowered.

8. Release and remove the front coil spring from the shock absorber.

To install:

9. Clean front spring seats.
10. Position front spring, with close coil at the top end, over the shock absorber and located in the cutout in the lower spring seat.
11. Ensure both front springs are correctly located in the spring seats, then raise the front axle. Install shock absorber–to–axle bolts and tighten to 33 ft. lbs. (45 Nm).
12. Ensure washer is in place on lower ball joint of each anti-roll bar link, then connect lower ball joints to axle. Tighten ball joint nuts to 74 ft. lbs. (100 Nm).
13. Install front road wheels. Tighten nuts to 103 ft. lbs. (140 Nm).
14. Remove safety stands and lower the vehicle.

REAR

1. Before servicing the vehicle, refer to the precautions in the beginning of this section.
2. Raise and support rear of vehicle, supporting vehicle under chassis.

**** WARNING**

Do not work on or under a vehicle supported only by a jack. Always use safety stands.

3. Remove rear road wheels.
4. Support weight of rear axle on a jack.
5. Remove bolt, on each side, securing each shock absorber to axle.

**** WARNING**

Ensure axle cannot move when shock absorber is disconnected. Shock absorbers limit downward motion of the axle. If axle is not restrained, disconnecting shock absorber will allow unrestricted movement which may cause personal injury or damage to vehicle or equipment.

6. Remove clip securing brake pipe to bracket.
7. Release ABS sensor lead from bracket.
8. Lower axle on jack and remove coil spring.

To install:

9. Clean coil spring seats.
10. Position coil springs in vehicle, with close coil at the top end.
11. Ensure the spring is correctly located on the spring seats and raise the axle. Install and tighten the shock absorber–to–axle bolt to 91 ft. lbs. (124 Nm).
12. Fit clip to secure brake pipe to bracket.
13. Fit ABS sensor lead to bracket.
14. Install road wheels and tighten nuts to 103 ft. lbs. (140 Nm).
15. Remove safety stands and jack and lower vehicle.

Air Springs

REMOVAL & INSTALLATION

Discovery (2001–04)

REAR

1. Before servicing the vehicle, refer to the precautions in the beginning of this section.
2. Using TestBook, depressurize SLS air system.

➡After depressurizing, about 15 psi (1 bar) of air pressure remains in the system.

3. Raise and support rear of vehicle.

**** WARNING**

Never work on or under a vehicle supported only by a jack. Always use safety stands.

4. Remove the road wheels.
5. Remove the 2 clips securing the air spring to the chassis.
6. Collapse the air spring and disconnect the pipe from the top of the air spring.

**** CAUTION**

Always fit plugs to open connections to prevent contamination.

7. Rotate air spring to unlock it from the axle and remove air spring.
8. Remove connector from pipe.

To install:

9. Clean mating faces of air spring, axle and chassis.

**** CAUTION**

Check air spring bag for any sign of damage. If bag is damaged, air spring assembly must be replaced.

10. Fit pipe connector to air spring.
11. Location air spring onto axle and rotate 90° to fully engage bayonet fitting.
12. Connect pipe to top of air spring.
13. Engage top locating pins of air spring in chassis and secure with clips.
14. Using TestBook, re-pressurized SLS system.

**** WARNING**

Eye protection must be worn during this procedure.

15. Install road wheels and tighten nuts to 103 ft. lbs. (140 Nm).
16. Remove safety stands and lower vehicle.

Range Rover (2001–02)

FRONT

1. Before servicing the vehicle, refer to the precautions in the beginning of this section.
2. Use TestBook to properly depressurize the self-leveling suspension (SLS) system.

3. Remove the road wheels.
4. Remove the wheel well liner.
5. Disconnect the air tube from the spring. Seal the openings.
6. Place supports under the front crossmember.
7. Remove the clips securing the air spring to the chassis.
8. Remove the bolt securing the air spring to the axle.
9. Remove the pin.
10. Raise the chassis slightly with a jack to provide clearance.
11. Remove the spring.

WARNING

Keep the chassis supported until the system is pressurized after repairs are complete. Never allow the chassis to rest on a deflated spring.

To install:

12. Clean the mating surfaces of the chassis, axle and spring.
13. Place the air spring on the axle and install the pin and bolt.
14. Lower the chassis just enough to connect the air spring.

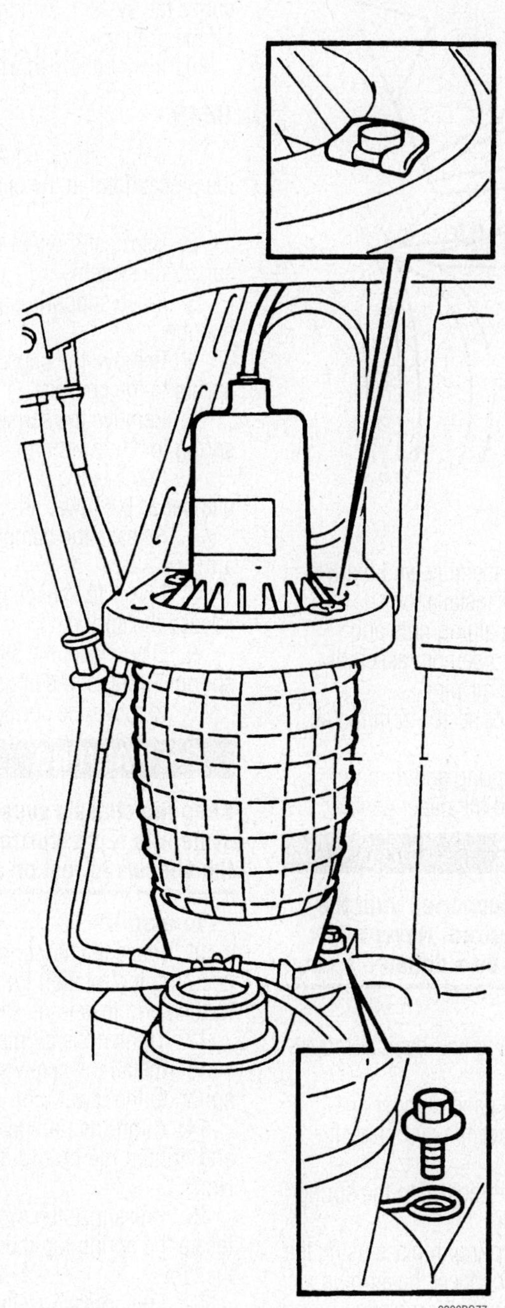

Front air spring mounting—Range Rover (2001–02)

15. Connect the pipe.
16. Repressurize the system.
17. Install the liner.

REAR

1. Before servicing the vehicle, refer to the precautions in the beginning of this section.
2. Use TestBook to properly depressurize the system.
3. Remove the wheel well liner.
4. Place supports under the rear crossmember.
5. Remove the clips securing the air spring to the chassis.
6. Remove the bolt securing the air spring to the axle.
7. Raise the chassis slightly with a jack to provide clearance.
8. Move the air spring away from the chassis.
9. Disconnect the air tube from the spring. Seal the openings.
10. Remove the spring.

WARNING

Keep the chassis supported until the system is repressurized. Never allow the chassis to rest on a deflated spring.

To install:

11. Clean the mating surfaces of the chassis, axle and spring.
12. Connect the pipe.
13. Connect the spring to the chassis.
14. Lower the chassis just enough to connect the air spring to the axle.
15. Place the air spring on the axle and install the clip.
16. Repressurize the system.
17. Install the liner.

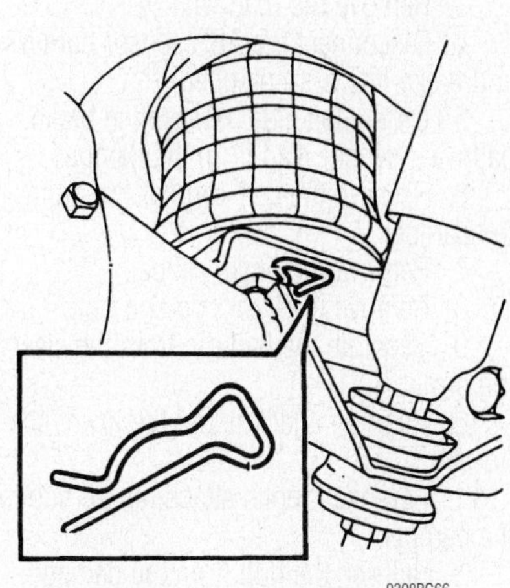

Rear air spring lower mounting clip—Range Rover (2001–02)

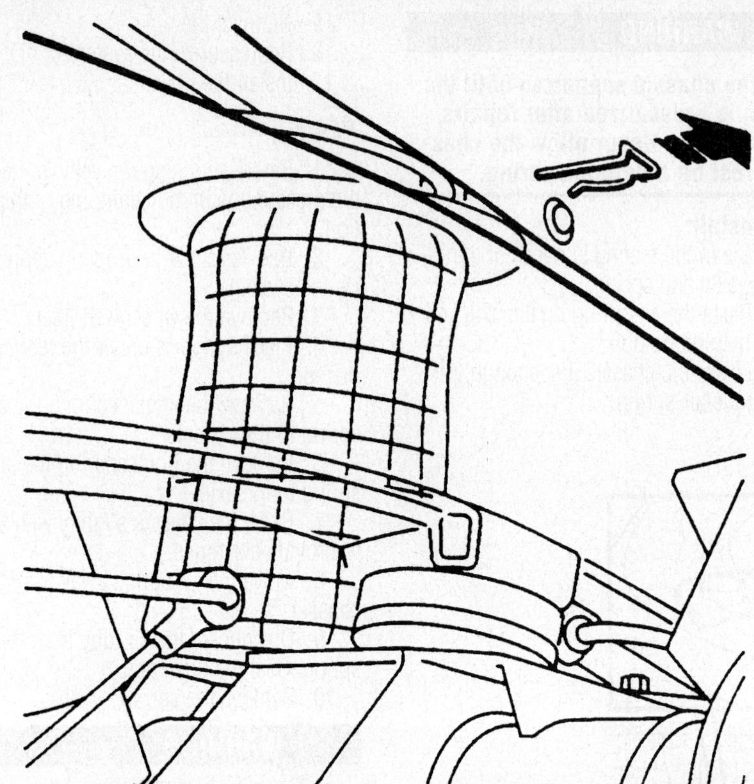

Rear air spring upper mounting clip—Range Rover (2001–02)

Range Rover (2003–04)

FRONT

1. Before servicing the vehicle, refer to the precautions in the beginning of this section.
2. Use TestBook/T4 to properly depressurize the self-leveling suspension (SLS) system.
3. Remove the road wheels.
4. Disconnect the ABS sensor harness and brake line from damper.
5. On the left side, release the brake pad wear sensor lead from the damper.
6. Remove the anti-roll bar link from the damper.
7. Remove the brake caliper.
8. Remove the track rod end nut.
9. Separate the ball pin from the steering arm.
10. Position a jack under the lower control arm.
11. Remove 2 bolts attaching the hub to the damper.
12. Release the hub from the damper.
13. From inside the engine compartment, remove the inboard nut attaching the spring and damper assembly, and mark the reference location for installation.
14. Remove 2 remaining nuts and release the spring and damper assembly enough to access the air pipe.
15. Remove the connector from the air spring.
16. Remove the spring and damper assembly and discard the paper gasket.

⁂ WARNING

Keep the chassis supported until the system is repressurized. Never allow the chassis to rest on a deflated spring.

To install:

17. Clean the spring and damper mounting locations.
18. If a new spring and damper are being installed, remove the air connector from the old spring.
19. Fit a new paper gasket to the spring and damper assembly.
20. Position the spring in place using the reference marks and connect the air pipe.
21. Install the mounting nuts and tighten the nuts to 42 ft. lbs. (56 Nm).
22. Install the hub to the damper. Tighten the bolts to 184 ft. lbs. (250 Nm).
23. Using a new nut, install the ball joint to the steering arm and tighten the nut to 59 ft. lbs. (80 Nm).
24. Install the brake caliper.
25. On the left side, install the brake pad wear sensor lead to the damper.
26. Connect the ABS sensor harness and brake line from damper.
27. Install the track rod end nut.
28. Install the anti-roll bar link to the damper. Tighten the nut to 74 ft. lbs. (100 Nm).
29. Use TestBook/T4 to properly pressurize the system, with the vehicle weight off the ground.
30. Install the road wheels.

REAR

1. Before servicing the vehicle, refer to the precautions in the beginning of this section.
2. Use TestBook/T4 to properly depressurize the system.
3. Place supports under the rear crossmember.
4. Remove the clips securing the air spring to the chassis.
5. Remove the screw securing the air spring to the lower arm.
6. Remove the bolt securing the damper to the lower arm.
7. Release the damper from the lower arm.
8. Lower the suspension arm and release the spring.
9. Disconnect the air tube from the spring. Seal the openings.
10. Remove the spring.

⁂ WARNING

Keep the chassis supported until the system is repressurized. Never allow the chassis to rest on a deflated spring.

To install:

11. Place the air spring on the axle, ensuring the rotation lock spigot is correctly located and install the clip.
12. Tighten the air pipe connection.
13. Install the screw securing the air spring to the lower arm.
14. Align the damper to the lower arm and tighten the bolt to 81 ft. lbs. (110 Nm).
15. Pressurize the system enough to locate the spring top mounting and install the clip.
16. Fully pressurize the system.
17. Install the road wheels.

Upper Ball Joint

REMOVAL & INSTALLATION

➡Each ball joint can be replaced up to 3 times before the yoke bore becomes over-sized. At each replacement, the yoke should be marked. Check for yellow paint marks about 0.5 inch (12mm) wide. If 3 marks are found, axle case must be replaced.

Discovery (2001–04)

1. Before servicing the vehicle, refer to the precautions in the beginning of this section.
2. Raise and support vehicle chassis with safety stands.
3. Support front axle on stands.
4. Remove 2 nuts securing anti-roll bar lower links to front axle. Use a 16mm wrench to prevent link joint from turning.
5. Remove 2 lower bolts securing each front shock absorber to axle.
6. Remove 8 bolts and remove crossmember.
7. Remove the bolt securing the brake hose and ABS sensor harness bracket to the axle.
8. Lower the front axle. Release the front spring from the shock absorber and remove the spring.
9. To tool LRT-54-008, fit LRT-54-008/4, LRT-54-008/5 and LRT-54-008/22 as illustrated.
10. Press upper ball joint from axle. When ram lead screw reach end of its stroke, retract lead screw and screw the ram further into the tool. Repeat operation until ball joint is freed from axle.
11. Dismantle the tool set.

To install:

12. Clean the upper ball joint location and surrounding area of axle yoke thoroughly.
13. Place a 0.5 inch (12mm) yellow paint stripe on the axle yoke, adjacent to the bore.
14. Place tool LRT-54-008 with LRT-54-021 (attached with screw), then fit tool LRT-54-008/7 to tool and position over ball joint and axle.
15. Align the tool assembly and press the ball joint into the axle yoke. Do not install nut at this time.

⁂ CAUTION

Damage to ball joint boot will result if tool is not correctly aligned during installation process.

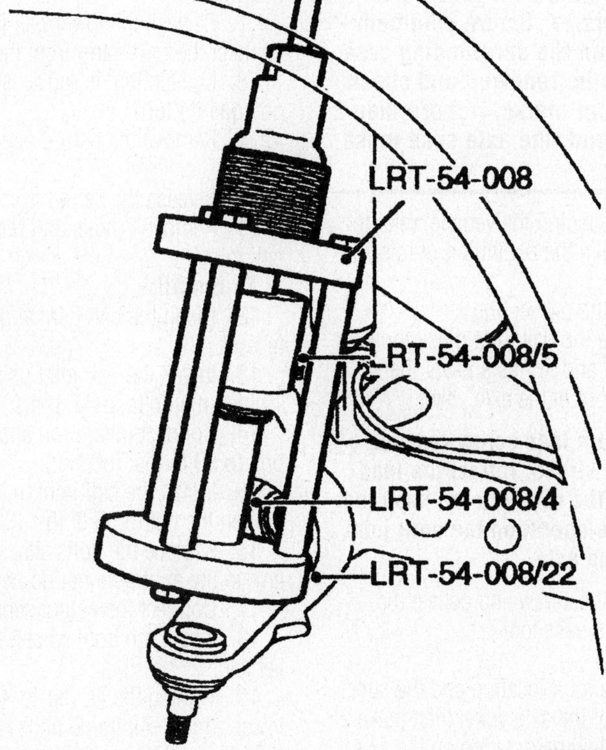

Tool assembly for removing upper ball joint—Discovery (2001–04)

16. Remove tools from axle yoke.
17. Clean the spring seats and install the coil springs over shock absorber. Tighten shock absorber bolts to 33 ft. lbs. (45 Nm).
18. Install crossmember and tighten bolts to 18 ft. lbs. (25 Nm).
19. Connect the anti-roll bar links to axle. Torque the nuts to 74 ft. lbs. (100 Nm).
20. Install steering knuckle. Tighten upper ball joint nut to 81 ft. lbs. (110 Nm)
21. The remainder of assembly is the reverse of disassembly.

Range Rover (2001–02)

1. Before servicing the vehicle, refer to the precautions in the beginning of this section.
2. Remove the swivel hub.
3. Using a ball joint forcing tool, such as LRT-54-008/4 and /5, remove the ball joint.

To install:

4. Clean the yoke thoroughly.
5. Place a yellow paint stripe on the yoke adjacent to the bore.
6. Place a forcing tool, such as LRT-54-008-8, on the yoke.
7. Place base LRT-54-008/7 on the yoke. Position the ball joint and align the tool.
8. Press the ball joint into the yoke.

Range Rover (2003–04)

REAR

1. Before servicing the vehicle, refer to the precautions in the beginning of this section.
2. Raise and support the vehicle.
3. Remove the wheels.
4. Disconnect the ABS sensor.
5. Remove the brake hose brackets.
6. Position a jack under the lower control arm.
7. Reference mark the eccentric cam to upper arm.

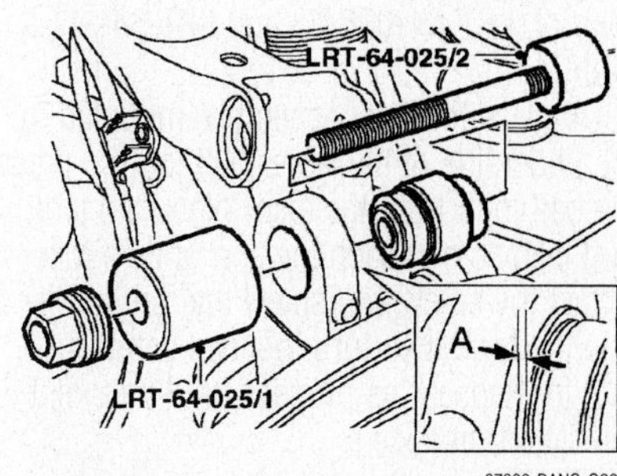

Rear upper ball joint mounting location—Range Rover (2003–04)

8. Remove nut attaching the hub to control arm, and disconnect the hub from the arm.

9. Using tool LRT-64-025, remove the ball joint.

To install:

10. Clean the ball joint and hub opening.

11. Using tool LRT-64-025, install the ball joint with the chamfer toward the rear.

12. Ensure that the ball joint protrudes from the machined face 0.030 inch (0.75 mm) as shown in dimension A.

13. Align the hub to the upper arm, install new bolt and using the reference mark align the eccentric cam and lightly tighten the bolt.

14. Connect the brake pipe.

15. Connect the ABS sensor.

16. Install the wheels.

17. Lower the vehicle.

18. Tighten the upper arm-to-hub nut to 121 ft. lbs. (165 Nm).

19. Check the wheel alignment,

Lower Ball Joint

REMOVAL & INSTALLATION

➡Each ball joint can be replaced up to 3 times before the yoke bore becomes over-sized. At each replacement, the yoke should be marked. Check for yellow paint marks about 0.5 inch (12mm) wide. If 3 marks are found, axle case must be replaced.

Freelander (2001–04)

The ball joints are not replaceable. If a ball joint becomes defective, the control arm must be replaced.

Discovery (2001–04)

1. Before servicing the vehicle, refer to the precautions in the beginning of this section.

2. Remove the steering knuckle.

3. On tool LRT-54-008, install LRT-54-008/22, LRT-54-008/23 and LRT-54-008/24.

4. Use this tool assembly, mounted to the underside of the lower ball joint to press the ball joint from the axle. When the tool ram lead screw reach the end of its stroke, retract the screw and screw the ram further into the tool, then turn the lead screw further in. Repeat this process until ball joint is released from axle.

5. Remove the tool set.

6. Remove the lower ball joint from the axle yoke.

To install:

7. Clean the ball joint location and surround area of axle yoke thoroughly.

8. Place a 0.5 inch (12mm) wide yellow paint stripe on the axle yoke adjacent to the lower ball joint bore.

9. Place a forcing tools, such as LRT-54-008 with LRT-54-022 and 008/14 onto the lower ball joint and position on the axle yoke.

10. Position the ball joint and ensure the tools are properly aligned.

11. Press the ball joint into the yoke.

⁂ CAUTION

Damage to ball joint boot will result if tools are not properly aligned.

12. Remove the tool set.

13. Position the knuckle on the ball studs. Torque the lower ball joint nut to 100 ft. lbs. (135 Nm).

Range Rover (2001–02)

➡Each ball joint can be replaced up to 3 times before the yoke bore becomes over-sized. At each replacement, the yoke should be marked. Factory-trained technicians mark the yoke with yellow paint at each replacement.

⁂ CAUTION

Each ball joint can be replaced up to 3 times before the axle yoke bore becomes oversize. Before commencing work, clean the surrounding area of the joint to be renewed and check for yellow paint marks. If more than 2 marks are found, the axle case must be renewed.

1. Before servicing the vehicle, refer to the precautions in the beginning of this section.

2. Remove the swivel hub.

3. Assemble the ball joint press tools LRT-54-008/10 and /11 as shown, then press the joint out of the axle yoke.

➡When the ram lead screw reaches the end of the stroke, retract the lead screw, screw the ram into the base tool and repeat the operation the until joint is free from the axle.

4. Remove the screw and collect the adapter from the base tool.

To install:

5. Clean the joint location and the surrounding area of the axle yoke, then make a 12mm wide yellow paint stripe on the axle yoke, adjacent to joint location.

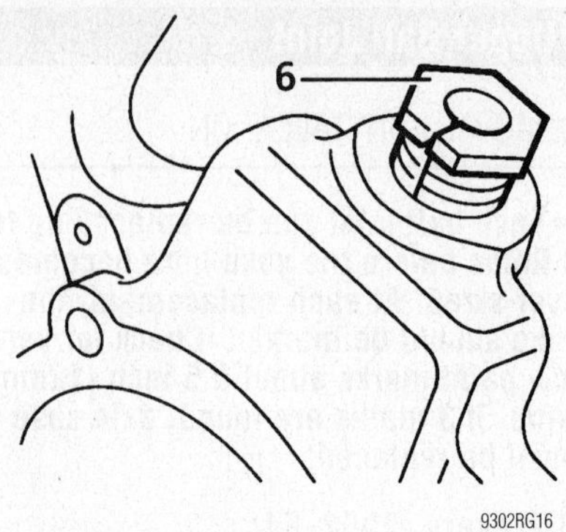

Be sure to install the adjustable collet (6) into the swivel hub—Range Rover (2001–02)

6. Align the tool assembly and press the joint into the axle yoke.

7. Remove the tools from the axle yoke.

8. Install the swivel hub onto the axle.

Range Rover (2003–04)

FRONT

1. Before servicing the vehicle, refer to the precautions in the beginning of this section.

2. Raise and support vehicle chassis with safety stands.

3. Remove the front wheels.

4. Loosen the lower arm-to-subframe bolt a quarter turn.

5. Remove the height sensor link.

6. Remove the lower arm ball joint-to-hub nut.

7. Release the lower ball joint and disconnect the lower arm from the hub.

8. Loosen the tie rod-to-subframe bolt one quarter turn.

9. Remove the tie rod end-to-ball joint nut.

10. Release the tie rod from the ball joint.

11. Remove 2 bolts and remove lower ball joint.

To install:

12. Clean the ball joint and mating surface.

13. Install the ball joint using new bolts. Tighten the bolts to 44 ft. lbs. (60 Nm).

14. Connect the tie rod and tighten the bolt to 59 ft. lbs. (80 Nm).

15. Install the ball joint to the hub and tighten the nut to 59 ft. lbs. (80 Nm).

16. Replace the bolts attaching the lower arm to the subframe but do not tighten.

17. Connect the height sensor link.

18. Install the front wheels and lower the vehicle.

19. Tighten the tie rod-to-subframe and lower arm-to-subframe bolts to 121 ft. lbs. (165 Nm), plus an additional 90 degrees.

20. Check the front wheel alignment.

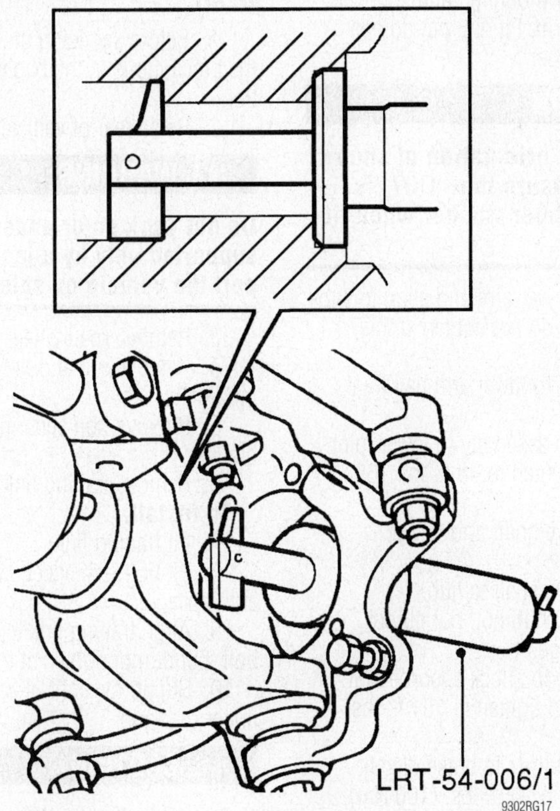

Install the tool into the axle casing with the 'TOP' mark facing out upwards—Range Rover (2001–02)

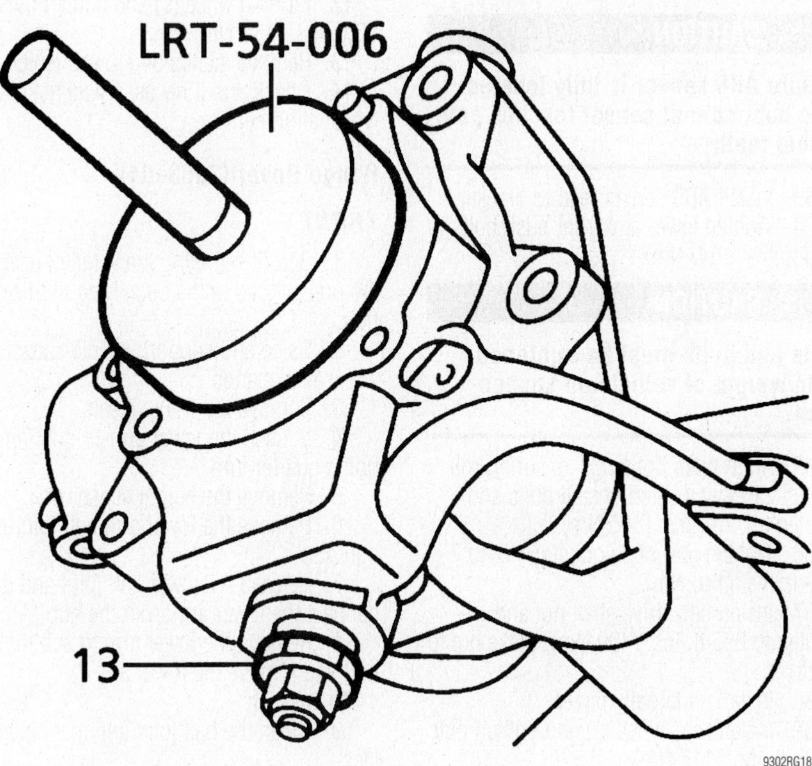

Using tool LTR-54-006/2 to adjust the height of the hub—Range Rover (2001–02)

Range Rover (2003–04)

REAR

1. Before servicing the vehicle, refer to the precautions in the beginning of this section.
2. Raise and support vehicle chassis with safety stands.
3. Remove the rear wheels.
4. Remove the tie rod.
5. Position a jack under the lower control arm.
6. Remove and discard the lower control arm-to-hub bolt.
7. Lower the jack and disconnect the hub from the arm.
8. Remove and discard the circlip.
9. Remove the lower brake shield bolt.
10. Using tool LRT-64-026, remove the ball joint.

To install:

11. Clean the ball joint and mating surface.
12. Install the ball joint with the circlip facing the rear.
13. Install the brake shield lower bolt.
14. Align the hub to the control arm and tighten the new bolt and nut to 184 ft. lbs. (250 Nm).
15. Install the tie rod.
16. Install the rear wheels.
17. Check the wheel alignment

Lower Control Arm

REMOVAL & INSTALLATION

Freelander (2001–04)

FRONT

1. Before servicing the vehicle, refer to the precautions in the beginning of this section.
2. Raise front of vehicle.

⁂ WARNING

Do not work on or under a vehicle supported only by a jack. Always support the vehicle on safety stands.

3. Remove road wheel.
4. Remove underbelly panel.
5. Release stake in drive shaft nuts.
6. With an assistant applying brakes, remove and discard drive shaft hub nut.
7. Remove clip securing brake hose to bracket on shock absorber.
8. Release ABS sensor harness and brake hose from shock absorber.
9. Release ABS sensor from hub.
10. Remove 2 bolts securing brake caliper to hub. Release caliper from hub and tie aside.

CAUTION

Do not allow caliper to hang on brake hose.

11. Remove 2 nuts and bolts securing hub to shock absorber.
12. Release drive shaft from hub.
13. Tie drive shaft aside.
14. Remove nut securing lower arm ball joint, discard nut.
15. Break taper joint using LRT-57-043.
16. Remove hub assembly.
17. Remove 2 bolts securing lower control arm rear bushing housing.
18. Remove bolt securing lower control arm front mounting.
19. Remove lower control arm.
20. Remove nut from rear mounting, remove snubber washer and remove mounting.

CAUTION

Note orientation of snubber washer.

To install:

21. Clean rear mounting mating faces.

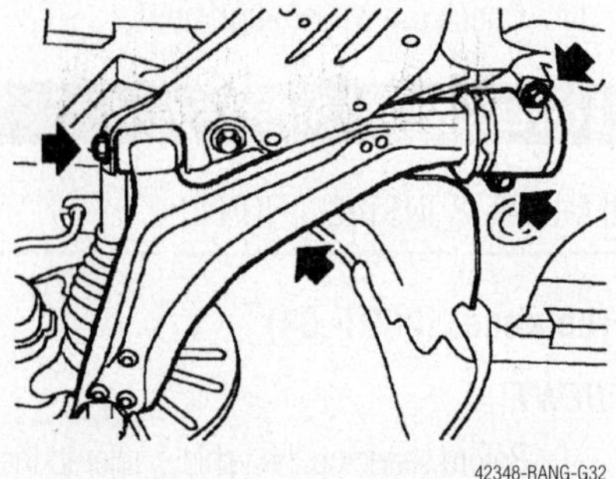

42348-RANG-G32

Showing the location of the bolts securing the lower control arm—Freelander (2001–04)

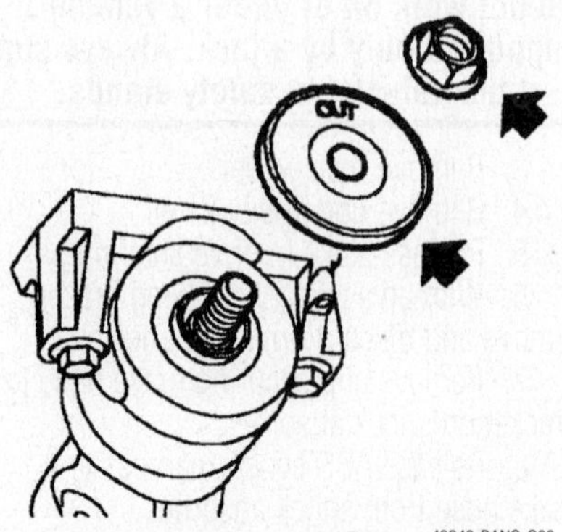

42348-RANG-G33

Removing the nuts and snubber washer from the rear lower control arm mounting—Freelander (2001–04)

22. Install rear mounting and snubber rubber to lower arm. Fit nut, but do not tighten at this stage.

CAUTION

Ensure correct orientation of snubber washer. Ensure that 'OUT' is visible on snubber washer when fitted.

23. Position lower arm and align to subframe fit bolt, but do not tighten at this stage.
24. Clean hub to lower arm ball joint mating faces.
25. Install hub assembly to lower joint, fit new nut and tighten to 48 ft. lbs. (65 Nm).
26. Clean drive shaft and flange splines.
27. Install drive shaft to hub.
28. Install new hub nut, but do not tighten at this stage.
29. Install hub to shock absorber. Fit nuts and bolts and tighten to 151 ft. lbs. (205 Nm).
30. Position caliper to brake disc fit bolts and tighten to 74 ft. lbs. (100 Nm).
31. Install brake hose to bracket on shock absorber and secure with clip.
32. Clean ABS sensor, smear sensor with an anti-seize grease and fit sensor to hub.

CAUTION

Ensure ABS sensor is fully located into hub, so that sensor touches pole wheel teeth.

33. Install ABS sensor lead to bracket.
34. Tighten lower arm front bush bolts to 140 ft. lbs. (190 Nm).

CAUTION

Nuts and bolts must be tightened with weight of vehicle on suspension.

35. Align bush housings, ensuring roll pin is correctly located. Install bolts and tighten to 77 ft. lbs. (105 Nm).
36. Tighten rear bush housing nut to 103 ft. lbs. (140 Nm).
37. Install new drive shaft nut and tighten to 295 ft. lbs. (400 Nm). Stake nut to shaft.
38. Install underbelly panel.
39. Install road wheel(s) and tighten nuts to 85 ft. lbs. (115 Nm).
40. Remove stands and lower vehicle.

REAR

1. Before servicing the vehicle, refer to the precautions in the beginning of this section.
2. Raise rear of vehicle.

WARNING

Do not work on or under a vehicle supported only by a jack. Always support the vehicle on safety stands.

3. Remove road wheel.
4. Remove nut and bolt securing trailing link to hub.
5. Remove bolt securing trailing link to bracket.
6. Remove trailing link.

To install:

7. Fit trailing link.
8. Fit bolt to bracket, but do not tighten at this stage.
9. Align trailing link to hub, fit nut and bolt, but do not tighten at this stage.
10. Support weight of vehicle with jack under rear hub.

CAUTION

Nuts and bolts must be tightened with weight of vehicle on suspension.

11. Tighten nuts and bolts to 89 ft. lbs. (120 Nm).
12. Fit road wheel(s) and tighten nuts to 85 ft. lbs. (115 Nm).
13. Remove stands and lower vehicle.
14. Check and if necessary adjust rear wheel alignment.

Range Rover (2003–04)

FRONT

1. Before servicing the vehicle, refer to the precautions in the beginning of this section.
2. Raise and support vehicle chassis with safety stands.
3. Remove the front wheels.
4. Loosen the lower arm-to-subframe bolt a quarter turn.
5. Remove the height sensor link.
6. Remove the lower arm ball joint-to-hub nut.
7. Release the lower ball joint and disconnect the lower arm from the hub.
8. Remove the lower arm to subframe bolt and remove the lower arm.

To install:

9. Clean the ball joint and mating surface.
10. Install the ball joint to the hub and tighten the nut to 59 ft. lbs. (80 Nm).

11. Connect the height sensor link.
12. Replace the bolts attaching the lower arm to the subframe but do not tighten.
13. Install the front wheels and lower the vehicle.
14. Tighten the tie rod-to-subframe and lower arm-to-subframe bolts to 121 ft. lbs. (165 Nm), plus an additional 90 degrees.
15. Check the front wheel alignment.

REAR

1. Before servicing the vehicle, refer to the precautions in the beginning of this section.
2. Raise and support vehicle chassis with safety stands.
3. Remove the lower sway bar link.
4. Remove the height sensor link from the lower arm.
5. Depressurize the air suspension system.
6. Remove the damper-to-lower arm bolt and disconnect the damper.
7. Remove the bolt attaching the air spring to the lower arm.
8. Remove and discard the lower arm-to-hub bolt.
9. Loosen 2 bolts attaching the lower arm to the subframe.
10. Disconnect the air spring spigot from the lower arm.
11. Remove the 2 bolts attaching the lower arm to the hub and remove the lower arm.

To install:

12. Clean the lower arm and pivot and subframe mating surfaces.
13. Position the lower arm, install the new bolts, but do not tighten.
14. Fit the air spring to the lower arm and tighten the bolt.
15. Align the damper to the lower arm and tighten the bolt to 81 ft. lbs. (110 Nm).
16. Tighten the lower arm-to-subframe bolts to 121 ft. lbs. (165 Nm).
17. Pressurize the air suspension system.
18. Install the height sensor link.
19. Install the sway bar link.
20. Lower the vehicle.

Bushing Replacement

REMOVAL & INSTALLATION

Freelander (2001–04)

LOWER CONTROL ARM FRONT BUSHING

1. Before servicing the vehicle, refer to the precautions in the beginning of this section.
2. Remove underbelly panel.
3. Raise front of vehicle.

✱✱ WARNING

Do not work on or under a vehicle supported only by a jack. Always support the vehicle on safety stands.

4. Remove 2 bolts securing rear bush housing to rear beam.
5. Release rear bush housing from body dowel.
6. Remove bolt securing lower arm front mounting.
7. Release lower arm front mounting from rear beam.
8. Using LRT-60-008 and adapters remove lower arm bush.

To install:

9. Ensure bush bore in hub is clean.
10. Using LRT-60-008 fit new bush into lower arm.
11. Align lower arm to rear beam and fit bolt, but do not tighten at this stage.
12. Align rear bush housing to body dowel and fit bolts, but do not tighten at this stage.
13. Tighten rear bush housing bolts to 77 ft. lbs. (105 Nm).

✱✱ CAUTION

Nuts and bolts must be tightened with the weight of the vehicle on the suspension.

14. Tighten lower arm front bush bolts to 140 ft. lbs. (190 Nm).
15. Fit underbelly panel.
16. Remove stands and lower vehicle.
17. Check and, if necessary, adjust wheel alignment.

LOWER CONTROL ARM REAR BUSHING

1. Before servicing the vehicle, refer to the precautions in the beginning of this section.
2. Raise front of vehicle.

✱✱ WARNING

Do not work on or under a vehicle supported only by a jack. Always support the vehicle on safety stands.

3. Remove nut securing lower control arm rear bush housing and remove snubber rubber.

✱✱ CAUTION

Note orientation of snubber washer.

4. Remove 2 bolts securing lower control arm rear bush housing.

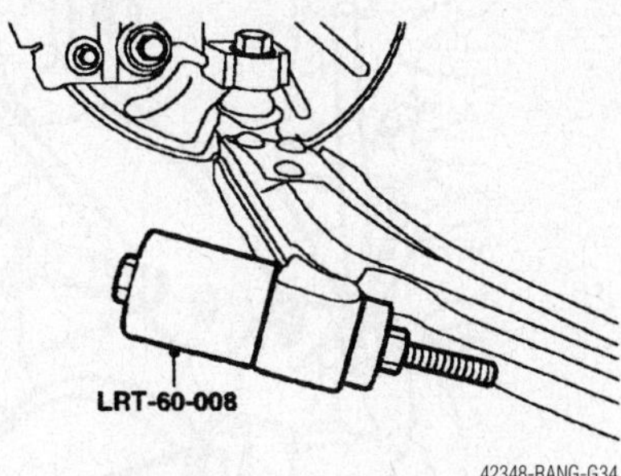

Using a bushing removal tool to remove the lower control arm front bushing—Freelander (2001–04)

5. Release bush locating pin from body and remove bush housing.

To install:

6. Clean lower control arm and bush housing mating faces.
7. Fit bush housing to lower arm and locate dowel into body.
8. Fit bolts securing lower arm rear bush housings and tighten to 77 ft. lbs. (105 Nm).
9. Fit snubber rubber and nut, but do not tighten at this stage.

✱✱ CAUTION

Ensure correct orientation of snubber washer. Ensure that 'OUT' is visible on snubber washer when fitted.

10. Tighten rear bush housing nut to 103 ft. lbs. (140 Nm).

✱✱ CAUTION

Nuts and bolts must be tightened with the weight of the vehicle on the suspension.

11. Remove stand and lower vehicle.
12. Check front wheel alignment.

Discovery (2001–04)

PANHARD ROD BUSHINGS

1. Before servicing the vehicle, refer to the precautions in the beginning of this section.
2. Raise and support front of vehicle.

✱✱ WARNING

Do not work on or under a vehicle supported only by a jack. Always support the vehicle on safety stands.

3. Remove 2 nuts and bolts securing panhard rod to axle and chassis. Remove panhard rod.
4. Using a suitable bushing removal tool, such as LRT-60-013 with 013/1 and 013/3, press bushings out of panhard rod.

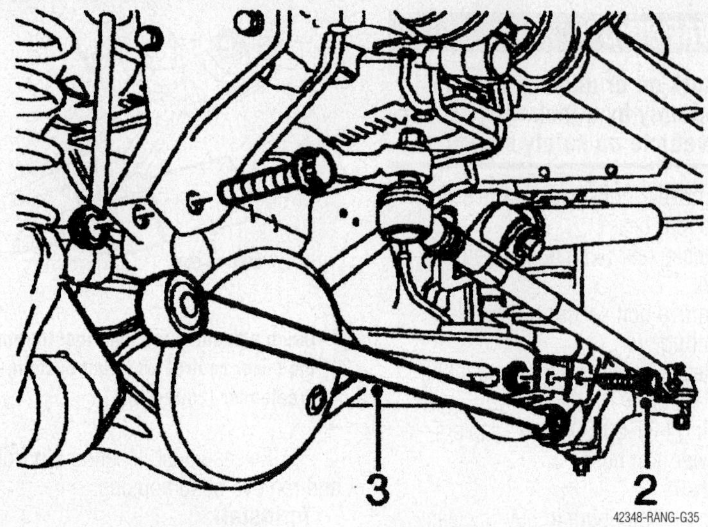

Showing the location of the panhard rod—Discovery (2000–04)

To install:

5. Clean bushing locations in panhard rod.

6. Use appropriate bushing installer, such as LRT-60-013 with 013/1 and 013/2, to press new bushings into panhard rod.

⁂ CAUTION

Ensure pressure is applied to outer edge of bushing only, not to rubber inner section.

7. Position panhard rod to axle and chassis. Install retaining bolts, but do not tighten at this time.

8. Remove safety stands and lower vehicle.

9. Tighten rod bolts to 170 ft. lbs. (230 Nm).

⁂ CAUTION

Bolts must be tightened with weight of vehicle on suspension.

RADIUS ARM BUSHINGS

1. Before servicing the vehicle, refer to the precautions in the beginning of this section.

2. Raise and support front of vehicle.

⁂ WARNING

Do not work on or under a vehicle supported only by a jack. Always support the vehicle on safety stands.

3. Remove front road wheel.

4. Remove and discard nut securing tie rod to steering knuckle.

5. Using a suitable ball joint removal tool, break taper of tie rod ball joint and release from steering knuckle. Position tie rod aside.

6. Remove and discard cable tie securing front axle breather tube.

7. Remove nut and bolt securing radius arm to chassis.

8. Remove 2 nuts securing radius arm to axle.

9. Remove rear bolt from radius arm.

10. Remove front bolt from radius arm and remove arm from axle.

11. Using LRT-60-011/1 with 011/2, press bushing out of rear end of radius arm.

12. With LRT-60-034/1, press out bushings from axle end of radius arm.

To install:

13. Clean and lubricate new bushings and radius arm mating faces.

14. Using LRT-60-011/1, 011/2 and 011/3, press new bushing into read end of radius arm.

15. Using LRT-60-034/1, press front radius arm bushing into LRT-60-034/2 until about 0.08 inch (2mm) of bushing is protruding from opposite end of tool.

16. Position tool over radius arm axle end and locate bushing to opening.

17. With assistance, press bushing into radius arm.

18. Position radius arm to vehicle and install retaining nuts and bolts, but do not tighten at this time.

⁂ CAUTION

Fasteners must be tightened with weight of vehicle on suspension.

19. Fit cable tie to secure axle breather tube.

20. Fit tie rod to steering knuckle. Install tie rod ball joint nut and tighten to 92 ft. lbs. (125 Nm).

21. Install road wheel and tighten nuts to 103 ft. lbs. (140 Nm).

22. Remove safety stands and lower vehicle.

23. Tighten radius arm nuts and bolts to 170 ft. lbs. (230 Nm).

ACE FRONT LONG ARM BUSHINGS

1. Before servicing the vehicle, refer to the precautions in the beginning of this section.

2. Remove ACE front actuator.

3. Remove securing nut and remove anti-roll bar link from torsion bar.

4. Restrain torsion bar and remove ACE long arm from torsion bar.

⁂ CAUTION

Short arm and torsion bar are supplied as an assembly and must not be separated.

5. Use a suitable drift to remove both halves of slipper bushings from ACE long arm.

6. Use suitable adapters to press actuator rod end bushings from long arm.

To install:

7. Clean bushing locations in ACE long arm.

8. Use suitable adapters to press new actuator rod end bushing into ACE long arm. Ensure hole in bushings is correctly aligned with hole in long arm.

9. Align slots in new slipper bushing halves with those in long arm. Carefully press both halves of slipper bushing into long arm. Ensure sealing rings on slipper bushing faces are not damaged.

10. Clean long arm and mating face on torsion bar.

11. Fit long arm to torsion bar. Restrain torsion bar and tighten bolt to 133 ft. lbs. (180 Nm).

12. Fit anti-roll bar link to long arm and tighten nut to 37 ft. lbs. (50 Nm).

13. Install ACE front actuator.

Range Rover (2003–04)

LOWER CONTROL ARM FRONT BUSHING

1. Before servicing the vehicle, refer to the precautions in the beginning of this section.

2. Raise front of vehicle.

⁂ WARNING

Do not work on or under a vehicle supported only by a jack. Always support the vehicle on safety stands.

3. Remove the lower control arm
4. Using LRT-60-031 and adapters remove lower arm bush.

To install:

5. Clean lower control arm and bush housing mating faces.
6. Using LRT-60-031 and adapters install lower arm bush.
7. Install the lower control arm.
8. Check front wheel alignment.

ACE Actuator

REMOVAL & INSTALLATION

Discovery (2001–04)

1. Before servicing the vehicle, refer to the precautions in the beginning of this section.

⁂ WARNING

The accelerated cornering enhancement (ACE) hydraulic system is extremely sensitive to any ingress of dirt and debris. The smallest amount could render the system unserviceable. It is imperative that the following precautions are taken:

- ACE components are thoroughly cleaned externally before any work begins.
- All open pipe and module ports are capped immediately.
- All fluid is stored in and added to system through clean containers.

2. Raise vehicle on ramp.
3. Remove right side road wheel.
4. Thoroughly clean ACE actuator and all connecting components.
5. Position container to collect fluid spillage.
6. Remove cap nuts securing fluid pipes to actuator. Disconnect pipes and discard sealing washers.

⁂ CAUTION

Always fit plugs to open connections to prevent contamination.

7. Remove 2 nuts securing anti-roll bar links to axle. Detach links from axle.
8. With assistance, remove 2 bolts and remove both clamp plates from torsion bar.
9. Remove mounting rubbers from torsion bar.
10. Remove torsion bar and actuator assembly.

⁂ CAUTION

While torsion bar is removed from vehicle, it must be stored without any load on anti-roll bar links, or ball joints and/or link rods could be damaged.

11. Use LRT-60-009 to remove nut securing ACE actuator to long arm.
12. Remove nut and bolt securing ACE actuator to short arm and remove actuator.

⁂ CAUTION

Short arm and torsion bar are supplied as an assembly and must not be separated.

To install:

13. Position ACE actuator to torsion bar. Use LRT-60-099 to tighten ACE actuator–to–long arm nut to 35 ft. lbs. (48 Nm). Tighten actuator–to–short arm nut and bolt to 133 ft. lbs. (180 Nm).
14. Ensure torsion bar mounting rubber mating faces are clean and free from damage, then fit to torsion bar.
15. With assistance, position torsion bar and ACE actuator assembly to vehicle. Fit clamp plates and tighten bolts to 33 ft. lbs. (45 Nm).
16. Ensure washer is in place on lower ball joint of each anti-roll bar link, then connect lower ball joint to axle and tighten nuts to 74 ft. lbs. (100 Nm).
17. Connect pipes to actuator with cap nuts and new sealing washers. Tighten cap nuts to 21 ft. lbs. (29 Nm). Ensure pipes are not under any tension or kinked.
18. Use TestBook to properly bleed ACE hydraulic system.
19. Fit right side road wheel and tighten nuts to 103 ft. lbs. (140 Nm).
20. Lower vehicle from ramp.

Wheel Bearings

ADJUSTMENT

The front and rear wheel bearings are not adjustable.

REMOVAL & INSTALLATION

Freelander (2001–04)

FRONT

1. Before servicing the vehicle, refer to the precautions in the beginning of this section.
2. Raise front of vehicle.

⁂ WARNING

Do not work on or under a vehicle supported only by a jack. Always support the vehicle on safety stands.

3. Remove road wheel.
4. Release stake in drive shaft nut.
5. With an assistant applying brakes, remove and discard drive shaft hub nut.
6. Remove clip securing right side brake hose to support bracket, release hose from bracket.

Release ABS sensor and pad wear sensor harnesses from bracket.

7. Remove 2 bolts securing brake caliper to hub. Release caliper from hub and tie aside.

⁂ CAUTION

Do not allow caliper to hang on brake hose.

8. Mark brake disc to hub relationship.
9. Remove 2 screws securing brake disc and remove brake disc.
10. Remove 2 nuts and bolts securing hub to shock absorber.
11. Release hub from shock absorber.
12. Release drive shaft from hub.
13. Restrain hub from rotating and remove nut from lower swivel joint.
14. Break taper joint using LRT-57-043.
15. Remove swivel hub.

➡Do not carry out further dismantling if component is removed for access only.

16. Remove 3 bolts securing brake disc shield.
17. Position hub assembly to press, support on tools LRT-54-017 and press out drive flange using tool LRT-54-014.
18. Remove brake disc shield.

➡Outer bearing track will remain on drive flange.

19. Remove bearing sealing plate from inner track.
20. Position drive flange in a vice.
21. Clamp both halves of a suitable bearing separator around inner track ensuring that inner lip fits in groove on inner track.

➡From 2002 model year, the groove in the inner track was deleted. To remove the inner track, clamp the separator around the bearing inner track surface.

22. Using tool LRT-99-500 and thrust pad LRT-54-014 withdraw inner track from drive flange.

23. Remove circlip from bearing.

24. Position hub to press and press out bearing using tool LRT-54-015 and LRT-54-017, discard bearing.

⁂ CAUTION

Never re-use existing bearing.

To install:

25. Clean hub and drive flange.

26. Support hub on tool LRT-54-016 and press in new bearing using LRT-54-015.

27. Fit circlip to hub.

28. Fit brake disc shield, fit bolts and tighten to 6.5 ft. lbs. (8.5 Nm).

29. Support bearing on tool LRT-54-015 and press drive flange into bearing using LRT-54-014.

30. Clean drive shaft.

31. Fit hub assembly to lower joint, fit new nut and tighten to 48 ft. lbs. (65 Nm).

32. Fit drive shaft to hub.

33. Fit hub to shock absorber, fit nuts and bolts and tighten to 151 ft. lbs. (205 Nm).

34. Clean brake disc to drive flange mating faces.

35. Fit disc to drive flange, align reference marks, fit screws and tighten to 4 ft. lbs. (5 Nm).

36. Clean mating faces of caliper and hub.

37. Position caliper to brake disc fit bolts and tighten to 74 ft. lbs. (100 Nm).

38. Fit brake hose to abutment bracket and fit clip.

39. Clean ABS sensor, smear sensor with an anti-seize grease and fit sensor to hub.

⁂ CAUTION

Ensure ABS sensor is fully located into hub, so that sensor touches pole wheel teeth.

40. Fit ABS sensor lead to bracket.

41. Fit new drive shaft nut and tighten to 295 ft. lbs. (400 Nm). Stake nut to shaft.

42. Stake drive shaft hub nut.

43. Fit road wheel(s) and tighten nuts to 85 ft. lbs. (115 Nm).

44. Remove stands and lower vehicle.

REAR

1. Before servicing the vehicle, refer to the precautions in the beginning of this section.

2. Raise rear of vehicle.

⁂ WARNING

Do not work on or under a vehicle supported only by a jack. Always support the vehicle on safety stands.

3. Remove road wheel.

4. With assistant depressing the brake pedal, remove and discard drive shaft nut.

5. Remove brake shoe assembly.

6. Using LRT-70-007 release handbrake cable from backplate and remove from backplate.

7. Remove clip securing brake hose to bracket on shock absorber.

8. Disconnect brake pipe union from wheel cylinder.

⁂ CAUTION

Always fit plugs to open connections to prevent contamination.

9. Release ABS sensor from hub.

10. Remove nut and bolt securing trailing link to hub.

11. Remove nut, bolt and washers securing transverse links to hub.

12. Remove 2 nuts and bolts securing hub to shock absorber.

13. Release shock absorber from hub.

14. Remove hub assembly from drive shaft.

➡Do not carry out further dismantling if component is removed for access only.

15. Position hub assembly to press, support on tools LRT-54-017 and press out drive flange using tool LRT-54-014.

➡Outer bearing track will remain on drive flange.

16. Remove bearing sealing plate from inner track.

17. Position drive flange in a vice.

18. Clamp both halves of a suitable bearing separator around inner track ensuring that inner lip fits in groove on inner track.

➡From 2002 model year, groove in the inner track was deleted. To remove the inner track, clamp the separator around the bearing inner track surface.

19. Using tool LRT-99-500 and thrust pad LRT-54-014 withdraw inner track from drive flange.

20. Install hub to vice and remove 4 bolts securing backplate to hub.

21. Remove backplate.

22. Remove circlip from bearing.

23. Position hub to press and press out bearing using tool LRT-54-015 and LRT-54-017, discard bearing.

⁂ CAUTION

Never re-use existing bearing.

To install:

24. Clean hub and drive flange.

25. Support hub on tool LRT-54-016 and press in new bearing using LRT-54-015.

26. Install circlip to hub.

27. Install hub to vice, fit backplate and tighten bolts to 33 ft. lbs. (45 Nm).

28. Support bearing on tool LRT-54-015 and press drive flange into bearing using LRT-54-014.

29. Install hub assembly to drive shaft.

30. Install hub to shock absorber and tighten bolts to 151 ft. lbs. (205 Nm).

⁂ CAUTION

Nuts and bolts must be tightened with weight of vehicle on suspension.

31. Install nut, bolt and washers securing transverse links to hub and tighten nut to 89 ft. lbs. (120 Nm).

32. Install trailing link to hub and tighten nut and bolt to 89 ft. lbs. (120 Nm).

➡Ensure that washers are fitted to both ends of bolts.

33. Clean ABS sensor, smear sensor with an anti-seize grease and fit sensor to hub.

⁂ CAUTION

Ensure ABS sensor is fully located into hub, so that sensor touches pole wheel teeth.

34. Install brake pipe to wheel cylinder and tighten union to 10 ft. lbs. (14 Nm).

35. Install clip securing brake pipe to bracket.

36. Install and secure handbrake cable to backplate.

37. Install brake shoes.

38. Install new drive shaft nut and tighten to 295 ft. lbs. (400 Nm). Stake nut to shaft.

39. Bleed brakes.

40. Install road wheel(s) and tighten nuts to 85 ft. lbs. (115 Nm).

41. Remove stands and lower vehicle.

Discovery (2001–02) and Range Rover (2001–02)

FRONT

1. Before servicing the vehicle, refer to the precautions in the beginning of this section.

2. Raise and safely support the vehicle and remove the front wheels.

3. Unbolt the brake caliper and position it aside, leaving the hose attached.
4. Remove the dust cap.
5. Remove the circlip and shim from driveshaft.
6. Remove the 5 bolts, then remove the driving member and joint washer.
7. Bend back the lockwasher tabs, then remove the locknut and washer.
8. Remove the hub adjusting nut.
9. Remove the spacing washer.
10. Remove the hub and brake disc assembly with bearings.
11. Remove the outer bearing.
12. Matchmark the hub to the brake disc, then remove the 5 hub–to–brake disk attaching bolts and separate the hub from the brake disc.
13. Remove the grease seal and inner bearing from the hub and discard the seal.
14. Using a punch and hammer, drive out the inner and outer bearing races.
15. Clean the hub of any old grease, and dry it.

To install:

16. Using a suitable bearing driver, install the inner and outer bearing races.
17. Pack the hub inner bearing with grease and install.
18. Using a suitable seal driver, install the seal until it is flush with the rear face of the hub. Apply grease between the seal lips.
19. Align the brake disc–to–hub matchmarks, apply Loctite® 270 to the mounting bolts and tighten to 54 ft. lbs. (73 Nm).
20. Grease and install the outer bearing to the hub.
21. Clean the stub axle and driveshaft
22. Install the hub assembly onto the axle and place the spacing washer .
23. Install the hub adjusting nut and tighten to 45 ft. lbs. (61 Nm).
24. Loosen the adjusting nut 90°, then tighten to 35 inch lbs. (4 Nm). This will give the required hub end-float of 0.0004 inch (0.010mm)
25. Install a new lockwasher and locknut.
26. Tighten the locknut to 45 ft. lbs. (61 Nm).
27. Fold the tab over the lockwasher to secure the adjusting nut and locknut.
28. Install driving member with a new joint washer and tighten the hub retaining bolts to 48 ft. lbs. (65 Nm).
29. Install the original driveshaft shim and secure the shim with a circlip.
30. Check the driveshaft end-play by mounting a dial gauge as shown.
31. Install a suitable bolt to the threaded

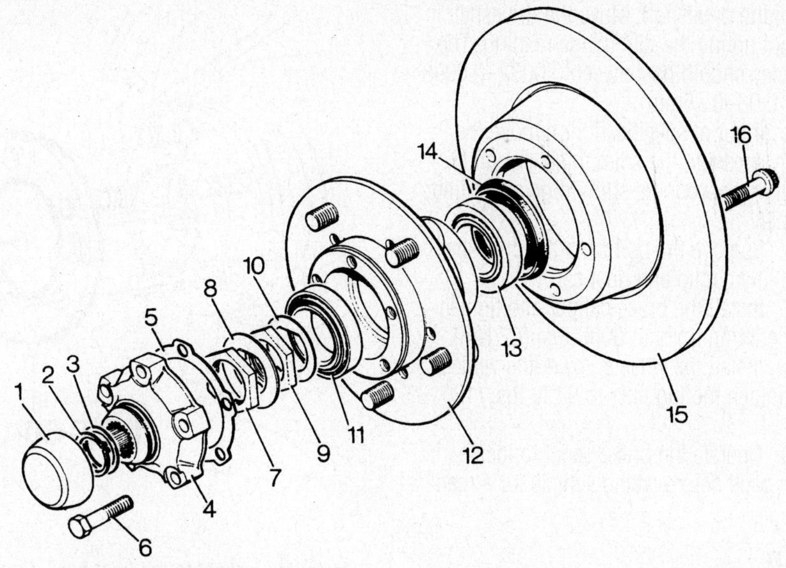

Exploded view of the front hub and related components—Discovery (2001–02)

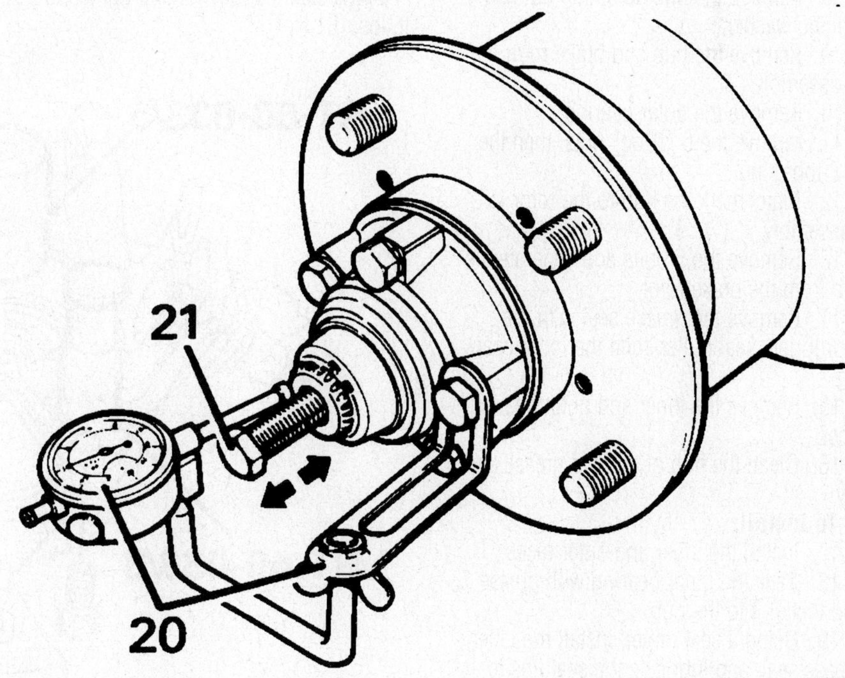

Install the dial indicator and bracket (20), then push and pull on bolt (21) to check the end-play—Discovery (2001–02)

end of the driveshaft. Move the driveshaft in and out noting the dial gauge reading. The end-play should be between 0.0032–0.0098 inch (0.08–0.25mm).

32. If out of specification, remove the circlip, measure the shim thickness and install an appropriate shim to give required end-play.

33. Remove the bolt from the driveshaft, install the circlip and dust cap.

34. Install the brake caliper and tighten the mounting bolts to 60 ft. lbs. (82 Nm).

35. Install the wheels, lower the vehicle and tighten the lug nuts to 93 ft. lbs. (126 Nm).

36. Operate the brake pedal to locate brake pads before taking vehicle for a road test.

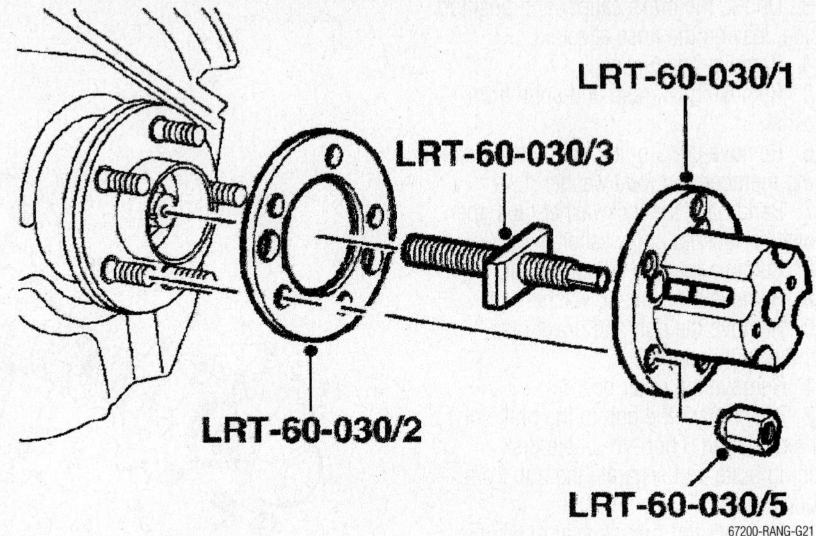

Installing tool set to the front hub—Range Rover (2003–04)

REAR

1. Before servicing the vehicle, refer to the precautions in the beginning of this section.

2. Raise and safely support the vehicle.

3. Remove the rear wheels.

4. Release the brake hose clip, then remove the caliper and position it aside taking care not to kink the brake hose.

5. Remove the 5 axle retaining bolts, then withdraw the axle shaft.

6. Remove the joint washer.

7. Straighten the lockwasher tabs, then remove the locknut and lockwasher.

8. Remove the hub adjusting nut and spacing washer.

9. Remove the hub and brake rotor as an assembly.

10. Remove the outer bearing.

11. remove the 5 Nyloc® nuts, then the ABS tone ring.

12. Matchmark the hub to the rotor for reassembly.

13. Remove the 5 bolts and separate the hub from the brake rotor.

14. Remove the grease seal with the appropriate seal puller, then the inner bearing.

15. Remove the inner and outer bearing races.

16. Clean the hub of any old grease and dry.

To install:

17. Install the inner and outer races.

18. Pack the inner bearing with grease and install it to the hub.

19. Using a seal driver, install the inner grease seal, and lubricate the seal lips to prevent premature wear.

20. Install the brake rotor to the hub, align the matchmarks, and apply Loctite® 270 or equivalent to the mounting bolts and tighten to 54 ft. lbs. (73 Nm).

21. Install the ABS tone ring and tighten the new Nyloc® nuts to 80 inch lbs. (9 Nm).

22. Pack the outer bearing with grease and install it to the hub.

23. Retract the ABS sensor slightly.

24. Install the hub and brake rotor assembly.

25. Install the spacing washer and nut. tighten the nut to 45 ft. lbs. (61 Nm), loosen ½ turn (90°), then tighten to 35 inch lbs. (4 Nm).

26. Install a new lockwasher.

27. Install the locknut and tighten to 45 ft. lbs. (61 Nm).

28. Fold the tab of the lockwasher over the locknut.

29. Using a new joint washer install the axle shaft to the hub, and tighten the 5 bolts to 48 ft. lbs. (65 Nm).

30. Install the brake caliper and tighten the mounting bolt to 60 ft. lbs. (82 Nm), and install the brake hose clip.

31. Reinstall the ABS sensor.

32. Install the wheels and tighten the nuts to 93 ft. lbs. (126 Nm), and operate the foot brake to seat the brake pads.

33. Check the fluid level in the axle.

34. Lower the vehicle and test drive.

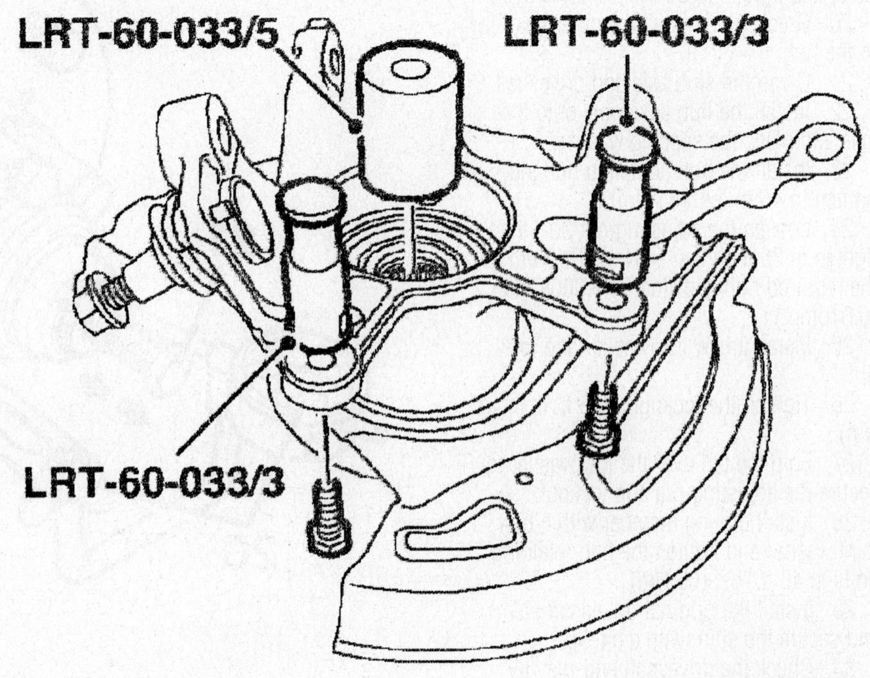

Installing bearing remover set to the front hub—Range Rover (2003–04)

Range Rover (2003–04)

FRONT

1. Before servicing the vehicle, refer to the precautions in the beginning of this section.
2. Raise and safely support the vehicle and remove the front wheels.
3. Remove the axle hub nut.
4. Unbolt the brake caliper and position it aside, leaving the hose attached.
5. Remove the brake rotor.
6. Disconnect the ABS sensor lead, then remove the ABS sensor from the hub.
7. Disconnect the height sensor link.
8. Loosen the lower arm-to-subframe bolt a quarter turn.
9. Remove the lower arm ball joint-to-hub nut.
10. Release the lower ball joint and disconnect the lower arm from the hub.
11. Loosen the tie rod-to-subframe bolt one quarter turn.
12. Remove the tie rod-to-ball joint nut.
13. Disconnect the tie rod from the ball joint.
14. Remove and discard the nut attaching the track rod ball joint to the steering arm.
15. Install an M14 nut to the ball pin so it is flush with the end of the pin.
16. Using a separator, disconnect the ball pin from the steering arm.
17. Remove the M14 nut.
18. Install tool set LRT-60-030 to the front hub.
19. Tighten the center screw on the tool to press the driveshaft from the flange.
20. Remove the tool set.
21. Remove the bolts attaching the hub to the damper and remove the hub.
22. Remove the brake shield from the hub.
23. Install mounting pins LRT-60-033/3 to the caliper mounting holes in the hub.
24. Fit adapter LRT-60-033/5 to the drive flange.
25. Fit plate LRT-6-033/2 to the mounting pin, then fit LRT-60-033/4 over the steering arm and fit to the plate.
26. Screw ram LRT-60-033/1 into the plate and press the drive flange from the hub.
27. Remove plate and pin from hub.
28. Remove and discard the circlip.
29. Fit adapter LRT-60-033/6 to the bearing. Reinstall the plate and the pin and press the bearing from the hub.
30. Remove the bearing race from the drive flange with a bearing separator.

To install:

31. Clean the hub and drive flange surfaces.
32. Position anvil LRT-60-033/10 on a hydraulic press bed.

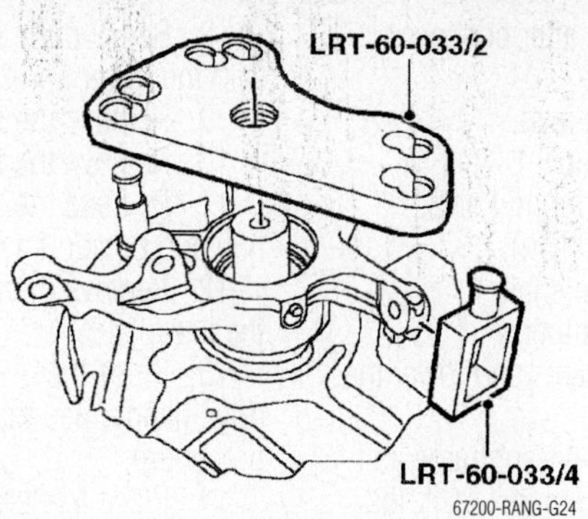

Installing remover plate to the front hub—Range Rover (2003–04)

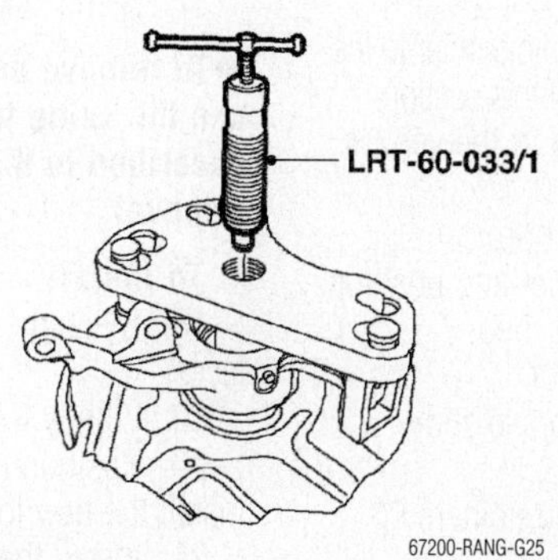

Installing bearing remover ram to the tool set—Range Rover (2003–04)

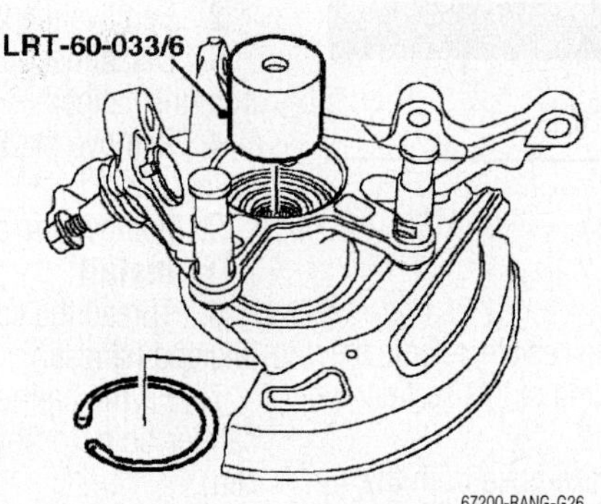

Installing bearing remover adapter to the tool set—Range Rover (2003–04)

33. Place the hub over the anvil.
34. Insert stud/bearing in hub about half way in.
35. Press bearing into the hub using adapter LRT-6-033/12.
36. Install a new circlip.
37. Fit bearing support LRT-60-033/11 to the anvil and press the drive flange into the bearing using adapter LRT-60-033/12.
38. Remove all tools
39. Install the brake shield.
40. Clean the hub-to-damper mating surfaces and the driveshaft and drive flange splines.
41. Lightly lubricate the splines.
42. Position the hub assembly and install the upper bolt.
43. Install the drive shaft to the hub, align the hub and install the lower bolt.
44. Tighten the upper and lower bolts to 184 ft. lbs. (250 Nm).
45. Using the same tools as in removal, install the driveshaft into the drive flange.
46. Using new nuts, install the ball joint to steering arm, tie rod to ball joint and ball joint to hub. Tighten all nuts to 59 ft. lbs. (80 Nm).
47. Install new bolts in lower arm to subframe and tie rod to subframe, but do not tighten.
48. Connect height sensor link.

49. Install ABS sensor after coating it with anti-sieze.
50. Attach ABS sensor lead.
51. Install the brake rotor.
52. Install a new axle hub nut and tighten to 311 ft. lbs. (420 Nm).
53. Stake the nut to the shaft.
54. Install the brake caliper.
55. Install the road wheels and lower the vehicle.
56. Tighten the tie rod-to-subframe and lower arm-to-subframe bolts to 121 ft. lbs. (165 Nm), plus an additional 90 degrees.
57. Check the front wheel alignment.

REAR

1. Before servicing the vehicle, refer to the precautions in the beginning of this section.
2. Raise and safely support the vehicle and remove the rear wheels.
3. Remove the axle hub nut.
4. Unbolt the brake caliper and position it aside, leaving the hose attached.
5. Remove the brake rotor.
6. Remove the parking brake shoe upper return spring.
7. Rotate the brake shoe retainers 90° and remove.
8. Remove the brake shoe assembly with the lower spring and adjuster.
9. Remove the shoe expander.
10. Remove the rear tie rod.
11. Remove the ABS sensor from the hub and position aside.
12. Remove the parking brake cable from the backplate.
13. Press the driveshaft out from the drive flange. See the Front Wheel Bearing procedure.
14. Place a reference mark to locate the eccentric cam to the upper arm.
15. Remove upper arm-to-hub bolt.
16. Remove lower arm-to-hub bolt.
17. Remove the hub.

➡To remove and replace the bearing, use the same tools and procedure as described in the Front Wheel Bearing removal.

To install:

18. Clean the hub and drive flange surfaces.
19. Lightly lubricate the splines.
20. Position the hub assembly and install the new lower bolt.
21. Install the drive shaft to the hub, align the hub and install the new upper lower bolt.
22. Align the eccentric cam to the reference mark and lightly tighten the bolt.
23. Tighten the lower bolt to 184 ft. lbs. (250 Nm).
24. Using the same tools as in removal, install the driveshaft into the drive flange.
25. Install the parking brake cable to the backplate.
26. Install the ABS sensor to the hub after coating it with anti-sieze.
27. Install the rear tie rod to ball joint nut and tighten to 121 ft. lbs. (165 Nm).
28. Install the shoe expander.
29. Install the brake shoe assembly with the lower spring and adjuster.
30. Install the parking brake shoe upper return spring.
31. Install the brake rotor.
32. Install a new axle hub nut and tighten the nut to 311 ft. lbs. (420 Nm).
33. Install the brake caliper.
34. Install the rear wheels and lower the vehicle.
35. Tighten the upper control arm to hub nut to 121 ft. lbs. (165 Nm).

BRAKES

Brake Caliper

REMOVAL & INSTALLATION

Freelander (2001–04)

FRONT

1. Before servicing the vehicle, refer to the precautions in the beginning of this section.
2. Remove the wheels.
3. Disconnect the brake hose from the caliper.
4. Remove the 2 bolts attaching the caliper.
5. Remove the caliper.

To install:

6. Thread the caliper onto the hydraulic line and hand-tighten it. Fit the caliper into place.
7. Torque the caliper bolts to bolts to 74 ft. lbs. (100 Nm).
8. Tighten the hydraulic line and bleed the brakes.

Discovery (2001–04)

FRONT AND REAR

1. Before servicing the vehicle, refer to the precautions in the beginning of this section.
2. Remove the wheels.
3. Disconnect the brake hose banjo bolt from the caliper.
4. Remove the 2 bolts attaching the caliper.
5. Remove the caliper.

To install:

6. Thread the caliper onto the hydraulic line and hand-tighten it.
7. Fit the caliper into place. Torque the caliper bolts to bolts to 22 ft. lbs. (30 Nm).
8. Tighten the hydraulic line banjo bolt to 24 ft. lbs. (32 Nm).
9. Bleed the brakes.

Range Rover (2001–04)

FRONT AND REAR

1. Before servicing the vehicle, refer to the precautions in the beginning of this section.
2. Remove the wheels.
3. Disconnect the brake hose from the caliper.
4. Remove the 2 bolts attaching the caliper.
5. Remove the caliper.

To install:

6. Thread the caliper onto the hydraulic line and hand-tighten it.
7. Fit the caliper into place. Torque the front caliper bolts to bolts to 81 ft. lbs. (110 Nm). Torque the rear caliper bolts to 48 ft. lbs. (65 Nm).
8. Tighten the hydraulic line.
9. Bleed the brakes.

Brake Pads

REMOVAL & INSTALLATION

Freelander (2001–04)

FRONT

1. Before servicing the vehicle, refer to the precautions in the beginning of this section.
2. Remove the front wheels.
3. Remove the lower guide pin bolt and rotate the caliper upward.
4. Remove the brake pads and retainers from the caliper brackets.

To install:

5. Compress the caliper piston into the bore.
6. Install the pad retainers to the caliper bracket.
7. Fit the pads into the caliper bracket.

➡The leading edge of the pads have a chamfer on them. The pads must be

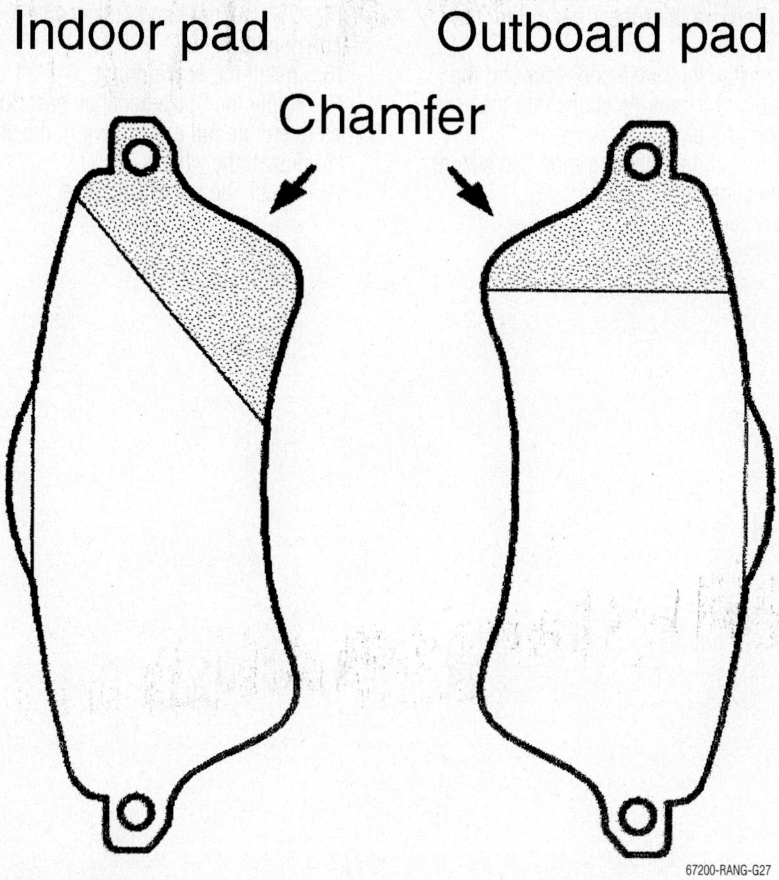

Front brake pad orientation—Freelander (2001–04)

correctly installed to the inside and outside of the caliper. The pads are not interchangeable.

8. Lower the caliper over the pads and bracket.
9. Ensure the flats on the guide pins locate with the lugs on the caliper housing.
10. Tighten the caliper mounting bolts to 20 ft. lbs. (27 Nm).
11. Install the wheels.
12. Depress the brake pedal several times to set the pad-to-disc clearance.
13. Check the brake fluid level.

Discovery (2001–04)

FRONT AND REAR

1. Before servicing the vehicle, refer to the precautions in the beginning of this section.
2. Remove the wheels.
3. Remove the lower guide pin bolt and rotate the caliper upward.
4. Remove the brake pads from the caliper. The shims are part of the pads.

To install:

5. Compress the caliper piston into the bore.
6. Fit the pads into the caliper bracket with the chamfer toward the leading edge of the rotor.
7. Lower the caliper over the pads.
8. Ensure the flats on the guide pins locate with the lugs on the caliper housing.
9. Tighten the caliper mounting bolts to 22 ft. lbs. (30 Nm).
10. Install the wheels.
11. Depress the brake pedal several times to set the pad-to-disc clearance.
12. Check the brake fluid level.

Range Rover (2001–04)

FRONT AND REAR

1. Before servicing the vehicle, refer to the precautions in the beginning of this section.
2. Remove the wheels.
3. On the left front, remove the pad wear sensor.
4. On the right rear, remove the pad wear sensor.
5. Remove the anti-rattle spring from the caliper housing.
6. Remove the 2 guide pin bolts and rotate the caliper upward.
7. Remove the brake pads from the caliper.

To install:

8. Compress the caliper piston into the bore.
9. Fit the pads into the caliper.
10. Install the caliper and tighten the mounting bolts to 24 ft. lbs. (33 Nm).
11. Install the anti-rattle springs.
12. Install the pad wear sensor.
13. Install the wheels.
14. Depress the brake pedal several times to set the pad-to-disc clearance.
15. Check the brake fluid level.

Brake Drums

REMOVAL & INSTALLATION

Freelander (2001–04)

REAR

1. Before servicing the vehicle, refer to the precautions in the beginning of this section.
2. Remove the rear wheels.
3. Remove 2 screws attaching the brake drum.
4. Remove the brake drum.

To install:

5. Apply Molykote 111 grease to the brake shoe and spring contact points.
6. Install the brake drum and tighten the screws.
7. Apply the brake pedal at least 30 times to ensure full adjustment of the shoes.
8. Install the wheels.
9. Check the brake fluid level.

Brake Shoes

REMOVAL & INSTALLATION

Freelander (2001–04)

REAR

1. Before servicing the vehicle, refer to the precautions in the beginning of this section.
2. Remove the rear wheels.
3. Remove the brake drum.
4. Depress the brake show clips and turn them to release, then remove the 2 clips and 2 retaining pins.
5. Release the hand brake cable from the brake shoe lever.
6. Ease one brake shoe out of the retaining groove, follower by the other shoe.
7. Maneuver the brake shoe assembly around the wheel cylinder and rear hub, then remove the brake shoes.
8. Place a rubber band around the wheel cylinder ends.

9. Remove the return springs, adjuster and the expander.

To install:

10. Install the adjuster and expander to the brake shoes, then fit the return springs.
11. Apply Molykote 111 grease to the brake shoe contact points.
12. Remove the rubber band from the wheel cylinder.
13. Install the brake shoes around the rear hub and locate the shoes into the retaining grove.
14. Install the retaining pins and secure with the clips.
15. Connect the hand brake cable to the brake shoe lever.
16. Install the brake drum.
17. Apply the brake pedal at least 30 times to ensure full adjustment of the shoes.
18. Install the wheels.
19. Check the brake fluid level.

MERCEDES-BENZ

C • CL • CLK • E • S • SL • SLK

7

SPECIFICATION CHARTS

ENGINE AND VEHICLE IDENTIFICATION CHART

Engine						
Code ①	Liters (cc)	Cu. In.	Cyl.	Fuel Sys.	Type	Eng. Mfg.
104.994	3.2 (3199)	195	6	HFM	DOHC	MB
111.961	1.8 (1790)	109	4	ME 2.0	DOHC	MB
111.973	2.3 (2295)	140	4	ME 2.1	DOHC	MB
111.974	2.3 (2295)	140	4	ME 2.1	DOHC	MB
112.912	2.6 (2597)	158	6	ME 2.0	SOHC	MB
112.920	2.8 (2799)	171	6	ME 2.0	SOHC	MB
112.940	3.2 (3199)	195	6	ME 2.0	SOHC	MB
112.941	3.2 (3199)	195	6	ME 2.0	SOHC	MB
112.946	3.2 (3199)	195	6	ME 2.0	SOHC	MB
112.947	3.2 (3199)	195	6	ME 2.0	SOHC	MB
112.960	3.2 (3199)	195	6	ME 2.0	SOHC	MB
112.961	3.2 (3199)	195	6	ME 2.0	SOHC	MB
113.940	4.3 (4266)	260	8	ME 2.0	SOHC	MB
113.941	4.3 (4266)	260	8	ME 2.0	SOHC	MB
113.943	4.3 (4266)	260	8	ME 2.0	SOHC	MB
113.960	5.0 (4973)	303	8	ME 2.0	SOHC	MB
113.961	5.0 (4973)	303	8	ME 2.0	SOHC	MB
113.967	5.0 (4973)	303	8	ME 2.0	SOHC	MB
113.968	5.0 (4973)	303	8	ME 2.0	SOHC	MB
113.991	5.5 (5439)	332	8	ME 2.0	SOHC	MB
113.992	5.5 (5439)	332	8	ME 2.0	SOHC	MB
113.998	5.5 (5439)	335	8	EFI	SOHC	MB
119.980	5.0 (4973)	303	8	ME-1	DOHC	MB
119.982	5.0 (4973)	303	8	ME-1	DOHC	MB
120.983	6.0 (5987)	365	12	ME-1	DOHC	MB

Model Year	
Code ②	Year
2	2002
3	2003
4	2004
5	2005

HFM: Hot Film Management (engine)

ME: Motronic Engine management

ME: Motronic Engine management

EFI: Electronic Fuel Injection

SFI: Sequential Fuel Injection

SOHC: Singe overhead camshaft

DOHC: Double overhead camshafts

① 8th digit of the VIN

② 10th digit of the VIN

06041_BENZ_C0001

ENGINE SPECIFICATIONS

Year	Engine Displacement Liters	Engine ID/VIN	Net Horsepower @ rpm	Net Torque@rpm (ft. lbs.)	Bore x Stroke (in.)	Com- pression Ratio	Oil Pressure @ rpm
2002	1.8	111.961 ①	189@5800	192@3700	3.23x3.35	8.7:1	43.5@3000
	2.3	111.973 ①	185@5300	200@25-4800	3.58x3.48	8.8:1	43.5@3000
	2.3	111.974	148@5500	162@4000	3.58x3.48	10.4:1	23.2-72.5@2000
	2.6	112.912	167@6000	177@4500	3.54x2.68	10.0:1	23.2-72.5@2000
	2.8	112.920	194@5800	195@30-4600	3.54x2.89	10.0:1	23.2-72.5@2000
	3.2	112.940	215@5500	229@30-4600	3.54x3.30	10.0:1	43.5@3000
	3.2	112.941	221@5500	232@30-4800	3.54x3.30	10.0:1	43.5@3000
	3.2	112.946	215@5700	228@3000	3.54x3.20	10.0:1	43.5@3000
	3.2	112.947	215@5700	228@3000	3.54x3.30	10.0:1	43.5@3000
	3.2	112.960 ①	349@6100	333@4400	3.54x3.30	9.0:1	43.5@3000
	3.2	112.961 ①	302@5600	339@2700	3.82x3.81	10.0:1	43.5@3000
	4.3	113.940	275@5750	295@30-4400	3.54x3.31	10.0:1	43.5@3000
	4.3	113.941	275@5750	295@2700	3.54x3.31	10.0:1	43.5@3000
	4.3	113.943	275@5750	295@3200	3.54x3.31	10.0:1	43.5@3000
	5.0	113.960	302@5500	339@2700	3.82x3.31	10.0:1	43.5@3000
	5.0	113.961	302@5500	339@2700	3.82x3.31	10.0:1	43.5@3000
	5.0	113.967	302@5500	339@2700	3.82x3.31	10.0:1	43.5@3000
	5.0	119.980	315@5600	347@3900	3.80x3.35	10.0:1	23.2-72.5@2000
	5.0	119.982	315@5600	345@3900	3.80x3.35	10.0:1	23.2-72.5@2000
	5.5	113.991 ①	493@6100	517@3400	3.82x3.60	9.0:1	43.5@3000
	5.5	113.992 ①	493@6100	517@3400	3.82x3.60	9.0:1	43.5@3000
	6.0	120.983	389@5200	420@3800	3.50x3.16	10.0:1	43.5@3000
2003	1.8	111.961 ①	189@5800	192@3700	3.23x3.35	8.7:1	43.5@3000
	2.3	111.973 ①	185@5300	200@25-4800	3.58x3.48	8.8:1	43.5@3000
	2.3	111.974	148@5500	162@4000	3.58x3.48	10.4:1	23.2-72.5@2000
	2.6	112.912	167@6000	177@4500	3.54x2.68	10.0:1	23.2-72.5@2000
	2.8	112.920	194@5800	195@30-4600	3.54x2.89	10.0:1	23.2-72.5@2000
	3.2	112.940	215@5500	229@30-4600	3.54x3.30	10.0:1	43.5@3000
	3.2	112.941	221@5500	232@30-4800	3.54x3.30	10.0:1	43.5@3000
	3.2	112.946	215@5700	228@3000	3.54x3.20	10.0:1	43.5@3000
	3.2	112.947	215@5700	228@3000	3.54x3.30	10.0:1	43.5@3000
	3.2	112.960 ①	349@6100	333@4400	3.54x3.30	9.0:1	43.5@3000

06041_BENZ_C0002

ENGINE SPECIFICATIONS

Year	Engine Displacement Liters	Engine ID/VIN	Net Horsepower @ rpm	Net Torque@rpm (ft. lbs.)	Bore x Stroke (in.)	Com-pression Ratio	Oil Pressure @ rpm
2003 (cont.)	3.2	112.961 ①	302@5600	339@2700	3.82x3.81	10.0:1	43.5@3000
	4.3	113.940	275@5750	295@30-4400	3.54x3.31	10.0:1	43.5@3000
	4.3	113.941	275@5750	295@2700	3.54x3.31	10.0:1	43.5@3000
	4.3	113.943	275@5750	295@3200	3.54x3.31	10.0:1	43.5@3000
	5.0	113.960	302@5500	339@2700	3.82x3.31	10.0:1	43.5@3000
	5.0	113.961	302@5500	339@2700	3.82x3.31	10.0:1	43.5@3000
	5.0	113.967	302@5500	339@2700	3.82x3.31	10.0:1	43.5@3000
	5.0	119.980	315@5600	347@3900	3.80x3.35	10.0:1	23.2-72.5@2000
	5.0	119.982	315@5600	345@3900	3.80x3.35	10.0:1	23.2-72.5@2000
	5.5	113.991 ①	493@6100	517@3400	3.82x3.60	9.0:1	43.5@3000
	5.5	113.992 ①	493@6100	517@3400	3.82x3.60	9.0:1	43.5@3000
	6.0	120.983	389@5200	420@3800	3.50x3.16	10.0:1	43.5@3000
2004	1.8	111.961 ①	189@5800	192@3700	3.23x3.35	8.7:1	43.5@3000
	2.3	111.973 ①	185@5300	200@25-4800	3.58x3.48	8.8:1	43.5@3000
	2.3	111.974	148@5500	162@4000	3.58x3.48	10.4:1	23.2-72.5@2000
	2.6	112.912	167@6000	177@4500	3.54x2.68	10.0:1	23.2-72.5@2000
	2.8	112.920	194@5800	195@30-4600	3.54x2.89	10.0:1	23.2-72.5@2000
	3.2	112.940	215@5500	229@30-4600	3.54x3.30	10.0:1	43.5@3000
	3.2	112.941	221@5500	232@30-4800	3.54x3.30	10.0:1	43.5@3000
	3.2	112.946	215@5700	228@3000	3.54x3.20	10.0:1	43.5@3000
	3.2	112.947	215@5700	228@3000	3.54x3.30	10.0:1	43.5@3000
	3.2	112.960 ①	349@6100	333@4400	3.54x3.30	9.0:1	43.5@3000
	3.2	112.961 ①	302@5600	339@2700	3.82x3.81	10.0:1	43.5@3000
	4.3	113.940	275@5750	295@30-4400	3.54x3.31	10.0:1	43.5@3000
	4.3	113.941	275@5750	295@2700	3.54x3.31	10.0:1	43.5@3000
	4.3	113.943	275@5750	295@3200	3.54x3.31	10.0:1	43.5@3000
	5.0	113.960	302@5500	339@2700	3.82x3.31	10.0:1	43.5@3000
	5.0	113.961	302@5500	339@2700	3.82x3.31	10.0:1	43.5@3000
	5.0	113.967	302@5500	339@2700	3.82x3.31	10.0:1	43.5@3000
	5.0	119.980	315@5600	347@3900	3.80x3.35	10.0:1	23.2-72.5@2000
	5.0	119.982	315@5600	345@3900	3.80x3.35	10.0:1	23.2-72.5@2000
	5.5	113.991 ①	493@6100	517@3400	3.82x3.60	9.0:1	43.5@3000
	5.5	113.992 ①	493@6100	517@3400	3.82x3.60	9.0:1	43.5@3000
	6.0	120.983	389@5200	420@3800	3.50x3.16	10.0:1	43.5@3000
2005	1.8	271.948	189@5800	192@3500	3.23x3.35	8.5:1	43.5@3000
	2.6	112.912	167@6000	177@4500	3.54x2.68	10.0:1	23.2-72.5@2000
	3.2	112.940	215@5500	229@30-4600	3.54x3.30	10.0:1	43.5@3000
	5.5	113.988	493@6100	517@3400	3.82x3.60	9.0:1	43.5@3000

① Supercharged

NA: Not Available

06041_BENZ_C0003

GASOLINE ENGINE TUNE-UP SPECIFICATIONS

Year	Engine Displacement Liters	Engine ID/VIN	Spark Plug Gap (in.)	Ignition Timing (deg.)		Fuel Pump (psi)	Idle Speed (rpm)		Valve Clearance	
				MT	AT		MT	AT	In.	Ex.
2002	1.8	111.961	0.031	①	①	②	700-800	700-800	HYD	HYD
	2.3	111.973	0.031	①	①	②	700-800	600-750	HYD	HYD
	2.3	111.974	0.031	—	①	②	—	650-750	HYD	HYD
	2.6	112.912	0.031	—	7-11B	②	—	650-750	HYD	HYD
	2.8	112.920	0.031	—	①	②	—	650-750	HYD	HYD
	3.2	112.940	0.031	—	①	②	—	600-750	HYD	HYD
	3.2	112.941	0.031	—	①	②	—	650-750	HYD	HYD
	3.2	112.946	0.031	—	①	②	—	650-750	HYD	HYD
	3.2	112.947	0.031	—	①	②	—	650-750	HYD	HYD
	3.2	112.960	0.031	—	①	②	—	600-750	HYD	HYD
	3.2	112.961	0.031	—	①	②	—	600-750	HYD	HYD
	4.3	113.940	0.031	—	①	②	700-800	700-800	HYD	HYD
	4.3	113.941	0.031	—	①	②	—	600-750	HYD	HYD
	4.3	113.943	0.031	①	①	②	—	650-750	HYD	HYD
	5.0	113.960	0.031	—	①	②	—	650-750	HYD	HYD
	5.0	113.961	0.031	—	7-11B	②	—	650-750	HYD	HYD
	5.0	113.967	0.031	—	①	②	—	600-750	HYD	HYD
	5.0	113.968	0.031	—	①	②	—	650-750	HYD	HYD
	5.5	113.991	0.031	—	①	②	—	650-750	HYD	HYD
	5.5	113.992	0.031	—	①	②	—	650-750	HYD	HYD
	5.0	119.980	0.031	—	①	②	—	600-750	HYD	HYD
	5.0	119.982	0.031	—	①	②	—	600-750	HYD	HYD
	6.0	120.983	0.031	—	①	②	—	600-750	HYD	HYD
2003	1.8	111.961	0.031	①	①	②	700-800	700-800	HYD	HYD
	2.3	111.973	0.031	①	①	②	700-800	600-750	HYD	HYD
	2.3	111.974	0.031	—	①	②	—	650-750	HYD	HYD
	2.6	112.912	0.031	—	7-11B	②	—	650-750	HYD	HYD
	2.8	112.920	0.031	—	①	②	—	650-750	HYD	HYD
	3.2	112.940	0.031	—	①	②	—	600-750	HYD	HYD
	3.2	112.941	0.031	—	①	②	—	650-750	HYD	HYD
	3.2	112.946	0.031	—	①	②	—	650-750	HYD	HYD
	3.2	112.947	0.031	—	①	②	—	650-750	HYD	HYD
	3.2	112.960	0.031	—	①	②	—	600-750	HYD	HYD

06041_BENZ_C0004

GASOLINE ENGINE TUNE-UP SPECIFICATIONS

Year	Engine Displacement Liters	Engine ID/VIN	Spark Plug Gap (in.)	Ignition Timing (deg.)		Fuel Pump (psi)	Idle Speed (rpm)		Valve Clearance	
				MT	AT		MT	AT	In.	Ex.
2003 (cont.)	3.2	112.961	0.031	—	①	②	—	600-750	HYD	HYD
	4.3	113.940	0.031	—	①	②	700-800	700-800	HYD	HYD
	4.3	113.941	0.031	—	①	②	—	600-750	HYD	HYD
	4.3	113.943	0.031	①	①	②	—	650-750	HYD	HYD
	5.0	113.960	0.031	—	①	②	—	650-750	HYD	HYD
	5.0	113.961	0.031	—	7-11B	②	—	650-750	HYD	HYD
	5.0	113.967	0.031	—	①	②	—	600-750	HYD	HYD
	5.0	113.968	0.031	—	①	②	—	650-750	HYD	HYD
	5.5	113.991	0.031	—	①	②	—	650-750	HYD	HYD
	5.5	113.992	0.031	—	①	②	—	650-750	HYD	HYD
	5.0	119.980	0.031	—	①	②	—	600-750	HYD	HYD
	5.0	119.982	0.031	—	①	②	—	600-750	HYD	HYD
	6.0	120.983	0.031	—	①	②	—	600-750	HYD	HYD
2004	1.8	111.961	0.031	①	①	②	700-800	700-800	HYD	HYD
	2.3	111.973	0.031	①	①	②	700-800	600-750	HYD	HYD
	2.3	111.974	0.031	—	①	②	—	650-750	HYD	HYD
	2.6	112.912	0.031	—	7-11B	②	—	650-750	HYD	HYD
	2.8	112.920	0.031	—	①	②	—	650-750	HYD	HYD
	3.2	112.940	0.031	—	①	②	—	600-750	HYD	HYD
	3.2	112.941	0.031	—	①	②	—	650-750	HYD	HYD
	3.2	112.946	0.031	—	①	②	—	650-750	HYD	HYD
	3.2	112.947	0.031	—	①	②	—	650-750	HYD	HYD
	3.2	112.960	0.031	—	①	②	—	600-750	HYD	HYD
	3.2	112.961	0.031	—	①	②	—	600-750	HYD	HYD
	4.3	113.940	0.031	—	①	②	700-800	700-800	HYD	HYD
	4.3	113.941	0.031	—	①	②	—	600-750	HYD	HYD
	4.3	113.943	0.031	①	①	②	—	650-750	HYD	HYD
	5.0	113.960	0.031	—	①	②	—	650-750	HYD	HYD
	5.0	113.961	0.031	—	7-11B	②	—	650-750	HYD	HYD
	5.0	113.967	0.031	—	①	②	—	600-750	HYD	HYD
	5.0	113.968	0.031	—	①	②	—	650-750	HYD	HYD
	5.5	113.991	0.031	—	①	②	—	650-750	HYD	HYD
	5.5	113.992	0.031	—	①	②	—	650-750	HYD	HYD
	5.0	119.980	0.031	—	①	②	—	600-750	HYD	HYD
	5.0	119.982	0.031	—	①	②	—	600-750	HYD	HYD
	6.0	120.983	0.031	—	①	②	—	600-750	HYD	HYD
2005	1.8	271.948	0.031	①	①	②	700-800	700-800	HYD	HYD
	2.6	112.912	0.031	—	7-11B	②	—	650-750	HYD	HYD
	3.2	112.940	0.031	—	①	②	—	600-750	HYD	HYD
	5.5	113.988	0.031	—	①	②	—	650-750	HYD	HYD

NOTE: The Vehicle Emission Control Information label often reflects specification changes made during production.
The label figures must be used if they differ from those in this chart.

B: Before top dead center

HYD: Hydraulic

① Timing controlled by engine control module. Adjustment is not possible.

② 46-52 psi without vacuum applied
53-61 psi with vacuum applied
36 psi retention after 30 minutes

06041_BENZ_C0005

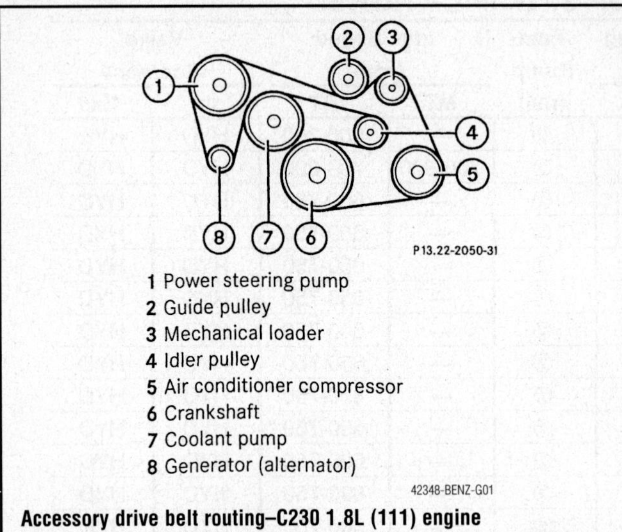

1 Power steering pump
2 Guide pulley
3 Mechanical loader
4 Idler pulley
5 Air conditioner compressor
6 Crankshaft
7 Coolant pump
8 Generator (alternator)

Accessory drive belt routing–C230 1.8L (111) engine

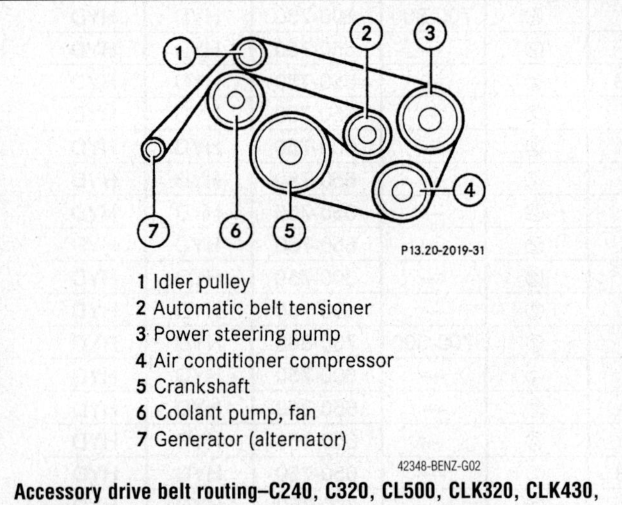

1 Idler pulley
2 Automatic belt tensioner
3 Power steering pump
4 Air conditioner compressor
5 Crankshaft
6 Coolant pump, fan
7 Generator (alternator)

Accessory drive belt routing–C240, C320, CL500, CLK320, CLK430, E Class and SL500 2.6L, 3.2L (112) 4.3L, 5.0L (113) engines

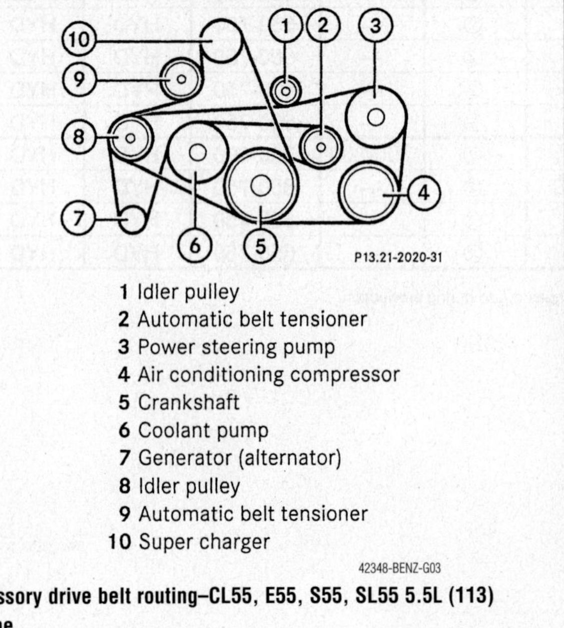

1 Idler pulley
2 Automatic belt tensioner
3 Power steering pump
4 Air conditioning compressor
5 Crankshaft
6 Coolant pump
7 Generator (alternator)
8 Idler pulley
9 Automatic belt tensioner
10 Super charger

Accessory drive belt routing–CL55, E55, S55, SL55 5.5L (113) engine

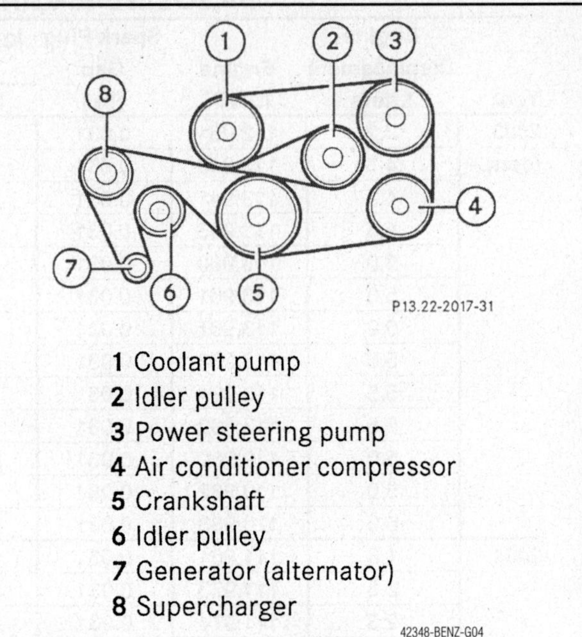

1 Coolant pump
2 Idler pulley
3 Power steering pump
4 Air conditioner compressor
5 Crankshaft
6 Idler pulley
7 Generator (alternator)
8 Supercharger

Accessory drive belt routing–SLK230 2.3L (111) engine

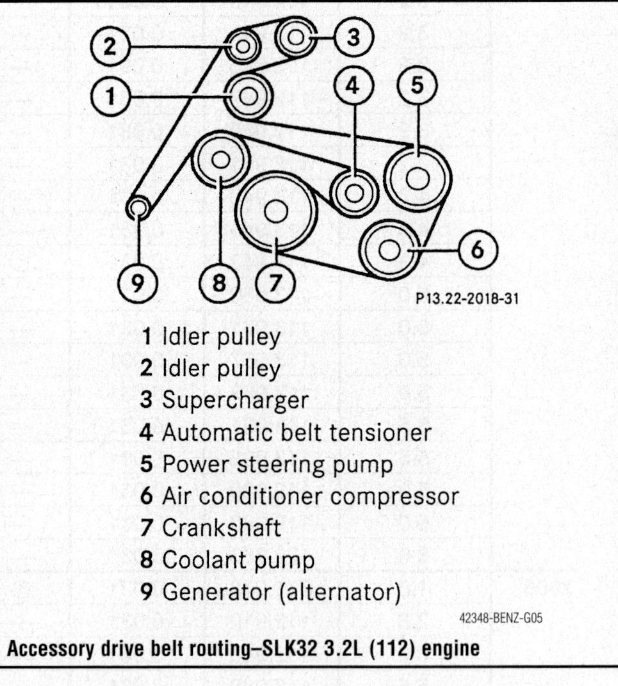

1 Idler pulley
2 Idler pulley
3 Supercharger
4 Automatic belt tensioner
5 Power steering pump
6 Air conditioner compressor
7 Crankshaft
8 Coolant pump
9 Generator (alternator)

Accessory drive belt routing–SLK32 3.2L (112) engine

CAPACITIES

Year	Model	Engine Displacement Liters	Engine ID/VIN	Engine Oil with Filter	Transmission (pts.) 5-Spd	Transmission (pts.) Auto.	Drive Axle Front (pts.)	Drive Axle Rear (pts.)	Fuel Tank (gal.)	Cooling System (qts.)
2002	C230	1.8	111.961	5.8	1.3	16.9	—	2.3	16.4	8.5
	C240	2.6	112.912	7.5	1.3	16.9	—	2.3	16.4	6.8
	C320	3.2	112.946	7.5	1.3	16.9	—	2.3	16.4	6.8
	C32	3.2	112.961	7.5	—	16.9	—	2.3	16.4	6.8
	E320	3.2	112.941	8.5	—	15.8	—	2.9	20.6	10.7
	E430	4.3	113.940	8.5	—	19.7	—	2.9	23.0	11.6
	E500	5.0	113.967	8.5	—	15.8	—	2.9	20.6	11.9
	S430	4.3	113.941	8.5	—	18.1	—	3.3	21.1	12.1
	S500	5.0	119.980	8.5	—	18.1	—	3.3	21.1	12.1
	S55	5.0	113.991	7.9	—	18.1	—	3.3	21.1	14.3
	CL500	5.0	113.960	8.5	—	19.2	—	3.4	23.3	12.1
	CL55	5.5	113.991	7.9	—	19.2	—	3.4	23.3	15.3
	SL500	5.0	113.961	7.9	—	18.1	—	2.9	21.1	12.2
	SL55	5.5	113.992	9	—	18.1	—	2.9	21.1	13.7
	CLK320	3.2	112.940	8.5	—	16.9	—	2.1	16.4	9.5
	CLK430	4.3	113.943	8.5	—	19.2	—	2.1	16.4	11.7
	CLK500	5.0	113.968	8.5	—	19.7	—	2.9	16.4	17.5
	CLK55	5.5	113.991	7.9	—	19.7	—	2.9	16.4	17.5
	SLK230	2.3	111.973	6.1	3.1	16.9	—	2.7	15.9	9.8
	SLK320	3.2	112.947	6.1	3.8	16.9	—	2.7	15.9	11.8
	SLK32	3.2	112.960	8.5	—	16.9	—	2.7	15.9	15.3
2003	C230	1.8	111.961	5.8	1.3	16.9	—	2.3	16.4	8.5
	C240	2.6	112.912	7.5	1.3	16.9	—	2.3	16.4	6.8
	C320	3.2	112.946	7.5	1.3	16.9	—	2.3	16.4	6.8
	C32	3.2	112.961	7.5	—	16.9	—	2.3	16.4	6.8
	E320	3.2	112.941	8.5	—	15.8	—	2.9	20.6	10.7
	E500	5.0	113.967	8.1	—	15.8	—	2.9	20.6	11.9
	S430	4.3	113.941	8.5	—	18.1	—	3.3	21.1	12.1
	S500	5.0	119.980	8.5	—	18.1	—	3.3	21.1	12.1
	S55	5.0	113.991	7.9	—	18.1	—	3.3	21.1	14.3
	CL500	5.0	113.960	8.5	—	19.2	—	3.4	23.3	12.1
	CL55	5.5	113.991	7.9	—	19.2	—	3.4	23.3	15.3
	SL500	5.0	113.961	7.9	—	18.1	—	2.9	21.1	12.2
	SL55	5.5	113.992	9	—	18.1	—	2.9	21.1	13.7
	CLK320	3.2	112.940	8.5	—	16.9	—	2.1	16.4	9.5
	CLK430	4.3	113.943	8.5	—	19.2	—	2.1	16.4	11.7
	CLK500	5.0	113.968	8.5	—	19.7	—	2.9	16.4	17.5
	CLK55	5.5	113.991	7.9	—	19.7	—	2.9	16.4	17.5
	SLK230	2.3	111.973	6.1	3.1	16.9	—	2.7	15.9	9.8
	SLK320	3.2	112.947	6.1	3.8	16.9	—	2.7	15.9	11.8
	SLK32	3.2	112.960	8.5	—	16.9	—	2.7	15.9	15.3

06041_BENZ_C0006

CAPACITIES

Year	Model	Engine Displacement Liters	Engine ID/VIN	Engine Oil with Filter	Transmission (pts.) 5-Spd	Transmission (pts.) Auto.	Drive Axle Front (pts.)	Drive Axle Rear (pts.)	Fuel Tank (gal.)	Cooling System (qts.)
2004	C230	1.8	111.961	5.8	1.3	16.9	—	2.3	16.4	8.5
	C240	2.6	112.912	7.5	1.3	16.9	—	2.3	16.4	6.8
	C320	3.2	112.946	7.5	1.3	16.9	—	2.3	16.4	6.8
	C32	3.2	112.961	7.5	—	16.9	—	2.3	16.4	6.8
	E320	3.2	112.941	8.5	—	15.8	—	2.9	20.6	10.7
	E500	5.0	113.967	8.1	—	15.8	—	2.9	20.6	11.9
	S430	4.3	113.941	8.5	—	18.1	—	3.3	21.1	12.1
	S500	5.0	119.980	8.5	—	18.1	—	3.3	21.1	12.1
	S55	5.0	113.991	7.9	—	18.1	—	3.3	21.1	14.3
	CL500	5.0	113.960	8.5	—	19.2	—	3.4	23.3	12.1
	CL55	5.5	113.991	7.9	—	19.2	—	3.4	23.3	15.3
	SL500	5.0	113.961	7.9	—	18.1	—	2.9	21.1	12.2
	SL55	5.5	113.992	9	—	18.1	—	2.9	21.1	13.7
	CLK320	3.2	112.940	8.5	—	16.9	—	2.1	16.4	9.5
	CLK430	4.3	113.943	8.5	—	19.2	—	2.1	16.4	11.7
	CLK500	5.0	113.968	8.5	—	19.7	—	2.9	16.4	17.5
	CLK55	5.5	113.991	7.9	—	19.7	—	2.9	16.4	17.5
	SLK230	2.3	111.973	6.1	3.1	16.9	—	2.7	15.9	9.8
	SLK320	3.2	112.947	6.1	3.8	16.9	—	2.7	15.9	11.8
	SLK32	3.2	112.960	8.5	—	16.9	—	2.7	15.9	15.3
2005	C230	1.8	271.948	5.8	1.3	16.9	—	2.3	16.4	8.5
	C240	2.6	112.912	7.5	1.3	16.9	—	2.3	16.4	6.8
	C320	3.2	112.940	7.5	1.3	16.9	—	2.3	16.4	6.8
	C55 AMG	5.5	113.988	7.9	—	19.7	—	2.9	16.4	17.5

NOTE: All capacities are approximate. Add fluid gradually and check to be sure a proper fluid level is obtained.

06041_BENZ_C0007

CRANKSHAFT AND CONNECTING ROD SPECIFICATIONS

All measurements are given in inches.

	Engine		Crankshaft				Connecting Rod		
Year	Displacement Liters	Engine ID/VIN	Main Brg. Journal Dia.	Main Brg. Oil Clearance	Shaft End-play	Thrust on No.	Journal Diameter	Oil Clearance	Side Clearance
2002	2.3	111.974	2.2815-2.2820	0.0012-0.0018	0.0039-0.0098	3	1.8872-1.8884	0.0008-0.0020	NA
	2.3	111.973	2.2815-2.2820	0.0012-0.0018	0.0039-0.0098	3	1.8872-1.8884	0.0008-0.0020	NA
	2.8	112.920	NA	NA	NA	NA	NA	NA	NA
	3.0	606.962	2.2811-2.2820	0.0012-0.0020	NA	5	NA	NA	NA
	3.2	112.941	NA	NA	NA	NA	NA	NA	NA
	3.2	104.994	2.2811-2.2820	0.0012-0.0018	0.0039-0.0098	5	1.8872-1.8884	0.0008-0.0020	NA
	3.2	112.940	NA	NA	NA	NA	NA	NA	NA
	4.2	119.981	2.5177-2.5183	0.0008-0.0018	0.0039-0.0087	3	1.8880-1.8884	0.0012-0.0022	0.0087-0.0149
	4.3	113.940	NA	NA	NA	NA	NA	NA	NA
	5.0	119.980	2.5177-2.5183	0.0008-0.0018	0.0039-0.0087	3	1.8880-1.8884	0.0012-0.0022	0.0087-0.0149
	5.0	119.982	2.5177-2.5183	0.0008-0.0018	0.0039-0.0087	3	1.8880-1.8884	0.0012-0.0022	0.0087-0.0149
	6.0	120.983	2.2815-2.2820	0.0010-0.0020	0.0039-0.0100	4	1.8880-1.8884	0.0008-0.0023	0.0088-0.0154
2003	2.3	111.974	2.2815-2.2820	0.0012-0.0018	0.0039-0.0098	3	1.8872-1.8884	0.0008-0.0020	NA
	2.3	111.973	2.2815-2.2820	0.0012-0.0018	0.0039-0.0098	3	1.8872-1.8884	0.0008-0.0020	NA
	2.8	112.920	NA	NA	NA	NA	NA	NA	NA
	3.0	606.962	2.2811-2.2820	0.0012-0.0020	NA	5	NA	NA	NA
	3.2	112.941	NA	NA	NA	NA	NA	NA	NA
	3.2	104.994	2.2811-2.2820	0.0012-0.0018	0.0039-0.0098	5	1.8872-1.8884	0.0008-0.0020	NA
	3.2	112.940	NA	NA	NA	NA	NA	NA	NA
	4.2	119.981	2.5177-2.5183	0.0008-0.0018	0.0039-0.0087	3	1.8880-1.8884	0.0012-0.0022	0.0087-0.0149
	4.3	113.940	NA	NA	NA	NA	NA	NA	NA
	5.0	119.980	2.5177-2.5183	0.0008-0.0018	0.0039-0.0087	3	1.8880-1.8884	0.0012-0.0022	0.0087-0.0149
	5.0	119.982	2.5177-2.5183	0.0008-0.0018	0.0039-0.0087	3	1.8880-1.8884	0.0012-0.0022	0.0087-0.0149
	6.0	120.983	2.2815-2.2820	0.0010-0.0020	0.0039-0.0100	4	1.8880-1.8884	0.0008-0.0023	0.0088-0.0154

06041_BENZ_C0008

CRANKSHAFT AND CONNECTING ROD SPECIFICATIONS

All measurements are given in inches.

	Engine		Crankshaft				Connecting Rod		
Year	Displacement Liters	Engine ID/VIN	Main Brg. Journal Dia.	Main Brg. Oil Clearance	Shaft End-play	Thrust on No.	Journal Diameter	Oil Clearance	Side Clearance
2004	2.3	111.974	2.2815-2.2820	0.0012-0.0018	0.0039-0.0098	3	1.8872-1.8884	0.0008-0.0020	NA
	2.3	111.973	2.2815-2.2820	0.0012-0.0018	0.0039-0.0098	3	1.8872-1.8884	0.0008-0.0020	NA
	2.8	112.920	NA	NA	NA	NA	NA	NA	NA
	3.0	606.962	2.2811-2.2820	0.0012-0.0020	NA	5	NA	NA	NA
	3.2	112.941	NA	NA	NA	NA	NA	NA	NA
	3.2	104.994	2.2811-2.2820	0.0012-0.0018	0.0039-0.0098	5	1.8872-1.8884	0.0008-0.0020	NA
	3.2	112.940	NA	NA	NA	NA	NA	NA	NA
	4.2	119.981	2.5177-2.5183	0.0008-0.0018	0.0039-0.0087	3	1.8880-1.8884	0.0012-0.0022	0.0087-0.0149
	4.3	113.940	NA	NA	NA	NA	NA	NA	NA
	5.0	119.980	2.5177-2.5183	0.0008-0.0018	0.0039-0.0087	3	1.8880-1.8884	0.0012-0.0022	0.0087-0.0149
	5.0	119.982	2.5177-2.5183	0.0008-0.0018	0.0039-0.0087	3	1.8880-1.8884	0.0012-0.0022	0.0087-0.0149
	6.0	120.983	2.2815-2.2820	0.0010-0.0020	0.0039-0.0100	4	1.8880-1.8884	0.0008-0.0023	0.0088-0.0154
2005	1.8	271.948	NA	NA	NA	NA	NA	NA	NA
	2.6	112.912	NA	NA	NA	NA	NA	NA	NA
	3.2	112.940	NA	NA	NA	NA	NA	NA	NA
	5.5	113.988	NA	NA	NA	NA	NA	NA	NA

NA: Not Available

06041_BENZ_C0009

PISTON AND RING SPECIFICATIONS

All measurements are given in inches.

Year	Engine Size Liters	Engine ID/VIN	Piston Clearance	Ring Gap Top Compression	Ring Gap Bottom Compression	Ring Gap Oil Control	Ring Side Clearance Top Compression	Ring Side Clearance Bottom Compression	Ring Side Clearance Oil Control
2002	2.3	111.974	NA	0.0012-0.0024	0.0004-0.0020	0.0008-0.0024	0.0012-0.0028	0.0008-0.0016	0.0004-0.0020
	2.3	111.973	NA	0.0012-0.0024	0.0004-0.0020	0.0008-0.0024	0.0012-0.0028	0.0008-0.0016	0.0004-0.0020
	2.8	112.920	NA	NA	NA	NA	NA	NA	NA
	3.0	606.962	NA	NA	NA	NA	NA	NA	NA
	3.2	112.941	NA	NA	NA	NA	NA	NA	NA
	3.2	104.994	0.0008-0.0016	0.0079-0.0157	0.0079-0.0157	0.0079-0.0177	0.0006-0.0020	0.0008-0.0016	0.0004-0.0018
	3.2	112.940	NA	NA	NA	NA	NA	NA	NA
	4.2	119.981	0.0002-0.0010	0.0020-0.0031	0.0004-0.0012	0.0004-0.0016	NA	NA	NA
	4.3	113.940	NA	NA	NA	NA	NA	NA	NA
	5.0	119.980	0.0002-0.0010	0.0020-0.0031	0.0004-0.0012	0.0004-0.0016	NA	NA	NA
	5.0	119.982	0.0002-0.0010	0.0020-0.0031	0.0004-0.0012	0.0004-0.0016	NA	NA	NA
	6.0	120.983	0.0001-0.0008	NA	NA	NA	NA	NA	NA
2003	2.3	111.974	NA	0.0012-0.0024	0.0004-0.0020	0.0008-0.0024	0.0012-0.0028	0.0008-0.0016	0.0004-0.0020
	2.3	111.973	NA	0.0012-0.0024	0.0004-0.0020	0.0008-0.0024	0.0012-0.0028	0.0008-0.0016	0.0004-0.0020
	2.8	112.920	NA	NA	NA	NA	NA	NA	NA
	3.0	606.962	NA	NA	NA	NA	NA	NA	NA
	3.2	112.941	NA	NA	NA	NA	NA	NA	NA
	3.2	104.994	0.0008-0.0016	0.0079-0.0157	0.0079-0.0157	0.0079-0.0177	0.0006-0.0020	0.0008-0.0016	0.0004-0.0018
	3.2	112.940	NA	NA	NA	NA	NA	NA	NA
	4.2	119.981	0.0002-0.0010	0.0020-0.0031	0.0004-0.0012	0.0004-0.0016	NA	NA	NA
	4.3	113.940	NA	NA	NA	NA	NA	NA	NA
	5.0	119.980	0.0002-0.0010	0.0020-0.0031	0.0004-0.0012	0.0004-0.0016	NA	NA	NA
	5.0	119.982	0.0002-0.0010	0.0020-0.0031	0.0004-0.0012	0.0004-0.0016	NA	NA	NA
	6.0	120.983	0.0001-0.0008	NA	NA	NA	NA	NA	NA

06041_BENZ_C0010

PISTON AND RING SPECIFICATIONS

All measurements are given in inches.

Year	Engine Size Liters	Engine ID/VIN	Piston Clearance	Ring Gap Top Compression	Ring Gap Bottom Compression	Ring Gap Oil Control	Ring Side Clearance Top Compression	Ring Side Clearance Bottom Compression	Ring Side Clearance Oil Control
2004	2.3	111.974	NA	0.0012-0.0024	0.0004-0.0020	0.0008-0.0024	0.0012-0.0028	0.0008-0.0016	0.0004-0.0020
	2.3	111.973	NA	0.0012-0.0024	0.0004-0.0020	0.0008-0.0024	0.0012-0.0028	0.0008-0.0016	0.0004-0.0020
	2.8	112.920	NA	NA	NA	NA	NA	NA	NA
	3.0	606.962	NA	NA	NA	NA	NA	NA	NA
	3.2	112.941	NA	NA	NA	NA	NA	NA	NA
	3.2	104.994	0.0008-0.0016	0.0079-0.0157	0.0079-0.0157	0.0079-0.0177	0.0006-0.0020	0.0008-0.0016	0.0004-0.0018
	3.2	112.940	NA	NA	NA	NA	NA	NA	NA
	4.2	119.981	0.0002-0.0010	0.0020-0.0031	0.0004-0.0012	0.0004-0.0016	NA	NA	NA
	4.3	113.940	NA	NA	NA	NA	NA	NA	NA
	5.0	119.980	0.0002-0.0010	0.0020-0.0031	0.0004-0.0012	0.0004-0.0016	NA	NA	NA
	5.0	119.982	0.0002-0.0010	0.0020-0.0031	0.0004-0.0012	0.0004-0.0016	NA	NA	NA
	6.0	120.983	0.0001-0.0008	NA	NA	NA	NA	NA	NA
2005	1.8	271.948	NA	NA	NA	NA	NA	NA	NA
	2.6	112.912	NA	NA	NA	NA	NA	NA	NA
	3.2	112.940	NA	NA	NA	NA	NA	NA	NA
	5.5	113.988	NA	NA	NA	NA	NA	NA	NA

NA: Not Available

06041_BENZ_C0011

TORQUE SPECIFICATIONS

All readings in ft. lbs.

Year	Engine Displacement Liters	Engine ID/VIN	Cylinder Head Bolts	Main Bearing Bolts	Rod Bearing Bolts	Crankshaft Damper Bolts	Flywheel Bolts	Manifold Intake	Manifold Exhaust	Spark Plugs	Lug Nut
2002	1.8	111.961	⑦	②	③	300	⑧	15	30	⑥	81
	2.3	111.973	⑦	②	③	300	⑧	15	30	⑥	81
	2.3	111.974	⑦	②	③	300	⑧	15	30	⑥	81
	2.6	112.912	⑨	⑩	⑪	⑫	⑬	15	12	21	100
	2.8	112.920	⑨	⑩	⑪	⑫	⑬	15	12	21	100
	3.2	112.940	⑨	⑩	⑪	⑫	⑬	15	12	21	100
	3.2	112.941	⑨	⑩	⑪	⑫	⑬	15	12	21	100
	3.2	112.946	⑨	⑩	⑪	⑫	⑬	15	12	21	100
	3.2	112.947	⑨	⑩	⑪	⑫	⑬	15	12	21	100
	3.2	112.96	⑨	⑩	⑪	⑫	⑬	15	12	21	100
	3.2	112.961	⑨	⑩	⑪	⑫	⑬	15	12	21	100
	4.3	113.940	⑨	⑩	⑪	⑫	⑬	15	12	21	100
	4.3	113.941	⑨	⑩	⑪	⑫	⑬	15	12	21	100
	4.3	113.943	⑨	⑩	⑪	⑫	⑬	15	12	21	100
	5.0	113.960	⑨	⑩	⑪	⑫	⑬	15	12	21	100
	5.0	113.961	⑨	⑩	⑪	⑫	⑬	15	12	21	100
	5.0	113.967	⑨	⑩	⑪	⑫	⑬	15	12	21	100
	5.0	119.980	⑧	⑦	⑬	400	⑬	18	22	⑥	81
	5.0	119.982	⑧	⑦	⑬	400	⑬	18	22	⑥	81
	5.5	113.991	⑨	⑩	⑪	⑫	⑬	15	12	21	100
	5.5	112.992	⑨	⑩	⑪	⑫	⑬	15	12	21	100
	6.0	120.983	⑤	⑦	⑬	400	⑨	18	30	⑥	81
2003	1.8	111.961	⑦	②	③	300	⑧	15	30	⑥	81
	2.3	111.973	⑦	②	③	300	⑧	15	30	⑥	81
	2.3	111.974	⑦	②	③	300	⑧	15	30	⑥	81
	2.6	112.912	⑨	⑩	⑪	⑫	⑬	15	12	21	100
	2.8	112.920	⑨	⑩	⑪	⑫	⑬	15	12	21	100
	3.2	112.940	⑨	⑩	⑪	⑫	⑬	15	12	21	100
	3.2	112.941	⑨	⑩	⑪	⑫	⑬	15	12	21	100
	3.2	112.946	⑨	⑩	⑪	⑫	⑬	15	12	21	100
	3.2	112.947	⑨	⑩	⑪	⑫	⑬	15	12	21	100
	3.2	112.96	⑨	⑩	⑪	⑫	⑬	15	12	21	100
	3.2	112.961	⑨	⑩	⑪	⑫	⑬	15	12	21	100
	4.3	113.940	⑨	⑩	⑪	⑫	⑬	15	12	21	100
	4.3	113.941	⑨	⑩	⑪	⑫	⑬	15	12	21	100
	4.3	113.943	⑨	⑩	⑪	⑫	⑬	15	12	21	100
	5.0	113.960	⑨	⑩	⑪	⑫	⑬	15	12	21	100
	5.0	113.961	⑨	⑩	⑪	⑫	⑬	15	12	21	100
	5.0	113.967	⑨	⑩	⑪	⑫	⑬	15	12	21	100
	5.0	119.980	⑧	⑦	⑬	400	⑬	18	22	⑥	81
	5.0	119.982	⑧	⑦	⑬	400	⑬	18	22	⑥	81
	5.5	113.991	⑨	⑩	⑪	⑫	⑬	15	12	21	100
	5.5	112.992	⑨	⑩	⑪	⑫	⑬	15	12	21	100
	6.0	120.983	⑤	⑦	⑬	400	⑨	18	30	⑥	81

06041_BENZ_C0012

TORQUE SPECIFICATIONS

All readings in ft. lbs.

Year	Engine Displacement Liters	Engine ID/VIN	Cylinder Head Bolts	Main Bearing Bolts	Rod Bearing Bolts	Crankshaft Damper Bolts	Flywheel Bolts	Manifold Intake	Manifold Exhaust	Spark Plugs	Lug Nut
2004	1.8	111.961	⑦	②	③	300	⑧	15	30	⑥	81
	2.3	111.973	⑦	②	③	300	⑧	15	30	⑥	81
	2.3	111.974	⑦	②	③	300	⑧	15	30	⑥	81
	2.6	112.912	⑨	⑩	⑪	⑫	⑬	15	12	21	100
	2.8	112.920	⑨	⑩	⑪	⑫	⑬	15	12	21	100
	3.2	112.940	⑨	⑩	⑪	⑫	⑬	15	12	21	100
	3.2	112.941	⑨	⑩	⑪	⑫	⑬	15	12	21	100
	3.2	112.946	⑨	⑩	⑪	⑫	⑬	15	12	21	100
	3.2	112.947	⑨	⑩	⑪	⑫	⑬	15	12	21	100
	3.2	112.96	⑨	⑩	⑪	⑫	⑬	15	12	21	100
	3.2	112.961	⑨	⑩	⑪	⑫	⑬	15	12	21	100
	4.3	113.940	⑨	⑩	⑪	⑫	⑬	15	12	21	100
	4.3	113.941	⑨	⑩	⑪	⑫	⑬	15	12	21	100
	4.3	113.943	⑨	⑩	⑪	⑫	⑬	15	12	21	100
	5.0	113.960	⑨	⑩	⑪	⑫	⑬	15	12	21	100
	5.0	113.961	⑨	⑩	⑪	⑫	⑬	15	12	21	100
	5.0	113.967	⑨	⑩	⑪	⑫	⑬	15	12	21	100
	5.0	119.980	⑧	⑦	⑬	400	⑬	18	22	⑥	81
	5.0	119.982	⑧	⑦	⑬	400	⑬	18	22	⑥	81
	5.5	113.991	⑨	⑩	⑪	⑫	⑬	15	12	21	100
	5.5	112.992	⑨	⑩	⑪	⑫	⑬	15	12	21	100
	6.0	120.983	⑤	⑦	⑬	400	⑨	18	30	⑥	81
2005	1.8	271.948	⑦	②	③	300	⑧	15	30	⑥	81
	2.6	112.912	⑨	⑩	⑪	⑫	⑬	15	12	21	100
	3.2	112.940	⑨	⑩	⑪	⑫	⑬	15	12	21	100
	5.5	113.988	⑨	⑩	⑪	⑫	⑬	15	12	21	100

① Not Used

② Step 1: 41 ft. lbs.
Step 2: plus 90-100 degrees

③ Step 1: 22 ft. lbs.
Step 2: plus 90-100 degrees

④ Not used
Step 2: plus 90 degrees

⑤ Step 1: 41 ft. lbs.
Step 2: plus 90 degrees
Step 3: plus 90 degrees

⑥ Spark plug with conical seat: 7.3-15 ft. lbs.
Spark plug with flat seat: 15-22 ft. lbs.

⑦ Step 1: 52 ft. lbs.
Step 2: plus 90 degrees
Step 3: plus 90 degrees

⑧ M8: 22 ft. lbs.
M10 bolts: 36.8 ft. lbs.

⑨ Step 1: 15 ft. lbs.
Step 2: 37 ft. lbs.
Step 3: 65 degrees
Step 4: 65 degrees

⑩ M8x40: 18 ft. lbs.
M8x75
Step 1: 10 ft. lbs.
Step 2: 90-100 degrees
M10x90
Step 1: 15 ft. lbs.
Step 2: 90-100 degrees

⑪ Step 1: 44 inch lbs.
Step 2: 18 ft. lbs
Step 3. 90 degrees

⑫ Step 1: 148 ft. lbs.
Step 2: 95 degrees

⑬ Step 1: 33 ft. lbs.
Step 2: 90 degrees

06041_BENZ_C0013

TIRE, WHEEL AND BALL JOINT SPECIFICATIONS

Year	Model	OEM Tires		Tire Pressures (psi)		Wheel	Ball Joint
		Standard	Optional	Front	Rear	Size	Inspection
2002	C Class	205/55R17	245/45R17	26	26	7-J	NS
	CL Class	Fr: 235/60R16	None	30	33	7.5-J	NS
		Rr: 235/60R16	None			8-J	
	CLK320/430	205/55R16	None	30	33	7-J	NS
	E320	225/55R16	None	30	33	8-J	NS
	E500	245/45R17	None	30	33	8.5-J	NS
	S430/500	225/55R17	None	30	33	7.5-J	NS
	S600	245/45R18	None	30	33	8-J	NS
	SL Class	255/45R17	None	30	33	8.5-J	NS
	SLK Class	225/45R17	None	30	33	7.5-J	NS
2003	C Class	205/55R17	245/45R17	26	26	7-J	NS
	CL Class	Fr: 235/60R16	None	30	33	7.5-J	NS
		Rr: 235/60R16	None			8-J	
	CLK320/430	205/55R16	None	30	33	7-J	NS
	E320	225/55R16	None	30	33	8-J	NS
	E500	245/45R17	None	30	33	8.5-J	NS
	S430/500	225/55R17	None	30	33	7.5-J	NS
	S600	245/45R18	None	30	33	8-J	NS
	SL Class	255/45R17	None	30	33	8.5-J	NS
	SLK Class	225/45R17	None	30	33	7.5-J	NS
2004	C Class	205/55R17	245/45R17	26	26	7-J	NS
	CL Class	Fr: 235/60R16	None	30	33	7.5-J	NS
		Rr: 235/60R16	None			8-J	
	CLK320/430	205/55R16	None	30	33	7-J	NS
	E320	225/55R16	None	30	33	8-J	NS
	E500	245/45R17	None	30	33	8.5-J	NS
	S430/500	225/55R17	None	30	33	7.5-J	NS
	S600	245/45R18	None	30	33	8-J	NS
	SL Class	255/45R17	None	30	33	8.5-J	NS
	SLK Class	225/45R17	None	30	33	7.5-J	NS
2005	C Class	Fr: 245/45R17	None	26	26	7.5-J	NS
		Rr: 245/40R17	None			8.5-J	

OEM: Original Equipment Manufacturer

PSI: Pounds Per Square Inch

Fr: Front

Rr: Rear

NS: Not specified by manufacturer

06041_BENZ_C0014

BRAKE SPECIFICATIONS

All measurements in inches unless noted

		Front Brake Disc			Rear Brake Disc			Minimum Lining Thickness		Brake Caliper	
Year	Model	Original Thickness	Minimum Thickness	Max. Runout	Original Thickness	Minimum Thickness	Max. Runout	Front	Rear	Bracket Bolts (ft. lbs.)	Mounting Bolts (ft. lbs.)
2002	C230	0.470	0.763	0.0047	0.354	0.287	0.0059	0.078	0.078	①	22-30
	C240	0.870	0.763	0.0047	0.350	0.287	0.0059	0.078	0.078	①	22-30
	C320	1.260	NA	NA	0.866	NA	NA	0.078	0.078	NA	NA
	C32	1.260	NA	NA	0.866	NA	NA	0.078	0.078	NA	NA
	E320	1.035	0.763	0.0047	0.400	0.337	0.0039	0.078	0.078	①	22-30
	E430	1.200	0.763	0.0047	0.480	0.385	0.0039	0.078	0.078	①	22-30
	E500	1.200	0.763	0.0047	0.480	0.385	0.0039	0.078	0.078	①	22-30
	S430	1.200	②	0.0031	0.480	0.763	0.0039	0.078	0.078	①	22-30
	S500	1.200	②	0.0031	0.480	0.763	0.0039	0.078	0.078	①	22-30
	S600	1.200	②	0.0031	0.480	0.763	0.0039	0.078	0.078	①	22-30
	CL500	1.200	②	0.0031	0.480	0.763	0.0039	0.078	0.078	①	22-30
	CL55	1.200	②	0.0031	0.480	0.763	0.0039	0.078	0.078	①	22-30
	CL600	1.200	②	0.0031	0.480	0.763	0.0039	0.078	0.078	①	22-30
	SL500	1.102	0.999	0.0047	0.354	0.287	0.0059	0.078	0.078	①	22-30
	SL55	1.102	0.999	0.0047	0.354	0.287	0.0059	0.078	0.078	①	22-30
	CLK320	1.100	0.999	0.0047	0.390	0.327	0.0039	0.078	0.078	①	22-30
	CLK430	1.100	0.999	0.0047	0.39	0.327	0.0039	0.078	0.078	①	22-30
	CLK500	1.100	0.999	0.0047	0.390	0.327	0.0039	0.078	0.078	①	22-30
	CLK55	1.100	0.999	0.0047	0.390	0.327	0.0039	0.078	0.078	①	22-30
	SLK230	0.980	0.763	0.0047	0.400	0.337	0.0039	0.078	0.078	①	22-30
	SLK320	0.980	0.763	0.0047	0.400	0.337	0.0039	0.078	0.078	①	22-30
2003	C230	0.470	0.763	0.0047	0.354	0.287	0.0059	0.078	0.078	①	22-30
	C240	0.870	0.763	0.0047	0.350	0.287	0.0059	0.078	0.078	①	22-30
	C320	1.260	NA	NA	0.866	NA	NA	0.078	0.078	NA	NA
	C32	1.260	NA	NA	0.866	NA	NA	0.078	0.078	NA	NA
	E320	1.035	0.763	0.0047	0.400	0.337	0.0039	0.078	0.078	①	22-30
	E500	1.200	0.763	0.0047	0.480	0.385	0.0039	0.078	0.078	①	22-30
	S430	1.200	②	0.0031	0.480	0.763	0.0039	0.078	0.078	①	22-30
	S500	1.200	②	0.0031	0.480	0.763	0.0039	0.078	0.078	①	22-30
	S600	1.200	②	0.0031	0.480	0.763	0.0039	0.078	0.078	①	22-30
	CL500	1.200	②	0.0031	0.480	0.763	0.0039	0.078	0.078	①	22-30
	CL55	1.200	②	0.0031	0.480	0.763	0.0039	0.078	0.078	①	22-30
	CL600	1.200	②	0.0031	0.480	0.763	0.0039	0.078	0.078	①	22-30
	SL500	1.102	0.999	0.0047	0.354	0.287	0.0059	0.078	0.078	①	22-30
	SL55	1.102	0.999	0.0047	0.354	0.287	0.0059	0.078	0.078	①	22-30
	CLK320	1.100	0.999	0.0047	0.390	0.327	0.0039	0.078	0.078	①	22-30
	CLK430	1.100	0.999	0.0047	0.390	0.327	0.0039	0.078	0.078	①	22-30
	CLK500	1.100	0.999	0.0047	0.390	0.327	0.0039	0.078	0.078	①	22-30
	CLK55	1.100	0.999	0.0047	0.390	0.327	0.0039	0.078	0.078	①	22-30
	SLK230	0.980	0.763	0.0047	0.400	0.337	0.0039	0.078	0.078	①	22-30
	SLK320	0.980	0.763	0.0047	0.400	0.337	0.0039	0.078	0.078	①	22-30

06041_BENZ_C0015

BRAKE SPECIFICATIONS

All measurements in inches unless noted

		Front Brake Disc			Rear Brake Disc			Minimum Lining Thickness		Brake Caliper	
Year	Model	Original Thickness	Minimum Thickness	Max. Runout	Original Thickness	Minimum Thickness	Max. Runout	Front	Rear	Bracket Bolts (ft. lbs.)	Mounting Bolts (ft. lbs.)
2004	C230	0.470	0.763	0.0047	0.354	0.287	0.0059	0.078	0.078	①	22-30
	C240	0.870	0.763	0.0047	0.350	0.287	0.0059	0.078	0.078	①	22-30
	C320	1.260	NA	NA	0.866	NA	NA	0.078	0.078	NA	NA
	C32	1.260	NA	NA	0.866	NA	NA	0.078	0.078	NA	NA
	E320	1.035	0.763	0.0047	0.400	0.337	0.0039	0.078	0.078	①	22-30
	E500	1.200	0.763	0.0047	0.480	0.385	0.0039	0.078	0.078	①	22-30
	S430	1.200	②	0.0031	0.480	0.763	0.0039	0.078	0.078	①	22-30
	S500	1.200	②	0.0031	0.480	0.763	0.0039	0.078	0.078	①	22-30
	S600	1.200	②	0.0031	0.480	0.763	0.0039	0.078	0.078	①	22-30
	CL500	1.200	②	0.0031	0.480	0.763	0.0039	0.078	0.078	①	22-30
	CL55	1.200	②	0.0031	0.480	0.763	0.0039	0.078	0.078	①	22-30
	CL600	1.200	②	0.0031	0.480	0.763	0.0039	0.078	0.078	①	22-30
	SL500	1.102	0.999	0.0047	0.354	0.287	0.0059	0.078	0.078	①	22-30
	SL55	1.102	0.999	0.0047	0.354	0.287	0.0059	0.078	0.078	①	22-30
	CLK320	1.100	0.999	0.0047	0.390	0.327	0.0039	0.078	0.078	①	22-30
	CLK430	1.100	0.999	0.0047	0.390	0.327	0.0039	0.078	0.078	①	22-30
	CLK500	1.100	0.999	0.0047	0.390	0.327	0.0039	0.078	0.078	①	22-30
	CLK55	1.100	0.999	0.0047	0.390	0.327	0.0039	0.078	0.078	①	22-30
	SLK230	0.980	0.763	0.0047	0.400	0.337	0.0039	0.078	0.078	①	22-30
	SLK320	0.980	0.763	0.0047	0.400	0.337	0.0039	0.078	0.078	①	22-30
2005	C230	0.470	0.763	0.0047	0.354	0.287	0.0059	0.078	0.078	①	22-30
	C240	0.870	0.763	0.0047	0.350	0.287	0.0059	0.078	0.078	①	22-30
	C320	1.260	NA	NA	0.866	NA	NA	0.078	0.078	NA	NA
	C55 AMG	NA	NA	NA	NA	NA	NA	NA	NA	NA	NA

NA: Not Available

① Front caliper: 85 ft. lbs.
Rear caliper: 38 ft. lbs.

② With 2 piston fixed caliper: 0.999
With 4 piston fixed caliper: 1.078

06041_BENZ_C0016

SCHEDULED MAINTENANCE INTERVALS

Mercedes Benz—C-Class, E-Class, S-Class, SLK, CLK, SL & CL

TO BE SERVICED	TYPE OF SERVICE	VEHICLE MILEAGE INTERVAL (x1000)															
		7.5	15	22.5	30	37.5	45	52.5	60	67.5	75	82.5	90	97.5	105	112.5	120
Accessory drive belt ①	S/I		✓		✓		✓		✓		✓		✓		✓		✓
Air filter element ②	R								✓								
Body for paint damage ③	S/I				✓				✓				✓				✓
Brake fluid ③	R				✓				✓				✓				✓
Charcoal filter ④	R								✓								✓
Driveshaft flex discs ②	S/I							✓							✓		
Engine coolant ⑤	R						✓						✓				✓
Engine oil & filter ①	R		✓		✓		✓		✓		✓		✓		✓		✓
Fuel filter ②	R								✓								
Sliding/pop-up roof	S/I	Clean slide rails and lubricate guide mechanism every 3 years.															
Spark plugs	R	Every 100,000 miles for all models except S-600 &CL-600; 60,000 for them.															
Steering gear bolts (tighten)	S/I							✓							✓		
Suspension & body structure ③	S/I				✓				✓				✓				✓
Underside of vehicle ②	S/I								✓								

R: Replace S/I: Service or Inspect, if needed

① Perform this at the mileage shown or once a year, whichever occurs first.

② Perform this at the mileage shown, or every 4 years, whichever occurs first.

③ Perform this at the mileage shown or every 2 years, whichever occurs first.

④ Replace at mileage interval or every 5 years, whichever comes first.

⑤ Replace at mileage or every 3 years regardless of mileage.

06041_BENZ_C0017

GASOLINE ENGINE REPAIR

➡Disconnecting the negative battery cable on some vehicles may interfere with the functions of the on board computer systems and may require the computer to undergo a relearning process, once the negative battery cable is reconnected.

Alternator

REMOVAL & INSTALLATION

2.8L, 3.2L, 3.6L (104), 1.8L, 2.3L (111), 2.6L, 2.8L, 3.2L (112), 4.3L, 5.0L and 5.5L (113) Engines

1. Before servicing the vehicle, refer to the Precautions Section.
2. Remove or disconnect the following:
   - Negative battery cable
   - Accessory drive belt
   - Engine under cover
   - Alternator electrical wires
3. Remove the alternator from the bottom.

To install:

4. Install the alternator assembly. Torque the mounting bolts to 31 ft. lbs. (42 Nm).
5. Install or connect the following:
   - Alternator electrical wires
   - Engine under cover
   - Accessory drive belt
   - Negative battery cable
6. Start the engine and check for proper operation.

5.0L (119) Engine

1. Before servicing the vehicle, refer to the Precautions Section.
2. Remove or disconnect the following:
   - Negative battery cable
   - Viscous fan coupling if equipped with ball-type alternator
   - Accessory drive belt
   - Alternator electrical connections
   - Alternator mounting bolts
   - Alternator

To install:

3. Install or connect the following:
   - Alternator assembly
   - Electrical connections
   - Accessory drive belt
   - Viscous fan coupling
   - Negative battery cable
4. Start the engine and check for proper operation.

6.0L (120) Engine

1. Before servicing the vehicle, refer to the Precautions Section.
2. Remove or disconnect the following:
   - Negative battery cable
   - Accessory drive belt
   - Alternator electrical connections
   - Automatic transmission coolant pipe
   - Engine under cover
   - Right torsion bar and spring leaf rockers
   - Alternator mounting bolts
3. Remove the alternator from the bottom.

To install:

4. Install or connect the following:
   - Alternator assembly
   - Right torsion bar and spring leaf rockers. Torque the bolts to 44 ft. lbs. (60 Nm).
   - Automatic transmission coolant pipe
   - Alternator electrical connections
   - Engine under cover
   - Negative battery cable
   - Accessory drive belt
5. Start the engine and check for proper operation.

Ignition Timing

ADJUSTMENT

All engines except the 5.0L (119) engine are equipped with a Distributorless Ignition System (DIS). No adjustment is necessary. The 5.0L timing is controlled by the Electronic Ignition with Anti-Knock Retard (EZL/ARK) control unit. No ignition timing adjustment is possible.

Engine Assembly

REMOVAL & INSTALLATION

2.8L, 3.2L and 3.6L (104) Engines

1. Before servicing the vehicle, refer to the Precautions Section.
2. Properly recover the air conditioning system refrigerant.
3. Properly relieve the fuel system pressure.
4. Drain the engine coolant and engine oil.
5. Remove or disconnect the following:
   - Negative battery cable
   - Exhaust
   - Steering damper
   - Automatic transmission cooling lines
   - Coolant expansion reservoir
   - Cooling fan and radiator
6. Cover the air conditioning condenser using a piece of sheet metal, plywood or plastic.
   - Charge air pipes
   - Air cleaner housing
   - Power steering lines
   - Power steering pump
   - Fuel lines
   - Engine wiring harness and vacuum lines
   - Throttle cable
   - Coolant hoses
   - Rear engine mount
   - Driveshaft
   - Park lock interlock cable
   - Automatic transmission control unit connector
   - Transmission linkage
   - Front engine mounts
7. Raise the engine sufficiently until bolts of air conditioning compressor are accessible
   - Air conditioning compressor
   - Engine assembly

To install:

8. Install or connect the following:
   - Engine assembly
   - Engine mounts. Torque the bolts to 41 ft. lbs. (55 Nm)
   - Air conditioning compressor
   - Transmission linkage
   - Automatic transmission control unit connector
   - Park lock interlock cable
   - Driveshaft
   - Coolant hoses
   - Throttle cable
   - Engine wiring harness and vacuum lines
   - Fuel lines
   - Power steering pump and lines
   - Air cleaner housing
   - Charge air pipes
9. Remove the air conditioning condenser guard
   - Cooling fan and radiator
   - Coolant expansion reservoir
   - Automatic transmission cooling lines
   - Steering damper
   - Exhaust
   - Negative battery cable
   - Coolant
   - Transmission fluid
   - Refrigerant

1.8L and 2.3L (111) Engines

1. Before servicing the vehicle, refer to the Precautions Section.
2. Properly recover the air conditioning system refrigerant.
3. Properly relieve the fuel system pressure.
4. Drain the engine coolant, engine and transmission oil.
5. Remove or disconnect the following:
   - Negative battery cable
   - Engine under cover
   - Fan shroud, clutch and fan assembly
   - Radiator
   - Heater hoses
   - Oil cooler lines
   - Mass Air Flow (MAF) sensor and air intake assembly
   - Charge air pipes if equipped with a supercharger
   - Accessory drive belts
6. Cover the air conditioning condenser using a piece of sheet metal, plywood or plastic.
   - Engine wiring harness, ground straps and electrical connections
   - Vacuum hoses
   - Fuel lines
7. Drain the power steering fluid from the reservoir.
   - Automatic transmission electrical connections, if equipped
   - Accelerator and transmission linkage
   - Power steering pump hoses
   - Air conditioning compressor and position it aside, leaving the hoses attached
   - Front halfshafts, if equipped with 4-MATIC
   - Slave cylinder clutch line, if equipped
   - Exhaust system
   - Driveshaft
8. Ensure that all electrical and mechanical connections are detached from the Automatic transmission
9. Support the engine and transmission. Disconnect the mounts, and lift the engine and transmission out of the vehicle as one assembly.

To install:

10. Position the engine/transmission assembly and install the mounts. Torque the bolts as follows:
    - Transmission mount-to-crossmember bolts: 18 ft. lbs. (25 Nm)
    - Transmission crossmember-to-body bolts: 30 ft. lbs. (40 Nm)
    - Transmission mount-to-transmission nut: 52 ft. lbs. (70 Nm)
    - Engine mount-to-subframe mounting bolt: 30 ft. lbs. (40 Nm)
11. Install or connect the following:
    - Driveshaft
    - Exhaust system
    - Slave cylinder clutch line
    - Front halfshafts
    - Air conditioning compressor
    - Power steering hoses
    - Accelerator and transmission linkage
    - Automatic transmission electrical connections
    - Fuel lines
    - Vacuum hoses
    - Engine wiring harness, ground straps and electrical connections
12. Remove the air conditioning condenser guard and fill the power steering reservoir.
    - Accessory drive belts
    - Charge air pipes
    - MAF sensor and air intake assembly
    - Radiator and coolant hoses
    - Oil cooler lines
    - Fan clutch, fan and shroud
    - Negative battery cable
    - Coolant
    - Engine oil
    - Transmission fluid
    - Refrigerant

2.6L, 2.8L, 3.2L (112), 4.3L (113) and 5.0L (119) Engines

1. Before servicing the vehicle, refer to the Precautions Section.
2. Properly relieve the fuel system pressure.
3. Drain the engine coolant and engine oil.
4. Remove or disconnect the following:
   - Negative battery cable
   - Engine under cover
   - Accessory drive belt
   - Air cleaner housing
   - Resonance pipe and body
   - Fan shroud, clutch and fan assembly
   - Mass Air Flow (MAF) sensor
5. Cover the air conditioning condenser using a piece of sheet metal, plywood or plastic.
6. Remove or disconnect the following:
   - Coolant hoses and radiator
   - Automatic transmission dipstick guide tube
7. Drain the power steering fluid and detach the lines from the pump. Plug the lines.
   - Automatic transmission fluid lines
   - Fuel lines
   - Wiring, vacuum and cable connections
   - Air conditioning compressor and position it aside with the hoses attached
   - Exhaust system and bracket
   - Driveshaft vibration damper
   - Driveshaft
   - Transmission linkage
   - Front wheels
   - Front halfshafts if equipped with 4-MATIC
   - Rear engine crossmember at transmission
   - Starter assembly
   - Front engine mounts
8. Support the transmission assembly and lift the engine out of the vehicle.

To install:

9. Install or connect the following:
   - Engine assembly
   - Front engine mounts. Torque the bolts to 41 ft. lbs. (55 Nm).
   - Starter assembly
   - Rear engine crossmember at transmission. Torque the bolts to 30 ft. lbs. (40 Nm).
   - Front halfshafts
   - Transmission linkage
   - Driveshaft
   - Driveshaft vibration damper
   - Exhaust system and bracket
   - Air conditioning compressor
   - Wiring, vacuum and cable connections
   - Fuel lines
   - Automatic transmission fluid lines and dipstick guide tube
   - Coolant hoses and radiator
   - Resonance pipe and body
   - Air cleaner housing
   - MAF sensor
10. Remove the air conditioning condenser guard and fill the power steering reservoir.
    - Fan shroud, clutch and fan assembly
    - Accessory drive belt
    - Negative battery cable
    - Coolant
    - Engine oil
    - Transmission fluid

5.0L and 5.5L (113) Engines

1. Before servicing the vehicle, refer to the Precautions Section.
2. Properly relieve the fuel system pressure.
3. Drain the engine coolant and engine oil.
4. Drain the transmission fluid on automatic transmissions.

5. Remove or disconnect the following:
   - Negative battery cable
   - Engine under cover
   - Accessory drive belt
   - Fan shroud, clutch and fan assembly
   - Coolant pipes and hoses
   - Air cleaner housing
   - Resonance pipe and body
   - Mass Air Flow (MAF) sensor
   - Radiator
6. Drain the power steering fluid and detach the lines from the pump and plug the lines.
   - Fuel lines
   - Wiring, vacuum and cable connections
7. Cover the air conditioning condenser using a piece of sheet metal, plywood or plastic.
   - Air conditioning compressor and position it aside with the hoses attached
   - Exhaust system and bracket
   - Clutch slave cylinder, if equipped
   - Driveshaft vibration damper
   - Driveshaft
   - Transmission linkage
   - Rear engine crossmember
   - Bolts from front engine mounts
8. Support the transmission assembly and lift the engine out of the vehicle.

To install:

9. Install or connect the following:
   - Engine assembly
   - Front engine mounts. Torque the bolts to 26 ft. lbs. (35 Nm).
   - Rear engine crossmember. Torque the bolts to 30 ft. lbs. (40 Nm).
   - Transmission linkage
   - Driveshaft. Torque the M10 bolts to 30 ft. lbs. (40 Nm), and M12 bolts to 44 ft. lbs. (60 Nm).
   - Driveshaft vibration damper
   - Clutch slave cylinder, if equipped
   - Exhaust system and bracket. Torque the bolts to 15 ft. lbs. (20 Nm).
   - Air conditioning compressor. Torque the bolts to 15 ft. lbs. (20 Nm).
   - Wiring, vacuum and cable connections
   - Fuel lines. Torque the bolts to 28 ft. lbs. (38 Nm).
   - Power steering pump. Torque the bolts to 31 ft. lbs. (42 Nm).
   - Radiator
   - MAF sensor
   - Resonance pipe and body
   - Air cleaner housing
   - Coolant pipes and hoses
   - Fan shroud, clutch and fan assembly
   - Accessory drive belt
   - Engine under cover
   - Negative battery cable
   - Coolant
   - Engine oil
   - Transmission fluid

6.0L (120) Engine

1. Before servicing the vehicle, refer to the Precautions Section.
2. Properly relieve the fuel system pressure.
3. Drain the engine coolant, and engine oil.
4. Remove or disconnect the following:
   - Negative battery cable
   - Engine under cover
   - Air cleaner
   - Radiator and coolant hoses
   - Automatic transmission fluid lines
5. Cover the air conditioning condenser using a piece of sheet metal, plywood or plastic.
   - Fan shroud, clutch and fan assembly
   - Accessory drive belts
   - Fuel lines
   - Wiring, vacuum and cable connections
   - Power steering pump
   - Oil cooler lines
   - Air conditioning compressor and position it aside with the lines attached
   - Oil pressure switch
   - Exhaust system
   - Starter
   - Driveshaft
   - Transmission linkage
   - Front and rear engine mounts
   - Rear engine crossmember
6. Support the transmission assembly and lift the engine out of the vehicle.

To install:

7. Position the engine assembly into the vehicle and install the mounts. Torque the bolts to 18 ft. lbs. (25 Nm).
8. Install or connect the following:
   - Rear engine crossmember. Torque the bolts to 30 ft. lbs. (40 Nm).
   - Transmission linkage
   - Driveshaft
   - Starter
   - Exhaust system
   - Oil pressure switch
   - Air conditioning compressor
   - Oil cooler lines
   - Power steering pump
   - Wiring, vacuum and cable connections
   - Fuel lines
   - Accessory drive belts
   - Fan shroud, clutch and fan assembly
9. Remove the air conditioning condenser guard and fill the power steering reservoir.
   - Automatic transmission fluid lines
   - Radiator and coolant hoses
   - Air cleaner
   - Negative battery cable
   - Coolant
   - Engine oil
   - Transmission fluid

Water Pump

REMOVAL & INSTALLATION

All Models

1. Before servicing the vehicle, refer to the Precautions Section.
2. Disconnect the negative battery cable and drain the cooling system.
3. Remove or disconnect the following:
   - Fan clutch, fan and shroud
   - Engine cover if equipped
   - Accessory drive belt and tensioner
   - Secondary air injection switchover valve if equipped
   - Power steering pump and position the pump aside, leaving the hoses attached
   - Water pump coolant hoses
   - Oil-to-water heat exchanger coolant hoses, if equipped
   - Belt pulley
   - Water pump mounting bolts
   - Water pump
4. Clean all gasket material from the sealing surfaces.

To install:

5. Install the water pump and gasket. Torque the bolts as follows:
   - 3.2L (104) and 5.0L (119) engines: 15 ft. lbs. (21 Nm)
   - 1.8L and 2.3L (111) engines: M6 bolts to 88 inch lbs. (10 Nm). M8 bolts to 18 ft. lbs. (25 Nm)
   - 2.6L, 2.8L and 3.2L (112) engines with M6 bolts to 88 inch lbs. (10 Nm) and M8 bolts to 15 ft. lbs. (20 Nm)
   - 6.0L (120) engine: M6 bolts to 88 inch lbs. M8 bolts to 15 ft. lbs. (21 Nm)
6. Install or connect the following:
   - Belt pulley and torque the mounting bolts to 88 inch lbs. (10 Nm)
   - Oil-to-water heat exchanger coolant hoses

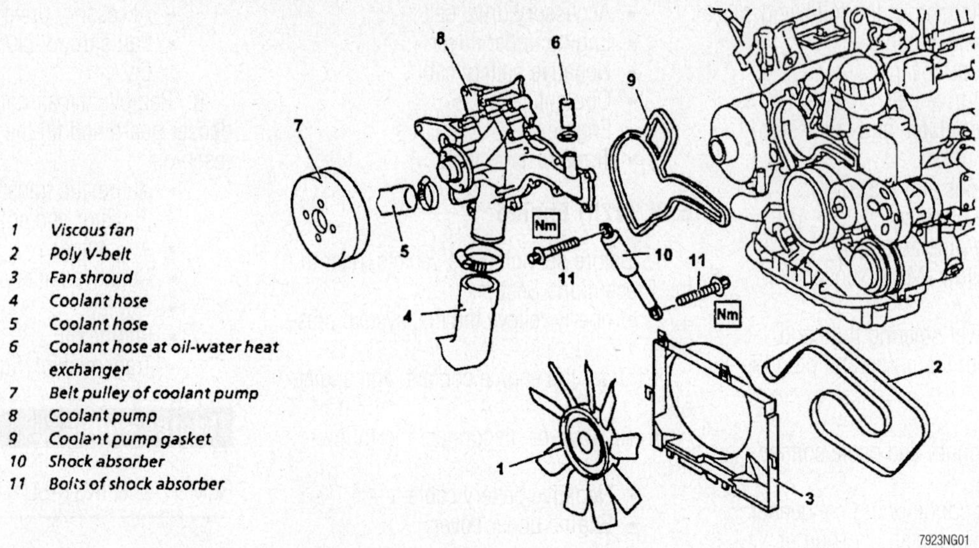

Exploded view of the water pump mounting—2.6L, 2.8L and 3.2L (112) and 4.3L, 5.0L and 5.5L (113) engines

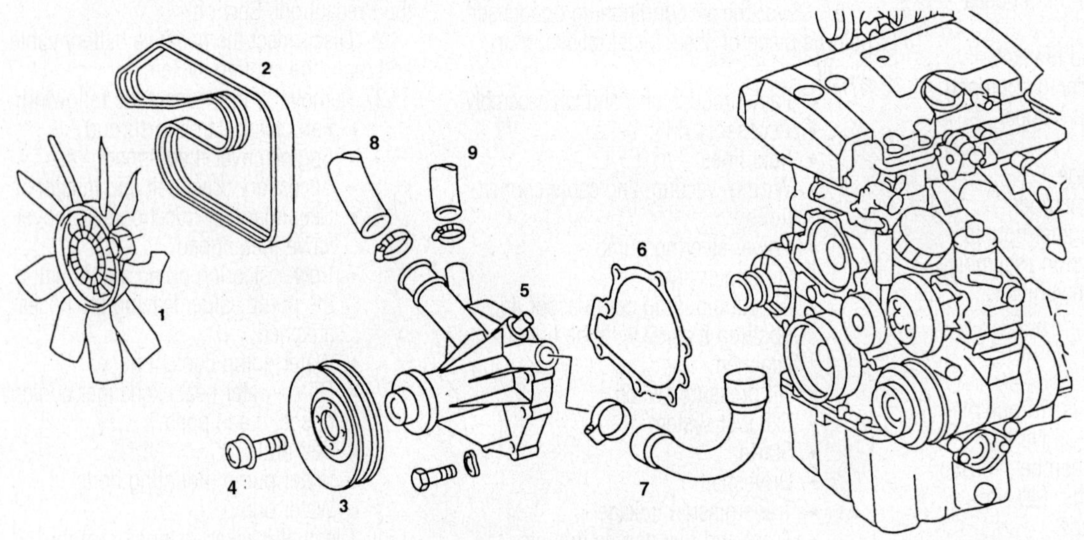

1. Viscous fan
2. Poly V-belt
3. Water pump pulley
4. Water pump pulley mounting
5. Water pump
6. Water pump gasket
7. Coolant hose
8. Heater hose
9. Heater hose

7923NG02

Exploded view of the water pump mounting—1.8L and 2.3L (111) engine

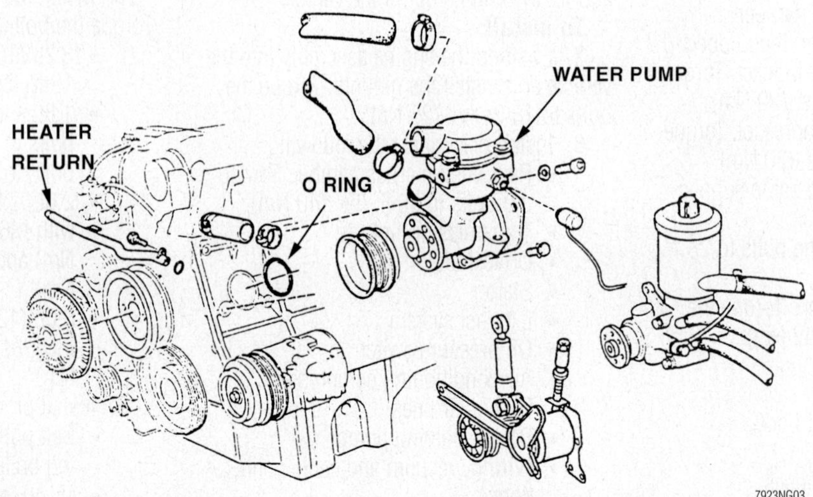

Exploded view of the water pump mounting—3.2L (104) engine in the SL

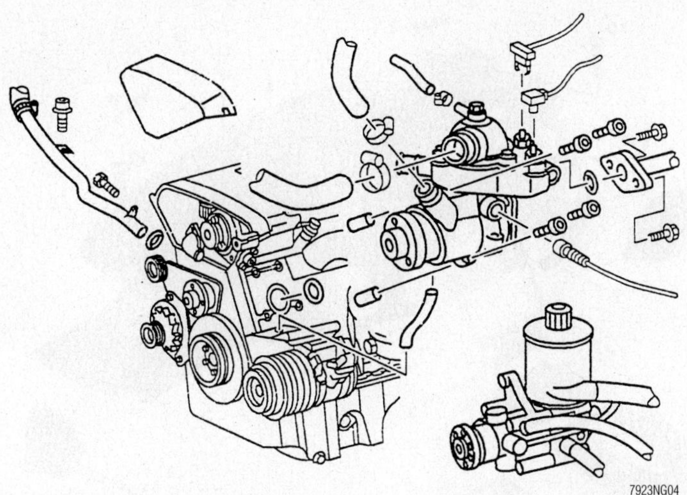

Exploded view of the water pump mounting—3.2L (104) engine in the C and E Class models

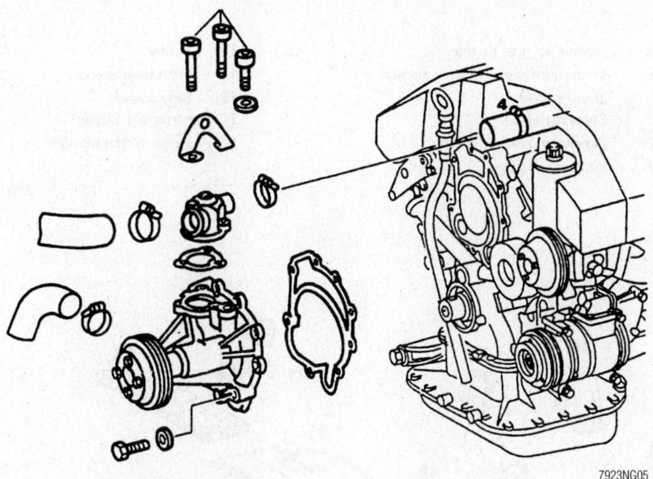

Exploded view of the water pump mounting—5.0L (119) engine

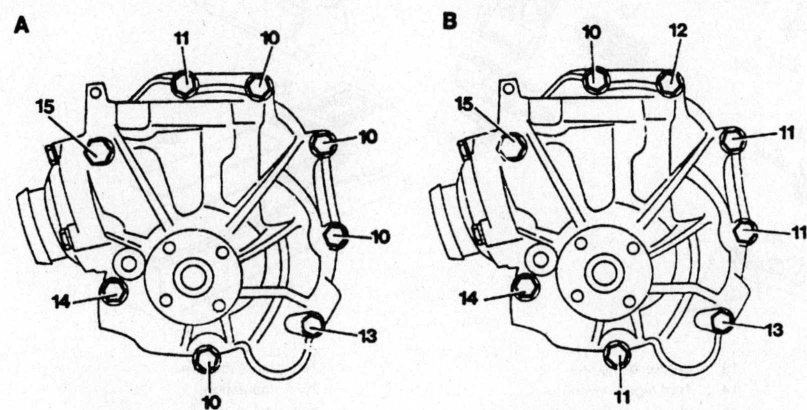

A	Engine 119.96	B	Engine 119.97/98
10	M8×60 bolt+washer	10	M8×60 bolt+washer
11	M8×65 bolt+washer	11	M8×65 bolt+washer
13	M8×85 bolt+washer	12	M8×75 bolt+washer
14	M8×90 bolt+washer (together with fan clutch carrier)	13	M8×85 bolt+washer
15	M8×135 bolt+washer	14	M8×90 bolt+washer (together with fan clutch carrier)
		15	M8×135 bolt+washer

7923NG06

Water pump mounting bolt identification—5.0L (119) engine

- Water pump coolant hoses
- Power steering pump
- Secondary air injection switchover valve
- Accessory drive belt and tensioner
- Fan clutch, fan and shroud
- Engine cover
- Coolant
- Negative battery cable

Heater Core

REMOVAL & INSTALLATION

C Class

1. Before servicing the vehicle, refer to the Precautions Section.
2. Disconnect the negative battery.
3. Drain the cooling system into a clean container for reuse.
4. Discharge and recovery the air conditioning refrigerant.
5. Remove or disconnect the following:
   - Steering wheel
   - Instrument cluster
   - Center air vent
   - Lock cover plate
   - Instrument panel to carrier screws
   - End panels
   - Passenger air bag screws
   - Air bag cover and air bag
   - A pillar trim
   - Fuse box
   - Instrument panel strut
   - Instrument panel screws
   - Electrical connectors
   - Instrument panel
   - Accelerator pedal
   - Left and right air ducts
   - Drier cartridge
   - Expansion valve line
   - Heater hoses
   - Drain hoses
   - Evaporator sensor connector
   - Heater core

To install:

6. To install, reverse the removal procedure.
7. Refill the cooling system.
8. Recharge the air conditioning system.
9. Connect the negative battery cable.
10. Operate the engine to normal operating temperatures; then, check the climate control operation and check for leaks.

E Class

1. Before servicing the vehicle, refer to the Precautions Section.
2. Disconnect the negative battery cable.

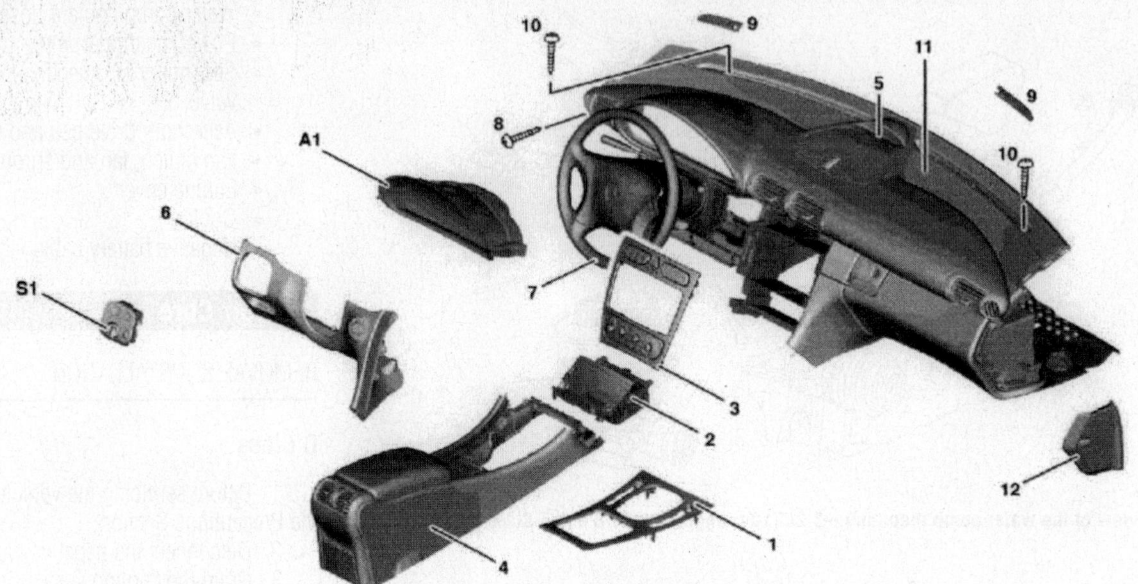

Illustrated on model 203.0

1 Cover on gearshift lever
2 Ashtray housing
3 Cover on center console
4 Center console
5 Center air nozzle, top
6 Instrument panel bottom section on driver's side
7 Steering wheel
8 Screw for side air nozzle
9 Lock cover plate
10 Screw
11 Instrument panel
12 Side cover
TO1 Instrument cluster
S1 Lamp switch module

42348-BENZ-G20

Exploded view of the instrument panel—C Class

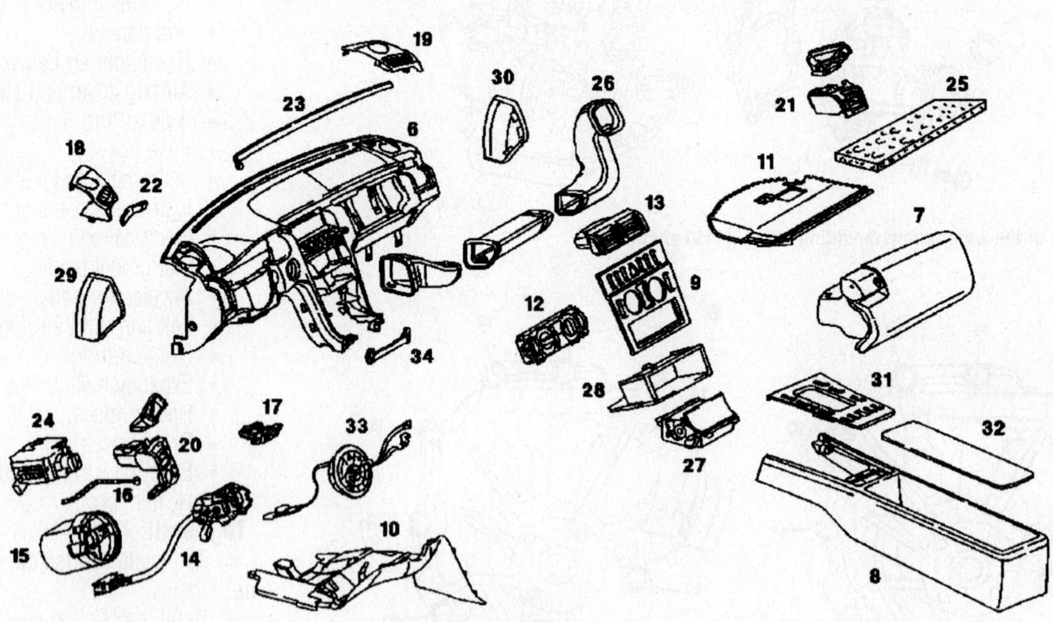

6 Instrument panel
7 Glove box
8 Center console
9 Cover for center console
10 Cover below instrument panel (left)
11 Cover below instrument panel (right)
12 Pushbutton control module for automatic heater or air conditioning
13 Center air nozzle
14 Turn signal switch
15 Steering column covering
16 Pull cable
17 Housing for release handle
18 Cover for loudspeaker (left)
19 Cover for loudspeaker (right)
20 Side air nozzle (left)
21 Side air nozzle (right)
22 Molding
23 Molding
24 Light module
25 Insulation
26 Air duct
27 Ashtray housing
28 Stowage compartment
29 Cover
30 Cover
31 Shift lever cover
32 Insert
33 Contact spiral
34 Guide

42348-BENZ-G21

Exploded view of the instrument panel—E Class

3. Drain the cooling system into a clean container for reuse.
4. Discharge and recovery the air conditioning refrigerant.
5. Remove or disconnect the following:
   - Hoses from the heater supply and return pipes. Using compressed air, blow the residual coolant from the heater core
   - Instrument panel lower covers
   - Steering wheel
   - Combination switch and jacket
   - Hood release
   - A pillar trims
   - Speaker covers
   - Wood moldings
   - Air vent nozzles
   - Lighting module
   - Glove box
   - Passenger air bag strap
   - Center console
   - Center storage compartment
   - Radio
   - A/C control push button panel
   - Sun sensor
   - Instrument panel
   - Air ducts
   - Floor coverings
   - Vacuum lines
   - Blower motor connector
   - Connectors and vent flaps
   - Heater core

To install:

6. To install, reverse the removal procedure.
7. Refill the cooling system.
8. Recharge the air conditioning system.
9. Connect the negative battery cable.
10. Operate the engine to normal operating temperatures; then, check the climate control operation and check for leaks.

CL and S Class

1. Before servicing the vehicle, refer to the Precautions Section.
2. Disconnect the negative battery cable.

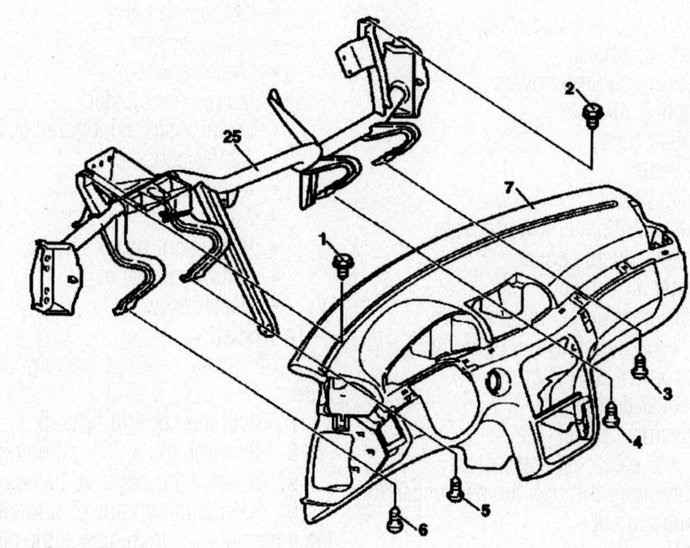

Instrument panel mounting to crossmember—CL and S Class

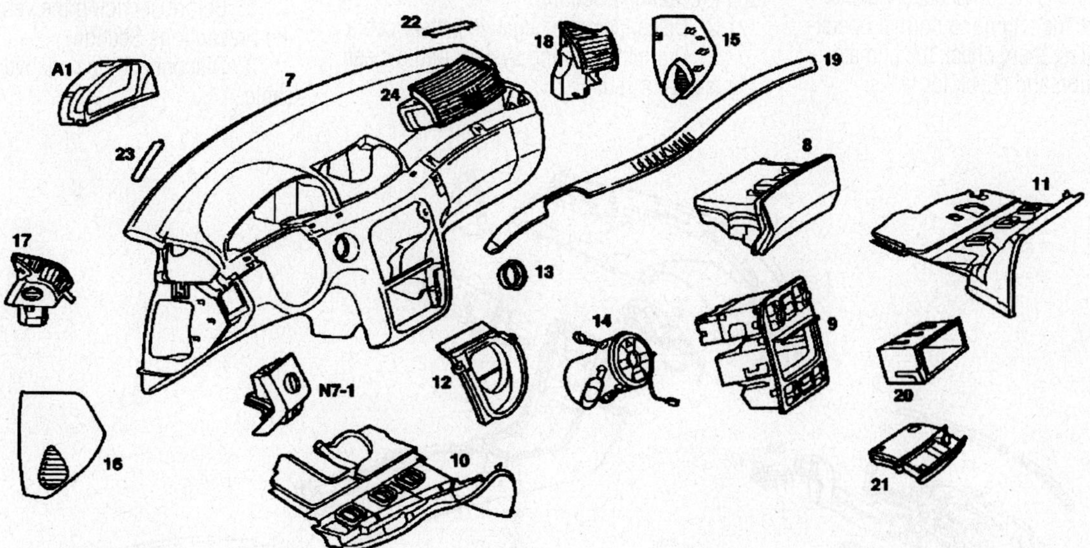

Exploded view of the instrument panel—CL and S Class

3. Drain the cooling system into a clean container for reuse.

4. Discharge and recovery the air conditioning refrigerant.

5. Remove or disconnect the following:

- Windshield wipers
- Cowl vents
- Expansion valve lines
- Heater hoses
- Ignition switch module
- Data link connector
- Steering wheel
- Steering column covers
- Lighting module
- A pillar trim
- Air vents
- Glove box
- Instrument panel lower covers
- Center console cover
- Center air nozzle
- Instrument panel upper covers
- Parktronic displays, if equipped
- Instrument panel screws
- Speaker connector
- Instrument panel
- All A/C electrical connectors
- Heater core through the passenger compartment

To install:

6. To install, reverse the removal procedure.

7. Refill the cooling system.

8. Recharge the air conditioning system.

9. Connect the negative battery cable.

10. Operate the engine to normal operating temperatures; then, check the climate control operation and check for leaks.

CLK

1. Before servicing the vehicle, refer to the Precautions Section.

2. Disconnect the negative battery cable.

3. Drain the cooling system into a clean container for reuse.

4. Discharge and recovery the air conditioning refrigerant.

5. Remove or disconnect the following:

- Expansion valve
- Heater hoses
- Steering wheel
- Instrument cluster
- Lower instrument panel cover
- Center console
- Air vents
- A pillar trim
- Instrument panel screws
- Instrument panel
- Heater core

To install:

6. To install, reverse the removal procedure.

7. Refill the cooling system.

8. Recharge the air conditioning system.

9. Connect the negative battery cable.

10. Operate the engine to normal operating temperatures; then, check the climate control operation and check for leaks.

SL

1. Before servicing the vehicle, refer to the Precautions Section.

2. Disconnect the negative battery cable.

3. Drain the cooling system into a clean container for reuse.

4. Discharge and recovery the air conditioning refrigerant.

5. Open the sunroof.

6. Place the steering wheel at its lowest position.

7. Move the front seats fully rearward

8. Remove or disconnect the following:

- Center console covers
- Center console screws
- Upper instrument panel bolts
- COMAND display module
- Radio opening
- Instrument panel lower bolts
- Driver side lower instrument panel cover
- Instrument cluster
- A pillar trims
- Electrical connectors and vacuum lines
- Speakers
- Instrument panel
- Heater core

To install:

9. To install, reverse the removal procedure.

10. Refill the cooling system.

11. Recharge the air conditioning system.

12. Connect the negative battery cable.

13. Operate the engine to normal operating temperatures; then, check the climate control operation and check for leaks.

SLK

1. Before servicing the vehicle, refer to the Precautions Section.

2. Disconnect the negative battery cable.

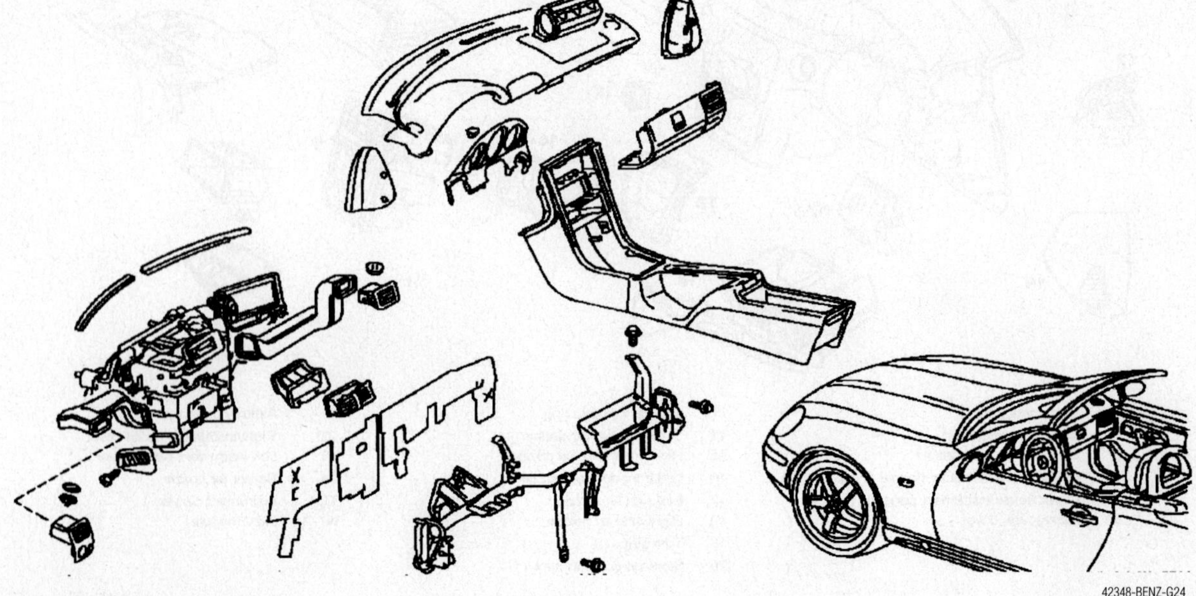

Exploded view of the instrument panel—SLK

3. Drain the cooling system into a clean container for reuse.
4. Discharge and recovery the air conditioning refrigerant.
5. Remove or disconnect the following:
- Heater hoses
- Expansion valve
- Center console
- A pillar trims
- A pillar clips
- Instrument cluster
- Upper instrument panel screws
- Instrument panel
- Electrical connectors
- Heater ducts
- Heater core

To install:

6. To install, reverse the removal procedure.
7. Refill the cooling system.
8. Recharge the air conditioning system.
9. Connect the negative battery cable.
10. Operate the engine to normal operating temperatures; then, check the climate control operation and check for leaks

Cylinder Head

REMOVAL & INSTALLATION

3.2L (104) Engine

1. Before servicing the vehicle, refer to the Precautions Section.
2. Properly relieve the fuel system pressure and drain the engine coolant.
3. Position the No. 1 cylinder at Top Dead Center (TDC).
4. Remove or disconnect the following:
- Negative battery cable
- Coolant hoses
- Upper timing cover
5. Matchmark the camshaft sprocket to the timing chain, and remove the pin (1) for the timing chain guide with a threaded extractor.
- Guide sprocket (left-hand thread)
- Bearing assembly
- Timing chain from the camshaft sprockets and wire the chain aside

➡Be sure the chain is securely wired so it will not slide into the engine.

- Vacuum hoses and electrical connectors
- Exhaust system
- Secondary air injection pipe, if equipped
- Dipstick tube support bracket
- Crankcase breather hose
- Fuel lines
- Dipstick tube
- Throttle, kickdown and cruise control cables

6. Loosen the head bolts in stages in the reverse order of the tightening sequence.

CAUTION

Never use a prybar between the head and the block.

7. Remove the cylinder head and clean all gasket material from the sealing surfaces. Be sure the cylinder head locating dowels are positioned correctly in the engine block.
8. Inspect length of the cylinder head bolt shaft. New bolt length is 6.30 inches (160mm), and the maximum permissible length is 6.44 inches (163.5mm). Replace bolts that measure greater than the maximum permissible length.

To install:

➡The head will not be watertight until the engine reaches operating temperature. Do not pressure test the cooling system until the engine has reached operating temperature.

9. Rotate the camshafts so that the bottom edge of the holes in the camshaft flange, are level with the top edge of the cylinder head.
10. Verify the TDC position of the No. 1 cylinder. Clean the head bolt threads, and then apply clean engine oil to the thread and head contact surfaces.
11. Install the head gasket and cylinder head.
12. Torque the head bolts using the illustrated sequence to 41 ft. lbs. (55 Nm) plus 90 degrees, then an additional 90 degrees.

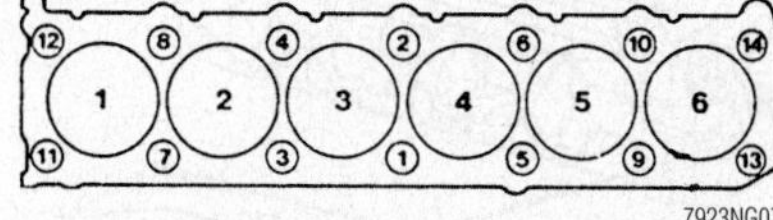

Cylinder head bolt tightening sequence—3.2L (104) engines

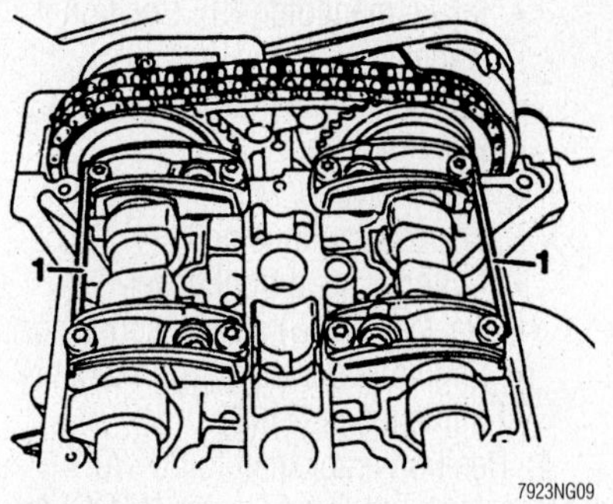

Verify the correct position of the camshafts for cylinder head installation using a 4mm pin (1)—3.2L (104) Engines

13. Install or connect the following:
- Timing chain and upper timing chain guide
- Timing chain guide pin
- Guide sprocket and bearing assembly. Torque the bolt to 26 ft. lbs. (35 Nm).
- Throttle, kickdown and cruise control cables
- Fuel lines
- Crankcase breather hose
- Oil dipstick tube
- Secondary air injection pipe, if equipped
- Exhaust system
- Vacuum hoses and electrical connectors
- Coolant hoses
- Coolant
- Negative battery cable

1.8L and 2.3L (111) Engines

1. Before servicing the vehicle, refer to the Precautions Section.
2. Drain the engine coolant.
3. Remove or disconnect the following:
- Negative battery cable
- Exhaust system and bracket at transmission
- Thermostat housing cover
- Air intake tubes
- Electrical connectors
- Coolant hose (3) at the rear of the cylinder head

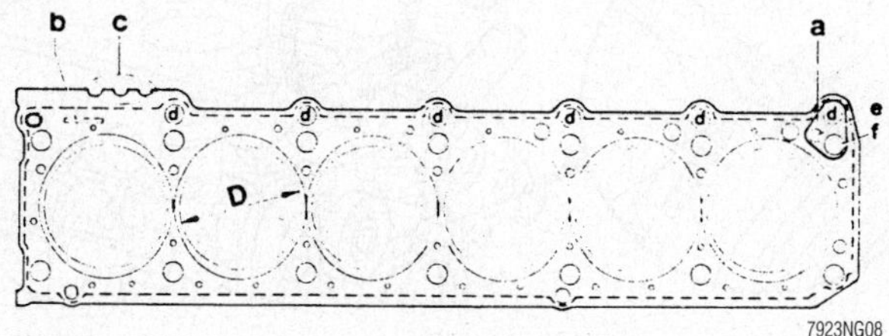

Cylinder head gasket identification—3.2L (104) Engines

- Intake manifold (19). Position it aside.
- Crankcase breather hose
- Manifold bracket (2)
- Valve cover
- Thermostat housing
- Cylinder head front cover
- Heated (HO_2S) sensor connector
- Engine and transmission dipstick guide tubes (7) (8)

4. Position crankshaft to 20–30 degrees After Top Dead Center (ATDC) for cylinder No. 1.
5. Lock the camshafts (75 and 78) with locking pins (01) and matchmark the timing chain to the sprockets.
6. Remove or disconnect the following:
   - Chain tensioner (14)
   - Chain guide rail in the cylinder head
   - Exhaust camshaft sprocket (76)
   - Intake camshaft sprocket (74)
7. Install the cylinder head removal bracket (16) and remove the guide rail pin (11), using a threaded extractor.
   - Timing chain housing bolts (A)
   - Cylinder head bolts in the reverse order of the illustrated tightening sequence

✲✲ CAUTION

Never use a prybar between the head and block.

8. Remove the cylinder head and clean all gasket material from the sealing surfaces. Be sure the cylinder head locating dowels are positioned correctly in the engine block.
9. Inspect length of the cylinder head bolt shaft. New bolt length is 102mm and the maximum permissible length is 105mm. Replace bolts that measure greater than the maximum permissible length.

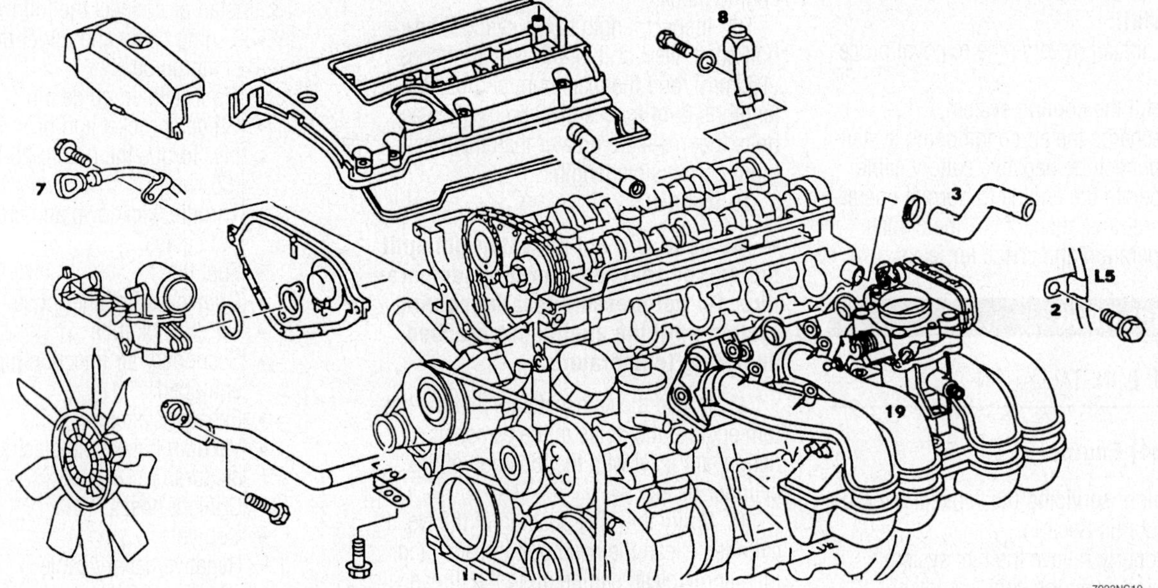

Exploded view of the cylinder head accessory components—1.8L and 2.3L (111) engines

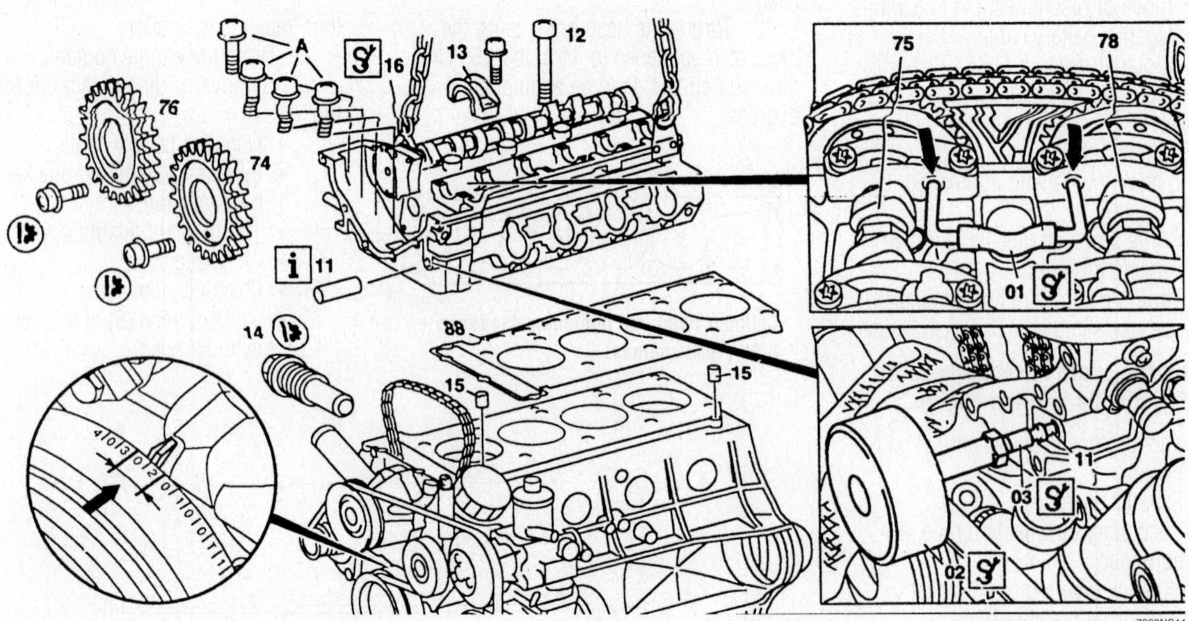

Exploded view of the cylinder head mounting—1.8L and 2.3L (111) engines

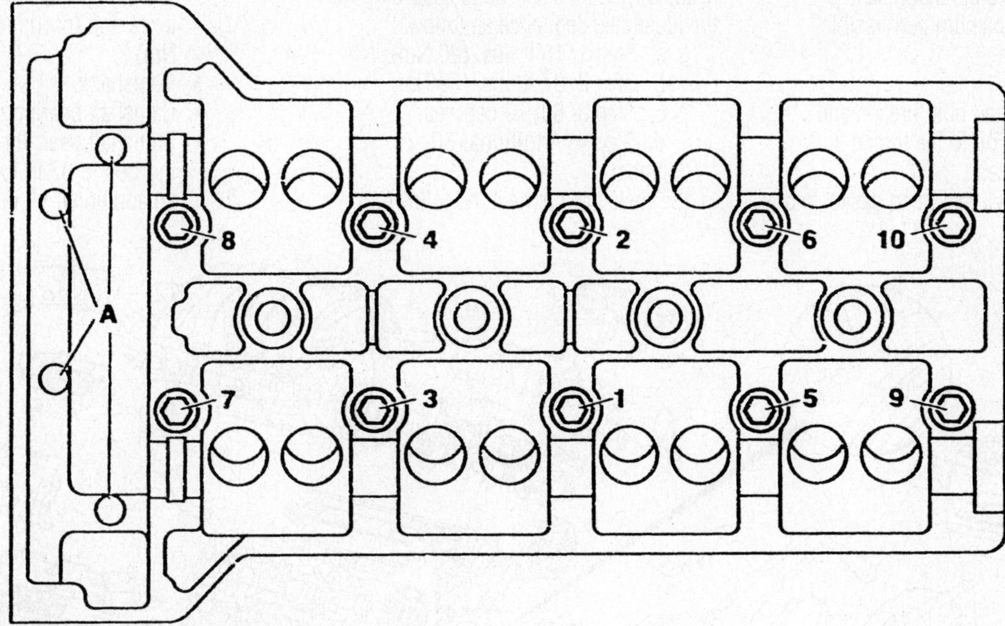

Cylinder head bolt tightening sequence—1.8L and 2.3L (111) engines

To install:

➡The head gasket is not watertight until the engine has reached operating temperature. Do not pressure test the cooling system until the engine has reached operating temperature.

10. Verify the Top Dead Center (TDC) position of the No. 1 cylinder. Clean the head bolt threads and apply clean engine oil to the thread and head contact surfaces.
11. Install or connect the following:
    - Head gasket and cylinder head. Torque the head bolts using the illustrated sequence to 41 ft. lbs. (55 Nm) plus 90 degrees, then an additional 90 degrees.
    - Timing chain housing bolts
    - Guide rail pin
12. Remove the cylinder head removal bracket.
    - Intake and exhaust camshaft sprockets
    - Guide rail
    - Chain tensioner
13. Remove the camshaft locking pins and verify timing mark alignment.
    - Engine oil and transmission dipstick guide tubes
    - HO_2S sensor
    - Cylinder head front cover
    - Thermostat housing
    - Valve cover
    - Cylinder head coolant hose
    - Crankcase breather hose
    - Intake manifold
    - Electrical connectors
    - Thermostat housing cover and air intake tubes
    - Exhaust system and bracket
    - Coolant
    - Negative battery cable

2.6L, 2.8L and 3.2L (112) and 4.3L, 5.0L and 5.5L (113) engines

1. Before servicing the vehicle, refer to the Precautions Section.
2. Properly relieve the fuel system pressure and drain the engine coolant. Place a guard plate behind the radiator/condenser to protect it from damage.
3. Remove or disconnect the following:
    - Negative battery cable
    - Fan clutch, fan and shroud
    - Engine cover
    - Air cleaner housing, resonance pipe and body
    - Fuel line
    - Ignition coils
    - Cylinder head covers

➡The intake manifold system must not be disassembled.

- Intake manifold
- Vacuum switchover valve
- Camshaft Position (CMP) sensor

4. Lock the Automatic belt tensioner by rotating the tensioner counterclockwise until a 5mm drift or pin fits through the tensioner, and then remove the serpentine belt.
5. Remove or disconnect the following:
    - Power steering pump and position it aside leaving the hoses attached
    - Serpentine belt
    - Heater hose at the firewall
    - Exhaust system
6. Rotate the engine clockwise to position the crankshaft 40 degrees After Top Dead Center (ATDC).

⁂ WARNING

The engine must not be rotated backwards.

7. Lock the camshafts using camshaft locking tools.
8. Remove or disconnect the following:
    - Generator
    - Timing chain tensioner
    - Camshaft gears. Attach to the chain with a cable tie.
    - Camshaft bearing bridges
    - Camshafts
    - Timing case-to-cylinder head bolts
    - Cylinder head bolts in the reverse order of the illustrated tightening sequence

⁂ CAUTION

Never use a prybar between the head and block.

9. Remove the cylinder head and clean all gasket material from the sealing surfaces. Be sure the cylinder head locating dowels are positioned in the engine block.
10. Inspect length of the cylinder head bolt shaft. New bolt length is 141.5mm and the maximum permissible length is

144.5mm. Replace bolts that measure greater than the maximum permissible length.

To install:

11. Clean the head bolt threads and apply clean engine oil to the thread and head contact surfaces.
12. Install the cylinder head gasket and head. Torque the head bolts according to the illustrated sequence as follows:
    a. Step 1: 15 ft. lbs. (20 Nm).
    b. Step 2: 37 ft. lbs. (50 Nm).
    c. Step 3: 60–70 degrees.
    d. Step 4: Additional 60–70 degrees.
13. Install or connect the following:
    - Timing case-to-cylinder head bolts and torque to 15 ft. lbs. (20 Nm)
    - Camshafts
    - Camshaft bearing bridges
    - Camshaft gear. Torque the mounting bolt to 37 ft. lbs. (50 Nm) plus an additional 90 degrees.

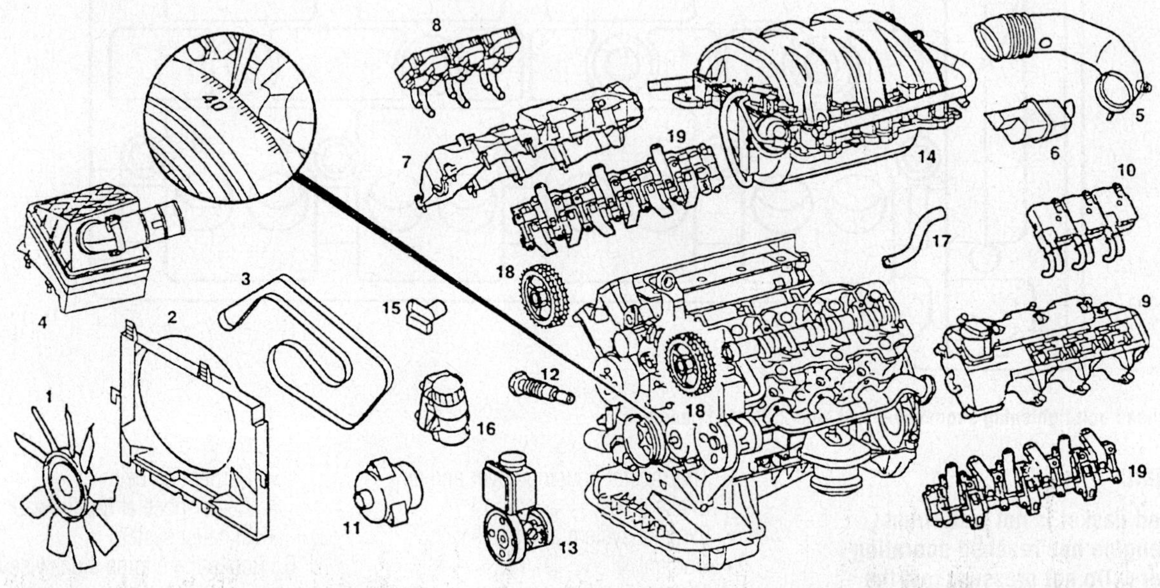

Exploded view of the cylinder head accessory components—2.6L, 2.8L and 3.2L (112) shown: 4.3L, 5.0L and 5.5L (113) engines similar

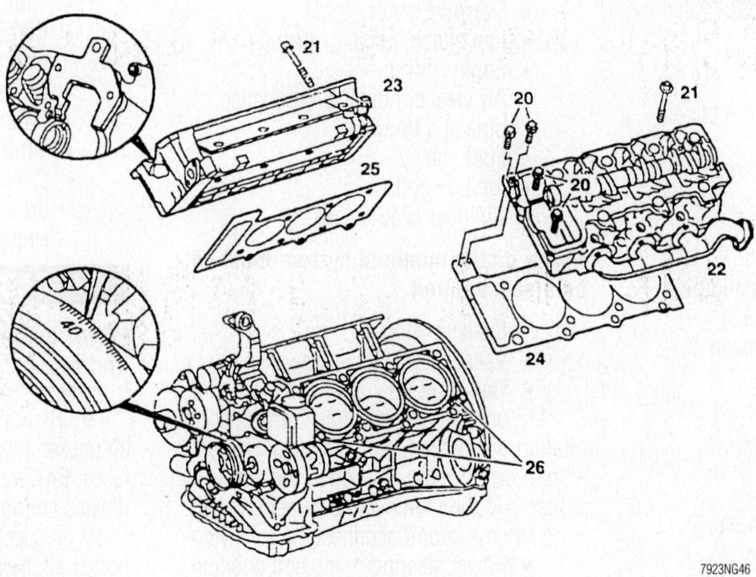

Exploded view of the cylinder head removal—2.6L, 2.8L and 3.2L (112) shown: 4.3L, 5.0L and 5.5L (113) engines similar

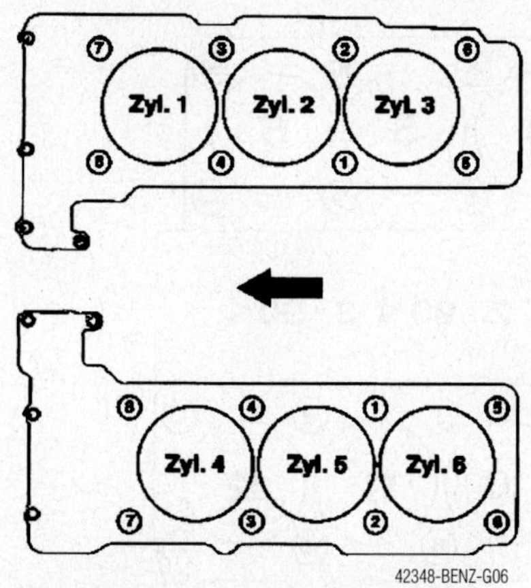

Cylinder head bolt tightening sequence—2.6L, 2.8L and 3.2L (112) engines

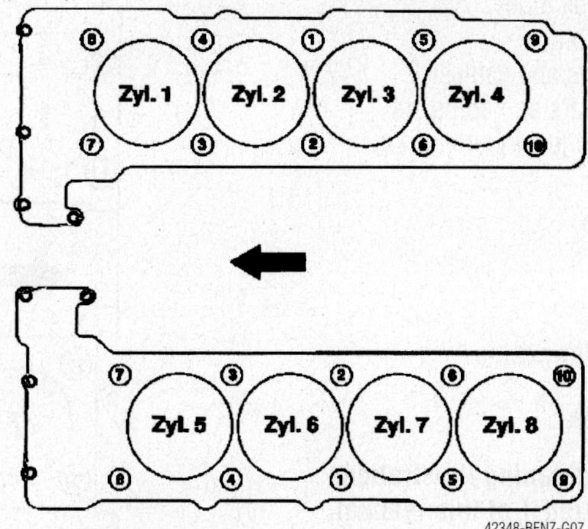

Cylinder head bolt tightening sequence—4.3L, 5.0L and 5.5L (113) engines

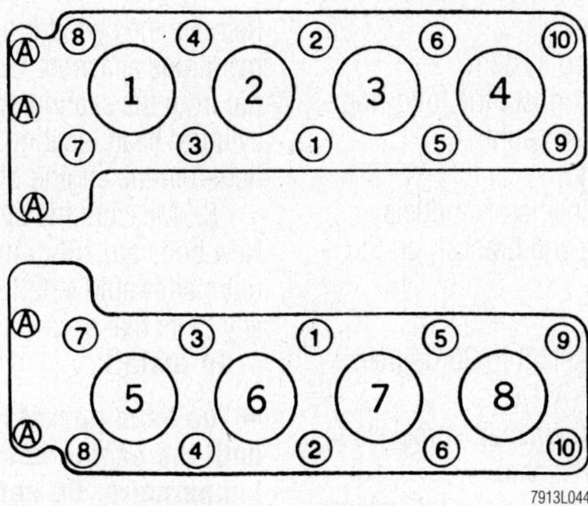

Cylinder head bolt tightening sequence—5.0L (119) Engine

- Timing chain tensioner with new gasket. Torque to 59 ft. lbs. (80 Nm).
- Generator

14. Remove the camshaft locking plates.
    - Exhaust system. Torque mounting nuts to 15 ft. lbs. (20 Nm).
    - Heater hose
    - Power steering pump
    - Serpentine belt
    - CMP sensor
    - Vacuum switchover valve
    - Intake manifold
    - Cylinder head covers. Torque the bolts to 88 inch lbs. (10 Nm).
    - Ignition coils. Torque the mounting bolts to 70 inch lbs. (8 Nm).
    - Fuel line
    - Air cleaner housing, resonance pipe and body
    - Engine cover
    - Fan clutch, fan and shroud
15. Remove the guard plate from the radiator/condenser.
16. Fill the cooling system, connect the negative battery cable, start the engine and check for leaks.

5.0L (119) Engine

1. Before servicing the vehicle, refer to the Precautions Section.
2. Drain the engine coolant.
3. Remove or disconnect the following:
    - Negative battery cable
    - Engine cover
    - Front covers
4. Crank the No. 1 piston of the engine to 45 degrees Before Top Dead Center (BTDC). Look for the **4/5** on the timing indicator.
5. Mark all 4 camshaft timing gears and timing chain with colored dots at about 11 o'clock for the right outer and left inner camshaft sprocket and 1 o'clock for the right inner and left outer camshaft sprocket.
6. Remove or disconnect the following:
    - Timing chain tensioner and top guide rails
    - Camshaft gears and adjusters
    - Electrical wiring, hoses and cables
    - Intake manifold
    - Dipstick guide tube
    - Exhaust pipes and manifolds
    - Cylinder head covers

➡Be sure the engine is cold before removing cylinder head bolts. On the "close-deck" crankcase, the cylinder head bolts have different lengths.

7. Loosen the cylinder head bolts in the reverse order of the illustrated tightening sequence.
8. Remove the cylinder head and clean all gasket material from the sealing surfaces. Be sure the cylinder head locating dowels are positioned in the engine block.
9. Inspect length of the cylinder head bolts. New bolt length is 160mm and the maximum permissible length is 163.5mm. Replace bolts that measure greater than the maximum permissible length.

To install:

➡The head gasket is not watertight until the engine has reached operating temperature. Do not pressure test the cooling system until the engine has reached operating temperature.

10. Clean the head bolt threads, and then apply clean engine oil to the thread and head contact surfaces.
11. Install the cylinder head gasket and head. Torque the bolts in sequence, as follows:
    a. Step 1: 41 ft. lbs. (55 Nm)
    b. Step 2: + 90 degrees
    c. Step 3: + additional 90 degrees
12. Torque the M8 bolts near the timing sprockets to 18 ft. lbs. (25 Nm).

➡Bolts that are screwed into the front of the cylinder head must be coated with sealant when installed.

13. Install or connect the following:
    - Cylinder head covers
    - Exhaust pipes and manifolds

- Dipstick guide tube
- Intake manifold
- Wiring, hoses and cables
- Camshaft gears and adjusters
- Timing chain tensioner and top guide rails
- Front covers
- Engine cover
- Coolant
- Engine oil
- Negative battery cable

6.0L (120) Engines

➡Cylinder head tightening illustrations are located in Section 1 of this manual. The illustrations follow the Torque Charts.

1. Before servicing the vehicle, refer to the Precautions Section.
2. Drain the cooling system.
3. Remove or disconnect the following:
   - Negative battery cable
   - Intake manifold
   - Engine assembly, on S models
   - Exhaust pipes and bracket, on SL models
   - Valve covers
4. Rotate the crankshaft to 30 degrees After Top Dead Center (ATDC).
5. Matchmark the camshaft sprockets to the timing chain as illustrated.
6. Remove or disconnect the following:
   - Timing chain tensioner
   - Timing chain from camshaft sprockets
   - Left head ventilation tube
   - Transmission dipstick tube
   - Oil dipstick tube
   - Timing chain guide rails
   - Secondary air injection lines
   - Exhaust manifold, on S models
   - Coolant hose at the rear of the head
7. Loosen the head bolts in stages, in the reverse of the tightening sequence.

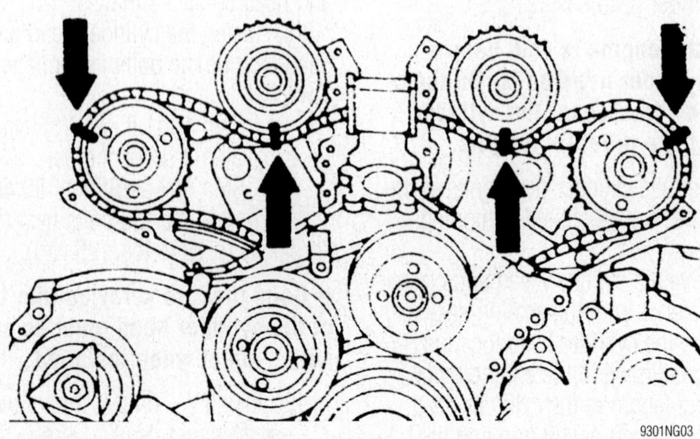

Matchmark the timing chain and sprockets for installation—6.0L (120) engines

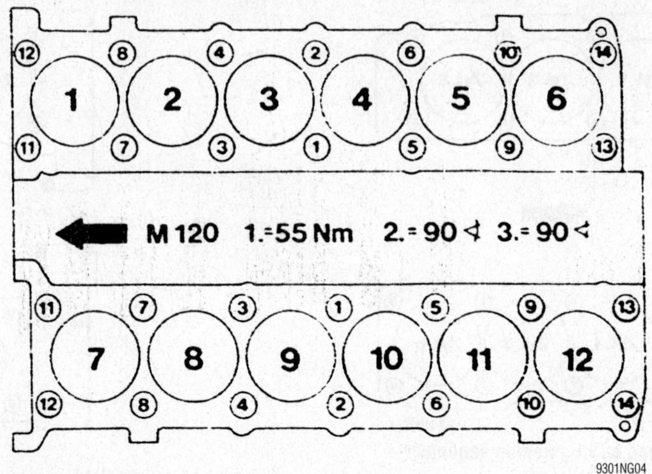

Cylinder head bolt tightening sequence—6.0L (120) Engine

Remove the cylinder heads with the exhaust manifolds attached. Clean all gasket material from the sealing surfaces. Be sure the cylinder head locating dowels are positioned in the engine block.

8. Measure the cylinder head bolts. New bolts are 168mm long, and the maximum allowable length is 171mm. Replace any bolts that exceed the maximum length.

To install:

➡The head gasket is not watertight until the engine has reached operating temperature. Do not pressure test the cooling system until the engine has reached operating temperature.

9. Install the cylinder head gasket and head. Torque the bolts in sequence, as follows:
   a. Step 1: 41 ft. lbs. (55Nm).
   b. Step 2: + 90 degrees.
   c. Step 3: + another 90 degrees.
10. Install or connect the following:
    - Coolant hose
    - Exhaust manifold, on S models
    - Secondary air injection lines
    - Timing chain guide rails
    - Oil dipstick tube
    - Transmission dipstick tube
    - Ventilation tube
11. Align the matchmarks and install the timing chain.
    - Timing chain tensioner. Verify the correct valve timing.
    - Valve covers
    - Engine assembly, on S models
    - Exhaust pipes to the manifolds and connect the exhaust pipe bracket to the transmission, on SL models
    - Intake manifold
    - Engine oil
    - Coolant
    - Negative battery cable

Rocker Arms/Shafts

REMOVAL & INSTALLATION

All engines except the 112 and 113 are not equipped with rocker arms/shafts. The camshaft(s) act directly on valve tappets. On the 112 and 113 engines, the rocker arm/shaft is part of the camshaft bearing cap assembly and is called the camshaft bearing bridge. This procedure is for removing and installing the camshaft bearing bridge.

2.6L, 2.8L and 3.2L (112) and 4.3L, 5.0L and 5.5L (113) engines

1. Before servicing the vehicle, refer to the Precautions Section.
2. Remove or disconnect the following:
   - Negative battery cable
   - Cylinder head cover
3. Rotate the engine clockwise to position the crankshaft 40 degrees After Top Dead Center (ATDC).

⁂ WARNING

The engine must not be rotated backward.

- Generator
- Timing chain tensioner

4. Cable tie the timing chain to the camshaft sprockets and remove the sprockets.
5. Loosen the camshaft bearing bridge bolts in reverse of the tightening sequence.

➡The camshaft bearing bridge must not be disassembled. If damage exists at the valve gear or at the top half of the camshaft bearing journal, the complete cylinder head should be replaced.

To install:

6. Lubricate the camshaft bearing journals.
7. Install or connect the following:
   - Camshaft bearing bridge. Torque the bolts to 11 ft. lbs. (15 Nm) in sequence as illustrated.
   - Camshaft sprockets
   - Timing chain tensioner with new gasket. Torque to 59 ft. lbs. (80 Nm).
   - Cylinder head covers. Torque to 88 inch lbs. (10 Nm).
8. Connect the negative battery cable, start the vehicle and check operation.

Supercharger

REMOVAL & INSTALLATION

1. Before servicing the vehicle, refer to the Precautions Section.
2. Drain the engine coolant.
3. Remove or disconnect the following:
   - Intake air assembly and plug supercharger openings
   - Sensor block cover
   - Electromagnetic clutch connector
   - Coolant hose, radiator to water pump
   - Coolant return hose
   - Fan and shroud
   - Supercharger drive belt
   - Supercharger pressure connections
   - Supercharger

To install:

➡Install the bottom securing bolt in the supercharger before installing.

4. Install or connect the following:
   - Supercharger. Torque the mounting bolts to 16 ft. lbs. (21 Nm).
   - Pressure connections with new seals
   - Drive belt
   - Fan and shroud
   - Coolant hoses
   - Electromagnetic clutch connector
   - Sensor block cover
5. Remove the plug from the inlet of the supercharger and install the intake air assembly.
6. Fill the cooling system, start the engine and check operation.

Intake Manifold

REMOVAL & INSTALLATION

3.2L (104) Engines

1. Before servicing the vehicle, refer to the Precautions Section.

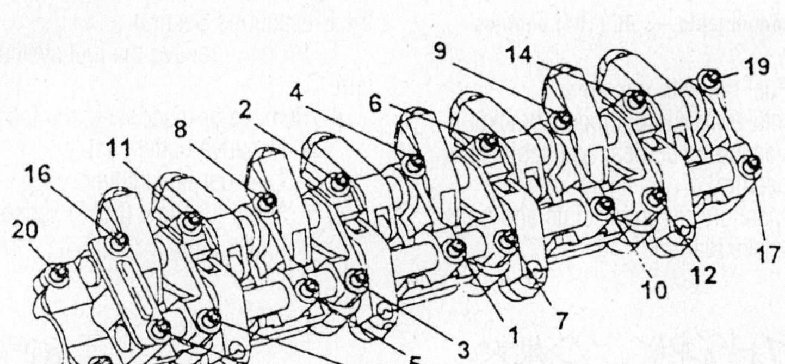

Camshaft bearing bridge tightening sequence—4.3L, 5.0L and 5.5L (113) engines

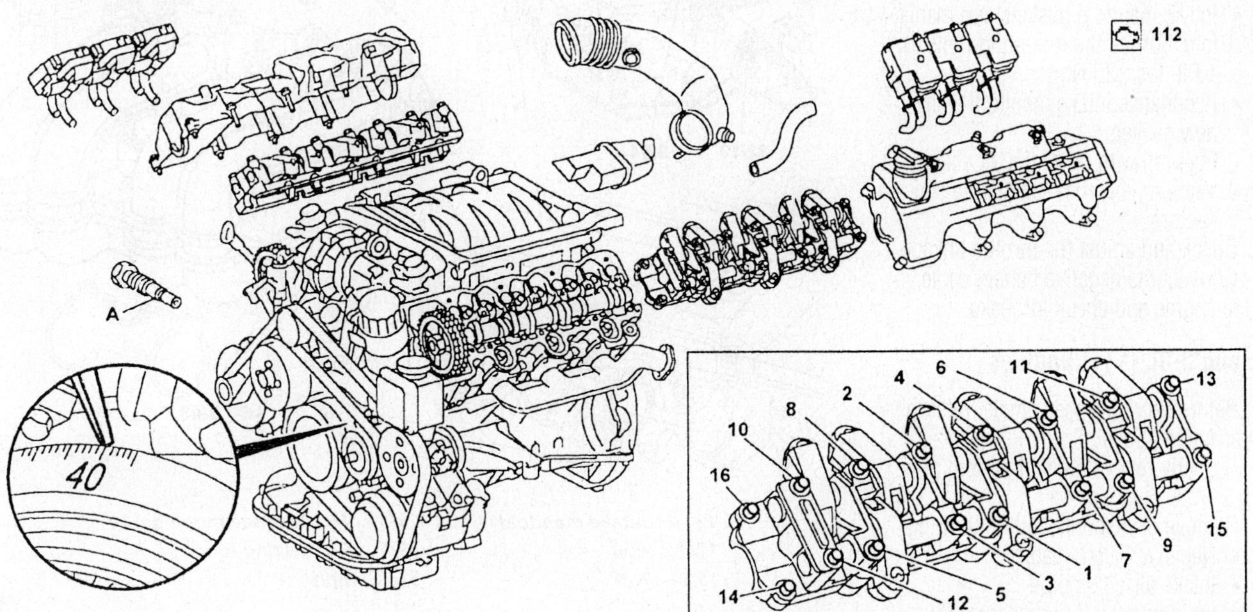

Camshaft bearing bridge tightening sequence—2.6L, 2.8L and 3.2L (112) engines

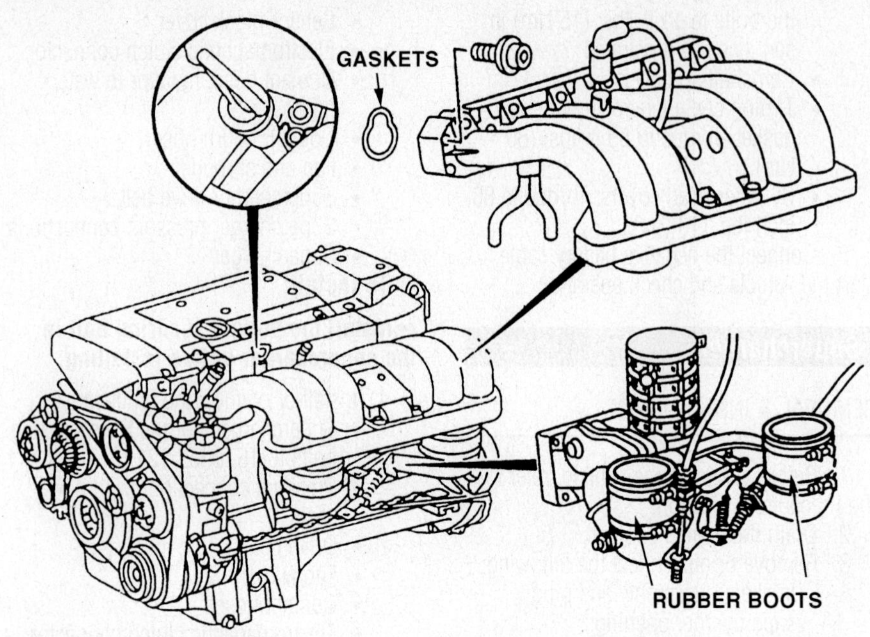

Exploded view of the intake manifold mounting and related components—3.2L (104) engines

2. Properly relieve the fuel system pressure.
3. Remove or disconnect the following:
   - Negative battery cable
   - Vacuum, electrical and cable connectors
   - Fuel rail with the injectors
   - Resonance intake manifold (throttle body). Clean the gasket surfaces.
   - Intake manifold

To install:

4. Clean all gasket material from the sealing surfaces.
5. Install or connect the following:
   - Intake manifold gaskets and manifold. Torque the mounting bolts to 18 ft. lbs. (25 Nm).
   - Resonance intake manifold with new gaskets
   - Fuel rail with new injector seals
   - Vacuum, electrical and cable connectors
6. Check and adjust the throttle linkage.
7. Connect the negative battery cable, start the engine and check for leaks.

1.8L and 2.3L (111) Engines

1. Before servicing the vehicle, refer to the Precautions Section.
2. Properly relieve the fuel system pressure.
3. Remove or disconnect the following:
   - Negative battery cable
   - Intake air cross pipe
   - Ignition coils if necessary
   - Power steering pump bracket mounting bolt, if necessary
   - Fuel rail with injectors
   - Intake manifold support bracket
   - Vacuum, electrical and cable connectors
   - Manifold mounting nuts and bolts
   - Intake manifold

To install:

4. Clean all gasket material from the sealing surfaces.
5. Install or connect the following:
   - Manifold gaskets and intake manifold. Torque the mounting bolts and nuts to 15 ft. lbs. (20 Nm).
   - Vacuum, electrical and cable connectors
   - Intake manifold support bracket
   - Fuel rail with new injector seals
   - Power steering pump bracket mounting
   - Ignition coils
   - Intake air cross pipe
6. Check and adjust the throttle linkage.
7. Connect the negative battery cable, start the engine and check for leaks.

2.6L, 2.8L and 3.2L (112) and 4.3L, 5.0L and 5.5L (113) engines

1. Before servicing the vehicle, refer to the Precautions Section.
2. Properly relieve the fuel system pressure.
3. Remove or disconnect the following:
   - Negative battery cable
   - Cylinder head cover
   - Mass Air Flow (MAF) sensor with intake pipe

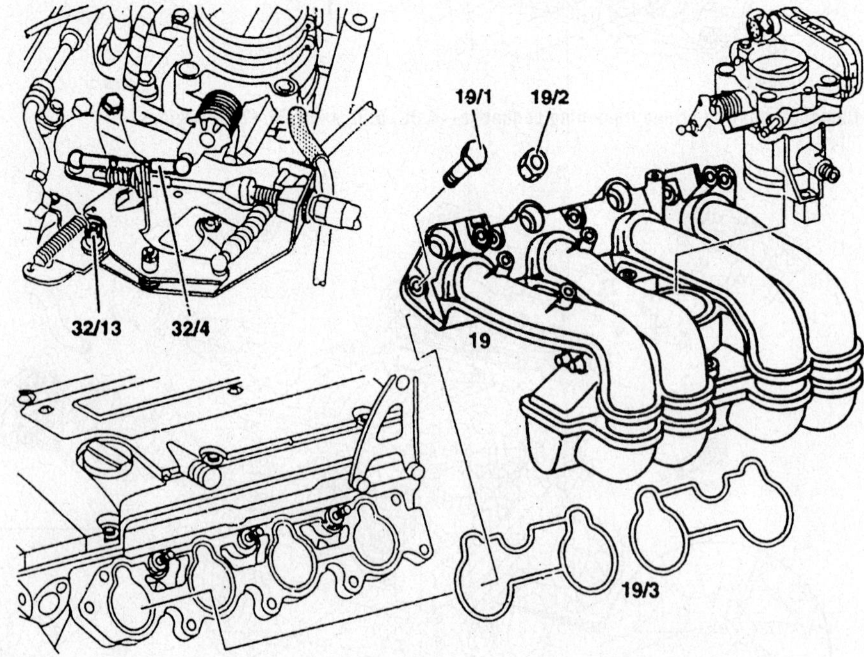

Exploded view of the intake manifold mounting and related components—1.8L and 2.3L (111) engines

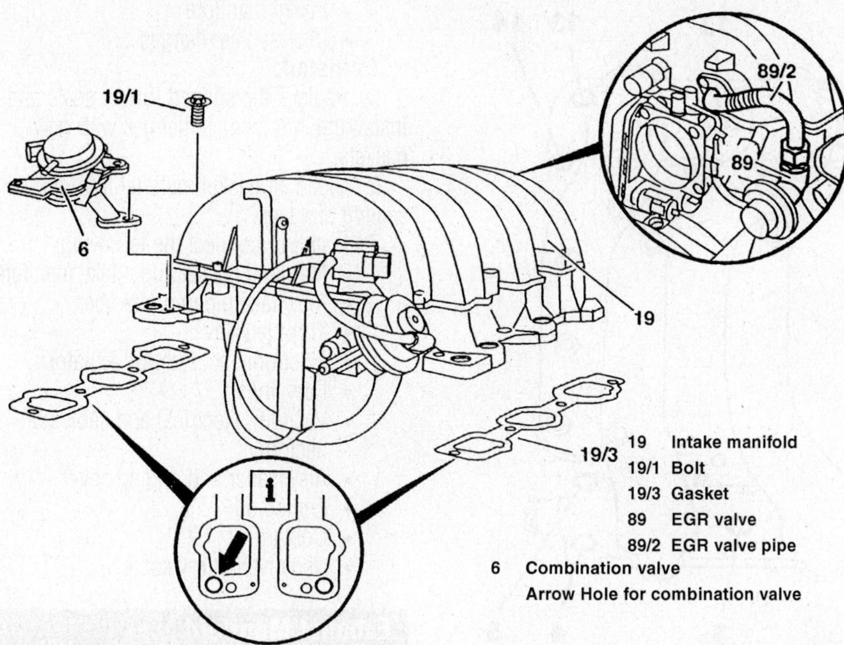

Exploded view of the intake manifold mounting and related components—2.6L, 2.8L and 3.2L (112) shown: 4.3L, 5.0L and 5.5L (113) engines similar

- Fuel rail with injectors
- Vacuum, electrical and cable connectors
- Exhaust gas Recirculation (EGR) valve
- Combination valve
- Manifold mounting bolts
- Intake manifold

To install:

1. Clean all gasket material from the sealing surfaces.
2. Install or connect the following:
   - New intake manifold gasket. Verify the secondary air injection passage opening in the gasket and install the intake manifold. Torque the mounting bolts to 15 ft. lbs. (20 Nm).
   - Combination valve. Torque the bolts to 15 ft. lbs. (20 Nm).
   - EGR valve
   - Vacuum, electrical and cable connectors
   - Fuel rail with new injector seals
   - MAF sensor with the air intake pipe
   - Cylinder head cover. Torque the bolts to 88 inch lbs. (10 Nm).
3. Reconnect the negative battery cable, start the vehicle and check for leaks.

5.0L (119) Engine

1. Before servicing the vehicle, refer to the Precautions Section.
2. Partially drain the coolant.
3. Properly relieve the fuel system pressure.
4. Remove or disconnect the following:
   - Negative battery cable
   - Air cleaner and engine cover
   - Vacuum and electrical connectors
   - Fuel supply (17/3) and return (17/2) lines
   - Guide element (27)
   - Throttle cable (30) at the control lever (9)
   - Control pressure cable (98)
   - Accelerator actuator (M16/1x1) connector and cable
   - Coolant hose (33) at the front of the manifold
   - Coolant hose at the rear of the manifold
   - Manifold mounting bolts
   - Intake manifold

To install:

5. Clean all gasket material from the sealing surfaces.
6. Install or connect the following:
   - Manifold gasket and intake manifold. Torque the bolts in sequence to 18 ft. lbs. (25 Nm).
   - Coolant hoses
   - Accelerator actuator connector and cable
   - Control pressure cable
   - Throttle and transmission cables
   - Guide element
   - Fuel lines
   - Vacuum, electrical and cable connectors
   - Air cleaner and engine cover
   - Engine oil
   - Coolant
   - Negative battery cable

6.0L (120) Engines

1. Before servicing the vehicle, refer to the Precautions Section.
2. Partially drain the coolant.
3. Properly relieve the fuel system pressure.

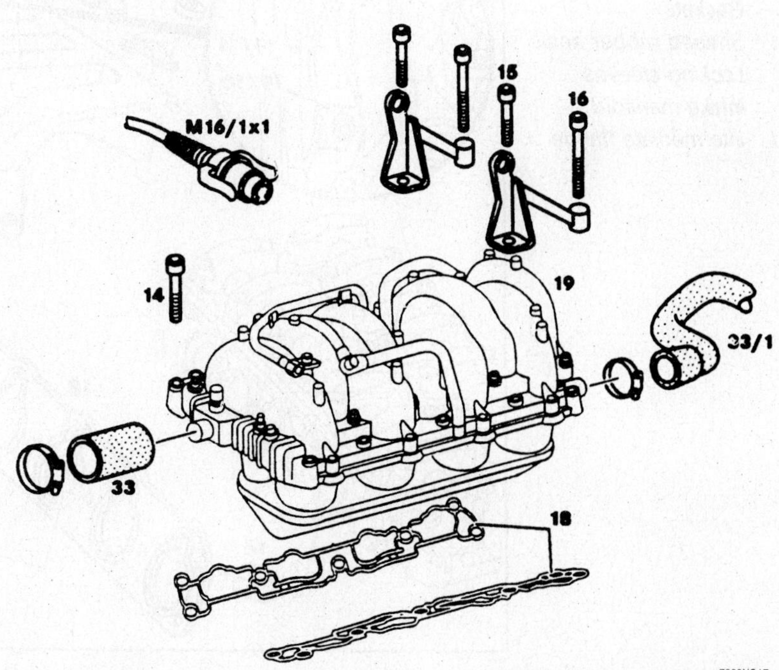

Exploded view of the intake manifold mounting and related components—5.0L (119) engine

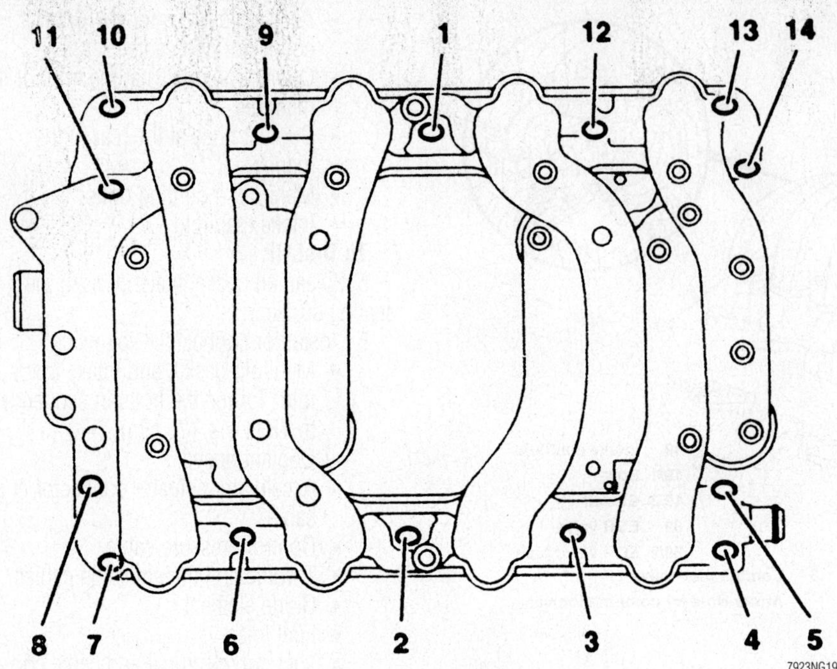

Intake manifold bolt tightening sequence—5.0L (119) engine

4. Remove or disconnect the following:
    - Negative battery cable
    - Air cleaner and engine cover
    - Vacuum, electrical and cable connectors
    - Fuel supply and return lines
    - Electronic accelerator actuator
    - Throttle body
    - Air inlet plenum chamber, if equipped
    - Manifold mounting bolts
    - Intake manifold
    - Intermediate flanges

To install:

5. Replace the shaped rubber seals and install the intermediate flanges with new gaskets.
6. Clean all gasket material from the sealing surfaces.
7. Install or connect the following:
    - Manifold gasket and intake manifold
    - Air intake plenum chamber
    - Throttle body
    - Electronic accelerator actuator
    - Fuel lines
    - Vacuum, electrical and cable connectors
    - Air cleaner and engine cover
    - Engine oil
    - Coolant
    - Negative battery cable

Exhaust Manifold

REMOVAL & INSTALLATION

Inline 4- and 6-Cylinder Engines

1. Before servicing the vehicle, refer to the Precautions Section.
2. Remove or disconnect the following:
    - Negative battery cable

14 *Bolts*
15 *Screws*
16 *Screws*
18 *Gaskets*
18/1 *Shaped rubber seals*
17 *Locking sleeves*
19 *Intake manifold*
19/1 *Intermediate flange*

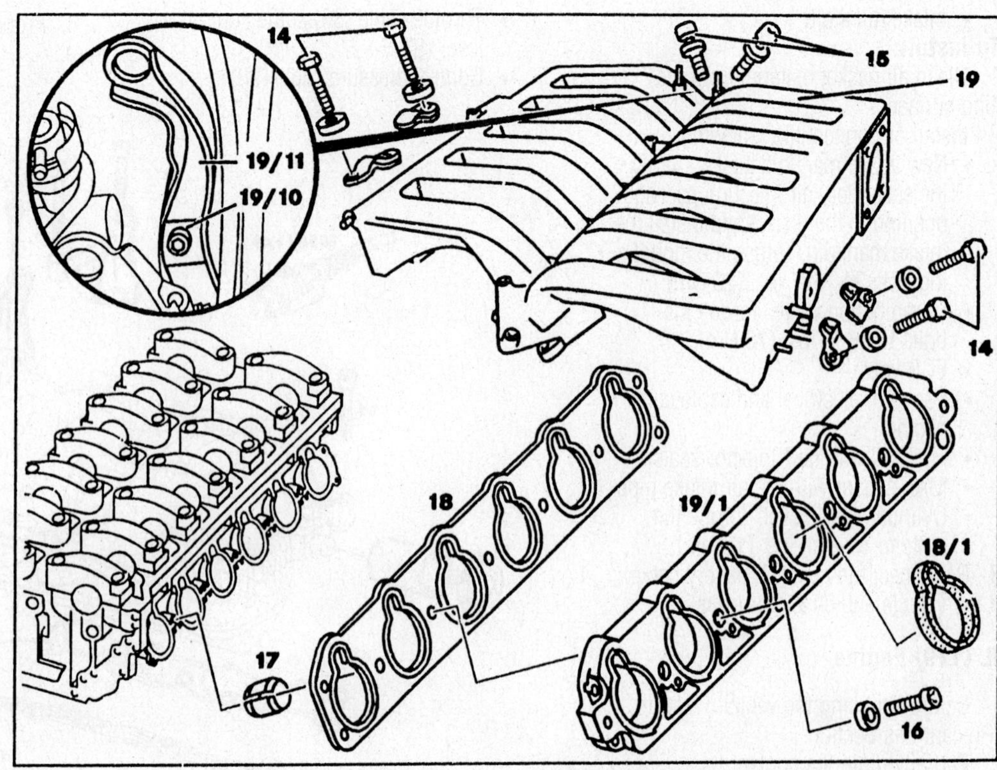

Exploded view of the intake manifold—6.0L (120) engines

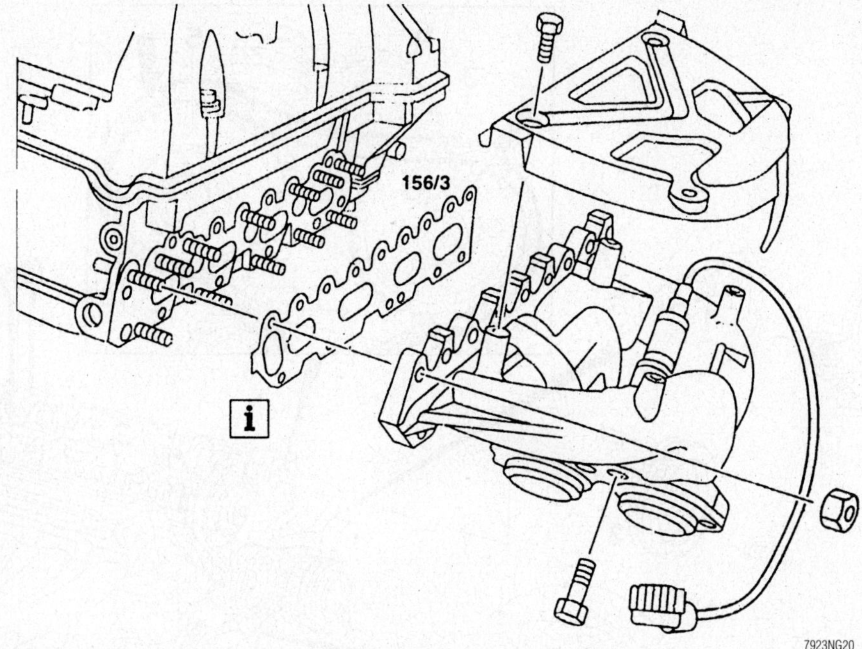

Exploded view of the mounting—4-cylinder shown, inline 6-cylinder similar

- Exhaust pipe
- Transmission exhaust support, if equipped
- Air injection tube, if equipped
- Manifold retaining bolts and manifold

To install:

3. Install or connect the following:
   - Manifold gasket and exhaust manifold. Torque the bolts in an even pattern.
   - Air injection tube
   - Transmission exhaust support
   - Exhaust pipe with new self-locking nuts. Torque to 25 ft. lbs. (34 Nm).
   - Negative battery cable

V6, V8 and 12-Cylinder Engines

1. Before servicing the vehicle, refer to the Precautions Section.
2. Properly relieve the fuel system pressure.
3. Remove or disconnect the following:
   - Negative battery cable
   - Engine undercover
   - Exhaust pipes
   - Fan shroud
   - Engine mount for the manifold being serviced
   - Fuel supply and return pipes, if removing left manifold
   - Self-locking bolts and exhaust manifold

To install:

4. Inspect the rivet nuts in the manifold and replace as needed.
5. Install or connect the following:
   - Manifold gasket and exhaust manifold. Torque the mounting bolts to 23 ft. lbs. (30 Nm).
   - Fuel lines, if removed
   - Engine mounts
   - Fan shroud
   - Exhaust pipes
   - Engine under cover
6. Connect the negative battery cable, start the engine and check for leaks.

Camshaft and Valve Lifters

REMOVAL & INSTALLATION

3.2L (104) Engine

1. Before servicing the vehicle, refer to the Precautions Section.
2. Disconnect the negative battery cable and position the No. 1 cylinder 30 degrees BTDC.
3. Remove or disconnect the following:
   - Valve cover
   - Timing chain tensioner (1)
   - Upper front cover (30) and top guide rail (42)
4. Matchmark the timing chain (6) to the camshaft sprockets (3 and 5).
   - Exhaust camshaft sprocket (3)
   - Timing chain from the intake camshaft (5)
5. Hold the camshafts so that the lobes of the cams of cylinder No. 2 press on the middle of the buckets. Unbolt the camshaft bearing caps 1, 4, 6 and 7 of the exhaust camshaft (14) and 8, 11, 13 and 14 of the intake camshaft (13). Loosen the remaining bearing cap bolts 1 turn at a time until there is no counterpressure.
6. Remove or disconnect the following:
   - Camshafts
   - Bucket tappets

To install:

7. Install the bucket tappets and oil the camshaft journals and lobes. Install the camshafts at 30 degrees Before Top Dead Center (BTDC) for cylinder No. 1. The lobes of the cams must be facing on the middle of the buckets for cylinder No. 2.
8. Install bearing caps 2, 3 and 5 for the exhaust camshaft and 9, 10 and 12 for the intake camshaft. Torque to 15 ft. lbs. (21 Nm). Hold the camshafts so that the lobes of the cams of cylinder No. 2 press on the middle of the buckets. Install the remaining bearing caps. Torque to 16 ft. lbs. (22 Nm).
9. Install or connect the following:
10. Timing chain
11. Exhaust camshaft sprocket
12. Align the matchmarks and install the sprocket. Torque to the following specification:
    - M7 13.5mm T30 Torx® bolt: 13 ft. lbs. (18 Nm)
    - M7 13.5mm T40 Torx® bolt: 16 ft. lbs. (22 Nm)
    - M7 13mm T30 Torx® bolt: 15 ft. lbs. (20 Nm) plus 60 degrees
13. Install or connect the following:
    - Front cover and top guide rail

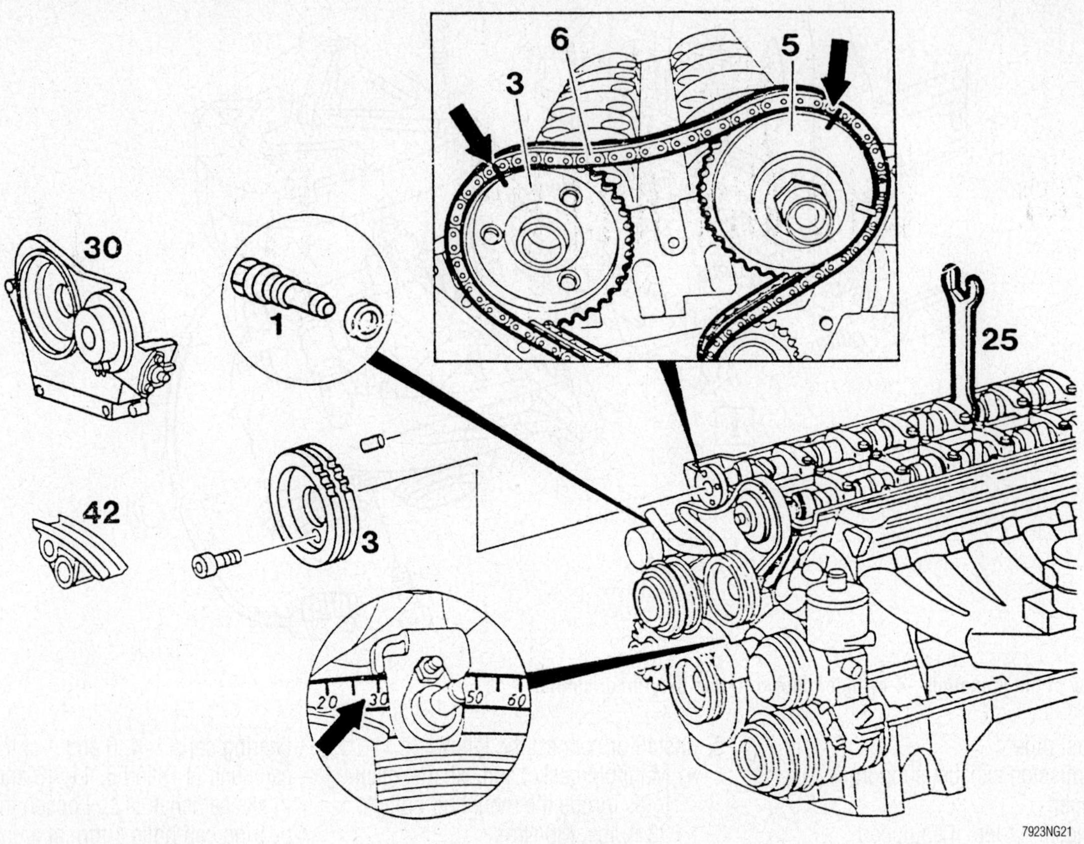

Position the crankshaft as shown to prevent piston-to-valve contact—3.2L (104) engines

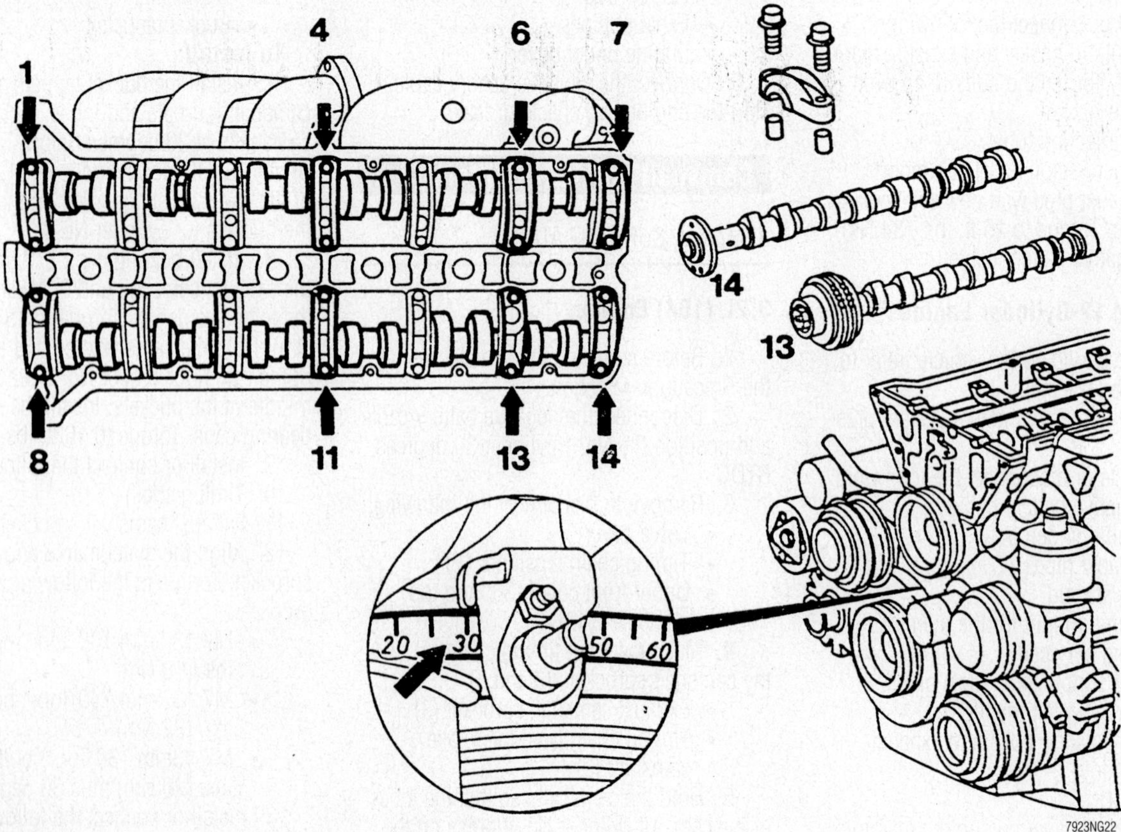

Remove the indicated bearing caps first—3.2L (104) engines

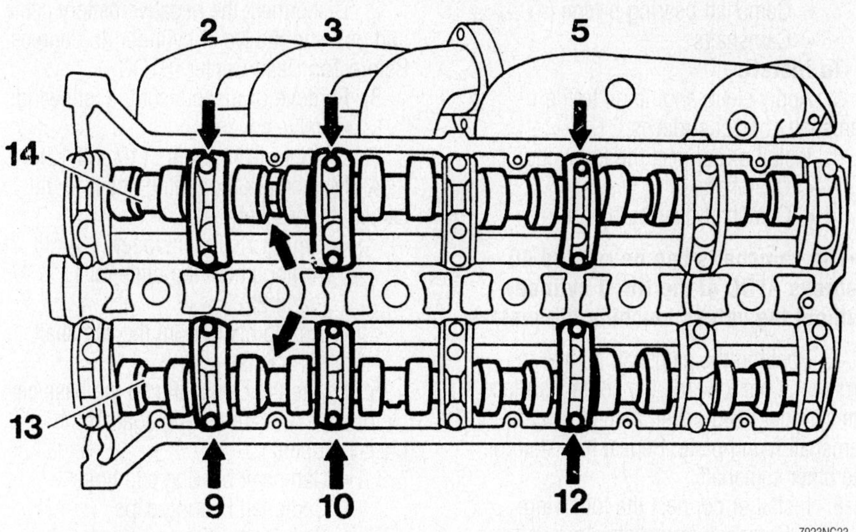

Be sure to install the indicated bearing caps first—3.2L (104) engines

- Timing chain tensioner
- Valve cover
- Negative battery cable

1.8L and 2.3L (111) Engines

1. Before servicing the vehicle, refer to the Precautions Section.
2. Disconnect the negative battery cable and position the No. 1 cylinder 30 degrees ATDC.
3. Remove or disconnect the following:
    - Valve cover
    - Front timing cover
4. Matchmark the timing chain (73) to the camshaft sprockets (75 and 76).
    - Timing chain tensioner (14)
    - Exhaust camshaft sprocket (76)
    - Timing chain from the intake camshaft (75)
5. Using a wrench (1), rotate the camshafts (77 and 78) so the base circle of the cams are resting against the bucket tappets.
    - Bearing caps, mark the bearing caps for reinstallation reference prior to removal
    - Camshafts
    - Bucket tappets

To install:

6. Install or connect the following:
    - Bucket tappets and oil the camshaft journals and lobes. Position the base circle of the camshafts so they are resting against the bucket tappets.

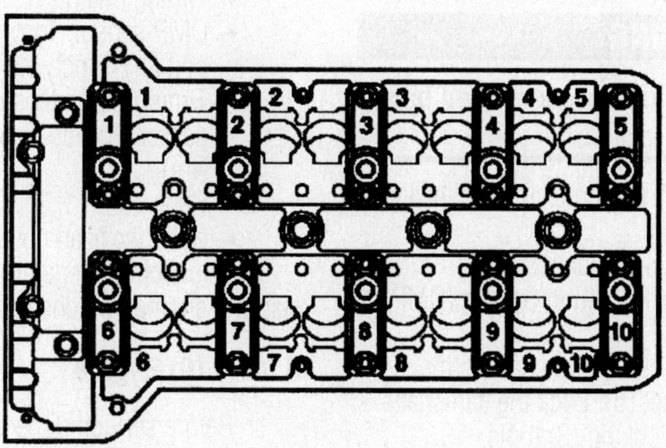

Correct positioning of camshaft bearing caps—1.8L and 2.3L (111) engines

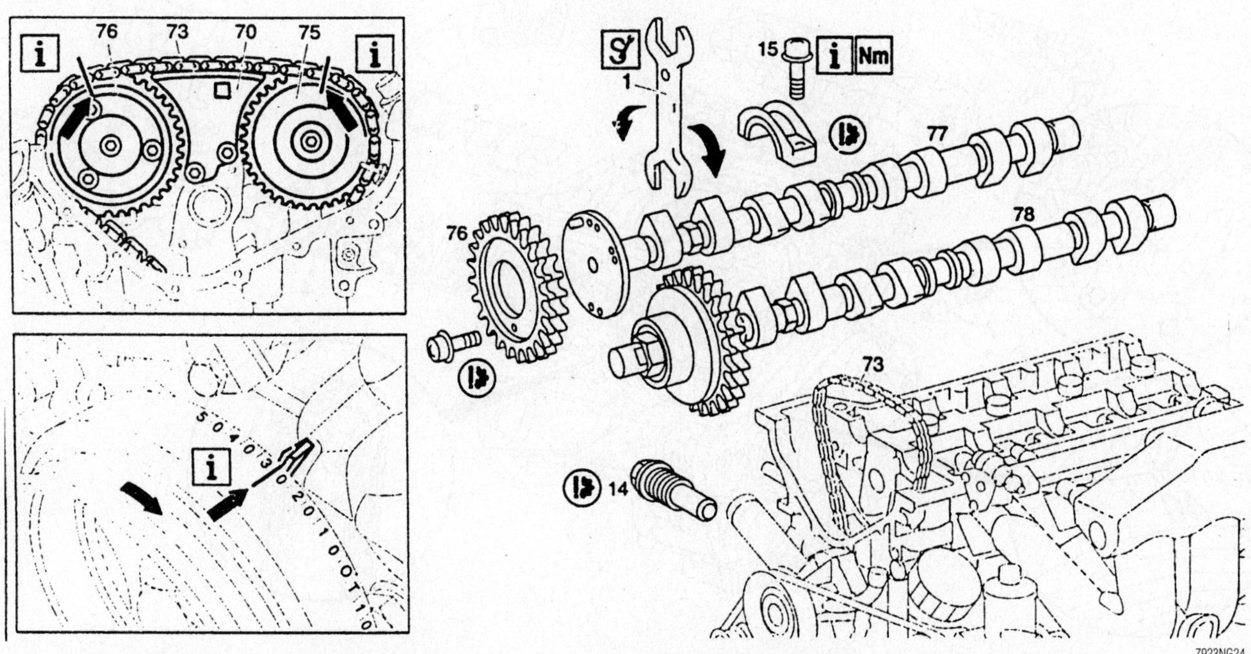

Exploded view of the camshaft mounting components and positioning—1.8L and 2.3L (111) engines

- Camshaft bearing caps in correct positions. Torque to 44 inch lbs. (5 Nm) plus 90 degrees.
- Timing chain
- Exhaust camshaft sprocket. Align matchmarks and torque to 15 ft. lbs. (20 Nm) plus 90 degrees.
- Front cover and top guide rail
- Timing chain tensioner
- Valve cover
- Negative battery cable

2.6L, 2.8L and 3.2L (112) and 4.3L, 5.0L and 5.5L (113) Engines

1. Before servicing the vehicle, refer to the Precautions Section.
2. Disconnect the negative battery cable and remove the cylinder head cover. Rotate the engine clockwise to position the crankshaft 40 degrees After Top Dead Center (ATDC).

⁂ WARNING

The engine must not be rotated backward.

3. Remove or disconnect the following:
   - Generator
   - Timing chain tensioner (2)
   - Camshaft Position Sensor (CMP) (1)
4. Cable-tie the timing chain to the camshaft sprocket (5). Lock the camshafts using camshaft locking (3) tools.
   - Camshaft gears
   - Camshaft bearing bridge (4)
   - Camshafts

To install:

5. Apply clean engine oil to the camshaft contact surfaces.
6. Install or connect the following:
   - Camshafts
   - Camshaft bearing bridge

➡The camshafts can be rotated 40 degrees ATDC of the No. 1 cylinder without the valves touching the pistons.

7. Position the camshaft so that the groove points towards the contact surface of the cylinder head cover, then attach the camshaft fixing plate. Repeat this step for the other camshaft.
8. Install or connect the following:
   - Camshaft sprockets. Torque the bolt to 37 ft. lbs. (50 Nm) plus 90–100 degrees.
   - Timing chain
   - CMP sensor. Torque the mounting bolt to 70 inch lbs. (8 Nm).
   - Timing chain tensioner with new gasket. Torque to 59 ft. lbs. (80 Nm).
   - Generator
   - Cylinder head cover
9. Connect the negative battery cable, start the engine and check operation.

5.0L (119) Engine

1. Before servicing the vehicle, refer to the Precautions Section.
2. Disconnect the negative battery cable and position the No. 1 cylinder 45 degrees Before Top Dead Center (BTDC).
3. Remove or disconnect the following:
   - Valve covers
   - Upper front covers (10 and 12)
4. Matchmark the timing chain to the camshaft sprockets.
   - Timing chain tensioner (13)
   - Timing chain top slide rails (1, 1a, 2 and 2a)
   - Timing chain from the camshaft sprockets
5. Rotate the camshafts so the base circle of the cams are resting against the bucket tappets.
   - Camshaft bearing cap bolts
   - Camshaft bearing caps
   - Camshafts with sprockets
   - Bucket tappets

To install:

6. Install or connect the following:
   - Bucket tappets and oil the camshaft journals and lobes
   - Left intake and exhaust camshafts
   - Timing chain on left cylinder head, aligning the matchmarks
   - Left side bearing caps. Torque to 88 inch lbs. (10 Nm).
   - Right intake and exhaust camshafts
   - Timing chain on right cylinder head, aligning the matchmarks
   - Right side bearing caps. Torque to 88 inch lbs. (10 Nm).
   - Timing chain top slide rails

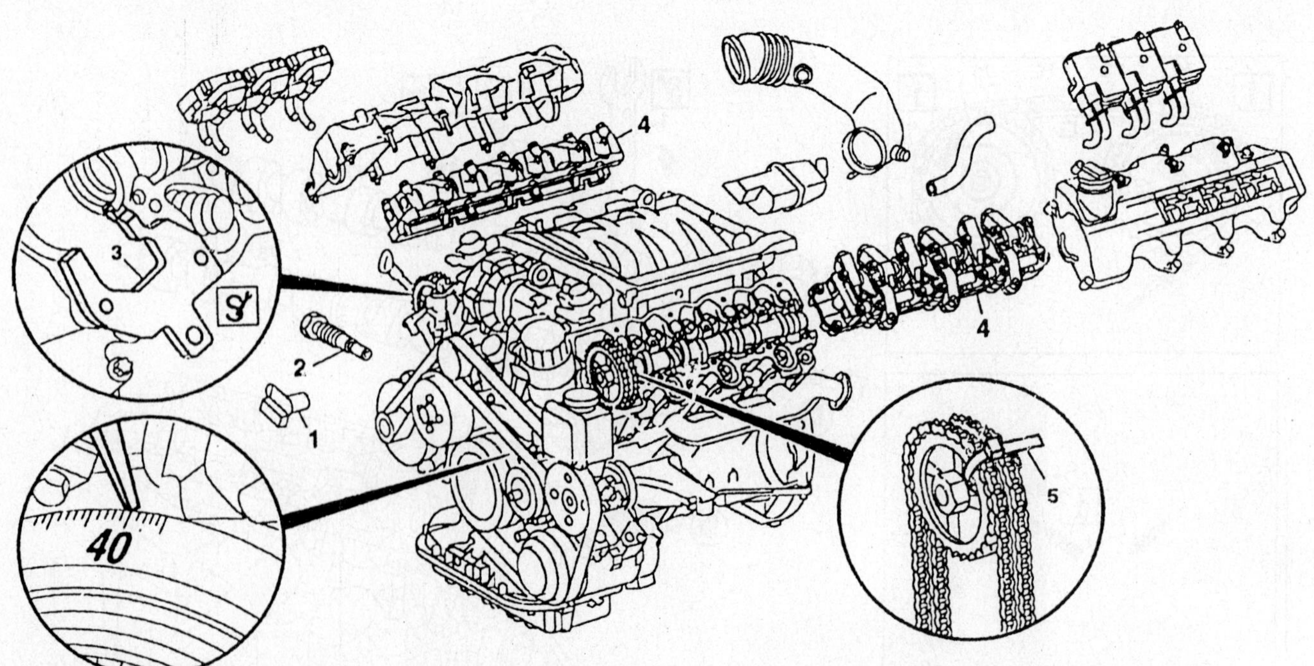

Exploded view of the camshaft mounting, showing related components—2.6L, 2.8L and 3.2L (112) and 4.3L, 5.0L and 5.5L (113) engines

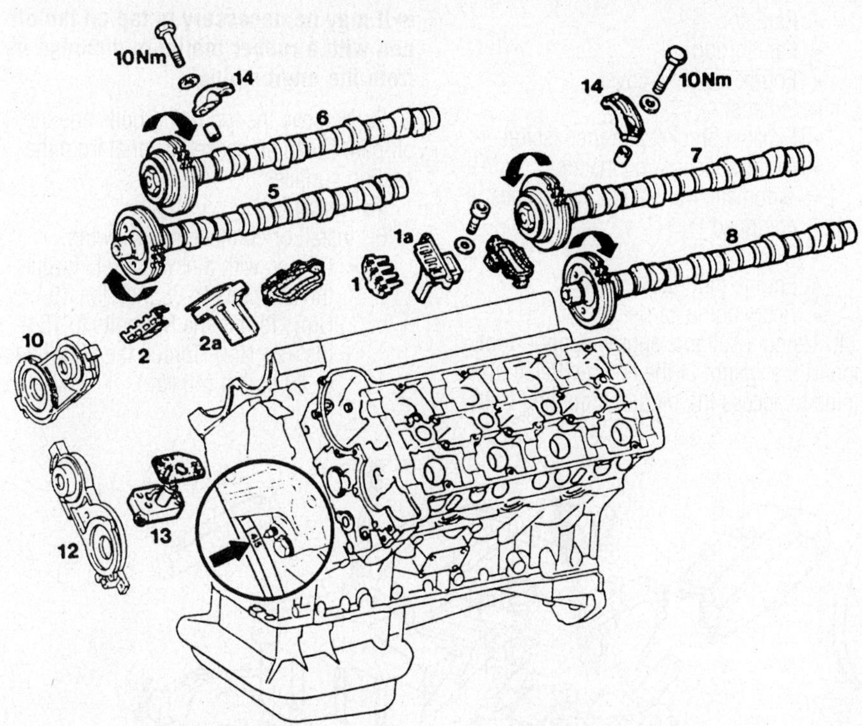

Exploded view of the camshaft mounting and related components—5.0L (119) engine

- Timing chain tensioner
- Upper front covers
- Valve covers
- Negative battery cable

6.0L (120) Engines

1. Before servicing the vehicle, refer to the Precautions Section.
2. Disconnect the negative battery cable and position the No. 1 cylinder 30 degrees ATDC.
3. Remove or disconnect the following:
   - Valve covers
   - Upper front covers and top guide rails
   - Timing chain tensioner
4. Matchmark the timing chain to the camshaft sprockets.
   - Exhaust camshaft sprockets
   - Timing chain from the intake camshaft sprockets
   - Bearing caps, except for caps 3, 4, 10, and 11 on the right head, and 17, 18, 24, and 25 on the left head
5. Loosen the bolts on the remaining bearing caps 1 turn at a time until the bearing caps can be removed.

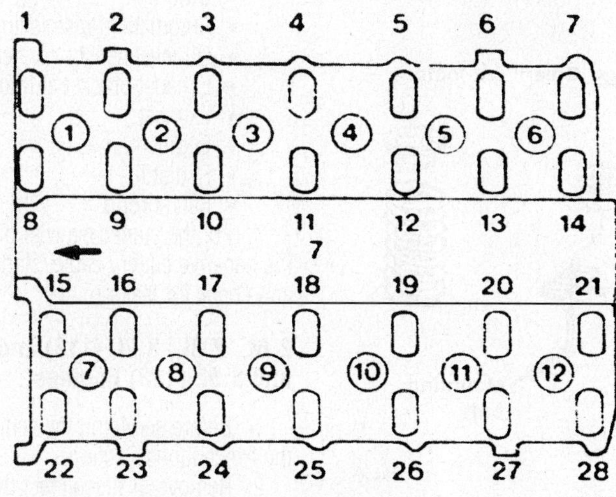

Cylinder and bearing cap identification—6.0 (120) engines

6. Remove or disconnect the following:
   - Camshafts
   - Bucket tappets

To install:

7. Install or connect the following:
   - Bucket tappets and oil the camshaft journals and lobes
   - Camshafts
   - Bearing caps 3, 4, 10, and 11 on the right head, and 17, 18, 24, and 25 on the left head. Torque the bolts in steps of 1 turn to 15 ft. lbs. (21 Nm).
   - Remaining bearing caps. Torque the bolts to 15 ft. lbs. (21 Nm).
   - Align the matchmarks and install the timing chain to the intake camshaft sprockets.
   - Exhaust camshaft sprockets. Torque the bolts to 16 ft. lbs. (22Nm).
   - Timing chain tensioner
   - Upper front covers top guide rails
   - Valve covers
   - Negative battery cable

Valve Lash

ADJUSTMENT

All Mercedes-Benz engines use hydraulic valve lifters. There is no provision for valve clearance adjustments.

Starter

REMOVAL & INSTALLATION

All Models

1. Before servicing the vehicle, refer to the Precautions Section.
2. Turn the steering wheel to full left position to gain access if necessary.
3. Remove or disconnect the following:
   - Negative ground cable
   - Engine under cover
   - Starter electrical connections
   - Rear engine mount, if necessary
   - Starter mounting bolts
   - Starter assembly

To install:

4. Install or connect the following:
   - Starter assembly. Torque the mounting bolts to 31 ft. lbs. (42 Nm).
   - Rear engine mount. Torque the bolts to 41 ft. lbs. (55 Nm).
   - Starter electrical connections
   - Negative ground cable
5. Start the engine and check for proper operation.

Oil Pan

REMOVAL & INSTALLATION

3.2L (104) and 1.8L and 2.3L (111) Engines

1. Before servicing the vehicle, refer to the Precautions Section.
2. Disconnect the negative battery cable and drain the engine oil.
3. Remove or disconnect the following:
   - Front torsion bar
   - Radiator
   - Fan shroud
   - Engine harness cover
   - Exhaust
   - Exhaust bracket at transmission
   - Oil level sensor connection
   - Automatic transmission oil pipes, if equipped
   - Drag link
   - Engine mounts
   - Rear engine carrier
4. Using a suitable engine support tool, support the weight of the engine. Raise the engine to access the oil pan mounting bolts.

➡It may be necessary to tap on the oil pan with a rubber mallet to dislodge it from the engine block.

5. Remove the mounting bolts and the oil pan. Clean all gasket material from the sealing surfaces.

To install:

6. Install or connect the following:
   - Oil pan with a new gasket. Torque the M6 bolts to 88 inch lbs. (9 Nm). Torque the M8 bolts to 15 ft. lbs. (21 Nm). Torque the M10 bolts to 30 ft. lbs. (40 Nm).

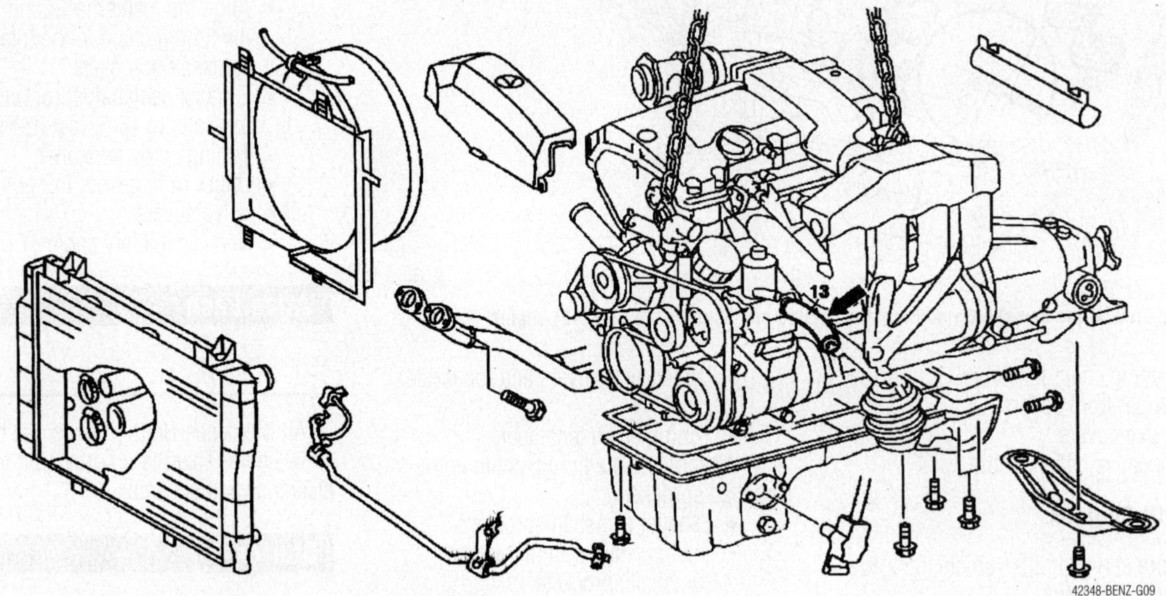

Exploded view of the oil pan mounting and related components—1.8L and 2.3L (111) engines

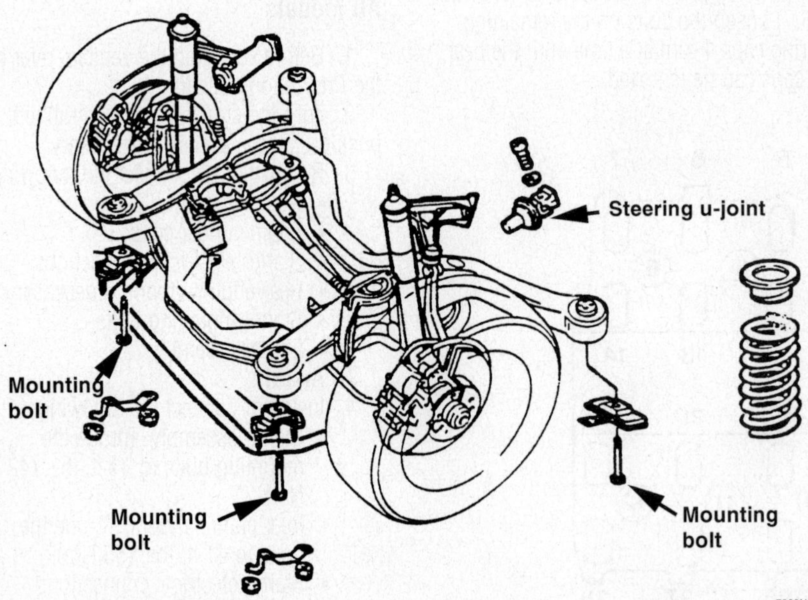

Lower the front subframe by removing the mounting bolts—except 3.2L (112) and 4.3L (113) engines

   - Rear engine carrier
   - Engine mounts. Torque the bolts to 41 ft. lbs. (55 Nm).
   - Front torsion bar
   - Drag link
   - Automatic transmission oil pipes
   - Oil level sensor connection
   - Exhaust bracket at transmission
   - Exhaust
   - Engine harness cover
   - Radiator
   - Fan shroud
7. Fill the crankcase with oil, connect the negative battery cable, start the engine and check for leaks.

2.6L, 2.8L, 3.2L (112) and 4.3L, 5.0L and 5.5L (113) Engines

1. Before servicing the vehicle, refer to the Precautions Section.
2. Remove or disconnect the following:
   - Negative battery cable
   - Engine oil

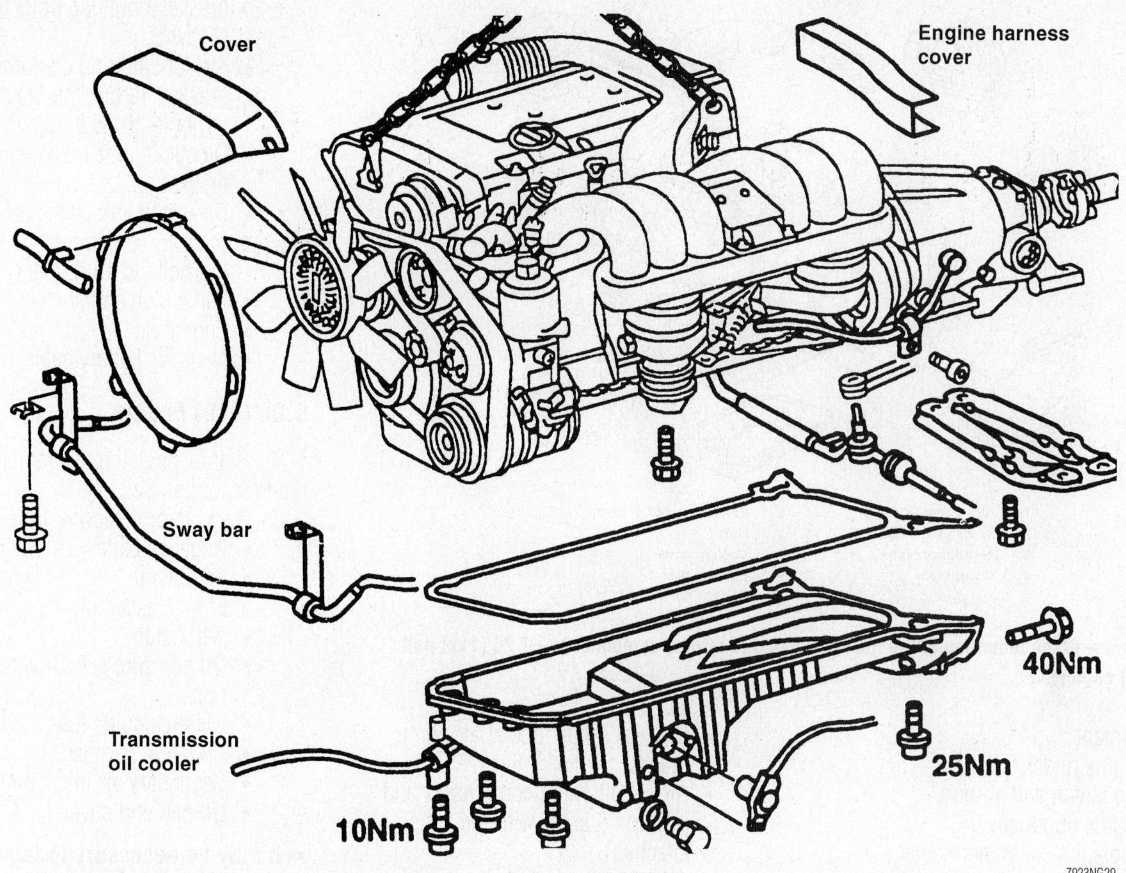

Exploded view of the oil pan mounting and related components—3.2L (104) engines

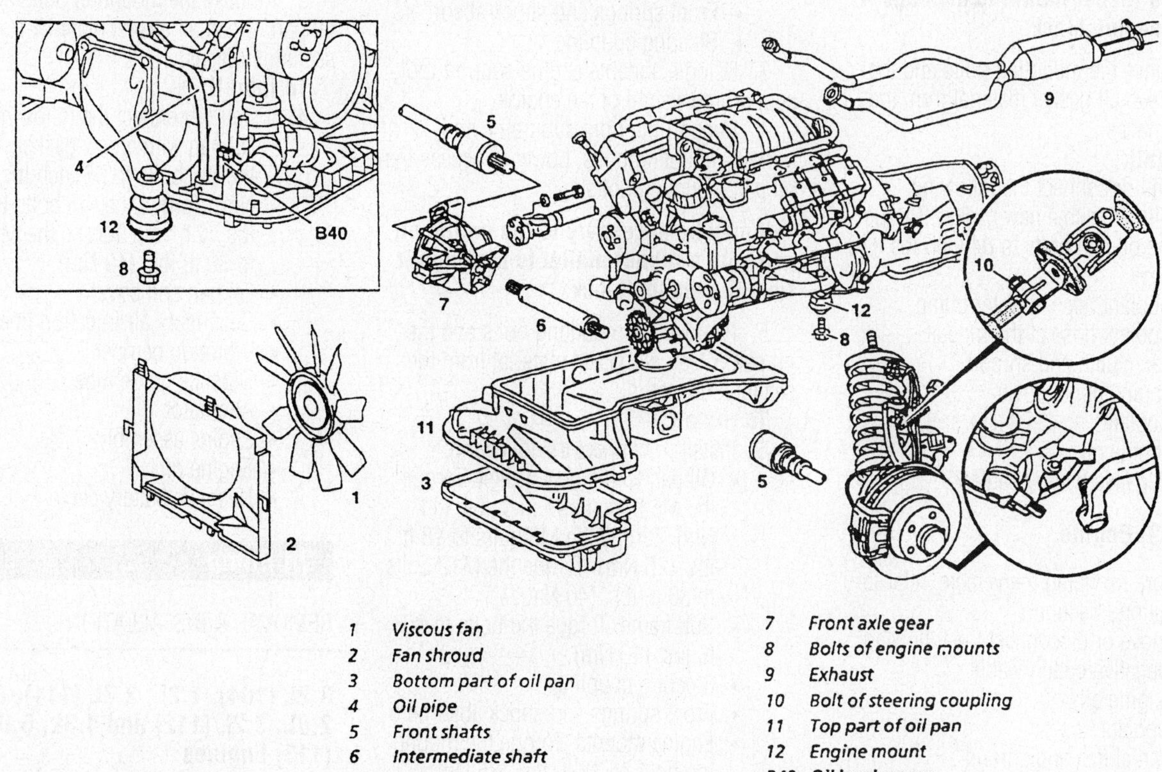

Exploded view of the mounting of the upper oil pan and related components—3.2L (112) and 4.3L (113) engines

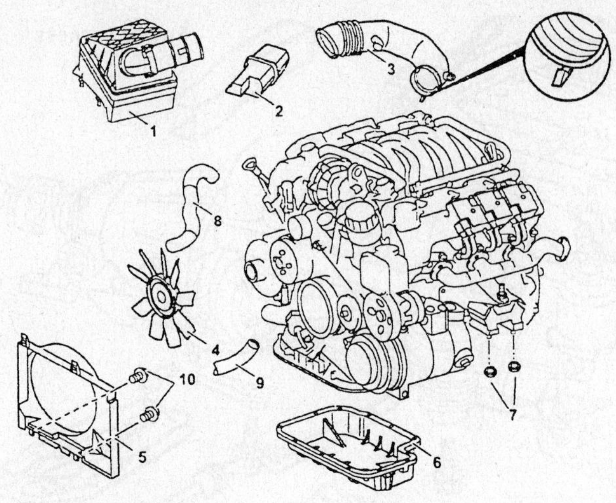

Exploded view of the mounting of the lower oil pan and related components—3.2L (112) and 4.3L (113) engines

- Coolant
- Engine under cover
- Fan clutch and shroud
- Engine upper cover
- Coolant hose at thermostat
- Coolant hose at water pump

➡It may be necessary to tap on the oil pan with a rubber mallet to dislodge it from the engine block.

3. Remove the mounting bolts and the oil pan. Clean all gasket material from the sealing surfaces.

To install:

4. Install or connect the following:
   - Oil pan with a new gasket. Torque the oil pan bolts to 84 inch lbs. (10 Nm).
   - Coolant hose at water pump
   - Coolant hose at thermostat
   - Fan clutch and shroud
   - Engine upper cover
   - Coolant
   - Engine oil
   - Negative battery cable

5.0L (119) Engine

1. Before servicing the vehicle, refer to the Precautions Section.
2. Remove or disconnect the following:
   - Negative battery cable
   - Engine oil
   - Coolant
   - Fan clutch and shroud
   - Drive belt and tensioner with bracket
   - Dipstick guide tube at air pump bracket
   - Air conditioning compressor and position it aside with the lines attached
   - Alternator and bracket
   - Dipstick guide tube
   - Engine mounts
   - Front springs and shock absorbers
   - Steering coupling
3. Using a suitable engine support tool, support the weight of the engine.
4. Support the front sub frame and remove mounting bolts. Lower to access the oil pan mounting bolts.

➡It may be necessary to tap on the oil pan with a rubber mallet to dislodge it from the engine block.

5. Remove the mounting bolts and the oil pan. Clean all gasket material from the sealing surfaces.

To install:

6. Install or connect the following:
   - Oil pan with a new gasket. Torque the M6 bolts to 97 inch lbs. (11 Nm). Torque the M8 bolts to 18 ft. lbs. (25 Nm). Torque the M10 bolts to 30 ft. lbs. (40 Nm).
   - Sub frame. Torque the bolts to 30 ft. lbs. (40 Nm).
   - Steering coupling
   - Front springs and shock absorbers
   - Engine mounts. Torque the mounting bolts to 18 ft. lbs. (25 Nm).
   - Dipstick guide tube at oil pump
   - Alternator and bracket. Torque the bracket mounting bolts to 15 ft. lbs. (21 Nm).
   - Air conditioning compressor and bracket. Torque the bracket mounting bolts to 15 ft. lbs. (21 Nm).
   - Dipstick guide tube at air pump bracket
   - Drive belt and tensioner with bracket. Torque the bracket mounting bolts to 15 ft. lbs. (21 Nm).
   - Fan clutch and shroud
   - Engine oil
   - Negative battery cable

6.0L (120) Engine

1. Before servicing the vehicle, refer to the Precautions Section.
2. Remove or disconnect the following:
   - Negative battery cable
   - Engine oil
   - Engine assembly
   - Alternator
   - Oil pressure sensor wire connection
   - Dipstick guide tube
   - Vibration damper
   - Secondary air injection line
   - Oil pan end cover

➡It may be necessary to tap on the oil pan with a rubber mallet to dislodge it from the engine block.

3. Remove the mounting bolts and the oil pan. Clean all gasket material from the sealing surfaces.

To install:

4. Install or connect the following:
   - Oil pan with a new gasket. Torque the M6 bolts to 88 inch lbs. (9 Nm). Torque the M8 bolts to 15 ft. lbs. (21 Nm). Torque the M10 bolts to 30 ft. lbs. (40 Nm).
   - Oil pan end cover
   - Secondary air injection line
   - Vibration damper
   - Dipstick guide tube
   - Alternator
   - Engine assembly
   - Engine oil
   - Negative battery cable

Oil Pump

REMOVAL & INSTALLATION

3.2L (104), 1.8L, 2.3L (111), 2.6L, 2.8L, 3.2L (112) and 4.3L, 5.0L, 5.5L (113) Engines

1. Before servicing the vehicle, refer to the Precautions Section.

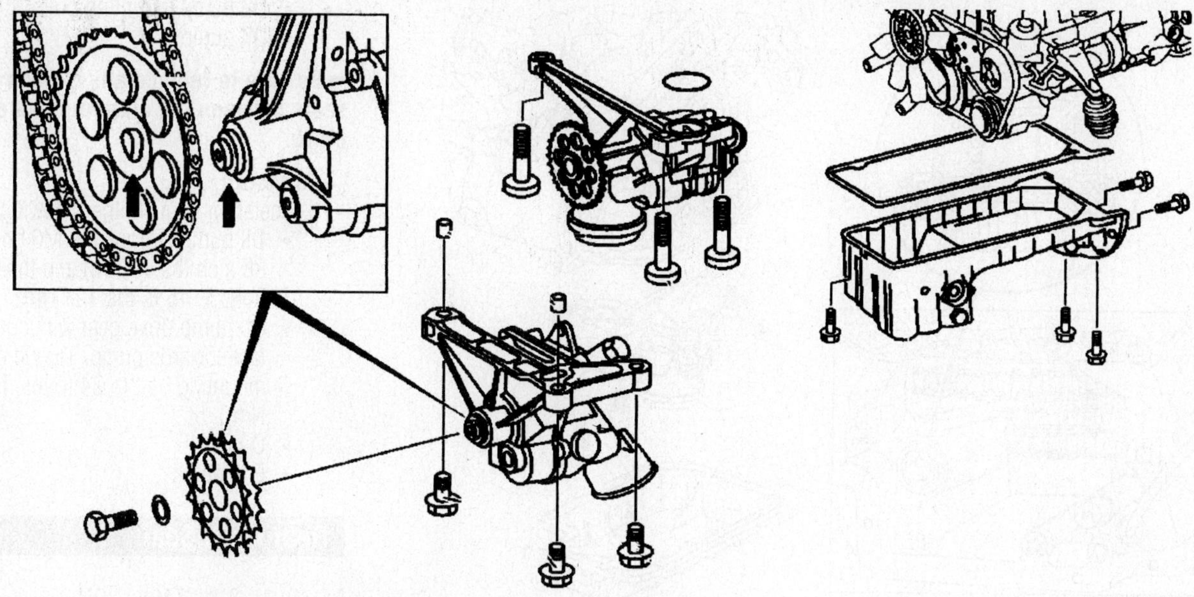

Exploded view of the oil pump mounting and related component—1.8L and 2.3L (111) engines

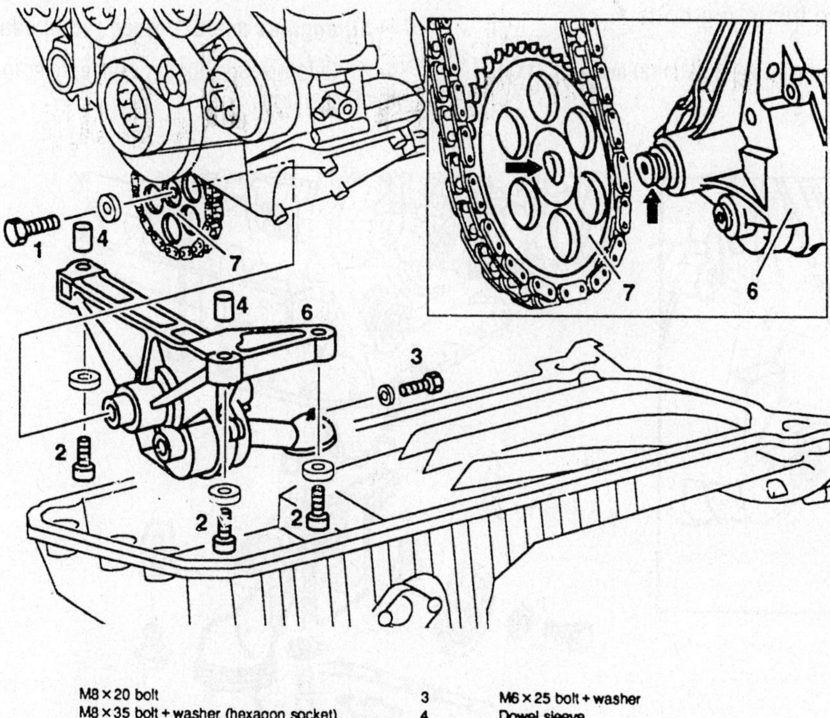

Exploded view of the oil pump mounting and related component—2.6L, 2.8L, 3.2L (112) and 4.3L, 5.0L, 5.5L (113) engines

2. Remove or disconnect the following:
   - Oil pan
   - Oil pump drive gear
   - Oil pump mounting bolts
   - Oil pump

To install:

➡Be sure to replace any O-rings or seals between the oil pump and engine block.

3. Install the oil pump and tighten the mounting bolts as follows:
   - M6 bolts: 88 inch lbs. (9 Nm)
   - M8 bolts: 15 ft. lbs. (21 Nm)
4. Install or connect the following:
   - Oil pump drive gear. Torque the bolt to 24 ft. lbs. (32 Nm).
   - Oil pan
   - Engine oil

5.0L (119) Engine

1. Before servicing the vehicle, refer to the Precautions Section.
2. Remove or disconnect the following:
   - Ground straps
   - Automatic transmission oil line bracket
   - Oil level control hose
   - Oil pan
   - Oil pump drive gear
   - Oil pump

➡Be sure to replace any O-rings or seals between the oil pump and engine block.

To install:

3. Install or connect the following:
   - Oil pump. Torque the mounting bolts to 15 ft. lbs. (21 Nm).
   - Oil pump drive gear. Torque the mounting bolt to 24 ft. lbs. (28 Nm).
   - Oil pan
   - Oil level control hose
   - Automatic transmission oil line bracket
   - Ground straps
   - Engine oil

6.0L (120) Engine

1. Before servicing the vehicle, refer to the Precautions Section.
2. Remove or disconnect the following:
   - Negative battery cable
   - Engine oil
   - Oil pan
   - Oil pump drive gear

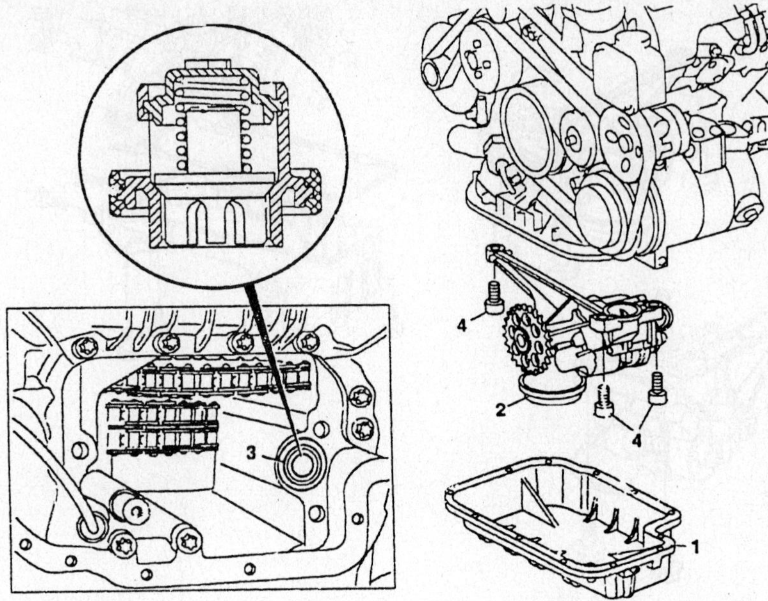

Exploded view of the oil pump mounting and related component—3.2L (112) and 4.3L (113) engines

- Oil pump mounting bolts
- Oil pump

➡**Be sure to replace any O-rings or seals between the oil pump and engine block.**

To install:

3. Install or connect the following:
   - Oil pump. Torque the M6 bolts to 88 inch lbs. (9 Nm) and the M8 bolts to 15 ft. lbs. (21 Nm).
   - Oil pump drive gear with curved face towards pump. Torque the mounting bolt to 24 ft. lbs. (28 Nm).
   - Oil pan
   - Engine oil

Rear Main Seal

REMOVAL & INSTALLATION

All Models

➡**All engines use a 1 piece radial seal.**

1. Before servicing the vehicle, refer to the Precautions Section.

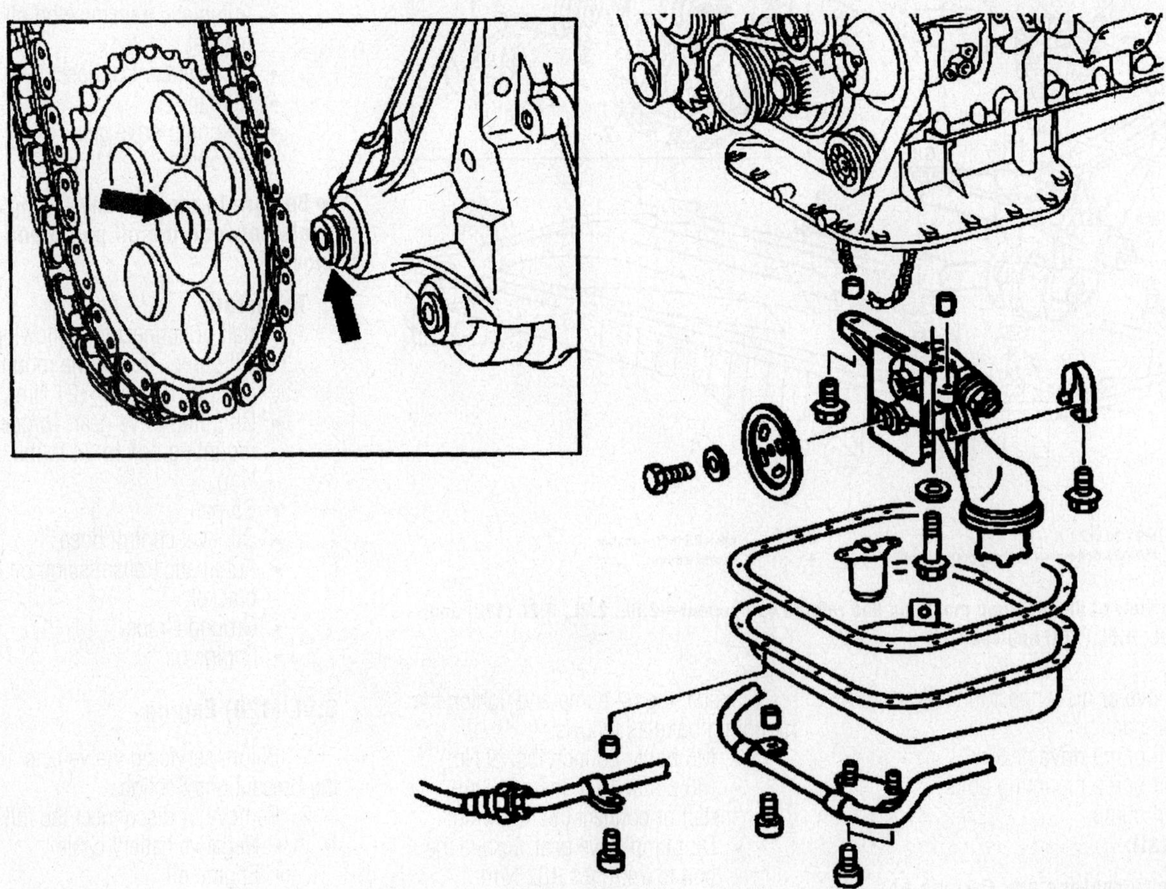

Exploded view of the oil pump mounting and related component—5.0L (119) engine

2. Remove or disconnect the following:
- Negative battery cable
- Transmission assembly
- Flywheel or flexplate
- Rear main seal using a suitable prying tool

To install:

3. Coat the inner lip of the seal with engine oil. Using a seal driver, install the seal into the retainer.
4. Install or connect the following:
- Flywheel or flexplate
- Transmission assembly
- Negative battery cable

Front Cover and Seal

REMOVAL & INSTALLATION

3.2L (104) Engine

1. Before servicing the vehicle, refer to the Precautions Section.
2. Remove or disconnect the following:
- Coolant
- Negative battery cable
- Alternator and bracket
- Engine cover, air cleaner and intake scoop
- Fan clutch, belt and pulleys
- Wiring and hoses

3. Remove the upper timing cover and mark the timing chain and sprockets.
- Timing chain and tensioner
- Accessories as needed, referring to the illustration for the specific model
- Crankshaft Position (CKP) sensor
- Crankshaft pulley
- Timing cover. Make sure to clean all gasket material from the sealing surfaces.

To install:

4. Install or connect the following:
- Timing cover with new gasket. Torque the M6 bolts to 84 inch lbs. (10 Nm). Torque the M8 bolts to 15 ft. lbs. (21 Nm).
- Crankshaft pulley
- CKP sensor
- Accessories
- Timing chain and tensioner
- Wiring and hoses
- Fan clutch, belt and pulleys
- Alternator and bracket
- Engine cover, air cleaner and intake scoop
- Negative battery cable
- Coolant

1.8L and 2.3L (111) Engines

1. Before servicing the vehicle, refer to the Precautions Section.
2. Remove or disconnect the following:
- Coolant
- Negative battery cable
- Left charge air pipe
- Oil filter
- Oil pan
- Fan clutch and shroud
- Accessory drive belt
- Air conditioning compressor and position it aside with the lines attached

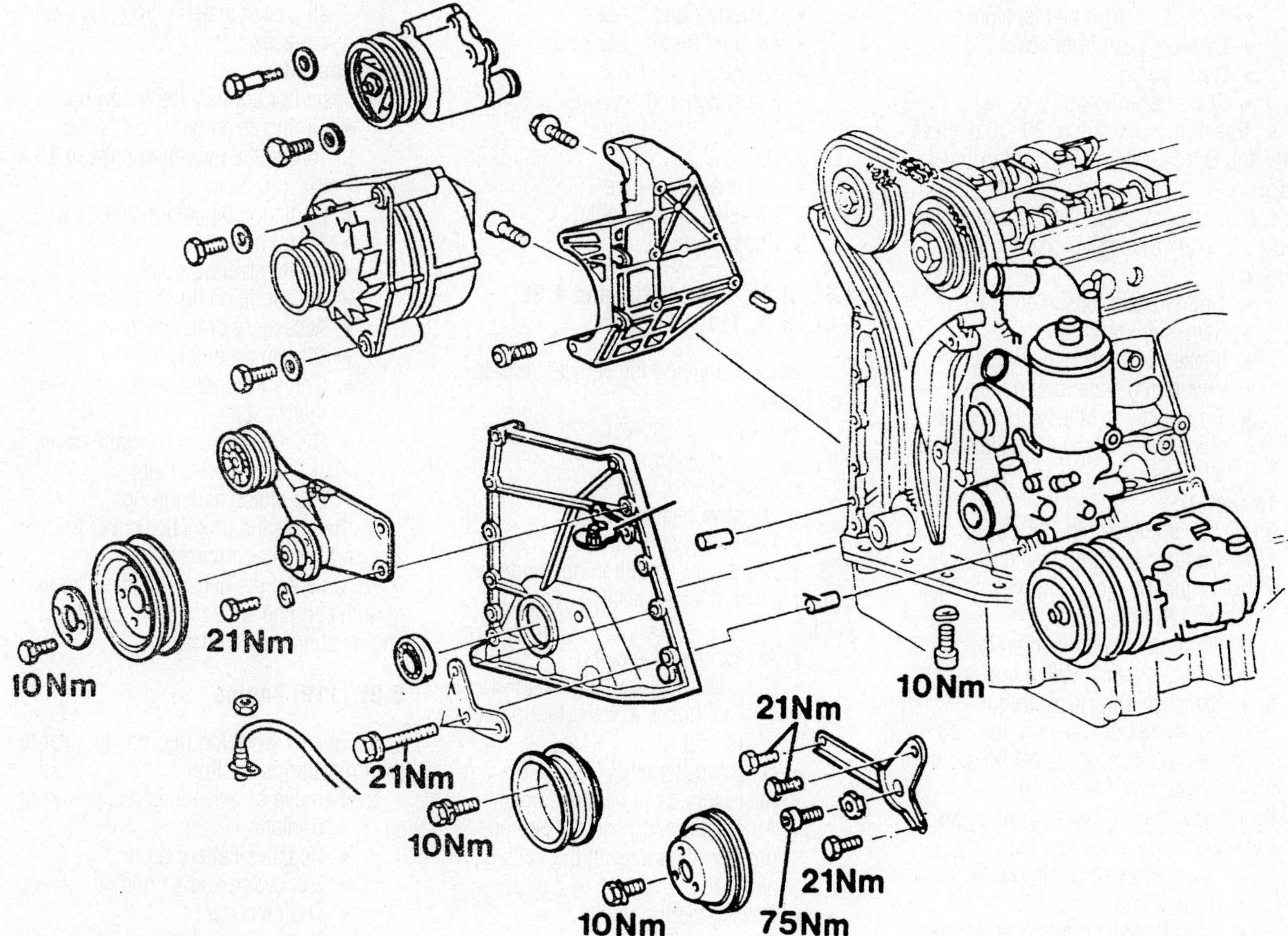

Exploded view of the lower timing cover mounting and related components—3.2L (104) engine

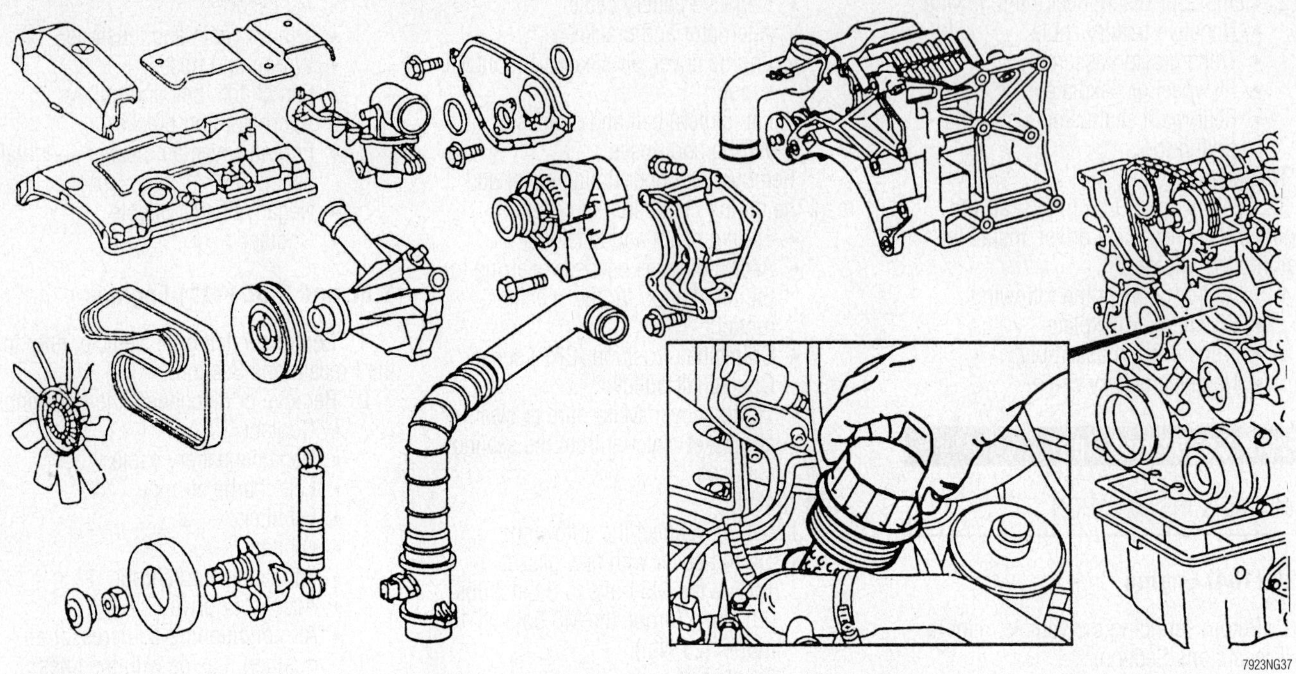

Exploded view of the lower timing cover mounting and related components—1.8L and 2.3L (111) engines

- Alternator and bracket
- Accessory drive belt tensioner
- Cylinder head front cover
- Water pump
- Power steering pump pulley

3. Position crankshaft to 20–30 degrees After Top Dead Center (ATDC) for cylinder No. 1.

4. Lock the camshafts with locking pins and matchmark the timing chain to the sprockets.

- Timing chain tensioner
- Camshaft sprockets
- Intake camshaft with adjuster
- Vibration damper and pulley
- Timing cover. Make sure to clean all gasket material from the sealing surfaces.

To install:

5. Install or connect the following:

- Timing cover with new gasket. Torque the mounting bolts to 18 ft. lbs. (25 Nm).
- Vibration damper and pulley
- Intake camshaft with adjuster
- Camshaft sprockets. Torque the mounting bolts to 15 ft. lbs. (20 Nm), plus an additional 90 degrees.
- Timing chain tensioner

6. Remove the camshaft locking pins and verify timing mark alignment.

- Power steering pump pulley
- Water pump
- Cylinder head front cover. Torque the mounting bolts to 18 ft. lbs. (25 Nm).
- Accessory drive belt tensioner
- Alternator and bracket
- Air conditioning compressor
- Accessory drive belt
- Fan clutch and shroud
- Oil pan
- Oil filter
- Left charge air pipe
- Negative battery cable
- Coolant

2.6L, 2.8L, 3.2L (112) and 4.3L, 5.0L, 5.5L (113) Engines

1. Before servicing the vehicle, refer to the Precautions Section.

2. Remove or disconnect the following:

- Coolant
- Negative battery cable
- Fan clutch and shroud

3. Cover the air conditioning condenser using a piece of sheet metal, plywood or plastic.

- Accessory drive belts
- Cylinder head and engine covers
- Coolant hoses at water pump and thermostat
- Oil pan assembly
- Accessory drive belt tensioner
- Air conditioning compressor and position it aside with the lines attached
- Power steering pump
- Air pump, if equipped
- Coolant temperature sensor wire
- Timing cover. Make sure to clean all gasket material from the sealing surfaces.

To install:

4. Install or connect the following:

- Timing cover with new gasket. Torque the mounting bolts to 15 ft. lbs. (20 Nm).
- Coolant temperature sensor wire
- Air pump
- Power steering pump
- Air conditioning compressor
- Accessory drive belt tensioner
- Oil pan assembly
- Coolant hoses at water pump and thermostat
- Cylinder head and engine cover
- Accessory drive belts
- Fan clutch and shroud

5. Remove the protective cover from the air conditioning compressor.

6. Connect the negative battery cable, refill the engine with coolant, start the engine and check for leaks.

5.0L (119) Engine

1. Before servicing the vehicle, refer to the Precautions Section.

2. Remove or disconnect the following:

- Coolant
- Negative battery cable
- Fan clutch and shroud
- Engine cover
- Upper timing covers

3. Position crankshaft to 40 degrees

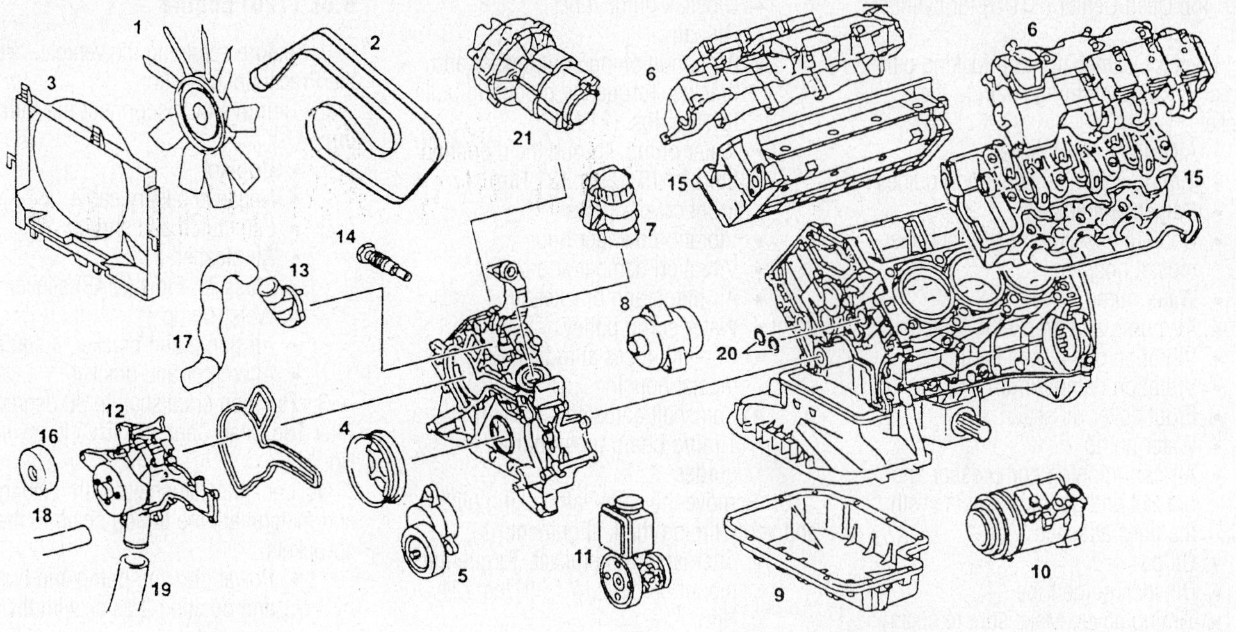

1 Viscous fan
2 Poly V-belt
3 Fan shroud
4 Vibration damper
5 Tensioning device
6 Cylinder head covers with ignition coils
7 Oil filter housing
8 Generator
9 Bottom part of oil pan
10 AC compressor
11 Power steering pump
12 Coolant pump
13 Thermostat housing
14 Chain tensioner
15 Cylinder heads
16 Guide pulley
17 Coolant hose to thermostat housing
18, 19 Coolant hoses to coolant pump
20 Timing case seals
21 Air pump

7923NG33

Exploded view of the lower timing cover mounting and related components—2.6L, 2.8L, 3.2L (112) and 4.3L, 5.0L, 5.5L (113) Engines

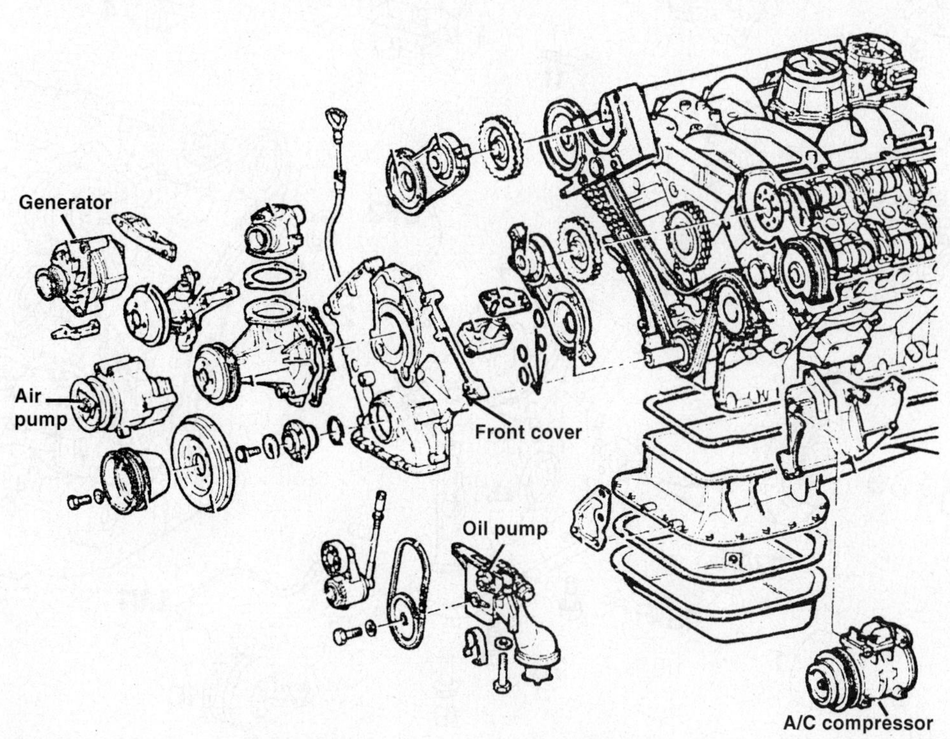

Exploded view of the lower timing cover mounting and related components—5.0L (119) engine

Before Top Dead Center (BTDC) for cylinder No. 1.

4. Lock the camshafts with locking pins and matchmark the timing chain to the sprockets.

- Alternator and bracket
- Timing chain tensioner and guides
- Camshaft adjusters
- Coolant hoses at radiator and thermostat housing
- Water pump pulley
- Air pump and bracket if equipped
- Vibration damper and pulley
- Vibration damper hub
- Front cover oil seal
- Water pump
- Air conditioning compressor and bracket and position it aside with the lines attached
- Oil pan
- Dipstick guide tube
- Timing cover. Make sure to clean all gasket material from the sealing surfaces. Replace the timing case cover O-rings.

To install:

5. Install or connect the following:

- Timing cover with new gasket. Torque the mounting bolts to 15 ft. lbs. (21 Nm).
- Dipstick guide tube
- Oil pan
- Air conditioning compressor and bracket. Torque the mounting bolts to 15 ft. lbs. (21 Nm).
- Water pump. Torque the mounting bolts to 15 ft. lbs. (21 Nm).
- Front cover oil seal
- Vibration damper hub
- Vibration damper and pulley
- Air pump and bracket
- Water pump pulley
- Coolant hoses at radiator and thermostat housing
- Camshaft adjusters
- Timing chain tensioner and guides

6. Remove the camshaft locking pins and verify timing mark alignment.

- Alternator and bracket. Torque the mounting bolts to 15 ft. lbs. (21 Nm).
- Upper timing covers. Torque the mounting bolts to 15 ft. lbs. (21 Nm).
- Engine cover
- Fan clutch and shroud
- Negative battery cable
- Coolant

6.0L (120) Engine

1. Before servicing the vehicle, refer to the Precautions Section.

2. Remove or disconnect the following:

- Coolant
- Negative battery cable
- Fan clutch and shroud
- Air cleaner
- Mass Air Flow (MAF) sensor
- Water pump
- Air pump and bracket, if equipped
- Alternator and bracket

3. Position crankshaft to 30 degrees After Top Dead Center (ATDC) for cylinder No. 1.

4. Lock the camshafts with locking pins and matchmark the timing chain to the sprockets.

- Power steering pump and bracket and position it aside with the lines attached
- Top Dead Center (TDC) sensors
- Secondary air pump air cleaner and bracket
- Timing chain tensioner
- Secondary air injection line, if equipped
- Oil pan bottom cover

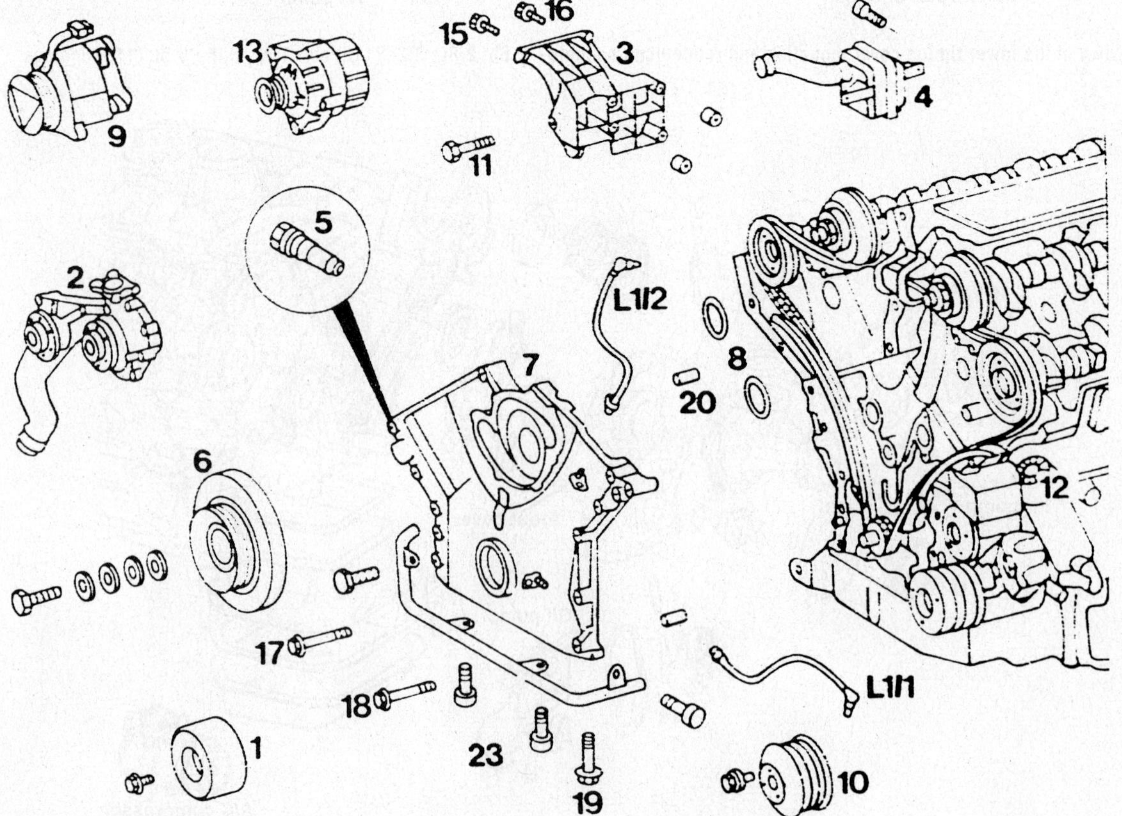

Exploded view of the lower timing cover mounting and related components—6.0L (120) engine

- Remove the timing cover. Make sure to clean all gasket material from the sealing surfaces. Replace the timing case cover O-rings.

To install:

5. Install or connect the following:
   - Timing cover with new gasket. Torque the M6 bolts to 88 inch lbs. (9 Nm), the M8 bolts to 15 ft. lbs. (21 Nm), and the M8 bolts to 18 ft. lbs. (25 Nm).
   - Oil pan bottom cover
   - Secondary air injection line
   - Timing chain tensioner
   - Secondary air pump air cleaner and bracket
   - TDC sensors
   - Power steering pump and bracket
6. Remove the camshaft locking pins and verify timing mark alignment.
   - Alternator and bracket
   - Air pump and bracket, if equipped
   - Water pump
   - MAF sensor
   - Air cleaner
   - Fan clutch and shroud
   - Negative battery cable
   - Coolant

Timing Chain

REPLACEMENT

All Models

➡An endless timing chain is used on production engines, but a split chain with a connecting link is used for service. The endless chain can be separated with a "chain breaker". Only 1 master link (connecting link) should be used on a chain.

1. Before servicing the vehicle, refer to the Precautions Section.
2. Remove or disconnect the following:
   - Negative battery cable
   - Spark plugs
   - Valve cover(s)
   - Engine timing covers
3. Clamp the chain to the camshaft gear and cover the opening of the timing chain case.
4. Separate the chain with a chain breaker.

To install:

5. Attach a new timing chain (5) to the old chain (4) with a master link (6), center plate (7) and end plate (8). Using a socket wrench on the crankshaft, slowly rotate the engine in the direction of normal rotation. Simultaneously, pull the old chain through until the master link is uppermost on the camshaft sprocket (9). Be sure to keep tension on the chain throughout this procedure.
6. Disconnect the old timing chain and connect the ends of the new chain with the master link. Insert the new connecting link from the rear so the lockwashers can be seen from the front.
7. Rotate the engine until the timing marks align. Check the valve timing.
8. Install or connect the following:
   - Engine timing covers
   - Valve covers
   - Spark plugs
   - Negative battery cable
9. Start the engine and check operation.

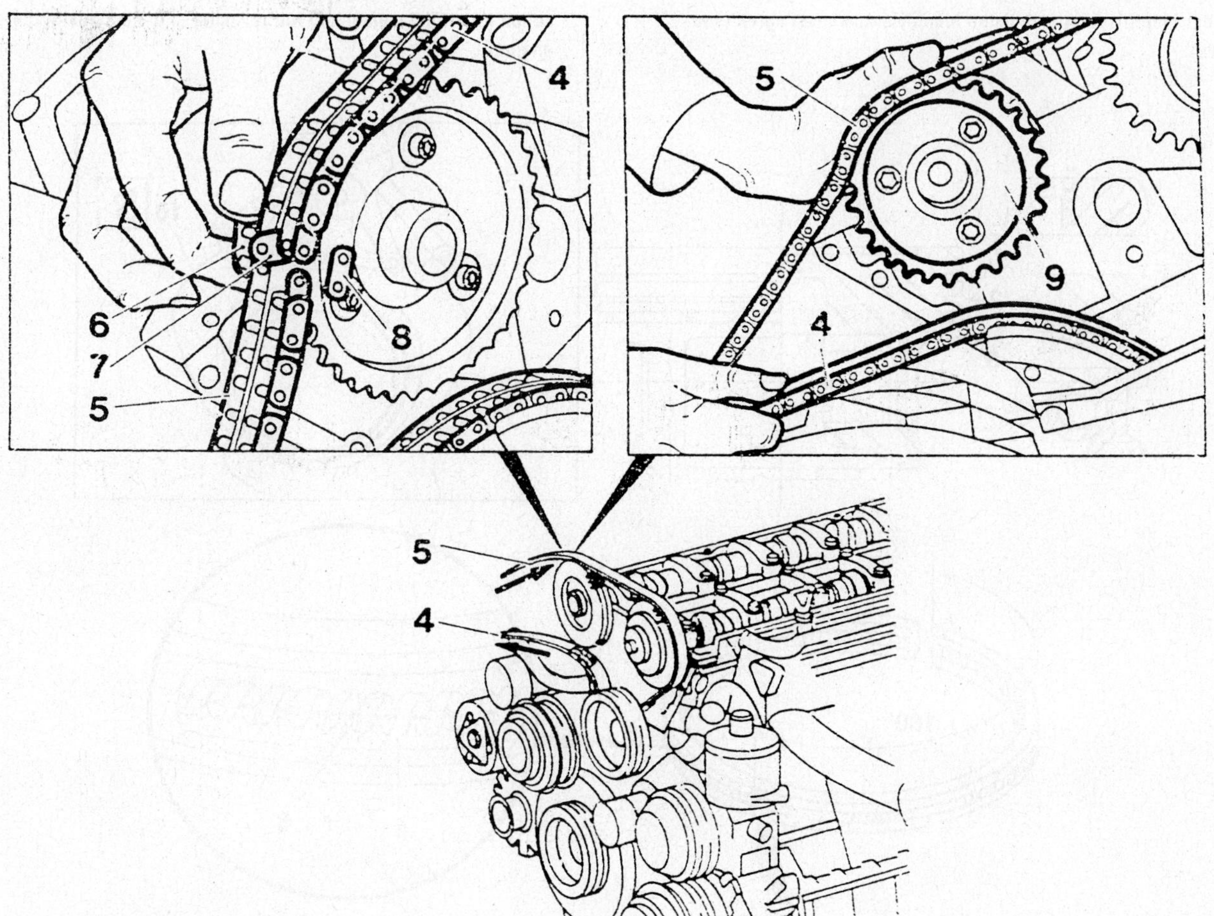

Replacing the timing chain—104 engine shown

Piston and Ring

POSITIONING

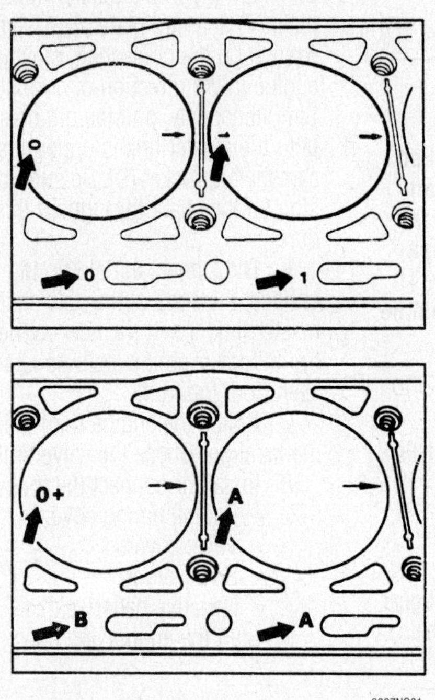

Cylinder identification and piston identifications must agree (see arrows)

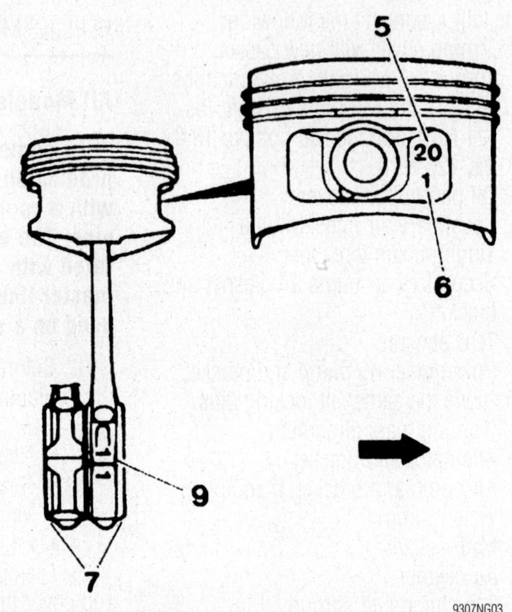

Install pistons so the code (5) next to the piston pin and group number (6) are pointing in the direction of travel, and the identification (9) on the connecting rod is facing to the outside of the engine.

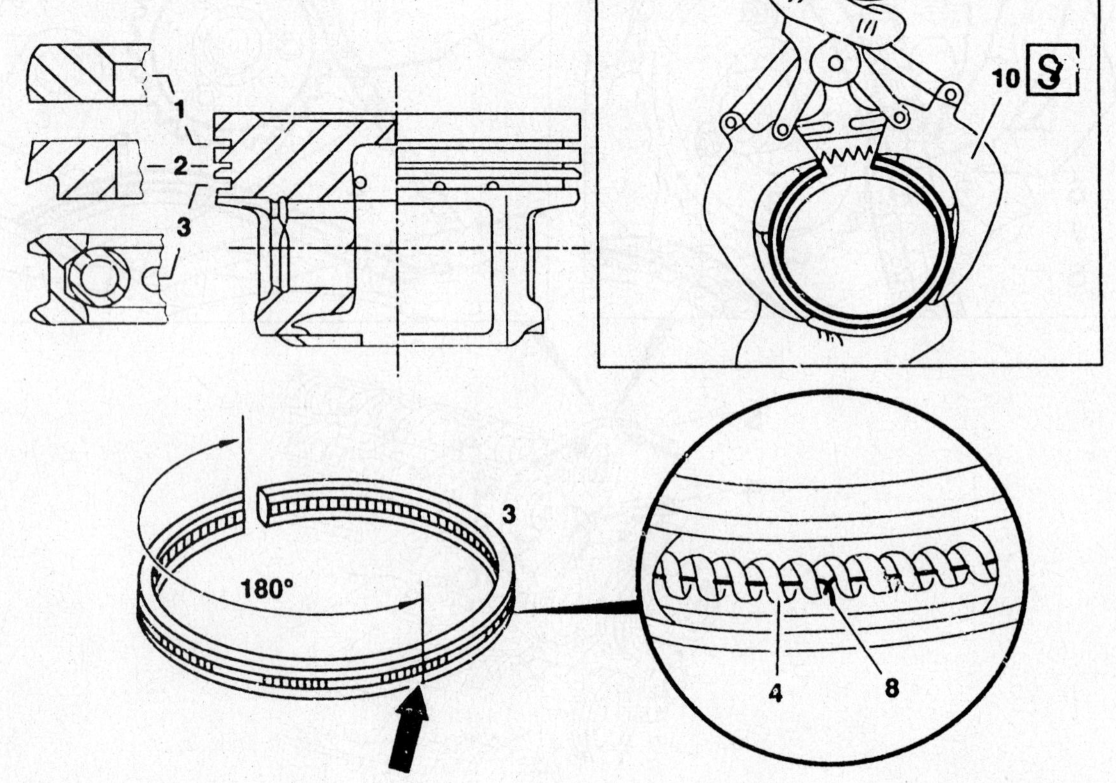

Ring positioning

DIESEL ENGINE REPAIR

➡Disconnecting the negative battery cable may interfere with the functions of the on board computer systems and may require the computer to undergo a relearning process, once the negative battery cable is reconnected.

Alternator

REMOVAL & INSTALLATION

1. Before servicing the vehicle, refer to the Precautions Section.
2. Remove or disconnect the following:
   - Negative battery cable
   - Accessory drive belt
   - Engine under cover
   - Alternator electrical wires
   - Alternator from the bottom of the vehicle

To install:

3. Install or connect the following:
   - Alternator. Torque the mounting bolts to 31 ft. lbs. (42 Nm).
   - Alternator electrical wires
   - Engine under cover
   - Accessory drive belt
   - Connect the negative battery cable
4. Start the engine and check for proper operation.

Engine Assembly

REMOVAL & INSTALLATION

➡In all cases, Mercedes-Benz engines and transmissions are removed as a unit.

1. Before servicing the vehicle, refer to the Precautions Section.
2. Remove or disconnect the following:
   - Coolant
   - Engine oil
   - Transmission fluid
   - Hood
   - Battery
   - Heater hoses and oil cooler lines
   - Fan shroud and radiator
   - Air cleaner
   - Fuel, vacuum and oil hoses
   - Fan clutch and fan
   - Accelerator linkage
   - Ground straps and electrical connections
   - Gearshift linkage
   - Exhaust pipes
3. Loosen the steering relay arm and move it aside, along with the center steering rod and hydraulic steering damper.
   - Hydraulic engine shock absorber
   - Transmission oil line connectors
   - Exhaust pipe bracket
4. Support the bell housing or place a cable sling under the oil pan to support the engine. On turbocharged models, disconnect the exhaust pipes at the turbocharger. Mark the position of the rear engine support and unbolt the 2 outer bolts. Remove the top bolt at the transmission and pull the support out.
   - Speedometer cable
   - Front driveshaft U-joint and push the driveshaft back and wire it aside
   - Engine mounts
5. Unbolt the power steering fluid reservoir and move it aside. Using a chain hoist and cable, lift the engine and transmission upward and outward.

To install:

6. Install or connect the following:
   - Engine and transmission assembly
   - Engine mounts. Torque the bolts to 30 ft. lbs. (40 Nm).
   - Front driveshaft U-joint
   - Speedometer cable
   - Exhaust pipes and bracket
   - Rear engine support
   - Exhaust pipe bracket
   - Exhaust pipes at the turbocharger, if equipped. Torque the bolts to 25 ft. lbs. (34 Nm).
   - Transmission oil line connectors
   - Hydraulic engine shock absorber
   - Steering relay arm to 15 ft. lbs. (20 Nm)
   - Steering damper to the steering linkage
   - Gearshift linkage and exhaust pipes. Torque the manifold bolts to 25 ft. lbs. (34 Nm).
   - Ground straps and electrical connections
   - Accelerator linkage
   - Fan clutch and fan
   - Fuel, vacuum and oil hoses
   - Air cleaner
   - Radiator and fan shroud
   - Heater hoses and oil cooler lines
   - Battery
   - Coolant
   - Engine oil
   - Transmission fluid

Water Pump

REMOVAL & INSTALLATION

1. Before servicing the vehicle, refer to the Precautions Section.
2. Remove or disconnect the following:
   - Negative battery cable
   - Fan clutch, fan and shroud

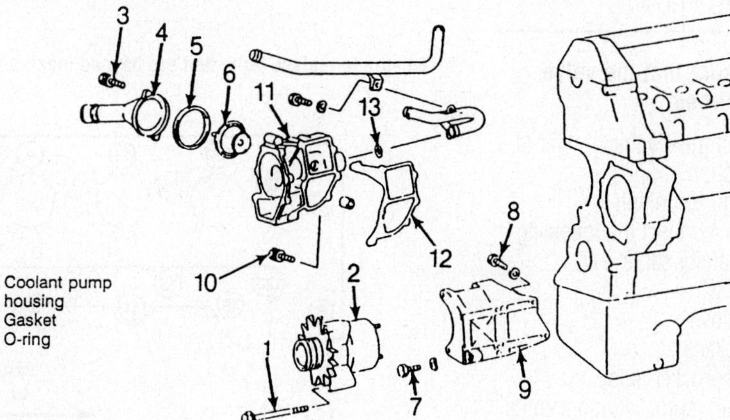

1. Collar screw—tightening torque 33 ft. lbs. (45 Nm)
2. Alternator
3. Screw—tightening torque 7 ft. lbs. (10 Nm)
4. Thermostat housing cap
5. Sealing ring
6. Thermostat
7. Screw—tightening torque 19 ft. lbs. (25 Nm)
8. Screw—tightening torque 19 ft. lbs. (25 Nm)
9. Carrier
10. Screw—tightening torque 7 ft. lbs. (10 Nm)
11. Coolant pump housing
12. Gasket
13. O-ring

Exploded view of the water pump housing assembly—Diesel engines

- Water pump pulley
- Hex nuts and magnet body
- Water pump housing

➡The magnet carrier is glued to the water pump housing and should not be pulled off.

To install:

3. Install or connect the following:
   - Water pump housing and gasket. Torque combination screws to 90 inch lbs. (10 Nm).
   - Magnet body and hex nuts
   - Water pump pulley. Torque the screws to 90 inch lbs. (10 Nm).
   - Fan clutch, fan and shroud
   - Coolant
   - Negative battery cable

Glow Plugs

REMOVAL & INSTALLATION

The glow plugs are located on the left side of the cylinder head.

1. Before servicing the vehicle, refer to the Precautions Section.
2. Remove or disconnect the following:
   - Negative battery cable
   - Glow plug electrical connectors
   - Glow plugs

To install:

3. Apply anti-seize compound to the glow plug threads.
4. Install or connect the following:
   - Glow plugs. Torque to 15 ft. lbs. (20 Nm).
   - Glow plug electrical connector. Torque the retaining nut to 35 inch lbs. (4 Nm).
   - Negative battery cable

Cylinder Head

REMOVAL & INSTALLATION

➡Use care to ensure that the valve timing is not disturbed.

1. Before servicing the vehicle, refer to the Precautions Section.
2. Drain the engine coolant.
3. Remove or disconnect the following:
   - Negative battery cable
   - Hoses and wires
   - Camshaft cover
   - Throttle linkage
   - Camshaft sprocket nuts
4. Mark the chain, sprocket and cam for ease of assembly. Remove the sprocket and chain and wire it aside. Unbolt the manifolds and exhaust header pipe and set them aside.

➡Be sure the chain is securely wired so it will not slide into the engine.

5. Remove or disconnect the following:
   - Cylinder head bolts in the reverse order of the illustrated sequence.
   - Cylinder head
6. Clean all gasket material from the sealing surfaces. Be sure the cylinder head locating dowels are positioned in the engine block.

⁂ WARNING

All Diesel engines utilize cylinder head stretch bolts. These bolts undergo a permanent stretch each time they are tightened. When a maximum length is reached, they must be replaced with new bolts. Before tightening the head bolts on these engines, refer to the illustration for the specific length of the bolts.

To install:

7. Install the cylinder head and gasket. Torque the cylinder head bolts in sequence as follows:
   a. Step 1: 11 ft. lbs. (15 Nm).
   b. Step 2: 20 ft. lbs. (35 Nm).
   c. Step 3: + 90 degrees.
   d. Step 5: wait 10 minutes.
   e. Step 6: + additional 90 degrees.
8. Install or connect the following:
   - Manifolds and exhaust header pipe. Torque the pipe fasteners to 24 ft. lbs. (34 Nm).
   - Sprocket and chain
   - Camshaft sprocket nuts
   - Trowel linkage
   - Camshaft cover
   - Hoses and wires
   - Coolant
   - Negative battery cable

Dimensions of cylinder head bolts and maximum permissible length (L)

New	Max. permissible length (L) in mm
M 10×80	83.6
M 10×102	105.6
M 10×115	118.6

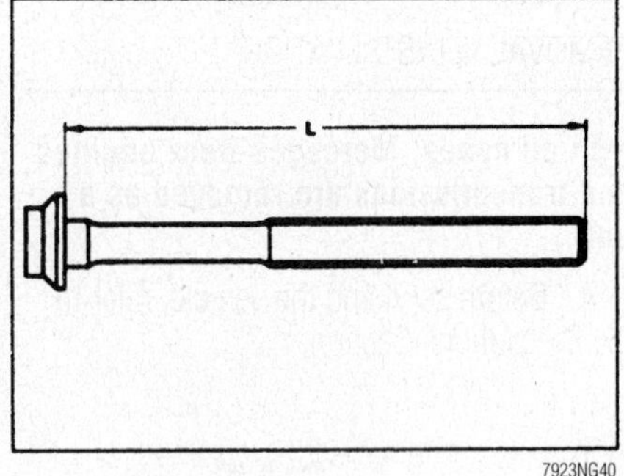

Be sure to replace bolts that are beyond maximum length—Diesel engine

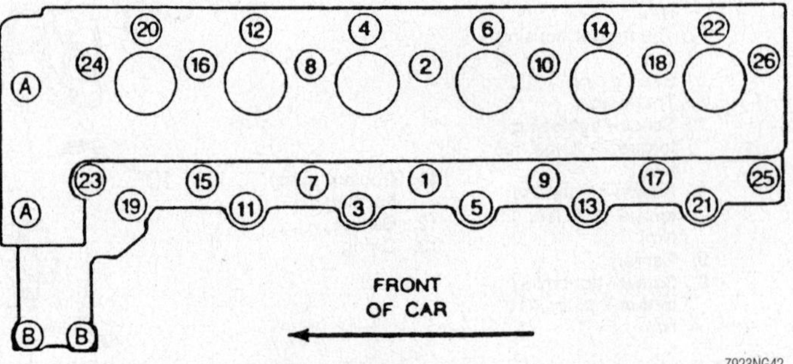

Cylinder head bolt tightening sequence—Diesel engine

Rocker Arms/Shafts

REMOVAL & INSTALLATION

The Diesel engine does not use rocker arms. The camshafts act directly on the hydraulic valve tappets.

Turbocharger

REMOVAL & INSTALLATION

1. Before servicing the vehicle, refer to the Precautions Section.
2. Remove or disconnect the following:
   - Negative battery cable
   - Air filter
   - Temperature switch electrical cable

- Air filter to compressor housing hose clamp
- Vacuum line and crankcase breather pipe
- Air intake duct
- Turbocharger oil line
- Air filter mounting bracket
- Turbocharger exhaust flange
- Transmission pipe bracket. Push the pipe rearward.
- Intermediate flange mounting bracket
- Turbocharger
- Intermediate flange and turbocharger oil return line

To install:

3. Install or connect the following:
   - Oil return line and intermediate flange
   - Flange gasket between the turbocharger and exhaust manifold with the reinforcing bead toward the exhaust manifold. Use only heat-proof nuts and bolts.
   - Intermediate flange mounting bracket
   - Transmission pipe bracket
   - Turbocharger exhaust flange
   - Air filter mounting bracket
   - Turbocharger oil line
   - Air intake duct
   - Vacuum line and crankcase breather pipe
   - Air filter-to-compressor housing hose clamp
   - Temperature switch electrical cable
   - Air filter
   - ¼ pint of engine oil through the oil supply into a new turbocharger
   - Negative battery cable

Intake Manifold

REMOVAL & INSTALLATION

1. Before servicing the vehicle, refer to the Precautions Section.
2. Remove or disconnect the following:
   - Negative battery cable
   - Engine ventilation tubes
   - Electrical wiring, cables and hoses
   - Manifold retaining bolts and manifold

To install:

3. Clean the gasket mating surfaces and install a new gasket.
4. Install or connect the following:
   - Manifold and bolts. Torque bolts to 18 ft. lbs. (25 Nm), working from the middle outward.
   - Electrical wiring, cables and hoses
   - Engine ventilation tubes
   - Negative battery cable

Exhaust Manifold

REMOVAL & INSTALLATION

1. Before servicing the vehicle, refer to the Precautions Section.
2. Remove or disconnect the following:
   - Negative battery cable
   - Exhaust pipe
   - Manifold retaining bolts and manifold

To install:

3. Clean the gasket mating surfaces and install a new gasket.
4. Install or connect the following:
   - Manifold and bolts. Torque bolts to 18 ft. lbs. (25 Nm), working from the middle outward.
   - Exhaust pipe
   - Negative battery cable

Camshaft and Valve Lifters

REMOVAL & INSTALLATION

1. Before servicing the vehicle, refer to the Precautions Section.
2. Remove or disconnect the following:
   - Negative battery cable
   - Cylinder head cover
3. Set crankshaft to Top Dead Center (TDC) of No. 1 cylinder. Remove the timing chain tensioner and mark the camshaft timing gears and timing chain in relation to each other.
4. Remove or disconnect the following:
   - Camshaft timing gears
   - Pressure oil pump, if equipped with level control and place aside leaving the lines connected
   - Camshaft bearing caps
   - Camshafts
   - Valve tappets using a lifter tool

To install:

5. Insert the circlip for axial locating into the cylinder head. Lubricate and install the valve tappets and camshafts.
6. Install or connect the following:
   - Camshaft bearing caps in reverse of the loosening sequence. Torque bearing caps to 18 ft. lbs. (25 Nm).
   - Camshaft timing gears. Torque the fastening screw to 48 ft. lbs. (65 Nm).
   - Timing chain tensioner
   - Pressure oil pump
   - Cylinder head cover
   - Negative battery cable

Valve Lash

ADJUSTMENT

Diesel engines use hydraulic valve lifters. There is no provision for valve clearance adjustments.

Starter

REMOVAL & INSTALLATION

1. Before servicing the vehicle, refer to the Precautions Section.
2. Remove or disconnect the following:
   - Negative battery cable
   - Starter wires and bracket
   - Starter mounting bolts
   - Starter assembly

To install:

3. Install or connect the following:
   - Starter assembly. Torque the mounting bolts to 31 ft. lbs. (42 Nm).
   - Starter wires and bracket
   - Negatives battery cable
4. Start the vehicle and check for proper operation.

Oil Pan

REMOVAL & INSTALLATION

1. Before servicing the vehicle, refer to the Precautions Section.
2. Disconnect the negative battery cable and drain the engine oil.
3. Using a suitable engine support tool, support the weight of the engine.
4. Support the front subframe and remove the mounting bolts. Lower to access the oil pan mounting bolts.
5. Remove the mounting bolts and the oil pan. Clean all gasket material from the sealing surfaces.

To install:

6. Install or connect the following:
   - Oil pan with new gasket
   - Oil pan mounting bolts in a crisscross pattern
   - Front subframe
   - Engine oil
   - Negative battery cable

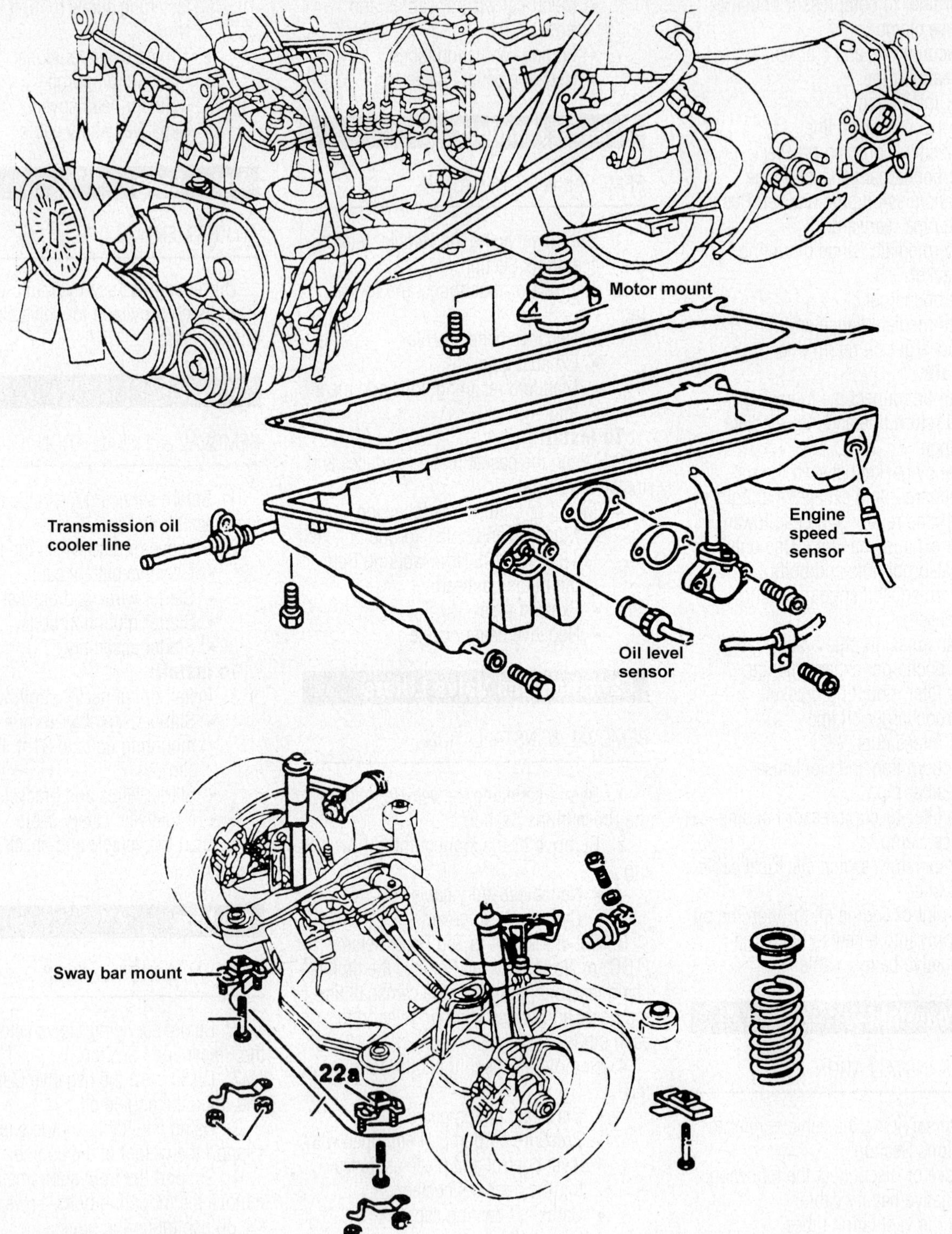

Exploded view of the oil pan mounting and related removal and installation components—Diesel engines

Oil Pump

REMOVAL & INSTALLATION

1. Before servicing the vehicle, refer to the Precautions Section.
2. Remove or disconnect the following:
   - Negative battery cable
   - Oil pan
   - Oil pump sprocket
   - Oil pump

To install:

3. Install or connect the following:
   - Oil pump. Torque to 18 ft. lbs. (25 Nm).
   - Sprocket and chain

➡Mount the sprocket so that the rise points toward the oil pump and so that the trochoid shape corresponds with that on the oil pump shaft.

   - Oil pan
   - Engine oil

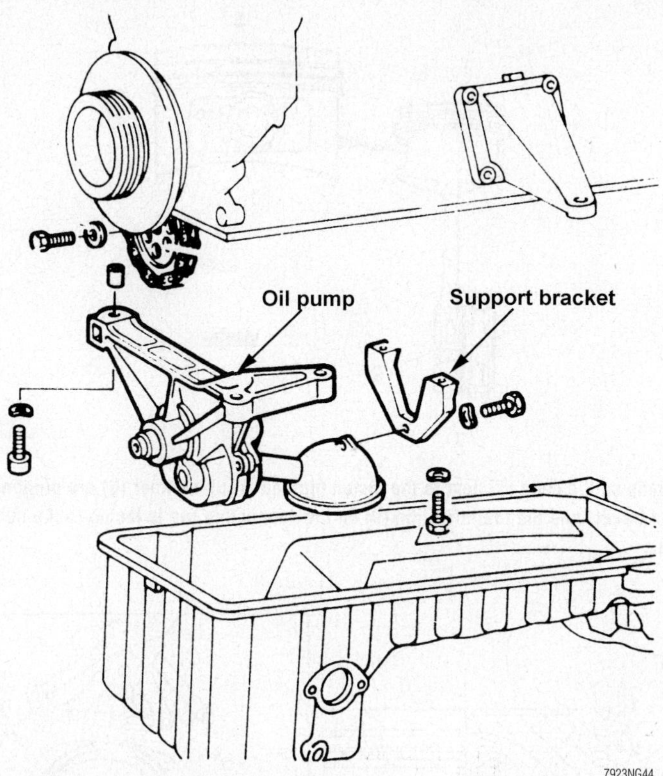

Exploded view of the oil pump mounting—Diesel engine

Rear Main Seal

REMOVAL & INSTALLATION

1. Before servicing the vehicle, refer to the Precautions Section.

➡**All engines use a 1-piece radial seal.**

2. Remove or disconnect the following:
   - Negative battery cable
   - Transmission assembly
   - Flywheel or flexplate
   - Rear main seal using a suitable prying tool

To install:

3. Coat the inner lip of the seal with engine oil.
4. Install or connect the following:
   - Seal into the retainer
   - Flywheel or flexplate
   - Transmission assembly
   - Negative battery cable

Timing Chain Cover and Seal

REMOVAL & INSTALLATION

1. Before servicing the vehicle, refer to the Precautions Section.
2. Remove or disconnect the following:
   - Coolant
   - Negative battery cable
   - Engine undercover
   - Serpentine belt
   - Fan shroud and radiator
   - Fan clutch and fan
   - Crankshaft pulley
   - Serpentine belt tensioner
   - Vacuum pump
   - Power steering pump
   - Self-leveling suspension hydraulic pump
   - Alternator
3. Cover the air conditioning condenser using a piece of sheet metal, plywood or plastic.
   - Cylinder head cover and charge air pipe
   - Oil dipstick tube
   - Top Dead Center (TDC) sensor, marking its position
4. Support the engine and remove the engine mounts. Raise the engine to gain access.
5. Remove the timing cover bolts and carefully remove the cover from the engine. Be careful not to damage the oil pan and cylinder head gaskets.

To install:

6. Install or connect the following:
   - Timing cover. Torque the bolts to 18 ft. lbs. (25 Nm).
   - Engine mounts
   - TDC sensor
   - Oil dipstick tube
   - Cylinder head cover and charge air pipe
7. Remove the air conditioning condenser guard
   - Alternator
   - Self-leveling suspension hydraulic pump
   - Power steering pump
   - Vacuum pump
   - Serpentine belt tensioner
   - Crankshaft pulley
   - Fan clutch and fan
   - Radiator and fan shroud
   - Serpentine belt
   - Engine undercover
   - Negative battery cable
   - Coolant

Timing Chain

REMOVAL & INSTALLATION

1. Before servicing the vehicle, refer to the Precautions Section.
2. Remove or disconnect the following:
   - Negative battery cable
   - Cylinder head cover
   - Injection nozzles
   - Chain tensioner
   - Fan and shroud
   - Engine timing cover
3. Clamp the chain to the camshaft gear and cover the opening of the timing chain case.
4. Separate the chain with a chain breaker.

To install:

5. Attach a new timing chain to the old chain with a master link. Using a socket wrench on the crankshaft, slowly rotate the engine in the direction of normal rotation. Simultaneously, pull the old chain through until the master link is uppermost on the camshaft sprocket. Be sure to keep tension on the chain throughout this procedure.

➡**Use only a rivet-type connecting link. Do not use connecting link that uses a retaining spring.**

6. Disconnect the old timing chain and connect the ends of the new chain with the master link. Insert the new connecting link from the rear so the lockwashers can be seen from the front.
7. Install the timing chain tensioner. Rotate the engine until the timing marks align. Check the valve timing.

8. Install or connect the following:
   - Engine timing cover. Torque the bolts to 18 ft. lbs. (25 Nm).
   - Cylinder head cover. Torque to 90 inch lbs. (10 Nm).
   - Fan and fan shroud
   - Injection nozzles
   - Negative battery cable

Piston and Ring

POSITIONING

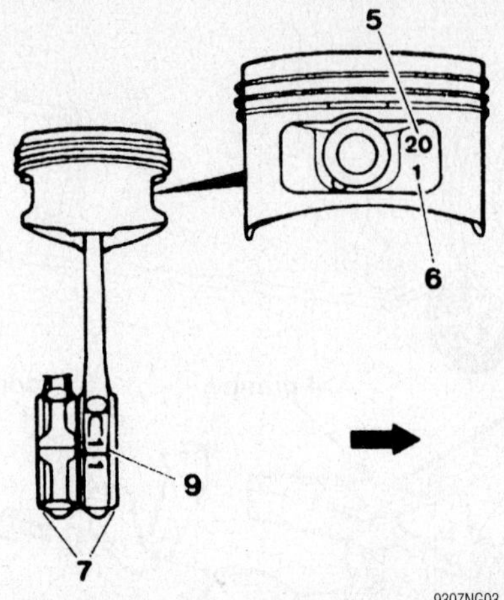

Install pistons so the code (5) next to the piston pin and group number (6) are pointing in the direction of travel, and the identification (9) on the connecting rod is facing to the outside of the engine.

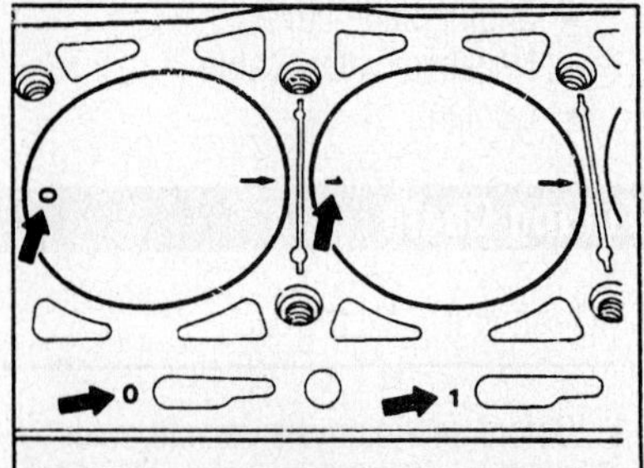

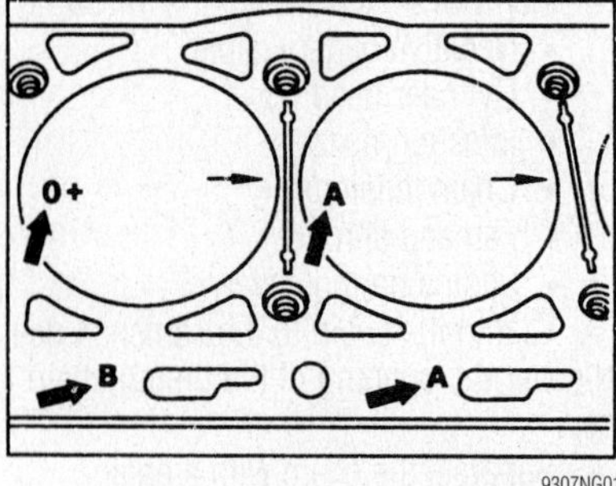

Cylinder identification and piston identifications must agree (see arrows)

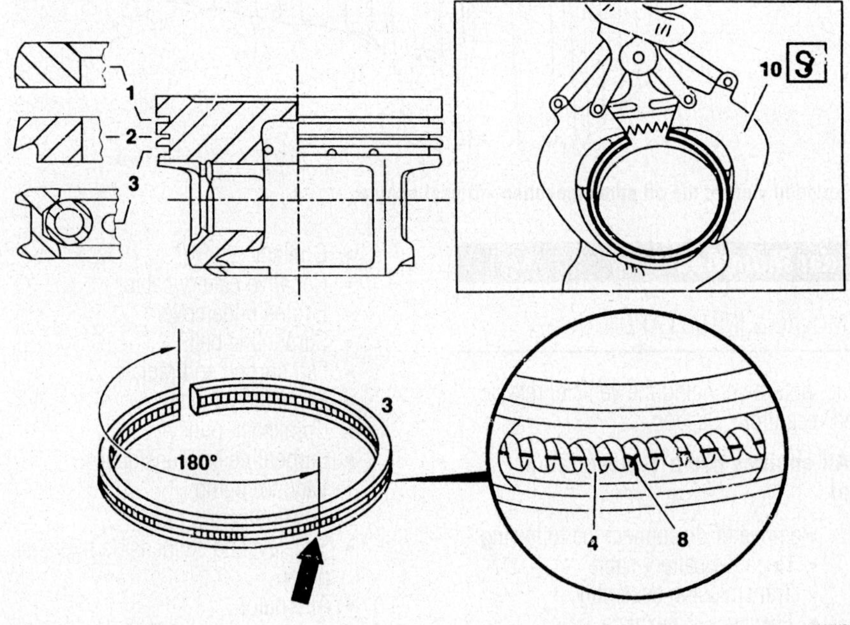

Ring positioning

GASOLINE FUEL SYSTEM

Fuel System Service Precautions

Safety is the most important factor when performing not only fuel system maintenance but also any type of maintenance. Failure to conduct maintenance and repairs in a safe manner may result in serious personal injury or death. Maintenance and testing of the vehicle's fuel system components can be accomplished safely and effectively by adhering to the following rules and guidelines.

- To avoid the possibility of fire and personal injury, always disconnect the negative battery cable unless the repair or test procedure requires that battery voltage be applied.
- Always relieve the fuel system pressure prior to disconnecting any fuel system component (injector, fuel rail, pressure regulator, etc.), fitting or fuel line connection. Exercise extreme caution whenever relieving fuel system pressure, to avoid exposing skin, face and eyes to fuel spray. Please be advised that fuel under pressure may penetrate the skin or any part of the body that it contacts.
- Always place a shop towel or cloth around the fitting or connection prior to loosening to absorb any excess fuel due to spillage. Ensure that all fuel spillage (should it occur) is quickly removed from engine surfaces. Ensure that all fuel soaked cloths or towels are deposited into a suitable waste container.
- Always keep a dry chemical (Class B) fire extinguisher near the work area.
- Do not allow fuel spray or fuel vapors to come into contact with a spark or open flame.
- Always use a back-up wrench when

loosening and tightening fuel line connection fittings. This will prevent unnecessary stress and torsion to fuel line piping.

- Always replace worn fuel fitting O-rings with new. Do not substitute fuel hose or equivalent, where fuel pipe is installed.

Fuel System Pressure

RELIEVING

1. Before servicing the vehicle, refer to the Precautions Section.
2. Disconnect the negative battery cable.
3. Connect a fuel pressure gauge with a pressure release valve to the service port on the fuel supply rail.
4. Place the fuel release tube into a container and open the valve.
5. Remove the fuel pressure gauge from the service port on the fuel supply rail.

Fuel Filter

REMOVAL & INSTALLATION

1. Before servicing the vehicle, refer to the Precautions Section.
2. Properly relieve the fuel system pressure.
3. Remove or disconnect the following:
   - Cover box
   - Gas cap
   - Fuel pump cover
   - Pressure hoses
   - Fuel filter
4. Remove the connecting plug from the old filter and install it on a new filter using a new gasket.

To install:

5. Install or connect the following:
   - Fuel filter
   - Attaching screws
   - Pressure hoses
   - Fuel pump cover
   - Gas cap
   - Cover box and check for proper sealing

Fuel Pump

REMOVAL & INSTALLATION

All Models

1. Before servicing the vehicle, refer to the Precautions Section.
2. Disconnect the negative battery cable and properly relieve the fuel system pressure.
3. Remove or disconnect the following:
   - Intake, outlet and bypass lines. Plug the lines.
   - Electrical leads
   - Fuel pump and vibration pads

To install:

➡Some models utilize 2 fuel pumps connected in series.

4. Install or connect the following:
   - Fuel pump and vibration pads
   - Electrical leads
   - Intake, outlet and bypass lines. Torque the cap nuts and banjo bolts to 18–22 ft. lbs. (25–30 Nm).

Fuel Injector

REMOVAL & INSTALLATION

All Models

1. Before servicing the vehicle, refer to the Precautions Section.
2. Properly relieve the fuel system pressure.
3. Remove or disconnect the following:
   - Negative battery cable
   - Brake servo vacuum hose
   - Pressure regulator vacuum hose
   - Positive Crankcase Ventilation (PCV) hose
   - Fuel feed pipe
   - Fuel system electrical connections
   - Left side ignition coil covers if necessary
   - Fuel rail with the injectors attached
   - Injector-to-fuel rail locking clip
   - Injectors

To install:

4. Install or connect the following:
   - Injectors with new O-rings on the fuel rail
   - Injector-to-fuel rail locking clip
   - Fuel rail/injector assembly
   - Left side ignition coil covers
   - Fuel system electrical connections
   - Fuel feed pipe
   - PCV hose
   - Pressure regulator vacuum hose
   - Brake servo vacuum hose
   - Negative battery cable

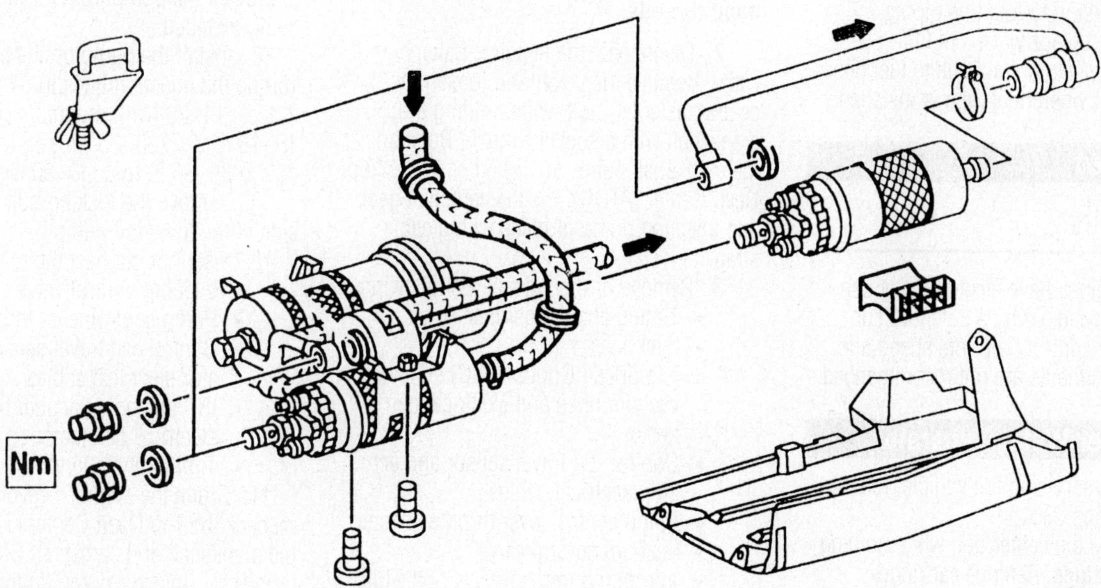

Typical fuel pumps and related components

DIESEL FUEL SYSTEM

Fuel System Service Precautions

Safety is the most important factor when performing not only fuel system maintenance but any type of maintenance. Failure to conduct maintenance and repairs in a safe manner may result in serious personal injury or death. Maintenance and testing of the vehicle's fuel system components can be accomplished safely and effectively by adhering to the following rules and guidelines.

- To avoid the possibility of fire and personal injury, always disconnect the negative battery cable unless the repair or test procedure requires that battery voltage be applied.
- Please be advised that fuel under pressure may penetrate the skin or any part of the body that it contacts.
- Always place a shop towel or cloth around the fitting or connection prior to loosening to absorb any excess fuel due to spillage. Ensure that all fuel spillage (should it occur) is quickly removed from engine surfaces. Ensure that all fuel soaked cloths or towels are deposited into a suitable waste container.
- Always keep a dry chemical (Class B) fire extinguisher near the work area.
- Do not allow fuel spray or fuel vapors to come into contact with a spark or open flame.
- Always use a back-up wrench when loosening and tightening fuel line connection fittings. This will prevent unnecessary stress and torsion to fuel line piping.
- Always replace worn fuel fitting O-rings with new. Do not substitute fuel hose or equivalent, where fuel pipe is installed.

Idle Speed

ADJUSTMENTS

These engines have electronically controlled idle speed, using a solenoid connected to the control unit. Idle speed and mixture adjustments are not recommended.

Fuel Filter/Water Separator

1. Before servicing the vehicle, refer to the Precautions Section.
2. Loosen the center bolt while holding the filter cartridge. Remove cartridge.
3. Lubricate the gasket on the new filter cartridge.

Fuel filter assembly—Diesel engines

4. Install the new cartridge, washer, O-ring and bolt.

➡These Diesel engines utilize a self-bleeding fuel pump, therefore the priming pump has been eliminated. There is no need to bleed the system.

Diesel Injection Pump

REMOVAL & INSTALLATION

1. Before servicing the vehicle, refer to the Precautions Section.

➡The central fastening bolt uses left-hand threads.

2. Disconnect the negative battery cable. Remove the cover and loosen the central fastening bolt while holding the crankshaft with a socket wrench. Position the crankshaft pulley at 15 degrees After Top Dead Center (ATDC). Fix the camshaft gear and injection pump in place with a cable strap.
3. Remove or disconnect the following:
   - Timing chain tensioner
   - Fuel lines
   - Idle Speed Control (ISC) solenoid vacuum lines and electrical connector
   - Control rod travel sensor and actuator electrical cables
   - Pump control lever throttle linkage
   - Vacuum control valve
   - Injection pump sprocket bolt while holding the crankshaft in place. The bolt is a left-hand thread.

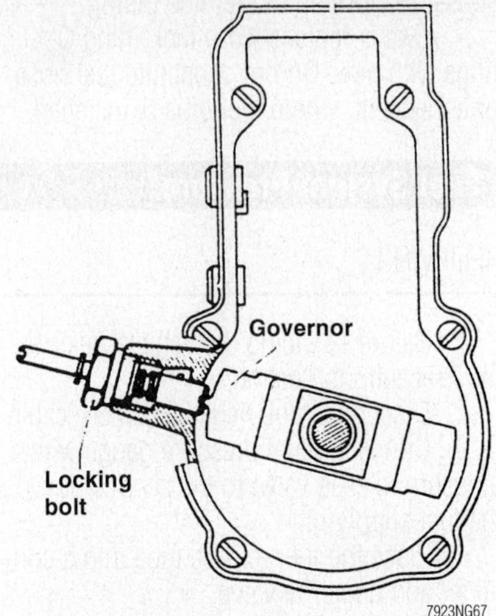

Install the locking bolt tool (023) to prevent the pump from turning during installation—Diesel engines

   - Sprocket bolt
   - Pump mounting bolts
   - Injection pump while holding the timing device in place

To install:

4. Be sure the crankshaft is still at 15 degrees ATDC.
5. Remove the timing plug from the pump.
6. Using special tool 601 589 00 08 00 (serrated wrench) turn the pump until the lug of the governor is visible through the hole.
7. Install the special tool 601 589 05 21 00 (locking bolt) in the timing plug hole to prevent the pump from turning while it is being installed.
8. Install the pump on the engine. Torque the injection lines to 84 inch lbs.–15 ft. lbs. (10–20 Nm), mounting bolts to 15–18 ft. lbs. (20–25 Nm) and the central bolt to 30–36 ft. lbs. (40–50 Nm).
9. Remove the locking bolt from the side of the injection pump.
10. Install or connect the following:
    - Vacuum control valve
    - Pump control lever throttle linkage
    - Control rod travel sensor and actuator electrical cables
    - ISC solenoid vacuum lines and electrical connector
    - Timing chain tensioner
11. Crank the engine 1 revolution and recheck the Top Dead Center (TDC) mark of the crankshaft and camshaft. Check the start of injector (injector pump timing) delivery with a digital tester.

Fuel Injector

REMOVAL & INSTALLATION

1. Before servicing the vehicle, refer to the Precautions Section.
2. Disconnect the negative battery cable and properly relieve the fuel system pressure.
3. Remove or disconnect the following:
   - Resonance intake pipe
   - Fuel lines
   - Injection pipes at nozzles
   - Injection nozzles

To install:

4. Install or connect the following:
   - Injectors with new O-rings. Torque to 30 ft. lbs. (40 Nm).
   - Injection pipes at nozzles. Torque to 17 ft. lbs. (23 Nm).
   - Fuel lines
   - Resonance intake pipe
   - Negative battery cable

Diesel Injection Timing

ADJUSTMENT

1. Before servicing the vehicle, refer to the Precautions Section.
2. Remove the screw plug from the side of the injection pump governor housing.
3. Install a position sensor into the hole and connect the tester to the vehicle battery and position sensor.
4. Turn the crankshaft by hand until lamps **A** and **B** on the tester light simultaneously. Take the reading on the crankshaft pulley.
5. Loosen the injection pump and turn the adjusting screw until the timing pointer is at 15 degrees After Top Dead Center (ATDC) and both lamps are ON.
6. Install the oil and plug. Torque the plug to 22 ft. lbs. (30 Nm).

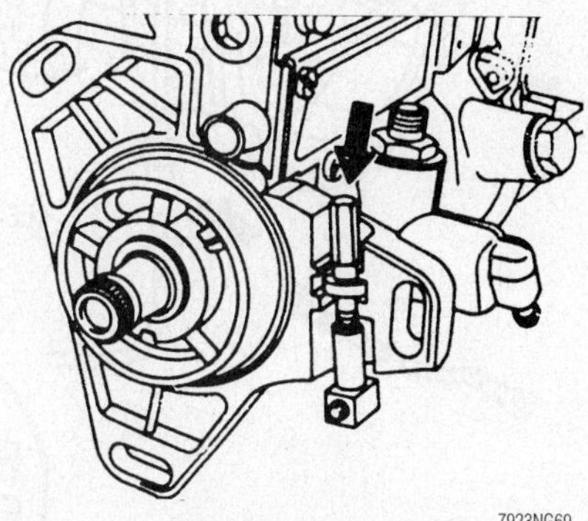

Turn the adjusting screw (arrow) to fine-tune the injection timing—Diesel engine

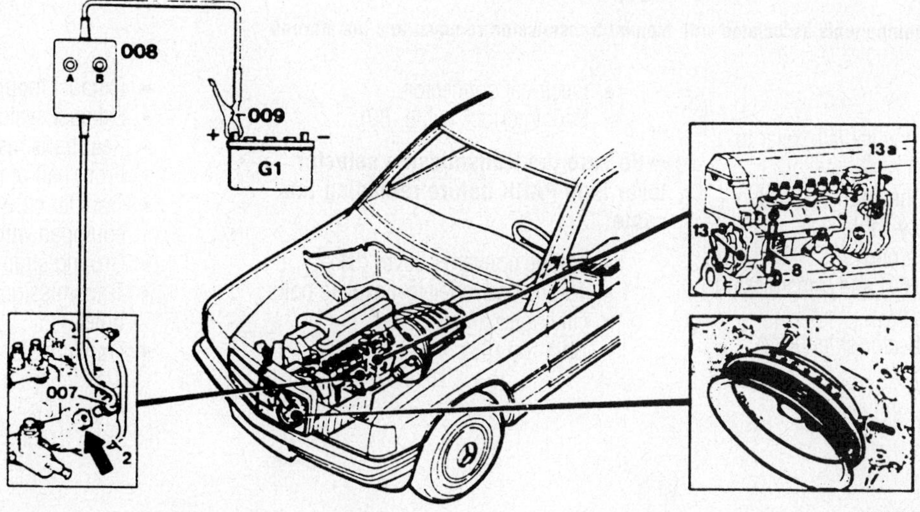

Remove the screw plug and install the injection timing position sensor

DRIVETRAIN

Transmission Assembly

REMOVAL & INSTALLATION

Manual

1. Before servicing the vehicle, refer to the Precautions Section.
2. Disconnect the negative battery cable.
3. Support the transmission with a jack and remove the transmission support assembly (19).
4. Remove or disconnect the following:
   - Exhaust support bracket (10) and U-bolt (11)
   - Heat shield (5)
   - Front driveshaft (71)
   - Exhaust system (60) at the rear mount. Use a piece of wire to suspend it.
   - Speedometer cable (9) or Vehicle Speed Sensor (VSS) (85a)
   - Clutch hydraulic line (arrow)
   - Shift linkage clips (32) and shift linkage
   - Transmission ground strap (27)
   - Transmission-to-engine mounting bolts
   - Transmission

To install:

5. Apply a light coating of grease to the splines of the input shaft (33).
6. Lift the transmission to the engine.
7. Install or connect the following:
   - Transmission mounting bolts
   - Ground strap
   - Shift linkage
   - Clutch hydraulic line
   - Speedometer cable or VSS
   - Exhaust system with new gaskets
   - Driveshaft
   - Heat shield
   - Exhaust support bracket and U-bolt
   - Transmission support
8. Check the transmission oil level and add if necessary.
9. Lower the vehicle to the floor and connect the negative battery cable. Bleed the hydraulic clutch system.

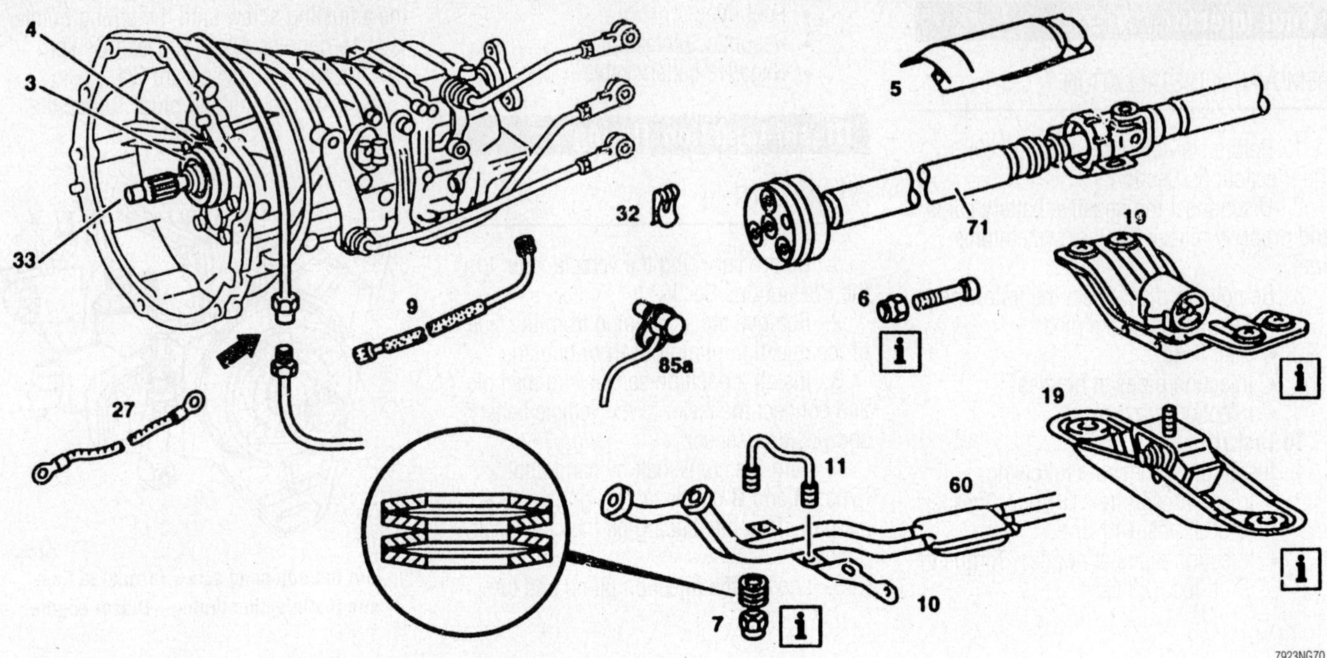

Exploded view of the components associated with manual transmission removal and installation

Automatic

1. Before servicing the vehicle, refer to the Precautions Section.
2. Remove or disconnect the following:
   - Negative battery cable
   - Radiator fan shroud
   - Dipstick tube (61) and transmission fluid
   - Electrical connector shield (62)
   - Electrical connector
   - Park interlock cable (80)

➡Be sure the transmission selector lever is in PARK before removing the cable.

   - Torque converter cover (81)
   - Torque converter-to-flexplate bolts
   - Oil cooler lines
   - Shift rod (63)
   - Exhaust mounting bracket (64)
   - Exhaust system (94)
   - Rear transmission mount (65)
   - Front half of rear driveshaft (66)
   - Transfer case driveshaft, if equipped with 4-MATIC
   - Ground strap (46)
   - Transmission-to-engine mounting bolts
   - Transmission

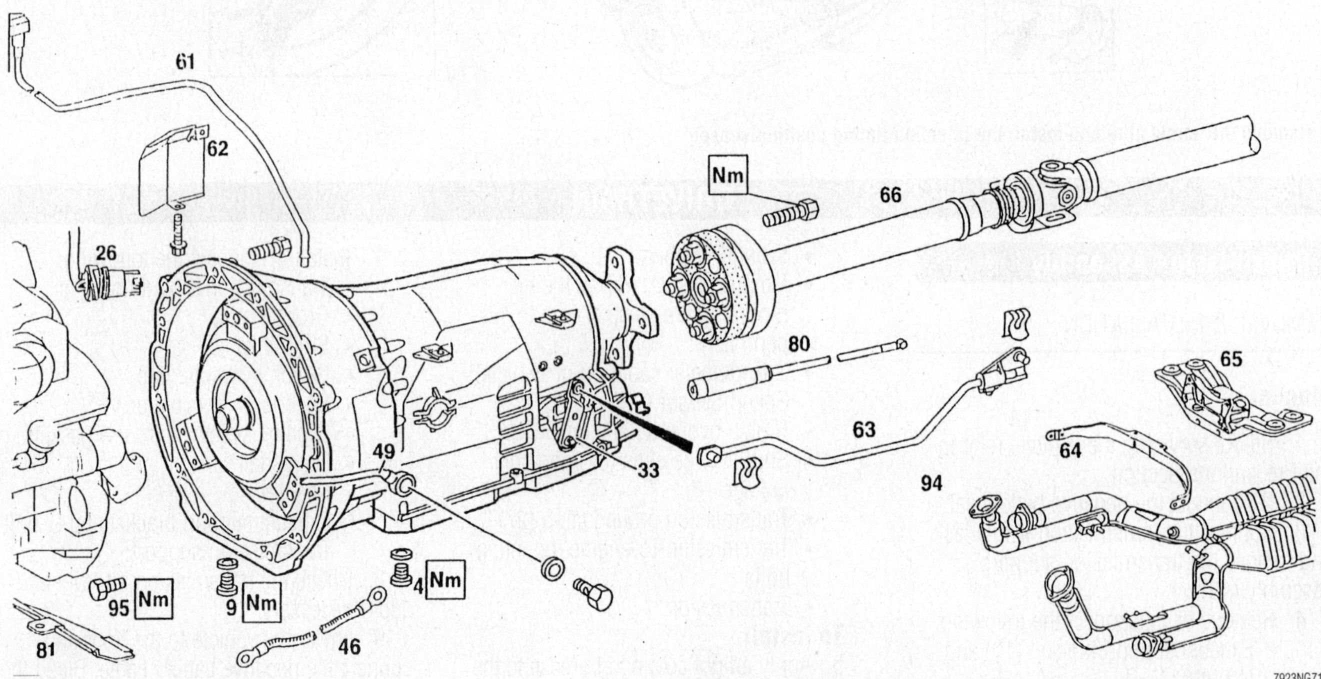

Exploded view of the components associated with automatic transmission removal and installation

To install:

3. Position the transmission on the engine and install the mounting bolts.
4. Install or connect the following:
    - Ground strap
    - Front half of rear driveshaft
    - Transfer case driveshaft, if equipped with 4-MATIC
    - Rear transmission mount
    - Exhaust system
    - Exhaust mounting bracket
    - Shift rod
    - Oil cooler lines
    - Torque converter-to-flexplate bolts to 31 ft. lbs. (42 Nm)
    - Torque converter cover
    - Park interlock cable
    - Electrical connector and shield
    - Transmission dipstick tube
    - Fan shroud
    - Negative battery cable
    - Transmission fluid

Clutch

ADJUSTMENTS

All clutch release mechanisms are hydraulic, no periodic adjustment is necessary.

REMOVAL & INSTALLATION

1. Before servicing the vehicle, refer to the Precautions Section.
2. Remove or disconnect the following:
    - Transmission assembly
    - Pressure plate (6) mounting bolts (9)
3. Inspect the pilot bearing, release bearing (7), release fork (8) and flywheel. Replace as necessary.

To install:

4. Deglaze the flywheel with coarse abrasive cloth.
5. Install or connect the following:
    - Clutch disc using a line-up tool (012)
    - Pressure plate. Torque the mounting bolts to 18 ft. lbs. (25 Nm).
6. Lightly grease the friction surfaces (arrows) on the release fork. Install the transmission assembly lower the vehicle and check operation.

Hydraulic Clutch System

BLEEDING

1. Before servicing the vehicle, refer to the Precautions Section.
2. Fill the reservoir with brake fluid. Do not allow brake fluid to contact painted surfaces.
3. Have a helper push the clutch pedal down and hold it.
4. Attach a clear plastic hose to the bleeder valve. Place the other end of the hose in a container to catch the brake fluid.
5. Open the bleeder valve on the release cylinder to expel air, then close the bleeder valve. Do not release the clutch pedal until the bleeder valve is closed.
6. Repeat the procedure until clear fluid flows out of the bleeder valve.

Transfer Case Assembly

REMOVAL & INSTALLATION

1. Before servicing the vehicle, refer to the Precautions Section.
2. Drain the fluid and remove the exhaust brackets. Suspend the exhaust with a piece of wire.
3. Remove or disconnect the following:
    - Driveshaft shield
    - Case bracket
    - Case vibration damper
    - Front half of rear driveshaft
4. Support the transmission and remove

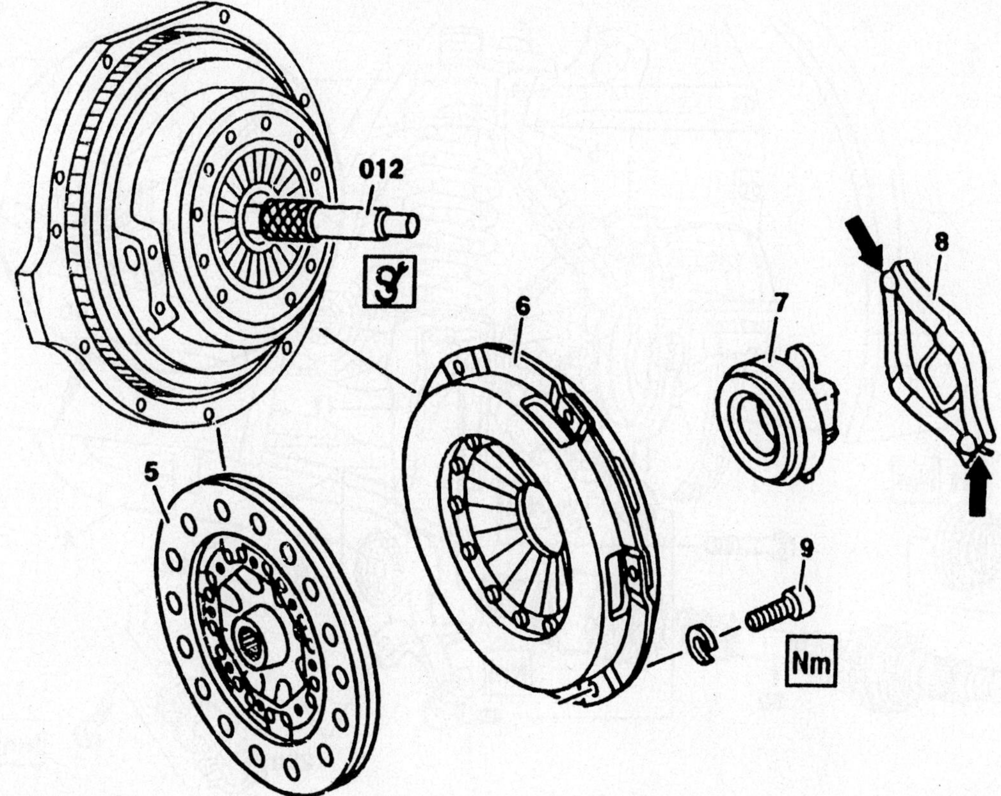

Exploded view of the clutch assembly

the engine support along with the rear engine mount. Lower the assembly slightly.

- Front driveshaft
- Transfer case bolts
- Transfer case and drain the residual oil

To install:

5. Install or connect the following:
   - Transfer case with new gasket. Torque the bolts to 22 ft. lbs. (30 Nm).
   - Rear engine support. Torque the bolts to 30 ft. lbs. (40 Nm).
   - Driveshafts. Torque the bolts to 18 ft. lbs. (25 Nm).
   - Case vibration damper
   - Driveshaft shield
   - Exhaust brackets

Halfshaft

REMOVAL & INSTALLATION

Front (4-Matic)

1. Before servicing the vehicle, refer to the Precautions Section.
2. Remove or disconnect the following:
   - Negative battery cable
   - Front wheel
   - 12-pointed axle nut (61)
   - Engine undercover
   - Brake hose bracket (arrow)
   - Stabilizer bar brackets (22b)
   - Headlight leveling linkage (80)
   - Lower ball joint (7) from the steering knuckle (5)
   - Shock absorber (11) from the lower control arm (4)
   - Lower control arm mounting bolts
   - Halfshaft

To install:

3. Install or connect the following:
   - Halfshaft
   - Lower ball joint
   - Shock absorber
   - Headlight leveling linkage
   - Stabilizer bar brackets
   - Brake hose bracket
   - Engine undercover
   - 12-pointed axle nut. Torque the nut to 162 ft. lbs. (220 Nm).
   - Front wheel
   - Negative battery cable

Rear

1. Before servicing the vehicle, refer to the Precautions Section.
2. Remove or disconnect the following:
   - Wheel
   - Center axle hold-down bolt (in hub)
   - Brake caliper. Suspend it with a wire.
3. Drain the differential oil and support the differential housing.
   - Differential housing cover
   - Axle lock ring
   - Axle shaft. If necessary, loosen the shock absorber.

To install:

4. Install or connect the following:
   - Axle shaft. Torque the nut to 148–175 ft. lbs. (200<\#208>240 Nm).
   - Axle lock ring
   - Differential housing cover
   - Brake caliper
   - Center axle hold down bolt
   - Wheel
5. Fill the axle with oil. New radial seal rings are used on all models. Lubricate the outside diameter of radial sealing rings with hypoid gear lubricant prior to installation.

➡Check end-play of the lock ring in the groove. If necessary, install a thicker lock ring or spacer to eliminate all end-play, while still allowing the lock ring to rotate. Do not allow the joints in the axle shaft to hang free or the joint bearing may be damaged.

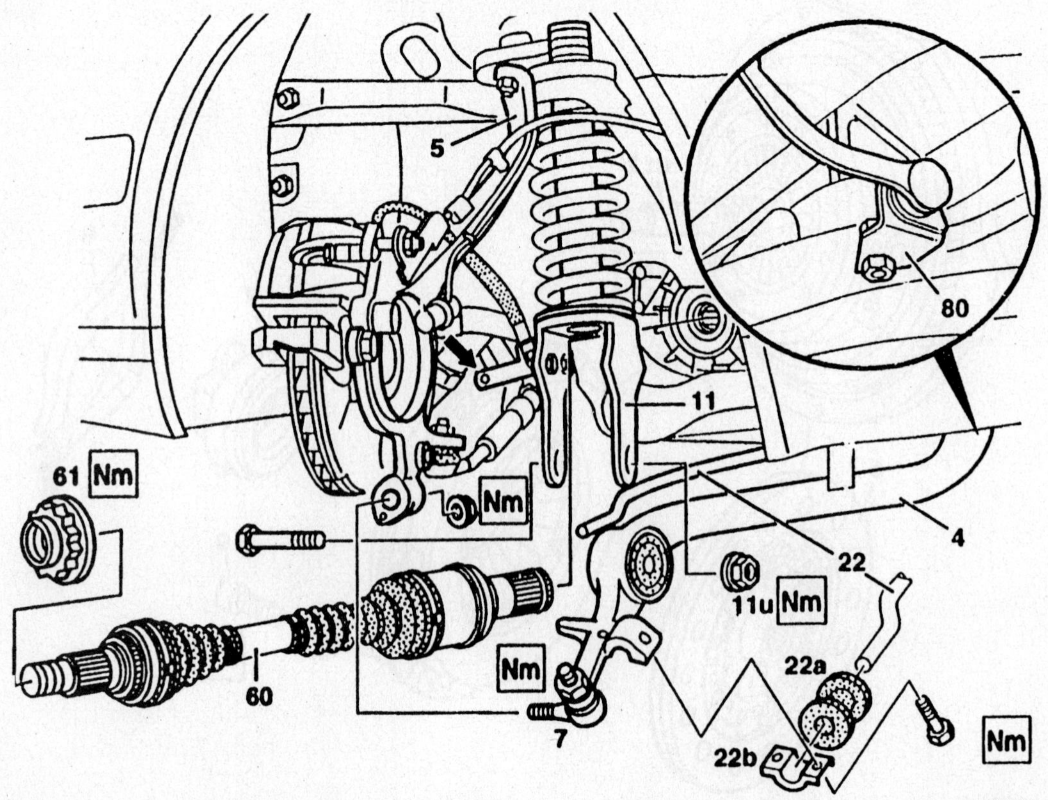

Exploded view of the front halfshaft mounted—4-Matic

CV-Joints

OVERHAUL

Front (4-Matic)

1. Before servicing the vehicle, refer to the Precautions Section.
2. Remove or disconnect the following:
   - Axle shaft
   - Rubber boot sealing clamp and CV-joint circlip
3. Using a soft metal drift or a piece of wood, tap the CV-joint off the shaft.
4. Use a rag to clean off the grease, inspect all components for wear and/or damage and replace as necessary.

To install:

5. Replace the rubber boot and install the CV-joint and circlip. Replace the grease with the appropriate CV-joint lubricant.
6. Install the axle assembly.

Rear

1. Before servicing the vehicle, refer to the Precautions Section.
2. Remove or disconnect the following:
   - Axle shaft
   - End cover and boot cap
   - Rubber boot sealing clamp
   - CV-joint locking ring
3. Remove the CV-joint from the axle using a press. Remove the rubber boot.
4. Use a rag to clean off the grease, inspect all components for wear and/or damage and replace as necessary.

To install:

5. Replace the rubber boot and install the CV-joint using a press. Replace the grease with the appropriate CV-joint lubricant.
6. Install or connect the following:
   - CV-joint locking ring
   - Rubber boot sealing clamp
   - End cover and boot cap
   - Axle shaft

Axle Shaft, Bearing and Seal

REMOVAL & INSTALLATION

Front (4-Matic)

1. Before servicing the vehicle, refer to the Precautions Section.
2. Remove or disconnect the following:
   - Steering knuckle
   - Axle shaft flange using an appropriate puller
   - Bearing circlip
   - Bearing using a appropriate removal tool

To install:

3. Install or connect the following:
   - Bearing using a appropriate installation tool
   - Bearing circlip
   - Axle shaft flange using an appropriate installation tool
   - Steering knuckle

Rear

1. Before servicing the vehicle, refer to the Precautions Section.
2. Remove or disconnect the following:
   - Axle shaft
   - Disc brake rotor
   - Axle shaft
   - Wheel bearing locking ring
   - Wheel bearing using an appropriate puller

To install:

3. Install or connect the following:
   - Wheel bearing using an appropriate installation tool
   - Wheel bearing locking ring
   - Axle shaft flange using an appropriate installation tool
   - Disc brake rotor
   - Axle shaft

Pinion Seal

REMOVAL & INSTALLATION

Front (4-Matic)

1. Before servicing the vehicle, refer to the Precautions Section.

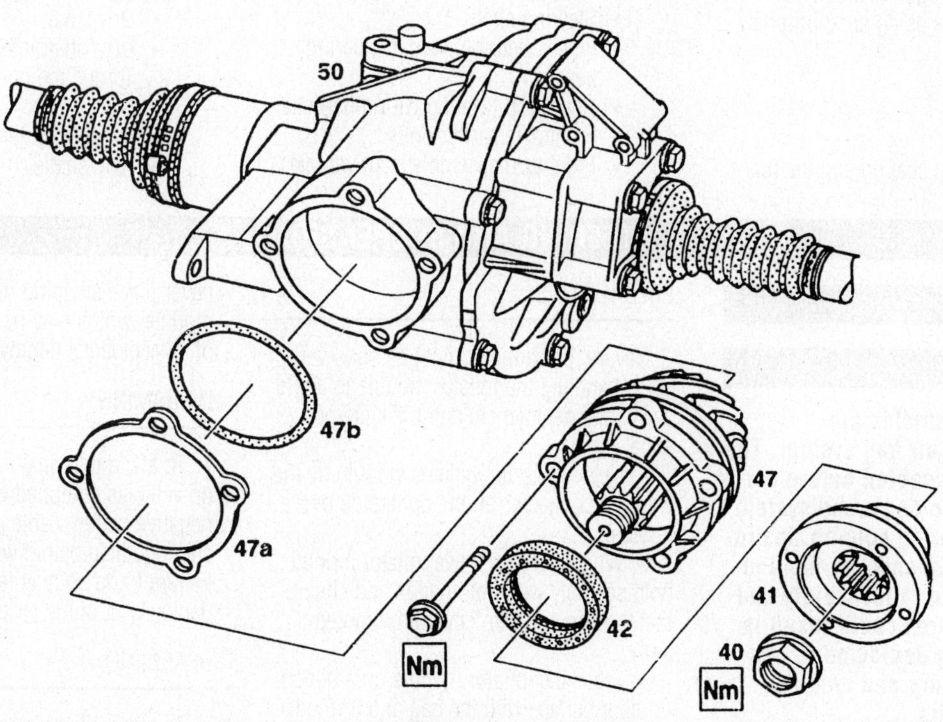

Exploded view of the components associated with the front drive pinion (4-MATIC)

2. Remove or disconnect the following:
   - Axle shaft
   - Drive pinion with bearing cover
3. Mark the bearing cover relative to the front axle drive housing.
4. Remove or disconnect the following:
   - Rubber O-ring
   - Pinion seal

To install:

5. Coat the pinion seal lip and the rubber O-ring with clean hypoid transmission oil.
6. Install or connect the following:
   - Pinion seal
   - O-ring
   - Bearing cover. Torque the bolts to 26 ft. lbs. (35 Nm).
   - Drive pinion and flange with new self-locking hexagon collar nut. Torque the nut to 350 ft. lbs. (475 Nm).
   - Axle shaft

Rear

1. Before servicing the vehicle, refer to the Precautions Section.
2. Remove or disconnect the following:
   - Exhaust system
   - Driveshaft
   - Fuel pump cover
   - Driveshaft intermediate bearing
   - Driveshaft at differential
   - Rear axle shafts at connecting flange
   - 12-point hexagon collared nut
   - Pinion flange using an appropriate puller
   - Rubber O-ring
   - Pinion seal

To install:

3. Coat the pinion seal lip and the rubber O-ring with clean hypoid transmission oil.
4. Install or connect the following:
   - Pinion seal
   - Rubber O-ring
   - Drive pinion and flange with new self-locking hexagon collared nut. Torque the nut to 133 ft. lbs. (180 Nm).
5. Install or connect the following:
   - Rear axle shafts
   - Driveshaft at differential
   - Driveshaft intermediate bearing. Torque the bolts to 22 ft. lbs. (30 Nm).
   - Fuel pump cover
   - Exhaust system

Axle Housing Assembly

REMOVAL & INSTALLATION

Front (4-Matic)

1. Before servicing the vehicle, refer to the Precautions Section.
2. Remove or disconnect the following:
   - Electrical connections
   - Axle shafts
   - Shock absorbers and front springs
   - Steering knuckles
3. Support the engine and remove the front mount. Drain the power steering fluid.
   - Feed line from reservoir, if equipped with level control
   - Steering shaft
   - Power steering lines at steering gear
   - Headlamp leveling wiring harness
   - Front axle carrier bolts
   - Front axle assembly from vehicle

To install:

4. Install or connect the following:
   - Axle assembly. Torque the carrier bolts to 96 ft. lbs. (130 Nm).
   - Headlamp leveling wiring harness
   - Power steering lines at steering gear
   - Steering shaft
   - Feed line from reservoir, if equipped with level control
   - Engine mount. Torque the bolt to 26 ft. lbs. (35 Nm).
   - Axle shafts
   - Steering knuckles
   - Shock absorbers and front springs
   - Electrical connections

Rear

1. Before servicing the vehicle, refer to the Precautions Section.
2. Drain the differential oil.
3. Remove or disconnect the following:
   - Axle shafts
   - Exhaust system
   - Driveshaft intermediate bearing
   - Driveshaft
4. Support the rear axle housing assembly
   - Axle housing suspension mountings
   - Axle housing assembly

To install:

5. Install or connect the following:
   - Axle housing assembly
   - Axle housing suspension mountings
   - Driveshaft
   - Driveshaft intermediate bearing. Torque the bolts to 22 ft. lbs. (30 Nm).
   - Exhaust system
   - Axle shafts

STEERING AND SUSPENSION

Air Bag

CAUTION

Mercedes-Benz vehicles are equipped with an air bag system. The system must be disabled before performing service on or around system components, steering column, instrument panel components, wiring and sensors. Failure to follow safety and disabling procedures could result in accidental air bag deployment, possible personal injury and unnecessary system repairs.

PRECAUTIONS

Several precautions must be observed when handling the inflator module to avoid accidental deployment and possible personal injury.

- Never carry the inflator module by the wires or connector on the underside of the module.
- When carrying a live inflator module, hold securely with both hands and ensure that the bag and trim cover are pointed away.
- Place the inflator module on a bench or other surface with the bag and trim cover facing up.
- With the inflator module on the bench, never place anything on or close to the module, which may be thrown in the event of an accidental deployment.

DISARMING

To avoid personal injury when working on vehicles equipped with an air bag, the negative battery cable must be disconnected and insulated before working on the system. Failure to do so may result in accidental deployment of the air bag.

REARMING

To rearm the air bag system, reattach the battery cable(s).

Power Steering Pump

REMOVAL & INSTALLATION

All Models

1. Before servicing the vehicle, refer to the Precautions Section.
2. Disconnect the negative battery cable.
3. Drain the power steering fluid.
4. Remove or disconnect the following:
   - Fan clutch, fan and shroud
   - Accessory drive belt
   - Power steering lines at pump
   - Oil cooler lines, if equipped
   - Power steering pump
   - Reservoir

To install:

5. Install or connect the following:
   - Reservoir
   - Power steering pump. Torque the mounting bolts to 18 ft. lbs. (25 Nm).
   - Power steering lines
   - Oil cooler lines, if equipped
   - Accessory drive belt
   - Fan clutch, fan and shroud
6. Refill the reservoir with fresh fluid. Start the engine and turn the steering wheel from lock-to-lock several times. The system will bleed Automatically. Check for leaks.

Rack and Pinion Steering Gear

REMOVAL & INSTALLATION

1. Before servicing the vehicle, refer to the Precautions Section.
2. Disconnect the negative battery cable.
3. Center the steering wheel and turn the key to the lock position. Remove the key from the ignition switch.
4. Drain the power steering reservoir.
5. Remove or disconnect the following:
   - Fluid level feed line, if equipped with level control
   - Tie rod ends
   - Pressure and return lines
   - Steering gear return line
   - Steering coupling heat shield
   - Steering coupling
   - Control valve connector, if equipped with speed sensitive steering
   - Steering gear retainer bolts (23g)
   - Steering gear

To install:

6. Install or connect the following:
   - Steering gear with new locking nuts. Torque to 36 ft. lbs. (50 Nm).
   - Control valve connector
   - Control valve coupling. Torque the bolt to 15 ft. lbs. (20 Nm).
   - Steering gear return line
   - Pressure and return lines with new O-rings. Torque the fittings to 30 ft. lbs. (40 Nm).
   - Heat shield
   - Tie rod ends with new nuts. Torque the nuts to 44 ft. lbs. (60 Nm).
   - Fluid level feed line
   - Negative battery cable
7. Refill the reservoir with fresh fluid. Start the engine and turn the steering wheel from lock-to-lock several times. The system will bleed Automatically. Check for leaks.
8. Check the front wheel alignment and adjust if necessary.

Worm and Sector Steering Gear

REMOVAL & INSTALLATION

1. Before servicing the vehicle, refer to the Precautions Section.
2. Disconnect the negative battery cable.
3. Center the steering wheel and turn the key to the lock position. Remove the key from the ignition switch.
4. Drain the power steering fluid reservoir
5. Remove or disconnect the following:
   - Trim panel if necessary
   - Drag link and tie rod

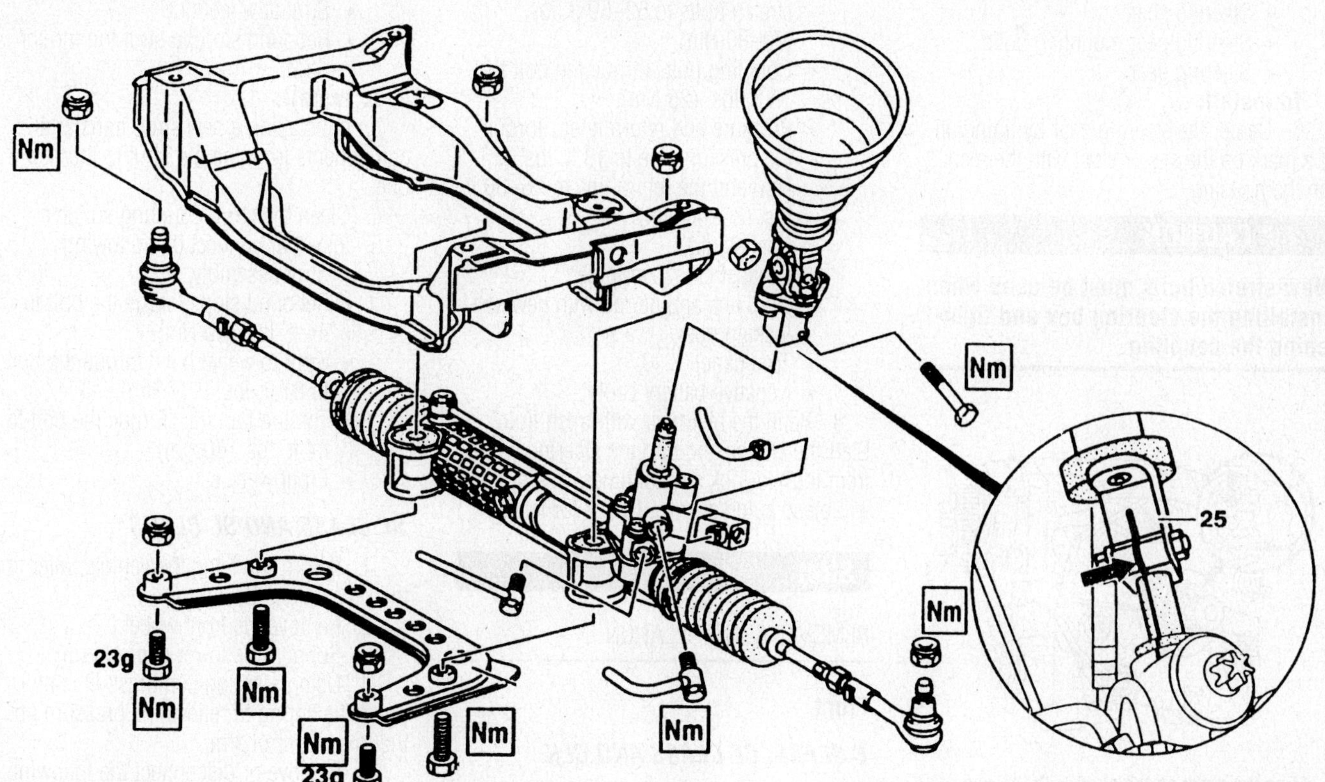

Exploded view of the rack and pinion steering gear mounting—E Class (210) shown, other models similar

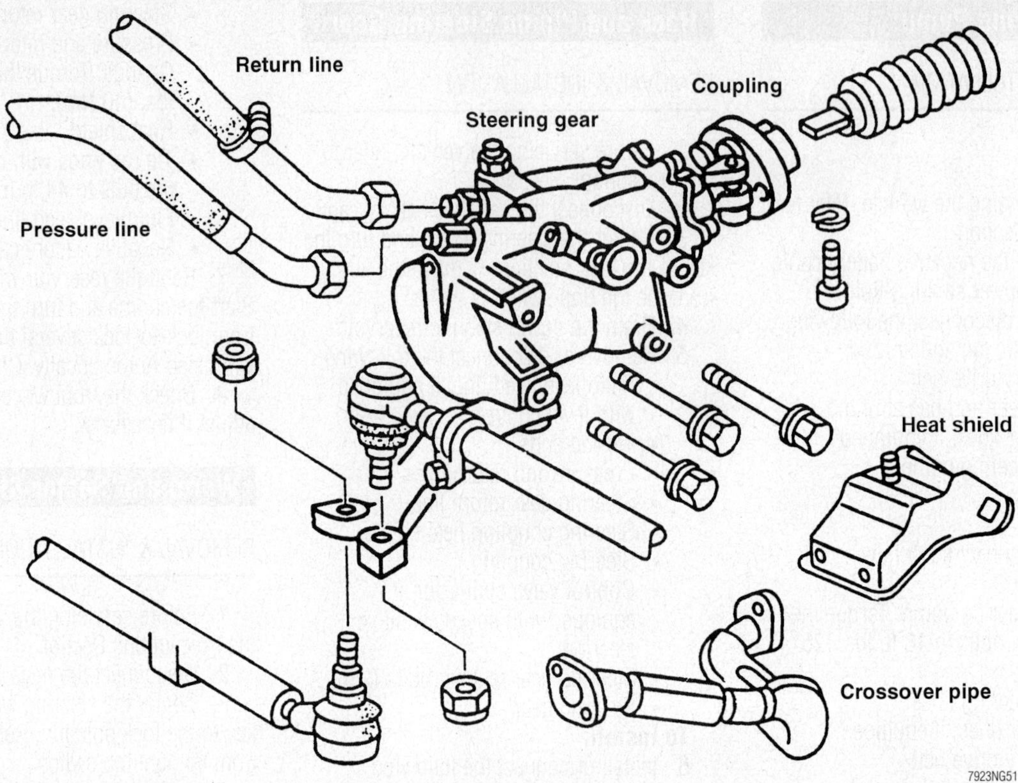

Exploded view of the worm and sector steering gear mounting

- Exhaust crossover pipe
- Steering gear heat shield
- Pressure and return lines
- Steering shaft
- Steering gear mounting bolts
- Steering gear

To install:

6. Center the steering gear by lining up the mark on the sector shaft with the mark on the housing.

CAUTION

New stretch bolts must be used when installing the steering box and tightening the coupling.

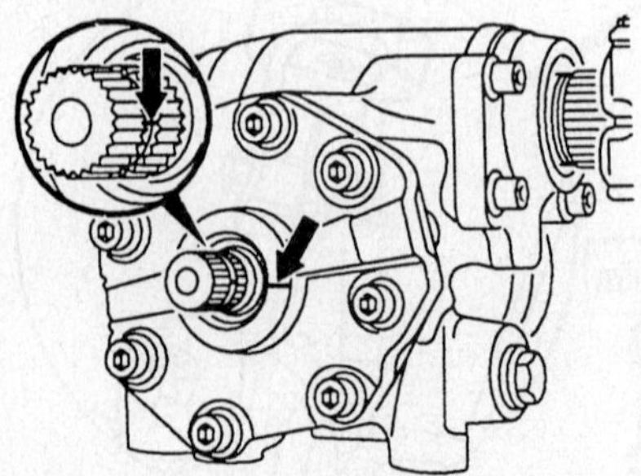

Align the mark on the shaft with the one on the housing to center the steering gear—worm and sector steering gear

7. Install or connect the following:
   - Steering gear
   - Steering shaft. Torque the new stretch bolts to 52–59 ft. lbs. (70–80 Nm).
   - Coupling bolt. Torque the bolt to 18 ft. lbs. (25 Nm).
   - Pressure and return lines. Torque the pressure line to 18 ft. lbs. (25 Nm) and the return line to 26–30 ft. lbs. (35–40 Nm).
   - Heat shield
   - Exhaust crossover pipe
   - Drag link and tie rod with new self-locking nuts
   - Trim panel
   - Negative battery cable
8. Refill the reservoir with fresh fluid. Start the engine and turn the steering wheel from lock-to-lock several times. The system will bleed automatically. Check for leaks.

Strut

REMOVAL & INSTALLATION

Front

C CLASS, CL CLASS AND CLK

1. Before servicing the vehicle, refer to the Precautions Section.
2. Remove or disconnect the following:
   - Front wheel
   - Torsion bar link rod at strut
   - Strut at wheel hub
   - Rebound stop on strut top mount
   - Strut

To install:

3. If installing a new strut, transfer all components from the old strut to the new one.
4. Clean the strut mounting surface.
5. Install or connect the following:
   - Strut assembly.
   - Rebound stop. Torque the bolt to 44 ft. lbs. (60 Nm).
   - Strut to wheel hub. Torque the bolt to 81 ft. lbs. (110 Nm).
   - Torsion bar link. Torque the bolt to 44 ft. lbs. (60 Nm).
   - Front wheels

SL CLASS AND SL CLASS

1. Before servicing the vehicle, refer to the Precautions Section.
2. Remove the front wheel.
3. Support the lower control arm.
4. Using a spring compressor, compress the spring to relieve the pressure on the lower control arm.
5. Remove or disconnect the following:
   - Upper mounting nut while holding the piston rod with a hex wrench

- ABS sensor wire
- Lower strut mounting bolts
- Strut assembly

⁂ WARNING

Do not allow the knuckle assembly to hang by the brake hose, ABS wire or brake pad sensor wire.

6. Remove the upper nut and remove the strut from the vehicle.

To install:

7. If installing a new strut, transfer all components from the old strut to the new one.
8. Clean the strut mounting surface of the knuckle.
9. Install or connect the following:
   - Strut assembly. Torque the lower bolts only until the heads make contact.
   - Upper mounting bolt with washer and new self-locking nut. Torque until the surface of the knuckle makes contact with the strut.
10. Torque the lower mounting bolts to 81 ft. lbs. (110 Nm), then tighten the upper bolt to 147 ft. lbs. (200 Nm).
11. Release the tension from the spring compressor. Be sure the upper and lower ends of the spring are properly seated.

Rear

S CLASS

1. Before servicing the vehicle, refer to the Precautions Section.
2. Relieve the pressure in the strut.
3. Unscrew the pressure line connection.
4. Remove the track control arm shield.
5. Remove or disconnect the following:
   - Upper strut mounting bolts

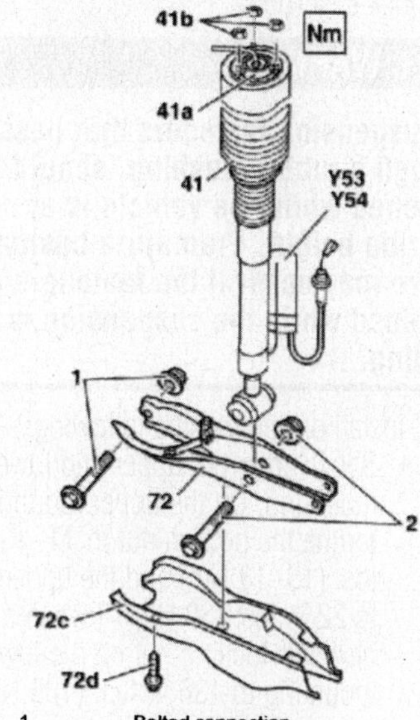

1	Bolted connection
2	Bolted connection
41	Rear suspension strut
41a	Pressure line connection
41b	Nuts (3 ea.)
72	Spring link
72c	Spring link cover
72d	Bolts (4 ea.)
Y53	Left rear axle damping valve unit
Y53x1	Damping valve unit connector
Y54	Right rear axle damping valve unit
Y54x1	Damping valve unit connector

42348-BENZ-G12

Rear strut assembly—S Class

   - Lower strut mounting bolt
   - Upper thrust arm bolt
   - Strut assembly

To install:

6. Install or connect the following:
   - Strut assembly. Torque the lower bolts to 81 ft. lbs. (110 Nm), torque the upper nuts 15 ft. lbs. (20 Nm).
   - Upper thrust arm bolt. Torque to 37 ft. lbs. (50 Nm) plus 90 °.
   - Track arm control shield
7. Pressurize the strut.

Shock Absorber

REMOVAL & INSTALLATION

Front

ALL MODELS—EXCEPT E CLASS (210) 4-MATIC

1. Before servicing the vehicle, refer to the Precautions Section.
2. Remove the front wheel.
3. Support the lower control arm with a jack. Raise the jack slightly to release the tension on the shock absorber.
4. Remove or disconnect the following:
   - Upper mounting
   - Lower mounting bolt
   - Shock absorber

To install:

5. Install or connect the following:
   - Shock absorber
   - Lower mounting bolt. Torque the bolt to 74 ft. lbs. (100 Nm).

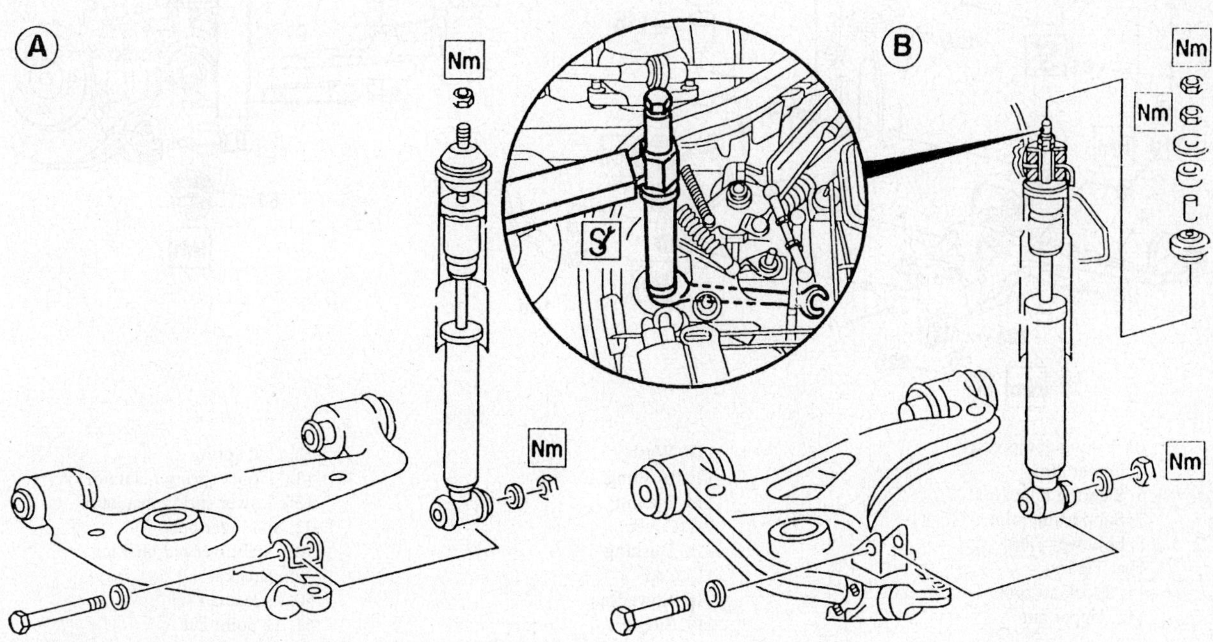

Front shock absorber mounting—S Class (A), C and E Class (B)

- Upper mounting. Torque the lower nut to 11–13 ft. lbs. (15–18 Nm) and the upper nut to 22 ft. lbs. (30 Nm).
- Front wheel

E CLASS (210) 4-MATIC

1. Before servicing the vehicle, refer to the Precautions Section.
2. Remove or disconnect the following:
   - Front wheel
   - Axle shaft 12-point nut
   - Brake hose bracket
   - Speed sensor and brake lining sensor wires
   - Tie rod end
   - Supporting joint
3. Pull the steering knuckle outward and move the front axle out of the axle shaft flange.
4. Separate the follower joint from the steering knuckle. Use wire to support the knuckle assembly.
   - Engine undercover
   - Axle shaft
   - Stabilizer bar bracket
   - Stabilizer bar end bushing
5. Compress the coil spring and remove the upper shock absorber mounting.
   - Lower shock absorber mounting
   - Shock absorber
   - Coil spring

To install:

6. Place the coil spring into position on the shock absorber.

⁂ WARNING

All suspension fasteners that pass through a rubber bushing, should be tightened while the vehicle is at normal ride height. Premature bushing failure may occur if the fasteners are tightened while the suspension is hanging.

7. Install or connect the following:
   - Shock absorber upper and lower mounting. On the upper mounting, torque the bottom nut to 11–13 ft. lbs. (15–18 Nm) and the upper nut to 22 ft. lbs. (30 Nm). Torque the new self-locking nut on the lower mounting to 136 ft. lbs. (185 Nm).
   - Stabilizer bar. Torque the bracket bolts to 15 ft. lbs. (20 Nm).
   - Axle shaft
   - Engine undercover

➡Ensure that the axle shaft is in the proper position before connecting the follower joint.

   - Follower joint with new self-locking nut. Torque nut to 33 ft. lbs. (45 Nm).
   - Supporting joint with new self-locking nut. Torque nut to 77 ft. lbs. (105 Nm).
   - Tie rod end with new self-locking nut. Torque nut to 44 ft. lbs. (60 Nm).
   - Speed and brake lining sensor wires
   - Brake hose bracket
   - New axle shaft 12-point nut. Torque to 162 ft. lbs. (220 Nm). Secure the nut with caulk (nut lock) so there is no gap between the groove and locking tab.
   - Front wheel

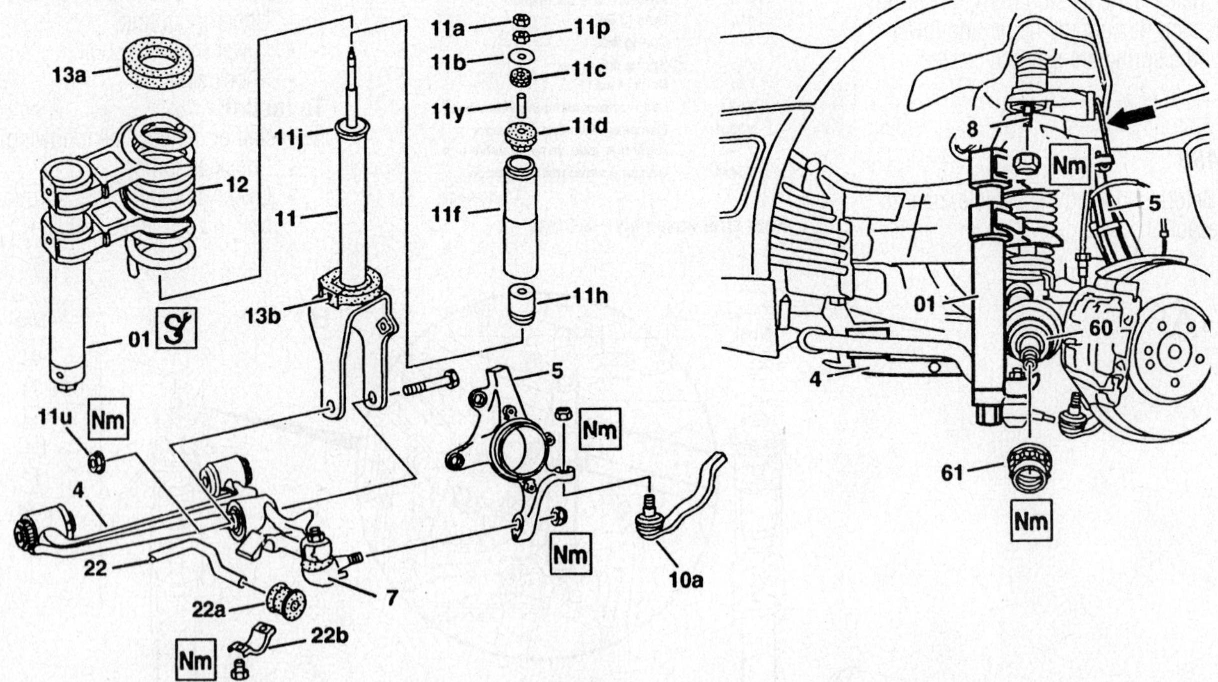

01. Spring compressor
4. Lower arm
5. Steering knuckle
7. Supporting joint
8. Follower joint
10a. Tie rod
11. Shock absorber
11a. Upper nut
11b. Washer
11c. Bushing
11d. Bushing
11f. Cover
11h. Bushing
11j. Cap
11p. Lower nut
11y. Sleeve
12. Coil spring
13a. Upper spring insulator
13b. Lower spring insulator
22. Stabilizer bar
22a. Stabilizer bar bushing
22b. Bracket
60. Axle shaft
61. 12 point nut

7923NG57

Exploded view of the front shock absorber/coil spring mounting—210 4-MATIC

Rear

ALL MODELS

1. Before servicing the vehicle, refer to the Precautions Section.
2. On models with Automatic level control (ALC), release the pressure from the system.
3. Support the lower control arm.

⁂ WARNING

Be sure that the piston rod does not turn when the removing the nut. The mount for the operating piston may loosen and release the gas and oil from the shock absorber.

4. Remove or disconnect the following:
   - Pressure lines, on ALC models
   - Upper shock absorber mounting nuts
   - Rollover bar switch, on convertible models
   - Lower control arm cover
   - Lower shock absorber mounting bolt
   - Shock absorber

To install:

5. Install or connect the following:
   - Shock absorber
   - Lower mounting bolt. Torque to 40 ft. lbs. (55 Nm).
   - Lower control arm cover
   - Rollover bar switch
   - Pressure lines, on ALC models. . Torque the banjo bolt to 18 ft. lbs. (25 Nm).
6. Lower the vehicle and install the spacer, upper bushing, washer and new self-locking nuts. Torque the bottom nut to

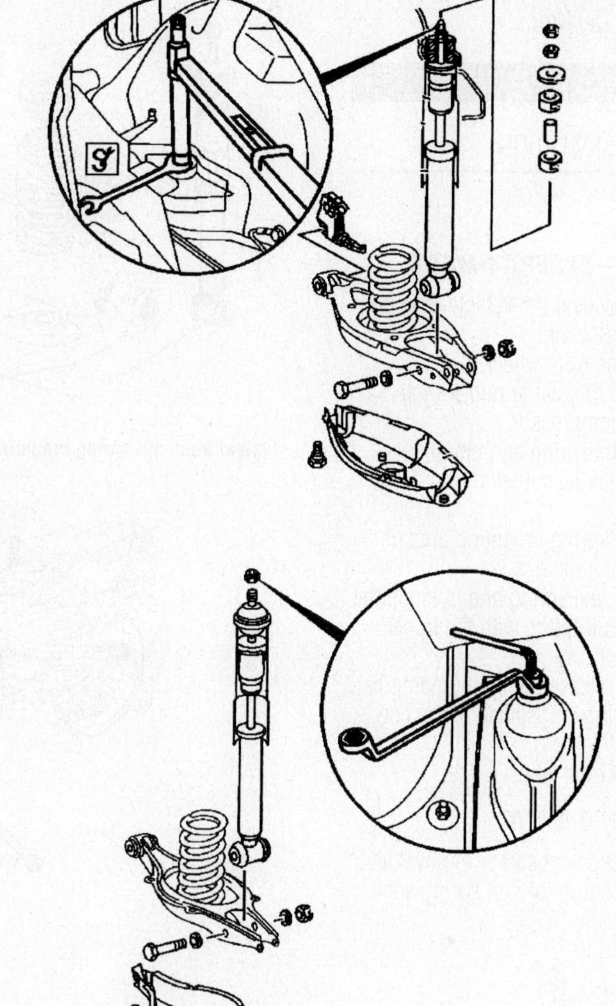

Exploded view of the rear shock absorber/coil spring mounting—Models without Automatic level Control

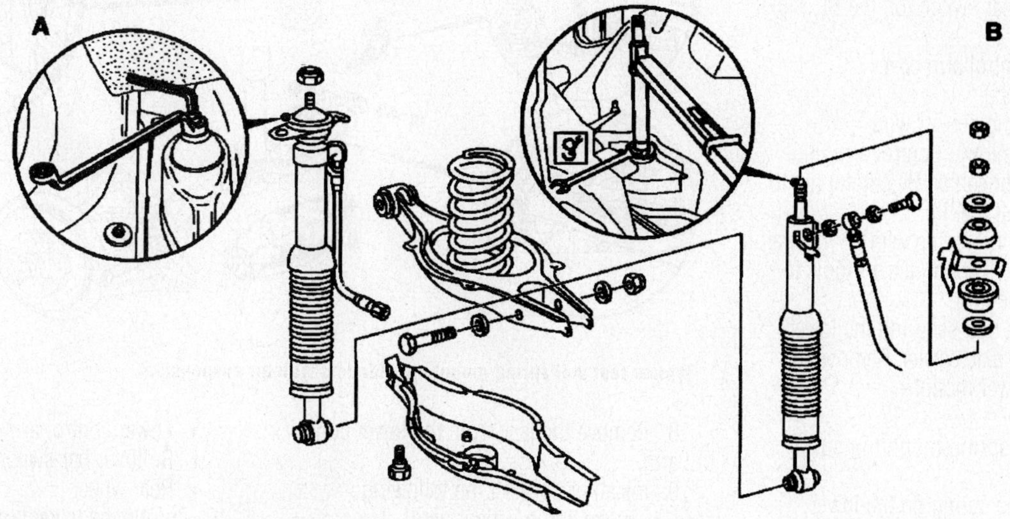

Exploded view of the rear shock absorber/coil spring mounting—Models with Automatic level Control—E Class (A), CLK (B)

11–13 ft. lbs. (15–18 Nm) and the upper nut to 22 ft. lbs. (30 Nm).

Coil Spring

REMOVAL & INSTALLATION

Front

ALL MODELS—EXCEPT 4-MATIC

1. Before servicing the vehicle, refer to the Precautions Section.
2. Remove the front wheel.
3. Compress the coil spring with an internal spring compressor.
4. Remove the spring and upper seat toward the front of the vehicle.

To install:

5. Clean the spring mounting area on the lower control arm.
6. Compress the spring and position the spring on the control arm with the upper seat.
7. Carefully release the spring compressor until the spring is installed in the correct position.
8. Install the front wheel.

E CLASS (210) 4-MATIC

Refer to the shock absorber removal and installation procedure for coil spring service.

Rear

ALL MODELS

1. Before servicing the vehicle, refer to the Precautions Section.
2. Remove or disconnect the following:
   - Rear wheel
   - Rollover bar switch, on the SL Class
   - Lower control arm cover
   - Cross brace
   - Brake lining sensor wire
3. Jack up the lower control arm until the axle shaft is horizontal. Be careful not to raise the vehicle off the lift.
4. Compress the spring with an internal spring compressor. Do not use air tools to compress the spring.
5. Loosen the bolt securing the lower control arm to the axle carrier. Remove the spring with the lower mount.

To install:

6. Clean the spring mounting surface.
7. Position the spring on the lower control arm and release the spring compressor.
8. Remove the jack from the lower control arm.
9. Install or connect the following:
   - Brake lining sensor wire
   - Cross brace
   - Lower control arm cover
   - Rollover bar switch
   - Rear wheel
10. Torque the lower control arm-to-axle carrier nut to 52 ft. lbs. (70 Nm).

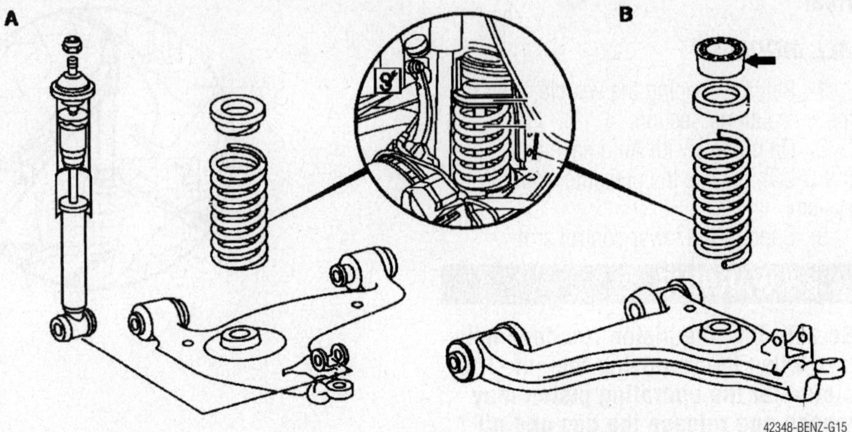

Typical front coil spring mountings

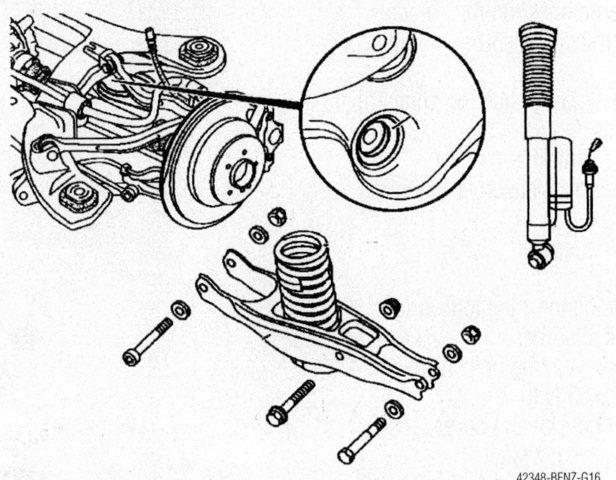

Typical rear coil spring mounting—Models without air suspension

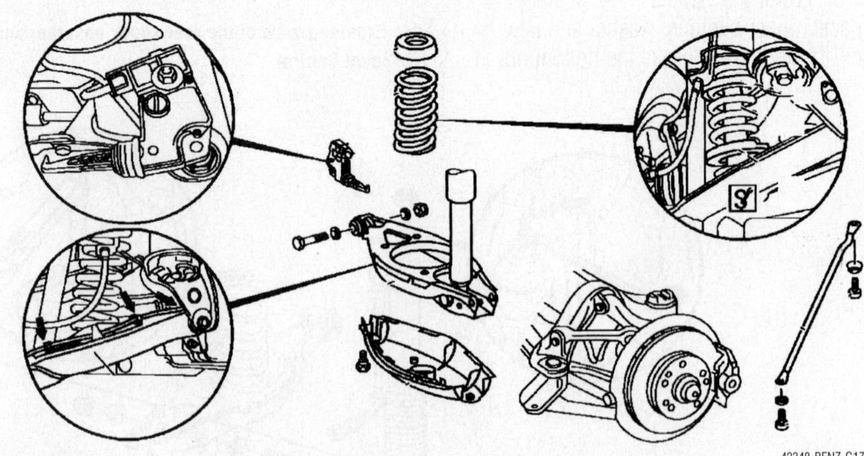

Typical rear coil spring mounting—Models with air suspension

Torsion Bar

REMOVAL & INSTALLATION

Front

ALL MODELS

1. Before servicing the vehicle, refer to the Precautions Section.
2. Remove or disconnect the following:
   - Torsion bar brackets
   - Spring swiveling lever and mounting plate, if equipped
   - Torsion bar

To install:

3. Install or connect the following:
   - Torsion bar and brackets with new bushings. Torque the bracket bolts as follows:
   - Control arm bracket self- locking nuts: 15 ft. lbs. (20 Nm)
   - Top spring swiveling lever bracket self-locking nuts: 30 ft. lbs. (40 Nm)
   - Bottom spring swiveling lever bracket self-locking nuts: 15 ft. lbs. (20 Nm)
   - Torsion bar-to-frame bracket self-locking bolts: 44 ft. lbs. (60 Nm)

Rear

ALL MODELS

1. Before servicing the vehicle, refer to the Precautions Section.
2. Properly relieve the fuel system pressure on SL Class vehicles.
3. Remove or disconnect the following:
   - Rear axle housing assembly
   - Torsion bar connecting rods
   - Torsion bar brackets
   - Fuel-tank-to-pump hose, on SL Class
   - Intermediate lever
   - Torsion bar

To install:

4. Install or connect the following:
   - Torsion bar and brackets with new bushings. Torque the bracket nuts to 15 ft. lbs. (20 Nm).
   - Intermediate lever. Torque the self-locking nuts to 84 inch lbs. (10 Nm).
   - Fuel-tank-to-pump hose
   - Torsion bar connecting rods. Torque the self-locking nuts to 30 ft. lbs. (40 Nm).
   - Rear axle housing assembly

Upper Ball Joint

REMOVAL & INSTALLATION

All Models

➡The upper ball joint is an integral part of the upper control arm. Control arm replacement is necessary if the ball joint becomes worn or damaged.

1. Before servicing the vehicle, refer to the Precautions Section.
2. Remove the front wheel and support the lower control arm with a jack.
3. Remove or disconnect the following:
   - Electrical connections, if equipped
   - Ball joint stud
   - Control arm mounting bolts
   - Bearing shell

To install:

4. Install or connect the following:
   - Control arm in the bearing shell. Torque the self-locking nut to 55 ft. lbs. (75 Nm) after the weight of the vehicle is on the suspension.
   - Control arm assembly using new micro-encapsulated bolts. Torque the bolts to 37 ft. lbs. (50 Nm).
   - Ball joint stud. Torque the nut to 88 ft. lbs. (120 Nm). Fill the gap in the clamp with a wax-like protectant.
   - Front wheel
5. Lower the vehicle to the floor and tighten the control arm through-bolt and nut to the specification.

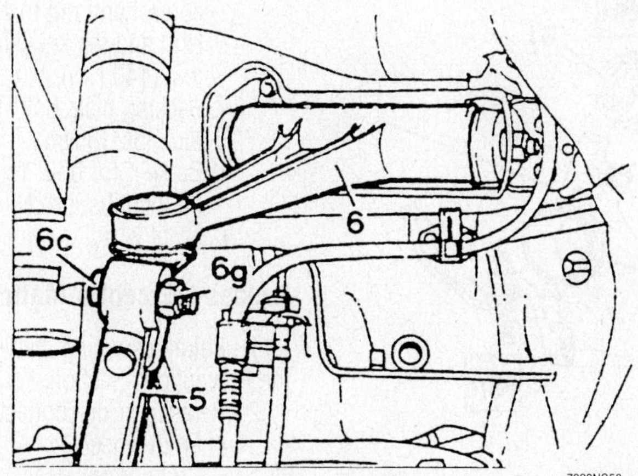

Steering knuckle (5), control arm (6), bolt (6c) and nut (6g)—C Class and S Class

Lower Ball Joint

REMOVAL & INSTALLATION

C Class, CLK, S Class and SLK 230 (170)

1. Before servicing the vehicle, refer to the Precautions Section.
2. Remove or disconnect the following:
   - Front wheel
   - Disc brake rotor securing bolt and backing plate bolt
   - Ball joint nut
   - Ball joint-to-steering knuckle nut
   - Ball joint

To install:

3. Install or connect the following:
   - Ball joint. Torque the control arm nut to 74 ft. lbs. (100 Nm) and the knuckle nut to 104 ft. lbs. (140 Nm).
   - Backing plate bolt. Torque to 72 inch lbs. (8 Nm).
   - Brake rotor bolt. Torque the bolt to 84 inch lbs. (10 Nm).
   - Front wheel

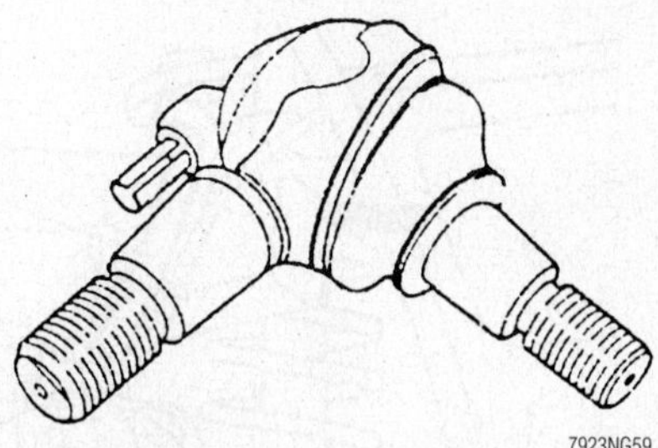

Common ball joint—C Class (202) and S Class (140)

E class—Except 4-Matic

➡The ball joint on vehicles with engines 104 and 119 cannot be pressed out of the control arm. They have been welded in place. In this case, replace the control arm assembly.

1. Before servicing the vehicle, refer to the Precautions Section.
2. Remove or disconnect the following:
   - Front wheel

- Coil spring
- Stabilizer bar and bracket
- Lower control arm-to-frame nuts and bolts
- Ball joint pinch bolt and nut
- Lower control arm
- Ball joint using a press

To install:

3. Install or connect the following:
   - Ball joint using a press
   - Control arm. Temporarily tighten the control arm-to-frame nuts.
   - Ball joint pinch bolt and nut. Torque the nut to 92 ft. lbs. (125 Nm).
   - Stabilizer bar bracket. Torque the bolts to 15 ft. lbs. (20 Nm).
   - Coil spring
   - Front wheel
4. Lower the vehicle to the floor. Torque the control arm-to-frame nuts to 88 ft. lbs. (120 Nm).

E Class 4-Matic

1. Before servicing the vehicle, refer to the Precautions Section.
2. Remove or disconnect the following:
   - Front wheel
   - Axle shaft 12-point nut (61)
   - Ball joint nut. Press the ball joint out of the steering knuckle (5).
3. Pull the steering knuckle outward and remove the axle shaft (60) from the flange.
4. Support the knuckle assembly by placing a block of wood (arrow) between the coil spring and the knuckle.
5. Position the axle shaft out of the way using a piece of wire.
6. Remove the nut and press the ball joint (7) out of the lower control arm (4).

To install:

7. Install or connect the following:
   - Ball joint. Torque a new self-locking nut to 77 ft. lbs. (105 Nm).
   - Axle shaft
   - Ball joint to the steering knuckle with a new self-locking nut. Torque the nut to 77 ft. lbs. (105 Nm).
   - 12-point axle nut. Torque to 162 ft. lbs. (220 Nm).
   - Front wheel

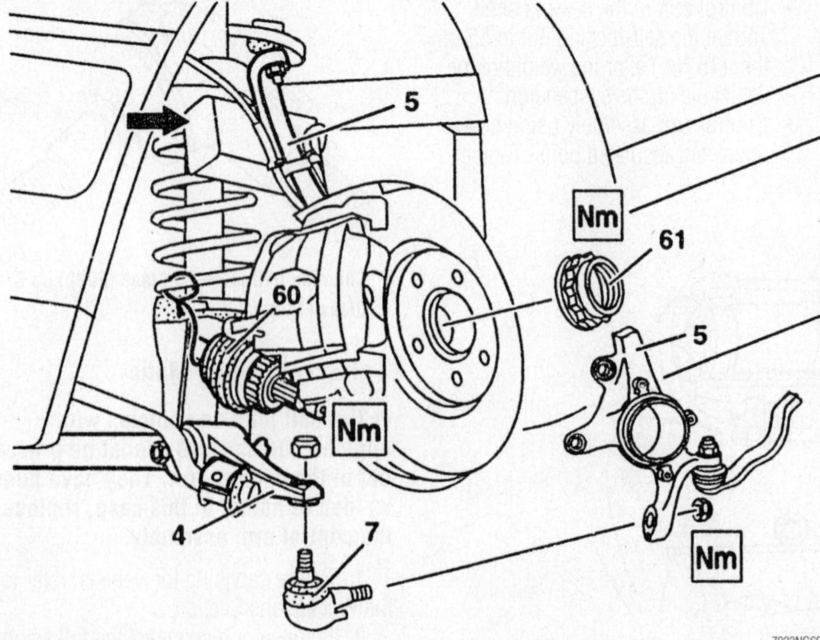

Lower ball joint mounting—E Class 4-Matic (210)

Upper Control Arm

REMOVAL & INSTALLATION

Front

1. Before servicing the vehicle, refer to the Precautions Section.
2. Remove the front wheel and support the lower control arm with a jack.
3. Remove or disconnect the following:
   - Electrical connections, if equipped
   - Ball joint stud
   - Control arm mounting bolts

To install:

4. Install or connect the following:
   - Control arm assembly using new bolts. Torque the bolts to 37 ft. lbs. (50 Nm).
   - Ball joint stud. Torque the nut to 88 ft. lbs. (120 Nm).
   - Electrical connections
5. Front wheel
6. Lower the vehicle to the floor and tighten the control arm bolt and nut.

CONTROL ARM BUSHING REPLACEMENT

Front

1. Before servicing the vehicle, refer to the Precautions Section.
2. Remove or disconnect the following:
   - Front wheel and support the lower control arm with a jack
   - Upper control arm
   - Steel spacer washers
   - Bushings from the control arm, using an appropriate removal tool

To install:

3. Install or connect the following:
   - Bushings using an appropriate installation tool
   - Steel spacer washers
   - Upper control arm

Front Lower Control Arm

REMOVAL & INSTALLATION

C Class, CLK, S Class and SLK 230

1. Before servicing the vehicle, refer to the Precautions Section.
2. Remove or disconnect the following:
   - Front wheel
   - Disc brake rotor securing bolt and backing plate bolt
   - Ball joint nut
   - Ball joint-to-steering knuckle nut
   - Control arm

To install:

3. Install or connect the following:
   - Control arm assembly. Torque the control arm nut to 74 ft. lbs. (100 Nm) and the knuckle nut to 104 ft. lbs. (140 Nm).
   - Backing plate bolt. Torque to 72 inch lbs. (8 Nm).
   - Brake rotor bolt. Torque the bolt to 84 inch lbs. (10 Nm).
   - Front wheel

E class—Except 4-Matic

1. Before servicing the vehicle, refer to the Precautions Section.
2. Remove or disconnect the following:
   - Front wheel
   - Coil spring
   - Stabilizer bar and bracket

- Lower control arm-to-frame nuts and bolts
- Ball joint pinch bolt and nut
- Lower control arm

To install:

3. Install or connect the following:
   - Control arm. Temporarily tighten the control arm-to-frame nuts.
   - Ball joint pinch bolt and nut. Torque the nut to 92 ft. lbs. (125 Nm).
   - Stabilizer bar bracket. Torque the bolts to 15 ft. lbs. (20 Nm).
   - Coil spring
   - Front wheel
4. Lower the vehicle to the floor. Torque the control arm-to-frame nuts to 88 ft. lbs. (120 Nm).

E Class 4-Matic

1. Before servicing the vehicle, refer to the Precautions Section.
2. Remove or disconnect the following:
   - Front wheel
   - Axle shaft 12-point nut
   - Ball joint nut and press the ball joint out of the steering knuckle
   - Axle shaft from the flange
3. Support the knuckle assembly by placing a block of wood between the coil spring and the knuckle.
4. Position the axle shaft out of the way using a piece of wire.
5. Remove the lower control arm assembly.

To install:

6. Install or connect the following:
   - Ball joint. Torque a new self-locking nut to 77 ft. lbs. (105 Nm).
   - Axle shaft
   - Ball joint to the steering knuckle with a new self-locking nut. Torque the nut to 77 ft. lbs. (105 Nm).
   - 12-point axle nut. Torque to 162 ft. lbs. (220 Nm).
   - Front wheel

FRONT CONTROL ARM BUSHING REPLACEMENT

1. Before servicing the vehicle, refer to the Precautions Section.
2. Remove or disconnect the following:
   - Lower control arm
3. Using appropriate removal tool, remove the bushings from the control arm.

To install:

4. Install the bushings using appropriate installation tool.
5. Install the lower control arm assembly

Rear Lower Control Arm

REMOVAL & INSTALLATION

1. Before servicing the vehicle, refer to the Precautions Section.
2. Support the lower control arm assembly.
3. Remove or disconnect the following:
   - Rear wheel
   - Electrical connections if equipped
   - Shock absorber
   - Lower control arm

To install:

4. Install or connect the following:
   - Lower control arm. Torque the bolts to 40 ft. lbs. (54 Nm).
   - Shock absorber. Torque the mounting nut to 52 ft. lbs. (70 Nm)
   - Electrical connections
   - Rear wheel

REAR CONTROL ARM BUSHING REPLACEMENT

1. Before servicing the vehicle, refer to the Precautions Section.
2. Remove or disconnect the following:
   - Rear wheel
   - Lower control arm
   - Rubber mount from the spring link using a press

To install:

3. Install or connect the following:
   - Rubber mounts using a press
   - Lower control arm
   - Rear wheel

Wheel Bearings

ADJUSTMENT

Front

ALL MODELS EXCEPT E CLASS 4-MATIC

1. Before servicing the vehicle, refer to the Precautions Section.
2. Remove the front wheel.
3. Install 1 wheel bolt (48a) in the flange on the opposite side of the locking bolt.
4. Push the brake pads into the caliper until they are clear of the rotor.
5. Remove the grease cap (9e).
6. Loosen the bolt (9i) and tighten the clamping nut (9d) while turning the rotor until all bearing play is eliminated.
7. Loosen the clamping nut until play is detectable.
8. Mount a dial indicator (022) on the hub with the measuring tip on the end of the spindle.
9. Turn the clamping nut to adjust the play while pulling and pushing on the rotor to measure the bearing play. Correct bearing play should be 0.0004–0.0008 inches (0.01–0.02mm).
10. Torque the bolt on the clamping nut to 96 inch lbs. (12 Nm) and recheck the

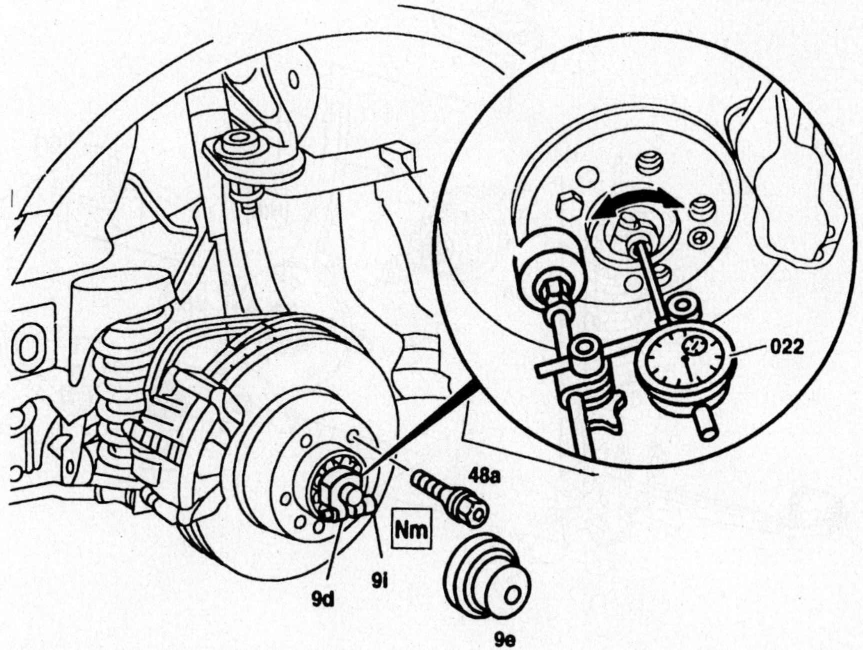

Mount a dial indicator on the rotor to measure the bearing play—except E Class 4-Matic

bearing play again. If the play is within specification, assemble the wheel. If not, repeat the procedure.

E CLASS 4-MATIC (210)

The front wheel bearings on the 4-Matic are not adjustable. If the bearings become loose or make noise, they must be replaced.

REMOVAL & INSTALLATION

Front

ALL MODELS EXCEPT E CLASS 4-MATIC

1. Before servicing the vehicle, refer to the Precautions Section.
2. Remove or disconnect the following:
   - Front wheel
   - Brake caliper and rotor
   - Grease cap
   - Clamping nut and washer
   - Wheel hub
   - Grease seal
   - Bearing

To install:

3. Install or connect the following:
   - Bearing
   - Grease seal
   - Wheel hub
   - Washer and clamping nut
4. Adjust bearing play. Torque the clamp bolt to 96 inch lbs. (12 Nm).
5. Fill the grease cap to the flared edge with high temperature roller bearing grease and install the cap.
   - Rotor and caliper
   - Front wheel

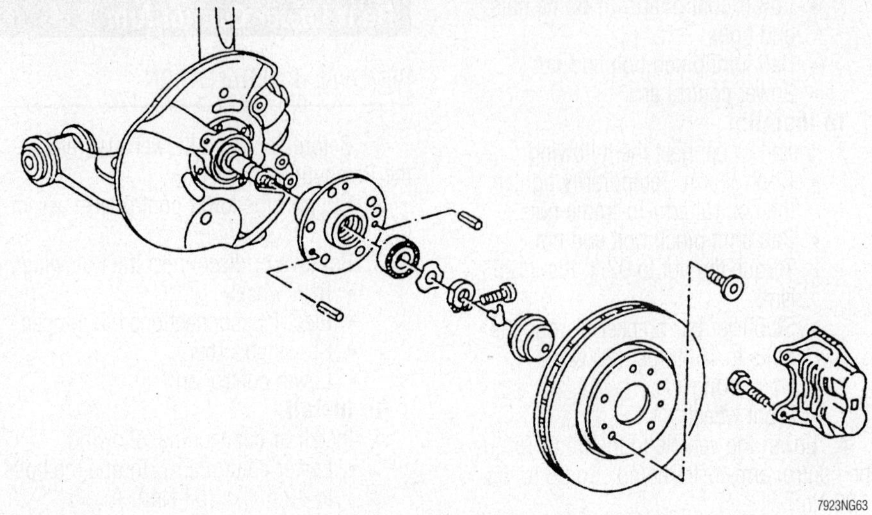

Exploded view of the front wheel bearing assembly—C Class, CLK 320, E Class, S Class and SLK 230

E CLASS 4-MATIC

1. Before servicing the vehicle, refer to the Precautions Section.
2. Remove or disconnect the following:
   - Front wheel
   - 12-point axle nut (61)
   - Front axle distributor connector
   - Speed and brake lining sensor wires and position out of the way
   - Brake disc (34)
   - Splash shield (35)
   - Speed sensor and bracket (5f)
   - Tie rod end (10a)
   - Ball joint (7)
   - Follower joint (8)
   - Steering knuckle
   - Axle flange
   - Snapring
   - Bearing assembly using a press

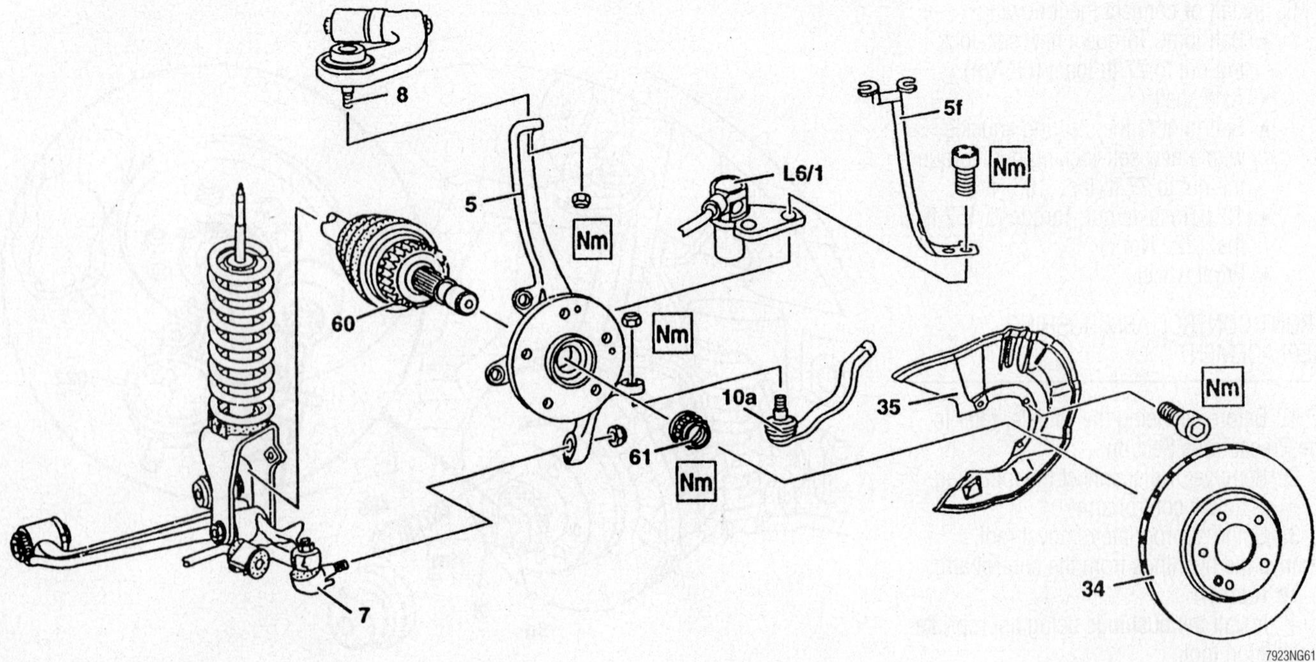

Steering knuckle and related components—E Class 4-Matic (210)

To install:

3. Press a new bearing into the knuckle and install the snapring.
4. Press the axle flange into the bearing.
5. Install or connect the following:
   - Steering knuckle
   - Follower joint. Torque the nut to 33 ft. lbs. (45 Nm).
   - Ball joint. Torque the nut to 77 ft. lbs. (105 Nm).
   - Tie rod end. Torque the nut to 16 ft. lbs. (22 Nm).
   - Speed sensor and bracket
   - Splash shield
   - Brake disc
   - Sensor wiring
   - Axle distributor connector
   - 12-point axle nut. Torque the nut to 162 ft. lbs. (220 Nm).
   - Front wheel

Rear

ALL MODELS

1. Before servicing the vehicle, refer to the Precautions Section.
2. Remove or disconnect the following:
   - Rear wheel
   - Exhaust if necessary
   - Cable guide rail
   - Level sensor control rod, if equipped
   - 12-point axle nut
   - Axle shaft to differential flange attaching bolts
   - Axle shaft
   - Brake disc
   - Parking brake shoes
   - Axle flange
   - Snap ring
   - Bearing using a puller
   - Axle shaft flange race using a puller

To install:

3. Press the new bearing (8) into the axle carrier.
4. Install or connect the following:
   - Snapring (9)
   - Axle flange (10)
   - Parking brake shoes
   - Brake disc
   - Axle shaft. Torque the M10 bolts to 52 ft. lbs. (70 Nm) and the M12 bolts to 74 ft. lbs. (100 Nm).
   - 12-point axle nut. Torque the nut to 236 ft. lbs. (320 Nm) on the 140 model or 162 ft. lbs. (220), on the remaining models.
   - Level sensor control rod
   - Cable guide
   - Exhaust system
   - Rear wheel

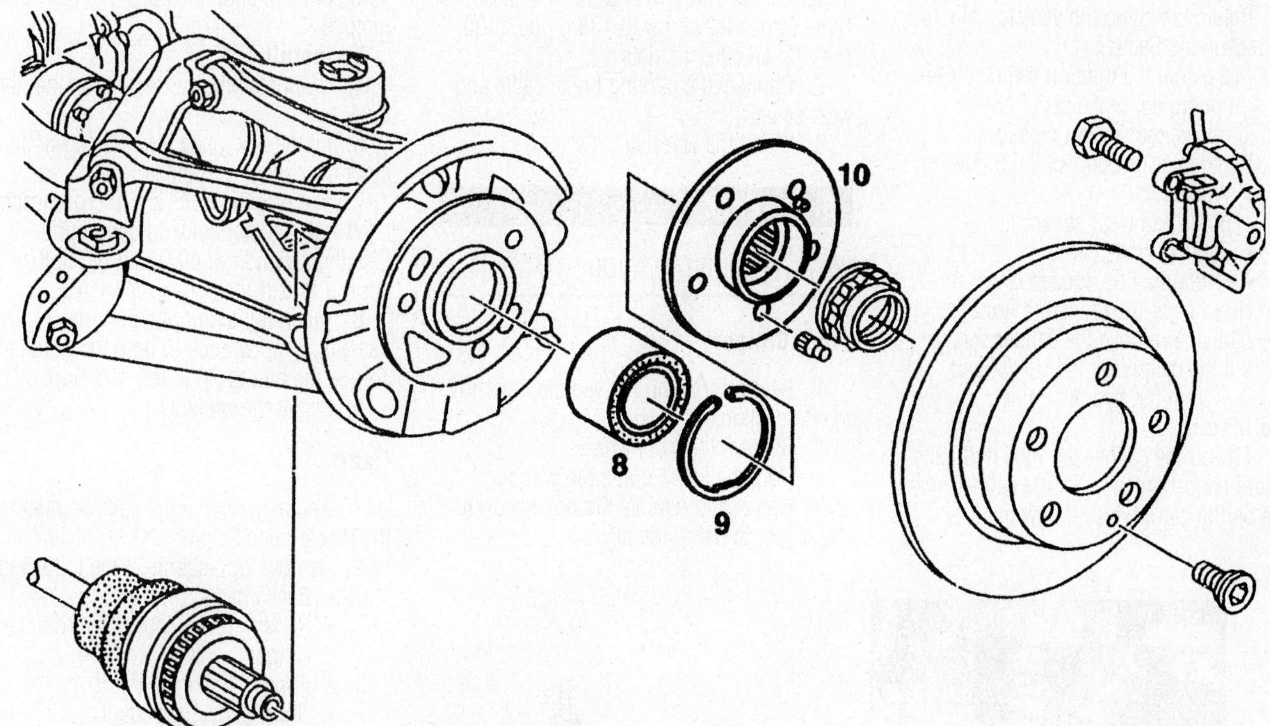

Exploded view of the rear wheel bearing assembly—all models

BRAKES

Brake Caliper

REMOVAL & INSTALLATION

Front

1. Before servicing the vehicle, refer to the Precautions Section.
2. Remove or disconnect the following:
   - Wheels
   - Brake pad wear sensor
   - Hydraulic line at the caliper, then remove the caliper from the carrier. Make sure to hold the pin with a back-up wrench when removing the caliper bolts.
   - Caliper from the hydraulic line, loosen only
3. The carrier can be removed by removing the 2 bolts.

To install:

4. If removed, install the carrier. Torque the bolts to 85 ft. lbs. (115 Nm).
5. Thread the caliper onto the hydraulic line and hand-tighten it. Fit the caliper into place on the carrier.
6. Torque the guide pin bolts to 18 ft. lbs. (25 Nm).
7. Tighten the hydraulic line and bleed the brakes.

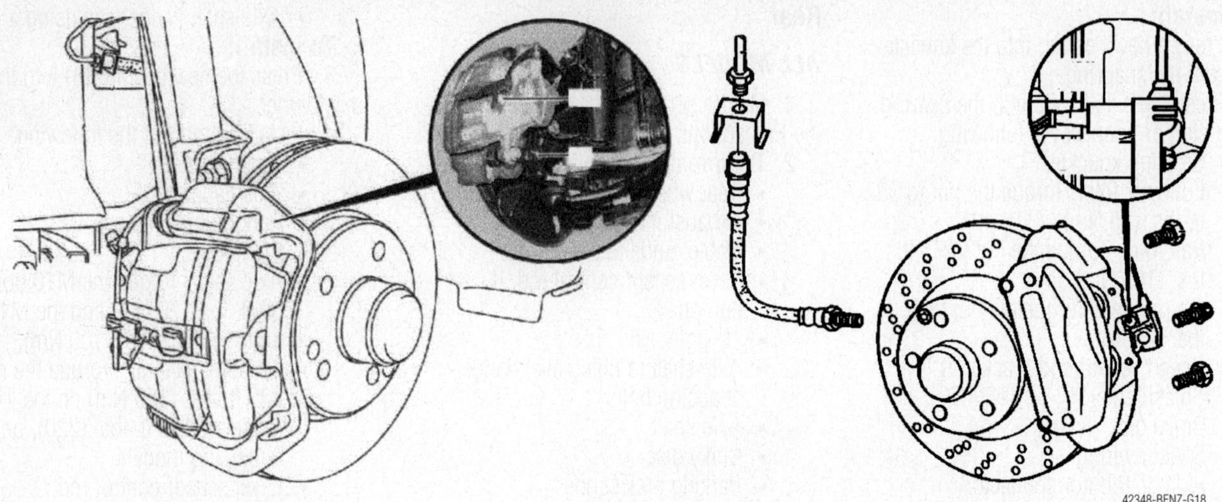

Typical front disc brake caliper mountings

Rear

1. Before servicing the vehicle, refer to the Precautions Section.
2. Make sure the ignition switch stays **OFF** and pump the brake pedal 25–35 times to relieve the system pressure.
3. Remove or disconnect the following:
   - Wheels
   - Brake pad wear sensor
   - Parking brake cable
   - Hydraulic line, loosen only
4. Use a back-up wrench to hold the guide pins and remove the caliper screw.
5. Lift the caliper off the carrier and unscrew it from the hydraulic line.

To install:

6. Thread the caliper onto the hydraulic line and hand-tighten it. Fit the caliper into place on the carrier. Install new micro-encapsulated screws. Torque the screw to 41 ft. lbs. (55 Nm), on C Class, E Class and CLK. Torque the screw to 133 ft. lbs. (180 Nm), on all other models.
7. Connect the parking brake cable and wear sensor.
8. Install the wheels.

Brake Pads

REMOVAL & INSTALLATION

Front Caliper

1. Before servicing the vehicle, refer to the Precautions Section.
2. Remove the front wheels.
3. Hold the lower guide pin with an open wrench and remove the bolt securing the caliper to the guide pin.
4. Pivot the caliper up on the upper guide pin and slide the pads straight out to remove.

To install:

5. Compress the caliper piston into the bore.
6. Fit the new pads into the carrier and pivot the caliper into place.
7. The original bolts are micro-encapsulated with a thread locking compound. Install a new bolt or clean the old bolt and apply a thread-locking compound.
8. When tightening the bolt, be sure to use a back-up wrench to hold the guide pin. Torque the bolt to 26 ft. lbs. (35 Nm).
9. Install the wheels.

Rear

1. Before servicing the vehicle, refer to the Precautions Section.
2. Remove or disconnect the following:
   - Rear wheels
   - Parking brake cable clip from the caliper
   - Parking brake cable
   - Brake pad wear sensor
   - Brake pad connector from the caliper
   - Brake pads

To install:

3. Retract the piston into the housing by rotating the piston clockwise.
4. Install or connect the following:
   - New brake pads onto the pad carrier
   - Brake pad connector
   - Brake pad wear sensor
   - Hand brake cable to the caliper.
5. Check the parking brake operation and adjust the cable if necessary.
6. Install the wheels.

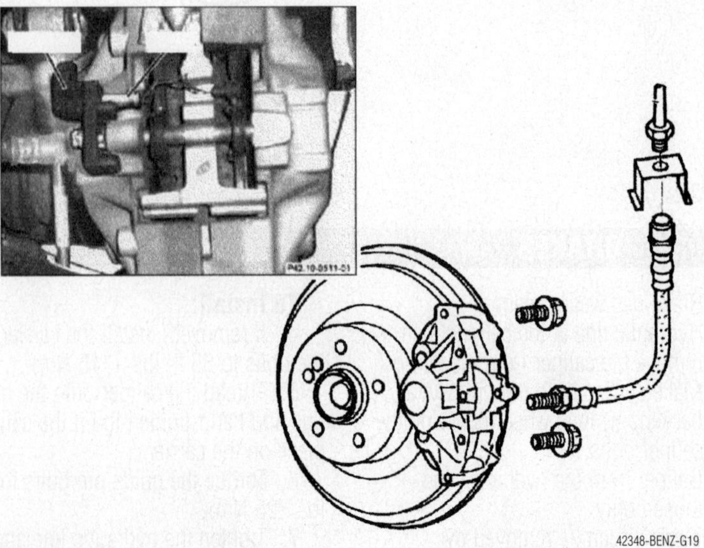

Typical rear disc brake caliper mounting

MERCEDES-BENZ

8

ML320 • ML350 • ML430 • ML500

SPECIFICATIONS AND MAINTENANCE CHARTS

ENGINE AND VEHICLE IDENTIFICATION CHART

Engine Code					
Code	Liters (cc)	Cu. In.	Cyl.	Fuel Sys.	Eng. Mfg.
M112	3.2 (3199)	195	6	SFI	MB
M112	3.5 (3500)	214	6	SFI	MB
M112	3.7 (3724)	227	6	SFI	MB
M113	5.0 (4966)	303	8	SFI	MB

Model Year	
Code ①	Year
3	2003
4	2004
5	2005
6	2006
7	2007

SFI: Sequential Fuel Injection

MB: Mercedes-Benz

① 10th digit of the VIN

06041_MCLS_C0001

GENERAL ENGINE SPECIFICATIONS

Year	Model	Engine Displacement Liters	Engine ID	Net Horsepower @ rpm	Net Torque @ rpm (ft. lbs.)	Bore x Stroke (in.)	Com-pression Ratio	Oil Pressure @ rpm
2003	ML320	3.2	M112	215@5600	233@3000	3.54x3.31	10.0:1	43.5@3000
	ML350	3.7	M112	241@5600	259@3000	3.81x3.31	10.0:1	43.5@3000
	ML500	5.0	M113	288@5600	325@2700	3.81x3.31	10.0:1	43.5@3000
2004	ML350	3.7	M112	241@5600	259@3000	3.81x3.31	10.0:1	43.5@3000
	ML500	5.0	M113	288@5600	325@2700	3.81x3.31	10.0:1	43.5@3000
2005	ML350	3.7	M112	241@5600	259@3000	3.81x3.31	10.0:1	43.5@3000
	ML500	5.0	M113	288@5600	325@2700	3.81x3.31	10.0:1	43.5@3000
2006	ML350	3.5	M112	268@6000	258@2400	3.39x3.66	10.7:1	43.5@3000
	ML500	5.0	M113	302@5600	339@2700	3.81x3.81	10.0:1	43.5@3000

06041_MCLS_C0002

ENGINE TUNE-UP SPECIFICATIONS

Year	Engine Displacement Liters	Engine ID	Spark Plug Gap (in.)	Ignition Timing (deg.)	Fuel Pump (psi)	Idle Speed (rpm)	Valve Clearance	
							In.	Ex.
2003	3.2	M112	0.032	①	55	700	②	②
	3.7	M112	0.032	①	55	700	②	②
	5.0	M113	0.032	①	55	700	②	②
2004	3.7	M112	NA	①	55	700	②	②
	5.0	M113	NA	①	55	700	②	②
2005	3.7	M112	NA	①	55	700	②	②
	5.0	M113	NA	①	55	700	②	②
2006	3.5	M112	NA	①	55	700	②	②
	5.0	M113	NA	①	55	700	②	②

NA: Information not available

① ECM controlled

② Hydraulic lash adjusters

06041_MCLS_C0003

1 Crankshaft
2 AC compressor
3 Power steering pump
4 Guide pulley 1
5 Generator
6 Coolant pump and fan
7 Tension pulley

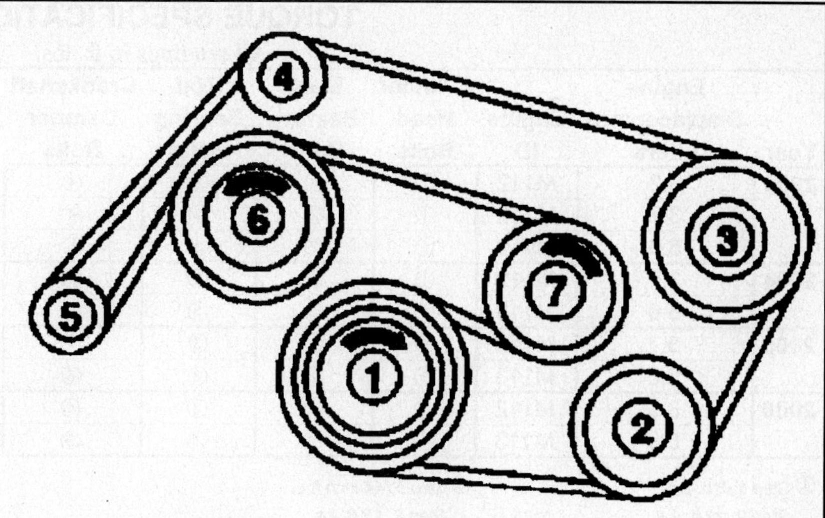

42348-MCLS-G02

Accessory drive belt routing—all engines

CAPACITIES

Year	Model	Engine Displacement Liters	Engine ID	Engine Oil with Filter (qts.)	Transmission (pts.)	Transfer Case (pts.)	Drive Axle Front (pts.)	Drive Axle Rear (pts.)	Fuel Tank (gal.)	Cooling System (qts.)
2003	ML320	3.2	M112	8.5	20	3.0	2.4	3.2	19.0	9.0
	ML350	3.7	M112	8.5	20	3.0	2.4	3.2	19.0	9.0
	ML430	5.0	M113	8.5	20	3.0	2.4	3.2	19.0	9.0
2004	ML350	3.7	M112	8.5	18.5	NA	NA	NA	NA	8.5
	ML500	5.0	M113	8.5	18.5	NA	NA	NA	NA	NA
2005	ML350	3.7	M112	8.5	18.5	NA	NA	NA	NA	8.5
	ML500	5.0	M113	8.5	18.5	NA	NA	NA	NA	NA
2006	ML350	3.5	M112	8.5	18.5	NA	NA	NA	NA	8.5
	ML500	5.0	M113	8.5	18.5	NA	NA	NA	NA	NA

NA: Information not available

Note: All capacities are approximate. Add fluid gradually and check to be sure a proper fluid level is obtained.

06041_MCLS_C0004

TORQUE SPECIFICATIONS

All readings in ft. lbs.

Year	Engine Displacement Liters	Engine ID	Cylinder Head Bolts	Main Bearing Bolts	Rod Bearing Bolts	Crankshaft Damper Bolts	Flywheel Bolts	Manifold Intake	Manifold Exhaust	Spark Plugs	Oil Pan Drain Plug
2003	3.2	M112	①	②	③	④	⑤	15	12	21	22
	3.7	M112	①	②	③	④	⑤	15	12	21	22
	5.0	M113	①	②	③	⑥	⑤	15	12	21	22
2004	3.7	M112	①	②	③	④	⑤	15	12	21	22
	5.0	M113	①	②	③	⑥	⑤	15	12	21	22
2005	3.7	M112	①	②	③	④	⑤	15	12	21	22
	5.0	M113	①	②	③	⑥	⑤	15	12	21	22
2006	3.5	M112	①	②	③	④	⑤	15	12	21	22
	5.0	M113	①	②	③	⑥	⑤	15	12	21	22

① Step 1: 8 ft. lbs.
Step 2: 22 ft. lbs.
Step 3: plus 90 degrees
Step 4: plus 90 degrees

② M8 vertical bolts
Step 1: 15 ft. lbs.
Step 2: plus 90 degrees
M8 side bolts: 22 ft. lbs.
M10 bolts
Step 1: 44 inch lbs.
Step 2: 22 ft. lbs.
Step 3: plus 90 degrees

③ Step 1: 44 inch lbs.
Step 2: 18 ft. lbs
Step 3. 90 degrees

④ Step 1: 148 ft. lbs.
Step 2: 95 degrees

⑤ Step 1: 33 ft. lbs.
Step 2: 90 degrees

⑥ Step 1: 148 ft. lbs.
Step 2: plus 90 degrees

06041_MCLS_C0005

TIRE, WHEEL AND BALL JOINT SPECIFICATIONS

Year	Model	OEM Tires Standard	OEM Tires Optional	Tire Pressures (psi) Front	Tire Pressures (psi) Rear	Wheel Size	Ball Joint Inspection	Lugnut Torque (ft. lbs.)
2002	ML320	255/65R16	None	32	32	8J	NS	111
	ML430	275/55R17	None	32	32	8.5J	NS	111
2003	ML320	255/60R17	275/55VR17	32	32	std.: 8J Opt.: 8.5J	NS	111
	ML350	255/60R17	275/55VR17	32	32	std.: 8J Opt.: 8.5J	NS	111
	ML500	275/55VR17	None	32	32	8.5J	NS	111
2004	ML350	P255/60HR17	P275/55VR17	①	①	std.: 8J Opt.: 8.5J	NS	111
	ML500	P275/55VR17	None	①	①	8.5J	NS	111
2005	ML350	P255/60HR17	P275/55VR17	①	①	std.: 8J Opt.: 8.5J	NS	111
	ML500	P275/55VR17	None	①	①	8.5J	NS	111
2006	ML350	P255/65HR17	NA	①	①	NA	NS	111
	ML500	P255/55R18	None	①	①	NA	NS	111

NA: Information not available
NS: Not Specified
① See placard on vehicle

06041_MCLS_C0006

BRAKE SPECIFICATIONS

All measurements in inches unless noted

Year	Model		Brake Disc Original Thickness	Brake Disc Minimum Thickness	Brake Disc Maximum Runout	Minimum Lining Thickness	Brake Caliper Bracket Bolts (ft. lbs.)	Brake Caliper Mounting Bolts (ft. lbs.)
2003	ML320	F	1.02	0.91	NA	①	NA	22
		R	0.59	0.49	NA	②	NA	17
	ML350	F	1.26	1.16	NA	①	NA	22
		R	0.55	0.47	NA	②	NA	17
	ML500	F	1.02	0.91	NA	①	NA	③
		R	0.59	0.49	NA	②	NA	④
2004	ML350	F	1.26	1.16	NA	①	NA	22
		R	0.55	0.47	NA	②	NA	17
	ML500	F	1.26	1.16	NA	①	NA	③
		R	0.55	0.47	NA	②	NA	④
2005	ML350	F	1.26	1.16	NA	①	NA	22
		R	0.55	0.47	NA	②	NA	17
	ML500	F	1.26	1.16	NA	①	NA	③
		R	0.55	0.47	NA	②	NA	④
2006	ML350	F	1.26	1.16	NA	①	NA	22
		R	0.55	0.47	NA	②	NA	17
	ML500	F	1.26	1.16	NA	①	NA	③
		R	0.55	0.47	NA	②	NA	④

NA: Information not available

① New: Inner pad w/backing plate 0.65 in.
Wear limit: 0.47 in
New: Outer pad w/backing plate 0.61 in.
Wear limit: 0.43 in.

② New: 0.61 in.
Wear limit: 0.43 in.

③ Fixed caliper: 133 ft. lbs.; floating caliper: 22 ft. lbs.

④ Fixed caliper: 85 ft. lbs.; floating caliper: 17 ft. lbs.

06041_MCLS_C0007

SCHEDULED MAINTENANCE INTERVALS

Mercedes Benz—ML320, ML350, ML500

TO BE SERVICED	TYPE OF SERVICE	VEHICLE MILEAGE INTERVAL (x1000)															
		7.5	15	22.5	30	37.5	45	52.5	60	67.5	75	82.5	90	97.5	105	112.5	120
Accessory drive belt ①	S/I		✓		✓		✓		✓		✓		✓		✓		✓
Air filter element ②	R								✓								
Body for paint damage ③	S/I				✓				✓				✓				✓
Brake fluid ③	R				✓				✓				✓				✓
Engine coolant ④	R						✓						✓				✓
Engine oil & filter ①	R		✓		✓		✓		✓		✓		✓		✓		✓
Fuel filter ②	R								✓								
Spark plugs ②	R	Every 100,000 miles															
Suspension & body structure ③	S/I				✓				✓				✓				✓
Underside of vehicle ②	S/I								✓								

R: Replace S/I: Service or Inspect, if needed

① Perform this at the mileage shown or once a year, whichever occurs first.

② Perform this at the mileage shown, or every 4 years, whichever occurs first.

③ Perform this at the mileage shown or every 2 years, whichever occurs first.

④ Perform this at the mileage shown or every 3 years, whichever occurs first.

06041_MCLS_C0008

ENGINE REPAIR

➡Disconnecting the negative battery cable on some vehicles may interfere with the functions of the on board computer systems and may require the computer to undergo a relearning process, once the negative battery cable is reconnected.

Alternator

REMOVAL & INSTALLATION

1. Before servicing the vehicle, refer to the Precautions Section.
2. Remove or disconnect the following:
   - Negative battery cable
   - Accessory drive belt
   - Right inner fender liner
   - Alternator electrical wires
   - Alternator

To install:

3. Install the alternator assembly. Torque the mounting bolts to 31 ft. lbs. (42 Nm).
4. Install or connect the following:
   - Alternator electrical wires
   - Right inner fender liner
   - Accessory drive belt
   - Connect the negative battery cable
5. Start the engine and check for proper operation.

Ignition Timing

ADJUSTMENT

The ignition timing is controlled by the Electronic Control Module (ECM) and is not adjustable.

Engine Assembly

REMOVAL & INSTALLATION

1. Before servicing the vehicle, refer to the Precautions Section.
2. Verify that the rear engine lifting eyes are correct. The left lifting eye is marked with a star and code number 04, and the right lifting eye is marked with a star and code number 02.
3. Remove or disconnect the following:
   - Negative battery cable
   - Engine undercover
   - Air baffle
   - Engine cooling fan and clutch, then the fan shroud

➡The fan clutch is equipped with right-hand thread.

4. Place a guard plate behind the radiator/condenser to protect it from damage during removal and installation.
   - Air cleaner housing
   - Resonance pipe and body
   - Coolant hoses from the water pump and thermostat housing
   - Radiator
   - Coolant expansion tank
   - Transmission dipstick tube from the cylinder head cover
   - Brake booster vacuum hose from the rear of the intake manifold
5. Drain the power steering fluid from the pump.
   - Hoses from the power steering pump, then plug the openings
   - Vacuum hose at the purge control valve
6. Relieve the fuel system pressure.
   - Fuel pipe
   - Heater hose from the rear of the cylinder head
   - Engine wiring harness
7. Lock the automatic belt tensioner by rotating the tensioner counterclockwise until a 5mm drift or pin fits through the tensioner.
   - Serpentine belt
   - Air conditioning compressor and position it aside, leaving the hoses attached
   - Left inner fender liner
   - Exhaust system from the manifolds
   - Cable for the park lock interlock
8. Matchmark the torque converter-to-ring gear.
   - Torque converter bolts
   - Left and right Oxygen (O_2S) sensors electrical connections
   - Starter
   - Engine-to-transmission mounting bolts. Leave the 2 upper bolts attached at this time.
   - Motor mounts from the front suspension support
   - Generator wiring harness
   - Air conditioning compressor electrical connector
   - Engine ground cable at the power steering pump
9. Attach the engine hoist to the lifting eyes, then raise the engine and support the transmission.

➡Be sure that the engine does not touch the body at the rear.

10. Remove the 2 remaining engine-to-transmission mounting bolts.
11. Slowly pull the engine out to the front and lift it out.

To install:

12. Install or connect the following:
    - Engine into the vehicle
    - Engine to the transmission
    - Upper 2 mounting bolts
    - Engine ground cable to the power steering pump
    - Motor mount bolts and tighten to 26 ft. lbs. (35 Nm)
13. Remove the engine hoist from the lifting eyes.
    - Remaining engine-to-transmission mounting bolts and tighten to 30 ft. lbs. (40 Nm).
    - Air conditioning compressor electrical connector
    - Generator wiring harness
    - Starter
    - Left and right O_2S sensors
14. Align the torque converter-to-ring gear matchmarks and tighten the bolts to 31 ft. lbs. (42 Nm).
    - Starter and tighten bolts to 31 ft. lbs. (42 Nm).
    - Cable for the park lock interlock
    - Exhaust system to the manifolds and tighten the mounting nuts to 15 ft. lbs. (20 Nm)
    - Air conditioning compressor
    - Serpentine belt and remove the locking pin
    - Engine wiring harness
    - Coolant hoses to the water pump, cylinder head and thermostat housing
    - Fuel pipe
    - Vacuum hose to the purge control valve
    - Power steering hoses to the pump and fill the reservoir
    - Brake booster vacuum hose to the intake manifold
    - Transmission dipstick tube to the cylinder head cover
    - Coolant expansion tank
    - Resonance pipe and body, then the air cleaner housing
15. Remove the radiator/condenser guard plate.
    - Fan shroud and fan
    - Engine undercover
16. Fill the engine with coolant and oil.
17. Connect the negative battery cable.
18. Read fault memory, encode the radio and normalize the power windows.

Water Pump

REMOVAL & INSTALLATION

1. Before servicing the vehicle, refer to the Precautions Section.
2. Remove or disconnect the following:
   - Negative battery cable
   - Engine cooling fan and clutch, then the fan shroud

➡The fan clutch is equipped with right-hand thread.

3. Drain the engine coolant.
   - Engine cover
4. Lock the automatic belt tensioner by rotating the tensioner counterclockwise until a 5mm drift or pin fits through the tensioner.
   - Serpentine belt
   - Coolant hoses from the water pump
   - Belt pulley
   - Water pump mounting bolts
   - Water pump
5. Clean and dry the gasket mating surface for the water pump.

To install:

6. Install or connect the following:
   - Water pump and gasket. Tighten the M6 bolts to 88 inch lbs. (10 Nm) and the M8 bolts to 177 inch lbs. (20 Nm).
   - Water pump belt pulley and tighten the mounting bolts to 88 inch lbs. (10 Nm)
   - Coolant hoses to the water pump
   - Serpentine belt and remove the locking pin
   - Engine cover
   - Fan shroud and fan
7. Fill the engine with coolant.
8. Connect the negative battery cable.
9. Read fault memory, encode the radio and normalize the power windows.
10. Start the vehicle and check for leaks.

Cylinder Head

REMOVAL & INSTALLATION

1. Before servicing the vehicle, refer to the Precautions Section.
2. Remove or disconnect the following:
   - Negative battery cable
3. Drain and recycle the engine coolant.
   - Engine cooling fan and clutch
   - Fan shroud

➡The fan clutch is equipped with right-hand thread.

4. Place a guard plate behind the radiator/condenser to protect it from damage during removal and installation.
   - Engine cover
   - Air cleaner housing, resonance pipe and body
5. Properly relieve the fuel system pressure.
   - Fuel line
   - Ignition coils
   - Cylinder head covers

➡The intake manifold system must not be disassembled.

   - Intake manifold
   - Vacuum switchover valve
   - Camshaft Position (CMP) sensor
6. Lock the automatic belt tensioner by rotating the tensioner counterclockwise until a 5mm drift or pin fits through the tensioner.
   - Serpentine belt
   - Heater hose at left cylinder head
   - Driver side fender liner
   - Exhaust support bracket
   - Power steering pump and position it aside leaving the hoses attached

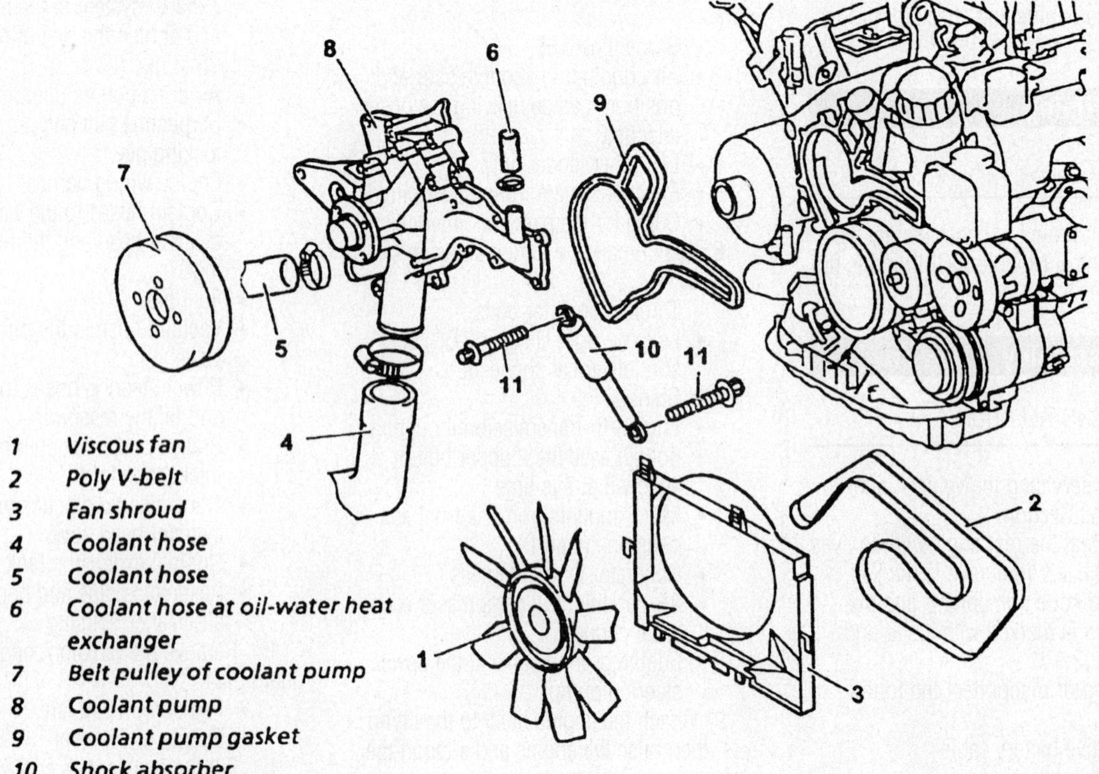

Exploded view of the water pump mounting and related components

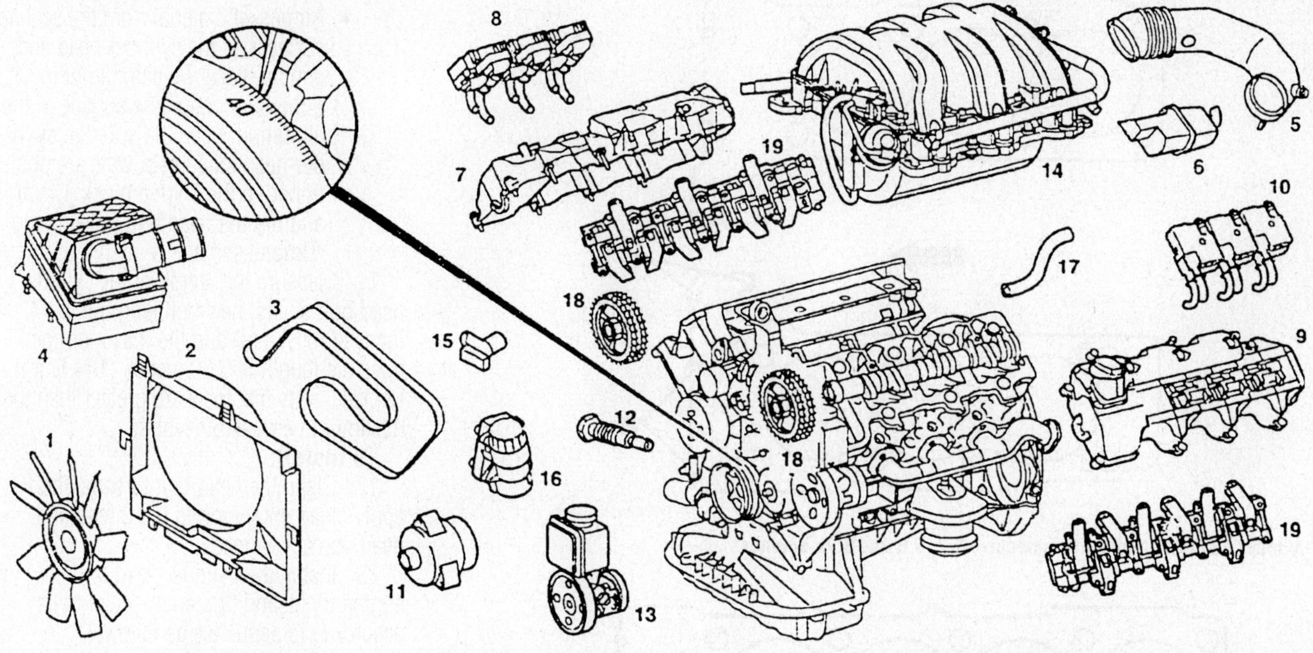

Exploded view of the cylinder head accessory components—3.2, 3.5 and 3.7L engines shown; 5.0L engine similar

- Exhaust system from the exhaust manifolds

7. Rotate the engine clockwise to position the crankshaft 40 degrees after top dead center.

WARNING

The engine must not be rotated backward.

8. Lock the camshafts using the Camshaft Locking tools 112 589 00 32 00 and 112 589 01 32 00.
9. Remove or disconnect the following:
   - Generator, then the timing chain tensioner
   - Oil filter housing and heat exchanger
   - Camshaft gears and attach them to the chain with a cable tie
   - Camshaft bearing bridges
   - Timing case-to-cylinder head bolts
10. Loosen and remove the cylinder head bolts in stages following the illustrated sequence.

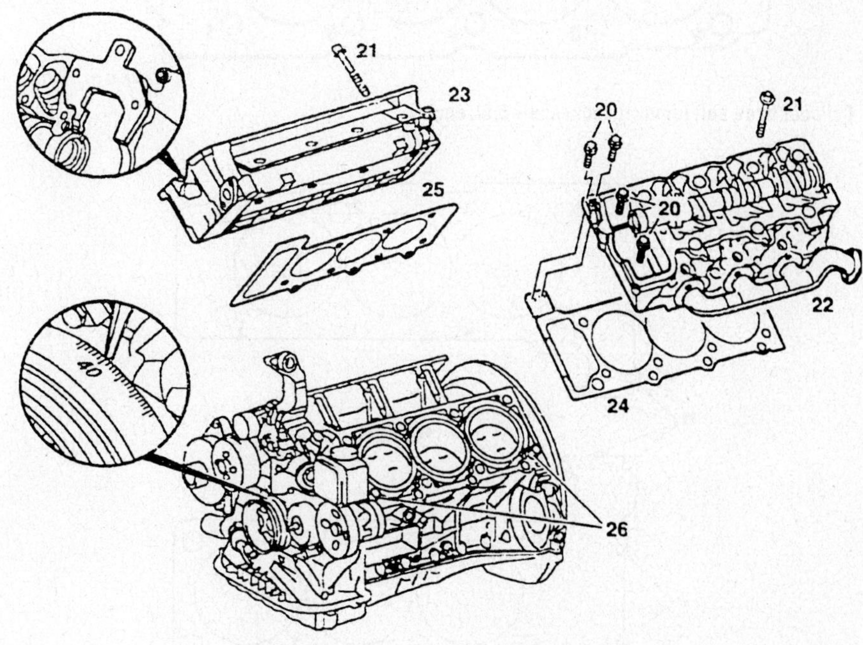

Exploded view of the cylinder head removal—3.2, 3.5 and 3.7L engines shown; 5.0L engine similar

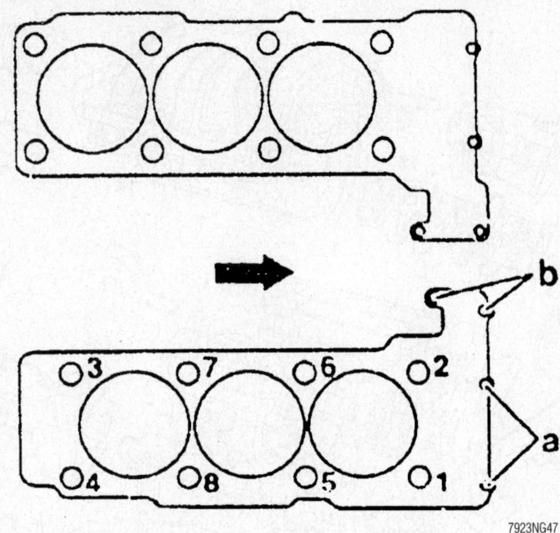

Cylinder head bolt removal sequence—3.2, 3.5 and 3.7L engines

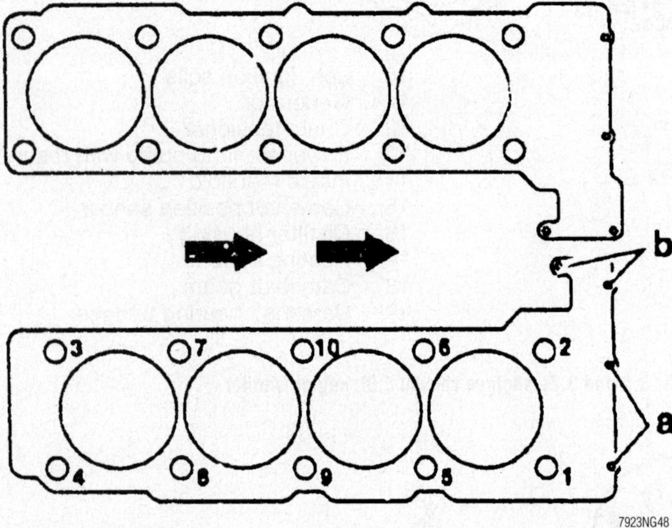

Cylinder head bolt removal sequence—5.0L engine

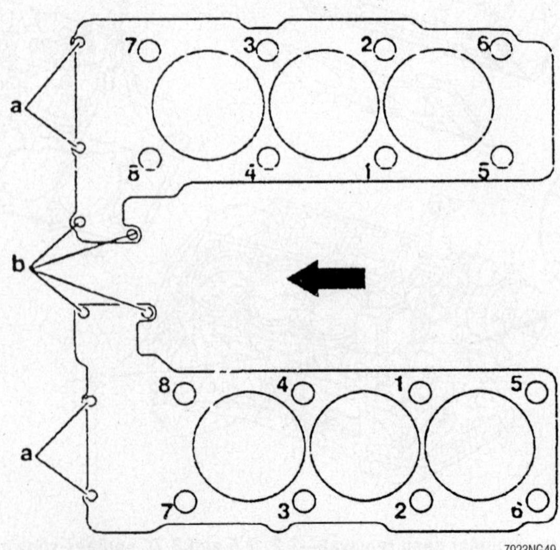

Cylinder head bolt tightening sequence—3.2, 3.5 and 3.7L engines

- Cylinder head off the engine block
- All gasket material from the sealing surfaces of the cylinder head and engine block. Be careful not to gouge or scratch the surface of the aluminum head. Be sure the cylinder head locating dowels are positioned in the engine block. Clean and dry the head bolt holes using compressed air.

11. Measure the length of the cylinder head bolt shafts, new bolt length is 5.57 inches (141.5mm) and the maximum permissible length is 5.69 inches (144.5mm). Replace bolts that measure greater than the maximum permissible length.

To install:

12. Clean the head bolt threads, then apply clean engine oil to the thread and head contact surfaces.
13. Install the cylinder head to the engine block and tighten the head bolts according to sequence as follows:
    a. Step 1. 96 inch lbs. (10 Nm).
    b. Step 2. 22 ft. lbs. (30 Nm).
    c. Step 3. 90 degrees.
    d. Step 4. additional 90 degrees.
14. Install or connect the following:
    - Timing case-to-cylinder head bolts and tighten to 15 ft. lbs. (20 Nm)
    - Camshaft bearing bridges
    - Camshaft gear and tighten the mounting bolt to 37 ft. lbs. (50 Nm) plus an additional 90 degrees. Check the basic camshaft position and adjust if necessary.
    - Timing chain tensioner with a new gasket and tighten to 59 ft. lbs. (80 Nm)
    - Generator
15. Remove the camshaft locking plates.
    - Exhaust system to the manifolds and tighten the mounting nuts to 15 ft. lbs. (20 Nm)
    - Heater hose to the cylinder head
    - Power steering pump
    - Serpentine belt and remove the locking pin
    - CMP sensor
    - Vacuum switchover valve
    - Intake manifold
    - Cylinder head covers and tighten the bolts to 89 inch lbs. (10 Nm)
    - Ignition coils and tighten the mounting bolts to 70 inch lbs. (8 Nm)
    - Fuel pipe
    - Air cleaner housing, resonance pipe and body
    - Engine cover
16. Remove the guard plate from the radiator/condenser.

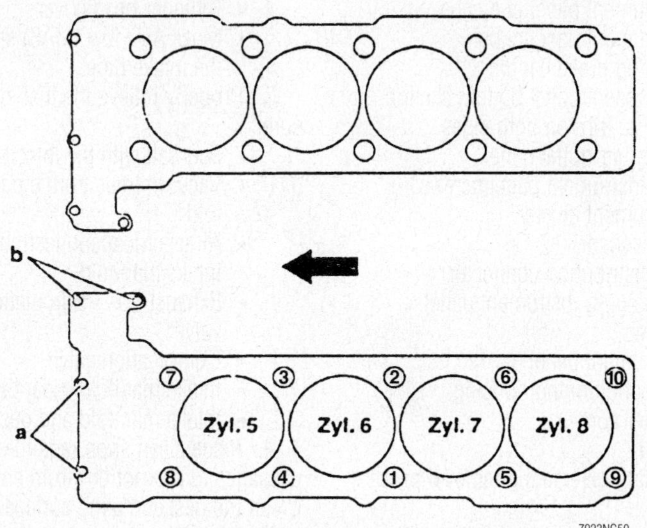

Cylinder head bolt tightening sequence—5.0L engine

- Fan shroud, then the cooling fan

17. Fill the engine with coolant.
18. Connect the negative battery cable.
19. Read fault memory, encode the radio and normalize the power windows.
20. Start the vehicle and check for leaks.

Rocker Arms/Shafts

The rocker arm/shaft is part of the camshaft bearing cap assembly and is called the camshaft bearing bridge. This procedure is for removing and installing the camshaft bearing bridge.

REMOVAL & INSTALLATION

1. Before servicing the vehicle, refer to the Precautions Section.
2. Remove or disconnect the following:
   - Negative battery cable
   - Cylinder head cover
3. Rotate the engine clockwise to position the crankshaft 40 degrees After Top Dead Center (ATDC).

✻✻ WARNING

The engine must not be rotated backward.

   - Generator
   - Timing chain tensioner
4. Cable tie the timing chain to the camshaft sprocket.
   - Camshaft bearing bridge bolts in the reverse order of installation, starting at 16 for the 3.2L engine, and 20 for the 4.3L engine

➡The camshaft bearing bridge must not be disassembled. If damage exists at the valve gear or at the top half of the camshaft bearing journal, the complete cylinder head should be replaced.

To install:

5. Lubricate the camshaft bearing journals.
6. Install or connect the following:
   - Camshaft bearing bridge and tighten the bolts to 88 inch lbs. (10 Nm) plus 90° in the reverse of the removal sequence.
7. Remove the cable ties from the camshaft sprockets.

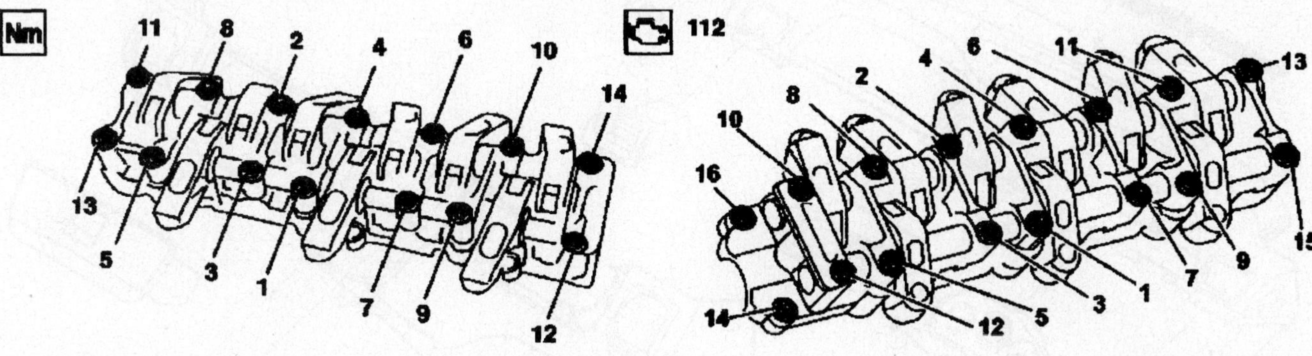

Camshaft bearing bridge bolt removal sequence—3.2L and 3.7L engines

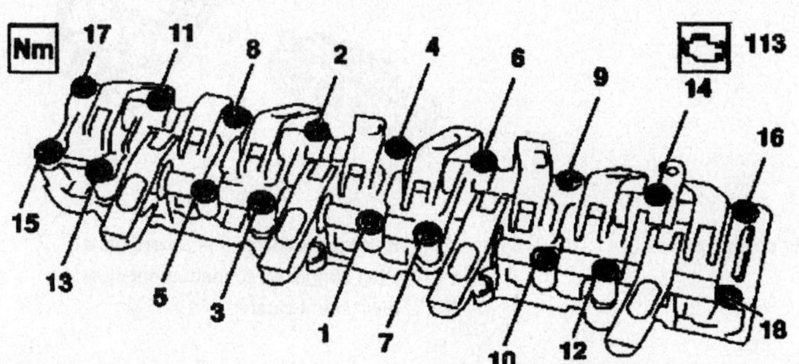

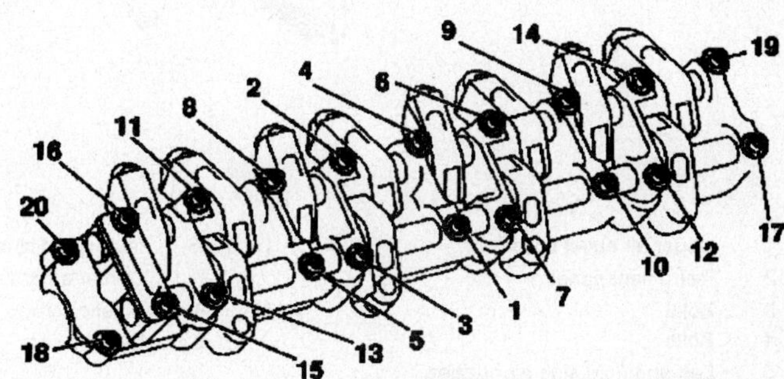

Camshaft bearing bridge bolt removal sequence —4.3L and 5.0L engine

- Timing chain tensioner with a new gasket and tighten to 59 ft. lbs. (80 Nm)
- Cylinder head covers and tighten the bolts to 88 inch lbs. (10 Nm)
- Negative battery cable

8. Read fault memory, encode the radio and normalize the power windows.
9. Start the vehicle and check for leaks.

Heater Core

REMOVAL & INSTALLATION

1. Before servicing the vehicle, refer to the Precautions Section.
2. Remove or disconnect the following:
   - Negative battery cable
   - Heater hoses
   - Upper and lower steering column covers
   - Combination switch
   - Instrument cluster trim
   - Instrument cluster screws
   - Footwell screws on both sides
   - Lower instrument panel screws
   - Instrument panel center section
   - A/C heater control panel
   - Center instrument panel screws
   - Glove box
   - Instrument panel end caps
   - Instrument panel clips
   - Parking brake handle
   - Instrument panel bottom section
   - A pillar trim on both sides
   - Upper air outlet panel
   - Top instrument panel screws
   - Instrument cluster
   - Sun sensor
   - Air distribution connector
   - Remaining instrument panels screws
   - Instrument panel
   - Air conditioning housing
   - Heater core

To install:

3. To install, reverse the removal procedure.
4. Connect the negative battery cable.
5. Fill the engine with coolant and check for proper heater operation.

Intake Manifold

REMOVAL & INSTALLATION

1. Before servicing the vehicle, refer to the Precautions Section.
2. Remove or disconnect the following:
   - Negative battery cable
   - Cylinder head cover
   - Mass Air Flow (MAF) sensor with the intake pipe
3. Properly relieve the fuel system pressure.
   - Fuel rail with the injectors
   - Vacuum lines from the intake manifold
   - All electrical connections to the intake manifold
   - Exhaust Gas Recirculation (EGR) valve
   - Combination valve
   - Intake manifold mounting bolts
   - Intake manifold and gaskets
4. Place clean shop rags into the intake passages to prevent dirt from entering. Clean the gasket mating surfaces.

To install:

5. Install the new gaskets and verify the secondary air injection passage opening in the gasket.
6. Remove the shop rags from the intake passages.
7. Install or connect the following:
   - Intake manifold to the engine and tighten the mounting bolts to 15 ft. lbs. (20 Nm)
   - Combination valve and tighten the bolts to 15 ft. lbs. (20 Nm)

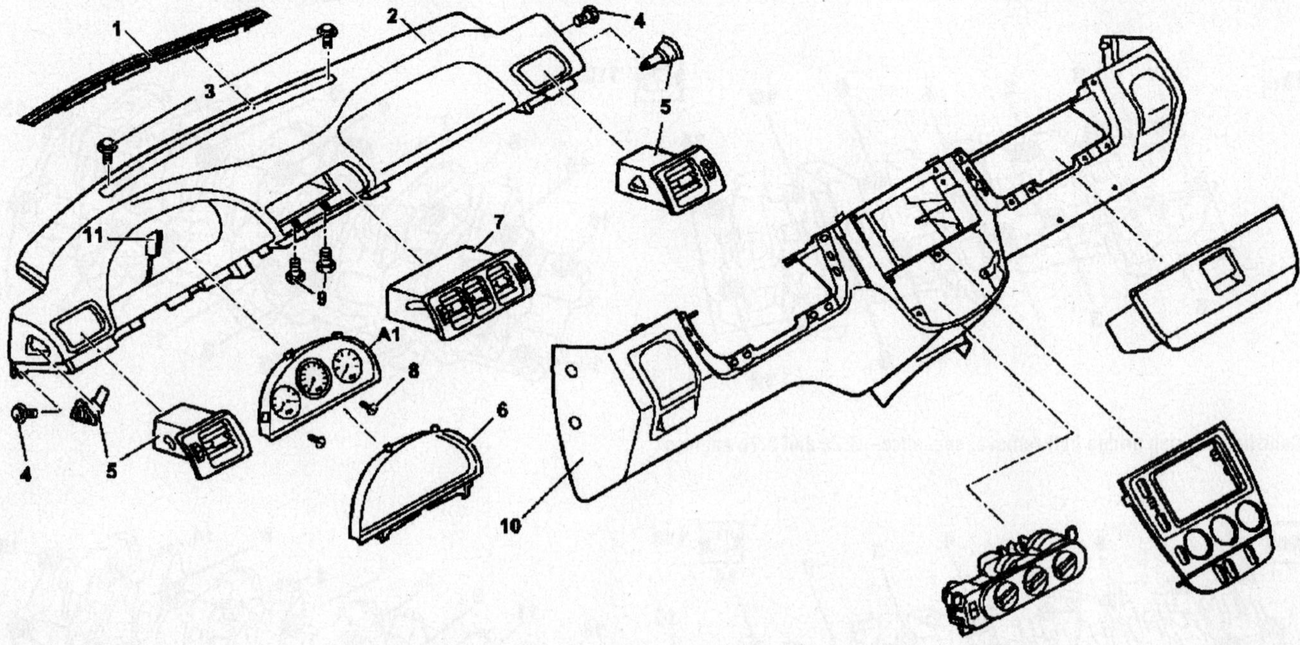

1 Upper air outlet cover
2 Instrument panel
3 Bolts
4 Bolts
5 Left and right side air nozzles
6 Instrument cluster cover frame
7 Center air nozzle
8 Plastic screws
9 Bolts
10 Bottom section of instrument panel
11 Plug connector to instrument cluster
A1 Instrument cluster

42348-MCLS-G05

Instrument panel assembly—

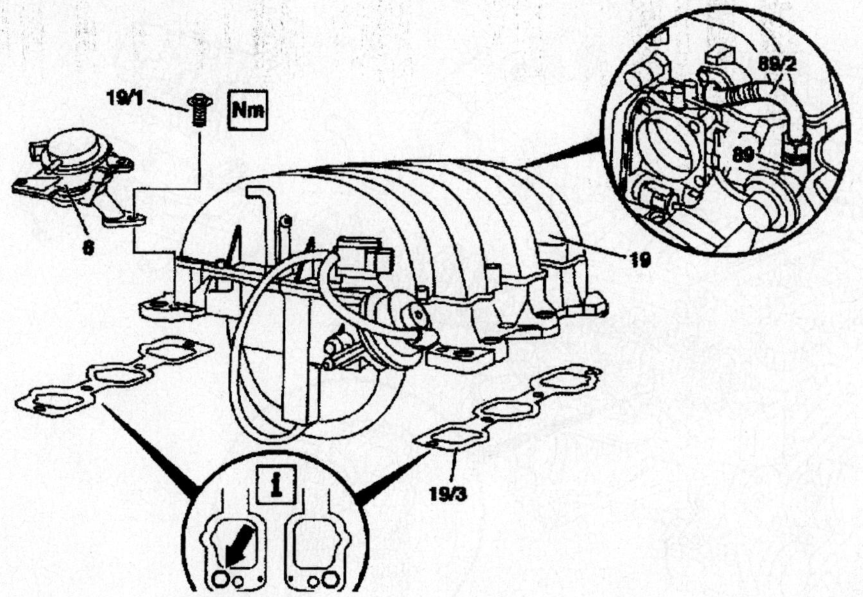

2/ 1	Bolt of cylinder head cover	19/ 3	Gasket
6	Combination valve	89	Exhaust gas recirculation valve
17/ 2	Feed line	89/ 2	Exhaust gas recirculation line
19	Intake manifold	Arrow	Hole for combination valve
19/ 1	Bolt		

42348-MCLS-G06

Exploded view of the intake manifold and related components—3.2L, 3.5L and 3.7L engines shown; 5.0L similar

- EGR valve
- Electrical connections to the intake manifold
- Vacuum lines to the manifold
- Fuel rail with the injectors
- MAF sensor with the air intake pipe
- Cylinder head cover and tighten the bolts to 88 inch lbs. (10 Nm)
- Negative battery cable

8. Read fault memory, encode the radio and normalize the power windows.
9. Start the vehicle and check for leaks.

Exhaust Manifold

REMOVAL & INSTALLATION

1. Before servicing the vehicle, refer to the Precautions Section.
2. Remove or disconnect the following:
   - Negative battery cable
   - Front wheels
   - Plastic inner fender liners
   - Exhaust manifold heat shields

➡If the bolts are difficult to remove or the threads show signs of damage, replace the rivet nuts in the manifold.

- Exhaust system-to-manifold flanged connection mounting bolts
- Front exhaust pipe at the transmission exhaust bracket
- Exhaust manifold-to-cylinder head mounting bolts
- Exhaust manifold

3. Clean the gasket mating surfaces.

To install:

4. Install or connect the following:
   - Exhaust manifold using new gaskets and nuts. Tighten the nuts to 12 ft. lbs. (16 Nm).
   - Front exhaust pipe to the transmission exhaust bracket
   - Exhaust system to the manifolds and tighten the mounting bolts to 15 ft. lbs. (20 Nm)
   - Exhaust manifold heat shields
   - Plastic inner fender liners
   - Front wheels and tighten the lug bolts to 110 ft. lbs. (150 Nm)
   - Negative battery cable
5. Read fault memory, encode the radio and normalize the power windows.
6. Start the vehicle and check for leaks.

Front Crankshaft Seal

REMOVAL & INSTALLATION

1. Before servicing the vehicle, refer to the Precautions Section.
2. Remove or disconnect the following:
   - Engine cooling fan and clutch
   - Fan shroud

➡The fan clutch is equipped with right-hand thread.

- Accessory drive belt
- Crankshaft pulley
- Crankshaft seal

To install:

3. Apply clean engine oil to the end of the crankshaft to ease installation.
4. Install or connect the following:
   - Crankshaft seal, using a suitable seal driver
   - Crankshaft pulley and tighten the bolt to 148 ft. lbs., then tighten another 95 degrees
   - Accessory drive belt
   - Fan shroud, engine cooling fan, and fan clutch

Camshaft

REMOVAL & INSTALLATION

1. Before servicing the vehicle, refer to the Precautions Section.
2. Remove or disconnect the following:
   - Negative battery cable
   - Cylinder head cover
3. Rotate the engine clockwise to position the crankshaft 40 degrees After Top Dead Center (ATDC).

⁂ WARNING

The engine must not be rotated backward.

- Generator
- Timing chain tensioner

4. Cable tie the timing chain to the camshaft sprocket.
   - Camshaft Position (CMP) sensor
5. Lock the camshafts using the Camshaft Locking tools 112–589–00–32–00 and 112–589–01–32–00.
   - Camshaft gears
   - Camshaft bearing bridge
   - Camshaft from the cylinder head

To install:

➡Be sure to install the correct camshaft for the corresponding cylinder head.

6. Apply clean engine oil to the camshaft contact surfaces.
7. Install or connect the following:
   - Camshaft
   - Camshaft bearing bridge

➡The camshafts can be rotated 40 degrees ATDC of the No. 1 cylinder without the valves touching the pistons.

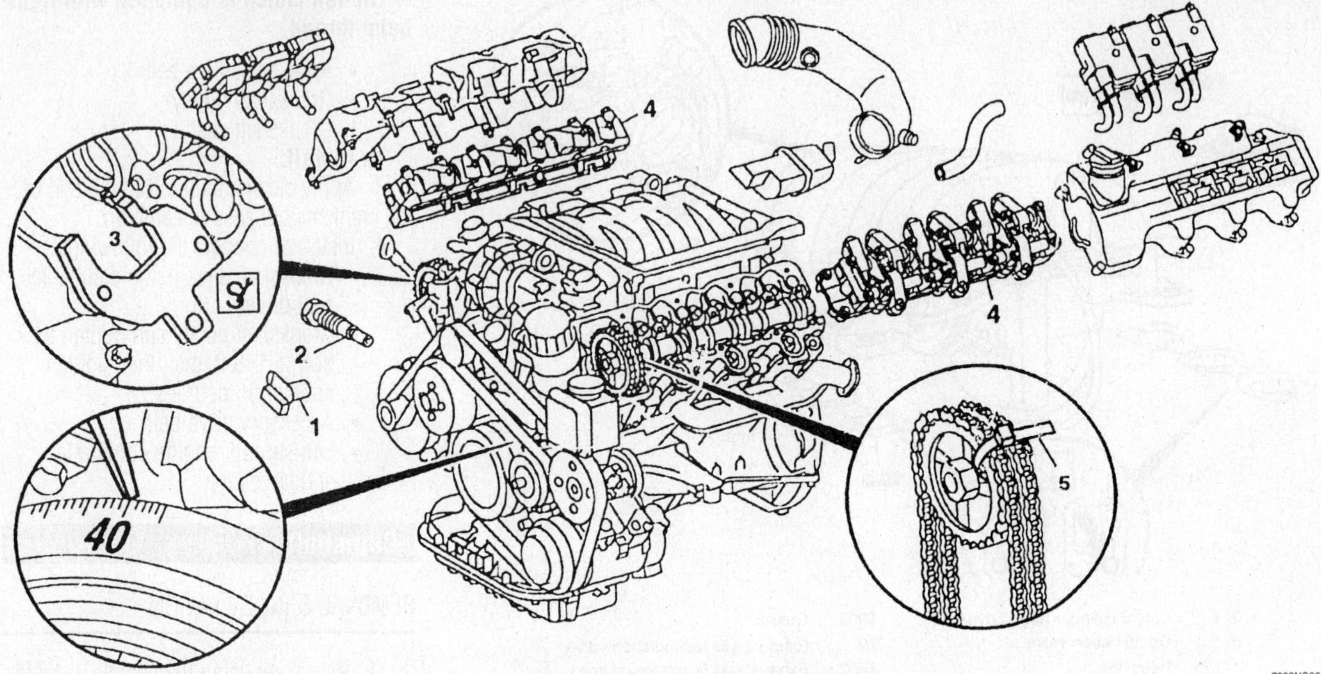

Exploded view of the camshaft removal and related components—3.2L, 3.5L and 3.7L engine shown; 5.0L engine similar

8. Position the camshaft so that the groove points centered towards the contact surface of the cylinder head cover, then attach the camshaft fixing plate. Repeat this step for the other camshaft.
   - Camshaft sprockets and tighten the attaching bolt to 37 ft. lbs. (50 Nm) plus 90–100 degrees
9. Remove the camshaft locking tools and the cable ties from the timing chain.
   - CMP sensor and tighten the mounting bolt to 70 inch lbs. (8 Nm)
   - Timing chain tensioner with a new gasket and tighten to 59 ft. lbs. (80 Nm)
   - Cylinder head cover
   - Negative battery cable
10. Read fault memory, encode the radio and normalize the power windows.

Starter Motor

REMOVAL & INSTALLATION

1. Before servicing the vehicle, refer to the Precautions Section.
2. Remove or disconnect the following:
   - Negative battery cable
   - Right side inner fender liner
   - Left engine mount shield
   - Starter electrical connectors
   - Starter

To install:

3. Install or connect the following:
   - Starter motor. Torque the mounting bolts to 31 ft. lbs. (42 Nm).
   - Starter electrical connectors
   - Engine mount shield and torque the bolts to 48 ft. lbs. (65 Nm)
   - Inner fender liner
   - Negative battery cable

Valve Lash

ADJUSTMENT

These vehicles are equipped with Hydraulic Lash Adjusters (HLA's), which do not require periodic adjustment.

Oil Pan

REMOVAL & INSTALLATION

1. Before servicing the vehicle, refer to the Precautions Section.
2. Disconnect the negative battery cable.
3. Drain the cooling system and engine oil.
4. Remove or disconnect the following:
   - Engine cover
   - Engine cooling fan, fan clutch, and fan shroud

➡The fan clutch is equipped with right-hand thread.

5. Attach a guard plate to protect the radiator.
   - Intake resonator and pipe
   - Upper and lower coolant hoses
   - Engine ground cable from the frame
   - Motor mounts at the front suspension supports
6. Attach an engine hoist and raise the engine. Be sure that the engine does not contact the firewall.
   - Lower oil pan

➡The upper pan bolts are different lengths and diameters. Note their locations for installation.

   - Upper pan toward the front of the vehicle. Rotate the crankshaft as necessary for clearance.

To install:

7. Clean the sealing surfaces and apply a bead of silicone sealant.

➡The pan must be installed within ten minutes after the sealant is applied.

8. Install or connect the following:
   - Upper pan. Tighten the 6mm bolts to 90 inch lbs. (10 Nm), and the 8mm bolts to 15 ft. lbs. (20 Nm).
9. In the same manner, apply a bead of silicone sealant and install the lower oil pan. Tighten the bolts to 90 inch lbs. (10 Nm).
10. Install or connect the following:
   - Motor mounts and remove the engine hoist

- Engine ground cable
- Upper and lower coolant hoses
- Intake resonator and pipe

11. Remove the radiator guard plate.
    - Fan shroud, engine cooling fan, and fan clutch
    - Engine cover
12. Fill the cooling system and crankcase to the correct levels.
13. Connect the negative battery cable.
14. Read fault memory, encode the radio and normalize the power windows.

Oil Pump

REMOVAL & INSTALLATION

1. Before servicing the vehicle, refer to the Precautions Section.
2. Remove the lower oil pan. Refer to the oil pan procedure in this section.
3. Push the chain tensioner back and remove the pump drive chain.
4. Unbolt and remove the oil pump.

To install:

5. Fill the oil pump with clean engine oil and install it to the engine. Tighten the bolts to 15 ft. lbs. (20 Nm).
6. Install the pump drive chain.
7. Install the lower oil pan. Refer to the oil pan procedure in this section.

Timing Chain, Sprockets, Tensioner and Front Cover

REMOVAL & INSTALLATION

1. Before servicing the vehicle, refer to the Precautions Section.
2. Disconnect the negative battery cable.
3. Drain the cooling system and engine oil.
4. Remove or disconnect the following:
    - Engine cover
    - Engine cooling fan, fan clutch, and fan shroud

➡The fan clutch is equipped with right-hand thread.

5. Place a guard plate behind the radiator/condenser to protect it from damage during removal and installation.
    - Accessory drive belt and tensioner
    - Cylinder head covers
    - Coolant hoses
    - A/C compressor and wire aside
    - Oil pan
    - Crankshaft pulley
    - Timing chain tensioner
    - Oil filter housing and heat exchanger
    - Water pump pulley

⁂ WARNING

The engine must not be rotated backward.

6. Rotate the crankshaft clockwise to align the crankshaft timing marks at 40° After Top Dead Center (ATDC). Insure that the grooves in the camshafts (6 in the illustration) align with the cylinder head cover mating surface on the intake side of the cylinder heads.

➡The timing chain cover bolts are different lengths and diameters. Note their locations for installation.

- Timing chain cover

7. Lock the camshafts using the Camshaft Locking tools 112–589–00–32–00 and 112–589–01–32–00.
    - Camshaft sprockets bolts
    - Timing chain

To install:

8. Align the copper plated links of the timing chain with the marks on the

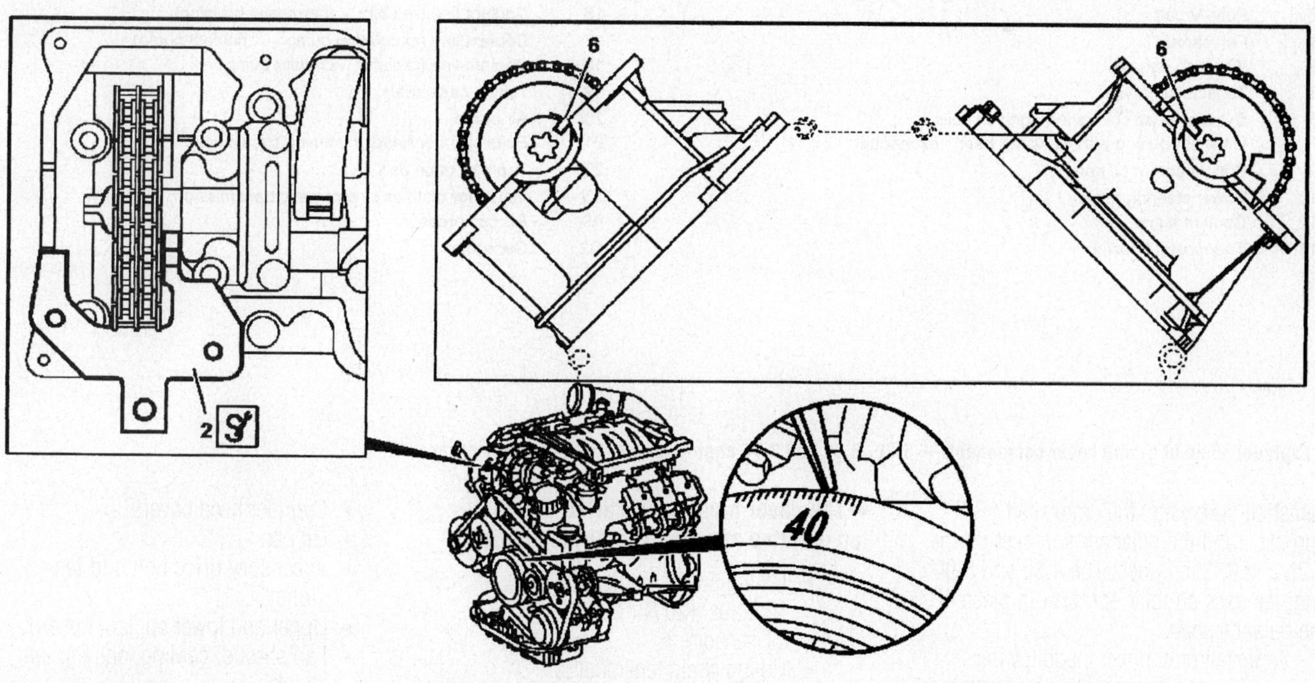

Camshaft timing marks

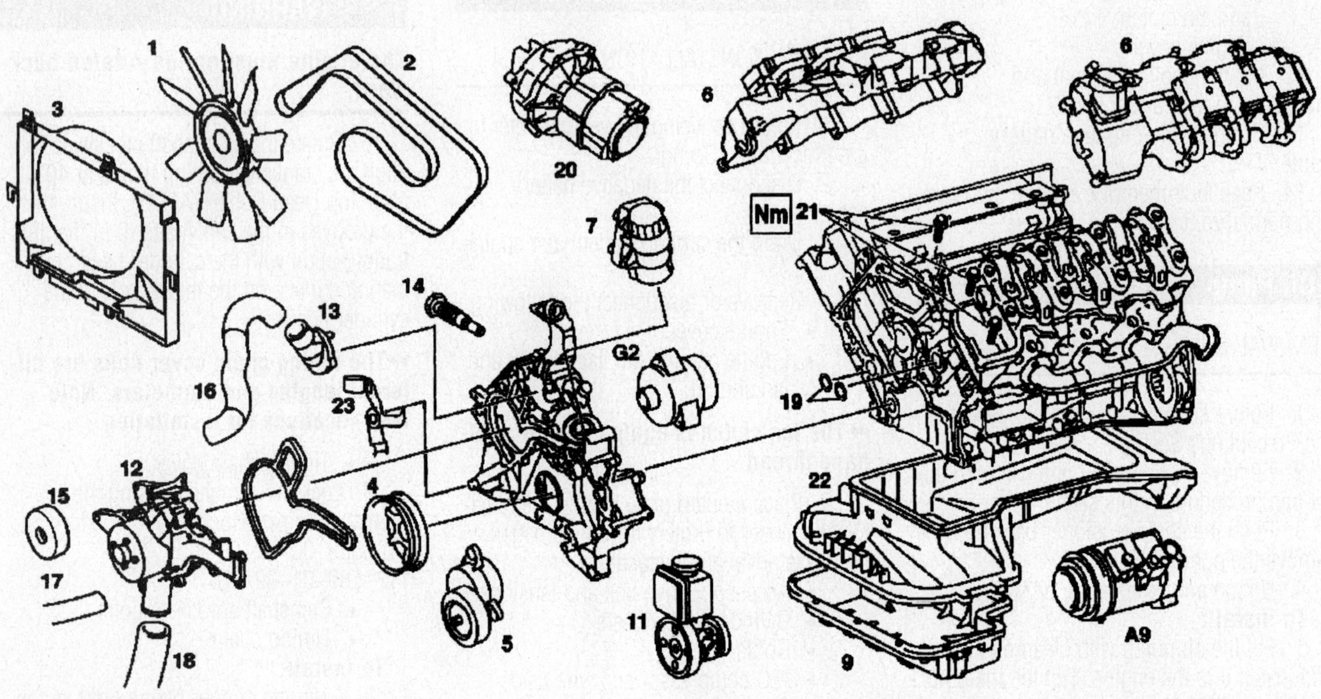

Shown on engine 112 in model 210

1	Viscous fan	14	Chain tensioner
2	Poly- V belt	15	Guide pulley
3	Fan shroud	16	Coolant line (radiator. thermostat housing)
4	Vibration damper	17	Coolant line (circulation pump . coolant pump)
5	Tensioning device	18	Coolant line (radiator. coolant pump)
6	Cylinder head covers with ignition coils	19	Timing case seals
7	Oil filter housing with oil- water heat exchanger	20	Air pump
9	Bottom part of oil pan	21	Bolts cylinder head . timing case cover
11	Power steering pump	22	Top part of oil pan
12	Coolant pump	23	Mount for oil lines in automatic transmission
13	Thermostat housing	A9	AC compressor
		G2	Generator

42348-MCLS-G08

Exploded view of timing cover components— 3.2L, 3.5L and 3.7L engine shown; 5.0L engine similar

camshaft sprockets, the crankshaft sprocket, and the balancer sprocket for the 3.2L and 3.7L engines. The 4.3L and 5.9L engines uses an idler sprocket in place of the balance shaft.

9. Install or connect the following:
   - Camshaft sprockets along with the timing chain. Tighten the attaching bolts to 37 ft. lbs. (50 Nm) plus 90–100 degrees.

10. Clean the sealing surfaces and apply a bead of silicone sealant to the timing chain cover.

➡The cover must be installed within ten minutes after the sealant is applied.

- Timing cover. Tighten the bolts to 15 ft. lbs. (20 Nm).
- CT sensor
- Timing chain tensioner
- Generator
- Crankshaft pulley and tighten the bolt to 148 ft. lbs., then tighten another 95 degrees
- Air pump, if equipped
- Air conditioning compressor and the power steering pump
- Cylinder head covers
- Oil pan
- Accessory drive belt and tensioner
- Upper and lower coolant hoses
- Fan shroud, cooling fan, and fan clutch

11. Fill the cooling system and the crankcase to the correct levels.
   - Engine cover
   - Negative battery cable

12. Read fault memory, encode the radio and normalize the power windows.

Piston and Ring

POSITIONING

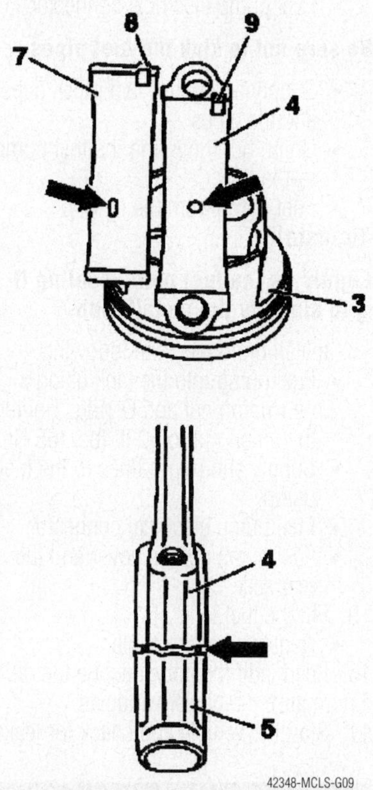

Cylinder identification and piston identifications must agree (see arrows) Install pistons so the code next to the piston pin and group number are pointing in the direction of travel, and the anti-twist lock (8) and the groove (9) are matched.

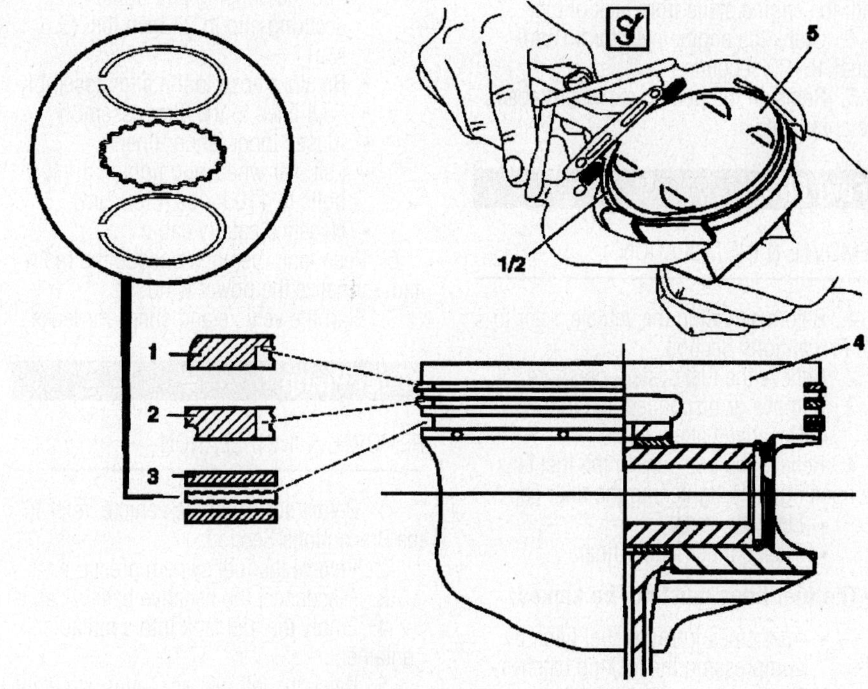

1 Compression ring (plain compression ring)
2 Compression ring (taper-faced hook scraper ring)
3 Oil scraper ring
4 Piston
5 Piston crown

42348-MCLS-G10

Ring positioning

FUEL SYSTEM

Fuel System Service Precautions

Safety is the most important factor when performing not only fuel system maintenance but any type of maintenance. Failure to conduct maintenance and repairs in a safe manner may result in serious personal injury or death. Maintenance and testing of the fuel system components can be accomplished safely and effectively by adhering to the following rules and guidelines.

- To avoid the possibility of fire and personal injury, always disconnect the negative battery cable unless the repair or test procedure requires that battery voltage be applied.
- Always relieve the fuel system pressure prior to disconnecting any fuel system component (injector, fuel rail, pressure regulator, etc.), fitting or fuel line connection. Exercise extreme caution whenever relieving fuel system pressure, to avoid exposing skin, face and eyes to fuel spray. Please be advised that fuel under pressure may penetrate the skin or any part of the body that it contacts.
- Always place a shop towel or cloth around the fitting or connection prior to loosening to absorb any excess fuel due to spillage. Ensure that all fuel spillage (should it occur) is quickly removed from engine surfaces. Ensure that all fuel soaked cloths or towels are deposited into a suitable waste container.
- Always keep a dry chemical (Class B) fire extinguisher near the work area.
- Do not allow fuel spray or fuel vapors to come into contact with a spark or open flame.
- Always use a back-up wrench when loosening and tightening fuel line connection fittings. This will prevent unnecessary stress and torsion to fuel line piping. Always follow the proper torque specifications.
- Always replace worn fuel fitting O-rings with new ones. Do not substitute fuel hose or equivalent, where fuel pipe is installed.

Fuel System Pressure

RELIEVING

1. Before servicing the vehicle, refer to the Precautions Section.
2. Locate the electric fuel pump fuse and remove it from the fuse box.

➡If the fuel pump fuse cannot be located, disconnect the vehicle wiring harness from the pump itself and perform the procedure.

3. Start the engine and allow it to idle until the engine stalls from lack of fuel.
4. Crank the engine over for an additional 15–20 seconds.
5. Reinstall the pump fuse when repairs are completed.

Fuel Filter

REMOVAL & INSTALLATION

1. Before servicing the vehicle, refer to the Precautions Section.
2. Relieve the fuel system pressure.
3. Remove or disconnect the following:
    - Negative battery cable
4. Relieve the pressure in the fuel tank by opening, then tightening the filler cap.
    - Left rear wheel
    - Plastic inner fender liner

➡The fuel lines must not be kinked.

- Fuel pipes from the fuel filter, by compressing the locking catches
- Breather hose from the filter assembly
- Filter securing clip
- Filter/pressure regulator

To install:

5. Install or connect the following:
    - New fuel filter/pressure regulator into the housing and tighten the securing clip to 27 inch lbs. (3 Nm)
    - Breather hose to the filter assembly
    - Fuel lines to the filter assembly
    - Plastic inner fender liner
    - Left rear wheel and tighten the lug bolts to 110 ft. lbs. (150 Nm)
    - Negative battery cable
6. Read fault memory, encode the radio and normalize the power windows.
7. Start the vehicle and check for leaks.

Fuel Pump

REMOVAL & INSTALLATION

1. Before servicing the vehicle, refer to the Precautions Section.
2. Relieve the fuel system pressure.
3. Disconnect the negative battery cable.
4. Empty the fuel tank into a suitable container.
5. Raise the left rear seat approximately 20 inches (50 cm).
6. Lift the carpeting to gain access to the fuel pump cover.
7. Remove or disconnect the following:
    - Fuel pump cover
    - Fuel pump electrical connector

➡Be sure not to kink the fuel pipes.

- Supply and return fuel pipes clips and the pipes
- Union nut mounting the fuel pump-to-the tank
- Fuel pump from the tank

To install:

➡Lightly oil the fuel pump sealing O-ring to simplify the installation.

8. Install or connect the following:
    - Fuel pump into the tank using a new union nut and O-ring. Tighten the union nut to 50 ft. lbs. (65 Nm).
    - Supply and return lines to the fuel pump
    - Fuel pump electrical connector
    - Fuel pump access cover and the rear seat
9. Fill the fuel tank.
    - Negative battery cable
10. Read fault memory, encode the radio and normalize the power windows.
11. Start the vehicle and check for leaks.

DRIVETRAIN

Transmission Assembly

REMOVAL & INSTALLATION

1. Before servicing the vehicle, refer to the Precautions Section.
2. Drain the transmission fluid.
3. Remove or disconnect the following:
    - Negative battery cable
    - Transfer case
    - Dipstick tube
    - Electrical connector shield
    - Electrical connector
    - Shift rod
    - Park interlock cable
    - Oxygen sensors
    - Transmission cooler lines
    - Torque converter cover
    - Torque converter-to-flexplate bolts
    - Rear transmission mount
    - Ground strap
    - Transmission-to-engine mounting bolts
    - Transmission

To install:

4. Position the transmission on the engine and install the mounting bolts.
5. Install or connect the following:
    - Ground strap
    - Rear transmission mount
    - Torque converter-to-flexplate bolts: tighten to 31 ft. lbs. (42 Nm)
    - Torque converter cover
    - Transmission cooler lines
    - Oxygen sensors
    - Park interlock cable
    - Shift rod
    - Electrical connector and shield
    - Transmission dipstick tube
    - Transfer case
    - Negative battery cable
    - Transmission fluid

Transfer Case Assembly

REMOVAL & INSTALLATION

1. Before servicing the vehicle, refer to the Precautions Section.
2. Drain the fluid and remove the exhaust brackets. Suspend the exhaust with a piece of wire.
3. Remove or disconnect the following:
    - Front half of rear driveshaft
    - Front driveshaft
    - Servomotor connector
4. Support the transmission and remove the engine support along with the rear engine mount. Lower the assembly slightly.
5. Remove or disconnect the following:
    - Transfer case bolts
6. Remove the transfer case and drain the residual oil.

To install:

7. Install or connect the following:
    - Transfer case with new gasket. Torque the bolts to 15 ft. lbs. (20 Nm)
    - Rear engine support. Torque the bolts to 30 ft. lbs. (40 Nm)
    - Front driveshaft. Torque the bolts to 37 ft. lbs. (50 Nm)
    - Front half of rear driveshaft. Torque the bolts to 30 ft. lbs. (40 Nm)
    - Servomotor connector
    - Exhaust brackets
8. Fill the transfer case with fluid.

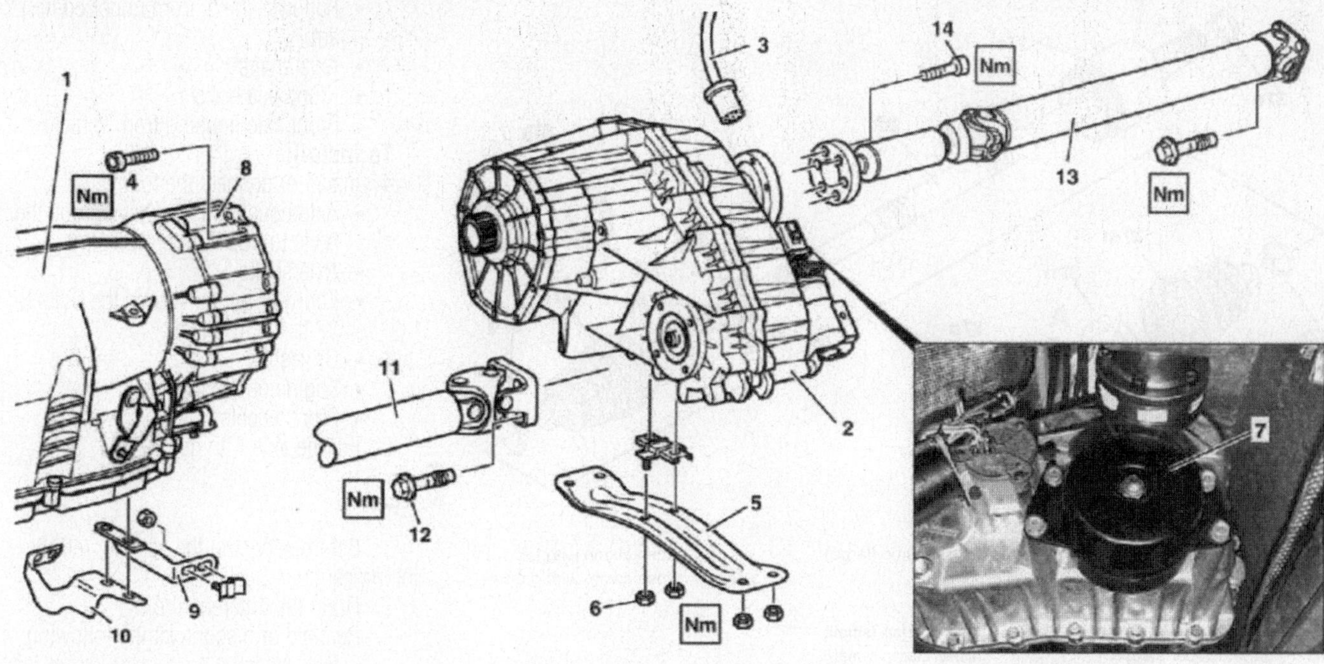

1	Transmission	6	Nut	10	Exhaust bracket
2	Transfer case	7	Vibration damper:	11	Propeller shaft
3	Connector	8	Adapter housing	12	Bolt
4	Bolt	9	Exhaust bracket	13	Propeller shaft
5	Engine support with rear engine mount			14	Bolt

42348-MCLS-G12

Exploded view of the components associated with transfer case removal and installation

Halfshaft

REMOVAL & INSTALLATION

Front

1. Before servicing the vehicle, refer to the Precautions Section.
2. Remove or disconnect the following:
   - Negative battery cable
   - Front wheel
   - Axle nut
   - Brake caliper
   - Track rod from steering knuckle
   - Follower joint from control arm
   - Halfshaft from flange
   - Halfshaft from axle gear
   - Halfshaft

To install:

3. Install or connect the following:
   - Halfshaft
   - Follower joint to control arm. Torque the nut to 37 ft. lbs. (50 Nm)
   - Track rod. Torque the nut to 41 ft. lbs. (55 Nm).
   - Brake caliper.
   - Axle nut. Torque the nut to 361 ft. lbs. (490 Nm)
   - Front wheel
   - Negative battery cable

Rear

1. Before servicing the vehicle, refer to the Precautions Section.
2. Remove or disconnect the following:
   - Wheel
   - Center axle hold-down bolt (in hub)
   - Wheel speed sensor
3. Push the follower joint out of carrier.
4. Press the track rod joint out of carrier.
5. Pull the axle shaft out of axle shaft flange.
6. Press the axle shaft out of axle carrier.

To install:

7. Install or connect the following:
   - Axle shaft. Torque the nut to 148–175 ft. lbs. (200–240 Nm)
   - Axle lock ring
   - Track rod joint. Torque the nut to 41 ft. lbs. 55 (Nm).
   - Follower joint. Torque the nut to 37 ft. lbs. (50 Nm).
   - Wheel speed sensor
   - Center axle hold down bolt. Torque the nut to 361 ft. lbs. (490 Nm).
   - Wheel

CV-Joints

OVERHAUL

Front

1. Before servicing the vehicle, refer to the Precautions Section.
2. Remove or disconnect the following:
   - Axle shaft
   - Rubber boot sealing clamp and CV-joint circlip
3. Using a soft metal drift or a piece of wood, tap the CV-joint off the shaft.
4. Use a rag to clean off the grease, inspect all components for wear and/or damage and replace as necessary.

To install:

5. Replace the rubber boot and install the CV-joint and circlip. Replace the grease with the appropriate CV-joint lubricant.
6. Install the axle assembly.

Rear

1. Before servicing the vehicle, refer to the Precautions Section.
2. Remove or disconnect the following:
   - Axle shaft
   - End cover and boot cap

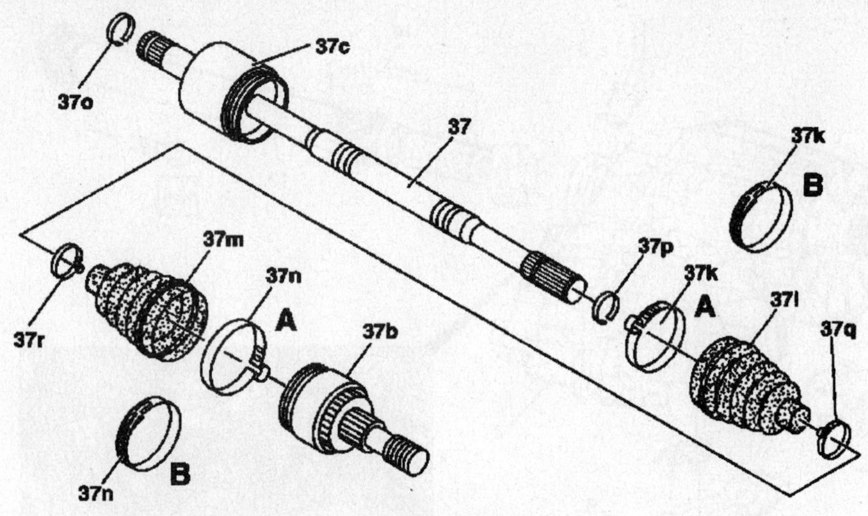

37	Remove/ install	37k	Hose clamp (large)	A	Hose clamp with tab
37b	Outer joint	37o	Circlip	B	Hose clamp with groove
37c	Inside joint	37p	Circlip		
37k	Hose clamp (large)	37q	Hose clamp (small)		
37l	Inner rubber sleeve	37r	Hose clamp (small)		
37m	Outer boot				

42348-MCLS-G13

Exploded view of the front and rear CV-joint assembly

- Rubber boot sealing clamp
- CV-joint locking ring

3. Remove the CV-joint from the axle using a press. Remove the rubber boot.
4. Use a rag to clean off the grease, inspect all components for wear and/or damage and replace as necessary.

To install:

5. Replace the rubber boot and install the CV-joint using a press. Replace the grease with the appropriate CV-joint lubricant.
6. Install or connect the following:
   - CV-joint locking ring
   - Rubber boot sealing clamp
   - End cover and boot cap
   - Axle shaft

Pinion Seal

REMOVAL & INSTALLATION

Front

1. Before servicing the vehicle, refer to the Precautions Section.
2. Remove or disconnect the following:
   - Driveshaft
3. Using torque wrench, loosen hex nut while measuring torque reading.
   - Drive flange
   - Pinion seal

To install:

4. Coat the pinion seal lip clean hypoid gear oil.
5. Install or connect the following:
   - Pinion seal
   - Drive flange
   - New self-locking hexagon collar nut. Torque the nut to measured torque.
   - Driveshaft

Axle Housing Assembly

REMOVAL & INSTALLATION

Front

1. Before servicing the vehicle, refer to the Precautions Section.
2. Drain the front axle oil.
3. Remove or disconnect the following:
   - Front wheels
   - Engine undercover
   - Driveshaft
   - Follower joints from upper control arm
   - Axle shafts
   - Front axle bolts
   - Front axle housing from vehicle

To install:

4. Install or connect the following:
   - Axle housing assembly. Torque the bolts to 100 ft. lbs. (135 Nm)
   - Axle shafts
   - Follower joints. Torque the bolts to 37 ft. lbs. (50 Nm).
   - Driveshaft
   - Engine undercover
   - Front wheels
5. Fill the axle with gear oil.

Rear

1. Before servicing the vehicle, refer to the Precautions Section.
2. Drain the differential oil.
3. Remove or disconnect the following:
   - Rear wheels
   - Axle shafts
   - Auxiliary fuel tank (if equipped)
   - Brake line brackets
   - Wheel speed sensors
   - Exhaust shield
   - Driveshaft
   - Strut at lower control arm
   - Tie rod
4. Support the rear axle housing assembly
5. Remove or disconnect the following:
   - Axle housing suspension mountings
   - Axle housing assembly

To install:

6. Install or connect the following:
   - Axle housing assembly. Torque the bolts to 148 ft. lbs. (200 Nm).
   - Tie rod. Torque the bolts to 37 ft. lbs. (50 Nm).
   - Strut at lower control arm. Torque the bolts to 63 ft. lbs. (85 Nm)
   - Driveshaft. Torque the bolts to 37 ft. lbs. (50 Nm).
   - Exhaust shield
   - Wheel speed sensors
   - Brake line brackets
   - Auxiliary fuel tank
   - Axle shafts
   - Rear wheels
7. Fill the rear axle with gear oil.

STEERING AND SUSPENSION

Air Bag

⁂ CAUTION

Mercedes-Benz vehicles are equipped with an air bag system. The system must be disabled before performing service on or around system components, steering column, instrument panel components, wiring and sensors. Failure to follow safety and disabling procedures could result in accidental air bag deployment, possible personal injury and unnecessary system repairs.

PRECAUTIONS

Several precautions must be observed when handling the inflator module to avoid accidental deployment and possible personal injury.

- Never carry the inflator module by the wires or connector on the underside of the module.
- When carrying a live inflator module, hold securely with both hands and ensure that the bag and trim cover are pointed away.
- Place the inflator module on a bench or other surface with the bag and trim cover facing up.
- With the inflator module on the bench, never place anything on or close to the module, which may be thrown in the event of an accidental deployment.

DISARMING

To avoid personal injury when working on vehicles equipped with an air bag, the negative battery cable must be disconnected and insulated before working on the system. Failure to do so may result in accidental deployment of the air bag.

REARMING

To rearm the air bag system, reattach the battery cable(s).

Power Steering Pump

REMOVAL & INSTALLATION

1. Before servicing the vehicle, refer to the Precautions Section.
2. Disconnect the negative battery cable.
3. Drain the power steering fluid.
4. Remove or disconnect the following:
   - Fan clutch, fan and shroud
   - Accessory drive belt
   - Power steering lines at pump
   - Oil cooler lines if equipped
   - Power steering pump
   - Reservoir

To install:

5. Install or connect the following:
   - Reservoir
   - Power steering pump. Torque the mounting bolts to 18 ft. lbs. (25 Nm)
   - Power steering lines
   - Oil cooler lines if equipped
   - Accessory drive belt
   - Fan clutch, fan and shroud
6. Refill the reservoir with fresh fluid. Start the engine and turn the steering wheel from lock-to-lock several times. The system will bleed automatically. Check for leaks.

Rack and Pinion Steering Gear

REMOVAL & INSTALLATION

1. Before servicing the vehicle, refer to the Precautions Section.

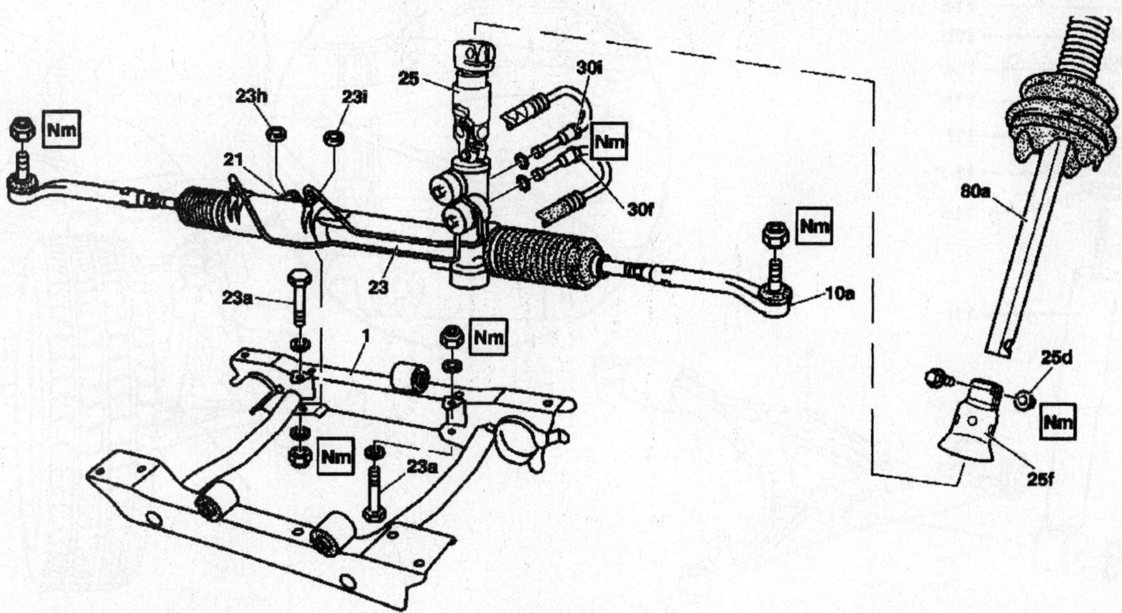

Shown on model 163.154 (vehicles without speed- sensitive power steering)

1 Front axle carrier
10a Tie rod joint
21 Rubber mount
23 Rack- and-pinion steering
23a Bolts
23h Shim
23i Shim
25 Steering coupling
25d Nut
25f Steering coupling shield
30f Return line
30i High- pressure expansion hose
80a Lower steering shaft

42348-MCLS-G14

Exploded view of the rack and pinion steering gear mounting

2. Disconnect the negative battery cable.
3. Center the steering wheel and turn the key to the lock position. Remove the key from the ignition switch.
4. Drain the power steering reservoir.
5. Remove or disconnect the following:
   - Front axle housing
   - Track rod
   - Pressure and return lines
   - Steering coupling heat shield
   - Steering coupling
   - Steering gear retainer bolts
   - Steering gear

To install:

   - Steering gear. Torque the bolts to 37 ft. lbs. (50 Nm).
   - Steering coupling. Torque the bolts to 19 ft. lbs. (26 Nm).
   - Steering coupling heat shield
   - Pressure and return lines. Torque the bolts to 11 ft. lbs. (15 Nm).
   - Track rod. Torque the bolts to 37 ft. lbs. (50 Nm).
   - Front axle housing

6. Refill the power steering reservoir with fresh fluid. Start the engine and turn the steering wheel from lock-to-lock several times. The system will bleed automatically. Check for leaks.

Shock Absorber

REMOVAL & INSTALLATION

Front

1. Before servicing the vehicle, refer to the Precautions Section.
2. Remove or disconnect the following:
   - Front wheel
   - Fender Liner
   - Upper mounting
   - Lower mounting bolt
   - Shock absorber

To install:

3. Install or connect the following:
   - Shock absorber
   - Lower mounting bolt. Torque the bolt to 96 ft. lbs. (130 Nm).
   - Upper mounting. Torque the nut to 22 ft. lbs. (30 Nm)
   - Fender liner
   - Front wheel

Rear

1. Before servicing the vehicle, refer to the Precautions Section.
2. Remove the upper strut mounting with the vehicle on the floor.
3. Remove or disconnect the following:
   - Rear wheel
   - Fender liner
   - Top shock mounting nuts
   - Lower shock mounting bolt
   - Support lower control am
   - Torsion bar link rod
   - Lower control arm
   - Shock/spring assembly
4. Compress the spring and remove the shock absorber.

To install:

5. Install or connect the following:
   - Shock absorber
   - Shock/spring assembly
   - Lower control arm. Torque to 100 ft. lbs. (135 Nm).
   - Torsion bar link rod. Torque to 21 ft. lbs. (28 Nm).
   - Lower strut mounting bolt. Torque to 63 ft. lbs. (85 Nm)
   - Top strut mounting nuts. Torque to 15 ft. lbs. (20 Nm)
   - Fender liner
   - Rear wheel

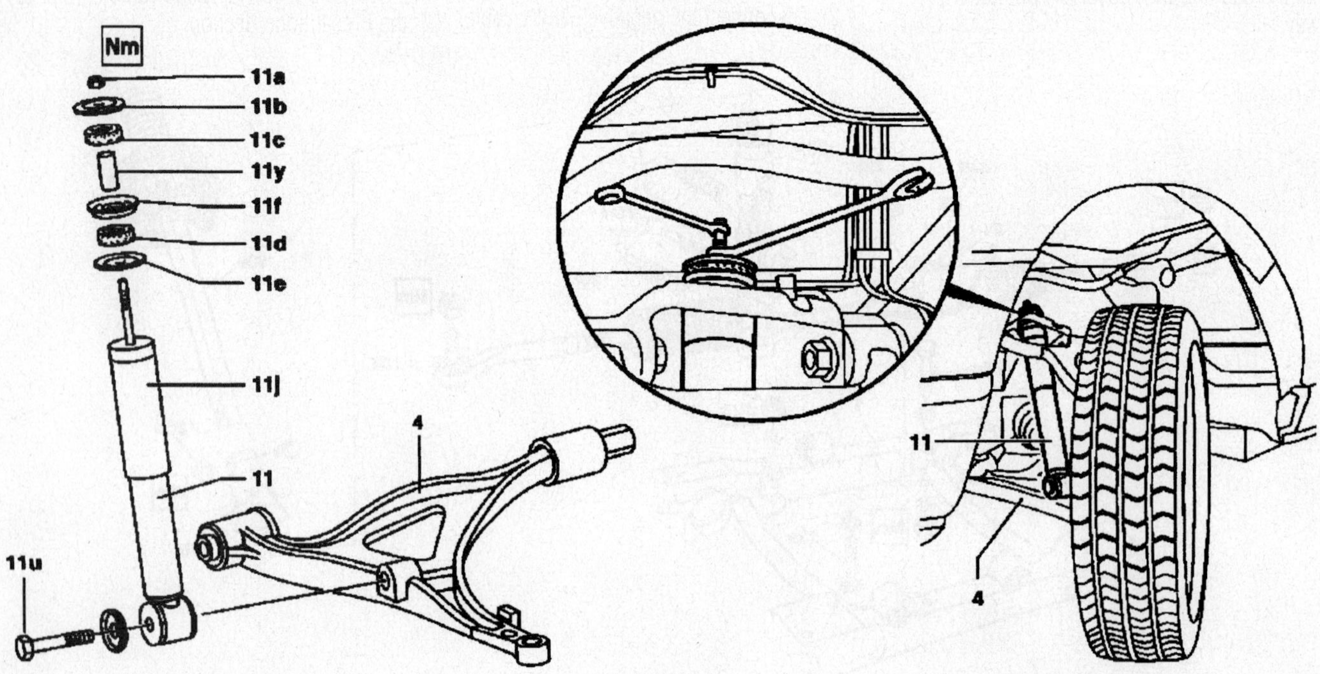

4	Lower transverse control arm	11c	Upper rubber mount	11j	Cover
11	Vibration damper	11d	Lower rubber mount	11u	Bolt
11a	Nut	11e	Plate	11y	Sleeve
11b	Plate	11f	Plate		

42348-MCLS-G15

Front shock absorber mounting

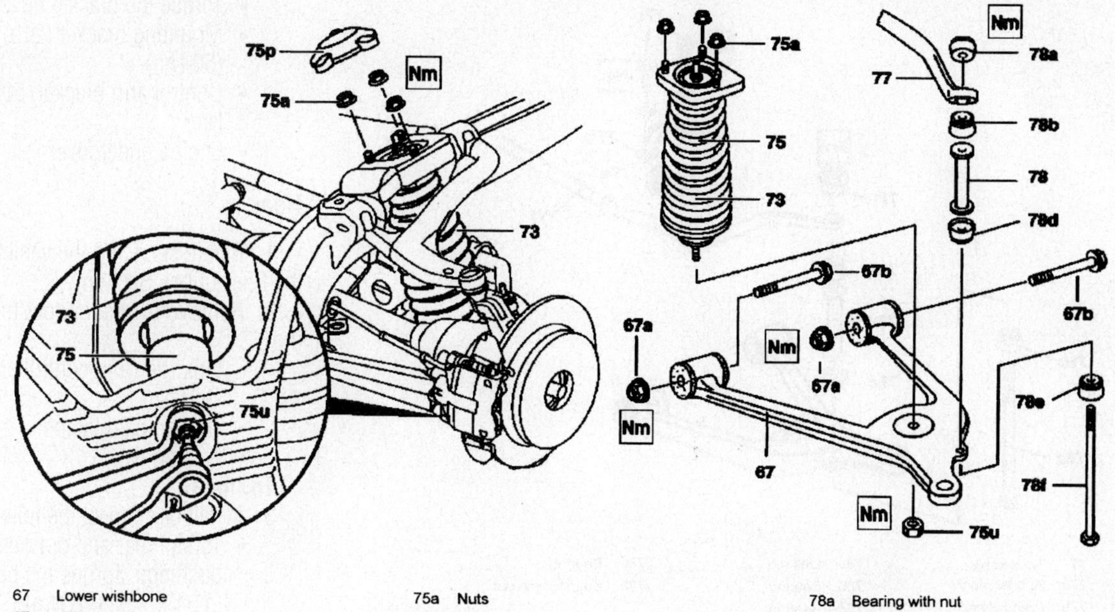

67	Lower wishbone	75a	Nuts	78a	Bearing with nut
67a	Nuts	75u	Nut	78b	Bearing
67b	Bolts	75p	Cover	78d	Bearing
73	Rear spring	77	Torsion bar	78e	Bearing
75	Vibration damper	78	Distance sleeve	78f	Link rod

42348-MCLS-G16

Rear shock absorber/spring mounting

Coil Spring

REMOVAL & INSTALLATION

Rear

Refer to the shock absorber removal and installation procedure for coil spring service.

Torsion Bar

REMOVAL & INSTALLATION

Front

1. Before servicing the vehicle, refer to the Precautions Section.
2. Remove or disconnect the following:
   - Engine undercover
   - Bracket from control arm
   - Mounting bracket (21b)
   - Torsion bar

To install:

3. Install or connect the following:
   - Torsion bar

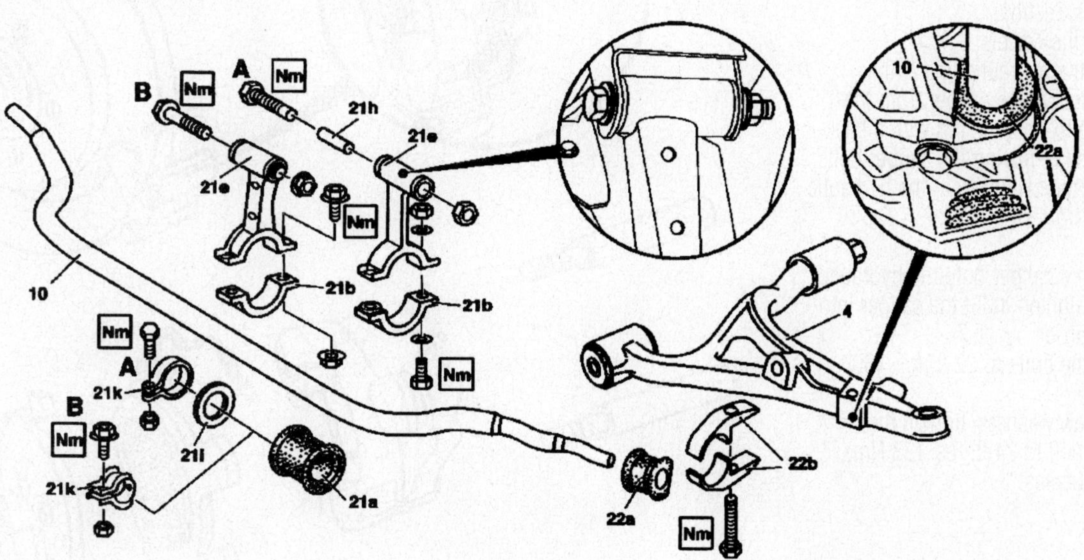

4	Lower transverse control arm	21e	Retaining bracket	22a	Rubber mount
10	Torsion bar	21h	Sleeve	22b	Bracket
21a	Rubber mount	21i	Washers	A	Version up to 31.07.98
21b	Retaining bracket	21k	Anti- shift device	B	Version as of 01.08.98

42348-MCLS-G17

Front torsion bar mounting

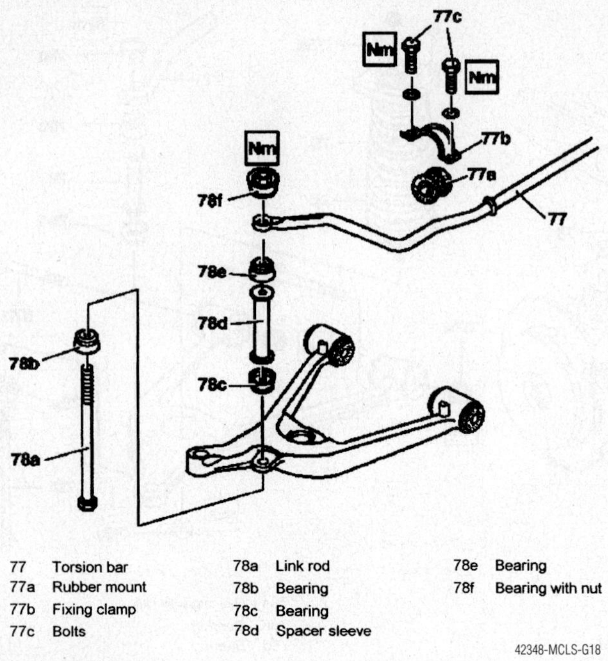

Rear torsion bar mounting

- Torque the bracket bolts as follows:
- Mounting bracket (21b): 37 ft. lbs. (50 Nm)
- Control arm bracket: 50 ft. lbs. (68 Nm)
- Engine undercover

Rear

1. Before servicing the vehicle, refer to the Precautions Section.
2. Remove or disconnect the following:
   - Link rod from control arm to torsion bar
   - Torsion bar clamps
   - Torsion bar

To install:

3. Install or connect the following:
   - Torsion bar and brackets with new bushings. Torque the bracket nuts to 21 ft. lbs. (28 Nm).
   - Link rod. Torque the bracket nuts to 15 ft. lbs. (21 Nm).

BRAKES

Brake Caliper

REMOVAL & INSTALLATION

Floating Caliper

FRONT

1. Before servicing the vehicle, refer to the Precautions Section.
2. Remove the wheels.
3. Loosen the hydraulic line at the caliper, then remove the caliper from the carrier. Be sure to hold the pin with a back-up wrench when removing the caliper bolts.
4. Remove the caliper from the hydraulic line.

To install:

5. Thread the caliper onto the hydraulic line and hand-tighten it. Fit the caliper into place on the carrier.
6. Torque the bolts to 22 ft. lbs. (30 Nm).
7. Using new washers, tighten the hydraulic line bolt to 24 ft. lbs. (33 Nm), and bleed the brakes.

REAR

1. Before servicing the vehicle, refer to the Precautions Section.
2. If equipped with ABS, make sure the ignition switch stays **OFF** and pump the brake pedal 25–35 times to relieve the system pressure.

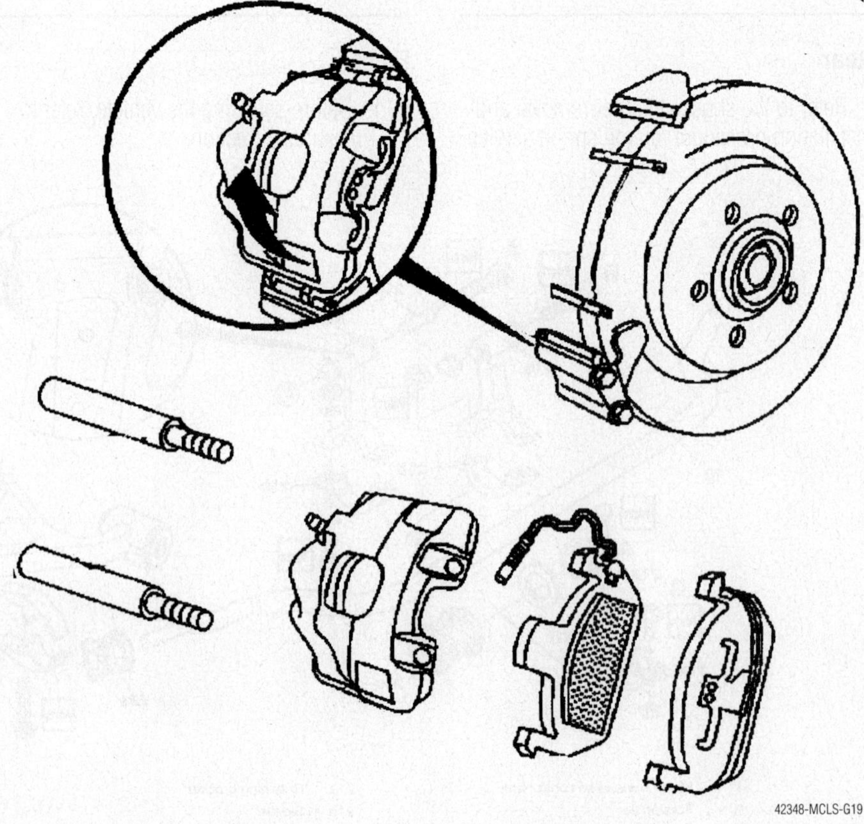

Front brake floating caliper

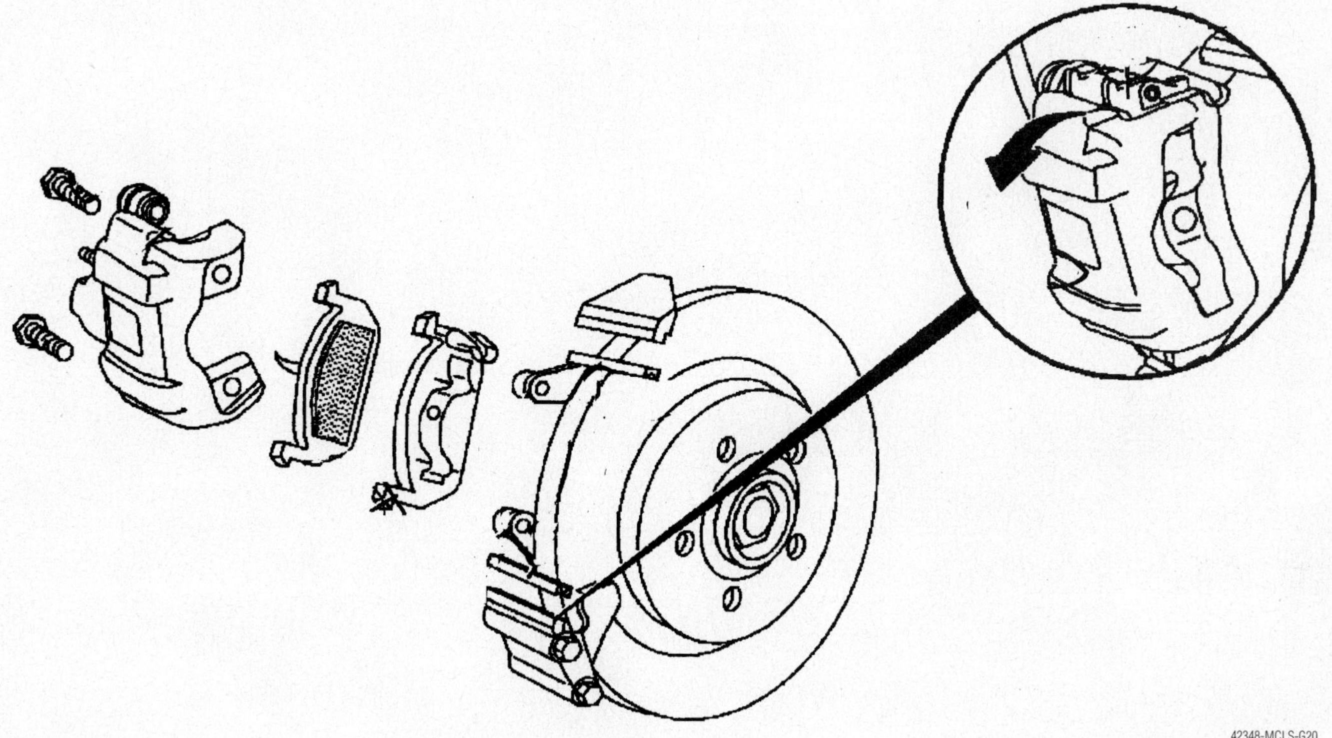

Rear brake caliper

3. Remove the wheels.
4. Disconnect the parking brake cable.
5. Loosen the hydraulic line.
6. Use a back-up wrench to hold the guide pins and remove the caliper bolts.
7. Lift the caliper off the carrier and unscrew it from the hydraulic line. Discard the washers.

To install:

8. Thread the caliper onto the hydraulic line and hand-tighten it. Fit the caliper into place on the carrier. Torque the bolts to 17 ft. lbs. (24 Nm).
9. Connect the brake line. Torque to 13 ft. lbs. (18 Nm).
10. Bleed the brakes.
11. Install the wheels.

Brake Pads

REMOVAL & INSTALLATION

Floating Caliper

FRONT

1. Before servicing the vehicle, refer to the Precautions Section.
2. Remove the front wheels.
3. Hold the lower guide pin with an open wrench and remove the bolt securing the caliper to the guide pin.
4. Pivot the caliper up on the upper guide pin and slide the pads straight out to remove them.

To install:

5. Compress the caliper piston into the bore.
6. Fit the new pads into the carrier and pivot the caliper into place.
7. The original bolts are micro-encapsulated with a thread locking compound. Install a new bolt or clean the old bolt and apply a thread-locking compound.
8. When tightening the bolt, be sure to use a back-up wrench to hold the guide pin. Torque the bolt to 22 ft. lbs. (30 Nm).
9. Install the wheels.

REAR

1. Before servicing the vehicle, refer to the Precautions Section.
2. Remove the rear wheels.

3. Remove the parking brake cable clip from the caliper. Disconnect the parking brake cable.

4. Hold the guide pin with a back-up wrench and remove the upper mounting bolt from the brake caliper.

5. Swing the caliper downward and remove the brake pads.

To install:

6. Retract the piston into the housing by rotating the piston clockwise.

7. Install the new brake pads onto the pad carrier.

8. Install the caliper to the pad carrier using a new self locking bolt or a thread locking compound and torque to 17 ft. lbs. (24 Nm).

9. Attach the hand brake cable to the caliper.

10. Check the parking brake operation and adjust the cable if necessary.

11. Install the wheels.

MINI

Mini Cooper • Mini Cooper S

9

SPECIFICATION CHARTS

ENGINE AND VEHICLE IDENTIFICATION

Engine							Model Year	
Code ①	Liters (cc)	Cu. In.	Cyl.	Fuel Sys.	Engine Type	Eng. Mfg.	Code ②	Year
W10	1.6 (1598)	97.5	4	NA	SOHC	BMW	2	2002
W11	1.6 (1598)	97.5	4	NA	SOHC	BMW	3	2003
							4	2004
							5	2005
							6	2006

NA: Not available

SOHC: Single Overhead Camshaft

① 8th position of VIN

② 10th position of VIN

06041_BMW_MINI_C0001

GENERAL ENGINE SPECIFICATIONS

Year	Model	Engine Displacement Liters (VIN)	Net Horsepower @ rpm	Net Torque @ rpm (ft. lbs.)	Bore x Stroke (in.)	Com-pression Ratio	Oil Pressure @ rpm
2002	R50	1.6 (W10/4)	115@6000	100@4500	3.03x3.38	10.6:1	25-80@3000
	R53	1.6 (W11/4)	163@6000	155@4000	3.03x3.38	8.3:1	NA
2003	R50	1.6 (W10/4)	115@6000	100@4500	3.03x3.38	10.6:1	25-80@3000
	R53	1.6 (W11/4)	163@6000	155@4000	3.03x3.38	8.3:1	NA
2004	R50	1.6 (W10/4)	115@6000	100@4500	3.03x3.38	10.6:1	25-80@3000
	R53	1.6 (W11/4)	163@6000	155@4000	3.03x3.38	8.3:1	NA
2005-06	R50	1.6 (W10/4)	115@6000	100@4500	3.03x3.38	10.6:1	25-80@3000
	R53	1.6 (W11/4)	163@6000	155@4000	3.03x3.38	8.3:1	NA

NA: Not available

06041_BMW_MINI_C0002

ENGINE TUNE-UP SPECIFICATIONS

Year	Engine Displacement Liters (VIN)	Spark Plug Gap (in.)	Ignition Timing (deg.)	Fuel Pump (psi)	Idle Speed (rpm)	Valve Clearance	
						In.	Ex.
2002	1.6 (W10/4)	NA	①	NA	②	HYD	HYD
	1.6 (W11/4)	NA	①	NA	②	HYD	HYD
2003	1.6 (W10/4)	NA	①	NA	②	HYD	HYD
	1.6 (W11/4)	NA	①	NA	②	HYD	HYD
2004	1.6 (W10/4)	NA	①	NA	②	HYD	HYD
	1.6 (W11/4)	NA	①	NA	②	HYD	HYD
2005-06	1.6 (W10/4)	NA	①	NA	②	HYD	HYD
	1.6 (W11/4)	NA	①	NA	②	HYD	HYD

NOTE: The Vehicle Emission Control Information label often reflects specification changes made during production. The label figures must be used if they differ from those in this chart.

NA: Not available

HYD: Hydraulic

① Ignition timing is regulated by the Electronic Control Module (ECM), and cannot be adjusted.

② Idle speed is controled by the Electronic Control Module (ECM), and cannot be adjusted.

06041_BMW_MINI_C0003

CAPACITIES

Year	Model	Engine Displacement Liters (VIN)	Engine Oil with Filter (qts.)	Automatic Transaxle (qts.)	Manual Transaxle (qts.)	Rear Drive Axle (pts.)	Fuel Tank (gal.)	Cooling System (qts.)
2002	R50	1.6 (W10/4)	4.8	—	2.1	NA	13.2	5.8
	R53	1.6 (W11/4)	4.8	—	1.6	NA	13.2	6.3
2003	R50	1.6 (W10/4)	4.8	—	2.1	NA	13.2	5.8
	R53	1.6 (W11/4)	4.8	—	1.6	NA	13.2	6.3
2004	R50	1.6 (W10/4)	4.8	—	2.1	NA	13.2	5.8
	R53	1.6 (W11/4)	4.8	—	1.6	NA	13.2	6.3
2005-06	R50	1.6 (W10/4)	4.8	—	2.1	NA	13.2	5.8
	R53	1.6 (W11/4)	4.8	—	1.6	NA	13.2	6.3

NOTE: All capacities are approximate. Add fluid gradually and check to be sure a proper fluid level is obtained.

NA: Not applicable to this model

06041_BMW_MINI_C0004

TORQUE SPECIFICATIONS

All readings in ft. lbs.

Year	Engine Displacement Liters (VIN)	Cylinder Head Bolts	Main Bearing Bolts	Rod Bearing Bolts	Crankshaft Damper Bolts	Flywheel Bolts	Manifold Intake	Manifold Exhaust	Spark Plugs	Oil Pan Drain Plug
2002	1.6 (W10/4)	①	②	15 ①	NA	59	19	18	20	18
	1.6 (W11/4)	①	②	15 ①	NA	66	19	18	20	18
2003	1.6 (W10/4)	①	②	15 ①	NA	59	19	18	20	18
	1.6 (W11/4)	①	②	15 ①	NA	66	19	18	20	18
2004	1.6 (W10/4)	①	②	15 ①	NA	59	19	18	20	18
	1.6 (W11/4)	①	②	15 ①	NA	66	19	18	20	18
2005-06	1.6 (W10/4)	①	②	15 ①	NA	59	19	18	20	18
	1.6 (W11/4)	①	②	15 ①	NA	66	19	18	20	18

NA: Not available

① Step 1: Bolts 1 through 10: 30 ft. lbs.

Step: 2 Bolts 1 through 10 tighten an additional 90 degrees

Step: 3 Bolts 11 through 12 21 ft. lbs.

② Inner (M10) bolts: 44 ft. lbs.

Outer (M8) bolts: 26 ft. lbs.

③ Plus an additional 90 degrees

06041_BMW_MINI_C0005

WHEEL ALIGNMENT

Year	Model		Caster Range (+/-Deg.)	Caster Preferred Setting (Deg.)	Camber Range (+/-Deg.)	Camber Preferred Setting (Deg.)	Toe-in (Deg.)
2002	R50	F	①	①	0.40	-0.50	0.18+/-0.05
		R	—	—	0.50	-1.50	0.24+/-0.08
	R53	F	①	①	0.40	-0.50	0.18+/-0.05
		R	—	—	0.50	-1.50	0.24+/-0.08
2003	R50	F	①	①	0.40	-0.50	0.18+/-0.05
		R	—	—	0.50	-1.50	0.24+/-0.08
	R53	F	①	①	0.40	-0.50	0.18+/-0.05
		R	—	—	0.50	-1.50	0.24+/-0.08
2004	R50	F	①	①	0.40	-0.50	0.18+/-0.05
		R	—	—	0.50	-1.50	0.24+/-0.08
	R53	F	①	①	0.40	-0.50	0.18+/-0.05
		R	—	—	0.50	-1.50	0.24+/-0.08
2005-06	R50	F	①	①	0.40	-0.50	0.18+/-0.05
		R	—	—	0.50	-1.50	0.24+/-0.08
	R53	F	①	①	0.40	-0.50	0.18+/-0.05
		R	—	—	0.50	-1.50	0.24+/-0.08

① With +/- 10<deg> steering angle: 4<deg>47' +/- 30'.
With +/- 20<deg> steering angle: 4<deg>55' +/- 30'.

② Inside wheel: 36<deg>10' (approx.).
Outside wheel: 30<deg>36' (approx.).

06041_BMW_MINI_C0006

TIRE, WHEEL AND BALL JOINT SPECIFICATIONS

Year	Model	OEM Tires Standard	OEM Tires Optional	Tire Pressures (psi) Front	Tire Pressures (psi) Rear	Wheel Size	Ball Joint Inspection	Lug Nuts
2002	R50	175/65R15	195/55R16	32	32	15x5.5	NA	89
	R53	195/55R16	205/45R17	32	32	16x6.5	NA	89
2003	R50	175/65R15	195/55R16	32	32	15x5.5	NA	89
	R53	195/55R16	205/45R17	32	32	16x6.5	NA	89
2004	R50	175/65R15	195/55R16	32	32	15x5.5	NA	89
	R53	195/55R16	205/45R17	32	32	16x6.5	NA	89
2005-06	R50	175/65R15	195/55R16	32	32	15x5.5	NA	89
	R53	195/55R16	205/45R17	32	32	16x6.5	NA	89

OEM: Original Equipment Manufacturer

PSI: Pounds Per Square Inch

NA: Not available

06041_BMW_MINI_C0007

BRAKE SPECIFICATIONS

All measurements in inches unless noted

Year	Model		Brake Disc: Original Thickness	Brake Disc: Minimum Thickness	Brake Disc: Maximum Run-out	Brake Drum Diameter: Original Inside Diameter	Brake Drum Diameter: Max. Wear Limit	Brake Drum Diameter: Maximum Machine Diameter	Min. Lining Thickness	Brake Caliper: Bracket Bolts (ft. lbs.)	Brake Caliper: Mounting Bolts (ft. lbs.)
2002	R50	F	NA	①	NA	NA	NA	NA	0.118	81	22-26
		R	NA	①	NA	NA	NA	NA	0.118	49	18-22
	R53	F	NA	①	NA	NA	NA	NA	0.118	81	22-26
		R	NA	①	NA	NA	NA	NA	0.118	49	18-22
2003	R50	F	NA	①	NA	NA	NA	NA	0.118	81	22-26
		R	NA	①	NA	NA	NA	NA	0.118	49	18-22
	R53	F	NA	①	NA	NA	NA	NA	0.118	81	22-26
		R	NA	①	NA	NA	NA	NA	0.118	49	18-22
2004	R50	F	NA	①	NA	NA	NA	NA	0.118	81	22-26
		R	NA	①	NA	NA	NA	NA	0.118	49	18-22
	R53	F	NA	①	NA	NA	NA	NA	0.118	81	22-26
		R	NA	①	NA	NA	NA	NA	0.118	49	18-22
2005-06	R50	F	NA	①	NA	NA	NA	NA	0.118	81	22-26
		R	NA	①	NA	NA	NA	NA	0.118	49	18-22
	R53	F	NA	①	NA	NA	NA	NA	0.118	81	22-26
		R	NA	①	NA	NA	NA	NA	0.118	49	18-22

F: Front

R: Rear

NA: Not available

① Minimum thickness is stamped in the brake disc shell

06041_BMW_MINI_C0008

ENGINE REPAIR

➡Disconnecting the negative battery cable on some vehicles may interfere with the functions of the on-board computer systems and may require the computer to undergo a relearning process.

Ignition Timing

ADJUSTMENT

The Digital Motor Electronics (DME) control, unit controls all ignition and fuel injection functions. Ignition timing is fully electronically controlled; there is no vacuum advance or manual adjustment. Ignition functions are calculated from internal maps and from the same sensors used for the fuel injection system. On vehicles with an automatic transmission, the control unit will retard ignition timing briefly when the transmission is about to shift up or down. For this reason, there is a data link between the DME control unit and the transmission control unit.

Since the ignition timing is controlled by the DME, checking and adjusting the timing is impossible. There is no method of setting dynamic or static timing.

Alternator

REMOVAL & INSTALLATION

➡When the battery is disconnected the radio code, on-board computer and clock settings will be lost. The radio code should be obtained before disconnecting the battery or radio. Once the battery has been reconnected, the radio will not function unless the code is keyed in.

Cooper Coupe

1. Before servicing the vehicle, refer to the precautions section.
2. Check for stored fault codes, then erase code memory.
3. Switch off ignition.
4. Disconnect negative battery cable.
5. Remove or disconnect the following:
   - Alternator drive belt
   - Electrical connections from alternator
   - Alternator mounting bolts
   - Alternator

Installing special tools to move modular front end for alternator access—Cooper S Coupe models

To install:

6. To install, reverse removal procedure.
7. Torque alternator mounting bolts to 18 ft. lbs. (25 Nm) and the power lead to the alternator stud to 7 ft. lbs. (10 Nm).
8. Check for any stored fault codes.
9. Clear fault code memory.

Cooper S Coupe

1. Before servicing the vehicle, refer to the precautions section.
2. Check for stored fault codes, the erase fault code memory.
3. Switch off ignition.
4. Disconnect the battery.
5. Remove or disconnect the following:
   - Alternator drive belt
   - Front bumper cover
6. Loosen the retainers for the modular front end and insert bracing tools, 11–8–401/2, to provide access room.
7. Detach the electrical connections from the alternator.
8. Remove the mounting bolts and remove the alternator.

To install:

9. To install, reverse removal procedure.
10. Torque alternator mounting bolts to 18 ft. lbs. (25 Nm) and the power lead to the alternator stud to 7 ft. lbs. (10 Nm).
11. Check for any stored fault codes.
12. Clear fault code memory.

Engine Assembly

REMOVAL & INSTALLATION

Cooper Coupe

1. Before servicing the vehicle, refer to the precautions section.
2. Disconnect the battery and battery container.
3. Properly relieve the fuel system pressure.
4. Remove or disconnect the following:
   - Front bumper cover
   - Bumper bracket
   - Air cleaner housing
   - Clutch cylinder
   - Top stabilizer bar bracket
   - Exhaust manifold with catalytic converter
   - Both driveshafts
   - Auxiliary drive belts
5. Drain the cooling system.
6. Remove the upper engine hood support and install a special support tool, 51–2–160.
7. Loosen the retainers for the modular front end and insert bracing tools, 11–8–401/2, to provide access room.
8. Remove or disconnect the following:
   - Coolant hoses
   - Heater hoses
   - Overflow hose

9. Release the fuel line (1), then plug the line opening.

⁂ CAUTION

Fuel may still be under pressure. Use caution when releasing any fuel lines. Prepare to catch fuel spillage.

10. Remove the starter heat shield.
11. Release the brake booster line connection at the booster by pressing downward on clip ring and pulling line upward.
12. Remove or disconnect the following:
    - Plug connector from fuse box
    - Fuse box (position aside)
    - Grounding cable from left spring strut
    - Circular connector (twist upper and lower halves in opposite directions)
    - Transmission shift cables from retaining clips and ball connections
    - Shift control housing
    - Engine stabilizer brace
    - Steering pump motor connector
    - A/C compressor connector and 2 retaining bolts (position compressor aside)
    - Both transmission brackets
13. Install a special holding tool, 11–8–352, on transmission as a lifting eye.
14. Install a special lifting tool, 11–8–351, on front of engine as a lifting eye.
15. Attach engine lifting equipment to lifting eyes.
16. Remove engine mount retaining nut and remove the engine.

To install:

17. Install in reverse of removal procedure.
18. Torque retainers to the following:
    - Engine mount retaining nut: 50 ft. lbs. (68 Nm)
    - Mount bracket to transmission bolts: 49 ft. lbs. (66 Nm)
    - A/C compressor mounting bolts: 18 ft. lbs. (25 Nm)

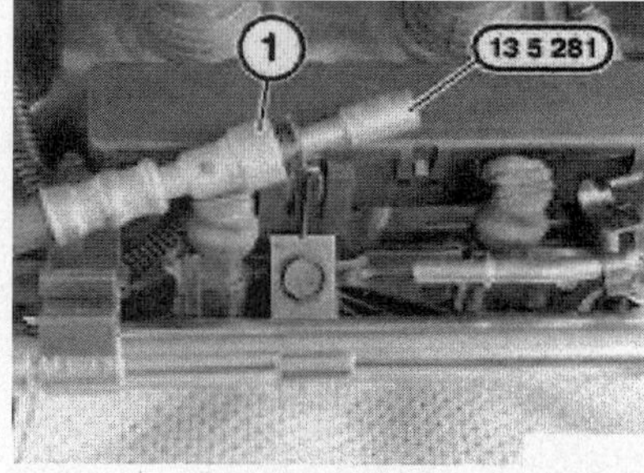

Releasing the fuel line—Cooper Coupe models

Showing location of engine stabilizer bar—Cooper Coupe models

Showing left transmission bracket—Cooper Coupe models

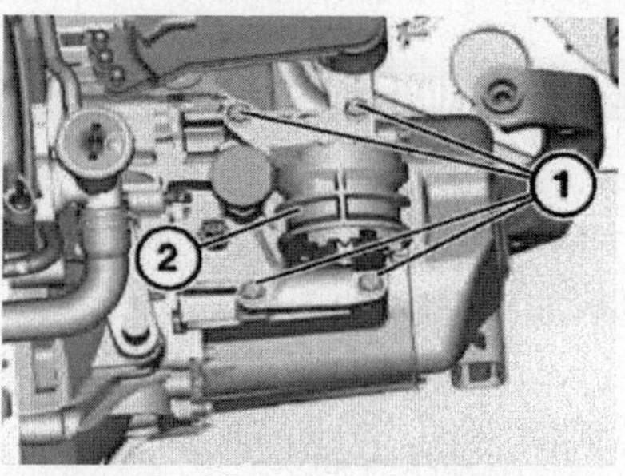

Showing right transmission bracket—Cooper Coupe models

- Engine stabilizer bar bolts: 74 ft. lbs. (100 Nm)
- Shift cable bracket bolt: 16 ft. lbs. (22 Nm)
- Ground cable at left strut: 7 ft. lbs. (9 Nm)

19. When engine is fully installed and assembled, refill transmission and cooling systems.

Cooper S Coupe

1. Before servicing the vehicle, refer to the precautions section.
2. Install the intercooler protector, 11–8–480 to ensure intercooler is not damaged during engine removal and installation.
3. Disconnect the battery and battery container.
4. Properly relieve the fuel system pressure.
5. Remove or disconnect the following:
    - Air cleaner housing
    - Auxiliary drive belt
6. Drain the transmission.
7. Remove the lower engine stabilizer bar bracket bolts and remove the bracket.
8. Disconnect the steering pump motor connector.
9. Remove the crush tubes as follows:
    a. Remove the bumper and bumper carrier.
    b. Remove the bolt securing the impact tube to the front end module.
    c. Remove the bolts retaining the impact tube to the chassis subframe.
    d. Remove the impact (crush) tube.
10. Loosen the retainers for the modular front end and insert bracing tools, 11–8–401/2, to provide access room.
11. Remove or disconnect the following:
    - Subframe
    - Driveshafts
    - Exhaust manifold
    - Starter motor heat shield
    - Starter motor connections
    - Oil pressure connector
    - Wiring harness from brackets on starter motor and cooling pipe (note routing for installation; move wiring harness aside)
    - Reverse lamp connector from transaxle
    - Intercooler
    - Throttle valve
    - Tank venting valve lower pipe from stabilizer bracket
    - Fuel line from fuel rail (quick-disconnect fitting); plug openings
    - Fusebox cover and connector from fusebox
    - Grounding cable from left spring strut
    - Circular connector (twist upper and lower halves in opposite directions)
    - Drain cooling system
    - Pipes from heater matrix
    - Expansion tank pipes at heater pipe junction
    - Upper radiator hose
    - Top engine stabilizer bracket
    - Clutch slave cylinder
    - Transmission shift cables from retaining clips and ball connections and bracket
    - A/C compressor connector and retaining bolts (position compressor aside)

- MAP sensor connector (near water housing)
- Coolant hose from water housing

12. Install a special lifting tools, 11–8–351 and 11–8–351, as a lifting eye.
13. Remove the supercharger inlet retaining bolt and release the hose clip.
14. Remove the supercharger inlet pipe to access and disconnect additional hose fittings.
15. Remove the brake booster pipe from the engine and the slave cylinder pipe from the transaxle (move aside).
16. Attach engine lifting equipment to lifting eyes.
17. Remove engine mount retaining nut and remove the engine.

42348-MINI-G06

Showing installed position of the intercooler protector—Cooper S Coupe models

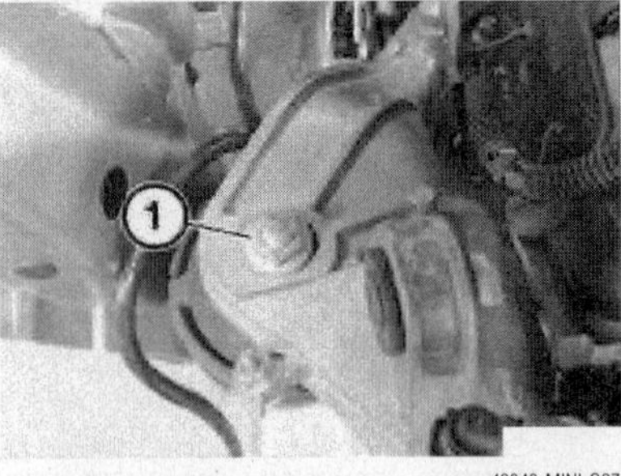

42348-MINI-G07

Identifying location of engine mount retaining nut—Cooper S Coupe models

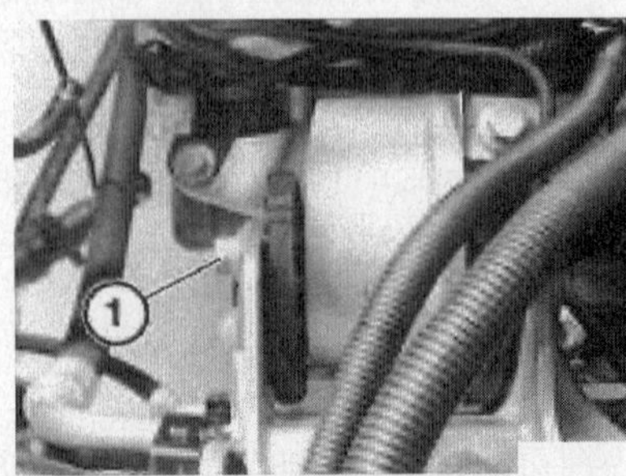

42348-MINI-G08

Identifying location of transaxle–to–upper mount retaining bolt—Cooper S Coupe models

18. Remove the transaxle–to–upper mount retaining bolt.
19. Lower and remove the engine.

To install:

20. Install in reverse of removal procedure.
21. Torque retainers to the following:
    - Engine mount retaining nut: 50 ft. lbs. (68 Nm)
    - Mount bracket to transmission bolts: 49 ft. lbs. (66 Nm)
    - Coolant pipe to cylinder head: 18 ft. lbs. (25 Nm)
    - A/C compressor mounting bolts: 18 ft. lbs. (25 Nm)
    - Ground cable at left strut: 7 ft. lbs. (9 Nm)
    - Starter motor to transaxle bolts: 63 ft. lbs. (85 Nm)
    - Alternator connection on starter: 10 ft. lbs. (14 Nm)
    - Starter motor heat shield bolts: 7 ft. lbs. (9 Nm)
    - Crush tube to Subframe bolts: (165 Nm)
    - Engine stabilizer bar bolts: 74 ft. lbs. (100 Nm)
22. When engine is fully installed and assembled, refill transmission and cooling systems.

Water Pump

REMOVAL & INSTALLATION

Cooper Coupe

1. Before servicing the vehicle, refer to the precautions section.
2. Disconnect the battery.
3. Drain the cooling system.
4. Remove the alternator drive belt and alternator.
5. Drain the cooling system. Drain plug is located on exhaust side of block, next to cylinder number 2.
6. Remove the lower modular front end (MFE) as follows:
    a. Discharge the A/C system.
    b. Remove the engine compartment under tray.
    c. Remove the front bumper assembly.
    d. Remove both front wheel well liners.
    e. Remove or disconnect the following:
    - Cooling fan connectors
    - A/C pipe from condenser (plug openings)
    - A/C pipe from A/C hose (plug openings)

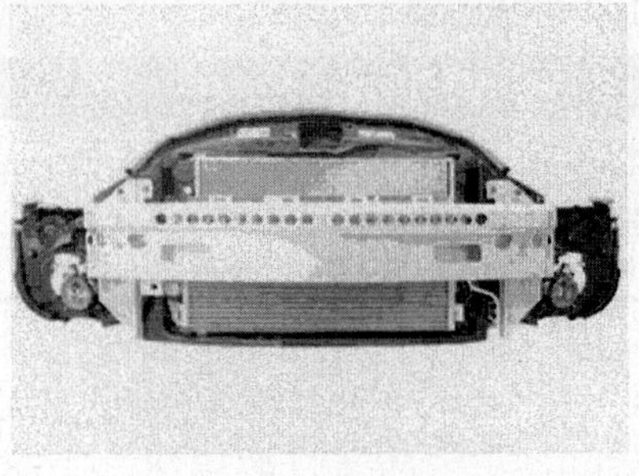
42348-MINI-G09

Showing the modular front end (MFE) assembly—Cooper Coupe

- ABS speed sensor connector from clip (both sides)
- Front fog lamp connector (both sides)
- Horn connector (both sides)
- Feed harness through MFE into wheel housing (both sides)
- Subframe crash tube bolts (both sides)
- Upper radiator hose
- Nuts securing MFE and bumper carrier assembly
- MFE

7. Insert bracing tools, 11–8–401/2, to provide access room.
8. Detach the hoses from the water pump.
9. Remove the bolts and remove the water pump.

To install:

10. Install or connect the following:
    - Water pump mounting bolts; torque to (30 Nm)
    - Hoses to water pump
    - Impact tubes; torque bolts to 74 ft. lbs. (100 Nm)
    - MFE
    - Alternator
    - Battery
11. Refill the cooling system. Start the engine and check for leaks.

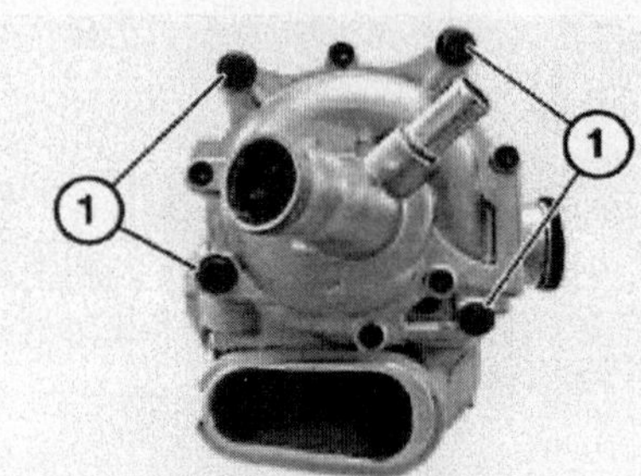

42348-MINI-G10

Showing the location of the water pump mounting bolts—Cooper S Coupe

42348-MINI-G11

Aligning water pump drive (1) to supercharger drive (2)—Cooper S Coupe

Cooper S Coupe

1. Before servicing the vehicle, refer to the precautions section.
2. Drain the cooling system.
3. Remove the negative battery cable.
4. Remove the supercharger
5. Remove the water pump bolts.
6. Remove the water pump.

To install:

7. Clean and remove any residual debris or material from the mounting surfaces for the water pump.
8. Align the water pump drive with the supercharger drive.
9. Install the water pump to the supercharger. Torque the bolts to 18 ft. lbs. (25 Nm).
10. Install a new sealing ring to the water pump and lubricate the seal.
11. Install the supercharger.
12. Reconnect the negative battery cable.
13. Fill and bleed the cooling system.
14. Start the vehicle, check for leaks and repair as necessary.

Heater Core

REMOVAL & INSTALLATION

Cooper Coupe & Cooper S Coupe

1. Before servicing the vehicle, refer to the precautions section.
2. Disconnect the negative battery cable.
3. Drain the cooling system.
4. Discharge and recover the A/C refrigerant.
5. Remove the instrument panel by removing or disconnecting the following:
   - Battery box
   - Intake filter housing
   - Heater hoses
   - A/C pipe from firewall fittings (plug openings)
   - Heater locating stud nut
   - Steering column from steering gear

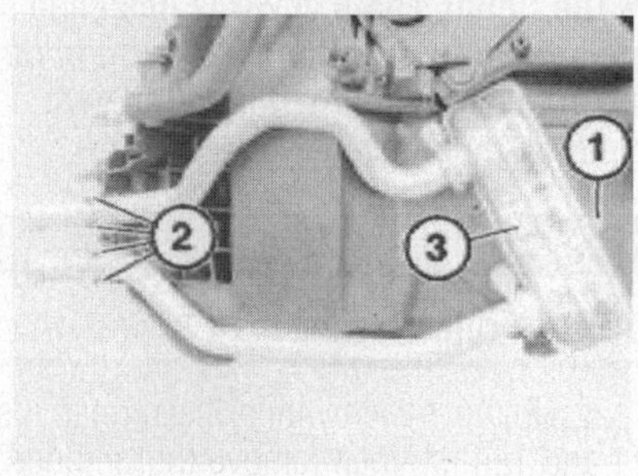

42348-MINI-G12

Removing heater pipes from heater housing

   - Steering column from rubber bellows
   - Left and right A-pillar trim
   - Radio
   - Cover from beneath center controls
   - Left and right kick panels
   - Connector behind left kick panel
   - Upper connector from blower control
   - Instrument panel end covers
   - Bolts behind end covers
   - Lower section of steering column trim panel
   - Electrical harness from instrument panel trim
   - Connectors on steering column
   - Heater connector
   - Lower bolts on instrument panel support (next to steering column lower end)
   - Instrument panel
6. Remove the cover panel from the heater core.
7. Remove the screw and pipes from the connector on the side of the heater housing.
8. Remove the heater core.

To install:

9. Install the heater core into the housing.
10. Connect the heater pipes on the side of the housing.
11. Install the heater housing side cover.
12. Install the instrument panel in reverse of the removal procedure.
13. Refill the cooling system.
14. Reconnect the negative battery cable.
15. Evacuate and recharge the A/C system.
16. Check the coolant level after starting the engine and running for several minutes.

Supercharger

REMOVAL & INSTALLATION

Cooper S Coupe

1. Before servicing the vehicle, refer to the precautions section.
2. Remove or disconnect the following:
   - Intercooler–to–intake manifold pipe
   - Supercharger outlet pipe
   - Auxiliary drive belt
   - Slacken and push out the modular front end (MFE)
   - Crush tubes
3. Drain the cooling system.
4. Remove or disconnect the following:
   - Alternator
   - Radiator hoses
   - Dipstick and move aside
   - Clamp from coolant pipe
   - Supercharger retaining bolts
   - Supercharger

To install:

5. Install new water pump sealing ring.
6. Install the supercharger (4) and torque the mounting M8 bolts to 17 ft. lbs. (25 Nm) and the M10 bolts to 33 ft. lbs. (45 Nm).
7. Install the coolant pipe clamp (2) and the dipstick (1).
8. Install or connect the following:
   - Radiator hoses
   - Supercharger intake pipe
   - Alternator; torque mounting bolts to 17 ft. lbs. (25 Nm).
   - Alternator connectors
   - Crush tubes
   - Modular front end
   - Auxiliary drive belt
   - Supercharger outlet pipe
   - Intercooler–to–intake manifold pipe
9. Refill cooling system.

Cylinder Head

REMOVAL & INSTALLATION

Cooper Coupe

1. Before servicing the vehicle, refer to the precautions section.
2. Drain the cooling system.
3. Remove or disconnect the following:
   - Wheel well liners
   - Battery and battery container
   - Vent hose and engine control DME connector from cylinder head
   - Fuel rail cover
   - Fuel injector wiring harness (move aside)
   - Heater hoses from cylinder head
   - Top hose from thermostat housing
   - Exhaust manifold from block
   - Spark plugs
   - Fuel line from fuel rail (plug openings)
   - Lines from stabilizer bar bracket

- Vacuum line to brake booster from intake manifold
- CMP sensor connector
- Dipstick

4. Remove the intake manifold bolts in reverse of the order as shown (start with bolt 5), then lift the manifold over the dipstick tube.
5. Disconnect the line from the filler neck to the expansion tank, then the engine wiring harness can be moved around the thermostat housing.
6. Tie back the intake manifold from the cylinder head.
7. Remove or disconnect the following:
   - Engine stabilizer bar bracket
   - Oxygen sensor plug connector
   - Holder for oxygen sensor plug from cylinder head
   - Coolant distributor pipe screw
   - Coolant temperature sensor connector
8. Support the engine with a trolley jack and rubber pad on the oil pan. Use caution so oil pan is not damaged.
9. Remove the engine carrier bolts and engine mount nut.
10. Remove the engine carrier.
11. Use a special tool, 11–8–200, and remove the hydraulic engine mount.
12. Mount a special engine holding tool, 11–8–370, on the cylinder block and body fixtures, as shown.
13. Remove the camshaft sensor.
14. Remove the plugs from the cover on each side of the camshaft.
15. Rotate the crankshaft until the triangular adjustment mark on the camshaft gear is at 12 o'clock. Apply a paint reference mark from camshaft and across timing chain for reassembly reference.

➡The brass-colored chain links are of no importance to the timing.

16. Install a special clamping fixture tool, 11–8–250, to camshaft gear. Slacken, but do not remove the center bolt from the camshaft gear.
17. Remove the wiring harness holder, timing chain tensioner and clamping fixture tool.
18. Remove center bolt from camshaft gear.
19. Remove the camshaft gear from the timing chain and secure the chain to prevent it from falling.
20. Remove the bolts from timing chain guides (through the plug openings).
21. Remove the clamping rail and timing chain guides.

➡The timing chain cover is designed so that the timing chain can remain on the crankshaft gear without any gear teeth being skipped.

⁂ CAUTION

DO NOT rotate crankshaft.

22. Remove the cylinder head retainers 11 and 12 first, then remove the cylinder head bolts in reverse of the order shown.
23. Remove the cylinder head.

To install:

24. Clean all sealing material from mating faces

⁂ CAUTION

There must be no oil in the cylinder head bolt holes in the block and timing case cover or there is a possibility of cracking and distorting torque values.

25. Install a new cylinder head gasket.
26. Position the cylinder head onto the block and install new cylinder head bolts (do not clean compound applied to new bolts).
27. Tighten cylinder head bolts, following the sequence shown for bolts 1 through 10, in 2 steps, to the following:
    a. Step 1: 30 ft. lbs. (40 Nm)
    b. Step 2: Additional 90 degrees
28. Tighten cylinder head retainers number 11 and 12 to 21 ft. lbs. (28 Nm).
29. Install or connect the following:
   - Clamping rail and timing chain guides
   - Chain guide bolts; torque to 21 ft. lbs. (28 Nm)
   - Timing chain onto camshaft gear
   - Center bolt in camshaft gear

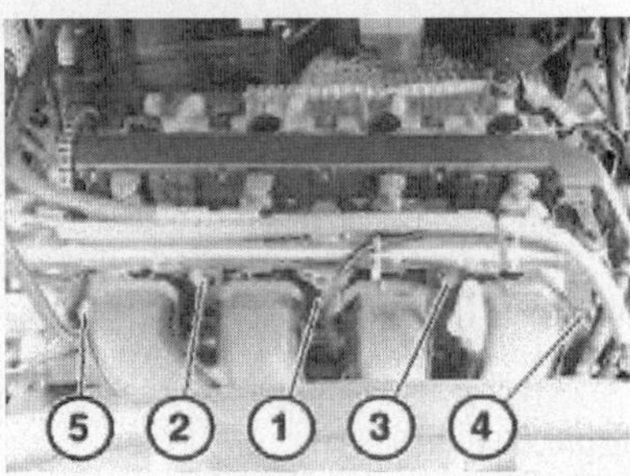

Showing intake manifold bolt tightening sequence—Cooper Coupe

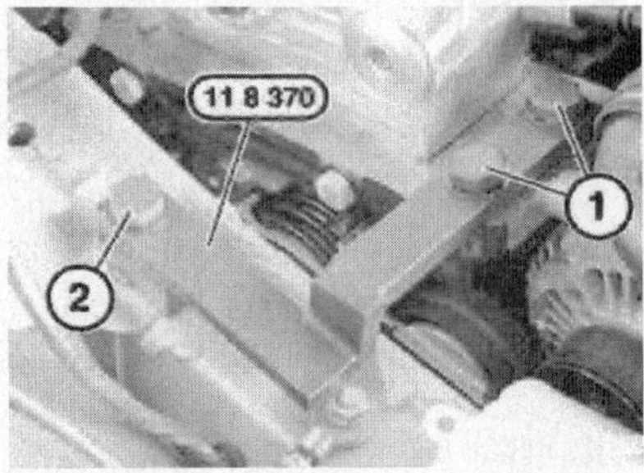

Mounting special holding tool on engine—Cooper Coupe

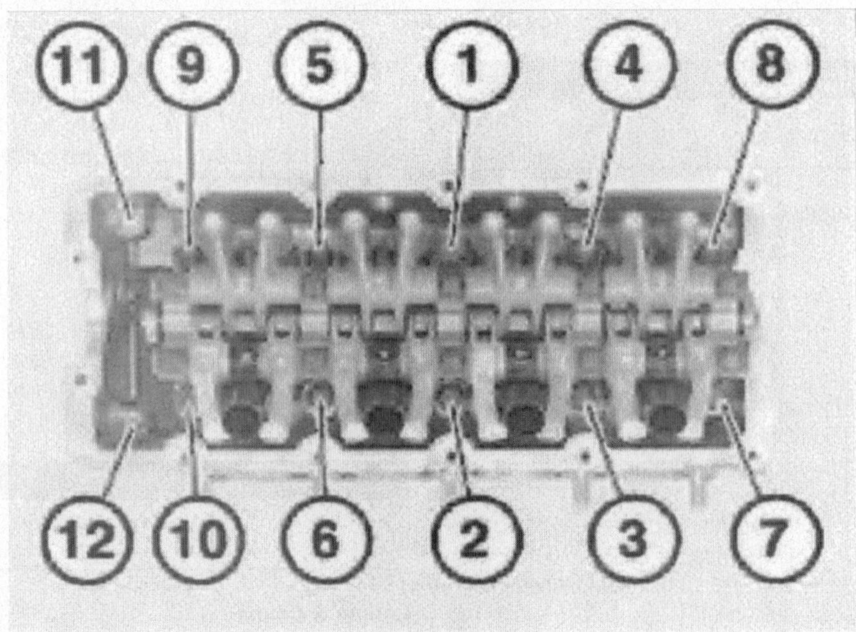

Cylinder head bolt tightening sequence—Cooper Coupe

42348-MINI-G16

Aligning camshaft and timing chain reference marks—Cooper Coupe

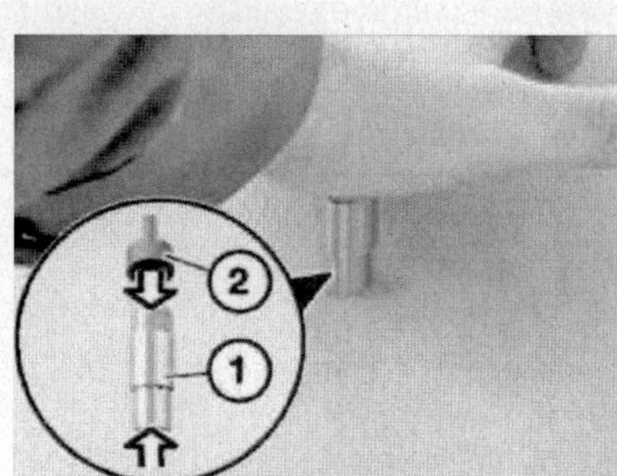

42348-MINI-G17

Reassembling timing chain tensioner clamping fixture—Cooper Coupe

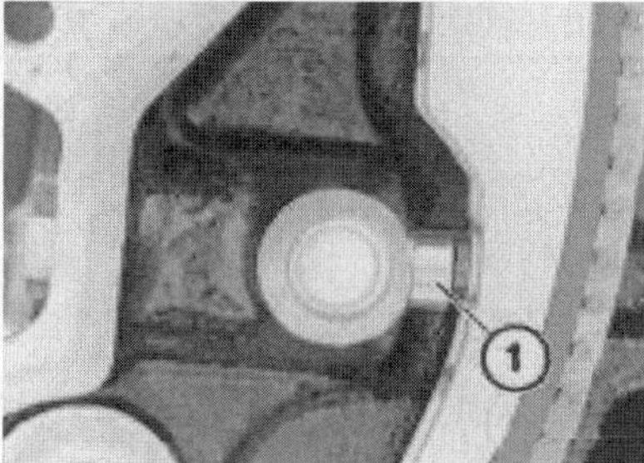

42348-MINI-G18

Showing the released position of the timing chain clamping fixture (with tension on the timing chain)—Cooper Coupe

30. Align the camshaft and timing chain paint marks made during removal.

31. Install camshaft gear holding special tool, 11–8–250, then torque center camshaft gear bolt to 75 ft. lbs. (102 Nm).

32. Move the timing chain tensioner into transition position. Place the timing chain tensioner clamping fixture (1) on a level surface and remove the cap (2).

33. Place palm of hand against the clamping fixture and exert continuous pressure until fixture is completely compressed. Replace clamping fixture cap. Position the clamping fixture in place.

34. Install the timing chain tensioner and torque screw plug to 46 ft. lbs. (63 Nm). Install the cable holder.

CAUTION

Timing chain tensioner is in the transition position. Ensure timing chain is correctly arranged inside the channel of the timing chain guides.

35. Use a prybar to lever the clamping rail until the timing chain tensioner applies tension to the timing chain (do not lever directly on the timing chain).

36. Exam the released position of the clamping fixture as shown.

37. Complete installation in reverse of the removal procedure.

38. Refill the cooling system.

Cooper S Coupe

1. Before servicing the vehicle, refer to the precautions section.

2. Disconnect or remove the following:
   - Battery
   - Intercooler
   - ECU connectors
   - Intake filter housing

3. Slacken the module front end (MFE) and install the extension tools, 11–8–401/2 to keep MFE extended for access.

4. Disconnect or remove the following:
   - Throttle assembly
   - Cylinder head cover
   - Fuel tank venting valve
   - Fuel line from fuel rail (quick-disconnect fitting); plug openings

CAUTION

Fuel system may be under pressure; be prepared to open line cautiously and ready to catch spilling fuel.

5. Release both pipes from the engine stabilizer bracket and move to one side.

6. Disconnect or remove the following:
   - Intake manifold
   - Supercharger outlet pipe
   - Engine stabilizer support bracket
   - Cap from coolant reservoir
   - Drain the cooling system
   - Oxygen sensor connector (1)
   - Oxygen sensor connector bracket from cylinder head (2)
   - Coolant rail support bolt (3)
   - Coolant sensor connector (4)
   - Coolant hoses (5)

7. Disconnect the camshaft (CMP) sensor connector.

8. Remove the dipstick.

9. Remove the exhaust heat shield and exhaust manifold bolts from cylinder head.

10. Remove the spark plugs.

11. Support the engine with a suitable jack.

12. Remove the engine mount support bracket.

13. Remove the engine hydra-mount, with special tool, 11–8–200.

14. Install a special engine retainer brace, 11–8–370, to cylinder block and engine mount chassis location.

15. Remove the CMP sensor.

16. Remove both plugs from the front of the cylinder head.

17. Remove both fender well liners.

18. Rotate the engine until the camshaft gear triangular timing mark is at the 12 o'clock position. Make a paint mark across the timing mark and timing chain for reassembly reference.

➡The copper colored link has no relation to timing. The design of the timing chain cover will allow the chain to stay on the crankshaft gear without skipping any teeth.

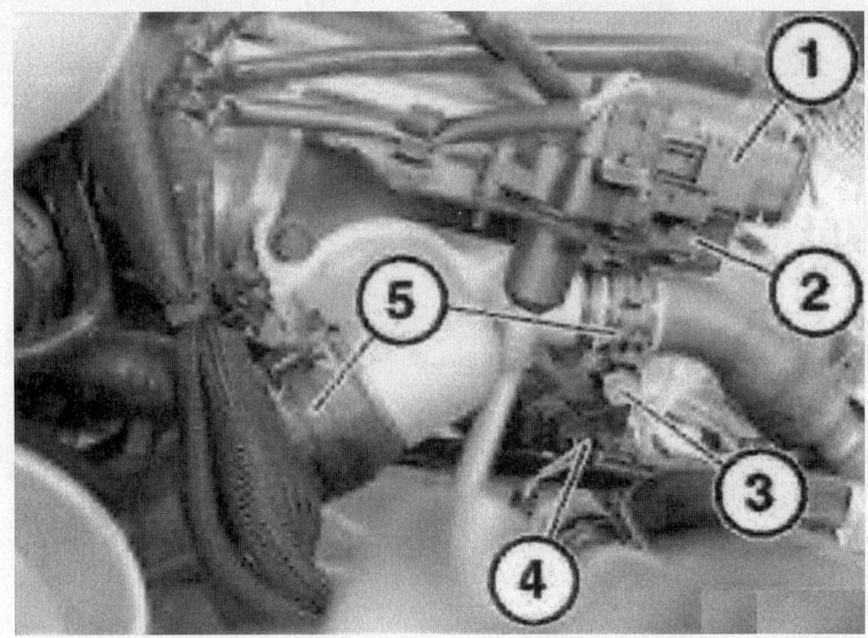

Showing components to remove from coolant housing area of engine—Cooper S Coupe

⁂ CAUTION

DO NOT rotate the engine with timing chain disconnected.

19. Install camshaft gear holding tool, 11–8–250, then slacken, but do not remove, the camshaft gear center bolt.
20. Remove the wiring harness holder, timing chain tensioner and clamping fixture tool.
21. Remove center bolt from camshaft gear.
22. Remove the camshaft gear from the timing chain and secure the chain to prevent it from falling.
23. Remove the bolts from timing chain guides (through the plug openings).
24. Remove the clamping rail and timing chain guides.

➡The timing chain cover is designed so that the timing chain can remain on the crankshaft gear without any gear teeth being skipped.

⁂ CAUTION

DO NOT rotate crankshaft.

25. Remove the cylinder head retainers 11 and 12 first, then remove the cylinder head bolts in reverse of the order shown.
26. Remove the cylinder head.

To install:

27. Clean all sealing material from mating faces

⁂ CAUTION

There must be no oil in the cylinder head bolt holes in the block and timing case cover or there is a possibility of cracking and distorting torque values.

28. Install a new cylinder head gasket.
29. Position the cylinder head onto the block and install new cylinder head bolts (do not clean compound applied to new bolts).
30. Tighten cylinder head bolts, following the sequence shown for bolts 1 through 10, in 2 steps, to the following:
    a. Step 1: 30 ft. lbs. (40 Nm)
    b. Step 2: Additional 90 degrees
31. Tighten cylinder head retainers number 11 and 12 to 21 ft. lbs. (28 Nm).
32. Install or connect the following:
    - Clamping rail and timing chain guides
    - Chain guide bolts; torque to 21 ft. lbs. (28 Nm)
    - Timing chain onto camshaft gear
    - Center bolt in camshaft gear
33. Align the camshaft and timing chain paint marks made during removal.
34. Install camshaft gear holding special tool, 11–8–250, then torque center camshaft gear bolt to 75 ft. lbs. (102 Nm).
35. Move the timing chain tensioner into transition position. Place the timing chain tensioner clamping fixture (1) on a level surface and remove the cap (2).
36. Place palm of hand against the clamping fixture and exert continuous pressure until fixture is completely compressed. Replace clamping fixture cap. Position the clamping fixture in place.
37. Install the timing chain tensioner and torque screw plug to 46 ft. lbs. (63 Nm). Install the cable holder.

⁂ CAUTION

Timing chain tensioner is in the transition position. Ensure timing chain is correctly arranged inside the channel of the timing chain guides.

38. Use a prybar to lever the clamping rail until the timing chain tensioner applies tension to the timing chain (do not lever directly on the timing chain).
39. Exam the released position of the clamping fixture as shown.
40. Complete installation in reverse of the removal procedure.
41. Refill the cooling system.

Intake Manifold

REMOVAL & INSTALLATION

Cooper Coupe

1. Before servicing the vehicle, refer to the precautions section.
2. Properly relieve the fuel system pressure.

⁂ CAUTION

Fuel system may still be under pressure; use caution when disconnecting any fuel system components. Be prepared to catch fuel spillage.

3. Remove or disconnect the following:
    - Battery
    - Air cleaner housing
    - Throttle assembly
    - Cover from fuel rail
    - Brake booster line from intake manifold (push attaching ring down to expose line)
    - Crankcase vent valve from inspection hole cover
    - Fuel line from fuel rail (quick-disconnect attachment)
    - Plug connector from intake air temperature/manifold air pressure sensor (TMAP)
    - Knock sensor plug from fuel rail wiring harness
    - Tank vent line and unclip at fuel rail
    - Vacuum line from intake manifold

- Retaining screws of fuel rail
- Injectors and fuel rail (plug all openings)
- Support or wire fuel rail assembly out of the way
- Dipstick
- Coolant line below intake manifold

4. Remove the intake manifold bolts, starting from the center and working outward in an alternating pattern.
5. Remove the intake manifold

To install:

6. Check all intake manifold gaskets and replace if necessary.
7. Install intake manifold and torque bolts to 19 ft. lbs. (26 Nm), in the sequence shown.
8. Install or connect the following:
    - Coolant line below intake manifold
    - Dipstick
    - Fuel rail and injectors
    - Intake manifold vacuum line
    - Tank venting line on fuel rail
    - Knock sensor plug to fuel rail wiring harness
    - Connector to intake air temperature/manifold air pressure sensor (TMAP)
    - Fuel line to fuel rail
    - Crankcase vent valve to inspection hole cover
    - Brake booster line to intake manifold
    - Fuel rail cover
    - Throttle assembly
    - Air cleaner housing
    - Battery

Cooper S Coupe

1. Before servicing the vehicle, refer to the precautions section.
2. Drain the cooling system
3. Properly relieve the fuel system pressure.

CAUTION

Fuel system may still be under pressure; use caution when disconnecting any fuel system components. Be prepared to catch fuel spillage.

4. Remove or disconnect the following:
    - Battery
    - Air cleaner housing
    - Throttle assembly
    - Fuel injector rail
    - Engine vent control valve
    - Coolant hose from upper connection at intake manifold
    - Knock sensor plug from manifold
    - Air intake sensor connector at manifold
    - Top tank vent valve line
    - TMAP sensor connector
5. Remove the intake manifold nuts, starting from the center and working outward in an alternating pattern.
6. Remove the intake manifold

To install:

7. Position the intake manifold and install the nuts. Torque the nuts in an alternating pattern working outward from the center to 19 ft. lbs. (26 Nm).
8. Install or connect the following:
    - TMAP sensor connector
    - Top tank vent valve line
    - Air intake sensor connector at manifold
    - Knock sensor plug from manifold
    - Coolant hose from upper connection at intake manifold
    - Engine vent control valve
    - Fuel injector rail
    - Throttle assembly
    - Air cleaner housing
    - Battery
9. Refill the cooling system.

Showing intake manifold bolt removal and tightening sequence—Cooper Coupe

Exhaust Manifold

REMOVAL & INSTALLATION

Cooper Coupe & Cooper S Coupe

1. Before servicing the vehicle, refer to the precautions section.
2. Remove or disconnect the following:
    - Exhaust system from manifold
    - Both oxygen sensor connectors
    - Heat shield
    - Exhaust manifold bolts
    - Exhaust manifold

To install:

3. Clean all mating faces.
4. Install new gaskets.
5. Position the exhaust manifold and torque retaining bolts to 18 ft. lbs. (24 Nm).
6. Install or connect the following:
    - Heat shield; torque bolts to 10 ft. lbs. (13 Nm)
    - Oxygen sensor connectors
    - Exhaust system to manifold with new nuts; torque to 44 ft. lbs. (60 Nm)

Camshaft

REMOVAL & INSTALLATION

Cooper Coupe & Cooper S Coupe

1. Before servicing the vehicle, refer to the precautions section.
    a. Remove or disconnect the following:
    - Battery
    - Spark plugs
    - Wheel well liners
    - Cylinder head cover
    - Left engine mount
    - Hydraulic chain tensioner
2. Remove bolts from rocker arm shafts in sequence shown.

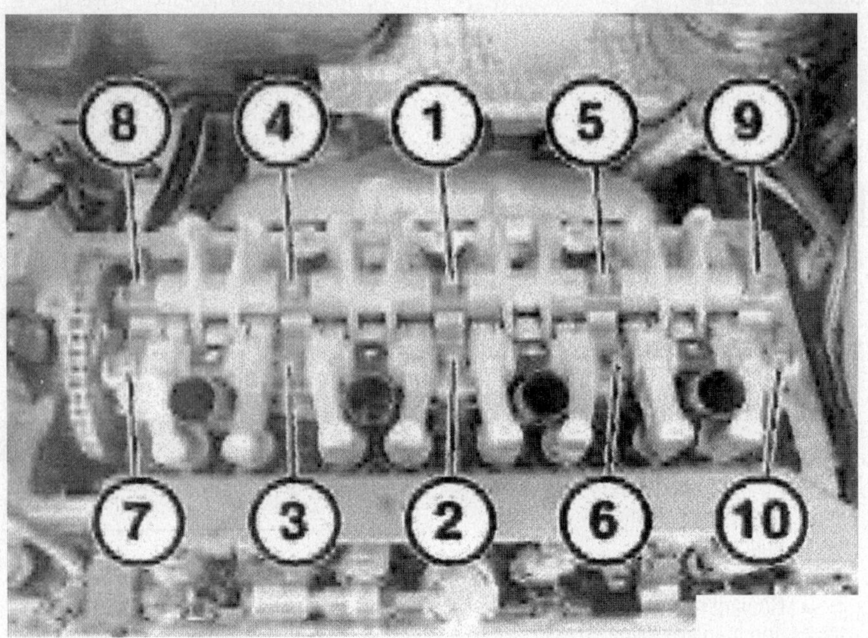

Showing rocker arm shaft bolt removal and tightening sequence—Cooper Coupe

3. Install a special engine holding tool, 11–8–370, onto cylinder block and fixture of engine mount.
4. Remove the camshaft (CMP) sensor connector and then the sensor.
5. Rotate the engine until the triangular adjustment mark on the camshaft gear is at the 12 o'clock position. Apply a paint mark across the adjustment mark and timing chain for reassembly reference. Also mark the vibration damper and timing case cover with a paint reference mark.

➡Brass color timing chain links are of no importance to chain timing.

6. Install a special locking tool, 11–8–250, onto camshaft gear and loosen, but do not remove, the camshaft gear center bolt. Remove the special tool.
7. Make sure that paint reference mark on camshaft gear and timing chain are aligned, then remove the camshaft gear center bolt.
8. Remove all the camshaft bearing caps and the camshaft. Be sure to keep bearing caps in same order and orientation as removed.

To install:

9. Check components for signs of wear or damage. Replace components as necessary.

➡If camshaft is replaced with a new unit, rocker arms must also be replaced.

⁂ CAUTION

Install bearing caps in same positions as removed.

10. Lubricate camshaft bearing journals and rocker arm rolling areas with clean engine oil.
11. Install timing chain to the camshaft gear.
12. Ensure that the timing reference paint marks are aligned.
13. Install the camshaft gear center bolt.
14. Install the camshaft gear locking tool, 11–8–250, then torque the center bolt to 75 ft. lbs. (102 Nm). Remove the locking tool.
15. Apply a thin coat of engine oil to the camshaft seal.
16. Remove the engine holding tool, 11–8–370.
17. Install or connect the following:
    - CMP sensor and connector
    - Rocker arm shafts; torque bolts first evenly by hand, then to 22 ft. lbs. (30 Nm) in sequence shown.
    - Hydraulic chain tensioner
18. Install the left engine (hydra) mount. Torque the bolts as follows:
    - M10x110 bolts: 41 ft. lbs. (56 Nm), then an additional 90 degrees
    - Other bolt: 74 ft. lbs. (100 Nm)
19. Install or connect the following:
    - Cylinder head cover
    - Wheel well liners
    - Spark plugs
    - Battery

Valve Lash

ADJUSTMENT

All engines are equipped with hydraulic valve lash adjusters. This design does not permit adjustments nor are adjustments possible.

Starter

REMOVAL & INSTALLATION

Cooper Coupe & Cooper S Coupe

1. Before servicing the vehicle, refer to the precautions section.
2. Remove or disconnect the following:
    - Battery
    - Exhaust system from manifold
    - Exhaust manifold
    - Heat shield from starter
    - Oxygen sensor cable from wire clip
    - Alternator connectors
    - Starter solenoid connectors
    - Starter

To install:

3. Install or connect the following:
    - Starter to transmission; torque mounting bolts to 63 ft. lbs. (85 Nm)
    - Alternator connector on starter; torque to 10 ft. lbs. (14 Nm)
    - Oxygen sensor cable to wire clip
    - Heat shield for starter; torque bolts to 7 ft. lbs. (9 Nm)
    - Exhaust manifold
    - Exhaust pipes to manifold
    - Battery

Oil Pan

REMOVAL & INSTALLATION

Cooper Coupe & Cooper S Coupe

1. Before servicing the vehicle, refer to the precautions section.
2. Remove or disconnect the following:
    - Battery
    - Alternator drive belt
    - Drain engine oil

⁂ CAUTION

Cover the alternator to prevent oil from dripping on it.

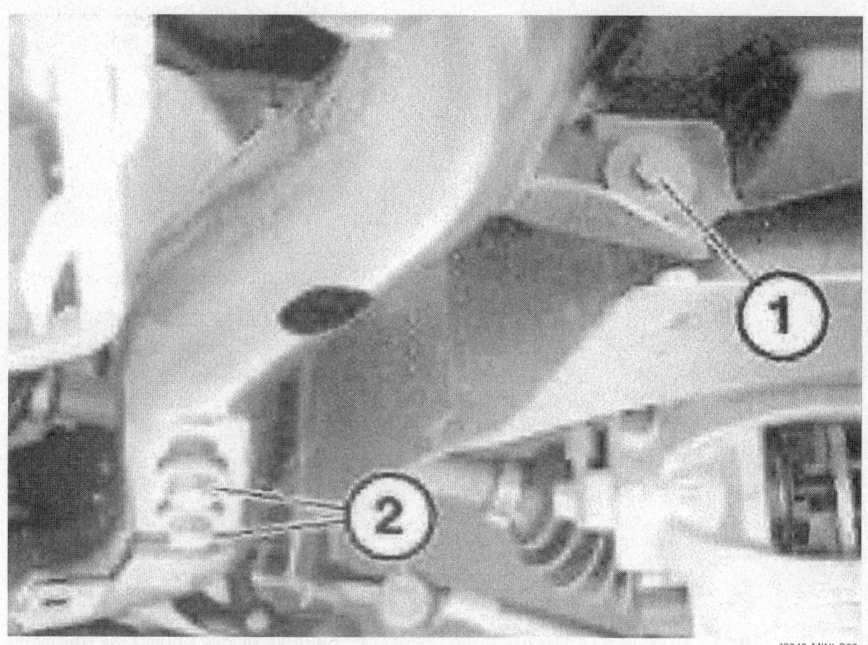

Showing location of impact tube retaining bolts

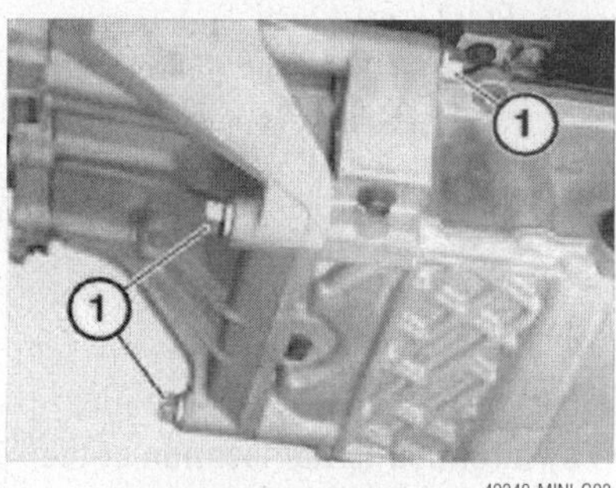

Showing location of the transaxle–to–oil pan bolts to remove

3. Remove the impact (crush) tube as follows:
   a. Remove the bumper and bumper carrier.
   b. Remove the bolt securing the impact tube to the front end module.
   c. Remove the impact tube.
4. Slacken retainers from modular front end (MFE) and push MFE outward and restrain with special tools 11–8–401/2.
5. Detach the A/C compressor connector.
6. Remove the compressor and lower out of the way. Secure it to MFE.
7. Unclip high-pressure A/C hose.
8. Remove or disconnect the following:
   - Lower engine stabilizer bar
   - Bracket from oil pan
   - 2 forward transaxle–to–oil pan bolts and upper bolts (1)
9. Remove the oil pan bolts in the specified sequence.

To install:

10. Ensure mating surfaces are clean.
11. Install a new gasket and position the oil pan in place.
12. Install and tighten the oil pan bolts to 23 ft. lbs. (31 Nm).
13. Install and tighten oil pan–to–transaxle bolts to 23 ft. lbs. (31 Nm).

➡**Shorter bolts go in lower locations.**

14. Install the lower stabilizer–to–oil pan holder and torque bolts to 74 ft. lbs. (100 Nm).
15. Install the lower engine stabilizer bar to oil pan and torque the bolts to 33 ft. lbs. (45 Nm).
16. Install the A/C compressor and torque the mounting bolts to 18 ft. lbs. (25 Nm).
17. Reconnect the A/C compressor connector.
18. Restore the MFE to normal position. Torque the bolts as follows:
    - M8x30 bolts: 16 ft. lbs. (22 Nm)
    - M6x16 bolts: 3 ft. lbs. (5 Nm)
19. Install the impact tube and torque the bolts to 74 ft. lbs. (100 Nm).
20. Install the bumper and bumper carrier. Torque nuts and bolts to 16 ft. lbs. (22 Nm).
21. Install or connect the following:
    - Splash guard
    - Alternator drive belt
    - Battery
22. Refill the engine oil. Start the engine and wait until the oil indicator lamp goes out. Switch the engine off and wait about 5 minutes, then recheck the oil level.

Oil Pump

REMOVAL & INSTALLATION

Cooper Coupe & Cooper S Coupe

1. Before servicing the vehicle, refer to the precautions section.

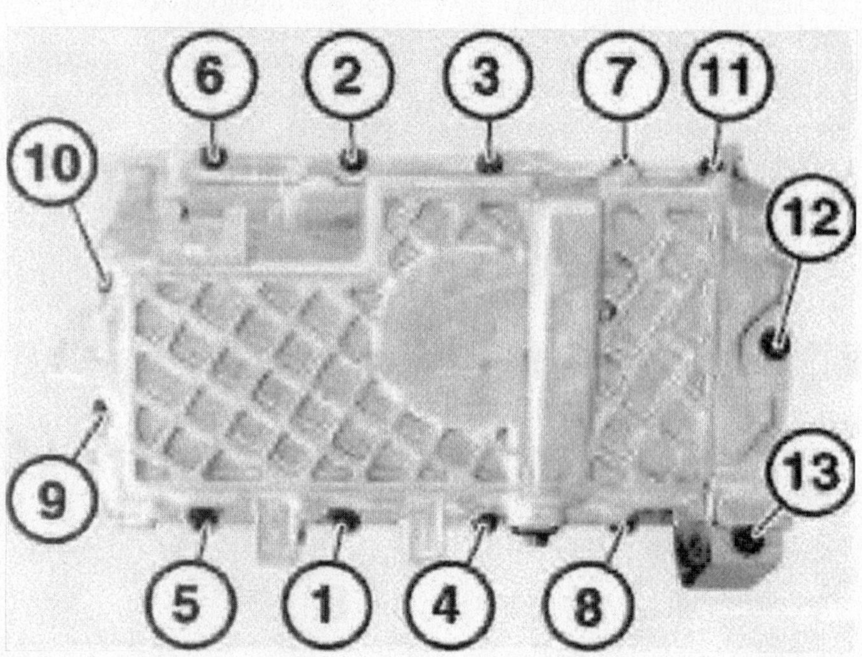

Oil pan bolt removal and installation sequence

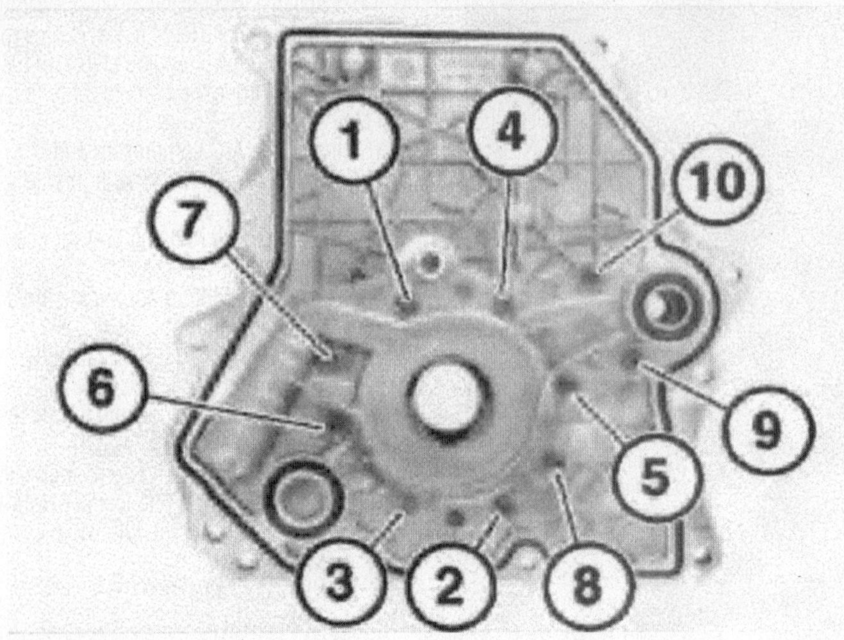

Showing oil pump cover bolt tightening sequence

2. Drain the engine oil.
3. Remove or disconnect the following:
   - Negative battery cable
   - Timing chain cover
   - Oil pump cover in reverse of order shown
   - Oil pump

To install:

4. Check oil pump gears, pressure relief valve and housing for signs of wear or damage.
5. Install or connect the following:
6. Fill the rotor cavity with clean engine oil before installing the oil pump.
   - Oil pump
   - Oil pump cover; torque bolts to 13 ft. lbs. (18 Nm)
   - Timing chain cover
   - Negative battery cable
7. Fill the engine with clean oil.
8. Start the vehicle and check for leaks, repair if necessary.

Rear Main Seal

REMOVAL & INSTALLATION

Cooper Coupe

The rear main bearing oil seal can be replaced after the transmission.

1. Before servicing the vehicle, refer to the precautions section.
2. Drain the transmission fluid.
3. Remove or disconnect the following:
   - Negative battery cable
   - Transmission
   - Clutch release bearing, bolts and guide tube (with seal)

➡Seal and guide tube are supplied as an assembly.

To install:

4. Tape input shaft splines to protect them during seal installation.
5. Coat the sealing lips of the new seal with oil.
6. Install or connect the following:
   - New seal into transaxle housing; torque guide sleeve bolts to 4 ft. lbs. (6 Nm)
   - Clutch release bearing
   - Transmission
   - Negative battery cable
7. Fill the transmission with new fluid.
8. Start the engine and check for oil leaks.
9. Check and top off all fluid levels.

Timing Chain, Sprockets, Front Cover and Seal

REMOVAL & INSTALLATION

Cooper Coupe & Cooper S Coupe

1. Before servicing the vehicle, refer to the precautions section.
2. Drain the engine oil.
3. Remove or disconnect the following:
   - Negative battery cable
   - Vibration damper
   - Alternator drive belt tensioner and belt
   - Front impact tube
4. Slacken the modular front end (MFE) retainers and restrain MFE outward with special tools 11–8–401/2.
5. Remove the water pump bolts and move the pump aside so front cover is accessible.
6. Remove timing chain cover bolts in reverse of sequence shown.

➡Pay attention to the location of the Torx and oval-head bolts.

Timing cover bolt tightening sequence

7. Remove O-ring seals and housing seal.

To install:

8. Clean all mating surfaces.
9. Install new timing chain cover seals.

➡If oil pump was removed, fill the rotor cavity with clean engine oil before installing the oil pump.

10. Install timing chain cover. Install and torque the bolts, in the sequence shown, as follows:
    - Torx bolts: 9 ft. lbs. (12 Nm)
    - Oval head bolts: 13 ft. lbs. (18 Nm)
    - M6 bolts: 9 ft. lbs. (12 Nm)
11. Install water pump and torque the mounting bolts to 41 ft. lbs. (56 Nm) plus an additional 90 degrees.
12. Restore MFE to its normal position.
13. Install the impact tube.
14. Install the belt tensioner, then install the auxiliary drive belt.
15. Install the vibration damper.
16. Connect the negative battery cable.
17. Refill the engine oil.

Piston and Ring Positioning

➡Offset position of ring end gaps by 120° from each other, but not above piston pin boss.

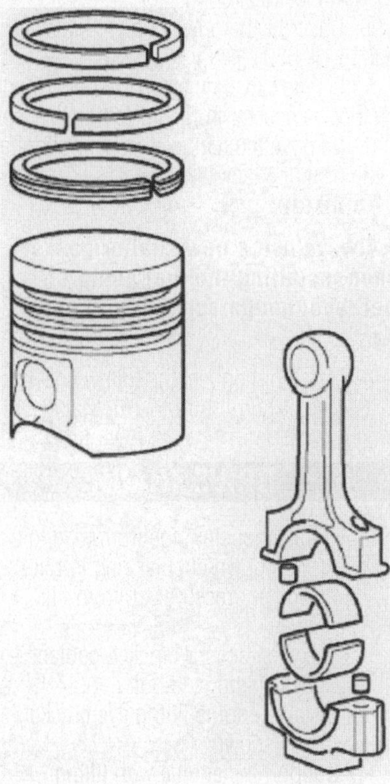

Showing orientation of piston, rings and connecting rod

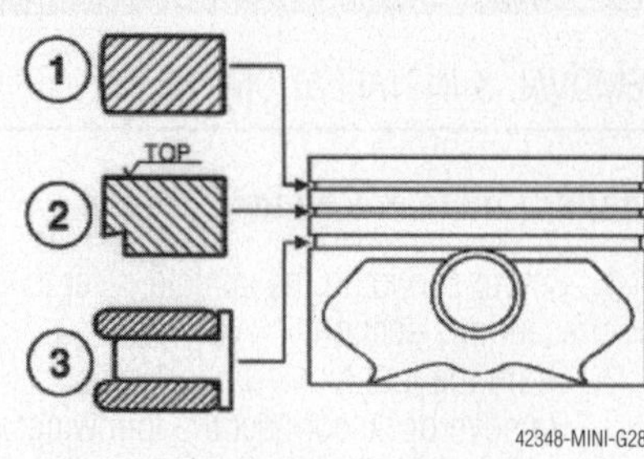

Compressor and oil ring locations

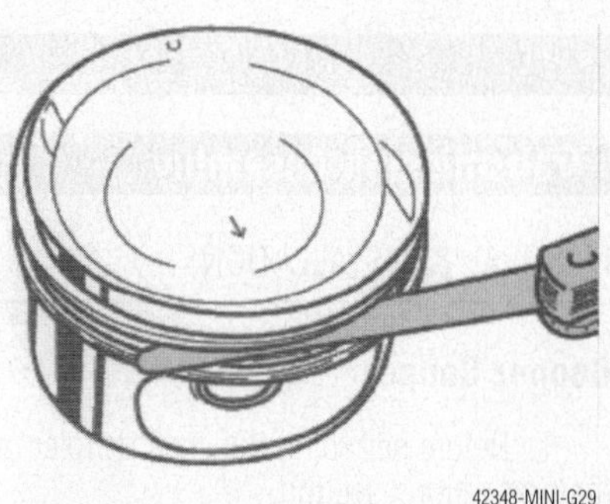

Showing piston positioning arrow pointing toward front of block

FUEL SYSTEM

Fuel System Service Precautions

Safety is the most important factor when performing not only fuel system maintenance but also any type of maintenance. Failure to conduct maintenance and repairs in a safe manner may result in serious personal injury or death. Maintenance and testing of the vehicle's fuel system components can be accomplished safely and effectively by adhering to the following rules and guidelines.

1. To avoid the possibility of fire and personal injury, always disconnect the negative battery cable unless the repair or test procedure requires that battery voltage be applied.
2. Always relieve the fuel system pressure prior to disconnecting any fuel system component (injector, fuel rail, pressure regulator, etc.), fitting or fuel line connection. Exercise extreme caution whenever relieving fuel system pressure, to avoid exposing skin, face and eyes to fuel spray. Fuel under pressure may penetrate the skin or any part of the body that it contacts.
3. Always place a shop towel or cloth around the fitting or connection prior to loosening to absorb any excess fuel due to spillage. Ensure that all fuel spillage (should it occur) is quickly removed from engine surfaces. Ensure that all fuel soaked cloths or towels are deposited into a suitable waste container.
4. Always keep a dry chemical (Class B) fire extinguisher near the work area.
5. Do not allow fuel spray or fuel vapors to come into contact with a spark or open flame.
6. Always use a back-up wrench when loosening and tightening fuel line connection fittings. This will prevent unnecessary stress and torsion to fuel line piping. Always follow the proper torque specifications.
7. Always replace worn fuel fitting O-rings with new. Do not substitute fuel hose or equivalent where fuel pipe is installed.

Fuel System Pressure

RELIEVING

1. Install special tool, 13–5–220.
2. Fit a suitable length of hose onto the special tool and route the hose into a fuel container.
3. Screw in check valve (1) of the special tool to release the fuel pressure from the injector rail.
4. Hold an absorbent cloth around the special tool and remove the hose and tool.

⁂ WARNING

Other parts of the fuel system may have some residual pressure. Always open fittings slowly and be prepared to catch any fuel.

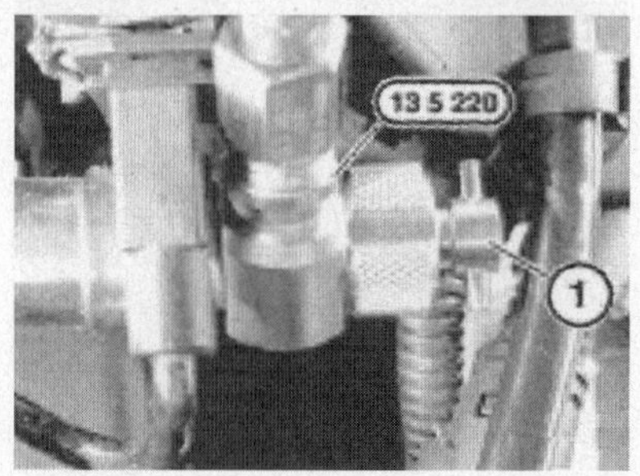

Using special tool to relieve fuel rail pressure

Fuel Pump

REMOVAL & INSTALLATION

Cooper Coupe & Cooper S Coupe

1. Before servicing the vehicle, refer to the precautions section.
2. Drain the fuel tank.
3. Remove or disconnect the following:
   - Negative battery cable
   - Rear seat
   - Fuel pump access plate (under trim panel)
   - Electrical connector
4. Unscrew the outer ring with special tool 16–1–020.
5. Lift the cap and detach the fuel line and electrical connector from the fuel level sensor.
6. Remove the fuel level sensor and fuel pump from the tank.

To install:

➡Always use a new seal or gasket when installing the fuel pump or fuel level gauge sending unit assembly.

7. Install or connect the following:
   - Fuel pump into the fuel tank taking care not to bend or damage the fuel sending unit assembly
   - New seal and torque the sealing ring, using tool No. 16-1-020, to 26 ft. lbs. (35 Nm).
   - Fuel gauge level sending unit electrical connector
   - Metal cover
   - Rear seat bench
   - Negative battery cable
8. Refill fuel tank.
9. Start the vehicle and check for leaks, repair if necessary.

DRIVE TRAIN

Transmission Assembly

REMOVAL & INSTALLATION

Cooper Coupe

1. Before servicing the vehicle, refer to the precautions section.
2. Remove or disconnect the following:
   - Negative battery cable
   - Battery and battery box
   - Manifold heat shield
   - Engine stabilizer (upper)
   - Fuel and vent pipes from bracket near stabilizer
   - Drain transaxle
   - Front left wheel well liner
   - Driveshafts with steering knuckle carrier
   - Lower stabilizer
   - Front subframe
   - Gearshift cables from ball joint attachment
   - Gearshift cable bracket
   - Clutch slave cylinder from transaxle
   - Reverse lamp connector from transaxle
   - Brake booster pipe from manifold (push down circular ring to release)
   - Coolant pressure cap from fill tower
   - Oxygen sensor bracket, coolant hose clamp, and bolt
3. Install a engine lifting eye bracket, 11–8–260 as shown.
4. Support the engine with lifting equipment.
5. Raise equipment enough to take weight of engine and transaxle
6. Remove the upper bracket retaining bolt (1), mount to transaxle bolts (2) and remove the transaxle mount.
7. Lower the engine about 1.5 inches (40mm).

⁂ CAUTION

DO NOT lower engine too much or exhaust system could be damaged. Also watch A/C pipe to compressor when lowering engine.

8. Remove or disconnect the following:
   - Starter heat shield
   - Oxygen sensor wiring from clip
   - Starter
   - Closure plate bracket around inner driveshaft opening
9. Support transaxle with suitable jack.
10. Remove the transaxle retaining bolts.
11. Remove the transaxle.

➡Shorter 2 bolts are located into oil pan.

To install:

12. Clean all mating surfaces.
13. Position the transaxle into the vehicle.
14. Install and torque the transaxle–to–engine housing bolts to 63 ft. lbs. (85 Nm).

➡2 shorter bolts go directly into oil pan.

15. Remove jack.
16. Install and torque closure plate bracket bolts to 7 ft. lbs. (9 Nm).
17. Install or connect the following:
   - Starter; torque bolts to 63 ft. lbs. (85 Nm)
   - Starter heat shield; torque bolts to 7 ft. lbs. (9 Nm)
   - Starter electrical connections
   - Oxygen sensor to clip near starter heat shield
18. Raise the engine back into normal position and install the transaxle mount. Torque the bolts as follows:
   - Mount bracket–to transaxle: 28 ft. lbs. (38 Nm)
   - Mount–to–upper bracket: 49 ft. lbs. (66 Nm)
19. Slowly release engine tension from lift equipment. Remove the equipment.
20. Remove the engine lifting eye bracket.
21. Install or connect the following:
   - Coolant hose, clamp and bolt

42348-MINI-G31

Installing engine lifting eye bracket—Cooper Coupe

42348-MINI-G32

Identifying the location of the transaxle mount—Cooper Coupe

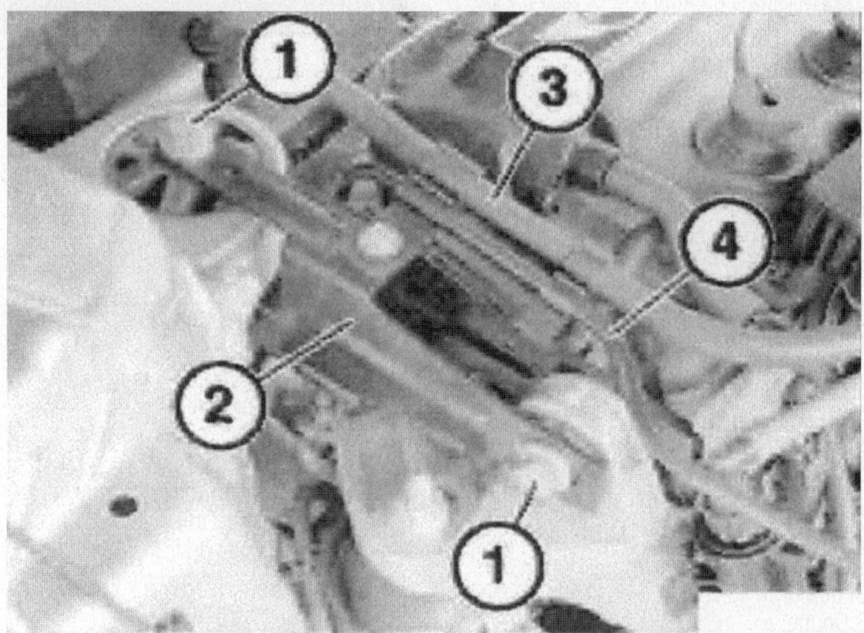

42348-MINI-G33

Showing the upper stabilizer bracket and bolt locations, plus the location of the fuel and vent pipes—Cooper Coupe

- Oxygen sensor bracket (near coolant hose)
- Brake booster pipe to manifold
- Reverse light switch connector
- Clutch slave cylinder; torque bolts to 18 ft. lbs. (24 Nm)
- Gearshift cable and bracket
- Front subframe
- Lower stabilizer bracket (2); torque bolts (1) to 74 ft. lbs. (100 Nm)
- Driveshafts and steering knuckle carrier
- Left wheel well liner

22. Refill the transaxle with proper oil.
23. Install the upper stabilizer bolts. Torque the bolts to 74 ft. lbs. (100 Nm).
24. Attach the fuel and vent pipes to the upper stabilizer.
25. Install the manifold heat shield.
26. Install and connect the battery.
27. Start the engine and check transaxle operation.

Cooper S Coupe

1. Before servicing the vehicle, refer to the precautions section.
2. Remove or disconnect the following:
   - Negative battery cable
   - Battery and battery box
   - Intake filter housing
   - Manifold heat shield
   - Engine stabilizer (upper)
   - Fuel and vent pipes from bracket near stabilizer
   - Drain transaxle
   - Front left wheel well liner
   - Driveshafts with steering knuckle carrier
   - Lower stabilizer
   - Crush tubes
   - Front subframe

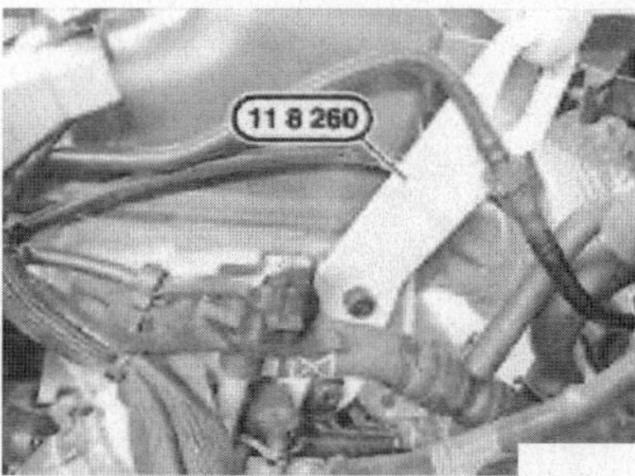

42348-MINI-G34

Installing engine lifting eye bracket—Cooper S Coupe

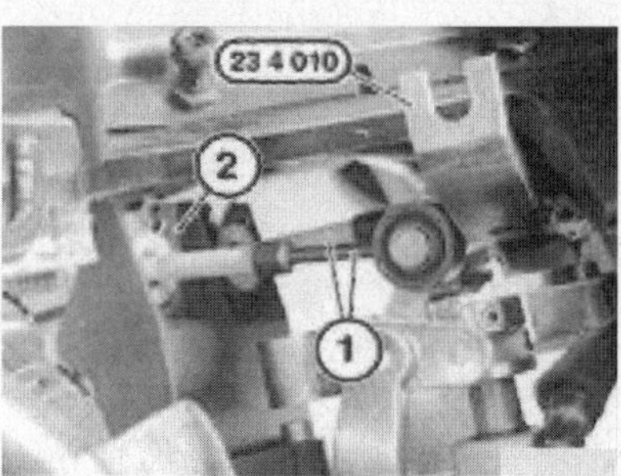

42348-MINI-G35

Disconnect the gearshift cables from the ball joints—Cooper S Coupe

   - Coolant expansion tank cap
   - Oxygen sensor bracket, coolant hose clamp, and bolt
3. Install an engine lifting eye bracket, 11–8–260 as shown.
4. Remove the gearshift cables from ball joint attachment, with special tool 23–4–010, then remove the gearshift cable mounting bracket.
5. Remove the clutch slave cylinder from the transaxle.
6. Disconnect the reverse light switch connector from the transaxle.
7. Open the hood to the full upright position and install strut extensions, 51–2–160, to hold the hood in this position.
8. Support the engine with lifting equipment.
9. Raise equipment enough to take weight of engine and transaxle.
10. Remove or disconnect the following:
    - Throttle housing
    - Supercharger intake hose
    - Detach other pipes by quick-fit couplings
    - Slave cylinder hose from transaxle and move aside
    - Closure plate around inner driveshaft opening
    - Starter heat shield
    - Oxygen sensor from clip on heat shield
    - Starter connections and move wiring harness aside
    - Starter
    - Transaxle mount
11. Lower the engine about 5 inches (135 mm).
12. Support transaxle with suitable jack.
13. Remove the transaxle retaining bolts.
14. Remove the transaxle.

To install:

15. Clean all mating surfaces.
16. Position the transaxle into the vehicle.
17. Install and torque the transaxle–to–engine housing bolts to 63 ft. lbs. (85 Nm).
18. Remove jack.
19. Raise the engine to normal position.
20. Install or connect the following:
    - Transaxle mount; torque bolts to 49 ft. lbs. (66 Nm)
    - Starter; torque bolts to 63 ft. lbs. (85 Nm)
    - Starter electrical connections and wiring harness
    - Oxygen sensor to clip near starter heat shield
    - Starter heat shield; torque bolts to 7 ft. lbs. (9 Nm)

- Closure plate bolts to 7 ft. lbs. (9 Nm).

21. Install the lower support bracket and torque the bolts as follows:
    - Mount bracket–to–transaxle: 28 ft. lbs. (38 Nm)
    - Mount–to–upper bracket: 49 ft. lbs. (66 Nm)
22. Install or connect the following:
    - MAP sensor
    - Slave cylinder hose to transaxle
    - Supercharger, pipes and hoses
    - Reverse light switch connector
    - Slave cylinder to transaxle; torque bolts to 18 ft. lbs. (24 Nm)
    - Gearshift cables and bracket
23. Slowly release engine tension from lift equipment. Remove the equipment.
24. Remove the engine lifting eye bracket.
25. Install or connect the following:
    - Coolant hose, clamp and bolt
    - Oxygen sensor bracket (near coolant hose)
26. Install the front subframe; torque bolts to 74 ft. lbs. (100 Nm)
27. Reinstall the MFE to its normal position. Torque the bolts as follows:
    - M8x30 bolts: 17 ft. lbs. (22 Nm)
    - M6x16 bolts: 3 ft. lbs. (5 Nm)
28. Install or connect the following:
    - Crush member–to–subframe: 74 ft. lbs. (100 Nm)
    - Lower stabilizer bracket (2); torque bolts (1) to 74 ft. lbs. (100 Nm)
    - Driveshafts and steering knuckle carrier
    - Left wheel well liner
29. Refill the transaxle with proper oil.
30. Install the upper stabilizer bolts. Torque the bolts to 74 ft. lbs. (100 Nm).
31. Attach the fuel and vent pipes to the upper stabilizer.
32. Install the manifold heat shield.
33. Install and connect the battery.
34. Start the engine and check transaxle operation.

Clutch

REMOVAL & INSTALLATION

Cooper Coupe & Cooper S Coupe

1. Before servicing the vehicle, refer to the precautions section.
2. Remove or disconnect the following:
3. Remove the transmission.
4. Using a holding tool to restrain or lock the crankshaft pulley in place (keep it from turning).
5. Slacken the pressure plate bolts evenly, in an alternating sequence, then remove all bolts.
6. Remove the pressure plate and disk.

To install:

7. Position the clutch disk onto the transmission input shaft and check for free movement.
8. Install the pressure plate and clutch disk onto the flywheel, using a special tool, 21–6–100 (Cooper Coupe) or 21–2–210 (Cooper S Coupe).
9. Install new pressure plate retaining bolts. Tighten them gradually and evenly, in an alternating pattern. Final torque setting is 15 ft. lbs. (20 Nm) for Cooper Coupe or to 17 ft. lbs. (23 Nm) for Cooper S Coupe.

➡During the tightening process, rotate the special holding tool. This will help to centralize the clutch disk.

10. Remove the special tool from the clutch.
11. Install the transmission.
12. Remove the holding tool from the crankshaft pulley.

Halfshafts

REMOVAL & INSTALLATION

Cooper Coupe & Cooper S Coupe

1. Before servicing the vehicle, refer to the precautions section.
2. Remove or disconnect the following:
    - Front wheel
    - Front wheel hub nut
    - Drain transaxle
    - Brake caliper from disc (tie out of way; hose connected)
    - Tie rod ball joint from steering knuckle
    - ABS sensor from steering knuckle
    - Control arm from steering knuckle
3. On right side driveshaft only, remove bolts holding the intermediate shaft housing to the bracket.
4. Pull the driveshaft from transaxle (discard snap ring)
5. Remove the bolt holding the steering knuckle to the McPherson strut, then lift the steering knuckle out with the driveshaft.

To install:

6. Install a new snap ring on the end of the driveshaft inner spline.
7. Install a special seal protector tool, 24–8–120, into side of transaxle.

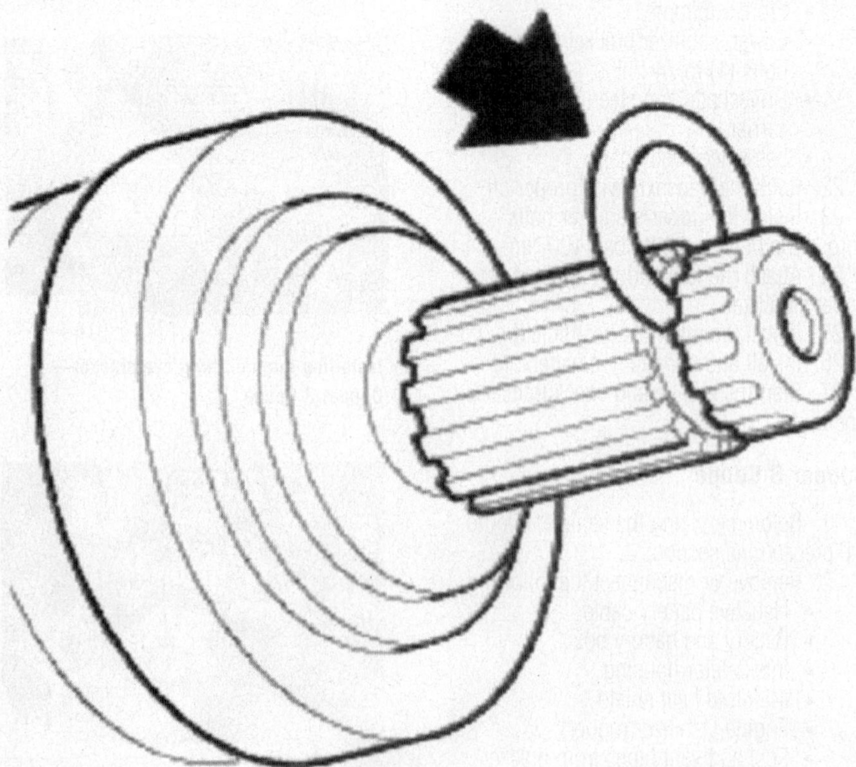

Installing a new snap ring on driveshaft inner spline

Showing special seal protector tool installed in transaxle

8. Position the driveshaft to the transaxle and insert into to seal. Pull on the special tool handle to remove once the driveshaft is in position.
9. Push in output shaft over the resistance of the retaining ring until it snaps in place.
10. Install the steering knuckle to the McPherson strut. Torque the retaining bolt to 60 ft. lbs. (81 Nm).
11. Install the intermediate shaft housing to the bracket. Torque the retaining bolts to 18 ft. lbs. (25 Nm).
12. Install or connect the following:
    - Control arm to steering knuckle; torque new nut to 41 ft. lbs. (56 Nm)
    - ABS sensor to steering knuckle; torque to 6 ft. lbs. (8 Nm)
    - Tie rod to steering knuckle; torque new ball joint nut to 38 ft. lbs. (52 Nm)
    - Brake caliper to disc; torque caliper guide bolts to 23 ft. lbs. (31 Nm)
    - Front wheel hub nut; torque new nut to 134 ft. lbs. (182 Nm)
    - Front wheel
13. Refill the transaxle.

CV Joints

REMOVAL & INSTALLATION

Cooper Coupe & Cooper S Coupe

1. Before servicing the vehicle, refer to the precautions section.
2. Remove the driveshaft.
3. Remove the bellows clamps.
4. Slide the bellows away from the inner CV joint.
5. Hold the shaft firmly and drive the inner CV joint off the shaft.
6. Remove the bellows.

To install:

7. Install a new bellows and seal onto the shaft.
8. Generously pack new joint with grease. Be sure the joint rests on the new snap ring on the shaft.
9. Press the snap ring into the shaft groove, then drive the CV joint onto the shaft.
10. Slide the bellows onto the joint and shaft and make sure the seal bearing of the bellows fits into the grooves on the shaft on one end and the grooves on the CV joint on the other end.
11. Install the bellows clamps.
12. Install the driveshaft.

STEERING AND SUSPENSION

Air Bag

✲✲ CAUTION

The vehicles are equipped with an air bag system. The system must be disarmed before performing service on, or around, system components, the steering column, instrument panel components, wiring and sensors. Failure to follow the safety precautions and the disarming procedure could result in accidental air bag deployment, possible personal injury and unnecessary system repairs.

PRECAUTIONS

Several precautions must be observed when handling the inflator module to avoid accidental deployment and possible personal injury.

1. Never carry the inflator module by the wires or connector on the underside of the module.
2. When carrying a live inflator module, hold securely with both hands, and ensure that the bag and trim cover are pointed away.
3. Place the inflator module on a bench or other surface with the bag and trim cover facing up.
4. With the inflator module on the bench, never place anything on or close to the module that may be thrown in the event of an accidental deployment.

DISARMING

1. Before servicing the vehicle, refer to the precautions section.
2. Place the ignition switch in the **OFF** position.
3. Disconnect the negative battery terminal and cover the battery terminal to prevent accidental contact.
4. Once the battery has been disconnected, wait for a short period of time to allow the capacitor in the control unit to discharge. Once the capacitor is discharged, a trigger pulse cannot be generated inadvertently.

REARMING

1. Before servicing the vehicle, refer to the precautions section.
2. Place the ignition switch in the **OFF** position.
3. Attach the sensors, the steering column connector and the seat belt tensioner connectors.
4. Connect the negative battery terminal.
5. Place the ignition switch in the **ON** position. Check that the SRS light illumi-

nates for 6 seconds and extinguishes. If it illuminates in any other pattern, check the components and their connections for proper operation and recheck operation of the warning light.

Power Rack and Pinion Steering Gear

REMOVAL & INSTALLATION

Cooper Coupe & Cooper S Coupe

1. Before servicing the vehicle, refer to the precautions section.

⁂ CAUTION

It is essential to maintain cleanliness of components, especially when removing hoses or otherwise opening the hydraulic system. Always plug all openings to seal against debris getting into the system.

2. Remove the front wheels.
3. Apply a reference mark on the steering tie rod with paint, for reassembly reference.
4. Remove the nuts on the left and right tie rod ends and separate the tie rod ends from the steering knuckle.
5. Release the nuts (1) at the bottom of the stabilizer bar.
6. Drain the steering fluid from the reservoir.
7. Remove the nut on the clamp at the lower end of the steering column (near the firewall).
8. Detach the high pressure line and low pressure line from the steering gear (banjo bolts). Plug the openings.
9. Remove the heat shield from the steering gear.
10. Detach the line bracket from the ends of the steering gear housing, then remove the 4 bolts holding the steering gear to the chassis.
11. Remove the steering gear.

To install:

12. Position the steering gear to the chassis mountings. Torque the retaining bolts to 41 ft. lbs. (56 Nm).
13. Install or connect the following:
    - Line bracket on end of steering gear
    - Heat shield; torque bolts to 14 ft. lbs. (19 Nm)
    - Power steering pipes to steering gear; torque high pressure pipe bolt to 25 ft. lbs. (34 Nm) and low pressure pipe bolt to 30 ft. lbs. (40 Nm)
    - Pinch bolt at steering gear to steering column; torque new nut to 16 ft. lbs. (22Nm)
    - New self-locking nuts of bottom of stabilizer bar; torque to 41 ft. lbs. (56 Nm)
    - Tie rod end ball joint nut to steering knuckle; torque new nut to 38 ft. lbs. (52 Nm)
14. Reset tie rod to steering gear by screwing connection in until paint marks align (this is an initial setting).
15. Bleed the power steering system.
16. Check and adjust alignment.

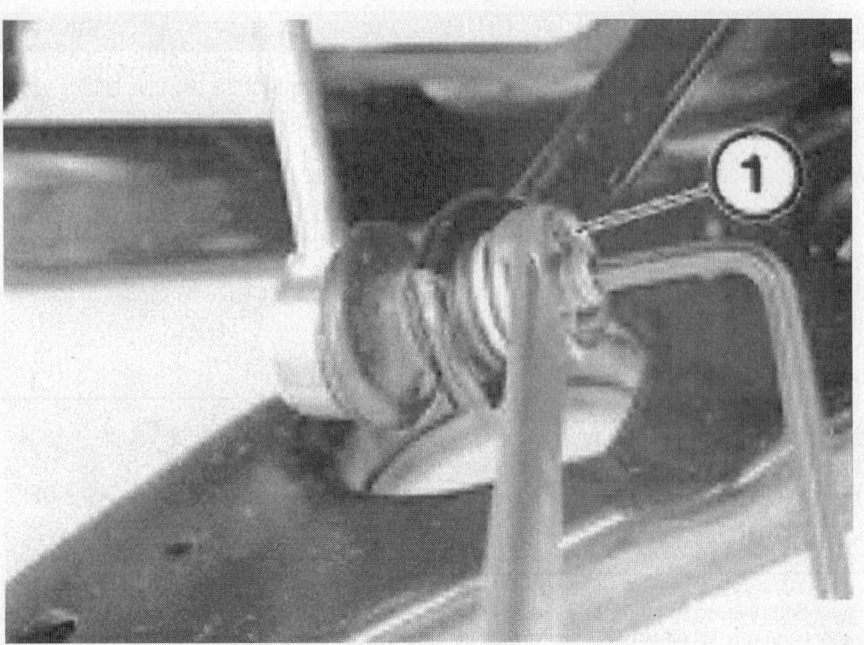

Releasing nuts on bottom of the stabilizer bar

BLEEDING

⁂ CAUTION

Thoroughly clean the reservoir and parts in immediate working area before removing the oil reservoir cap. No dirt must enter the system.

1. Check power steering fluid level and top up as needed.
2. Start the engine and turn the steering wheel 2 times to the left and right.
3. Stop the engine and check the fluid level. Adjust if needed.
4. Repeat this process if the fluid level went down significantly or if presence of bubbles is still noted.

Strut

REMOVAL & INSTALLATION

Cooper Coupe & Cooper S Coupe

1. Before servicing the vehicle, refer to the precautions section.
2. Remove or disconnect the following:
    - Tire and wheel assembly
    - ABS cable and brake hose from retainers on strut bracket
    - Brake caliper from disk and tie out of way
    - Nut from stabilizer end.
    - Tie rod end ball joint from steering knuckle
    - Transverse link ball joint from steering knuckle
3. With steering knuckle supported, remove the clamping screw from the lower end of the McPherson strut and detach the steering knuckle from the strut.
4. Remove the 3 upper strut retaining nuts (on top of strut tower).
5. Remove the strut assembly.

To install:

6. Install or connect the following:
    - Strut assembly
    - 3 new nuts on strut tower; torque to 25 ft. lbs. (34 Nm)
7. Fit the support bracket into the gap and press the steering knuckle upward until the bolt fits in the bracket hole (lower end of strut). Torque the bolt to 60 ft. lbs. (81 Nm).
8. Remove the steering knuckle support.

9. Install or connect the following:
   - Transverse link to steering knuckle; torque new self-locking nut to 41 ft. lbs. (56 Nm)
   - Tie rod end to steering knuckle; torque new self-locking nut to 38 ft. lbs. (52 Nm)
   - New self-locking nut on stabilizer end; torque to 41 ft. lbs. (56 Nm)
   - Brake caliper; torque guide pin bolts to 23 ft. lbs. (31 Nm)
   - ABS cable and brake hose to strut retainers
   - Front wheel

Shock Absorber

REMOVAL & INSTALLATION

Cooper Coupe & Cooper S Coupe

1. Before servicing the vehicle, refer to the precautions section.
2. Place a jack or other support to relieve the load on the trailing arm.
3. Remove the lower shock absorber mounting bolt.
4. Release the ABS sensor and brake hose from the retainers on the shock absorber.
5. Remove the 2 upper retaining bolts for the shock absorber.
6. Support the shock absorber and remove it from the vehicle.

To install:

7. Position the shock absorber and install the upper mounting bolts. Torque the bolts to 41 ft. lbs. (56 Nm).
8. Be sure that the rubber grommets for the ABS sensor and brake hose are correctly installed.
9. Install the ABS sensor and brake hose to the shock absorber retainers.
10. Install the lower shock absorber bolt and torque to 103 ft. lbs. (140 Nm).
11. Remove the jack from under the trailing arm.

Coil Spring

REMOVAL & INSTALLATION

Cooper Coupe & Cooper S Coupe

FRONT

⁂ CAUTION

This procedure calls for the spring to be compressed. A compressed spring has high potential energy and if released suddenly can cause severe damage and personal injury.

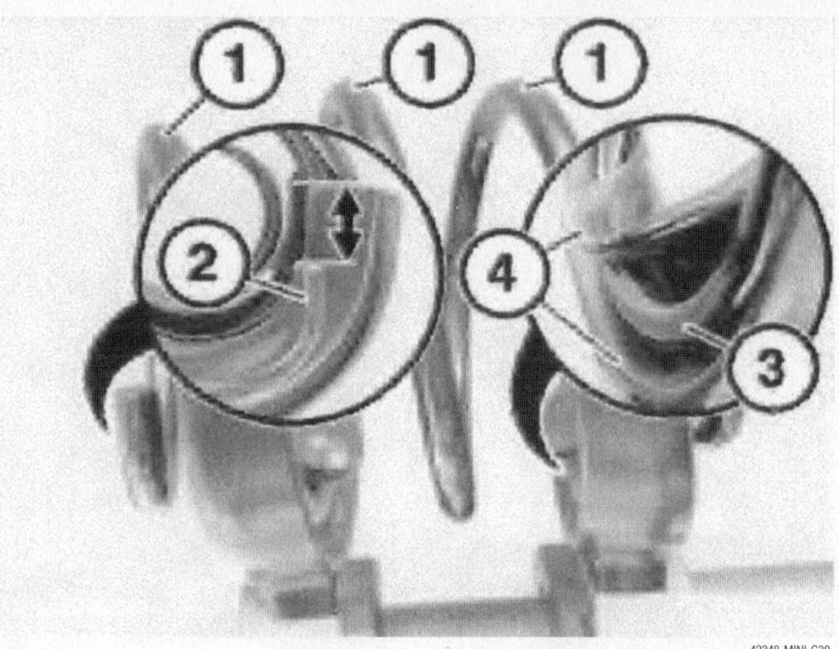

Showing the proper position of the coil spring in the spring retainer

➡Springs with identical color code must be used in pairs (color code is on the end of the spring coil).

1. Before servicing the vehicle, refer to the precautions section.
2. Remove the strut from the vehicle and mount in a vise using a strut holder. This will prevent damage to the strut tube
3. Using a proper spring compressor, 31–3–341 with 31–3–355, compress the spring until the lock pins on the coil spring holding tools are heard and felt to lock in place.
4. Only tighten the coil springs until the stress on the thrust bearing is relieved.
5. Remove the retaining nut and the coil spring and strut assembly.
6. Slowly release the compression of the spring.

To install:

7. Check the condition of the spring pad and replace it, if necessary.
8. Be sure the spring pad fits over the tongue on the lower spring seat.
9. Check the protective sleeve on top of the upper spring seat. During installation, make sure the tabs of the sleeve fit correctly over the trim.
10. Be sure spring seats are fit from the chamfered side of the special tools and that the lock pins are heard and felt to lock in position.
11. Recheck the fit of the spring seats.
12. There must be 3 spring coils between the spring retainers when properly positioned. The end of the spring must be located under end of spring retainer. Coil spring must lie completely in the recess when tensioned in the spring retainer.
13. Insert the spring strut into the coil spring. Mount the protective sleeve, auxiliary spring, and upper spring seat.
14. Screw the self-locking nut onto the piston rod.
15. Release the coil spring until it is fully resting on the lower spring plate.

➡The end of the coil spring must be aligned correctly at the rubber seal of the spring seat.

16. Fully tighten the new self-locking nut to 47 ft. lbs. (64 Nm).
17. Fully release the coil spring and remove the spring compressor tool.
18. Install the strut assembly.

Front Subframe (Axle Carrier)

REMOVAL & INSTALLATION

1. Before servicing the vehicle, refer to the precautions section.
2. Remove or disconnect the following:
   - Front wheels
   - Front bumper and bumper carrier
   - Impact tube
   - Bulkhead nuts and clamp nut holding reservoir tank to bulkhead

42348-MINI-G40

Illustrating how the power steering reservoir should be held after the subframe is removed

- Tie rod ends from steering knuckle
- Left and right stabilizer rods from stabilizer bar
- Lower control arm ball joints from steering knuckle
- Power steering pump electrical plug
- Steering column shaft from steering gear
- Lower engine stabilizer bracket

3. Position a jack at the jacking point on chassis subframe.
4. Remove or disconnect the following:
   - Center bolts for left and right side of subframe at vehicle body
   - Bolts holding left and right retaining bushing housing on body
   - All bolts from rear end of subframe to body
5. Lower the subframe and power steering reservoir through the engine compartment.
6. Remove the cable assembly between the subframe and steering gear (if equipped).
7. Use a 0.16 inch (4mm) rod to hold the power steering reservoir vertically, as shown.
8. Remove or disconnect the following:
   - Stabilizer bar
   - Lower control arms from subframe
   - Retaining clips for lower covering from subframe
   - Steering gear heat shield
   - Power steering line and power steering gear from subframe
   - Power steering pump from subframe
   - Stone guard

To install:

9. Install or connect the following:
   - New retaining clips for ABS wiring harness to subframe
   - Stone guard
   - Power steering pump to subframe; torque mounting bracket bolts to 14 ft. lbs. (19 Nm)
   - Steering gear to subframe; torque mounting bolts to 41 ft. lbs. (56 Nm)
   - Power steering line bracket
   - Steering gear heat shield
   - Bottom covering to subframe clips
   - Lower control arms to subframe; torque mounting bolts 74 ft. lbs. (100 Nm)
   - Stabilizer bar to subframe; torque mounting bolts to 122 ft. lbs. (165 Nm)
10. Position the subframe and power steering reservoir into position. Install the rear subframe bolts. Torque bolts to 71 ft. lbs. (100 Nm).
11. Install, but do not tighten, bolts holding left and right retaining bushing housing to body.
12. Install the center bolts for the front left and right side subframe mountings to body. Torque bolts to 74 ft. lbs. (100 Nm).
13. Fully tighten the left and right bushing housing bolt to 44 ft. lbs. (59 Nm), plus an additional 90 degrees, then an additional 15 degrees.
14. Remove the jack from the subframe jacking point.
15. Install or connect the following:
    - Lower engine stabilizer; torque bolts to 74 ft. lbs. (100 Nm)
    - Steering gear to lower end of steering column; tighten new pinch bolt nut to 16 ft. lbs. (22 Nm)
    - High voltage connector to power steering pump
    - Lower control arm ball joints to steering knuckle; torque new nuts to 41 ft. lbs. (56 Nm)
    - Left and right stabilizer rods to stabilizer bar; torque new nuts to 41 ft. lbs. (56 Nm)
    - Tie rod end ball joints to steering knuckle; torque new nuts to 38 ft. lbs. (52 Nm)
    - Power steering reservoir clamp bolt and mounting bolts to bulkhead; torque nuts to 14 ft. lbs. (19 Nm)
    - Impact tube
    - Bumper carrier and bumper
    - Front wheels
16. Check and adjust wheel alignment.

Lower Ball Joint

REMOVAL & INSTALLATION

➡Vehicle uses ball joints on lower control arm and on tie rod ends. Lower ball joints can be separated from steering knuckle after raising front end of vehicle. Complete removal of lower control arm requires removal of subframe.

Lower Control Arm

REMOVAL & INSTALLATION

➡See "Subframe."

Wheel Hub & Bearings

ADJUSTMENT

Wheel bearings can not be adjusted and must be replaced as a unit and never be reused once removed.

REMOVAL & INSTALLATION

Front

1. Before servicing the vehicle, refer to the precautions section.
2. Remove or disconnect the following:
   - Front wheels

- Hub nut
- Brake caliper
- Rotor
- ABS sensor
- Wheel hub from steering knuckle and driveshaft
- Wheel bearings

To install:

3. Install or connect the following:
   - Wheel bearing into hub
   - Wheel hub to steering knuckle and driveshaft; torque bolts to 41 ft. lbs. (56 Nm)
   - ABS sensor; torque to 6 ft. lbs. (8 Nm)
   - Brake rotor and caliper
   - New flanged hub nut; torque to 134 ft. lbs. (182 Nm)
4. Stake flanged nut into groove of thread on driveshaft.

Rear

1. Before servicing the vehicle, refer to the precautions section.
2. Remove or disconnect the following:
   - Wheel assembly
   - Brake caliper (tie to body without strain on hose)
   - ABS sensor from trailing arm
   - Stabilizer bar, if equipped
   - Brake rotor
   - Wheel hub from trailing arm
   - Wheel bearing from hub

To install:

3. Install or connect the following:
   - Wheel bearing
   - Wheel hub to trailing arm; torque bolts to 41 ft. lbs. (56 Nm)
   - Brake rotor; torque bolts to 20 ft. lbs. (27 Nm)
   - Stabilizer bar, if equipped; torque new nuts to 41 ft. lbs. (56 Nm)
   - ABS sensor
   - Brake caliper; torque guide pin bolts to 48 ft. lbs. (65 Nm)
   - Wheel assembly

BRAKES

Brake Caliper

REMOVAL & INSTALLATION

Front

1. Before servicing the vehicle, refer to the precautions section.
2. Apply the brake pedal slightly with a brake clamp.
3. Remove or disconnect the following:
   - Wheel assembly
   - Brake pipe from caliper
   - Retaining spring across caliper
   - Plastic plugs over guide pin bolts
   - Guide pin bolts
   - Caliper from rotor, pulling off toward rear

To install:

4. Position the caliper into place.
5. Clean but do not grease guide pin bolts
6. Install the torque the guide pin bolts to 23 ft. lbs. (31 Nm).
7. Replace the plastic plugs over the guide pin bolts.
8. Install the retainer spring.
9. Install the brake pipe and banjo bolt to the caliper; torque to 30 ft. lbs. (40 Nm).
10. Remove the brake clamp.
11. Install the front wheels.
12. Bleed the brakes.

Rear

1. Before servicing the vehicle, refer to the precautions section.
2. Apply the brake pedal slightly with a brake clamp.
3. Remove or disconnect the following:
   - Wheel assembly
   - Retaining spring from bottom, then top
   - Plastic plugs from guide pin bolts
   - Handbrake cable from handbrake lever and at rear caliper
   - Brake hose from caliper
   - Caliper guide bolts
   - Rear caliper; remove toward rear

To install:

4. Position the caliper into place.
5. Install the torque the guide bolts to 21 ft. lbs. (28 Nm).
6. Attach brake hose to caliper at torque banjo bolt, with new seals, to 33 ft. lbs. (45 Nm).
7. Install the handbrake cable to caliper and to handbrake.
8. Install the plastic plugs over the guide pin bolts.
9. Install the retaining spring at the top, then at the bottom.
10. Remove the brake clamp.
11. Install the rear wheels.
12. Adjust the parking brake.
13. Bleed the brakes.

Disk Brake Pads

REMOVAL & INSTALLATION

Front

1. Before servicing the vehicle, refer to the precautions section.
2. Remove the front wheels.
3. Remove the disk pad retaining spring from the caliper, from bottom and then from the top.
4. Remove the plastic plugs over the caliper guide pin bolts.
5. Remove the guide pin bolts and the calipers from the rotor.
6. Press the piston back into caliper.

⁂ CAUTION

Watch the brake fluid level in reservoir during this procedure.

7. Remove the outer brake pad (inner pad is held in place with a spring in the piston).

To install:

8. Check piston dust sleeves for damage; replace if needed.
9. Clean all mating surfaces.
10. Apply anti-squeak compound to all mounting surfaces.
11. Install calipers; torque bolts to 21 ft. lbs. (31 Nm). Install the plastic plugs.
12. Reposition retaining spring at the top, then at the bottom.
13. Install front wheels.
14. Fully depress brake pedal several times to set proper contact of pads with rotor.
15. Check fluid level and bleed brake system, if necessary.

Rear

1. Before servicing the vehicle, refer to the precautions section.

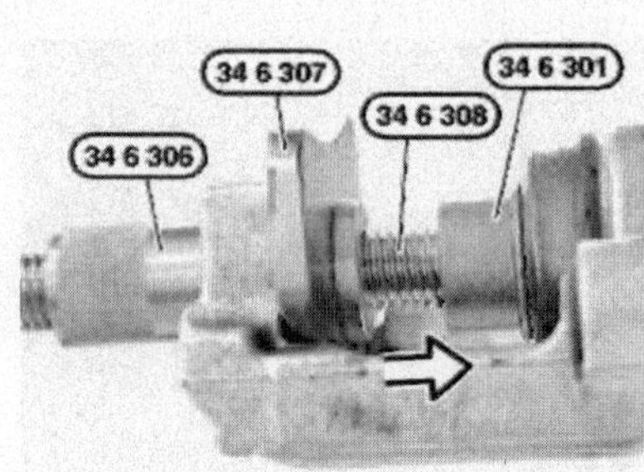

Using special tools to push piston into caliper for removal of disk pads

2. Remove the rear wheels.

3. Remove the retaining spring from the top and then from the bottom.

4. Remove the plastic plugs from the inside of the caliper.

5. Remove the caliper guide pin bolts and remove the caliper from the rotor.

6. Use special tools, 34–6–301, 34–6–306/7/8, force piston back into caliper, as shown.

7. Remove the disk pads.

To install:

8. Check condition of dust sleeve on piston; replace if needed.

9. Clean all contact surfaces.

10. Apply anti-squeak compound to all mounting surfaces.

11. Install the new disk pads evenly in their mounted positions.

12. Clean caliper guide pin bolts; do not apply grease.

13. Install calipers and guide pin bolts. Torque bolts to 21 ft. lbs. (28 Nm).

14. Install the plastic plugs.

15. Install the retaining spring at the bottom and then at the top.

16. Install rear wheels.

17. Fully depress brake pedal several times to set proper contact of pads with rotor.

18. Bleed brake system, if necessary.

SAAB

9-3 • 9-5

10

SPECIFICATIONS AND MAINTENANCE CHARTS

ENGINE AND VEHICLE IDENTIFICATION CHART

Engine						
Code ①	Liters (cc)	Cu. In.	Cyl.	Fuel Sys.	Engine Type	Eng. Mfg.
E	2.3 (2290)	140	I4	MFI-Turbo	DOHC	Saab
G	2.3 (2290)	140	I4	MFI-Turbo	DOHC	Saab
H	2.0 (1985)	121	I4	MFI-Turbo	DOHC	Saab
K	2.0 (1985)	121	I4	MFI-Turbo	DOHC	Saab
S	2.0 (1985)	121	I4	MFI-Turbo	DOHC	Saab
Y	2.0 (1985)	121	I4	MFI-Turbo	DOHC	Saab
U	2.8 (2792)	170	V6	MFI-Turbo	DOHC	Saab
Z	3.0 (2962)	180	V6	MFI-Turbo	DOHC	Saab

Model Year	
Code ②	Year
2	2002
3	2003
4	2004
5	2005
6	2006

MFI: Multiport Fuel Injection

① 8th position of VIN

② 10th position of VIN

06041_SAAB_C01

GENERAL ENGINE SPECIFICATIONS

Year	Model	Engine Displacement Liters	Engine ID/VIN	Net Horsepower @ rpm	Net Torque @ rpm (ft. lbs.)	Bore x Stroke (in.)	Com-pression Ratio	Oil Pressure @ rpm
2002	9-3	2.0	B205L/H	185@5500	192@2100	3.54x3.07	8.8:1	36@2000
	9-3	2.0	B205R/K	205@5500	203@2200	3.54x3.07	8.8:1	36@2000
	9-3	2.3	B235R/G	230@5500	288@2500	3.54x3.54	9.3:1	36@2000
	9-5	2.3	B235E/E	185@5500	192@1800	3.54x3.54	9.3:1	36@2000
	9-5	2.3	B235R/G	250@5200	295@1800	3.54x3.54	9.3:1	36@2000
	9-5	3.0	B308E/Z	200@5100	230@2500	3.38x3.35	9.5:1	NS
2003	9-3	2.0	B207L/Y	185@5500	192@2100	3.54x3.07	8.8:1	36@2000
	9-3	2.0	B207R/S	205@5500	203@2200	3.54x3.07	8.8:1	36@2000
	9-5	2.3	B235E/E	185@5500	192@1800	3.54x3.54	9.3:1	36@2000
	9-5	2.3	B235R/G	250@5200	295@1800	3.54x3.54	9.3:1	36@2000
	9-5	3.0	B308E/Z	200@5100	230@2500	3.38x3.35	9.5:1	NS
2004	9-3	2.0	B207L/Y	185@5500	185@5500	3.39x3.39	9.5:1	29@1000
	9-3	2.0	B207R/S	125@5500	125@5500	3.39x3.39	9.5:1	29@1000
	9-5	2.3	B235E/E	185@5300	207@1800	3.54x3.54	9.3:1	36@2000
	9-5	2.3	B235R/G	250@5300	258@1900	3.54x3.54	9.3:1	36@2000
2005	9-3	2.0	B207L/Y	185@5500	185@5500	3.39x3.39	9.5:1	29@1000
	9-3	2.0	B207R/S	125@5500	125@5500	3.39x3.39	9.5:1	29@1000
	9-5	2.3	B235E/E	185@5300	207@1800	3.54x3.54	9.3:1	36@2000
	9-5	2.3	B235R/G	250@5300	258@1900	3.54x3.54	9.3:1	36@2000
2006	9-3	2.0	B207L/Y	210@5500	221@2500	3.38x3.38	9.5:1	29@1000
	9-3	2.0	B207R/S	210@5500	221@2500	3.38x3.38	9.5:1	29@1000
	9-3	2.8	B284E/U	250@5500	258@2000	3.50x2.94	9.5:1	138@2000
	9-5	2.3	B235E/E	260@5300	258@1900	3.54x3.54	9.3:1	36@2000
	9-5	2.3	B235R/G	260@5300	258@1900	3.54x3.54	9.3:1	36@2000

NS: Not specified by manufacturer

06041_SAAB_C02

TUNE-UP SPECIFICATIONS

Year	Engine Displacement Liters	Engine ID/VIN	Spark Plugs Gap (in.)	Ignition Timing (deg.) MT	Ignition Timing (deg.) AT	Fuel Pump (psi)	Idle Speed (rpm) MT	Idle Speed (rpm) AT	Valve Clearance In.	Valve Clearance Ex.
2002	2.0	B205L/H	0.035-0.039	①	①	43 ②	900	900	HYD	HYD
	2.0	B205R/K	0.035-0.039	①	①	43 ②	900	900	HYD	HYD
	2.3	B235E/E	0.039-0.043	①	①	43 ②	825	825	HYD	HYD
	2.3	B235R/G	0.035-0.039	①	①	43 ②	825	825	HYD	HYD
	3.0	B308E/Z	0.039-0.043	①	①	43 ②	700	700	HYD	HYD
2003	2.0	B207L/Y	0.035-0.039	①	①	43 ②	900	900	HYD	HYD
	2.0	B207R/S	0.035-0.039	①	①	43 ②	900	900	HYD	HYD
	2.3	B235E/E	0.039-0.043	①	①	43 ②	825	825	HYD	HYD
	2.3	B235R/G	0.035-0.039	①	①	43 ②	825	825	HYD	HYD
	3.0	B308E/Z	0.039-0.043	①	①	43 ②	700	700	HYD	HYD
2004	2.0	B207L/Y	0.035-0.039	①	①	43 ②	720	720	HYD	HYD
	2.0	B207R/S	0.035-0.039	①	①	43 ②	850	720	HYD	HYD
	2.3	B235E/E	0.039-0.043	①	①	43 ②	825	825 ③	HYD	HYD
	2.3	B235R/G	0.035-0.039	①	①	43 ②	825	825 ③	HYD	HYD
2005	2.0	B207L/Y	0.035-0.039	①	①	43 ②	720	720	HYD	HYD
	2.0	B207R/S	0.035-0.039	①	①	43 ②	850	720	HYD	HYD
	2.3	B235E/E	0.039-0.043	①	①	43 ②	825	825 ③	HYD	HYD
	2.3	B235R/G	0.035-0.039	①	①	43 ②	825	825 ③	HYD	HYD
2006	2.0	B207L/Y	0.035-0.039	①	①	43 ②	720	720	HYD	HYD
	2.0	B207R/S	0.035-0.039	①	①	43 ②	850	720	HYD	HYD
	2.3	B235E/E	0.039-0.043	①	①	43 ②	825	825 ③	HYD	HYD
	2.3	B235R/G	0.035-0.039	①	①	43 ②	825	825 ③	HYD	HYD
	2.8	B284E/U	0.035-0.039	①	①	43 ②	700	700	HYD	HYD

NOTE: The Vehicle Emission Control Information label often reflects specification changes made during production. The label figures must be used if they differ from figures in this chart.

HYD: Hydraulic

① Pre-programmed in ECU and cannot be adjusted

② With engine warm and operating at 2000 rpm.

③ Idle speed given is with transmission in Neutral; with engine in Drive: 860 rpm.

06041_SAAB_C03

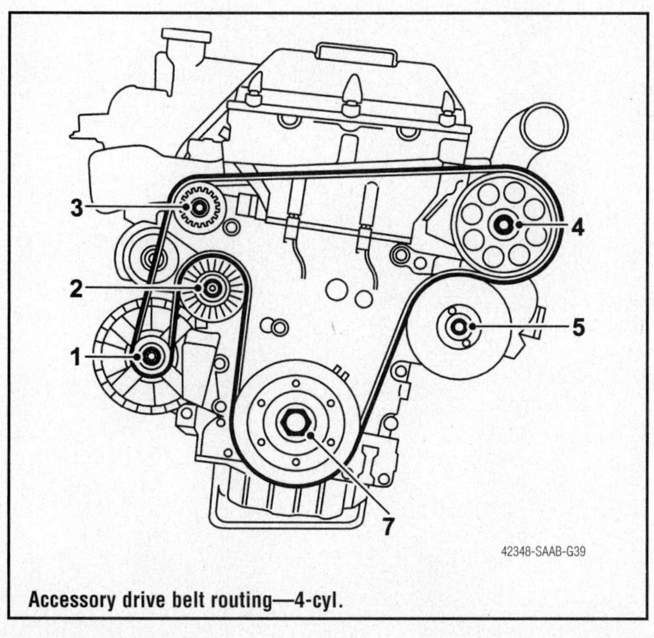

Accessory drive belt routing—4-cyl.

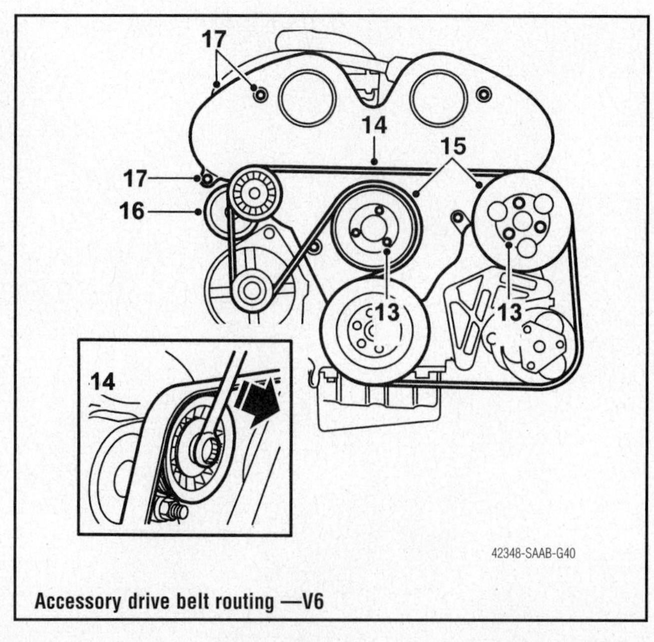

Accessory drive belt routing —V6

CAPACITIES

Year	Model	Engine Displacement Liters	Engine ID/VIN	Engine Oil with Filter (qts.)	Transmission (pts.) 5-Spd	6-Spd	Auto.	Transfer Case (pts.)	Drive Axle Front (pts.)	Rear (pts.)	Fuel Tank (gal.)	Cooling System (qts.)
2002	9-3	2.0	B205L/H	4.3	3.8	—	6.8	NA	NA	NA	16.3 ①	7.8
	9-3	2.0	B205R/K	4.3	3.8	—	6.8	NA	NA	NA	16.3 ①	7.8
	9-3	2.3	B235R/G	4.3	3.8	—	6.8	NA	NA	NA	18.5	7.8
	9-5	2.3	B235E/E	4.1	3.8	—	6.8	NA	NA	NA	18.5	7.8
	9-5	2.3	B235R/G	4.1	3.8	—	6.8	NA	NA	NA	18.5	7.8
	9-5	3.0	B308E/Z	4.6	3.2	—	6.8	NA	NA	NA	18.5	7.6
2003	9-3	2.0	B207L/Y	4.3	3.8	—	6.8	NA	NA	NA	16.3 ①	7.8
	9-3	2.0	B207R/S	4.3	3.8	—	6.8	NA	NA	NA	16.3 ①	7.8
	9-5	2.3	B235E/E	4.1	3.8	—	6.8	NA	NA	NA	18.5	7.8
	9-5	2.3	B235R/G	4.1	3.8	—	6.8	NA	NA	NA	18.5	7.8
	9-5	3.0	B308E/Z	4.6	3.2	—	6.8	NA	NA	NA	18.5	7.6
2004	9-3	2.0	B207L/Y	4.3	3.8	3.8	6.8	NA	NA	NA	16.3 ①	7.8
	9-3	2.0	B207R/S	4.3	3.8	3.8	6.8	NA	NA	NA	16.3 ①	7.8
	9-5	2.3	B235E/E	4.1	3.8	—	6.8	NA	NA	NA	18.5	7.8
	9-5	2.3	B235R/G	4.1	3.8	—	6.8	NA	NA	NA	18.5	7.8
2005	9-3	2.0	B207L/Y	4.3	3.8	3.8	6.8	NA	NA	NA	16.3 ①	7.8
	9-3	2.0	B207R/S	4.3	3.8	3.8	6.8	NA	NA	NA	16.3 ①	7.8
	9-5	2.3	B235E/E	4.1	3.8	—	6.8	NA	NA	NA	18.5	7.8
	9-5	2.3	B235R/G	4.1	3.8	—	6.8	NA	NA	NA	18.5	7.8
2006	9-3	2.0	B207L/Y	4.3	3.8	3.8	6.8	NA	NA	NA	16.4	7.8
	9-3	2.0	B207R/S	4.3	3.8	3.8	6.8	NA	NA	NA	16.4	7.8
	9-3	2.8	B284E/U	4.6	3.8	3.8	6.8	NA	NA	NA	16.4	10.0
	9-5	2.3	B235E/E	4.1	3.8	—	6.8	NA	NA	NA	18.5	7.8
	9-5	2.3	B235R/G	4.1	3.8	—	6.8	NA	NA	NA	18.5	7.8

NOTE: All capacities are approximate. Add fluid gradually and check to be sure a proper fluid level is obtained.

NA: Not available.

① Sedan fuel tank capacity; For convertible: 17.0 gal.

06041_SAAB_C04

VALVE SPECIFICATIONS

Year	Engine Displacement Liters	Engine ID/VIN	Seat Angle (deg.)	Face Angle (deg.)	Spring Test Pressure (lbs. @ in.)	Spring Installed Height (in.)	Stem-to-Guide Clearance (in.) Intake	Stem-to-Guide Clearance (in.) Exhaust	Stem Diameter (in.) Intake	Stem Diameter (in.) Exhaust
2002	2.0	B205L/H	45	44.5	138-150@ 1.25	1.48	0.0067	0.0087	0.1957-0.1963	0.1949-0.1955
	2.0	B205R/K	45	44.5	138-150@ 1.25	1.48	0.0067	0.0087	0.1957-0.1963	0.1949-0.1955
	2.3	B235E/E	45	①	138-150@ 1.25	1.48	0.0067	0.0087	0.1957-0.1963	0.1949-0.1955 ②
	2.3	B235R/G	45	①	138-150@ 1.25	1.48	0.0067	0.0087	0.1957-0.1963	0.1949-0.1955 ②
	3.0	B308E/Z	45	45.3	142-153@ 0.95	NS	NS	NS	0.2344-0.2350	0.2341-0.2346
2003	2.0	B207L/Y	45	44.5	138-150@ 1.25	1.48	0.0067	0.0087	0.1957-0.1963	0.1949-0.1955
	2.0	B207R/S	45	44.5	138-150@ 1.25	1.48	0.0067	0.0087	0.1957-0.1963	0.1949-0.1955
	2.3	B235E/E	45	①	138-150@ 1.25	1.48	0.0067	0.0087	0.1957-0.1963	0.1949-0.1955 ②
	2.3	B235R/G	45	①	138-150@ 1.25	1.48	0.0067	0.0087	0.1957-0.1963	0.1949-0.1955 ②
	3.0	B308E/Z	45	45.3	142-153@ 0.95	NS	NS	NS	0.2344-0.2350	0.2341-0.2346
2004	2.0	B207L/Y	45	①	NS	0.89-1.28	0.0012-0.0022	0.0016-0.0026	0.2344-0.2350	0.2341-0.2346
	2.0	B207R/S	45	①	NS	0.89-1.28	0.0012-0.0022	0.0016-0.0026	0.1957-0.1963	0.1949-0.1955
	2.3	B235E/E	45	①	138-150@ 1.25	1.48	0.0067	0.0087	0.1957-0.1963	0.1949-0.1955 ②
	2.3	B235R/G	45	①	138-150@ 1.25	1.48	0.0067	0.0087	0.1957-0.1963	0.1949-0.1955 ②
2005	2.0	B207L/Y	45	①	NS	0.89-1.28	0.0012-0.0022	0.0016-0.0026	0.2344-0.2350	0.2341-0.2346
	2.0	B207R/S	45	①	NS	0.89-1.28	0.0012-0.0022	0.0016-0.0026	0.1957-0.1963	0.1949-0.1955
	2.3	B235E/E	45	①	138-150@ 1.25	1.48	0.0067	0.0087	0.1957-0.1963	0.1949-0.1955 ②
	2.3	B235R/G	45	①	138-150@ 1.25	1.48	0.0067	0.0087	0.1957-0.1963	0.1949-0.1955 ②
2006	2.0	B207L/Y	45	①	NS	0.89-1.28	0.0012-0.0022	0.0016-0.0026	0.2344-0.2350	0.2341-0.2346
	2.0	B207R/S	45	①	NS	0.89-1.28	0.0012-0.0022	0.0016-0.0026	0.1957-0.1963	0.1949-0.1955
	2.8	B284E/U	45	NS	NS	1.38	NS	NS	NS	NS
	2.3	B235E/E	45	①	138-150@ 1.25	1.48	0.0067	0.0087	0.1957-0.1963	0.1949-0.1955 ②
	2.3	B235R/G	45	①	138-150@ 1.25	1.48	0.0067	0.0087	0.1957-0.1963	0.1949-0.1955 ②

NS - Information Not Specified by manufacturer.

① Exhaust: 44.5

Intake: 45.3

② If equipped with Nimonic engine: 0.1956-0.1961 inches.

06041_SAAB_C05

CRANKSHAFT AND CONNECTING ROD SPECIFICATIONS

All measurements are given in inches.

	Engine		Crankshaft				Connecting Rod		
Year	Displacement Liters	Engine ID/VIN	Main Brg. Journal Dia.	Main Brg. Oil Clearance	Shaft End-play	Thrust on No.	Journal Diameter	Oil Clearance	Side Clearance
2002	2.0	B205L/H	2.283	0.0006-0.0024	0.003-0.013	NS	2.046	0.0008-0.0027	NS
	2.0	B205R/K	2.283	0.0006-0.0024	0.003-0.013	NS	2.046	0.0008-0.0027	NS
	2.3	B235R/G	2.283	0.0006-0.0024	0.003-0.013	NS	2.046	0.0008-0.0027	NS
	2.3	B235E/E	2.283	0.0006-0.0024	0.003-0.013	NS	2.046	0.0008-0.0027	NS
	3.0	B308E/Z	2.676	0.0006-0.0017	0.004-0.030	NS	2.126	0.0005-0.0024	NS
2003	2.0	B207L/Y	2.283	0.0006-0.0024	0.003-0.013	NS	2.046	0.0008-0.0027	NS
	2.0	B207R/S	2.283	0.0006-0.0024	0.003-0.013	NS	2.046	0.0008-0.0027	NS
	2.3	B235R/G	2.283	0.0006-0.0024	0.003-0.013	NS	2.046	0.0008-0.0027	NS
	2.3	B235E/E	2.283	0.0006-0.0024	0.003-0.013	NS	2.046	0.0008-0.0027	NS
	3.0	B308E/Z	2.676	0.0006-0.0017	0.004-0.030	NS	2.126	0.0005-0.0024	NS
2004	2.0	B207L/Y	2.205	0.0012-0.0026	0.001-0.003	NS	1.929	0.0012-0.0028	NS
	2.0	B207R/S	2.205	0.0012-0.0026	0.001-0.003	NS	1.929	0.0012-0.0028	NS
	2.3	B235R/G	2.283	0.0006-0.0024	0.003-0.013	NS	2.046	0.0008-0.0027	NS
	2.3	B235E/E	2.283	0.0006-0.0024	0.003-0.013	NS	2.046	0.0008-0.0027	NS
2005	2.0	B207L/Y	2.205	0.0012-0.0026	0.001-0.003	NS	1.929	0.0012-0.0028	NS
	2.0	B207R/S	2.205	0.0012-0.0026	0.001-0.003	NS	1.929	0.0012-0.0028	NS
	2.3	B235R/G	2.283	0.0006-0.0024	0.003-0.013	NS	2.046	0.0008-0.0027	NS
	2.3	B235E/E	2.283	0.0006-0.0024	0.003-0.013	NS	2.046	0.0008-0.0027	NS
2006	2.0	B207L/Y	2.205	0.0012-0.0026	0.001-0.003	NS	1.929	0.0012-0.0028	NS
	2.0	B207R/S	2.205	0.0012-0.0026	0.001-0.003	NS	1.929	0.0012-0.0028	NS
	2.8	B284E/U	2.868	2.3622-3.9370	0.001-0.003	NS	2.347	2.7559-3.9370	NS
	2.3	B235R/G	2.283	0.0006-0.0024	0.003-0.013	NS	2.046	0.0008-0.0027	NS
	2.3	B235E/E	2.283	0.0006-0.0024	0.003-0.013	NS	2.046	0.0008-0.0027	NS

NS: Not specified by manufacturer.

06041_SAAB_C06

PISTON AND RING SPECIFICATIONS

All measurements are given in inches.

Year	Engine Displacement Liters	Engine ID/VIN	Piston Clearance	Ring Gap: Top Compression	Ring Gap: Bottom Compression	Ring Gap: Oil Control	Ring Side Clearance: Top Compression	Ring Side Clearance: Bottom Compression	Ring Side Clearance: Oil Control
2002	2.0	B205L/H	0.001-0.002	0.012-0.019	0.012-0.019	0.030-0.039	0.0014-0.0031	0.0016-0.0030	0.0985
	2.0	B205R/K	0.001-0.002	0.012-0.019	0.012-0.019	0.030-0.039	0.0014-0.0031	0.0016-0.0030	0.0985
	2.3	B235R/G	0.001-0.002	0.012-0.019	0.012-0.019	0.030-0.039	0.0014-0.0031	0.0016-0.0030	0.0985
	2.3	B235E/E	0.001-0.002	0.012-0.019	0.012-0.019	0.030-0.039	0.0014-0.0031	0.0016-0.0030	0.0985
	3.0	B308E/Z	0.001-0.002	0.012-0.019	0.012-0.019	0.016-0.055	NS	NS	NS
2003	2.0	B207L/Y	0.001-0.002	0.012-0.019	0.012-0.019	0.030-0.039	0.0014-0.0031	0.0016-0.0030	0.0985
	2.0	B207R/S	0.001-0.002	0.012-0.019	0.012-0.019	0.030-0.039	0.0014-0.0031	0.0016-0.0030	0.0985
	2.3	B235R/G	0.001-0.002	0.012-0.019	0.012-0.019	0.030-0.039	0.0014-0.0031	0.0016-0.0030	0.0985
	2.3	B235E/E	0.001-0.002	0.012-0.019	0.012-0.019	0.030-0.039	0.0014-0.0031	0.0016-0.0030	0.0985
	3.0	B308E/Z	0.001-0.002	0.012-0.019	0.012-0.019	0.016-0.055	NS	NS	NS
2004	2.0	B207L/Y	0.0005-0.0018	0.006-0.014	0.016-0.024	0.010-0.030	0.0010-0.0030	0.0010-0.0020	0.002-0.006
	2.0	B207R/S	0.0005-0.0018	0.006-0.014	0.016-0.024	0.010-0.030	0.0010-0.0030	0.0010-0.0020	0.002-0.006
	2.3	B235R/G	0.001-0.002	0.012-0.019	0.012-0.019	0.030-0.039	0.0014-0.0031	0.0016-0.0030	0.0985
	2.3	B235E/E	0.001-0.002	0.012-0.019	0.012-0.019	0.030-0.039	0.0014-0.0031	0.0016-0.0030	0.0985
2005	2.0	B207L/Y	0.0005-0.0018	0.006-0.014	0.016-0.024	0.010-0.030	0.0010-0.0030	0.0010-0.0020	0.002-0.006
	2.0	B207R/S	0.0005-0.0018	0.006-0.014	0.016-0.024	0.010-0.030	0.0010-0.0030	0.0010-0.0020	0.002-0.006
	2.3	B235R/G	0.001-0.002	0.012-0.019	0.012-0.019	0.030-0.039	0.0014-0.0031	0.0016-0.0030	0.0985
	2.3	B235E/E	0.001-0.002	0.012-0.019	0.012-0.019	0.030-0.039	0.0014-0.0031	0.0016-0.0030	0.0985
2006	2.0	B207L/Y	0.0005-0.0018	0.006-0.014	0.016-0.024	0.010-0.030	0.0010-0.0030	0.0010-0.0020	0.002-0.006
	2.0	B207R/S	0.0005-0.0018	0.006-0.014	0.016-0.024	0.010-0.030	0.0010-0.0030	0.0010-0.0020	0.002-0.006
	2.8	B284E/U	NS	NS	NS	NS	NS	NS	NS
	2.3	B235R/G	0.001-0.002	0.012-0.019	0.012-0.019	0.030-0.039	0.0014-0.0031	0.0016-0.0030	0.0985
	2.3	B235E/E	0.001-0.002	0.012-0.019	0.012-0.019	0.030-0.039	0.0014-0.0031	0.0016-0.0030	0.0985

NS: Not specified by manufacturer

06041_SAAB_C07

TORQUE SPECIFICATIONS

All readings in ft. lbs.

Year	Engine Displacement Liters	Engine ID/VIN	Cylinder Head Bolts	Main Bearing Bolts	Rod Bearing Bolts	Crankshaft Damper Bolts	Flywheel Bolts	Manifold Intake	Manifold Exhaust	Spark Plugs	Oil Pan Drain Plug
2002	2.0	B205L/H	①	②	②	130	③	18	13	21	19
	2.0	B205R/K	①	②	②	130	③	18	13	21	19
	2.3	B235R/G	①	②	②	130	③	18	13	21	19
	2.3	B235E/E	①	②	②	130	③	18	13	21	19
	3.0	B308E/Z	④	⑤	⑥	230	⑦	15	15	19	19
2003	2.0	B207L/Y	①	②	②	130	③	18	13	21	19
	2.0	B207R/S	①	②	②	130	③	18	13	21	19
	2.3	B235R/G	①	②	②	130	③	18	13	21	19
	2.3	B235E/E	①	②	②	130	③	18	13	21	19
	3.0	B308E/Z	④	⑤	⑥	230	⑦	15	15	19	19
2004	2.0	B207L/Y	⑧	NS	⑨	⑩	⑪	7	⑫	21	19
	2.0	B207R/S	⑧	NS	⑨	⑩	⑪	7	⑫	21	19
	2.3	B235R/G	①	②	②	130	③	18	13	21	19
	2.3	B235E/E	①	②	②	130	③	18	13	21	19
2005	2.0	B207L/Y	⑧	NS	⑨	⑩	⑪	7	⑫	21	19
	2.0	B207R/S	⑧	NS	⑨	⑩	⑪	7	⑫	21	19
	2.3	B235R/G	①	②	②	130	③	18	13	21	19
	2.3	B235E/E	①	②	②	130	③	18	13	21	19
2006	2.0	B207L/Y	⑧	NS	⑨	⑩	⑪	7	⑫	21	19
	2.0	B205R/K	⑧	NS	⑨	⑩	⑪	7	⑫	21	19
	2.3	B235R/G	①	②	②	130	③	18	13	21	19
	2.3	B235E/E	①	②	②	130	③	18	13	21	19
	2.8	B284E/U	⑬	⑭	⑮	⑯	⑰	17	15	21	19

NS: Not specified by manufacturer

① Step 1: 29.5 ft. lbs.
Step 2: 44 ft. lbs.
Step 3: Tighten each bolt an additional 90 degrees

② Step 1: 15 ft. lbs.
Step 2: Tighten each bolt an additional 70 degrees

③ Step 1: 15 ft. lbs.
Step 2: Tighten each bolt an additional 50 degrees

④ Step 1: 18 ft. lbs.
Step 2: Tighten each bolt an additional 90 degrees
Step 3: Tighten each bolt an additional 90 degrees
Step 4: Tighten each bolt an additional 90 degrees

⑤ Step 1: 37 ft. lbs.
Step 2: Tighten each bolt an additional 60 degrees
Step 3: Tighten each bolt an additional 15 degrees

⑥ Step 1: 26 ft. lbs.
Step 2: Tighten each bolt an additional 45 degrees
Step 3: Tighten each bolt an additional 15 degrees

⑦ Step 1: 48 ft. lbs.
Step 2: Tighten each bolt an additional 30 degrees

⑧ Step 1: 22 ft. lbs.
Step 2: Tighten each bolt an additional 75 degrees
Step 3: Tighten each bolt an additional 80 degrees

⑨ Step 1: 18 ft. lbs.
Step 2: Tighten each bolt an additional 100 degree

⑩ Step 1: 74 ft. lbs.
Step 2: Tighten each bolt an additional 75 degrees

⑪ Step 1: 39 ft. lbs.
Step 2: Tighten each bolt an additional 25 degrees

⑫ Step 1: 18 ft. lbs.
Step 2: 24 ft. lbs.

⑬ Step 1: Tighten M8 bolts to 11 ft. lbs.
Step 2: Tighten each M8 bolt an additional 60 degrees
Step 3: Tighten M11 bolts to 33 ft. lbs.
Step 4: Tighten each M11 bolt an additional 120 degrees

⑭ Main bearing inner screw: 15 ft. lbs. plus 80 degrees
Main bearing outer screw: 11 ft. lbs. plus 110 degrees
Main bearing side screw: 22 ft. lbs. plus 60 degrees

⑮ Step 1: 22 ft. lbs.
Step 2: Loosen back to 0
Step 3: 18 ft. lbs.
Step 4: Tighten each bolt an additional 110 degrees

⑯ Step 1: 74 ft. lbs.
Step 2: Tighten an additional 150 degrees

⑰ Step 1: 22 ft. lbs.
Step 2: Tighten each bolt an additional 45 degrees

06041_SAAB_C08

WHEEL ALIGNMENT

Year	Model		Caster Range (+/-Deg.)	Caster Preferred Setting (Deg.)	Camber Range (+/-Deg.)	Camber Preferred Setting (Deg.)	Toe-in (in.)
2002	9-3	F	0.70	+1.85	0.80	-0.80	0.39 ①
		R	NS	NS	0.30	-1.70	0.39 ①
	9-5	F	0.50	+2.90	0.50	-0.80	②
		R	NS	NS	0.25	-0.80	③
2003	9-3	F	0.70	+1.85	0.80	-0.80	④
		R	NS	NS	0.30	-0.70	④
	9-5	F	0.50	+2.90	0.50	-0.80	②
		R	NS	NS	0.25	-0.80	③
2004	9-3	F	0.50	+2.90	0.50	-0.80	④
		R	NS	NS	0.30	-0.70	④
	9-5	F	0.50	+2.90	0.50	-0.80	⑤
		R	NS	NS	0.25	⑥	⑦
2005	9-3	F	0.50	+2.90	0.50	-0.80	④
		R	NS	NS	0.30	-0.70	④
	9-5	F	0.50	+2.90	0.50	-0.80	⑤
		R	NS	NS	0.25	⑥	⑦
2006	9-3	F	0.50	+2.90	0.50	-0.80	④
		R	NS	NS	0.30	-0.70	④
	9-5	F	0.50	+2.90	0.50	-0.80	⑤
		R	NS	NS	0.25	⑥	⑦

NS: Not specified by manufacturer

① Range +/- 0.79 inches

② With 15 inch wheels: 0.039 inch
With 16 inch wheels: 0.43 inch
With 17 inch wheels: 0.43 inch
With 17 inch wheels (sport chassis) : 0.47 inch

③ With 15 inch wheels: 0.059 inch
With 16 inch wheels: 0.63 inch
With 17 inch wheels: 0.67 inch
With 17 inch wheels (sport chassis) : 0.67 inch

④ With 16 inch wheels: 0.034 inch (+/-0.011 inch)
With 17 inch wheels: 0.037 inch (+/- 0.012 inch)
With 18 inch wheels : 0.041 inch (+/- 0.013 inch)

⑤ With 16 inch wheels: 0.43 inch
With 17 inch wheels: 0.43 inch
With 17 inch wheels (sport chassis) : 0.47 inch

⑥ With 16 inch wheels: -0.80 degrees
With 17 inch wheels: -0.80 degrees
With 17 inch wheels (sport chassis): -1.10 degrees

⑦ With 16 inch wheels: 0.63 inch
With 17 inch wheels: 0.67 inch
With 17 inch wheels (sport chassis) : 0.67 inch

06041_SAAB_C09

TIRE, WHEEL AND BALL JOINT SPECIFICATIONS

Year	Model	OEM Tires Standard	OEM Tires Optional	Tire Pressures (psi) Front	Tire Pressures (psi) Rear	Wheel Size	Ball Joint Inspection	Wheel Lug Torque (ft. lbs.)
2002	9-3	195/60R15	None	32	32	6J	NP	81
	9-3	205/50R16	None	33	33	6J	NP	81
	9-3 Viggen	215/45R17	None	38	35	7J	NP	81
	9-5	205/65R15	None	30	30	6.5J	NP	80
	9-5	215/55R15	None	32	32	6.5J	NP	80
	9-5	225/45R17	None	32	32	7J	NP	80
2003	9-3 Sedan	215/60R15	None	32	32	6J	NP	81
	9-3 Sedan	215/55R16	None	33	33	6J	NP	81
	9-3 Sedan	215/50R17	225/45R17 ①	38	35	7J	NP	81
	9-3 Conv.	195/60R15	None	32	32	6.5J	NP	81
	9-3 Conv.	205/55R16	None	33	33	6.5J	NP	81
	9-3 Conv.	215/45R17	None	35	35	7J	NP	81
	9-5	205/65R15	None	33	30	6J	NP	80
	9-5	215/55R15	None	35	32	6.5J	NP	80
	9-5	225/45R17	None	35	33	7J	NP	80
2004	9-3 Sedan	215/60R15	None	32	32	6J	NP	81
	9-3 Sedan	215/55R16	None	33	33	6J	NP	81
	9-3 Sedan	215/50R17	225/45R17 ①	38	35	7J	NP	81
	9-3 Conv.	195/60R15	None	32	32	6.5J	NP	81
	9-3 Conv.	205/55R16	None	33	33	6.5J	NP	81
	9-3 Conv.	215/45R17	None	35	35	7J	NP	81
	9-5 Linear	215/55VR16	None	35	32	6J	NP	80
	9-5 Arc	215/55VR16	225/45R17	35	32	6J	NP	80
	9-5 Aero	225/45R17	None	35	33	7J	NP	80
2005	9-3 Sedan	215/60R15	None	32	32	6J	NP	81
	9-3 Sedan	215/55R16	None	33	33	6J	NP	81
	9-3 Sedan	215/50R17	225/45R17 ①	38	35	7J	NP	81
	9-3 Conv.	195/60R15	None	32	32	6.5J	NP	81
	9-3 Conv.	205/55R16	None	33	33	6.5J	NP	81
	9-3 Conv.	215/45R17	None	35	35	7J	NP	81
	9-5 Linear	215/55VR16	None	35	32	6J	NP	80
	9-5 Arc	215/55VR16	225/45R17	35	32	6J	NP	80
	9-5 Aero	225/45R17	None	35	33	7J	NP	80
2006	9-3 Sedan	215/55HR16	None	35	35	6.5J	NP	81
	9-3 Aero Sedan	235/45VR17	None	35	35	7.5J	NP	81
	9-3 Conv.	215/55HR16	None	35	35	6.5J	NP	81
	9-3 Aero Conv.	235/45VR17	None	35	35	7.5J	NP	81
	9-3 SportCombi	215/55HR16	None	35	35	6.5J	NP	81
	9-3 SportCombi	235/45VR17	None	35	35	7.5J	NP	81
	9-5 Sedan	235/45VR17	None	35	35	7.5J	NP	80
	9-5 SportCombi	235/45VR17	None	35	35	7.5J	NP	80

NP: No play visible upon inspection

① With optional 18 inch wheels: 225/45R18

06041_SAAB_C10

BRAKE SPECIFICATIONS

All measurements in inches unless noted

			Brake Disc			Minimum Lining Thickness		Brake Caliper	
Year	Model		Original Thickness	Minimum Thickness	Maximum Runout	Front	Rear	Bracket Bolts (ft. lbs.)	Mounting Bolts (ft. lbs.)
2002	9-3	F	0.980	0.870	0.003	0.200	NA	81	①
		R	0.390	0.310	0.003	NA	0.200	59	NA
	9-5	F	0.980	0.870	0.003	0.200	NA	86	21
		R	0.470	0.320	0.003	NA	0.160	59	21
2003	9-3	F	②	②	0.003	0.550	NA	③	21
		R	④	④	0.003	NA	0.433	⑤	21
	9-5	F	0.980	0.870	0.003	0.200	NA	⑥	21
		R	0.390	0.320	0.003	NA	0.160	59	21
2004	9-3	F	②	②	0.003	0.550	NA	③	21
		R	④	④	0.003	NA	0.433	⑤	21
	9-5	F	0.980	0.870	0.003	0.200	NA	⑥	21
		R	0.390	0.320	0.003	NA	0.160	59	21
2005	9-3	F	②	②	0.003	0.550	NA	③	21
		R	④	④	0.003	NA	0.433	⑤	21
	9-5	F	0.980	0.870	0.003	0.200	NA	⑥	21
		R	0.390	0.320	0.003	NA	0.160	59	21
2006	9-3	F	②	②	0.003	0.550	NA	③	21
		R	④	④	0.003	NA	0.433	⑤	21
	9-5	F	0.980	0.870	0.003	0.200	NA	⑥	21
		R	0.390	0.320	0.003	NA	0.160	59	21

NA: Not applicable

① Step 1: 103 ft. lbs.
Step 2: Tighten bolts an additional 45 degrees

② Level 1: Original thickness: 0.980 inch; minimum thickness: 0.870 inch
Levels 2, 3 and 4: Original thickness: 1.100 inch; minimum thickness: 0.980 inch

③ Step 1: 155 ft. lbs.
Step 2: Tighten bolts an additional 30 degrees

④ Levels 1 and 4: Original thickness: 0.470 inch; minimum thickness: 0.390 inch
Levels 2 and 3: Original thickness: 0.790 inch; minimum thickness: 0.730 inch

⑤ Step 1: 96 ft. lbs.
Step 2: Tighten bolts an additional 45 degrees

⑥ Except Viggen models: 18 ft. lbs.
Viggen models: 21 ft. lbs.

06041_SAAB_C11

SCHEDULED MAINTENANCE INTERVALS

SAAB—9-3 & 9-5

TO BE SERVICED	TYPE OF SERVICE	VEHICLE MILEAGE INTERVAL (x1000)												
		5	10	15	20	25	30	35	40	45	50	55	60	65
Engine oil and filter	R	✓	✓	✓	✓	✓	✓	✓	✓	✓	✓	✓	✓	✓
Battery electrolyte	S/I	✓		✓		✓		✓		✓		✓		✓
Brake fluid	S/I	✓		✓		✓		✓		✓		✓		✓
Brake lines and hoses	S/I	✓		✓		✓		✓		✓		✓		✓
Brake pads & discs	S/I	✓		✓		✓		✓		✓		✓		✓
Drive belts	S/I	✓		✓		✓		✓		✓		✓		✓
Engine coolant strength	S/I	✓		✓		✓		✓		✓		✓		✓
Exhaust system	S/I	✓		✓		✓		✓		✓		✓		✓
Final drive oil level (900 A/T)	S/I	✓		✓		✓		✓		✓		✓		✓
Gear oil	S/I	✓		✓		✓		✓		✓		✓		✓
Outer & inner drive joint boots	S/I	✓		✓		✓		✓		✓		✓		✓
Rotate tires (front to rear)	S/I	✓		✓		✓		✓		✓		✓		✓
Automatic transmission fluid & filter	R	✓						✓						✓
Air cleaner element	R							✓						✓
Engine coolant	R							✓						✓
Spark plugs	R							✓						✓
Ventilation air filter	R							✓						✓
Ball joint clearance	S/I							✓						✓
Engine cooling system, hoses & cap	S/I	✓						✓						✓
Front wheel alignment	S/I							✓						✓
Fuel lines	S/I							✓						✓
Shock absorbers & bushings	S/I							✓						✓
Fuel filter	R													✓
Power steering fluid	R							✓						✓
Crankcase ventilation & vacuum lines	S/I													✓
Distributor cap & rotor	S/I													✓
EVAP system	S/I													✓
Front suspension, rear axle mountings	S/I	✓												
Parking brake adjustment	S/I	✓												
Spark plug wires	S/I													✓

R: Replace S/I: Inspect and service, if needed

FREQUENT OPERATION MAINTENANCE (SEVERE SERVICE)

If a vehicle is operated under any of the following conditions it is considered severe service:

- Extremely dusty areas.

- 50% or more of the vehicle operation is in 32°C (90°F) or higher temperatures, or constant operation in temperatures below 0°C (32°F).

- Prolonged idling (vehicle operation in stop and go traffic).

- Frequent short running periods (engine does not warm to normal operating temperatures).

- Police, taxi, delivery usage or trailer towing usage.

Oil and oil filter: change every 2500 miles.

Air cleaner element: service or inspect every 15,000 miles.

06041_SAAB_C12

ENGINE REPAIR

➡Disconnecting the negative battery cable on some vehicles may interfere with the functions of the on board computer systems and may require the computer to undergo a relearning process, once the negative battery cable is reconnected.

➡Disconnecting the negative battery cable will deprogram the radio system security code. Be sure to get the code from the customer before starting service, as it will have to be re-entered upon completion of service in order to operate the radio.

Alternator

REMOVAL

9-3

2002 MODEL

1. Before servicing the vehicle, refer to the Precautions Section.
2. Remove or disconnect the following:
   - Negative battery cable
   - Air cleaner
   - Air intake hose to throttle body
   - Drive belt tensioner
   - Right front wheel
   - Belt drive cover from under vehicle
   - Alternator electrical connections
   - Alternator retaining bolts
   - Exhaust manifold to exhaust pipe nuts and rubber hanger on front pipe
   - Push catalytic converter carefully aside.
   - Alternator

2003–06 MODEL (EXCEPT 2.8L ENGINE)

1. Before servicing the vehicle, refer to the Precautions Section.
2. Remove or disconnect the following:
   - Negative battery cable
   - Lower engine cover
   - Drain coolant
   - Plastic nuts from wheel well liner
   - Drive belt (through wheel well liner)
   - Upper engine cover
   - Connector(s) from ECM
   - ECM from mounting
   - Air hose from air filter housing
   - Secondary air injection hose
   - Air filter housing cover and air filter
   - Pressure sensor connector
   - Air filter housing
   - Radiator hose
   - Vent hose from air hose to throttle body
   - Loosen fastener in camshaft cover
   - Bend vent hose aside
   - Alternator electrical connections
   - Alternator retaining bolts
   - Alternator (lift upward)

2.8L ENGINE

1. Before servicing the vehicle, refer to the Precautions Section.
2. Remove the battery cover.
3. Disconnect the negative battery cable.
4. Raise and safely support the front of the vehicle.
5. Remove the front right wheel.
6. Remove the right wheel well liner.
7. Relieve the belt tensioner and unhook the belt. Use a ½ inch drawbar handle.
8. Remove the bolt from the steering swivel member, lower the suspension arm and place a 83 95 238 Wedge between the suspension arm and the anti-roll bar.
9. Tap the drive shaft out of the intermediate shaft with a brass drift and large hammer. Move aside the drive shaft. Use protective goggles to protect against flying splinters of metal.
10. Remove the alternator electrical connections.
11. Remove the alternator retaining bolts.
12. Remove the alternator.

9-5

2.3L ENGINE

1. Before servicing the vehicle, refer to the Precautions Section.
2. Remove or disconnect the following:
   - Negative battery cable
   - Intake manifold shield
   - Crankcase ventilation solenoid and constant pressure valves
   - Belt tensioner with special tool 83 95 254
   - Alternator upper mounting bolt
   - Right side engine mount nut
   - Rear engine mount nut
   - Front exhaust pipe

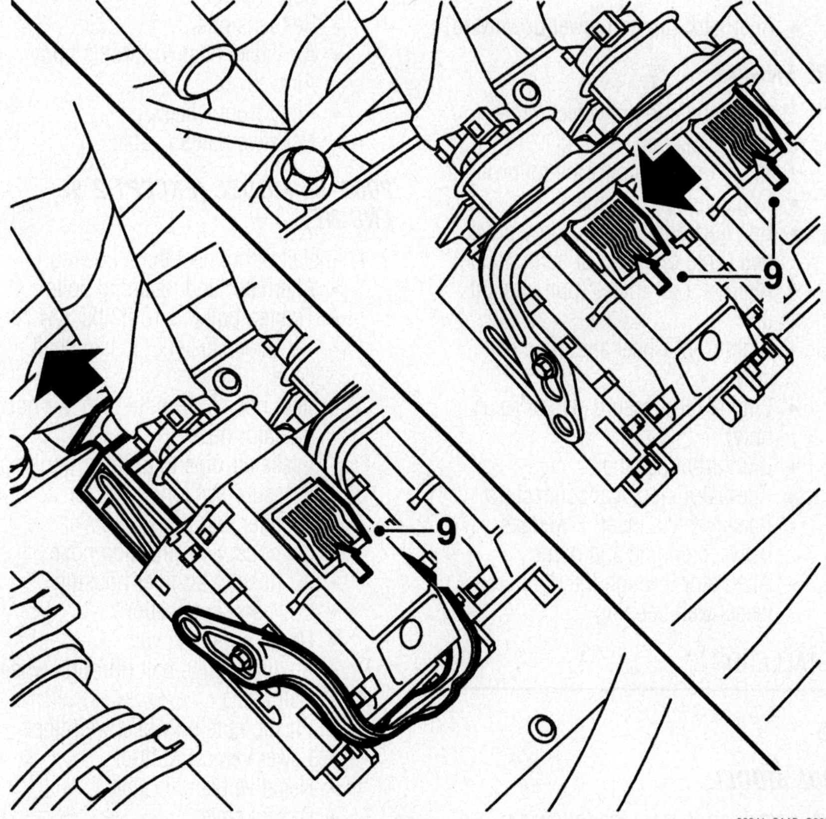

Showing location of ECM connectors—9-3 2003–06

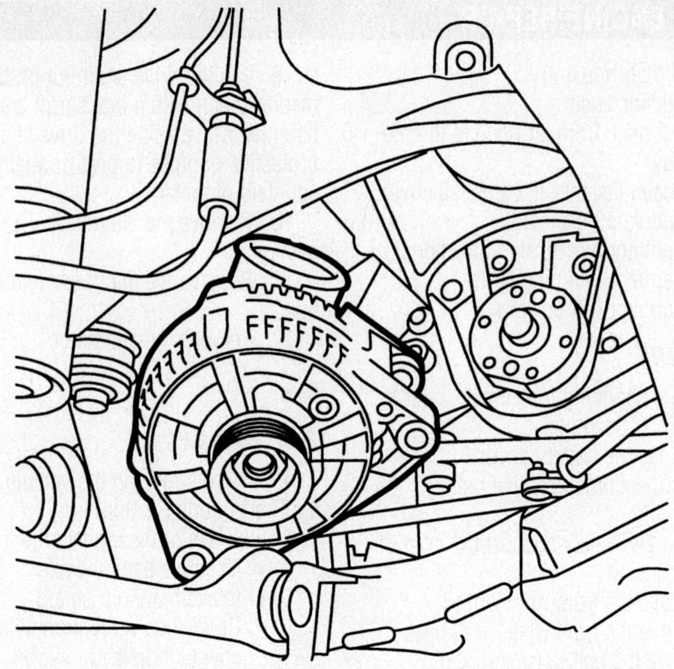

Remove the alternator through the right front wheel well opening—3.0L Engine

- Hose between the oil trap and the oil pan
- Alternator electrical connectors
- Alternator lower bolt and insert a bar between the gearbox and subframe
- Alternator (tip and lower downward)

3.0L ENGINE

1. Before servicing the vehicle, refer to the Precautions Section.
2. Remove or disconnect the following:
   - Negative battery cable
   - Belt tensioner and auxiliary drive belt (note direction of belt rotation)
   - Alternator air intake (pull straight up)
   - Right front wheel and drive belt cover
   - Engine oil pressure sensor (plug hole)
   - Right front halfshaft
   - Alternator electrical connectors
   - Raise vehicle insert 2 wedges between engine and frame
   - Alternator through the right side wheel well opening

INSTALLATION

9-3

2002 MODEL

1. Install or connect the following:
   - Alternator
   - Catalytic converter mounting bolt
   - Exhaust manifold to exhaust pipe mounting hardware
   - Alternator retaining bolts
   - Alternator electrical connections
   - Belt drive cover
   - Belt tensioner
   - Air intake hose to throttle body
   - Air cleaner
   - Right front wheel
   - Negative battery cable

2003–06 MODEL (EXCEPT 2.8L ENGINE)

1. Install or connect the following:
   - Alternator and retaining bolts
   - Tighten bolts to 18 ft. lbs. (24 Nm)
   - Vent hose fastener to camshaft cover
   - Vent hose to throttle body air hose
   - Radiator hose
   - Intake air pipe to air filter housing
   - Pressure sensor connector
   - Air filter and housing cover
   - Secondary air injection hose
   - Air hose to air filter housing
   - ECM and connections
   - Upper engine cover
   - Auxiliary drive belt (through wheel well liner)
   - Plastic nuts on wheel well liner
   - Lower vehicle to floor
   - Negative battery cable
   - Battery cover
   - Refill cooling system
2. Reset the clock and radio, if necessary.

2.8L ENGINE

1. Install the alternator. Torque the retaining bolts to 18 ft. lbs. (24 Nm).
2. Connect the alternator electrical connections.
3. Grease the drive shaft splines and make sure the rubber seal is mounted on the shaft.
4. Fit the drive shaft to the intermediate shaft. Make sure the lock ring clicks into place.
5. Remove the wedge and connect the ball joint to the steering swivel member with a nut and bolt. The pin must be visible above the steering swivel member before the bolt is fitted. Tighten the nut to 37 ft. lbs. (50 Nm).

➡Press up the pin carefully. The groove in the pin must be visible in the screw hole in the spindle housing. If the rubber gaiter is pressed down, it will not seal properly to the swivel pin.

6. Relieve the belt tensioner, use a ´ inch drawbar handle and mount the belt.
7. Install the right wheel well liner.
8. Install the front wheel.
9. Lower the vehicle.
10. Connect the negative battery cable.
11. Fit the battery cover.
    - Reset clock and radio

9-5

2.3L ENGINE

1. Install or connect the following:

➡To make it easier to refit alternator, tap bushings until they are level with the mountings.

   - Insert bar between gearbox and subframe.
   - Alternator and loosely install the lower retaining bolt
   - Alternator electrical connectors
   - Hose between the oil trap and oil pan
   - Front exhaust pipe
   - Upper retaining bolts and torque the bolts to 33 ft. lbs. (45 Nm)
   - Auxiliary drive belt and tensioner
   - Crankcase ventilation solenoid and constant pressure valves
   - Right side engine mount and torque the nut to 37 ft. lbs. (50 Nm)
   - Rear engine mount and torque the nut to 19 ft. lbs. (25 Nm)
   - Intake manifold shield
   - Negative battery cable

3.0L ENGINE

1. Install or connect the following:
   - Position alternator and bolts
   - Alternator electrical connectors
   - Wedges between engine and frame (remove)
   - Engine oil pressure sensor
   - Right front halfshaft
   - Wheel housing drive belt cover
   - Right front wheel
   - Belt tensioner and torque the bolt to 30 ft. lbs. (40 Nm)
   - Alternator air intake
   - V-belt
   - Negative battery cable
   - Reset clock and radio

Ignition Timing

ADJUSTMENT

Saab vehicles are equipped with a Distributorless Ignition System (DIS). Ignition timing is controlled by the ECM. No adjustment is necessary or possible.

Engine Assembly

REMOVAL & INSTALLATION

9-3

2002 MODEL

➡The engine and transmission are removed together as an assembly.

1. Before servicing the vehicle, refer to the Precautions Section.
2. With all 4 wheels on the ground, loosen the axle hub nuts.
3. Relieve the fuel system pressure.
4. Drain the engine coolant.
5. Remove or disconnect the following:
   - Battery
   - Air cleaner assembly and attached hoses
   - Resonator
   - Throttle cable and move it to one side
   - Cruise control wiring harness and cables from the throttle body
   - Cruise control unit retaining nuts
   - Cruise control unit
   - Fuel lines at their connections at the front of the fuel injection manifold and on the fuel pressure regulator
   - Turbocharger pressure line from the turbocharger compressor, if equipped
   - Intercooler/throttle housing, if equipped
   - Vacuum hose from the secondary injection
   - Brake booster vacuum hose from the intake manifold.
   - Boost pressure control unit, if equipped
   - Drive belt using a ⅜ ratchet extension
   - Pressure and return pipe from the steering servo pump and plug to prevent oil escaping from the pipe.
6. Disconnect the cooling system hoses at the following connections:
   a. Heater control valve
   b. Expansion tank

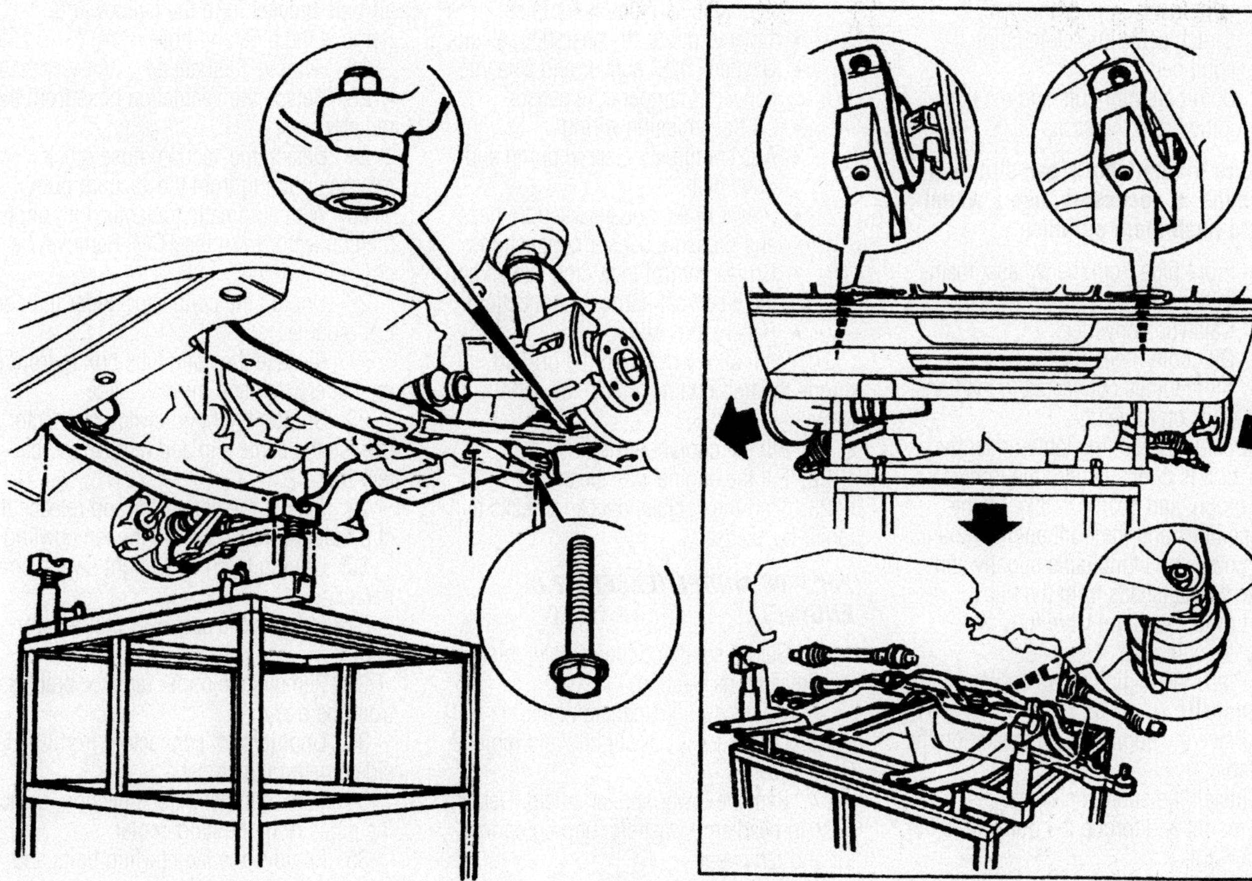

Lowering the entire powertrain assembly—9-3 models with 2.0L and 2.3L engines

c. Bottom of the radiator
d. Thermostat housing
- A/C compressor and its mounting bracket without disconnecting any hoses. Place them on the filter housing for the heater system. Secure the alternator so it will not drop or become damaged.
- Positive battery cable from the clips holding it to the body
- Ground cable from the transmission
- Ignition cable
- Electrical connections from the ignition coil
- Wiring connectors from the transmission
- O_2S connector
- Catalytic converter temperature sensor connector, if equipped

7. Inside the vehicle, pull back carpet under glove box to gain access to the central locking system and disconnect the control module. Feed the wires to the engine compartment through the grommet.

8. Remove or disconnect the following:
- Hub nuts and both wheels
- Front splash shield
- Tapered pin from the gearshift rod joint located under the vehicle
- Clutch cable and clutch pipe if equipped
- Both ball joint nuts and the ball joints from the struts

➡**Be sure the halfshafts can slide out of the hubs. If necessary, use a wheel puller to push them out now.**

- Front pipe from the exhaust manifold
- Catalytic converter
- Oil cooler lines
- All the lines connected to the transmission housing

9. Position the lifting table under the vehicle so it is directly under the front engine mount and gearbox. Take off the subframe bolts and the front engine mounts.

10. Lower the lifting table slightly and separate the halfshafts from the hubs.

11. Lower the lifting table fully and remove the subframe.

12. Place the engine on an engine stand.

To install:

13. Place the engine/transmission on the lifting table.

14. Install the subframe on the rear engine mount and torque the bolts to 35 ft. lbs. (47 Nm).

15. Install the halfshafts into the hubs as the engine is raised into the engine compartment.

16. Install the engine mount and subframe bolts and torque them in sequence as follows:

a. Front bolts: 85 ft. lbs. (115 Nm).
b. Middle subframe bolts: 141 ft. lbs. (192 Nm).
c. Rear subframe bolts: 90 ft. lbs. (122 Nm).
d. Rear engine mount: 54 ft. lbs. (73 Nm).

17. Install or connect the following:
- All oil cooler and hydraulic lines
- All charge air cooler lines between the turbocharger and the cooler
- Front exhaust pipe to the exhaust manifold
- Catalytic converter to the rear exhaust section
- Engine shield to the underside of the vehicle
- Hub nuts and wheels and torque the wheel nuts to 89 ft. lbs. (121 Nm) but do not tighten the hub nuts yet

18. Feed the wiring through the grommet and reconnect the central locking system control module.

19. Install or connect the following:
- Clutch cable or bleed the clutch hydraulic cylinder, if equipped
- Tapered pin to the gearshift selector
- Oxygen (O_2S) sensor and catalytic converter temperature sensor
- All transmission wiring
- A/C compressor, servo pump and drive belt
- Power brake booster vacuum hose and pressure sensor connections
- Cruise control including cables and the electrical harness connections
- Battery and cables

20. With all 4 wheels on the ground torque the hub nuts to 215 ft. lbs. (290 Nm).

21. Fill the coolant system.

22. Fill the engine with clean oil.

23. Start the vehicle, check for leaks and repair if necessary.

2003–06 MODEL (EXCEPT 2.8L ENGINE)

1. Before servicing the vehicle, refer to the Precautions Section.
2. Open cap on expansion tank.
3. On lift, raise car slightly and remove front wheels.
4. Remove lower spoiler shield. Detach hose to headlamp washers, unplug connector and remove it.
5. Drain cooling system.
6. Remove wheel well liners.
7. Remove cover by the gearbox.
8. Detach the hose for the headlamp washers and drain washer fluid reservoir.
9. Lower the car slightly and remove the front bumper.
10. Remove the cover on the left side and the hose between the charge air pipe and cooler.
11. Remove the lower charge air pipe bracket from the fan cowling.
12. Remove cover on right side and hose between charge air cooler and pipe.
13. Remove receiver-drier bolts from the radiator.
14. Remove the lower seal between the charge air cooler and the radiator.
15. Lower the car and remove the upper engine cover.
16. Remove the battery cover and battery cooler pipe.
17. Remove the hood release cable coupling by first undoing the clip in the body and then separating the quick-release couple using a screwdriver.
18. Remove the upper radiator support.
19. Unplug the pressure-temperature sensor connector.
20. Disconnect the bypass valve hose.
21. Detach the charge air hose from the throttle body and release the upper charge air pipe bracket from the fan cowling. Remove the pipe and hose.
22. Remove the battery.
23. Detach the ventilation hose from the radiator.
24. Detach the vacuum hose quick-release coupling from the vacuum pump.
25. With automatic transmission, unplug the connector from the TCM. Remove the control module.
26. Unplug the connector under the control module.
27. Release the main fuse box in front of the battery tray and move it aside.
28. Unplug the hood switch connector, release the cable clip and remove the battery tray.
29. Detach the connector and release the clip holding the cables on the fan cowling.
30. Unplug the reverse light switch connector.
31. Remove the seal over the radiator core.
32. Remove the upper radiator brackets from the body.
33. Unplug both connectors from the left side structural member.
34. With automatic transmission, detach the pipes from the fluid cooler.
35. Remove the fan cowling bolts and loosen cowling slightly.
36. Secure the condenser and charge air cooler to the body with cable ties.

37. Remove the fan cowling. Carefully detach A/C pipes from retaining clip on right side structural member.

38. With receptacle under radiator, remove lower radiator hose.

39. Remove upper radiator hose.

40. Remove the radiator.

41. Unplug the A/C pressure sensor connector.

42. Detach the hoses from the turbocharger inlet pipe and remove the pipe.

43. Unplug the mass air flow (MAF) sensor connector. Cap the turbocharger inlet.

44. Remove the air cleaner cover and the air cleaner.

45. Detach the intake hose and remove the air cleaner casing.

46. Remove the windshield washer filler pipe.

47. Remove 2 mounting screws from inside the main fuse box.

48. Disconnect the positive battery cable connection.

49. Undo the engine harness connector retaining screw in the main fuse box.

50. Remove the engine harness clamp from the body and the ground cables.

51. Bend up the engine harness and secure it to the engine with a suitable strap.

52. Unplug the coolant level connector.

53. Detach the expansion tank from the body and secure it to the engine.

54. Undo coolant hose quick-release couplings while trapping any coolant spill. Bend hose aside and secure it to the engine.

55. Detach cables from gearbox. Carefully bend then aside and secure to expansion tank bracket with cable ties.

56. With pliers, pinch off clutch hose and disconnect quick-release coupling from clutch slave cylinder.

57. Release any pressure in the fuel system by carefully pressing the service valve needle. Collect any fuel spillage.

58. Detach both fuel lines from fuel rail while gripping the lower nut. Plug the fuel line openings.

59. Disconnect quick-release coupling for ventilation line and detach fuel lines and vent lines from clips on camshaft cover. Bend up vent hose from intake manifold and place it on camshaft cover.

60. Be sure steering wheel and front wheels are in the straight-ahead position.

WARNING

To prevent contact roller for airbag from breaking due to twisting, steering wheel should be fixed in position, such as taping it to the instrument panel.

61. Detach the steering shaft from the steering gear.

62. Raise the car. Mark the rotation direction of the auxiliary drive belt, then remove it.

63. Unplug the A/C compressor connector and remove the compressor retaining bolts.

64. Remove the power steering pipe clamp bolts from the subframe.

65. Slightly loosen rear torsion arm bolts.

66. Fit an engine centering tool kit, 83 96 152, to subframe, as illustrated. Adjust the centering tool bolts so adjuster screw slots are completely visible.

➡Because of narrow tolerances on the driveshafts, centering tool must be used to carefully fit the powertrain to the subframe and body during installation.

67. Lower the car and re-install the radiator support member.

68. Attach a strap, 83 95 212, to radiator support, lower the strap and wind it an extra turn around the A/C compressor to relieve any strain.

69. Remove left and right side engine mounts.

70. Raise the car and disconnect the driveshaft from the hub by removing the center nuts. It may be necessary to use a driveshaft puller, 89 96 951, or a brass drift and mallet.

71. Remove outboard steering link nuts and separate links from steering swivel member.

72. Grip anti-roll bar link flats with a wrench, then disconnect the anti-roll bar links.

73. Undo the lower swivel joints from the steering swivel members and lower the suspension arms.

74. Disconnect the driveshafts and move them away.

75. Measure a distance of about 8 inches (87 mm) from front end of muffler and cut the exhaust pipe between the muffler and flexpipe.

76. Detach front pipe from the catalytic converter.

77. Disconnect the ground cable from the gearbox and the connector for the angle sensor (if equipped).

78. Position a subframe-to-engine centering tool, 83 96 145, on engine lift (70).

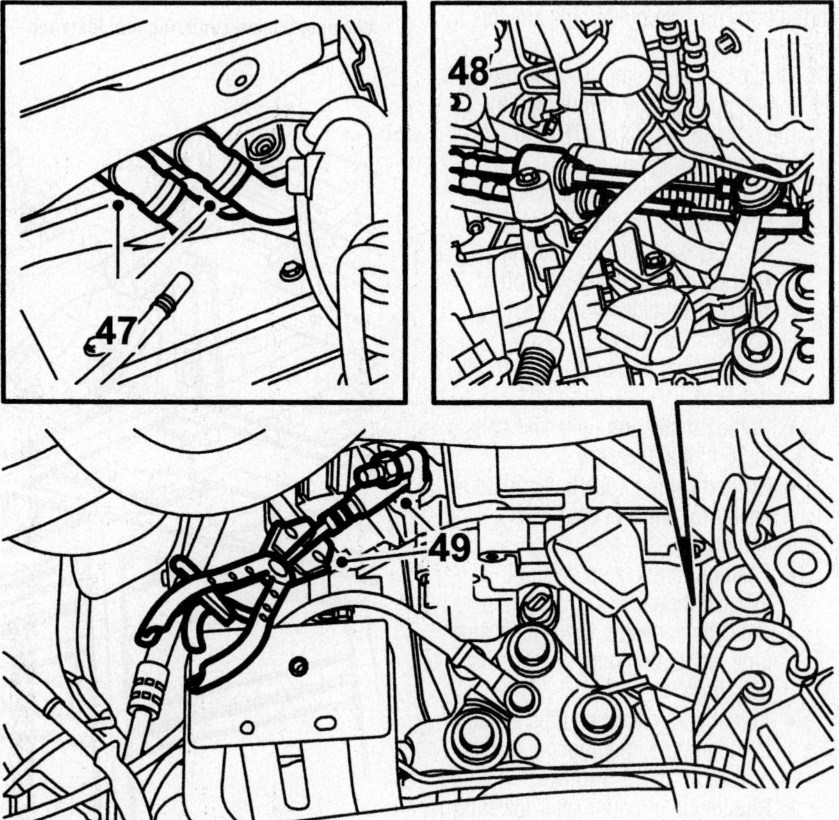

Showing location of coolant quick-release connections, gearbox cables and clutch hose—9-3 2003–06

➡Ensure adjuster screws on height adjusters, 83 95 170, are in their lowest position.

79. Position the engine lift, the raise and insert the guide pins in the subframe reference holes. Adjust the lifting pillars with the height adjusters so they rest evenly on the subframe.

80. Check with the body guide pins that the subframe is positioned correctly in relation to the body.

81. Raise engine lift slightly to ensure stable contact, then remove the subframe bolts in the body (72).

82. Carefully lower the engine and transmission assembly on the engine lift, checking that no components a caught or damaged during removal.

To install:

83. Position the engine lift under the vehicle, moving the engine and transmission in position.

84. Lift the powertrain until the inboard driveshaft U-joints are level with the wheel hubs, then insert the driveshafts into the hubs.

85. Lift the powertrain a little higher and position the lower swivel joints to steering swivel members. If necessary, adjust the engine lift screws to provide even contact with the body. Ensure guide pins are correctly positioned.

86. Install the subframe-to-body bolts. Raise the engine lift fully and tighten the subframe bolts to 55 ft. lbs. (75 Nm), plus and additional 90 degrees.

87. Install or connect the following:

- Lower swivel joint bolts to 37 ft. lbs. (50 Nm); ensure steering knuckle stub is visible on top of steering knuckle housing before bolt is installed.
- Anti-roll bar link bolts to 47 ft. lbs. (64 Nm)
- Outboard steering links and bolts to 26 ft. lbs. (35 Nm)
- Ground cable to gearbox and connector for angle sensor (if equipped).
- Position joint clamp on front exhaust pipe
- Catalytic converter to front exhaust pipe, tightening nuts to 18 ft. lbs. (25 Nm)
- Adjust joint clamp so pipe ends are in the middle, then tighten clamp nuts to 30 ft. lbs. (40 Nm)
- Right engine mount, after lowering car: right mount to 52 ft. lbs. (70 Nm), plus an additional 60 degrees

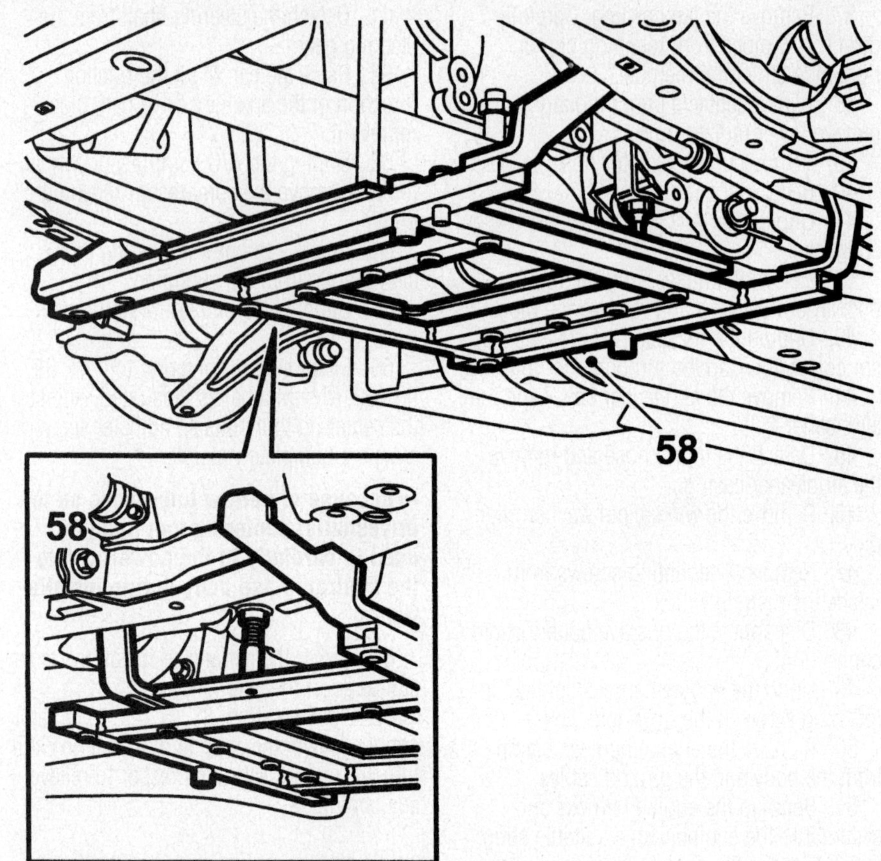

Showing engine centering tool installed—9-3 2003-06

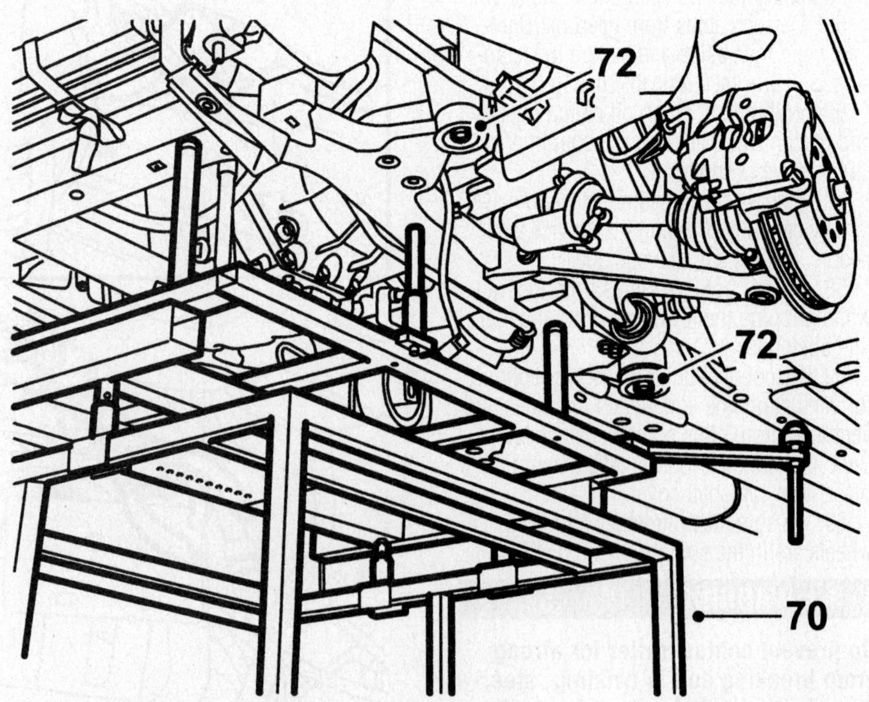

Showing engine lift equipment installed—9-3 2003-06

- Left engine mount to 52 ft. lbs. (70 Nm), plus an additional 45 degrees
- Right engine pad-to-body to 30 ft. lbs. (40 Nm), plus an additional 60 degrees
- Release strap from radiator member and A/C compressor
- Raise car and remove engine centering kit
- Rear torsion arm bolts to 52 ft. lbs. (70 Nm), plus an additional 90 degrees
- Bolts to power steering pipe clamps on subframe
- A/C compressor and connector
- Auxiliary drive belt (in proper direction of rotation)
- Steering shaft to steering gear. Use new clamp bolt and tighten to 20 ft. lbs. (27 Nm)

88. Lower the car and remove the restraint from steering wheel
89. Connect the quick-release coupling on the vent line and press fuel lines and vent lines into clips on camshaft cover.
90. Unplug fuel lines and connect to fuel rail while gripping lower nut. Use new seals.
91. Connect the quick-release coupling to the clutch slave cylinder and release the pliers.
92. Connection at clutch slave cylinder must be mounted at the correct angle, as illustrated.
93. Install or connect the following:
    - Cables to gearbox
    - Coolant hoses at quick-release couplings
    - Expansion tank to body
    - Coolant level sensor connector
    - Ground cables and clamp for engine harness to body
    - Retaining screw for engine harness connector in main fuse box
    - Positive cable to battery
    - 2 retaining screws for main fuse box and box cover
    - Windshield washer filler pipe
    - Air cleaner housing, intake hose and air cleaner cover

⁂ CAUTION

To reduce risk of hoses coming loose from delivery side of turbocharger, hoses and connecting pieces must be cleaned thoroughly with a cleaning agent, 30 15 815. Use new hose clamps.

- Mass airflow sensor connector
- Turbocharger inlet pipe and hoses to inlet pipe
- Connector for A/C pressure sensor

94. Install the radiator. Cut off the cable ties.
95. Be sure the condenser and charge air cooler hooks are positioned correctly and fit the radiator retaining bolts.
96. Install the upper and lower radiator hoses.
97. Position the fan cowling and press the A/C pipes into the retaining clips. For automatic transmission vehicles, connect the oil pipes from the transmission to the radiator.
98. Install or connect the following:
    - Both connectors on left side structural member
    - Upper radiator brackets to body
    - Seal over radiator core
    - Reverse light switch connector
    - Connector and clip securing cables on fan cowling
    - Battery tray and cable clip
    - Hood switch connector
    - Main fuse box in front of battery tray
    - Connector to control module
    - Cover and connector to transmission control module (with automatic transmission)
    - Vacuum hose coupling to vacuum pump
    - Ventilation hose to radiator
    - Battery
    - Charge air hose to throttle body and upper charge air pipe bracket on fan cowling
    - Hose to bypass valve and connector to pressure/temperature sensor
    - Upper radiator support
    - Hood cable quick-release coupling; secure it to body with new clips
    - Battery cover and battery cooler pipes
99. Raise the car and install the lower seal between the charge air cooler and the radiator.
100. Install the receiver-drier bolts to the radiator.
101. Install the hose between the charge air cooler and charge air pip and fit cover on right side.
102. Install the charge air pipe lower bracket on the fan cowling (left side).
103. Install the charge air hose between the charge air pip and charge air cooler and install cover (left side).
104. Lower the car slightly and install the front bumper.
105. Install or connect the following:
    - Hose to headlamp washers
    - Cover on gearbox
    - Inner wheel well liners and both front wheel housings
    - Lower spoiler shield and hose to headlamp washers
    - Headlamp washer connector
    - Driveshafts and tighten center hub nuts to 170 ft. lbs. (230 Nm)
    - Front wheels and tighten lug nuts to 81 ft. lbs. (110 Nm)
106. Refill the engine oil, cooling system and washer fluids.
107. Bleed the cooling system, if necessary.
108. Bleed the clutch system (manual transmission).
109. Restore all electrical functions.
110. Install upper engine cover.

2.8L ENGINE

1. Before servicing the vehicle, refer to the Precautions Section.
2. Disconnect the negative battery cable.
3. Drain the oil and coolant from the engine and the fluid from the gearbox.
4. Remove the alternator.
5. Remove the right-hand drive shaft with intermediate shaft.
6. Remove the bearing anchorage of the intermediate shaft.
7. Remove the engine mounting from the engine.
8. Remove the drive belt.
9. Remove the power steering pump. Let the servo hoses remain on the pump.
10. Unplug the control module connectors and disconnect the ground cable.
11. Unhook the power steering reservoir and servo hose clips.
12. Remove the A/C bracket and the turbocharger delivery pipe mounting.
13. Remove the power steering pump bracket.
14. Remove the secondary air injection pump complete with connectors, hoses and mountings.
15. Detach the turbocharger delivery hose from the turbocharger.
16. Detach the coolant hoses from the oil cooler. Collect any fluid spill.
17. Detach the coolant pipe with hoses from the thermostat housing.
18. Remove the starter motor.
19. If equipped with automatic transaxle, remove the bolts holding the torque converter in place.
20. Remove the wiring harness between the starter motor and alternator as well as the one from the rear of the intake manifold. Release the two clips from the back of the engine and the cable duct on the gearbox.

21. If equipped with manual transaxle, remove the cable mounting from the gearbox.

22. If equipped with automatic transaxle, remove the clamp and ground cable from the plate bracket and unplug the connector from the gear selector position sensor/control module.

23. Detach the 7-pin connector and the cable duct from the intake manifold.

24. Remove the upper and lower heat shields from the turbocharger.

25. Unplug the connectors of the heated oxygen sensors.

26. Detach the upper oxygen sensor from the catalytic converter.

27. Remove the catalytic converter with mounting.

28. Remove the upper mounting of the catalytic converter from the engine.

29. Remove the coolant pipe between the heat exchanger and thermostat housing.

30. Unplug the connectors and remove the engine control module with mounting.

31. Connect 83 92 409 Lifting yoke to the engine and unload with an engine lift.

32. Remove the bolts holding the torque arms to the subframe.

33. Raise the engine from the subframe.

34. Remove the 4 lower bolts of the transaxle.

35. Place the engine on the floor.

36. Remove the transaxle bolts and lift away the transaxle.

37. If equipped with automatic transaxle, press the torque converter against the gearbox.

38. If equipped with automatic transaxle, fit 87 92 574 Holder to hold the torque converter in place.

39. Transfer the crankcase ventilation hose.

To install:

➡Read the engine designation (L or E) and the serial number on the old engine and stamp in the equivalent on the new one.

40. If equipped with automatic transaxle, turn the torque converter so that the bolt holes align with the holes on the driver plate and turn the engine so that the oval hole in the driver is centered at the starter motor opening. Fit 87 92 574 Holder to hold the torque converter in place during fitting. Lubricate the torque converter guide pin with grease.

41. Make sure the two guide sleeves are fitted on the engine and apply grease to the guide sleeves.

42. Install the transaxle on the engine. If equipped with automatic transaxle, remove 87 92 574 Holder just before the gearbox is in place.

43. Tighten the upper bolts between the engine and transaxle to 14 ft. lbs. (19 Nm). If equipped with automatic transaxle, install the plate bracket for the ground cable.

44. Connect 83 92 409 Lifting yoke, raise the engine with an engine lift and place it on the subframe.

45. Install the bolts that secure the torque arms to the subframe. Tighten the rear torque arm bolts to 59 ft. lbs. (80 Nm), and the front torque arm bolts to 44 ft. lbs. (60 Nm) plus an additional 90°.

46. Fit the lower bolts between the engine and transaxle.

47. If equipped with automatic transaxle, perform the following:

   a. Press the torque converter against the driver plate. Install the plug.

   b. Apply a thread locking adhesive to the bolts holding the torque converter to the drive plate. Use the original bolts with corresponding washers. The torque converter will be ruined if the bolts that are used are too long.

   c. Turn the crankshaft clockwise with the pulley and tighten the bolts one at a time to 46 ft. lbs. (62 Nm), starting with the oval hole in the drive plate.

48. Fit the coolant pipe between the heat exchanger and thermostat housing. Use a new O-ring coated with acid-free vaseline.

49. Position the wiring harness.

50. If equipped with automatic transaxle, plug in the connector of the gear selector position sensor/control module. Install the clamp and ground cable to the plate bracket.

51. If equipped with manual transaxle, install the cable mounting to the transaxle.

52. Connect the two clips on the rear of the engine and fit the cable duct on the transaxle.

53. Fit the engine control module with mounting.

54. Fit the cable duct of the intake manifold and plug in the 7-pin connector.

55. Install the starter.

56. Fit the coolant pipe to the thermostat housing. Use a new O-ring coated with acid-free vaseline.

57. Attach the coolant hoses to the oil cooler.

58. Attach the turbocharger delivery hose to the turbocharger.

59. Install the secondary air injection pump complete with connectors, hoses and mountings.

60. Install the upper mounting of the catalytic converter to the engine and tighten the M10 size bolt to 30 ft. lbs. (42 Nm).

61. Install the catalytic converter with bracket, using 90 513 210 Universal paste on the nut threads. Install the upper heated oxygen sensor. Tighten the flange-to-turbocharger bolts to 18 ft. lbs. (25 Nm), mounting-to-oil sump bolts to 14 ft. lbs. (19 Nm), and the mounting-to-exhaust flange bolts to 30 ft. lbs. (40 Nm).

62. Plug in the connectors of the heated oxygen sensors.

63. Fit the upper and lower heat shields to the turbocharger.

64. Fit the power steering pump bracket.

65. Install the A/C bracket and the turbocharger delivery pipe mounting. Tighten the A/C bracket bolts to 28 ft. lbs. (38 Nm).

66. Hook on the power steering fluid reservoir; secure the hose with the clip.

67. Plug in the engine control module connectors and connect the ground cable.

68. Install the power steering pump and tighten the mounting bolts to 14 ft. lbs. (19 Nm).

69. Fit the drive belt.

70. Install the engine mounting to the engine and tighten the mounting bolts to 69 ft. lbs. (93 Nm).

71. Fit the bearing anchorage of the intermediate shaft and tighten the bolts to 18 ft. lbs. (24 Nm).

72. Install the alternator bracket. Tighten the M8 size bolt to 16 ft. lbs. (22 Nm), and the M10 size bolt to 28 ft. lbs. (38 Nm)

73. Install the alternator.

74. Install the right-hand drive shaft with intermediate shaft.

75. Fill the engine with coolant

76. Fill the engine with clean oil.

77. Fill the transmission to the proper level.

78. Connect the negative battery cable.

79. Start the vehicle, check for leaks and repair if necessary.

80. Bleed the cooling system, if necessary.

81. Bleed the clutch system (manual transmission).

82. Restore all electrical functions.

9-5

2.3L ENGINE

1. Before servicing the vehicle, refer to the Precautions Section.

2. Properly relieve the fuel system pressure.

3. Drain the engine coolant.

4. Drain the engine oil.

5. Remove or disconnect the following:

   - Battery and tray
   - Ground cable
   - Breather hose from under the fuse

box and cables to the gearbox, automatic transmission only
- Reverse light switch and remove the lower engine cover, manual transmission only
- Steering column locking bolt and move the steering column upward
- Throttle cable
- Evaporative Emissions (EVAP) purge valve vacuum hose from the intake manifold
- Brake servo vacuum hose
- Fuel connections and plug the lines
- Positive battery cable at the distribution terminal
- Negative battery cable at the gearbox
- Clutch slave cylinder hose, manual transmission only
- Selector rod after placing the vehicle in 3rd gear, manual transmission only
- Selector lever cable, automatic transmission only
- Upper radiator hose
- Bypass valve vacuum hose
- Hose between the throttle body and charge air cooler
- Pressure/temperature sensor electrical connector
- Bypass valve and intake manifold
- Radiator fan electrical connector
- Upper radiator hose to the oil cooler, if equipped
- Fan cowling
- Grille
- Hose between the turbocharger and charge air cooler
- MAF sensor
- Lower radiator from the water pump
- Heat exchanger hoses
- Vacuum hose from the bypass, if equipped with ACC
- Radiator
- Engine wire harness and bracket
- Wiper arms and covers
- Control module electrical connectors
- Hub nuts and raise the vehicle
- Both front wheels. Drive a wedge between the gearbox and subframe, and between the oil pan and subframe
- A/C compressor retaining bolts
- Air filter housing retaining nuts and air hose. Lower the vehicle
- Condenser cooler, charge air cooler and A/C compressor and suspend them to the radiator member
- A/C compressor connector
- Quick release coupling to the wastegate
- Right side engine mount and yoke
- Belt tensioner and belt
- Left side engine mount
- Power steering reservoir and hoses
- Separate the quick release couplings to the gearbox oil cooler, if equipped
- Steering swivel joint
- Hardware securing the anti-roll bar supports to the strut
- Outer steering links
- Halfshafts from the hub
- Oil cooler
- Separate the exhaust system behind the catalytic converter
- Temperature sensor and place a lifting trolley under the subframe
- Subframe supporting plate bolts

6. Lower the trolley slightly and unhook the oil cooler hose from the bracket
7. Move the trolley to the rear to allow access for the A/C compressor and remove the powertrain assembly.

To install:

8. Position the trolley under the vehicle with the engine in position.
9. Install or connect the following:
   - Driveshafts to the steering swivel joint and torque the bolts to 22 ft. lbs. (30 Nm) plus 90 degrees
   - Subframe and supporting plate. Torque the subframe bolts to 84 ft. lbs. (115 Nm) and the supporting plate to 48 ft. lbs. (65 Nm). Move the trolley away from the vehicle.
   - Exhaust pipe and temperature sensor and torque the bolts to 18 ft. lbs. (25 Nm)
   - Air cleaner
   - Outer steering swivel members to the anti-roll bar support
   - Hub nuts but do not tighten them
   - A/C compressor and air filter housing and torque the bolt to 35 ft. lbs. (47 Nm)
   - Steering gear into position. Torque the locking nut to 19 ft. lbs. (25 Nm).
   - Left side engine mount
   - Right side engine mount and yoke and remove the wedges. Torque the mounting bolts to 36 ft. lbs. (50 Nm)
   - Selector lever cable, if equipped
   - Selector rod on manual tranaxles and remove the locking pins
   - Bleeder hose under the fuse box
   - Hoses to the heat exchanger
   - Coolant hose to the expansion tank
   - Engine wire harness to the bulkhead partition
   - Engine control module
   - Wiper arms and cover plate
   - Ground cable to the gearbox
   - Positive cable to the positive terminal
   - Connector with electrical leads to the gear box, automatic transmission only
   - Reverse light connector, manual transmission only
   - Quick release couplings to the automatic transmission fluid cooler
   - Fuel connections and rubber protectors
   - Vacuum hose for the brake servo and intake manifold
   - EVAP canister purge valve vacuum hose
   - Throttle cable
   - Hose between the turbocharger and charge air cooler
   - Engine oil cooler and cover
   - Radiator fans
   - Front grille
   - Upper and lower radiator hoses
   - Power steering reservoir
   - Power steering pump pipes
   - Quick release coupling for the turbocharger wastegate vacuum hose
   - A/C compressor electrical connectors
   - MAF sensor connector
   - Bypass valve and intake manifold
   - Pressure pipe and charge air cooler hose
   - Pressure/temperature sensor connector
   - Front wheels and torque the hub nuts to 215 ft. lbs. (290 Nm)
   - Battery tray and battery
   - Intake manifold cover and torque the bolts to 16 ft. lbs. (22 Nm)
   - Shut off valve vacuum hose
10. Fill the engine with coolant
11. Fill the engine with clean oil.
12. Fill the transmission to the proper level.
13. Start the vehicle, check for leaks and repair if necessary.

3.0L ENGINE

1. Before servicing the vehicle, refer to the Precautions Section.
2. Properly relieve the fuel system pressure by removing fuel pump fuse while engine is running. Switch ignition off when engine has stopped, then re-install the fuse.
3. Drain the engine coolant.
4. Drain the engine oil.
5. Remove or disconnect the following:
   - Unbolt the steering column inside the vehicle
   - Engine and battery cover

- Battery and tray
- Automatic transmission electrical connectors
- Ground cable from the gearbox
- Gearbox vent hose
- Selector lever cable
- Positive cable from the junction box
- Throttle cable from the throttle body
- Turbocharger pressure hose from the throttle body
- Turbocharger pressure sensor electrical connectors
- Intake Air Temperature (IAT) sensor electrical connector
- Ground cable from the pressure hose pipe
- Turbocharger pressure hose from the charge air cooler
- Lower spoiler sections
- Air cleaner rubber mounts and install wedges between the oil pan to subframe and the gearbox to the subframe
- Rear Heated Oxygen Sensor (HO_2S), if equipped
- Front exhaust system after the catalytic converter
- Anti-roll bar link rods at the arms
- Outer tie rod ends from the strut
- Lower steering swivel joint bolts from the link arms
- Vacuum hose from the charge air bypass valve
- Turbocharger pressure hose with the pipe and valve
- Crankcase breather pipe from the turbocharger intake pipe
- Mass Air Flow (MAF) sensor and intake hose
- Turbocharger intake pipe
- Radiator fan
- Wiper arms and cover
- Engine control modules rubber seal and connector
- Engine wire harness cover
- Engine wire harness and place it on top of the engine
- Vacuum hose from the shut off valve
- Brake servo vacuum hose
- Vent hose from the throttle body
- Fuel rail connections
- Power steering fluid reservoir
- Upper, lower and expansion tank radiator hoses
- Automatic transmission hydraulic hoses
- Radiator
- Heat exchanger hose connections
- Hose between the turbocharger outlet and charge air cooler
- Front grille
- Charge air cooler and condenser
- A/C compressor electrical connectors
- A/C compressor pipes from the subframe
- V-belt tension
- Right side engine mount and bracket
- Left side engine mount and place a lifting trolley under the vehicle and raise it into position
- Subframe and triangular stiffener bolts and lower the powertrain slightly
- Halfshafts from the hub
- A/C compressor retaining bolts
- Engine assembly

To install:

6. Position the lifting trolley under the vehicle.
7. Install or connect the following:
   - A/C compressor
   - Halfshafts into the hubs
   - Alternator cooling air duct
   - Subframe and triangular stiffener and torque the subframe bolts to 74 ft. lbs. (100 Nm), plus an additional 45 degrees
   - Torque the triangular plate bolts to 46 ft. lbs. (63 Nm).
   - Left side engine mount
   - Right side engine mount and bracket
   - V-belt
   - A/C compressor pipes to the subframe
   - Front grille
   - Hose between the turbocharger outlet and charge air cooler
   - A/C compressor electrical connectors
   - Radiator
   - Automatic transmission hydraulic hoses
   - Upper and lower radiator hoses and the expansion tank hose (check for proper routing of each hose)
   - Cabin heat exchanger hoses
   - Power steering reservoir
   - Fuel lines
   - Vent hose to the throttle body
   - Vacuum hose for the brake servo to the intake manifold
   - Shut off valve vacuum hose
   - Engine wire harness connector and cover
   - Engine control module and rubber seal
   - Wiper arms and cover plate
   - Radiator fan
   - Turbocharger intake pipe and hose
   - MAF sensor and vent hose
   - Crankcase breather pipe to the turbocharger intake pipe and torque the fasteners to 18 ft. lbs. (24 Nm)
   - Exhaust manifold heat shield and torque the bolts to 18 ft. lbs. (24 Nm)
   - Turbocharger pressure hose with the charge air bypass pipe and valve
   - Lower steering swivel joint and torque the lower bolts to 44 ft. lbs. (60 Nm), plus an additional 30 degrees, and bolt on steering swivel member to 37 ft. lbs. (50 Nm)
   - Link rods to the anti-roll bar
   - Outer tie rod ends to the strut
   - Front exhaust system and torque the bolts to 16 ft. lbs. (22 Nm)
   - Air cleaner rubber mounts to the subframe
   - HO_2S electrical connector
   - Lower spoiler sections
   - Turbocharger pressure hose to the charge air cooler
   - Turbocharger pressure sensor and IAT sensor connectors
   - Ground cable for the pressure hose connecting pipe
   - Turbocharger pressure hose to the throttle body
   - Throttle cable
   - Positive cable to the junction terminal
   - Selector lever cable to the gear actuator, tighten to 15 ft. lbs. (20 Nm)
   - Ground cable to the gearbox
   - Gearbox vent hose
   - Ground cable to the left side wheel housing
   - Front wheels and hub nuts and torque the nuts to 81 ft. lbs. (110 Nm); torque hub center nut to 170 ft. lbs. (230 Nm)
   - Battery tray and battery
   - Battery cables and engine cover
   - Automatic transmission wiring harness
   - Steering column locking bolt and torque the bolt to 18 ft. lbs. (25 Nm)
8. Fill the engine with coolant
9. Fill the engine with clean oil.
10. Fill the transmission to the proper level.
11. Start the vehicle, check for leaks and repair if necessary.

Water Pump

REMOVAL & INSTALLATION

9-3

2002 MODEL

1. Before servicing the vehicle, refer to the Precautions Section.
2. Loosen the expansion tank pressure cap.
3. Raise the car and remove the lower front cover.
4. Drain the coolant.
5. Detach the coolant pipe from the turbocharger.
6. Lower the car.
7. Disconnect the negative battery cable.
8. For the turbocharger, remove or disconnect the following:
   - Air hose between air collector casing and turbocharger intake pipe
   - Tie securing main engine harnesses
   - Tension from auxiliary drive belt tensioner
   - Auxiliary drive belt from power steering pump and water pump pulleys
   - Engine top cover
   - Crankcase vent pipe from turbocharger inlet pipe and camshaft cover (bend aside)

➡Keep pipe seals and washers in a safe place.

   - Solenoid valve connector and hoses
   - Engine lifting eye
   - Bypass pipe with valve from turbocharger intake pipe
   - Exhaust manifold heat shield
   - Vent hose from T-connection
   - Turbocharger intake pipe; plug openings in turbocharger
9. Remove the power steering pump bolt on right side behind the belt pulley.
10. Remove the upper power steering pump bolt.
11. Lift up the pump and remove the lower bolt and 2 upper bolts on the bracket.
12. Move the power steering pump and bracket aside and secure with a cable tie.
13. Remove hose clips and detach hoses from water pump inlet.
14. Detach the 2 longitudinal coolant pipes from cylinder block.
15. Disconnect coolant pipes from water pump. Keep the O-rings and move pipes aside.
16. Loosen water pump slightly.
17. Detach the coolant pipe from the water pump. Retain the seals.
18. Remove the 3 water pump bolts.
19. Carefully wiggle the water pump away from its mounting and connecting piece in cylinder block.
20. Remove the connecting piece from the cylinder block.

To install:

21. Clean the sealing surfaces and use a new gasket.
22. Install or connect the following:
   - Connecting piece to cylinder block
   - Coat seals with sealant Gleitmo 805 30-06 442.
   - Water pump loosely to mounting
   - Coolant pipe between water pump and turbocharger with new washers and seals
   - Coolant pipe to water pump to 15 ft. lbs. (20 Nm)
   - 2 longitudinal pipes to water pump with new O-rings coated with non-acidic Vaseline
   - 3 water pump bolts to 16 ft. lbs. (22 Nm)
   - 2 longitudinal pipes to 7 ft. lbs. (10 Nm)
   - Power steering pump and bracket, tightening bolts to 18 ft. lbs. (24 Nm)
   - Hoses and clips on water pump inlet
   - Turbocharger inlet pipe, but do not tighten (coat O-ring with non-acidic Vaseline)
   - Engine lifting eye
   - Solenoid valve and hoses
   - Turbocharger intake pipe V-clamp
   - Vent hose to T-connection
   - Exhaust manifold heat shield
   - Bypass pipe and valve, with new O-ring, to turbocharger inlet pipe
   - Solenoid valve connector
   - Crankcase ventilation pipe to turbocharger inlet pipe, with new copper seals, to 18 ft. lbs. (24 Nm)
   - Crankcase vent pipe to camshaft cover clip
   - Throttle body cover
   - Coolant line to turbocharger to 18 ft. lbs. (24 Nm) from under the car
   - Auxiliary drive belt in all pulleys
   - Release tension from drive belt tensioner
   - Main wiring harness secured with cable ties
   - Air hose between air collector casing and turbocharger inlet pipe
23. Refill the cooling system.
24. Bleed the cooling system as follows:
   a. Ensure A/C is off.
   b. Fill system to MAX level. Fit pressure cap.
   c. Start engine and warm to normal operating temperature, varying engine speeds until cooling fan starts.
   d. Switch off engine and top coolant up to MAX, if necessary.
25. Check for any leaks.

2003–06 MODEL (EXCEPT 2.8L ENGINE)

1. Before servicing the vehicle, refer to the Precautions Section.
2. Loosen the expansion tank pressure cap.
3. Remove or disconnect the following:
   - Connector for oxygen sensors and temperature sensor
   - Connector from MAF sensor
   - Cover on air cleaner housing
   - Air filter
   - Air cleaner housing

CAUTION

Use caution when removing air cleaner housing to ensure A/C pressure sensor is not damaged.

   - Turbocharger inlet hose (plug turbocharger opening)
   - Upper bolt for water pump in timing cover and engine block
   - Turbocharger heat shield
   - Upper oxygen sensor
   - Catalytic converter-to-turbocharger nuts
   - Turbocharger oil delivery pipe bolt
4. Raise the car and remove the right front wheel.
5. Remove the front spoiler shield.
6. Remove the right fender well liner.
7. Drain the cooling system.
8. Remove the lower water pump bolt in the timing cover and engine block.
9. Remove plug from water pump and drain coolant.
10. Mark direction of rotation of the auxiliary drive belt.
11. Relieve belt tension and remove the belt.
12. Remove or disconnect the following:
   - Turbocharger lower pressure pipe
   - Front exhaust pipe
   - Steering gear heat shield
   - Catalytic converter heat shield
   - Temperature sensor
   - Lower oxygen sensor
   - Catalytic converter and bracket from engine block
   - Turbocharger coolant pipe from

thermostat housing and turbocharger
- Thermostat housing bolts

13. Remove the water pump cover from the timing cover (34).
14. Install water pump holder tool, B207, to secure the balancer shaft pinion (35).
15. Remove the pinion bolts (36). Use a magnetic socket to remove the bolts from the holder tool.
16. Remove the bolts on the back of the water pump.
17. Move aside the thermostat housing, with the pipe, and remove the water pump.
18. Remove the water pipe from the thermostat housing.

To install:

19. Clean water pipe fittings. Install new O-rings, lightly coated with Vaseline.
20. Clean the thermostat housing connections, fit a new seal and install the water pipe.
21. Install the stud to the water pump driver using a holding tool, 83 96 103.
22. Install the water pump, with a new seal coated lightly with Vaseline. Guide the stud through one of the holes in the balancer shaft pinion.
23. Install the 2 rear water pump bolts and the lower bolt through the timing cover.
24. Lower the car.
25. Torque the water pump bolts to 15 ft. lbs. (20 Nm) for short bolts and 18 ft. lbs. (25 Nm) for the long bolts.
26. Install the coolant pipe to the water pump and install the thermostat.
27. Fit the coolant pipe between the turbocharger and thermostat with a new seal. Tighten pipe to 15 ft. lbs. (20 Nm).
28. Lift catalytic converter and position its bracket to studs on the engine block. Position the catalytic converter to the turbocharger connection.
29. Insert the catalytic converter bracket bolts into the engine block and new nuts on the studs.
30. Lower the car. Lubricate the turbocharger studs with screw thread paste 30 20 971.
31. Install the new nuts so they are firmly against the turbocharger, then torque to 7 ft. lbs. (10 Nm).
32. Raise the car.
33. Tighten the catalytic converter nut and bolts one last time to 19 ft. lbs. (25 Nm).
34. Lubricate the lower oxygen sensor threads with screw thread paste and install the sensor and tighten to 30 ft. lbs. (40 Nm).

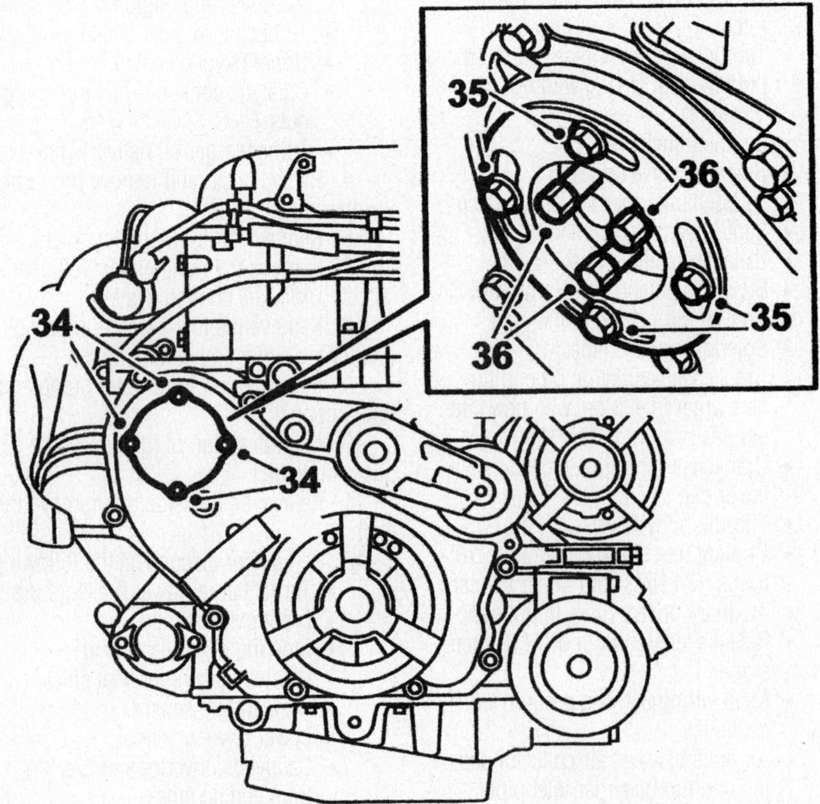

Showing water pump holding tool for securing balancer shaft pinion—2.0L and 2.3L engines

⁂ CAUTION

Be sure oxygen sensor cables are not twisted or damaged.

35. Lubricate the temperature sensor threads with screw thread paste and install the sensor and tighten to 33 ft. lbs. (45 Nm).
36. Lift the oxygen sensor and temperature sensor cables to check for proper routing.
37. Install the heat shields on the catalytic converter and steering gear.
38. Clean the catalytic converter joint at pipe connection with front muffler. Install clamp and position front pipe, with new gasket, mounting nuts and rubber mountings. Apply thread paste to studs.
39. Position joint clamp so pipe ends are in middle of the clamp. Torque joint at catalytic converter to 19 ft. lbs. (25 Nm) and clamp nuts to 30 ft. lbs. (40 Nm).
40. Remove the stud and install bolts to the balancer shaft pinion, using a 10mm magnetic socket.

➡It may be necessary to remove the fixing tool for access before finally tightening the bolts.

41. Install the timing cover with a new gasket.
42. Install the auxiliary drive belt over the pulleys, making sure it is in proper direction of rotation.
43. Install the lower turbocharger delivery pipe.
44. Close the radiator drain and lower the car.
45. Apply thread paste and install upper oxygen sensor. Tighten it to 30 ft. lbs. (40 Nm).
46. Plug in and secure the oxygen sensor and temperature sensor connectors.
47. Install the turbocharger heat shield. Tighten bolts to 15 ft. lbs. (20 Nm).
48. Install the air cleaner assembly and hoses, including to the turbocharger.
49. Plug in the MAF sensor.
50. Fill cooling system and bleed system by starting and running the engine at varying speeds until the cooling fan comes on. Then, check and top off the coolant.
51. Install remaining components and check for any signs of leaks.

2.8L ENGINE

1. Before servicing the vehicle, refer to the Precautions Section.
2. Disconnect the negative battery cable.
3. Take the cap off the expansion tank to release any pressure.

4. Raise and safely support the vehicle.

5. Undo the right-hand side of the spoiler shield. Place a suitable receptacle under the radiator. Connect a hose to the radiator and drain the coolant. Close the cock and refit the spoiler shield.

6. Remove the chassis reinforcement from the front subframe (convertible model).

7. Fit 83 96 145 Centering tool, subframe—engine and 83 96 152 Centering tool, power train with 32 025 059 Support. First remove the two long supports. Relieve the load on the engine.

8. Lower the vehicle.

9. Disconnect the turbocharger intake hose.

10. Remove the air cleaner casing cover.

11. Remove the air cleaner. Detach the intake hose and remove the air cleaner casing.

12. Remove the engine pad from the body and bracket.

13. Remove the engine bracket. The lower hole is slotted.

14. Loosen the bolts of the water pump pulley slightly.

15. Unload the drive belt tensioner using a ½ inch puller and remove the belt from the water pump pulley.

16. Remove the pulley.

17. Remove the water pump.

To install:

18. Clean the sealing surfaces.

19. Install the water pump and tighten the bolts to 7 ft. lbs. (10 Nm).

20. Install the pulley.

21. Install the drive belt.

22. Tighten the bolts on the pulley to 9 ft. lbs. (12 Nm).

23. Install the engine bracket. Tighten the M10 size bolts to 31 ft. lbs. (42 Nm), and the M12 size bolts to 69 ft. lbs. (93 Nm).

24. Install the engine pad to the body and bracket using 32 025 063 Special box spanner. Tighten the bolts to the body to 37 ft. lbs. (50 Nm), plus an additional 180°. Tighten the bolts to the engine mounting to 33 ft. lbs. (45 Nm), plus an additional 90°.

25. Install the air cleaner casing and connect the intake hose. Install the air cleaner.

26. Install the air cleaner casing cover.

27. Install the turbocharger inlet hose.

28. Raise and safely support the vehicle.

29. Remove 83 96 145 Centering tool, subframe—engine and 83 96 152 Centering tool, power train with 32 025 059 Support.

30. Install the chassis reinforcement to the front subframe (convertible model).

31. Install the spoiler shield.

32. Lower the vehicle.

33. Connect the negative battery cable.

34. Fill cooling system and bleed system by starting and running the engine at varying speeds until the cooling fan comes on.

35. Stop the engine and top off the coolant.

36. Install remaining components and check for any signs of leaks.

9-5

2.3L ENGINE

1. Before servicing the vehicle, refer to the Precautions Section.

2. Open cap of expansion tank and release any pressure.

3. Raise the car and remove the lower front cover.

4. Drain the cooling system.

5. Lower the car and remove or disconnect the following:
   - Negative battery cable
   - Mass Air Flow (MAF) sensor electrical connector
   - MAF sensor and air hose
   - V-belt tension with tool 83 95 254
   - V-belt from the power steering pump and water pump
   - Crankcase breather pipe
   - Camshaft cover
   - Boost pressure control valve connector
   - Engine lifting eye
   - Turbocharger wastegate valve hoses, loosen only
   - Bypass valve and pipe
   - Exhaust manifold heat shield
   - Quick release coupling on the vent hose with tool 83 95 261
   - Turbocharger intake pipe (plug turbocharger opening)
   - Power steering pump and move it aside
   - Water pump inlet hoses
   - 2 longitudinal coolant pipes from the engine block and water pump
   - Water pump and connecting piece

To install:

6. Clean all mating surfaces and connecting piece seals. Replace the seals if necessary

7. Install or connect the following:
   - Connecting piece to the engine block
   - Water pump and torque the bolts to 16 ft. lbs. (22 Nm)
   - Coolant pipe to the water pump and turbocharger. Torque the water pump bolt to 15 ft. lbs. (20 Nm) and the turbocharger bolt to 18 ft. lbs. (24 Nm)

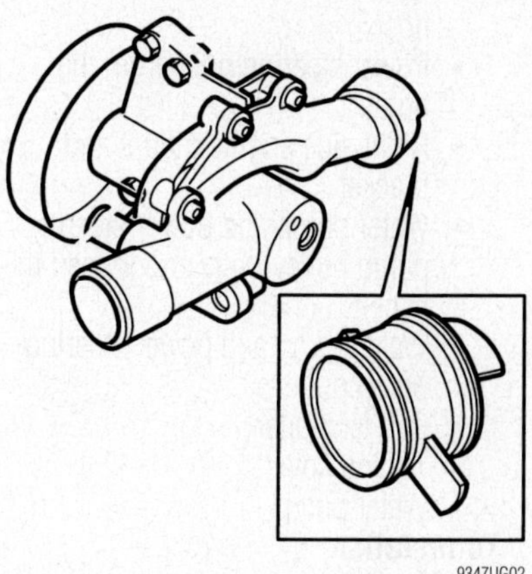

Remove the water pump and engine block connecting piece

   - Longitudinal pipes to the water pump and torque the bolts to 7 ft. lbs. (10 Nm)
   - Water pump inlet hoses
   - Power steering pump and torque the bolts to 18 ft. lbs. (24 Nm)
   - Turbocharger intake pipe
   - Turbocharger vent hose
   - Exhaust manifold heat shield and torque the bolts to 15 ft. lbs. (20 Nm)
   - Bypass pipe and valve and torque the bolts to 6 ft. lbs. (8 Nm)
   - Turbocharger wastegate valve hoses
   - Engine lifting eye
   - Boost pressure control valve connector
   - Crankcase breather pipe fastened to turbocharger inlet pipe and camshaft cover, torque to 18 ft. lbs. (24 Nm)
   - V-belt
   - MAF sensor and air hose
   - Negative battery cable

8. Fill the cooling system.

9. Start the vehicle, check for leaks and repair if necessary.

3.0L ENGINE

1. Before servicing the vehicle, refer to the Precautions Section.

2. Open cap on expansion tank and drain off any pressure.

3. Raise the car and remove the lower front cover.

4. Drain the cooling system.

5. Lower the car and remove or disconnect the following:
   - Negative battery cable
   - Lower front cover and raise the engine until the right front mount is unloaded

- Mass Air Flow (MAF) sensor and hose
- Power steering hose from the holder
- Right side engine mount and bracket
- Water pump and power steering pump pulley bolts and loosen the V-belt
- Water pump and power steering pump pulleys
- Belt tensioner
- Timing cover
- Water pump

To install:

6. Clean all mating surfaces.
7. Install or connect the following:
   - Water pump with a new O-ring and torque the bolts to 18 ft. lbs. (24 Nm)
   - Timing cover
   - Belt tensioner and torque the bolts to 30 ft. lbs. (40 Nm)
   - Water pump and power steering pump pulleys. Do not tighten the bolts
   - V-belt and torque the water pump pulley bolts to 6 ft. lbs. (8 Nm) and the power steering pump pulley bolts to 15 ft. lbs. (20 Nm)
   - Right side engine mount and bracket
   - Power steering hose to the holder
   - MAF sensor
   - Lower front cover
   - Negative battery cable
8. Fill the cooling system.
9. Start the vehicle, run at varying speeds until cooling fan comes on. Top up the coolant. Let the engine run until the cooling fan comes on 3 more times.
10. Stop the engine and top up the coolant.
11. Check for leaks and repair if necessary.

Heater Core

REMOVAL & INSTALLATION

9-3

2002 MODEL

1. Before servicing the vehicle, refer to the Precautions Section.
2. Disconnect the negative battery cable.
3. Drain the cooling system into a clean container for reuse.
4. Pinch heater hoses by firewall connections with hose clamp pliers.
5. Disconnect the heater hoses at the firewall in the engine compartment. Plug the ends of the fittings on the valve.
6. Blow any remaining coolant from the heater core with compressed air.
7. Remove the glove compartment retaining screws, bolt, quick-release pin and bracket catch. Pull the glove compartment out partway to disconnect the lamp; the remove the glove compartment.
8. If equipped with automatic climate control (ACC), remove ACC unit.
9. Remove or disconnect the following:
   - Glove box
   - Panels from both side of center console
   - Cover around ignition
   - Rear ashtray
   - Rear air vents and cover
   - Bolts holding rear part of center console
   - Window switches
   - Center console
10. Remove the floor air ducts in front of the heater box.
11. Open the heater box housing.
12. Remove the clips securing the hoses to the heater box.
13. Remove toggle clips on side of heater box, pull down on pipes and lift out the heater core.

To install:

14. Install the heater core and connect the pipes with clips.
15. Close the heater box.
16. Install the toggle clips at the sides of the heater core case.
17. Install or connect the following:
    - Install the center console
    - Window switches
    - Rear part of center console
    - Rear air vent and cover
    - Rear ashtray
    - Cover around ignition
    - Glove box
18. Install the ACC unit, if removed.
19. Connect the heater hoses at the firewall in the engine compartment. Remove pinching pliers.
20. Refill the cooling system.
21. Connect the negative battery cable.
22. Operate the engine to normal operating temperatures; then, check the climate control operation and check for leaks.

2003–06 MODELS

1. Before servicing the vehicle, refer to the Precautions Section.
2. Disconnect the negative battery cable.
3. Drain the cooling system into a clean container for reuse.
4. Disconnect the hoses from the heater core fittings at the firewall by pulling up the clasp and pulling the hoses forward.
5. Blow any remaining coolant from the heater core with compressed air.
6. Remove or disconnect the following:
   - Glove box
   - Front side console panel on passenger side.
   - Passenger sound shield.
   - ACC or MCC module.
   - Ashtray
   - Instrument panel right side end cover
   - Passenger side A-pillar trim
   - Nuts holding dashboard on passenger side (through end opening)
   - 4 nuts holding instrument panel (through center console opening)
7. Remove heater core cover.
8. Remove the pipe clamps from the heater core.
9. Pull the pipes out from the heater core.
10. Carefully pull of the instrument panel on the passenger side and pull out the heater core.

To install:

11. Place O-rings and insert the heater core into position.
12. Reposition instrument panel.
13. Install pipes into heater core and install clamps.
14. Install heater core cover.
15. Install or connect the following:
    - Nuts in center of instrument panel
    - Nuts on end of instrument panel
    - Instrument panel end cover
    - A-pillar trim
    - Ashtray
    - ACC or MCC module
    - Passenger sound shield
    - Floor console side panel
    - Glove box
    - Heater hoses at firewall
16. Refill the cooling system.
17. Start and run the engine at varying speeds until the cooling fan comes on. Top up the coolant, then continue to run the engine until the cooling fan comes on 3 more times.
18. Stop the engine and check coolant level.
19. Check for leaks.

9-5

1. Before servicing the vehicle, refer to the Precautions Section.
2. Disconnect the negative battery cable.

View of the heater core removal—9-5

3. Drain the cooling system into a clean container for reuse.
4. Disconnect the heater hoses at the firewall in the engine compartment. Plug the ends of the fittings on the valve.
5. Blow any remaining coolant from the heater core with compressed air.
6. Remove the glove compartment retaining screws, bolt, quick-release pin and bracket catch. Pull the glove compartment out partway to disconnect the lamp; the remove the glove compartment.
7. Remove the trim from the side of the center console.
8. Remove the air vent from the air duct. Do not remove the seal.
9. Detach the pad connector for the piping to the heater core.
10. Remove the 4 heater core housing bolts.
11. Remove the heater core.

To install:

12. Install the heater core and 4 mounting bolts.
13. Replace and lubricate the O-rings with synthetic Vaseline. Place O-rings on heater core pipes.
14. Connect the pipes to the heater core and screw on the pad connector.
15. Fit the air vent to the air duct.
16. Replace the side trim on the center console.
17. Install the glove compartment.
18. Connect the heater core hoses to the engine bulkhead connections.
19. Refill the cooling system.
20. Run engine at varying speeds until the cooling fan comes on.
21. Open the expansion tank cap and top up the coolant.
22. Let the engine run until the cooling fan has started 3 more times.
23. Stop the engine and top up the coolant.
24. Check for leaks.

Rocker Arm (Valve) Cover

REMOVAL & INSTALLATION

2.0L Engine

1. Before servicing the vehicle, refer to the Precautions Section.
2. Remove or disconnect the following:
   - Negative battery cable
   - Engine cover
   - Crankcase breather hose from the rocker arm cover
   - Turbocharger delivery pipe
   - Front wiring harness channel from the cylinder head
   - Fuel lines from the rocker arm cover and engine mount
   - Rocker arm cover and discard the gasket
3. Clean all mating surfaces of any residual gasket material.

To install:

4. Install or connect the following:
   - Rocker arm cover with a new gasket and torque the bolts to 72 inch lbs. (8 Nm)
   - Fuel lines to the fasteners
   - Wiring harness channel to the cylinder head
   - Turbocharger delivery pipe
   - Crankcase breather hose
   - Engine cover
   - Negative battery cable

2.3L Engines

1. Before servicing the vehicle, refer to the Precautions Section.
2. Remove or disconnect the following:
   - Negative battery cable
   - Ignition discharge module/ignition cables from the rocker arm cover
   - Turbocharger pipes, if necessary for access
   - Spark plugs
   - Rocker arm cover and discard the gasket
3. Clean all mating surfaces of any residual gasket material.

To install:

4. Install or connect the following:
   - Rocker arm cover with a new gasket and torque the bolts, in sequence, to 11 ft. lbs. (15 Nm)
   - Spark plugs and torque to 21 ft. lbs. (28 Nm)
   - Turbocharger pipes, if removed
   - Ignition discharge module/ignition cables and torque to 96 inch lbs. (11 Nm)
   - Negative battery cable

2.8L Engines

REAR COVER

1. Before servicing the vehicle, refer to the Precautions Section.
2. Disconnect the negative battery cable.
3. Remove the heat shield between the engine and turbo.
4. Remove the bleeder pipe.
5. Release the quick-coupling of the secondary air pipe and put the pipe aside.
6. Remove the power steering fluid reservoir's bolt.
7. Unplug the connectors from ignition coils 2, 4 and 6.
8. Remove the left grounding point.
9. Raise the heat shield and unplug the coolant temperature sensor connector.
10. Release the right ground cable from the wiring harness.
11. Remove the wiring harness from the camshaft cover.
12. Unplug the connectors from the camshaft position sensor and the camshaft solenoid valve.
13. Raise the wiring harness slightly.
14. Remove the ground cable of the engine control module.

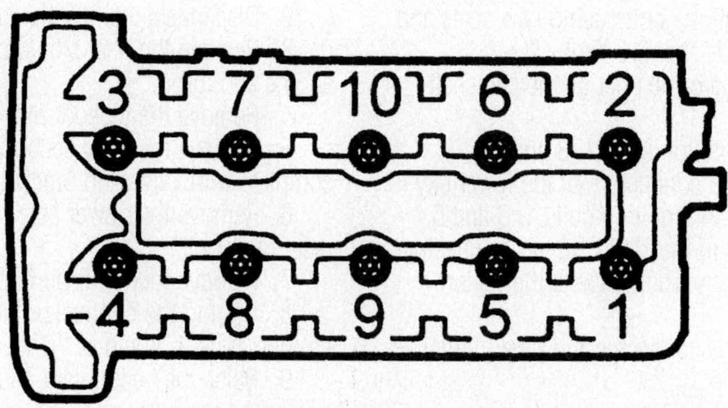

Torque the rocker arm cover bolts as illustrated

15. Unplug the engine control module connectors (608A and 608B).
16. Detach the cable duct from the intake manifold.
17. Unplug the atmospheric pressure sensor connectors.
18. Detach the corner bracket and move it aside.
19. Remove the bolt holding the injector bracket in place and unplug the connector.
20. Detach the brake vacuum hose from the intake manifold.
21. Detach the crankcase ventilation from the intake manifold.
22. Unplug the fuel pressure sensor connector.
23. Detach the fuel line, wiping up any fuel spill with a cloth.
24. Remove the intake manifold.
25. Unplug the connectors from the camshaft position sensor and the camshaft solenoid valve.
26. Raise the wiring harness slightly.
27. Remove the connector and pipe from the fuel bleeder valve. Lift away the valve.
28. Detach the vacuum pipe from the venturi valve.
29. Unplug the 7-pin contact connector.
30. Unplug the lower oxygen sensor connector and the connectors of ignition coils 1, 3 and 5.
31. Remove ignition coils 1, 3 and 5.
32. Remove the rear cable duct from the camshaft cover. Use an angled screwdriver to release the 3 clips from the cable duct. Press the cable duct back toward the bulkhead.
33. Remove the camshaft cover bolts.
34. Raise the wiring harness and servo reservoir. Remove the camshaft cover.
35. Remove the camshaft cover gasket and seals around the spark plug holes.

To install:

36. Fit new seals on the camshaft cover.
37. Add a spot of sealant to each mating surface between the cylinder head and timing cover.
38. Position the camshaft cover.
39. Install the bolts using new seals and tighten to 7 ft. lbs. (10 Nm).
40. Fit the cable duct on the camshaft cover.
41. Fit ignition coils 1, 3 and 5.
42. Fit the connectors of the lower oxygen sensor and ignition coils 1, 3 and 5.
43. Plug in the 7-pin contact connector.
44. Fit the vacuum pipe to the venturi valve.
45. Fit the connector and pipe to the fuel bleeder valve.
46. Fit the wiring harness to the camshaft cover.
47. Plug in the connectors of the camshaft position sensor and the camshaft solenoid valve.
48. Install the intake manifold.
49. Fit the fuel line.
50. Fit the fuel pressure sensor connector.
51. Fit the crankcase ventilation to the intake manifold.
52. Fit the brake vacuum hose to the intake manifold.
53. Plug in the injector connectors and fit the bracket.
54. Reposition the corner bracket and fit the bolts.
55. Plug in the atmospheric pressure sensor connector.
56. Fit the cable duct to the intake manifold.
57. Plug in the engine control module connectors (608A and 608B).
58. Fit the ground cable of the engine control module.
59. Plug in the connectors of the camshaft position sensor and the camshaft solenoid valve.
60. Fit the wiring harness to the camshaft cover using new cable ties.
61. Secure the ground cable to the wiring harness.
62. Plug in the coolant temperature sensor connector and press down the heat shield.
63. Fit ignition coils 2, 4 and 6. Plug in the connectors.
64. Fit the left grounding point.
65. Fit the power steering fluid reservoir's bolt and tighten to 7 ft. lbs. (10 Nm).
66. Fit the secondary air pipe and check that the quick-coupling engages.
67. Fit the bleeder pipe and tighten fasteners to 7 ft. lbs. (10 Nm).
68. Install the heat shield between the engine and turbo.
69. Connect the negative battery cable.

FRONT COVER

1. Before servicing the vehicle, refer to the Precautions Section.
2. Disconnect the negative battery cable.
3. Remove the heat shield between the engine and turbo.
4. Remove the bleeder pipe.
5. Remove the front secondary air pipe complete with valve and bracket.
6. Remove the power steering fluid reservoir's bolt.
7. Remove the left grounding point.
8. Unplug the connectors and remove ignition coils 2, 4 and 6.
9. Raise the heat shield and unplug the coolant temperature sensor connector.
10. Release the right ground cable from the wiring harness.
11. Remove the wiring harness from the camshaft cover.
12. Unplug the connectors from the camshaft position sensor and the camshaft solenoid valve.
13. Raise the wiring harness slightly.
14. Remove the ground cable of the engine control module.
15. Unplug the engine control module connectors (608A and 608B).
16. Detach the cable duct from the intake manifold.
17. Unplug the atmospheric pressure sensor connector.
18. Detach the corner bracket and move it aside.
19. Remove the bolt holding the injector bracket in place and unplug the connector.
20. Detach the brake vacuum hose from the intake manifold.
21. Detach the crankcase ventilation from the intake manifold.
22. Unplug the fuel pressure sensor connector.
23. Detach the fuel line, wiping up any fuel spill with a cloth.
24. Remove the intake manifold.
25. Loosen the lifting eye. Remove one of the bolts and fold the lifting eye down.
26. Remove the camshaft cover bolts and seals.
27. Remove the camshaft cover.
28. Remove the camshaft cover gasket and seals around the spark plug holes.

To install:

29. Fit new seals on the camshaft cover.
30. Add a spot of sealant to each mating surface between the cylinder head and timing cover.
31. Position the camshaft cover.
32. Install the bolts using new seals and tighten to 7 ft. lbs. (10 Nm).
33. Fold the eye up and fit the bolt. Tighten the bolt to 48 ft. lbs. (65 Nm).
34. Install the intake manifold.
35. Fit the fuel line.
36. Fit the fuel pressure sensor connector.
37. Fit the crankcase ventilation to the intake manifold.
38. Fit the brake vacuum hose to the intake manifold.
39. Plug in the injector connectors and fit the bracket.
40. Reposition the corner bracket and fit the bolts.
41. Plug in the atmospheric pressure sensor connector.
42. Fit the cable duct to the intake manifold.
43. Plug in the engine control module connectors (608A and 608B).

44. Fit the ground cable of the engine control module.

45. Plug in the connectors of the camshaft position sensor and the camshaft solenoid valve.

46. Fit the wiring harness to the camshaft cover using new cable ties.

47. Secure the ground cable to the wiring harness.

48. Plug in the coolant temperature sensor connector and press down the heat shield.

49. Fit ignition coils 2, 4 and 6. Plug in the connectors.

50. Fit the left grounding point.

51. Fit the power steering fluid reservoir's bolt and tighten to 7 ft. lbs. (10 Nm).

52. Fit the secondary air pipe complete with valve and bracket. Use a new gasket. Tighten the bolts to the exhaust manifold to 16 ft. lbs. (22 Nm), and the bracket bolts to 48 ft. lbs. (65 Nm).

53. Fit the bleeder pipe and tighten fasteners to 7 ft. lbs. (10 Nm).

54. Install the heat shield between the engine and turbo.

55. Connect the negative battery cable.

56. Connect the negative battery cable.

3.0L Engines

1. Before servicing the vehicle, refer to the Precautions Section.
2. Remove or disconnect the following:
   - Negative battery cable
   - Crankcase breather hose and vacuum control unit
   - Temperature sensor wiring
   - Rocker arm (camshaft) cover and discard the gasket
3. Clean all mating surfaces of any residual gasket material.

To install:

4. Install or connect the following:
   - Rocker arm (camshaft) cover with a new gasket and torque the bolts to 16 ft. lbs. (22 Nm)
   - Temperature sensor wiring
   - Crankcase breather hose and vacuum control unit
   - Negative battery cable

Cylinder Head

REMOVAL & INSTALLATION

**** CAUTION**

The fuel system pressure must be relieved before disconnecting any fuel lines. Failure to do so may result in personal injury.

9-3

2002 MODEL

1. Before servicing the vehicle, refer to the Precautions Section.
2. Raise the car and remove the turbocharger support.
3. Remove the belt circuit cover.
4. Drain the cooling system.
5. Remove or disconnect the following:
   - Hose clamp between turbocharger inlet pipe and connecting pipe
   - EVAP canister purge valve
   - MAF sensor connector and wiring
   - Air intake box
   - Engine cover
   - Exhaust manifold heat shield
   - Turbocharger delivery pipe
   - Bypass valve vacuum hose
   - Pressure/temperature sensor hose
   - Venturi hose from charge air pipe (manual transmission)
   - Signal hose from turbocharger wastegate
   - Turbocharger delivery hose (plug openings)
   - Cable ties on wiring harness at strut tower
   - Auxiliary drive belt

**** CAUTION**

DO NOT pry too hard on tensioner or it could be damaged at its end position.

   - Crankcase vent pipe (move aside)
   - EVAP connector and purge hose
   - Engine lifting eye
   - Turbocharger inlet pipe (leaving control valve)
   - Power steering pump with bracket (position aside)
   - Vacuum unit
   - Turbocharger bolts at exhaust manifold
   - Connector for ignition discharge module, turbo pressure sensor and coolant temperature sensor
   - Cylinder head coolant hose
   - Coolant pipe from turbocharger to pressure sensor
   - Banjo connector at cylinder head
   - Thermostat hose
   - Coolant pipe from water pump to cylinder block
   - Upper throttle body hose at coolant pipe
   - Dipstick tube (plug hole in block)
   - Crankcase vent hose (24)
   - Vacuum hose from camshaft cover (24)
   - Ties to camshaft cover at non-return valve (24)
   - Cover at accelerator connection (25)
   - Grounding connections on wiring harness duct (26)
   - Fuel lines at quick-disconnect coupling (27)
   - Vacuum hoses at throttle body, intake manifold and brake servo
   - Connectors for oxygen sensor, timing sensor, limp-home solenoid, throttle body, MAP sensor and injectors
   - Tie holding wiring harness to starter solenoid (position harness channel aside)
   - Top bolt on intake manifold stay
   - Ignition discharge module
   - Spark plugs
6. Remove the camshaft cover and "zero" the engine by turning the crankshaft in rotating direction until mark on belt pulley is aligned with timing cover mark. Be sure camshafts are aligned with their set marks.
7. Remove the bolts on the camshaft sprockets. Use a wrench on rear end of camshaft as a counter-stay.

**** WARNING**

DO NOT allow camshaft to turn during this procedure or damage to valves and pistons may occur.

8. Remove the idler sprocket bolt, then insert a 27mm socket on the chain tensioner. Remove the tensioner by removing the plug, pin and spring, then remove the chain tensioner.
9. Remove the sprocket and restrain the chain guides together.
10. Remove the cylinder head bolts in sequence as illustrated.
11. Lift off cylinder head with intake and exhaust manifolds.

To install:

12. Thoroughly clean all mating surfaces and bolt holes of oil and residue. Wipe combustion chambers to be sure no contamination enters the head or block.
13. Install new gaskets and mount the exhaust manifold and turbocharger to the cylinder head.
14. Apply a 2mm thick by 10-20mm long bead of Loctite 518 to inner part of upper contact surface of timing cover against cylinder head.
15. Turn the crankshaft by 45 degrees to lower the pistons.
16. Wrap a rubber band around the

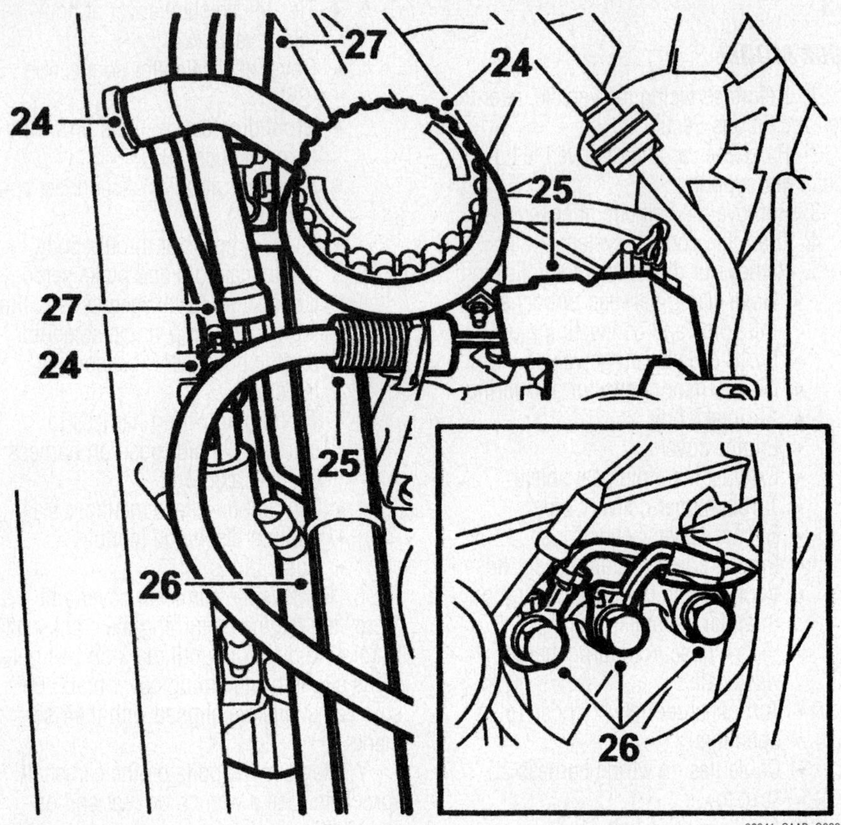

Disconnecting components near throttle cable—9-3 2.0L engine 2002

chain guides and install the cylinder head with a new gasket.

17. Verify that cylinder head is correctly seated on the locating sleeves and that the timing chain runs freely.

18. Tighten the cylinder head bolts, in sequence shown, in 3 steps as follows:

a. Step 1: 30 ft. lbs. (40 Nm)
b. Step 2: 44 ft. lbs. (60 Nm)
c. Step 3: Tighten each bolt, in sequence, an additional 90 degrees.

19. Install 2 screws between timing cover and cylinder head and torque to 18 ft. lbs. (24 Nm).

20. Align the **0** mark on the crankshaft with the timing marks on the camshafts.

21. Remove the rubber band and fit the camshaft sprockets and camshaft timing chain. Start with the exhaust camshaft. Do not tighten the bolts at this time.

22. Prepare the chain tensioner for mounting by pressing down the catch and pressing in the chain tensioner.

23. Install the chain tensioner with a 27mm socket. If necessary, change the tensioner gasket. Torque the bolt to 47 ft. lbs. (63 Nm).

24. Install or connect the following:

- Chain tensioner plug with push rod and spring torqued to 17 ft. lbs. (22 Nm)
- Idler sprocket
- Top bolt on intake manifold stay

25. Make sure chain is positioned correctly on the chain guards. Rotate the crankshaft 2 revolutions and check the setting of the flywheel and camshaft timing marks.

26. Tighten the camshaft sprocket bolts to 47 ft. lbs. (63 Nm).

27. Clean the camshaft cover sealing surface with a safe solvent.

28. Apply soap to the opening the camshaft cover and install the cover.

29. Tighten the bolt located furthest to the front at the timing chain end. Continue with this sequence around the outside and inside. Tighten bolts to 11 ft. lbs. (15 Nm).

30. Install the spark plugs to 21 ft. lbs. (28 Nm).

31. Apply sealant 30 19 312 to rubber seal on the ignition discharge module and install the module.

32. Install the auxiliary drive belt loosely on the pulleys.

33. Install the wiring harness channel and fasten the cable tie to the starter solenoid.

34. Install or connect the following:

- Connectors for oxygen sensor, timing sensor, limp-home solenoid, throttle body, MAP sensor and fuel injectors
- Vacuum hoses to throttle body
- Hose between intake manifold and brake servo
- 3 grounding connections
- Fuel lines with couplings and seals
- Accelerator cable and cover
- Hoses for crankcase vent and camshaft cover
- Dipstick tube
- Wiring harness ties on camshaft cover by non-return valve
- Lower and upper hoses to throttle body and each coolant pipe
- Coolant pipe to water pump
- Hose from charge air cooler to thermostat housing
- Turbocharger cooler pip with banjo coupling (to thermostat housing, with new gaskets) and tightened to 19 ft. lbs. (25 Nm)

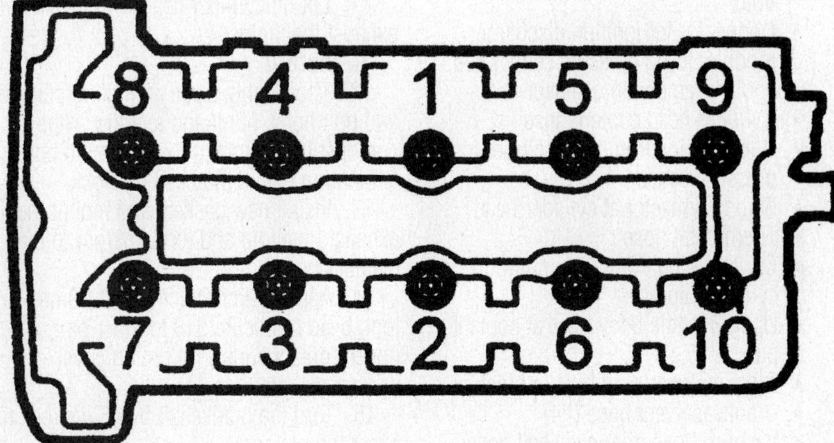

Remove the cylinder head bolts in sequence—2.3L engine 9-3 and 9-5

- Coolant hoses to cylinder head and thermostat housing
- Connectors to turbocharger pressure sensor, coolant temperature sensor, and ignition discharge module
- Exhaust manifold to turbocharger with Molykote on the studs
- Vacuum unit to turbocharger
- Power steering pump and bracket, with belt on pulley, then release tensioner
- Turbocharger inlet pipe with bracket
- V-clamp
- Control valve on bracket with connector
- EVAP purge hose
- Crankcase vent pipe with banjo bolt tightened to 18 ft. lbs. (24 Nm)
- Harness ties to strut tower
- Heat shield on exhaust manifold
- Charge air pipe
- Pressure/temperature sensor connector
- Hose to bypass valve
- Venturi hose to charge air pipe (manual transmission)
- 3 nuts holding rubber bushings

⁂ WARNING

To prevent hoses on delivery side of turbocharger from coming loose under pressure, thoroughly clean all mating areas on inside of hoses and pipes before attaching.

- Connecting hose to turbocharger inlet pipe
- MAF sensor cable to clamps on top of air collection box
- Evap canister purge valve to air collection box
- Negative battery cable
- Engine cover

35. Raise the car and ensure coolant drain plug is tight. Also tighten bottom bolts on the intake manifold stay and the turbocharger stay.
36. Visually check that auxiliary drive belt is properly positioned on all pulleys.
37. Install the air shields and lower the car
38. Refill the cooling system
39. With the A/C off, ensure coolant is at MAX level. Start the engine and run at varying speeds until the cooling fan comes on.
40. Stop the engine. Open the expansion tank cap and top up the coolant.
41. Start the vehicle and run it until the thermostat opens. Turn off the engine and check coolant level.

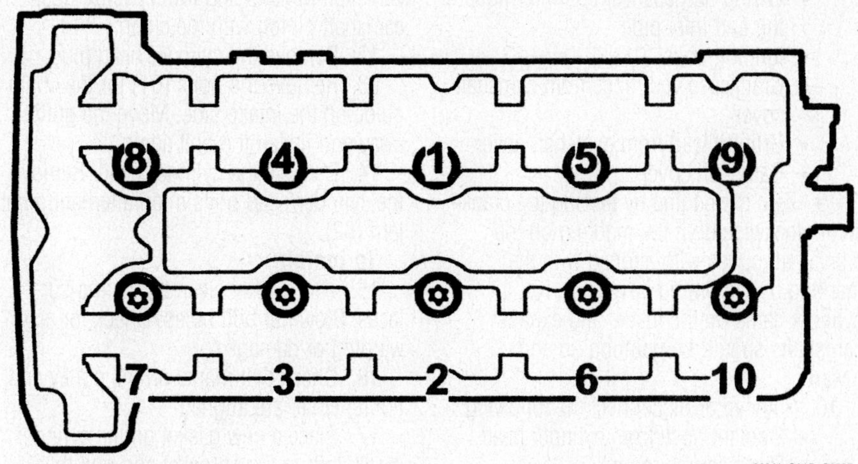

Cylinder head bolt tightening sequence—9-3 and 9-5 4-cylinder engines 2002

42. Check for leaks and repair if necessary.
43. Reset clock and radio code.

2003–06 MODEL (EXCEPT 2.8L ENGINE)

1. Before servicing the vehicle, refer to the Precautions Section.
2. Remove or disconnect the following:
   - Cap from expansion tank
   - Oxygen sensor connector
   - Temperature sensor connector
   - Heat shield over turbocharger
   - Turbocharger oil delivery pipe
   - Upper oxygen sensor
   - Catalytic converter connection from turbocharger
   - Right front wheel and fenderwell liner
   - Lower spoiler shield
   - Hose and connector from headlamp washers
   - Drain the cooling system
   - Lower turbocharger pressure pipe
   - Front exhaust pipe from muffler (cut about 3.5 inches (87 mm) from the end of the muffler)
   - Front pipe from catalytic converter
   - Steering gear heat shield
   - Catalytic converter heat shield
   - Temperature sensor
   - Lower oxygen sensor
   - Bracket from catalytic converter and block
   - Catalytic converter
   - Drain remaining coolant
   - Turbocharger oil return pipe
   - Stay from intake manifold and cylinder block
   - Starter positive cable
   - Power steering pipe clamps on subframe
   - Rear torsion bar bolts (loosen slightly)
3. Fit an engine centering tool kit, 83 96 152, to subframe, as illustrated. Adjust the centering tool bolts so adjuster screw slots are completely visible.
4. Lower the car and remove the upper engine cover.
5. Remove or disconnect the following:
   - Turbocharger delivery hose from throttle body and pipe
   - Turbocharger inlet pipe (plug openings)
   - MAF sensor connector
   - Air cleaner assembly
6. Release any fuel pressure by carefully pressing the service valve needle. Collect spilled fuel.
7. Detach both fuel lines for the fuel rail while gripping the lower nut. Plug the fuel line openings.
8. Remove or disconnect the following:
   - Quick-release coupling for ventilation line
   - Fuel lines and vent lines from camshaft cover (bend up hose and out of the way)
   - Ignition coil cover
   - Ignition coil connectors
   - Turbocharger solenoid connector
   - Connectors for A/C pressure sensor connector, coolant temperature sensor, engine control module, bypass solenoid valve, throttle body, MAP sensor, atmospheric pressure sensor, IDM module
   - Ground connection on ECM
   - Hose from turbocharger wastegate and solenoid valve
   - Wiring harness duct and bracket from camshaft cover (bend wiring harness aside)

- Wiring harness from oil filter housing and inlet pipe
- Ignition coils
- Crankcase vent hose from camshaft cover
- Ground lead from camshaft cover
- Camshaft cover

9. Zero the engine by turning the crankshaft clockwise until the marking on the crankshaft pulley with aligned with the marking on the timing cover. The No. 1 cylinder cams on the intake and exhaust camshafts should be pointing up and inward.
10. Remove or disconnect the following:
    - Coolant hose from cylinder head
    - Right engine mount
    - IDM module
    - Couplings from vacuum pump
    - Power steering pump
    - EVAP canister purge valve and bracket
    - Throttle body
    - ECM
    - Fuel injector connectors
    - Dipstick tube bolt
    - Vacuum hose and purge hose from intake manifold
    - Vacuum hose from fuel pressure regulator
    - Intake manifold
    - Timing chain tensioner
    - Camshaft pinions (hold camshaft flats with wrench)
11. Fit a cable tie to the exhaust camshaft pinion and chain. Remove the camshaft pinions and lower the exhaust camshaft pinion with the chain.
12. Remove the cylinder head plug.
13. Remove the bolts (61) for the chain guide on the intake side. Move the guide away and install the bolt again.
14. Remove the cylinder head. Remove the cylinder head bolts in an alternating pattern (62).

To install:

15. Thoroughly clean all mating surfaces. Blow out bolt holes. Check for any warping or damage.
16. Check that marks on the pulley and timing cover are aligned.
17. Place a new gasket on the cylinder block. Lift up the sprocket and pull the chain guides together to lower the cylinder head.
18. Remove the exhaust camshaft and intake camshaft No. 2 bearing caps. Fit an adjustment tools, 83 96 046, to the camshafts (5). Tighten retaining bolts.
19. Install cylinder head bolts, in sequence as illustrated. DO NOT drop bolts into place or threads will be damaged. Tighten cylinder head bolts in 3 steps as follows:
    a. Step 1: 22 ft. lbs. (30 Nm)
    b. Step 2: Each bolt, in sequence, an additional 75 degrees
    c. Step 3: Each bolt, in sequence, an additional 80 degrees
20. Install and tighten transmission section bolts to 26 ft. lbs. (35 Nm).
21. Remove the bolt for the chain guide. Put the guide in place an install and tighten the bolt.
22. Install the plug on the front of the cylinder head, with a new sealing washer. Tighten the plug to 18 ft. lbs. (25 Nm).
23. Fit the pinion on the exhaust camshaft and intake camshaft without tightening. Verify that the engine is still "zeroed".
24. Install the chain tensioner reset with a new seal. The groove on the end of the chain tensioner should be vertical when installed. Torque tensioner reset to 55 ft. lbs. (75 Nm).
25. Torque the camshaft pinions while gripping the respective camshaft with a wrench on the flats. Torque to 63 ft. lbs. (85 Nm), plus an additional 30 degrees. Remove the cable tie.
26. Remove the camshaft adjustment tool and fit the camshaft No. 2 bearing caps. Torque the bolts to 6 ft. lbs. (8 Nm).
27. Turn the engine over twice and zero it so the markings are aligned. Remove camshaft bearing caps No. 2 and No. 7 and install that camshaft adjustment tool, 83 96 046, to check that camshaft setting is correct. Remove the tools and install the bearing caps.
28. Clean the sealing surfaces on the intake manifold and cylinder head. Install the intake manifold with a new gasket. Torque the nuts to 84 inch lbs. (10 Nm).
29. Install or connect the following:
    - Vacuum hose to fuel pressure regulator
    - Dipstick tube bolt
    - Vacuum hose and purge hose to intake manifold
    - Fuel injector connectors
    - Throttle body and ECM (leaving upper right screw for ground cable)
    - Bracket and coupling to EVAP canister purge valve
    - Power steering pump, with a new seal, to 16 ft. lbs. (22 Nm)
    - Hose to vacuum pump
    - IDM module
    - Coolant hose to pipe on cylinder head
    - Right side engine mount: body bolts to 30 ft. lbs. (40 Nm), plus an additional 60 degrees and bracket bolts to 52 ft. lbs. (70 Nm), plus an additional 60 degrees
    - Remaining coolant hose to cylinder head pipe
    - Camshaft cover, with a new seal, to 84 inch lbs. (10 Nm)
    - Ground cable to camshaft cover
    - Crankcase vent hose to camshaft cover

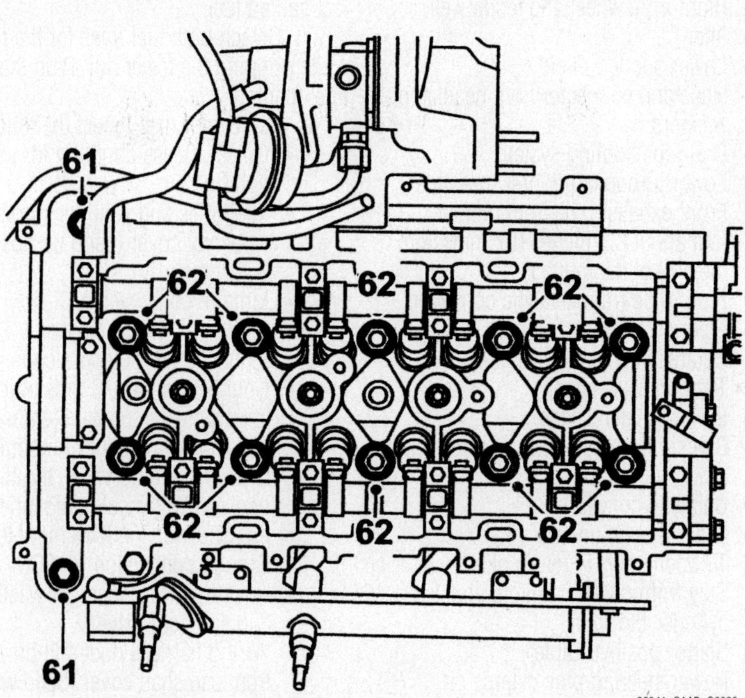

Showing cylinder head bolt locations—9-3 4-cylinder engines

- Ignition coils
- Wiring harness to intake manifold and oil filter housing
- Wiring harness duct and bracket to camshaft cover
- Hose to turbocharger wastegate and solenoid valve
- Connectors for A/C pressure sensor, coolant temperature sensor, ECM, wastegate solenoid, throttle body, MAP sensor, barometric pressure sensor, IDM module
- Ground cable to ECM
- Ignition module connectors (ensure cables are not pinched by cover)
- Turbocharger solenoid valve connectors with cable clips
- Cover over ignition coils
- Fuel lines and vent lines into clips on camshaft cover
- Coupling for vent line
- Both fuel lines, with new seals, to fuel rail while gripping lower nut to 84 inch lbs. (10 Nm)
- Air cleaner assembly

⁂ WARNING

To prevent hoses on delivery side of turbocharger from coming loose under pressure, thoroughly clean all mating areas on inside of hoses and pipes before attaching.

- Air cleaner cover
- MAF sensor connector
- Turbocharger inlet pipe to air cleaner
- Hoses to inlet pipe
- Turbocharger hose to throttle body and turbo delivery pipe

30. Raise the car and remove the engine centering tool from the subframe.

31. Torque the rear torsion arm bolts to 52 ft. lbs. (70 Nm), plus an additional 90 degrees.

32. Install or connect the following:
- Bolts on power steering pipe clamps on subframe
- Positive cable to starter
- Stay between intake manifold and cylinder block to 16 ft. lbs. (22 Nm)
- Oil return line to turbocharger with new gasket and O-ring; torque to 11 ft. lbs. (15 Nm)
- Stay between exhaust manifold and cylinder block to 16 ft. lbs. (22 Nm)
- New seals on turbocharger coolant pipe; torque to 30 ft. lbs. (40 Nm)
- Catalytic converter and bracket to studs on engine block and to turbocharger connection
- Catalytic converter bracket bolts into engine block and new nuts on studs

33. Lower the car and install the catalytic converter with new nuts so it is evenly positioned to the turbocharger. Use thread paste on the studs.

34. Raise the car and tighten the catalytic converter bracket nuts and bolts to 18 ft. lbs. (25 Nm).

35. Sparingly apply thread paste to the lower oxygen sensor threads and install it. Be sure wiring is not twisted or damaged. Torque oxygen sensor to 30 ft. lbs. (40 Nm).

36. Install or connect the following:
- Catalytic converter heat shield
- Steering gear heat shield
- Front exhaust pipe to catalytic converter with new gasket and nuts torqued to 18 ft. lbs. (25 Nm)
- Joining clamp centered on exhaust pipe at cut; torque nuts to 30 ft. lbs. (40 Nm)
- Lower turbocharger delivery pipe
- Headlamp washer hose and connector
- Lower spoiler shield
- Chassis reinforcement front support frame, if equipped
- Right front fenderwell liner and front wheel
- Upper oxygen sensor, with thread paste sparingly, torqued to 30 ft. lbs. (40 Nm)
- Turbocharger oil delivery pipe, with new seals, to 16 ft. lbs. (22 Nm)
- Turbocharger heat shield
- Negative battery cable and battery cover

37. Check engine oil level and top as needed. Fill cooling system.

38. With the A/C off, ensure coolant is at MAX level. Start the engine and run at varying speeds until the cooling fan comes on.

39. Stop the engine. Open the expansion tank cap and top up the coolant.

40. Start the vehicle and run it until the thermostat opens. Turn off the engine and check coolant level.

41. Check for leaks and repair if necessary.

42. Install upper engine cover.

43. Reset clock and radio and other electrical accessories as needed.

44. Use scan tool to reprogram ECM. Follow tool instructions.

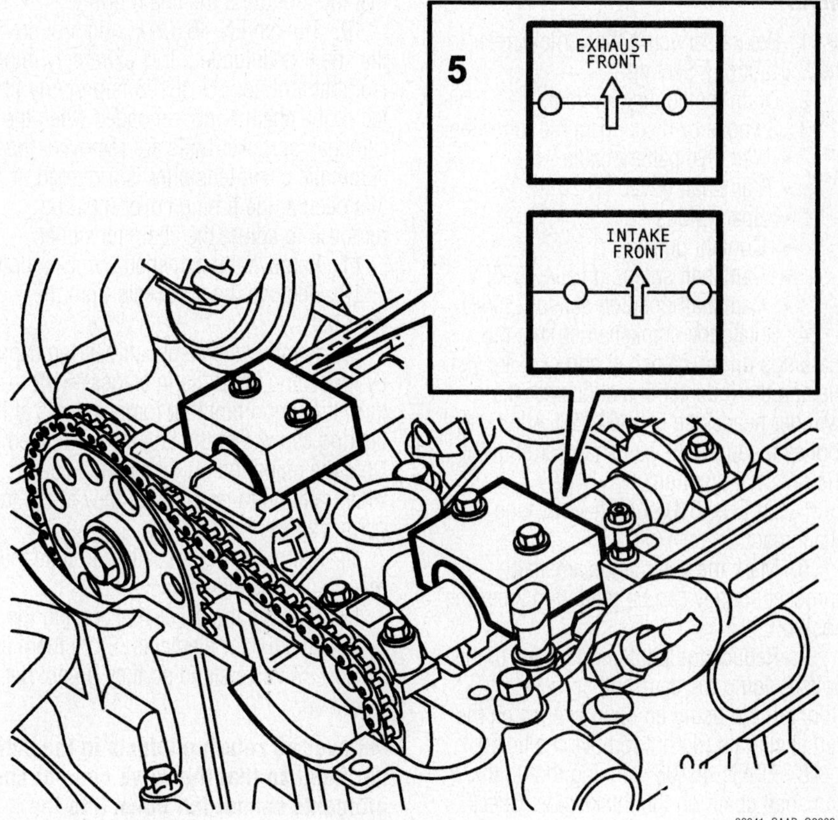

Using camshaft adjusting tools—9-3 4-cylinder engine

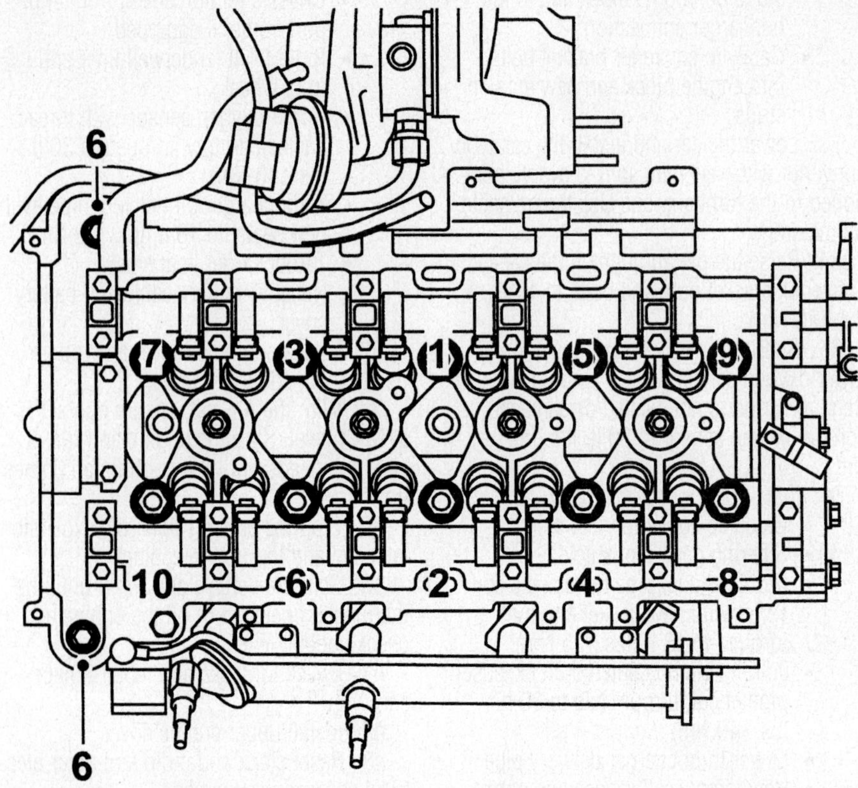

Cylinder head bolt tightening sequence—9-3 4-cylinder engine 2003–06

2.8L ENGINE—FRONT CYLINDER HEAD

1. Before servicing the vehicle, refer to the Precautions Section.
2. Drain the cooling system.
3. Remove or disconnect the following:
    - Negative battery cable
    - Camshaft cover
    - Spark plugs
    - Coolant port
    - Camshaft solenoid valve (695F)
    - Camshaft position sensor (555F)
4. Rotate the crankshaft so that the markings on the camshaft sprocket are visible and the indentations align with the cylinder head. Use EN-46105-1 Adjustment tool as a gauge (the camshafts are now in their neutral position).
5. Fit EN-46106 Flywheel locking attachment.
6. Mark the chain and camshaft sprocket so they can be refitted in the same position.
7. Reduce the tightening torque of the bolts holding the camshaft sprockets. Counterhold using an open wrench on the camshaft flats (do not remove the bolts).
8. Fit EN-46108-1 Fixing tool to the camshaft chain on the intake side. Check that the tool grips the chain firmly.
9. Fit EN-46108-2 Fixing tool to the camshaft chain on the exhaust side. Check that the tool grips the chain firmly.
10. Tighten EN-46108 Fixing tool so that the chain is unloaded. It is extremely important that tool no. EN-46108 is properly fitted. If the chain is not unloaded when the camshaft sprocket bolts are removed, the automatic chain tensioner is triggered. If this occurs, the timing cover must be removed to adjust the chain tensioner.
11. Remove the camshaft sprocket bolts.
12. Remove the sprockets from the camshafts.
13. Remove the camshaft bearing caps by loosening the bolts in stages—1/2 to 1 turn. It is important that removal ends at the bearing cap where the tappets are loaded. Note the markings on the bearing caps when refitting to ensure that they are in the correct spots.
14. Lift out the camshafts and note the rear marking on the camshafts.
15. Remove the roller rockers and the hydraulic adjuster elements. Store them in 83 93 787 Valve stand so they do not get mixed up.

➡Cover all return oil ducts in the cylinder head so that the valve cone or contaminants cannot fall down into the engine.

16. Fit EN-46110-eu Valve spring tool.
17. Fit 83 94 173 Air nipple for spark plug hole. Connect compressed air and put the piston and valves under pressure.
18. Press down the valve plate with EN-46110-eu Valve spring tool.
19. Remove the valve cone with a magnet. The compressed air must not be disconnected when the valve cone is removed. Do not press on the valve stem either as counterpressure in the cylinder could reduce and the valve could fall down into the cylinder.
20. Lift out the spring plate and valve spring.
21. Lift out the valve guide seal using 83 94 157 Pliers, valve guide seal.
22. Remove the cylinder head bolts in an alternating sequence by loosening them ¼ turn then ½ turn
23. Remove the cylinder head.

To install:

24. Clean all mating surfaces of any residual gasket material.
25. Install the cylinder head. Torque the new bolts in an alternating sequence as follows:
    a. M8 size bolts torque 11 ft. lbs. (15 Nm).
    b. Plus an additional 60 degrees.
    c. M11 size bolts torque 33 ft. lbs. (45 Nm).
    d. Plus an additional 120 degrees.
26. Clean all parts and check contact surfaces and bearing surfaces for wear.
27. Fit the valve guide seal using 83 94 157 Pliers, valve guide seal.
28. Fit the valve spring and spring plate.
29. Press down the valve plate with EN-46110-eu Valve spring tool.
30. Fit the valve cone using EN-46117 Fitting tool.
31. Unload the valve and remove EN-46110-eu Valve spring tool.
32. Disconnect the compressed air and remove the air nipple.
33. Position the camshafts and use EN-46105 Adjustment tool to check that the camshafts are mounted at the correct angle.
34. Fit the bearing caps as marked on the caps. Start where the cams point down and load the tappets. Tighten the bolts in stages—1/2 to 1 turn. Tighten to 7 ft. lbs. (10 Nm).
35. Fit the sprockets with chain on the camshafts. Tighten the bolts by hand.
36. Check that the marks on the chain match the markings on the sprockets.
37. Remove EN-46108-2 and EN-46108-1 Fixing tools
38. Tighten the bolts of both camshaft sprockets, counterholding with an open

spanner on the camshaft flats to 48 ft. lbs. (65 Nm).

39. Remove EN-46106 Flywheel locking attachment.

40. Install or connect the following:

- Camshaft position sensor (555F) using a new gasket and tighten the mounting bolt to 7 ft. lbs. (10 Nm)
- Camshaft solenoid valve (695F) using a new seal and tighten to 7 ft. lbs. (10 Nm)
- Coolant port using a new seal and gasket. Tighten the mounting bolts to 7 ft. lbs. (10 Nm)
- Spark plugs
- Camshaft cover.
- Negative battery cable

41. Fill the cooling system to the proper level.

42. Start the vehicle, check for leaks and repair if necessary.

2.8L ENGINE—REAR CYLINDER HEAD

1. Before servicing the vehicle, refer to the Precautions Section.

2. Drain the cooling system.

3. Remove or disconnect the following:

- Camshaft cover
- Spark plugs
- Camshaft solenoid valve (695R)
- Camshaft position sensor (555R)

4. Rotate the crankshaft so that the markings on the camshaft sprocket are visible and the indentations align with the cylinder head. Use EN-46105 Adjustment tool as a gauge (the camshafts are now in their neutral position).

5. Fit EN-46106 Flywheel locking attachment.

6. Mark the chain and camshaft sprocket so they can be refitted in the same position.

7. Slightly loosen the bolts holding the camshaft sprockets. Counterhold using an open wrench on the camshaft flats (do not remove the bolts).

8. Fit EN-46108-2 Fixing tool to the camshaft chain on the intake side. Check that the tool grips the chain firmly.

9. Fit EN-46108-1 Fixing tool to the camshaft chain on the exhaust side. Check that the tool grips the chain firmly.

10. Tighten EN-46108 Fixing tool so that the chain is unloaded. It is extremely important that tool no. EN-46108 is properly fitted. If the chain is not unloaded when the camshaft sprocket bolts are removed, the automatic chain tensioner is triggered. If this occurs, the timing cover must be removed to adjust the chain tensioner.

11. Remove the camshaft sprocket bolts.

12. Remove the sprockets from the camshafts.

13. Remove the camshaft bearing caps by loosening the bolts in stages—1/2 to 1 turn. It is important that removal ends at the bearing cap where the tappets are loaded. Note the markings on the bearing caps when refitting to ensure that they are in the correct spots.

14. Lift out the camshafts and note the rear marking on the camshafts.

15. Remove the roller rockers and the hydraulic adjuster elements. Store them in 83 93 787 Valve stand so they do not get mixed up.

➡Cover all return oil ducts in the cylinder head so that the valve cone or contaminants cannot fall down into the engine.

16. Fit EN-46110-eu Valve spring tool.

17. Fit 83 94 173 Air nipple for spark plug hole. Connect compressed air and put the piston and valves under pressure.

18. Press down the valve plate with EN-46110-eu Valve spring tool.

19. Remove the valve cone with a magnet. The compressed air must not be disconnected when the valve cone is removed. Do not press on the valve stem either as counterpressure in the cylinder could reduce and the valve could fall down into the cylinder.

20. Lift out the spring plate and valve spring.

21. Lift out the valve guide seal using 83 94 157 Pliers, valve guide seal.

22. Remove the cylinder head bolts in an alternating sequence by loosening them ¼ turn then ½ turn

23. Remove the cylinder head.

To install:

24. Clean all mating surfaces of any residual gasket material.

25. Install the cylinder head. Torque the new bolts in an alternating sequence as follows:

a. M8 size bolts torque 11 ft. lbs. (15 Nm).

b. Plus an additional 60 degrees.

c. M11 size bolts torque 33 ft. lbs. (45 Nm).

d. Plus an additional 120 degrees.

26. Clean all parts and check contact surfaces and bearing surfaces for wear.

27. Fit the valve guide seal using 83 94 157 Pliers, valve guide seal.

28. Fit the valve spring and spring plate.

29. Press down the valve plate with EN-46110-eu Valve spring tool.

30. Fit the valve cone using EN-46117 Fitting tool.

31. Unload the valve and remove EN-46110-eu Valve spring tool.

32. Disconnect the compressed air and remove the air nipple.

33. Position the camshafts and use EN-46105-1 Adjustment tool to check that the camshafts are mounted at the correct angle.

34. Fit the bearing caps as marked on the caps. Start where the cams point down and load the tappets. Tighten the bolts in stages—1/2 to 1 turn. Tighten to 7 ft. lbs. (10 Nm).

35. Fit the sprockets with chain on the camshafts. Tighten the bolts by hand.

36. Check that the marks on the chain match the markings on the sprockets.

37. Remove EN-46108-2 and EN-46108-1 Fixing tools.

38. Tighten the bolts of both camshaft sprockets, counterholding with an open spanner on the camshaft flats to 48 ft. lbs. (65 Nm).

39. Remove EN-46106 Flywheel locking attachment.

40. Install or connect the following:

- Camshaft position sensor (555R) using a new gasket and tighten the mounting bolt to 7 ft. lbs. (10 Nm)
- Camshaft solenoid valve (695R) using a new seal and tighten to 7 ft. lbs. (10 Nm)
- Spark plugs
- Camshaft cover.
- Negative battery cable

41. Fill the cooling system to the proper level.

42. Start the vehicle, check for leaks and repair if necessary.

9-5

2.3L ENGINE

1. Before servicing the vehicle, refer to the Precautions Section.

2. Run engine at idle speed. Remove fuse No 19 from fuse panel. When engine stops, turn off ignition. Re-install the fuse.

3. Remove the battery cover and intake manifold cover.

4. Drain the cooling system.

5. Remove or disconnect the following:

- Battery and intake manifold cover
- Negative battery cable
- Lower engine cover
- Right front wheel
- Steering servo pipe from the subframe and install a wedge between the oil pan and subframe on the right hand side
- Right front engine mount and bracket
- Mass Air Flow (MAF) sensor

- Crankcase breather pipe
- Front lifting eye
- Charge air pipe and bypass valve
- Pressure/temperature sensor connector
- Bypass valve vacuum hose
- V-belt
- Belt tensioner
- Alternator (move the bracket aside)
- Temperature sensor and ignition discharge module connectors
- Coolant hoses from the cylinder head
- Throttle body lever cover
- Throttle cable and dipstick tube
- Fuel hoses
- Crankcase breather nipple from the timing cover
- Intake manifold steady bar
- Turbocharger steady bar
- Turbocharger and exhaust manifold heat shield
- Servo pump and suspend it on the radiator crossmember
- Lower screw from the steering servo pump bracket
- Bracket for the ignition discharge module's connector from the rear lifting eye
- Heat exchanger pipe bracket
- Intake manifold and partition and move it rearward
- Ignition discharge module and spark plugs
- Camshaft cover and align the pulley marks with the timing cover and make certain that the camshafts are in line with their timing marks
- Camshaft sprocket bolts. Make certain that the camshaft does not turn
- Idler sprocket bolt and remove the chain tensioner
- Camshaft sprockets and place a rubber band between the chain guides
- Cylinder head by first removing the bolts from the timing cover

To install:

6. Clean all mating surfaces.
7. Install or connect the following:
   - Inner gasket for the camshaft cover partition
   - Cylinder head gasket and turn the crankshaft 45 degrees in the rotational direction of the engine to lower pistons

8. Bind chain guides with strap or rubber band and install the new cylinder head gasket and position cylinder head. Make sure chain is not trapped.

9. Install the cylinder head bolts. Torque the bolts in sequence shown as follows:
   a. 30 ft. lbs. (40 Nm).
   b. 44 ft. lbs. (60 Nm).
   c. Plus an additional 90 degrees.
10. Install or connect the following:
    - 2 bolts between the timing cover and cylinder head and torque the bolts to 16 ft. lbs. (22 Nm). Make certain that the camshafts are aligned with the timing marks and reset the crankshaft to the **O** mark.
    - Camshaft sprocket and chain starting with the exhaust camshaft. Do not tighten the bolts
    - Timing chain tensioner and torque the bolts to 47 ft. lbs. (63 Nm)
    - Timing chain tensioner plug, push rod and spring and torque the bolt to 16 ft. lbs. (22 Nm)
    - Torque the idler pulley and camshaft sprocket bolts to 47 ft. lbs. (63 Nm). Rotate the crankshaft 2 revolutions and make check the settings of the crankshaft pulley and camshafts
    - Camshaft cover after lightly coating the opening with clean oil and torque the bolts, in sequence from the front at the timing chain end and working around, to 11 ft. lbs. (15 Nm)
    - Spark plugs and torque the plugs to 21 ft. lbs. (28 Nm)
    - Ignition discharge module and torque the bolt to 8 ft. lbs. (11 Nm)
    - Intake manifold and intermediate partition. Remove the straps and torque the bolts to 16 ft. lbs. (22 Nm).
    - Alternator bracket and make certain that the adjuster sleeve is tapped out slightly
    - Belt tensioner and tighten the ignition discharge module connector bracket
    - Coolant hoses to the cylinder head and install the bolt securing the pipe to the thermostat housing cover
    - Ignition discharge module electrical connector
    - Temperature sensor electrical connector
    - Turbocharger retaining nuts between the exhaust manifold and turbocharger and torque the bolts to 19 ft. lbs. (25 Nm)
    - Turbocharger heat shield
    - Dipstick tube
    - Turbocharger and intake manifold steady bars
    - Fuel hoses and rubber protectors
    - Crankcase ventilation nipple on the valve cover
    - Throttle cable and adjust as necessary
    - Ground connections to the cylinder head
    - Throttle body cover
    - Servo pump and install the V-belt. Torque the pulley bolts to 15 ft. lbs. (20 Nm)
    - Lifting eye
    - Solenoid valve connector
    - Charge air pipe and bypass valve
    - Bypass valve vacuum hose and pressure/temperature sensor connector
    - MAF sensor
    - Right side engine mount and bracket. Torque the bolts to 39 ft. lbs. (47 Nm) and the nuts to 78 ft. lbs. (105 Nm). Remove the wedge and secure the servo pump pipe to the subframe
    - Right front wheel
    - Crankcase breather pipe and torque the banjo bolt to 18 ft. lbs. (24 Nm)
    - Lower engine cover
    - Negative battery cable
    - Intake manifold and battery cover
11. Fill the cooling system and check all fluid levels.
12. Start the vehicle, check for leaks and repair if necessary.

3.0L ENGINE—FRONT CYLINDER HEAD

1. Before servicing the vehicle, refer to the Precautions Section.
2. Remove the vent hose and battery.
3. Remove the lower engine cover.
4. Drain the cooling system.
5. Remove or disconnect the following:
   - Turbocharger bracket and install wedges between the oil pan and subframe and gearbox and subframe
   - Turbocharger delivery pipe from the charge air cooler
   - Mass Air Flow (MAF) sensor and hose
   - Power steering hose from the holder
   - Right side engine mount and bracket
   - V-belt tension
   - Water pump and power steering pump pulleys
   - Belt tensioner and alternator air intake
   - Timing cover
   - Throttle cable

- Upper intake manifold vacuum hoses and pressure sensor connector
- Vacuum hose and water hoses from the throttle body
- Turbocharger delivery pipe with the bypass pipe and valve from the throttle body
- Pressure/temperature sensor connector
- Throttle body connector and purge valve hose from the throttle body
- Ignition discharge module connector
- Fuel injector electrical connector
- Fuel lines
- Upper intake manifold and move it aside
- Middle intake manifold
- Lower intake manifold
- Coolant bridge and move it aside
- Crankshaft pulley
- Tensioner pulley and adjusting rollers and mark the rotation direction of the belt
- Lower adjusting roller and washer
- Timing belt. Refer to the appropriate section for the timing belt removal and installation.
- Adjuster roller bolts
- Ignition discharge module
- Camshaft cover and make sure the O-rings do not fall into the engine. Rotate the engine 60 degrees Before Top Dead Center (BTDC) and note the marking
- Camshaft sprockets
- Water pump
- Inner timing cover
- Front exhaust manifold heat shield
- Hose connections for the front SAI valve
- SAI pipe and valve from the exhaust manifold
- Crankcase ventilation pipe from between the turbocharger intake pipe and the cylinder head
- Turbocharger intake pipe and outlet hose
- Water hose from the thermostat housing
- Front exhaust manifold from the turbocharger
- Turbocharger water and oil pipes
- Exhaust camshaft by removing the bolts in stages
- Bearing caps where the tappets are under load
- Stuffing box
- Cylinder head bolts in the proper sequence by loosening them ¼ turn then ½ turn

To install:

6. Clean all mating surfaces of any residual gasket material.
7. Make certain that the crankshaft is at 60° BTDC.
8. Be sure that the proper gasket is used and is marked **OPEN/TOP**
9. Install or connect the following:
   - Tappets
10. Cylinder head. Torque the new bolts in sequence as follows:
    a. 19 ft. lbs. (25 Nm).
    b. Plus an additional 90 degrees.
    c. Plus an additional 90 degrees.
    d. Plus an additional 90 degrees.
11. Install or connect the following:
    - Exhaust camshaft into position after lubricating it with clean engine oil. The locating pins for the intake camshaft should be in line with the inside bolts for the bearing caps. The guide pin for the exhaust camshaft should point straight up from the plane of the cylinder head
    - Stuffing box and bearing caps. Torque the bolts in ½ to 1 turn to 6 ft. lbs. (8 Nm).
    - Camshafts sprockets in relation to the locating pins. The number in the sprocket hub should match the
    - Rear timing cover
    - Camshaft cover and make certain that the O-rings are in place.
    - Lubricate with a soap solution and place tool (10)-81-52-381 in the corners at the camshaft bearing caps. Torque the bolts to 6 ft. lbs. (8 Nm).
    - Ignition discharge module
    - Turbocharger oil pipes and torque the oil pipe to 15 ft. lbs. (20 Nm); the oil pipe to filter adapter to 19 ft. lbs. (25 Nm) and the oil return pipe to 10 ft. lbs. (15 Nm)
    - Turbocharger water pipes. Torque the water pipes to 19 ft. lbs. (25 Nm) and the cylinder head retaining bolt to 10 ft. lbs. (15 Nm)
    - Water hose to the thermostat housing
    - Front exhaust pipe to the turbocharger and torque the bolts to 18 ft. lbs. (24 Nm)
    - Hose between the turbocharger and charge air cooler
    - Crankcase breather pipe to the turbocharger intake pipe and torque the bolt to 18 ft. lbs. (24 Nm)
    - Turbocharger intake pipe and torque the V-clamp to 2 ft. lbs. (3 Nm) and the U clamp to 10 ft. lbs. (15 Nm)
    - SAI pipe with new gaskets to the front exhaust manifold
    - SAI hose to the valve
    - Crankcase ventilation pipe between the cylinder head turbocharger intake pipe and torque to 18 ft. lbs. (24 Nm)
    - Water pump and torque the bolts to 19 ft. lbs. (25 Nm)
    - Bracket with the tensioner roller and the upper adjusting roller
12. Install locking tool KM-800-1 on No. 1 and 2 camshaft sprockets and tool KM-800-2 on No. 3 and 4 camshaft sprockets
13. Rotate the crankshaft to just before the 0 degree mark and install tool KM-800–10. Carefully rotate the crankshaft in the direction of the engine until the arm is lying against the water pump flange.
14. Install the camshaft belt to the markings on the belt and the marked direction of rotation. Loosely install the lower tensioner pulley and make certain that the washer is in place.
15. Adjust the tensioner pulley counter-

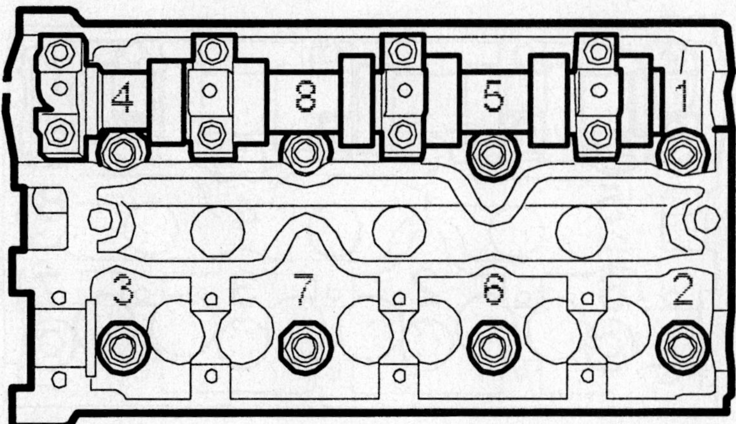

Remove the cylinder head bolts as shown

clockwise by hand and make certain that the markings are aligned properly.

16. Tighten the center adjuster roller bolts lightly. Adjust the lower adjuster roller counterclockwise until a tension of 275–300 Nm is reached. When properly tightened, remove the belt tension meter.

17. Make certain that the adjusting nut o0n the upper adjuster roller rotates freely. Adjust the tensioning pulley until the lines are aligned. Torque the fastener to 15 ft. lbs. (20 Nm).

18. Remove the locking tool from camshafts No. 1–2 and install tool KM-800-20. Adjust the upper roller counterclockwise with tool 83-94-983 until the markings on the camshaft sprocket are aligned with the markings on tool KM-800-20. Torque the fastener to 30 ft. lbs. (40 Nm).

19. Remove all locking tools.

20. Rotate the engine in the proper direction 2 revolutions until just before the **0** mark. Install tool KM-800-10 on the crankshaft.

21. Rotate the crankshaft in the proper direction until the arm lies against water pump and tighten it.

22. Install tool KM-800-20 and make certain that the markings on the camshaft sprocket are opposite the markings on the tool and that the edge of the belt is aligned with the edge of the sprocket.

23. Install or connect the following:

- Coolant bridge after lubricate the bolts and torque them to 22 ft. lbs. (30 Nm)
- Lower intake manifold with a new gasket and torque the bolts to 15 ft. lbs. (20 Nm)
- Middle intake manifold and torque the bolts to 15 ft. lbs. (20 Nm)
- Upper intake manifold with a new gasket and torque the bolts to 15 ft. lbs. (20 Nm)
- Fuel lines
- Ignition discharge module and fuel injector connectors
- Turbocharger delivery pipe to the throttle body
- Turbocharger delivery pipe to the charge air cooler and tighten all clamps
- Pressure/temperature sensor electrical connector
- Bypass pipe and valve
- Water hoses and vacuum hose to the throttle body
- Throttle body connector and attach the hoses to the purge valve. Torque the bypass pipe to 6 ft. lbs. (8 Nm).
- Vacuum hoses and pressure/temperature sensor connectors to the intake manifold
- Throttle cable
- Crankshaft pulley and torque the bolts to 15 ft. lbs. (20 Nm)
- Timing cover and torque the bolts to 6 ft. lbs. (8 Nm)
- Belt tensioner and alternator air intake
- Water pump and power steering pump pulleys
- V-belt and torque the water pump pulley bolts to 6 ft. lbs. (8 Nm) and the power steering pump pulley bolts to 15 ft. lbs. (20 Nm)
- Right side engine mount and bracket and torque the bolts to 46 ft. lbs. (63 Nm)
- Power steering hose into the holder
- MAF sensor and hose and remove the wedges
- Turbocharger bracket
- Battery and vent hose
- Lower engine cover
- Negative battery cable
- Engine cover

24. Fill the cooling system to the proper level.

25. Start the vehicle, check for leaks and repair if necessary.

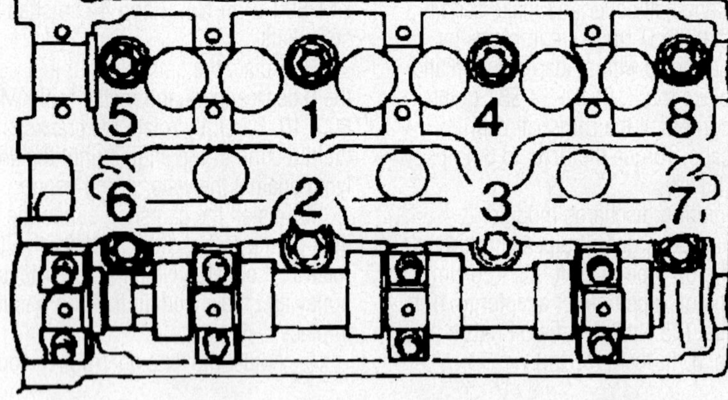

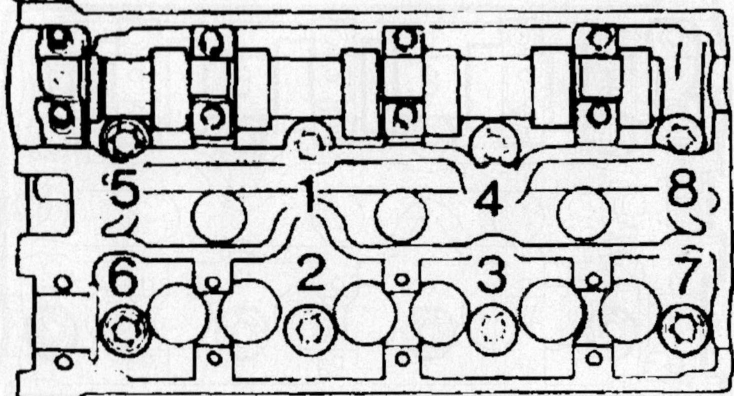

Cylinder head bolt torque sequence—6-cylinder engine

3.0L ENGINE—REAR CYLINDER HEAD

1. Before servicing the vehicle, refer to the Precautions Section.
2. Drain the cooling system.
3. Remove or disconnect the following:

- Engine and battery cover
- Lower engine cover and install wedges between the oil pan and subframe and the gearbox and subframe
- Mass Air Flow (MAF) sensor and hose
- Power steering hose from the holder
- Right side engine mount and bracket
- Water pump and power steering pump pulley bolts
- Tension from the V-belt
- Water pump and power steering pump pulleys
- Belt tensioner and alternator air intake
- Timing cover
- Throttle cable
- Vacuum hoses and pressure sensor

connector from the upper intake manifold
- Oxygen sensor connector from the holder on the rear of the SAI valve
- SAI pipe from the cylinder head and exhaust manifold connections
- Water hoses and vacuum hose from the throttle body
- Turbocharger delivery hose with the bypass pipe
- Throttle body connector and hose to the purge valve
- Fuel lines
- Upper intake manifold and move it aside
- Ignition discharge module and fuel injector connectors
- Middle intake manifold
- Lower intake manifold
- Coolant bridge
- Crankshaft pulley

➡Make certain that the markings on the camshaft sprockets and timing cover are aligned along with the crankshaft.

- Tensioner pulley and adjusting rollers. Mark the rotation of the belt
- Lower adjuster roller and washer
- Timing belt. Refer to the appropriate section for the removal and installation procedure.
- Bracket with the tensioning roller and the upper adjuster roller
- Ignition discharge module
- Camshaft cover with the O-rings. Rotate the crankshaft 60° BTDC
- Camshaft sprockets
- Inner timing cover
- Separate the exhaust pipes
- Exhaust camshaft
- Stuffing box
- Tappets and place them in the proper order for installation
- Cylinder head bolts in the proper sequence by loosening them ¼ turn then ½ turn

To install:

4. Clean all mating surfaces of any residual gasket material.
5. Make certain that the crankshaft is at 60° BTDC.
6. Be sure that the proper gasket is used and is marked **OPEN/TOP**
7. Install the tappets.
8. Install the cylinder head. Torque the new bolts in sequence as follows:
   a. 19 ft. lbs. (25 Nm).
   b. Plus an additional 90 degrees.
   c. Plus an additional 90 degrees.
   d. Plus an additional 90 degrees.
9. Install or connect the following:
   - Exhaust camshaft into position after lubricating it with clean engine oil. The locating pins for the intake camshaft should be in line with the inside bolts for the bearing caps. The guide pin for the exhaust camshaft should point straight up from the plane of the cylinder head
   - Stuffing box and bearing caps. Torque the bolts in ½ to 1 turn to 6 ft. lbs. (8 Nm).
   - Camshafts sprockets in relation to the locating pins. The number in the sprocket hub should match the number on the camshaft. Torque the bolt to 37 ft. lbs. (50 Nm) plus 60 degrees.
   - Rear timing cover
   - Camshaft cover and make certain that the O-rings are in place.
   - Lubricate with a soap solution and place tool (10), 81 52 381, in the corners at the camshaft bearing caps. Torque the bolts to 6 ft. lbs. (8 Nm).
   - Ignition discharge module
   - Hose to the SAI pipe and install the pipe with new gaskets to the exhaust manifold and cylinder head
   - Oxygen sensor connector in the holder
   - Front and rear exhaust pipes and torque the bolts to 30 ft. lbs. (40 Nm)
   - Water pump and torque the bolts to 19 ft. lbs. (25 Nm)
   - Bracket with the tensioning roller and the upper adjuster roller
10. Install locking tool KM-800-1 on No. 1 and 2 camshaft sprockets and tool KM-800-2 on No. 3 and 4 camshaft sprockets
11. Rotate the crankshaft to just before the **0** degree mark and install tool KM-800-10. Carefully rotate the crankshaft in the direction of the engine until the arm is lying against the water pump flange.
12. Install the camshaft belt to the markings on the belt and the marked direction of rotation. Loosely install the lower tensioner pulley and make certain that the washer is in place.
13. Adjust the tensioner pulley counterclockwise by hand and make certain that the markings are aligned properly.
14. Tighten the center adjuster roller bolts lightly. Adjust the lower adjuster roller counterclockwise until a tension of 275–300 Nm is reached. When properly tightened, remove the belt tension meter.
15. Make certain that the adjusting nut on the upper adjuster roller rotates freely. Adjust the tensioning pulley until the lines are aligned. Torque the fastener to 15 ft. lbs. (20 Nm).
16. Remove the locking tool from camshafts No. 1-2 and install tool KM-800-20. Adjust the upper roller counterclockwise with tool 83-94-983 until the markings on the camshaft sprocket are aligned with the markings on tool KM-800-20. Torque the fastener to 30 ft. lbs. (40 Nm).
17. Remove all locking tools.
18. Rotate the engine in the proper direction 2 revolutions until just before the **0** mark. Install tool KM-800-10 on the crankshaft.
19. Rotate the crankshaft in the proper direction until the arm lies against water pump and tighten it.
20. Install tool KM-800-20 and make certain that the markings on the camshaft sprocket are opposite the markings on the tool and that the edge of the belt is aligned with the edge of the sprocket.
21. Install or connect the following:
   - Coolant bridge after lubricating the bolts and torque them to 22 ft. lbs. (30 Nm)
   - Lower intake manifold with a new gasket and torque the bolts to 15 ft. lbs. (20 Nm)
   - Middle intake manifold and torque the bolts to 15 ft. lbs. (20 Nm)
   - Upper intake manifold with a new gasket and torque the bolts to 15 ft. lbs. (20 Nm)
   - Fuel lines
   - Ignition discharge module and fuel injector connectors
   - Turbocharger delivery pipe to the throttle body
   - Bypass pipe and valve
   - Water hoses and vacuum hose to the throttle body
   - Throttle body connector and attach the hoses to the purge valve and torque the bypass pipe to 6 ft. lbs. (8 Nm)
   - Vacuum hoses and pressure/temperature sensor connectors to the intake manifold
   - Throttle cable
   - Turbocharger delivery pipe to the charge air cooler
   - Vacuum hose and pressure/temperature sensor connector to the upper intake manifold
   - Turbocharger bypass pipe and valve
   - Crankcase pulley and torque the bolts to 15 ft. lbs. (20 Nm)
   - Timing cover and torque the bolts to 6 ft. lbs. (8 Nm)

- Belt tensioner and alternator air intake
- Water pump and power steering pump pulleys
- V-belt. Torque the water pump pulley bolts to 6 ft. lbs. (8 Nm) and the power steering pump pulley bolts to 15 ft. lbs. (20 Nm).
- Right side engine mount and bracket and torque the bolts to 46 ft. lbs. (63 Nm)
- Power steering hose into the holder
- MAF sensor and hose. Remove the wedges
- Turbocharger bracket
- Battery and vent hose
- Lower engine cover
- Negative battery cable
- Engine cover

22. Fill the cooling system to the proper level.
23. Start the vehicle, check for leaks and repair if necessary.

Turbocharger

REMOVAL & INSTALLATION

9-3

2002 MODEL

1. Before servicing the vehicle, refer to the Precautions Section.
2. Drain the cooling system.
3. Remove or disconnect the following:
    - Negative battery cable
    - Turbocharger stay and front exhaust pipe stay
    - Coolant pipe from turbocharger
    - Oil return pipe from the turbocharger (plug openings)
    - Oil pipe from the oil filter housing
    - Turbocharger outlet hose from charge air cooler (plug openings)
    - Hose between MAF sensor and turbocharger inlet pipe
    - Bypass pipe with valve (keep O-ring)
    - Exhaust manifold heat shield
    - Solenoid valve connector and hoses
    - Crankcase vent pipe from turbocharger inlet pipe and camshaft cover (move pipe aside)
    - EVAP purge hose from T-connector on solenoid valve
    - Engine lifting eye
    - Turbocharger inlet pipe
    - Top 2 nuts on joint between turbocharger and front exhaust pipe (move exhaust pipe aside)

⁂ CAUTION

DO NOT bend bellows on exhaust pipe more than 5 degrees or bellows will be permanently deformed.

- Wastegate (2 bolts and 1 clip)
- Oil pipe from turbocharger
- Transverse coolant pipe from cylinder head and from pressure sensor mounting
- Coolant hose from water pump and turbocharger
- Turbocharger-to-exhaust manifold nuts
- Turbocharger (lower from car)

To install:

4. Install or connect the following:
    - Turbocharger (from below) with a new gasket and torque the nuts to 18 ft. lbs. (24 Nm) and fill the turbocharger interchamber passage with engine oil

⁂ WARNING

It is very important that there is oil in the turbocharger when the engine is started to avoid damage to the unit.

- Coolant pipe with new banjo bolt and new gaskets to cylinder head and pressure sensor mounting; torque to 18 ft. lbs. (25 Nm)
- Turbocharger coolant pipe from water pump
- Wastegate (2 bolts and 1 clip)
- Exhaust pipe to joint mount with turbocharger; torque nuts, in alternating pattern, to 18 ft. lbs. (24 Nm)

⁂ CAUTION

DO NOT allow exhaust pipe flange to contact turbocharger flange.

- Turbocharger inlet pipe with O-rings; do not tighten V-clamp
- Hoses to solenoid valve
- Engine lifting eye
- Tighten inlet pipe V-clamp
- EVAP purge hose to T-connection
- Crankcase vent pipe to turbocharger inlet pipe with new sealing washers torqued to 18 ft. lbs. (24 Nm)
- Fasten crankcase vent pipe to camshaft cover
- Exhaust manifold heat shield
- Bypass pipe and valve, with new O-ring, to inlet pipe; torque to 6 ft. lbs. (8 Nm)

⁂ WARNING

To reduce risk of turbocharger hoses coming loose under pressure, be sure fittings and inside of hoses are cleaned before connections are made.

- Connecting hose between MAF sensor and turbocharger inlet pipe
- Solenoid valve connector
- Hose connecting the charge air cooler with the turbocharger
- Oil pipe from turbocharger to oil filter adapter, with new copper washers, torqued to 18 ft. lbs. (25 Nm)

➡Make sure oil pipe connection is straight when tightened.

- Oil return pipe from turbocharger to cylinder block
- Coolant pipe of turbocharger with new sealing washers and torque to 18 ft. lbs. (25 Nm)
- Bottom nut on joint between turbocharger and front exhaust pipe torqued to 18 ft. lbs. (25 Nm)
- Negative battery cable

5. Fill cooling system and bleed by starting engine (A/C off) and running at varying speeds until the cooling fan comes on. Turn engine off and top up coolant in expansion tank.
6. Check for leaks.

2003–06 MODELS (EXCEPT 2.8L ENGINE)

1. Before servicing the vehicle, refer to the Precautions Section.
2. Remove expansion tank cap to release any pressure.
3. Remove or disconnect the following:
    - Upper oxygen sensor and temperature sensor connectors
    - Crankcase vent hose from camshaft cover and turbocharger inlet (B207L)
    - Air cleaner housing and turbo inlet hose
    - Hose to crankcase ventilation and secondary air injection (SAI) hose (B205R)
    - Turbocharger heat shield
    - Solenoid valve hoses from turbocharger (B205R)
    - Pipe connection for crankcase ventilation (B205R)
    - Locking clip from turbocharger control arm
    - Vacuum unit from turbocharger

- Hose from turbocharger bypass (B205R)
- Upper oxygen sensor
- Catalytic converter connection from turbocharger
- Front spoiler shield
- Drain the cooling system (radiator and water pump)
- Coolant pipe from turbocharger and cylinder head (grip inner connection so it does not come off turbo)
- Oil pipe from turbocharger
- Lower pressure pipe from turbocharger
- Front exhaust pipe from catalytic converter
- Steering gear heat shield
- Catalytic converter heat shield
- Temperature sensor
- Lower oxygen sensor
- Catalytic converter retainer from engine block
- Catalytic converter
- Coolant pipe from turbocharger and thermostat housing
- Oil return pipe from turbocharger and engine block (collect O-ring)
- Stay between engine block and turbocharger (B205R)
- Oil pressure pipe from engine block
- Turbocharger from bottom of car

To install:

4. Install a new gasket for turbocharger.
5. Coat turbocharger stud bolts and exhaust manifold stud bolts with thread paste.
6. Raise and position turbocharger to engine. Tighten mounting stud bolts to 18 ft. lbs. (24 Nm).
7. Install or connect the following:
   - Oil return pipe to turbocharger, with new gasket and O-ring; torque to 11 ft. lbs. (15 Nm).
   - Coolant pipe between turbocharger and thermostat housing, with new sealing washers; torque to 15 ft. lbs. (20 Nm)
   - Bolt, with new sealing washers, for turbocharger oil delivery pipe on engine block; do not tighten
   - Stay between turbocharger and engine block torqued to 24 ft. lbs. (32 Nm) for bolt in turbocharger and 35 ft. lbs. (48 Nm) for M10 bolt
   - Fill turbocharger oil delivery connection with clean engine oil while turning turbine wheel
   - Oil delivery pipe to turbocharger and tighten connection, with new sealing washers; torque to 21 ft. lbs. (28 Nm)
   - Coolant pipe to turbocharger and pipe mounting on cylinder head
   - Hose to bypass valve
   - Oil delivery pipe connection on engine block (from under car); torque to 21 ft. lbs. (28 Nm)
8. Position the catalytic converter with new nuts on studs to 7 ft. lbs. (10 Nm). Ensure converter rests against turbocharger.
9. Raise the car and tighten the catalytic converter nuts to 18 ft. lbs. (25 Nm) and bolts to 16 ft. lbs. (22 Nm) a last time.
10. Lightly coat threads of lower oxygen sensor with thread paste and install sensor. Torque it to 30 ft. lbs. (40 Nm). Be sure oxygen sensor cables are not twisted or damaged.
11. Lightly coat threads of temperature sensor with thread paste and install sensor. Tighten to 33 ft. lbs. (45 Nm). Check to be sure all wiring from oxygen sensor and temperature sensor is not twisted or damaged.
12. Install or connect the following:
    - Catalytic converter heat shield
    - Steering gear heat shield
    - Catalytic converter and pipe to front muffler; torque joint of converter to 19 ft. lbs. (25 Nm) and joint clamp nuts to 30 ft. lbs. (40 Nm)
    - Turbocharger delivery pipe
    - Radiator drain cock
    - Front spoiler shield
    - Upper oxygen sensor (threads coated lightly with thread paste); torque sensor to 30 ft. lbs. (40 Nm)
    - Oxygen sensor and temperature sensor connectors
    - Vacuum unit to turbocharger
    - Control rod to control arm
    - Pipe for crankcase ventilation (B205R)
    - Control valve hoses to turbocharger
    - Turbocharger heat shield

⁂ WARNING

To reduce the risk that turbocharger hoses will come loose under pressure, thoroughly clean all fittings and inside of hoses before connecting.

- Air cleaner housing and turbocharger inlet hose
- Crankcase vent hose to camshaft cover and turbocharger and hose from solenoid valve
- Negative battery cable

13. Fill cooling system and bleed by starting engine (A/C off) and running at varying speeds until the cooling fan comes on. Turn engine off and top up coolant in expansion tank.
14. Check for leaks.

2.8L ENGINE

1. Before servicing the vehicle, refer to the Precautions Section.
2. Remove lower front cover and drain the cooling system.
3. Remove or disconnect the following:
   - Negative battery cable
   - Battery and engine covers
   - Battery cable
   - Expansion tank cap
4. Raise and safely support the vehicle.
5. Remove the chassis reinforcement, front sub-frame (convertible)
6. Remove the front nuts of the rear catalytic converter and move aside. Fix in position with 83 95 212 Strap.
7. Remove the mounting from the front catalytic converter and oil sump using 32 025 061 Crow foot wrench.
8. Lower the vehicle.
9. Remove the battery and battery cover bottom.
10. Detach the breather hoses from the coolant reservoir.
11. Detach the breather pipe from the turbocharger and heat shield.
12. Detach and raise the expansion tank slightly.
13. Detach the lower hose of the coolant reservoir, unplug the level sensor conenctor and lift away the coolant reservoir.
14. Remove the upper heat shield from the turbocharger.
15. Remove the bolt of the ventilation pipe mounting.
16. Unplug the upper oxygen sensor connector and remove the cable fixings. Move the cable up.
17. Carefully raise the heat shield slightly. Use an oxygen sensor socket to remove the oxygen sensor.
18. Remove the turbocharger's lower heat shield.
19. Remove the bolt of the catalytic converter mounting.
20. Remove the catalytic converter nuts to the turbocharger. Lubricate the studs and nuts with rust removal oil and carefully loosen.
21. Remove the upper radiator member (B284E).
22. Unplug the mass air flow sensor connector. Detach the secondary air pipe, the delivery hose from the turbocharger intake manifold and the crankcase ventilation pipes.
23. Detach the turbocharger intake mani-

fold from the air cleaner casing cover and the turbocharger.

24. Remove the bypass valve and unplug the connector.

25. Remove the upper radiator mounting.

26. Unplug the radiator fan connectors and detach the wiring harness from the fan cowling.

27. Remove the bolts of the lower turbocharger delivery pipe and the fan cowling.

28. Press the radiator assembly forward and remove the fan cowling.

29. Detach the intake manifold of the secondary air injection pump from the pump.

30. Detach the discharge pipe of the secondary air injection pump from the pump.

31. Unplug the intake air sensor connector (688).

32. Detach the upper turbocharger delivery pipe.

33. Remove the screw holding the dipstick tube.

34. Remove the heat shield of the front exhaust manifold. Lubricate the studs and nuts with rust removal oil and carefully loosen.

35. Detach the intermediate exhaust pipe from the turbocharger and the front exhaust manifold.

36. Detach the vacuum hoses and crankcase ventilation pipe from the turbocharger.

37. Detach the lower turbocharger delivery hose from the turbocharger.

38. Detach the oil delivery pipe from the engine.

39. Remove the oil return pipe from the turbocharger.

40. Remove the oil delivery pipe's banjo screws from the turbocharger.

41. Remove the coolant pipe's banjo screw from the turbocharger.

42. Remove the nuts of the rear intermediate exhaust pipe.

43. Remove the bolt holding the turbocharger to the bracket. Use an impact socket and a long puller.

44. Remove the turbocharger.

To install:

➡It is essential that there is oil in the turbocharger bearing housing when the engine is started, especially after a new turbocharger has been fitted.

45. Fill the turbocharger's oil inlet with oil from an oilcan. Rotate the compressor wheel by hand a number times to ensure that oil enters the bearings.

46. Position the turbocharger and install the bolt that holds the turbocharger to the bracket. Tighten the turbocharger-to-bracket bolt to 48 ft. lbs. (65 Nm).

47. Install the nuts of the rear intermediate exhaust pipe. Tighten alternately to 22 ft. lbs. (30 Nm). The conical side of the nut should face the centre exhaust pipe.

48. Install the coolant pipe's banjo screw to the turbocharger. Use new gaskets and tighten the banjo screw to 22 ft. lbs. (30 Nm).

49. Install the oil delivery pipe's banjo screws to the turbochager using new gaskets. Tighten the banjo screws to 22 ft. lbs. (30 Nm).

50. Connect the oil return pipe to the turbochager using new gaskets. Tighten the fasteners to 10 ft. lbs. (14 Nm).

51. Attach the oil delivery pipe to the engine using new gaskets. Tighten the fasteners to 22 ft. lbs. (30 Nm).

52. To reduce the risk of hoses mounted on the delivery side of the turbocharger coming loose due to low friction at high air pressure, the hoses and connecting pieces must be cleaned thoroughly before fitting. Use a rag dampened with 93 160 907 Motip Dupli cleaning agent to wipe clean inside the ends of the hoses. Clean the connecting pieces as well. If hose clips are rusty or damaged, they must be replaced so the correct clamping force is maintained.

53. Attach the lower turbocharger delivery hose to the turbocharger.

54. Attach the vacuum hoses and crankcase ventilation pipes to the turbocharger.

55. Attach the intermediate exhaust pipe to the front exhaust manifold and the turbocharger. Use a new gasket and coat the studs with 90 513 210 Universal paste. The conical side of the nut should face the centre exhaust pipe.

56. Tighten the nuts first on the exhaust manifold and then on the turbocharger to 22 ft. lbs. (30 Nm). Tightening order 11a, 11b, 11c, and 11d alternatingly on the turbocharger. The surfaces should align with one another when the nuts are fitted. Check using a feeler gauge. If the surfaces are not flat or the pipe is deformed, the centre exhaust pipe and/or exhaust manifold must be replaced.

57. Install the heat shield of the front exhaust manifold and tighten the fasteners to 7 ft. lbs. (10 Nm).

58. Install the screw holding the dipstick tube and tighten to 7 ft. lbs. (10 Nm).

59. Connect the upper turbocharger delivery pipe.

60. Plug in the intake air sensor connector (688).

61. Connect the discharge pipe of the secondary air injection pump.

62. Connect the secondary air injection pump's intake manifold to the pump.

63. Press the radiator assembly forward and position the fan cowling.

64. Install the bolts of the lower turbocharger delivery pipe and the fan cowling.

65. Plug in the radiator fan connectors and attach the wiring harness to the fan cowling.

66. Fit the upper radiator mounting.

67. Fit the bypass valve and plug in the connector.

68. Attach the turbocharger intake manifold to the turbocharger and the air cleaner casing cover.

69. Plug in the mass air flow sensor connector. Attach the secondary air pipe, the delivery hose to the turbocharger intake manifold and the crankcase ventilation pipes.

70. Install the upper radiator member (B284E).

71. Install the catalytic converter nuts to the turbocharger. Coat the studs with 90 513 210 Universal paste. When installing the front catalytic converter to the turbocharger, tighten the nuts in a crosswise pattern to avoid tightening the flange crooked. Tighten the nuts to 18 ft. lbs. (25 Nm).

72. Install the bolt of the catalytic converter mounting.

73. Install the turbocharger's lower heat shield.

74. Lightly coat the threads of the front oxygen sensor with 90 513 210 Universal paste and install using the proper socket. Tighten the oxygen sensor to 33 ft. lbs. (45 Nm).

75. Carefully press down the heat shield.

76. Secure the cable and plug in the connector.

77. Fit the bolt of the ventilation pipe mounting.

78. Fit the upper heat shield to the turbocharger.

79. Attach the lower hose of the coolant reservoir and plug in the level sensor connector.

80. Install the coolant reservoir in the mounting.

81. Attach the breather pipe to the turbocharger and heat shield. Use a new gasket. Tighten the banjo screw to 22 ft. lbs. (30 Nm).

82. Attach the breather hoses to the coolant reservoir.

83. Install battery cover bottom.

84. Install the battery.

85. Raise and safely support the vehicle.

86. Install the mounting on the front catalytic converter and oil sump.
87. Install the rear catalytic converter using a new gasket. Tighten the bolts to 18 ft. lbs. (25 Nm).
88. Install the chassis reinforcement, front sub-frame (convertible).
89. Lower the vehicle.
90. Refill the cooling system. Bleed and pressure test the cooling system.
91. Check the oil level and top off as necessary.
92. Connect the negative battery cable.
93. Install the battery cover and the engine cover.
94. Remove the wing covers and restore car electrical functions.
95. Start the engine and check for any exhaust, oil and coolant leakage.

9-5

2.3L ENGINE

1. Before servicing the vehicle, refer to the Precautions Section.
2. Remove lower front cover and drain the cooling system.
3. Remove or disconnect the following:
   - Negative battery cable
   - Turbocharger stays
   - Loosen the oil return pipe from the turbocharger to the engine
   - Oil pipe between the oil filter adapter and the turbocharger
   - Bypass pipe and valve
   - Control valve connector
   - Mass Air Flow (MAF) sensor electrical connector
   - Loosen hoses to turbocharger and diaphragm housing
   - Crankcase breather pipe
   - Engine lifting eye
   - MAF sensor and air hose
   - Exhaust manifold heat shield
   - Quick-release coupling on the vent hose
   - Intake manifold V-clamp
   - Intake manifold
   - Hose between the turbocharger and the charge air cooler
   - Exhaust system from the turbocharger
   - Oil pipe from the oil filter adapter and copper washers
   - Coolant pipe to the turbocharger
   - Coolant return pipe from the cylinder head and the pressure sensor bracket
   - Turbocharger

To install:

4. Install or connect the following:
   - Turbocharger and torque the 2 upper bolts to 18 ft. lbs. (24 Nm)
   - Coolant return pipe with new copper washers and torque the nipple to 26 ft. lbs. (35 Nm), the banjo bolt to 19 ft. lbs. (25 Nm), the coolant return pipe to 19 ft. lbs. (25 Nm) and the coolant return pipe clamp to 7 ft. lbs. (9 Nm)
   - Turbocharger oil pipe with new copper washers and torque the banjo bolt to 15 ft. lbs. (20 Nm)
   - Turbocharger intake pipe with a new O-ring and loosen the adjusting screw on the intake manifold
   - Torque the V clamp between the turbocharger to intake pipe to 30 inch lbs. (4 Nm) and the intake manifold and adjusting screw to 18 ft. lbs. (24 Nm).
   - Hoses to the turbocharger and diaphragm housing
   - Charge air cooler inlet hose and torque the bolts to 72 inch lbs. (8 Nm)
   - Heat shield and torque the bolts to 15 ft. lbs. (20 Nm)
   - MAF sensor and air hose
   - Engine lifting eye
   - Crankcase breather pipe and torque the bolts to 18 ft. lbs. (24 Nm)
   - Charge air bypass valve and pipe with a new O-ring and torque the bolts to 6 ft. lbs. (8 Nm)
   - Lower nut to the turbo charger and torque the stud to 16 ft. lbs. (22 Nm) and the lock nut to 18 ft. lbs. (24 Nm)
   - Turbocharger oil pipe with new copper washers and torque the bolt to 19 ft. lbs. (25 Nm)
   - Turbocharger stays and torque to 18 ft. lbs. (24 Nm)
   - Oil return pipe between the turbocharger and engine and torque the bolts to 10 ft. lbs. (15 Nm)
   - Lower engine cover
   - Negative battery cable
5. Fill the engine with coolant.
6. Start the vehicle and run at varying speeds until radiator fan comes on. Stop the engine and top up the coolant and engine oil as needed.
7. Check for leaks and repair if necessary.

3.0L ENGINE

1. Before servicing the vehicle, refer to the Precautions Section.
2. Remove the upper engine cover.
3. Drain the cooling system.
4. Remove or disconnect the following:
   - Negative battery cable

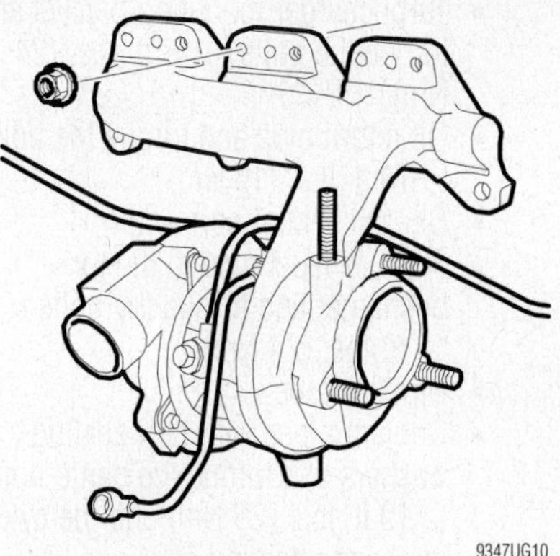

Remove the turbocharger and exhaust manifold as an assembly

   - Turbocharger pressure hose with the bypass pipe and valve (plug opening)
   - Exhaust manifold heat shield
   - Crankcase breather pipe (bend pipe aside)
   - Turbocharger intake pipe (bend pipe aside)
   - Turbocharger outlet hose to the charge air cooler
   - Radiator fan and clamp off the coolant hoses to the turbocharger
   - Coolant pipes
   - Dipstick tube
   - Front pipe from the turbocharger
   - Oxygen sensor connector
   - Separate front and rear exhaust systems
   - Front pipe from the rear exhaust manifold
   - Turbocharger oil return pipe
   - Turbocharger mounting bracket
   - Oil pipe from oil filter adapter
   - Front exhaust manifold and turbocharger as an assembly

To install:

5. Fill the turbo inlet with oil. Rotate the compressor wheel by hand several times to assure the bearings are well lubricated.
6. Install or connect the following:
   - Turbocharger and exhaust manifold with a new gasket and torque the bolts to 15 ft. lbs. (20 Nm)
   - Front pipe to the exhaust manifold with a new gasket and torque the bolts to 30 ft. lbs. (40 Nm)
   - Pipe joint between the front and rear exhaust systems and torque the bolts to 16 ft. lbs. (22 Nm)
   - Oxygen sensor wiring harness (secure with cable tie)
   - Oil pipe to the oil filter adapter with new gaskets and sealing washers and torque the bolts to 15 ft. lbs. (20 Nm)

- Turbocharger mounting bracket and torque the bolts to 18 ft. lbs. (24 Nm)
- Oil return pipe and torque the bolts to 10 ft. lbs. (15 Nm)
- Oxygen sensor connector
- Front exhaust pipe to the turbocharger and torque the bolts to 18 ft. lbs. (24 Nm)
- Dipstick tube
- Coolant pipes with new sealing washers and torque the banjo bolt to 19 ft. lbs. (25 Nm) and the cylinder head retaining screw to 10 ft. lbs. (15 Nm)
- Radiator fan and unclamp the coolant hoses
- Hose between the turbocharger and charge air cooler and torque the fasteners to 72 inch lbs. (8 Nm)
- Turbocharger intake pipe and torque V-clamp to 24 inch lbs. (3 Nm) and U-clamp to 10 ft. lbs. (15 Nm)
- Crankcase breather pipe and torque the banjo bolt to 18 ft. lbs. (24 Nm)
- Exhaust manifold heat shield and torque stud to 16 ft. lbs. (22 Nm) and lock nut to 18 ft. lbs. (24 Nm)
- Turbocharger pressure hose with the bypass pipe and valve and torque to 72 inch lbs. (8 Nm)
- Negative battery cable
- Engine upper cover

7. Fill the cooling system to the proper level.
8. Start the vehicle, check for leaks and repair if necessary.

Intake Manifold

REMOVAL & INSTALLATION

2.0L and 2.3L Engines

CAUTION

The fuel injection system remains under pressure, even after the engine has been turned OFF. The fuel system pressure must be relieved before disconnecting any fuel lines. Failure to do so may result in fire or personal injury.

1. Before servicing the vehicle, refer to the Precautions Section.
2. Drain the cooling system.
3. Remove or disconnect the following:
   - Negative battery cable
   - Rubber elbow running between the throttle housing and the turbocharger, if equipped
   - Throttle position sensor connector
   - Hoses at the throttle housing
   - Throttle housing
   - Oil filler pipe bracket at the manifold. Position it out of the way
   - All hoses and lines attached to the manifold. Label them prior to removal
   - AIC valve
   - Fuel line from the pressure regulator
   - Banjo fitting connecting the fuel line to the fuel rail
   - Fuel line and pulsator
   - Each fuel injector electrical lead
   - Temperature sensor
   - Ground wires at the manifold
   - Harness assembly from underneath the manifold
   - EGR pipe and all connectors
   - Intake manifold

To install:

4. Scrape off any excess gasket material
5. Install or connect the following;
   - Intake manifold with a new gasket and torque the bolts in a crisscross pattern to 16 ft. lbs. (22 Nm)
   - Wire harness
   - EGR pipe
   - Ground wires
   - Temperature sensor
   - Injector leads
   - Fuel line to the pressure regulator
   - Fuel line/pulsator to the fuel rail. Secure it with a plastic tie
   - Oil filler pipe bracket to the manifold
   - AIC valve
   - Throttle housing
   - Rubber elbow between the turbocharger and the intake manifold, if equipped
   - Negative battery cable
6. Fill the cooling system.
7. Start the vehicle, check for leaks and repair if necessary.

2.8L Engine

CAUTION

The fuel injection system remains under pressure, even after the engine has been turned OFF. The fuel system pressure must be relieved before disconnecting any fuel lines. Failure to do so may result in fire or personal injury.

1. Before servicing the vehicle, refer to the Precautions Section.
2. Drain the cooling system.
3. Remove or disconnect the following:
   - Negative battery cable
   - Engine cover
   - Power steering fluid reservoir bolt and place reservoir aside
   - Connector of intake air sensor (688)
   - Upper charge air pipe
   - Throttle body
   - Ground cable of the control module
   - Engine control module connectors (608A and 608B)
   - Engine breather pipe
   - Cable duct and move it aside
   - Atmospheric pressure sensor bracket
   - Injectors' wiring harness bracket
   - Brake servo pipe from intake manifold
   - Crankcase ventilation pipes from intake manifold
   - Upper intake manifold bolts
   - Upper intake manifold and gaskets
   - Injectors' wiring harness from engine

➡Plug the intake ducts and blow the intake manifold clean.

   - Fuel pressure sensor connector.
4. Release any fuel pressure by pressing in the needle valve on the fuel rail. Catch any fuel spill with a cloth.
   - Fuel line from fuel rail using 83 95 261 Fuel line tool
   - Lower intake manifold bolts
   - Lower intake manifold and gaskets

To install:

5. Plug the ducts and clean the sealing surfaces. Remove the plugs.
6. Install new gaskets and position the lower intake manifold. Install and tighten the bolts to 17 ft. lbs. (23 Nm).
7. Remove or disconnect the following:
   - Fuel line to fuel rail
   - Fuel pressure sensor connector
   - Injectors' wiring harness to the engine
8. Clean the sealing surfaces and fit new gaskets. Carefully position the upper intake manifold.
9. Install and tighten the upper intake manifold bolts to 17 ft. lbs. (23 Nm).
   - Crankcase ventilation pipes to intake manifold and fit bolt
   - Brake servo pipe to intake manifold
   - Injectors' wiring harness bracket
   - Atmospheric pressure sensor bracket
   - Cable duct.
   - Engine breather pipe

- Engine control module connectors (608A and 608B)
- Ground cable of the control module
- Throttle body. Tighten throttle body bolts to 7 ft. lbs. (9 Nm).
- Charge air pipe
- Intake air sensor (688)
- Power steering fluid reservoir and fit bolt
- Engine cover
- Negative battery cable

10. Fill the cooling system.
11. Start the vehicle, check for leaks and repair if necessary.

3.0L Engine

1. Before servicing the vehicle, refer to the Precautions Section.
2. Remove or disconnect the following:
   - Negative battery cable
   - Engine covers
   - Cruise control cable
   - Throttle cable
   - Control rod
   - Intake tube from the Mass Air Flow (MAF) sensor assembly
   - Vacuum hoses and electrical connectors from MAF assembly
   - MAF assembly
   - Hoses and wiring from the upper intake manifold. Label prior to removal.
   - Upper intake manifold
   - Idle Air Control (IAC) valve
   - Fuel pressure regulator hose
   - Harness from under the throttle plate housing
   - Injector electrical harness
   - Camshaft Position (CMP) sensor connector
   - Fuel line. Plug to prevent contamination
   - Lower intake manifold

To install:

3. Install or connect the following:
   - Lower intake manifold with new gaskets and torque the bolts to 15 ft. lbs. (20 Nm)
   - Fuel lines
   - Injector wiring
   - CMP sensor
   - Upper intake manifold with new gaskets
   - Throttle plate housing and torque the bolts to 15 ft. lbs. (20 Nm)
   - Wiring and hoses
   - IAC valve
   - Hose to the fuel pressure regulator
   - MAF assembly and hoses
   - Throttle control rod
   - Cruise control cable
   - Engine covers
   - Negative battery cable

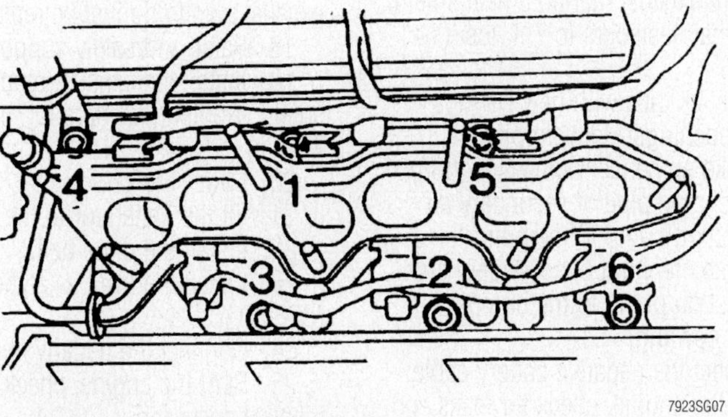

Intake manifold bolt torque sequence—3.0L engine

Exhaust Manifold

REMOVAL & INSTALLATION

2.0L Engine

1. Before servicing the vehicle, refer to the Precautions Section.
2. Drain the cooling system.
3. Remove or disconnect the following:
   - Negative battery cable
   - Turbocharger and relieve tension on the belt tensioner
   - Drive belt
   - Steering servo pump and bracket
   - Exhaust manifold and discard the gasket
4. Clean all mating surfaces of any residual gasket material.

To install:

5. Install or connect the following:
   - Exhaust manifold with a new gasket and torque the bolts to 19 ft. lbs. (25 Nm)
   - Steering servo pump and bracket
   - Drive belt
   - Turbocharger
   - Negative battery cable
6. Fill the cooling system.
7. Start the vehicle, check for leaks and repair if necessary.

2.3L Engine

1. Before servicing the vehicle, refer to the Precautions Section.
2. Drain the cooling system.
3. Remove or disconnect the following:
   - Negative battery cable
   - Lower front cover
   - Turbocharger-to-block mounting bolt
   - Turbocharger upper mounting bolt
   - Turbocharger coolant pipe
   - Turbocharger return hose
   - Oil pipe from the turbocharger to the oil filter adapter
   - Bypass pipe and valve
   - Exhaust manifold heat shield
   - Solenoid connector
   - MAF sensor-to-turbocharger connecting hose
   - Crankcase breather pipe (bend pipe aside)
   - Solenoid valve hoses (note positions)
   - Engine lifting eye
   - Turbocharger intake pipe from V-clamp (plug turbocharger inlet opening)
   - 2 upper nuts between turbocharger and front exhaust pipe (move exhaust aside)
   - Turbocharger coolant return pipe from cylinder head and pressure sensor mount
   - Wastegate from turbocharger
   - Nuts holding turbocharger to exhaust manifold (slightly lower turbocharger)
   - Relieve auxiliary drive belt tensioner
   - Auxiliary drive belt from power steering pump pulley
   - Through-bolts, lower bolt and 2 upper bolts holding power steering pump (lift pump and bracket aside)
   - Exhaust manifold

To install:

4. Install or connect the following:
   - Exhaust manifold with a new gasket and torque the bolts to 19 ft. lbs. (25 Nm)
   - Power steering pump and bracket. Torque the upper left mounting bolt to 18 ft. lbs. (24 Nm) first then torque the remaining bolts to the same specification
   - Auxiliary drive belt to power steering pump pulley and tighten it with the tensioner

➡Fill the turbo inlet with oil. Rotate the compressor wheel by hand several times to assure the bearings are well lubricated.

- Turbocharger to the exhaust manifold and torque the lock nuts to 18 ft. lbs. (24 Nm) and the studs to 16 ft. lbs. (22 Nm)
- Front exhaust pipe to the turbocharger and torque the bolts to 18 ft. lbs. (24 Nm)
- Wastegate
- Coolant return pipe to turbocharger with new copper washers and torque the bolts to 19 ft. lbs. (25 Nm)
- Coolant return pipe to cylinder head and torque bolts to 19 ft. lbs. (25 Nm)
- Turbocharger intake pipe with a new lubricated O-ring to the power steering pump bracket. Torque the bolt and adjusting screw to 18 ft. lbs. (24 Nm).
- Coolant pipe to turbocharger, with new sealing washers, and torque to 19 ft. lbs. (25 Nm)
- Oil pipe from turbo to oil filter adapter, with new washers, and torque to 19 ft. lbs. (25 Nm)
- Turbocharger-to-block upper bolt then remaining bolt torqued to 18 ft. lbs. (24 Nm)
- Lower nut on joint between turbocharger and exhaust pipe
- Intake pipe with lifting eye to turbocharger
- Solenoid valve hoses
- Solenoid valve connector
- Crankcase-to-camshaft cover and turbo intake pipe and torque the bolts to 18 ft. lbs. (24 Nm)
- MAF sensor-to-turbocharger intake pipe hose
- Bypass hose and valve with new O-ring
- Exhaust manifold heat shield
- Negative battery cable

5. Fill the cooling system and bleed by running engine (A/C off), at varying speeds, until cooling fan comes on. Stop the engine and top up the coolant.
6. Check and adjust level of engine oil
7. Start the vehicle, check for leaks and repair if necessary.

2.8L Engine

FRONT

1. Before servicing the vehicle, refer to the Precautions Section.
2. Disconnect the negative battery cable.
3. Remove or disconnect the following:
   - Fan cowling
   - Intake manifold of secondary air injection pump from pump
   - Discharge pipe of secondary air injection pump from pump
   - Intake air sensor connector (688)
   - Turbocharger delivery pipe
   - Dipstick tube
   - Front exhaust manifold heat shield
   - Intermediate exhaust pipe from turbocharger and front exhaust manifold
   - Secondary air pipe bolts to raise it slightly
4. Lubricate the studs and nuts with rust removal oil and carefully loosen.
5. Remove the exhaust manifold.

To install:

6. Clean the contact surfaces from gasket residue and soot.
7. Install the exhaust manifold using new bolts and a new gasket. Tighten the bolts to 15 ft. lbs. (20 Nm).
8. Install the secondary air pipe using a new gasket and tighten the bolts to 17 ft. lbs. (23 Nm).
9. Connect the intermediate exhaust pipe to the front exhaust manifold and the turbocharger. Use a new gasket and coat the studs with 90 513 210 Universal paste. Install all nuts.

➡The conical side of the nut should face the centre exhaust pipe.

10. Tighten the mounting nuts first on the exhaust manifold and then on the turbocharger to 22 ft. lbs. (30 Nm). Tighten in an alternating fashion on the turbocharger. The surfaces should align with one another when the nuts are fitted. Check using a feeler gauge. If the surfaces are not flat or the pipe is deformed, the centre exhaust pipe and/or exhaust manifold must be replaced.
11. Install or connect the following:
    - Front exhaust manifold heat shield. Tighten fasteners to 7 ft. lbs. (10 Nm).
    - Dipstick tube with new O-rings
    - Turbocharger delivery pipe
    - Intake air sensor connector (688)
    - Discharge pipe of secondary air injection pump to the pump
    - Intake manifold of secondary air injection pump to the pump
    - Fan cowling
12. Connect the negative battery cable.
13. Start the engine, check for leaks and repair if necessary.

REAR

1. Before servicing the vehicle, refer to the Precautions Section.
2. Disconnect the negative battery cable.
3. Remove or disconnect the following:
   - Front catalytic converter (B284E)
   - Drive belt
   - Alternator
4. Raise and safely support the vehicle.
5. Remove the exhaust manifold heat shield.
6. Lower the vehicle.
7. Remove the intermediate exhaust pipe.
8. Raise and safely support the vehicle.
9. Detach the rear secondary air pipe between the exhaust manifold and the check valve.
10. Remove the exhaust manifold.

To install:

11. Clean the contact surfaces from gasket residue and soot.
12. Install new gaskets. Coat the studs with 90 513 210 Universal paste.
13. Install the exhaust manifold using new bolts and a new gasket. Tighten the bolts to 15 ft. lbs. (20 Nm).
14. Attach the rear secondary air pipe between the exhaust manifold and the check valve.
15. Lower the vehicle.
16. Connect the intermediate exhaust pipe to the front exhaust manifold and the turbocharger. Use a new gasket and coat the studs with 90 513 210 Universal paste. Install all nuts.

➡The conical side of the nut should face the centre exhaust pipe.

17. Tighten the mounting nuts first on the exhaust manifold and then on the turbocharger to 22 ft. lbs. (30 Nm). Tighten in an alternating fashion on the turbocharger. The surfaces should align with one another when the nuts are fitted. Check using a feeler gauge. If the surfaces are not flat or the pipe is deformed, the centre exhaust pipe and/or exhaust manifold must be replaced.
18. Raise and safely support the vehicle.
19. Install the exhaust manifold heat shield. Tighten fasteners to 7 ft. lbs. (10 Nm).
20. Lower the vehicle.
21. Fit the generator to the bracket.
22. Install the drive belt.
23. Install the front catalytic converter (B284E).
24. Connect the negative battery cable.
25. Start the engine, check for leaks and repair if necessary.

3.0L Engine

FRONT

1. Before servicing the vehicle, refer to the Precautions Section.
2. Remove or disconnect the following:
   - Upper engine cover
   - Turbocharger bypass pipe and valve
   - Temperature/pressure sensor connector
   - Turbocharger delivery pipe from the throttle body
   - Turbocharger delivery pipe from the charge air cooler (plug charge air opening)
   - Exhaust manifold heat shield
   - SAI valve hose and nut on valve bracket
   - Battery and vent hose
   - Front SAI pipe bolt and pipe from exhaust manifold
   - Crankcase ventilation pipe (between turbo inlet pipe and cylinder head)
   - Oil trap hose
   - Turbocharger intake pipe
   - Turbocharger outlet hose from charge air cooler
   - Fan motor connectors
   - Radiator fan
   - Pinch off inlet and return water hoses to turbocharger
   - Turbocharger water pipes
   - Dipstick tube
   - Front pipe from the turbocharger
   - Oxygen sensor connectors (cut cable ties, note routing of sensor wiring)
   - Separate pipe joint between front and rear exhaust systems
   - Front pipe from rear exhaust manifold (lower pipe away from front exhaust)
   - Turbocharger oil return pipe
   - Turbocharger mounting bracket
   - Oil pipe from the oil filter adapter
   - Front exhaust manifold with the turbocharger and discard the gasket
3. Clean all mating surfaces of any residual gasket material.

To install:

4. Install or connect the following:
   - Front exhaust manifold assembly with a new gasket and torque the bolts to 15 ft. lbs. (20 Nm)
   - Front exhaust pipe to rear exhaust manifold with a new gasket and torque the bolts to 30 ft. lbs. (40 Nm)
   - Pipe joint between front and rear exhaust systems; torque to 16 ft. lbs. (22 Nm)
   - Secure oxygen sensor wiring with cable ties (note original routing of wiring)
   - Oil pipe to the oil filter adapter and torque the fastener to 15 ft. lbs. (20 Nm)
   - Turbocharger mounting bracket and torque the bolts to 18 ft. lbs. (25 Nm)
   - Turbocharger oil return pipe and torque the bolt to 10 ft. lbs. (15 Nm)
   - Oxygen sensor connector
   - Front exhaust pipe to the turbocharger and torque the bolts to 18 ft. lbs. (25 Nm)
   - Dipstick tube
   - Turbocharger water return pipe and torque the banjo bolt to 18 ft. lbs. (25 Nm) and the retaining screw to 10 ft. lbs. (15 Nm)
   - Remove pinching pliers from turbocharger water hoses
   - Radiator fan; torque the bolts to 72 inch lbs. (8 Nm)
   - Radiator fan connectors
   - Turbocharger-to-charge air cooler hose and torque to 6 ft. lbs. (8 Nm)
   - Turbocharger intake pipe and torque U-clamp to 24 inch lbs. (3 Nm) and V-clamp to 10 ft. lbs. (15 Nm)
   - MAF sensor
   - Crankcase ventilation pipe (between cylinder head and turbo inlet pipe)
   - Oil trap hose
   - SAI pipe and valve
   - Turbocharger delivery pipe to the throttle body and charge air cooler
   - Turbocharger delivery pipe to charge air cooler and throttle body
   - Temperature/pressure sensor connector
   - Exhaust manifold heat shield and torque the bolts to 18 ft. lbs. (25 Nm)
   - Turbocharger by pass pipe and valve and torque to 72 inch lbs. (8 Nm)
   - Battery and vent hose
   - Upper engine cover
   - Negative battery cable
5. Start the engine, check for leaks and repair if necessary.

REAR

1. Before servicing the vehicle, refer to the Precautions Section.
2. Remove or disconnect the following:
   - Negative battery cable
   - Upper engine cover
   - Air intake to alternator
   - Oxygen sensor connectors from holder on rear SAI valve
   - SAI pipe from cylinder head and connections from exhaust manifold
   - SAI valve and pipe
   - Outer nuts from the rear exhaust manifold
   - Front exhaust pipe from rear exhaust
   - 2 lower nuts from the rear exhaust manifold
   - 3 upper nuts from the rear exhaust manifold
   - Rear exhaust manifold and discard the gasket
3. Clean all mating surfaces of any residual gasket material.

To install:

4. Install or connect the following:
   - Rear exhaust manifold with a new gasket and torque all the bolts to 15 ft. lbs. (20 Nm)
   - Front exhaust pipe and torque the turbocharger locknuts to 18 ft. lbs. (25 Nm) and rear exhaust manifold bolts to 30 ft. lbs. (40 Nm)
   - SAI pipe and valve
   - Oxygen sensor connectors to SAI holder
   - Air intake to the alternator
   - Upper engine cover
   - Negative battery cable
5. Start the vehicle, check for leaks and repair if necessary.

Front Crankshaft Seal

REMOVAL & INSTALLATION

2.0L Engine

1. Before servicing the vehicle, refer to the Precautions Section.
2. Remove or disconnect the following:
   - Negative battery cable
   - Right front wheel and inner cover
   - Drive belt tension and belt from crankshaft pulley
   - Protective plate and install flywheel locking tool, 83 94 868, on the ring gear
   - Crankshaft pulley
   - Front crankshaft seal

To install:

3. Install or connect the following:
   - New crankshaft seal using special tool 83 94 876
   - Crankshaft pulley and torque to 130 ft. lbs. (175 Nm)

- Drive belt on crankshaft pulley and reset tensioner
- Inner cover and right front wheel
- Negative battery cable

2.3L Engine

2002 MODEL

1. Before servicing the vehicle, refer to the Precautions Section.
2. Remove or disconnect the following:
   - Negative battery cable
   - Power steering pump pipe from subframe
   - Lower spoiler sections
   - Insert a wedge between pan and sub frame
   - Right front engine attachment and mount
   - Belt tension
   - Flywheel cover plate and install flywheel locking tool 83 94 868
   - Crankshaft pulley
3. Remove the crankshaft seal.

To install:

4. Install the crankshaft seal.
5. Install or connect the following:
   - Crankshaft pulley and torque the bolt to 130 ft. lbs. (175 Nm)
   - Remove flywheel locking tool
   - Flywheel cover plate
6. Drive in wedge at subframe to raise engine slightly, then install auxiliary drive belt to all pulleys and restore belt tension.
7. Install or connect the following:
   - Right side engine attachment and mount and torque to 37 ft. lbs. (50 Nm)
   - Wedge from subframe
   - Install power steering pump pipe to subframe
   - Belt circuit cover
   - Lower spoiler sections
   - Negative battery cable

2003–06 MODELS

1. Before servicing the vehicle, refer to the Precautions Section.
2. Remove or disconnect the following:
   - Right front wheel
   - Right fenderwell liner
   - Auxiliary drive belt (mark direction of rotation)
   - Crankshaft pulley, using holding tools 83 95 360 (shaft only) and 83 96 210
   - Crankshaft seal, using seal remover 87 91 360

To install:

3. Clean all mating and sealing surfaces
4. Position the protective collar, 83 96 202, onto the crankshaft. Lube the new sealing ring with non-acidic Vaseline and position it on the fitting tool
5. Position the fitting tool on the crankshaft. Screw in the sealing ring, using the crankshaft pulley bolt until it is flush with timing cover. Remove the tool.
6. Install the crankshaft pulley, with a new bolt. Use pulley holding tool 83 95 360 and 83 96 210 to restrain the pulley. Pry down the engine slightly and insert the bolt. Torque the pulley bolt to 74 ft. lbs. (100 Nm), plus an additional 75 degrees. Bend the metal lug back into place.
7. Install or connect the following:
   - Auxiliary drive belt
   - Restore tensioner position
   - Inner fenderwell liner
   - Front wheel; torque lug nuts to 81 ft. lbs. (110 Nm)
   - Check oil level

2.8L ENGINE

1. Before servicing the vehicle, refer to the Precautions Section.
2. Disconnect the negative battery cable.
3. Raise and safely support the vehicle.
4. Remove the right front wheel.
5. Remove the right wing liner.
6. Mark the belt's direction of rotation. Unload the belt tensioner. Use a ´ inch puller and remove the poly-V-belt from the crankshaft pulley.
7. Remove the bolt holding the crankshaft pulley using 83 95 360 Holding tool, crankshaft pulley (handle only) and EN-47981 Holding tool, pulley.

➡The pulley is pressed onto the crankshaft.

8. Remove the pulley using EN-47982 Puller and EN-47981 Holding tool, pulley.
9. Remove the front crankshaft seal using 87 91 360 Extractor.

To install:

10. Clean the sealing surfaces.
11. Lubricate the new seal with non-acidic Vaseline.
12. Install the seal using EN-47635 Fitting tool and a plastic mallet.
13. Clean the sealing surface of the pulley.
14. Press the pulley onto the crankshaft using KM-J 41998 Fitting tool.
15. Install the bolt using 83 95 360 Holding tool, crankshaft pulley (handle only) and EN-47981 Holding tool, pulley. Tighten the crankshaft pulley bolt to 74 ft. lbs. (100 Nm), plus an additional 150°.
16. Install the drive belt.
17. Install the right wing liner.
18. Install the right front wheel.
19. Lower the vehicle.
20. Connect the negative battery cable.

3.0L Engine

1. Before servicing the vehicle, refer to the Precautions Section.
2. Remove the lower spoiler sections.
3. Install 2 wedges between the oil pan and subframe and between the gearbox and subframe.
4. Remove or disconnect the following:
   - MAF sensor with hose
   - Power steering hose from holder
   - Right side engine attachment with bracket and engine mount
   - Auxiliary drive belt
   - Water pump pulley and power steering pump pulley
   - Belt tensioner
   - Timing cover
   - Crankshaft pulley
5. Zero the engine by rotating crankshaft until markings on camshaft sprockets (14) and the timing cover are aligned. Use camshaft sprocket locking tool KM-800-1/2 and crankshaft locking tool KM-800-10, when marks are aligned, to hold this position (14).
6. Remove timing belt tensioning pulley (15) and adjusting roller (15).
7. Mark timing belt adjacent to all timing marks for installation reference.
8. Remove locking tool from crankshaft. Ensure washer behind adjusting roller is also removed.
9. Remove the timing belt (16).
10. Rotate the crankshaft to 60° BTDC.
11. From under the car, install counterstay, 83 95 063, to hold crankshaft from turning, and remove the crankshaft timing belt sprocket and spacer ring.
12. Use a screwdriver to pry out the front crankshaft seal. Use care not to scratch the sealing surface.

To install:

13. Install crankshaft protector sleeve, 83 94 942, onto crankshaft with taper outward (1-insert).
14. Position spacer ring on back of crankshaft timing belt sprocket (2). Lubricate new crankshaft seal (1) and tap into place, using outer part of the tool.
15. Use the counterstay 83 95 063 to hold sprocket while tightening into place with bolt. Remove counterstay. Tighten crankshaft sprocket bolt to 185 ft. lbs. (250 Nm), plus an additional 45 degrees.
16. Position camshaft locking tool KM-800-1 on No. 1 and No. 2 camshaft sprockets. Position camshaft locking took

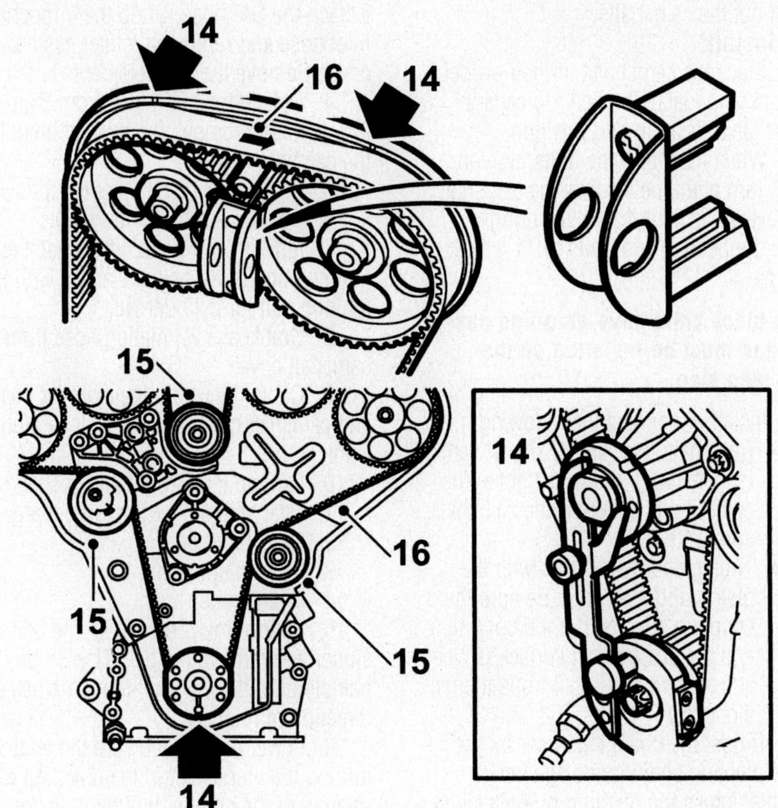

Installing locking tools prior to timing belt removal—9-5 3.0L engines

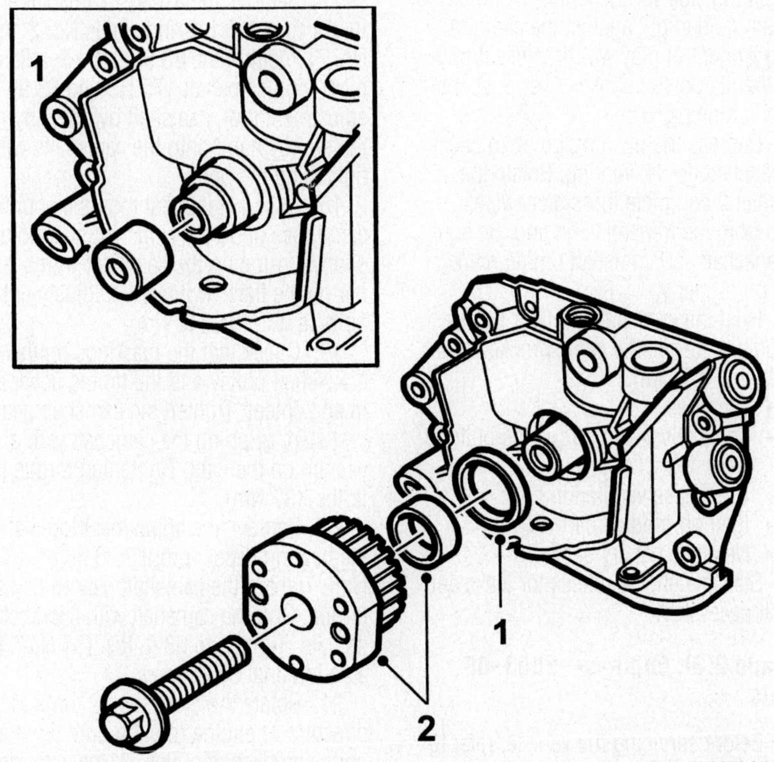

Using crankshaft seal remover/installer tool—9-5 3.0L engines

KM-800-2 on No. 3 and No. 4 camshaft sprockets.

17. Rotate crankshaft to just before zero degree mark and install holding tool KM-800-10 between camshaft sprockets.

18. Carefully rotate crankshaft in normal rotation direction until arm is lying against the water pump flange, then remove the holding tool.

19. Install the timing belt, using marks made during removal. Be sure belt is in same direction of rotation as removed.

20. Use a setting tool KM-800-30 to keep belt in place.

21. Install lower tensioning pulley loosely. Be sure washer is in place.

22. Adjust tensioning pulley by hand to belt does not jump. Adjust belt counter-clockwise. Check that timing reference marks on belt still align with timing marks on camshaft sprockets and crankshaft. Remove belt setting tool.

23. Use a belt tension meter 83 93 985 to measure timing belt tension. If adjustment is needed, tighten center bolts of adjuster rollers lightly. Adjust lower roller counterclockwise until tension of 203-222 ft. lbs. (275–300 Nm) is obtained. Tighten roller bolts to 30 ft. lbs. (40 Nm).

24. Check that adjusting nut on upper adjusting roller rotates freely. Adjust tensioning pulley with a 5mm Allen key until the 2 lines on the tensioning pulley correspond with each other. Tighten to 15 ft. lbs. (20 Nm).

25. Remove the locking tool for camshaft sprockets No. 1 and No. 2. Fit tool KM-800-20, then adjust upper adjusting roller counterclockwise with tool 83 94 983 until markings on camshaft sprockets are aligned with corresponding markings on tool KM-800-20. Use a counterstay 83 94 983. Tighten roller to 30 ft. lbs. (40 Nm).

26. Remove all locking tools and KM-800-20.

27. Rotate engine 2 revolutions in direction of rotation until just before the zero mark. Fit locking tool KM-800-10 on crankshaft.

28. Carefully rotate crankshaft in direction of rotation until arm is lying against water pump, and tighten it.

29. Fit tool KM-800-20 and ensure markings on camshaft sprockets are opposite markings on the tool. Ensure edge of the timing belt is aligned with the edge of the sprocket.

30. Install or connect the following:

- Crankshaft pulley. Torque bolts to 15 ft. lbs. (20 Nm)
- Timing cover. Torque bolts to 6 ft. lbs. (8 Nm)

- Auxiliary drive belt tensioner
- Alternator air intake
- Water pump pulley (install bolts loosely)
- Power steering pump pulley (install bolts loosely)
- Auxiliary drive belt
- Tighten water pump pulley bolts to 6 ft. lbs. (8 Nm)
- Tighten power steering pump pulley bolts to 15 ft. lbs. (20 Nm)
- Right side engine attachment with bracket and engine mount. Torque bolt to body to 47 ft. lbs. (63 Nm)
- Power steering hose into holder
- MAF sensor with hose

31. Carefully remove the 2 wedges from the subframe.
32. Install the lower spoiler sections.
33. Start engine and check operations and for leaks.

Camshaft

REMOVAL & INSTALLATION

2.0L and 2.3L Engines—2002 Models

1. Before servicing the vehicle, refer to the Precautions Section.
2. Remove or disconnect the following:
   - Negative battery cable
   - Ignition module cartridge
   - Valve cover
   - Crankcase ventilation hose
3. Position the crankshaft for TDC. The **0** mark on the flywheel should align with the timing mark on the bell housing end plate. The camshafts should be lined up with their respective timing marks.
4. Remove or disconnect the following:
   - Oil pipes
   - Center bolts securing the camshaft sprockets.

***** WARNING**

Use a proper holding tool to hold the camshafts from rotating. Always keep the camshafts in their correct basic setting. If the setting of the crankshaft or camshafts is altered at this stage the valves can be damaged.

   - Timing chain tensioner
   - Camshaft sprockets
5. Mark the bearing cap positions and relation to the front of the engine. The caps must be installed in their original location.
6. Loosen the camshaft bearing cap bolts 1 turn at a time to avoid uneven valve spring pressure on the camshafts. When all bolts are loose, remove the bearing caps and lift out the camshafts.

To install:

7. Place the camshafts in their proper positions and install the bearing caps in their original location and position.
8. When installing the bolts, tighten them 1 turn at a time to draw the camshaft down evenly against the valve springs. Torque the bearing cap bolts to 11 ft. lbs. (15 Nm).

➡The black bolts have an oiling passage and must be installed on the spark plug side.

9. Install or connect the following:
   - Camshaft sprockets. Exhaust cam is installed first. Hand-tighten the center bolts securing the camshaft sprockets
   - Timing chain tensioner with the piston under tension. Be sure the copper gasket is in good condition and that the sealing surface is clean and free from burrs. Torque the tensioner to 47 ft. lbs. (63 Nm).
10. Trigger the chain tensioner by pressing the pivoting chain guide against it. Thereafter, press the pivoting guide against the chain to give the chain its basic tension. Check that the chain tensioner maintains tension on the chain when the pressure on the chain guide is released and that the basic setting stop for the tensioner holds the chain guide tight against the chain. A limited amount of play will be present until the hydraulic pressure takes over once the engine is running.
11. Depress the pivoting guide to check that the tensioner is working. Rotate the crankshaft 2 complete turns clockwise, viewed from the transmission end. Be sure the crankshaft and camshaft timing marks still align properly.
12. Hold the camshafts in their proper position and torque the cam sprocket bolts to 49 ft. lbs. (67 Nm).
   - Oil pipes
   - Valve cover and torque the bolts to 11 ft. lbs. (14 Nm)
   - Crankcase ventilation hose
   - Ignition module cartridge
   - Negative battery cable
13. Start the engine, check for leaks and repair if necessary.

2.0L and 2.3L Engines—2003–06 Models

1. Before servicing the vehicle, refer to the Precautions Section.
2. Remove the upper engine cover.
3. Unplug the MAF sensor connector. Detach the SAI hose. Undo the turbocharger inlet hose and remove the filter casing cover. Remove the filter element.
4. Detach the fuel pipes from their snap fasteners and remove the bracket bolts from the camshaft cover.
5. Remove or disconnect the following:
6. Cover over the ignition coils
7. Ignition coil bolts, lift up each ignition coil and move aside, starting with the ignition coil for cylinder No. 1
8. Crankcase ventilation hose from the camshaft cover
9. Cable duct from the cylinder head and camshaft cover and move it carefully aside
10. Ground lead from the camshaft cover
11. Heat shield over the turbocharger
12. Camshaft cover
13. Right-hand wheel
14. Right fender liner
15. Relieve the pressure on the belt tensioner and remove the belt. Use 83 96 095 belt circuit relieving tool. Mark the belt's direction of rotation.
16. Lower the car and zero the engine by turning the crankshaft in the direction of rotation of the engine until the marking on the crankshaft pulley is aligned with the marking on the timing cover. The cam lobes on cylinder No. 1 intake and exhaust camshafts should be pointing up and in.
17. Remove the exhaust camshaft and intake camshaft bearing shells No. 2 and No. 7. Position the 83 96 046 kit, adjustment tool, camshaft 175 HP and 83 96 079 adjustment tool, camshaft by pressing down the tool by hand onto the camshafts without tightening the bolts.
18. If the adjustment tool does not fit the cam lobes undo the camshaft gear bolts slightly, gripping the camshaft with a spanner on the flats. Tighten the adjustment tool bolts to 7 ft. lbs. (10 Nm).
19. Check that the markings on the crankshaft pulley and the timing cover are in agreement. Tighten the camshaft gear a first step, gripping the camshaft with a wrench on the flats. Tightening torque is 22 ft. lbs. (30 Nm).
20. Remove the adjustment tools and fit the bearing caps. Torque to 6 ft. lbs. (8 Nm). Tighten the camshaft gear to its final torque. Grip the camshaft with a wrench on its flats. Tighten to 63 ft. lbs. (85 Nm), plus an additional 30 degrees.
21. Rotate the crankshaft 2 turns in the direction of engine rotation until the mark on the crankshaft pulley agrees with the mark on the timing cover. Remove the bear-

ing caps and fit 83 96 046 kit, adjustment tool (175 HP engines) and 83 96 079 adjustment tool (B207, 200 HP) again to check the camshaft setting. Remove the adjustment tool and refit the bearing caps. Torque to 6 ft. lbs. (8 Nm).

To install:

22. Raise the car, relieve the belt tensioner and fit the belt in the marked direction of rotation. Fit the right fender liner. Install the front wheels.
23. Lower the car and fit the camshaft cover with new seals. Take care not to disturb the position of the seals when fitting. Torque cover bolts to 8 ft. lbs. (10 Nm).
24. Install or connect the following:
    - Heat shield over turbocharger (make sure clip under heat shield snaps onto holder)
    - Ground cable to camshaft cover
    - Cable duct on camshaft cover
    - Crankcase ventilation hose to the camshaft cover
    - Ignition coils (start with coil for cylinder No. 4); torque to 6 ft. lbs. (8 Nm)
    - Cover over the ignition coils
    - Bolts for bracket to camshaft cover; fit fuel pipes in snap fasteners
    - Filter element
    - Cover on air cleaner casing
    - MAF sensor connector
    - SAI hose
    - Turbocharger inlet hose to the air cleaner
    - Upper engine cover

2.8L Engine

1. Before servicing the vehicle, refer to the Precautions Section.
2. Disconnect the negative battery cable.
3. Drain the cooling system.
4. Remove the camshaft covers.
5. Remove the spark plugs.
6. Remove the coolant port.
7. Remove the camshaft solenoid valve.
8. Remove the camshaft position sensor.
9. Rotate the crankshaft so that the markings on the camshaft sprocket are visible and the indentations align with the cylinder head. Use EN-46105-1 Adjustment tool as a gauge (the camshafts are now in their neutral position).
10. Fit EN-46106 Flywheel locking attachment.
11. Mark the chain and camshaft sprocket so they can be refitted in the same position.
12. Reduce the tightening torque of the bolts holding the camshaft sprockets. Counterhold using an open wrench on the camshaft flats (do not remove the bolts).
13. Fit EN-46108-1 Fixing tool to the camshaft chain on the intake side. Check that the tool grips the chain firmly.
14. Fit EN-46108-2 Fixing tool to the camshaft chain on the exhaust side. Check that the tool grips the chain firmly.
15. Tighten EN-46108 Fixing tool so that the chain is unloaded.

➡It is extremely important that tool no. EN-46108 is properly fitted. If the chain is not unloaded when the camshaft sprocket bolts are removed, the automatic chain tensioner is triggered. If this occurs, the timing cover must be removed to adjust the chain tensioner.

16. Remove the camshaft sprocket bolts.
17. Remove the sprockets from the camshafts.
18. Remove the camshaft bearing caps by loosening the bolts in stages—1/2 to 1 turn. It is important that removal ends at the bearing cap where the tappets are loaded. Note the markings on the bearing caps when refitting to ensure that they are in the correct spots.
19. Lift out the camshafts and note the rear marking on the camshafts.
20. Remove the roller rockers and the hydraulic adjuster elements. Store them in order so they do not get mixed up.
21. Cover all return oil ducts in the cylinder head so that the valve cone or contaminants cannot fall down into the engine.
22. Fit EN-46110-eu Valve spring tool.
23. Fit 83 94 173 Air nipple for spark plug hole. Connect compressed air and put the piston and valves under pressure.
24. Press down the valve plate with EN-46110-eu Valve spring tool.
25. Remove the valve cone with a magnet. The compressed air must not be disconnected when the valve cone is removed. Do not press on the valve stem either as counterpressure in the cylinder could reduce and the valve could fall down into the cylinder.
26. Lift out the spring plate and valve spring.
27. Lift out the valve guide seal using 83 94 157 Pliers, valve guide seal.

To install:

28. Clean all parts and check contact surfaces and bearing surfaces for wear.
29. Fit the valve guide seal using 83 94 157 Pliers, valve guide seal.
30. Fit the valve spring and spring plate.
31. Press down the valve plate with EN-46110-eu Valve spring tool.
32. Fit the valve cone using EN-46117 Fitting tool.
33. Unload the valve and remove EN-46110-eu Valve spring tool.
34. Disconnect the compressed air and remove the air nipple.
35. Position the camshafts and use EN-46105 Adjustment tool to check that the camshafts are mounted at the correct angle.
36. Install the bearing caps in order (markings on caps). Start where the cams point down and load the tappets. Tighten the bolts in stages—1/2 to 1 turn. Tighten to 7 ft. lbs. (10 Nm).
37. Install the sprockets with chain on the camshafts. Tighten the bolts by hand.
38. Check that the marks on the chain match the markings on the sprockets.
39. Remove EN-46108-2 and EN-46108-1 Fixing tools.
40. Tighten the bolts of both camshaft sprockets, counterholding with an open wrench on the camshaft flats. Tighten to 48 ft. lbs. (65 Nm).
41. Remove EN-46106 Flywheel locking attachment.
42. Install the camshaft position sensor (555F) using a new gasket. Torque to 7 ft. lbs. (10 Nm).
43. Install the camshaft solenoid valve (695F) using a new seal. Torque to 7 ft. lbs. (10 Nm).
44. Install the coolant port using a new seal and gasket.
45. Install the spark plugs.
46. Install the camshaft covers.
47. Refill the cooling system.
48. Connect the negative battery cable.
49. Start the engine, check for leaks and repair if necessary.

3.0L Engine

⁂ WARNING

To avoid damage to the valves, DO NOT rotate the camshafts. The crankshaft may only be turned between 0 TDC and 60° BTDC when the camshafts are locked in position with the appropriate locking tool.

1. Drain the cooling system.
2. Before servicing the vehicle, refer to the Precautions Section.
3. Remove or disconnect the following:
    - Negative battery cable
    - Front exhaust pipe from the exhaust manifolds
    - Bracket for the check valves
    - Lower center air deflector
    - Top engine covers

- Cruise control cable and throttle cable
- Throttle control rod from the bracket
- Bracket mounting bolt and set the bracket with cables attached to the side
- Clamps on the air intake pipes at the intake manifold and the mass air flow meter
- Air intake pipes. Raise them slightly
- Vacuum hoses and electrical connections and label them for identification
- Pipes with resonator attached
- Intake plenum bolts
- Idle Air Control (IAC) valve connector
- Fuel pressure regulator hose
- Wiring harness from under the throttle body
- Throttle position indicator
- Ignition coil
- TCS connectors
- Intake plenum and plug the intake runners
- Fuel injector
- CPS sensor connectors
- Fuel line connections
- Center intake manifold and fuel rails and set them aside
- Spark plug wires
- Ignition coil. Bend the ignition coil aside and disconnect the ignition coil bracket
- Lifting eye
- Heat shield over the exhaust manifold
- Resonator bracket and secondary air injection pipe from the exhaust manifold
- Power steering reservoir
- Torque arm engine mount connection
- Torque arm
- Power steering line clamp from the torque arm engine mount
- Engine mount
- Hose from the coolant expansion tank
- Upper alternator air intake
- Power steering pump pulley
- Water pump pulley
- Six outer crankshaft pulley bolts

➡When removing the crankshaft pulley, remove the 6 outer bolts only, DO NOT remove the center bolt.

- Drive belt and tensioner
- Power steering pump
- Timing cover
- Crankshaft pulley
- Timing belt, tensioner, and camshaft sprockets.

4. Rotate the crankshaft back to 60° Before Top Dead Center (BTDC), to prevent damage to the valves.
5. Remove or disconnect the following:
   - Valve cover. Be sure the O-rings stay in position and do not fall into the engine
   - Bearing caps. Note the position markings on the bearing caps and loosen the bearing cap bolts in stages of ½ to 1 turn at a time

➡The bearing caps located where valve tappets are compressed should be removed last.

   - Camshafts

To install:

6. Be sure the crankshaft is still positioned at 60° BTDC.
7. Be sure all gasket contact surfaces are clean.

➡The camshaft bearing caps are marked L1-l8 for the Front or Left bank and R1-r8 for the Rear or Right bank and the camshaft bearing seats in each head are numbered 1–8.

8. Thoroughly lubricate the camshafts and install the camshafts with new front camshaft gaskets. Be sure the locating pins are properly positioned. Install the bearing caps in their proper location and position. Tighten the bearing cap bolts in sequence ½ to 1 turn at a time, to a torque of 72 inch lbs. (8 Nm).
9. Install or connect the following:
   - Timing covers
10. Check to be sure that the camshaft locating pins are in the proper position. Check the locating pins, if they are hollow, replace them with solid pins.

➡The locating pins of both camshafts 1 & 2 should be point towards the inboard bolts of the camshaft bearing caps. The locating pin for the intake camshaft No, 3 should be pointing downwards, in line with the inboard bolt of the bearing caps. The locating pin for the exhaust No. 4 camshaft should be pointing upwards, in line with the edge of the camshaft sensor.

   - Camshaft sprockets
   - Tensioner and timing belt.
   - Crankcase ventilation housing
11. Be sure the valve cover O-rings are still in position and clean. Lubricate the O-rings with soapy water, apply (81-52-381) sealer at the corners of the large end bearing caps.
   - Valve cover
   - Timing cover
   - Crankshaft pulley and torque the bolts to 15 ft. lbs. (20 Nm)
   - Water pump pulley
   - Power steering pump pulley
   - Drive belt tensioner, and drive belt
   - Upper alternator air intake
   - Torque arm engine mount
   - Power steering line clamp
   - Torque arm and bolt the connection to the torque arm engine mount
   - Upper hose to the coolant expansion tank
   - Power steering reservoir
   - Lifting eye
   - Secondary air injection pipe to the exhaust manifold
   - Exhaust manifold heat shield
   - Resonator bracket
   - Ignition coil bracket
   - Ignition coil
   - Spark plugs and torque them to 19 ft. lbs. (25 Nm)
   - Spark plug wires
   - Center intake manifold with fuel rail
   - Center intake manifold and torque the bolts to 15 ft. lbs. (20 Nm)
   - Fuel lines
   - CKP sensor
   - Fuel injector connectors
   - Intake plenum into position and connect the TCS throttle body and torque to 15 ft. lbs. (20 Nm)
   - Throttle position indicator connector
   - Ignition coil connector
   - TCS connector
   - Wiring harness
   - Fuel pressure regulator hose
   - IAC valve connector
   - Labeled electrical and vacuum connections on the intake manifold
   - Air intake pipes, with resonator attached
   - Vacuum hoses and electrical connection to air intake
   - Air intake pipes to the mass airflow meter and intake manifold
   - Throttle control bracket with cables
   - Throttle control rod
   - Throttle cable and cruise control cable. Adjust the kick-down cable and the throttle cable
   - Upper engine covers
   - Lower center air deflector
   - Bracket for the check valves
   - Front exhaust pipe to the exhaust manifold
   - Negative battery cable
12. Fill the cooling system.
13. Start the vehicle, check for leaks and repair if necessary.

Valve Lash

ADJUSTMENT

2.0L and 2.3L Engines

1. Check and adjust the valve clearance relative to the working range of the tappet.
2. Place the clearance gauge across two of the camshaft bearing seats with the depth gauge against the end of the valve stem.
3. Verify that the maximum gauge depth of 0.81 inch (20.5mm) reaches down to the end of the valve stem as seen by the valve clearance gauge not bottoming against the bearing seat closest to the depth gauge.
4. Then, check that the minimum depth 0.77 inch (19.5mm) does not reach the end of the valve stem. Correct valve position should be between the depth measurement minimum and maximum values.
5. If the valve position deviates from the given measurements, adjustments are done on the valve stem or valve seat.
6. The valve stem is shortened if measurement is below the minimum value and the valve seat is machined if the maximum value is exceeded. When adjusting the valve position, set this at a nominal value of 0.80 inch (20.2mm).

The hydraulic cam followers used in Saab engines do not require adjusting. The cam followers keep the valve clearance within 18.75–20.8mm. However, if the cam followers are making excessive noise or are diagnosed to be defective, perform the following procedure:

7. Before servicing the vehicle, refer to the precautions in the beginning of this section.
8. Disconnect the negative battery cable.
9. If a cam follower is noisy, it can be found by removing the valve cover and, using a screwdriver, gently pushing down on each cam follower until the defective follower(s) is found by exhibiting a spongy feeling.
10. Replace the defective cam follower(s); first removing the camshaft(s).
11. Reinstall the camshaft(s) and the valve cover.

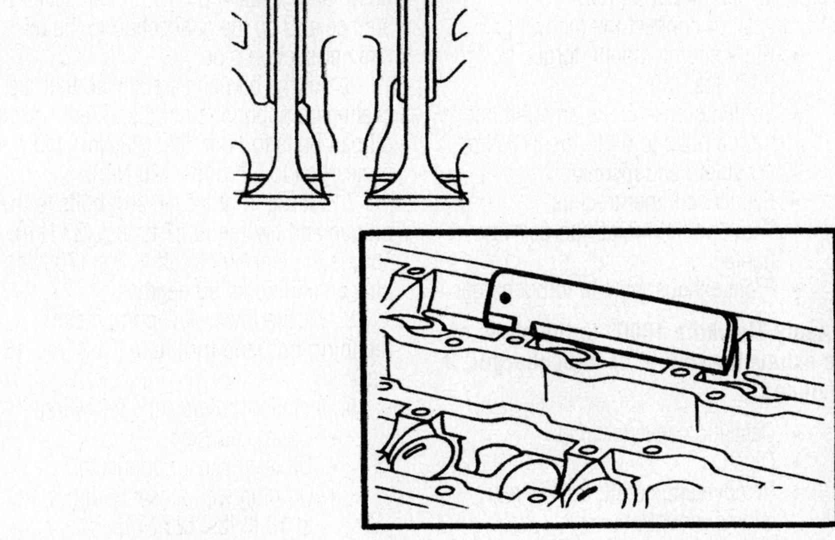

Checking valve clearance—2.0L engines 2002

Starter Motor

REMOVAL & INSTALLATION

2.0L and 2.3L Engines

1. Before servicing the vehicle, refer to the Precautions Section.
2. Remove or disconnect the following:
   - Negative battery cable
   - Upper mounting bolt
   - Starter motor electrical connections
   - Lower mounting bolt
   - Starter motor

To install:

3. Install or connect the following:
   - Starter motor
   - Lower mounting bolt
   - Starter electrical connections
   - Upper mounting bolt
   - Negative battery cable

2.8L Engines

1. Before servicing the vehicle, refer to the Precautions Section.
2. Remove the battery cover.
3. Disconnect the negative battery cable.
4. Raise and safely support the vehicle.
5. Remove the secondary air pump.
6. Disconnect the starter motor connecting cables.
7. Remove the starter motor.

To install:

8. Install the starter motor and tighten the mounting bolts to 35 ft. lbs. (47 Nm).
9. Connect the cables to the starter motor.
10. Install the secondary air pump.
11. Lower the vehicle.
12. Connect the negative battery cable.
13. Install the battery cover.
14. Restore all electrical functions.

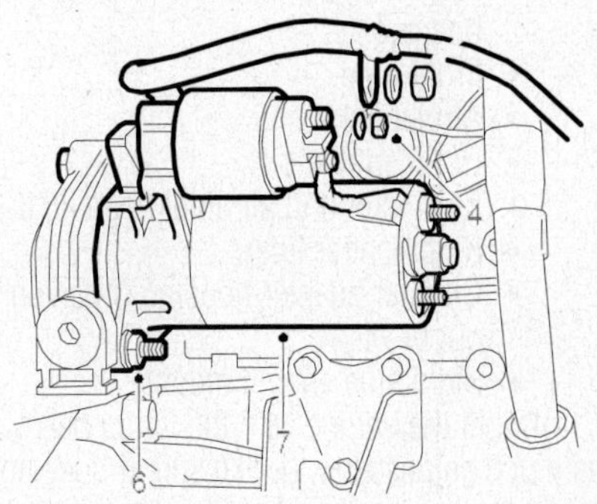

Remove the starter motor

3.0L Engine

1. Before servicing the vehicle, refer to the Precautions Section.
2. Remove or disconnect the following:
   - Negative battery cable
   - Alternator
   - Electrical connections at the starter
   - Upper retaining bolt. Access is through the right front wheel housing using a ratchet and extension
   - Lower retaining bolt
   - Starter

To install:

3. Install or connect the following:
   - Starter motor
   - Lower retaining bolt
   - Upper retaining bolt
   - Electrical connections at starter
   - Alternator
   - Negative battery cable

Oil Pan

REMOVAL & INSTALLATION

9-3

2002 MODEL

1. Before servicing the vehicle, refer to the Precautions Section.
2. Drain the engine oil.
3. Remove or disconnect the following:
   - Negative battery cable
   - Engine cover
   - Dipstick tube
   - Install engine lifting beam 83 94 850 and slightly raise the engine after taking center bolt from rear engine mount
   - Exhaust manifold heat shield and bypass tube
   - Oxygen sensor connector and sensor
   - 2 top nuts on exhaust pipe
   - Front exhaust pipe at the exhaust manifold and intermediate pipe
   - Front exhaust pipe from the turbocharger
   - Front pipe
   - Air spoilers
   - Drain engine oil
   - Resonator and air shield behind it
   - Right fender liner
   - Oil filter adapter housing (position out of way)
   - Right side engine mount
4. Lift the engine with the lifting beam as much as possible. (Make sure it does not touch the engine mounting bracket and that the belt pulley goes free.)
5. Raise the car and remove the bolts on the right side of the subframe. Place wedges between the chassis and the subframe.
6. Remove the access panel for the flywheel and the oil trap return hose.
7. Unscrew the oil pan bolts and remove the pan.

To install:

8. Make sure there are no impurities in the oil pan and clean the sealing surfaces on the oil pan, oil filter adapter and engine block. Use benzene. Check the O-rings on the oil pipe to the oil filter adapter and replace with new ones, if necessary.
9. Apply an even bead of flange sealant, part no. 93 21 795, to the sealing surface of the oil pan.
10. Make sure the pipe to the filter adapter is positioned correctly. Carefully lift up the oil pan at a slightly anti-clockwise angle. Make sure the sealant does not scrape off. Carefully turn the pan clockwise and place it in the correct position.
11. Install or connect the following:
    - Oil pan and torque the bolts to 16 ft. lbs. (22 Nm)
    - Return hose from oil trap
12. Apply flange sealant 93 21 795 to the sealing surfaces of the oil filter adapter. Apply engine oil to the oil pipe O-ring and fit the adapter housing. Use new washers on the oil pipe banjo screw. Tighten the screw by hand first to prevent it from entering crookedly.
13. Install the flywheel access panel.
14. Install the sub frame and tighten the sub frame bolts as follows:
    a. Front bolts: 85 ft. lbs. (115 Nm).
    b. Center bolts: 140 ft. lbs. (190 Nm).
    c. Rear bolts: 81 ft. lbs. (110 Nm) plus 75° additional torque.
15. Install or connect the following:
    - Right engine mount; torque bolt to 29 ft. lbs. (39 Nm)
    - Engine bracket bolts on subframe; torque bolts to 54 ft. lbs. (73 Nm)
    - Air shield and resonator
    - Fenderwell liner to clips
    - Rear engine mount nut; do not tighten
    - Front exhaust pipe to turbocharger

➡Apply Molykote 1000® to the studs of the exhaust manifold or turbocharger, if equipped).

   - Catalytic converter joint
   - Dipstick
   - Remove engine lift; tighten rear engine mount nut
   - 2 top nuts on exhaust pipe
   - Oxygen sensor and connector
   - Bypass pipe and heat shield
   - Negative battery cable
16. Fill the engine with oil.
17. Start the vehicle, check for leaks and repair if necessary.

2003–06 MODELS (EXCEPT 2.8L ENGINE)

1. Before servicing the vehicle, refer to the Precautions Section.
2. Remove or disconnect the following:
   - Negative battery cable.
   - Front right wheel
   - Right fender liner
   - Drain the engine oil
   - Fit oil plug with a new seal and lower the car
   - Upper engine cover
   - SAI pump inlet hose from the pump and air cleaner
   - Dipstick and dipstick pipe retaining screw
   - Connector from mounting on dipstick (undo clip and pull up dipstick from oil pan)
   - Oil level sensor connector
   - Lower charge air pipe
   - Lower A/C compressor retaining bolt
   - Rear torque arm bolt and remove the front bolts
   - Oil pan retaining bolts
3. Place a screwdriver between the oil pan and the timing cover by the A/C compressor and carefully pry the oil pan loose.

To install:

4. Clean the sealing surfaces on the engine block and the pan. Remove any impurities in the oil pan.
5. Apply a 2mm thick bead of 83 95 691 Flange sealant on the oil pan sealing surface and on the connection to the oil strainer suction pipe.
6. Fit the oil pan carefully so that the sealing compound is not disturbed. Torque oil pan bolts to 16 ft. lbs. (22 Nm) and bolts in gearbox to 52 ft. lbs. (70 Nm).
7. Fit the front torque arm bolts to the oil pan and tighten to 27 ft. lbs. (37 Nm). Torque the rear bolt to 52 ft. lbs. (70 Nm), plus an additional 90 degrees.
8. Fit the lower A/C compressor retaining bolt and torque to 18 ft. lbs. (24 Nm).
9. Install or connect the following:
   - Charge air pipe
   - Oil level sensor connector
   - Oil plug with a new O-ring; torque to 18 ft. lbs. (25 Nm)
   - New fastening clip on connector
   - Right fender liner

- Right front wheel; torque wheel nuts to 81 ft. lbs. (110 Nm)
- Remove connector fastening clip and fit dipstick pipe with new O-rings lubricated with 30 15 286 Vaseline
- Connector clip to the dipstick; plug in connector
- Ventilation line clip to dipstick
- Dipstick
- SAI hose to pump and filter element

10. Fill engine with the specified engine oil.

11. Start the engine and idle. Switch off the engine and wait 2-5 minutes. Check the oil level and adjust as necessary.

12. Install the upper engine cover.

2.8L ENGINE

1. Before servicing the vehicle, refer to the Precautions Section.
2. Disconnect the negative battery cable.
3. Drain the cooling system.
4. Drain the engine oil.
5. Remove or disconnect the following:
   - Engine assembly
   - Camshaft covers
   - Timing cover (refer to timing chain procedure)
   - Oil cooler assembly and gasket from cylinder block
   - Oil pan. Carefully remove with a pry bar.
6. Clean sealing surfaces of gasket residue and the like.

To install:

7. Apply a 2mm thick bead of flange sealant on the oil pan's sealing surface and install a new seal on the oil pan intake pipe.
8. Install the oil pan. Tighten the M6 size bolts to 7 ft. lbs. (10 Nm), and the M8 size bolts to 18 ft. lbs. (25 Nm).
9. Install a new gasket and fit the oil cooler to the cylinder block. Tighten the mounting bolts to 18 ft. lbs. (25 Nm).
10. Install or connect the following:
    - Timing cover (refer to timing chain procedure)
    - Camshaft covers
    - Engine assembly
11. Fill the engine cooling system.
12. Fill the engine with clean oil.
13. Connect the negative battery cable.
14. Start the vehicle, check for leaks and repair if necessary.

9-5

2.3L ENGINE

1. Before servicing the vehicle, refer to the Precautions Section.

Exploded view of the oil pan and splash guard 2.3L engine—9-5

2. Drain the engine oil.
3. Remove or disconnect the following:
   - Negative battery cable
   - Upper engine cover
   - Dipstick
   - Oxygen sensor cables
   - Turbocharger bypass pipe
   - Exhaust manifold heat shield
   - Lower engine cover
   - Front exhaust pipe
   - Gearbox cover plate
   - Crankcase breather hose from the oil pan
   - Oil pan

To install:

4. Transfer the splash guard and pie to a new oil pan, if replacing the pan.
5. Clean all mating surfaces of gasket material.
6. Apply an even bead of flange sealant to the mating surface on the oil pan.
7. Install or connect the following:
   - Oil pan and make certain that the pipe to the oil filter adapter is properly positioned in the oil pan and torque the oil pan bolts evenly to 16 ft. lbs. (22 Nm)
   - Crankcase ventilation hose
   - Gearbox cover plate
   - Front exhaust pipe
   - O_2S sensor cables
   - Exhaust manifold heat shield
   - Turbocharger bypass pipe
   - Dipstick
   - Upper and lower engine covers
   - Negative battery cable
8. Fill the engine with clean oil.
9. Start the vehicle, check for leaks and repair if necessary.

3.0L ENGINE

1. Before servicing the vehicle, refer to the Precautions Section.
2. Drain the engine oil.

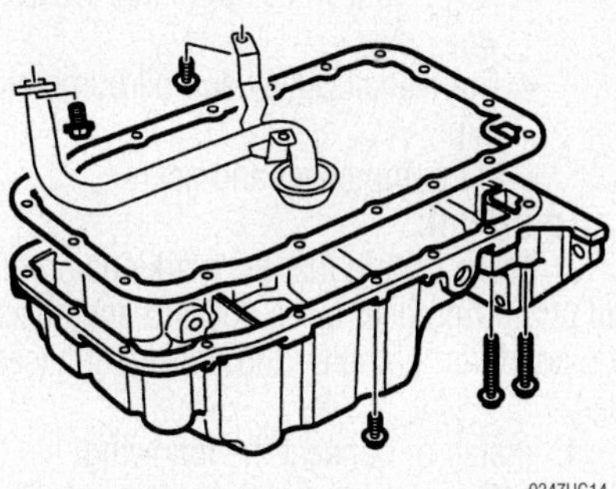

Exploded view of the oil pan, suction pipe and the antislosh baffle—3.0L engine

3. Remove or disconnect the following:
   - Negative battery cable
   - Oxygen sensor cables
   - Lower rear spoiler section
   - Front exhaust pipe
   - Oil pan
4. If replacing the oil pan, remove the following components:
   a. Oil suction pipe and O-ring.
   b. Antislosh baffle.

To install:

5. Clean all mating surfaces of any gasket material.
6. Install or connect the following:
   - Antislosh baffle to the oil pan and torque the bolts to 6 ft. lbs. (8 Nm)
   - Oil suction pipe with new O-rings and torque the bolts to 6 ft. lbs. (8 Nm)
   - Oil pan with a new gasket and torque the bolts to 10 ft. lbs. (15 Nm)
   - Front exhaust pipe and torque the front bolts to 18 ft. lbs. (24 Nm) and the exhaust manifold bolts to 30 ft. lbs. (40 Nm)
   - Oxygen sensor cables
   - Lower rear spoiler section
   - Negative battery cable
7. Fill the engine with clean oil.
8. Start the vehicle, check for leaks and repair if necessary.

Oil Pump

REMOVAL & INSTALLATION

9-3

2.0L AND 2.3L ENGINES

1. Before servicing the vehicle, refer to the Precautions Section.
2. Remove or disconnect the following:
   - Negative battery cable
   - Air cleaner and relieve the belt tension

- Right front wheel and inner wheel well cover
- Crankshaft pulley and oil pump circlip
- Oil pump cover and gears

To install:

3. Make certain that the markings on the oil pump ring gear faces outward and check the condition of the O-ring, replace if necessary.
4. Install or connect the following:
   - Oil pump gears and cover
   - Oil pump circlip with the chamfer facing outward and the opening facing downward
   - Crankshaft pulley and torque the bolt to 130 ft. lbs. (175 Nm)
   - Right front inner wheel well and front tire
   - Air cleaner and tension the drive belt
   - Negative battery cable

2.8L ENGINE

1. Before servicing the vehicle, refer to the Precautions Section.
2. Disconnect the negative battery cable.
3. Drain the cooling system.
4. Drain the engine oil.
5. Remove the timing chain, sprockets and front cover.
6. Remove the oil pump.

To install:

7. Install the oil pump against the flat surface at the crankshaft.
8. Install the oil pump bolts and tighten to 17 ft. lbs. (23 Nm).
9. Install the timing chain, sprockets and front cover.
10. Fill the cooling system to the proper level.
11. Fill the engine with clean oil.
12. Connect the negative battery cable.
13. Start the vehicle, check for leaks and repair if necessary.

9-5

2.3L ENGINE

1. Before servicing the vehicle, refer to the Precautions Section.
2. Remove or disconnect the following:
   - Negative battery cable
   - Intake manifold cover
   - Right front wheel
   - Power steering pump from the subframe. Install a wedge between the oil pan and the subframe and the gearbox and the subframe
   - Right side engine mount and bracket and relieve the belt tension
   - Flywheel cover plate
   - Crankshaft pulley
   - Oil pan circlip
   - Oil pump cover and remove the oil pump

To install:

3. Clean the oil pan circlip.
4. Make certain that the marking on the oil pump ring gear face out and the pump gear with flange are facing in.
5. Lubricate the O-ring with vaseline.
6. Install or connect the following:
   - Oil pump gears
   - Oil pump cover using the locating arrows
   - Oil pan circlip with the chamfer facing outward and the opening downward
   - Crankshaft pulley and torque the bolts to 130 ft. lbs. (175 Nm)
   - Flywheel cover plate
   - Drive belt
   - Right side engine mount and bracket and torque the bolts to 37 ft. lbs. (50 Nm)
   - Power steering pump pie to the holder
   - Right front wheel
   - Negative battery cable
7. Start the vehicle, check for leaks and repair if necessary.

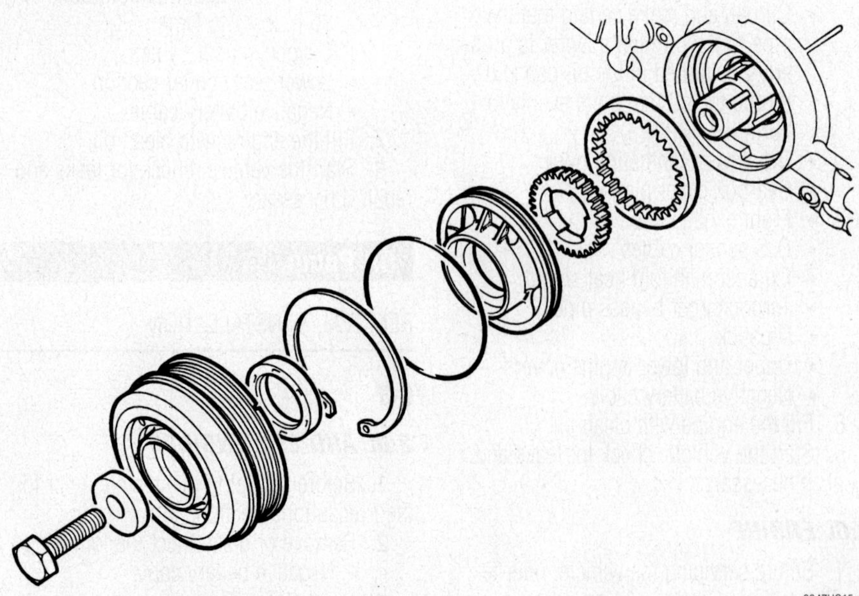

Exploded view of the oil pump assembly—2.3L Engine

3.0L ENGINE

1. Before servicing the vehicle, refer to the Precautions Section.
2. Drain the cooling system.
3. Drain the engine oil.
4. Remove or disconnect the following:
   - Negative battery cable
   - Engine cover
   - Right front wheel
   - Lower spoiler section and install a lifting eye on the front cylinder head and suspend the engine.
   - Mass Air Flow (MAF) sensor and hose
   - Power steering hose from the holder
   - Right side engine mount and bracket
   - Front exhaust manifold heat shield
   - Crankcase breather pipe from the turbocharger intake pipe
   - Turbocharger intake pipe and vent valve
   - Water pump and power steering pump pulley bolts
   - Tension from the auxiliary drive belt
   - Water pump and power steering pump pulleys
   - Belt tensioner and alternator air intake
   - Timing cover
   - Crankshaft pulley and make certain that the markings on the camshaft sprockets and timing cover are aligned with the crankshaft marking
5. Install locking tool KM-800-1 and KM-800-2 for the camshaft sprockets and KM-800-10 for the crankshaft.
   - Tensioning pulley and adjusting rollers. Mark the direction of rotation for the belt and remove the crankshaft tool.

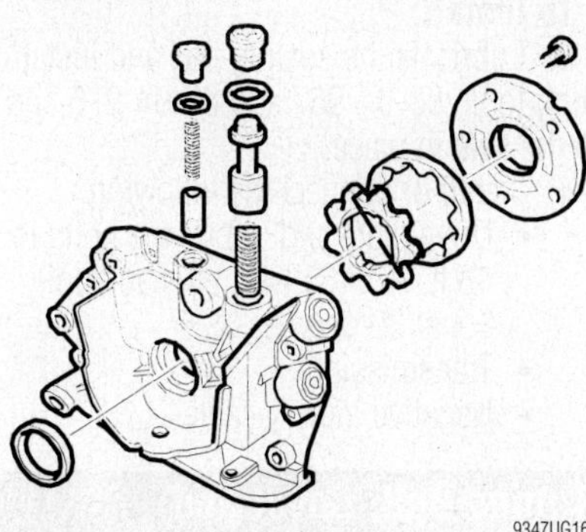

Exploded view of the 9-5 3.0L oil pump assembly

- Lower adjusting roller and washer. Rotate the crankshaft to 60° BTDC
- Camshaft sprockets
- Water pump
- Inner timing cover
- Crankshaft timing belt pulley
- Alternator and oil pressure switch connector
- Front exhaust pipe
- Oil pan and strainer
- A/C compressor and bracket and move them aside
- Oil pump housing
- Cover and remove the 2 pump impellers

To install:

6. Install or connect the following:
   - Oil pump impellers and cover. Torque the bolts to 53 inch lbs. (6 Nm). Make certain that the marks on the impellers are properly aligned.
   - Oil pump housing with a new gasket and torque the bolts to 53 inch lbs. (6 Nm)
   - A/C compressor and bracket
   - Right front wheel
   - Oil strainer and oil pan and torque the oil strainer bolts to 6 ft. lbs. (8 Nm) and the oil pan bolts to 10 ft. lbs. (15 Nm)
   - Front exhaust pipe and torque the front bolt to 16 ft. lbs. (22 Nm) and the front pipe to manifold bolt to 30 ft. lbs. (40 Nm)
   - Alternator and oil pressure switch connector
   - Crankshaft timing belt pulley and torque the bolts to 185 ft. lbs. (250 Nm) plus 45 degrees
   - Water pump and torque the bolts to 19 ft. lbs. (25 Nm)
   - Upper and lower adjusting rollers
   - Camshaft sprockets and torque the bolts to 37 ft. lbs. (50 Nm) plus 60 degrees
   - Locking tools between the camshaft sprockets to lock the camshafts of both heads in position.

7. Rotate the crankshaft forward to just before **0** TDC and install a crankshaft locking tool on the crankshaft. Carefully rotate the engine until the arm of the tool is against the water pump flange. Be sure the crankshaft is at **0** TDC and all timing marks are aligned. Remove the locking tool.

8. If reusing the belt, install the timing belt according to its marked direction of rotation and timing marks. Adjust the tensioning roller loosely by hand to prevent the belt from slipping out of the cogs. Always adjust counterclockwise.

9. Measure the belt tension.

10. Tighten the center bolts of the adjusting roller lightly. Adjust the adjusting rollers counterclockwise. Begin with the lower roller and adjust it to a belt tension of 275–300 Nm.

➡Adjustment of the belt tension is only a preparatory measure and must not be used as a check when the belt is finally adjusted.

11. Continue to carry out the adjustment by means of the tensioning roller, mark against mark. Remove the locking tool for camshaft sprockets 1 and 2. Carry out the final adjustment with the upper center adjusting roller until camshaft sprocket No. 2 moves 0.04–0.08 in. (1–2mm) forward.

12. Remove the locking tool for camshaft sprockets 3 and 4 and also remove the crankshaft locking tool.

13. Torque the tensioning roller to 15 ft. lbs. (20 Nm); the upper adjusting roller to 30 ft. lbs. (40 Nm) and the lower adjusting roller to 15 ft. lbs. (20 Nm).

14. Turn the engine over 2 revolutions to the **0** mark and refit the locking tool on the crankshaft. Check that the markings on the camshaft sprockets are in alignment with the markings on the timing cover. Check the positioning by installing the 2 camshaft locking tools, which should fit, and also by fitting tool 83-94-926 on camshaft sprockets 1 and 2 and 3 and 4. Also check the tensioning roller to ensure that the marks are still in alignment.

15. Install or connect the following:
   - Crankshaft pulley and torque the bolts to 15 ft. lbs. (20 Nm)
   - Timing belt cover and torque the bolts to 72 inch lbs. (8 Nm)
   - Belt tensioner and alternator air intake
   - Water pump and power steering pump pulleys
   - V-belt and torque the water pump pulley bolts to 72 inch lbs. (8 Nm) and the power steering pulley bolts to 15 ft. lbs. (20 Nm)
   - Right side engine mount and bracket and torque the bolts to 47 ft. lbs. (63 Nm)
   - Power steering hose in the holder
   - MAF sensor and hose and remove the engine lifting support
   - Turbocharger intake pipe
   - Crankcase breather pipe and torque the bolts to 18 ft. lbs. (24 Nm)
   - Front exhaust manifold heat shield and torque the bolts to 18 ft. lbs. (24 Nm)
   - Negative battery cable
   - Lower spoiler section
   - Engine cover

16. Fill the cooling system to the proper level.

17. Fill the engine with clean oil.

18. Start the vehicle, check for leaks and repair if necessary.

Rear Main Seal

REMOVAL & INSTALLATION

2.0L and 2.3L Engines

1. Before servicing the vehicle, refer to the Precautions Section.

2. Disconnect the negative battery cable.

3. Remove the transmission.

4. Remove the pressure plate and clutch drive plate.

5. Remove the drive plate or the flywheel.

6. Carefully pry out the seal using a screwdriver.

To install:

7. Clean the sealing surfaces of gasket residue.

8. Position the rear crankshaft seal protective collar 83 94 975 on the crankshaft. Lubricate the new sealing ring with non-acidic Vaseline and position it on the fitting tool.

9. Position the fitting tool on the crankshaft. Drive in the sealing ring with the shaft and the narrow guide sleeve until it is flush with the timing cover. Remove the tool.

10. Fit the flywheel or drive plate with new bolts.

11. Install the pressure plate and the clutch drive plate.

12. Install the transmission.

13. Connect the negative battery cable.

14. Check the oil level in the engine. Top up as necessary.

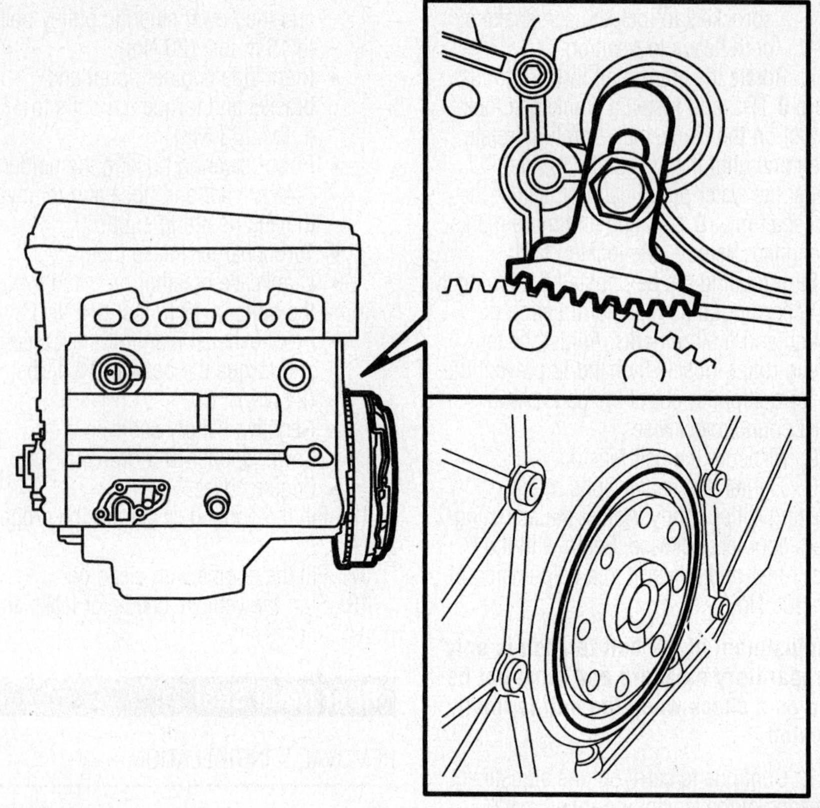

Rear main seal on 2.3L engine

2.8L Engines

1. Before servicing the vehicle, refer to the Precautions Section.
2. Disconnect the negative battery cable.
3. Remove the transaxle.
4. Remove the flywheel.
5. Remove the crankshaft seal as follows:
   a. Remove the bolts.
   b. Prize off the crankshaft seal using a suitable tool from the cylinder block and the oil sump. The crankshaft sealing ring is only supplied complete with housing.
   c. Clean the sealing surfaces and oil sump grooves.

To install:

6. Apply sealing compound and fit the special tool as follows:
   a. Apply an approx. 3mm thick bead of sealing compound on the oil sump's grooves.
   b. Press gently so that the sealing string becomes flat.
   c. Apply sealing compound on the edges to the left and right of the sealing surface.
   d. Install EN-47839 Fitting tool onto the crankshaft.
   e. Tighten the bolts.

➡Check that the oil sump's grooves are filled with sealing compound and that the sealing string is applied consistently across the whole width of the oil sump's sealing surface. The component may only be touched on its metal housing—the sealing ring must not be touched!

7. Apply an approx. 3mm thick bead of sealing compound
8. Install the crankshaft seal as follows:
   a. Press the crankshaft seal into the cylinder block over EN-47839 Fitting tool.
   b. Install the bolts and tighten to 7 ft. lbs. (10 Nm).
9. Remove EN-47839 Fitting tool.
10. Install the flywheel with new bolts.
11. Install the transaxle.
12. Connect the negative battery cable.

3.0L Engines

1. Before servicing the vehicle, refer to the Precautions Section.
2. Remove or disconnect the following:
   - Negative battery cable
   - Transmission
   - Drive plate pry out the rear main seal

To install:

3. Lubricate the sealing lips and install fitting tools 83-94-967 and 83-94-975 and tap the seal in place.
4. Install or connect the following:
   - Drive plate and torque the bolts to 48 ft. lbs. (65 Nm) plus an additional 30 degrees
   - Transmission
   - Negative battery cable

Timing Belt, Cover and Crankshaft Seal

REMOVAL & INSTALLATION

3.0L Engine

1. Before servicing the vehicle, refer to the Precautions Section.
2. Remove the negative battery cable.
3. Remove the lower spoiler sections.
4. Insert 2 wedges, 83 95 238, between oil sump and subframe and between the gearbox and subframe.
5. Remove or disconnect the following:
   - MAF sensor with hose
   - Power steering hose from holder
   - Right-hand engine attachment with bracket and engine mount
   - Retaining bolts from water pump pulley and power steering pump pulley
   - Release tension on auxiliary drive belt
   - Water pump pulley and power steering pump pulley
   - Belt tensioner and generator air intake
   - Timing cover
   - Crankshaft pulley
6. Zero the engine. The markings on the camshaft sprockets and timing cover should be aligned. Use adjustment tools for cam sprockets and timing belt, KM-800-20. The crankshaft markings should also be aligned. Fit locking tools, KM-800-1 and KM-800-2, for the camshaft sprockets and locking tool, KM-800-10, for the crankshaft.
7. Remove the tensioning pulley and adjusting rollers. Use counterstay tool, 83 94 983, on the adjusting rollers.
8. Mark the direction of rotation on the belt. The belt can also be marked with the respective camshaft markings and crankshaft marking to facilitate reinstallation.
9. Remove locking tool from the crankshaft. Let camshaft locking tools remain.
10. Remove the lower adjusting roller. Ensure the washer behind the roller is also removed.

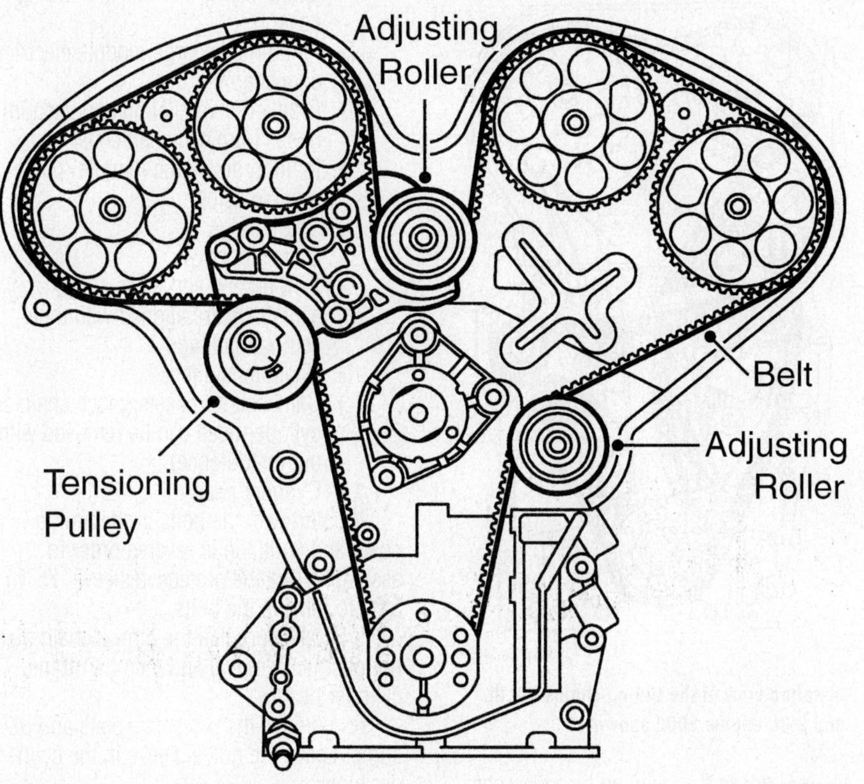

Showing routing of timing belt—9-5 3.0L engines

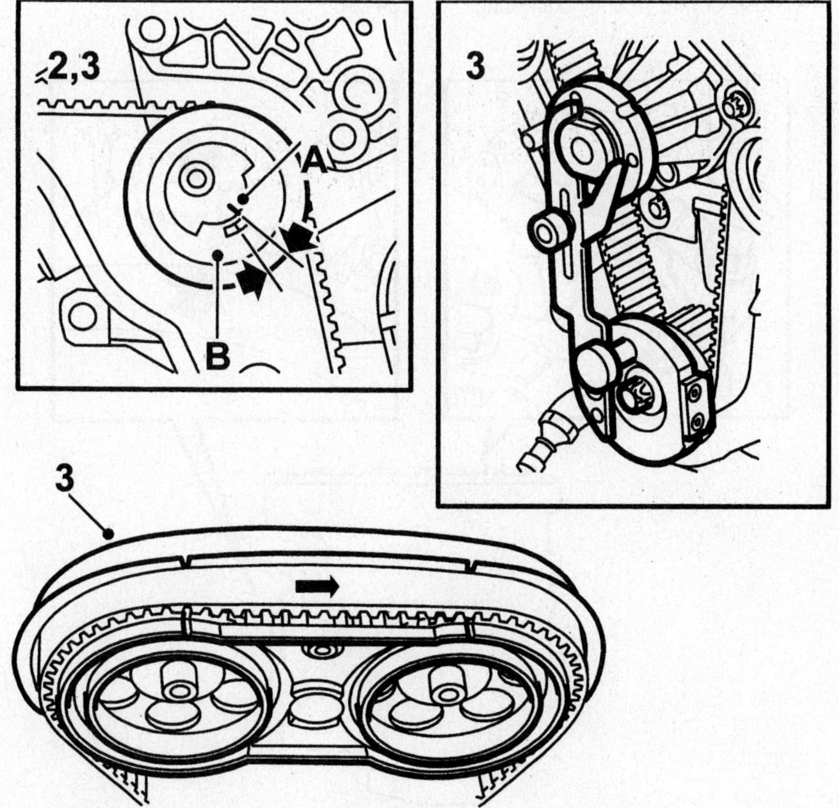

Timing belt is installed noting direction of rotation, using a counterstay device, and properly setting position of tensioning pulley—9-5 3.0L

11. Remove the timing belt.

To install:

12. Position the camshaft belt in the correct direction of rotation. Use adjustment tools for cam sprockets and timing belt, KM-800-30, to hold the belt in place.
13. Fit the lower tensioning pulley loosely. Make sure the washer is in place.
14. Lightly adjust the tensioning pulley by hand so the belt does not jump. Adjust counterclockwise.
15. Make sure the markings on the camshaft belt and the corresponding markings on the camshaft sprockets and crankshaft are aligned. Remove the tools, KM-800-30.
16. Install a locking tool, KM-800-10, and position a piece of camshaft belt, fitting a belt tension meter, 83 93 985, to measure the belt tension.

CAUTION

Adjustment of belt tension is only a preparatory measure and must not be used as a check when the belt is finally adjusted

17. Tighten the center bolts of the adjuster rollers lightly. Adjust the lower adjuster roller counterclockwise until a tension of 239–260 ft. lbs. (275–300 Nm) is obtained.
18. Use tool 83 94 983 to tighten the adjuster roller bolt. Use a counterstay tool, 83 94 983. Check the belt tension again and adjust as necessary. Torque adjuster roller bolt to 30 ft. lbs. (40 Nm).
19. Remove the belt tension meter and piece of the toothed belt.

➡Adjustment of belt tension is only a preparatory measure and must not be used as a check when the belt is finally adjusted.

20. Ensure the adjusting nut on the upper adjusting roller rotates freely and is not tight. Adjust the tensioning pulley with a 5mm Allen key until the 2 lines on the tensioning pulley correspond with each other. Tighten tensioning pulley bolt to 15 ft. lbs. (20 Nm).
21. Remove the locking tool for camshaft sprockets. Fit tool KM-800-20. Then, adjust the upper adjusting roller counterclockwise with tool 83 94 983 until the markings on the camshaft sprocket are aligned with the corresponding markings on KM-800-20. Tighten camshaft sprocket bolts, using a counterstay, 83 94 983. Tighten to 30 ft. lbs. (40 Nm).
22. Remove KM-800-20 and the other locking tools.

23. Rotate the engine 2 revolutions in the direction of rotation until just before the zero mark and fit a locking tool, KM-800-10, on the crankshaft.

24. Carefully rotate the crankshaft in the direction of rotation of the engine until the arm is lying against the water pump and tighten it.

25. Fit gauge tool KM-800-20 and ensure the markings on the camshaft sprocket are opposite the markings on the tool and that the edge of the belt is aligned with the edge of the sprocket.

26. Raise the car and install the crankshaft pulley, torquing the bolts to 15 ft. lbs. (20 Nm).

27. Lower the car.

28. Install or connect the following:

- Timing cover, tightening to 6 ft. lbs. (8 Nm)
- Auxiliary drive belt tensioner and the generator air intake
- Water pump pulley and the power steering pump pulley. Use Loctite, part No. 74 96 292. Do not tighten the bolts.
- Auxiliary drive belt
- Water pump pulley bolts to 6 ft. lbs. (8 Nm).
- Power steering pump pulley bolts to 15 ft. lbs. (20 Nm)
- Right-hand engine attachment with bracket and engine mount. Torque bolts to 47 ft. lbs. (63 Nm)
- Power steering hose in holder
- MAF sensor with hose
- Lower the engine by carefully removing the wedges
- Lower spoiler sections

Timing Chain, Sprockets, Front Cover and Seal

REMOVAL & INSTALLATION

9-3

2002 MODEL

1. Before servicing the vehicle, refer to the precautions in the beginning of this section.

2. Drain the engine oil, disconnect the cables to the starter motor, generator and engine oil pressure sensor.

3. Remove or disconnect the following:

- Oil filler pipe
- Intake manifold stay
- Crankcase ventilation pipe with intake manifold and nuts on turbocharger

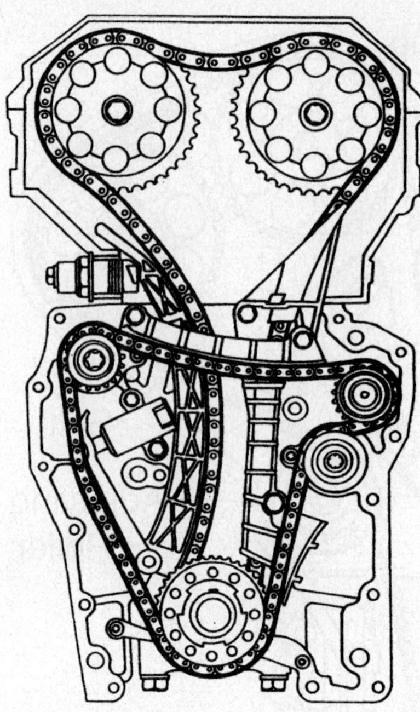

Installed view of the timing chains—2.0L and 2.3L engine 2002 shown

- Loosen the bolts on turbocharger stay slightly
- Water pipe for turbocharger from thermostat housing
- Hose for throttle body preheating
- and unscrew thermostat housing cover
- Ignition discharge module electrical connection
- Crankcase ventilation and vacuum hoses from camshaft cover
- Ignition discharge module (place module aside)
- Spark plugs
- Camshaft cover
- Zero engine and make sure camshafts are aligned with their setting markings
- Chain tensioner
- Camshaft sprockets (place chain so cylinder head can be removed without interference)
- Cylinder head

4. Start with the bolts in the timing cover and continue in reverse order to assembly. Use the protective sleeve, 75 19 531, to retrieve the bolts.

5. Make sure that the timing chain dies not obstruct removal and then lift off the cylinder head.

6. Remove the protective plate and oil sump. Leave the guide sleeve in the cylinder block.

7. Remove or disconnect the following:

8. Remove the crankshaft pulley. Use the flywheel locking attachment, part no. 83 94 868.

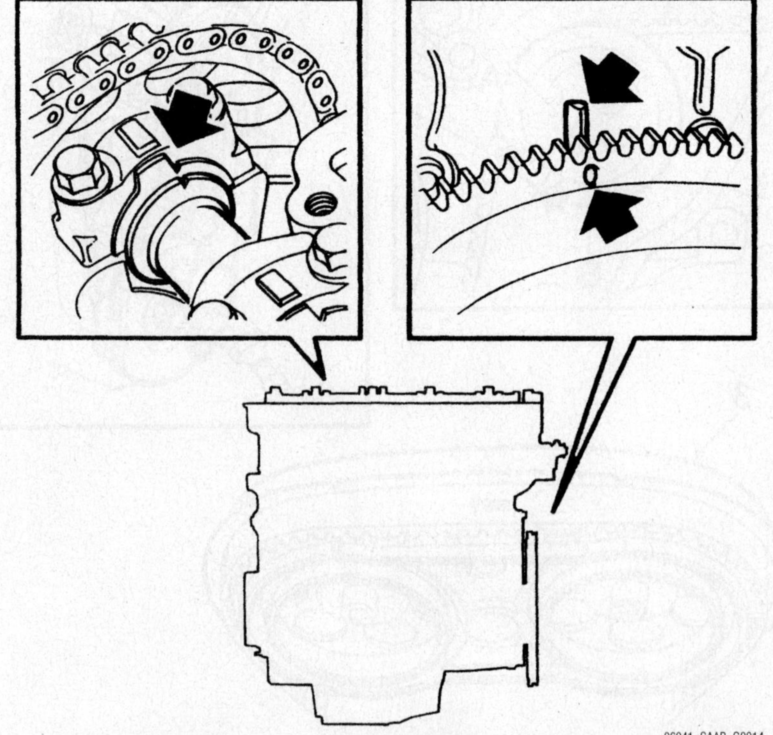

Be sure to align the crankshaft and camshaft timing marks, as shown—2.0L and 2.3L engines 2002 shown

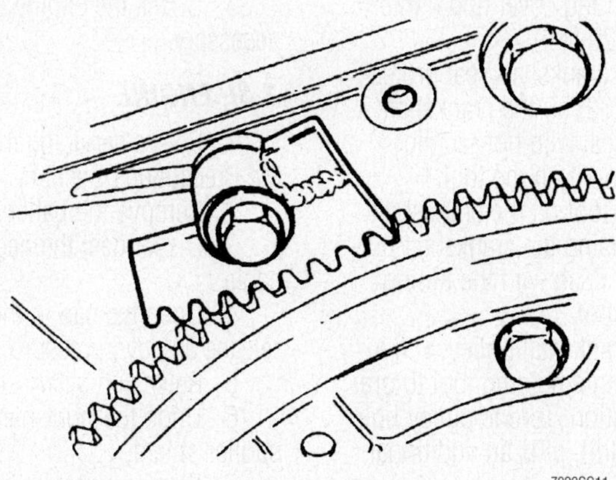

Secure the engine in position using a flywheel locking tool—2.0L and 2.3L engines

9. Remove the water pump and the sleeve with O-rings.

10. Remove the belt tensioner, generator and the bolt for the support bearing bracket in the timing cover.

11. Remove all the bolts on the timing cover, carefully tap the cover off and remove it.

To install:

12. Thoroughly remove all remains of sealant on all surfaces. Wash clean with benzene.

13. Apply a bead of flange sealant, 93 21 795, about 0.04 inch (1 mm) wide on the center of the sealing surfaces.

14. Fit the timing cover and torque to 16 ft. lbs. (22 Nm).

15. Install the bolt for support bearing bracket in the timing cover and fit the generator and belt tensioner.

16. Install water pump. Use new O-rings if needed.

17. Fit the crankshaft belt pulley. Use the flywheel locking attachment, part no. 83 94 868. Torque pulley bolt to 130 ft. lbs. (175 Nm).

18. Verify that oil pan is clean. Clean mating surfaces. Apply an even bead of flange sealant, 93 21 795, to the sealing surface of the oil sump and fit the oil pan. Torque pan bolts to 16 ft. lbs. (22 Nm).

19. Install the oil pan protective plate.

20. Rotate the crankshaft 45 degrees and fit the cylinder head with a new gasket. Make sure it is correctly seated on the guide sleeves and that the chain runs freely.

21. Install and tighten the cylinder head bolts according to bolt tightening sequence and to specified torque.

22. Install the 2 bolts between the timing cover and cylinder head. Tighten to 18 ft. lbs. (24 Nm).

23. Make sure the camshafts are aligned with their setting marks and rotate the crankshaft back to its 0-marking.

24. Starting with the intake camshaft, install the camshaft sprockets and chain.

➡**Do not tighten the bolts at this time.**

25. Prepare the chain tensioner for mounting by pressing down the catch and pressing in the chain tensioner. Install the chain tensioner, with a 27mm socket, and torque to 47 ft. lbs. (63 Nm).

26. Install the chain tensioner plug with push rod and spring. Torque the plug to 17 ft. lbs. (23 Nm).

27. Be sure the chain is positioned correctly on the chain guards. Turn the crankshaft 2 revolutions and check the setting of the flywheel and camshafts.

28. Tighten the camshaft sprocket bolts to 47 ft. lbs. (63 Nm).

29. Clean the sealing surface of the camshaft cover with benzene.

30. Apply soap to the opening in the camshaft cover and fit the cover, starting at the opening. Then, tighten the bolt located furthest to the front at the timing chain end. Continue all the way round the outside and inside. Torque the bolts to 11 ft. lbs. (15 Nm).

31. Install the spark plugs to 21 ft. lbs. (28 Nm). Install the ignition discharge module. Tighten to 8 ft. lbs. (11 Nm).

32. Connect the ignition discharge module electrical connection and the crankcase ventilation and vacuum hoses.

33. Install the thermostat housing cover and connect the hose for the throttle body preheating.

34. Tighten the lock nuts on the turbocharger to 19 ft. lbs. (25 Nm).

17

42348-SAAB-G19

Loosen the bolts from the timing chain front cover, starting with the last bolt and working in reverse. To tighten, start with the bolt hole in the opening of the cover and work around in sequence —2.0L and 2.3L engines

35. Tighten the bolts on the turbocharger stay. Start with the one in the cylinder block. Torque to 16 ft. lbs. (22 Nm).

36. Fit the water pipe to the thermostat housing and tighten bolts to 19 ft. lbs. (25 Nm).

37. Fit the crankcase ventilation pipe together with the intake manifold (turbo). Tighten the turbo/intake manifold bolts to 6 ft. lbs. (8 Nm) and the crankcase vent pip to 18 ft. lbs. (24 Nm).

38. Tighten the bolts on the intake manifold stay, starting with the one in the cylinder block and fit the oil filler pipe. Tighten stay-to-block bolts to 18 ft. lbs. (24 Nm). Tighten the upper mounting bolts to 7 ft. lbs. (10 Nm). Tighten the lower mounting bolts to 16 ft. lbs. (22 Nm).

39. Be sure the oil plug is tight and connect the cables to the engine oil pressure sensor, generator and starter motor.

2003–06 MODELS (EXCEPT 2.8L ENGINE)

1. Before servicing the vehicle, refer to the Precautions Section.
2. Raise the car and remove the front right wheel.
3. Remove the right wing liner.
4. Relieve the belt tensioner, using 83 96 095, and remove the belt. Mark the direction of rotation of the belt.
5. Remove the belt tensioner.
6. Lower the car and remove the upper engine cover.
7. Unplug the mass air flow sensor connector and detach the inlet hose from the air cleaner casing cover. Remove the cover and the air filter. Remove the SAI hose.
8. Detach the inlet hose and remove the air cleaner casing. Unplug the connector for the A/C pressure sensor.
9. Remove the turbocharger heat shield.
10. Install the lifting eye kit 83 96 178.
11. Fit the lifting beam, 83 94 850, and holder, 83 95 287. Take the weight off the engine and remove the bolts in the engine bracket.
12. Lower the engine slightly.
13. Raise the car and remove the crankshaft pulley using a holding tool.
14. Carefully remove the timing cover.
15. Cut away the timing cover gasket around the engine mounting and remove the gasket.
16. Remove the crankshaft seal.

To install:

17. Clean all the sealing surfaces.
18. Cut off the part of the gasket that is around the engine mounting and position the new timing cover gasket.
19. Install the timing cover and tighten bolts to 15 ft. lbs. (20 Nm).
20. Position the crankshaft seal protective sleeve, 83 96 202, on the crankshaft. Lubricate the new seal with non-acidic Vaseline and position it on the tool.
21. Position the tool on the crankshaft. Screw in the seal, using the crankshaft pulley bolt so that it is flush with the timing cover. Remove the tool.
22. Install the crankshaft pulley with a new bolt, while using a holding tool to prevent crankshaft rotation. Torque pulley bolt to 74 ft. lbs. (100 Nm), plus an additional 75 degrees.
23. Lower the car and raise the engine with the lifting beam until it rests against the engine mount.
24. Fit the bolts to the engine bracket. Remove the lifting beam with holder. Tighten the bolts to 52 ft. lbs. (70 Nm), plus an additional 60 degrees.
25. Remove the lifting eye kit.
26. Install the turbocharger heat shield.
27. Install the air cleaner casing and connect the intake hose. Plug in the A/C pressure sensor connector.
28. Install the filter element and the air cleaner casing cover. Connect the intake hose and the MAF sensor connector. Fit the SAI hose.
29. Install the upper engine cover.
30. Raise the car and fit the belt tensioner, tightening the bolt to 37 ft. lbs. (50 Nm).
31. Relieve the belt tensioner with the belt circuit relieving tool, 83 96 095 and install the belt in the marked direction of rotation. Make sure the belt is located correctly on all the pulleys.
32. Install the right fender liner and the front wheel.
33. Check the engine oil level. Top up as necessary.

2.8L ENGINE

1. Before servicing the vehicle, refer to the Precautions Section.
2. Remove the battery cover.
3. Disconnect the negative battery cable.
4. Open the cap to the expansion tank, release any overpressure.
5. Raise and safely support the vehicle.
6. Undo the right-hand side of the spoiler shield.
7. Place a suitable receptacle under the radiator. Connect a hose to the radiator and drain the coolant. Close the cock and refit the spoiler shield.
8. Lower the vehicle.
9. Remove the upper section of the air cleaner. When removing the air cleaner casing cover, be careful not to damage the brake vacuum pump sensor.
10. Remove the lower section of the air filter.
11. Unplug the mass air flow sensor connector. Detach the secondary air pipe, the delivery hose from the turbocharger intake manifold and the crankcase ventilation pipes. Take care when releasing the locking mechanism on the connector so as not to damage the connector. Pull the halves straight apart to avoid bending the pins.
12. Detach the turbocharger intake manifold from the air filter housing cover and the turbocharger.
13. Remove the upper engine cover.
14. Remove the charge air pipe by first removing the clamps and the charge pressure sensor connectors. Detach the power steering line.

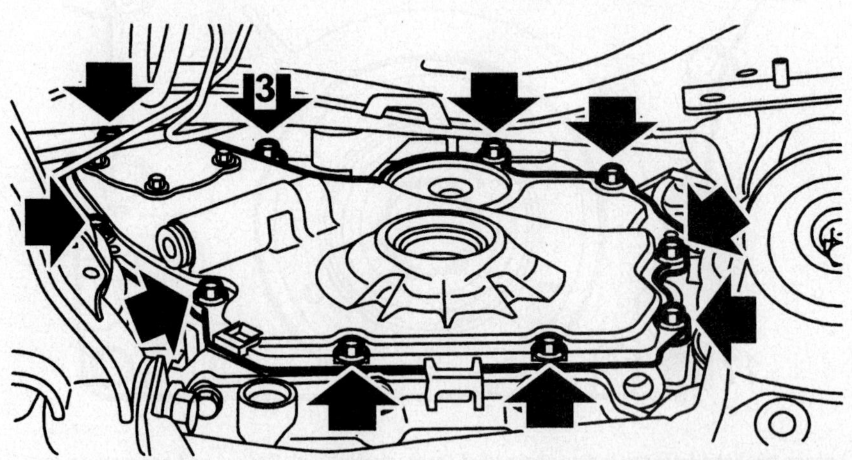

Showing the location of timing cover bolts (start loosening or tightening with the bolt shown with the number)—9-3 2.0L engine 2003–06

15. Remove the coolant return line by first removing the wiring harnesses.

16. Remove the connector and the ground connection to the engine control module.

17. Remove the engine control module bolts.

18. Remove the wiring harnesses holders by removing the cable duct, the connectors and the nut. Unplug the connector to the intake manifold pressure sensor. Remove the bolt, cable tie and connector.

19. Undo the quick coupling and remove the vacuum line from the upper section of the intake manifold.

20. Remove the crankcase ventilation lines by removing the bolt and removing the holder.

21. Remove the holder to the injector's connector.

22. Remove the throttle body connector and remove the upper section of the intake manifold.

23. Undo the quick couplings and remove the bleeder valve plus lines.

24. Place a collecting pan under the vehicle.

25. Detach the fuel line, wiping up any fuel spill with a cloth.

26. Remove the power steering reservoir by removing the bolt, return hose and clamp. Seal the hose and reservoir. Turn the reservoir to the side.

27. Unplug the connectors to the fuel pressure sensor.

28. Remove the wiring harnesses from the cylinder head for cylinders 1-3-5. Unplug the camshaft position sensor connector, the connector to the camshaft setting valve and the holder.

29. Remove the wiring harnesses from the cylinder head for cylinders 2-4-6. Unplug the camshaft position sensor connector, the connector to the camshaft setting valve and the holder.

30. Remove the wiring harness for the injectors by first unplugging the injectors' connectors.

31. Remove the fuel rail with injectors.

32. Remove the lower section of the intake manifold. Seal the inlet ducts to the cylinder head with lintless rags.

33. Remove the ignition coils for cylinders 2-4-6 by first unplugging the connectors.

34. Remove the wiring harness holder from the camshaft cover for cylinders 1-3-5.

35. Remove the ignition coils for cylinders 1-3-5 by first unplugging the connectors.

36. Remove the ground cable from the wiring harnesses by carefully cutting open the tape strip (arrows). Place the ground cable to one side.

37. Remove the wiring harnesses plus connectors. Remove the cable tie.

38. Remove the battery cover bottom section.

39. Loosen the expansion tank from the mounting.

40. Remove the turbocharger heat shield panel.

41. Remove the crankcase ventilation line from the turbocharger.

42. Remove the crankcase ventilation check valve from the camshaft cover for cylinders 1-3-5.

43. Undo the quick coupling and remove the secondary air system's valve for cylinders 2-4-6.

44. Undo the upper quick coupling by removing the bolt and removing the secondary air system's delivery line.

45. Remove the turbocharger diaphragm unit by undoing the bolt and pressing it out.

46. Remove the secondary air system's delivery line for cylinders 1-3-5.

47. Remove the secondary air system's delivery line for cylinders 2-4-6.

48. Remove the ventilation line to the oil dipstick's guide tube by undoing the quick coupling.

49. Remove the engine's wiring harness from cylinders 2-4-6 by unplugging the connectors, and by removing the head shield and the ground cable from the cylinder head. Fold the wiring harnesses to the side.

50. Remove the engine's transport lug by removing the bolts.

51. Remove the camshaft cover for cylinders 2-4-6 by removing the bolts.

52. Remove the camshaft cover for cylinders 1-3-5 by removing the bolts.

53. Raise the car to half height.

54. Remove the front wheels.

55. Remove the right wing liner.

56. Remove the power steering hydraulic line from the holder on the front axle beam.

57. Remove the right engine mounting.

58. Relieve the load on the belt tensioner. Use a " inch puller handle and remove the driver belt from the A/C compressor pulley.

59. Remove the bolt holding the crankshaft pulley using 83 95 360 Holding tool, crankshaft pulley (handle only) and EN-47981 Holding tool, pulley.

60. Remove the pulley using EN-47982 Puller and EN-47981 Holding tool, pulley.

61. Remove the transaxle housing at the bottom by removing the bolts.

62. Remove the wiring harness holder from the bottom of the transaxle housing by removing the bolt.

63. Lower the vehicle.

64. Remove the coolant line from the coolant outlet pipe by removing the bolt.

65. Remove the coolant outlet pipe by removing the coolant hose, remove the clamp and remove the bolts.

66. Remove the coolant pump pulley by removing the bolts. Hold firm with a screwdriver.

67. Remove the coolant pump by removing the bolts.

68. Remove the power steering pump at the top by removing the bolted joint.

69. Raise and safely support the vehicle.

70. Remove the power steering pump at the bottom by removing the bolt. Move the power steering pump towards the front of the car

71. Lower the vehicle.

72. Remove the drive belt tensioner by removing the bolt.

73. Remove the camshaft setting valve for cylinders 1-3-5 by removing the bolt.

74. Remove the camshaft setting valve for cylinders 2-4-6 by removing the bolt.

75. Remove the upper bolts of the transaxle housing.

76. Remove the timing cover. Carefully pry the cover loose with a pry bar and a screw that is carefully screwed in (to press the cover loose).

77. Remove the starter motor's wiring harness holder from the wiring harnesses.

➡There is no marking to check valve timing on the camshafts and cylinder heads. The timing chains have silver links that are to align with the marks on the sprockets. The links of the primary chain are marked with yellow.

78. Turn the crankshaft so that the cams of cylinder 1 point upward and inward at an angle, thus aligning the installation marks on the crankshaft sprocket and the oil pump. When the engine is approx. 15 degrees ATDC, use EN-46111 Sleeve to turn the engine.

79. Remove the timing chain tensioner of the rear cylinder bank.

80. Remove the chain tensioner gasket.

➡The exhaust camshaft turns somewhat when tension is relieved from the chain.

81. Remove the guides and lift off the chain.

82. Remove the chain tensioner of the primary circuit and remove the gasket.

83. Remove the guides and lift off the chain.

84. Remove the rear primary sprocket.

85. Remove the timing chain tensioner of the front cylinder bank and remove the gasket.

86. Remove the guides and lift off the front primary sprocket. Lift off the chain.

87. Remove the crankshaft sprocket, counterholding with a spanner on the camshaft flats.

88. Remove the crankshaft sprocket, counterholding with a fixed wrench on the camshaft flats.

⁂ CAUTION

The camshafts must not be rotated as the valves may collide with the pistons or other valves and be damaged.

To install:

89. Fit the crankshaft sprocket, counterholding with a fixed wrench on the camshaft flats. Tighten the bolt to 48 ft. lbs. (65 Nm).

90. Install the sprocket on the crankshaft. The marking should be visible. The intake sprocket is variable.

91. Check that the installation marks on the crankshaft sprocket and the oil pump are aligned. Use EN-46111 Sleeve to adjust the position of the crankshaft. The crankshaft must not be rotated. Otherwise the pistons could damage the valves.

92. Install the camshaft drive of the front cylinder bank. Position the chain on the camshaft sprockets so that the silver links align with the marks on the sprockets. The front cylinder bank sprockets are marked with an L.

93. Install the primary sprocket so that the lower silver link is visible through the hole in the primary sprocket and tighten the bolt to 48 ft. lbs. (65 Nm).

94. Install the chain guides and tighten the fasteners to 18 ft. lbs. (25 Nm).

95. Restore the chain tensioner to its previous position by removing the piston and turning the spring clockwise until it engages in the inner position. Use a screwdriver. Refit the piston in the tensioner and check that the groove in the piston is vertical so that is fits with the tensioner guide.

96. Install the chain tensioner using a new gasket and tighten the fasteners to 18 ft. lbs. (25 Nm).

97. Trigger/activate the chain tensioner by carefully pressing on the timing chain or guide.

98. Install the rear primary sprocket. Tighten the bolt to 48 ft. lbs. (65 Nm).

99. Install the primary chain so that the yellow links align with the marks on the primary sprockets and crankshaft sprocket.

100. Install the front chain guide and tighten the fasteners to 7 ft. lbs. (10 Nm).

101. Install the upper guide and tighten the fasteners to 18 ft. lbs. (25 Nm).

102. Restore the chain tensioner to its previous position by removing the piston and turning the spring clockwise until it engages in the inner position. Use a screwdriver. Refit the piston in the tensioner. Check that the groove in the piston is vertical so that is fits with the tensioner guide.

103. Install the chain tensioner using a new gasket and tighten the fasteners to 18 ft. lbs. (25 Nm).

104. Trigger/activate the chain tensioner by carefully pressing on the timing chain or guide.

105. Position the chain on the camshaft sprockets so that the silver links align with the marks on the sprockets. The rear cylinder bank sprockets are marked with an R.

106. Rotate the crankshaft approx. " turn approx. 100° ATDC so that the lower silver link aligns with the marking on the primary sprocket.

107. Install the chain guides and tighten the fasteners to 18 ft. lbs. (25 Nm).

108. Restore the chain tensioner to its previous position by removing the piston and turning the spring clockwise until it engages in the inner position. Use a screwdriver. Refit the piston in the tensioner. Check that the groove in the piston is vertical so that is fits with the tensioner guide.

109. Install the chain tensioner using a new gasket and tighten the fasteners to 18 ft. lbs. (25 Nm). Trigger/activate the chain tensioner by carefully pressing on the timing chain or guide.

110. Rotate the crankshaft one turn to check and ensure that all chain tensioners have been triggered and that the chain is correctly seated in the guide.

111. Clean the sealing surfaces.

112. Fit EN-46109-2 Centering tool in the engine block and cylinder head.

113. Apply 93 165 267 Flange sealant. Apply an approx. 3mm thick bead on the sealing surfaces of the transaxle housing.

114. Install the transaxle housing with help from an assistant. Check that the sealing bead remains undamaged during the alignment of the transaxle housing.

115. Install the upper bolts for the transmission housing and tighten to 17 ft. lbs. (23 Nm). Remove the centering tools.

116. Secure the starter motor's wiring harness holder on the wiring harnesses.

117. Raise and safely support the vehicle.

118. Fit the lower bolts of the transaxle housing and tighten to 17 ft. lbs. (23 Nm).

119. Install the lower bolt to the power steering pump and tighten to 16 ft. lbs. (22 Nm).

120. Install the pulley to the crankshaft using J-41998 Fitting tool to press on the pulley.

121. Install the pulley bolt using 83 95 360 Holding tool, crankshaft pulley (handle only) and EN-47981 Pulley holding tool. Tighten the pulley bolt to 74 ft. lbs. (100 Nm), plus an additional 150°.

122. Secure the starter motor's wiring harness holder at the bottom. Tighten the bolt.

123. Lower the vehicle.

124. Install the camshaft setting solenoid valves Tighten to 8 ft. lbs. (10 Nm).

125. Install the belt tensioner and the bolt. Tighten the bolt to 16 ft. lbs. (22 Nm).

126. Secure the starter motor's wiring harness holder at the top. Tighten the bolt.

127. Secure the power steering pump at the top. Tighten the fasteners to 16 ft. lbs. (22 Nm).

128. Install the mounting bolts to the engine pad's attaching plate and tighten to 32 ft. lbs. (44 Nm).

129. Install the coolant pump. Use a new gasket and tighten the bolts to 8 ft. lbs. (10 Nm).

130. Install the coolant pump's belt pulley. Counterhold with a screwdriver. Tighten the bolts to 9 ft. lbs. (12 Nm).

131. Install the coolant outlet pipe. Use a new gasket and tighten the bolts to 9 ft. lbs. (12 Nm).

132. Install the coolant hose and secure the clamp.

133. Unload the belt tensioner. Use a 1/4 inch puller and fit the drive belt.

134. Install the right engine mounting.

135. Install the holder to the power steering line and tighten the bolt to 6 ft. lbs. (8 Nm).

136. Fit the coolant line to the coolant outlet pipe. Use a new sealing ring. Tighten to 6 ft. lbs. (8 Nm).

137. Raise and safely support the vehicle.

138. Secure the power steering hydraulic line in the holder on the front axle beam.

139. Lower the vehicle to half height.

140. Install the right-hand wing liner.

141. Install the wheels.

142. Clean the camshaft cover sealing surfaces.

143. Replace the camshaft cover gasket for cylinders 2-4-6.

144. Apply an approx. 3mm thick bead of sealing compound on the separating points

of the cylinder head and transmission housing.

145. Install the camshaft cover for cylinders 2-4-6 and tighten the bolts to 8 ft. lbs. (10 Nm).

146. Install the engine's transport lug and tighten the bolt to 48 ft. lbs. (65 Nm).

147. Clean the camshaft cover sealing surfaces.

148. Replace the camshaft cover gasket for cylinders 1-3-5. Use new gaskets.

149. Apply an approx. 3mm thick bead of sealing compound on the separating points of the cylinder head and transmission housing.

150. Install the camshaft cover for cylinders 1-3-5 and tighten the bolts to 8 ft. lbs. (10 Nm).

151. Install the engine wiring harnesses to the cylinder head for cylinders 2-4-6 by plugging in the connectors on the cylinder head for cylinders 2-4-6. Fit the heat shield and fit the ground cable on the cylinder head for cylinders 2-4-6. Fit the bolts. Secure the wiring harnesses. Secure the engine control module's ground cable in the wiring harnesses. Fit the cable ties.

152. Take care when plugging in the connector so as not to damage or press out the pins/sleeves in the connector.

153. Install the secondary air system line for cylinders 2-4-6. Fit the exhaust pipe and tighten the bolts to 8 ft. lbs. (10 Nm).

154. Install the ventilation line to the oil dipstick's guide tube. Connect the quick coupling.

155. Install the turbocharger diaphragm unit by fitting the bolts.

156. Install the secondary air system's delivery line. Connect the quick couplings.

157. Install the secondary air system's check valve for cylinders 2-4-6. Use a new gasket. Tighten the bolts and connect the quick coupling.

158. Install the secondary air system's delivery line and bolt.

159. Install the check valve to the crankcase ventilation on the camshaft cover for cylinders 1-3-5.

160. Install the crankcase ventilation line on the turbocharger and tighten the bolt to 8 ft. lbs. (10 Nm).

161. Install the turbocharger's upper heat shield panel. Fit the nut and tighten to 8 ft. lbs. (10 Nm).

162. Install the expansion tank in the holder.

163. Install the battery cover bottom section.

164. Install the charge pressure regulator's wiring harness. Plug in the connector. Fit the cable tie.

165. Install the ignition coils for cylinders 2-4-6. Plug in the connectors and fit the bolts. Tighten the bolts to 8 ft. lbs. (10 Nm).

166. Install the wiring harness holder on the camshaft cover for cylinders 1-3-5. Secure in the holder.

167. Install the ignition coils for cylinders 1-3-5. Plug in the connectors and fit the bolts. Tighten the bolts to 8 ft. lbs. (10 Nm).

168. Install the lower section of the intake manifold with a new seal. Remove the rags from the inlet ducts.

169. Install the fuel rail with injectors and tighten the bolts to 8 ft. lbs. (10 Nm).

170. Install the injectors' wiring harness. Plug in the connectors.

171. Plug in the connectors to the fuel pressure sensor.

172. Install the wiring harness to the camshaft guide for cylinders 2-4-6. Plug in the connectors and secure in the holder.

173. Connect the wiring harness to the camshaft guide for cylinders 1-3-5. Plug in the connectors and secure in the holder.

174. Install the power steering reservoir. Fit the return hose. Secure the clamp and tighten to 7 ft. lbs. (9 Nm).

175. Connect the fuel delivery line.

176. Fit the bleeder valve plus lines. Connect the quick couplings. Plug in the connectors to the bleeder valve.

177. Fit the upper section of the intake manifold. Plug in the throttle body's connector. Use new gaskets and tighten the fasteners to 17 ft. lbs. (23 Nm).

178. Fit the holder to the injector's connector.

179. Fit the lines for the crankcase ventilation.

180. Fit the vacuum line on the upper section of the intake manifold. Connect the quick coupling.

181. Fit the wiring harnesses' holders. Fit the bolt. Plug in the connectors for the intake manifold's pressure sensor. Fit the cable cables. Secure the connectors. Fit the cable duct. Plug in the connectors. Fit the nut.

182. Install the engine control module. Fit the bolts and tighten to 8 ft. lbs. (10 Nm).

183. Fit the ground connection.

184. Plug in the engine control module connector.

185. Install the coolant return line. Fit the clamps and tighten the bolts to 8 ft. lbs. (10 Nm).

186. Install the charge air pipe. Install the clamp.

187. Plug in the charge pressure sensor connector. Secure the power steering line with clips.

188. Install the air intake pipe.

189. Install the air filter housing with the mass air flow sensor as follows:

a. Plug in the mass air flow sensor connector.

b. Secure the cable ties.

c. Fit the secondary air system's intake manifold. Secure the line.

d. Connect the quick coupling.

190. Install the upper engine cover. Remove the oil filler opening's screw cap. Fit the protective cover. Fit the oil filler opening's screw cap.

191. Fill and bleed the cooling system.

192. Connect the negative battery cable and install the battery cover.

9-5

2.3L ENGINE

1. Before servicing the vehicle, refer to the Precautions Section.
2. Remove or disconnect the following:
   - Dipstick
   - Idler pulley
   - Power steering pump and bracket with the lifting eye
   - Water pump and the sleeve with O-rings
   - Protective plate and oil pan (leave guide sleeve in cylinder block)
   - Crankshaft pulley (use locking segment 83 94 868 on flywheel)
   - Crankcase breather hose from the oil pan
   - Locating pins in the timing cover by cutting an internal thread in them using a ⅜ inch UNC thread tap and withdraw them with sliding hammer 83 90 270.
   - Timing cover retaining bolts
3. Pull the timing cover away, starting at the bottom, then lift the cover outwards/downwards to avoid damaging the gasket at the cylinder head.

To install:

4. Thoroughly remove all remains of sealant on all surfaces.
5. Apply a bead of Loctite 518 about 0.40 inch (1mm) thick along the middle of the sealing surfaces.
6. Position the timing cover, carefully turning it into position. Fit the retaining bolts, but do not tighten them. Tap the locating pins in place.
7. Now, tighten the timing cover bolts to 16 ft. lbs. (22 Nm).
8. Install new water pump O-rings
9. Clean the hole in the block and tighten the sleeve until its large wing is

pointing horizontally towards the flywheel end.

10. Lubricate the O-rings and fit the water pump.

11. Install the crankshaft pulley, using a locking segment, 83 94 868, on the flywheel. Tighten to 130 ft. lbs. (175 Nm).

12. Be sure oil pan is clean. Also clean mating surfaces.

13. Apply an even bead of Loctite 518 on the oil pan sealing surface and position the pan in place. Tighten bolts to 16 ft. lbs. (22 Nm).

14. Install the oil pan protective plate.

15. Plug in the oil level sensor connector and press the cable back into its clamps.

16. Check that the oil plug is properly tightened and connect the cables to the oil pressure sensor, generator, and starter motor. Tighten to 19 ft. lbs. (25 Nm).

Piston and Ring

POSITIONING

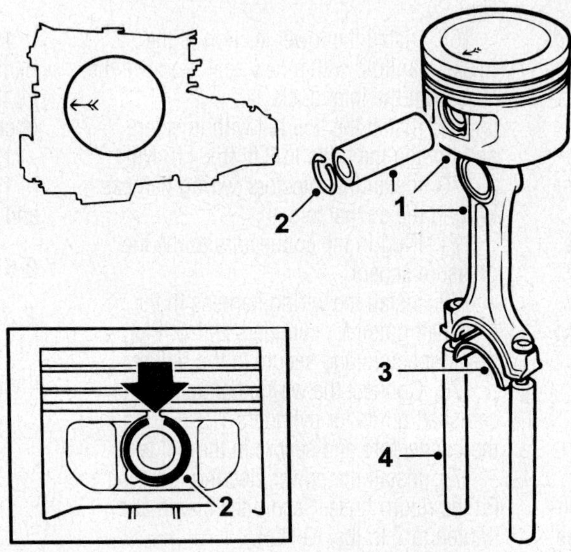

Piston and connecting rod assembly positioning—2.0L and 2.3L engines

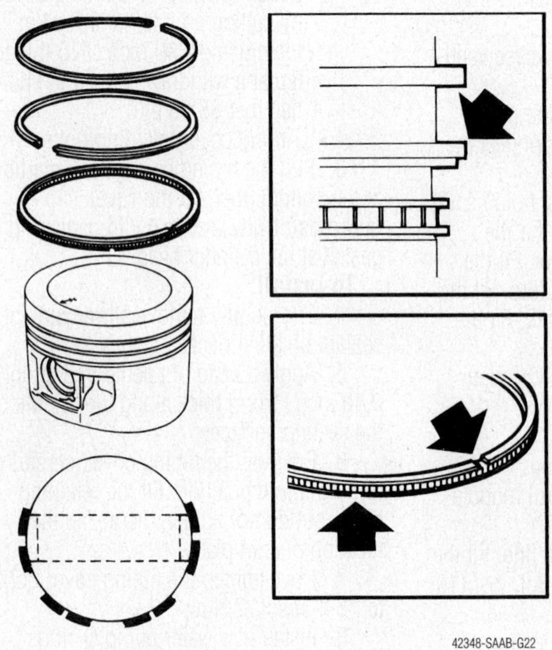

Piston ring positioning—2.0L and 2.3L engines

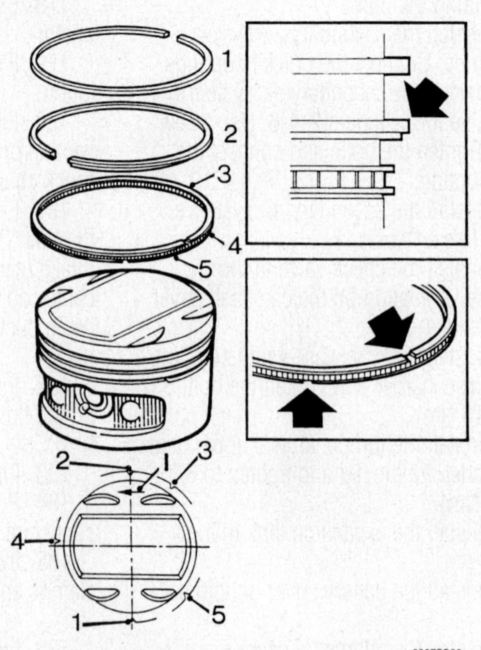

3.0L engine—piston and connecting rod assembly positioning

FUEL SYSTEM

Fuel System Service Precautions

Safety is the most important factor when performing not only fuel system maintenance but any type of maintenance. Failure to conduct maintenance and repairs in a safe manner may result in serious personal injury or death. Maintenance and testing of the vehicle's fuel system components can be accomplished safely and effectively by adhering to the following rules and guidelines.

1. To avoid the possibility of fire and personal injury, always disconnect the negative battery cable unless the repair or test procedure requires that battery voltage be applied.
2. Always relieve the fuel system pressure prior to disconnecting any fuel system component (injector, fuel rail, pressure regulator, etc.), fitting or fuel line connection. Exercise extreme caution whenever relieving fuel system pressure to avoid exposing skin, face, and eyes to fuel spray. Please be advised that fuel under pressure may penetrate the skin or any part of the body that it contacts.
3. Always place a shop towel or cloth around the fitting or connection prior to loosening to absorb any excess fuel due to spillage. Ensure that all fuel spillage (should it occur) is quickly removed from engine surfaces. Ensure that all fuel soaked cloths or towels are deposited into a suitable waste container.
4. Always keep a dry chemical (Class B) fire extinguisher near the work area.
5. Do not allow fuel spray or fuel vapors to come into contact with a spark or open flame.
6. Always use a back-up wrench when loosening and tightening fuel line connection fittings. This will prevent unnecessary stress and torsion to fuel line piping. Always follow the proper torque specifications.
7. Always replace worn fuel fitting O-rings with new. Do not substitute fuel hose where fuel pipe is installed.

Fuel System Pressure

RELIEVING

9-3 and 9-5

⁂ CAUTION

The fuel injection system remains under pressure, even after the engine has been turned OFF. The fuel system pressure must be relieved before disconnecting any fuel lines. Failure to do so may result in fire and/or personal injury.

1. Before servicing the vehicle, refer to the Precautions Section.
2. Remove fuel pump fuse from fuse panel while the engine is running.
3. Switch off the ignition once the engine has stopped.
4. Reinstall fuel pump fuse.

Fuel Filter

REMOVAL & INSTALLATION

9-3 (2002 Models) and 9-5 (2002–05 Models)

⁂ CAUTION

The fuel injection system remains under pressure, even after the engine has been turned OFF. The fuel system pressure must be relieved before disconnecting any fuel lines. Failure to do so may result in fire and/or personal injury.

1. Before servicing the vehicle, refer to the Precautions Section.
2. Properly relieve the fuel system pressure.
3. Be sure that the ignition switch is in the **OFF** position.
4. Locate the fuel filter, which is mounted under the vehicle and forward of the fuel tank.
5. Remove the fuel filter splash guard, if equipped.
6. Drain the fuel from the filter before removing it from the vehicle.
7. Thoroughly clean the area around the banjo fittings before continuing.
8. Remove or disconnect the banjo fittings on both sides of the fuel filter. Contain the fuel with the shop towel. Properly dispose of the towel once the job is complete.

To install:

9. Install or connect the following:
   - New fuel filter. Make sure that the arrow is pointing in the correct direction of fuel flow
   - Banjo fittings to the filter with new sealing washers and torque to 16 ft. lbs. (21 Nm)
10. Install the fuel filter splash guard, if equipped.
11. Start the vehicle, check for leaks and repair if necessary.

9-3 (2003–06 Models) and 9-5 (2006 Models)

SAAB 9-3 for 2003–06 and 9-5 for 2006 have the fuel filter mounted on the fuel pump/sending unit located in the fuel tank.

Fuel Pump

REMOVAL & INSTALLATION

9-3

⁂ CAUTION

The fuel injection system remains under pressure, even after the engine has been tuned OFF. The fuel system pressure must be relieved before disconnecting any fuel lines. Failure to do so may result in fire and/or personal injury.

1. Before servicing the vehicle, refer to the Precautions Section.
2. Remove or disconnect the following:
   - Negative battery cable
   - Fuel into a suitable container
   - Rubber hoses from the tank. Plug the openings
   - Fuel filter clamp
   - Metal straps holding fuel tank. Support the tank from below with a pole jack or other suitable device.
3. Carefully lower the tank, right side first, until the top is visible.
4. Remove or disconnect the following:
   - Two wiring connectors to pump
   - Fuel pressure and return lines
   - The fuel tank
   - Retaining ring
5. Lift the pump about 2 inches, then rotate clockwise about 80° and remove

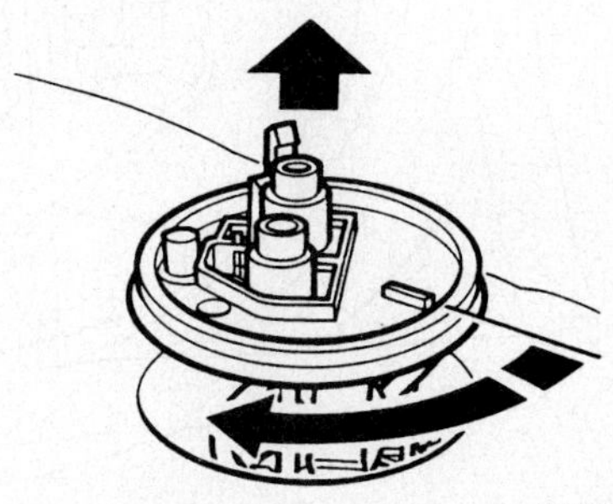
9301SG03

Removing the fuel pump

To install:

6. Install or connect the following:
   - Fuel pump and new O-ring
   - Retaining ring and torque the ring to 55 ft. lbs. (75 Nm)
7. Raise fuel tank and support with appropriate stands.
   - Fuel lines
   - Electrical connectors
   - Metal straps
   - Rubber hoses to tank
   - Fuel filter clamp
8. Lower the car and refill the tank.
9. Reconnect the negative battery cable and check that the system is operating properly and that there are no fuel leaks.

9-5

CAUTION

The fuel injection system remains under pressure, even after the engine has been tuned OFF. The fuel system pressure must be relieved before disconnecting any fuel lines. Failure to do so may result in fire and/or personal injury.

1. Before servicing the vehicle, refer to the Precautions Section.
2. Properly relieve the fuel system pressure.
3. Raise the rear seat cushions and fold the carpeting out of the way.
4. Remove or disconnect the following:
   - Negative battery cable
   - Fuel pump cover
   - Upper connector from the fuel pump
5. Carefully loosen the check valves and fuel lines from the pump. Move the yellow hooks to one side. The check valves are connected to the pump with quick release couplings.
6. The white check valve is the delivery side of the valve and marked "Pressure". The black check valve is the return side of the valve and is marked "Return" on the fuel pump.
   - Screw ring with tool 83-94-462 and lift the fuel pump until the upper section is slightly above the fuel tank
7. Rotate the fuel pump approximately 80 degrees and remove the pump from the fuel tank.

To install:

8. Clean the sealing surfaces on the fuel pump and fuel tank.
9. Install or connect the following:
   - New O-ring in the groove on the fuel tank and carefully lower the fuel pump into position. Make certain that the marks on the fuel pump and tank are opposite one another and press the pump into position.
   - Screw ring and torque the ring to 55 ft. lbs. (75 Nm)
   - Check valves to the fuel pump with new O-rings. Make certain that the valves are properly positioned
   - Fuel lines and upper connector
   - Negative battery cable
10. Start the vehicle, check for leaks and proper operation of the fuel pump and repair if necessary.
11. Install the fuel pump cover.
12. Reposition the carpeting and the rear seat cushions.

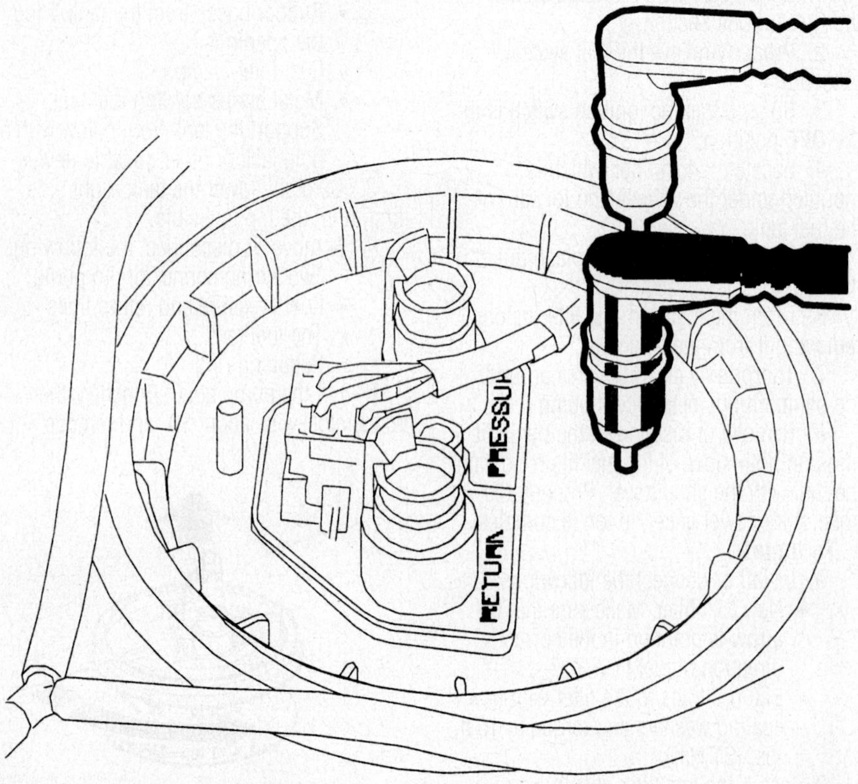

Remove the Pressure and Return valves from the top of the fuel pump

Fuel Injector

REMOVAL & INSTALLATION

9-3

2.0L & 2.3L ENGINES

CAUTION

The fuel injection system remains under pressure, even after the engine has been tuned OFF. The fuel system pressure must be relieved before disconnecting any fuel lines. Failure to do so may result in fire and/or personal injury.

1. Before servicing the vehicle, refer to the Precautions Section.
2. Properly relieve the fuel system pressure.
3. Disconnect the negative battery cable.
4. Detach the crankcase vent hoses.
5. Remove or disconnect the following:
   - Fuel line connections from fuel rail
   - Dipstick and filler pipe (plug pipe opening)
   - Cover over throttle valve
   - Accelerator wire from throttle spindle
   - Accelerator wire with holder (bend it to one side)
   - Cable channel (put it to one side)
   - Cable tie securing check valve to camshaft cover
   - Injector connectors

WARNING

Secure each injector clip with a narrow cable tie. Make sure they are positioned correctly and tighten them. Otherwise, the injectors may come loose from the fuel rail while being removed.

⁂ CAUTION

To prevent entry of foreign matter, regularly blow around the injectors with compressed air.

6. Undo both rail retaining screws.
7. Detach the pressure regulator vacuum hose.
8. Carefully pry off the rail using 2 crowbars or similar tool and lift out the rail.
9. Plug the nozzle holes.

To install:

10. Remove the plugs in the nozzle holes and lubricate the injector O-rings.
11. Carefully lift the rail into position and attach with 2 screws.
12. Install or connect the following:
    - Pressure regulator vacuum hose
    - Remove the cable ties round the injectors.
    - Plug in injector connectors
    - Cable channel and fit the 3 screws
    - Check valve to camshaft cover with cable tie
    - Throttle cable and bracket (adjust the throttle cable)
    - Throttle valve cover
    - Dipstick and filler pipe
    - Crankcase ventilation hoses
    - Fuel connections to fuel rail
    - Upper engine cover
    - Negative battery cable

2.8L ENGINE

⁂ CAUTION

The fuel injection system remains under pressure, even after the engine has been tuned OFF. The fuel system pressure must be relieved before disconnecting any fuel lines. Failure to do so may result in fire and/or personal injury.

1. Before servicing the vehicle, refer to the Precautions Section.
2. Properly relieve the fuel system pressure.
3. Disconnect the negative battery cable.
4. Remove the upper intake manifold.
5. Remove the intake manifold gaskets.
6. Unplug the injector connectors using a small screwdriver.
7. Detach the injectors' wiring harness.
8. Unplug the fuel pressure sensor connector.
9. Release any fuel pressure by pressing in the needle valve on the fuel rail. Catch any fuel spill with a cloth.
10. Detach the fuel line from the fuel rail using 83 95 261 Fuel line tool. Catch any fuel spill.
11. Remove the bolts. Remove the fuel rail with injectors from the engine.
12. Remove the clips from the valves.
13. Remove the valves from the fuel rail.

To install:

14. Install the injectors to the fuel rail using new seals coated with acid-free vaseline. Fit the clips on the injectors and press them into the fuel rail.
15. Install the fuel rail with injectors to the engine. Fit the bolts and tighten to 7 ft. lbs. (10 Nm).
16. Connect the fuel line to the fuel rail.
17. Plug in the fuel pressure sensor connector.
18. Connect the injectors' wiring harness.
19. Plug in the injector connectors.
20. Install the intake manifold gaskets.
21. Install the upper intake manifold.
22. Connect the negative battery cable.
23. Start the vehicle, check for leaks and repair if necessary.

9-5

2.3L ENGINE

⁂ CAUTION

The fuel injection system remains under pressure, even after the engine has been tuned OFF. The fuel system pressure must be relieved before disconnecting any fuel lines. Failure to do so may result in fire and/or personal injury.

1. Before servicing the vehicle, refer to the Precautions Section.
2. Properly relieve the fuel system pressure.
3. Remove or disconnect the following:
    - Negative battery cable
    - Upper engine cover
    - Crankcase ventilation hoses
    - Dipstick with the oil filler pipe
    - Cover and detach the throttle cable from the spindle
    - Turbocharger delivery pipe
    - Fuel injector connectors
    - Ignition discharge module
    - Manifold Absolute Pressure (MAP) sensor
    - Boost pressure control valve
    - Turbocharger pressure sensor and slacken the upper bolt on the wiring holder
    - Fuel rail retaining bolts and cable ties
    - Fuel lines
    - Pressure regulator vacuum hose
    - Fuel rail with the injectors
    - Fuel injector locking clips
    - Fuel injectors

To install:

4. Lubricate the O-rings on the fuel injectors
5. Install or connect the following:
    - Fuel injectors to the fuel rail and make certain that the injectors are fitted to the proper cables
    - Fuel rail and torque the bolts to 6 ft. lbs. (8 Nm)
    - Vacuum hose to the pressure regulator
    - Fuel injector hoses
    - Screws for the wiring holder
    - Throttle cable and cover and adjust if necessary
    - Dipstick and filler pipe
    - Crankcase ventilation hoses
    - Upper engine cover
    - Negative battery cable
6. Start the vehicle, check for leaks and repair if necessary.

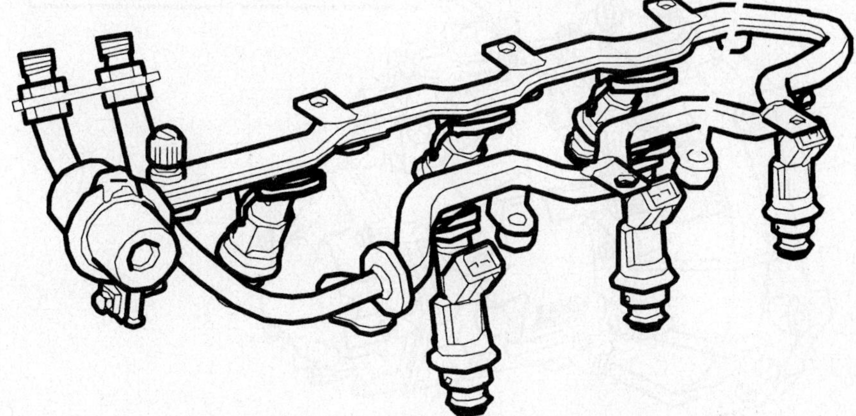

Remove the fuel rail and injectors as an assembly

3.0L ENGINE

CAUTION

The fuel injection system remains under pressure, even after the engine has been tuned OFF. The fuel system pressure must be relieved before disconnecting any fuel lines. Failure to do so may result in fire and/or personal injury.

1. Before servicing the vehicle, refer to the Precautions Section.
2. Properly relieve the fuel system pressure.
3. Remove or disconnect the following:
   - Negative battery cable
   - Engine cover
   - Turbocharger delivery hose from the throttle body
   - Throttle body from the intake manifold
   - Upper intake manifold
   - Fuel rail wiring
   - Vacuum hose from the fuel pressure regulator
   - Fuel rail and injectors as an assembly

To install:

4. Install or connect the following:
   - New O-rings on the injectors
   - Fuel rail and injectors as an assembly and torque the bolts to 6 ft. lbs. (8 Nm)
   - Fuel pressure regulator vacuum hose
   - Fuel lines and wiring
   - Upper intake manifold with a new gasket and torque the bolts to 15 ft. lbs. (20 Nm)
   - Throttle body
   - Throttle cable and adjust if needed
   - Turbocharger delivery hose
   - Engine cover
   - Negative battery cable
5. Start the vehicle, check for leaks and repair if necessary.

DRIVE TRAIN

Manual Transmission

REMOVAL & INSTALLATION

9-3 2002 Model

1. Before servicing the vehicle, refer to the Precautions Section.
2. Drain the transmission fluid.
3. Remove the cover over the intake manifold.

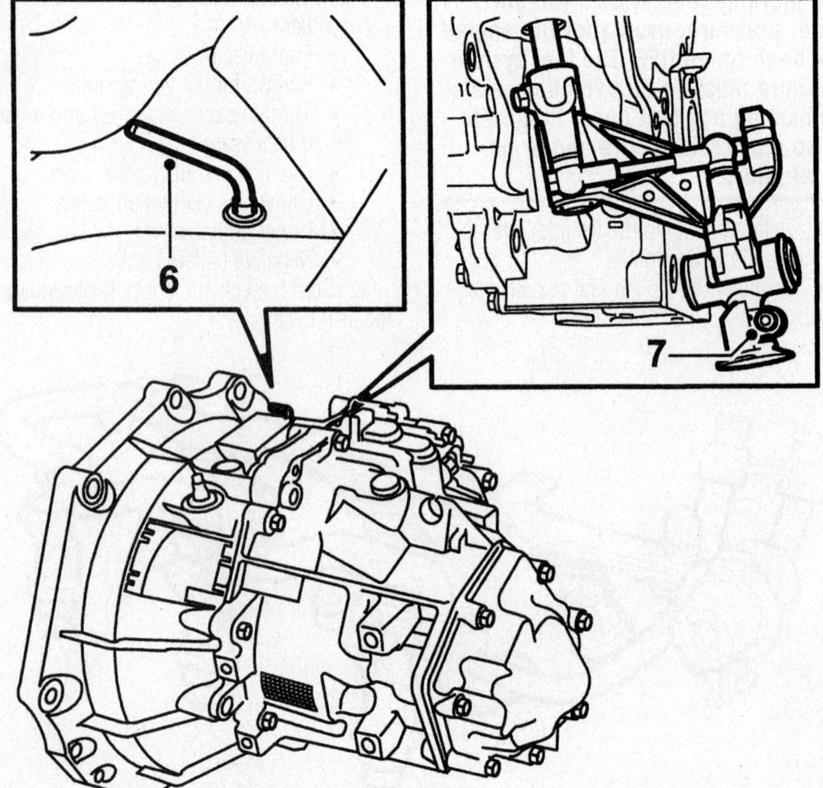

Engage 4th gear with a locking pin and remove selector rod clamp—9-3

4. If necessary, remove the resonator, together with hoses and mass air flow sensor.
5. Remove the battery.
6. Engage 4th gear and remove the plastic plug from the gearbox. Lock 4th gear in the gearbox by fitting a locking pin, 87 92 335.
7. Loosen the clamp securing the selector rod to the linkage in the gearbox from above.
8. Lift up the selector lever boot. Engage 3rd gear so that the selector rod disengages from the linkage and fit the locking pin, 87 92 335, in the gear lever housing.
9. Depress the clutch pedal about 2 inches (50mm) with a brake bracket. This prevents the brake fluid from flowing out of the reservoir.
10. Remove or disconnect the following:
    - Locking clip on the slave cylinder (disconnect delivery pipe)
    - Reverse light connector
    - Positive battery cable connection from transmission housing
    - 3 upper gearbox bolts
    - Connector(s) for the oxygen sensor(s) (release from clamps)
11. Slightly lift the engine and gearbox with lifting equipment, 83 94 850. Position the lifting device on the wheel housings so that it rests on the edges of the inner fenders.
12. Remove or disconnect the following:
    - Front wheels
    - Any clamps on oxygen sensor cable
    - Front exhaust system (first undo flange on turbo and then undo rear mounting)

CAUTION

It is important that the exhaust system is removed in this way as the flexible bellows may other wise be damaged. Be also careful with the oxygen sensor(s).

- Air shields and belt circuit cover in right-hand wheel housing
- End piece, suspension arm on both sides (use a puller, 89 96 696)

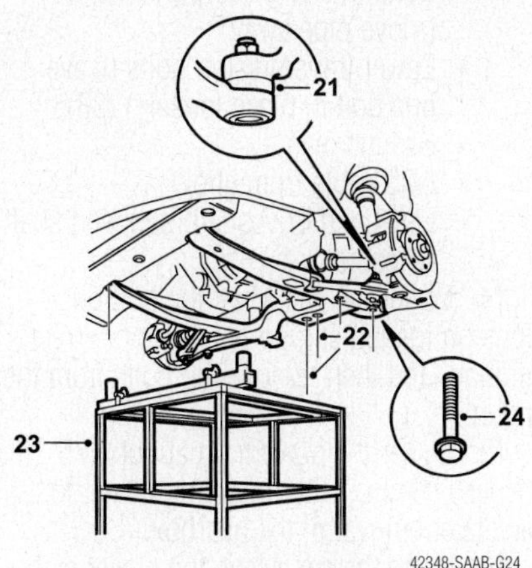

Using lifting trolley to support power train for transmission removal—9-3

- 2 nuts on rear engine mount and 2 center bolts on subframe (22)

13. Position complete lifting trolley (23) with front and rear holders with 83 94 801 parent fixture under the subframe. Raise the lifting trolley against the subframe.

14. Remove the remaining bolts (24) securing the subframe (note washers on the rear mountings).

15. Lower the trolley lift and push it to one side.

16. Remove or disconnect the following:

- Oil drain plug and drain oil from gearbox (screw in drain plug)
- Left-hand driveshaft (use removal tool for driveshafts, 89 96 654)
- Suspend driveshaft and insert plug, 87 92 244, in gearbox
- Splash plate by flywheel
- Ground leads from gearbox
- Bolts in engine/gearbox mating face accessible from below (except front bolt)
- 2 left-hand engine mount bolts at body
- Engine mount from gearbox

17. Lower the car and lower the whole drive unit slightly.

18. Raise the car and install a holding tool, 87 92 608, for a single column lift on a single column jack. Adjust the lifting tool so that it is centered in line with the gearbox center and mating face. Connect the lifting tool to the gearbox.

19. Remove the last bolt between the engine and the gearbox.

20. Remove the rear engine mount bracket from the gearbox.

21. Pull out and lower the gearbox. Lift the gearbox from the pillar jack with and engine lift and undo the lifting tool from the gearbox.

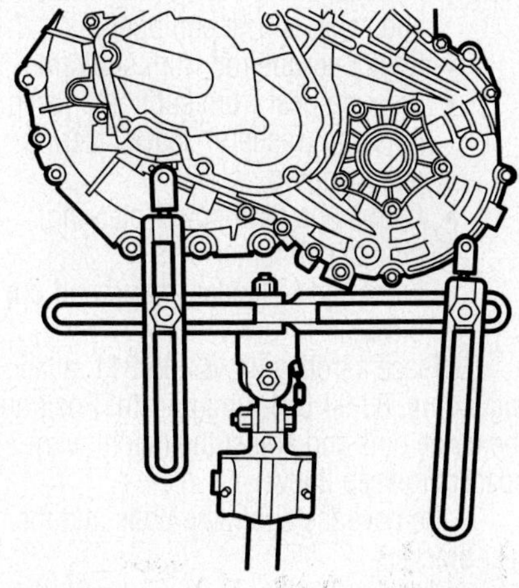

Using single column lift and jack to remove manual transmission—9-3

To install:

22. Check that the selector lever in the car is in 3rd gear and that it is locked with the locking pin, 87 92 335, for the gear lever housing. The locking pin ring should be down in the hole.

23. Grease the spines of the primary shaft.

➡Check that the two locating sleeves are on the engine.

24. Lubricate the sleeves with a thin coating of anti-corrosion oil.

25. If the slave cylinder is new or has been emptied of brake fluid, bleed the slave cylinder.

26. Install a protective collar, 83 95 162, in the right-hand shaft seal in the gearbox to protect the seal when the gearbox is mounted.

➡Always change the shaft seal in the gearbox. Lubricate with gearbox oil.

27. Attach a holding device, 87 92 608, for a single column lift. Make sure the lifting tool center is in line with the gearbox center and mating surface.

28. Fit the gearbox, with lifting tool, onto a pillar jack

29. Push in the gearbox until about. 0.8 inch (20mm) remain, remove the protective collar from the shaft seal, then push in the gearbox completely. Turn the engine shaft, if needed, so that the gearbox can be fitted.

30. Install or connect the following:

- Rear engine mount on gearbox
- Bottom bolts between the engine and gearbox to 50 ft. lbs. (70 Nm)
- Left-hand engine mount on gearbox and torque to 46 ft. lbs. (62 Nm)
- Lift drive unit with jack and tighten left-hand engine mount on the body to 46 ft. lbs. (62 Nm)
- Remove lifting tool from gearbox and remove jack
- Ground leads to gearbox
- Splash plate by flywheel

31. Install a protective collar, 83 95 162, onto left-hand shaft seal.

➡Make sure the driveshaft is clean and then position it in the tool.

32. Install the driver in the gearbox until about 0.8 inch (20mm) remains and withdraw the tool before the shaft's sealing surface reaches the shaft seal.

33. Insert the shaft until the locking ring snaps in place.

34. Carefully raise the subframe. Make sure the washers on the rear mounts are in place.

35. Adjust the subframe so that the screw holes correspond with the body.

36. Tighten the 4 corner bolts on the subframe as follows:

a. Front: 85 ft. lbs. (115 Nm)

b. Rear: 81 ft. lbs. (110 Nm), plus an additional 75 degrees

37. Lower the lifting trolley and move it away.

38. Tighten the remaining bolts on the subframe to 141 ft. lbs. (190 Nm).

39. Tighten the nuts on the rear engine mount to 37 ft. lbs. (50 Nm).

40. Position the suspension arm end piece on both sides and tighten the nuts to 55 ft. lbs. (75 Nm).

41. Reinstall the air shields and belt circuit cover in the right-hand wheel housing.

42. Install the front exhaust system and put back the clamps for the oxygen sensor cable.

43. Install the front wheels and tighten nuts to 81 ft. lbs. (110 Nm).

44. Remove the lifting equipment and tighten the upper bolts between the engine and gearbox to 50 ft. lbs. (70 Nm).

45. Install or connect the following:

- Oxygen sensor(s)
- Positive cable attachment on engine
- Lead to reverse light switch
- Clutch delivery pipe to gearbox with snap ring
- Remove locking pin from gear lever housing, engage 4th gear and insert locking pin again.
- Clamp on selector rod to 15 ft. lbs. (20 Nm)
- Gear lever boot

46. Make an initial bleed of the clutch as follows:

a. Take locking pin out of gearbox and put back plastic plug

b. Remove the brake bracket from the clutch pedal.

c. Attach a hose to the bleed nipple, open the nipple and let a little brake fluid run out.

d. Tighten the nipple and test the clutch.

e. Check that there are no system leaks.

f. Fit and connect the battery.

g. Install the engine cover or the resonator with hoses and the MAF sensor.

9-3 2003–06

5-SPEED & 6-SPEED

1. Before servicing the vehicle, refer to the Precautions Section.
2. Remove or disconnect the following:
    - Upper engine cover and battery cover
    - Battery (and cooling hose, if equipped)
    - Cable clamp under battery tray
    - Hood switch connector
    - Battery tray
    - Ground cable from engine bracket and reverse light switch connection (7)
    - Gear cables from the transmission by pulling locking sleeves back, lifting up cables from levers, and carefully moving them aside (8)
    - Pinch off clutch hose and undo quick-release coupling from clutch slave cylinder (9)
    - Coolant reservoir (6-speed)
    - Upper gearbox bolts
3. Install lift eyes, 83 96 178, to the engine and lifting beam 83 94 850 with holder 83 95 287 holder.
4. Remove the upper bumper shell mountings.
5. Suspend 2 straps in place above the radiator core so they are accessible from below. Remove or disconnect the following:
    - Front wheels
    - Front spoiler shield
    - Hose from the headlamp washers and unplug and remove connector
    - Secure radiator core with the straps
    - Radiator brackets from subframe
    - Front part of exhaust pipe by cutting 3.43 inches (87mm) from front end of muffler
    - Engine torque rod from subframe (6-speed)
    - Suspension arm from the steering swivel member on both sides
    - Anti-roll bar from link arm.
    - Headlamp angle sensor connector and cable clip (if equipped)
    - Engine torque rod from subframe
    - Steering gear from subframe (keep nuts and washers); leave steering gear hanging
    - Clamps holding power steering pipe to subframe
    - Front part of fender liner (bend out of way)
6. Place a trolley lift, 83 95 311, attaching fixture, 83 94 801, underneath. Position the guide pins and adjust the height with spacers to keep it level.
7. Remove the subframe bolts and the rear brackets.
8. Lower the subframe slightly.
9. Remove the anti-roll bar from the subframe, while protecting the steering gear boot.
10. Fit bolts loosely in the link arms so the anti-roll bar is kept hanging in place.
11. Pull out the steering arm ball joints from the steering swivel members.
12. Lower the subframe.
13. Remove or disconnect the following:
    - Torque rod bracket (remove torque rod bolt for easier fitting)
    - Gearbox oil (refit the oil plug)
    - Bolts in charge air pipe bracket (move pipe away)
    - Lower transmission bolts (leave one bolt in place loosely) (38)
    - Ground lead
    - ABS cable from clip
    - Left-hand driveshaft (suspend shaft by cables)
14. Lower the car. Mark the bolt positions on the left engine bracket for correct refitting, and then remove the bolts from the mount.
15. Lower the powertrain about 2.75 inches (70mm), with the lifting beam, to facilitate removal of the gearbox.
16. Raise the car. Move the power steering cooling coil aside and suspend it with a strap.
17. Install single-column lift holder, 87 92 608, onto a column jack. Adjust and secure the tool in the transmission. Use bolt of 8.8 grade that are about 0.8 inch (20mm) longer than the bolts that were removed.
18. Remove the last bolt, pull out the transmission and lower it.
19. Lift the transmission down from the column jack with an engine lift and chain.
20. Remove the lifting tool from the transmission.

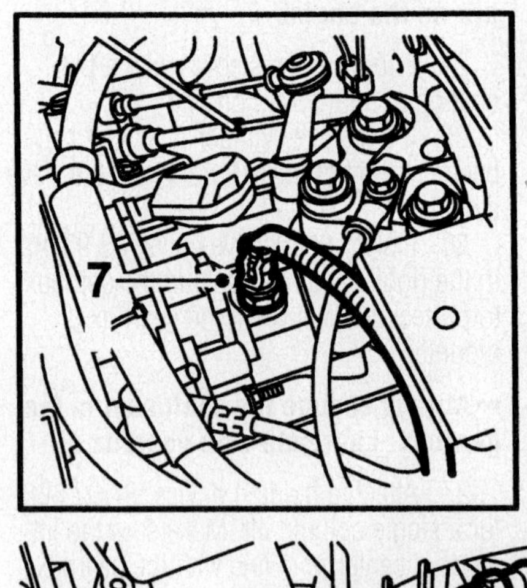

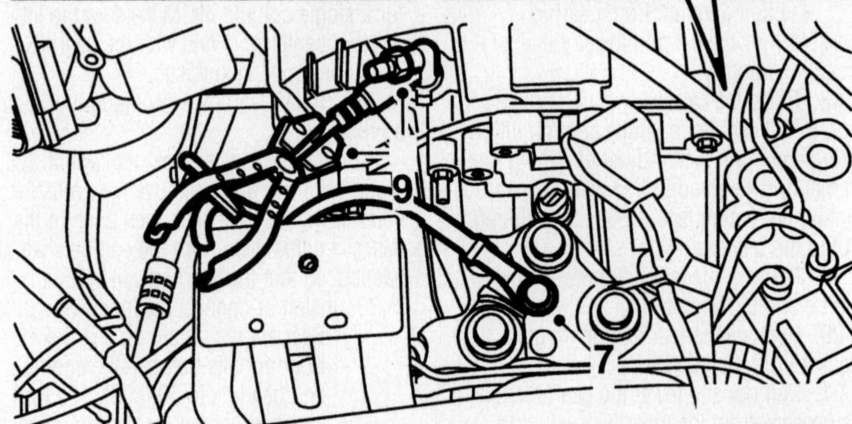

Disconnecting reverse light switch, removing gearbox cables, and pinching off clutch hoses—9-3 with 5-speed manual transmission

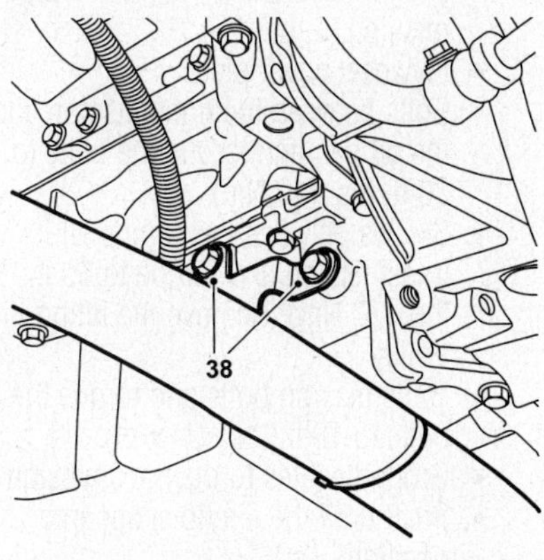

Showing the location of the lower transmission bolts—9-3 with 5-speed manual transmission

To install:

21. Bleed the slave cylinder, if necessary.
22. Lubricate the primary shaft splines.
23. Lubricate the guide sleeves on the engine with anti-corrosion agent.
24. Install a protective collar, 83 95 162, in the right shaft seal in the transmission. This is done to protect the seal while the transmission is being fitted. Lubricate the seal.

⁂ CAUTION

Always replace the shaft seals in the gearbox. Lubricate with gearbox oil.

25. Fit the gearbox, with the lifting tool, on the column jack.
26. Slide in the gearbox until approx. 0.8 inch (20mm) are remaining and remove the tool.
27. Push in the rest of the gearbox. Turn the crankshaft if necessary to get the gearbox in place.
28. Tighten all bolts, except the top one between the engine and the gearbox, to 30 ft. lbs. (40 Nm) for M10 bolts, and 52 ft. lbs. (70 Nm) for M12 bolts.
29. Install the charge air pipe bracket.
30. Remove the lifting tool from the gearbox and move the jack out of the way.
31. Reinstall the gearbox bolts and torque to 18 ft. lbs. (24 Nm).
32. Undo the straps and position the power steering cooling coil. Make sure it is fitted on the correct side of the rubber seal.
33. Lower the car. Lift in place the powertrain with the lifting beam until it meets the engine mounting.
34. Install the bracket for the engine mount according to the marks made earlier. Torque the bolts to 52 ft. lbs. (70 Nm), plus an additional 45 degrees.
35. Remove the holder and the lifting beam.
36. Install or connect the following:
    - 3 upper gearbox bolts to 52 ft. lbs. (70 Nm).
    - Protective collar, 83 95 162, in left driveshaft seal (remove strap, ensure driveshaft is clean, lubricate it and then align it with tool).
    - Driver in gearbox until approx.0.8 inch (20mm) are remaining and pull out tool before sealing surface of shaft reaches shaft seal; push in rest of driveshaft into gearbox until circlip clicks in.
    - ABS cable to the clip
    - Ground cable
    - Refill gearbox oil until level with plug hole.
    - Filler plug to 37 ft. lbs. (50 Nm) on 5-speed, or to 7 ft. lbs. (10 Nm) on 6-speed
    - Torque rod bracket, torque bolts to 52 ft. lbs. (70 Nm), plus an additional 90 degrees
37. Position the subframe using the fixture. Also use a 3/8 inch square extension, 82 93 102, to guide the subframe. Lift until the anti-roll bar can be fitted to the subframe.
38. Remove the link arm nuts and fit the anti-roll bar to the subframe. Torque bolts to 13 ft. lbs. (18 Nm).
39. Install the suspension arm ball joints to the steering swivel member and torque bolts to 37 ft. lbs. (50 Nm).

⁂ CAUTION

Ensure that the steering knuckle stub is visible on the top of the steering knuckle housing before the bolt is fitted.

40. Check that the guide pins fit into the reference holes, adjust the subframe until the guide pins go in easily and fit the subframe bolts and brackets. Torque as follows:
    - Subframe bolts (except rear): 52 ft. lbs. (70 Nm), plus as an additional 90 degrees
    - Rear subframe bolts: 66 ft. lbs. (90 Nm), plus an additional 45 degrees
41. Install the anti-roll bar link arms and torque bolts to 47 ft. lbs. (64 Nm).
42. Move the lift and jig away.
43. Fit the torque rods to the subframe and torque bolts to 52 ft. lbs. (70 Nm), plus an additional 90 degrees.
44. Set up the engine centering tool kit, 83 96 152, and check that the engine is located correctly in relation to the subframe. Remove the centering tool.
45. Install or connect the following:
    - Radiator brackets to subframe; remove straps and torque bracket bolts to 35 ft. lbs. (47 Nm)
    - Clamps holding power steering pipe to subframe
    - Steering gear to subframe, torquing bolts to 37 ft. lbs. (50 Nm), plus an additional 60 degrees
    - Joint clamp on front exhaust pipe and fit it to catalytic converter; use a new gasket and new nuts, tightening to 18 ft. lbs. (25 Nm)
    - Joint clamp where pipe was cut (place pipe ends in middle; torque joint clamp nuts to 30 ft. lbs. (40 Nm)
    - Connector and cable clip to the headlamp angle sensor (if equipped)
    - Coolant reservoir (6-speed)
    - Left-hand side cover
    - Fender liner section
    - Connector and hose to headlamp washer
    - Front spoiler shield
    - Front wheels
    - Bumper shell upper mountings
    - Quick-release coupling to clutch slave cylinder (remove hose pinch-off pliers); ensure connection is in correct position
    - Bleed the clutch
    - Gear cables to gearbox
    - Reverse light switch connection and ground cable to engine mount; torque bolt to 13 ft. lbs. (18 Nm)
    - Battery tray and hood switch connector
    - Battery (and cooling hose, if equipped)
    - Battery cover
    - Upper engine cover
46. Check the gear positions and adjust if necessary as described below:
    a. Engage 4th gear on the gearbox.
    b. Undo the gear lever cover with rubber mat and lift up the cover.
    c. Undo the cable adjusters using a screwdriver.
    d. Secure the gear lever in its adjusted position by lifting the adjusting sleeve while pressing in the two catches. Use a pair of pliers. Let the catch engage in the adjusting position. Check that 4th gear is engaged in the gearbox.
    e. Press on the gear adjusters with a screwdriver.
    f. Lift up the adjuster sleeve to its normal position.

g. Check the shifting positions.

47. Fit the gear lever cover and rubber mat.

9-5

1. Before servicing the vehicle, refer to the Precautions Section.
2. Drain the transmission.
3. Remove or disconnect the following:
   - Front grille (2001)
   - Intake manifold cover
   - Battery cover
   - Battery and tray
   - Main fuse box (2002-04)
4. Place the vehicle in 4th gear. Remove the plastic plug from the gearbox, and lock 4th gear in position with tool 87 92 335.
5. Remove the clip from the selector rod.
6. Raise gear selector boot. Engage 3rd gear so the selector rod disengages from the linkage. Install locking pin 87 92 335 in the lever housing.
7. Clamp brake hose or depress brake pedal about 2 inches (50mm) with brake clamp to prevent brake fluid from overflowing reservoir.
8. Remove or disconnect the following:
   - Clip from slave cylinder and disconnect delivery line
   - Reverse light connector
   - 3 upper gearbox bolts and install locating studs in engine upper, outer guide holes
   - Oxygen sensor
   - Selector linkage from the gearbox
   - Rear engine mount nut
   - 3 bolts from rear engine pad (loosen but do not remove)
   - Slightly lift engine and gearbox with a lifting beam on engine
   - Both front wheels, wheelwell covers, and under covers
   - Headlamp leveling sensor brackets (if equipped); bend sensors aside
   - Front exhaust pipe
   - Stay between the engine mount and engine
   - Rear engine mount and pad
   - Bolt securing torque arm to subframe (if equipped)
   - Steering gear bolts
   - Clips securing steering servo line to subframe
   - A/C lines from the subframe
   - Engine oil cooler from charge air cooler
   - Place strap around radiator and crossmember to retain radiator
   - Bolts securing outer ball joints to steering swivel
   - Loosen upper ball joint (both sides) from anti-roll bar
   - Flywheel cover plate
   - Bolts between gearbox and oil pan (B235R engines)
   - Bolts securing the rear support plate to subframe and place a lifting trolley under subframe
   - Subframe and front mounts for the steering servo delivery line
   - Gearbox oil
   - Left driveshaft
   - Ground cables from the gearbox
   - Left engine pad
   - Lower transmission so it clears structural member
   - Transmission mounting
9. Connect a lifting tool to the transmission
10. Remove the transmission from the vehicle

To install:

11. Lubricate the primary shaft splines.
12. Install a protective collar in the right shaft seal in the transmission.
13. Position the lifting trolley under the vehicle
14. Install or connect the following:
    - Transmission and turn the engine shaft if needed to install the transmission
    - Flywheel cover plate
    - Bolts between the transmission and the oil pan and torque the bolts to 30 ft. lbs. (40 Nm)
    - Screws between the engine and transmission and torque to 50 ft. lbs. (70 Nm). Remove the lifting trolley
    - Transmission bolts and torque the bolts to 18 ft. lbs. (24 Nm)
    - Ground cables to the transmission
    - Shaft until the circlip snaps into position
    - Transmission mountings and torque the bolts to 30 ft. lbs. (40 Nm)
    - Clutch delivery line and clips
    - Left engine pad. Torque the engine pad-to-transmission bolts to 62 ft. lbs. (85 Nm) and the engine pad to body bolts to 45 ft. lbs. (60 Nm)
    - Reverse light electrical connector
15. Position the lifting trolley under the vehicle with the subframe properly aligned.
    - Front ball joints and tighten the steering servo delivery lines
    - Radiator journals to the subframe

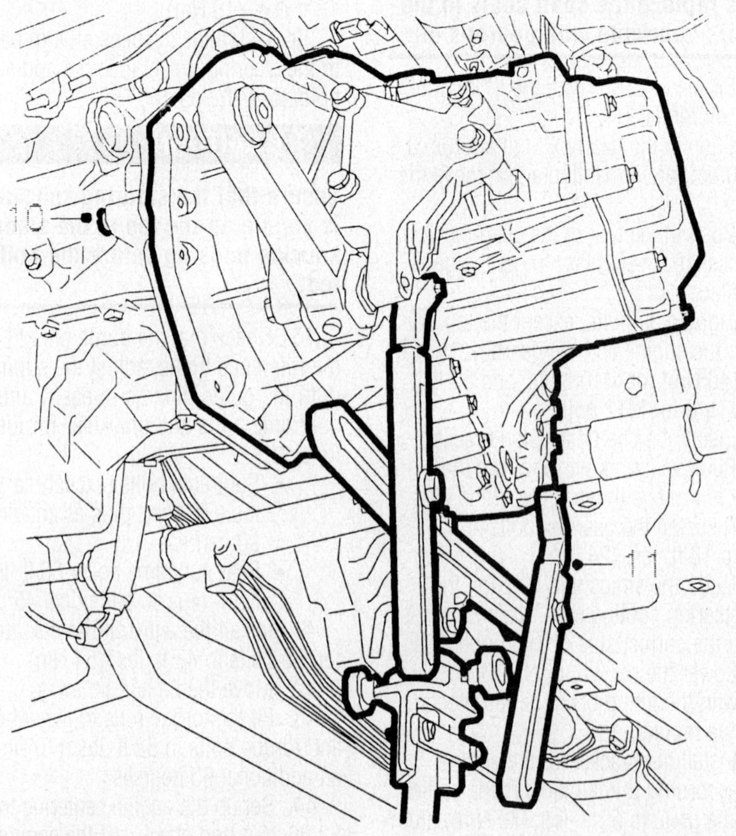

Install a lifting tool to the transmission

- A/C lines into the brackets on the subframe
- Screws and rear support plate for the subframe. Move the lifting trolley aside. Torque the subframe bolts to 74 ft. lbs. (100 Nm) plus an additional 45 degrees
- Torque the support plate bolts to 44 ft. lbs. (60 Nm)
- Torque rod-to-subframe bolts and torque the bolts to 22 ft. lbs. (30 Nm)
- Engine oil cooler
- Rear attaching clamps for the steering servo line to the subframe
- Steering gear and torque the bolts to 66 ft. lbs. (90 Nm)
- Outer ball joints to the steering swivel member and torque the bolts to 64 ft. lbs. 85 Nm)
- Anti-roll bar stay and torque the fastener to 64 ft. lbs. (85 Nm)
- Cover plate between the engine and transmission
- Rear engine pad and engine mount and torque the bolts to 50 ft. lbs. (70 Nm)
- Engine mount stay and torque the fastener to 16 ft. lbs. (22 Nm)
- Front exhaust pipe and torque the flange to the turbocharger to 19 ft. lbs. (25 Nm)
- Engine stay and torque the fastener to 16 ft. lbs. (22 Nm)
- Inner wheel well covers and remove the lifting devise from the top of the engine
- Upper engine bolts and torque the bolts to 50 ft. lbs. (70 Nm)
- Front torque arm to transmission and torque the arm to 34 ft. lbs. (47 Nm)
- Selector linkage to the transmission and torque the bolt to 18 ft. lbs. (24 Nm)
- Gear linkage to the selector rod. Place the vehicle in 4th gear and secure with a locking pin. Torque the clamp on the linakge to 16 ft. lbs. (22 Nm)
- Oxygen sensor connectors
- Battery and tray. Connect the battery cables
- Battery cover
- Front wheels
- Throttle body cover
- Grille

16. Fill and bleed the clutch system.
17. Fill the transmission fluid to the proper level.
18. Start the vehicle, check for leaks and proper operation. Repair if necessary.

Automatic Transmission

REMOVAL & INSTALLATION

9-3

2002 MODEL

1. Before servicing the vehicle, refer to the Precautions Section.
2. Drain the transmission.
3. Remove or disconnect the following:
    - Intake manifold cover or resonator with hoses and MAF sensor
    - Battery
    - Ground strap from the transmission housing
    - Dipstick and sleeve (plug opening in block)
    - Vent hose from the transmission housing
    - Positive cable routing straps
    - Gearshift selector lever from housing
    - Electrical harness on transmission housing
    - Oxygen sensor connectors and securing straps
    - 3 upper bolts on gearbox
    - Remove strain from powertrain with engine hoist
    - Front wheels
    - Front exhaust pipes
    - Cooler lines from the transmission (plug all openings)
    - Air shields and cover in right wheel housing
    - End pieces from suspension arms on both sides
    - 2 nuts from rear engine mount and 2 middle bolts from subframe
4. Position lifting trolley, 83 94 793, with fixture, 83 94 801, and front and rear holders, 83 94 819 and 83 90 827 under subframe.
5. Remove the remaining bolts securing the subframe. Note position of washers on rear mount.
6. Lower the lifting trolley assembly, with subframe and move it out of the way.
7. Detach the left driveshaft from the gearbox by releasing at the splines, then pulling straight out.
8. Suspend the driveshaft with a securing strap.
9. Remove or disconnect the following:
10. Remove the flywheel splash plate.
11. Remove the torque converter-to-driveplate bolts (rotate crankshaft to access bolts).
12. Remove the rubber plug and install a torque converter holding tool, 87 92 277, into opening.

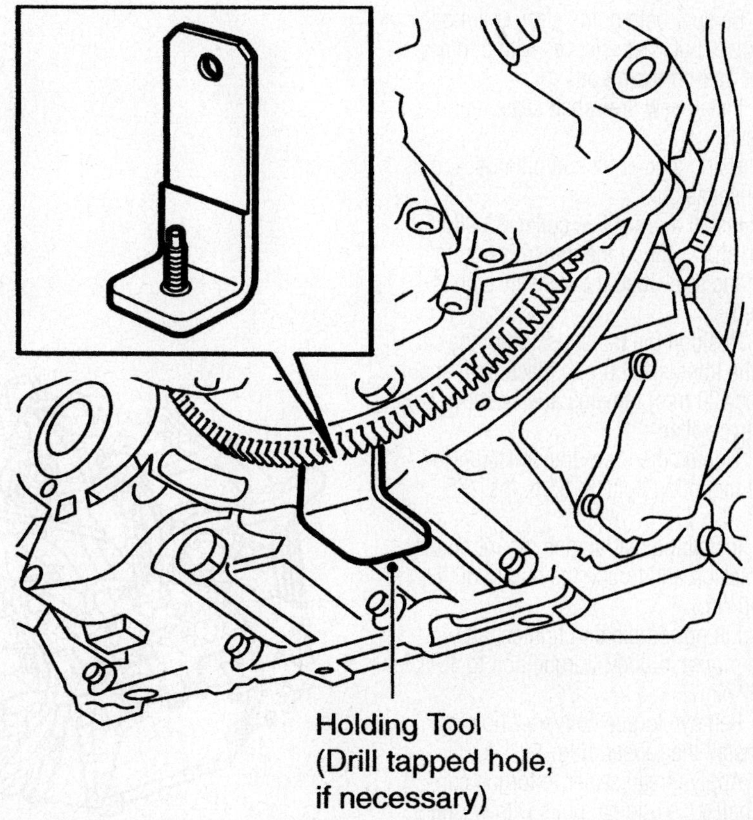

Installing torque converter holding tool—9-3 2002 shown

➡Some transmissions may not have a hole in which to fit the tool. Therefore, drill a 5mm hole and thread it was an M6 tap.

13. Remove the rear engine-to-transmission bolt.
14. Remove the front engine mount bolts.
15. Lower the car, then lower the engine and transmission assembly with the lifting equipment until there is a slight amount of play between the body and from transmission mount (play can be seen from left wheel housing).
16. Raise the car. Install a lifting tool,87 92 608, onto a pillar jack. Adjust the tool so it is centered with lifting lug on top of the transmission. Tighten it on the transmission.
17. Remove the front bolt between the engine and transmission.
18. Pull out and lower the transmission, using care with the pillar jack.
19. Lift transmission from pillar jack with engine lift.

To install:

➡If installing a new transmission, transfer torque converter holding tool to new assembly.

20. Turn torque converter so bolt holes line up with drive plate holes.
21. Be sure both guide sleeves are on the engine. Lubricate sleeves with a thin layer of anti-corrosion oil.
22. Install new driveshaft seals. Lubricate seals.
23. Using a hoist lift and pillar jack to lift the transmission.
24. Install a protective collar, 83 95 162, in right shaft seal on the transmission to protect the seal when the transmission is installed.
25. Position the transmission in the car. Push the transmission into place until about 0.8 inch (20 mm) remains, then remove the protective collar.
26. Tighten the lower transmission-to-engine mounting bolts to 55 ft. lbs. (75 Nm).
27. Install the rear engine mount on the transmission and torque the bolt to 30 ft. lbs. (40 Nm).
28. Lift powertrain and tighten left engine mount to body. Torque bolt to 46 ft. lbs. (62 Nm).
29. Remove torque converter holding tool. Install the rubber plug.
30. Apply thread sealer to torque converter bolt. Use original bolts with washers. Longer bolts will ruin the torque converter. Install all 6 bolts, then tighten the bolts to 44 ft. lbs. (60 Nm).
31. Install the flywheel splash guard.
32. Install a protective collar (16), 83 95 162, in the sealing ring prior to installing the driveshaft.
33. Make sure the driveshaft is clean. Install driveshaft, inserting all except about 0.8 inch (20 mm) into the transmission. Remove the protective collar tool before the shaft sealing surface reaches the shaft seal.
34. Push the driveshaft fully in until the circlip clicks, indicating engagement.
35. Reattach transmission oil cooler lines. If necessary, clean oil cooling system first. Make sure lines do not contact the subframe after installation. Use new lubricated seals. Tighten fittings to 20 ft. lbs. (27 Nm).
36. Carefully raise the subframe. Ensure washers on rear mounts are in place, then position suspension arm end pieces, but do not tighten at this time.
37. Adjust subframe so the bolt holes are aligned with the body.
38. Tighten the 4 corner subframe bolts as follows:
    - Front bolts: 85 ft. lbs. (115 Nm)
    - Rear bolts: 81 ft. lbs. (110 Nm), plus an additional 75 degrees.
39. Lower the lifting trolley and move out of the way.
40. Tighten the remaining subframe bolts to 141 ft. lbs. (190 Nm).
41. Tighten the rear engine mount nuts to 37 ft. lbs. (50 Nm).
42. Tighten the suspension arm end pieces on both sides. Tighten nuts to 55 ft. lbs. (75 Nm).
43. Install or connect the following:
    - Ground lead to transmission
    - Air shields and lining in right wheel housing
    - Front exhaust pipe; torque exhaust flange nuts to 30 ft. lbs. (40 Nm)
    - Install clips for oxygen sensor wiring.
    - Front wheels; torque wheel nuts to 81 ft. lbs. (110 Nm)
    - Remove engine hoist and tighten top bolts between engine and transmission to 55 ft. lbs. (75 Nm)
    - Oxygen sensor (attach clips)
    - Positive cable mount on engine
    - Connectors for transmission (behind battery)
    - Dipstick tube
    - Transmission vent tube
    - Shift cable to bracket on transmission (install clamp)
    - Shift cable to selector lever
    - Intake manifold cover, or resonator, with hoses and MAF sensor

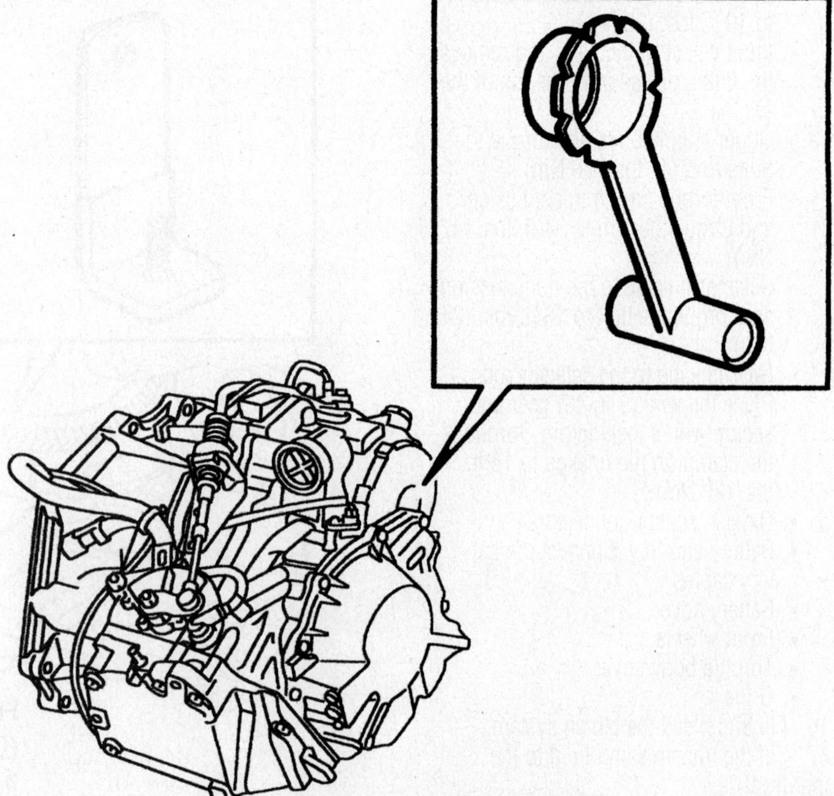

Showing protective collar for using when installing the driveshaft—9-3 2002 shown

44. Check positions of selector lever and adjust if necessary.

45. Connect the battery negative cable to ground on transmission

46. Install the battery.

47. Refill the transmission with Dexron III®.

48. Test drive the car at varying speeds and loads. Check for diagnostic codes.

2003–06 MODELS

1. Before servicing the vehicle, refer to the Precautions Section.

2. Remove or disconnect the following:
   - Upper engine cover
   - Battery cover
   - Battery with coolant pipe, bonnet switch and fuse box
   - Control module connector
   - Battery tray
   - Connector under control module
   - Ventilation hose from transmission
   - Gear cable from selector lever arm (selector lever must be in N)
   - Cable from the cable retainer (pull back locking sleeve)
   - Ground cable from transmission casing
   - Upper bolts on transmission casing
   - Expansion tank from its holder
   - Install engine lifting eyes. Attach a lifting beam and hoist to lifting eyes.
   - Dipstick

3. With the hoist, take weight off the engine and transmission.

4. Remove the upper bumper shell mountings.

5. Attach 2 straps above the radiator core so they are accessible from below (these will be used to restrain the radiator).

6. Remove or disconnect the following:
   - Front wheels
   - Lower front cover, then secure radiator core with straps
   - Radiator brackets from subframe
   - Exhaust pipe (cut about 3.5 inches (87mm) from front of muffler)
   - Engine torque rod from subframe
   - Bolt and nut holding suspension arm to steering swivel member on both sides
   - Anti-roll bar from link arm
   - Connector and cable clip of headlamp angle sensor (if equipped)
   - Steering gear from subframe (retain nuts and washers); leave steering gear hanging
   - Power steering pipe from subframe
   - Front part of fender liner (bend it out of the way)

7. Position a trolley lift with jig underneath. Remove the subframe bolts and the rear brackets. Lower the subframe slightly.

8. Separate the steering arm ball joints from the steering swivel members.

9. Remove the anti-roll bar from the subframe (protect the steering gear boot). Install the bolts loosely in the link arms so the anti-roll bar is kept hanging in place.

10. Lower the subframe.

11. Remove the torque rod bracket. Undo the bolt for the torque rod for easier fitting.

12. Remove or disconnect the following:
    - Starter motor
    - Torque converter from drive plate (6 bolts; turn crankshaft to access bolts)
    - Undo plug and press torque converter towards transmission; install holding tool
    - Transmission fluid
    - Oil cooling hoses; plug holes and hoses
    - Bolts in charge air pipe bracket and move pipe aside
    - Lower transmission bolts; leave one bolt loosely in place
    - Ground lead from transmission
    - ABS cable from clip
    - Left driveshaft; suspend driveshaft by means of a cable tie

13. Lower the car, mark the location of the bolts on the left engine mount with a marker pen to ensure correct refitting. Remove the engine mount bracket.

14. Lower the powertrain with the lifting beam to facilitate removal of the transmission.

15. Undo the clip securing the power steering hose.

16. Raise the car and move the power steering cooling coil aside and restrain with a strap.

17. Install a holding device, 87 92 608, onto a column jack. Adjust and tighten the tool on the transmission. Use a bolt with 8.8 grade that is approx. 0.8 inch (20mm) longer than the removed bolt as well as a bolt (8.8 grade) with nut on the rear lifting eye.

18. Remove the remaining bolts securing the transmission to the engine.

19. Remove the last bolt, pull out the transmission and lower it.

20. Lift the transmission down from the column jack with an engine hoist.

To install:

➡If installing a new transmission, transfer the torque converter holder to the new transmission from the old one.

21. Turn the torque converter so that the bolt holes line up with the holes in the drive plate. Install torque converter holder during bolt installation.

22. Make sure the 2 guide sleeves are on the engine and apply anti-corrosion agent to the sleeves.

23. Install new driveshaft seals (lubricate them before installation).

24. Install the holder, 87 92 608, onto the column jack. Secure the tool to the transmission as for removal.

25. Install a protective collar, 83 95 162, in the right shaft seal in the transmission. This is done to protect the seal while the transmission is being fitted. Lubricate the seal.

26. Place the transmission in position. Slide in the gearbox until about 0.8 inch (20mm) are remaining and remove the tool.

27. Push in the transmission until it is against the mating face. Tighten the bolts between the engine and the transmission that are accessible from below to 52 ft. lbs. (70 Nm).

28. Remove the lifting tool from the transmission and remove the jack. Fit the bolt into the transmission and torque it to 16 ft. lbs. (22 Nm).

29. Undo the strap and position the power steering cooling coil in place. Make sure it is fitted on the correct side of the rubber seal.

30. Lower the car and install the mounting for the engine bracket onto the transmission. Lift up the unit and install the bolts for the engine bracket as marked earlier. Tighten bolts to 69 ft. lbs. (93 Nm).

31. Remove the lifting beam and undo the lifting eyes.

32. Install the expansion tank in its holder.

33. Fit the remaining upper bolts in the transmission and torque to 52 ft. lbs.(70 Nm).

34. Fit Install a protective collar, 83 95 162, into the sealing ring. Lubricate the sealing ring.

35. Make sure the driveshaft is clean and align it with the tool. Slide in the gearbox until about 0.8 inch (20mm) are remaining and remove the tool. Push in the rest of the shaft until the circlip clicks in.

36. Attach the ABS cable to the clip.

37. Raise the car and remove the torque converter holding tool. Press the torque converter against the drive plate. Fit the plug.

38. Apply thread lock to the torque converter-to-drive plate bolts. Use the original bolts and washers. Using longer bolts will damage the torque converter. Fit the 6 bolts without tightening.

39. Rotate the engine clockwise and tighten the bolts once they are all in place. Torque bolts to 22 ft. lbs. (30 Nm).

40. Fit the starter motor with electrical connections. Torque bolts to 35 ft. lbs. (47 Nm).

41. Install the charge air pipe bracket. Torque the bolts to 15 ft. lbs. (20 Nm).

42. If original transmission is installed, change oil cooler hose seals. Fit new seals lubricated with Vaseline on the oil cooler hoses. Lubricate the pipes and fit the oil cooler hoses. The pipes will penetrate the seals when installed.

CAUTION

Ensure that the surfaces in the gearbox gasket opening are not damaged.

43. Install or connect the following:
- Ground cable to transmission; torque to 7 ft. lbs. (10 Nm)
- Bracket for torque arm on transmission; torque to 69 ft. lbs. (93 Nm).
- Place subframe on trolley lift with jig; lift the frame
- Power steering pipe
- Suspension arm ball joints in steering swivel member
- Anti-roll bar to subframe (opening of rubber bushes must face front); torque to 47 ft. lbs. (64 Nm)
- Subframe into position
- Guide pins in fixture to holes in body
- Subframe bolts and rear brackets
- Subframe against body so bushings do not rotate when bolts are tightened (remove lift); torque to 55 ft. lbs. (75 Nm), plus an additional 90 degrees and to 60 ft. lbs. (90 Nm), plus an additional 45 degrees for rear bracket bolts
- Steering gear bolts, washers and nuts to 37 ft. lbs. (50 Nm), plus an additional 60 degrees
- Engine torque rod to subframe; torque to 52 ft. lbs. (70 Nm), plus an additional 90 degrees
- Suspension arms to steering swivel members; torque to 37 ft. lbs. (50 Nm)

CAUTION

Ensure that the steering knuckle stub is visible on the top of the steering knuckle housing before the bolt is fitted.

- Radiator brackets to subframe; torque to 35 ft. lbs. (47 Nm); remove the straps

44. Set up the engine centering kit, 83 96 152, and check that engine is located correctly in relation to the subframe. Remove the centering tool.

45. Fit the power steering pipe to the subframe. Insert all bolts first and then tighten.

46. Install a joint clamp on the front pipe and fit the front exhaust pipe to the catalytic converter. Use a new gasket and new nuts. Torque nuts to 18 ft. lbs. (25 Nm). Adjust the joint clamp so that the pipe ends are in the middle. Tighten the joint clamp nuts to 30 ft. lbs. (40 Nm).

47. Connect or install the following:
- Fender liner section on both sides
- Front wheels
- Lower front cover
- Upper bumper shell mountings
- Cable to retainer (by pulling back the locking sleeve)
- Gear cable to selector lever arm

48. Carry out selector lever cable adjustment as follows:

a. Engage P with the gear selector on the transmission.

b. Rock the car until the parking lock engages.

c. Press down the locking brace by the selector lever housing securing the adjuster.

d. Check the shifting positions.

e. Fit the plastic cover for the selector lever and the rubber mat in the storage compartment.

49. Install or connect the following:
- Transmission ventilation hose (lightly lubricated)
- Connectors under control module
- Battery tray
- Control module connector
- Hood switch
- Cooling pipe
- Main fuse box
- Battery, cables and cover
- Upper engine cover
- Dipstick
- Transmission fluid

50. Test drive the car with varying engine loads and speeds. Check for trouble codes. Also check the position of the steering wheel when driving straight ahead on a level road. Adjust if needed.

9-5

1. Before servicing the vehicle, refer to the Precautions Section.

2. Remove or disconnect the following:
- Intake manifold cover
- Battery and tray
- MAXI fuse board
- 16 pin and 10 pin wire connectors
- Transmission breather hose
- Gear selector arm from the transmission (place lever in L position)
- Shifting cable
- Cable channel from the engine and gear case (3.0L engine only)
- 3 upper bolts from the transmission
- Dipstick tube (plug hole in block)
- Oxygen sensor connectors
- Rear engine mount nut
- Top 2 bolts from the rear engine cushion, loosen only

3. Install a lifting beam to the engine and relieve the weight on the engine and transmission.

4. Remove or disconnect the following:
- Both front wheels
- Lower engine cover
- Front exhaust system
- Rear engine bracket and pad
- Steering gear bolts
- Rear clamps securing the power steering delivery pipe to the subframe
- A/C pipes from the subframe holder
- Air cleaner casing from the subframe
- Engine oil cooler from the charge air cooler (2.3L engine only)

5. Attach a strap around the radiator and crossmember to restrain the radiator.

6. Remove or disconnect the following:
- Outer ball joint to steering knuckle bolts
- Upper anti-roll bar ball joints
- Torque arm from the subframe
- Rear support plates

7. Position a lifting trolley with a holder under the vehicle. Align the trolley to the subframe.

8. Remove or disconnect the following:
- Remaining bolts from the subframe and slightly lower the subframe
- Power steering delivery pipe clamps. Lower the lifting trolley and move the subframe aside
- Splash plate (2.3L engine only)
- Plug covering the torque converter bolts (3.0L engine only)
- Bolts securing the torque converter to the drive plate. Rotate the plate and pulley together

➡The crankshaft must be rotated to gain access for all the bolts.

9. Press the torque converter against the transmission to keep the converter in place during transmission removal.

10. Remove or disconnect the following:

- Ground leads
- Torque arm and bracket
- Transmission fluid
- Oil cooler inlet and outlet hoses (plug openings)
- Left driveshaft and suspend it
- Left side engine pad; lower transmission to clear structural member
- Transmission bracket
- Transmission from the engine assembly using a single-column jack
- Transmission from the vehicle

To install:

11. Rotate the torque converter so that the bolt holes align with the drive plate.
12. Make certain that the 2 guide sleeves are on the engine.
13. Lubricate and install new driveshaft seals.
14. Raise the transmission into position under the vehicle.
15. Install the transmission and torque the bottom bolts between the engine and transmission to 55 ft. lbs. (74 Nm) and the bolts between the transmission and oil pan to 34 ft. lbs. (47 Nm)
16. Remove the lifting beam from the transmission .
17. Install or connect the following:
    - Transmission bolts and torque the bolts to 18 ft. lbs. (24 Nm)
    - Engine pad mount to the transmission and torque the bolts to 62 ft. lbs. (84 Nm)
    - Bolts securing the engine pad to the body and torque the bolts to 46 ft. lbs. (63 Nm)
    - Protective sleeve in driveshaft seal
    - Driveshaft and make certain that the circlip snaps into position (remove protective sleeve just before driveshaft is fully into transmission)
    - Torque converter to drive plate and torque the bolts to 22 ft. lbs. (30 Nm)
    - Splash plate (2.3L engine only)
    - Plug covering the torque converter bolts (3.0L engine only)
    - Cooler hoses and torque the fasteners to 20 ft. lbs. (27 Nm)
    - Torque arm bracket
    - Ground cable to the bracket and raise the subframe into position
    - Power steering delivery pipe clamps
    - Outer ball joints and A/C pipes
    - Subframe and rear support plates and torque the subframe bolts to 74 ft. lbs. (100 Nm) plus 45 degrees and the support plate bolts to 44 ft. lbs. (60 Nm)
    - Outer ball joints to the steering knuckles and torque the bolts to 63 ft. lbs. (85 Nm)
    - Anti-roll bar link and torque the fastener to 68 ft. lbs. (92 Nm)
    - Steering gear and torque the bolts to 66 ft. lbs. (90 Nm)
    - Engine oil cooler and air filter housing (2.3L engine only)
    - Rear engine cushion
    - Rear engine mount; torque bolts to 55 ft. lbs. (70 Nm)
    - Oil pan-to-transmission bolts; torque to 34 ft. lbs. (47 Nm) (3.0L engine only)
    - Torque arm-to-subframe; torque bolts to 19 ft. lbs. (25 Nm)
    - Front exhaust system; torque bolts to 18 ft. lbs. (24 Nm) on 2.3L engine. On the 3.0L engine torque turbo bolts to 18 ft. lbs. (24 Nm) and exhaust manifold bolts to 30 ft. lbs. (40 Nm).
    - Oxygen sensor wiring connectors and remove lifting beam
    - Torque rear engine cushion bolts to 18 ft. lbs. (24 Nm) and nut to 33 ft. lbs. (45 Nm).
    - Engine to transmission; torque bolts to 55 ft. lbs. (70 Nm)
    - Cable channel to the transmission (3.0L engine only)
    - Dipstick tube
    - Transmission breather hose
    - Shifter cable to bracket
    - Shifter cable to selector lever; adjust if necessary
    - Torque rod and torque bolt to 34 ft. lbs. (47 Nm)
    - MAXI fuse board
    - Battery tray and battery
    - Battery cables and cover
    - Intake manifold cover
    - Front wheels
18. Fill the transmission to the proper level.
19. Start the vehicle, check for leaks and repair if necessary.

Clutch

ADJUSTMENT

The clutch cable utilized in Saab vehicles is self-adjusting. Although it does not require periodic adjustments, when the clutch cable is first reinstalled in the vehicle, a small procedure will insure that the self-adjuster is properly functioning.

1. Check the functioning of the clutch cable by moving the clutch lever forward in the car's direction of travel. The balancing spring should, then be compressed and the length of the clutch cable cover reduced. When the clutch lever is released, the cover should regain its original length. Repeat 3 or 4 times.
2. Hold the balancing spring tightly and give it a small jerk to remove any free-play.

REMOVAL & INSTALLATION

9-3

1. Before servicing the vehicle, refer to the Precautions Section.
2. Remove or disconnect the following:
    - Transmission assembly and lock the flywheel in position
    - Pressure plate retaining nuts
    - Pressure plate
    - Clutch disc

To install:

3. Install or connect the following:
    - Clutch disc with pressure plate and torque the nuts to 16 ft. lbs. (22 Nm)
    - Transmission
4. To help with installation use a guide pin to line up the transmission.
5. Fill the transmission with fluid.
    - Clutch cable
6. Test drive the vehicle.

9-5

1. Before servicing the vehicle, refer to the Precautions Section.
2. Remove or disconnect the following:
    - Transmission and install flywheel tool 83 94 868
    - Pressure plate
    - Clutch assembly

To install:

3. Install or connect the following:
    - Driven plate and drive plate on the flywheel. Hand tighten the bolts
    - Center the drive plate and torque the bolts to 16 ft. lbs. (22 Nm). Remove the flywheel locking tool
    - Transmission assembly
4. Fill the transmission with clean fluid to the proper level.

Hydraulic Clutch System

BLEEDING

1. Before servicing the vehicle, refer to the precautions in the beginning of this section.
2. Connect a hose to the slave cylinder bleeder valve. Place the other end of the

hose in a suitable jar partially filled with brake fluid.

3. Fill the master cylinder with brake fluid.

4. Open the bleeder valve on the slave cylinder ½ turn.

5. Place a cooling system pressure tester gauge over the opening of the master cylinder.

6. Pump the tester until all air is expelled from the hydraulic clutch system.

7. Close the slave cylinder bleeder valve.

8. Check that all air was removed from the system and the clutch is functioning properly. Adjust the fluid level, as required.

Halfshaft

REMOVAL & INSTALLATION

9-3

1. Before servicing the vehicle, refer to the Precautions Section.

2. Remove or disconnect the following:
   - Negative battery cable
   - Lower engine cover (2003–06 only)
   - Wheel
   - Hub center nut
   - Ball joint
   - Sway bar hardware and rubber bushing

3. Push down on the lower control arm. Using a rubber mallet, tap the halfshaft out of the hub.

4. Move the strut to one side. Be extremely careful not to stretch or break the Antilock Braking System (ABS) sensor cables or brake hoses. Place a drain pan under the transmission to catch any fluid spillage.

5. On left side halfshaft, remove the power steering pressure line from clamps.

6. Remove or disconnect the following:
   - Halfshaft

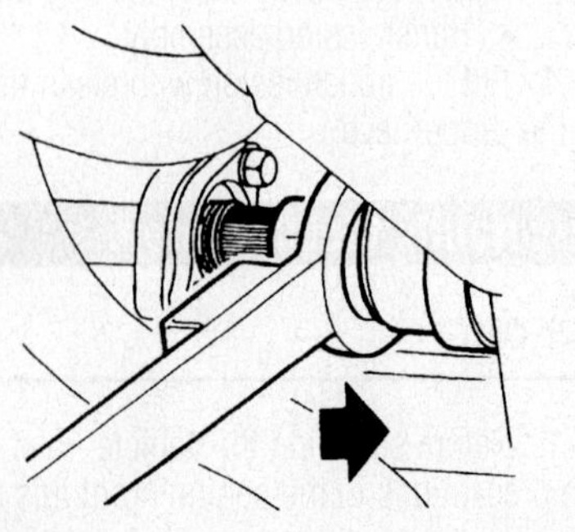

Removing the halfshaft joint from the intermediate shaft—9-3

a. Remove the inner halfshaft joint from the transmission on the left-hand side

b. Remove the halfshaft joint from the intermediate shaft on the right-hand side.

To install:

7. Install or connect the following:
   - Halfshaft
   - On left side, power steering line in clips
   - Hub and new center nut onto halfshaft. Do not tighten yet.
   - Ball joint and torque the ball joint nut to 55 ft. lbs. (75 Nm)
   - Sway bar to the lower control arm. Install the rubber bushing, a new washer and new retaining nut and torque the nut to 89 inch lbs. (10 Nm)
   - Wheels
   - Negative battery cable

8. Tighten the hub center nut to 214 ft. lbs. (290 Nm).

9. Check the transmission fluid level and top off if necessary.

9-5

1. Before servicing the vehicle, refer to the Precautions Section.

2. Remove or disconnect the following:
   - Negative battery cable
   - Front wheel
   - Hub nut
   - Headlamp position sensor fasteners (move fastener aside), if equipped
   - Halfshaft from the transmission, left side
   - Cover and knock out the halfshaft from the intermediate shaft, right side only
   - Antilock Brake System (ABS) cable fasteners and clips
   - Ball joint from the steering swivel and lower the suspension arm
   - Strut
   - Halfshaft from the hub by tapping lightly on it

To install:

3. Install or connect the following:
   - Halfshaft to the hub and hand tighten the hub nut
   - Halfshaft to the transmission and make certain that the circlip snaps into position
   - Halfshaft to the intermediate shaft and make certain that the circlip snaps into position
   - Ball joint to the steering swivel and torque the fastener to 74 ft. lbs. (100 Nm)
   - ABS cable to the clips
   - Front wheel and torque the hub nut to 123 ft. lbs. (170 Nm) plus an additional 45 degrees
   - Negative battery cable

4. Check and fill the transmission fluid to the proper level.

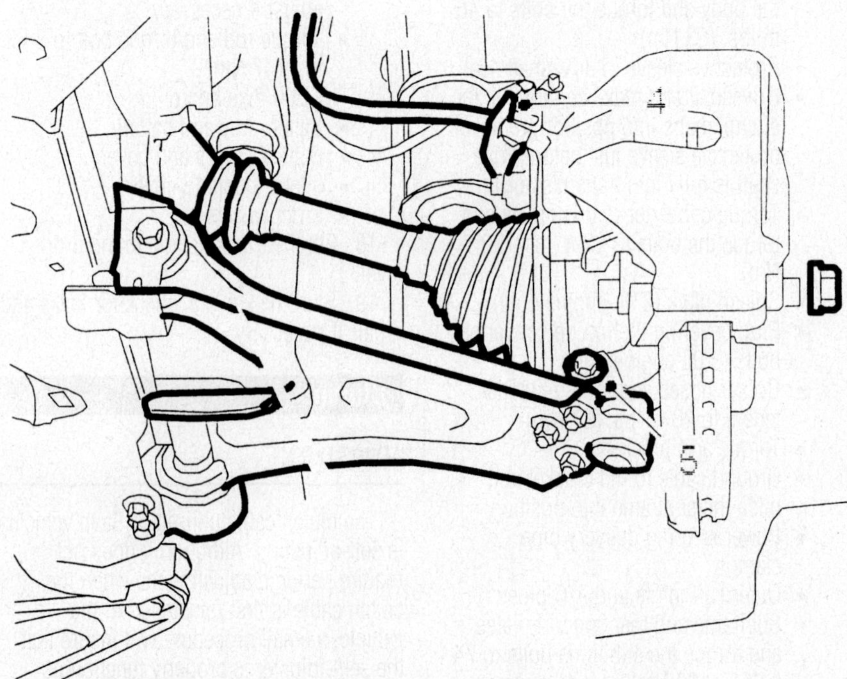

Remove the right side halfshaft from the transmission—9-5

CV-Joints

REMOVAL & INSTALLATION

9-3

INNER CV-JOINTS

1. Remove the driveshafts (1).
2. Clean, remove the circlip (2) and pull out the CV-joint tripod (3) from the shaft. Use a suitable 3-armed puller.
3. On models with B235R engines, bend out the plastic protector to remove the CV-joint from the housing.
4. Pull the rubber boot (4) from the driveshaft.
5. Wipe the CV-joint components and driveshaft clean.

To install:

6. Install the new rubber boot to the driveshaft and fit a new clamp.
7. Fit the CV-joint tripod. Make sure that its beveled edge is positioned inward on the shaft.
8. Install and tighten the circlip.
9. Fill the joint with new grease.
10. Join the driveshaft and CV-joint housing.
11. On models with B235R engines, bend back the plastic protector and make sure no protruding sharp edges can damage the rubber boot.
12. Thread on the rubber boot and fit a new clamp.
13. Install the driveshafts.

OUTER CV-JOINTS

➡This procedure applies to 9-5 models as well.

1. Make sure the shaft and boot are clean. Fasten the drive shaft in a vice.
2. Loosen the clamps (2) on the outer CV-joint (1) and press the rubber boot (3) up the shaft. Clean the joint of any grease.
3. Expand the circlip (4) and tap off the CV-joint from the shaft with a hammer and a brass drift.
4. Turn the bearing so that the steel balls can be removed.
5. Turn the inner steel ball cage so that it can be removed from the outer cage.
6. Lift out the outer cage.
7. Clean all parts.

To install:

8. Fit a new rubber boot on shaft.
9. Place the outer cage in the CV-joint.
10. Place the inner cage in the outer cage and turn it so that the steel balls can be positioned.
11. Press in each ball.
12. Fill the CV-joint with new grease.
13. Press the driveshaft into the CV-joint.
14. Place the new rubber boot in the correct position and fit new clamps.

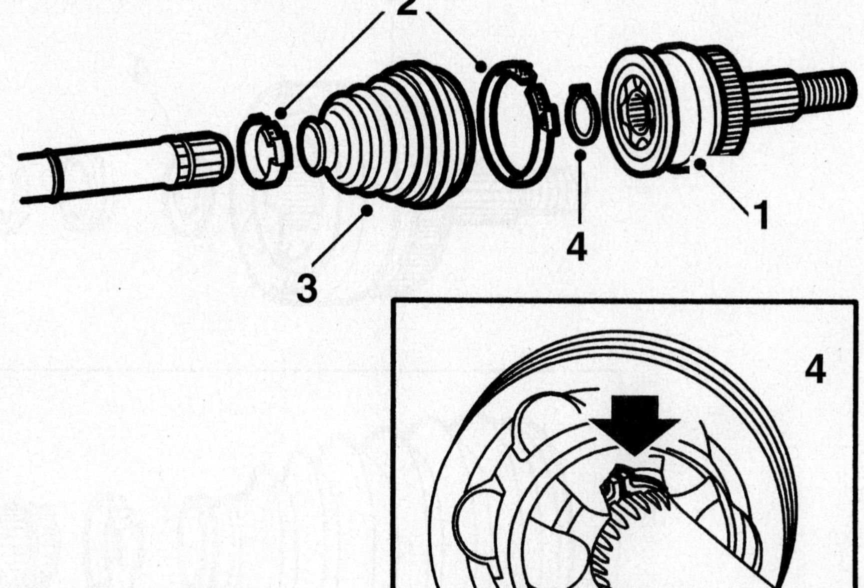

Exploded view of outer CV-joint assembly—9-3 (except models with B235R engines) and 9-5

9-5

INNER CV-JOINTS

1. Clean the outside of the CV-joint assembly (1). Remove the boot clips (2). Remove the plate end (not shown) and lift off the CV-joint assembly (1).

➡The plate does not need to be refitted (only for production purposes).

2. Wipe and remove the circlip (4) and pull the tripod (3) from the shaft. Use a suitable 3-arm puller.
3. Pull the rubber boot (5) from the driveshaft.
4. Clean the universal joint, tripod and driveshaft thoroughly.

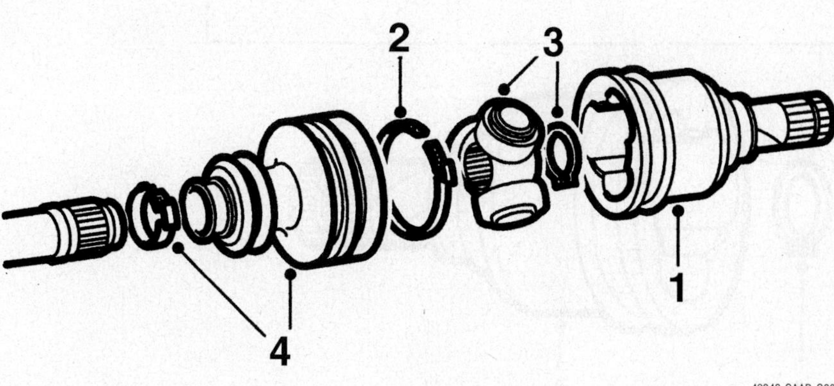

Exploded view of the inner CV-joint assembly—9-3

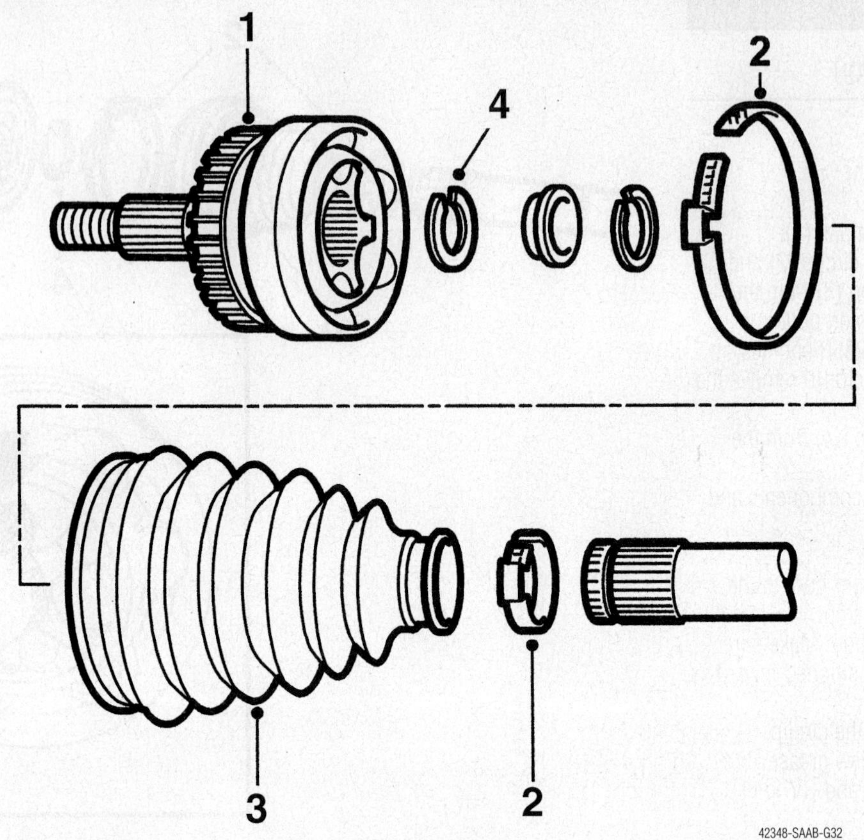

Exploded view of outer CV-joint assembly—9-3 (models with B235R engines)

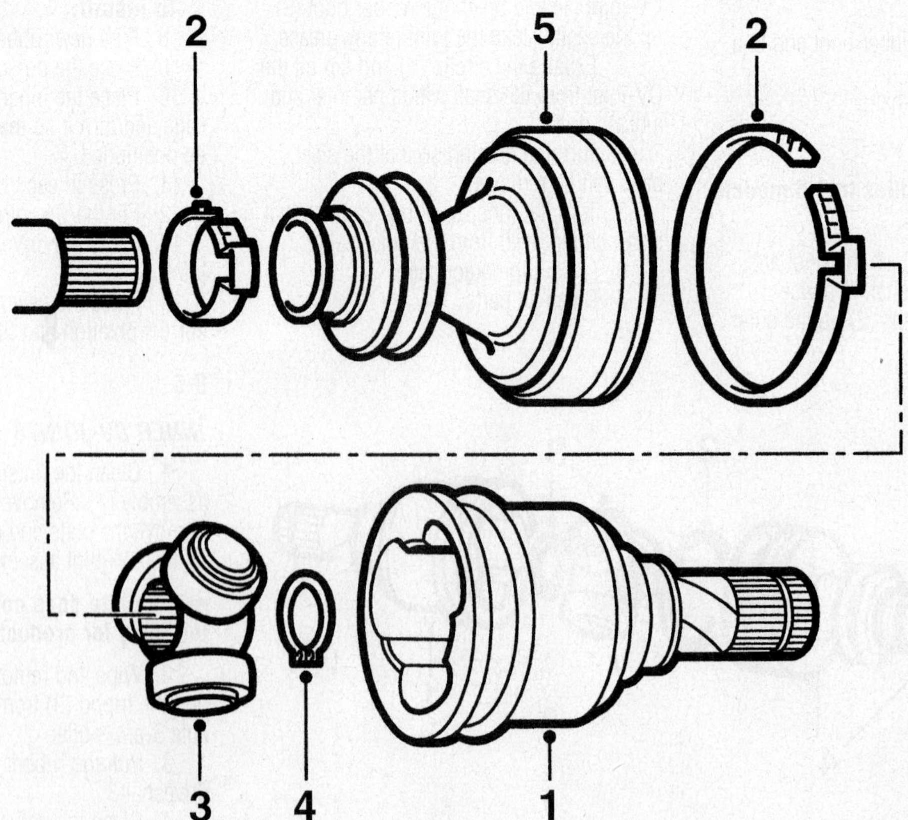

Exploded view of inner CV-joint assembly—9-5

To install:

5. Fit the new rubber boot (5) to the driveshaft and fit a new small clip.
6. Fit the tripod (3). Check that the chamfered edge on the tripod faces in towards the driveshaft.
7. Install a new circlip (4).
8. Pack the joint with new grease
9. Join the driveshaft and CV-joint housing. Bend back the plate, if it has not been removed.

➡**The plate does not need to be refitted (only for production purposes).**

10. Thread on the rubber boot and fit a new large clip.

STEERING

Air Bag

✻✻ CAUTION

Air bag system must be disabled before performing service on or around system components, steering column, instrument panel components, wiring and sensors. Failure to follow safety and disabling procedures could result in accidental air bag deployment, possible personal injury and unnecessary system repairs.

PRECAUTIONS

Several precautions must be observed when handling the inflator module to avoid accidental deployment and possible personal injury.

- Never carry the inflator module by the wires or connector on the underside of the module.
- When carrying a live inflator module, hold securely with both hands, and ensure that the bag and trim cover are pointed away.
- Place the inflator module on a bench or other surface with the bag and trim cover facing up.
- With the inflator module on the bench, never place anything on or close to the module which may be thrown in the event of an accidental deployment.

DISARMING

All Models

✻✻ CAUTION

The air bag system must be disarmed before performing service around air bag system components or system wiring. Failure to do so may cause accidental deployment of the air bag, resulting in unnecessary air bag system repairs and/or personal injury.

Always disconnect the battery cables (negative cable first) and wait 20 minutes prior to performing service around air bag system components or system wiring.

✻✻ CAUTION

Do not use any diagnostic instruments that are battery powered, such as buzzers, ohmmeters or diode testers, to diagnose faults in the steering wheel or electronic control unit. Using such devices may trigger the air bag. Also, ensure that the battery cables cannot accidentally come into contact with the battery terminals.

REARMING

To rearm the air bag system, reconnect the battery cables.

Rack and Pinion Steering Gear

REMOVAL & INSTALLATION

9-3

2002 MODEL

1. Before servicing the vehicle, refer to the Precautions Section.
2. Slightly lift the front assembly.
3. Remove the brace between the MacPherson struts.
4. Undo the main fuse box and put it to one side.
5. Remove the driver side lower dashboard panel (with the data link connector).
6. Remove the steering shaft lock bolt.

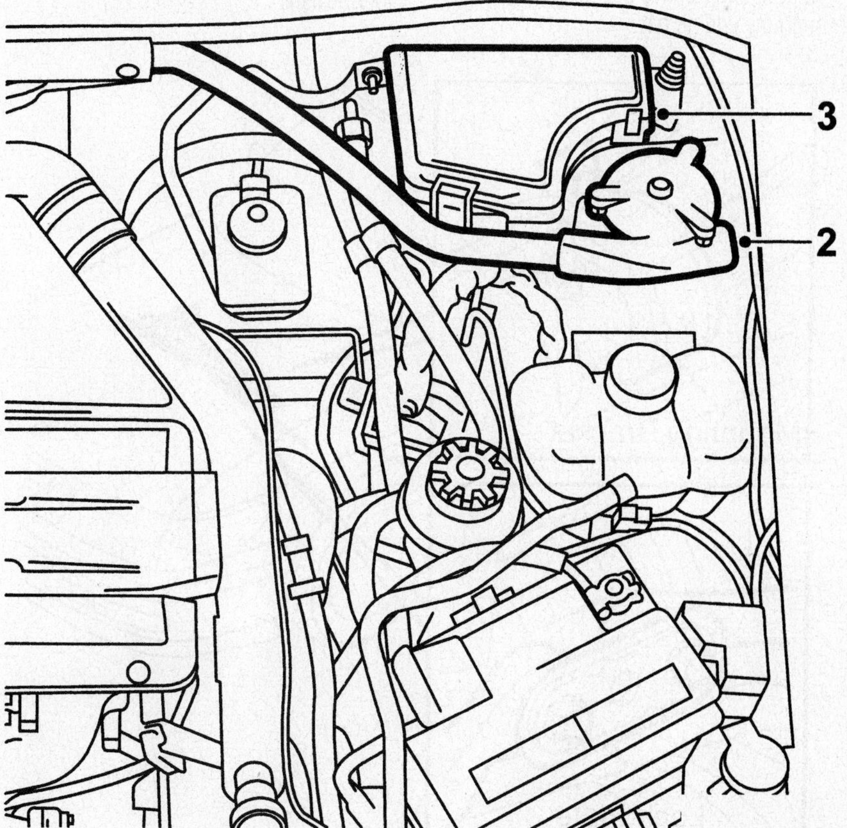

Identifying the MacPherson strut brace and location of the fuse box—9-3

✲✲ WARNING

Secure the steering wheel to avoid damaging the contact roller (coil spring). If the basic position of the contact roller is disturbed, the spiral conductor in the roller will be damaged when the steering wheel is turned full-lock. The airbag will then fail to function, causing risk for injury in the event of a collision.

7. Turn the steering wheel to the straight-ahead position.
8. Pull the steering shaft from the pinion shaft.
9. Cut the cable tie between the return hose and the pressure hose and separate the two hoses.
10. Detach the return hose from the steering gear and plug it. Pinch the hose with pliers. Bend the hose to one side.
11. Remove the delivery pipe from the steering gear. Plug the hole in the steering gear.
12. Remove the lock plate and separate the steering arm from the steering gear.
13. Remove the right and left steering gear mounting brackets.
14. Remove the bulkhead seal.
15. Lift and remove the steering gear up through the engine bay.

To install:

16. If installing a new steering gear, transfer the return pipe to the new unit.
17. Fit the rubber bushing (lubricated with petroleum jelly) to the left mounting bracket.
18. Position the steering gear by lowering it down through the engine bay.
19. Install the right mounting bracket, with the rubber bushing loosely installed to the bulkhead.
20. Install the bulkhead seal. Position the groove in the seal against the groove in the steering gear.
21. Install the left mounting bracket first. Make sure the smooth surface of the steering gear is against the bulkhead partition. Tighten the bolts to 18 ft. lbs. (24 Nm) after applying thread lock sealant.

✲✲ WARNING

It is extremely important that the steering gear mounting brackets are fitted in the correct order. Otherwise, the steering gear can be fitted in such a position that the angle to the steering column fitting is incorrect.

22. Tighten the right mounting bracket to 18 ft. lbs. (24 Nm) after applying thread lock sealant.

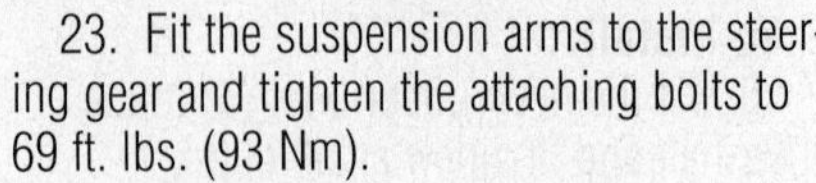

23. Fit the suspension arms to the steering gear and tighten the attaching bolts to 69 ft. lbs. (93 Nm).

➡A new lock plate must be fitted each time the steering gear has been removed from the car.

24. Remove the plug and attach the delivery pipe to the steering gear and tighten it.
25. Remove the plug from the return hose and attach it to the steering gear.
26. Install a new cable tie between the return hose and the pressure hose and fix them together.
27. Install the main fuse box.
28. Install the brace between the MacPherson struts.
29. Release the steering wheel and fit the steering shaft to the pinion shaft. Tighten the lock bolt to 18 ft. lbs. (24 Nm).

✲✲ WARNING

If the steering wheel position has been altered, the coil spring must be adjusted.

30. Adjust coil spring position as follows:
    a. Check that the road wheels are set straight ahead.
    b. Remove air bag module and steering wheel.
    c. Turn the coil spring clockwise until you feel stiff resistance.
    d. Turn the coil spring back 2 complete turns, plus the distance that remains until the electric leads are standing straight up.

➡If a new coil spring is being fitted, no adjustment to the central position is needed. The coil comes ready-set to the central position, being held there by a transport arrester.

    e. Check that the leads connected to the coil spring are not twisted.
    f. Install the steering wheel and air bag module.
31. Lower the car and turn the wheels to point straight ahead.
32. Fill the hydraulic fluid reservoir. Start the engine, turn the steering wheel to the right and left lock positions 2-3 times.
33. Turn off the engine and correct the fluid level. At 68°F (20°C), it should be in between the MAX and MIN markings on the dipstick in the hydraulic fluid reservoir.
34. Check for leaks at the control valve couplings.
35. Make sure the steering wheel is positioned "straight ahead" when the car is steering straight ahead on a level road.

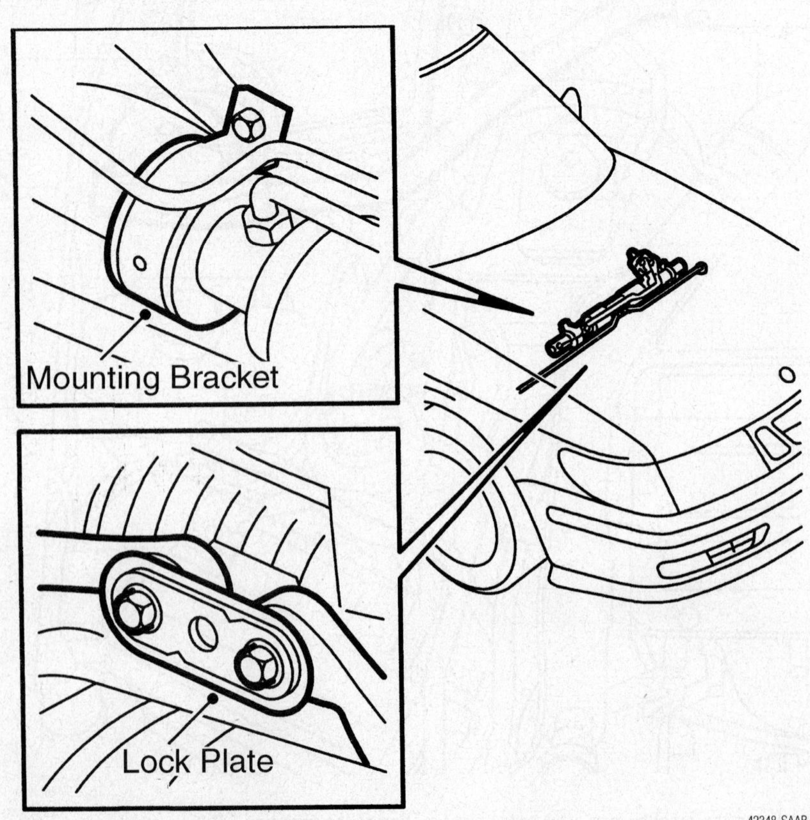

Showing the steering gear lock plate and mounting brackets—9-3

2003–06 MODELS

1. Before servicing the vehicle, refer to the Precautions Section.
2. Clamp the return hose using suitable pinch-off pliers.
3. Position the steering wheel and the wheels so that they face straight ahead. Use woven tape to secure the steering wheel to the dashboard.
4. Raise the car and remove the front wheels.
5. Remove the nuts securing the track rod ends to the steering swivel members. Remove the track rod.
6. Remove the anti-roll bar links from the anti-roll bar.
7. Remove the steering shaft joint from the steering gear.
8. Detach the delivery line and the return line from the steering gear. Plug the lines.
9. Remove the heat shield from the steering gear.
10. Remove the steering gear bolts, nuts and washers.
11. Twist up the anti-roll bar as high as possible and lift out the steering gear through the wheel housing on the passenger side.

To install:

12. Lift the steering gear into place through the wheel housing on the passenger side.
13. Install the steering gear to the subframe and tighten mounting bolts to 37 ft. lbs. (50 Nm), plus an additional 60 degrees.
14. Install the heat shield to the steering gear.
15. Attach the delivery line and return line to the steering gear. Fit new O-rings. Tighten fittings to 21 ft. lbs. (28 Nm).
16. Twist the anti-roll bar down into position. Install the anti-roll bar links to the anti-roll bar. Hold with a thin 17mm open wrench so that the ball joint does not turn. Torque to 47 ft. lbs. (64 Nm).
17. Fit the track rod ends to the steering swivel members. Torque nuts to 26 ft. lbs. (35 Nm).
18. Install the steering shaft joint to the steering gear. Torque pinch bolt to 20 ft. lbs. (27 Nm).
19. Lower the car and install the wheels.
20. Remove the hose pinch-off pliers and fill steering reservoir with power steering fluid.
21. Remove the tape used to prevent the steering wheel from moving.
22. Bleed the power steering system and check for leaks.
23. Carry out a wheel alignment .

9-5

1. Before servicing the vehicle, refer to the Precautions Section.

⁂ CAUTION

Carefully clean area around all fittings for steering gear and hoses. Immediately plug openings.

2. Drain the power steering reservoir.
3. For 4-cylinder models only, remove or disconnect the following:
   - Power steering reservoir and move it aside
   - Return hose from the power steering reservoir (place end of hose in large container)
4. On 4-cylinder models, with wheels clear of floor, start the engine and turn steering wheel from lock-to-lock until flow of power steering fluid slows down into container. Turn engine off (do not allow pump to run dry).
5. For all engines, set wheels in straight-ahead position and secure steering wheel so it will not turn.
6. Remove or disconnect the following:
   - Steering column shaft from the steering gear
   - Intake manifold cover (4-cylinder)
   - Rear engine cushion to the subframe
   - Engine cushion to the engine mount
   - Turbocharger bypass pipe, bend pipe aside and plug opening (V6)
   - Loosen, but do not remove, exhaust pipe from flange (V6)
   - Attach engine lift and hoist
   - Both front wheels
   - Track rod lock nut
   - Track rod end from the steering swivel
   - Track rod end from the track rod (count number of turns)
   - Reinforcement from the rear attaching point on the subframe
   - Exhaust pipe between the catalytic converter and the silencer (cut pipe)
   - Exhaust pipe at rear exhaust manifold (V6)
   - Subframe center attaching point, lower the subframe
   - Engine cushion (4-cylinder)
   - Delivery and return pipes from the valve body
   - Steering gear retaining bolts
   - Steering gear through the passenger side wheel well housing

To install:

7. Turn pinion shaft until rack is in center position.
8. Install or connect the following:
   - Steering gear into position. Hand tighten the bolts at this time
   - Valve body seal after lubricating
   - Delivery and return pipes to the valve body, with new sealing rings; do not torque the bolts
   - Delivery pipe to steering gear with clamp
   - Steering gear bolts, with bracket on right side, but do not torque bolts
   - Torque power steering pipes to steering gear to 25 ft. lbs. (30 Nm)
   - Return hose and oxygen sensor cables (V6) to delivery pipe
   - Torque steering gear retaining bolts to 70 ft. lbs. (95 Nm)
   - Engine cushion; install bolt only snug
   - Raise the subframe and torque the center attaching bolts to 75 ft. lbs. (100 Nm) plus 45 degrees
   - Exhaust system between the catalytic

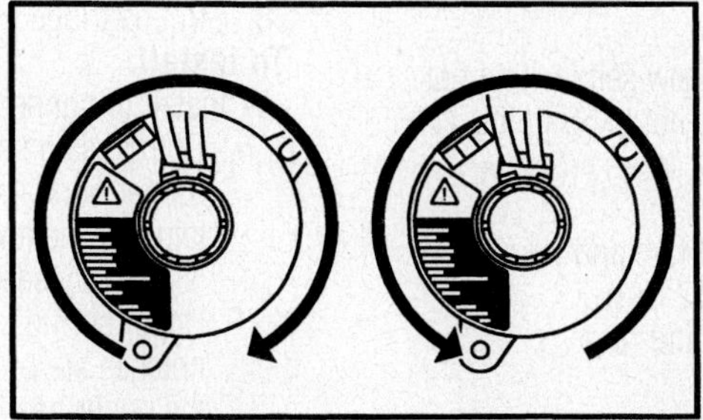

1. Turn clockwise until resistance is felt.
2. Turn counter-clockwise 2 full turns, plus more until wires are straight up.

42348-SAAB-G36

Illustrating adjustment of coil spring in steering wheel—9-3

converter and the silencer; torque clamp bolts to 30 ft. lbs. (40 Nm)
- Subframe rear attaching points to the reinforcement and torque the bolts to 75 ft. lbs. (100 Nm) plus 45 degrees
- Reinforcement at subframe rear attachment point; torque to 50 ft. lbs. (65 Nm)
- Track rod ends to the track rods same number of turns as removed. Hand tighten the lock nuts
- Track rod ends on the steering swivel members and torque the lock nuts to 45 ft. lbs. (60 Nm)
- Both front wheels
- Rear engine cushion to the sub-frame and torque the bolts to 20 ft. lbs. (25 Nm)
- Rear engine cushion to engine mount and torque the bolts to 35 ft. lbs. (50 Nm)
- Intake manifold cover
- Steering column shaft with the steering gear and torque the bolts to 25 ft. lbs. (30 Nm)
- Release steering wheel
- Return hose to the power steering reservoir (4-cylinder)
- Power steering fluid reservoir (4-cylinder)

9. Fill and bleed the power steering system by starting engine turning steering wheel lock-to-lock 2–3 times. Turn the engine off and top up the fluid.
10. Check the toe-in and adjust if necessary.
11. Tighten the track rod lock nuts to 55 ft. lbs. (70 Nm).

FRONT SUSPENSION

Strut

REMOVAL & INSTALLATION

9-3

1. Before servicing the vehicle, refer to the Precautions Section.
2. Slacken center nut on top end of strut while vehicle is resting on the ground.
3. Remove or disconnect the following:
   - Negative battery cable
   - Front wheel
   - Hub nut
   - Wheel speed sensor
   - Caliper. Be sure the caliper is properly supported and not hanging from the brake hose
   - Rotor and backing plate
   - Tie rod end
   - Sway bar hardware and bushing
   - Ball joint
   - Three upper strut mounting bolts
   - Strut housing

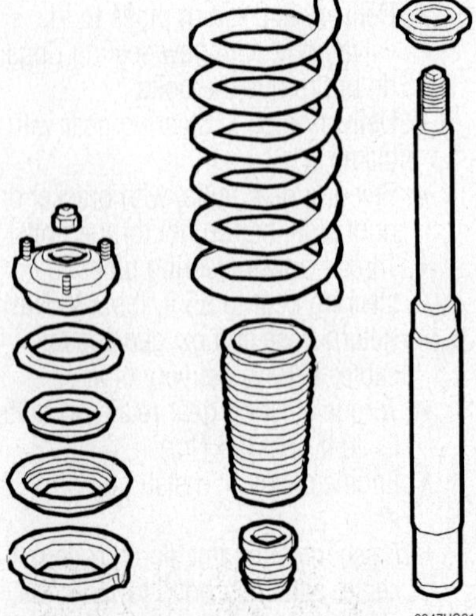

Exploded view of the strut assembly

   - Spring and shock absorber cartridge
4. Place the strut in a vise and compress the spring.
5. Remove the self-locking nut from the bearing plate and discard the nut.
6. Remove the coil spring, bellows and rubber bumper.
7. Remove the upper flange nut.
8. Remove the strut cartridge from the housing.

To install:

9. Install or connect the following:
   - New strut cartridge and torque the spanner nut to 159 ft. lbs. (215 Nm)
   - Coil spring. Be sure the end of the spring is up against the spring stop
   - Upper spring seat. Make sure the notches are properly aligned. Place the upper bearing assembly on the upper seat and secure it using a new self-locking nut
10. Remove the spring compressor and be sure the coil spring stays properly seated.
11. Install or connect the following:
   - Strut assembly and torque the upper mounting bolts 13 ft. lbs. (18 Nm)
   - Ball joint and new self locking nut and torque the nut to 55 ft. lbs. (75 Nm)
   - Tie rod end
   - Sway bar hardware and torque to 96 inch lbs. (12 Nm)
   - Backing plate and rotor
   - Caliper
   - Wheel speed sensor
   - Hub nut. Do not tighten yet.
   - Front wheel
   - Negative battery cable
   - Hub nut and torque the nut to 125 ft. lbs. (170 Nm) plus an additional 45 degrees
12. Pump the brake pedal to position the brake caliper piston.

CAUTION

Do not attempt to move the vehicle until a firm pedal is obtained.

13. Check front wheel alignment.

9-5

1. Before servicing the vehicle, refer to the Precautions Section.
2. Remove or disconnect the following:
   - Negative battery cable
   - Front wheel
   - Nut from the anti-roll bar link
   - Steering swivel member from the strut and move the Antilock Braking System (ABS) sensor cable and brake hose aside
   - Upper retaining bolts from the strut
   - Strut assembly
3. Place the strut in a vise.
4. Compress the spring.
5. Remove the self locking nut.
6. Coil spring, bellows and rubber snubber.
7. Upper flange nut
8. Strut cartridge from the housing

To install:

9. Install or connect the following:
   - Compression stop, spring seat, rubber boot and spring. Make certain that the lower end of the spring is properly seated
   - Upper spring seat. Make sure the notches are properly aligned. Place the upper bearing assembly on the upper seat and secure it using a new self-locking nut and torque the nut to 55 ft. lbs. (75 Nm)
10. Remove the spring compressor and be sure the coil spring stays properly seated.
11. Install or connect the following:

- Strut assembly and torque the bolts to 13 ft. lbs. (18 Nm)
- Press the steering swivel member inward and torque the bolts to 75 ft. lbs. (100 Nm) plus 45 degrees
- Anti roll bar link and torque the nut to 70 ft. lbs. (95 Nm)
- ABS sensor cable and brake hose
- Front wheel
- Negative battery cable

Coil Spring

REMOVAL & INSTALLATION

The front coil spring removal and installation procedure is covered in the strut removal and installation procedure.

Lower Ball Joint

REMOVAL & INSTALLATION

9-3

➡The ball joint cannot be removed from the control arm. To replace the ball joint, the control arm must be replaced.

1. Before servicing the vehicle, refer to the Precautions Section.
2. Raise the vehicle.
3. Remove or disconnect the following:
   - Wheel
   - Sway bar link bolt
   - Ball joint out of the steering knuckle
   - Retaining nut at the support arm
   - Retaining bolt at the subframe

To install:

4. Install or connect the following:
   - Arm and bolt at the subframe and torque the retaining bolt at the subframe to 85 ft. lbs. (115 Nm)
   - Bolt at the support arm and torque the bolt to 68 ft. lbs. (92 Nm)
   - Sway bar link and torque the link to 89 inch lbs. (10 Nm)
   - Ball joint and torque to 55 ft. lbs. (75 Nm)
   - Wheel
5. Check front wheel alignment.

9-5

1. Before servicing the vehicle, refer to the Precautions Section.
2. Remove or disconnect the following:
   - Wheel
   - Transverse link from the longitudinal link and place a spacer between the upper and lower links
   - Press the brake pistons back
   - Brake line bracket from the longitudinal link
   - Brake caliper bolts and support the caliper with wire ties
   - Antilock Braking System (ABS) sensor connector, if equipped
   - Hub retaining nuts
   - Unhook the handbrake return spring and cable
   - Hub and brake disc
   - Press out the ball joint with tool 89 96 761

To install:

3. Install or connect the following:
   - Press in the ball joint with tool 89 96 761
   - Hook the handbrake return spring and cable to the hub and brake disc
   - Hub on the back plate and secure it to the longitudinal link with new nuts and torque to 40 ft. lbs. (50 Nm) plus 30 degrees
   - ABS sensor connector
   - Brake caliper
   - Brake line bracket to the longitudinal link and torque the bolt to 70 ft. lbs. (90 Nm) plus 60 degrees

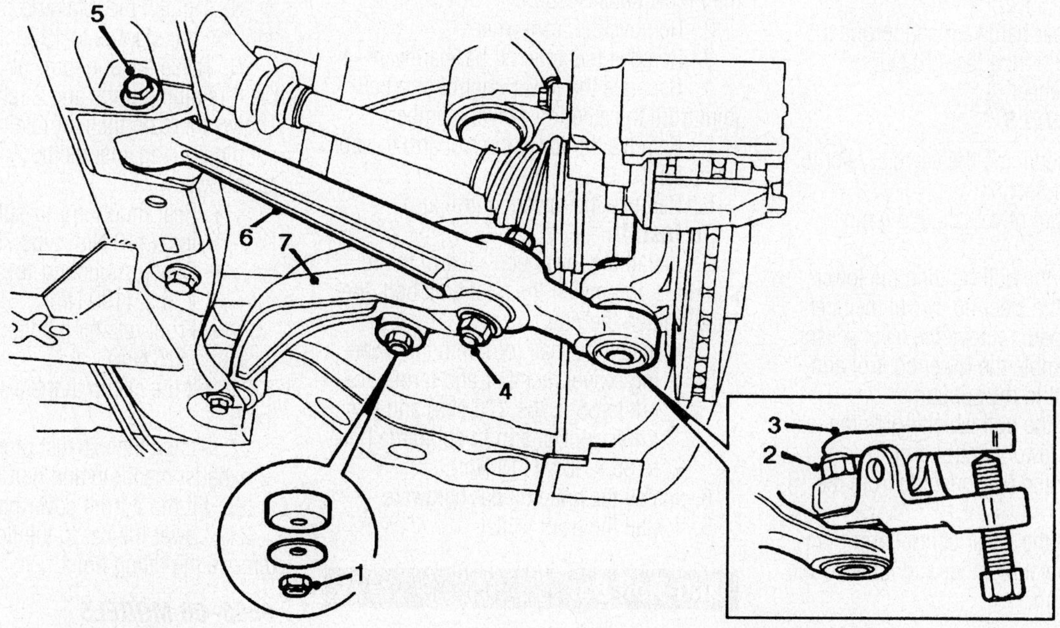

Lower control arm connection points—9-3

- Wheel

4. Depress the brake pedal and verify proper operation of the braking system.

Lower Control Arm

REMOVAL & INSTALLATION

9-3

2002 MODEL

1. Before servicing the vehicle, refer to the Precautions Section.
2. Remove or disconnect the following:
   - Front wheel
   - Sway bar hardware
   - Ball joint
   - Support arm to lower control arm nut
   - Lower control arm to subframe bolt
   - Lower control arm

To install:

3. Install or connect the following:
   - Lower control arm and subframe bolt and hand tighten the bolt
   - Support arm to lower control arm using a new nut and torque the nut to 68 ft. lbs. (92 Nm)
   - Ball joint and torque the nut to 55 ft. lbs. (75 Nm) and the lower control arm to subframe bolt to 85 ft. lbs. (115 Nm)
   - Sway bar hardware and torque the nut to 89 inch lbs. (10 Nm)
   - Front wheel

2003–06 MODELS

4. Before servicing the vehicle, refer to the Precautions Section.
5. Raise front of vehicle and remove front wheels.
6. Remove the bolt holding the lower control arm to the steering swivel member.
7. If equipped, remove the level sensor.
8. Disassemble the lower control arm front attachment to the subframe.
9. Slacken the bolt which holds the rear bushing to the lower control arm.
10. Remove the rear bushing from the subframe.
11. Remove the lower control arm from the steering swivel member and lift away the lower control arm.

To install:

12. Lift up the lower control arm and fit the bolt for the front bushing.
13. Fit the rear bushing to the subframe. Torque bolt to 48 ft. lbs. (65 Nm), plus 90 degrees.
14. Connect the lower control arm to the steering swivel member with a new nut. Hold the bolt with a wrench so that the bolt does not rotate. Torque nut to 37 ft. lbs. (50 Nm).

**** WARNING**

Press up the pin carefully. The groove in the pin must be visible in the screw hole in the spindle housing. If the rubber gaiter is pressed down, it will not seal properly to the swivel pin.

15. Remove the anti-roll bar link from the anti-roll bar and lift the lower control arm up to normal position.
16. Tighten the front lower control arm attachment to 48 ft. lbs. (65 Nm) plus 90 degrees.
17. Tighten the bolt holding the rear bush to the lower control arm. Torque the bolt to 30 ft. lbs. (40 Nm) plus 30 degrees.
18. Reconnect the anti-roll bar link and torque bolt to 47 ft. lbs. (64 Nm).
19. If equipped, fit the level sensor.
20. Fit the wheel.
21. Lower the car to the floor.
22. Carry out 4-wheel checking and alignment.

9-5

1. Before servicing the vehicle, refer to the Precautions Section.
2. Remove the front wheel.
3. Remove the anti-roll bar hardware.
4. Separate the lower control arm ball joint from the steering swivel member.
5. Remove the lower control arm to subframe bolt.
6. Remove the lower control arm.

To install:

7. Install the lower control arm to the subframe and install the subframe bolt, then hand tighten the bolt.
   - Install the ball joint into the steering swivel member and torque the nut to 55 ft. lbs. (75 Nm) and the lower control arm to subframe bolt to 85 ft. lbs. (115 Nm).
8. Install the anti-roll bar hardware.
9. Install the front wheel.

Anti-Roll Bar

REMOVAL & INSTALLATION

9-3

2002 MODEL

1. Before servicing the vehicle, refer to the Precautions Section.
2. Suspend the engine using an engine hoist.
3. Raise the car and remove the front wheels.
4. Remove the 2 front cover panels.
5. Detach the exhaust pipe at the turbo or exhaust manifold.
6. Remove the exhaust pipe at the joint.
7. Place a stand under the subframe.
8. Slacken the 2 front subframe retaining bolts slightly (7).
9. Remove the 4 retaining bolts and the 2 retaining nuts at the rear of the subframe (8).
10. Lower the subframe at the rear.
11. Unscrew the 2 nuts securing the anti-roll bar (10).
12. Remove the anti-roll bar's two retaining clamps (11).
13. Remove the anti-roll bar.

To install:

14. Lubricate the bushings with Molykote 33 or other grease of similar type.
15. Position the anti-roll bar on the subframe.
16. Fit the rubber bushings on left-hand and right-hand sides.
17. Position the retaining clamps and screw the bolts in place finger-tight.
18. Fit the rubber bushings and retaining nuts for the anti-roll bar links. Torque the nuts to 7 ft. lbs. (10 Nm).
19. Tighten the retaining bolts for the anti-roll bar clamps to 19 ft. lbs. (26 Nm).
20. Raise the subframe at the rear and fit the 4 retaining bolts and 2 retaining nuts. The washers on the rear retaining bolts should abut against the body. Torque these as follows:
    - Rear mounting-to-subframe: 81 ft. lbs. (110 Nm) plus 75 degrees
    - Center mounting-to-subframe: 140 ft. lbs. (190 Nm)
    - Front retaining bolts: 85 ft. lbs. (115 Nm)
21. Fit the clamp at the exhaust pipe joint.
22. Attach the exhaust pipe to the turbo or exhaust manifold and bolt it in place.
23. Fit the 2 front cover panels.
24. Lower the car to the floor and remove the lifting hoist.

2003–06 MODELS

1. Before servicing the vehicle, refer to the Precautions Section.
2. Place the wheels in a straight forward position and secure steering wheel in this position.
3. Remove or disconnect the following:
   - Front left wheel
   - Rear nut from the left fender liner

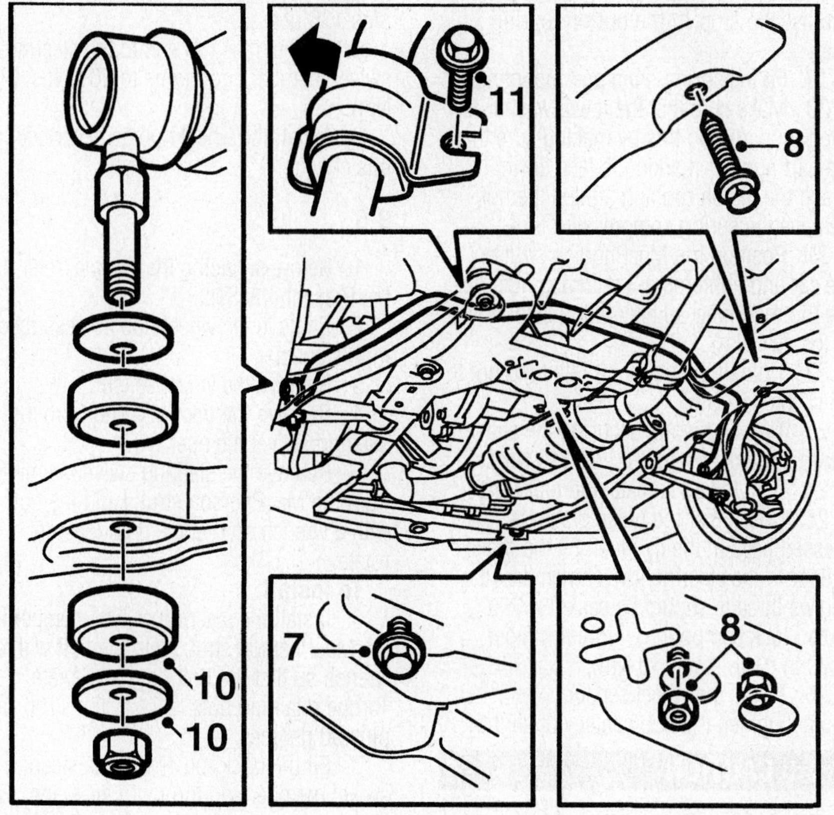

Identifying components to remove before removing the anti-roll bar—9-3

- Spoiler shield
- Cut exhaust pipe at right angles about 3.4 inches (87mm) in front of front muffler
- Front part of exhaust pipe; clean cut section of burrs and dirt
- Front section of exhaust system
- Steering shaft from steering gear at pinch bolt
- Heat shield from steering gear
- Track rod from steering swivel member on left side
- Engine torque rod from subframe
- Rear subframe bolts and stays; slacken front bolts several turns; hold down subframe with wedges
- Anti-roll bar links from anti-roll bar
- Anti-roll bar from subframe; pull out to left between bulkhead partition and subframe

To install:

4. Position the anti-roll bar in place between the bulkhead partition and the subframe.
5. Lubricate the anti-roll bar bushings.
6. Attach the anti-roll bar to the subframe. Be sure the opening of the rubber bushings faces forward. Torque fasteners to 13 ft. lbs. (18 Nm).
7. Lift up the subframe to the body. Enter the bolts for the subframe and stays on both sides.
8. Position the trolley lift with jig underneath and tighten the bolts on both sides to 55 ft. lbs. (75 Nm) plus 135 degrees.
9. Lower and pull trolley lift out of the way.
10. Tighten the rear stay bolts to 66 ft lbs. (90 Nm) plus 45 degrees.
11. Install or connect the following:
    - Heat shield to steering gear
    - Engine torque rod; torque to 52 ft. lbs. (70 Nm), plus 90 degrees
    - Anti-roll bar links on both sides; torque to 47 ft. lbs. (64 Nm)
    - Left-hand track rod to steering swivel member; torque to 26 ft. lbs. (35 Nm)
    - Rear fender liner edge
    - Steering shaft to steering gear; be sure groove in steering gear shaft is set in correct position so bolt fits in the groove; torque to 20 ft. lbs. (27 Nm)
    - Front part of exhaust pipe and fit a splice; torque clamps to 30 ft. lbs. (40 Nm)
    - Front wheel
    - Release steering wheel fixing

9-5

1. Before servicing the vehicle, refer to the Precautions Section.
2. Remove or disconnect the following:
    - Cover over intake manifold
    - Engine cushion from engine mount and subframe
3. Position an engine hoist and chain and fit the hook in the engine's rear lifting eyebolt.
4. Raise the car and remove the front wheels.
5. Remove or disconnect the following:
    - Reinforcement at subframe rear mounting point
    - Catalytic converter from muffler at joint
    - 2 steering gear retaining bolts
    - Bolts from subframe center attachment point; lower rear of subframe
    - Anti-roll bar from links
    - Bolts from clamps securing anti-roll bar to subframe
6. Withdraw the anti-roll bar through the wheel housing on the passenger side.

To install:

7. Lift the anti-roll bar into position through the wheel housing on the passenger side. Make sure that it does not catch on any hoses or cables in the engine bay.
8. Lubricate the bushes with Molykote 33 and fit them with the opening facing rearwards.
9. Install or connect the following:
    - Bolts in clamps securing anti-roll bar to subframe; torque to 20 ft. lbs. (25 Nm).
    - Nuts securing anti-roll bar to the links; torque to 65 ft. lbs. (90 Nm)
    - Subframe rear bolts at center attachment points; torque to 75 ft. lbs. (100 Nm) plus 45 degrees
    - Joint between catalytic converter and muffler
    - Subframe at rear attachment points with the reinforcement; torque to 75 ft. lbs. (100 Nm) plus 45 degrees
    - Bolts securing reinforcement to subframe; torque to 50 ft. lbs. (65 Nm)
    - Steering gear retaining bolts; torque to 70 ft. lbs. (95 Nm)
    - Front wheels
10. Lower the engine onto the rear engine cushion and remove the lifting beam.
11. Tighten the bolts securing the engine cushion to the subframe to 20 ft. lbs. (25 Nm).
12. Tighten the bolts securing the rear

engine cushion to the engine mount to 35 ft. lbs. (50 Nm).

13. Refit the cover over the intake manifold.

Steering Swivel Member

REMOVAL & INSTALLATION

9-3

2002 MODEL

➡The steering swivel member is also referred to as the steering knuckle.

1. Before servicing the vehicle, refer to the Precautions Section.
2. Slacken the wheel hub center-nut when all 4 wheels of the car are on the floor.
3. Raise the car and remove the wheel.
4. Unscrew the hub center-nut completely.
5. Remove the wheel sensor.
6. Use a pair of slip-joint pliers to press in the brake piston. Remove the caliper from the steering swivel member and suspend it in the wheel housing by means of a cable tie.
7. Remove the brake disc and the cover plate. Slacken the tie-rod end nut slightly. Press out the tie-rod end bolt. Undo the nut and lift the tie-rod end bolt out of the hole.
8. Unscrew the nut securing the anti-roll bar to the lower control arm.
9. Undo the nut on the outer ball joint. Press the ball joint out of the steering swivel member. Unscrew the nut. This nut is of self-locking type and must not be reused.
10. Lift the plastic cover off the top MacPherson strut mounting and unscrew the 3 nuts.
11. Lift off the MacPherson strut and place it in a vise. Compress the spring with a proper spring compression tool.
12. Grip the strut piston rod and remove the nut. The nut is of self-locking type and must not be reused.
13. Remove the bearing and upper spring seat.
14. Remove the spring, rubber boot and compression stop.
15. Remove the shock absorber from the MacPherson strut.

To install:

16. Inspect the gaiter and bushes for damage.
17. Install and tighten the shock absorber to 159 ft. lbs. (215 Nm).
18. Install the spring and zinc washer. The lower end of the spring should abut against the stop in the bottom spring seat.
19. Fit the spring shim and mounting.
20. Make sure that the lower washer is fitted properly. Do this by making sure that the part number marking is face down. Grasp the piston rod and tighten the nut. Remove the spring compression tool.
21. Position the MacPherson strut on the car and tighten the 3 retaining nuts on the top mounting, in alternating steps to 18 ft. lbs. (24 Nm).
22. Tighten the ball joint retaining nut to 55 ft. lbs. (75 Nm).
23. Tighten the nut securing the anti-roll bar to the link arm to 7 ft. lbs. (10 Nm).
24. Refit the back plate and brake disc.
25. Check that the brake piston is pressed back in the cylinder. Fit the brake caliper to the steering swivel member (the screws threads should be sealed with a thread lock compound). Tighten caliper bolts to 81 ft. lbs. (110 Nm).
26. Install the wheel sensor.
27. Tighten the hub center-nut slightly.

⁂ CAUTION

A new hub center-nut should always be fitted if it has been loosened, as the clamping force of the lock embossing is reduced when refitted. Use nut with top groove.

28. Install the wheel.
29. Lower the car and tighten the hub center-nut when all four wheels of the car are on the floor to 126 ft. lbs. (170 Nm) plus 45 degrees.
30. Tighten the nut on the upper strut bearing to 55 ft. lbs. (75 Nm). Fit the plastic cover over the nuts.
31. Pump the brake pedal to advance the brake pistons in the calipers.

2003–06 MODELS

1. Before servicing the vehicle, refer to the Precautions Section.
2. Remove the wheel hub as described in this chapter.
3. Remove the track rod end from the steering swivel member.
4. Remove the steering swivel member from the MacPherson strut. Hold the bolt with a wrench so that the bolt does not rotate.

To install:

5. Install the steering swivel member to the MacPherson strut. Hold the bolt with a wrench so that the bolt does not rotate. Torque retaining nuts to 59 ft. lbs. (80 Nm) plus 135 degrees.
6. Fit the track rod end to the steering swivel member and torque to 26 ft. lbs. (35 Nm).
7. Install the wheel hub as described in this chapter.

9-5

1. Before servicing the vehicle, refer to the Precautions Section.
2. Remove the wheel hub as described in this chapter.
3. Remove the wheel sensor.
4. Remove the track rod end from the steering swivel member.
5. Remove the steering swivel member from the MacPherson strut. Hold the bolt with a wrench so that the bolt does not rotate.

To install:

6. Install the steering swivel member to the MacPherson strut. Hold the bolt with a wrench so that the bolt does not rotate. Torque retaining nuts to 74 ft. lbs. (100 Nm) plus 90 degrees.
7. Fit the track rod end to the steering swivel member and torque to 26 ft. lbs. (35 Nm).
8. Install the wheel sensor.
9. Install the wheel hub as described in this chapter.

Wheel Bearings

ADJUSTMENT

The wheel bearings found in all Saab vehicles are sealed units requiring no adjustment.

Wheel Hub

REMOVAL & INSTALLATION

9-3 2002 Model

1. Before servicing the vehicle, refer to the Precautions Section.
2. Remove the MacPherson strut as described in this chapter.
3. Place 2 pieces of flat or square bar under the MacPherson strut and press off the front wheel hub (3) from the wheel bearing (2). Use a wheel bearing drift, 89 96 704.
4. Remove the circlips from the MacPherson strut (4).

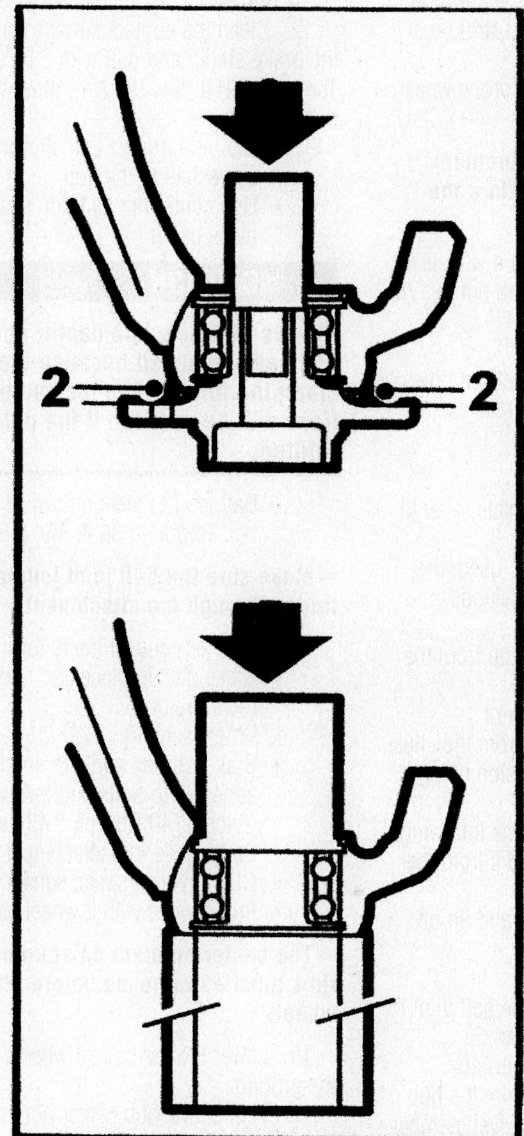

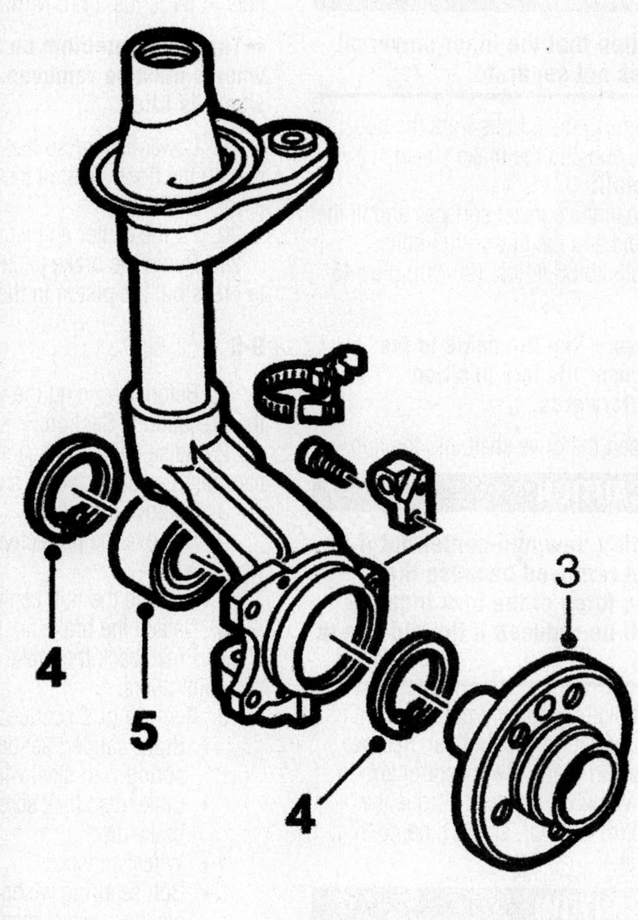

Exploded view of the front wheel hub—9-3 2002 shown

5. Remove the wheel bearing (5) from the steering swivel member with the wheel bearing drift.

To install:

6. Fit the outer circlip on the steering swivel member. The opening in the circlip should face downwards.
7. Using the wheel bearing drift, 89 96 704, press in the wheel bearing until it abuts with the circlip.
8. Lubricate the bearing seat and outer circumference of the bearing with Molykote G Rapid Plus grease.
9. Fit the inner circlip on the steering swivel member. The opening in the circlip should face downwards.
10. Install the wheel hub into the wheel bearing, using the wheel bearing drift, 89 96 704.
11. Install the MacPherson strut as described in this chapter.

9-3 2003–06 Models

1. Before servicing the vehicle, refer to the Precautions Section.
2. Slightly loosen the wheel hub nut while all 4 wheels are on the ground.
3. Raise the car and remove the wheel.
4. Remove the hub nut.
5. Press in the driveshaft so that it loosens from the hub; use hub drive shaft puller 89 96 951.
6. Press in the piston by prying as follows:
   - Level 1 and 2: Pry against the brake pad using a screwdriver.
   - Level 3: Pry between the brake disc and brake caliper using a screwdriver.
7. Remove or disconnect the following:
   - Brake caliper; suspend it from the spring
   - Brake disc lock bolt and lift off the brake disc
   - Wheel sensor connection
   - Bolt securing lower control arm ball joint to steering swivel member
   - Pull out the ball joint pin
8. Place a wedge between the lower control arm and the anti-roll bar in order to hold down the lower control arm.

9. Pull the drive shaft out from the hub and lift the hub and steering swivel member from the shaft.

CAUTION

Use caution that the inner universal joint does not separate.

10. Remove the 3 bolts from the hub. Lift off the hub and the brake shield.

To install:

11. Clean the contact surfaces and fit the brake shield and the hub with 3 bolts. Torque bolts to 66 ft. lbs. (90 Nm) plus 45 degrees.

➡Make sure that the cable to the wheel sensor fits into position upwards/forwards.

12. Insert the drive shaft into the hub.

CAUTION

Always fit a new hub-center nut if it has been removed because the clamping force of the lock indentations will be reduced if the old one is refitted.

13. Fit the hub nut and pull the shaft to the correct position with the nut. Remove the wedge and fit the lower control arm to the steering swivel member. Torque the lower control arm nut, at this time, to 37 ft. lbs. (50 Nm).

CAUTION

Check that the spindle on the ball joint protrudes above the steering swivel member attachment.

14. Clean the brake disc surfaces thoroughly.
15. Install the brake disc into place and tighten the brake disc bolts to 5 ft. lbs. (7 Nm).
16. Connect the wheel sensor and fit the connector into the holder.
17. Install the brake caliper and torque caliper bolts to 155 ft. lbs. (210 Nm) plus 30 degrees.
18. Install the front wheel. Torque wheel nuts to 84 ft. lbs. (110 Nm).

➡The center emblem on aluminum wheels must be removed before the wheel is fitted.

19. Lower the car so that all 4 wheels rest on the floor. Tighten the hub nut to 170 ft. lbs. (230 Nm).
20. Fit the center emblem.
21. Pump the brake pedal several times to press out the piston in the brake caliper.

9-5

1. Before servicing the vehicle, refer to the Precautions Section.
2. Raise the car. If a jack is used, the car must be supported on axle stands.
3. Remove the wheel.
4. Remove the protective cup from the wheel hub.
5. Remove the hub center nut.
6. Detach the brake hose from the clips.
7. Press back the brake piston using slip-joint pliers.
8. Remove or disconnect the following:
   - Brake caliper; suspend it from the spring with steel wire
   - Brake disc lock screw and lift off brake disc
   - Wheel sensor
   - Bolt securing wishbone ball joint to steering swivel member
9. Pull out the ball joint journal.
10. Tap out the driveshaft from the hub and lift the hub and steering swivel member from the shaft.

CAUTION

Take care so that the inboard universal joint does not separate.

11. Remove the 3 hub bolts. Lift off the hub and the disc back-plate.

To install:

12. Clean the contact surfaces and fit the brake shield and hub with 3 bolts. Torque to 66 ft. lbs. (90 Nm) plus 45 degrees.
13. Install or connect the following:
   - Driveshaft to the hub
   - Hub center nut and pull shaft in place using nut

CAUTION

Always fit a new hub center nut if it has been removed because the clamping force of the lock indentations will be reduced if the old one is refitted.

   - Ball joint to steering swivel member. Torque to 36 ft. lbs. (49 Nm).

➡Make sure the ball joint journal protrudes through the attachment.

14. Install or connect the following:
   - Brake disc in place and tighten its retaining bolt
   - Wheel sensor
   - Brake caliper, apply thread locking adhesive to bolts and torque to 103 ft. lbs. (140 Nm) plus 45 degrees.
   - Brake hose into attachment and clip
   - Clip to wheel speed sensor cable
   - Front wheel with 2 wheel bolts.

➡The center emblem on aluminum rims must be removed before fitting the wheel.

15. Lower the car so that wheels are on the ground.
16. Tighten the hub center nut to 170 ft. lbs. (230 Nm).
17. Raise the car and remove the wheel. Fit the center emblem.
18. Fit the protective cup to the wheel hub.
19. Install the front wheel.
20. Depress the brake to press out the piston in the brake caliper.

REAR SUSPENSION

Shock Absorber

REMOVAL & INSTALLATION

9-3 2002 Model

1. Before servicing the vehicle, refer to the Precautions Section.
2. Remove or disconnect the following:
   - Upper mounting bolt in trunk. Cut out a flap in the carpeting to access the mounting bolt and bushings.
   - Nut, washer and bushing from the mounting point
   - Rear wheel
   - Lower shock mounting bolt
   - Shock

To install:

3. Install or connect the following:
   - Shock to the upper mounting. Be sure to install the lower part of the bushing on the shock
   - Lower shock mounting bolt and torque the bolt to 46 ft. lbs. (62 Nm)
   - Rear wheel
   - Upper bushing, washer and nut and torque to 15 ft. lbs. (20 Nm)
   - Carpeting

9-3 2003–06 Models

1. Before servicing the vehicle, refer to the Precautions Section.
2. Raise the car and remove the rear wheels.
3. Relieve the weight on the lower suspension arm with a jack.
4. Clean, lubricate and remove the bolts which holds the shock absorber to the steering swivel member.
5. Remove the shock absorber bracket from the body and lift the shock absorber out.
6. Remove the bracket from the shock absorber.

To install:

7. Install the bracket to the shock absorber. Torque bracket bolts to 20 ft. lbs. (27 Nm).
8. Fit the shock absorber bracket to the body and torque to 39 ft. lbs. (53 Nm).
9. Install the shock absorber to the steering swivel member. Torque retaining bolts to 111 ft. lbs. (150 Nm).
10. Install the rear wheels. Torque wheel nuts to 81 ft. lbs. (110 Nm).

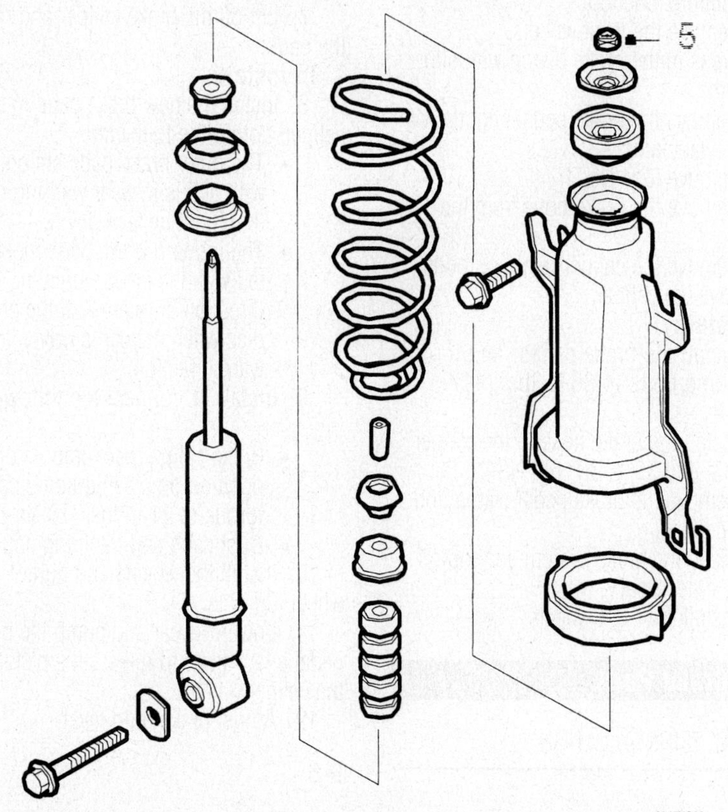

Exploded view of the shock absorber and spring assembly

9-5

1. Before servicing the vehicle, refer to the Precautions Section.
2. Remove or disconnect the following:
   - Negative battery cable
   - Rear wheel
   - Spring bracket lower bolts and loosen the upper bolts
   - Lower shock absorber retaining bolt
   - Shock absorber and spring assembly and loosen the lock nut
3. Press the spring bracket down to relieve the load. Remove the center nut, washer and bushing. If necessary, use a spring compressor tool 88 18 791.
4. Remove the shock absorber.

To install:

5. Install or connect the following:
   - Spring, spacer ring and bracket on the shock absorber
   - Press the bracket down to relieve the shock absorber load and install the bushing and washer
   - New locknut and torque the nut to 15 ft. lbs. (20 Nm)
   - Shock absorber and torque the bolts to 40 ft. lbs. (55 Nm)
   - Lower mounting bolt on the rear axle and torque the bolt to 140 ft. lbs. (190 Nm)
   - Rear wheel
   - Negative battery cable

Coil Spring

REMOVAL & INSTALLATION

9-3 2002 Model

1. Before servicing the vehicle, refer to the Precautions Section.
2. Raise the vehicle and safely support it on jackstands. Do not place the stands under the rear axle assembly.
3. Remove or disconnect the following:
   - Rear wheels
4. Place a floor jack under the lower control arm and raise it upward slightly.

➡If the same spring is to be reinstalled, mark the rear of the spring to be sure it is reinstalled in the proper position.

5. Remove the lower mounting bolt from the shock absorber
6. Slowly lower the floor jack until the lower arm is relaxed.
7. Use a prybar to bring the lower arm down far enough to remove the spring.

To install:

8. Install or connect the following:
   - Spring with rubber cushions
   - Lower shock mount and torque to 46 ft. lbs. (62 Nm)
   - Rear Wheels
9. Lower the vehicle and road test.

9-3 2003–06 Models

1. Before servicing the vehicle, refer to the Precautions Section.
2. Raise the vehicle and safely support it on jackstands. Do not place the stands under the rear axle assembly.
3. Remove the wheel.
4. Remove the spring from the brake caliper.
5. Remove the protective covers.
6. Remove and move aside the hydraulic body.
7. Remove the outer brake pad.
8. Compress the spring and lift out the spring. Remove the spring from the compression tool.

To install:

9. Compress the spring and place the spring support in the spring.
10. Lift the spring into position. Remove the spring compressor.
11. Remove the inner brake pad and screw in the brake piston.
12. Fit the brake pads.
13. Make sure the springs on the inner pad rest in the piston groove.
14. Fit the hydraulic body. Torque the bolts to 21 ft. lbs. (28 Nm).
15. Install the protective covers.
16. Fit the spring to the brake caliper.
17. Install the rear wheel.
18. Depress the brake pedal several times to press out the self adjustment of the brake pistons and the parking brake.

9-5

Refer to the shock absorber removal and installation procedure for the rear coil spring.

Wheel Bearings

ADJUSTMENT

The wheel bearings found in all Saab vehicles are sealed units requiring no adjustment.

FRONT BRAKES

Brake Caliper

REMOVAL & INSTALLATION

9-3

1. Before servicing the vehicle, refer to the Precautions Section.
2. Remove the wheels.
3. Press caliper piston back into caliper.
4. Use a brake pedal clamp to hold the pedal down.
5. On 2003–06 models, remove fuse No. 6 from electrical center.
6. Disconnect the brake hose from the brake caliper.
7. Remove the brake caliper mounting bolts and remove the brake caliper.

To install:

8. Fit the brake caliper onto the strut. Apply thread-locking compound to the mounting bolts and tighten the caliper mounting bolts to the following:
   - 2002: 81 ft. lbs. (110 Nm)
   - 2003–06: 155 ft. lbs. (210 Nm) plus 30 degrees
9. Install the brake hose onto the caliper, with new seals, if equipped. Torque connection to 30 ft. lbs. (40 Nm).

⁂ CAUTION

Be careful to install the brake hose in its original position.

10. Release the brake pedal clamp.
11. Reinstall fuse No. 6 into electrical center.
12. Bleed the brake system and fill the fluid to the proper level.
13. Install the wheels.

9-5

1. Before servicing the vehicle, refer to the Precautions Section.
2. Remove the front wheels.
3. Press in the brake piston with slip joint pliers.
4. Depress the brake pedal slightly with a brake pedal clamp.
5. Remove fuse No. 1.
6. Remove the brake hose from the caliper.
7. Remove the caliper retaining bolts and remove the caliper.

To install:

8. Install the brake caliper and torque the retaining bolts to 86 ft. lbs. (117 Nm).
9. Install the brake hose to the caliper and torque it to 29 ft. lbs. (40 Nm).
10. Remove the brake pedal clamp and replace the fuse
11. Bleed the brake system and top off the fluid, if necessary.
12. Install the front wheels.

Disc Brake Pads

REMOVAL & INSTALLATION

9-3

1. Before servicing the vehicle, refer to the Precautions Section.
2. Remove the wheels.
3. Push caliper piston back.
4. Remove the retaining clip from the brake caliper.
5. Remove the caliper guide pins.
6. Remove caliper and suspend it from the strut.
7. Lift off the brake caliper and remove the pads.

To install:

8. Install the new brake pads in the caliper, noting the following:
   - The outer brake pads are equipped with acoustic wear warning devices that must be face down.
   - The inboard brake pads must be fitted with the arrows pointing in the direction of rotation of the brake disc when the car is driven in a forward direction.
9. Install or connect the following:
   - Brake caliper back into its original position on clean guide pins; torque to 21 ft. lbs. (28 Nm)
   - Dust caps and retaining clip
10. Install the wheels and tighten the wheel lug nuts.
11. Lower the car and pump the brake pedal a few times to release the piston into the caliper.
12. Adjust brake fluid level.

9-5

1. Before servicing the vehicle, refer to the Precautions Section.
2. Remove the front wheels.

3. Press back the caliper piston and remove the clip.
4. Remove the caliper guide pins.
5. Remove the caliper and suspend it from the strut.
6. Remove the brake pads.

To install:

7. Install the brake pads in the caliper.
8. Install the brake caliper and torque the bolts to 21 ft. lbs. (28 Nm).
9. Install the brake clip depress the brake pedal to force out the pistons.
10. Install the front wheels.
11. Check and top off the brake fluid, if necessary.

REAR BRAKES

Brake Caliper

REMOVAL & INSTALLATION

9-3

2002 MODEL

1. Before servicing the vehicle, refer to the Precautions Section.
2. Remove the wheels.
3. Press caliper piston back into the caliper.
4. Use a brake pedal clamp to slightly depress brake pedal.
5. Remove the brake pipe from the caliper and plug the openings.
6. Remove the locking spring and caliper pins.
7. Remove the brake pads.
8. Remove 2 caliper bolts and remove the caliper.

To install:

9. Position the caliper and install retaining bolts. Torque the bolts to 59 ft. lbs. (80 Nm).
10. Connect the brake pipe and tighten to 10 ft. lbs. (14 Nm).
11. Install the brake pads, caliper pins and locking spring.
12. Remove brake pedal clamp.
13. Install the wheels and tighten the lug nuts.
14. Pump brake pedal several times to release the piston.

2003–06 MODELS

1. Before servicing the vehicle, refer to the Precautions Section.
2. Remove fuse No. 6 from the instrument panel electrical center.
3. Depress the brake pedal slightly using a brake pedal clamp.
4. Loosen the handbrake adjustment so that the cables can be unhooked from the brake caliper.
5. Raise the car and remove the rear wheel.
6. Remove or disconnect the following:
   - Brake hose and handbrake cable from caliper body
   - Carrier from brake caliper
   - Protective covers and remove caliper body
   - Brake pads
   - Carrier from steering swivel member.

To install:

7. Clean the contact surfaces between the steering swivel member and the brake caliper carrier.
8. Install or connect the following:
   - Brake caliper carrier to steering swivel member, 74 96 268 thread lock adhesive to bolts and torque to 96 ft. lbs. (130 Nm) plus 45 degrees.
   - Screw in brake piston, if necessary
   - Brake pads

➡Make sure the springs on the inner pad enter the groove in the piston.

   - Caliper hydraulic body; torque bolts to 21 ft. lbs. (28 Nm)
   - Protective covers
   - Retaining spring to the brake caliper
   - Brake hose, with new seals, and torque connection to 30 ft. lbs. (40 Nm)
9. Remove the brake pedal clamp and refit fuse No. 6 in the instrument panel electrical center.
10. Bleed the brake system.
11. Adjust the handbrake.
12. Install the rear wheels.
13. Lower the car.
14. Depress the brake pedal repeatedly in order to press out the brake piston and the handbrake self adjustment.
15. Check the operation of the foot brake and handbrake.

9-5

1. Before servicing the vehicle, refer to the Precautions Section.
2. Remove the rear wheels.
3. Remove the brake pads.
4. Depress the brake pedal slightly with a brake pedal clamp.
5. Remove fuse No. 1.
6. Remove the clip from the brake pedal pipe, swinging caliper only.
7. Remove the hose from the brake pedal pipe.
8. Remove the caliper guide pins and remove the hydraulic body, swinging caliper only.
9. Remove the brake hose from the caliper.
10. Remove the caliper.

To install:

11. Install the brake lines to the hydraulic body, swinging caliper only. Torque the line to 32 ft. lbs. (43 Nm).
12. Install the hydraulic body and torque the bolts to 21 ft. lbs. (28 Nm).
13. Install the caliper and torque the bolts to 59 ft. lbs. (80 Nm), fixed caliper only.
14. Install the brakes lines and torque them to 11 ft. lbs. (16 Nm), fixed caliper only.
15. Install the brake hose to the pipe and install the clip, swinging caliper only.
16. Install the brake pads.
17. Remove the brake pedal clamp and replace the fuse.
18. Bleed the brake system and top off the fluid, if necessary.
19. Install the rear wheels.

Disc Brake Pads

REMOVAL & INSTALLATION

9-3

1. Before servicing the vehicle, refer to the Precautions Section.
2. Remove the rear wheels.
3. Compress the caliper piston into the caliper bore.
4. Remove the brake pad retaining pins and lock spring.
5. Remove the brake pads.

To install:

6. Install the brake pads into the brake caliper.
7. Install the lock spring and the brake pad retaining pins.

8. Install the wheels and tighten the lug nuts. Torque the lug nuts to 81 ft. lbs. (102 Nm).
9. Adjust brake fluid level.

9-5

1. Before servicing the vehicle, refer to the Precautions Section.
2. Remove the rear wheels.
3. Press back the piston and remove the caliper clip.
4. Remove the caliper guide pins.
5. Remove the brake caliper.
6. Remove the brake pads.

To install:

7. Install the brake pads.
8. Install the caliper and torque the pins to 21 ft. lbs. (28 Nm).
9. Install the caliper clip and depress the brake pedal to force out the pistons.
10. Install the rear wheels.
11. Check and top off the brake fluid, if necessary.

VOLKSWAGEN

Cabrio • Golf • GTI • Jetta • New Beetle • Passat • Rabbit

11

SPECIFICATIONS AND MAINTENANCE CHARTS

ENGINE AND VEHICLE IDENTIFICATION

Engine						
Code ①	Liters (cc)	Cu. In.	Cyl.	Fuel Sys.	Engine Type	Eng. Mfg.
AAA	2.8 (2792)	170	6	Motronic	DOHC	Volkswagen
AAZ	1.9 (1896)	116	4	Diesel	SOHC	Volkswagen
ABA	2.0 (1984)	121	4	Motronic	SOHC	Volkswagen
AEB	1.8 (1781)	109	4	Motronic	DOHC	Volkswagen
AEG	2.0 (1984)	121	4	Motronic	SOHC	Volkswagen
AHA	2.8 (2771)	170	6	Motronic	DOHC	Volkswagen
AHH	1.9 (1896)	116	4	Diesel	SOHC	Volkswagen
ALH	1.9 (1896)	116	4	Diesel	SOHC	Volkswagen
ATQ	2.8 (2771)	170	6	Motronic	DOHC	Volkswagen
ATW	1.8 (1781)	109	4	Motronic	DOHC	Volkswagen
AVH	2.0 (1984)	121	4	Motronic	SOHC	Volkswagen
AWM	1.8 (1781)	109	4	Motronic	DOHC	Volkswagen
AWP	1.8 (1781)	109	4	Motronic	DOHC	Volkswagen
AWV	1.8 (1781)	109	4	Motronic	DOHC	Volkswagen
AWW	1.8 (1781)	109	4	Motronic	DOHC	Volkswagen
AZG	2.0 (1984)	121	4	Motronic	SOHC	Volkswagen
BDC	2.0 (1984)	121	4	Motronic	SOHC	Volkswagen
BDF	2.8 (2771)	170	6	Motronic	DOHC	Volkswagen
BDP	4.0 (3999)	244	8	Motronic	DOHC	Volkswagen
BEV	2.0 (1984)	121	4	Motronic	SOHC	Volkswagen
BEW	1.9 (1896)	116	4	Diesel	SOHC	Volkswagen
BGD	2.0 (1984)	121	4	Motronic	SOHC	Volkswagen
BGP	2.5 (2480)	151	5	Motronic	DOHC	Volkswagen
BGQ	2.5 (2480)	151	5	Motronic	DOHC	Volkswagen
BKF	1.8 (1781)	109	4	Motronic	DOHC	Volkswagen
BLV	3.6 (3597)	219	6	Motronic	DOHC	Volkswagen
BPR	2.5 (2480)	151	5	Motronic	DOHC	Volkswagen
BPS	2.5 (2480)	151	5	Motronic	DOHC	Volkswagen
BPY	2.0 (1984)	121	4	Motronic	DOHC	Volkswagen
BRM	1.9 (1896)	116	4	Diesel	SOHC	Volkswagen

Model Year	
Code ②	Year
2	2002
3	2003
4	2004
5	2005
6	2006

DOHC: Double Overhead Camshafts

SOHC: Single Overhead Camshaft

① Located on the vehicle data plate

② 10th digit of VIN

06041_VWMC_C0001

GENERAL ENGINE SPECIFICATIONS

Year	Model	Engine Displacement Liters	Engine ID/VIN	Net Horsepower @ rpm	Net Torque@rpm (ft. lbs.)	Bore x Stroke (in.)	Com-pression Ratio	Oil Pressure @ rpm
2002	Cabrio	2.0	ABA	115@5400	122@3200	3.25x3.65	10.0:1	29@2000
	Golf	1.9	AAZ	90@3750	155@1900	3.13x3.76	19.5:1	29@2000
	Golf	2.0	ABA	115@5200	122@2600	3.25x3.65	10.0:1	29@2000
	GTI	2.0	ABA	115@5200	122@2600	3.25x3.65	10.0:1	29@2000
	GTI	2.8	AAA	174@5800	181@3200	3.19x3.56	10.0:1	29@2000
	Jetta	1.9	AAZ	90@3750	155@1900	3.13x3.76	19.5:1	29@2000
	Jetta	2.0	ABA	115@5500	122@2600	3.25x3.65	10.5:1	29@2000
	Jetta	2.8	AAA	174@5800	181@3200	3.19x3.56	10.0:1	29@2000
	New Beetle	1.9	ALH	90@3750	154@1900	3.13x3.76	19.5:1	29@2000
	New Beetle	2.0	AEG	115@5400	125@2400	3.25x3.65	10.0:1	29@2000
	Passat	1.8	AEB	150@5700	155@3200	3.19x3.40	9.5:1	29@2000
	Passat	1.9	AHH	90@3750	154@1900	3.13x3.76	19.5:1	29@2000
	Passat	2.8	AHA	190@6000	206@3200	3.25x3.40	10.6:1	29@2000
2003	Convertible	1.8	AWV	150@5700	155@3200	3.19x3.40	9.5:1	29@2000
	Convertible	2.0	BDC	115@5400	125@2400	3.25x3.65	10.0:1	29@2000
	Golf	1.9	ALH	90@3750	154@1900	3.13x3.76	19.5:1	29@2000
	Golf	2.0	AEG	115@5400	125@2400	3.25x3.65	10.0:1	29@2000
	Golf	2.0	AVH	115@5400	125@2400	3.25x3.65	10.0:1	29@2000
	Golf	2.0	AZG	115@5400	125@2400	3.25x3.65	10.0:1	29@2000
	GTI	1.8	AWW	150@5700	155@3200	3.19x3.40	9.5:1	29@2000
	GTI	1.8	AWP	177@5500	173@3500	3.19x3.40	9.5:1	29@2000
	GTI	2.8	BDF	197@6200	195@3200	3.19x3.55	10.75:1	29@2000
	Jetta	1.8	AWP	177@5500	173@3500	3.19x3.40	9.5:1	29@2000
	Jetta	1.8	AWW	150@5700	155@3200	3.19x3.40	9.5:1	29@2000
	Jetta	1.9	ALH	90@3750	154@1900	3.13x3.76	19.5:1	29@2000
	Jetta	2.0	AEG	115@5400	125@2400	3.25x3.65	10.0:1	29@2000
	Jetta	2.0	AVH	115@5400	125@2400	3.25x3.65	10.0:1	29@2000
	Jetta	2.0	AZG	115@5400	125@2400	3.25x3.65	10.0:1	29@2000
	Jetta	2.8	BDF	197@6200	195@3200	3.19x3.55	10.75:1	29@2000
	New Beetle	1.8	AWV	150@5700	155@3200	3.19x3.40	9.5:1	29@2000
	New Beetle	1.8	AWP	177@5500	173@3500	3.19x3.40	9.5:1	29@2000
	New Beetle	1.9	ALH	90@3750	154@1900	3.13x3.76	19.5:1	29@2000
	New Beetle	2.0	AEG	115@5400	125@2400	3.25x3.65	10.0:1	29@2000
	New Beetle	2.0	AVH	115@5400	125@2400	3.25x3.65	10.0:1	29@2000
	New Beetle	2.0	AZG	115@5400	125@2400	3.25x3.65	10.0:1	29@2000
	New Beetle	2.0	BDC	115@5400	125@2400	3.25x3.65	10.0:1	29@2000
	Passat	1.8	AEB	150@5700	155@3200	3.19x3.40	9.5:1	29@2000
	Passat	1.8	ATW	150@5700	155@3200	3.19x3.40	9.5:1	29@2000
	Passat	1.8	AWM	168@5900	166@3500	3.19x3.40	9.3:1	29@2000
	Passat	2.8	AHA	190@6000	206@3200	3.25x3.40	10.6:1	29@2000
	Passat	2.8	ATQ	190@6000	206@3200	3.25x3.40	10.6:1	29@2000
	Passat	4.0	BDP	271@6000	273@3250	3.30x3.55	10.5:1	29@2000

06041_VWMC_C0002

GENERAL ENGINE SPECIFICATIONS

Year	Model	Engine Displacement Liters	Engine ID/VIN	Net Horsepower @ rpm	Net Torque@rpm (ft. lbs.)	Bore x Stroke (in.)	Com-pression Ratio	Oil Pressure @ rpm
2004	Convertible	1.8	AWV	150@5700	155@3200	3.19x3.40	9.5:1	29@2000
	Convertible	1.8	BKF	150@5700	155@3200	3.19x3.40	9.5:1	29@2000
	Convertible	2.0	BDC	115@5400	125@2400	3.25x3.65	10.0:1	29@2000
	Convertible	2.0	BEV	115@5400	125@2400	3.25x3.65	10.0:1	29@2000
	Convertible	2.0	BGD	115@5400	125@2400	3.25x3.65	10.0:1	29@2000
	Golf	1.9	ALH	90@3750	154@1900	3.13x3.76	19.5:1	29@2000
	Golf	2.0	AVH	115@5400	125@2400	3.25x3.65	10.0:1	29@2000
	Golf	2.0	AZG	115@5400	125@2400	3.25x3.65	10.0:1	29@2000
	GTI	1.8	AWW	150@5700	155@3200	3.19x3.40	9.5:1	29@2000
	GTI	1.8	AWP	177@5500	173@3500	3.19x3.40	9.5:1	29@2000
	GTI	2.8	BDF	197@6200	195@3200	3.19x3.55	10.75:1	29@2000
	Jetta	1.8	AWP	177@5500	173@3500	3.19x3.40	9.5:1	29@2000
	Jetta	1.8	AWW	150@5700	155@3200	3.19x3.40	9.5:1	29@2000
	Jetta	1.9	ALH	90@3750	154@1900	3.13x3.76	19.5:1	29@2000
	Jetta	2.0	AVH	115@5400	125@2400	3.25x3.65	10.0:1	29@2000
	Jetta	2.0	AZG	115@5400	125@2400	3.25x3.65	10.0:1	29@2000
	Jetta	2.8	BDF	197@6200	195@3200	3.19x3.55	10.75:1	29@2000
	New Beetle	1.8	AWV	150@5700	155@3200	3.19x3.40	9.5:1	29@2000
	New Beetle	1.8	AWP	177@5500	173@3500	3.19x3.40	9.5:1	29@2000
	New Beetle	1.9	ALH	90@3750	154@1900	3.13x3.76	19.5:1	29@2000
	New Beetle	2.0	AVH	115@5400	125@2400	3.25x3.65	10.0:1	29@2000
	New Beetle	2.0	AZG	115@5400	125@2400	3.25x3.65	10.0:1	29@2000
	New Beetle	2.0	BEV	115@5400	125@2400	3.25x3.65	10.0:1	29@2000
	Passat	1.8	AEB	150@5700	155@3200	3.19x3.40	9.5:1	29@2000
	Passat	1.8	ATW	150@5700	155@3200	3.19x3.40	9.5:1	29@2000
	Passat	1.8	AWM	168@5900	166@3500	3.19x3.40	9.3:1	29@2000
	Passat	2.8	AHA	190@6000	206@3200	3.25x3.40	10.6:1	29@2000
	Passat	2.8	ATQ	190@6000	206@3200	3.25x3.40	10.6:1	29@2000
	Passat	4.0	BDP	271@6000	273@3250	3.30x3.55	10.5:1	29@2000
2005	Convertible	1.8	BKF	150@5700	155@3200	3.19x3.40	9.5:1	29@2000
	Convertible	2.0	BEV	115@5400	125@2400	3.25x3.65	10.0:1	29@2000
	Convertible	2.0	BGD	115@5400	125@2400	3.25x3.65	10.0:1	29@2000
	Golf	1.9	ALH	90@3750	154@1900	3.13x3.76	19.5:1	29@2000
	Golf	2.0	AVH	115@5400	125@2400	3.25x3.65	10.0:1	29@2000
	Golf	2.0	AZG	115@5400	125@2400	3.25x3.65	10.0:1	29@2000
	GTI	1.8	AWW	150@5700	155@3200	3.19x3.40	9.5:1	29@2000
	GTI	1.8	AWP	177@5500	173@3500	3.19x3.40	9.5:1	29@2000
	GTI	2.8	BDF	197@6200	195@3200	3.19x3.55	10.75:1	29@2000
	Jetta	1.9	BRM	100@4000	177@1800	3.13x3.76	18.5:1	29@2000
	Jetta	2.0	BPY	200@5100	207@1800	3.20x3.70	10.3:1	29@2000
	Jetta	2.5	BGP	150@5000	170@3750	3.25x3.65	9.5:1	29@2000
	Jetta	2.5	BGQ	150@5000	170@3750	3.25x3.65	9.5:1	29@2000
	New Beetle	1.8	AWV	150@5700	155@3200	3.19x3.40	9.5:1	29@2000
	New Beetle	1.8	AWP	177@5500	173@3500	3.19x3.40	9.5:1	29@2000
	New Beetle	1.9	BEW	100@4000	177@1800	3.13x3.76	19.0:1	29@2000
	New Beetle	2.0	AVH	115@5400	125@2400	3.25x3.65	10.0:1	29@2000
	New Beetle	2.0	AZG	115@5400	125@2400	3.25x3.65	10.0:1	29@2000

06041_VWMC_C0003

GENERAL ENGINE SPECIFICATIONS

Year	Model	Engine Displacement Liters	Engine ID/VIN	Net Horsepower @ rpm	Net Torque@rpm (ft. lbs.)	Bore x Stroke (in.)	Com-pression Ratio	Oil Pressure @ rpm
2005	New Beetle	2.0	BEV	115@5400	125@2400	3.25x3.65	10.0:1	29@2000
	Passat	1.8	AEB	150@5700	155@3200	3.19x3.40	9.5:1	29@2000
	Passat	1.8	ATW	150@5700	155@3200	3.19x3.40	9.5:1	29@2000
	Passat	1.8	AWM	168@5900	166@3500	3.19x3.40	9.3:1	29@2000
	Passat	2.8	AHA	190@6000	206@3200	3.25x3.40	10.6:1	29@2000
	Passat	2.8	ATQ	190@6000	206@3200	3.25x3.40	10.6:1	29@2000
	Passat	4.0	BDP	271@6000	273@3250	3.30x3.55	10.5:1	29@2000
2006	Convertible	2.5	BPR	150@5000	170@3750	3.25x3.65	9.5:1	29@2000
	Convertible	2.5	BPS	150@5000	170@3750	3.25x3.65	9.5:1	29@2000
	Rabbit	2.5	BGP	150@5000	170@3750	3.25x3.65	9.5:1	29@2000
	Rabbit	2.5	BGQ	150@5000	170@3750	3.25x3.65	9.5:1	29@2000
	GTI	2.0	BPY	200@5100	207@1800	3.20x3.70	10.3:1	29@2000
	Jetta	1.9	BRM	100@4000	177@1800	3.13x3.76	18.5:1	29@2000
	Jetta	2.0	BPY	200@5100	207@1800	3.20x3.70	10.3:1	29@2000
	Jetta	2.5	BGP	150@5000	170@3750	3.25x3.65	9.5:1	29@2000
	Jetta	2.5	BGQ	150@5000	170@3750	3.25x3.65	9.5:1	29@2000
	New Beetle	1.9	BEW	100@4000	177@1800	3.13x3.76	19.0:1	29@2000
	New Beetle	2.5	BPR	150@5000	170@3750	3.25x3.65	9.5:1	29@2000
	New Beetle	2.5	BPS	150@5000	170@3750	3.25x3.65	9.5:1	29@2000
	Passat	2.0	BPY	200@5100	207@1800	3.20x3.70	10.3:1	29@2000
	Passat	3.6	BLV	280@6200	265@2750	3.50x3.80	12.0:1	29@2000

06041_VWMC_C0004

GASOLINE ENGINE TUNE-UP SPECIFICATIONS

Year	Engine Displacement Liters	Engine ID/VIN	Spark Plug Gap (in.)	Ignition Timing (deg.) A		Fuel Pump (psi)	Idle Speed (rpm)		Valve Clearance	
				MT	AT		MT	AT	Intake	Exhaust
2002	1.8 (1781)	AEB	0.035-0.043	5-7B ①	5-7B ①	52 ②	820-920	820-920	HYD	HYD
	2.0 (1984)	ABA	0.024	5-7B ①	5-7B ①	44 ②	800-880	800-880	HYD	HYD
	2.0 (1984)	AEG	0.035-0.043	5-7B ①	5-7B ①	52 ②	760-880	760-880	HYD	HYD
	2.8 (2792)	AAA	0.028	5-7B ①	5-7B ①	58 ②	650-750	650-750	HYD	HYD
	2.8 (2771)	AHA	0.063	5-7B ①	5-7B ①	52 ②	620-740	620-740	HYD	HYD
2003	1.8 (1781)	AEB	0.035-0.043	5-7B ①	5-7B ①	52 ②	820-920	820-920	HYD	HYD
	1.8 (1781)	ATW	0.035-0.043	5-7B ①	5-7B ①	52 ②	820-920	820-920	HYD	HYD
	1.8 (1781)	AWM	0.035-0.043	5-7B ①	5-7B ①	50 ②	740-920	740-920	HYD	HYD
	1.8 (1781)	AWP	0.035-0.043	5-7B ①	5-7B ①	36 ②	760-810	700-750	HYD	HYD
	1.8 (1781)	AWV	0.035-0.043	5-7B ①	5-7B ①	52 ②	760-810	700-750	HYD	HYD
	1.8 (1781)	AWW	0.035-0.043	5-7B ①	5-7B ①	52 ②	820-920	820-920	HYD	HYD
	2.0 (1984)	AEG	0.035-0.043	5-7B ①	5-7B ①	52 ②	760-880	760-880	HYD	HYD
	2.0 (1984)	AVH	0.035-0.043	5-7B ①	5-7B ①	52 ②	730-830	790-890	HYD	HYD
	2.0 (1984)	AZG	0.035-0.043	5-7B ①	5-7B ①	52 ②	730-830	790-890	HYD	HYD
	2.0 (1984)	BDC	0.035-0.043	5-7B ①	5-7B ①	36 ②	760-880	760-880	HYD	HYD
	2.8 (2771)	AHA	0.063	5-7B ①	5-7B ①	52 ②	620-740	620-740	HYD	HYD
	2.8 (2771)	ATQ	0.063	5-7B ①	5-7B ①	52 ②	620-740	620-740	HYD	HYD
	2.8 (2771)	BDF	0.035-0.043	5-7B ①	5-7B ①	36 ②	640-720	640-720	HYD	HYD
	4.0 (3999)	BDP	0.031	5-7B ①	5-7B ①	50 ②	620-740	620-740	HYD	HYD
2004	1.8 (1781)	AEB	0.035-0.043	5-7B ①	5-7B ①	52 ②	820-920	820-920	HYD	HYD
	1.8 (1781)	ATW	0.035-0.043	5-7B ①	5-7B ①	52 ②	820-920	820-920	HYD	HYD
	1.8 (1781)	AWM	0.035-0.043	5-7B ①	5-7B ①	50 ②	740-920	740-920	HYD	HYD
	1.8 (1781)	AWP	0.035-0.043	5-7B ①	5-7B ①	36 ②	760-810	700-750	HYD	HYD
	1.8 (1781)	AWV	0.035-0.043	5-7B ①	5-7B ①	52 ②	760-810	700-750	HYD	HYD
	1.8 (1781)	AWW	0.035-0.043	5-7B ①	5-7B ①	52 ②	820-920	820-920	HYD	HYD
	1.8 (1781)	BKF	0.035-0.043	5-7B ①	5-7B ①	52 ②	820-920	820-920	HYD	HYD
	2.0 (1984)	AVH	0.035-0.043	5-7B ①	5-7B ①	52 ②	730-830	790-890	HYD	HYD
	2.0 (1984)	AZG	0.035-0.043	5-7B ①	5-7B ①	52 ②	730-830	790-890	HYD	HYD
	2.0 (1984)	BDC	0.035-0.043	5-7B ①	5-7B ①	36 ②	760-880	760-880	HYD	HYD
	2.0 (1984)	BEV	0.039-0.043	5-7B ①	5-7B ①	44 ②	730-830	730-830	HYD	HYD
	2.0 (1984)	BGD	0.039-0.043	5-7B ①	5-7B ①	44 ②	730-830	730-830	HYD	HYD
	2.8 (2771)	AHA	0.063	5-7B ①	5-7B ①	52 ②	620-740	620-740	HYD	HYD
	2.8 (2771)	ATQ	0.063	5-7B ①	5-7B ①	52 ②	620-740	620-740	HYD	HYD
	2.8 (2771)	BDF	0.035-0.043	5-7B ①	5-7B ①	36 ②	640-720	640-720	HYD	HYD
	4.0 (3999)	BDP	0.031	5-7B ①	5-7B ①	50 ②	620-740	620-740	HYD	HYD

06041_VWMC_C0005

GASOLINE ENGINE TUNE-UP SPECIFICATIONS

Year	Engine Displacement Liters	Engine ID/VIN	Spark Plug Gap (in.)	Ignition Timing (deg.) A MT	Ignition Timing (deg.) A AT	Fuel Pump (psi)	Idle Speed (rpm) MT	Idle Speed (rpm) AT	Valve Clearance Intake	Valve Clearance Exhaust
2005	1.8 (1781)	AEB	0.035-0.043	5-7B ①	5-7B ①	52 ②	820-920	820-920	HYD	HYD
	1.8 (1781)	ATW	0.035-0.043	5-7B ①	5-7B ①	52 ②	820-920	820-920	HYD	HYD
	1.8 (1781)	AWM	0.035-0.043	5-7B ①	5-7B ①	50 ②	740-920	740-920	HYD	HYD
	1.8 (1781)	AWP	0.035-0.043	5-7B ①	5-7B ①	36 ②	760-810	700-750	HYD	HYD
	1.8 (1781)	AWV	0.035-0.043	5-7B ①	5-7B ①	52 ②	760-810	700-750	HYD	HYD
	1.8 (1781)	AWW	0.035-0.043	5-7B ①	5-7B ①	52 ②	820-920	820-920	HYD	HYD
	1.8 (1781)	BKF	0.035-0.043	5-7B ①	5-7B ①	52 ②	820-920	820-920	HYD	HYD
	2.0 (1984)	AVH	0.035-0.043	5-7B ①	5-7B ①	52 ②	730-830	790-890	HYD	HYD
	2.0 (1984)	AZG	0.035-0.043	5-7B ①	5-7B ①	52 ②	730-830	790-890	HYD	HYD
	2.0 (1984)	BEV	0.039-0.043	5-7B ①	5-7B ①	44 ②	730-830	730-830	HYD	HYD
	2.0 (1984)	BGD	0.039-0.043	5-7B ①	5-7B ①	44 ②	730-830	730-830	HYD	HYD
	2.0 (1984)	BPY	0.027-0.031	5-7B ①	5-7B ①	58 ②	620-800	620-800	HYD	HYD
	2.5 (2480)	BGP	0.039-0.043	5-7B ①	5-7B ①	58 ②	680-780	680-780	HYD	HYD
	2.5 (2480)	BGQ	0.039-0.043	5-7B ①	5-7B ①	58 ②	680-780	680-780	HYD	HYD
	2.8 (2771)	AHA	0.063	5-7B ①	5-7B ①	52 ②	620-740	620-740	HYD	HYD
	2.8 (2771)	ATQ	0.063	5-7B ①	5-7B ①	52 ②	620-740	620-740	HYD	HYD
	2.8 (2771)	BDF	0.035-0.043	5-7B ①	5-7B ①	36 ②	640-720	640-720	HYD	HYD
	4.0 (3999)	BDP	0.031	5-7B ①	5-7B ①	50 ②	620-740	620-740	HYD	HYD
2006	2.0 (1984)	BPY	0.027-0.031	5-7B ①	5-7B ①	58 ②	620-800	620-800	HYD	HYD
	2.5 (2480)	BGP	0.039-0.043	5-7B ①	5-7B ①	58 ②	680-780	680-780	HYD	HYD
	2.5 (2480)	BGQ	0.039-0.043	5-7B ①	5-7B ①	58 ②	680-780	680-780	HYD	HYD
	2.5 (2480)	BPR	0.039-0.043	5-7B ①	5-7B ①	58 ②	680-780	680-780	HYD	HYD
	2.5 (2480)	BPS	0.039-0.043	5-7B ①	5-7B ①	58 ②	680-780	680-780	HYD	HYD
	3.6 (3597)	BLV	0.043	5-7B ①	5-7B ①	44 ②	640-760	640-760	HYD	HYD

Note: The Vehicle Emission Control Information label reflects specification changes made during production.

The label figures must be used if they differ from those in this chart.

B: Before top dead center.

HYD: Hydraulic

① Specifications for reference only. The ignition timing is controlled bt the ECM and is not adjustable.

② System pressure at idle.

06041_VWMC_C0006

DIESEL ENGINE TUNE-UP SPECIFICATIONS

Year	Engine Displacement cu. in. (cc)	Engine ID/VIN	Valve Clearance Intake (in.)	Valve Clearance Exhaust (in.)	Intake Valve Opens (deg.)	Start of Injection Stroke @ TDC (cyl 1) (in.)	Injection Nozzle Pressure (psi) New	Injection Nozzle Pressure (psi) Used	Idle Speed (rpm)	Cranking Compression Pressure (psi)
2002	1.9 (1896)	AAZ	HYD	HYD	6	①	2175-2291	2030	870-930	NA
	1.9 (1896)	AHH	HYD	HYD	8-14	NA	2175-2291	2030	870-930	NA
	1.9 (1896)	ALH	HYD	HYD	8-14	NA	2175-2291	2030	870-930	NA
2003	1.9 (1896)	ALH	HYD	HYD	8-14	NA	2175-2291	2030	870-930	NA
2004	1.9 (1896)	ALH	HYD	HYD	8-14	NA	2175-2291	2030	870-930	NA
2005	1.9 (1896)	ALH	HYD	HYD	8-14	NA	2175-2291	2030	870-930	NA
	1.9 (1896)	BEW	HYD	HYD	8-14	NA	2175-2291	2030	870-930	NA
	1.9 (1896)	BRM	HYD	HYD	8-14	NA	2175-2291	2030	870-930	NA
2006	1.9 (1896)	BEW	HYD	HYD	8-14	NA	2175-2291	2030	870-930	NA
	1.9 (1896)	BRM	HYD	HYD	8-14	NA	2175-2291	2030	870-930	NA

Note: The Vehicle Emission Control Information label often reflects specification changes made during production.
The label figures must be used if they differ from those in this chart

HYD: Hydraulic

B: Before top dead center

NA: Not Available

① Checking: 0.0287 - 0.0342 in.
Adjusting: 0.0307 - 0.0323 in.

06041_VWMC_C0007

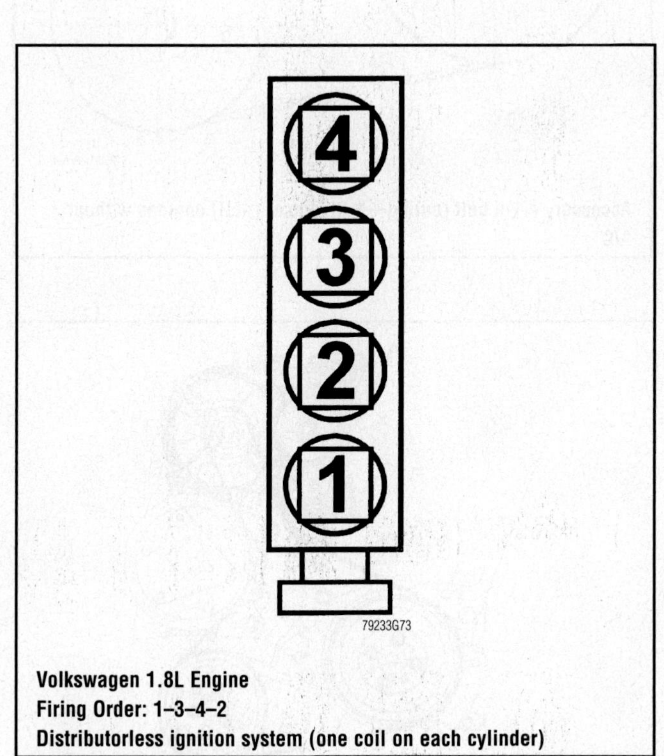

Volkswagen 1.8L Engine
Firing Order: 1–3–4–2
Distributorless ignition system (one coil on each cylinder)

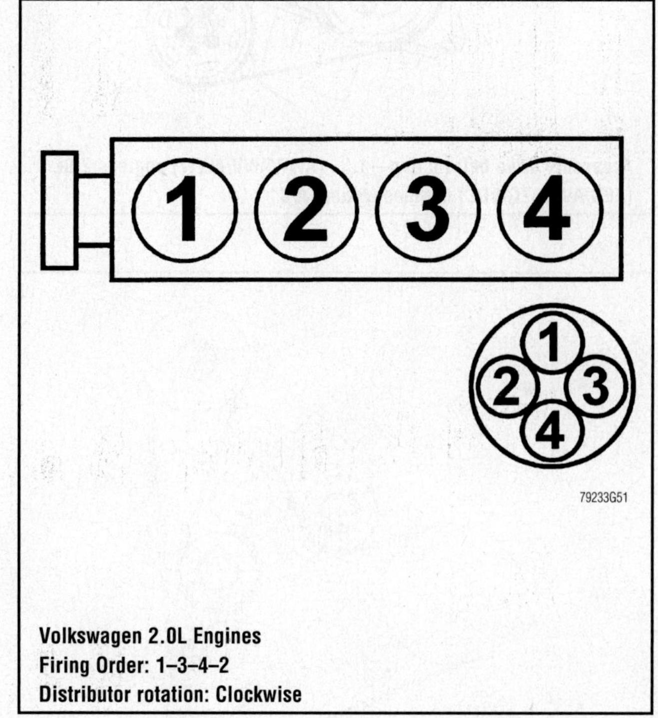

Volkswagen 2.0L Engines
Firing Order: 1–3–4–2
Distributor rotation: Clockwise

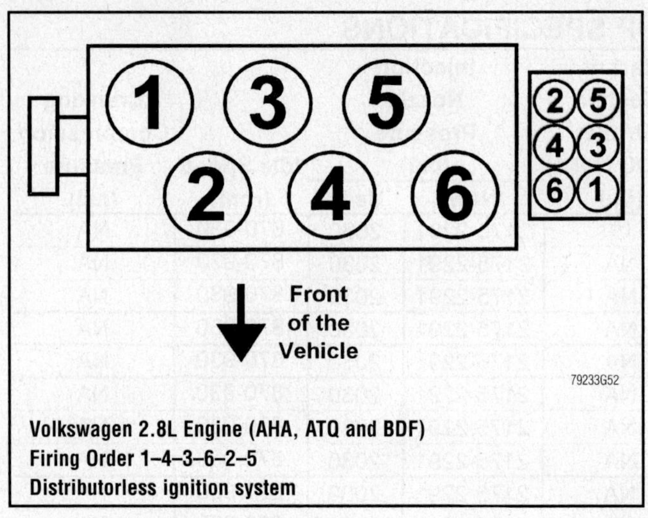

Volkswagen 2.8L Engine (AHA, ATQ and BDF)
Firing Order 1–4–3–6–2–5
Distributorless ignition system

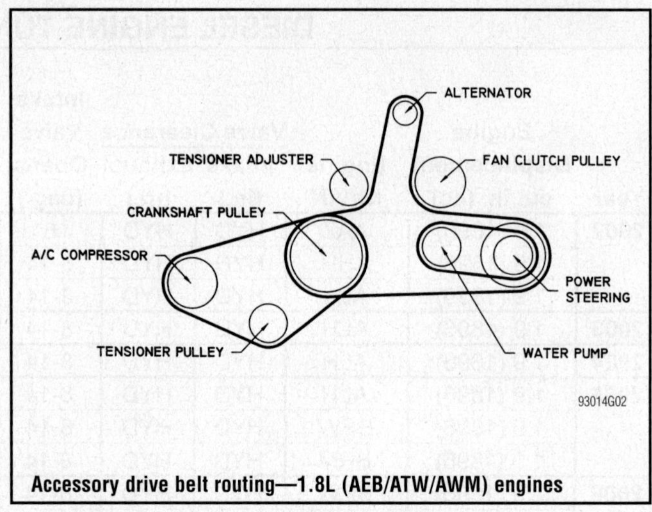

Accessory drive belt routing—1.8L (AEB/ATW/AWM) engines

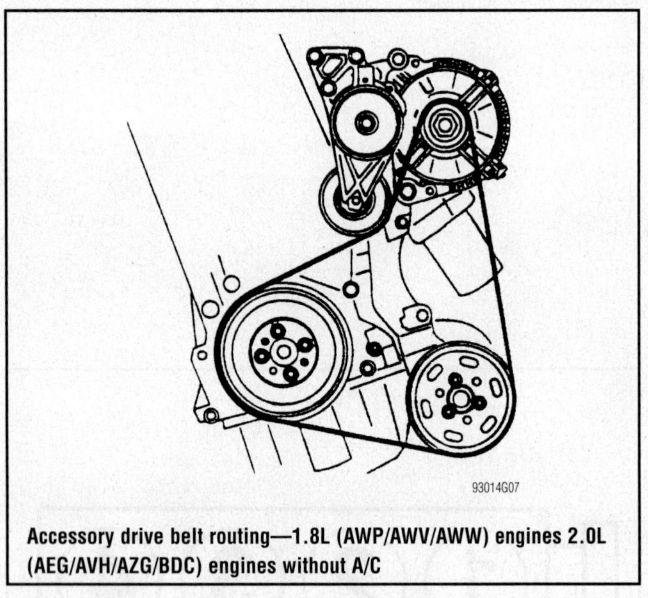

Accessory drive belt routing—1.8L (AWP/AWV/AWW) engines 2.0L (AEG/AVH/AZG/BDC) engines without A/C

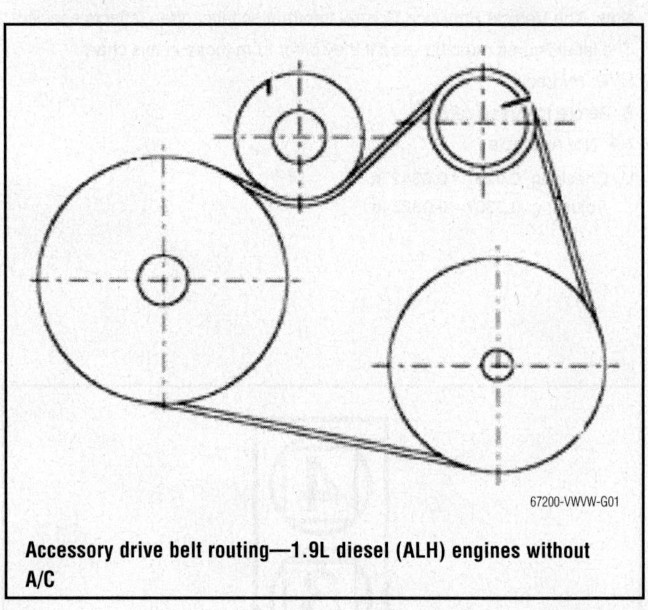

Accessory drive belt routing—1.9L diesel (ALH) engines without A/C

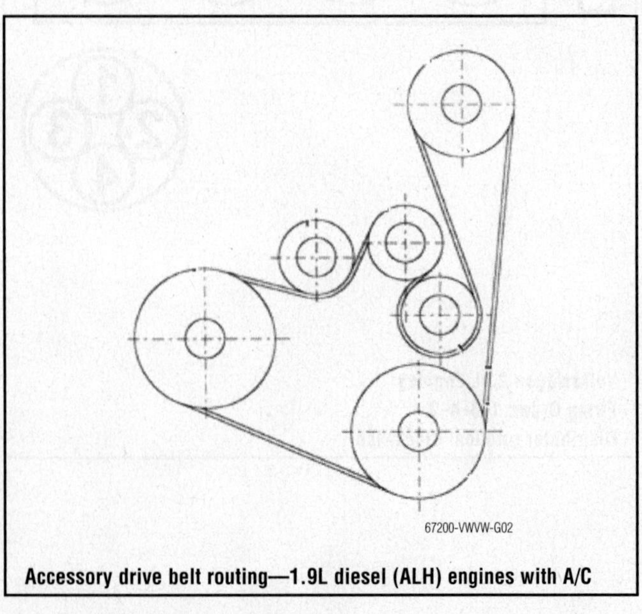

Accessory drive belt routing—1.9L diesel (ALH) engines with A/C

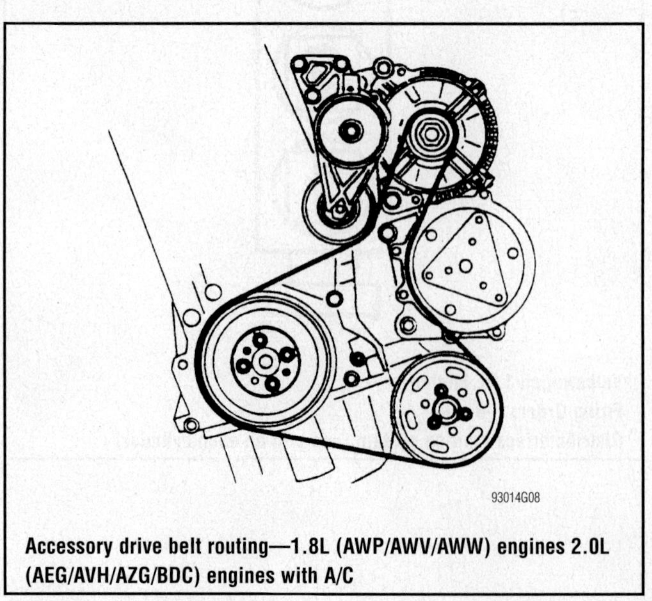

Accessory drive belt routing—1.8L (AWP/AWV/AWW) engines 2.0L (AEG/AVH/AZG/BDC) engines with A/C

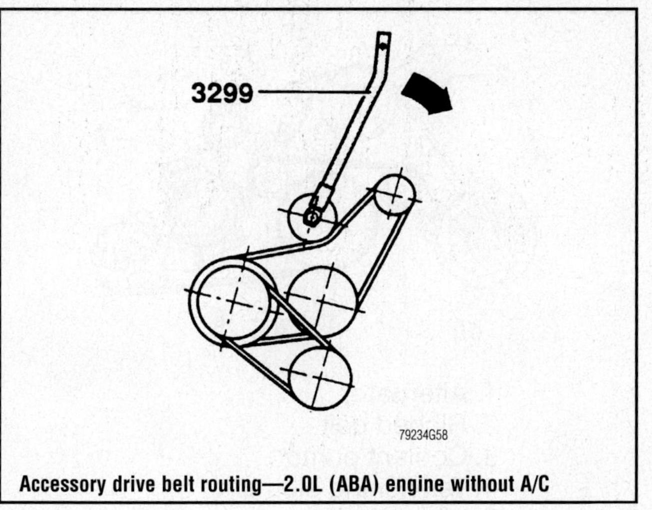

Accessory drive belt routing—2.0L (ABA) engine without A/C

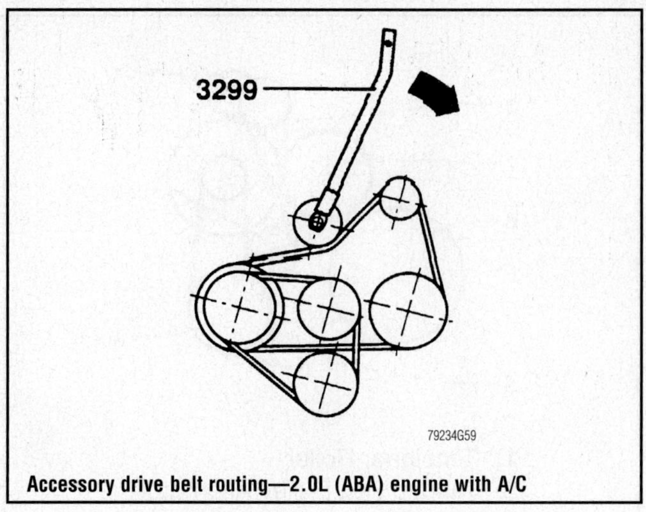

Accessory drive belt routing—2.0L (ABA) engine with A/C

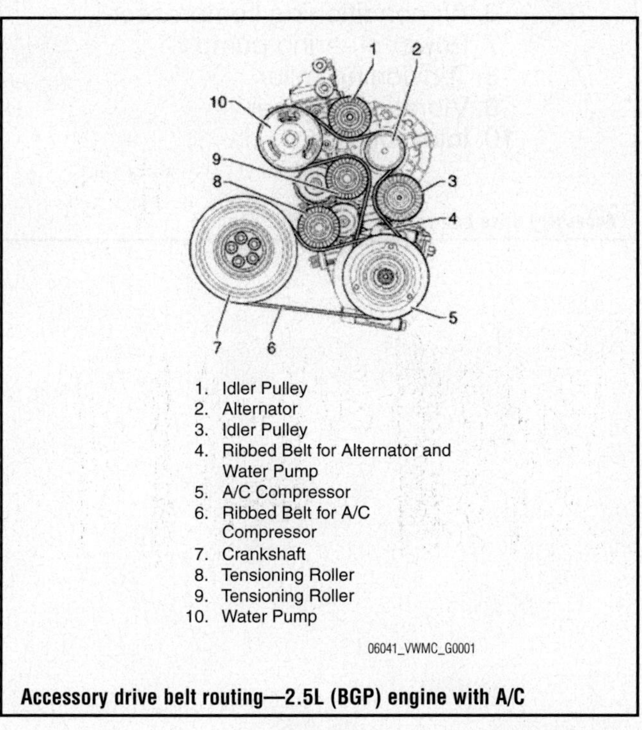

Accessory drive belt routing—2.5L (BGP) engine with A/C

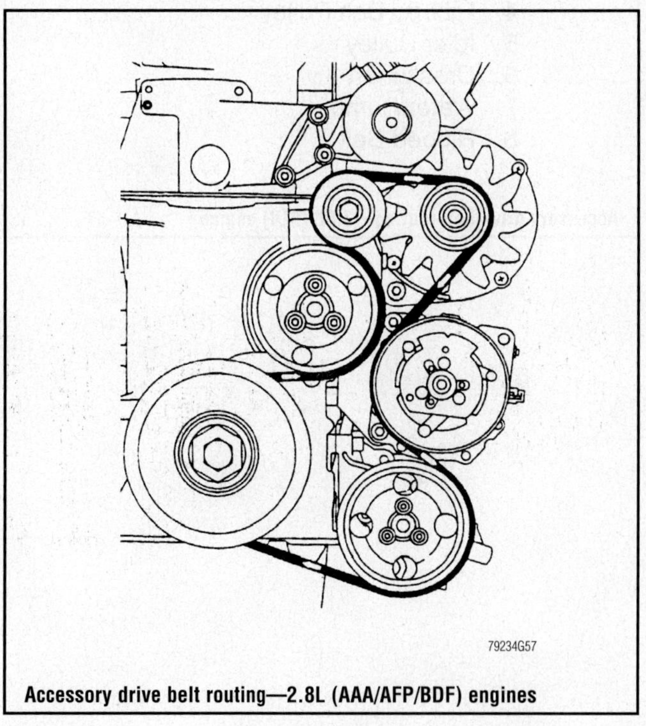

Accessory drive belt routing—2.8L (AAA/AFP/BDF) engines

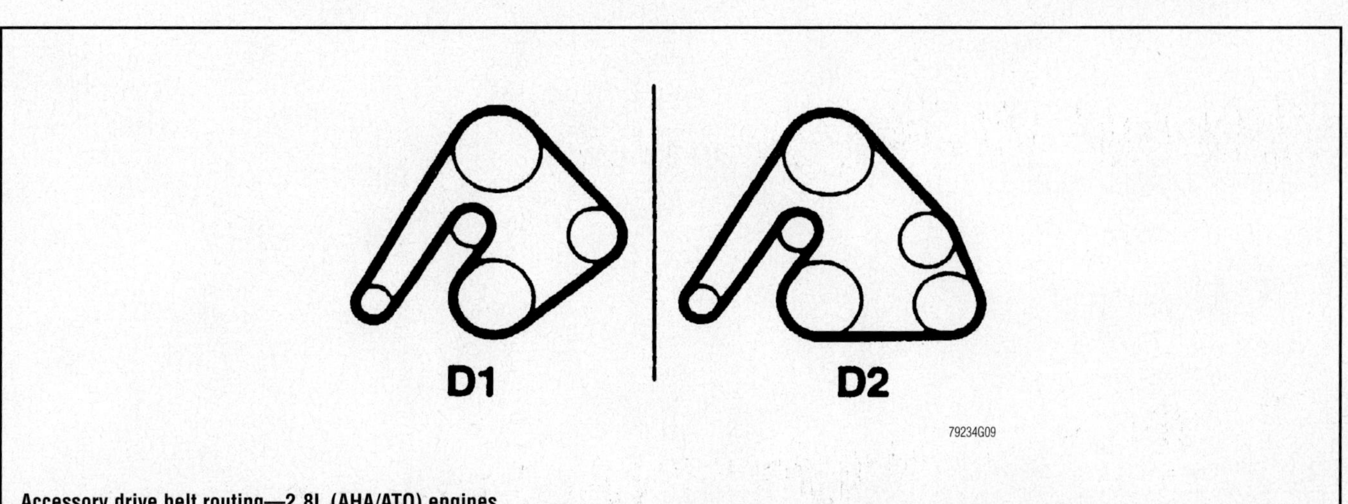

Accessory drive belt routing—2.8L (AHA/ATQ) engines

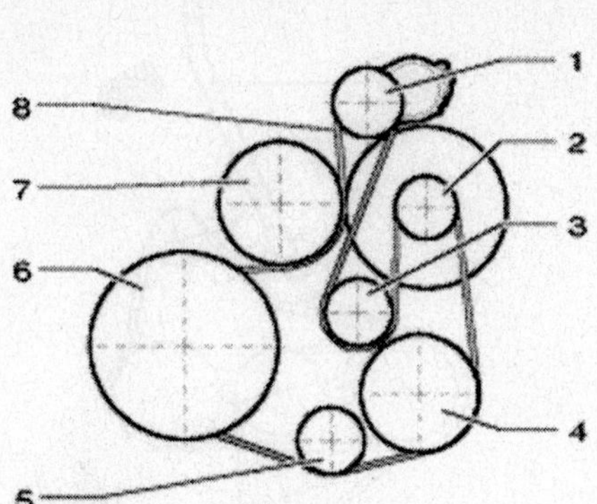

1. Tensioner Roller
2. Ribbed Belt Pulley Alternator
3. Idler Pulley
4. Ribbed Belt Pulley
5. Idler Pulley
6. Crankshaft
7. Water Pump
8. Ribbed Belt

06041_VWMC_G0002

Accessory drive belt routing—3.6L (BLV) engine

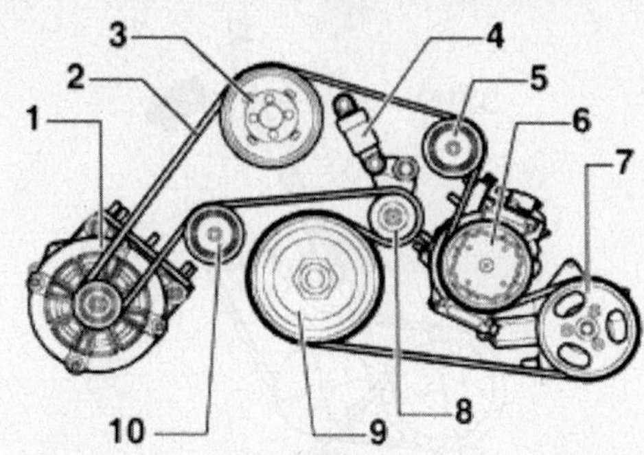

1. Alternator
2. Ribbed belt
3. Coolant pump
4. Tensioning element
5. Idler roller
6. Air conditioning compressor
7. Power steering pump
8. Tensioning roller
9. Vibration damper
10. Idler roller

42348-VWVW-G01

Accessory drive belt routing—4.0L (BDP) engine

CAPACITIES

Year	Model	Engine Displacement Liters	Engine ID/VIN	Engine Oil with Filter	Transmission Manual (pts.)	Transmission Auto. (qts.)	Final Drive (pts.)	Fuel Tank (gal.)	Cooling System (qts.)
2002	Cabrio	2.0	ABA	4.8	4.2	6.4	—	14.5	6.7
	Golf	1.9	AAZ	4.8	4.2	6.4	—	14.6	5.5
	Golf	2.0	ABA	4.2	4.2	6.4	—	14.6	5.8
	GTI	2.0	ABA	4.8	4.2	6.4	—	14.5	6.7
	GTI	2.8	AAA	5.8	4.2	—	—	14.5	8.5
	Jetta	1.9	AAZ	4.8	4.2	6.4	—	14.5	5.5
	Jetta	2.0	ABA	4.2	4.2	6.4	—	14.5	5.8
	Jetta	2.8	AAA	5.8	4.2	6.4	—	14.5	8.5
	New Beetle	1.9	ALH	4.8	4.2	6.8	①	14.5	6.7
	New Beetle	2.0	AEG	4.2	4.2	6.8	①	14.5	6.7
	Passat	1.8	AEB	4.3	4.8	5.5	①	16.4	7.4
	Passat	1.9	AHH	3.7	4.8	7.4	—	NA	NA
	Passat	2.8	AHA	5.8	4.8	5.5	①	16.4	8.5
2003	Convertible	1.8	AWV	4.8	4.2	5.5	①	14.5	5.2
	Convertible	2.0	BDC	4.2	4.2	6.4	—	14.5	5.2
	Golf	1.9	ALH	4.8	4.2	6.4	—	14.5	6.3
	Golf	2.0	AEG	4.8	4.2	6.4	—	14.5	5.2
	Golf	2.0	AVH	4.1	4.2	—	—	14.5	5.2
	Golf	2.0	AZG	4.1	4.2	6.4	—	14.5	5.2
	GTI	1.8	AWW	4.6	4.2	6.4	—	14.5	5.2
	GTI	1.8	AWP	4.8	4.2	6.4	—	14.5	5.2
	GTI	2.8	BDF	4.1	4.2	6.8	①	14.5	9.7
	Jetta	1.8	AWP	4.8	4.2	6.8	①	14.5	5.2
	Jetta	1.8	AWW	4.6	4.8	5.5	①	14.5	5.2
	Jetta	1.9	ALH	4.8	4.8	7.4	—	14.5	6.3
	Jetta	2.0	AEG	4.1	4.2	6.4	—	14.5	5.2
	Jetta	2.0	AVH	4.1	4.2	—	—	14.5	5.2
	Jetta	2.0	AZG	4.1	4.2	6.4	—	14.5	5.2
	Jetta	2.8	BDF	4.1	4.8	7.4	—	14.5	9.7
	New Beetle	1.8	AWV	4.8	4.2	5.5	①	14.5	5.2
	New Beetle	1.8	AWP	4.8	4.2	6.4	①	14.5	5.2
	New Beetle	1.9	ALH	4.8	4.2	6.4	—	14.5	6.3
	New Beetle	2.0	AEG	4.1	4.2	6.4	—	14.5	5.2
	New Beetle	2.0	AVH	4.1	4.2	6.4	—	14.5	5.2
	New Beetle	2.0	AZG	4.1	4.2	—	—	14.5	5.2
	New Beetle	2.0	BDC	4.2	4.2	6.4	—	14.5	5.2
	Passat	1.8	AEB	3.9	4.2	6.4	—	16.4	7.4
	Passat	1.8	ATW	3.9	4.2	6.4	—	16.4	7.4
	Passat	1.8	AWM	3.7	4.2	6.8	①	16.4	7.4
	Passat	2.8	AHA	6.5	4.2	6.8	①	16.4	8.5
	Passat	2.8	ATQ	6.5	4.8	5.5	①	16.4	8.5
	Passat	4.0	BDP	NA	4.8	5.5	—	21.1	9.4

06041_VWMC_C0008

CAPACITIES

Year	Model	Engine Displacement Liters	Engine ID/VIN	Engine Oil with Filter	Transmission		Final Drive (pts.)	Fuel Tank (gal.)	Cooling System (qts.)
					Manual (pts.)	Auto. (qts.)			
2004	Convertible	1.8	AWV	4.8	4.2	7.4	—	14.5	5.2
	Convertible	1.8	BKF	4.8	4.2	7.4	—	14.5	5.2
	Convertible	2.0	BDC	4.2	4.2	7.4	—	14.5	5.2
	Convertible	2.0	BEV	4.4	4.2	7.4	—	14.5	5.2
	Convertible	2.0	BGD	4.2	4.2	7.4	—	14.5	5.2
	Golf	1.9	ALH	4.8	4.2	6.4	—	14.5	6.3
	Golf	2.0	AVH	4.1	4.2	—	—	14.5	5.2
	Golf	2.0	AZG	4.1	4.2	6.4	—	14.5	5.2
	GTI	1.8	AWW	4.6	4.2	6.4	—	14.5	5.2
	GTI	1.8	AWP	4.8	4.2	6.4	—	14.5	5.2
	GTI	2.8	BDF	4.1	4.2	6.8	①	14.5	9.7
	Jetta	1.8	AWP	4.8	4.2	6.8	①	14.5	5.2
	Jetta	1.8	AWW	4.6	4.8	5.5	①	14.5	5.2
	Jetta	1.9	ALH	4.8	4.8	7.4	—	14.5	6.3
	Jetta	2.0	AVH	4.1	4.2	—	—	14.5	5.2
	Jetta	2.0	AZG	4.1	4.2	6.4	—	14.5	5.2
	Jetta	2.8	BDF	4.1	4.8	7.4	—	14.5	9.7
	New Beetle	1.8	AWV	4.8	4.2	5.6	①	14.5	5.2
	New Beetle	1.8	AWP	4.8	4.2	5.6	①	14.5	5.2
	New Beetle	1.9	ALH	4.8	4.2	6.8	—	14.5	6.3
	New Beetle	2.0	AVH	4.1	4.2	7.4	—	14.5	5.2
	New Beetle	2.0	AZG	4.1	4.2	7.4	—	14.5	5.2
	New Beetle	2.0	BEV	4.4	4.2	7.4	—	14.5	5.2
	Passat	1.8	AEB	3.9	4.2	3.2	—	16.4	7.4
	Passat	1.8	ATW	3.9	4.2	3.2	—	16.4	7.4
	Passat	1.8	AWM	3.7	4.2	3.2	①	16.4	7.4
	Passat	2.8	AHA	6.5	4.2	3.2	①	16.4	8.5
	Passat	2.8	ATQ	6.5	4.8	3.2	①	16.4	8.5
	Passat	4.0	BDP	NA	4.8	3.2	—	21.1	9.4
2005	Convertible	1.8	BKF	4.8	4.2	7.4	—	14.5	5.2
	Convertible	2.0	BEV	4.4	4.2	7.4	—	14.5	5.2
	Convertible	2.0	BGD	4.2	4.2	7.4	—	14.5	5.2
	Golf	1.9	ALH	4.8	4.2	6.4	—	14.5	6.3
	Golf	2.0	AVH	4.1	4.2	—	—	14.5	5.2
	Golf	2.0	AZG	4.1	4.2	6.4	—	14.5	5.2
	GTI	1.8	AWW	4.6	4.2	6.4	—	14.5	5.2
	GTI	1.8	AWP	4.8	4.2	6.4	—	14.5	5.2
	GTI	2.8	BDF	4.1	4.2	6.8	①	14.5	9.7
	Jetta	1.9	BRM	5.1	4.8	5.5	—	14.5	8.7
	Jetta	2.0	BPY	4.4	4.8	5.5	—	14.5	7.6
	Jetta	2.5	BGP	6.3	4.2	7.4	—	14.5	10.0
	Jetta	2.5	BGQ	6.3	4.2	7.4	—	14.5	10.0
	New Beetle	1.8	AWV	4.8	4.2	5.6	①	14.5	5.2
	New Beetle	1.8	AWP	4.8	4.2	5.6	①	14.5	5.2
	New Beetle	1.9	BEW	4.8	4.2	6.8	—	14.5	5.3
	New Beetle	2.0	AVH	4.1	4.2	7.4	—	14.5	5.2
	New Beetle	2.0	AZG	4.1	4.2	7.4	—	14.5	5.2
	New Beetle	2.0	BEV	4.4	4.2	7.4	—	14.5	5.2

06041_VWMC_C0009

CAPACITIES

Year	Model	Engine Displacement Liters	Engine ID/VIN	Engine Oil with Filter	Transmission Manual (pts.)	Transmission Auto. (qts.)	Final Drive (pts.)	Fuel Tank (gal.)	Cooling System (qts.)
2005	Passat	1.8	AEB	3.9	4.2	6.4	—	16.4	7.4
	Passat	1.8	ATW	3.9	4.2	6.4	—	16.4	7.4
	Passat	1.8	AWM	3.7	4.2	6.8	①	16.4	7.4
	Passat	2.8	AHA	6.5	4.2	6.8	①	16.4	8.5
	Passat	2.8	ATQ	6.5	4.8	5.5	①	16.4	8.5
	Passat	4.0	BDP	NA	4.8	5.5	—	21.1	9.4
2006	Convertible	2.5	BPR	6.3	4.2	②	—	14.5	10.0
	Convertible	2.5	BPS	6.3	4.2	②	—	14.5	10.0
	Rabbit	2.5	BGP	6.3	—	5.5	—	14.5	10.0
	Rabbit	2.5	BGQ	6.3	③	5.5	—	14.5	10.0
	GTI	2.0	BPY	4.4	4.8	5.5	—	14.5	7.6
	Jetta	1.9	BRM	5.1	4.8	5.5	—	14.5	8.7
	Jetta	2.0	BPY	4.4	4.8	5.5	—	14.5	7.6
	Jetta	2.5	BGP	6.3	4.2	7.4	—	14.5	10.0
	Jetta	2.5	BGQ	6.3	4.2	7.4	—	14.5	10.0
	New Beetle	1.9	BEW	4.8	4.2	②	—	14.5	5.3
	New Beetle	2.5	BPR	6.3	4.2	②	—	14.5	10.0
	New Beetle	2.5	BPS	6.3	4.2	②	—	14.5	10.0
	Passat	2.0	BPY	4.4	4.8	7.4	—	18.5	7.6
	Passat	3.6	BLV	5.8	4.8	7.4	—	18.5	8.5

Note: All capacities are approximate. Add fluid gradually and check often to avoid overfilling.

NA: Not Available

① For models equipped with an automatic transmission: 1.6 pints

② For models equipped with the 02E Direct Shift automatic transmission: 4.8 quarts
For models equipped with the 09G automatic transmission: 7.4 quarts

③ For models equipped with the 5-speed manual transmission: 3.6 pints
For models equipped with the 6-speed manual transmission: 4.9 pints

06041_VWMC_C0010

VALVE SPECIFICATIONS

Year	Engine Displacement Liters	Engine ID/VIN	Seat Angle (deg.)	Face Angle (deg.)	Spring Test Pressure (lbs. @ in.)	Spring Installed Height (in.)	Stem-to-Guide Clearance (in.)		Stem Diameter (in.)	
							Intake	Exhaust	Intake	Exhaust
2002	1.8	AEB	45	45	NA	NA	0.039	0.051	0.2348	0.2340
	1.9	AAZ	45	45	NA	NA	0.039	0.051	0.2741	0.2734
	1.9	AHH	45	45	NA	NA	0.039	0.051	0.2741	0.2734
	1.9	ALH	45	45	NA	NA	0.039	0.051	0.2741	0.2734
	2.0	ABA	45	45	NA	NA	0.039	0.051	0.2744	0.2736
	2.0	AEG	45	45	NA	NA	0.039	0.051	0.2724	0.2724
	2.8	AAA	45	45	NA	NA	0.039	0.051	0.2744	0.2736
	2.8	AHA	45	45	NA	NA	0.039	0.051	0.2350	0.2343
2003	1.8	AEB	45	45	NA	NA	0.031	0.031	0.2346	0.2346
	1.8	ATW	45	45	NA	NA	0.031	0.031	0.2346	0.2346
	1.8	AWM	45	45	NA	NA	0.031	0.031	0.2346	0.2338
	1.8	AWP	45	45	NA	NA	0.031	0.031	0.2346	0.2338
	1.8	AWV	45	45	NA	NA	0.031	0.031	0.2346	0.2338
	1.8	AWW	45	45	NA	NA	0.031	0.031	0.2346	0.2338
	1.9	ALH	45	45	NA	NA	0.039	0.051	0.2741	0.2734
	2.0	AEG	45	45	NA	NA	0.039	0.051	0.2724	0.2724
	2.0	AVH	45	45	NA	NA	0.039	0.051	0.2748	0.2740
	2.0	AZG	45	45	NA	NA	0.039	0.051	0.2748	0.2740
	2.0	BDC	45	45	NA	NA	0.039	0.051	0.2748	0.2740
	2.8	AHA	45	45	NA	NA	0.031	0.031	0.2350	0.2343
	2.8	ATQ	45	45	NA	NA	0.031	0.031	0.2350	0.2343
	2.8	BDF	45	45	NA	NA	0.031	0.031	0.2346	0.2338
	4.0	BDP	45	45	NA	NA	0.031	0.031	0.2346	0.2338
2004	1.8	AEB	45	45	NA	NA	0.031	0.031	0.2346	0.2346
	1.8	ATW	45	45	NA	NA	0.031	0.031	0.2346	0.2346
	1.8	AWM	45	45	NA	NA	0.031	0.031	0.2346	0.2338
	1.8	AWP	45	45	NA	NA	0.031	0.031	0.2346	0.2338
	1.8	AWV	45	45	NA	NA	0.031	0.031	0.2346	0.2338
	1.8	AWW	45	45	NA	NA	0.031	0.031	0.2346	0.2338
	1.9	ALH	45	45	NA	NA	0.039	0.051	0.2741	0.2734
	2.0	AEG	45	45	NA	NA	0.039	0.051	0.2724	0.2724
	2.0	AVH	45	45	NA	NA	0.039	0.051	0.2748	0.2740
	2.0	AZG	45	45	NA	NA	0.039	0.051	0.2748	0.2740
	2.0	BDC	45	45	NA	NA	0.039	0.051	0.2748	0.2740
	2.8	AHA	45	45	NA	NA	0.031	0.031	0.2350	0.2343
	2.8	ATQ	45	45	NA	NA	0.031	0.031	0.2350	0.2343
	2.8	BDF	45	45	NA	NA	0.031	0.031	0.2346	0.2338
	4.0	BDP	45	45	NA	NA	0.031	0.031	0.2346	0.2338

06041_VWMC_C0011

VALVE SPECIFICATIONS

Year	Engine Displacement Liters	Engine ID/VIN	Seat Angle (deg.)	Face Angle (deg.)	Spring Test Pressure (lbs. @ in.)	Spring Installed Height (in.)	Stem-to-Guide Clearance (in.)		Stem Diameter (in.)	
							Intake	Exhaust	Intake	Exhaust
2005	1.8	AEB	45	45	NA	NA	0.031	0.031	0.2346	0.2346
	1.8	ATW	45	45	NA	NA	0.031	0.031	0.2346	0.2346
	1.8	AWM	45	45	NA	NA	0.031	0.031	0.2346	0.2338
	1.8	AWP	45	45	NA	NA	0.031	0.031	0.2346	0.2338
	1.8	AWV	45	45	NA	NA	0.031	0.031	0.2346	0.2338
	1.8	AWW	45	45	NA	NA	0.031	0.031	0.2346	0.2338
	1.9	ALH	45	45	NA	NA	0.039	0.051	0.2741	0.2734
	1.9	BEW	45	45	NA	NA	0.039	0.051	0.2741	0.2734
	1.9	BRM	45	45	NA	NA	0.039	0.051	0.2741	0.2734
	2.0	AVH	45	45	NA	NA	0.039	0.051	0.2748	0.2740
	2.0	AZG	45	45	NA	NA	0.039	0.051	0.2748	0.2740
	2.0	BDC	45	45	NA	NA	0.039	0.051	0.2748	0.2740
	2.8	AHA	45	45	NA	NA	0.031	0.031	0.2350	0.2343
	2.8	ATQ	45	45	NA	NA	0.031	0.031	0.2350	0.2343
	2.8	BDF	45	45	NA	NA	0.031	0.031	0.2346	0.2338
	4.0	BDP	45	45	NA	NA	0.031	0.031	0.2346	0.2338
2006	1.9	BEW	45	45	NA	NA	0.039	0.051	0.2741	0.2734
	1.9	BRM	45	45	NA	NA	0.039	0.051	0.2741	0.2734
	2.0	BPY	45	45	NA	NA	0.039	0.051	0.2748	0.2740
	2.5	BGP	45	45	NA	NA	NA	NA	NA	NA
	2.5	BGQ	45	45	NA	NA	NA	NA	NA	NA
	2.5	BPR	45	45	NA	NA	NA	NA	NA	NA
	2.5	BPS	45	45	NA	NA	NA	NA	NA	NA
	3.6	BLV	45	45	NA	NA	NA	NA	NA	NA

NA: Not Available

06041_VWMC_C0012

CRANKSHAFT AND CONNECTING ROD SPECIFICATIONS

All measurements are given in inches.

Year	Engine Displacement Liters	Engine ID/VIN	Crankshaft Main Brg. Journal Dia.	Crankshaft Main Brg. Oil Clearance	Crankshaft Shaft End-play	Crankshaft Thrust on No.	Connecting Rod Journal Diameter	Connecting Rod Oil Clearance	Connecting Rod Side Clearance
2002	1.8	AEB	2.1260	0.0008-0.0024	0.0028-0.0066	3	1.8819	0.0004-0.0024	0.0020-0.0122
	1.9	AAZ	2.1600	0.0012-0.0032	0.0028-0.0066	3	1.8819	0.0012-0.0032	0.0028-0.0066
	1.9	AHH	2.1260	0.0012-0.0032	0.0028-0.0066	3	1.8819	0.0012-0.0032	0.0028-0.0066
	1.9	ALH	2.1260	0.0012-0.0032	0.0028-0.0066	3	1.8819	0.0012-0.0032	0.0028-0.0066
	2.0	ABA	2.1260	0.0008-0.0024	0.0028-0.0066	3	1.8819	0.0004-0.0024	0.0020-0.0122
	2.0	AEG	2.1260	0.0008-0.0024	0.0028-0.0066	3	1.8819	0.0004-0.0024	0.0020-0.0122
	2.8	AAA	2.3606-2.3613	0.0008-0.0024	0.0028-0.0091	5	2.1243-2.1251	0.0004-0.0024	0.0020-0.0122
	2.8	AHA	2.5590	0.0007-0.0018	0.0028-0.0091	3	2.1260	0.0006-0.0024	0.0020-0.0122
2003	1.8	AEB	2.1260	0.0008-0.0016	0.0028-0.0090	3	1.8819	0.0004-0.0024	0.0020-0.0122
	1.8	ATW	2.1260	0.0008-0.0016	0.0028-0.0090	3	1.8819	0.0004-0.0024	0.0020-0.0122
	1.8	AWM	2.1260	0.0008-0.0016	0.0028-0.0090	3	1.8819	0.0008-0.0016	0.0020-0.0122
	1.8	AWP	2.1260	0.0008-0.0016	0.0028-0.0090	3	1.8819	0.0008-0.0016	0.0020-0.0122
	1.8	AWV	2.1260	0.0008-0.0016	0.0028-0.0090	3	1.8819	0.0008-0.0016	0.0020-0.0122
	1.8	AWW	2.1260	0.0008-0.0016	0.0028-0.0090	3	1.8819	0.0008-0.0016	0.0020-0.0122
	2.0	AEG	2.1260	0.0008-0.0016	0.0028-0.0090	3	1.8819	0.0008-0.0016	0.0020-0.0122
	2.0	AVH	2.1260	0.0008-0.0016	0.0028-0.0090	3	1.8819	0.0008-0.0016	0.0020-0.0122
	2.0	AZG	2.1260	0.0008-0.0016	0.0028-0.0090	3	1.8819	0.0008-0.0016	0.0020-0.0122

06041_VWMC_C0013

CRANKSHAFT AND CONNECTING ROD SPECIFICATIONS

All measurements are given in inches.

	Engine		Crankshaft				Connecting Rod		
Year	Displacement Liters	Engine ID/VIN	Main Brg. Journal Dia.	Main Brg. Oil Clearance	Shaft End-play	Thrust on No.	Journal Diameter	Oil Clearance	Side Clearance
2003	2.0	BDC	2.1260	0.0004-0.0016	0.0028-0.0090	3	1.8819	0.0004-0.0020	0.0040-0.0137
	1.9	ALH	2.1260	0.0012-0.0032	0.0028-0.0066	3	1.8819	0.0012-0.0032	0.0028-0.0066
	2.8	AHA	2.5590	0.0007-0.0018	0.0028-0.0090	3	2.1570	0.0006-0.0025	0.0020-0.0122
	2.8	ATQ	2.5590	0.0007-0.0018	0.0028-0.0090	3	2.1570	0.0006-0.0025	0.0020-0.0122
	2.8	BDF	2.3606-2.3613	0.0008-0.0024	0.0028-0.0091	5	2.1243-2.1251	0.0008-0.0028	0.0020-0.0137
	4.0	BDP	NA	NA	NA	NA	NA	NA	NA
2004	1.8	AEB	2.1260	0.0008-0.0016	0.0028-0.0090	3	1.8819	0.0004-0.0024	0.0020-0.0122
	1.8	ATW	2.1260	0.0008-0.0016	0.0028-0.0090	3	1.8819	0.0004-0.0024	0.0020-0.0122
	1.8	AWM	2.1260	0.0008-0.0016	0.0028-0.0090	3	1.8819	0.0008-0.0016	0.0020-0.0122
	1.8	AWP	2.1260	0.0008-0.0016	0.0028-0.0090	3	1.8819	0.0008-0.0016	0.0020-0.0122
	1.8	AWV	2.1260	0.0008-0.0016	0.0028-0.0090	3	1.8819	0.0008-0.0016	0.0020-0.0122
	1.8	AWW	2.1260	0.0008-0.0016	0.0028-0.0090	3	1.8819	0.0008-0.0016	0.0020-0.0122
	2.0	AEG	2.1260	0.0008-0.0016	0.0028-0.0090	3	1.8819	0.0008-0.0016	0.0020-0.0122
	2.0	AVH	2.1260	0.0008-0.0016	0.0028-0.0090	3	1.8819	0.0008-0.0016	0.0020-0.0122
	2.0	AZG	2.1260	0.0008-0.0016	0.0028-0.0090	3	1.8819	0.0008-0.0016	0.0020-0.0122
	2.0	BDC	2.1260	0.0004-0.0016	0.0028-0.0090	3	1.8819	0.0004-0.0020	0.0040-0.0137
	1.9	ALH	2.1260	0.0012-0.0032	0.0028-0.0066	3	1.8819	0.0012-0.0032	0.0028-0.0066
	2.8	AHA	2.5590	0.0007-0.0018	0.0028-0.0090	3	2.1570	0.0006-0.0025	0.0020-0.0122
	2.8	ATQ	2.5590	0.0007-0.0018	0.0028-0.0090	3	2.1570	0.0006-0.0025	0.0020-0.0122
	2.8	BDF	2.3606-2.3613	0.0008-0.0024	0.0028-0.0091	5	2.1243-2.1251	0.0008-0.0028	0.0020-0.0137
	4.0	BDP	NA	NA	NA	NA	NA	NA	NA
2005	1.8	AEB	2.1260	0.0008-0.0016	0.0028-0.0090	3	1.8819	0.0004-0.0024	0.0020-0.0122
	1.8	ATW	2.1260	0.0008-0.0016	0.0028-0.0090	3	1.8819	0.0004-0.0024	0.0020-0.0122
	1.8	AWM	2.1260	0.0008-0.0016	0.0028-0.0090	3	1.8819	0.0008-0.0016	0.0020-0.0122
	1.8	AWP	2.1260	0.0008-0.0016	0.0028-0.0090	3	1.8819	0.0008-0.0016	0.0020-0.0122

06041_VWMC_C0014

CRANKSHAFT AND CONNECTING ROD SPECIFICATIONS

All measurements are given in inches.

Year	Engine Displacement Liters	Engine ID/VIN	Crankshaft Main Brg. Journal Dia.	Crankshaft Main Brg. Oil Clearance	Crankshaft Shaft End-play	Crankshaft Thrust on No.	Connecting Rod Journal Diameter	Connecting Rod Oil Clearance	Connecting Rod Side Clearance
2005	1.8	AWV	2.1260	0.0008-0.0016	0.0028-0.0090	3	1.8819	0.0008-0.0016	0.0020-0.0122
	1.8	AWW	2.1260	0.0008-0.0016	0.0028-0.0090	3	1.8819	0.0008-0.0016	0.0020-0.0122
	2.0	AEG	2.1260	0.0008-0.0016	0.0028-0.0090	3	1.8819	0.0008-0.0016	0.0020-0.0122
	2.0	AVH	2.1260	0.0008-0.0016	0.0028-0.0090	3	1.8819	0.0008-0.0016	0.0020-0.0122
	2.0	AZG	2.1260	0.0008-0.0016	0.0028-0.0090	3	1.8819	0.0008-0.0016	0.0020-0.0122
	2.0	BDC	2.1260	0.0004-0.0016	0.0028-0.0090	3	1.8819	0.0004-0.0020	0.0040-0.0137
	1.9	ALH	2.1260	0.0012-0.0032	0.0028-0.0066	3	1.8819	0.0012-0.0032	0.0028-0.0066
	2.8	AHA	2.5590	0.0007-0.0018	0.0028-0.0090	3	2.1570	0.0006-0.0025	0.0020-0.0122
	2.8	ATQ	2.5590	0.0007-0.0018	0.0028-0.0090	3	2.1570	0.0006-0.0025	0.0020-0.0122
	2.8	BDF	2.3606-2.3613	0.0008-0.0024	0.0028-0.0091	5	2.1243-2.1251	0.0008-0.0028	0.0020-0.0137
	4.0	BDP	NA	NA	NA	NA	NA	NA	NA
2006	1.9	BEW	2.1260	0.0012-0.0032	0.0028-0.0066	3	1.8819	0.0012-0.0032	0.0028-0.0066
	1.9	BRM	2.1260	0.0012-0.0032	0.0028-0.0066	3	1.8819	0.0012-0.0032	0.0028-0.0066
	2.0	BPY	2.1260	0.0008-0.0016	0.0028-0.0090	3	1.8819	0.0008-0.0016	0.0020-0.0122
	2.5	BGP	2.2834	0.0009-0.0017	0.0028-0.0083	3	1.8819	0.0008-0.0023	0.0040-0.0137
	2.5	BGQ	2.2834	0.0009-0.0017	0.0028-0.0083	3	1.8819	0.0008-0.0023	0.0040-0.0137
	2.5	BPR	2.2834	0.0009-0.0017	0.0028-0.0083	3	1.8819	0.0008-0.0023	0.0040-0.0137
	2.5	BPS	2.2834	0.0009-0.0017	0.0028-0.0083	3	1.8819	0.0008-0.0023	0.0040-0.0137
	3.6	BLV	NA	NA	NA	NA	NA	NA	NA

NA: Not Available

06041_VWMC_C0015

PISTON AND RING SPECIFICATIONS

All measurements are given in inches.

	Engine			Ring Gap			Ring Side Clearance		
Year	Displacement Liters	Engine ID/VIN	Piston Clearance	Top Compression	Bottom Compression	Oil Control	Top Compression	Bottom Compression	Oil Control
2002	1.8	AEB	0.0450	0.008-0.016	0.008-0.016	0.010-0.020	0.0035-0.0047	0.0020-0.0031	0.0012-0.0024
	1.9	AAZ	0.0012	0.008-0.016	0.008-0.016	0.010-0.020	0.0035-0.0047	0.0020-0.0031	0.0012-0.0024
	1.9	AHH	0.0400	0.008-0.016	0.008-0.016	0.010-0.020	0.0035-0.0047	0.0020-0.0031	0.0012-0.0024
	1.9	ALH	0.0400	0.008-0.016	0.008-0.016	0.010-0.020	0.0035-0.0047	0.0020-0.0031	0.0012-0.0024
	2.0	ABA	0.0010	0.008-0.016	0.008-0.016	0.010-0.020	0.0008-0.0020	0.0008-0.0020	0.0008-0.0020
	2.0	AEG	0.0450	0.008-0.016	0.008-0.016	0.010-0.020	0.0035-0.0047	0.0020-0.0031	0.0012-0.0024
	2.8	AAA	0.0010	0.008-0.016	0.008-0.016	0.010-0.020	0.0008-0.0020	0.0008-0.0020	0.0008-0.0019
	2.8	AHA	0.0012	0.014-0.020	0.020-0.028	0.010-0.020	0.0008-0.0031	0.0008-0.0031	0.0008-0.0031
2003	1.8	AEB	0.0018	0.008-0.016	0.008-0.016	0.010-0.020	0.0023-0.0037	0.0023-0.0037	0.0012-0.0024
	1.8	ATW	0.0018	0.008-0.016	0.008-0.016	0.010-0.020	0.0023-0.0037	0.0023-0.0037	0.0012-0.0024
	1.8	AWM	0.0018	0.008-0.016	0.008-0.016	0.010-0.020	0.0023-0.0037	0.0023-0.0037	0.0012-0.0024
	1.8	AWP	0.0018	0.008-0.016	0.008-0.016	0.010-0.020	0.0023-0.0037	0.0023-0.0037	0.0012-0.0024
	1.8	AWV	0.0018	0.008-0.016	0.008-0.016	0.010-0.020	0.0023-0.0037	0.0023-0.0037	0.0012-0.0024
	1.8	AWW	0.0018	0.008-0.016	0.008-0.016	0.010-0.020	0.0023-0.0037	0.0023-0.0037	0.0012-0.0024
	1.9	ALH	0.0400	0.008-0.016	0.008-0.016	0.010-0.020	0.0035-0.0047	0.0020-0.0031	0.0012-0.0024
	2.0	AEG	0.0018	0.008-0.016	0.008-0.016	0.010-0.020	0.0023-0.0037	0.0023-0.0037	0.0012-0.0024

06041_VWMC_C0016

PISTON AND RING SPECIFICATIONS

All measurements are given in inches.

Year	Engine Displacement Liters	Engine ID/VIN	Piston Clearance	Ring Gap Top Compression	Ring Gap Bottom Compression	Ring Gap Oil Control	Ring Side Clearance Top Compression	Ring Side Clearance Bottom Compression	Ring Side Clearance Oil Control
2003	2.0	AVH	0.0018	0.008-0.016	0.008-0.016	0.010-0.020	0.0023-0.0037	0.0023-0.0037	0.0012-0.0024
	2.0	AZG	0.0018	0.008-0.016	0.008-0.016	0.010-0.020	0.0023-0.0037	0.0023-0.0037	0.0012-0.0024
	2.0	BDC	0.0018	0.008-0.016	0.008-0.016	0.010-0.020	0.0023-0.0037	0.0023-0.0037	0.0012-0.0024
	2.8	AHA	0.0010	0.014-0.020	0.020-0.028	0.010-0.020	0.0008-0.0031	0.0008-0.0031	0.0008-0.0031
	2.8	ATQ	0.0010	0.014-0.020	0.020-0.028	0.010-0.020	0.0008-0.0031	0.0008-0.0031	0.0008-0.0031
	2.8	BDF	0.0010	0.014-0.020	0.020-0.028	0.010-0.020	0.0008-0.0031	0.0008-0.0031	0.0008-0.0031
	4.0	BDP	NA	NA	NA	NA	NA	NA	NA
2004	1.8	AEB	0.0018	0.008-0.016	0.008-0.016	0.010-0.020	0.0023-0.0037	0.0023-0.0037	0.0012-0.0024
	1.8	ATW	0.0018	0.008-0.016	0.008-0.016	0.010-0.020	0.0023-0.0037	0.0023-0.0037	0.0012-0.0024
	1.8	AWM	0.0018	0.008-0.016	0.008-0.016	0.010-0.020	0.0023-0.0037	0.0023-0.0037	0.0012-0.0024
	1.8	AWP	0.0018	0.008-0.016	0.008-0.016	0.010-0.020	0.0023-0.0037	0.0023-0.0037	0.0012-0.0024
	1.8	AWV	0.0018	0.008-0.016	0.008-0.016	0.010-0.020	0.0023-0.0037	0.0023-0.0037	0.0012-0.0024
	1.8	AWW	0.0018	0.008-0.016	0.008-0.016	0.010-0.020	0.0023-0.0037	0.0023-0.0037	0.0012-0.0024
	1.9	ALH	0.0400	0.008-0.016	0.008-0.016	0.010-0.020	0.0035-0.0047	0.0020-0.0031	0.0012-0.0024
	2.0	AEG	0.0018	0.008-0.016	0.008-0.016	0.010-0.020	0.0023-0.0037	0.0023-0.0037	0.0012-0.0024
	2.0	AVH	0.0018	0.008-0.016	0.008-0.016	0.010-0.020	0.0023-0.0037	0.0023-0.0037	0.0012-0.0024
	2.0	AZG	0.0018	0.008-0.016	0.008-0.016	0.010-0.020	0.0023-0.0037	0.0023-0.0037	0.0012-0.0024
	2.0	BDC	0.0018	0.008-0.016	0.008-0.016	0.010-0.020	0.0023-0.0037	0.0023-0.0037	0.0012-0.0024
	2.8	AHA	0.0010	0.014-0.020	0.020-0.028	0.010-0.020	0.0008-0.0031	0.0008-0.0031	0.0008-0.0031
	2.8	ATQ	0.0010	0.014-0.020	0.020-0.028	0.010-0.020	0.0008-0.0031	0.0008-0.0031	0.0008-0.0031
	2.8	BDF	0.0010	0.014-0.020	0.020-0.028	0.010-0.020	0.0008-0.0031	0.0008-0.0031	0.0008-0.0031
	4.0	BDP	NA	NA	NA	NA	NA	NA	NA
2005	1.8	AEB	0.0018	0.008-0.016	0.008-0.016	0.010-0.020	0.0023-0.0037	0.0023-0.0037	0.0012-0.0024
	1.8	ATW	0.0018	0.008-0.016	0.008-0.016	0.010-0.020	0.0023-0.0037	0.0023-0.0037	0.0012-0.0024
	1.8	AWM	0.0018	0.008-0.016	0.008-0.016	0.010-0.020	0.0023-0.0037	0.0023-0.0037	0.0012-0.0024

06041_VWMC_C0017

PISTON AND RING SPECIFICATIONS

All measurements are given in inches.

Year	Engine Displacement Liters	Engine ID/VIN	Piston Clearance	Ring Gap Top Compression	Ring Gap Bottom Compression	Ring Gap Oil Control	Ring Side Clearance Top Compression	Ring Side Clearance Bottom Compression	Ring Side Clearance Oil Control
2005	1.8	AWP	0.0018	0.008-0.016	0.008-0.016	0.010-0.020	0.0023-0.0037	0.0023-0.0037	0.0012-0.0024
	1.8	AWV	0.0018	0.008-0.016	0.008-0.016	0.010-0.020	0.0023-0.0037	0.0023-0.0037	0.0012-0.0024
	1.8	AWW	0.0018	0.008-0.016	0.008-0.016	0.010-0.020	0.0023-0.0037	0.0023-0.0037	0.0012-0.0024
	1.9	ALH	0.0400	0.008-0.016	0.008-0.016	0.010-0.020	0.0035-0.0047	0.0020-0.0031	0.0012-0.0024
	2.0	AEG	0.0018	0.008-0.016	0.008-0.016	0.010-0.020	0.0023-0.0037	0.0023-0.0037	0.0012-0.0024
	2.0	AVH	0.0018	0.008-0.016	0.008-0.016	0.010-0.020	0.0023-0.0037	0.0023-0.0037	0.0012-0.0024
	2.0	AZG	0.0018	0.008-0.016	0.008-0.016	0.010-0.020	0.0023-0.0037	0.0023-0.0037	0.0012-0.0024
	2.0	BDC	0.0018	0.008-0.016	0.008-0.016	0.010-0.020	0.0023-0.0037	0.0023-0.0037	0.0012-0.0024
	2.8	AHA	0.0010	0.014-0.020	0.020-0.028	0.010-0.020	0.0008-0.0031	0.0008-0.0031	0.0008-0.0031
	2.8	ATQ	0.0010	0.014-0.020	0.020-0.028	0.010-0.020	0.0008-0.0031	0.0008-0.0031	0.0008-0.0031
	2.8	BDF	0.0010	0.014-0.020	0.020-0.028	0.010-0.020	0.0008-0.0031	0.0008-0.0031	0.0008-0.0031
	4.0	BDP	NA	NA	NA	NA	NA	NA	NA
2006	1.9	BEW	0.0400	0.008-0.016	0.008-0.016	0.010-0.020	0.0035-0.0047	0.0020-0.0031	0.0012-0.0024
	1.9	BRM	0.0400	0.008-0.016	0.008-0.016	0.010-0.020	0.0035-0.0047	0.0020-0.0031	0.0012-0.0024
	2.0	BPY	0.0018	0.008-0.016	0.008-0.016	0.010-0.020	0.0023-0.0037	0.0023-0.0037	0.0012-0.0024
	2.5	BGP	0.0018	0.008-0.016	0.008-0.016	0.010-0.020	0.0023-0.0035	0.0023-0.0035	0.0012-0.0024
	2.5	BGQ	0.0018	0.008-0.016	0.008-0.016	0.010-0.020	0.0023-0.0035	0.0023-0.0035	0.0012-0.0024
	2.5	BPR	0.0018	0.008-0.016	0.008-0.016	0.010-0.020	0.0023-0.0035	0.0023-0.0035	0.0012-0.0024
	2.5	BPS	0.0018	0.008-0.016	0.008-0.016	0.010-0.020	0.0023-0.0035	0.0023-0.0035	0.0012-0.0024
	3.6	BLV	NA	NA	NA	NA	NA	NA	NA

NA: Not Available

06041_VWMC_C0018

TORQUE SPECIFICATIONS

All readings in ft. lbs.

Year	Engine Displacement Liters	Engine ID/VIN	Cylinder Head Bolts	Main Bearing Bolts	Rod Bearing Bolts	Crankshaft Damper Bolt	Flywheel Bolts	Manifold Intake	Manifold Exhaust	Spark Plugs	Oil Pan Drain Plug
2002	1.8	AEB	①	②	③	④	⑤	18	18	22	22
	1.9	AAZ	①	②	③	④	⑤	18	18	—	22
	1.9	AHH	①	②	③	④	⑤	18	18	—	22
	1.9	ALH	①	②	③	⑥	⑤	18	18	—	22
	2.0	ABA	①	②	⑦	④	⑤	15	15	22	22
	2.0	AEG	⑧	②	③	④	⑨	18	18	22	22
	2.8	AAA	①	⑩	⑦	④	⑤	15	15	22	22
	2.8	AHA	①	⑪	③	⑫	⑬	18	18	22	22
2003	1.8	AEB	①	②	③	④	⑤	18	18	22	22
	1.8	ATW	①	②	③	④	⑤	18	18	22	22
	1.8	AWM	①	②	③	④	⑨	18	18	22	22
	1.8	AWP	⑧	②	⑦	④	⑤	15	15	22	22
	1.8	AWV	⑧	②	③	④	⑤	18	18	22	22
	1.8	AWW	①	②	③	⑥	⑤	18	18	22	22
	1.9	ALH	①	②	③	⑥	⑤	18	18	—	22
	2.0	AEG	⑧	⑩	⑦	④	⑤	15	15	22	22
	2.0	AVH	⑧	⑪	③	⑫	⑬	18	18	22	22
	2.0	AZG	⑧	②	③	④	⑤	18	18	22	22
	2.0	BDC	⑧	②	③	④	⑤	18	18	22	22
	2.8	AHA	⑪	②	③	④	⑤	18	18	22	22
	2.8	ATQ	⑪	②	③	⑥	⑤	18	18	22	22
	2.8	BDF	⑭	⑩	⑦	⑮	⑤	15	15	22	22
	4.0	BDP	⑭	NA	NA	⑮	⑪	NA	NA	22	22
2004	1.8	AEB	①	②	③	④	⑤	18	18	22	22
	1.8	ATW	①	②	③	④	⑤	18	18	22	22
	1.8	AWM	①	②	③	④	⑨	18	18	22	22
	1.8	AWP	⑧	②	⑦	④	⑤	15	15	22	22
	1.8	AWV	⑧	②	③	④	⑤	18	18	22	22
	1.8	AWW	⑧	②	③	④	⑤	18	18	22	22
	1.9	ALH	①	②	③	⑥	⑤	18	18	—	22
	2.0	AEG	⑧	⑩	⑦	④	⑤	15	15	22	22
	2.0	AVH	⑧	⑪	③	⑫	⑬	18	18	22	22
	2.0	AZG	⑧	②	③	④	⑤	18	18	22	22
	2.0	BDC	⑧	②	③	④	⑤	18	18	22	22
	2.8	AHA	⑪	②	③	④	⑤	18	18	22	22
	2.8	ATQ	⑪	②	③	⑥	⑤	18	18	22	22
	2.8	BDF	⑭	⑩	⑦	⑮	⑤	15	15	22	22
	4.0	BDP	⑭	NA	NA	⑮	⑪	NA	NA	22	22
2005	1.8	AEB	①	②	③	④	⑤	18	18	22	22
	1.8	ATW	①	②	③	④	⑤	18	18	22	22
	1.8	AWM	①	②	③	④	⑨	18	18	22	22
	1.8	AWP	⑧	②	⑦	④	⑤	15	15	22	22
	1.8	AWV	⑧	②	③	④	⑤	18	18	22	22
	1.8	AWW	⑧	②	③	④	⑤	18	18	22	22
	1.8	BKF	⑧	②	⑦	④	⑤	15	15	22	22
	1.9	ALH	①	②	③	⑥	⑤	18	18	—	22
	1.9	BEW	①	②	③	⑥	⑤	18	18	—	22
	1.9	BRM	①	②	③	⑥	⑤	18	18	—	22
	2.0	AVH	⑧	⑪	③	⑫	⑬	18	18	22	22

06041_VWMC_C0019

TORQUE SPECIFICATIONS

All readings in ft. lbs.

Year	Engine Displacement Liters	Engine ID/VIN	Cylinder Head Bolts	Main Bearing Bolts	Rod Bearing Bolts	Crankshaft Damper Bolt	Flywheel Bolts	Manifold		Spark Plugs	Oil Pan Drain Plug
								Intake	Exhaust		
2005	2.0	AZG	⑧	②	③	④	⑤	18	18	22	22
	2.0	BEV	⑧	②	③	④	⑤	18	18	22	22
	2.0	BGD	⑧	②	③	④	⑤	18	18	22	22
	2.0	BPY	⑯	②	③	④	⑤	7	18	22	22
	2.5	BGP	⑰	⑱	⑲	48	⑳	6	30	22	22
	2.5	BGQ	⑰	⑱	⑲	48	⑳	6	30	22	22
	2.8	AHA	⑪	②	③	④	⑤	18	18	22	22
	2.8	ATQ	⑪	②	③	⑥	⑤	18	18	22	22
	2.8	BDF	⑭	⑩	⑦	⑮	⑤	15	15	22	22
	4.0	BDP	⑭	NA	NA	⑮	⑪	NA	NA	22	22
2006	1.9	BEW	①	②	③	⑥	⑤	18	18	—	22
	1.9	BRM	①	②	③	⑥	⑤	18	18	—	22
	2.0	BPY	⑯	②	③	④	⑤	7	18	22	22
	2.5	BGP	⑰	⑱	⑲	48	⑳	6	30	22	22
	2.5	BGQ	⑰	⑱	⑲	48	⑳	6	30	22	22
	2.5	BPR	⑰	⑱	⑲	48	⑳	6	30	22	22
	2.5	BPS	⑰	⑱	⑲	48	⑳	6	30	22	22
	3.6	BLV	㉑	㉒	㉓	7	⑳	NA	30	22	22

① Torque in four steps: (use new bolts).
Step 1: 30 ft. lbs.
Step 2: 44 ft. lbs.
Step 3: plus 90 degrees
Step 4: plus 90 degrees

② 48 ft. lbs., plus 90 degrees. Use new bolts.

③ Torque to 22 ft. lbs. plus 90 degrees

④ Step 1: 66 ft. lbs.
Step 2: plus 90 degrees

⑤ 44 ft. lbs. plus 90 degrees;
except Passat GLX: 52 ft. lbs. plus 90 degrees.

⑥ 88 ft. lbs. plus 90 degrees. Always replace bolt.

⑦ 22 ft. lbs. plus 90 degrees. Use new bolts.

⑧ Torque in three steps. Use new bolts.
Step 1: 30 ft. lbs.
Step 2: plus 90 degrees
Step 3: plus 90 degrees

⑨ 30 ft. lbs. plus 90 degrees. Use new bolt.

⑩ 22 ft. lbs. plus 180 degrees. Use new bolts.

⑪ 44 ft.lbs. plus 180 degrees

⑫ Step 1: 148 ft. lbs.
Step 2: plus 90 degrees

⑫ Step 1: 148 ft. lbs.
Step 2: plus 90 degrees

⑬ Driveplate: 44 ft.lbs. plus 90 degrees
Flywheel: 44 ft. lbs. plus 180 degrees

⑭ Torque in four steps: (use new bolts).
Step 1: 22 ft. lbs.
Step 2: 37 ft. lbs.
Step 3: plus 90 degrees
Step 4: plus 90 degrees

⑮ 73 ft. lbs. plus 90 degrees

⑯ Torque in three steps: (use new bolts).
Step 1: 28 ft. lbs.
Step 2: plus 90 degrees
Step 3: plus 90 degrees

⑰ Torque in four steps: (use new bolts).
Step 1: 30 ft. lbs.
Step 2: plus 90 degrees
Step 3: plus 90 degrees
Step 4: bolt 13 to 88 inch lbs. (10 Nm)

⑱ Torque in two steps: (use new bolts).
Step 1: 30 ft. lbs.
Step 3: plus 90 degrees

⑲ Torque in two steps: (use new bolts).
Step 1: 22 ft. lbs.
Step 3: plus 90 degrees

⑳ Torque in two steps: (use new bolts).
Step 1: 41 ft. lbs.
Step 3: plus 90 degrees

㉑ Torque in three steps: (use new bolts).
Step 1: 10 ft. lbs.
Step 2: 22 ft. lbs.
Step 3: plus 180 degrees

㉒ Torque in two steps: (use new bolts).
Step 1: 21 ft. lbs.
Step 3: plus 180 degrees

㉓ Torque in two steps: (use new bolts).
Step 1: 21 ft. lbs.
Step 3: plus 90 degrees

06041_VWMC_C0020

WHEEL ALIGNMENT

			Caster		Camber		
Year	Model		Range (+/-Deg.)	Preferred Setting (Deg.)	Range (+/-Deg.)	Preferred Setting (Deg.)	Toe-in (in.)
2002	Cabrio						
	1.8L, 1.9L	F	0.50	+1.75	0.33	-0.50	0+/-0.16
	2.0L	F	0.50	+1.50	0.33	-0.58	0+/-0.16
	2.8L	F	0.50	+3.27	0.33	-0.50	0+/-0.16
		R	—	—	0.16	-1.50	0.33+/-0.16
	Golf						
	Standard	F	0.50	+7.67	0.50	-0.50	0+/-0.16
		R	—	—	0.16	-1.45	0.33+/-0.16
	Sport	F	0.50	+7.83	0.50	-0.58	0+/-0.16
		R	—	—	0.16	-1.58	0.42+/-0.16
	Jetta						
	Standard	F	0.50	+7.67	0.50	-0.50	0+/-0.16
		R	—	—	0.16	-1.45	0.33+/-0.16
	Sport	F	0.50	+7.83	0.50	-0.58	0+/-0.16
		R	—	—	0.16	-1.45	0.42+/-0.16
	New Beetle						
	Standard	F	0.50	+7.67	0.50	-0.50	0+/-0.16
		R	—	—	0.16	-1.45	0.33+/-0.16
	Sport	F	0.50	+7.83	0.50	-0.55	0+/-0.16
		R	—	—	0.16	-1.45	0.42+/-0.16
	Passat FWD						
	Standard	F	NA	NA	0.42	-0.42	0.16+/-0.08
		R	—	—	0.33	-1.50	0.33+/-0.16
	Sport	F	NA	NA	0.42	-0.67	0.16+/-0.08
		R	—	—	0.33	-1.50	0.47+/-0.16
	Heavy Duty	F	NA	NA	0.42	-0.25	0.16+/-0.08
		R	—	—	0.33	-1.50	0.18+/-0.16
	Passat AWD						
	Standard	F	NA	NA	0.42	-0.42	0.16+/-0.08
		R	—	—	0.50	-0.67	0.27+/-0.16
	Sport	F	NA	NA	0.42	-0.67	0.16+/-0.08
		R	—	—	0.50	-0.67	0.27+/-0.16
	Heavy Duty	F	NA	NA	0.42	-0.25	0.16+/-0.08
		R	—	—	0.50	-0.67	0.27+/-0.16

06041_VWMC_C0021

WHEEL ALIGNMENT

Year	Model		Caster Range (+/-Deg.)	Caster Preferred Setting (Deg.)	Camber Range (+/-Deg.)	Camber Preferred Setting (Deg.)	Toe-in (in.)
2003	Convertible						
	Standard	F	0.50	+7.67	0.50	-0.50	0+/-0.16
		R	—	—	0.16	-1.45	0.33+/-0.16
	Sport	F	0.50	+7.83	0.50	-0.55	0+/-0.16
		R	—	—	0.16	-1.45	0.42+/-0.16
	Golf & GTI						
	Standard	F	0.50	+7.67	0.50	-0.50	0+/-0.16
		R	—	—	0.16	-1.45	0.33+/-0.16
	Sport	F	0.50	+7.83	0.50	-0.58	0+/-0.16
		R	—	—	0.16	-1.58	0.42+/-0.16
	Jetta						
	Standard	F	0.50	+7.67	0.50	-0.50	0+/-0.16
		R	—	—	0.16	-1.45	0.33+/-0.16
	Sport	F	0.50	+7.83	0.50	-0.58	0+/-0.16
		R	—	—	0.16	-1.45	0.42+/-0.16
	New Beetle						
	Standard	F	0.50	+7.67	0.50	-0.50	0+/-0.16
		R	—	—	0.16	-1.45	0.33+/-0.16
	Sport	F	0.50	+7.83	0.50	-0.55	0+/-0.16
		R	—	—	0.16	-1.45	0.42+/-0.16
	Passat FWD						
	Standard	F	NA	NA	0.42	-0.42	0.16+/-0.08
		R	—	—	0.33	-1.50	0.33+/-0.16
	Sport	F	NA	NA	0.42	-0.67	0.16+/-0.08
		R	—	—	0.33	-1.50	0.47+/-0.16
	Heavy Duty	F	NA	NA	0.42	-0.25	0.16+/-0.08
		R	—	—	0.33	-1.50	0.18+/-0.16
	Passat AWD						
	Standard	F	NA	NA	0.42	-0.42	0.16+/-0.08
		R	—	—	0.50	-0.67	0.27+/-0.16
	Sport	F	NA	NA	0.42	-0.67	0.16+/-0.08
		R	—	—	0.50	-0.67	0.27+/-0.16
	Heavy Duty	F	NA	NA	0.42	-0.25	0.16+/-0.08
		R	—	—	0.50	-0.67	0.27+/-0.16
2004	Convertible						
	Standard	F	0.50	+7.67	0.50	-0.50	0+/-0.16
		R	—	—	0.16	-1.45	0.33+/-0.16
	Sport	F	0.50	+7.83	0.50	-0.55	0+/-0.16
		R	—	—	0.16	-1.45	0.42+/-0.16

06041_VWMC_C0022

WHEEL ALIGNMENT

Year	Model		Caster Range (+/-Deg.)	Caster Preferred Setting (Deg.)	Camber Range (+/-Deg.)	Camber Preferred Setting (Deg.)	Toe-in (in.)
2004	Golf & GTI						
	Standard	F	0.50	+7.67	0.50	-0.50	0+/-0.16
		R	—	—	0.16	-1.45	0.33+/-0.16
	Sport	F	0.50	+7.83	0.50	-0.58	0+/-0.16
		R	—	—	0.16	-1.58	0.42+/-0.16
	Jetta						
	Standard	F	0.50	+7.67	0.50	-0.50	0+/-0.16
		R	—	—	0.16	-1.45	0.33+/-0.16
	Sport	F	0.50	+7.83	0.50	-0.58	0+/-0.16
		R	—	—	0.16	-1.45	0.42+/-0.16
	New Beetle						
	Standard	F	0.50	+7.67	0.50	-0.50	0+/-0.16
		R	—	—	0.16	-1.45	0.33+/-0.16
	Sport	F	0.50	+7.83	0.50	-0.55	0+/-0.16
		R	—	—	0.16	-1.45	0.42+/-0.16
	Passat FWD						
	Standard	F	NA	NA	0.42	-0.42	0.16+/-0.08
		R	—	—	0.33	-1.50	0.33+/-0.16
	Sport	F	NA	NA	0.42	-0.67	0.16+/-0.08
		R	—	—	0.33	-1.50	0.47+/-0.16
	Heavy Duty	F	NA	NA	0.42	-0.25	0.16+/-0.08
		R	—	—	0.33	-1.50	0.18+/-0.16
	Passat AWD						
	Standard	F	NA	NA	0.42	-0.42	0.16+/-0.08
		R	—	—	0.50	-0.67	0.27+/-0.16
	Sport	F	NA	NA	0.42	-0.67	0.16+/-0.08
		R	—	—	0.50	-0.67	0.27+/-0.16
	Heavy Duty	F	NA	NA	0.42	-0.25	0.16+/-0.08
		R	—	—	0.50	-0.67	0.27+/-0.16
2005	Convertible						
	Standard	F	0.50	+7.67	0.50	-0.50	0+/-0.16
		R	—	—	0.16	-1.45	0.33+/-0.16
	Sport	F	0.50	+7.83	0.50	-0.55	0+/-0.16
		R	—	—	0.16	-1.45	0.42+/-0.16
	Golf & GTI						
	Standard	F	0.50	+7.67	0.50	-0.50	0+/-0.16
		R	—	—	0.16	-1.45	0.33+/-0.16
	Sport	F	0.50	+7.83	0.50	-0.58	0+/-0.16
		R	—	—	0.16	-1.58	0.42+/-0.16
	Jetta						
	Standard	F	0.50	+7.67	0.50	-0.50	0+/-0.16
		R	—	—	0.16	-1.45	0.33+/-0.16
	Sport	F	0.50	+7.83	0.50	-0.58	0+/-0.16
		R	—	—	0.16	-1.45	0.42+/-0.16
	New Beetle						
	Standard	F	0.50	+7.67	0.50	-0.50	0+/-0.16
		R	—	—	0.16	-1.45	0.33+/-0.16
	Sport	F	0.50	+7.83	0.50	-0.55	0+/-0.16
		R	—	—	0.16	-1.45	0.42+/-0.16

06041_VWMC_C0023

WHEEL ALIGNMENT

Year	Model		Caster Range (+/-Deg.)	Caster Preferred Setting (Deg.)	Camber Range (+/-Deg.)	Camber Preferred Setting (Deg.)	Toe-in (in.)
2005	Passat FWD						
	Standard	F	NA	NA	0.42	-0.42	0.16+/-0.08
		R	—	—	0.33	-1.50	0.33+/-0.16
	Sport	F	NA	NA	0.42	-0.67	0.16+/-0.08
		R	—	—	0.33	-1.50	0.47+/-0.16
	Heavy Duty	F	NA	NA	0.42	-0.25	0.16+/-0.08
		R	—	—	0.33	-1.50	0.18+/-0.16
	Passat AWD						
	Standard	F	NA	NA	0.42	-0.42	0.16+/-0.08
		R	—	—	0.50	-0.67	0.27+/-0.16
	Sport	F	NA	NA	0.42	-0.67	0.16+/-0.08
		R	—	—	0.50	-0.67	0.27+/-0.16
	Heavy Duty	F	NA	NA	0.42	-0.25	0.16+/-0.08
		R	—	—	0.50	-0.67	0.27+/-0.16
2006	Convertible						
	Standard	F	0.50	+7.67	0.50	-0.50	0+/-0.16
		R	—	—	0.16	-1.45	0.33+/-0.16
	Sport	F	0.50	+7.83	0.50	-0.55	0+/-0.16
		R	—	—	0.16	-1.45	0.42+/-0.16
	Rabbit, Jetta						
	Standard	F	0.50	+7.67	0.50	-0.50	0+/-0.16
		R	—	—	0.16	-1.45	0.33+/-0.16
	Sport	F	0.50	+7.83	0.50	-0.58	0+/-0.16
		R	—	—	0.16	-1.45	0.42+/-0.16
	New Beetle						
	Standard	F	0.50	+7.67	0.50	-0.50	0+/-0.16
		R	—	—	0.16	-1.45	0.33+/-0.16
	Sport	F	0.50	+7.83	0.50	-0.55	0+/-0.16
		R	—	—	0.16	-1.45	0.42+/-0.16
	Passat FWD						
	Standard	F	NA	NA	0.42	-0.42	0.16+/-0.08
		R	—	—	0.33	-1.50	0.33+/-0.16
	Sport	F	NA	NA	0.42	-0.67	0.16+/-0.08
		R	—	—	0.33	-1.50	0.47+/-0.16
	Heavy Duty	F	NA	NA	0.42	-0.25	0.16+/-0.08
		R	—	—	0.33	-1.50	0.18+/-0.16
	Passat AWD						
	Standard	F	NA	NA	0.42	-0.42	0.16+/-0.08
		R	—	—	0.50	-0.67	0.27+/-0.16
	Sport	F	NA	NA	0.42	-0.67	0.16+/-0.08
		R	—	—	0.50	-0.67	0.27+/-0.16
	Heavy Duty	F	NA	NA	0.42	-0.25	0.16+/-0.08
		R	—	—	0.50	-0.67	0.27+/-0.16

NA: Not Available

06041_VWMC_C0024

TIRE, WHEEL AND BALL JOINT SPECIFICATIONS

		OEM Tires		Tire Pressures (psi)		Wheel	Lug Nut Torque
Year	Model	Standard	Optional	Front	Rear	Size	Ft. Lbs.
2002	Cabrio	P195/60HR14	None	30	30	6-J	89
	GTI	205/50HR15	None	32	32	6.5-J	89
	Jetta GLX	205/50VR15	None	28	26	6.5-J	89
	Jetta GL, GLS	185/60HR14	195/60HR14	33	30	6-J	89
	New Beetle	205/55R16	None	28	28	6.5-J	89
	Passat	195/65R15	None	32	32	6.5-J	89
2003	Convertible	P195/60HR14	None	30	30	6-J	89
	Golf	P195/65HR15	None	32	32	6.5-J	89
	GTI	205/50HR15	None	32	32	6.5-J	89
	Jetta GLX	205/55HR16	None	28	26	6.5-J	89
	Jetta GL, GLS	195/65HR15	195/60HR14	33	30	6-J	89
	New Beetle	205/55R16	P225/45HR17	28	28	6.5-J	89
	Passat	195/65R15	205/55HR16	32	32	6.5-J	89
2004	Convertible	205/55HR16	P225/45HR17	28	28	6.5-J	89
	Golf	P195/65HR15	None	32	32	6.5-J	89
	GTI	205/50HR15	None	32	32	6.5-J	89
	Jetta GLX	205/55HR16	None	28	26	6.5-J	89
	Jetta GL, GLS	195/65HR15	195/60HR14	33	30	6-J	89
	New Beetle	205/55HR16	P225/45HR17	28	28	6.5-J	89
	Passat	195/65R15	205/55HR16	32	32	6.5-J	89
2005	Convertible	205/55HR16	P225/45HR17	28	28	6.5-J	89
	Golf	P195/65HR15	None	32	32	6.5-J	89
	GTI	205/50HR15	None	32	32	6.5-J	89
	Jetta Value Ed.	195/65HR15	None	28	28	6-J	89
	Jetta 2.5, TDI	205/55HR16	None	28	28	6.5-J	89
	Jetta 2.0T	205/55HR16	P225/45HR17	28	28	6.5-J	89
	Jetta GLI	225/45WR17	225/45HR17	28	28	7-J	89
	New Beetle	205/55HR16	P225/45HR17	28	28	6.5-J	89
	Passat	195/65R15	205/55HR16	32	32	6.5-J	89
2006	Convertible	205/55HR16	P225/45HR17	28	28	6.5-J	89
	Rabbit	195/65HR15	205/55HR16	32	32	6-J	89
	GTI	225/45R17	None	32	32	6.5-J	89
	Jetta Value Ed.	195/65HR15	None	28	28	6-J	89
	Jetta 2.5, TDI	205/55HR16	None	28	28	6.5-J	89
	Jetta 2.0T	205/55HR16	P225/45HR17	28	28	6.5-J	89
	Jetta GLI	225/45WR17	225/45HR17	28	28	7-J	89
	New Beetle	205/55HR16	P225/45HR17	28	28	6.5-J	89
	Passat 2.0T	215/55HR16	235/45HR17	32	32	7-J	89
	Passat 3.6L	235/45HR17	None	32	32	7-J	89

OEM: Original Equipment Manufacturer
PSI: Pounds Per Square Inch
STD: Standard
OPT: Optional

06041_VWMC_C0025

BRAKE SPECIFICATIONS

All measurements in inches unless noted

			Brake Disc			Drum Diameter			Brake Caliper	
Year	Model		Original Thickness	Minimum Thickness	Maximum Run-out	Original Inside Diameter	Maximum Machine Diameter	Minimum Lining Thickness	Bracket Bolts (ft. lbs.)	Mounting Bolts (ft. lbs.)
2002	Cabrio	F	0.790	0.709	0.002	—	—	0.28	18-26	92
		R	0.390	0.315	0.002	7.87	7.91	0.275 ①	22	41
	Golf ②	F	0.790	0.709	0.002	—	—	0.28	18-26	92
		R	0.390	0.315	0.002	7.87	7.91	0.275 ①	22	41
	Golf ③	F	0.870	0.787	0.002	—	—	0.28	18-26	92
		R	0.390	0.315	0.002	7.87	7.91	0.28	22	41
	GTI ④	F	0.790	0.709	0.002	—	—	0.28	18-26	92
		R	0.390	0.315	0.002	7.87	7.91	0.28	22	41
	GTI ⑤	F	0.870	0.787	0.002	—	—	0.28	18-26	92
		R	0.390	0.315	0.002	7.87	7.91	0.28	22	41
	Jetta ⑥	F	0.790	0.709	0.002	—	—	0.28	18-26	92
		R	0.390	0.315	0.002	7.87	7.91	0.275 ①	22	41
	Jetta ⑦	F	0.870	0.787	0.002	—	—	0.28	18-26	92
		R	0.390	0.315	0.002	7.87	7.91	0.28	22	41
	New Beetle	F	0.790	0.950	0.002	—	—	0.27	18	92
		R	0.390	0.315	0.002	9.06	9.09	0.275 ①	26	48
	Passat	F	0.98 ⑧	0.900 ⑨	0.002	—	—	0.28	22	89
		R	0.393	0.314	0.002	NA	NA	0.27	22	70
2003	Convertible	F	0.790	0.950	0.002	—	—	0.27	18	92
		R	0.390	0.315	0.002	—	—	0.275 ①	26	48
	Golf ②	F	0.790	0.709	0.002	—	—	0.28	18-26	92
		R	0.390	0.315	0.002	—	—	0.275 ①	22	41
	Golf ③	F	0.870	0.787	0.002	—	—	0.28	18-26	92
		R	0.390	0.315	0.002	—	—	0.28	22	41
	GTI ④	F	0.790	0.709	0.002	—	—	0.28	18-26	92
		R	0.390	0.315	0.002	—	—	0.28	22	41
	GTI ⑤	F	0.870	0.787	0.002	—	—	0.28	18-26	92
		R	0.390	0.315	0.002	—	—	0.28	22	41
	Jetta ⑥	F	0.790	0.709	0.002	—	—	0.28	18-26	92
		R	0.390	0.315	0.002	—	—	0.275 ①	22	41
	Jetta ⑦	F	0.870	0.787	0.002	—	—	0.28	18-26	92
		R	0.390	0.315	0.002	—	—	0.28	22	41
	New Beetle	F	0.790	0.950	0.002	—	—	0.27	18	92
		R	0.390	0.315	0.002	—	—	0.275 ①	26	48
	Passat	F	0.98 ⑧	0.900 ⑨	0.002	—	—	0.28	22	89
		R	0.393 ⑧	0.314 ⑨	0.002	—	—	0.27	22	70
2004	Convertible	F	0.790	0.950	0.002	—	—	0.27	18	92
		R	0.390	0.315	0.002	—	—	0.275 ①	26	48
	Golf ②	F	0.790	0.709	0.002	—	—	0.28	18-26	92
		R	0.390	0.315	0.002	—	—	0.275 ①	22	41
	Golf ③	F	0.870	0.787	0.002	—	—	0.28	18-26	92
		R	0.390	0.315	0.002	—	—	0.28	22	41
	GTI ④	F	0.790	0.709	0.002	—	—	0.28	18-26	92
		R	0.390	0.315	0.002	—	—	0.28	22	41
	GTI ⑤	F	0.870	0.787	0.002	—	—	0.28	18-26	92
		R	0.390	0.315	0.002	—	—	0.28	22	41

06041_VWMC_C0026

BRAKE SPECIFICATIONS

All measurements in inches unless noted

			Brake Disc			Drum Diameter			Brake Caliper	
Year	Model		Original Thickness	Minimum Thickness	Maximum Run-out	Original Inside Diameter	Maximum Machine Diameter	Minimum Lining Thickness	Bracket Bolts (ft. lbs.)	Mounting Bolts (ft. lbs.)
2004	Jetta ⑥	F	0.790	0.709	0.002	—	—	0.28	18-26	92
		R	0.390	0.315	0.002	—	—	0.275 ①	22	41
	Jetta ⑦	F	0.870	0.787	0.002	—	—	0.28	18-26	92
		R	0.390	0.315	0.002	—	—	0.28	22	41
	New Beetle	F	0.790	0.950	0.002	—	—	0.27	18	92
		R	0.390	0.315	0.002	—	—	0.275 ①	26	48
	Passat	F	0.98 ⑧	0.900 ⑨	0.002	—	—	0.28	22	89
		R	0.393 ⑧	0.314 ⑨	0.002	—	—	0.27	22	70
2005	Convertible	F	0.790	0.950	0.002	—	—	0.27	18	92
		R	0.390	0.315	0.002	—	—	0.275 ①	26	48
	Golf ②	F	0.790	0.709	0.002	—	—	0.28	18-26	92
		R	0.390	0.315	0.002	—	—	0.275 ①	22	41
	Golf ③	F	0.870	0.787	0.002	—	—	0.28	18-26	92
		R	0.390	0.315	0.002	—	—	0.28	22	41
	GTI ④	F	0.790	0.709	0.002	—	—	0.28	18-26	92
		R	0.390	0.315	0.002	—	—	0.28	22	41
	GTI ⑤	F	0.870	0.787	0.002	—	—	0.28	18-26	92
		R	0.390	0.315	0.002	—	—	0.28	22	41
	Jetta ⑥	F	0.790	0.709	0.002	—	—	0.28	18-26	92
		R	0.390	0.315	0.002	—	—	0.275 ①	22	41
	Jetta ⑦	F	0.870	0.787	0.002	—	—	0.28	18-26	92
		R	0.390	0.315	0.002	—	—	0.28	22	41
	New Beetle	F	0.790	0.950	0.002	—	—	0.27	18	92
		R	0.390	0.315	0.002	—	—	0.275 ①	26	48
	Passat	F	0.98 ⑧	0.900 ⑨	0.002	—	—	0.28	22	89
		R	0.393 ⑧	0.314 ⑨	0.002	—	—	0.27	22	70
2006	Convertible	F	0.790	0.950	0.002	—	—	0.27	18	92
		R	0.390	0.315	0.002	—	—	0.275 ①	26	48
	Rabbit	F	0.790	0.709	0.002	—	—	0.28	18-26	92
		R	0.390	0.315	0.002	—	—	0.28	22	41
	GTI	F	0.870	0.787	0.002	—	—	0.28	18-26	92
		R	0.390	0.315	0.002	—	—	0.28	22	41
	Jetta ⑥	F	0.790	0.709	0.002	—	—	0.28	18-26	92
		R	0.390	0.315	0.002	—	—	0.275 ①	22	41
	Jetta ⑦	F	0.870	0.787	0.002	—	—	0.28	18-26	92
		R	0.390	0.315	0.002	—	—	0.28	22	41
	New Beetle	F	0.790	0.950	0.002	—	—	0.27	18	92
		R	0.390	0.315	0.002	—	—	0.275 ①	26	48
	Passat	F	0.98 ⑧	0.900 ⑨	0.002	—	—	0.28	22	89
		R	0.393 ⑧	0.314 ⑨	0.002	—	—	0.27	22	70

F: Front
R: Rear
NA: Not Available
① VR6 Model
② GLX Model
③ GL model
④ 2.0L Engine
⑤ 2.8L Engine
⑥ ABS equipped models have a diameter of 0.937 in.
⑦ Models equipped with drum brakes have a lining limit of 0.098 in.
⑧ Lucas caliper: 0.87 in.
⑨ Lucas caliper: 0.78 in.

06041_VWMC_C0027

SCHEDULED MAINTENANCE INTERVALS
VOLKSWAGEN—CABRIOLET, GOLF, GTI, JETTA, NEW BEETLE, PASSAT & RABBIT

TO BE SERVICED	TYPE OF SERVICE	VEHICLE MILEAGE INTERVAL (x1000)												
		7.5	15	22.5	30	37.5	45	52.5	60	67.5	75	82.5	90	97.5
Engine oil & filter	R	✓	✓	✓	✓	✓	✓	✓	✓	✓	✓	✓	✓	✓
Brake pad thickness	R	✓	✓	✓	✓	✓	✓	✓	✓	✓	✓	✓	✓	✓
A/T final drive fluid level	S/I		✓		✓		✓		✓		✓		✓	
Battery	S/I		✓		✓		✓		✓		✓		✓	
Brake system	S/I		✓		✓		✓		✓		✓		✓	
Cooling system	S/I		✓		✓		✓		✓		✓		✓	
Driveshaft boots	S/I		✓		✓		✓		✓		✓		✓	
Engine (check for leaks)	S/I		✓		✓		✓		✓		✓		✓	
Engine coolant level	S/I		✓		✓		✓		✓		✓		✓	
Engine exhaust	S/I		✓		✓		✓		✓		✓		✓	
Fuel system	S/I		✓		✓		✓		✓		✓		✓	
Idle speed (gasoline)	S/I		✓		✓		✓		✓		✓		✓	
Idle speed (diesel)	S/I				✓				✓				✓	
Intake air system	S/I		✓		✓		✓		✓		✓		✓	
OBD system - check for codes	S/I		✓		✓		✓		✓		✓		✓	
Power steering fluid level	S/I		✓		✓		✓		✓		✓		✓	
Steering system	S/I		✓		✓		✓		✓		✓		✓	
Timing belt (diesel)	S/I				✓				✓				✓	
Transaxle fluid level	S/I		✓		✓		✓		✓		✓		✓	
Water separator (diesel)	S/I		✓		✓		✓		✓		✓		✓	
Air filter element	R				✓				✓				✓	
Engine coolant	R				✓				✓				✓	
Fuel filter (diesel)	R				✓				✓				✓	
Spark plugs	R								✓					
Passenger compartment air filter	R				✓				✓				✓	
Drive belts	S/I				✓				✓				✓	
Dust seals on ball joints, tie rod ends & tie rods	S/I				✓				✓				✓	
Brake fluid ①	R													

R: Replace S/I: Service or Inspect

① Replace every two years regardless of mileage.

FREQUENT OPERATION MAINTENANCE (SEVERE SERVICE)

If a vehicle is operated under any of the following conditions it is considered severe service:

- Extremely dusty areas.

- 50% or more of the vehicle operation is in 32°C (90°F) or higher temperatures, or constant operation in temperatures below 0°C (32°F).

- Prolonged idling (vehicle operation in stop and go traffic).

- Frequent short running periods (engine does not warm to normal operating temperatures).

- Police, taxi, delivery usage or trailer towing usage.

Oil & oil filter change: change every 3750 miles.

Air filter element: service or inspect every 15,000 miles.

Automatic transaxle fluid & filter: replace every 30,000 miles.

06041_VWMC_C0028

GASOLINE ENGINE REPAIR

Distributor

All engines in this section are equipped with a Distributorless Ignition System (DIS).

Alternator

REMOVAL & INSTALLATION

Before purchasing a replacement alternator, read the specification plate on the housing. The number 14V will appear to indicate maximum voltage rating. On the same line will be two more digits followed by the letter **A**. This is the maximum amperage output. Be sure to purchase an alternator with the same rating. The regulator can be replaced without removing the alternator.

All Engines (except 3.6L and 2.5L Engines)

1. Before servicing the vehicle, refer to the Precautions Section.
2. Disconnect the negative battery cable.
3. Disconnect the wiring from the alternator. If necessary, mark the wires with tape or other means to ensure they are connected properly upon installation.
4. Loosen the belt tension, and remove the drive belt from the alternator. On VR6 (AAA) engines only, remove the belt tensioner from the cylinder head.
5. Remove the alternator adjustment bolt, followed by the pivot bolts.
6. Carefully lift the alternator from the bracket.

To install:

7. Hold the alternator in position on the mounting bracket, and install the pivot bolt. On later engines with automatic belt tensioners, install the upper mounting bolt.
8. Install (but do not tighten) the alternator adjustment bolt (earlier models only).
9. Place the alternator drive belt on the pulley.
10. Adjust belt tension and tighten the mounting and adjustment bolts as necessary.
11. Connect the wiring to the alternator. If tape was used to identify the wires, make sure it is removed once the wires are connected.
12. Connect the negative battery cable.

3.6L Engine

1. Before servicing the vehicle, refer to the Precautions Section.
2. Disconnect the negative battery cable.
3. Remove noise insulation.
4. Remove forward section of wheel housing liner.
5. Mark running direction of ribbed belt.
6. Release tension on ribbed belt by swinging tensioning element using a 16mm open-end wrench.
7. Lock tensioner using Mandrel T10060 A and remove ribbed belt.
8. Remove a/c compressor (without opening refrigerant circuit) and secure to body using wire.
9. Remove nut from harness retainer.

Alternator mounting detail—2.0L engines

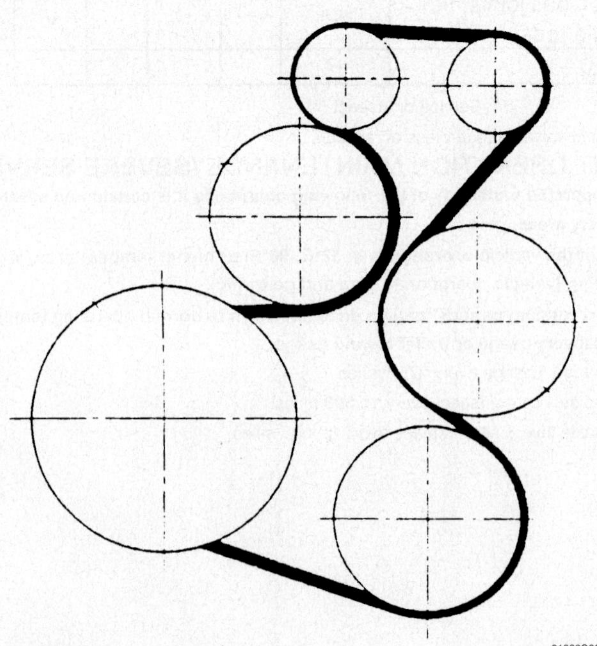

Alternator belt mounting detail—VR6 engines

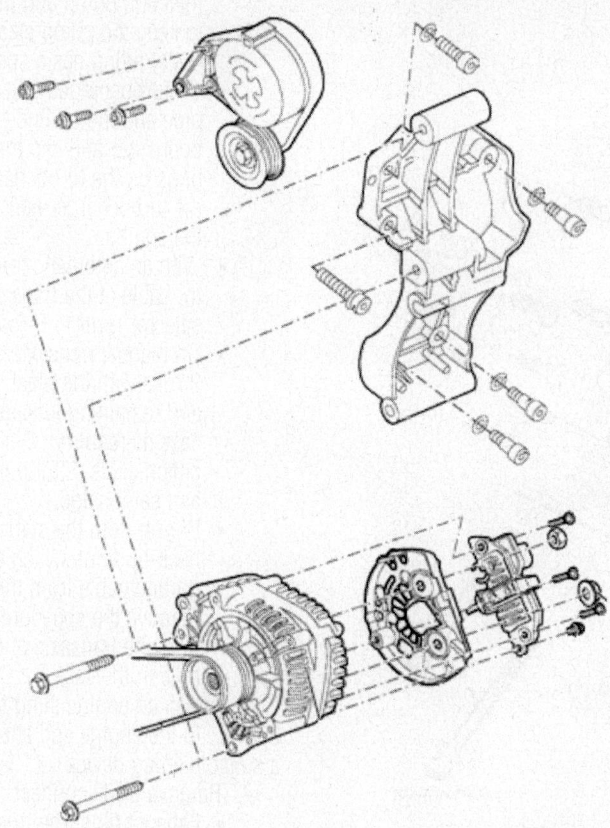

Alternator mounting detail—6 cylinder engines (except VR6 and Passat)

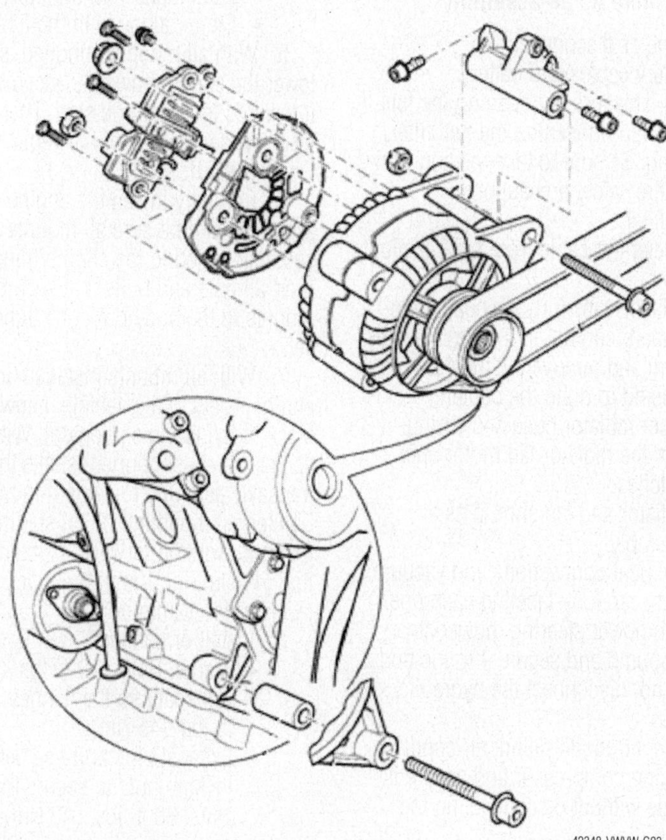

Alternator mounting detail—6 cylinder engine (Passat)

10. Disconnect terminal 61 electrical connection.
11. Remove cap, open retainer and remove harness.
12. Remove nut at positive (B+) cable and set aside cable.
13. Remove bolt and remove tensioner roller.
14. Remove bolts and remove alternator from engine.

To install:

15. Install in reverse order of removal, noting the following:
    a. Drive threaded sleeves out of the alternator housing approximately 4mm towards the rear of the alternator.
    b. Screw wire retainer firmly to rear side of alternator in 9 o'clock position.
    c. Tighten the alternator mounting bolts to 16 ft. lbs. (23 Nm).
    d. Install ribbed belt according to routing diagram.
16. Connect the negative battery cable.
17. Start engine and verify the belt is running properly.
18. Turn off engine.

2.5L Engine

1. Before servicing the vehicle, refer to the Precautions Section.
2. Disconnect the negative battery cable.
3. Remove screws for air intake to air filter/engine cover from lock carrier.

➡Air intake and engine cover component layout may differ between vehicles.

4. Disconnect the Mass Airflow (MAF) sensor (if necessary).
5. Loosen spring clamp using pliers VAS 5024 and remove intake air hose from MAF (if necessary).
6. Pull engine cover upwards and remove.
7. Bring lock carrier into service position.
8. Remove ribbed belt.
9. Disconnect terminal DF electrical connection and remove protective cap.
10. Remove B+-wire from alternator.
11. Remove upper idler pulley, lower idler pulley and tensioning roller for ribbed belt pulley for alternator and water pump.
12. Remove bolts, harness retainer and alternator.

To install:

13. Install in reverse order of removal, noting the following:
    a. Drive threaded sleeves out of the alternator housing approximately 4mm towards the rear of the alternator.

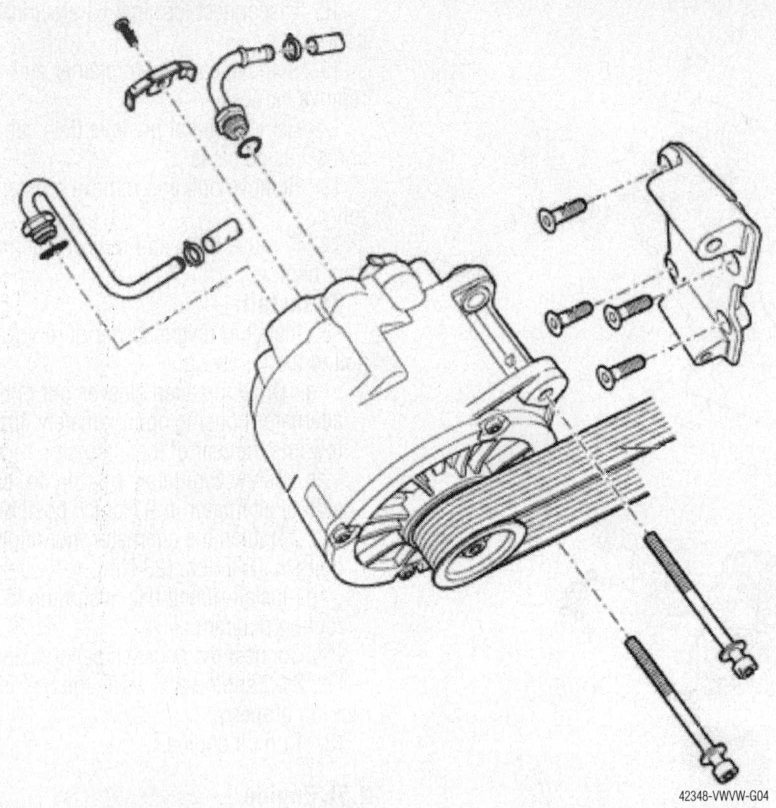

Alternator mounting detail—V8 engine (Passat)

b. Screw wire retainer firmly to rear side of alternator in 9 o'clock position.

c. Tighten the alternator mounting bolts to 18 ft. lbs. (25 Nm).

d. Install ribbed belt according to routing diagram.

14. Connect the negative battery cable.

15. Start engine and verify the belt is running properly.

16. Turn off engine.

Ignition Timing

ADJUSTMENT

➡The ignition timing is controlled by the engine control module and is not adjustable. However the timing can be monitored on a scan tool connected to the DLC in the vehicle. No specification has been given by the manufacturer.

Engine Assembly

REMOVAL & INSTALLATION

Cabrio, Golf, GTI, Rabbit and Jetta (Except 2.8L BDF Engine)

1. Before servicing the vehicle, refer to the Precautions Section.

➡The engine and transaxle are lifted from the vehicle as an assembly.

2. Remove or disconnect:
   - Battery cables and battery.
   - Fuel pressure, by opening the fuel filler cap, loosening the fuel filter fitting. Be sure to take the appropriate fire safety precautions.
   - Air filter
   - Accelerator cable from the injection pump.
   - Radiator cap. Turn the heater temperature control all the way towards warm and remove the thermostat housing to drain the coolant.
   - Upper radiator hose and wiring from the radiator fan motor and switches.
   - Radiator and fan shroud as an assembly.
   - Electrical connections and vacuum lines, carefully labeling each one.
   - With power steering, power steering pump and secure it to the body. Do not disconnect the hydraulic lines.
   - With air conditioning, air conditioning compressor and secure it aside without disconnecting the lines.
   - Fuel inlet and outlet lines from the injection pump and plug the holes to keep the pump clean. Note the outlet fitting has a special orifice.
   - On turbocharged engines, exhaust pipe and the oil lines from the turbocharger and cap the oil line fittings on the turbocharger. Unbolt the turbocharger and lift it out of the engine.
   - With an automatic transaxle, selector cable at the transaxle, with selector lever in P (park).
   - On manual transaxle shift linkage, 2 rods with the plastic socket ends and remaining linkage from the case as required. Disconnect the clutch cable, lift it out of the case and set it aside.
   - Wiring from the starter and the back-up light switch and the ground cable from the transaxle. Remove the speedometer cable from the transaxle and plug the hole in the case.

3. Attach an engine sling tool VW-2024A , to the engine and attach the sling to a suitable lifting device.

4. Remove or disconnect:
   - Exhaust pipe from the manifold or turbocharger.
   - Halfshafts from the flanges and hang them from the body with wire.
   - Starter along with the front mount.

5. With all mounts unbolted, slightly lower the engine/transaxle assembly and tilt it towards the transaxle side. Then, carefully lift the assembly out of the vehicle.

To install:

6. Carefully install the engine/transaxle assembly and be sure all mounts are securely bolted to the engine/transaxle. Start all nuts and bolts that secure the mounts to the body but don't tighten them yet.

7. With all mounts installed and the engine safely in the vehicle, allow some slack in the lifting equipment. With the vehicle safely supported, shake the engine/transaxle as a unit to settle it in the mounts. Tighten all mounting bolts, starting at the rear and working forward. Tighten to 33 ft. lbs. (41 Nm) for 10mm bolts or 54 ft. lbs. (73 Nm) for 12mm bolts.

8. Install or connect:
   - Starter. Bolts: 33 ft. lbs. (45 Nm).
   - Halfshafts to the flanges. Bolts: 33 ft. lbs. (45 Nm).
   - Exhaust pipe and use new self-locking nuts to secure the flange. Nuts: 30 ft. lbs. (40 Nm). If equipped with spring clamps, the clamps can be used again.

- Shift linkage and the clutch cable, if equipped.
- Fuel injector lines and tighten to 18 ft. lbs. (25 Nm). Be careful not to over-tighten the line nuts. If a line is damaged or clogged, replace all lines as a set.
- Inlet and outlet lines to the injector pump, using new gaskets. Note the special outlet fitting has the word "OUT" printed on the top.
- Air conditioning compressor and/or power steering pump, if equipped. Install and adjust the drive belts.
- Wiring and vacuum hoses.
- Radiator, fan and heater hoses. Use a new O-ring on the thermostat and tighten the thermostat housing bolts to 84 inch lbs. (10 Nm).
- Coolant
- Air filter.
- Battery and battery cables.

GTI, Jetta (2.8L BDF Engine)

➡Tag all hoses and wiring during removal, to use as reference during reassembly.

1. Before servicing the vehicle, refer to the Precautions Section.

➡The engine and transaxle are lifted from the vehicle as an assembly.

2. Remove or disconnect:
   - Battery cables and battery.
   - Fuel pressure, by opening the fuel filler cap, loosening the fuel filter fitting. Be sure to take the appropriate fire safety precautions.
   - Front engine cover
   - Ignition coil connectors
   - Air cleaner and intake hose
   - Throttle valve vacuum hose
   - Fuel return and supply lines
   - Engine undercovers
   - Radiator grille
3. Lock the carrier into service position as follows:
   a. Remove the front bumper.
   b. Tag and remove any wiring or connector that would inhibit locking the carrier.
   c. Disconnect Bowden cable from lock.
   d. Remove the 2 No. 2 bolts.
   e. Remove the 4 No. 7 bolts.
   f. Remove the 4 No. 8 bolts.
   g. Install Guide Rods 3411.
   h. Remove the remaining bolts and pull the lock carrier out to the stop.
   i. To secure the lock carrier, install the appropriate M6 bolts into the rear of the lock carrier and fender.

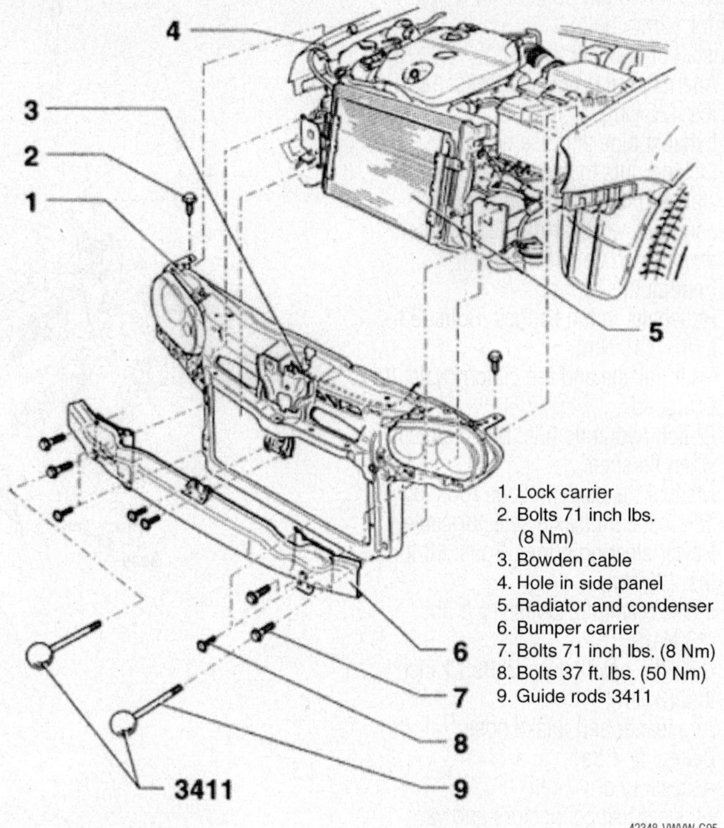

Moving the lock carrier into the service position—2.8L (BDF) engine GTI and Jetta

4. Remove or disconnect:
   - Cylinder head cover.
   - Intake manifold.
   - Accessory drive belt.
   - Power steering pump and secure it to the body. Do not disconnect the hydraulic lines.
   - Air conditioning compressor and secure it to the body.
   - On manual transaxle shift linkage, 2 rods with the plastic socket ends and remaining linkage from the case as required. Disconnect the clutch cable, lift it out of the case and set it aside.
   - Clutch hydraulic line and cap opening.
   - With an automatic transaxle, selector cable at the transaxle, with selector lever in P (park).
   - All remaining vacuum hoses and electrical connectors.
   - Drain coolant.
   - Coolant hoses from radiator and engine.
   - Halfshafts from the flanges and hang them from the body with wire.
   - Pendulum support under vehicle.
   - Exhaust pipe from the manifold.
   - Alternator and bracket.
   - Air injection pump bracket from block and oil pan.
5. Install engine bracket 3395 to cylinder block and tighten securing nuts.
6. Install engine transmission jack to bracket tool 3395.
7. With all mounts unbolted, slightly lower the engine/transaxle assembly and tilt it towards the transaxle side. Then, carefully lift the assembly out of the vehicle.

To install:

➡Be sure that the centering sleeves for the engine-to-transaxle are correctly installed in the cylinder block.

8. Carefully install the engine/transaxle assembly and be sure all mounts are securely bolted to the engine/transaxle. Start all nuts and bolts that secure the mounts to the body but don't tighten them yet.
9. With all mounts installed and the engine safely in the vehicle, allow some slack in the lifting equipment. With the vehicle safely supported, shake the engine/transaxle as a unit to settle it in the mounts. Tighten all mounting bolts, starting at the rear and working forward. Tighten to 11 ft. lbs. (15 Nm) for 7mm bolts, tighten to 18 ft. lbs. (25 Nm) for 8mm bolts, tighten to 33 ft.

lbs. (41 Nm) for 10mm bolts or 44 ft. lbs. (60 Nm) for 12mm bolts.

10. Install or connect:
- Alternator and bracket. Bolts: 18 ft. lbs. (25 Nm).
- Exhaust pipe and use new self-locking nuts to secure the flange. Nuts: 30 ft. lbs. (40 Nm). If equipped with spring clamps, the clamps can be used again.
- Pendulum support.
- Halfshafts to the flanges. Bolts: 33 ft. lbs. (45 Nm).
- Shift linkage and the clutch cable, if equipped.
- Clutch hydraulic line. Bleed clutch when finished.
- Manual transaxle linkage rods, or automatic transaxle selector cable.
- Power steering pump. Bolts: 18 ft. lbs. (25 Nm).
- Intake manifold. Bolts: 10 ft. lbs. (13 Nm).
- Cylinder head cover. Bolts 71 inch lbs. (8 Nm).
- Air cleaner and intake hose.
- Battery and tray.
- Accessory drive belt.
- All electrical connectors and vacuum lines.

11. The completion of the installation procedure is the reverse of the removal.

12. If equipped with an automatic transaxle, check the ATF level

13. Fill the engine with coolant

14. Fully close all power windows, operate all window switches for at least 1 second in the close direction to activate the one-touch opening/closing function

15. Connect the negative battery cable.

➡**DTCs are stored when harness connectors are detached.**

16. Read the DTCs and clear the fault codes.

Passat

1. Before servicing the vehicle, refer to the Precautions Section.

➡**Tag all hoses and wiring during removal, to use as reference during reassembly.**

2. Lock the carrier into service position as follows:

a. Remove the front bumper.

b. Tag and remove any wiring or connector that would inhibit locking the carrier.

c. Remove the 3 quick-release screws on the front noise insulation panel.

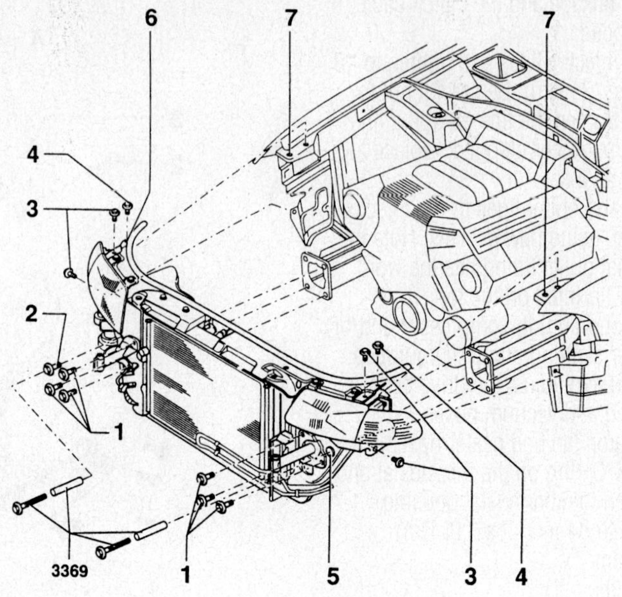

1. Bolts 33 ft. lbs. (45 Nm)
2. Bolts 33 ft. lbs. (45 Nm)
3. Bolts 7 ft. lbs. (10 Nm)
4. Bolts 7 ft. lbs. (10 Nm)
5. Bore for support tool
6. Lock carrier bore
7. Fender bore

7923CG12

Moving the lock carrier into the service position—Passat

d. Unbolt the air guide between the lock carrier and the air filter.

e. If installed, remove the retaining clamps for the wiring harness at the left side of the radiator frame.

f. Remove the No. 2 bolts and install Support tool 3369.

g. Remove the remaining bolts and pull the lock carrier out to the stop.

h. To secure the lock carrier, install the appropriate M6 bolts into the rear of the lock carrier and fender.

3. Position the wipers to the vertical position.

4. Properly relieve the fuel system pressure.

5. Remove or disconnect:
- Negative battery cable.
- Engine under cover.
- Coolant.
- Front bumper.
- Power steering cooling coil from the radiator, leaving it connected and hanging.
- If equipped, transaxle oil cooling lines.
- Electric cooling fan thermal switch at the lower left of the radiator.
- Air intake duct and assembly.
- Headlight height adjuster wiring harness.
- Turn signal bulbs from the light housing.
- Coolant hose from the radiator at the upper coolant pipe.
- Hood release cable at the carrier lock.
- Power steering fluid reservoir cap/dipstick.
- Wiring harness for the ABS hydraulic unit.
- Horn electrical connectors.
- Air guides at the left and right sides of the radiator.
- Air conditioning condenser retaining fasteners.
- Air conditioning low pressure switch.
- Condenser, and position it over the fender.
- Green harness connector from the air conditioning compressor magnetic clutch.
- Engine covers.
- Wiring harness connectors for the wastegate bypass regulator valve, the EVAP canister purge regulator valve, the power output stage, and the MAF sensor.

- ECL warning switch.
- Coolant hoses at the expansion tank, then remove the tank and position it aside.
- With cruise control, actuating rod from the throttle valve control module, then remove the vacuum hose from the vacuum unit.
- Accelerator pedal cable from the throttle valve control module.
- Hose for the Leak Detection Pump (LDP).
- Fuel supply and return lines.
- Brake booster vacuum hose.
- Vacuum hose for the EVAP canister purge regulator valve.
- ECM retaining bracket.
- Wiring harness to the ECM.
- With an automatic transaxle, the kickdown switch connector.
- Heated oxygen sensor wiring harness.
- Ground connection at the plenum chamber.
- Heater hoses from the heater core.
- VSS from the transaxle and position it aside.
- With a manual transaxle, back-up light switch from the transaxle.
- Engine driven cooling fan.
- Accessory drive belts.
- Air conditioning compressor from the mounting bracket and position it aside using wire.
- Power steering pump and position it aside leaving the hoses attached.

⁂ WARNING

The flexpipe at the front exhaust pipe must not be bent more than 10°, otherwise it may be damaged.

- Catalytic converter from the turbocharger.
- Starter, and the ground strap at the right engine mount.
- With an automatic transaxle, 3 torque converter-to-driveplate mounting bolts through the opening left by the starter.

6. Loosen the upper nuts for the left and right engine mounts, matchmark the threaded bolt and centering sleeves at the bottom of the left and right engine mounts, then remove the mounting nuts.
7. Remove or disconnect:
   - Lower engine-to-transaxle mounting bolts.
   - With an automatic transaxle, ATF cooler line bracket form the left side of the engine.
   - Upper nuts from the engine mounts.
8. Position an Engine Support Bridge 10-222A to the bolted flanges of the fenders with the spindle facing forward.
9. Attach the Engine Support Adapter 3147 or equivalent to a bolt hole in the transaxle bell housing.
10. Connect the Engine Support Adapter 3147 and the Engine Support Bridge 10-222A using Adapter 2024A/1 and Extension 2024A/2.
11. Attach an engine sling between the engine and the hoist.
12. Remove the upper engine-to-transaxle mounting bolts.
13. Separate the engine from the transaxle, then slowly lift the engine up and out the front of the engine compartment.
14. If equipped with an automatic transaxle, secure the torque converter to prevent it from falling out.

To install:

➡Be sure that the centering sleeves for the engine-to-transaxle are correctly installed in the cylinder block.

15. Verify that the intermediate plate is over the centering sleeves.
16. Install or connect the following:
    - Engine into the vehicle
    - Upper engine-to-transaxle mounting bolts
17. Lower the engine into position, then remove the engine sling and hoist and the transaxle support apparatus from the vehicle.
18. Install or connect:
    - Engine mounting fasteners without any tension or pre-load.
    - With an automatic transaxle, ATF cooler line bracket to the left side of the engine.
    - Lower engine-to-transaxle mounting bolts. M12 bolts: 48 ft. lbs. (65 Nm), M10 bolts: 33 ft. lbs. (45 Nm)
    - Engine mounting nuts/bolts: 18 ft. lbs. (25 Nm).
    - With an automatic transaxle, driveplate-to-torque converter mounting bolts through the starter opening. Bolts: 63 ft. lbs. (85 Nm).
    - Starter and attach the ground strap to the right engine mount.
    - Catalytic converter to the turbocharger. Mounting bolts: 22 ft. lbs. (30 Nm).
    - Power steering pump, the air conditioning compressor, and the engine cooling fan, then the accessory drive belts.
    - With a manual transaxle, back-up light switch to the transaxle.
    - VSS to the transaxle.
    - Heater hoses to the heater core.
    - Ground connection at the plenum chamber.
    - Heated oxygen sensor wiring harness.
    - With an automatic transaxle, kick-down switch connector.
    - Wiring harness to the ECM.
    - ECM retaining bracket and cover the E-box.
    - Vacuum hose for the EVAP canister purge regulator valve.
    - Fuel supply and return lines.
    - Brake booster vacuum hose.
19. The completion of the installation procedure is the reverse of the removal.
20. If equipped with an automatic transaxle, check the ATF level
21. Fill the engine with coolant
22. Fully close all power windows, operate all window switches for at least 1 second in the close direction to activate the one-touch opening/closing function
23. Connect the negative battery cable.

➡DTCs are stored when harness connectors are detached.

24. Read the DTCs and clear the fault codes.

New Beetle and New Beetle Convertible

The engine and transaxle are removed from under the vehicle as an assembly.

1. Before servicing the vehicle, refer to the Precautions Section.
2. Remove or disconnect:
   - Negative battery cable.
   - Engine cover.
   - Power steering reservoir from the battery support leaving the hoses attached, and positioning it aside.
   - Battery and bracket.
   - Fuel supply and return lines.
   - Air cleaner and intake air duct.
   - Throttle cable.
   - With a manual transaxle, transaxle range selector mechanism and the clutch slave cylinder.
   - With an automatic transaxle, range selector lever cable at the transaxle.
   - Engine undercover.
   - Coolant.
   - Accessory drive belt.
   - If necessary, right side cooling fan.
   - Power steering pump and bracket and position it aside leaving the hoses connected.

- Air conditioner compressor and position it aside leaving the hoses connected.
- Coolant, vacuum and intake hoses.
- Secondary air injection pump and bracket.
- All wires from the transaxle, starter and generator, position the wires out of the way.
- Any remaining wires or connectors that would interfere with the engine removal.
- Engine pendulum support.
- Right-hand halfshaft and disconnect the left-hand halfshaft at the transaxle.
- Front exhaust pipe from the manifold.

3. Install Engine Bracket T10012, or equivalent, in Engine/transaxle Jack VAG 1383 A or equivalent.

4. Remove the bracket for the coolant hose under the engine block.

5. Attach the Engine Bracket T10012 to the engine block using the threaded holes at the corners of the engine block.

6. Raise the engine/transaxle jack to relieve the tension on the mounts.

7. Disconnect the engine and transaxle mounts from inside the engine compartment.

8. Lower the engine/transaxle assembly from the vehicle, be sure to guide the power steering pressure hose past the transaxle.

To install:

9. Raise the engine/transaxle assembly into the vehicle, be sure to guide the power steering hose around the transaxle.

10. Using new bolts, connect the engine/transaxle mounts following the accopanying illustration.

11. Remove the engine/transaxle jack and Engine Bracket T10012.

12. Install or connect:
- Bracket for the coolant hose under the engine block.
- Front exhaust pipe to the manifold.
- Left halfshaft and the right halfshaft.
- Engine pendulum support. Bolts (1) and (2): 30 ft. lbs. (40 Nm) plus 90 degrees, and bolts (3): 15 ft. lbs. (20 Nm) plus 90 degrees.
- Any wiring that was removed.
- Secondary air injection pump and bracket.
- Any hoses that were removed.
- Air conditioner compressor.
- Power steering bracket and pump.
- If removed, right side cooling fan.
- Accessory drive belt and fill the cooling system.
- With an automatic transaxle, the transaxle range selector cable.
- With a manual transaxle, clutch slave cylinder and connect the range selector lever.
- Throttle cable.
- Intake air duct and air cleaner assembly.
- Fuel supply and return lines.
- Battery bracket.
- Power steering reservoir.
- Engine oil.
- Battery cables.

13. Check and clear any DTCs, then match the ECM to the TCM.

14. Install the engine covers.

Engine / transmission mounts

Tightening torques

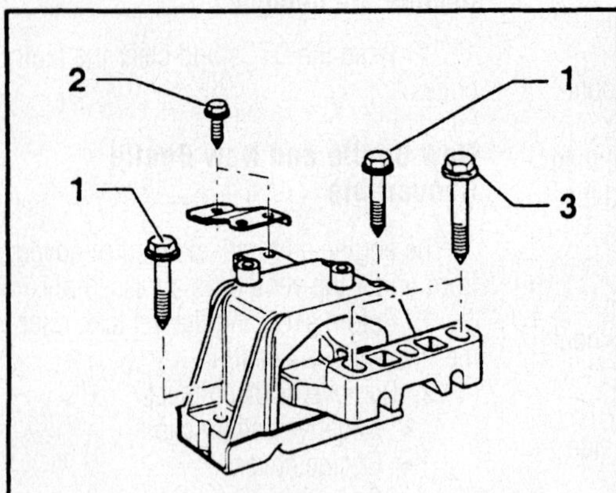

◄ Front assembly mounting

1 -	Mount to body [1]	40 Nm (30 ft lb) + 90° (1/4 turn)
2 -	Mount/bracket to body	25 Nm (18 ft lb)
3 -	Mount to engine bracket [1]	60 Nm (44 ft lb) + 90° (1/4 turn)

[1] Replace bolts

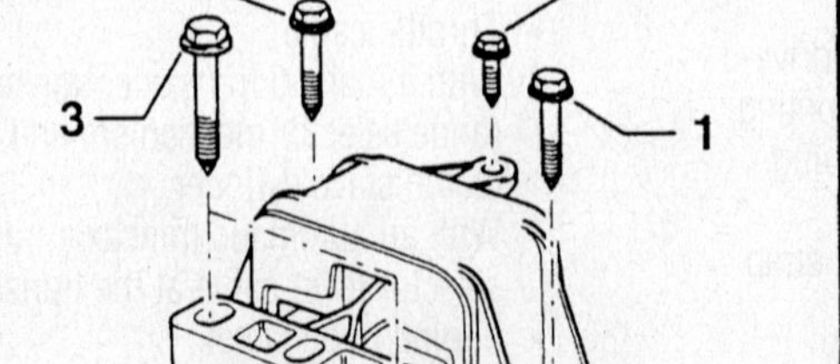

◄ Rear right assembly mounting

1 -	Mount to body [1]	40 Nm (30 ft lb) + 90° (1/4 turn)
2 -	Mount to body	25 Nm (18 ft lb)
3 -	Bearing on transmission console [1]	60 Nm (44 ft lb) + 90° (1/4 turn)

[1] Replace bolts

9301WG02

Engine mount fastener locations—New Beetle

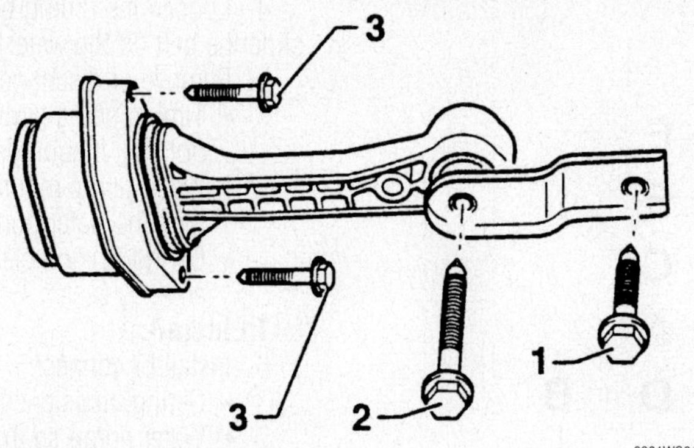

Pendulum support mounting bolt locations—New Beetle

Water Pump

REMOVAL & INSTALLATION

1.8L Engine

➡**The coolant pump is bolted to the brackets for the generator, power steering pump, and cooling fan.**

1. Before servicing the vehicle, refer to the Precautions Section.
2. Position the lock carrier into the service position as follows:
   a. Remove or disconnect:
   - Front bumper.
   - Any wiring or connector that would inhibit locking the carrier.
   - 3 quick-release screws on the front noise insulation panel.
   - Air guide between the lock carrier and the air filter.
   - Retaining clamps for the wiring harness at the left side of the radiator frame.
   - No. 2 bolts and install Support tool 3369 or equivalent.
   - Remaining bolts and pull the lock carrier out to the stop.

   b. To secure the lock carrier, install the appropriate M6 bolts into the rear of the lock carrier and fender.
3. Remove or disconnect
   - Negative battery cable.
   - Accessory drive belt, then the engine driven cooling fan.
   - Coolant.
   - Clamps for the coolant hoses at the water pump.
   - Intake air duct between the intake manifold and the charge air cooler.
   - Generator mounting bolts and slide it forward.
   - Wiring from the generator once it is removed.
4. Unbolt the following supports and brackets for the generator, power steering pump, and engine cooling fan:
   a. Intake manifold support
   b. Support for the engine torque bracket
   c. Brace to the cylinder block (remove completely)
5. Remove or disconnect:
   - Brackets for the generator, power steering pump, and engine cooling fan, positioning the brackets to the left side using a piece of wire
   - Coolant hoses from the pump and thermostat housing
   - Coolant pump housing from the timing belt cover
   - Coolant pump mounting bolts, then the pump
   - Impeller housing from the pump housing

To install:

6. Using a new gasket, mount the new coolant pump to the pump housing. Mounting bolts: 84 inch lbs. (10 Nm).
7. Using a new gasket and O-ring, install the coolant pump. Mounting bolts in alphabetical sequence: 18 ft. lbs. (25 Nm).
8. Tighten the coolant pump housing to the timing belt cover to 84 inch lbs. (10 Nm).
9. Install or connect:
   - Coolant hoses to the pump and thermostat housing.
   - Brackets that were removed. Mounting bolts: 18 ft. lbs. (25 Nm).
   - Wires to the generator, then install the generator.
   - Air intake duct between the intake manifold and the charge air cooler.
10. The remaining steps are the reverse of the removal.
11. Fill the engine with coolant
12. Fully close all power windows to stop, operate all window switches for at least 1 second in the close direction to activate the one-touch opening/closing function
13. After installing the lock carrier, check the wiring for proper routing near the cooling fan

2.0L (ABA) Engine

1. Before servicing the vehicle, refer to the Precautions Section.
2. Remove or disconnect:
   - Coolant
   - Thermostat housing from under the water pump housing.
3. Loosen but don't remove the bolts holding the pulley to the water pump.
4. Remove or disconnect:
   - Timing belt cover.

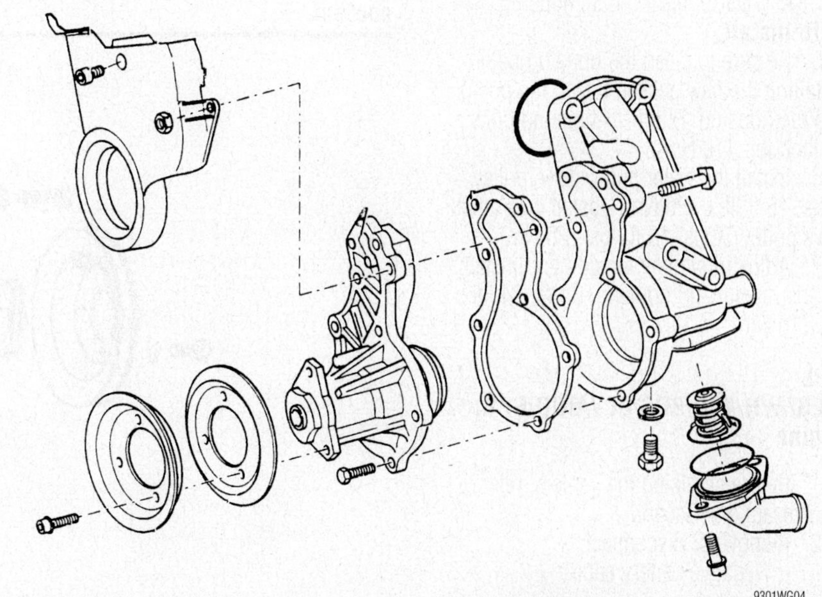

Exploded view of the water pump, housing and related components—1.8L engine

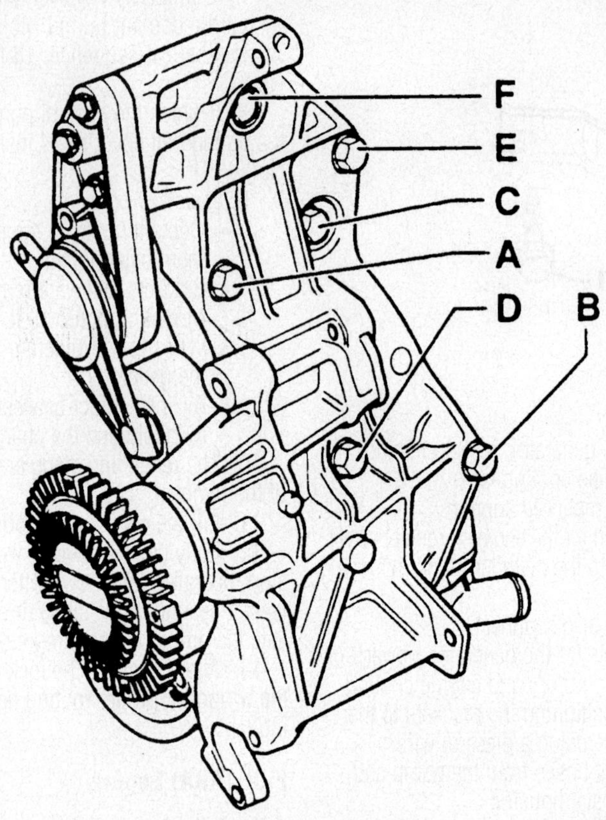

Assembly bracket with water pump mounting bolt tightening sequence—1.8L engine

- Alternator and/or steering pump as required to remove the water pump drive belt.
- Water pump pulley. On some vehicles, the crankshaft pulley must also be removed by removing the bolts holding the pulley to the timing belt sprocket.
- Water pump from its housing.

To install:

5. Be sure to clean the housing before installing the new gasket. Install the pump into the housing. Pump-to-housing bolts: 84 inch lbs. (10 Nm).
6. Install the water pump drive pulley. Bolts: 15 ft. lbs. (20 Nm). Install crankshaft drive pulley. Bolts: 15 ft. lbs. (20 Nm).
7. Adjust drive belt tension and install the thermostat and housing. Bolts: 84 inch lbs. (10 Nm).

2.0L (AEG/AVH/AZG/BDC/BEV/BGD/BPY) Engine

1. Before servicing the vehicle, refer to the Precautions Section.
2. Remove or disconnect:
   - Negative battery cable.
   - Coolant.
   - Accessory drive belt and tensioner.
   - Upper and center timing belt covers.
3. Position the engine so that the No. 1 cylinder is at TDC.

⁂ WARNING

Cover the timing belt to protect it from being contaminated with coolant.

4. Loosen the timing belt tension and slide the belt off the water pump sprocket.
5. Remove or disconnect:
   - Timing belt guard (2) mounting bolt (1), if equipped.
   - Water pump mounting bolts (5), then the water pump (4).
   - O-ring (3) and clean the seating area.

To install:

6. Install or connect:
   - O-ring, moistened with coolant.
   - Water pump so that the plug in the housing faces downward. Mounting bolts: 11 ft. lbs. (15 Nm).
   - Timing belt guard, if equipped. Mounting bolt: 15 ft. lbs. (20 Nm).
   - Timing belt.
   - Upper and center timing belt covers.
   - Accessory drive belt tensioner. Mounting bolt: 18 ft. lbs. (25 Nm).
   - Accessory drive belt.
   - Coolant.
   - Negative battery cable.
7. Check and clear any DTCs, then match the ECM to the TCM.

2.8L (AAA) Engine

1. Before servicing the vehicle, refer to the Precautions Section.
2. Remove or disconnect:
   - Negative battery cable.
   - Front exhaust pipe from the catalytic converter.
   - Coolant.
   - Accessory drive belt.
   - Air intake duct.
   - Ignition wires from the coils and unclip them from the retainers.

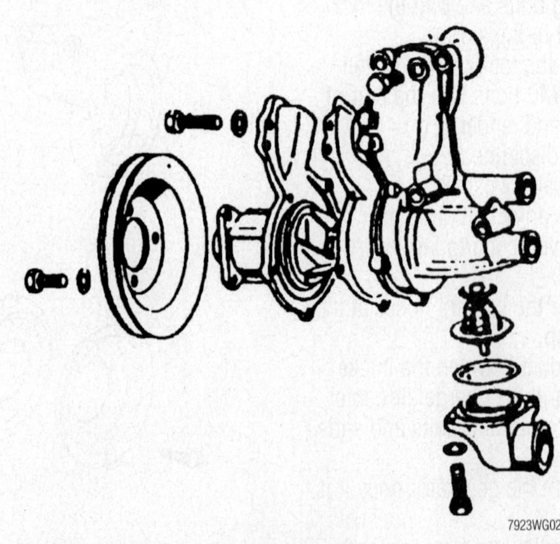

Water pump and thermostat housing—2.0L (ABA) engine

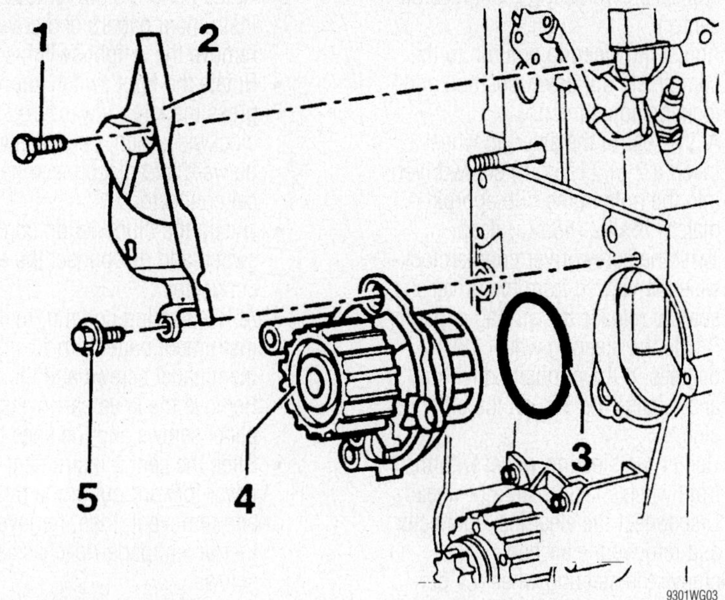

Exploded view of the water pump mounting—2.0L (AEG/AVH/AZG/BDF) engine

- Ignition wire guide above coil assembly.
- Vacuum hose from the fuel pressure regulator.
- IAT sensor from the upper intake manifold.

3. Without disconnecting the hoses, remove and place the coolant expansion tank to the side.
4. Install an engine support fixture to the lifting eyes on the left and right sides of the cylinder head. Lift the engine slightly to remove the weight from the mounts.
5. Remove or disconnect:
   - Right and left rear engine/transaxle mount center bolts.
   - Front engine mounting center bolts.
6. Carefully raise the engine to gain access to the water pump pulley mounting bolts.
7. Remove the water pump pulley using wrench VAG 1590 or equivalent. Modify the wrench as shown if necessary to fit the bolt.
8. Remove the mounting bolts and the water pump.

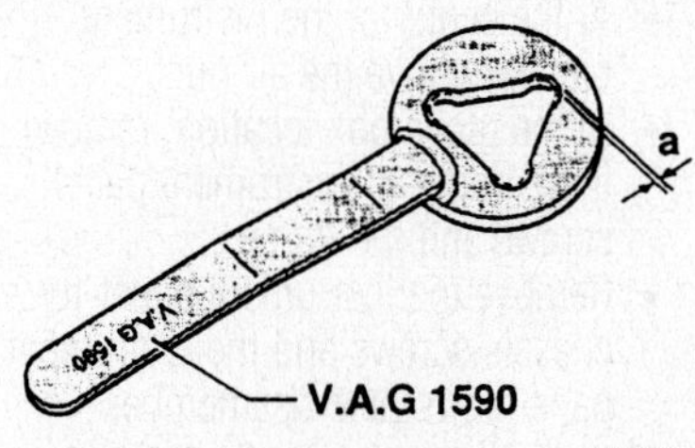

Modify wrench VAG 1590 or equivalent to fit the water pump pulley bolt as needed

To install:

9. Install or connect:
   - Water pump, using a new O-ring. Mounting bolts: 15 ft. lbs. (20 Nm).
   - Water pump pulley. Bolt: 18 ft. lbs. (25 Nm).
   - Engine/transaxle mount bolts. Mounting bolts: 44 ft. lbs. (60 Nm). Tighten the front and right rear mounts first, then the left rear mount.
   - Expansion tank. Bolts: 84 inch lbs. (10 Nm).
   - IAT sensor in the upper intake manifold.
   - Vacuum hose to the fuel pressure regulator.
   - Ignition wires and the wire guide.
   - Air duct and accessory drive belt.
   - Coolant.

2.8L (AHA/ATQ) Engine

1. Before servicing the vehicle, refer to the Precautions Section.
2. Disconnect the negative battery cable.
3. Lock the carrier into service position as follows:
   a. Remove or disconnect:
   - Front bumper.
   - Any wiring or connector that would inhibit locking the carrier.
   - 3 quick-release screws on the front noise insulation panel.
   - Air guide between the lock carrier and the air filter.
   - Retaining clamps for the wiring harness at the left side of the radiator frame.
   - No. 2 bolts and install Support 3369 or equivalent.
   - Remaining bolts and pull the lock carrier out to the stop.

   b. To secure the lock carrier, install the appropriate M6 bolts into the rear of the lock carrier and fender.
4. Remove or disconnect:
   - Accessory drive belt.
   - Timing belt.
   - Coolant.
   - Timing belt tensioner and idler rollers.
   - Water pump mounting bolts, then the pump.

To install:

5. Install or connect:
   - New gasket on the water pump flange.
   - Water pump. Mounting bolts: 10 ft. lbs. (15 Nm).
   - Timing belt idler roller. Mounting bolt: 33 ft. lbs. (45 Nm).
   - Timing belt tensioner roller. Mounting bolt: 15 ft. lbs. (20 Nm).
   - Timing belt and accessory drive belt.
   - Coolant.
6. Return the lock carrier to the normal position.
7. Connect the negative battery cable.

2.8L (BDF) Engine

1. Before servicing the vehicle, refer to the Precautions Section.
2. Disconnect the negative battery cable.
3. Remove the accessory drive belt.
4. Drain the coolant.
5. Using a socket through the holes in the pump pulley, remove the bolts and remove water pump.

To install:

6. Install or connect:
   - New O ring on the water pump.
   - Water pump. Mounting bolts: 15 ft. lbs. (20 Nm).
   - Accessory drive belt.
   - Coolant.

3.6L (BLV) Engine

1. Before servicing the vehicle, refer to the Precautions Section.
2. Disconnect the negative battery cable.
3. Remove sound insulation.
4. Drain the coolant.
5. Remove the accessory drive belt.
6. Assemble engine support bridge 10—222 A with frame 10—222 A /1 and

support engine in installation position at right lifting eyelet.

7. Remove bolts for subframe mount on engine side from engine mount.

8. Lower engine on engine support bridge 10—222 A only far enough that water pump can be removed. Make sure when doing this that pendulum support is not overloaded.

9. Remove belt pulley of water pump. To do so, counterhold belt pulley using a drift (standard).

10. Remove water pump mounting bolts and remove water pump.

To install:

11. Installation is performed in reverse order of removal. Note the following:

a. Moisten new O-ring with coolant.

b. Insert water pump.

c. Tighten mounting bolts for water pump to 6 ft. lbs. (8 Nm).

d. Install belt pulley for water pump. Tightening bolts to 14 ft. lbs. (20 Nm).

e. Install drive belt.

12. Refill the cooling system.

13. Connect the negative battery cable.

4.0L (BDP) Engine

1. Before servicing the vehicle, refer to the Precautions Section.
2. Disconnect the negative battery cable.
3. Remove the accessory drive belt.
4. Drain the coolant.
5. Remove the water pump bolts and the water pump.

To install:

6. Install or connect:
   - New O ring on the water pump.
   - Water pump. Mounting bolts: 15 ft. lbs. (20 Nm).
   - Accessory drive belt.
   - Coolant.

Heater Core

REMOVAL & INSTALLATION

New Beetle

➡Be sure to obtain the anti-theft code for the radio.

1. Disconnect the negative battery cable.
2. Drain the cooling system into a clean container for reuse.
3. Remove the driver's side air bag by performing the following procedure:
   - Release the steering column adjustment.
   - Rotate the steering wheel until the spokes are vertical (90 degrees off center).
   - Adjust the steering column to the completely out down position and lock the adjustment.
   - At the rear of the steering wheel, insert a 7 in. (175mm) screwdriver into the hole in the hub approximately 1¾ in. (45mm) deep.
   - Twist the screwdriver counterclockwise (as viewed from the driver's seat) to release the air bag clip.
   - Rotate the steering wheel 180 degrees in the opposite direction and release the clip on the other side.
   - Center the steering wheel with the front wheels facing straight-ahead.
   - Disconnect the electrical connector and remove the air bag.
4. Remove the steering wheel by performing the following procedure:
   - Place the front wheels in the straight-ahead position.
   - Disconnect the horn's electrical connector.
   - Remove the steering wheel-to-steering column bolt and discard it.
   - Remove the steering wheel from the steering column.
5. Remove the instrument panel by performing the following procedure:
   - Remove the upper steering column cover screws and the cover.
   - Remove the lower steering column cover screws, release the steering wheel height adjustment and remove the lower steering column switch trim.
   - Remove the combination switch-to-steering column bolt, disconnect the electrical connectors and remove the combination switch.
   - At the driver's side, remove the lower instrument panel cover screws and the cover.
   - At the passenger's side of instrument panel, remove the side cover.
   - At the center of the instrument panel, remove both switch assembly covers, the 4 switch assembly-to-panel screws and the switch assembly.
   - Remove the 4 heater/ventilator control unit-to-instrument panel screws and move the control unit under the instrument panel with the cables attached.
   - Remove the 6 glove box-to-instrument panel screws, disconnect the glove box light connector and remove the glove box.
   - At the driver's side, remove the instrument panel's end cover and remove the 2 light switch screws.
   - Rotate the light switch knob to **0**, press the knob inward, rotate it clockwise; then, pull the switch outward and disconnect the electrical connector.
   - Unclip the illumination control switch and disconnect the electrical connector.
   - At the steering column, remove the instrument panel trim-to-instrument panel screws and the trim.
   - Remove the knee bar-to-instrument panel screws and the knee bar.
   - Slide the center instrument panel cover forward out of the retainer and remove it; then, remove the 3 instrument panel-to-chassis screws.
   - At the upper center of the instrument panel, lift both (left and right) plenum panel covers from the front and rear edge clips; then, move the covers toward the center and out of the A-pillar retainers.
   - Unclip and remove the instrument cluster trim.
   - Remove the 2 instrument cluster-to-instrument panel screws, disconnect the multi-pin electrical connector and remove the instrument cluster.
   - Slide the Radio Release tool No. 3344 into the slots at both sides of the radio until they engage; then, pull the radio out of the instrument panel and disconnect the electrical connectors.

➡To remove the Radio Release tool(s) No. 3344, press the locating tabs (located on both sides of the radio) inward and remove them.

   - If not equipped with a radio, remove the center instrument panel's trim cover.
   - Remove the 2 center instrument panel-to-instrument panel screws and the center panel.
   - At the center of the instrument panel, remove the air duct.
   - At the glove box location, remove the 2 frame-to-instrument panel screws and the frame.
   - Remove the instrument panel-to-chassis screws and the instrument panel from the crossmember.
6. Loosen the instrument panel crossmember-to-chassis bolts and lift the crossmember upward.

Releasing the driver's side air bag—New Beetle

7. Disconnect the heater hoses from the heater core.

8. If not equipped with air conditioning, perform the following procedure:

- Partially remove the heater/ventilation housing assembly.
- Depress the retainer clips and remove the heater core from the heater housing.

9. If equipped with air conditioning, perform the following procedure:

- Discharge and recover the air conditioning system refrigerant.
- Disconnect the refrigerant lines from the evaporator core, discard the O-rings and plug the openings to prevent contamination.
- Remove the ventilation ducts from the heater/air conditioning housing assembly.
- Remove the heater core from the heater/air conditioning housing assembly.

To install:

10. If equipped with air conditioning, perform the following procedure:

- Install the heater core to the heater/air conditioning housing assembly.

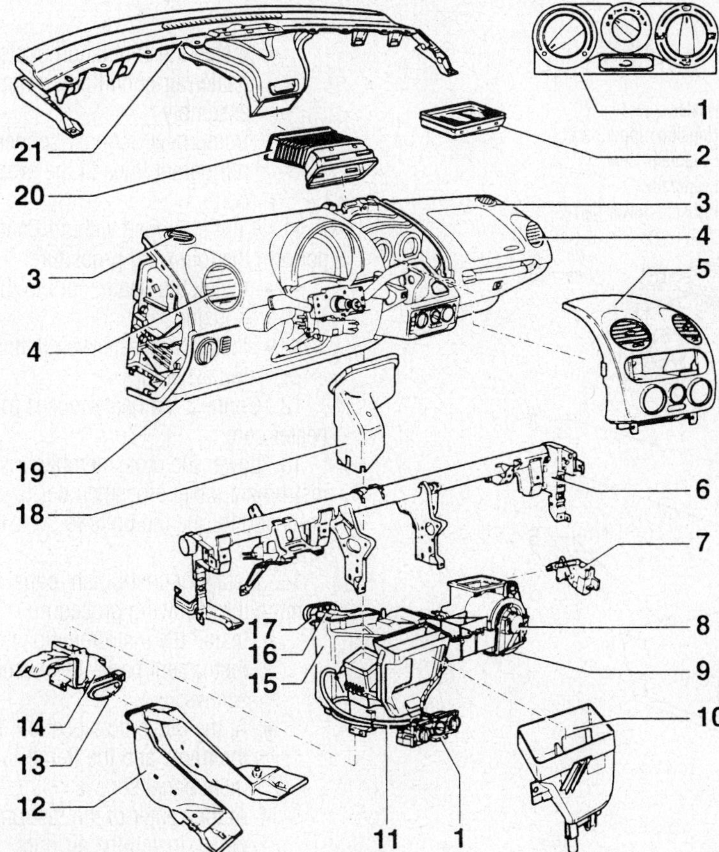

1. Heating and ventilation controls
2. Dust and pollen filter
3. Side window air outlet
4. Side air outlets
5. Center air outlet console
6. Instrument panel cross member
7. Servo motor for fresh/recirculating air door
8. Fresh air blower
9. Fresh air blower series resistance with fuse
10. Intermediate duct
11. Cables
12. Rear footwell air outlet duct
13. Gasket
14. Connecting duct
15. Heating and ventilation unit
16. Heater core
17. Heater core/bulkhead seal
18. Defroster duct
19. Instrument panel
20. Center air outlet duct
21. Defroster air outlet panel

Exploded view of the instrument panel, crossmember and related components—New Beetle

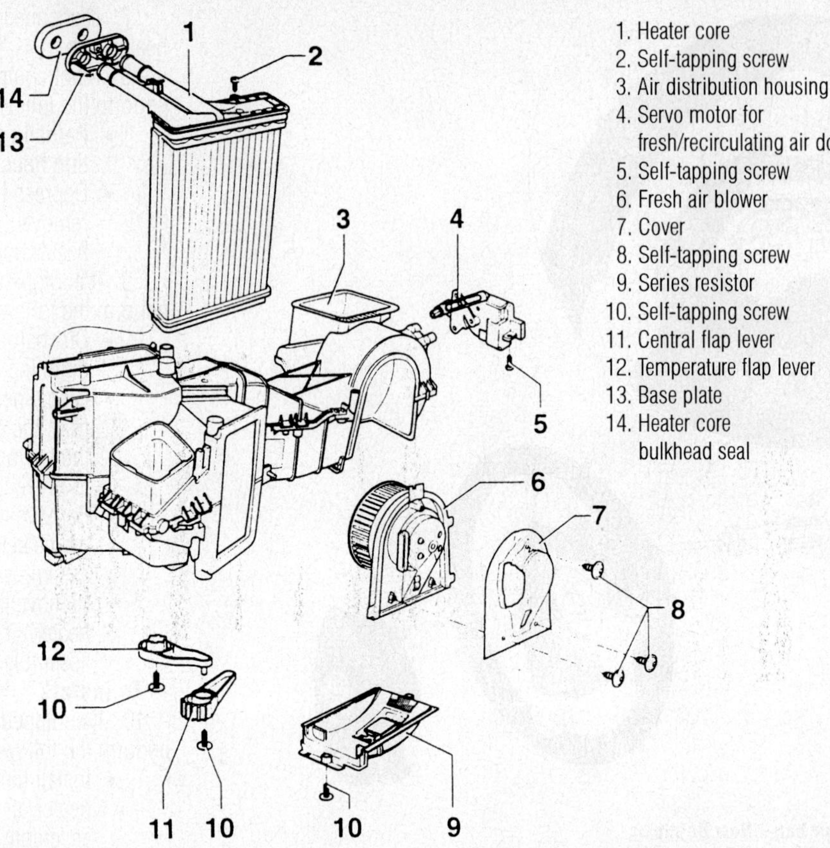

Exploded view of the heater core, heater/ventilation housing and related components—New Beetle without air conditioning

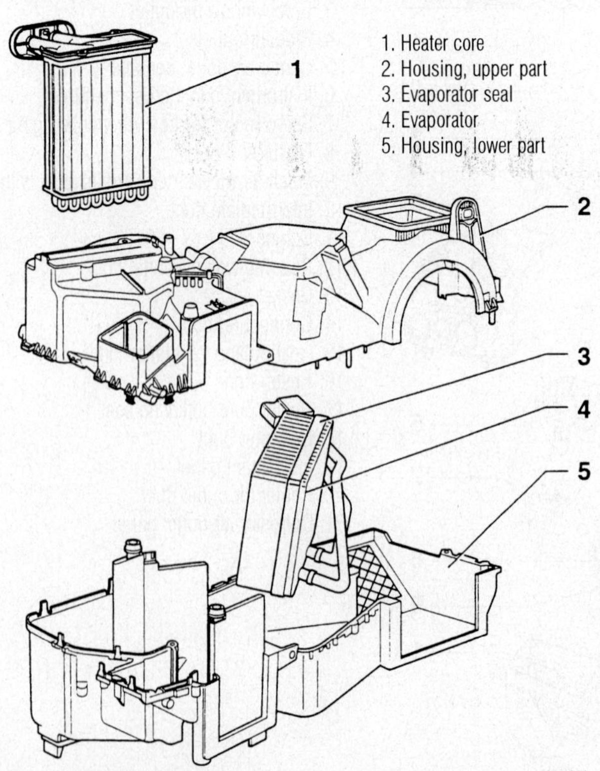

Exploded view of the heater core, heater/air conditioning housing and related components–New Beetle with air conditioning

- Install the ventilation ducts to the heater/air conditioning housing assembly.
- Using new O-rings, connect the refrigerant lines to the evaporator core.

11. If not equipped with air conditioning, perform the following procedure:
    - Install the heater core to the heater housing.
    - Install the heater/ventilation housing assembly.
12. Connect the heater hoses to the heater core.
13. Lower the crossmember, install the instrument panel crossmember-to-chassis bolts and torque the bolts to 18 ft. lbs. (25 Nm).
14. Install the instrument panel by performing the following procedure:
    - Install the instrument panel and the instrument panel-to-chassis screws.
    - At the glove box location, install the frame and the frame-to-instrument panel screws.
    - At the center of the instrument panel, install the air duct.
    - Install the center instrument panel

and the 2 center panel-to-instrument panel screws.

- If not equipped with a radio, install the center instrument panel's trim cover.
- Connect the electrical connectors and slide the radio into the instrument panel.
- Install the instrument cluster, connect the multi-pin electrical connector and install the 2 instrument cluster-to-instrument panel screws.
- Install the instrument cluster trim.
- Install both (left and right) plenum panel covers.
- Install the 3 instrument panel-to-chassis screws and the center instrument panel cover.
- Install the knee bar and the knee bar-to-instrument panel screws.
- At the steering column, install the instrument panel trim and the trim-to-instrument panel screws.
- Connect the electrical connector and install the illumination control switch.
- Connect the light switch electrical connector and install the switch.
- At the driver's side, install the 2 light switch screws and the instrument panel's end cover.
- Install the glove box, connect the glove box light connector and install the 6 glove box-to-instrument panel screws.
- Install the heater/ventilator control unit and the 4 heater/ventilator control unit-to-instrument panel screws.
- At the center of the instrument panel, install the switch assembly, the 4 switch assembly-to-panel screws and both switch assembly covers.
- At the passenger's side of instrument panel, install the side cover.
- At the driver's side, install the lower instrument panel cover and the cover screws.
- Install the combination switch, connect the electrical connectors and install the combination switch-to-steering column bolt.
- Install the lower steering column switch trim and the lower steering column cover screws.
- Install the upper steering column cover and the cover screws.

15. Install the steering wheel by performing the following procedure:
    - Install the steering wheel to the steering column.
    - Install the new steering wheel-to-steering column bolt and torque it to 30 ft. lbs. (40 Nm).
    - Connect the horn's electrical connector.
16. Connect the electrical connector and install the air bag.
17. Refill the cooling system.
18. Connect the negative battery cable.
19. Evacuate, charge and leak test the air conditioning system refrigerant.
20. Operate the engine to normal operating temperatures; then, check the climate control operation and check for leaks.

Golf

➡**Be sure to obtain the anti-theft code for the radio.**

1. Disconnect the negative battery cable.
2. Drain the cooling system into a clean container for reuse.
3. Remove the driver's side air bag by performing the following procedure:
    - Release the steering column adjustment.
    - Rotate the steering wheel until the spokes are vertical (90 degrees off center).
    - Adjust the steering column to the completely out down position and lock the adjustment.
    - At the rear of the steering wheel, insert a 7 in. (175mm) screwdriver into the hole in the hub approximately 1¾ in. (45mm) deep.
    - Raise the screwdriver (as viewed from the driver's seat) to release the air bag clip.
    - Rotate the steering wheel 180 degrees in the opposite direction and release the clip on the other side.
    - Center the steering wheel with the front wheels facing straight-ahead.
    - Disconnect the electrical connector and remove the air bag.
4. Remove the steering wheel by performing the following procedure:
    - Place the front wheels in the straight-ahead position.
    - Remove the steering wheel-to-steering column bolt.

➡**The Multi-point socket head bolt can be used up to 5 times; be sure to place a punch mark on it each time it is reused.**

    - Remove the steering wheel from the steering column.
5. Remove the center console-to-chassis bolts and the console.
6. Remove the instrument panel by performing the following procedure:
    - Remove the upper steering column cover screws and the cover.
    - Remove the 3 lower steering column cover screws, release the steering wheel height adjustment and remove the lower steering column switch trim.
    - Remove the combination switch-to-steering column bolt, disconnect the electrical connectors and remove the combination switch.
    - At the passenger's side of instrument panel, remove the side cover.
    - Remove the 7 glove box-to-instrument panel screws, disconnect the glove box light connector and remove the glove box.
    - At the driver's side, remove the both lower instrument panel cover screws and the covers.
    - Remove the DLC-to-bracket screws.
    - At the driver's side, remove the 6 sound damper panel-to-instrument panel screws and the panel.
    - Slide the Radio Release tool No. 3316 into the slots at both sides of the radio until they engage; then, pull the radio out of the instrument panel and disconnect the electrical connectors.

➡**To remove the Radio Release tool(s) No. 3316, press the locating tabs (located on both sides of the radio) inward and remove them.**

    - At the center of the instrument pane (above the radio), unclip and remove the switches and disconnect the electrical connectors.
    - Unclip and remove the Climatronic control (heater control) trim.
    - Remove the 4 Climatronic control (heater control)-to-instrument panel screws and the control; then, disconnect the electrical connector.
    - Remove the 5 center reinforcement-to-chassis screws and the reinforcement.
    - At the driver's side, remove the 3 footwell cover screws and the cover.
    - Remove the 2 instrument cluster-to-instrument panel screws, disconnect the multi-pin electrical connector and remove the instrument cluster.
    - At the upper center of the instru-

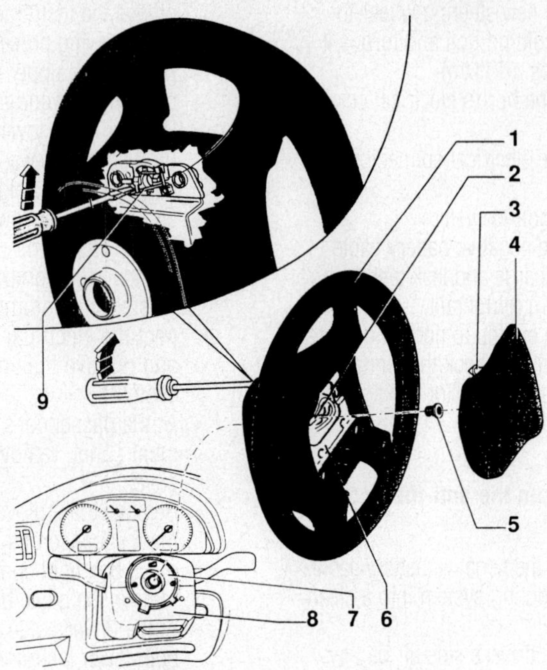

Releasing the driver's side air bag—Golf

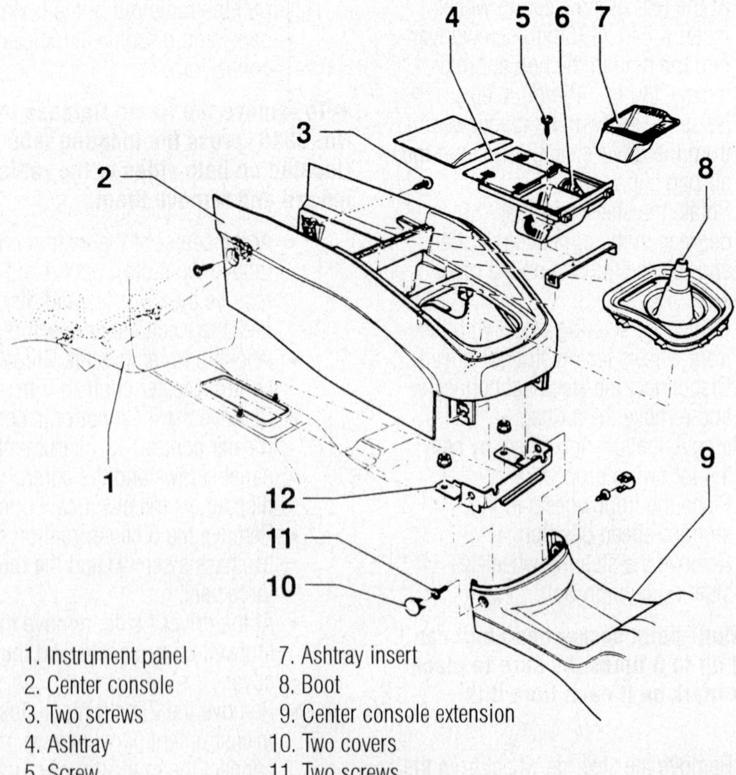

Exploded view of the center console and related components—Golf

ment panel, unclip the photo sensor and disconnect the electrical connector.

- Rotate the light switch knob to **0**, press the knob inward, rotate it clockwise; then, pull the switch outward and disconnect the electrical connector.
- Unclip the illumination control switch and disconnect the electrical connector.
- Remove the instrument panel-to-chassis screws and the instrument panel from the crossmember.

7. Loosen the instrument panel crossmember-to-chassis bolts and lift the crossmember upward.
8. Disconnect the heater hoses from the heater core.
9. If not equipped with air conditioning, perform the following procedure:
   - Partially remove the heater/ventilation housing assembly.
   - Depress the retainer clips and remove the heater core from the heater housing.

➡If the retainer clips break, use a self-tapping screw to hold the heater core in place.

10. If equipped with air conditioning, perform the following procedure:
    - Discharge and recover the air conditioning system refrigerant.
    - Disconnect the refrigerant lines from the evaporator core, discard the O-rings and plug the openings to prevent contamination.
    - Remove the ventilation ducts from the heater/air conditioning housing assembly.
    - Remove the heater core from the heater/air conditioning housing assembly.

To install:

11. If equipped with air conditioning, perform the following procedure:
    - Install the heater core to the heater/air conditioning housing assembly.
    - Install the ventilation ducts to the heater/air conditioning housing assembly.
    - Using new O-rings, connect the refrigerant lines to the evaporator core.
12. If not equipped with air conditioning, perform the following procedure:
    - Install the heater core to the heater housing.
    - Install the heater/ventilation housing assembly.

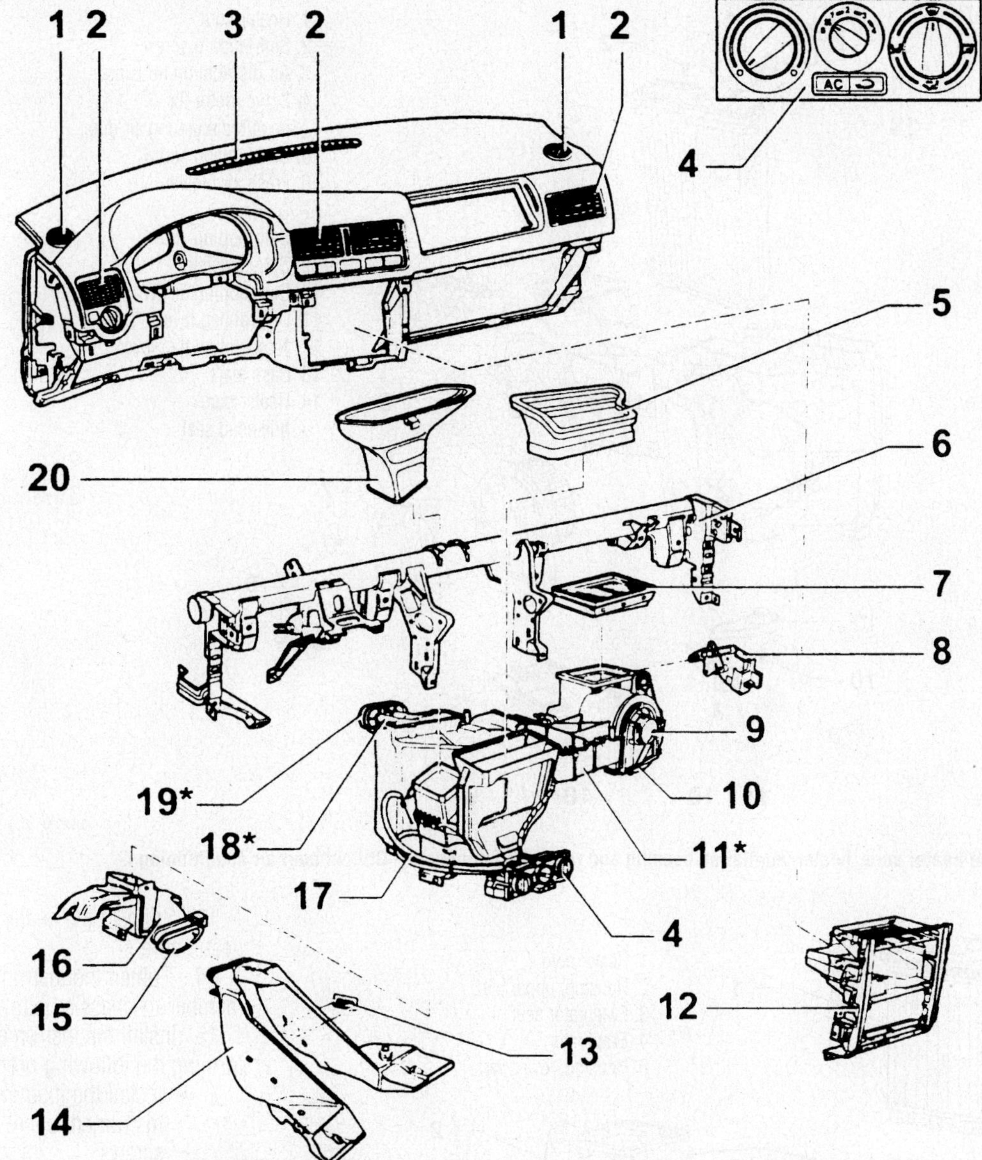

1. Side window air outlet
2. Side air outlets
3. Defroster air outlet
4. Heating and A/C controls
5. Intermediate duct
6. Instrument panel cross member
7. Dust and pollen filter
8. Servo motor for fresh/recirculated air door
9. Fresh air blower
10. Fresh air blower series resistance with fuse
11. Heating and A/C unit
12. Center trim
13. Footwell air outlet
14. Rear duct
15. Gasket
16 Connecting duct
17. Cables
18. Heater core
19. Heater core/bulkhead seal
20. Defroster duct

93112GS0

Exploded view of the instrument panel, crossmember and related components—Golf

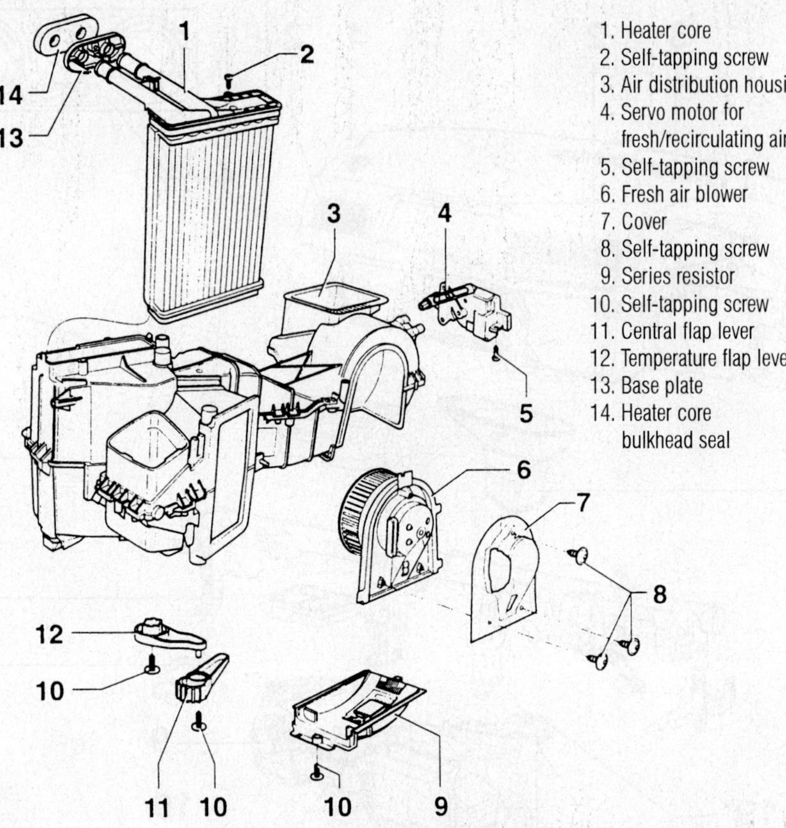

Exploded view of the heater core, heater/ventilation housing and related components—Golf without air conditioning

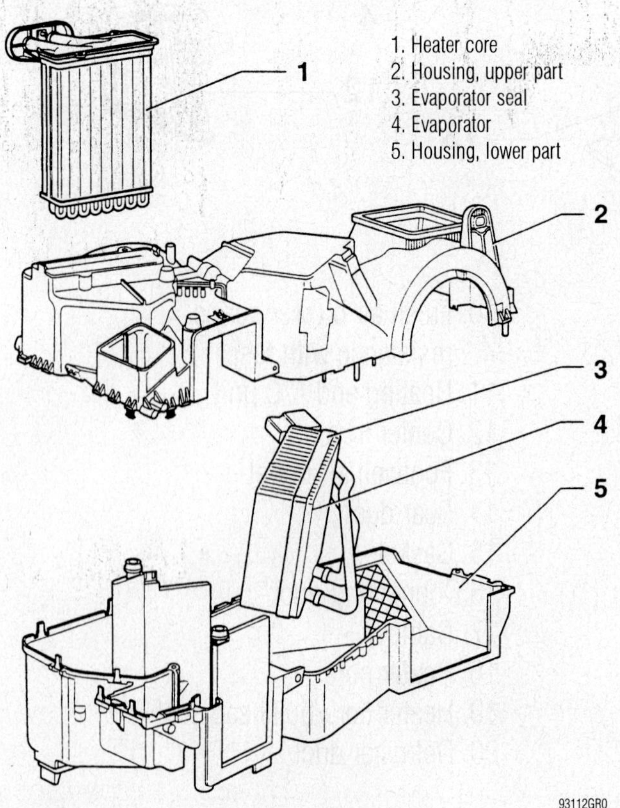

Exploded view of the heater core, heater/air conditioning housing and related components—Golf with air conditioning

13. Connect the heater hoses to the heater core.

14. Tighten the instrument panel cross-member-to-chassis bolts.

15. Install the instrument panel by performing the following procedure:

- Install the instrument panel and the instrument panel-to-chassis screws.
- Connect the electrical connectors; then, install the light switch knob and the illumination control switch.
- At the upper center of the instrument panel, connect the electrical connector and install the photo sensor.
- Install the instrument cluster, connect the multi-pin electrical connector and install the 2 instrument cluster-to-instrument panel screws.
- At the driver's side, install the footwell cover and the 3 cover screws.
- Install the center reinforcement and the 5 reinforcement-to-chassis screws.
- Connect the electrical connector; then, install the Climatronic control (heater control) and the 4 control-to-instrument panel screws.

- Install the Climatronic control (heater control) trim.
- At the center of the instrument pane (above the radio), Install the switches and connect the electrical connectors.
- Connect the electrical connectors and install the radio.
- At the driver's side, install the sound damper panel and the 6 panel-to-instrument panel screws.
- Install the DLC-to-bracket screws.
- At the driver's side, install the both lower instrument panel covers and the cover screws.
- Install the glove box, connect the glove box light connector and install the 7 glove box-to-instrument panel screws.
- At the passenger's side of instrument panel, install the side cover.
- Install the combination switch, connect the electrical connectors and install the combination switch-to-steering column bolt.
- Install the lower steering column cover, the 3 lower steering column cover screws and the lower steering column switch trim.
- Install the upper steering column cover and the cover screws.

16. Install the center console and the console-to-chassis bolts.
17. Install the steering wheel and the steering wheel-to-steering column bolt. Torque the multi-point socket head bolt to 44 ft. lbs. (60 Nm).

➡The Multi-point socket head bolt can be used up to 5 times; be sure to place a punch mark on it each time it is reused.

18. Connect the electrical connector and install the air bag.
19. Refill the cooling system.
20. Connect the negative battery cable.
21. Evacuate, charge and leak test the air conditioning system.
22. Operate the engine to normal operating temperatures; then, check the climate control operation and check for leaks.

2002–05 GTI and 2002–04 Jetta

➡Be sure to obtain the anti-theft code for the radio.

1. Disconnect the negative battery cable.
2. Drain the cooling system into a clean container for reuse.
3. Remove the driver's side air bag by performing the following procedure:
   - Release the steering column adjustment.
   - Rotate the steering wheel until the spokes are vertical (90 degrees off center).
   - Adjust the steering column to the completely out down position and lock the adjustment.
   - At the rear of the steering wheel, insert a 7 in. (175mm) screwdriver into the hole in the hub approximately 1¾ in. (45mm) deep.
   - Raise the screwdriver (as viewed from the driver's seat) to release the air bag clip.
   - Rotate the steering wheel 180 degrees in the opposite direction and release the clip on the other side.
   - Center the steering wheel with the front wheels facing straight-ahead.
   - Disconnect the electrical connector and remove the air bag.
4. Remove the steering wheel by performing the following procedure:
   - Place the front wheels in the straight-ahead position.
   - Remove the steering wheel-to-steering column bolt.

➡The Multi-point socket head bolt can be used up to 5 times; be sure to place a punch mark on it each time it is reused.

   - Remove the steering wheel from the steering column.
5. Remove the center console-to-chassis bolts and the console.
6. Remove the instrument panel by performing the following procedure:
   - Remove the upper steering column cover screws and the cover.
   - Remove the 3 lower steering column cover screws, release the steering wheel height adjustment and remove the lower steering column switch trim.
   - Remove the combination switch-to-steering column bolt, disconnect the electrical connectors and remove the combination switch.
   - At the passenger's side of instrument panel, remove the side cover.
   - Remove the 7 glove box-to-instrument panel screws, disconnect the glove box light connector and remove the glove box.
   - At the driver's side, remove the both lower instrument panel cover screws and the covers.
   - Remove the DLC-to-bracket screws.
   - At the driver's side, remove the 6 sound damper panel-to-instrument panel screws and the panel.
   - Slide the Radio Release tool No. 3316 into the slots at both sides of the radio until they engage; then, pull the radio out of the instrument panel and disconnect the electrical connectors.

➡To remove the Radio Release tool(s) No. 3316, press the locating tabs (located on both sides of the radio) inward and remove them.

   - At the center of the instrument pane (above the radio), unclip and remove the switches and disconnect the electrical connectors.
   - Unclip and remove the Climatronic control (heater control) trim.
   - Remove the 4 Climatronic control (heater control)-to-instrument panel screws and the control; then, disconnect the electrical connector.
   - Remove the 5 center reinforcement-to-chassis screws and the reinforcement.
   - At the driver's side, remove the 3 footwell cover screws and the cover.
   - Remove the 2 instrument cluster-to-instrument panel screws, disconnect the multi-pin electrical connector and remove the instrument cluster.
   - At the upper center of the instrument panel, unclip the photo sensor and disconnect the electrical connector.
   - Rotate the light switch knob to **0**, press the knob inward, rotate it clockwise; then, pull the switch outward and disconnect the electrical connector.
   - Unclip the illumination control switch and disconnect the electrical connector.
   - Remove the instrument panel-to-chassis screws and the instrument panel from the crossmember.
7. Loosen the instrument panel crossmember-to-chassis bolts and lift the crossmember upward.
8. Disconnect the heater hoses from the heater core.
9. If not equipped with air conditioning, perform the following procedure:
   - Partially remove the heater/ventilation housing assembly.
   - Depress the retainer clips and remove the heater core from the heater housing.

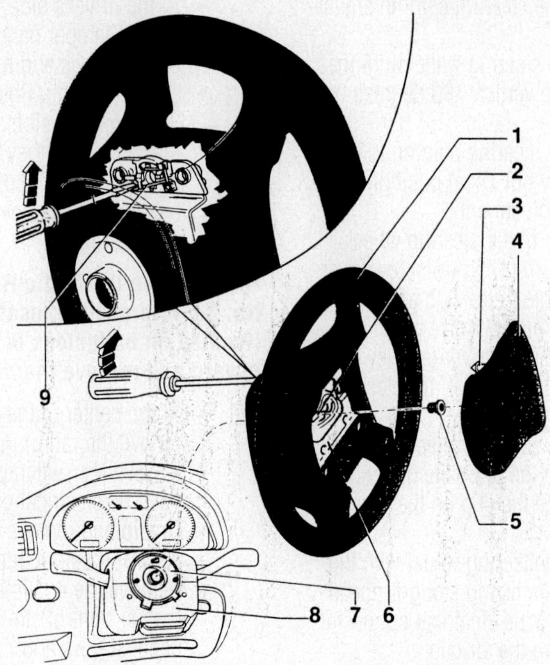

1. Steering wheel
2. Connector
3. Locking lug
4. Airbag unit (with airbag igniter, driver's side)
5. Multi-point socket head bolt
6. Securing plate
7. Coil spring for airbag/return spring with slip ring
8. Trim
9. Clip

93112GS8

Releasing the driver's side air bag—GTI and Jetta

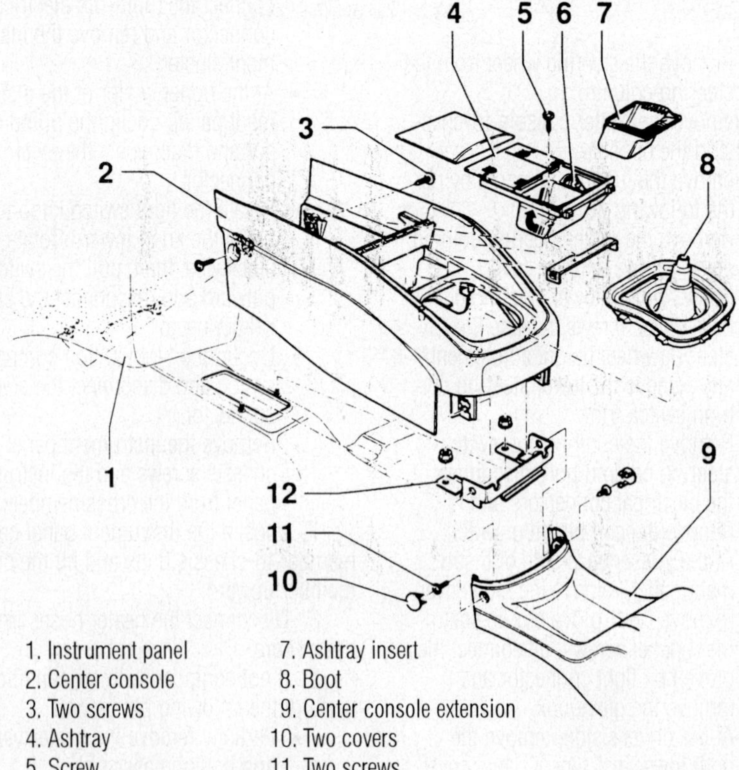

1. Instrument panel
2. Center console
3. Two screws
4. Ashtray
5. Screw
6. Cigarette lighter
7. Ashtray insert
8. Boot
9. Center console extension
10. Two covers
11. Two screws
12. Mounting bracket

93112GS9

Exploded view of the center console and related components— GTI and Jetta

➡If the retainer clips break, use a self-tapping screw to hold the heater core in place.

10. If equipped with air conditioning, perform the following procedure:
    - Discharge and recover the air conditioning system refrigerant.
    - Disconnect the refrigerant lines from the evaporator core, discard the O-rings and plug the openings to prevent contamination.
    - Remove the ventilation ducts from the heater/air conditioning housing assembly.
    - Remove the heater core from the heater/air conditioning housing assembly.

To install:

11. If equipped with air conditioning, perform the following procedure:
    - Install the heater core to the heater/air conditioning housing assembly.
    - Install the ventilation ducts to the heater/air conditioning housing assembly.
    - Using new O-rings, connect the refrigerant lines to the evaporator core.
12. If not equipped with air conditioning, perform the following procedure:
    - Install the heater core to the heater housing.
    - Install the heater/ventilation housing assembly.
13. Connect the heater hoses to the heater core.
14. Tighten the instrument panel cross-member-to-chassis bolts.
15. Install the instrument panel by performing the following procedure:
    - Install the instrument panel and the instrument panel-to-chassis screws.
    - Connect the electrical connectors; then, install the light switch knob and the illumination control switch.
    - At the upper center of the instrument panel, connect the electrical connector and install the photo sensor.
    - Install the instrument cluster, connect the multi-pin electrical connector and install the 2 instrument cluster-to-instrument panel screws.
    - At the driver's side, install the footwell cover and the 3 cover screws.
    - Install the center reinforcement and the 5 reinforcement-to-chassis screws.

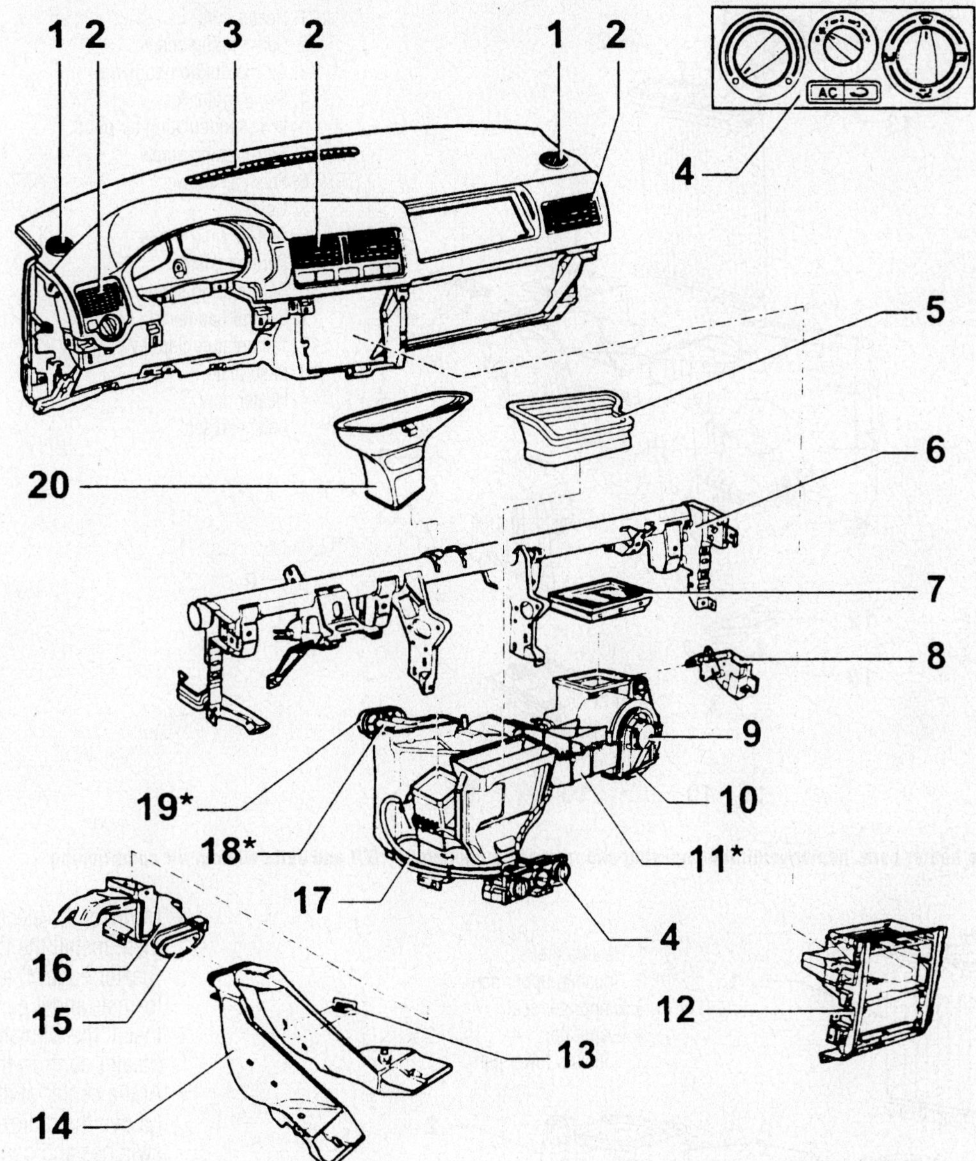

1. Side window air outlet
2. Side air outlets
3. Defroster air outlet
4. Heating and A/C controls
5. Intermediate duct
6. Instrument panel cross member
7. Dust and pollen filter
8. Servo motor for fresh/recirculated air door
9. Fresh air blower
10. Fresh air blower series resistance with fuse
11. Heating and A/C unit
12. Center trim
13. Footwell air outlet
14. Rear duct
15. Gasket
16 Connecting duct
17. Cables
18. Heater core
19. Heater core/bulkhead seal
20. Defroster duct

93112GS0

Exploded view of the instrument panel, crossmember and related components— GTI and Jetta

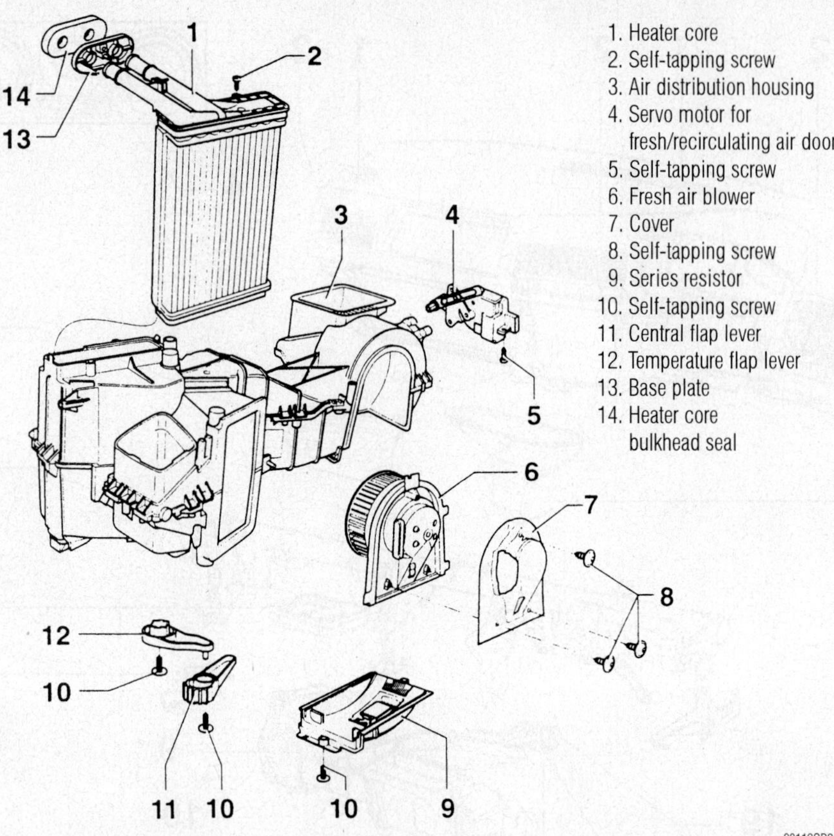

Exploded view of the heater core, heater/ventilation housing and related components— GTI and Jetta without air conditioning

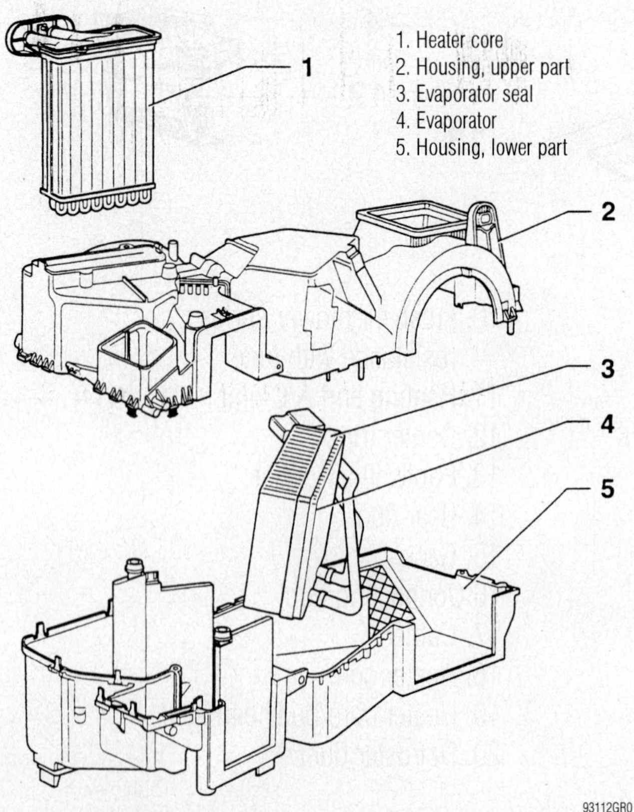

Exploded view of the heater core, heater/air conditioning housing and related components— GTI and Jetta with air conditioning

- Connect the electrical connector; then, install the Climatronic control (heater control) and the 4 control-to-instrument panel screws.
- Install the Climatronic control (heater control) trim.
- At the center of the instrument pane (above the radio), Install the switches and connect the electrical connectors.
- Connect the electrical connectors and install the radio.
- At the driver's side, install the sound damper panel and the 6 panel-to-instrument panel screws.
- Install the DLC-to-bracket screws.
- At the driver's side, install the both lower instrument panel covers and the cover screws.
- Install the glove box, connect the glove box light connector and install the 7 glove box-to-instrument panel screws.
- At the passenger's side of instrument panel, install the side cover.
- Install the combination switch, connect the electrical connectors and install the combination switch-to-steering column bolt.

- Install the lower steering column cover, the 3 lower steering column cover screws and the lower steering column switch trim.
- Install the upper steering column cover and the cover screws.

16. Install the center console and the console-to-chassis bolts.
17. Install the steering wheel and the steering wheel-to-steering column bolt. Torque the multi-point socket head bolt to 44 ft. lbs. (60 Nm).

➡The Multi-point socket head bolt can be used up to 5 times; be sure to place a punch mark on it each time it is reused.

18. Connect the electrical connector and install the air bag.
19. Refill the cooling system.
20. Connect the negative battery cable.
21. Evacuate, charge and leak test the air conditioning system.
22. Operate the engine to normal operating temperatures; then, check the climate control operation and check for leaks.

Rabbit, 2006 GTI and 2005–06 Jetta

1. Before servicing the vehicle, refer to the Precautions Section.

➡Be sure to obtain the anti-theft code for the radio.

2. Disconnect the negative battery cable.
3. Remove facing wall in plenum chamber.
4. Place Drip tray for VAS6100 VAS 6208 under the engine.
5. Clamp off coolant hoses using hose clamps, up to 1.6 inches (40mm).
6. Disconnect coolant hoses to heater core.
7. Connect a section of hose onto upper connection of heater core.
8. Hold suitable container under the lower connection.
9. Using a compressed air gun, carefully blow coolant from heater core into container.
10. Remove bolt (6mm socket head) from connection flange between heater core connections. This allows the coolant pipes to move for removing the heater core.
11. Remove driver side footwell trim and left footwell air vent.
12. Remove screws and remove heater core trim.
13. Cover floor carpet in area below heater core with leak-proof foil and absorbent paper.
14. Open hose clamps and disconnect coolant pipes from heater core.
15. Remove heater core from heating and ventilation unit.

To install:

16. Install in reverse order of removal, noting the following:
    a. Check seals installed on heater core, only install a heater core with undamaged seals. An incorrectly glued seal can roll up into heating and ventilation unit when sliding in the heater core.

➡If seal is damaged or installed incorrectly, cold air can flow past the heater core.

    b. Push heater core into the heating and ventilation unit.
    c. Coat sealing rings with coolant before installing.
    d. Insert sealing rings into connection on heater core.
    e. Ensure sealing rings are installed with the chamfered end facing the heater core.
    f. If hose clamps are deformed, replace them.
    g. Connect coolant pipes to heater core.
    h. Hose clamps must be able to be twisted slightly when installing onto the coolant pipes. Tighten hose clamps to 1.5 ft. lbs. (2 Nm).
    i. Check seating of both clamps after tightening screws, clamps must enclose the flange on heater core and coolant pipe completely and must not touch other components.
    j. Thread bolt into connection flange and while doing this ensure bolt is actually threaded into the mounting point intended for it.
    k. Check position of grommet in bulkhead for proper seating.
    l. Seal flange for coolant pipes to heater core and for expansion valve (to evaporator, only in vehicles with A/C system) at grommet pass-through with silicon adhesive sealant if necessary (to prevent water from penetrating).

➡Sealing rings must always be replaced. If hose clamps are deformed, replace them.

17. After replacing heater core, coolant must be replaced completely.
18. Check coolant circuit for leaks, pay particular attention between coolant hoses and heater core.
19. Connect the negative battery cable.
20. Evacuate, charge and leak test the air conditioning system refrigerant.
21. Operate the engine to normal operating temperatures; then, check the climate control operation and check for leaks.

Passat

➡Be sure to obtain the anti-theft code for the radio.

1. Disconnect the negative battery cable.
2. Drain the cooling system into a clean container for reuse.
3. Disconnect the heater hoses from the heater core.
4. Discharge and recover the air conditioning system refrigerant.
5. Remove the driver's side air bag by performing the following procedure:
    - Release the steering column adjustment.
    - Rotate the steering wheel until the spokes are vertical (90 degrees off center).
    - Adjust the steering column to the completely out down position and lock the adjustment.
    - At the rear of the steering wheel, insert a 7 in. (175mm) screwdriver into the hole in the hub approximately 1¾ in. (45mm) deep.
    - Raise the screwdriver (as viewed from the driver's seat) to release the air bag clip.
    - Rotate the steering wheel 180 degrees in the opposite direction and release the clip on the other side.
    - Center the steering wheel with the front wheels facing straight-ahead.
    - Disconnect the electrical connector and remove the air bag.
6. Remove the steering wheel by performing the following procedure:
    - Place the front wheels in the straight-ahead position.
    - Remove the steering wheel-to-steering column bolt.

➡Two kinds of bolts are used: Standard (disconnected shortly after production) and Multi-point socket head. If equipped with a standard bolt, discard it. If equipped with a multi-point socket head bolt, it can be used up to 5 times; be sure to place a punch mark on it each time it is reused.

    - Remove the steering wheel from the steering column.
7. Remove the instrument panel by performing the following procedure:
    - Remove the 2 upper steering column cover screws and the cover.

- Remove the 4 lower steering column cover screws/bolt, release the steering wheel height adjustment and remove the lower steering column switch trim.
- Remove the combination switch-to-steering column bolt, disconnect the electrical connectors and remove the combination switch.
- At the driver's side, remove the lower "A" pillar trim bolt cover, the "A" pillar trim-to-chassis bolts and the trim.
- At the driver's side, remove the instrument panel's end cover.
- Remove the 2 stowage compartment-to-instrument panel bolts, the 2 instrument panel-to-frame bolts and the stowage compartment.
- Disconnect the electrical connectors from the light switch and the headlight brightness control.
- At the steering column, slide the cover upward and place a screwdriver handle between the cover the steering column to secure it.
- Remove the 4 cover-to-instrument panel screws and the cover.
- Between the steering column and the instrument cluster, remove the 2 instrument panel insert screws and the panel.
- Disconnect the electrical connectors.
- At both sides of the steering column, remove the 2 instrument panel-to-frame bolts.
- Slide the Radio Release tool No. 3316 into the slots at both sides of the radio until they engage; then, pull the radio out of the instrument panel and disconnect the electrical connectors.

➡To remove the Radio Release tool(s) No. 3316, press the locating tabs (located on both sides of the radio) inward and remove them.

- At the center of the instrument panel, unclip and remove the control cluster trim.
- Remove the 8 center instrument panel cover-to-instrument panel bolts and the center panel; then, disconnect the electrical connector.
- At the top of the center instrument panel cover opening, remove the 5 instrument panel-to-bracket bolts.
- Under the left side of the glove box, remove the small piece of trim bolt and the trim.
- Under the right side of the glove box, remove the lower "A" pillar trim bolt cover, the "A" pillar trim-to-chassis bolts and the trim.
- Remove the 2 lower glove box bolts and open the glove box.
- Remove the 5 glove box-to-instrument panel screws, disconnect the glove box light connector and remove the glove box.
- At the passenger's side of instrument panel, remove the side cover.
- Remove the 4 instrument panel-to-bracket bolts and disconnect the electrical connector.
- Remove the instrument cluster.

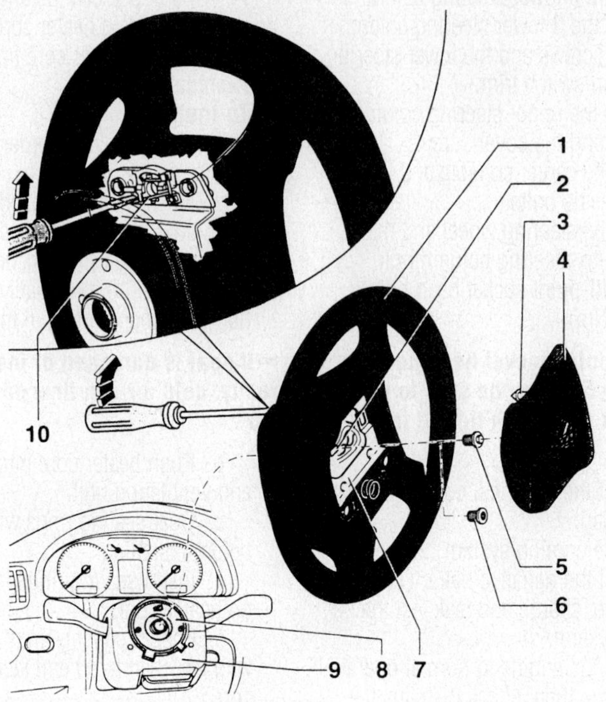

1. Steering wheel
2. Connector
3. Locking lug
4. Airbag unit
5. Bolt
6. Multi-point socket head bolt
7. Securing plate
8. Spiral spring connector with slip ring
9. Trim panel
10. Clip

93112GT1

Releasing the driver's side air bag—Passat

8. Remove the ventilation ducts from the heater/air conditioning housing assembly.
9. Remove the heater/air conditioning housing assembly-to-chassis fasteners and the assembly.
10. Remove the heater core.

To install:

11. Install the heater core.
12. Install the heater/air conditioning housing assembly and the assembly-to-chassis fasteners.
13. Install the ventilation ducts to the heater/air conditioning housing assembly.
14. Install the instrument panel by performing the following procedure:
    - Install the instrument cluster.
    - Install the 4 instrument panel-to-bracket bolts and connect the electrical connector.
    - At the passenger's side of instrument panel, install the side cover.
    - Install the glove box, connect the glove box light connector and install the 5 glove box-to-instrument panel screws.
    - Install the glove box and open the 2 lower glove box bolts.
    - Under the right side of the glove box, install the lower "A" pillar trim, the "A" pillar trim-to-chassis bolts and the trim bolt cover.
    - Under the left side of the glove box, install the small piece of trim and the bolt.
    - At the top of the center instrument panel cover opening, install the 5 instrument panel-to-bracket bolts.
    - Install the center instrument panel cover and the 8 center panel cover-to-instrument panel bolts; then, connect the electrical connector.
    - At the center of the instrument panel, install the control cluster trim.

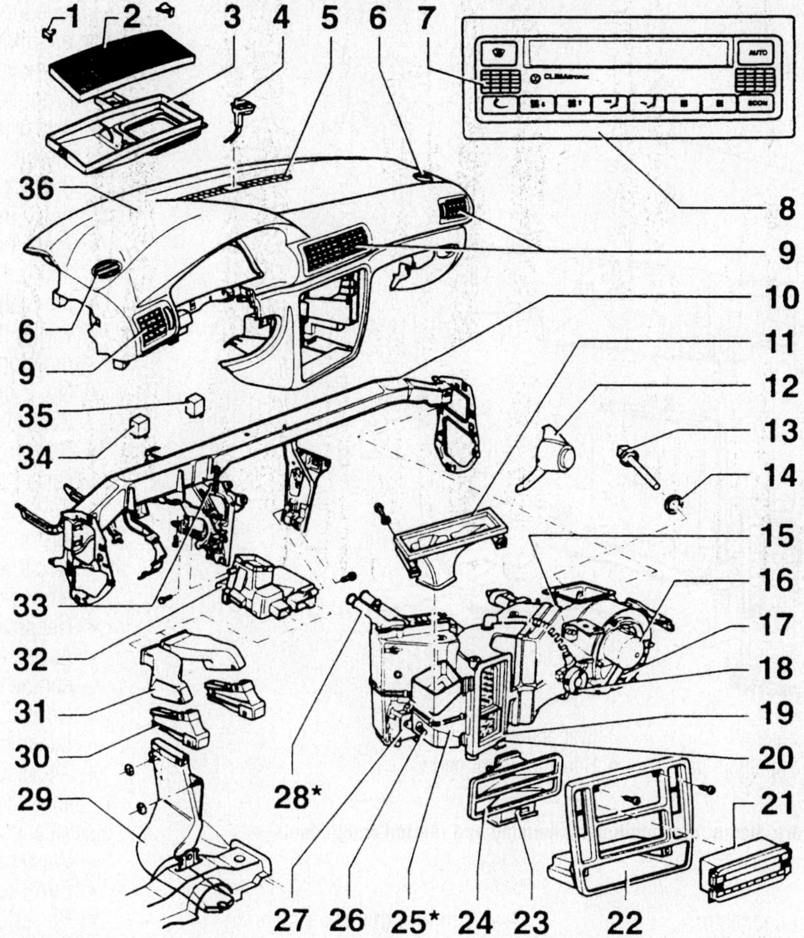

1. Clip
2. Dust and pollen filter
3. Air intake housing
4. Sunlight photo sensor
5. Defroster air outlet
6. Side window air outlet
7. Instrument panel interior temperature sensor
8. A/C control head
9. Center air outlet
10. Instrument panel cross member
11. Defroster duct
12. Evaporator water drain guide
13. Fresh air intake duct temperature sensor
14. Seal
15. Air flow flap motor
16. Fresh air blower
17. Control module for fresh air blower
18. Air outlet for glove box
19. Temperature regulator flap motor
20. Central air flap motor
21. Climatronic control unit
22. Center trim
23. Sender for outlet temperature, center
24. Center air outlet duct
25. Heater/evaporator unit
26. Sender for outlet temperature, floor outlet
27. Footwell/defroster flap motor
28. Heater core
29. Left rear duct
30. Lower duct
31. Upper duct
32. Footwell air outlet
33. Instrument panel cross member and left side support securing bolt
34. Relay
35. A/C clutch relay
36. Instrument panel

93112GT2

Exploded view of the instrument panel, crossmember and related components—Passat

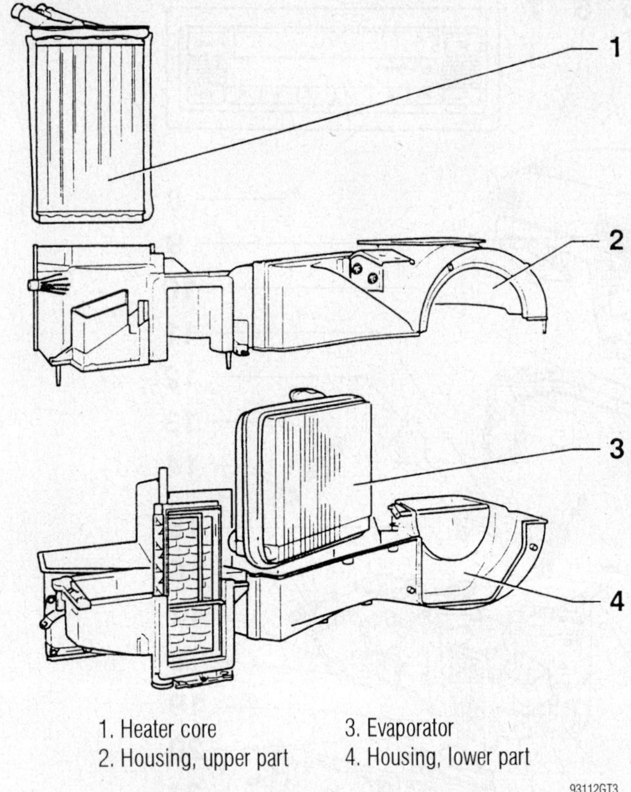

Exploded view of the heater core, heater/air conditioning housing and related components—Passat

- Connect the electrical connectors and install the radio.
- At both sides of the steering column, install the 2 instrument panel-to-frame bolts.
- Connect the electrical connectors.
- Between the steering column and the instrument cluster, install the instrument panel insert and the 2 panel screws.
- At the steering column, install the cover and the 4 cover-to-instrument panel screws.
- Connect the electrical connectors to the light switch and the headlight brightness control.
- Install the stowage compartment, the 2 instrument panel-to-frame bolts and the 2 stowage compartment-to-instrument panel bolts.
- At the driver's side, install the instrument panel's end cover.
- At the driver's side, install the lower "A" pillar trim, the "A" pillar trim-to-chassis bolts and the trim bolt cover.
- Install the combination switch, connect the electrical connectors and install the combination switch-to-steering column bolt.
- Install the lower steering column cover trim and the lower steering column switch trim screws/bolt.
- Install the upper steering column cover and the 2 cover screws.

15. Install the steering wheel and the steering wheel-to-steering column bolt. Torque the standard bolt to 55 ft. lbs. (75 Nm) or the multi-point socket head bolt to 44 ft. lbs. (60 Nm).
16. Connect the electrical connector and install the air bag.
17. Connect the heater hoses to the heater core.
18. Refill the cooling system.
19. Connect the negative battery cable.
20. Evacuate, charge and leak test the air conditioning system refrigerant.
21. Operate the engine to normal operating temperatures; then, check the climate control operation and check for leaks.

Cylinder Head

REMOVAL & INSTALLATION

1.8L Engine

1. Before servicing the vehicle, refer to the Precautions Section.
2. Place the lock carrier into the service position as follows:
   a. Remove or disconnect:
   - Front bumper.
   - Any wiring or connector that would inhibit locking the carrier.
   - 3 quick-release screws on the front noise insulation panel.
   - Air guide between the lock carrier and the air filter.
   - If installed, retaining clamps for the wiring harness at the left side of the radiator frame.
   - No. 2 bolts and install Support 3369 or equivalent.
   - Remaining bolts and pull the lock carrier out to the stop.

   b. To secure the lock carrier, install the appropriate M6 bolts into the rear of the lock carrier and fender.
3. Remove or disconnect:
   - Negative battery cable.
   - Accessory drive belt, then the engine driven cooling fan.
   - Coolant.
   - Intake manifold.
   - Accessory drive belts.
4. Label and detach the following lines and electrical connectors:
   - Wastegate bypass regulator valve
   - EVAP canister purge regulator valve
   - Power outage stage
   - MAF sensor
5. Remove or disconnect:
   - Air cleaner housing.
   - ECT and the Temperature II sensor harness connector.
   - All connections from the cylinder head and position them aside.
   - Crankcase breather line.
   - Fuel supply and return lines.
   - Oil supply line at the cylinder head.
   - Exhaust manifold heat shield.
   - Turbocharger from the exhaust manifold.
   - Coolant hose to the heat exchanger at the rear of the cylinder head.
   - Upper timing belt cover.
6. Turn the crankshaft, in the direction of rotation (clockwise), until the No. 1 cylinder is at TDC.
7. Using a T45 Torx® wrench, loosen the timing belt tensioner, and remove the belt from the camshaft gear.
8. Remove:
   - The Torx® bolt and swing the tensioner assembly bracket forward.
   - Valve cover.
   - Cylinder head bolts in sequence, as shown.
   - Cylinder head, then clean the gasket mating surfaces.

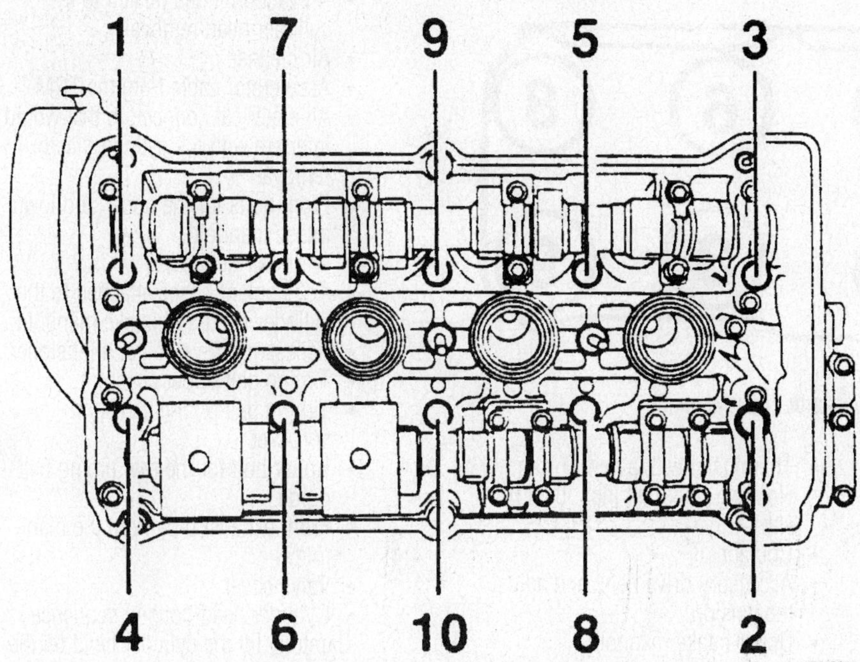

Cylinder head bolt removal sequence—1.8L engine

9. Clean and dry out the cylinder head bolt holes.

To install:

WARNING

Always replace the cylinder head bolts. Always replace self-locking nuts, bolts, gaskets and O-rings. Do not remove the new head gasket from the package until immediately before installing.

10. Before installing the cylinder head, be sure NO pistons are at TDC.

11. Loosen the turbocharger support bracket to reduce the likelihood of any tension while installing the cylinder head.

12. Install or connect:
- Head gasket with the part number visible from the intake side.
- Cylinder head.
- New cylinder head bolts and tighten by hand.

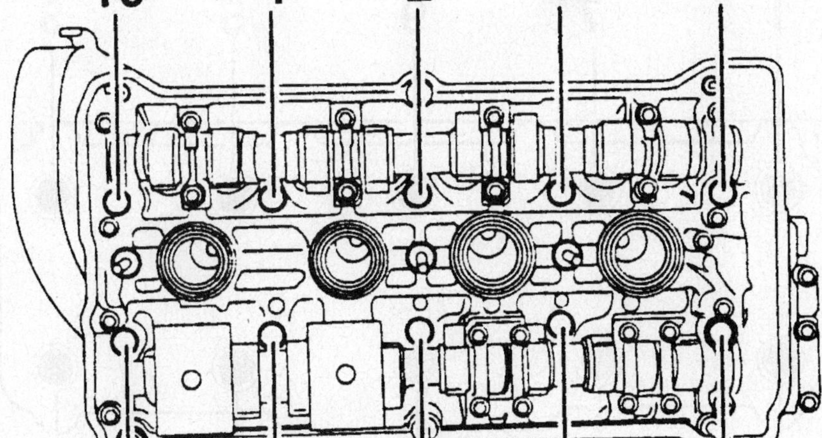

Cylinder head bolt tightening sequence—1.8L engine

13. Tighten the new cylinder head bolts in sequence in 2 steps:
   a. Tighten all of the bolts to 44 ft. lbs. (60 Nm).
   b. Tighten all of the bolts an additional ½ turn (180°).

➡It is not necessary to retighten the cylinder head bolts.

14. Using new gaskets, install the turbocharger to the exhaust manifold , coat the bolts with Hot Bolt Paste G 052 112 A3 (or equivalent). Mounting bolts: 26 ft. lbs. (35 Nm). Turbo support bracket mounting bolts: 33 ft. lbs. (40 Nm).

15. Install or connect:
- Valve cover.
- Timing belt.
- Accessory drive belts.
- Exhaust manifold heat shield.
- Oil supply lines to the cylinder head and tighten the retaining straps to 15 ft. lbs. (20 Nm).
- Crankcase breather.
- Fuel supply and return lines.
- Any items removed during disassembly.
- Coolant temperature sensors, and the air cleaner housing.
- Coolant.
- Negative battery cable.

16. Fully close all power windows, operate all window switches for at least 1 second in the close direction to activate the one-touch opening/closing function

➡DTCs are stored when harness connectors are detached.

17. Read the DTCs and clear the fault codes.

2.0L (ABA) Engine

1. Before servicing the vehicle, refer to the Precautions Section.
2. Remove or disconnect:
- Negative battery cable.
- Coolant.
- Throttle cable.
- All wiring and vacuum lines from the intake manifold.
- Upper intake manifold.
- Fuel supply and return lines.
- Radiator and heater hoses from the cylinder head.
- Wiring for oil pressure and temperature sensors.
- Distributor cap and wires.
- Exhaust pipe from the exhaust manifold.
- If equipped, EGR pipe from the exhaust manifold.

Tighten the cylinder head bolts in the sequence shown—2.0L (ABA) engine

- Accessory drive belts and any accessory that is bolted to the head.
- Cylinder head cover, timing belt cover and belt.

3. Loosen the cylinder head bolts in the reverse of the tightening sequence.
4. Remove the bolts and lift the head straight off.

To install:

5. Before reinstalling the head, check the flatness of the head and block in both width and length, then diagonally from each corner.
6. Install the new cylinder head gasket with the word TOP or OBEN facing upward; do not use any sealing compound.
7. To align the cylinder head, install Guide Pins from tool 3070, or equivalent, into the holes for cylinder head bolts 8 and 10.
8. Install:
   - Cylinder head.
   - Cylinder head bolts, except 8 and 10, by hand.
9. Remove the Guide Pins using tool 3070 or equivalent, then install head bolts 8 and 10.
10. Tighten the head bolts in sequence in the following steps:
    - 30 ft. lbs. (40 Nm)
    - 44 ft. lbs. (60 Nm)
    - + 90°
    - + an additional 90°
11. Install or connect:
    - Camshaft drive belt and adjust the tension.
    - Exhaust pipe to the manifold. Use new gaskets and self-locking nuts. Nuts: 18 ft. lbs. (25 Nm).
    - EGR pipe, if equipped.
    - Wiring to the oil pressure and temperature sensors.
    - Ignition system components.
    - Radiator and heater hoses.
    - Throttle cable and all wiring and vacuum lines.
    - Thermostat with a new O-ring. Housing bolts: 84 inch lbs. (10 Nm).
    - Coolant.
    - Accessory drive belts and adjust the tension.
    - Upper intake manifold.
    - All wiring and vacuum lines disconnected from the intake manifold. Connect the throttle cable.
    - Fuel supply and return lines.
    - Negative battery cables.

2.0L (AEG/AVH/AZG/BDC/BEV/BGD) Engine

1. Before servicing the vehicle, refer to the Precautions Section.
2. Remove or disconnect:
   - Negative battery cable.
   - Engine cover.
   - Coolant.
   - Fuel supply and return lines.
   - All vacuum lines related to the cylinder head removal.
   - Air cleaner.
   - Accelerator cable from the TCM.
   - All electrical connectors that would interfere with the cylinder head removal.
   - Front bolts for the upper and lower intake manifold.
   - Warm air deflector.
   - Any coolant hoses attached to the cylinder head and intake manifold.
   - Accessory drive belt and tensioner.
   - Timing belt upper cover.
   - Timing belt off the camshaft sprocket.
   - Upper bolt for the rear timing belt guide.
   - Front exhaust pipe from the manifold.
   - Valve cover.
   - Cylinder head bolts in sequence.
3. Carefully lift the cylinder head off the engine block.
4. Clean the gasket sealing surfaces.

To install:

5. Install:
   - Guide Pins 3450/2A, or equivalent, into the head bolt holes as shown.

➡The head gasket part number must be readable when installed on the engine block.

   - Head gasket.
   - Cylinder head and tighten the 8 remaining head bolts by hand.
6. Remove the Guide Pins 3450/2A using the Pin Removal tool 3450/3, or

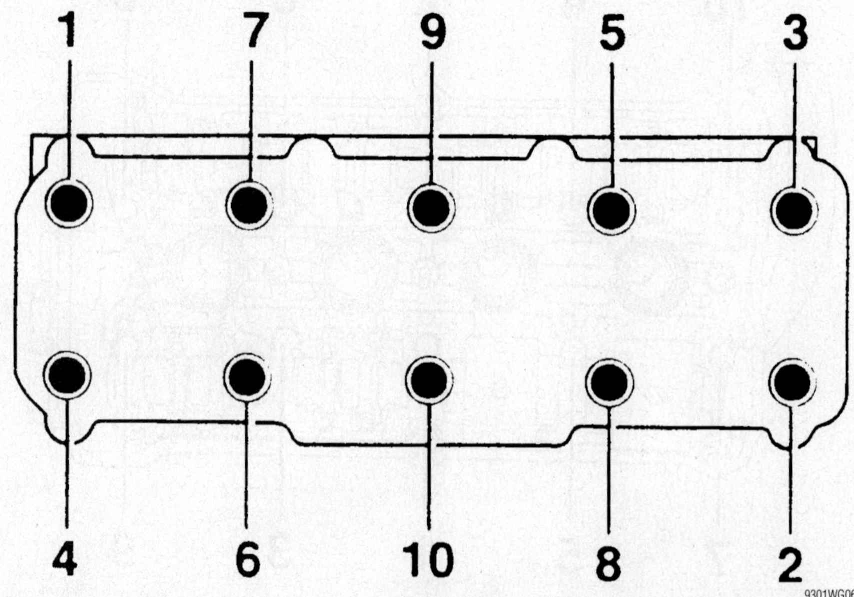

Cylinder head bolt removal sequence—2.0L (AEG) engine

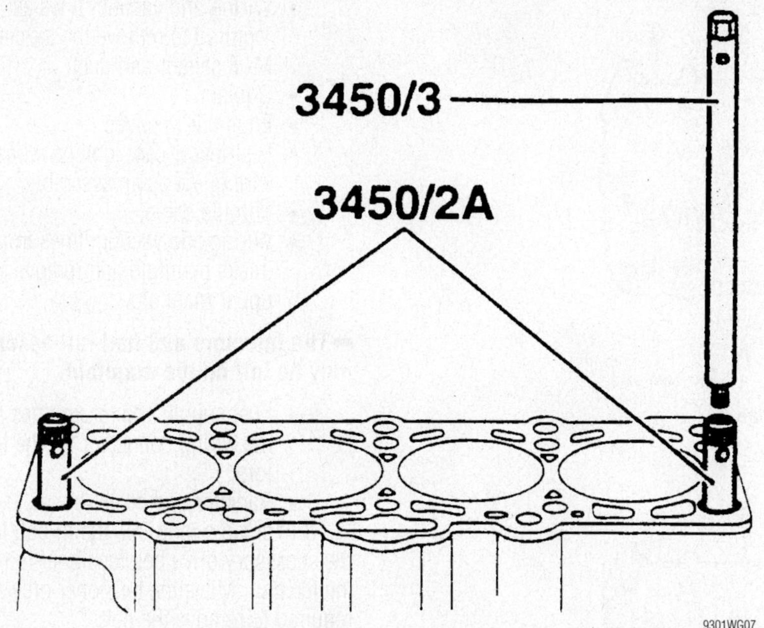

To properly install the head gasket and cylinder head, install the tools as shown—2.0L (AEG) engine

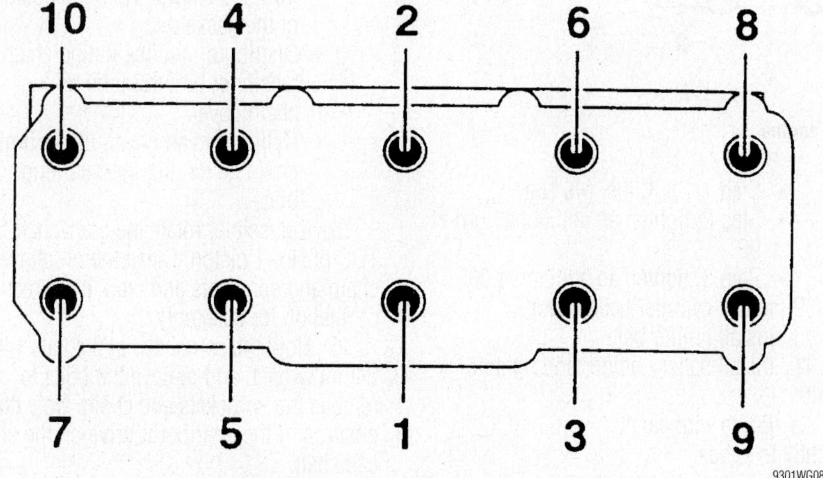

Cylinder head bolt torque sequence—2.0L (AEG) engine

equivalent and install the last 2 head bolts and tighten by hand.

7. Tighten all the cylinder head bolts in sequence as follows:
   - 30 ft. lbs. (40 Nm)
   - + 90°
   - + an additional 90°.
8. Install or connect:
   - Upper bolt for the rear timing belt cover and tighten to 15 ft. lbs. (20 Nm).
   - Timing belt and covers.
   - Front exhaust pipe to the manifold.
   - Accessory drive belt tensioner and belt.
   - Valve cover.
   - Coolant hoses that were removed.
   - Warm air deflector. Mounting bolts: 18 ft. lbs. (25 Nm).
   - Any electrical connectors that were removed.
   - Throttle cable to the TCM.
   - Air cleaner and any vacuum hoses that were removed.
   - Coolant and engine oil.
   - Negative battery cable.
9. Check and clear and DTCs, then match the ECM to the TCM.

2.0L (BPY) Engine

1. Before servicing the vehicle, refer to the Precautions Section.
2. Disconnect the negative battery cable.

➡All cable ties opened or cut during engine removal must be reinstalled at the same locations during installation.

3. Perform oil change if engine oil is contaminated.
4. Determine whether a coded radio is installed. In this case, determine anti-theft coding.
5. Remove plenum chamber cover.
6. Remove intake manifold.
7. Drain coolant.
8. Remove sound insulation at right.
9. Remove catalytic converter with front exhaust pipe.
10. Remove charge air pipe.
11. Pull off/disconnect all lines from turbocharger and set aside.
12. Remove oil supply line and coolant supply line from turbocharger.
13. Remove oil return line from turbocharger.
14. Remove support for turbocharger.
15. Remove battery.
16. Open cover of E-box and remove wire.
17. Remove battery holder.
18. Disconnect coolant hoses.
19. Remove Ground (GND) wire.
20. Disconnect connector and water hoses.
21. Pull off/disconnect all electrical wires from cylinder head and set aside.
22. Remove cable bracket and set wiring harness aside.
23. Remove coolant flange.
24. Remove drive belt.
25. Remove Locking Pin T10060A from drive belt tensioner.
26. Remove cylinder head cover.
27. Remove timing belt.
28. Follow sequence for loosening cylinder head bolts.

➡Verify that all hose and line connections to engine, transmission and body have been disconnected.

29. Remove cylinder head.

To install:

➡Replace cylinder head bolts. Always replace self-locking nuts and bolts during assembly work. Always replace bolts that are secured with tightening torque as well as O-rings and gaskets.

30. After installing a replacement cylinder head with camshaft installed, oil contact surfaces between roller cam followers and cam lubricating surfaces after installing cylinder head.

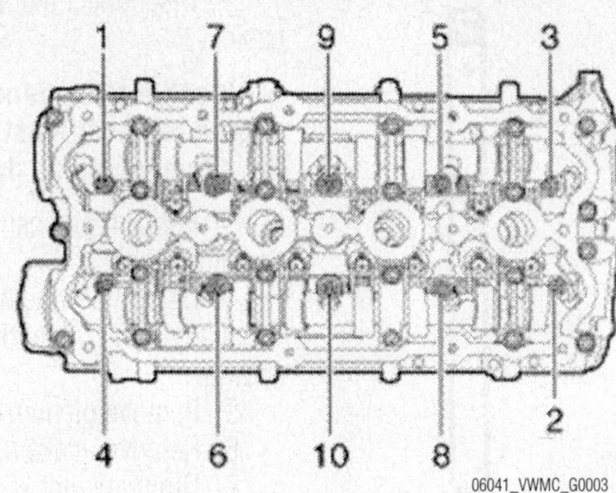

Cylinder head bolt loosening sequence—2.0L (BPY) engine

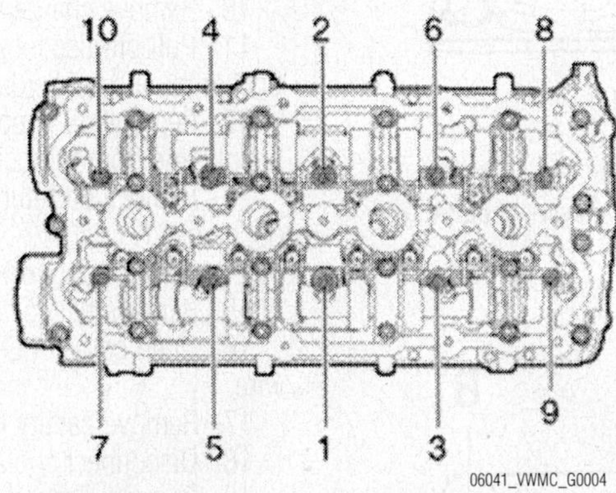

Cylinder head bolt tightening sequence—2.0L (BPY) engine

➡Do not remove new cylinder head gasket from package until immediately before installing. Handle new gasket with extra care. Damage in silicon layer and recessed area will cause leaks. Holes in the engine block for cylinder head bolts must be free of oil and coolant. Stuff clean cloths into cylinders so that no dirt or abrasive powder can get between cylinder wall and piston.

31. Carefully clean sealing surfaces of cylinder head and cylinder block. Avoid introducing scratches or scoring (do not use sandpaper with grit below 100).
32. Thoroughly remove all sanding and grinding residue and remove cloths.
33. If crankshaft was turned: Set number 1 piston at TDC position and slightly turn crankshaft backward.
34. Position new cylinder head gasket. Text (replacement part number) must be visible.
35. Note alignment sleeves in cylinder block.
36. Install cylinder head.
37. Insert new cylinder head bolts and tighten them hand-tight.
38. Tighten cylinder head bolts in the sequence shown, to the following steps:
    - Step 1, 28 ft. lbs. (40 Nm)
    - Step 2, tighten an additional turn of 90°
    - Step 3, tighten an additional 90°
39. Install cylinder head cover.
40. Install timing belt.
41. Install battery holder and tighten bolts.
42. Route wire carefully and bolt it tightly to E-box.
43. The rest of installation is in reverse order of removal.
44. Install battery.
45. Refill with new coolant.
46. Connect the negative battery cable.

2.8L (AAA) Engine

This procedure requires special tool 3268 or equivalent. This is a setting tool that holds the camshafts in the correct position for installing the timing chains. Before removing the cylinder head, be sure new bolts are available. The cylinder head bolts are made to stretch and cannot be used again.

1. Before servicing the vehicle, refer to the Precautions Section.
2. Remove or disconnect:
    - Battery cables and battery.
    - Wiring and vacuum lines as required to remove the air cleaner, MAF sensor and duct.
    - Coolant.
    - Engine trim cover.
    - Distributor cap, ignition wires and wire guide as an assembly.
    - Throttle cable.
    - Wiring and vacuum lines from the intake manifold and remove the upper manifold.

➡The injectors and fuel rail assembly may be left on the manifold.

    - Fuel supply and return lines and the wiring connector for the injectors.
    - Radiator and heater hoses.
3. Thread a long 8 x 10mm bolt into the accessory drive belt tensioner to release the tension. Move the tensioner only as required to remove the belt.
4. Remove or disconnect:
    - Alternator and belt tensioner.
    - Heat shield and the bolts to disconnect the 2 piece exhaust manifold from the engine. Note the position of the gaskets.
    - Distributor and the timing chain tensioner bolt from the timing chain cover.
    - Cylinder head cover, upper timing chain cover and the retaining plate.
5. If possible, rotate the crankshaft to TDC of No. 1 piston. Clean the oil off the chain and sprockets and mark the direction of rotation for assembly.
6. Hold the camshafts at the flats with a 24mm wrench and remove the bolts to remove the sprockets and chain. Note the position of the distributor drive on the short camshaft.

⁂ WARNING

Do not use the setting tool to hold the camshafts when removing or installing the sprocket bolts. The camshafts and the tool will be damaged.

7. Carefully check to be sure all necessary wires, hoses and brackets and components have been removed.
8. Loosen the cylinder head bolts in the reverse of the torque sequence. Remove and discard the bolts.
9. Remove the cylinder head.

To install:

10. Carefully clean the old gasket material from the head and the block. Before reinstalling the head, check the flatness of

the head and block in both width and length, then diagonally from each corner. Maximum allowable distortion is 0.004 in. (0.1mm).

11. If the new head gasket already has sealant in the small holes at the timing chain end, remove the sealant. Apply a silicone sealer to the timing chain end and install the gasket onto the block with the word TOP or OBEN facing up.

12. Fit the cylinder head over the locating dowels and set the head onto the engine. Install new bolts and hand-tighten them.

13. Tighten the bolts in sequence as described in the following steps:
- 30 ft. lbs. (40 Nm)
- 44 ft. lbs. (60 Nm)
- + 90°
- + an additional 90°

14. Be sure the crankshaft is at TDC on No. 1 piston. Install the setting tool to lock the camshafts in place, then install the timing chain and sprockets. Be sure they are positioned to rotate in the original direction.

15. Hold the camshaft with a 24mm wrench and install the sprocket bolt. Be sure the distributor drive is correctly positioned. Bolts: 74 ft. lbs. (100 Nm).

16. Install the tensioner shoe and temporarily install the upper timing chain cover. Install the tensioner bolt and remove the setting tool. Rotate the crankshaft 4 full turns and stop at TDC of No. 1 piston. The setting tool should fit into the camshafts.

17. Remove the tensioner bolt and upper timing chain cover again. Apply new sealant as required, install the cover. Bolts: 82 inch lbs. (10 Nm). Tensioner bolt: 15 ft. lbs. (20 Nm).

18. Install or connect:
- Cylinder head cover.
- Intake and exhaust manifolds, using new gaskets. Nuts and bolts: 18 ft. lbs. (25 Nm).
- Alternator belt and adjust tension.
- Accessory drive belt and adjust tension.
- Radiator and heater hoses.
- Injectors, fuel rail assembly and manifold. Connect the fuel supply and return lines. Connect the wiring connector for the injectors.
- Upper manifold. Bolts: 18 ft. lbs. (25 Nm).
- Wiring and vacuum lines disconnected from the intake manifold.
- Throttle cable.
- Distributor cap, ignition wires and wire guide as an assembly.
- Trim cover.
- Battery cables and battery.
- Coolant.

2.8L (AHA/ATQ) Engine

➡This procedure is for removing the left cylinder head, the right cylinder head service is similar.

1. Before servicing the vehicle, refer to the Precautions Section.

2. Place the lock carrier into the service position as follows:

a. Remove or disconnect:
- Front bumper.
- Any wiring or connector that would inhibit locking the carrier.
- 3 quick-release screws on the front noise insulation panel.
- Air guide between the lock carrier and the air filter.
- If installed, retaining clamps for the wiring harness at the left side of the radiator frame.
- No. 2 bolts and install Support 3369 or equivalent.
- Remaining bolts and pull the lock carrier out to the stop.

b. To secure the lock carrier, install the appropriate M6 bolts into the rear of the lock carrier and fender.

3. Remove or disconnect:
- Negative battery cable.
- Engine cover and under cover.
- Accessory drive belt.
- Timing belt.
- Left and right front exhaust pipes from the manifold.
- Exhaust system from the transaxle bracket.
- Coolant.
- Coolant expansion tank.
- Crankshaft housing ventilation hose from the valve cover.
- Intake air duct between the MAF sensor and elbow.
- Intake Manifold Tuning (IMT) valve and IAT sensor electrical connectors.
- Vacuum hoses from the cruise control diaphragm and fuel pressure regulator.
- Cruise control vacuum diaphragm and throttle control linkage from the throttle body.
- ECT and CMP sensor electrical connector.
- Tie wraps for the wiring harnesses and position the wires to the rear.
- Any remaining connectors and hoses that would interfere with the cylinder head removal.
- Fuel supply and return lines.
- 4 bolts mounting the fuel rail to the intake manifold.
- Fuel rail from the intake manifold.

4. Install Camshaft Locator Bar 3391, or equivalent, onto the camshaft sprocket.

5. Loosen the 2 bolts for the camshaft and back out approximately 5 turns.

6. Remove the Camshaft Locating Bar 3391.

7. Using Camshaft Gear Puller T40001, or equivalent, remove the camshaft sprocket.

8. Remove or disconnect:
- Spark plug connectors.

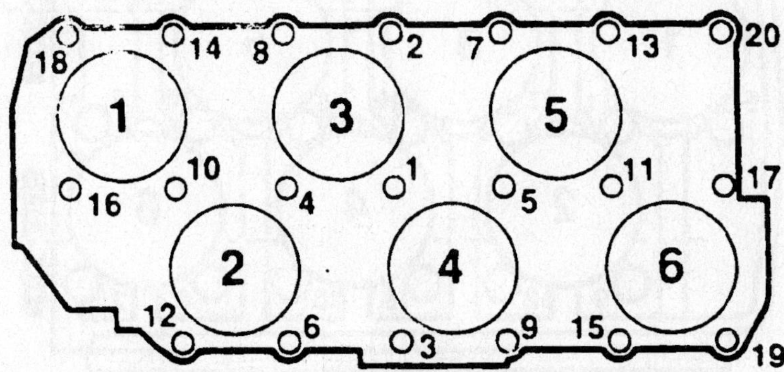

Be sure to tighten the cylinder head bolts in the sequence shown—2.8L (AAA) engine

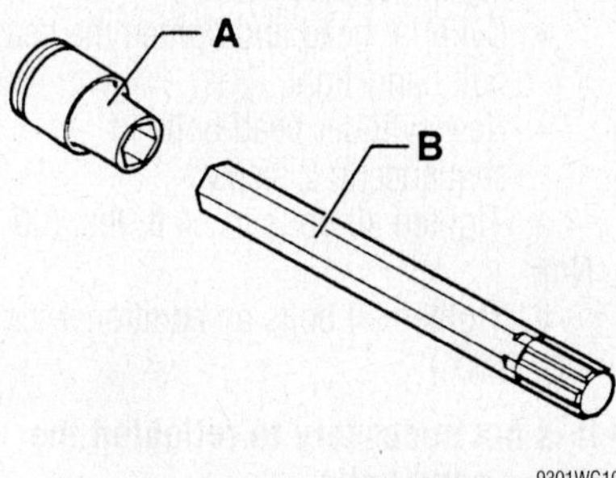

To remove and install the cylinder head bolts, use the Polydrive Special tool 3452 with a 10mm socket—2.8L (AHA/ATQ) engine

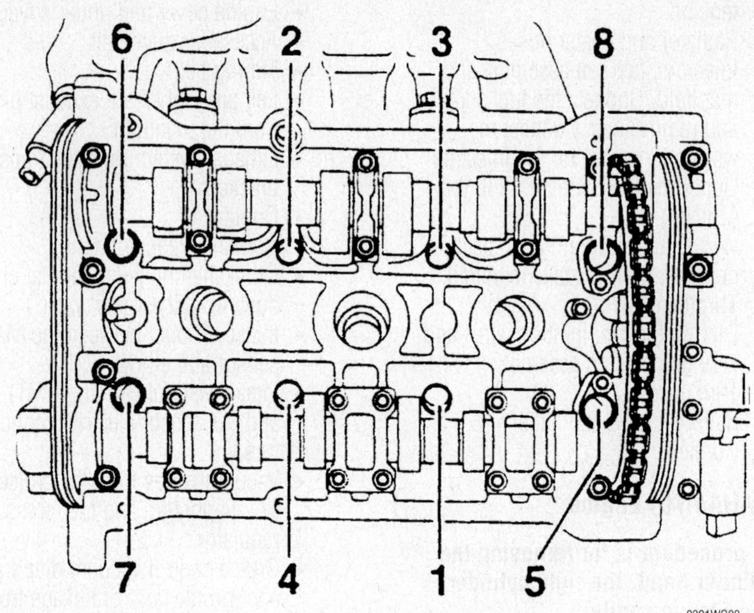

Cylinder head bolt torque sequence—2.8L (AHA/ATQ) engine

- Ignition coil bracket and position it aside with the coils.
- Intake manifold.
- Timing belt rear cover.
- Coolant hoses related to the cylinder head service.
- Valve cover.

9. Using Polydrive Special tool 3452, or equivalent, loosen, then remove the cylinder head bolts in the reverse order of the tightening sequence.

To install:

➡The bolt holes in the engine block MUST be clean and free of debris and fluid. Always replace the cylinder head bolts.

10. Set the crankshaft and camshaft to TDC for cylinder No. 3.
11. Install or connect:
    - Head gasket, making sure that the part No. or the word "OBEN" is facing the cylinder head.
    - Cylinder head and tighten the head bolt hand-tight.
    - New cylinder head bolts in sequence in 2 steps:

    a. Tighten all bolts to 44 ft. lbs. (60 Nm).

    b. Tighten all bolts an additional ½ turn (180°).

➡It is not necessary to retighten the cylinder head bolts.

- Timing belt rear cover.
- Any coolant hoses that were removed.
- Intake manifold.
- Ignition coil bracket with the coil attached.
- Camshaft sprocket. Retaining bolt: 41 ft. lbs. (55 Nm).
- Camshaft Locating Bar 3391, tighten the 2 bolts for the camshafts, then remove the tool.
- Fuel rail to the intake manifold.
- Fuel supply and return lines.
- All hoses, vacuum lines and electrical connectors.
- Any tie wraps that were cut during the removal steps.
- Exhaust system.
- Coolant expansion tank and fill the cooling system.

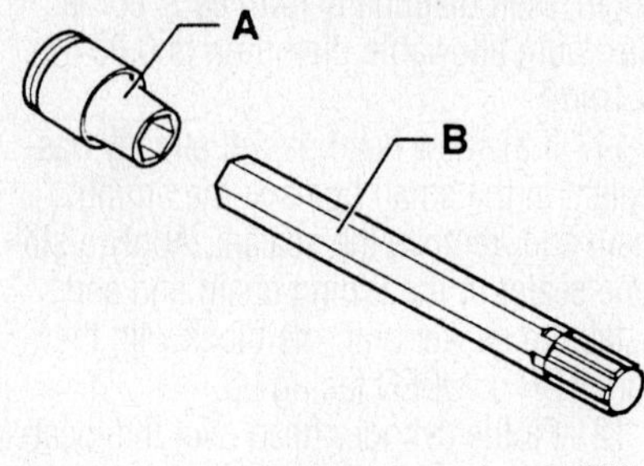

To remove and install the cylinder head bolts, use the Polydrive Special tool 3452 with a 10mm socket—2.8L (BDF) engine

- Timing belt.
- Accessory drive belt.

12. Return the lock carrier to the normal position.
13. Connect the negative battery cable.
14. Install the engine cover and under cover.

2.8L (BDF) Engine

1. Before servicing the vehicle, refer to the Precautions Section.
2. Disconnect the negative battery cable.
3. Remove the engine top cover.
4. Remove the engine insulation cover.
5. Drain the engine coolant into an approved container.
6. Disconnect the coolant hoses at the radiator.
7. Disconnect the connectors to the ignition coils with power output stage and move the wiring harness clear.
8. Remove the ignition coils using T10094 extractor.

    a. Remove or disconnect:

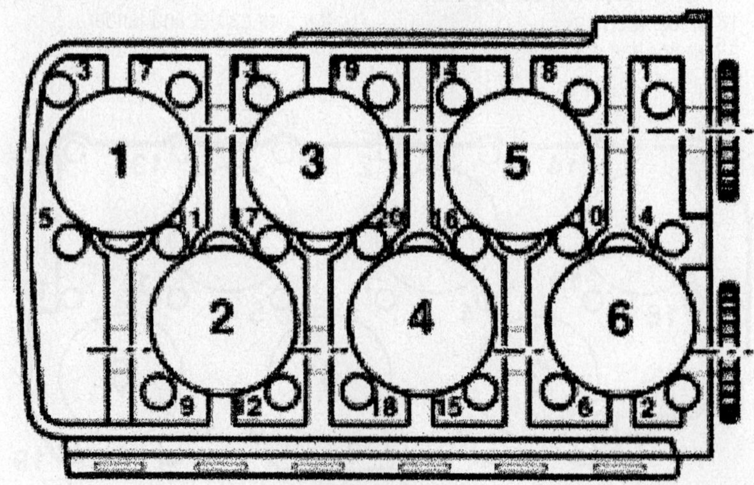

Cylinder head bolt removal sequence—2.8L (BDF) engine

- Connector from throttle valve control module.
- Coolant hoses from throttle valve control module and plug hose ends.
- All vacuum lines at cylinder head.
- Center support between intake manifold and heat shield/exhaust manifold.
- Pressure hose at combination valve and connecting hose at cylinder head cover.
- Unclip pressure hose and all other lines from retainers at intake manifold and at cylinder head cover.
- Intake manifold supports at both sides at bracket from heat shield/exhaust manifold.
- Secondary Air Injection (AIR) pump motor.
- Dipstick and dipstick tube.
- Intake manifold with throttle valve control module.
- Cylinder head cover.
- Fuel supply and return lines.
- Coolant temperature sensor.
- All remaining harness connectors and vacuum lines at cylinder head.
- Thermostat housing.
- Exhaust pipe from exhaust manifold.
- Timing chain.

9. Using Polydrive Special tool 3452, or equivalent, loosen, then remove the cylinder head bolts in the proper sequence.
10. Remove the cylinder head.

To install:

➡The bolt holes in the engine block MUST be clean and free of debris and fluid. Always replace the cylinder head bolts.

11. Set the crankshaft and camshaft to TDC for cylinder No. 1.
12. Install or connect:
    - Head gasket, making sure that the part No. faces up.

➡Make sure dowel sleeves in cylinder block bore 12 and 20 are inserted and the cylinder head gasket is secured. The longer cylinder head bolts go into the center bores.

- Cylinder head and tighten the head bolt hand-tight.
- New cylinder head bolts in reverse of the removal sequence in 4 steps:

a. Tighten all bolts to 22 ft. lbs. (30 Nm).
b. Tighten all bolts to 37 ft. lbs. (50 Nm).
c. Tighten all bolts an additional ½ turn (180°), using 2 steps of 90° each.

➡It is not necessary to retighten the cylinder head bolts.

13. Installation is the reverse of the removal procedure using the following torque values.
    - Camshaft roller chain adjusting bolt: 30 ft. lbs. (40 Nm).
    - Camshaft adjuster: 88 inch lbs. (10 Nm).
    - Drive belt tensioner bolt: 88 inch lbs. (10 Nm).
    - Exhaust pipe and exhaust manifold nuts: M8 18 ft. lbs. (25 Nm), M10 30 ft. lbs. (40 Nm).
    - Intake manifold bolts: 18 ft. lbs. (25 Nm).
    - Thermostat housing bolt: 88 inch lbs. (10 Nm).
    - Valve cover bolts: 88 inch lbs. (10 Nm).
14. Connect the negative battery cable.

4.0L Engine

➡This procedure is for removing the left cylinder head, the right cylinder head service is similar.

1. Before servicing the vehicle, refer to the Precautions Section.
2. Disconnect the negative battery cable.
3. Remove the engine top cover.
4. Remove the engine insulation cover.
5. Drain the engine coolant into an approved container.
6. Disconnect the coolant hoses at the radiator.
7. Remove or disconnect:
    - Upper intake manifold.
8. Disconnect the connectors to the ignition coils with power output stage and move the wiring harness clear.
9. Remove the ignition coils using T10094 extractor.
10. Remove or disconnect:
    - Cylinder head cover
    - Fuel rail with injectors.
    - Lower intake manifold spacer.

➡Cover intake ports to prevent debris from entering.

- Coolant pipes.
- Intake pipes.

11. Turn the crankshaft until no. 1 cylinder is at TDC. Mark on crankshaft damper must align with mark on cylinder block joint.
12. Remove the camshaft chain tensioner.
13. Remove the timing chain upper cover.
14. Remove the timing chain lower cover.
15. Mark the timing chain rotation direction for installation.
16. Remove the exhaust camshaft adjuster bolts.
17. Remove the intake camshaft adjuster bolts.

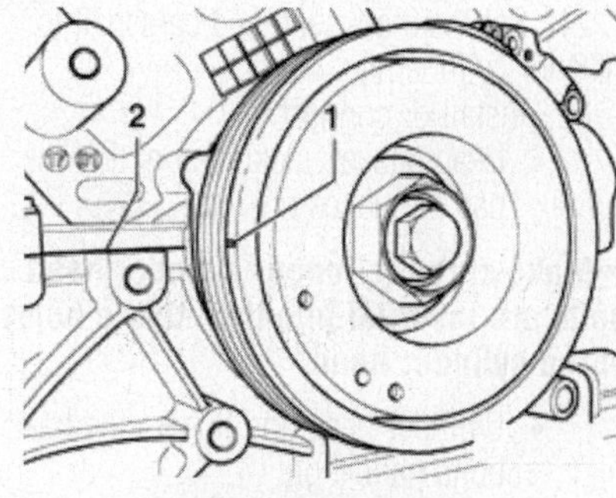

Locating TDC alignment marks—4.0L engine

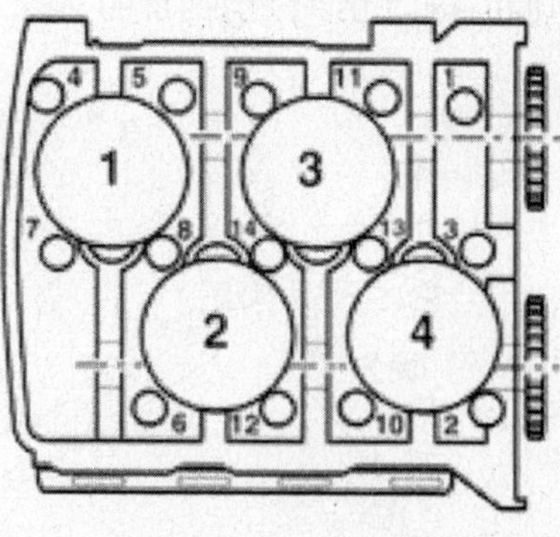

Cylinder head bolt removal sequence bank 1—4.0L engine

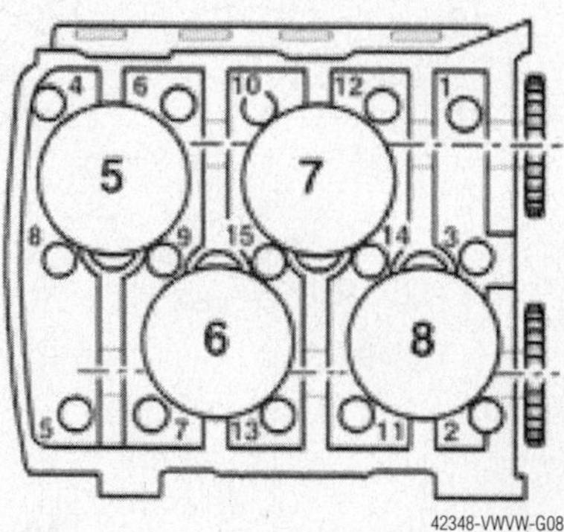

Cylinder head bolt removal sequence bank 2—4.0L engine

18. Remove the camshaft adjusters together with the timing chain.
19. Remove the timing chain guide rail.
20. Remove the cylinder head bolts in the proper sequence, and remove the cylinder head.

To install:

➡The bolt holes in the engine block MUST be clean and free of debris and fluid. Always replace the cylinder head bolts.

21. Set the crankshaft and camshaft to TDC for cylinder No. 1.
22. Install or connect:
    - Head gasket, making sure that the part No. faces up.

➡Make sure the longer cylinder head bolts are inserted into the middle holes of the cylinder head.

- Cylinder head and tighten the head bolts hand-tight.
- New cylinder head bolts in reverse of the removal sequence in 4 steps:

a. Tighten all bolts to 22 ft. lbs. (30 Nm).
b. Tighten all bolts to 37 ft. lbs. (50 Nm).
c. Tighten all bolts an additional ½ turn (180°), using 2 steps of 90° each.

➡It is not necessary to retighten the cylinder head bolts.

23. Installation is the reverse of the removal procedure using the following torque values.
    - Timing chain guide rail bolt: 88 inch lbs. (10 Nm).
    - Camshaft adjuster bolt: 44 ft. lbs. (60 Nm) plus 90°.
    - Lower timing chain cover bolts: 15 ft. lbs. (20 Nm).
    - Upper timing chain cover bolts: 88 inch lbs. (10 Nm).
    - Timing chain tensioner bolts: 88 inch lbs. (10 Nm).
    - Intake manifold bolts: 88 inch lbs. (10 Nm).
    - Fuel rail bolts: 88 inch lbs. (10 Nm).
24. Connect the negative battery cable.

Turbocharger

REMOVAL & INSTALLATION

1.8L Engine

1. Before servicing the vehicle, refer to the Precautions Section.
2. Remove or disconnect:
    - Negative battery cable
    - Engine undercover, and unbolt the air conditioning compressor
    - Turbocharger support bracket
    - Oil return line at the turbocharger
    - Air hoses from the turbocharger
    - Oil feed line at the turbocharger
    - Hose for the boost pressure regulation valve vacuum diaphragm
    - Bracket for the coolant supply line at the boost pressure regulation valve vacuum diaphragm, and using Clamp 3094 or equivalent, pinch off the coolant supply hose
    - Intake air duct between the cowl and the air cleaner housing
    - Air cleaner housing cover
3. Label and detach the following lines and electrical connectors:
    - Wastegate bypass regulator valve
    - EVAP canister purge regulator valve
    - Power outage stage
    - MAF sensor
4. Remove or disconnect:
    - Air cleaner housing and the engine cover
    - Crankcase breather hose at the valve cover
    - Oil supply line at the turbocharger
    - Heat shield, and sleeve from the coolant return hose, and using

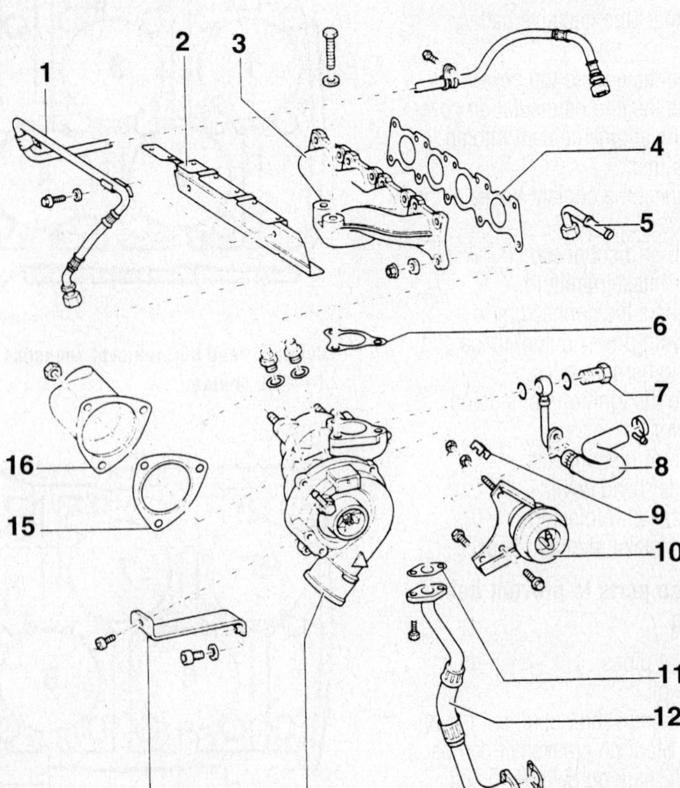

1. Oil supply line
2. Heat shield
3. Exhaust manifold
4. Exhaust manifold gasket
5. Coolant return line
6. Exhaust manifold-to-turbo gasket
7. Banjo bolt
8. Coolant supply hose
9. Fuse
10. Vacuum diaphragm for the wastegate
11. Gasket
12. Oil return line
13. Turbocharger
14. Support bracket
15. Gasket
16. Three Way Catalytic Converter (TWC)

Exploded view of the turbocharger and related components—1.8L engine (AEB/ATW/AWV)

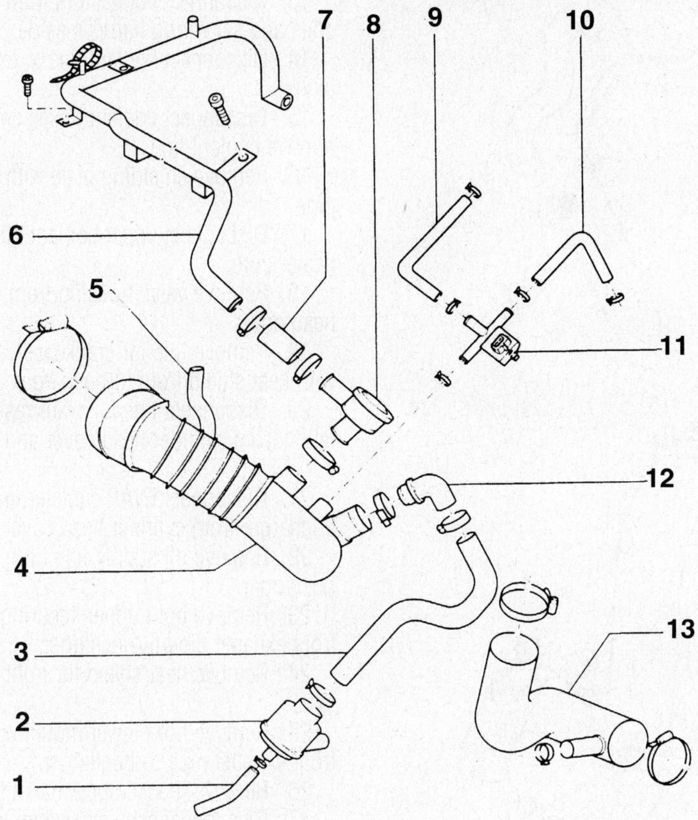

Exploded view of the hoses related to the turbocharger—1.8L engine (AEB/ATW/AWV)

Clamp 3094 or equivalent, pinch off the coolant return hose, then remove the hose.

**** WARNING**

The exhaust flexpipe may be damaged if bent more than 10 degrees.

- TWC from the turbocharger
- Turbocharger from the exhaust manifold
- Coolant supply banjo fitting
- Turbocharger

To install:

5. Install or connect:
   - Turbocharger
   - Coolant supply banjo fitting and tighten to 18 ft. lbs. (25 Nm)
   - Turbocharger to the exhaust manifold, using new gaskets. Mounting bolts, coating the bolts with Hot Bolt Paste G 052 112 A3 (or equivalent): 26 ft. lbs. (35 Nm). Turbocharger support bracket mounting bolts: 33 ft. lbs. (45 Nm).
   - TWC to the turbo
   - Coolant supply hose
   - Sleeve and heat shield to the return hose
   - Oil return hose

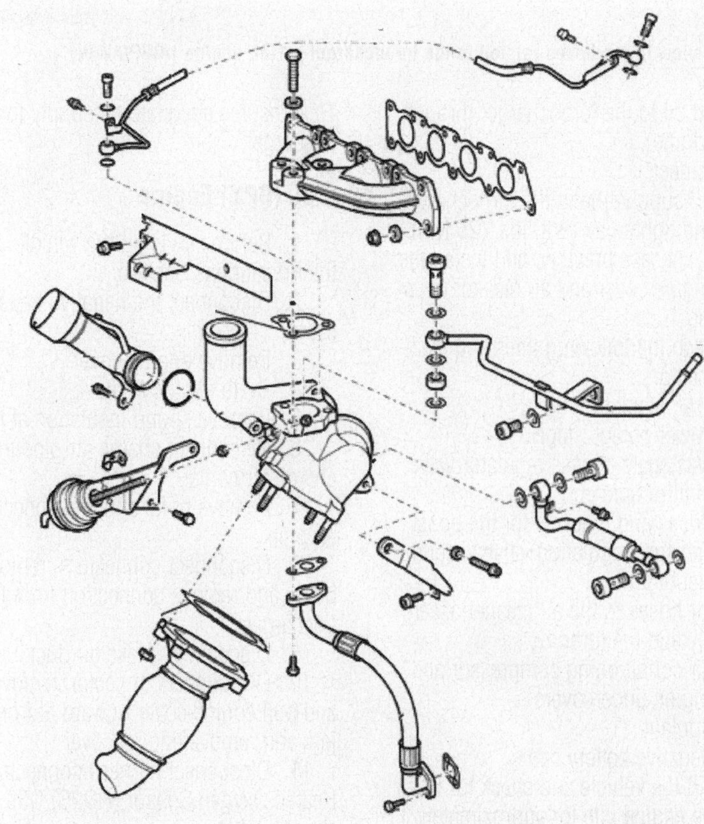

Exploded view of the turbocharger and related components—1.8L engine (AWP/AWW)

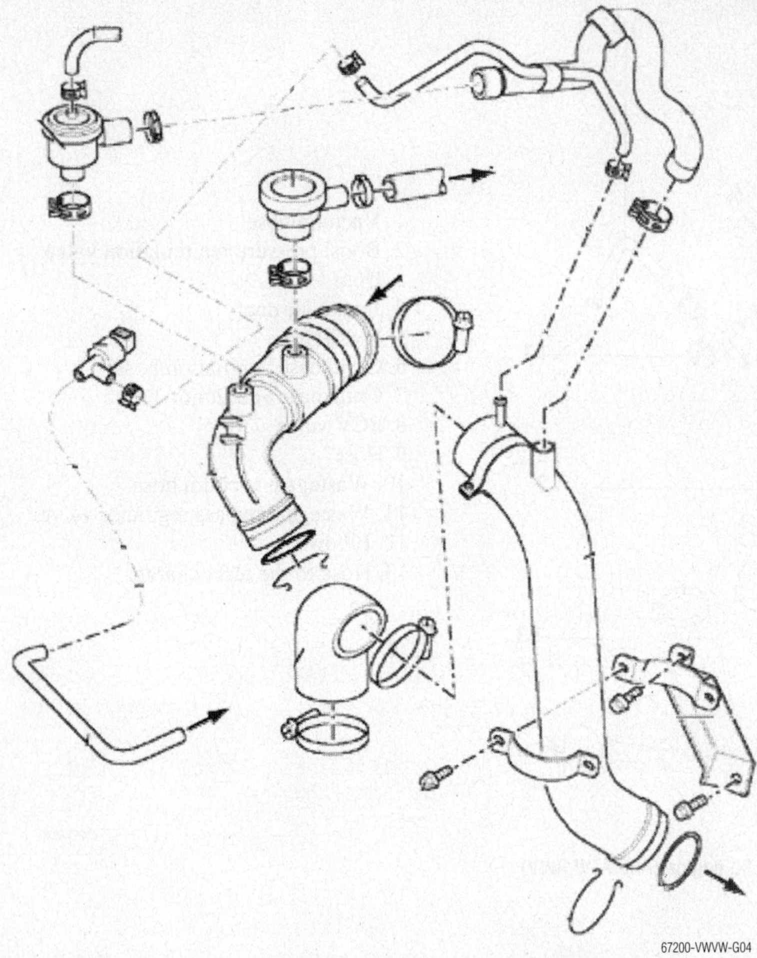

Exploded view of the hoses related to the turbocharger—1.8L engine (AWP/AWW)

6. Add oil to the turbocharger through the oil feed line.
7. Connect:
   - Oil supply line to the turbocharger and tighten to 18 ft. lbs. (25 Nm).
   - Crankcase breather, and install the engine cover and air cleaner housing.
8. Attach the following lines and electrical connectors:
   - MAF sensor
   - Power outage stage
   - Wastegate bypass regulator valve
9. Install or connect:
   - Hoses and brackets for the boost pressure regulation valve vacuum diaphragm.
   - Air hoses to the air cleaner assembly and the turbo
   - Air conditioning compressor and engine undercovers
   - Coolant
   - Negative battery cable
10. Start the vehicle and check for leaks, then let the engine idle for approximately 1 minute without increasing the engine speed. This ensures adequate oil supply to the turbocharger.

2.0L (BPY) Engine

1. Before servicing the vehicle, refer to the Precautions Section.
2. Disconnect the negative battery cable.
3. Remove engine cover.
4. Drain engine coolant.
5. Remove sound insulation at right.
6. Disconnect charge air pipe from charge air cooler.
7. Remove bolts and disconnect charge air pipe.
8. Disconnect connectors, remove bolts, and remove connection from turbocharger.
9. Disconnect intake air duct.
10. Disconnect connector, remove bolts, and pull engine cover upward at front with a jerk and remove engine cover.
11. Disconnect harness connector for Heated Oxygen Sensor (HO2S) G39 and expose cable.
12. Remove plenum chamber bulkhead.
13. Disconnect connectors from ignition coils and set wiring harness aside.
14. Disconnect coolant line to expansion tank.
15. Disconnect coolant hoses and remove coolant line.
16. Remove shielding plate with coolant pipe.
17. Disconnect upper coolant hose to heater core.
18. Remove valve housing from cylinder head cover.
19. Remove line for crankcase ventilation with heat shield from turbocharger.
20. Disconnect line for crankcase ventilation from cylinder head cover and remove it.
21. Disconnect EVAP canister line to turbocharger from cylinder head cover.
22. Remove oil supply line from turbocharger.
23. Remove both upper securing nuts of front exhaust pipe/turbocharger.
24. Remove heat shield for right drive axle.
25. Remove both lower mounting nuts of front exhaust pipe/turbocharger.
26. Remove bracket for exhaust system.
27. Disconnect exhaust system at double clamp.
28. Disconnect front exhaust pipe from turbocharger and slide it toward rear slightly.

➡The catalytic converter is not removed and the oxygen sensor connector is not disconnected.

29. Remove oil supply line and coolant supply line from turbocharger.
30. Disconnect oil return line from turbocharger.
31. Remove bolts and remove support for turbocharger.

➡The following work step is necessary to obtain some more room between cylinder head and bulkhead.

32. Remove bolts and pull engine approximately 20mm toward rear using tensioning strap T10038.
33. Disconnect coolant pipe.

➡Retaining strip nuts do not have to be loosened.

34. Remove bolt for coolant pipe and remove upper nuts.
35. Remove turbocharger/exhaust manifold.

To install:

36. When installing the turbocharger, please note the following torque specifications:

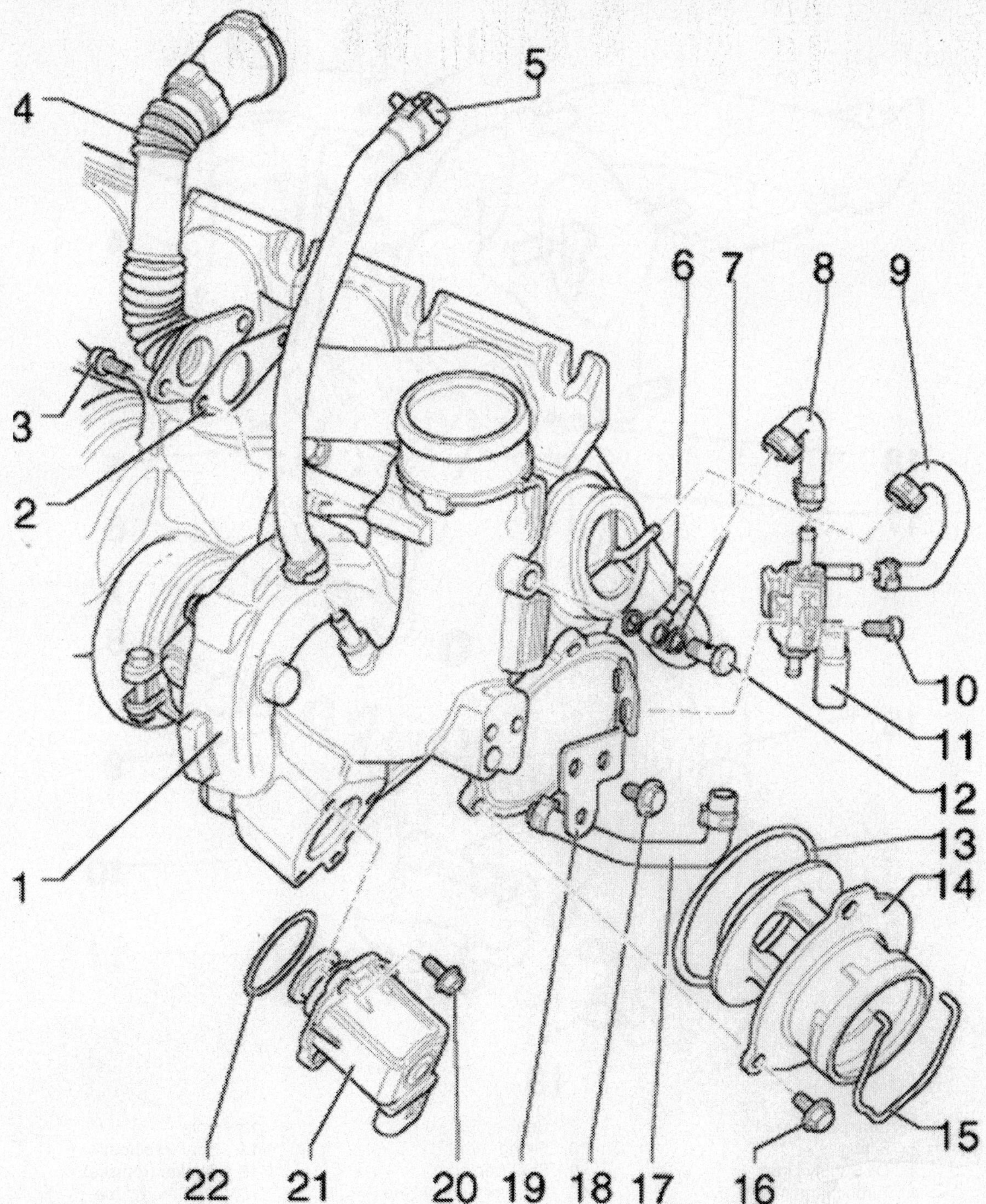

1. Turbocharger
2. Gasket (replace)
3. Screw
4. Crankcase Ventilation Line
5. Connection Hose
6. Ring Connection
7. Seal (replace)
8. Hose
9. Hose
10. Screw
11. Wastegate Bypass Regulator Valve N75
12. Fitting
13. Seal (replace)
14. Connection
15. Locking Clip
16. Bolts
17. Hose
18. Bolts
19. Bracket
20. Bolts
21. Turbocharger Recirculating Valve N249
22. Seal (replace)

06041_VWMC_G0005

Exploded view of the turbocharger and related components—2.0L engine (BPY)

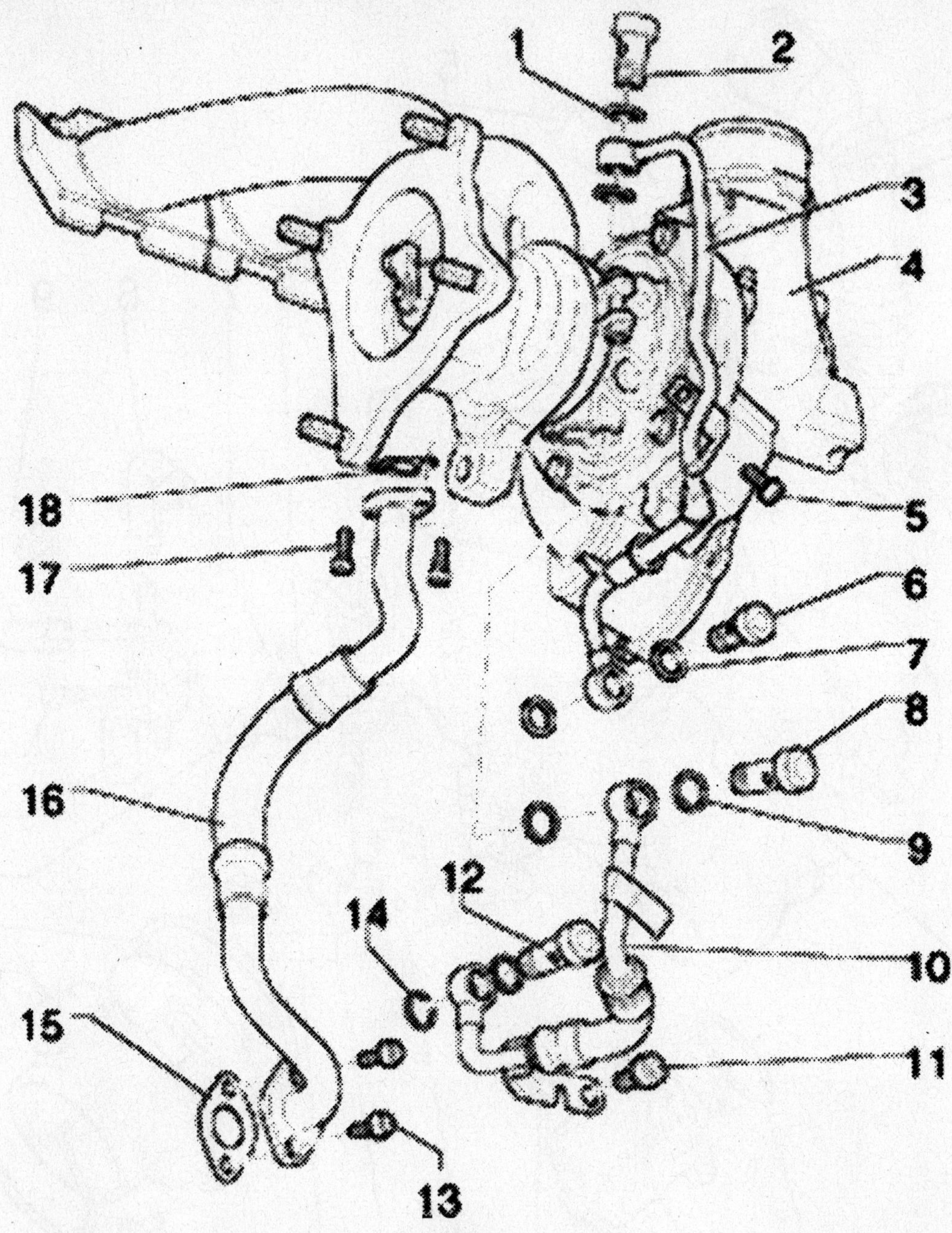

1. Seal (replace)
2. Fitting
3. Oil Supply Line
4. Turbocharger
5. Bolt
6. Fitting
7. Seal
8. Fitting
9. Seal (replace)
10. Coolant Supply Line
11. Bolt
12. Fitting
13. Bolt
14. Seal (replace)
15. Gasket (replace)
16. Oil Return Line
17. Screws
18. Gasket (replace)

06041_VWMC_G0006

Exploded view of the hoses and lines related to the turbocharger—2.0L engine (BPY)

- Exhaust manifold/turbocharger-to-cylinder head with new nuts: 15 ft. lbs. (20 Nm)
- Oil supply line-to-exhaust turbocharger: 21 ft. lbs. (30 Nm)
- Turbocharger bracket-to-engine block: 21 ft. lbs. (30 Nm)
- Turbocharger bracket-to-turbocharger: 21 ft. lbs. (30 Nm)
- Right charge air pipe-to-oil pan: 7 ft. lbs. (10 Nm)
- Insert with hot bolt paste G 052 112 A3

37. Installation is performed in reverse order of removal. Note the following:
    - Always replace seals, gaskets and self-locking nuts
    - Add oil to turbocharger through oil feed line connecting piece
    - After installing turbocharger, let

engine idle for approx. 1 minute to ensure adequate oil supply to the turbocharger
- Hose unions and hoses for charge air system must be free of oil and grease before installing
- Secure all hose connections with hose clamps

38. Install exhaust system and align it free of tension.
39. When assembling charge air pipes with harness connector, make sure that securing clip engages audibly into retaining tab.
40. Refill cooling system.

Intake Manifold

REMOVAL & INSTALLATION

1. Before servicing the vehicle, refer to the Precautions Section.
2. Remove or disconnect:
   - Negative battery cable.
   - Engine cover as required.
   - Air duct from the throttle body and disconnect the accelerator cable.
   - Vacuum and coolant hoses as required.
   - Any remaining wiring as required.
   - Fuel supply and return lines.
   - Intake manifold support brackets.
   - If equipped, EGR pipe.
   - If equipped, bolts to separate the upper intake manifold from the lower intake manifold.
   - Bolts and remove the manifold from the cylinder head.

To install:

3. Install the intake manifold(s) to the cylinder head with a new gasket(s). Tighten the bolts as follows:
   - 1.8L, 2.0L (BPY) and 2.8L (AHA/ATQ) engines: 7 ft. lbs. (10 Nm)

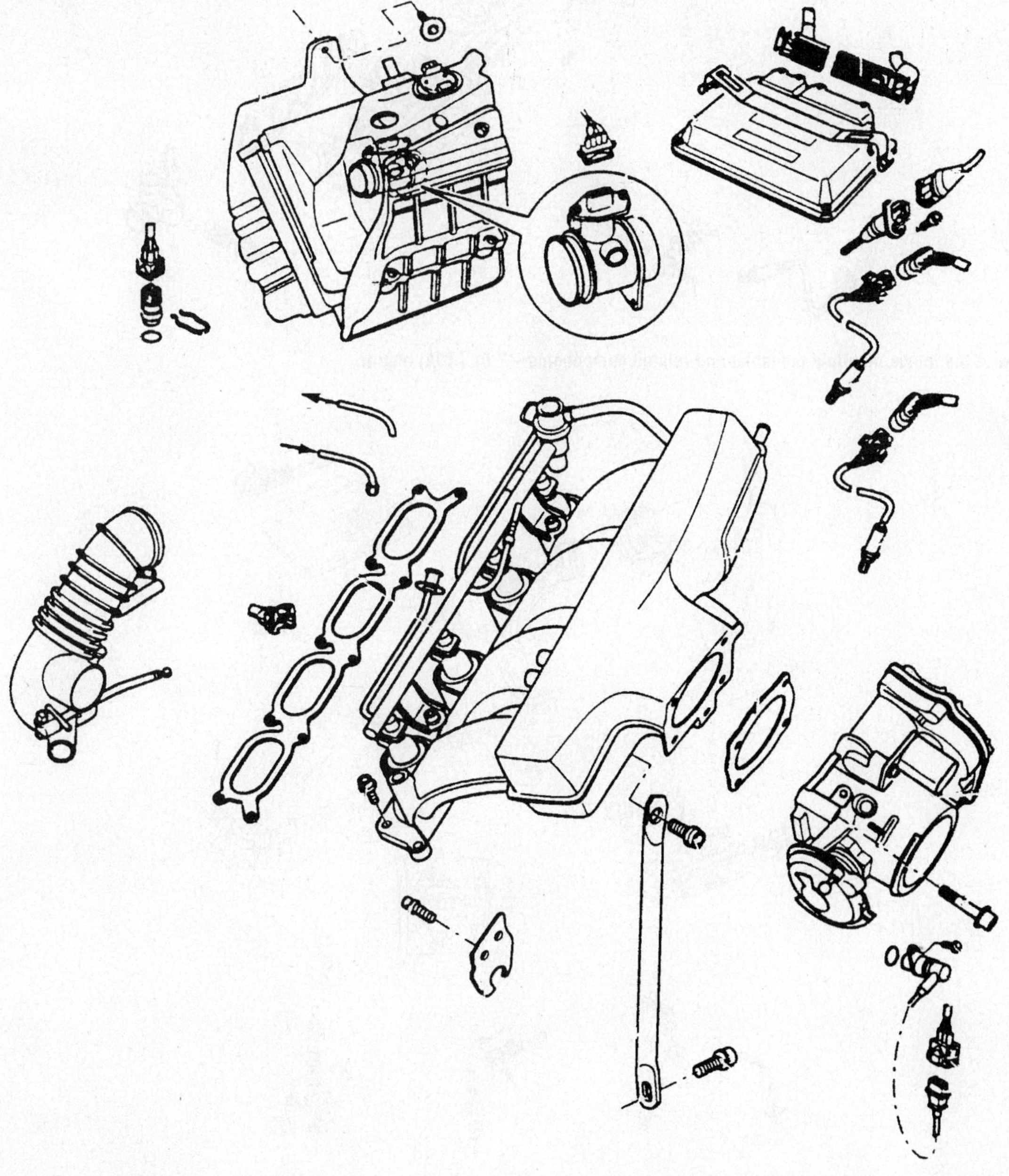

Exploded view of the intake manifold mounting and related components—1.8L engine

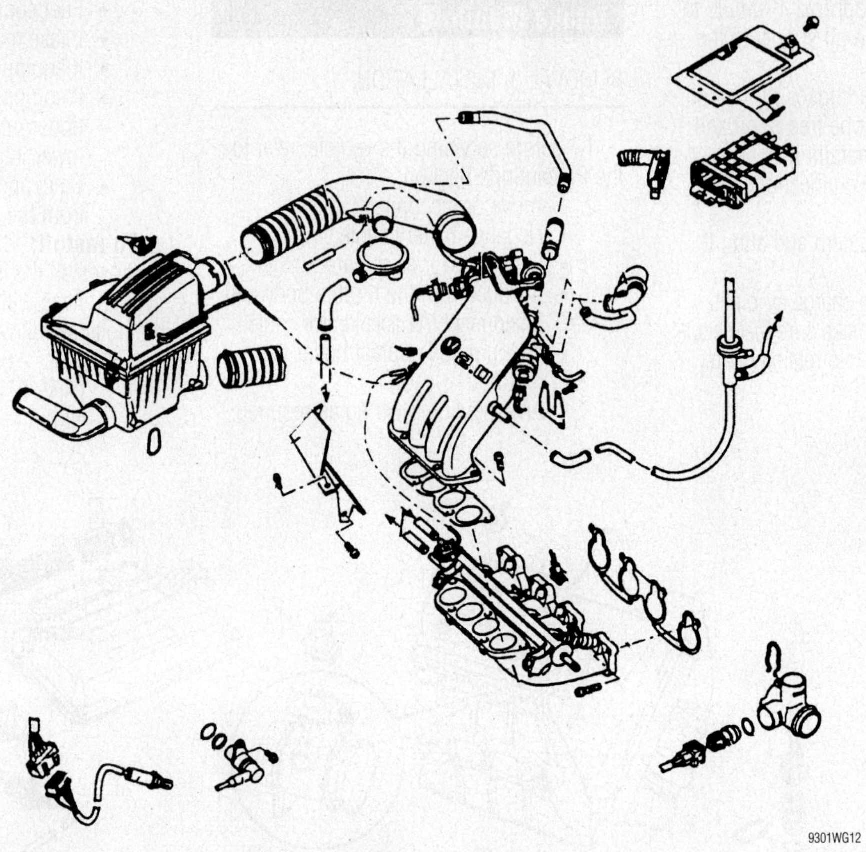

Exploded view of the intake manifold mounting and related components—2.0L (ABA) engine

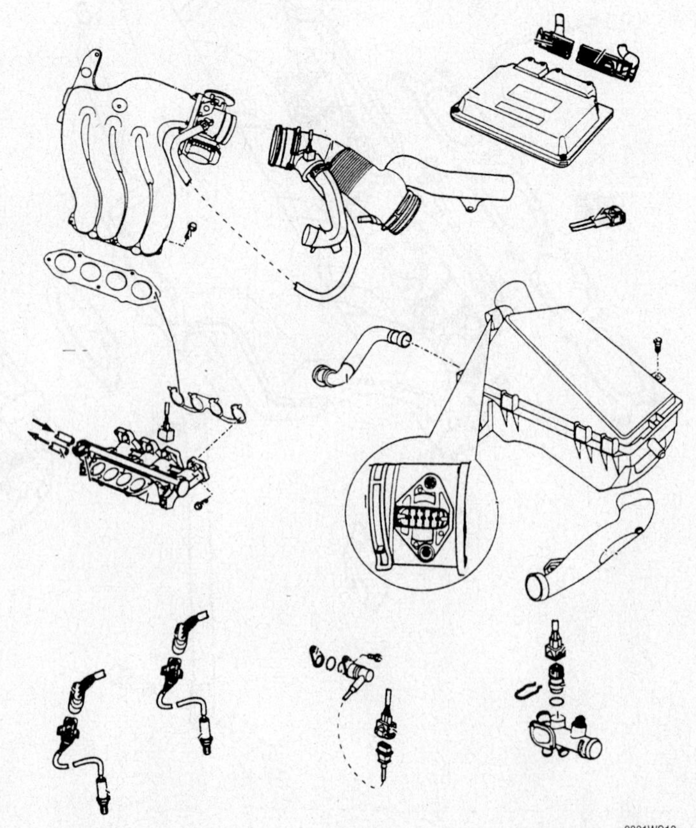

Exploded view of the intake manifold mounting and related components—2.0L (AEG) engine

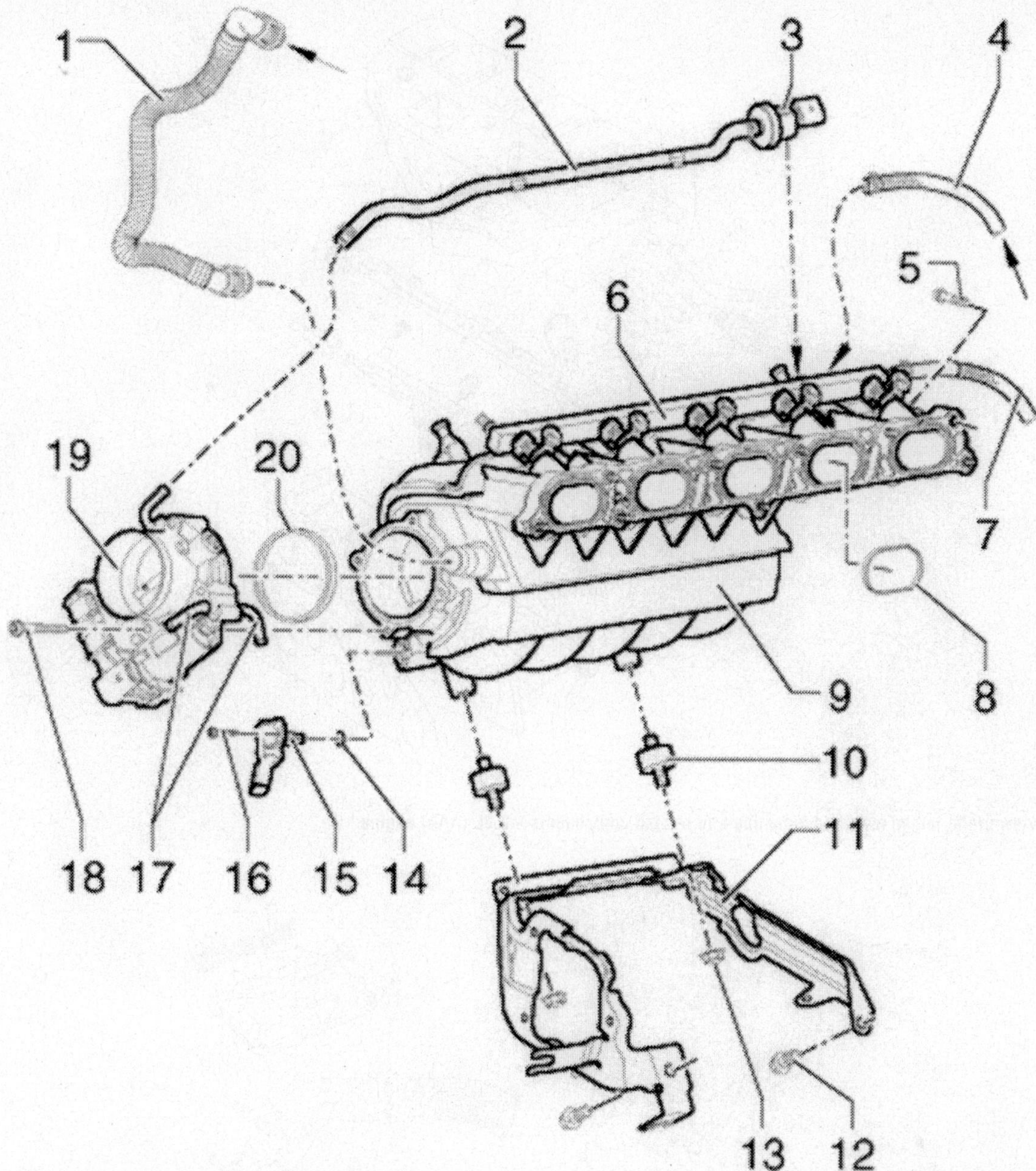

1. Bleeder Hose for Crankcase Ventilation
2. Ventilation Hose
3. EVAP canister purge regulator valve N80
4. Vacuum Hose
5. Mounting Bolts
6. Fuel Rail
7. Fuel Supply Line
8. Gasket (replace)
9. Intake Manifold
10. Rubber Bushing
11. Intake Manifold Support
12. Screw
13. Nut
14. O-ring (replace if damaged)
15. MAP Sensor G71
16. Screw
17. Coolant Conections
18. Screw
19. Throttle Valve Control Module J338
20. Oil Seal (replace if damaged)

06041_VWMC_G0007

Exploded view of the intake manifold mounting and related components—2.5L (BGP/BGQ/BPR/BPS) engine

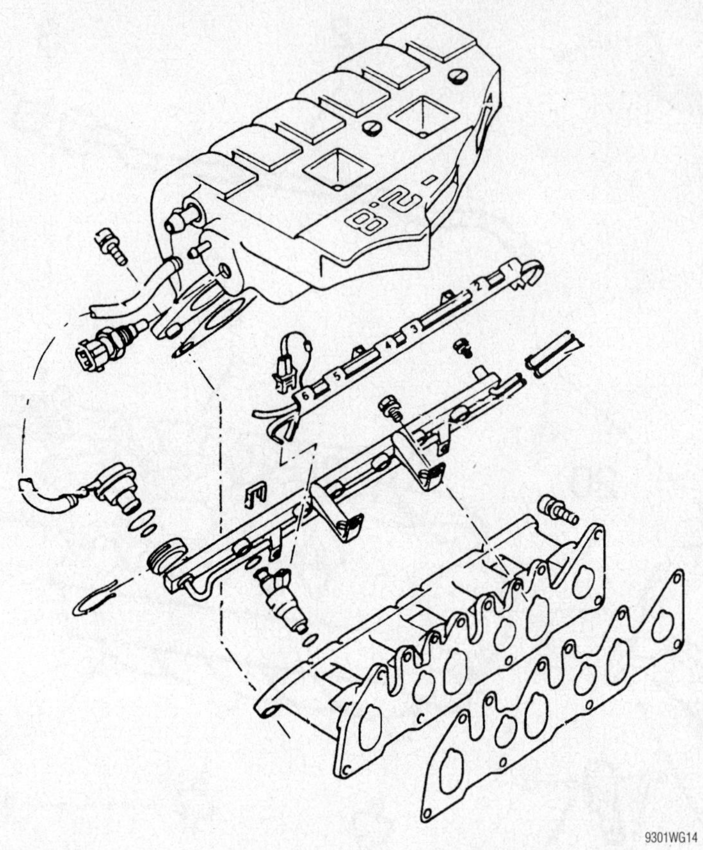

Exploded view of the intake manifold mounting and related components—2.8L (AAA) engine

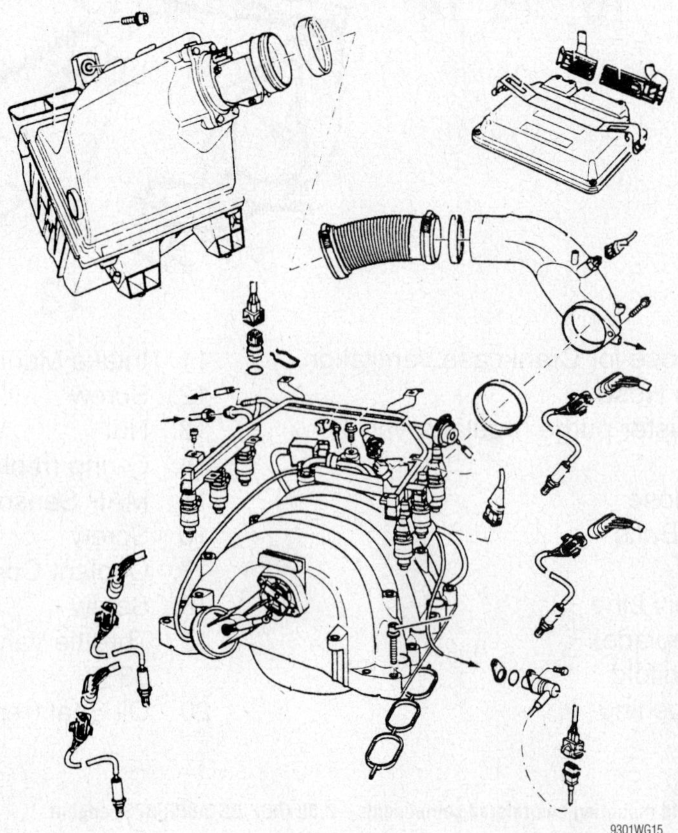

Exploded view of the intake manifold mounting and related components—2.8L (AHA) engine

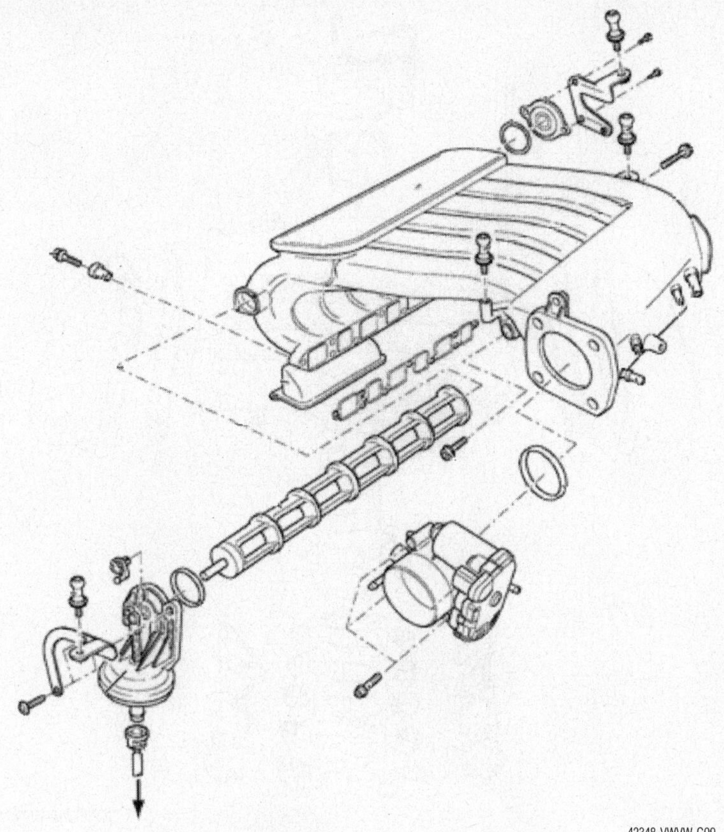

Exploded view of the intake manifold mounting and related components—2.8L (BDF) engine

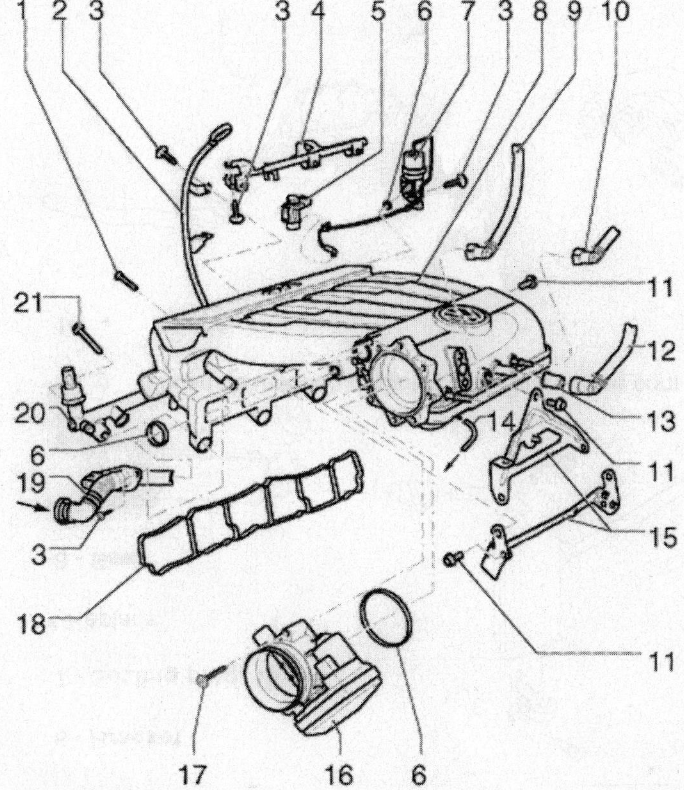

1. Screw
2. Oil Dipstick Guide tube
3. Screw
4. Bracket
5. Intake Manifold Runner Control (IMRC) Valve N316
6. Seal (replace)
7. Vacuum Actuator
8. Intake Manifold
9. EVAP Canister Purge Regulator Balve N80
10. Leak Detection Pump (LDP) V144
11. Screw
12. Suction Jet Pump Hose
13. Sealing Grommet
14. Connection to IMRC Valve N316
15. Intake Manifold Support
16. Throttle Valve Control Module J338
17. Throttle Valve Control Module Fasteners
18. Gasket (replace)
19. Crankcase Ventilation Connection
20. Fuel Distributor
21. Ruel Rail Fasteners

06041_VWMC_G0008

Exploded view of the intake manifold mounting and related components—3.6L (BLV) engine

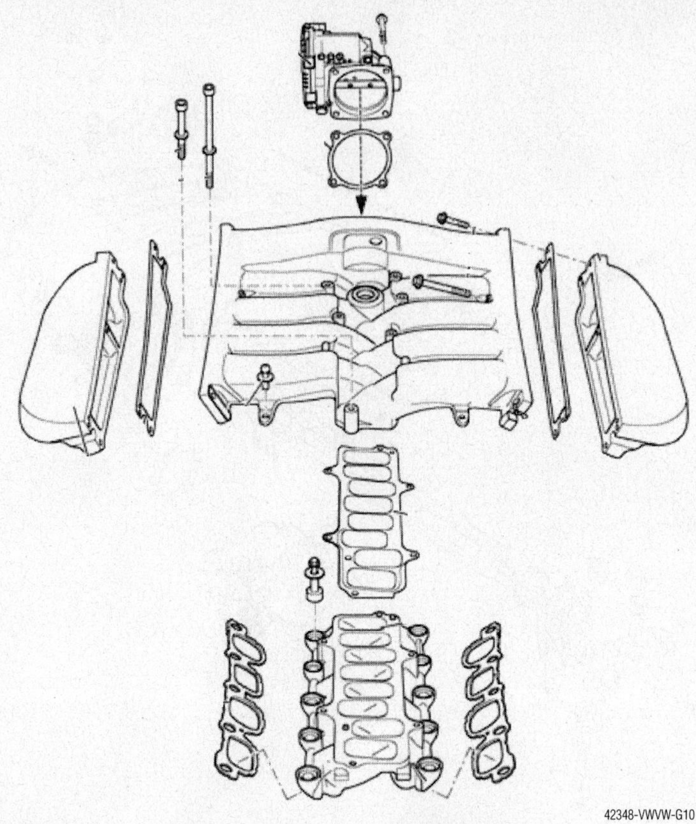

Exploded view of the intake manifold mounting and related components—4.0L (BDP) engine

- 2.0L (ABA/AEG/AVH/AZG/BDC/BEV/BGD) engine upper and lower mounting bolts: 15 ft. lbs. (20 Nm)
- 2.5L (BGP/BGQ/BPR/BPS) engines: 6 ft. lbs. (9 Nm)
- 2.8L (AAA) engine upper and lower mounting bolts: 18 ft. lbs. (25 Nm)
- 2.8L (BDF) engines: 10 ft. lbs. (13 Nm)
- 3.6L (BLV) and 4.0L (BDP) engines: 71 inch lbs. (8 Nm)

4. On 6-cylinder engines, if the fuel injectors were removed, examine the injector O-rings and replace as required. Install the injectors and rail.

5. Install or connect:
   - Fuel system hoses or the injectors to protect the system.
   - All vacuum hoses and wiring.
   - If equipped, EGR pipe.
   - Throttle cable.
   - Remaining components.

Exhaust Manifold

REMOVAL & INSTALLATION

1. Before servicing the vehicle, refer to the Precautions Section.
2. Remove or disconnect:

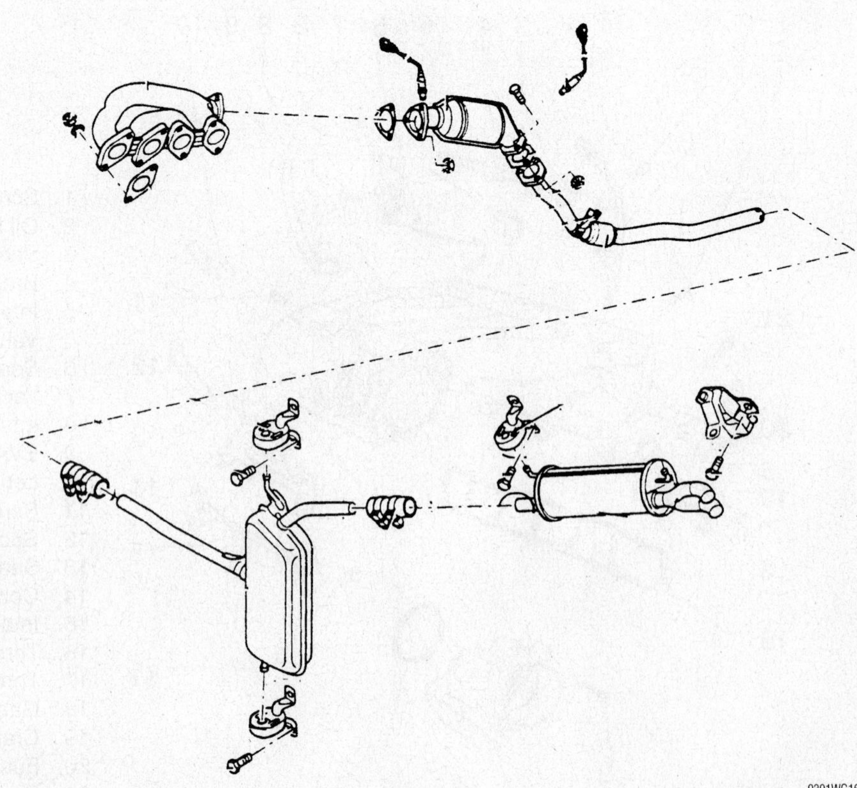

Exploded view of the exhaust manifold mounting and related components—1.8L engine

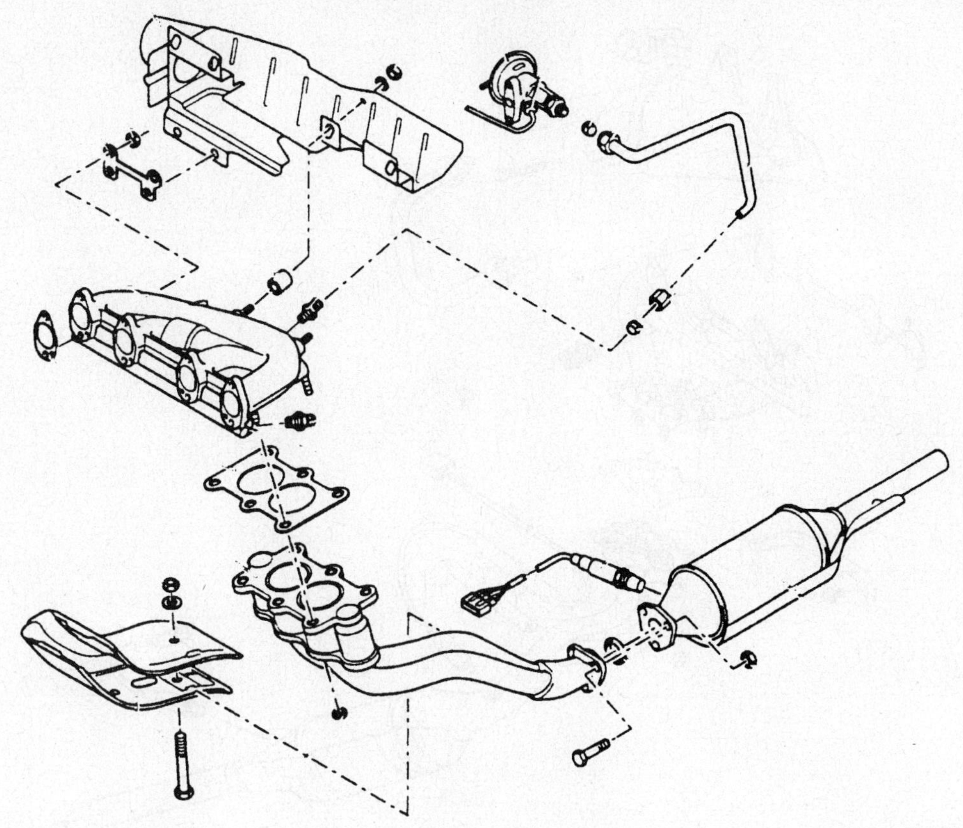

Exploded view of the exhaust manifold mounting and related components—2.0L (ABA) engine

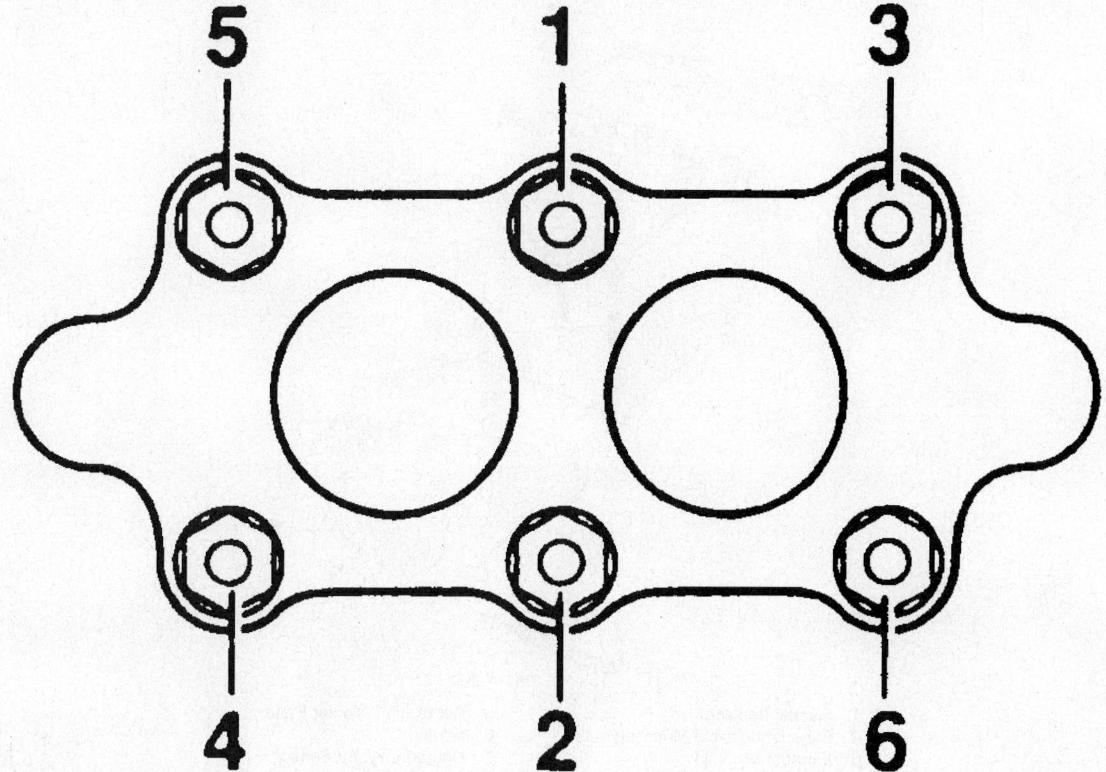

Exhaust pipe-to-manifold torque sequence—2.0L (ABA) engine

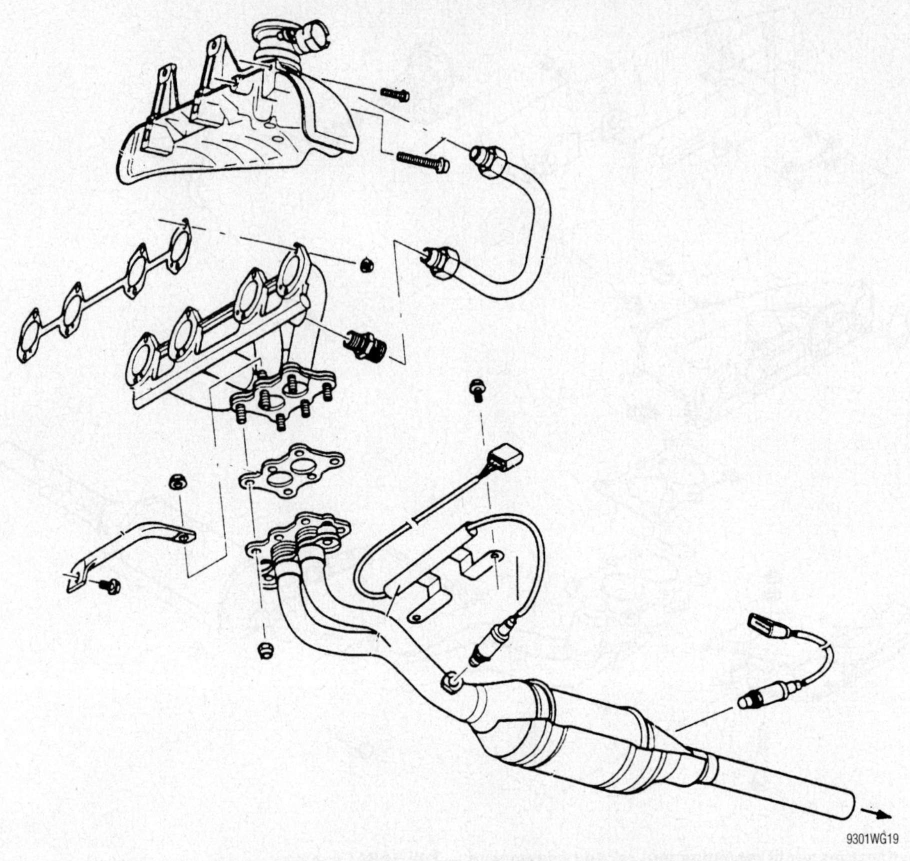

Exploded view of the exhaust manifold mounting and related components—2.0L (AEG/AVH/AZG/BDC) engine

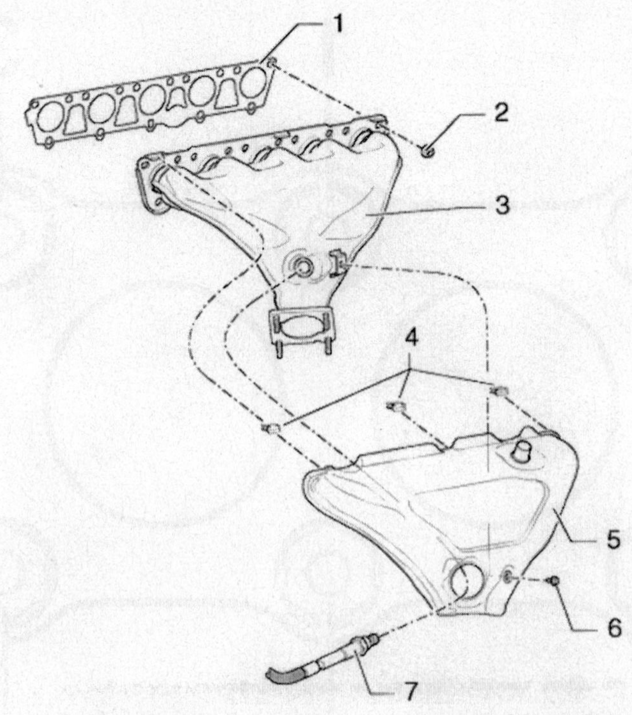

Exploded view of the exhaust manifold mounting and related components—2.5L (BGP/BGQ/BPR/BPS) engine

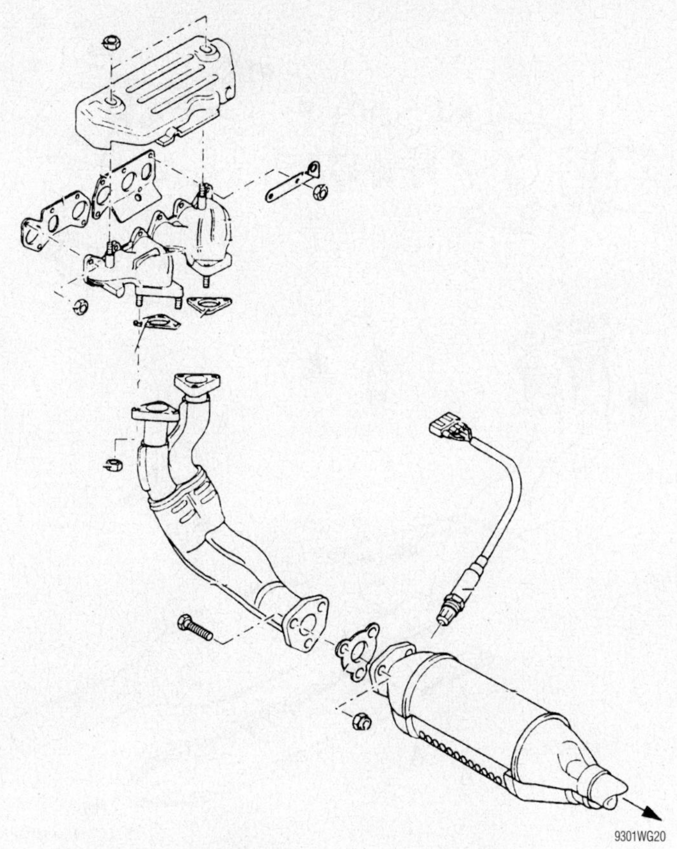

Exploded view of the exhaust manifold mounting and related components—2.8L (AAA) engine

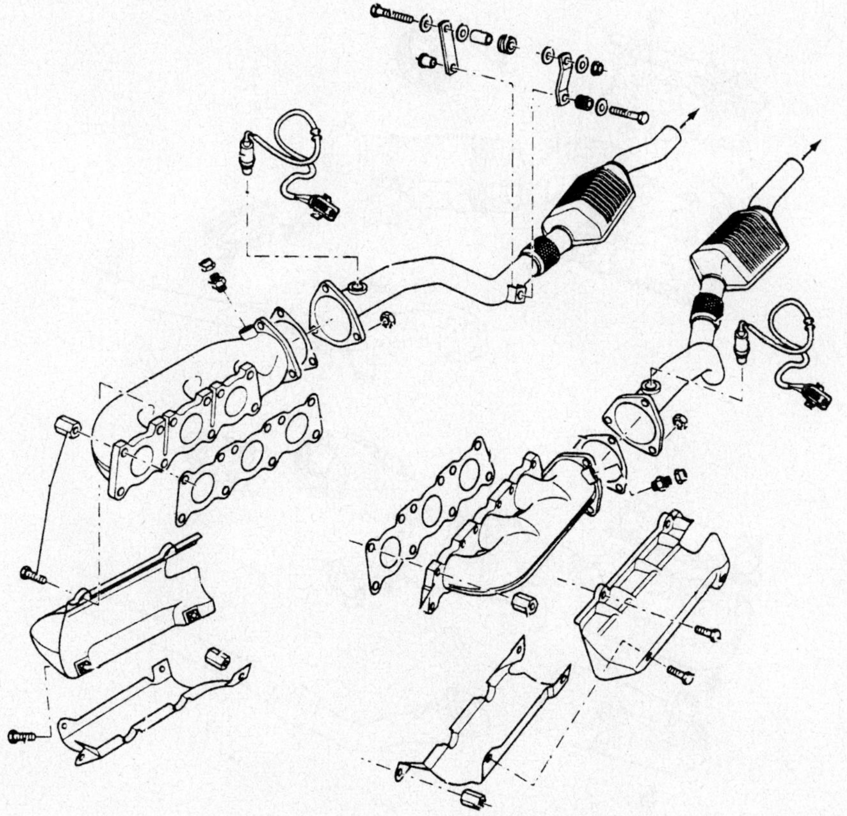

Exploded view of the exhaust manifold mounting and related components—2.8L (AHA/ATQ) engine

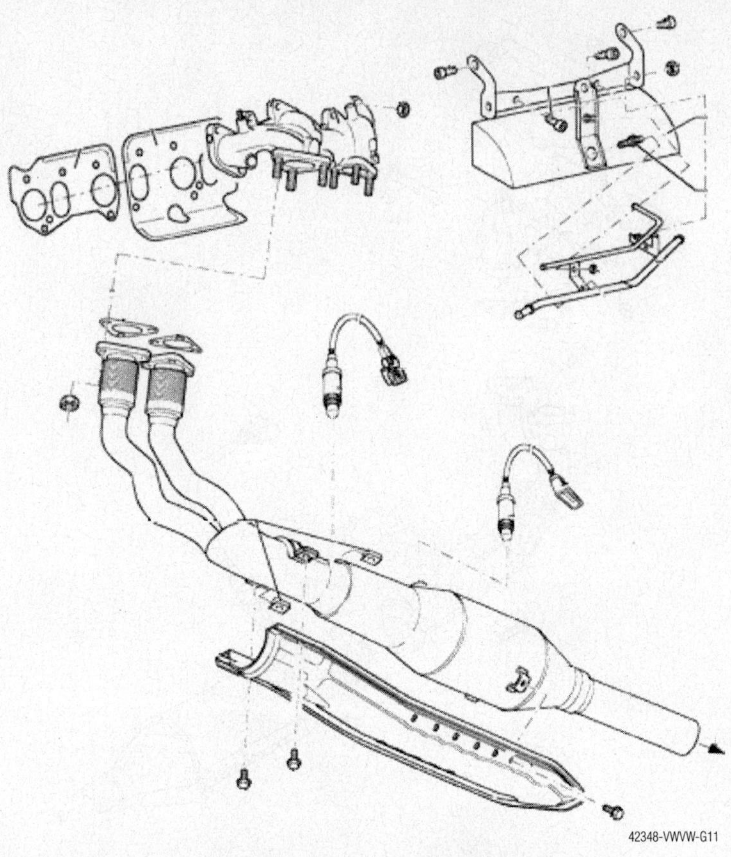

Exploded view of the exhaust manifold mounting and related components—2.8L (BDF) engine

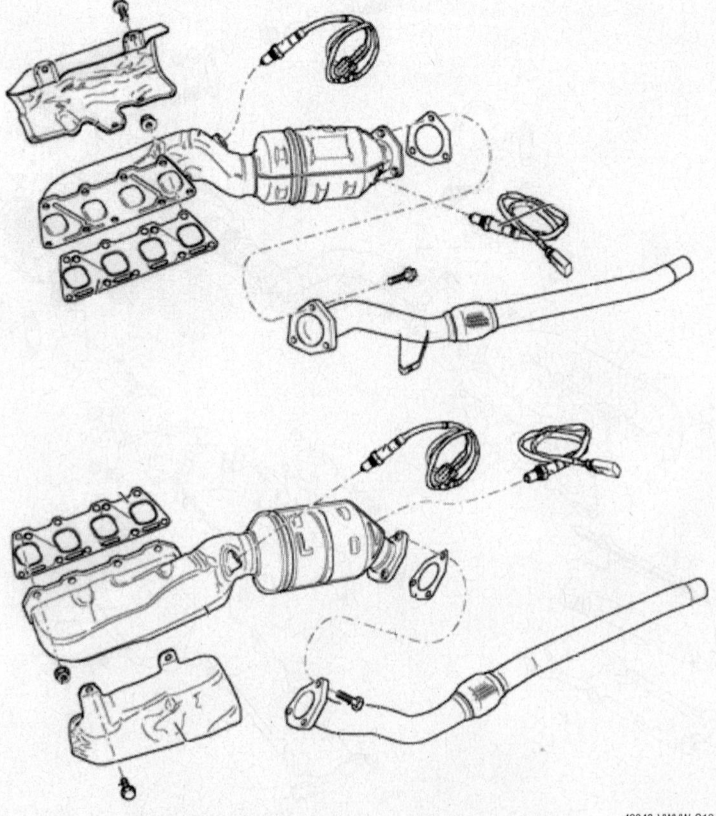

Exploded view of the exhaust manifold mounting and related components—4.0L (BDP) engine

- O_2 sensor wiring and remove any heat shields that may be in the way.
- Exhaust support brackets as necessary.
- Front exhaust pipe from the manifold or turbocharger.
- If equipped, the turbocharger.
- Self-locking nuts or bolts and the manifold.

To install:

3. Installation is the reverse of removal. Use new gaskets and self-locking nuts. Nuts: 18 ft. lbs. (25 Nm).

➡Coat stud bolts on cylinder head with hot bolt paste.

4. If the exhaust pipe is bolted to the manifold, install a new gasket and use new self-locking nuts. Nuts: 30 ft. lbs. (40 Nm).
5. Check and clear any DTCs.

Front Crankshaft Seal

REMOVAL & INSTALLATION

4-Cylinder Engines

1. Before servicing the vehicle, refer to the Precautions Section.
2. Remove or disconnect:
   - Negative battery cable.
   - Accessory drive belts.
   - Timing belt.
3. Hold the crankshaft sprocket with tool 3099 and remove the center bolt and sprocket.
4. Unscrew the inner part of Oil Seal Extractor 2085 or equivalent out of the outer part about 2 turns.
5. Install the socket head bolt into the crankshaft to guide the tool.
6. Apply oil to Oil Seal Extractor 2085 or equivalent. Apply firm pressure and screw the tool into the oil seal as far as it will go.
7. Loosen the knurled screw and turn the inner part against the crankshaft until the seal is removed.
8. Remove the socket head bolt.

To install:

9. Install the guide sleeve from tool 3083 on the end of the crankshaft.
10. Apply oil to the lip of the seal and slide the seal over the guide sleeve.
11. Press the seal in using the thrust sleeve from tool 3083 and a socket head bolt.
12. Remove the tools and install the crankshaft sprocket. Bolt: 66 ft. lbs. (90 Nm) plus ¼ turn.
13. Install or connect:
    - Timing belt.
    - Accessory drive belt and remaining components.
    - Negative battery cable.

5-Cylinder Engines

1. Before servicing the vehicle, refer to the Precautions Section.
2. Remove drive belt for A/C compressor.
3. Lock crankshaft in place and remove belt pulley from crankshaft.
4. Remove tensioning element.
5. Remove bolts and begin pressing off sealing flange at alignment sleeves using a suitable screwdriver.
6. Use removal wedge 3409 to support screwdriver so that sealing surface of cylinder block will not be damaged.

➡The sealing flange will be damaged when removing.

7. Press off sealing flange completely.

➡After removing sealing flange, clean trim removal wedge 3409 which is intended for the removal of interior equipment parts.

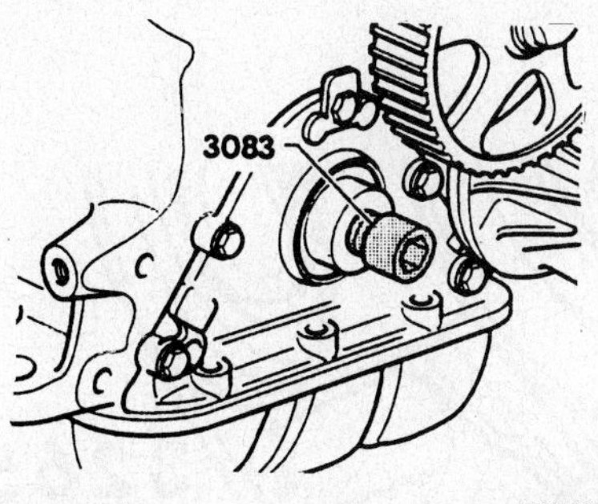

Install the socket head bolt (3083) in the crankshaft to guide the Seal Extractor—4-cylinder engines

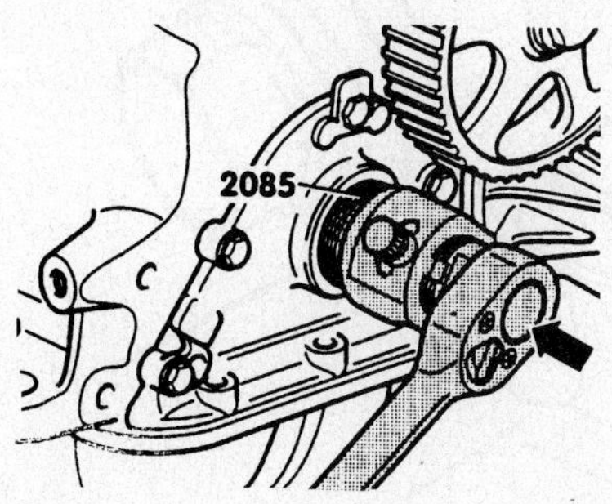

Screw the Seal Extractor (2085) into the oil seal while applying pressure—4-cylinder engines

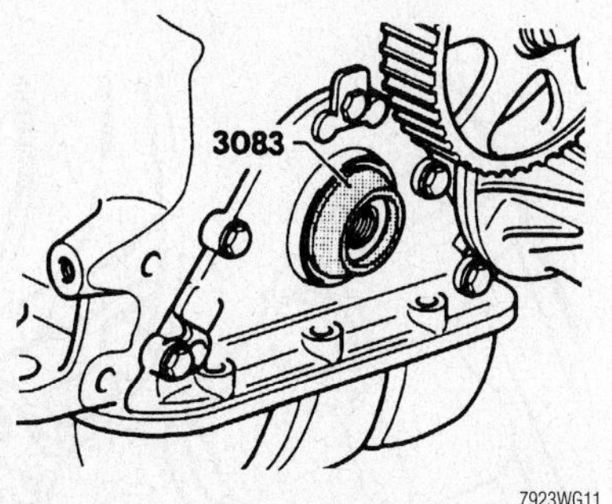

Install the guide sleeve (3083) on the crankshaft—4-cylinder engines

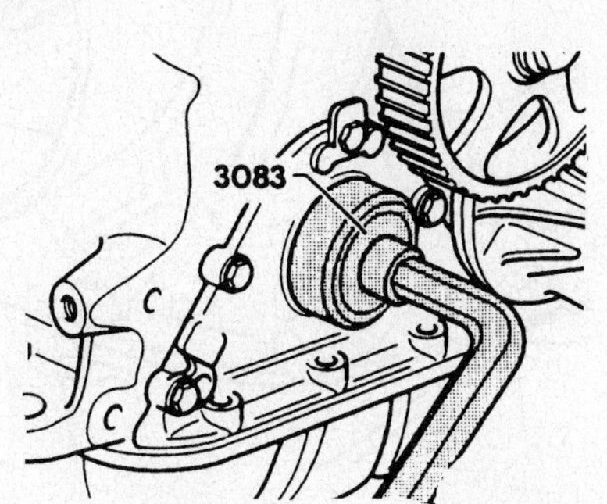

Tighten the socket head bolt to press the seal into place—4-cylinder engines

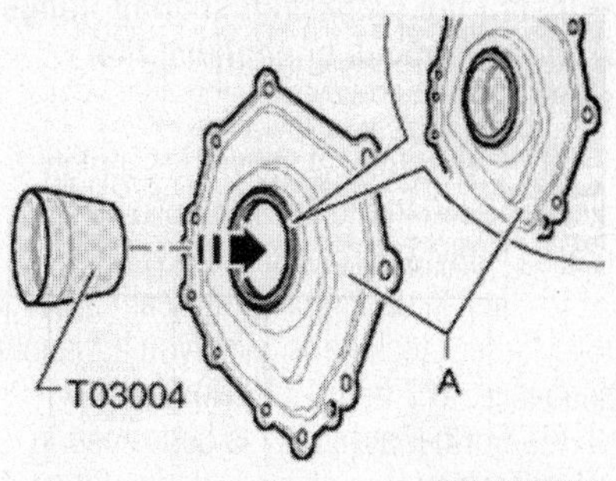

Spread sealing lip of sealing flange with assembly sleeve T03004—5-cylinder engines

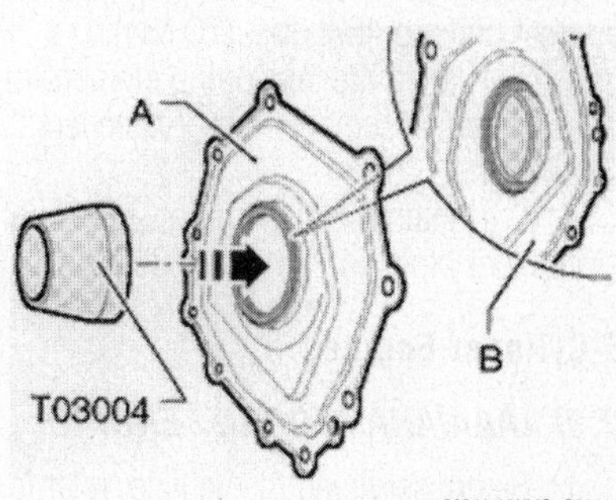

After a short while, remove assembly sleeve T03004, rotate it 180° and insert it into seal—5-cylinder engines

To install:

8. While wearing safety goggles, remove sealant residue from the cylinder block, e.g. with a rotating plastic brush. Make sure that no sealant residue enters the engine.
9. Clean sealing surfaces of cylinder block and end of crankshaft; they must be free of oil and grease.

➡Do not oil or grease the sealing lip of the sealing flange! To prevent the sealing lip of the sealing flange from inverting, follow the work steps exactly.

10. Spread sealing lip of sealing flange, as shown, with assembly sleeve T03004. The surface **A** is the outer side.
11. After a short while, remove assembly sleeve T03004, rotate it 180° and insert it into seal. The assembly sleeve T03004 must protrude approx. 0.12 inch (3mm) on the inner side **B**. The surface **A** is the outer side. The surface **B** is the inner side (sealing surface).
12. Cut tube nozzle at front marking (nozzle: approx. 0.08 inch (2mm) diameter).
13. Apply a 0.10–0.12 inch (2.5–3mm) width x 0.04 inch (1mm) height bead of sealant into the outer groove of the mating surface of the sealing flange.

➡Install the crankshaft sealing flange within 5 minutes of applying the sealant.

14. Set sealing flange with assembly sleeve T03004 onto end of crankshaft and press evenly onto cylinder block.
15. Install and tighten the sealing flange-to-cylinder block bolts, evenly in a diagonal sequence, to 7 ft. lbs. (10 Nm).
16. Further assembly is performed in reverse order of removal. Note the following:
    a. Remove locking pin T40069 from rear of cylinder block and screw in locking bolt (30 Nm).
    b. Tighten the belt pulley-to-crankshaft bolts to 48 ft. lbs. (70 Nm).
    c. Tighten the tensioning element-to-accessory assembly bracket fasteners to 24 ft. lbs. (35 Nm).
    d. Tighten the locking bolt-to-rear of cylinder block to 21 ft. lbs. (30 Nm).

6-Cylinder Engines

2.8L (AAA/AHA/ATQ/BDF) ENGINE

1. Before servicing the vehicle, refer to the Precautions Section.
2. Remove or disconnect:
    - Timing belt.

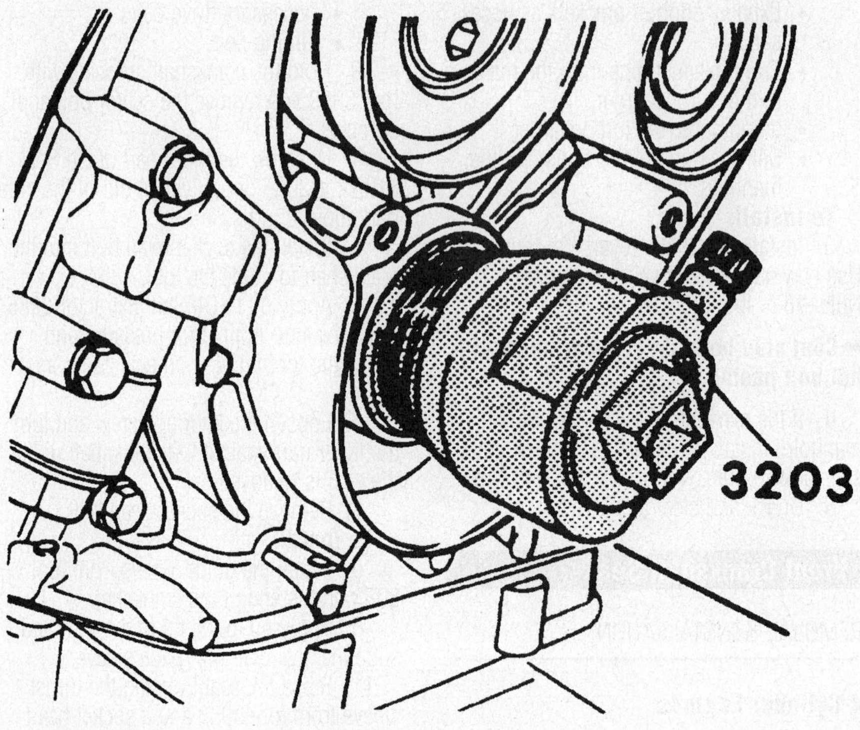

Removing the seal using Seal Remover 3203—6-cylinder engines

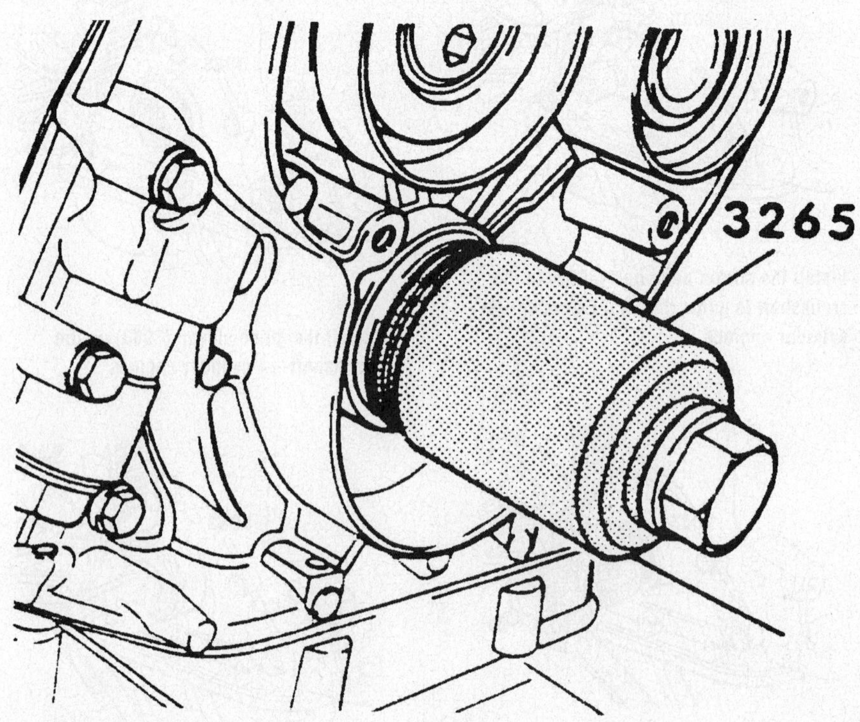

Installing the seal using the seal installer and the crankshaft center bolt—6-cylinder engines

- Timing belt sprocket from the crankshaft.
- Seal with Seal Remover 3203 or equivalent.

3. Clean the running and sealing surfaces.

To install:

4. Slide the seal over the Installing Sleeve 3202 or equivalent.
5. Press the oil seal flush with Seal Installer 3265, or equivalent, and the center crankshaft bolt.
6. Install the timing belt sprocket. Center crankshaft bolt: 66 ft. lbs. (90 Nm) plus ¼ turn.
7. Install the timing belt and all remaining components.

3.6L (BLV) ENGINE

1. Before servicing the vehicle, refer to the Precautions Section.
2. Remove or disconnect:
   - Drive belt
   - Vibration damper using counter-holder tool T10069
   - Oil pan
   - Crankshaft seal
   - Sealant residue on sealing surfaces

To install:

3. Remove oil residue at crankshaft pin using a clean rag.
4. Cut tube nozzle at front marking (nozzle approx. 3mm diameter).
5. Apply a bead of sealant D 176 501 of approx. 0.08–0.12 inch (2–3mm) on clean sealing flange sealing surface.

⁂ CAUTION

Cover seal using a clean rag before applying sealant bead. Sealant bead must not be thicker than 0.08–0.12 inch (2–3mm), otherwise excess sealant could get into oil pan and clog oil intake pipe strainer.

➡Note expiration date for sealant. The sealing flange must be installed within 5 minutes after application of silicon sealant.

6. Insert guide sleeve T10215/1 at front onto crankshaft pin.
7. Slide sealing flange with seal carefully over guide sleeve.
8. Tighten bolt sealing flange on cylinder crankshaft housing to 7 ft. lbs. (10 Nm).
9. Install oil pan.
10. Install vibration damper and secure using counter-holder tool T10069. Tighten new mounting bolt to 69 ft. lbs. (100 Nm) plus an additional turn of 90°.
11. Install drive belt and all remaining components.

8-Cylinder Engines

1. Before servicing the vehicle, refer to the Precautions Section.
2. Remove or disconnect:
   - Timing belt.
   - Timing belt sprocket from the crankshaft.
   - Pry out seal with Remover lever 681 or equivalent.

3. Clean the running and sealing surfaces.

To install:

4. Slide the seal over the Installing Sleeve 10124/1 or equivalent.
5. Press the oil seal flush with Seal Installer 10124/2, or equivalent, and the center crankshaft bolt.
6. Install the timing belt sprocket. Center crankshaft bolt: 74 ft. lbs. (100 Nm) plus ¼ turn.
7. Install the timing belt and all remaining components.

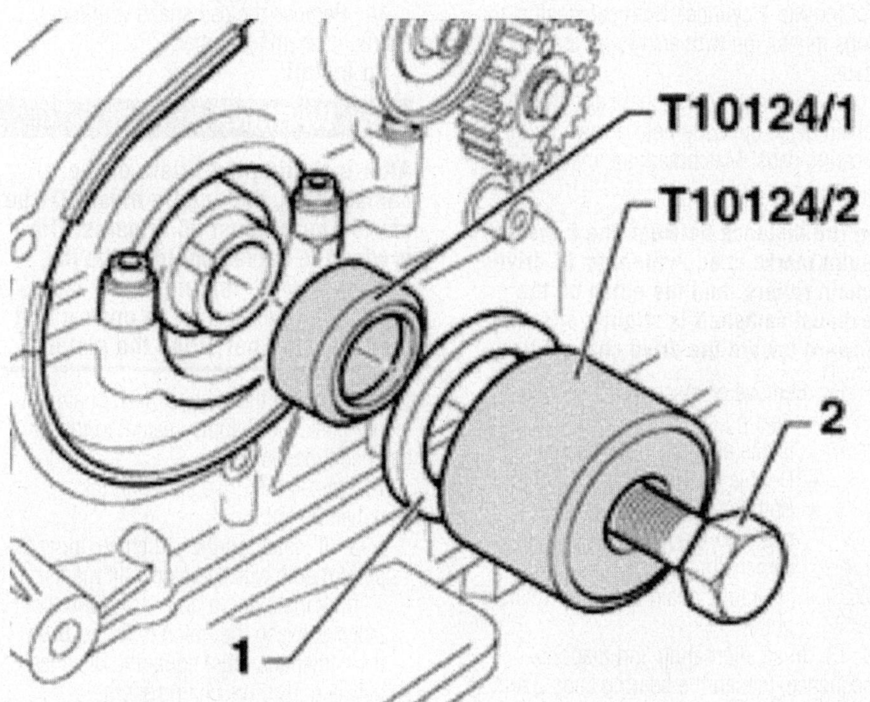

Installing the seal using the seal installer and the crankshaft center bolt—8-cylinder engines

Camshaft and Valve Lifters

REMOVAL & INSTALLATION

1.8L Engine

1. Place the lock carrier into the service position as follows:
   a. Remove or disconnect:
      - Front bumper.
      - Any wiring or connector that would inhibit locking the carrier.
      - 3 quick-release screws on the front noise insulation panel.
      - Air guide between the lock carrier and the air filter.
      - If installed, retaining clamps for the wiring harness at the left side of the radiator frame.
      - No. 2 bolts and install Support 3369 or equivalent.
      - Remaining bolts and pull the lock carrier out to the stop.

   b. To secure the lock carrier, install the appropriate M6 bolts into the rear of the lock carrier and fender.
2. Remove or disconnect:
   - Negative battery cable.
   - Accessory drive belts.
   - Engine covers.
   - Timing belt upper cover.
3. Turn the crankshaft, in the direction of rotation (clockwise), until the No. 1 cylinder is at TDC.
4. Using a T45 Torx® wrench, loosen the timing belt tensioner.
5. Push down on the tensioner, and remove the belt from the camshaft gear.
6. Remove or disconnect:
   - Torx® bolt and swing the tensioner assembly bracket forward.
   - Valve cover.

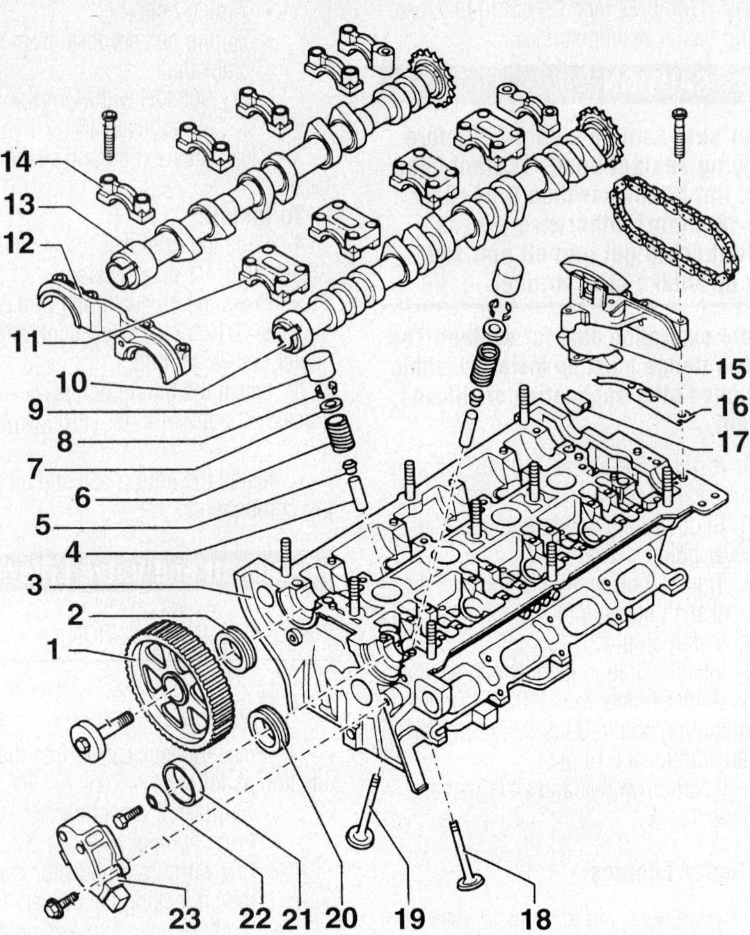

1. Camshaft gear
2. Oil seal
3. Cylinder head
4. Valve guide
5. Valve stem oil seal
6. Valve spring
7. Valve spring retainer
8. Valve keeper
9. Valve lifter
10. Intake camshaft
11. Intake camshaft bearing cap
12. Double bearing cap
13. Exhaust camshaft
14. Exhaust camshaft bearing cap
15. Hydraulic chain tensioner
16. Rubber/metal seal
17. Gasket
18. Exhaust valve
19. Intake valve
20. Oil seal
21. Shutter wheel for CMP sensor
22. Washer
23. Camshaft Position Sensor (CMP)

7923WG14

Exploded view of the camshaft and related components—1.8L engine

7. Using Retainer 3036 or equivalent, loosen the cam gear retaining bolt.
8. Remove:
   - Camshaft gear.
   - Housing for CMP sensor and shutter wheel.
9. Secure the hydraulic chain tensioner with Bracket-Tensioner 3366 or equivalent.
10. Verify that the camshafts are at TDC for the No. 1 cylinder. Both camshaft markings must align with arrows on the bearing caps.

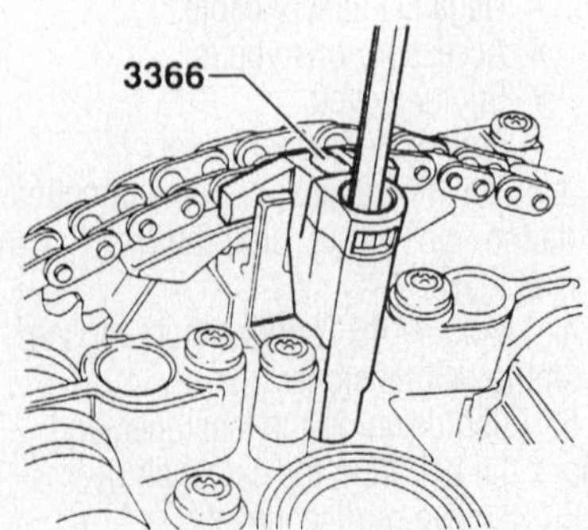

Over-tightening will damage the chain tensioner (3366)—1.8L engine

11. Clean the drive chain and the cam chain gears opposite both arrows on the bearing caps. Matchmark the installed position using paint.

➡The distance between the 2 arrows/paint marks is equivalent to 16 drive chain rollers, and the notch on the exhaust camshaft is slightly offset inward toward the drive chain roller.

12. Remove or disconnect:
    - Bearing caps 3 and 5 from the intake and exhaust camshafts.
    - Double bearing cap.
    - Both bearing caps from the chain gears on the intake and exhaust camshafts.
    - Hydraulic chain tensioner retaining bolts.
13. In an alternating and diagonal sequence, loosen the bearing caps 2 and 4 of the intake and exhaust manifold, then remove.
14. Remove the camshafts with the hydraulic chain tensioner.

To install:

WARNING

After installing the lifters or the camshaft(s), the engine must NOT be started for at least 30 minutes. Otherwise the valves could strike the pistons. Rotate the engine by hand, at least 2 revolutions, to ensure that the valves do not strike the pistons.

15. Replace the rubber/metal chain tensioner gasket and apply sealant to the hatched area, as shown.
16. Install the drive chain on the camshaft as follows:
    a. If installing the old chain, align the paint marks with the camshaft marks.
    b. If installing a new chain, the distance between the notches **A** and **B** on the camshafts must equal the distance between 16 drive chain rollers.
17. Slide the hydraulic chain tensioner between the drive chain.

To ensure proper installation, matchmark the chain-to-camshaft position—1.8L engine

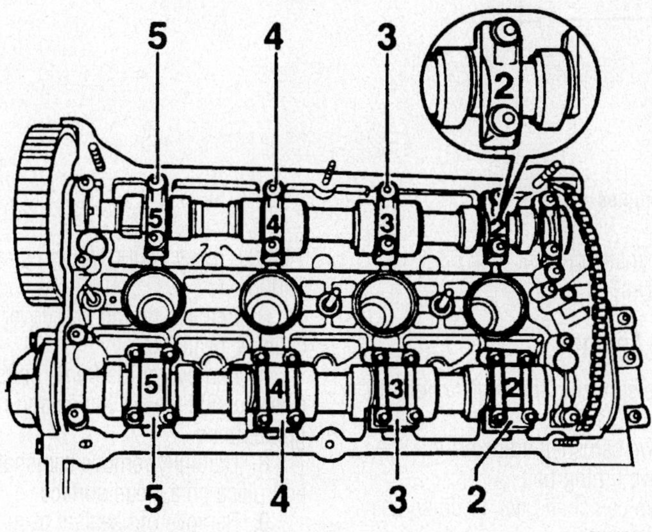

Camshaft bearing cap identification—1.8L engine

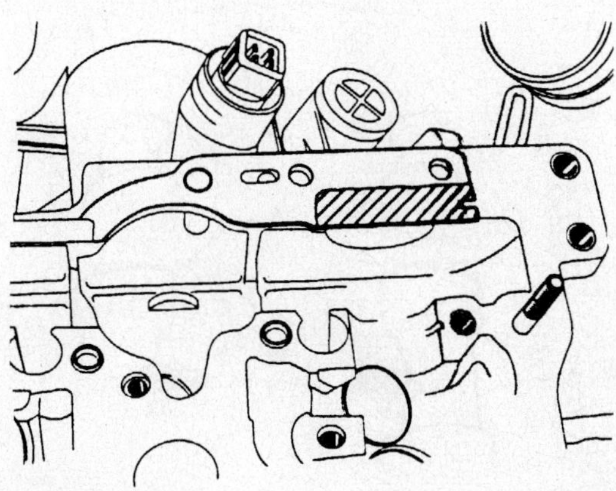

To ensure a proper seal, be sure to apply sealant to the hatched area—1.8L engine

18. Install the camshafts with the chain tensioner into the cylinder head.
19. Oil the camshaft contact surfaces.

➡When installing the bearing caps, verify the markings on the caps are readable from the intake side of the cylinder head.

20. Tighten the bearing caps 2 and 4 of the intake and exhaust camshafts in an alternating diagonal sequence to 84 inch lbs. (10 Nm).
21. Install both bearing caps on the chain sprockets of the intake and exhaust camshafts and tighten to 84 inch lbs. (10 Nm).
22. Verify the correct positions of the camshafts.
23. Remove the Bracket-Tensioner 3366.
24. Lightly coat the cylinder head mating surface of the double bearing cap with sealant, then install.
25. Install:
   - Remaining bearing caps: 84 inch lbs. (10 Nm).
   - Camshaft gear. Retaining bolt: 48 ft. lbs. (65 Nm).
   - CMP shutter wheel and housing cover.
   - Valve cover.
26. Align the camshaft gear and the vibration damper with the TDC markings.
27. Install or connect:
   - Timing belt.
   - Accessory drive belts, then the engine cover.
   - Lock carrier.
   - Negative battery cable.

2.0L SOHC Engines

1. Before servicing the vehicle, refer to the Precautions Section.
2. Remove or disconnect:
   - Negative battery cable.
   - Timing belt cover(s), the timing belt
   - Camshaft sprocket and cylinder head cover.
3. Number the bearing caps from front-to-back. Scribe an arrow pointing towards the front of the engine. The caps are offset and must be installed correctly. Factory numbers on the caps are not always on the same side.
4. Remove or disconnect:
   - Front and rear bearing caps. Loosen the remaining bearing cap nuts diagonally, in several steps, starting from the outside caps near the ends of the head and working toward the center.
   - Bearing caps and the camshaft.

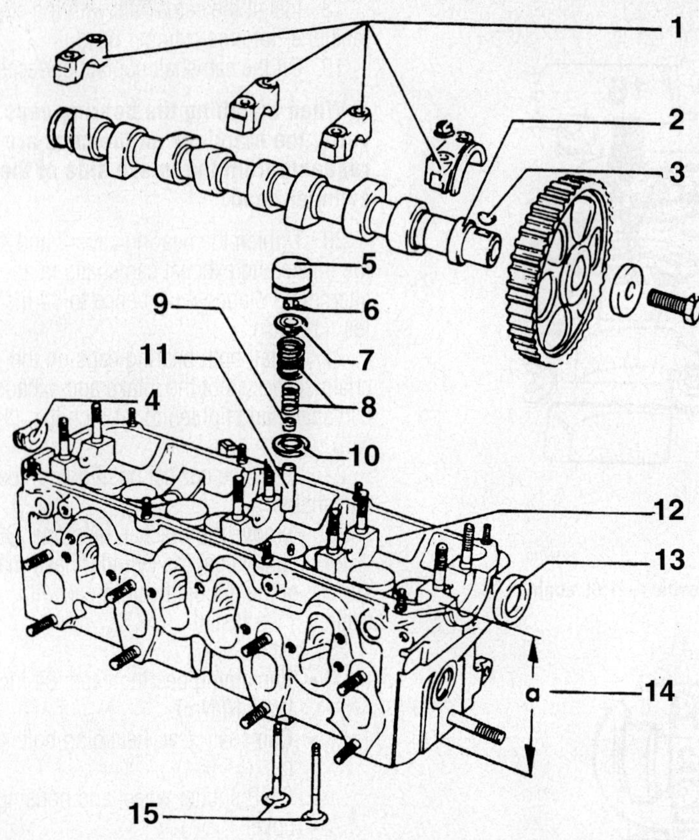

1. Bearing cap
2. Camshaft
3. Woodruff key
4. End cap
5. Valve lifter
6. Valve keeper
7. Upper spring seat
8. Valve spring
9. Valve stem oil seal
10. Lower spring seat
11. Valve guide
12. Cylinder head
13. Oil Seal
14. Cylinder head machining dimension
15. Valves

Exploded view of the camshaft and related components—2.0L engines

- If required, lifters from the valves. Keep them in order so they can be installed in their original positions. Place them in a bath of oil or place them upside down to prevent air from entering them.

To install:

5. Install or connect:
   - New oil seal and end-plug in the cylinder head. Lubricate the camshaft bearing journals and lobes and set the camshaft in place.
   - Bearing caps in the correct position with the arrow pointing towards the front of the engine. Tighten the cap nuts diagonally and in several steps until they are tightened to 15 ft. lbs. (20 Nm). Do not over-tighten.
   - Drive sprocket. Bolt: 58 ft. lbs. (80 Nm).
6. Align the timing marks, install the timing belt and adjust the tension.
7. On engines with hydraulic lifters, wait at least ½ hour after installing the camshaft before starting the engine to allow the lifters to leak down. Observe the following values:
   - Camshaft shaft end-play: 0.006 in. (0.15mm)
   - Bearing cap bolts: 15 ft. lbs. (20 Nm)
   - Camshaft sprocket bolt: 58 ft. lbs. (80 Nm)

2.0L DOHC Engines

1. Before servicing the vehicle, refer to the Precautions Section.
2. Remove camshaft adjuster.
3. Remove timing belt.
4. Loosen camshaft gear, to do so use retainer 3036.
5. Pull off camshaft gear using puller T40001 , claw T40001/6 and claw T40001/7.
6. Remove timing belt cover at rear of cylinder head.
7. Remove guide frame bolts evenly from outside toward inside and remove guide frame.
8. Carefully remove camshaft upward and place on a clean surface.
9. Remove old sealant from guide frame groove as well as from sealing surfaces.

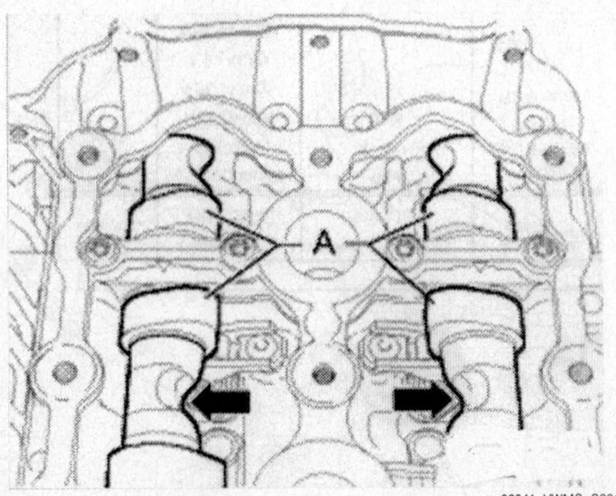

Cams of cylinder 4 must face each other and recesses must point to each other—2.0L DOHC engine

10. Prevent dirt and adhesive residue from entering cylinder head.

To install:

➡Sealing surfaces must be completely free of oil and grease. Pistons must not be at Top Dead Center (TDC). Make sure that all roller cam followers properly rest on valve shaft tips. Oil journal surfaces of camshafts.

11. Carefully insert camshafts into camshaft bearings of cylinder head. Cams of cylinder 4 must face each other. Recesses must point to each other.
12. Apply an even, slightly raised bead of sealant D 188 800 A1 into clean grooves of guide frame. Apply sealant uniformly on sealing surface. Note expiration date for sealant. Sealant must not be applied too thick. Wipe off excess sealant if necessary using a lint-free cloth.

➡Attaching and bolting the guide frame should be performed without interruption because the sealant begins to harden immediately as soon as it contacts the sealing surfaces.

13. Place guide frame onto cylinder head.
14. Gently tighten new bolts from inside working toward outside in several stages.
15. Tighten bolts in sequence to 6 ft. lbs. (8 Nm).
16. Drive in cap using pressure piece 3334 approx. 0.04–0.08 inch (1–2mm).
17. Install camshaft seal.
18. Install rear timing belt cover.
19. Insert fitting key into camshaft.
20. Install camshaft gear. To tighten bolt, hold camshaft gear using retainer 3036. Tightening torque 45 ft. lbs. (65 Nm).
21. When turning camshaft, pistons may not be at TDC for any cylinder. Valves and/or pistons may be damaged.
22. Be sure fitting key is properly seated.
23. Install timing belt.
24. Install camshaft adjuster.

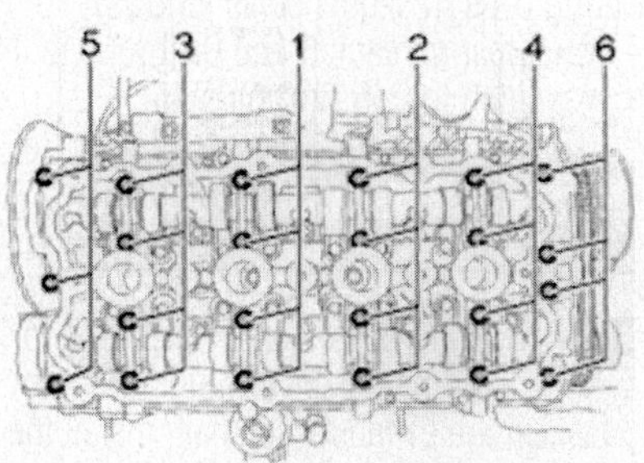

Tighten bolts in sequence shown—2.0L DOHC engine

25. Install cylinder head cover.
26. The rest of installation is in reverse order of removal.

2.8L (AAA) Engine

This procedure requires special tool 3268 or equivalent. This is a setting tool that holds the camshafts in the correct position for installing the timing chains.

1. Before servicing the vehicle, refer to the Precautions Section.
2. Remove or disconnect:
   - Distributor cap, wires and wire guide as an assembly.
   - Upper intake manifold.
   - Cylinder head cover.
   - Timing chain tensioner bolt and the upper timing chain cover.
3. Rotate the crankshaft to TDC of No. 1 piston.
4. Mark the direction of travel on the upper camshaft drive chain. Remove the tensioner shoe and the chain.
5. Hold the camshafts at the flats with a 24mm wrench and remove the bolts to remove the sprockets. Note the position of the distributor drive on the short camshaft.

✲✲ WARNING

Do not use the setting tool to hold the camshafts when removing or installing the sprocket bolts. The camshafts or the tool will be damaged.

6. On the long camshaft, remove the end bearing caps. Loosen the center cap nuts in a diagonal pattern 2 turns at a time until the valve springs are relieved. Remove the camshaft.
7. On the short camshaft, remove the center bearing cap and loosen the nuts on the end-caps in a diagonal pattern 2 turns at a time. When the valve springs are relieved, remove the camshaft.

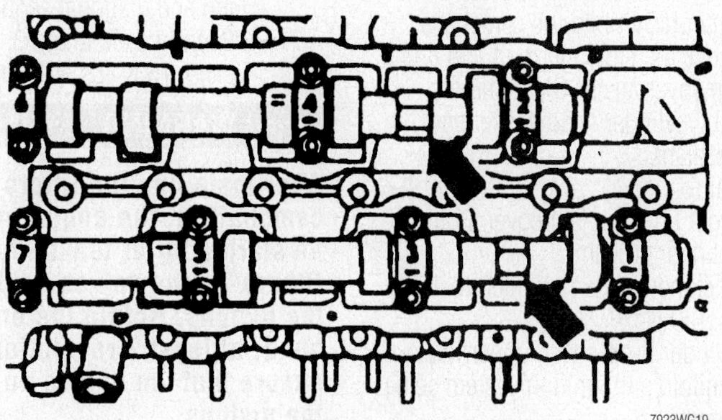

Hold the camshafts with an open end wrench on the flats (arrows) when removing or installing the sprocket bolts

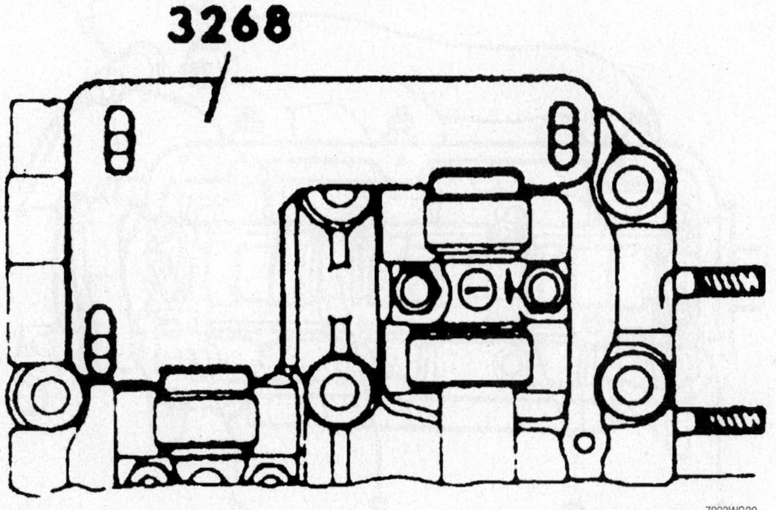

The camshaft setting tool holds the camshafts in place when installing the chain—do not use this tool to loosen or tighten the sprocket bolts

To install:

8. Lubricate the long camshaft and the cylinder head bearing surfaces and set the camshaft in place. Install bearing caps 3 and 5 and tighten the bolts 2 turns at a time in a diagonal pattern to draw the camshaft down against the valve springs.

9. Install the other bearing caps. Nuts: 15 ft. lbs. (20 Nm).

10. Repeat the process with the short camshaft, using bearing caps 2 and 6 to draw the camshaft down against the springs.

11. Hold the camshaft with a 24mm wrench and install the sprockets. Be sure the distributor drive is correctly positioned. Bolts: 74 ft. lbs. (100 Nm).

12. Be sure the crankshaft is at TDC on No. 1 piston. Install the setting tool and install the timing chain.

13. Install the tensioner shoe and temporarily install the upper timing chain cover. Install the tensioner bolt and remove the setting tool. Rotate the crankshaft 4 full turns and stop at TDC of No. 1 piston. The setting tool should fit into the camshafts.

14. Remove the tensioner bolt and upper timing chain cover again. Clean the old sealant off the cylinder head gasket and apply new sealant.

15. Install:

- Upper timing chain cover. Bolts: 82 inch lbs. (10 Nm).
- Tensioner bolt and tighten to 15 ft. lbs. (20 Nm).
- Cylinder head cover, upper intake manifold and ignition system components.

2.8L (AHA/ATQ) Engine

1. Remove or disconnect:

- Timing belt.
- Valve cover(s).
- On the left cylinder head, Camshaft Position (CMP) sensor.
- On the right cylinder head, plug/cover on the head.
- Camshaft timing belt sprocket.

➡DO NOT allow the bearing caps to become mixed up.

- Camshaft bearing caps 2 and 3.

2. Gradually and evenly, loosen the nuts for the camshaft bearing caps 1 and 4, in a diagonal sequence.

3. Remove the camshaft and lift out the valve lifters. If it is to be reused, it must go in the bore from which it was removed.

To install:

4. Install or connect:

- Lifters into their respective bore.
- Bearing caps 1 and 4 in a alternating and diagonal sequence.
- Bearing caps 2 and 3. All bearing caps: 15 ft. lbs. (20 Nm).

⁂ WARNING

After installing the lifters or the camshaft(s), the engine must NOT be started for at least 30 minutes. Otherwise the valves could strike the pistons. Rotate the engine by hand, at least 2 revolutions, to ensure that the valves do not strike the pistons.

- Camshaft timing belt sprocket. Mounting bolt: 52 ft. lbs. (71 Nm).
- On the right cylinder head, plug/cover on the head.
- On the left cylinder head, CMP sensor. Mounting bolts: 89 inch lbs. (10 Nm).
- Valve cover.
- Timing belt.

2.8L (BDF) Engine

1. Remove or disconnect:

- Valve cover.
- Intake manifold.
- Timing chain.
- Four bolts and control housing from cylinder head.

➡DO NOT allow the bearing caps to become mixed up.

2. On the intake camshaft, remove the bearing caps as follows:

- Bearing caps 1 and 13.
- Bearing caps 3 and 11.
- Bearing cap 7.
- Bearing caps 5 and 9 alternating in a diagonal sequence.

3. On the exhaust camshaft, remove the bearing caps as follows:

- Bearing caps 2 and 14.
- Bearing cap 4 and 12.
- Bearing cap 8.
- Bearing caps 6 and 10 alternating in a diagonal sequence.

4. Remove the camshaft and lift out the valve lifters.

To install:

5. Install or connect:

6. Install the lifters into their respective bore.

7. Ensure cam lobes for cylinder no 1 are pointing upward and that the pistons are not at TDC.

8. Oil the camshaft journal surfaces.

9. Insert the respective camshafts into their mountings.

10. On the intake camshaft, install the bearing caps in sequence as follows:

- Bearing caps 5 and 9: Torque to 44 inch lbs. (5 Nm) plus 45°.
- Bearing caps 1 and 13: Torque to 44 inch lbs. (5 Nm) plus 45°.
- Bearing cap 7: Torque to 44 inch lbs. (5 Nm) plus 45°.
- Bearing caps 3 and 11: Torque to 44 inch lbs. (5 Nm) plus 45°.

11. On the exhaust camshaft, install the bearing caps in sequence as follows:

- Bearing caps 6 and 10: Torque to 44 inch lbs. (5 Nm) plus 45°.

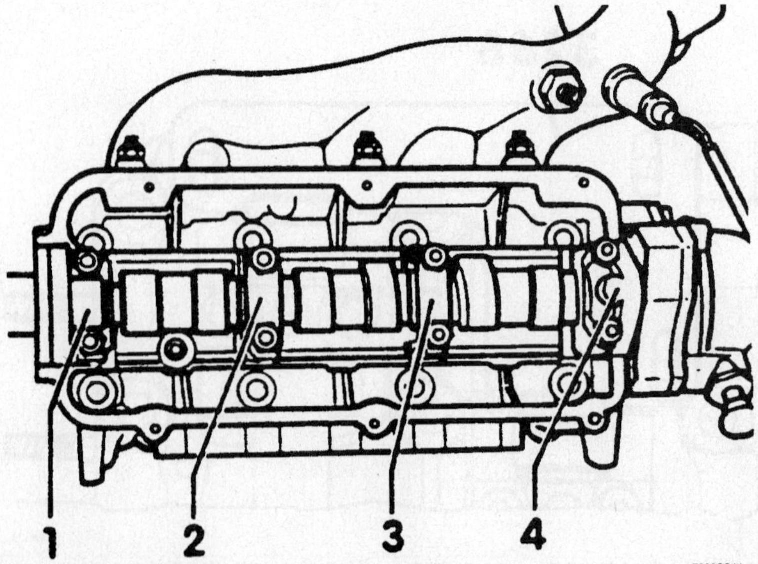

Camshaft bearing cap identification—2.8L (AHA) engine

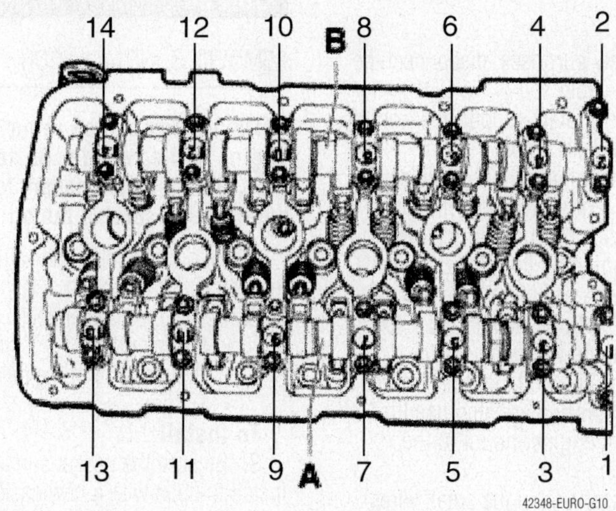

Camshaft bearing cap identification—2.8L engine (BDF)

- Bearing caps 2 and 14: Torque to 44 inch lbs. (5 Nm) plus 45°.
- Bearing cap 8: Torque to 44 inch lbs. (5 Nm) plus 45°.
- Bearing caps 4 and 12: Torque to 44 inch lbs. (5 Nm) plus 45°.

12. Adjust the camshafts in cylinder head to TDC on cylinder no. 1.
13. Lightly oil the sealing ring contact surfaces of camshaft before installing control housing.
14. Lightly oil camshafts and slowly slide control housing over camshaft sealing rings.
15. Install control housing to cylinder head and tighten to 72 inch lbs. (8 Nm).
16. Install the timing chain and adjust the valve timing.
17. To complete installation, reverse the removal procedure.

4.0L (BDP) Engine

1. On both cylinder heads, remove or disconnect:
   - Valve cover.
   - Intake manifold.
   - Timing chain.
   - Control housing from cylinder head.

CYLINDER BANK NO. 1

➡DO NOT allow the bearing caps to become mixed up.

1. Remove camshafts as follows:
2. On the intake camshaft, remove the bearing caps as follows:
   - Bearing caps 1 and 9.
   - Bearing cap 5.
   - Bearing caps 3 and 7 alternating in a diagonal sequence.
3. On the exhaust camshaft, remove the bearing caps as follows:
   - Bearing caps 2 and 10.
   - Bearing cap 6.
   - Bearing caps 4 and 8 alternating in a diagonal sequence.
4. Remove the camshaft and lift out the valve lifters.

CYLINDER BANK NO. 2

➡DO NOT allow the bearing caps to become mixed up.

1. Remove camshafts as follows:
2. On the intake camshaft, remove the bearing caps as follows:
   - Bearing caps 21 and 29.
   - Bearing cap 25.
   - Bearing caps 23 and 27 alternating in a diagonal sequence.
3. On the exhaust camshaft, remove the bearing caps as follows:

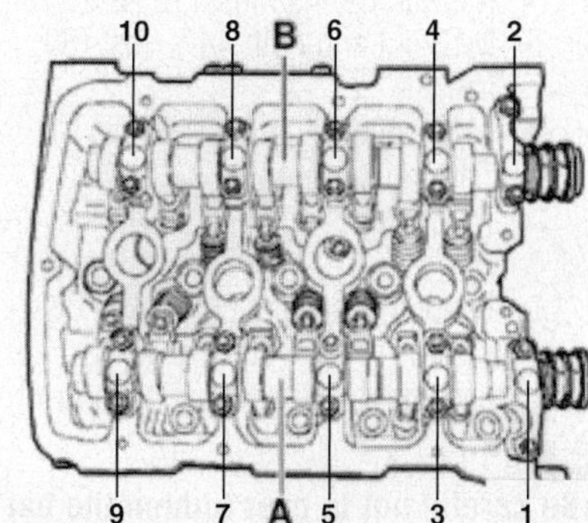

Camshaft bearing cap identification cylinder bank 1—4.0L engine

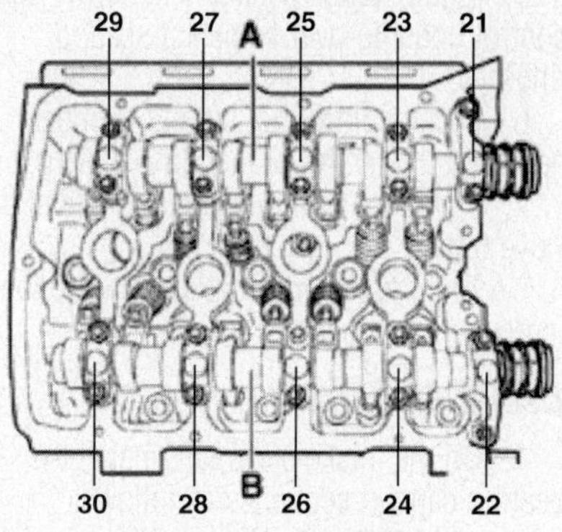

Camshaft bearing cap identification cylinder bank 2—4.0L engine

- Bearing caps 22 and 30.
- Bearing cap 26.
- Bearing caps 24 and 28 alternating in a diagonal sequence.

4. Remove the camshaft and lift out the valve lifters.

To install:

5. On both cylinder heads, install or connect:
6. Install the lifters into their respective bore.
7. Ensure cam lobes for cylinder no 1 are pointing upward and that the pistons are not at TDC.
8. Oil the camshaft journal surfaces.
9. Insert the respective camshafts into their mountings.

CYLINDER BANK NO. 1

1. On the intake camshaft, install the bearing caps in sequence as follows:
   - Bearing caps 3 and 7 diagonally: Torque to 44 inch lbs. (5 Nm) plus 45°.
   - Bearing cap 5: Torque to 44 inch lbs. (5 Nm) plus 45°.
   - Bearing caps 1 and 9: Torque to 44 inch lbs. (5 Nm) plus 45°.
2. On the exhaust camshaft, install the bearing caps in sequence as follows:
   - Bearing caps 4 and 8 diagonally: Torque to 44 inch lbs. (5 Nm) plus 45°.
   - Bearing cap 6: Torque to 44 inch lbs. (5 Nm) plus 45°.
   - Bearing caps 2 and 10: Torque to 44 inch lbs. (5 Nm) plus 45°.
3. Adjust the camshafts in cylinder head to TDC on cylinder no. 1.
4. Lightly oil the sealing ring contact surfaces of camshaft before installing control housing.

5. Lightly oil camshafts and slowly slide control housing over camshaft sealing rings.
6. Install control housing to cylinder head and tighten to 72 inch lbs. (8 Nm).
7. Install the timing chain and adjust the valve timing.
8. To complete installation, reverse the removal procedure.

CYLINDER BANK NO. 2

1. On the intake camshaft, install the bearing caps in sequence as follows:
   - Bearing caps 23 and 27 diagonally: Torque to 44 inch lbs. (5 Nm) plus 45°.
   - Bearing cap 25: Torque to 44 inch lbs. (5 Nm) plus 45°.
   - Bearing caps 21 and 29: Torque to 44 inch lbs. (5 Nm) plus 45°.
2. On the exhaust camshaft, install the bearing caps in sequence as follows:
   - Bearing caps 24 and 28 diagonally: Torque to 44 inch lbs. (5 Nm) plus 45°.
   - Bearing cap 26: Torque to 44 inch lbs. (5 Nm) plus 45°.
   - Bearing caps 22 and 20: Torque to 44 inch lbs. (5 Nm) plus 45°.
3. Adjust the camshafts in cylinder heads to TDC on cylinder no. 1.
4. Lightly oil the sealing ring contact surfaces of camshaft before installing control housing.
5. Lightly oil camshafts and slowly slide control housing over camshaft sealing rings.
6. Install control housing to cylinder head and tighten to 72 inch lbs. (8 Nm).
7. Install the timing chain and adjust the valve timing.
8. To complete installation, reverse the removal procedure.

Valve Lash

ADJUSTMENT

All engines are equipped with hydraulic valve lash adjusters. No periodic valve lash adjustment is necessary.

Starter

REMOVAL & INSTALLATION

➡On A1 and A2 platform vehicles equipped with 010 automatic transaxle, access to the starter motor is limited. Although not necessary, it is recommended that the intake and exhaust manifolds be removed to access the starter.

1. For safety purposes, disconnect the battery ground cable.
2. Raise and safely support the front of the vehicle with jackstands.
3. On some vehicles (A3 vehicles, for example) may require the use of a floor jack (with a block of wood on the chock) to support the engine by the oil pan. Do NOT jack the car by the oil pan under any circumstances! The floor jack is ONLY to support the engine while the starter is being removed. The bolts that secure the starter to the transaxle are also used to secure the front engine mount to the transaxle.
4. If necessary, label the small wires before disconnecting them.
5. Disconnect the large cable, which is the positive battery cable, from the solenoid.
6. If equipped, remove the bracket that secures the starter to the engine.
7. Remove the starter mounting bolts, while supporting the weight of the starter.
8. Pull the starter straight out from the transaxle.

To install:

➡On vehicles with a manual transaxle, there is a bushing where the starter shaft fits into the bell housing. If the shaft or bushing are worn or if the starter has been jamming, the bushing should be replaced. There is a special bushing removal tool available but a small inside bearing removal tool is usually sufficient.

9. Install the starter into the transaxle.
10. Tighten the starter mounting bolts as follows:
    a. All vehicles except 010 transaxle:
    - M8 nut: 89 inch lbs. (10 Nm)
    - M10 nut and bolt: 44 ft. lbs. (60 Nm)
    - M12 bolt: 33 ft. lbs. (45 Nm)

    b. 010 transaxle only:
    - Mounting flange bolt: 15 ft. lbs. (20 Nm)
    - Mounting bracket bolt: 18 ft. lbs. (25 Nm)
11. Attach the electrical connections to the starter.

➡Be careful not to over tighten the battery cable connection. The metal is soft and the threads will strip easily.

12. Lower the vehicle from the jackstands.
13. Connect the negative battery cable.

Oil Pan

REMOVAL & INSTALLATION

➡The oil pan can be removed with the engine in the vehicle. On some vehicles, it may be necessary to lower the subframe to service the oil pan.

1. Before servicing the vehicle, refer to the Precautions Section.
2. Remove or disconnect:
   - Engine oil.
   - Bolts retaining the oil pan.
   - Oil pan.

To install:

3. Be sure the gasket surface is flat and install the pan with a new gasket.
4. Tighten the retaining bolts in a crisscross pattern as follows:
   - 1.8L and 2.8L (AAA) engine: 11 ft. lbs. (15 Nm)
   - 2.0L (AEG/AVH/AZG/BDC/BEV/BGD) engines: 11 ft. lbs. (15 Nm)
   - 2.0L (ABA) engine: 15 ft. lbs. (20 Nm)
   - 2.0L (BPY) engine: oil pan-to-transmission bolts, 28 ft. lbs. (40 Nm); oil pan-to-cylinder block bolts, 10 ft. lbs. (15 Nm)
   - 2.5L (BGP/BGQ/BPR/BPS) engine: oil pan lower-to-oil pan upper bolts, 7 ft. lbs. (10 Nm); oil pan upper-to-cylinder block bolts, 17 ft. lbs. (25 Nm)
   - 2.8L (AHA/ATQ) engine: 84 inch lbs. (10 Nm)
   - 2.8L (BDF) engine: screws to 9 ft. lbs. (12 Nm)
   - 3.6L (BLV) engine: oil pan-to-transmission bolts, 31 ft. lbs. (45 Nm); oil pan-to-cylinder block bolts, 8 ft. lbs. (12 Nm)
   - 4.0L (BDP) engine: 71 inch lbs. (8 Nm)
5. Refill the engine with oil. Start the engine and check for leaks.

Oil Pump

REMOVAL & INSTALLATION

1.8L (Except AWP/AWW/BKF), 2.0L (ABA), and 2.8L (AAA) Engines

1. Before servicing the vehicle, refer to the Precautions Section.
2. Remove or disconnect:
   - Oil pan
   - Baffle plate, if equipped
   - Mounting bolts

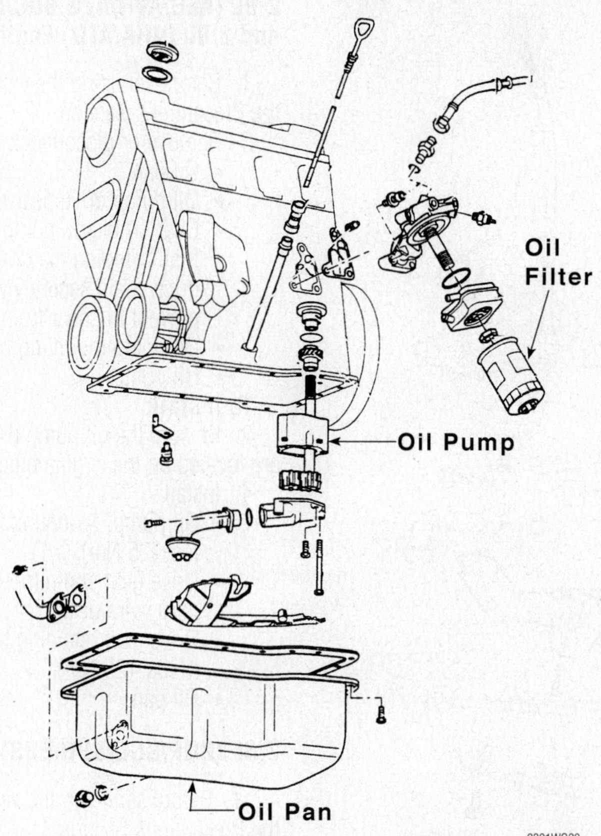

Exploded view of the lubrication system components—1.8L engine (except AWP/AWW), 2.0L (ABA) engine similar

3. Lower the pump from the engine.
4. Install the oil pump in the reverse order of removal.
5. Observe the following values:
   - Oil pump bottom cover bolts: 84 inch lbs. (10 Nm)
   - Oil pump suction foot bolts: 84 inch lbs. (10 Nm)
   - Oil pump retaining bolts: 18 ft. lbs. (25 Nm)

1.8L (AWP/AWW/BKF) Engines

1. Before servicing the vehicle, refer to the Precautions Section.
2. Remove or disconnect:
   - Oil pan
   - Baffle plate
   - Oil pump drive chain
   - Drive sprocket
   - Suction pipe
   - Mounting bolts
3. Lower the pump from the engine.
4. Install the oil pump in the reverse order of removal.
5. Observe the following values:
   - Oil pump and baffle plate bolts: 11 ft. lbs. (15 Nm)
   - Oil pump suction pipe bolts: 11 ft. lbs. (15 Nm)
   - Oil pump sprocket bolt: 18 ft. lbs. (25 Nm)

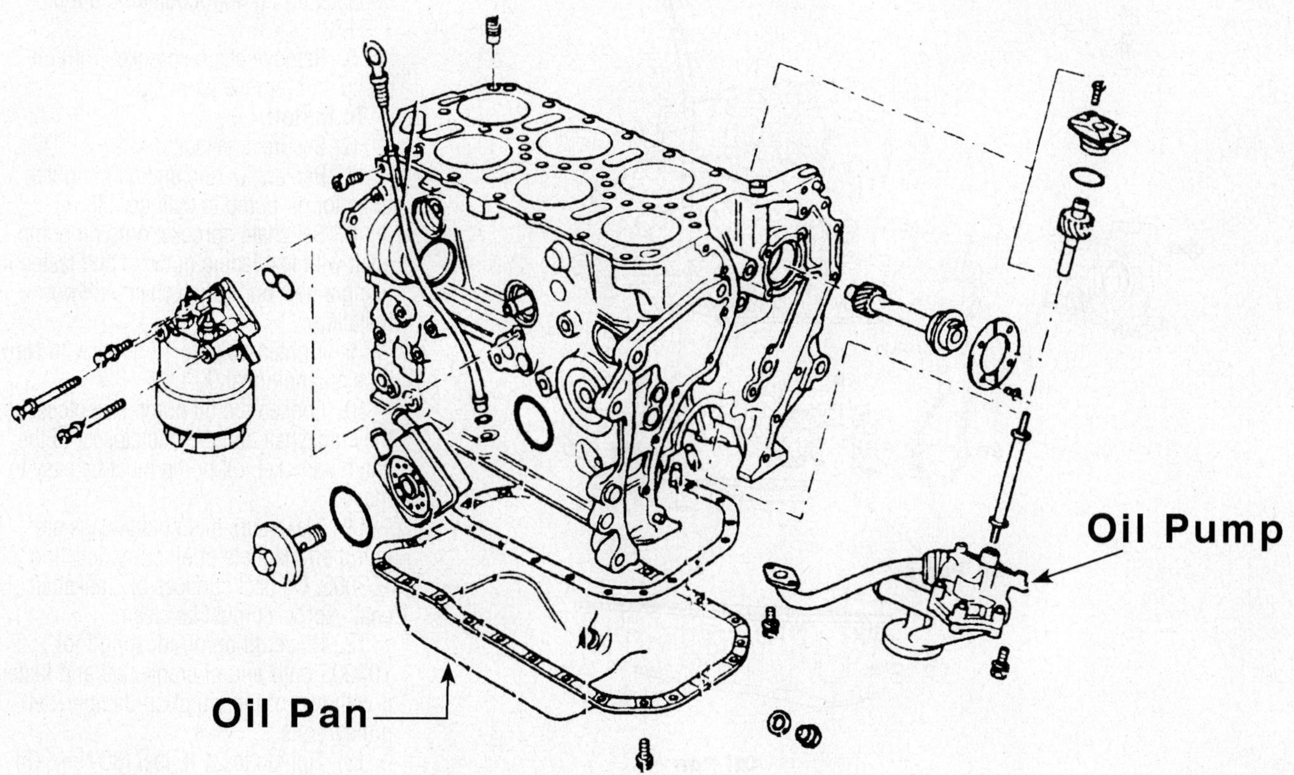

Exploded view of the lubrication system components—2.8L (AAA) engine

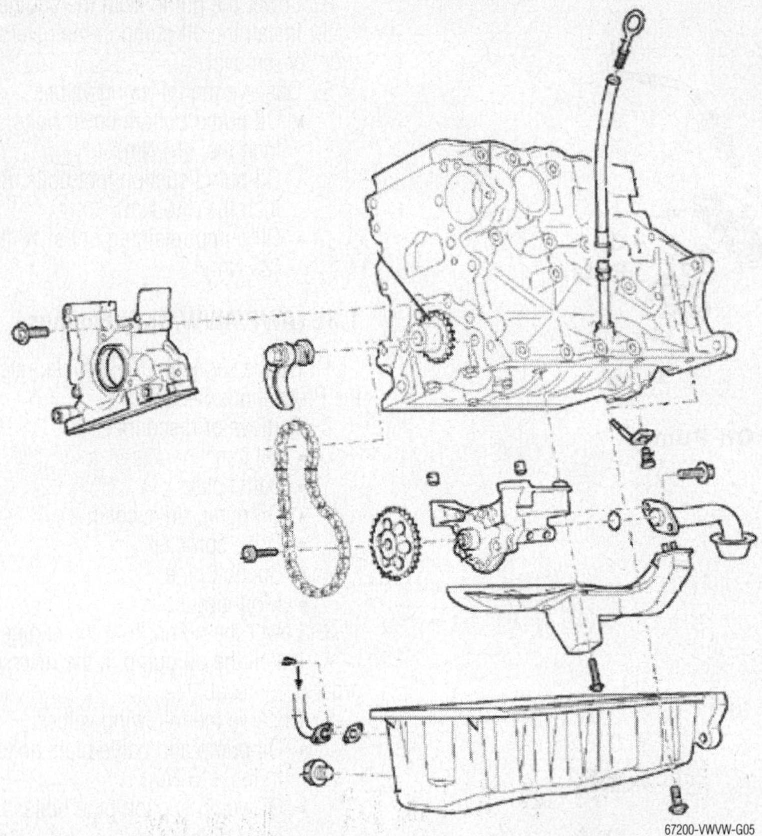

Exploded view of the lubrication system components—1.8L engine (AWP/AWW)

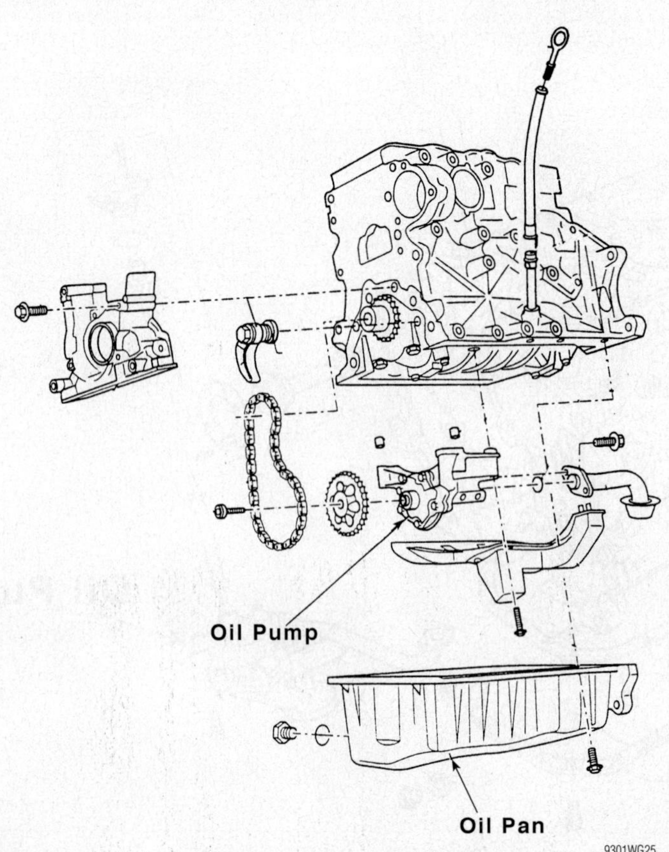

Exploded view of the lubrication system components—2.0L (AEG/AVH/AZG/BDC) engine

2.0L (AEG/AVH/AZG/BDC/BEV/BGD) and 2.8L (AHA/ATQ) Engines

1. Before servicing the vehicle, refer to the Precautions Section.
2. Remove or disconnect:
   - Oil pan.
   - Oil pump drive gear retaining bolt.
   - Gear off the oil pump driveshaft, first fastening the drive gear to the drive chain securely with wire to prevent the gear from falling.
   - Oil pump mounting bolts.
   - Oil pump.

To install:

3. Be sure the oil pump dowel sleeves are located on the engine block.
4. Install:
   - Oil pump. Mounting bolts: 18 ft. lbs. (25 Nm).
   - Drive gear and remove the string, cord, wire or tie.
   - Drive gear retaining bolt. Bolt: 18 ft. lbs. (25 Nm).
   - Oil pan.

2.5L (BGP/BGQ/BPR/BPS) Engine

1. Before servicing the vehicle, refer to the Precautions Section.
2. Remove oil pan.
3. Tension chain tensioner, secure it with locking pin T10115 and remove it.
4. Remove bolts and hold chain sprocket tightly with counterhold tool T10172.
5. Remove chain sprocket from oil pump and remove oil pump.

To install:

6. Secure crankshaft.
7. Replace O-ring and hand tighten bolts for oil pump to cylinder block.
8. Set chain sprocket onto oil pump shaft with text facing outward and fasten it using a new bolt (drive chain still not installed).
9. Tighten bolts to 14 ft. lbs. (20 Nm), plus an additional 90°.
10. Loosen the oil pump bolt closest to the crankshaft and the one closest to the chain sprocket; oil pump must be easy to move.
11. Make sure that no shavings are found on magnets of oil pump locating tool T03005. Contact surfaces of crankshaft and chain sprocket must be clean.
12. Place oil pump locating tool T03005 onto end of crankshaft and fasten it with two of the vibration damper/belt pulley bolts.
13. Tighten to 21 ft. lbs. (30 Nm). Oil pump is pulled into place by magnets.

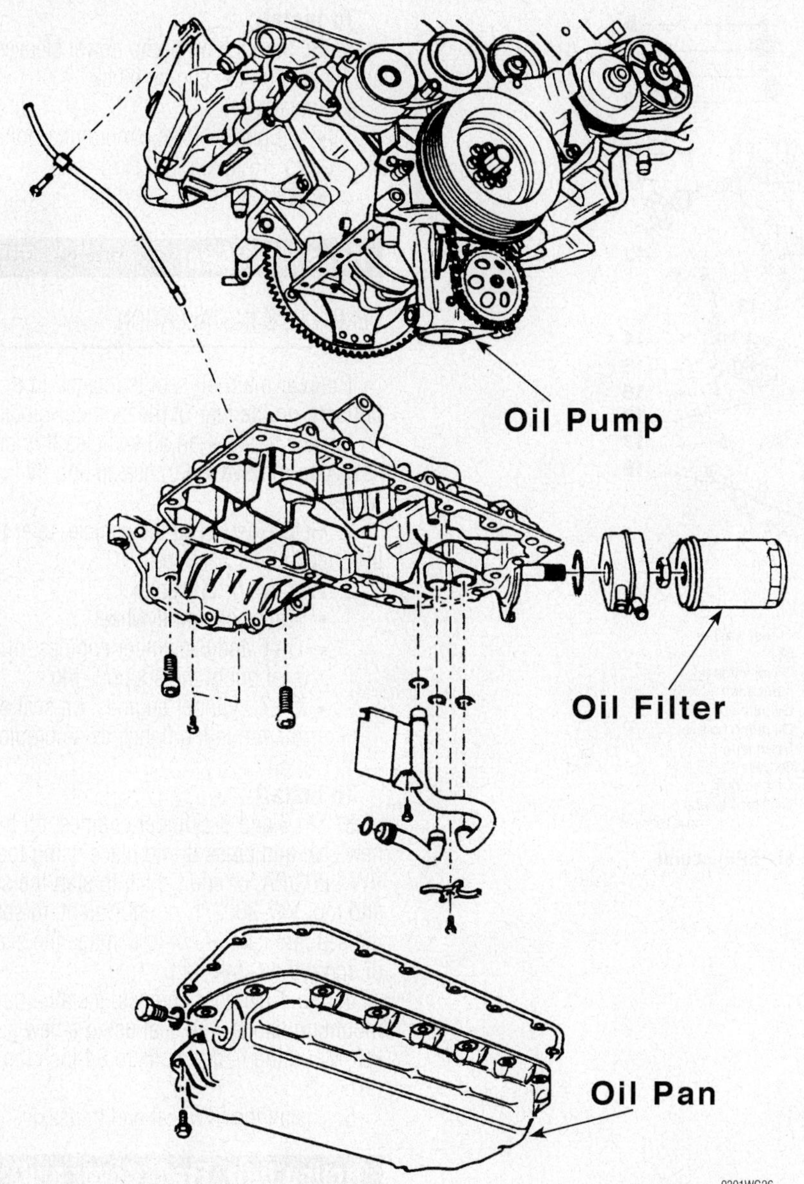

Exploded view of the lubrication system components—2.8L (AHA/ATQ) engine

14. Remove locking pin T40069.
15. Push crankshaft, as for checking axial bearing play, toward pulley drive.

➡This work step is important to guarantee the correct positioning of the chain sprockets to one another.

16. In this position, first tighten the bolt closest to the crankshaft, the one closest to the chain sprocket, and then the remaining bolt to 17 ft. lbs. (25 Nm).
17. Install locking pin T40069 again. Crankshaft may only be moved minimally around TDC point. Otherwise, there is a risk that valves will contact piston.
18. Remove oil pump locating tool T03005.
19. If a new oil pump is installed, fill oil pump with some clean engine oil via suction pipe and turn pump shaft a several times.
20. Place drive chain onto chain sprocket of oil pump.
21. Install oil pan.
22. Install guide rail, release tension on chain tensioner and pull out locking pin T10115.

➡Make sure that the drive chain is correctly positioned in the guide rail and in the tensioning rail.

23. Adjust valve timing.
24. Further assembly is performed in reverse order of removal. Note the following:
    a. Remove locking pin T40069 from rear of cylinder block and screw in locking bolt. Tighten to 21 ft. lbs. (30 Nm).
    b. Refill the engine cooling system.

2.8L (BDF) Engine

1. Before servicing the vehicle, refer to the Precautions Section.
2. Remove or disconnect:
    - Oil pan.
    - Oil pump mounting bolts.
    - Oil pump.

To install:

3. Be sure the oil pump dowel sleeves are located on the engine block.
4. Install:
    - Oil pump. Torque mounting bolts to: 18 ft. lbs. (25 Nm).
    - Oil pan.

3.6L (BLV) Engine

1. Before servicing the vehicle, refer to the Precautions Section.
2. Remove or disconnect:
    - Engine assembly
    - Transmission
    - Cover piece on cylinder head
    - Drive plate and sealing flange from engine
    - Chain tensioner and camshaft roller chain from front oil pump chain sprocket
    - Chain sprockets with roller chain from oil pump
3. Remove mounting bolts of oil pump.
4. Screw a threaded bolt M10x30 (standard) into oil pump shaft.
5. Screw extractor T10055 with adapter T10055/3 onto threaded bolt.
6. Pull oil pump out of cylinder block using light knocking motions.

To install:

7. Installation is performed in reverse of removal. Note the following:
    a. O-ring as well as mounting bolts for oil pump and chain sprockets must always be replaced.
    b. Tighten oil pump with new mounting bolts. Tightening torque: 6 ft. lbs. (8 Nm)
    c. Tighten chain sprockets with a new mounting bolt. Tightening torque: 28 ft. lbs. (40 Nm), plus an additional 90°
    d. Adjusting valve timing.

4.0L (BDP) Engine

1. Before servicing the vehicle, refer to the Precautions Section.
2. Remove or disconnect:
    - Oil pan.
    - Oil pump mounting bolts.

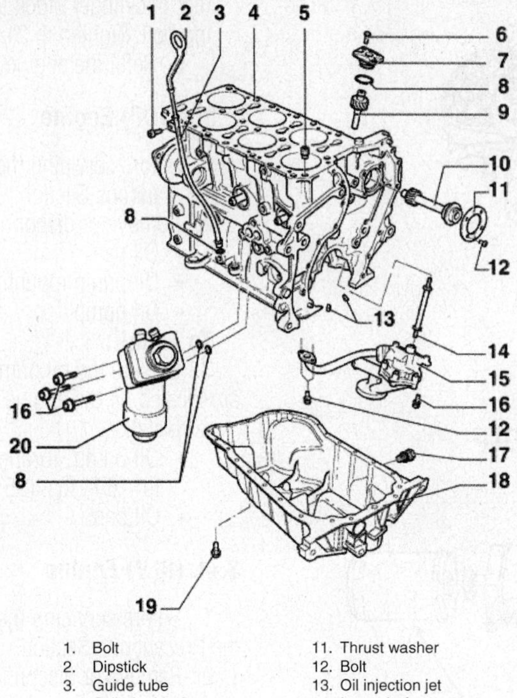

Exploded view of the lubrication system components—2.8L (BDF) engine

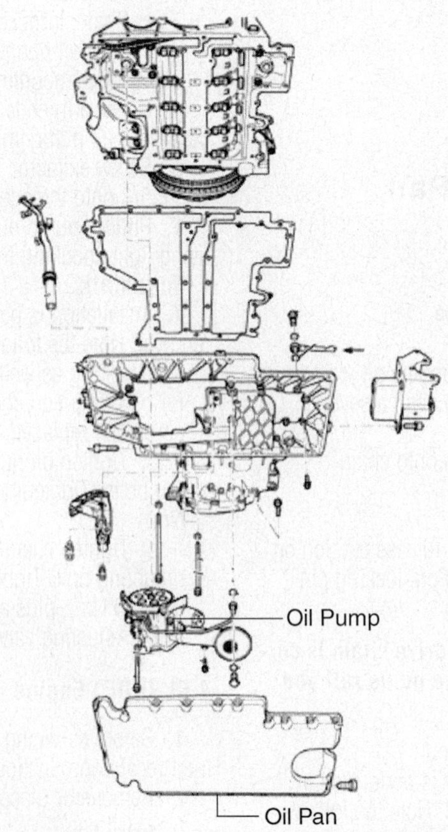

Exploded view of the lubrication system components—4.0L (BDP) engine

- Oil pump.

To install:

3. Be sure the oil pump dowel sleeves are located on the engine block.
4. Install:
   - Oil pump. Torque mounting bolts to: 15 ft. lbs. (20 Nm).
   - Oil pan.

Rear Main Seal

REMOVAL & INSTALLATION

The rear main oil seal is located in a housing on the rear of the cylinder block. To replace the seal on all vehicles it is necessary to remove the transaxle and flywheel.

1. Before servicing the vehicle, refer to the Precautions Section.
2. Remove or disconnect:
   - Transaxle and flywheel.
   - On 6 and 8-cylinder engines, old seal out of the support ring.
   - On 4-cylinder engines, oil seal with the mounting flange as a complete unit.

To install:

3. On 6 and 8-cylinder engines, oil the new seal and press it into place using tool VW-2003/2A, or equivalent, to start the seal and tool VW-2003/1, or equivalent, to seat the seal. Be careful not to damage the seal or score the crankshaft.
4. On 4-cylinder engines, install a new mounting flange with seal using a new gasket. Mounting flange bolts to 84 inch lbs. (10 Nm).
5. Install the flywheel and transaxle.

Timing Belt, Front Cover and Seal

REMOVAL & INSTALLATION

2.0L (ABA) Engine

➡Do not turn the engine or camshaft with the camshaft drive belt removed. The pistons will contact the valves and cause internal engine damage.

1. Disconnect the negative battery cable and remove the accessory drive belts, crankshaft pulley and the timing belt cover(s).
2. Temporarily reinstall the crankshaft pulley bolt, if removed, and turn the crankshaft to Top Dead Center (TDC) of No. 1 piston. The mark on the camshaft sprocket should be aligned with the mark on the

inner drive belt cover, if equipped, or the edge of the cylinder head.

3. On 8-valve engines, the notch on the crankshaft pulley should align with the dot on the intermediate shaft sprocket. With the distributor cap removed, the rotor should be pointing toward the No. 1 mark on the rim of the distributor housing.

4. Loosen the locknut on the tensioner pulley and turn the tensioner counterclockwise to relieve the tension on the timing belt.

5. Slide the timing belt off the sprockets.

To install:

6. Install the new timing belt and tension the belt so that it can be twisted 90 degrees at the middle of its longest section, between the camshaft and intermediate sprockets.

7. Recheck the alignment of the timing marks, if correct, turn the engine 2 full revolutions to return to TDC of No. 1 piston. Recheck belt tension and timing marks. Readjust as required. Tighten the tensioner nut to 33 ft. lbs. (45 Nm).

8. Reinstall the belt cover and accessory drive belts.

➡When running the engine, there will be a growling noise that rises and falls with engine speed if the belt is too tight.

1.8L (AEB/ATW/AWM/AWV) 4-Cylinder Engines

1. Disconnect the negative battery cable.

2. Remove the necessary components to gain access to the front of the engine.

3. Place an open-end wrench on the alternator belt tensioner and rotate it clockwise toward the alternator to release the belt's tension. Remove the alternator serpentine drive belt and release the tensioner.

➡If necessary to lock the alternator tensioner in position, align the housing holes and insert an Allen wrench into the holes to secure its movement.

4. Using a 5 x 60mm bolt, secure the viscous fan pulley. Using a hex wrench, remove the viscous fan-to-pulley bolts. Remove the viscous fan assembly.

5. Remove the upper timing belt cover.

➡If reusing the timing belt, mark its rotational direction so it may be installed in its original position.

6. Using the center bolt, rotate the crankshaft in the direction of engine rotation to position the No. 1 cylinder at Top Dead Center (TDC) of its compression stroke.

7. Remove the damper pulley-to-crankshaft bolts and the damper.

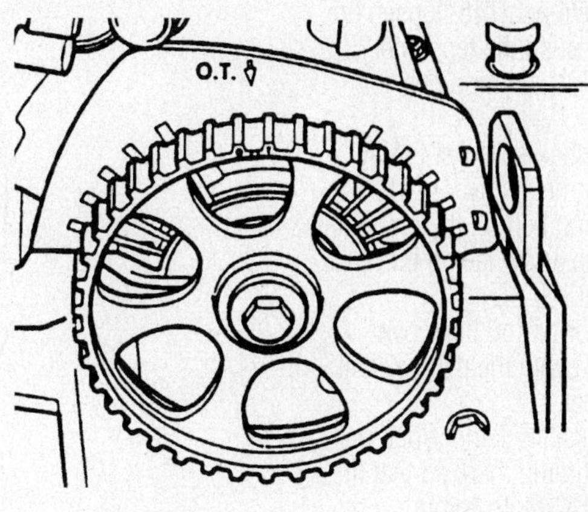

Camshaft timing belt sprocket TDC alignment mark—Volkswagen 1.8L (ACC) and 2.0L (ABA) engines

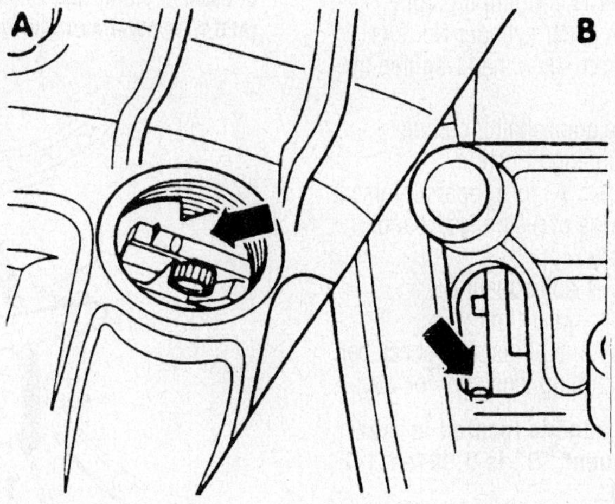

Align the flywheel (A) or driveplate (B) as shown for TDC alignment for cylinder No. 1—Volkswagen 1.8L (ACC) and 2.0L (ABA) engines

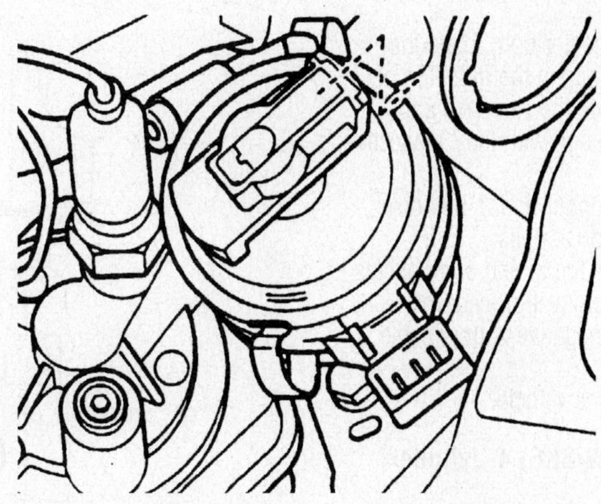

When the No. 1 cylinder is at TDC, the ignition rotor should face the notch in the distributor housing—Volkswagen 2.0L (ABA) engine

8. Remove the lower timing belt cover.
9. Using a Torx Wrench T45, loosen the timing belt tensioner, push the tensioner downward and remove the timing belt.

To install:

10. Align the camshaft sprocket timing mark with the cylinder head cover mark.
11. Install the timing belt on the crankshaft sprocket with the arrow facing the rotational direction.
12. Install the lower timing belt cover.
13. Using a bolt, secure the damper/belt pulley on the crankshaft.
14. Align the crankshaft damper/belt pulley with the housing timing mark so that the No. 1 cylinder is at TDC of its compression stroke.
15. Install the timing belt on the camshaft sprocket and belt tensioner.
16. Using a 2-pin Spanner Matra V159 Wrench, lift (turn clockwise) the timing belt tensioner cylinder No. 1 until it is fully extended and tensioner cylinder No. 2 is raised approx. 1mm; then, hand-tighten the mounting bolt.
17. Rotate the crankshaft 2 complete rotation in the running direction.
18. Inspect area "A" for proper alignment with the upper edge of piston No. 2 and adjust if necessary.

- Area "A"—adjustment OK
- Area "B"—wear limit
- Area "C"—re-adjust and check belt drive including tensioner for wear.

➡If the piston edge is located in area "A", measurement "D" is 0.984–1.142 in. (25–29mm).

19. After adjustment has been verified, secure the tensioner with a 2-pin Spanner Matra V159 Wrench and tighten the mounting bolt.
20. Complete the damper to crankshaft installation.
21. Using the center bolt, rotate the crankshaft 2 rotations in the direction of engine rotation until the camshaft and crankshaft marks align with their respective reference points.
22. Install the upper timing belt cover.
23. Install the drive belts.
24. Replace the remaining components by reversing the removal procedures.
25. Install the negative battery cable last.
26. Test drive the vehicle.

1.8L (AWP/AWW/BKF) 4-Cylinder Engines

1. Disconnect the negative battery cable.

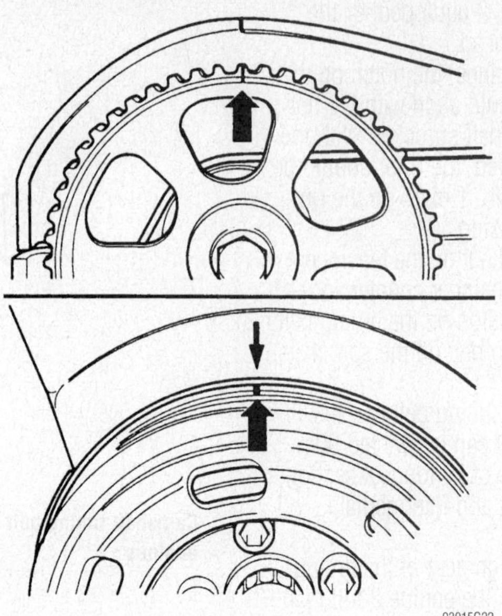

Crankshaft pulley and camshaft sprocket alignment locations—Volkswagen 1.8L (AEB/ATW/AWM/AWP/AWV/AWW) 4-Cyl engine

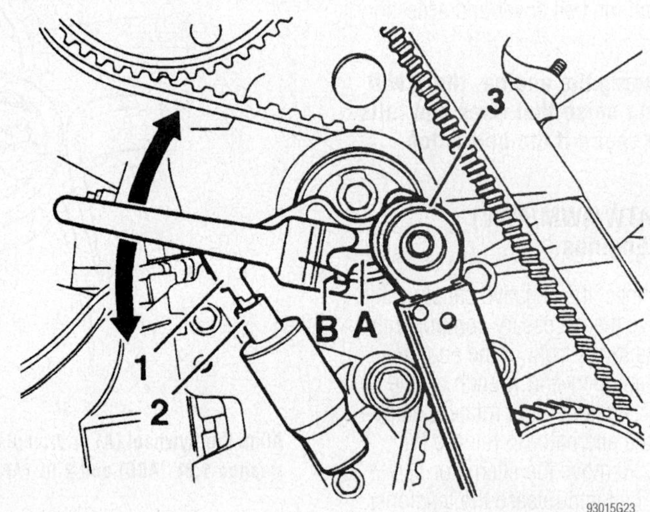

Timing belt tension adjustment— Volkswagen 1.8L (AEB/ATW/AWM/AWV) 4-Cyl engine

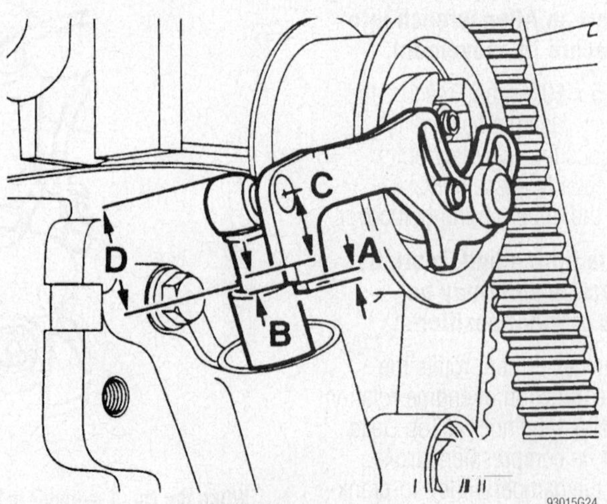

Timing belt tension wear limits— Volkswagen 1.8L (AEB/ATW/AWM/AWV) 4-Cyl engine

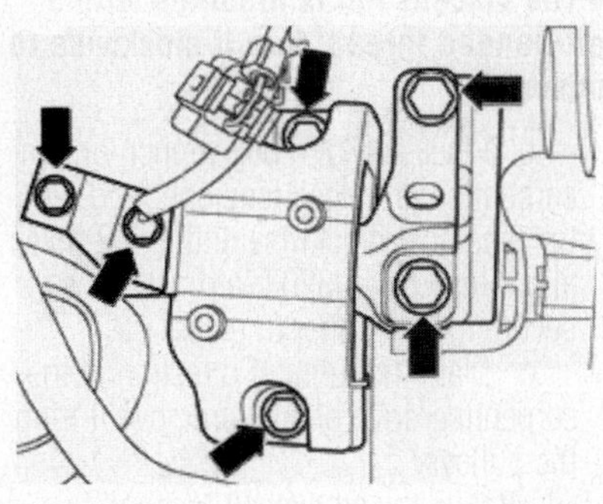

Removing engine and bolt mounts bolts—1.8L (AWP/AWW) 4-Cyl engine

2. Remove the engine appearance cover and splash shield.
3. Remove the right side headlight assembly.
4. Remove the intake pipe between the turbocharger and the charge air cooler.
5. Place an open-end wrench on the alternator belt tensioner and rotate it clockwise toward the alternator to release the belt's tension. Remove the alternator serpentine drive belt and release the tensioner.

➡If necessary to lock the alternator tensioner in position, align the housing holes and insert an Allen wrench into the holes to secure its movement.

6. Using the center bolt, rotate the crankshaft in the direction of engine rotation to position the No. 1 cylinder at Top Dead Center (TDC) of its compression stroke.
7. Remove the upper timing belt cover.

➡If reusing the timing belt, mark its rotational direction so it may be installed in its original position.

8. Use an engine hoist and take the weight off the engine.
9. Remove the engine mount and body mount bolts and remove the engine mount from the cylinder block.
10. Remove the damper pulley-to-crankshaft bolts and the damper.
11. Remove the center and lower timing belt covers.
12. Place tensioning bolt T10092 into the timing belt tensioner and tension the element so the tensioner piston can be locked with a locking pin.
13. Remove the timing belt.
14. turn the crankshaft back slightly so the pistons are not at TDC.

To install:

15. Align the camshaft sprocket timing mark with the cylinder head cover mark.
16. Install the timing belt on the crankshaft sprocket with the arrow facing the rotational direction.
17. Install the engine mount and body mount bolts to the cylinder block and tighten to 33 ft. lbs. (45 Nm).
18. Install the lower timing belt cover and tighten the bolts to 89 inch lbs. (10 Nm).
19. Using a bolt, secure the damper/belt pulley on the crankshaft.
20. Using the center bolt, rotate the crankshaft in the direction of engine rotation to position the No. 1 cylinder at Top Dead Center (TDC) of its compression stroke.
21. Install the timing belt on the camshaft sprocket, water pump and belt tensioner.
22. Remove locking pin and tensioning tool from the belt tensioner.
23. Rotate the crankshaft 2 complete rotation in the running direction.
24. Check that the crankshaft and camshaft timing marks are still aligned.
25. Install the crankshaft pulley bolts and tighten to 18 ft. lbs. (25 Nm).
26. Install the center timing belt cover and tighten the bolts to 89 inch lbs. (10 Nm).
27. Install the upper timing belt cover.
28. Install the drive belts.
29. Replace the remaining components by reversing the removal procedures.
30. Install the negative battery cable last.
31. Test drive the vehicle.

2.0L (AEG/AVH/AZG/BDC/BEV/BGD) 4-Cylinder Engine

1. Disconnect the negative battery cable.
2. Remove the necessary components to gain access to the front of the engine.
3. Remove the serpentine drive belt by performing the following procedure:
   a. Using an open-end wrench, rotate the serpentine drive belt tensioner clockwise to relieve the belt tension. Using tool 3090 or a drift awl, lock the tensioner in place.
   b. Remove the serpentine drive belt.
4. Remove the drive tensioner from the engine.
5. Rotate the crankshaft to position the No. 1 cylinder on Top Dead Center (TDC) of its compression stroke.
6. Remove the upper timing belt cover.

➡If the camshaft sprocket is not aligned with its timing mark, rotate the crankshaft 1 complete turn until the crankshaft and camshaft timing marks align.

7. Install the Engine Support tool 10-222A with Leg tools and support the weight of the engine.
8. From the front of the engine, remove the engine mount and bracket which will interfere with the timing belt removal.
9. Mark the rotational direction of the timing belt for reinstallation purposes.
10. Remove the vibration damper-to-crankshaft sprocket bolts and the damper.
11. Remove the center and lower timing belt covers.
12. Release the timing belt tensioner and remove the timing belt.

To install:

13. Make sure that the camshaft and crankshaft timing marks are aligned.
14. Install the timing belt, with the arrow pointing in the direction of rotation, on the crankshaft sprocket, the water pump sprocket and the camshaft sprocket.
15. Install the lower and center timing belt covers.
16. Install the vibration damper and torque the damper-to-crankshaft sprocket bolts to 18 ft. lbs. (25 Nm).
17. Using the 2-Hole Tensioning tool T-10020, rotate the timing belt tensioner counterclockwise until notch **1** and indicator **2** align. Then, tighten the timing belt tensioner nut to 15 ft. lbs. (20 Nm).
18. Rotate the crankshaft 2 complete revolutions and recheck the timing marks.
19. Install the upper timing belt cover.
20. Install the engine bracket and torque the bolts to 33 ft. lbs. (45 Nm).
21. Install the engine mount and torque the mount-to-bracket bolts to 30 ft. lbs. (40 Nm) plus 90 degree turn (¼ turn), the mount-to-body bolts to 18 ft. lbs. (25 Nm) and the mount-to-engine bracket bolts to 44 ft. lbs. (60 Nm) plus 90 degree turn (¼ turn).
22. Install the timing belt tensioner and torque the bolts to 18 ft. lbs. (25 Nm).
23. Complete the installation by reversing the removal procedures.
24. Test drive the vehicle.

2.8L (AHA/ATQ) V6 Engines

1. Disconnect the negative battery cable.
2. Remove the necessary components to gain access to the front of the engine.
3. Remove the serpentine drive belt by performing the following procedure:
   a. Using Spanner Wrench No. 3212, secure the viscous fan pulley. Using an Open-end Spanner Wrench 3212, remove the viscous fan bearing housing by turning it clockwise.

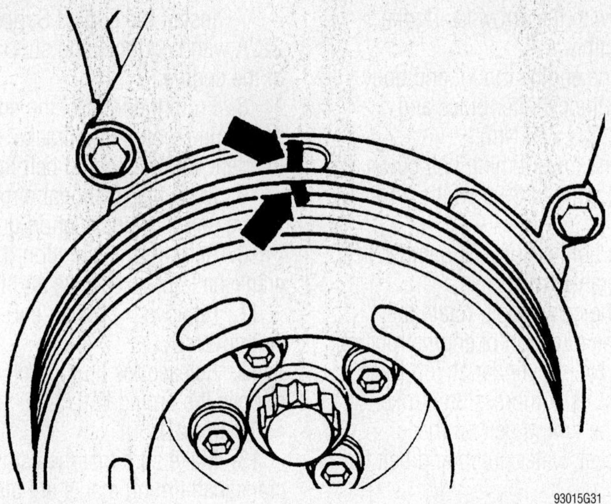

View of crankshaft damper aligned with the timing mark—Volkswagen 2.0L (AEG/AVH/AZG/BDC) 4-Cyl. Engine

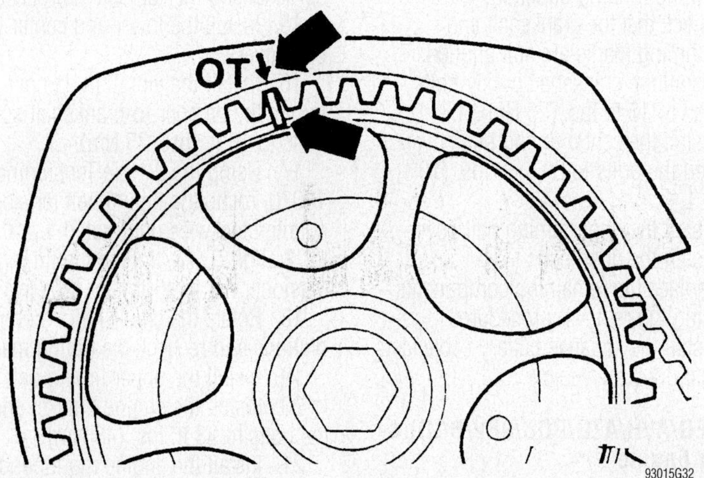

View of camshaft damper aligned with the timing mark—Volkswagen 2.0L (AEG/AVH/AZG/BDC) 4-Cyl. Engine

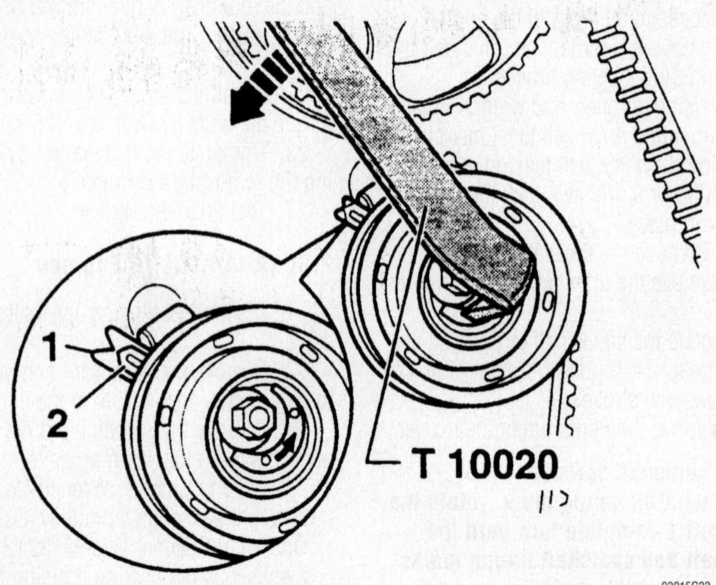

Adjusting the timing belt tensioner—Volkswagen 2.0L (AEG/AVH/AZG/BDC) 4-Cyl. engine

➡The viscous fan is mounted with a left-handed thread; turn it clockwise to loosen it.

b. Place a 17mm box wrench on the serpentine drive belt tensioner and turn the tensioner clockwise until the 2 holes align; insert drift 3204 into the holes to secure the tensioner in place.

c. Mark the running direction of the serpentine drive belt and remove it from the pulleys.

4. Rotate the crankshaft by hand to align the crankshaft pulley mark with the arrow on the engine housing and the large hole in each camshaft sprocket must face inward and must align; this should be Top Dead Center (TDC) of the No. 1 cylinder's compression stroke. If these conditions are not correct, rotate the crankshaft one complete revolution and realign.

5. On the left side of the cylinder block near the crankshaft, remove the sealing plug.

6. Insert Crankshaft Holder tool No. 3242 into the sealing plug hole to secure the crankshaft.

7. Using a 8mm Allen® wrench, rotate the timing belt tensioner roller clockwise until the tensioner is compressed; then, insert a 2mm spring pin through the tensioner housing and tensioner plunger to secure it in place. When the plunger is secure, release the wrench tension.

8. Remove the dampener-to-crankshaft bolts and the damper.

➡It is not necessary to remove the center bolt when removing the crankshaft damper.

9. Remove the serpentine belt idler and the crankshaft damper guard.

10. Mark the running direction of the timing belt and remove it from the pulleys.

To install:

11. Make sure that the camshaft pulleys and the crankshaft pulley are in alignment with TDC of the No. 1 cylinder's compression stroke.

12. Install the timing belt; make sure the timing belt is installed in the correct running direction from which it was removed.

13. Using a 8mm Allen® wrench, rotate the timing belt tensioner roller clockwise until the tensioner is compressed; then, remove the 2mm spring pin from the tensioner housing. Slowly, release the tensioner's spring pressure to put pressure on the timing belt.

14. Install the crankshaft damper guard and the serpentine belt idler pulley; torque the idler pulley bolts to 33 ft. lbs. (45 Nm).

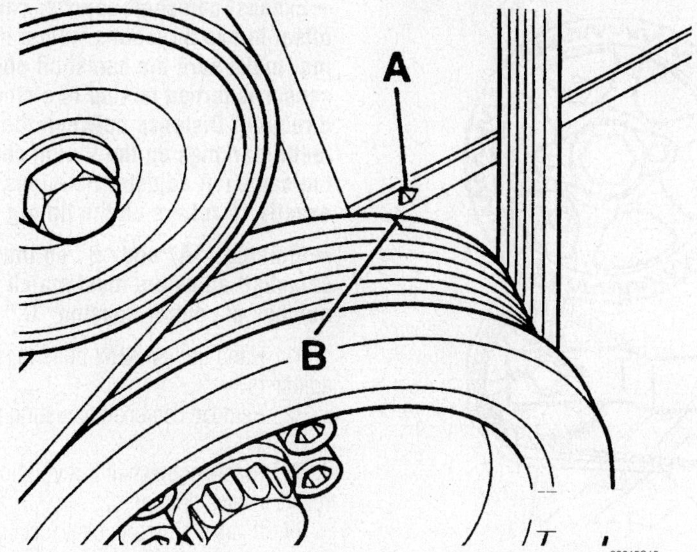

Crankshaft pulley alignment location for TDC—Volkswagen 2.8L (AHA/ATQ) V6 engine

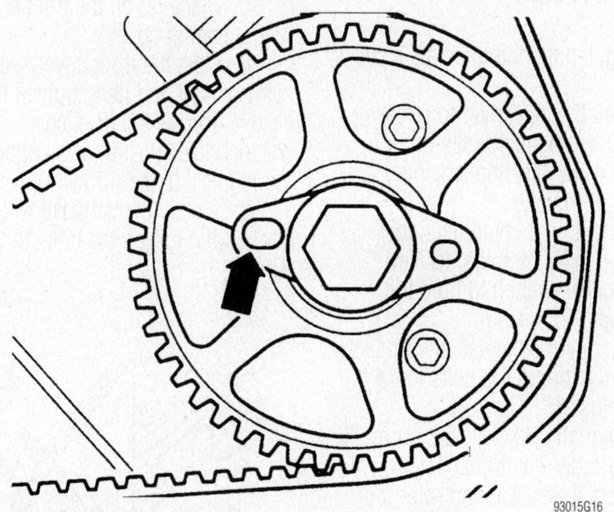

Left camshaft sprocket alignment position for TDC; right camshaft position is similar—Volkswagen 2.8L (AHA/ATQ) V6 engine

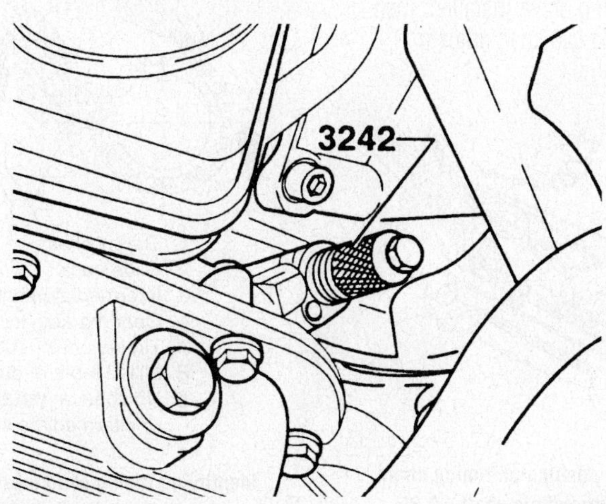

View of crankshaft holding tool installed—Volkswagen 2.8L (AHA/ATQ) V6 engine

15. Install the crankshaft damper and torque the damper-to-crankshaft bolts to 15 ft. lbs. If the damper-to-crankshaft center bolt was removed, torque it to 147 ft. lbs. (200 Nm) plus 180° (½ turn).
16. Remove the Crankshaft Holder tool No. 3242 and install the sealing plug.
17. Replace the remaining components by reversing the removal procedures.
18. Refill the cooling system and the automatic transaxle. Connect the electrical connectors. Install the negative battery cable last.
19. Test drive the vehicle.

2.8L (BDF) Engines

1. Before servicing the vehicle, refer to the Precautions Section.
2. Disconnect the negative battery cable.
3. Remove engine covers at the front of the engine compartment and underneath the vehicle.
4. Mark the rotation direction of the drive belt.
5. Insert an M8x50 bolt into the threaded hole of the belt tensioner roller and tighten just enough until tension is off belt and drive belt can be removed.
6. Remove the cylinder head cover.
7. Remove the necessary components to gain access to the front of the engine.
8. Place the crankshaft at TDC for no. 1 cylinder.
9. Mark the camshaft timing chain rotation direction with a paint mark.
10. Remove the camshaft roller chain tensioner.
11. Remove the 8 mounting bolts and remove the chain front cover
12. Disconnect the front exhaust pipe from the exhaust manifold.
13. Check that timing is at TDC by looking down between timing chain sprockets of the intermediate shaft. A groove should be visible between the chain edges.

➡If the groove cannot be seen, turn crankshaft one more revolution until groove can be seen.

14. Insert camshaft gauge T10068 into both camshaft grooves.

➡If the gauge cannot be inserted, rotate crankshaft about .20 inches (5 mm) above TDC position until gauge can be inserted.

➡Markings "A" and "B" on the camshaft adjusters must match the notches of control housing "C."

15. Remove the camshaft adjusters together with the timing chain from the camshafts.

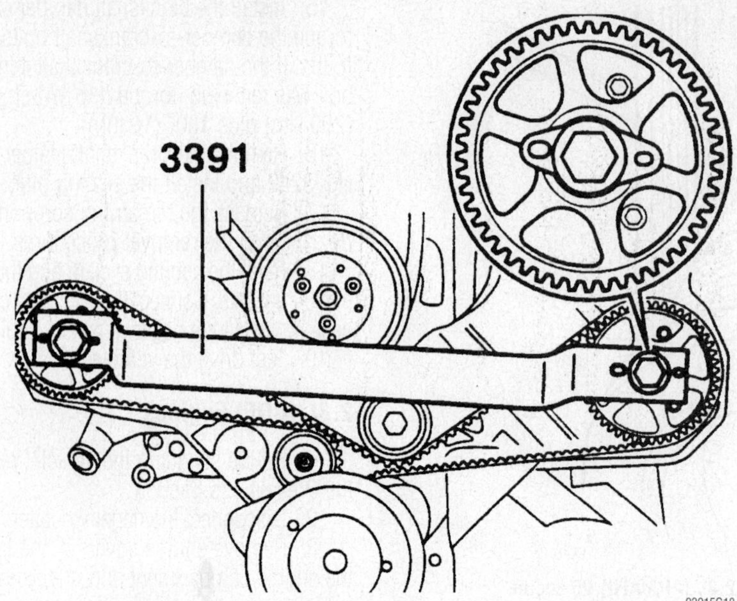

View of camshaft locator bar installed—Volkswagen 2.8L (AHA/ATQ) V6 engine

16. Remove the two bolts from the glide track located at lower left below camshaft sprocket.

17. Remove the timing chain from the camshaft adjusters and place chain on its side.

To install:

18. Adjust the milled tooth of the drive pinion to line up with the bearing separating gap at TDC on cylinder no. 1.

19. Install the glide track bolts, without the collar, and torque to 89 inch lbs. (10 Nm).

20. Install the timing chains and the intermediate shaft gears (if removed), ensuring that the timing chain rotation direction is correct.

21. Check that the marking on the gear matches with the notches on the intermediate shaft thrust washer.

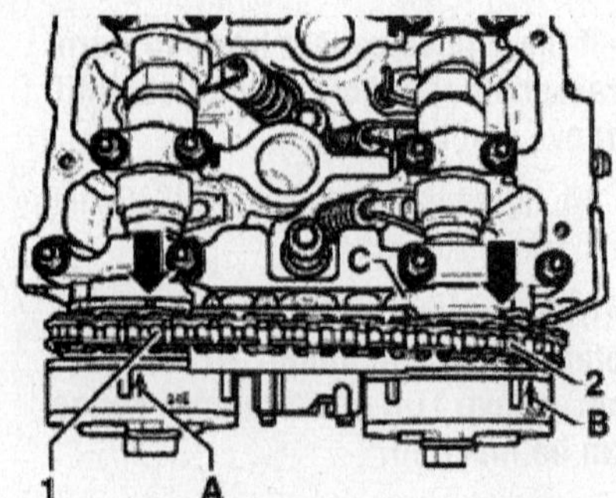

Lining up camshaft adjusters to remove timing chain—2.8L (BDF) engine

22. Hand tighten the intermediate shaft gear bolts.

23. Use a small screwdriver to release the tension of the locking splines in the chain tensioner and press the tensioner track against the tensioner.

24. Install the timing chain tensioner bolts and torque to 72 inch lbs. (8 Nm).

25. Use a socket wrench to hold the crankshaft damper in position.

26. With the intermediate shaft sprockets lined up correctly, torque the bolts to 44 ft. lbs. (60 Nm), plus 90°.

27. Check that all sprockets are correctly lined up and cylinder no. 1 is at TDC.

28. Insert camshaft adjusting ruler tool T10068 into camshaft grooves.

29. Guide the timing chain between the tensioning track and the glide track.

30. Install the intake camshaft adjuster first with the timing chain installed, then install the exhaust camshaft adjuster.

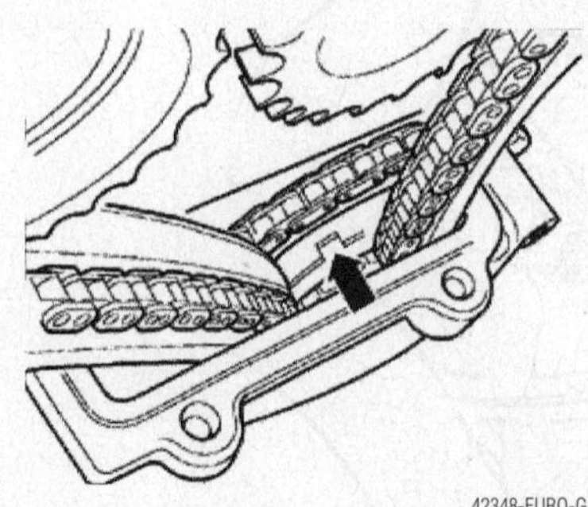

Identifying TDC position on timing chain sprockets of intermediate shaft—2.8L (BDF) engine

➡Exhaust camshaft adjuster can be offset in two directions. When installing, make sure the camshaft position sensor is turned to stop in a clockwise direction. Distance between the milled teeth markings on the timing chain at the camshaft adjuster markings is exactly 16 rollers on the timing chain.

➡Markings "A" and "B" on the camshaft adjusters must match the notches of control housing "C."

31. Hand tighten NEW bolts for camshaft adjusters.

32. Remove camshaft adjusting ruler T10068.

33. Rotate crankshaft 2 revolutions and recheck valve timing.

34. If valve timing is okay, secure camshafts in position and torque camshaft adjuster bolts to 44 ft. lbs. (60 Nm), plus 90°.

35. Place oil on the front cover O ring and install in cover.

36. Coat the front cover sealing surface with sealant and hand tighten the bolts.

37. Torque the front cover M6 bolts to 72 inch lbs. (8 Nm), and tighten the M8 bolts to 18 ft. lbs. (25 Nm).

38. Install the camshaft roller chain tensioner and torque the bolts to 30 ft. lbs. (40 Nm).

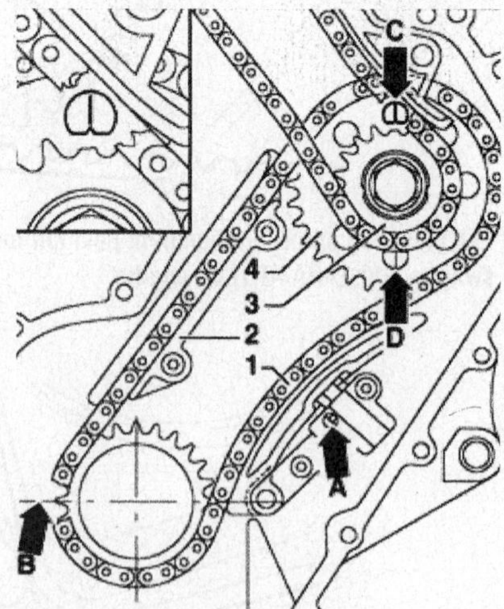

Identifying timing chain positioning on crankshaft and intermediate shaft sprockets—2.8L (BDF) engine

39. Rotate crankshaft 2 revolutions and recheck valve timing.

40. Replace the remaining components by reversing the removal procedures.

41. Reconnect the negative battery cable.

42. Test drive the vehicle.

4.0L (BDP) Engines

➡The following procedure is for cylinder bank no. 1. Cylinder bank no. 2 is similar.

1. Before servicing the vehicle, refer to the Precautions Section.

2. Disconnect the negative battery cable.

3. Remove the engine top cover.

4. Remove the engine insulation cover.

5. Drain the engine coolant into an approved container.

6. Disconnect the coolant hoses at the radiator.

7. Remove or disconnect:

- Upper intake manifold.

8. Disconnect the connectors to the ignition coils with power output stage and move the wiring harness clear.

9. Remove the ignition coils using T10094 extractor.

10. Remove or disconnect:

- Cylinder head cover
- Fuel rail with injectors.
- Lower intake manifold spacer.

➡Cover intake ports to prevent debris from entering.

- Coolant pipes.
- Intake pipes.

11. Turn the crankshaft until no. 1 cylinder is at TDC. Mark on crankshaft damper must align with mark on cylinder block joint.

12. Remove the camshaft chain tensioner.

13. Remove the timing chain upper cover.

1. TDC mark on pulley
2. Cylinder block joint

42348-VWVW-G06

Locating TDC alignment marks—4.0L engine

14. Remove the timing chain lower cover.

15. Mark the timing chain rotation direction for installation.

16. Remove the exhaust camshaft adjuster bolts.

17. Remove the intake camshaft adjuster bolts.

18. Remove the camshaft adjusters together with the timing chain.

19. Remove the timing chain guide rail.

To install:

20. Set the crankshaft and camshafts to TDC for cylinder No. 1.

21. Secure the crankshaft from turning.

22. Install camshaft adjuster guide T10068 into horizontal slots on camshafts to lock them in place

23. On cylinder bank no. 1, install the timing chain as follows:

24. On the intake camshaft adjuster, turn the Hall sensor wheel clockwise to its stop.

25. On the exhaust camshaft adjuster, turn the Hall sensor wheel counterclockwise to its stop.

26. Install the timing chain and camshaft adjuster as a unit.

27. Copper link "A" of timing chain must be positioned on the top of the intermediate shaft.

28. Copper link of timing chain must align with "B1" arrow on intake camshaft adjuster

29. Copper link of timing chain must align with "B1" arrow on exhaust camshaft adjuster

30. Tighten NEW camshaft adjuster bolts hand tight.

31. Install the guide rail and tensioning rail.

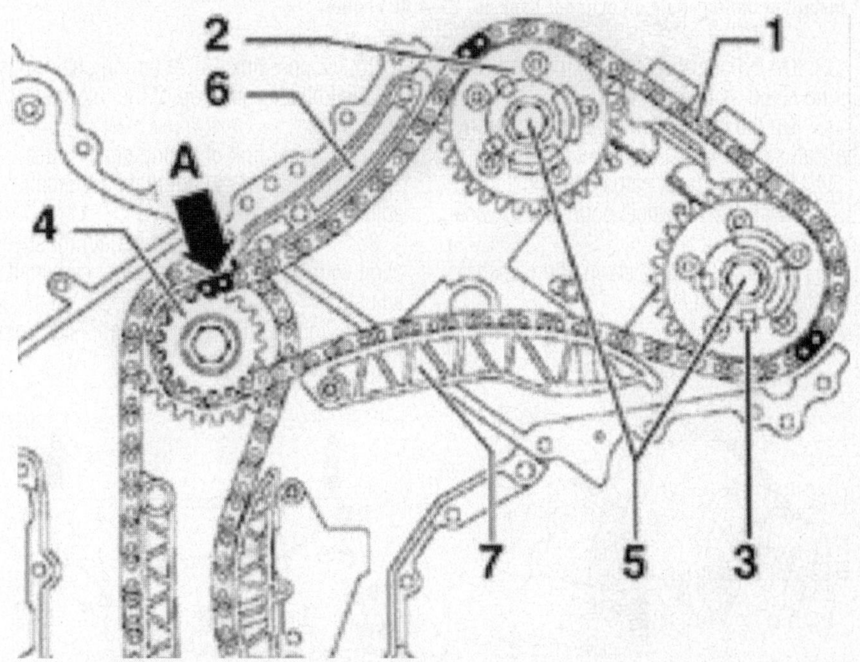

42348-VWVW-G17

Installing timing chain on cylinder bank no. 1—4.0L engine

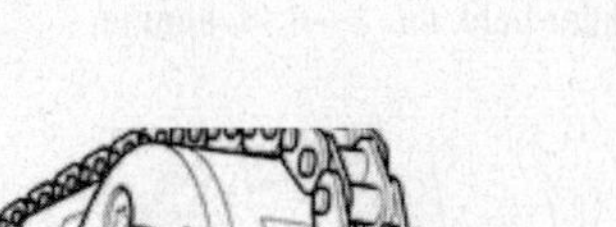

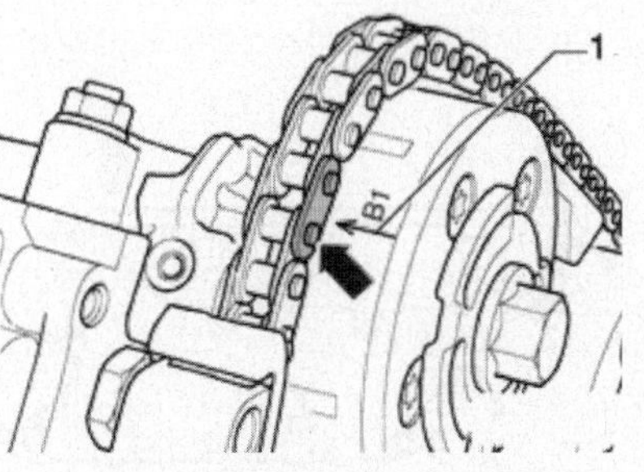

42348-VWVW-G18

Locating copper timing chain link with arrow on intake camshaft adjuster for cylinder bank no. 1—4.0L engine

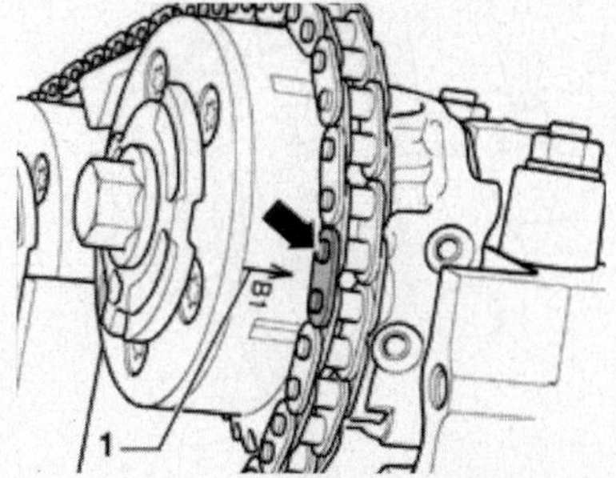

42348-VWVW-G19

Locating copper timing chain link with arrow on exhaust camshaft adjuster for cylinder bank no. 1—4.0L engine

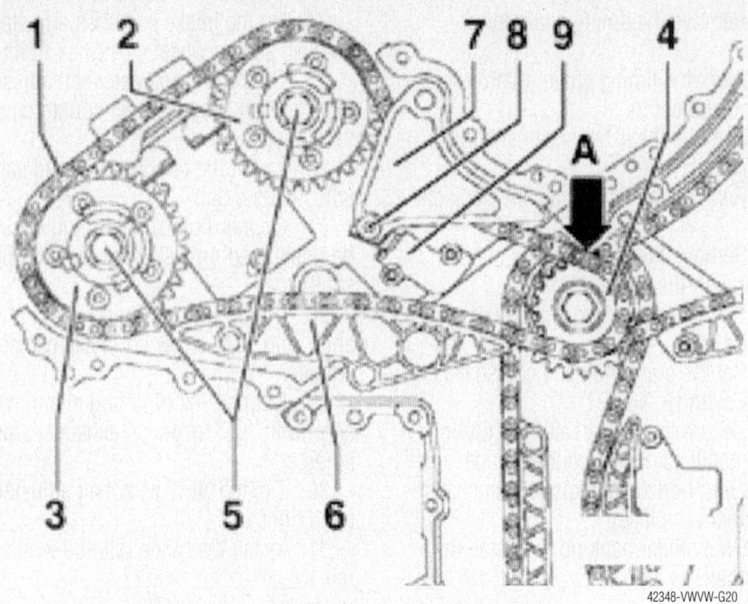

Installing timing chain on cylinder bank no. 2—4.0L engine

32. On cylinder bank no. 2, install the timing chain as follows:

33. On the intake camshaft adjuster, turn the Hall sensor wheel clockwise to its stop.

34. On the exhaust camshaft adjuster, turn the Hall sensor wheel counterclockwise to its stop.

35. Install the timing chain and camshaft adjusters as a unit.

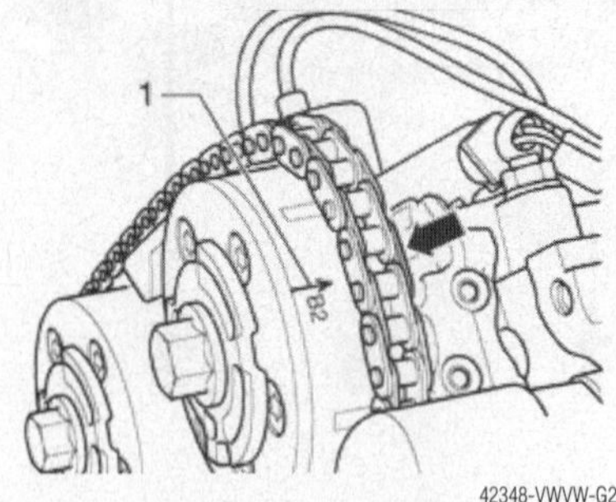

Locating copper timing chain link with arrow on intake camshaft adjuster for cylinder bank no. 2—4.0L engine

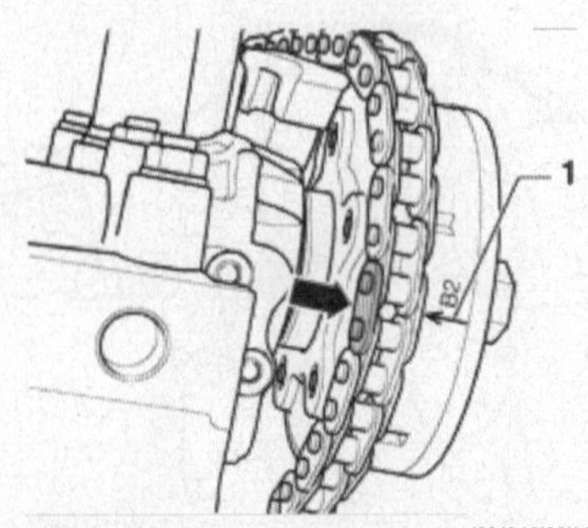

Locating copper timing chain link with arrow on exhaust camshaft adjuster for cylinder bank no. 1—4.0L engine

36. Copper link "A" of timing chain must be positioned on the top of the intermediate shaft.

37. Copper link of timing chain must align with "B2" arrow on intake camshaft adjuster

38. Copper link of timing chain must align with "B2" arrow on exhaust camshaft adjuster

39. Tighten NEW camshaft adjuster bolts hand tight.

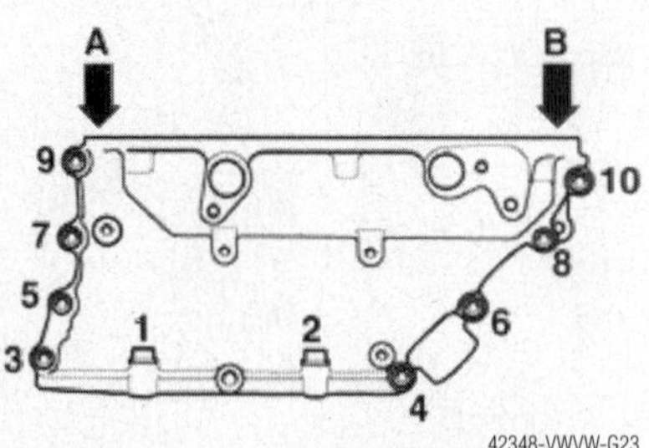

Tightening sequence for upper timing chain cover for cylinder bank no. 1—4.0L engine

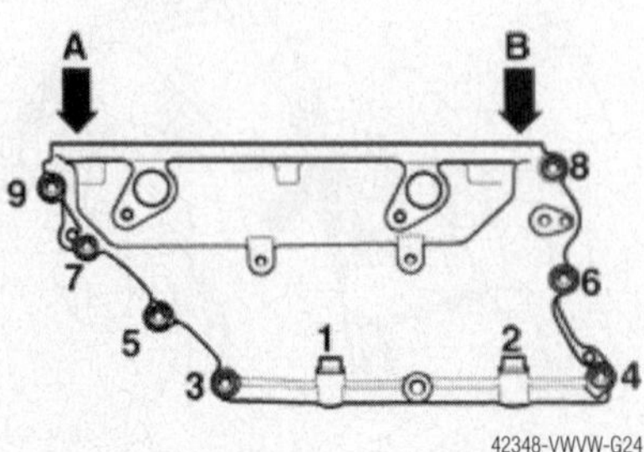

Tightening sequence for upper timing chain cover for cylinder bank no. 2—4.0L engine

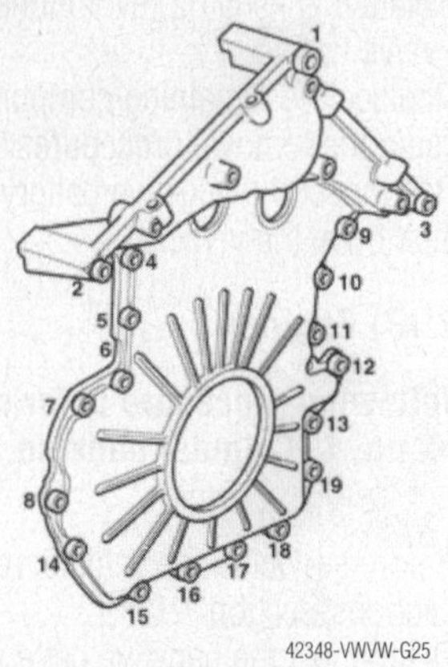

Tightening sequence for lower timing chain cover—4.0L engine

40. Install the guide rail and tensioning rail.

41. Install the chain tensioner limit stop and tighten to 15 ft. lbs. (20 Nm).

42. Install the chain tensioner and tighten to 71 inch lbs. (10 Nm).

43. On both cylinder banks, proceed as follows:

44. Remove camshaft adjuster guide T10068 from both heads.

45. Hold crankshaft from turning and install NEW intermediate sprocket bolt and tighten to 44 ft. lbs. (60 Nm).

46. Hold camshafts in place with a 32 mm open end wrench, and tighten NEW camshaft adjuster bolts to 44 ft. lbs. (60 Nm).

47. Install lower timing chain cover and tighten bolts in sequence.

48. Install sealant to cylinder bank no. 1 upper timing belt cover and tighten bolts in sequence to 71 inch lbs. (10 Nm).

49. Install sealant to cylinder bank no. 2 upper timing belt cover and tighten bolts in sequence to 71 inch lbs. (10 Nm).

50. Replace the remaining components by reversing the removal procedures.

51. Reconnect the negative battery cable.

52. Test drive the vehicle.

Piston and Ring Positioning

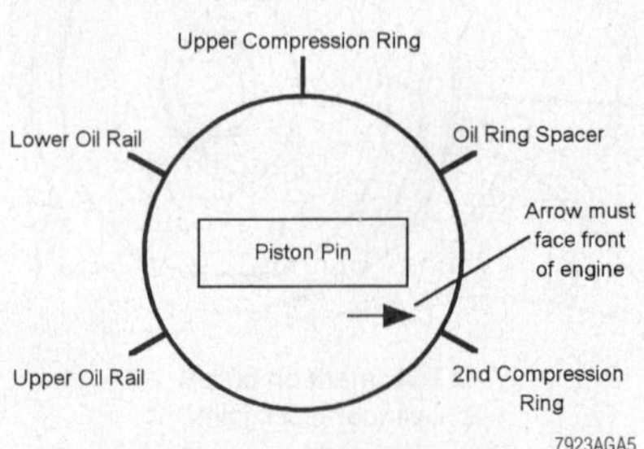

1.8L (ABE), and 2.0L (ABA) engines—piston ring end-gap spacing

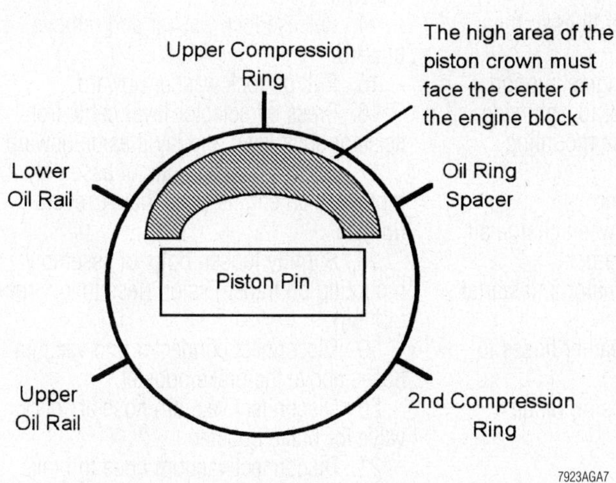

2.8L (AAA) engine—piston ring end-gap spacing

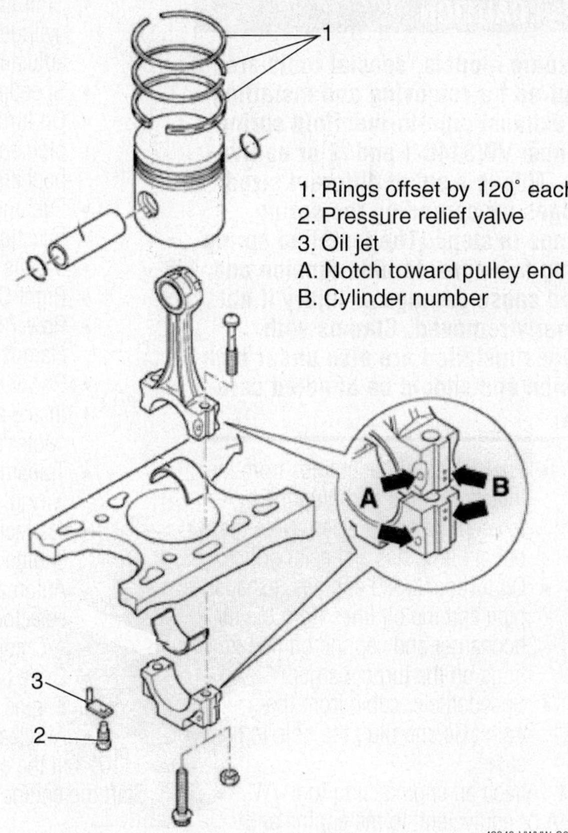

Piston and connecting rod installation—all other engines

DIESEL ENGINE REPAIR

Alternator

REMOVAL & INSTALLATION

1. Disconnect the negative battery cable.
2. Remove the engine undercover, if equipped.
3. Disconnect the wiring from the alternator. If necessary, mark the wires with tape or other means to ensure they are connected properly upon installation.
4. Loosen the belt tension, and remove the drive belt from the alternator.
5. Remove the alternator adjustment bolt, followed by the pivot bolts.
6. Carefully lift the alternator from the bracket.

To install:

7. Hold the alternator in position on the mounting bracket, and install the pivot bolt. On later engines with automatic belt tensioners, install the upper mounting bolt.
8. Install (but do not tighten) the alternator adjustment bolt (earlier models only).
9. Place the alternator drive belt on the pulley.
10. Adjust belt tension and tighten the mounting and adjustment bolts to 18 ft. lbs. (25 Nm).
11. Connect the wiring to the alternator. If tape was used to identify the wires, make sure it is removed once the wires are connected.
12. Install the engine undercover, if equipped.
13. Connect the negative battery cable.

Engine Assembly

REMOVAL & INSTALLATION

Golf and 2002–04 Jetta

1. Before servicing the vehicle, refer to the Precautions Section.
2. The engine and transaxle are removed from the vehicle as an assembly.
3. Remove or disconnect:
   - Battery cables and battery.
   - Fuel pressure. Open the fuel filler cap to relieve tank pressure, then loosen the fuel filter fitting. Be sure to take the appropriate fire safety precautions.
   - Engine top cover
   - Right driveshaft
   - Disconnect left driveshaft at transmission
   - Radiator hoses
   - 4 bolts and front engine support
   - Air cleaner
   - Engine undercover
   - Drive belt
   - A/C compressor and wire aside
   - Automatic transmission range selector lever
   - All vacuum and breather hoses from engine
   - Transmission, alternator and starter wiring
   - Intake air pipes between charge air cooler and turbocharger
   - Power steering pump
   - Power steering hose mounting clamps
   - Right CV joint cover from right side
   - Front exhaust pipe and catalytic converter

⁂ CAUTION

On some models, special tools are required for removing and installing the exhaust pipe-to-manifold spring clamps: VW3140/1 and /2 or equivalent. This is a set of different sized wedges for spreading the spring clamps in steps. The installed spring clamp has considerable tension and could cause damage or injury if not properly removed. Clamps with wedges installed are also under high tension and should be handled carefully.

- Fuel inlet and outlet lines from the injection pump and plug the holes to keep the pump clean. Note the outlet fitting has a special orifice.
- On turbocharged engines, exhaust pipe and the oil lines from the turbocharger and cap the oil line fittings on the turbocharger.
- Speedometer cable from the transaxle and plug the hole in the case.

4. Attach an engine sling tool VW-2024A or equivalent, to the engine and attach the sling to a suitable lifting device.
5. Remove or disconnect:
   - Engine mount from engine bracket
   - Engine transmission mount from transmission bracket
6. With all mounts unbolted, slightly lower the engine/transaxle assembly and tilt it towards the transaxle side. Then, carefully lower the assembly out of the vehicle.

To install:

7. Carefully install the engine/transaxle assembly and be sure all mounts are securely bolted to the engine/transaxle. Start all nuts and bolts that secure the mounts to the body but don't tighten them yet.
8. With all mounts installed and the engine safely in the vehicle, allow some slack in the lifting equipment. With the vehicle safely supported, shake the engine/transaxle as a unit to settle it in the mounts.Bolts, starting at the rear and working forward: 30 ft. lbs. (40 Nm) for 10mm bolts or 44 ft. lbs. (60 Nm) for 12mm bolts.
9. Install or connect:
   - Starter. Bolts: 50 ft. lbs. (65 Nm).
   - Halfshafts to the flanges. Bolts: 30 ft. lbs. (40 Nm).
   - Exhaust pipe and use new self-locking nuts to secure the flange. Nuts: 18 ft. lbs. (25 Nm). If equipped with spring clamps, the clamps can be used again.
   - Shift linkage and the clutch slave cylinder. Make any necessary adjustments.
   - Speedometer cable
   - On turbocharged engines, exhaust pipe and the oil lines to the turbocharger
   - Fuel inlet and outlet lines to the injection pump
   - 4 bolts and front engine support
   - Right CV joint cover to right side
   - Power steering hose mounting clamps
   - Power steering pump
   - Intake air pipes between charge air cooler and turbocharger
   - Transmission, alternator and starter wiring
   - All vacuum and breather hoses to engine
   - Automatic transmission range selector lever
   - A/C compressor
   - Drive belt
   - Engine undercover
   - Air cleaner
10. Fill the engine with coolant and oil. Start the engine and check for leaks.

2005–06 Jetta

1. Before servicing the vehicle, refer to the Precautions Section.
2. The engine and transaxle are removed from the vehicle as an assembly.
3. Disconnect the negative battery cable.
4. Remove engine cover. To do this, jerk front of engine cover upward and then pull it forward out of rear retainer.
5. Remove bulkhead plenum chamber.
6. Remove air filter housing with Mass Air Flow (MAF) Sensor G70 and connecting pipe.
7. Disconnect connector at Mass Air Flow (MAF) Sensor G70.
8. Disconnect breather hose, as well as air guide hoses.
9. Remove bolt and air filter housing.
10. Remove battery.
11. Remove battery carrier.

⁂ CAUTION

When touching metal parts, it is possible that an electrostatic discharge could occur. The reason for this is the electrostatic charge of the human body. This charge can lead to functional problems by touching electrical components of the transmission and selector lever mechanism.

12. Touch a grounded object, e.g. a water pipe or a lifting platform before working on Mechatronic! Do not make direct contact with connector terminals.
13. Turn bayonet connection of the connector toward left and pull connector off transmission.
14. Release lock washer and remove upward.
15. Pull off lock washer upward.
16. Press off selector lever cable from selector shaft lever and lay it aside upward.
17. Slightly loosen bolts of assembly mounting on engine (less than 1 revolution).
18. Slightly loosen bolts of assembly mounting on transmission (less than 1 revolution).
19. Disconnect connector and vacuum hoses above the brake booster.
20. Disconnect vacuum hose at check valve for brake booster.
21. Disconnect vacuum hose to brake booster at tandem pump.
22. Disconnect coolant hoses to heater core at bulkhead.
23. Remove Ground (GND) wire.
24. Disconnect wires on starter and free up cables.
25. Disconnect connector at Engine Coolant Temperature (ECT) Sensor G62.
26. Pull off central connector for pump injector unit by first pulling the release pin and opening the knurled nut. Then free up wiring harness.
27. Disconnect connector at Fuel Temperature Sensor G81 and free up wiring.
28. Remove coolant hose between engine and radiator.
29. Pull off hoses and free them up. Remove air duct. To do this, lift retaining clips slightly.
30. Disconnect coolant hose.
31. Disconnect fuel supply line and fuel return line. To do this, pull release buttons.
32. Disconnect coolant hose from coolant expansion tank.
33. Disconnect connector at Intake Flap Motor V157.
34. Remove connector for Camshaft Position (CMP) Sensor G40 from bracket and disconnect it.
35. Disconnect connector at oil pressure switch F1.
36. Unscrew oil pressure switch.
37. Disconnect the connector at Engine Speed (RPM) Sensor G28 with assembly tool T10118.

➡To release the connector without the assembly tool T10118 , push the connector against the Engine Speed (RPM)

Sensor G28 and simultaneously lift up on release tab with a thin wire hook.

38. Free up cable and disconnect connector strip from glow plugs.
39. Free up wiring harnesses to body.
40. Remove sound insulation.
41. Move lock carrier into service position.

⁂ CAUTION

To prevent damage to condenser and also to the refrigerant lines/hoses, make sure that the lines and hoses are not stretched, kinked or bent.

42. To facilitate removal and installation of the engine without having to open refrigerant circuit:
    a. Remove ribbed belt.
    b. Remove A/C compressor.
    c. Secure A/C compressor to lock carrier so that refrigerant lines/hoses are not stressed.
43. Remove alternator.
44. Remove radiator.
45. Remove left air duct hose, by lifting retaining clips slightly.
46. Remove nut and wiring bracket from threaded stud.
47. Remove right air duct hose, by lifting retaining clips slightly.
48. Disconnect connector at Engine Coolant Temperature (ECT) Sensor (on Radiator) G83.
49. Remove bolt for front air guide tube.
50. Remove right and left drive axle at transmission.
51. Remove front exhaust pipe.
52. Release connector from engine control module (ECM) and remove.
53. Open various retainers of cable guide.
54. Remove engine wiring harness from cable guide on longitudinal member and lay it aside on top of engine.
55. Remove right air guide tube. To do this, lift retaining clips slightly.
56. Remove air guide hose from turbocharger. To do this, lift retaining clip slightly.
57. Pull off/disconnect all electrical connections, as necessary, from engine and transmission and lay them aside.
58. Disconnect all connections, coolant lines, vacuum hoses and intake hoses from engine.
59. First remove bolt at frame, then remove the 2 bolts from the pendulum support at the transmission.
60. Set Engine/Trans. Support T10012 into engine/transmission jack V.A.G 1383 A.
61. Fasten Engine/Trans. Support T10012 to cylinder block with bolt M10 x 25/8.8 to approximately 28 ft. lbs. (40 Nm).
62. Lift engine and transmission slightly using engine/transmission jack V.A.G 1383 A.

➡Use ladder VAS 5085 to remove securing bolts.

63. From above, unbolt assembly mounting from engine bracket on engine side.
64. From above, unbolt assembly mounting from transmission bracket on transmission side.

➡Verify that all hose and line connections between engine, transmission and body have been disconnected.

65. While lowering, carefully guide engine/transmission assembly in order to prevent damages.
66. Carefully lower engine with transmission downward.
67. Secure engine to assembly stand using holding fixture VW 540 on holding fixture VW 313 when performing repair work.
68. Move engine/transmission jack V.A.G 1383 A to a work bench.
69. Lower engine/transmission assembly so that transmission makes contact.
70. Remove connecting bolts for engine/transmission.
71. Press transmission off from engine.
72. Hook in engine sling 2024 A and lift engine using shop crane V.A.G 1202 A out of engine/transmission jack V.A.G 1383 A.

➡The peg positions on the carrying strap marked with 1 through 4 face the belt pulley. The bores in the hole rail are counted from the hook.

73. Secure engine using holding fixture VW 540 on holding fixture VW 313.

To install:

74. Install the engine-to-transaxle attaching hardware and tighten as follows:
    - M6 fasteners: 84 inch lbs. (10 Nm)
    - M8 fasteners: 15 ft. lbs. (20 Nm)
    - M10 fasteners: 33 ft. lbs. (45 Nm)
    - M12 fasteners: 44 ft. lbs. (60 Nm)
75. Install the engine into the vehicle.
76. The remaining steps of the installation procedure is the reverse of the removal, keep in mind the following items:
    - During installation, all cable ties must be re-installed at the same location
    - Secure all hose connections with hose clamps
    - During assembly, replace self-locking nuts and bolts
    - Always replace bolts that are tightened to torque as well as O-rings and gaskets
    - Make sure alignment sleeves for centering engine/transmission are installed in cylinder block. Install if necessary.
    - Make sure that the intermediate plate is hooked in at sealing flange and is pushed onto alignment sleeves
    - Align engine mounts so that they are free of tension, via shaking

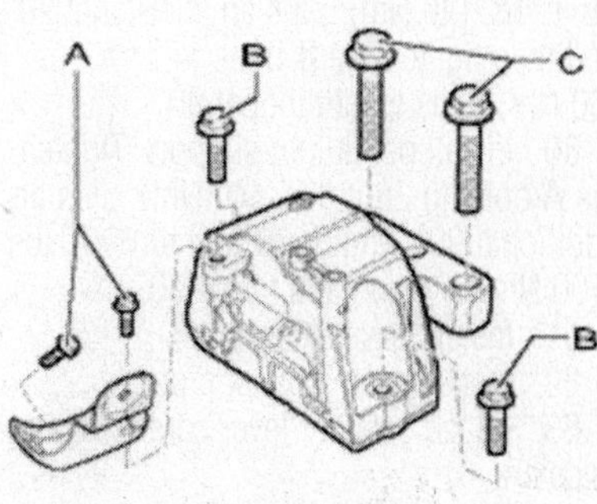

Engine mounting and mounting bolts—2005–06 Jetta with 1.9L engine

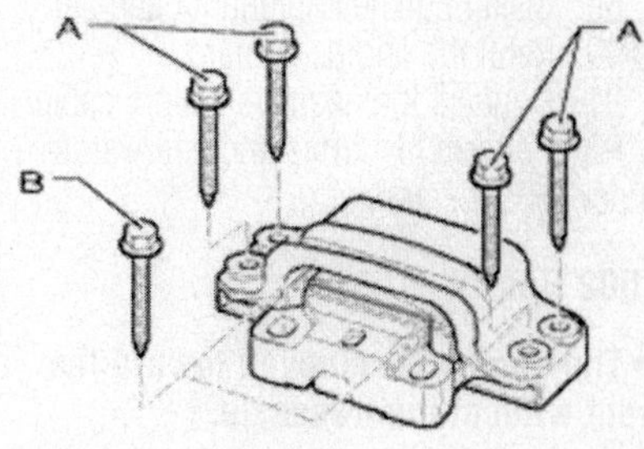

Transmission assembly mounting and mounting bolts—2005–06 Jetta with 1.9L engine

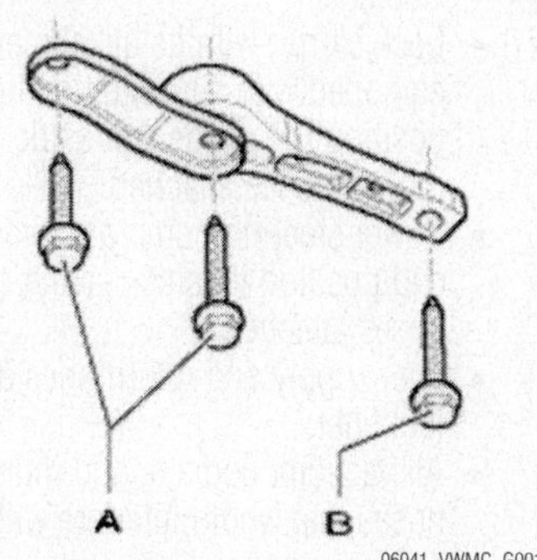

Pendulum support and mounting bolts—2005–06 Jetta with 1.9L engine

motions. If necessary, loosen engine mounts on body as well

77. Make sure that there is a 135mm clearance between engine mounting support and right longitudinal member and that the casting edge of the engine support is positioned parallel to the support arm. The dimension must be the same at top and bottom.

78. After installation of the engine mounting, tighten the **A** bolts to 14 ft. lbs. (20 Nm), plus an additional 90°; tighten the **B** bolts to 28 ft. lbs. (40 Nm), plus an additional 90°, and then tighten the **C** bolts to 41 ft. lbs. (60 Nm), plus an additional 90°.

79. After installation of the transmission assembly mounting, tighten the **A** bolts to 28 ft. lbs. (40 Nm), plus an additional 90°, and then tighten the **B** bolts to 41 ft. lbs. (60 Nm), plus an additional 90°.

80. Install pendulum support. Tighten the **A** bolts to 28 ft. lbs. (40 Nm), plus an additional 90°, and the **B** bolt to 69 ft. lbs. (100 Nm), plus an additional 90°.

81. Install drive axles.

82. Install front exhaust pipe.

83. Attach selector lever cable to transmission.

84. Install alternator.

85. Install A/C compressor.

86. Install ribbed belt.

87. Install sound insulation.

88. Install Engine Control Module (ECM).

89. Install bulkhead plenum chamber.

90. Refill the engine coolant.

91. Connect the negative battery cable.

92. Perform test drive and afterwards check DTC memory.

2002 Passat

➡The engine is removed toward the front without the transaxle.

1. Before servicing the vehicle, refer to the Precautions Section.
2. Remove or disconnect:
   - Negative battery cable.
   - Engine cover and under cover.
   - Coolant.
   - Lock carrier with its attachments. If equipped with air conditioning, position the condenser aside leaving the hoses attached.
   - Power steering pump and bracket, and position it aside leaving the hoses attached.
   - Fuel supply and return lines at the fuel filter.
   - All vacuum, coolant, and intake air hoses that would interfere with engine removal.
   - Any electrical connections that would interfere with engine removal.
   - Starter.
   - Front exhaust pipe from the turbocharger.
   - Engine-to-transaxle mounting bolts.
3. Unbolt:
   - Upper left and right-hand motor mounts.
   - Air conditioning compressor and position it aside leaving the hoses attached.
4. Using Engine Support Bracket 10–222A (or equivalent) and Transaxle Support Bracket 3147 (or equivalent) support the transaxle.
5. Attach an engine sling and hoist.
6. Raise the engine off the mounts and separate the engine from the transaxle.
7. Slowly lift the engine out forward.

To install:

8. Verify that the centering dowels are installed in the engine block, install if necessary.
9. Install the engine into the vehicle.
10. Remove the engine hoist and sling.
11. Remove the transaxle support apparatus.
12. Install the engine-to-transaxle and motor mount attaching hardware and tighten as follows:
    - M6 fasteners: 84 inch lbs. (10 Nm)
    - M8 fasteners: 15 ft. lbs. (20 Nm)
    - M10 fasteners: 33 ft. lbs. (45 Nm)
    - M12 fasteners: 44 ft. lbs. (60 Nm)
13. Connect the front exhaust pipe to the turbocharger.
14. Install the starter.
15. The remaining steps of the installation procedure is the reverse of the removal, keep in mind the following items:
    - Be sure all electrical connections are made before connecting the battery.
    - Install any tie-wraps that were removed.
    - Be sure to fill the coolant and check the oil level.
    - Check the operation of the clutch.
    - Install the lock carrier with its attachments.
16. Connect the negative battery cable.
17. Install the engine cover and undercover.
18. Read and clear any DTCs.

New Beetle

The engine and transaxle are removed from beneath the vehicle as an assembly.

1. Before servicing the vehicle, refer to the Precautions Section.
2. Remove or disconnect:
   - Negative battery cable.
   - Engine cover.
   - Power steering reservoir from the battery support leaving the hoses attached, and position it aside.
   - Battery and bracket.
   - Fuel supply and return lines at the fuel filter.
   - Air cleaner and intake air duct.
   - Throttle cable.
   - With a manual transaxle, transaxle range selector mechanism and the clutch slave cylinder.
   - With an automatic transaxle, range selector lever cable at the transaxle.
   - Engine undercover.
   - Coolant.
   - Accessory drive belt.
   - If necessary, right side cooling fan.
   - Power steering pump and bracket and position it aside leaving the hoses connected.
   - Air conditioner compressor and position it aside leaving the hoses connected.
   - Any coolant, vacuum and intake hoses.
   - All wires from the transaxle, starter and generator; position the wires out of the way.
   - Any remaining wires or connectors that would interfere with the engine removal.
   - Engine pendulum support.
   - Right-hand halfshaft
   - Left-hand halfshaft at the transaxle.
   - Front exhaust pipe from the manifold.
3. Install Engine Bracket T10012, or equivalent, in the Engine/Transaxle Jack VAG 1383 A or equivalent.
4. Remove the bracket for the coolant hose under the engine block.
5. Attach the Engine Bracket T10012 to the engine block using the threaded holes at the corners of the engine block.
6. Raise the engine/transaxle jack to relieve the tension on the mounts.
7. Disconnect the engine and transaxle mounts from inside the engine compartment.
8. Lower the engine/transaxle assembly from the vehicle, be sure to guide the power steering pressure hose past the transaxle.

To install:

9. Raise the engine/transaxle assembly into the vehicle, be sure to guide the power steering hose around the transaxle.
10. Using new bolts, connect the engine/transaxle mounts following the accompanying illustration.

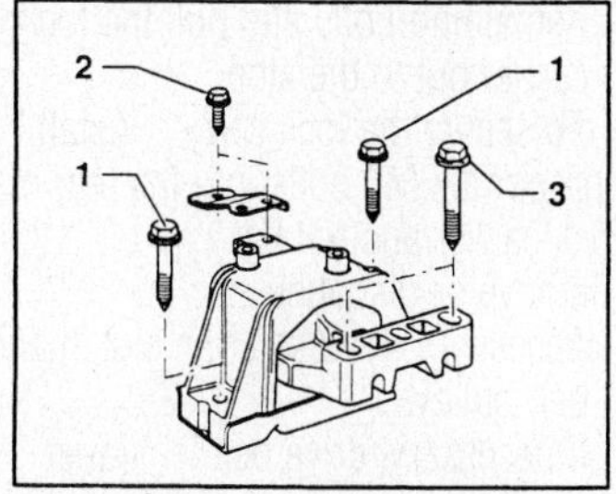

Engine / transmission mounts

Tightening torques

Front assembly mounting

1 -	Mount to body [1]	40 Nm (30 ft lb) + 90° (1/4 turn)
2 -	Mount/bracket to body	25 Nm (18 ft lb)
3 -	Mount to engine bracket [1]	60 Nm (44 ft lb) + 90° (1/4 turn)

[1] Replace bolts

Rear right assembly mounting

1 -	Mount to body [1]	40 Nm (30 ft lb) + 90° (1/4 turn)
2 -	Mount to body	25 Nm (18 ft lb)
3 -	Bearing on transmission console [1]	60 Nm (44 ft lb) + 90° (1/4 turn)

[1] Replace bolts

9301WG02

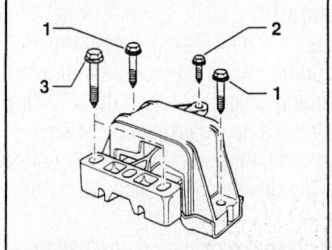

Exploded view of the engine mounts—New Beetle

11. Remove the engine/transaxle jack and Engine Bracket T10012.
12. Install or connect:
    - Bracket for the coolant hose under the engine block.
    - Front exhaust pipe to the manifold.
    - Left halfshaft and right halfshaft.
    - Engine pendulum support. Bolts (1) and (2) to 30 ft. lbs. (40 Nm) plus 90 degrees, and bolts (3) to 15 ft. lbs. (20 Nm) plus 90 degrees.
    - Any wiring that was removed.
    - Any hoses that were removed.
    - Air conditioner compressor.
    - Power steering bracket and pump.
    - If removed, right side cooling fan.
    - Accessory drive belt and fill the cooling system.
    - With an automatic transaxle, transaxle range selector cable.
    - With a manual transaxle, clutch slave cylinder and connect the range selector lever.
    - Throttle cable.
    - Intake air duct and air cleaner assembly.
    - Fuel supply and return lines to the filter housing.
    - Battery bracket.
    - Power steering reservoir.
    - Battery cables.
    - Engine covers.

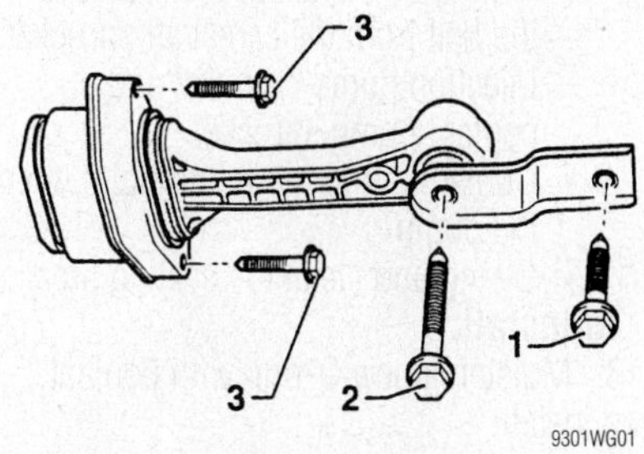

Pendulum support mounting bolt locations—New Beetle

Water Pump

REMOVAL & INSTALLATION

1.9L (AAZ) Engine

On some Diesel engines, the belt tension is adjusted with shims between the outer and inner halves of the pulley. On others, the alternator swivels to adjust belt tension.

1. Before servicing the vehicle, refer to the Precautions Section.
2. Drain the coolant.
3. Loosen but don't remove the bolts holding the pulley to the water pump.
4. With a movable alternator, loosen the alternator and remove the drive belt.
5. Remove the water pump pulley and water pump.

To install:

6. Installation is the reverse of removal. Be sure to clean the pump housing before installing the new gasket. Tighten the following:
    - Water pump-to-housing—84 inch lbs. (10 Nm)
    - Water pump drive pulley—15 ft. lbs. (20 Nm)
    - Thermostat housing—84 inch lbs. (10 Nm)
    - Alternator mounting bolts—18 ft. lbs. (25 Nm)

1.9L (AHH) Engine

1. Before servicing the vehicle, refer to the Precautions Section.
2. Remove or disconnect:
    - Negative battery cable
    - Engine cover and under cover
    - Coolant
3. Position the lock carrier into the service position, as follows:
    a. Remove or disconnect:
    - Front bumper
    - Any wiring or connector that would inhibit locking the carrier
    - 3 quick-release screws on the front noise insulation panel
    - Air guide between the lock carrier and the air filter
    - If installed, retaining clamps for the wiring harness at the left side of the radiator frame

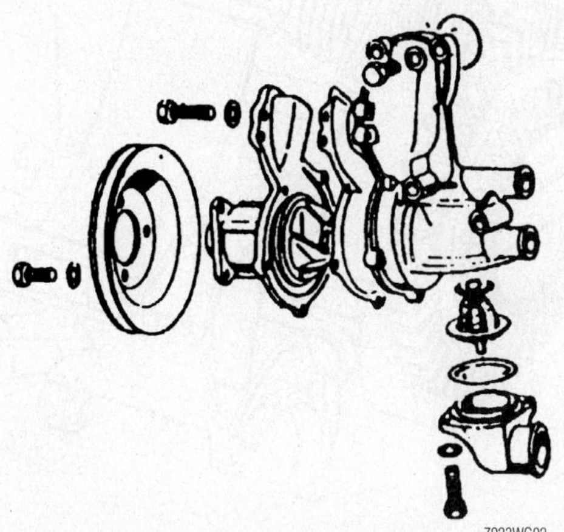

Exploded view of the water pump assembly—1.9L (AAZ) engine

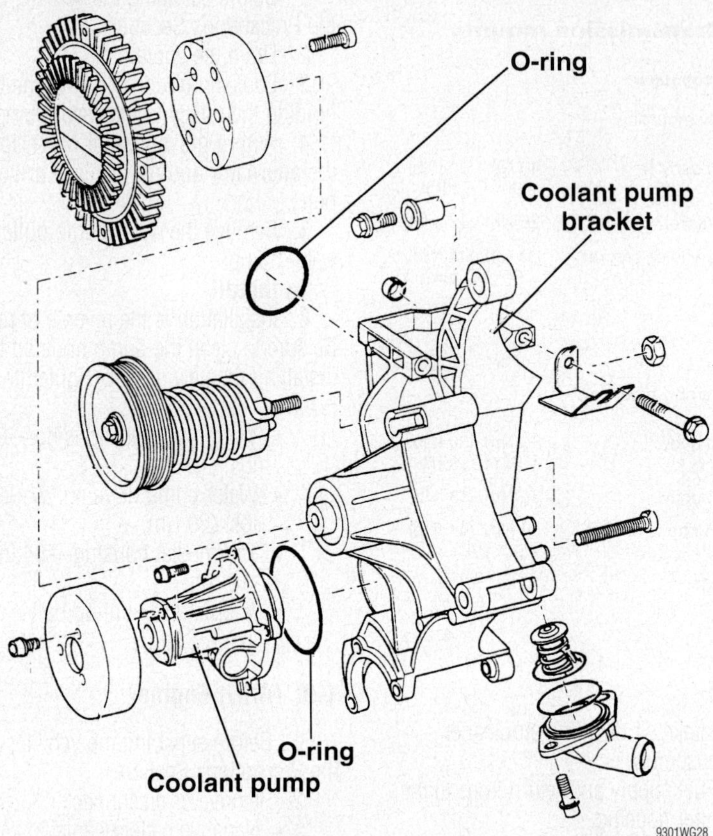

Exploded view of the coolant pump, bracket and related components—1.9L (AHH) engine

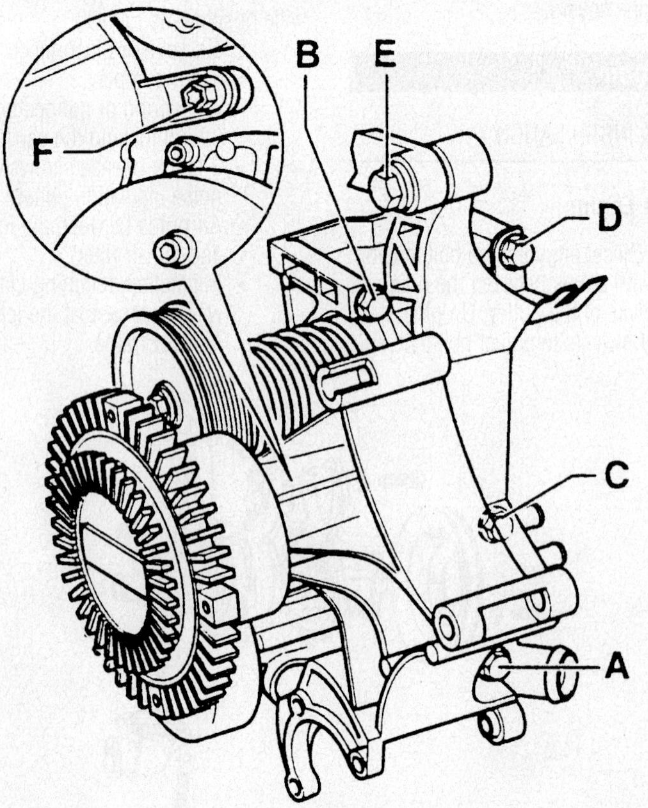

Coolant pump bracket torque sequence—1.9L (AHH) engine

- No. 2 bolts and install Support tool 3369 or equivalent
- Remaining bolts and pull the lock carrier out to the stop

b. To secure the lock carrier, install the appropriate M6 bolts into the rear of the lock carrier and fender

4. Remove or disconnect:
   - Accessory drive belt, fan clutch and belt pulleys
   - If necessary, drive belt tensioner
   - Coolant pump assembly bracket
   - Coolant pump from the bracket

To install:

5. Using a new O-ring, install the coolant pump onto the bracket and tighten the mounting bolts to 84 inch lbs. (10 Nm).
6. Using a new gasket and O-ring, install the coolant pump bracket. Mounting bolts, in alphabetical sequence: 18 ft. lbs. (25 Nm).
7. Tighten the coolant pump housing to the timing belt cover mounting bolt (F) to 84 inch lbs. (10 Nm).
8. If removed, install the drive belt tensioner.
9. Install the belt pulleys, fan clutch and accessory drive belt.
10. Return the lock carrier into the normal position.
11. Install or connect:
    - Coolant
    - Negative battery cable
    - Engine cover and under cover

1.9L (ALH/BEW/BRM) Engine

1. Before servicing the vehicle, refer to the Precautions Section.
2. Remove or disconnect:
   - Negative battery cable
   - Engine cover and under cover
   - Coolant
   - Accessory drive belt
   - Fuel supply and return lines at the filter
   - Hose between the charge air cooler and the intake manifold
   - Fuel filter and bracket
   - Timing belt upper and center covers
   - Timing belt tension, then remove the belt from the camshaft sprocket, injection pump sprocket and coolant pump sprocket
   - Coolant pump mounting bolts, then the pump
   - O-ring and clean the seating area

To install:

3. Moisten a new O-ring with coolant, then install.
4. Install or connect:
   - Water pump, so that the plug in the

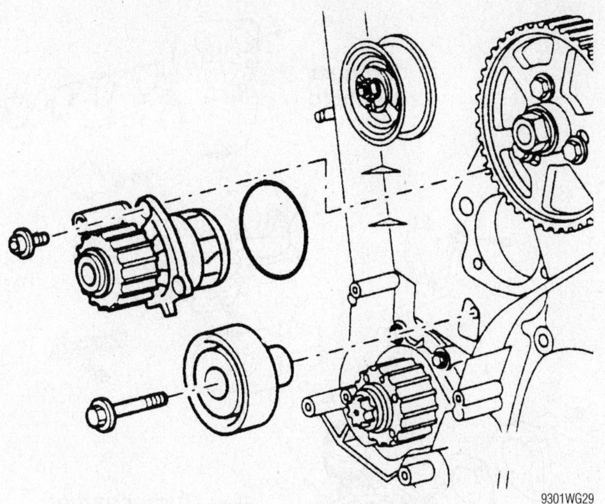

Exploded view of the water pump mounting—1.9L (ALH) engine

housing faces downward. Mounting bolts: 11 ft. lbs. (15 Nm).
- Timing belt.
- Upper and center timing belt covers.
- Fuel filter and bracket.
- Fuel lines.
- Hose between the charge air cooler and the intake manifold.
- Accessory drive belt tensioner and tighten the new bolt to 30 ft. lbs. (40 Nm), plus an additional 90°.
- Accessory drive belt.
- Coolant.
- Negative battery cable.

Diesel Glow Plugs

REMOVAL & INSTALLATION

1. Before servicing the vehicle, refer to the Precautions Section.
2. Remove the bus bar connecting the glow plugs and determine which plugs need replacement.
3. Remove the defective plugs.
4. When installing new plugs, tighten them to 11 ft. lbs. (15 Nm).

➡Diesel glow plugs have an air gap much like a spark plug to prevent overheating of the plug. Over-tightening the glow plug will close the gap and cause the plug to burn out.

Cylinder Head

REMOVAL & INSTALLATION

➡The cylinder head bolts on all diesel vehicles are stretch bolts and must be replaced when removed.

1. Before servicing the vehicle, refer to the Precautions Section.
2. Remove or disconnect:
   - Battery ground cable.
   - Accessory drive belt.
   - Coolant.
   - Fuel lines from the injectors and the pump as an assembly. Put the lines where they will stay clean; protect the injector and pump fittings with caps.
   - Radiator and heater hoses.
   - All vacuum and electrical connections and carefully label for installation.
   - Front exhaust pipe from the turbocharger, then remove the turbocharger from the manifold.
   - Cylinder head cover and timing belt.
3. Turn the engine to TDC of No. 1

Measure piston pop-up to determine the required head gasket thickness—1.9L diesel engine

cylinder, if possible, and remove the camshaft drive belt.

4. Remove the head bolts in the reverse order of installation sequence and lift the head out of the vehicle.

To install:

5. On these engines, the pistons actually project above the deck of the block. If the crankshaft and pistons are not to be removed, examine the old head gasket to see how many notches are on the edge near the oil return hole, between cylinders Nos. 2 and 3. Replace the gasket with the same thickness.
6. If the pistons were removed or if the old gasket is not available, the piston height (pop up) must be measured to select the proper head gasket. Use a dial indicator or caliper to obtain the measurement.

Pop-up measurements:
- 0.036–0.039 in. (0.91–1.00mm): 1 notch
- 0.040–0.043 in. (1.01–1.10mm): 2 notches

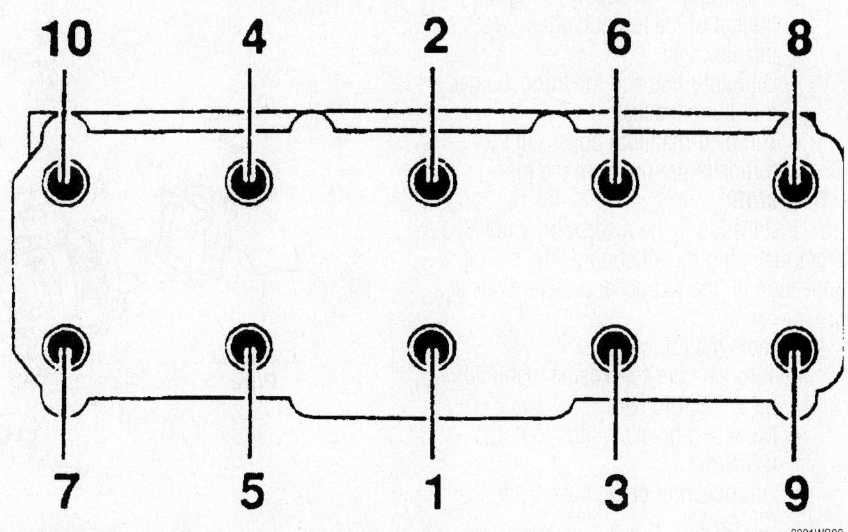

Cylinder head bolt torque sequence—Diesel engines

- 0.044–0.047 in. (1.11–1.20mm): 3 notches

7. Install the new cylinder head gasket with the word TOP or OBEN facing upward. Do not use any sealing compound.
8. Turn the crankshaft to TDC of No. 1 cylinder, then back about ¼ turn to bring all pistons about even.
9. Carefully lower the head on and install new head bolts into No. 8 and 10 first. These holes are smaller and will properly locate the gasket and cylinder head.
10. Install the remaining bolts and tighten in the proper sequence in 3 steps: 30 ft. lbs. (40 Nm), 44 ft. lbs. (60 Nm), then a full ½ turn more. 2 quarter turns are allowed.
11. Installation of the remaining parts is the reverse of removal, be sure to change the oil and filter. Install the camshaft drive belt and set injection pump timing.
12. Install the fuel injector lines and tighten to 18 ft. lbs. (25 Nm). Be careful not to over tighten the line nuts. If a line is damaged or clogged, replace all lines as a set.
13. Connect the negative battery cable.

Turbocharger

REMOVAL & INSTALLATION

1. Before servicing the vehicle, refer to the Precautions Section.
2. Remove or disconnect:
   - Negative battery cable.
   - Exhaust pipe from the turbocharger outlet.
   - Oil supply line and bracket, after cleaning the oil supply fitting on the top of the turbocharger.
   - Inlet air hose.
   - Oil return line and the turbocharger mounting bracket.
   - Turbo-to-manifold bolts. Lift the turbocharger out from the top.

To install:

3. Installation is the reverse of removal. Before installing the oil supply line, fill the connection on the turbocharger with engine oil.
4. Tighten the following:
   - Turbocharger-to-exhaust manifold: 33 ft. lbs. (45 Nm)
   - Mounting bracket nuts: 18 ft. lbs. (25 Nm)
   - Turbocharger outlet nuts: 18 ft. lbs. (25 Nm)
   - Oil return line: 22 ft. lbs. (30 Nm)

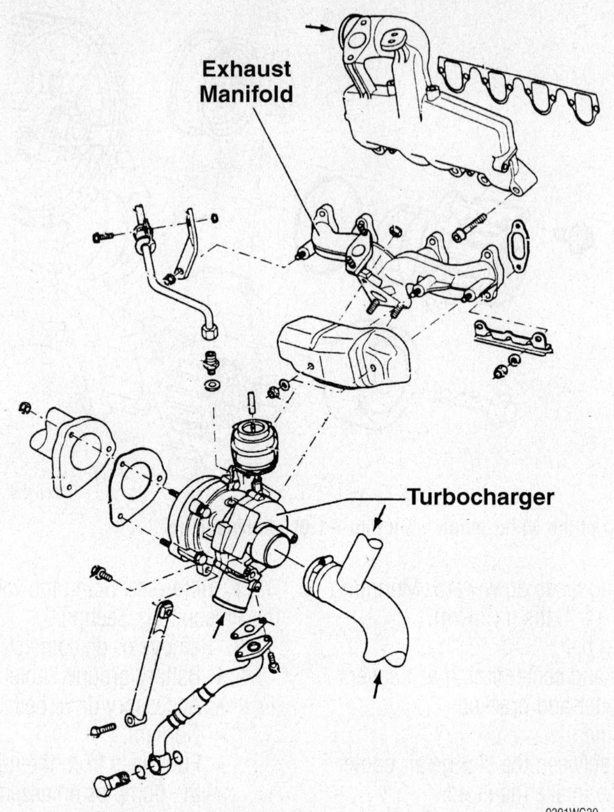

Exploded view of the intake and exhaust manifold, turbocharger and related components—1.9L (AHH) engine

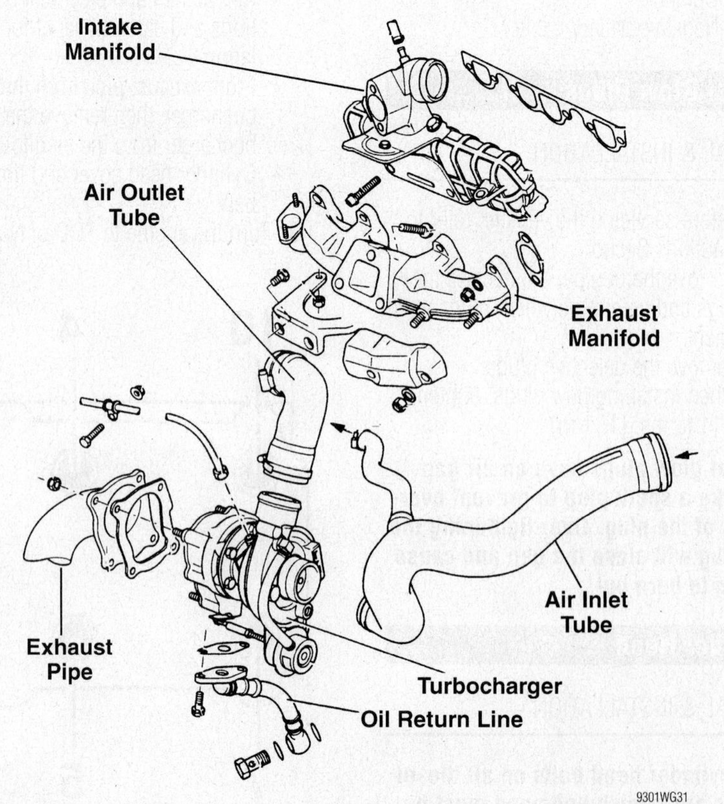

Exploded view of the intake and exhaust manifold, turbocharger and related components—1.9L (AAA) engine

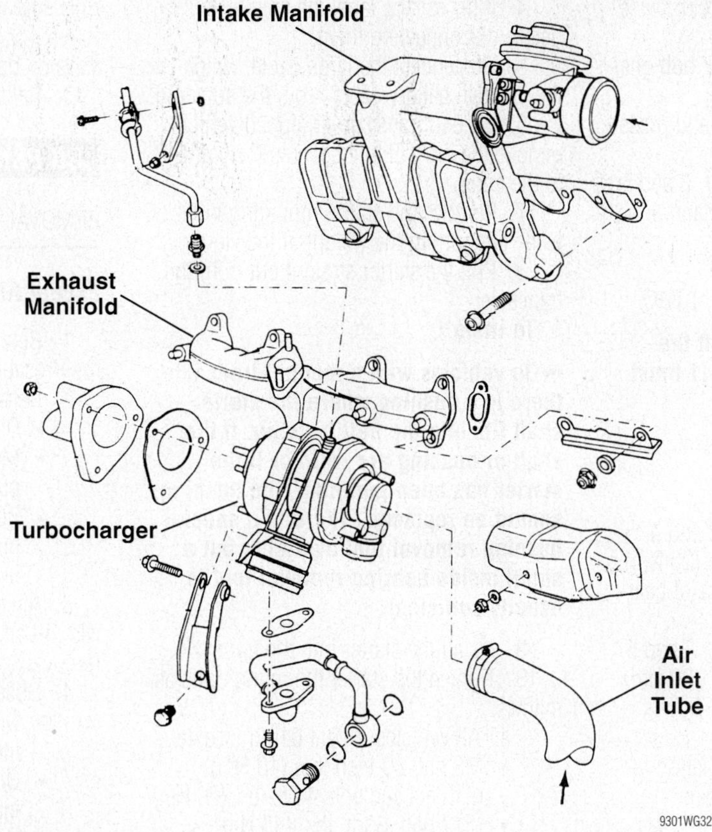

Exploded view of the intake and exhaust manifold, turbocharger and related components—1.9L (ALH) engine

Intake Manifold

REMOVAL & INSTALLATION

1. Before servicing the vehicle, refer to the Precautions Section.
2. Label and detach any hoses and electrical connections related to intake manifold service.
3. Disconnect the air inlet hose.
4. Remove the intake manifold mounting bolts.
5. Installation is the reverse of removal. Use a new gasket and tighten the bolts to 18 ft. lbs. (25 Nm).

Exhaust Manifold

REMOVAL & INSTALLATION

1. Before servicing the vehicle, refer to the Precautions Section.
2. Remove or disconnect:
   - Negative battery cable and any heat shields that may be in the way.
   - Exhaust pipe from the turbocharger outlet.
   - Turbocharger oil lines and the turbocharger.
   - Manifold locking nuts
3. Lift the manifold off the head.
4. Installation is the reverse of removal. Use new gaskets and locking nuts and tighten to 18 ft. lbs. (25 Nm).

Camshaft and Valve Lifters

REMOVAL & INSTALLATION

AAZ and AHH Engines

1. Before servicing the vehicle, refer to the Precautions Section.
2. Remove or disconnect:
   - Negative battery cable.
   - Timing belt cover(s), the timing belt, cylinder head cover and the camshaft sprocket.
3. Number the bearing caps from front-to-back. If the cap does not already have one, scribe an arrow pointing towards the front of the engine. The caps are offset and must be installed correctly. Factory numbers on the caps are not always on the same side.
4. Remove:
   - Front and rear bearing caps. Loosen the remaining bearing cap nuts a little at a time to avoid bending the camshaft. Start from the outside caps near the ends of the head and work toward the center.
   - Bearing caps and the camshaft.

To install:

5. Install a new oil seal and end-plug in the cylinder head. Lubricate the camshaft bearing journals and lobes and set the camshaft in place.
6. Install the bearing caps in the correct position with the arrow pointing towards the front of the engine. Cap nuts, diagonally and in several steps: 15 ft. lbs. (20 Nm). Do not over tighten. Camshaft shaft end-play should be about 0.006 in. (0.15mm).
7. Install the drive sprocket and timing belt. Wait at least ½ hour after installing the camshaft before starting the engine to allow the lifters to leak down.

ALH/BEW/BRM Engines

1. Before servicing the vehicle, refer to the Precautions Section.
2. Remove or disconnect:
   - Negative battery cable
   - Fuel lines from fuel filter
   - Right headlight
   - Connecting line between charge air cooler and intake manifold
   - Upper timing belt cover
   - Valve cover
   - Brake booster vacuum pump
   - Timing belt from camshaft sprocket

- Rotate crankshaft counterclockwise slightly
- Loosen camshaft sprocket bolt one turn
- Use 2 jaw puller T40001 and press camshaft sprocket off

3. Remove bearing caps no. 1, 3 and 5.
4. Loosen bearing caps no. 2 and 4 in steps and diagonally.

To install:

5. Ensure the pistons are not at TDC.

➡When installing the camshaft the camshaft lobes for cylinder no.1 must point upward.

6. Lubricate the bearing camshaft surface with oil.
7. Install the camshaft.
8. Tighten bearing caps no. 2 and 4 alternately and diagonally to 15 ft. (20 Nm).
9. Lightly coat the contact face of bearing cap no. 1 with sealant.
10. Tighten bearing caps no. 1, 3 and 5 alternately and diagonally to 15 ft. (20 Nm).
11. Lightly tap bearing cap no. 5 to fit the camshaft.
12. The remainder of the installation is the reverse of the removal procedure.

Valve Lash

ADJUSTMENT

All vehicles have hydraulic valve lifters and require no adjustment. On these vehicles there will be a sticker under the hood indicating hydraulic lifters.

Starter

REMOVAL & INSTALLATION

➡On A1 and A2 platform vehicles equipped with 010 automatic transaxle, access to the starter motor is limited. Although not necessary, it is recommended that the intake and exhaust manifolds be removed to access the starter.

1. For safety purposes, disconnect the battery ground cable.
2. Raise and safely support the front of the vehicle with jackstands.
3. For A3 vehicles, use a floor jack (with a block of wood on the chock) to support the engine by the oil pan. Do NOT jack the car by the oil pan under any circumstances! The floor jack is ONLY to support the engine while the starter is being removed. The bolts that secure the starter to the transaxle are also used to secure the front engine mount to the transaxle.
4. If necessary, label the small wires before disconnecting them.
5. Disconnect the large cable, which is the positive battery cable, from the solenoid.
6. On 010 transaxle-equipped vehicles, remove the bracket that secures the starter to the engine.
7. Remove the starter mounting bolts, while supporting the weight of the starter.
8. Pull the starter straight out from the transaxle.

To install:

➡On vehicles with a manual transaxle, there is a bushing where the starter shaft fits into the bell housing. If the shaft or bushing are worn or if the starter has been jamming, the bushing should be replaced. There is a special bushing removal tool available but a small inside bearing removal tool is usually sufficient.

9. Install the starter into the transaxle.
10. Tighten the starter mounting bolts as follows:
    a. All vehicles except 010 transaxle:
    - M8 nut: 89 inch lbs. (10 Nm)
    - M10 nut and bolt: 44 ft. lbs. (60 Nm)
    - M12 bolt: 33 ft. lbs. (45 Nm)

    b. 010 transaxle only:
    - Mounting flange bolt: 15 ft. lbs. (20 Nm)
    - Mounting bracket bolt: 18 ft. lbs. (25 Nm)
11. Attach the electrical connections to the starter.

➡Be careful not to over tighten the battery cable connection. The metal is soft and the threads will strip easily.

12. Lower the vehicle from the jackstands.
13. Connect the negative battery cable.

Oil Pan

REMOVAL & INSTALLATION

The oil pan can be removed with the engine in the vehicle.

1. Before servicing the vehicle, refer to the Precautions Section.
2. Remove or disconnect:
    - Engine oil.
    - Bolts retaining the oil pan.
    - Oil pan.

To install:

3. Be sure the gasket surface is flat and install the pan with a new gasket.
4. On all except ALH engines, tighten the retaining bolts in a crisscross pattern to 15 ft. lbs. (20 Nm). Do not over-tighten. On ALH engine, tighten the oil pan bolts to 11 ft. lbs. (15 Nm), and the oil pan-to-transmission bolts to 33 ft. lbs. 45 Nm).
5. Refill the engine with oil.

Oil Pump

REMOVAL & INSTALLATION

Except ALH Engines

1. Before servicing the vehicle, refer to the Precautions Section.
2. Remove or disconnect:
    - Oil pan.
    - Mounting bolts and lower the pump from the engine.
    - Bottom cover and disassemble the pump. The pressure relief valve is in the bottom cover.
3. After reassembling the pump, prime it with oil and install it in the reverse order of removal.
4. Observe the following values:
    - Oil pump bottom cover bolts: 84 inch lbs. (10 Nm)
    - Oil pump suction foot bolts: 84 inch lbs. (10 Nm)
    - Oil pump retaining bolts: 18 ft. lbs. (25 Nm)

ALH Engines

1. Before servicing the vehicle, refer to the Precautions Section.
2. Remove or disconnect:
    - Oil pan
    - Baffle plate
    - Oil pump drive chain
    - Drive sprocket
    - Suction pipe
    - Mounting bolts
3. Lower the pump from the engine.
4. Install the oil pump in the reverse order of removal.
5. Observe the following values:
    - Oil pump and baffle plate bolts: 11 ft. lbs. (15 Nm)
    - Oil pump suction pipe bolts: 11 ft. lbs. (15 Nm)
    - Oil pump sprocket bolt: 18 ft. lbs. (25 Nm)

Rear Main Seal

REMOVAL & INSTALLATION

The rear main oil seal is located in a housing on the rear of the cylinder block. To replace the seal on all vehicles it is necessary to remove the transaxle and flywheel.

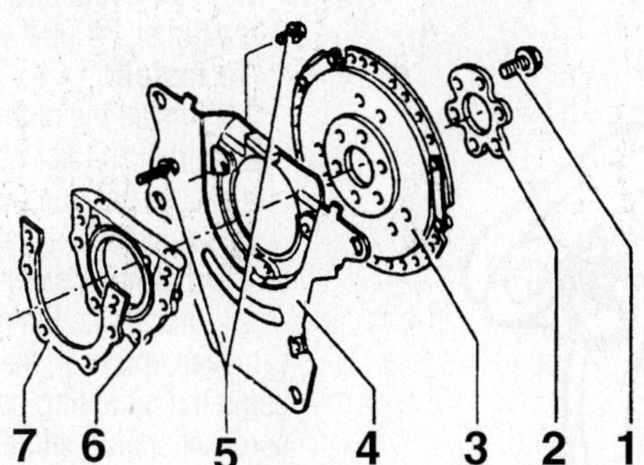

Rear main oil seal and related components—1.9L diesel engine

1. Before servicing the vehicle, refer to the Precautions Section.
2. Remove the transaxle and flywheel.
3. Remove the oil seal with the mounting flange as a complete unit.

To install:

4. Install a new mounting flange with seal using a new gasket. Mounting flange bolts: 11 ft. lbs. (15 Nm) on ALH engines or 84 inch lbs. (10 Nm) on all other engines.
5. Install the flywheel and transaxle.

Timing Belt, Front Cover and Seal

REMOVAL & INSTALLATION

Except ALH Engines

Some special tools are required to perform this procedure properly. A flat bar, VW tool 2065A is used to secure the camshaft in position. A pin, VW tool 2064 is used to fix the pump position while the timing belt is removed. The camshaft and pump work against spring pressure and will move out of position when the timing belt is removed. It is not difficult to find substitutes but do not remove the timing belt without these tools.

WARNING

Do not turn the engine or camshaft with the timing belt removed. The pistons will contact the valves and cause internal engine damage.

1. Disconnect the negative battery cable and remove the accessory drive belts, crankshaft pulley and the timing belt cover(s). Remove the camshaft cover and rubber plug at the back end of the camshaft.
2. Temporarily reinstall the crankshaft pulley bolt and turn the crankshaft to Top Dead Center (TDC) of No. 1 piston. The mark on the camshaft sprocket should be aligned with the mark on the inner timing belt cover or the edge of the cylinder head.
3. With the engine at TDC, insert the bar into the slot at the back of the camshaft. The bar rests on the cylinder head to will hold the camshaft in position.
4. Insert the pin into the injection pump drive sprocket to hold the pump in position.
5. Loosen the locknut on the tensioner pulley and turn the tensioner counterclockwise to relieve the tension on the timing belt. Slide the timing belt from the sprockets.

To install:

6. Install the new timing belt and adjust the tension so the belt can be twisted 45 degrees at the halfway point between the camshaft and pump sprockets. Tighten the tensioner nut to 33 ft. lbs. (45 Nm).
7. Remove the holding tools.
8. Turn the engine 2 full revolutions to return to TDC for the No. 1 cylinder. Recheck belt tension and timing mark alignment, readjust as required.
9. Install the belt cover and accessory drive belts.

➡If the belt is too tight, there will be a growling noise that rises and falls with engine speed.

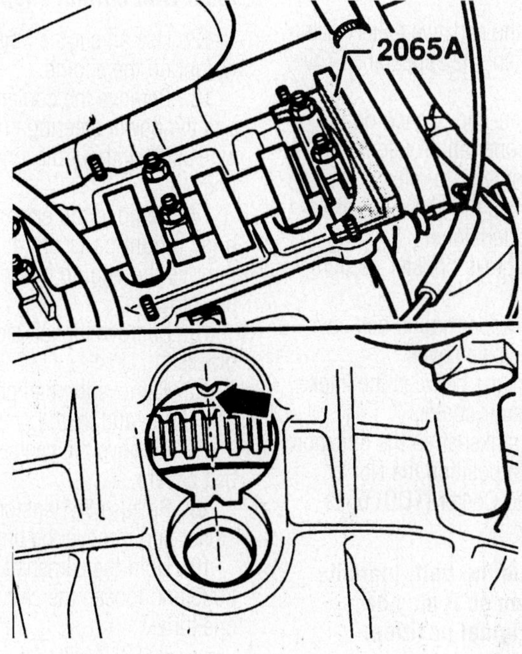

Use the VW tool to lock the camshaft at TDC for timing belt replacement—Volkswagen 1.9L Diesel engines

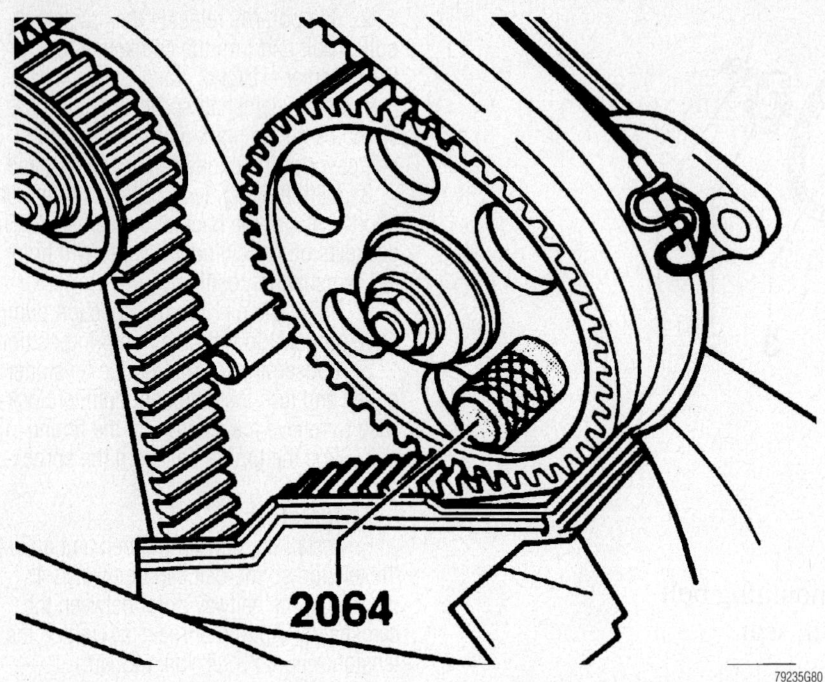

Locking the injection pump with the VW tool as shown—Volkswagen 1.9L Diesel engines

ALH Engines

Some special tools are required to perform this procedure properly. A camshaft locking bar, VW tool 3418 is used to secure the camshaft in position. A pin, VW tool 3359 is used to fix the injection pump position while the timing belt is removed. Camshaft sprocket holding tool 3036 is used to lock the camshaft sprocket. Puller T40001 is needed to remove the camshaft sprocket.

1. Disconnect the negative battery cable.
2. Remove the engine appearance cover and splash shield.
3. Place an open-end wrench on the alternator belt tensioner and rotate it clockwise toward the alternator to release the belt's tension. Remove the alternator drive belt and release the tensioner.
4. Remove the brake booster vacuum pump.
5. Remove the upper timing belt cover.
6. Remove the valve cover.
7. Remove the tube between the intake pipe and the charge air cooler.
8. Rotate the crankshaft in the direction of engine rotation to position the No. 1 cylinder at Top Dead Center (TDC) of its compression stroke.

➡If reusing the timing belt, mark its rotational direction so it may be installed in its original position.

9. Install camshaft locking bar onto the camshaft and cylinder head so it is equally balanced between both legs of the tool.
10. Lock the injection pump in place using pin 3359.
11. Loosen the injection pump sprocket bolts in a diagonal sequence.

➡DO NOT loosen the pump sprocket hub bolt or the injection pump basic setting will be altered and cannot be reset with normal shop equipment.

12. Use an engine hoist and take the weight off the engine.
13. Remove the coolant reservoir tank and the power steering fluid reservoir and wire aside without disconnect the fluid lines.
14. Remove the engine mount and body mount bolts on the right side and remove the engine mount from the cylinder block.
15. Remove the left side engine mount bracket.
16. Remove the damper pulley-to-crankshaft bolts and the damper.
17. Remove the center and lower timing belt covers.
18. Release the tension on the belt tensioner and remove the timing belt.
19. Hold the camshaft sprocket with tool 3036 and loosen the camshaft sprocket bolt one turn.
20. Remove the camshaft sprocket using 2-jaw puller T40001.

To install:

21. Install the timing belt onto the crankshaft sprocket, idler pulley, injection pump sprocket, water pump and tensioner.
22. Adjust the injection pump sprocket so the mounting slots are centered.
23. Install the timing belt on the camshaft sprocket, then install the sprocket using the mounting bolt. The camshaft sprocket should still turn.
24. On manual transmission models, tension the timing belt using tool T10020 on the tensioner roller eccentric and turn it clockwise until the notch and the raised mark are aligned as shown.
25. On automatic transmission models, tension the timing belt using an Allen wrench on the tensioner roller eccentric and turn it clockwise until the notch and the pointer are aligned as shown.
26. Using new bolts, tighten the injection pump sprocket bolts to 15 ft. lbs. (20 Nm).

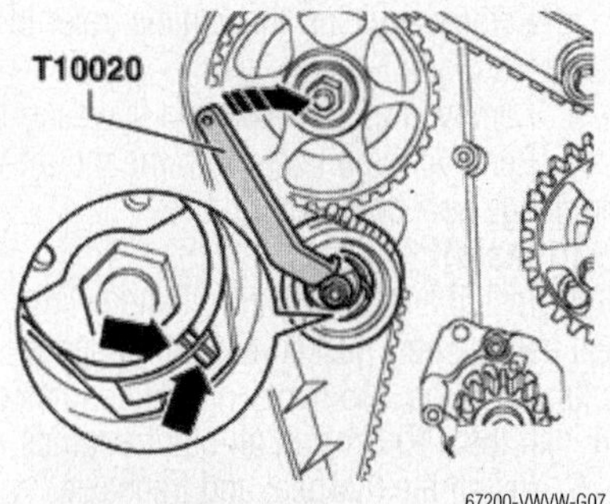

Aligning the timing belt tensioner marks on manual transmission models—1.9L (ALH) Diesel engines

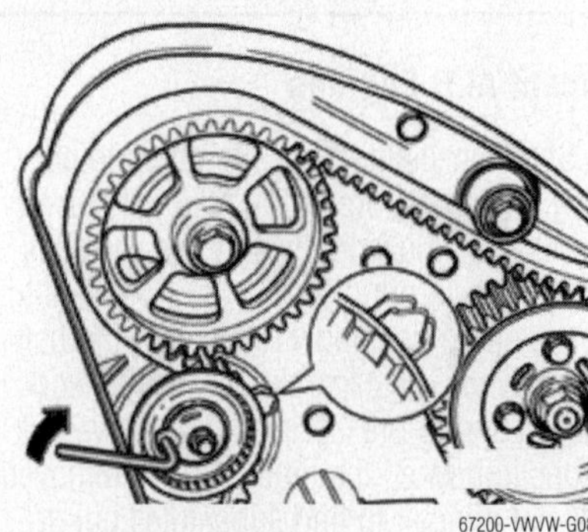

Aligning the timing belt tensioner marks on automatic transmission models—1.9L (ALH) Diesel engines

➡After dynamically testing the start of injection, retighten the sprocket bolts an additional 90°.

27. Remove the camshaft locking tool and the injection pump sprocket locking pin. holding tools.
28. Turn the engine 2 full revolutions to return to TDC for the No. 1 cylinder. Recheck belt tension and timing mark alignment, readjust as required.
29. Install the left side engine mount bracket and tighten the bolts to 33 ft. lbs. (45 Nm).
30. Install the center and lower timing belt covers.
31. Install the crankshaft damper and tighten the bolts to 89 inch lbs. (10 Nm), plus an additional 90°.
32. Install the engine mount and body mount bolts to the cylinder block and tighten to 33 ft. lbs. (45 Nm).
33. The remainder of the installation is the reverse of the removal procedure.

Piston and Ring Positioning

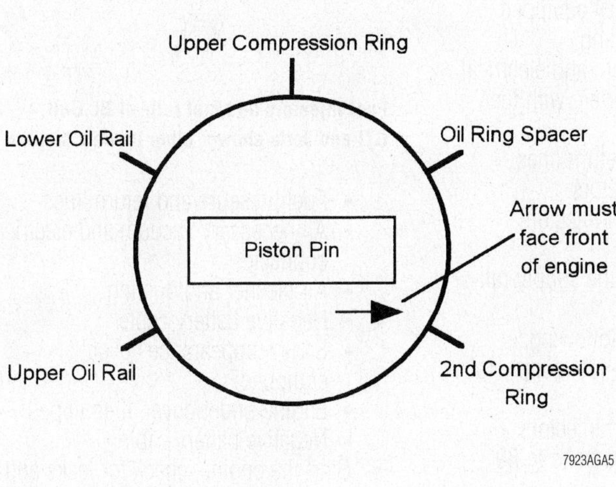

Diesel engines—piston ring end-gap spacing

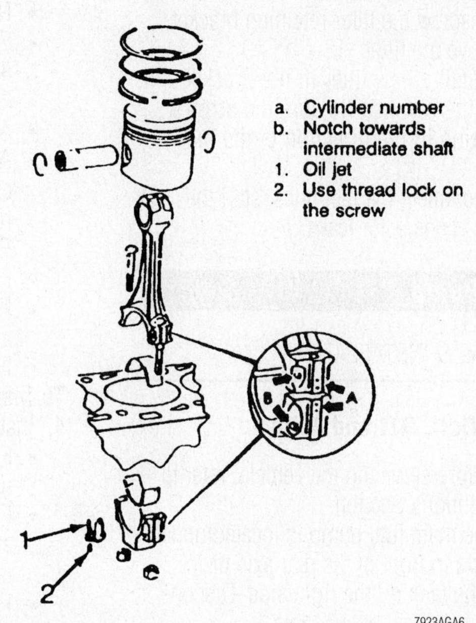

Piston and connecting rod installation

GASOLINE FUEL SYSTEM

Fuel System Service Precautions

Whenever working on or around gasoline or the fuel delivery system, heed the following precautions:

- Do not allow fuel spray or fuel vapors to come into contact with a heating element or open flame. Do not smoke while working on the fuel system.
- Always disconnect the negative battery cable unless the repair or test procedure requires that battery voltage be applied.
- Always relieve the fuel system pressure prior to disconnecting any fitting or fuel line connection.
- To control fuel spray when relieving system pressure, place a shop towel around the fitting prior to loosening to catch the spray. Ensure that all fuel spillage is quickly wiped up and that all fuel soaked rags are deposited into a proper fire safety container.
- Always keep a dry chemical (Class B) fire extinguisher near the work area.
- Always use a back-up wrench when loosening and tightening fuel line fittings.
- Do not re-use fuel system gaskets and O-rings, replace with new ones. Do not substitute fuel hose where fuel pipe is installed.

➡Before servicing the vehicle, also refer to the precautions in the beginning of this section.

Fuel System Pressure

RELIEVING

The fuel injection system operates under high pressure. This makes it necessary to first relieve the system of pressure before servicing. The pressurized fuel, when released, may ignite or cause personal injury.

1. Before servicing the vehicle, refer to the Precautions Section.
2. Remove or disconnect:
   - Power to the fuel pump by removing the relay or the fuel pump fuse. Check the list on the fuse box lid to be sure. The fuse can be removed to stop the fuel pump from running. With the engine operating at idle, wait until the engine stalls from fuel starvation.
   - Negative battery cable.
3. Carefully loosen the fuel line on the control pressure regulator or component to be serviced.
4. Wrap a clean rag around the connection, while loosening, to catch any fuel.
5. After service is complete, discard the fuel soaked rag in the proper manner and reconnect the negative battery cable, relay or fuses.

Fuel Filter

REMOVAL & INSTALLATION

Most vehicles use a fuel filter mounted under the vehicle, below the fuel tank. An arrow should be on the filter indicating fuel flow direction. Install with arrow pointing to engine. Use care not to mix up fuel supply or return lines. Fuel pressure applied to the return side of the system will cause damage.

In addition, some vehicles use a filter in the engine compartment near the fuel distributor. If equipped, use the following procedure:

1. Before servicing the vehicle, refer to the Precautions Section.
2. Make certain to follow the precautions and relieve fuel the pressure.
3. Disconnect the fuel lines leading into and out of the fuel filter.
4. Unscrew the filter retaining bracket and remove the filter.
5. Install a new filter in the bracket and reattach the bracket. Be sure the arrows are pointing in the direction of the fuel flow.
6. Reconnect the fuel lines, start the engine and check for leaks.

Fuel Pump

REMOVAL & INSTALLATION

Except Golf, GTI and Jetta

1. Before servicing the vehicle, refer to the Precautions Section.
2. The main fuel pump is located under the vehicle in front of the rear axle or in front of the tank on the right side. Disconnect the negative battery cable.
3. Remove or disconnect:
   - Electrical connector.
   - Fuel system pressure.
   - Mounting bolts
   - Fuel pump.
4. Installation is the reverse of removal. Be sure to use new sealing rings and/or gaskets.

Golf, GTI and Jetta

1. Before servicing the vehicle, refer to the Precautions Section.
2. Relieve the fuel system pressure.
3. Disconnect the negative battery cable.
4. Remove or disconnect:
   - Cover under rear seat
   - Electrical connector
   - Fuel supply and return lines
5. Using spanner tool 3217, remove the fuel pump flange nut.
6. Remove the fuel pump unit and seal.
7. Installation is the reverse of removal. Be sure to use new sealing rings and/or gaskets. Tighten the fuel pump flange nut to 55 ft. lbs. (75 Nm) and make sure the markings on the flange nut match the marking on the fuel tank.

Fuel Injectors

REMOVAL & INSTALLATION

1. Before servicing the vehicle, refer to the Precautions Section.
2. Relieve the fuel system pressure.
3. Remove or disconnect the following:
   - Negative battery cable
   - Engine appearance cover, if equipped
   - Engine under cover, if equipped
   - Air cleaner and ducting
   - All necessary vacuum and electrical connectors that interfere with fuel rail removal
   - Fuel pressure and return lines
   - Fuel injector connectors
   - Fuel supply rail with injectors attached
   - Fuel injectors from the supply rail

To install:

4. Install or connect the following:
   - Fuel injectors to the fuel supply rail using new O-rings
   - Fuel supply rail with injectors attached. Tighten the bolts to 89 inch lbs. (10 Nm)
   - Fuel injector connectors

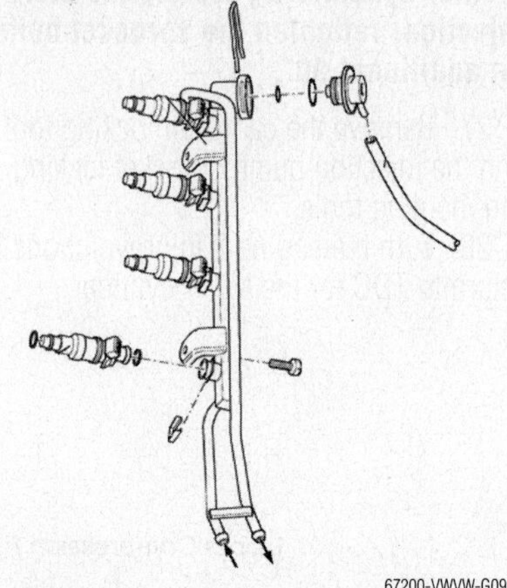

Fuel injectors and fuel rail—1.8L Golf, GTI and Jetta shown: other models similar

   - Fuel pressure and return lines
   - All necessary vacuum and electrical connectors
   - Air cleaner and ducting
   - Negative battery cable
   - Engine appearance cover, if equipped
   - Engine under cover, if equipped
   - Negative battery cable
5. Start the engine, check for leaks and repair if necessary.

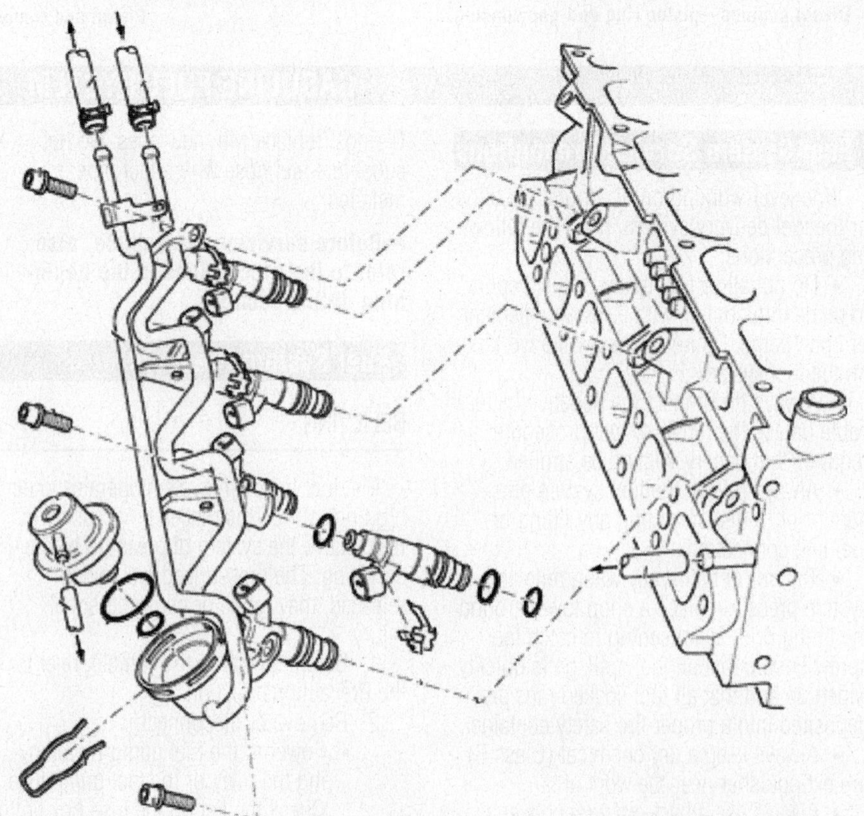

Fuel injectors and fuel rail—2.8L Golf, GTI and Jetta shown: other models similar

DIESEL FUEL SYSTEM

Fuel System Service Precautions

Although Diesel fuel is not as flammable as gasoline, whenever working on or around diesel fuel or the fuel delivery system, heed the following precautions:

- Do not allow fuel spray or fuel vapors to come into contact with a heating element or open flame. Do not smoke while working on the fuel system.
- Always disconnect the negative battery cable unless the repair or test procedure requires that battery voltage be applied.
- Always relieve the fuel system pressure prior to disconnecting any fitting or fuel line connection.
- To control fuel spray when relieving system pressure, place a shop towel around the fitting prior to loosening to catch the spray. Ensure that all fuel spillage is quickly wiped up and that all fuel soaked rags are deposited into a proper fire safety container.
- Always keep a dry chemical (Class B) fire extinguisher near the work area.
- Always use a back-up wrench when loosening and tightening fuel line fittings.
- Do not re-use fuel system gaskets and O-rings. Replace with new ones. Do not substitute fuel hose where fuel pipe is installed.

Fuel Filter/Water Separator

DRAINING WATER

Although diesel fuel and water do not readily mix, fuel does tend to entrap moisture from the air each time it is moved from one container to another. Eventually every diesel fuel system collects enough water to become a potential hazard. Fortunately, when it's allowed to settle, the water will always drop to the bottom of the tank or filter housing. Some diesel fuel filters are equipped with a water drain; a bolt or petcock at the bottom of the housing.

At The Water Separator

1. Before servicing the vehicle, refer to the Precautions Section.
2. Remove the fuel filler cap.
3. At the separator, connect a hose from the separator drain to a catch pan.
4. Open the drain valve (3 turns) and drain the separator until a steady stream of fuel flows from the separator, then close the valve.

➡Don't forget to install the filler cap.

At The Filter

1. Before servicing the vehicle, refer to the Precautions Section.
2. If the filter is equipped with a water drain at the bottom, place a pan under the drain to catch the water and fuel.
3. If equipped, loosen the vent bolt on the filter base. If there is no vent, loosen the return line at the pump (the line not connected to the filter).
4. Loosen the bolt or valve. When fuel flows in a clean stream, close the drain and tighten the vent or return line.

REMOVAL & INSTALLATION

⁂ WARNING

Do not allow diesel fuel to contact the coolant hoses. If this happens, wipe it off and wash the hoses with soap and water immediately.

1. Before servicing the vehicle, refer to the Precautions Section.
2. Remove or disconnect:
   - Fuel pressure and return lines
   - Retaining clip
   - Regulator valve and O-rings
   - Filter assembly

To install:

3. Install the filter.
4. Install the regulator valve and new O-rings
5. Install the retaining clip.
6. Connect the fuel pressure and return lines and secure them with clamps.

Diesel Injection Pump

REMOVAL & INSTALLATION

Except ALH Engines

➡Special tools are required for injection pump installation. Do not remove the pump without these tools on hand.

1. Before servicing the vehicle, refer to the Precautions Section.
2. Remove or disconnect:
   - Negative battery cable
   - Air cleaner
   - Cylinder head cover

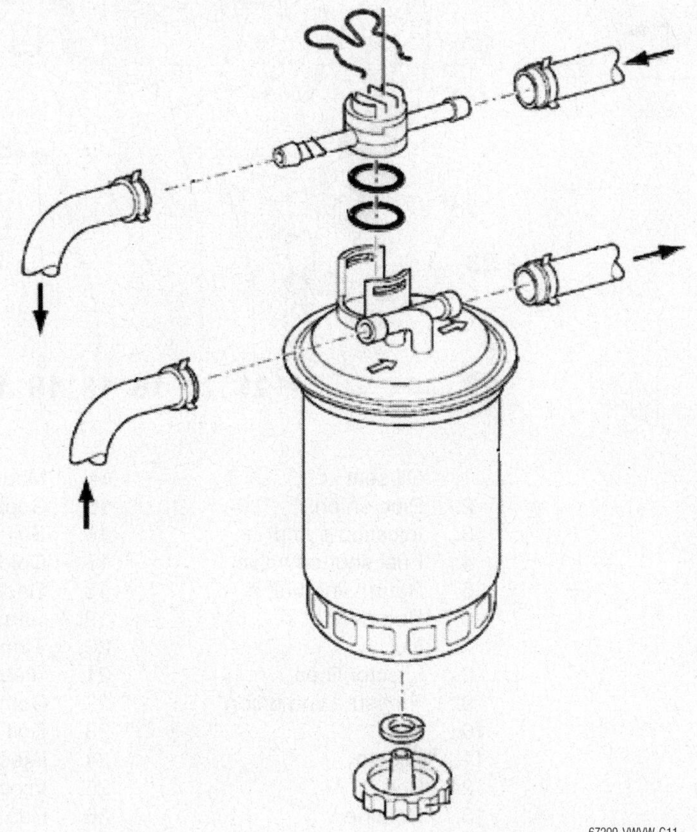

Fuel filter assembly—1.9L (ALH) engine shown; other engines similar

- Timing belt cover.

3. Turn the engine to TDC of No. 1 cylinder and insert a setting bar into the slot on the rear of the camshaft, VW tool 2065A or equivalent, to hold the camshaft in place. Remove the timing belt. Be careful to not turn the engine while the belt is removed.
4. Loosen the pump drive sprocket nut but don't remove it yet. Install a puller on the sprocket and apply moderate tension.
5. Rap the puller bolt with light hammer taps until the sprocket jumps off the tapered shaft, then remove the puller and sprocket. Be careful not to lose the Woodruff key.
6. Hold the pump fittings with a wrench

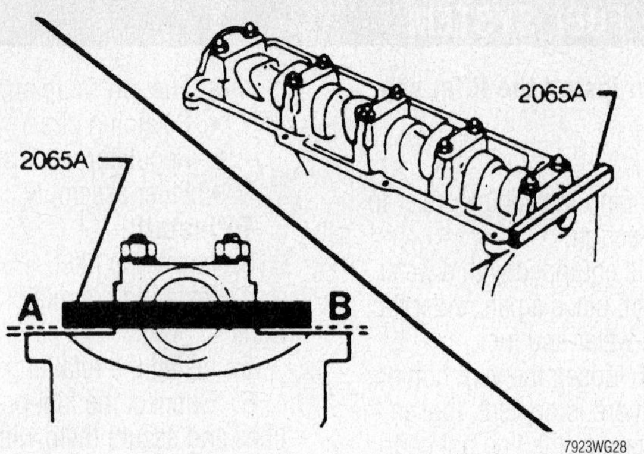

Install the bar to hold the camshaft in position during diesel injection pump service

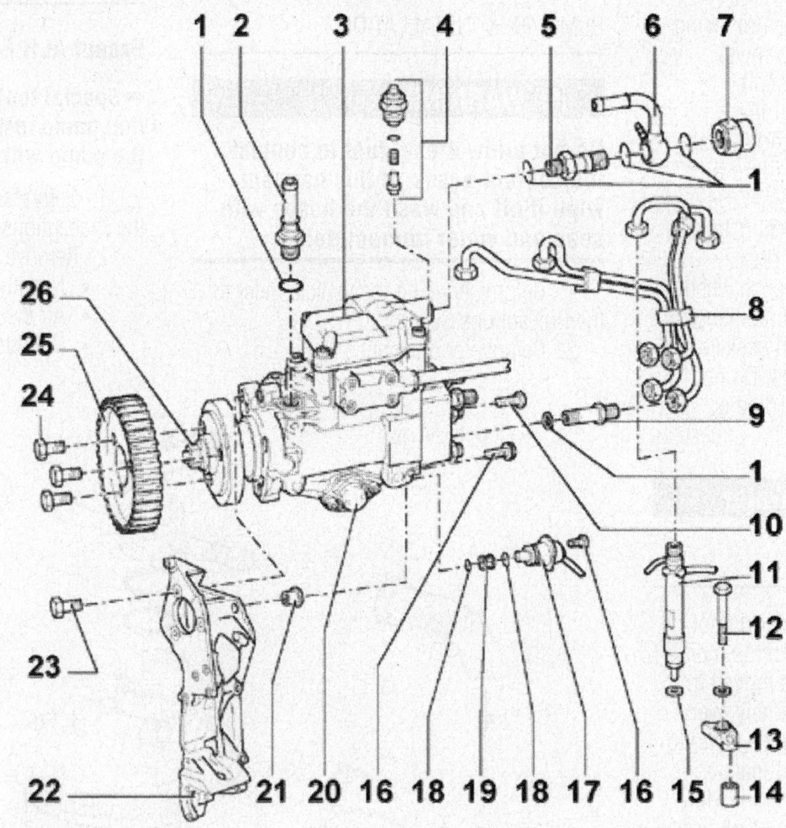

1. Oil seal
2. Pipe union
3. Injection pump
4. Fuel shut-off valve
5. Return line union
6. Return line
7. Nut
8. Injector lines
9. Pressure line union
10. Bolt
11. Injector
12. Bolt
13. Retainer
14. Mounting
15. Copper washer
16. Bolt
17. Cold start injector
18. Gasket
19. Strainer
20. Timing control cover
21. Sleeve nut
22. Compact bracket
23. Bolt
24. Injection pump sprocket bolts
25. Injection pump sprocket
26. Hub nut

67200-VWVW-G12

Exploded view of injection pump assembly—1.9L ALH engine

and using a line wrench, remove the injection lines from the pump. Cap the pump fittings to keep dirt out. It may be easier to remove the lines from the injectors also and set them aside as an assembly. Cap the injector fittings to keep dirt out.

7. Disconnect the control cables, fuel solenoid wire and fuel supply and return lines.

8. Remove the pump mounting bolts and lift the pump from the vehicle.

To install:

9. When reinstalling, align the marks on the top of the mounting flange and the pump. Mounting bolts: 18 ft. lbs. (25 Nm).

10. Install the Woodruff key and sprocket. Nut: 33 ft. lbs. (45 Nm).

11. When reinstalling the supply and return lines, be sure the fitting marked OUT is used for the return line. This fitting has an orifice and must be in the correct place. Use new gaskets.

12. Turn the pump sprocket so the mark aligns with the mark on the side of the mounting flange and insert a pin through the hole in the sprocket to hold it in place.

13. Install the camshaft drive sprocket and belt and set the belt tension. Tension the drive belt by turning the tensioner pulley clockwise until the belt can be flexed ½ in. (13mm) between the camshaft and the pump sprockets. Remove the pin.

14. Remove the camshaft holding bar. Turn the engine through 2 full turns, return to TDC of the No. 1 cylinder and recheck the belt tension and camshaft timing.

15. Reinstall the injection lines, wiring and control cables. Line nuts: 18 ft. lbs. (25 Nm).

ALH Engines

➡Special tools are required for injection pump installation. Do not remove the pump without these tools on hand.

1. Before servicing the vehicle, refer to the Precautions Section.

2. Remove or disconnect:
   - Negative battery cable
   - Timing belt from camshaft and injection pump sprockets

➡The vibration damper and center and lower timing belt covers can remained installed.

   - Fuel lines at fuel pump
   - Adjuster connector from retainer
   - Injection pump sprocket bolts

➡DO NOT loosen the pump sprocket center bolt or the basic pump setting will be disturbed.

   - Compact bracket bolts
   - Injection pump rear support bolt
   - Injection pump

To install:

3. Install the injection pump into the compact bracket, then tighten the rear support bolt first.

4. Insert the front bolts and tighten to 18 ft. lbs. (25 Nm).

5. Install the injection pump sprocket to the hub and lightly tighten the bolts.

6. Lock the injection pump sprocket with holding tool 3359.

7. Align the elongated holes of the sprocket install the timing belt and tighten the sprocket bolts to 15 ft. lbs. (20 Nm) plus 90°.

8. The remainder of the installation is the reverse of the removal procedures, noting the following torque specifications:
   a. Tighten the injector lines to 18 ft. lbs. (25 Nm).
   b. Tighten the fuel line fittings to 18 ft. lbs. (25 Nm).
   c. Tighten the pressure line union to 33 ft. lbs. (45 Nm).
   d. Tighten the compact bracket bolts to 18 ft. lbs. (25 Nm).

Injectors

REMOVAL & INSTALLATION

1. Relieve the fuel system pressure.

2. Remove the fuel pipe by unscrewing the union nuts.

3. Remove the fuel pipes at the pump and at the injectors.

4. Remove the fuel injection lines as an assembly.

5. Unscrew the injector from the cylinder head.

To install:

➡Replace the heat shields at each injector hole every time a new injector is installed.

6. Screw the new injector into the cylinder head and tighten it.

7. Install the fuel line assembly.

DRIVE TRAIN

Transaxle Assembly

REMOVAL & INSTALLATION

Manual

CABRIO, GOLF, GTI, RABBIT AND JETTA

1. Before servicing the vehicle, refer to the Precautions Section.

2. Remove or disconnect:
   - Negative battery cable.
   - Back-up light switch connector and the speedometer cable from the transaxle; plug the speedometer cable hole.
   - Upper engine-to-transaxle bolts.
   - 3 right side engine mount bolts, between engine and firewall.
   - Shift linkage as follows: Pry open the ball joint ends and remove the shift and relay shaft rods.
   - Center bolt from the left transaxle mount.
   - Front wheels. Connect the engine sling tool VW-10–222A or equivalent, to the loop in the cylinder head and just take the weight of the engine off the mounts. On 16V engine, the idle stabilizer valve must be removed to attach the tool. Do not try to support the engine from below.
   - Oil from the transaxle.
   - Left inner fender liner.
   - Halfshafts from the inner drive flanges and hang them from the body.
   - Clutch cover plate and the small plate behind the right halfshaft flange.
   - Starter and front engine mount.
   - Clutch cable and remove it from the transaxle housing.
   - Remaining transaxle mount bolts and mounts.

3. Place a jack under the transaxle and remove the last bolts holding it to the engine. Carefully pry the transaxle away from the engine and lower it from the vehicle.

To install:

4. Coat the input shaft lightly with molybdenum grease and carefully fit the transaxle in place. If necessary, put the transaxle in any gear and turn an output flange to align the input shaft spline with the clutch spline.

5. Install the engine-to-transaxle bolts. Bolts: 55 ft. lbs. (75 Nm).

6. When installing the mounts to the transaxle, tighten the rear bracket-to-engine

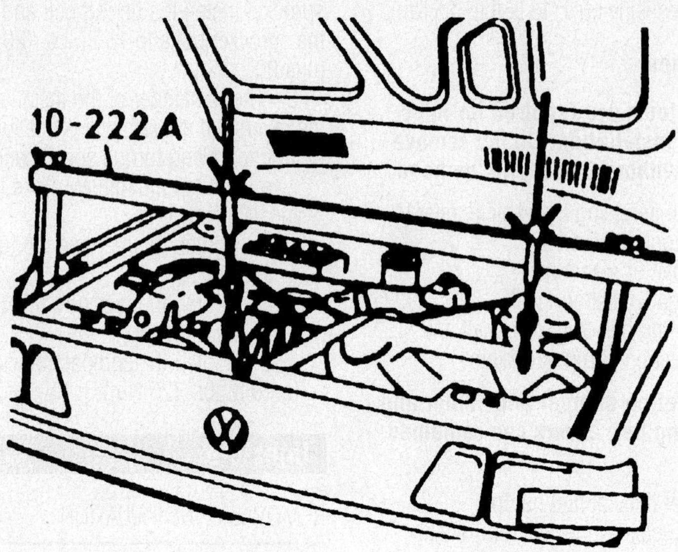

Supporting the engine to remove the transaxle

bolts and the transaxle support bolts to 18 ft. lbs. (25 Nm). Tighten the left bracket-to-transaxle bolts to 25 ft. lbs. (35 Nm) and the remaining mounting bolts to 44 ft. lbs. (60 Nm). Install but do not tighten the bolts that go into the rubber mounts.

7. Install the starter and front mount.

8. With all mounts installed and the transaxle safely in the vehicle, allow some slack in the lifting equipment. With the vehicle safely supported, shake the engine/transaxle as a unit to settle it in the mounts. Tighten all mounting bolts, starting at the rear and working forward. Tighten the bolts that go into the rubber mounts to 44 ft. lbs. (60 Nm).

9. Remove or disconnect:
- Halfshafts. Bolts: 33 ft. lbs. (45 Nm).
- Clutch cover plates.
- Shift linkage and clutch cable and adjust as required.
- Inner fender and complete the remaining installation.
- Transaxle with oil.

NEW BEETLE

1. Before servicing the vehicle, refer to the Precautions Section.

2. Remove or disconnect:
- Engine cover and undercover.
- Negative, then the positive battery cable.
- Power steering reservoir from the battery tray and position it aside leaving the hoses attached.
- Battery and tray.
- Air cleaner and intake air hose.
- Connector for the reverse light switch and VSS.
- Gear selector cable from the transaxle.
- Slave cylinder and position it side leaving the hydraulic hose attached.
- Cable retainer near the starter.
- Upper starter mounting bolt.
- Ground strap at the engine-to-transaxle top mounting bolt.
- Upper engine-to-transaxle mounting bolts.

3. Reposition any wiring or hoses that would interfere with installing an engine support tool.

4. Install a suitable engine support tool.

5. Remove or disconnect:
- Starter.
- Halfshafts at the transaxle, after turning the wheels to the left lock, and tie them up out of the way.
- Flywheel cover plates.
- Front exhaust pipe from the manifold.
- Pendulum and transaxle support.
- Bolts for the transaxle mount.
- If necessary, right side cooling fan.

6. Install a suitable transaxle jack and support the weight of the transaxle.

7. Remove the lower engine-to-transaxle mounting bolts and lower the transaxle out of the vehicle.

To install:

8. Clean the input shaft and lubricate lightly with grease.

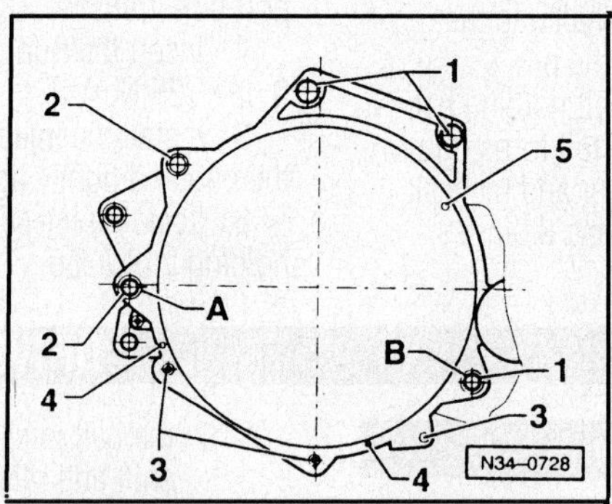

Pos.	Bolt	Nm (ft lb)
1	M12 x 55	80 (59)
2[1)]	M12 x 150	80 (59)
3[2)]	M10 x 50	45 (33)
4[3)]	M7 x 12	10 (7)
5[4)]	M7 x 12	10 (7)

[1)] Also starter to transmission

[2)] Only on engines with an aluminium oil pan,

[3)] Large cover plate for flywheel only on engines with sheet steel oil pan (painted black)

[4)] Small cover plate for flywheel

Pos. A + B = Dowel sleeves

9301WG33

Engine-to-transaxle (automatic and manual) mounting bolt torque specification and locations—New Beetle

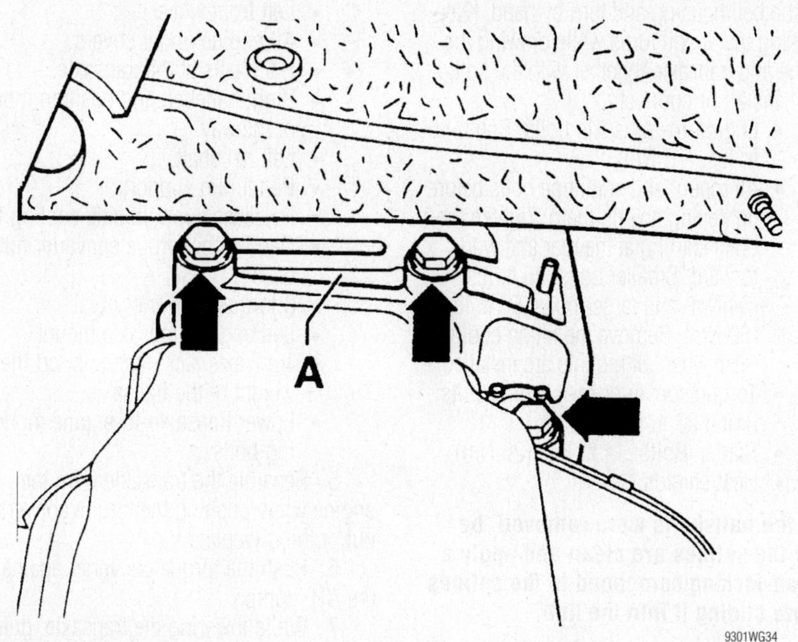

When installing the transaxle mount (A), be sure to install new bolts (arrows)—New Beetle

9. Raise the transaxle into the vehicle.
10. Install the lower engine-to-transaxle mounting bolts.
11. Remove the transaxle jack.
12. Install or connect:
   - Upper engine-to-transaxle mounting bolts.
   - Transaxle mount, using new bolts. Mounting bolt: 30 ft. lbs. (40 Nm) plus 90°.
   - Pendulum and transaxle supports.
   - Exhaust system.
   - Install the flywheel cover plates.
   - Halfshafts to the transaxle.
   - Starter and clutch slave cylinder.
13. The remaining steps of the installation are the reverse of the removal, keeping in mind the following items:
   - Attach any ground cables that were removed
   - Remove the engine support bracket
   - Be sure all connectors are attached
   - Check the transaxle
   - Install engine covers
   - Read and clear any DTCs

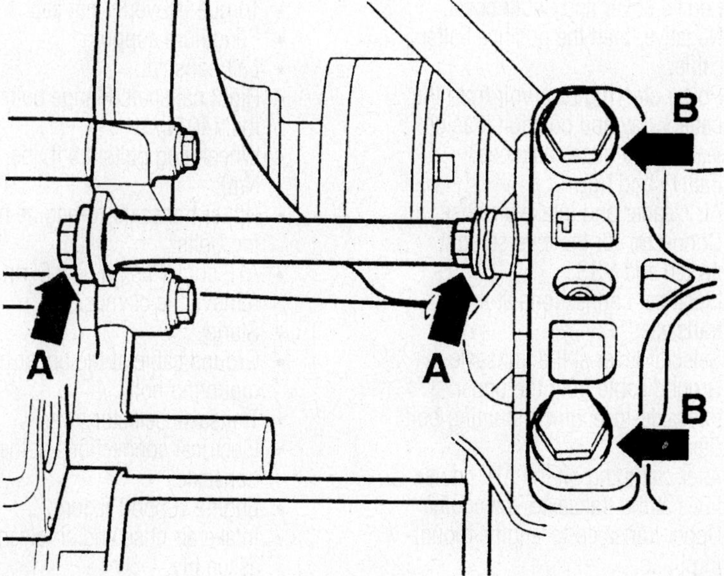

Be sure to tighten bolts (A) to 18 ft. lbs. (25 Nm) and bolts (B) to 44 ft. lbs. (60 Nm) plus 90°—New Beetle

PASSAT

1. Before servicing the vehicle, refer to the Precautions Section.
2. Remove or disconnect:
   - Negative battery cable.
   - Engine undercover.
   - Front exhaust pipes from the engine. Loosen the U-bolt and push to the rear.
   - Starter.
   - Shift rod from the transaxle.
   - Speed sensor and left back-up light connectors from the transaxle.
3. Support the transaxle with a jack.
4. Remove:
   - Right and left transaxle mounts.
   - Right and left halfshaft from the transaxle.
   - Halfshaft shield.
   - Transaxle-to-engine mounting bolts.
5. Pry the transaxle from the engine and lower it about 6 in. (13cm) to access the slave cylinder.
6. Remove the slave cylinder with bracket without disconnecting the fluid line.
7. Lower and remove the transaxle assembly.

To install:

8. Clean the input shaft and apply a thin film of No. 000 100 high-performance grease or equivalent to the splines.
9. Lubricate the plunger contact surface on the release lever with Dow Corning® CU-7439 Plus Copper Paste or equivalent.
10. Raise the transaxle into position and install the slave cylinder. Mounting bolts: 18 ft. lbs. (25 Nm).
11. Install or connect:
   - Transaxle-to-engine bolts. M8 bolts: 18 ft. lbs. (25 Nm), M10 bolts: 33 ft. lbs. (45 Nm) and the M12 bolts: 48 ft. lbs. (65 Nm).
   - Transaxle mounts. Mounting bolts: 30 ft. lbs. (40 Nm).
   - Halfshafts. M8 bolts: 33 ft. lbs. (45 Nm) and M10 bolts: 59 ft. lbs. (80 Nm).
   - Halfshaft shield.
   - Shift rod. Bolts: 15 ft. lbs. (20 Nm).
   - Starter.
   - Exhaust system.
   - Engine undercover.
   - Negative battery cable.

Automatic

CABRIO, GOLF, GTI, RABBIT AND JETTA

1. Before servicing the vehicle, refer to the Precautions Section.

2. Remove or disconnect:
   - Battery and the speedometer drive and plug the hole in the transaxle.
   - On the Golf and Jetta models, front axle nuts.

CAUTION

When loosening or tightening an axle nut, be sure the vehicle is on the ground. Axle nut torque is high enough that attempting to loosen it may cause the vehicle to fall.

   - Remove the front wheels. Connect the engine sling tool VW-10–222A or equivalent, to the cylinder head and just take the weight of the engine off the mounts. On 16V engine, the idle stabilizer valve must be removed to attach the tool. Do not try to support the engine from below.
   - Driver's side rear transaxle mount and support bracket.
   - On the Golf and Jetta models, front mount bolts from the transaxle and from the body and mount as a complete assembly.
   - Selector and accelerator cables from the transaxle lever but leave them attached to the bracket. Remove the bracket assembly to save the adjustment.
   - Halfshafts from the drive flanges. On the Golf and Jetta models, the shafts must be removed, which may require separating the ball joints from the wheel bearing housing to gain the necessary clearance. Remove the ball joint clamping bolt.
   - Heat shield and brackets and remove the starter. On the Cabrio models, the front mount comes off with the starter.
   - Torque converter-to-flywheel bolts, turning the engine as needed.
   - Remaining transaxle mounts, on the Golf and Jetta models, the subframe bolts and allow the subframe to hang free.

3. Support the transaxle with a jack and remove the remaining engine-to-transaxle bolts. Be careful to secure the torque converter so it does not fall out of the transaxle.

4. Carefully lower the transaxle from the vehicle.

To install:

5. When reinstalling, be sure the torque converter is fully seated on the pump shaft splines. The converter should be recessed into the bell housing and turn by hand. Keep checking that it still turns while drawing the engine and transaxle together with the bolts.

6. Install or connect:
   - Engine-to-transaxle bolts. Bolts: 55 ft. lbs. (75 Nm).
   - All mount and subframe bolts before tightening any on them. Tighten the bolts starting at the rear and work forward. Smaller bolts: 25 ft. lbs. (34 Nm) and larger bolts: 58 ft. lbs. (80 Nm). Remove the lifting equipment when all mounts are installed.
   - Torque converter-to-flywheel bolts. Bolts: 26 ft. lbs. (35 Nm).
   - Starter. Bolts: 14 ft. lbs. (20 Nm).
   - Heat shields.

➡If the halfshafts were removed, be sure the splines are clean and apply a thread-locking compound to the splines before sliding it into the hub.

   - Halfshafts to the drive flanges. Bolts: 37 ft. lbs. (50 Nm). Install new axle nuts but do not fully tighten them until the vehicle is on the ground.
   - If removed, fit the ball joints to the control arm. Clamping bolt: 37 ft. lbs. (50 Nm).
   - Shift linkage. Adjust as required.

7. When assembly is complete and the vehicle is on its wheels, tighten the axle nuts to 195 ft. lbs. (265 Nm).

NEW BEETLE

1. Before servicing the vehicle, refer to the Precautions Section.

2. Remove or disconnect:
   - Engine cover and under cover.
   - Negative, then the positive battery cable.
   - Power steering reservoir from the battery tray and position it aside leaving the hoses attached.
   - Battery and tray.
   - Air cleaner and intake air hose.
   - Connector for the reverse light switch and VSS.
   - Electrical connections at the transaxle.
   - Selector lever at the transaxle.
   - Ground cable from the upper transaxle-to-engine mounting bolt.
   - Starter.
   - After clamping off the ATF cooler lines at the transaxle, the cooler.
   - Upper transaxle-to-engine mounting bolts.

3. Install a suitable engine support fixture and slightly lift the engine.

4. Remove or disconnect:
   - Left front wheel.
   - All engine under covers
   - Halfshafts at the transaxle.
   - Right halfshaft and position it out of the way.
   - Left halfshaft.
   - Pendulum support.
   - If necessary, right side cooling fan.
   - Cap for the torque converter nut cover.
   - 3 torque converter nuts.
   - Left engine/transaxle mount.
   - Transaxle jack, then support the weight of the transaxle.
   - Lower transaxle-to-engine mounting bolts.

5. Separate the transaxle from the engine while pushing the torque converter out of the driveplate.

6. Push the torque converter against the ATF pump.

7. While lowering the transaxle, guide the power steering hoses past the transaxle.

8. Secure the torque converter to prevent it from falling out.

To install:

9. When reinstalling, be sure the torque converter is fully seated on the pump shaft splines. The converter should be recessed into the bell housing and turn by hand.

10. Install or connect:
   - Transaxle into the vehicle.
   - Lower transaxle-to-engine mounting bolts.
   - Left engine/transaxle mount, then remove the transaxle jack.
   - Torque converter-to-flexplate mounting nuts. Nuts: 44 ft. lbs. (60 Nm).
   - Torque converter nut cap.
   - Pendulum support.
   - Left halfshaft.
   - Right halfshaft. Flange bolts: 30 ft. lbs. (40 Nm).
   - Wheel. Lug bolts: 89 ft. lbs. (120 Nm).
   - Upper transaxle-to-engine mounting bolts.
   - ATF cooler using new O-rings and remove the clamps.
   - Starter.
   - Ground cable at the upper transaxle mounting bolt.
   - Transaxle selector lever.
   - Electrical connections at the transaxle.
   - Engine support fixture.
   - Intake air hose and air cleaner assembly.
   - Battery tray and battery.
   - Power steering reservoir to the battery.

- Battery cables.
- All covers that were removed.

PASSAT

1. Before servicing the vehicle, refer to the Precautions Section.
2. Remove or disconnect:
   - Battery.
   - Wiring from the transaxle.
   - Upper engine-to-transaxle bolts.
   - Front wheels. Connect the engine sling tool VW-10–222A or equivalent, to the cylinder head and just take the weight of the engine off the mounts. The idle stabilizer valve must be removed to attach the tool. Do not try to support the engine from below.
   - Shift cable.
   - Hoses at the transaxle cooler.
   - Starter and the engine's left and right mounts.
   - Skid plate
   - Halfshafts from the drive flanges. Hang them from the body with wire.
   - Torque converter plate and turn the engine as needed to remove the torque converter-to-flywheel bolts.
   - Remaining transaxle mounts and lower the hoist slightly.
3. Support the transaxle with a jack and remove the remaining engine-to-transaxle bolts. Be careful to secure the torque converter so it does not fall out of the transaxle.
4. Carefully lower the transaxle out of the vehicle.

To install:

5. Fit the transaxle into the vehicle and be sure the guide pins fit properly between the engine and transaxle. Install the bolts and tighten the 12mm bolts to 59 ft. lbs. (80 Nm), the 10mm bolts to 44 ft. lbs. (60 Nm).
6. Install or connect:
   - Transaxle mounts. Bolts: 44 ft. lbs. (60 Nm). Left side bracket-to-transaxle bolts: 18 ft. lbs. (25 Nm).
   - Torque converter bolts: 44 ft. lbs. (60 Nm).
   - Halfshafts. Bolts: 33 ft. lbs. (45 Nm).
   - Shift linkage. Adjust as required.
7. Install the remaining parts and check the fluid level in the transaxle.

Clutch

ADJUSTMENT

All vehicles use a hydraulic clutch release mechanism. No free-play adjustment is required or possible. If the clutch does not release or engage properly, bleed the system before moving on to more extensive repairs.

REMOVAL & INSTALLATION

1. Before servicing the vehicle, refer to the Precautions Section.
2. Remove the transaxle.
3. Matchmark the flywheel and pressure plate if the pressure plate is going to be reused.
4. Gradually loosen the pressure plate bolts 1–2 turns at a time in a crisscross pattern to prevent distortion.
5. Remove the pressure plate and disc.
6. Check the clutch disc for uneven or excessive lining wear. Examine the pressure plate for cracking, scorching or scoring. Replace any questionable components.

To install:

7. Install the clutch disc and pressure plate with the springs on the disc towards the plate. Use an alignment tool to keep the clutch disc centered.
8. Gradually tighten the pressure plate-to-flywheel bolts in a crisscross pattern. Bolts: 18 ft. lbs. (24 Nm).
9. Install the clutch release bearing.
10. Install the transaxle.

Hydraulic Clutch System

BLEEDING

1. Before servicing the vehicle, refer to the Precautions Section.
2. The clutch and brakes share the same reservoir. Clean all dirt and grease from the cap to be sure no foreign substances enter the system.
3. Remove the cap and diaphragm and fill the reservoir to the top with the approved DOT 3 or 4 brake fluid. Fully loosen the bleed screw which is in the slave cylinder body next to the inlet connection.
4. At this point bubbles of air will appear at the bleed screw outlet. When the slave cylinder is full and a steady stream of fluid comes out of the slave cylinder bleeder, tighten the bleed screw.
5. Refill the reservoir and cap it. Exert a light load of about 20 lbs. (9 kg) to the slave cylinder piston by pushing the release lever towards the cylinder and loosen the bleed screw. Maintain a constant light load; fluid and any air that is left will be expelled through the bleed port. Tighten the bleed screw when a steady flow of fluid with no air is being expelled.
6. Fill the reservoir fluid level back to normal capacity, if necessary repeat Step 4.
7. Exert a light load to the release lever but do not open the bleeder screw as the piston in the slave cylinder will move slowly down the bore. Repeat this operation 2–3 times; the fluid movement will force any air left in the system into the reservoir. The hydraulic system should now be fully bled.
8. Check the operation of the clutch hydraulic system and repeat this procedure, if necessary.

Halfshaft

REMOVAL & INSTALLATION

Except Passat

⁂ CAUTION

When loosening or tightening axle nuts, be sure the vehicle is on the ground. Axle nut torque is high enough that attempting to loosen it may cause the vehicle to fall.

1. Before servicing the vehicle, refer to the Precautions Section.
2. Remove or disconnect:
   - Front axle nut.
   - Front wheels.
   - Socket head bolts retaining the halfshaft to the transaxle flange.
3. Matchmark the installed position of the ball joint, then detach the ball joint from the control arm.
4. Remove the transaxle side of the halfshaft from the drive flange and secure it out of the way. Do not let it hang unsupported.
5. Push the halfshaft out of the hub. A wheel puller may be required.

To install:

6. Fit the halfshaft to the drive flange and install the bolts. It is not necessary to tighten them yet.
7. Apply a thread-locking compound to the outer ¼ in. (6mm) of the spline. Slip the spline through the hub and loosely install a new axle nut.
8. Assemble the front suspension, being careful to align the matchmarks.
   a. On New Beetle models, tighten the ball joint bolts to 15 ft. lbs. (20 Nm) plus 90°.
   b. On Cabrio, Golf, GTI, Jetta and Passat models, tighten the ball joint clamping bolt to 33 ft. lbs. (45 Nm).
9. Install the wheel and hold it to keep the axle from turning. Inner axle bolts: 33 ft. lbs. (45 Nm).

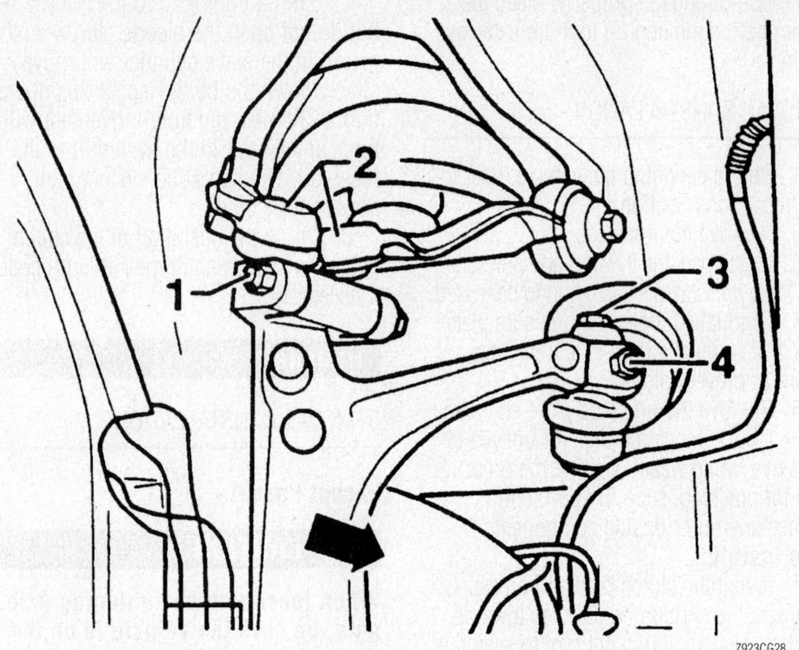

Loosen nut (1), remove the hex bolt, and pull both arms (2) upward and out—Passat

10. With the vehicle on the ground, tighten the axle nut as follows:
    - New Beetle—221 ft. lbs. (300 Nm), loosen 1 turn, tighten to 37 ft. lbs. (50 Nm) then an additional 30°
    - Cabrio, Golf, GTI, Jetta and Passat—66 ft. lbs. (90 Nm) plus 45°
11. Check and adjust the front wheel alignment.

PASSAT

CAUTION

When loosening or tightening axle nut or bolt, be sure the vehicle is on the ground. Axle nut torque is high enough that attempting to loosen it may cause the vehicle to fall.

1. Remove the hub cap or center cap.
2. Loosen the hex collar bolt.
3. Remove or disconnect:
    - Front wheels.
    - Halfshaft-to-transaxle flange bolts, then the hex collar bolt.
    - ABS wheel speed sensor cable from the brake caliper bracket.
4. Slide the ABS speed sensor partly out of its mount.
5. Remove nut/bolt No. 1, as shown, then pull both arms up and out of the swing arm.

➡The slots in the swing arm must not be widened. Do not loosen the bolts No. 3 and 4, otherwise the axle geometry must be checked.

6. Tilt the swing arm out and to the rear of the vehicle, then remove the halfshaft.

To install:

7. Install or connect:
    - Halfshaft into the wheel hub.
    - Swing arm bolt. Bolt: 30 ft. lbs. (40 Nm).
    - Halfshaft-to-transaxle flange. M8 bolts: 30 ft. lbs. (40 Nm) and M10 bolts: 57 ft. lbs. (77 Nm).
    - ABS wheel speed sensor, and the sensor cable into the caliper bracket.
8. With the wheels installed and the vehicle on the ground, tighten the axle bolt as follows: M14 bolt 85 ft. lbs. (115 Nm) plus an additional ¼ (90°) turn, M16 bolt 140 ft. lbs. (190 Nm) plus an additional ¼ (90°) turn.

CV-JOINT AND BOOT OVERHAUL

The constant velocity joints (CV-joints) can be disassembled for cleaning and inspection but they cannot be repaired. All parts are machined to a matched tolerance and the entire CV-joint must be replaced as a unit. On Golf, GTI, Rabbit and Jetta, the CV-joints are different on the left and right sides and cannot be interchanged.

1. Raise and safely support the vehicle and remove the halfshaft.
2. Pry open and remove the boot clamps with a pair of wire cutters.
3. With the halfshaft securely clamped in a vise, the outer CV-joint can be removed by sharply rapping out on the joint with a plastic hammer. The joint will snap off of the circlip and slide off the axle.
4. To remove the inner joint, remove the circlip from the center and slide the joint off the axle.
5. Both boots can be removed after removing the CV-joint

To install:

6. Always replace both circlips and make sure the CV-joint is clean before installation. Wrap a piece of black electrical tape around the shaft splines and slip the inner clamp and the boot onto the shaft.
7. Remove the tape and install the dished washer with the concave side out so it acts as a spring pushing the CV-joint out. On the outer joint, install the thrust washer and a new circlip.
8. To install the outer joint, place it onto the spline and carefully tap straight in on the end with a plastic hammer. The joint will click into place over the circlip.
9. To install the inner joint, slide it onto the spline and push in enough to allow the circlip to fit into the groove in the axle shaft.
10. Pack the CV-joint with special CV-joint grease. DO NOT use any other type of grease.
11. Pack any remaining grease into the boot and install the clamps on the outer boot.

STEERING

Air Bag

⁂ CAUTION

Some vehicles are equipped with an air bag system. The system must be disabled before performing service on or around system components, steering column, instrument panel components, wiring and sensors. Failure to follow safety and disabling procedures could result in accidental air bag deployment, possible personal injury and unnecessary system repairs.

PRECAUTIONS

Several precautions must be observed when handling the inflator module to avoid accidental deployment and possible personal injury.

- Never carry the inflator module by the wires or connector on the underside of the module.
- When carrying a live inflator module, hold securely with both hands, and ensure that the bag and trim cover are pointed away.
- Place the inflator module on a bench or other surface with the bag and trim cover facing up.
- With the inflator module on the bench, never place anything on or close to the module which may be thrown in the event of an accidental deployment.

1. Before servicing the vehicle, also refer to the precautions in the beginning of this section.

DISARMING

To avoid personal injury when working on vehicles equipped with an air bag, the negative battery cable must be disconnected before working on the system. Failure to do so may result in deployment of the air bag.

1. Before servicing the vehicle, refer to the precautions in the beginning of this section.
2. Turn the ignition switch to the **LOCK** position.
3. Disconnect the negative battery cable.
4. Wait 10 minutes for the battery back-up power to discharge.

Rack and Pinion Steering Gear

REMOVAL & INSTALLATION

Except Passat

1. Before servicing the vehicle, refer to the Precautions Section.
2. Remove or disconnect:
   - Front wheels and disengage both tie rod ends.
   - Low pressure (suction) hose from the pump and drain the system into a catch pan.
   - At the steering column, the boot clamp, push the boot towards the body and remove the clamp bolt from the universal joint.
   - On Cabrio models, exhaust manifold and shift linkage bracket.
   - Rack mounting clamp nuts and clamps.
3. At this point on some vehicles, the rack cannot be removed from the body.

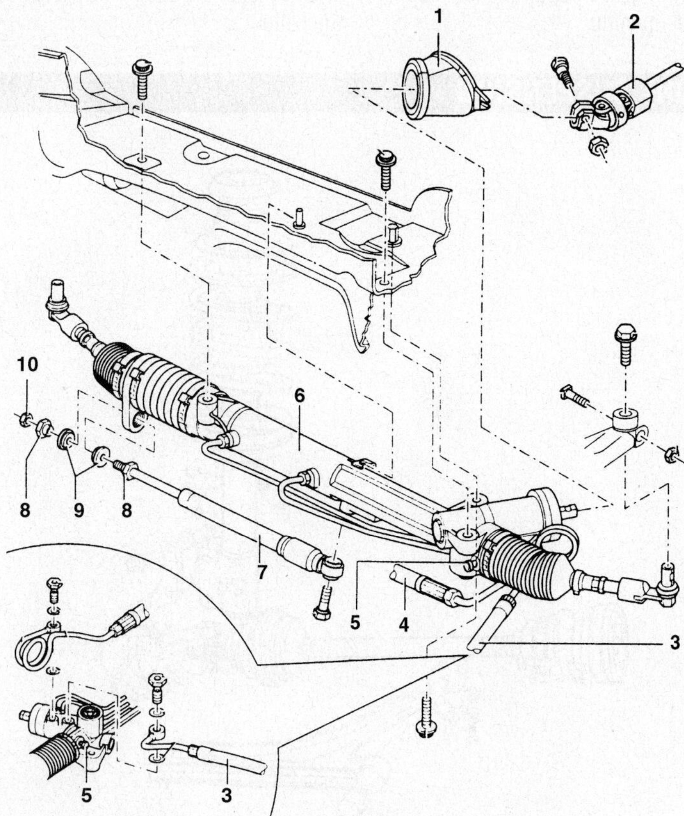

Exploded view of the steering gear mounting—Passat

Support the engine/transaxle and remove the subframe bolts to allow the rack to move towards the rear. On Cabrio models, remove the transaxle mount and bracket.

4. Disconnect the power steering hydraulic lines and remove the rack toward the right.

To install:

5. Be sure the mounting bushings are in good condition. Fit the rack assembly into place and tighten the clamp nuts as follows:
   - Except New Beetle: 22 ft. lbs. (32 Nm)
   - New Beetle: 15 ft. lbs. (20 Nm) plus 90°
6. Install any subframe bolts that were removed.
7. Connect the hydraulic lines and install the steering column universal joint bolt.
8. Fill the system with new fluid and run the engine to check for leaks and bleed the system.

Passat

1. Before servicing the vehicle, refer to the Precautions Section.
2. Remove or disconnect:
   - Battery, then the battery box.
   - Bolt at the steering column U-joint.
3. Release the eccentric by turning the T50 Torx® bolt clockwise, then remove the bolt.
4. Before removing the steering column form the steering gear, secure the steering column with safety wire.

⁂ WARNING

Be sure to lock the steering wheel, otherwise the air bag unit coil spring may be damaged.

5. Lock the steering wheel in the center position and do not move during the repairs.

➡The splines between the top and bottom part of the steering column must not be separated.

6. Move the U-joint down and out of the way.
7. Using Hose Clamps 3094 or equivalent, pinch off the suction and return lines to the steering gear.
8. Remove or disconnect:
   - Front wheels.
   - Left and right tie rods.
   - Tie rod opening cover.

➡Place a drip tray under the vehicle to catch any residual power steering fluid.

   - Banjo bolts for the steering gear suction and return hydraulic hoses.
   - Steering gear mounting bolts.
   - Steering gear through the left side wheel opening.

To install:

9. Remove the screw plug to lock the steering gear in the center position with Locking tool VAG 1907 or equivalent, and tighten to 13 ft. lbs. (18 Nm).
10. Insert the steering gear into the vehicle through the left side wheel opening.
11. Hand-tighten mounting bolts 1 and 2.
12. Install bolt 3 and tighten to 48 ft. lbs. (65 Nm), then tighten bolts 1 and 2 to 48 ft. lbs. (65 Nm).
13. Using new sealing gaskets, install the return hose. Banjo bolt: 37 ft. lbs. (50 Nm). Suction hose banjo bolt: 30 ft. lbs. (40 Nm).
14. Connect the left and right tie rods. Mounting through-bolt: 33 ft. lbs. (45 Nm).
15. Install the tie rod opening cover.
16. Fit the J-joint to the steering gear, then insert the Torx® adjusting bolt by turning it clockwise.
17. Remove the Locking tool VAG 1907, then reinstall the screw plug and tighten to 13 ft. lbs. (18 Nm).
18. Tighten the adjusting bolt nut to 30 ft. lbs. (40 Nm).
19. Remove the steering wheel lock.
20. Remove the hose clamps 3094, and check the hydraulic fluid.
21. Install the battery tray, then connect the battery.

FRONT SUSPENSION

Strut

REMOVAL & INSTALLATION

Except Passat

The upper strut-to-steering knuckle bolt may have an eccentric washer for adjusting wheel camber. Use a wire brush to clean the area and use a cold chisel to mark a fine line on the washer and the strut together. This matchmark may be enough to preserve the front wheel camber adjustment. It will at least be accurate enough to allow driving the vehicle until a proper front wheel alignment can be performed. If there is no eccentric washer, a new bolt and eccentric washer can be substituted. The parts are available through the dealer.

A special tool is required to remove the upper strut nut. If necessary, it can be made by cutting away part of a 22mm socket.

1. Before servicing the vehicle, refer to the Precautions Section.
2. Remove or disconnect:

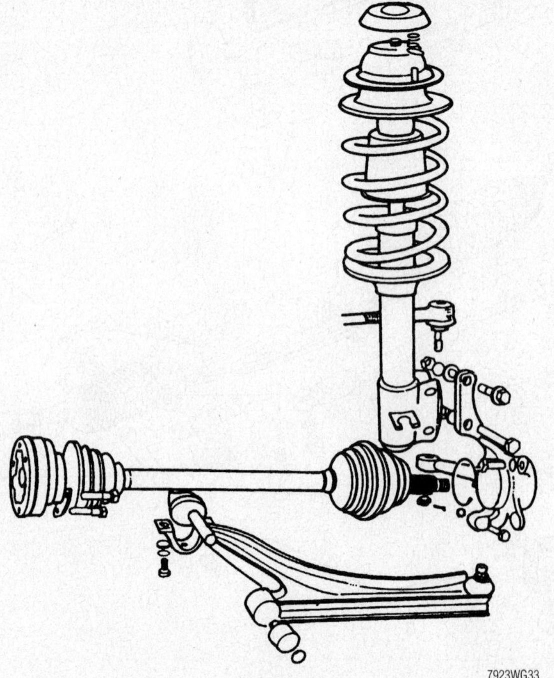

The MacPherson strut is mounted between the steering knuckle and body

Cut away socket for removing the upper strut nut

- Front wheels.
- If equipped, ABS wheel speed sensor.
- Brake line from the strut and remove the caliper. DO NOT let the caliper hang by the hydraulic line, hang it from the body with wire.
- On the New Beetle, strut-to-wheel bearing housing pinch bolt and, slightly, spread the joint.

3. Clean and matchmark the position of the strut-to-steering knuckle bolt.
4. Remove the bolts and push the steering knuckle down away from the strut. Support the knuckle so it is not hanging on the outer CV-joint.

➡On the New Beetle, it may be necessary to remove the wiper arms and external plenum chamber to access the upper strut mounting.

5. Use a hex wrench to hold the shock absorber rod and use the cut away socket to remove the upper nut. Lower the strut from the vehicle.

To install:

6. Place the strut into the fender and install the nuts. Install a new center nut and tighten it to 44 ft. lbs. (60 Nm).
7. On the New Beetle, tighten the pinch bolt to 37 ft. lbs. (50 Nm) plus 90°.
8. On the Cabrio, Golf, Jetta, fit the knuckle into the strut and install the bolts. Be sure the matchmarks are aligned and install the nuts.

➡On the Cabrio, Golf, Jetta, the strut-to-knuckle bolts are 2 different wrench sizes. Tighten the bolts to 70 ft. lbs. (95 Nm).

9. Install the brake caliper. Bolts: 44 ft. lbs. (60 Nm).
10. Install the wheel and align the front wheels.

Passat

1. Before servicing the vehicle, refer to the Precautions Section.
2. Remove or disconnect:
   - Front wheels.
   - Rubber grommets from the plenum chamber.
   - Upper strut-to-body mounting nuts.
   - ABS wheel speed sensor wire from the bracket at the brake caliper.
   - Upper control arm pinchbolt, then lift out both upper control links upward and out.
   - Guide link ball joint by swiveling the wheel bearing housing aside.
   - Lower strut mounting bolt.

➡When removing the strut be sure not to damage the CV joint boot.

3. Remove the strut downward.

To install:

⁂ WARNING

The bonded rubber bushing can only turned to a limited extent. The bolted connections between the suspension strut and the lower track control links should therefore only be tightened when the vehicle is standing on the ground.

4. Install or connect:
   - Strut into the vehicle and position it so that the hole in the spring plate faces the middle of the vehicle.
   - Lower mounting bolt. Bolt: 66 ft. lbs. (90 Nm).
   - Upper control links to the wheel bearing housing. Pinchbolt: 30 ft. lbs. (40 Nm).

➡It may be necessary to hold the ball joint stud with a 4mm hex wrench.

   - Ball joint nut. Nut: 74 ft. lbs. (100 Nm).
   - ABS wheel speed sensor wire into the holder at the brake caliper.
   - Upper strut-to-body mounting nuts. Nuts: 15 ft. lbs. (20 Nm).
   - Rubber grommets into the plenum chamber.
   - Wheels. Lug bolts to 89 ft. lbs. (120 Nm).

Coil Spring

REMOVAL & INSTALLATION

1. Before servicing the vehicle, refer to the Precautions Section.
2. Remove the strut from the vehicle.
3. Clamp the Spring Compressor VAG 1752/2 or equivalent in a vise.
4. Install the strut into the spring compressor.

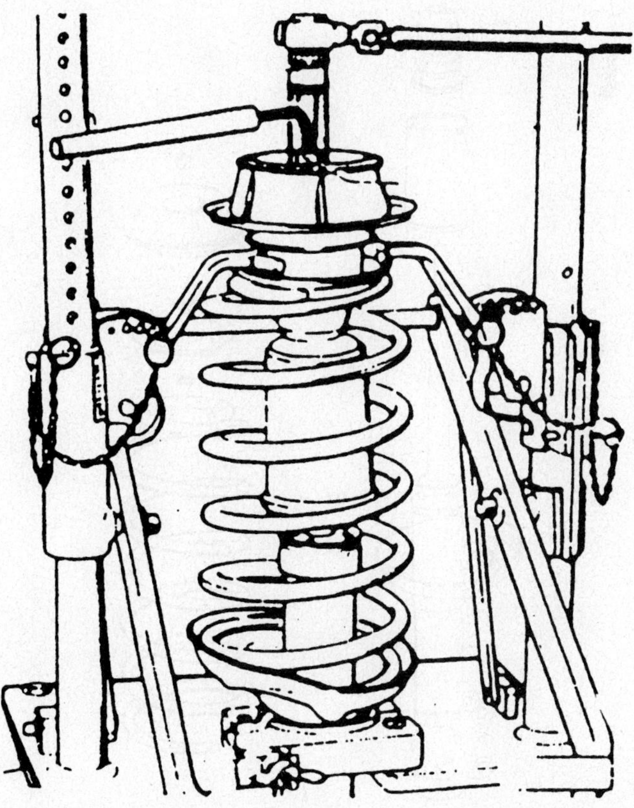

Compress the coil spring before removing the upper strut rod nut

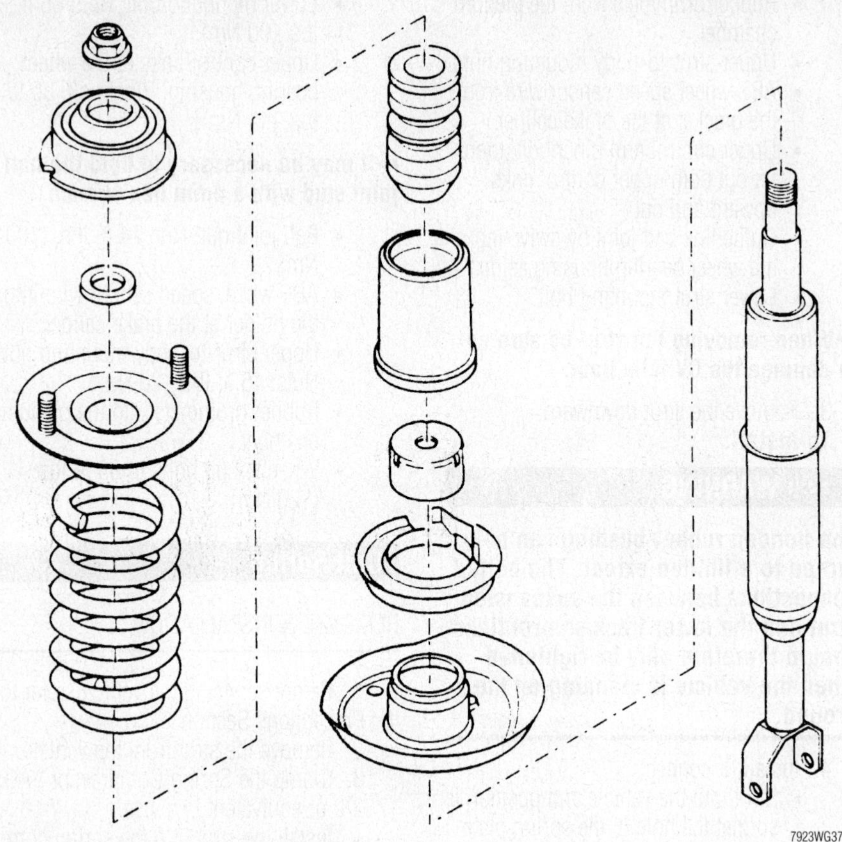

Exploded view of the front strut—Passat

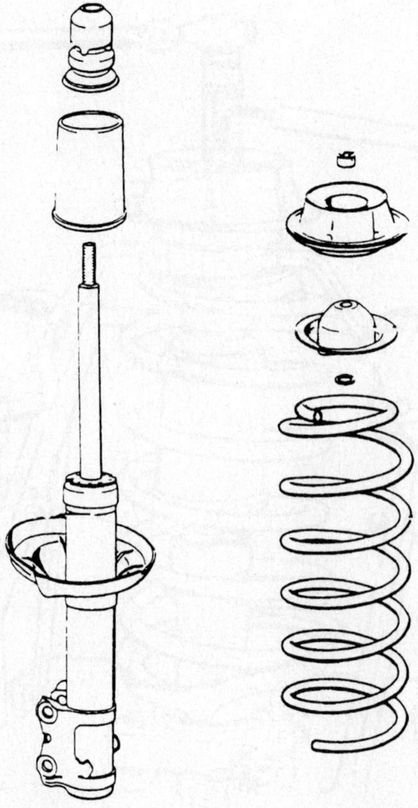

Exploded view of the front strut—except Passat

5. Pry off the mounting bolt cap.
6. Compress the coil spring and remove the self-locking nut from the piston rod.
7. Matchmark the position of the spring retainer and spring mount.
8. Remove:
   - Spring seat and related components noting the order of removal.
   - Strut from the spring compressor.
   - Spring out of the compressor.

To install:

9. Install the new spring into the compressor.
10. Compress the spring and insert the strut through the spring.
11. Install the spring seat and related components in the reverse order as they were removed and aligning the matchmarks.
12. Install a new self-locking nut.
13. Reinstall the mounting bolt cap.
14. Release the spring compressor and install the strut into the vehicle.

Upper Ball Joint

REMOVAL & INSTALLATION

Passat

The Passat front suspension is equipped with 2 separate upper ball joints that are not replaceable, the upper link (front or rear) must be replaced as follows:

1. Before servicing the vehicle, refer to the Precautions Section.
2. Remove or disconnect:
   - Front wheels.
   - Clip No. 1 as shown. The clip does not have to be replaced.
   - Pinchbolt and pull both control arms upward and out.
3. Cover the steering gear boot.
4. Remove or disconnect:
   - Guide link ball joint and press off the joint.
   - ABS wheel speed sensor wire front the bracket on the brake caliper.
5. Support the suspension from excessive rebound travel.
6. Remove or disconnect:
   - Lower strut mounting bolt and swing the wheel bearing housing aside.
   - Rubber grommets from the plenum chamber.
   - Upper strut-to-body mounting nuts.
   - Strut together with the mounting bracket.
7. Clamp the strut in a vise with the protective jaw covers.

8. Remove:
   - Upper link bolts and detach both of the links.
   - Bracket-to-strut mounting nuts, then separate.

To install:

9. Position the brackets and links as shown, and tighten the bracket-to-strut mounting nuts to 15 ft. lbs. (20 Nm).
10. Align the links as shown, then tighten to 37 ft. lbs. (50 Nm) plus ¼ turn (90°).
11. Install or connect:
   - Strut with mounting bracket into the vehicle. Upper strut-to-body mounting nuts: 48 ft. lbs. (75 Nm).
   - Lower strut mounting bolt: 66 ft. lbs. (90 Nm).
   - Nut on the ball joint: 74 ft. lbs. (100 Nm).
   - Upper links to the wheel bearing housing. Pinchbolt: 30 ft. lbs. (40 Nm).
   - ABS wiring to the brake caliper bracket.
   - Wheel.

Lower Ball Joint

REMOVAL & INSTALLATION

1. Before servicing the vehicle, refer to the Precautions Section.
2. Remove:
   - Front wheels.
   - Ball joint clamping bolt.
   - Ball joint from the steering knuckle, by prying the lower control arm down.
   - Ball joint-to-lower control arm retaining nuts and bolts or drill out the rivets with a ¼ in. (6mm) drill.
   - Ball joint.

To install:

3. Install the new ball joint in the reverse order of removal. If no parts were installed other than the ball joint, no camber adjustment is necessary. Tighten the 2 control arm-to-ball joint bolts to 18 ft. lbs. (25 Nm) and the ball joint clamping bolt to 37 ft. lbs. (50 Nm).

Lower Control Arm

REMOVAL & INSTALLATION

Cabrio, Golf and Jetta

When removing the driver's side control arm on Golf and Jetta equipped with an automatic transaxle, it may be necessary to lift the engine/transaxle. First support the engine from above or below. Remove the front left engine mounting nut and bolt, remove the rear mount and raise the engine to expose the front control arm bolt.

1. Raise and safely support the vehicle and remove the wheels.
2. Remove the ball joint clamping bolt and pry the control arm down.
3. Remove the rubber bushings to unfasten the stabilizer bar.
4. Remove the control arm mounting bolts and remove the control arm.

To install:

5. Installation is the reverse of removal. Tighten the following components:

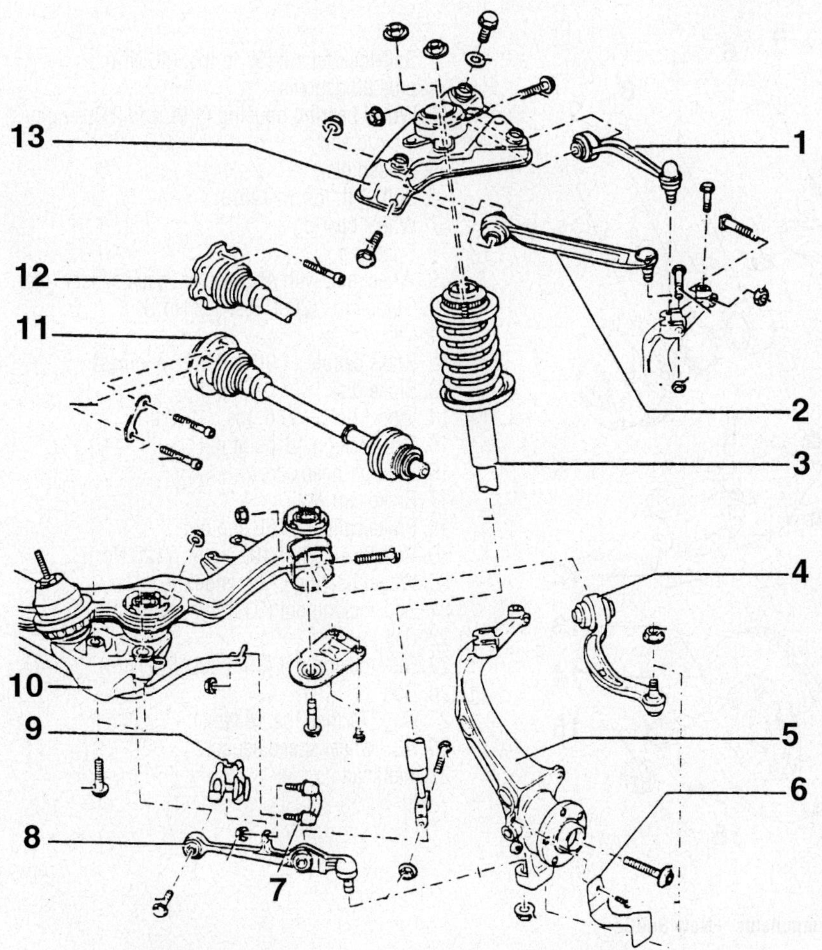

1. Upper link, rear
2. Upper link, front
3. Suspension strut
4. Guide link
5. Wheel bearing housing
6. Splash shield
7. Connecting link
8. Lower track control link
9. Clamp
10. Subframe
11. Halfshaft w/CV joint
12. Halfshaft w/triple-rotor joint
13. Mounting bracket

Exploded view of the front suspension—Passat

- Cabriolet control arm bushing bolts—50 ft. lbs. (68 Nm).
- Golf and Jetta front bushing bolts—96 ft. lbs. (130 Nm), rear bolts—59 ft. lbs. (80 Nm).
- Stabilizer bar link rods—18 ft. lbs. (25 Nm).
- Stabilizer bar bushing clamp bolts—32 ft. lbs. (43 Nm).
- Ball joint clamping bolt—37 ft. lbs. (50 Nm).

BUSHING REPLACEMENT

1. Remove the control arm.
2. Position the control arm on a press. Carefully push the bushing out of the control arm using the press.

To install:

3. Lightly lubricate, then position the bushing on the control arm. On Golf and Jetta models, align one arrow with the dimple on the control arm (the kidney shaped opening in the bushing must face the center of the vehicle when the control arm is installed).
4. Carefully press the bushing into the control arm.
5. Install the control arm.

Wheel Bearings

ADJUSTMENT

The front wheel bearings are sealed, no adjustment is necessary or possible.

REMOVAL & INSTALLATION

Except Passat

➡The hub and bearing are pressed into the knuckle and the bearing cannot be reused once the hub has been removed.

1. Before servicing the vehicle, refer to the Precautions Section.
2. With the vehicle on the ground, remove the front axle nut.
3. Remove the steering knuckle.
4. Clamp the upper knuckle-to-strut bolt boss in a vice.
5. Install the special press tool onto the hub as shown and press the hub out of the bearing.
6. If the inner bearing race stayed on the hub, clamp the hub in a vise and use a bearing puller to remove it.
7. On the knuckle, remove the splash shield and internal snaprings from the bearing housing.
8. After removing the snapring, the same press tool can be used to push the bearing out of the knuckle.
9. Clean the bearing housing and hub with a wire brush and inspect all parts. Replace parts that have been distorted or discolored from heat. If the hub is not

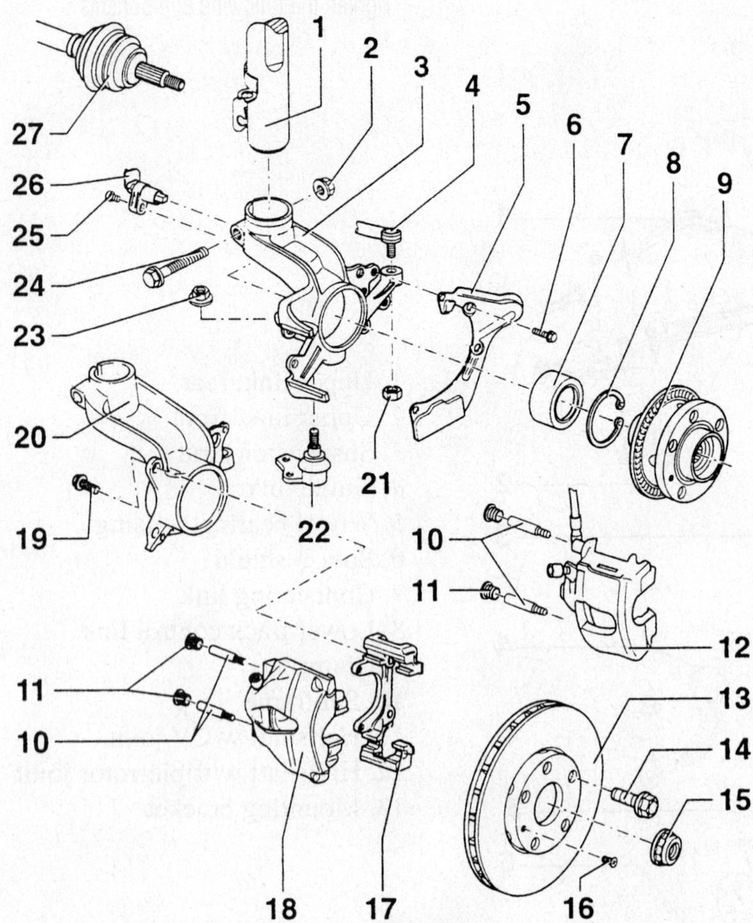

1. Strut
2. Self-locking nut (37 ft. lbs. (50 Nm) plus 90 degrees)
3. Wheel bearing housing (1.9L and 2.0L engines)
4. Tie rod end
5. Splash plate
6. Bolt (7 ft. lbs. (10 Nm))
7. Wheel bearing
8. Snapring
9. Wheel hub with ABS wheel speed sensor rotor
10. Guide pins (20 ft. lbs. (28 Nm))
11. Cap
12. Brake caliper (1.9L and 2.0L engines)
13. Brake disk
14. Wheel bolts (89 ft. lbs. (120 Nm))
15. Self-locking 12-point nut
16. Phillips-head screw
17. Brake carrier
18. Brake caliper (1.8L engine)
19. Self-locking bolt (92 ft. lbs. (125 Nm))
20. Wheel bearing (1.8L engine)
21. Self-locking nut (33 ft. lbs. (45 Nm))
22. Ball joint
23. Self-locking nut (33 ft. lbs. (45 Nm))
24. Bolt
25. Bolt (71 inch lbs. (8 Nm))
26. ABS wheel speed sensor
27. Halfshaft

9301WG36

Exploded view of the front wheel bearing and related components—New Beetle

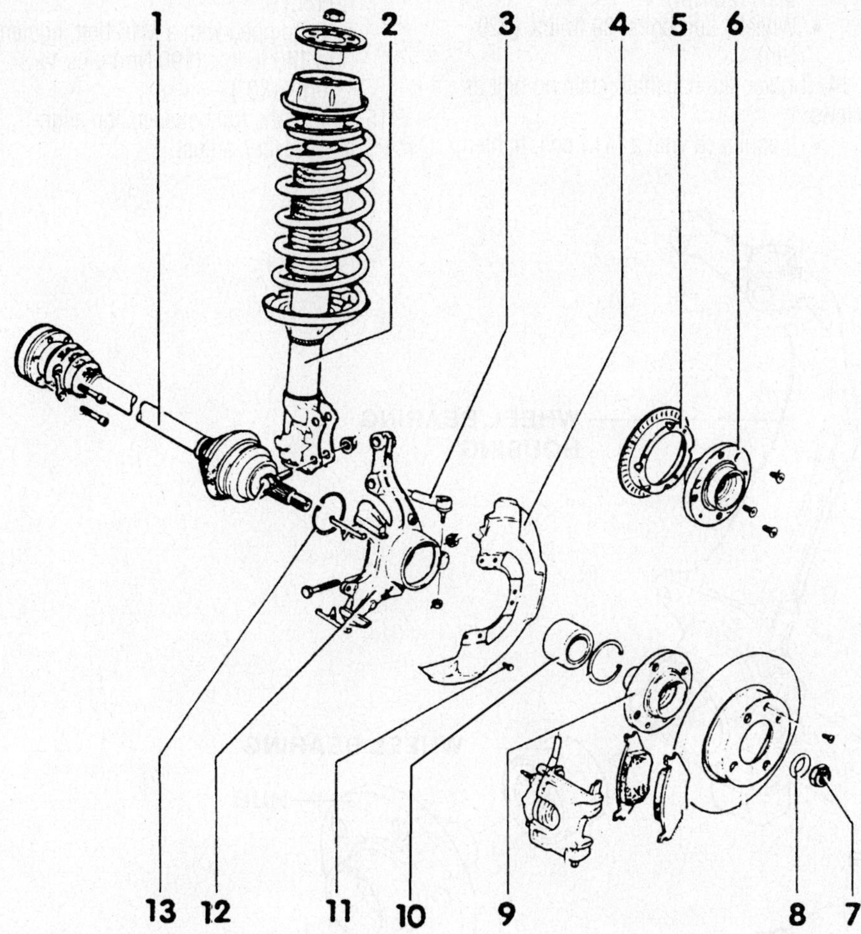

1. Halfshaft
2. Strut
3. Tie rod end
4. Splash plate
5. ABS wheel speed sensor
6. Wheel hub w/ABS
7. Axle nut (195 ft. lbs. (265 Nm))
8. Washer
9. Wheel hub wo/ABS
10. Snapring
11. Wheel bearing
12. Wheel bearing housing
13. Snapring

Exploded view of the front wheel bearing and related components—Cabrio, Golf, GTI and Jetta

absolutely prefect where it contacts the inner bearing race, the new bearing will fail quickly.

To install:

10. The new bearing is pressed in from the hub side using a regular arbor press. Install the snapring and support the steering knuckle on the press.
11. Using the old bearing as a press tool, press the new bearing into the housing up against the snapring. Be sure the press tool contacts only the outer race of the bearing.
12. Install the outer snapring and splash shield. If removed, install the speed sensor rotor onto the hub.
13. Support the inner race on the press and press the hub into the bearing. Be sure the inner race is supported or the bearing fail quickly.
14. Install the steering knuckle and be sure to tighten the axle nut correctly before allowing the vehicle to roll.

Passat

1. Before servicing the vehicle, refer to the Precautions Section.
2. Loosen the halfshaft retaining bolt.
3. Remove or disconnect:
   - Front wheel.
   - ABS wheel speed sensor.
   - Caliper bracket mounting bolts, then the rotor.
   - Brake splash guard.
4. Loosen the mounting nuts for the lower guide and track links.
5. Remove or disconnect:
   - Tie rod end from the wheel bearing housing.
   - Mounting nuts for the lower guide and track links and press out the joints.
   - Upper control arm pinchbolt and disconnect the arms.
   - Wheel bearing housing.
6. Place the wheel bearing housing on a press.
7. Drive out the hub with the wheel bearing.
8. Using a bearing separator and press, drive hub out of the bearing.

To install:

9. Press the new wheel bearing into the bearing housing using the appropriate bearing driver.
10. Press the hub into the wheel bearing using the appropriate bearing driver.
11. Install the wheel bearing housing.
12. Slide the CV-joint through the wheel hub and hand-tighten the new nut.
13. Install or connect:
    - Lower track control and guide link. New self-locking nut: 74 ft. lbs. (100 Nm).
    - Both of the upper link ball joints into the wheel bearing. Pinchbolt: 30 ft. lbs. (40 Nm).

- Tie rod end. New self-locking nut: 37 ft. lbs. (50 Nm). Bolt: 44 inch lbs. (5 Nm).
- ABS wheel speed sensor.
- Brake splash guard. Retaining bolts: 84 inch lbs. (10 Nm).
- Brake rotor.
- Brake caliper. Retaining bolt: 89 ft. lbs. (120 Nm).
- Wheels. Lug bolts: 89 ft. lbs. (120 Nm).

14. Tighten the halfshaft retaining bolt as follows:

- If equipped with a M14 bolt, tighten to 85 ft. lbs. (115 Nm) plus ½ turn (180°)
- If equipped with a M16 bolt, tighten to 140 ft. lbs. (190 Nm) plus ½ turn (180°)

15. Check the front suspension alignment, if necessary, adjust.

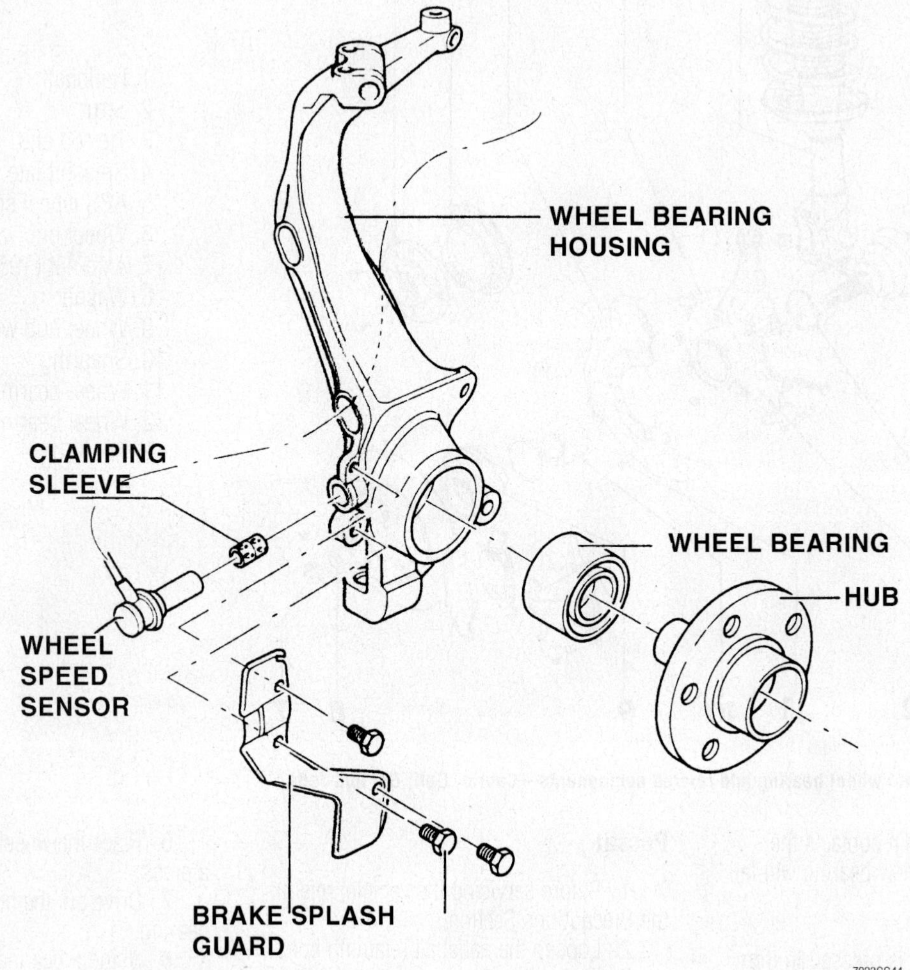

Exploded view of the front wheel bearing housing—Passat

REAR SUSPENSION

Strut

REMOVAL & INSTALLATION

⁂ WARNING

Do not remove both suspension struts at the same time or the axle beam will be hanging on the brake lines.

1. Before servicing the vehicle, refer to the Precautions Section.
2. On Cabrio, Golf, Jetta, perform the following:
   a. Working inside the vehicle, remove the cap from the top shock mount and remove the top nut, washer and rubber bushings.
   b. Remove the second nut.
   c. Slowly lift the vehicle and safely support it. Do not place supports under the axle beam.
3. On the New Beetle and Passat, raise the vehicle and remove the upper mounting bolts through the wheel opening.
4. Unbolt the strut from the axle and carefully remove the strut and spring from the vehicle. It may be necessary to press the axle down slightly.

To install:

5. Install the shock on the axle assembly. Do not tighten the nut until the vehicle is on the floor at normal riding height.
6. On the Cabrio, Golf, Jetta, perform the following:
   a. Install the upper end of the strut to the body. Lower nut: 11 ft. lbs. (15 Nm) and Upper nut: 18 ft. lbs. (25 Nm).
   b. Install the wheel and lower the vehicle to the floor.
   c. Tighten the lower strut mounting nut to 52 ft. lbs. (70 Nm).
7. On the New Beetle and Passat, perform the following:
   a. Install the upper end of the strut to the body and tighten the attaching bolt to 55 ft. lbs. (75 Nm).
   b. Install the wheel.
   c. Tighten the lower strut mounting nut to 44 ft. lbs. (60 Nm).

Coil Spring

REMOVAL & INSTALLATION

Cabrio, Golf, 2002–05 GTI, 2002–04 Jetta, New Beetle and Passat

1. Before servicing the vehicle, refer to the Precautions Section.
2. Remove the strut from the vehicle.
3. Clamp the Spring Compressor VAG 1752/2 or equivalent in a vise.
4. Install the strut into the spring compressor.
5. Pry off the mounting bolt cap.
6. Compress the coil spring and remove the self-locking nut from the piston rod.
7. Matchmark the position of the spring retainer and spring mount.
8. Remove:
   - Spring seat and related components noting the order of removal.
   - Strut from the spring compressor.
   - Spring out of the compressor.

To install:

9. Install the new spring into the compressor.
10. Compress the spring and insert the strut through the spring.
11. Install the spring seat and related components in the reverse order as they were removed and aligning the matchmarks.

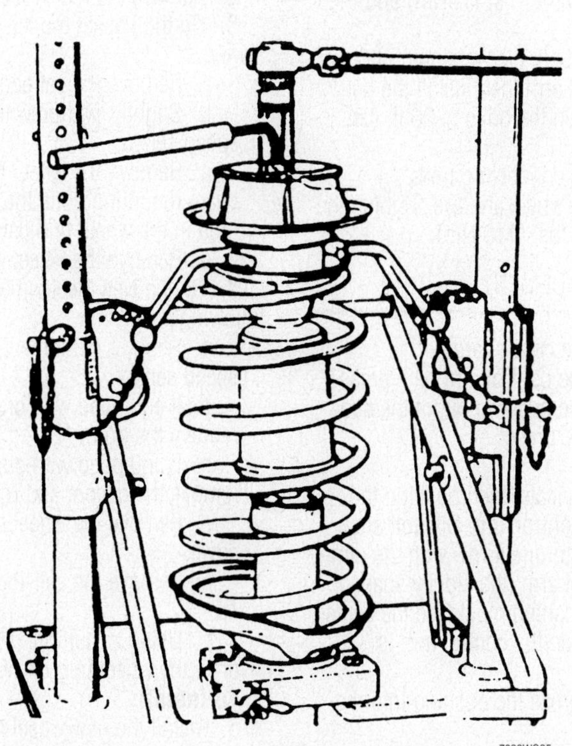

Compress the coil spring before removing the upper strut rod nut

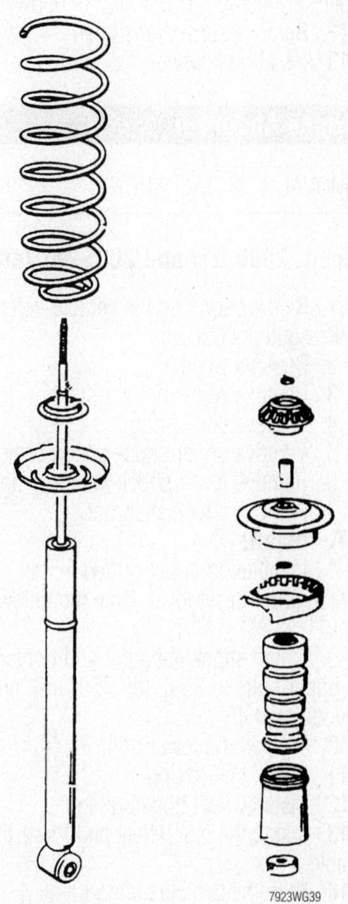

Exploded view of the rear strut—except Passat

12. Install a new self-locking nut.
13. Reinstall the mounting bolt cap.
14. Release the spring compressor and install the strut into the vehicle.

Rabbit, 2006 GTI and 2005–06 Jetta

1. Before servicing the vehicle, refer to the Precautions Section.
2. Remove wheel.
3. Insert spring compressor V.A.G 1752/3A.
4. Use wrench or ratchet for tightening spring compressor.
5. Compress coil spring until it can be removed.
6. Remove spring.

To install:

➡Note installation position.

7. The spring start must touch bottom of spring seat.
8. Install spring together with spring seat.
9. Lower spring support has a pin.
10. Insert this pin into hole of lower transverse link, then insert upper spring support into upper end of spring.
11. Release tension of spring, guiding upper spring support onto tab of body.
12. Remove spring tensioner.
13. Install rear wheel.

Shock Absorber

REMOVAL & INSTALLATION

Rabbit, 2006 GTI and 2005–06 Jetta

1. Before servicing the vehicle, refer to the Precautions Section.
2. Remove wheel.
3. Remove wheel housing liner.
4. Remove coil spring.
5. Remove upper shock absorber bolts.
6. Remove lower shock absorber bolt.
7. Remove shock absorber.

To install:

8. Installation is performed in the reverse order of removal. Note the following:
9. Install shock absorber and tighten the upper bolts to 35 ft. lbs. (50 Nm), plus an additional 45°.
10. Tighten the lower bolt.
11. Install coil spring.
12. Install wheel housing liner.
13. Install the rear wheel and lower the vehicle.
14. With the vehicle sitting on all 4 wheels, tighten the lower shock absorber bolt to 17 ft. lbs. (25 Nm).

Lower Control Arm

REMOVAL & INSTALLATION

Cabrio, Golf, GTI and Jetta

1. Chock the rear wheels and then lift the vehicle supporting it on jack stands.
2. Remove the front wheel.
3. Base suspension models require the removal of the connecting link to the control arm.
4. Plus suspension models require the removal of the nut that attaches the stabilizer bar to the control arm.
5. Separate the link from the arm.
6. Matchmark the correct installed position of the ball joint into the control arm.
7. Remove the ball joint.
8. Remove the pivot bolt from the control arm.
9. Remove the rear control arm mounting bolt.
10. Pull the rear control arm from the vehicle.

To install:

11. Install the control arm and slide the ball joint into it.
12. Push the control arm pivot bolt through the control arm. Also install the rear mounting bolt. Tighten the pivot bolt to 37 ft. lbs. (50Nm) and then turn it another 90°. The rear mounting bolt is to be tightened to 52 ft. lbs. (70Nm) and then turn it 90°.
13. Align the ball joint retaining plate on top of the control arm then install the ball joint bolts. Tighten the bolts to 26 ft. lbs. (35Nm).
14. Install the connecting links.
15. Install the wheel and tire. Tighten the lug nuts to 81 ft. lbs. (110 Nm).

BUSHING REPLACEMENT

1. Remove the control arm.
2. Position the control arm on a press. Carefully push the bushing out of the control arm using the press.

To install:

3. Lightly lubricate, then position the bushing on the control arm. On Golf and Jetta models, align one arrow with the dimple on the control arm (the kidney shaped opening in the bushing must face the center of the vehicle when the control arm is installed).
4. Carefully press the bushing into the control arm.
5. Install the control arm.

Wheel Bearings

ADJUSTMENT

The New Beetle, Passat, 2006 Rabbit/GTI and 2005–06 Jetta are equipped with non-adjustable wheel bearings, no adjustment is possible or required.

1. Before servicing the vehicle, refer to the Precautions Section.
2. Remove the grease cap.
3. Remove the cotter pin and the locking nut.
4. While turning the wheel, so the wheel bearing does not jam, tighten the adjusting nut firmly.
5. Back the nut off slightly. The nut is properly adjusted when it is possible to pry the thrust washer side to side with some drag by using finger pressure on the tool.
6. Install the locking nut and a new cotter pin. When installing the cap, be sure it is securely in place.

REMOVAL & INSTALLATION

New Beetle and Passat

The Beetle and Passat are equipped with a sealed bearing. The wheel bearing, wheel hub and wheel speed sensor are replaced as an assembly.

1. Before servicing the vehicle, refer to the Precautions Section.
2. Remove the rear wheels.
3. On the Passat model, perform the following:
   a. Remove the caliper and rotor.
   b. Slightly, withdraw the ABS wheel speed sensor.
   c. Remove the wheel hub-to-axle beam mounting bolts through the openings in the wheel hub flange.
   d. Remove the wheel hub.
4. On the New Beetle model, perform the following:
   a. Slightly, withdraw the ABS wheel speed sensor.
   b. If equipped with drum brakes, remove the drum.
   c. If equipped with disk brakes, remove the caliper and rotor.
   d. Remove the wheel hub center dust cover.
   e. Remove the self-locking 12-point nut.
   f. Using a multiple jaw puller, withdraw the wheel hub off of the stub axle.

To install:

5. Install the new wheel hub and tighten the mounting fasteners as follows:

- Passat—44 ft. lbs. (60 Nm)
- New Beetle—129 ft. lbs. (175 Nm)

6. Install:
   - Drum or caliper and rotor.
   - If equipped, center dust cap.
   - If removed, ABS wheel speed sensor.
   - Wheels.
7. If necessary, check and adjust the wheel alignment.

Cabrio, Golf, 2002–05 GTI and 2002–04 Jetta

1. Before servicing the vehicle, refer to the Precautions Section.
2. Remove the rear wheels.
3. On drum brakes, insert a small pry-tool through a wheel bolt hole and push up on the adjusting wedge to slacken the rear brake adjustment.
4. Remove or disconnect:
   - On disc brakes, the caliper.
   - Grease cap, cotter pin, locking ring, axle nut and thrust washer. Carefully remove the bearing and put all these parts where they will stay clean.

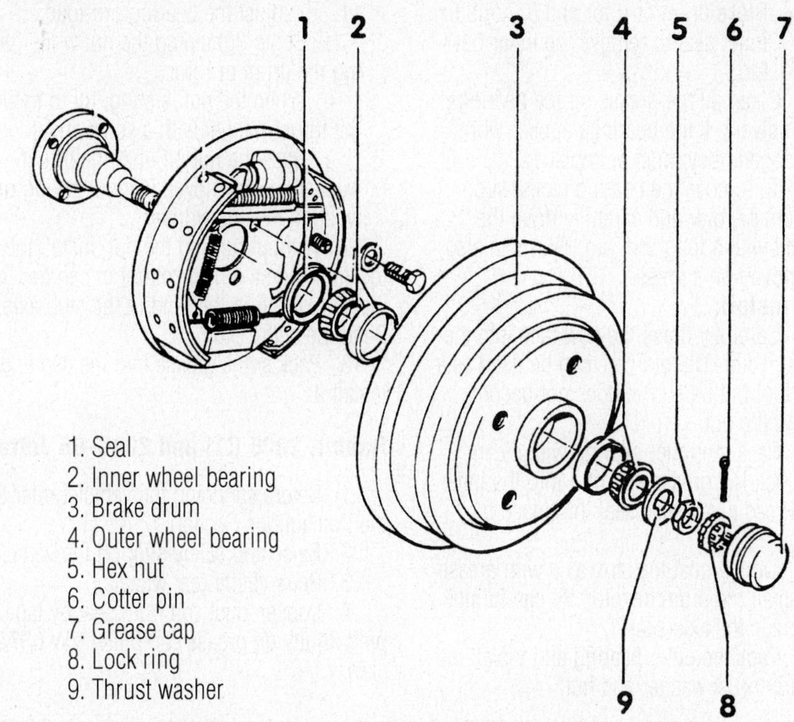

Exploded view of the rear wheel bearing—Cabrio, Golf, GTI and Jetta w/drum brakes

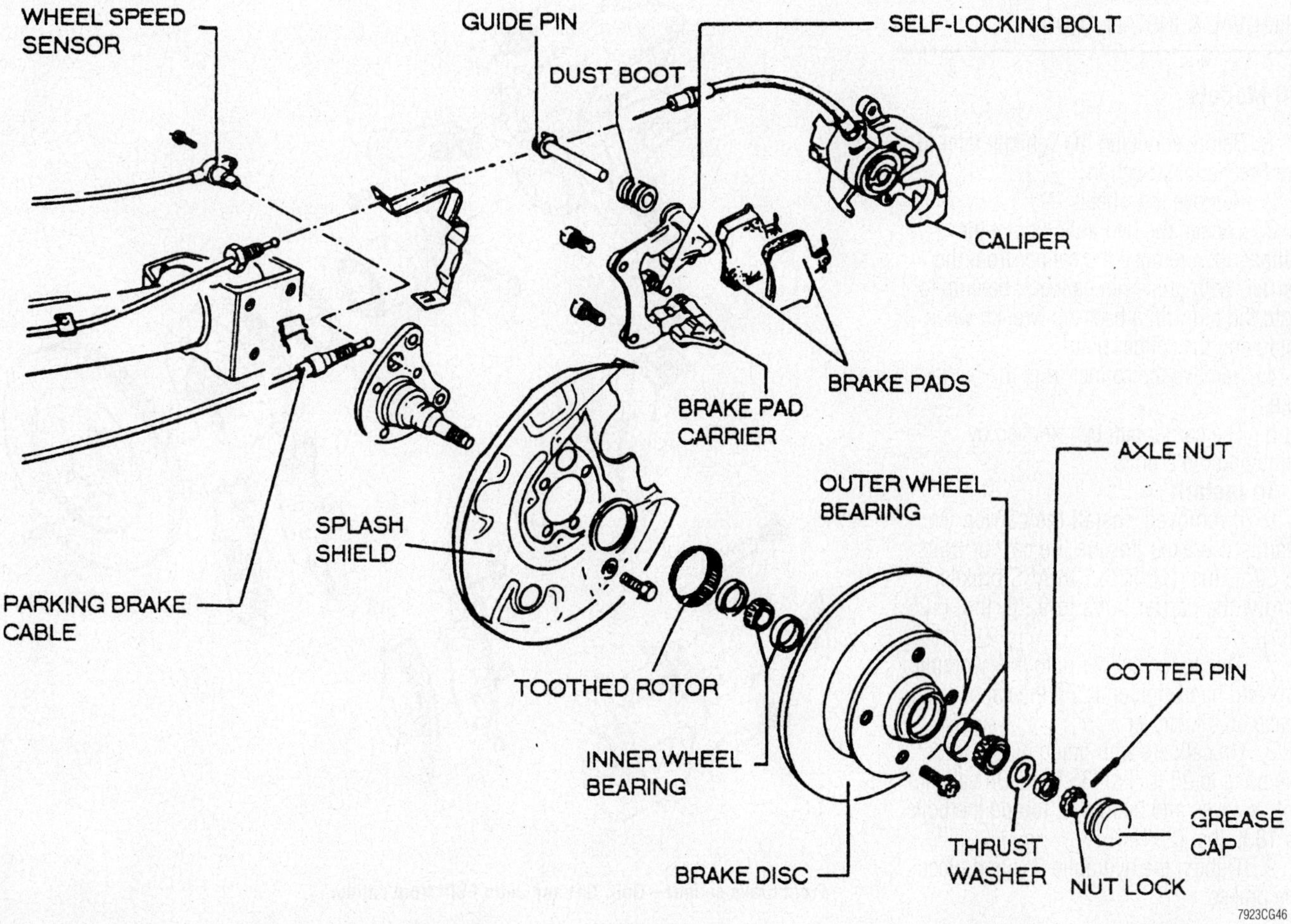

WHEEL SPEED SENSOR
GUIDE PIN
SELF-LOCKING BOLT
DUST BOOT
CALIPER
BRAKE PADS
BRAKE PAD CARRIER
AXLE NUT
OUTER WHEEL BEARING
SPLASH SHIELD
PARKING BRAKE CABLE
TOOTHED ROTOR
COTTER PIN
INNER WHEEL BEARING
GREASE CAP
THRUST WASHER
BRAKE DISC
NUT LOCK
7923CG46

Exploded view of the rear wheel bearing—Cabrio, Golf, GTI AND Jetta w/disc brakes

- Brake drum or rotor and pry out the inner seal to remove the inner bearing.

5. Clean all the grease off the bearings using solvent. If the bearings appear worn or damaged, they must be replaced.
6. To remove the bearing races, support the drum or rotor and carefully drive the race out with a long drift pin. They can also be removed on a press.

To install:

7. Carefully press the new race into the drum or rotor. The old race can be used as a press tool but be sure it does not become stuck in the hub.
8. Pack the inner bearing with clean wheel bearing grease and fit it into the inner race. Press a new axle seal into place by hand.
9. Lightly coat the stub axle with grease and install the drum or rotor. Be careful not to damage the axle seal.
10. Pack the outer bearing and install the bearing, thrust washer and nut.
11. To adjust the bearing pre-load:
    a. Begin tightening the nut while turning the drum or rotor.
    b. When the nut is snug, try to move the thrust washer with a screwdriver.
    c. Back the nut off until the thrust washer can be moved without prying or twisting the screwdriver.
12. Without turning the nut, install the locking ring so a new cotter pin can be installed through the hold in the stub axle. Bend the cotter pin.
13. Pack some grease into the cap and install it.

Rabbit, 2006 GTI and 2005–06 Jetta

1. Before servicing the vehicle, refer to the Precautions Section.
2. Raise and safely support the vehicle.
3. Remove the rear wheels.
4. Loosen dust cap from seat by tapping lightly on grease cap puller VW 637/2 claw.
5. Pull off dust cap.
6. Remove brake carrier with brake caliper and hang on body with tie wire.
7. Remove Phillips-head screw for brake disc and take out brake disc.
8. Remove socket head bolt using bit XZN 18 T10162.
9. Pull wheel hubs/wheel bearing unit off stub axle.

To install:

10. Carefully slide wheel hubs/wheel bearing unit onto stub axle.
11. Make sure that wheel hubs/wheel bearing unit do not cant.
12. Use new socket head bolt and tighten to 124 ft. lbs. (180 Nm), plus an additional 180°.
13. Install brake disc and caliper.
14. Drive dust cap on using installation sleeve 3241/4. Always replace dust caps.
15. Install rear wheels.
16. Lower the vehicle.

FRONT BRAKES

Brake Caliper

REMOVAL & INSTALLATION

All Models

1. Before servicing the vehicle, refer to the Precautions Section.
2. Remove the wheels.
3. Loosen the hydraulic line at the caliper, then remove the caliper from the carrier. With guide pin calipers, be sure to hold the pin with a back-up wrench when removing the caliper bolts.
4. Remove the caliper from the hydraulic line.
5. The carrier can be removed by removing the 2 bolts.

To install:

6. If removed, install the carrier. On standard brakes, torque the carrier bolts to 52 ft. lbs. (70 Nm). On ABS brakes, torque the carrier bolts to 92 ft. lbs. (125 Nm).
7. Thread the caliper onto the hydraulic line and hand-tighten it. Fit the caliper into place on the carrier.
8. On calipers with guide pins, torque the bolts to 25 ft. lbs. (35 Nm). On calipers with sleeves and bushings, torque the bolts to 18 ft. lbs. (25 Nm).
9. Tighten the hydraulic line and bleed the brakes.

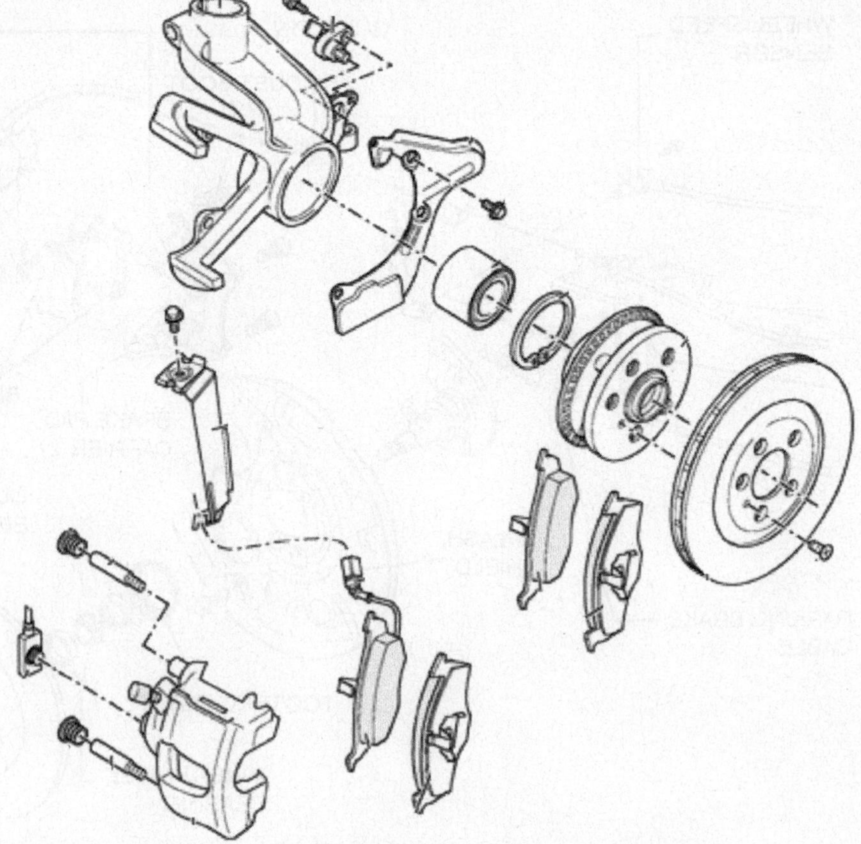

Front brake caliper—Golf, GTI and Jetta FSIII front caliper

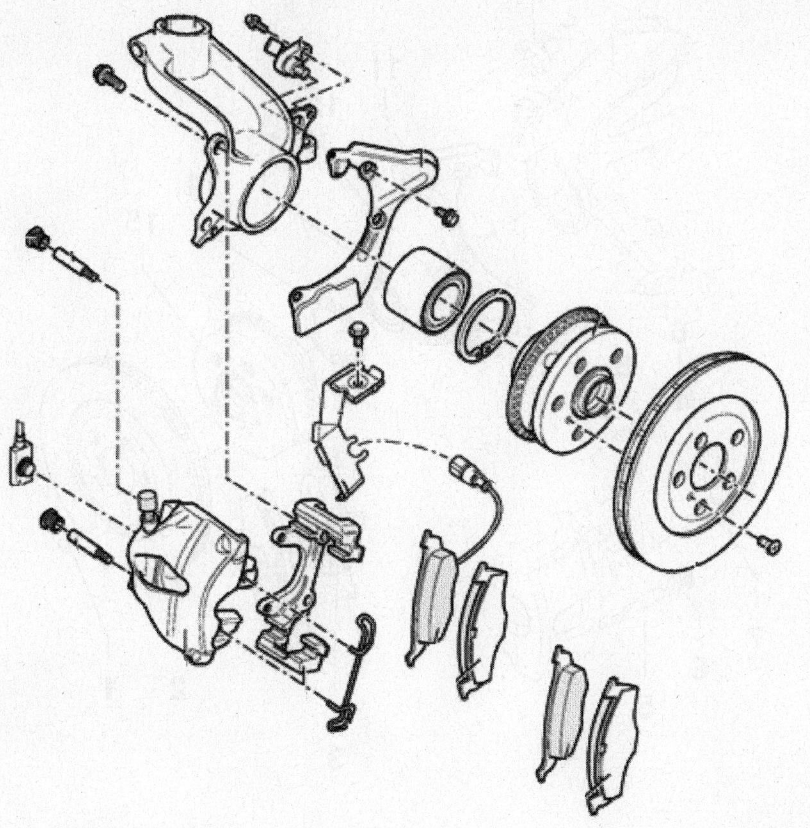

Front brake caliper—Golf, GTI and Jetta FN3 front caliper

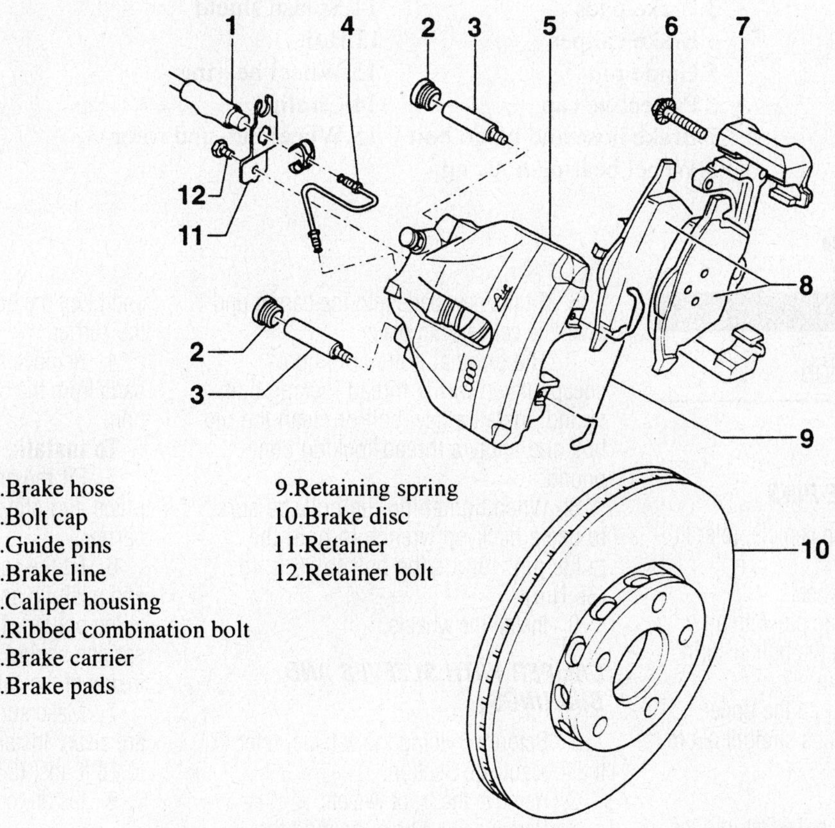

Front brake caliper—Passat

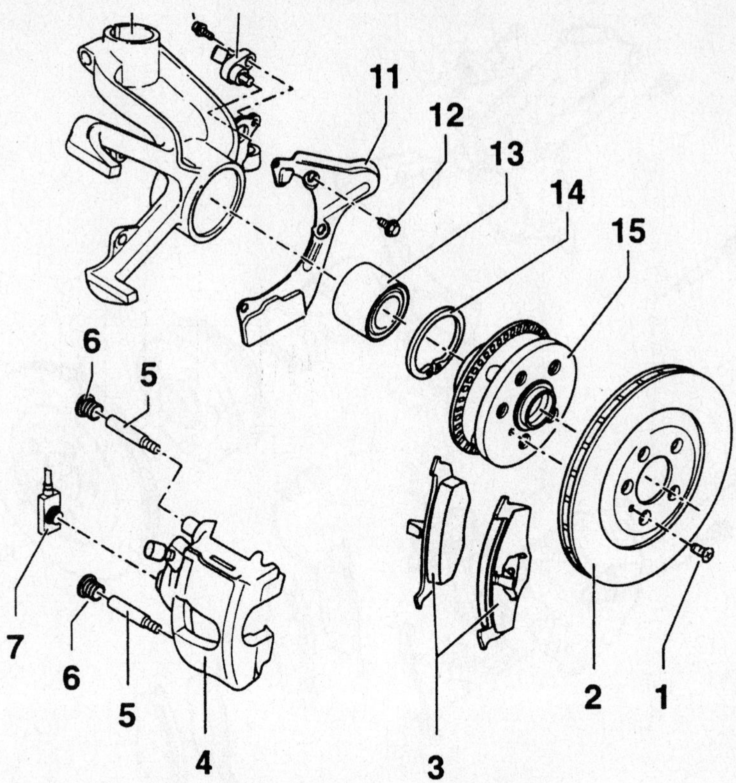

1.Screw
2.Brake disc
3.Brake pads
4.Brake caliper
5.Guide pin
6.Protective cap
7.Brake hose and banjo bolt
8.Wheel bearing housing
9.Socket head bolt
10.ABS sensor
11.Splash shield
12.Bolt
13.Wheel bearing
14.Circlip
15.Wheel hub and rotor

93016G73

Front caliper—New Beetle

Brake Pads

REMOVAL & INSTALLATION

All Models

CALIPER WITH GUIDE PINS

1. Before servicing the vehicle, refer to the Precautions Section.
2. Remove the front wheels.
3. Hold the lower guide pin with an open wrench and remove the bolt securing the caliper to the guide pin.
4. Pivot the caliper up on the upper guide pin and slide the pads straight out to remove them.

To install:

5. Compress the caliper piston into the bore.
6. Fit the new pads into the carrier and pivot the caliper into place.
7. The original bolts are micro-encapsulated with a thread locking compound. Install a new bolt or clean the old bolt and apply a thread-locking compound.
8. When tightening the bolt, be sure to use a back-up wrench to hold the guide pin. Torque the bolt to 26 ft. lbs. (35 Nm).
9. Install the wheels.

CALIPER WITH SLEEVES AND BUSHINGS

1. Before servicing the vehicle, refer to the Precautions Section.
2. Remove the front wheels.
3. Remove the 2 bolts holding the caliper to the carrier. Push the caliper up and pivot the bottom of the caliper out of the carrier.
4. Remove the anti-rattle springs and the pads from the carrier and note their location.

To install:

5. Fit the anti-rattle springs into place and slide the new pads onto the carrier.
6. Fit the caliper into place at the top and push up so it can be pivoted into place at the bottom. The tabs on the anti-rattle springs should be pushing against the inside of the caliper.
7. Make sure the caliper mounting bolts are clean. Install the bolts and torque them to 18 ft. lbs. (25 Nm).
8. Install the wheels.

REAR BRAKES

Brake Caliper

REMOVAL & INSTALLATION

All Models

1. Before servicing the vehicle, refer to the Precautions Section.
2. If equipped with ABS, make sure the ignition switch stays **OFF** and pump the brake pedal 25–35 times to relieve the system pressure.
3. Remove the wheels.
4. Disconnect the parking brake cable.
5. Loosen the hydraulic line.
6. Use a back-up wrench to hold the guide pins and remove the caliper bolts.
7. Lift the caliper off the carrier and unscrew it from the hydraulic line.

To install:

8. Thread the caliper onto the hydraulic line and hand-tighten it. Fit the caliper into place on the carrier. Torque the bolts to 26 ft. lbs. (35 Nm).
9. Bleed the brakes.
10. Install the wheels.

Brake Pads

REMOVAL & INSTALLATION

All Models

1. Before servicing the vehicle, refer to the Precautions Section.
2. Remove the rear wheels.
3. Remove the parking brake cable clip from the caliper. Disconnect the parking brake cable.
4. Hold the guide pin with a back-up wrench and remove the upper mounting bolt from the brake caliper.
5. Swing the caliper downward and remove the brake pads.

To install:

6. Retract the piston into the housing by rotating the piston clockwise.
7. Install the new brake pads onto the pad carrier.
8. Install the caliper to the pad carrier using a new self locking bolt or a thread locking compound and torque to 26 ft. lbs. (35 Nm).
9. Attach the hand brake cable to the caliper.
10. Check the parking brake operation and adjust the cable if necessary.
11. Install the wheels.

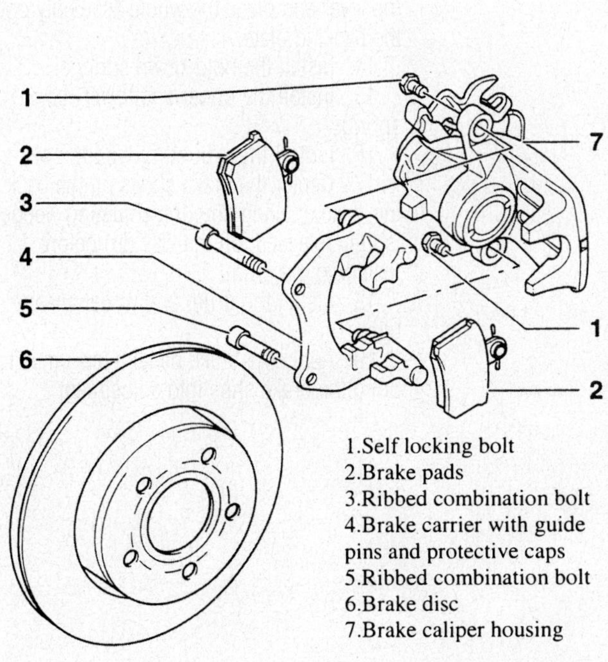

Rear disc brakes—Passat

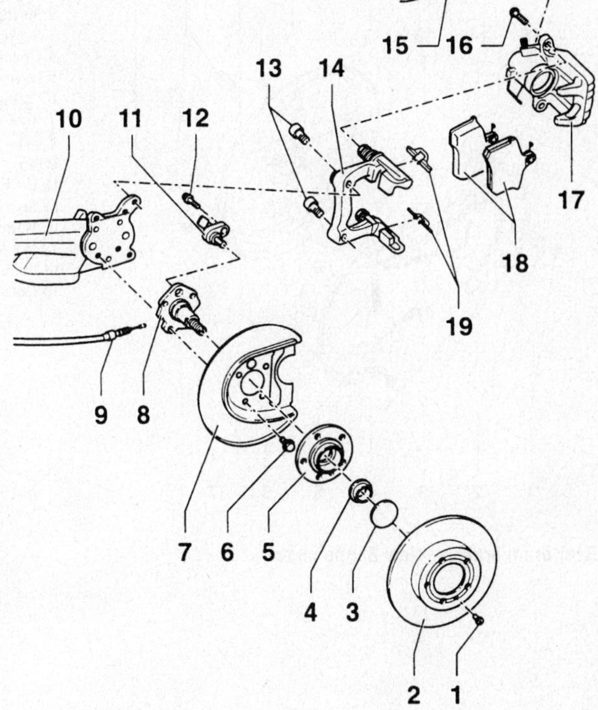

Rear disc brakes—Golf, GTI, Jetta and New Beetle

Brake Drums

REMOVAL & INSTALLATION

All Models

1. Before servicing the vehicle, refer to the Precautions Section.
2. Remove the rear wheels.
3. Insert a small pry tool through a wheel bolt hole and push up on the adjusting wedge to slacken the rear brake adjustment.
4. Remove the grease cap, cotter pin, locking ring, axle nut, thrust washer, and outer bearing.
5. Remove the drum.

To install:

6. Install the drum. Install the outer bearing, washer, and nut.
7. Install the lock ring and cotter pin. Install the grease cap.
8. Install the wheels.

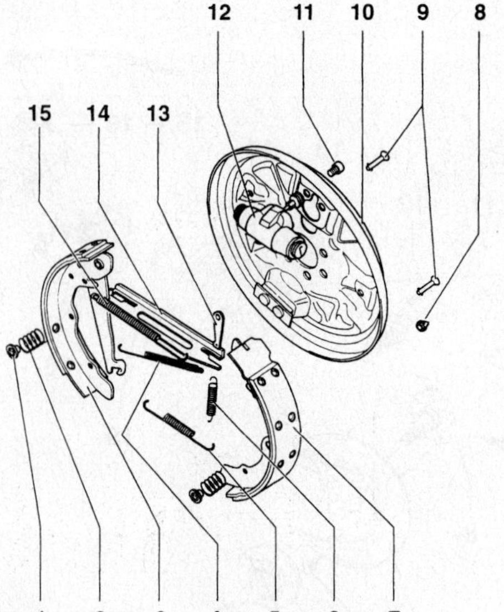

Rear drum brakes—New Beetle shown

Brake Shoes

REMOVAL & INSTALLATION

All Models

1. Before servicing the vehicle, refer to the Precautions Section.
2. Remove the rear wheels.
3. Remove the rear brake drum.
4. Remove the brake shoe hold-down spring retainers.
5. Remove the shoes from the back plate by pulling first 1 shoe, then the other against the upper spring and from its wheel cylinder slot. Detach the parking brake cable from the brake lever. Remove the brake shoe assembly from the vehicle.
6. Clamp the pushrod that holds the shoes apart at the top in a vise and begin removing the springs. Start with the lower return spring, adjusting wedge spring, upper return spring and then the tensioning spring and adjusting wedge.
7. On most vehicles, the parking brake lever must be removed from the old shoes and reused.

To install:

8. Clean the back plate and lubricate the shoe contact points with a suitable brake lubricant.
9. With the push rod clamped in a vise, attach the front brake shoe and tensioning spring.
10. Insert the adjusting wedge between the front shoe and pushrod so its lug is pointing toward the backing plate.
11. Remove the parking brake lever from the old shoe and attach it onto the new rear brake shoe.
12. Put the rear brake shoe and parking brake lever assembly onto the pushrod and hook up the spring.
13. Connect the parking brake cable to the lever and place the whole assembly onto the backing plate.
14. Install the hold-down springs.
15. Install the upper and lower return springs.
16. Install the adjusting wedge spring.
17. Center the brake shoes on the backing plate, making sure the adjusting wedge is fully released (all the way up) before installing the drum.
18. Install the drum and wheel assembly.
19. Apply the brake pedal a few times to bring the brake shoe into adjustment.

VOLKSWAGEN

12

Eurovan

SPECIFICATIONS AND MAINTENANCE CHARTS

ENGINE AND VEHICLE IDENTIFICATION

Engine							Model Year	
Code	Liters (cc)	Cu. In.	Cyl.	Fuel Sys.	Engine Type	Eng. Mfg.	Code ①	Year
AXK	2.8 (2792)	170	6	MFI	DOHC	Volkswagen	2	2002
							3	2003

MFI: Multi-point Fuel Injection

DOHC: Dual Overhead Camshaft

① 10th digit of the Vehicle Identification Number (VIN)

06041_EURO_C0001

GENERAL ENGINE SPECIFICATIONS

Year	Model	Engine Displacement Liters	Engine ID	Net Horsepower @ rpm	Net Torque @ rpm (ft. lbs.)	Bore x Stroke (in.)	Com-pression Ratio	Oil Pressure @ rpm
2002	Eurovan	2.8	AXK	201@6200	181@2500	3.19x3.55	10.7:1	44-80@2000
2003	Eurovan	2.8	AXK	201@6200	181@2500	3.19x3.55	10.7:1	44-80@2000

06041_EURO_C0002

ENGINE TUNE-UP SPECIFICATIONS

Year	Engine Displacement Liters	Engine ID	Spark Plug Gap (in.)	Ignition Timing (deg.)		Fuel Pump (psi)	Idle Speed (rpm)		Valve Clearance	
				MT	AT		MT	AT	In.	Ex.
2002	2.8	AXK	0.039	—	①	36	—	650-750	HYD	HYD
2003	2.8	AXK	0.039	—	①	36	—	650-750	HYD	HYD

NOTE: The Vehicle Emission Control Information label reflects specification changes made during production and must be used if difernent from this chart.

NOTE: Fuel pump pressure specifications with the fuel pressure regulator vacuum hose attached.

HYD: Hydraulic

① The basic setting is controlled by the ECU and is not adjustable

06041_EURO_C0003

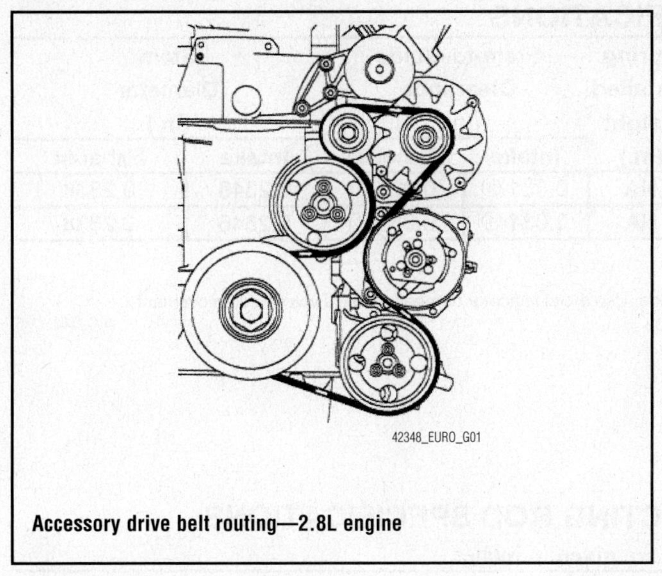

Accessory drive belt routing—2.8L engine

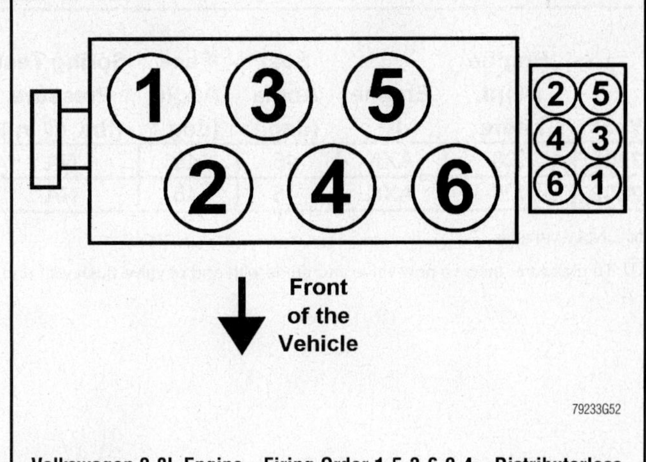

Volkswagen 2.8L Engine—Firing Order 1-5-3-6-2-4—Distributorless ignition system

CAPACITIES

Year	Model	Engine Displacement Liters	Engine ID	Engine Oil with Filter	Transmission (pts.) 5-Spd	Transmission (pts.) Auto	Drive Axle Front (pts.)	Drive Axle Rear (pts.)	Fuel Tank (gal.)	Cooling System (qts.)
2002	Eurovan	2.8	AXK	5.9	—	5.4	NA	NA	21.1	9.7
2003	Eurovan	2.8	AXK	5.9	—	5.4	NA	NA	21.1	9.7

NA: Information not available

06041_EURO_C0004

VALVE SPECIFICATIONS

Year	Engine Displ. Liters	Engine ID	Seat Angle (deg.)	Face Angle (deg.)	Spring Test Pressure (lbs. @ in.)	Spring Installed Height (in.)	Stem-to-Guide Clearance (in.)		Stem Diameter (in.)	
							Intake	Exhaust	Intake	Exhaust
2002	2.8	AXK	45	45	NA	NA	0.031 ①	0.031 ①	0.2346	0.2338
2003	2.8	AXK	45	45	NA	NA	0.031 ①	0.031 ①	0.2346	0.2338

NA: Not Available

① To measure: Insert a new valve into guide with end of valve flush with end of guide. Use a dial indicator to measure axial valve head movement.

06041_EURO_C0005

CRANKSHAFT AND CONNECTING ROD SPECIFICATIONS

All measurements are given in inches.

Year	Engine Displ. Liters	Engine ID	Crankshaft Main Brg. Journal Dia.	Crankshaft Main Brg. Oil Clearance	Crankshaft Shaft End-play	Crankshaft Thrust on No.	Connecting Rod Journal Diameter	Connecting Rod Oil Clearance	Connecting Rod Side Clearance
2002	2.8	AXK	2.3605-2.3613	0.0007-0.0023	0.0027-0.0090	5	2.1243-2.1251	0.0007-0.0027	0.0019-0.0122
2003	2.8	AXK	2.3605-2.3613	0.0007-0.0023	0.0027-0.0090	5	2.1243-2.1251	0.0007-0.0027	0.0019-0.0122

06041_EURO_C0006

PISTON AND RING SPECIFICATIONS

All measurements are given in inches

Year	Engine Displ. Liters	Engine ID	Piston Clearance	Ring Gap Top Compression	Ring Gap Bottom Compression	Ring Gap Oil Control	Ring Side Clearance Top Compression	Ring Side Clearance Bottom Compression	Ring Side Clearance Oil Control
2002	2.8	AXK	0.0010-0.0040	0.0078-0.0157	0.0078-0.0157-	0.0098-0.0197	0.0016-0.0035	0.0011-0.0023	0.0008-0.0023
2003	2.8	AXK	0.0010-0.0040	0.0078-0.0157	0.0078-0.0157-	0.0098-0.0197	0.0016-0.0035	0.0011-0.0023	0.0008-0.0023

06041_EURO_C0007

TORQUE SPECIFICATIONS

All readings in ft. lbs.

Year	Engine Displacement Liters	Engine ID	Cylinder Head Bolts	Main Bearing Bolts	Rod Bearing Bolts	Crankshaft Damper Bolts	Flywheel Bolts	Manifold Intake	Manifold Exhaust	Spark Plugs	Oil Drain Plug
2002	2.8	AXK	①	②	③	④	⑤	7	18	18	22
2003	2.8	AXK	①	②	③	④	⑤	7	18	18	22

① Step 1: 22 ft. lbs.
Step 2: 37 ft. lbs.
Step 3: 90 degrees
Step 4: 90 degrees

② Step 1: 22 ft. lbs.
Step 2: 180 degrees
Step 2: 180 degrees

③ Step 1: 22 ft. lbs.
Step 2: 90 degrees

④ Step 1: 74 ft. lbs.
Step 2: 90 degrees

⑤ Step 1: 45 ft. lbs.
Step 2: 90 degrees
Step 2: 90 degrees
Step 3: 90 degrees

06041_EURO_C0008

WHEEL ALIGNMENT

Year	Model		Caster Range (+/-Deg.)	Caster Preferred Setting (Deg.)	Camber Range (+/-Deg.)	Camber Preferred Setting (Deg.)	Toe-in (in.)
2002	Eurovan	F	—	—	0.33	-0.66	0.16 +/- 0.16
		R	—	—	0.33	+0.50	0.00 +/- 0.50
2003	Eurovan	F	—	—	0.33	-0.66	0.16 +/- 0.16
		R	—	—	0.33	+0.50	0.00 +/- 0.50

06041_EURO_C0009

TIRE, WHEEL AND BALL JOINT SPECIFICATIONS

Year	Model	OEM Tires Standard	OEM Tires Optional	Tire Pressures (psi) Front	Tire Pressures (psi) Rear	Wheel Size	Ball Joint Inspection	Wheel Lug Nut Torque (ft. lbs.)
2002	Eurovan	225/60HR16	—	34	34	7-J	NS	125
2003	Eurovan	225/60HR16	—	34	34	7-J	NS	125

NS: Not specified by the manufacturer
OEM: Original Equipment Manufacturer
PSI: Pounds Per Square Inch

06041_EURO_C0010

BRAKE SPECIFICATIONS

All measurements in inches unless noted

			Brake Disc			Minimum Lining Thickness		Brake Caliper	
Year	Model		Original Thickness	Minimum Thickness	Maximum Runout	Front	Rear	Bracket Bolts (ft. lbs.)	Mounting Bolts (ft. lbs.)
2002	Eurovan	F	1.023	0.866	0.002	0.080	—	—	18
		R	0.472	0.452	0.002	—	0.080	—	35
2003	Eurovan	F	1.023	0.866	0.002	0.080	—	—	18
		R	0.472	0.452	0.002	—	0.080	—	35

06041_EURO_C0011

SCHEDULED MAINTENANCE INTERVALS

Volkswagen—Eurovan

		VEHICLE MILEAGE INTERVAL (x1000)										
TO BE SERVICED	TYPE OF SERVICE	5	10	20	30	40	50	60	70	80	90	100
Engine oil & filter ①	R	✓	✓	✓	✓	✓	✓	✓	✓	✓	✓	✓
Automatic shiftlock operation	S/I		✓	✓	✓	✓	✓	✓	✓	✓	✓	✓
Cooling system	S/I			✓		✓		✓		✓		✓
Passenger compartment air filter	R			✓		✓		✓		✓		✓
Automatic transmission fluid, filter & final drive	S/I			✓		✓		✓		✓		✓
Battery electrolyte level	S/I			✓		✓		✓		✓		✓
Brake system (brake pads & fluid level)	S/I		✓	✓	✓	✓	✓	✓	✓	✓	✓	✓
Drive axle shaft boots	S/I			✓		✓		✓		✓		✓
Engine (check for leaks)	S/I			✓		✓		✓		✓		✓
Exhaust system	S/I			✓		✓		✓		✓		✓
Idle speed	S/I											
Manual transmission fluid	S/I											
OBD System check for codes	S/I			✓		✓		✓		✓		✓
V-belts	S/I			✓		✓				✓		✓
Air cleaner element	R					✓				✓		
Spark plugs	R					✓				✓		
Power steering fluid level	S/I					✓				✓		
Automatic transmission fluid	R					✓				✓		
Timing belt	R											
Brake Fluid ②	R					✓				✓		
Front axle dust seals on ball joints & tie rod ends	S/I					✓				✓		✓
Poly-ribbed belt	S/I					✓				✓		
Rotate tires	S/I		✓	✓	✓	✓	✓	✓	✓	✓	✓	✓

R: Replace   S/I: Service or Inspect

① Reset service interval display, if equipped.

② Replace every 2 years regardless of mileage.

FREQUENT OPERATION MAINTENANCE (SEVERE SERVICE)

If a vehicle is operated under any of the following conditions it is considered severe service:

- Extremely dusty areas.

- 50% or more of the vehicle operation is in 32°C (90°F) or higher temperatures, or constant operation in temperatures below 0°C (32°F).

- Prolonged idling (vehicle operation in stop and go traffic).

- Frequent short running periods (engine does not warm to normal operating temperatures).

- Police, taxi, delivery usage or trailer towing usage.

Oil & oil filter: change every 5000 miles.

06041_EURO_C0012

ENGINE REPAIR

Distributor

The engines in this section are equipped with a Distributorless Ignition System (DIS).

Alternator

REMOVAL & INSTALLATION

1. Before servicing the vehicle, refer to the Precautions Section.
2. Disconnect the negative battery cable.
3. Remove the engine undercover.
4. Disconnect the wiring from the alternator. If necessary, mark the wires with tape or other means to ensure they are connected properly upon installation.
5. Loosen the belt tension, and remove the drive belt from the alternator.
6. Remove the alternator adjustment bolt, followed by the pivot bolt.
7. Carefully lift the alternator from the bracket.

To install:

8. Hold the alternator in position on the mounting bracket, and install the pivot bolt.
9. Install (but do not tighten) the alternator adjustment bolt.
10. Place the alternator drive belt on the pulley.
11. Adjust belt tension and tighten the mounting and adjustment bolts as necessary.
12. Connect the wiring to the alternator. If tape was used to identify the wires, make sure it is removed once the wires are connected.
13. Connect the negative battery cable.

Ignition Timing

ADJUSTMENT

➡The ignition timing is controlled by the engine control module and is not adjustable. However the timing can be monitored on a scan tool connected to the DLC in the vehicle.

Engine Assembly

REMOVAL & INSTALLATION

1. Before servicing the vehicle, refer to the Precautions Section.

➡The engine and transaxle are lowered from the vehicle as an assembly.

2. Relieve fuel pressure, by opening the fuel filler cap. Be sure to take the appropriate fire safety precautions.
3. Remove or disconnect:
   - Battery cover.
   - Battery cables and battery.
   - Front engine cover.
   - Ignition coils cover.
   - Ignition coil connectors.
   - Mass air flow (MAF) sensor hose.
   - Throttle valve control module.
   - Upper radiator hose and wiring from the radiator fan motor and switches.
   - Radiator and cooling fan.
   - Intake hose to combination valve.
   - Secondary air injection pump.
   - Vacuum and vent hoses carefully labeling each one.
   - Manifold heat shield brace.
   - Electrical connections, carefully labeling each one.
   - Oxygen sensor connectors on subframe.
   - Fuel supply and return lines.
   - Remaining coolant hoses.
   - Power steering pump and secure it to the body. Do not disconnect the hydraulic lines.
   - Vehicle speed sensor at transaxle.
   - Front exhaust pipe.
   - Exhaust manifold.
   - Catalytic converter.
   - Air conditioning compressor and secure it aside without disconnecting the lines.
   - Halfshafts from the flanges and hang them from the body with wire.
4. Attach engine support tool 3408 on cylinder block.
5. Attach an engine/transaxle jack tool VAG-1383A into support tool.
6. Disconnect lower engine support and subframe and transaxle.
7. Remove all bolts and left and right subframes.

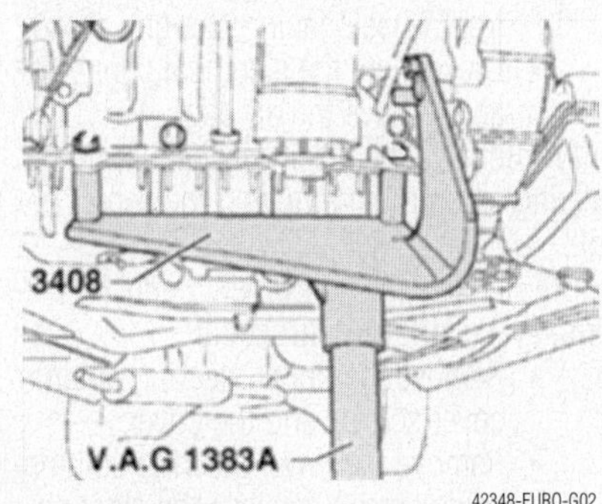

Attaching engine support tool 3408

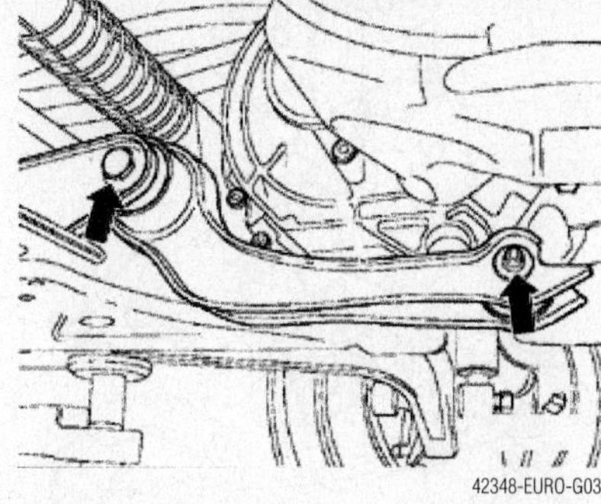

Removing engine support at subframe and transaxle

8. Carefully lower the engine/transaxle assembly and remove from vehicle.

To install:

9. Start all nuts and bolts that secure the subframe mounts to the body but don't tighten them yet.

➡When installing the right subframe mount, make sure the recess in engine bracket engages into the pin of the mount.

10. Carefully install the engine/transaxle.
11. With all mounts installed and the engine safely in the vehicle, allow some slack in the lifting equipment. With the vehicle safely supported, shake the engine/transaxle as a unit to settle it in the mounts.
12. Torque the subframe bolts as follows:
    - Right subframe nuts and bolt "A": 41 ft. lbs. (55 Nm); bolt "B": 37 ft. lbs. (50 Nm), plus 90°; bolt "C": 73 inch lbs. (10 Nm)
    - Left subframe nuts "A": 41 ft. lbs. (55 Nm); bolts "B" and "C": 37 ft. lbs. (50 Nm), plus 90°; bolts "D": 22 ft. lbs. (30 Nm).
13. Torque the engine support bolts as follows:

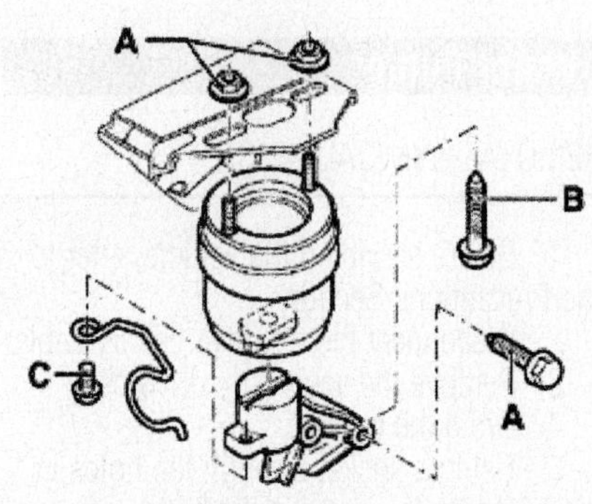

Identifying right subframe mounting bolts

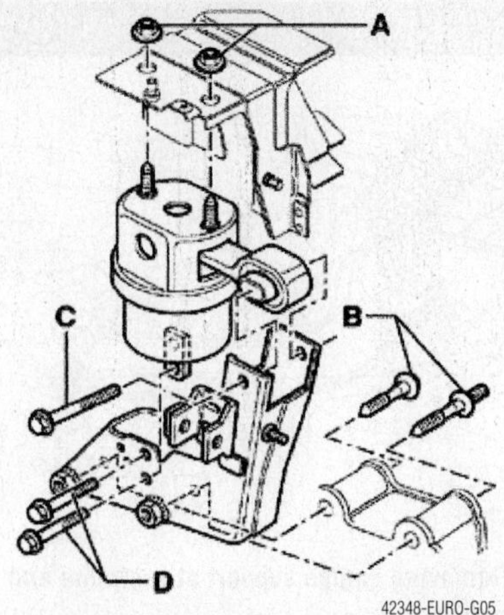

Identifying left subframe mounting bolts

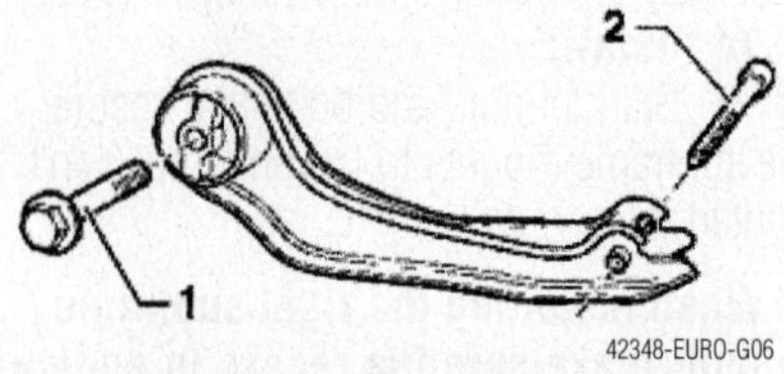

Identifying engine support mounting bolts

- Support to subframe bolt "1": 148 ft. lbs. (200 Nm)
- Support to transmission bolt "2": 59 ft. lbs. (80 Nm) plus 90°.

14. The remaining installation is the reverse of the removal procedure, using the following torque values.
    - Starter bolts: 33 ft. lbs. (45 Nm).
    - Halfshafts to flange bolts: 59 ft. lbs. (80 Nm).
    - Halfshafts to wheel hub bolts: 111 ft. lbs. (150 Nm) plus 90°.
    - Exhaust pipe to the catalytic converter. 18 ft. lbs. (25 Nm).
    - Exhaust pipe to the exhaust manifold bolts: 30 ft. lbs. (40 Nm).

➡DTCs are stored when harness connectors are detached.

15. Read the DTCs and clear the fault codes.

Water Pump

REMOVAL & INSTALLATION

1. Before servicing the vehicle, refer to the Precautions Section.
2. Disconnect the negative battery cable.
3. Remove the accessory drive belt.
4. Drain the coolant.
5. Using a socket through the holes in the pump pulley, remove the bolts and remove water pump.

To install:

6. Install or connect:
    - New O ring on the water pump.
    - Water pump. Mounting bolts: 15 ft. lbs. (20 Nm).
    - Accessory drive belt.
    - Coolant.

Heater Core

REMOVAL & INSTALLATION

➡Be sure to obtain the anti-theft code for the radio.

1. Disconnect the negative battery cable.
2. From inside the engine compartment, remove the dust/pollen filter.
3. Drain the cooling system into a clean container for reuse.
4. Disconnect the heater hoses from heater core.
5. Remove the right side air intake duct at instrument panel.
6. Release the retaining tabs and pull out air vent housing for air intake duct.
7. Remove the radio trim panel.
8. Remove the hex bolts on both sides of the passenger's side air bag unit and remove air bag.
9. Remove the hex bolt from the air duct and cross panel support in engine compartment behind vacuum reservoir.
10. Remove the driver side airbag by performing the following procedure:
    - Place the steering wheel so front tires are facing straight-ahead.
    - At the rear of the steering wheel, insert a 7 in. (175mm) screwdriver into the hole in the hub approximately 1¾ in. (45mm) deep.
    - Pull the screwdriver upward to release the air bag clip.
    - Rotate the steering wheel 180 degrees in the opposite direction and release the clip on the other side.
    - Center the steering wheel with the front wheels facing straight-ahead.
    - Disconnect the electrical connector and remove the air bag.
11. Remove the steering wheel from the steering column by removing the center hex bolt.
12. Remove the instrument panel by performing the following procedure:
    - Remove the upper steering column cover screws and the cover.
    - Remove the lower steering column cover screws, release the steering wheel height adjustment and remove the lower steering column switch trim.
    - Remove the combination switch-to-steering column bolt, disconnect the electrical connectors and remove the combination switch.
    - Remove the 2 screws at the rear of the instrument cluster trim cover, press in the clips and remove the trim cover.
    - Remove the 2 screws at the side of the speedometer cluster cover, pull the cover forward and disconnect the electrical connector.
    - Below the instrument cluster, remove the fresh air vent and the vent outlet.
    - Insert a thin feeler gauge above the headlight dimmer switch and push to the right to release the locks and then remove the dimmer switch.
    - Remove the left and right center air outlets and the ashtray.
    - Starting at the ashtray opening, carefully loosen the center instrument panel trim panel and remove the panel.
    - Insert radio release tool 3316 into unlocking slots on side of radio until engaged.
    - Remove radio from instrument panel by pulling out on tool.
    - Disconnect radio connectors.
    - Remove the A/C control head trim panel, remove 4 screws and the A/C control panel.
    - Remove the heater control trim panel, the heater rotary knobs and remove the heater control panel.
    - Disconnect all electrical connectors from the center instrument panel.
    - At the driver's side, remove the fuse/relay panel cover and the storage tray.
    - At the driver's side, remove the footwell air duct screws and the air duct.
    - At the driver's side, remove the A-pillar trim panel.
    - At the driver's side, remove the photo sensor from the top of the instrument panel.
    - Disconnect the radio speaker connectors from both sides of the instrument panel.
    - On both sides of the instrument panel, remove the instrument panel end trim cover and remove the hex screws.
    - Remove the instrument panel assembly from the instrument panel carrier.

- Remove the 3 heater core mounting bolts and remove the heater core.

To install:

13. Install the heater core to the heater housing.
14. Connect the heater hoses to the heater core.
15. Install the instrument panel by reversing the removal procedure.
16. Torque the steering wheel nut to 37 ft. lbs. (50 Nm). Torque the driver's side air bag mounting screws to 44 inch lbs. (5 Nm).

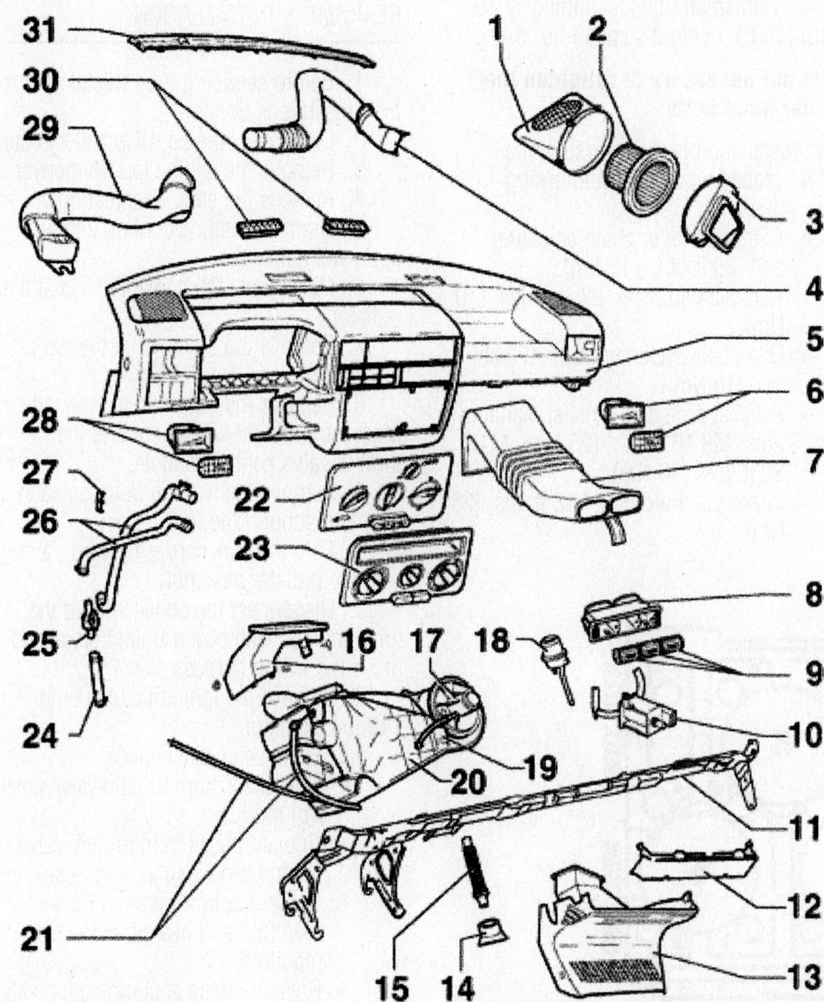

1. Filter housing
2. Dust/pollen filter
3. Air intake duct
4. Right air duct
5. Instrument panel
6. Right air outlet
7. Center air duct
8. Center air outlet hose
9. Air guide
10. Recirculation two-way valve
11. Instrument panel carrier
12. Trim panel
13. Footwell air outlet
14. Water drain valve
15. Drain pipe
16. Defroster duct
17. Fresh air blower
18. Recirculation flap vacuum unit
19. Blower motor fuse
20. Heater core
21. Cables
22. A/C-heater controls (old)
23. A/C-heater controls (new)
24. Coolant hose
25. Heater control valve
26. Coolant hose
27. Clip
28. Left air outlet
29. Left air duct
30. Defroster vent
31. Air intake grille

42348-EURO-G07

Exploded view of the instrument panel, and related components—Eurovan

Cylinder Head

REMOVAL & INSTALLATION

2.8L Engine

1. Before servicing the vehicle, refer to the Precautions Section.
2. Disconnect the negative battery cable.
3. Remove the engine top front cover.
4. Remove the engine insulation cover.
5. Drain the engine coolant into an approved container.
6. Disconnect the coolant hoses at the radiator.
7. Remove the 3 screws at the top of the front grille.
8. Remove the headlight washer spray tube, if equipped.
9. Carefully pry away the trim frame starting below the headlight.
10. Remove the front grille and the trim frame as a unit.
11. Remove the intake hose between Mass Air Flow (MAF) sensor and the throttle valve control module.
12. Remove the two top bolts on each side that support the lock carrier.
13. Fold the lock carrier forward , along with the radiator assembly.
14. Disconnect the connectors to the ignition coils with power output stage and move the wiring harness clear.
15. Remove the ignition coils using T10094 extractor.
    a. Remove or disconnect:
    - Connector from throttle valve control module.
    - Coolant hoses from throttle valve control module and plug hose ends.
    - All vacuum lines at cylinder head.
    - Center support between intake manifold and heat shield/exhaust manifold.
    - Pressure hose at combination valve and connecting hose at cylinder head cover.
    - Unclip pressure hose and all other lines from retainers at intake manifold and at cylinder head cover.
    - Intake manifold supports at both sides at bracket from heat shield/exhaust manifold.
    - Secondary Air Injection (AIR) pump motor.
    - Dipstick and dipstick tube.
    - Intake manifold with throttle valve control module.
    - Cylinder head cover.

- Fuel supply and return lines.
- Coolant temperature sensor.
- All remaining harness connectors and vacuum lines at cylinder head.
- Thermostat housing.
- Exhaust pipe from exhaust manifold.
- Timing chain.

16. Using Polydrive Special tool 3452, or equivalent, loosen, then remove the cylinder head bolts in the proper sequence.

To install:

➡The bolt holes in the engine block MUST be clean and free of debris and fluid. Always replace the cylinder head bolts.

17. Set the crankshaft and camshaft to TDC for cylinder No. 1.

18. Install or connect:

- Head gasket, making sure that the part No. faces up.

➡Make sure dowel sleeves in cylinder block bore 12 and 20 are inserted and the cylinder head gasket is secured. The longer cylinder head bolts go into the center bores.

- Cylinder head and tighten the head bolt hand-tight.
- New cylinder head bolts in sequence in 4 steps:

a. Tighten all bolts to 22 ft. lbs. (30 Nm).

b. Tighten all bolts to 37 ft. lbs. (50 Nm).

c. Tighten all bolts an additional ½ turn (180°), using 2 steps of 90° each.

➡It is not necessary to retighten the cylinder head bolts.

19. Installation is the reverse of the removal procedure using the following torque values.

- Camshaft roller chain adjusting bolt: 30 ft. lbs. (40 Nm).
- Camshaft adjuster: 88 inch lbs. (10 Nm).
- Drive belt tensioner bolt: 88 inch lbs. (10 Nm).
- Exhaust pipe and exhaust manifold nuts: M8 18 ft. lbs. (25 Nm), M10 30 ft. lbs. (40 Nm).
- Intake manifold bolts: 18 ft. lbs. (25 Nm).

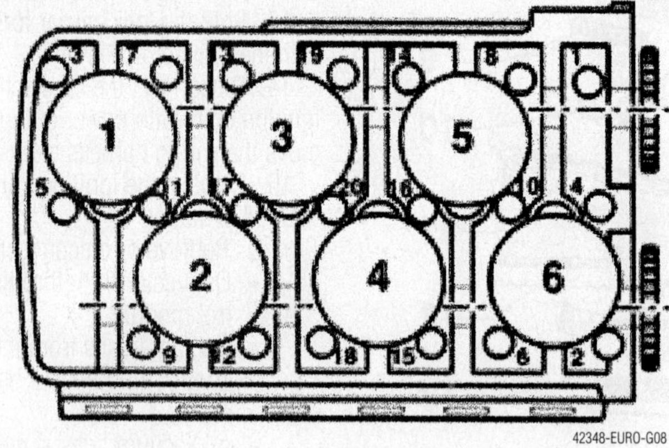

Cylinder head bolt removal sequence—2.8L engine

Cylinder head bolt tightening sequence—2.8L engine

- Thermostat housing bolt: 88 inch lbs. (10 Nm).
- Valve cover bolts: 88 inch lbs. (10 Nm).

20. Connect the negative battery cable.

Intake Manifold

REMOVAL & INSTALLATION

1. Before servicing the vehicle, refer to the Precautions Section.
2. Disconnect the negative battery cable.
3. Remove the engine top front cover.
4. Remove the engine insulation cover.
5. Drain the engine coolant into an approved container.
6. Disconnect the coolant hoses at the radiator.
7. Remove the 3 screws at the top of the front grille.
8. Remove the intake hose between Mass Air Flow (MAF) sensor and the throttle valve control module.
9. Remove the two top bolts on each side that support the lock carrier.
10. Fold the lock carrier forward , along with the radiator assembly.
11. Disconnect the connectors to the ignition coils with power output stage and move the wiring harness clear.
12. Remove the ignition coils using T10094 extractor.

a. Remove or disconnect:

- Connector from throttle valve control module..
- Coolant hoses from throttle valve control module and plug hose ends.
- Center support between intake manifold and heat shield/exhaust manifold.
- Pressure hose at combination valve and connecting hose at cylinder head cover.
- Unclip pressure hose and all other lines from retainers at intake manifold and at cylinder head cover.
- Intake manifold supports at both sides at bracket from heat shield/exhaust manifold.
- Secondary Air Injection (AIR) pump motor.
- Dipstick and dipstick tube.
- Intake manifold with throttle valve control module.

To install:

13. Install the intake manifold(s) to the cylinder head with a new gasket(s). Tighten the bolts to 18 ft. lbs. (25 Nm)
14. Reverse the removal procedure to install remaining components.

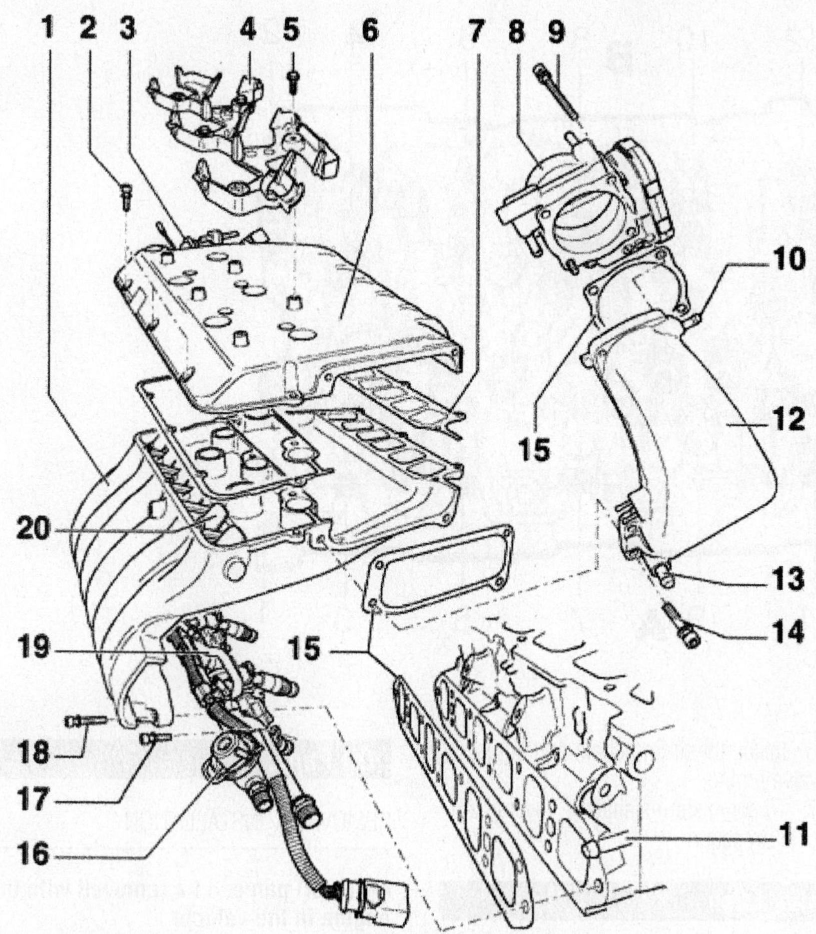

1. Intake manifold
2. Valve cover bolt
3. Vacuum element
4. Wiring harness
5. Harness bolt
6. Valve cover
7. Gasket
8. Throttle valve control module
9. Throttle valve bolt
10. Vacuum connection
11. Vacuum connection
12. Intake manifold sleeve
13. Vacuum connection
14. Manifold sleeve bolt
15. Gasket
16. Fuel pressure regulator
17. Fuel pressure regulator bolt
18. Intake manifold bolt
19. Fuel rail
20. Change over flaps

42348-EURO-G09

Exploded view of the intake manifold mounting and related components—2.8L engine

Exhaust Manifold

REMOVAL & INSTALLATION

1. Before servicing the vehicle, refer to the Precautions Section.
2. Remove or disconnect:
   - O_2 sensor wiring and remove any heat shields that may be in the way.
   - Exhaust support brackets as necessary.
   - Front exhaust pipe from the manifold.
   - Self-locking nuts or bolts and the manifold.

To install:

3. Installation is the reverse of removal. Use new gaskets and self-locking nuts. Torque as follows: M8 bolts to 18 ft. lbs. (25 Nm), M10 bolts to 30 ft. lbs. (40 Nm).
4. Check and clear any DTCs.

Front Crankshaft Seal

REMOVAL & INSTALLATION

1. Before servicing the vehicle, refer to the Precautions Section.
2. Remove or disconnect:
   - Accessory drive belt.

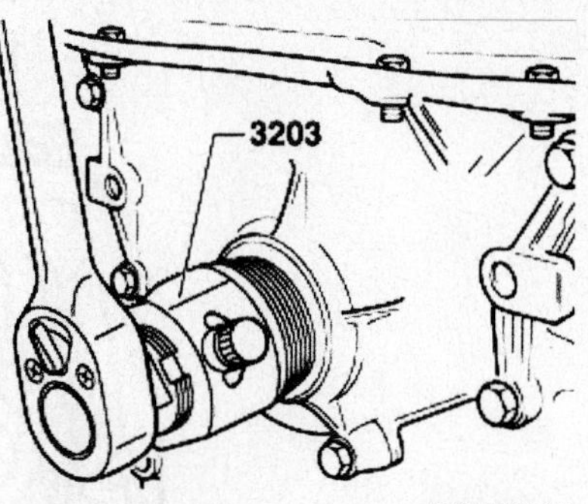

Removing the seal using Seal Remover 3203—2.8L engine

   - Crankshaft balancer.
   - Seal with Seal Remover 3203 or equivalent.
3. Clean the running and sealing surfaces.

To install:

4. Press the oil seal flush with Seal Installer 3266, or equivalent.
5. Install the crankshaft balancer. Torque the NEW crankshaft bolt: 73 ft. lbs. (100 Nm) plus 1/4 turn.
6. Install the accessory drive belt.

Camshaft and Valve Lifters

REMOVAL & INSTALLATION

1. Remove or disconnect:
   - Valve cover.
   - Intake manifold.
   - Timing chain.
   - Four bolts and control housing from cylinder head.

➡DO NOT allow the bearing caps to become mixed up.

2. On the intake camshaft, remove the bearing caps as follows:
   - Bearing caps 1 and 13.
   - Bearing cap 7.
   - Bearing caps 3 and 11.
   - Bearing caps 5 and 9 alternating in a diagonal sequence.
3. On the exhaust camshaft, remove the bearing caps as follows:
   - Bearing caps 2 and 14.
   - Bearing cap 4 and 12.
   - Bearing cap 8.
   - Bearing caps 6 and 10 alternating in a diagonal sequence.
4. Remove the camshaft and lift out the valve lifters.

To install:

5. Install or connect:
6. Install the lifters into their respective bore.

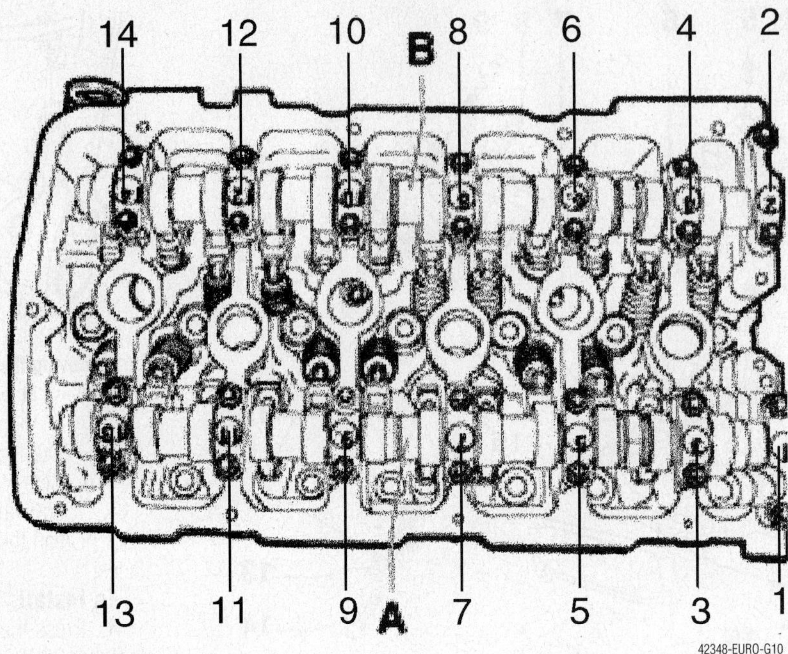

Camshaft bearing cap identification—2.8L engine

7. Ensure cam lobes for cylinder no 1 are pointing upward and that the pistons are not at TDC.
8. Oil the camshaft journal surfaces.
9. Insert the respective camshafts into their mountings.
10. On the intake camshaft, install the bearing caps in sequence as follows:
   - Bearing caps 5 and 9: Torque to 44 inch lbs. (5 Nm) plus 45°.
   - Bearing caps 1 and 13: Torque to 44 inch lbs. (5 Nm) plus 45°.
   - Bearing cap 7: Torque to 44 inch lbs. (5 Nm) plus 45°.
   - Bearing caps 3 and 11: Torque to 44 inch lbs. (5 Nm) plus 45°.
11. On the exhaust camshaft, install the bearing caps in sequence as follows:
   - Bearing caps 6 and 10: Torque to 44 inch lbs. (5 Nm) plus 45°.
   - Bearing caps 2 and 14: Torque to 44 inch lbs. (5 Nm) plus 45°.
   - Bearing cap 8: Torque to 44 inch lbs. (5 Nm) plus 45°.
   - Bearing caps 4 and 12: Torque to 44 inch lbs. (5 Nm) plus 45°.
12. Adjust the camshafts in cylinder head to TDC on cylinder no. 1.
13. Lightly oil the sealing ring contact surfaces of camshaft before installing control housing.
14. Lightly oil camshafts and slowly slide control housing over camshaft sealing rings.
15. Install control housing to cylinder head and tighten to 72 inch lbs. (8 Nm).
16. Install the timing chain and adjust the valve timing.
17. To complete installation, reverse the removal procedure.

Valve Lash

ADJUSTMENT

All engines are equipped with hydraulic valve lash adjusters. No periodic valve lash adjustment is necessary.

Starter

REMOVAL & INSTALLATION

1. Disconnect the battery ground cable.
2. Raise and safely support the front of the vehicle with jackstands.
3. If necessary, label the small wires before disconnecting them.
4. Disconnect the large cable, which is the positive battery cable, from the solenoid.
5. Remove the starter mounting bolts, while supporting the weight of the starter.
6. Pull the starter straight out from the transaxle.

To install:

7. Install the starter into the transaxle.
8. Attach the electrical connections to the starter.
9. Lower the vehicle from the jackstands.
10. Connect the negative battery cable.

Oil Pan

REMOVAL & INSTALLATION

➡The oil pan can be removed with the engine in the vehicle

1. Before servicing the vehicle, refer to the Precautions Section.
2. Drain the engine oil.
3. Remove or disconnect:
   - Oil pan bolts.
   - Oil pan.

To install:

4. Be sure the gasket surface is flat and install the pan with a new gasket.
5. Tighten the oil pan screws to 9 ft. lbs. (12 Nm).
6. Tighten bolts between oil pan and transmission to 33 ft. lbs. (45 Nm).
7. Refill the engine with oil. Start the engine and check for leaks.

Oil Pump

REMOVAL & INSTALLATION

1. Before servicing the vehicle, refer to the Precautions Section.
2. Remove or disconnect:
   - Oil pan.
   - Oil pump mounting bolts.
   - Oil pump.

To install:

3. Be sure the oil pump dowel sleeves are located on the engine block.

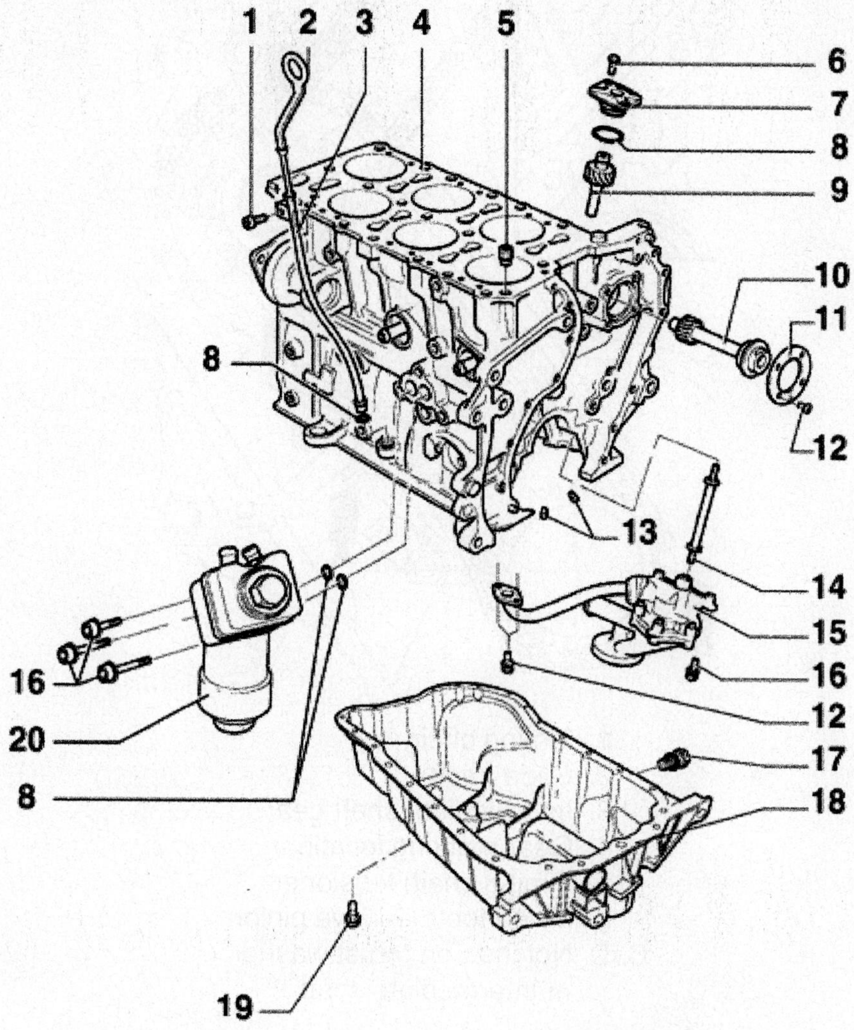

1. Bolt
2. Dipstick
3. Guide tube
4. Cylinder block
5. Oil check valve
6. Oil pump drive cover bolt
7. Oil pump drive cover
8. O-ring
9. Oil pump drive
10. Intermediate shaft
11. Thrust washer
12. Bolt
13. Oil injection jet
14. Input shaft
15. Oil pump
16. Oil pump bolt
17. Drain plug
18. Oil pan
19. Oil pan bolt
20. Oil filter housing

42348-EURO-G11

Exploded view of the lubrication system components—2.8L engine

4. Install:
   - Oil pump. Torque mounting bolts to: 18 ft. lbs. (25 Nm).
   - Oil pan.

Rear Main Seal

REMOVAL & INSTALLATION

The rear main oil seal is located in a housing on the rear of the cylinder block. To replace the seal on all vehicles it is necessary to remove the transaxle and flywheel.

1. Before servicing the vehicle, refer to the Precautions Section.
2. Remove or disconnect:
   - Transaxle and flywheel.
   - Old seal out of the support ring.

To install:

3. Oil the new seal and press it into place using tool VW-2003/2A, or equivalent, to start the seal and tool VW-2003/1, or equivalent, to seat the seal. Be careful not to damage the seal or score the crankshaft.
4. Install the flywheel and transaxle.

Timing Chain, Front Cover and Seal

REMOVAL & INSTALLATION

1. Before servicing the vehicle, refer to the Precautions Section.
2. Disconnect the negative battery cable.
3. Remove engine covers at the front of the engine compartment and underneath the vehicle.
4. Mark the rotation direction of the drive belt.
5. Insert an M8x50 bolt into the threaded hole of the belt tensioner roller and tighten just enough until tension is off belt and drive belt can be removed.
6. Remove the cylinder head cover.
7. Remove the necessary components to gain access to the front of the engine.
8. Place the crankshaft at TDC for no. 1 cylinder.
9. Mark the camshaft timing chain rotation direction with a paint mark.
10. Remove the camshaft roller chain tensioner.
11. Remove the 8 mounting bolts and remove the chain front cover
12. Disconnect the front exhaust pipe from the exhaust manifold.
13. Check that timing is at TDC by looking down between timing chain sprockets of the intermediate shaft. A groove should be visible between the chain edges.

➡If the groove cannot be seen, turn crankshaft one more revolution until groove can be seen.

14. Insert camshaft gauge T10068 into both camshaft grooves.

➡If the gauge cannot be inserted, rotate crankshaft about .20 inches (5 mm) above TDC position until gauge can be inserted.

➡Markings "A" and "B" on the camshaft adjusters must match the notches of control housing "C."

15. Remove the camshaft adjusters together with the timing chain from the camshafts.
16. Remove the two bolts from the glide track located at lower left below camshaft sprocket.
17. Remove the timing chain from the camshaft adjusters and place chain on its side.

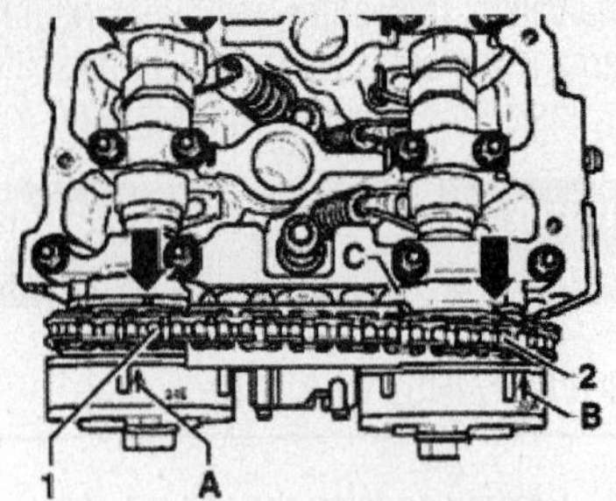

Lining up camshaft adjusters to remove timing chain—2.8L engine

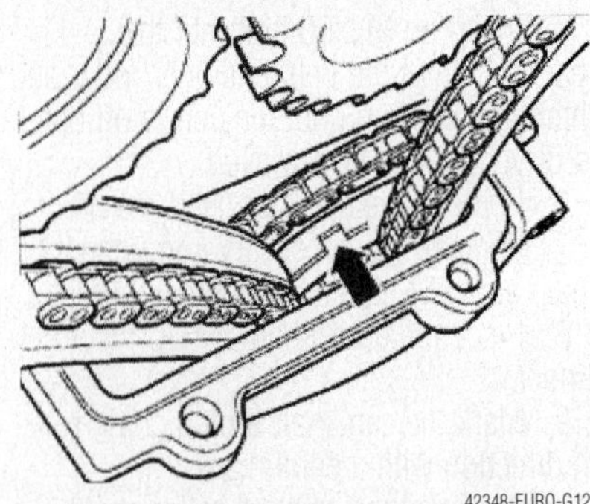

Identifying TDC position on timing chain sprockets of intermediate shaft—2.8L engine

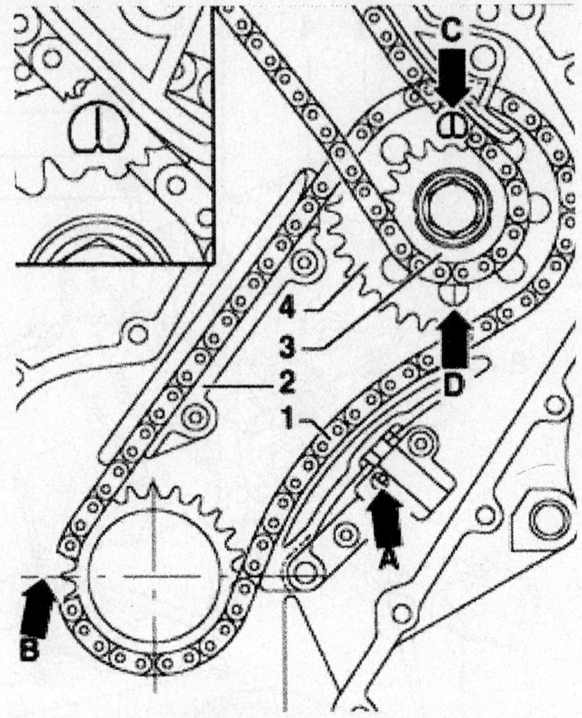

Identifying timing chain positioning on crankshaft and intermediate shaft sprockets—2.8L engine

To install:

18. Adjust the milled tooth of the drive pinion to line up with the bearing separating gap at TDC on cylinder no. 1.
19. Install the glide track bolts, without the collar, and torque to 89 inch lbs. (10 Nm).
20. Install the timing chains and the intermediate shaft gears (if removed), ensuring that the timing chain rotation direction is correct.
21. Check that the marking on the gear matches with the notches on the intermediate shaft thrust washer.
22. Hand tighten the intermediate shaft gear bolts.
23. Use a small screwdriver to release the tension of the locking splines in the chain tensioner and press the tensioner track against the tensioner.
24. Install the timing chain tensioner bolts and torque to 72 inch lbs. (8 Nm).
25. Use a socket wrench to hold the crankshaft damper in position.
26. With the intermediate shaft sprockets lined up correctly, torque the bolts to 44 ft. lbs. (60 Nm), plus 90°.
27. Check that all sprockets are correctly lined up and cylinder no. 1 is at TDC.
28. Insert camshaft adjusting ruler tool T10068 into camshaft grooves.
29. Guide the timing chain between the tensioning track and the glide track.
30. Install the intake camshaft adjuster first with the timing chain installed, then install the exhaust camshaft adjuster.

➡Exhaust camshaft adjuster can be offset in two directions. When installing, make sure the camshaft position sensor is turned to stop in a clockwise direction. Distance between the milled teeth markings on the timing chain at the camshaft adjuster markings is exactly 16 rollers on the timing chain.

➡Markings "A" and "B" on the camshaft adjusters must match the notches of control housing "C."

31. Hand tighten NEW bolts for camshaft adjusters.
32. Remove camshaft adjusting ruler T10068.
33. Rotate crankshaft 2 revolutions and recheck valve timing.
34. If valve timing is okay, secure camshafts in position and torque camshaft adjuster bolts to 44 ft. lbs. (60 Nm), plus 90°.
35. Place oil on the front cover O ring and install in cover.
36. Coat the front cover sealing surface with sealant and hand tighten the bolts.
37. Torque the front cover M6 bolts to 72 inch lbs. (8 Nm), and tighten the M8 bolts to 18 ft. lbs. (25 Nm).
38. Install the camshaft roller chain tensioner and torque the bolts to 30 ft. lbs. (40 Nm).
39. Rotate crankshaft 2 revolutions and recheck valve timing.
40. Replace the remaining components by reversing the removal procedures.
41. Reconnect the negative battery cable.
42. Test drive the vehicle.

Piston and Ring Positioning

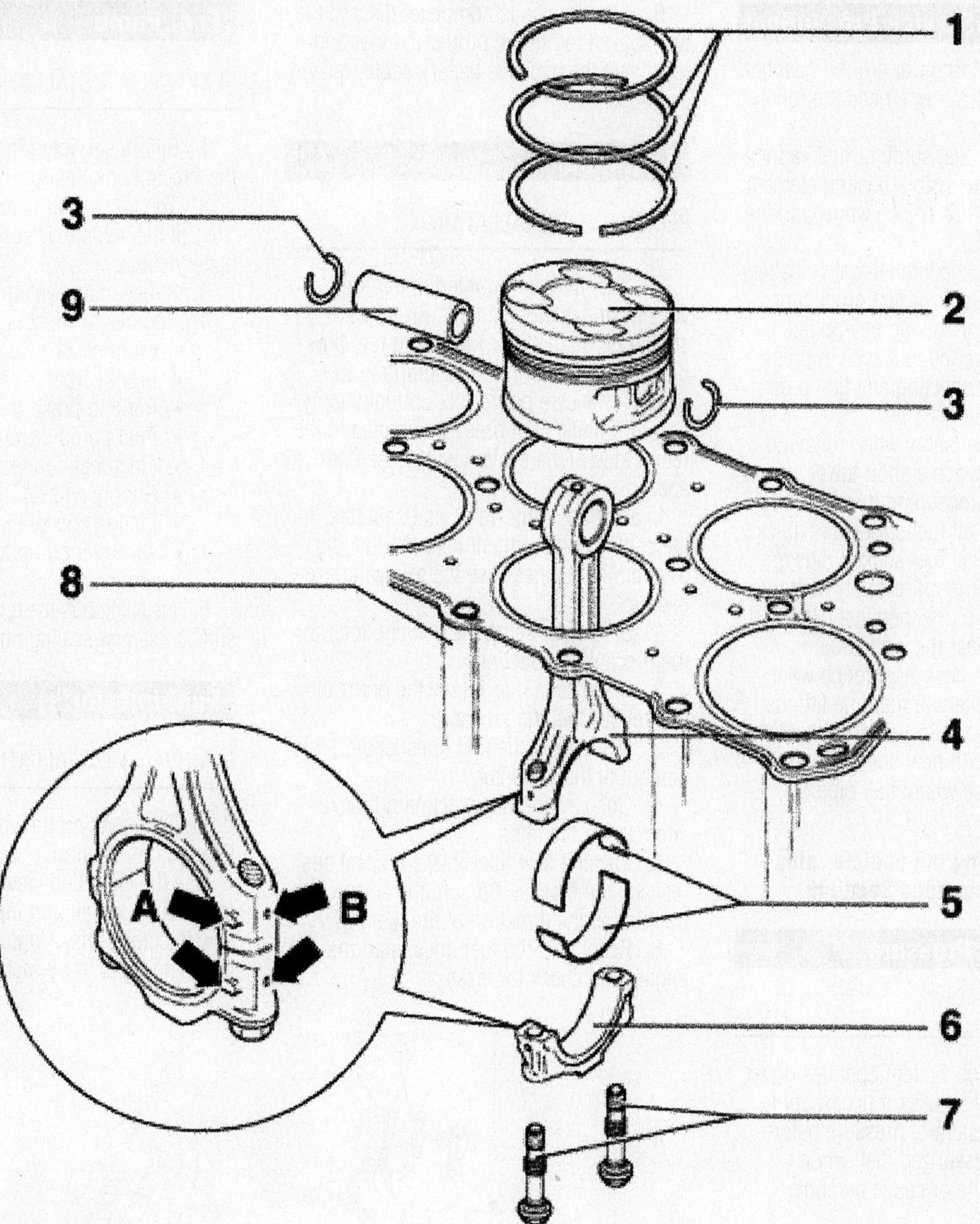

1. Piston rings. Offset each ring by 120°
2. Piston
3. Snap ring
4. Connecting rod
5. Bearing
6. Connecting rod bearing cap
7. Bearing cap bolts
8. Block
9. Pinstons pins

A. / B. Connecting rod match marks

42348-EURO-G15

Piston and connecting rod installation—2.8L engine

FUEL SYSTEM

Fuel System Service Precautions

Whenever working on or around gasoline or the fuel delivery system, heed the following precautions:

- Do not allow fuel spray or fuel vapors to come into contact with a heating element or open flame. Do not smoke while working on the fuel system.
- Always disconnect the negative battery cable unless the repair or test procedure requires that battery voltage be applied.
- Always relieve the fuel system pressure prior to disconnecting any fitting or fuel line connection.
- To control fuel spray when relieving system pressure, place a shop towel around the fitting prior to loosening to catch the spray. Ensure that all fuel spillage is quickly wiped up and that all fuel soaked rags are deposited into a proper fire safety container.
- Always keep a dry chemical (Class B) fire extinguisher near the work area.
- Always use a back-up wrench when loosening and tightening fuel line fittings.
- Do not re-use fuel system gaskets and O-rings, replace with new ones. Do not substitute fuel hose where fuel pipe is installed.

➡Before servicing the vehicle, also refer to the Precautions Section.

Fuel System Pressure

RELIEVING

The fuel injection system operates under high pressure. This makes it necessary to first relieve the system of pressure before servicing. The pressurized fuel, when released, may ignite or cause personal injury.

1. Before servicing the vehicle, refer to the Precautions Section.
2. Remove or disconnect:
   - Power to the fuel pump by removing the relay or the fuel pump fuse. Check the list on the fuse box lid to be sure. The fuse can be removed to stop the fuel pump from running. With the engine operating at idle, wait until the engine stalls from fuel starvation.
   - Negative battery cable.
3. Carefully loosen the fuel line on the control pressure regulator or component to be serviced.
4. Wrap a clean rag around the connection, while loosening, to catch any fuel.
5. After service is complete, discard the fuel soaked rag in the proper manner and reconnect the negative battery cable, relay or fuses.

Fuel Filter

REMOVAL & INSTALLATION

The fuel filter is mounted under the vehicle, above the fuel tank. An arrow should be on the filter indicating fuel flow direction. Install with arrow pointing to engine. Use care not to mix up fuel supply or return lines. Fuel pressure applied to the return side of the system will cause damage.

In addition, some vehicles use a filter in the engine compartment near the fuel distributor. If equipped, use the following procedure:

1. Before servicing the vehicle, refer to the Precautions Section.
2. Make certain to follow the precautions and relieve fuel the pressure.
3. Disconnect the fuel lines leading into and out of the fuel filter.
4. Unscrew the filter retaining bracket and remove the filter.
5. Install a new filter in the bracket and reattach the bracket. Be sure the arrows are pointing in the direction of the fuel flow.
6. Reconnect the fuel lines, start the engine and check for leaks.

Fuel Pump

REMOVAL & INSTALLATION

1. Before servicing the vehicle, refer to the Precautions Section.
2. The fuel pump is located under the front of the vehicle. Disconnect the negative battery cable.
3. Relieve fuel system pressure.
4. Remove or disconnect:
   - Front seats
   - Front carpet
   - Parking brake lever
   - Fuel pump cover
   - Electrical connector
   - Supply and return lines
   - Flange and sealing ring
   - Turn fuel pump to the left and remove
5. Installation is the reverse of removal. Be sure to use new sealing rings and/or gaskets.

Fuel Injector(s)

REMOVAL & INSTALLATION

Before servicing the vehicle, refer to the Precautions Section.

Use the exploded view to remove and install the fuel rail and injectors. Always use new injector seals. Tighten the fuel rail attaching bolts to 88 inch lbs. (10 Nm).

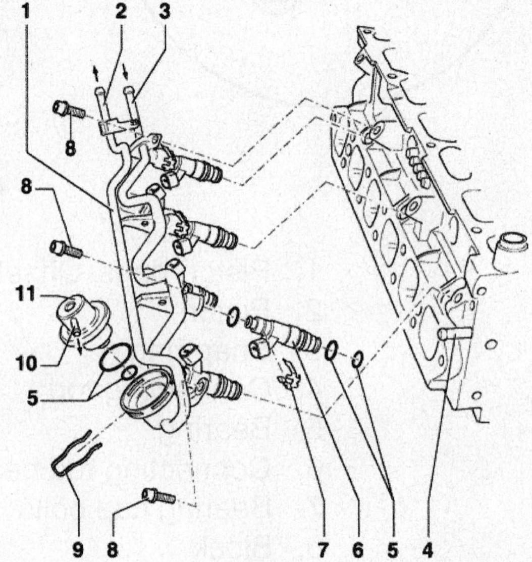

Exploded view of fuel rail and injectors—2.8L engine

Transaxle Assembly

REMOVAL & INSTALLATION

Automatic

1. Before servicing the vehicle, refer to the Precautions Section.
2. Relieve fuel pressure, by opening the fuel filler cap. Be sure to take the appropriate fire safety precautions.
3. Drain the engine coolant into an approved container.
4. Remove the charge air cooler from the lock carrier.
5. Disconnect the coolant hoses at the radiator.
6. Remove the 3 screws at the top of the front grille.
7. Remove the headlight washer spray tube, if equipped.
8. Carefully pry away the trim frame starting below the headlight.
9. Remove the front grille and the trim frame as a unit.
10. Remove the intake hose between Mass Air Flow (MAF) sensor and the throttle valve control module.
11. Remove the two top bolts on each side that support the lock carrier.
12. Fold the lock carrier forward , along with the radiator assembly.
13. Remove left side drive axle bolt while vehicle is on ground.
14. Raise and support vehicle and remove left side wheel.
15. Remove drive axle bolt.

➡Measure distance from bottom of torsion bar nut to the end of torsion bar threaded rod and record.

16. Remove the torsion bar nut.
17. Remove or disconnect:
    - Transmission speed sensor connector.
    - Upper engine to transmission bolts.
    - Engine undercover.
    - All electrical connections from transmission.
    - Starter.
    - Drive axles from transmission flange shafts.
    - Raise right drive axle and wire aside.
    - Flanged shaft bearing from transmission.
    - Right flange shaft from intermediate shaft.
    - Front exhaust pipe at exhaust manifold.

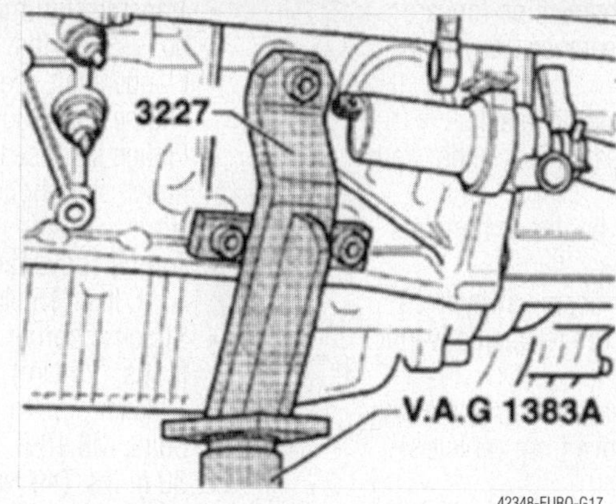

Installing engine/transmission support tools

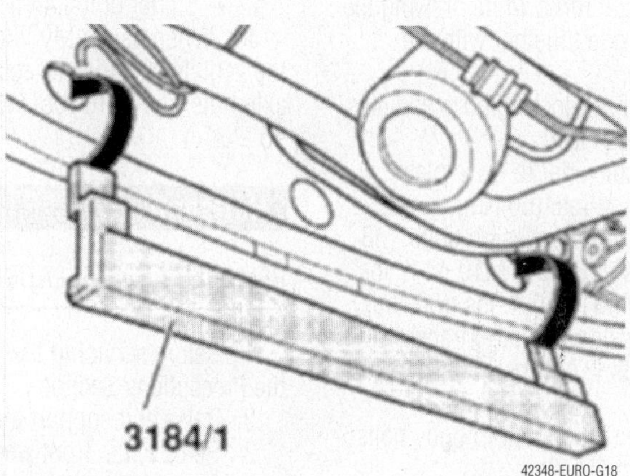

Installing auxiliary support tool in cross panel

    - Cover plate at torque converter flange.
    - Torque converter nut through cover plate opening.
    - Left side wheel bearing housing/lower swivel joint bolted connection.
    - Left side shock absorber/stabilizer bar coupling rod bolt.
    - Left side shock absorber auxiliary heater end pipe, if equipped.
18. Install engine support tool 3148 to support engine/transmission.
19. Install engine support tool 3227 to engine block.
20. Install auxiliary support tool 3184/1 in holes in cross panel.
21. Install support tool 3184 into auxiliary support tool 3184/4 and into subframe.
22. Install spindle "C" into hole on engine mount.
23. Align support tool so that slide "B" is resting against right side tube.
24. Tighten angle plates to secure support tool to subframe and auxiliary support.

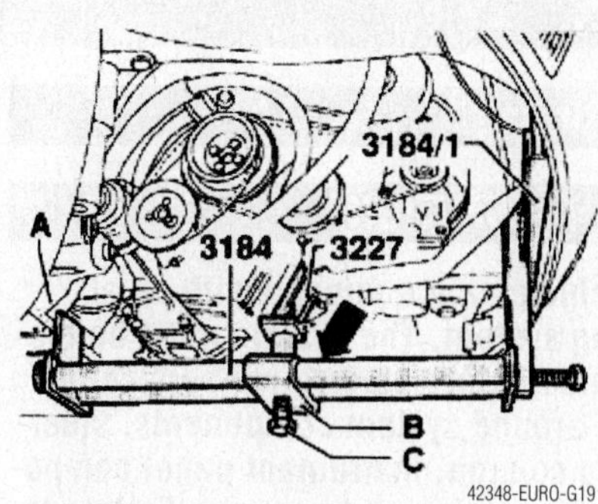

Installing support tools into subframe

25. Relieve subframe tension by raising engine/transmission using spindle "C".
26. Remove 2 bolts and remove transmission support.
27. Remove 2 nuts and unbolt transmission mount from body
28. Lower engine and remove 2 bolts from transmission mount to transmission.
29. Carefully swivel engine/transmission forward and lower, ensuring that selector lever cable, multi-function switch and drive axles are not damaged.

30. Clamp off ATF lines at cooler and then remove lines.

31. Tilt engine/transmission forward again until there is clearance away from crossmember.

32. Remove left side drive axle.

33. Place transmission jack under transmission and secure.

34. Remove lower engine to transmission bolts.

35. Remove transmission away from engine enough to secure torque converter from falling out.

36. Carefully lower transmission away from engine and remove from vehicle.

To install:

37. When reinstalling, be sure the torque converter drive pins engage in the ATF pump inner wheel recesses. Keep checking that it still turns while drawing the engine and transaxle together with the bolts.

38. Replace all self-locking bolts and nuts.

39. Install or connect:

- Torque converter to drive plate bolts: 44 ft. lbs. (60 Nm).
- Engine-to-transmission bolts. M8 17 ft. lbs. (23 Nm), M10 44 ft. lbs. (60 Nm), M12 59 ft. lbs. (80 Nm).
- Transmission mount to transmission bolts: 37 ft. lbs. (50 Nm) plus 90°.
- Transmission mount to body bolts: 41 ft. lbs. (55 Nm).
- Transmission mount to bearing housing bolts: 18 ft. lbs. (25 Nm).
- Transmission mount to engine bolts: 30 ft. lbs. (40 Nm).
- Flange shaft bearing carrier to engine bolts: 30 ft. lbs. (40 Nm).
- Flange shaft bearing carrier to flange shaft bolts: 18 ft. lbs. (25 Nm).
- Torque converter cover plate bolts: 11 ft. lbs. (15 Nm).
- Shock absorber coupling bolt: 18 ft. lbs. (25 Nm).
- Exhaust pipe to exhaust manifold bolts: M8 18 ft. lbs. (25 Nm), M10 30 ft. lbs. (40 Nm).
- Drive axles to transmission flange bolts: M10 59 ft. lbs. (80 Nm), M12 74 ft. lbs. (100 Nm).
- Starter bolts: 44 ft. lbs. (60 Nm).

40. When assembly is complete and the vehicle is on its wheels, tighten the axle nuts to 111 ft. lbs. (150 Nm) plus 90°.

Halfshaft

REMOVAL & INSTALLATION

1. Before servicing the vehicle, refer to the Precautions Section.
2. Raise and support the vehicle.
3. Remove the front wheels.
4. Remove the engine undercover.
5. Relieve the tension on the torsion bar.
6. Remove or disconnect:
   - Front axle nut.
   - Lower shock absorber mounting.
   - Wheel bearing housing from lower ball joint.
   - Multi-function switch connector.
   - Drive axle from transmission flange.
   - Transmission support.
   - Halfshaft.

To install:

7. Reverse the removal procedure to install. Always using NEW bolts and using the following torque specifications:
   - Drive axle to transmission flange bolts: M10 59 ft. lbs. (80 Nm), M12 74 ft. lbs. (100 Nm).
   - Wheel bearing housing to ball joint: 66 ft. lbs. (90 Nm) plus 90°.
   - Shock absorber to control arm: 118 ft. lbs. (160 Nm).
   - Drive axle to wheel hub: 111 ft. lbs. (150 Nm).
   - Transmission support: 59 ft. lbs. (80 Nm) plus 90 °.

CV-JOINT AND BOOT OVERHAUL

The constant velocity joints (CV-joints) can be disassembled for cleaning and inspection but they cannot be repaired. All parts are machined to a matched tolerance and the entire CV-joint must be replaced as a unit.

STEERING AND SUSPENSION

Air Bag

⁂ CAUTION

Vehicles are equipped with an air bag system. The system must be disabled before performing service on or around system components, steering column, instrument panel components, wiring and sensors. Failure to follow safety and disabling procedures could result in accidental air bag deployment, possible personal injury and unnecessary system repairs.

PRECAUTIONS

Several precautions must be observed when handling the inflator module to avoid accidental deployment and possible personal injury.

- Never carry the inflator module by the wires or connector on the underside of the module.
- When carrying a live inflator module, hold securely with both hands, and ensure that the bag and trim cover are pointed away.
- Place the inflator module on a bench or other surface with the bag and trim cover facing up.
- With the inflator module on the bench, never place anything on or close the module which may be thrown in t he event of an accidental deployment.

1. Before servicing the vehicle, also refer to the Precautions Section.

DISARMING

To avoid personal injury when working on vehicles equipped with an air bag, the negative battery cable must be disconnected before working on the system. Failure to do so may result in deployment of the air bag.

1. Before servicing the vehicle, refer to the Precautions Section.
2. Turn the ignition switch to the **LOCK** position.
3. Disconnect the negative battery cable.
4. Wait 10 minutes for the battery back-up power to discharge.

Rack and Pinion Steering Gear

REMOVAL & INSTALLATION

Before servicing the vehicle, refer to the Precautions Section.

Use the exploded view to remove and install the power steering gear.

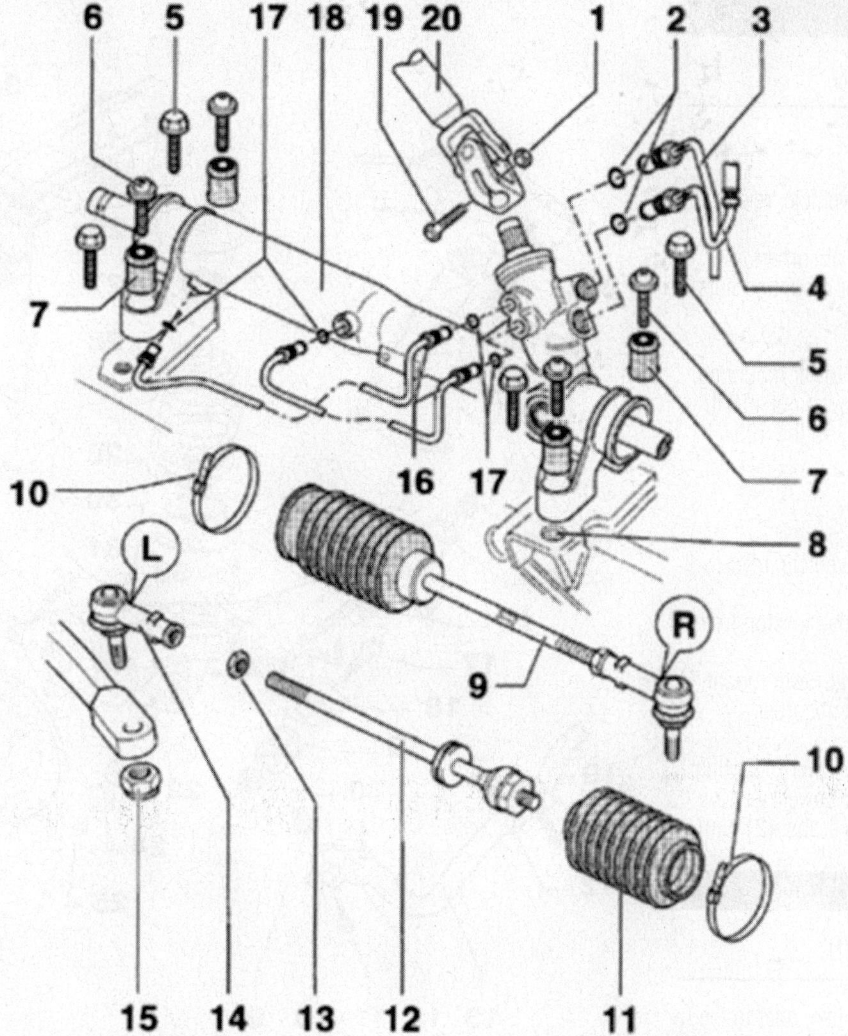

1. Self-locking hex nut—22 ft. lbs. (30 Nm) plus 90°
2. Sealing ring
3. Pressure line—22 ft. lbs. (30 Nm) plus 90°
4. Return line—22 ft. lbs. (30 Nm) plus 90°
5. Hex bolt—44 ft. lbs. (60 Nm)
6. Torx bolt—44 ft. lbs. (60 Nm)
7. Rubber bearing
8. Subframe threaded bore
9. Right tie rod—52 ft. lbs. (70 Nm)
10. Cable ties
11. Boot
12. Left tie rod—52 ft. lbs. (70 Nm)
13. Hex nut—41 ft. lbs. (55 Nm)
14. Tie rod end
15. Hex nut—48 ft. lbs. (65 Nm)
16. Control line
17. Sealing ring
18. Power steering gear
19. Hex bolt
20. Lower intermediate steering shaft with universal joint

42348_EURO_G20

Exploded view of the steering gear assembly

Shock Absorber

REMOVAL & INSTALLATION

Front

1. Before servicing the vehicle, refer to the Precautions Section.
2. To remove the shock absorbers, remove the upper and lower mounting nuts and remove shock absorber.

To install:

3. Install the upper and lower mounting nuts a torque as follows: Lower nut 118 ft. lbs. (160 Nm), upper nut 18 ft. lbs. (25 Nm).

Rear

1. Before servicing the vehicle, refer to the Precautions Section.
2. Raise vehicle to remove tension from coil springs.
3. Remove the upper and lower mounting nuts and remove shock absorber.

To install:

4. Install the upper and lower mounting nuts and torque as follows: Lower nut 18 ft. lbs. (25 Nm), upper nut 15 ft. lbs. (21 Nm).

Coil Spring

REMOVAL & INSTALLATION

Before servicing the vehicle, refer to the Precautions Section.

Use the exploded view to remove and install the rear oil springs.

Upper Ball Joint

REMOVAL & INSTALLATION

1. Before servicing the vehicle, refer to the Precautions Section.
2. Relieve the tension on the torsion bar.
3. Remove or disconnect:
   - Front wheels.
   - Wheel hub bolt.
   - Brake caliper and suspend aside.
   - Lower shock absorber bolt.
   - Stabilizer bar coupling.

➡Mark the position of the camber eccentric bushing to the wheel housing.

4. Remove or disconnect:
   - Upper ball joint nut.
   - Tie rod from steering knuckle.
   - Eccentric washer bolt from wheel bearing housing.

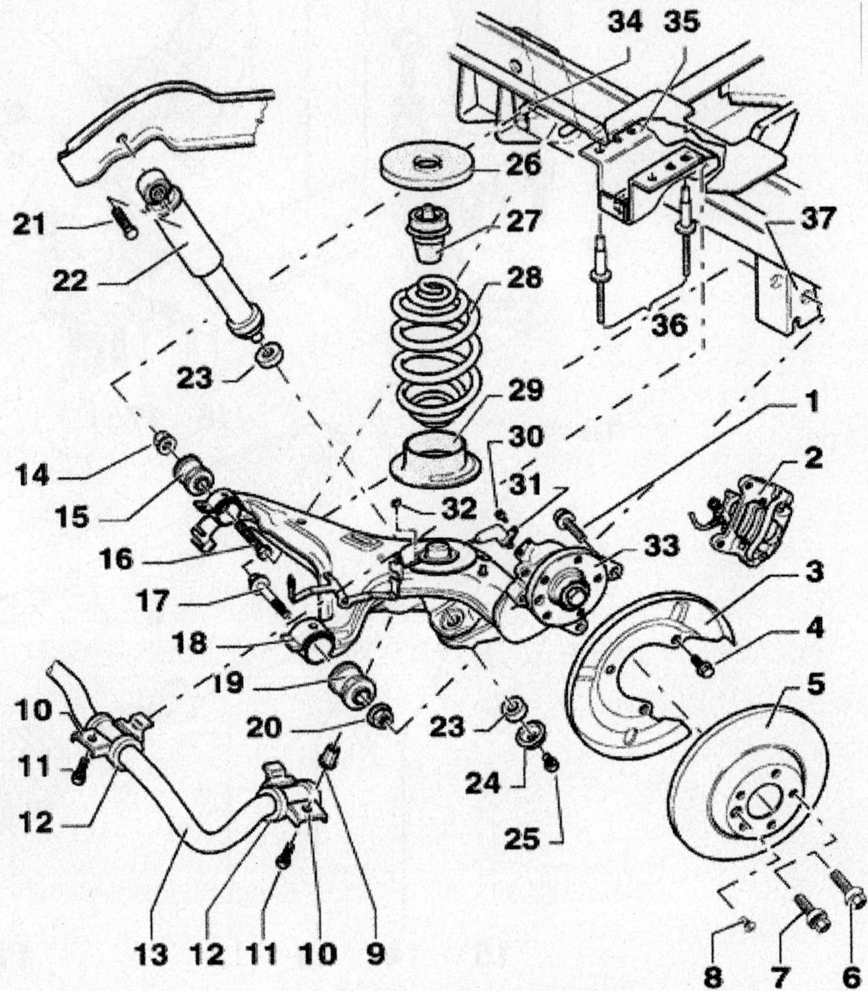

1. Hex bolt—125 ft. lbs. (170 Nm)
2. Brake caliper
3. Backing plate
4. Hex bolt
5. Brake disc
6. Wheel bolt
7. Wheel bolt
8. Socket head bolt— 11 ft. lbs. (15 Nm)
9. Pop rivet
10. Clamp
11. Hex bolt—22 ft. lbs. (30 Nm)
12. Rubber bearing
13. Stabilizer bar
14. Nut—74 ft. lbs. (100 Nm)
15. Inner rubber bushing
16. Hex bolt
17. Hex bolt
18. Control arm
19. Outer rubber bushing
20. Nut—118 ft. lbs. (160 Nm)
21. Hex bolt—74 ft. lbs. (100 Nm)
22. Shock absorber
23. Rubber bearing
24. Washer
25. Hex nut—33 ft. lbs. (45 Nm)
26. Upper backing plate
27. Stop buffer
28. Coil spring
29. Cap
30. Socket head screw
31. ABS sensor
32. Shock absorber buffer
33. Wheel bearing
34. Inner mounting bracket
35. Mounting bracket
36. Pop rivet
37. Outer mounting bracket

42348-EURO-G21

Exploded view of rear suspension

- Wheel bearing housing.
- Eccentric bushing.

5. Using a puller, press out upper ball joint.

To install:

6. Reverse the removal procedure to install. Torque the ball joint nut to 44 ft. lbs. (60 Nm), plus 90°.

Lower Ball Joint

REMOVAL & INSTALLATION

1. Before servicing the vehicle, refer to the Precautions Section.
2. Relieve the tension on the torsion bar.
3. Remove:
   - Front wheels.
   - Stabilizer bar coupling.
   - Ball joint to wheel housing connection.
   - Ball joint.

To install:

4. Install the new ball joint in the reverse order of removal. If no parts were installed other than the ball joint, no camber adjustment is necessary. Torque the ball joint bolts to 44 ft. lbs. (60 Nm), plus 90°.

Upper and Lower Control Arm

REMOVAL & INSTALLATION

➡The upper control arm can only be removed with the vehicle subframe removed first.

BUSHING REPLACEMENT

1. Remove the control arm.
2. Position the control arm on a press. Carefully push the bushing out of the control arm using the press.

To install:

3. Lightly lubricate, then position the bushing on the control arm.
4. Carefully press the bushing into the control arm.
5. Install the control arm.

Wheel Bearings

ADJUSTMENT

Front And Rear

The wheel bearings are sealed; no adjustment is necessary or possible.

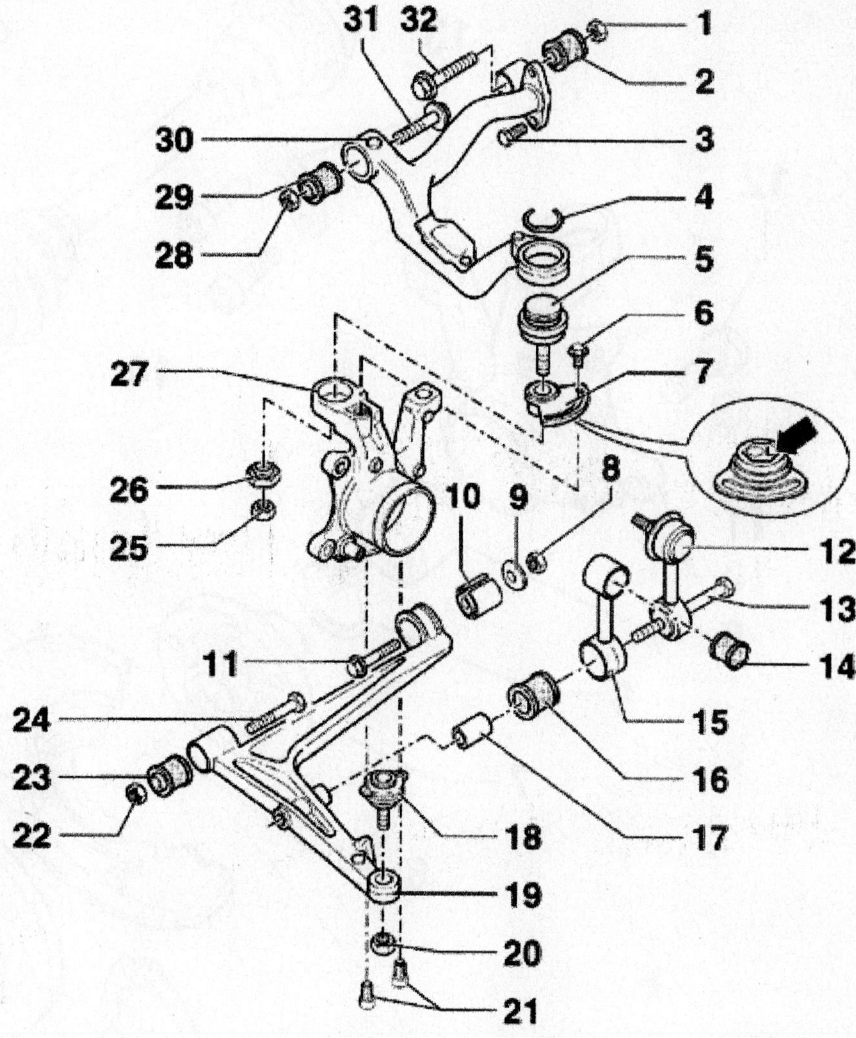

1. Hex nut—74 ft. lbs. (100 Nm)
2. Upper control arm rear bearing
3. Hex bolt
4. Snap ring—replace
5. Upper ball joint
6. Hex bolt—44 ft. lbs. (60 Nm)
7. Washer
8. Eccentric bushing
9. Hex nut—96 ft. lbs. (130 Nm)
10. Eccentric washer
11. Lower control arm rear bearing
12. Eccentric bolt
13. Coupling link
14. Hex bolt
15. Rubber bearing
16. Coupling link
17. Rubber bearing
18. Socket
19. Lower ball joint
20. Lower control arm
21. Self-locking nut—44 ft. lbs. (60 Nm) plus 90°
22. Socket head bolt—66 ft. lbs. (90 Nm) plus 90°
23. Hex nut—118 ft. lbs. (160 Nm)
24. Lower control arm front bearing
25. Hex bolt
26. Self-locking nut—44 ft. lbs. (60 Nm) plus 90°
27. Eccentric washer
28. Wheel bearing housing
29. Hex nut—74 ft. lbs. (100 Nm)
30. Upper control arm front bearing
31. Upper control arm
32. Hex bolt
33. Hex bolt

42348-EURO-G22

Exploded view of the front suspension assembly—Eurovan

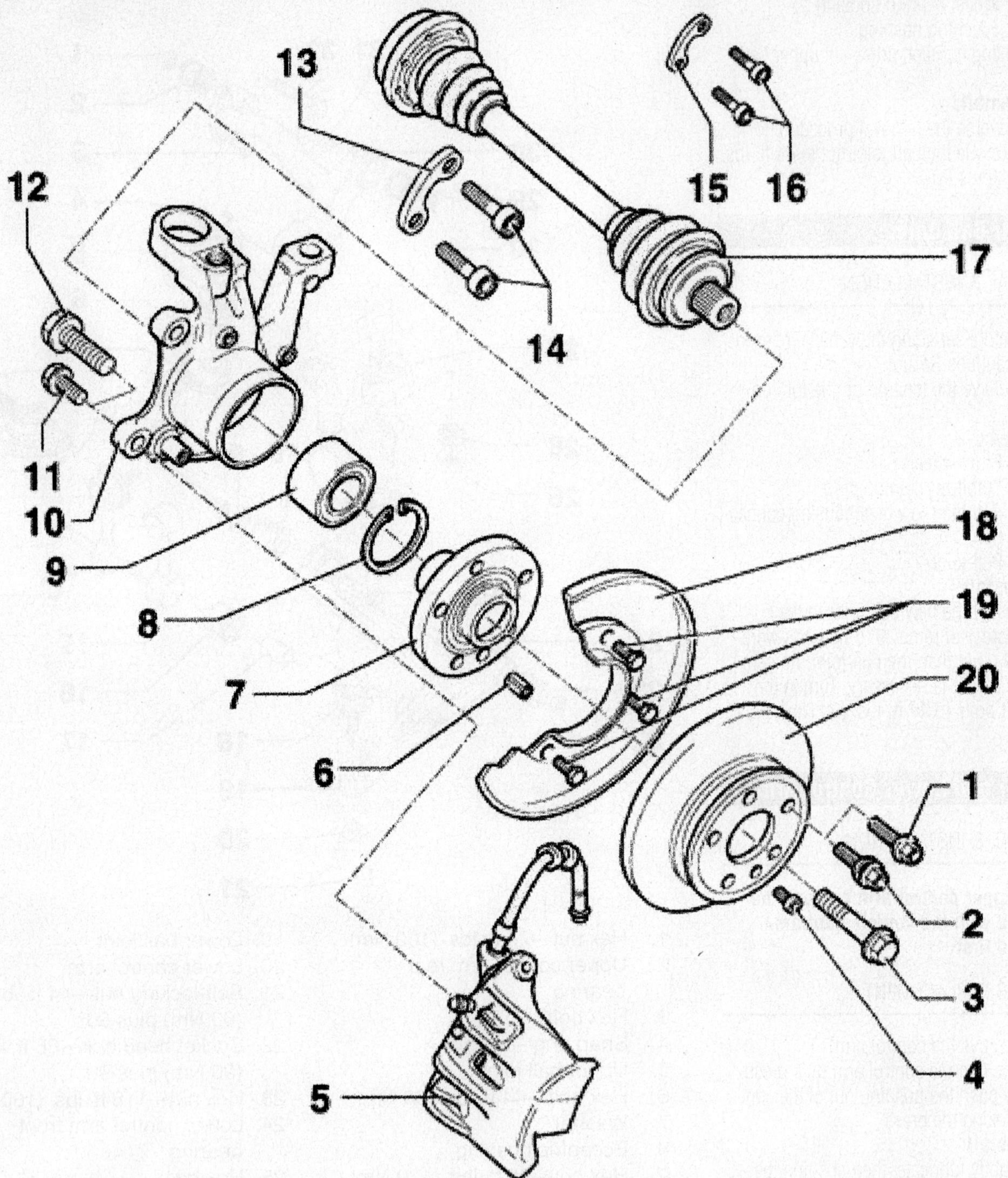

1. Wheel bolt (old)
2. Wheel bolt (new)
3. Hex bolt
4. Socket head bolt
5. Brake caliper
6. Spring dowel sleeve
7. Wheel hub
8. Snap ring
9. Wheel bearing
10. Wheel bearing housing
11. Hex bolt
12. Hex bolt
13. Backing plate
14. Socket head bolts
15. Backing plate
16. Socket head bolts
17. Drive axle
18. Backing plate
19. Hex bolt
20. Brake disc

42348-EURO-G23

Exploded view of the front wheel bearing housing

REMOVAL & INSTALLATION

Front

1. Before servicing the vehicle, refer to the Precautions Section.
2. Relieve the tension on the torsion bar.
3. Loosen the halfshaft retaining bolt.
4. Remove or disconnect:
   - Front wheel.
   - Lower bolt for stabilizer bar/coupling rod.
   - Caliper bracket mounting bolts and wire caliper aside.
   - Brake rotor and backing plate.
   - Tie rod end from the wheel bearing housing.
   - Threaded connection for upper ball joint.

➡**Mark position of eccentric bushing to wheel bearing housing or the front axle must be adjusted on the wheel alignment stand.**

   - Wheel bearing housing.
5. Place the wheel bearing housing on a press.
6. Drive out the hub with the wheel bearing.
7. Using a bearing separator and press, drive hub out of the bearing.

To install:

8. Press the new wheel bearing into the bearing housing using the appropriate bearing driver.
9. Press the hub into the wheel bearing using the appropriate bearing driver.
10. Install the wheel bearing housing. Torque M16 bolts to 207 ft. lbs. (280 Nm), torque M18 bolts to 262 ft. lbs. (355 Nm).
11. Reverse the removal procedure to install remaining components, using the following torque values:
   - Eccentric bushing nut: 96 ft. lbs. (130 Nm).
   - Upper ball joint nut: New self-locking nut: 44 ft. lbs. (60 Nm) plus 90°.
   - Tie rod end. New self-locking nut: 37 ft. lbs. (50 Nm). Bolt: 44 inch lbs. (5 Nm).
   - ABS wheel speed sensor.
   - Brake backing plate. Retaining bolts: 84 inch lbs. (10 Nm).
   - Brake rotor.
   - Brake caliper. Retaining bolt: 15 ft. lbs. (25 Nm).
   - Wheel lug bolts: 125 ft. lbs. (170 Nm).
   - Halfshaft retaining bolt: 111 ft. lbs. (150 Nm) plus 90°

Rear

1. Before servicing the vehicle, refer to the Precautions Section.
2. Loosen the drive axle bolt with vehicle resting on its wheels.
3. Raise and support vehicle and remove the rear wheels.
4. Remove or disconnect:
   - Wheel bearing threaded connection.
   - ABS speed sensor.
   - Parking brake cable at caliper.
   - Brake caliper and suspend out of the way.
   - Brake rotor and backing plate.
5. Using a puller, press the wheel hub out of the control arm.
6. Using a puller, press the wheel bearing out of the control arm.
7. Place the wheel hub in a vise and press out the inner bearing race.

To install:

8. Carefully press the new race into the wheel hub.
9. Press the new wheel bearing into the control arm until it reaches the stop.
10. Press the wheel hub into the control arm until it reaches the stop.
11. Reverse the removal procedure to install remaining components, using the following torque values:
   - Brake backing plate. Retaining bolts: 84 inch lbs. (10 Nm).
   - Brake rotor.
   - Brake caliper. Retaining bolt: 15 ft. lbs. (25 Nm).
   - Parking brake cable.
   - ABS wheel speed sensor.
   - Wheel lug bolts: 125 ft. lbs. (170 Nm).
   - Drive axle retaining bolt: 111 ft. lbs. (150 Nm) plus 90°.

BRAKES

Brake Caliper

REMOVAL & INSTALLATION

Front

1. Before servicing the vehicle, refer to the Precautions Section.
2. Remove the wheels.
3. Loosen the hydraulic line at the caliper, then remove the caliper from the carrier. Make sure to hold the pin with a back-up wrench when removing the caliper bolts.
4. Remove the caliper from the hydraulic line.
5. The carrier can be removed by removing the 2 bolts.

To install:

6. If removed, install the carrier.
7. Thread the caliper onto the hydraulic line and hand-tighten it. Fit the caliper into place on the carrier.
8. Torque the guide pin bolts to 18 ft. lbs. (25 Nm).
9. Tighten the hydraulic line and bleed the brakes.

Rear

1. Before servicing the vehicle, refer to the Precautions Section.
2. Make sure the ignition switch stays **OFF** and pump the brake pedal 25–35 times to relieve the system pressure.
3. Remove the wheels.
4. Disconnect the parking brake cable.
5. Loosen the hydraulic line.
6. Use a back-up wrench to hold the guide pins and remove the caliper bolts.
7. Lift the caliper off the carrier and unscrew it from the hydraulic line.

To install:

8. Thread the caliper onto the hydraulic line and hand-tighten it. Fit the caliper into place on the carrier. Torque the bolts to 26 ft. lbs. (35 Nm).
9. Bleed the brakes.
10. Install the wheels.

Brake Pads

REMOVAL & INSTALLATION

Front

1. Before servicing the vehicle, refer to the Precautions Section.
2. Remove the front wheels.
3. Hold the lower guide pin with an open wrench and remove the bolt securing the caliper to the guide pin.
4. Pivot the caliper up on the upper guide pin and slide the pads straight out to remove.

To install:

5. Compress the caliper piston into the bore.

6. Fit the new pads into the carrier and pivot the caliper into place.

7. The original bolts are micro-encapsulated with a thread locking compound. Install a new bolt or clean the old bolt and apply a thread-locking compound.

8. When tightening the bolt, be sure to use a back-up wrench to hold the guide pin. Torque the bolt to 26 ft. lbs. (35 Nm).

9. Install the wheels.

Rear

1. Before servicing the vehicle, refer to the Precautions Section.

2. Remove the rear wheels.

3. Remove the parking brake cable clip from the caliper. Disconnect the parking brake cable.

4. Hold the guide pin with a back-up wrench and remove the upper mounting bolt from the brake caliper.

5. Swing the caliper downward and remove the brake pads.

To install:

6. Retract the piston into the housing by rotating the piston clockwise.

7. Install the new brake pads onto the pad carrier.

8. Install the caliper to the pad carrier using a new self locking bolt or a thread locking compound and torque to 26 ft. lbs. (35 Nm).

9. Attach the hand brake cable to the caliper.

10. Check the parking brake operation and adjust the cable if necessary.

11. Install the wheels.

VOLKSWAGEN 13

Touareg

SPECIFICATIONS AND MAINTENANCE CHARTS

ENGINE AND VEHICLE IDENTIFICATION

Engine							Model Year	
Code ①	Liters (cc)	Cu. In.	Cyl.	Fuel Sys.	Engine Type	Eng. Mfg.	Code ②	Year
BAA	3.2 (3189)	195	6	Motronic	DOHC	Volkswagen	4	2004
BMX	3.2 (3189)	195	6	Motronic	DOHC	Volkswagen	5	2005
BHX	4.2 (4172)	255	8	Motronic	DOHC	Volkswagen	6	2006

DOHC: Double Overhead Camshafts

① Located on the vehicle data plate

② 10th digit of VIN

06041_TOUA_C0001

GENERAL ENGINE SPECIFICATIONS

Year	Model	Engine Displacement Liters	Engine ID	Net Horsepower @ rpm	Net Torque@rpm (ft. lbs.)	Bore x Stroke (in.)	Com-pression Ratio	Oil Pressure @ rpm
2004	Touareg	3.2	BAA	220@5400	225@3200	3.31x3.78	11.3:1	44@2000
	Touareg	4.2	BHX	310@6200	302@3000	3.33x3.66	11.0:1	44@2000
2005	Touareg	3.2	BMX	240@6000	229@3200	3.31x3.78	10.9:1	44@2000
	Touareg	4.2	BHX	310@6200	302@3000	3.33x3.66	11.0:1	44@2000
2006	Touareg	3.2	BMX	240@6000	229@3200	3.31x3.78	10.9:1	44@2000
	Touareg	4.2	BHX	310@6200	302@3000	3.33x3.66	11.0:1	44@2000

06041_TOUA_C0002

GASOLINE ENGINE TUNE-UP SPECIFICATIONS

Year	Engine Displacement Liters	Engine ID	Spark Plug Gap (in.)	Ignition Timing (deg.)		Fuel Pump (psi)	Idle Speed (rpm)		Valve Clearance	
				MT	AT		MT	AT	Intake	Exhaust
2004	3.2 (3189)	BAA	0.043	—	①	58 ②	—	650-750	HYD	HYD
	4.2 (4172)	BHX	0.043	—	①	58 ②	—	670-730	HYD	HYD
2005	3.2 (3189)	BMX	0.043	—	①	58 ②	—	650-750	HYD	HYD
	4.2 (4172)	BHX	0.043	—	①	58 ②	—	670-730	HYD	HYD
2006	3.2 (3189)	BMX	0.043	—	①	58 ②	—	650-750	HYD	HYD
	4.2 (4172)	BHX	0.043	—	①	58 ②	—	670-730	HYD	HYD

Note: The Vehicle Emission Control Information label reflects specification changes made during production.

The label figures must be used if they differ from those in this chart.

HYD: Hydraulic

NA: Information not available

① The ignition timing is controlled by the ECM and is not adjustable.

② System pressure at idle.

06041_TOUA_C0003

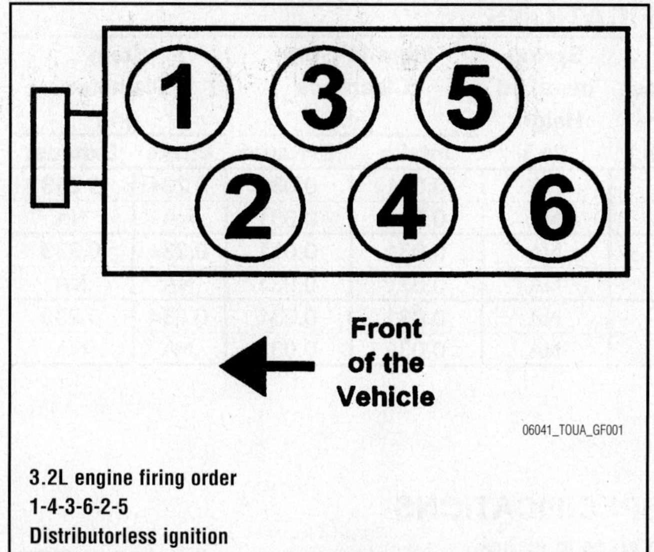

3.2L engine firing order
1-4-3-6-2-5
Distributorless ignition

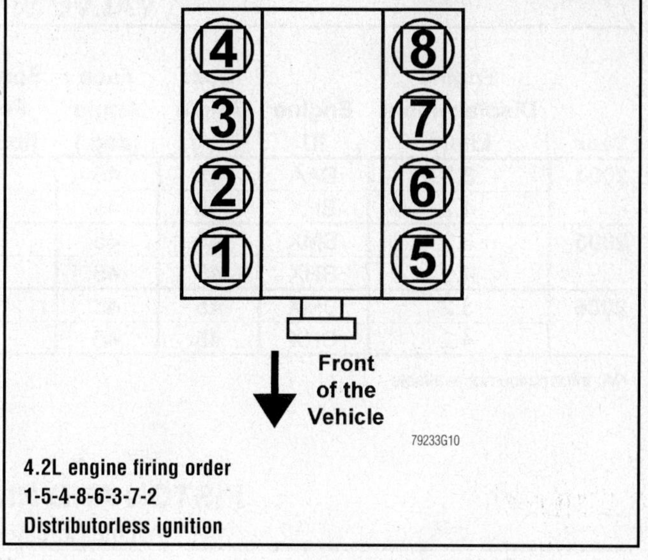

4.2L engine firing order
1-5-4-8-6-3-7-2
Distributorless ignition

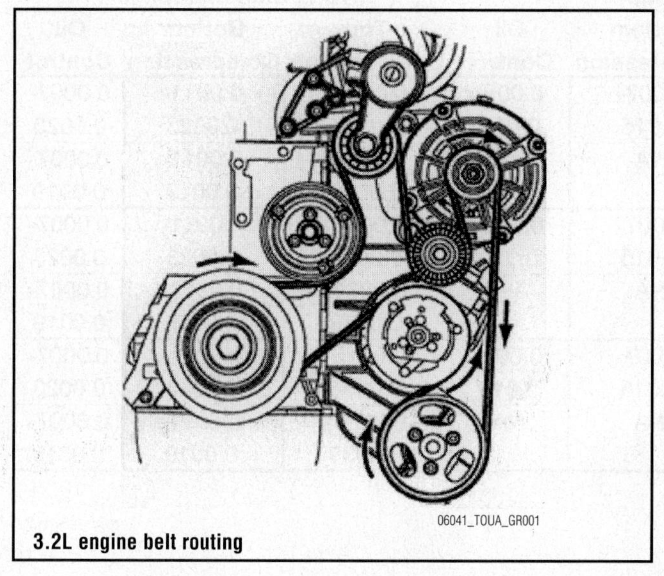

3.2L engine belt routing

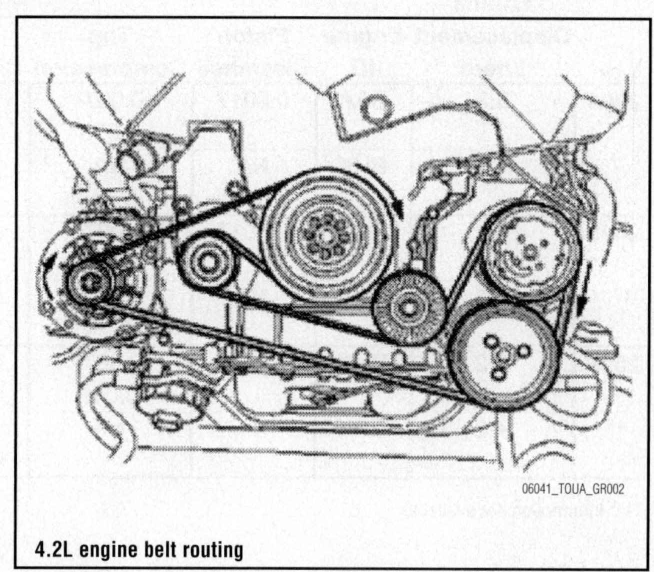

4.2L engine belt routing

CAPACITIES

Year	Model	Engine Displacement Liters	Engine ID	Engine Oil with Filter	Transmission (pts)		Final Drive (pts.)	Fuel Tank (gal.)	Cooling System (qts.)
					Manual	Auto.			
2004	Touareg	3.2	BAA	6.7	NA	9.5	—	26.4	9.5
	Touareg	4.2	BHX	8.0	NA	9.5	—	26.4	9.5
2005	Touareg	3.2	BMX	6.7	NA	9.5	—	26.4	9.5
	Touareg	4.2	BHX	8.0	NA	9.5	—	26.4	9.5
2006	Touareg	3.2	BMX	6.7	NA	9.5	—	26.4	9.5
	Touareg	4.2	BHX	8.0	NA	9.5	—	26.4	9.5

Note: All capacities are approximate. Add fluid gradually and check often to avoid overfilling.

NA: Information not available

06041_TOUA_C0004

VALVE SPECIFICATIONS

Year	Engine Displacement Liters	Engine ID	Seat Angle (deg.)	Face Angle (deg.)	Spring Test Pressure (lbs. @ in.)	Spring Installed Height (in.)	Stem-to-Guide Clearance (in.) Intake	Stem-to-Guide Clearance (in.) Exhaust	Stem Diameter (in.) Intake	Stem Diameter (in.) Exhaust
2004	3.2	BAA	45	45	NA	NA	0.031	0.031	0.234	0.233
	4.2	BHX	45	45	NA	NA	0.031	0.031	NA	NA
2005	3.2	BMX	45	45	NA	NA	0.031	0.031	0.234	0.233
	4.2	BHX	45	45	NA	NA	0.031	0.031	NA	NA
2006	3.2	BMX	45	45	NA	NA	0.031	0.031	0.234	0.233
	4.2	BHX	45	45	NA	NA	0.031	0.031	NA	NA

NA: Information not available

06041_TOUA_C0005

PISTON AND RING SPECIFICATIONS

All measurements are given in inches.

Year	Engine Displacement Liters	Engine ID	Piston Clearance	Ring Gap Top Compression	Ring Gap Bottom Compression	Ring Gap Oil Control	Ring Side Clearance Top Compression	Ring Side Clearance Bottom Compression	Ring Side Clearance Oil Control
2004	3.2	BAA	0.0017	0.007-0.015	0.007-0.015	0.009-0.019	0.0015-0.0035	0.0011-0.0023	0.0007-0.0023
	4.2	BHX	NA	NA	NA	NA	0.0031-0.0039	0.0011-0.0019	0.0007-0.0019
2005	3.2	BMX	0.0017	0.007-0.015	0.007-0.015	0.009-0.019	0.0015-0.0035	0.0011-0.0023	0.0007-0.0023
	4.2	BHX	NA	NA	NA	NA	0.0031-0.0039	0.0011-0.0019	0.0007-0.0019
2006	3.2	BMX	0.0017	0.007-0.015	0.007-0.015	0.009-0.019	0.0015-0.0035	0.0011-0.0023	0.0007-0.0023
	4.2	BHX	NA	NA	NA	NA	0.0031-0.0039	0.0011-0.0019	0.0007-0.0019

NA: Information not available

06041_TOUA_C0006

CRANKSHAFT AND CONNECTING ROD SPECIFICATIONS

All measurements are given in inches.

Year	Engine Displacement Liters	Engine ID	Crankshaft Main Brg. Journal Dia.	Crankshaft Main Brg. Oil Clearance	Crankshaft Shaft End-play	Crankshaft Thrust on No.	Connecting Rod Journal Diameter	Connecting Rod Oil Clearance	Connecting Rod Side Clearance
2004	3.2	BAA	2.3605	0.0007-0.0023	0.0027-0.0090	5	2.1243	0.0007-0.0027	0.0019-0.0122
	4.2	BHX	NA	NA	NA	NA	NA	NA	NA
2005	3.2	BMX	2.3605	0.0007-0.0023	0.0027-0.0090	5	2.1243	0.0007-0.0027	0.0019-0.0122
	4.2	BHX	NA	NA	NA	NA	NA	NA	NA
2006	3.2	BMX	2.3605	0.0007-0.0023	0.0027-0.0090	5	2.1243	0.0007-0.0027	0.0019-0.0122
	4.2	BHX	NA	NA	NA	NA	NA	NA	NA

NA: Information not available

06041_TOUA_C0007

TORQUE SPECIFICATIONS

All readings in ft. lbs.

Year	Engine Displacement Liters	Engine ID	Cylinder Head Bolts	Main Bearing Bolts	Rod Bearing Bolts	Crankshaft Damper Bolt	Flywheel Bolts	Manifold Intake	Manifold Exhaust	Spark Plugs	Oil Pan Drain Plug
2004	3.2	BAA	①	②	③	④	⑤	10	⑥	15	22
	4.2	BHX	⑦	NA	NA	⑧	⑨	7	18	22	37
2005	3.2	BMX	①	②	③	④	⑤	10	⑥	15	22
	4.2	BHX	⑦	NA	NA	⑧	⑨	7	18	22	37
2006	3.2	BMX	①	②	③	④	⑤	10	⑥	15	22
	4.2	BHX	⑦	NA	NA	⑧	⑨	7	18	22	37

NA: Information not available

① Torque in four steps: (use new bolts)
- Step 1: 22 ft. lbs.
- Step 2: 37 ft. lbs.
- Step 3: plus 90 degrees
- Step 4: plus 90 degrees

② Use new bolts: 22 ft. lbs. plus 180 degrees

③ Use new bolts: 22 ft. lbs. plus 90 degrees

④ Use new bolts: 74 ft. lbs. plus 90 degrees

⑤ Use new bolts: 44 ft. lbs. plus 90 degrees

⑥ M8 nuts: 18 ft. lbs
M10 nuts: 30 ft. lbs.

⑦ Torque in four steps: (use new bolts)
- Step 1: 26 ft. lbs.
- Step 2: 44 ft. lbs.
- Step 3: plus 90 degrees
- Step 4: plus 90 degrees

⑧ Use new bolts: 148 ft. lbs. plus 180 degrees

⑨ Use new bolts: 22 ft. lbs. plus 90 degrees

06041_TOUA_C0008

WHEEL ALIGNMENT

Year	Model		Caster Range (+/-Deg.)	Caster Preferred Setting (Deg.)	Camber Range (+/-Deg.)	Camber Preferred Setting (Deg.)	Toe-in (Deg.)
2004	1BA Strut	F	0.16	8.16	0.33	0.00	0.04 +/-0.04
		R	—	—	0.33	1.00	0.16 +/-0.08
	1BE Strut	F	0.16	8.50	0.33	0.00	0.04 +/-0.04
		R	—	—	0.33	1.00	0.16 +/-0.08
	1BK Strut	F	0.16	8.58	0.33	-0.16	0.08 +/-0.04
		R	—	—	0.33	1.00	0.16 +/-0.08
2005	1BA Strut	F	0.16	8.16	0.33	0.00	0.04 +/-0.04
		R	—	—	0.33	1.00	0.16 +/-0.08
	1BE Strut	F	0.16	8.50	0.33	0.00	0.04 +/-0.04
		R	—	—	0.33	1.00	0.16 +/-0.08
	1BK Strut	F	0.16	8.58	0.33	-0.16	0.08 +/-0.04
		R	—	—	0.33	1.00	0.16 +/-0.08
2006	1BA Strut	F	0.16	8.16	0.33	0.00	0.04 +/-0.04
		R	—	—	0.33	1.00	0.16 +/-0.08
	1BE Strut	F	0.16	8.50	0.33	0.00	0.04 +/-0.04
		R	—	—	0.33	1.00	0.16 +/-0.08
	1BK Strut	F	0.16	8.58	0.33	-0.16	0.08 +/-0.04
		R	—	—	0.33	1.00	0.16 +/-0.08

06041_TOUA_C0009

TIRE, WHEEL AND BALL JOINT SPECIFICATIONS

Year	Model	OEM Tires		Tire Pressures (psi)		Wheel	Lug Nut Torque
		Standard	Optional	Front	Rear	Size	Ft. Lbs.
2004	Touareg 3.2L	P235/65R17	NA	36	38	NA	118
	Touareg 3.2L	P255/60R17	NA	36	38	NA	118
	Touareg 3.2L	P255/55R17	NA	38	42	NA	118
	Touareg 4.2L	P255/60R17	NA	36	38	NA	118
	Touareg 4.2L	P255/55R18	NA	38	42	NA	118
	Touareg 4.2L	P275/45R19	NA	39	44	NA	118
2005	Touareg 3.2L	P235/65R17	NA	36	38	NA	118
	Touareg 3.2L	P255/60R17	NA	36	38	NA	118
	Touareg 3.2L	P255/55R17	NA	38	42	NA	118
	Touareg 4.2L	P255/60R17	NA	36	38	NA	118
	Touareg 4.2L	P255/55R18	NA	38	42	NA	118
	Touareg 4.2L	P275/45R19	NA	39	44	NA	118
2006	Touareg 3.2L	P235/65R17	NA	36	38	NA	118
	Touareg 3.2L	P255/60R17	NA	36	38	NA	118
	Touareg 3.2L	P255/55R17	NA	38	42	NA	118
	Touareg 4.2L	P255/60R17	NA	36	38	NA	118
	Touareg 4.2L	P255/55R18	NA	38	42	NA	118
	Touareg 4.2L	P275/45R19	NA	39	44	NA	118

NA: Information not available
OEM: Original Equipment Manufacturer
PSI: Pounds Per Square Inch

06041_TOUA_C0010

BRAKE SPECIFICATIONS

All measurements in inches unless noted

Year	Model		Brake Disc			Minimum	Brake Caliper	
			Original Thickness	Minimum Thickness	Maximum Run-out	Lining Thickness	Guide Pins (ft. lbs.)	Mounting Bolts (ft. lbs.)
2004	Touareg	F	①	②	NA	0.07	22	199
		R	③	④	NA	0.07	22	133
2005	Touareg	F	①	②	NA	0.07	22	199
		R	③	④	NA	0.07	22	133
2006	Touareg	F	①	②	NA	0.07	22	199
		R	③	④	NA	0.07	22	133

NA: Information not available

① 16" FN3: 1.16
17" Brembo: 1.25
18" Brembo: 1.33

② 16" FN3: 1.00
17" Brembo: 1.18
18" Brembo: 1.25

③ 16" FN44: 0.86
17" Brembo: 1.10

④ 16" FN44: 0.70
17" Brembo: 1.02

06041_TOUA_C0011

SCHEDULED MAINTENANCE INTERVALS
VOLKSWAGEN TOUAREG

TO BE SERVICED	TYPE OF SERVICE	VEHICLE MILEAGE INTERVAL (x1000)												
		5	10	20	30	40	45	50	60	70	75	80	90	100
Engine oil & filter	R	✓	✓	✓	✓	✓		✓	✓	✓		✓	✓	✓
Brake pad thickness	S/I		✓	✓	✓	✓		✓	✓	✓		✓	✓	✓
Dust seals on ball joints, tie rod ends & tie rods	S/I			✓		✓			✓			✓		✓
Battery	S/I		✓	✓	✓	✓		✓	✓	✓		✓	✓	✓
Brake system	S/I		✓	✓	✓	✓		✓	✓	✓		✓	✓	✓
Cooling system	S/I			✓		✓			✓			✓		✓
Driveshaft boots	S/I			✓		✓			✓			✓		✓
Engine (check for leaks)	S/I			✓		✓			✓			✓		✓
Engine exhaust	S/I			✓		✓			✓			✓		✓
Power steering fluid level	S/I			✓		✓			✓			✓		✓
Underbody	S/I						✓					✓		
Timing belt (V8)	R											✓		
Transmission fluid level	S/I			✓		✓			✓			✓		✓
Air filter element	R					✓						✓		
Spark plugs (V6)	R					✓						✓		
Spark plugs (V8)	R								✓					
Passenger compartment air filter	R			✓		✓			✓			✓		✓
Drive belts	S/I											✓		
Brake fluid ①	R													

R: Replace S/I: Service or Inspect

① Replace every two years regardless of mileage.

FREQUENT OPERATION MAINTENANCE (SEVERE SERVICE)

If a vehicle is operated under any of the following conditions it is considered severe service:

- Extremely dusty areas.

- 50% or more of the vehicle operation is in 32°C (90°F) or higher temperatures, or constant operation in temperatures below 0°C (32°F).

- Prolonged idling (vehicle operation in stop and go traffic).

- Frequent short running periods (engine does not warm to normal operating temperatures).

- Police, taxi, delivery usage or trailer towing usage.

Oil & oil filter change: change every 3750 miles.

Air filter element: service or inspect every 15,000 miles.

Automatic transaxle fluid & filter: replace every 30,000 miles.

06041_TOUA_C0012

ENGINE REPAIR

➡Disconnecting the negative battery cable on some vehicles may interfere with the functions of the on board computer system. The computer may undergo a relearning process once the negative battery cable is reconnected.

Battery

REMOVAL & INSTALLATION

➡In order to prevent damaging the trim panel under the drivers seat during removal, release of the trim panel locking tabs must be performed using the following procedure in the sequence given. Locking tabs (A) are engaged into a transverse member under the seat. Locking tabs (B) engage in slots located ahead of the seat. Retaining tabs (C) are clipped into side trim of seat

1. Before servicing the vehicle, refer to the Precautions Section.
2. Move the driver's seat to the rearmost and highest position.
3. Release locking tabs (A) by pressing down and rearward on the rear edge of the trim panel.
4. Release locking tabs (B) and locating tabs (C) by pushing the front edge of the panel downward and pulling panel away from seat mounting frame.
5. Pry upwards on the cover of the front seat spindle to remove it.
6. Pull the front spindle trim from the seat track by pulling it out.
7. Move the seat fully forward and to the highest position.
8. Pry the rear seat spindle covers upwards to remove them.
9. Pull off the rear spindle trim from the seat tracks.
10. Carefully pry off the seat console side trim and pull it up and out.
11. Remove the front bolts securing the seat to the floor then tilt the seat and frame to the rear.
12. Remove the bolt securing the air duct to the battery box.
13. Release the clips and remove the battery box cover.
14. Disconnect the negative battery cable first, then the positive.

➡Failure to disconnect the cables in the proper order will cause the main battery switch E74 to deactivate and may damage electrical system components.

15. Remove the vent hose and battery retaining clamps, then lift the battery out.

To install:

16. Installation is the reverse of removal. Observe the following tightening torques:
    - Battery hold down bracket bolt: 17 ft. lbs. (23 Nm)
    - Battery retainer bracket bolt: 7 ft. lbs. (9 Nm)
    - Battery terminal clamp nut: 7 ft. lbs. (9 Nm)
    - Seat frame to body bolts: 33 ft. lbs. (45 Nm)

➡Be sure to connect the positive cable first, then the negative.

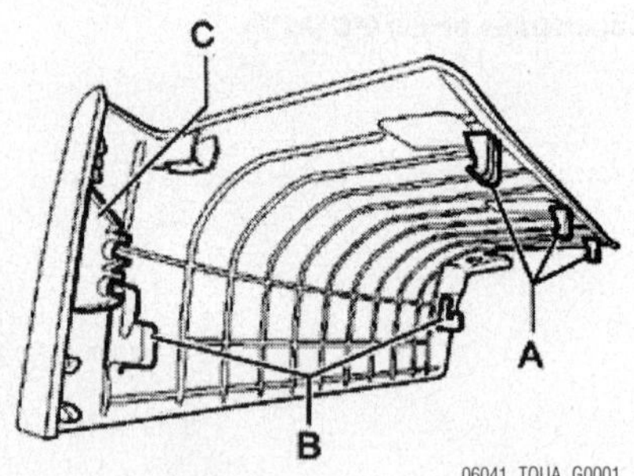

Location of the locking and retaining tabs

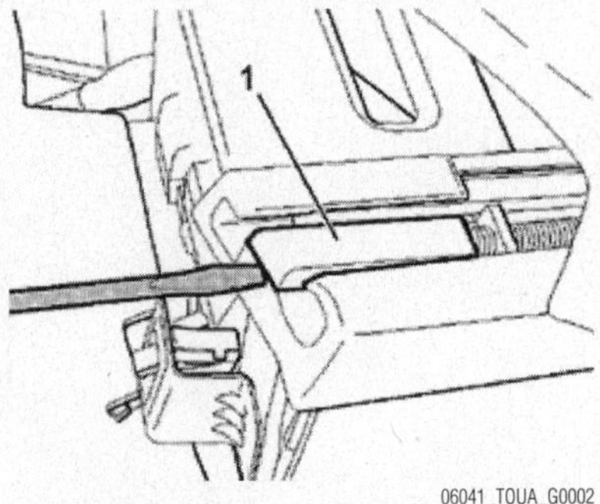

Remove the spindle covers by prying upwards

Distributor

All engines in this section are equipped with a distributorless ignition system.

Alternator

REMOVAL & INSTALLATION

3.2L Engines

1. Before servicing the vehicle, refer to the Precautions Section.
2. Disconnect the battery.
3. Unplug the electrical connector from the back of the alternator.
4. Remove the cap from the positive terminal on the alternator and remove the nut securing the cable.
5. On water cooled alternators, clamp the coolant hoses shut. Remove the bolt securing the coolant pipes to the alternator and pull it away from the alternator.
6. On air cooled alternators, remove the air guide over the radiator.
7. Mark the direction of rotation on the belt.
8. Screw a M8x50 bolt into the threaded hole of the tensioner to release tension on the belt. Screw the bolt in only far enough to allow removal of the belt.
9. Remove the 3 bolts securing the tensioner, then remove it together with the belt.
10. Remove the 2 bolts securing the alternator, then remove it from the engine.

To install:

11. Drive the threaded sleeves out of the alternator housing 4mm towards the rear of the alternator.
12. Remaining installation is the reverse of removal. Observe the following tightening torques:
    - Alternator bracket and mounting bolts: 15 ft. lbs. (20 Nm)

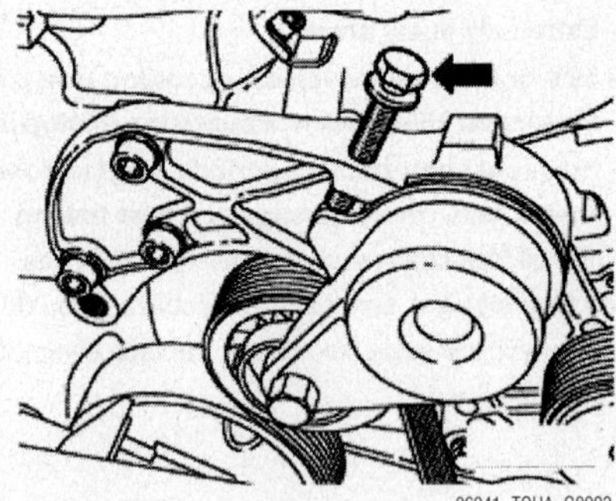

Use a M8x50 bolt to release tension on the belt—3.2L engines

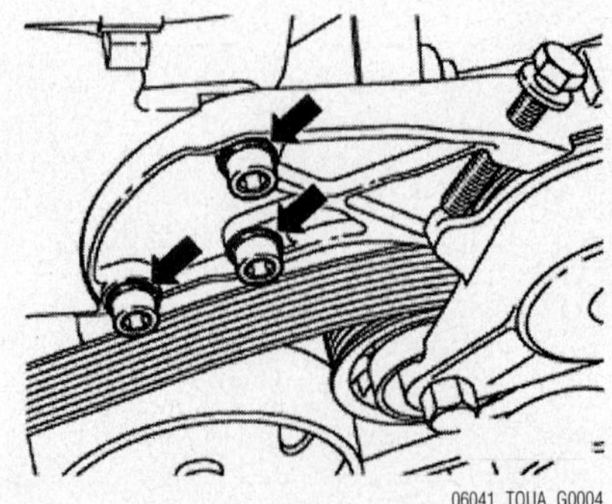

Tensioner retaining bolts—3.2L engines

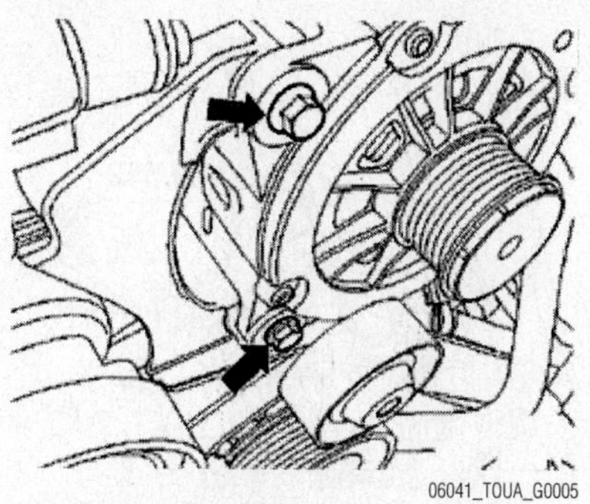

Alternator mounting bolts—3.2L engines

- Tensioner mounting bolts: 15 ft. lbs. (20 Nm)
- Coolant pipe: 7 ft. lbs. (9 Nm)
- Positive terminal nut: 11 ft. lbs. (15 Nm)

➡Be sure to reinstall the belt in the proper direction of rotation and to remove the belt tension release bolt. On water cooled alternators, check the coolant level.

4.2L Engines

1. Before servicing the vehicle, refer to the Precautions Section.
2. Disconnect the battery.
3. Remove the engine assembly.
4. Mark the direction of rotation on the belt.
5. Rotate the tensioner using a 19mm wrench and secure it in the released position using tool T10060A or a similar sized pin punch. Remove the belt.
6. Unplug the electrical connector from the back of the alternator.
7. Remove the cap from the positive terminal on the alternator and remove the nut securing the cable.
8. Remove the bolt securing the coolant lines on the alternator, then pull the lines away from it.
9. Remove the 2 bolts securing the alternator to the engine.

To install:

10. Installation is the reverse of removal. Observe the following tightening torques:
    - Alternator bracket and mounting bolts: 15 ft. lbs. (20 Nm)
    - Tensioner mounting bolts: 15 ft. lbs. (20 Nm)
    - Coolant pipe: 7 ft. lbs. (9 Nm)
    - Positive terminal nut: 11 ft. lbs. (15 Nm)

➡Be sure to reinstall the belt in the proper direction of rotation and to remove the belt tension release tool. Use new O-rings on the coolant pipes and check the coolant level.

Ignition Timing

ADJUSTMENT

➡The ignition timing is controlled by the engine control module and is not adjustable. However the timing can be monitored on a scan tool connected to the DLC in the vehicle. No specification has been given by the manufacturer.

Engine Assembly

REMOVAL & INSTALLATION

➡The engine is lowered from the vehicle together with the front driveline, transmission and transfer case. All tie wraps that are loosened or removed in order to remove the engine must be replaced or reinstalled to the same place when the engine is reinstalled. Tag all hoses and wiring during removal to use as reference during reassembly.

3.2L Engines

1. Before servicing the vehicle, refer to the Precautions Section.
2. Disconnect the battery.
3. Remove the left and right wiper arms.
4. Remove the engine covers (A, B and C) and plenum cover (D).
5. Relieve the fuel system pressure.
6. Disconnect the fuel supply line.
7. Disconnect the breather line to the EVAP purge valve.
8. Unplug the small connector from the Engine Control Module (ECM) and the wiring harness ground connection.
9. Unplug the wiring harness connectors from the left plenum fuse box. Remove the harness from the plenum chamber and place it on the engine.
10. Disconnect the harness from the starter and alternator, place it on the engine.
11. Disconnect the vacuum hose from the intake manifold.
12. Pull the pressure hose from the secondary air injection pump.
13. Unplug the oxygen sensor wiring harness and place it on the engine.
14. If equipped, unplug the connecting line to the air suspension compressor at the air filter.
15. Remove the top part of the air filter upwards and out.
16. Remove the transmission breather line from the air filter housing.

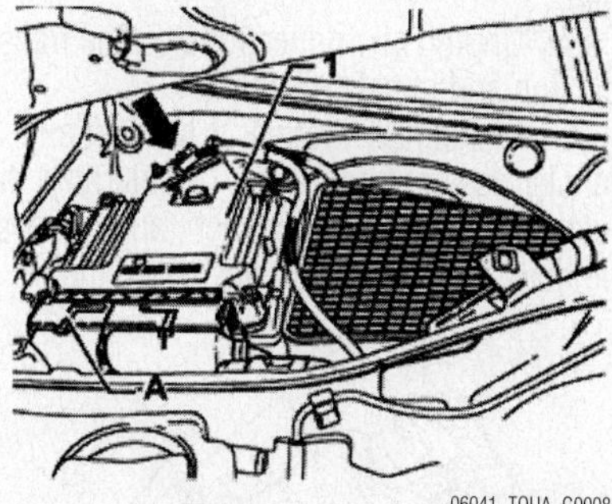

Unplug connector (A) from the ECM (1), and the ground connection (arrow)—3.2L and 4.2L engines

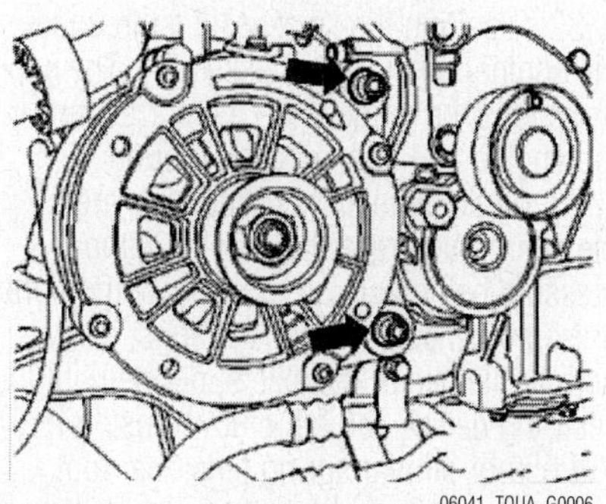

Alternator mounting bolts—4.2L engines

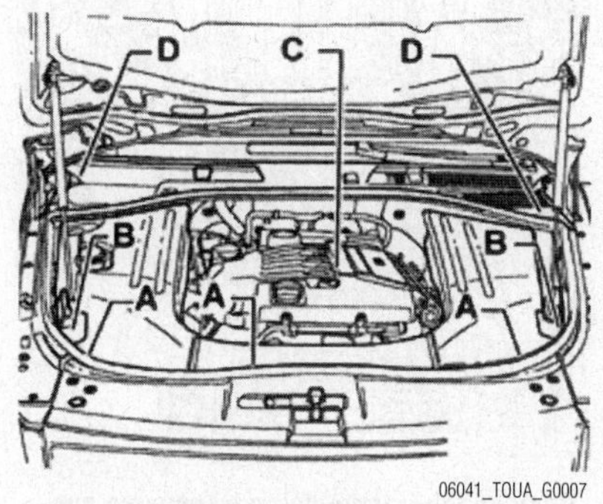

Engine cover identification—3.2L engines

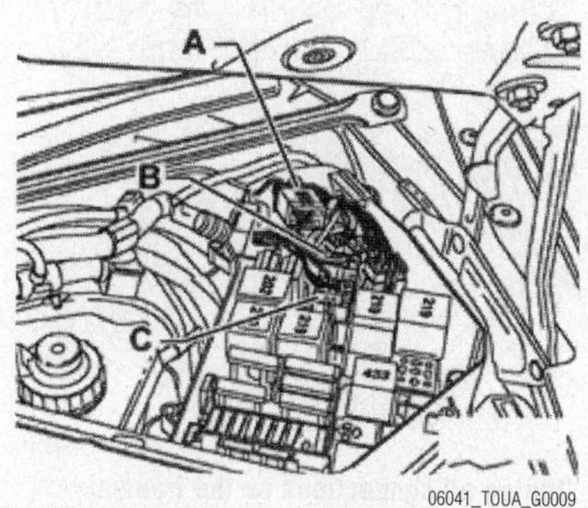

Unplug connectors (A), (B) and (C) from the fuse box—3.2L and 4.2L engines

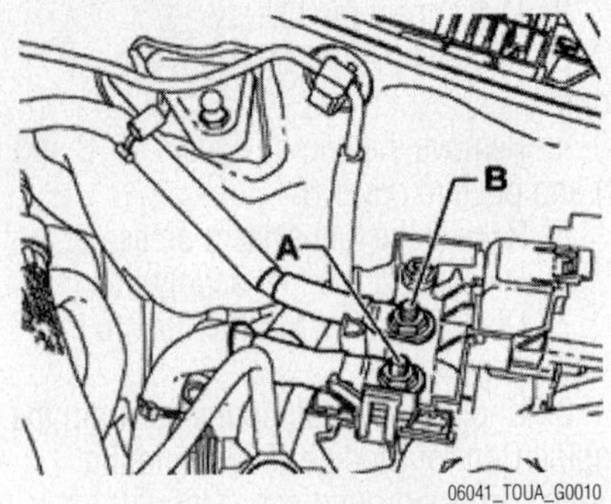

Disconnect the starter (A) and alternator (B) harness—3.2L engines

17. Discharge and recover the air conditioning system refrigerant.

18. Remove the sound insulation tray.

19. Drain the cooling system. Pull the upper hose from the radiator and the lower hose off the coolant pipe.

20. Disconnect the lines from the transmission cooler and power steering cooler; catch any fluid that runs out.

21. Loosen the double clamp between the catalytic converter and the center muffler and push the clamp forward.

22. Support the rear muffler, remove the rear mountings and lower the exhaust system.

23. Remove the heat shield from the steering shaft and loosen the universal joint.

24. Unplug all connections on the transmission and transfer case.

25. Matchmark the rear driveshaft and transfer case output flange. Counterhold the nuts with an open end wrench, and remove the bolts securing the driveshaft to the flange. Secure the driveshaft to prevent it from flexing.

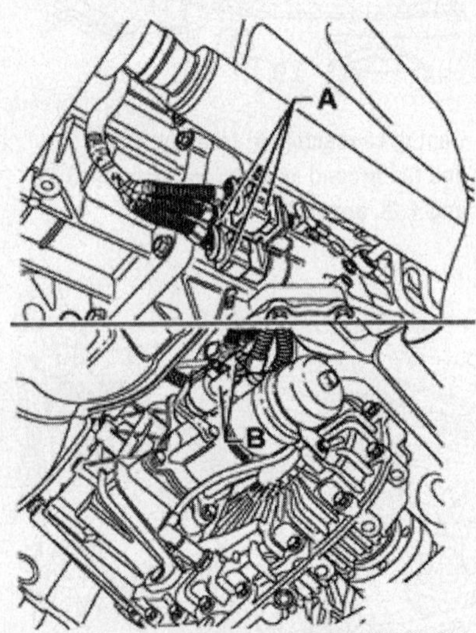

Unplug all connections on the transmission (A) and transfer case (B)—3.2L and 4.2L engines

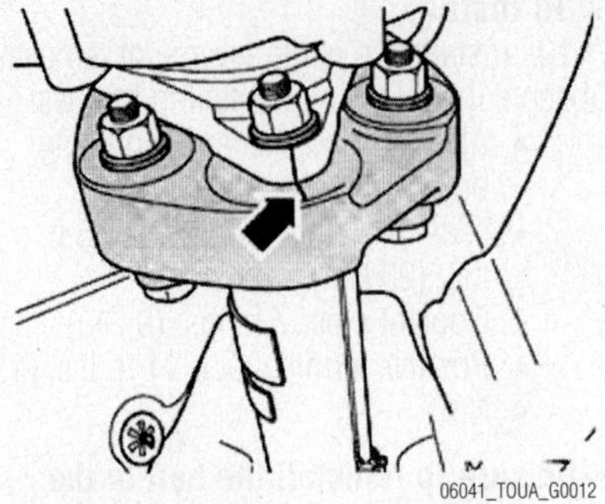

Matchmark the driveshaft and transfer case flanges—3.2L and 4.2L engines

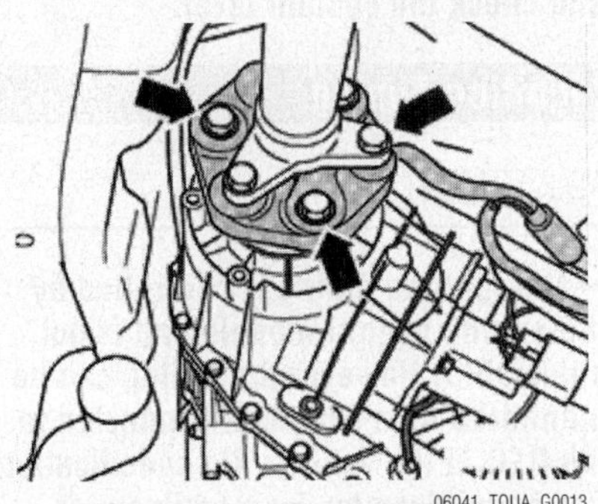

Remove the fasteners securing the driveshaft to the transfer case flange—3.2L and 4.2L engines

26. Remove the front wheels and wheel house liners.

27. Disconnect and plug the brake lines from the brake hoses. Catch any fluid that runs out.

28. Label and separate connectors between the body and front axle. If equipped, disconnect the lines for the air suspension at the strut.

29. Remove the transmission transverse support

30. Use tool VW 552 or similar spring compressor to secure the struts, then remove the upper bolts securing the struts to each side of the vehicle.

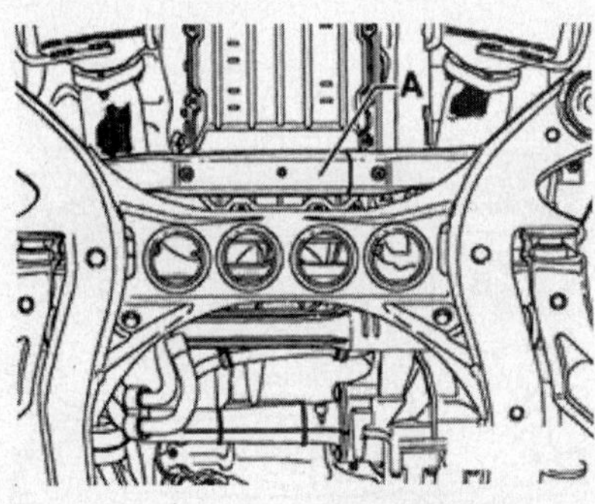

Remove the transmission transverse support—3.2L and 4.2L engines

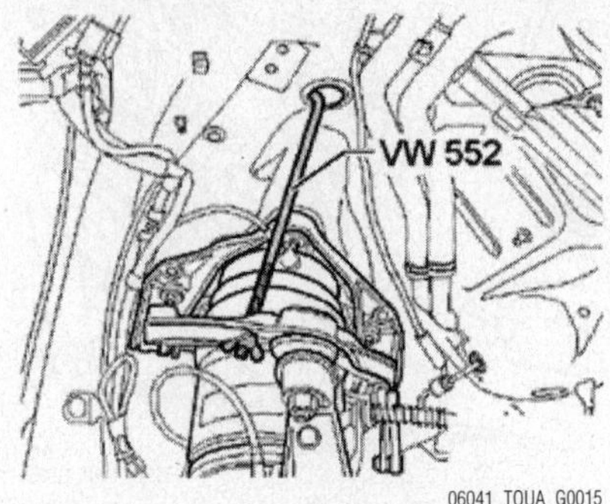

Secure the strut as shown—3.2L engines

➡Use of tool VAS6131 or a similar scissor lift table providing support of the subframe and engine/driveline assembly is recommended. This allows for proper alignment of the subframe and engine/driveline assembly during removal and installation.

31. Prepare tool VAS6131 or similar scissor lift as follows:
   a. Attach pulley end engine supports VAS6131/6-1 and VAS6131/6-2, or similar. Secure the bolts at the following positions:
      - Left subframe support: A5, B4 and B5
      - Right subframe support: G4, H4 and H5
   b. Attach axle supports VAS6131/6-3 and VAS6131/6-4, or similar. Secure the bolts at the following positions:
      - Left suspension support: A6 and C6
      - Right suspension support: F6 and H6
   c. Rotate the turntables of the axle supports downward.
   d. Place axle supports VAS6131/6-5 (or similar) for the subframe and VAS6131/6-6 (or similar) for the transmission mount at the appropriate positions on the scissor lift table.

32. Position the scissor lift table under the engine/transmission assembly. The supports must be guided into the corresponding engine and transmission mounts.

33. Make sure all four supports are in the appropriate receptacle holes without pressure being applied to them. Using light pressure beneath the control arms, screw both turntables of the axle supports until the securing device can be removed using tool VW 552 or similar spring tensioner tool.

34. Guide the subframe supports into their respective receptacles, correcting the

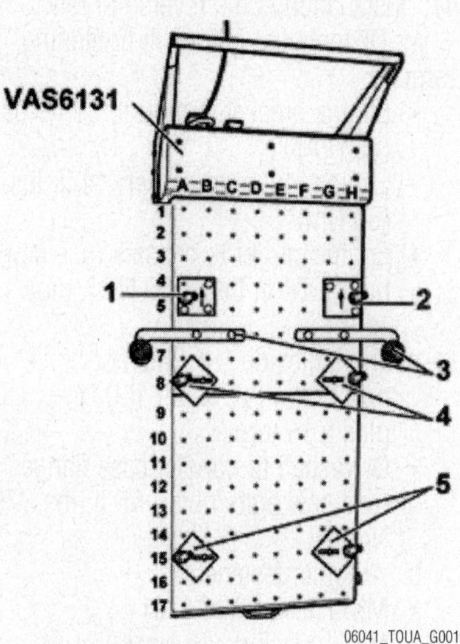

The carrier supports must be positioned on the scissor table as shown—3.2L and 4.2L engines

Guide the supports into the corresponding mounts—3.2L and 4.2L engines

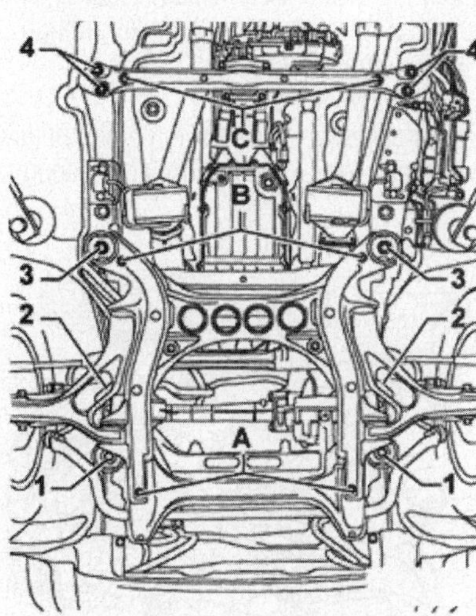

Loosen subframe bolts (1, 2 and 3), as well as transmission transverse support (4)—3.2L and 4.2L engines

support height as necessary with the knurled nuts.

35. Loosen subframe bolts (1, 2 and 3), as well as transmission transverse support (4).

➡Anytime the subframe is loosened, it will be necessary to perform a wheel alignment on the vehicle.

36. Slowly lower the engine/transmission assembly, constantly observing clearance and checking for items that may not have been disconnected.

To install:

➡When installing the drivetrain assembly back into the vehicle, make sure to bring the mounts to rest against the body using the scissor lift table. Do not let the bolts draw up the drivetrain assembly as the threaded inserts in the body could become damaged.

37. Installation is the reverse of removal.

a. Observe the following tightening torques:

- Engine bracket to mount: 55 ft. lbs. (75 Nm)
- Engine mount to carrier: 44 ft. lbs. (60 Nm)
- Engine carrier to chassis (use new bolts): 74 ft. lbs. (100 Nm), plus a ½ turn
- Transmission crossmember (use new bolts): 37 ft. lbs. (50 Nm), plus a ¼ turn
- Driveshaft to transfer case flange (use new bolts/nuts): 55 ft. lbs. (75 Nm)

b. All other fasteners:

- M6: 7 ft. lbs. (10 Nm)
- M7: 11 ft. lbs. (15 Nm)
- M8: 18 ft. lbs. (25 Nm)
- M10: 30 ft. lbs. (40 Nm)
- M12: 44 ft. lbs. (60 Nm)

38. Make sure all lines, hoses and wiring harnesses are properly routed. Refill or top off all fluids and recharge the A/C system. Perform a wheel alignment.

4.2L Engines

1. Before servicing the vehicle, refer to the Precautions Section.
2. Disconnect the battery.
3. Remove the left and right wiper arms.
4. Remove the engine covers (A and B) and plenum covers (C).
5. Relieve the fuel system pressure.
6. Disconnect the fuel supply line.
7. Disconnect the breather line to the EVAP purge valve.
8. Unplug the small connector from the

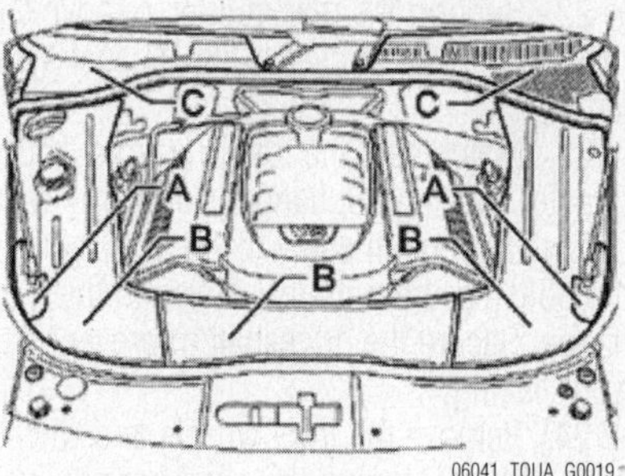

Engine cover identification—4.2L engines

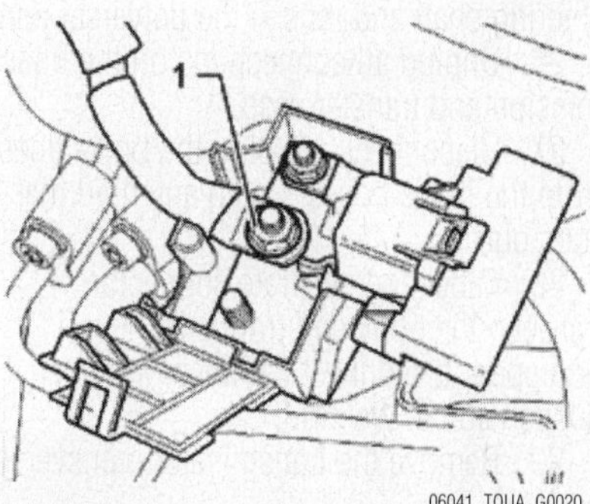

Disconnect the starter (1) harness—4.2L engines

Engine Control Module (ECM) and the wiring harness ground connection.

9. Unplug the wiring harness connectors from the left plenum fuse box. Remove the harness from the plenum chamber and place it on the engine.
10. Disconnect the harness from the starter, place it on the engine.
11. Unplug the left Mass Air Flow (MAF) sensor connector. Remove the intake hose and left air filter housing.
12. Unplug the oxygen sensor wiring harness and place it on the engine.
13. Remove the right intake hose leading to the throttle body.
14. Pull the intake hose from the secondary air injection pump on the air filter cover. Remove the cover and air filter.
15. If equipped, unplug the connecting line to the air suspension compressor at the air filter.
16. Remove the transmission breather line from the air filter housing.
17. Remove the bracket for the right engine cover.
18. Pull the lower part of the air filter housing from the air duct and remove it from the vehicle.
19. Discharge and recover the air conditioning system refrigerant.
20. Remove the sound insulation tray.
21. Loosen the double clamp between the catalytic converter and the center muffler and push the clamp forward.

22. Support the rear muffler, remove the rear mountings and lower the exhaust system.

23. Matchmark the rear driveshaft and transfer case output flange. Counterhold the nuts with an open end wrench, and remove the bolts securing the driveshaft to the flange. Secure the driveshaft to prevent it from flexing.

24. Remove the front wheels and wheel house liners. Unplug the hose from the vacuum reservoir in the left wheel housing.

25. Remove the heat shield from the steering shaft and loosen the universal joint.

26. Unplug all connections on the transmission and transfer case.

27. Disconnect and plug the brake lines from the brake hoses. Catch any fluid that runs out.

28. Label and separate connectors between the body and front axle. If equipped, disconnect the lines for the air suspension at the strut.

29. Remove the transmission transverse support

30. Drain the cooling system. Pull the upper hose from the radiator and the lower hose off the coolant pipe. Disconnect the hoses from the heater core in the left wheel housing.

31. Disconnect the lines from the transmission cooler and power steering cooler; catch any fluid that runs out.

32. Disconnect the ground harness (1) from the longmember. Pull the coolant hose (3) and unfasten the coolant pump and bracket (2).

33. From inside the plenum area, remove the 2 least accessible bolts from the suspension strut on each side of the vehicle.

➡Use of tool VAS6131 or a similar scissor lift table providing support of the subframe and engine/driveline assembly is recommended. This allows for proper alignment of the subframe and engine/driveline assembly during removal and installation.

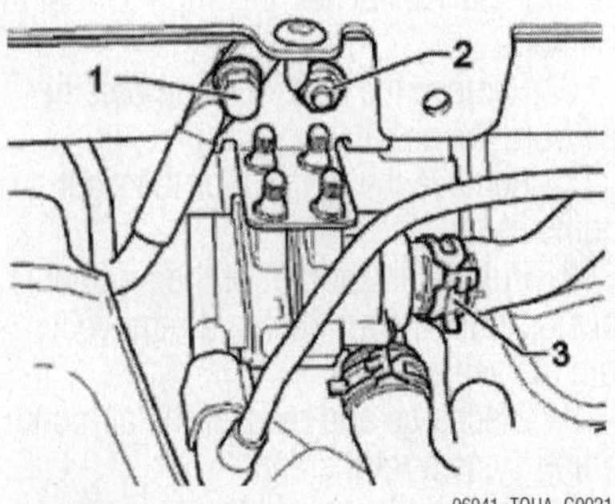

Ground harness (1), coolant pump and bracket (2), and coolant hose (3)—4.2L engines

34. Prepare tool VAS6131 or similar scissor lift as follows:

a. Attach pulley end engine supports VAS6131/6-1 and VAS6131/6-2, or similar. Secure the bolts at the following positions:

- Left subframe support: A5, B4 and B5
- Right subframe support: G4, H4 and H5

b. Attach axle supports VAS6131/6-3 and VAS6131/6-4, or similar. Secure the bolts at the following positions:

- Left suspension support: A6 and C6
- Right suspension support: F6 and H6

c. Rotate the turntables of the axle supports downward.

d. Place axle supports VAS6131/6-5 (or similar) for the subframe and VAS6131/6-6 (or similar) for the transmission mount at the appropriate positions on the scissor lift table.

35. Position the scissor lift table under the engine/transmission assembly. The supports must be guided into the corresponding engine and transmission mounts.

36. Make sure all four supports are in the appropriate receptacle holes without pressure being applied to them. Extend both turntables of the axle supports to slightly preload the control arms.

37. Guide the subframe supports into their respective receptacles, correcting the support height as necessary with the knurled nuts.

38. Loosen subframe bolts (1, 2 and 3), as well as transmission transverse support (4).

➡Anytime the subframe is loosened, it will be necessary to perform a wheel alignment on the vehicle.

39. Remove the third bolt securing each strut.

40. Slowly lower the engine/transmission assembly, constantly observing clearance and checking for items that may not have been disconnected.

To install:

➡When installing the drivetrain assembly back into the vehicle, make sure to bring the mounts to rest against the body using the scissor lift table. Do not let the bolts draw up the drivetrain assembly as the threaded inserts in the body could become damaged.

41. Installation is the reverse of removal.

a. Observe the following tightening torques:

- Engine bracket to mount: 55 ft. lbs. (75 Nm)
- Engine mount to carrier: 44 ft. lbs. (60 Nm)
- Engine carrier to chassis (use new bolts): 74 ft. lbs. (100 Nm), plus a ½ turn
- Transmission crossmember (use new bolts): 37 ft. lbs. (50 Nm), plus a ¼ turn
- Driveshaft to transfer case flange (use new bolts/nuts): 55 ft. lbs. (75 Nm)

b. All other fasteners:

- M6: 7 ft. lbs. (10 Nm)
- M7: 11 ft. lbs. (15 Nm)
- M8: 18 ft. lbs. (25 Nm)
- M10: 30 ft. lbs. (40 Nm)
- M12: 44 ft. lbs. (60 Nm)

42. Make sure all lines, hoses and wiring harnesses are properly routed. Refill or top off all fluids and recharge the A/C system. Perform a wheel alignment.

Water Pump

REMOVAL & INSTALLATION

3.2L Engines

1. Before servicing the vehicle, refer to the Precautions Section.
2. Drain the cooling system.
3. Mark the direction of rotation on the belt.
4. Screw a M8x50 bolt into the threaded hole of the tensioner to release tension on the belt. Screw the bolt in only far enough to allow removal of the belt.
5. Remove the belt.
6. If necessary to remove the belt pulley, use wrench VAG 1590 or similar to secure

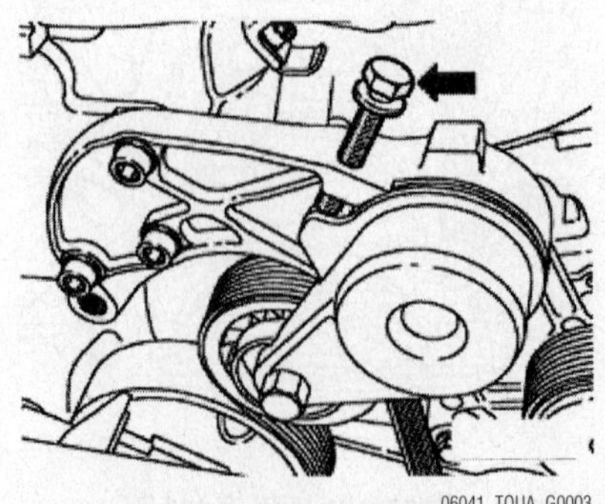

Use a M8x50 bolt to release tension on the belt—3.2L engines

the pulley, then remove the 3 bolts securing it to the pump.

7. Remove the 3 pump bolts. If the pulley is still installed, the bolts can be accessed through the holes in the pulley.

To install:

8. Installation is the reverse of removal. Note the following:

- Moisten a new O-ring with coolant.
- Tighten the mounting bolts to 5 ft. lbs. (8 Nm)

4.2L Engines

1. Before servicing the vehicle, refer to the Precautions Section.
2. Drain the cooling system.
3. Remove the upper coolant hose at the thermostat housing.
4. Unfasten the coolant pipe from the fan mount, then loosen the spring clamps and rotate it upwards with the spacers.
5. Unplug the cooling fan connector, loosen the harness clamps and lay the harness aside.
6. Pull off the ring securing the rubber mounting from both vacuum pumps, then lay then pumps aside.
7. Release the tabs securing the fan assembly to the radiator, then pull the assembly up and out.
8. Mark the direction of rotation on the belt.
9. Rotate the tensioner using a 19mm wrench and secure it in the released position using tool T10060A or a similar sized pin punch. Remove the belt.
10. Remove the timing belt.
11. Remove the 2 bolts from the timing belt cover (1), then remove the bolts securing the water pump.

To install:

12. Installation is the reverse of removal. Note the following:

- Be sure to use a new seal with the coolant pump.

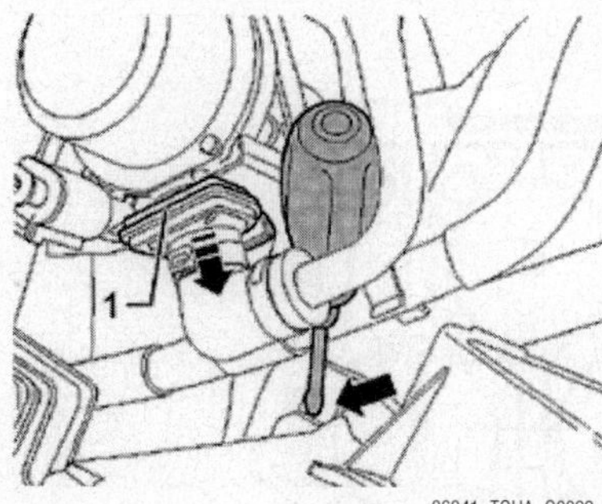

Remove the rubber mounting from the vacuum pump (1). The tabs on the fan can be released with a screwdriver (arrow)—4.2L engines

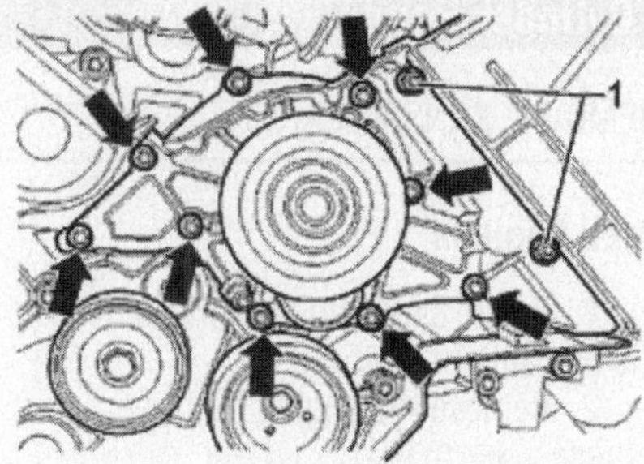

Timing belt cover bolts (1) and pump mounting bolt locations (arrows)—4.2L engines

- Tighten the mounting bolts to 10 ft. lbs. (14 Nm)

Heater Core

REMOVAL & INSTALLATION

1. Before servicing the vehicle, refer to the Precautions Section.
2. Remove the screws and disengage the clips securing the driver's side lower footwell panel, then remove the panel.
3. Remove the screws and disengage the clips securing the driver's side lower dashboard panel. Remove the panel and disconnect the Bowden cable from the parking brake.
4. Disconnect the rotary light switch and headlight range regulator harness connectors.
5. Remove the Access/Start control module as follows:
   a. Push the locking hooks of bracket outward (arrows B) and pull the bracket forward (arrow A).
   b. Pull retainer (1) in direction of arrow (A).
   c. Unplug electrical connection (2) in direction of arrow (B).
   d. Unplug electrical connection (3) in direction of arrow (C).

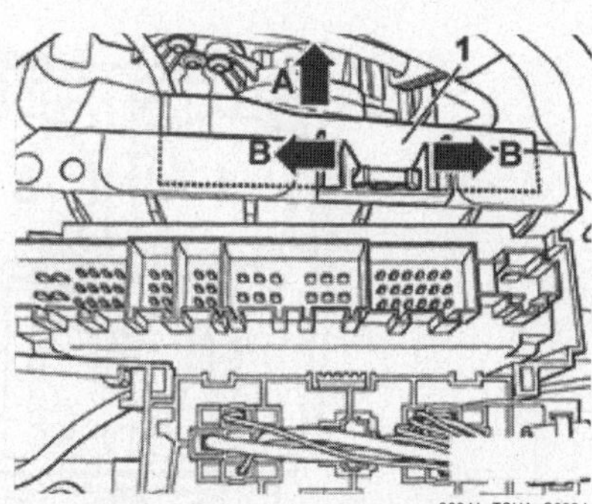

Push the locking hooks of bracket outward (B) and pull the bracket forward (A)

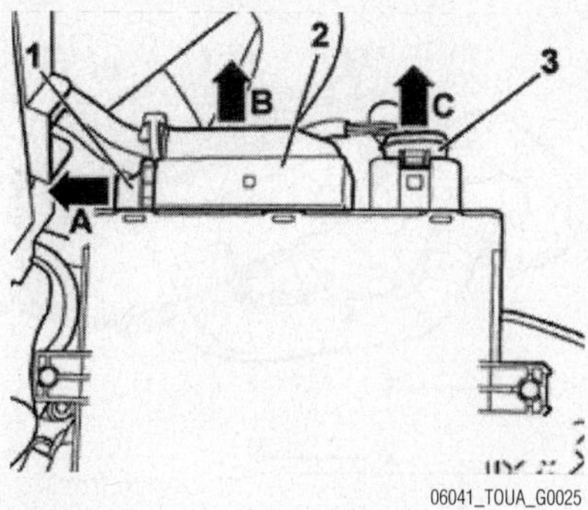

Access/Start control module electrical connections

   e. Remove the module from the bracket.
6. Remove the relay carriers as follows:
   a. Unplug the electrical connection from the brake pedal switch.
   b. Loosen the screws at the left and right of the brake pedal switch bracket, then remove brake pedal switch bracket.
   c. Press the retaining tabs together to remove the relay carriers with the connected wires out of bracket.
   d. Remove the relay carriers.
7. Remove the electrical control module as follows:
   a. Release the retainers and unplug the electrical connections from the control module located above the accelerator pedal assembly.
   b. Remove the control module with the wires still connected to rear side out of the bracket.
   c. Release the retainers and unplug the electrical connections from the module.
   d. Remove the control module.
8. Drain the cooling system.
9. Remove the right heater core cover by loosening bolt (1) and removing bolt (2). Unclip the cover from mounting points (3).
10. Pull the carpet off of the center tunnel.

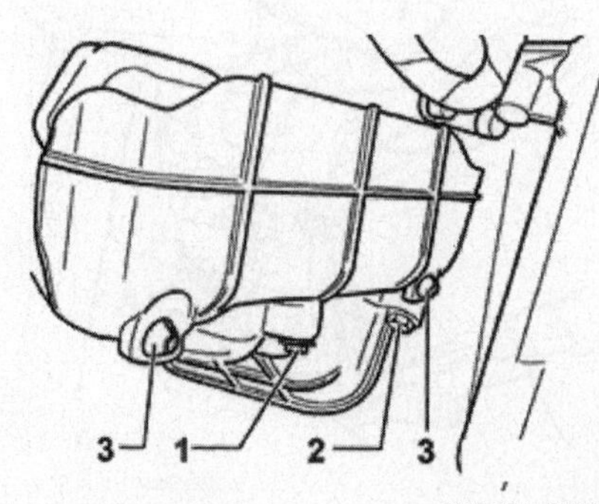

Right heater core cover fasteners

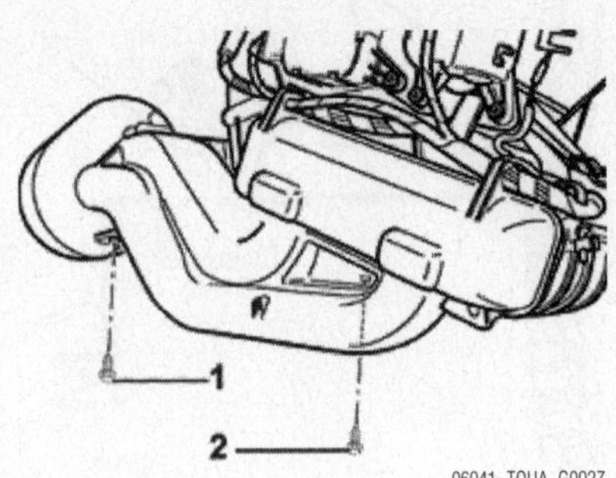

Left heater core cover fasteners

11. Remove the 2 screws securing the left heater core cover, then unclip the cover. Note that one of the screws may be hidden by the heater core cover.
12. Remove the footwell vent.
13. Cover the carpet in the area under the heater core with water absorbing materials.
14. Pull the retaining clips from the coolant hoses.
15. Slowly pull the coolant lines off of heater core and collect any coolant that leaks out.

➡A retaining pin is located under the heater core on the left-hand side. This connects the lower heater core cover to the heating and A/C unit.

16. Lift the retaining pin out from the locking device and pull the retaining pin out. Fold the cover downward and take heater core out to left.

To install:

17. Installation is the reverse of removal. Note the following:
    - Always replace the O-rings.
    - When installing the retaining clips, ensure they are seated securely on the heater core.

Cylinder Head

REMOVAL & INSTALLATION

3.2L Engines

1. Before servicing the vehicle, refer to the Precautions Section.
2. Remove the engine from the vehicle.
3. Set the crankshaft to TDC of cylinder 1 by turning the vibration damper securing bolt in the direction of engine rotation.
4. Screw a M8x50 bolt into the threaded hole of the tensioner to release tension on the belt. Screw the bolt in only far enough to allow removal of the belt.
5. Remove the belt.
6. Remove the intake manifold.
7. At front of cylinder head cover, unbolt the wiring bracket for the knock sensor and oil level thermal sensor. Unplug the fuel injector connectors.
8. Remove the cylinder head cover.
9. Unbolt the connecting line from the intake manifold support and unplug the connector for the knock sensor.
10. Unclip the wire harness from the retainer on cylinder head.
11. Unbolt the connecting line from the combination valve and pull off the coolant hose below the combination valve on the cylinder head.

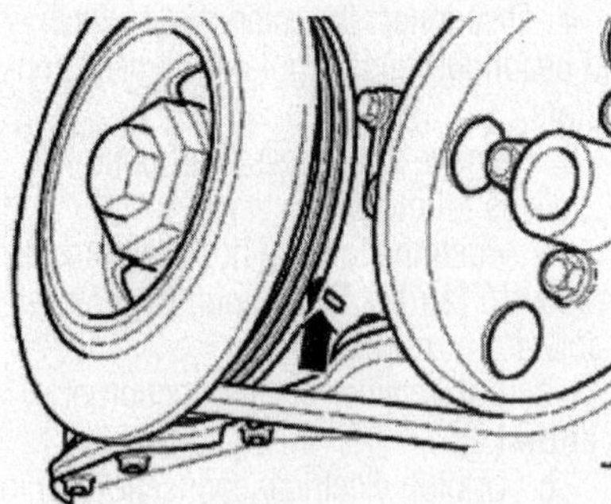

TDC mark located on the vibration damper—3.2L engines

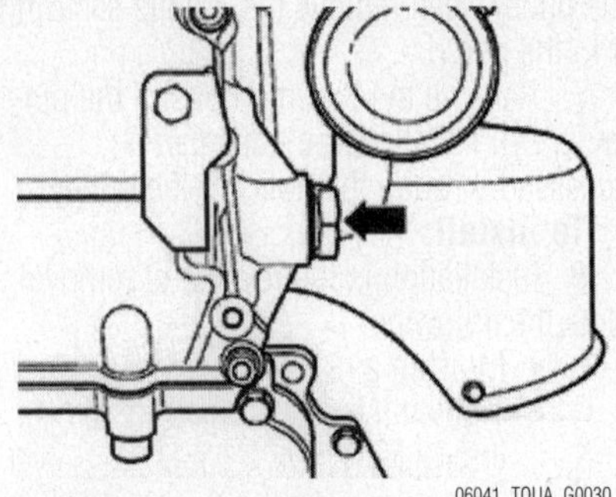

Timing chain tensioner is located in the front timing chain cover—3.2L engines

12. Unbolt the exhaust manifold from the cylinder head.
13. Unplug the connectors from the camshaft adjustment valves.
14. Unbolt the securing clamps for the wiring harness at the bracket.
15. Remove the thermostat housing.
16. Remove the camshaft roller chain tensioner.
17. Remove the timing chain cover. Note that the cover can be removed only after the thermostat housing has been removed.
18. Remove the bolts for the camshaft adjuster.

➡When loosening or tightening the camshaft adjuster, counter-hold using a 32 mm open end wrench on the camshaft flats. Make sure the camshaft is not locked using a holding bar tool.

19. Remove the camshaft adjuster together with the roller chain from the intake camshaft.
20. Remove the exhaust camshaft adjuster.
21. Remove the timing chain guide rail securing bolts and remove the guide rail. Place the chain aside.
22. First loosen the cylinder head bolts

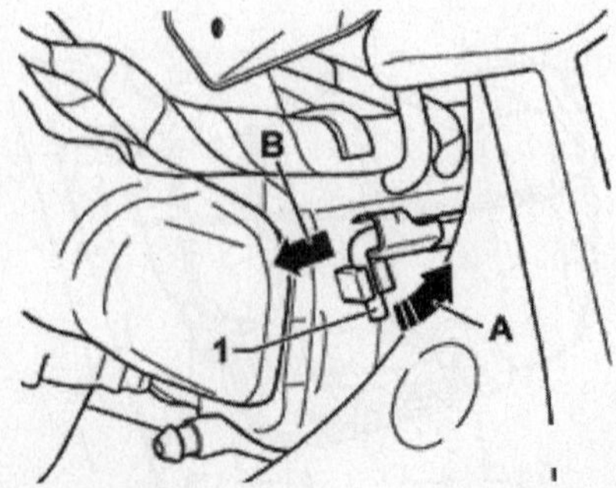

Lift the retaining pin (1) upwards (A), then slide it out (B)

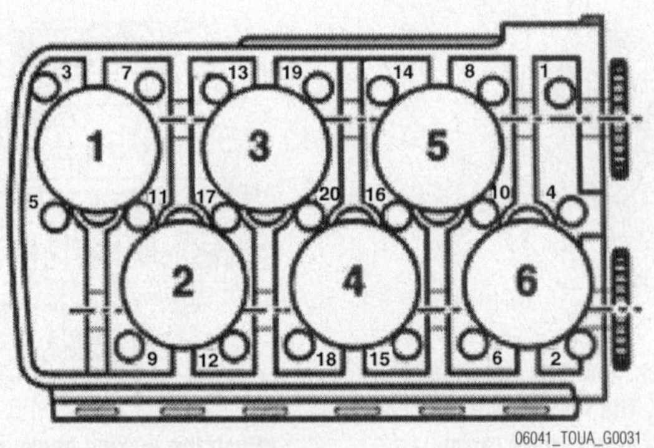

Cylinder head bolt removal sequence—3.2L engines

in sequence starting from the outside to the inside, then remove them completely.

➡For Polydrive® cylinder head bolts, use Polydrive® key 3452 or equivalent.

23. Carefully lift off the cylinder head with a workshop crane. Place a clean cloth in cylinders so that no dirt or emery cloth particles can get in between cylinder wall and piston. Also prevent dirt and emery cloth particles from getting into the coolant.

To install:

24. Carefully clean the cylinder head and cylinder block sealing surfaces. Ensure that no scoring or scratches are formed. Abrasive paper grade must not be less than 100.

25. Clean all threaded holes for the cylinder head bolts. Carefully remove any metal particles, emery remains and cloths.

26. If the piston for cylinder 1 is not at TDC, set it to the TDC marks by turning the vibration damper securing bolt in the direction of engine rotation. Guide the camshaft roller chain by hand with aid of an assistant while doing this.

27. Install a new cylinder head gasket; handle the new gasket with care to prevent damage.

28. Install alignment bushings into cylinder block holes 12 and 20 and that cylinder head gasket is seated.

29. Position the camshafts in the cylinder head to TDC of cylinder 1.

30. Lock the camshafts using holding bar T10068 or equivalent. It must engage in both shaft grooves.

31. Lower the cylinder head onto the block; insert new cylinder head bolts and tighten by hand.

➡The longer cylinder head bolts must be inserted in the middle holes of cylinder head.

32. Tighten the cylinder head bolts in the proper sequence starting from the inside to the outside:

- Step 1: 22 ft. lbs. (30 Nm)

Cylinder head bolt installation sequence—3.2L engines

- Step 2: 37 ft. lbs. (50 Nm)
- Step 3: Tighten all bolts a ¼ turn
- Step 4: Tighten all bolts again an additional ¼ turn

33. The remaining installation is the reverse of the removal procedure. Refer to the timing chain procedures in this section to check and adjust the valve timing. Observe the following additional torque specifications:

- Cylinder head cover: 7 ft. lbs. (10 Nm)
- All other fasteners: Refer to the appropriate procedures in this section.

4.2L Engines

LEFT SIDE

1. Before servicing the vehicle, refer to the Precautions Section.
2. Remove the engine from the vehicle.
3. Remove the front exhaust pipe from the exhaust manifold.
4. Remove the bolt securing the oil dipstick guide tube to the cylinder head.
5. Label and unplug the electrical connections from the front of the cylinder head.
6. Remove the timing belt and camshaft gear.

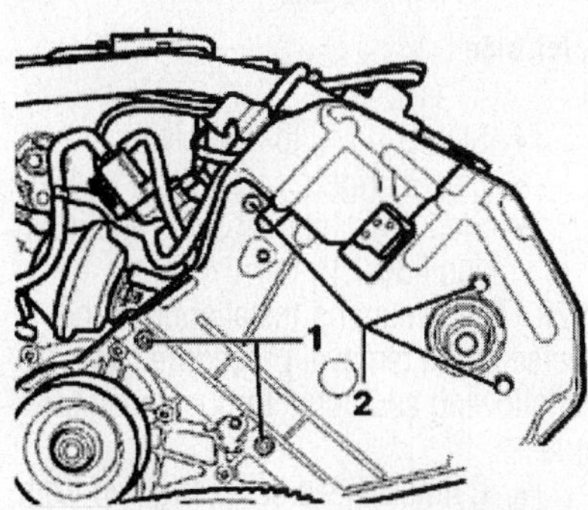

Remove the bolts securing the rear timing belt cover (1 and 2)—4.2L engines, left side

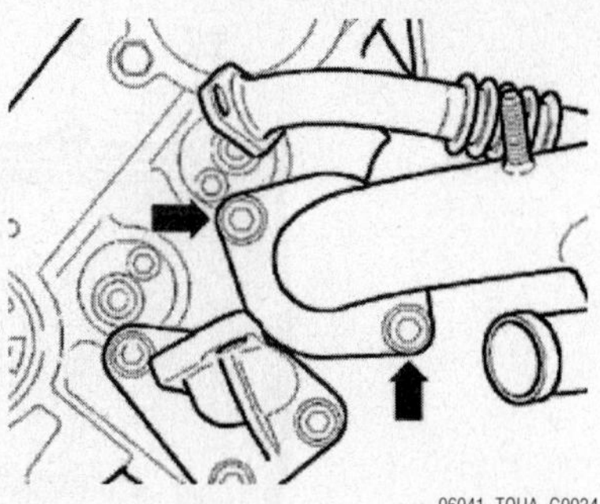

Remove the bolts securing the coolant pipe located below the combination valve—4.2L engines

7. Remove the bolts securing the rear timing belt cover, then remove the cover.
8. Unplug the electrical connectors from the back of the cylinder heads.
9. Unplug the vacuum hoses, then remove the combination valves from the end of the cylinder heads.
10. Remove the bolts from the coolant pipes on the end of the cylinder heads.
11. Unplug the knock sensor connectors.
12. Unplug the connectors from all of the fuel injectors.
13. Remove the coolant pipe assembly from the cylinder heads.
14. Remove the intake manifold.
15. Remove the bracket from below the ignition coils.
16. Pull off the connectors from the ignition coils, then remove the coils from the cylinder head cover.
17. Disconnect the crankcase ventilation hoses.
18. Loosen the bolts in the sequence from 12 to 1, then remove the cylinder head cover.
19. Loosen the cylinder head socket head bolts in the given sequence, starting from the outside to the inside. Then remove the bolts completely.
20. Carefully lift the cylinder head off. Place a clean cloth in cylinders so that no dirt or emery cloth particles can get in between cylinder wall and piston. Also prevent dirt and emery cloth particles from getting into the coolant.

To install:

21. Carefully clean the cylinder head and cylinder block sealing surfaces. Ensure that no scoring or scratches are formed. Abrasive paper grade must not be less than 100.
22. Clean all threaded holes for the cylinder head bolts. Carefully remove any metal particles, emery remains and cloths.
23. Install a new cylinder head gasket; handle the new gasket with care to prevent damage.

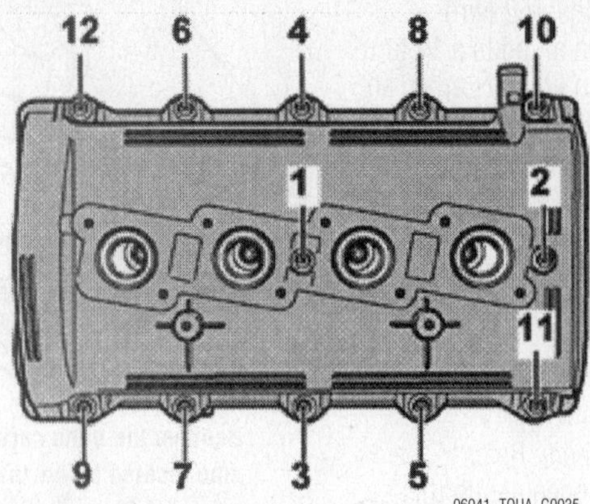

Cylinder head cover bolt identification—4.2L engines, left side

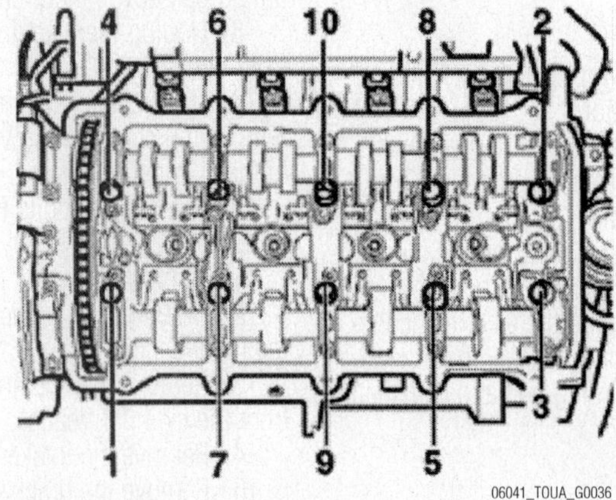

Cylinder head bolt removal sequence—4.2L engines, left side

24. Ensure that alignment bushings are inserted into the cylinder block and that cylinder head gasket is seated.

25. Lower the cylinder head onto the block; insert new cylinder head bolts and tighten by hand.

26. Tighten the cylinder head bolts in the proper sequence starting from the inside to the outside:

- Step 1: 26 ft. lbs. (35 Nm)
- Step 2: 44 ft. lbs. (60 Nm)
- Step 3: Tighten all bolts a ¼ turn
- Step 4: Tighten all bolts again an additional ¼ turn

27. The remaining installation is the reverse of the removal procedure. Observe the following additional torque specifications:

- Cylinder head cover (tighten from 1 to 12): 7 ft. lbs. (10 Nm)

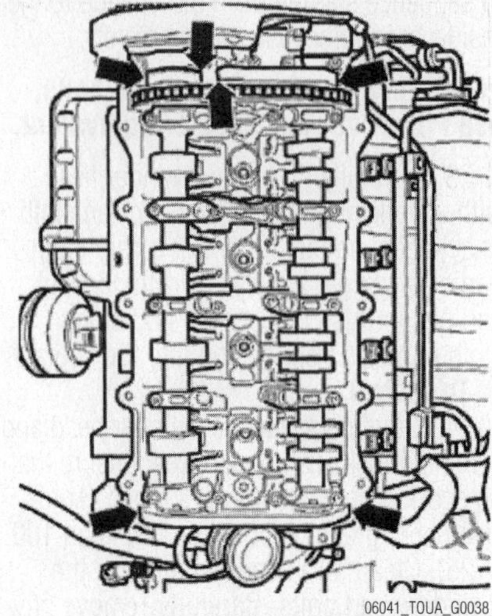

Apply sealant to these points on the cylinder head—4.2L engines, left side

- All other fasteners: Refer to the appropriate procedures in this section.

RIGHT SIDE

1. Before servicing the vehicle, refer to the Precautions Section.
2. Remove the engine from the vehicle.
3. Remove the front exhaust pipe from the exhaust manifold.
4. Label and unplug the electrical connections from the front of the cylinder head.
5. Remove the timing belt and camshaft gear.
6. Remove the idler roller.
7. Remove the bolts securing the rear timing belt cover, then remove the cover.
8. Unplug the knock sensor connectors.
9. Unplug the connectors from all of the fuel injectors.
10. Unplug the electrical connector and vacuum hose, then remove the combination valve from the end of the cylinder head.

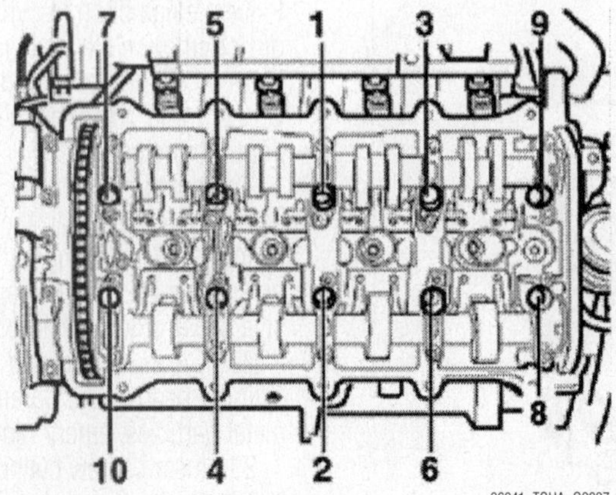

Cylinder head bolt tightening sequence—4.2L engines, left side

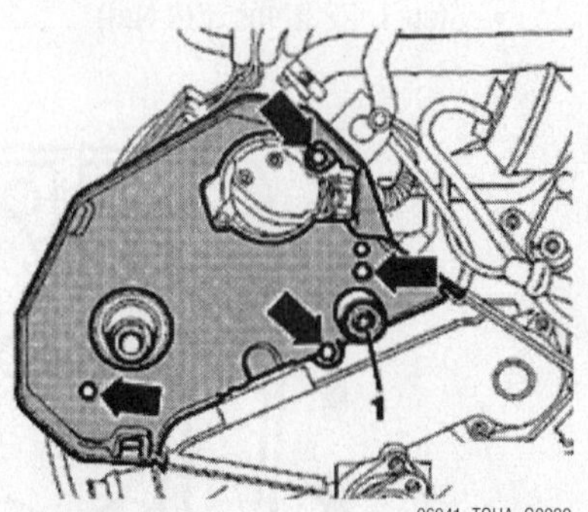

Remove the bolts securing the rear timing belt cover (arrows) and the idler roller (1)—4.2L engines, right side

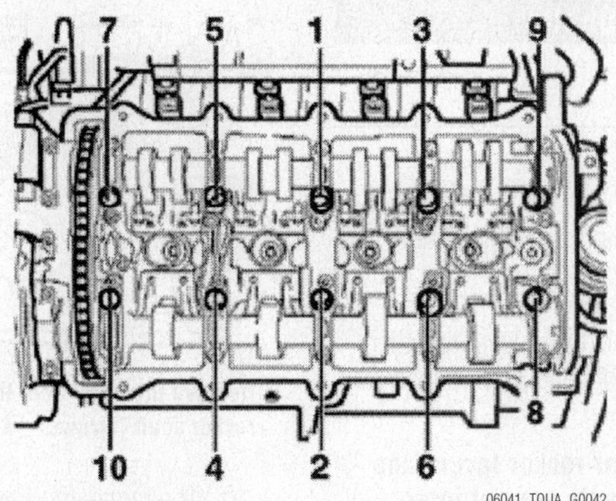

Cylinder head bolt tightening sequence—4.2L engines, right side

11. Remove the bolts from the coolant pipe on the end of the cylinder head. Detach the pipe from the head.

12. Remove the intake manifold.

13. Pull off the connectors from the ignition coils, then remove the coils from the cylinder head cover.

14. Disconnect the line for the secondary air injection system from the cylinder head cover.

15. Disconnect the crankcase ventilation hose.

16. Loosen the bolts in the sequence from 12 through 1, then remove the cylinder head cover.

17. Loosen the cylinder head socket head bolts in the given sequence, starting from the outside to the inside. Then remove the bolts completely.

18. Carefully lift the cylinder head off. Place a clean cloth in cylinders so that no dirt or emery cloth particles can get in between cylinder wall and piston. Also prevent dirt and emery cloth particles from getting into the coolant.

To install:

19. Carefully clean the cylinder head and cylinder block sealing surfaces. Ensure that no scoring or scratches are formed. Abrasive paper grade must not be less than 100.

20. Clean all threaded holes for the cylinder head bolts. Carefully remove any metal particles, emery remains and cloths.

21. Install a new cylinder head gasket; handle the new gasket with care to prevent damage.

22. Ensure that alignment bushings are inserted into the cylinder block and that cylinder head gasket is seated.

23. Lower the cylinder head onto the block; insert new cylinder head bolts and tighten by hand.

24. Tighten the cylinder head bolts in the proper sequence starting from the inside to the outside:

- Step 1: 26 ft. lbs. (35 Nm)
- Step 2: 44 ft. lbs. (60 Nm)
- Step 3: Tighten all bolts a ¼ turn
- Step 4: Tighten all bolts again an additional ¼ turn

25. The remaining installation is the reverse of the removal procedure. Observe the following additional torque specifications:

- Cylinder head cover: 7 ft. lbs. (10 Nm)
- All other fasteners: Refer to the appropriate procedures in this section.

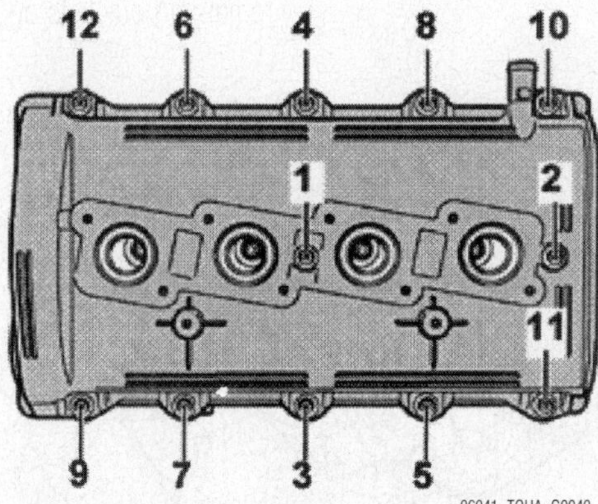

Cylinder head cover bolt identification—4.2L engines, right side

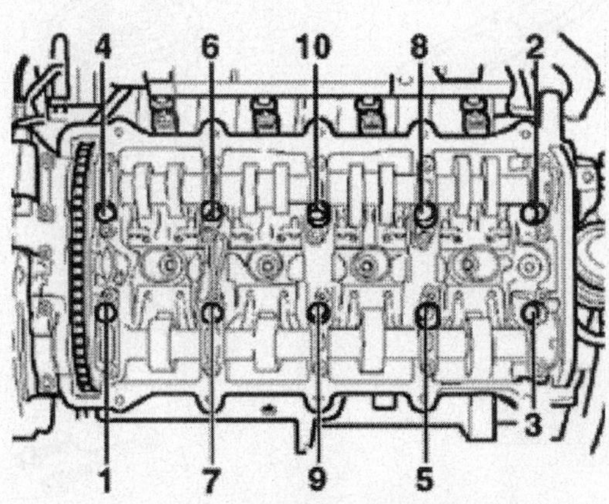
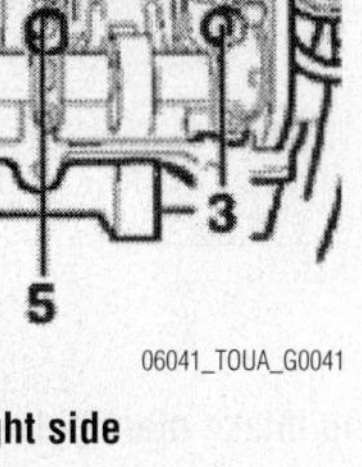

Cylinder head bolt removal sequence—4.2L engines, right side

06041_TOUA_G0043

Apply sealant to these points on the cylinder head—4.2L engines, right side

Rocker Arms/Shafts

REMOVAL & INSTALLATION

3.2L Engines

1. Before servicing the vehicle, refer to the Precautions Section.
2. Remove the camshafts.
3. Remove roller rocker lever together with support elements and place on a clean surface.

➡Ensure the roller rocker levers and the support elements are not interchanged. Number them as you remove them.

4. Installation is the reverse of removal. Be sure to fully seat the rocker levers in the support elements.

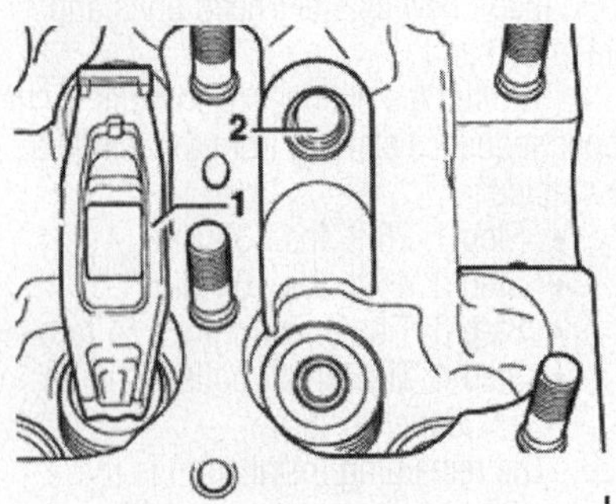

Rocker lever (1) and support element (2)—3.2L engines

4.2L Engines

1. Before servicing the vehicle, refer to the Precautions Section.
2. Remove the engine from the vehicle.
3. Remove the camshafts and camshaft adjusters from the cylinder head.
4. Mark the position of the roller rocker levers and roller rocker lever shafts to aid with installation.
5. Remove the bolts from the end of the rocker shaft.
6. Thread a M6 bolt into the holes of the rocker shafts, then pull the shafts out of the cylinder head.
7. Remove the rockers from the head.

To install:

➡Always replace the O-rings of the rocker shaft when installing it.

8. Oil the bearing areas of the roller rocker levers. Slide the shaft into the head and rocker levers.
9. Tighten the bolts of the rocker shaft to 7 ft. lbs. (10 Nm).

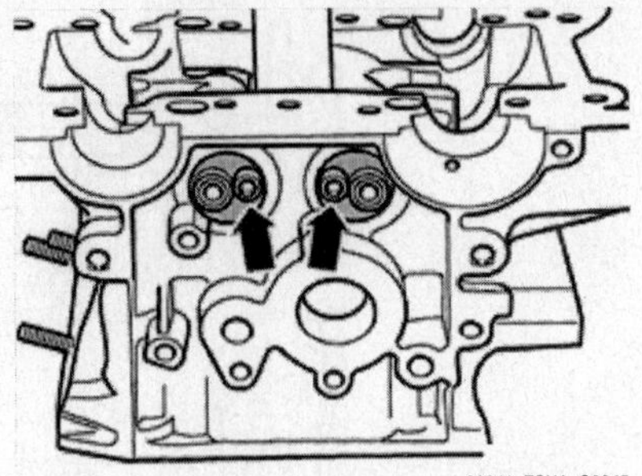

Remove the bolts from the end of the rocker shaft (arrows)—4.2L engines

10. The remaining installation is the reverse of removal.

Intake Manifold

REMOVAL & INSTALLATION

3.2L Engines

1. Before servicing the vehicle, refer to the Precautions Section.
2. Remove the ignition coils from the spark plug holes.
3. Remove the intake hose between the Mass Air Flow (MAF) sensor and the throttle body.

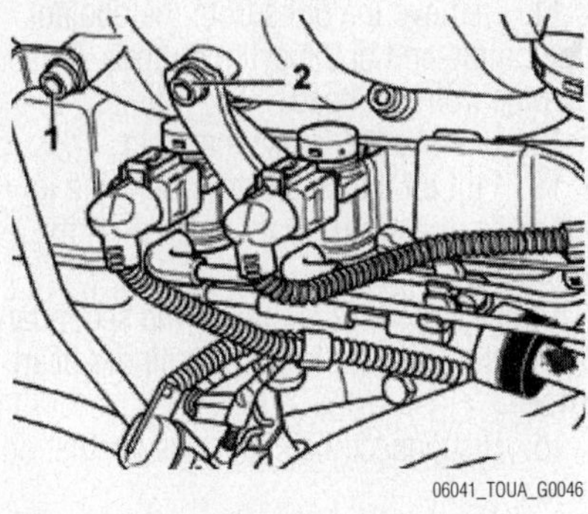

Remove the bolts from the two remaining brackets on the intake manifold—3.2L engines

4. Unplug the 6-pin connector from the throttle body, as well as the 2-pin connector from the EVAP canister purge valve and crankcase breather valve.
5. Label and remove the vacuum hoses from the intake manifold.
6. Remove the securing bolt from the intake manifold support.
7. Unbolt the intake manifold supports from right and left side of the intake manifold.
8. Disconnect the crankcase breather hose from the cylinder head cover.
9. Remove the bolts from the two remaining brackets on the intake manifold.

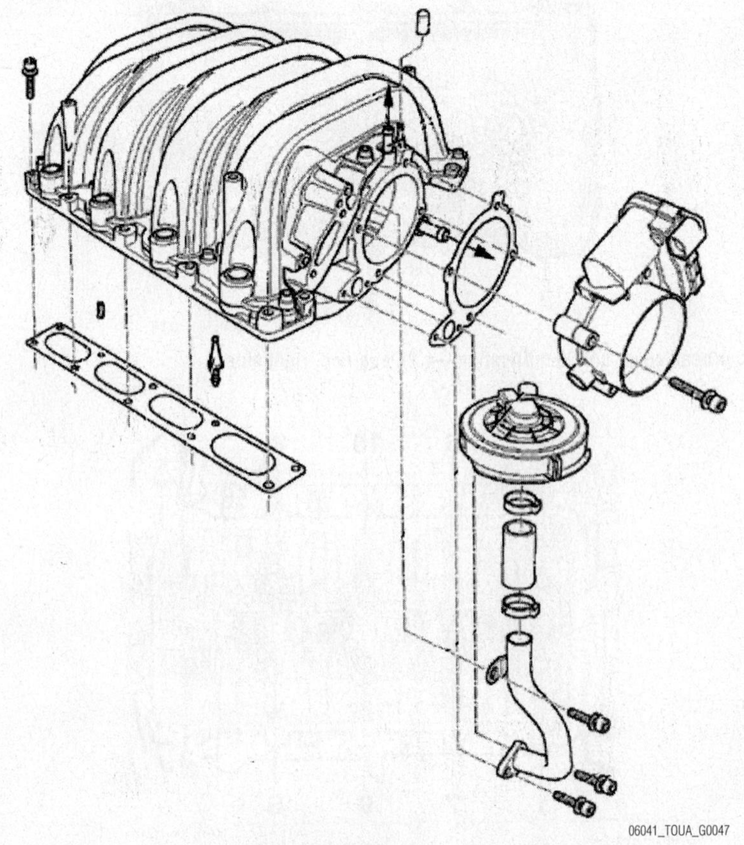

Exploded view of the intake manifold—4.2L engines

10. Pull off the vacuum hose from the intake manifold vacuum actuator.
11. Unbolt the coolant pipe on the front of the cylinder head.
12. Remove the intake manifold securing bolts.
13. Remove the manifold and place it on a suitable surface so that the vacuum actuator is not damaged.

➡Seal the intake ports in the intake manifold and in cylinder head with clean cloths.

14. Installation is the reverse of removal. Observe the following torque specifications:
- Intake manifold: 10 ft. lbs. (13 Nm)
- Intake manifold supports: 15 ft. lbs. (20 Nm)
- Oil dipstick: 5 ft. lbs. (8 Nm)

4.2L Engines

1. Before servicing the vehicle, refer to the Precautions Section.
2. Relieve the fuel system pressure.
3. Remove the intake hose from the throttle body.
4. Remove or disconnect any components necessary to ease access to the intake manifold.
5. Disconnect the fuel supply line.
6. Unplug the injector wiring harness and set aside.
7. Remove the bolts securing the intake manifold.
8. Remove the manifold and place it on a suitable surface so that it is not damaged.

➡Seal the intake ports with clean cloths to prevent entry of foreign objects.

9. Installation is the reverse of removal. Observe the following:
- Set new gaskets in place, then install the alignment cones and pins. Tighten to 26 inch lbs. (3 Nm)
- Tighten the intake manifold bolts to 7 ft. lbs. (10 Nm)

Exhaust Manifold

REMOVAL & INSTALLATION

3.2L Engines

1. Before servicing the vehicle, refer to the Precautions Section.
2. Remove the intake manifold.
3. Unbolt the heat shield with the intake manifold support from the cylinder head.
4. Unplug the oxygen sensor connectors. Unclip the wires from the harness guides.

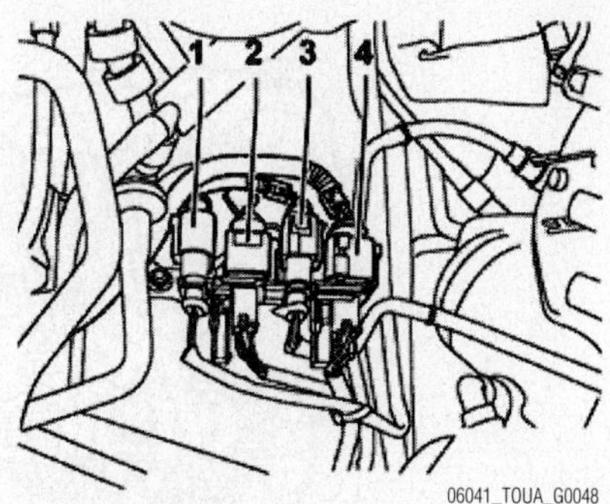

Unplug the oxygen sensor connectors—3.2L engines

5. Remove the oxygen sensors from the catalytic converters.
6. Raise the vehicle.
7. Remove the securing nuts from the main catalytic converter flange.
8. Unbolt the support to the transmission.
9. Unbolt the flange to the exhaust manifold. Remove the exhaust manifold.

To install:

➡When installing the exhaust pipes, ensure that the flange connection after the catalytic converter seals tightly. Leaks in this area allow ambient air to reach the oxygen sensors after the catalytic converter, resulting in false readings.

10. Installation is the reverse of removal. Observe the following:
- Adjust the exhaust system so that there is sufficient clearance to the transmission and subframe.
- Use new exhaust manifold gaskets
- M8 securing nuts: 18 ft. lbs. (25 Nm)
- M10 securing nuts: 30 ft. lbs. (40 Nm)
- Heat shield center bolt: 15 ft. lbs. (20 Nm)
- Other heat shield bolts: 17 ft. lbs. (23 Nm)

4.2L Engines

1. Before servicing the vehicle, refer to the Precautions Section.
2. Remove the engine from the vehicle.
3. To remove the exhaust manifold from the left cylinder bank, unfasten the bracket for the oil dipstick guide tube and disconnect the refrigerant line. Close off the connections with a clean rag.
4. Remove the nuts from the exhaust flange and exhaust manifold. Remove the manifold.

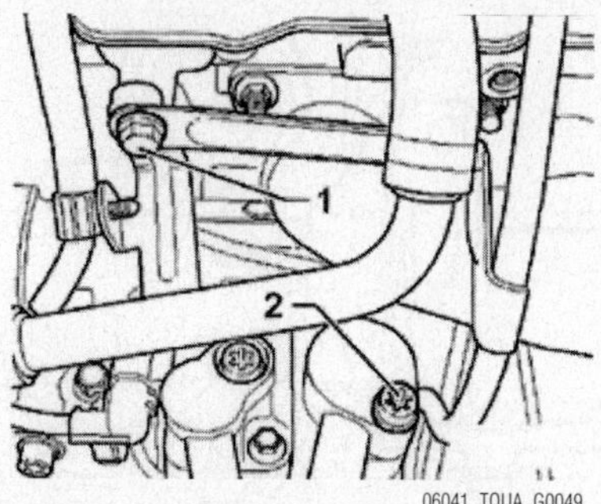

On the left exhaust manifold, remove the dipstick tube bolt (1) and refrigerant line bolt (2)—4.2L engines

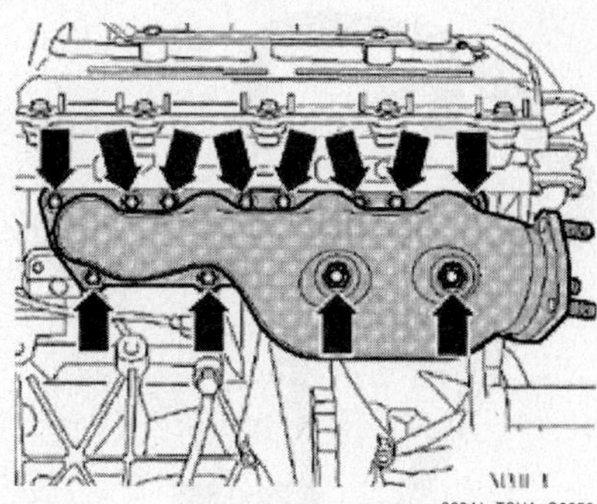

Remove the nuts securing the exhaust manifold—4.2L engines

5. Installation is the reverse of removal. Observe the following:
- Use new exhaust manifold gaskets
- Exhaust manifold securing nuts: 18 ft. lbs. (25 Nm)
- M8 exhaust flange securing nuts: 18 ft. lbs. (25 Nm)
- M10 exhaust flange securing nuts: 30 ft. lbs. (40 Nm)

Camshaft and Valve Lifters

REMOVAL & INSTALLATION

3.2L Engines

1. Before servicing the vehicle, refer to the Precautions Section.
2. Remove the engine assembly and separate it from the transmission.
3. Set crankshaft to TDC of cylinder 1 by turning the vibration damper securing bolt in the direction of engine rotation. Align the mark on the vibration damper.
4. Remove intake manifold and cylinder head cover.
5. Remove the thermostat housing.

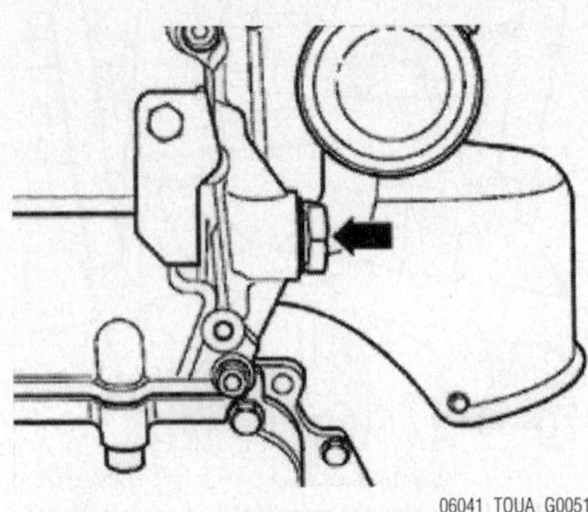
06041_TOUA_G0051

Remove the timing chain tensioner—3.2L engines

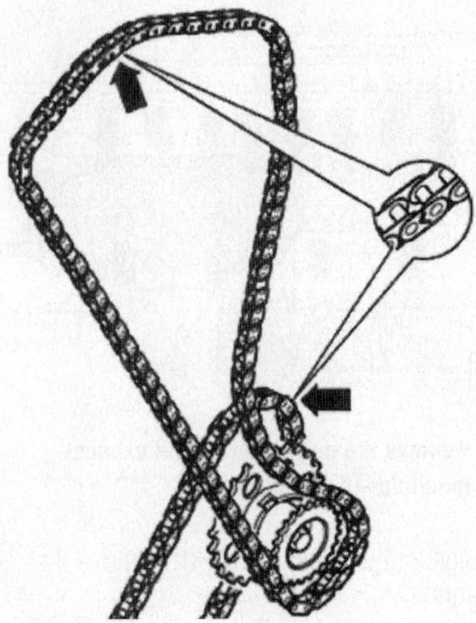
06041_TOUA_G0052

Mark the position of the chains and an arrow pointing in the direction of rotation—3.2L engines

6. Remove the upper timing chain tensioner.

7. Remove the upper timing chain cover. Note that the cover has a hidden bolt that can be removed only after the thermostat housing has been removed.

8. Mark the position of the timing chain before removing with paint. Also mark an arrow pointing in the direction of rotation. Do not mark chain with a center punch or similar means!

9. Remove the exhaust camshaft adjuster. Counter-hold the camshaft using a 32 mm open end wrench on the camshaft flats.

10. Remove camshaft adjuster together with the timing chain from the intake camshaft. Place the chain aside.

11. Remove the control housing from the cylinder head. Pull the control housing carefully off of the camshaft seals.

12. Remove the intake camshaft:

a. First remove bearing caps 1 and 13.

b. Remove bearing caps 3 and 11.

c. Remove bearing cap 7.

d. Loosen bearing caps 5 and 9 alternately and diagonally, then remove the caps.

13. Remove the exhaust camshaft:

a. First remove bearing caps 2 and 14.

b. Remove bearing caps 4 and 12.

c. Remove bearing cap 8.

d. Loosen bearing caps 6 and 10 alternately and diagonally, then remove the caps.

14. Carefully remove the camshafts and place them on a clean surface.

To install:

➡When installing the camshafts, the cam lobes for cylinder 1 must point upward.

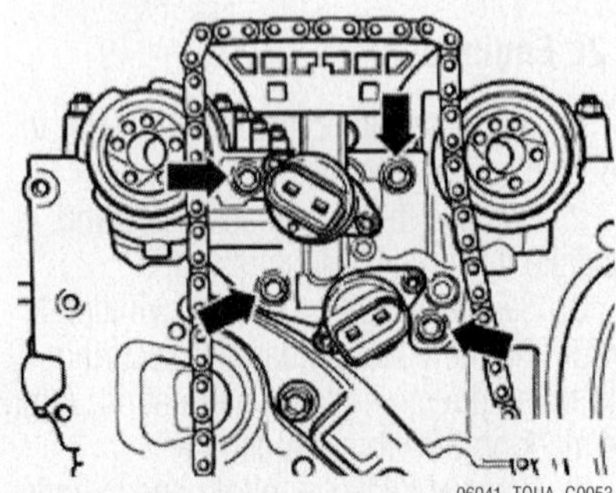
06041_TOUA_G0053

Remove the bolts securing the control housing (arrows)—3.2L engines

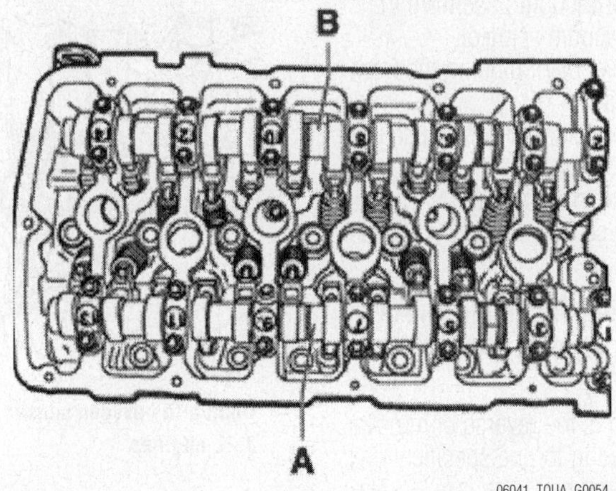

06041_TOUA_G0054

Intake camshaft (A) and exhaust (B)—3.2L engines

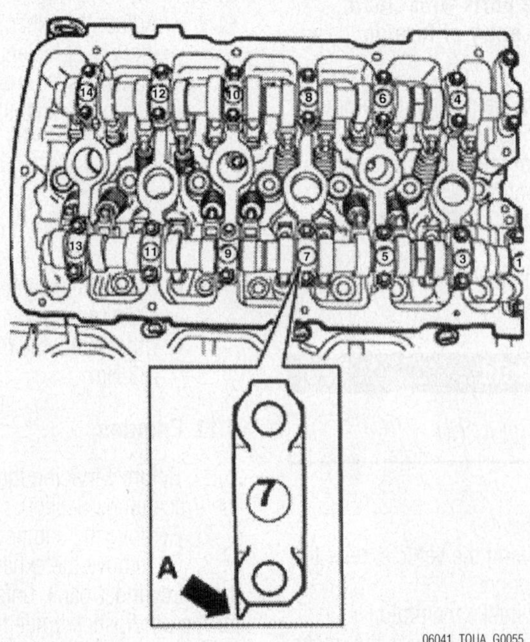

06041_TOUA_G0055

Bearing cap identification. The numbers on the cap should be legible when viewing from the intake side—3.2L engines

15. Lightly coat the contact surfaces of bearing caps 7 and 8 with grease G 052 723 A2 or equivalent before installing.

16. Ensure that all roller rocker levers seat properly on valve stem ends and are clipped onto their respective support elements.

17. Oil the wear surfaces of both camshafts.

18. Place camshafts carefully in camshaft bearings of cylinder head. Observe proper camshaft identification.

➡Observe the installed position of the bearing caps. Points of the intake and exhaust camshaft bearing caps face outwards. The writing on the caps should be legible when viewed from the intake side.

19. Install the intake camshaft:

a. Tighten bearing caps 5 and 9 alternately and diagonally to 44 inch lbs. (5 Nm), then an addition ⅛ turn further.

b. Install bearing caps 1 and 13 and tighten to 44 inch lbs. (5 Nm), then an additional ⅛ turn further.

c. Install bearing cap 7 and tighten to 44 inch lbs. (5 Nm), then an additional ⅛ turn further.

d. Install bearing caps 3 and 11 and tighten to 44 inch lbs. (5 Nm), then an additional ⅛ turn further.

20. Install the exhaust camshaft:

a. Tighten bearing caps 6 and 10 alternately and diagonally to 44 inch lbs. (5 Nm), then an addition ⅛ turn further.

b. Install bearing caps 2 and 14 and tighten to 44 inch lbs. (5 Nm), then an addition ⅛ turn further.

c. Install bearing cap 8 and tighten to 44 inch lbs. (5 Nm), then an addition ⅛ turn further.

d. Install bearing caps 4 and 12 and tighten to 44 inch lbs. (5 Nm), then an addition ⅛ turn further.

21. Position the camshafts to TDC of cylinder 1.

22. Install camshaft locking bar T10068 or similar. It must engage in both shaft grooves.

23. Check the control housing screen for contamination. Lightly lubricate the surfaces of the control housing contacting the camshaft seals. Lightly lubricate contact surfaces of camshaft seals and slowly push control housing over camshaft seals. Tighten the fasteners to 71 inch lbs. (8 Nm).

24. The remaining installation is the reverse of removal. Refer to the timing chain procedures for valve timing adjustment.

4.2L Engines

1. Before servicing the vehicle, refer to the Precautions Section.

2. Remove the timing belt and camshaft gear.

3. Remove the right or left cylinder head cover as appropriate. Refer to the cylinder head removal procedure for the proper bolt removal/installation sequence of the cover(s).

4. For the left cylinder head unscrew bolts (1) and (2), then remove rear timing belt cover.

5. For the right cylinder head, remove the idler roller (1). Unscrew the bolts (arrows) and remove rear timing belt cover.

6. Unplug the connector from the Hall

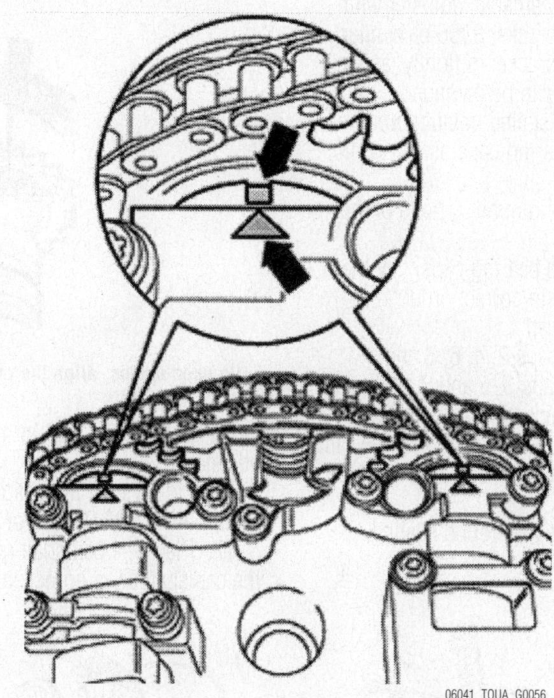

Align the marks on the camshafts with the bearing caps—4.2L engines

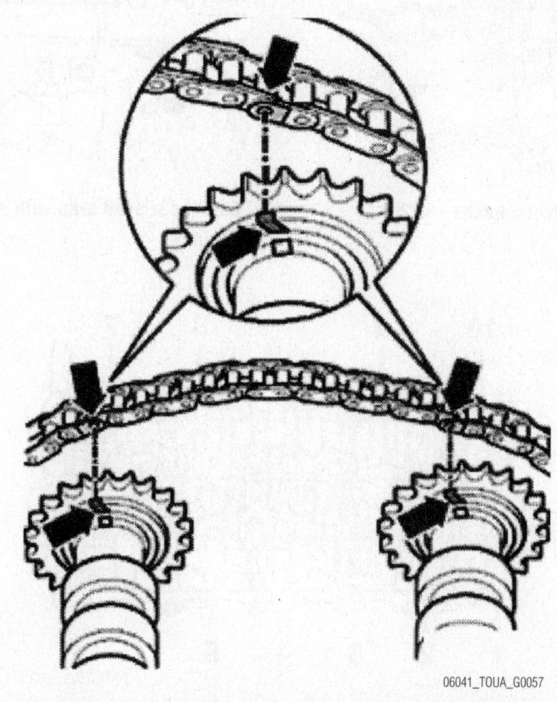

Mark the position and direction of rotation of the roller chain—4.2L engines

sender and remove the Hall sender housing with the cover and cone.

7. Verify the TDC position of the camshafts. The markings on the camshafts must be aligned with both arrows on the bearing cap.

8. Mark the roller chain position with paint and with an arrow pointing in direction of rotation. Do not mark the chain with a center punch or similar means!

9. Secure the camshaft adjuster with chain adjustment bracket 3366 or similar. Do not fasten the bracket to tightly, as the camshaft adjuster can be damaged.

10. Mark the installed position and sequence of all bearing caps as shown in the illustration.

11. Remove the mounting bolts of the camshaft adjuster.

12. Remove the bearing caps. Lay them on a clean surface in correct order:

a. Remove cap 1.

b. Remove caps 2, 4, 6, 8 and 9.

c. Loosen caps 3, 5 and 7 alternately in a diagonal sequence, then remove.

13. Remove both camshafts with the camshaft adjusters and place on a clean surface.

To install:

14. Replace the half-round sealing plugs.

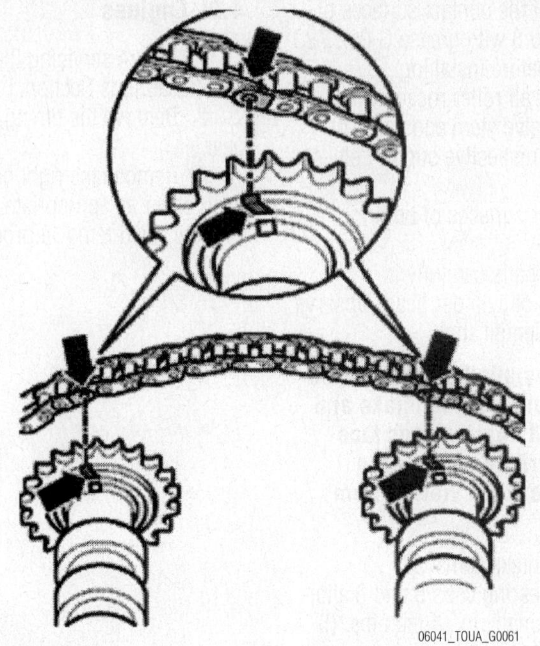

On used chains, align the markings made during removal—4.2L engines

15. Replace the seal for the camshaft adjuster.

16. Lightly coat the shaded area with sealant AMV 188 001 02 or equivalent.

17. Place the camshaft roller chain onto the camshaft drive sprockets as follows:

a. When re-using the camshaft roller chain: Match the markings made during removal.

b. When using a new camshaft roller chain: The distance between notches (A) and (B) must be 16 rollers of the drive chain. Refer to the illustration.

18. Push the camshaft adjuster in between the camshaft roller chain.

19. Install the camshafts with the roller chain and camshaft adjuster into the cylinder head.

20. Oil the journal surfaces of the camshafts.

➡Alignment bushings for the bearing caps and camshaft adjusters must be installed in cylinder head.

21. Install bearing caps 3, 5 and 7. Using new bolts, tighten them alternately in a diagonal sequence to 44 inch lbs. (5 Nm), plus an additional ¼ turn.

22. Tighten the camshaft adjuster to 44 inch lbs. (5 Nm) plus an additional ¼ turn.

23. Install the chain tensioner.

24. Verify the TDC position of the camshafts. The markings on the camshafts must be aligned with both arrows on the bearing cap. If necessary, turn the camshaft slightly back and forth to align them.

25. Lightly coat double bearing cap 6 and bearing cap 1 with grease AMV 188 001 02 or equivalent. Install the bearing caps and tighten to 44 inch lbs. (5 Nm), plus an additional ¼ turn.

26. Install the remaining bearing caps

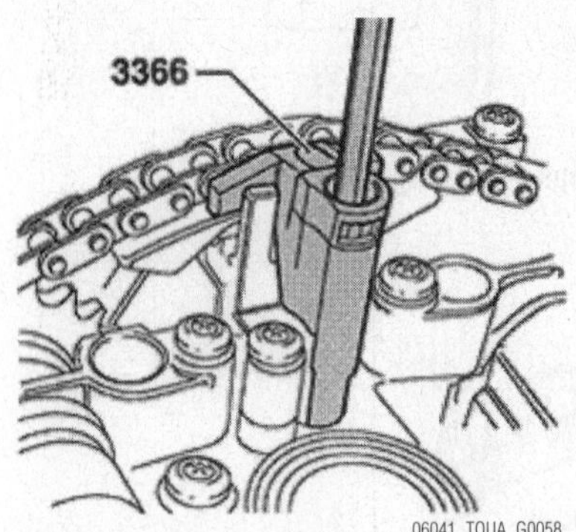

Secure the camshaft adjuster—4.2L engines

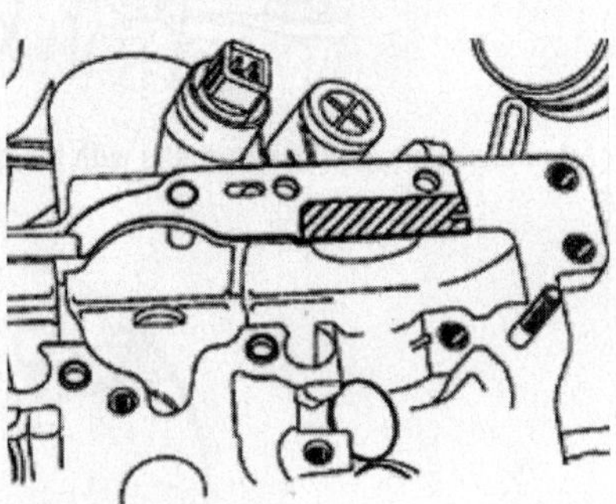

Coat the shaded area with sealant—4.2L engines

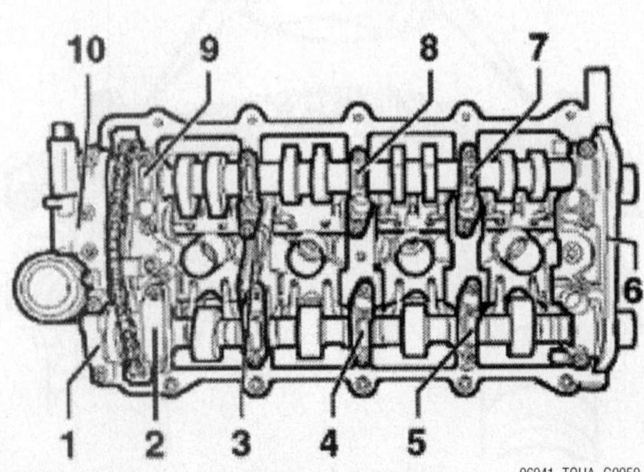

Mark the bearing caps as shown—4.2L engines

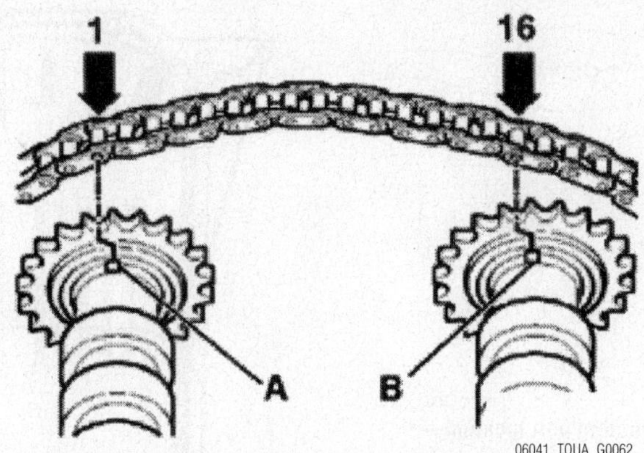

On new chains, count 16 links between (A) and (B)—4.2L engines

and tighten them to 44 inch lbs. (5 Nm), plus an additional ¼ turn.

27. If necessary, replace the camshaft seals and the seals for the Hall sender. Carefully drive in the sealing cover

28. The remainder of installation is the reverse of removal. Note the following:

- Refer to the cylinder head removal procedure for the proper bolt removal/installation sequence of the cylinder head cover(s).
- After installing the camshafts, the engine may not be started for approximately 30 minutes. The hydraulic valve adjusters must bleed down, otherwise the valves will crash into the pistons.
- Carefully rotate the crankshaft by hand at least 2 full revolutions before starting to be sure that valves do not strike the pistons.

Valve Lash

ADJUSTMENT

All engines in this section are equipped with hydraulic lash adjusters. No adjustment is possible.

Starter

REMOVAL & INSTALLATION

3.2L Engines

1. Before servicing the vehicle, refer to the Precautions Section.
2. Remove the engine from the vehicle.
3. Open the snaps and remove the starter solenoid heat shield mat.

➡When removing the nuts from the starter solenoid terminals, always counter-hold the terminal stud in order to avoid damage to the starter solenoid.

4. Pry off the caps from the nuts at the positive (B+) terminal and at terminal 50 on the starter, if present. Remove the nuts.

➡When removing the nuts from the starter solenoid terminals, always counterhold the terminal stud in order to avoid damage to starter solenoid.

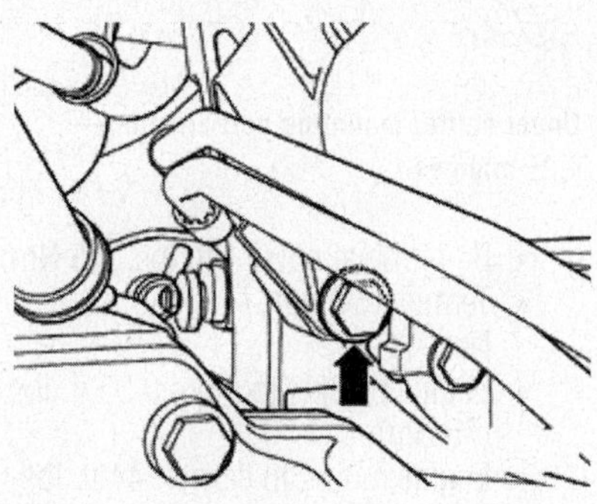

Remove the lower starter bolt on the engine side—3.2L engines

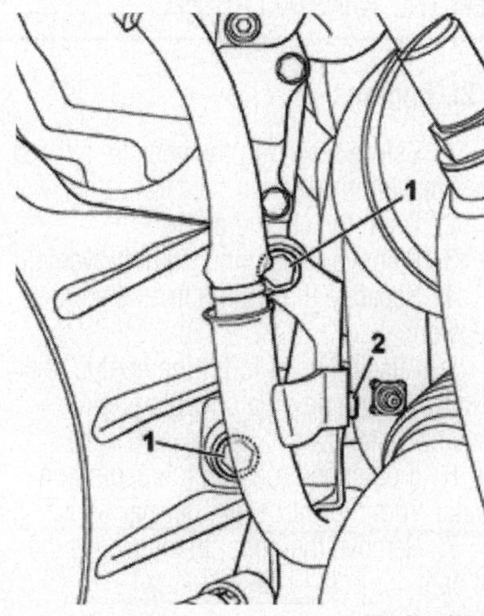

Remove the bolt (2) for the cable retainer and nuts (1) —3.2L engines

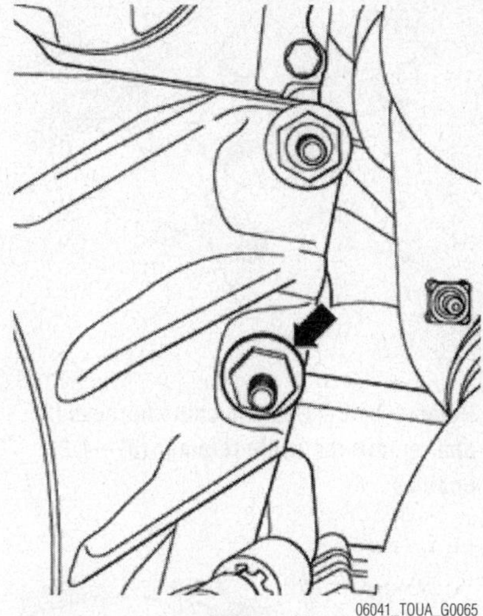

Remove the upper starter bolt—3.2L engines

5. Remove the lower starter bolt on the engine side.
6. Remove the bolt (2) for the cable retainer and set it aside. Remove the nuts (1)
7. Remove the bolt from the exhaust system bracket from the engine side and remove the bracket.
8. Remove the bolt for the coolant pipe on the side of the oil pan.
9. Remove the upper starter bolt on the transmission side.
10. Remove the starter.

To install:

11. Installation is the reverse of removal. Note the following torque specifications:
    - Starter mounting bolts: 55 ft. lbs. (75 Nm)
    - B+ terminal nut: 11 ft. lbs. (15 Nm)
    - Terminal 50 nut: 71 inch lbs. (8 Nm)

4.2L Engines

1. Before servicing the vehicle, refer to the Precautions Section.
2. Remove the engine from the vehicle.
3. Remove the exhaust manifold.
4. Remove bolt (1). Unclip cable harness (2) and remove the cable retainer (3) from mounting.
5. Remove the nuts and bolts from the engine mount (arrows).
6. Lift the engine with a hoist to take the weight off the mount. Remove the engine mount.
7. Remove the three bolts for the engine mount bracket at engine block, then remove the bracket.

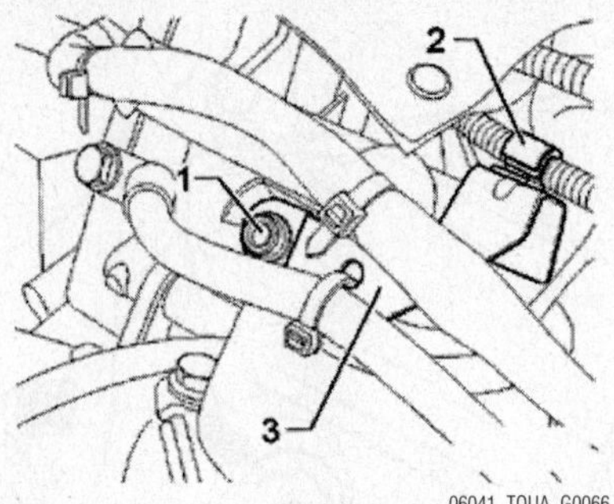

Remove bolt (1), unclip cable harness (2) and remove the cable retainer (3)—4.2L engines

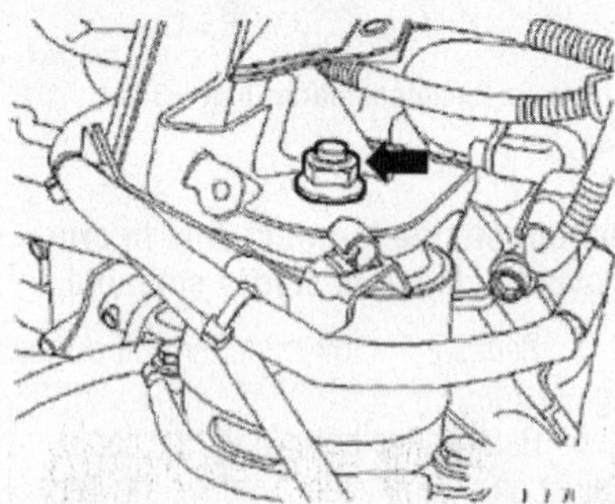

Remove the nut from the mount—4.2L engines

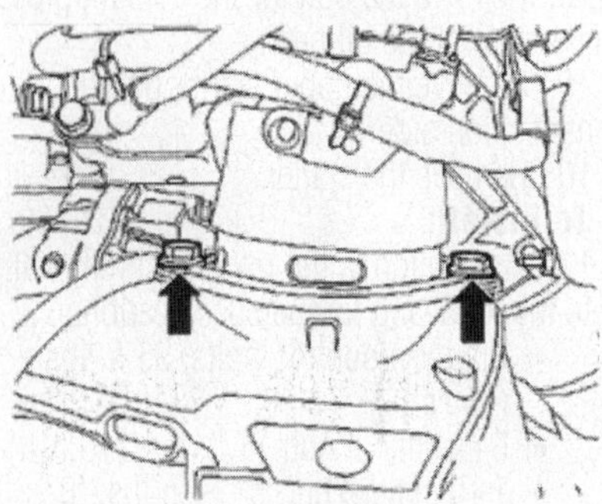

Remove the bolts securing the mount to the carrier—4.2L engines

8. Pry off the caps from the nuts at the positive (B+) terminal and at terminal 50 on the starter, if present. Remove the nuts.

➡When removing the nuts from the starter solenoid terminals, always counterhold the terminal stud in order to avoid damage to starter solenoid.

9. Remove the bolts securing the starter, then remove starter.

To install:

10. Installation is the reverse of removal. Note the following torque specifications:

- Starter mounting bolts: 55 ft. lbs. (75 Nm)

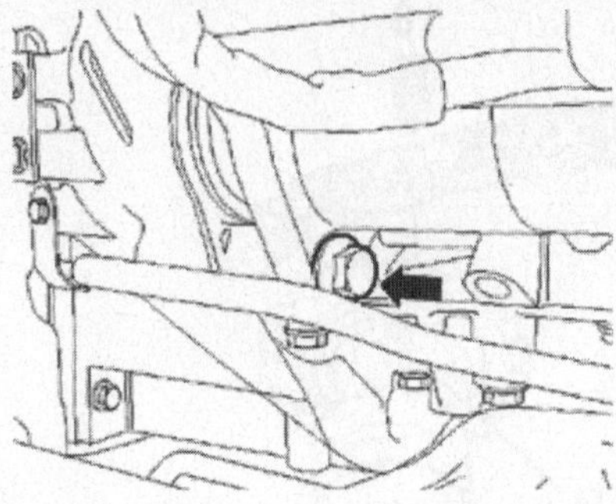

Lower starter mounting bolt location—4.2L engines

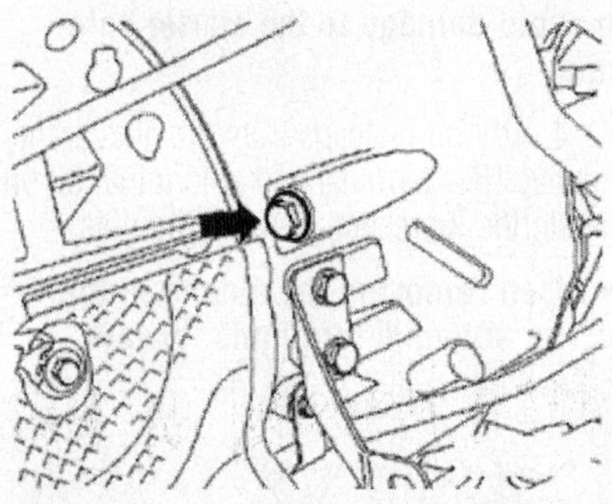

Upper starter mounting bolt location—4.2L engines

- B+ terminal nut: 11 ft. lbs. (15 Nm)
- Terminal 50 nut: 71 inch lbs. (8 Nm)
- Engine bracket to mount: 55 ft. lbs. (75 Nm)
- Engine mount to carrier: 44 ft. lbs. (60 Nm)

Oil Pan

REMOVAL & INSTALLATION

3.2L Engines

1. Before servicing the vehicle, refer to the Precautions Section.
2. Drain the engine oil.
3. Remove the engine from the vehicle.
4. Separate the engine from the transmission.
5. Attach a hoist to lift the engine, then secure the engine on a suitable engine assembly stand.
6. If equipped, unplug the 3-pin connector from the Oil Level Thermal Sensor.
7. Remove the bolts, then remove the oil pan.

➡Loosen the oil pan with light blows of a rubber mallet if necessary.

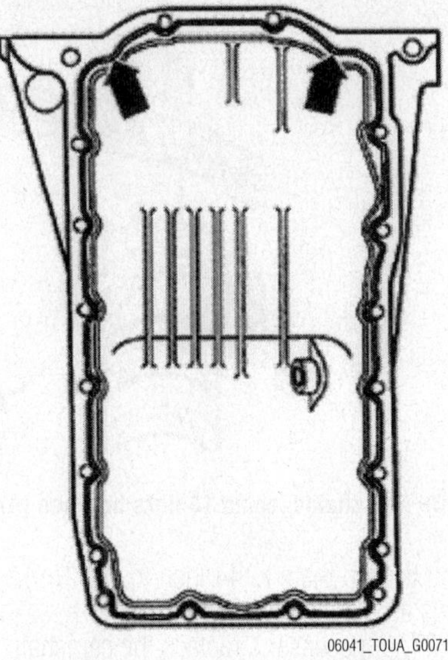

Apply the sealant as shown—3.2L engines

8. Remove the sealant residue from cylinder block and oil pan. Sealing surfaces must be free of oil and grease.

To install:

9. Apply silicone sealant D 176 404 A2 or equivalent as shown. Sealing compound bead must be 2 to 3 mm thick and run on the inside of bolt holes.

➡The sealing compound bead must not be thicker otherwise excess sealant will enter the oil pan and may block the oil suction pipe strainer.

10. Install oil pan immediately and tighten all oil pan bolts lightly. Tighten the oil pan bolts to 9 ft. lbs. (12 Nm).

➡After installing the oil pan, the sealant must be allowed to dry for 30 minutes. Only after then may the engine oil be replenished.

11. The remaining installation is the reverse of removal.

4.2L Engines

1. Before servicing the vehicle, refer to the Precautions Section.
2. Drain the engine oil.
3. Remove the engine from the vehicle.
4. Separate the engine from the transmission.
5. Attach a hoist to lift the engine, then secure the engine on a suitable engine assembly stand.
6. Remove the right engine mount.
7. Unscrew the oil filter housing.
8. At the left cylinder head, unfasten the bracket from the oil dipstick guide tube and pull the guide tube out.

9. Unfasten the bracket of the coolant drain pipe at the upper oil pan. If necessary, loosen the banjo bolt slightly.

10. Remove the bolts, then remove the lower and upper part of oil pan.

➡Loosen the oil pan with light blows of a rubber mallet if necessary.

11. Remove the sealant residue from cylinder block and oil pan. Sealing surfaces must be free of oil and grease.

To install:

12. Apply silicone sealant D 176 404 A2 or equivalent as shown to the upper oil pan. Sealant bead must be about 1.5 mm thick and run on the inside of bolt holes.

➡The sealing compound bead must not be thicker otherwise excess sealant will enter the oil pan and may block the oil suction pipe strainer.

13. Install the upper oil pan immediately.

a. First tighten all oil pan bolts lightly in a diagonal sequence.

b. Tighten oil pan bolts (1) to 11 ft. lbs. (15 Nm) and bolts (2) to 16 ft. lbs. (22 Nm) in a diagonal sequence.

14. Apply silicone sealant D 176 404 A2 or equivalent as shown to the lower oil pan. Sealant bead must be about 1.5 mm thick and run on the inside of bolt holes.

15. Install the lower oil pan immediately.

16. First tighten all oil pan bolts lightly in a diagonal sequence. Then tighten again in a diagonal sequence to 7 ft. lbs. (10 Nm).

17. Set the oil dipstick guide tube in place. Fasten the coolant drain pipe to upper part of oil pan to 7 ft. lbs. (10 Nm) and to the cylinder block to 22 ft. lbs. (30 Nm).

➡After installing the oil pan, the sealant must be allowed to dry for 30 minutes. Only after then may the engine oil be replenished.

18. The remaining installation is the reverse of removal.

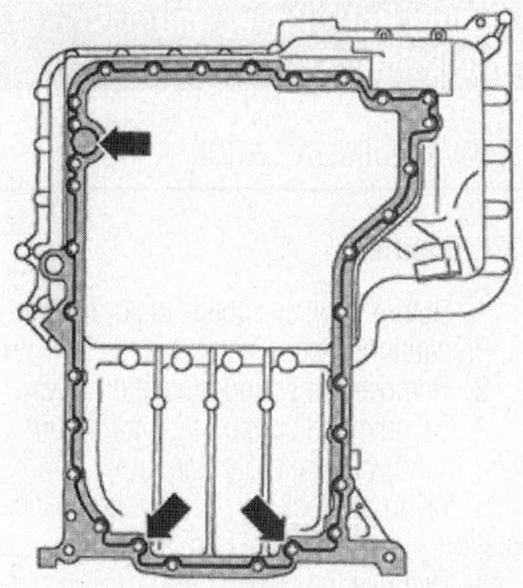

Apply sealant to the upper oil pan as shown—4.2L engines

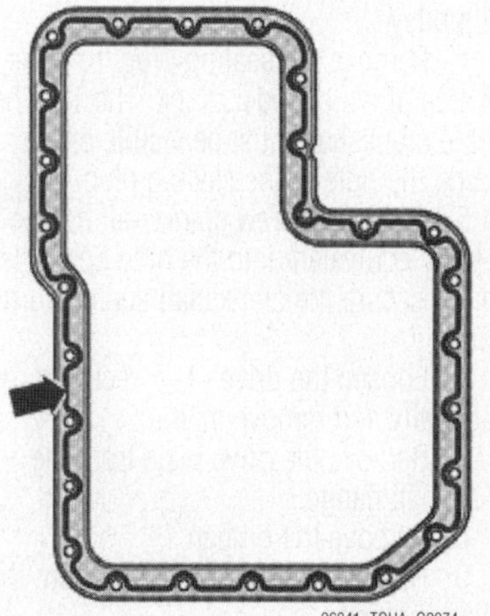

Apply sealant to the lower oil pan as shown—4.2L engines

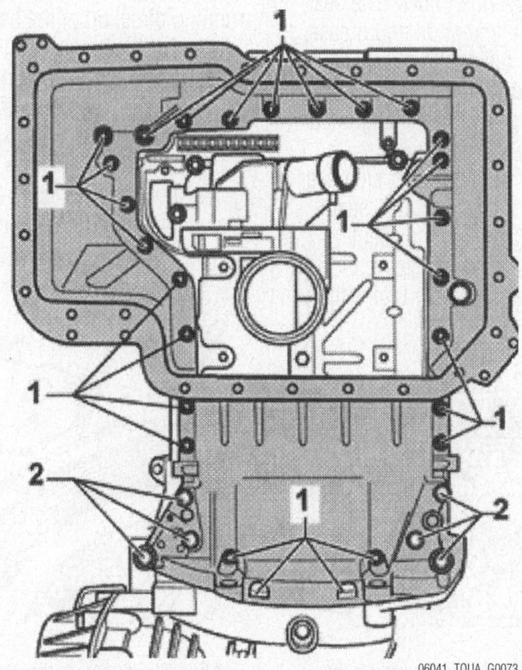

Upper oil pan bolt identification—4.2L engines

Oil Pump

REMOVAL & INSTALLATION

3.2L Engines

1. Before servicing the vehicle, refer to the Precautions Section.

2. Remove the oil pan.

3. Remove the bolts securing the oil pump and the oil pipe.

4. Remove the oil pump; be careful not to let the oil pump input shaft fall.

To install:

5. Installation is the reverse of removal. Observe the following torque specifications:

- Oil pump: 17 ft. lbs. (23 Nm)
- Oil pipe: 71 inch lbs. (8 Nm)

4.2L Engines

1. Before servicing the vehicle, refer to the Precautions Section.

2. Remove the lower and upper oil pan.

3. Compress the chain tensioner and secure it with locking pin tool T40011 or equivalent.

4. Unbolt the chain sprocket from the oil pump. Note that the text on the sprocket faces outward.

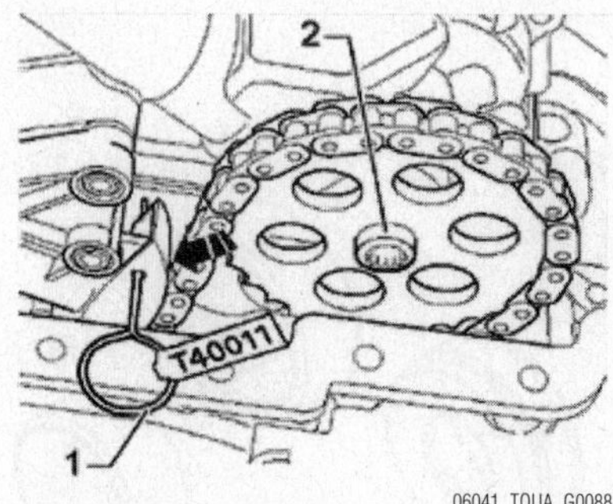

Compress the tensioner in the direction of the arrow and install the locking pin—4.2L engines

5. Unscrew the three bolts from the oil pump and the oil intake tube.
6. Remove the oil pump.

To install:

7. Installation is the reverse of removal. Observe the following torque specifications:
    - Oil pump: 22 ft. lbs. (30 Nm)
    - Oil pipe to oil pump: 7 ft. lbs. (10 Nm)
    - Chain sprocket bolt: 25 ft. lbs. (34 Nm)

Rear Main Seal

REMOVAL & INSTALLATION

3.2L Engines

1. Before servicing the vehicle, refer to the Precautions Section.
2. Remove the engine. Separate the engine from the transmission.
3. Lock the vibration damper with tool T10069 or equivalent.
4. Loosen the drive plate securing bolts diagonally and remove them.
5. Remove the drive plate from the crankshaft flange.
6. Place removal hook T20143/2 or equivalent behind the sealing lip of the oil seal. Pry out the oil seal.

To install:

7. Pull the seal with the outer side over sleeve T10122/1 or equivalent onto seal assembly tool T10122/2 or equivalent.
8. Separate the assembly sleeves.
9. Place the seal assembly tool with the seal onto the crankshaft flange. Do not lubricate the seal, install it dry.
10. Drive the seal into the sealing flange. Using thrust piece T10122/3 or equivalent.
11. Place the drive plate on the crankshaft flange. Using at least 3 old securing bolts, tighten them to 22 ft. lbs. (30 Nm).
12. Measure dimension (a) using a depth gauge. It should be 21–23 mm. If not, remove the drive plate and install an appropriate sized shim.

➡Only one shim of the appropriate thickness may be used.

13. Insert new securing bolts and tighten by hand (be sure to replace the 3 old bolts). Tighten the bolts to 44 ft. lbs. (60 Nm) plus an additional ¼ turn.
14. The remaining instillation is the reverse of removal.

4.2L Engines

1. Before servicing the vehicle, refer to the Precautions Section.
2. Remove the engine. Separate the engine from the transmission.
3. Set the crankshaft to Top Dead Center (TDC). The marking on the timing belt cover must be aligned with the notch on the belt pulley.
4. Remove the sealing plug from the left side of the cylinder block. The TDC hole for the crankshaft must be visible or felt behind the hole of the sealing plug.
5. Carefully screw crankshaft holder 3242 or equivalent into the hole up to the stop to secure the crankshaft against turning.
6. Loosen the drive plate securing bolts diagonally and remove them.
7. Remove the drive plate from the crankshaft flange.
8. Remove the oil pan.
9. Remove the sealing flange from the cylinder block.
10. Remove the rear main seal.

To install:

11. Clean the sealing surfaces of the sealing flange and cylinder block. The sealing surfaces must be free of oil and grease.
12. Wipe any oil residue from end of crankshaft with a clean cloth.
13. Place a new seal onto the alignment bushings.
14. Pull the seal with the outer side over sleeve T10122/1 or equivalent onto seal assembly tool T10122/2 or equivalent.
15. Separate the assembly sleeves.
16. Place the seal assembly tool with the seal onto the crankshaft flange. Do not lubricate the seal, install it dry.
17. Secure the sealing flange to the cylinder block. Tighten to 7 ft. lbs. (10 Nm).
18. Install the oil pan.
19. Place the shim (3.4 mm thickness) on the end of the crankshaft.
20. Install the drive plate with the washer (1.5 mm thickness).
21. Use new securing bolts and tighten by hand. Then tighten to 22 ft. lbs. (30 Nm) plus an additional ¼ turn.
22. Remove the crankshaft holder. Install the sealing plug with new seal. Tighten to 26 ft. lbs. (35 Nm).
23. The remaining installation is the reverse of removal.

Timing Belt, Cover and Crankshaft Seal

REMOVAL & INSTALLATION

4.2L Engines

1. Before servicing the vehicle, refer to the Precautions Section.
2. Remove the sound insulation tray.
3. Remove the cooling fan and mount (refer to the water pump procedure).
4. Bring the lock carrier into the service position.
5. Remove the ribbed belt.
6. Remove the fasteners from the left and right timing belt covers. Remove the covers.
7. If reusing the timing belt, mark the running direction of the belt.
8. Set the crankshaft to Top Dead Center (TDC). The marking on the timing belt cover must be aligned with the notch on the belt pulley.
9. Check the position of the camshaft gears. The large holes on the securing plates must align on the inside.

T10122 assembly tools

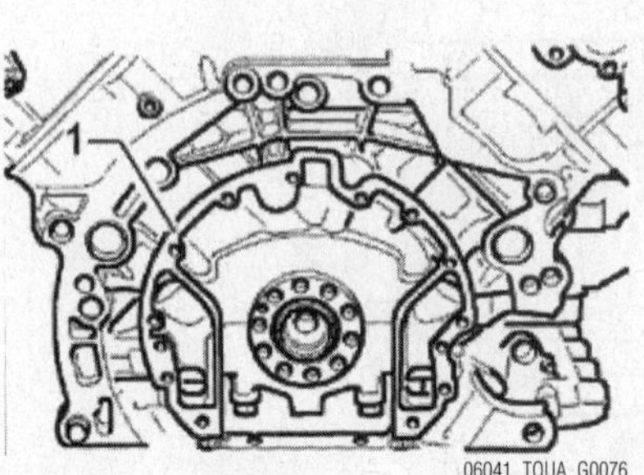

Place a new seal (1) for the flange on the alignment bushings—4.2L engines

Align the marks on the crankshaft—4.2L engines

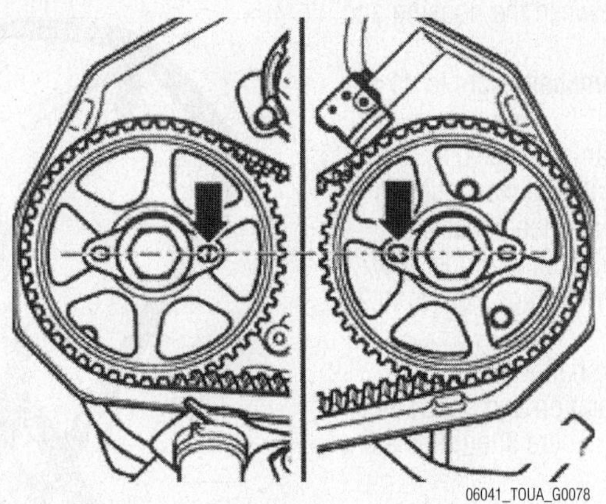

The large holes (arrows) must align on the inside—4.2L engines

➡If the large holes are positioned on the outer side of the timing belt gears, the crankshaft must be turned one rotation further in the direction of engine rotation.

10. Remove the sealing plug from the left side of the cylinder block. The TDC hole for the crankshaft must be visible or felt behind the hole of the sealing plug.
11. Carefully screw crankshaft holder 3242 or equivalent into the hole up to the stop to secure the crankshaft against turning.
12. Remove the center timing belt guard.
13. Remove the 8 bolts securing the vibration damper.
14. Remove the bolt from the oil filter housing and then remove the bolts for the timing belt tensioner cover. Remove the cover.

➡The timing belt tensioner is oil-dampened. Therefore it can only be compressed slowly.

15. Use locking pin T40011 or equivalent to secure the tensioner.
16. If necessary, align the holes in the housing and the tensioner piston using a pair of needle nose pliers or a thin piece of wire, before tensioning.
17. Turn the tensioning lever of the tensioning roller (1) in the direction of the arrow. When the lever pushes the tensioner together so that the holes in the housing and piston align, use locking pin T40011 or equivalent to lock the piston.
18. Place camshaft locking tool T40005 or equivalent into the securing plates of the camshafts and loosen the bolts about 5 turns.
19. Remove the camshaft locking tool.
20. Remove the camshaft gears from the cone using puller T40001 and T40001/2 or equivalent.
21. Loosen the tensioning roller and remove the timing belt.

To install:

➡On used timing belts, observe the marking for the direction of rotation.

22. Set the timing belt onto the timing belt crankshaft gear first, then onto the idler roller of the timing belt tensioner, then onto the tensioning roller. Now, set it onto the camshaft gears and the coolant pump belt pulley.

➡The camshaft gears must just be able to be twisted on the camshaft cones.

23. Install camshaft locking tool T40005 or equivalent onto the camshaft gears again.

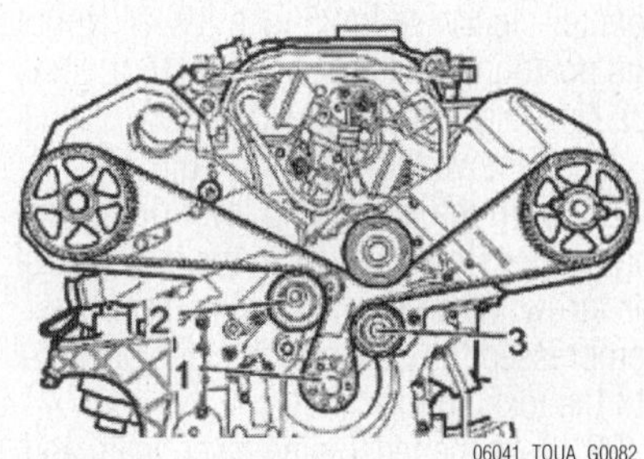

Timing belt routing—4.2L engines

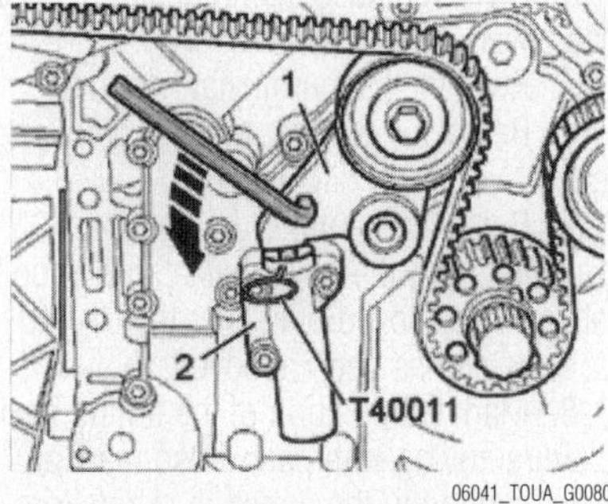

Turn the tensioning lever of the tensioning roller (1) in the direction of the arrow—4.2L engines

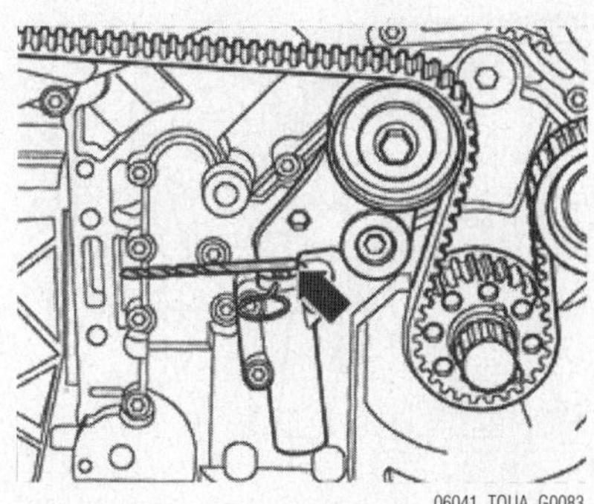

Set a 5 mm drill as shown (arrow)—4.2L engines

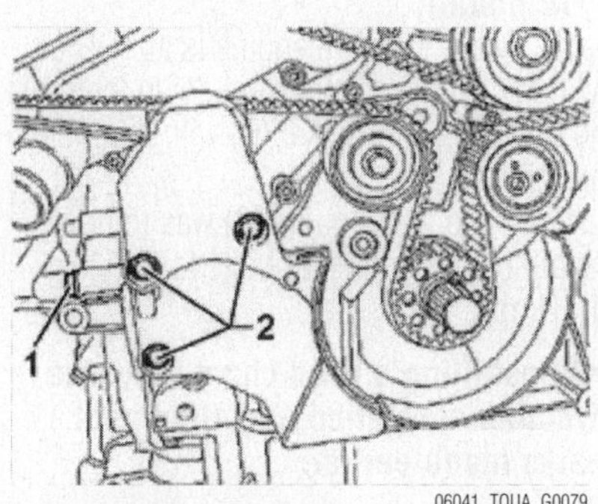

Remove the bolt (1) from the filter housing and the tensioner cover (2)—4.2L engines

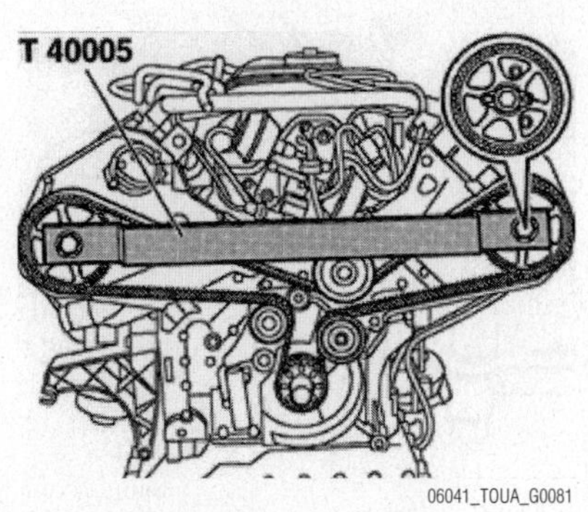

Camshaft locking tool—4.2L engines

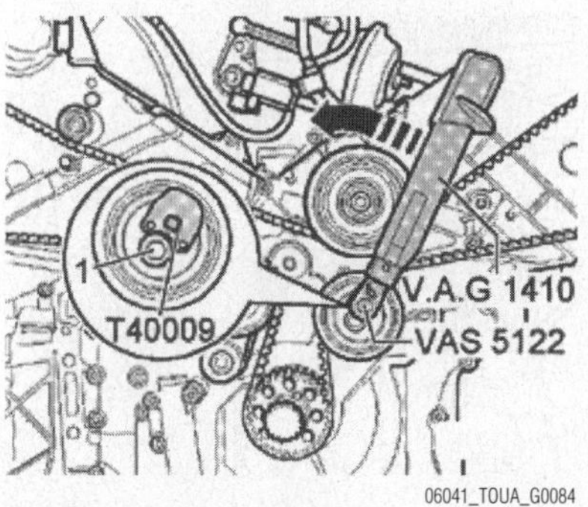

Rotate the tensioning roller counter-clockwise—4.2L engines

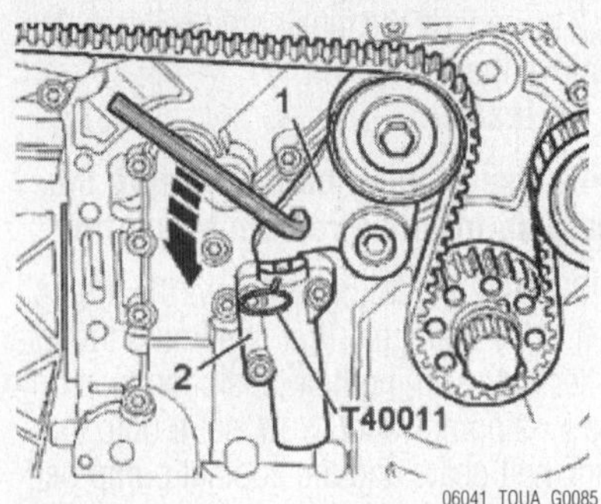

Rotate the tensioner lever as shown. Remove the locking pin when the piston is compressed

24. Set a 5 mm drill bit between the tensioner lever and the piston of the tensioner.
25. Pretension the still loosened tensioning roller using a torque wrench. Rotate counter-clockwise to 35 inch lbs. (4 Nm). In this position, tighten the bolt to 16 ft. lbs. (22 Nm). Refer to the illustration.
26. Remove the 5 mm drill bit.
27. Turn the tensioning lever of the timing belt tensioning roller in the direction of the arrow. When the tensioning lever has compressed the piston in the tensioner, pull out the locking pin.
28. Turn the tensioning lever again and place a 7 mm drill between the housing and the tensioning lever.
29. Tighten the camshaft gears to 41 ft. lbs. (55 Nm).
30. Remove the camshaft locking tool.
31. Remove the drill that was inserted between the housing and tensioner.
32. Remove the crankshaft holder tool and screw in the sealing plug. Tighten to 22 ft. lbs. (30 Nm).
33. Turn the crankshaft twice in the direction of engine rotation and check the gap of dimension (a). There should be a 5 mm gap.
34. The remaining installation is the reverse of removal.

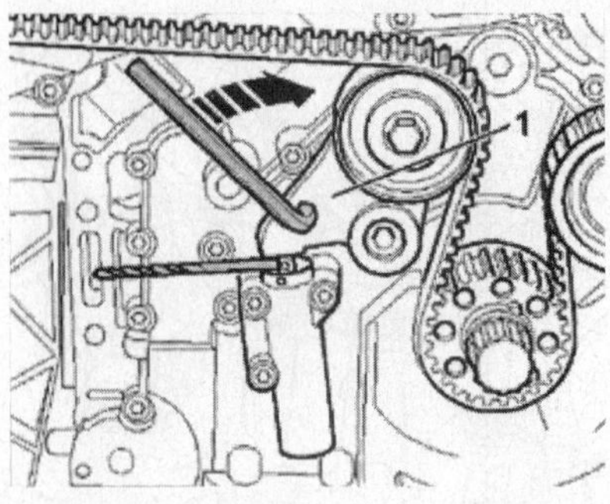

Rotate the lever again and place a 7 mm drill as shown—4.2L engines

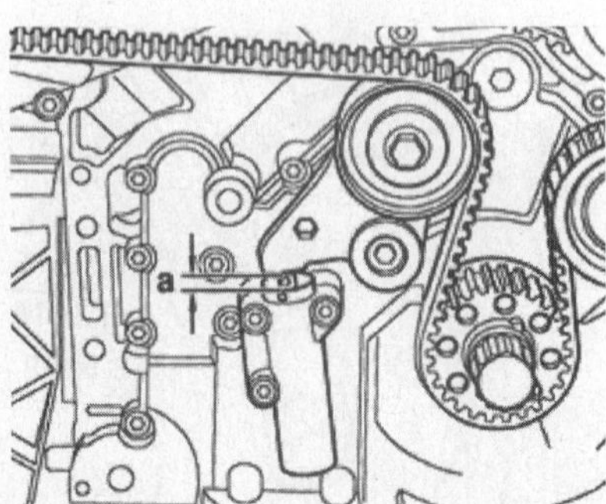

Dimension (a) should be 5 mm—4.2L engines

Timing Chain, Sprockets, Rear Covers and Seal

REMOVAL & INSTALLATION

3.2L Engines

1. Before servicing the vehicle, refer to the Precautions Section.
2. Remove the oil pan.
3. Set crankshaft to TDC of cylinder 1 by turning the vibration damper securing bolt in the direction of engine rotation. Align the mark on the vibration damper.
4. Remove intake manifold and cylinder head cover.
5. Remove the thermostat housing.
6. Remove the upper timing chain tensioner.
7. Remove the upper timing chain cover. Note that the cover has a hidden bolt that can be removed only after the thermostat housing has been removed.
8. Mark the position of the timing chain before removing with paint. Also mark an arrow pointing in the direction of rotation. Do not mark chain with a center punch or similar means!

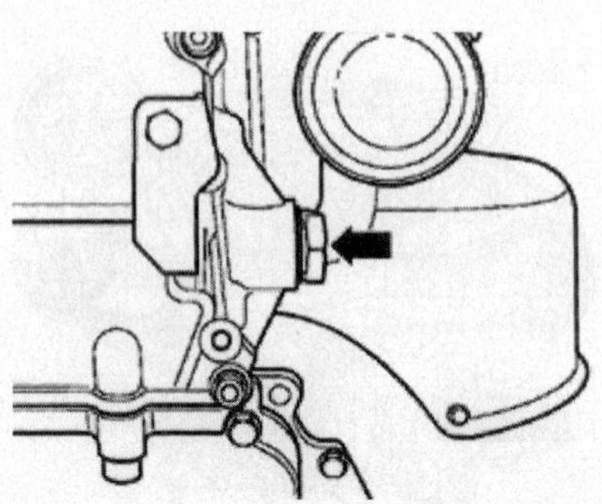

Remove the timing chain tensioner—3.2L engines

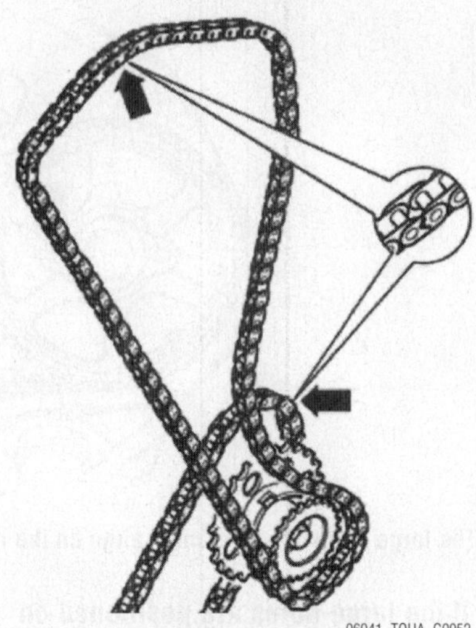

Mark the position of the chains and an arrow pointing in the direction of rotation—3.2L engines

9. Remove the exhaust camshaft adjuster. Counter-hold the camshaft using a 32 mm open end wrench on the camshaft flats.
10. Remove camshaft adjuster together with the timing chain from the intake camshaft. Place the chain aside.
11. Remove the drive plate (refer to the Rear Main Seal procedure).
12. Remove the bolts securing the lower timing chain cover.
13. The upper (camshaft) timing chain can now be removed.

➡**Be careful not to damage the cylinder head gasket.**

14. Mark the position of the lower timing chain before removing with paint. Also mark an arrow pointing in the direction of rotation. Do not mark chain with a center punch or similar means!
15. Remove the bolt securing the sprockets to the intermediate shaft.
16. Release the lower timing chain tensioner, then remove the chain together with the intermediate shaft sprockets.

To install:

17. Make sure the engine is at TDC of cylinder 1. Align the ground down tooth of the drive chain sprocket (B) with the mounting joint.
18. If the lower guide (2) was removed, install it and tighten the bolts to 7 ft. lbs. (10 Nm).

➡**If installing a used chain, note the direction of rotation and alignment marks made earlier.**

19. Install the timing chain (1) and both chain sprockets (3) and (4). The marking on the timing chain sprocket (4) must align

Timing chain components and timing mark identification—3.2L engines

with notch (C) or (D) on the thrust washer of the intermediate shaft.

➡Make sure that the chain runs completely straight in the guide rail from the crankshaft to the intermediate shaft.

20. Hand tighten the intermediate shaft bolt. Use a new bolt.
21. Release the locking splines of the lower chain tensioner (A) with a small screwdriver while the tensioning rail is pressed against chain tensioner. If removed, the chain tensioner should be installed in this position and tightened to 71 inch lbs. (8 Nm).
22. Lock the vibration damper with tool T10069 or equivalent.
23. Tighten the new intermediate shaft sprockets securing bolt to 44 ft. lbs. (60 Nm), plus an additional ¼ turn.
24. Remove the vibration damper counter holder tool.
25. Check the position of crankshaft (B) to intermediate shaft (C) and (D) once again. The engine must still be at TDC of cylinder 1.
26. Position the camshafts to TDC of cylinder 1.
27. Install camshaft locking bar T10068 or similar. It must engage in both shaft grooves
28. Place the camshaft timing chain on the intermediate shaft chain sprocket. Route the chain between the tensioning rail and guide rail towards the control housing.
29. Install the intake camshaft adjuster first. Turn the sensor wheel (1) in the camshaft adjuster to the right onto its stop and bolt it on in this position. The marking 24E must align with the notch in the control housing.

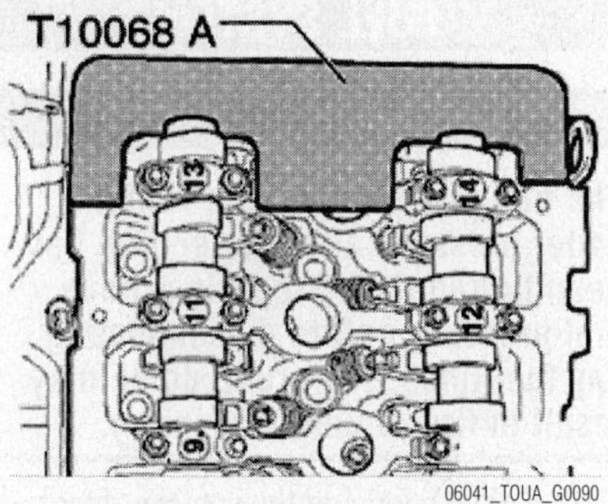

Lock the position of the camshafts as shown—3.2L engines

30. Place the camshaft timing chain over the sprocket in this position.

➡If the camshaft adjuster is to be fastened to the camshaft, the adjuster with the corresponding chain sprocket must be turned to the left. In any case, the marking 24E (intake) or 32A (exhaust) must be aligned to the notch in the control housing.

31. Count exactly 16 rollers of the timing chain to the right from the 24E marking of the camshaft adjuster. Insert the tooth of the exhaust camshaft adjuster with the marking 32A at that point.
32. There must now be exactly 16 rollers between those two teeth. Turn the sensor wheel (1) in the camshaft adjuster to the right onto its stop and bolt it on in this position.
33. Marks (A) and (B) on the camshaft adjusters must align with notches (arrows) on the control housing (C).
34. Remove the camshaft locking bar.
35. Using new bolts for the camshaft adjusters, tighten them by hand.
36. Check the timing adjustment:
    a. Press the tensioner plate of upper timing chain against the chain.
    b. Turn the crankshaft over twice in direction of engine rotation and check the timing marks.

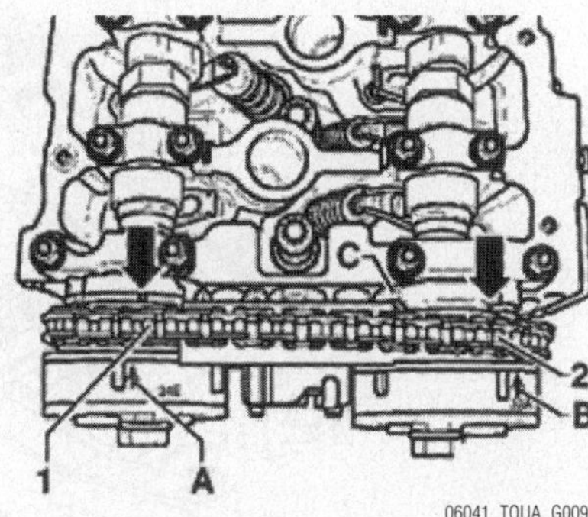

The timing marks (A) and (B) must align with the notches (arrows)—3.2L engines

➡When turning the crankshaft, the tensioning plate must be pressed against the chain by hand at the chain tensioner, so that the chain does not jump over a tooth.

37. If the marks do not align, repeat the valve timing adjustment.
38. If the marks align, hold the respective camshaft tightly with a 32 mm open end wrench. Tighten the intake and exhaust camshaft adjusters to 44 ft. lbs. (60 Nm), plus and additional ¼ turn.
39. Remove the rear main seal from the lower timing cover.
40. Coat the sealing surface of the lower timing with sealant AMV 176 501 or equivalent.
41. Align the cover with the dowel pins in the cylinder block. Tighten the bolts to 71 inch lbs. (8 Nm).
42. Install a new rear main seal.
43. Lubricate a new oil channel seal and install it into the upper timing cover.
44. Fill the 3mm holes in the cylinder head gasket with sealant AMV 174 004 01 (or equivalent). Coat the sealing surface on the upper timing cover with sealant AMV 176 501 (or equivalent). Install the cover immediately after applying the sealant.
45. Insert all the securing bolts with corresponding brackets and tighten lightly. Then tighten the M8 bolts to 17 ft. lbs. (23 Nm) and the M6 bolts to 71 inch lbs. (8 Nm).
46. Install the camshaft timing chain tensioner and tighten to 30 ft. lbs. (40 Nm).
47. Turn the crankshaft over twice in direction of engine rotation and check the timing marks. If the marks do not align, repeat the valve timing adjustment.
48. Install the thermostat housing. Tighten the bolts to 71 inch lbs. (8 Nm).
49. The remaining installation is the reverse of removal.

Piston and Ring

POSITIONING

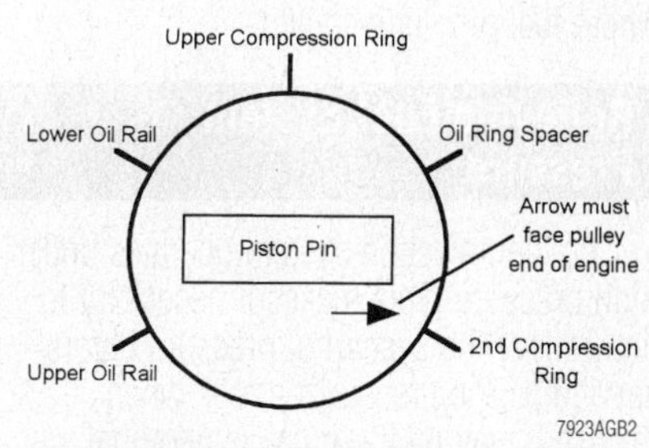

Piston and piston ring positioning—3.2L and 4.2L engines

FUEL SYSTEM

Fuel System Service Precautions

Safety is the most important factor when performing not only fuel system maintenance but also any type of maintenance. Failure to conduct maintenance and repairs in a safe manner may result in serious personal injury or death. Maintenance and testing of the vehicle's fuel system components can be accomplished safely and effectively by adhering to the following rules and guidelines.

- To avoid the possibility of fire and personal injury, always disconnect the negative battery cable unless the repair or test procedure requires that battery voltage be applied.
- Always relieve the fuel system pressure prior to disconnecting any fuel system component (injector, fuel rail, pressure regulator, etc.), fitting or fuel line connection. Exercise extreme caution whenever relieving fuel system pressure, to avoid exposing skin, face and eyes to fuel spray. Please be advised that fuel under pressure may penetrate the skin or any part of the body that it contacts.
- Always place a shop towel or cloth around the fitting or connection prior to loosening to absorb any excess fuel due to spillage. Ensure that all fuel spillage is quickly removed from engine surfaces. Ensure that all fuel soaked cloths or towels are deposited into a suitable waste container.
- Always keep a dry chemical (Class B) fire extinguisher near the work area.
- Do not allow fuel spray or fuel vapors to come into contact with a spark or open flame.
- Always use a back-up wrench when loosening and tightening fuel line connection fittings. This will prevent unnecessary stress and torsion to fuel line piping. Always follow the proper torque specifications.
- Always replace worn fuel fitting O-rings with new. Do not substitute fuel hose where fuel pipe is installed.

Relieving Fuel System Pressure

The fuel injection system operates under high pressure. This makes it necessary to first relieve the system of pressure before servicing. The pressurized fuel, when released, may ignite or cause personal injury.

** CAUTION

The fuel injection system remains under pressure after the engine has been turned OFF. Properly relieve fuel pressure before disconnecting any fuel lines. Failure to do so may result in fire or personal injury.

1. Before servicing the vehicle, refer to the precautions in the beginning of this section.

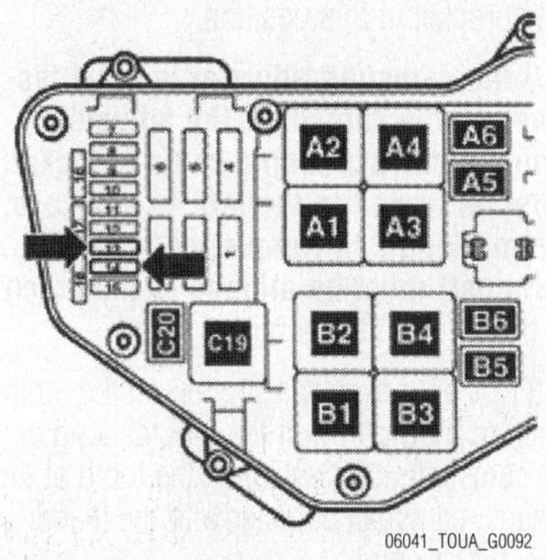

Remove fuses 13 and 14 (arrows)

2. Remove the fuel pump fuses 13 and 14 in the left plenum chamber. The fuse can be removed to stop the fuel pump from running. With the engine operating at idle, wait until the engine stalls from fuel starvation.
3. Switch the ignition **OFF** and remove the negative battery cable.
4. Slowly and carefully open the fuel tank filler cap for a brief moment and reinstall.
5. Carefully loosen the fuel line on the control pressure regulator or component to be serviced.
6. Wrap a clean rag around the connection, while loosening, to catch any fuel.
7. After service is complete, discard the fuel soaked rag in the proper manner and reconnect negative battery cable and fuses.

Fuel Filter

REMOVAL & INSTALLATION

1. Before servicing the vehicle, refer to the Precautions Section.

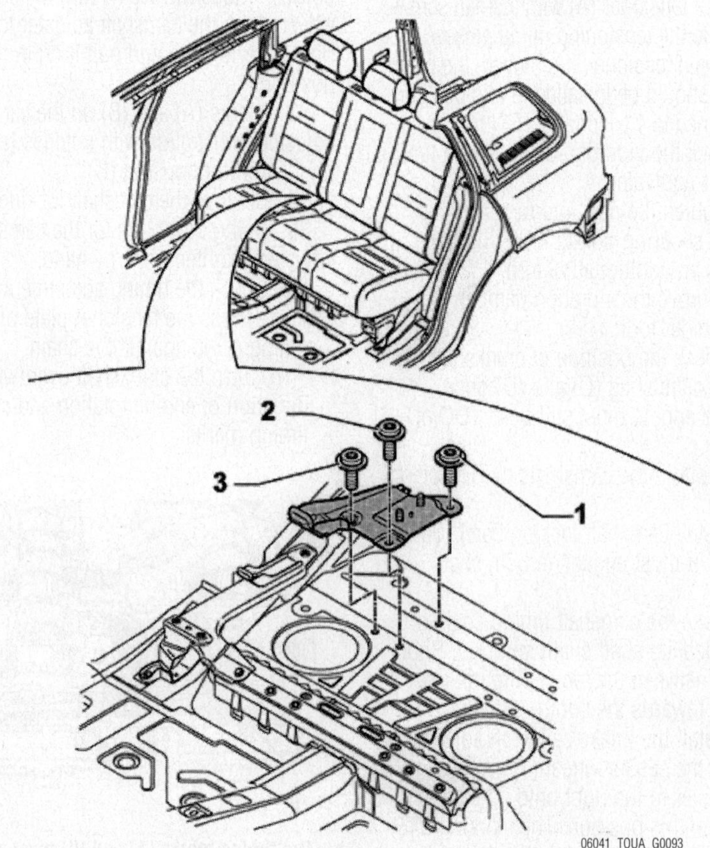

The seat backrest support is secured with 3 bolts. When installing, tighten them in this order

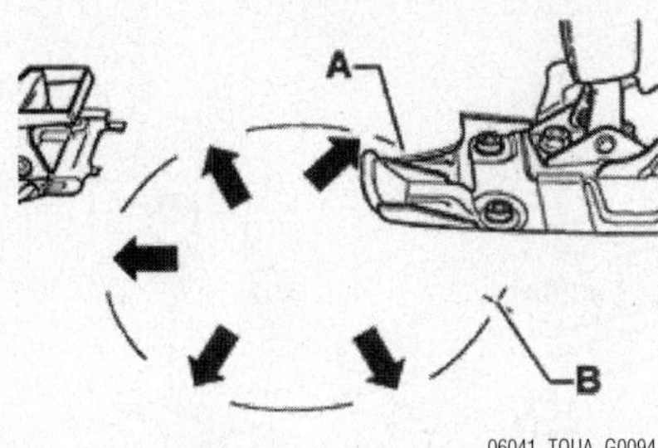

Cut the carpet in the area shown (arrows)

2. Relieve the fuel system pressure.
3. Drain the fuel tank.
4. The fuel filter housing in the fuel tank can be accessed from under the rear seat. Cut open the carpet in the pre-cut area.
5. Remove the 4 nuts from the filter housing cover. It may be necessary to remove the seat backrest support.
6. Unplug any hose and electrical connections.
7. Remove the locking ring. Tool T10202 or similar can be used.
8. Pull the filter housing out and empty it of fuel.
9. Matchmark the position of the filter cover.
10. Disconnect the ground wire and remove the screws from the filter cover.
11. Remove the filter from the housing.

To install:

12. Installation is the reverse of removal. Observe the following torque specifications:
   - Fuel filter cover: 7 ft. lbs. (10 Nm)
   - Locking ring: 107 ft. lbs. (145 Nm)
   - Rear seat bracket: 15 ft. lbs. (20 Nm), tighten bolts in sequence of 1, 2 then 3

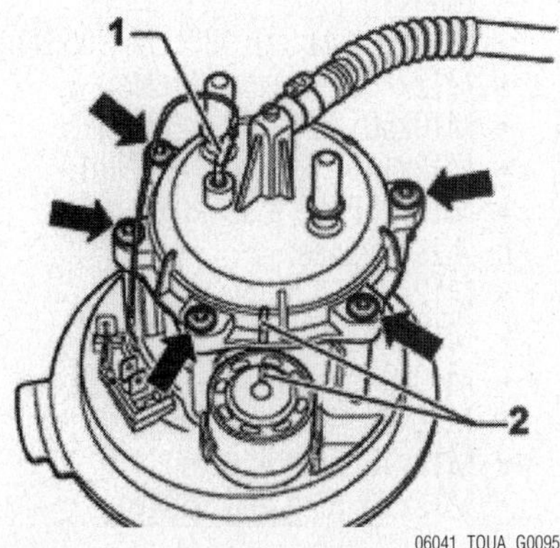

Filter housing cover bolts (arrows)

Fuel Pump

REMOVAL & INSTALLATION

1. Before servicing the vehicle, refer to the Precautions Section.
2. Drain the fuel tank.
3. Relieve the fuel system pressure.
4. The fuel pumps in the fuel tank can be accessed from under the rear seat. Cut open the carpet in both pre-cut areas.
5. Remove the 4 nuts from the housing covers. It may be necessary to remove the seat backrest support.
6. Unplug any hose and electrical connections.
7. Remove the locking rings. Tool T10202 or similar can be used.
8. Remove both sending unit flanges.
9. Unclip the black filler hose from both fuel delivering units in the tank.
10. Pull the fuel supply line off both delivery units. Remove both delivery units by twisting them quarter of a turn to the left. Be careful when removing them as fuel may run out.

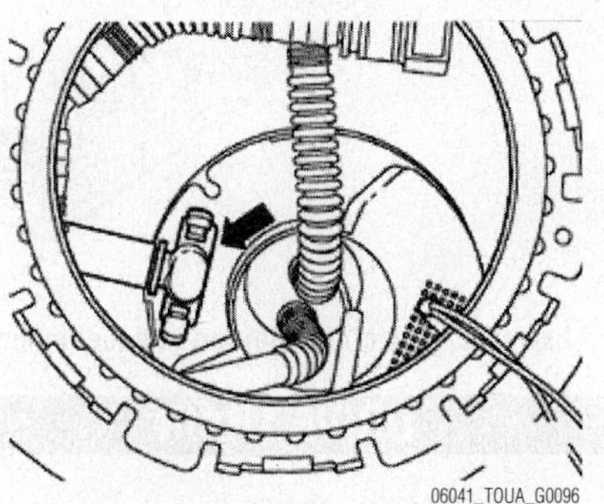

Unclip the fill hose

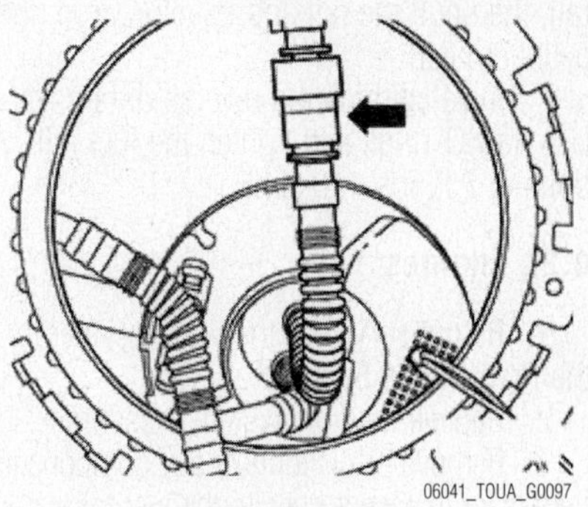

Disconnect the supply line

11. Unclip the fuel pumps on the bottom of each side, and remove them with a slight turn.

To install:

12. Installation is the reverse of removal. Observe the following torque specifications:
   - Fuel filter cover: 7 ft. lbs. (10 Nm)
   - Locking ring: 107 ft. lbs. (145 Nm)
   - Rear seat bracket: 15 ft. lbs. (20 Nm), tighten bolts in sequence of 1, 2 then 3

Fuel Rail

REMOVAL & INSTALLATION

3.2L Engines

1. Before servicing the vehicle, refer to the Precautions Section.
2. Relieve the fuel system pressure.
3. Disconnect the fuel supply line.
4. Remove the intake manifold.
5. Unplug the injector wiring harness and set aside.
6. Remove the bolts securing the fuel

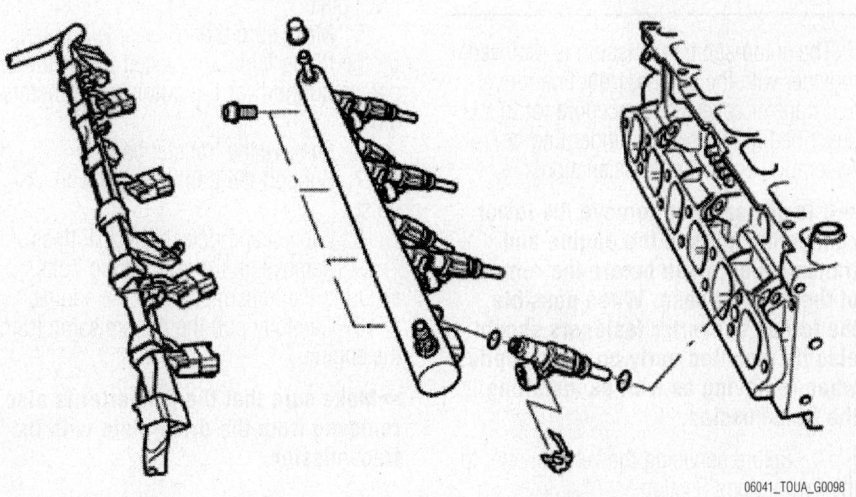

Exploded view of the fuel rail and fuel injector assembly—3.2L engines

rail, then pull the rail and injectors from the cylinder head.

7. Installation is the reverse of removal. Use new O-rings and tighten the fuel rail bolts to 7 ft. lbs. (10 Nm).

4.2L Engines

1. Before servicing the vehicle, refer to the Precautions Section.
2. Relieve the fuel system pressure.
3. Remove or disconnect any components necessary to ease access to the fuel rail.
4. Disconnect the fuel supply line.
5. Unplug the injector wiring harness and set aside.
6. Remove the bolts securing the fuel rail, then pull the rail and injectors from the cylinder head.
7. Installation is the reverse of removal. Use new O-rings and tighten the fuel rail bolts to 7 ft. lbs. (10 Nm).

Fuel Injectors

REMOVAL & INSTALLATION

1. Before servicing the vehicle, refer to the Precautions Section.
2. Relieve the fuel system pressure.
3. Remove the fuel rail.
4. Remove the clips securing the injectors to the rail, then pull the injector from its socket in the rail.
5. Installation is the reverse of removal. Lightly coat new O-rings with clean engine oil to ease installation.

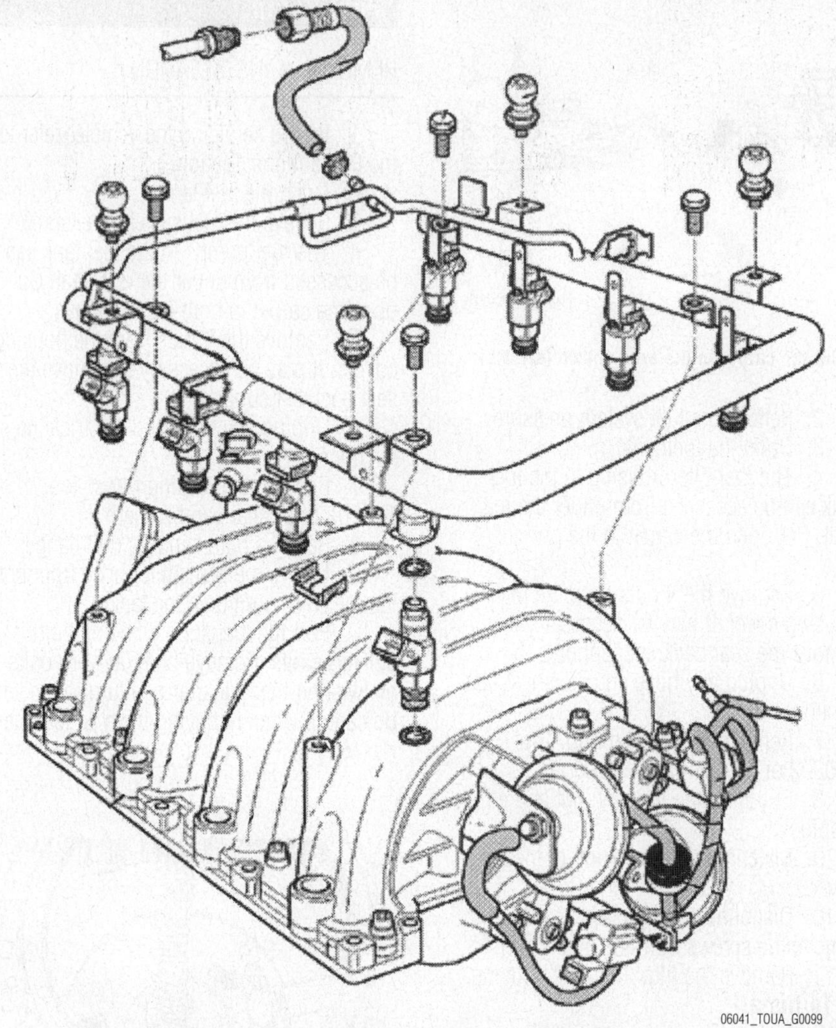

Exploded view of the fuel rail and fuel injector assembly—4.2L engines

DRIVE TRAIN

Transmission

REMOVAL & INSTALLATION

The automatic transmission is removed together with the engine, front final drive and transfer case. The procedure for this is described in this section under Engine Assembly removal and installation.

➡It is advisable to remove the lower connecting bolts of the engine and transmission, even before the removal of the transfer case. When possible, the torque converter fasteners should also be mounted early on. This applies when removing as well as installing the transmission.

1. Before servicing the vehicle, refer to the Precautions Section.
2. Remove the engine.
3. Remove the starter motor.
4. Unbolt the torque converter from the drive plate.
5. Make sure that at least two plates on the lifting table are placed under the oil pan as support for the automatic transmission.
6. Remove the transfer case.
7. Support the transmission with a hoist.
8. Loosen and detach the ATF lines.
9. Remove all the connecting bolts securing the transmission to the engine.
10. Carefully pull the transmission from the engine.

➡Make sure that the converter is also removed from the drive plate with the transmission.

To install:

11. Installation is the reverse of removal. Observe the following torque specifications (refer to the illustrations for bolt identification):

a. 3.2L engines:
- M12x60 (except starter): 59 ft. lbs. (80 Nm)
- M12x60 (starter): 48 ft. lbs. (65 Nm)
- M12x70: 59 ft. lbs. (80 Nm)
- M10x50: 33 ft. lbs. (45 Nm)
- M12x140: 59 ft. lbs. (80 Nm)
- M12x155: 59 ft. lbs. (80 Nm)

b. 4.2L engines:
- M12x60: 59 ft. lbs. (80 Nm)
- M12x70: 59 ft. lbs. (80 Nm)
- M10x205: 33 ft. lbs. (45 Nm)
- M12x80: 48 ft. lbs. (65 Nm)
- M12x45: 59 ft. lbs. (80 Nm)
- M12x70: 59 ft. lbs. (80 Nm)

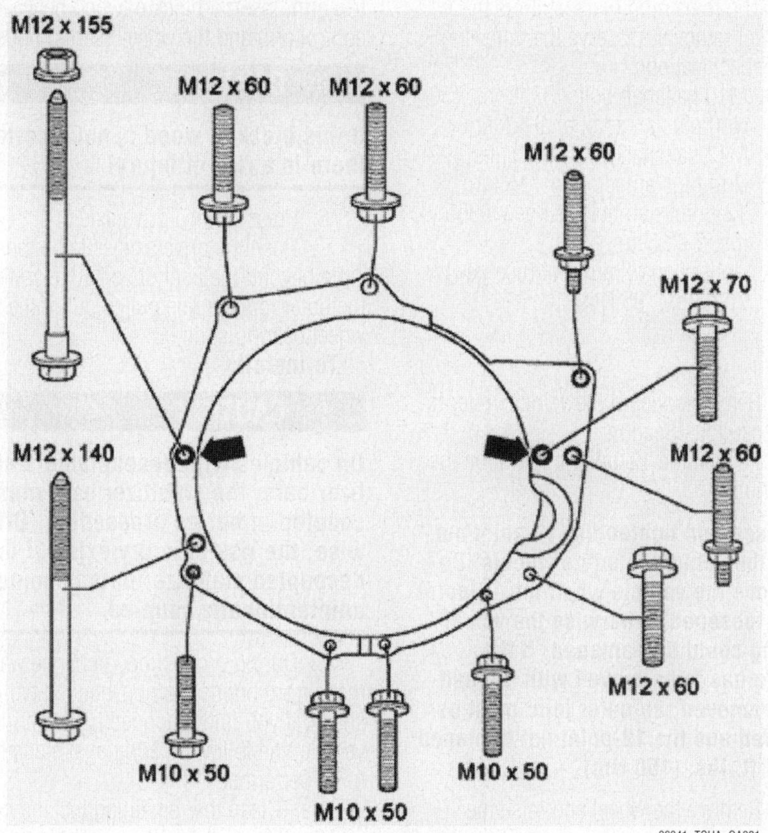

3.2L engine bolt identification

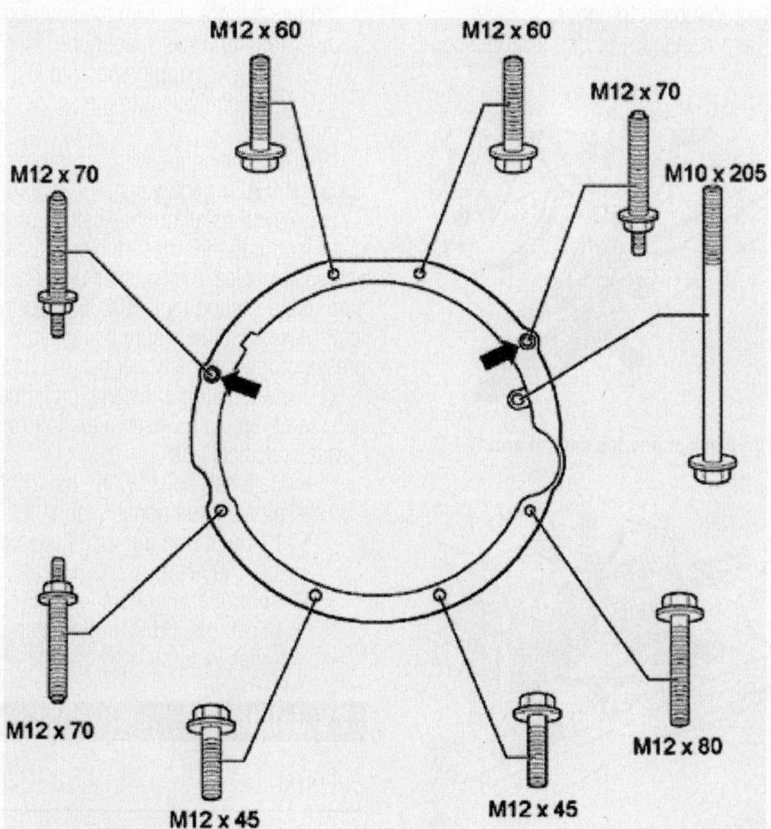

4.2L engine bolt identification

Transfer Case Assembly

REMOVAL & INSTALLATION

The transfer case is removed together with the engine, front final drive and transmission. The procedure for this is described in this section under Engine Assembly removal and installation.

1. Before servicing the vehicle, refer to the Precautions Section.
2. Remove the engine.
3. Remove the driveshaft between the transfer case and the front final drive. Counterhold the driveshaft using tool T10172 or equivalent to ease removal of the bolts.
4. Support the transfer case with a hoist.
5. Remove the bolts, then separate the transfer case and set it aside.

To install:

6. Check if the alignment sleeves for centering the transmission/transfer case are installed in the transmission; install if necessary.
7. Always replace the O-ring seal between the transfer case and transmission; lubricate it lightly.
8. Lubricate the transfer case input shaft and transmission output shaft with grease G 000 100 or equivalent.
9. Push the transfer case completely onto the transmission, ensuring that transfer case input shaft splines are centered as they are guided onto the transmission output shaft.

➡If the splines are correctly positioned, the transfer case will slide to a stop against the transmission. Do not use the bolts to pull the transfer case onto the transmission!

10. The remaining installation is the reverse of removal. Observe the following torque specifications:
    - Transfer case mounting bolts (use new): ft. lbs. (45 Nm)
    - Transfer case damper: ft. lbs. (32 Nm)

Halfshaft

REMOVAL & INSTALLATION

Front

1. Before servicing the vehicle, refer to the Precautions Section.
2. Loosen the 12-point hub nut on the wheel hub.

➡Loosen and tighten the 12-point nut when the vehicle is on its wheels. Do not move the vehicle when the 12-point nut is loosened, otherwise the wheel bearing could be damaged. If the vehicle has to be moved with the half-shaft removed, an outer joint must be installed and the 12-point nut tightened to 111 ft. lbs. (150 Nm).

3. Remove the wheel and raise the vehicle.
4. Remove the halfshaft bolts from the front final drive. Use socket T10099/1 or equivalent to loosen the bolts.
5. Press the tie rod from the steering knuckle.
6. Press the drive axle out from the wheel bearing hub.
7. Support the lower control arm/steering knuckle assembly.
8. Press out the upper control arm from the steering knuckle.
9. Loosen the strut lower control arm bolt.
10. Remove the brake line retainer from the steering knuckle.
11. Disconnect the link from the stabilizer bar.
12. Pull out the bolt from the lower control arm.
13. Lower the steering knuckle assembly until you are able to remove drive axle.

To install:

⁂ CAUTION

On vehicles with decouplable stabilizer bars, the stabilizer bars must be coupled in before proceeding. Otherwise, the risk of injury exists if the decoupled stabilizer bars become unintentionally coupled.

14. During production, a change was made from bonded to non-bonded half-shafts. If you find a bonded halfshaft in the vehicle, the following additional steps have to be performed:
    a. Clean the remaining locking compound from the splines on the outer joint and on the wheel hub.
    b. The surface should be clean metal; remove the grease from the spline on the outer joint and the wheel hub.
    c. Apply locking compound D 154 000 A1 or equivalent to the spline on the outer joint.
15. When installing the shafts, guide the outer joint into the wheel hub splines as far as possible. Seat the halfshaft as far as the stop using seating tool T10206 or equivalent.
16. The remaining installation is the reverse of removal. Observe the following torque specifications:
    - M10 halfshaft bolts: 37 ft. lbs. (50 Nm) plus an additional ¼ turn
    - M12 halfshaft bolts: 66 ft. lbs. (90 Nm) plus an additional ¼ turn
    - 12-point hub nut (use new): 369 ft. lbs. (500 Nm)
    - Upper control arm nut (use new): 70 ft. lbs. (95 Nm)

Rear

1. Before servicing the vehicle, refer to the Precautions Section.
2. Loosen the 12-point hub nut on the wheel hub.

➡Loosen and tighten the 12-point nut when the vehicle is on its wheels. Do not move the vehicle when the 12-point nut is loosened, otherwise the wheel bearing could be damaged. If the vehicle has to be moved with the half-shaft removed, an outer joint must be installed and the 12-point nut tightened to 111 ft. lbs. (150 Nm).

3. Remove the wheel and raise the vehicle.
4. Remove the halfshaft bolts from the rear final drive. Use socket T10099 or equivalent to loosen the bolts.
5. Remove the lower bolt from the control arm. Insert a block of wood between the control arm and the wheel bearing housing.

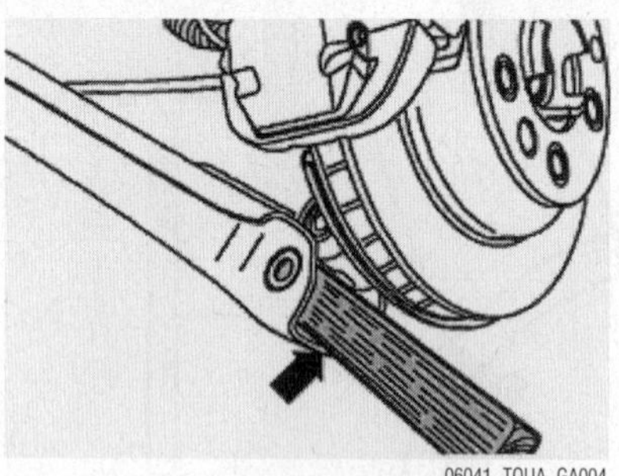

Insert a block of into the control arm

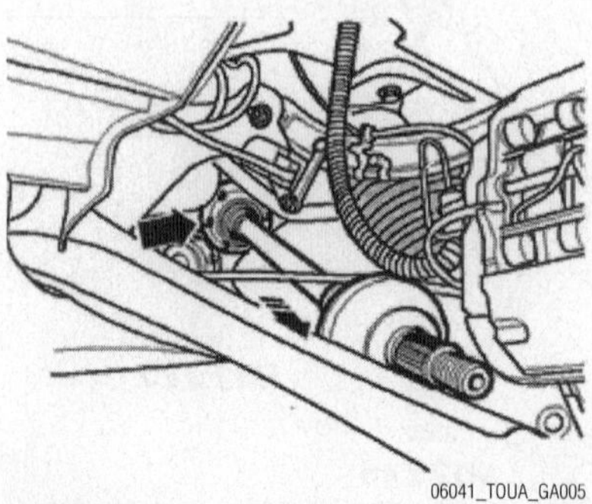

Pull the halfshaft out in the direction shown

⁂ CAUTION

If this block of wood is not inserted, there is a risk of injury!

6. Press out the drive axle.
7. Push the drive axle as far as possible beyond the final drive flange so that the outer joint can be pulled out of the wheel bearing spline.

To install:

⁂ CAUTION

On vehicles with decouplable stabilizer bars, the stabilizer bars must be coupled in before proceeding. Otherwise, the risk of injury exists if the decoupled stabilizer bars become unintentionally coupled.

8. During production, a change was made from bonded to non-bonded half-shafts. If you find a bonded halfshaft in the vehicle, the following additional steps have to be performed:
    a. Clean the remaining locking compound from the splines on the outer joint and on the wheel hub.
    b. The surface should be clean metal; remove the grease from the spline on the outer joint and the wheel hub.
    c. Apply locking compound D 154 000 A1 or equivalent to the spline on the outer joint.
9. Coat the spline with assembly paste G 052 109 A2 or equivalent.
10. When installing the shafts, guide the outer joint into the wheel hub splines as far as possible. Seat the halfshaft as far as the stop using seating tool T10206 or equivalent. Make absolutely sure that you are pressing only on the wheel bearing housing.
11. The remaining installation is the reverse of removal. Observe the following torque specifications:
    - Halfshaft bolts: 37 ft. lbs. (50 Nm) plus an additional ¼ turn
    - 12-point hub nut (use new): 369 ft. lbs. (500 Nm)
    - Control arm bolt and nut (use new): 111 ft. lbs. (150 Nm) plus an additional ¼ turn

CV-Joints

OVERHAUL

1. Before servicing the vehicle, refer to the Precautions Section.

2. Remove both clamps from the outer constant velocity joint boot. Pull the constant velocity joint boot off.

3. Drive the outer constant velocity joint off of the axle using firm blows with a plastic faced hammer. The outer constant velocity joint cannot be rebuilt and must be replaced if necessary.

4. Remove both clamps from the inner constant velocity joint boot. Pull the boot off.

5. Drive off the cover for the inner constant velocity joint. Remove the circlip using circlip pliers.

6. Press the inner constant velocity joint from the axle.

7. Before disassembling the joint, mark the ball hub position in relation to ball cage and housing.

8. Swivel the ball hub and ball cage. Remove the balls one after another.

9. Lift out the cage together with the hub.

10. Pivot the segment of the hub into an opening in the cage. Tilt the hub out of the cage.

11. Check the stub axle, hub, cage and balls for small indentations or pitting and signs of seizure. Excessive backlash in the joint will be noticed as a knock, in such cases the joint must be replaced. Polished areas and ball track marks are normal; it is not necessary to replace the joint.

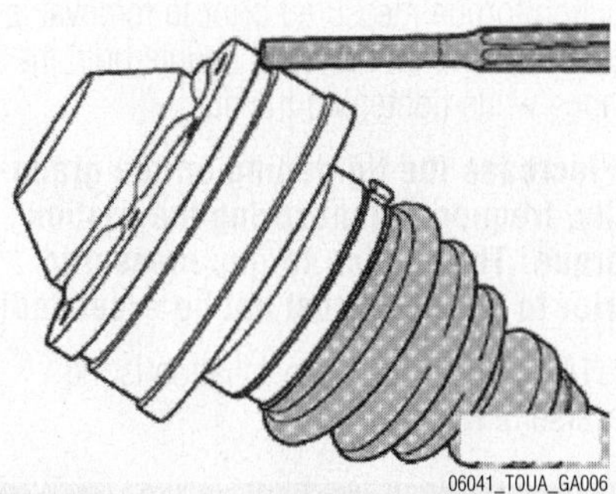

Drive the cover off of the inner joint

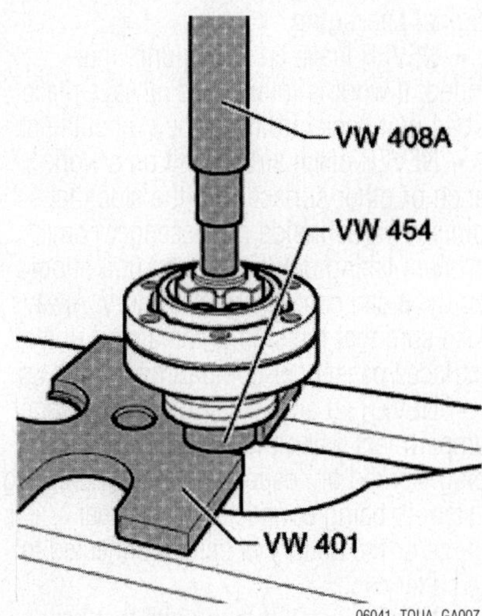

Press the inner joint from the shaft as shown

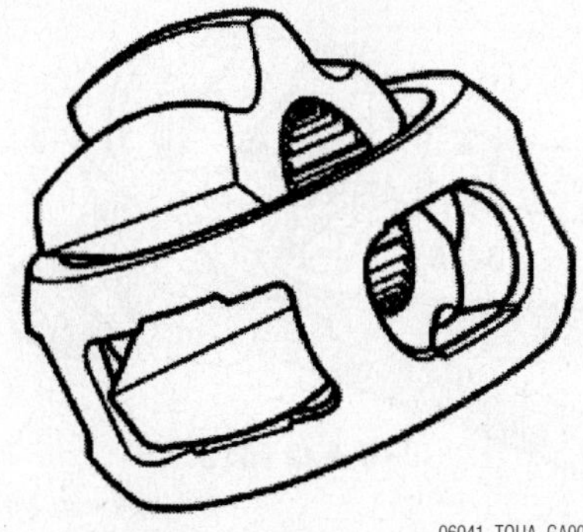

Pivot the segment into an opening, then tilt hub out of the cage

To install:

12. Press half of the total amount of grease into the joint body. Total per joint is 1.4 oz. (60 grams).

13. Install the cage with the hub in the joint housing.

14. When installing, make sure that the large chamfer on the inner diameter of the ball hub and the large inner diameter of ball cage face toward the drive axle.

15. Press in the balls one after another from opposite sides until the original position of the hub, cage and joint housing is restored.

16. Pack the remaining grease into a new joint boot and install the boot.

17. Install a new cover onto the inner constant velocity joint.

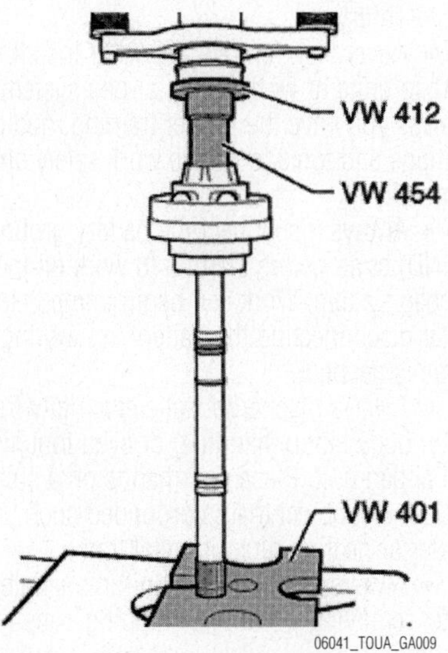

Pressing the inner joint onto the shaft

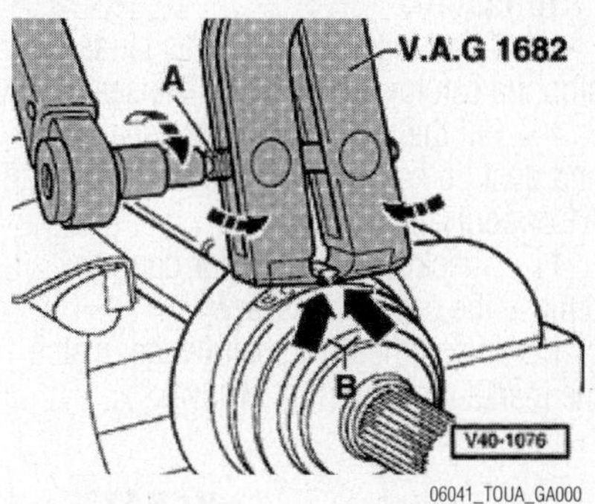

Use tool VAG 1682 or similar to install the boot clamps

18. Press the inner constant velocity joint onto the axle.

19. Install a new circlip onto the axle using circlip pliers.

20. Install boot clamp tool VAG 1682 or equivalent as shown in illustration. Make sure that the jaws of the tension clamp seat in the corners of the clip.

➡Make sure the threads of the clamp are not tight and dirty. If necessary lubricate with grease.

21. Tighten the hose clamp to 18 ft. lbs. (25 Nm). Do not bend the tensioning clamp.

Front Differential Pinion Seal

REMOVAL & INSTALLATION

The front differential/final drive is removed together with the engine, transmission and transfer case. The procedure for this is described in this section under Engine Assembly removal and installation.

1. Before servicing the vehicle, refer to the Precautions Section.

2. Remove the bolts securing the driveshaft and halfshafts to the front final drive.

3. Remove the front final drive and place it on a workbench.

4. Rotate the drive pinion nut with a beam type torque wrench. Measure friction torque before loosening the drive pinion nut. Note the friction torque value. After the seal is replaced, friction torque must be restored to this value.

5. Remove the hex nut securing the drive shaft flange. Counterhold the flange using tool 3435 or equivalent to ease removal.

6. Mark the installation position of the driveshaft flange relative to the drive pinion.

7. Pull off the driveshaft flange. If necessary, use a slide hammer.

8. Pull out the sealing ring. Clean the threads on the drive pinion.

To install:

9. Drive in a new seal in as far as the stop; be careful not to distort the seal.
10. Fill the area between the sealing lip and dust lip halfway with grease G 052 128 A1 or equivalent.
11. Check the dust cap for damage, replace the cap if necessary.
12. Install the driveshaft flange, match the marks made during removal.

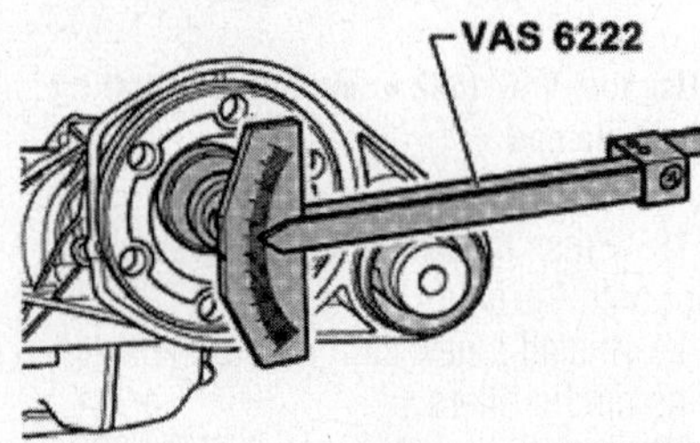

Measure the friction torque with a beam type torque wrench

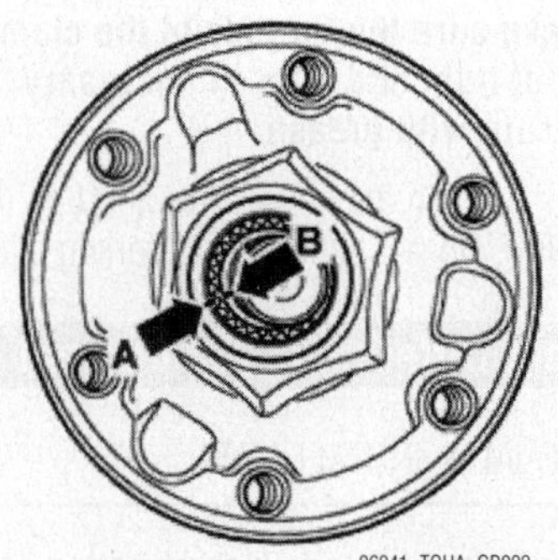

Mark the installation position of the driveshaft flange relative to the drive pinion

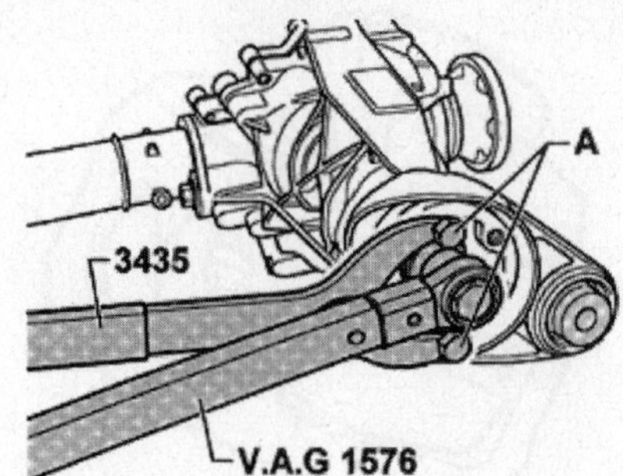

Have an assistant counterhold the flange while tightening the nut

13. Coat the threads of a new nut with Loctite® 648.
14. Tighten the drive pinion nut until the friction torque measured prior to removal is attained. Have an assistant counterhold the flange while tightening the nut.

➡Increase the tightening torque gradually, frequently measuring the friction torque. The friction torque measured prior to removal must not be exceeded!

15. The remaining installation is the reverse of removal.

Rear Differential Pinion Seal

REMOVAL & INSTALLATION

1. Before servicing the vehicle, refer to the Precautions Section.
2. Disconnect the rear driveshaft and the halfshafts from the rear differential/final drive.
3. Rotate the drive pinion nut with a beam type torque wrench. Measure friction torque before loosening the drive pinion nut. Note the friction torque value. After the seal is replaced, friction torque must be restored to this value.
4. Remove the hex nut securing the drive shaft flange. Counterhold the flange using tool 3435 or equivalent to ease removal.
5. Mark the installation position of the driveshaft flange relative to the drive pinion.
6. Pull off the driveshaft flange. If necessary, use a slide hammer.
7. Pull out the sealing ring. Clean the threads on the drive pinion.

To install:

8. Drive in a new seal in as far as the stop; be careful not to distort the seal.
9. Fill the area between the sealing lip and dust lip halfway with grease G 052 128 A1 or equivalent.
10. Check the dust cap for damage, replace the cap if necessary.
11. Install the driveshaft flange, match the marks made during removal.
12. Coat the threads of a new nut with Loctite® 648.
13. Tighten the drive pinion nut until the friction torque measured prior to removal is attained. Have an assistant counterhold the flange while tightening the nut.

➡Increase the tightening torque gradually, frequently measuring the friction torque. The friction torque measured prior to removal must not be exceeded!

14. The remaining installation is the reverse of removal.

STEERING

Air Bag

⁂ CAUTION

Some vehicles are equipped with an air bag system. The system must be disabled before performing service on or around system components, steering column, instrument panel components, wiring and sensors. Failure to follow safety and disabling procedures could result in accidental air bag deployment, possible personal injury and unnecessary system repairs.

PRECAUTIONS

Several precautions must be observed when handling the inflator module to avoid accidental deployment and possible personal injury.

- Never test, remove, repair or install airbag units or work on the airbag system unless you have the proper training, qualifications and tools to do the work safely and correctly
- Always disconnect the battery ground (GND) strap before starting to work on the airbag system. Work can begin immediately after disconnecting the battery, no waiting time is needed.
- Always discharge static electricity from your body before handling or even touching an airbag unit. Place your hands on a suitable metal object that is grounded such as a water or heating pipe or metal frame.
- Always install airbag unit immediately after removing it from its shipping container.
- If installation work is interrupted, immediately return the airbag unit to its original packaging.
- NEVER leave an airbag unit unattended. If work is interrupted, always place airbag unit back into its shipping container.
- NEVER place airbag unit on a workbench or other surface with the side that points towards inside of passenger compartment facing down. If airbag unit should deploy, it can cause personal injury. Always make sure that the side of the airbag unit that faces passenger compartment faces up.
- NEVER let anyone sit in the passenger compartment when the airbag system is being worked on, especially when the airbag system is being connected to a power source or the battery is being connected to the battery.
- NEVER install airbag units that have dropped onto a hard surface, or which have signs of damage.

• Always put airbag units that have not deployed or have not completely deployed in the airbag shipping container and mark the outside of the container so that others will know that there is still a possibility of airbag deployment.

• Always store and handle airbag units according to all applicable legal requirements.

1. Before servicing the vehicle, also refer to the precautions in the beginning of this Section.

DISARMING

To avoid personal injury when working on vehicles equipped with an air bag, the negative battery cable must be disconnected before working on the system. Failure to do so may result in deployment of the air bag.

1. Before servicing the vehicle, refer to the precautions in the beginning of this Section.
2. Turn the ignition switch to the **LOCK** position.
3. Disconnect the negative battery cable. Shield the cable by wrapping electrical tape around it. Work can begin immediately after disconnecting the battery, no waiting time is needed.

Power Steering Gear

REMOVAL & INSTALLATION

➡The steering angle sensor must be set using diagnostic tool VAS5051 after installation of the steering gear.

1. Before servicing the vehicle, refer to the precautions in the beginning of this Section.
2. Raise the vehicle and remove the front wheels.
3. Clamp off the power steering intake and return line using hose clamps.
4. Remove the nuts, and press the tie rods from the steering knuckle.
5. Remove the left tie rod ball joint from the tie rod. Mark it's position before removing.
6. Remove the retainers for the hydraulic lines. Disconnect the lines from the steering gear.
7. Turn the steering to full left lock.
8. If installed, remove the guard plate from the steering gear.
9. Remove the bolt from the steering gear universal joint, then remove the universal joint from the steering gear.
10. Remove the heat shield.

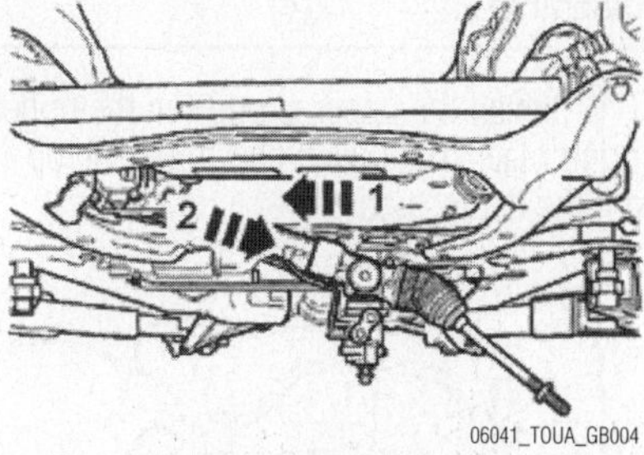

Remove the steering gear as shown

11. Slide the steering gear to the right side of the vehicle.
12. Swing the left tie rod downward, then with the steering gear moved to the left side of the vehicle, remove it downward.

To install:

13. Installation is the reverse of removal. Bleed the steering system and perform a wheel alignment. Note the following torque specifications:
    • Hydraulic lines: 22 ft. lbs. (30 Nm)
    • Steering gear universal joint (use new): 30 ft. lbs. (40 Nm) plus an additional 1/4 turn
    • Steering gear shield: 7 ft. lbs. (10 Nm)
    • Tie rod nuts (use new): 66 ft. lbs. (90 Nm)

FRONT SUSPENSION

Strut

REMOVAL & INSTALLATION

1. Before servicing the vehicle, refer to the precautions in the beginning of this Section.
2. Remove the wheel and raise the vehicle.
3. Remove the nuts securing the suspension strut to the body.
4. Remove the halfshaft.
5. Make sure the lower control arm/steering knuckle assembly is properly supported.
6. Pull out bolt securing the strut to the lower control arm and remove suspension strut.

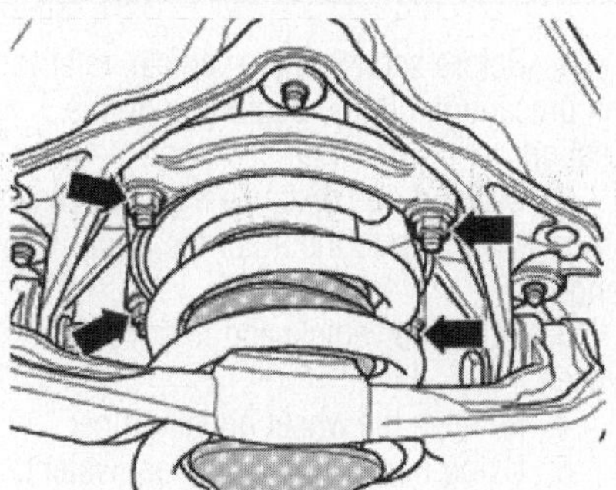

Remove the nuts securing the strut to the body

To install:

7. Installation is the reverse of removal. Perform a wheel alignment and observe the following torque specifications (use new hardware):
    • Suspension strut to body nuts: 22 ft. lbs. (30 Nm)
    • Suspension strut to lower control arm: 111 ft. lbs. (150 Nm) plus an additional 1/4 turn
    • Upper control arm to steering knuckle: 70 ft. lbs. (95 Nm)
    • Tie rod to steering knuckle: 66 ft. lbs. (90 Nm)

DISASSEMBLY

➡The shock absorber portion of the strut is not rebuildable and must be replaced if necessary.

1. Before servicing the vehicle, refer to the precautions in the beginning of this Section.
2. Compress the coil spring using VAS 6046 or equivalent until the upper spring plate is no longer under load.
3. Remove the hex nut from the piston rod.
4. Remove the individual components of the suspension strut.

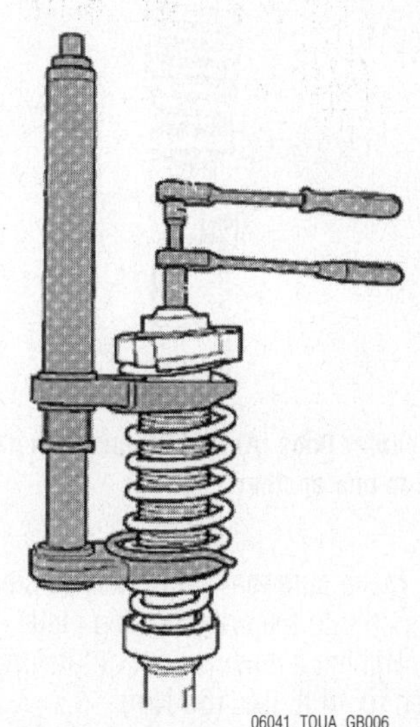

Compress the spring, then remove the hex nut from the piston rod

ASSEMBLY

1. Install the upper support in the upper spring plate. The raised part must line up with the depression.

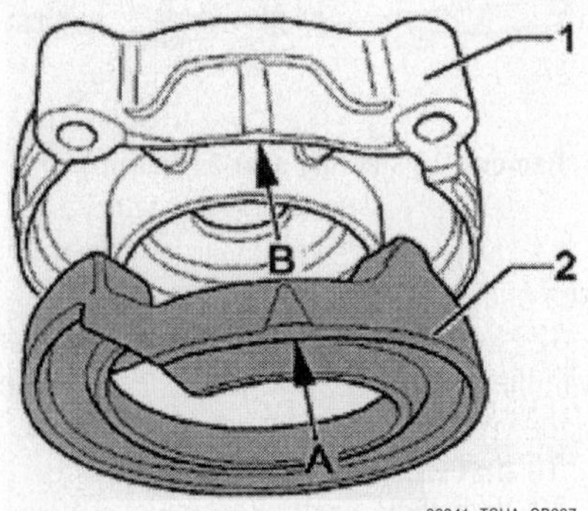

Install the upper support as shown. Notch (A) must line up with groove (B)

2. Make sure the coil spring is properly seated in the spring holder.
3. The upper spring plate must be set on the spring so that the raised part is located opposite the attachment for the link.
4. The center lines (A) and (B) must run parallel to one another. The end of the spring must lie against the stop on the support.

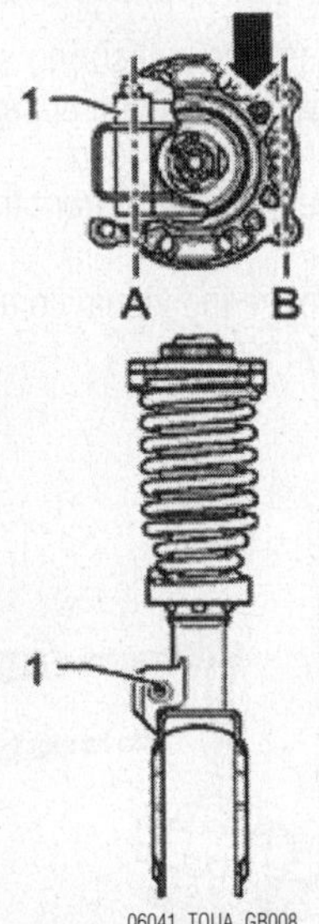

The center lines (A) and (B) must run parallel to one another

5. Make sure that the protective sleeve is clipped into the upper spring plate.
6. Tighten a new hex nut on piston rod. Tighten to 44 ft. lbs. (60 Nm).
7. Release the tension on spring compressor, then remove it from the coil spring.

Stabilizer (Sway) Bar

REMOVAL & INSTALLATION

**** CAUTION**

On vehicles with decouplable stabilizer bars, the stabilizer bars must be coupled in before proceeding. Otherwise, the risk of injury exists if the decoupled stabilizer bars become unintentionally coupled.

➡The hydraulic system pressure must first be released using diagnostic tool VAS 5051. After installation, the system must be bled using this tool as well.

1. Before servicing the vehicle, refer to the precautions in the beginning of this Section.
2. Remove the underbody cover.
3. Disconnect the hydraulic lines and unplug the electrical connection. Label the lines before removing so they are not switched when installing.

➡When removing or installing the hydraulic lines, counterhold them with an open-end wrench.

4. Remove the stabilizer bar clamps and mark the installed location of the stabilizer bar mount on the bar.

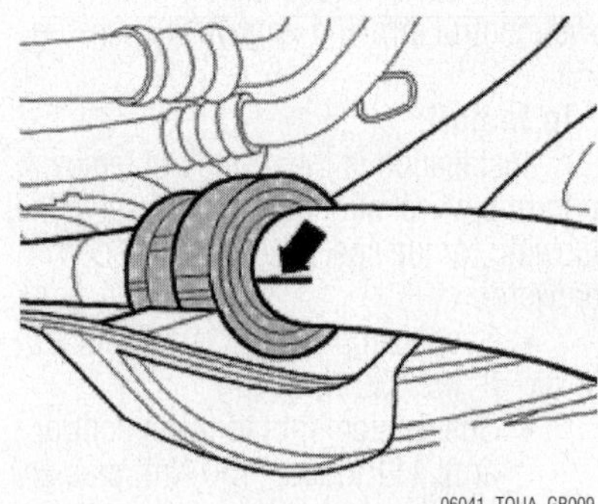

Mark the installed position of the bar

5. Remove the stabilizer bar from the connecting links.

To install:

➡The larger outside diameter of the stabilizer bar mount halves face toward the outside of vehicle.

6. Installation is the reverse of removal. Align the markings made during removal. Observe the following torque specifications (use new hardware):
    - Coupling rod to stabilizer bar: 81 ft. lbs. (110 Nm)
    - Stabilizer bar to subframe: 44 ft. lbs. (60 Nm)
    - Hydraulic lines: 11 ft. lbs. (15 Nm)

Upper and Lower Ball Joints

REMOVAL & INSTALLATION

The ball joints are not replaceable. If necessary, the entire control arm must be replaced.

Upper Control Arm

REMOVAL & INSTALLATION

The ball joint and bushings on the upper control arm are not replaceable. If necessary, the entire control arm must be replaced.

1. Before servicing the vehicle, refer to the precautions in the beginning of this Section.
2. Remove the suspension strut.
3. Loosen the self locking nuts from the sides of the control arm.
4. Remove the mounting bracket from the body, then remove it together with the control arm.

To install:

5. Installation is the reverse of removal. The ball joint pin must be clean of oil or grease. Observe the following torque specifications (use new hardware):
    - Bracket mount bolt (M10x37): 37 ft. lbs. (50 Nm) plus an additional ¼ turn
    - Control arm bolts/nuts: 37 ft. lbs. (50 Nm) plus an additional ¼ turn
    - Ball joint nut: 70 ft. lbs. (95 Nm)

Lower Control Arm

REMOVAL & INSTALLATION

1. Before servicing the vehicle, refer to the precautions in the beginning of this Section.
2. Measure the distance from the center of the wheel hub to the lip of the front fender.
3. Raise the vehicle and remove the wheel.
4. Remove the wheel housing liner.
5. Using hooks VW 552 or equivalent, secure the upper control arm to the upper opening of the wheel housing.
6. Raise the lower control arm slightly

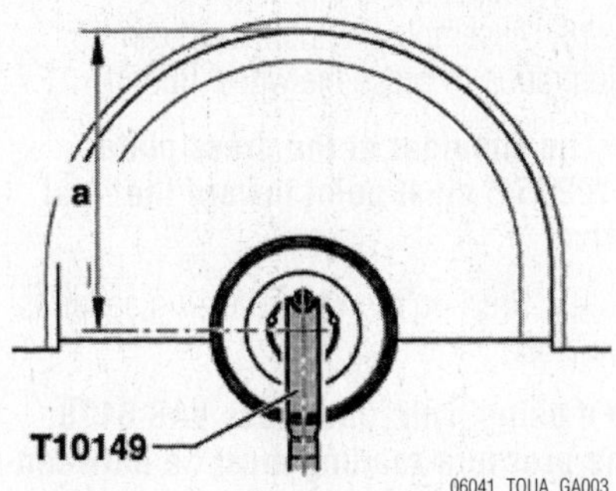

Measure distance (A). The lower strut nut must be tightened in this position

to preload it. This prevents damage to the stud of the ball joint.

7. Remove the suspension strut from the lower control arm.

8. Remove the ball joint nut from the lower control arm and press off the ball stud using a ball joint puller.

9. Remove the nuts, then remove the lower control arm from the subframe and steering knuckle.

To install:

10. Installation is the reverse of removal. When installing the strut, raise the control arm to the measurement taken in the beginning of the procedure. The nut must be tightened when in this position. Perform a wheel alignment and observe the following torque specifications (use new hardware):

- Lower control arm to subframe: 133 ft. lbs. (180 Nm)
- Suspension strut to lower control arm: 111 ft. lbs. (150 Nm) plus an additional ¼ turn
- Lower control arm to steering knuckle: 77 ft. lbs. (105 Nm)

LOWER CONTROL ARM BUSHING REPLACEMENT

1. Before servicing the vehicle, refer to the precautions in the beginning of this Section.

2. Remove lower control arm from the vehicle.

3. Press out bonded rubber bushing using a hydraulic press. Use the appropriate sized press pieces and plates.

4. When installing, note the following:

a. For the camber adjustment bolt bushing, press in bonded rubber bushing up to the stop. Then, press the center of the bonded rubber bushing in further so that the outside lip of the bushing will contact the control arm when installed.

b. For the caster adjustment bolt bushing, use a felt-tipped pen to draw an

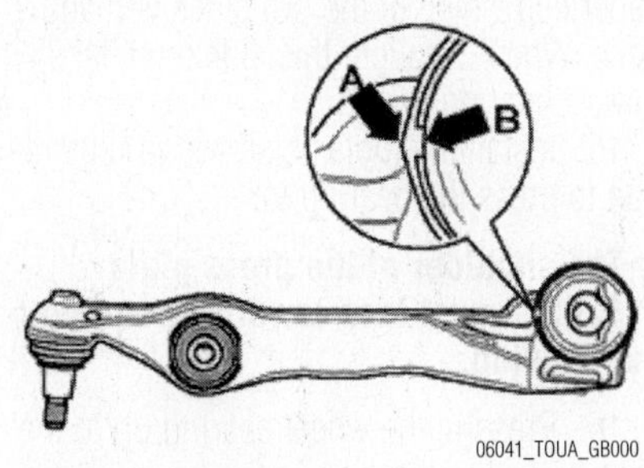

On the caster bushing, align the marks when installing

extension of the relief notch on the outer diameter of the bonded rubber bushing. Align the bushing according to the marking and press it in up to the stop.

c. For the suspension strut bushing, press in the bushing so that it is flush with the control arm.

5. Install the lower control arm and perform a wheel alignment.

Wheel Bearings

REMOVAL & INSTALLATION

1. Before servicing the vehicle, refer to the precautions in the beginning of this Section.

2. Remove the halfshaft.

3. Attach the upper control arm to the steering knuckle.

4. Secure the brake rotor with a wheel bolt.

5. Remove the brake caliper and secure it with wire out of the way.

6. Remove the wheel bolt and the brake rotor. Remove the disc shield.

7. Unclip the wheel speed sensor wire from the retainer.

8. Remove the bolt and the ABS wheel speed sensor.

⁂ CAUTION

On vehicles with decouplable stabilizer bars, the stabilizer bars must be coupled in before proceeding. Otherwise, the risk of injury exists if the decoupled stabilizer bars become unintentionally coupled.

9. The following tools or their commercially available equivalents can be used to remove and install the hub and bearing:

- Hydraulic press VAS 6178
- Threaded rod M20T10205/8-1
- Threaded nut M20T10205/8-2
- Cup T10205/2
- Puller T10205/1
- Press plates/supports T10205/3, 4, 6, 7, 9 and 10
- Mounting tube T10205/5

10. Attach the puller to the wheel hub using wheel bolts. Pull the wheel hub out.

11. Remove the circlip.

12. Assemble the tools as shown in illustration. Pull the wheel bearing out.

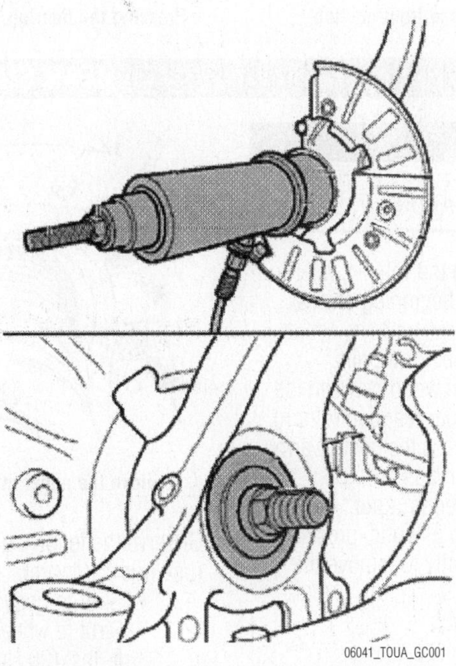

Pulling the bearing out of the housing

➡The shoulder of press plate T10205/6 must point toward the final drive.

13. To pull the inner race off the wheel hub, install the separating tool behind the inner race.

➡The chamfers on the jaws face toward the inner race.

⁂ CAUTION

To prevent injury, hold the tool when pulling the bearing inner race off so that the race faces downward.

To install:

14. Make sure that the rubberized ABS sensor ring faces towards the final drive. If no rubberized ring is visible, check using a paper clip which of the two sides is magnetic. When installed, this side must face toward final drive.

15. Install the tools as shown in illustration to press the bearing in.

➡The shoulder of the press plate T10205/9 must face toward wheel bearing housing.

16. Press in the wheel bearing up to the stop.

➡If using hydraulic press VAS 6178, the pressure reading must be between 100 and 190 bar shortly before the end of pressing in. The maximum pressure should not exceed 310 bar.

17. Install the circlip.

18. Assemble the tools as shown in illustration to press the wheel hub in.

➡The shoulder of the press plate T10205/6 must point toward the final drive.

19. Press in the wheel bearing as far as the stop.

➡If using hydraulic press VAS 6178, the pressure reading must be between 30 and 100 bar shortly before the end of pressing in. The maximum pressure should not exceed 140 bar.

20. The remaining installation is the reverse of removal.

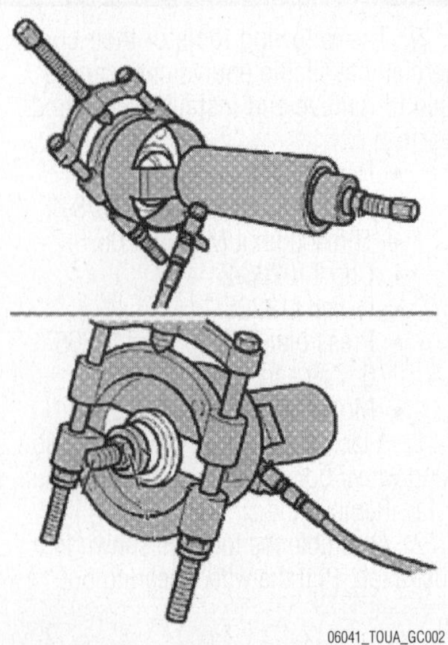

Removing the inner race from the hub

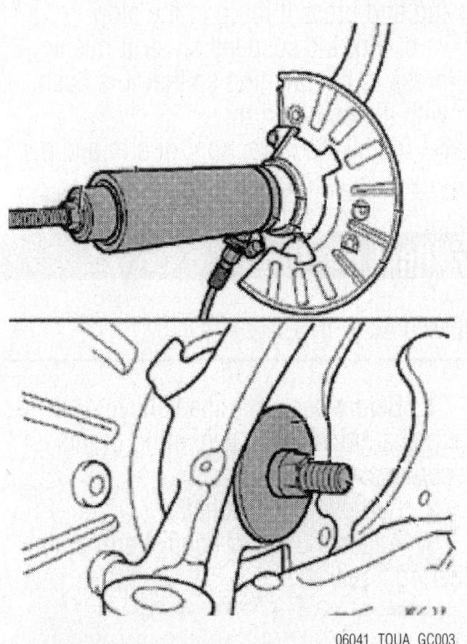

Pressing the bearing into the housing

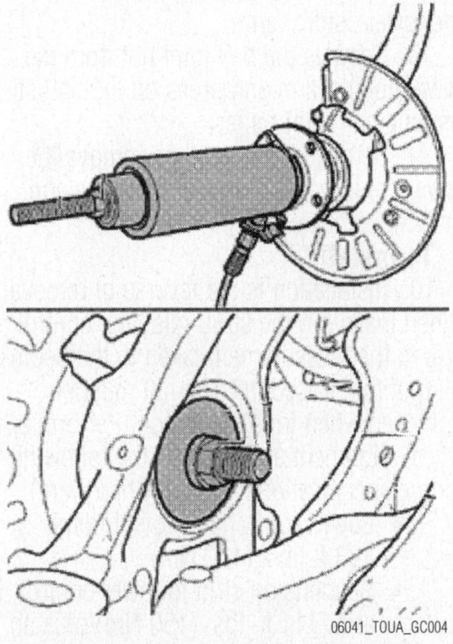

Pressing the hub into the housing

REAR SUSPENSION

Strut

REMOVAL & INSTALLATION

1. Before servicing the vehicle, refer to the precautions in the beginning of this Section.
2. Remove the upper strut bolts.
3. Insert a length of wood between the underbody and the upper rear control arm. This will be used to press the control arm down to remove the strut.
4. Remove the lower strut nut.
5. Using the length of wood, press control arm down sufficiently to remove the bolt.
6. Remove the strut.

To install:

7. Installation is the reverse of removal.

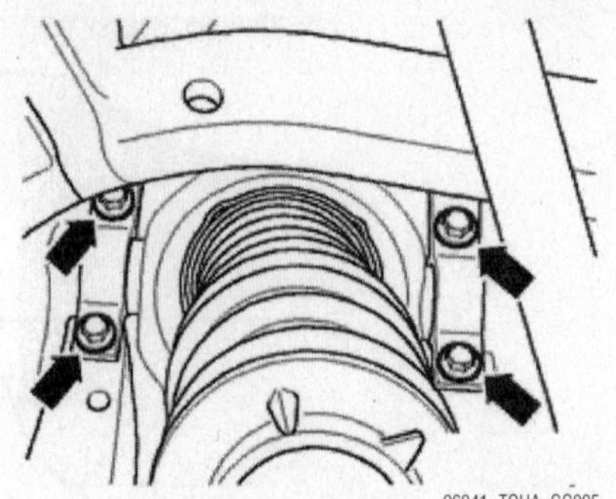

Remove the upper strut bolts

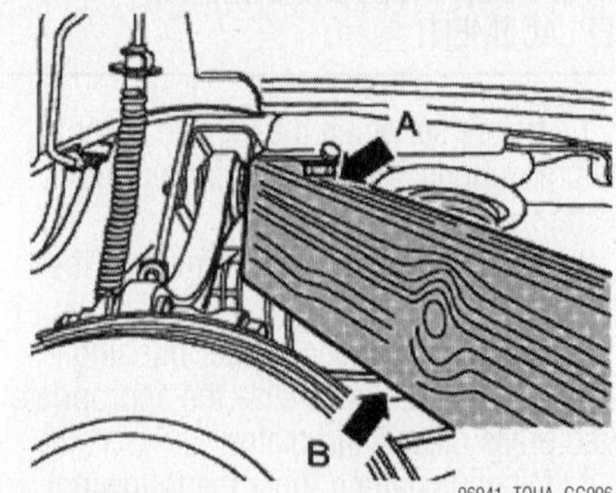

Insert the length of wood as shown

Observe the following torque specifications (use new hardware):

- Strut to body: 44 ft. lbs. (60 Nm)
- Strut to wheel bearing housing: 66 ft. lbs. (90 Nm) plus an additional ¼ turn

DISASSEMBLY

➡The shock absorber portion of the strut is not rebuildable and must be replaced if necessary.

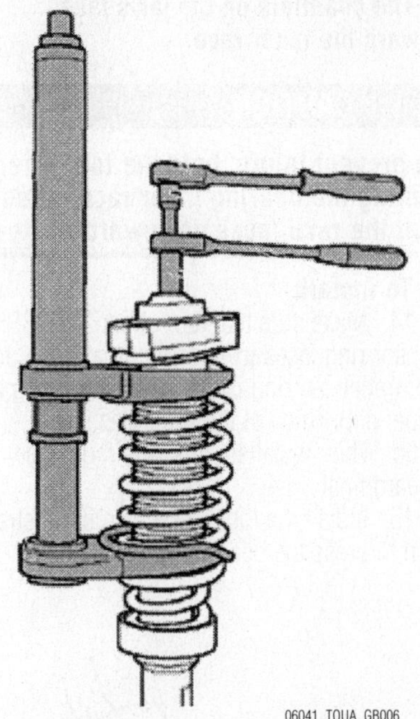

Compress the spring, then remove the hex nut from the piston rod

1. Before servicing the vehicle, refer to the precautions in the beginning of this Section.
2. Compress the coil spring using VAS 6046 or equivalent until the upper spring plate is no longer under load.
3. Remove the hex nut from the piston rod.
4. Remove the individual components of the suspension strut.

ASSEMBLY

1. Install the upper support in the upper spring plate. The raised part must line up with the depression.
2. Make sure the coil spring is properly seated in the spring holder.
3. The upper spring plate must be set on the spring so that the raised part is located opposite the attachment for the link.

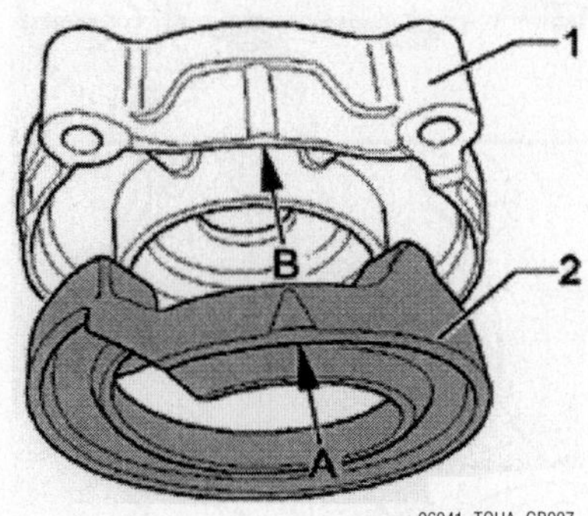

Install the upper support as shown. Notch (A) must line up with groove (B)

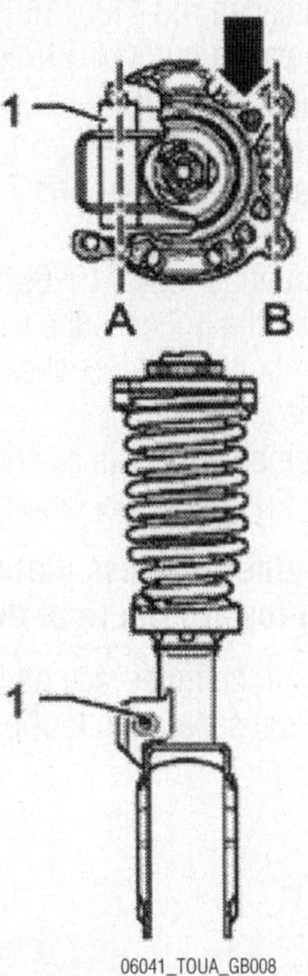

The center lines (A) and (B) must run parallel to one another

4. The center lines (A) and (B) must run parallel to one another. The end of the spring must lie against the stop on the support.
5. Make sure that the protective sleeve is clipped into the upper spring plate.
6. Tighten a new hex nut on piston rod. Tighten to 44 ft. lbs. (60 Nm).
7. Release the tension on spring compressor, then remove it from the coil spring.

Stabilizer (Sway) Bar

REMOVAL & INSTALLATION

✲✲ CAUTION

On vehicles with decouplable stabilizer bars, the stabilizer bars must be coupled in before proceeding. Otherwise, the risk of injury exists if the decoupled stabilizer bars become unintentionally coupled.

➡The hydraulic system pressure must first be released using diagnostic tool VAS 5051. After installation, the system must be bled using this tool as well.

1. Before servicing the vehicle, refer to the precautions in the beginning of this Section.
2. Remove the left rear wheel housing liner.

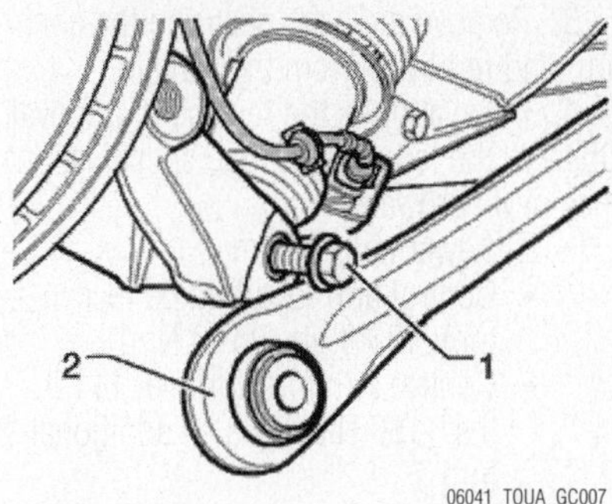

Back out bolt (1) to lock the tie rod (2)

3. Disconnect the hydraulic lines and unplug the electrical connection. Label the lines before removing so they are not switched when installing.

➡When removing or installing the hydraulic lines, counterhold them with an open-end wrench.

4. Remove the connecting links.
5. Disconnect the tie rods from both wheel bearing housings.
6. Back out bolt to lock the tie rod.
7. Remove the stabilizer bar clamps and mark the installed location of the stabilizer bar mount on the bar.
8. Remove the stabilizer bar.

To install:

➡The larger outside diameter of the stabilizer bar mount halves face toward the front of vehicle.

9. Installation is the reverse of removal. Align the markings made during removal. Observe the following torque specifications (use new hardware):
   - Coupling rod to stabilizer bar: 74 ft. lbs. (100 Nm)
   - Stabilizer bar to subframe: 37 ft. lbs. (50 Nm)
   - Connecting link to wheel bearing housing: 74 ft. lbs. (100 Nm)
   - Tie rod to wheel bearing housing: 111 ft. lbs. (150 Nm) plus an additional ¼ turn
   - Hydraulic lines: 11 ft. lbs. (15 Nm)

Control Arms

REMOVAL & INSTALLATION

The bushings of the control arms are not replaceable. If necessary, the entire control arm must be replaced

1. Before servicing the vehicle, refer to the precautions in the beginning of this Section.
2. Raise the vehicle and remove the wheels.

3. Remove the bolts securing the control arm and remove it from the vehicle.

4. Installation is the reverse of removal. Observe the following torque specifications (use new hardware):

a. Lower control arm:
- Control arm to subframe (eccentric): 133 ft. lbs. (180 Nm)
- Control arm to subframe: 111 ft. lbs. (150 Nm) plus an additional ¼ turn
- Control arm to wheel bearing housing: 111 ft. lbs. (150 Nm) plus an additional ¼ turn

b. Upper rear control arm:
- Control arm to subframe: 66 ft. lbs. (90 Nm) plus an additional ¼ turn
- Control arm to wheel bearing housing: 111 ft. lbs. (150 Nm) plus an additional ¼

c. Upper front control arm:
- Control arm to wheel bearing housing: 111 ft. lbs. (150 Nm) plus an additional ¼
- Control arm to subframe: 133 ft. lbs. (180 Nm)

Wheel Bearings

REMOVAL & INSTALLATION

1. Before servicing the vehicle, refer to the precautions in the beginning of this Section.

2. Remove the halfshaft.

3. Secure the brake rotor with a wheel bolt.

4. Remove the brake caliper and secure it with wire out of the way.

5. Remove the wheel bolt and the brake rotor. Remove the disc shield.

6. Unclip the wheel speed sensor wire from the retainer.

7. Remove the bolt and the ABS wheel speed sensor.

8. Unhook the parking brake cable from wheel bearing housing and remove parking brake shoes.

CAUTION

On vehicles with decouplable stabilizer bars, the stabilizer bars must be coupled in before proceeding. Otherwise, the risk of injury exists if the decoupled stabilizer bars become unintentionally coupled.

9. The following tools or their commercially available equivalents can be used to remove and install the hub and bearing:
- Hydraulic press VAS 6178
- Threaded rod M20T10205/8-1
- Threaded nut M20T10205/8-2
- Cup T10205/2
- Puller T10205/1
- Press plates/supports T10205/3, 4, 6, 7, 9 and 10
- Mounting tube T10205/5

10. Attach the puller to the wheel hub using wheel bolts. Pull the wheel hub out.

11. Remove the circlip.

12. Assemble the tools as shown in illustration. Pull the wheel bearing out.

➡The shoulder of press plate T10205/6 must point toward the final drive.

13. To pull the inner race off the wheel hub, install the separating tool behind the inner race.

➡The chamfers on the jaws face toward the inner race.

CAUTION

To prevent injury, hold the tool when pulling the bearing inner race off so that the race faces downward.

To install:

14. Make sure that the rubberized ABS sensor ring faces towards the final drive. If no rubberized ring is visible, check using a paper clip which of the two sides is magnetic. When installed, this side must face toward final drive.

15. Install the tools as shown in illustration to press the bearing in.

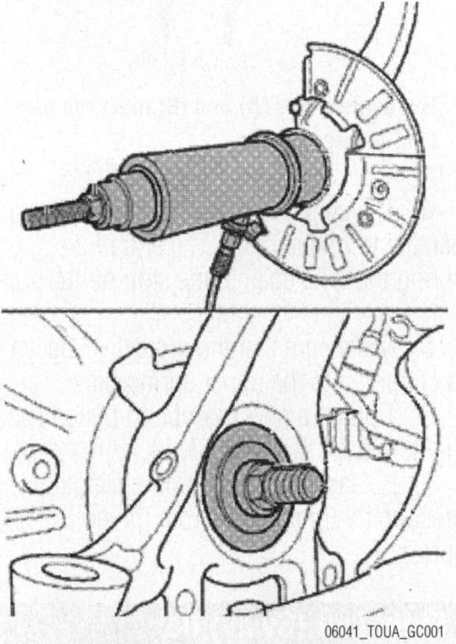

Pulling the bearing out of the housing

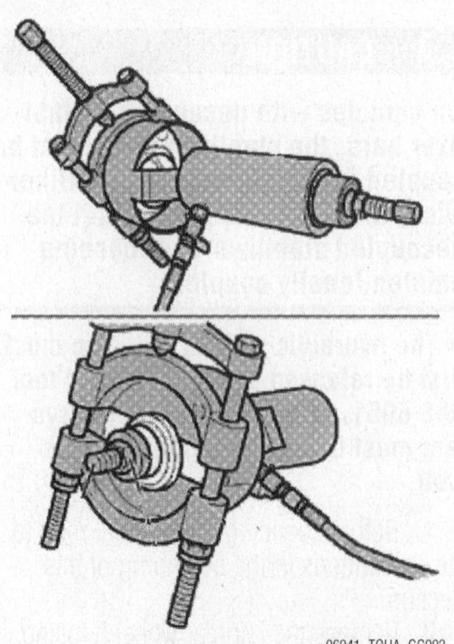

Removing the inner race from the hub

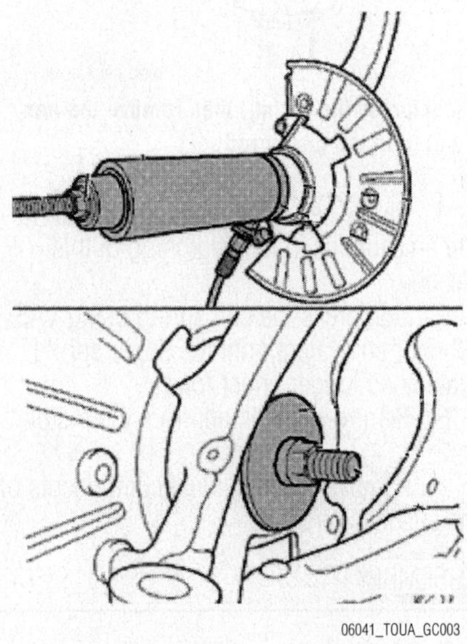

Pressing the bearing into the housing

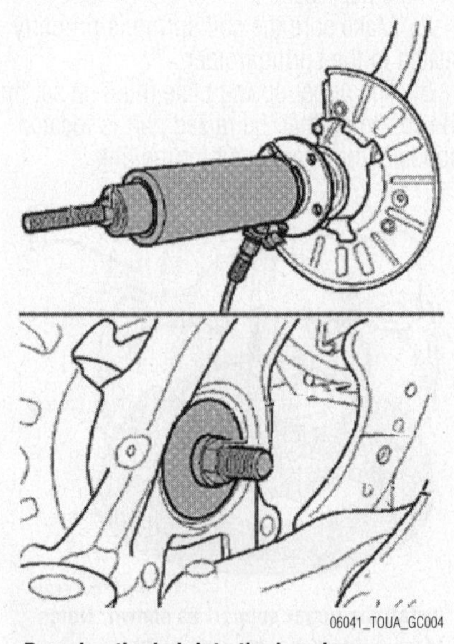

Pressing the hub into the housing

➡The shoulder of the press plate T10205/9 must face toward wheel bearing housing.

16. Press in the wheel bearing up to the stop.

➡If using hydraulic press VAS 6178, the pressure reading must be between 100 and 190 bar shortly before the end of pressing in. The maximum pressure should not exceed 310 bar.

17. Install the circlip.
18. Assemble the tools as shown in illustration to press the wheel hub in.

➡The shoulder of the press plate T10205/6 must point toward the final drive.

19. Press in the wheel bearing as far as the stop.

➡If using hydraulic press VAS 6178, the pressure reading must be between 30 and 100 bar shortly before the end of pressing in. The maximum pressure should not exceed 140 bar.

20. The remaining installation is the reverse of removal.

BRAKES

Brake Caliper

REMOVAL & INSTALLATION

1. Before servicing the vehicle, refer to the precautions in the beginning of this Section.
2. Raise the vehicle and remove the wheels.
3. Press the retaining spring in until it can be pressed out of the bore.
4. Turn the retaining spring clockwise until it can be removed from the upper bore.
5. Unplug the connector for the brake pad wear indicator.
6. Remove the dust cap and lift off the wire for the brake pad wear indicator.
7. Remove the brake pad wear indicator wire from the retainer on the caliper.
8. Attach the hose from a bleeder bottle onto the valve of the brake caliper and then open the bleeder valve.
9. Lightly preload the brake pedal using tool V.A.G 1869/2 or equivalent.
10. Close the bleeder valve and remove bleeder bottle.
11. Disconnect and plug the brake line.
12. Remove the caps, then remove both guide pins from the brake caliper.
13. Remove the brake caliper from the carrier.
14. If necessary, remove the brake pads from the caliper.

To install:

15. Push the piston back in its bore with piston resetting tool T10145-1 or equivalent.
16. If removed, install the brake pads.
17. Attach the brake caliper to the carrier with the guide pins. Install both caps.
18. Connect the brake line to the brake caliper.
19. Remove the brake pedal preloading tool.
20. Install the retaining spring into the upper bore and turn it downwards.
21. Set the retaining spring on the bottom of the carrier.
22. First press the retaining spring in, then at the same time into the bore of brake caliper.
23. The remaining installation is the reverse of removal. Bleed the brake system and check the brake fluid level. Observe the following torque specifications:
    - Guide pins to carrier: 22 ft. lbs. (30 Nm)
    - Front brake line to brake caliper: 10 ft. lbs. (14 Nm)
    - Rear brake line banjo fitting bolt: 28 ft. lbs. (38 Nm)

➡Depress the brake pedal firmly several times with vehicle stationary so that the brake pads are properly seated in their normal operating position.

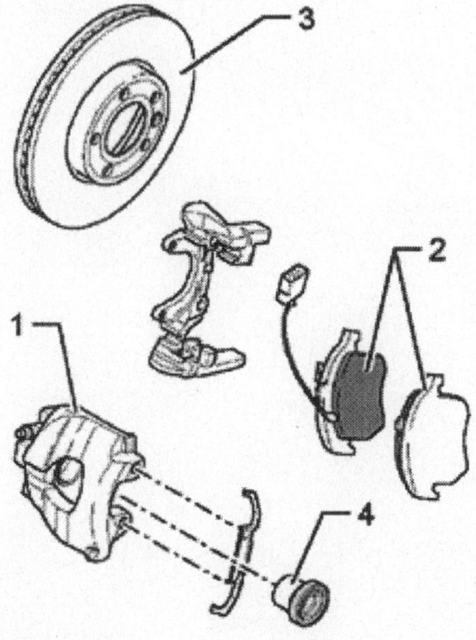

Exploded view of the brake assembly

Disc Brake Pads

REMOVAL & INSTALLATION

1. Before servicing the vehicle, refer to the precautions in the beginning of this Section.
2. Raise the vehicle and remove the wheels.
3. Press the retaining spring in until it can be pressed out of the bore.
4. Turn the retaining spring clockwise until it can be removed from the upper bore.
5. Unplug the connector for the brake pad wear indicator.
6. Remove the dust cap and lift off the wire for the brake pad wear indicator.
7. Remove the brake pad wear indicator wire from the retainer on the caliper.
8. Remove the caps, then remove both guide pins from the brake caliper.
9. Remove the brake caliper from the carrier. Suspend it with a wire so that the weight of the caliper does not cause any stress or damage to the brake line.
10. Remove the brake pads.

To install:

11. Clean the brake caliper with mineral spirits of any adhesive residue. The surface must be dry and clean.
12. Before pressing the piston back in its bore to insert the new brake pads, draw off brake fluid from the reservoir with a bleeder bottle.
13. Push the piston back in its bore with piston resetting tool T10145-1 or equivalent.

➡The inner (piston side) brake pads are directional and specific to each side of the vehicle. They are designated by markings that show direction of travel.

14. Install the brake pad by inserting the retaining springs on pad into the piston.
15. Remove the protective foil from the backing plate on the outer brake pad.
16. Set the outer brake pad into the carrier.

➡When installing the brake caliper, be careful not to let the brake pad adhere itself to the brake caliper before the correct installed position has been achieved.

17. Attach the brake caliper to the carrier with the guide pins. Install both caps.
18. Connect the brake line to the brake caliper.
19. Remove the brake pedal preloading tool.
20. Install the retaining spring into the upper bore and turn it downwards.
21. Set the retaining spring on the bottom of the carrier.
22. First press the retaining spring in, then at the same time into the bore of brake caliper.
23. The remaining installation is the reverse of removal. Bleed the brake system and check the brake fluid level. Tighten the guide pins to 22 ft. lbs. (30 Nm).

➡Depress the brake pedal firmly several times with vehicle stationary so that the brake pads are properly seated in their normal operating position.

14

VOLVO

C70 • S40 • S60 • S80 • V40 • V50 • V70 • V70XC • XC70

SPECIFICATIONS AND MAINTENANCE CHARTS

ENGINE AND VEHICLE IDENTIFICATION CHART

Engine						
Code	Liters (cc)	Cu. in.	Cyl.	Fuel Sys.	Type	Eng. Mfg.
B4204T3	1.9 (1948)	119	4	EFI	DOHC	Volvo
B4204T4	1.9 (1948)	119	4	EFI	DOHC	Volvo
B5234T3	2.3 (2319)	144	5	EFI	DOHC	Volvo
B5234T9	2.3 (2319)	142	5	EFI	DOHC	Volvo
B5244S	2.4 (2435)	149	5	EFI	DOHC	Volvo
B5244T3	2.4 (2435)	149	5	EFI	DOHC	Volvo
B5244T7	2.4 (2435)	149	5	EFI	DOHC	Volvo
B5254T	2.5 (2521)	154	5	EFI	DOHC	Volvo
B5254T3	2.5 (2521)	154	5	EFI	DOHC	Volvo
B5254T4	2.5 (2521)	154	5	EFI	DOHC	Volvo
B6294S2	2.9 (2922)	178	6	EFI	DOHC	Volvo
B6294T	2.9 (2922)	178	6	EFI	DOHC	Volvo

EFI: Electronic Fuel Injection

DOHC: Double Overhead Camshafts

Model Year	
Code	Year
2	2002
3	2003
4	2004
5	2005
6	2006

06041_VOLV_C0001

GENERAL ENGINE SPECIFICATIONS

Year	Model	Engine Displ. Liters	Engine ID	Net Horsepower @ rpm	Net Torque @ rpm (ft. lbs.)	Bore x Stroke (in.)	Com-pression Ratio	Oil Pressure @ rpm
2002	S40	1.9	B4204T3	160@5250	177@1800	NA	10.3:1	NA
	S60	2.4	B5244S	168@5700	170@4500	3.27 x 3.54	10.3:1	51@4000
	S60 2.4T	2.4	B5244T3	197@6000	210@1800	3.27 x 3.54	9.0:1	51@4000
	S60 T5	2.3	B5234T3	247@5220	241@2520	3.19 x 3.54	8.5:1	51@4000
	S60 AWD	2.4	B5244T3	197@6000	210@1800	3.27 x 3.54	9.0:1	51@4000
	C70 HPT Coupe	2.3	B5234T3	247@5220	241@2520	3.19 x 3.54	8.5:1	51@4000
	C70 LPT Conv	2.4	B5244T3	197@6000	210@1800	3.27 x 3.54	9.0:1	51@4000
	C70 HPT Conv	2.3	B5234T3	247@5220	241@2520	3.19 x 3.54	8.5:1	51@4000
	S80 2.9	2.9	B6294S2	197@6000	277@2100	3.27 x 3.54	10.3:1	51@4000
	S80 T6	2.9	B6294T	268@5400	206@4200	3.19 x 3.54	8.5:1	51@4000
	V40	1.9	B4204T3	160@5250	177@1800	NA	10.3:1	NA
	V70 2.4	2.4	B5244S	168@5700	170@4500	3.27 x 3.54	10.3:1	51@4000
	V70 2.4T	2.4	B5244T3	197@6000	210@1800	3.27 x 3.54	9.0:1	51@4000
	V70 T5	2.3	B5234T3	247@5220	241@2520	3.19 x 3.54	8.5:1	51@4000
	V70 AWD	2.4	B5244T3	197@6000	210@1800	3.27 x 3.54	9.0:1	51@4000
	XC70 AWD	2.4	B5244T3	197@6000	210@1800	3.27 x 3.54	9.0:1	51@4000
2003	S40	1.9	B4204T3	190@5500	177@1800	NA	9.0:1	NA
	S60	2.4	B5244S	168@5700	170@4500	3.27 x 3.54	10.3:1	51@4000
	S60 2.4T	2.4	B5244T3	197@6000	210@1800	3.27 x 3.54	9.0:1	51@4000
	S60 T5	2.3	B5234T3	247@5220	241@2520	3.19 x 3.54	8.5:1	51@4000
	S60 AWD	2.5	B5254T	208@5000	235@1500	NA	9.5:1	NA
	C70 LPT	2.4	B5244T7	197@5700	210@1800	3.27 x 3.54	9.0:1	NA
	C70 HPT	2.3	B5234T9	242@5400	243@2400	3.19 x 3.54	8.5:1	NA
	S80 2.9	2.9	B6294S2	197@6000	277@2100	3.27 x 3.54	10.3:1	51@4000
	S80 T6	2.9	B6294T	268@5400	206@4200	3.19 x 3.54	8.5:1	51@4000
	V40	1.9	B4204T3	160@5250	177@1800	NA	10.3:1	NA
	V70 2.4	2.4	B5244S	168@5700	170@4500	3.27 x 3.54	10.3:1	51@4000
	V70 2.4T	2.4	B5244T3	197@6000	210@1800	3.27 x 3.54	9.0:1	51@4000
	V70 T5	2.3	B5234T3	247@5220	241@2520	3.19 x 3.54	8.5:1	51@4000
	V70 AWD	2.5	B5254T	208@5000	235@1500	NA	9.5:1	NA
	XC70 AWD	2.5	B5254T	208@5000	235@1500	NA	9.5:1	NA
2004	S40	1.9	B4204T4	190@5500	177@1800	NA	9.0:1	NA
	S60 2.4	2.4	B5244S	168@6000	166@4500	NA	NA	NA
	S60 2.5T	2.5	B5254T	208@5000	236@1500	NA	NA	NA
	S60 2.5T AWD-R	2.5	B5254T	208@5000	236@1500	NA	NA	NA
	S60 T5	2.3	B5234T3	247@5200	243@2400	3.19 x 3.54	8.5:1	NA
	C70 LPT	2.4	B5244T7	197@5700	210@1800	3.27 x 3.54	9.0:1	NA
	C70 HPT	2.3	B5234T9	242@5400	243@2400	3.19 x 3.54	8.5:1	NA
	S80 2.9	2.9	B6294S2	197@6000	277@2100	3.27 x 3.54	10.3:1	51@4000
	S80 T6	2.9	B6294T	268@5400	206@4200	3.19 x 3.54	8.5:1	51@4000
	V40	1.9	B4204T4	190@5500	177@1800	NA	9.0:1	NA
	V70 2.4	2.4	B5244S	168@6000	166@4500	NA	NA	NA
	V70 2.5T	2.5	B5254T	208@5000	236@1500	NA	NA	NA
	V70 T5	2.3	B5234T3	247@5200	243@2400	3.19 x 3.54	8.5:1	NA
	V70 AWD	2.5	B5254T	208@5000	236@1500	NA	NA	NA
	XC70 AWD	2.5	B5254T	208@5000	236@1500	NA	NA	NA

NA: Information not available

06041_VOLV_C0002

GENERAL ENGINE SPECIFICATIONS

Year	Model	Engine Displ. Liters	Engine ID	Net Horsepower @ rpm	Net Torque @ rpm (ft. lbs.)	Bore x Stroke (in.)	Com-pression Ratio	Oil Pressure @ rpm
2005	S40 2.4i	2.4	B5244S	169@6000	170@4400	3.24 x 3.54	10.3:1	NA
	S40 T5	2.5	B5254T3	218@5000	236@4800	3.27 x 3.67	9.0:1	51@4000
	S60 2.4	2.4	B5244S	165@6000	166@4500	NA	NA	NA
	S60 2.5T	2.5	B5254T	208@5000	236@1500	NA	NA	NA
	S60 R	2.5	B5254T4	300@5500	295@1950	NA	8.5:1	NA
	S60 T5	2.4	B5244T	257@5500	258@2100	NA	NA	NA
	S80 2.5T	2.5	B5254T	208@5000	236@1500	3.27 x 3.67	9.0:1	NA
	S80 T6	2.9	B6294T	268@5200	280@1800	3.28 x 3.56	8.7:1	51@4000
	V50 2.4i	2.4	B5244S	168@6000	170@4400	3.27 x 3.54	10.3:1	NA
	V50 T5	2.5	B5254T3	218@5000	236@4800	3.27 x 3.67	9.0:1	51@4000
	V70 2.4	2.4	B5244S	165@6000	166@4500	3.27 x 3.54	10.3:1	NA
	V70 2.5T	2.5	B5254T2	208@5000	236@1500	NA	8.5:1	51@4000
	V70 R	2.5	B5254T4	300@5500	295@1950	3.27 x 3.67	8.5:1	NA
	XC70	2.5	B5254T2	208@5000	236@1500	NA	NA	51@4000
2006	C70 T5	2.5	B5254T3	218@5000	236@4800	3.27 x 3.67	9.0:1	51@4000
	S40 2.4i	2.4	B5244S	168@6000	170@4400	3.24 x 3.54	10.3:1	51@4000
	S40 T5	2.5	B5254T3	218@5000	236@4800	3.27 x 3.67	9.0:1	51@4000
	S60 2.5T	2.5	B5254T	208@5000	236@1500	3.27 x 3.67	9.0:1	51@4000
	S60 R	2.5	B5254T4	300@5500	295@1950	3.27 x 3.67	8.5:1	51@4000
	S60 T5	2.4	B5244T	257@5500	258@2100	3.20 x 3.67	8.5:1	51@4000
	S80 2.5T	2.5	B5254T	208@5000	236@1500	3.27 x 3.67	9.0:1	51@4000
	V50 2.4i	2.4	B5244S	168@6000	170@4400	3.27 x 3.54	10.3:1	51@4000
	V50 T5	2.5	B5254T3	218@5000	236@4800	3.27 x 3.67	9.0:1	51@4000
	V70 2.4	2.4	B5244S	165@6000	166@4500	3.27 x 3.54	10.3:1	51@4000
	V70 2.5T	2.5	B5254T2	208@5000	236@1500	3.27 x 3.67	8.5:1	51@4000
	V70 R	2.5	B5254T4	300@5500	295@1950	3.27 x 3.67	8.5:1	51@4000
	XC70	2.5	B5254T2	208@5000	236@1500	3.27 x 3.67	8.5:1	51@4000

NA: Information not available

06041_VOLV_C0003

ENGINE TUNE-UP SPECIFICATIONS

	Engine Displacement Liters	Engine ID	Spark Plug Gap (in.)	Ignition Timing (deg.)		Fuel Pump (psi)	Idle Speed (rpm)		Valve Clearance (in.)	
				MT	AT		MT	AT	In.	Ex.
2002	1.9	B4204T3	0.030	NA	NA	32-45	NA	NA	HYD	HYD
	2.3	B5234T3	0.030	NA	NA	55	670	670	HYD	HYD
	2.4	B5244T3	0.030	NA	NA	55	670	670	HYD	HYD
	2.4	B5244S	0.030	NA	NA	55	670	670	HYD	HYD
	2.9	B6294S2	0.030	NA	NA	55	650	650	HYD	HYD
	2.8	B6294T	0.030	NA	NA	55	650	650	HYD	HYD
2003	1.9	B4204T3	0.030	NA	NA	32-45	NA	NA	HYD	HYD
	2.3	B5234T3	0.030	NA	NA	55	670	670	HYD	HYD
	2.4	B5244T3	0.030	NA	NA	55	670	670	HYD	HYD
	2.4	B5244S	0.030	NA	NA	55	670	670	HYD	HYD
	2.5	B5254T	0.030	NA	NA	55	850	850	HYD	HYD
	2.3	B5234T9	0.030	NA	NA	55	670	670	HYD	HYD
	2.4	B5244T7	0.030	NA	NA	55	670	670	HYD	HYD
	2.9	B6294T	0.030	NA	NA	55	650	650	HYD	HYD
	2.9	B6294S2	0.030	NA	NA	55	650	650	HYD	HYD
2004	1.9	B4204T3	0.030	NA	NA	32-45	NA	NA	HYD	HYD
	2.3	B5234T3	0.030	NA	NA	55	670	670	HYD	HYD
	2.4	B5244S	0.030	NA	NA	55	670	670	HYD	HYD
	2.5	B5254T	0.030	NA	NA	55	850	850	HYD	HYD
	2.3	B5234T9	0.030	NA	NA	55	670	670	HYD	HYD
	2.4	B5244T7	0.030	NA	NA	55	670	670	HYD	HYD
	2.9	B6294S2	0.030	NA	NA	55	650	650	HYD	HYD
	2.9	B6294T	0.030	NA	NA	55	650	650	HYD	HYD
2005	2.4	B5244S	0.030	NA	NA	NA	670	670	HYD	HYD
	2.5	B5254T3	0.030	NA	NA	NA	NA	NA	HYD	HYD
	2.5	B5254T	0.030	NA	NA	NA	NA	NA	HYD	HYD
	2.5	B5254T4	0.030	NA	NA	NA	720	720	HYD	HYD
	2.4	B5244T	0.030	NA	NA	NA	670	670	HYD	HYD
	2.9	B6294T	0.030	NA	NA	55	NA	650	HYD	HYD
	2.5	B5254T2	0.030	NA	NA	NA	850	850	HYD	HYD
2006	2.4	B5244S	NA	NA	NA	NA	720	720	HYD	HYD
	2.4	B5244T	NA	NA	NA	NA	770	770	HYD	HYD
	2.5	B5254T	NA	NA	NA	NA	670	NA	HYD	HYD
	2.5	B5254T2	NA	NA	NA	NA	670	NA	HYD	HYD
	2.5	B5254T3	NA	NA	NA	NA	NA	NA	HYD	HYD
	2.5	B5254T4	NA	NA	NA	NA	NA	NA	HYD	HYD

HYD: Hydraulic lash adjusters

B: Before top dead center

NA: Information not available

06041_VOLV_C0004

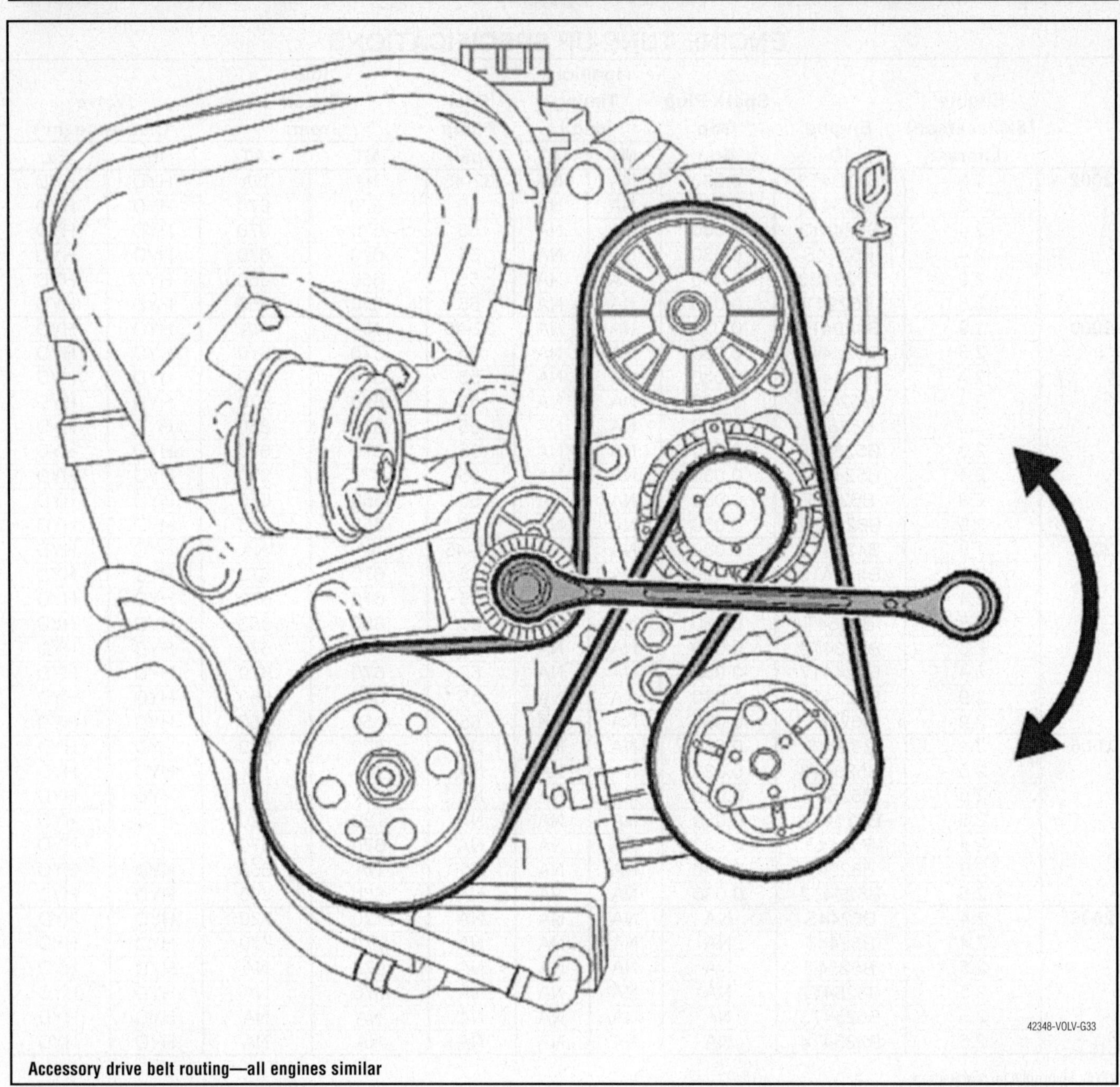

Accessory drive belt routing—all engines similar

CAPACITIES

Year	Model	Engine Displacement Liters	Engine ID	Engine Oil with Filter (qts.) ①	Transmission (qts.)		Transfer Case (pts.)	Drive Axle		Fuel Tank (gal.)	Cooling System (qts.)
					5-Spd	Auto.		Front (pts.)	Rear (pts.)		
2002	S40	1.9	B4204T3	5.7	—	7.5	—	—	—	NA	6.6
	S60	2.4	B5244S	5.8	2.2	②	—	—	—	NA	NA
	S60 2.4T	2.4	B5244T3	6.1	—	②	—	—	—	NA	NA
	S60 T5	2.3	B5234T3	6.1	2.2	②	—	—	—	NA	NA
	S60 AWD	2.4	B5244T3	6.1	—	②	—	—	—	NA	NA
	C70 LPT	2.4	B5244T3	6.2	—	②	—	—	—	18.0	NA
	C70 HPT	2.3	B5234T3	6.2	2.2	②	—	—	—	18.0	NA
	S80 2.9	2.9	B6294S2	7.3	—	②	—	—	—	NA	NA
	S80 T6	2.8	B6294T	7.3	—	②	—	—	—	NA	NA
	V40	1.9	B4204T3	5.7	—	②	—	—	—	NA	NA
	V70 2.4	2.4	B5244S	6.1	2.2	②	—	—	—	NA	NA
	V70 2.4T	2.4	B5244T3	6.1	—	②	—	—	—	NA	NA
	V70 T5	2.3	B5234T3	6.1	2.2	②	—	—	—	NA	NA
	V70 AWD	2.4	B5244T3	6.1	—	NA	—	—	NA	NA	NA
	XC70 AWD	2.4	B5244T3	6.1	—	NA	—	—	NA	NA	NA
2003	S40	1.9	B4204T3	5.7	—	8.0	—	—	—	16	6.0
	S60	2.4	B5244S	5.8	2.2	7.9	—	—	—	18.5	8.5
	S60 2.4T	2.4	B5244T3	6.1	2.2	7.9	—	—	—	18.5	9.3
	S60 T5	2.3	B5234T3	6.1	2.2	7.9	—	—	—	18.5	9.3
	S60 AWD	2.3	B5254T	6.1	2.2	7.9	—	—	—	19	9.3
	C70 LPT	2.4	B5244T7	6.2	—	8.0	—	—	—	17.9	7.4
	C70 HPT	2.3	B5234T9	6.2	2.2	8.0	—	—	—	17.9	7.4
	S80 2.9	2.9	B6294S2	7.3	—	13.4	—	—	—	21.1	9.3
	S80 T6	2.8	B6294T	7.3	—	13.4	—	—	—	21.1	10.1
	V40	1.9	B4204T3	5.7	—	8.0	—	—	—	16	6.0
	V70 2.4	2.4	B5244S	6.1	2.2	7.9	—	—	—	18.5	NA
	V70 2.4T	2.4	B5244T3	6.1	—	7.9	—	—	—	18.5	NA
	V70 T5	2.3	B5234T3	6.1	2.2	7.9	—	—	—	18.5	NA
	V70 AWD	2.3	B5254T	6.1	—	7.9	—	—	NA	19	NA
	XC70 AWD	2.3	B5254T	6.1	—	7.9	—	—	NA	19	NA
2004	S40	1.9	B4204T3	5.7	—	8.0	—	—	—	16	6.0
	S60	2.4	B5244S	5.8	2.2	7.9	—	—	—	18.5	8.5
	S60 2.4T	2.4	B5244T3	6.1	2.2	7.9	—	—	—	18.5	9.3
	S60 T5	2.3	B5234T3	6.1	2.2	7.9	—	—	—	18.5	9.3
	S60 AWD	2.3	B5254T	6.1	2.2	7.9		—	—	19	9.3
	C70 LPT	2.4	B5244T7	6.2	—	8.0	—	—	—	17.9	7.4
	C70 HPT	2.3	B5234T9	6.2	2.2	8.0	—	—	—	17.9	7.4
	S80 2.9	2.9	B6294S2	7.3	—	13.4	—	—	—	21.1	9.3
	S80 T6	2.8	B6294T	7.3	—	13.4	—	—	—	21.1	10.1
	V40	1.9	B4204T3	5.7	—	8.0	—	—	—	16	6.0
	V70 2.4	2.4	B5244S	6.1	2.2	7.9	—	—	—	18.5	NA
	V70 2.5T	2.4	B5244T3	6.1	—	7.9	—	—	—	18.5	NA
	V70 T5	2.3	B5234T3	6.1	2.2	7.9	—	—	—	18.5	NA
	V70 AWD	2.3	B5254T	6.1	—	7.9	NA	—	NA	19	NA
	XC70 AWD	2.3	B5254T	6.1	—	7.9	NA	—	NA	19	NA

NA: Information not available

NOTE: All capacities are approximate. Add fluid gradualy and check to be sure a proper fluid level is obtained.

① On turbocharged engines, add 0.7 US qts. if the cooler is drained

② M56: 2.2 qts.
M66: 2.1 qts.

06041_VOLV_C0005

CAPACITIES

Year	Model	Engine Displacement Liters	Engine ID	Engine Oil with Filter (qts.) ①	Transmission (qts.)		Transfer Case (pts.)	Drive Axle		Fuel Tank (gal.)	Cooling System (qts.)
					5-Spd	Auto.		Front (pts.)	Rear (pts.)		
2005	S40 2.4i	2.4	B5244S	5.8	3.7	13.5	—	—	—	NA	NA
	S40 T5	2.5	B5254T3	NA	3.5	13.5	—	—	—	NA	NA
	S60 2.4	2.4	B5244S	5.8	3.7	12.5	—	—	—	NA	NA
	S60 2.5T	2.5	B5254T	NA	3.7	12.5	—	—	—	NA	NA
	S60 R	2.5	B5254T4	5.8	3.5	12.5	—	—	1.2	NA	NA
	S60 T5	2.4	B5244T	5.8	3.5	12.5	—	—	—	NA	NA
	S80 2.5T	2.5	B5254T2	NA	—	12.5	—	—	—	NA	NA
	S80 T6	2.9	B6294T	NA	—	20.9	—	—	—	NA	NA
	V50 2.4i	2.4	B5244S	5.8	3.7	13.5	—	—	—	NA	NA
	V50 T5	2.5	B5254T3	NA	3.5	13.5	—	—	—	NA	NA
	V70 2.4	2.4	B5244S	5.8	3.7	12.5	—	—	—	NA	NA
	V70 2.5T	2.5	B5254T2	NA	—	12.5	—	—	—	NA	NA
	V70 R	2.5	B5254T4	5.8	—	12.5	—	—	1.2	NA	NA
	XC70	2.5	B5254T2	NA	—	12.5	NA	—	—	NA	NA
2006	C70 T5	2.5	B5254T3	6.1	②	8.2	—	—	—	15.9	④
	S40 2.4i	2.4	B5244S	6.1	②	8.2	—	—	—	15.9	④
	S40 T5	2.5	B5254T3	6.1	②	8.2	—	—	—	15.9	④
	S60 2.5T	2.5	B5254T	6.1	—	③	—	—	—	18	8.4
	S60 R	2.5	B5254T4	5.8	②	③	—	—	—	18	8.4
	S60 T5	2.4	B5244T	6.1	②	③	—	—	—	18	8.4
	S80 2.5T	2.5	B5254T	6.1	—	③	—	—	—	18	8.4
	V50 2.4i	2.4	B5244S	6.1	②	8.2	—	—	—	15.9	④
	V50 T5	2.5	B5254T3	6.1	②	8.2	—	—	—	15.9	④
	V70 2.4	2.4	B5244S	6.1	②	③	—	—	—	18.5	9.3
	V70 2.5T	2.5	B5254T2	6.1	—	③	—	—	—	18	8.5
	V70 R	2.5	B5254T4	5.8	②	③	—	—	—	18	8.4
	XC70	2.5	B5254T2	6.1	—	③	—	—	—	18	8.5

NA: Information not available

NOTE: All capacities are approximate. Add fluid gradualy and check to be sure a proper fluid level is obtained.

① On turbocharged engines, add 0.7 US qts. if the cooler is drained

② M56: 2.2 qts.
M66: 2.1 qts.

③ AW55: 7.9 qts.
TF-80SC: 7.5 qts.

④ Automatic: 9 qts.
Manual: 8.5 qts.

06041_VOLV_C0006

VALVE SPECIFICATIONS

Year	Engine Displacement Liters	Engine ID	Seat Angle (deg.)	Face Angle (deg.)	Spring Test Pressure (lbs. @ in.)	Spring Installed Height (in.)	Stem-to-Guide Clearance (in.)		Stem Diameter (in.)	
							Intake	Exhaust	Intake	Exhaust
2002	1.9	B4204T3	NA	NA	NA	NA	NA	NA	NA	NA
	2.3	B5234T3	45	44.5	①	NA	0.0012-0.0024	0.0015-0.0028	0.2738-0.2750	0.2734-0.2746
	2.4	B5244T3	45	44.5	①	NA	0.0012-0.0024	0.0015-0.0028	0.2738-0.2750	0.2734-0.2746
	2.4	B5244S	45	44.5	①	NA	0.0012-0.0024	0.0015-0.0028	0.2738-0.2750	0.2734-0.2746
	2.9	B6294S2	44.5	44.5	NA	NA	NA	NA	0.2344-0.2350	0.2344-0.2350
	2.9	B6294T	44.5	44.5	NA	NA	NA	NA	0.2344-0.2350	0.2344-0.2350
2003	1.9	B4204T3	NA	NA	NA	NA	NA	NA	NA	NA
	2.3	B5234T3	45	44.5	①	NA	0.0012-0.0024	0.0015-0.0028	0.2738-0.2750	0.2734-0.2746
	2.4	B5244T3	45	44.5	①	NA	0.0012-0.0024	0.0015-0.0028	0.2738-0.2750	0.2734-0.2746
	2.4	B5244S	45	44.5	①	NA	0.0012-0.0024	0.0015-0.0028	0.2738-0.2750	0.2734-0.2746
	2.5	B5254T	45	44.5	①	NA	0.0012-0.0024	0.0015-0.0028	0.2738-0.2750	0.2734-0.2746
	2.3	B5234T9	45	44.5	①	NA	0.0012-0.0024	0.0015-0.0028	0.2738-0.2750	0.2734-0.2746
	2.4	B5244T7	45	44.5	①	NA	0.0012-0.0024	0.0015-0.0028	0.2738-0.2750	0.2734-0.2746
	2.9	B6294S2	44.5	44.5	NA	NA	NA	NA	0.2344-0.2350	0.2344-0.2350
	2.9	B6294T	44.5	44.5	NA	NA	NA	NA	0.2344-0.2350	0.2344-0.2350
2004	1.9	B4204T3	NA	NA	NA	NA	NA	NA	NA	NA
	2.3	B5234T3	45	44.5	①	NA	0.0012-0.0024	0.0015-0.0028	0.2738-0.2750	0.2734-0.2746
	2.4	B5244S	45	44.5	①	NA	0.0012-0.0024	0.0015-0.0028	0.2738-0.2750	0.2734-0.2746
	2.5	B5254T	45	44.5	①	NA	0.0012-0.0024	0.0015-0.0028	0.2738-0.2750	0.2734-0.2746
	2.3	B5234T9	45	44.5	①	NA	0.0012-0.0024	0.0015-0.0028	0.2738-0.2750	0.2734-0.2746
	2.4	B5244T7	45	44.5	①	NA	0.0012-0.0024	0.0015-0.0028	0.2738-0.2750	0.2734-0.2746
	2.9	B6294S2	44.5	44.5	NA	NA	NA	NA	0.2344-0.2350	0.2344-0.2350
	2.9	B6294T	44.5	44.5	NA	NA	NA	NA	0.2344-0.2350	0.2344-0.2350

NA: Information not available

① Intake valve: 150@1.00
Exhaust valve: 61@1.34

06041_VOLV_C0007

VALVE SPECIFICATIONS

Year	Engine Displacement Liters	Engine ID	Seat Angle (deg.)	Face Angle (deg.)	Spring Test Pressure (lbs. @ in.)	Spring Installed Height (in.)	Stem-to-Guide Clearance (in.)		Stem Diameter (in.)	
							Intake	Exhaust	Intake	Exhaust
2005	2.4	B5244S	45	44.5	NA	NA	0.0012-0.0024	0.0012-0.0024	NA	0.2344-0.2350
	2.5	B5254T3	45	44.5	NA	NA	0.0012-0.0024	0.0012-0.0024	NA	0.2344-0.2350
	2.5	B5254T	45	44.5	NA	NA	0.0012-0.0024	0.0012-0.0024	NA	0.2344-0.2350
	2.5	B5254T4	45	44.5	NA	NA	0.0012-0.0024	0.0012-0.0024	NA	0.2344-0.2350
	2.5	B5254T2	45	44.5	NA	NA	0.0012-0.0024	0.0012-0.0024	NA	0.2344-0.2350
	2.4	B5244T	45	44.5	NA	NA	0.0012-0.0024	0.0012-0.0024	NA	0.2344-0.2350
	2.9	B6294T	45	44.5	NA	NA	0.0012-0.0024	0.0012-0.0024	NA	0.2344-0.2350
2006	2.4	B5244S	45	44.5	NA	NA	0.0012-0.0024	0.0012-0.0024	NA	NA
	2.4	B5244T	45	44.5	NA	NA	0.0012-0.0024	0.0012-0.0024	NA	NA
	2.5	B5254T	45	44.5	NA	NA	0.0012-0.0024	0.0012-0.0024	0.2344-0.2356	0.2735-0.2746
	2.5	B5254T2	45	44.5	NA	NA	0.0012-0.0024	0.0012-0.0024	0.2344-0.2356	0.2735-0.2746
	2.5	B5254T3	45	44.5	NA	NA	0.0012-0.0024	0.0012-0.0024	0.2344-0.2356	0.2735-0.2746
	2.5	B5254T4	45	44.5	NA	NA	0.0012-0.0024	0.0012-0.0024	0.2344-0.2356	0.2735-0.2746

NA: Information not available

06041_VOLV_C0008

CRANKSHAFT AND CONNECTING ROD SPECIFICATIONS

All measurements are given in inches.

	Engine		Crankshaft				Connecting Rod		
Year	Displacement Liters	Engine ID	Main Brg. Journal Dia.	Main Brg. Oil Clearance	Shaft End-play	Thrust on No.	Journal Diameter	Oil Clearance	Side Clearance
2002	1.9	B4204T3	NA	NA	NA	NA	NA	NA	NA
	2.3	B5234T3	2.5589 - 2.5597	NA	0.003-0.007	NA	1.9679-1.9685	NA	0.006-0.018
	2.4	B5244T3	2.5589 - 2.5597	NA	0.003-0.007	NA	NA	NA	NA
	2.4	B5244S	2.5589 - 2.5597	NA	0.003-0.007	NA	NA	NA	NA
	2.9	B6294S2	2.5589 - 2.5597	NA	0.003-0.007	NA	NA	NA	NA
	2.9	B6294T	2.5589 - 2.5597	NA	0.003-0.007	NA	NA	NA	NA
2003	1.9	B4204T3	NA	NA	NA	NA	NA	NA	NA
	2.3	B5234T3	2.5589 - 2.5597	NA	0.003-0.007	NA	1.9679-1.9685	NA	0.006-0.018
	2.4	B5244T3	2.5589 - 2.5597	NA	0.003-0.007	NA	NA	NA	NA
	2.4	B5244S	2.5589 - 2.5597	NA	0.003-0.007	NA	NA	NA	NA
	2.5	B5254T	2.5589 - 2.5597	NA	0.003-0.007	NA	NA	NA	NA
	2.3	B5234T9	2.5589 - 2.5597	NA	0.003-0.007	NA	NA	NA	NA
	2.4	B5244T7	2.5589 - 2.5597	NA	0.003-0.007	NA	NA	NA	NA
	2.9	B6294S2	2.5589 - 2.5597	NA	0.003-0.007	NA	NA	NA	NA
	2.9	B6294T	2.5589 - 2.5597	NA	0.003-0.007	NA	NA	NA	NA
2004	1.9	B4204T3	NA	NA	NA	NA	NA	NA	NA
	2.3	B5234T3	2.5589 - 2.5597	NA	0.003-0.007	NA	1.9679-1.9685	NA	0.006-0.018
	2.4	B5244S	2.5589 - 2.5597	NA	0.003-0.007	NA	NA	NA	NA
	2.5	B5254T	2.5589 - 2.5597	NA	0.003-0.007	NA	NA	NA	NA
	2.3	B5234T9	2.5589 - 2.5597	NA	0.003-0.007	NA	NA	NA	NA
	2.4	B5244T7	2.5589 - 2.5597	NA	0.003-0.007	NA	NA	NA	NA
	2.9	B6294S2	2.5589 - 2.5597	NA	0.003-0.007	NA	NA	NA	NA
	2.9	B6294T	2.5589 - 2.5597	NA	0.003-0.007	NA	NA	NA	NA

NA: Information not available

06041_VOLV_C0009

CRANKSHAFT AND CONNECTING ROD SPECIFICATIONS

All measurements are given in inches.

	Engine		Crankshaft				Connecting Rod		
Year	Displacement Liters	Engine ID	Main Brg. Journal Dia.	Main Brg. Oil Clearance	Shaft End-play	Thrust on No.	Journal Diameter	Oil Clearance	Side Clearance
2005	2.4	B5244S	2.5584 - 2.5592	NA	0.003- 0.007	NA	NA	NA	NA
	2.5	B5254T3	2.5584 - 2.5592	NA	0.003- 0.007	NA	NA	NA	NA
	2.5	B5254T	2.5584 - 2.5592	NA	0.003- 0.007	NA	NA	NA	NA
	2.5	B5254T4	2.5584 - 2.5592	NA	0.003- 0.007	NA	NA	NA	NA
	2.5	B5254T2	2.5584 - 2.5592	NA	0.003- 0.007	NA	NA	NA	NA
	2.4	B5244T	2.5584 - 2.5592	NA	0.003- 0.007	NA	NA	NA	NA
	2.9	B6294T	2.5584 - 2.5592	NA	0.003- 0.007	NA	NA	NA	NA
2006	2.4	B5244S	NA	NA	NA	NA	NA	NA	NA
	2.4	B5244T	NA	NA	NA	NA	NA	NA	NA
	2.5	B5254T	2.5590- 2.5596	NA	0.0030- 0.0070	NA	NA	NA	NA
	2.5	B5254T2	2.5590- 2.5596	NA	0.0030- 0.0070	NA	NA	NA	NA
	2.5	B5254T3	2.5590- 2.5596	NA	0.0030- 0.0070	NA	NA	NA	NA
	2.5	B5254T4	2.5590- 2.5596	NA	0.0030- 0.0070	NA	NA	NA	NA

NA: Information not available

06041_VOLV_C0010

PISTON AND RING SPECIFICATIONS

All measurements are given in inches.

Year	Engine Displ. Liters	Engine ID	Piston Clearance	Ring Gap Top Compression	Ring Gap Bottom Compression	Ring Gap Oil Control	Ring Side Clearance Top Compression	Ring Side Clearance Bottom Compression	Ring Side Clearance Oil Control
2002	1.9	B4204T3	NA	NA	NA	NA	NA	NA	NA
	2.3	B5234T3	0.0004-0.0012	0.047	0.059	0.118	0.0020-0.0033	0.0011-0.0026	0.0008-0.0022
	2.4	B5244T3	0.0004-0.0012	0.047	0.059	0.118	0.0020-0.0033	0.0011-0.0026	0.0008-0.0022
	2.4	B5244S	0.0004-0.0012	0.047	0.059	0.118	0.0020-0.0033	0.0011-0.0026	0.0008-0.0022
	2.9	B6294S2	0.0004-0.0012	0.047	0.059	NA	0.0012-0.0028	0.0012-0.0028	0.0015-0.0056
	2.9	B6294T	0.0004-0.0012	0.047	0.059	NA	0.0012-0.0028	0.0012-0.0028	0.0015-0.0056
2003	1.9	B4204T3	NA	NA	NA	NA	NA	NA	NA
	2.3	B5234T3	0.0004-0.0012	0.047	0.059	0.118	0.0020-0.0033	0.0011-0.0026	0.0008-0.0022
	2.4	B5244T3	0.0004-0.0012	0.047	0.059	0.118	0.0020-0.0033	0.0011-0.0026	0.0008-0.0022
	2.4	B5244S	0.0004-0.0012	0.047	0.059	0.118	0.0020-0.0033	0.0011-0.0026	0.0008-0.0022
	2.5	B5254T	0.0004-0.0012	0.047	0.059	NA	0.0012-0.0028	0.0012-0.0028	0.0015-0.0056
	2.3	B5234T9	0.0004-0.0012	0.047	0.059	NA	0.0012-0.0028	0.0012-0.0028	0.0015-0.0056
	2.4	B5244T7	0.0004-0.0012	0.047	0.059	NA	0.0012-0.0028	0.0012-0.0028	0.0015-0.0056
	2.9	B6294S2	0.0004-0.0012	0.047	0.059	NA	0.0012-0.0028	0.0012-0.0028	0.0015-0.0056
	2.9	B6294T	0.0004-0.0012	0.047	0.059	NA	0.0012-0.0028	0.0012-0.0028	0.0015-0.0056
2004	1.9	B4204T3	NA	NA	NA	NA	NA	NA	NA
	2.3	B5234T3	0.0004-0.0012	0.047	0.059	0.118	0.0020-0.0033	0.0011-0.0026	0.0008-0.0022
	2.4	B5244S	0.0004-0.0012	0.047	0.059	0.118	0.0020-0.0033	0.0011-0.0026	0.0008-0.0022
	2.5	B5254T	0.0004-0.0012	0.047	0.059	NA	0.0012-0.0028	0.0012-0.0028	0.0015-0.0056
	2.3	B5234T9	0.0004-0.0012	0.047	0.059	NA	0.0012-0.0028	0.0012-0.0028	0.0015-0.0056
	2.4	B5244T7	0.0004-0.0012	0.047	0.059	NA	0.0012-0.0028	0.0012-0.0028	0.0015-0.0056
	2.9	B6294S2	0.0004-0.0012	0.047	0.059	NA	0.0012-0.0028	0.0012-0.0028	0.0015-0.0056
	2.9	B6294T	0.0004-0.0012	0.047	0.059	NA	0.0012-0.0028	0.0012-0.0028	0.0015-0.0056

NA: Information not available

06041_VOLV_C0011

PISTON AND RING SPECIFICATIONS

All measurements are given in inches.

Year	Engine Displ. Liters	Engine ID	Piston Clearance	Ring Gap			Ring Side Clearance		
				Top Compression	Bottom Compression	Oil Control	Top Compression	Bottom Compression	Oil Control
2005	2.4	B5244S	0.0004-0.0012	0.047	0.059	NA	0.0012-0.0028	0.0012-0.0028	0.0015-0.0056
	2.5	B5254T3	0.0004-0.0012	NA	NA	NA	0.0012-0.0028	0.0012-0.0028	0.0015-0.0056
	2.5	B5254T	0.0004-0.0012	NA	NA	NA	0.0012-0.0028	0.0012-0.0028	0.0015-0.0056
	2.5	B5254T4	0.0004-0.0012	NA	NA	NA	0.0012-0.0028	0.0012-0.0028	0.0015-0.0056
	2.5	B5254T2	0.0004-0.0012	NA	NA	NA	0.0012-0.0028	0.0012-0.0028	0.0015-0.0056
	2.4	B5244T	0.0004-0.0012	0.047	0.059	NA	0.0012-0.0028	0.0012-0.0028	0.0015-0.0056
	2.9	B6294T	0.0004-0.0012	0.047	0.059	NA	0.0012-0.0028	0.0012-0.0028	0.0015-0.0056
2006	2.4	B5244S	0.0004-0.0012	NA	NA	NA	0.0012-0.0028	0.0012-0.0028	0.0015-0.0056
	2.4	B5244T	0.0004-0.0012	NA	NA	NA	NA	NA	NA
	2.5	B5254T	0.0004-0.0012	NA	NA	NA	0.0012-0.0028	0.0012-0.0028	0.0015-0.0056
	2.5	B5254T2	0.0004-0.0012	NA	NA	NA	0.0012-0.0028	0.0012-0.0028	0.0015-0.0056
	2.5	B5254T3	0.0004-0.0012	NA	NA	NA	0.0012-0.0028	0.0012-0.0028	0.0015-0.0056
	2.5	B5254T4	0.0004-0.0012	NA	NA	NA	0.0012-0.0028	0.0012-0.0028	0.0015-0.0056

NA: Information not available

06041_VOLV_C0012

TORQUE SPECIFICATIONS

All readings in ft. lbs.

Year	Engine Displacement Liters	Engine ID	Cylinder Head Bolts	Main Bearing Bolts	Rod Bearing Bolts	Crankshaft Damper Bolts	Flywheel Bolts	Manifold Intake	Manifold Exhaust	Spark Plugs	Lug Nut
2002	1.9	B4204T3	①	②	③	132-134	④	14	18	22	81
	2.3	B5234T3	①	②	③	132-134	④	14	18	22	81
	2.4	B5244T3	①	②	③	132-134	④	14	18	22	81
	2.4	B5244S	①	②	③	132-134	④	14	18	22	81
	2.9	B6294S2	①	②	③	220-222	④	14	18	22	81
	2.9	B6294T	①	②	③	220-222	④	14	18	22	81
2003	1.9	B4204T3	①	②	③	132-134	④	14	18	22	81
	2.3	B5234T3	①	②	③	132-134	④	14	18	22	81
	2.4	B5244T3	①	②	③	132-134	④	14	18	22	81
	2.4	B5244S	①	②	③	132-134	④	14	18	22	81
	2.5	B5254T	①	②	③	132-134	④	14	18	22	81
	2.3	B5234T9	①	②	③	132-134	④	14	18	22	81
	2.4	B5244T7	①	②	③	132-134	④	14	18	22	81
	2.9	B6294S2	①	②	③	220-222	④	14	18	22	81
	2.9	B6294T	①	②	③	220-222	④	14	18	22	81
2004	1.9	B4204T3	①	②	③	132-134	④	14	18	22	81
	2.3	B5234T3	①	②	③	132-134	④	14	18	22	81
	2.4	B5244S	①	②	③	132-134	④	14	18	22	81
	2.5	B5254T	①	②	③	132-134	④	14	18	22	81

① Step 1: 15 ft. lbs.
Step 2: 44 ft. lbs.
Step 23: Plus 130 degrees

② Step 1: M10 bolts:15 ft. lbs.
Step 2: M10 bolts: 33 ft. lb
Step 3: M8 bolts: 18 ft. lbs
Step 4: M7 bolts: 13 ft. lbs.
Step 5: M10 bolts: Plus 90 degrees

③ Step 1: 15 ft. lbs.
Step 2: Plus 90 degrees

④ Step 1: 33 ft. lbs.
Step 2: Plus 65 degrees

06041_VOLV_C0013

TORQUE SPECIFICATIONS

All readings in ft. lbs.

Year	Engine Displacement Liters	Engine ID	Cylinder Head Bolts	Main Bearing Bolts	Rod Bearing Bolts	Crankshaft Damper Bolts	Flywheel Bolts	Manifold		Spark Plugs	Lug Nut
								Intake	Exhaust		
2004 cont'd	2.3	B5234T9	①	②	③	132-134	④	14	18	22	81
	2.4	B5244T7	①	②	③	132-134	④	14	18	22	81
	2.9	B6294S2	①	②	③	220-222	④	14	18	22	81
	2.9	B6294T	①	②	③	220-222	④	14	18	22	81
2005	2.4	B5244S	①	②	③	132	④	14	18	22	103
	2.5	B5254T3	①	②	③	132	④	14	18	22	103
	2.5	B5254T	①	②	③	132	④	14	18	22	103
	2.5	B5254T4	①	②	③	132	④	14	18	22	103
	2.5	B5254T2	①	②	③	132	④	14	18	22	103
	2.4	B5244T	①	②	③	132	④	14	18	22	103
	2.9	B6294T	①	②	③	221	④	15	15	22	103
2006	2.4	B5244S	①	②	③	132	④	15	19	22	66
	2.4	B5244T	①	②	③	132	④	15	19	22	66
	2.5	B5254T	①	②	③	132	④	15	19	22	66
	2.5	B5254T2	①	②	③	132	④	15	19	22	66
	2.5	B5254T3	①	②	③	132	④	15	19	22	66
	2.5	B5254T4	①	②	③	132	④	15	19	22	66

① Step 1: 15 ft. lbs.
Step 2: 44 ft. lbs.
Step 23: Plus 130 degrees

② Step 1: M10 bolts:15 ft. lbs.
Step 2: M10 bolts: 33 ft. lbs.
Step 3: M8 bolts: 18 ft. lbs.
Step 4: M7 bolts: 13 ft. lbs.
Step 5: M10 bolts: Plus 90 degrees

③ Step 1: 15 ft. lbs.
Step 2: Plus 90 degrees

④ Step 1: 33 ft. lbs.
Step 2: Plus 65 degrees

06041_VOLV_C0014

WHEEL ALIGNMENT

Year	Model		Caster Range (+/-Deg.)	Caster Preferred Setting (Deg.)	Camber Range (+/-Deg.)	Camber Preferred Setting (Deg.)	Toe-in (in.)
2002	S40	F	1.00	+4.1	0.50	-0.16	0.08+/-0.05①
		R	—	—	0.25	-0.90	0.15+/-0.05 ①
	S60	F	1.00	+4.00	0.90	-0.30	0.10+/-0.10
		R	—	—	1.00	0	0.20+/-0.20
	C70	F	1.00	+3.35	1.00	0	0.14+/-0.04
		R	—	—	0.50	-1.00	0.07+/-0.-5
	S80	F	1.00	+4.00	0.90	-0.30	0.10+/-0.10
		R	—	—	1.00	0	0.20+/-0.20
	V40	F	1.00	+4.00	0.50	-0.16	0.15+/-0.05 ①
		R	—	—	0.25	-0.90	0.15+/-0.05 ①
	V70	F	1.00	+4.00	0.90	-0.30	0.10+/-0.10
		R	—	—	1.00	0	0.20+/-0.20
	V70 AWD	F	1.00	+4.00	0.90	-0.30	0.10+/-0.10
		R	—	—	1.00	0	0.20+/-0.20
	V70 XC	F	1.00	+5.00	0.90	-0.30	0.20+/-0.20
		R	—	—	1.00	+0.30	0.20+/-0.20
2003	S40	F	1.00	+4.1	0.50	-0.16	0.08+/-0.05①
		R	—	—	0.25	-0.90	0.15+/-0.05 ①
	S60	F	1.00	+4.00	0.90	-0.30	0.10+/-0.10
		R	—	—	1.00	0	0.20+/-0.20
	C70	F	1.00	+3.35	1.00	0	0.14+/-0.04
		R	—	—	0.50	-1.00	0.07+/-0.-5
	S80	F	1.00	+4.00	0.90	-0.30	0.10+/-0.10
		R	—	—	1.00	0	0.20+/-0.20
	V40	F	1.00	+4.00	0.50	-0.16	0.15+/-0.05 ①
		R			0.25	-0.90	0.15+/-0.05 ①
	V70	F	1.00	+4.00	0.90	-0.30	0.10+/-0.10
		R	—	—	1.00	0	0.20+/-0.20
	V70 AWD	F	1.00	+4.00	0.90	-0.30	0.10+/-0.10
		R	—	—	1.00	0	0.20+/-0.20
	V70 XC	F	1.00	+5.00	0.90	-0.30	0.20+/-0.20
		R	—	—	1.00	+0.30	0.20+/-0.20
2004	S40	F	1.00	+4.1	0.50	-0.16	0.08+/-0.05①
		R			0.25	-0.90	0.15+/-0.05 ①
	S60	F	1.00	+4.00	0.90	-0.30	0.10+/-0.10
		R	—	—	1.00	0	0.20+/-0.20
	C70	F	1.00	+3.35	1.00	0	0.14+/-0.04
		R	—	—	0.50	-1.00	0.07+/-0.-5
	S80	F	1.00	+4.00	0.90	-0.30	0.10+/-0.10
		R	—	—	1.00	0	0.20+/-0.20
	V40	F	1.00	+4.00	0.50	-0.16	0.15+/-0.05 ①
		R	—	—	0.25	-0.90	0.15+/-0.05 ①
	V70	F	1.00	+4.00	0.90	-0.30	0.10+/-0.10
		R	—	—	1.00	0	0.20+/-0.20
	V70 AWD	F	1.00	+4.00	0.90	-0.30	0.10+/-0.10
		R	—	—	1.00	0	0.20+/-0.20
	V70 XC	F	1.00	+5.00	0.90	-0.30	0.20+/-0.10
		R	—	—	1.00	+0.30	0.20+/-0.20

F: Front

R: Rear

① Measurement is in degrees

06041_VOLV_C0015

WHEEL ALIGNMENT

Year	Model		Caster Range (+/-Deg.)	Caster Preferred Setting (Deg.)	Camber Range (+/-Deg.)	Camber Preferred Setting (Deg.)	Toe-in (in.)
2005	S40	F	1.00	+3.60	0.70	-0.60	0.20+/-0.10
		R	—	—	0.35	-1.50	0.30+/-0.10
	S60	F	1.00	+4.00	0.90	-0.30	0.10+/-0.10
		R	—	—	1.00	0	0.20+/-0.20
	S80	F	1.00	+4.00	0.90	-0.30	0.10+/-0.10
		R	—	—	1.00	0	0.20+/-0.20
	V50	F	1.00	+3.60	0.70	-0.60	0.20+/-0.10
		R	—	—	0.35	-1.50	0.30+/-0.10
	V70	F	1.00	+4.00	0.90	-0.30	0.10+/-0.10
		R	—	—	1.00	0	0.20+/-0.20
	XC70	F	1.00	+5.00	0.90	-0.30	0.20+/-0.10
		R	—	—	1.00	+0.30	0.20+/-0.20
2006	C70	F	1.50	+3.60	0.70	-0.60	0.20+/-0.10
		R	—	—	1.00	-1.50	0.30+/-0.10
	S40	F	1.50	+3.60	0.70	-0.60	0.20+/-0.10
		R	—	—	1.00	-1.50	0.30+/-0.10
	S60	F	1.00	+4.00	0.90	-0.30	0.10+/-0.10
		R	—	—	1.00	0	0.20+/-0.20
	S80	F	1.00	+4.00	0.90	-0.30	0.10+/-0.10
		R	—	—	1.00	0	0.20+/-0.20
	V50	F	1.50	+3.60	0.70	-0.60	0.20+/-0.10
		R	—	—	1.00	-1.50	0.30+/-0.10
	V70	F	1.00	+4.00	0.90	-0.30	0.10+/-0.10
		R	—	—	1.00	0	0.20+/-0.20
	XC70	F	1.00	+5.00	0.90	-0.30	0.20+/-0.10
		R	—	—	1.00	+0.30	0.20+/-0.20

F: Front

R: Rear

① Measurement is in degrees

06041_VOLV_C0016

TIRE, WHEEL AND BALL JOINT SPECIFICATIONS

		OEM Tires		Tire Pressures (psi)		Wheel	Ball Joint
Year	Model	Standard	Optional	Front	Rear	Size	Inspection
2002	S40	195/60R15	None	①	①	NA	NA
	S60	195/65R15	205/55R16 215/55R16	①	①	NA	NA
	C70 Conv	205/55R16	225/50R16	①	①	NA	NA
	C70 Coupe	225/45R17	None	①	①	NA	NA
	S80	215/55R16	225/55R16	①	①	NA	NA
	V40	195/60R15	None	①	①	NA	NA
	V70	195/65R15	205/55R16 215/55R16	①	①	NA	NA
	XC70	215/65R16	None	①	①	NA	NA
2003	S40	195/60R15	None	①	①	NA	NA
	S60	195/65R15	205/55R16 215/55R16	①	①	NA	NA
	C70 Conv	205/55R16	225/50R16	①	①	NA	NA
	C70 Coupe	225/45R17	None	①	①	NA	NA
	S80	215/55R16	225/50R17	①	①	NA	NA
	V40	195/60R15	None	①	①	NA	NA
	V70	195/65R15	205/55R16 215/55R16	①	①	NA	NA
	XC70	215/65R16	None	①	①	NA	NA
2004	S40	195/60R15	215/50R16	①	①	NA	NA
	S60	195/65R15	205/55R16 215/55R16 235/45R17	①	①	NA	NA
	C70 Conv	205/55R16	225/50R16	①	①	NA	NA
	S80	215/55R16	225/50R17	①	①	NA	NA
	V40	195/60R15	215/50R16	①	①	NA	NA
	V70	195/65R15	205/55R16 215/55R16 235/45R17	①	①	NA	NA
	XC70	215/65R16	None	①	①	NA	NA

OEM: Original Equipment Manufacturer

PSI: Pounds Per Square Inch

STD: Standard

NA: Information not available

OPT: Optional

① See the tire placard on the vehicle

06041_VOLV_C0017

TIRE, WHEEL AND BALL JOINT SPECIFICATIONS

		OEM Tires		Tire Pressures (psi)		Wheel	Ball Joint
Year	Model	Standard	Optional	Front	Rear	Size	Inspection
2005	S40	205/55R16	205/50R17	①	①	NA	NA
	S60	195/65R15	215/55R16 235/45ZR17 235/40ZR18	①	①	NA	NA
	S80	215/55R16	225/50R17 235/45R17 235/40R18	①	①	NA	NA
	V50	205/55R16	205/50R17	①	①	NA	NA
	V70	195/65R15	205/55R16 235/45R17	①	①	NA	NA
	XC70	215/65R16	None	①	①	NA	NA
2006	C70	235/45R17	None	①	①	NA	NA
	S40	205/55R16	205/50R17	①	①	NA	NA
	S60	205/55R16	215/55R16 235/45ZR17	①	①	NA	NA
	S80	225/50R17	235/45R17 235/40R18	①	①	NA	NA
	V50	205/55R16	205/50R17	①	①	NA	NA
	V70	195/65R15	205/55R16 235/45R17	①	①	NA	NA
	XC70	215/65R16	None	①	①	NA	NA

OEM: Original Equipment Manufacturer
PSI: Pounds Per Square Inch
STD: Standard
NA: Information not available
OPT: Optional
① See the tire placard on the vehicle

06041_VOLV_C0018

BRAKE SPECIFICATIONS

All measurements in inches unless noted

Year	Model		Brake Disc Original Thickness	Brake Disc Minimum Thickness	Brake Disc Maximum Runout	Minimum Lining Thickness	Brake Caliper Bracket bolts (ft. lbs.)	Brake Caliper Mounting bolts (ft. lbs.)
2002	S40	F	0.870	0.847	0.001	0.079	74	26
		R	0.350	0.331	0.001	0.079	41	24
	S60	F	①	②	0.001	0.118	74	22
		R	0.421	0.397	0.001	0.118	44	22
	C70	F	0.937	0.906	0.001	0.118	74	22
		R	0.937	0.906	0.001	0.079	60	45
	S80	F	①	②	0.002	0.118	74	22
		R	0.421	0.397	0.002	0.118	44	22
	V40	F	0.870	0.847	0.002	0.079	74	26
		R	0.350	0.331	0.002	0.079	41	24
	V70	F	①	②	0.002	0.118	74	22
		R	0.421	0.397	0.002	0.118	44	22
	V70 XC	F	①	②	0.002	0.118	74	22
		R	0.421	0.397	0.002	0.118	44	22
2003	S40	F	0.870	0.847	0.001	0.079	74	26
		R	0.350	0.331	0.001	0.079	41	24
	S60	F	①	②	0.001	0.118	74	22
		R	0.421	0.397	0.001	0.118	44	22
	C70	F	0.937	0.906	0.001	0.118	74	22
		R	0.937	0.906	0.001	0.079	60	45
	S80	F	①	②	0.002	0.118	74	22
		R	0.421	0.397	0.002	0.118	44	22
	V40	F	0.870	0.847	0.002	0.079	74	26
		R	0.350	0.331	0.002	0.079	41	24
	V70	F	①	②	0.002	0.118	74	22
		R	0.421	0.397	0.002	0.118	44	22
	XC70	F	①	②	0.002	0.118	74	22
		R	0.421	0.397	0.002	0.118	44	22
2004	S40	F	0.870	0.847	0.001	0.079	74	26
		R	0.350	0.331	0.001	0.079	41	24
	S60	F	①	②	0.001	0.118	74	22
		R	0.421	0.397	0.001	0.118	44	22
	C70	F	0.937	0.906	0.001	0.118	74	22
		R	0.937	0.906	0.001	0.079	60	45
	S80	F	①	②	0.002	0.118	74	22
		R	0.421	0.397	0.002	0.118	44	22
	V40	F	0.870	0.847	0.002	0.079	74	26
		R	0.350	0.331	0.002	0.079	41	24
	V70	F	①	②	0.002	0.118	74	22
		R	0.421	0.397	0.002	0.118	44	22
	XC70	F	1.100	0.980	0.002	0.118	74	22
		R	0.421	0.397	0.002	0.118	44	22

NA: Information not available
F: Front
R: Rear

① 15 inch wheel - 0.937
16 inch wheel -1.016
② 15 inch wheel - 0.906
16 inch wheel - 0.984

06041_VOLV_C0019

BRAKE SPECIFICATIONS

All measurements in inches unless noted

			Brake Disc				Brake Caliper	
Year	Model		Original Thickness	Minimum Thickness	Maximum Runout	Minimum Lining Thickness	Bracket bolts (ft. lbs.)	Mounting bolts (ft. lbs.)
2005	S40	F	0.937	0.906	0.001	0.118	52	22
		R	0.385	0.354	0.001	0.118	52	22
	S60	F	①	②	0.001	0.118	74	22
		R	0.394	0.421	0.002	0.118	44	22
	S80	F	①	②	0.002	0.118	74	22
		R	0.421	0.397	0.002	0.118	44	22
	V50	F	0.937	0.906	0.003	0.118	52	22
		R	0.385	0.354	0.001	0.118	52	22
	V70	F	①	②	0.001	0.118	74	22
		R	0.394	0.421	0.002	0.118	44	22
	XC70	F	1.100	0.980	0.001	0.118	74	22
		R	0.394	0.421	0.002	0.118	44	22
2006	C70	F	NA	NA	0.003	0.118	88	22
		R	NA	NA	0.002	0.118	52	22
	S40	F	1.000	0.905	0.001	0.118	88	20
		R	0.400	0.350	0.002	0.118	52	20
	S60	F	①	②	0.001	0.118	74	22
		R	0.394	0.421	0.002	0.118	44	22
	S80	F	①	②	0.002	0.118	74	22
		R	0.421	0.397	0.002	0.118	44	22
	V50	F	0.937	0.906	0.001	0.118	88	20
		R	0.385	0.354	0.002	0.118	52	20
	V70	F	①	②	0.001	0.118	74	22
		R	0.394	0.421	0.002	0.118	44	22
	XC70	F	1.100	0.980	0.001	0.118	74	20
		R	0.394	0.421	0.002	0.118	44	20

NA: Information not available

F: Front

R: Rear

① 15 inch wheel - 0.937
16 inch wheel -1.016

② 15 inch wheel - 0.906
16 inch wheel - 0.984

06041_VOLV_C0020

SCHEDULED MAINTENANCE INTERVALS
VOLVO

TO BE SERVICED	TYPE OF SERVICE	VEHICLE MILEAGE INTERVAL (x1000)									
		7.5	15	22.5	30	37.5	45	52.5	60	67.5	75
Engine oil & filter	R	✓	✓	✓	✓	✓	✓	✓	✓	✓	✓
Automatic transmission fluid	S/I	✓	✓	✓	✓	✓	✓	✓	✓	✓	✓
Brake pads & parking brake	I	✓	✓	✓	✓	✓	✓	✓	✓	✓	✓
Cabin air filter	R		✓		✓		✓		✓		✓
Engine coolant	S/I		✓		✓		✓		✓		✓
Air cleaner filter	R					✓					✓
Spark plugs	R								✓		
Accessory drive belt	R										✓
Fuel lines	I								✓	✓	✓
Exhaust system	S/I							✓			
Check suspension	S/I								✓	✓	✓
Brake fluid ①	R										

R: Replace S/I: Service or Inspect

① Replace every 2 years or 30,000 miles, whichever comes first under normal conditions, more frequently in mountainous areas or moist climates.

FREQUENT OPERATION MAINTENANCE (SEVERE SERVICE)

If a vehicle is operated under any of the following conditions it is considered severe service:

- Extremely dusty areas.

- 50% or more of the vehicle operation is in 32°C (90°F) or higher temperatures, or constant operation in temperatures below 0°C (32°F).

- Prolonged idling (vehicle operation in stop and go traffic).

- Frequent short running periods (engine does not warm to normal operating temperatures).

- Police, taxi, delivery usage or trailer towing usage.

Oil & oil filter: change every 5000 miles.

Air filter element: service or inspect every 15,000 miles.

06041_VOLV_C0021

ENGINE REPAIR

➡Disconnecting the negative battery cable on some vehicles may interfere with the functions of the on board computer system. The computer may undergo a relearning process once the negative battery cable is reconnected.

Distributor

REMOVAL & INSTALLATION

These models use a Distributorless Ignition System (DIS) controlled by the Powertrain Control Module (PCM).

Alternator

REMOVAL

4-Cylinder Engines

1. Before servicing the vehicle, refer to the Precautions Section.

➡Make a note of the anti-theft radio code

2. Remove or disconnect the following:
    - Engine splash guards, if equipped
    - Negative battery cable
    - Drive belt
    - Power steering pump
    - Mounting bracket
    - Connector and wiring
    - Upper mounting screw
    - Upper screws for the air conditioning (A/C) compressor. Move the screws approximately 30 mm forward.
    - Lower screws for the air conditioning (A/C) compressor
    - Alternator

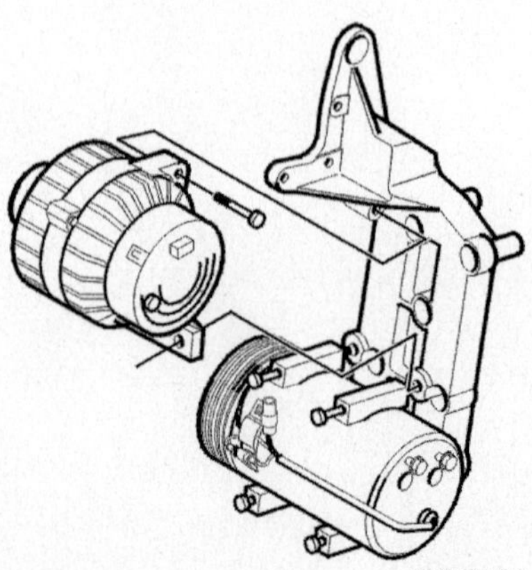

Typical alternator mounting

5-Cylinder Engines 2002–2004 Models

1. Before servicing the vehicle, refer to the Precautions Section.
2. Remove or disconnect the following:
    - Negative battery cable
    - Air intake between the radiator and the air cleaner housing, if equipped
    - Charge air pipe over the engine, if equipped
    - Engine cooling fan
    - Charge air pipe from cooling system
    - Accessory drive belt
    - Screw and nut that run through the upper mounting
    - Lower screws
    - Connector and wiring
    - Alternator

5-Cylinder Engines—2005-06 Models, except S40, V50 and C70

1. Before servicing the vehicle, refer to the Precautions Section.
2. Remove or disconnect the following:
    - Negative battery cable
    - Accessory drive belt
    - Power steering pump
    - Turbo hose from air conditioning assembly
    - Radiator hose from radiator
    - Mounting bolts common to alternator and compress
    - Alternator connector and wiring
    - Alternator mounting bolts
    - Alternator

5-Cylinder Engines—2005-06 S40, V50 and C70 models

1. Before servicing the vehicle, refer to the Precautions Section.
2. Remove or disconnect the following:
    - Negative battery cable
    - Accessory drive belt
    - Air cleaner assembly
    - Oil dipstick assembly
    - Throttle body
    - Bolts common to the alternator and compressor
    - Alternator connector and wiring
    - Alternator mounting bolts
    - Alternator

6-Cylinder Engines—2002–2004 Models

1. Before servicing the vehicle, refer to the Precautions Section.
2. Remove or disconnect the following:
    - Negative battery cable
    - Accessory drive belt
    - Power steering pump
    - Upper mounting bolts
    - Bolts common to the alternator and compressor
    - Lower bolts on the alternator and the upper bolts on the compressor
    - Connector and wiring
    - Alternator

6-Cylinder Engines—2005 Models

1. Before servicing the vehicle, refer to the Precautions Section.
2. Remove or disconnect the following:
    - Negative battery cable
    - Air cleaner assembly
    - Power steering pump
    - Turbo hose from air conditioning assembly
    - Radiator hose from the radiator
    - Accessory drive belt
    - Alternator connector and wiring
    - Mounting bolts common the alternator and compressor
    - Alternator mounting bolts
    - Alternator

INSTALLATION

4-Cylinder Engines

Install or connect the following:
- Alternator
- Upper bolt. Do not tighten.
- Mounting bracket
- Two bolts for the air conditioning compressor. Do not tighten.
- Tighten the screws to 18 ft. lbs. (25 Nm)
- Connector and wiring
- Power steering pump
- Drive belt
- Negative battery cable
- Engine splash guards, if equipped

5-Cylinder Engines—2005-06 Models, except S40 and V50

Install or connect the following:
- Alternator, and tighten mounting bolts and bolts common to the alternator and compressor to 18 ft. lbs. (24 Nm)
- Alternator connector and wiring
- Radiator hose to radiator
- Turbo hose to air conditioning assembly
- Power steering pump

- Accessory drive belt
- Negative battery cable

5-Cylinder Engines—2005-06 S40/V50/C70 Models

Install or connect the following:

- Alternator, and tighten mounting bolts and bolts common to the alternator and compressor to 18 ft. lbs. (24 Nm)
- Alternator connector and wiring
- Throttle body
- Oil dipstick assembly
- Air cleaner assembly
- Accessory drive belt
- Negative battery cable

6-Cylinder Engines—2002–2004 Models

Install or connect the following:

- Alternator
- Connector and wiring
- Bolt in the upper mounting. Do not tighten
- Bolts common to the alternator and compressor (the lower bolts on the alternator and the upper screws on the compressor). Do not tighten
- Tighten the lower bolts for the compressor to 18 ft. lbs. (24 Nm)
- Tighten the bolts common to the alternator and the compressor to 18 ft. lbs. (24 Nm)
- Tighten the screw and nut that run through the upper mounting to 18 ft. lbs. (24 Nm)
- Power steering pump
- Accessory drive belt
- Negative battery cable

6-Cylinder Engines—2005 Models

Install or connect the following:

- Alternator, and tighten mounting bolts and bolts common to the alternator and compressor to 18 ft. lbs. (24 Nm)
- Alternator connector and wiring
- Accessory drive belt
- Radiator hose to radiator
- Turbo hose to air conditioning assembly
- Power steering pump
- Negative battery cable

Ignition Timing

ADJUSTMENT

All engines are equipped with a Motronic 4.3 or 4.4 control system. Manual adjustment of the ignition timing is not possible.

Engine Assembly

REMOVAL & INSTALLATION

4-Cylinder Engines

1. Before servicing the vehicle, refer to the Precautions Section.
2. Drain the cooling system.
3. Remove or disconnect the following:
   - Left-hand side drive shaft nut
   - Hub cap and nut
   - Air conditioning hose from the fan shroud
   - Front and rear splash guards under the engine, if equipped
   - Lateral strut
   - Front-rear member
   - Down pipe from the exhaust manifold
   - Rear engine mounting bracket
   - Front engine mounting
   - Drive belt
   - A/C compressor
   - Oil pressure sensor and alternator connectors
   - Right-hand drive shaft from the transmission
   - Left-hand side drive shaft assembly
   - Transmission oil pipes
4. Drain the fuel injection system.
5. Remove or disconnect the following:
   - Battery and tray
   - Air cleaner
   - Gear selector cables
   - Vacuum hose from the canister purge valve
   - Hoses from the turbocharger
   - Throttle cable
   - Engine speed (ES) sensor
   - Camshaft position (CMP) sensor
   - Knock sensor (KS)
   - Brake servo hose from the power brake booster
   - Hose from the valve for the air cleaner
   - Fuel line from the engine
   - Engine coolant temperature (ECT) sensor
   - Injectors
   - Throttle position (TP) sensor
   - Idle air control (IAC) valve
   - Cable guide
   - Tail lamp switch
   - Upper radiator hose
   - Lower radiator hose on the engine side
   - Heater hoses
   - Power steering pump
6. Place the hood in the service position.
7. Lift the engine and transmission out of the vehicle.

To install:

8. Lower the drivetrain into the vehicle.
9. Install or connect the following:
   - Gear selector cables
   - Transmission side engine mount. Tighten bolt to 33 ft. lbs. (45 Nm).
   - Left and right side engine mounts. Tighten bolts to 72 ft. lbs. (98 Nm).
   - Fuel injectors and wiring
   - Ignition coil wiring
   - Cable mount on the engine. Tighten to 84 inch lbs. (10 Nm).
   - PCV valve
   - Throttle position (TP) sensor
   - Idle air control (IAC) valve
   - Fuel supply line
   - Upper charge air hose from the turbocharger, if equipped
   - A/C compressor. Tighten bracket mounting bolts to 18 ft. lbs. (25 Nm).
   - Alternator
   - A/C compressor sensor
   - Oil pressure sensor
   - Right-hand drive shaft in the transmission
   - Rear engine mount. Tighten to 29 ft. lbs. (40 Nm).
   - Inlet hose to the charge air cooler (CAC). Tighten mounting bolts to 37 ft. lbs. (50 Nm).
   - Front-rear member. Tighten bolts to 51 ft. lbs. (69 Nm).

➡Position the rubber elements in the correct positions. The position of the mounting points for the front-rear member must be the same at both the front and the rear.

10. Install or connect the following:
    - Front exhaust pipe on the exhaust manifold. Tighten the nuts to 18 ft. lbs. (25 Nm).
    - Oil pipes for automatic transmission, if equipped
    - Left-hand side drive shaft
    - Lateral strut. Tighten bolts to 37 ft. lbs. (50 Nm).
    - Engine splash guards, if equipped
    - Ground lead to the transmission
    - Brake servo hose to the power brake booster
    - Wiring for the starter
    - Coolant hoses
    - Cap nut on the oil cooler pipe connector on the transmission. Tighten to 22 ft. lbs. (30 Nm).
    - Heater hoses

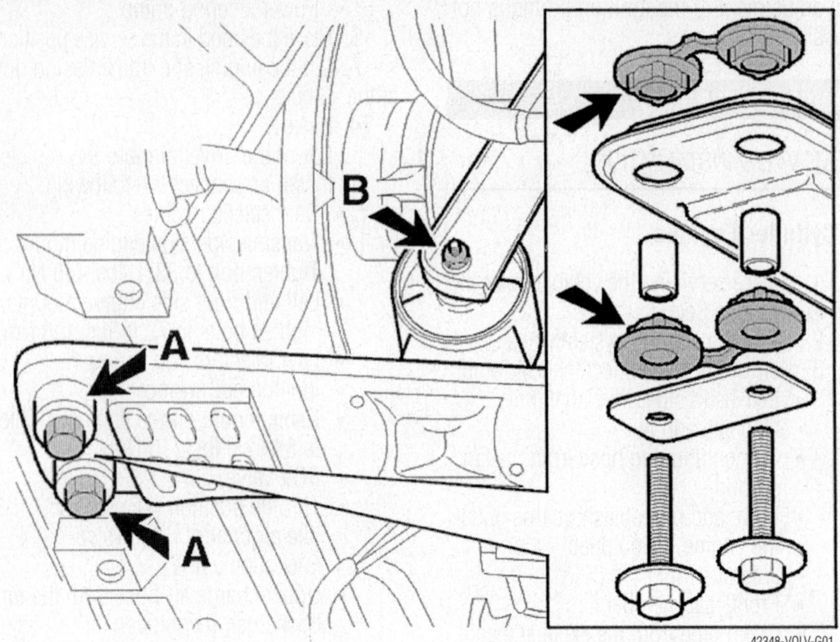

Position of the rubber elements for the front-rear member

- Air cleaner (ACL)
- Battery and tray
- Air intake hose
- Idle hose to the idle air control (IAC) valve
- Power steering pump
- Drive belt

11. Fill the cooling system.
12. Fill the crankcase to the correct level.
13. Start the engine and check for leaks.

5-Cylinder Engines—2002–2004 Models

1. Before servicing the vehicle, refer to the Precautions Section.
2. Drain the cooling system.
3. Remove or disconnect the following:
   - Battery
   - Cross member between the suspension turrets
   - Air cleaner (ACL) assembly
   - Charge air pipe
   - Ground strips at the top of the cylinder head
   - Engine wiring
   - Control modules
   - Body ground cable
   - Engine cooling fan and connectors
   - Brake vacuum hose from the ejector valve
   - Canister purge (CP) valve
   - Upper oil hose for the transmission at the connection to the radiator, if equipped
   - Pressure sensor and temperature sensor
   - Rubber hose at the intake for the charge air cooler (CAC)
   - Brake vacuum hose at the intake manifold
   - Plastic pipe between the throttle body (TB) and the charge air cooler (CAC)
   - Lower oil hose for the transmission at the connection to the radiator, if equipped
   - Transmission cable
4. Drain the fuel line.
5. Raise and support the vehicle.
6. Remove or disconnect the following:
   - Front wheels
   - Splash guards, if equipped
7. Drain the cooling system.
8. Remove or disconnect the following:
   - Connector from the engine block heater
   - Drive shafts
   - Front exhaust pipe
   - Radiator
   - A/C compressor
   - Heat deflector plate above the turbocharger, if equipped
   - Hoses for the heater unit
   - Bleed hose for the radiator at the expansion tank
   - ABS line and clips and the power steering hose
   - Steering shaft
   - Three-way catalytic converter, AWD only
   - Propeller shaft from the transmission and drive distribution box, AWD only
   - Sub-frame bracket bolts
   - Bolts for the sub-frame
   - Control arms from the ball joint pinion
9. Lower the engine and transmission unit out of the vehicle.

To install:

10. Raise engine and transmission into vehicle. Stop when sub-frame is approximately 6 inches (15 cm) below the frame.
11. Install or connect the following:
    - Drive shaft to the hub. Install the center screws. Do not tighten yet.
    - Ball joint to the control arm. Install the nuts. Do not tighten yet
12. Ensure the engine block heater (if equipped), A/C compressor, tie rod, gear selector cable, and steering shaft joint are in the correct position.
13. Install a new bolt in the lower steering shaft joint. Tighten to 15 ft. lbs. (20 Nm).
14. Lift the engine and transmission unit up to the frame members. Tighten sub-frame bolts to 77 ft. lbs. (105 Nm) plus 120 degrees. Tighten sub frame brackets to 37 ft. lbs. (50 Nm).
15. Install or connect the following:
    - Propeller shaft, AWD only
    - Three-way catalytic converter, AWD only
    - Front exhaust pipe
    - Front exhaust pipe on the turbocharger, if equipped. Tighten nuts to 22 ft. lbs. (30 Nm).
    - Fuel lines
    - Engine block heater, if equipped
    - A/C compressor
    - Right ground cable
    - Accessory belt
    - Control modules
    - Expansion tank
    - Servo oil reservoir
    - Heat deflector plate above the turbocharger, if equipped
    - Ground strips on the top of the cylinder head
    - Engine wiring
    - Transmission cable
    - Lower engine coolant hose to the radiator
    - Lower oil cooler hose for the transmission to the radiator, if equipped
    - Charge air pipe between the throttle body and charge air cooler. Do not tighten the clamp yet.
    - Engine cooling fan.
    - Charge air pipe to the connection at the charge air cooler. Tighten both clamps.
    - Canister purge (CP) valve
    - Brake vacuum hoses
    - Upper oil cooler hose for the transmission to the radiator, if equipped

- Charge air pipe above the engine. Tighten all clamps.
- Upper coolant hose
- Ventilation pipe between the control module and fan shroud
- Air cleaner assembly
- Mass air flow (MAF) sensor
- Support brace between the suspension turrets
- Pressure switch connector
- Heated oxygen (HO_2S) sensor connectors
- Splash guards, if equipped
- Front wheels
- Negative battery cable

16. Fill the cooling system.
17. Fill the crankcase to the correct level.
18. Start the engine and check for leaks.

5-Cylinder Engines—2005-06 Models, except S40 and V50

1. Before servicing the vehicle, refer to the Precautions Section.
2. Remove or disconnect the following:
   - Negative battery cable
   - Expansion tank cap
   - Air pre-heating hose
   - Cross-member between suspension turrets
   - Engine positive lead from left suspension turret
3. Relieve the fuel system pressure.
4. Drain the cooling system.
5. Remove or disconnect the following:
   - Fuel line and EVAP line at the right rear sub-frame mounting
   - Halfshafts

➡Install sealing plugs in the holes where the halfshafts were positioned.

   - Bolt for lower torque rod in the transmission
   - Lower bolts for the front and rear engine mounts
   - Heated oxygen sensors (HO_2S) connectors and wiring
6. Separate the exhaust system as follows:
   a. Remove the nuts where the manifold and front exhaust pipe divide.
   b. Remove the bolts holding the Side Impact Protection System (SIPS) member to the bodywork.
   c. Press the brake pipe out of the holders on the SIPS member.
   d. Press the exhaust system backwards to remove it from the manifold.
7. Remove or disconnect the following:
   - Mass air flow (MAF) sensor connector
   - Air assisted control valve connector
   - Air intake assembly
   - Preheat control vacuum hose
   - Ground strip between engine and cowl
   - Heater hoses
   - Brake vacuum hose from the intake manifold
   - Engine cooling fan connectors from the fan shroud
   - Canister purge (CP) valve
   - Manifold absolute pressure (MAP) sensor
   - EVAP canister vacuum hose
   - Engine cooling fan and fan shroud
   - Lower radiator hose
   - Transmission ground lead
8. The following applies to vehicles equipped with manual transmissions:
   a. Matchmark and remove the mechanical transmission cables
   b. Remove the rubber mounted section of gear selector cable
   c. Plug the hose between the brake fluid reservoir and clutch master cylinder using hose pliers.
   d. Disconnect the clutch cable
9. The following applies to vehicles equipped with automatic transmissions:
   a. Remove the transmission mission cooling hoses from the radiator.
   b. Remove the mechanical gear selector cable
10. Remove or disconnect the following:
   - Transmission air filter bracket
   - Upper radiator hose
   - Coolant expansion tank
   - Accessory drive belt
   - Temperature sensor cable from power steering pump
   - Three bolts from power steering pump

➡Leave the power steering pump in place.

   - Connector for the magnetic clutch for the A/C compressor
   - A/C compressor and hang it on the front cover plate
11. Remove the control module as follows:
   a. Remove the cover from the module box.
   b. Insert special tool 999-5722 around the control module as far as it will go.
   c. Raise the tool and carefully pull up the control module.
   d. Remove the intermediate section of the control module box.
   e. Pry out the connect plate from the lower section of the module box.
   f. Disconnect the engine and transmission wiring connectors and place on top of engine.
12. Lift up the power steering pump and place to one side.
13. Using lifting yoke 999-2810, position a hook between the engine lifting eyelet at the auxiliary equipment bracket and the upper torque rod bracket.
14. Install lifting eyelet 999-5464 in the exhaust manifold.

67200-VOLV-G116

Removing the control module from the box

67200-VOLV-G103

Lifting yoke assembly installed

15. Install mounting bracket 999-5428 to the hooks and eyelet and carefully lift the engine assembly out of the vehicle.

To install:

16. Carefully lower the engine assembly into the vehicle and tighten the front engine mount nut.
17. Install the control module as follows:
    a. Connect the engine and transmission wiring connectors to the connector plate.
    b. Press the connect plate into place in the lower section of the module box.
    c. Install the intermediate section of the module box, aligning the ventilation pipe at the rear edge of the box.
    d. Install the module box into place in the fender liner
    e. Carefully press the control modules down and lock into place using special tool 999-5722.
    f. Press the cover into place.
18. Install or connect the following:
    - A/C compressor and tighten mounting bolts to 18 ft. lbs. (25 Nm)
    - Connector for the magnetic clutch in the A/C compressor
    - Engine and transmission wiring on the sub-frame
    - Power steering pump and tighten bolts to 18 ft. lbs. (25 Nm)
    - Coolant expansion tank and coolant hoses
    - Upper radiator hose
    - Accessory drive belt
    - Transmission air filter bracket
    - Transmission ground lead
19. The following applies to cars with manual transmissions:
    a. Install the rubber mounted section of the gear selector cable and mechanical transmission cables.
    b. Connect the clutch cable using a new seal, remove the hose pliers and bleed the clutch.
20. The following applies to cars with automatic transmissions:
    a. Install the transmission cooling pipes to the radiator
    b. Install the mechanical gear selector cable.
21. Install or connect the following:
    - Lower radiator hose
    - Engine cooling fan assembly with shroud
    - Air intake assembly
    - CP valve and EVAP canister vacuum hose
    - MAP sensor
    - Engine cooling fan wiring connectors on the shroud
    - Brake vacuum hose on the intake manifold
    - Heater hoses
    - Group strip between the engine and cowl
    - Air preheat hose
    - Preheat control vacuum hose
    - Front and rear engine mounting bolts and tighten to 37 ft. lbs. (50 Nm)
    - Lower torque rod transmission bolt, using new bolt and tighten to 26 ft. lbs. (35 Nm) plus 90 degrees.
22. Connect the exhaust system as follows:
    a. Lubricate the studs at the manifold using paste 116-14-08 or equivalent.
    b. Install, but do not tighten, the front exhaust pipe on the manifold using new gasket and nuts.
    c. Install the bolts securing the SIPS member to the body and brake pipe in the holders.
    d. Tighten the nuts for the manifold and bolts securing the SIPS member to 18 ft. lbs. (25 Nm).
23. Install or connect the following:
    - HO_2S sensor connectors and wiring
    - Halfshafts
    - Positive lead for the engine or left suspension turret
    - Cross-member between suspension turrets, and tighten the bolts at the turrets to 37 ft. lbs. (50 Nm) and the bolts at the engine bracket to 59 ft. lbs. (80 Nm)
    - Negative battery cable
24. Fill the cooling system.
25. Fill the crankcase to the correct level.
26. Start the engine and check for leaks.

2005-06 S40, V50 and C70 models

1. Before servicing the vehicle, refer to the Precautions Section.
2. Relieve the fuel system pressure.
3. Remove or disconnect the following:
    - Negative battery cable
    - Battery and battery box
    - Air intake assembly
    - Oil dipstick assembly
    - Cable harness for fuse box on left hand fender liner and clamp
    - Ground lead in left-hand suspension turret
    - Positive cable on the starter and cable clamp
    - Ground lead on the front of the transmission
    - Steering shaft joint bolt, and release the steering shaft from the steering gear
4. Separate the quick-release fuel pressure line connector using special tool 999-5666
5. Separate the quick-release connector on the canister purge (CP) valve by pressing the connector together.
6. Disconnect the heated oxygen sensors (HO_2S) connectors
7. Lower the rear section of the engine.
8. Drain the cooling system.

9. Remove or disconnect the following:
   - Halfshafts
   - Lower torque rod
   - Delivery and return lines from the power steering pump and seal the openings
   - Power steering lines from the mountings on the engine and sub-frame
   - Catalytic converter from the sub-frame
   - Rear HO_2S connector and cable harness

10. Support the sub-frame with a jack with universal plate 999-5972 or equivalent.

11. Remove the following:
    - Bolts for the front edge of the sub-frame
    - Bolts for rear support brackets of the sub-frame
    - Bolts for rear edge of the sub-frame

12. Carefully lower the sub-frame

13. Remove or disconnect the following:
    - Side impact protection system (SIPS) member
    - Catalytic converter
    - Accessory drive belt
    - Bolts for A/C compressor and hang the compressor on the radiator member
    - Heater hoses
    - Coolant hoses from radiator
    - Gear selector cable using special tool 999-7077
    - Clutch cable at the firewall, if equipped with manual transmission
    - Oil hoses for transmission, if equipped with automatic transmission
    - Transmission coolant hoses, if equipped with automatic transmission

14. Position a retaining strap over the engine and either side of the engine.

15. Raise the vehicle and position a lifting table against the oil pan and use post hoist support arms to support the engine and transmission.

16. Secure the engine and transmission in the lifting table with the retaining straps.

17. Open the cock for the hydraulic modulator for the lifting table and lower the vehicle at the same time as opening the lifting table compressed air unit. Stop the car immediately before the lifting table reaches its lower position.

18. Close the cock for the hydraulic modulator for the lifting table.

19. Using the lifting table, lift up the engine assembly slightly to take the slack of the engine mounts.

20. Remove or disconnect the following:

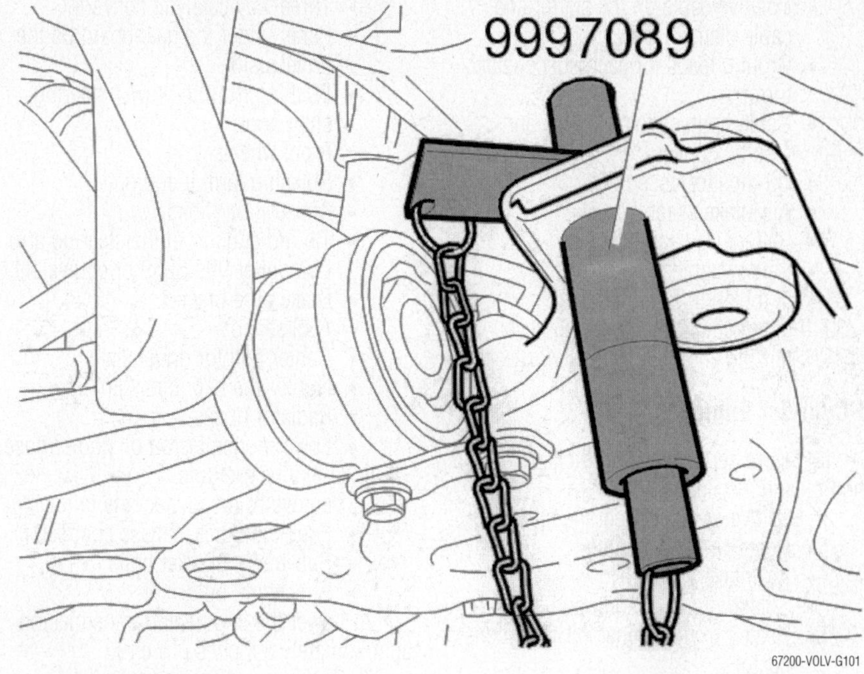

Sub-frame adjustment tool

- Ground lead in the right-hand suspension turret
- Deflection limiter
- Right hand engine mount
- Bolts for the engine mount on the transmission
- Raise the vehicle to free the engine assembly from the vehicle.

To install:

21. Place the engine assembly on a lifting table and raise the vehicle enough to clear the engine assembly.

22. Lower the vehicle over the engine assembly and stop lowering the vehicle just before the engine assembly is in position. Raise the engine the last bit to put it in position to install the engine mounts.

23. Install or connect the following:
    - Front engine mount
    - Bolt for engine mount on transmission and tighten to 26 ft. lbs. (35 Nm) plus 60 degrees
    - Deflection limiter and right-hand engine mount and tighten to 59 ft. lbs. (80 Nm)

24. Disconnect the restraining straps around the engine and remove the lifting table.

25. Install or connect the following:
    - Catalytic converter using new gaskets and nuts
    - SIPS member
    - Lower torque rod

26. Raise the sub-frame so adjustment tools 999-7089 are lined up with the holes in the side members.

27. Tighten the rear sub-frame bolts to 85 ft. lbs. (115 Nm).

28. Tighten the front sub-frame bolts to 85 ft. lbs. (115 Nm).

29. Tighten the support brackets for the rear sub-frame mounts.

30. Install or connect the following:
    - Rear HO_2S connector and cable harness
    - Halfshafts
    - A/C compressor
    - Accessory drive belt
    - Power steering lines and tighten as follows:

    a. Return line to 13 ft. lbs. (18 Nm)

    b. Delivery line to 22 ft. lbs. (30 Nm)

    - Front wheels
    - Engine splash guard, if equipped
    - Coolant hoses to radiator
    - Heater hoses
    - Fuel pressure line
    - EVAP canister hose
    - Oil hoses for transmission, if equipped with automatic transmission
    - Transmission coolant hoses, if equipped with automatic transmission
    - Clutch cable, if equipped with manual transmission
    - Gear selector cable
    - Steering shaft on the steering gear and tighten the bolt to 18 ft. lbs. (24 Nm)
    - Ground lead on transmission

- Positive cable on the starter and cable clamp
- Ground leads for each suspension turret
- Cable harness for fuse box and fuse box cover
- Oil dipstick assembly
- Air intake assembly
- Battery box and battery
- Battery cables

31. Fill the engine with coolant.
32. Fill the crankcase to correct level.
33. Start the engine and check for leaks.

6-Cylinder Engines

1. Before servicing the vehicle, refer to the Precautions Section.
2. Remove or disconnect the following:
   - Negative battery cable
   - Air intake assembly
   - Splash guard, if equipped
   - Heated oxygen sensor (HO_2S) connectors
3. Drain the cooling system.
4. Remove or disconnect the following:
   - Engine stabilizer brace mounting bolt
   - Cross stay between the suspension turrets
   - Charge air pipe, if equipped
   - Intake pipe, if equipped
   - Heat deflector plates
   - Heated oxygen (HO_2S) sensors
   - Negative ground cable
   - Control module box and intermediate section
   - Engine and transmission wiring
   - Upper torque rod bolt
   - Brake vacuum hose from the intake manifold
   - Transmission cable
   - Expansion tank
   - Heating system hoses
   - Radiator hoses
   - Upper oil cooler hose to the transmission
   - Canister purge (CP) valve
   - Intake hose between the turbo pipe and the charge air cooler, if equipped
   - Intake manifold from the charge air cooler
   - Fan shroud
   - Accessory drive belt
   - A/C compressor
   - Front exhaust pipe
   - Exhaust system at the flange behind the rear three-way catalytic converter
   - Center Side Impact Protection System (SIPS) member
   - Front Side Impact Protection System (SIPS) bolts
   - Three-way catalytic converter
   - Panel under the dashboard on the driver's side
   - Bushing from the lower steering shaft boot
   - Front wheels
   - Splash guard, if equipped.
   - Anti-roll bar links
   - Tie rod ends from the steering arm. Use puller 999 5259 or equivalent.
   - Brake pipe bracket
   - ABS sensor
   - Center bolt for drive shafts
   - Nut for the ball joint / link arm.
   - Radiator hoses
   - Lower transmission oil cooler hose
5. Drain the fuel line.
6. Remove or disconnect the following:
   - Fuel line quick-release connector
   - Sub-frame bracket bolts
   - Sub-frame bolts
7. Lower the engine and transmission approximately 6 inches (15 cm).
8. Remove or disconnect the following:
   - Control arms from the ball joint pinion
   - Drive shafts from the hub

➡Do not pull the drive shafts, they do not have axial locks.

9. Lower the engine and transmission unit out of the vehicle.

To install:

10. Raise engine and transmission into vehicle. Stop when sub-frame is approximately 6 inches (15 cm) below the frame.
11. Install or connect the following:
    - Drive shaft on the hub
    - Ball joint to the control arm
12. Guide in the lower steering shaft in the boot.
13. Lift the engine and transmission unit up to the frame members. Tighten the left hand side of the sub-frame first to 77 ft. lbs. (105 Nm) plus 120 degrees. Use the same procedure for the right hand side.
14. Install or connect the following:
    - Sub-frame brackets. Tighten to 37 ft. lbs. (50 Nm).
    - Three-way catalytic converter. Tighten to
    - Front exhaust pipe
    - Exhaust system to the rear exhaust pipe flange and turbocharger
    - Bolts for the front Side Impact Protection System (SIPS) member
    - Center Side Impact Protection System (SIPS) member
    - Rear heated oxygen (HO_2sensor connectors
    - Lower transmission oil cooler hose
    - Lower radiator hose
    - Ball joint/control arm. Tighten to 59 ft. lbs. (80 Nm).
    - Tie rods. Tighten to 52 ft. lbs. (70 Nm).
    - Anti-roll bar links. Tighten to 37 ft. lbs. (50 Nm).
    - Center bolts for drive shafts. Tighten to 37 ft. lbs. (50 Nm)
    - ABS sensors. Tighten to 7 ft. lbs. (10 Nm).
    - Splash guards, if equipped
    - Front wheels
    - Bearing in the lower steering shaft boot
    - Steering shaft joint. Tighten 22 ft. lbs. (30 Nm).
    - Dashboard panel
15. Tighten the three-way catalytic converters (TWC) to the turbocharger (TC) to 18 ft. lbs. (24 Nm).
16. Install or connect the following:
    - Upper heated oxygen (HO_2) sensors. Tighten to 33 ft. lbs. (45 Nm).
    - Heat deflector plates
    - Intake and charge air pipes
    - Engine stabilizer brace mounting brace
    - A/C compressor. Tighten to 18 ft. lbs. (25 Nm).
    - Accessory belt
    - Hose between the charge air cooler and the turbocharger pipe, if equipped
    - Upper radiator hose
    - Engine and transmission wiring
    - Control module intermediate section and box
    - Negative ground cable
    - Expansion tank
    - Servo reservoir
    - Fan shroud
    - Engine cooling fan connector
    - CP valve
    - Charge air pipe to the charge air cooler, if equipped
    - Upper transmission oil cooler hose
    - Transmission cable
    - Air intake assembly
    - Battery cables
17. Fill the cooling system.
18. Fill the crankcase to the correct level.
19. Start the engine and check for leaks.

Heater Core

REMOVAL & INSTALLATION

V70 and 2002–2004 C70

1. Before servicing the vehicle, refer to the Precautions Section.

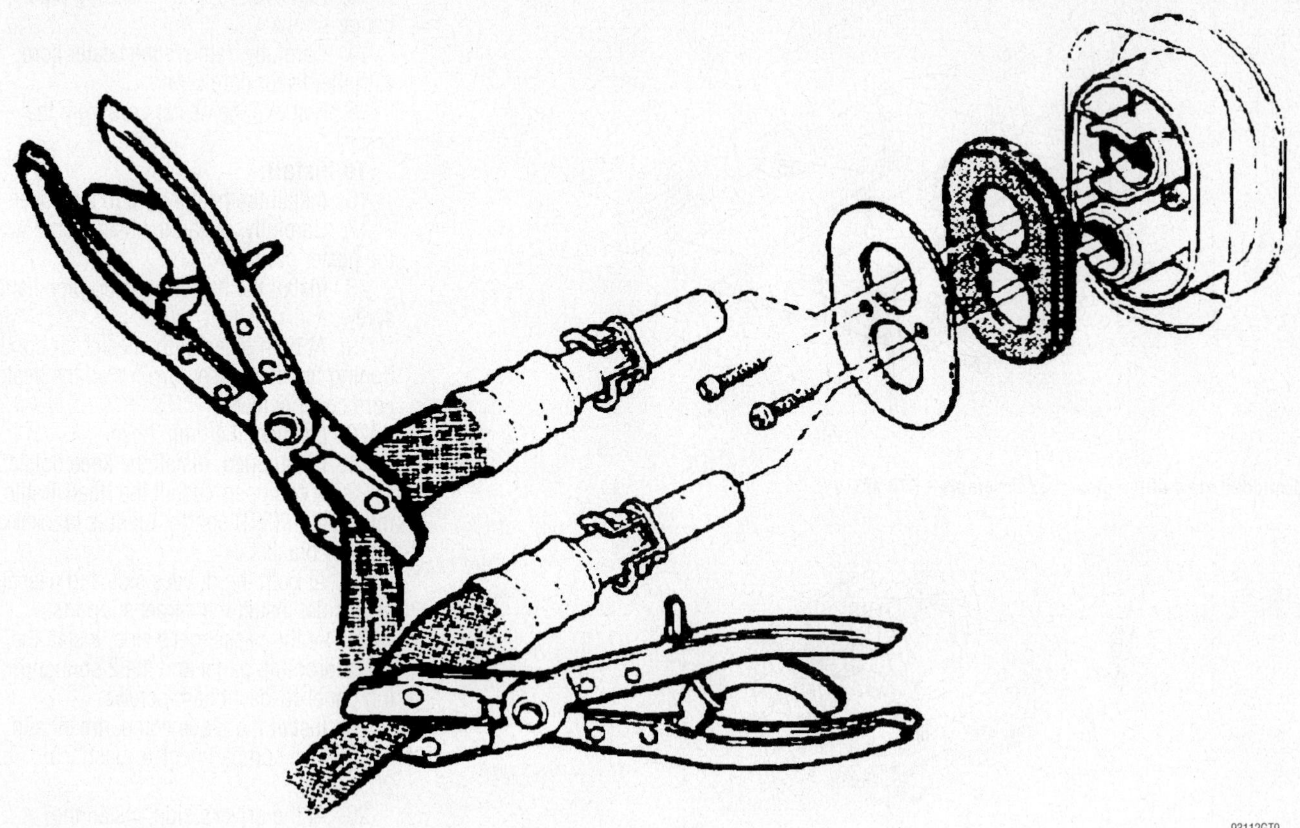

Exploded view of the heater hose connections—C70 and V70

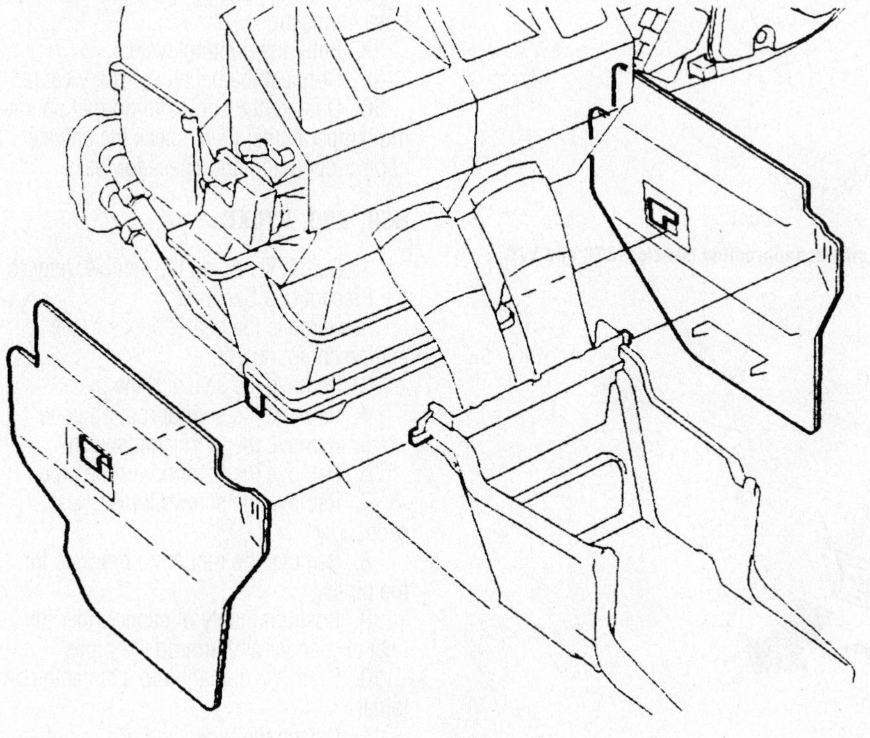

Exploded view of the carpet supports—C70 and V70

2. Disconnect the negative battery cable.

3. Drain the cooling system into a clean container for reuse.

4. Using tools No. 155 8957, disconnect the heater hoses from the heater core by pressing in the hose connections, squeezing the locking catches and pulling out the hoses. Discard the O-rings.

5. At the driver's side, remove the soundproofing panel-to-dashboard Torx 20 screw and the soundproofing panel.

6. Open the glove box door; then, remove the 4 glove compartment-to-dashboard Torx 20 screws and the glove compartment.

7. At the passenger's side, remove the 2 soundproofing panel-to-dashboard screws and the soundproofing panel.

8. At both the driver's side and passenger's side, remove the carpet supports.

9. If equipped, remove the Road Traffic Information (RTI) control module bracket or booster bracket.

10. If equipped, remove the knee bolster.

11. Disconnect the drain hose and fold it out of the way.

12. At both sides of the heater/air conditioning housing assembly, remove the heater core cover screws.

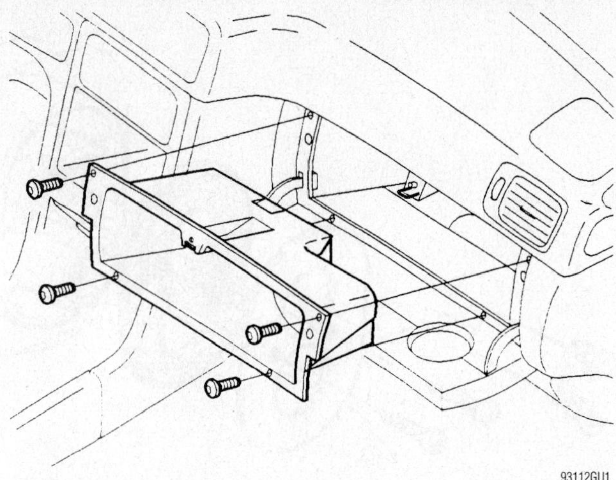

93112GU1

Exploded view of the glove box assembly—C70 and V70

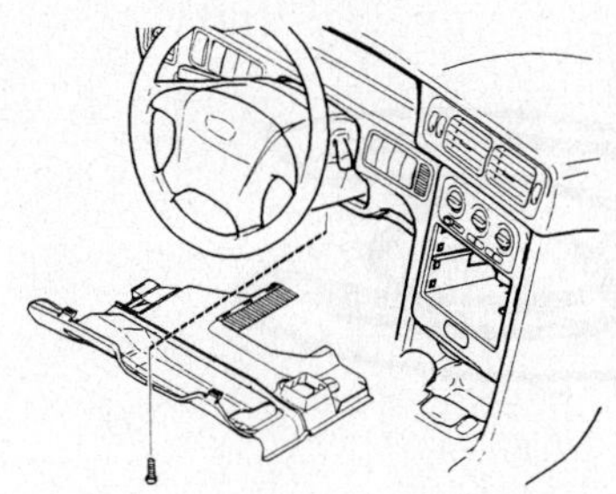

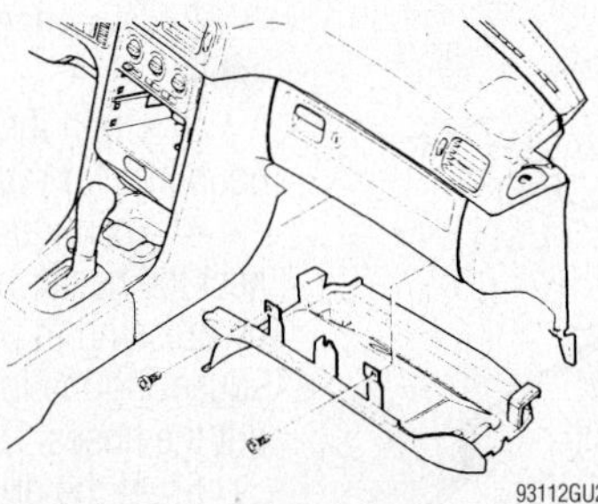

93112GU2

Exploded view of the driver's side and passenger's side soundproofing panels—C70 and V70

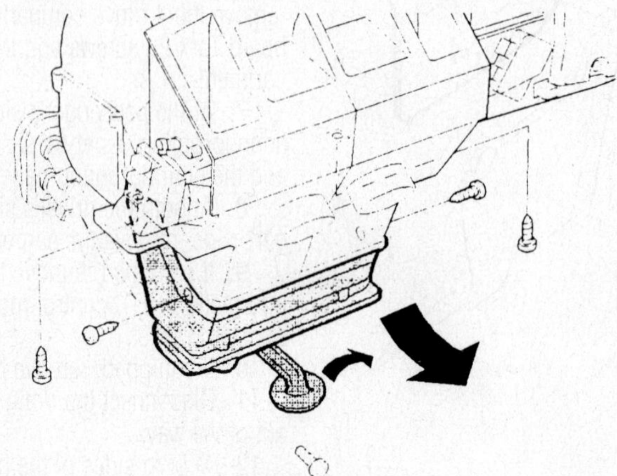

93112GU3

View of the heater core cover—C70 and V70

13. Remove the heater housing pipe flange screw.
14. Carefully, remove the heater core with the heater core cover.
15. Remove the heater core from the cover.

To install:

16. Install the heater core to the cover.
17. Carefully, install the heater core with the heater core cover.
18. Install the heater housing pipe flange screw.
19. At both sides of the heater/air conditioning housing assembly, install the heater core cover screws.
20. Connect the drain hose.
21. If equipped, install the knee bolster.
22. If equipped, install the Road Traffic Information (RTI) control module bracket or booster bracket.
23. At both the driver's side and passenger's side, install the carpet supports.
24. At the passenger's side, install the soundproofing panel and the 2 soundproofing panel-to-dashboard screws.
25. Install the glove compartment and the 4 glove compartment-to-dashboard Torx 20 screws.
26. At the driver's side, install the soundproofing panel and the soundproofing panel-to-dashboard Torx 20 screw.
27. Connect the heater hoses to the heater core by inserting the hoses, squeezing the locking catches and pressing in the hose connections.
28. Refill the cooling system.
29. Connect the negative battery cable.
30. Operate the engine to normal operating temperatures; then, check the climate control operation and check for leaks.

S60, S80, V70XC

1. Before servicing the vehicle, refer to the Precautions Section.
2. Remove the heater hoses in the engine compartment.
3. Remove the seal and the plate.
4. Remove the soundproofing panel.
5. Remove the stop lamp switch.
6. Remove the tie strap for the pipes
7. Remove the screws for the heat exchanger
8. Remove the mounting brackets for the pipes.
9. Position plenty of paper under the heat exchanger and around the pipes
10. Drain the coolant into a suitable container.
11. Detach the pipes and position them out of the way
12. Pull out the heat exchanger.

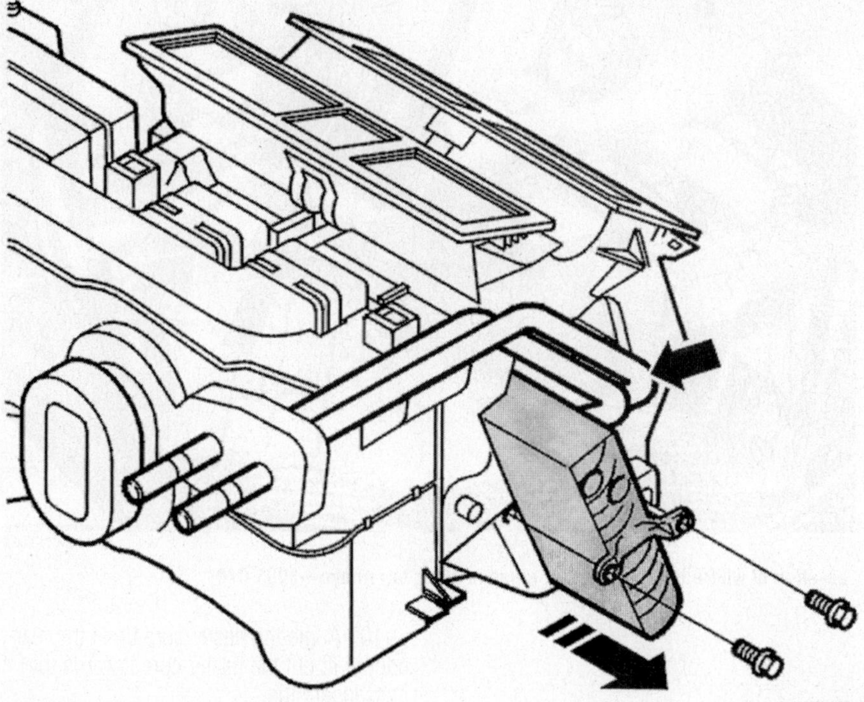

Heater core removal—S60, S80, V70XC

To install:

➡**Always use new O-rings**

13. Install the heat exchanger carefully
14. Screw the heat exchanger into place
15. Install the upper pipes
16. Install the locking plate using round-nosed pliers
17. Press in the pipes using a screwdriver. Press the catch using pliers
18. Install the lower pipes
19. Install the locking plate using round-nosed pliers
20. Press in the pipe using a screwdriver. Press in the locking bracket using round-nosed pliers
21. Install the tie strap for the pipes
22. Install the seal and the plate
23. Install the heating hoses in the engine compartment
24. Start the engine and the heating system
25. Fill the cooling system with coolant
26. Check that the pipe connectors do not leak
27. Install the soundproofing panel.

S40, V40

1. Before servicing the vehicle, refer to the Precautions Section.
2. In the engine compartment:
   a. Remove the top splashguard.
   b. Install two hose clips on the hoses to the heat exchanger. Remove the hoses.
   c. Remove the air conditioning pipe in the engine compartment from the support.
3. In the passenger compartment:
   a. Disconnect the wiring and the connectors located on the center support.
   b. Remove the mounting bolts from the center support. Remove the support.
   c. Remove the control panel.
   d. Disconnect all the connectors.
   e. Remove the power unit
4. Cars with air conditioning remove the air conditioning pipe in the engine compartment from the support.
5. Remove the bolts and the nut from the power unit. Pull the power unit backwards as far as possible
6. Remove the upper section from the housing.
7. Remove the section that connects with the floor heater.
8. Press the pipes down. Remove the lower section.
9. Remove the four mounting bolts.
10. Pull off the housing. Press the floor heater down. Remove the housing.
11. Remove the control lever on the valve side.
12. Remove the mounting and the clips around it.
13. Carefully separate the components.
14. Replace damaged components if necessary.

➡**Do not disassemble damper motors when adjusting.**

15. Replace the refrigerant temperature sensor.
16. Replace the heat exchanger.
17. Check the seals (also on the dampers).
18. Install the control mechanism last.
19. Install in reverse order.

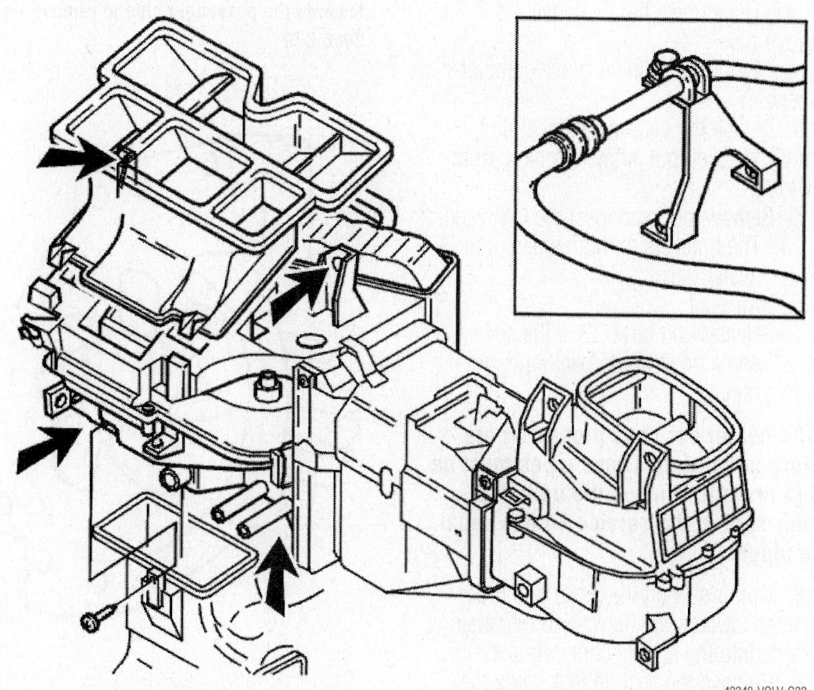

Heater unit—S40, V40

20. Check the setting of the damper motors.

2006 C70

1. Before servicing the vehicle, refer to the Precautions Section.
2. Drain the cooling system.
3. Remove or disconnect the following:
   - Negative battery cable
   - Expansion tank cover
   - Engine splash guard
   - Radiator hoses
4. Remove the center console as follows:
   a. Remove the parking brake boot.
   b. Remove the panel inside the center console.
   c. Using a plastic weatherstrip tool, carefully pry off the radio panel.
   d. Remove the infotainment control module (ICM).
   e. Remove the center console.
   f. Remove the cigarette lighter/electrical socket panel
   g. Remove the panel undercover/armrest.
   h. Lift the rear edge of the center console. Then lift the center console forwards and upwards to remove.
5. Remove the glove compartment as follows:
   a. Remove the dash panel end cover.
   b. Remove the air duct clip.
   c. Pull the glove compartment straight back into the vehicle to release the 2 hidden clips in the top edge.
   d. Disconnect the 2 connectors in the glove box.
   e. Remove the inner glove compartment.
6. Detach the central electrical unit from the bracket, but allow it to hang from its wiring.
7. Remove or disconnect the following:
   - The brackets on both sides of the heater unit
   - Air duct
   - Heater unit cover. Pull the cover at angle downwards/backwards to remove.

➡The heater core has pipes that are factory integrated. These pipes must be cut in order to remove the unit. The heater core in the service kit has separate pipes.

8. Cut or saw off the pipes and press the pipes towards the firewall so that they protrude into the engine compartment.
9. Remove the section that holds the heater core and evaporator pipes apart.

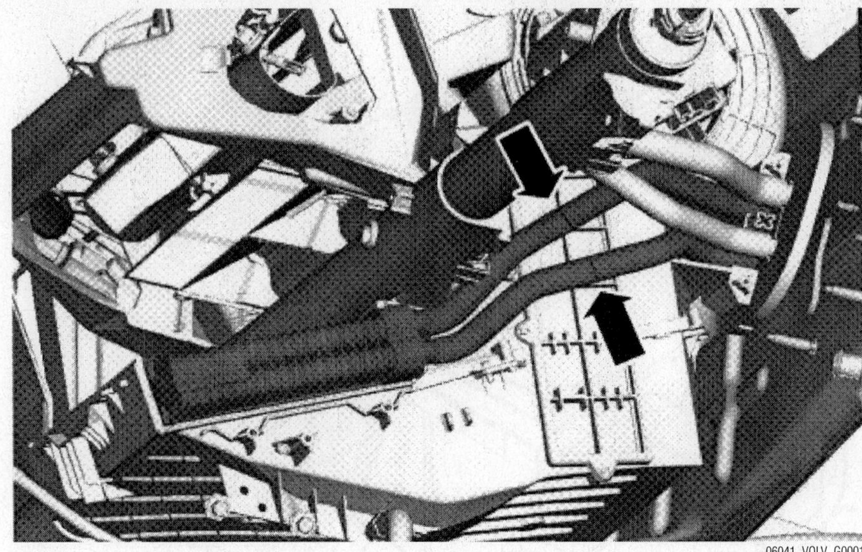

Location of where to cut the heater pipes in order to remove—2006 C70

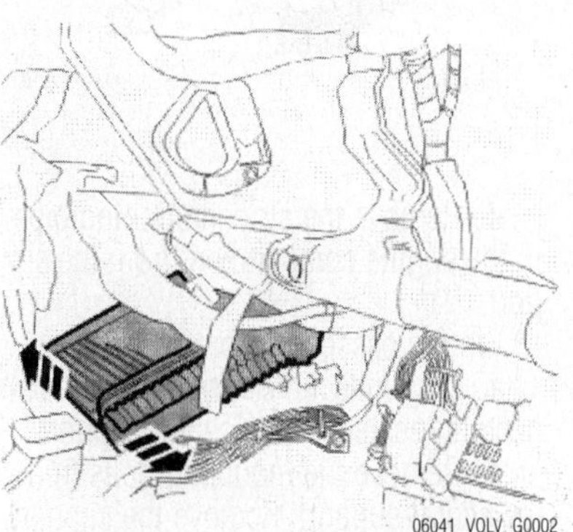

Angle the heater core up and lift out towards the passenger side to remove—2006 C70

10. Angle the heater core up at the rear edge. Lift out the heater core towards the passenger side.
11. Installation is the reverse order of removal.
12. Refill the engine with coolant.
13. Start the engine and check for leaks.

Water Pump

REMOVAL & INSTALLATION

4-Cylinder Engines

1. Before servicing the vehicle, refer to the Precautions Section.
2. Drain the cooling system.

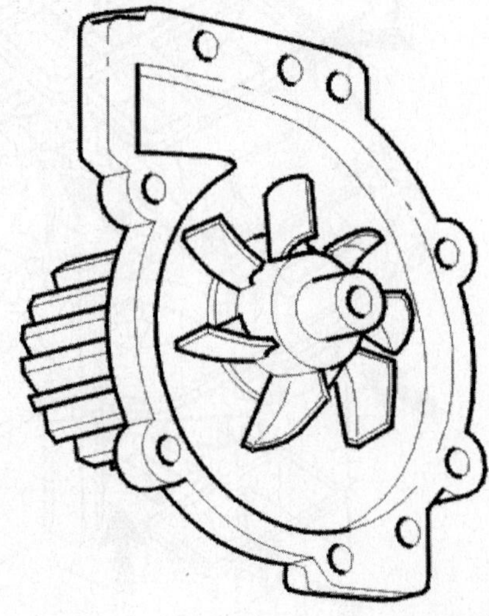

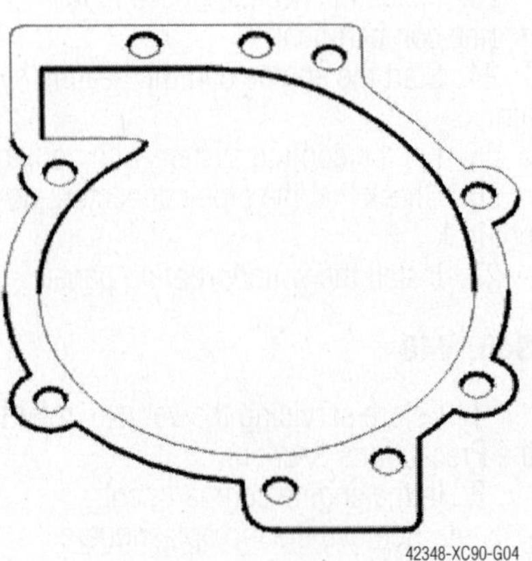

Water pump and gasket

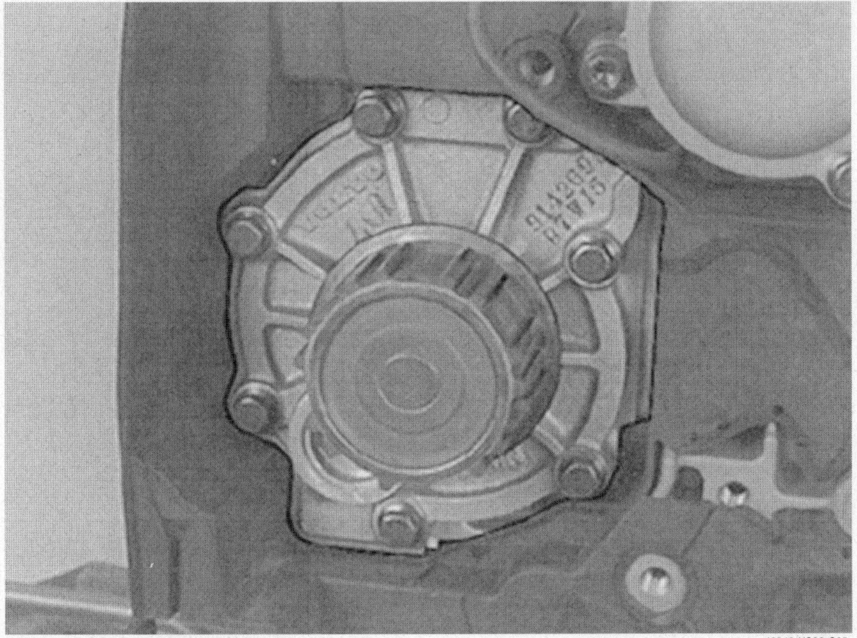
42348-XC90-G05

Water pump installation

3. Remove or disconnect the following:
    - Negative battery cable
    - Accessory belt
    - Timing belt cover
    - Timing belt
    - Water pump

To install:

4. Install or connect the following:
    - Water pump with a new gasket. Tighten bolts to 13 ft. lbs. (17 Nm).
    - Timing belt
    - Timing belt cover. Tighten bolts to 9 ft. lbs. (12 Nm).
    - Accessory belt
5. Fill the cooling system.
6. Start the engine and check for leaks.

5-Cylinder Engines, except 2005-06 S40/V50/C70

1. Before servicing the vehicle, refer to the Precautions Section.
2. Drain the cooling system.
3. Remove or disconnect the following:
    - Negative battery cable
    - Engine stabilizer brace
    - Servo reservoir
    - Expansion tank
    - Accessory belt
    - Front timing belt cover
    - Front wheel
    - Inner fender panel
    - Timing belt
    - Water pump

To install:

4. Install or connect the following:
    - Water pump with a new gasket. Tighten bolts to 13 ft. lbs. (17 Nm).
    - Timing belt
    - Front timing belt cover. Tighten bolts to 9 ft. lbs. (12 Nm).
    - Accessory belt
    - Servo reservoir
    - Expansion tank
    - Engine stabilizer brace. Tighten the bolts at suspension turrets to 37 ft. lbs. (50 Nm). Tighten bolt for engine bracket to 59 ft. lbs. (80 Nm).
    - Inner fender panel
    - Front wheel
5. Fill the cooling system.
6. Start the engine and check for leaks.

2005-06 S40/V50/C70 Models

1. Before servicing the vehicle, refer to the Precautions Section.
2. Drain the cooling system.
3. Remove the timing belt.
4. Securely attach a lifting device to the engine and raise it slightly to access the top two bolts of the water pump.
5. Lower the engine and remove the remaining bolts.
6. Gently tap the pump wheel with a mallet to remove.

To install:

7. Using a new pump and gasket, tighten the bolts alternately to 13 ft. lbs. (17 Nm).
8. Install the timing belt.
9. Fill the cooling system.
10. Start the engine and check for leaks.

6-Cylinder Engines

1. Before servicing the vehicle, refer to the Precautions Section.
2. Drain the cooling system.
3. Remove or disconnect the following:
    - Negative battery cable
    - Timing belt
    - Water pump

To install:

4. Install or connect the following:
    - Water pump with a new gasket. Tighten bolts to 13 ft. lbs. (17 Nm).
    - Timing belt
5. Fill the cooling system.
6. Start the engine and check for leaks.

Cylinder Head

REMOVAL & INSTALLATION

4-Cylinder Engines

1. Before servicing the vehicle, refer to the Precautions Section.
2. Drain the cooling system.
3. Remove or disconnect the following:
4. Relieve the fuel system.
    - Negative battery cable
    - Exhaust manifold
    - Starter motor
    - Injector cover
    - Throttle body cover
    - Upper timing cover
    - Front timing cover
    - Throttle cable from the bracket
    - Vacuum hose for the brake servo
    - Turbocharger control valve from the air intake assembly
    - Mass air flow (MAF) sensor
    - EVAP hose
    - Upper section of the air intake assembly
    - Accessory drive belt
    - Camshaft position (CMP) sensor
    - Engine coolant temperature sensor
    - Ignition coils and ignition cables
    - Rear cover for the camshaft
    - Camshaft position (CMP) sensor housing
    - Intake manifold
    - Timing belt
    - Water pump inlet pipe
    - Camshafts
    - Valve lifters
    - Cylinder head

To install:

5. Install the cylinder head with a new gasket. Tighten the bolts in the following sequence:
    a. Step 1: 15 ft. lbs. (20 Nm)
    b. Step 2: 44 ft. lbs. (60 Nm)

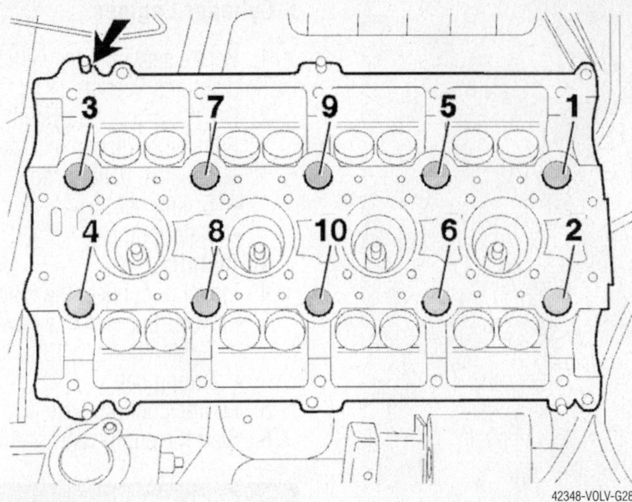

Cylinder head loosening sequence—4-Cylinder Engine

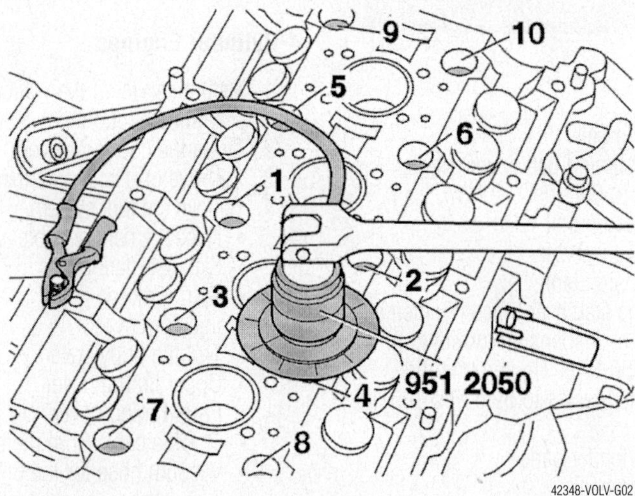

Cylinder head torque sequence—4-Cylinder Engine

c. Step 3: Plus 130 degrees

6. Install or connect the following:
   - Valve lifters
   - Camshafts
   - Water pump inlet pipe
   - Timing belt
   - Accessory drive belt
   - CMP sensor housing
   - Intake manifold
   - Ignition coils and ignition cables
   - CMP sensor
   - EVAP hose
   - MAF sensor
   - Turbocharger control valve from the air intake assembly
   - Brake servo vacuum hose
   - Throttle cable
   - Front timing cover
   - Upper timing cover
   - Exhaust manifold
   - Starter motor
   - Negative battery cable

7. Fill the cooling system.

8. Start the engine and check for leaks.

5-Cylinder Engines

1. Before servicing the vehicle, refer to the Precautions Section.
2. Drain the cooling system.
3. Remove or disconnect the following:
   - Negative battery cable
   - Engine appearance cover
   - Air intake assembly
   - Battery
   - Expansion tank
   - Right front wheel
4. Separate the exhaust system from the intake manifold.
5. Relieve the fuel pressure.
6. Remove or disconnect the following:
   - Engine stabilizer brace
   - Upper engine coolant hose
   - Mass air flow (MAF) sensor
   - Fuel rail
   - Ignition coils
   - Intake manifold
   - Accessory drive belt
   - Timing belt
   - Turbocharger, if equipped
   - Intake pipe for water pump
   - Camshafts
   - Cylinder head

To install:

7. Install the cylinder head with a new gasket. Tighten the bolts in the following sequence:
   a. Step 1: 15 ft. lbs. (20 Nm)
   b. Step 2: 44 ft. lbs. (60 Nm)
   c. Step 3: Plus 130 degrees

Cylinder head loosening sequence—5-cylinder engine

Cylinder head torque sequence—5-cylinder engines

8. Install or connect the following:
   - Camshafts
   - Intake pipe for water pump
   - Turbocharger, if equipped
   - Timing belt
   - Accessory drive belt
   - Expansion tank
   - Upper engine stabilizer brace
   - Intake manifold
   - MAF sensor
   - Ignition coils
   - Fuel rail
   - Upper engine coolant hose
   - Battery
   - Air intake assembly
9. Reassemble the exhaust system.
10. Install the right front wheel.
11. Connect the negative battery cable.
12. Fill the cooling system.
13. Start the engine and check for leaks.

6-Cylinder Engine

1. Before servicing the vehicle, refer to the Precautions Section.
2. Drain the cooling system.
3. Remove or disconnect the following:
   - Negative battery cable
   - Right axle half shaft
   - Nuts where the exhaust system connects with the turbocharger exhaust bends
   - First mounting bracket for the exhaust system downstream of the three-way catalytic converter (TWC) from the car body
   - Front Side Impact Protection System (SIPS) member from the car body
4. Separate the exhaust system from the intake manifold.
5. Remove or disconnect the following:
   - Engine stabilizer brace
   - Intake and charge air pipes above the engine
   - Upper engine coolant hose
   - Air intake assembly.
   - Accessory drive belt
   - Upper timing cover
   - Front timing cover.
   - Cover over the ignition coils.
6. Relieve the fuel pressure.
7. Remove or disconnect the following:
   - Intake manifold
   - Turbochargers, if equipped
   - Exhaust manifolds
   - Timing belt
   - Thermostat housing
   - Camshafts
   - Valve lifters

➡Mark up and position the lifters so that their original positions can be established.

   - Intake manifold
   - Coolant outlet pipe
   - Cylinder head

To install:

8. Install the cylinder head with a new gasket. Tighten the bolts in the following sequence:
   a. Step 1: 15 ft. lbs. (20 Nm)
   b. Step 2: 44 ft. lbs. (60 Nm)
   c. Step 3: Plus 130 degrees
9. Install or connect the following:
   - Valve lifters
   - Camshafts
   - Thermostat housing
   - Timing belt
   - Fuel rail
   - Intake pipe for coolant pump
   - Exhaust manifolds
   - Turbochargers, if equipped

Cylinder head loosening sequence—6-cylinder engine

Cylinder head torque sequence—6-cylinder engines

- Intake manifold
- Front timing cover
- Cover over ignition coils
- Upper timing cover
- Air intake assembly
- Intake and charge air pipes above the engine
- Upper engine coolant hose
- Engine stabilizer brace
- Assemble the exhaust assembly using new fasteners
- Front SIPS member
- Right axle half shaft
- Wheels
- Negative battery cable

10. Fill the cooling system.
11. Start the engine and check for leaks.

Turbocharger

REMOVAL & INSTALLATION

4-Cylinder Engines

1. Before servicing the vehicle, refer to the Precautions Section.
2. Remove or disconnect the following:
    - Air intake assembly
    - Rear engine cover
    - Heat shield from the intake manifold
    - Air intake hoses
    - Upper charge air pipe
3. Drain the cooling system.
4. Remove or disconnect the following:
    - Heat shield from the firewall
    - Turbo coolant hoses
    - Heated oxygen (HO_2S) sensors
    - Front exhaust pipe
    - Turbo oil lines
    - Hoses from the turbo control valve
    - Three-way catalytic converter
    - Turbocharger

To install:

5. Install or connect the following:
    - Turbo oil lines with new gasket
    - Hoses in the turbo control valve
    - Three-way catalytic converter
    - Turbocharger and tighten the fasteners to 33 ft. lbs. (45 Nm)
    - Turbo coolant hoses
    - Heated oxygen (HO_2S) sensors
    - Front exhaust pipe. Tighten fasters to 18 ft. lbs. (25 Nm).
    - Upper charge air pipe
    - Heat shield on firewall
    - Air intake hoses
    - Heat shield on intake manifold
    - Air intake assembly
    - Engine cover
6. Fill the cooling system.
7. Start the engine and check for leaks.

5-Cylinder engines, except 2005-06 S40/V50/C70 models

1. Before servicing the vehicle, refer to the Precautions Section.
2. Drain the cooling system.
3. Remove or disconnect the following:
    - Negative battery cable
    - Engine stabilizer brace
    - Heat shield
    - Air intake hose
    - Upper charge air pipe
    - Turbo coolant hose
    - Splash guard
    - Turbo oil lines
    - Catalytic converter
    - Right hand halfshaft
    - Drive shaft
    - Bevel gear
    - Turbocharger hoses
    - Turbocharger and exhaust manifold

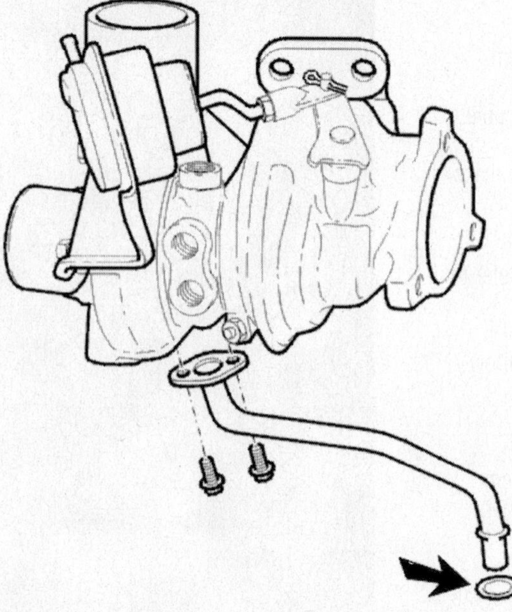

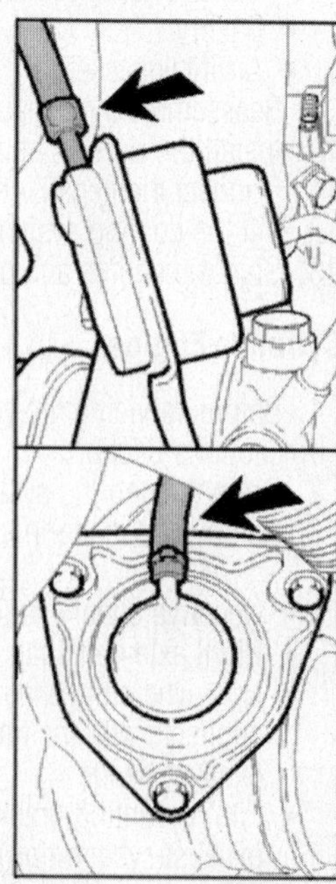

4-Cylinder engine turbocharger removal

To install:

4. Install the turbo oil lines. Tighten to 19 ft. lbs. (26 Nm).
5. Install the turbocharger and tighten the fasteners as follows:
   - Lower nuts where the turbocharger is connected to the manifold to 18 ft. lbs. (25 Nm)
   - Lower nut where the front exhaust pipe is connected to the turbocharger to 22 ft. lbs. (30 Nm)
   - Upper nuts where the turbocharger is connected to the manifold to 18 ft. lbs. (25 Nm)
   - Upper nuts where the front exhaust pipe is connected to the turbocharger to 22 ft. lbs. (30 Nm)
6. Install or connect the following:
   - Bevel gear
   - Drive shaft
   - Halfshaft
   - Catalytic converter
   - Turbo coolant hoses. Tighten to 19 ft. lbs. (26 Nm).
   - Air intake hose
   - Turbo oil lines. Tighten to 28 ft. lbs. (38 Nm).
   - Splash guard
   - Upper charge air pipe
   - Heat shield
   - Engine stabilizer brace. Tighten the bolts at the suspension turrets to 37 ft. lbs. (50 Nm). Tighten the engine bracket bolt to 59 ft. lbs. (80 Nm).
7. Fill the cooling system.
8. Start the engine and check for leaks.

2005-06 S40/V50/C70 Models

1. Before servicing the vehicle, refer to the Precautions Section.
2. Drain the cooling system.
3. Remove or disconnect the following:
   - EVAP canister hose
   - Air intake hose
   - Front heated oxygen sensor (HO_2S) connector
   - Upper timing belt cover
   - Heat deflector plate
   - Crankcase ventilation (CCV) hose from turbocharger (TC) intake
   - Lower charge air pipe
   - Bevel gear
   - Catalytic converter
   - Oil pressure pipe bolt on the TC and place the pipe to one side
   - Nuts and washers at the junction between the cylinder head and exhaust manifold
   - Oil return line and seal the opening in the cylinder block
   - Right halfshaft
   - Manifold and turbocharger as a unit

To install:

4. Install or connect the following:
   - Turbocharger and manifold with new gasket, but do not tighten
   - Oil return line with a new O-ring. Tighten nuts to 9 ft. lbs. (12 Nm)
   - Oil pressure pipe with new gaskets and tighten to 19 ft. lbs (26 Nm)
5. Tighten the mounting nuts on the manifold to 27 ft. lbs. (37 Nm).
6. Install or connect the following:
   - CCV hose to TC intake, using a new clamp
   - Heat deflector plate
   - Bevel gear
   - Catalytic converter
   - Front HO_2S connector
   - Halfshaft
   - Lower charge air pipe
   - Upper timing belt cover
   - Air intake hose
   - EVAP canister hose
7. Fill the cooling system.
8. Start the engine and check for leaks.

6-Cylinder Engines

1. Before servicing the vehicle, refer to the Precautions Section.
2. Remove or disconnect the following:
   - Engine stabilizer brace
   - Air charge pipe
   - Air intake pipe
   - Heat deflector plate
3. Drain the cooling system
4. Remove or disconnect the following:
   - Rear Side Impact Protection System (SIPS) member
   - Heated oxygen (HO_2S) wiring connectors
   - Catalytic converters
   - Right half shaft
   - Turbo coolant pipes
   - Turbo oil lines
   - Turbocharger

To install:

5. Install or connect the following:
   - Turbocharger and tighten the fasteners to 18 ft. lbs. (25 Nm)
   - Turbo oil lines. Tighten to 28 ft. lbs. (38 ft. lbs).
   - Turbo coolant pipes
   - Catalytic converters with new gaskets. Tighten fasteners to 18 ft. lbs. (25 Nm).
   - Rear SIPS member. Tighten to 18 ft. lbs. (25 Nm).
   - Heated oxygen (HO_2S) wiring connectors
   - Right half shaft

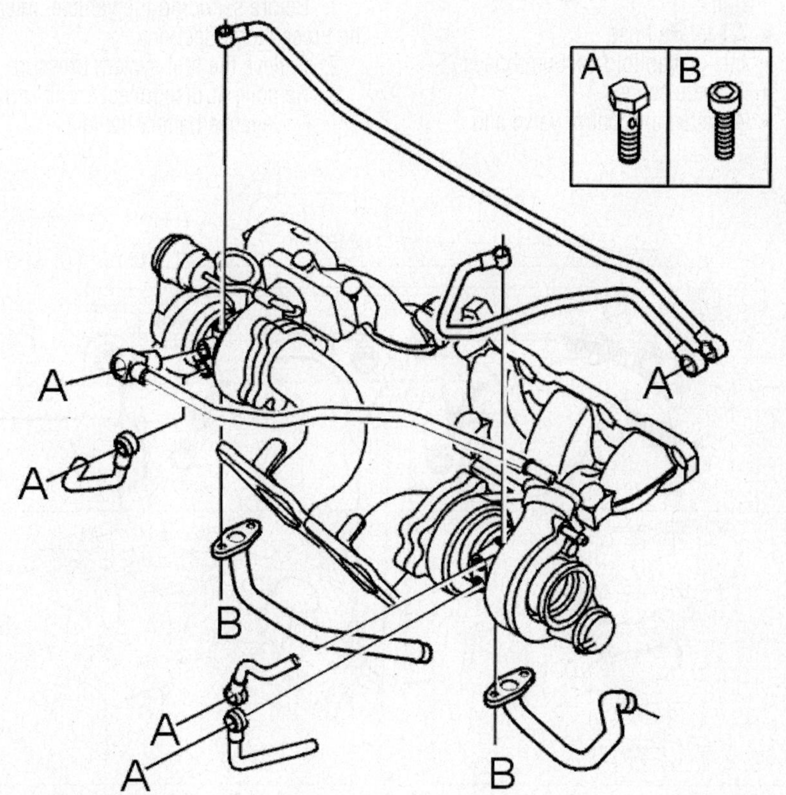

Mounting bolt locations for 6-cylinder turbo

- Heat deflector plate. Tighten to 18 ft. lbs. (25 Nm).
- Air intake pipe
- Air charge pipe
- Engine stabilizer brace. Tighten the bolts at the suspension turrets to 37 ft. lbs. (50 Nm). Tighten the engine bracket bolt to 59 ft. lbs. (80 Nm).

6. Fill the cooling system.
7. Start the engine and check for leaks.

Intake Manifold

REMOVAL & INSTALLATION

4-Cylinder Engines

1. Before servicing the vehicle, refer to the Precautions Section.
2. Relieve fuel system pressure.
3. Remove or disconnect the following:
   - Negative battery cable
   - Ignition coil cover
   - Crankcase ventilation hose
   - Throttle body cover
   - Injector connectors
   - Throttle body vacuum hose
   - Quick-release connector between the fuel line and the fuel rail using tool 951 2666 or equivalent.
   - Fuel rail and injectors as a single unit
   - Air intake hose
   - Intake manifold vacuum hoses
   - Throttle cable
   - Assisted air control valve and throttle position (TP) switch connectors
   - Brake servo hose
   - Intake manifold support bracket
   - Intake manifold

To install:

4. Install or connect the following:
   - Intake manifold with new gasket. Tighten the bolts to 15 ft. lbs. (20 Nm).
   - Intake manifold support bracket
   - Brake servo hose
   - Assisted air control valve and throttle position (TP) switch connectors
   - Throttle cable
   - Intake manifold vacuum hoses
   - Air intake hose
   - Fuel rail and injectors. Tighten fuel rail bolts to 7 ft. lbs. (10 Nm).
   - Fuel line
   - Throttle body vacuum hose
   - Crankcase ventilation hose
   - Injector connectors and protective cover
   - Throttle body cover
   - Ignition coil cover
   - Negative battery cable
5. Start the engine and check for leaks.

4-cylinder intake manifold bolts

5-Cylinder Engines, except 2005-06 S40/V50/C70 models

1. Before servicing the vehicle, refer to the Precautions Section.
2. Relieve the fuel system pressure.
3. Remove or disconnect the following:
   - Negative battery cable
   - Ignition coil cover
   - Crankcase ventilation hose
   - Air intake hose
   - Injector connectors
   - Charge air pipe, if equipped
   - Quick-release connector between the fuel line and the fuel rail using tool 951 2666 or equivalent.
   - Fuel rail and injectors as a single unit
   - Intake manifold vacuum hoses
   - Dipstick tube bracket
   - Intake manifold support bracket
   - Throttle body (TB) connector
   - Intake manifold

To install:

4. Install or connect the following:
   - Intake manifold with new gasket. Tighten the bolts to 15 ft. lbs. (20 Nm).
   - Intake manifold support bracket
   - Fuel rail and injectors. Tighten fuel rail bolts to 7 ft. lbs. (10 Nm).
   - Injector connectors and protective cover
   - Fuel line
   - Crankcase ventilation hose
   - Intake manifold vacuum hoses
   - Charge air pipe, if equipped
   - Air intake hose
   - Ignition coil cover
   - TB connector
   - Dipstick tube bracket
   - Negative battery cable
5. Start the engine and check for leaks.

2005-06 S40/V50/C70 models

1. Before servicing the vehicle, refer to the Precautions Section.
2. Relieve the fuel system pressure.
3. Remove or disconnect the following:
   - Negative battery cable
   - Air intake assembly
   - Front 6 bolts for upper intake manifold and 2 rear bolts
   - EVAP canister hose
   - Brake vacuum hose
   - Throttle body (TB) connector
   - Fuel pressure sensor connector
   - Upper intake manifold
   - Fuel injector connectors
   - Fuel pressure sensor in fuel rail
   - Fuel injectors and fuel rail as a unit
   - Lower intake manifold mounting bolts
   - Lower intake manifold

To install:

4. Install or connect the following:
   - Lower intake manifold using a new gasket. Tighten 2 lower bolts to 14 ft. lbs. (19 Nm)

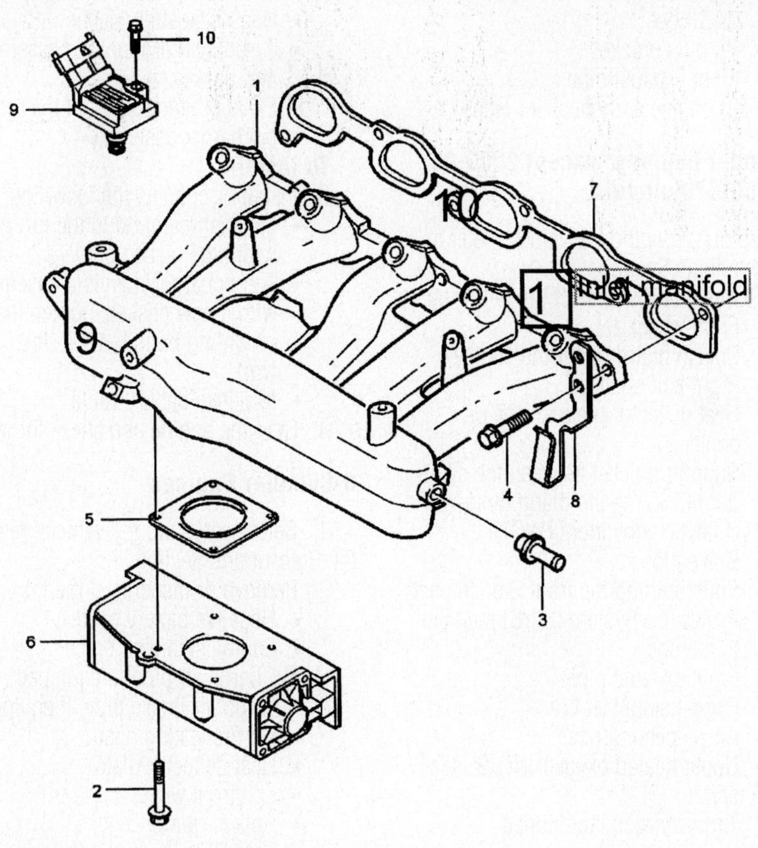

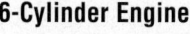

1 Inlet manifold
2 Flange screw
3 Nipple
4 Flange screw
5 Gasket
6 Throttle body
7 Gasket
8 Bracket
9 Map sensor
10 Six point socket screw

42348-XC90-G14

Intake manifold—5-cylinder engine

➡Route CCV hose up between cylinders 2 and 3.

- Fuel rail with fuel injectors
- Fuel injector connectors
- Fuel pressure sensor connector
- Upper intake manifold with new gasket. Tighten 6 upper bolts starting with the center and working outward to 7 ft. lbs. (10 Nm)
- EVAP canister hose
- Brake vacuum hose
- Air intake assembly
- Negative battery cable

5. Start engine and check for leaks.

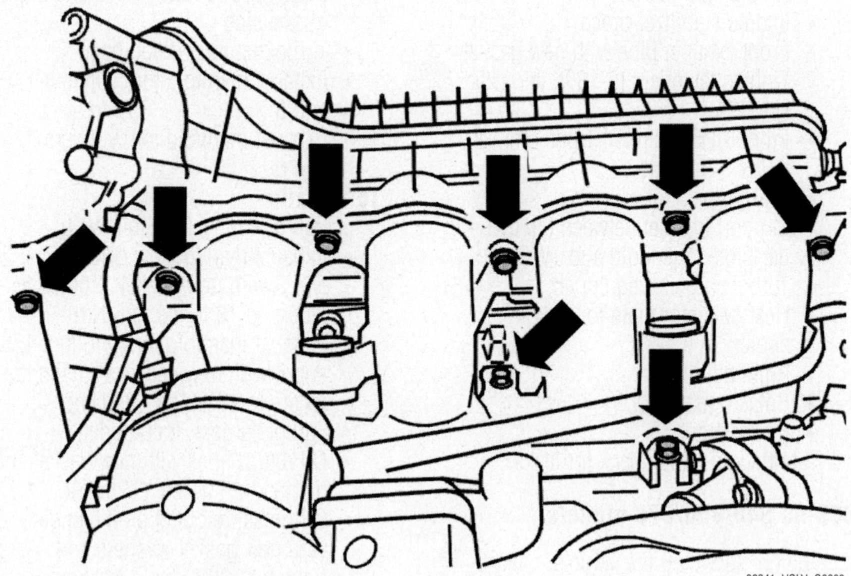

Intake manifold six upper and two lower mounting bolts—2005–06 S40/V50/C70

6-Cylinder Engine

1. Before servicing the vehicle, refer to the Precautions Section.
2. Relieve the fuel system pressure.
3. Remove or disconnect the following:
   - Negative battery cable
   - Charge air pipe, if equipped
   - Charge air cooler (CAC) hose, if equipped
   - Air intake hose
   - Injector connectors
   - Fuel rail and injectors as a single unit
   - Quick-release connector between the fuel line and the fuel rail using tool 951 2666 or equivalent
   - Dipstick tube bracket
   - Crankcase ventilation hose
   - Canister purge (CP) valve hose
   - Engine cooling fan shroud, if equipped
   - Engine cooling fan connector, if equipped
   - Throttle body
   - Intake manifold vacuum hoses
   - Upper intake manifold bolts
   - Outer intake manifold bolts from lower row
4. Slacken off the remaining bolts and remove the intake manifold.

5. Remove the remaining bolts and gasket.

To install:

6. Install or connect the following:
   - Intake manifold with new gasket. Tighten bolts to 15 ft. lbs. (20 Nm).
   - Throttle body, with new gasket.
   - Intake manifold vacuum hoses
   - Engine cooling fan connectors, if equipped
   - Engine cooling fan shroud, if equipped
   - Fuel rail and injectors. Tighten fuel rail bolts to 84 inch lbs. (10 Nm).
   - Injector connectors
   - CP valve hose
   - Crankcase ventilation hose
   - Dipstick tube bracket
   - Air intake hose
   - CAC hose, if equipped
   - Charge air pipe, if equipped
   - Negative battery cable
7. Start engine and check for leaks.

Exhaust Manifold

REMOVAL & INSTALLATION

4-Cylinder Engine

1. Before servicing the vehicle, refer to the Precautions Section.
2. Remove or disconnect the following:
   - Heat deflector plate
   - Exhaust manifold bracket
   - Catalytic converter
   - Turbocharger
   - Rear engine cover
   - Front exhaust pipe
   - Oil pipe bracket
   - Upper oil pipe connection
   - Nuts for the exhaust manifold
   - Air cleaner hose
   - Charge air cooler (CAC) hose
   - Lower oil pipe
   - Exhaust manifold

To install:

3. Install or connect the following:
   - Exhaust manifold with new gaskets. Tighten bolts to 18 ft. lbs. (25 Nm).
   - Exhaust manifold support bracket. Tighten to 17 ft. lbs. (24 Nm).
   - Lower oil pipe
   - Turbocharger
   - Catalytic converter
   - Upper oil pipe connection. Tighten to 17 ft. lbs. (24 Nm).
   - Heat deflector plate
   - Air cleaner hose
   - CAC hose
   - Oil pipe bracket
   - Front exhaust pipe
4. Start engine and check for leaks.

5-Cylinder Engines, except 2005-06 S40/V50/C70 models

1. Before servicing the vehicle, refer to the Precautions Section.
2. Remove or disconnect the following:
   - Front wheel
   - Splash guard under the engine
   - Right half shaft.
   - Heat deflector plate for the oil cooler
   - Support bracket between the cylinder block, manifold and three-way catalytic converter (TWC)
   - Brake pipe
   - Bolts holding the front Side Impact Protection System (SIPS) member to the body
   - Front exhaust pipe
   - Engine stabilizer brace
   - Air preheating hose
   - Upper heated oxygen (HO_2S) sensor
   - Turbocharger, if equipped
   - Upper heat deflector plate over the manifold
   - Exhaust manifold with the TWC

To install:

3. Install or connect the following:
   - Manifold with new gasket and TWC. Tighten fasteners to 18 ft. lbs. (25 Nm).
   - Heat deflector plate over the manifold
   - Upper heated oxygen (HO_2S) sensor
   - Air preheating hose
   - Engine stabilizer brace
   - Front exhaust pipe with new gasket. Tighten fasteners to 18 ft. lbs. (25 Nm)
   - Front SIPS member. Tighten to 18 ft. lbs. (25 Nm).
   - Brake pipe
   - Support bracket between the cylinder block, manifold and TWC
   - Turbocharger, if equipped
   - Heat deflector plate for the oil cooler
   - Right half shaft
   - Splash guard
   - Front wheel
4. Start engine and check for leaks.

2005-06 S40/V50/C70 models

1. Before servicing the vehicle, refer to the Precautions Section.
2. Remove or disconnect the following:
   - Negative battery cable
   - Turbocharger/manifold assembly and gasket
   - Exhaust manifold from the turbocharger assembly

To install:

3. Install or connect the following:
   - Exhaust manifold to the turbocharger assembly
   - Turbocharger/manifold assembly with a new gasket. Tighten the mounting bolts to 27 ft. lbs. (37 Nm).
   - Negative battery cable
4. Start the engine and check for leaks

6-Cylinder Engines

1. Before servicing the vehicle, refer to the Precautions Section.
2. Remove or disconnect the following:
   - Negative battery cable
   - Engine stabilizer brace
   - Charge air pipe, if equipped
   - Turbo air intake pipe, if equipped
   - Air preheating hose
   - Heat deflector plate
   - Left front wheel
   - Splash guard
   - Rear Side Impact Protection System (SIPS) member, if equipped
   - Catalytic converters
3. Drain the cooling system.
4. Remove or disconnect the following:
   - Right hand half shaft
   - Bolts for coolant hoses and place to one side
   - Bolts for oil pressure outlet on the engine
   - Bolts for oil return hoses and place to one side
   - Turbochargers, if equipped
   - Exhaust manifold for cylinders 1, 2 and 3
   - Exhaust manifold for cylinders 4, 5 and 6

To install:

5. Install or connect the following:
   - Exhaust manifold for cylinders 1, 2 and 3 with new gasket. Tighten fasteners to 18 ft. lbs. (25 Nm).
   - Exhaust manifold for cylinders 4, 5 and 6 with new gasket. Tighten fasteners to 18 ft. lbs. (25 Nm).
   - Turbochargers, if equipped
   - Oil return lines with new gasket and tighten to 9 ft. lbs. (12 Nm)
   - Oil pressure outlet on the engine with new gasket washers and tighten to 28 ft. lbs. (38 Nm)
   - Coolant hoses with new gasket

washers and tighten to 19 ft. lbs. (26 Nm).
- Catalytic converters. Tighten fasteners to 18 ft. lbs. (25 Nm).
- Rear SIPS member. Tighten to 18 ft. lbs. (25 Nm).
- Splashguard
- Front wheel
- Heat deflector plate. Tighten bolts to 18 ft. lbs. (25 Nm).
- Air preheating hose
- Engine stabilizer brace. Tighten the bolts at the suspension turrets to 37 ft. lbs. (50 Nm). Tighten the engine bracket bolt to 59 ft. lbs. (80 Nm).
- Turbo air intake pipe, if equipped
- Charge air pipe, if equipped
- Negative battery cable

6. Start engine and check for leaks.

Front Crankshaft Seal

REMOVAL & INSTALLATION

5-Cylinder Engines, except 2005-06 S40/V50/C70 Models

1. Before servicing the vehicle, refer to the Precautions Section.
2. Remove or disconnect the following:
   - Engine stabilizer brace
   - Upper timing belt cover
   - Servo reservoir
   - Expansion tank
   - Accessory drive belt
   - Front timing belt cover
   - Front wheel
   - Timing belt
   - Vibration damper
   - Crankshaft timing gear pulley
   - Front crankshaft seal

To install:

3. Install or connect the following:
   - Front crankshaft seal
   - Crankshaft timing gear pulley
   - Vibration damper and tighten nut to 133 ft. lbs. (180 Nm)
   - Timing belt
   - Front wheel
   - Front timing belt cover
   - Upper timing belt cover
   - Accessory drive belt
   - Expansion tank
   - Servo reservoir
   - Engine stabilizer brace. Tighten the bolts at the suspension turrets to 37 ft. lbs. (50 Nm). Tighten the engine bracket bolt to 59 ft. lbs. (80 Nm).
4. Start the engine and check for leaks.

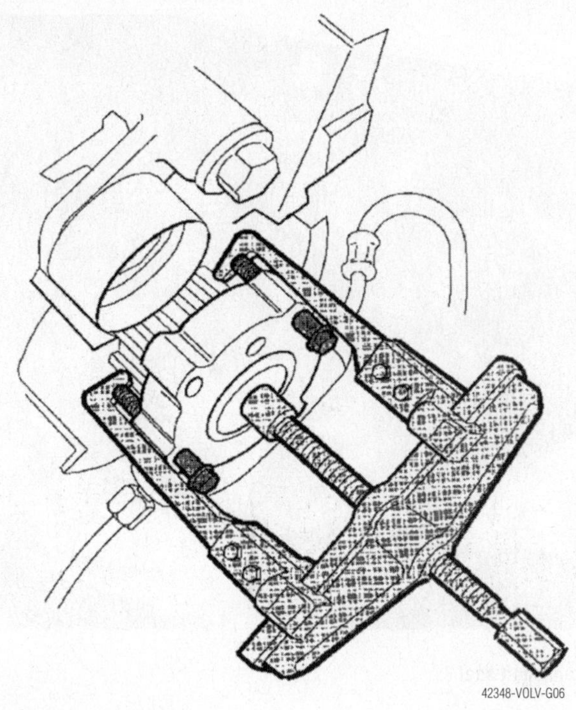
42348-VOLV-G06

Removing crankshaft timing gear pulley

2005-06 S40/V50/C70 Models

1. Before servicing the vehicle, refer to the Precautions Section.
2. Remove or disconnect the following:
   - Timing belt
   - Crankshaft pulley using special tool 999-5304
   - Front crankshaft seal using special tool 999-5919 or equivalent groove-cut chisel

To install:

3. Lubricate the seal mating surfaces with clean engine oil.
4. Install or connect the following:
   - Front crankshaft seal using special tool 999-5455
   - Crankshaft pulley
   - Timing belt

67200-VOLV-G112

Crankshaft pulley removal tool

Installing the crankshaft seal

6-Cylinder Engine

1. Before servicing the vehicle, refer to the Precautions Section.
2. Remove or disconnect the following:
   - Timing belt
   - Timing gear pulley
   - Front crankshaft seal

To install:

3. Install or connect the following:
   - Front crankshaft seal
   - Timing gear pulley
   - Timing belt

Camshaft and Valve Lifters

REMOVAL & INSTALLATION

1. Before servicing the vehicle, refer to the Precautions Section.
2. Remove or disconnect the following:
   - Negative battery cable
   - Accessory drive belts
   - Front cover
   - Timing belt
   - Ignition coil cover
   - Camshaft Position (CMP) sensor or distributor, as equipped
   - Switch holder and shield at left rear of the engine
   - Ignition coils
   - Camshaft sprockets
   - Cylinder head cover
   - Camshafts and lifters

➡Keep all valvetrain components in order for assembly.

To install:

3. Lubricate the valve lifters and camshaft bearing positions and lobes with clean engine oil.
4. Install the valve lifters in their original positions.
5. Install the intake camshaft. Make sure that the groove in the rear edge of the camshaft extends across an imagined center line.
6. Install the exhaust camshaft. Make sure that the groove in the rear edge of the camshaft extends across an imagined center line.
7. Tighten the camshaft cover bolts alternately, keeping it parallel to the cylinder head using the press tools.
8. Using Special Tool 0951-2767 or equivalent, apply liquid gasket to the mating surface of the cylinder head cover.

✱✱ WARNING

Ensure no liquid gasket material gets in the oil ducts of the cylinder head cover.

9. Position the Press Tool 999-5454 in the spark plug holes. Fit the two M6 bolts under each press tool to set the right height.
10. Crosswise tighten the cylinder head cover parallel to the cylinder head with the press tool. Install all the bolts and tighten from the middle out to the edges. Tighten the bolts to 13 ft. lbs. (17 Nm). Remove the press tool and bolts under the tool.
   - Camshaft sprockets and tighten the bolts to 15 ft. lbs. (20 Nm)
   - Ignition coils
   - Switch holder and shield at left rear of the engine
   - CMP sensor or distributor, as equipped
   - Ignition coil cover
   - Timing belt
   - Front cover
   - Accessory drive belts
   - Negative battery cable

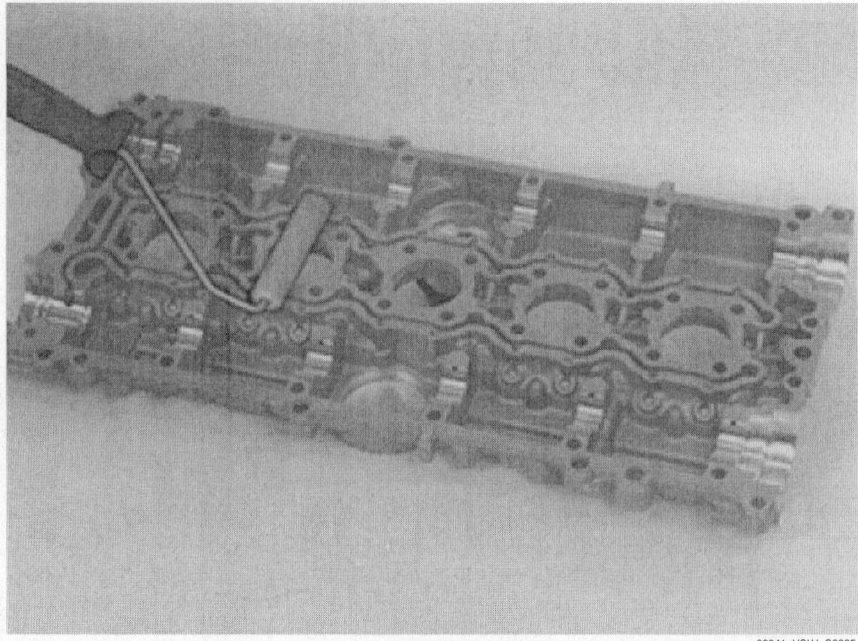

Using Special Tool 0951-2767 to apply the liquid gasket to the cylinder head cover.

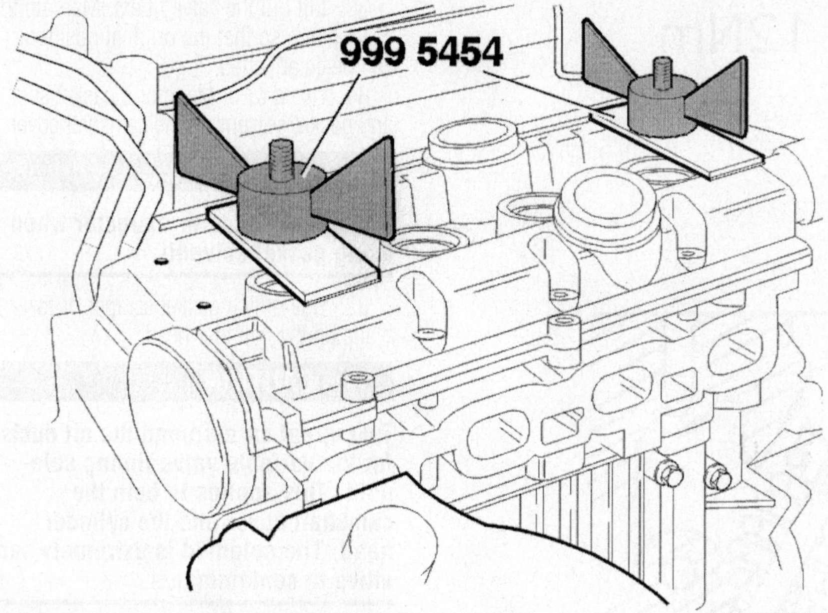

Using Special Tool 999-5454 to install the cylinder head cover.

Valve Lash

ADJUSTMENT

All engines covered use hydraulic lash adjusters. No adjustment is necessary. However, when the camshaft is replaced, the following procedures must be followed.

1. Remove the cable from the battery negative terminal.
2. Remove:
   - The cross stay between the suspension turrets
   - The ground strip from the cylinder head
   - The upper engine stabilizer brace
   - The cover in the cylinder head at the rear of the exhaust camshaft
   - The crankcase ventilation hose from the top of the camshaft cover
   - The radiator breather tube from the expansion tank. Install lock grip pliers.
3. Lift up the brake fluid reservoir.
4. Disconnect the ABS sensor connector.
5. Place the brake fluid reservoir over the engine.

⁂ WARNING

Ensure that no fluid is spilled on the engine. It is extremely flammable!

6. Disconnect the connector for the level sensor in the expansion tank.
7. Lift up and place the expansion tank on top of the engine.
8. Remove:
   - The auxiliary's belt
   - The front timing cover.
9. Position the upper timing cover.
10. Turn the crankshaft clockwise until the markings on the crankshaft belt pulley and the timing belt pulley are aligned with the markings on the oil pump and the upper timing cover.
11. Remove the upper timing cover.

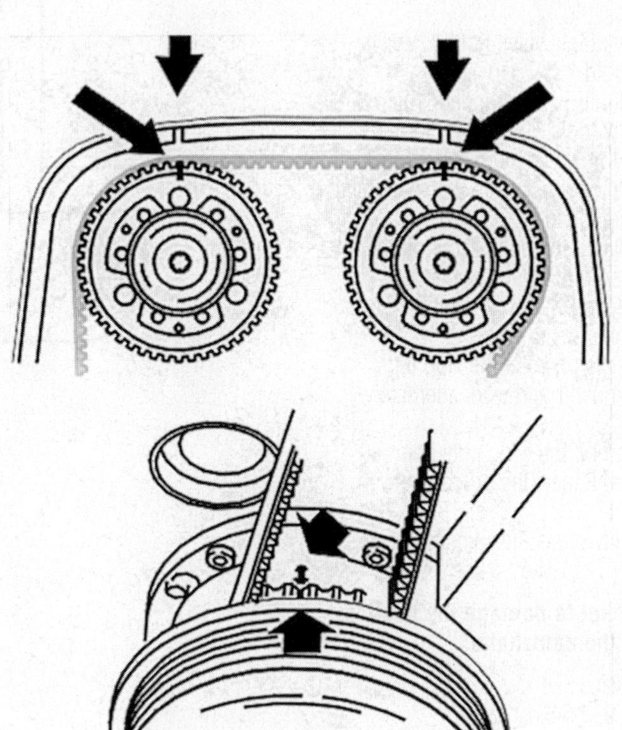

Setting the timing marks

⁂ CAUTION

Crankshaft and camshafts must not be turned more than is stated in the method description. If the shafts are turned in any other way the valves may be damaged.

12. Slacken off the center bolt for the belt tensioner slightly.
13. Hold the center bolt still. Turn the tensioner eccentric clockwise to 10 o'clock using a 6 mm Allen key.
14. Remove the timing belt from the camshaft pulleys.
15. Install camshaft adjustment tool 999 5452 at the rear of the camshafts.
16. Check that the bolts securing the adjustment tool to the camshafts and the bolt holding the tool together are well tightened
17. Remove: (timing gear pulley with variable valve timing unit)
    - The plug at the front of the variable valve timing unit
    - The center bolt in the variable valve timing unit. Carefully pull out the variable valve timing unit with the timing gear pulley.
18. Remove: (timing gear pulley without variable valve timing unit)
    - The bolts securing the timing gear pulley on the camshaft
    - The timing gear pulley.

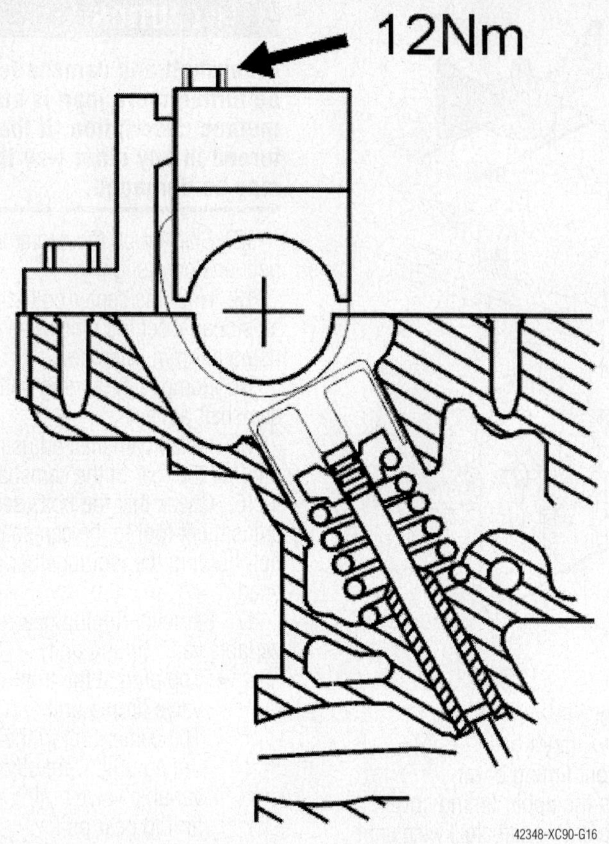

Camshaft press installed

19. Remove tool 999 5452.
20. Reinstall the expansion tank and the brake fluid reservoir at the fender liner.
21. Remove:
    - The variable valve timing (VVT) solenoid
    - spark plugs for cylinders 1 and 5.
22. Install 2 tools 999 5454. Leave a 2–3 mm gap to the camshaft cover.
23. Ensure that the bolt in the spark plug well is fully tightened.
24. Remove all the bolts securing the camshaft cover to the cylinder head.
25. Use pliers 999 5670 to lift the cover from the cylinder head.
26. Install the pliers at the stop lugs. Start with cylinder 1 and work alternately backward.
27. Slacken off the wing nuts approximately 2 turns. Repeat the procedure with the pliers.
28. Carefully press the camshaft seals free.

➡Take care not to damage the sealing surfaces on the camshafts.

29. Remove:
    - Tool 999 5454
    - The camshaft cover
    - The camshafts.

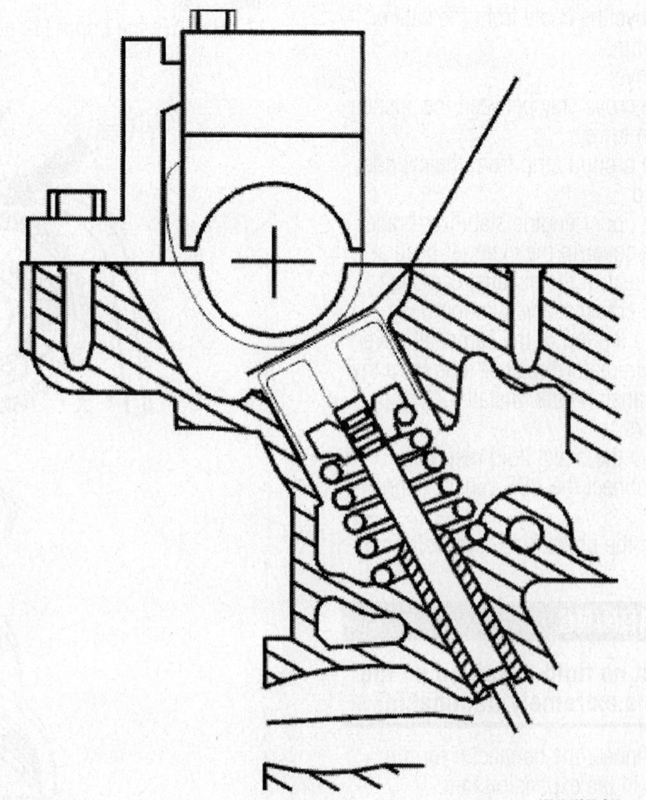

Checking the valve clearance

30. Lift out the valve lifters. Mark up the valve lifters so that the original positions can be established.
31. Use a razor blade or a gasket scraper and gasket solvent on the camshaft cover.

✻✻ WARNING

Use a fume hood or extractor when using gasket solvent!

32. Use only a gasket scraper or razor blade on the cylinder head.

✻✻ CAUTION

Take great care around the oil ducts for the variable valve timing solenoid. This applies to both the camshaft cover and the cylinder head. The solenoid is extremely sensitive to contaminants.

33. Dry and blow all surfaces clean.
34. Carefully tap the end of the valve stem to ensure that the valve is correctly located in the seat.
35. Use a plastic, aluminum or brass drift to protect the valve and the surface of the valve lifter.
36. The sound made by tapping reveals if the valve is correctly seated.

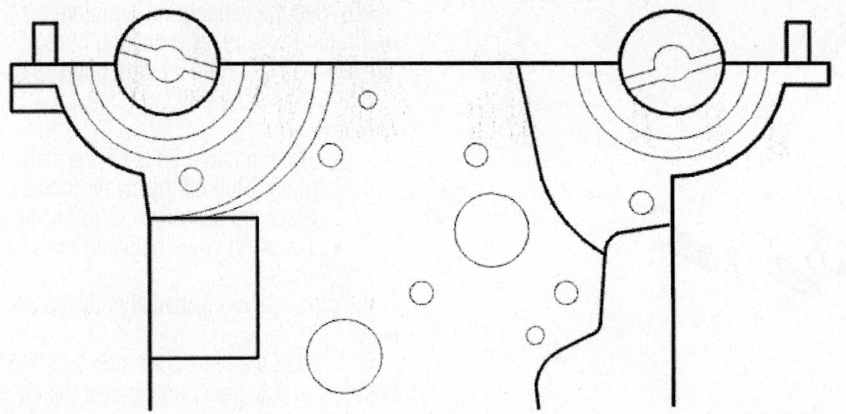

Camshafts properly positioned

37. Install both the valve lifters for the inlet valves at cylinder 1.

38. Check the notes made earlier. Select new valve lifters if necessary.

➡Only install two valve lifters. The valve lifters should be placed at the same cylinder!

39. Other valve lifters are available as replacement part/replacement part kits.

40. The valve clearance on a cold engine, approximately 68°F (20°C) should be:

- Inlet valve: 0.008 +/- 0.001 inch (0.20 +/- 0.03 mm).
- Exhaust valve: 0.016 inch +/- 0.001 inch (0.40 +/- 0.03 mm).

➡The tolerances are less at setting! When checking the valve clearance through the plug hole the tolerances are larger.

41. Position the intake camshaft. Ensure that the lobes at cylinder 1 point upwards.

42. Apply a little oil to the cam lobe and the upper side of the valve lifter to facilitate later measurement.

43. Install the lower section of camshaft press 999 5765 at the inlet valves for cylinder 1.

44. Tighten the tool against the cylinder head. Tighten to 13 ft. lbs. (17 Nm).

45. Turn the camshaft until it stops against the camshaft press.

46. Install the upper section of the camshaft press.

47. Tighten the bolt which tensions the camshaft. Tighten to 84 inch lbs. (12 Nm).

➡Measurements should only be taken on a cold engine. A suitable temperature is approximately 68°F (20°C).

48. Using a feeler gauge, press with a finger so that the feeler gauge lies parallel to the upper side of the valve lifter.

49. Move the feeler gauge sideways when taking the reading in order to obtain as accurate a measurement as possible.

50. The valve clearance measured on cold engines (approximately 68°F (20°C) should be:

- Inlet valve: 0.008 inch +/- 0.001 inch (0.20 +/- 0.03 mm).
- Exhaust valve: 0.016 inch +/- 0.001 inch 0.40 +/- 0.03 mm.

➡The tolerances are less at setting! When checking the valve clearance through the plug hole the tolerances are larger.

51. Differences in valve clearance for different engines/ambient temperatures:

- -0.0004 inch (- 0.01 mm) at 59°F (15°C)
- +0.0004 inch (+ 0.01 mm) at 77°F (25°C)
- +0.0008 inch (+ 0.02 mm) at 86°F (30°C)
- +0.0012 inch (+ 0.03 mm) at 95°F (35°C)
- +0.0016 inch (+ 0.04 mm) at 113°F (45°C)

52. Correcting measured clearance:

a. Lift out the upper section of the press tool.

b. Lift out the camshaft.

c. Adjust the play by replacing the valve lifters.

d. Other valve lifters are available as replacement part / replacement part kits.

e. Reinstall the camshaft and the upper section of the press tool. Tighten to 84 inch lbs. (12 Nm).

f. Take a new measurement.

53. When the correct valve clearance is reached, remove:

- The press tool 999 5765
- The camshaft
- The valve lifters.

54. Carefully mark the valve lifters so that exact reinstallation can be carried out.
For example:
Intake side: I1, I2, I3........I10.
Exhaust side: A1, A2, A3........A10.

55. Repeat the procedure for measuring the valve clearance for all cylinders on both the intake and exhaust sides.

56. Lubricate the valve guide wells.

57. Install all the valve lifters.

58. Lubricate the camshaft bearing seats and the upper sides of the valve lifters.

59. Position the intake camshaft. Ensure that the groove at the rear edge of the camshaft is above an imaginary center line.

60. Position the exhaust camshaft. Ensure that the groove at the rear edge of the camshaft is below an imaginary center line.

61. Wipe the oil film off the mating surfaces on the camshaft cover and cylinder head.

62. Install new O-rings around the spark plug wells at the cylinder head.

63. Apply liquid gasket 1161 059 on the camshaft cover. Use roller 951 2767.

➡The surface must be completely covered without any excess.

⁂ CAUTION

Ensure that no liquid gasket gets in to the oil ducts.

64. Lubricate the camshaft lobes, the camshaft bearing surfaces and the valve lifters.

65. Install the camshaft cover.

66. Install press tool 999 5454 (2x).

67. Tighten the camshaft cover bolts alternately, keeping it parallel to the cylinder head using the press tools.

68. Install all the bolts. Tighten the bolts from the middle and outwards.

69. Remove the press tool 999 5454

70. Install:

- The variable valve timing (VVT) solenoid. Use a new gasket
- The spark plugs. Tighten to 22 ft. lbs. (30 Nm)
- The plugs for the test holes. Tighten to 15 ft. lbs. (20 Nm)
- The crankcase ventilation hose to the top of the camshaft cover
- The ignition coils according to the earlier marking
- The ground terminals between the ignition coils.

71. To clean the shaft journal and mating surface, use emery cloth.

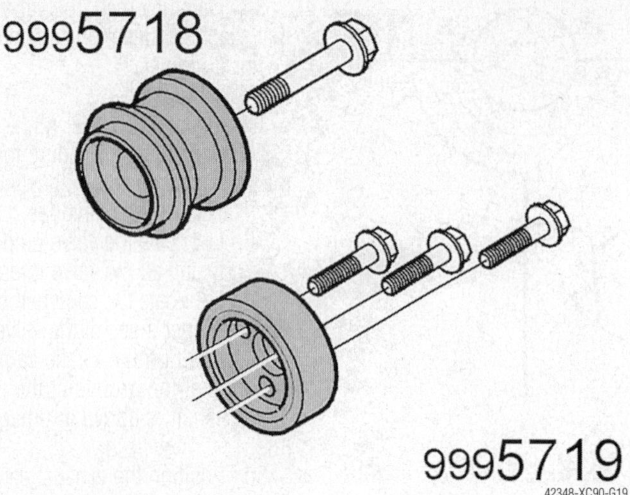

Front seal installation tools

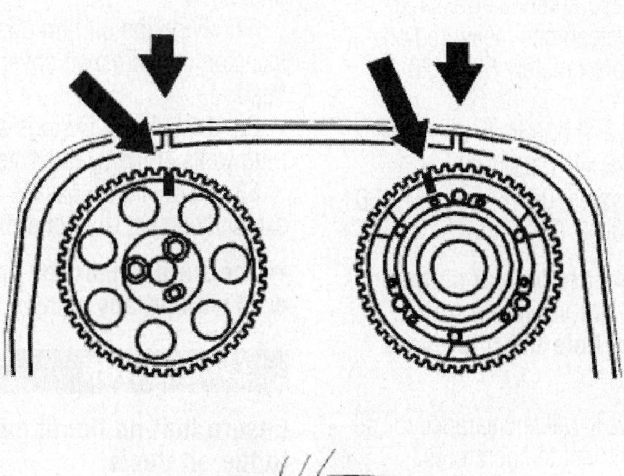

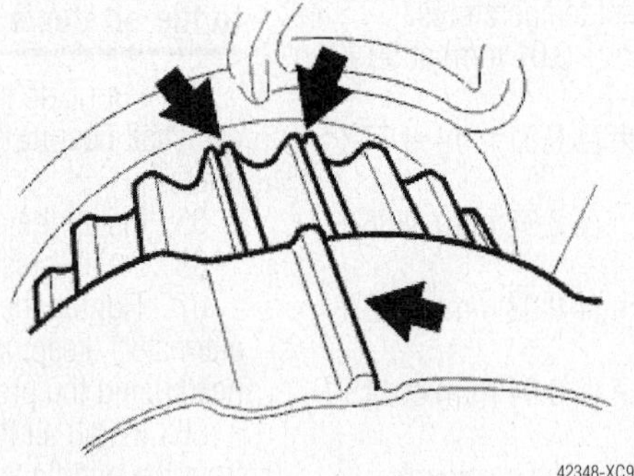

Timing marks, with variable timing units

➡When cleaning work around the shaft journal, not in and out. It is essential that any residue from the emery cloth and any other contaminants are completely removed before the new sealing ring is installed.

72. Use drift 999 5450.
73. Lubricate the surface of the seal that the camshaft rotates against.
74. Press in the seal until the drift bottoms out.

➡If there are wear grooves on the camshaft, the seal can be pressed in a further 2 mm by reversing the sleeve.

75. Install camshaft adjustment tool 999 5452 at the rear of the camshafts.
76. Check that the bolts securing the adjustment tool to the
77. Lift up and position the brake fluid reservoir and the expansion tank on top of the engine.
78. Use drift 999 5718 on camshafts with variable valve timing units. Use drift 999 5719 on camshafts without variable valve timing.
79. Use new seals and lubricate the surface of the seal that the camshaft rotates against.
80. Use the variable valve timing unit/timing belt pulley mounting bolts. Tighten the bolts until the drift bottoms out.
81. Remove the drift.
82. Remove:
   - The mounting bolts for the starter motor. Pull off the starter motor. Place the starter motor to one side
   - The blind cover plug and the seal washer.
83. Turn the crankshaft slightly clockwise.
84. Install the crankshaft stop 999 5451 Ensure that it bottoms out against the cylinder block.
85. Turn the crankshaft counter-clockwise until it stops against the crankshaft stop.
86. Check that the marking on the crankshaft timing gear pulley corresponds with the marking on the oil pump.

➡The purpose of the section is to ensure that the VVT unit is correctly positioned and to reset the camshaft timing gear pulley to the correct position using the markings made at the factory. This is to ensure that the conditions are correct for any later fault-tracing.

87. Slacken off, but do not remove the bolts which secure the timing gear pulley to the variable valve timing unit.
88. Press the variable valve timing unit/timing gear onto the camshaft.
89. Install the center bolt which secures the variable valve timing unit to the camshaft. Tighten slightly.
90. Turn the variable valve timing unit counter-clockwise as far as it will go.
91. Remove the center bolt.
92. Position the upper timing cover.
93. Turn the timing gear pulley clockwise until the bolts at the oval holes are in the limit position.
94. Continue turning clockwise until the timing gear pulley marking is 1 tooth before the marking on the upper timing cover.

⁂ CAUTION

Do not turn counter-clockwise during this procedure.

95. Check that the timing gear pulley is still in its limit position at the oval holes.
96. Tighten the center bolt for the variable valve timing (VVT) unit. Tighten to 88 ft. lbs. (120 Nm). Check that the variable valve timing unit does not rotate when tightening.
97. Install the center bolt. Tighten to 26 ft. lbs. (35 Nm).

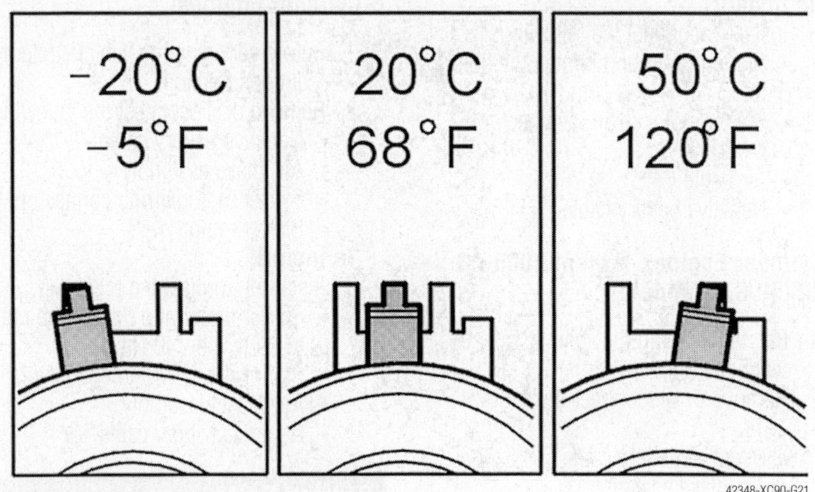

Tensioning indicator at different temperatures

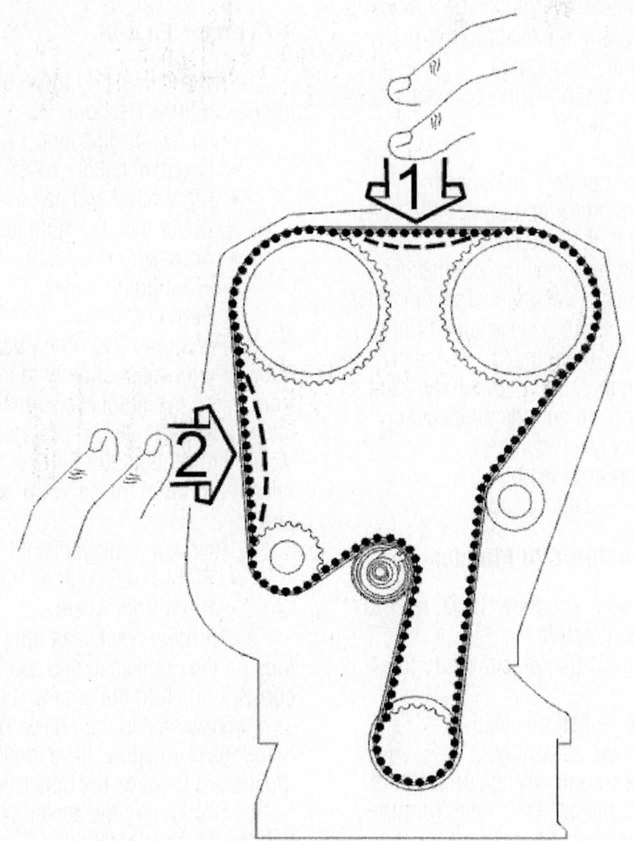

Checking belt tension

98. Install the timing gear pulley. Install the bolts.

99. Install two bolts without tightening. Allow the third bolt to protrude.

100. Adjust the timing gear pulleys so that the markings on the timing gear pulleys / upper timing cover correspond.

101. Tighten the center bolt on the belt tensioner. Tighten to 44 inch lbs. (5 Nm).

102. Turn the variable valve timing unit clockwise to the stop.

103. Hold it secure in the limit position.

104. Install the belt in the following order:

- Crankshaft
- The idler pulley
- Intake cam
- Exhaust cam
- Water pump
- Belt tensioner

➡Adjust the timing gear pulleys so that the bolts are not at a limit position in the oval holes. Also check that the markings correspond.

➡This adjustment is always carried out on a cold engine. A suitable temperature is approximately 68°F (20°C).

At higher temperatures (with the engine at operating temperature or a high outside temperature for example) the indicator is further to the right.

The illustration shows the position of the indicator when aligning the timing belt tensioner at different temperatures.

105. Hold the center bolt secure and turn the belt tensioner eccentric counter-clockwise until the tensioner indicator passes the marked position.

➡Check that the variable valve timing unit is in its limit position.

106. Tighten the three bolts at the intake camshaft timing gear pulley. Tighten to 74 inch lbs. (10 Nm).

107. Tighten the three bolts at the exhaust camshaft timing gear pulley. Tighten to 15 ft. lbs. (20 Nm).

108. Turn the eccentric on the belt tensioner back so that the indicator reaches the marked position in the center of the window. Remember to hold the center bolt secure at the same time.

109. Hold the eccentric secure and tighten the center bolt. Tighten to 15 ft. lbs. (20 Nm).

110. Check that the indicator is in the correct position.

111. Remove camshaft adjustment tool 999 5451 and crankshaft stop 999 5452.

112. Install the plug with a new blind cover plug. Tighten to 29 ft. lbs. (40 Nm).

113. Press the timing belt to check that the indicator on the tensioner moves easily.

114. Turn the crankshaft two turns. Check that the markings on the crankshaft timing gear pulley and the camshaft timing gear pulley match up with the markings on the oil pump and upper timing cover respectively.

115. Check that the indicator on the belt tensioner is within the marked position.

116. Remove the upper timing cover.

117. Install:

- front timing cover
- The auxiliary's belt
- The expansion tank
- The servo oil reservoir

- The bleed hose for the expansion tank

118. Close the clamp and check that the hoses lie correctly.

119. Connect the connectors for the:
- The ABS sensor by the right suspension turret
- The coolant level sensor in the expansion tank

120. Install:
- The starter motor
- The plastic nuts for the cover in the fender liner
- The right front wheel

121. Install:
- The trigger wheel.

➡Ensure that the trigger wheel is correctly positioned against the camshaft.

- The camshaft position sensor housing
- The cover at the rear of the exhaust camshaft
- The bolts holding both the ground strips from the firewall to the cylinder head
- The upper engine stabilizer brace
- The cover over the ignition coils
- The upper timing cover
- The crankcase ventilation hose on the inlet hose
- The inlet hose between the air cleaner and throttle body
- The engine stabilizer brace between the suspension turrets. Secure the servo hose at the right mounting for the engine stabilizer brace in the bodywork
- The cable to the battery negative terminal.

122. Check the level in the expansion tank and the brake fluid reservoir.

123. Test drive the engine until the thermostat opens and check for any leakage.

Starter Motor

REMOVAL & INSTALLATION

4-Cylinder Engines

1. Before servicing the vehicle, refer to the Precautions Section.
2. Remove or disconnect the following:
- Negative battery cable
- Air baffle
- Charge air cooler (CAC) pipe
- Splash guard
- Starter motor wiring connectors
- Starter motor

To install:

3. Install or connect the following:
- Starter motor and tighten the bolts to 37 ft. lbs. (50 Nm)
- Starter motor wiring connectors
- CAC pipe
- Air baffle
- Negative battery cable

5-Cylinder Engines, except 2005-06 S40/V50/C70 Models

1. Before servicing the vehicle, refer to the Precautions Section.
2. Remove or disconnect the following:
- Negative battery cable
- Air intake assembly
- Intake manifold between the front cover plate and the air cleaner assembly, if equipped
- Hose between the charge air cooler (CAC) and the electronic throttle module, if equipped
- Starter motor wiring connectors
- Starter motor

To install:

3. Install or connect the following:
- Starter motor and tighten the bolts to 29 ft. lbs. (40 Nm)
- Starter motor wiring connectors
- Hose between the charge air cooler (CAC) and the electronic throttle module, if equipped
- Intake manifold between the front cover plate and the air cleaner assembly, if equipped
- Air intake assembly
- Negative battery cable

2005-06 S40/V50/C70 Models

1. Before servicing the vehicle, refer to the Precautions Section.
2. Remove or disconnect the following:
- Negative battery cable
- Air intake assembly
- Starter motor wiring connections

3. Remove the coolant hose from the bracket and slacken the nuts off for the hose bracket. Bend the bracket to one side to allow access for bolts to the starter motor.
4. Remove the start motor.

To install:

5. Install or connect the following:
- Starter motor
- Start motor wiring connections
- Coolant hose bracket
- Air intake assembly
- Negative battery cable

6-Cylinder Engines

1. Before servicing the vehicle, refer to the Precautions Section.
2. Remove or disconnect the following:
- Negative battery cable
- Air intake assembly
- Starter motor wiring connectors
- Starter motor

To install:

3. Install or connect the following:
- Starter motor and tighten the bolts to 29 ft. lbs. (40 Nm)
- Starter motor wiring connectors
- Air intake assembly
- Negative battery cable

Timing Belt

REMOVAL & INSTALLATION

4-Cylinder Engine

1. Before servicing the vehicle, refer to the Precautions Section.
2. Remove or disconnect the following:
- Negative battery cable
- Right hand engine cover
- Cover over the right headlamp
- Accessory drive belt
- Front timing cover
- Servo reservoir

3. Position Volvo lifting beam 999 5006 or equivalent slightly in front of the front lifting eyelet for the engine. Use Volvo lifting arm 999 5383 or equivalent and Volvo lifting hook 999 5460 or equivalent to raise the front of the engine a few millimeters.
4. Remove or disconnect the following:
- Right-hand engine mount
- Right front wheel

5. Turn the crankshaft until the markings on the crankshaft and camshaft pulley correspond. Turn the crankshaft a further ¼ turn clockwise and then back again until the markings correspond. The markings are illustrated. Remove the upper timing cover.
6. Slacken off the center screw for the belt tensioner slightly.
7. Turn the tensioner clockwise to 10 o'clock.
8. Remove or disconnect the following:
- Air baffles
- Lower belt cover
- Timing belt

To install:

9. Turn all the pulleys listening for bearing noise. Check to see that the contact surfaces are clean and smooth. If the tensioner pulley lever or idler is seized, replace it.

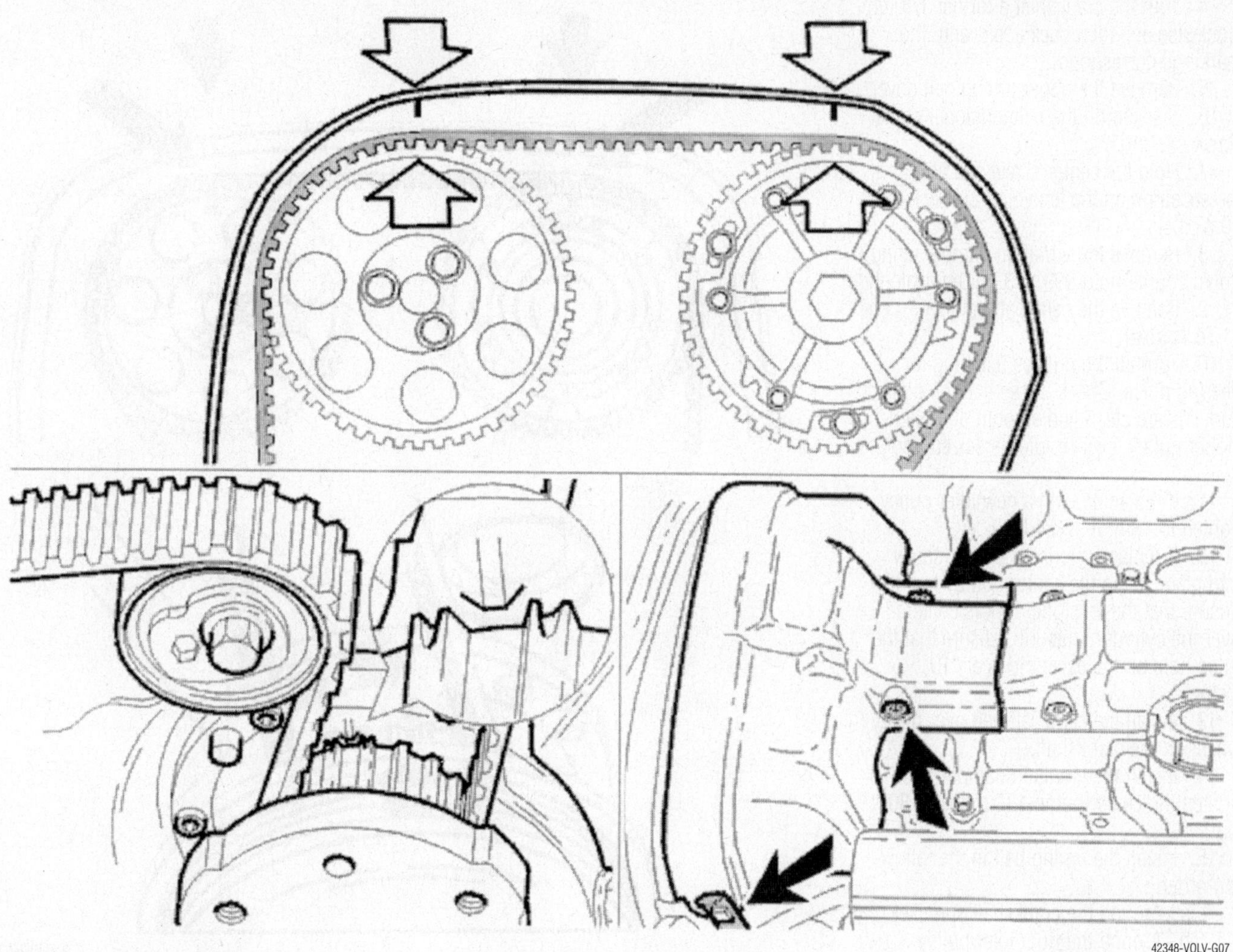

Timing belt alignment markings

10. If replacing with a new idler pulley, tighten to 18 ft. lbs. (24 Nm).

11. If replacing tension pulley, screw into place using the center bolt by hand. Ensure that the tensioner fork is centered over the cylinder block rib. Ensure that the Allen hole on the eccentric is at "10 o'clock".

12. Install the timing belt in the following order:

a. Around the crankshaft sprocket
b. Around the idler pulley
c. Around the camshaft sprockets
d. Around the water pump
e. Onto the belt tensioner

13. Carefully turn the crankshaft clockwise until the timing belt is tensioned. The belt must be tensioned between the intake camshaft pulley, the idler pulley and the crankshaft

14. Hold the belt tensioner center screw secure. Turn the belt tensioner eccentric counter-clockwise until the tensioner indicator passes the marked position.

15. Then turn the eccentric back so that the indicator reaches the marked position in the center of the window and tighten the center bolt to 15 ft. lbs. (20 Nm).

16. Press the belt to check that the indicator on the tensioner moves easily.

17. Install or connect the following:

- Lower belt guard
- Air baffles
- Upper timing cover

18. Turn the crankshaft two turns. Check that the markings on the crankshaft and camshaft pulley correspond.

➡Check that the indicator on the belt tensioner is within the marked area.

19. Install or connect the following:

- Lower timing cover
- Front timing cover and tighten to 9 ft. lbs. (12 Nm)
- Accessory drive belt
- Engine mount and tighten to 49 ft. lbs. (67 Nm)
- Servo reservoir. Tighten bolts to 37 ft. lbs. (50 Nm).
- Front wheel
- Right hand engine cover
- Cover over the right headlamp
- Negative battery cable

5-Cylinder Engines, except 2005-06 S40/V50/C70 Models

1. Before servicing the vehicle, refer to the Precautions Section.

2. Remove or disconnect the following:

- Negative battery cable
- Engine stabilizer brace
- Servo reservoir
- Expansion tank
- Accessory drive belt
- Front timing belt cover
- Right front wheel
- Inner fender liner

3. Turn the crankshaft clockwise until the markings on the crankshaft and camshaft pulley correspond.

4. Turn the crankshaft a further 1/4 turn clockwise and then back again until the markings correspond.

5. Remove the upper timing belt cover.

6. Slacken off the belt tensioner center screw slightly.

7. Hold the center screw still and turn the eccentric on the tensioner clockwise to 10 o'clock.

8. Remove the vibration damper using Volvo counterhold 999 5433 or equivalent.

9. Remove the camshaft belt.

To install:

10. Turn all the pulleys listening for bearing noise. Check to see that the contact surfaces are clean and smooth. If the tensioner pulley lever or idler is seized, replace it.

11. If replacing with a new idler pulley, tighten to 18 ft. lbs. (24 Nm).

12. If replacing tension pulley, screw into place using the center bolt by hand. Ensure that the tensioner fork is centered over the cylinder block rib. Ensure that the Allen hole on the eccentric is at "10 o'clock".

13. Install the camshaft belt over pulley wheel on the crankshaft.

14. Install the vibration damper and tighten the center nut to 133 ft. lbs. (180 Nm).

15. Install the timing belt in the following order:

    a. Around the crankshaft sprocket.
    b. Around the right idler pulley
    c. Around the camshaft sprockets
    d. Around the water pump
    e. Onto the tensioner pulley

16. Carefully turn the crankshaft clockwise until the timing belt is tensioned. The belt must be tensioned between the intake camshaft pulley, the idler pulley and the crankshaft

17. Hold the belt tensioner center screw secure. Turn the belt tensioner eccentric counter-clockwise until the tensioner indicator passes the marked position.

18. Then turn the belt tensioner eccentric back so that the indicator reaches the marked position in the center of the window and tighten the center bolt to 15 ft. lbs. (20 Nm).

19. Press the belt to check that the indicator on the tensioner moves easily.

20. Install or connect the following:

- Front timing belt cover and tighten to 9 ft. lbs. (12 Nm)
- Upper timing belt cover
- Accessory drive belt
- Servo reservoir
- Expansion tank
- Engine stabilizer brace. Tighten the

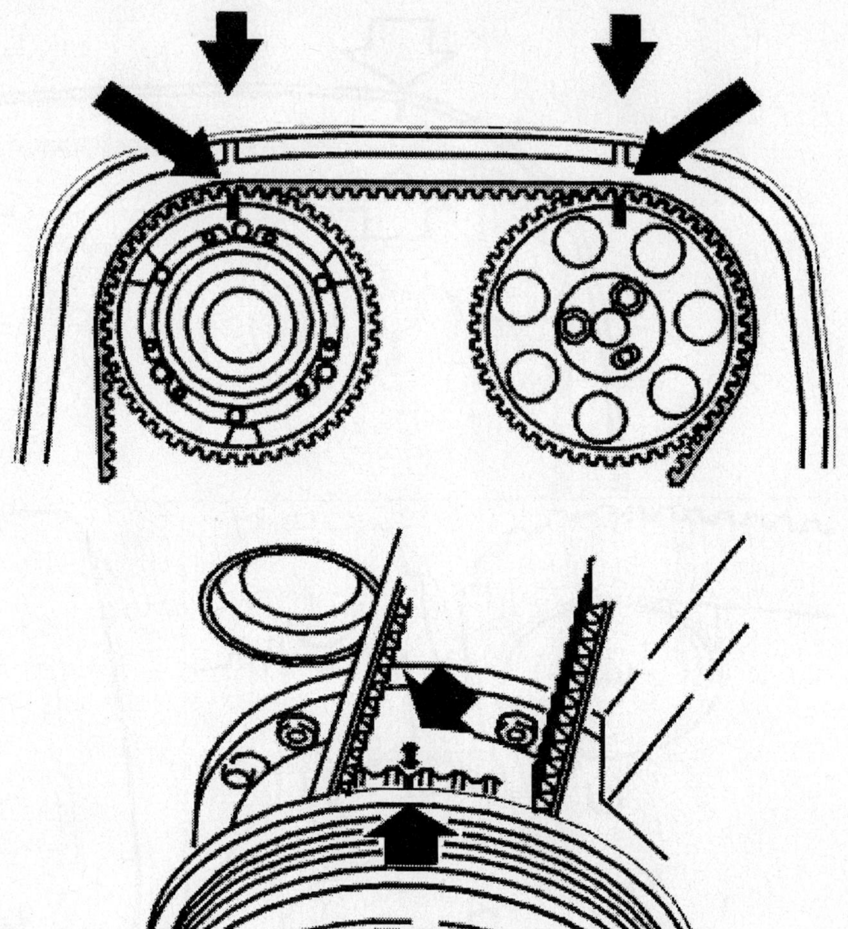

Timing mark alignment

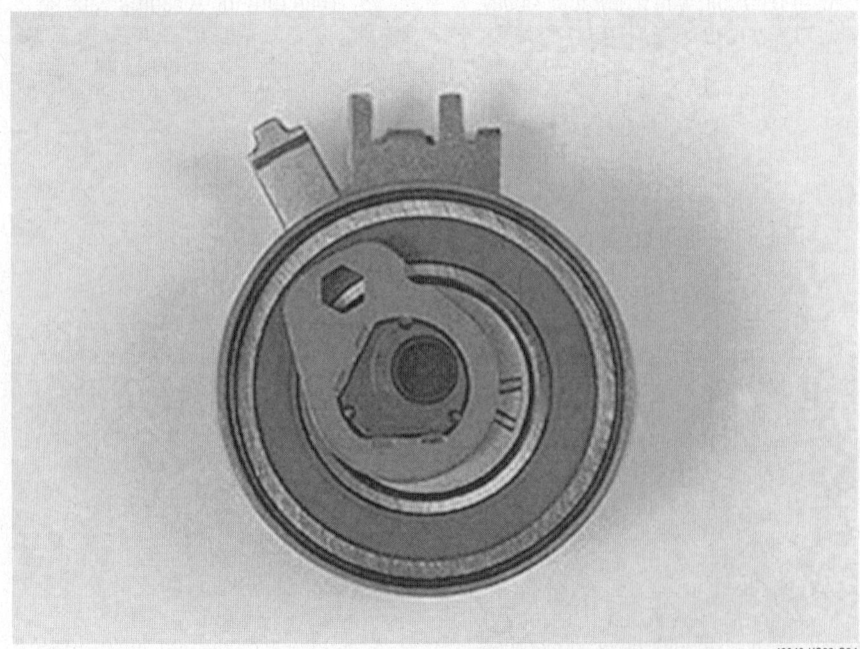

Belt tensioner

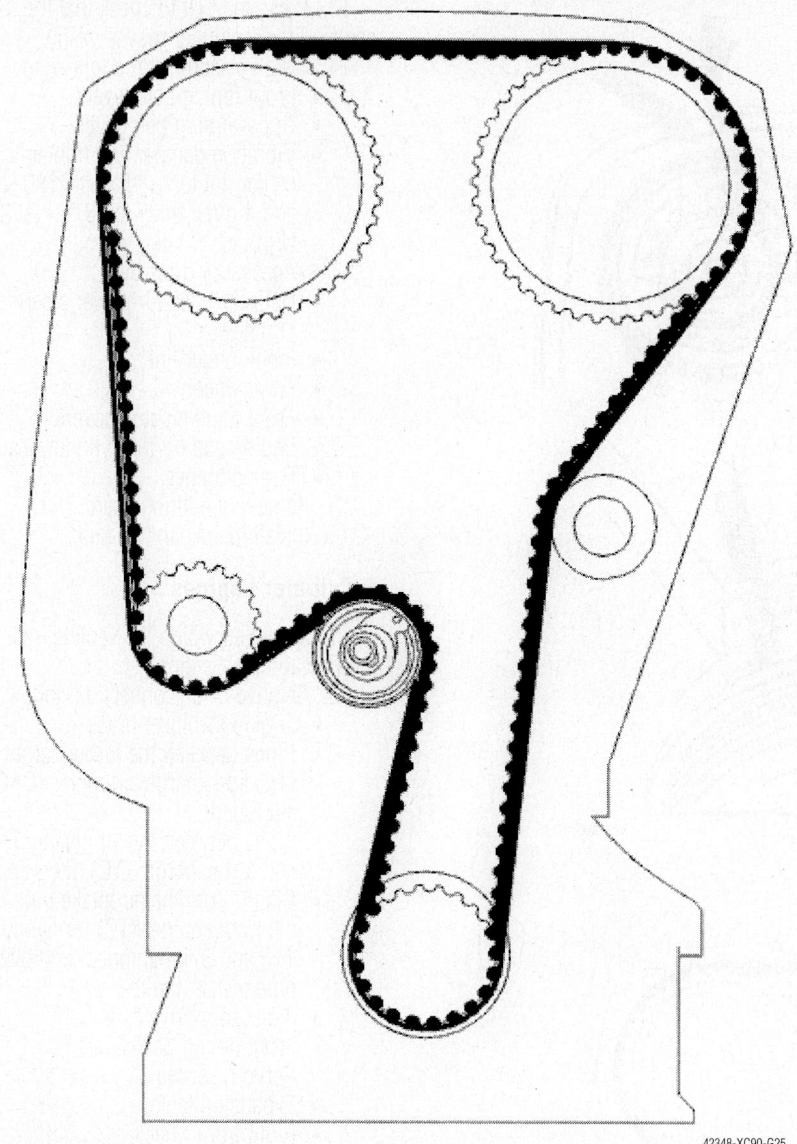

Timing belt installed

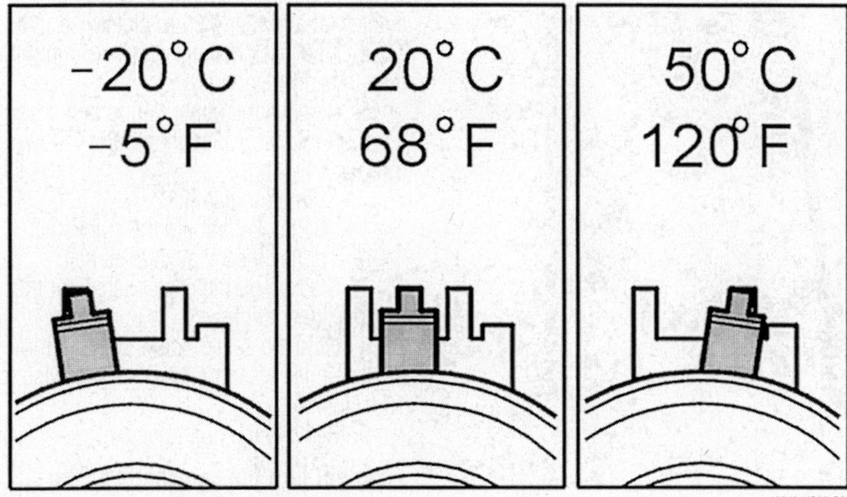

Tension indicator at different temperatures

bolts at the suspension turrets to 37 ft. lbs. (50 Nm). Tighten the engine bracket bolt to 59 ft. lbs. (80 Nm).
- Inner fender liner
- Front wheel
- Negative battery cable

2005-06 S40 and V50 Models

1. Before servicing the vehicle, refer to the Precautions Section.
2. Lift the expansion tank and place it on top of the engine.
3. Install lifting beam 99-7103 or equivalent and tighten.
4. Remove or disconnect the following:
   - Engine limiter
   - Engine pad on the cylinder head
   - Front transmission covers
   - Right hand front wheel
   - Inner fender liner
   - Delivery line from the power steering pump
   - Accessory drive belt
   - Vibration damper
   - Upper timing belt cover
5. Lower the engine slightly until the belt tensioner bolts can be accessed from beneath.
6. Turn the crankshaft clockwise until the markings on the crankshaft and camshaft pulleys correspond.
7. Slacken off the center bolt for the belt tensioner slightly and hold the center bolt still. Turn the tensioner clockwise to 10 o'clock.
8. Remove the timing belt.

To install:

9. Install the timing belt in the following order:
   a. Around the crankshaft
   b. Around the idler pulley
   c. Around the intake camshaft pulley
   d. Around the exhaust camshaft pulley
   e. Around the water pump
   f. Onto the belt tensioner
10. Carefully turn the crankshaft clockwise until the timing belt is tensioned. The belt must be tensioned between the intake camshaft pulley, the idler pulley and the crankshaft
11. Hold the belt tensioner center screw secure. Turn the belt tensioner eccentric counter-clockwise until the tensioner indicator passes the marked position.
12. Then turn the belt tensioner eccentric back so that the indicator reaches the marked position in the center of the window and tighten the center bolt to 15 ft. lbs. (20 Nm).

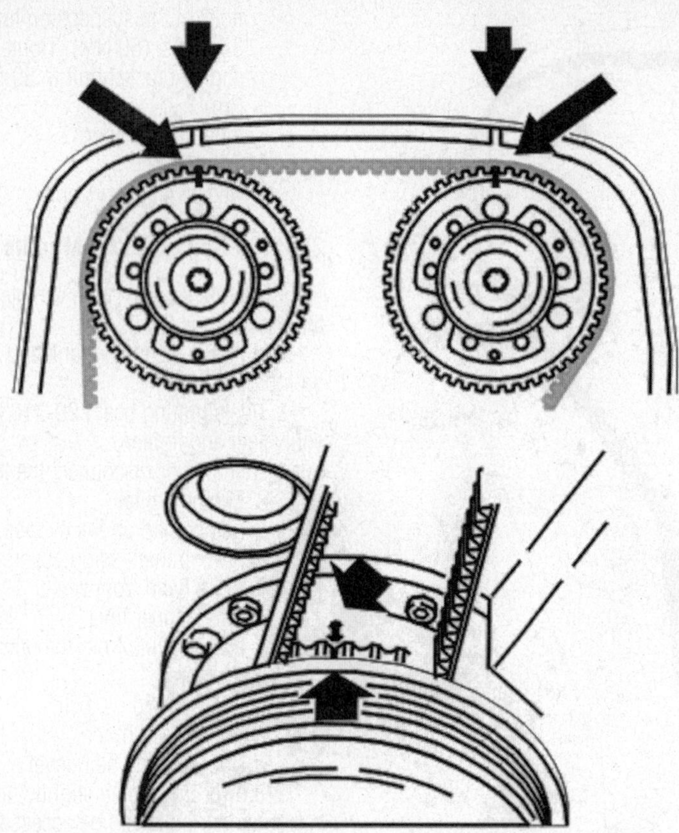

Identifying the timing marks

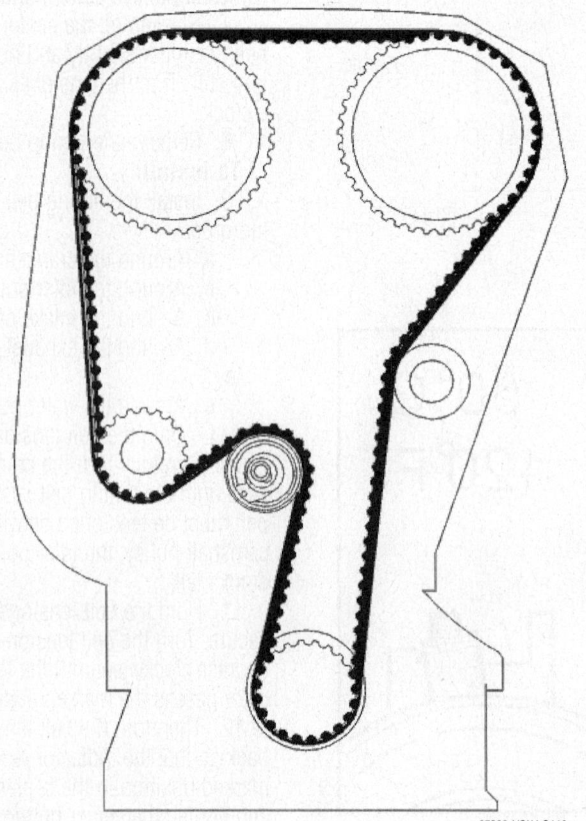

2005–06 S40/V50/C70 timing belt routing

13. Press the belt to check that the indicator on the tensioner moves easily.
14. Install or connect the following:
    - Front timing belt cover
    - Upper timing belt cover
    - Vibration damper and tighten the center nut to 133 ft. lbs. (180 Nm) and 4 outer bolts to 18 ft. lbs. (25 Nm) plus 60 degrees.
    - Accessory drive belt
    - Delivery hose for power steering pump
    - Inner fender liner
    - Front wheel
    - Front transmission covers
    - Engine pad on the cylinder head
    - Engine limiter
15. Remove the lifting beam.
16. Install the expansion tank.

6-Cylinder Engines

1. Before servicing the vehicle, refer to the Precautions Section.
2. Remove or disconnect the following:
    - Engine stabilizer brace
    - Pipes between the turbocharger (TC) and charge air cooler (CAC), if equipped
    - Pipes between the air cleaner (ACL) and turbocharger (TC), if equipped
    - Clamp from the air intake pipe for the turbocharger (TC) for cylinders 1, 2 and 3, if equipped, and twist pipe toward the rear
    - Accessory drive belt
    - Front timing belt cover.
    - Servo reservoir
    - Expansion tank
    - Right front wheel
    - Inner fender liner
3. Turn the crankshaft clockwise until the markings on the crankshaft and camshaft pulleys correspond. Turn the crankshaft a further ¼ turn clockwise.
4. Turn back counter-clockwise until the markings correspond.
5. Remove the upper timing belt cover.
6. Remove or disconnect the following:
    - Vibration damper
    - Belt cover behind the crankshaft pulley for the auxiliaries belt
7. Slacken off the center screw for the belt tensioner slightly.
8. Hold the center screw still and turn the tensioner eccentric counter-clockwise to 10 o'clock.
9. Remove the timing belt.

➡**Do not turn the camshafts or the crankshaft when the timing belt has been removed.**

To install:

10. Turn all the pulleys listening for bearing noise. Check to see that the contact surfaces are clean and smooth. If the tensioner pulley lever or idler is seized, replace it.
11. If replacing with a new idler pulley, tighten to 18 ft. lbs. (24 Nm).
12. If replacing tension pulley, screw into place using the center bolt by hand. Ensure that the tensioner fork is centered over the cylinder block rib. Ensure that the Allen hole on the eccentric is at “10 o'clock”.
13. Install the camshaft belt over pulley wheel on the crankshaft.
14. Install the timing belt in the following order:
    a. Around the crankshaft sprocket.
    b. Around the right idler pulley
    c. Around the camshaft sprockets
    d. Around the water pump
    e. Onto the tensioner pulley
15. Carefully turn the crankshaft clockwise until the timing belt is tensioned. The belt must be tensioned between the intake camshaft pulley, the idler pulley and the crankshaft
16. Hold the belt tensioner center screw secure. Turn the belt tensioner eccentric counter-clockwise until the tensioner indicator passes the marked position.
17. Then turn the eccentric back so that the indicator reaches the marked position in the center of the window and tighten the center bolt to 18 ft. lbs. (25 Nm).
18. Press the belt to check that the indicator on the tensioner moves easily.
19. Install or connect the following:
    - Accessory drive belt
    - Vibration damper. Tighten the bolts to 26 ft. lbs. (35 Nm) plus 50 degrees
    - Inner fender liner
    - Front wheel
    - Front timing belt cover
    - Expansion tank
    - Servo reservoir
    - Clamp for the air intake pipe
    - Pipes between the air cleaner (ACL) and turbocharger (TC), if equipped
    - Pipes between the turbocharger (TC) and charge air cooler (CAC), if equipped
    - Engine stabilizer brace. Tighten the bolts at the suspension turrets to 37 ft. lbs. (50 Nm). Tighten the engine bracket bolt to 59 ft. lbs. (80 Nm).

Oil Pan

REMOVAL & INSTALLATION

5-Cylinder Engines, except 2005-06 S40/V50/C70 Models

1. Before servicing the vehicle, refer to the Precautions Section.
2. Drain the engine oil
3. Remove or disconnect the following:
    - Oil dipstick tube
    - Splashguard
    - Oil filter
    - Oil cooler
    - Oil pan

To install:

4. Install or connect the following:
    - New O-rings for oil pan
    - Oil pan and tighten the bolts to 13 ft. lbs. (17 Nm)
    - Oil cooler with new O-rings
    - New oil filter
    - Oil dipstick tube
    - Splashguard

2005-06 S40/V50/C70 models

1. Before servicing the vehicle, refer to the Precautions Section.
2. Drain the engine oil.
3. Remove or disconnect the following:
    - Air intake assembly
    - Oil dipstick assembly
    - Right hand front wheel
    - Inner fender liner
    - Engine splashguard
4. Remove the bolts for the oil cooling system and hang the system off to the side.
5. Remove or disconnect the following:
    - Lower charge air pipe, if equipped
    - Pressure hose and return hose from power steering pump
    - Power steering hose bracket and move to one side
    - Accessory drive belts
    - A/C compressor
    - Oil pan

To install:

6. Install or connect the following:
    - New O-rings for oil pan
    - Oil pan and tighten the bolts as follows:

a. Tighten bolts 1, 2, 3 and 4 to 2 ft. lbs. (3 Nm)

b. Tighten bolts 5 to 18 ft. lbs. (25 Nm) and then to 35 ft. lbs. (45 Nm)

c. Tighten all bolts to 13 ft. lbs. (17 Nm). Start at the transmission and work forward in pairs.

- Oil cooling system using new O-rings
- A/C compressor
- Accessory drive belts
- Oil pressure and return hoses for the power steering pump
- Engine splashguard
- Inner fender liner
- Wheel
- Oil dipstick assembly
- Air intake pipe

7. Start engine and check for leaks.

6-Cylinder Engine

1. Drain the engine oil
2. Remove or disconnect the following:
    - Oil dipstick tube
    - Turbo charge air pipe, if equipped
    - Turbo air intake pipe, if equipped
3. Raise and support the engine.
4. Remove or disconnect the following:
    - Lower engine cover
    - Front air baffle
    - Oil filter
    - Front wheel

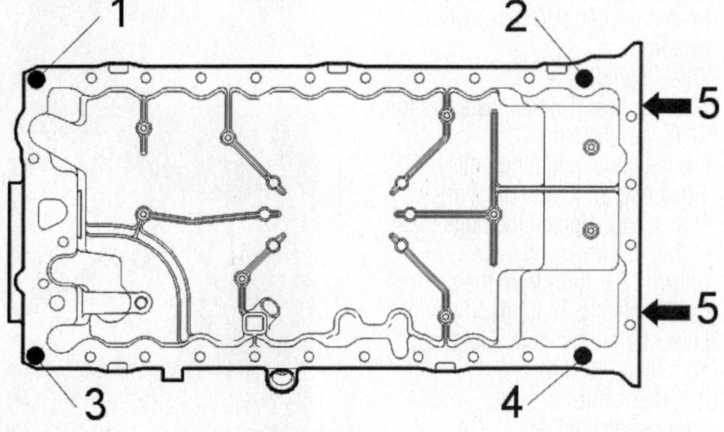

Oil pan bolt identification—2005-06 S40/V50/C70 models

- Lower ball joints nuts
- Anti-roll bar links from the anti-roll bar
- Right-hand axle shaft center bolt
- Steering gear
- Front engine mounting bolts
- Torque rod including the mounting in the sub-frame
- Wheel spindles
- Oil pipe for the steering gear from the snap fasteners along the sub-frame
- EVAP canister hose
- Fuel line for the engine block heater, if equipped
- Negative ground cable bracket
- Heated oxygen (HO_2) sensor connectors

5. Lower the sub-frame slightly.
6. Remove the following:
   - Oil cooler
   - Oil pan

To install:

7. Install or connect the following:
   - New O-rings
   - Oil pan and tighten the bolts as follows:

a. Tighten bolts 1, 2, 3 and 4 to 2 ft. lbs. (3 Nm)

b. Tighten bolts 5 to 18 ft. lbs. (25 Nm) and then to 35 ft. lbs. (45 Nm)

c. Tighten all bolts to 13 ft. lbs. (17 Nm). Start at the transmission and work forward in pairs.

- Oil cooler with new O-rings

8. Tighten the sub-frame bolts to 77 ft. lbs. (105 Nm) plus 120 degrees.
9. Install or connect the following:
   - Oil pipe for the steering gear from the snap fasteners along the sub-frame
   - EVAP canister hose
   - Fuel line for the engine block heater, if equipped
   - Negative ground cable bracket
   - Heated oxygen (HO_2) sensor connectors
   - Wheel spindles
   - Steering gear. Tighten fasteners to 37 ft. lbs. (50 Nm).
   - Front engine mounting bolts. Tighten to 37 ft. lbs. (50 Nm).
   - Torque rod. Tighten the bolts to 37 ft. lbs. (50 Nm).
   - Anti-roll bar links from the anti-roll bar. Tighten to 37 ft. lbs. (50 Nm).
   - Lower ball joints nuts. Tighten to 59 ft. lbs. (80 Nm).
   - Front air baffle
   - Lower engine cover
   - Engine mounts
   - Oil dipstick tube

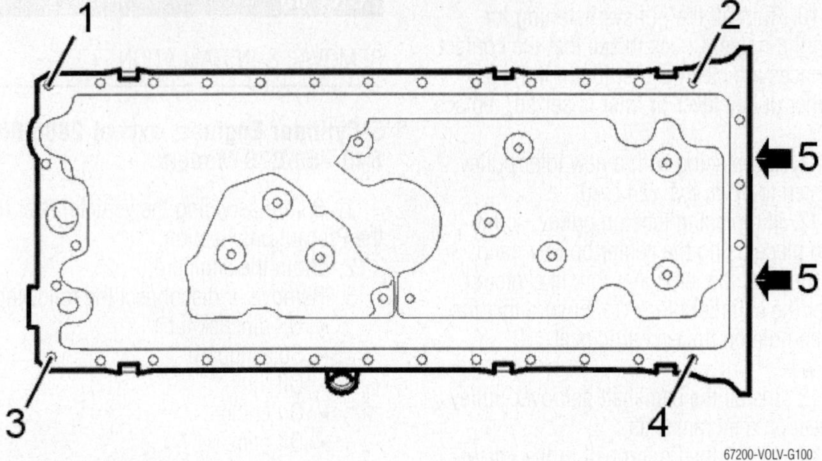

6-Cylinder oil pan bolts

- Turbo air intake pipe, if equipped
- Turbo charge air pipe, if equipped

10. Start the engine and check for leaks.

Oil Pump

REMOVAL & INSTALLATION

4-Cylinder Engine

1. Before servicing the vehicle, refer to the Precautions Section.
2. Remove or disconnect the following:
   - Timing belt
   - Vibration damper
   - Crankshaft timing gear pulley
   - Front crankshaft seal
   - Oil pump

To install:

3. Install the oil pump using special tool 999-5455. Use the oil pump bolts to guide the pump. Use the crankshaft nut to press the pump in until it is seated fully. Tighten the oil pump bolts to 84 inch lbs. (10 Nm).
4. Install or connect the following:
   - Front crankshaft seal
   - Crankshaft timing gear pulley
   - Vibration damper and tighten the center nut to 133 ft. lbs. (180 Nm)
   - Timing belt
5. Start the engine and check for leaks.

5-Cylinder Engines, except 2005-06 S40/V50/C70 Models

1. Before servicing the vehicle, refer to the Precautions Section.

Use special tool to install the oil pump.

42348-XC90-G30

Oil pump installation

2. Remove or disconnect the following:
   - Negative battery cable
   - Engine stabilizer brace
   - Servo reservoir
   - Expansion tank
   - Accessory drive belt
   - Front timing belt cover
   - Right front wheel
   - Upper camshaft cover
   - Timing belt
   - Vibration damper
   - Crankshaft timing gear pulley
   - Front crankshaft seal
   - Oil pump

To install:

3. Install the oil pump using special tool 999-5455. Use the oil pump bolts to guide the pump. Use the crankshaft nut to press the pump in until it is seated fully. Tighten the oil pump bolts to 84 inch lbs. (10 Nm).
4. Install or connect the following:
   - Front crankshaft seal
   - Crankshaft timing gear pulley
   - Vibration damper and tighten the center nut to 133 ft. lbs. (180 Nm)
   - Timing belt
   - Upper camshaft cover
   - Right front wheel
   - Front timing belt cover
   - Accessory drive belt
   - Expansion tank
   - Servo reservoir
   - Engine stabilizer brace. Tighten the bolts at the suspension turrets to 37 ft. lbs. (50 Nm). Tighten the engine bracket bolt to 59 ft. lbs. (80 Nm).
   - Negative battery cable
5. Start the engine and check for leaks.

6-Cylinder Engine and 2005-06 S40/V50/C70 Models

1. Before servicing the vehicle, refer to the Precautions Section.
2. Remove or disconnect the following:
   - Timing belt
   - Timing gear pulley
   - Front crankshaft seal
   - Belt guard from oil pump, if equipped
   - Oil pump

To install:

3. Install the oil pump with a new gasket using special tool 999-5455. Use the oil pump bolts to guide the pump. Use the crankshaft nut to press the pump in until it is seated fully. Tighten the oil pump bolts to 84 inch lbs. (10 Nm).
4. Install or connect the following:
   - Front crankshaft seal
   - Timing gear pulley
   - Timing belt
5. Start the engine and check for leaks.

Rear Main Seal

REMOVAL & INSTALLATION

1. Before servicing the vehicle, refer to the Precautions Section.
2. Remove or disconnect the following:
   - Negative battery cable
   - Transmission
   - Clutch, if equipped
   - Flexplate or flywheel
   - Rear main seal

To install:

3. Install the rear main seal using Special Tool 999-7174 or equivalent, so that the seal is flush with the cylinder block.

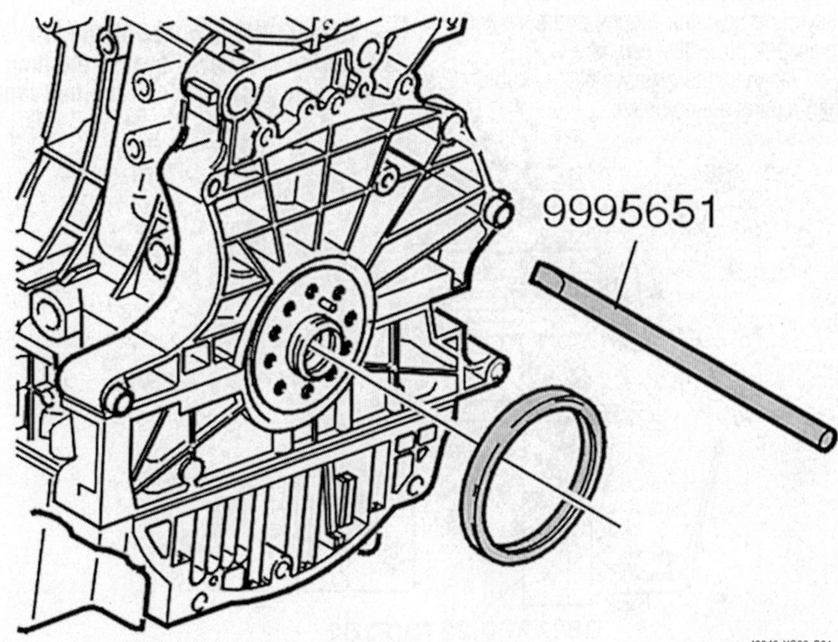

42348-XC90-G31

Rear main seal removal

➡The felt side of the seal should face outward.

4. Install or connect the following:
   - Flexplate or flywheel. Tighten the bolts to 33 ft. lbs. (45 Nm) plus 65 degrees.
   - Clutch, if equipped
   - Transmission
   - Negative battery cable
5. Start the engine and check for leaks.

Piston and Ring

POSITIONING

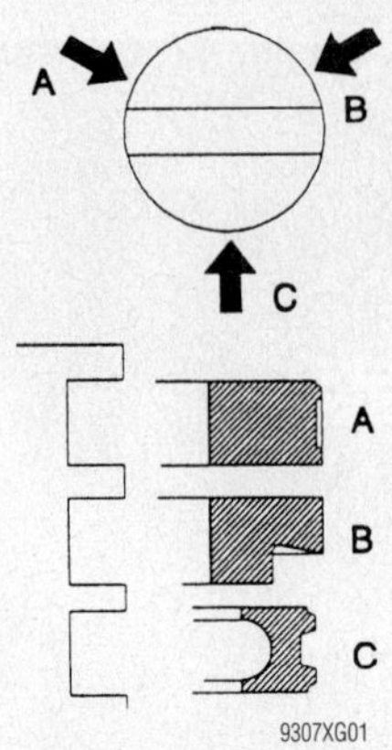

Ring identification and end-gap spacing—all engines

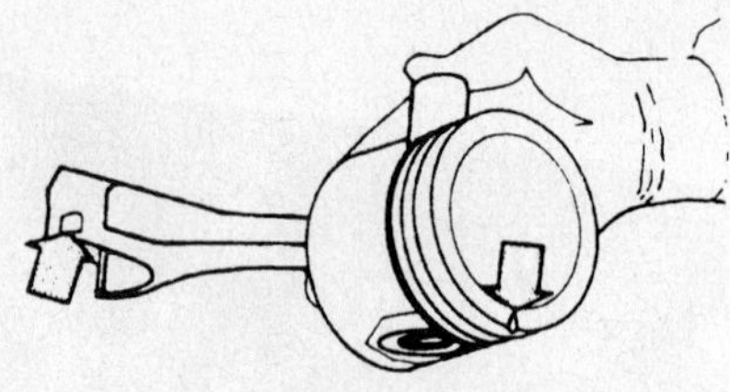

Piston and rod positioning—all engines The notch on the piston crown faces the front of the engine

FUEL SYSTEM

Fuel System Pressure

RELIEVING

All Models Except 2005-06 S40/V50/C70

1. Before servicing the vehicle, refer to the Precautions Section.
2. Disconnect the negative battery cable.
3. Remove protective cap from the valve on the fuel rail.
4. Connect a hose to adapter 999-5484 and place the other end in a clean container.
5. Connect the adapter in the locked or closed position to the valve on the fuel rail.
6. Unlock or open the adapter valve.
7. After the fuel system pressure is relieved, remove the adapter and hose and replace the protective cap.
8. Connect the negative battery cable when repairs are complete.

2005-06 S40/V50/C70 Models

1. Before servicing the vehicle, refer to the Precautions Section.
2. Remove the fuse for the fuel pump.
3. Start the engine and allow it to idle until is stops.
4. Turn the engine using the starter motor for approximately 5 seconds until the pressure in the fuel rail has been released.

Fuel Filter

REMOVAL & INSTALLATION

All Models, except 2005-06 S40/V50/C70 Models

➡The fuel filter is either on the left side of the firewall or next to the fuel pump near the left side of the fuel tank.

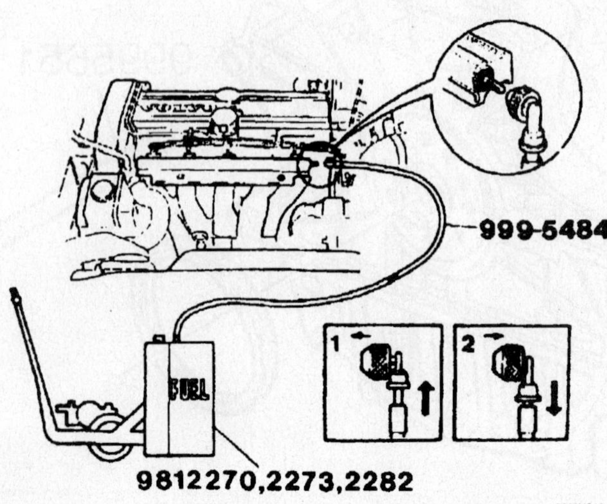

Connecting the adapter and fuel drainage unit—5-cylinder engine shown

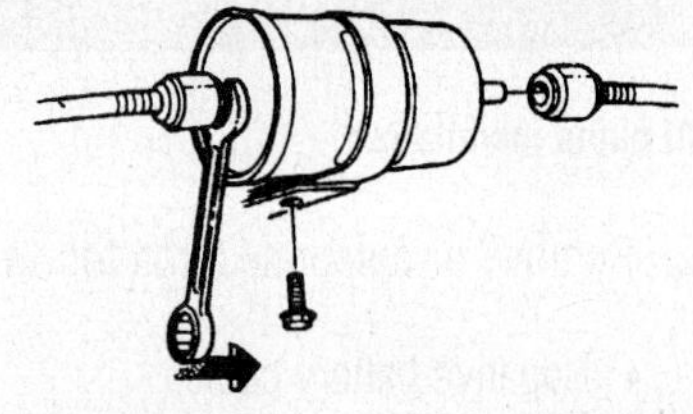

Using an open end wrench to push the quick disconnect coupler sleeves back—5-cylinder models

1. Before servicing the vehicle, refer to the Precautions Section.
2. Relieve the fuel system pressure.
3. Remove or disconnect the following:
   - Negative battery cable
   - Fuel filler cap
   - Fuel lines from the fuel filter
   - Fuel filter from the bracket

To install:

4. Install or connect the following:
   - Fuel filter to the bracket
   - Fuel lines to the fuel filter
   - Fuel filler cap
   - Negative battery cable
5. Start the engine and check for leaks.

2005-06 S40/V50/C70 Models

On these models, the fuel filter is part of the fuel pump assembly.

Fuel Pump

REMOVAL & INSTALLATION

4-Cylinder Engine

1. Before servicing the vehicle, refer to the Precautions Section.

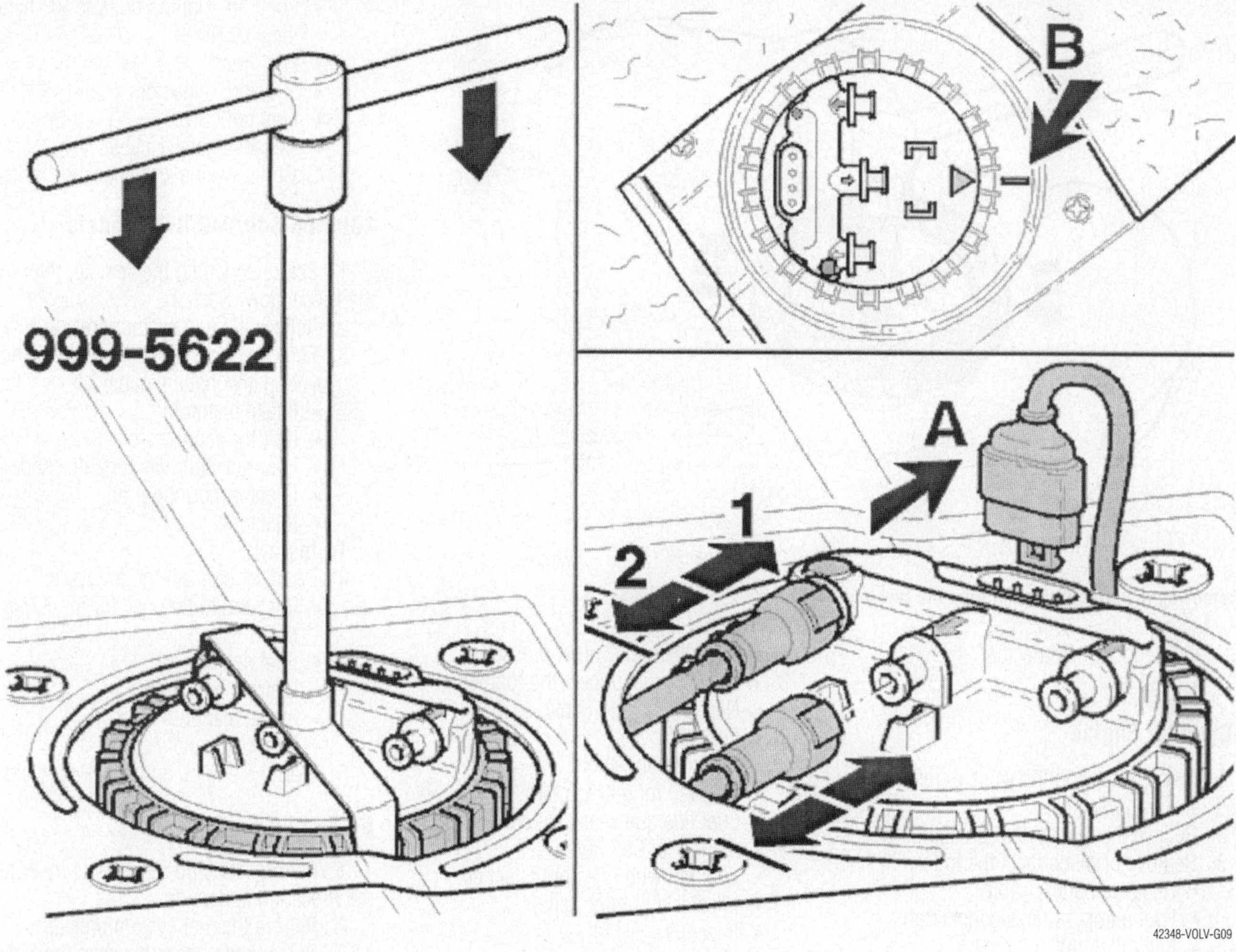

Installing the cap nut for the fuel pump.

2. Relieve the fuel system pressure.
3. Tilt the rear seat forward and remove or fold back the trunk compartment carpet.
4. Remove or disconnect the following:
   - Protective cover
   - Fuel pump harness connector. Label the connections for the fuel lines.
   - Fuel lines by pressing in the catches and pulling out the lines
   - Cap nut using special tool 999 5622
   - Pump unit

➡**Replace the fuel pump *immediately* or reinstall the cap nut temporarily because the threaded collar swells.**

To install:

5. Install or connect the following:
   - Pump unit. Ensure that the arrow points towards the marking on the fuel tank
   - Cap nut using special tool 999 5622 and tighten to 37 ft. lbs. (50 Nm)
   - Fuel pump harness connector
   - Fuel lines
   - Protective cover
6. Start engine and check for leaks.

5-Cylinder Engine, except 2005-06 S40/V50/C70 Models

1. Before servicing the vehicle, refer to the Precautions Section.
2. Relieve the fuel system pressure.
3. Tilt the rear seat forward and remove, remove rear seat cushion or fold back the trunk compartment carpet over the right-hand wheel well panel.
4. Remove or disconnect the following:
   - Protective cover
   - Fuel pump harness connector
   - Fuel supply and return lines using tool 999 5666

➡**The supply line is marked with yellow tape, corresponding to the color marking on the pump.**

   - Fitting nut using wrench 999 5485
   - Pump unit and rubber gasket

To install:

5. Install or connect the following:
   - Fuel pump with new gasket
   - Fitting nut using wrench 999 5485 and tighten to 29 ft. lbs. (40 Nm)
   - Fuel supply and return lines
   - Fuel pump harness connector
   - Protective cover
6. Start engine and check for leaks.

2005-06 S40/V50/C70 Models

1. Before servicing the vehicle, refer to the Precautions Section.
2. Relieve the fuel system pressure.
3. Remove or disconnect the following:
   - Fuel tank
   - Fuel pump connector
   - Large plastic fitting nut using special tool 999-7093
   - Fuel pump

To install:

4. Install or connect the following:
   - Fuel pump with new O-ring
   - Large plastic fitting nut using special tool 999-7093

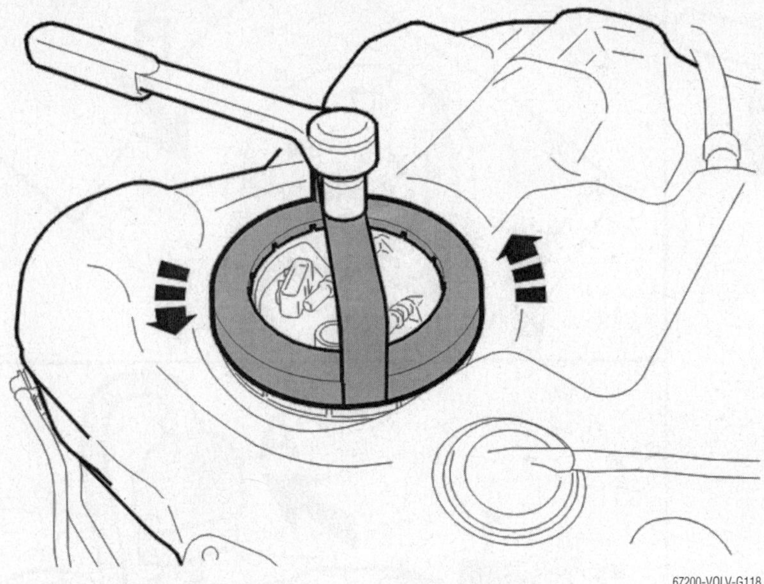
67200-VOLV-G118

Removing the large fitting nut from the fuel pump assembly

- Fuel pump connector
- Fuel tank

6-Cylinder Engine

1. Before servicing the vehicle, refer to the Precautions Section.
2. Relieve the fuel system pressure.
3. Remove or disconnect the following:
   - Negative battery cable
   - Fuel pump harness connections
   - Fuel fill and vent hoses
   - Fuel supply and return lines
4. Loosen the lockring at the top of the fuel tank and remove the sending unit with the transfer pump attached. Note the direction of the float in the tank.
5. Remove the transfer pump from the sending unit.

To install:

6. Install the transfer pump on the sending unit. Install the sending unit in the fuel tank and tighten the lockring.
7. Install or connect the following:
   - Fuel supply and return lines
   - Fuel fill and vent hoses
   - Fuel pump harness connections
   - Negative battery cable
8. Start the engine and check for leaks.

Fuel Injector

REMOVAL & INSTALLATION

4 and 5-Cylinder Engines, except 2005-06 S40/V50/C70 Models

1. Before servicing the vehicle, refer to the Precautions Section.
2. Relieve the fuel system pressure.
3. Remove or disconnect the following:
   - Negative battery cable
   - Fuel rail cover
   - Injector connectors
   - Fuel rail mounting bolts
   - Fuel line quick-release connector using tool 951 2666
   - Fuel rail with injectors attached
   - Injector mounting rail
   - Injectors

To install:

4. Install or connect the following:
   - Injectors with new O-ring seals
   - Injector mounting rail
   - Fuel rail. Tighten bolts to 84 inch lbs. (10 Nm)
   - Fuel lines
   - Injector connectors
   - Fuel rail cover
   - Negative battery cable
5. Start the engine and check for leaks.

2005-06 S40/V40/C70 Models

1. Before servicing the vehicle, refer to the Precautions Section.
2. Relieve the fuel system pressure.
3. Remove or disconnect the following:
   - Negative battery cable
   - Intake manifold
   - Injector connectors
   - Fuel rail with injectors attached
   - Injector mounting rail
   - Injectors

To install:

4. Install or connect the following:
   - Injectors to the injector mounting rail
   - Fuel rail
   - Injector connectors
   - Intake manifold
   - Negative battery cable
5. Start the engine and check for leaks.

6-Cylinder Engines

1. Before servicing the vehicle, refer to the Precautions Section.
2. Relieve the fuel system pressure.
3. Remove the hoses between the turbocharger/charge air cooler and air cleaner/turbocharger on turbocharged engines.

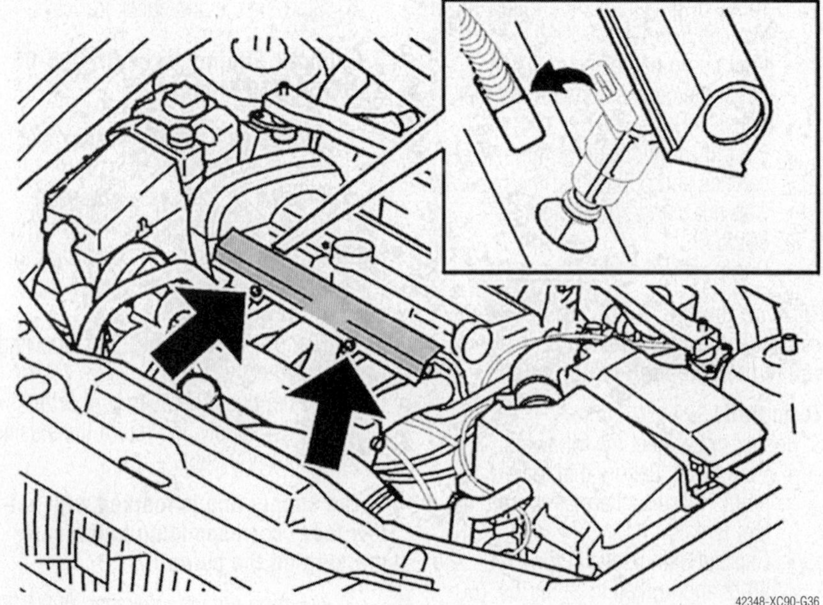
42348-XC90-G36

Injector removal

4. Remove the upper section of intake manifold on non-turbo engines.

5. Remove or disconnect the following:
- Injector connectors
- Fuel rail
- Fuel line quick-release connectors using tool 999 5666
- Injector mounting rail
- Injectors

To install:

6. Install or connect the following:
- Injector
- Injector mounting rail
- Fuel rail and tighten mounting bolts to 33 ft. lbs. (45 Nm)
- Injector connectors

7. Install the upper section of the intake manifold with a new gasket on non-turbo engines.

8. Install the hoses for the charge air cooler and air cleaner on turbocharged engines.

9. Start engine and check for leaks.

DRIVE TRAIN

Transaxle Assembly

REMOVAL & INSTALLATION

4-Cylinder Engine

1. Before servicing the vehicle, refer to the Precautions Section.
2. Drain the transaxle fluid.
3. Remove or disconnect the following:
- Engine cover
- Battery and tray
- Air intake assembly
- Air intake hose
- Gear selector cable
- Transaxle ground lead
- Vehicle speed (VSS) sensor

4. Raise and support the engine.
5. Remove or disconnect the following:
- Heat deflector plate
- Rear engine bracket
- Front engine bracket
- Axle half shafts
- Transaxle oil pipes
- Oil pipe clamp

6. Install transaxle fixture 999 5463 on the transaxle jack and install the fixture on the transmission. At the same time align the support plate 999 5463-1 on the raise the jack so it making light contact.
7. Remove or disconnect the following:
- 5 bolts between the transaxle and the engine and starter
- Left-hand engine mount
- Remaining 7 bolts between the engine and transaxle

8. Carefully detach the transaxle from the engine and lower the engine slightly.
9. Turn the transaxle slightly and remove it.

To install:

10. Install or connect the following:
- Transaxle and tighten the bolts to 37 ft. lbs. (50 Nm)
- Transaxle mount and tighten to the bolts to 37 ft. lbs. (50 Nm)

11. Remove the transaxle fixture and jack from the transmission.
12. Using new bolts, tighten the 2 bolts for the transaxle and the starter motor to 37 ft. lbs. (50 Nm). Tighten the remaining transaxle to engine bolts to 37 ft. lbs (50 Nm).
13. Align the torque converter against the carrier plate by turning engine with the nut for the crankshaft pulley while turning the torque converter. Tighten the bolts to 26 ft. lbs. (35 Nm).
14. Install or connect the following:
- Front engine bracket and tighten to 49 ft. lbs. (67 Nm)
- Rear engine bracket and tighten to 49 ft. lbs. (67 Nm)
- Axle half shafts
- Heat deflector plate
- Oil pipe clamp
- Transaxle oil pipes
- Transaxle ground lead
- VSS sensor
- Gear select cable
- Air intake hose
- Air intake assembly
- Battery and tray
- Engine cover

15. Fill the transaxle to the correct level.
16. Start the engine and check for leaks.

Using the transaxle fixture

5-Cylinder Engine, except 2005-06 S40/V50/C70 Models

MANUAL

1. Before servicing the vehicle, refer to the Precautions Section.
2. Drain the transaxle fluid.
3. Remove or disconnect the following:
- Negative battery cable
- Air intake assembly
- Upper charge air pipe
- Upper torque rod
- Engine stabilizer brace
- Intake pipe from the charge air cooler (CAC)
- Coolant hose from the heater unit
- Transmission cables
- Reverse lamp switch connector
- Cable conduit
- Starter motor
- Ignition setting sender

4. Raise and support the engine.
5. Remove or disconnect the following:

- Engine splash guards
- Anti-roll bar link at the spring strut
- Wheel spindle from the control arm
- Drive shaft from the wheel spindle
- Sealing ring for the CV joint housing
- Drive shaft
- Steering gear sub-frame bolts
- Rear engine pad
- Torque rod mounting bolts from the transaxle
- Cross members
- Heated oxygen (HO_2S) sensor cable bracket
- Left half shaft from the transaxle
- Bevel gear, if equipped
- Rear engine mount
- Inner fender liner

6. Install transaxle fixture 999 5463 on transaxle jack, using the torque rod mounting bolts to hold it in place. At the same time, fit tool 999 5463-2 support plate and raise the jack so that it is making light contact.
7. Remove the transaxle flange bolts.
8. Remove the transaxle.

To install:

9. Install or connect the following:
   - Transaxle and tighten the bolts to 37 ft. lbs. (50 Nm)
   - Rear engine mount and tighten the bolts to 37 ft. lbs. (50 Nm)
   - Bevel gear, if equipped
   - Sealing ring for the CV joint housing
   - Drive shaft on the transaxle
   - Wheel spindle on the control arm
   - Steering gear sub-frame bolts using new lock nuts. Tighten the bolts to 37 ft. lbs. (50 Nm).
   - Rear engine pad and tighten the bolt to 37 ft. lbs. (50 Nm)
   - Heated oxygen (HO_2S) sensor cable bracket
   - Cross members and tighten the bolts to 18 ft. lbs. (25 Nm)
   - Torque rod mounting bolts using new bolts and tighten to 37 ft. lbs. (50 Nm)
   - Fender inner liner
   - Starter motor and tighten the bolts to 29 ft. lbs. (40 Nm)
   - Ignition setting sender and tighten to 88 inch lbs. (10 Nm)
   - Cable conduit
   - Reverse lamp switch connector
   - Transaxle cables
   - Coolant hose on the heater unit
   - Intake pipe from the charge air cooler (CAC)
   - Engine stabilizer brace
   - Torque rod
   - Upper charge air pipe
   - Air intake assembly
   - Negative battery cable
10. Fill the transaxle to the correct level.

AUTOMATIC

1. Before servicing the vehicle, refer to the Precautions Section.
2. Drain the transaxle fluid.
3. Disconnect the negative battery cable.
4. Remove or disconnect the following on turbocharged engines only:
   - Torque rod
   - Air intake hose
   - Air intake assembly
   - Charge air (CAC) pipe
5. Remove or disconnect the following on naturally aspirated engines only:
   - Upper engine stabilizer brace
   - Air intake assembly with the intake pipe and preheating hose
   - Vacuum hoses from the intake pipe for the control valves
   - Engine ground cable
6. Remove or disconnect the following:
   - Gear selector cable and mount.
   - Transaxle oil pipe
   - Air intake housing bracket
   - Heated oxygen (HO_2S) sensor connector bracket
   - Starter motor
   - Turbo coolant hoses, if equipped
7. Raise and support engine.
8. Remove or disconnect the following:
   - Air baffle
   - Engine splash guard
   - Sub-frame cable duct
   - Parking heater, if equipped
   - Cross members
   - Exhaust bracket rubber mounts
   - Torque rod mounting bolts
   - Steering gear bolt
   - Power steering line
   - ABS line from the mounting in the spring strut
   - Drive shaft bolt
   - Anti-roll bar link
   - Inner fender liner
   - Sub-frame support plate bolts
   - Control arm from the ball joint using tool 999 7076
   - Spring strut
   - Drive shaft
   - Bevel gear, AWD only
   - Rear engine mount
   - Torque converter bolts
   - Transaxle oil pipe
   - Transaxle ground cable
9. Lower the left side of the engine and transmission to the lifting hook stop
10. Install transaxle fixture 999 5463 on transaxle jack, using the torque rod mounting bolts to hold it in place. At the same time, fit tool 999 5463-2 support plate and raise the jack so that it is making light contact.
11. Remove the transaxle flange bolts.
12. Remove the transaxle.

To install:

13. Install or connect the following:
    - Transaxle and tighten the bolts to 37 ft. lbs. (50 Nm)
    - Torque converter and tighten the bolts to 26 ft. lbs. (35 Nm)
    - Rear engine mount

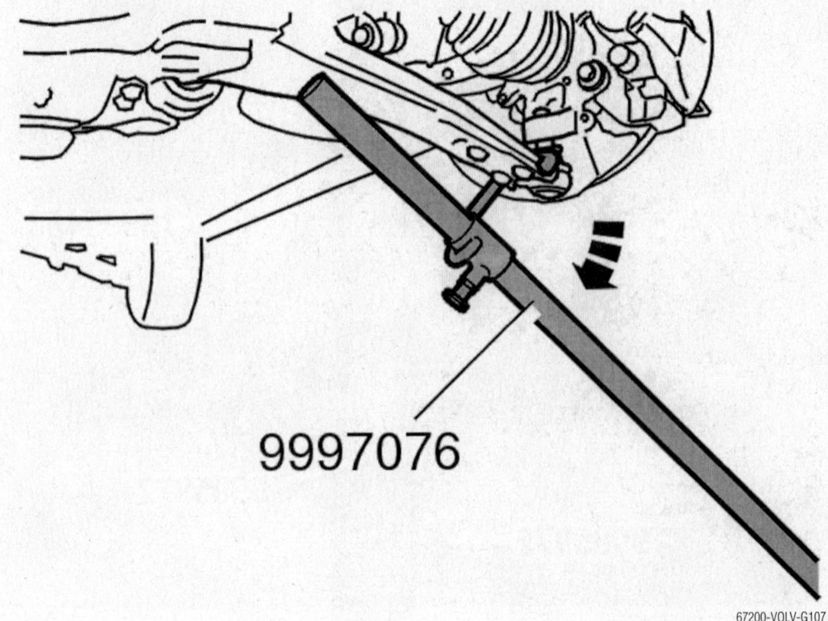

Remove the ball joint from the control arm.

- Bevel gear, AWD only
- Transaxle oil pipe using new O-ring
- Transaxle ground cable
- Control arm
- Spring strut
- Drive shafts
- Subframe using new bolts
- Power steering line
- Cross members
- Anti-roll bar link using new lock nut
- ABS line in the mounting in the spring strut
- Inner fender liner
- Sub-frame cable duct
- Parking heater, if equipped
- Splash guard
- Turbo coolant hoses, if equipped
- Starter motor
- Heated oxygen (HO_2S) sensor connector bracket
- Air intake housing bracket
- Transaxle oil pipe
- Gear selector cable and mount

14. Install or connect the following on turbocharged engines only:
    - CAC pipe
    - Air intake hose
    - Air intake assembly
    - Torque rod
    - Engine ground cable
15. Install or connect the following on naturally aspirated engines only:
    - Air intake assembly with the intake pipe and preheating hose
    - Vacuum hoses for the intake manifold from the control valves
    - Engine ground cable
    - Upper engine stabilizer brace
16. Connect the negative battery cable
17. Fill the transaxle to the correct level.

2005-06 S40/V50/C70 Models

MANUAL

1. Drain the transaxle fluid.

➡Leave the key in the ignition switch.

2. Remove or disconnect the following:
    - Negative battery cable
    - Battery box
    - Front wheels
    - Engine splash guard
    - Halfshafts
    - Anti-roll bar
    - Steering gear and hang securely out of the way
    - Sub-frame
    - Transmission ground lead
    - Torque rod mounting
    - Air intake assembly
    - Transmission cables
    - Coolant hose bracket
    - Clutch cable
    - Gear selector cable bracket
3. Install lifting beam 999-7103 and lifting hook 999-7102 to the engine.
4. Remove the left hand engine mount and lower the engine slightly.
5. Remove or disconnect the following:
    - Engine mount bracket
    - Starter motor
    - Top three mounting bolts between the engine and the transmission
    - Reverse lamp switch
6. Install transaxle fixture 999 5463 on transaxle jack, using the torque rod mounting bolts to hold it in place. At the same time, fit tool 999 5463-2 support plate and raise the jack so that it is making light contact.
7. Remove the remaining mounting bolts between the engine and transmission.
8. Remove the transmission.

To install:

9. Install or connect the following:
    - Transmission and tighten the lower seven mounting bolts to 37 ft. lbs. (50 Nm)
    - Reverse lamp switch
    - Engine mount bracket and tighten to 59 ft. lbs. (80 Nm)
    - Starter motor
    - Top three mounting bolts between the engine and transmission and tighten to 37 ft. lbs. (50 Nm)
10. Raise the transmission slight and install the left hand engine mount bolt and tighten to 96 ft. lbs. (130 Nm).
11. Install or connect the following:
    - Gear selector cable bracket and tighten to 18 ft. lbs. (24 Nm)
    - Clutch cable
    - Coolant hose bracket
    - Transmission cables and adjust if necessary
    - Air intake assembly
    - Transmission ground lead
    - Torque rod mounting and tighten to 37 ft. lbs. (50 Nm)
    - Sub-frame
    - Anti-roll bar
    - Steering gear
    - Halfshafts
    - Engine splash guard
    - Front wheels
    - Battery box and battery
12. Fill the transmission to the correct level.
13. Start the engine and check for leaks.

AUTOMATIC

1. Drain the transaxle fluid.

➡Leave the key in the ignition switch.

2. Remove or disconnect the following:
    - Negative battery cable
    - Battery box
    - Front wheels
    - Engine splash guard
    - Halfshafts
    - Anti-roll bar
    - Steering gear and hang securely out of the way
    - Sub-frame
    - Oil line, and plug the line and hole with plastic plugs
    - Transmission ground lead
    - Torque rod mounting
    - Torx® bolts from the torque converter
    - Air intake assembly
    - Coolant hose bracket
    - Gear selector cable bracket
    - Starter motor
    - Top four bolts between the engine and transmission
3. Install lifting beam 999-7103 and lifting hook 999-7102 to the engine.
4. Remove the left hand engine mount and lower the engine slightly.
5. Remove or disconnect the following:
    - Engine mount bracket and tighten to 37 ft. lbs. (50 Nm)
6. Install transaxle fixture 999 5463 on transaxle jack, using the torque rod mounting bolts to hold it in place. At the same time, fit tool 999 5463-2 support plate and raise the jack so that it is making light contact.
7. Remove the remaining mounting bolts between the engine and transmission.
8. Remove the transmission.

To install:

9. Install or connect the following:
    - Transmission and tighten the lower six mounting bolts to 37 ft. lbs. (50 Nm)
    - Engine mount bracket and tighten to 37 ft. lbs. (50 Nm)
10. Raise the transmission slight and install the left hand engine mount bolt and tighten to 96 ft. lbs. (130 Nm).
11. Install or connect the following:
    - Top four mounting bolts between the engine and transmission and tighten to 37 ft. lbs. (50 Nm)
    - Starter motor
    - Coolant hose bracket
    - Gear selector cable bracket and tighten to 18 ft. lbs. (24 Nm)
    - Air intake assembly
    - Torx® bolts for the torque converter
    - Oil line
    - Transmission ground lead
    - Torque rod mounting and tighten to 37 ft. lbs. (50 Nm)

- Sub-frame
- Anti-roll bar
- Steering gear
- Halfshafts
- Engine splash guard
- Front wheels
- Battery box and battery

12. Fill the transmission to the correct level.
13. Start the engine and check for leaks.

6-Cylinder Engine

1. Drain the transaxle fluid.
2. Remove or disconnect the following:
   - Negative battery cable
   - Air intake assembly
   - Charge air cooler (CAC) pipe, if equipped
   - Upper engine stabilizer brace
   - Charge air hose and clamp
   - Gear selector cable
   - Heater hose
   - Clamp holding oil pipe to power steering
   - Boost pressure turbo pipe, if equipped
   - Transmission dipstick tube assembly
3. Install lifting fixture 999-7018 on the left hand side of the engine.
4. Remove or disconnect the following:
   - Top three bolts between the engine and transmission
   - Ignition setting sender
   - Starter motor
   - Positive cable from engine
   - Torx® bolts for torque converter
5. Install lifting beam 999-5716 and lifting hook 999-5460 on the engine.
6. Raise the vehicle and remove or disconnect the following:
   - Front wheels
   - Splash guard
   - Parking heater, if equipped
   - Lower ball joints
   - Control arms
   - Subframe support plates
   - Right axle half shaft
   - Steering gear
   - Cross members
   - Heated oxygen (HO_2S) sensor and bracket
   - Left axle half shaft
   - Rear engine mount bolt
7. Lower the sub-frame approximately 5.5 in. (140mm) and secure the steering gear.

➡The steering shaft must not slide out of the rubber boot.

8. Remove or disconnect the following:

67200-VOLV-G102

Lifting beam installed on engine

   - Rear engine mount
   - Transmission oil pipes
9. Lower the left side of the engine and transaxle assembly slightly.
10. Install fixture 999 5463 on fixture 999 5972. Install this assembly on the transaxle jack. Install fixture 999 5735 on the entire assembly.
11. Install the transaxle jack against the transaxle and secure the side mountings.
12. Remove the transmission flange bolts.
13. Remove the transmission.

To install:

14. Install or connect the following:
   - Transmission and tighten crosswise the bolts to 37 ft. lbs. (50 Nm)

➡If necessary, turn the engine so the stator shaft is correctly aligned in the torque converter.

   - Splash guard above the transmission
   - HO_2S sensor and bracket and tighten to 18 ft. lbs. (25 Nm)
15. Raise the engine and transmission up to the horizontal position.
16. Install or connect the following:
   - Transmission oil pipes and tighten to 22 ft. lbs. (30 Nm)
   - Rear engine mount and tighten the bolts to 37 ft. lbs. (50 Nm)
   - Rear engine pad and tighten the bolt to 37 ft. lbs. (50 Nm)
   - Sub-frame, using new bolts. Tighten the frame bolts to 79 ft. lbs. (105 Nm) plus 120 degrees. Tighten the bracket bolts to 38 ft. lbs. (50 Nm).
   - Left axle half shaft
   - Right axle half shaft
   - Cross members
   - Steering gear
   - Lower ball joints
   - Control arms
   - Parking heater, if equipped
   - Torx® bolts for torque converter to 26 ft. lbs. (35 Nm)
   - Clamp for power steering oil pipe
   - Charge air hose and clamp
   - Top three bolts between the engine and transmission and tighten to 37 ft. lbs. (50 Nm)
   - Ignition setting sensor and tighten to 84 inch lbs. (10 Nm)
   - Positive cable to the engine
   - Starter motor and tighten to 30 ft. lbs. (40 Nm)
   - Transmission dipstick tube assembly
   - Splash guard
   - Boost pressure turbo pipe, if equipped
   - Heater hose
   - Gear selector cable
   - Upper engine stabilizer brace and tighten the bolts to 38 ft. lbs. (50 Nm)
   - CAC pipe, if equipped
   - Air intake assembly
   - Negative battery cable
17. Fill the transmission to the correct level

Clutch

ADJUSTMENT

The vehicles covered are equipped with a hydraulic clutch system that is self-adjusting.

REMOVAL & INSTALLATION

1. Remove the transmission.
2. Block the flywheel using gear sector 999-5112.

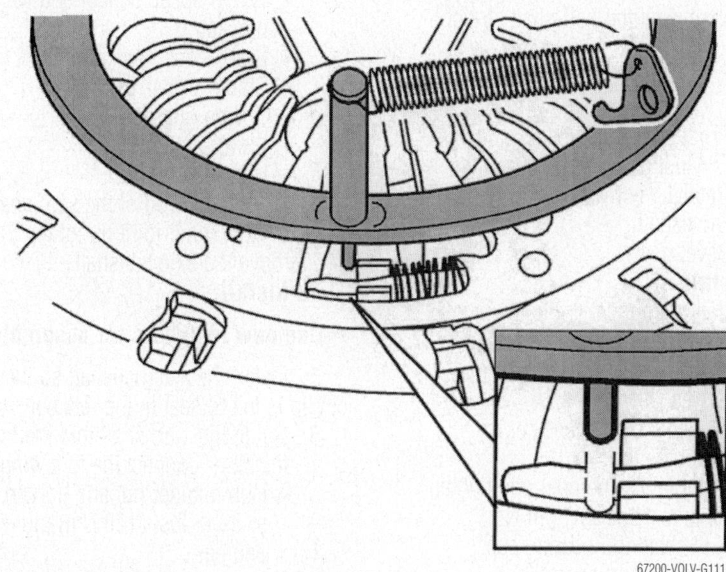

Secure the clutch automatic adjustment

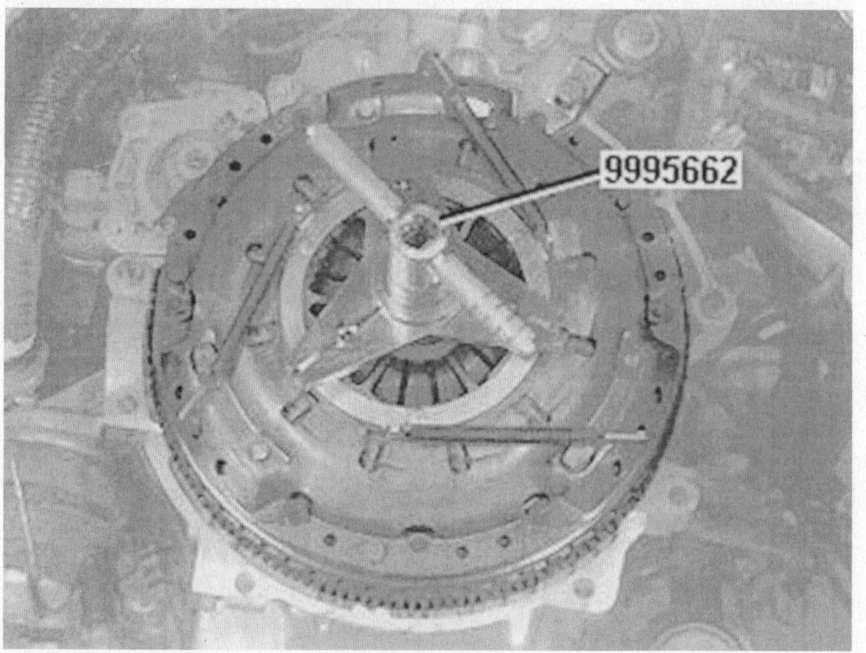

Install the compression tool onto the clutch to release the load.

3. Secure the clutch automatic adjustment with counterhold 999-5677.
4. Install compression tool 999-5662 on the clutch. Center the tool and tighten the screw to make light contact with the diaphragm spring.
5. Compress the clutch so the automatic adjuster is not under load. Screw in the compression tool until the diaphragm spring has pressed the pressure plate to a free position.
6. Remove the pressure plate and clutch disk.

To install:

7. Install compression tool 999-5662 on the clutch. Center the tool and tighten the screw to make light contact against the diaphragm spring, then compress the clutch.
8. Ensure the clutch automatic adjustment is secured with counterhold 999-5677.
9. Install the clutch, ensure the clutch is in contact with the flywheel all around its circumference, and tighten the six bolts crosswise and tighten to 18 ft. lbs. (25 Nm).
10. Remove the compression tool and gear sector.
11. Install the transmission.

Hydraulic Clutch System

BLEEDING

**** CAUTION**

Use only DOT 4+ brake fluid.

1. Before servicing the vehicle, refer to the Precautions Section.
2. Depress the clutch pedal a few times to purge the air bubbles in the master cylinder.
3. Connect a hose from a drain bottle to the nipple on the slave cylinder.
4. While the clutch pedal is depressed to the floor, open the bleed nipple.
5. Hold the pedal to the floor to allow brake fluid and air bubble to exit through the hose. Close the nipple.
6. Repeat this procedure until no air bubbles are visible in the escaping fluid.
7. Pump the clutch pedal a few times to build pressure in the system.
8. Check the fluid reservoir. The fluid level should not be above the MAX level.

Final Drive Assembly

REMOVAL & INSTALLATION

1. Before servicing the vehicle, refer to the Precautions Section.
2. Remove or disconnect the following:
   - Rear wheels
   - Rear exhaust system assembly

➡Support the front section of the exhaust system so that there is no strain on the exhaust manifold.

   - Driveshaft
   - Control arms
   - Cables for the parking brake
   - Anti-roll bar

- Active on Demand Coupling (AOC) connector

3. Install fixture 999 5972 on the jack and support the final drive assembly.
4. Remove the final drive assembly.

To install:

5. Install or connect the following:
   - Final drive assembly
   - AOC connector
   - Anti-roll bar
   - Control arms
   - Cable for the parking brake
   - Driveshaft
   - Rear exhaust system assembly
   - Rear wheels

Bevel Gear

REMOVAL & INSTALLATION

2005-06 S40 and V50 Models

1. Remove or disconnect the following:
   - Right half shaft
   - Charge air hose under engine
   - Heated oxygen sensor (HO_2S) wiring
   - Lower torque rod
   - Driveshaft CV joint
   - Bevel gear bracket attached to engine block
   - Bevel gear heat shield
   - Bevel gear

To install:

2. Install or connect the following:
   - Bevel gear and tighten bolts to 37 ft. lbs. (50 Nm)
   - Heat shield and tighten bolts to 18 ft. lbs. (24 Nm)
   - Bevel gear bracket. Tighten bolts into bevel gear to 37 ft. lbs. (50 Nm) and bolts into engine block to 18 ft. lbs. (24 Nm).
   - Driveshaft CV joint
   - Lower torque rod
   - HO_2S wiring
   - Charge air hose
   - Right half shaft

S60, V70, S80 Models

➡Ensure the parking brake is not applied and the gear selector is in position P.

1. Remove or disconnect the following:
   - Engine splash guard
   - Right front wheel
   - Bolt for the half shaft and wheel hub
   - Link for the anti-roll bar in the spring strut
   - Nut for the ball joint/link arm. Counterhold the ball joint pinion with a Torx® wrench so the boot does not twist.
2. Position a jack under the right-hand side of the sub-frame. Remove the bolts from the support bracket for the right-hand side of the sub-frame.
3. Slacken off, but do not remove, the remaining bolts for the sub-frame.
4. Unhook the control arm from the ball joint.
5. Press out the wheel spindle.
6. Detach the half shaft from the hub, remove the bolts from the bearing cap for the half-shaft. Pull the half-shaft straight out.
7. Remove or disconnect the following:
   - Front cross-member
   - Exhaust pipe
   - Catalytic converter
   - Driveshaft
   - Bevel gear

To install:

8. Install or connect the following:
   - Bevel gear, and tighten bolts to 37 ft. lbs. (50 Nm)
   - Driveshaft
   - Catalytic converter
   - Exhaust pipe
   - Front cross-member and tighten bolts to 18 ft. lbs. (24 Nm)
   - Half shaft and bearing cap.
   - Ball joint and control arm
   - Anti-roll bar link
   - Engine splash guard
   - Wheel
9. Fill the bevel to the correct level.

Halfshaft

REMOVAL & INSTALLATION

Front

1. Before servicing the vehicle, refer to the Precautions Section.
2. Remove or disconnect the following:
   - Front wheel
   - Wheel speed sensor and wiring bracket
   - Brake hose bracket
   - Stabilizer bar link
   - Splash guards
   - Lower ball joint
   - Hub retainer nut
3. Pull the hub off of the stub shaft.
4. Pry the inner joint out of the transaxle and remove the axle halfshaft.

To install:

➡Use new fasteners for assembly.

5. Install the axle halfshaft so that the circlip is felt to seat in the retainer groove.
6. Guide the stub shaft into the hub.
7. Install or connect the following:
   - Hub retainer nut and tighten the nut to 89 ft. lbs. (120 Nm) plus 60 degrees
   - Lower ball joint

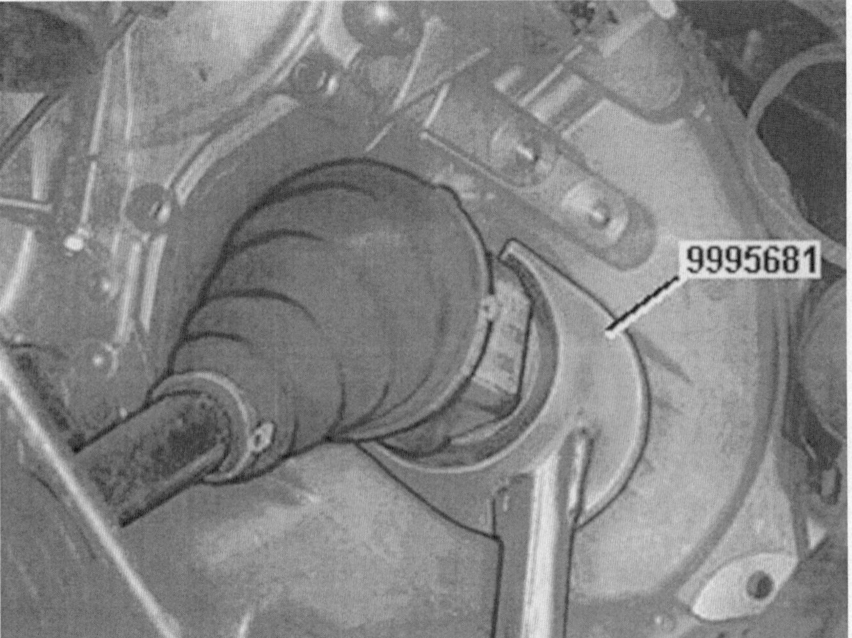

67200-VOLV-G106

Prying the half shaft from the differential

- Splash guards
- Stabilizer bar link
- Brake hose bracket
- Wheel speed sensor and wiring bracket
- Front wheel

Rear

1. Before servicing the vehicle, refer to the Precautions Section.
2. Remove or disconnect the following:
   - Rear wheel
   - ABS sensor
   - Hub retainer nut
   - Brake caliper and hang caliper securely from the spring
3. Remove the bolts between the spring strut and wheel spindle. Angle the hub out and remove the drive shaft from the hub.
4. Using special tool 999-5681 or equivalent, pry in the inner joint out of the final drive assembly and remove the half-shaft.

To install:

➡Use new fasteners for assembly.

5. Fit the inner joint stub shaft into the final drive assembly and ensure the half shaft snap ring engages correctly in the groove on the differential gear.
6. Align the half shaft in the hub and spring strut in the wheel spindle and tighten new lock nuts to 77 ft. lbs. (105 Nm).
7. Install or connect the following:
   - Brake caliper
   - Hub retainer nut
   - ABS sensor
   - Rear wheel

CV-Joints

OVERHAUL

Inner CV-Joint

The inner CV-joint is serviced with the axle shaft as an assembly. The inner CV-joint boot can be serviced by removing the outer CV-joint.

Outer CV-Joint

1. Remove the half shaft from the vehicle and secure in a vise.
2. Remove the boot clamps and cut the boot off of the shaft.
3. Measure and record the distance of the tripod joint in relation to the CV joint housing.
4. Using a brass drift on suitable equivalent, tap the CV-joint housing off of the shaft.
5. Matchmark the tripod joint and remove the snap ring.
6. Matchmark and remove the vibration damper from the halfshaft, if equipped with automatic transmission.

⁂ WARNING

Do not separate the outer joint from the halfshaft.

To install:

7. Clean the outer joint and check for damage.
8. Install the vibration damper, if equipped with automatic transmission.
9. Using Special Tool 999-5519, tap the tripod joint onto the halfshaft.
10. Install the snap ring to the tripod joint.

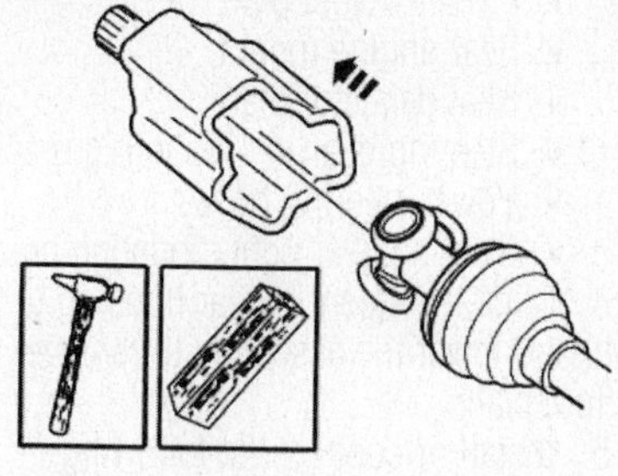

06041_VOLV_G0006

Tap the CV-joint housing from the shaft.

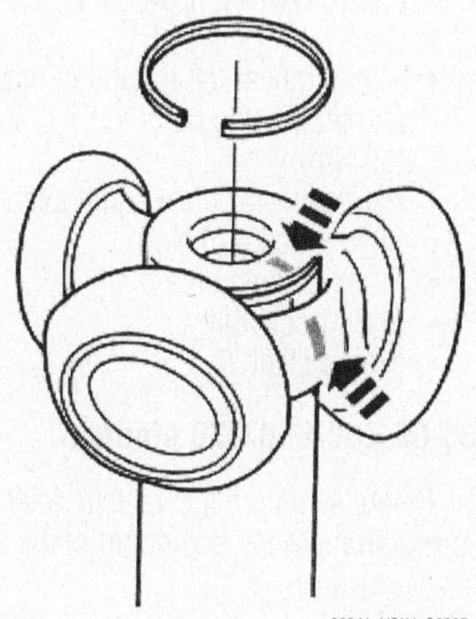

06041_VOLV_G0007

Matchmark the joint and remove the snap ring.

11. Lubricate the rollers of the tripod joint with grease.
12. Install the CV-joint housing.
13. Carefully insert a suitable prying device between the half shaft and boot to vent any over pressure in the boot.
14. Install new clamps onto the boot and tighten with Special Tool 951-2619.

STEERING AND SUSPENSION

Air Bag

⁂ CAUTION

Some vehicles are equipped with an air bag system. The system must be disarmed before performing service on, or around, system components, the steering column, instrument panel components, wiring and sensors. Failure to follow the safety precautions and the disarming procedure could result in accidental air bag deployment, possible injury and unnecessary system repairs.

DISARMING

1. Before servicing the vehicle, refer to the Precautions Section.
2. Disconnect the negative battery cable.
3. Wait at least 1 minute before working on the vehicle. The air bag system is designed to retain enough power to deploy the air bag for a short time after the battery has been disconnected.
4. After repairs are complete, connect the negative battery cable. Turn the ignition switch to the **ON** position and check the SRS light for proper operation.

Rack and Pinion Steering Gear

REMOVAL & INSTALLATION

2002–2004 S40/V40/C70 Models

1. Before servicing the vehicle, refer to the Precautions Section.
2. Separate the steering gear from the steering column by lifting the floor carpet below the steering column and remove the pinion shaft bolt.
3. Remove or disconnect the following:
   - Right front wheel
   - Lower ball joints
   - Center cross member

- Front exhaust pipe
- Rear engine mount
- Heat deflector plate
- Steering gear left mounting bracket
- Power steering hoses
- Steering gear right mounting bracket

4. Slide the steering gear housing to the right and lower the assembly to remove.

To install:

5. Install or connect the following:
   - Steering gear
   - Steering gear right mounting bracket and tighten the bolts to 51 ft. lbs. (69 Nm)
   - Power steering hoses with new O-rings
   - Steering gear left mounting bracket and tighten the bolts to 51 ft. lbs. (69 Nm)
   - Heat deflector plate and tighten to 88 inch lbs. (10 Nm)
   - Front exhaust pipe
   - Right front wheel
   - Pinion shaft

2005-06 S40/V50/C70 Models

1. Before servicing the vehicle, refer to the precautions in the beginning of the section.

➡Ensure the steering wheel is set to 'straight ahead' position and remove the key so the steering wheel lock engages.

2. Remove the left front wheel and measure and record the length of the tie rod in relation to the steering gear housing.
3. Remove or disconnect the following:
   - Left side soundproofing panel
   - Left front wheel
   - Steering shaft from the steering gear
   - Sub-frame
   - Powering steering hoses from the steering gear
4. Slacken off the nut for the outer steering arm. Press the steering arm out of the wheel spindle using special tool 951-2945 or equivalent.

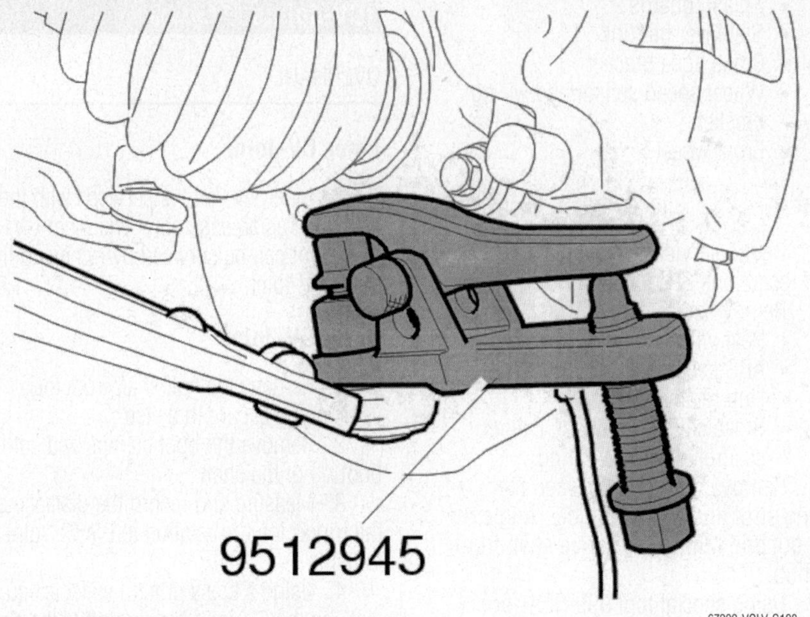

Pull the steering arm from the wheel spindle.

5. Remove the steering gear.

To install:

6. Install or connect the following:
   - Steering gear
   - Outer steering arm, tighten nuts to 37 ft. lbs. (50 Nm)
   - Power steering hoses using new O-rings
   - Sub-frame
   - Steering shaft to the steering gear
   - Wheel
   - Soundproofing panel
7. Fill and bleed the power steering system.

S60, S80, V70 and 2002–04 C70 Models

1. Before servicing the vehicle, refer to the Precautions Section.

➡Do not turn the front wheels so that the steering gear position is changed. The SRS-system contact reel can be damaged.

2. Raise and support the engine.
3. Remove or disconnect the following:
   - Front wheels
   - SRS connector, if equipped
   - Disconnect the connector cable from the clips along the ABS cable.
   - Push down the connector and cable to the steering gear.
   - Outer tie rod ends
   - Splash guard
   - Exhaust pipe bracket
4. Support the subframe with a jack
5. Remove the side and rear subframe bolts and loosen the front subframe bolts.
6. Remove the heated oxygen (HO_2S) sensor bracket, 6-cylinder engines only
7. Lower the rear edge of the sub-frame approximately 3.5 inches (90 mm). Ensure that the steering gear screws are released from the sub-frame.
8. Remove or disconnect the following:
   - Rear engine mount
   - Power steering hoses
   - Steering shaft coupler
9. Slide the steering gear to the left and lower the assembly to remove.

To install:

10. Slide the steering gear in from the left hand side.
11. Install or connect the following:
    - Power steering hoses using new O-rings.
    - Steering shaft coupler and tighten to 15 ft. lbs. (20 Nm)
    - Rear engine mount and tighten to 37 ft. lbs. (50 Nm)
    - Subframe. Use new bolts and tighten them to 77 ft. lbs. (105 Nm) plus 120 degrees.
    - Subframe brackets and tighten to 37 ft. lbs. (50 Nm)
    - Heated oxygen (HO_2S) sensor bracket, 6-cylinder engines only
    - Exhaust pipe bracket and tighten to 18 ft. lbs. (25 Nm)
    - Outer tie rod ends
    - Splash guards
    - SRS connector, if equipped
    - Front wheels

Strut

REMOVAL & INSTALLATION

Front

2002–2004 S40 AND V40 MODELS

1. Remove or disconnect the following:
   - Wheel
   - Brake hose
   - ABS cable

- Anti-roll bar link
- Wheel spindle
- Strut upper mount nuts
- Strut assembly

To install:

2. Install or connect the following:
   - Strut assembly and tighten the upper mount nuts to 33 ft. lbs. (45 Nm)
   - Wheel spindle and tighten the bolts to 66 ft. lbs. (90 Nm)
   - Anti-roll bar link and tighten the bolts to 33 ft. lbs. (45 Nm)
   - ABS cable
   - Brake hose
   - Wheel
3. Check the wheel alignment and adjust as necessary.

2005-06 S40/V50/C70 MODELS

1. Remove or disconnect the following:
   - Wheel
   - Half shaft retainer bolt
   - Ball joint from the wheel spindle
   - Brake caliper
   - Half shaft from the wheel hub
   - Anti-roll bar link
2. Remove the lower strut mounting bolt and carefully tap the wheel spindle down until it release from the strut assembly.
3. Remove the three top strut mounting bolts and remove the strut assembly.

To install:

4. Install or connect the following:
   - Strut assembly using new fasteners
   - Wheel spindle to strut assembly and tighten to 66 ft. lbs. (90 Nm)
   - Half shaft into the wheel hub
   - Ball joint to the wheel spindle and tighten to 52 ft. lbs. (70 Nm)
   - Half shaft retainer bolt
   - Brake caliper
   - Anti-roll bar link
   - Wheel

S60, S80, V70 AND 2002–04 C70 MODELS

1. Remove or disconnect the following:
   - Wheel
   - Anti-roll bar link
   - ABS sensor
   - Bolts holding the strut in the wheel spindle
   - Strut upper mount nuts
   - Strut assembly

To install:

➡Use new fasteners for assembly.

2. Install or connect the following:
   - Strut assembly and tighten the upper mount nuts to 18 ft. lbs. (25 Nm)
   - Strut assembly onto the wheel spindle and tighten to 77 ft. lbs. (105 Nm) plus 90 degrees
   - Anti-roll bar link and tighten to 37 ft. lbs. (50 Nm)
   - ABS sensor
   - Wheel

Shock Absorber

REMOVAL & INSTALLATION

Front

1. Before servicing the vehicle, refer to the Precautions Section.
2. Remove the strut assembly
3. Install spring compressor 951-2911 or suitable equivalent on the coil spring.
4. Remove the shock absorber mounting nut using socket 999-5500 using a Torx® as a counterhold.
5. Remove the shock absorber.

To install:

6. If replacing, transfer the ABS line holder to the new shock absorber.
7. Install spring compressor 951-2911 or suitable equivalent on the coil spring.
8. Install or connect the following:
   - Shock absorber and tighten the upper shock nut to 52 ft. lbs. (70 Nm)
   - Strut assembly
9. Check the wheel alignment and adjust as necessary.

Rear

2002–2004 S40 AND V40 MODELS

1. Before servicing the vehicle, refer to the Precautions Section.
2. Remove the side access panel in the trunk compartment
3. Remove or disconnect the following:
   - Anti-roll bar
   - Wheel
   - Lower shock absorber mounting nut
   - Lower trailing arm bolt
   - Upper shock absorber nut
   - Shock absorber

To install:

4. Install or connect the following:
   - Shock absorber and tighten the upper mounting nut to 37 ft. lbs. (50 Nm)
   - Lower trailing arm bolt using a new nut.
   - Lower shock absorber mounting nut using a new nut.
   - Wheel
   - Anti-roll bar

2005-06 S40 /V50/C70 MODELS

1. Before servicing the vehicle, refer to the Precautions Section.
2. Remove or disconnect the following:
   - Wheel
   - Upper shock absorber mounting bolts
   - Lower shock absorber mounting bolt
   - Shock absorber

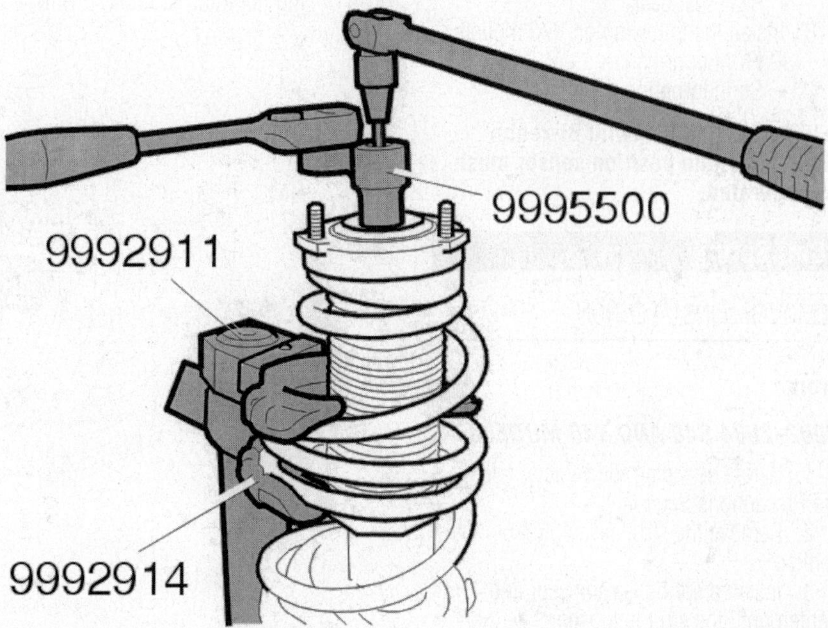

Removing the shock mounting nut

To install:

3. Install or connect the following:
   - Shock absorber
   - Upper shock absorber bolts and tighten to 18 ft. lbs. (24 Nm)
   - Lower shock absorber bolt
   - Wheel

S60, S80 AND V70 MODELS

1. Before servicing the vehicle, refer to the Precautions Section.
2. The following applies to S60 and S80 models:
   a. Fold the rear seat back forward
   b. Fold out the front edge of the side panel in the cargo compartment
   c. Remove the soundproofing panels
3. The following applies to V70 models:
   a. Remove the front floor hatch and sound proofing panels.
4. Remove the upper shock absorber mounting nut using socket 999-5500 and a Torx® as a counterhold.
5. Remove the lower shock absorber mounting bolt and remove the shock absorber.

To install:

6. Install or connect the following:
   - Shock absorber and tighten the lower bolt to 59 ft. lbs. (80 Nm)
   - Upper shock absorber nut using new nut and tighten to 44 ft. lbs. (60 Nm)
7. Install the following on S60 and S80 models:
   - Soundproofing panels
   - Side panel in the cargo compartment
   - Rear seat back
8. Install the following on V70 models
   - Front floor hatch
   - Soundproofing panels

➡If car is equipped with Bi-xenon headlamps, the position sensor must be calibrated.

Coil Spring

REMOVAL & INSTALLATION

Front

2002–2004 S40 AND V40 MODELS

1. Before servicing the vehicle, refer to the Precautions Section.
2. Remove the strut assembly from the vehicle.
3. Install a spring compressor and tighten until the strut is loose inside the spring.
4. Remove or disconnect the following:
   - Upper strut mount
   - Spring retainer
   - Coil spring

To install:

5. Install or connect the following:
   - Coil spring
   - Spring retainer
   - Upper strut mount. Tighten the nut to 52 ft. lbs. (70 Nm).
6. Remove the spring compressor.
7. Install the strut to the vehicle.
8. Check the wheel alignment and adjust as necessary.

S60, C70, S80, V70 AND 2005-06 S40 AND V50 MODELS

1. Before servicing the vehicle, refer to the Precautions Section.
2. Remove the strut assembly
3. Install spring compressor 951-2911 or suitable equivalent on the coil spring.
4. Remove the shock absorber mounting nut using socket 999-5500 using a Torx® as a counterhold.
5. Remove or disconnect the following:
   - Shock absorber
   - Spring seat
   - Bump stop
   - Coil spring

To install:

6. Install spring compressor 951-2911 or suitable equivalent on the coil spring.
7. Install or connect the following:
   - Bump stop
   - Coil spring
   - Spring seat
   - Shock absorber and tighten mounting nut to 52 ft. lbs. (70 Nm)
8. Remove the spring compressor.
9. Install the strut into the vehicle.
10. Check the wheel alignment and adjust as necessary.

Rear

2002–2004 S40 AND V40 MODELS

1. Remove the side access panels from inside the trunk compartment.
2. Remove or disconnect the following:
   - Wheel
   - Anti-roll bar
   - Shock absorber
3. Install a spring compressor and tighten until the shock absorber is loose inside the spring.
4. Remove or disconnect the following:
   - Upper shock nut
   - Upper spring cap
   - Coil spring

To install:

5. Pull out the shock absorber piston as far as possible.
6. Install or connect the following:
   - Coil spring
   - Upper spring cap
   - Upper shock nut and tighten to 22 ft. lbs. (30 Nm)
   - Shock absorber
   - Anti-roll bar
   - Wheel
7. Reinstall the side access panels inside the trunk compartment.

2005-06 S40/V50/C70 MODELS

1. Before servicing the vehicle, refer to the Precautions Section.
2. Remove the rear wheel

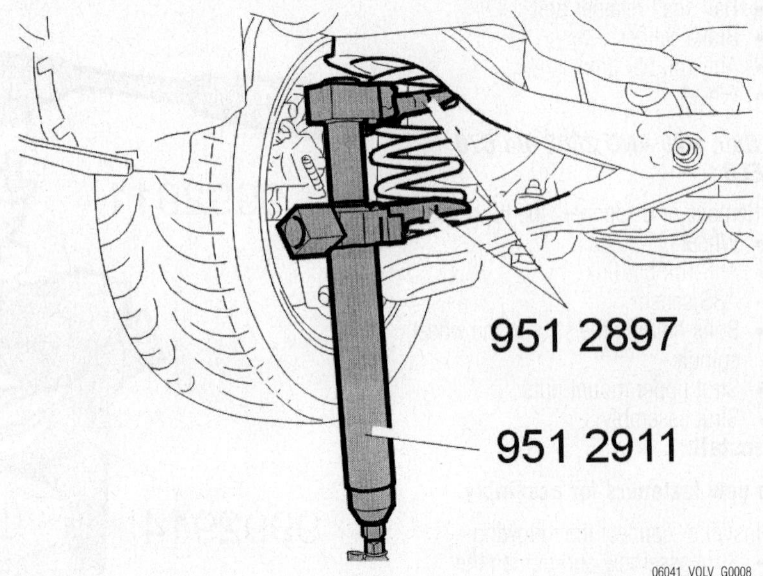

Using spring compressor 951-2911 to remove the rear spring—2005–06 S40/V50/C70

3. For right-hand side spring only:
a. Place a transmission jack under the control arm to relieve the tension on the shock absorber.
b. Remove the shock absorber lower mounting bolt.
4. Install spring compressor 951-2911 or suitable equivalent to compress the spring.
5. Remove the spring.

To install:

6. Install spring compressor 951-2911 or suitable equivalent to compress the spring.
7. Install the spring.

➡**Ensure that the spring engages in the control arm grooves.**

8. For right side spring only:
a. Place a transmission jack under the control arm so the lower shock absorber mounting bolt can be installed.

S60, S80 AND V70 MODELS

1. Before servicing the vehicle, refer to the Precautions Section.
2. Install plate 999 7079 to secure the spring.
3. Install a retaining strap between the control arms.
4. Remove or disconnect the following:
   - Brake caliper
   - Shock absorber
5. Install a spring compressor and tighten until spring tension is relieved.
6. Remove or disconnect the following:
   - Grommet
   - Striker plate 999 7079
   - Coil spring

To install:

7. Install or connect the following:
   - Coil spring
   - Striker plate 999 7079 to secure the spring
   - Grommet
8. Remove the spring compressor.
9. Install or connect the following:
   - Shock absorber
   - Brake caliper
10. Remove striker plate 999 7079

2002–04 C70 MODELS

1. Remove the access panel located under the carpet in the trunk compartment.
2. Raise the rear suspension to remove the load from the rear springs.
3. Remove or disconnect the following:
   - Strut upper mount nuts
   - Shock absorber
   - Spring mounting nut
   - Coil spring

To install:

4. Install or connect the following:
   - Coil spring and tighten the mounting nut to 29 ft. lbs. (40 Nm)
   - Shock absorber
   - Strut upper mount nuts
5. Reinstall access panels and trunk compartment carpet.

Lower Ball Joint

REMOVAL & INSTALLATION

V70XC Models

1. Remove the stub axle
2. Place the stub axle in a soft jawed vise
3. Remove the two bolts holding the ball joint in the stub axle.
4. Tap out the ball joint alternately. Use a drift.

To install:

5. Clean the stub axle mating surfaces of the ball joint.
6. Center the ball joint against the stub axle. Use the bolt (but do not tighten the bolt yet)

➡**Use new bolts.**

7. Tap the ball joint fully into the stub axle. Tap alternately with a drift.
8. Tighten the bolts. Tighten to 37 ft. lbs. (50 Nm).
9. Install the stub axle

V70, S60, S80

1. Remove the front wheels.
2. Remove the nut from the control arm. Use a Torx wrench as a counterhold.

⁂ WARNING

Ensure that the tension strap is correctly secured in the control arms.

3. Pull down the control arms (1) using a tension strap (2).
4. Release the spring strut from the control arm.
5. Remove the bolt from the halfshaft. Use a screwdriver as a counterhold on the brake disc.
6. Remove the halfshaft from the hub. Knock the halfshaft out using a brass drift.
7. Push the spring strut to one side.
8. Remove the bolts from the ball joint.
9. Remove the rubber seal (1).

Installing the inner sleeve (P/N 999 5781-1)

10. Install the inner sleeve so that both halves hook securely in the flange (see arrow) on the ball joint.
11. Turn the inner sleeve so that the slit is at right angles to the bolt holes.

Installing the outer sleeve (P/N 999 5781-2)

12. Install the outer sleeve around the inner sleeve halves. Tighten the bolt (1).
13. Install an M12 nut on the ball joint bolt (2).

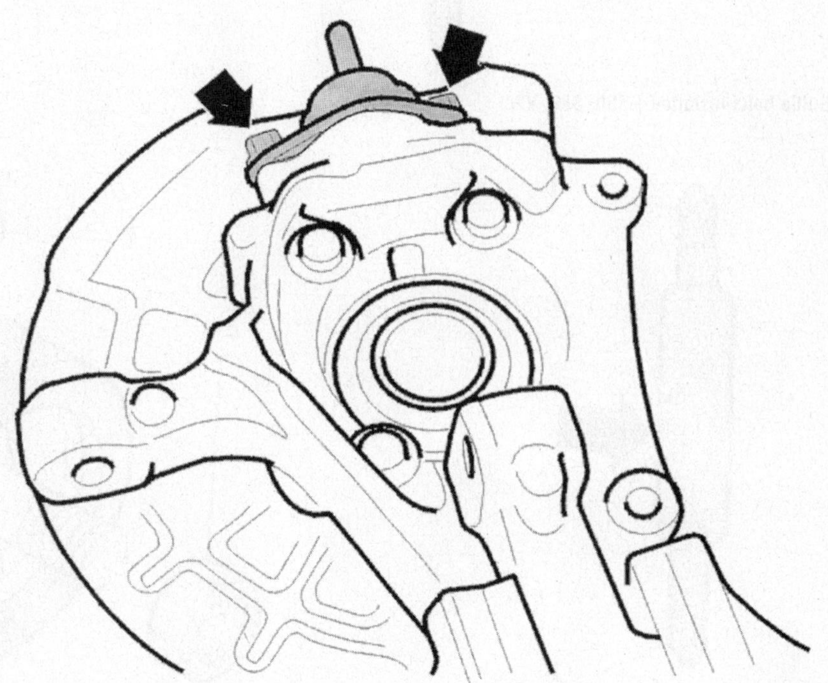

Ball joint removal—V70XC

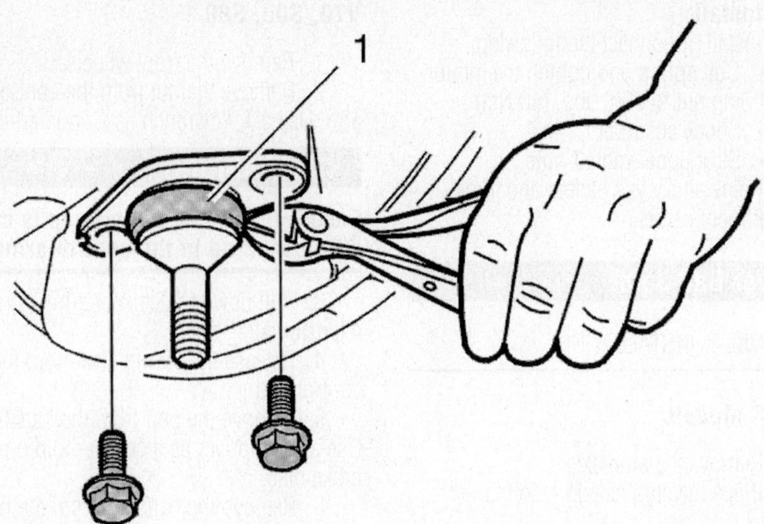

Removing rubber seal—S60, S80, V70

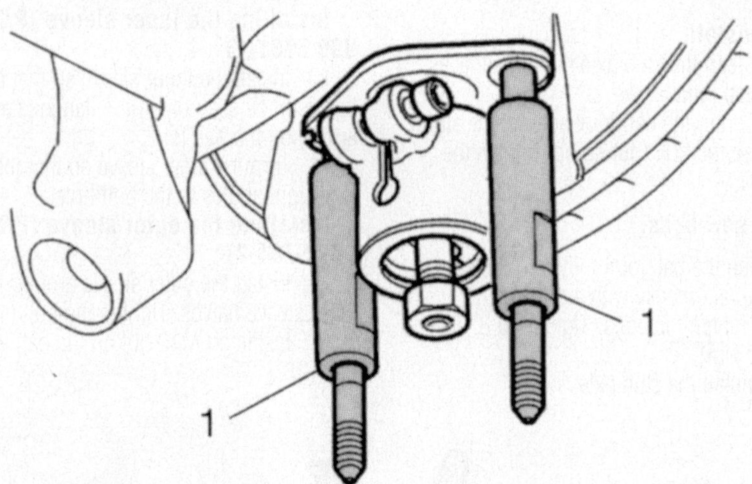

Guide bolts installed—S60, S80, V70

14. Screw on the nut so that the ball joint bolt is just protruding from the nut.

Installing the guide bolts (P/N 999 5781-3)

➡The guide bolts have a wrench grip (see arrow) which is asymmetrically positioned. 1. Aluminum wheel spindle: Use the bolt (1) with the greater distance from the wrench grip 2. Steel wheel spindle: Use the bolt (2) with the shortest distance from the wrench grip.

15. Screw in the guide bolts (1) fully.
16. Install the support (P/N 999 5781) using 3 wheel studs.
17. Screw the guide bolts (1) down until they are in contact with the support.
18. Position the supplementary support (2) (P/N 999 5781-4) on the hydraulic press (1).
19. Screw the connector (3) (P/N 999 5781-5) into place on the hydraulic press.
20. Install the connector on the nut on the ball joint.
21. Press off the ball joint.
22. Remove the tool and the ball joint.

⁂ WARNING

Use protective goggles.

23. Clean the ball joint seat and the mating surfaces of the ball joint to the wheel spindle. Use a rotary wire brush.
24. Lubricate the seat using wheel bearing grease.

To install:

25. Loosely install the new ball joint using the guide bolts.

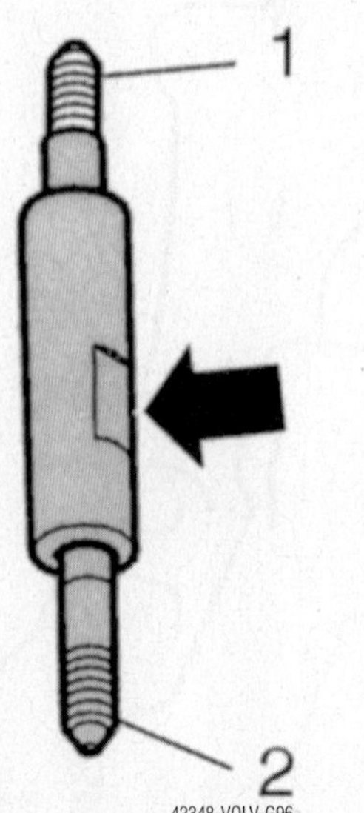

Guide bolt identification—S60, S80, V70

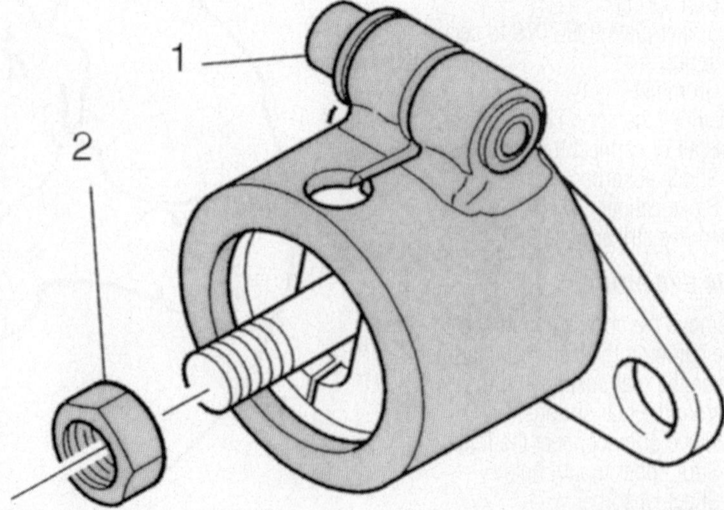

Installing outer sleeve—S60, S80, V70

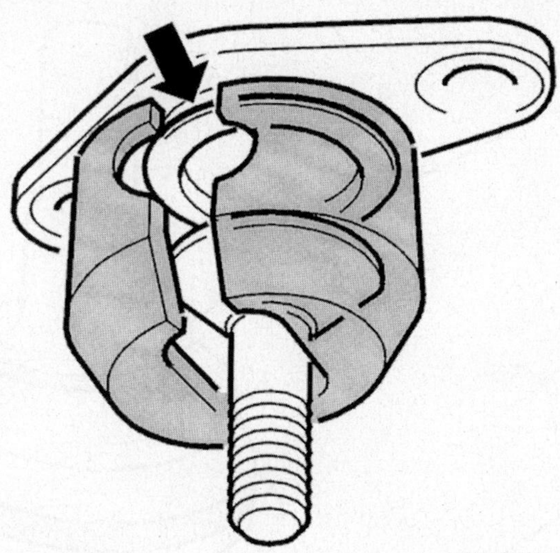

Installing inner sleeve—S60, S80, V70

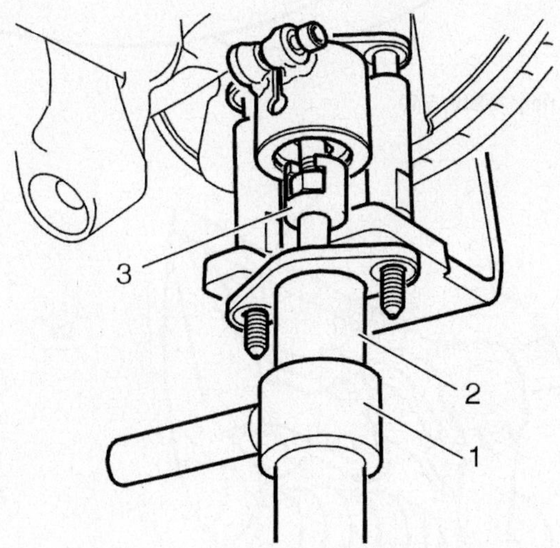

Ball joint removal tool installed—S60, S80, V70

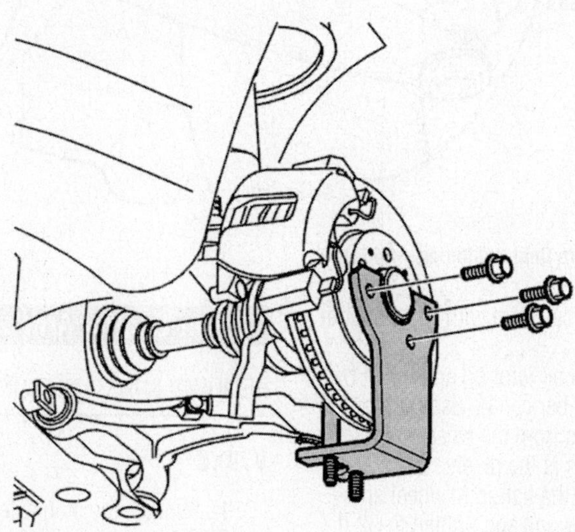

Support tool installed—S60, S80, V70

**** WARNING**

Leave the protective cap for the ball joint in position to prevent damaging the rubber seal.

26. Press the ball joint up towards the seat using the impact drift (1) (P/N 999 5796). Check that the ball joint is centered in the seat.
27. Knock in the ball joint using a copper mallet.
28. Tighten the ball joint using new bolts. Tighten to 37 ft. lbs. (50 Nm)
29. Remove the protective cap (1).
30. Install the halfshaft in the hub. Install a new bolt.
31. Align the ball joints (1) in the control arm.

**** CAUTION**

Take care when releasing the tension strap.

32. Slowly release the control arms (1) by releasing the tension strap (2).
33. Install a new nut.

➡The ball joint pinion must not rotate. Use a Torx wrench as a counterhold so that the rubber boot is not damaged. Tighten to 59 ft. lbs. (80 Nm).

34. Install the wheels. Tighten to 103 ft. lbs. (140 Nm).

V40, V50 and S40

1. Raise the car.
2. Slacken off the ball joint nut as much as possible.

➡Clean the ball joint thread using a wire brush. Use rust penetrator. If the ball joint releases in the cone, position a transmission jack under the control arm. Carefully lift (not more than so the front spring is in the resting position) until the cone jams. Remove the nut.

3. Remove the stud in the front edge
4. Remove the two bolts at the rear edge. Move the wheel to one side. Pull out the control arm from the cross member.

➡Ensure that the halfshaft is not pulled out of the transmission.

5. Slacken off the nut positioned on the ball joint as much as possible. Press the rubber down.

**** WARNING**

Secure the tool to prevent damage.

6. Press out the ball joint.
7. Remove the nut and draw out the control arm.

To install:

⁎⁎ CAUTION

Replace the ball joint rubber ring if it is worn or damaged.

8. Clean the area around the ball joint.
9. Remove the old rubber ring.
10. Lubricate the new ring and install it.
11. Ensure that the locking springs are correctly installed.
12. Install the control arm with the ball joint on the wheel spindle.
13. Install the control arm in the cross member. Install the bolt and nut.
14. Install the two bolts in the rear bushing.
15. Tighten all bolts and nuts by hand.
16. Post-tighten the two rear bolts. Tighten to 68 ft. lbs. (90 Nm).
17. Press the ball joint all the way down. Tighten to 51 ft. lbs. (67 Nm).
18. Remove the support and rock the car a few times so that the bushings come into the correct position.
19. Tighten the nut. Tighten to 68 ft. lbs. (90 Nm) + 60 deg.
20. Check the front mounting and wheel alignment.

C70

1. Remove the wheel
2. Remove the halfshaft bolt. Use a counterhold.
3. Detach the halfshaft end at the hub Tap in the shaft end approximately 10-15 mm. Use a rubber or copper mallet.
4. Blow clean around the ball joint with compressed air.
5. Generously spray the ball joint and the bolt and nut retaining the ball joint using rust penetrator.

➡Do not insert a screwdriver or punch between the ears on the wheel spindle.

6. If the bolt has rusted solid, slacken off the nut a few turns. Knock the bolt so that it loosens.
7. Remove the bolt and the nut.
8. Spray the ball joint with rust penetrator again.

➡The inner diameter of the socket must be larger than the outer diameter of the cover. This is to prevent damage to the cover.

9. Tap the ball joint off the wheel spindle. Use a socket as a spacer.

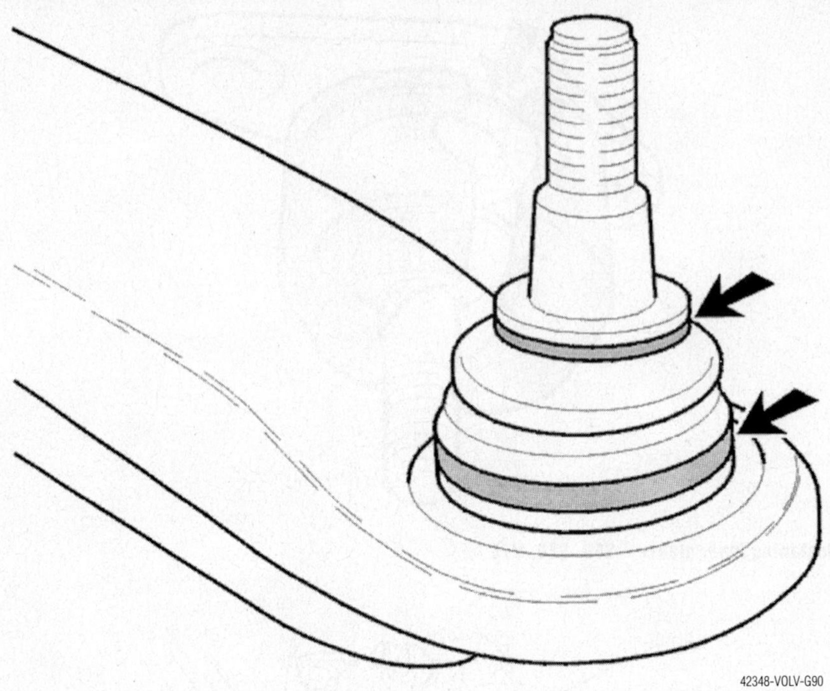

Inspect the rubber rings—S40, V40

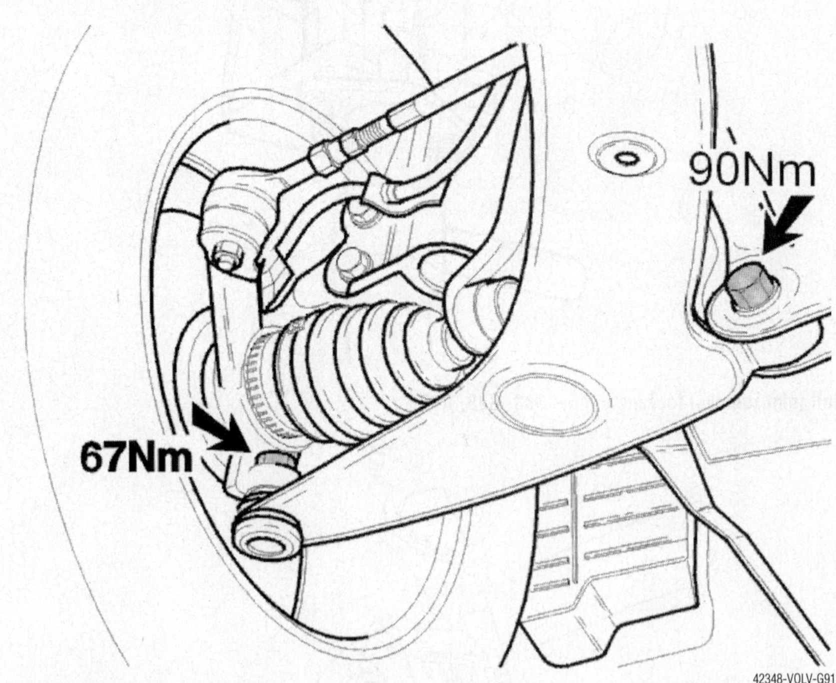

Lower control arm final tightening—S40, V40

10. Spray the ball joint with rust penetrator again.
11. Move the ball joint up and down by hand (or a jimmy bar can be used) until the ball joint detaches from the wheel spindle.
12. Installation is the reverse order of removal. Replace the ball joint wheel spindle nut with a new one and tighten to 52 ft. lbs. (70 Nm).

Lower Control Arm

REMOVAL & INSTALLATION

V70XC

1. Remove the stub axle.
2. Remove the splash guard under the engine.

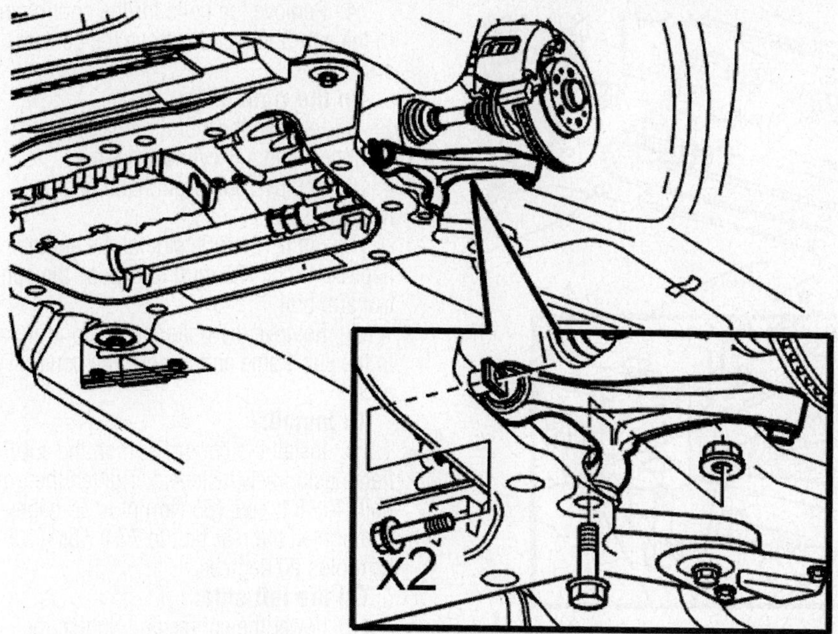

Lower control arm removal—V70XC

On the left side:

3. Remove the bolts and the nut from the control arm in the sub-frame
4. Remove the control arm.

On the right side:

5. Remove the hose from the EVAP canister and the fuel line from the engine block heater (where applicable). Move the hose and fuel line to one side.

➡When removing the rear bolt on the front mounting from the control arm, the engine must be raised slightly using a prybar

6. Remove the bolts and nut from the control arm in the sub-frame
7. remove the control arm.

To install:

8. Lift the engine slightly when installing the right control arm. When installing the rear bolt in the front mounting for the control arm, see removing the right front control arm.
9. Install the control arm on the sub-frame. Use new bolts and a new nut.
10. Tighten the front bolts. Tighten to 49 ft. lbs. (65 Nm). Angle-tighten 90 deg.
11. Tighten the rear bolt and nut. Tighten to 80 ft. lbs. (105 Nm). Angle-tighten 90 deg.
12. Install the stub axle.
13. Reinstall the splash guard under the engine.
14. Install the hose for the EVAP canister and the fuel line for the engine block heater (if applicable) in their mountings on the sub-frame.

V70 and S60

1. Remove the front wheels
2. Remove the nut on the ball joint. Use a Torx wrench as a counterhold so that the rubber boot is not damaged.
3. Remove the splash guard under the engine.
4. Install the retaining strap to bend the control arms down.

⁂ WARNING

Ensure that the tension strap is correctly secured in the control arms.

5. Pull down the control arms using a tension strap.
6. Release the spring strut from the control arm
7. Remove the tensioner strap.
8. Remove the bolt from the halfshaft. Use a counterhold on the brake disc
9. Remove the halfshaft from the hub. Knock the halfshaft out using a brass drift
10. Push the spring strut to one side.

On the right side:

11. Remove the hose from the EVAP canister and the fuel line from the engine block heater (where applicable) from their mountings. Move the hose and fuel line to one side.
12. Lift the engine approximately 10 mm using a mobile jack so that the front 2 bolts on the control arm can be removed
13. Position a mobile jack with a universal plate under the oil pan on the right side of the engine.
14. Remove the bolts and the nut from the control arm in the sub-frame
15. Remove the control arm.

On the left side:

16. Lift the engine approximately 15 mm using a mobile jack so that the front bolts on the control arm can be removed.
17. Position a mobile jack with a univer-

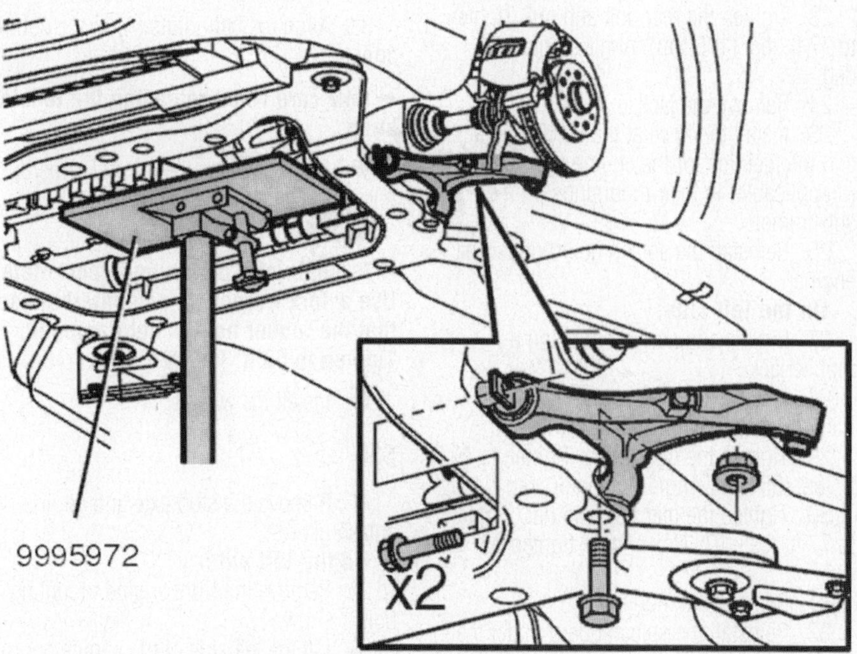

Left lower control arm removal—S60, S80, V70

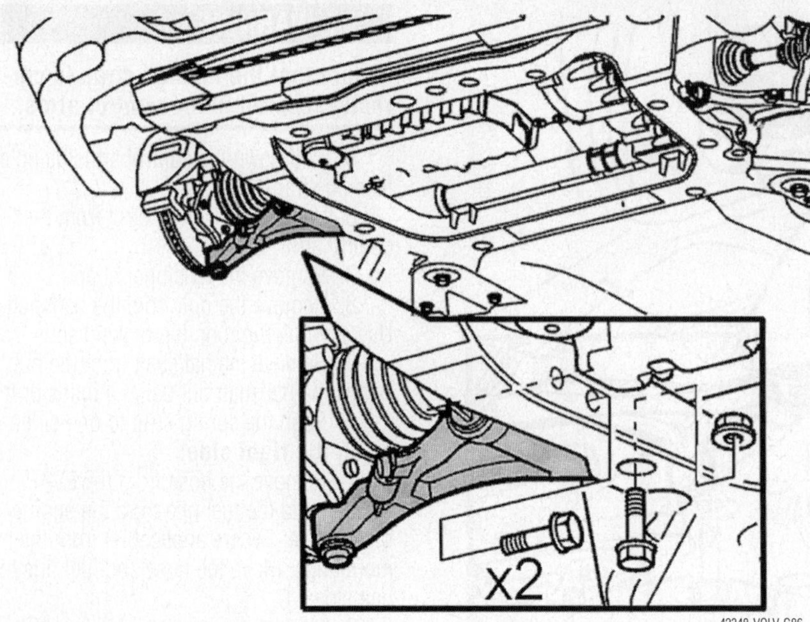

Right lower control arm removal—S60, S80, V70

sal plate under the oil pan on the left side of the engine.

18. Remove the bolts and the nuts from the control arm in the sub-frame

19. Remove the control arm.

To install:

On the right side:

20. For the right-hand control arm, lift the engine 10 mm using a mobile jack.

21. Install the control arm on the sub-frame. Use new bolts and a new nut

22. Tighten the front bolts. Tighten to 49 ft. lbs. (65 Nm). Angle-tighten 90 deg.

23. Tighten the rear bolt and nut. Tighten to 77 ft. lbs. (105 Nm). Angle-tighten 90 deg.

24. Remove the jack.

25. Install the hose at the EVAP canister and the fuel line for the engine block heater (if applicable) in their mountings on the sub-frame.

26. Reinstall the splash guard under the engine.

On the left side:

27. Lift the engine 15 mm using a mobile jack.

28. Install the control arm on the sub-frame. Use new bolts and new nuts

29. Tighten the front bolts. Tighten to 49 ft. lbs. (65 Nm). Angle-tighten 90 deg.

30. Tighten the rear bolt and nut. Tighten to 77 ft. lbs. (105 Nm). Angle-tighten 90 deg.

31. Remove the mobile jack

32. Reinstall the splash guard under the engine.

33. Turn in the wheel spindle and locate the halfshaft in the hub.

➡Check that the seal is not worn or damaged. Replace if necessary. Ensure that the sealing ring is correctly positioned on the halfshaft.

34. Clean the splines on the halfshaft.

35. Install the halfshaft in the hub.

36. Install a new bolt. Use a counterhold. Tighten to 27 ft. lbs. (35 Nm) plus 90 deg.

37. Install the retaining strap to bend the control arms down

➡Ensure that the tension strap is correctly secured in the control arms.

38. Align the ball joints (1) in the control arm.

➡Take care when releasing the tension strap.

39. Slowly release the control arms by releasing the tension strap.

40. Install a new nut

➡The ball joint pinion must not rotate. Use a Torx wrench as a counterhold so that the rubber boot is not damaged. Tighten to 59 ft. lbs. (80 Nm).

41. Install the wheels.

S80

1. Remove the stub axle and engine splash guard.

On the left side:

2. Remove the front engine mounting bolt.

3. Lift the left side of the engine approx. 1 inch (25 mm) to expose the control arm bolt.

4. Remove the bolts for the control arm in the sub-frame and remove the control arm.

On the right side:

5. Remove the vibration damper, if equipped with a 6-cylinder engine.

6. Remove both right hand engine mounting bolts.

7. Lift up the ride side of the engine approx. 1 inch (25 mm) to expose the control arm bolt.

8. Remove the bolts for the control arm in the sub-frame and remove the control arm.

To install:

9. Install the control arm on the sub-frame using new fasteners. Tighten the front bolts to 48 ft. lbs. (65 Nm) plus 90 degrees and tighten the rear bolt to 77 ft. lbs. (105 Nm) plus 90 degrees.

On the left side:

10. Lower the engine and tighten the front engine mounting bolt to 37 ft. lbs. (50 Nm).

On the right side:

11. Lower the engine and tighten the right engine mounting bolts to 26 ft. lbs. (35 Nm) plus 90 degrees.

12. Install the vibration damper using new bolts and tighten to 26 ft. lbs. (35 Nm) plus 50 degrees.

Applies to either side:

13. Install the stub axle and engine splash guard.

C70

1. Remove the wheel

2. Remove the halfshaft bolt. Use a counterhold.

3. Detach the halfshaft end at the hub Tap in the shaft end approximately 10-15 mm. Use a rubber or copper mallet.

4. Blow clean around the ball joint with compressed air.

5. Generously spray the ball joint and the bolt and nut retaining the ball joint using rust penetrator.

➡Do not insert a screwdriver or punch between the ears on the wheel spindle.

6. If the bolt has rusted solid, slacken off the nut a few turns. Knock the bolt so that it loosens.

7. Remove the bolt and the nut.

8. Spray the ball joint with rust penetrator again.

➡The inner diameter of the socket must be larger than the outer diameter of the cover. This is to prevent damage to the cover.

9. Tap the ball joint off the wheel spindle. Use a socket as a spacer.
10. Spray the ball joint with rust penetrator again.
11. Move the ball joint up and down by hand (or a jimmy bar can be used) until the ball joint detaches from the wheel spindle.
12. Withdraw the halfshaft from the hub.
13. Twist the spring strut to one side. Hang the halfshaft so that the control arm is not in contact with the halfshafts.
14. Remove the two bolts from the engine pad on the right side.
15. Remove the bolt (1) from the rear engine pad
16. Remove the bolt (3) from the front engine pad
17. Remove the bolt and nut (2) from the torque rod.

On the left side:

18. Position a jack on the torque rod mounting in the transmission.

➡Use a wooden block as a shim.

19. Remove the four bolts from the control arm
20. Remove the control arm.

On the right side:

21. Position a jack against the oil pan.

➡Use a wooden block as a shim.

22. Lift the engine to access the control arm bolts.
23. Remove the four bolts (1) from the control arm
24. Remove the control arm.

To install:

On the left side:

25. Install the control arm on the subframe. Use new bolts. Tighten to 49 ft. lbs. (65 Nm). Angle-tighten 90 deg.
26. Lower and remove the jack.

On the right side:

27. Install the control arm on the subframe. Use new bolts. Tighten to 49 ft. lbs. (65 Nm). Angle-tighten 90 deg.
28. Lower and remove the mobile jack.
29. Install the two bolts (1) for the engine pad on the right side. Tighten to 27 ft. lbs. (35 Nm). Angle-tighten 90 deg.

➡Use new bolts.

30. Install the bolts (1, 3). Tighten to 37 ft. lbs. (50 Nm).
31. Install the bolt and nut (2) for the torque rod. Tighten to 30 ft. lbs. (40 Nm). Angle-tighten 40 deg.

➡Use a new bolt and nut.

32. Clean the hub and the halfshaft splines.
33. Fold out and twist the spring strut and locate the halfshaft in the hub. Use an adjustable wrench to hold the control arm down.
34. Install a new halfshaft bolt. Tighten to 27 ft. lbs. (35 Nm). Angle-tighten 90 deg. Counterhold the brake disc.

➡Lubricate the threads and flange on the bolt.

35. Check that the groove on the ball joint spindle lines up with the bolt hole in the wheel spindle.

➡Press the ball joint upwards so that the conical top is visible above the bolted joint on the wheel spindle. Use a new locknut in the wheel spindle bolt. Tighten to 38 ft. lbs. (50 Nm).

36. Install the wheel.

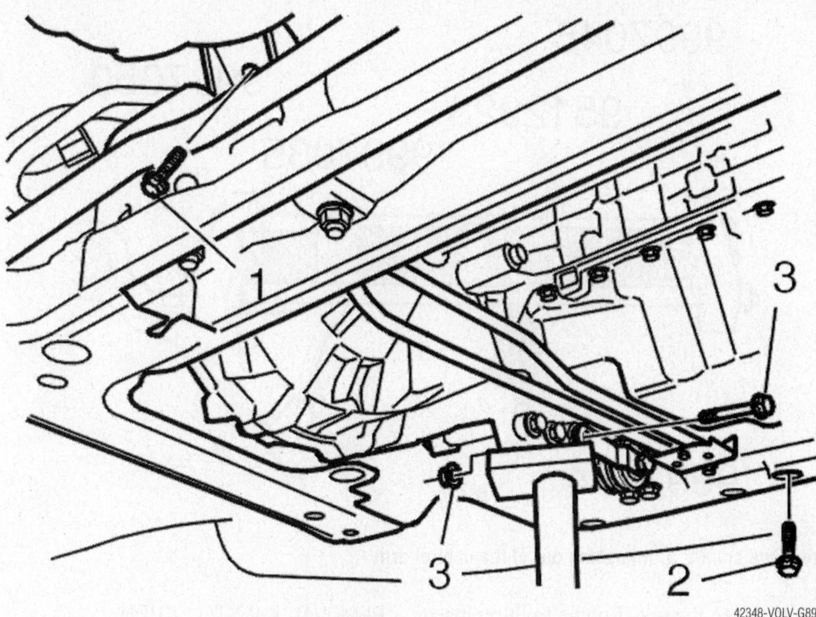

Support the torque rod with a jack—S40, V40

V40, V50 and S40

1. Raise the car.
2. Slacken off the ball joint nut as much as possible.

➡Clean the ball joint thread using a wire brush. Use rust penetrator. If the ball joint releases in the cone, position a transmission jack under the control arm. Carefully lift (not more than so the front spring is in the resting position) until the cone jams. Remove the nut.

3. Remove the stud in the front edge
4. Remove the two bolts at the rear edge.
5. Move the wheel to one side. Pull out the control arm from the cross member.

➡Ensure that the halfshaft is not pulled out of the transmission.

6. Slacken off the nut positioned on the ball joint as much as possible. Press the rubber down.

⁂ WARNING

Secure the tool to prevent damage.

7. Press out the ball joint.
8. Remove the nut and draw out the control arm.

To install:

⁂ CAUTION

Replace the ball joint rubber ring if it is worn or damaged.

9. Clean the area around the ball joint.
10. Remove the old rubber ring.
11. Lubricate the new ring and install it.
12. Ensure that the locking springs are correctly installed.
13. Install the control arm with the ball joint on the wheel spindle.
14. Install the control arm in the cross member. Install the bolt and nut.
15. Install the two bolts in the rear bushing.
16. Tighten all bolts and nuts by hand.
17. Post-tighten the two rear bolts. Tighten to 68 ft. lbs. (90 Nm).
18. Press the ball joint all the way down. Tighten to 51 ft. lbs. (67 Nm).
19. Remove the support and rock the car a few times so that the bushings come into the correct position.

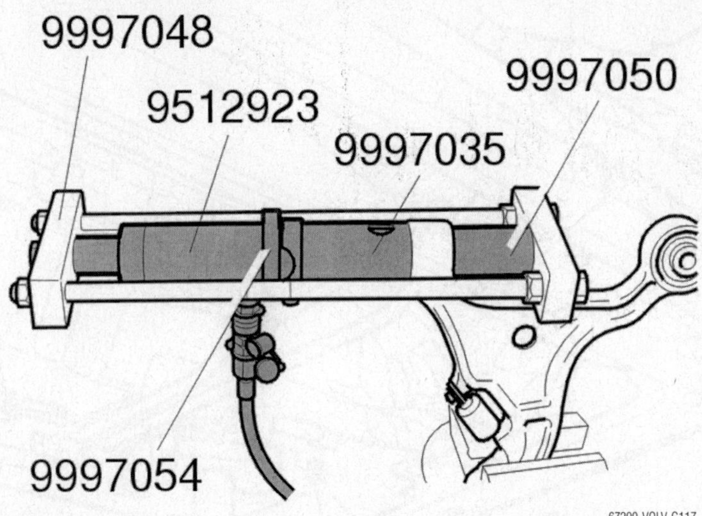

Press the control arm bushing out of the control arm

20. Tighten the nut. Tighten to 68 ft. lbs. (90 Nm) + 60 deg.
21. Check the front mounting and wheel alignment.

CONTROL ARM BUSHING REPLACEMENT

1. Remove the bushing. Use a press. Lubricate the tool with soap water to facilitate removal.
2. Cut the bushing in the necessary places so that the tool can be correctly installed.
3. Press the bushing out of the control arm.
4. Using a press, press the new bushing into the control arm.

➡Do not press in more than the measured bushing position.

⁂ CAUTION

Caution! Ensure that the bushing is at the correct angle in the control arm.

Wheel Bearing

ADJUSTMENT

Front

The front wheel bearings are not adjustable. If the lateral run-out on the hub with the disc removed exceeds 0.0007 in. (0.020mm), the hub must be replaced.

Rear

The rear wheel bearings are sealed, pressed-in units, and no adjustment is possible.

REMOVAL & INSTALLATION

Front

S60, S80 AND 2002–04 C70 W/STEEL SPINDLE

1. Remove the wheel
2. Remove the ABS sensor.
3. Detach the ABS sensor cable from the spring strut.
4. Remove both mounting bolts from the brake caliper.
5. Hang up the brake caliper, on the spring strut for example.

➡Take care not to damage the brake hose while working.

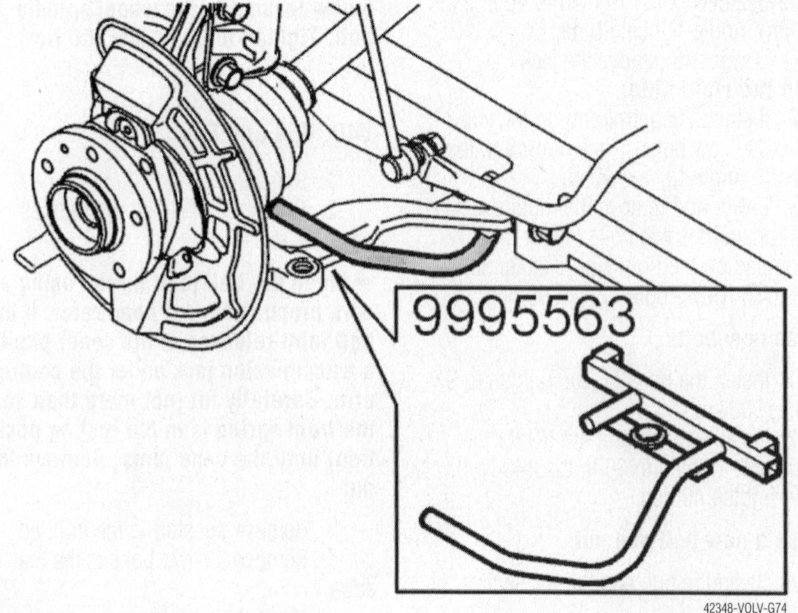

Securing the control arm—2002–04 C70

6. Remove the bolt holding the halfshaft. Counterhold the brake disc.
7. Remove the locating pin from the brake disc
8. Remove the brake disc.
9. Tap the end of the halfshaft into the hub approximately 10–15 mm. Use a brass drift.
10. Remove the upper link from the anti-roll bar from the spring strut.
11. Remove the steering arm from the wheel spindle.
12. Clean around the ball joint with compressed air.
13. Spray the area around the ball joint with plenty of rust penetrator. Spray the bolt and the nut holding the ball joint with rust penetrator. If the bolt has rusted solid, slacken of the nut a few turns. Knock the bolt so that it loosens.
14. Remove the nut and bolt.
15. Spray the ball joint with rust penetrator again.

➡Do not insert a punch between the ears on the wheel spindle.

16. Tap the ball joint off the wheel spindle. Use a socket as a spacer.

➡The inner diameter of the sleeve must be larger than the outer diameter of the cover. This is to prevent damage to the cover.

17. Spray the ball joint with rust penetrator again.
18. Move the ball joint up and down by hand (alternatively use a jimmy bar) until the ball joint detaches from the wheel spindle.

19. Install protective socket 999 5562 on the ball joint.

20. Secure the control arm in the depressed position using tool 999 5563.

21. Remove the halfshaft from the hub.

22. Turn the spring strut a half turn and lock it by positioning the steering arm on the control arm.

➡Take care not to damage the brake hose.

23. Hang up the halfshaft on the spring strut.

24. Remove the wheel bearing unit.

25. Check the wheel spindle mating surfaces against the wheel bearing unit. Also check the bolt head mating surfaces on the wheel spindle.

To install:

26. Lubricate all the mating surfaces sparingly.

27. Install the wheel bearing unit. Use new bolts.

28. Lightly lubricate the outer threads and tighten crosswise. Tighten to 49 ft. lbs. (65 Nm). Angle-tighten the bolts for the wheel bearing unit crosswise. Tighten an additional 60 deg.

29. Check the halfshaft for damage to the mating surfaces with the wheel bearing.

30. Check the splines.

31. Clean and check the pulse wheel. Check the shape of the cogs.

32. Lubricate the splines and mating surfaces with the wheel bearing sparingly.

33. Remove the halfshaft from where it was hung earlier. Turn the spring strut a half turn to disconnect it from the control arm.

➡Take care not to damage the brake hose.

34. Install the halfshaft in the hub by hand. It must be possible to move the halfshaft easily backward and forward when it is in position.

35. Lubricate a new halfshaft bolt. Screw it in a few turns by hand.

36. Clean off any rust from the ball joint and wheel spindle.

37. Lubricate the ball joint pinion. Lubricate the wheel spindle.

38. Install the ball joint in the wheel spindle

➡Press the ball joint upwards so that the conical top is visible above the bolted joint on the wheel spindle.

39. Install the bolt and a new nut. Tighten to 38 ft. lbs. (50 Nm).

40. Check the dust boot on the anti-roll bar link for damage.

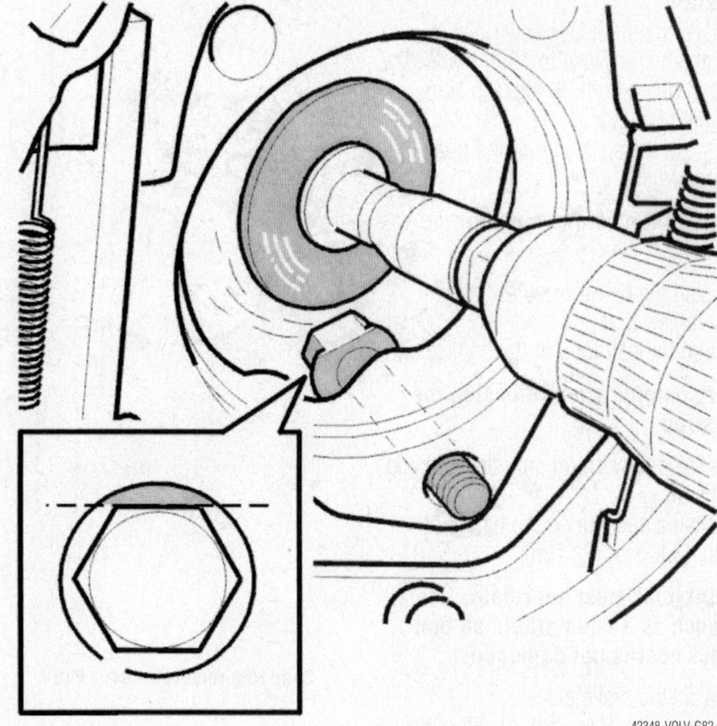

Grind the bolt flange as shown—V70

41. Check the threads on the link for damage.

42. Lubricate the threads.

43. Install the upper link of the anti-roll bar on the spring strut. Use a new lock nut. Tighten to 38 ft. lbs. (50 Nm).

44. Install the tie rod on the wheel spindle. Use a new lock nut. Tighten to 53 ft. lbs. (70 Nm).

45. Check that the brake disc mating surfaces with the wheel bearing unit are smooth and even.

46. Install the brake disc

47. Install the locating pin for the brake disc. Tighten to 72 inch lbs. (8 Nm).

48. Check the mating surfaces of the brake caliper and wheel spindle.

49. Clean and grease the surfaces of the steering limiter.

50. Install the brake caliper. Use new bolts. Tighten to 76 ft. lbs. (100 Nm).

51. Tighten the center bolt for the halfshaft to 27 ft. lbs. (35 Nm) plus 90 degrees. Counterhold the brake disc.

52. Clean and check that the ABS sensor mating surface on the wheel spindle is true.

53. Apply a very small amount of grease to the mating surface.

54. Install the ABS sensor cable in the holder on the spring strut.

55. Install the ABS sensor. Lubricate the bolt and tighten it. Tighten to 84 inch lbs. (10 Nm).

56. Install the wheel

V70, V70XC, S80 W/ALUMINUM SPINDLE

1. Remove the wheel.
2. Remove the brake caliper and limiter bolts.
3. Hang the caliper up using a piece of wire.
4. Remove the halfshaft bolt. Use a screwdriver as a counterhold on the brake disc.
5. Remove the locating pin holding the brake disc.
6. Remove the brake disc.
7. Slacken off the end of the halfshaft in the hub by knocking the halfshaft into the hub approximately 10–15 mm.
8. Use a rubber or copper mallet.
9. Remove the nut holding the control arm. Use a Torx wrench as a counterhold.

⁂ WARNING

Ensure that the tension strap is correctly secured in the control arms.

10. Pull down the control arm (1) using the tension strap (2)
11. Release the spring strut from the control arm.
12. Remove the halfshaft. Knock the halfshaft out using a brass drift. Hang the halfshaft as illustrated
13. Secure the spring strut in position
14. Remove the 4 bolts from the hub.

To install:

15. Install the hub. Use new bolts
16. Tighten crosswise to 15 ft. lbs. (20 Nm). Then tighten to 49 ft. lbs. (65 Nm). Angle-tighten 60 deg.
17. Turn in the spring strut and position the halfshaft in the hub.

➡Check that the seal is not worn or damaged.

18. Install the bolt loosely on the halfshaft. Use a new bolt
19. Align the ball joint in the control arm

➡Be very careful when releasing the tension strap.

20. Release the control arm slowly using the tension strap.
21. Install a new nut on the ball joint. Tighten to 43 ft. lbs. (80 Nm).

➡The ball joint must not rotate. Use a Torx wrench as a counterhold so that the rubber boot is not damaged.

22. Install the brake disc.
23. Tighten the brake disc locating pin. Tighten to 72 inch lbs. (8 Nm).

➡Ensure that the brake disc and wheel rim hub mating surfaces are clean.

24. Tighten the new halfshaft bolt.

V40 AND 2002–2004 S40

1. Slacken off the center nut from the halfshaft.
2. Remove the wheel.
3. Remove the lock clip. Remove the brake hose from the bracket.
4. Remove the two bolts from the brake caliper holder and hang them up to prevent damage to the brake hose.
5. Remove the brake disc.
6. Remove the nut from the halfshaft.
7. Clean thoroughly around the sensor.
8. Remove the ABS cable from the bracket.
9. Remove the bolt. Carefully remove the sensor from the wheel spindle with a twisting movement.
10. Remove the halfshaft nut.
11. Remove the tie rod ball joint.
12. Remove slacken off the wheel spindle nut as much as possible.
13. Remove the two wheel spindle bolts.
14. Use a universal puller to withdraw the halfshaft from the wheel spindle.
15. Remove the nut from the ball joint.

⁂ WARNING

Secure the tool by tying it into place to prevent damage.

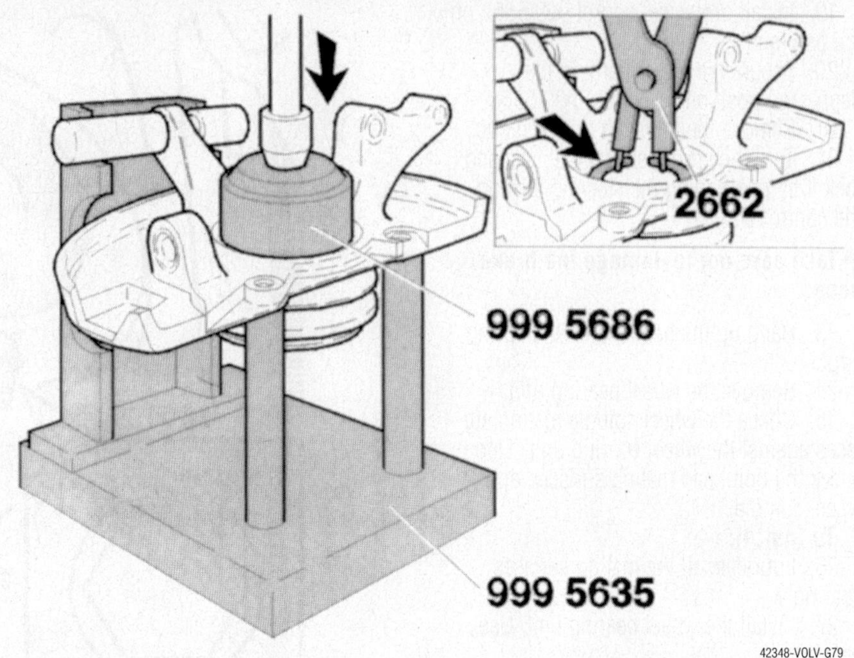

Snap ring removal—S40, V40

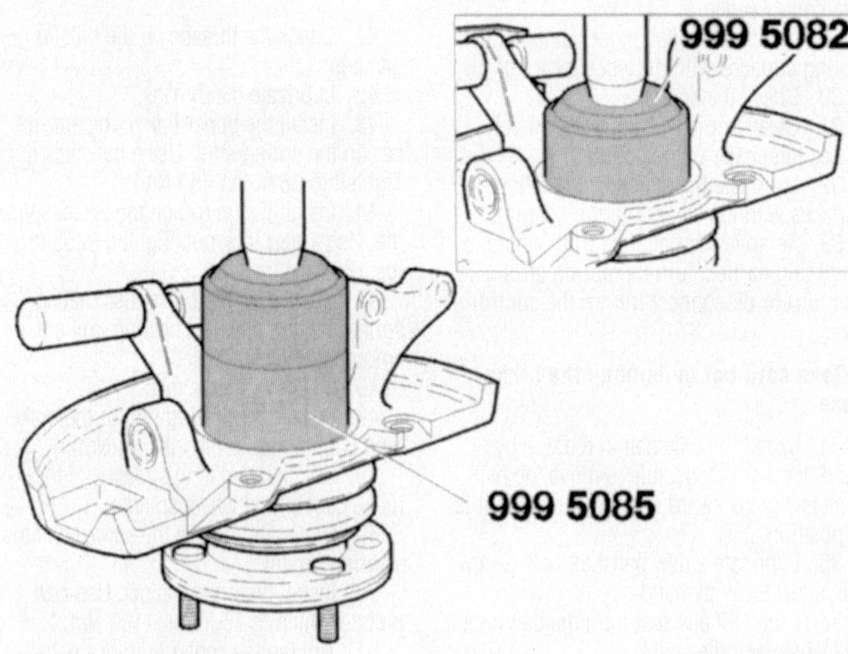

Hub bearing removal—S40, V40

16. Press out the ball joint.
17. Remove the wheel spindle.
18. Clean the hub.
19. Screw together special tools 5635-1 and 5635-2 in the conical opening in the end of the tie rod.
20. Install the wheel spindle with the hub downwards in special tool 999 5635.
21. Press the hub out of the bearing.

➡the inner ring can be left in place on the hub.

22. Remove the inner sealing ring
23. Remove the snap ring.
24. Turn the wheel spindle over in the tool.
25. Press the hub out of the wheel spindle.

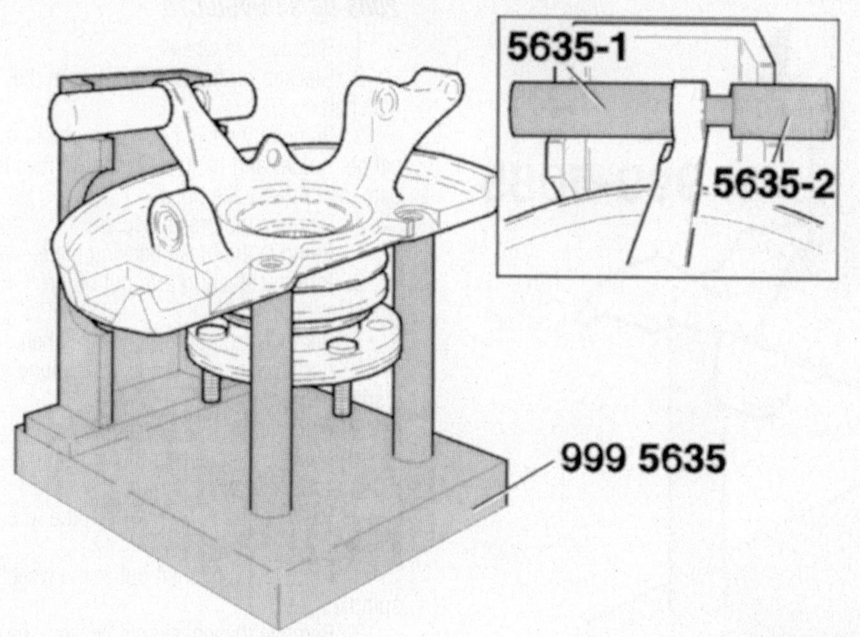

Tool 5635—S40, V40

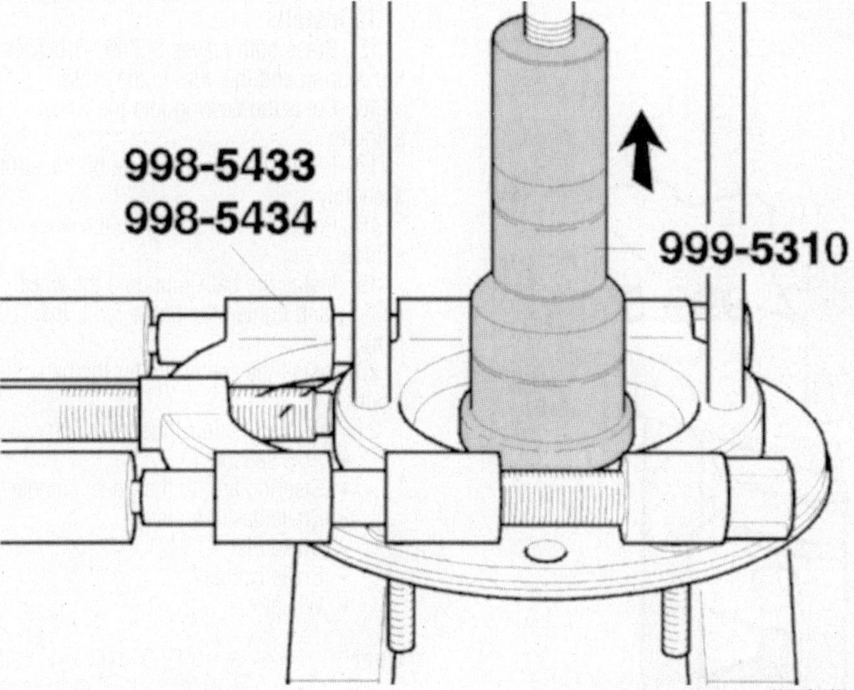

Tool assembly—S40, V40

➡Move the inner ring to the other side of the bearing if necessary.

26. Connect special tool 998 5433 and 998 5434.
27. Connect special tool 999 5310 to the hub. Pull off the bearing cap.
28. Remove the sealing ring.
29. Connect special tool 998 5433 and 998 5434.
30. Connect special tool 999 5310 to the hub. Pull off the bearing cap.
31. Remove the sealing ring.

➡When installing a new wheel spindle, a new splash guard must be installed on the wheel spindle.

32. Secure it in three places with a screwdriver.
33. Press out the wheel studs. Pull the wheel stud into place with a wheel nut and a spacer.

To install:

34. Install the wheel spindle in the special tool 999 5635.
35. Install the bearing with the text turned inwards.
36. Apply locking fluid to the wheel spindle
37. Guide the bearing and press it into place.
38. Install a new snap ring.
39. Position the hub under a press.

➡Avoid damaging the wheel studs.

40. Press the wheel spindle over the hub.
41. Clean the control arm. Position the wheel spindle over the ball joint. Tighten the lock nut as much as possible. Press the halfshaft inwards. Install the halfshaft in the wheel spindle.
42. Install the wheel spindle in the spring strut. Install the two bolts.
43. Install the nut and the snap ring in the halfshaft as far in as possible.

➡Cars with turbocharged engines have only a nut.

➡Oil the turbo splines.

44. Install two new nuts on the wheel spindle bolts.
45. Press the wheel spindle to the bottom. Tighten the wheel spindle bolts. Tighten to 68 ft. lbs. (90 Nm).
46. Tighten the ball joint nut. Tighten to 51 ft. lbs. (67 Nm).
47. Install the tie rod ball joint on the wheel spindle
48. Clean the hub. Install the brake disc.
49. Install the brake caliper bracket on the wheel spindle. Tighten to 76 ft. lbs. (100 Nm).
50. Install the brake hose in the bracket. Install the snap ring.
51. Remove any dirt from the sensor seat so that the sensor is in the correct position in relation to the pulse wheel.
52. Use compressed air to blow off any dirt on the pulse wheel.
53. Lightly lubricate the sensor and install it. Tighten to 19 ft. lbs. (25 Nm).
54. Ensure that the brake disc and wheel rim mating surfaces are clean.
55. Install the wheel.
56. Tighten the new nut for the halfshaft. Tighten to 91 ft. lbs. (120 Nm) +60 deg.
57. Install the hub cap.

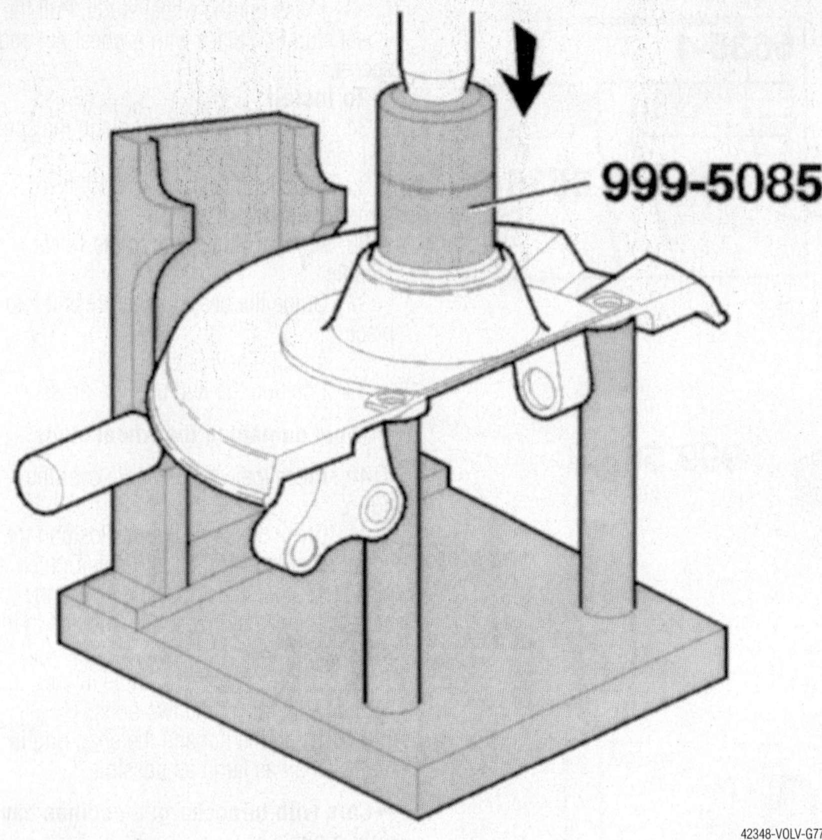

Seal removal—S40, V40

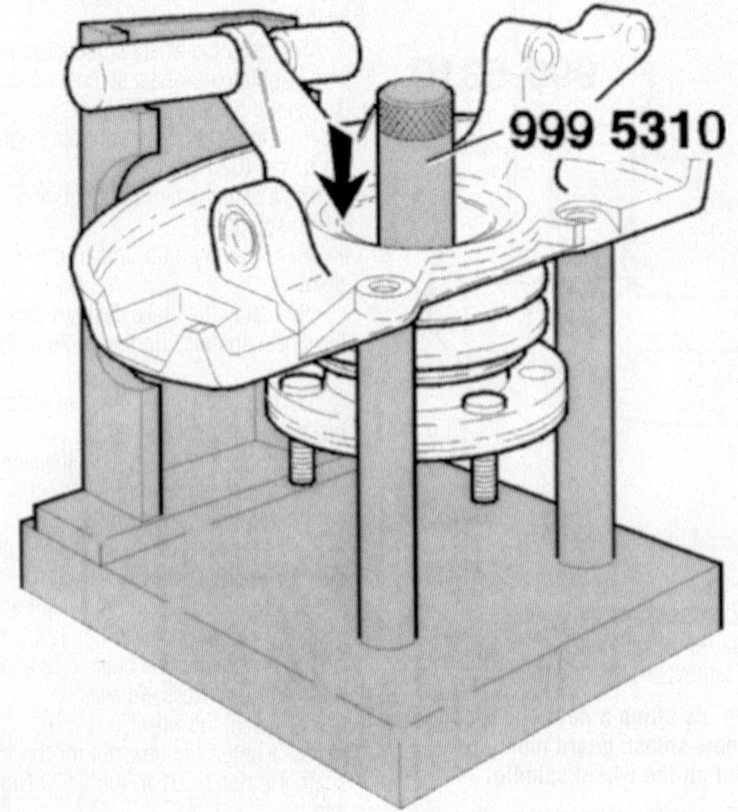

Tool 5310 installed—S40, V40

2005-06 S40/V50/C70

1. Remove the wheel.
2. Slacken off the center nut from the halfshaft.
3. Remove the two bolts from the brake caliper holder and hang them up to prevent damage to the brake hose.
4. Remove the brake disc.
5. Remove the brake backing plate.
6. Pull the steering arm out of the wheel spindle.
7. Remove the nut from the halfshaft.
8. Remove the ABS sensor and hang in a suitable position.
9. Remove the tie rod ball joint.
10. Remove slacken off the wheel spindle nut as much as possible.
11. Remove the ball-joint from the wheel spindle.
12. Press the half shaft out of the wheel spindle.
13. Remove the wheel spindle from the spring strut.
14. Secure the wheel spindle in a hydraulic press using socket 999-7090 and drift 999-1801 to press out the bearing.

To install:

15. Press both halves of 7090-1 between the bearing and the hub on the press.
16. Press the bearing into the wheel spindle.
17. Install the wheel spindle on the strut assembly.
18. Press the half shaft into the wheel spindle.
19. Install the ball-joint onto the wheel spindle and tighten the nut to 52 ft. lbs. (70 Nm).
20. Install the center nut for the half shaft.
21. Install or connect the following:
    - ABS sensor
    - Steering arm to the wheel spindle
    - Brake backing plate
    - Brake disc
    - Brake caliper
    - Wheel

Rear

V70XC MODELS

1. Remove the rear wheel.
2. Remove the bolt from the halfshaft/wheel hub.
3. Remove the brake caliper mounting bolts. Hang the caliper up with a piece of wire
4. Remove the locating pin from the brake disc
5. Remove the brake disc
6. Remove the ABS-cable holder

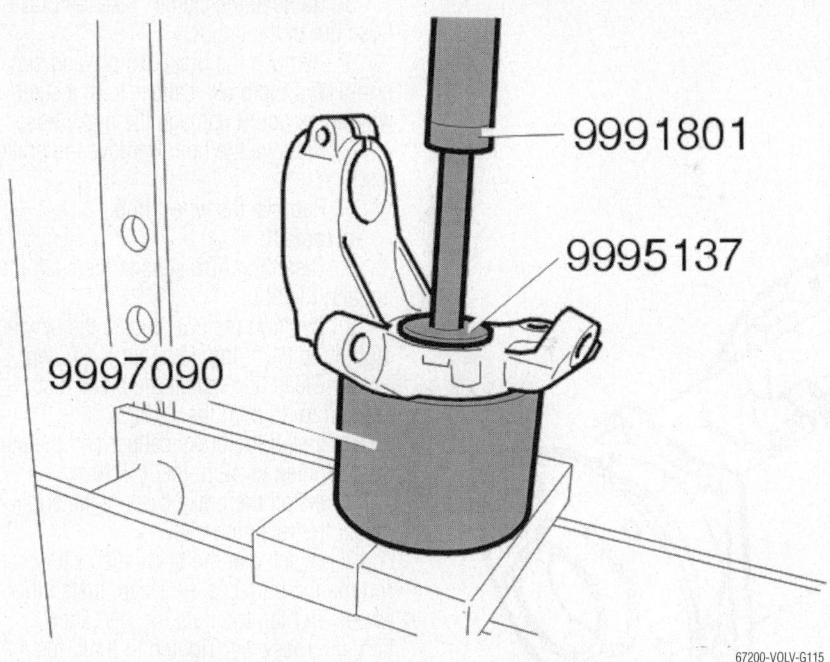

Securing the wheel spindle in a hydraulic press

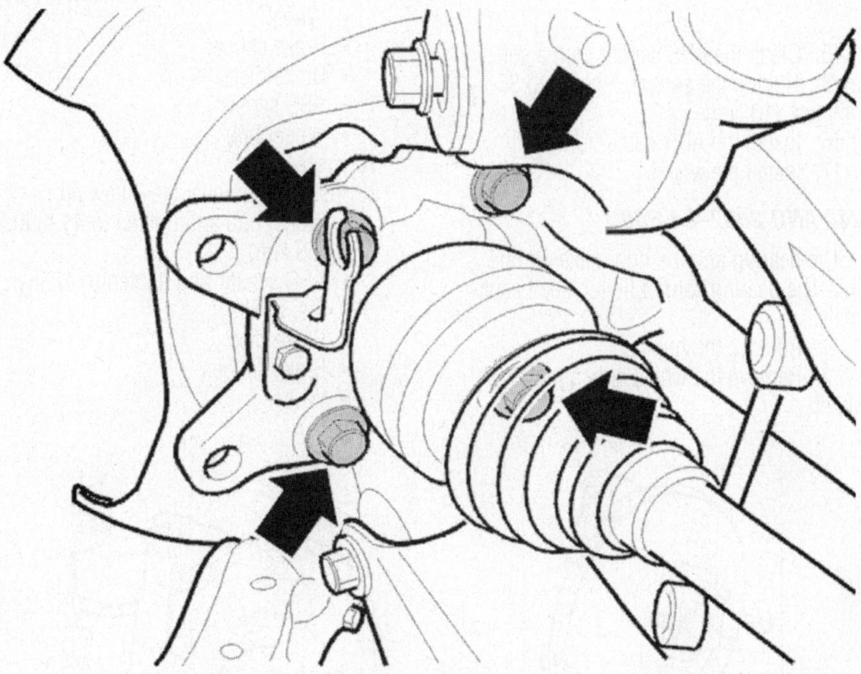

Rear hub bolts—V70XC

7. Remove the ABS sensor. Hang up the sensor using a piece of wire.

8. Remove the spring strut from the lower control arm.

➡the spring strut is slightly compressed.

9. Remove the anti-roll bar from the upper control arm.

10. Use a Torx wrench as a counterhold so that the rubber boot is not damaged.

11. Remove the tie rod from the wheel spindle. Angle the tie rod upwards so that the inner mounting of the control arm is exposed.

12. Remove the bolt holding the lower mounting from the control arm to the sub-frame. Bend the control arm downwards.

13. Remove the bolts from rear wheel hub.

14. Remove the rear wheel hub from the wheel spindle and halfshaft.

➡Clean the splines on the halfshaft before installing the new wheel hub. Lubricate the splines using oil.

To install:

15. Install the rear wheel hub. Use new bolts and torque wrench with ring wrench kit. Tighten crosswise. Tighten to 15 ft. (20 Nm). Then tighten to 49 ft. lbs. (65 Nm). Finally angle-tighten 60°.

16. Install the control arm to the sub-frame. Do not tighten yet.

17. Install the tie rod on the wheel spindle. Do not tighten yet.

18. Install the anti-roll bar link to the upper control arm. Tighten to 58 ft. lbs. (80 Nm). Use a Torx wrench as a counterhold so that the boot is not damaged.

19. Press the suspension up to its normal position.

20. Install the lower control arm to the sub-frame. Tighten to 58 ft. lbs. (80 Nm).

21. Install the tie rod to the wheel spindle. Tighten to 58 ft. lbs. (80 Nm).

22. Remove the mobile jack.

23. Install the spring strut on the lower control arm. Use a larger jimmy bar to bend the control arm down and align the spring strut against the control arm.

24. Install the lower bolt for the spring strut. Tighten to 58 ft. lbs. (80 Nm).

25. Install the brake disc.

26. Install the locating pin for the brake disc. Tighten the locating pin. Tighten to 86 inch lbs. (10 Nm).

27. Install the brake caliper. Use new bolts. Tighten to 46 ft. lbs. (60 Nm).

28. Install the ABS sensor.

➡Ensure that the sensor seat in the wheel spindle is completely clean. Clean the ABS sensor with a soft brush. Tighten the sensor. Tighten to 86 inch lbs. (10 Nm).

29. Install the ABS-cable holder.

30. Install the bolt for the halfshaft. Use a new bolt. Tighten to 27 ft. lbs. (35 Nm) plus 90 deg.

31. Install the wheel.

V70, S80, S60

1. Remove the wheel.

2. Remove ABS sensor and the cable from the wheel spindle.

3. Remove the brake caliper mounting bolts. Hang up the ABS line and the brake caliper. Use a piece of wire.

4. Remove the locating pin from the brake disc.

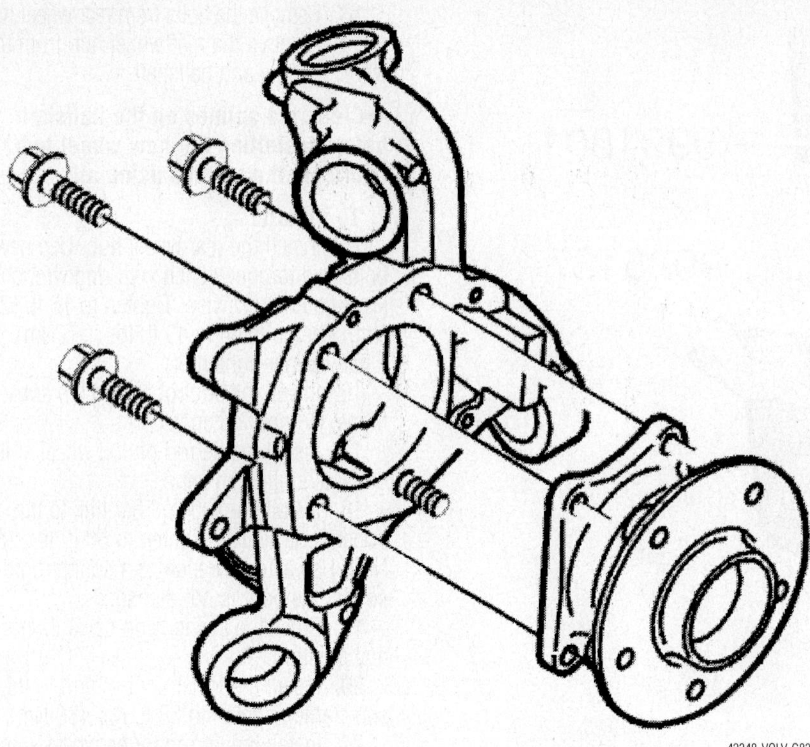

Rear hub removal—S60, S80, V70

5. Remove the brake disc.

6. Remove the holder from the ABS line on the wheel spindle.

7. Press up the control arm slightly. Use a mobile jack.

8. Remove the bolts from rear wheel hub.

9. Remove the rear wheel hub.

➡The flange on the lower rear bolt must be ground off as illustrated before it can be removed from the wheel spindle.

To install:

10. Install the rear wheel hub. Use new bolts and torque wrench with ring wrench kit. Tighten crosswise. Tighten to 15 ft. lbs. (20 Nm). Then tighten to 49 ft. lbs. (65 Nm). Finally angle-tighten 60 deg.

➡The bolt without a flange has a washer which is included in the kit. Position the bolt in the rear lower bolt hole.

11. Install the brake disc.

12. Install the locating pin for the brake disc. Tighten the locating pin. Tighten to 86 inch lbs. (10 Nm).

13. Install the brake caliper. Use new bolts. Tighten to 46 ft. lbs. (60 Nm).

14. Install the ABS sensor.

➡Ensure that the sensor seat in the wheel spindle is completely clean.

15. Clean the ABS sensor with a soft brush. Tighten the sensor. Tighten to 86 inch lbs. (10 Nm).

16. Install the ABS-cable holder.

17. Install the wheel.

V40 AND 2002–04 S40

The bearing and the hub make up one unit. The bearing cannot be replaced separately.

1. Remove the hub cap.

2. Remove the wheel nuts and the wheel.

3. Remove the clamp. Take the brake hose out of the support.

4. Remove the brake caliper and support. Hang the brake caliper from a steel wire so as not to damage the brake hose.

5. Remove the bolt. Remove the brake disc.

6. Remove the wheel hub.

To install:

7. Clean the ABS sensor, the shaft journal and the hub.

8. Position the hub. Install the new nut. Tighten to 91 ft. lbs. (120 Nm) +30 deg.

9. Clean and install the brake disc. Tighten to 46 inch lbs. (5 Nm).

10. Install the brake caliper and the support. Tighten to 42 ft. lbs. (55 Nm).

11. Install the brake hose in the support. Install the retaining clip.

12. Check that the brake disc and wheel rim mating surfaces are clean. Install the wheel. Tighten the nuts loosely. Then tighten crosswise. Tighten to 84 ft. lbs. (110 Nm).

2005-06 S40/V50/C70

1. Remove or disconnect the following:
   - Wheel
   - Brake caliper
   - Brake disc
   - ABS sensor
   - Wheel hub

To install:

2. Install or connect the following:
   - Wheel hub and tighten to 41 ft. lbs. (55 Nm)
   - ABS sensor and tighten to 42 inch lbs. (5 Nm)
   - Brake disc
   - Brake caliper
   - Wheel

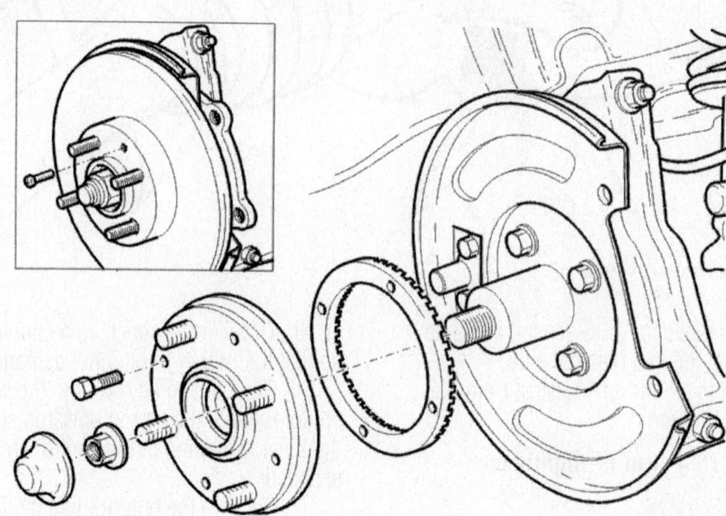

Rear hub removal—S40, V40

BRAKES

Brake Caliper

REMOVAL & INSTALLATION

2002–04 C70 Series

FRONT

1. Before servicing the vehicle, refer to the Precautions Section.
2. Remove the wheels.
3. Disconnect the ABS wires.
4. Remove the spring.
5. Remove the caliper bolts, lift the caliper off and unscrew the caliper from the hose.

To install:

6. Grease the caliper bolts with lithium grease and insert them into the sleeves.
7. Screw the caliper onto the brake hose.
8. Install the caliper and tighten the caliper bolts to 22 ft. lbs. (30 Nm).
9. If loosened, tighten the brake hose to 13 ft. lbs. (18 Nm).
10. Fill the master cylinder and bleed the brake system.
11. Connect the ABS wires.
12. Install the wheels.

REAR

1. Before servicing the vehicle, refer to the Precautions Section.
2. Remove the wheels.
3. Disconnect the ABS wires.
4. Remove the caliper mounting bolts and lift the caliper off.

To install:

5. Install the caliper and new mounting bolts. Tighten the bolts to 44 ft. lbs. (60 Nm).
6. Connect the brake line to the caliper and tighten to 10 ft. lbs. (14 Nm).
7. Fill the brake master cylinder and bleed the brake system.
8. Connect the ABS wires.
9. Install the wheels.

S60, S80, V70XC, 2002–04 V70 Series

FRONT

1. Before servicing the vehicle, refer to the Precautions Section.
2. Remove the wheels.
3. Disconnect the ABS wires.
4. Remove the spring.
5. Remove the caliper bolts, lift the caliper off and unscrew the caliper from the hose.

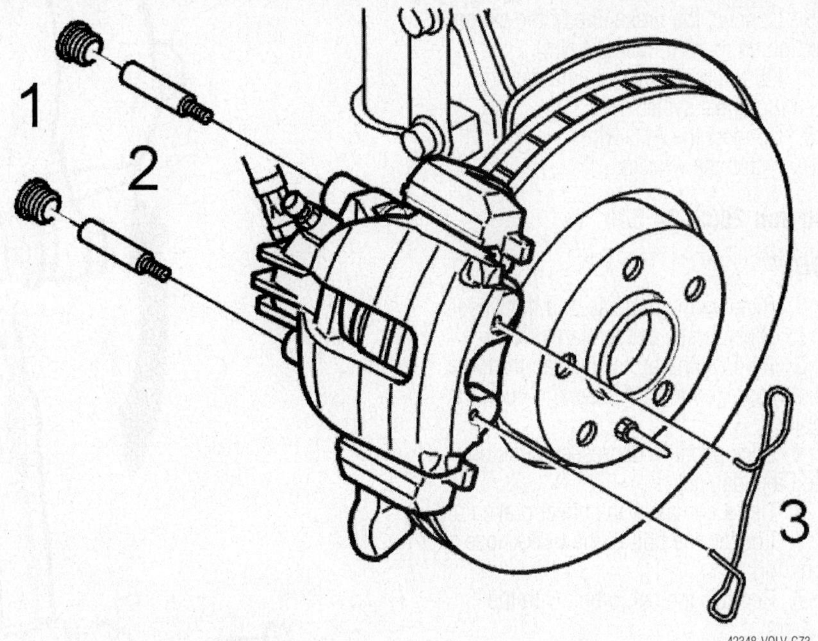

Front caliper removal—S60, V70, V70XC, S80

To install:

6. Grease the caliper bolts with lithium grease and insert them into the sleeves.
7. Screw the caliper onto the brake hose.
8. Install the caliper and tighten the caliper bolts to 22 ft. lbs. (30 Nm).
9. If loosened, tighten the brake hose to 13 ft. lbs. (18 Nm).
10. Fill the master cylinder and bleed the brake system.
11. Connect the ABS wires.
12. Install the wheels.

REAR

1. Before servicing the vehicle, refer to the Precautions Section.
2. Remove the wheels.
3. Disconnect the ABS wires.
4. Remove the caliper mounting bolts and lift the caliper off.

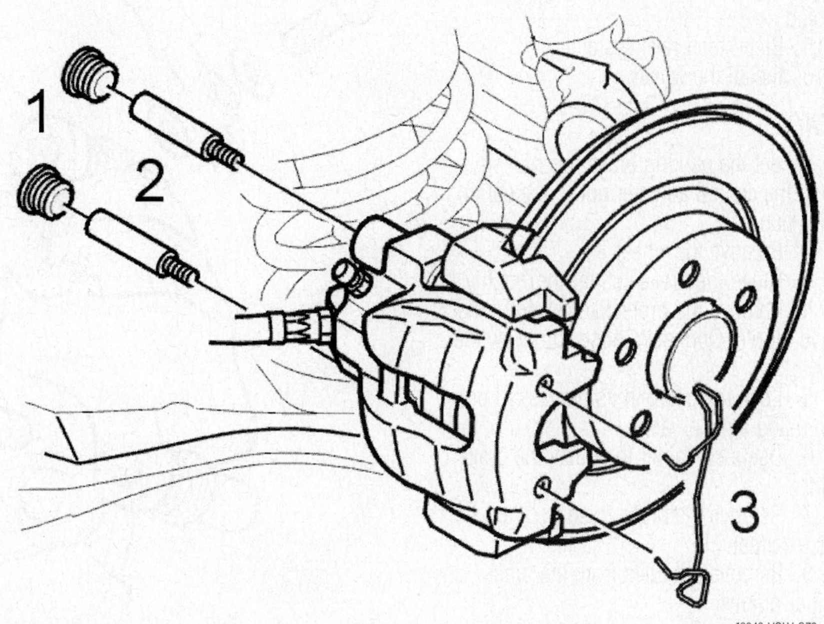

Rear caliper removal—S60, V70, V70XC, S80

To install:

5. Install the caliper and new mounting bolts. Tighten the bolts to 22 ft. lbs. (30 Nm).
6. Connect the brake line to the caliper and tighten to 10 ft. lbs. (14 Nm).
7. Fill the brake master cylinder and bleed the brake system.
8. Connect the ABS wires.
9. Install the wheels.

V40 and 2002–04 S40

FRONT

1. Remove the hub cap and the wheel.
2. Clean brake caliper carefully.
3. Remove the protective cap from the bleed nipple, connect a hose and open the nipple.
4. Lock pedal in depressed position using a pedal jack.
5. Use a container to collect brake fluid.
6. Loosen the bolt of the brake hose mounting.
7. Remove the banjo bolt with the washers.
8. Remove the two bolts from the brake caliper guide pin.
9. Remove the brake caliper from the holder.
10. Remove the brake pads.
11. Remove the remaining brake fluid.

To install:

12. Install the brake pads, paying attention to the wear-indicator.
13. Install the brake caliper to the guide pins. Tighten to 27 ft. lbs. (35 Nm).
14. Install the brake hose. Use new washers. Tightening torque 11 ft. lbs. (15 Nm). Make sure that the brake hose is not twisted.
15. Bleed the brake system.
16. Install the wheel

REAR

1. Set the parking brake mechanism using the central adjuster nut in the parking brake handle.
2. Remove the wheel.
3. Clean the brake caliper thoroughly.
4. Remove the protective cap from the bleed nipple. Connect a hose. Unscrew the nipple.
5. Lock the pedal in its depressed position using a pedal jack.
6. Use a container to collect the brake fluid.
7. Remove the brake hose from the brake caliper
8. Remove the bolts from the brake caliper guides
9. Remove the brake caliper from the holder, and the brake pads

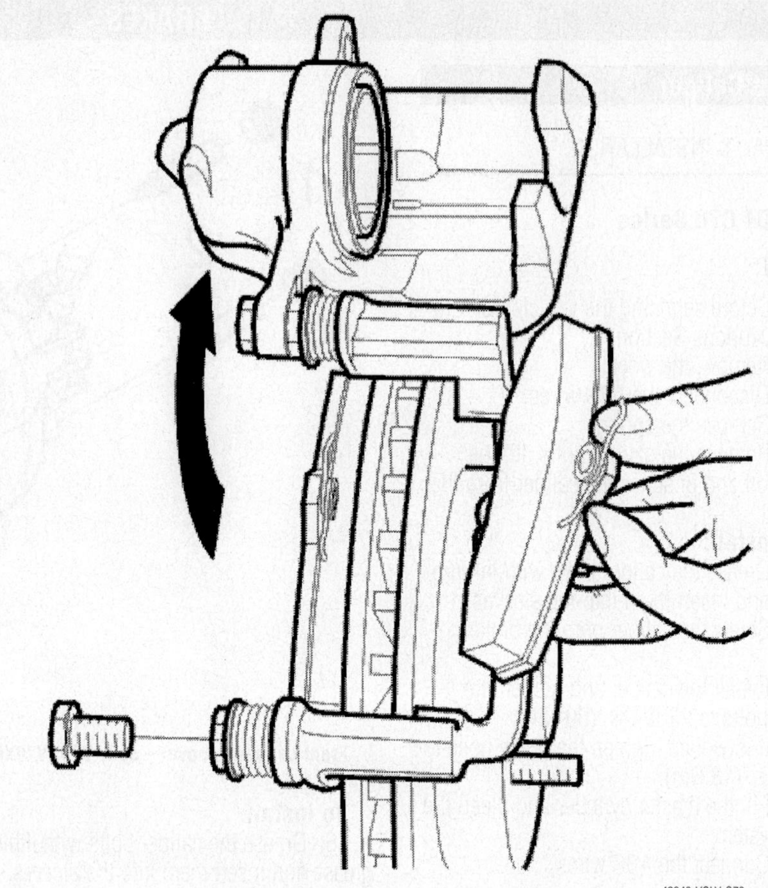

Front caliper removal—2002–04 S40, V40

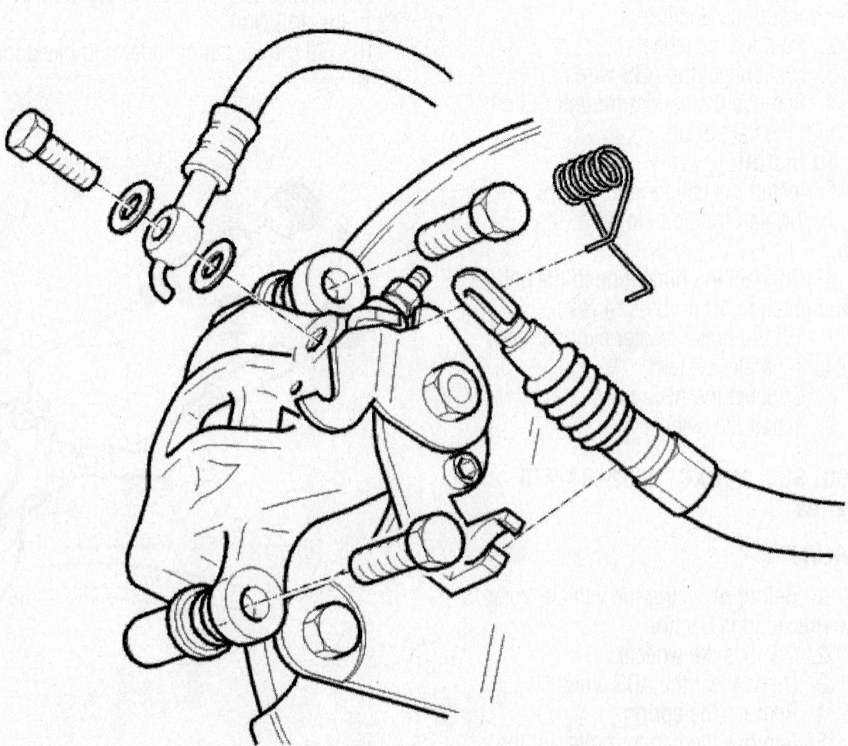

Rear caliper removal—2002–04 S40, V40

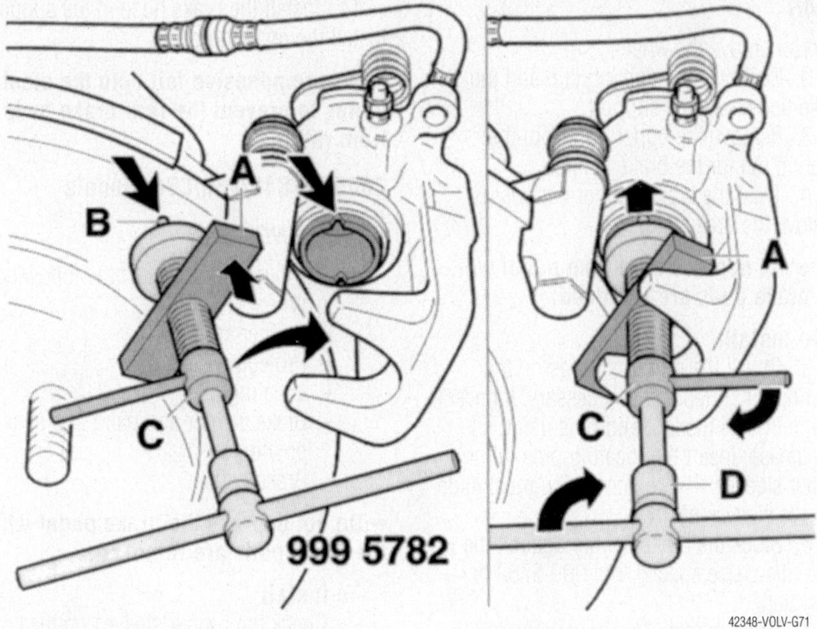

Rotating the rear caliper—2002–04 S40, V40

10. Remove the parking brake cable from the support. First pull the rubber sleeve up around the parking brake cable
11. Remove the parking brake cable spring (if installed)
12. Remove the inner mechanical cable from the parking brake handle
13. Remove the remaining brake fluid.

To install:

14. Install the brake pads.
15. Install the parking brake cable.
16. Install the brake caliper on the holder. Install the bolt. Tighten to 33 Nm. Install the brake hose. Use new seals. Tightening torque 29 Nm.
17. Ensure that the brake hose is not twisted.
18. Bleed the brake system.
19. Install the wheel. Depress the brake pedal a few times. Check the brake fluid level.
20. Check the parking brake setting.

2005-06 S40/V50/C70

FRONT

1. Remove the wheel.
2. Secure the pedal in the depressed position using a pedal jack.
3. Remove the protective cap from the bleed nipple, connect a hose and open the nipple.
4. Use a container to collect brake fluid.
5. Slacken off the brake hose half a turn
6. Remove the brake pads.
7. Remove the brake caliper
8. Unscrew the caliper from the brake hose.

To install:

9. Install the brake hose in the brake caliper, but do not tighten.
10. Install the brake pads.
11. Tighten the brake hose to 17 ft. lbs. (23 Nm).
12. Bleed the brake system.
13. Install the wheel. Depress the brake pedal a few times. Check the brake fluid level.

REAR

1. Remove the wheel.
2. Secure the pedal in the depressed position using a pedal jack.
3. Remove the protective cap from the bleed nipple, connect a hose and open the nipple.
4. Use a container to collect brake fluid.
5. Slacken off the brake hose half a turn
6. Remove the brake pads.
7. Remove the brake caliper
8. Remove the parking brake cable.
9. Unscrew the caliper from the brake hose.

To install:

10. Install the brake hose in the brake caliper, but do not tighten.
11. Install the parking brake cable.
12. Install the brake pads.
13. Tighten the brake hose to 17 ft. lbs. (23 Nm).
14. Remove the pedal jack and bleed the brake system.
15. Install the wheel. Depress the brake pedal a few times. Check the brake fluid level.

Brake Pads

REMOVAL & INSTALLATION

2002–04 C70

FRONT

1. Before servicing the vehicle, refer to the Precautions Section.
2. Remove the front wheels.
3. Remove the spring.
4. Remove the pad retaining pins.
5. Remove caliper from the carrier.
6. Remove the brake pads.

To install:

7. Press the piston back into the caliper cylinder using a suitable tool.
8. Lubricate the caliper guide pins with silicone grease.
9. Insert the brake pads and slide the caliper on over them.
10. Tighten the guide pins to 22 ft. lbs. (30 Nm) and replace the dust caps.
11. Install the wheels.

REAR

1. Before servicing the vehicle, refer to the Precautions Section.
2. Remove the wheels.
3. Remove the pad retaining pins with a 3mm punch, and the spring.
4. Remove the brake pads and shims.

To install:

5. Press the pistons back into their housing using a suitable tool.
6. Grease the pad shims on both sides with a thin layer of silicone grease.
7. Install the shims on the pads,
8. Install the pads in the caliper.
9. Install the retaining pins and spring.
10. Install the wheels.

V70, V70XC, S60, S80 Series

FRONT

1. Before servicing the vehicle, refer to the Precautions Section.
2. Remove the front wheels.
3. Remove the spring.
4. Remove the pad retaining pins.
5. Remove caliper from the carrier.
6. Remove the brake pads.

To install:

7. Press the piston back into the caliper cylinder using a suitable tool.
8. Lubricate the caliper guide pins with silicone grease.
9. Insert the brake pads and slide the caliper on over them.

10. Tighten the guide pins to 22 ft. lbs. (30 Nm) and replace the dust caps.
11. Install the wheels.

REAR

1. Remove the securing spring (3) carefully so that it does not deform
2. Remove the protective caps (1) from the two locating pins (2)
3. Remove the locating pins, use hex socket 7 mm.
4. Remove the brake caliper from the holder
5. Remove the brake pads.
6. Hang brake caliper from a steel wire from the spring so as not to damage brake hose.
7. Press piston back into cylinder on brake caliper.
8. Check that the dust cover is correctly positioned.

To install:

9. Install new brake pads
10. Install the brake caliper.
11. Check the rubber sleeves of the locating pins. Replace if necessary.
12. Lubricate the locating pins with silicone brake grease. Insert the locating pins (2) into the rubber sleeves. The pins should slide into the sleeves easily.
13. Tighten locating pins. Tighten to 22 ft. lbs. (30 Nm). Install protective caps (1).
14. Install the retaining spring (3).
15. Depress the brake pedal a few times.
16. Top up brake fluid if necessary.
17. Adjust parking brake as required

2002–2004 V40, S40

FRONT

1. Remove the hub cap and the wheel
2. Remove the bolt from the lower locating pin from the brake caliper.
3. Twist the brake caliper upwards. Remove the brake pads.
4. Remove the wear indicator.

➡Do not depress the brake pedal while the brake pads are removed.

To install:

5. Press the piston into the brake caliper. Check that no brake fluid comes out of the brake fluid reservoir. Check that dirt guard is correctly positioned.
6. Press down the wear indicators on the brake pads.
7. Install the brake pads and twist the brake caliper down. Check that the bolts for the brake pads fit in the brake caliper.
8. Secure the brake caliper. Tighten to 27 ft. lbs. (35Nm).

REAR

1. Remove the wheel.
2. Remove the retaining clip and the brake hose from the support.
3. Remove the upper bolt from the locating pin for the brake caliper.
4. Twist the brake caliper backwards. Remove the brake pads.

➡Do not depress the brake pedal while the brake pads are removed.

To install:

5. Check the rubber sleeves on the locating pins. Replace if necessary. If necessary, lubricate the locating pins using silicon grease. Insert the locating pins in the rubber sleeves. The locating pins must slide into the sleeves easily.
6. Block the brake caliper against the brake disc. Use special tool 999 5782 or equivalent.
7. Turn the plate (A) forward over the shaft and insert the tool
8. Install the lugs (B) in the cut-out in the brake piston
9. Turn the shaft (C) so that the plate (A) presses against the brake caliper
10. Turn the axle (D) to the right and press the piston back while turning the shaft (C) to the right so that the plate (A) is in contact with the brake caliper
11. Turn the brake piston back fully and remove the tool.
12. Install the wear indicator on the under side of the inner installed brake pads. Install the brake pads.
13. Position the brake caliper. Install the new locking bolt. Tighten to 25 ft. lbs. (33 Nm).

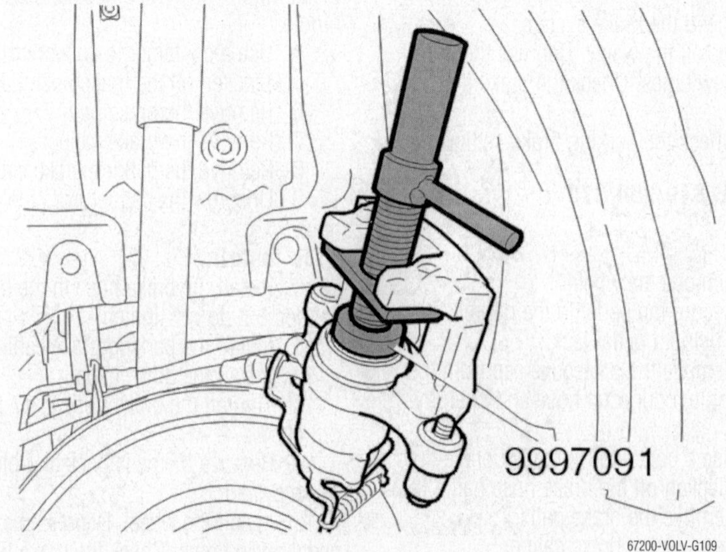

Compressing the piston into the cylinder

14. Install the brake hose in the support. Install the retaining clip.

➡Secure adhesive foil onto the steel plates to prevent the rear brake pads from rattling.

2005-06 S40/V50/C70 Models

FRONT AND REAR

1. Remove or disconnect the following:
   - Wheel
   - Caliper spring
   - Protective caps
   - Locating pins
   - Brake caliper and hang in a suitable position
   - Brake pads

➡Do not depress the brake pedal while the brake pads are removed.

To install:

2. Check the rubber sleeves on the locating pins. Replace if necessary. If necessary, lubricate the locating pins using silicon grease. Insert the locating pins in the rubber sleeves. The locating pins must slide into the sleeves easily.
3. Press the piston back into the cylinder using special tool 999-7091 or equivalent.
4. Install or connect the following:
   - Brake pads
   - Brake caliper
   - Locating pins
   - Protective caps
   - Caliper spring
   - Wheel

VOLVO

XC90

15

SPECIFICATIONS AND MAINTENANCE CHARTS

ENGINE AND VEHICLE IDENTIFICATION

Engine						
Code ①	Liters (cc)	Cu. In.	Cyl.	Fuel Sys.	Engine	Eng. Mfg.
B5254T2	2.5 (2435)	151	5	MFI	DOHC	Volvo
B6294T	2.9 (2922)	178	6	EFI	DOHC	Volvo
B8444S	4.4 (4414)	269	8	EFI	DOHC	Volvo

Model Year	
Code	Year
3	2003
4	2004
5	2005
6	2006

MFI: Multi-port Fuel Injection

EFI: Electronic Fuel Injection

DOHC: Double Overhead Camshafts

06041_XC90_C0001

GENERAL ENGINE SPECIFICATIONS

Year	Model	Engine Displacement Liters	Engine ID	Net Horsepower @ rpm	Net Torque @ rpm (ft. lbs.)	Bore x Stroke (in.)	Com-pression Ratio	Oil Pressure @ rpm
2003	XC90	2.5	B5254T2	208@5700	208@1500	3.27 x 3.54	8.5:1	43@2000
		2.9	B6294T	268@5200	280@1800	3.27 x 3.54	10.0:1	50@4000
2004	XC90	2.5	B5254T2	208@5700	208@1500	3.27 x 3.54	8.5:1	43@2000
		2.9	B6294T	268@5200	280@1800	3.27 x 3.54	10.0:1	50@4000
2005	XC90	2.5	B5254T2	208@5700	208@1500	3.27 x 3.54	8.5:1	43@2000
		2.9	B6294T	268@5200	280@1800	3.27 x 3.54	10.0:1	50@4000
		4.4	B4444S	311@5500	325@3900	3.70 x 3.13	10.4:1	68-109@2000
2006	XC90	2.5	B5254T2	208@5700	208@1500	3.27 x 3.54	8.5:1	43@2000
		4.4	B4444S	311@5500	325@3900	3.70 x 3.13	10.4:1	68-109@2000

NA: Information not available

06041_XC90_C0002

ENGINE TUNE-UP SPECIFICATIONS

Year	Engine Displacement Liters	Engine ID	Spark Plug Gap (in.)	Ignition Timing	Fuel Pump (psi)	Idle Speed RPM	Valve Clearance (in.) In.	Valve Clearance (in.) Ex.
2003	2.5	B5254T2	0.030	10B	51 ①	850	HYD	HYD
	2.9	B6294T	NA	NA	NA	650	HYD	HYD
2004	2.5	B5254T2	0.030	10B	51 ①	850	HYD	HYD
	2.9	B6294T	NA	NA	NA	650	HYD	HYD
2005	2.5	B5254T2	0.030	10B	51 ①	850	HYD	HYD
	2.9	B6294T	NA	NA	NA	650	HYD	HYD
	4.4	B8444S	NA	NA	NA	675	HYD	HYD
2006	2.5	B5254T2	0.030	10B	51 ①	850	HYD	HYD
	4.4	B8444S	NA	NA	NA	675	HYD	HYD

NA: Not available

NOTE: The Vehicle Emission Control Information label often reflects specification changes made during production. The label figures must be used if they differ from those in this chart.

B: Before top dead center

HYD: Hydraulic

① At idle

06041_XC90_C0003

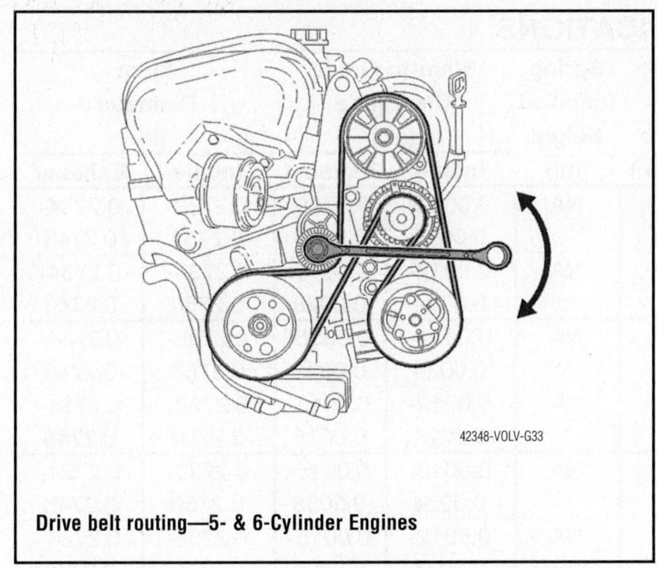

Drive belt routing—5- & 6-Cylinder Engines

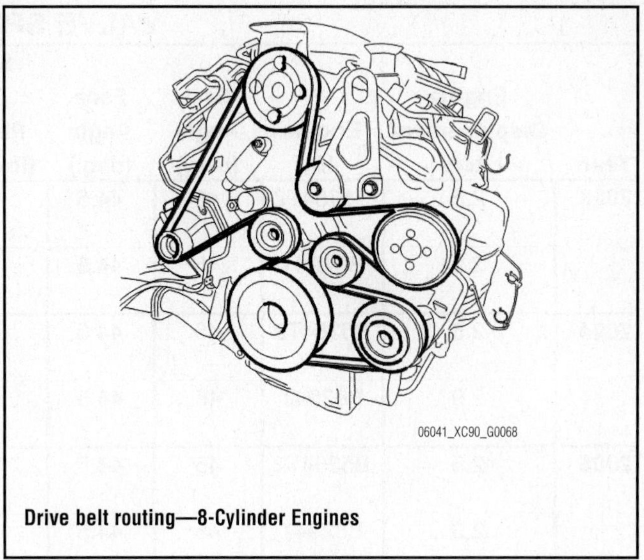

Drive belt routing—8-Cylinder Engines

CAPACITIES

Year	Model	Engine Displacement Liters	Engine ID	Engine Oil with Filter (qts.)	Transmission (pts.)	Transfer Case (pts.)	Drive Axle Front (pts.)	Drive Axle Rear (pts.)	Fuel Tank (gal.)	Cooling System (qts.)
2003	XC90	2.5	B5254T2	6.9 ①	15.0	②	—	NA	19.0	NA
		2.9	B6294T	7.6 ①	NA	②	—	NA	19.0	NA
2004	XC90	2.5	B5254T2	6.9 ①	15.0	②	—	NA	19.0	NA
		2.9	B6294T	7.6 ①	NA	②	—	NA	19.0	NA
2005	XC90	2.5	B5254T2	6.9 ①	15.0	②	—	NA	19.0	NA
		2.9	B6294T	7.6 ①	NA	②	—	NA	19.0	NA
		4.4	B8444S	7.0	③	②	—	NA	21.1	NA
2006	XC90	2.5	B5254T2	6.9 ①	15.0	②	—	NA	19.0	NA
		4.4	B8444S	7.0	③	②	—	NA	21.1	NA

NA: Information not available

NOTE: All capacities are approximate. Add fluid gradually and check to be sure a proper fluid level is obtained.

① Amount to fill after engine reconditioning, including filter

② Included in transmission amount

③ AW55-51: drain & refill, 6.4 pts.
GM 4T65: drain & refill, 8.4 pts.
TF-80SC (AWD) drain & refill, 8.4 pts.

06041_XC90_C0004

VALVE SPECIFICATIONS

Year	Engine Displacement Liters	Engine ID	Seat Angle (deg.)	Face Angle (deg.)	Spring Test Pressure (lbs. @ in.)	Spring Installed Height (in.)	Stem-to-Guide Clearance (in.)		Stem Diameter (in.)	
							Intake	Exhaust	Intake	Exhaust
2003	2.5	B5254T2	45	44.5	NA	NA	0.0012-0.0024	0.0015-0.0028	0.2738-0.2750	0.2734-0.2746
	2.9	B6294T	45	44.5	NA	NA	0.0012-0.0024	0.0015-0.0028	0.2738-0.2750	0.2734-0.2746
2004	2.5	B5254T2	45	44.5	NA	NA	0.0012-0.0024	0.0015-0.0028	0.2738-0.2750	0.2734-0.2746
	2.9	B6294T	45	44.5	NA	NA	0.0012-0.0024	0.0015-0.0028	0.2738-0.2750	0.2734-0.2746
2005	2.5	B5254T2	45	44.5	NA	NA	0.0012-0.0024	0.0015-0.0028	0.2738-0.2750	0.2734-0.2746
	2.9	B6294T	45	44.5	NA	NA	0.0012-0.0024	0.0015-0.0028	0.2738-0.2750	0.2734-0.2746
	4.4	B8444S	NA	NA	NA	NA	NA	NA	NA	NA
2006	2.5	B5254T2	45	44.5	NA	NA	0.0012-0.0024	0.0015-0.0028	0.2738-0.2750	0.2734-0.2746
	4.4	B8444S	NA	NA	NA	NA	NA	NA	NA	NA

NA: Information not available

06041_XC90_C0005

CRANKSHAFT AND CONNECTING ROD SPECIFICATIONS

All measurements are given in inches.

Year	Engine Displacement Liters	Engine ID	Crankshaft: Main Brg. Journal Dia.	Crankshaft: Main Brg. Oil Clearance	Crankshaft: Shaft End-play	Crankshaft: Thrust on No.	Connecting Rod: Journal Diameter	Connecting Rod: Oil Clearance	Connecting Rod: Side Clearance
2003	2.5	B5254T2	2.5584-2.5592	0.0007-0.0017	0.003-0.007	5	1.9679-1.9685	NA	0.006-0.018
	2.9	B6294T	2.5584-2.5592	0.0007-0.0017	0.003-0.007	5	1.9679-1.9685	NA	0.006-0.018
2004	2.5	B5254T2	2.5584-2.5592	0.0007-0.0017	0.003-0.007	5	1.9679-1.9685	NA	0.006-0.018
	2.9	B6294T	2.5584-2.5592	0.0007-0.0017	0.003-0.007	5	1.9679-1.9685	NA	0.006-0.018
2005	2.5	B5254T2	2.5584-2.5592	0.0007-0.0017	0.003-0.007	5	1.9679-1.9685	NA	0.006-0.018
	2.9	B6294T	2.5584-2.5592	0.0007-0.0017	0.003-0.007	5	1.9679-1.9685	NA	0.006-0.018
	4.4	B8444S	NA	NA	NA	NA	NA	NA	NA
2006	2.5	B5254T2	2.5584-2.5592	0.0007-0.0017	0.003-0.007	5	1.9679-1.9685	NA	0.006-0.018
	4.4	B8444S	NA	NA	NA	NA	NA	NA	NA

NA: Information not available

06041_XC90_C0006

PISTON AND RING SPECIFICATIONS

All measurements are given in inches.

Year	Engine Displacement Liters	Engine ID	Piston Clearance	Ring Gap Top Comp.	Ring Gap Bottom Comp.	Ring Gap Oil Control	Ring Side Clearance Top Comp.	Ring Side Clearance Bottom Comp.	Ring Side Clearance Oil Control
2003	2.5	B5254T/2	0.0004-0.0012	0.047	0.069	0.118	0.0020-0.0033	0.0011-0.0026	0.0008-0.0022
	2.9	B6249T	0.0004-0.0012	0.047	0.069	0.118	0.0020-0.0033	0.0011-0.0026	0.0008-0.0022
2004	2.5	B5254T/2	0.0004-0.0012	0.047	0.069	0.118	0.0020-0.0033	0.0011-0.0026	0.0008-0.0022
	2.9	B6249T	0.0004-0.0012	0.047	0.069	0.118	0.0020-0.0033	0.0011-0.0026	0.0008-0.0022
2005	2.5	B5254T/2	0.0004-0.0012	0.047	0.069	0.118	0.0020-0.0033	0.0011-0.0026	0.0008-0.0022
	2.9	B6249T	0.0004-0.0012	0.047	0.069	0.118	0.0020-0.0033	0.0011-0.0026	0.0008-0.0022
	4.4	B8444S	NA	NA	NA	NA	NA	NA	NA
2006	2.5	B5254T/2	0.0004-0.0012	0.047	0.069	0.118	0.0020-0.0033	0.0011-0.0026	0.0008-0.0022
	4.4	B8444S	NA	NA	NA	NA	NA	NA	NA

NA: Information not available

06041_XC90_C0007

TORQUE SPECIFICATIONS

All readings in ft. lbs.

Year	Engine Displacement Liters	Engine ID	Cylinder Head Bolts	Main Bearing Bolts	Rod Bearing Bolts	Crankshaft Damper Bolts	Driveplate Bolts	Manifold Intake	Manifold Exhaust	Spark Plugs	Oil Pan Drain Plug
2003	2.5	B5254T/2	①	②	③	133	④	15	18	20	28
	2.9	B6249T	①	②	⑤	219	④	15	18	22	28
2004	2.5	B5254T/2	①	②	③	133	④	15	18	20	28
	2.9	B6249T	①	②	⑤	219	④	15	18	22	28
2005	2.5	B5254T/2	①	②	③	133	④	15	18	20	28
	2.9	B6249T	①	②	⑤	222	④	15	18	22	28
	4.4	B8444S	NA	NA	NA	⑥	NA	14	18	18	36
2006	2.5	B5254T/2	①	②	③	133	④	15	18	20	28
	4.4	B8444S	NA	NA	NA	⑥	NA	14	18	18	36

NA :Information not available

① Step 1: 14 ft. lbs.
Step 2: 43 ft. lbs.
Step 3: Plus 130 degrees

② Tighten cylinder block, intermediate section, in stages:
Step 1: M10 bolts: 15 ft. lbs. (20mm)
Step 2: M10 bolts: 33 ft. lbs. (45mm)
Step 3: M8 bolts: 18 ft. lbs. (25mm)
Step 4: M7 bolts: 13 ft. lbs. (17mm)
Step 5: M10 bolts: Plus 90 degrees

③ Connecting rod with treated toothed surface between the connecting rod and cap, screw with shoulder:
Step 1: 14 ft. lbs.
Step 2: plus 90 degrees
Connecting rod with fracture surface between the connecting rod and cap, fully threaded screw:
Step 1: 22 ft. lbs.
Step 2: Plus 90 degrees

④ Step 1: 33 ft. lbs.
Step 2: Plus 65 degrees

⑤ 15 ft. lbs., plus 90 degrees

⑥ Step 1: 89 ft. lbs.
Step 2: Back off at least 1 full turn
Step 3: 37 ft. lbs.
Step 4: Plus 90 degrees

06041_XC90_C0008

WHEEL ALIGNMENT

Year	Model		Caster Range (+/-Deg.)	Caster Preferred Setting (Deg.)	Camber Range (+/-Deg.)	Camber Preferred Setting (Deg.)	Toe-in (in.)
2003	XC90	F	1.00	+5.70	0.90	-0.33	-0.20+/-0.10
		R	—	—	1.00	-0.04	0.30+/-0.20
2004	XC90	F	1.00	+5.70	0.90	-0.33	-0.20+/-0.10
		R	—	—	1.00	-0.04	0.30+/-0.20
2005	XC90	F	1.00	+5.70	0.90	-0.33	-0.20+/-0.10
		R	—	—	1.00	-0.04	0.30+/-0.20
2006	XC90	F	1.00	+5.70	0.90	-0.33	-0.20+/-0.10
		R	—	—	1.00	-0.04	0.25+/-0.15

06041_XC90_C0009

TIRE, WHEEL AND BALL JOINT SPECIFICATIONS

		OEM Tires		Tire Pressures (psi)		Wheel	Ball Joint	Lugnut Torque
Year	Model	Standard	Optional	Front	Rear	Size	Inspection	(ft. lbs.)
2003	XC90	P225/70R16	P235/65R17	①	①	7-JJ	NA	102
2004	XC90	P235/65R17	none	①	①	7-JJ	NA	102
2005	XC90	P235/65R17	none	①	①	7-JJ	NA	102
2006	XC90	P235/65R17	none	①	①	7-JJ	NA	103

OEM: Original Equipment Manufacturer

PSI: Pounds Per Square Inch

NA: Information not available

① See placard on vehicle

06041_XC90_C0010

BRAKE SPECIFICATIONS

All measurements in inches unless noted

			Brake Disc			Minimum	Brake Caliper	
			Original	Minimum	Maximum	Lining	Bracket Bolts	Mounting Bolts
Year	Model		Thickness	Thickness	Runout	Thickness	(ft. lbs.)	(ft. lbs.)
2003	XC90	F	NA	0.984	0.002	0.12	①	①
		R	NA	0.669	0.002	0.12	②	②
2004	XC90	F	NA	0.984	0.002	0.12	①	①
		R	NA	0.669	0.002	0.12	②	②
2005	XC90	F	NA	0.984	0.002	0.12	①	①
		R	NA	0.669	0.002	0.12	②	②
2006	XC90	F	NA	NA	NA	NA	①	①
		R	NA	NA	NA	NA	②	②

NA: Not available

① Caliper holder (steel spindle): 77 ft. lbs. plus 60 degrees

Caliper sliding pin: 20 ft. lbs.

② Caliper to wheel bearing housing: 26 ft. lbs. plus 60 degrees

06041_XC90_C0011

SCHEDULED MAINTENANCE INTERVALS
VOLVO XC90

TO BE SERVICED	TYPE OF SERVICE	VEHICLE MILEAGE INTERVAL (x1000)									
		7.5	15	22.5	30	37.5	45	52.5	60	67.5	75
Engine oil & filter	R	✓	✓	✓	✓	✓	✓	✓	✓	✓	✓
Automatic transmission fluid	S/I	✓	✓	✓	✓	✓	✓	✓	✓	✓	✓
Brake pads & parking brake	I	✓	✓	✓	✓	✓	✓	✓	✓	✓	✓
Cabin air filter	R		✓		✓		✓		✓		✓
Engine coolant	S/I		✓		✓		✓		✓		✓
Air cleaner filter	R					✓					✓
Spark plugs	R								✓		
Accessory drive belt	R										✓
Fuel lines	I								✓	✓	✓
Exhaust system	S/I							✓			
Check suspension	S/I								✓	✓	✓
Brake fluid ①	R										

R: Replace S/I: Service or Inspect

① Replace every 2 years or 30,000 miles, whichever comes first under normal conditions, more frequently in mountainous areas or moist climates.

FREQUENT OPERATION MAINTENANCE (SEVERE SERVICE)

If a vehicle is operated under any of the following conditions it is considered severe service:

- Extremely dusty areas.

- 50% or more of the vehicle operation is in 32°C (90°F) or higher, or constant operation below 0°C (32°F).

- Prolonged idling (vehicle operation in stop and go traffic).

- Frequent short running periods (engine does not warm to normal operating temperatures).

- Police, taxi, delivery usage or trailer towing usage.

Oil & oil filter: change every 5000 miles.

Air filter element: service or inspect every 15,000 miles.

06041_XC90_C0012

ENGINE REPAIR

➡Disconnecting the negative battery cable on some vehicles may interfere with the functions of the on board computer system. The computer may undergo a relearning process once the negative battery cable is reconnected.

Alternator

REMOVAL & INSTALLATION

5-Cylinder Engines

1. Before servicing the vehicle, refer to the Precautions Section.
2. Remove or disconnect the following:
   - Negative battery cable
   - Accessory drive belt
   - Power steering pump
   - Turbocharger hose
   - Alternator harness connectors
   - Alternator

To install:

Install or connect the following:

- Alternator
- Alternator harness connectors
- Turbocharger hose
- Power steering pump
- Accessory drive belt
- Negative battery cable

6-Cylinder Engines

1. Disconnect the battery negative lead.
2. Raise the car.
3. Remove the lower cover plate.
4. Drain the coolant.
5. Install the lower splash guard.
6. Lower the car.
7. Remove the upper engine coolant hose.
8. Remove the auxiliary belt.
9. Remove the power steering pump.
10. Disconnect the turbocharger hose from the cooling system. Push the hose to one side.
11. Remove the right-hand headlamp.
12. Detach the bumper cover on the right-hand side.
13. Remove the bolts from the front panel on the right-hand side as illustrated.
14. Remove the bolt that runs through the upper mounting and the lower bolts.

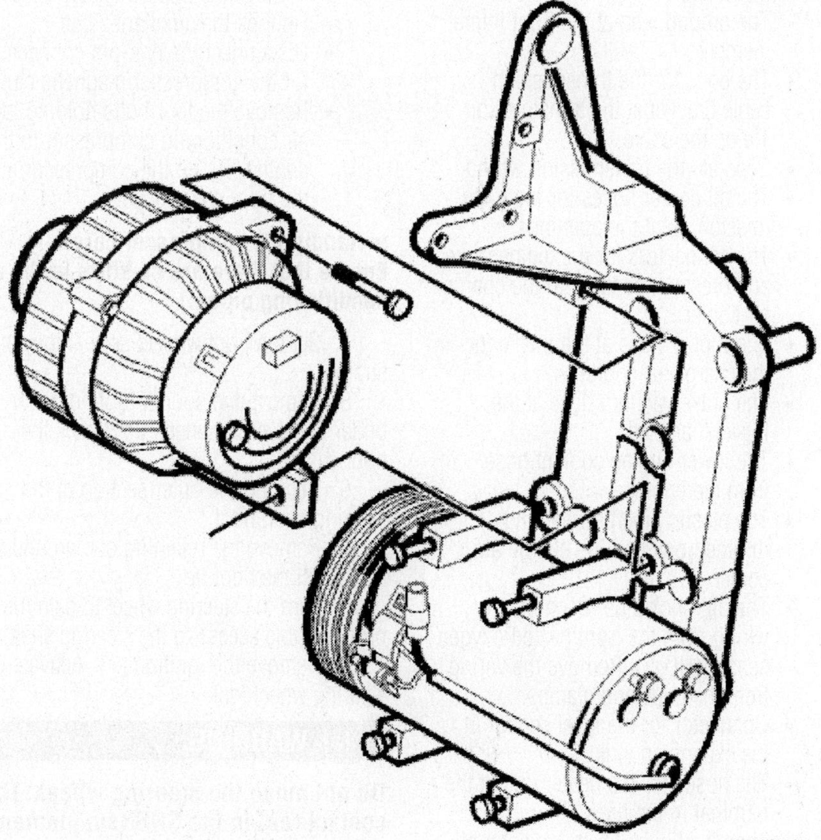

Alternator mounting

15. With AC, slacken off the lower bolts for the compressor, to provide clearance.
16. Disconnect the battery positive lead.
17. Disconnect the connector.
18. Remove the alternator .
19. Turn the alternator into position.
20. Carefully pry the front panel forwards.
21. Pull the alternator upwards and out.

To install:

22. Position the alternator. Install the bolt through the upper mounting. Do not tighten.
23. Connect the cable and the connector.
24. Install the lower bolts for the alternator (upper bolts for the compressor).
25. With AC, tighten the 2 lower bolts for the compressor.
26. Tighten the 3 bolts for the alternator.
27. Install:
    - The front panel
    - The bumper cover
    - The right-hand headlamp
    - The turbocharger hose
    - The power steering pump
    - The auxiliaries belt
    - The engine coolant hose. Top up the coolant
    - The battery negative lead

8-Cylinder Engines

1. Disconnect the battery negative cable.
2. Remove the auxiliary drive belt.
3. Detach the rear ignition coil cover on the right-hand side. Pull straight up and fold it aside.
4. Remove the right hand halfshaft.
5. Remove the cover to the wheel arch
6. Remove the plastic nuts.
7. Removing the fuel line brackets
8. Remove the bolts and nuts.
9. Move the fuel lines to the side.
10. Lift the generator out through the wheel arch.

To install:

11. Lift the generator into position.
12. Install the bolts and connect the battery cable.
13. Install the fuel lines
14. Install the cover to the wheel arch
15. Install the right-hand halfshaft
16. Install the fuel pipe
17. Press in the ignition coil cover.
18. Install the auxiliary belt.
19. Connect the battery negative cable.

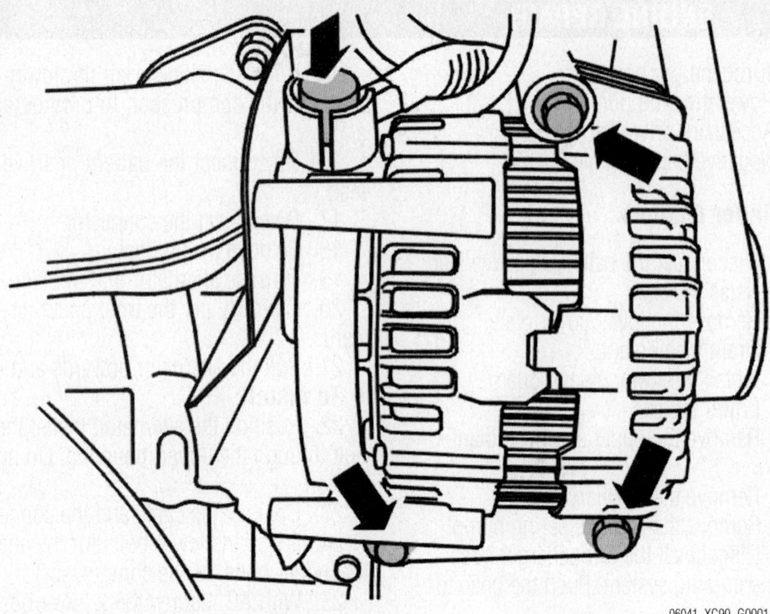

Alternator mounting—8-cylinder engines

Ignition Timing

ADJUSTMENT

Manual adjustment of the ignition timing is not possible.

Engine Assembly

REMOVAL & INSTALLATION

5-Cylinder Engine

1. Before servicing the vehicle, refer to the Precautions Section.
2. Drain the cooling system.
3. Remove or disconnect the following:
   - The battery negative lead
   - The cross stay between the suspension turrets
   - The turbocharger control valve from the air cleaner
   - Mass air flow sensor connector
   - The brake vacuum hose from the air cleaner
   - The air cleaner assembly. Seal the openings in the intake pipe for the turbocharger
   - The plastic charge air pipe above the engine. Seal the opening in the turbocharger
   - The ground strip at the top/rear of the cylinder head.
4. Position a container under the radiator drain cock. Drain the coolant. Close the cock.
5. Drain the fuel line.
6. Remove or disconnect the following:
   - The cover over the central electrical unit
   - The connector for the engine wiring from the central electrical unit
   - The positive lead for the starter motor from the terminal block
   - The ground lead at the front frame member
   - The bolts for the transmission cable bracket at the transmission. Tie up the cable
   - Separate the transmission wiring
   - The oil cooler hoses for the transmission. Seal the openings.
   - The connectors for the temperature and pressure sensors in the charge air cooler
   - The rubber hose at the intake for the charge air cooler
   - The brake vacuum hose at the intake manifold
   - The lower engine coolant hose from the engine
   - The plastic pipe between the throttle body and the charge air cooler
   - The dip stick pipe. Separate the wiring from the front heated oxygen sensor (HO_2S). Remove the wiring from the clips and clamps.
   - Connector for the level sensor in the expansion tank
   - The hoses for the heater unit at the terminal in the firewall
   - The bleed hose for the radiator at the expansion tank
   - The ABS line and clips from the power steering hose. Lift up the expansion tank and servo oil reservoir. Place them on top of the engine.
   - Control module unit cover
   - The control modules
   - The surround for the control module box.
   - The red retaining clip from the connector using a screwdriver
   - The connector by releasing the catches at the front edge using pliers. Pry out the connector using a screwdriver. Lift up the connector with the protection underneath and put it to one side.
   - Remove the ground lead from the right fender liner. Pull out the wiring between the front cover plate and the air conditioning pipe.
   - The right headlamp by pulling up the two lock facings. Disconnect the wiring and put the wiring to one side
   - The connector on the air conditioning receiver drier
   - The gray connector on the fan shroud. Remove the wiring from the clips and clamps on the fan shroud.
   - Remove the upper radiator hose
   - Unhook the auxiliaries belt
   - Disconnect the one-pin connector for the compressor magnetic clutch
   - Remove the four bolts holding the air conditioning compressor to the engine. Tie up the compressor in the front cover plate.

➡Handle the compressor carefully. Ensure that there are no kinks in the air conditioning pipes.

7. Set the front wheels so that they are straight
8. Remove the soundproofing panel under the steering wheel. Fold back the floor carpet
9. Bend up the outer section of the steering shaft boot
10. Remove the retaining clip around the boot. Pull the boot up
11. Turn the steering wheel to gain the best possible access to the steering shaft nut
12. Remove the ignition key. Activate the steering wheel lock.

⁂ WARNING

Do not move the steering wheel. The contact reel in the SRS (supplemental restraint system) could be damaged.

13. Remove or disconnect the following:
    - The bolt securing the mounting for the steering shaft in the joint
    - The joint from the steering shaft by pressing the shaft upwards
14. Raise the car.
15. Remove or disconnect the following:
    - The left and right front wheels
    - The splash guard under the engine.
    - The 6 bolts. Lift up and slide the plate forward
    - The protective plate.

➡There is a guide sleeve on the upper mounting for the plate. This must be lifted slightly so that the panel can be slid forward.

16. Remove the tie rod from the wheel spindle on the left and right sides.

17. On one side, measure the length of the tie rod in relation to the steering gear housing. Note the measurement.

18. Remove on the left and right sides:
    - The link rod from the anti-roll bar
    - The sound insulation cover for the halfshaft
    - The nut holding the control arm to the ball joint
    - The center bolt and the end of the halfshaft.

➡If an engine block heater is installed, mark the position of the fuel lines.

19. Separate the quick-release connectors for the fuel pressure line and engine block heater (option).

20. Pull apart the EVAP line at the rubber joint behind the rear right mounting for the sub-frame.

21. Remove the three-way catalytic converter.

22. Remove the propeller shaft.

23. Use a post hoist with a lifting table and stud guides. Position the lifting table against the sub-frame.

➡Ensure that the lifting table is centered under the sub-frame for optimal weight distribution. Check that the lifting table is correctly applied against the sub-frame.

24. Remove or disconnect the following:
    - The bumper cover on the left hand side. Use a weatherstrip tool
    - The M8 nut at the top of the left-hand fender liner for the engine coolant heater
    - The connector for the engine coolant heater. Press up the engine coolant heater. Lift it out of its upper mounting.
    - The bolts for the sub-frame brackets

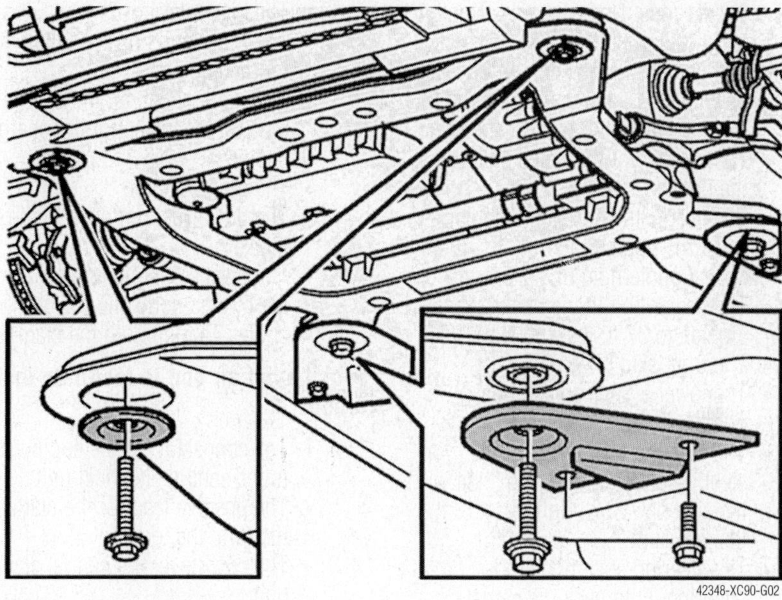

Sub-frame mounting brackets

- The bolts for the sub-frame. Lower the engine and transmission approximately 15 cm so that the halfshafts describe a horizontal line between the hub and transmission.

⁂ WARNING

Ensure that no hoses, wiring or anything else catch in the engine and transmission unit when it is lowered.

- The control arms from the ball joint pinion

25. Hold the constant velocity joint. Pull the halfshaft off the hub

26. Let the halfshafts rest against the control arms

27. Turn the engine block heater out of its position

28. Check that the air conditioning compressor is hanging free. Continue lowering the engine and transmission unit.

To install:

29. Raise the engine and the transmission unit. Use a lifting table with locating pins. Stop when the sub-frame is approximately 15 cm below the frame members.

30. Install, on the left and right sides:
    - The halfshaft to the hub. Install the center bolts. Do not tighten yet!
    - The ball joint to the control arm. Install the nuts. Do not tighten yet!

➡Ensure that the mating surfaces on the ball joints and control arms are clean. Twist the engine block heater (option) into position. Check that the air conditioning compressor is correctly positioned.

31. Check that the tie rod is in the correct position. Check that the transmission cable is routed between the steering shaft and the firewall.

32. Align the steering shaft joint with the hole in the firewall.

33. Lift the engine and transmission unit up to the frame members.

➡Ensure that no hoses, wiring or anything else catches.

34. Install or connect the following:
    - New bolt in the sub-frame. Lubricate the bolts.
    - The washers at the front edge and the support plates at the rear edge of the sub-frame
    - Engine coolant heater left front (option).

35. Tighten the bolts for the sub-frame. Tighten to 77 ft. lbs. (105Nm) plus 120 degrees. Start on the left hand side of the sub-frame. Continue with the right hand side.

36. Tighten the bolts for the brackets at the rear of the sub-frame. Tighten to 37 ft. lbs. (50 Nm).

37. Remove the lifting table.

38. Connect the quick-release connectors and fuel line to the engine and engine block heater (option). Press the EVAP line together at the rubber joint.

39. Install and tighten the M8 nut in the upper mounting for the engine coolant heater. Connect the connector. Install the front bumper cover.

40. Install or connect the following:
    - The propeller shaft
    - The three-way catalytic converter (TWC)

- The tie rods. Use new nuts. Tighten to 51 ft. lbs. (70 Nm)
- The link rod to the anti-roll bar. Use new nuts. Tighten to 37 ft. lbs. (50 Nm). Counterhold using Torx 27
- The bracket for the brake pipe to the body
- The nut for the ball joint and control arm. Tighten to 58 ft. lbs. (80 Nm). Counterhold using Torx 40.
- The center bolt in the halfshaft. Tighten to 37 ft. lbs. (50 Nm).
- The front skid plate
- The lower engine splash guard
- The front wheels
- The steering shaft joint on the steering gear and steering shaft. Install a new bolt. Tighten to 18 ft. lbs. (25 Nm).
- The steering shaft boot and snapring. Fold back the carpet
- Soundproofing panel.
- The compressor

➡Handle the compressor carefully. Ensure that there are no kinks in the air conditioning (A/C) pipes. Connect the one-pin connector to the compressor magnetic clutch.

- The connector on the air conditioning receiver/drier
- The gray connector for the fan shroud
- The wiring in the clips and clamps on the fan shroud
- The connectors for the pressure and temperature sensors on the charge air cooler
- The right-hand headlamp using two lock facings.
- The engine wiring to the control module box.
- The ground lead to the ground terminal at the right fender edge in the engine compartment
- The auxiliary belt
- The intermediate piece for the control module box
- The control modules
- The cover for the control module box
- Expansion tank. Connect the radiator breather tube to the expansion tank. Tighten the clamp. Connect the connector for the level sensor in the expansion tank
- Servo oil reservoir. Check that the hoses are correctly positioned.

41. Clamp the ABS cable to the power steering hose.

42. Install or connect the following:

- The transmission cable
- The oil cooler hoses
- The lower and upper coolant hoses for the engine
- The rubber hose on the charge air pipe on the intake for the charge air cooler
- The dip stick pipe
- The brake vacuum hose to the intake
- The plastic charge air pipe between the throttle body and the charge air cooler. Tighten both the clamps.

➡Use a hot air gun to facilitate installation.

- The connector for the engine wiring to the central electrical unit
- The positive lead for the starter motor to the terminal block
- The cover over the central electrical unit
- The ground lead between the engine and body in the front frame member.
- The ground strip at the top/rear of the cylinder head
- The plastic charge air pipe over the engine
- Air cleaner assembly
- The inlet hose for the turbocharger to the air cleaner
- Mass air flow sensor connector
- The turbocharger control valve on the air cleaner
- The brake vacuum hose in the clamp on the air cleaner housing
- The intake pipe between the front cover plate and the air cleaner
- The engine stabilizer brace between the suspension turrets
- The battery negative lead.
- Coolant
- Engine oil
- Power steering fluid.

43. Start the engine

44. Warm up the engine until the thermostat opens

45. Check for leakage

46. Check the oil and coolant levels. Adjust if necessary

47. Check the transmission oil level.

6-Cylinder Engine

1. Remove the cable from the battery negative terminal.
2. The expansion tank cap.
3. Drain the refrigerant.

➡There is a guide sleeve on the upper mounting for the plate. This must be lifted slightly so that the panel can be slid forward.

4. Remove the 6 bolts.
5. Lift up and slide the plate forward.
6. Remove the protective plate.
7. Remove the splash guard under the engine
8. Disconnect the connectors for the heated oxygen sensors (HO_2S). Remove the wiring from the clips and tie straps.
9. Connect a hose to the engine coolant drain cock.
10. Drain the coolant into a clean container.
11. Remove:
    - The bolt for the engine stabilizer brace mounting on the engine
    - The cross stay between the suspension turrets
    - The plastic charge air pipe over the engine
    - The charge air pipes which connect to the turbocharger. Seal the openings
    - The air cleaner.

➡Note the position of the probes to make it easier to reinstall them. If they are mixed up, diagnostic trouble codes (DTCs) may be stored.

12. Slide the wiring for the front heated oxygen sensors (HO_2S) upwards, Unscrew the heated oxygen sensors (HO_2S) from the three-way catalytic converter.
    - Remove the heat deflector plates from the turbocharger
    - Remove the nuts from the flange between the turbocharger and the three-way catalytic converter.

➡Rustproof the studs.

13. Separate the connections. Unhook the three-way catalytic converter.
14. Remove the ground strip between the engine and bodywork which is positioned above the turbocharger for cylinders 1, 2 and 3.
15. Remove both the hoses for the heater unit on the firewall.
16. Remove:
    - The battery lead from the terminal block at the fusebox
    - The connector in the fuse box.
    - The rear radiator hose at the transmission
    - The gear selector cable from the transmission
    - The connector for the input speed sensor on the transmission
    - The connector on the transmission
    - The vacuum hose on the intake for the power brake booster
    - The connector on the sensor in the charge air cooler

- The ground lead from the engine to the car body (detach in the car body).

17. Remove the charge air pipe on the intake between the throttle module and the charge air cooler.

➡If necessary use a hot air gun to loosen the pipe.

18. Detach the upper radiator hoses on the engine.
19. Lift the power steering reservoir. Place it on top of the engine.

➡Ensure that no oil leaks out of the reservoir through the ventilation holes in the cover.

20. Remove the coolant expansion tank. Slacken off the hose clamp on the hose to the engine. Disconnect the connector.
21. Remove the radiator breather tube by the expansion tank. Put the expansion tank to one side.
22. Remove:
    - The cover from the distribution box
    - The control module
    - The surround from the distribution box.
23. Pry out the connector plate from the lower section of the control module box. Pull out the engine and transmission wiring connectors from the plate.
24. Detach the ground lead on the fender liner.

➡Seal the openings.

25. Separate the climate control unit pipe at the frame member inside the central electrical unit.
26. Remove the right-hand headlamp by pulling the 2 lock facings on the lamp straight up. Then carefully pull the headlamp forward. Disconnect the connector and lift out the headlamp.
27. Disconnect the connector from the sensor on the accumulator. Pull the wiring through to the engine compartment.
28. Set the front wheels so that they are straight.
29. Fold back the floor carpet.
30. Bend up the outer section of the steering shaft boot.
31. Remove the retaining clip around the boot.
32. Pull the boot up.
33. Turn the steering wheel to gain the best possible access to the steering shaft nut.
34. Remove the ignition key. Activate the steering wheel lock.

⁂ WARNING

Do not move the steering wheel. The contact reel in the SRS (supplemental restraint system) could be damaged.

35. Remove:
    - The bolt securing the mounting for the steering shaft in the joint
    - The joint from the steering shaft by pressing the shaft upwards.

➡Clean and spray the exposed threads with rust-proofing agent before beginning removal.

- The front wheel
- The tie rod ends from the steering arm.
- The bolt holding the halfshaft to the hub
- The nut for the ball joint / link arm
- The nut for the anti-roll bar.

➡Counterhold using Torx 40 so that the ball joint boot does not turn when the nut is removed.

⁂ WARNING

The steering shaft must not be turned. The contact reel in the SRS system could be damaged.

36. On one side, measure the length of the tie rod in relation to the steering gear housing. Note the measurement.
37. Disconnect the brake pipe from the clips on the front SIPS member.
38. Remove the 4 bolts for the front SIPS member.
39. Remove the member.
40. Unhook the three-way catalytic converter (from the rubber mounting.
41. Remove:
    - The tie straps from the wiring for the rear probe. Separate the probe wiring
    - The nut for the flange between the three-way catalytic converter and exhaust system
    - The propeller shaft.
42. Drain the fuel line.
43. Disconnect the fuel line quick-release connector at the engine.

➡Seal the openings against dirt.

44. Place the cable to one side.
45. Separate the EVAP canister cable at the rear right sub-frame mounting.
46. Remove the transmission oil cooling hoses from the transmission. Seal the openings.

⁂ CAUTION

Ensure that the lifting table is properly centered under the engine.

47. Position a lifting table under the sub-frame. Secure the table against the sub-frame.

⁂ CAUTION

Do not damage the brake pipe that runs across the transmission tunnel.

48. Remove:
    - The sub-frame brackets bolts
    - The bolts for the sub-frame. Lower the engine and transmission approximately 15 cm so that the halfshafts describe a horizontal line between the hub and transmission.

➡Take great care that no hoses, wiring or anything else catches in the engine/ transmission unit when it is lowered.

49. Unhook the control arms from the ball joint pinion.
50. Remove the halfshafts from the wheel hubs.

➡Do not pull the halfshafts. They do not have axial locks.

⁂ CAUTION

Do not damage the ABS wiring. Do not damage the brake pipe.

51. Let the halfshafts rest against the control arms. Lower the three-way catalytic converter.
52. Continue lowering until the hose for the compressor in the climate control unit is accessible. Detach the inner hose for the climate control unit. Seal the openings. Continue lowering the unit.

To install:

53. Lift the engine and transmission unit. Use a lifting table.
54. Stop the lift approximately 15 cm before the sub-frame reaches the frame members.
55. Install (on both sides) the halfshaft on the hub.
56. Lift the three-way catalytic converter up into position. Connect the exhaust system to the rear flange. Use a new gasket. Install new nuts loosely.

➡The three-way catalytic converter cannot be raised after the engine has been raised to its correct position.

57. Continue raising the engine until it is only approximately 50 mm from its correct position.

➡**Guide the lower steering shaft into the hole in the firewall.**

58. Align the ball joint in the control arm.
59. Install the inner pipe on the compressor in the climate control unit. Use a new O-ring.
60. Lift the sub-frame until it is against the frame members.
61. Install new bolts in the sub-frame. Lubricate the bolt threads.
62. Install the washers in the front edge and support plates at the rear edge of the sub-frame.
63. First tighten the bolts on the left side of the sub-frame. Tighten to 47 ft. lbs. (65 Nm) plus 60 degrees.
64. Then tighten the bolts on the right-hand side of the sub-frame to the same values.
65. Tighten the brackets. Tighten to 37 ft. lbs. (50 Nm).
66. Remove the lifting table.
67. Tighten the ball joint and control arm. Tighten to 58 ft. lbs. (80 Nm). Counterhold using Torx 40 so that the ball joint boot does not rotate.
68. Install:
- The anti-roll bar links. Use new locknuts. Tighten to 37 ft. lbs. (50 Nm).
- New center bolts for the halfshafts. Tighten to 37 ft. lbs. (50 Nm).

69. Install the tie rod using the marking made during removal. Use new locknuts. Tighten to 51 ft. lbs. (70 Nm).
70. Connect the fuel line to the engine.
71. Connect the evaporative emission system (EVAP) line at the right rear sub-frame mounting.
72. Install the oil cooler hoses on the transmission.
73. Install the propeller shaft.
74. Lower the car.
75. Pull up and align the three-way catalytic converter (TWC) against the turbocharger (TC).
76. Lubricate the threads using copper paste 116 1408, or equivalent. Install new nuts loosely.
77. Lower the car.
78. Install the front SIPS member. Press the brake pipe into place in the member.
79. Hook the rubber mounting for the three-way catalytic converter (TWC) in the bracket on the sub-frame.
80. Connect the connector to the rear probe. Secure the wiring as before.
81. Tighten the screwed joint between the three-way catalytic converter (TWC) and the exhaust system. Tighten to 18 ft. lbs. (24 Nm).
82. Install:
- The steering shaft joint on the steering gear and steering shaft
- A new bolt. Tighten to 18 ft. lbs. (25 Nm)
- The steering shaft boot and snap-ring. Fold back the carpet.

83. Connect the engine and transmission wiring connectors to the control module box base plate.
84. Press the base plate into place in the lower section of the control module box.
- The climate control unit pipe at the frame member inside the control unit box. Use a new O-ring
- The cable from the engine wiring for the sensor on the receiver to the right of the radiator
- The front right headlamp

85. Install the intermediate section. Align the ventilation pipe in the rear edge of the control module box.
86. Press the control modules down carefully. Lock into place using mounting bracket 999 5722, or equivalent.
87. Press the cover into place. Screw the ground lead into place on the fender liner.
- The connector for the expansion tank from the engine compartment wiring
- The hose from the engine to the expansion tank
- The breather tube from the radiator
- The expansion tank
- The servo reservoir
- The upper radiator hose.
- The plastic pipe between the charge air cooler and the throttle module.

➡**Use a hot-air gun to facilitate installation of the pipe.**

- The connector for the sensor on the charge air cooler
- The ground lead from the engine to the car body
- The vacuum hose to the intake and power brake booster
- The connectors for the transmissions
- The mechanical transmission cable to the transmission
- The rear radiator hose
- The connector to the fuse box
- The battery lead to the terminal block on the fuse box.
- The radiator hoses on the heater unit. They are coded and will only connect to the correct sleeve connector
- The ground strip between the engine and the car body

88. Tighten the three-way catalytic converter to the turbocharger. Use new nuts. Tighten to 18 ft. lbs. (24 Nm).
89. Install the upper heated oxygen sensors using the marks made during removal. Lubricate the threads. Use copper paste 116 1408, or equivalent. Tighten to 33 ft. lbs. (45 Nm).
90. Lower the wiring down to the probes. Secure the wiring in the clips and clamps as before. Connect the wiring.
91. Install the heat deflector plates.
92. Install:
- The charge air pipes on the turbocharger
- The plastic charge air pipe over the engine
- Air cleaner assembly
- Mass air flow sensor connector
- The turbocharger control valve on the air cleaner
- The intake manifold to the air cleaner
- The engine stabilizer brace between the suspension turrets.

93. Tighten the bolts on the suspension turrets. Tighten to 37 ft. lbs. (50 Nm). Tighten the bolt in the torque bracket. Tighten to 58 ft. lbs. (80 Nm).
94. Install the battery negative lead.
95. Fill with coolant. The coolant volume is approximately 7.2 liters. Top up the engine oil.
96. Start the engine.
97. Warm up the engine until the thermostat opens.
98. Check for leakage.
99. Check the oil and coolant levels. Adjust if necessary.
100. Check the transmission oil level.
101. Install:
- The splash guard under the engine
- The front skid plate on the sub-frame
- The front wheels

8-Cylinder Engine

1. Remove the battery negative lead.
2. Remove the expansion tank cap.

➡**There is a guide sleeve on the upper mounting for the plate. This must be lifted slightly so that the panel can be slid forward.**

3. Remove the 6 bolts. Lift and slide the protective plate forward. Remove the protective plate.
4. Remove the 6 bolts and the lower engine cover.

5. Drain coolant.
6. Remove the 4 bolts at the suspension turrets, the bolts of the upper engine mount on the engine and the upper cross stay.
7. Remove the air cleaner cover
8. Remove the 2 connectors from the engine control module (ECM); clip the band clamp to the cable harness, remove the connector from the mass air flow sensor.
9. Remove the brake vacuum hose from the vacuum pump.
10. Remove the 2 bolts from the fresh air hose in the front plate.
11. Remove the fresh air hose.
12. Remove the air cleaner by pulling it straight up.
13. Remove the positive lead from the electrical distribution box.
14. Remove the ground cable between the engine and body (loosen in the body).
15. Remove the connector at the electrical distribution box from the brake vacuum valve. Detach the vacuum hose from the valve that leads to the brake vacuum amplifier.
16. Remove the selector lever cable using a break fork to break the cable loose from its ball.
17. Remove the transmission control module (TCM) connector.
18. Remove the lower line of the transmission oil cooler.
19. Remove the upper line of the transmission oil cooler.

➨Seal the openings and collect the oil from the oil cooler in a receptacle. Do not reuse the drained oil.

20. Detach the radiator hoses at the quick-release coupling toward the firewall.
21. Remove the upper radiator hose from the distributor housing.
22. Remove the lower radiator hose from the thermostat housing.
23. Remove the upper engine covers.
24. Remove the upper engine covers by pulling them straight up.
25. Remove the 2 bolts on the cover over the servo pump.
26. Remove the cover.
27. Remove the bolt on the washer fluid filler pipe; bend out the pipe.
28. Remove the return line for servo oil from the clips and clamps.
29. Lift up the servo oil reservoir.
30. Remove the 2 coolant hoses from the expansion tank.
31. Remove the 3 ground cables from the body at the expansion tank.
32. Remove the connector from the electric cooling fan.
33. Position the oil reservoir on the engine.

➨Make sure no oil runs out. Make sure that the oil return line is on the correct side of the upper coolant hose that runs to the engine.

34. Remove the auxiliary drive belt.
35. Remove the ground cable bolt between the engine and body. It is located beneath the expansion tank.
36. Raise the vehicle.
37. Remove the front wheels.
38. Remove the ball joint from the link arm (on both sides of the vehicle)
   a. Remove the halfshaft bolt. Counterhold with a prybar in the brake disc cooling channels.
   b. Remove the rubber sleeve in the wheel hub.
   c. Remove the nut between the ball joint and link arm.
   d. Remove the nut of the anti-roll bar link in the anti-roll bar.

➨Counterhold with a Torx wrench so that the ball joint gaiters do not turn when the nuts are removed.

39. Mark the position of the track rod.
40. Remove the track rod end from the guide arm, using a puller.
41. Detach the ball joint pin from the link arm.
42. Fit the cap nut, 7062-1, on the ball joint pin.
43. Tighten the nut toward the link arm. Then turn it back one rotation to create play between the cap nut and the link arm. Position the puller on the link arm and detach the ball joint from the link arm.
44. Remove the nuts at intermediate pipe joints and the intermediate pipes; scrap the gaskets.

➨Mark the position of the carrier on the driveshaft.

45. Remove the driveshaft bolts.

➨Do not move the steering wheel. The contact reel in the SRS (supplemental restraint system) could be damaged.

46. Set the front wheels in the straight-ahead position.
47. Fold the floor carpet out of the way.
48. Bend up the outer section of the steering shaft boot.
49. Remove the retaining clip around the boot.
50. Pull the boot up.
51. Turn the steering wheel to gain the best possible access to the steering shaft nut.
52. Remove the ignition key and activate the steering wheel lock.
53. Remove the bolt securing the mounting for the steering shaft in the joint
54. Remove the joint from the steering shaft by pressing the shaft upwards.
55. Detach the connector (blue) from the steering gear.
56. Drain the fuel line.
57. Detach the quick-release coupling of the fuel line at the right rear sub-frame mount.
58. Detach the fuel return line.

➨Seal the openings against dirt.

Detach the connector from the oil level sensor in the oil sump and the oil pressure sensor.

59. Remove the A/C compressor.
60. Pull out the cable harness and bend it aside.

➨Ensure that the lifting table is properly centered under the engine.

61. Position a lifting table under the sub-frame. Secure the table against the sub-frame.
62. Remove the connector from the circulation pump.
63. Remove the 4 M10 bolts in the rear edge of the sub-frame brackets.
64. Remove the 4 bolts in the sub-frame. Lower the engine/transmission unit approx. 15 cm so that the halfshafts create a horizontal line between the hub and the transmission.

➨Take great care that no hoses, wiring or anything else catches in the engine/transmission unit when it is lowered.

65. Tap the link arms off of the ball joint pin. Release the drive shafts from the wheel hubs by pressing the drive shafts into the hub.
66. Remove the 3 A/C compressor bolts.
67. Lift out the A/C compressor and suspend it in a suitable manner.
68. Continue lowering the engine/transmission unit.

To install:

⁂ WARNING

Ensure that no hoses, wiring or other components catch or become trapped during lifting.

➨Make sure that the mating surfaces on the ball joints and control arms are clean.

69. Raise the engine/transmission unit using a lifting table. Stop approx. 15 cm before the sub-frame reaches the frame members.

70. Install the lower radiator hose in the mount on the sub-frame.

71. Install the A/C compressor; tighten 3 bolts

72. Attach the connectors to the A/C compressor, the oil pressure sensor, and the oil level sensor; secure the cable harness in the clips and clamps.

73. Install the drive shaft in the hub on both sides.

74. Continue raising the engine to approx. 50 mm before it makes contact.

➡**Guide the lower steering shaft into the hole in the firewall.**

75. Guide the ball joint to the link arm.

76. Raise the sub-frame toward the frame members until it makes contact.

77. Install new bolts on the sub-frame. Oil the bolt threads.

78. Install the washers in the front edge and the support plates in the rear edge of the sub-frame.

79. First tighten the bolts on the left side of the sub-frame. Tighten the M10 bolts to 26 ft. lbs. (35 Nm) plus 60 deg. Tighten the M8 bolts to 15 ft. lbs. (20 Nm) plus 60 deg.

80. Then tighten the bolts on the right side of the sub-frame to the same values.

81. Tighten the bolts of the support plates.

82. Remove the lifting table.

83. Tighten the ball joint and link arm. Counterhold with Torx 40 so that the ball joint gaiter does not rotate.

84. Install the anti-roll bar links. Use new lock nuts.

85. Install new center bolts and a rubber sleeve on the drive shafts.

86. Install the track rods following the marks made during removal. Use new lock nuts.

➡**Remove the seals.**

87. Connect the fuel lines.

88. Attaching the steering gear connector

89. Attach the connector to the steering gear.

90. Install the anti-roll bar.

91. Install the driveshaft.

92. Install the intermediate pipes using new gaskets.

93. Install the nuts at intermediate pipe joints.

94. Install the gear selector cable in the mount on the transmission and press the cable into the ball.

95. Install the transmission control module (TCM) connector.

96. Install the upper radiator hose on the distributor housing; tighten the clamp.

97. Install the lower radiator hose on the thermostat housing; tighten the clamp.

98. Connect the radiator hoses at the firewall.

99. Remove the seals.

100. Install the lower line to the transmission oil cooler as marked; secure with a quick-acting lock.

101. Install the upper line to the transmission oil cooler as marked; secure with a quick-acting lock.

102. Install the positive lead in the electrical distribution box.

103. Install the ground cable in the body (between the engine and body)

104. Install the connector at the electrical distribution box.

105. Install the connector to the brake vacuum valve and the vacuum hose on the valve leading to the brake vacuum amplifier.

106. Install the auxiliary belt.

107. Installing the ground cable.

108. Install the servo oil reservoir.

109. Install the cover over the servo pump.

110. Position the upper engine covers and press them into the retainers.

111. Install the upper cross stay.

112. Install the lower engine cover.

113. Install the protective plate.

114. Install the front wheels.

115. Install the battery cable.

➡**Make sure the outlet hose of filling equipment is not blocked as this could damage the cooling system.**

116. Fill with coolant.

117. Check the oil level in the engine and transmission. Start the engine and allow it to warm to normal operating temperature. Check for leaks and proper function. Fill up fluids as necessary.

Heater Core

REMOVAL & INSTALLATION

1. Before servicing the vehicle, refer to the Precautions Section.
2. Drain the cooling system.
3. Remove:
   - The hoses from the heater core pipes in the engine compartment
   - The seal and the plate
   - The soundproofing panel
   - The bolt for the heat exchanger
   - The mounting brackets for the pipes

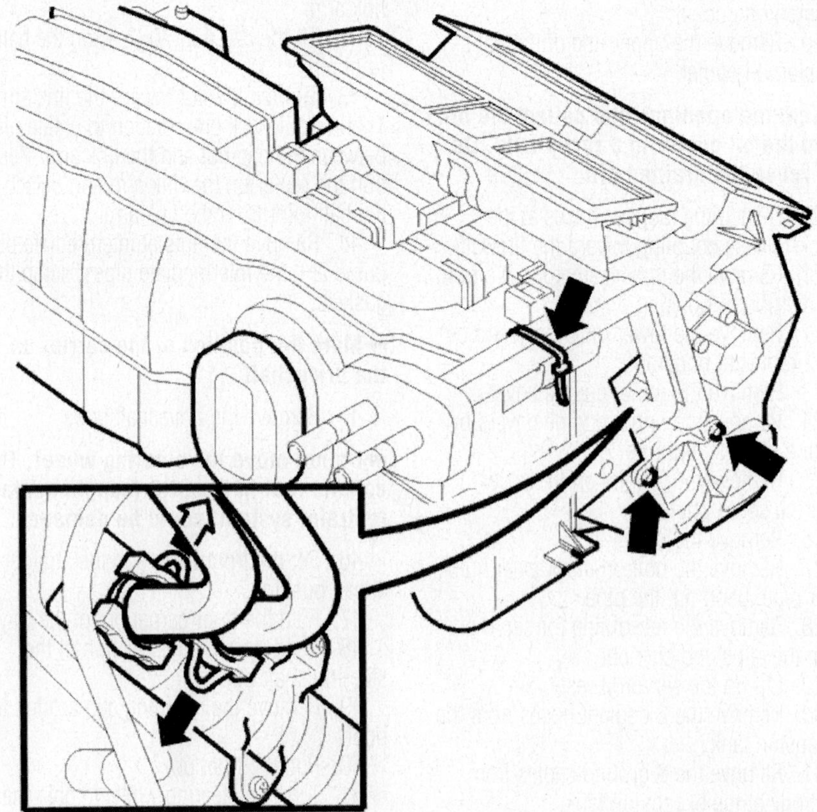
42348-XC90-G03

Heater core

4. Position plenty of paper under the heat exchanger and around the pipes.
5. Drain the coolant into a suitable container.
6. Detach the pipes and move them out of the way.
7. Pull out the heat exchanger.

To install:

8. Install:
  - The upper pipe
  - The striker plate using round-nosed pliers.
  - The catch using a pair of pliers.
  - The lower pipe
  - The striker plate using round-nosed pliers.
  - The striker plate using round-nosed pliers.
  - The seal and the plate
  - The heating loops in the engine compartment.
9. Start the engine and the heating system.
10. Fill the cooling system.
11. Check that the pipe connectors do not leak.
12. Install the soundproofing panel.

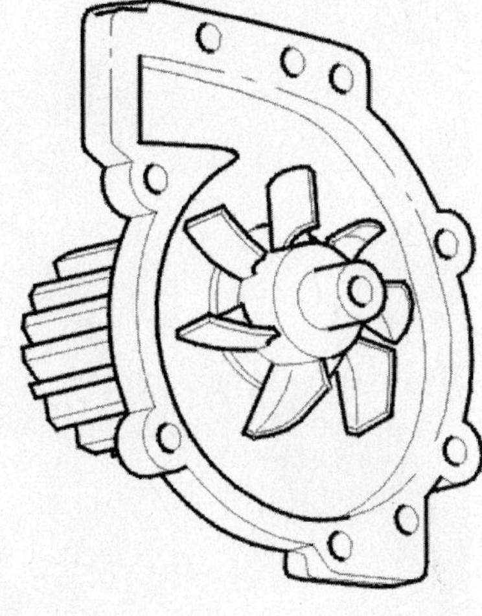

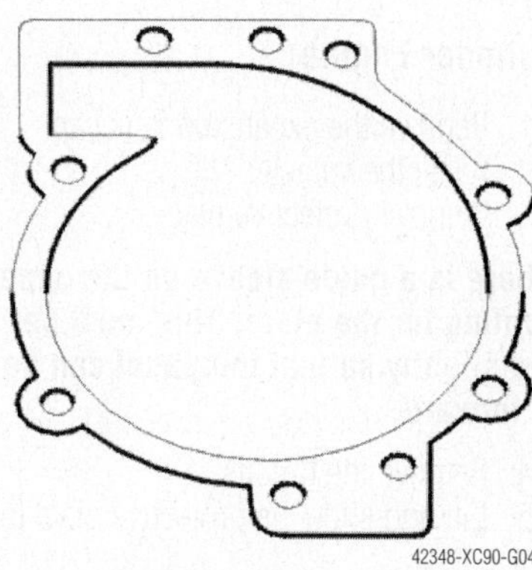

Water pump and gasket

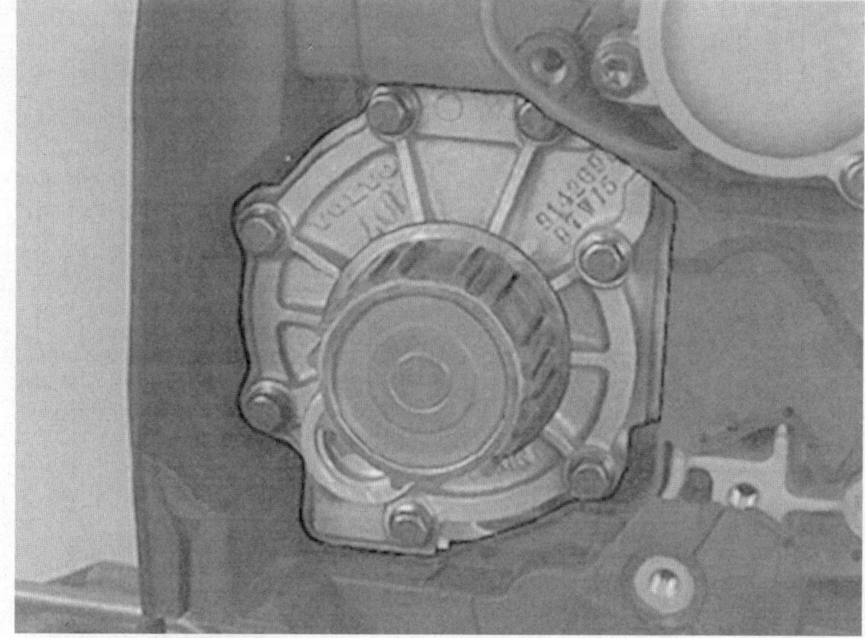

Water pump installation

Water Pump

REMOVAL & INSTALLATION

5-Cylinder Engine

1. Before servicing the vehicle, refer to the Precautions Section.
2. Drain the cooling system.
3. Remove or disconnect the following:
  - Negative battery cable
  - Spark plug cover
  - Fuel line clips
  - Expansion tank
  - Accessory drive belts
  - Vibration damper guard
  - Front wheel
  - Inner fender panel
  - Front timing cover
  - Timing belt
  - Water pump

To install:

4. Install or connect the following:
  - Water pump with a new gasket. Tighten the bolts to 15 ft. lbs. (20 Nm).
  - Timing belt
  - Front timing cover
  - Inner fender panel
  - Front wheel
  - Vibration damper guard
  - Accessory drive belts
  - Expansion tank
  - Fuel line clips
  - Spark plug cover
  - Negative battery cable
5. Fill the cooling system.
6. Start the engine and check for leaks.

6-Cylinder Engine

1. Before servicing the vehicle, refer to the Precautions Section.
2. Drain the cooling system.
3. Raise the car.
4. Remove the lower engine splash guard.
5. Open the engine nipple. Drain the coolant into a container. Close the nipple.
6. Remove the timing belt.
7. Remove the bolts holding the pump at the cylinder block.
8. Tap the pump wheel using the shaft of a hammer. Remove the pump.

To install:

9. Carefully clean the gasket face.
10. Install a new gasket and the coolant pump. Tighten crosswise. Tighten to 13 ft. lbs. (17 Nm).
11. Install the timing belt.
12. Remove the lock grip pliers from the lower coolant hose.
13. Install the lower engine splash guard.

14. Remove the lock grip pliers from the upper coolant hose.
15. Fill coolant and check the level.
16. Check the power steering fluid level.

8-Cylinder Engine

1. Remove the expansion tank cap.
2. Raise the vehicle.
3. Remove protective plate

➡There is a guide sleeve on the upper mounting for the plate. This must be lifted slightly so that the panel can be slid forward.

4. Remove the 6 bolts.
5. Lift and slide the protective plate forward.
6. Remove the protective plate.
7. Draining coolant
8. Open the cock on the lower left edge of the radiator.
9. Drain the coolant into a clean receptacle.
10. Close the cock.
11. Install the protective plate.
12. Loosen the 4 bolts of the coolant pump belt pulley a few turns.
13. Remove the belt.
14. Remove the 4 bolts of the belt pulley.
15. Remove the 7 bolts of the coolant pump and the coolant pump. Discard the gasket.

To install:

16. Clean the gasket surface for the coolant pump.
17. Install a new gasket and the coolant pump. Tighten the M6 bolts to 89 inch lbs. (10 Nm).
18. Position the belt pulley but do not tighten the bolts.
19. Install the auxiliary belt.
20. Tighten the 4 M6 bolts of the belt pulley to 89 inch lbs. (10 Nm).
21. Fill with coolant.
22. Start the engine and allow it to warm to normal operating temperature. Check for leaks and proper function. Fill up fluid as necessary.

Water pump—8-cylinder engines

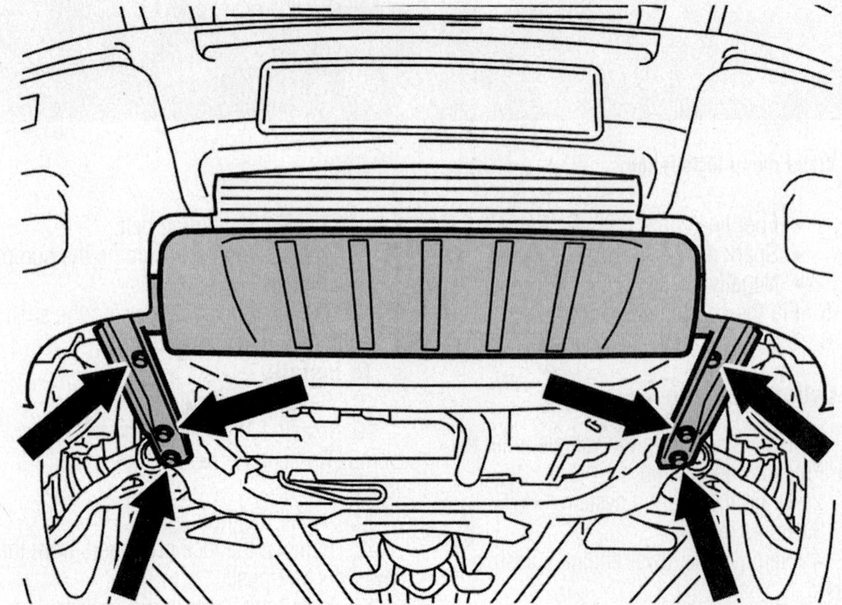

Protective plate—8-cylinder engines

Cylinder Head

REMOVAL & INSTALLATION

5-Cylinder Engines

1. Before servicing the vehicle, refer to the Precautions Section.
2. Drain the cooling system.
3. Remove or disconnect the following:
   - Negative battery cable
   - Air intake assembly
   - Accessory drive belts
   - Front cover
   - Timing belt
   - Exhaust manifold
   - Fuel supply manifold covers
   - Fuel line retainers
4. Install fuel injector holders on the injectors.
   - Fuel pressure regulator vacuum hose
   - Fuel supply manifold with injectors attached
   - Ignition coils
   - Cooling fan
   - Intake manifold with turbocharger attached
   - Upper radiator hose
   - Camshaft sprockets
   - Rear timing cover
   - Camshaft Position (CMP) sensor, if equipped
   - Extension arm and brackets
   - Cylinder head cover
   - Camshafts

Cylinder head loosening sequence—5-cylinder engine

Cylinder head torque sequence—5-cylinder engines

- Coolant pipe
- Cylinder head

To install:

5. Install the cylinder head with a new gasket. Tighten the bolts in sequence as follows:
    a. Step 1: 15 ft. lbs. (20 Nm)
    b. Step 2: 44 ft. lbs. (60 Nm)
    c. Step 3: Plus 130 degrees
6. Install or connect the following:
    - Coolant pipe
    - Camshafts
    - Upper cylinder head. Tighten the bolts to 13 ft. lbs. (17 Nm).
    - Extension arm and brackets
    - CMP sensor, if equipped
    - Distributor, if equipped
    - Rear timing cover
    - Camshaft sprockets
    - Upper radiator hose
    - Intake manifold
    - Cooling fan
    - Fuel supply manifold with injectors attached and remove the injector holders
    - Fuel pressure regulator vacuum hose
    - Fuel line retainers
    - Fuel supply manifold covers
    - Exhaust manifold
    - Timing belt
    - Front cover
    - Accessory drive belts
    - Air intake assembly
    - Negative battery cable
7. Fill the cooling system.
8. Start the engine and check for leaks.

6-Cylinder Engine

1. Before servicing the vehicle, refer to the Precautions Section.
2. Disconnect the battery negative lead.
3. Remove the radiator breather tube at the expansion tank
4. Install lock grip pliers
5. Install the cover from the expansion tank.
6. Lift up the servo reservoir and place it on top of the engine.

➡Ensure that the oil does not leak from the ventilation hole in the filler cap.

7. Seal the hose between the expansion tank and the radiator.
8. Disconnect the hose at the tank. Lift up the expansion tank and place it on top of engine.
9. Remove the turbocharger.
10. Remove:
    - The ground strip between the cylinder head and the engine
    - The upper engine coolant hose between the engine and radiator
    - The solenoid valve (turbocharger control system) from the air cleaner
    - The air cleaner intake from the connection on the front cover plate
    - Air cleaner assembly
    - Torque rod for cross stay
    - The upper timing cover
    - The cover over the ignition coils
    - The front timing cover.
11. Relieve the load from the belt tensioner. Remove the auxiliary belt.
12. Drain the fuel line.
13. Remove:
    - The brake vacuum hose from the intake manifold
    - The charge air pipe from the electronic throttle module
    - The dip stick pipe from the intake manifold
    - The pipe bolt for the crankcase

ventilation connection at the intake manifold
- The protective cover over the nozzle connectors
- The fuel rail mounting bolts.

14. Spray penetrating oil around the injector connectors on the intake manifold. Gently work the fuel rail and injectors loose.
15. Separate the fuel line quick-release connector under the intake manifold. Use tool 951 2666, or equivalent. Carefully put the fuel rail to one side.
16. Remove the EVAP hose from the intake manifold.
17. Remove:
    - The camshaft position sensor housing and trigger wheels
    - The pipe bolt at the rear edge of the cylinder head for the water heated crankcase ventilation
    - The clamps for the high tension wiring and positive cable.
    - The bolt that holds the inlet pipe for the coolant pump on the cylinder block
    - The two bolts for the connector between the coolant pipe and the bypass channel. Carefully turn the pipe to remove it from the cylinder block.
18. Remove the nuts from the cover on the fender liner.
19. Position the upper timing cover.
20. Turn the crankshaft clockwise until the markings on the crankshaft and camshafts correspond.
21. Remove the upper timing cover.
22. Remove:
    - The variable valve timing solenoid connectors
    - The connector for the coolant temperature sensor
    - The bolts for the ignition coils and the two ground terminals.

➡Mark up the ignition coils before removal.

23. Lift up and place the ignition coils and wiring to one side.
24. Slacken off the center bolt for the belt tensioner slightly.
25. Hold the center bolt still. Turn the tensioner eccentric clockwise to 10 o'clock using a (6 mm) Allen key.
26. Remove the timing belt.
27. Install tool 999 5452, or equivalent, at the rear of the camshafts.
28. Remove the center plugs for the variable valve timing units.

➡Collect any oil spills using paper under the plugs.

42348-XC90-G11

Tool 9995452 installed

29. Remove the variable valve timing unit center bolts.
30. Pull or work off the variable valve timing units and timing gear pulley.
31. Remove tool 999 5452.
32. Remove:
    - The bolt for the thermostat housing cable duct
    - Bolts holding the thermostat housing to the cylinder head. Lift out the thermostat housing, gasket and cable duct
    - The bolt holding the inner timing cover to the cylinder head.
    - Spark plugs for cylinders 2 and 5. Install tool 999 5454, or equivalent. Leave a 2-3 mm gap to the camshaft cover. Ensure that the bolt in the spark plug well is fully tightened

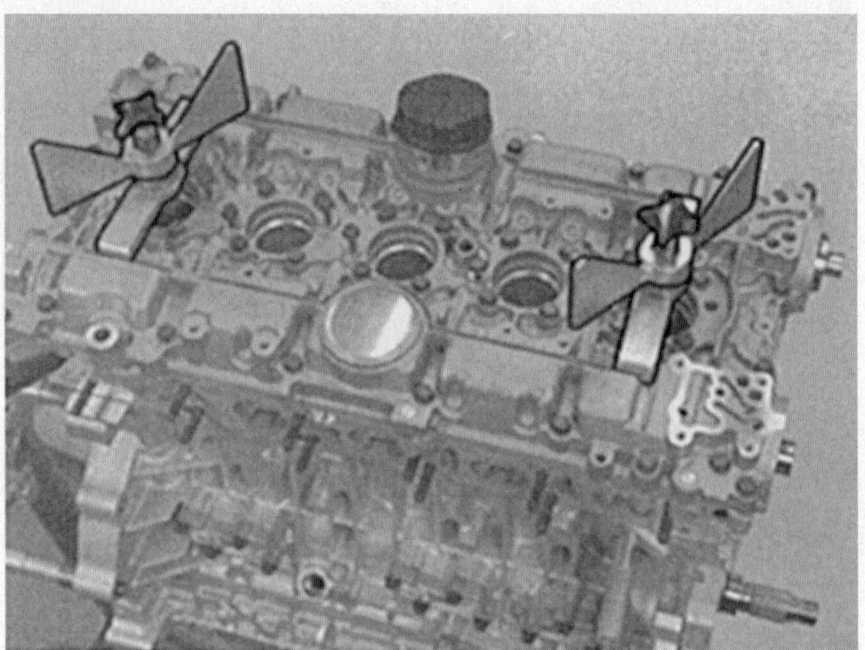

42348-XC90-G10

Tool 9995454 installed

- Variable valve timing solenoids
- The bolts holding the camshaft cover to the cylinder head.

33. Use to lift the cover from the cylinder head. Install the pliers at the stop lugs.
34. Start with cylinder 1 and work alternately backward.
35. Slacken off the wing nuts approximately 2 turns.
36. Repeat the procedure with the pliers.
37. Carefully press out the front and rear camshaft seals.
38. Remove the seals.
39. Remove:
    - Tools 999 5454
    - The camshaft cover
    - The camshafts.

✲✲ CAUTION

Mark up and position the lifters so that their original positions can be established.

40. Lift out the valve lifters. Use a suction cup.
41. Position the valve lifters on a piece of paper to drain.
42. Remove the bolts holding the cylinder head on the cylinder block.
43. Start at the sides and work alternately towards the center.

➡Do not damage the mating surfaces.

44. Remove:
    - The intake manifold and gasket
    - The coolant outlet pipe.

Cylinder head torque sequence—6-cylinder engines

To install:

45. Clean the gasket surfaces of:
    - The manifold. Check that the studs are tightened
    - The coolant bypass channel
    - The coolant outlet pipe
    - The intake manifold
    - The cylinder block
    - Camshaft cover.
46. Blow the oil ducts clean.
47. Remove the starter motor mounting bolts. Pull out the starter motor and place to one side.
48. Remove the blind cover plug.
49. Turn the crankshaft clockwise slightly. Install the crankshaft stop 999 5451, or equivalent. Ensure that it bottoms out against the cylinder block.
50. Turn the crankshaft counter-clockwise until it stops against the crankshaft stop.
51. Check that the marking on the crankshaft timing gear pulley corresponds with the marking on the oil pump.
52. Install a new cylinder head gasket.
53. Install the cylinder head.
54. Use new bolts. Lubricate and install all the bolts.
55. Tighten the bolts in the order illustrated. See the torque chart at the beginning of this chapter.
56. When installing a new cylinder head, the valve clearance must be set

✲✲ CAUTION

Make sure that the lifters are in the same position as before.

57. Lubricate the valve lifter wells and the lifters. Install all the valve lifters.
58. Lubricate the camshaft bearing positions.
59. Install the intake camshaft. Ensure that the groove at the back of the camshaft is above an imaginary center line.
60. Position the exhaust camshaft.

Cylinder head loosening sequence—6-cylinder engine

Ensure that the groove at the back of the camshaft is below an imaginary center line.

➡Ensure that no liquid gasket gets into the oil ducts.

61. Install new O-rings around the spark plug wells at the cylinder head.
62. Apply liquid gasket 11 61,059, or equivalent, to the camshaft cover. Use a roller. The surface must be completely covered without any excess.
63. Lubricate the camshaft lobes, the camshaft bearing surfaces and the valve lifters.
64. Install the camshaft cover. Install 2 press tools 999 5454, or equivalent.
65. Tighten the camshaft cover bolts alternately, keeping it parallel to the cylinder head using the press tools.
66. Install all the bolts and tighten.
67. Install the variable valve timing solenoids using new gaskets.
68. Remove the press tools.
69. Install the spark plugs.
70. Lubricate the surface of the seal that the camshaft rotates against.
71. Using a drift, press in the seal until the drift bottoms out.

➡If there are grooves worn into the camshaft, the seal can be pressed in a further 2mm by reversing the sleeve on the tool.

72. Using camshaft tool 999 5452, or equivalent, separate the camshaft adjustment tool into two units.

✲✲ CAUTION

The camshafts must not be turned more than is listed in the text below. The valves may be damaged.

73. Install the exhaust camshaft section of the adjustment tool. Screw into the rear edge of the camshaft.
74. Carefully turn the camshaft adjustment tool clockwise (as viewed from the back of the engine) until the camshaft adjustment tool is parallel with the join between the cylinder head and camshaft cover.
75. Install the intake camshaft section of the adjustment tool.
76. Screw into the rear edge of the camshaft.
77. Carefully turn the camshaft adjustment tool counter-clockwise until it is in contact with the exhaust camshaft section of the camshaft adjustment tool.
78. Screw the adjustment tool sections together into one.
79. Check that the bolts retaining the

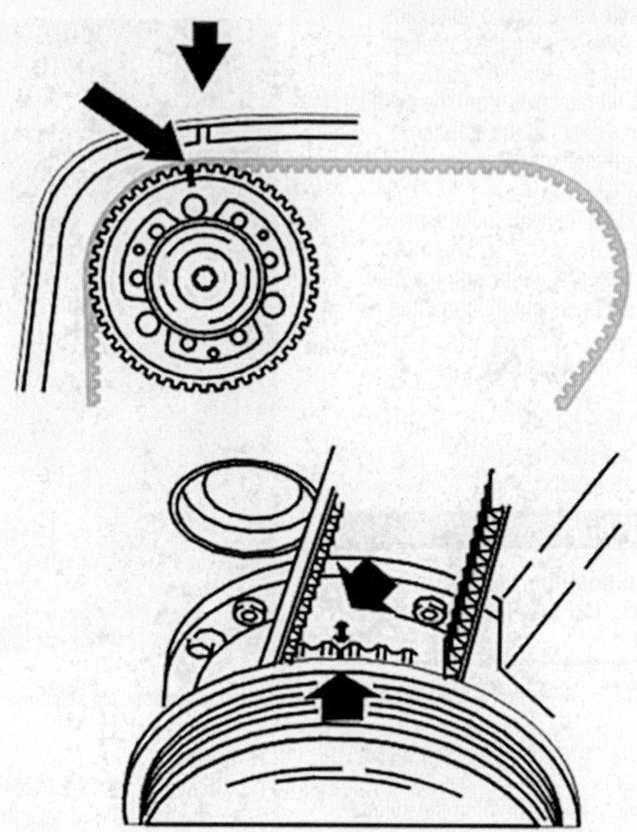

Allen hole positioned at 10:00 o'clock

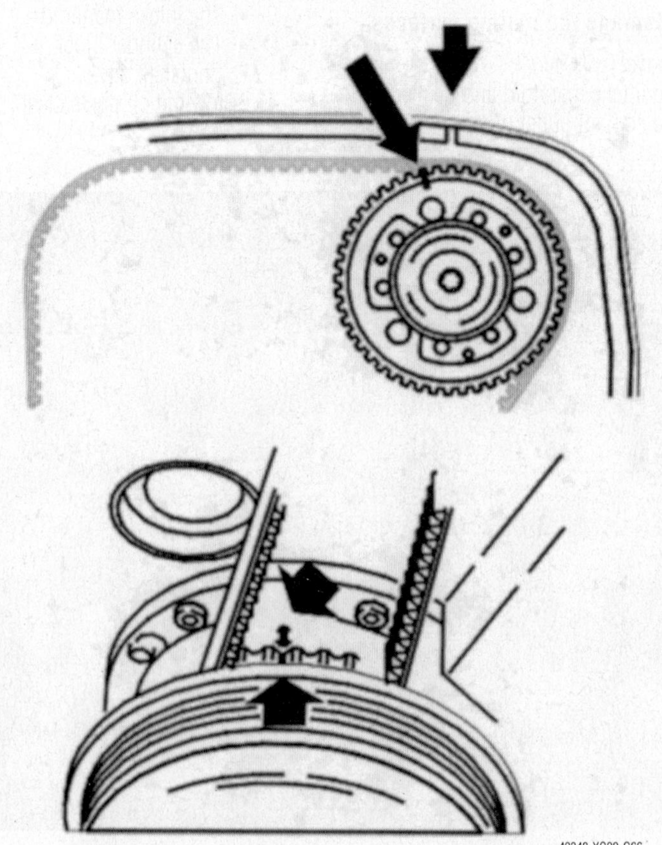

Installing the VVT unit

adjustment tool to the camshafts and the bolts holding the tool together are well tightened.

80. Lubricate the surface of the seal that the camshaft rotates against.

81. Use the mounting bolts for the variable valve timing unit.

82. Using drift 999 5718, or equivalent, tighten the bolts until the drift bottoms out.

83. Install:

- The bolt holding the inner timing cover to the cylinder head
- The thermostat housing and cable duct. Use a new gasket
- The bolt for the thermostat housing cable duct.

84. Spin the idler pulley and listen for noise. If replacing with a new idler pulley, tighten to 18 ft. lbs. (24 Nm).

85. Spin the tension pulley and listen for noise. When replacing, bolt the tension pulley into place using the center bolt. Screw in the center bolt by hand.

86. Ensure that the tensioner fork is centered over the cylinder block rib/bracket. Ensure that the Allen hole on the eccentric is at 10 o'clock.

87. Slacken off, but do not remove the bolts which secure the timing gear pulley to the VVT unit.

88. Press the variable valve timing unit and timing gear onto the camshaft.

89. Install the center bolt which secures the variable valve timing unit to the camshaft. Tighten slightly.

90. Turn the variable valve timing unit counter-clockwise as far as it will go. Slacken off the center bolt.

91. Position the upper timing cover.

92. Turn the timing gear pulley clockwise until the bolts at the oval holes are in the limit position. Continue turning clockwise until the timing gear pulley marking is 1 cog before the marking on the upper timing cover.

➡Do not turn counter clockwise during this procedure.

93. Check that the timing gear pulley is still in the limit position in the oval holes.

94. Tighten the center bolt in the VVT unit to 88 ft. lbs. (120 Nm). Check that the variable valve timing unit does not rotate when tightening.

95. Install and tighten the center plug.

96. Turn the variable valve timing unit clockwise to the limit position. Turn the timing gear pulley so that the markings correspond.

97. Slacken off, but do not remove the bolts which secure the timing gear pulley to the VVT unit.

98. Press the variable valve timing unit/timing gear onto the camshaft.

99. Install the center bolt which secures the variable valve timing unit to the camshaft. Tighten slightly.

100. Turn the variable valve timing unit counter-clockwise as far as it will go.

101. Slacken off the center bolt.

102. Turn the timing gear pulley clockwise until the bolts at the oval holes are in the limit position. Continue turning clockwise until the timing gear pulley marking is 2 cogs before the marking on the upper timing cover.

➡Do not turn counter clockwise during this procedure.

103. Check that the timing gear pulley is still in the limit position in the oval holes.

104. Tighten the center bolt in the VVT unit to 88 ft. lbs. (120 Nm). Check that the variable valve timing unit does not rotate when tightening.

105. Install and tighten the center plug.

106. Turn the variable valve timing unit clockwise to the limit position.

107. Turn the timing gear pulley so that the markings correspond.

108. Tighten the timing belt tensioner center bolt. Tighten to 44 inch lbs. (5 Nm).

109. Install the timing belt in the following order:

- Crankshaft
- The idler pulley
- Intake cam
- Exhaust cam
- Water pump
- Tension pulley.

110. The variable valve timing units does not have a return spring and is easily dislodged when reinstalling the timing belt. Check that the markings correspond.

➡Adjust the timing gear pulleys so that the bolts are not at the limit positions in any of the oval holes.

111. This adjustment is to be made with a cold engine. A suitable temperature is approximately 20°C/68°F.

112. At higher temperatures (with the engine at operating temperature or a high outside temperature for example) the indicator is further to the right.

113. The illustration shows the position of the indicator at different engine temperatures.

114. Hold the center bolt secure and turn the belt tensioner eccentric clockwise until the tensioner indicator passes the marked position.

115. Turn the eccentric back so that the indicator reaches the marked position in the center of the window.

116. Remember to hold the center bolt secure at the same time.

117. Hold the eccentric secure and tighten the center bolt. Tighten to 18 ft. lbs. (25 Nm).

118. Check that the indicator is in the correct position.

119. Check:

- That the timing gear pulley is still in the limit position in the oval holes

42348-XC90-G12

Camshaft adjustment tools installed

- The timing gear pulley so that the markings line up with the upper timing cover.

120. Tighten the bolts for the timing gear pulleys.

121. Remove:

- Camshaft adjustment tools 999 5451
- Crankshaft stop 999 5452

122. Install the plug with a new blind cover plug. Tighten to 30 ft. lbs. (40 Nm).

123. Press the timing belt to check that the indicator on the tensioner moves easily.

124. Position the upper timing cover.

125. Turn the crankshaft two turns. Check that the markings on the crankshaft timing gear pulley and the camshaft timing gear pulley match up with the markings on the oil pump and upper timing cover respectively.

126. Check that the indicator on the belt tensioner is within the marked position.

127. Remove the upper timing cover.

128. Install:

- The trigger wheel.

⁂ CAUTION

The trigger wheel can only be installed in one way

- The camshaft position sensor housing
- The pipe bolt for the water heated crankcase ventilation at the rear of the cylinder head. Use new seal washers. Tighten to 19 ft. lbs. (26 Nm).
- The ignition coils according to the earlier markings
- The connector for the engine coolant temperature sensor
- The connectors for the variable valve timing solenoids
- The bolts for the 2 ground terminals between ignition coils 1 and 2 and 5 and 6.
- The starter motor
- The throttle body connector to the engine wiring
- The charge air pipe to the throttle body. Tighten the clamp
- The brake vacuum hose to the intake manifold
- The pipe bolts at the underside of the intake pipe for the water heated crankcase ventilation. Tighten to 19 ft. lbs. (26 Nm)
- The dip stick pipe onto the intake manifold.
- The fuel rail. Use new bolts. Tighten to 88 inch lbs. (10 Nm). Press the quick-release connector for the fuel pressure line until it clicks
- The clamps that hold the injector wiring and the positive cable to the rear edge of the cylinder head
- The protective cover over the nozzle connectors
- The upper engine coolant hose to the cylinder head and radiator. Tighten the clamp.
- The inlet pipe for the coolant pump to the cylinder head. Use a new gasket
- The bolt which secures the coolant pump inlet pipe to the cylinder block
- The turbocharger.
- The bracket for the torque rod
- The upper bolt securing the automatic transmission dip stick pipe to the engine
- The ground strip between the right frame member and the cylinder head
- The cable duct on the inside of the right suspension turret
- The front timing cover.
- Expansion tank. Connect the connector to the coolant level sensor
- The hose to the expansion tank from the radiator. Secure the clamp and remove the lock grip pliers. Connect the connector to the ABS sensor
- The servo reservoir. Ensure that the hoses are correctly positioned.
- The upper timing cover
- The cover over the ignition coils
- The plastic intake pipe over the engine. Adjust and tighten the intake pipe at the turbocharger. Tighten the clamps and mounting bolts on the plastic pipe
- The plastic charge air pipe over the engine. Tighten the clamps and mounting bolts
- The hose between the charge air cooler and the plastic charge air pipe over the engine. Tighten the clamps
- The crankcase ventilation hose to the PTC resistor in the plastic intake pipe over the engine. Use a new clamp
- The connector for the PTC resistor
- The protective cover over the PTC resistor
- The EVAP hose to the plastic intake pipe
- The air cleaner assembly
- The plastic intake pipe to the mass air flow sensor. Tighten the clamp
- The connector to the mass air flow sensor. Clamp the wiring at the front of the air cleaner
- The turbocharger control valve to the mounting on the air cleaner
- The blue marked hose and the turbocharger control system to the intake manifold
- The red marked hose and turbocharger control system to the charge air pipe
- The yellow marked hose and turbocharger control system to the pressure regulators
- The cross stay between the suspension turrets
- The suspension turret covers. Clamp the servo hose to the right mounting on the engine stabilizer brace
- The battery negative lead.

129. Replace the oil and oil filter if necessary.

130. Check:

- The engine oil level
- The servo oil level
- The coolant level.
- The engine for leaks
- The oil level in the transmission
- The coolant level.

Turbocharger

REMOVAL & INSTALLATION

5-Cylinder Engines

1. Before servicing the vehicle, refer to the Precautions Section.
2. Drain the cooling system.
3. Remove or disconnect the following:
   - Negative battery cable
   - Exhaust manifold heat shield
   - Upper air charge pipe
   - Inner heat shield
   - Air intake hose
   - Turbo coolant hoses
   - Turbo oil lines
   - Exhaust front pipe and bracket
   - Red boost pressure hose
   - White bypass valve hose
   - Yellow pressure regulator hose
   - Turbocharger

To install:

4. Install new pin bolts with thread locking compound. Tighten them to 15 ft. lbs. (20 Nm).
5. Install or connect the following:
   - Turbocharger and tighten the fasteners to 18 ft. lbs. (25 Nm) on

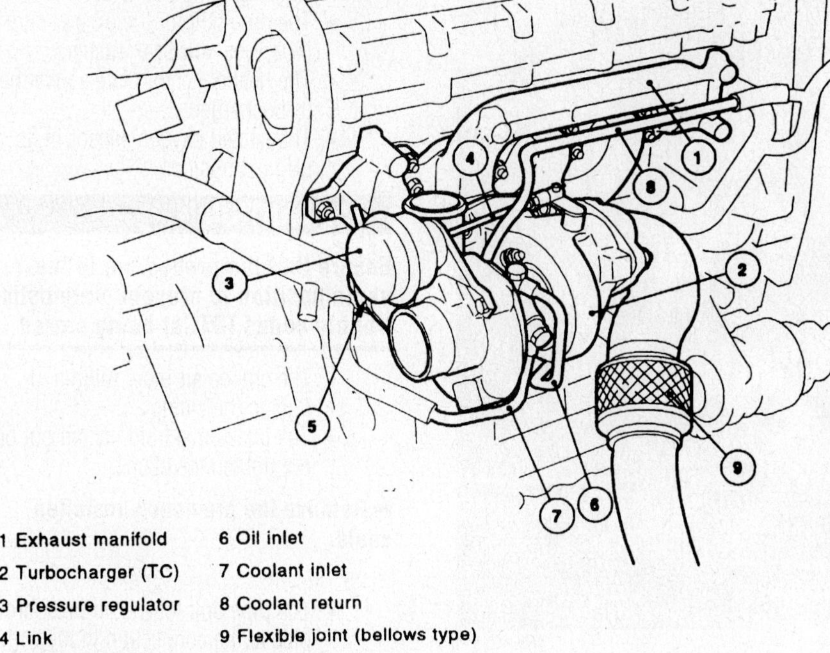

Turbocharger and exhaust manifold component identification—5-cylinder engines

models without a gasket; 27 ft. lbs. (37 Nm) on models with a gasket.

- Yellow pressure regulator hose
- White bypass valve hose
- Red boost pressure hose
- Exhaust front pipe and bracket
- Turbo oil lines
- Turbo coolant hoses
- Air intake hose
- Inner heat shield
- Upper air charge pipe
- Exhaust manifold heat shield
- Negative battery cable

6. Fill the cooling system.
7. Start the engine and check for leaks.

6-Cylinder Engines

➡The turbocharger for cylinders 1, 2 and 3 is designated turbo 1. The turbo for cylinders 4, 5 and 6 is designated turbocharger 2.

1. Before servicing the vehicle, refer to the Precautions Section.
2. Remove:
   - The splash guard under the engine
   - The right front wheel and the mudguard at the halfshaft on the right-hand side
   - The bolt for the fuel line in the oil pan
   - The support bracket for the bevel gear in the cylinder block
   - The pipe bolt for the turbocharger oil pressure pipe by the cylinder block.
   - The connectors for the front heated oxygen sensors. Remove the wiring from the clips and clamps
   - The brake pipe from the clips on the front SIPS member
   - The front SIPS member
   - The three-way catalytic converter from the rubber mountings.
3. Connect a hose to the engine coolant drain cock.
4. Drain the coolant into a clean container. Close the cock.
5. Remove:
   - The bolt for the engine stabilizer brace mounting on the engine
   - The engine stabilizer brace between the suspension turrets
   - The plastic charge air pipe. Place to one side
   - The plastic intake pipe. Put it to one side.
6. Remove from turbo 1:
   - The intake and charge air pipes. Seal the openings
   - The heated oxygen sensor.

⁂ CAUTION

Note the location of the heated oxygen sensor

- The heat deflector plate
- The 3 nuts for the three-way catalytic converter

➡Use rust solvent

- The pipe bolt for the oil pressure pipe
- The oil pressure pipe
- The pipe bolts for the coolant pipe. Move the upper coolant pipe and hose to one side

➡Position a container underneath for the coolant.

- Clamp the control pipe for the pressure regulator.

7. Remove from turbo 2:
   - The heated oxygen sensor
   - The heat deflector plate
   - The 3 nuts for the three-way catalytic converter

➡Use rust solvent

- The upper bolt for the automatic transmission dip stick pipe
- The intake and charge air pipes. Seal the openings
- The hose from the pressure regulator
- The clamps between the coolant pump intake pipe and the turbocharger oil pressure pipe.

8. Raise the car.
9. Remove:
   - The oil pressure pipe for turbo 1
   - The hose from the pressure regulator.
10. Lower the car.
11. Remove:
    - The pipe bolts for the coolant pipe at turbo 2. Place the lower pipe to one side
    - move the upper coolant pipe and hose from turbo 2 to one side
    - The air control pipe for the pressure regulators, 2 bolts
    - The nuts and washers for the manifold mounting at cylinders 1, 2 and 3.

➡Use rust solvent or similar. Lift off the manifold/turbocharger.

12. Remove from turbo 2:
    - The bolts for the oil return line
    - The pipe bolt for the oil pressure pipe
    - The nuts and washers where the manifold is connected to the cylinder head for cylinders 4, 5 and 6. Carefully lift off the turbocharger and manifold so that the oil pressure pipe is exposed

➡Do not damage the oil pressure pipe. Lift up and remove the turbocharger and manifold.

42348-XC90-G13

Turbocharger mounting—6-cylinder engine

To install:

13. Remove the old gaskets
14. Clean the gasket faces thoroughly
15. Ensure that all the studs are tightened in the cylinder head. Tighten to 15 ft. lbs. (20 Nm).
16. Installing the manifold and turbo 2
    - Install new gaskets
    - Lubricate the studs at the exhaust ports and at the flange on the turbocharger flange to the three-way catalytic converter using paste P/N 1161408, or equivalent.
    - Lower and install the exhaust manifold and turbocharger.
17. Install:
    - The washers and new nuts that hold the manifold onto the cylinder head.
    - The pipe bolt holding the oil pressure pipe to the turbocharger. Use new seal washers. Do not tighten
    - The bolts for the oil return line.
    - Lower coolant pipe. Use new gaskets.
18. Installing the manifold and turbo 1
    - Install new gaskets
    - Lubricate the studs at the exhaust ports and at the flange on the turbocharger flange to the three-way catalytic converter using paste P/N 1161408, or equivalent.
19. Install:
    - The washers and new nuts that hold the manifold onto the cylinder head.
    - The pipe for pressure regulator control
    - The upper coolant pipe and hose to turbo 2. Use new seal washers.
    - The upper coolant pipe and hose to turbo 1. Use new seal washers.
20. Install the three-way catalytic converter on the turbocharger. Use new nuts. Lubricate the studs using copper paste 116 1408, or equivalent.
21. Installing components at turbo 1
    - The oil return line. Use a new gasket
    - The hose for the pressure regulator
    - The oil pressure pipe
    - The pipe bolt for the connection between the oil pressure pipes and the cylinder block. Use new seal washers. Do not tighten.
22. Installing components at turbo 2
23. Lower the car. Tighten the pipe bolt for the oil pressure pipe.
    - The charge air pipe. Align and tighten the clamp
    - The intake manifold. Align and tighten the clamp

➡Remove the previously installed seals

    - The upper bolt securing the automatic transmission dip stick pipe to the engine. Tighten the lower bolt
    - The heat deflector plate above the turbocharger
    - The heated oxygen sensor
    - The clamps between the coolant pump intake pipe and the turbocharger oil pressure pipe
24. Installing components at turbo 1
    - The pipe bolt for the oil pressure pipe. Use new seal washers
    - The heat deflector plate above the turbocharger
    - The heated oxygen sensor in its original position

⁂ CAUTION

Ensure that the probes are in the same position to prevent diagnostic trouble codes (DTCs) being stored

    - The charge air pipe. Align and tighten the clamp
    - The intake manifold. Install but do not tighten the clamp

➡Remove the previously installed seals.

25. Install:
    - The pipe bolt for the oil pressure pipe at the connection in the cylinder block
    - The support bracket for the bevel gear. First screw 4 M10 bolts in to the cylinder block to contact. Then tighten 2 M10 bolts in the bevel gear. Then tighten the M10 bolts in the cylinder block.
    - The bolt for the fuel line in the oil pan
    - The splash guard at the halfshaft
    - Hook the three-way catalytic converter into the rubber mounting
    - The front SIPS member. Press the brake pipe into place. Connect the wiring for the heated oxygen sensors. Secure the wiring as before
    - The right front wheel
    - The plastic intake pipe
    - The plastic charge air pipe
    - The vacuum hoses for the purge valve and turbocharger control valve
    - The engine stabilizer brace
26. Fill with coolant.
27. Warm up the engine until the thermostat opens.
28. Check:
    - The engine for leaks
    - The coolant level
29. Install the lower engine cover.

Intake Manifold

REMOVAL & INSTALLATION

5-Cylinder Engines

1. Disconnect the battery negative lead.
2. Remove the fuel injection system.

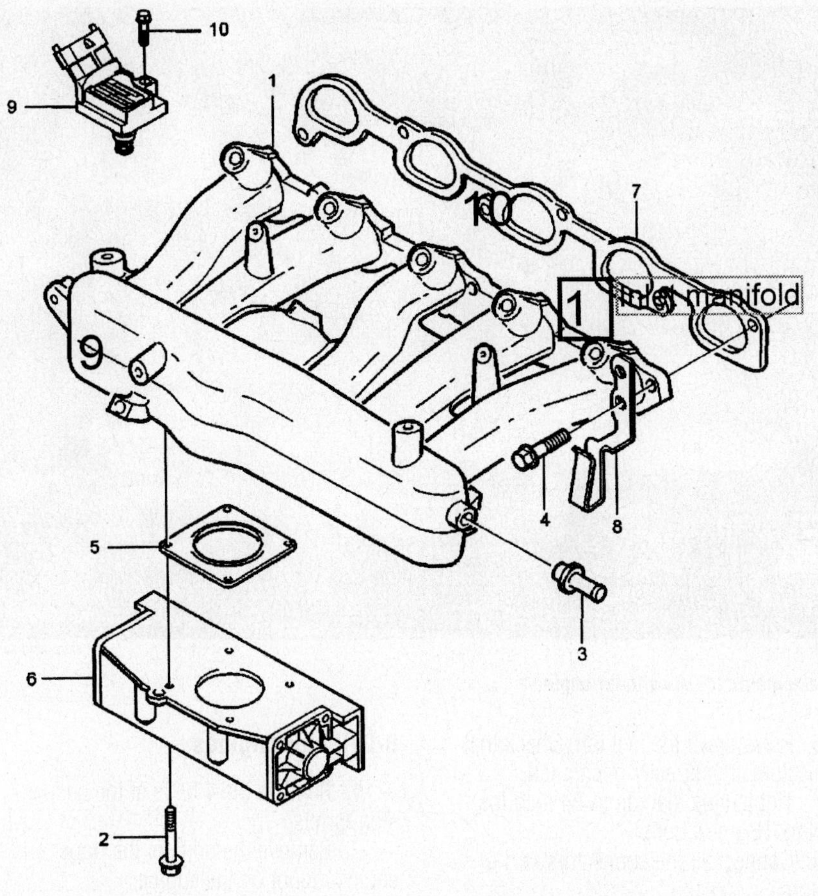

1 Inlet manifold
2 Flange screw
3 Nipple
4 Flange screw
5 Gasket
6 Throttle body
7 Gasket
8 Bracket
9 Map sensor
10 Six point socket screw

Intake manifold—5-cylinder engine

3. Drain the fuel injection system.
4. Remove or disconnect:
   - The charge air pipe over the engine
   - The cover plate over the connectors from the injectors
   - The fuel rail mounting bolts. Spray penetrating oil around the injector nozzle at the terminal on the intake manifold. Gently work the fuel rail and injector nozzles loose. Separate the fuel line quick-release connector.
   - Separate the throttle body connector
   - The plastic charge air pipe between the throttle body and the charge air cooler
   - Vacuum hoses from the intake manifold.
   - The oil filler cap
   - The cover over the ignition coils
   - The hose from the flame trap to the cam cover at the terminal in the cam cover.
   - The dip stick pipe

5. Slacken off the lower bolts for the intake manifold a few turns.

6. Remove the mounting bolts in the upper row and the outer bolts in the lower row.

7. Remove the intake manifold by lifting it approximately 20 mm.

8. Remove the nipple for the water heated crankcase ventilation.

9. Allow the crankcase ventilation hose to run through the intake manifold without disconnecting it from the flame trap.

10. Transfer components when replacing the intake manifold

11. Transfer the electronic throttle module. Use a new gasket.

To install:

12. Install or connect:
   - A new gasket held in position by the lower bolts for the intake manifold. Do not forget the clamp for the fuel line at the bolt in the lower row.
   - The nipple for the water heated crankcase ventilation
   - The intake manifold. Do not forget the crankcase ventilation hose. The hose must be inserted up through the gap between the second and third ducts
   - The 3 upper bolts. Tighten all the bolts starting from the center to 14 ft. lbs. (19Nm).
   - The fuel rail. Press together the quick-release connector. Use new bolts.
   - Injector connectors
   - The protective cover over the fuel rail
   - The vacuum hoses according to the earlier markings
   - The plastic charge air pipe between the throttle body and the charge air cooler
   - The connectors to the throttle body
   - The dip stick in the intake manifold.

6-Cylinder Engines

1. Before servicing the vehicle, refer to the Precautions Section.

2. Remove:
   - The battery negative lead
   - Plastic charge air pipe and hose to charge air cooler
   - The intake pipe between the air cleaner and the turbocharger

3. Remove the intake pipe for the air cleaner housing from the connection above radiator. Twist the pipe up.

4. Drain the fuel line.
5. Remove:
   - The protective cover over the injector connectors
   - Injector connectors
   - The fuel rail mounting bolts

➡**Handle the injectors carefully to avoid damaging the nozzles and needles.**

6. Spray universal oil or similar around the injector terminals on the intake manifold.
7. Gently work the fuel rail and injectors loose.
8. Detach the fuel line from the nozzle pipe by pressing the quick-release connector.
9. Remove:
   - The dip stick pipe mounting at the intake manifold
   - The nipple for the crankcase ventilation under the intake
10. Disconnect the throttle body connector. Remove the plastic charge air pipe from the throttle body.
11. Remove the upper bolts.
12. Remove the outer bolts from the lower row.
13. Slacken off the remaining bolts.
14. Lift up and remove the intake manifold.
15. Remove the remaining bolts and gasket.

To install:

16. Transfer the throttle body with a new gasket. Tighten to 84 inch lbs. (10 Nm).
17. Ensure that the gasket faces are clean.
18. Install a new gasket with the two centermost bolts in the lower row.
19. Lower the intake over the 2 bolts.
20. Install the remaining bolts.
21. Tighten all bolts. Tighten to 15 ft. lbs. (20 Nm).
22. Connect the throttle body connector.
23. Install:
    - The vacuum hoses according to the earlier markings
    - The nipple for the crankcase ventilation under the intake manifold
    - The dip stick mounting bolt to the intake manifold. Tighten to 18 ft. lbs. (25 Nm)
    - The plastic charge air pipe between the throttle body and the charge air cooler
24. Lubricate the injector rings with petroleum jelly.
25. Hold the fuel rail. Press the fuel line quick-release connector together until it clicks.

Intake manifold—6-cylinder engine

26. Press down the fuel rail. Check that all injectors are correctly positioned.
27. Tighten the fuel rail to 84 inch lbs. (10 Nm). Use new bolts.
28. Connect all the connectors to the injectors.
29. Install the protective cover over the connectors to the nozzles.
30. Install:
    - Intake manifold between air cleaner and turbocharger
    - Plastic charge air pipe and hose to charge air cooler
    - The air cleaner intake pipe
    - The battery negative lead

8-Cylinder Engines

1. Remove the 4 bolts at the suspension turrets
2. Remove the bolts of the upper engine mount on the engine.
3. Remove the upper cross stay.
4. Remove the air cleaner cover by pulling it straight up.
5. Remove the 2 connectors from the engine control module (ECM); clip the band clamp to the cable harness.
6. Remove the connector from the mass air flow sensor.
7. Remove the brake vacuum hose from the vacuum pump.

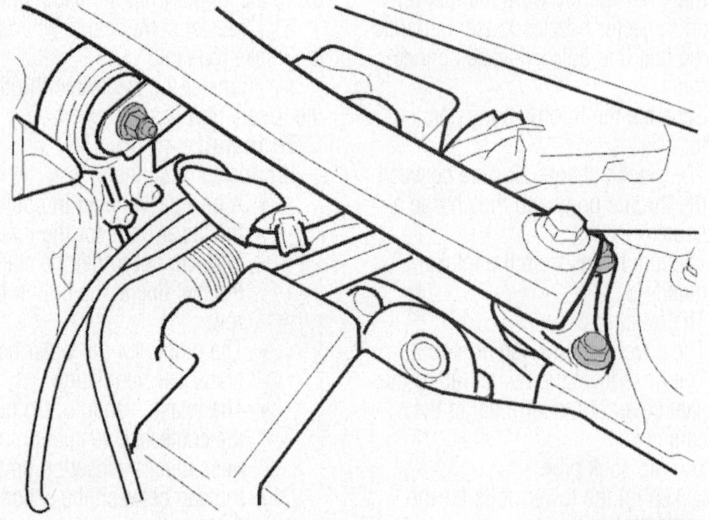

Upper cross stay—8-cylinder engines

8. Remove the 2 bolts of the fresh air hose in the front plate.
9. Remove the fresh air hose.
10. Remove the air cleaner by pulling it straight up.
11. Remove the upper engine covers by pulling them straight up.
12. Remove the cover over the servo pump.
13. Detach the connector of the pressure and temperature sensor.
14. Remove the 2 vacuum hoses of the check valve of the distribution damper.

➡There is a guide sleeve on the upper mounting for the plate. This must be lifted slightly so that the panel can be slid forward.

15. Remove the 6 bolts.
16. Lift and slide the protective plate forward.
17. Remove the protective plate.
18. Open the cock on the lower left edge of the radiator.
19. Drain the coolant into a clean receptacle.
20. Close the cock.
21. Install the protective plate.
22. Install the 6 bolts.
23. Lower the vehicle.
24. Remove the 4 bolts and the upper engine mount.
25. Remove the brake vacuum hose from the intake manifold by releasing the clamp.
26. Remove the intake hose between the air cleaner and throttle body by releasing the 2 clamps.
27. Remove the 2 nuts on the throttle body.
28. Remove the 2 bolts on the throttle body.
29. Position aside the throttle body.
30. Detach the 2 coolant hoses, one from the throttle body and one from the coolant distribution pipe.
31. Remove the hose clip on the vacuum hose.
32. Remove the vacuum hose.
33. Remove the 2 M6 bolts on the non-return valve for crankcase ventilation
34. Remove the non-return valve for crankcase ventilation.
35. Remove the 21 bolts of the upper intake manifold.
36. Remove the upper intake manifold. Discard the gaskets.
37. Drain the fuel system.
38. Remove the crankcase ventilation hose between the cylinder heads by releasing the 2 clamps.
39. Remove the 3 bolts from the fuel rail.

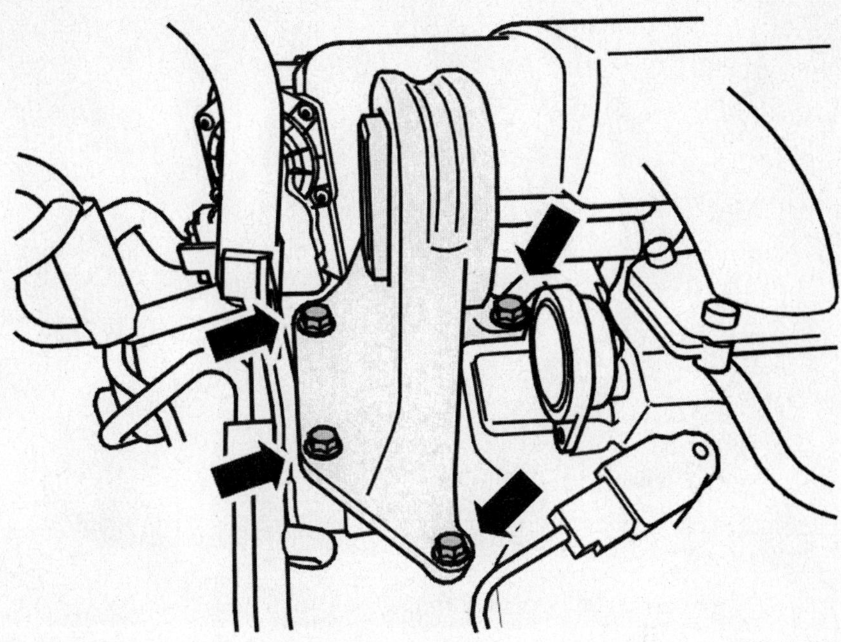

Upper engine mount—8-cylinder engine

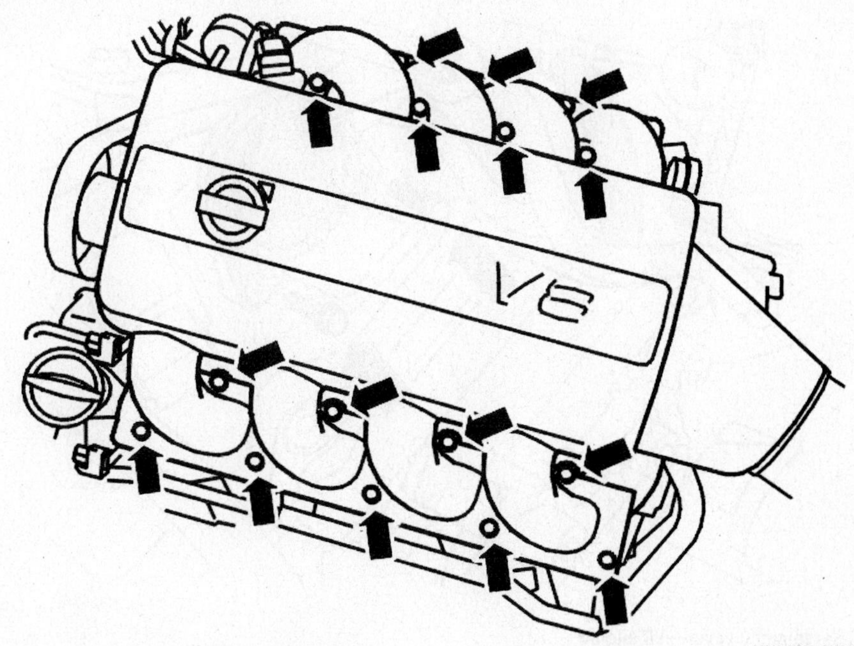

Upper intake manifold bolts—V8 engine

40. Remove the 2 bolts on the fuel pressure sensor.
41. Lift the fuel rail assembly with injector straight up and position it aside.
42. Remove the 12 bolts and the lower intake manifold. Discard the gaskets.

To install:

43. Clean the gasket surfaces.
44. Install new gaskets.
45. Install the lower intake manifold. Tighten the 12 bolts to 14 ft. lbs. (19 Nm).
46. Lubricate the O-rings on the injectors with Vaseline or the like.
47. Install the fuel rail assembly with injectors.
48. Install the 3 bolts of the fuel rail.
49. Install the 2 bolts on the fuel pressure sensor.
50. Install the crankcase ventilation hose and 2 clamps.
51. Connect the fuel supply line and secure the line in the clamp.

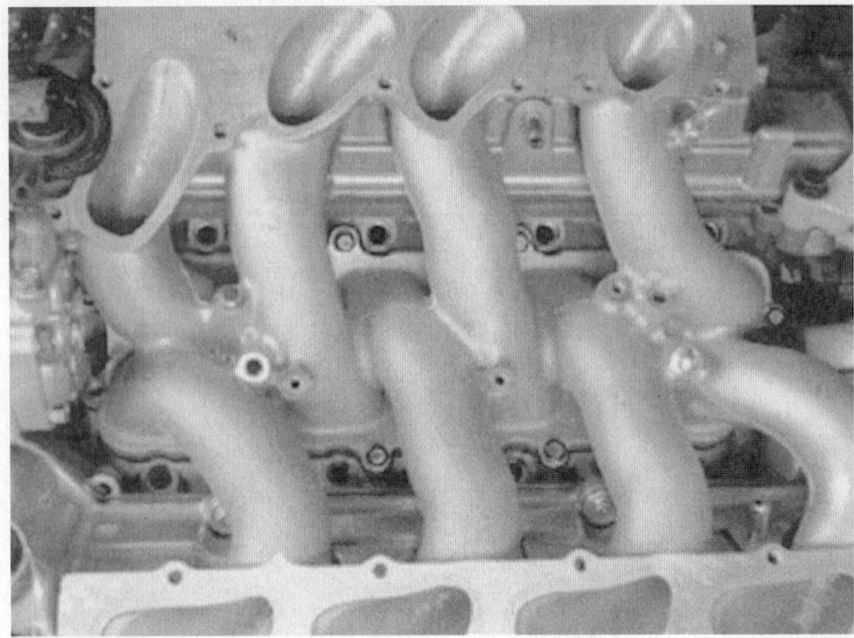

Lower intake manifold bolts—V8 engine

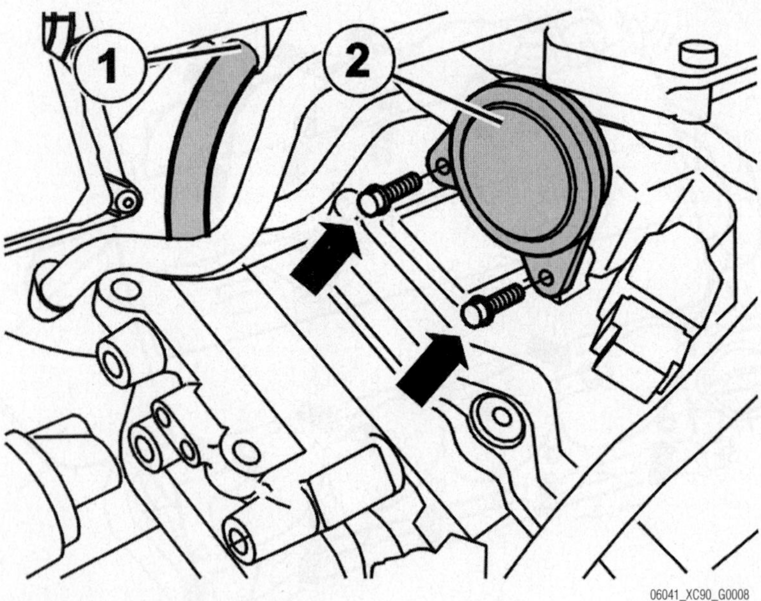

Servo pump cover—V8 engine

**** WARNING**

Make sure there is an audible click during connection.

52. Install new gaskets.
53. Install the upper intake manifold. Tighten the 21 bolts to 14 ft. lbs. (19 Nm).
54. Install the hoses to the intake manifold.
55. Install the non-return valve for crankcase ventilation.
56. Install the connector to the pressure and temperature sensor.
57. Install the 2 vacuum hoses of the check valve of the distribution damper.
58. Install the upper engine mount. Tighten the 4 M10 bolts.
59. Install the cover over the servo pump
60. Position the engine covers and press them into the retainers.
61. Install the upper cross stay between the suspension turrets. Tighten the 4 bolts.
62. Install the bolt of the cross stay mount on the engine.
63. Install the air cleaner.
64. Install the brake vacuum hose on the brake pump.
65. Install the fresh air hose.
66. Install the 2 bolts of the fresh air hose in the front plate.
67. Install the 2 connectors to the engine control module (ECM).
68. Install the connector to the mass air flow sensor.
69. Install the band clamp on the cable harness in the air cleaner housing.
70. Install the cover over the air cleaner.
71. Fill with coolant.
72. Start the engine and allow it to warm to normal operating temperature. Check for leaks and proper function. Fill up fluids as necessary.

Exhaust Manifold

REMOVAL & INSTALLATION

5-Cylinder Engines

1. Before servicing the vehicle, refer to the Precautions Section.
2. Remove or disconnect the following:
   - Negative battery cable
   - Exhaust manifold heat shield
   - Exhaust front pipe
   - Turbocharger
   - Exhaust manifold

To install:

3. Install or connect the following:
   - Exhaust manifold and tighten the fasteners to 18 ft. lbs. (25 Nm)
   - Turbocharger, if equipped
   - Exhaust front pipe and tighten the fasteners to 44 ft. lbs. (60 Nm)
   - Exhaust manifold heat shield and tighten the fasteners to 86 inch lbs. (10 Nm)
4. Loosen the joint at the catalytic converter and re-tighten to 18 ft. lbs. (25 Nm). This is necessary to prevent stresses in the system.
5. Connect the negative battery cable.
6. Start the engine and check for leaks.

6-Cylinder Engines

See the procedures under Turbocharger Removal and Installation.

8-Cylinder Engine

1. Remove the front wheels on both sides.
2. Remove the bolts to the torque rod.

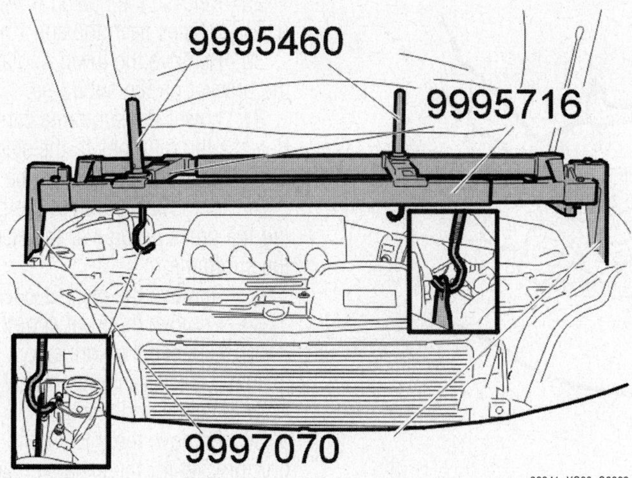

Lifting tool—V8 engine

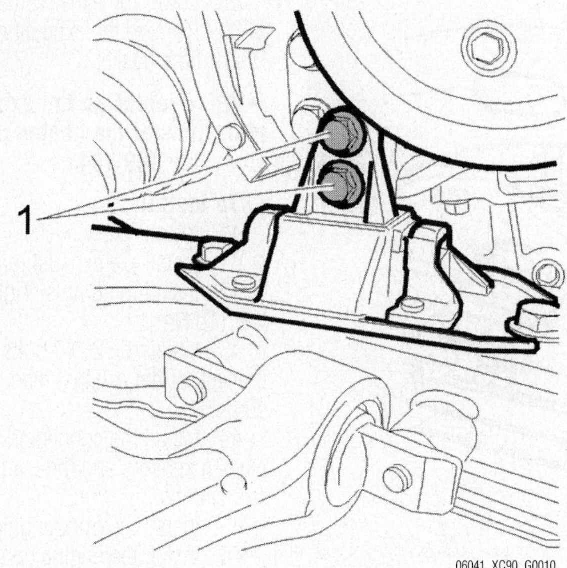

Right hand engine pad—V8 engine

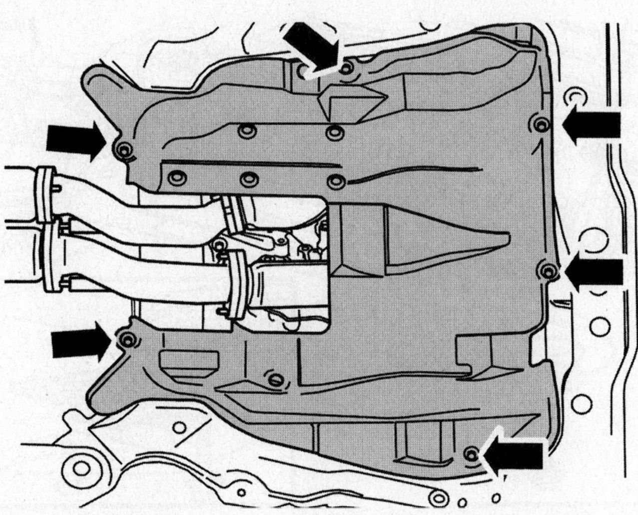

Engine undercover—V8 engine

3. Remove the oil filler cap.
4. Remove the upper front engine casing by pulling it straight up.

WARNING

When lifting the engine ensure that the lifting eye does not scratch the torque rod. Tape the torque rod and keep it away from your eyes.

5. Install the oil filler cap.
6. Install the lifting tool: 999 5716 converted with kit 999 7070 and the lifting hooks: 999 5460 . Raise the engine slightly so that the engine pads are unloaded.
7. Remove the wing liner
8. Remove the bolts for the right-hand engine pad

➡There is a guide sleeve on the upper mounting for the plate. This must be lifted slightly so that the panel can be slid forward.

9. Raise and move the protective plate forward. Remove the protective plate.
10. Remove the protective cover under the engine
11. Remove the gearbox bolts to the lower torque rod.
12. Remove the exhaust pipes between front and rear catalytic converters
13. Remove the lower bolt to the engine pad.
14. Remove the anti-roll bar links from the anti-roll bar on both sides. Use: 999 5500 and a Torx wrench as counterhold to avoid damaging the rubber gaiter.
15. Remove the halfshaft
16. Remove the link arm from the wheel spindle on both sides
17. Detach the ball joint pin from the control arm.
18. Fit the cap nut, 7062-1, for the ball joint pin. Tighten the nut against the link arm. Then turn back one turn so that clearance is created between the cap nut and link arm.
19. Place the puller on the link arm and loosen the ball joint from the link arm.
20. Detach the link arm from the ball joint pin.
21. Pull the control arm down. Use: 999 7076 and move the wheel spindle to the side, and install the protective sleeve 999 5562 on the ball joint pin at the same time. Keep it in position using the ball joint nut.
22. Remove the steering gear from the subframe.

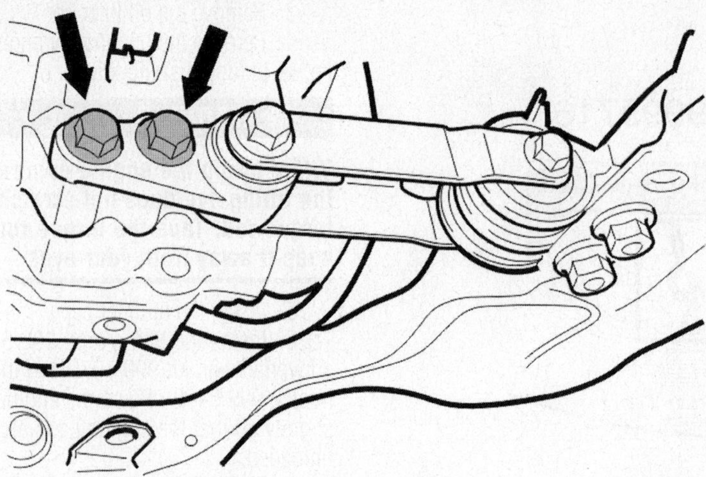

Gearbox bolts to the lower torque rod—V8 engine

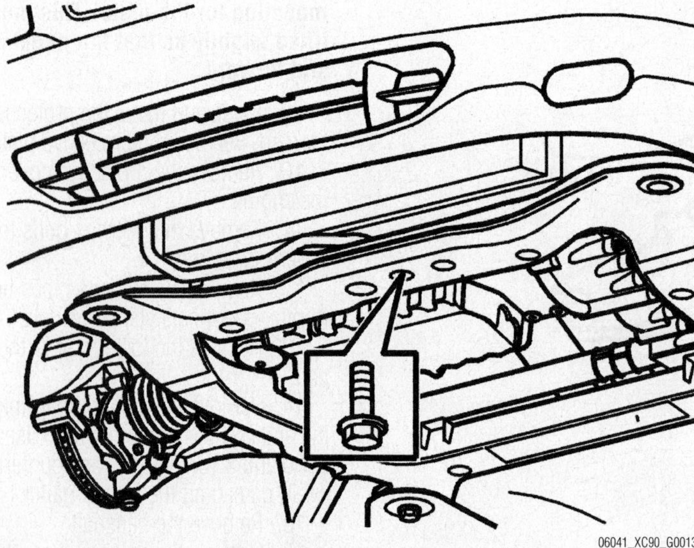

Remove the lower bolt to the engine pad—V8 engine

⁂ WARNING

Ensure that no pipes or cables are under load or are trapped when lowering the sub-frame.

23. Release the 2 x heated oxygen sensor (HO2S) connectors from the rear edge of the subframe; the heated oxygen sensor (HO2S) connector from the front edge of the subframe.
24. Remove the bolt to the ground cable bracket on the front edge of the subframe.
25. Position a lifting table under the subframe.
26. Remove the bolts to the subframe's brackets at the rear edge.
27. Remove the bolts to the subframe in the front and rear edge.
28. Remove the subframe's brackets.
29. Release the heated oxygen sensor (HO_2S) cables from the anti-roll bar.
30. Remove the lower coolant hose from the bracket on the subframe.
31. Lower the subframe carefully. Release the oil pipes to the steering gear from the clips on the subframe.
32. Remove the bolt to the bracket holding the pressure pipe to the steering gear on the subframe.
33. Remove the electric cooling fan.
34. Remove the front engine pad
35. Remove the connectors from the heated oxygen sensors and detach the lead from the clips and clamps.
36. Remove the 2 M6 bolts of the support bracket for the exhaust manifold in the oil sump.
37. Remove 8 M8 nuts at the exhaust manifold flange toward the cylinder head.
38. Lower the exhaust manifold.
39. Remove the exhaust manifold gasket. Discard the gasket.

➡When replacing the exhaust manifold, transfer the heated oxygen sensors. Use 999 5543.

To install:

40. Install a new gasket.
41. Install the exhaust manifold.
42 Install the 8 nuts. Tighten to 84 inch lbs. (10 Nm).
43. Install the 2 M6 bolts of the support bracket for the exhaust manifold in the oil sump.
44. Install the connectors to the heated oxygen sensors and the lead in the clips and clamps.
45. Install the front engine pad.
46. Install the engine cooling fan.

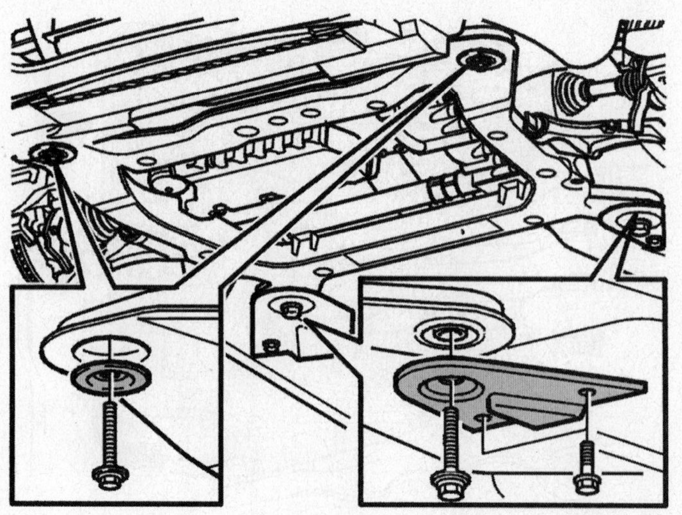

Sub-frame brackets—V8 engine

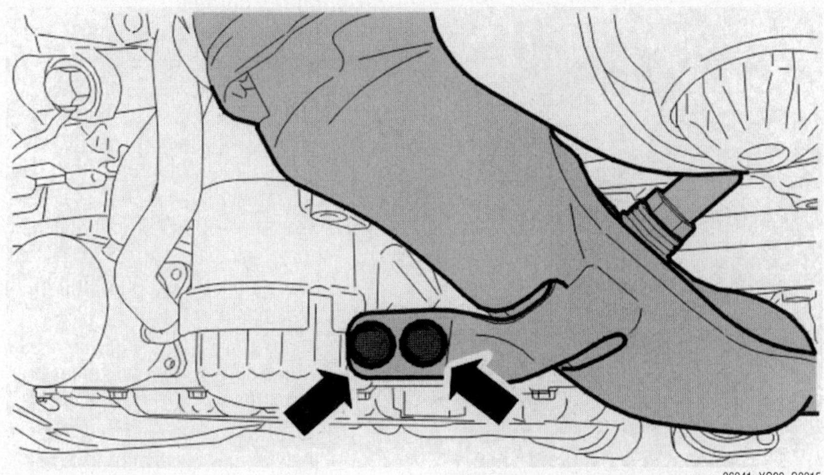

Exhaust manifold support bracket bolts—V8 engine

Exhaust manifold nuts—V8 engine

➡Ensure that no pipes or cables are loaded or trapped when raising the sub-frame.

47. Raise the subframe.
48. Install:
   a. The oil pipes to the steering gear in the clips on the subframe
   b. The bolt, M6, to the bracket holding the pressure pipe to the steering gear on the subframe
   c. The lower coolant hose in the bracket on the sub-frame
   d. The bolts and washers at the front edge and the brackets at the rear edge.
   e. The bolts to the brackets.
   f. The heated oxygen sensor (HO2S) cables on the anti-roll bar.
49. Remove the lifting table.
50. Install:
   a. The 2 x heated oxygen sensor (HO2S) connectors on the rear edge of the subframe
   b. The heated oxygen sensor (HO2S) connector on the front edge of the sub-frame
   c. The M6 bolt to the ground cable mounting on the front edge of the sub-frame.
51. Install the steering gear on the sub-frame.
52. Install the lower M10 bolt for the engine pad.
53. Install the gearbox bolts to the lower torque rod.
54. Install the exhaust pipes between front and rear catalytic converters.
55. Installing the link arm on both sides.
56. Install the halfshaft.
57. Install the protective cover under the engine.
58. Install the protective plate.
59. Move the skid plate up into the bumper cover. Press forward until the mounting brackets enter the sub-frame. Install the bolts.
60. Fit the bolts to the right-hand engine pad.
61. Install the wing liner.
62. Remove the lifting tool 999 5716, kit 999 7070 and lifting hooks 999 5460.
63. Fit the bolts to the torque rod.
64. Install the 2 x M10 bolts, by the suspension turret on both sides.
65. Install the front wheels on both sides.
66. Remove the oil filler cap.
67. Install the upper front engine casing and the oil filler cap. Carry out wheel alignment in accordance with.
68. Start the engine and check for leaks and proper function.

Front Crankshaft Seal

REMOVAL & INSTALLATION

5- and 6-Cylinder Engines

1. Before servicing the vehicle, refer to the Precautions Section.
2. Remove or disconnect the following:
   - Negative battery cable
   - Fuel line clips
   - Coolant recovery tank
   - Accessory drive belts

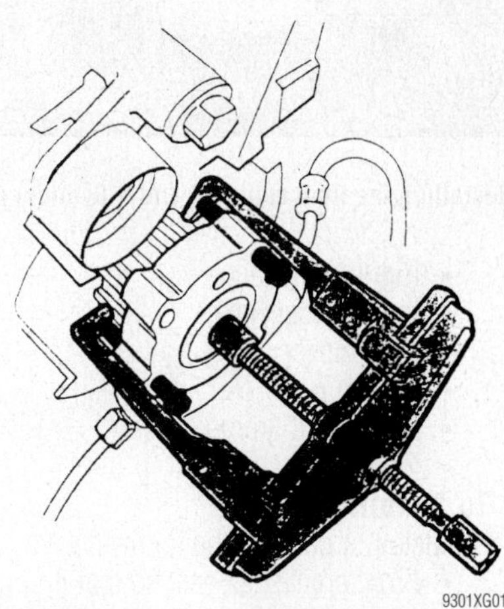

Removing the crankshaft timing belt sprocket—5- and 6-Cylinder Engines

67200-XC90-G02

Removing the front seal—5- and 6-Cylinder Engines

67200-XC90-G03

Installing the front seal—5- and 6-Cylinder Engines

- Right front wheel
- Inner fender liner
- Front cover
- Timing belt
- Crankshaft timing sprocket
- Front crankshaft seal

To install:

3. Install or connect the following:
   - Front crankshaft seal so that it is flush with the oil pump housing
   - Crankshaft timing sprocket and tighten the nut to 133 ft. lbs. (180 Nm)
   - Timing belt
   - Front cover
   - Inner fender liner
   - Right front wheel
   - Accessory drive belts
   - Coolant recovery tank
   - Fuel line clips
   - Negative battery cable
4. Start the engine and check for leaks.

8-Cylinder Engines

1. Remove the auxiliaries belt.
2. Remove the right front wheel.
3. Remove 1 plastic nut of the wheel arch liner. Bend aside the wheel arch liner.
4. Install counterhold 999 7196. Undo the center bolt.
5. Insert bolt 999 7198 in the center hole of the crankshaft.
6. Install puller 999 5304. Pull off the vibration damper.
7. Remove the seal using puller 999 5069. Screw 999 7198 in the center hole of the crankshaft.

To install:

8. Install a new seal with drift 999 7197. Pull the drift in to the stop with the old center bolt. Remove the drift.
9. Clean the surface for the sealing ring on the vibration damper with cloth 951 1024.
10. Install the vibration damper. Install bolt 999 7198 in the center hole of the crankshaft.
11. Install counterhold 999 7196, drift 999 7198 and press the vibration damper in to the stop using the old center bolt. Tighten to 74 ft. lbs. (100 Nm).
12. Remove the center bolt and discard it.
13. Install tools 999 7198 and 999 7198.
14. Insert a new bolt at the vibration damper and tighten:
    - Step 1: 88 ft. lbs. (120 Nm)
    - Step 2: back off at least 1 turn
    - Step 3: 37 ft. lbs. (50 Nm)
    - Step 4: plus 90 degrees
15. Remove counterhold 999 7196.
16. Install the wheel arch liner; tighten the plastic nut.
17. Install the right front wheel.
18. Lower the vehicle.
19. Installing the auxiliary belt.
20. Start the engine and check for leaks and proper function.

06041_XC90_G0017

Removing the damper—V8 engine

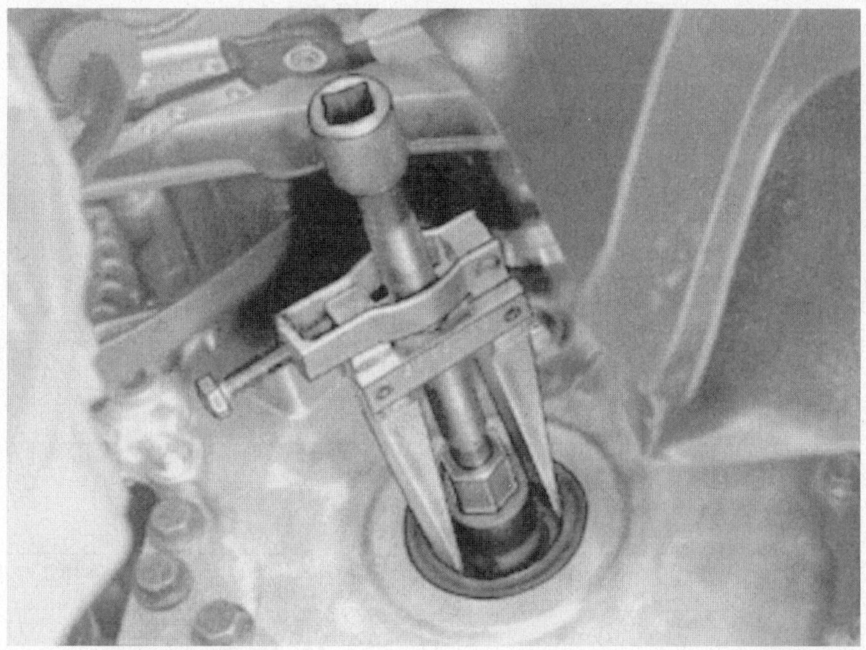

Removing the seal—V8 engine

Camshaft and Valve Lifters

REMOVAL & INSTALLATION

5- and 6-Cylinder Engines

1. Before servicing the vehicle, refer to the Precautions Section.
2. Remove or disconnect the following:
   - Negative battery cable
   - Accessory drive belts
   - Front cover
   - Timing belt
   - Ignition coil cover
   - Camshaft Position (CMP) sensor or distributor, as equipped
   - Switch holder and shield at left rear of the engine
   - Ignition coils
   - Camshaft sprockets
   - Cylinder head cover
   - Camshafts
   - Hydraulic lash adjusters

➡**Keep all valvetrain components in order for assembly.**

To install:

3. Install or connect the following:
   - Hydraulic lash adjusters
   - Camshafts
   - Cylinder head cover and tighten the bolts to 13 ft. lbs. (17 Nm)
   - Camshaft sprockets and tighten the bolts to 15 ft. lbs. (20 Nm)
   - Ignition coils
   - Switch holder and shield at left rear of the engine
   - CMP sensor or distributor, as equipped
   - Ignition coil cover
   - Timing belt
   - Front cover
   - Accessory drive belts
   - Negative battery cable

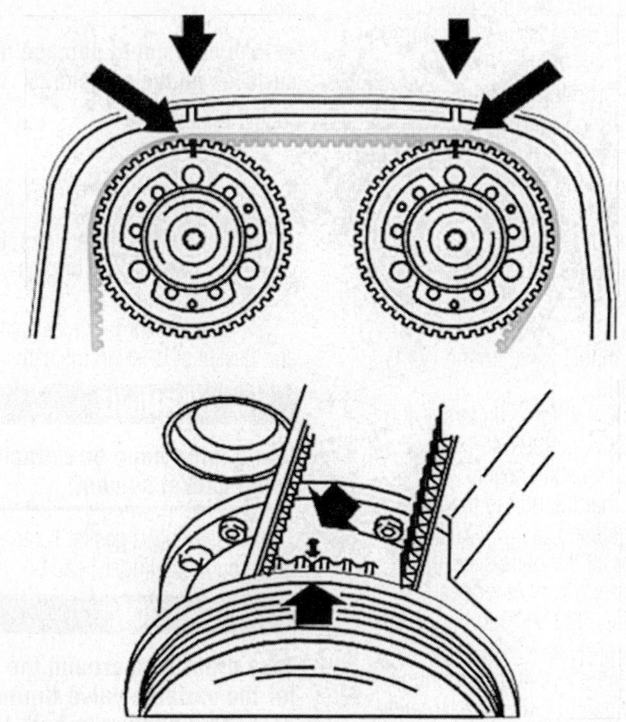

Setting the timing marks—5- and 6-Cylinder Engines

Valve Adjustment

5- and 6-Cylinder Engines

1. Remove the cable from the battery negative terminal.
2. Remove:
   - The cross stay between the suspension turrets
   - The ground strip from the cylinder head
   - The upper engine stabilizer brace
   - The cover in the cylinder head at the rear of the exhaust camshaft
   - The crankcase ventilation hose from the top of the camshaft cover
   - The radiator breather tube from the expansion tank. Install lock grip pliers.
3. Lift up the brake fluid reservoir.
4. Disconnect the ABS sensor connector.
5. Place the brake fluid reservoir over the engine.

✲✲ WARNING

Ensure that no fluid is spilled on the engine. It is extremely flammable!

6. Disconnect the connector for the level sensor in the expansion tank.
7. Lift up and place the expansion tank on top of the engine.

8. Remove:
   - The auxiliaries belt
   - The front timing cover.
9. Position the upper timing cover.
10. Turn the crankshaft clockwise until the markings on the crankshaft belt pulley and the timing belt pulley are aligned with the markings on the oil pump and the upper timing cover.
11. Remove the upper timing cover.

CAUTION

Crankshaft and camshafts must not be turned more than is stated in the method description. If the shafts are turned in any other way the valves may be damaged.

12. Slacken off the center bolt for the belt tensioner slightly.
13. Hold the center bolt still. Turn the tensioner eccentric clockwise to 10 o'clock using a 6 mm Allen key.
14. Remove the timing belt from the camshaft pulleys.
15. Install camshaft adjustment tool 999 5452 at the rear of the camshafts.
16. Check that the bolts securing the adjustment tool to the camshafts and the bolt holding the tool together are well tightened
17. Remove: (timing gear pulley with variable valve timing unit)
   - The plug at the front of the variable valve timing unit
   - The center bolt in the variable valve timing unit. Carefully pull out the variable valve timing unit with the timing gear pulley.
18. Remove: (timing gear pulley without variable valve timing unit)
   - The bolts securing the timing gear pulley on the camshaft
   - The timing gear pulley.
19. Remove tool 999 5452.
20. Reinstall the expansion tank and the brake fluid reservoir at the fender liner.
21. Remove:
   - The variable valve timing (VVT) solenoid
   - Spark plugs for cylinders 1 and 5.
22. Install 2 tools 999 5454. Leave a 2-3 mm gap to the camshaft cover.
23. Ensure that the bolt in the spark plug well is fully tightened.
24. Remove all the bolts securing the camshaft cover to the cylinder head.
25. Use pliers 999 5670 to lift the cover from the cylinder head.
26. Install the pliers at the stop lugs. Start with cylinder 1 and work alternately backward.

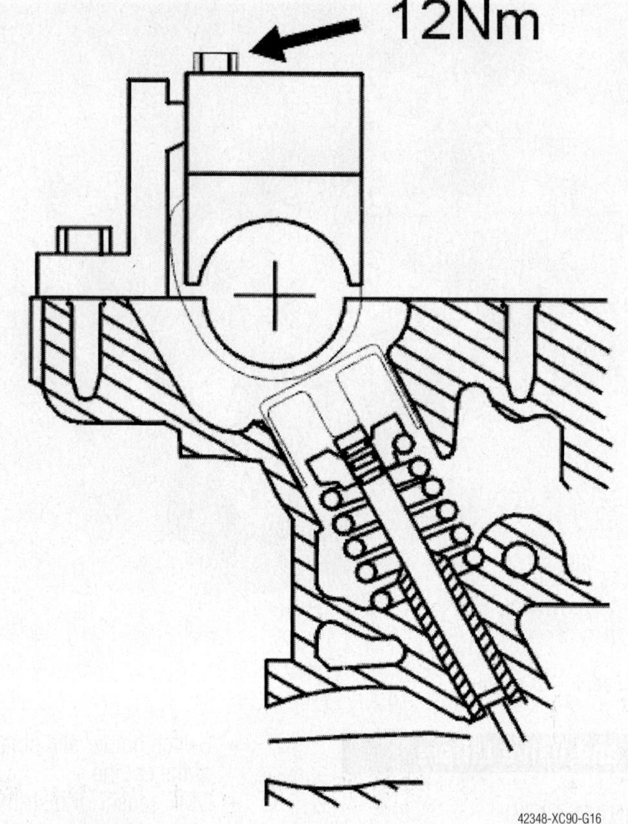

Camshaft press installed—5- and 6-Cylinder Engines

27. Slacken off the wing nuts approximately 2 turns. Repeat the procedure with the pliers.
28. Carefully press the camshaft seals free.

➡Take care not to damage the sealing surfaces on the camshafts.

29. Remove:
   - Tool 999 5454
   - The camshaft cover
   - The camshafts.
30. Lift out the valve lifters. Mark up the valve lifters so that the original positions can be established.
31. Use a razor blade or a gasket scraper and gasket solvent on the camshaft cover.

WARNING

Use a fume hood or extractor when using gasket solvent!

32. Use only a gasket scraper or razor blade on the cylinder head.

CAUTION

Take great care around the oil ducts for the variable valve timing solenoid. This applies to both the camshaft cover and the cylinder head. The solenoid is extremely sensitive to contaminants.

33. Dry and blow all surfaces clean.
34. Carefully tap the end of the valve stem to ensure that the valve is correctly located in the seat.
35. Use a plastic, aluminum or brass drift to protect the valve and the surface of the valve lifter.
36. The sound made by tapping reveals if the valve is correctly seated.
37. Install both the valve lifters for the inlet valves at cylinder 1.
38. Check the notes made earlier. Select new valve lifters if necessary.

➡Only install two valve lifters. The valve lifters should be placed at the same cylinder!

39. Other valve lifters are available as replacement part / replacement part kits.
40. The valve clearance on a cold engine (approximately 20° C) should be:
   - Inlet valve: 0.20 ± 0.03 mm.
   - Exhaust valve: 0.40 ± 0.03 mm.

➡The tolerances are less at setting! When checking the valve clearance through the plug hole the tolerances are larger.

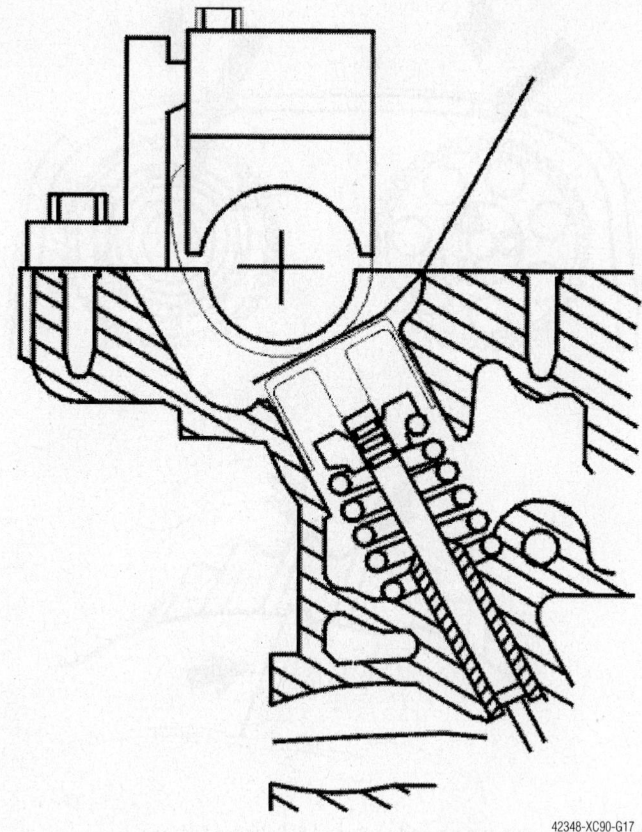

Checking the valve clearance—5- and 6-Cylinder Engines

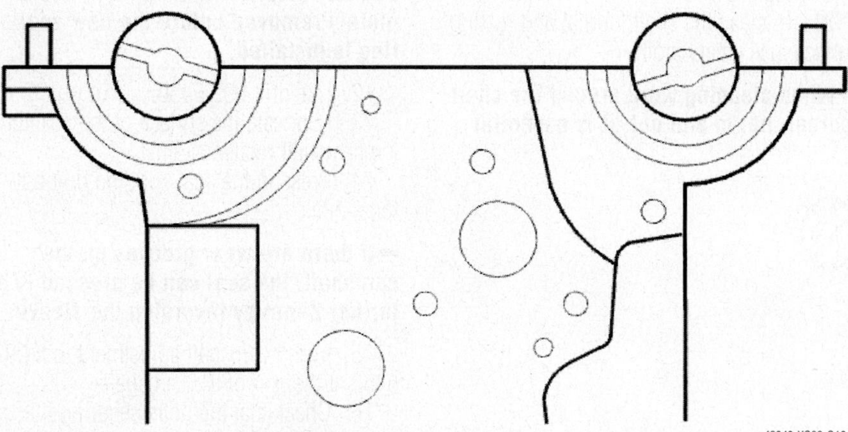

Camshafts properly positioned—5- and 6-Cylinder Engines

41. Position the intake camshaft. Ensure that the lobes at cylinder 1 point upwards.

42. Apply a little oil to the cam lobe and the upper side of the valve lifter to facilitate later measurement.

43. Install the lower section of camshaft press 999 5765 at the inlet valves for cylinder 1.

44. Tighten the tool against the cylinder head. Tighten to 17 Nm.

45. Turn the camshaft until it stops against the camshaft press.

46. Install the upper section of the camshaft press.

47. Tighten the bolt which tensions the camshaft. Tighten to 84 inch lbs. (12 Nm).

➡Measurements should only be taken on a cold engine. A suitable temperature is approximately 68°F (20°C).

48. Using a feeler gauge, press with a finger so that the feeler gauge lies parallel to the upper side of the valve lifter.

49. Move the feeler gauge sideways when taking the reading in order to obtain as accurate a measurement as possible.

50. The valve clearance measured on cold engines (approximately 68°F (20°C) should be:

- Inlet valve: 0.008 inch +/- 0.001 inch (0.20 +/- 0.03 mm).
- Exhaust valve: 0.016 inch +/- 0.001 inch 0.40 +/- 0.03 mm.

➡The tolerances are less at setting! When checking the valve clearance through the plug hole the tolerances are larger.

51. Differences in valve clearance for different engines/ambient temperatures:

- -0.0004 inch (- 0.01 mm) at 59°F (15°C)
- +0.0004 inch (+ 0.01 mm) at 77°F (25°C)
- +0.0008 inch (+ 0.02 mm) at 86°F (30°C)
- +0.0012 inch (+ 0.03 mm) at 95°F (35°C)
- +0.0016 inch (+ 0.04 mm) at 113°F (45°C)

52. Correcting measured clearance:

a. Lift out the upper section of the press tool.

b. Lift out the camshaft.

c. Adjust the play by replacing the valve lifters.

d. Other valve lifters are available as replacement part / replacement part kits.

e. Reinstall the camshaft and the upper section of the press tool. Tighten to 84 inch lbs. (12 Nm).

f. Take a new measurement.

53. When the correct valve clearance is reached, remove:

- The press tool 999 5765
- The camshaft
- The valve lifters.

54. Carefully mark the valve lifters so that exact reinstallation can be carried out.

For example:

Intake side: I1, I2, I3........I10.

Exhaust side: A1, A2, A3........A10.

55. Repeat the procedure for measuring the valve clearance for all cylinders on both the intake and exhaust sides.

56. Lubricate the valve guide wells.

57. Install all the valve lifters.

58. Lubricate the camshaft bearing seats and the upper sides of the valve lifters.

59. Position the intake camshaft. Ensure that the groove at the rear edge of the camshaft is above an imaginary center line.

60. Position the exhaust camshaft. Ensure that the groove at the rear edge of the camshaft is below an imaginary center line.

61. Wipe the oil film off the mating surfaces on the camshaft cover and cylinder head.

62. Install new O-rings around the spark plug wells at the cylinder head.

63. Apply liquid gasket 1161 059 on the camshaft cover. Use roller 951 2767.

➡**The surface must be completely covered without any excess.**

⁂ CAUTION

Ensure that no liquid gasket gets in to the oil ducts.

64. Lubricate the camshaft lobes, the camshaft bearing surfaces and the valve lifters.

65. Install the camshaft cover.

66. Install press tool 999 5454 (2x).

67. Tighten the camshaft cover bolts alternately, keeping it parallel to the cylinder head using the press tools.

68. Install all the bolts. Tighten the bolts from the middle and outwards.

69. Remove the press tool 999 5454

70. Install:

- The variable valve timing (VVT) solenoid. Use a new gasket
- The spark plugs. Tighten to 22 ft. lbs. (30 Nm)
- The plugs for the test holes. Tighten to 15 ft. lbs. (20 Nm)
- The crankcase ventilation hose to the top of the camshaft cover
- The ignition coils according to the earlier marking

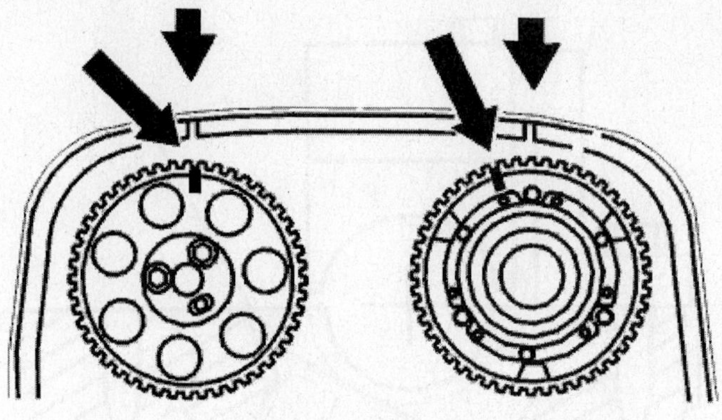

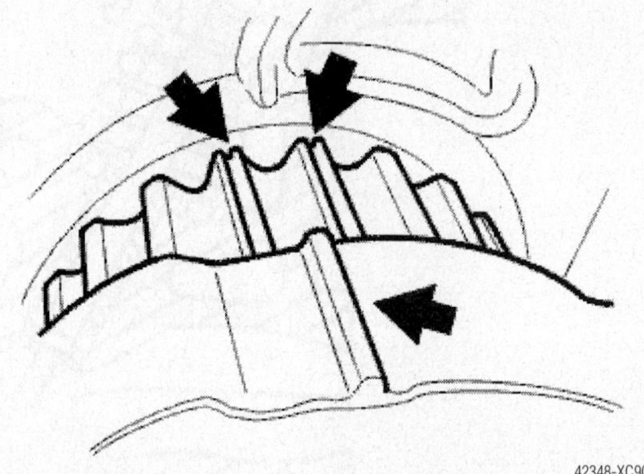

Timing marks, with variable timing units—5- and 6-Cylinder Engines

- The ground terminals between the ignition coils.

71. To clean the shaft journal and mating surface, use emery cloth.

➡**When cleaning work around the shaft journal, not in and out. It is essential that any residue from the emery cloth and any other contaminants are completely removed before the new sealing ring is installed.**

72. Use drift 999 5450.

73. Lubricate the surface of the seal that the camshaft rotates against.

74. Press in the seal until the drift bottoms out.

➡**If there are wear grooves on the camshaft, the seal can be pressed in a further 2 mm by reversing the sleeve.**

75. Install camshaft adjustment tool 999 5452 at the rear of the camshafts.

76. Check that the bolts securing the adjustment tool to the

77. Lift up and position the brake fluid reservoir and the expansion tank on top of the engine.

78. Use drift 999 5718 on camshafts with variable valve timing units. Use drift 999 5719 on camshafts without variable valve timing.

79. Use new seals and lubricate the surface of the seal that the camshaft rotates against.

80. Use the variable valve timing unit/timing belt pulley mounting bolts. Tighten the bolts until the drift bottoms out.

81. Remove the drift.

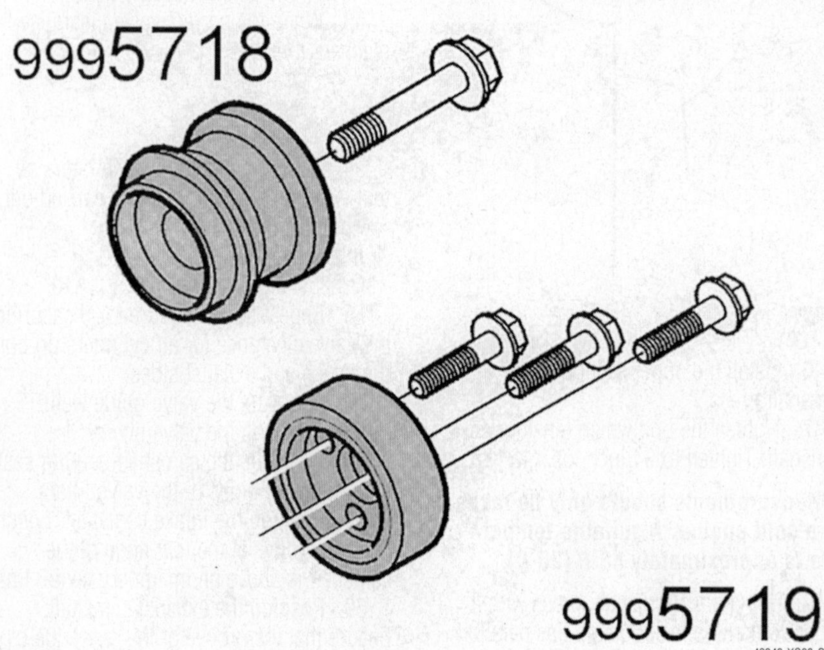

Front seal installation tools—5- and 6-Cylinder Engines

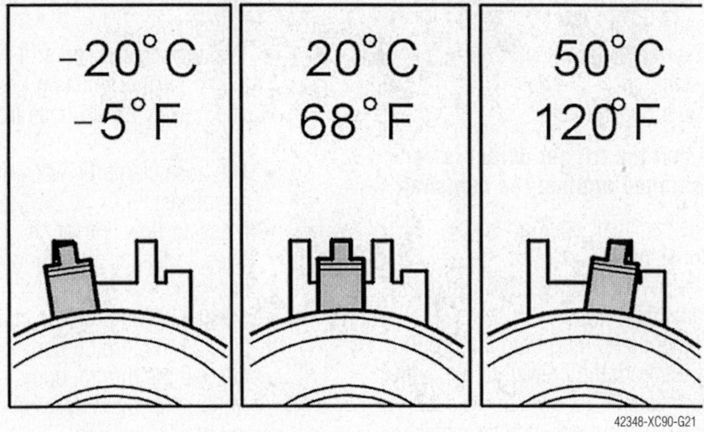

Tensioning indicator at different temperatures—5- and 6-Cylinder Engines

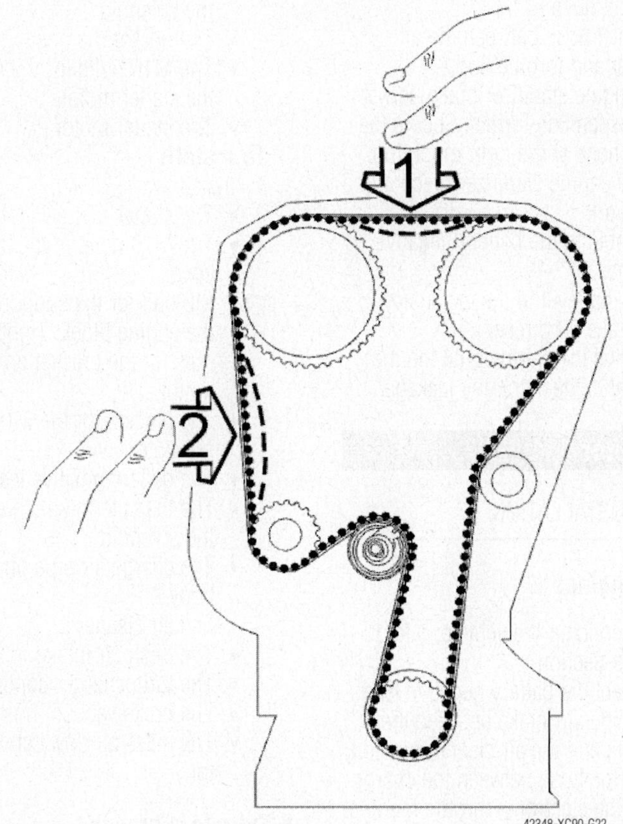

Checking belt tension—5- and 6-Cylinder Engines

82. Remove:
    - The mounting bolts for the starter motor. Pull off the starter motor. Place the starter motor to one side
    - The blind cover plug and the seal washer.

83. Turn the crankshaft slightly clockwise.

84. Install the crankshaft stop 999 5451 Ensure that it bottoms out against the cylinder block.

85. Turn the crankshaft counter-clockwise until it stops against the crankshaft stop.

86. Check that the marking on the crankshaft timing gear pulley corresponds with the marking on the oil pump.

➡**The purpose of the section is to ensure that the VVT unit is correctly positioned and to reset the camshaft timing gear pulley to the correct position using the markings made at the factory. This is to ensure that the conditions are correct for any later fault-tracing.**

87. Slacken off, but do not remove the bolts which secure the timing gear pulley to the variable valve timing unit.

88. Press the variable valve timing unit/timing gear onto the camshaft.

89. Install the center bolt which secures the variable valve timing unit to the camshaft. Tighten slightly.

90. Turn the variable valve timing unit counter-clockwise as far as it will go.

91. Remove the center bolt.

92. Position the upper timing cover.

93. Turn the timing gear pulley clockwise until the bolts at the oval holes are in the limit position.

94. Continue turning clockwise until the timing gear pulley marking is 1 tooth before the marking on the upper timing cover.

CAUTION

Do not turn counter-clockwise during this procedure.

95. Check that the timing gear pulley is still in its limit position at the oval holes.

96. Tighten the center bolt for the variable valve timing (VVT) unit. Tighten to 88 ft. lbs. (120 Nm). Check that the variable valve timing unit does not rotate when tightening.

97. Install the center bolt. Tighten to 26 ft. lbs. (35 Nm).

98. Install the timing gear pulley. Install the bolts.

99. Install two bolts without tightening. Allow the third bolt to protrude.

100. Adjust the timing gear pulleys so that the markings on the timing gear pulleys / upper timing cover correspond.

101. Tighten the center bolt on the belt tensioner. Tighten to 44 inch lbs. (5 Nm).

102. Turn the variable valve timing unit clockwise to the stop.

103. Hold it secure in the limit position.

104. Install the belt in the following order:
    - Crankshaft
    - The idler pulley
    - Intake cam
    - Exhaust cam
    - Water pump
    - Belt tensioner

➡**Adjust the timing gear pulleys so that the bolts are not at a limit position in the oval holes. Also check that the markings correspond.**

➡**This adjustment is always carried out on a cold engine. A suitable temperature is approximately 68°F (20°C).**

At higher temperatures (with the engine at operating temperature or a high outside temperature for example) the indicator is further to the right.

The illustration shows the position of the indicator when aligning the timing belt tensioner at different temperatures.

105. Hold the center bolt secure and turn the belt tensioner eccentric counter-clockwise until the tensioner indicator passes the marked position.

➡Check that the variable valve timing unit is in its limit position.

106. Tighten the three bolts at the intake camshaft timing gear pulley. Tighten to 74 inch lbs. (10 Nm).

107. Tighten the three bolts at the exhaust camshaft timing gear pulley. Tighten to 15 ft. lbs. (20 Nm).

108. Turn the eccentric on the belt tensioner back so that the indicator reaches the marked position in the center of the window. Remember to hold the center bolt secure at the same time.

109. Hold the eccentric secure and tighten the center bolt. Tighten to 15 ft. lbs. (20 Nm).

110. Check that the indicator is in the correct position.

111. Remove camshaft adjustment tool 999 5451 and crankshaft stop 999 5452.

112. Install the plug with a new blind cover plug. Tighten to 29 ft. lbs. (40 Nm).

113. Press the timing belt to check that the indicator on the tensioner moves easily.

114. Turn the crankshaft two turns. Check that the markings on the crankshaft timing gear pulley and the camshaft timing gear pulley match up with the markings on the oil pump and upper timing cover respectively.

115. Check that the indicator on the belt tensioner is within the marked position.

116. Remove the upper timing cover.

117. Install:

- Front timing cover
- The auxiliaries belt
- The expansion tank
- The servo oil reservoir
- The bleed hose for the expansion tank

118. Close the clamp and check that the hoses lie correctly.

119. Connect the connectors for the:

- The ABS sensor by the right suspension turret
- The coolant level sensor in the expansion tank

120. Install:

- The starter motor
- The plastic nuts for the cover in the fender liner
- The right front wheel

121. Install:

- The trigger wheel.

➡Ensure that the trigger wheel is correctly positioned against the camshaft.

- The camshaft position sensor housing
- The cover at the rear of the exhaust camshaft
- The bolts holding both the ground strips from the firewall to the cylinder head
- The upper engine stabilizer brace
- The cover over the ignition coils
- The upper timing cover
- The crankcase ventilation hose on the inlet hose
- The inlet hose between the air cleaner and throttle body
- The engine stabilizer brace between the suspension turrets. Secure the servo hose at the right mounting for the engine stabilizer brace in the bodywork
- The cable to the battery negative terminal.

122. Check the level in the expansion tank and the brake fluid reservoir.

123. Test drive the engine until the thermostat opens and check for any leakage.

Starter Motor

REMOVAL & INSTALLATION

5-Cylinder Engines

1. Before servicing the vehicle, refer to the Precautions Section.
2. Disconnect the battery negative lead.
3. Remove the air intake between the front cover plate and the air cleaner housing
4. Remove the hose between the charge air cooler and the electronic throttle module.
5. Disconnect the connector and the positive battery lead from the starter motor.
6. Remove the 2 M10 bolts securing the starter motor to the transmission.
7. Remove the starter motor.
8. Installation is the reverse of removal.

6-Cylinder Engines

1. Before servicing the vehicle, refer to the Precautions Section.
2. Disconnect the battery negative lead.
3. Remove:
   - The intake pipe between the front cover plate and the air cleaner
   - The connector from the mass air flow sensor
   - The turbocharger control valve from the air cleaner housing
   - The brake vacuum hose from the air cleaner
   - The hose clamp between the fresh air intakes
   - Mass air flow sensor
   - The air cleaner by pulling it straight up.
   - The hose clamp on the charge air pipe on the throttle body. Pull the pipe off the throttle body
   - M8 bolt for the bracket in the cylinder block
   - M8 nuts in the bracket for the wiring and the engine coolant hose
   - The blade terminal connector on the solenoid
   - The nut for the positive lead
   - The M10 bolts for the mounting for the starter motor
   - The starter motor.

To install:

4. Install:
   - The starter motor
   - The M10 bolts in the transmission cover
   - M8 bolt for the support bracket on the engine block. Tighten. The M6 nuts for the bracket on the starter motor
   - The bracket for the wiring. Use 2 M8 nuts.
   - The battery positive lead
   - The blade terminal connector on the solenoid
   - The charge air pipe on the throttle body
   - The air cleaner
   - The fresh air intake
   - The turbocharger control valve
   - The brake vacuum hose
   - The mass air flow sensor connector.

8-Cylinder Engines

1. Remove:

   a. The 4 bolts at the suspension turrets

   b. The bolts of the upper engine mount on the engine

   c. The upper cross stay.

   d. The battery negative cable.

   e. The expansion tank cap.

2. Raise the vehicle.
3. Remove the protective plate (skid plate).

➡There is a guide sleeve on the upper mounting for the plate. This must be

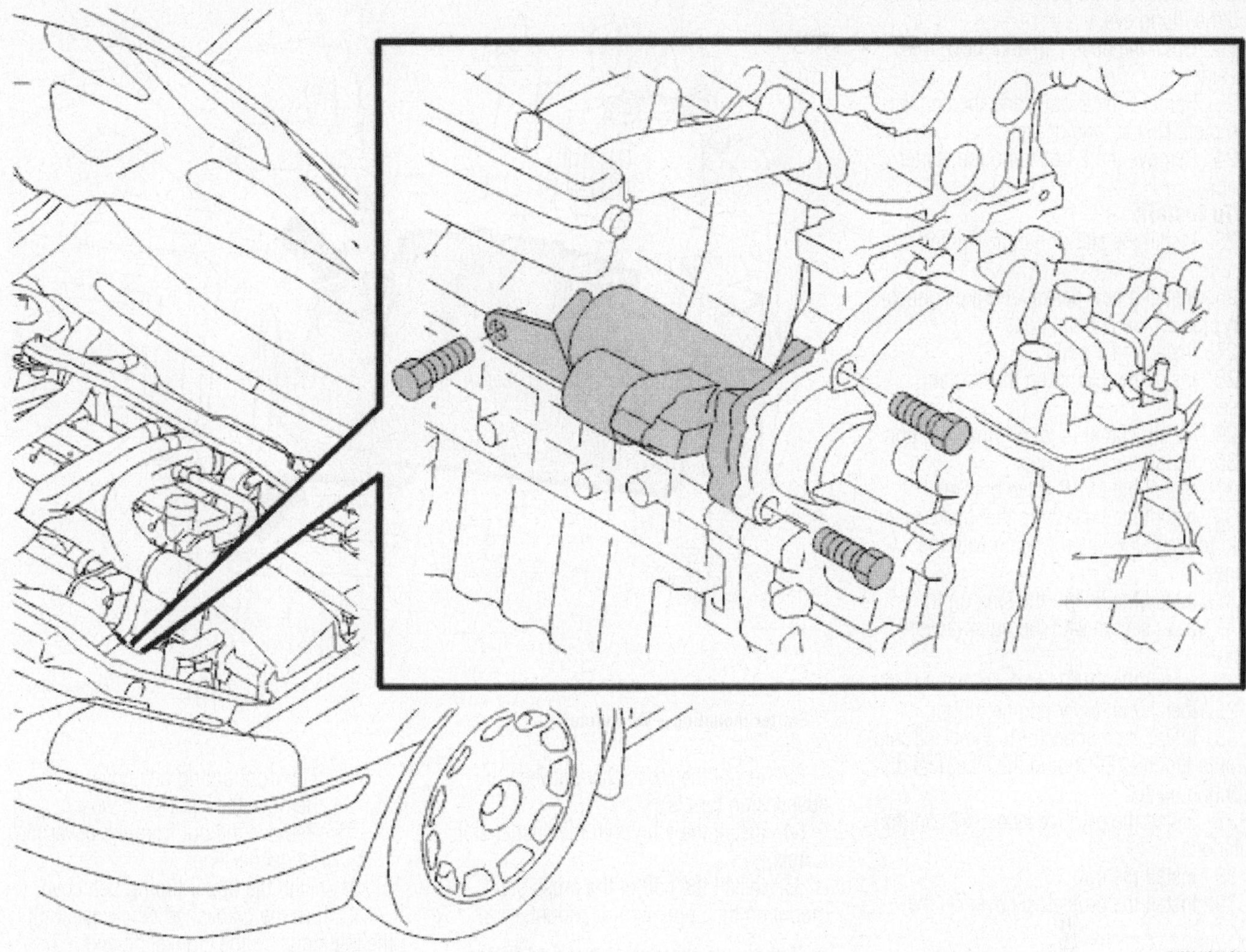

Starter mounting—5- and 6-cylinder engines

lifted slightly so that the panel can be slid forward.

4. Draining coolant
5. Open the cock on the lower left edge of the radiator.
6. Drain the coolant into a clean receptacle.
7. Close the cock.
8. Installing the protective plate.
9. Install the protective plate.
10. Lower the vehicle.
11. Remove the air cleaner cover by pulling it straight up.
12. Remove:
    a. The 2 connectors from the engine control module (ECM); clip the band clamp to the cable harness.
    b. The connector from the mass air flow sensor.
13. Remove:
    a. The 2 bolts of the fresh air hose in the front plate.
    b. The fresh air hose.
    c. The hose clips of the intake manifold between the mass air flow sensor and the electronic throttle module (ETM).
    d. The air cleaner by pulling it straight up.
14. Remove:
    a. The connector to the brake vacuum valve.
    b. The vacuum hose from the vacuum pump by pressing in the quick-action lock.
    c. The vacuum hose from the brake vacuum valve.
    d. The vacuum hose from the intake manifold.
    e. The brake vacuum from the retainer.
15. Detach the radiator hoses at the quick-release coupling toward the firewall.
16. Detach the coolant hose from the throttle body by undoing the clamp.
17. Bend aside the coolant hoses.
18. Remove:
    a. The connector to the electronic throttle module (ETM) and the electrical distribution center.
    b. The protective cap from the starter motor solenoid.
    c. The nut of the positive leads on the starter motor solenoid.
    d. The positive leads.
    e. The control current lead from the starter motor solenoid.
19. Remove:
    a. The 4 bolts.
    b. The upper engine mount.
20. Remove:
    a. The EVAP valve from the bracket.
    b. The line from the EVAP valve to the tank by releasing the quick-action lock.
    c. The hose from the intake manifold to the EVAP valve by releasing the hose clip.
    d. The 2 bolts of the EVAP valve bracket.
    e. The EVAP valve bracket.

21. Remove the 3 bolts of the lifting eye and the lifting eye.
22. Open the hose clip and move it to the side.
23. Remove the 2 bolts and the connection pipe. Discard the gasket.
24. Remove the 2 bolts and the starter motor.

To install:

25. Install the starter motor. Tighten the 2 bolts.
26. Install a new gasket and the connection pipe on the coolant hose.
27. Install the 2 bolts.
28. Install the clamp on the coolant hose.
29. Fit the clamp in its original position.
30. Install the lifting eye.
31. Install the EVAP valve bracket.
32. Install the hose from the intake manifold to the EVAP valve; secure with the clamp.
33. Install the line to the tank on the EVAP valve; secure with the quick-action lock.
34. Install the EVAP valve on the bracket.
35. Install the upper engine mount.
36. Install the connector to the electronic throttle module (ETM) and the electrical distribution center.
37. Install the positive leads, x 2, on the solenoid.
38. Install the nut.
39. Install the protective cover on the nut.
40. Install the control current lead on the solenoid.
41. Install the 2 coolant hoses on the connections at the firewall; secure with the quick-action lock.
42. Install the coolant hose on the throttle body; secure with the clamp.
43. Install the air cleaner on the left frame member.
44. Install the intake hose on the throttle body; secure with the clamp.
45. Install the connector to the mass air flow sensor.
46. Install the vacuum hose to the vacuum pump; secure with the quick-action lock.
47. Install the vacuum hose on the brake vacuum valve.
48. Install the vacuum hose on the intake manifold; secure with the clamp.
49. Install the brake vacuum valve in the retainer.
50. Install the connector to the brake vacuum valve.
51. Install the fresh air intake manifold.
52. Install the engine control module (ECM) connectors.

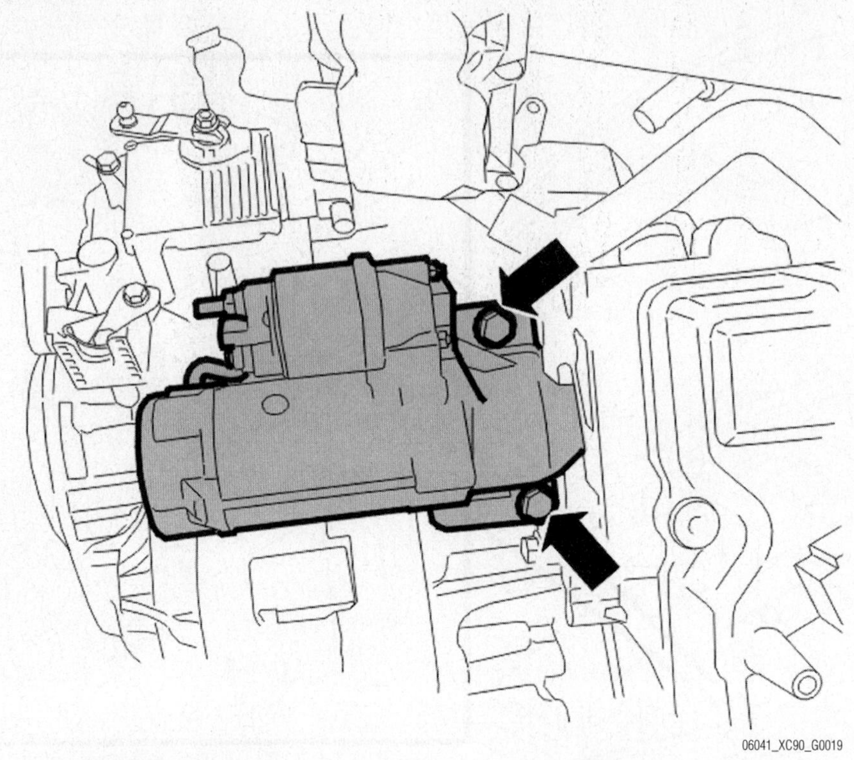

Starter mounting—V8 engine

53. Install the cross stay between the suspension turrets.
54. Install the 4 bolts in the suspension turrets.
55. Install the bolt of the cross stay mount on the upper engine mount.

➡**Make sure the outlet hose of filling equipment is not blocked as this could damage the cooling system.**

56. Fill with coolant.
57. Start the engine and allow it to warm to normal operating temperature. Check for leaks and proper function. Fill up fluid as necessary.

Timing Belt, Front Cover and Seal

REMOVAL & INSTALLATION

5-Cylinder Engines

1. Before servicing the vehicle, refer to the Precautions Section.
2. Remove or disconnect the following:
    - The cross stay between the suspension turrets
    - The upper timing belt cover
    - The servo reservoir and the expansion tank. Lift up and place on top of the engine.
    - The auxiliary drive belt
    - The front timing belt cover.
    - Remove the right front wheel
    - Remove the nut from the cover in the fender liner
3. Install the upper timing belt cover
4. Turn the crankshaft clockwise until the markings on the crankshaft and camshaft pulley correspond
5. Turn the crankshaft a further ¼ turn clockwise and then back again until the markings correspond. The markings are illustrated
6. Remove the upper timing belt cover.
7. Back off the center bolt of the belt tensioner slightly.
8. Hold the center bolt still. Turn the tensioner eccentric clockwise, using a 6 mm Allen key, to 10 o'clock.
9. Remove the timing belt from the tension pulley, camshaft pulley and water pump.
10. Remove the vibration damper. Use a counterhold. Work the vibration damper loose.
11. Remove the timing belt.
12. Spin the idler pulley and listen for noise. If replacing with a new idler pulley, tighten to 18 ft. lbs. (24 Nm).
13. Spin the tension pulley and listen for noise. When replacing, bolt the tension pulley into place using the center bolt. Screw in the center bolt by hand.
14. Ensure that the tensioner fork is centered over the cylinder block rib.

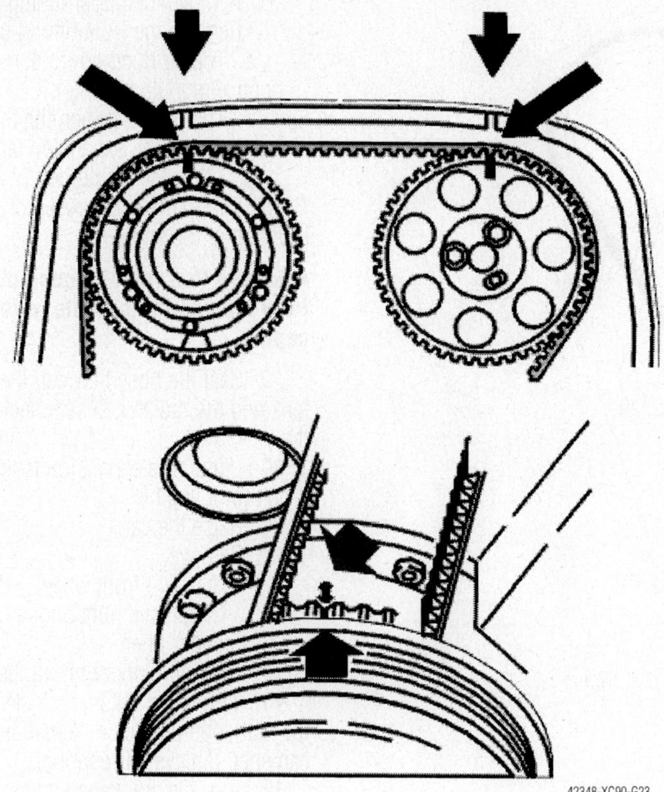
42348-XC90-G23

Timing mark alignment

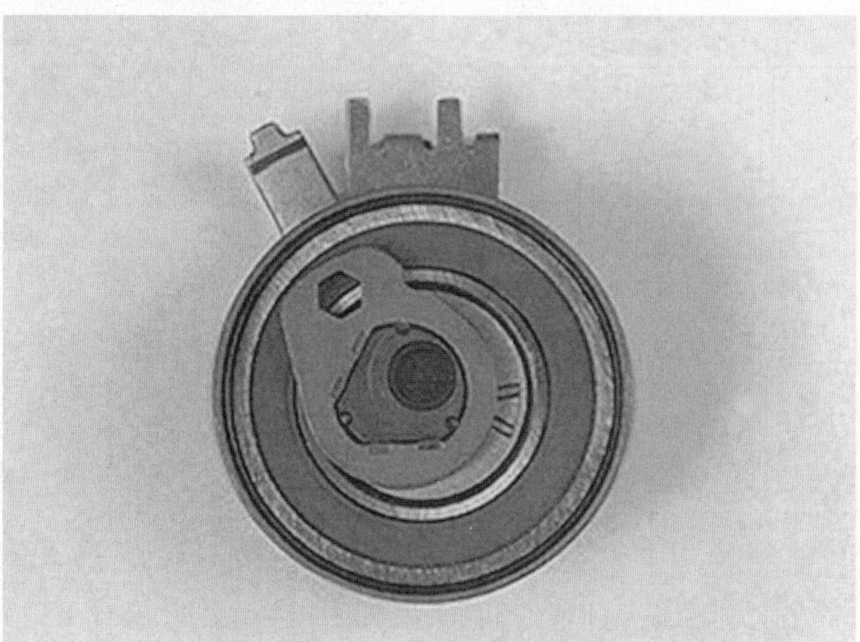
42348-XC90-G24

Belt tensioner—5-cylinder engine

15. Ensure that the Allen hole on the eccentric is at "10 o'clock".

To install:

16. Install the timing belt over the pulley on the crankshaft.

17. Install the vibration damper. Tighten the center nut. Tighten to 133 ft. lbs. (180 Nm).

18. Remove the counterhold and install new bolts. Tighten to 18 ft. lbs. (25 Nm) and angle-tighten 30 degrees.

19. Install or connect the following:

- Crankshaft
- The idler pulley
- Intake camshaft pulley
- Exhaust camshaft pulley
- Water pump
- Belt tensioner.

➡The following adjustment is carried out on a cold engine. A suitable temperature is approximately 68°F. At higher temperatures, for example with the engine at operating temperature or at higher ambient temperature, the needle is further to the right. The illustration shows the position of the indicator when aligning the timing belt tensioner at different temperatures.

20. Tension the timing belt as follows:
    - Turn the crankshaft clockwise carefully until the timing belt is tensioned. The belt must be tensioned between the intake camshaft pulley, the idler pulley and the crankshaft
    - Hold the center bolt on the belt tensioner secure. Turn the belt tensioner eccentric counter-clockwise until the tensioner indicator passes the marked position. Then turn the eccentric back so that the indicator reaches the marked position in the center of the window
    - Hold the eccentric secure and tighten the center bolt. Tighten to 15 ft. lbs. (20 Nm). Check that the indicator is in the correct position.
21. To check the alignment:
    - Press the belt to check that the indicator on the tensioner moves easily
    - Install the upper timing belt cover
    - Turn the crankshaft 2 turns. Check that the markings on the crankshaft and camshaft pulley correspond
    - Check that the indicator on the belt tensioner is within the marked area.
22. Install or connect the following:
    - The front timing belt cover. Tighten to 105 inch lbs. (12 Nm).
    - The upper timing belt cover
    - install the auxiliaries belt
    - The servo reservoir
    - The expansion tank.

➡Ensure that the hoses are correctly positioned.

- The engine stabilizer brace. Tighten the bolts at the suspension turrets. Tighten to 37 ft. lbs. (50 Nm). Tighten the engine bracket bolt. Tighten to 58 ft. lbs. (80 Nm).
- The cover in the fender liner
- The front wheel.

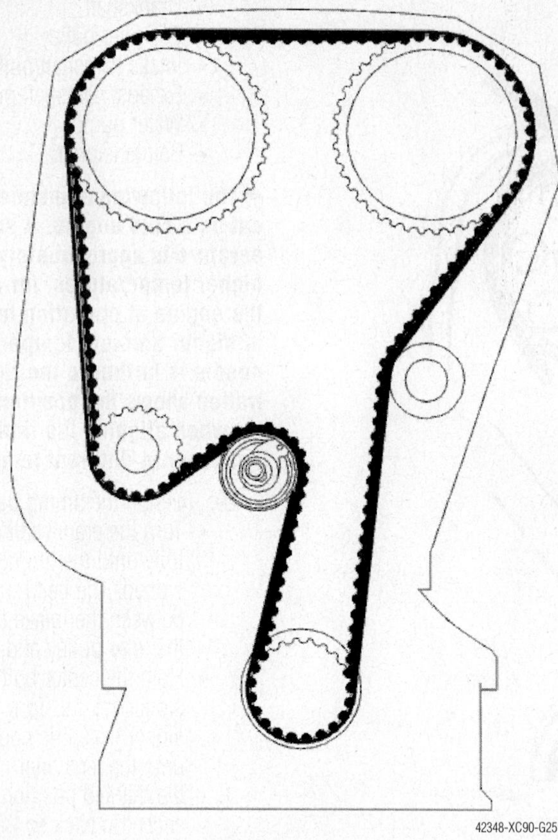

Timing belt installed

Tension indicator at different temperatures

6-Cylinder Engines

1. Before servicing the vehicle, refer to the Precautions Section.
2. Remove:
   - The bolt holding the engine stabilizer brace to the bracket on the engine
   - The bolts holding the engine stabilizer brace to the suspension turrets
   - The engine stabilizer brace.
3. Remove the plastic pipes between the turbocharger and charge air cooler and between the air cleaner and turbocharger. Put them to one side.
4. Remove the clamp from the intake manifold for the turbocharger for cylinders 1, 2 and 3. Turn the upper section of the pipe towards the firewall.
5. Relieve the load from the belt tensioner. Remove the auxiliary belt.
6. Remove the upper timing belt cover.
7. Remove the front timing belt cover.
   a. The air duct, cover, surround and control module box
   b. The hose between the thermostat housing and the expansion tank
   c. The front timing belt cover.
8. Lift up the servo reservoir and place it on top of the engine.

➡Ensure that the oil does not leak from the ventilation hole in the filler cap.

9. Seal the hose between the expansion tank and the radiator. Disconnect the hose at the tank.
10. Lift up the expansion tank and place it on top of engine.
11. Raise the car.
12. Remove:
    - The right front wheel
    - The plastic nuts on the cover in the fender liner.
13. Install the upper timing belt cover.
14. Turn the crankshaft clockwise until the markings on the crankshaft and camshaft pulleys correspond.
15. Turn the crankshaft a further ¼ turn clockwise.
16. Then turn back counter-clockwise until the markings correspond.
17. Remove the upper timing belt cover.
18. Remove the 4 vibration damper bolts. Counterhold the crankshaft central nut.
19. Remove:
    - The vibration damper
    - The auxiliaries belt
    - The belt cover behind the crankshaft pulley for the auxiliary belt.
20. Spray universal oil or similar around the rubber sleeve on the underside of the oil pump.
21. Remove the rubber sleeve.
22. Slacken off the center bolt for the belt tensioner slightly.
23. Hold the center bolt still. Turn the tensioner eccentric counter-clockwise using a 6 mm Allen key. Turn to 10 o'clock.
24. Unhook and remove the timing belt.Caution! Do not turn the camshafts or the crankshaft when the timing belt has been removed.
25. Spin the idler pulley and listen for noise. If replacing with a new idler pulley, tighten to 17 ft. lbs. (24 Nm).
26. Spin the tension pulley and listen for noise. When replacing, bolt the tension pulley into place using the center bolt.

To install:

27. Screw in the center bolt by hand.
28. Ensure that the tensioner fork is centered over the cylinder block rib/bracket.

29. Ensure that the Allen hole on the eccentric is at 10 o'clock.

30. Install the belt as follows:

- Crankshaft
- The idler pulley
- Intake cam
- Exhaust cam
- Water pump
- Tension pulley

➡Belt tension adjustment is carried out on a cold engine. A suitable temperature is approximately 20°C/68°F.

31. At higher temperatures (with the engine at operating temperature or a high outside temperature for example) the indicator is further to the right.

32. The illustration shows the position of the indicator when aligning the timing belt tensioner at different temperatures.

33. Carefully turn the crankshaft clockwise until the timing belt is tensioned. The belt must be in tension between the intake camshaft pulley, the idler pulley and the crankshaft.

34. Hold the belt tensioner center bolt secure. Turn the belt tensioner eccentric clockwise until the tensioner indicator passes the marked position.

35. Then turn the eccentric back so that the indicator reaches the marked position in the center of the window.

36. Hold the eccentric secure. Tighten the center bolt to 18 ft. lbs. (25 Nm).

37. Check that the indicator is in the correct position.

38. Press the belt to check that the indicator on the tensioner moves easily.

39. Install the upper timing belt cover.

40. Turn the crankshaft 2 turns. Check that the markings on the crankshaft and camshaft pulley correspond.

41. Check that the indicator on the belt tensioner is within the marked area.

42. Install the auxiliary belt around the pulley on the crankshaft.

43. Install:

- The rubber sleeve on the underside of the oil pump
- The vibration damper. Use new bolts. Tighten and angle tighten. See the torque chart at the beginning of this chapter. Use the crankshaft center nut as a counterhold
- The plastic nuts for the cover in the right-hand fender liner
- The right front wheel.

44. Lower the car.

45. Remove the front timing belt cover.

46. Tension the auxiliary belt.

47. Install the front cover:

a. The front timing belt cover

b. The hose between the thermostat housing and the expansion tank

c. The surround, cover and air ducts for the control module box.

48. Install:

- The expansion tank
- The hose between the expansion tank and the radiator. Remove the lock grip pliers
- The servo reservoir
- The upper timing belt cover.

49. Twist the intake pipe for the turbocharger for cylinders 1, 2 and 3 into position. Tighten the clamp.

50. Install the plastic hoses between the turbocharger and charge air cooler and between the air cleaner and turbocharger.

51. Tighten the hose clamps.

52. Install:

- The engine stabilizer brace. Tighten to 50 Nm
- The bolt holding the engine stabilizer brace to the bracket on the engine. Tighten to 80 Nm.

53. Check:

- Coolant level
- The servo fluid level.

Timing Chain Cover and Chain

REMOVAL & INSTALLATION

8-Cylinder Engines

1. Disconnect the battery negative cable.

2. Raise and support the vehicle safely.

3. Remove the valve covers.

4. Remove the auxiliaries belt/tensioner.

5. Remove the right front wheel.

6. Removing vibration damper protective cover nuts.

7. Push the protection towards the vibration damper.

8. Remove the center bolt, use: 999 7196

9. Install 999 7198 in the center hole on the crankshaft.

10. Pull off the vibration damper.

11. Remove the alternator.

12. Remove right engine pad

13. Remove the right-hand engine pad/bracket.

14. Remove the power steering pump. Place the power steering pump to one side.

15. Remove the bolt and clamp for the power steering hose and place to one side.

16. Remove the pulley for the water pump.

17. Remove the engine bracket and idler pulley.

18. Remove the 25 bolts and the front cover. Discard the gaskets.

➡Cover the hole in oil pan using a piece of paper.

19. Use position setting tool 999 7235 to align the position on the crankshaft with the guide on the engine block.

20. Install position setting tool: 999 7236.

21. Remove the position setting tool: 999 7235.

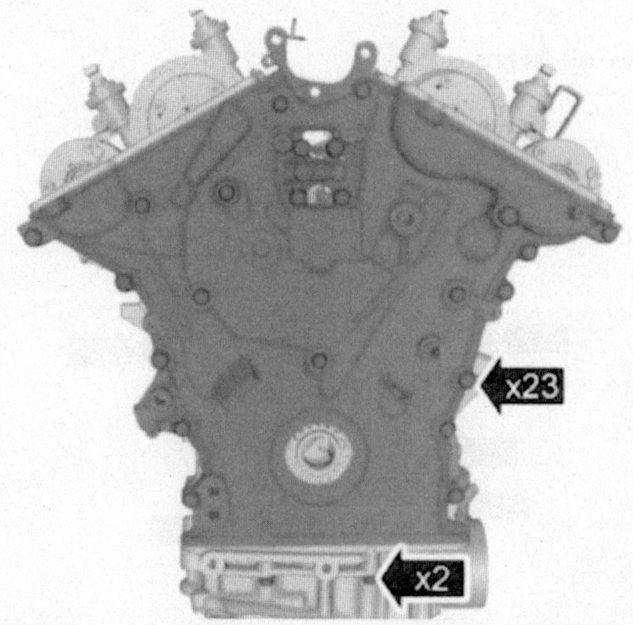

Front cover bolts—V8

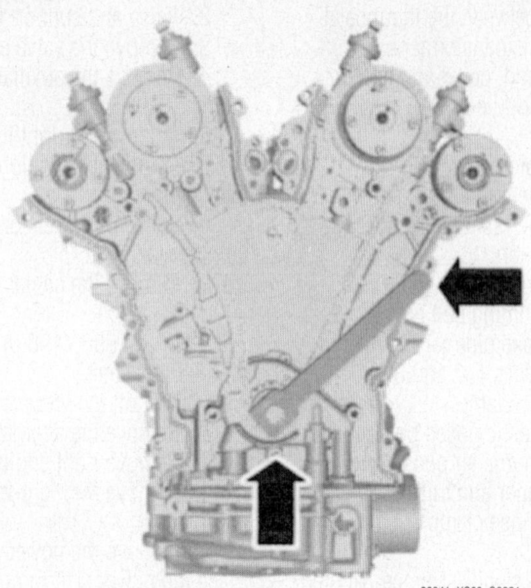

Position setting tool 999 7235—V8 engine

Position setting tool 999 7236—V8 engine

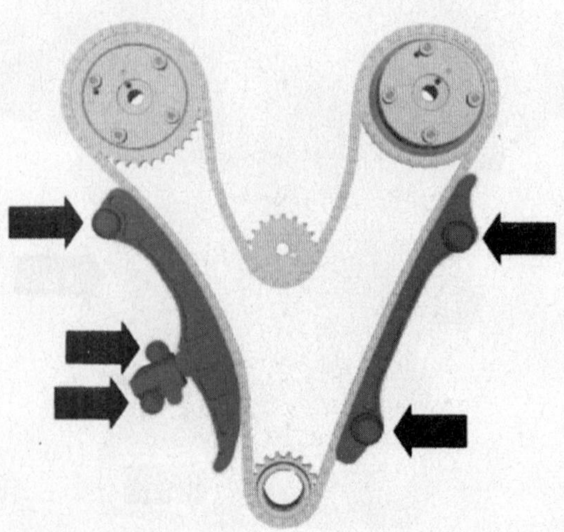

Timing chain tensioners—V8

22. Lock the chain tensioner using a drift.
23. Remove the 2 M6 bolts for the chain tensioner
24. Remove the chain tensioner
25. Remove the 2 rails
26. Remove the camshaft chain.

To install:

➡Adjust the color marking on the camshaft chain to the marking on the gear.

27. Install the camshaft chain and both rails.
28. Install the chain tensioner.
29. Remove the drift from the chain tensioner.
30. Minimize the slack of the chain by turning the crankshaft slightly.

➡The front cover must be cleaned before installing. Use a razor blade or plastic scraper.

31. Apply liquid gasket 307 57050 as illustrated. The bead diameter must be greater than 1 mm.

➡Install the front plate within 5 minutes. (curing time for liquid gasket).

32. Fold the front panel in at the lower edge.
33. Install the 23 M8 bolts for the front side.
34. Install the 2 M8 bolts for the oil sump.
35. Install the engine bracket and the idler pulley.

➡The washer must be correctly installed.

36. Install the pulley for the water pump.
37. Install the power steering pump.
38. Install the hose clamp and the bolt.
39. Install the front bolt for the alternator.
40. Install lifting beam: 999 5716, converted with: 999 7070 and 999 5460.
41. Remove tool 999 5550.
42. Install the right-hand engine pad/bracket.

➡The crankshaft flange must be cleaned.

43. Clean the surface for the sealing ring on the vibration damper with cloth 951 1024.

➡If necessary, replace the front crankshaft seal.

44. Install tool 999 7196, drift: 999 7198 and press in the vibration damper to its limit. Position using the old center bolt.

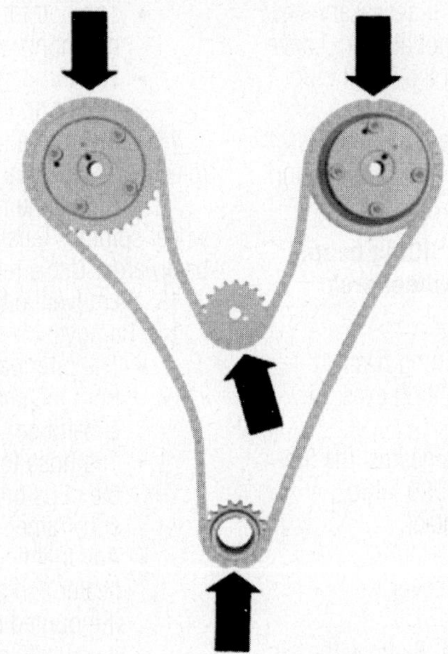

Timing chain alignment marks—V8

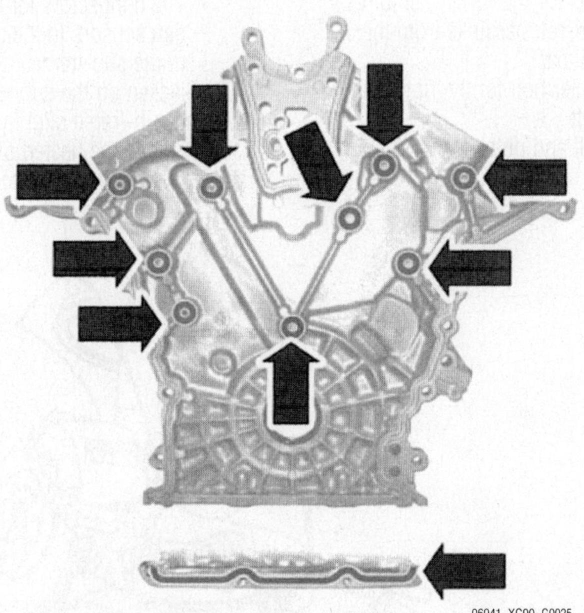

Liquid gasket application points—V8

45. Remove the center bolt.
46. Insert a new bolt for the vibration damper and tighten. See Front Seal Removal & Installation.
47. Remove tool 999 7196.
48. Install the protective cover.
49. Install the valve covers.
50. Install the auxiliaries belt/tensioner.
51. Install the right front wheel.
52. Start the engine and run the engine to normal operating temperature. Check the function of the engine. Check for leakage. Fill with fluids if necessary.

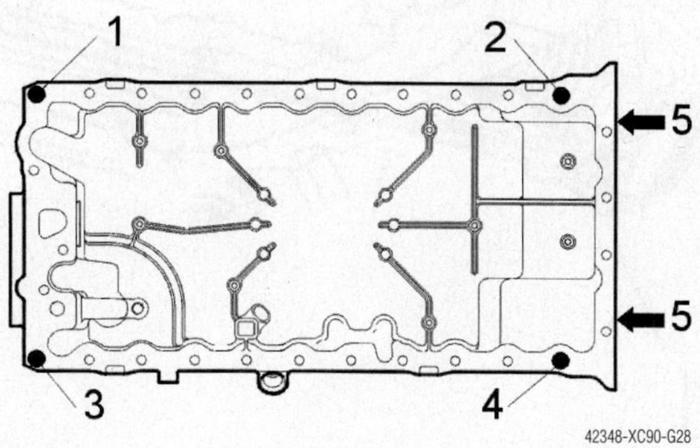

5-cylinder oil pan

Oil Pan

REMOVAL & INSTALLATION

5-Cylinder Engines

1. Before servicing the vehicle, refer to the Precautions Section.
2. Remove the oil dipstick and its pipe.
3. Remove the splashguard under the engine.
4. Drain the engine oil and remove the oil filter.
5. Release the oil cooler from the oil pan. Hang up at the rear.
6. Remove the bolt from the bracket for the fuel line.
7. Removing the oil pan
8. Back off all bolts holding the oil pan.
9. Remove all bolts except for four. It is recommended that the four bolts in the corners of the oil pan are left in place.
10. Carefully tap the oil pan with a rubber mallet until the joint and its liquid gasket releases.
11. Remove the four remaining bolts.
12. Remove the oil pan.

To install:

13. Apply liquid gasket 1161 059-9, or equivalent, to the oil pan.
14. Install new O-rings.
15. Position the oil pan. Secure it loosely with a few bolts.
16. Install the remaining bolts loosely.
17. Press the oil pan against the transmission. Tighten bolts 1, 2, 3 and 4 to 24 inch lbs. (3 Nm). Tighten bolts 5 to 18 ft. lbs.(25 Nm); then tighten to 35 ft. lbs. (48 Nm).
18. Tighten all bolts in the oil pan flange to 12 ft. lbs. (17 Nm). Start at the transmission end and continue forwards in pairs.
19. Install the bolt for the bracket for the fuel line.
20. Connect the oil cooler to the oil pan. Use new O-rings. Reinstall the pipe on the sub-frame.

21. Install a new oil filter.
22. Install the oil drain plug with a new gasket.
23. Install the oil dipstick and its pipe. Use a new O-ring.

➡Check that the O-ring is correctly positioned.

24. Fill with engine oil. Run the engine to operating temperature.
25. Check for oil leaks from the oil pan or oil cooler.
26. Install the splashguard under the engine.
27. Check the oil level. Top up if required.

6-Cylinder

1. Before servicing the vehicle, refer to the Precautions Section.
2. Remove the oil dipstick and pipe.
3. Remove:
   - The connector and hose from the crankcase ventilation terminal in the plastic intake pipe
   - The hose from the plastic charge air pipe
   - The bolts for the charge air pipe in the intake pipe
   - The hose clamps from the plastic pipe at the rear of the engine
   - The plastic pipes. If necessary, heat carefully using a hot air gun. Leave the pipes lying on the spark plug cover
   - The cover over the ignition coils
   - The 2 bolts at the engine mounting
4. Install fixture 999 5717.

➡Position the stand for lifting beam 999 5716 on the upper wheel arch members.

5. Then position the lifting beam directly above the engine lifting eyes on both sides.
6. Connect to the hole nearest the fire-wall. Connect lifting hook 999 5460.
7. Tighten to light contact.
8. Remove:
   - The lower engine cover
   - The front air baffle
9. Drain the engine oil. Remove the oil filter.
10. Remove:
    - The front wheel
    - The nuts for the lower ball joints
    - The anti-roll bar links from the anti-roll bar
    - The center bolt for the right-hand halfshaft
    - The bolt and nuts for the steering gear
    - The bolt for the front engine mounting
    - The torque rod including the mounting in the sub-frame
11. Tension the trailing arms together to release the spindles. Use a retaining strap.
12. Unhook the right halfshaft from the wheel spindle. Tension the spring strut backwards. Use a retaining strap.
13. Remove the bolt for the cable duct.
14. Remove:
    - The oil pipe for the steering gear from the snap fasteners along the sub-frame
    - The hose for the EVAP canister in the clips on the upper side of the sub-frame
    - Any fuel line for the engine block heater and any engine block heater if mounted on the left hand side of the sub-frame
    - The air duct from the clip
    - The bolt for the ground cable bracket
    - The connectors for the heated oxygen sensors in the clips at the rear of the sub-frame
15. Slacken off the subframe bolts. Lower the sub-frame slightly.
16. Remove the heated oxygen sensor bracket on the reverse of the sub-frame.

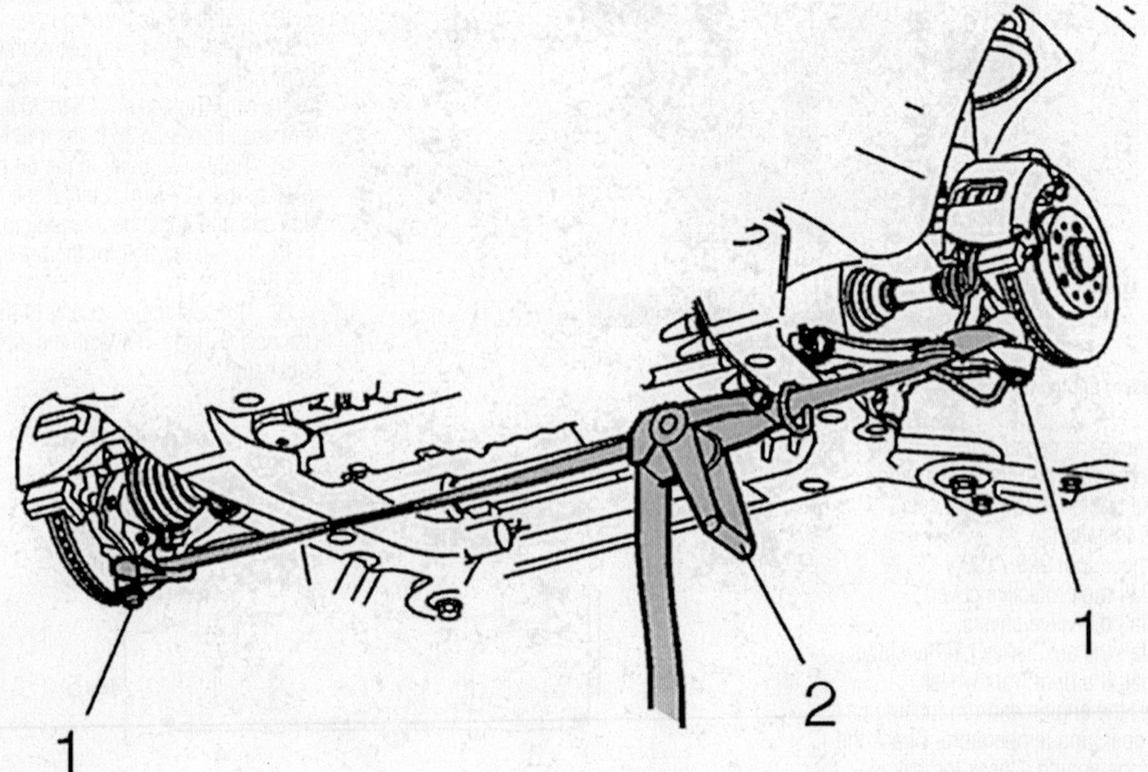

Tensioning the trailing arms

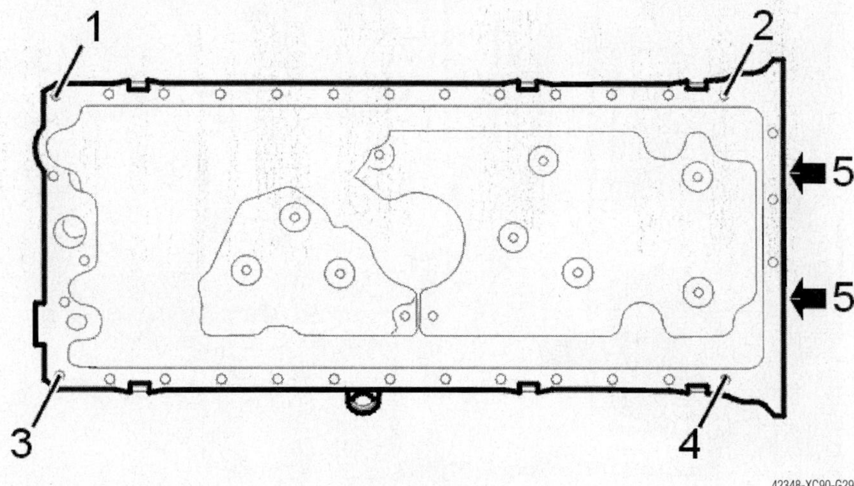

6-cylinder oil pan

17. Unbolt the bolts for the sub-frame approximately 2.5 cm on the right-hand side.

18. Fully remove the bolts on the left-hand side.

19. Angle the sub-frame down on the left-hand side.

20. Ensure that the bolts for the steering gear release from the frame.

21. Disconnect the oil cooler from the oil pan. Hang the oil cooler from a suitable place.

22. Remove the bolt from the bracket for the fuel line.

23. Remove the bolt for the bracket for the fuel line on the auxiliary bracket.

24. Slacken off all the bolts holding the oil pan.

25. Remove all but four bolts. It is recommended that the four bolts in the corners of the pan are left in place.

26. Carefully tap the pan using a rubber mallet until the joint and its liquid gasket releases.

27. Remove:
- The four remaining bolts
- The oil pan

28. Clean the gasket surfaces on the oil pan and cylinder block.

WARNING

Use a fume hood or extractor when using gasket solvent. Note! Also clean the gasket faces for the bolts for the torque rod.

29. Apply liquid gasket 116 1059-9 to the oil pan.

➡Note! Also apply liquid gasket around the holes for the bolts for the torque rod.

To install:

30. Install:
- new O-rings
- The oil pan. Secure the oil pan loosely using a few bolts
- The remaining bolts loosely

31. Press the oil pan against the transmission. Tighten the bolts (1), (2), (3) and (4). Tighten to 26 inch lbs. (3 Nm).

32. Tighten the bolts 5. First tighten to 18 ft. lbs. (25 Nm). Then tighten to 35 ft. lbs. (48 Nm).

33. Tighten all bolts in the oil pan flange. Tighten to 12 ft. lbs. (17 Nm). Start at the transmission and continue forwards in pairs.

34. Connect the oil cooler to the oil pan. Use new O-rings.

35. Install the bolt for the bracket for the fuel line.

36. Install:
- A new oil filter
- The oil plug with a new gasket

37. Lift the sub-frame. Use a mobile jack.

38. Install:
- The bracket for the heated oxygen sensors in the rear edge of the sub-frame
- New bolt in the sub-frame. Lubricate the bolts
- The washers at the front edge and the support plates at the rear edge of the sub-frame.

39. Tighten the bolts for the sub-frame. Tighten to 77 ft. lbs. (105 Nm). Angle-tighten 120°.

40. Start on the left hand side of the sub-frame. Continue with the right hand side.

41. Tighten the bolts for the brackets at the rear of the sub-frame. Tighten to 37 ft. lbs. (50 Nm).

42. Install:
- The bolt for the cable and air duct
- The oil pipe for the steering gear in the snap fasteners along the sub-frame
- The hose for the EVAP canister in the clip on the upper side of the sub-frame
- The engine block heater and the fuel line to the engine block heater if applicable
- The air duct in the clips
- The bolt for the ground cable bracket
- The connectors for the heated oxygen sensors in the clips at the rear of the sub-frame

43. Slacken off the retaining straps for the spring strut. Thread the halfshaft into the wheel spindle.

44. Carefully release the retaining strap for trailing arms. At the same time guide the lower ball joints into the trailing arms.

45. Install new nuts on the ball joints. Tighten to 58 ft. lbs. (80 Nm).

46. Counterhold using a Torx wrench so that the ball joint boot is not damaged.

47. Install the anti-roll bar links. Use new nuts. Tighten to 37 ft. lbs. (50 Nm).

48. Counterhold using a Torx wrench so that the boot is not damaged.

49. Install:
- The bolt and nuts for the steering gear. Use new nuts and new bolts. Tighten to 37 ft. lbs. (50 Nm)
- The front engine pad. Tighten to 37 ft. lbs. (50 Nm)
- The torque rod. Tighten the bolts through the oil pan. Tighten to 37 ft. lbs. (50 Nm). Tighten the nuts in the sub-frame. Tighten to 47 ft. lbs. (65 Nm)
- The center bolt for the halfshaft. Use a new bolt. Counterhold using a prybar in the brake disc vents. Tighten to 37 ft. lbs. (50 Nm)
- The front air baffle
- The front wheel
- The lifting beam
- The lifting hooks
- The lifting fixture from the engine. Install the bolts in the engine mounting
- The cover over the ignition coils
- The oil dipstick and pipe. Use a new O-ring. Check that the O-ring is correctly positioned
- The plastic pipes. Tighten the hose clamps
- The bolts for the charge air pipe in the intake pipe
- The connector and hose to the

crankcase ventilation terminal in the plastic intake pipe
- The hose to the plastic charge air pipe

50. Fill with new engine oil.
51. Start the engine and check for oil leakage.
52. Stop the engine. Give the oil time to run down into the oil pan. Then check the oil level.
53. Top up if necessary but do not overfill.
54. Wipe the engine compartment clean.
55. Install the lower engine cover.

8-Cylinder Engines

1. Drain the engine oil, remove the oil filter.
2. Remove the exhaust manifolds.
3. Remove the right-hand engine mounting.
4. Remove the 1 M10 bolts for the bracket.
5. Remove the 4 M10 bolts for the transmission.
6. Remove the oil cooler. Discard the O-ring.
7. Remove the 20 M6 bolts.
8. Remove the oil level sensor.
9. Remove the 17 x M8 bolts and the 1 x M6 bolt securing the intermediate section.
10. Carefully tap the side of the oil pan using a rubber mallet. Lift out the oil pan.
11. Discard the O-rings.

To install:

➡Clean the surface, using a razor blade or plastic scraper.

12. Apply liquid gasket 307 57050 as illustrated.

➡Minimum 1 mm bead width.

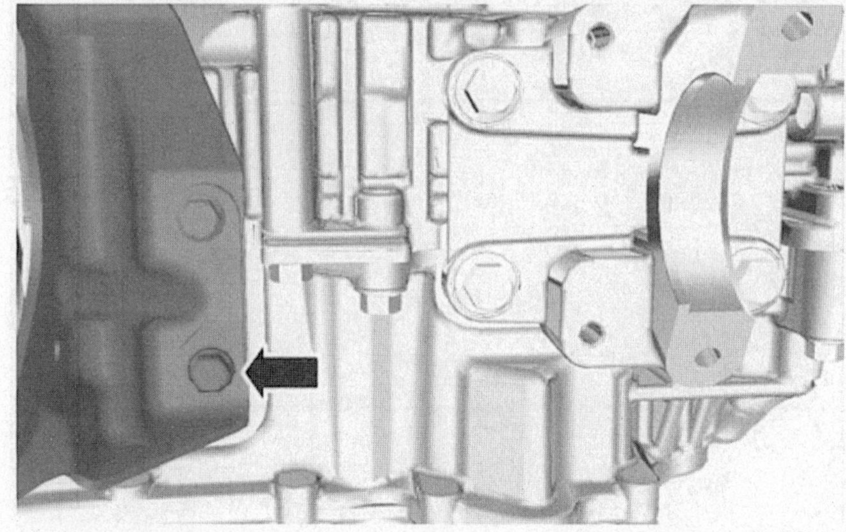

M10 bracket bolt—V8

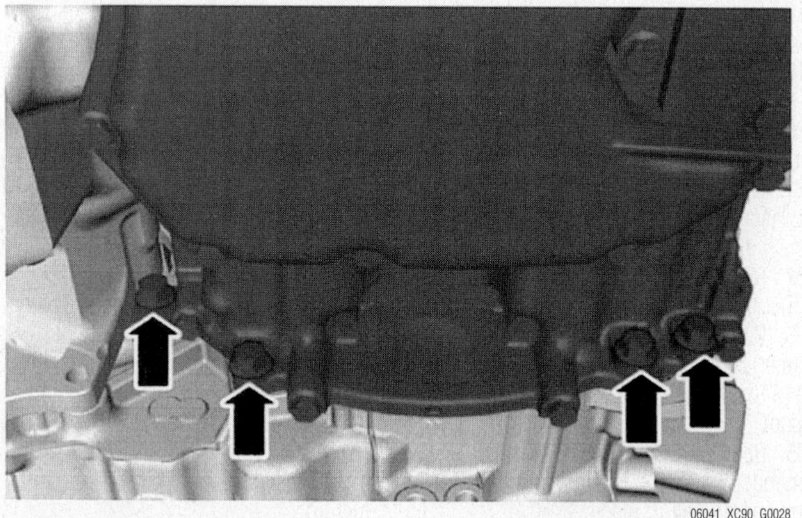

Transmission bolts—V8

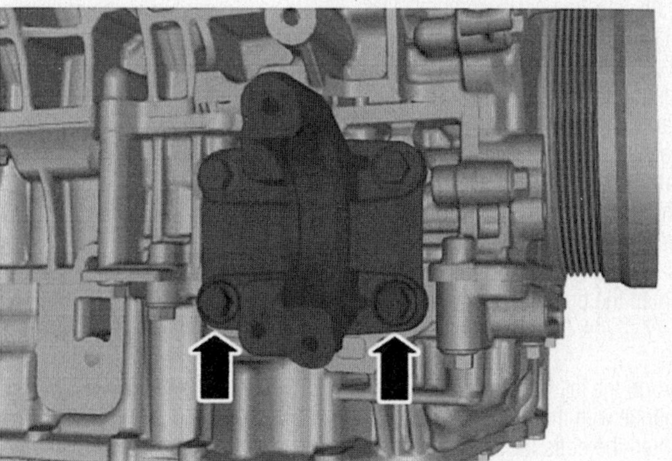

Right side engine mount—V8

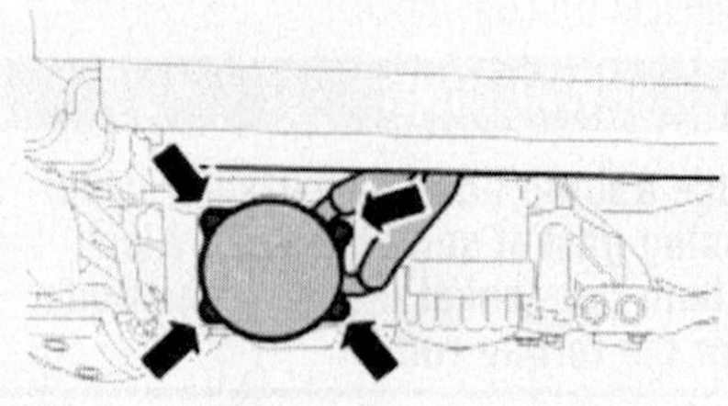

Oil cooler—V8

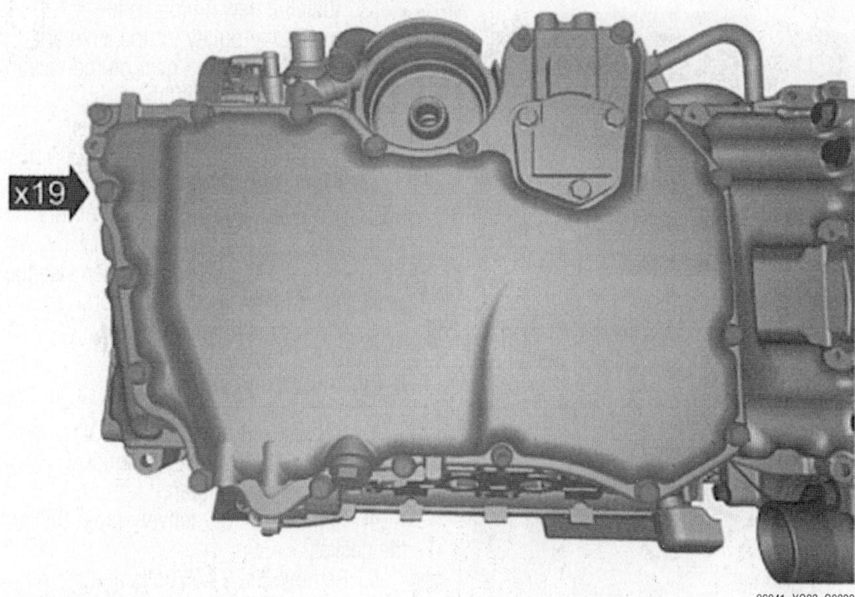

Oil pan bolts—V8

Intermediate section bolts—V8

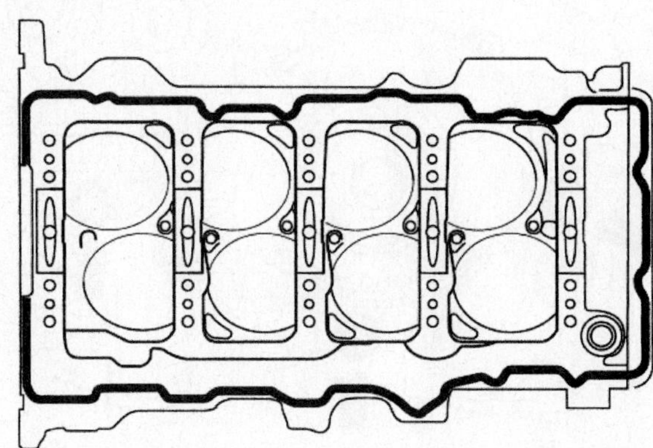

Liquid gasket application—V8

13. Install the oil pan within 5 minutes (curing time for liquid gasket).
14. Install the oil pan, use new O-rings
15. Install the 17 x M8 bolts.
16. Install the bolt securing the intermediate section.

➡**Clean the surface, using a razor blade or plastic scraper.**

17. Apply liquid gasket 307 57050 as illustrated.

➡**Minimum 1 mm bead width.**

18. Install the oil pan within 5 minutes (cure time for liquid gasket)
19. Install the 20 M6 bolts.
20. Install the oil level sensor.
21. Install the oil filter, using a new O-ring.
22. Install the oil cooler. Use new O-ring
23. The remainder of installation is the reverse of removal.
24. Top up oil.

Oil Pump

REMOVAL & INSTALLATION

5-Cylinder Engines

1. Before servicing the vehicle, refer to the Precautions Section.
2. Remove or disconnect the following:
   - Negative battery cable
   - Fuel line clips
   - Coolant recovery tank
   - Accessory drive belts
   - Right front wheel
   - Inner fender liner
   - Front cover
   - Timing belt
   - Crankshaft timing sprocket
   - Front crankshaft seal
   - Oil pump

To install:

3. Install the oil pump using special tool 999-5455. Use the oil pump bolts to guide the pump. Use the crankshaft nut to press the pump in until it is seated fully. Tighten the oil pump bolts to 84 inch lbs. (10 Nm).
4. Remove the crankshaft nut and the press tool.
5. Install or connect the following:
   - Front crankshaft seal
   - Crankshaft timing sprocket and tighten the nut to 133 ft. lbs. (180 Nm)
   - Timing belt
   - Front cover
   - Inner fender liner
   - Right front wheel

Oil pump installation

- Accessory drive belts
- Coolant recovery tank
- Fuel line clips
- Negative battery cable

6. Start the engine and check for leaks.

6-Cylinder Engines

1. Before servicing the vehicle, refer to the Precautions Section.
2. Remove the timing belt.
3. Check the tensioner and idler pulleys.
4. Carefully pull crankshaft timing gear pulley free. Remove the four oil pump bolts.
5. Carefully pry upward diagonally between the stop lugs and the cylinder block.
6. Lift out the oil pump.

To install:

7. Install a new gasket.
8. Carefully insert the oil pump over the end of the crankshaft.

➡The sealing ring in the oil pump is very easy to damage if not installed correctly.

9. Install four new bolts as a guide.
10. Pull in the oil pump with tool 999 5455, or equivalent, and the crankshaft center nut.
11. Tighten the oil pump bolt. Tighten to 84 inch lbs. (10 Nm).
12. Install the timing gear pulley. Carefully tap alternately around the timing gear pulley until reaching its end position.

➡The timing gear pulley can only be in one position on the crankshaft end splines.

13. Install a new timing belt.
14. Install the pulley for the auxiliary belt. Locate the steering gear on the locating pin in the timing gear pulley.
15. Install the auxiliary belt.
16. Install counterhold 999 5433.
17. Tighten crankshaft center nut. See the torque chart at the beginning of this chapter.
18. Install the vibration damper. Use the crankshaft center nut as a counterhold.
19. Check the engine oil level.

8-Cylinder Engines

1. Remove the timing chain.
2. Install the front engine bracket.
3. Remove the oil pan.
4. Remove the oil delivery pipe. Discard the gasket.
5. Remove the 3 M7 bolts and the oil pump. Discard the o-ring and the gasket for the oil delivery pipe.

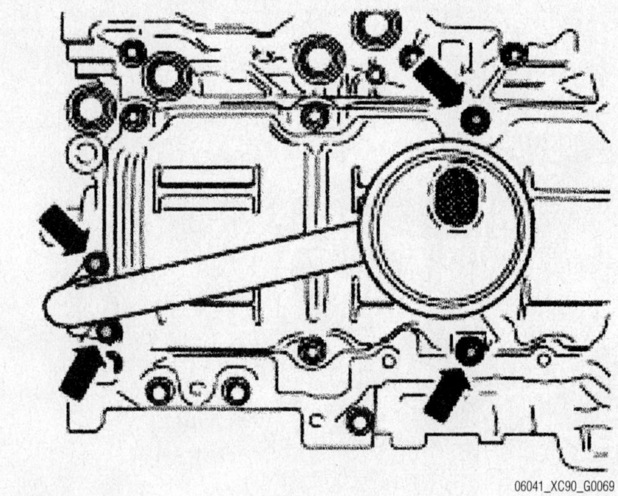

Oil delivery pipe—V8

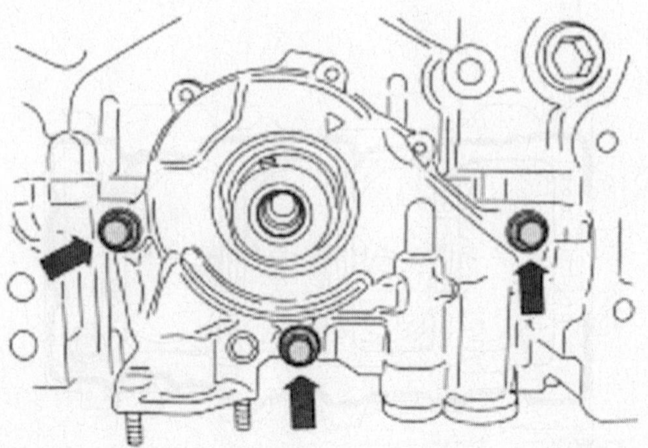

Oil pump mounting—V8

To install:

6. Install a new O-ring and the oil pump.
7. Install the oil delivery pipe using a new gasket.
8. Install the oil pan.
9. Install the timing chain.

Rear Main Seal

REMOVAL & INSTALLATION

5- and 6-Cylinder Engines

1. Before servicing the vehicle, refer to the Precautions Section.
2. Remove or disconnect the following:
   - Negative battery cable
   - Transmission
   - Flexplate
   - Rear main seal

To install:

3. Install the rear main seal so that it is flush with the cylinder block.
4. Install or connect the following:
   - Flexplate. Tighten the bolts to 33 ft. lbs. (45 Nm) plus 65 degrees.
   - Transmission
   - Negative battery cable
5. Start the engine and check for leaks.

8-Cylinder Engines

1. Remove the transmission according.
2. Lock the carrier plate using an M10 bolt.
3. Remove the carrier plate.

⁂ WARNING

Do not damage the mating surfaces of the crankshaft and cylinder block.

4. Remove the old sealing ring using puller 999 5651.
5. Clean around the crankshaft flange and bolt holes to remove any thread sealant residue. If necessary, wash and dry away any oil from the engine's rear engine block plane and the transmission clutch housing. Blow clean with compressed air.

⁂ WARNING

When cleaning work around the flange, not in and out. No grinding tools or similar may be used. It is extremely important that there is no thread sealant residue or other dirt. This could cause leakage.

➡To clean the crankshaft, use emery cloth such as 9511024.

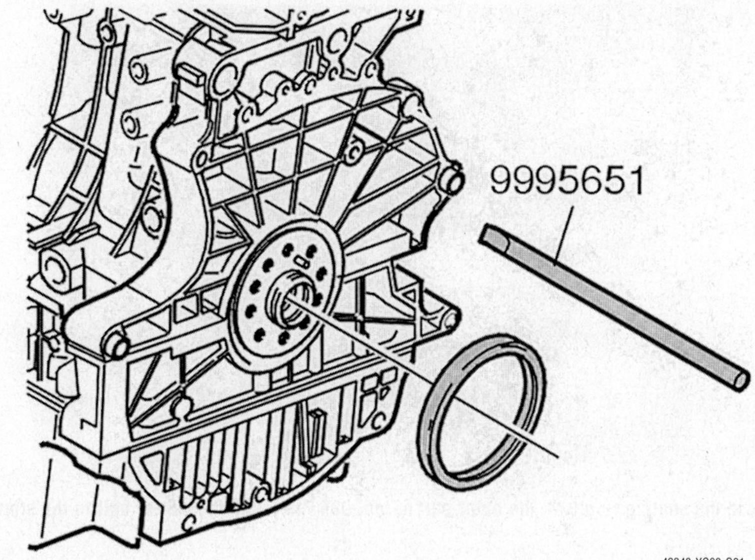

Rear main seal removal

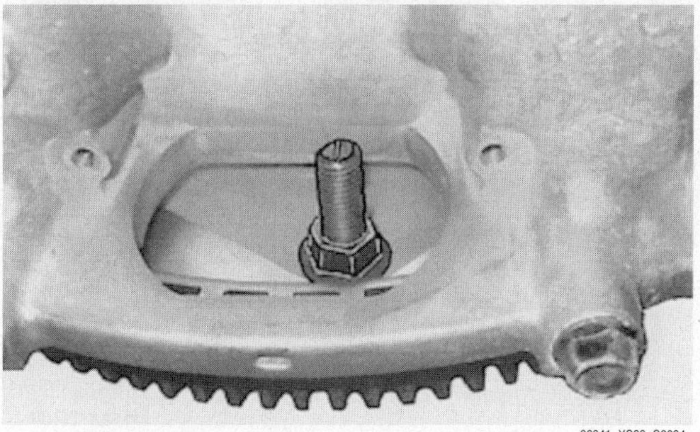

Lock the carrier plate using an M10 bolt—V8

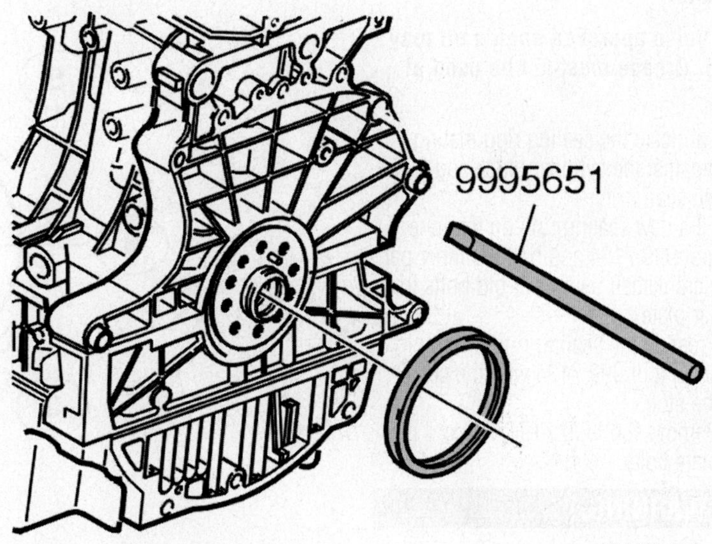

Remove the old sealing ring using puller 999 5651—V8

Press in the sealing ring with the outer part of tool 999 7194 with the center bolt to the stop—V8

Fit a new sealing ring on the inner part of tool 999 7194 and screw the inner part onto the crankshaft using two old bolts from the carrier plate—V8

To install:

➡Only Volvo approved engine oil may be used. Grease must not be used at all.

6. Lubricate the sealing ring mating surface against the cylinder block and the lips of the seal.
7. Fit a new sealing ring on the inner part of tool 999 7194 and bolt the inner part onto the crankshaft using two old bolts from the carrier plate.
8. Press in the sealing ring with the outer part of tool 999 7194 with the center bolt to the stop.
9. Remove tool 999 7194. Discard the carrier plate bolts.

⁂ WARNING

Ensure that the mating surfaces on the crankshaft and carrier plate are completely clean. The carrier plate must lie against the crankshaft flange before the bolts are installed. There is a risk that excess locking fluid on the bolts will come between the mating surfaces.

10. Brush off all the old locking fluid from the two old bolts. Use a wire brush. Mark the cleaned bolts using a marker pen.
11. Install the carrier plate, tighten it in towards the crankshaft with the two cleaned and marked bolts.

➡Check that the locating pin on the crankshaft is positioned opposite the hole in the carrier plate.

12. Remove the two old marked bolts.
13. Install the remaining new bolts.
14. Remove the locking bolt.
15. Install the transmission.

Piston and Ring

POSITIONING

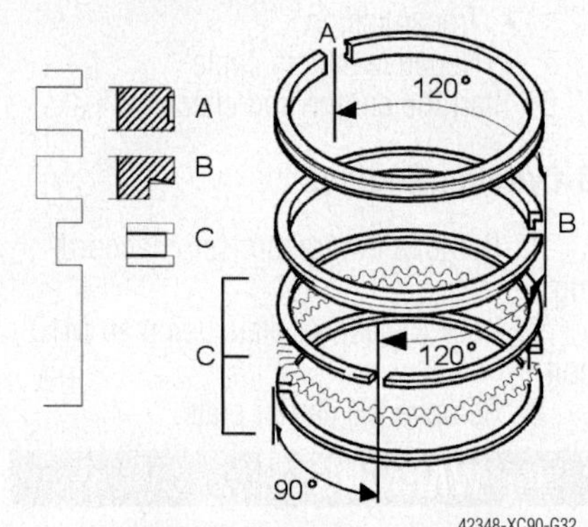

Ring identification and end-gap spacing—5- and 6-cylinder engines

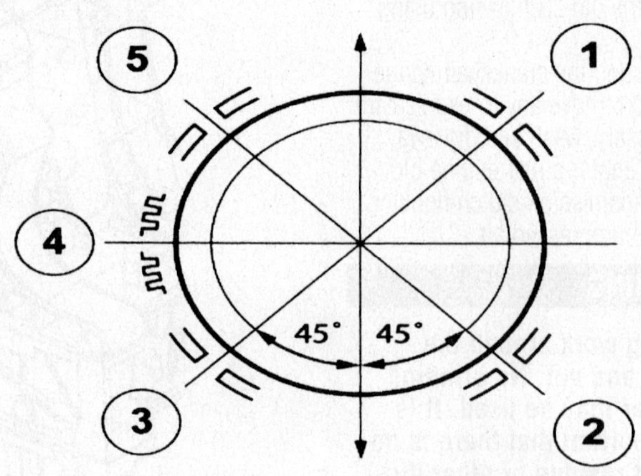

Ring identification and end-gap spacing—8-cylinder engines

FUEL SYSTEM

Fuel System Service Precautions

Safety is the most important factor when performing not only fuel system maintenance but any type of maintenance. Failure to conduct maintenance and repairs in a safe manner may result in serious personal injury or death. Maintenance and testing of the vehicle fuel system components can be accomplished safely and effectively by adhering to the following rules and guidelines.

- To avoid the possibility of fire and personal injury, always disconnect the negative battery cable unless the repair or test procedure requires that battery voltage be applied.
- Always relieve the fuel system pressure prior to disconnecting any fuel system component (injector, fuel rail, pressure regulator, etc.), fitting or fuel line connection. Exercise extreme caution whenever relieving fuel system pressure to avoid exposing skin, face and eyes to fuel spray. Please be advised that fuel under pressure may penetrate the skin or any part of the body that it contacts.
- Always place a shop towel or cloth around the fitting or connection prior to loosening to absorb any excess fuel due to spillage. Ensure that all fuel spillage (should it occur) is quickly removed from engine surfaces. Ensure that all fuel soaked cloths or towels are deposited into a suitable waste container.
- Always keep a dry chemical (Class B) fire extinguisher near the work area.
- Do not allow fuel spray or fuel vapors to come into contact with a spark or open flame.
- Always use a back-up wrench when loosening and tightening fuel line connection fittings. This will prevent unnecessary stress and torsion to fuel line piping. Always follow the proper tighten specifications.
- Always replace worn fuel fitting O-rings with new. Do not substitute fuel hose or equivalent, where fuel pipe is installed.

Fuel System Pressure

RELIEVING

5- and 6- Cylinder Engines

1. Before servicing the vehicle, refer to the Precautions Section.
2. Disconnect the negative battery cable.
3. Remove protective cap from the valve on the fuel rail.
4. Connect a hose to adapter 999-5484 and place the other end in a clean container.
5. Connect the adapter in the locked or closed position to the valve on the fuel rail.
6. Unlock or open the adapter valve.
7. After the fuel system pressure is relieved, remove the adapter and hose and replace the protective cap.
8. Connect the negative battery cable when repairs are complete.

8-Cylinder Engines

1. Remove the cover on the valve which is positioned on the fuel rail.
2. Connect adapter 999 5484 to fuel pressure relief unit 981 2270, 981 2273 and 981 2282.
3. Connect the adapter to the valve on the fuel rail in the locked position (figure 1, valve closed).
4. Unlock the adapter (illustration 2, valve open).
5. Raise the vehicle.

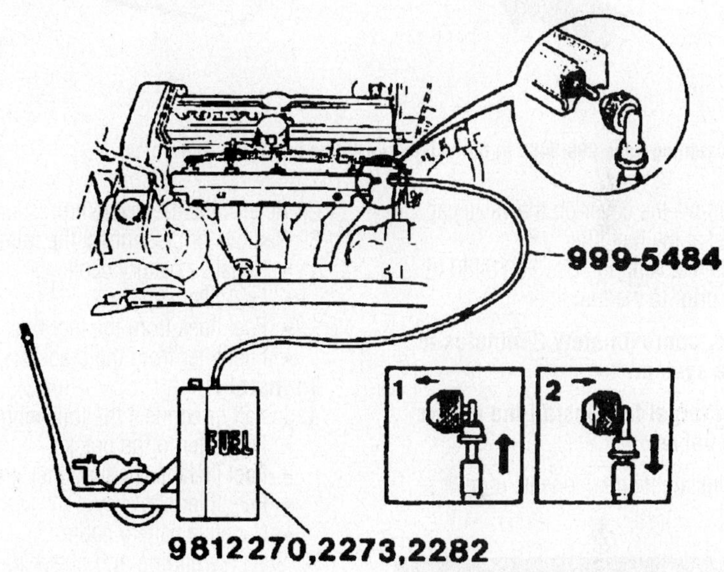

Connecting the adapter and fuel drainage unit—5-cylinder engine shown

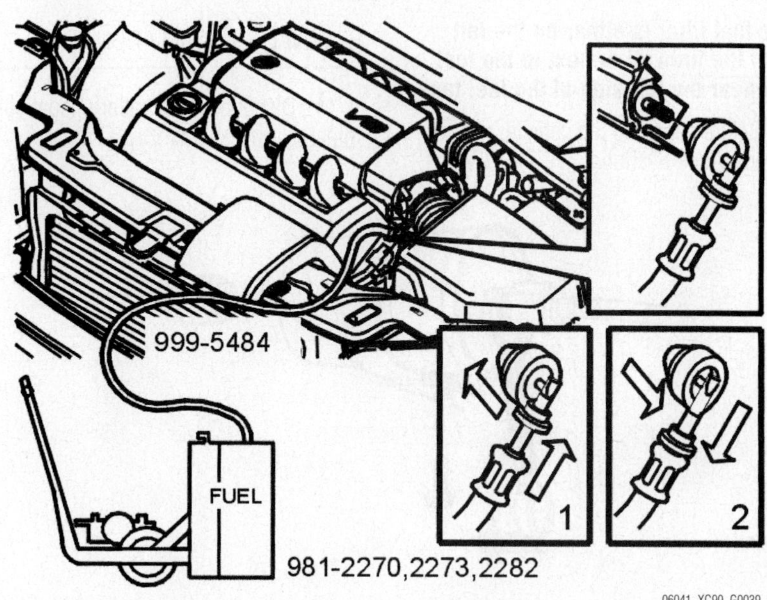

Connect adapter 999 5484 to fuel draining unit 981 2270, 981 2273 and 981 2282—V8

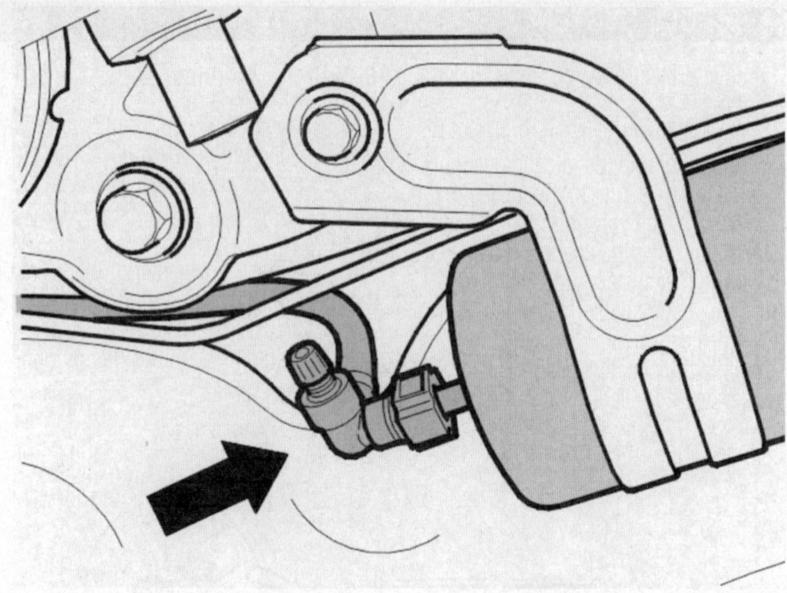

Connect venting hose 999 5480 to the valve prior to the fuel filter—V8

6. Remove the cover on the valve cap positioned at the fuel filter.

7. Connect venting hose 999 5480 to the valve prior to the fuel filter.

➡**It takes approximately 2 minutes to drain the system.**

➡**Do not forget to reinstall the covers over the valves.**

8. Reinstall the components in reverse order.

Fuel Filter

REMOVAL & INSTALLATION

➡**The fuel filter is either on the left side of the firewall or next to the fuel pump near the left side of the fuel tank.**

1. Before servicing the vehicle, refer to the Precautions Section.

2. Relieve the fuel system pressure.

3. Remove or disconnect the following:
   - Negative battery cable
   - Fuel filler cap
   - Fuel lines from the fuel filter
   - Fuel filter from the bracket

To install:

4. Install or connect the following:
   - Fuel filter to the bracket
   - Fuel lines to the fuel filter
   - Fuel filler cap
   - Negative battery cable

5. Start the engine and check for leaks.

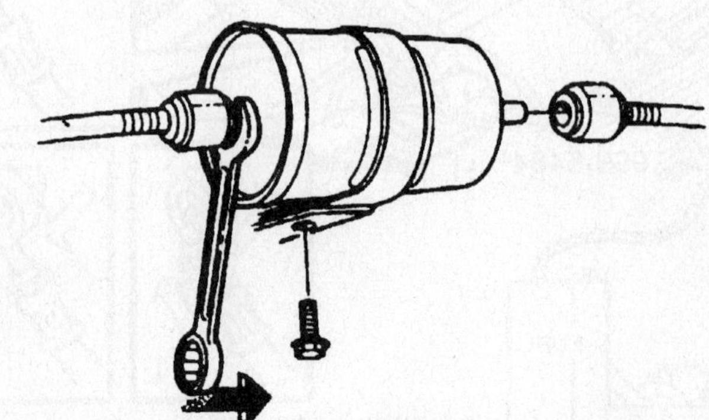

Using an open end wrench to push the quick disconnect coupler sleeves back—5 cylinder models shown

Fuel Pump

REMOVAL & INSTALLATION

1. Disconnect the battery negative terminal.

✲✲ WARNING

The use of a fresh air mask is recommended.

2. Insert the heavy duty hose 1.3 meters into the fuel filler pipe, measured from the edge of the opening of the filler pipe.
3. Pump until air comes out.
4. Remove the outer rear seats.
5. Remove the center rear seat.
6. Fold the carpet back.
7. Remove the ventilation pipe
8. Remove the insulation block
9. Remove the insulation block
10. Remove the 6 seat frame bolts
11. Remove the 2 seat frame bolts
12. Raise the rear edge of the seat frame approximately 25 mm. Position a block between the frame and the transmission tunnel.
13. Cover the area around the cover and between the doors with absorbent material.

✲✲ CAUTION

Ensure that there is no risk of dirt getting into the tank.

14. Remove the covers (1 and 2) over the fuel tank units.
15. Disconnect the level sensor connector (3).
16. Remove the hoses from the right-hand level sensor.
17. Clean the area around the fuel tank unit thoroughly. Open the right-hand level sensor. Insert the heavy duty hose and suck out the remains from the bottom of the right-hand side of the fuel tank.
18. Connect hose 999 5721 to the heavy duty hose. Suck out the remains from the bottom of the level sensor reservoir.
19. Carefully remove the overflow pipe from the reservoir.
20. Remove the thin hose 999 5721 from the heavy duty hose. Connect the thin hose to the overflow pipe. Use a hose clamp.
21. Pump until a lot of air comes out.
22. Open the left-hand level sensor. Insert the heavy duty hose and suck out the remains from the bottom of the left-hand side of the fuel tank.

To install:

23. Place the overflow pipe approximately half a centimeter to the left-hand side.
24. Grip around the right level sensor reservoir. Pinch the level sensor reservoir so that the float is held in its lowest position.

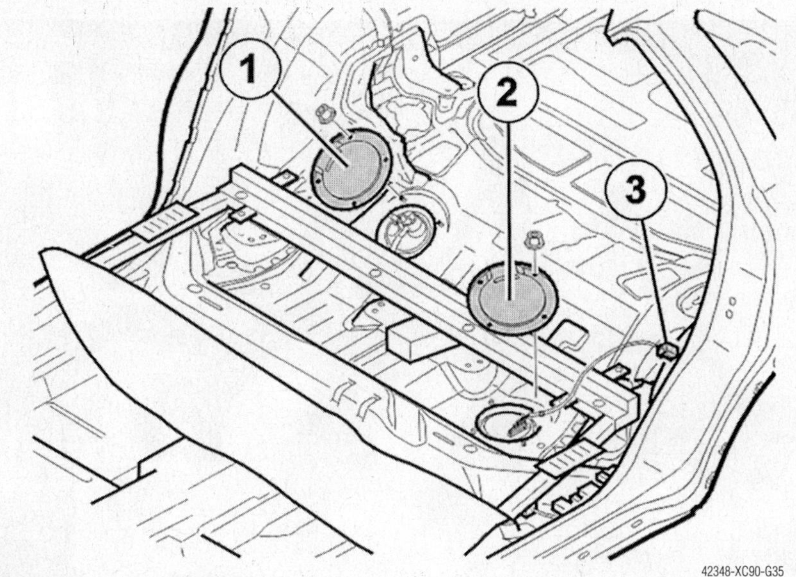

Accessing the level sensors

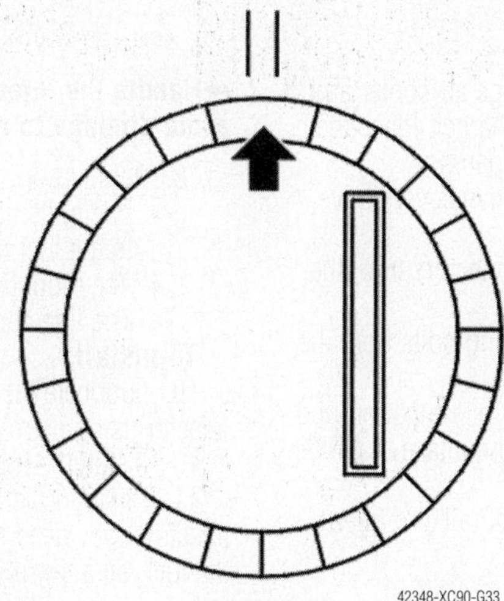

Left side sensor installation

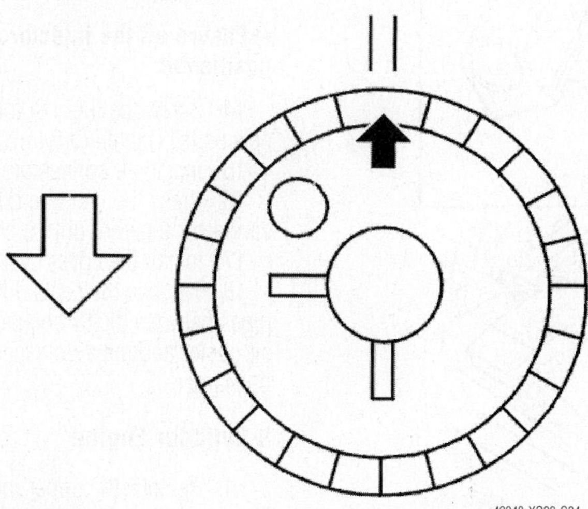

Right side sensor installation

25. Lower the reservoir carefully while rotating it backwards around the front-rear shaft on the float.

26. Install the overflow pipe on the reservoir.

27. On the left-hand level sensor:
   a. Install a new O-ring.
   b. Press the left level sensor down so that the row of protruding cables runs along the car. Check that the O-ring is not trapped.
   c. Tighten the bolt. Tighten to 44 ft. lbs. (60 Nm).
   d. Position the wiring in the cut-out in the rubber ring

➡Check that the arrow on the cover is between the markings on the tank. If the arrow is outside the markings the fuel gauge sensor will give an incorrect value and the float may catch against the inner wall of the tank. If not: Slacken off the bolt and adjust the position of the sensor. Tighten as above.

28. On the right-hand level sensor:
   a. Install a new O-ring.
   b. Press the right-hand level sensor down so that the fuel line connections are pointing forwards and to the right.
   c. Check that the O-ring is not trapped.
   d. Tighten the bolt. Tighten to 44 ft. lbs. (60 Nm).
   e. Install the hoses on the level sensor.
   f. Position the wiring in the cut-out in the rubber ring.

➡Check that the arrow on the cover is between the markings on the tank. If the arrow is outside the markings the fuel gauge sensor will give an incorrect value and the float may catch against the inner wall of the tank. If not: Slacken off the bolt and adjust the position of the sensor. Tighten as above.

29. Install the covers over the level sensors. Tighten the bolts.

30. Connect the wiring and the connectors.

31. Remove the block from under the seat frame

32. Install the 8 seat frame bolts. Tighten to 37 ft. lbs. (50 Nm).

33. Install the 2 seat frame bolts. Tighten to 18 ft. lbs. (24 Nm).

34. Fold the carpet down

35. Install the center rear seat.

36. Install the outer rear seat.

Fuel Injector

REMOVAL & INSTALLATION

5-Cylinder Engines

1. Before servicing the vehicle, refer to the Precautions Section.
2. Remove or disconnect the following:
   - Negative battery cable
   - Injector cover
   - Accelerator cable
3. Install fuel injector holders on the injectors.
   - Fuel pressure regulator vacuum hose
   - Fuel supply manifold with injectors attached
4. Remove the injector holders and separate the injectors from the fuel supply manifold.

To install:

5. Install or connect the following:
   - Injectors to the fuel supply manifold with new O-ring seals
   - Injector holders
   - Fuel supply manifold with injectors attached. Remove the injector holders.
   - Fuel pressure regulator vacuum hose
   - Accelerator cable
   - Injector cover
   - Negative battery cable
6. Start the engine and check for leaks.

6-Cylinder Engines

1. Drain the fuel injection system.
2. Remove the plastic hoses between the turbocharger and charge air cooler and the air cleaner and turbocharger. Place them to one side and seal the openings.
3. Remove the protective cover above the nozzles
4. Disconnect the connectors from the injectors
5. Remove the mounting bolts from the fuel rail
6. Spray universal oil or similar around the injector terminals on the intake manifold.
7. Gently work the fuel rail and injectors loose.

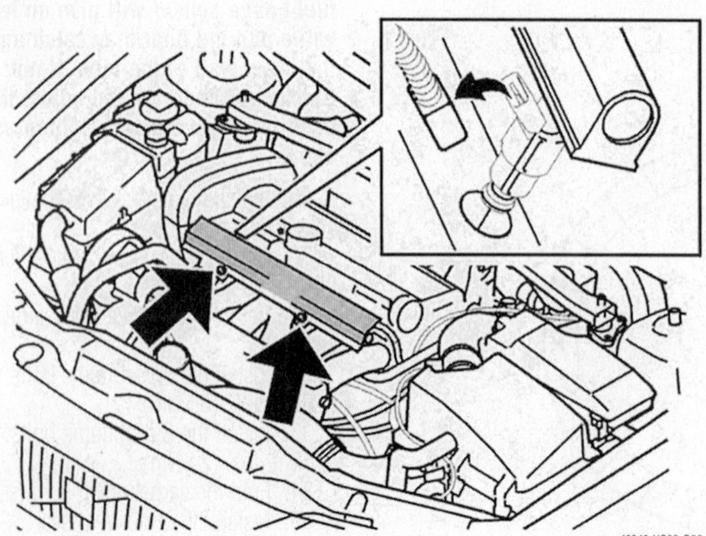

Injector removal

➡Handle the injectors carefully to avoid damage to nozzles and needles.

8. Remove:
   - The bolts holding the mounting rail to the fuel rail
   - The mounting rail
   - The injector.

To install:

9. Lubricate the O-ring for the new injector using petroleum jelly.
10. Install the injector.
11. Position the mounting rail. The injectors may need to be pulled away from the fuel rail a few millimeters to install the rail
12. Tighten the mounting rail to the fuel rail
13. Press down the fuel rail.

➡Ensure all the injectors are correctly positioned.

14. Screw the fuel rail into place. Use new bolts. Tighten to 84 inch lbs. (10 Nm)
15. Install all connectors on the injectors
16. Press the fuel line quick-release connector together until it clicks
17. Install the cover over the nozzles
18. Remove the seals. Install the plastic pipes between the turbocharger and charge air cooler and the air cleaner and turbocharger.

8-Cylinder Engine

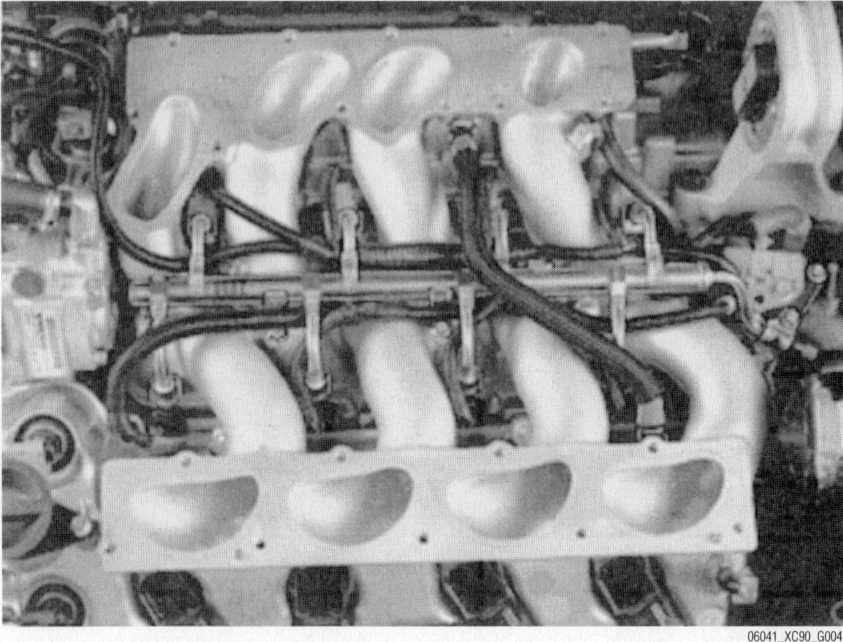

Fuel rail assembly—V8

1. Remove the upper intake manifold. See the Intake Manifold R&I procedures.
2. Remove the crankcase ventilation

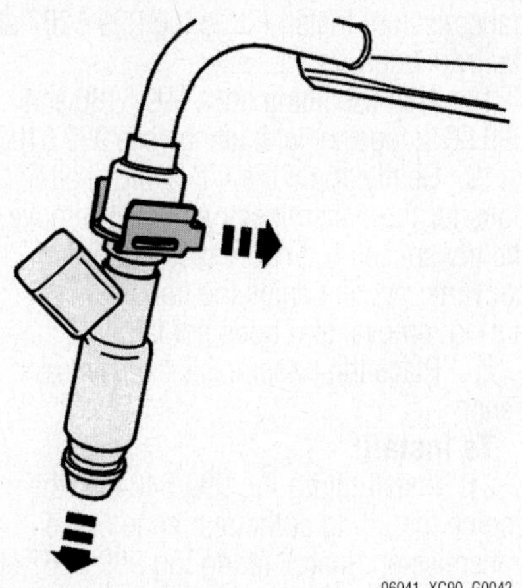

Injector removal—V8

hose between the cylinder heads by releasing the 2 clamps.

3. Remove the 3 bolts from the fuel rail
4. Remove the 2 bolts on the fuel pressure sensor.
5. Lift the fuel rail assembly with injector straight up and bend it aside.
6. Unplug the connector of the injector.
7. Remove the circlip of the injector; pull the clip laterally.
8. Remove the injector; jiggle and pull it straight out.

To install:

➡Lubricate the O-rings on the injectors with Vaseline or the like.

9. Install the injector.
10. Install the circlip.
11. Plug in the connector to the injector.
12. Install the fuel rail assembly with injector.
13. Install the 3 bolts of the fuel rail.
14. Install the 2 bolts on the fuel pressure sensor.
15. Install the crankcase ventilation hose and 2 clamps.

✻✻ WARNING

Make sure there is an audible click during connection.

16. Connect the fuel supply line and secure the line in the clamp.
17. Install the upper intake manifold.

DRIVE TRAIN

Transaxle Assembly

REMOVAL & INSTALLATION

5-Cylinder Engine

1. Before servicing the vehicle, refer to the Precautions Section.
2. Drain the transaxle fluid.
3. Attach a support fixture to the engine lifting eyes.
4. Remove or disconnect the following:
   - Battery and tray
   - Air intake assembly
   - Turbo control valve
   - Turbocharger inlet tube
   - Gear select cable
   - Transaxle harness connector
   - Transaxle ground cable
   - Heated Oxygen (HO_2S) sensor connector
   - Transaxle oil cooler lines
   - Transaxle dipstick tube
   - Exhaust Gas Recirculation (EGR) valve hoses, if equipped
   - Starter motor
   - Coolant recovery tank
   - Torque rod extension arm
   - Front wheels
   - Wheel speed sensors
   - Brake line brackets
   - Wheel speed sensor wiring brackets
   - Inner fender liners
   - Axle halfshafts
   - Transfer case, if equipped
   - Splash guards
   - Lower ball joints
   - Stabilizer bar links
   - Evaporative Emissions (EVAP) canister and hoses
   - Exhaust front pipe
   - Oil line bracket
   - Steering gear engine mount
   - Steering gear mounting nuts
   - Vehicle Speed (VSS) sensor connector
   - Transaxle mount
   - Torque converter
5. Loosen the 2 right side subframe-to-body bolts approximately ½ inch. Support the subframe with a jack and remove the subframe-to-body bolts on the left side.
6. Lower the jack and let the frame hang down from the right side bolts.
7. Tie the left side of the steering gear to the left side frame rail for support. Remove the steering gear engine mount.
8. Lower the engine and transaxle with the lifting hook.
9. Install transaxle fixture 5463 on the transaxle jack, using the torque rod mounting bolts to hold it in place. At the same time, fit tool 5463-1 support plate and raise the jack so that it is making light contact.
10. Remove the transaxle flange bolts.
11. Remove the transaxle.

To install:

➡Use new fasteners where indicated.

12. Install or connect the following:
    - Transaxle and tighten the bolts to 37 ft. lbs. (50 Nm)
    - Torque converter. Use new bolts and tighten them to 22 ft. lbs. (30 Nm).
    - Rear transaxle mount and torque the bolts to 37 ft. lbs. (50 Nm)
    - HO_2S sensor connector
    - VSS sensor connector
13. Install the subframe using new bolts. Starting on the left side, lift the frame with a jack. Mount the support brackets on both

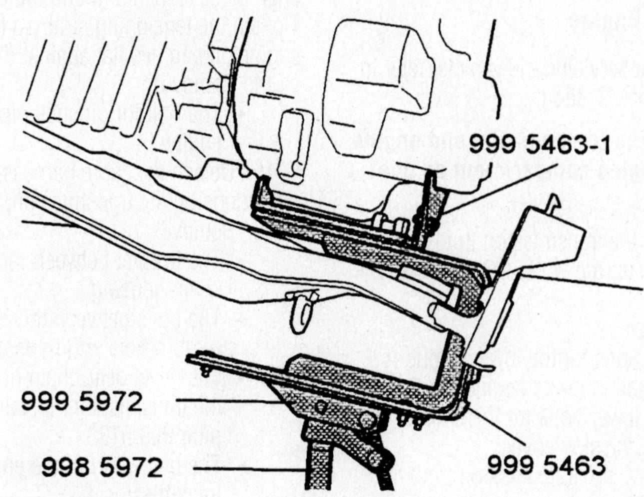

Using the transaxle fixture

sides. Tighten the frame bolts to 78 ft. lbs. (105 Nm) plus 120 degrees. Tighten the bracket bolts to 37 ft. lbs. (50 Nm). Repeat the procedure for the right side. Remove the subframe jack.

14. Install or connect the following:
   - Steering gear. Use new mounting nuts and tighten them to 37 ft. lbs. (50 Nm).
   - Steering gear engine mount and tighten the bolts to 37 ft. lbs. (50 Nm)
   - Oil line bracket
   - Torque rod extension arm and tighten the bolts to 26 ft. lbs. (35 Nm) plus 40 degrees
   - Exhaust front pipe
   - EVAP canister and hoses
   - Stabilizer bar links
   - Lower ball joints
   - Splash guards
   - Transfer case, if equipped
   - Axle halfshafts
   - Inner fender liners
   - Wheel speed sensor wiring brackets
   - Brake line brackets
   - Wheel speed sensors
   - Front wheels
   - Coolant recovery tank
   - Starter motor
   - EGR valve hoses, if equipped
   - Transaxle dipstick tube
   - Transaxle oil cooler lines
   - Transaxle ground cable
   - Transaxle harness connector
   - Gear select cable
   - Turbocharger inlet tube
   - Turbo control valve
   - Air intake assembly
   - Battery and tray

15. Fill the transaxle to the correct level.

6-Cylinder Engine

1. Before servicing the vehicle, refer to the Precautions Section.

➡Remove the transmission and engine from the engine compartment as one unit.

2. Drain the transmission fluid at a suitable time during work. Use an oil suction unit.
3. Remove:
   - The bolts for the lower torque rod mounting in the engine
   - The lower bolts for the engine transmission cover.
4. Remove the transmission from the engine. See the Engine removal instructions earlier in this chapter.
5. Remove:
   - The halfshafts. Install sealing plug on the transmission. Install a plug on the chain housing
   - The bolt for the engine pad and steering gear
   - The cable harness from the bracket on the steering gear.
   - The bolt for the engine pad/subframe
   - The bolts for the cable duct/subframe.
6. Detach the starter motor from the transmission cover and the engine to access the carrier plate/torque converter bolts.
7. Remove:
   - The torque converter bolts.
   - The vibration damper. Use the crankshaft nut as a counterhold
   - The bolts between the engine mounting in the engine and the sub-frame
   - The drive belt
   - The power steering pump from the engine.
8. Move the power steering pump with hoses and fluid reservoir to one side.
9. Install lifting lug 999 7018, or equivalent, at the rear edge of the engine. Use lifting lugs 999 5185 and 999 5186 together with lifting lug 999 2810.
10. Attach the lifting yoke and adjust for optimum balance. Raise the engine/transmission from the sub-frame.
11. Place the unit on a bench with a strong top surface and a stable base.
12. Place wooden or hard rubber blocks, approximately 30 mm thick, between the bench top and the entire flat surface of the engine oil pan.
13. Place a tensioning strap between the engine intake manifold and the bench. Tighten the tensioning strap so that the engine oil pan lies flat against the blocks.
14. Remove:
   - The bolt for dipstick pipe in the engine.
15. Detach the cable harness from the mountings on the transmission.
16. Remove:
   - The bracket between the engine and chain housing
   - The bolts between the chain housing/brackets and transmission
   - The bevel gear/chain housing from the transmission. Install sealing plug 999 5733
   - The bracket from the end of the transmission.
17. Install lifting lug 999 5464 on the engine mounting at the rear edge of the transmission. Install lifting lug 999 5267 at the front edge.
18. Also use lifting hook 999 5186 and 999 5642 together with lifting yoke 999 5100.
19. Gently apply the lift. Remove all bolts for the transmission/engine. Remove the transmission. Ensure that the torque converter accompanies the transmission during removal and does not fall out.
20. Place the transmission on a workbench.

To install:

21. Install lifting lug 999 5464 on the engine mounting at the rear edge of the transmission. Install lifting lug 999 5267 at the front edge.
22. Also use lifting hook 999 5186 and 999 5642 together with lifting yoke 999 5100.
23. Ensure that the torque converter does not slide from its axle.
24. Adjust the transmission against the guide sleeves on the engine. Install the 7 M10 bolts. Tighten alternately to a light contact. Tighten. (2 more bolts will be installed and tightened later.)
25. Remove the lifting tools from the transmission.
26. Install:
   - The bracket on the end of the transmission. Tighten to 37 ft. lbs. (50 Nm)
   - The bolt for the upper bracket. Tighten to ft. lbs. (50 Nm)
27. Tighten the bolts for the bracket in the cylinder block: 4 M10 bolts. Tighten to ft. lbs. (50 Nm). 1 M8 bolt. Tighten to 18 ft. lbs. (25 Nm).
28. Install:
   - The bevel gear/chain housing on the transmission. Lightly tighten the 2 M10 bolts alternately. Tighten
   - The bracket between the engine and the chain housing. Lightly tighten the 4 M10 bolts to the cylinder block
   - The 2 M10 bolts to the chain housing/bracket. Tighten.
29. Tighten the 4 M10 bolts for the bracket/cylinder block. Tighten.
30. Install:
   - The 1 M8 bolt for the dip stick pipe in the engine. Tighten
   - The cable harness to the mountings on the transmission.
31. Install lifting lug 999 7018 at the rear edge of the engine. Use lifting lugs 999 5185 and 999 5186 together with lifting lug 999 2810 .
32. Apply the lifting yoke and adjust for optimum balance. Raise the engine/transmission to the sub-frame.

33. Install:
- The 2 M10 bolts between the engine mounting in the engine and the sub-frame. Use new bolts.
- The drive belt
- The power steering pump on the engine. Install the 3 M8 bolts. Tighten
- The vibration damper using 4 M10 new bolts. Tighten. Use the crankshaft nut as a counterhold.

34. Install:
- The 6 M8 torque converter bolts. Use new bolts.

35. Loosely insert all bolts. Tighten the bolts until the heads are in contact with the carrier plate.

36. Use tool 999 5734 to turn the engine and as a counterhold.

37. Install:
- 1 M10 bolt for the engine pad/subframe.
- The bolt for the cable duct/sub-frame
- The starter motor on the transmission cover and engine. Install the 2 M10 bolts in the transmission cover. Install 1 M8 bolt in the bracket. Tighten.
- The halfshafts. Check that the snap rings for the halfshafts are in the grooves by pulling the inner constant velocity joint
- 1 M10 bolt in the steering gear engine pad. Tighten
- The cable harness to the bracket on the steering gear.

38. Check the oil level in the chain housing.

39. Tighten the level plug.

40. Check the oil level in the bevel gear.

41. Tighten the level plug.

42. Top up the oil in the transmission via the dip stick pipe.

43. Install:
- The 4 M10 bolts for the lower torque rod mounting in the engine. Tighten
- The 2 M10 lower bolts for the engine transmission cover. Tighten.

44. To install the engine/transmission, see the Engine instructions earlier in this chapter.

45. Check the oil level.

Final Drive

REMOVAL & INSTALLATION

1. Remove the halfshafts on both sides.

2. Remove the rear section of the exhaust system and the spare wheel.

3. Support the rear edge of the three-way catalytic converter so as not to put strain on the exhaust manifold.

4. Remove the propeller shaft from the final drive. Remove the member for the center bearing.

5. Tie up the shaft.

6. Use the rear cross member as support.

7. Remove one bolt and slacken off the other. Turn out the member.

8. Remove the connector from the Differential Electronic Module.

9. Detach the cable harness from the final drive housing and the mounting in the rear axle member.

10. Fold the cable harness to one side to prevent damage when removing the final drive.

11. Remove:
- The cover for the rear axle member
- The bolt for the front mounting for the final drive housing.

12. Slacken off the 2 rear bolts for the rear axle member/final drive housing. Leave the bolts in place by a few threads.

13. Bring mobile jack 998 5972 into light contact with the front section of the rear axle member.

14. Remove the bolts from the front mounting in the body for the rear axle member, on both sides.

15. Carefully lower the frame. Remove the jack.

16. Support the final drive housing and remove the 2 loose bolts.

17. Lift out the final drive housing from the rear axle member. Position the final drive housing on a workbench.

To install:

18. Install the final drive housing in the rear axle member. Loosely secure the final drive housing to the rear bushings in the frame using bolts.

19. Bring mobile jack 998 5972 into light contact with the front section of the rear axle member.

20. Install the bolts in the front mounting for the rear axle member in the body on both sides.

21. Install the 2 M12 bolts for the bracket/body. Tighten.

22. Install the 1 M12 bolt for the bracket/rear axle member/body.

23. Remove the mobile jack.

24. Install:
- 1 M12 front bolt for the final drive housing. Tighten.
- The cover for the rear axle member using 4 M12 bolts. Tighten.
- The 2 rear M12 bolts for the sub-frame/final drive housing. Tighten.

25. Connect the connector to the control module for the differential electronic module (DEM). Install the cable harness in the mounting in the rear axle member.

26. Install the propeller shaft on the final drive.

27. Install the member for the center bearing.

28. Install the halfshafts on both sides. Check that the gasket for the flanged joint

42348-XC90-G37

Removing the final drive

between the front and rear exhaust pipe is not damaged. Replace if necessary.

29. Install the rear cross member. Use 2 x M8 bolts. Tighten.

30. Fill the Active on Demand Coupling with oil. Check the oil level. The front level plug is for the Active on Demand Coupling.

31. Fill the final drive with oil. Check the oil level. The rear level plug is for the final drive.

32. Tighten both level plugs.

Halfshaft

REMOVAL & INSTALLATION

Front

See the appropriate steps under "Transmission Removal & Installation".

Rear

5- & 6-CYLINDER MODELS

1. Raise the car.
2. Remove the wheel.
3. Push up the rear suspension using tensioner tool 999 5659, or equivalent, 500 mm (19.7 in.) between the fender edge and the center of the wheel hub.
4. Remove:
   - The parking brake cable from the mounting in the tie rod
   - The bolt for the halfshaft/wheel hub (1)
   - The bolt for the tie rod mounting (2)
   - The bolt for the control arm mounting (3)
   - The nut for the anti-roll bar link mounting (4). Counterhold using a Torx wrench so that the boot does not rotate.
   - The bolt for the lateral link mounting (5).
5. Install a floor jack the wheel spindle and raise carefully.

➡Pull out the halfshaft from the wheel hub whilst raising the wheel spindle. Angle the shaft past the wheel spindle.

6. Release the tensioner so that the control arm hangs in the shock absorber.
7. Using a prybar, position the tip between the constant velocity joint housing and the final drive housing.
8. Tap firmly so that the snap ring for the halfshaft releases. Pull out the halfshaft. Do not damage the halfshaft seal.
9. Install sealing plug.
10. Lubricate the halfshaft mating surface using P/N 116 1329-6, or equivalent.

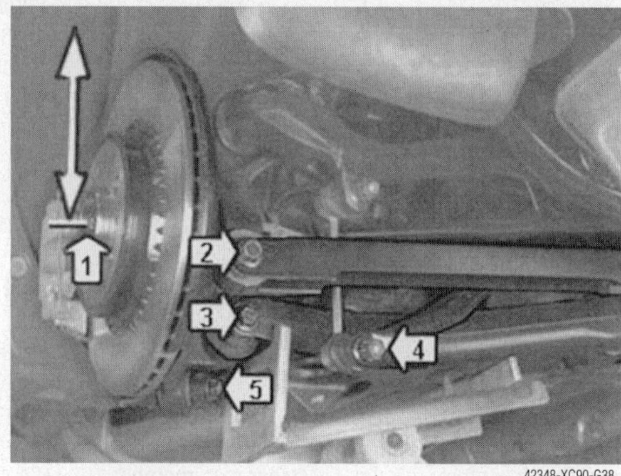

Remove these components for rear halfshaft removal—5- & 6-cylinder models

11. Press the halfshaft into the final drive housing. Check that the snap ring is in the groove in the final drive.
12. Check by pulling the constant velocity joint on the halfshaft.
13. Lubricate the halfshaft splines using a small amount of oil.
14. Position the halfshaft in the wheel hub. Carefully lower the mobile jack.
15. Remove the mobile jack.
16. Press up the control arm and tensioner to the previous specified measurement so that the bolts can be installed.

➡After installing the bolts, press tensioner 999 5659, or equivalent, up to the normal position. Tighten the bolts.

To install:

17. Install:
   - 1 M12 bolt for the lateral link mounting (5). Tighten
   - 1 M12 nut for the lateral link mounting (4). Tighten. Counterhold using a Torx wrench so that the boot does not rotate
   - 1 M12 bolt for the lateral link mounting (3). Tighten to 37 ft. lbs. (50 Nm).
   - 1 M12 bolt for the tie rod mounting (2). Tighten.
   - 1 M10 bolt to the halfshaft/wheel hub (1). Tighten to 26 ft. lbs. (35 Nm), plus 90 degrees.
   - The parking brake cable and mounting for the tie rod.
18. Remove the tensioner.
19. Install the wheel.

Driveshaft

REMOVAL & INSTALLATION

5- & 6-Cylinder Engines

1. Remove:
   - The front cross member from the body. Remove the brake pipe from the mountings in the member
   - The rear cross member
   - The rear exhaust pipe. Support the

Removing the front driveshaft bolts

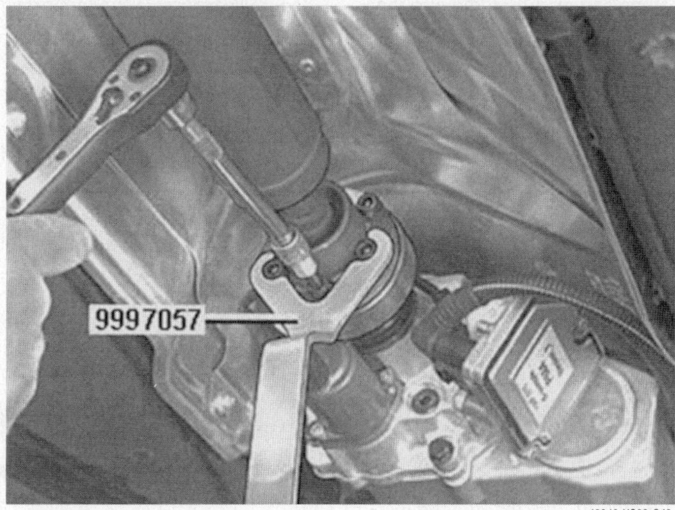

Removing rear driveshaft bolts—5-cyl. engines without vibration damper

Removing rear driveshaft bolts—5-cyl. engines with vibration damper

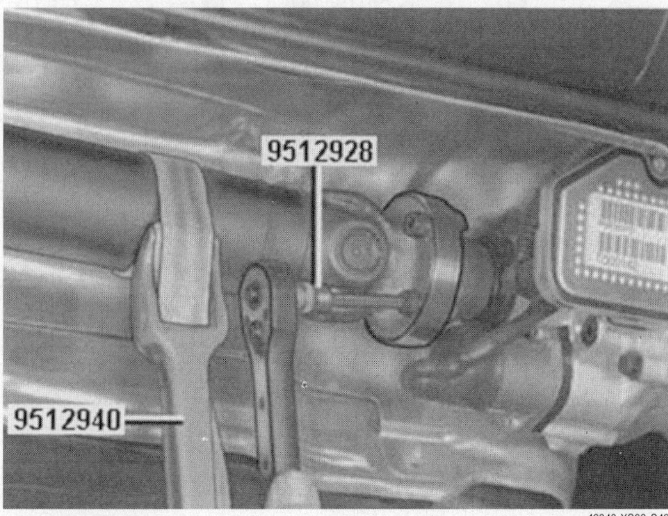

Removing rear driveshaft bolts—6-cyl. engines

front section of the exhaust pipe/three-way catalytic converter (TWC) so as not to put strain on the exhaust manifold.

2. Remove the bolts from the propeller shaft.

3. The following applies only to cars with 5-cylinder engines.

a. Remove the bolts from the propeller shaft. Use a counterhold.

b. Leave one bolt in place by a few threads for safety.

4. The following applies to cars with 5-cylinder engines and with a vibration damper on the flange for the Active on Demand Coupling.

a. Remove the bolts from the propeller shaft. Use a counterhold.

b. Leave one bolt in place by a few threads for safety.

5. The following applies only to cars with 6-cylinder engines.

a. Remove the bolts from the propeller shaft. Use a counterhold and socket.

b. Leave one bolt in place by a few threads for safety.

6. Remove:

- The bolts for the bearing for the propeller shaft in the center bearing member
- The bolts for the center bearing member. Let the member rest on the exhaust system.

7. Press the propeller shaft joints together. Detach the front joint from the flange for the bevel gear and move it to the side.

8. Remove:

- The bolt from the rear joint and detach it from the flange
- The propeller shaft.

To install:

9. Check carefully that the mating surfaces on the propeller shaft and flange are clean.

10. Align the propeller shaft joints with the flanges. Loosely install the bolts in the center member and the center bearing. Tighten the flange bolts. Step 1, 84 inch lbs. (10 Nm); step 2, 18 ft. lbs. (25 Nm).

➡Use new bolts.

11. Tighten the 4 M8 bolts for the support for the center bearing. Tighten the 2 M8 bolts for the propeller shaft center bearing. Tighten all bolts to 22 ft. lbs. (30 Nm).

12. Install:

- The front cross member to the body using 4 M8 bolts. Tighten
- The brake pipe in the mountings in the member
- The rear exhaust pipe assembly using 2 M8 bolts. Tighten

➡**Check that the gasket on the flanged joint for the exhaust pipe is intact.**

- The rear cross member, 2 M8 nuts.

CV-Joints

OVERHAUL

Inner CV-Joint

The inner CV-joint is serviced with the axle shaft as an assembly. The inner CV-joint boot can be serviced by removing the outer CV-joint.

Outer CV-Joint

1. Before servicing the vehicle, refer to the Precautions Section.
2. Remove the halfshaft from the vehicle.
3. Remove the grease boot clamps and push the boot away from the joint.
4. Expand the inner race circlip and pull the CV-joint off of the axle shaft.
5. Disassemble the inner race, cage and balls for cleaning and inspection.

To install:

6. Assemble the inner race, cage and balls into the outer joint housing.
7. Expand the circlip and push the joint on to the axle shaft.
8. Fill the outer race and the grease boot with CV-joint grease and tighten the boot clamps.
9. Install the axle halfshaft.

STEERING AND SUSPENSION

Air Bag

⁂ CAUTION

Some vehicles are equipped with an air bag system. The system must be disarmed before performing service on, or around, system components, the steering column, instrument panel components, wiring and sensors. Failure to follow the safety precautions and the disarming procedure could result in accidental air bag deployment, possible injury and unnecessary system repairs.

PRECAUTIONS

Several precautions must be observed when handling the inflator module to avoid accidental deployment and possible personal injury.

- Never carry the inflator module by the wires or connector on the underside of the module.
- When carrying a live inflator module, hold securely with both hands, and ensure that the bag and trim cover are pointed away.
- Place the inflator module on a bench or other surface with the bag and trim cover facing up.
- With the inflator module on the bench, never place anything on or close to the module which may be thrown in the event of an accidental deployment.

DISARMING

1. Before servicing the vehicle, refer to the Precautions Section.
2. Disconnect the negative battery cable.
3. Wait at least 1 minute before working on the vehicle. The air bag system is designed to retain enough power to deploy the air bag for a short time after the battery has been disconnected.
4. After repairs are complete, connect the negative battery cable. Turn the ignition switch to the **ON** position and check the SRS light for proper operation.

Rack and Pinion Steering Gear

REMOVAL & INSTALLATION

5- & 6-Cylinder Engines

1. Set the steering wheel to the straight-ahead position. Remove the key so that the steering wheel lock engages.

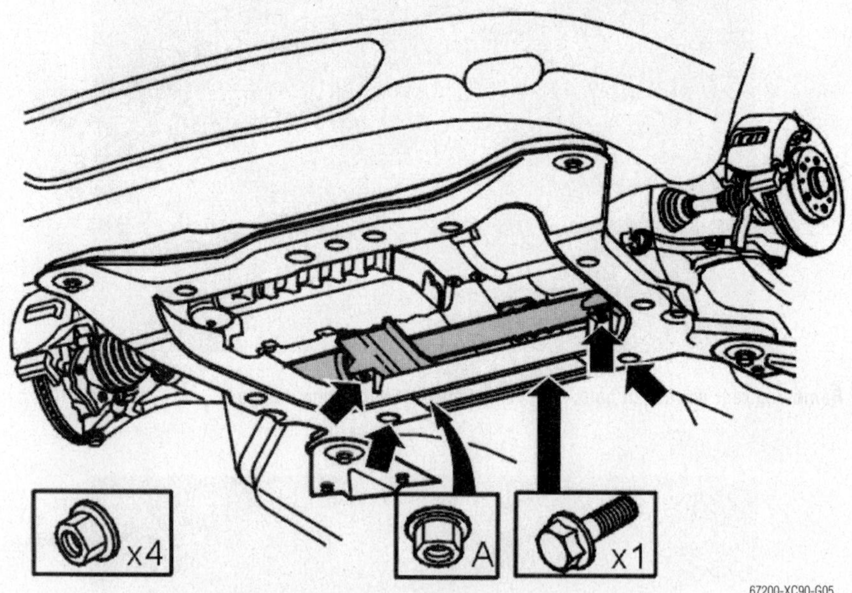

Steering gear mounting bolts—5- & 6-cylinder engines

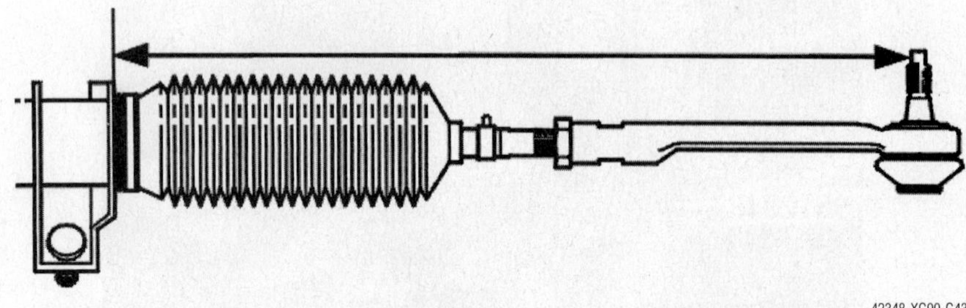

Measuring tie-rod length in relation to the steering gear housing—5- & 6-cylinder engines

Draining the power steering reservoir—V8

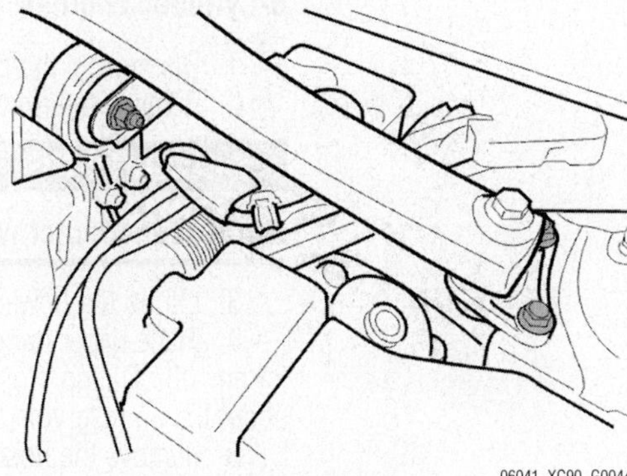

Remove the bolts at the suspension turrets on both sides

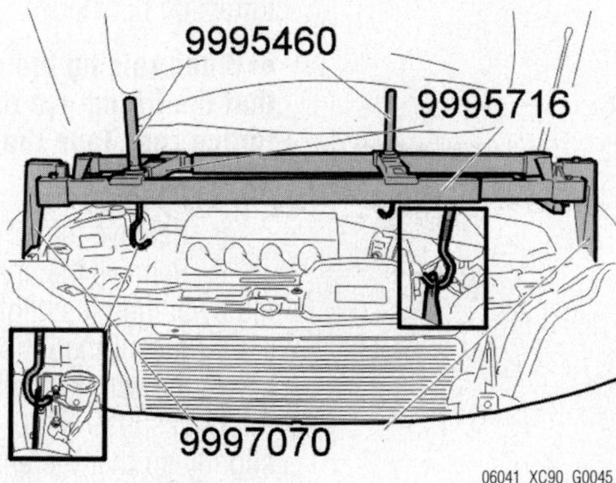

Installing the lifting tool

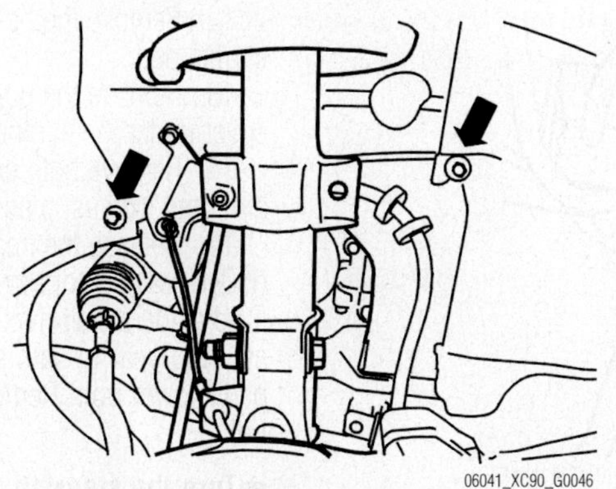

Removing the wing liner

2. Remove the battery lead.

3. Install the lock grip pliers on the feeder hose and the return hose for the power steering pump.

4. Remove the bolt on the steering shaft joint mounting in the steering gear from the engine compartment.

5. Remove the sub-frame.

6. Remove:

- The front wheels
- The tie rod ends from the steering arms.

7. Counterhold using a fixed wrench so that the boot is not damaged.

➡**On one side, measure the length of the tie rod in relation to the steering gear housing. Note the measurement.**

8. Place a container under the steering gear.

9. Remove:

- The return lines from the steering gear and drain the fluid
- The delivery line from the steering gear and drain the fluid.

10. Remove the clips holding the same lines.

11. Remove the steering gear.

To install:

12. Measure and adjust according to the measurements made earlier between the tie rod end and the steering gear housing on the new steering gear.

13. Install the steering gear. Torque the steering gear fasteners to 37 ft. lbs. (50 Nm).

14. Install the lines according to the noted routing.

15. Tighten the return line and the delivery line to the steering gear. Tighten to 18 ft. lbs. (25 Nm).

➡**Use a new O-ring on the tube at the steering gear.**

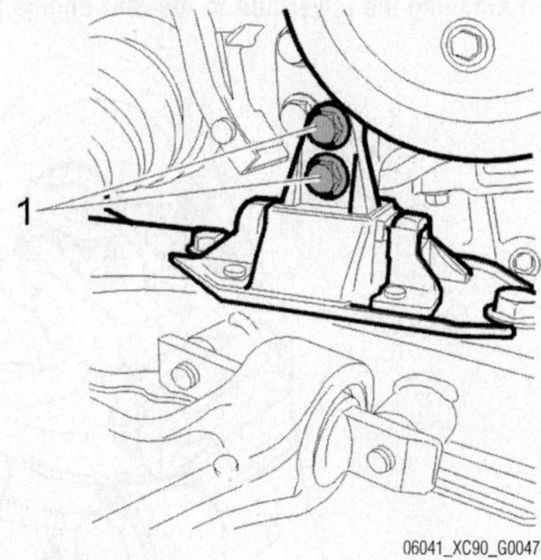

Removing the bolts to the engine pad, right

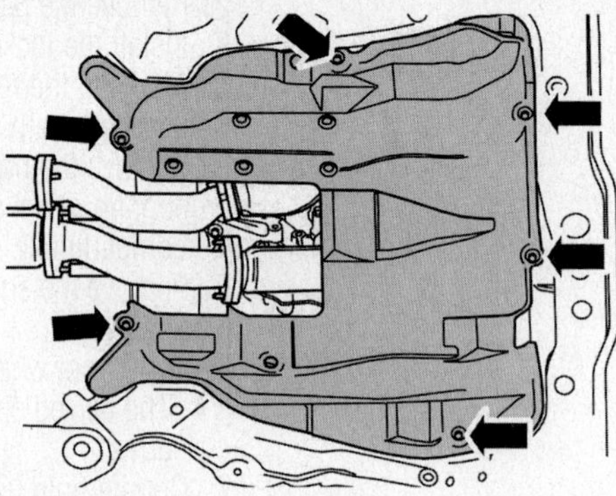

Removing the protective cover under the engine

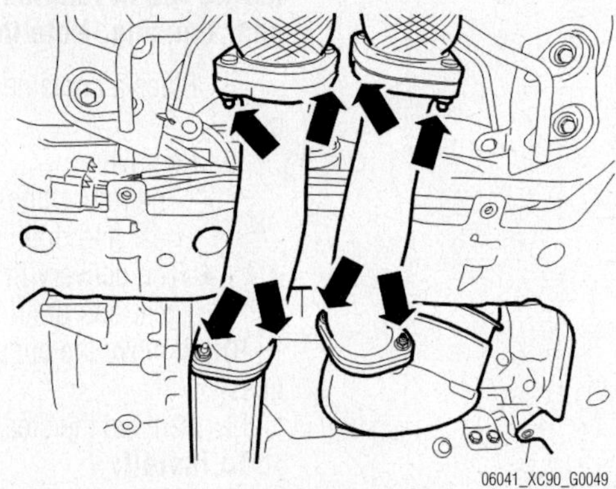

Removing the exhaust pipes between front and rear catalytic converters

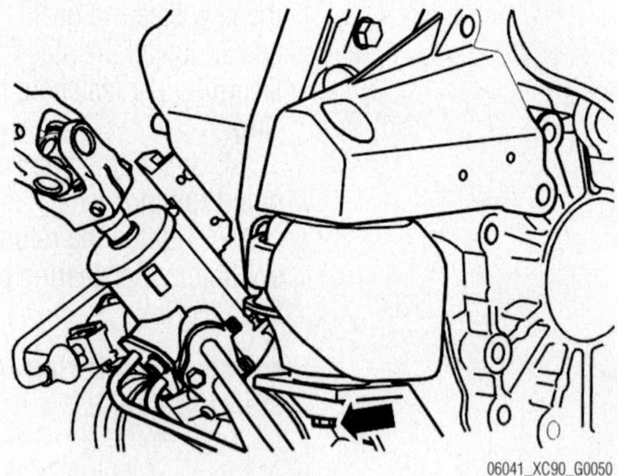

Removing the lower bolt to the rear engine pad

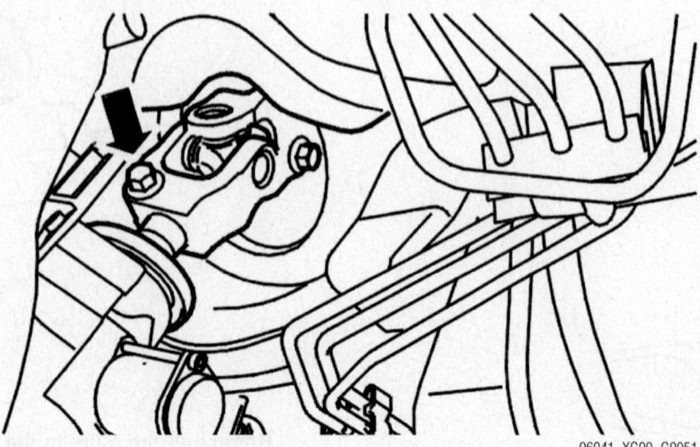

Removing the steering shaft joint, lower

16. Install the clips holding the same lines.
17. Install the tie rod ends on the steering arms. Use new nuts. Tighten to 58 ft. lbs. (80 Nm). Counterhold using a fixed wrench so that the boot is not damaged.
18. Install the sub-frame.
19. Expose the steering wheel lock. Turn the steering wheel so that the bolt in the mounting for the steering shaft joint points straight ahead.
20. Tighten the lower bolt on the steering shaft joint mounting in the steering gear from the engine compartment. Tighten to 18 ft. lbs. (25 Nm). Use a new bolt.

8-Cylinder Engines

1. Disconnect the battery.
2. Remove the front wheels.

WARNING

Avoid skin contact with the oil.

3. Lift up the power steering reservoir.
4. Place paper under the oil reservoir for any oil spillage. Position a container under the oil reservoir.
5. Remove the hose with the clamp from the oil reservoir.
6. Drain the oil.
7. Remove the bolts at the suspension turrets on both sides.

➡When raising the engine make sure that the lifting eye does not scratch the torque rod. Tape the torque rod at the lifting eye.

8. Remove the oil filler cap.
9. Remove the upper front engine casing by pulling it straight up.
10. Install the lifting tool: 999 5716 converted with kit: 999 7070 and the lifting hooks: 999 5460. Raise the engine slightly so that the engine pads are unloaded.
11. Remove the plastic nuts and the wing liner.
12. Remove the bolts (1) from the right engine pad.
13. Remove the bolts and the protective cover under the engine.
14. Remove both exhaust pipes' nuts and both exhaust pipes.
15. Release the heated oxygen sensor (HO_2S) connectors from the subframe. Fold aside the protective plate under the steering gear to gain access to the engine pad's lower bolt. Remove the engine pad's bolt.

➡Turn the steering wheel until optimum access to the bolt is achieved.

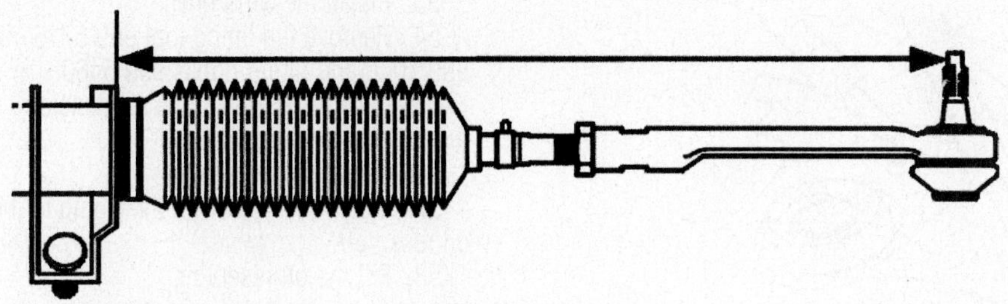

On one side, measure the length of the track rod in relation to the steering gear housing. Make a note of the measurement

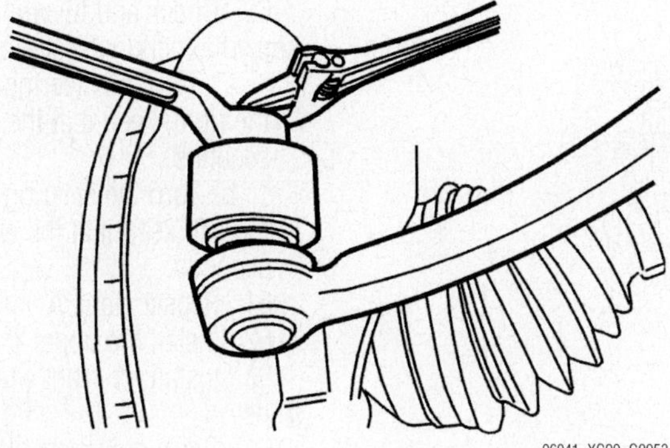

Removing the track rod ends

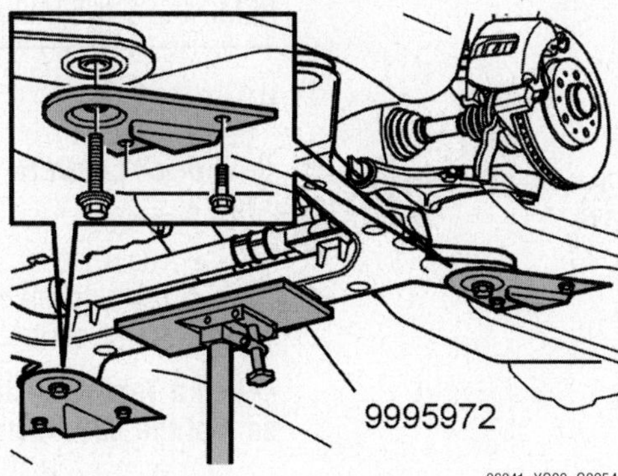

Lowering the subframe at the rear edge

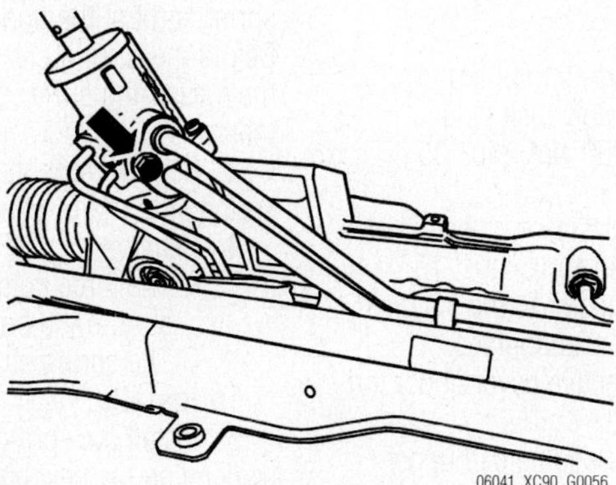

Removing the pressure pipe and return pipe from the steering gear

Lock the steering wheel and take out the ignition key, so that the steering wheel lock is activated.

16. On one side, measure the length of the track rod in relation to the steering gear housing. Make a note of the measurement.
17. Remove the bolt to the lower steering shaft joint.
18. Remove the connector to the solenoid.
19. Make sure that the wheels are in the straight forward position. Lock the steering wheel by removing the ignition key. Do not turn the front wheels so that the position of the steering gear is changed. This could damage the SRS system's contact reel.
20. Remove the track rod ends. Counter-hold with a wrench so that the gaiter is not damaged.
21. Press the lower steering shaft joint up from the steering gear. Make sure that no pipes or lines are loaded or trapped when lowering.
22. Position a mobile jack at the rear end of the subframe.
23. Remove the bolts to the subframe's brackets at the rear edge.
24. Remove the subframe's brackets.
25. Release the heated oxygen sensor (HO_2S) lines from the anti-roll bar.
26. Lower the subframe carefully.
27. Remove the bolt to the bracket holding the pipe to the subframe
28. Remove the bolt to the bracket holding the pipe to the subframe.
29. Release the pipes from the clips.
30. Place a container under the steering gear.
31. Remove the pipes' bolt on the valve housing and the pipes.
32. Remove the bolt holding the protective plate on the steering gear.
33. Remove the bolt holding the protective plate to the subframe.
34. Remove the protective plate.
35. Remove the nuts (1) and the bolt (2).
36. Remove the steering gear through the left-hand side of the vehicle.

To install:

37. Install the steering gear. Torque the nuts and bolt to 37 ft. lbs. (50 Nm).
38. Install the metal protective cover.
39. Install the oil pipes to the steering gear. Torque the pressure line to 18 ft. lbs. (25 Nm).
40. Raise the subframe's rear edge using a mobile jack.
41. Install the bolts with brackets. Torque to 74 ft. lbs. (100 Nm).
42. Install the front edge bolts to the brackets. Torque to:

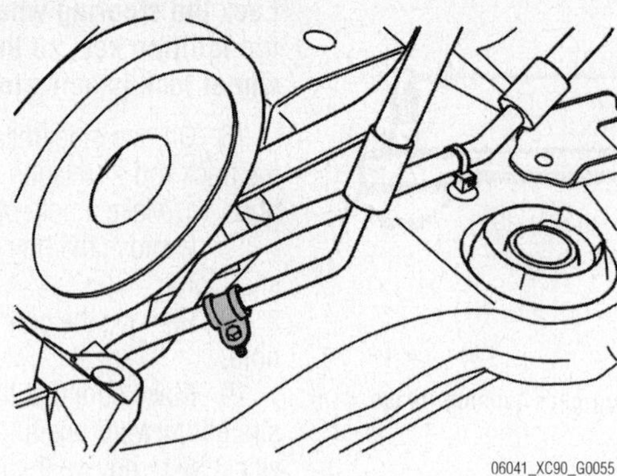

Remove the bolts to the subframe in the rear edge

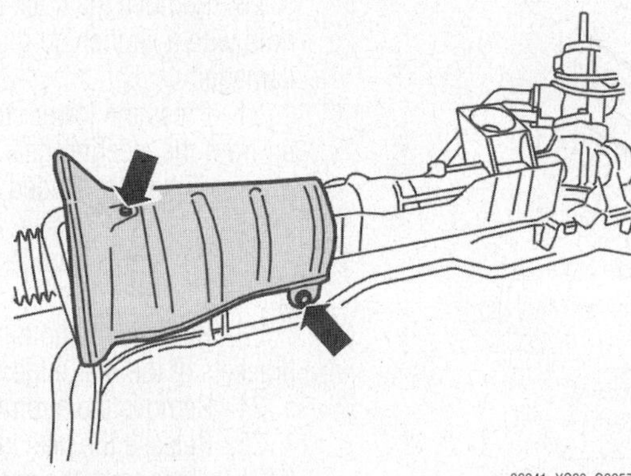

Removing the metal protective cover

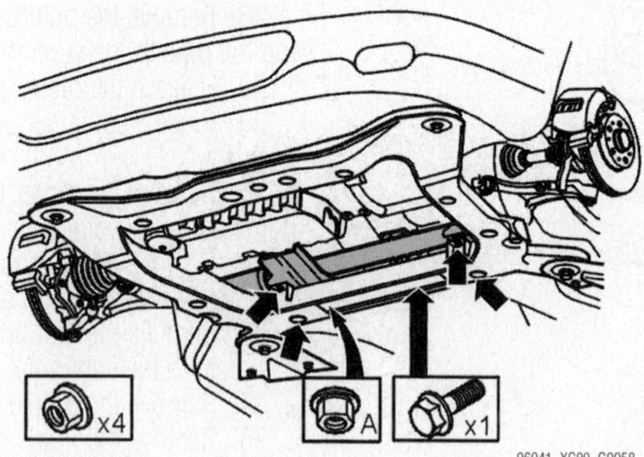

Removing the steering gear

- Step 1: 77 ft. lbs. (105 Nm)
- Step 2: plus 120 degrees

43. Install the heated oxygen sensor lines on the anti-roll bar.

➡Unscrew the upper nut in order to facilitate fitting the lower bolt.

44. Install the lower bolt to the engine pad. Torque to 37 ft. lbs. (50 Nm).
45. Connect the heated oxygen sensor connectors on the subframe.
46. Refit the protective plate under the steering gear.
47. Install the steering arms to the wheels spindles. Use new lock nuts. Tighten to 37 ft. lbs. (50 Nm) plus 35 degrees.
48. Install the lower steering shaft joint. Torque to 18 ft. lbs. (25 Nm).
49. Install the connector to the solenoid.
50. Connect both exhaust pipes.
51. Install the protective cover under the engine.
52. Install the bolts engine pad upper bolts. Torque to 37 ft. lbs. (50 Nm).
53. Install the wing liner.
54. Remove the lifting tool 999 5716, kit 999 7070 and lifting hooks 999 5460.
55. Install the bolts at the suspension turrets on both sides.
56. Connect the battery.
57. Install the hose with the clamp to the oil reservoir
58. Fill the oil reservoir.

➡For the correct oil grade use Volvo power steering fluid, part number 116 1529-1.

59. Turn the steering wheel between both end positions several times.
60. Check and fill with oil as necessary. Apply the parking brake. Start the engine.
    a. Turn the steering wheel slowly to the right, keep it at the end position for 2 seconds.
    b. Turn the steering wheel slowly to the left, keep it at the end position for 2 seconds.
61. Repeat steps A and B 10 times.
62. Install the upper front engine casing.
63. Install the front wheels on both sides.
64. Carry out wheel alignment.

Strut

REMOVAL & INSTALLATION

Front

5- & 6- CYLINDER

1. Remove:
    - The wheel
    - The anti-roll bar link from the spring strut.

➡Use a Torx wrench as a counterhold so that the boot is not damaged.

- The ABS line from the spring strut
- The ABS sensor. Hang up the sensor using a piece of wire.

2. Measure over the wheel spindle and spring strut at the upper bolt as illustrated. Before measuring clean off any dirt from the measuring surfaces. Note the measurement.
3. Remove both the bolts retaining the spring strut in the wheel spindle.
4. Remove:
    - The three nuts holding the spring strut in the bodywork
    - The spring strut.

To install:

5. Install the spring strut in the bodywork using the new nuts. Tighten to 18 ft. lbs. (25 Nm).

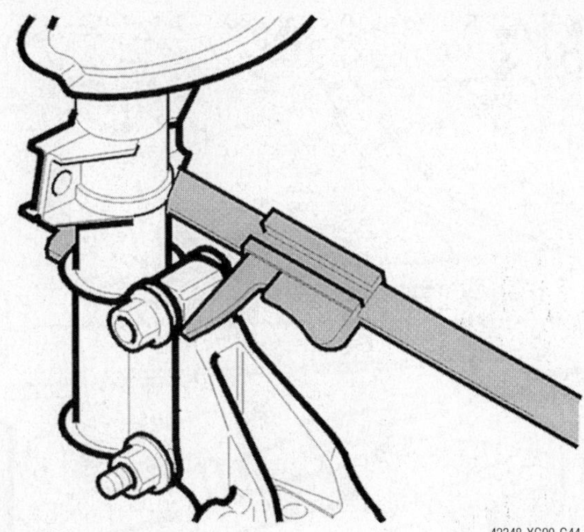
42348-XC90-G44
Measuring over the spindle and strut

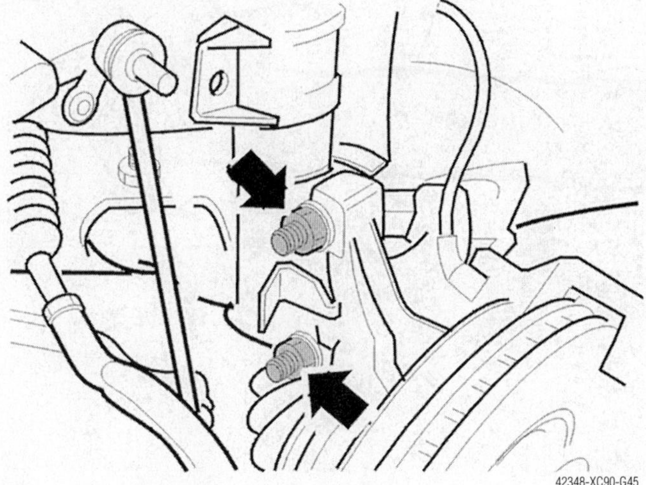
42348-XC90-G45
Removing the lower strut bolts

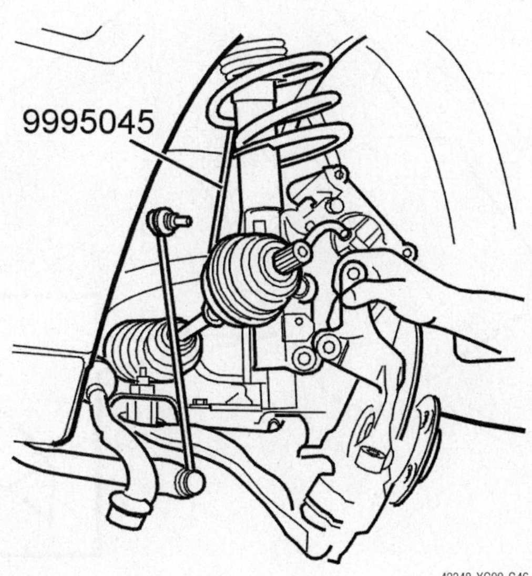

42348-XC90-G46
Removing the spindle

6. Install the spring strut on the wheel spindle. Use new bolts and new nuts.
7. Adjust the spring strut and wheel spindle to the measured value.
8. Tighten to 77 ft. lbs. (105 Nm), plus 75 degrees.
9. Install:
   - The anti-roll bar link to the spring strut. Use a new nut. Tighten to 37 ft. lbs. (50 Nm). Counterhold using a Torx wrench to prevent damage to the boot
   - The ABS sensor. Tighten to 84 inch lbs. (10 Nm).

➡Ensure that the ABS sensor seat in the wheel spindle is absolutely clean. Clean the ABS sensor with a soft brush.

   - The ABS sensor cable in the bracket
   - The wheel.

Shock Absorber

REMOVAL & INSTALLATION

Rear

1. Raise the vehicle.
2. Remove the wheels.
3. Install tensioner 999 5659 as follows:
   a. Remove the parking brake cable mountings from the sub-frame.
   b. Install the tensioner.

➡The tensioner must be positioned as far out as possible on the tensioner plates to provide the correct lifting force. It is vital that the screwed joints for the rubber bushings are tightened in the normal position (in the same position as when the car is on the ground and has three people in the car and a full fuel tank). The tensioner relieves the pressure on the components in the suspension when removing and installing.

   c. The tool is secured to the sub-frame with mounting 999 7061. This means that the lift stability is not affected during work.
   d. Install the thread bar on the tensioner 999 5659.
   e. Lift up the tensioner with the threaded bar 999 5659.
   f. Insert the threaded bar in the mounting 999 7061 from underneath.
   g. Insert the locking washer with the handle in from the side in mounting 999 7061 under the threaded bolt.

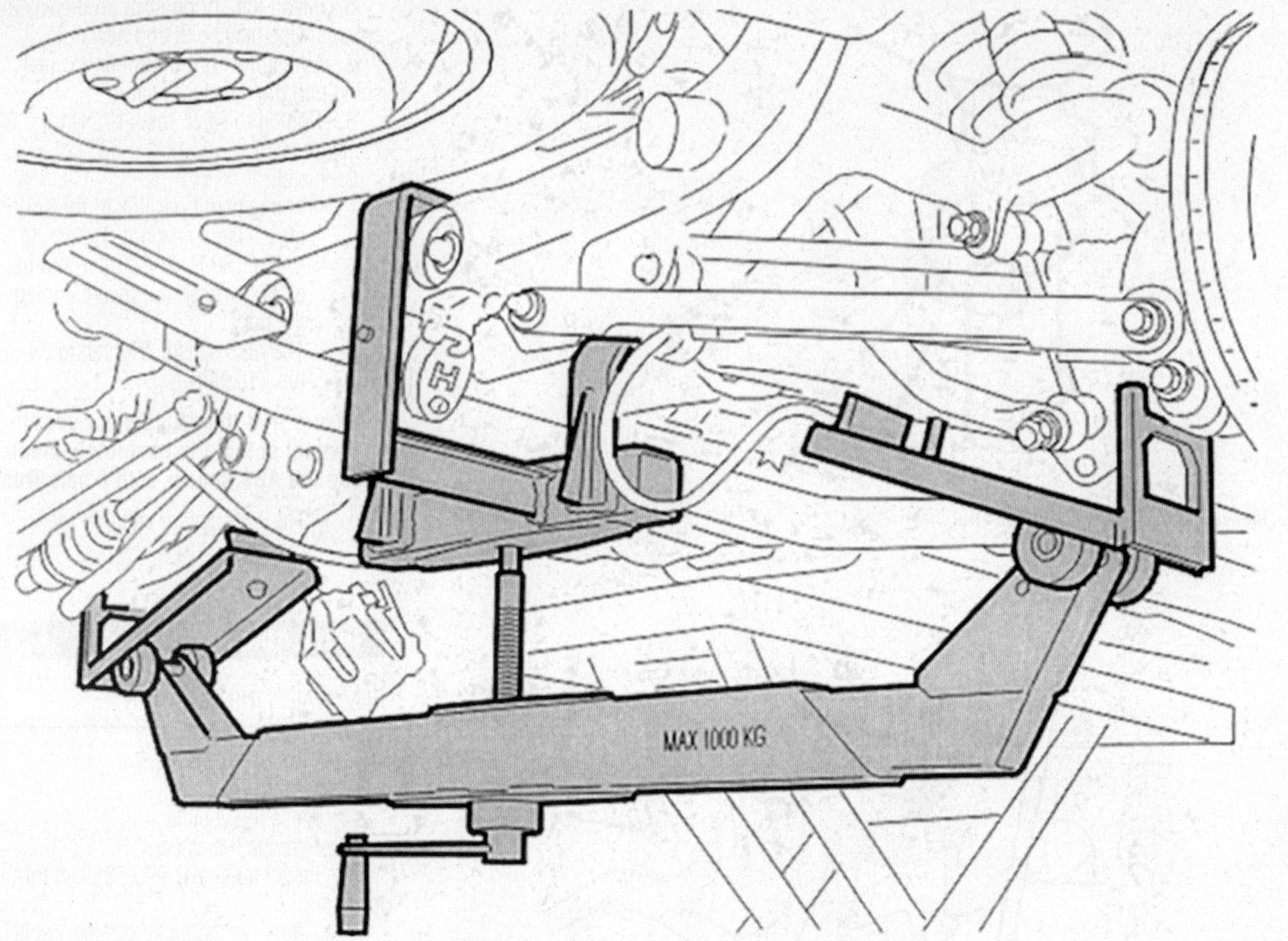

MAX 1000 KG
42348-XC90-G47

Tensioner installed

h. Lock the bolt in the locking washer. Align the locking washer in the mounting.

➨**The locking washer has a locating pin which must be aligned with the bolt.**

i. Install the 2 bolts (5740-36) from kit 998 9761 in the control arm holes.

j. Install the rails 5740-1 modified according to (WG-276) on the control arms.

k. Ensure that the rollers lie against the rails on both sides of the control arms.

➨**The tensioner must be positioned as far out as possible on the tensioner plates to provide the correct lifting force.**

l. Raise the control arms to their normal position using the tensioner. The normal position of the rear suspension is 453 mm from the fender edge to the center of the wheel.

4. With 5 seats:

42348-XC90-G48

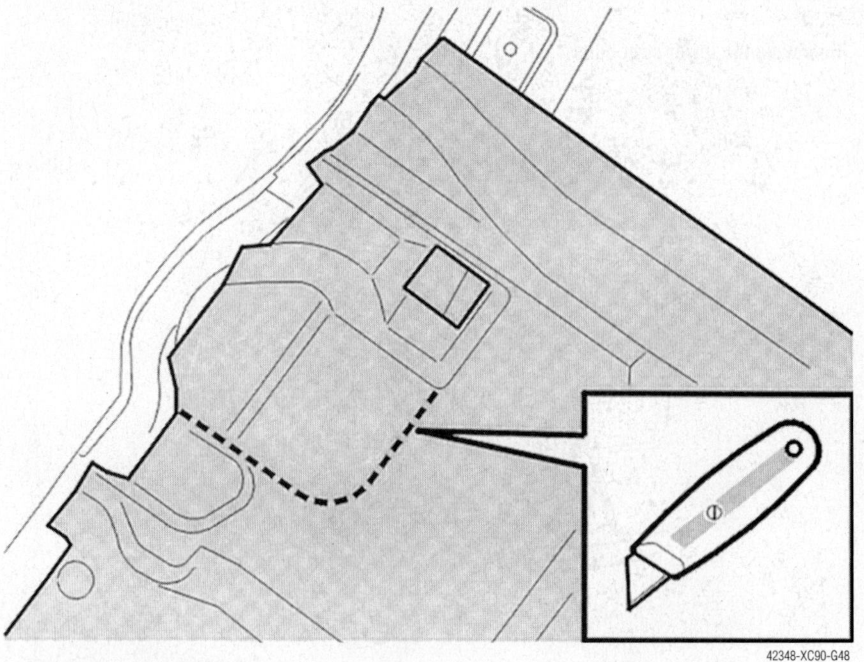

Cut the sound proofing

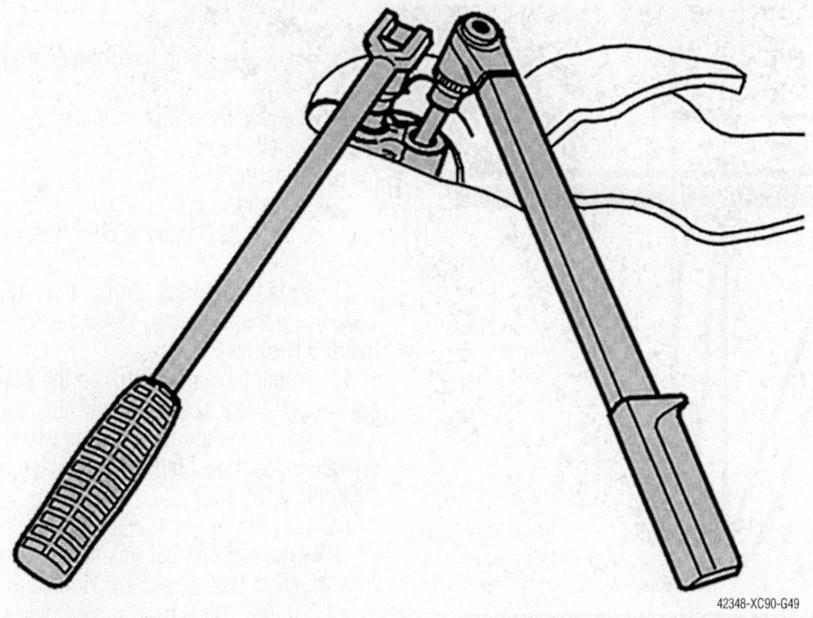

Removing the upper shock absorber mounting nut

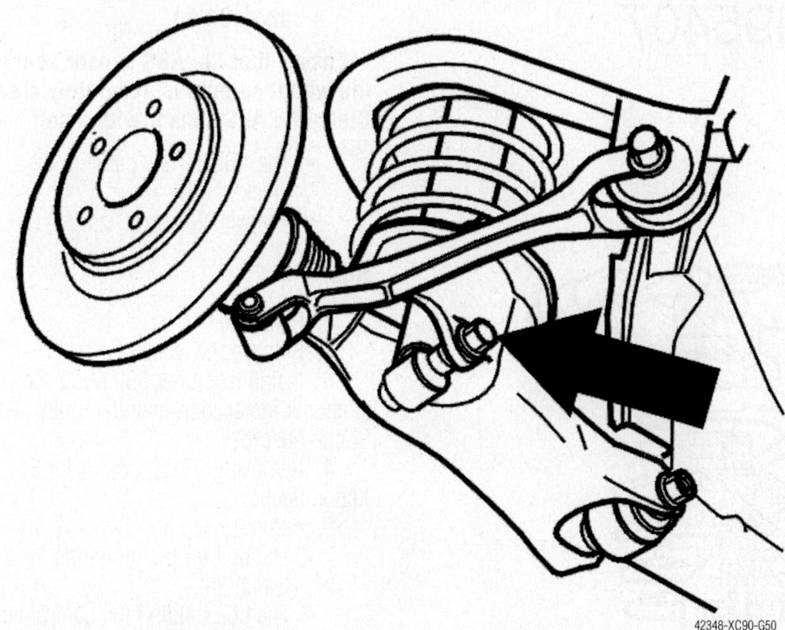

Removing the lower shock absorber mounting bolt

a. Remove:
- The front floor hatch
- The form-molded floor mat.

5. Fold the soundproofing over the shock absorber mounting out of the way.
6. With 7 seats:

a. Slide the seat cushion back on the rear row of seats.

b. Lift up the carpet.

c. Cut up the soundproofing as illustrated.

7. Fold up the soundproofing. Remove the cover on the splash guard.
8. Remove the shock absorber nut. Use socket 999 5500 and a Torx-wrench as a counterhold.
9. Remove:
   - The bolt in the lower mounting for the shock absorber
   - The shock absorber.

To install:

10. Insert the new shock absorber through the control arm to the upper mounting.
11. Install the new nut loosely.

➡Ensure that the bushing seats correctly in the opening in the rear suspension.

12. Fix the shock absorber in the lower mounting. Tighten the upper nut for the shock absorber. Tighten to 44 ft. lbs. (60 Nm). Tighten the lower bolt to 58 ft. lbs. (80 Nm).
13. Reinstall the carpet.
14. With 5 seats: Reinstall the front floor hatch.
15. With 7 seats: Slide the seat cushion forward.
16. Detach the tensioner, pull out the fork and lower the tensioner using the threaded bar.
17. Remove:
    - Mounting 999 5659 from the sub-frame
    - The rails 5740-1 from the control arms.
18. Install the parking brake cable mountings in the sub-frame.
19. Install the wheel.

➡If the car is equipped with Bi-Xenon lamps, the position sensor must be calibrated.

Coil Spring

REMOVAL & INSTALLATION

Front

5- & 6- CYLINDER ENGINE

1. Secure the strut in a vise.
2. Remove:
   - The nut for the shock absorber bearing.
   - Washer
   - Mounting.
3. Take the load off the spring seat. Use hydraulic tool 951 2911 together with 951 2914.
4. Remove:
   - The fixing nut on the shock absorber.
   - The spring seat
   - The spring
   - The rubber bump stop and gaiter.
5. Check that the spring strut bearing (1), spring seat (2) and rubber bump stop with boot (3) are free of damage. Replace if necessary.

To install:

6. Compress the new spring to a length of approximately 260 mm. Use spring clamps.
7. Install:
   - The bump stop with the boot

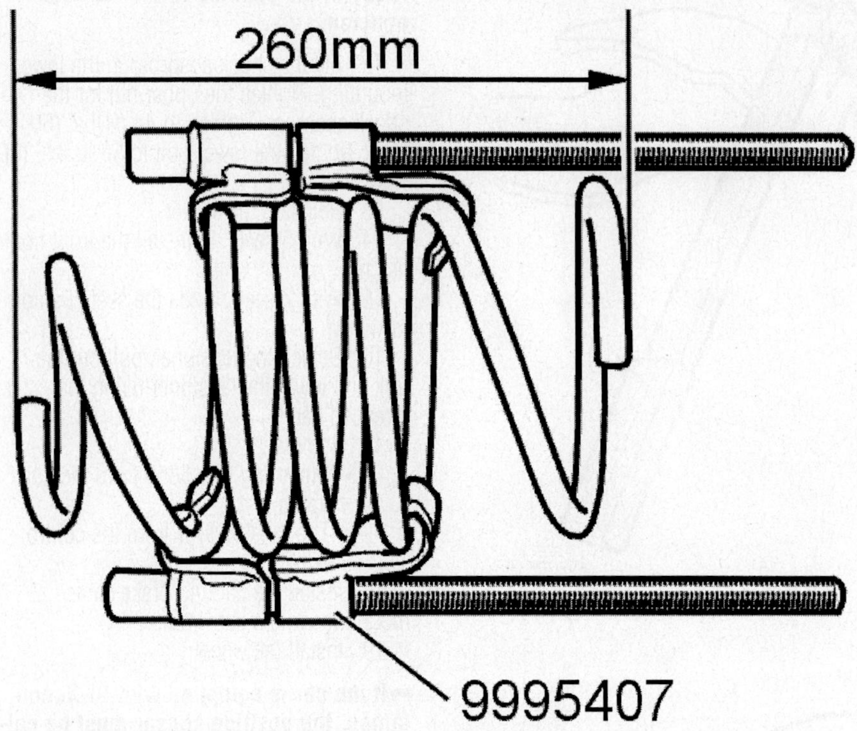

Compressing the spring

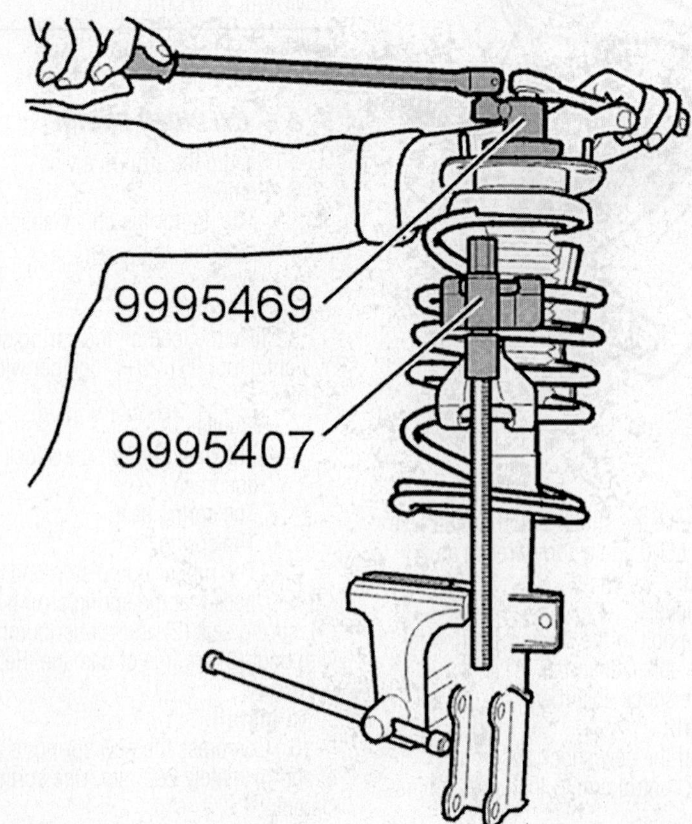

Installing the spring on the strut

- The spring
- The spring seat.

8. Install the fixing nut. Tighten to 51 ft. lbs. (70 Nm).
9. Remove the spring clamps.
10. Install:
    - Bearing
    - Washer
    - The nut. Tighten to 51 ft. lbs. (70 Nm).
11. Install the spring strut in the bodywork using the new nuts. Tighten to 18 ft. lbs. (25 Nm).
12. Install the spring strut on the wheel spindle. Use new bolts and new nuts.
13. Adjust the spring strut and wheel spindle to the measured value. Tighten to 77 ft. lbs. (105 Nm), plus 90 degrees.
14. Install:
    - The anti-roll bar link to the spring strut. Use a new nut. Tighten to 37 ft. lbs. (50 Nm). Counterhold using a Torx wrench to prevent damage to the boot
    - The ABS sensor. Tighten to 84 inch lbs. (10 Nm).

➡Ensure that the ABS sensor seat in the wheel spindle is absolutely clean. Clean the ABS sensor with a soft brush.

- The ABS sensor cable in the bracket
- The wheel.

Rear

1. Raise the car.
2. Remove the wheels.
3. Install tensioner 999 5659. See the tensioner installation material under "Rear Shock Absorber".
4. Remove position sensor for Bi-Xenon lamps.
5. Remove:
    - The bolt for the mounting for the lateral link
    - The bolt holding the control arm in the wheel spindle.
6. Lower the control arm with the tensioner. The spring is now unloaded.
7. Lower the tensioner. Twist the tensioner so that it is positioned along the length of the car.
8. Press the control arm down by hand. Remove the spring.

To install:

9. Install the new spring.
10. Check that the spring is correctly installed in the lower spring seat.
11. Lift the removed control arm by hand. Position the tensioner in the rails.

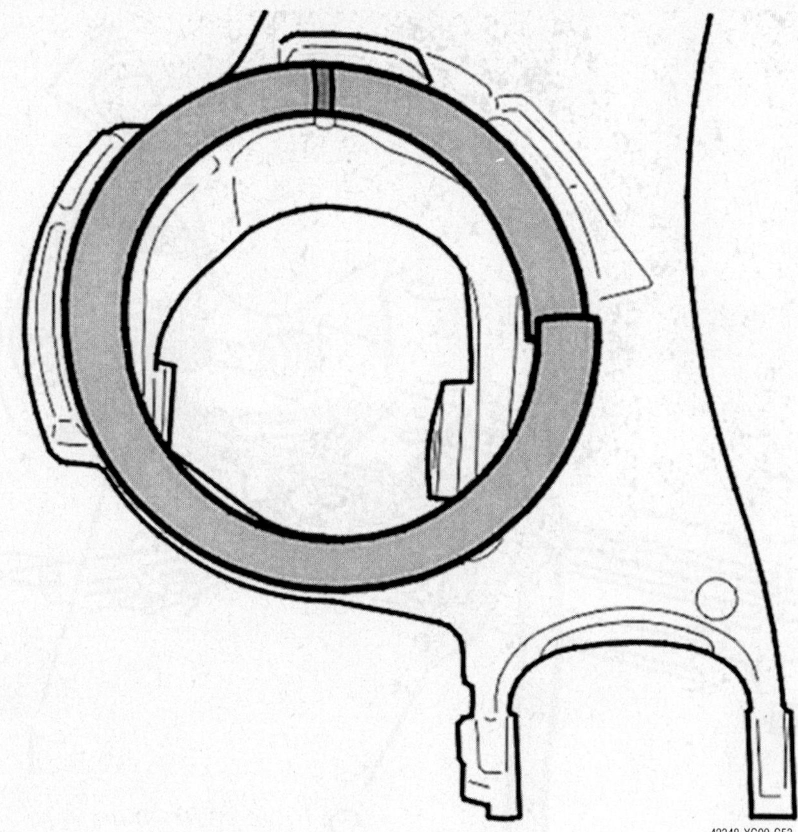

Spring correctly installed

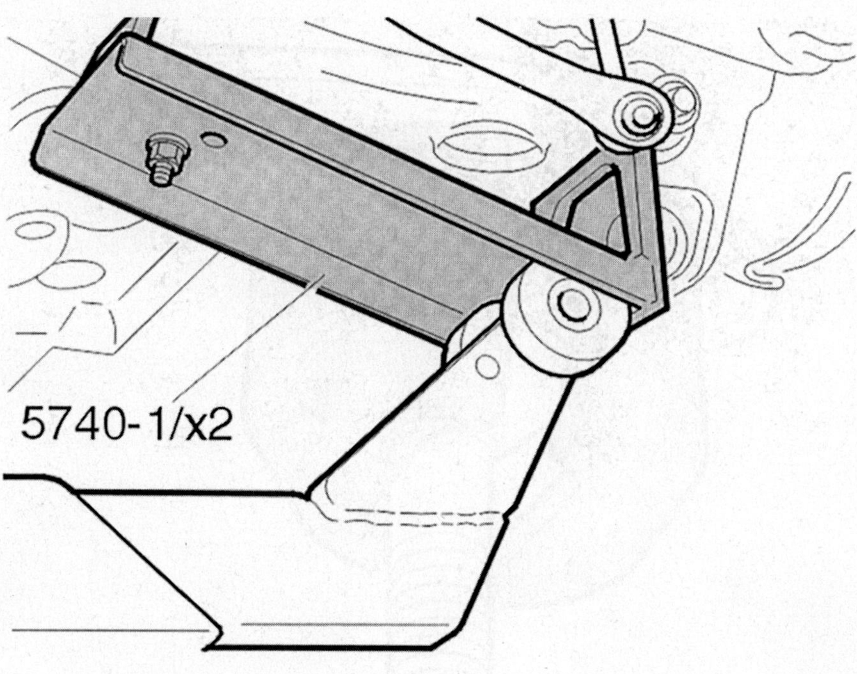

Tensioner positioned on control arm

➡**The tensioner must be positioned as far out as possible on the tensioner plates to achieve the correct lifting force.**

12. Lift the control arms so that the bolt for the control arm and lateral link can be installed. Align the bolt. Tighten by hand. Do not tighten yet.
13. Lower the tensioner. Install the bolt in the lower mounting for the shock absorber. Do not tighten yet.
14. Tension the control arms to the normal position.
15. Tighten the bolts for the control arm, lateral link and shock absorber. M12
16. Install the position sensor for Bi-Xenon lamps. .
17. Remove the tensioner.
18. Install the wheel.

➡**If the car is equipped with Bi-Xenon lamps, the position sensor must be calibrated.**

Lower Ball Joint

REMOVAL & INSTALLATION

2003–04

1. Remove the wheel.
2. Install cap nut 7062-1 on the ball joint pinion.
3. Tighten the nut against the control arm.
4. Then turn back one turn to form a gap between the cap nut and the control arm.
5. Position the extractor on the control arm. Detach the ball joint from the control arm.
6. Pull down the control arms using a tension strap.
7. Release the spring strut from the control arm.
8. Remove the bolt for the halfshaft. Use a prybar as a counterhold on the brake disc.
9. Remove the halfshaft from the hub. Knock the halfshaft out using a brass drift.
10. Push the spring strut to one side.
11. Remove the bolts for the ball joint.
12. Remove the rubber seal.
13. Install the inner sleeve so that both halves hook securely in the flange (see arrow) on the ball joint.
14. Turn the inner sleeve so that the slit is at right angles to the bolt holes.
15. Install the outer sleeve around the inner sleeve halves. Tighten the bolt.
16. Install an M12 nut on the ball joint bolt.

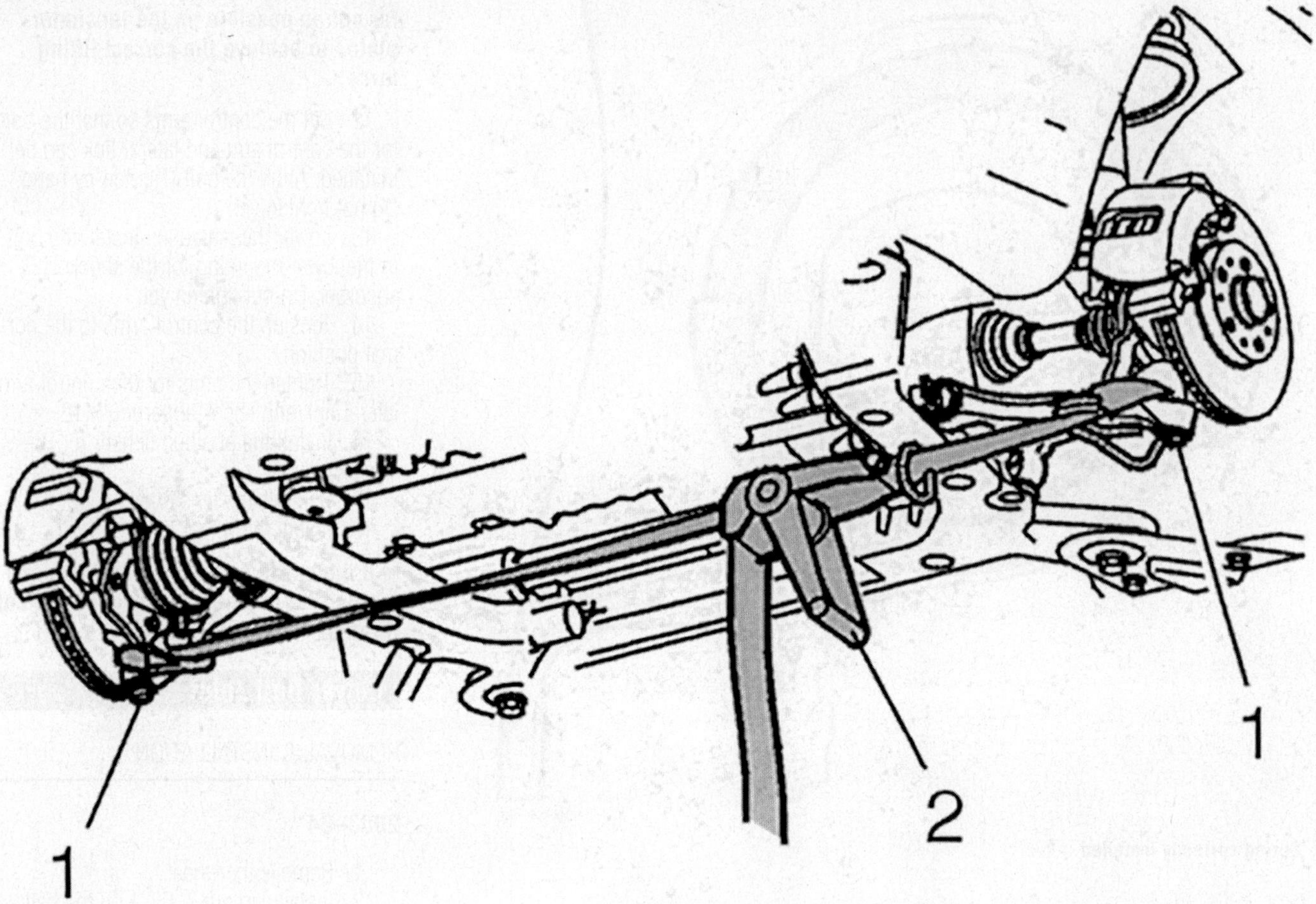

42348-XC90-G55

Securing the control arms—2003–04

17. Screw on the nut so that the ball joint bolt is just protruding from the nut.
18. Using Tool 999 5781-3:
    a. The guide bolts have a wrench grip (see arrow) which is asymmetrically positioned.
    b. Aluminum wheel spindle: Use the bolt (1) with the greater distance to the wrench grip.
    c. Steel wheel spindle: Use the bolt (2) with the shortest distance from the wrench grip.
    d. Screw in the guide bolts fully.
19. Using Tool 999 5781:
    a. Install the support using 3 wheel studs.
    b. Screw the guide bolts down until they are in contact with the support.
    c. Position supplementary support 999 5781-4 on the hydraulic press.
    d. Screw the connector 999 5781-5 into place on the hydraulic press.
    e. Install the connector on the nut on the ball joint.
20. Press off the ball joint.
21. Remove the tool and the ball joint.

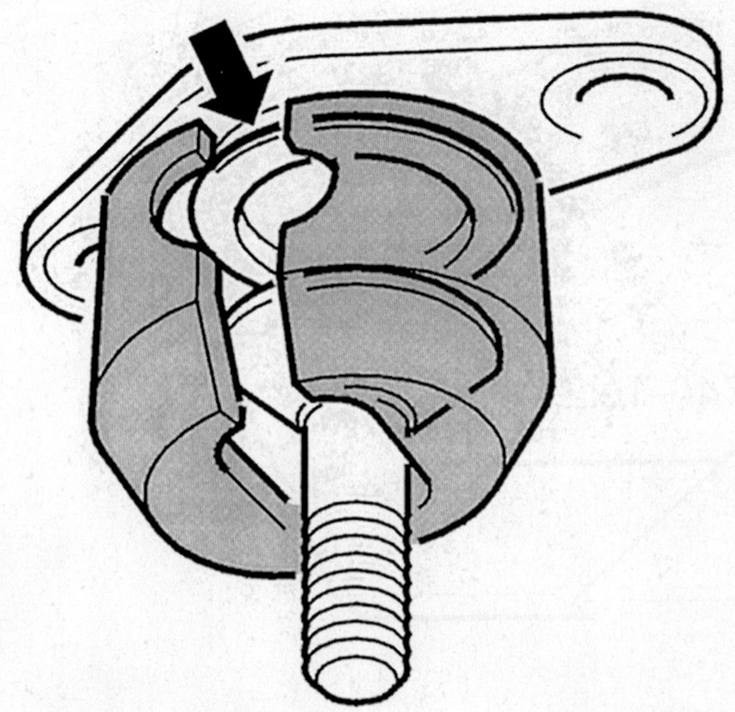

Inner sleeve installation—2003–04

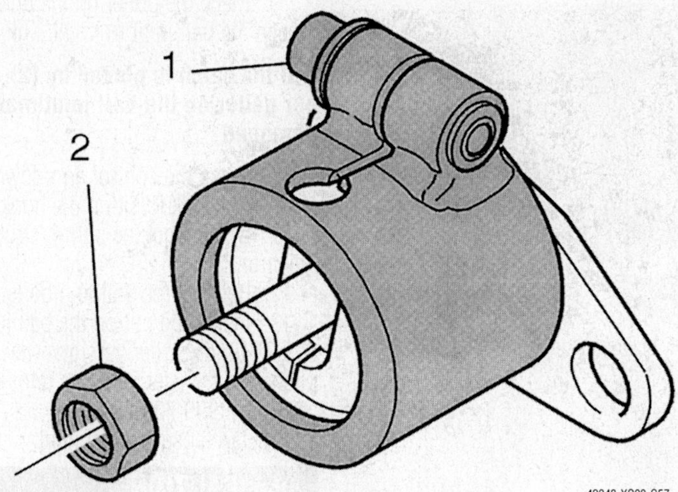

Outer sleeve installation—2003–04

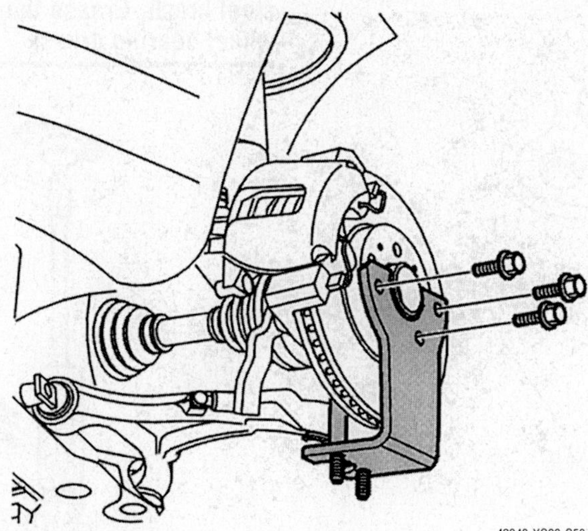

Installing the support—2003–04

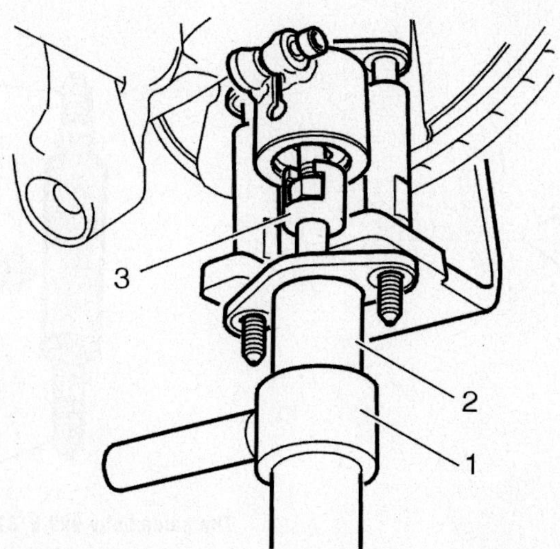

Removing the ball joint—2003–04

To install:

22. Clean the ball joint seat and the mating surfaces for the ball joint to the wheel spindle. Use a rotary wire brush.
23. Lubricate the seat using wheel bearing grease.
24. Loosely install the new ball joint using the guide bolts.

➡Leave the protective cap for the ball joint in position to prevent damaging the rubber seal.

25. Press the ball joint up towards the seat using impact drift 999 5796. Check that the ball joint is centered in the seat.
26. Knock in the ball joint using a copper mallet.
27. Tighten the ball joint using new bolts. Tighten to 37 ft. lbs. (50 Nm).
28. Install the halfshaft in the hub.
29. Install a new bolt. Use a prybar as a counterhold. Tighten to 37 ft. lbs. (50 Nm).
30. Align the ball joints (1) in the control arm.

⁂ CAUTION

Take care when releasing the tension strap.

31. Release the control arms by releasing the tension strap.
32. Install a new nut.

⁂ WARNING

The ball joint pinion must not rotate. Use a Torx wrench as a counterhold so that the rubber boot is not damaged.

33. Tighten to 58 ft. lbs. (80 Nm).
34. Install the wheels.

2005–06

1. Remove the front wheels.
2. Remove the bolt for the drive shaft. Use a prybar in the brake disc as counterhold.
3. Remove the drive shaft from the hub by tapping out the drive shaft with a brass drift.
4. Fold aside the spring strut.
5. Loosen the ball joint pin from the link arm. Use tool 999 7062.
6. Install the cap nut for the ball joint pinion. For earlier versions (deep thread pitch) use 999 7062 and for later versions (closer thread pitch) use 999 7231.
7. Tighten the nut against the link arm.
8. Then turn back one turn so that clearance is created between the cap nut and link arm.

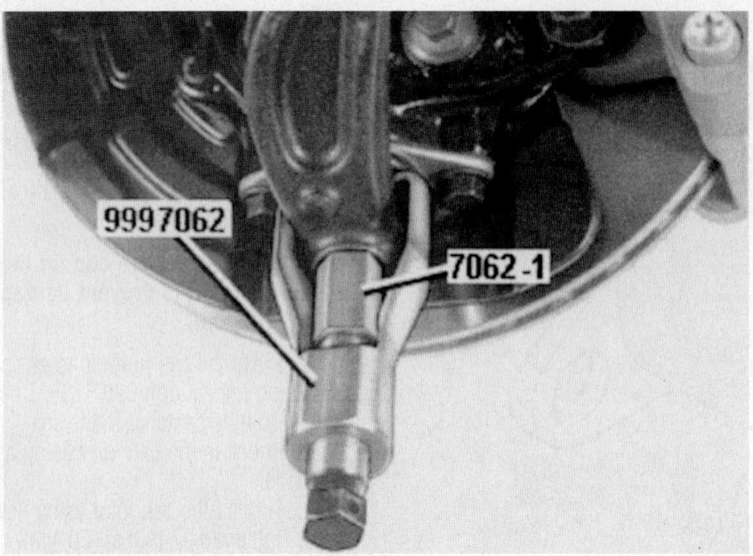

06041_XC90_G0059

Loosen the ball joint pin from the link arm—2005–06

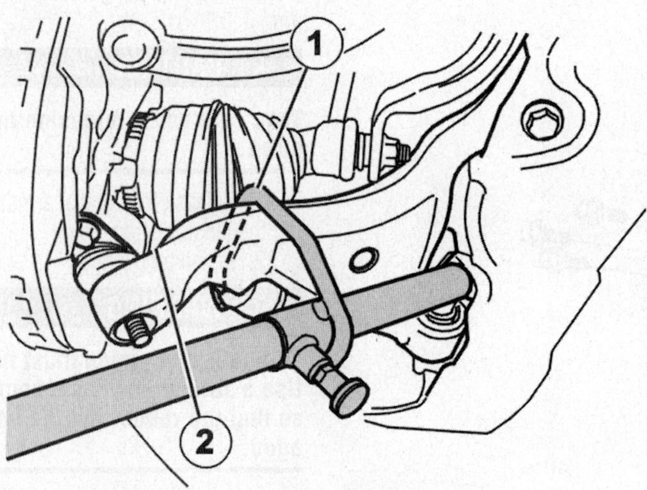

06041_XC90_G0060

Press the control arm down using lever 999 7076—2005–06

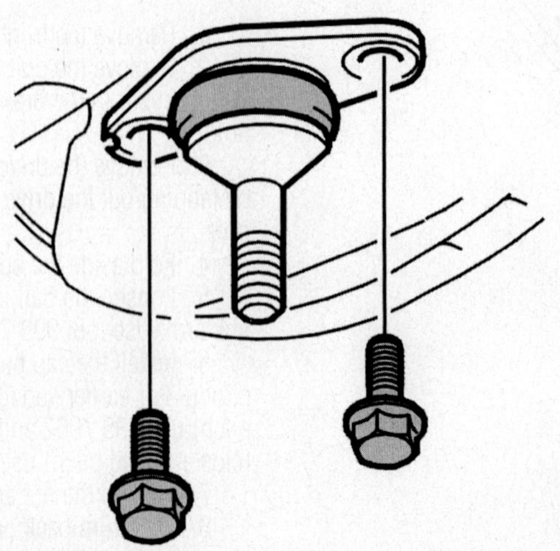
06041_XC90_G0061

Ball joint bolts—2005–06

9. Place the puller on the link arm and loosen the ball joint from the link arm.

➡If the catch is placed by (2), the rubber gaiter on the ball joint may be damaged.

10. Press the control arm down using lever 999 7076. Position the hook at (1).
11. Disconnect the spring strut from the link arm.
12. Remove the ball join bolts.
13. Install the nut on the ball joint. Use 999 7149 for earlier versions (deep thread pitch) and for later versions (closer thread pitch) use 999 7230.
14. Tap loose the ball joint.

CAUTION

Always use safety glasses. Clean both in the seat for the ball joint and the contact surfaces with a rotating steel brush. Grease the seat with wheel bearing grease.

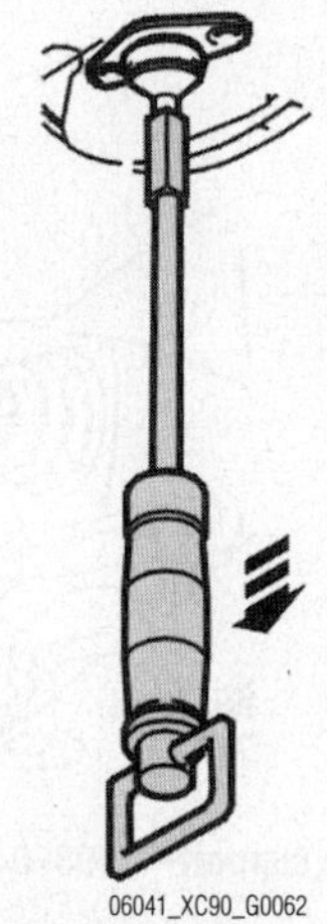
06041_XC90_G0062

Ball joint removal—2005–06

06041_XC90_G0063

The guide bolts 999 5781 have a wrench grip (see arrow) which is asymmetrically positioned—2005–06

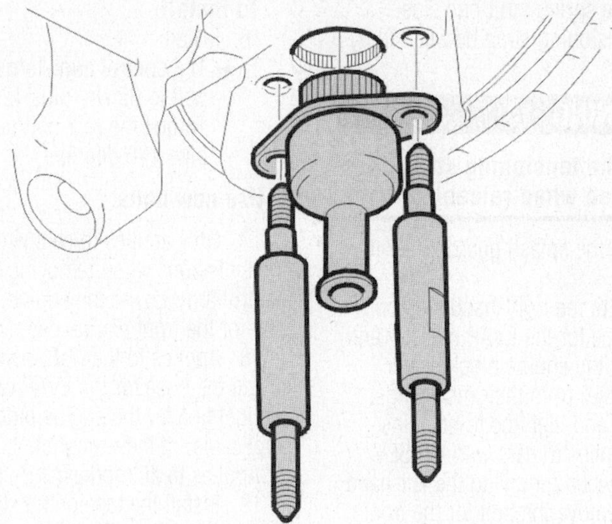

Install the new ball joint loosely using the guide bolts—2005–06

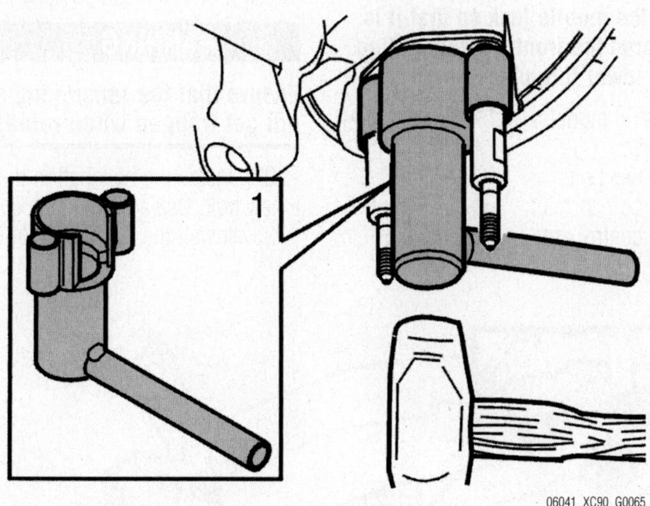

Drive in the ball joint with a copper hammer—2005–06

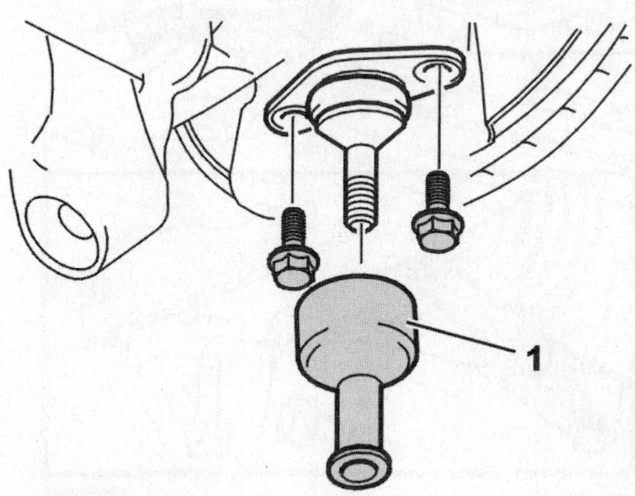

Tighten down the ball joint with the new bolts—2005–06

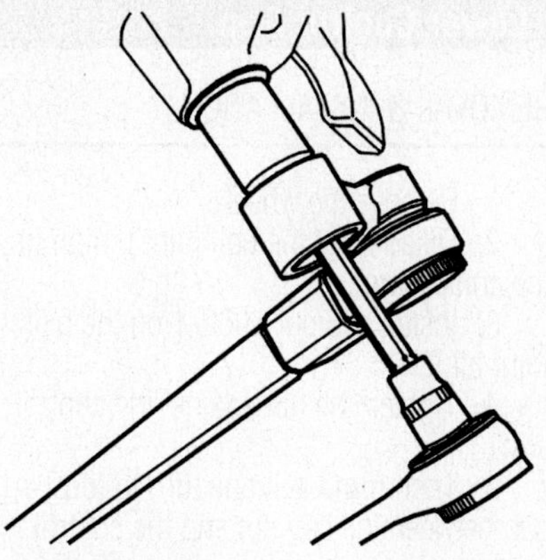

Tightening the stud nut—2005–06

To install:

➡The guide bolts 999 5781 have a wrench grip (see arrow) which is asymmetrically positioned.

➡Use the bolt (2) with the shortest distance from the wrench grip.

➡Leave the ball joint's protective cup on to avoid damaging the rubber gaiter.

15. Install the new ball joint loosely using the guide bolts.
16. Press up the ball joint against the seat with press tool 999 5796.
17. Check that the ball joint is centered in the seat.
18. Drive in the ball joint with a copper hammer.
19. Remove the guide bolts.
20. Tighten down the ball joint with the new bolts. Tighten to 74 ft. lbs. (100 Nm).
21. Remove the protective cup.
22. Install the halfshaft.

WARNING

If the catch is placed by (2), the rubber gaiter on the ball joint may be damaged.

23. Press down the link with lever 999 7076.
24. Place the catch by (1).
25. Install the spring strut in the link arm.

➡The ball joint pin may not rotate. Use a bit Torx as counterhold in order to prevent damaging the rubber gaiter.

26. Use 999 5500 as adapter for the torque wrench.
27. Install the nut. Tighten according to 37 ft. lbs. (50 Nm) plus 35 degrees.
28. Install the wheels.

Lower Control Arm

REMOVAL & INSTALLATION

1. Remove the wheel.
2. Disconnect the ball joint pinion from the control arm.
3. Install cap nut 7062-1 on the ball joint pinion.
4. Tighten the nut against the control arm.
5. Then turn back one turn to form a gap between the cap nut and the control arm.
6. Position the extractor on the control arm. Detach the ball joint from the control arm.
7. Pull down the control arms using a tension strap.
8. Release the spring strut from the control arm.

***** WARNING**

Ensure that the tension strap is correctly secured in the control arms.

9. Remove the bolt for the halfshaft. Use a prybar as a counterhold on the brake disc.
10. Remove the halfshaft from the hub. Knock the halfshaft out using a brass drift.
11. Push the spring strut one side. Remove the tensioning strap between the control arms.

***** WARNING**

Ensure that the tensioning strap does not get trapped when releasing.

12. Remove the splash guard under the engine.
13. Applies to the right-hand side only: Remove the hose for the EVAP canister and the fuel line for the engine block heater (where applicable) from their mountings. Move the hose and fuel line to one side.
14. Only applies to cars with 4T65EV transmissions when removing the left-hand control arm: Remove the bolt for the front engine pad. Lift the engine approximately 10mm using a mobile jack to remove the rear bolt on the front mounting for the control arm.

➡Position the mobile jack so that it is raised towards the front right mounting bolt for the lower torque rod bracket.

Applies to all models:

15. Remove:
    - The two bolts
    - The bolt and nut
    - The control arm.

To install:

16. Install:
    - The control arm. Torque the front bolt to 65 Nm plus 120 degrees. Torque the rear bolt/nut to 105 Nm, plus 120 degrees

➡Use new bolts.

17. Only applies to cars with 4T65EV transmissions when removing the left-hand control arm: Lower the engine. Install the bolt for the front engine pad. M10.
18. Applies to the right-hand side only: Install the hose for the EVAP canister and the fuel line for the engine block heater (if applicable) in their mountings.

Applies to all models:

19. Install the tensioning strap between the control arms and tighten.
20. Insert the halfshaft in the hub.
21. Position the spring strut over the ball joint. Release the tensioning strap so that the control arm meets the ball joint.

***** WARNING**

Ensure that the tensioning strap does not get trapped when releasing.

22. Install the halfshaft in the hub. Install a new bolt. Use a prybar as a counterhold.
23. Install the upper ball joint nut.

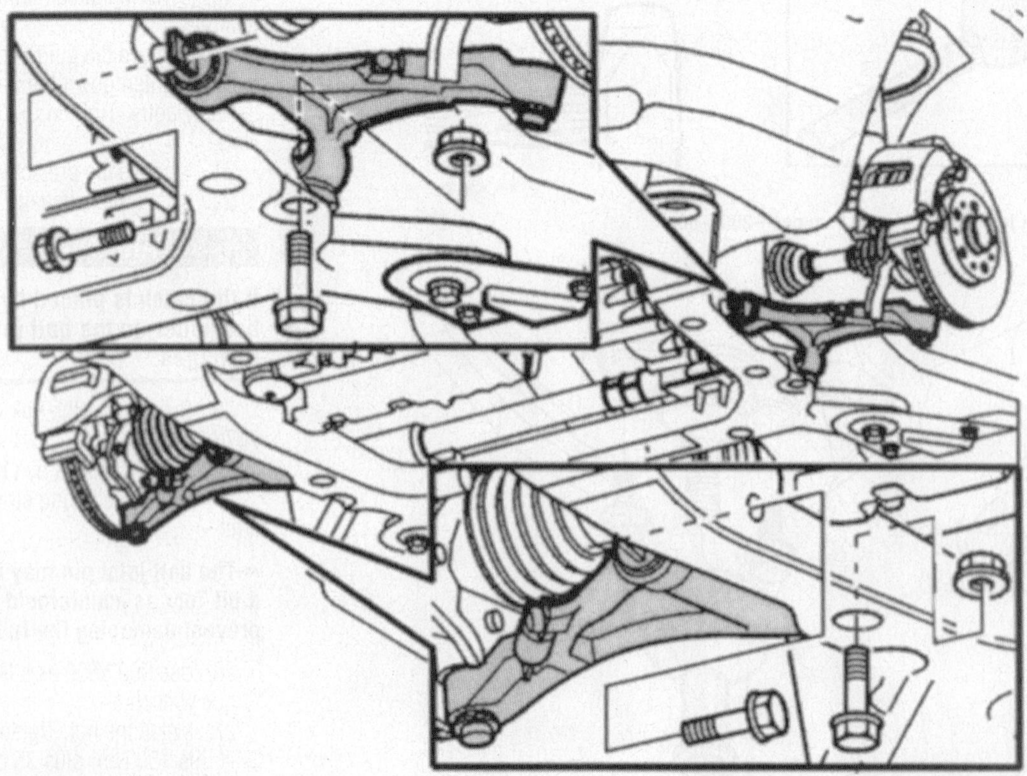

Removing the lower control arm

24. Install the splashguard under the engine.
25. Install the wheels.

CONTROL ARM BUSHING REPLACEMENT

The control arm bushings are serviced with the control arm as an assembly.

Wheel Bearing

ADJUSTMENT

The front and rear wheel bearings are not adjustable

REMOVAL & INSTALLATION

Front

1. Remove the wheel.
2. Remove:
   - The two bolts
   - The brake caliper
3. Hang up the caliper in a suitable position.
4. Remove the halfshaft bolt. Use a pry-bar as a counterhold on the brake disc.
5. Detach the end of the halfshaft in the hub by knocking the halfshaft into the hub approximately 10–15 mm. Use a rubber or copper mallet.
6. Remove:
   - The bolt
   - The brake disc
7. Remove the nut.

➡Counterhold using the wrench grip so that the bolt does not rotate and damage the rubber boot.

8. Remove the nut. Pull the sway bar link out of its mounting.

➡Counterhold using a Torx wrench so that the boot for the link is not damaged.

9. Remove the ABS cable harness from its brackets.
10. Remove the nut from the ball joint on both sides. Counterhold using a Torx wrench so that the ball joint boot is not damaged.
11. Press out the ball joints.
12. Pull down the control arm using a tension strap.
13. Release the spring strut from the control arm and halfshaft.

⁂ WARNING

Ensure that the tension strap is correctly secured in the control arms.

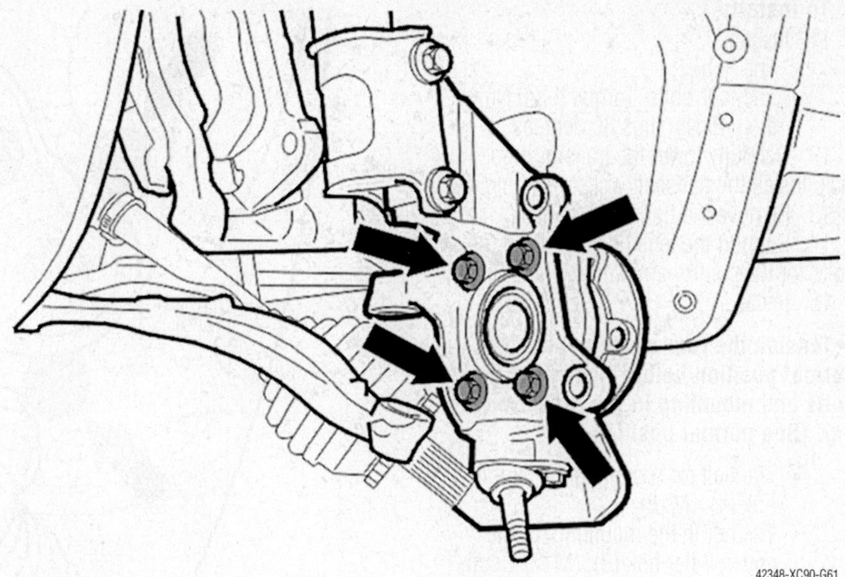

Front hub bolts

14. Remove:
    - The four bolts
    - The hub

To install:

15. Install:
    - The hub
    - The four bolts. For 5- and 6-cylinder models, torque to 15 ft. lbs. (20 Nm), then 44 ft. lbs. (60 Nm), then an additional 60 degrees. For 8-cylinder models, torque to 15 ft. lbs. (20 Nm), then 33 ft. lbs. (45 Nm), then an additional 60 degrees.
16. Install the halfshaft in the hub. Align the ball joint against the control arm.
17. Release the tensioning strap between the control arms.
18. Install a nut on the ball joint.

➡Counterhold using the Torx wrench so as not to damage the boot.

19. Remove the tensioner strap.
20. Install the ABS cable harness in the brackets.
21. Install:
    - The anti-roll bar link
    - The nut (M12). Torque to 44 ft. lbs. (60 Nm)
    - The steering arm
    - The nut. Torque to 50 ft. lbs. (70 Nm)
    - The brake disc
    - The bolt.
22. Install a new halfshaft bolt.
23. Install:
    - The brake caliper
    - The two bolts
24. Install the wheel.

Rear

1. Lock the self-adjuster unit for the parking brake.
2. Remove:
   - The rear wheel
   - The bolt for the halfshaft.
3. Remove the wheel.
4. Remove:
   - The brake caliper mounting bolts. Hang the caliper up using a piece of wire
   - The brake disc locating pin/bolt
   - The brake disc.
5. Install tensioner 999 5659. See the procedure under Shock Absorber R&I.
6. Support the rear suspension 500 mm up between the fender edge and the center of the wheel.
7. Remove:
   - The bolt in the mounting for the tie rod (1)
   - The bolt in the mounting for the control arm (2)
   - The nut in the mounting for the anti-roll bar link (3). Counterhold using a wrench so that the boot does not rotate
   - The bolt for the mounting for the lateral link (4).
8. Position a transmission jack under the wheel spindle and carefully raise.

➡Pull the halfshaft from the hub whilst raising the wheel spindle.

9. Release the tensioner so that the control arm hangs in the shock absorber.
10. Remove:
    - The four bolts
    - The hub.

To install:

11. Install:
    - The hub
    - The four bolts. Torque to 20 Nm, then 45 Nm plus 60 degrees
12. Carefully lower the transmission jack. Install the halfshaft whilst lowering.
13. Remove the transmission jack.
14. Support the tensioner to install the bolts for the control arm and lateral link.
15. Install:

➡Tension the rear suspension to the normal position before tightening the bolts and mounting in the rubber bushing. (See normal position.)

- The bolt for mounting the lateral link (4). (M12).
- The nut in the mounting for the anti-roll bar link (3). (M12) Counterhold using a wrench so that the boot does not rotate Torque to 58 ft. lbs. (80 Nm)
- The bolt for the mounting for the control arm (2) (M12)
- The bolt for the mounting for the tie rod (1). (M12).

16. Remove tensioner 999 5659.

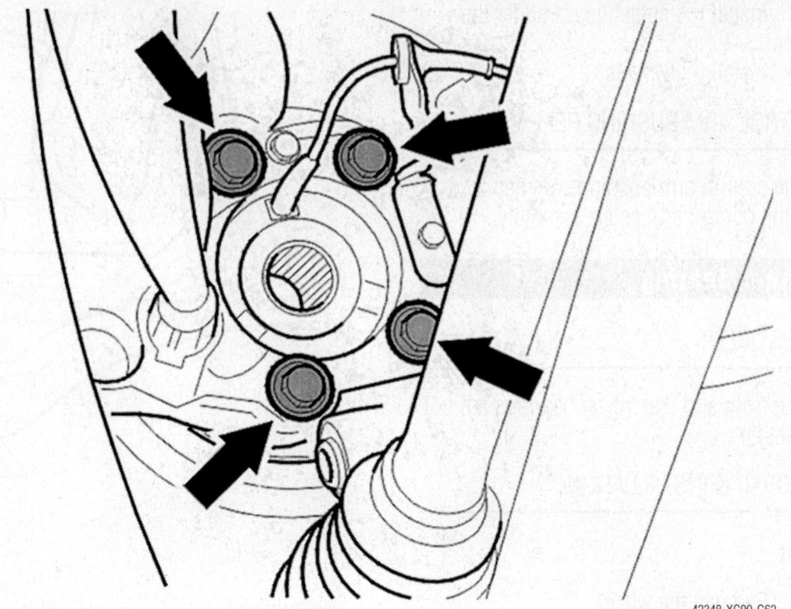

Rear hub bolts

17. Install:
    - The brake disc
    - The brake disc locating pin/bolt (M6)
    - The brake caliper
    - The brake caliper mounting bolts.
    - The bolt for the halfshaft (M10)
    - The rear wheel.
18. Activate the self adjuster for the parking brake.

BRAKES

Caliper

REMOVAL & INSTALLATION

Front

1. Remove the wheel.
2. If the caliper is being replaced:
   a. Secure the pedal in the depressed position. Use a pedal jack.
   b. Clean the brake caliper thoroughly.
   c. Remove the protective cap from the bleed nipple.
   d. Install a plastic hose on the nipple.
   e. Open the bleed nipple. Collect the brake fluid in a container. Shut the bleed nipple.
   f. Slacken the off brake hose approximately half a turn.
3. Remove the retaining spring carefully so as not to deform it.
4. Remove:
   - The protective caps from the two locating pins
   - The locating pins
   - The brake caliper from the holder
   - The brake pads.
5. Unscrew the brake caliper from the brake hose.
6. Drain the remaining brake fluid.
7. Insert the brake caliper on brake hose (Do not tighten)

To install:

8. Install:
   - brake pads
   - brake caliper in the holder.
9. Lubricate the locating pins using grease caliper grease.
10. Insert the locating pins into the rubber sleeves. The pins should slide into the sleeves easily.
11. Tighten the locating pins to 20 ft. lbs. (27.5 Nm).
12. Install the protective caps.
13. Install the securing spring.
14. Tighten the brake hose to 13 ft. lbs. (18 Nm).

➡The brake hose must not be twisted.

15. Fill and bleed the brake system.

Rear

1. Remove the wheel.
2. If the caliper is being replaced:
   a. Secure the pedal in the depressed position. Use a pedal jack.
   b. Clean the brake caliper thoroughly.
   c. Remove the protective cap from the bleed nipple.
   d. Install a plastic hose on the nipple.
   e. Open the bleed nipple. Collect the brake fluid in a container. Shut the bleed nipple.
   f. Slacken the off brake hose approximately half a turn.
3. Remove the brake caliper and brake pads. See the Pad R&I procedure.
4. Unscrew the brake caliper from the brake hose.
5. Drain the remaining brake fluid.

To install:

6. Insert the brake caliper on the hose and hand tighten.
7. Clean and install the brake pads and brake calipers.
8. Tighten the locating pins to 20 ft. lbs. (27.5 Nm) plus 60 degrees.
9. Tighten the brake hose to 13 ft. lbs. (18 Nm).

➡The brake hose must not be twisted.

10. Fill and bleed the brake system.
11. Depress the brake pedal a few times. Check the level of the brake fluid reservoir.

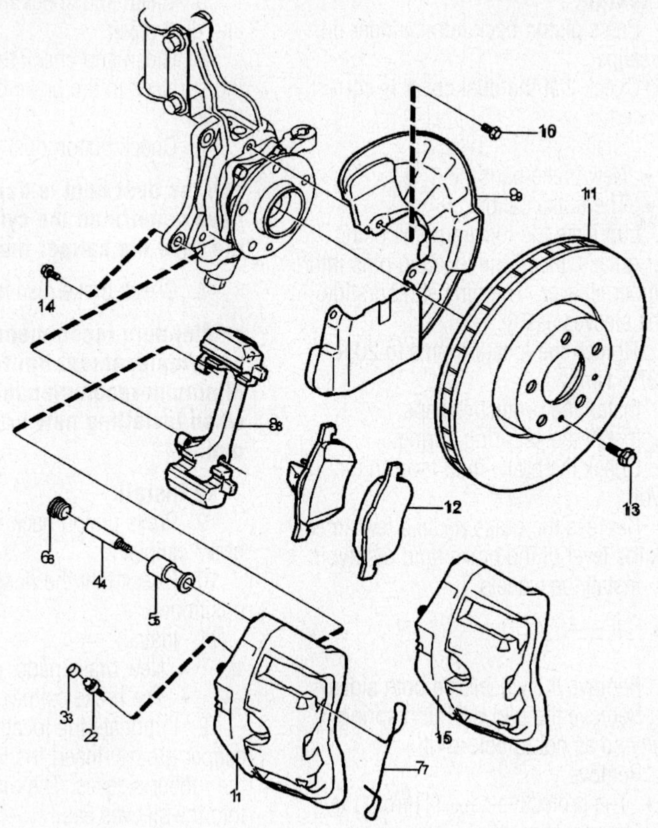

1 Brake caliper, exch left
Brake caliper, exch left
2 • Bleeder screw
3 • Protection
4 • Guide pin upper
• Guide pin
5 • Rubber bellows upper
• Bushing
6 • Protecting cover
7 • Spring
8 • Brace
• Brace
9 Protecting plate
Protecting plate
10 Flange screw
11 Brake disc
Brake disc
12 Brake pad kit

Brake pad kit
13 Hexagon screw
14 Flange screw
15 Brake caliper, exch left
Housing left

42348-XC90-G63

Front brake exploded view

1 Brake caliper, exch
2 • Brace
3 • Spring
4 • Seal
5 • Bolt
6 • Plug
7 Flange screw
8 Bearing housing, l.h.
9 Rear wheel hub
10 Flange screw
11 Gasket
12 Flange screw
13 Protecting plate
14 Screw
15 Brake disc
16 Brake pad kit

Service kits Rear wheel brake

17 Bleeder screw
18 Protection
19 Brake caliper, exch left

42348-XC90-G64

Rear brake exploded view

Brake Pads

REMOVAL & INSTALLATION

Front

1. Remove the wheels on both sides.
2. Remove the pad retaining spring carefully so as not to deform it.
3. Remove:
   - The protective caps (1) from the two locating pins (2)
   - The locating pins, use hex socket 7 mm.
   - The brake caliper from the holder
   - Brake pads.
4. Hang brake caliper from a steel wire from the front spring so as not to damage brake hose.

➡Do not depress the brake pedal while the brake pads are removed.

5. Clean and check the brake caliper and dust cover.
6. Clean and check the brake pad mating surfaces in the brake caliper and caliper holder.
7. Check piston dust boot.

➡If the dust boot is damaged dirt may have penetrated the cylinder. If this is the case the caliper must be replaced.

8. Check brake disc friction surfaces.

➡Minimum recommended disc thickness/replacement limit is 25.0 mm. Minimum recommended disc thickness when installing new brake pads is 25.8 mm.

To install:

9. Press piston back into cylinder on brake caliper.
10. Check that the dust cover is correctly positioned.
11. Install:
    - New brake pads
    - The brake caliper.
12. Lubricate the locating pins using caliper grease. Insert the locating pins into the rubber sleeves. The pins should slide into the sleeves easily.
13. Tighten the locating pins to 20 ft. lbs. (27.5 Nm).
14. Install the protective caps.
15. Install the securing spring.
16. Check the brake fluid level in the reservoir.
17. Depress the brake pedal a few times. Check the level of the brake fluid reservoir.
18. Install the wheels.

Rear

1. Remove the wheels on both sides.
2. Remove the pad retaining spring carefully so as not to deform it.
3. Remove:
   - The protective caps (1) from the two locating pins (2)
   - The locating pins, use hex socket 7 mm.
   - The brake caliper from the holder
   - Brake pads.
4. Hang brake caliper from a steel wire from the front spring so as not to damage brake hose.

➡Do not depress the brake pedal while the brake pads are removed.

5. Clean and check the brake caliper and dust cover.
6. Clean and check the brake pad mating surfaces in the brake caliper and caliper holder.
7. Check piston dust boot.

➡If the dust boot is damaged dirt may have penetrated the cylinder. If this is the case the caliper must be replaced.

8. Check brake disc friction surfaces.

➡Minimum recommended disc thickness/replacement limit is 17.0 mm. Minimum recommended disc thickness when installing new brake pads is 17.7 mm.

To install:

9. Press piston back into cylinder on brake caliper.
10. Check that the dust cover is correctly positioned.
11. Install:
    - New brake pads
    - The brake caliper.
12. Lubricate the locating pins using caliper grease. Insert the locating pins into the rubber sleeves. The pins should slide into the sleeves easily.
13. Tighten the locating pins to 20 ft. lbs. (27.5 Nm) plus 60 degrees.
14. Install the protective caps.
15. Install the securing spring.
16. Check the brake fluid level in the reservoir.
17. Depress the brake pedal a few times. Check the level of the brake fluid reservoir.
18. Install the wheels.

Commonly Used Abbreviations

2

2WD Two Wheel Drive

4

4WD Four Wheel Drive

A

A/C Air Conditioning
ABDC After Bottom Dead Center
ABS Anti-lock Brakes
AC Alternating Current
ACL Air cleaner
ACT Air Charge Temperature
AIR Secondary Air Injection
ALCL Assembly Line Communications Link
ALDL Assembly Line Diagnostic Link
AT Automatic Transaxle/Transmission
ATDC After Top Dead Center
ATF Automatic Transmission Fluid
ATS Air Temperature Sensor
AWD All Wheel Drive

B

BAP Barometric Absolute Pressure
BARO Barometric Pressure
BBDC Before Bottom Dead Center
BCM Body Control Module
BDC Bottom Dead Center
BPT Backpressure Transducer
BTDC Before Top Dead Center
BVSV Bimetallic Vacuum Switching Valve

C

CAC Charge Air Cooler
CARB California Air Resources Board
CAT Catalytic Converter
CCC Computer Command Control
CCCC Computer Controlled Catalytic Converter
CCCI Computer Controlled Coil Ignition
CCD Computer Controlled Dwell
CDI Capacitor Discharge Ignition
CEC Computerized Engine Control
CFI Continuous Fuel Injection
CIS Continuous Injection System
CIS-E Continuous Injection System - Electronic
CKP Crankshaft Position
CL Closed Loop
CMP Camshaft Position
CPP Clutch Pedal Position
CTOX Continuous Trap Oxidizer System
CTP Closed Throttle Position
CVC Constant Vacuum Control
CYL Cylinder

D

DBC Dual Bed Catalyst
DC Direct Current
DFI Direct Fuel Injection
DIS Distributorless Ignition System
DLC Data Link Connector
DMM Digital Multimeter
DOHC Double Overhead Camshaft
DRB Diagnostic Readout Box
DTC Diagnostic Trouble Code
DTM Diagnostic Test Mode
DVOM Digital Volt/Ohmmeter

E

EBCM Electronic Brake Control Module
ECM Engine Control Module
ECT Engine Coolant Temperature
ECU Engine Control Unit or Electronic Control Unit
EDIS Electronic Distributorless Ignition System
EEC Electronic Engine Control
EEPROM Electrically Erasable Programmable Read Only Memory
EFE Early Fuel Evaporation
EGR Exhaust Gas Recirculation
EGRT Exhaust Gas Recirculation Temperature
EGRVC EGR Valve Control
EPROM Erasable Programmable Read Only Memory
EVAP Evaporative Emissions
EVP EGR Valve Position

F

FBC Feedback Carburetor
FEEPROM Flash Electrically Erasable Programmable Read Only Memory
FF Flexible Fuel
FI Fuel Injection
FT Fuel Trim
FWD Front Wheel Drive

G

GND Ground

H

HAC High Altitude Compensation
HEGO Heated Exhaust Gas Oxygen sensor
HEI High Energy Ignition
HO2 Sensor Heated Oxygen Sensor

I

IAC Idle Air Control
IAT Intake Air Temperature
ICM Ignition Control Module
IFI Indirect Fuel Injection
IFS Inertia Fuel Shutoff
ISC Idle Speed Control
IVSV Idle Vacuum Switching Valve

Commonly Used Abbreviations

K

KOEO	Key On, Engine Off
KOER	Key ON, Engine Running
KS	Knock Sensor

M

MAF	Mass Air Flow
MAP	Manifold Absolute Pressure
MAT	Manifold Air Temperature
MC	Mixture Control
MDP	Manifold Differential Pressure
MFI	Multiport Fuel Injection
MIL	Malfunction Indicator Lamp or Maintenance
MST	Manifold Surface Temperature
MVZ	Manifold Vacuum Zone

N

NVRAM	Nonvolatile Random Access Memory

O

O2 Sensor	Oxygen Sensor
OBD	On-Board Diagnostic
OC	Oxidation Catalyst
OHC	Overhead Camshaft
OL	Open Loop

P

P/S	Power Steering
PAIR	Pulsed Secondary Air Injection
PCM	Powertrain Control Module
PCS	Purge Control Solenoid
PCV	Positive Crankcase Ventilation
PIP	Profile Ignition Pick-up
PNP	Park/Neutral Position
PROM	Programmable Read Only Memory
PSP	Power Steering Pressure
PTO	Power Take-Off
PTOX	Periodic Trap Oxidizer System

R

RABS	Rear Anti-lock Brake System
RAM	Random Access Memory
ROM	Read Only Memory
RPM	Revolutions Per Minute
RWAL	Rear Wheel Anti-lock Brakes
RWD	Rear Wheel Drive

S

SBC	Single Bed Converter
SBEC	Single Board Engine Controller
SC	Supercharger
SCB	Supercharger Bypass
SFI	Sequential Multiport Fuel Injection
SIR	Supplemental Inflatible Restraint
SOHC	Single Overhead Camshaft
SPL	Smoke Puff Limiter
SPOUT	Spark Output
SRI	Service Reminder Indicator
SRS	Supplemental Restraint System
SRT	System Readiness Test
SSI	Solid State Ignition
ST	Scan Tool
STO	Self-Test Output

T

TAC	Thermostatic Air Clearner
TBI	Throttle Body Fuel Injection
TC	Turbocharger
TCC	Torque Converter Clutch
TCM	Transmission Control Module
TDC	Top Dead Center
TFI	Thick Film Ignition
TP	Throttle Position
TR Sensor	Transaxle/Transmission Range Sensor
TVV	Thermal Vacuum Valve
TWC	Three-way Catalytic Converter

V

VAF	Volume Air Flow, or Vane Air Flow
VAPS	Variable Assist Power Steering
VRV	Vacuum Regulator Valve
VSS	Vehicle Speed Sensor
VSV	Vacuum Switching Valve

W

WOT	Wide Open Throttle
WU-TWC	Warm Up Three-way Catalytic Converter

ENGLISH TO METRIC CONVERSION: TORQUE

To convert foot-pounds (ft. lbs.) to Newton-meters (Nm), multiply the number of ft. lbs. by 1.36
To convert Newton-meters (Nm) to foot-pounds (ft. lbs.), multiply the number of Nm by 0.7376

ft. lbs.	Nm	ft. lbs.	Nm	ft. lbs.	Nm	ft. lbs.	Nm
0.1	0.1	34	46.2	76	103.4	118	160.5
0.2	0.3	35	47.6	77	104.7	119	161.8
0.3	0.4	36	49.0	78	106.1	120	163.2
0.4	0.5	37	50.3	79	107.4	121	164.6
0.5	0.7	38	51.7	80	108.8	122	165.9
0.6	0.8	39	53.0	81	110.2	123	167.3
0.7	1.0	40	54.4	82	111.5	124	168.6
0.8	1.1	41	55.8	83	112.9	125	170.0
0.9	1.2	42	57.1	84	114.2	126	171.4
1	1.4	43	58.5	85	115.6	127	172.7
2	2.7	44	59.8	86	117.0	128	174.1
3	4.1	45	61.2	87	118.3	129	175.4
4	5.4	46	62.6	88	119.7	130	176.8
5	6.8	47	63.9	89	121.0	131	178.2
6	8.2	48	65.3	90	122.4	132	179.5
7	9.5	49	66.6	91	123.8	133	180.9
8	10.9	50	68.0	92	125.1	134	182.2
9	12.2	51	69.4	93	126.5	135	183.6
10	13.6	52	70.7	94	127.8	136	185.0
11	15.0	53	72.1	95	129.2	137	186.3
12	16.3	54	73.4	96	130.6	138	187.7
13	17.7	55	74.8	97	131.9	139	189.0
14	19.0	56	76.2	98	133.3	140	190.4
15	20.4	57	77.5	99	134.6	141	191.8
16	21.8	58	78.9	100	136.0	142	193.1
17	23.1	59	80.2	101	137.4	143	194.5
18	24.5	60	81.6	102	138.7	144	195.8
19	25.8	61	83.0	103	140.1	145	197.2
20	27.2	62	84.3	104	141.4	146	198.6
21	28.6	63	85.7	105	142.8	147	199.9
22	29.9	64	87.0	106	144.2	148	201.3
23	31.3	65	88.4	107	145.5	149	202.6
24	32.6	66	89.8	108	146.9	150	204.0
25	34.0	67	91.1	109	148.2	151	205.4
26	35.4	68	92.5	110	149.6	152	206.7
27	36.7	69	93.8	111	151.0	153	208.1
28	38.1	70	95.2	112	152.3	154	209.4
29	39.4	71	96.6	113	153.7	155	210.8
30	40.8	72	97.9	114	155.0	156	212.2
31	42.2	73	99.3	115	156.4	157	213.5
32	43.5	74	100.6	116	157.8	158	214.9
33	44.9	75	102.0	117	159.1	159	216.2

METRIC TO ENGLISH CONVERSION: TORQUE

To convert foot-pounds (ft. lbs.) to Newton-meters (Nm), multiply the number of ft. lbs. by 1.36
To convert Newton-meters (Nm) to foot-pounds (ft. lbs.), multiply the number of Nm by 0.7376

Nm	ft. lbs.	Nm	ft. lbs.	Nm	ft. lbs.	Nm	ft. lbs.	Nm	ft. lbs.
0.1	0.1	34	25.0	76	55.9	118	86.8	160	117.6
0.2	0.1	35	25.7	77	56.6	119	87.5	161	118.4
0.3	0.2	36	26.5	78	57.4	120	88.2	162	119.1
0.4	0.3	37	27.2	79	58.1	121	89.0	163	119.9
0.5	0.4	38	27.9	80	58.8	122	89.7	164	120.6
0.6	0.4	39	28.7	81	59.6	123	90.4	165	121.3
0.7	0.5	40	29.4	82	60.3	124	91.2	166	122.1
0.8	0.6	41	30.1	83	61.0	125	91.9	167	122.8
0.9	0.7	42	30.9	84	61.8	126	92.6	168	123.5
1	0.7	43	31.6	85	62.5	127	93.4	169	124.3
2	1.5	44	32.4	86	63.2	128	94.1	170	125.0
3	2.2	45	33.1	87	64.0	129	94.9	171	125.7
4	2.9	46	33.8	88	64.7	130	95.6	172	126.5
5	3.7	47	34.6	89	65.4	131	96.3	173	127.2
6	4.4	48	35.3	90	66.2	132	97.1	174	127.9
7	5.1	49	36.0	91	66.9	133	97.8	175	128.7
8	5.9	50	36.8	92	67.6	134	98.5	176	129.4
9	6.6	51	37.5	93	68.4	135	99.3	177	130.1
10	7.4	52	38.2	94	69.1	136	100.0	178	130.9
11	8.1	53	39.0	95	69.9	137	100.7	179	131.6
12	8.8	54	39.7	96	70.6	138	101.5	180	132.4
13	9.6	55	40.4	97	71.3	139	102.2	181	133.1
14	10.3	56	41.2	98	72.1	140	102.9	182	133.8
15	11.0	57	41.9	99	72.8	141	103.7	183	134.6
16	11.8	58	42.6	100	73.5	142	104.4	184	135.3
17	12.5	59	43.4	101	74.3	143	105.1	185	136.0
18	13.2	60	44.1	102	75.0	144	105.9	186	136.8
19	14.0	61	44.9	103	75.7	145	106.6	187	137.5
20	14.7	62	45.6	104	76.5	146	107.4	188	138.2
21	15.4	63	46.3	105	77.2	147	108.1	189	139.0
22	16.2	64	47.1	106	77.9	148	108.8	190	139.7
23	16.9	65	47.8	107	78.7	149	109.6	191	140.4
24	17.6	66	48.5	108	79.4	150	110.3	192	141.2
25	18.4	67	49.3	109	80.1	151	111.0	193	141.9
26	19.1	68	50.0	110	80.9	152	111.8	194	142.6
27	19.9	69	50.7	111	81.6	153	112.5	195	143.4
28	20.6	70	51.5	112	82.4	154	113.2	196	144.1
29	21.3	71	52.2	113	83.1	155	114.0	197	144.9
30	22.1	72	52.9	114	83.8	156	114.7	198	145.6
31	22.8	73	53.7	115	84.6	157	115.4	199	146.3
32	23.5	74	54.4	116	85.3	158	116.2	200	147.1
33	24.3	75	55.1	117	86.0	159	116.9	201	147.8

ENGLISH/METRIC CONVERSION: TEMPERATURE

To convert Fahrenheit (F°) to Celsius (C°), take F° temperature and subtract 32, multiply the result by 5 and divide the result by 9
To convert Celsius (C°) to Fahrenheit (F°), take C° temperature and multiply it by 9, divide the result by 5 and add 32

F°	C°	F°	C°	C°	F°	C°	F°
-40	-40.0	150	65.6	-38	-36.4	46	114.8
-35	-37.2	155	68.3	-36	-32.8	48	118.4
-30	-34.4	160	71.1	-34	-29.2	50	122
-25	-31.7	165	73.9	-32	-25.6	52	125.6
-20	-28.9	170	76.7	-30	-22	54	129.2
-15	-26.1	175	79.4	-28	-18.4	56	132.8
-10	-23.3	180	82.2	-26	-14.8	58	136.4
-5	-20.6	185	85.0	-24	-11.2	60	140
0	-17.8	190	87.8	-22	-7.6	62	143.6
1	-17.2	195	90.6	-20	-4	64	147.2
2	-16.7	200	93.3	-18	-0.4	66	150.8
3	-16.1	205	96.1	-16	3.2	68	154.4
4	-15.6	210	98.9	-14	6.8	70	158
5	-15.0	212	100.0	-12	10.4	72	161.6
10	-12.2	215	101.7	-10	14	74	165.2
15	-9.4	220	104.4	-8	17.6	76	168.8
20	-6.7	225	107.2	-6	21.2	78	172.4
25	-3.9	230	110.0	-4	24.8	80	176
30	-1.1	235	112.8	-2	28.4	82	179.6
35	1.7	240	115.6	0	32	84	183.2
40	4.4	245	118.3	2	35.6	86	186.8
45	7.2	250	121.1	4	39.2	88	190.4
50	10.0	255	123.9	6	42.8	90	194
55	12.8	260	126.7	8	46.4	92	197.6
60	15.6	265	129.4	10	50	94	201.2
65	18.3	270	132.2	12	53.6	96	204.8
70	21.1	275	135.0	14	57.2	98	208.4
75	23.9	280	137.8	16	60.8	100	212
80	26.7	285	140.6	18	64.4	102	215.6
85	29.4	290	143.3	20	68	104	219.2
90	32.2	295	146.1	22	71.6	106	222.8
95	35.0	300	148.9	24	75.2	108	226.4
100	37.8	305	151.7	26	78.8	110	230
105	40.6	310	154.4	28	82.4	112	233.6
110	43.3	315	157.2	30	86	114	237.2
115	46.1	320	160.0	32	89.6	116	240.8
120	48.9	325	162.8	34	93.2	118	244.4
125	51.7	330	165.6	36	96.8	120	248
130	54.4	335	168.3	38	100.4	122	251.6
135	57.2	340	171.1	40	104	124	255.2
140	60.0	345	173.9	42	107.6	126	258.8
145	62.8	350	176.7	44	111.2	128	262.4

LENGTH CONVERSION

To convert inches (in.) to millimeters (mm), multiply the number of inches by 25.4
To convert millimeters (mm) to inches (in.), multiply the number of millimeters by 0.04

Inches	Millimeters	Inches	Millimeters	Inches	Millimeters	Inches	Millimeters
0.0001	0.00254	0.005	0.1270	0.09	2.286	4	101.6
0.0002	0.00508	0.006	0.1524	0.1	2.54	5	127.0
0.0003	0.00762	0.007	0.1778	0.2	5.08	6	152.4
0.0004	0.01016	0.008	0.2032	0.3	7.62	7	177.8
0.0005	0.01270	0.009	0.2286	0.4	10.16	8	203.2
0.0006	0.01524	0.01	0.254	0.5	12.70	9	228.6
0.0007	0.01778	0.02	0.508	0.6	15.24	10	254.0
0.0008	0.02032	0.03	0.762	0.7	17.78	11	279.4
0.0009	0.02286	0.04	1.016	0.8	20.32	12	304.8
0.001	0.0254	0.05	1.270	0.9	22.86	13	330.2
0.002	0.0508	0.06	1.524	1	25.4	14	355.6
0.003	0.0762	0.07	1.778	2	50.8	15	381.0
0.004	0.1016	0.08	2.032	3	76.2	16	406.4

ENGLISH/METRIC CONVERSION: LENGTH

To convert inches (in.) to millimeters (mm), multiply the number of inches by 25.4
To convert millimeters (mm) to inches (in.), multiply the number of millimeters by 0.04

Inches		Millimeters	Inches		Millimeters	Inches		Millimeters
Fraction	Decimal	Decimal	Fraction	Decimal	Decimal	Fraction	Decimal	Decimal
1/64	0.016	0.397	11/32	0.344	8.731	11/16	0.688	17.463
1/32	0.031	0.794	23/64	0.359	9.128	45/64	0.703	17.859
3/64	0.047	1.191	3/8	0.375	9.525	23/32	0.719	18.256
1/16	0.063	1.588	25/64	0.391	9.922	47/64	0.734	18.653
5/64	0.078	1.984	13/32	0.406	10.319	3/4	0.750	19.050
3/32	0.094	2.381	27/64	0.422	10.716	49/64	0.766	19.447
7/64	0.109	2.778	7/16	0.438	11.113	25/32	0.781	19.844
1/8	0.125	3.175	29/64	0.453	11.509	51/64	0.797	20.241
9/64	0.141	3.572	15/32	0.469	11.906	13/16	0.813	20.638
5/32	0.156	3.969	31/64	0.484	12.303	53/64	0.828	21.034
11/64	0.172	4.366	1/2	0.500	12.700	27/32	0.844	21.431
3/16	0.188	4.763	33/64	0.516	13.097	55/64	0.859	21.828
13/64	0.203	5.159	17/32	0.531	13.494	7/8	0.875	22.225
7/32	0.219	5.556	35/64	0.547	13.891	57/64	0.891	22.622
15/64	0.234	5.953	9/16	0.563	14.288	29/32	0.906	23.019
1/4	0.250	6.350	37/64	0.578	14.684	59/64	0.922	23.416
17/64	0.266	6.747	19/32	0.594	15.081	15/16	0.938	23.813
9/32	0.281	7.144	39/64	0.609	15.478	61/64	0.953	24.209
19/64	0.297	7.541	5/8	0.625	15.875	31/32	0.969	24.606
5/16	0.313	7.938	41/64	0.641	16.272	63/64	0.984	25.003
21/64	0.328	8.334	21/32	0.656	16.669	1/1	1.000	25.400
			43/64	0.672	17.066